ENGINEERING MECHANICS

STATICS

PWS Series in Engineering

Anderson, *Thermodynamics*
Askeland, *The Science and Engineering of Materials, Third Edition*
Borse, *FORTRAN 77 and Numerical Methods for Engineers, Second Edition*
Bolluyt / Stewart / Oladipupo, *Modeling for Design Using SilverScreen*
Clements, *68000 Family Assembly Language*
Clements, *Microprocessor Systems Design, Second Edition*
Clements, *Principles of Computer Hardware, Second Edition*
Das, *Principles of Foundation Engineering, Third Edition*
Das, *Principles of Geotechnical Engineering, Third Edition*
Das, *Principles of Soil Dynamics*
Duff / Ross, *Freehand Sketching for Engineering Design*
El-Wakil / Askeland, *Materials Science and Engineering Lab Manual*
Fleischer, *Introduction to Engineering Economy*
Gere / Timoshenko, *Mechanics of Materials, Third Edition*
Glover / Sarma, *Power System Analysis and Design, Second Edition*
Janna, *Design of Fluid Thermal Systems*
Janna, *Introduction to Fluid Mechanics, Third Edition*
Kassimali, *Structural Analysis*
Keedy, *An Introduction to CAD Using CADKEY 5 and 6, Third Edition*
Keedy / Teske, *Engineering Design Using CADKEY 5 and 6*
Knight, *The Finite Element Method in Mechanical Design*
Knight, *A Finite Element Method Primer for Mechanical Design*
Logan, *A First Course in the Finite Element Method, Second Edition*
McDonald, *Continuum Mechanics*
McGill / King, *Engineering Mechanics: Statics, Third Edition*
McGill / King, *Engineering Mechanics: An Introduction to Dynamics, Third Edition*
McGill / King, *Engineering Mechanics: Statics and An Introduction to Dynamics, Third Edition*
Meissner, *Fortran 90*
Raines, *Software for Mechanics of Materials*
Ray, *Environmental Engineering*
Reed-Hil / Abbaschian, *Physical Metallurgy Principles, Third Edition*
Reynolds, *Unit Operations and Processes in Environmental Engineering*
Russ, *CD-ROM for Materials Science*
Schmidt / Wong, *Fundamentals of Surveying, Third Edition*
Segui, *Fundamentals of Structural Steel Design*
Segui, *LRFD Steel Design*
Shen / Kong, *Applied Electromagnetism, Second Edition*
Sule, *Manufacturing Facilities, Second Edition*
Vardeman, *Statistics for Engineering Problem Solving*
Weinman, *VAX FORTRAN, Second Edition*
Weinman, *FORTRAN for Scientists and Engineers*
Wempner, *Mechanics of Solids*
Wolff, *Spreadsheet Applications in Geotechnical Engineering*
Zirkel / Berlinger, *Understanding FORTRAN 77 and 90*

Third Edition

ENGINEERING MECHANICS

STATICS

DAVID J. MCGILL AND WILTON W. KING
Georgia Institute of Technology

PWS Publishing Company
Boston

ITP An International Thomson Publishing Company

Boston • Albany • Bonn • Cincinnati • Detroit • London • Madrid
Melbourne • Mexico City • New York • Paris • San Francisco
Singapore • Tokyo • Toronto • Washington

PWS PUBLISHING COMPANY
20 Park Plaza, Boston, Massachusetts 02116-4324

This book is printed on recycled, acid-free paper

I(T)P™
International Thomson Publishing
The trademark ITP is used under license.

For more information, contact:

PWS Publishing Company
20 Park Plaza
Boston, MA 02116

International Thomson Publishing Europe
Berkshire House I68-I73
High Holborn
London WC1V7AA
England

Thomas Nelson Australia
102 Dodds Street
South Melbourne, 3205
Victoria, Australia

Nelson Canada
1120 Birchmount Road
Scarborough, Ontario
Canada M1K 5G4

International Thomson Editores
Campos Eliseos 385, Piso 7
Col. Polanco
11560 Mexico D.F., Mexico

International Thomson Publishing GmbH
Konigswinterer Strasse 418
53227 Bonn, Germany

International Thomson Publishing Asia
221 Henderson Road
#05-10 Henderson Building
Singapore 0315

International Thomson Publishing Japan
Hirakawacho Kyowa Building, 31
2-2-1 Hirakawacho
Chiyoda-ku, Tokyo 102
Japan

Library of Congress Cataloging-in-Publication Data

McGill, David J., 1939–
Engineering mechanics, statics / David J. McGill and Wilton W. King.
— 3rd ed.
p. cm.
Includes index
ISBN 0-534-93393-9
1. Mechanics, Applied 2. Statics I. King, Wilton W., . II. Title
TA350.M385 1994
620.1'03 — dc20 94-33845
CIP

Printed and bound in the United States of America
95 96 97 98 99 — 10 9 8 7 6 5 4 3 2 1

Sponsoring Editor: Jonathan Plant
Editorial Assistant: Cynthia Harris
Developmental Editor: Mary Thomas
Production Coordinator: Kirby Lozyniak
Marketing Manager: Nathan Wilbur
Manufacturing Coordinator: Marcia Locke
Production: York Production Services

Cover Designer: Julie Gecha
Interior Designer: York Production Services
Compositor: Progressive Information Technologies
Cover Photo: © Grant Faint. Courtesy of the Image Bank
Cover Printer: John P. Pow Company, Inc.
Text Printer: Quebecor Printing, Hawkins

To the memory of Robert W. Shreeves,
friend and colleague

CONTENTS

PREFACE xi

1 INTRODUCTION 1

1.1 Engineering Mechanics, 2
1.2 The Primitives, 2
1.3 Basic Laws, 3
1.4 Units and Dimensions, 5
1.5 Problem Solving and Accuracy of Solutions, 8

2 FORCES AND PARTICLE EQUILIBRIUM 11

2.1 Introduction, 12
2.2 Forces, 12
2.3 Equilibrium of a Particle, 24
2.4 Equilibrium of a System of Particles, 33
Summary, 45
Review Questions, 46

3 THE MOMENT OF A FORCE; RESULTANTS 47

3.1 Introduction, 48
3.2 Moment of a Force About a Point, 49
3.3 Moment of a Force About a Line, 59
3.4 The Couple, 75
3.5 Laws of Equilibrium: Relationship Between Sums of Moments, 83
3.6 Equipollence of Force Systems, 87
3.7 The Force-and-Couple Resultant of a System of Forces, 99
3.8 The Simplest Resultant of a Force System, 106
3.9 Distributed Force Systems, 129
Summary, 148
Review Questions, 150

4 ANALYSIS OF GENERAL EQUILIBRIUM PROBLEMS 152

4.1 Introduction, 153
4.2 The Free-Body Diagram, 156

4.3 Fundamental Applications of the Equilibrium Equations, 169
4.4 Applications of the Equations of Equilibrium to Interacting Bodies or Parts of a Structure, 220
Summary, 271
Review Questions, 272

5 Structural Applications 273
5.1 Introduction, 274
I Trusses, 275
5.2 Definition of a Truss; Examples of Trusses, 275
5.3 The Method of Joints (or Pins), 281
5.4 Shortcuts and Rigidity / Determinacy Results, 291
5.5 The Method of Sections, 300
5.6 Space Trusses, 318
5.7 A Brief Introduction to the Mechanics of Deformable Bodies, 327
II Systems Containing Multiforce Members, 348
5.8 Axial and Shear Forces and Bending Moments, 348
5.9 Beams / Shear and Moment Distributions, 358
5.10 Differential Relationships Between $q(x)$, $V(x)$, and $M(x)$ in a Beam / Shear and Bending Moment Diagrams, 371
III Cables, 387
5.11 Parabolic and Catenary Cables, 387
5.12 Cables Under Concentrated Loads, 397
Summary, 408
Review Questions, 409

6 Friction 411
6.1 Introduction, 412
6.2 Laws, Coefficients, and Basic Applications of Coulomb Friction, 412
6.3 Special Applications of Coulomb Friction, 454
Summary, 475
Review Questions, 477

7 Centroids and Mass Centers 478
7.1 Introduction, 479
7.2 Centroids of Lines, Areas, and Volumes, 480
7.3 The Method of Composite Parts, 495
7.4 Center of Mass, 511
7.5 The Theorems of Pappus, 517
Summary, 525
Review Questions, 526

8 Inertia Properties of Plane Areas 528
8.1 Introduction, 529
8.2 Moments of Inertia of a Plane Area, 529
8.3 The Polar Moment of Inertia of a Plane Area, 534

8.4 The Parallel-Axis Theorem (or Transfer Theorem) for Moments of Inertia / The Radius of Gyration, 536
8.5 The Method of Composite Areas, 543
8.6 Products of Inertia of Plane Areas, 554
8.7 The Parallel-Axis Theorem for Products of Inertia, 558
8.8 Moments and Products of Inertia with Respect to Rotated Axes Through a Point / Mohr's Circle, 561
Summary, 572
Review Questions, 573

9 SPECIAL TOPICS 575
9.1 Introduction, 576
9.2 The Principle of Virtual Work, 576
9.3 Hydrostatic Pressure on Submerged Bodies, 598
Review Questions, 621

APPENDIX CONTENTS 623
Appendix A VECTORS 624
A.1 Vectors: Addition, Subtraction, and Multiplication by a Scalar, 624
A.2 Unit Vectors and Orthogonal Components, 625
A.3 Scalar (Dot) Product, 626
A.4 Vector (Cross) Product, 628
A.5 Scalar and Vector Triple Products, 631
Appendix B TABLES RELATING TO UNITS 636
Appendix C MOMENTS AND PRODUCTS OF INERTIA OF AREAS 639
Appendix D EXAMPLES OF NUMERICAL ANALYSIS / THE NEWTON-RAPHSON METHOD 644
Appendix E EQUILIBRIUM: A BODY AND ITS PARTS 646
Appendix F ANSWERS TO ODD-NUMBERED PROBLEMS 648
INDEX 659

PREFACE

Statics is the first book in a two-volume set on basic mechanics. It is a text for standard courses in statics as found in most colleges of engineering. This text includes more material than is normally covered in such a course because we have attempted to include most traditional special applications from which schools and instructors make selections to augment the core subject matter.

In the writing of this text we have followed one basic guideline—to write the book in the same way we teach the course. To this end, we have written many explanatory footnotes and included frequent questions interspersed throughout the chapters. The answer to each question is provided on the same page (instead of at the end of the chapter as in previous editions). These questions are the same kind as the ones we ask in class; to make the most of them, treat them as serious homework as you read, and look at the answers only after you have your own answer in mind. The questions are intended to encourage thinking about tricky points and to emphasize the basic principles of the subject.

In addition to the text questions, a set of approximately one dozen review questions and answers are included at the end of each chapter. These true-false questions are designed for both classroom discussion and for student review. Homework problems of varying degrees of difficulty appear at the end of every major section. There are over 1,300 of these exercises, and the answers to the odd-numbered ones constitute Appendix F in the back of the book.

There are a number of reasons (besides carelessness) why it may be difficult to get the correct answer to a homework problem in statics or dynamics on the first try. The problem may require an unusual amount of thinking and insight; it may contain tedious calculations; or it may challenge the student's advanced mathematics skills. We have placed an asterisk beside especially difficult problems falling into one or more of these categories.

Statics is characterized by only two basic equations, but these equations are applied in a wide variety of circumstances. Thus it is imperative that students develop a feel for realistically modeling an engineering situation. Consequently, we have included a large number of actual engineering problems among the examples and exercises. Being aware of the assumptions and accompanying limitations of the model and of the solution method can be developed only by sweating over many problems outside the classroom. Only in this way can students develop the insight and creativity needed to solve engineering problems.

Some examples and problems are presented in SI (Système International) metric units, whereas others use traditional United States engineering system units. Whereas the United States is slowly and painfully converting to SI units, our consulting activities make it clear that much engineering work is still being performed using traditional units. Most United States engineers still tend to think in pounds instead of newtons and in feet instead of meters. We believe students will become much better engineers, scientists, and scholars if they are thoroughly familiar with both systems.

In Chapter 1 we introduce engineering mechanics and its primitives, and we set forth the basic laws of statics. This chapter also covers units and dimensions as well as techniques of problem solving and the importance of accuracy.

In this edition, we have split the concepts of forces and moments into separate chapters. Chapter 2, new to the third edition, introduces the reader to the force, to its vector representation, and to a group of equilibrium problems that can be solved without the need for summing moments.

In Chapter 3, the moment of a force is defined and covered in detail, after which the pair of equilibrium equations for the finite-sized body are presented. The reader is then prepared for the second half of the chapter, which deals with equipollence and with resultants of discrete and distributed force systems.

The heart of the book is Chapter 4, in which we analyze equilibrium problems. The chapter begins with the free-body diagram—crucial to successful analysis of problems in statics and dynamics. We then examine the equilibrium of a single body and expand that study to interacting bodies and to parts of a structure.

In Chapter 5 we extend our study to structures of four common types: trusses, frames, beams, and cables. In preparation for later courses in strength of materials or deformable bodies, we include a section on shear and moment diagrams. The studies of Chapter 5 differ from most of those in Chapter 4 in that the bodies are routinely "cut" (on paper) in order to determine their important internal force distributions.

Though friction forces may sometimes act on the bodies studied in Chapter 4, the special nature of these forces was not elaborated on there. This detailed study is done in Chapter 6, which deals exclusively with Coulomb (or dry friction). This chapter also includes fundamental problems and applications of dry friction along with special applications such as the friction on a flexible flat belt wrapped around a cylindrical surface.

Chapters 7 and 8, although not statics per se, treat topics often covered in statics courses. Chapter 7 includes the topics of centroids (of lines, areas, and volumes) and of centers of mass. Chapter 8 follows with a study of inertia properties of areas—a necessary background for studies of the strength and deflection of beams in courses on the mechanics of deformable solids. To this end we include a closing section on Mohr's circle for principal axes and principal moments of inertia of areas. Mohr's circle is also useful in studies of stress and strain as well as in studies of moments of inertia of masses (the latter of which is covered in our dynamics volume).

Finally, Chapter 9 includes two special topics in statics. The first is the principle of virtual work, a very powerful method in mechanics and an elegant alternative to the equations of equilibrium. The second is "fluid statics," or the statics of submerged bodies subjected to hydrostatic fluid pressure.

We thank Meghan Root for cheerfully typing all our third edition changes. And for their many useful suggestions, we are grateful to our third edition reviewers:

William Bickford
Arizona State University

Donald E. Carlson
University of Illinois at Urbana

Robert L. Collins
University of Louisville

John Dickerson
University of South Carolina

John F. Ely
North Carolina State University

Laurence Jacobs
Georgia Institute of Technology

Seymour Lampert
University of Southern California

Vincent WoSang Lee
University of Southern California

Joseph Longuski
Purdue University

Robert G. Oakberg
Montana State University

Joseph E. Parnarelli
University of Nebraska

Mario P. Rivera
Union College

Wallace S. Venable
West Virginia University

Carl Vilmann
Michigan Technological University

We also thank our colleagues at *Georgia Institute of Technology:* Larry Jacobs, Charles Ueng, Wan-Lee Yin, Don Berghaus, Jianmin Qu, Al Ferri, Dewey Hodges, Manohar Kamat, and Alan Larson, for helpful comments since the second edition.

We are grateful to the following professors, who each responded to a questionnaire we personally sent out in 1991: Don Carlson, *University of Illinois;* Patrick MacDonald and John Ely, *North Carolina State University;* Vincent Lee, *University of Southern California;* Charles Krousgrill, *Purdue University;* Samuel Sutcliffe, *Tufts University;* Larry Malvern and Martin Eisenberg, *University of Florida;* John Dickerson, *University of South Caro-*

lina; Bill Bickford, *Arizona State University;* James Wilson, *Duke University;* Mario Rivera, *Union College;* and Larry Jacobs, *Georgia Institute of Technology.* Their comments were also invaluable.

Special appreciation is expressed to our insightful editor, Jonathan Plant, and to the following individuals involved in the smooth production of this third edition: Mary Thomas and Kirby Lozyniak of *PWS Publishing Company,* and Tamra Winters of *York Production Services.*

David J. McGill
Wilton W. King

We are pleased to introduce to this edition a new set of model-based problems. These problems, presented in a full-color insert bound into the book, introduce students to the process of building three-dimensional models from commonly found objects in order to observe as well as calculate mechanical behavior. Many students beginning their engineering education lack a hands-on, intuitive feel for this behavior, and these specially designed problems can help build confidence in their observational and analytical abilities.

We wish to acknowledge the following contributors to the model-based problems insert:

David J. McGill and Wilton W. King for their initial conception and presentation of the model-based problem idea in *Dynamics Model Problems* written to accompany *Engineering Mechanics: An Introduction to Dynamics,* Third Edition.

David Barnett, *Stanford University,* Mario P. Rivera, *Union College;* Robert G. Oakberg, *Montana State University;* John F. Ely, *North Carolina State University;* Carl Vilmann, *Michigan Technological University;* Robert L. Collins, *University of Louisville;* Nicholas P. Jones, *Johns Hopkins University* and William B. Bickford, *Arizona State University* for their evaluations of McGill and King's *Dynamics Model Problems.*

We thank Mario P. Rivera, Robert G. Oakberg, and John F. Ely, for developing additional model problems for the insert. And a very special thanks to Michael K. Wells, *Montana State University,* for developing and editing the final text of the insert, and for providing an introduction and additional problems.

PWS Publishing Company

ENGINEERING MECHANICS

STATICS

1 Introduction

1.1 Engineering Mechanics
1.2 The Primitives
1.3 Basic Laws
1.4 Units and Dimensions
1.5 Problem Solving and Accuracy of Solutions

1.1 Engineering Mechanics

Two things that are basic to understanding the physical world and universe in which we live are (a) the motions of bodies and (b) their mechanical interactions. Engineering mechanics provides the basic principles by which these motions and interactions are described, related, and predicted.

There are many diverse applications of mechanics, which begin in most undergraduate engineering curricula with studies of statics, dynamics, mechanics of materials, and fluid mechanics. Applications of the principles learned in these studies have led to solutions of such problems as:

1. The invention and continuing refinement of the bicycle, the automobile, the airplane, the rocket, and machines for manufacturing processes.
2. The description of the motions of the planets and of artificial satellites.
3. The description of the flows of fluids that allow motion and flight to occur.
4. The determination of the stresses (intensities of forces) produced in machines and structures under load.
5. The control of undesirable vibrations that would otherwise cause discomfort in vehicles and buildings.

In solving problems such as these, mathematical models are created and analyzed. It will be important for students to learn to bridge the gap between problems of the real world and the mathematical models used to describe them. This, too, is part of mechanics — being able to visualize the actual problem and then to come up with a realistic and workable model of it. Proficiency will come only from the experience of comparing the predictions of mathematical models with observations of the physical world for large numbers of problems. The reader will find that there are not a great number of basic ideas and principles in mechanics, but they provide powerful tools for engineering analysis if they are thoroughly understood.

In the first part of this introductory mechanics text, we shall be considering bodies at rest in an inertial (or Newtonian) reference frame; a body in this situation is said to be in equilibrium. Statics is the study of the equilibrium interactions (forces) of a body with its surroundings. In another study, called dynamics, we explore the relation between motions and forces, especially in circumstances in which the body may be idealized as rigid.

1.2 The Primitives

There are several concepts that are *primitives* in the study of mechanics.

Space We shall be using ordinary Euclidean three-dimensional geometry to describe the positions of points on the bodies in which we are interested, and, by extension, the regions occupied by these bodies. The coordinate axes used in locating the points will be locked into a reference frame, which is itself no more or less than a rigid body (one for which the distance between any two points is constant).

Time Time will be measured in the usual way. It is, of course, the measure used to identify the chronology of events. Time will not really enter the picture in statics; it becomes important when the bodies are no longer at rest, but are instead moving in the reference frame.

Force Force is the action of one body upon another, most easily visualized as a push or pull. A force acting on a body tends to accelerate it in the direction of the force.

Mass The resistance of a body to motion is measured by its mass and by the distribution of that mass. Mass per unit volume, called density, is a fundamental material property. Mass is a factor in the gravitational attraction of one body to another. It is this manifestation of mass that we shall encounter in statics.

1.3 Basic Laws

When Isaac Newton first set down the basic laws or principles upon which mechanics has come to be based, he wrote them for a particle. This is a piece of material sufficiently small that we need not distinguish its material points as to locations (or velocities or accelerations). Therefore, we could actually consider the Earth and Moon as particles for some applications such as the analysis of celestial orbits (as Newton did).

Newton published a treatise called *The Principia* in 1687, in which certain principles governing the motion of a particle were developed. These have come to be known as **Newton's Laws of Motion** and are commonly expressed today as follows:

1. In the absence of external forces, a particle has constant velocity (which means it either remains at rest or travels in a straight line at constant speed).
2. If a force acts on a particle, it will be accelerated in the direction of the force, with an acceleration magnitude proportional to that of the force.
3. The two forces exerted on a pair of particles by each other are equal in magnitude, opposite in direction, and collinear along the line joining the two particles.

We must recognize that the laws will not apply when velocities approach the speed of light, when relativistic effects become important. Neither will Newton's Laws apply at a spatial scale smaller than that of atoms. It is also important to understand that what we are really doing is hypothesizing the existence of certain special frames of reference in which the laws are valid. These frames are called Newtonian, or inertial.

This poses a chicken-and-egg problem where one tries to reason which comes first — the inertial frame or the three laws. It is true that the laws hold only in inertial frames, but also that inertial frames are those in which the laws hold, so that neither is of any value without the other. To establish that a frame is inertial requires numerous comparisons of the predictions of the laws of motion with experimental observations. Such comparisons have failed to provide any contradiction of the assertion that a frame containing the mass center of the solar system and having fixed orientation relative to the "fixed" stars is inertial. For this reason many writers refer to this frame of reference as "fixed" or "absolute." While the earth, which moves and turns relative to this standard, is not an inertial frame, it closely enough approximates one for the analysis of most earth-bound engineering problems.

An important extension of Newton's Laws was made in the 18th century by the Swiss mathematician Leonhard Euler. The extension was the postulation of two vector laws of motion for the finite-sized body. These laws **(Euler's Laws),** again valid only in inertial frames, are expressible as:

1. The resultant of the external forces on a body is at all times equal to the time derivative of its momentum.
2. The resultant moment of these external forces about a fixed point is equal to the time derivative of the body's moment of momentum about that point.

Euler's Laws allow us to study the motions (or the special case in which the motions vanish) of bodies, whether or not they are particles. The first law yields the motion of the mass center, and the second leads to the orientational, or rotational, motion of a rigid body. It can be shown (see Appendix E) that an "action-reaction" principle (equivalent to Newton's Third Law) follows from these two laws of Euler.

Another contribution by Isaac Newton which is of monumental importance in mechanics is his **Law of Gravitation,** which expresses the gravitational attraction between two particles in terms of their masses (m_1 and m_2) and the distance (r) between them. The magnitude (F) of the force on either particle is given by

$$F = \frac{Gm_1m_2}{r^2}$$

where G is the universal gravitation constant. For a small body (particle) being attracted by the earth, the force is given approximately by an equation of the same form,

$$F = \frac{GMm}{r^2}$$

where now M is the mass of the earth, m is the mass of the particle, and r the distance from the particle to the center of the earth. If the particle is near the earth's surface, r is approximately the radius, r_e, of the earth and to good approximation

$$F = \left(\frac{GM}{r_e^2}\right) m = mg$$

The symbol g is called the strength of the gravitational field or the gravitational acceleration, since this is the free-fall acceleration of a body near the surface of the earth. Although g varies slightly from place to place on the earth, we shall, unless otherwise noted, use the nominal values of 32.2 lb/slug (or ft/sec^2) and 9.81 N/kg (or m/s^2). The force, mg, that the earth exerts on the body is called the weight of the body.

1.4 Units and Dimensions

The numerical value assigned to a physical entity expresses the relationship of that entity to certain standards of measurement called **units.** There is currently an international set of standards called the International System (SI) of Units. This is a descendant of the MKS metric system. In the SI system the unit of time is the **second** (s), the unit of length is the **meter** (m), and the unit of mass is the **kilogram** (kg). These independent (or *basic*) units are defined by physical entities or phenomena: the second is defined by the period of a radiation occurring in atomic physics, and the meter is defined by the wavelength of a different radiation. One kilogram is defined to be the mass of a certain piece of material that is stored in France. Any other SI units we shall need are *derived* from these three basic units. The unit of force, the **newton** (N), is derived by way of Newton's Second Law, so that, for example, one newton is the force required to give a mass of one kilogram an acceleration of one meter per second per second, or 1 N = 1 kg · m/s^2.

Until very recently almost all engineers in the United States have used a different system (sometimes called the British gravitational or U.S. system) in which the basic units are the **second** (sec) for time, the **foot** (ft) for length, and the **pound** (lb) for force. The pound is the weight, at a standard gravitational condition (location) of a certain body of material that is stored in the United States. In this system the unit of mass is derived and is the **slug,** one slug being the mass that is accelerated one foot per second per second by a force of one pound, or 1 slug = 1 lb-sec^2/ft. For the foreseeable future, United States engineers will find it desirable to be as comfortable as possible with both the U.S. and SI systems; for that reason we have used both sets of units in examples and problems throughout this book.

We next give a brief discussion of unit conversion. The conversion of units is very quickly and efficiently accomplished by multiplying by equivalent fractions until the desired units are achieved. For example, suppose we wish to know how many newton · meters (N · m) of torque are equivalent to 1 lb-ft; since we know there to be 3.281 ft per m and

4.448 N per lb,

$$1 \text{ lb-ft} = 1 \cancel{\text{lb-ft}} \left(\frac{1 \text{ m}}{3.281 \cancel{\text{ft}}}\right)\left(\frac{4.448 \text{ N}}{1 \cancel{\text{lb}}}\right) = 1.356 \text{ N} \cdot \text{m}$$

Note that if the undesired units don't cancel, the fraction is erroneously upside-down. For a second example, let us find how many slugs of mass there are in a kilogram:

$$1 \text{ kg} = 1 \frac{\cancel{\text{N}} \cdot \text{s}^2}{\cancel{\text{m}}} \left(\frac{1 \text{ lb}}{4.448 \cancel{\text{N}}}\right)\left(\frac{1 \cancel{\text{m}}}{3.281 \text{ ft}}\right)$$

$$= 0.06852 \frac{\text{lb-sec}^2}{\text{ft}} \qquad \text{or} \qquad 0.06852 \text{ slug}$$

Inversely, 1 slug = 14.59 kg. A table of units and conversion factors may be found in Appendix B.

It is a source of some confusion that sometimes there is used a unit of mass called the pound, or pound mass, which is the mass whose weight is one pound of force at standard gravitational conditions. Also, the term kilogram has sometimes been used for a unit of force, particularly in Europe. Grocery shoppers in the U.S. are exposed to this confusion by the fact that packages are marked as to weight (or is it mass?) both in pounds and in kilograms. Throughout this book, without exception, *the pound is a unit of force and the kilogram is a unit of mass.*

The reader is no doubt already aware of the care that must be exercised in numerical calculations using different units. For example, if two lengths are to be summed in which one length is 2 feet and the other is 6 inches, the simple sum of the measures, $2 + 6 = 8$, does not provide a measure of the desired length. It is also true that we may not add or equate the numerical measures of different types of entities; thus it makes no sense to attempt to add a mass to a length. These are said to have different dimensions, a **dimension** being the name assigned to the *kind* of measurement standard involved, as contrasted with the choice of a particular measurement standard (unit). In science and engineering we attempt to develop equations expressing the relationships among various physical entities in a physical phenomenon. We express these equations in symbolic form so that they are valid regardless of the particular choice of system of units; nonetheless, they must be *dimensionally consistent.*

To aid in verification of dimensional consistency, we assign some common symbols for basic dimensions: L for length, M for mass, F for force, and T for time. Just as there are derived units of measure, there are derived dimensions; thus the dimension of velocity or speed is L/T and the dimension of acceleration is L/T^2. In SI units, force is derived from L, M, and T; we have, dimensionally, $F = ML/T^2$. In U.S. units, mass is derived from L, F, and T; hence, dimensionally, $M = FT^2/L$. Some things are dimensionless. An example of this is the radian measure of an angle. Since the measure is defined by the ratio of two lengths, the numerical value is thus independent of the choice of unit of length. Arguments of transcendental functions must always be dimensionless.

To check an expression for dimensional consistency, we replace each symbol for a physical quantity by the symbol (or symbols) for its dimension. We likewise replace any dimensionless quantity by unity. The dimension symbols in each separate term of an equation must combine to yield the same dimension for each term. The following examples illustrate this process:

1. The distance, d, of a runner from the finish line of a race has been derived to be (for an interval of constant acceleration)

$$d = d_0 - v_0 t - \frac{1}{2} a t^2$$

where t is time, d_0 is the distance at $t = 0$, v_0 is the speed at $t = 0$, and a is the constant acceleration. Substituting the dimension symbols in each term

$$L = L - \frac{L}{T}(T) - \frac{L}{T^2}(T^2)$$

where the equality sign and the minus signs serve only the purpose of identifying the terms under consideration. Since each term has the dimension of length (L) the equation is dimensionally consistent.

2. A square plate is supported by a pair of ropes; suppose that a student deduces that the force, P, exerted by one rope is

$$P = mg(2\ell + 3\ell^2)$$

where ℓ is the length of a side of the plate, m is the mass of the plate, and g is the acceleration of gravity. If, as is intended here, every length appearing in the problem is a multiple (or fraction) of ℓ, then a student must immediately conclude that the analysis is in error since the dimension of 2ℓ is L and the dimension of $3\ell^2$ is L^2; thus they cannot be added.

A second student analyzing the problem concludes that

$$P = \frac{1}{2} mg\ell$$

This student also must conclude that the analysis is in error since the dimension of P is F while the dimension of $mg\ell$ is FL.

A third student analyzing this problem concludes that

$$P = \frac{1}{2} mg$$

This solution may be in error, but at least it satisfies the requirement of dimensional consistency.

3. Analyzing the dynamics of a rotating plate with edge lengths a and b, a student finds the angular speed, ω (the dimension is $1/T$, and typical units are rad/sec), at a certain instant to be

$$\omega = 5g/(a + b^2)$$

which cannot be true since the denominator is dimensionally inconsistent (adding an L to an L^2).

A second student obtains

$$\omega = 5g/(a + b)$$

Noting that g, the acceleration of gravity, has dimension L/T^2, we test the dimensional consistency of the result by writing

$$\frac{1}{T} \stackrel{?}{=} \frac{L/T^2}{L} = \frac{1}{T^2}$$

which demonstrates that this result is not dimensionally consistent either.

A third student obtains

$$\omega = 5\sqrt{g/(a + b)}$$

which is dimensionally consistent since the dimension of $\sqrt{g/(a + b)}$ is $\sqrt{1/T^2} = 1/T$.

4. A student's analysis of vibrations of an airplane wing yields the displacement, u (its dimension is L), of a certain point to be

$$u = Ae^{-\alpha t} \sin \beta t$$

where t is time. For this equation to be dimensionally consistent, (αt) and (βt) must be dimensionless; therefore, α and β must each have the dimension $1/T$. Moreover, the dimension of A must be length (L).

These examples illustrate a compelling reason for expressing the solutions to problems in terms of symbols so that *any system of units can be used.* When that is done it is relatively easy to check the dimensional consistency of the proposed expressions. With a solution in terms of symbols, we can also examine limiting cases of the parameters to check the solution itself. Sometimes we can even undertake to optimize a solution quantity with respect to one or more of the parameters.

1.5 Problem Solving and Accuracy of Solutions

In Chapters 2 and 3 we shall undertake a study of the two vectors of prime importance in statics: forces and moments. These vectors will be used to develop the concept of the resultant of a force/couple system. We shall then be ready in Chapter 4 to solve general equilibrium problems. At that time we shall give a detailed discussion of problem solving, emphasizing one of the most useful concepts in mechanics — the free-body diagram. Until we reach that point in our study, however, it is important that the student/reader do the following with the problems in the first three chapters:

1. Read the problem carefully, digest the physical meaning, and list the "givens" and the "requireds."

2. Sketch any diagrams that might be helpful.
3. Carry out the calculations, using only as many digits as the least accurate number in the given data.*
4. Look over your answers. See if they make sense, and draw and state all the conclusions you can from them.

In the examples, unless stated otherwise, we shall retain three significant digits (unless one or more digits are lost through additions or subtractions; for example, $90.2 - 90.1 = 0.1$). If, say, a length ℓ is given in the data to be 2 ft, it will be assumed throughout the example that ℓ is actually 2.00 ft.

In the next two chapters (indeed, throughout most of the rest of the book), we shall be using vectors to represent the three entities commonly known as force, moment, and position. In Appendix A we offer a review of vectors, and we encourage all student readers to glance through this appendix at this time and to study any unfamiliar topics.

* For instance, if $g = 32.2$ ft/sec² or 9.81 m/s² is used in a calculation, it is ridiculous to give an answer to four significant digits.

PROBLEMS ▶ Chapter 1

1.1 Describe a physical problem in which we already know the configuration (location) of a body at rest, and are interested in knowing the forces that keep it there.

1.2 Describe a physical problem in which we know at least one of the forces acting on a body at rest and are interested in knowing its configuration.

1.3 Explain why velocity and energy are not primitives in the study of mechanics.

1.4 A dyne is one gram · centimeter/s². How many dynes are there in one pound?

1.5 How many kilometers are there in one mile?

1.6 What is the weight in newtons of a 2500-pound automobile?

1.7 The Btu (British thermal unit) is a unit of energy used in thermodynamic calculations. There are 778 ft-lb in one Btu. How many joules are there in one Btu? (One joule = 1 N · m of energy.)

1.8 Determine which of the terms in the following equation is dimensionally inconsistent with all the others:

$$mg \cos\theta - N = \frac{mv^2}{r} + \frac{mr^2}{t^2}$$

where m = mass, g = gravitational acceleration, N = force, v = velocity, r = radius, and t = time.

1.9 With the same symbols as in the preceding problem, is the equation $v = \sqrt{2gr}$ dimensionally consistent?

1.10 Suppose that a certain (fictitious) quantity has dimension $L^2 M^3/T^4$, and that one quix = 1 m² · kg³/s⁴. The corresponding unit in U.S. units is a quax. How many quix in a quax?

1.11 Determine the units of the universal gravitation constant G, using the fact that the gravity force is expressible as Gm_1m_2/r^2. Roughly calculate the value of G using your own weight and mass, and the fact that for the earth, (a) radius ≈ 3960 miles, and (b) average specific gravity = 5.51. (The specific gravity of a material is the ratio of its density to that of water.)

1.12 In studying the flow of a fluid around an object (such as an airplane wing), there arises a quantity called the drag coefficient, which is defined as $C_F = F/(\frac{1}{2}\rho v^2 L^2)$. In this equation, F is the force on the object in the direction of the flow, ρ is the mass density of the fluid, v is its velocity, and L is a characteristic length of the object. Show that C_F is a dimensionless parameter.

1.13 The unit of stress in the SI system of units is the Pascal (Pa). If a material has an elastic modulus of 30×10^6 psi (pounds per square inch), what is its elastic modulus in (a) Pascals? (b) kilopascals (kPa)? (c) megapascals (MPa)? (d) gigapascals (GPa)? Note: 1 kPa = 1000 Pa, 1 MPa = 10^6 Pa, 1 GPa = 10^9 Pa.

1.14 A spring has a modulus of 30 lb/ft. What is its modulus in N/m?

1.15 If the measure of a quantity is known to three significant figures, what is the maximum percentage of uncertainty?

1.16 A rectangular parallelepiped has sides of lengths 2.00 m, 3.00 m, and 4.00 m. Another has sides 2.02 m, 3.03 m, and 4.04 m. Find the difference in their volumes. To how many significant figures is the difference known?

* **1.17** Each edge of a cube is increased in length by 0.002%. What is the percent increase in volume of the cube? Notice the number of significant figures required to calculate the change in volume if it is done by calculating a numerical value for the new volume and then subtracting the old. Repeat the problem for an increase of 2×10^{-6} %. Try to find a way to avoid this "small difference of large numbers" problem.

---

* Asterisks identify the more difficult problems.

2 Forces and Particle Equilibrium

2.1 **Introduction**

2.2 **Forces**

Forces as Vectors

Unit Vectors and Orthogonal (Mutually Perpendicular) Components

Dot Product to Find Components

2.3 **Equilibrium of a Particle**

The Free-Body Diagram

Analysis of Equilibrium

2.4 **Equilibrium of a System of Particles**

Summary

Review Questions

2.1 Introduction

To understand the subject of statics, one must master two fundamental concepts: (a) forces, and (b) moments of forces (i.e., turning effects of forces). In this chapter, we will study the first of these, the force. The reader will see that a force is just what one thinks it is — a "push" or "pull," but that with the help of vectors we can actually represent forces in a simple mathematical form.

Having learned how to express forces as vectors in Section 2.2, we will then (in Section 2.3) learn to use the force vectors in equilibrium equations to solve some simple engineering problems involving a single particle. In the process, we will also encounter a critical concept called the "free-body diagram," which is a sketch of the object to be analyzed, showing clearly all externally-applied forces acting upon it.

Finally, in Section 2.4, we expand our study of forces to include the case of more than one particle in equilibrium.

2.2 Forces

A **force** is a mechanical action exerted by one physical body on another. Very simply it is what we perceive as a "push" or a "pull." A force has *magnitude* which is given numerical value in a system of units as described in Section 1.4. Most commonly the unit would be the pound (lb) or the newton (N). A force also has *direction,* which we can describe using the tools of geometry such as angles.

It is usually natural to think of the action of a force on a body to be distributed over a surface or a volume. Push on a table with your finger and the action is distributed over the fingerprint. Sometimes, however, the action is sufficiently localized that it makes sense to characterize its place of application by a single point on the body; this is called the **point of application** of the force. For the time being we shall adopt this viewpoint, leaving the details of the "fingerprint" until later.

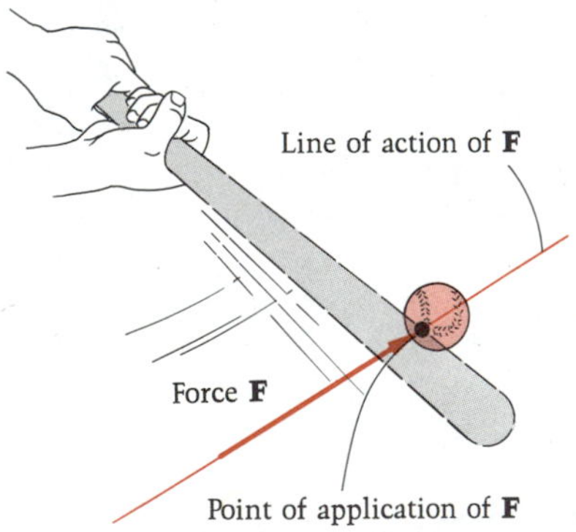

Figure 2.1 Line of action of a force.

The **line of action** of a force is the line in space that passes through the point of application and has the same direction as the force. This concept is illustrated in Figure 2.1 where the force exerted on a ball by a bat is shown. We shall see that the line of action of a force plays a central role in mechanics.

Forces as Vectors A **vector*** is the mathematical entity by which a force is represented. The magnitude-and-direction qualities of a vector can be displayed graphically by an "arrow" — that is, a directed line segment whose length is proportional to the magnitude of the vector. A defining property of vectors is that they satisfy the **parallelogram law of addi-**

* See Appendix A for a more formal treatment of vector algebra.

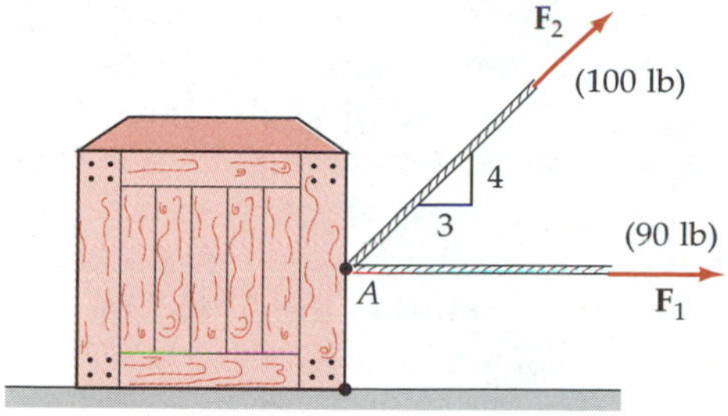

Figure 2.2

tion. To illustrate this law, first consider Figure 2.2 where is shown a crate and the vectors representing two forces exerted on it by cables. Note that we use bold type to denote vectors. Figure 2.3(a) shows the application of the parallelogram law to form $\mathbf{F}_1$ and $\mathbf{F}_2$, their sum being labeled $\mathbf{F}_3$. It's possible to find the magnitude of $\mathbf{F}_3$ by laying out the parallelogram with a ruler and then measuring the diagonal. Or we can observe that the 4-to-3 slope of the line of action of $\mathbf{F}_2$ means that it makes with the horizontal and vertical the angles of a 3-4-5 triangle. Thus in Figure 2.3(a)

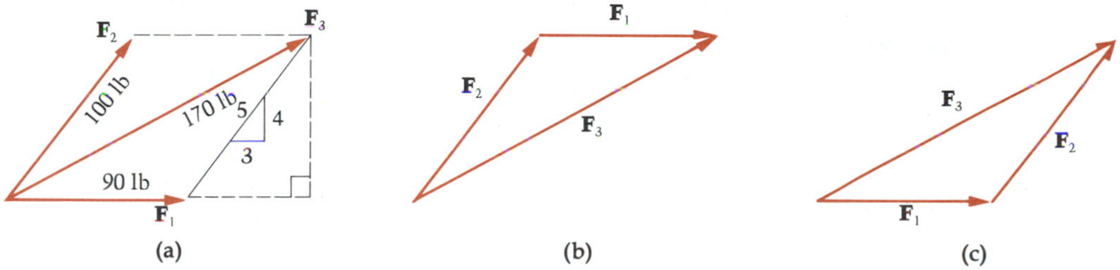

Figure 2.3

we can see a right triangle whose base is of "length"

$$90 + \frac{3}{5}(100) = 150 \text{ lb}$$

and whose "height" is

$$\frac{4}{5}(100) = 80 \text{ lb}$$

The magnitude of $\mathbf{F}_3$ is then the "hypotenuse" of this triangle and is given by

$$\sqrt{(150)^2 + (80)^2} = 170 \text{ lb.}$$

Equivalent head-to-tail representations of the sum of the forces are shown in Figure 2.3(b) and (c).

The legitimacy of representing forces by vectors rests upon the experimental evidence that the effect of two forces simultaneously applied (and having a common point of application) is the same as the effect that arises from a single force, related to the first two by the parallelogram law. However, it is also important to point out that the *mathematical* operation of summing the two vectors representing the forces does not depend in any way upon the forces having a common point of application; only their magnitudes and directions come into play.

In our first example we further illustrate the addition of forces by the parallelogram law.

EXAMPLE 2.1

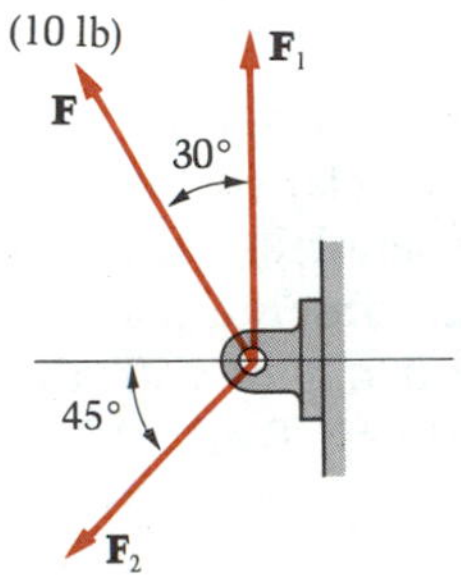

Figure E2.1a

Find the magnitudes of forces $\mathbf{F}_1$ and $\mathbf{F}_2$ having the directions shown in Figure E2.1a so that their sum is the force **F**, whose magnitude is 10 lb. All three forces lie in the plane of the paper.

Solution

The head-to-tail version of the parallelogram law of addition is shown in the sketch.

One way to solve this problem is by the Law of Sines; that is, using the triangle in Figure E2.1b,

$$\frac{\sin 45°}{10} = \frac{\sin 30°}{|\mathbf{F}_2|} = \frac{\sin 105°}{|\mathbf{F}_1|}$$

where $|\mathbf{F}_1|$ and $|\mathbf{F}_2|$ denote the magnitudes of forces $\mathbf{F}_1$ and $\mathbf{F}_2$. Thus

$$|\mathbf{F}_2| = \frac{10(0.5)}{0.707} = 7.07 \text{ lb}$$

$$|\mathbf{F}_1| = \frac{10}{0.707}(0.966) = 13.7 \text{ lb}$$

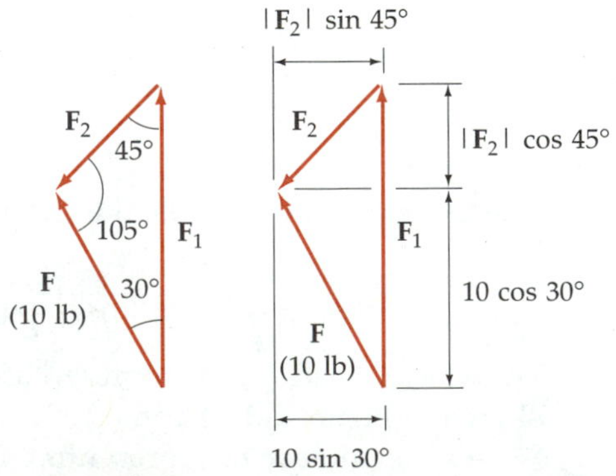

Figure E2.1b

An alternative approach is to observe that:

1. The horizontal projection of $|\mathbf{F}_2|$ must be equal to the horizontal projection of 10; that is,

$$|\mathbf{F}_2| \sin 45° = 10 \sin 30°$$

$$|\mathbf{F}_2| = \frac{10(0.5)}{0.707} = 7.07 \text{ lb}$$

2. $|\mathbf{F}_1|$ is the sum of the vertical projections of 10 and $|\mathbf{F}_2|$; that is,

$$\begin{aligned} |\mathbf{F}_1| &= 10 \cos 30° + |\mathbf{F}_2| \cos 45° \\ &= 10(0.866) + (7.07)(0.707) \\ &= 8.66 + 5 \\ &= 13.7 \text{ lb} \end{aligned}$$

This second approach is closely associated with the concept of orthogonal components of a force, which will be discussed after this example.

A third approach that could be used for this problem is graphic. That is, we could use a scale, a straightedge, and a protractor to draw the "force triangle" shown. The student is encouraged to do this and then to think about the effects of measurement errors on the accuracy of a solution by this method.

Unit Vectors and Orthogonal (Mutually Perpendicular) Components The directionality of a vector is easily communicated by the graphical means so far shown in this section when we are working in a plane (two-dimensional space). This is often awkward in three-dimensional space and, besides, it's useful to have some formal tools which can be used for mathematical manipulations. **Unit vectors** are the *direction indicators* of vector algebra. Such a vector, as the name suggests, has a *magnitude of unity* and it is *dimensionless.* It is a way of labeling some preassigned direction relative, of course, to some physical reference body. Suppose we let x, y, and z be mutually perpendicular axes, or reference directions, and we let $\hat{\mathbf{i}}$, $\hat{\mathbf{j}}$, and $\hat{\mathbf{k}}$ be dimensionless unit vectors* parallel, respectively, to those directions as shown in Figure 2.4. Two applications of the parallelogram law (using the shaded plane first) allow us to decompose the force **F** into three mutually perpendicular parts written $F_x\hat{\mathbf{i}}$, $F_y\hat{\mathbf{j}}$, and $F_z\hat{\mathbf{k}}$ so that, as suggested by Figure 2.4,

$$\mathbf{F} = F_x\hat{\mathbf{i}} + F_y\hat{\mathbf{i}} + F_z\hat{\mathbf{k}} \tag{2.1}$$

$F_x\hat{\mathbf{i}}$, $F_y\hat{\mathbf{j}}$, and $F_z\hat{\mathbf{k}}$ are called orthogonal (or rectangular) **vector components of F**, and F_x, F_y, and F_z are called the corresponding **scalar compo-**

* In this book a caret, or "hat," over a bold lower-case letter signifies that the vector is a unit vector. All unit vectors that we use are dimensionless. Throughout the book, the unit vectors ($\hat{\mathbf{i}}$, $\hat{\mathbf{j}}$, $\hat{\mathbf{k}}$) are always parallel, respectively, to the assigned directions of (x, y, z).

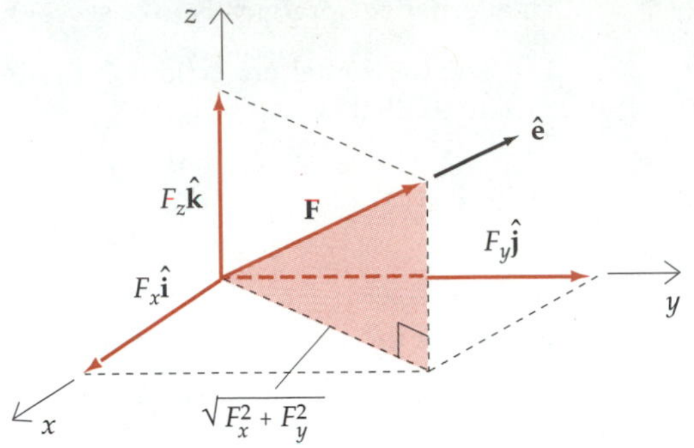

Figure 2.4

nents.* Referring to Figure 2.4, we see that the magnitude, $|\mathbf{F}|$, of $\mathbf{F}$ is given by

$$|\mathbf{F}| = \sqrt{(\sqrt{F_x^2 + F_y^2})^2 + F_z^2} = \sqrt{F_x^2 + F_y^2 + F_z^2} \tag{2.2}$$

where $\sqrt{F_x^2 + F_y^2}$ is itself the magnitude of the component of $\mathbf{F}$ in the xy plane. (Thus the components of a force need not be associated with coordinate directions.) We can speak of the component in a plane, or normal to a plane, or along a skewed line, and so on.

Sometimes we shall need to write a force as the product of its magnitude $|\mathbf{F}|$ and a unit vector $\hat{\mathbf{e}}$ (as shown in Figure 2.4) in its direction:

$$\mathbf{F} = |\mathbf{F}|\hat{\mathbf{e}} \tag{2.3}$$

Both Equations (2.1) and (2.3) are very important in the study of statics. It is also important to realize that the scalar components of $\hat{\mathbf{e}}$ are the cosines of the angles (or direction cosines) that $\mathbf{F}$ makes with the positive x, y, and z axes:

$$\begin{aligned}\mathbf{F} &= |\mathbf{F}|(e_x\hat{\mathbf{i}} + e_y\hat{\mathbf{j}} + e_z\hat{\mathbf{k}}) \\ &= |\mathbf{F}|[(\cos\theta_x)\hat{\mathbf{i}} + (\cos\theta_y)\hat{\mathbf{j}} + (\cos\theta_z)\hat{\mathbf{k}}]\end{aligned}$$

A comment about notation as it relates to figures is in order here. Sometimes the figures show an arrow labeled with a bold letter denoting a vector. The purpose of this is to display a vector pictorially, usually to depict some general relationship. At other times the figures show an arrow labeled with a scalar. In these instances we are communicating that *the vector in question is expressed by the scalar multiplying a unit vector in the direction of the arrow.* The examples that follow illustrate the use of this "code."

* Sometimes in this book we refer to "components" without an adjective; in such instances it should be clear from the context which orthogonal components, scalar or vector, are intended.

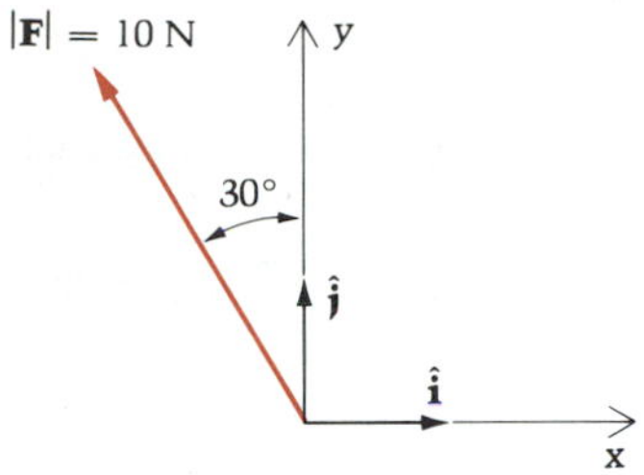

Figure E2.2a

Example 2.2

A force **F** of magnitude 10 newtons (N) is depicted in Figure E2.2a. Express the vector in component form using the reference directions x, y, and z.

Solution

$$\theta_x = 90° + 30° = 120°$$
$$\theta_y = 30°$$
$$\theta_z = 90°$$

Therefore,

$$\cos\theta_x = \cos 120° = -0.5$$
$$\cos\theta_y = \cos 30° = 0.866$$
$$\cos\theta_z = \cos 90° = 0$$

and so

$$F_x = 10(-0.5) = -5 \text{ N}$$
$$F_y = 10(0.866) = 8.66 \text{ N}$$
$$F_z = 10(0) = 0$$

The force is therefore expressible as:

$$\begin{aligned}\mathbf{F} &= F_x\hat{\mathbf{i}} + F_y\hat{\mathbf{j}} + F_z\hat{\mathbf{k}}\\ &= -5\hat{\mathbf{i}} + 8.66\hat{\mathbf{j}} \text{ N}\end{aligned}$$

The same result may be obtained by decomposing **F** as shown in Figure E2.2b. Thus we see that, because a unit vector to the left is $(-\hat{\mathbf{i}})$ and a unit vector upward is $\hat{\mathbf{j}}$,

$$\begin{aligned}\mathbf{F} &= 10(0.5)(-\hat{\mathbf{i}}) + 10(0.866)\hat{\mathbf{j}}\\ &= -5\hat{\mathbf{i}} + 8.66\hat{\mathbf{j}} \text{ N}\end{aligned}$$

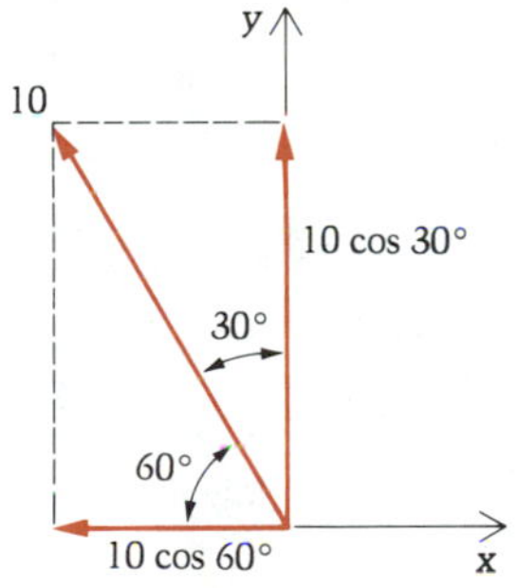

Figure E2.2b

Consequently, by picking off the coefficients of $\hat{\mathbf{i}}$, $\hat{\mathbf{j}}$, and $\hat{\mathbf{k}}$,

$$F_x = -5 \text{ N}$$
$$F_y = 8.66 \text{ N}$$
$$F_z = 0 \text{ N}$$

are seen again to be the scalar components.

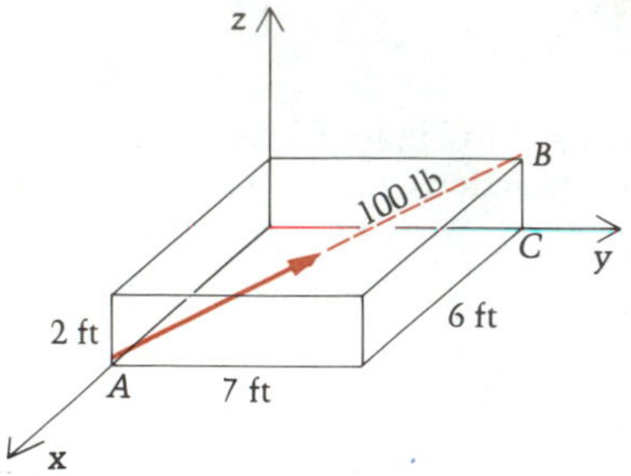

Figure E2.3

EXAMPLE 2.3

Find the components of the force of 100 lb having as its line of action the diagonal of the rectangular solid as shown in Figure E2.3.

Solution

A vector in the direction of the force is the vector from A to B*:

$$\mathbf{r}_{AB} = -6\hat{\mathbf{i}} + 7\hat{\mathbf{j}} + 2\hat{\mathbf{k}} \text{ ft}$$

The unit vector parallel to $\mathbf{r}_{AB}$ is $\mathbf{r}_{AB}/|\mathbf{r}_{AB}|$, or $\hat{\mathbf{e}}_{AB}$:

$$\hat{\mathbf{e}}_{AB} = \frac{-6\hat{\mathbf{i}} + 7\hat{\mathbf{j}} + 2\hat{\mathbf{k}}}{\sqrt{6^2 + 7^2 + 2^2}} = -0.636\hat{\mathbf{i}} + 0.742\hat{\mathbf{j}} + 0.212\hat{\mathbf{k}}$$

The reader should note that $\hat{\mathbf{e}}_{AB}$ has unit magnitude and is dimensionless. Now, writing the force as a vector in the form of its magnitude times the unit vector in its direction, we have

$$\mathbf{F} = 100(-0.636\hat{\mathbf{i}} + 0.742\hat{\mathbf{j}} + 0.212\hat{\mathbf{k}}) \text{ lb}$$

$$\mathbf{F} = -63.6\hat{\mathbf{i}} + 74.2\hat{\mathbf{j}} + 21.2\hat{\mathbf{k}} \text{ lb}$$

and so the scalar components of $\mathbf{F} = F_x\hat{\mathbf{i}} + F_y\hat{\mathbf{j}} + F_z\hat{\mathbf{k}}$ are

$$F_x = -63.6 \text{ lb}$$

$$F_y = 74.2 \text{ lb}$$

$$F_z = 21.2 \text{ lb}$$

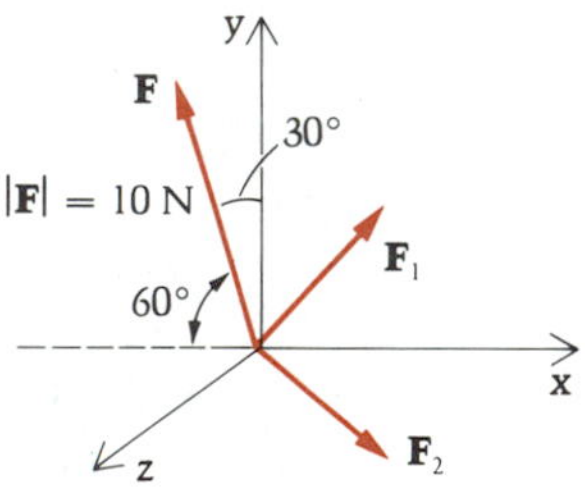

Figure E2.4

EXAMPLE 2.4

Express the force of Example 2.2, $\mathbf{F} = -5\hat{\mathbf{i}} + 8.66\hat{\mathbf{j}}$ N, as the sum of one force making equal angles with x, y, and z and a second force whose direction is in the xz plane. (See Figure E2.4.)

Solution

The first force, $\mathbf{F}_1$, may be expressed as

$$\mathbf{F}_1 = F_1\hat{\mathbf{e}}_1$$

where $\hat{\mathbf{e}}_1$ is a unit vector in the preassigned direction.

To find $\hat{\mathbf{e}}_1$, we write a vector in the direction making equal angles with x, y, and z, such as $\hat{\mathbf{i}} + \hat{\mathbf{j}} + \hat{\mathbf{k}}$, and then divide it by its magnitude:

$$\hat{\mathbf{e}}_1 = \frac{\hat{\mathbf{i}} + \hat{\mathbf{j}} + \hat{\mathbf{k}}}{\sqrt{1^2 + 1^2 + 1^2}}$$

$$\hat{\mathbf{e}}_1 = \frac{1}{\sqrt{3}}\hat{\mathbf{i}} + \frac{1}{\sqrt{3}}\hat{\mathbf{j}} + \frac{1}{\sqrt{3}}\hat{\mathbf{k}}$$

Now the scalar F_1 may be positive or negative, but, since $\hat{\mathbf{e}}_1$ is a unit vector, the absolute value of F_1 is the magnitude, $|\mathbf{F}_1|$, of $\mathbf{F}_1$.

---

* To get a vector from point A to point B, we merely start at A and write down "what we have to do" to get to B; thus, in this case, we travel $-6\hat{\mathbf{i}}$ to get to the origin plus $7\hat{\mathbf{j}}$ to get to C, plus $2\mathbf{k}$ to move finally up to B.

The only thing we know about the second force is that it has no component in the y direction. Therefore, without any loss of generality we may express $\mathbf{F}_2$ as

$$\mathbf{F}_2 = F_{2x}\hat{\mathbf{i}} + F_{2z}\hat{\mathbf{k}}$$

Now, setting the sum of $\mathbf{F}_1$ and $\mathbf{F}_2$ equal to $\mathbf{F}$:

$$\mathbf{F}_1 + \mathbf{F}_2 = -5\hat{\mathbf{i}} + 8.66\hat{\mathbf{j}}$$

Therefore

$$F_1(0.577\hat{\mathbf{i}} + 0.577\hat{\mathbf{j}} + 0.577\hat{\mathbf{k}}) + F_{2x}\hat{\mathbf{i}} + F_{2z}\hat{\mathbf{k}} = -5\hat{\mathbf{i}} + 8.66\hat{\mathbf{j}}$$

Equating the respective coefficients of $\hat{\mathbf{i}}$, $\hat{\mathbf{j}}$, and $\hat{\mathbf{k}}$, we have

$$\hat{\mathbf{i}}:\quad 0.577\,F_1 + F_{2x} = -5$$
$$\hat{\mathbf{j}}:\quad 0.577\,F_1 = 8.66$$
$$\hat{\mathbf{k}}:\quad 0.577\,F_1 + F_{2z} = 0$$

from which

$$F_1 = 8.66/0.577 = 15.0 \text{ N}$$
$$F_{2z} = -0.577\,F_1 = -8.66 \text{ N}$$
$$F_{2x} = -5 - 0.577\,F_1 = -5 - 8.66 = -13.7 \text{ N}$$

Therefore, the required forces are

$$\mathbf{F}_1 = 8.66\hat{\mathbf{i}} + 8.66\hat{\mathbf{j}} + 8.66\hat{\mathbf{k}} \text{ N}$$
$$\mathbf{F}_2 = -13.7\hat{\mathbf{i}} - 8.66\hat{\mathbf{k}} \text{ N}$$

Throughout this book, we have inserted questions for the reader to think about. The answer to each may be found below the Example or at the bottom of the page. The first question follows:

Question 2.1 Are there sources of force other than those from direct pushes or pulls — that is, those involving physical contact?

Dot Product to Find Components As the reader is perhaps aware, the **dot product** (or **scalar product**) of two vectors can be used to find the orthogonal component of one of them in the direction of the other. The dot product of two vectors $\mathbf{F}$ and $\mathbf{Q}$ is defined by

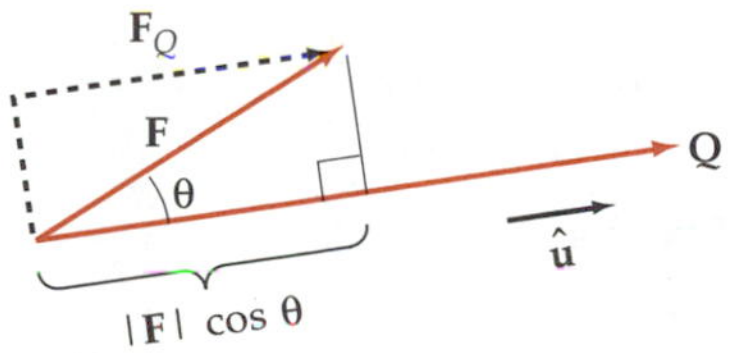

Figure 2.5

$$\mathbf{F} \cdot \mathbf{Q} = |\mathbf{F}||\mathbf{Q}|\cos\theta = \mathbf{Q} \cdot \mathbf{F} \tag{2.4}$$

where θ is the angle between $\mathbf{F}$ and $\mathbf{Q}$ in their plane. Thus if $\mathbf{F}$ represents a force and we wish to find its component (see Figure 2.5) in the direction of $\mathbf{Q}$, we just dot $\mathbf{F}$ with the unit vector $\hat{\mathbf{u}}$ in the direction of $\mathbf{Q}$, which is $\hat{\mathbf{u}} = \mathbf{Q}/|\mathbf{Q}|$, and obtain

$$\mathbf{F} \cdot \hat{\mathbf{u}} = |\mathbf{F}|(1)\cos\theta$$

Answer 2.1 Yes. Gravity and electromagnetic forces are two such examples.

which is the desired projection. Therefore the vector rectangular component of **F** in the direction of **Q** is

$$(\mathbf{F} \cdot \hat{\mathbf{u}})\hat{\mathbf{u}} = \mathbf{F}_Q \tag{2.5}$$

We now develop a useful expression for the dot product of two vectors when they are expressed in component form. We have:

$$\begin{aligned}\mathbf{F} \cdot \mathbf{Q} &= (F_x\hat{\mathbf{i}} + F_y\hat{\mathbf{i}} + F_z\hat{\mathbf{k}}) \cdot (Q_x\hat{\mathbf{i}} + Q_y\hat{\mathbf{i}} + Q_z\hat{\mathbf{k}}) \\ &= F_xQ_x(\hat{\mathbf{i}} \cdot \hat{\mathbf{i}}) + F_xQ_y(\hat{\mathbf{i}} \cdot \hat{\mathbf{j}}) + F_xQ_z(\hat{\mathbf{i}} \cdot \hat{\mathbf{k}}) \\ &\quad + F_yQ_x(\hat{\mathbf{j}} \cdot \hat{\mathbf{i}}) + F_yQ_y(\hat{\mathbf{j}} \cdot \hat{\mathbf{j}}) + F_yQ_z(\hat{\mathbf{j}} \cdot \hat{\mathbf{k}}) \\ &\quad + F_zQ_x(\hat{\mathbf{k}} \cdot \hat{\mathbf{i}}) + F_zQ_y(\hat{\mathbf{k}} \cdot \hat{\mathbf{j}}) + F_zQ_z(\hat{\mathbf{k}} \cdot \hat{\mathbf{k}})\end{aligned}$$

But Equation (2.4) gives

$$\hat{\mathbf{i}} \cdot \hat{\mathbf{i}} = \hat{\mathbf{j}} \cdot \hat{\mathbf{j}} = \hat{\mathbf{k}} \cdot \hat{\mathbf{k}} = (1)(1)\cos 0° = 1$$

and

$$\hat{\mathbf{i}} \cdot \hat{\mathbf{j}} = \hat{\mathbf{j}} \cdot \hat{\mathbf{k}} = \hat{\mathbf{i}} \cdot \hat{\mathbf{k}} = (1)(1)\cos 90° = 0$$

Therefore

$$\mathbf{F} \cdot \mathbf{Q} = F_xQ_x + F_yQ_y + F_zQ_z$$

We shall now use the dot product in an example to find a component of a force.

EXAMPLE 2.5

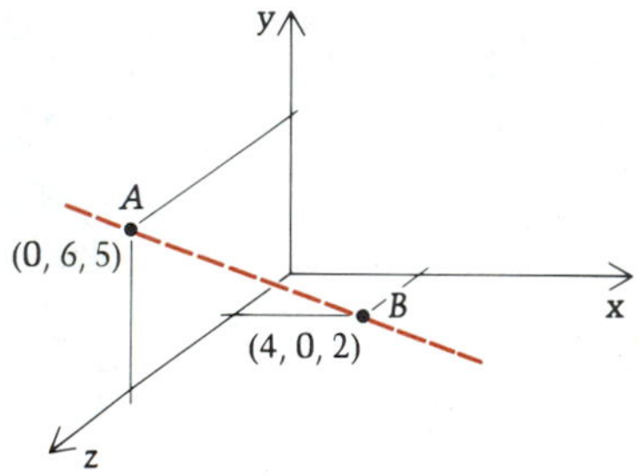

Figure E2.5

Given the forces

$$\mathbf{F}_1 = 2\hat{\mathbf{i}} + 3\hat{\mathbf{j}} - 4\hat{\mathbf{k}}\ \text{N}$$

and

$$\mathbf{F}_2 = \hat{\mathbf{i}} - 2\hat{\mathbf{j}} + 5\hat{\mathbf{k}}\ \text{N}$$

find the component of $\mathbf{F}_1 + \mathbf{F}_2$ in the direction of the line through the points whose rectangular coordinates are (0, 6, 5,) and (4, 0, 2) m as shown in Figure E2.5.

Solution

$$\begin{aligned}\mathbf{F}_1 + \mathbf{F}_2 &= (2 + 1)\hat{\mathbf{i}} + (3 - 2)\hat{\mathbf{j}} + (-4 + 5)\hat{\mathbf{k}} \\ &= 3\hat{\mathbf{i}} + \hat{\mathbf{j}} + \hat{\mathbf{k}}\ \text{N}\end{aligned}$$

We next construct the directed line segment (vector whose dimension is length) from point *A* to point *B*.

$$\mathbf{r}_{AB} = (4 - 0)\hat{\mathbf{i}} + (0 - 6)\hat{\mathbf{j}} + (2 - 5)\hat{\mathbf{k}} = 4\hat{\mathbf{i}} - 6\hat{\mathbf{j}} - 3\hat{\mathbf{k}}\ \text{m}$$

A unit vector $\hat{\mathbf{e}}_{AB}$ in the direction of the line is then

$$\begin{aligned}\hat{\mathbf{e}}_{AB} &= \frac{\mathbf{r}_{AB}}{|\mathbf{r}_{AB}|} = \frac{4\hat{\mathbf{i}} - 6\hat{\mathbf{j}} - 3\hat{\mathbf{k}}}{\sqrt{(4)^2 + (6)^2 + (3)^2}} = \frac{1}{\sqrt{61}}(4\hat{\mathbf{i}} - 6\hat{\mathbf{j}} - 3\hat{\mathbf{k}}) \\ &= 0.512\hat{\mathbf{i}} - 0.768\hat{\mathbf{j}} - 0.384\hat{\mathbf{k}}\end{aligned}$$

Therefore the scalar component of $\mathbf{F}_1 + \mathbf{F}_2$ associated with the direction of $\hat{\mathbf{e}}_{AB}$ is

$$(\mathbf{F}_1 + \mathbf{F}_2) \cdot \hat{\mathbf{e}}_{AB} = 3(0.512) + (1)(-0.768) + 1(-0.384) = 0.384 \text{ N}$$

The vector component along the line AB is

$$0.384\,\hat{\mathbf{e}}_{AB} = 0.384(0.512\hat{\mathbf{i}} - 0.768\hat{\mathbf{j}} - 0.384\hat{\mathbf{k}}) = 0.197\hat{\mathbf{i}} - 0.295\hat{\mathbf{j}} - 0.147\hat{\mathbf{k}} \text{ N}$$

Had we begun the analysis by forming

$$\hat{\mathbf{e}}_{BA} = \frac{\mathbf{r}_{BA}}{|\mathbf{r}_{BA}|} = -\hat{\mathbf{e}}_{AB}$$

we would have found the *scalar* component, $(\mathbf{F}_1 + \mathbf{F}_2) \cdot \hat{\mathbf{e}}_{BA}$, to be -0.384 N, but the *vector* component along AB is of course the same, because

$$[(\mathbf{F}_1 + \mathbf{F}_2) \cdot \hat{\mathbf{e}}_{AB}]\hat{\mathbf{e}}_{AB} = [(\mathbf{F}_1 + \mathbf{F}_2) \cdot \hat{\mathbf{e}}_{BA}]\hat{\mathbf{e}}_{BA}$$

PROBLEMS ► Section 2.2

2.1 Which force has the largest magnitude?

$$\mathbf{F}_1 = 2\hat{\mathbf{i}} + 3\hat{\mathbf{j}} + 6\hat{\mathbf{k}} \text{ N}$$
$$\mathbf{F}_2 = 9\hat{\mathbf{j}} \text{ N}$$
$$\mathbf{F}_3 = 3\hat{\mathbf{i}} - 7\hat{\mathbf{j}} + \sqrt{7}\hat{\mathbf{k}} \text{ N}$$

2.2 If $\mathbf{F}_1 = 5\hat{\mathbf{i}} + 6\hat{\mathbf{j}}$ lb and $\mathbf{F}_2 = 2\hat{\mathbf{i}} - 3\hat{\mathbf{j}} - 4\hat{\mathbf{k}}$ lb, find $\mathbf{F}_3$ so that the sum of the three forces is zero.

2.3 Prove that the sum of the magnitudes of two forces $\mathbf{F}_1$ and $\mathbf{F}_2$ is greater than or equal to the magnitude of their sum.

2.4 Express the 238-N force $\mathbf{F}$ in Figure P2.4 as a vector. Write it (a) as a magnitude times a unit vector in its direction; (b) in terms of its components parallel to $\hat{\mathbf{i}}$ and $\hat{\mathbf{j}}$.

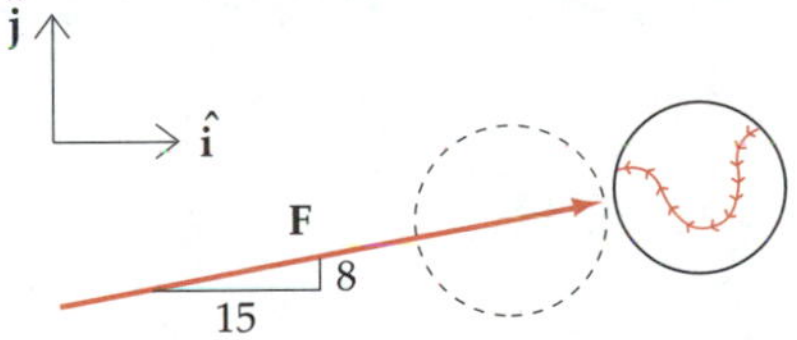

Figure P2.4

2.5 The orthogonal components of a certain force are

40 N in the positive x direction
50 N in the positive y direction
60 N in the negative z direction

a. What is the magnitude of the force?
b. What are its direction cosines?

2.6 A force has a magnitude of 100 lb and direction cosines $l = 0.7$, $m = -0.2$, $n = ?$ relative to an xyz frame of reference. Determine the orthogonal components of the force.

2.7 What is the unit vector in the direction of the force $6000\hat{\mathbf{i}} - 6000\hat{\mathbf{j}} + 7000\hat{\mathbf{k}}$ lb?

2.8 Find a force along $\hat{\mathbf{e}} = 0.8\hat{\mathbf{i}} - 0.6\hat{\mathbf{j}}$ and another force normal to $\hat{\mathbf{e}}$ that add up to the force $\mathbf{F} = 5\hat{\mathbf{i}} - 10\hat{\mathbf{j}} + 3\hat{\mathbf{k}}$ N.

2.9 Determine the component of the force in Problem 2.6 along a line having the direction cosines (−0.3, 0.1, 0.9487).

2.10 Obtain the dot product of the two vectors $\mathbf{F} = 10\hat{\mathbf{i}} + 6\hat{\mathbf{j}} - 3\hat{\mathbf{k}}$ lb and $\mathbf{B} = 6\hat{\mathbf{i}} - 2\hat{\mathbf{j}}$ ft.

2.11 Given the vectors $\mathbf{A} = 2\hat{\mathbf{i}} - 4\hat{\mathbf{j}}$ lb, $\mathbf{B} = 3\hat{\mathbf{j}} - 48\hat{\mathbf{k}}$ lb, and $\mathbf{C} = 3\hat{\mathbf{i}}$ (dimensionless), determine $\mathbf{C}(\mathbf{A} \cdot \mathbf{C}) + \mathbf{B}$.

2.12 Express the 500-N force making equal angles with x, y, and z (see Figure P2.12): (a) as a magnitude multiplied by a unit vector; (b) in terms of its orthogonal vector components.

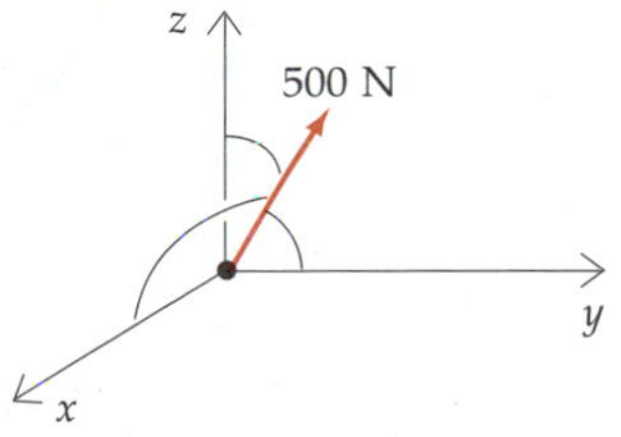

Figure P2.12

2.13 Find a unit vector in the xy plane that is perpendicular to the force $3\hat{\mathbf{i}} - 4\hat{\mathbf{j}} + 12\hat{\mathbf{k}}$ N.

2.14 Find the direction cosines of the force $\mathbf{F} = 30\hat{\mathbf{i}} + 40\hat{\mathbf{j}} - 120\hat{\mathbf{k}}$ lb, and use them to determine the angles the force makes with the coordinate axes (x, y, z).

2.15 For what value of F_y are the vectors $\mathbf{F}_1 = 3\hat{\mathbf{i}} + F_y\hat{\mathbf{j}} + 15\hat{\mathbf{k}}$ N and $\mathbf{F}_2 = 7\hat{\mathbf{i}} - 2\hat{\mathbf{j}} + 3\hat{\mathbf{k}}$ N orthogonal?

2.16 A force is given by $\mathbf{F} = 20\hat{\mathbf{i}} - 60\hat{\mathbf{j}} + 90\hat{\mathbf{k}}$ N. Find its magnitude and the angles it forms with the coordinate axes.

2.17 Given the forces $\mathbf{F}_1 = 6\hat{\mathbf{i}} + 10\hat{\mathbf{j}} + 16\hat{\mathbf{k}}$ lb, $\mathbf{F}_2 = 2\hat{\mathbf{i}} - 3\hat{\mathbf{j}}$ lb, and $\mathbf{F}_3$ = a third force in the xy plane at an inclination of 45° to both the negative y and positive x axes. The magnitude of $\mathbf{F}_3$ is 25 lb. Find (a) $\mathbf{F}_1 + \mathbf{F}_2 + \mathbf{F}_3$ and (b) $\mathbf{F}_1 - 2\mathbf{F}_2 + 3\mathbf{F}_3$.

2.18 If $\mathbf{F}_1 = 2\hat{\mathbf{i}} + 4\hat{\mathbf{j}}$ kN (kilonewton), $\mathbf{F}_2 = \hat{\mathbf{i}} - 2\hat{\mathbf{k}}$ kN, $\mathbf{F}_3 = \hat{\mathbf{i}} + \hat{\mathbf{j}} - 7\hat{\mathbf{k}}$ kN, and $\mathbf{F}_4 = 2\hat{\mathbf{i}} - 9\hat{\mathbf{j}} + 3\hat{\mathbf{k}}$ kN, determine scalars a, b, and c such that $\mathbf{F}_4 = a\mathbf{F}_1 + b\mathbf{F}_2 + c\mathbf{F}_3$.

2.19 If the tension in the guy wire AB in Figure P2.19 is 10 kN, find the tension in the other guy wire BC if the sum of the two tension forces exerted on the column at B is known to be vertical. Then find the sum of the two forces.

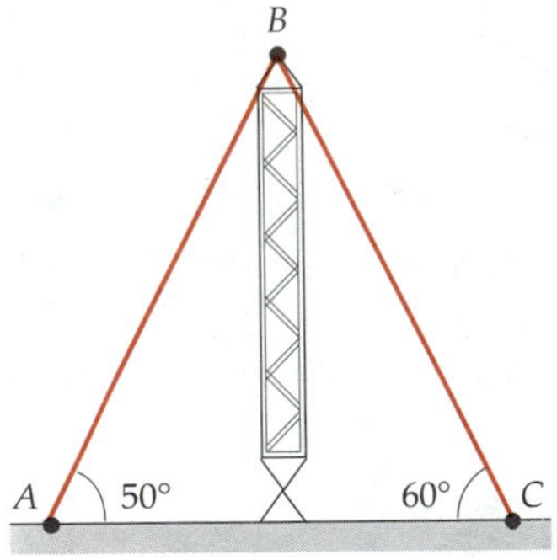

Figure P2.19

2.20 Find the angle between the forces $\mathbf{F}_1 = 2\hat{\mathbf{i}} + \hat{\mathbf{j}} - \hat{\mathbf{k}}$ N and $\mathbf{F}_2 = 5\hat{\mathbf{i}} - 6\hat{\mathbf{j}} + 8\hat{\mathbf{k}}$ lb.

2.21 Express the 20 kN force acting on the beam in Figure P2.21 in terms of its axial (parallel to the beam's axis, x) and transverse (normal to the beam's axis, y) components.

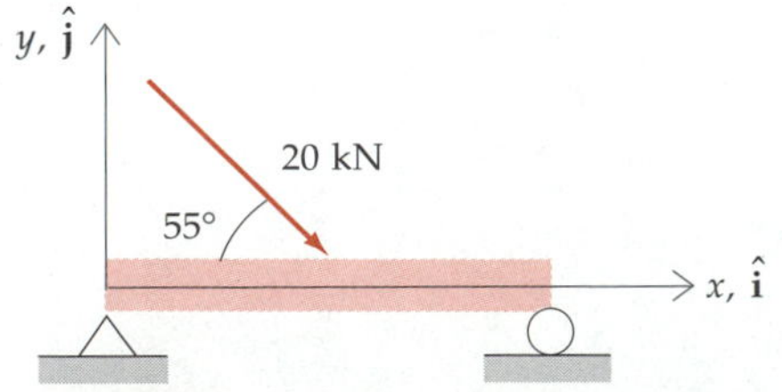

Figure P2.21

2.22 Show that if a, b, and c are nonvanishing scalars, and if $a\mathbf{F}_1 + b\mathbf{F}_2 + c\mathbf{F}_3 = \mathbf{0}$, then the three forces $\mathbf{F}_1$, $\mathbf{F}_2$, and $\mathbf{F}_3$ have lines of action in parallel planes.

2.23 Prove that $(\mathbf{A} \cdot \mathbf{B})^2$ is never greater than $|\mathbf{A}|^2|\mathbf{B}|^2$.

2.24 The cable BCA in Figure P2.24 passes smoothly through a hole at the end of the strut at C, and ties to the ground at A and to the end of the pole at B. If the tension in the cable is 800 lb, what is the force exerted by the cable onto the pole at B?

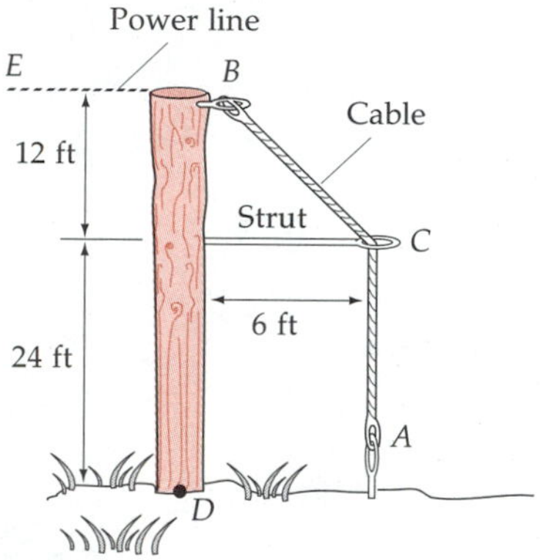

Figure P2.24

2.25 Resolve the force $\mathbf{F}$ into a part perpendicular to AB and a part parallel to BC. (See Figure P2.25.)

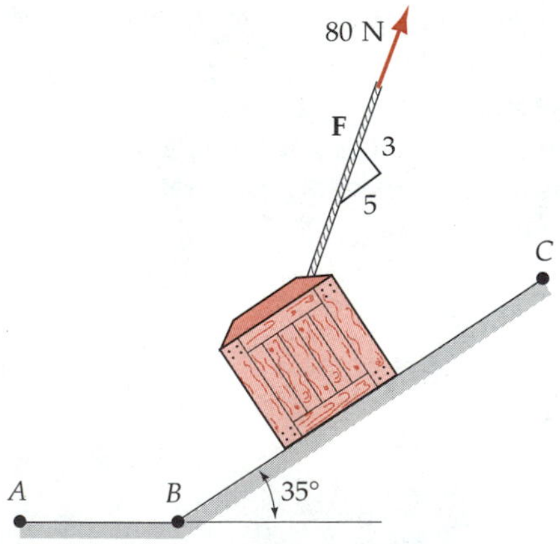

Figure P2.25

2.26 What are the orthogonal components of the 100-N force shown in Figure P2.26? What are the direction cosines associated with this force?

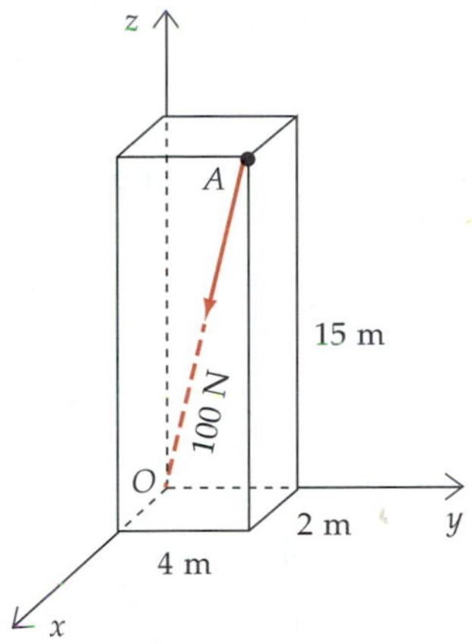

Figure P2.26

2.27 Express the force in Figure P2.27 (a) as a magnitude times a unit vector and (b) in terms of its components.

2.28 The x component of the force **P** in Figure P2.28 is 140 N to the left. Find **P**.

2.29 Show that the component of the downward force W in Figure P2.29 that is:

a. perpendicular to the inclined plane is $W \cos \theta$ (toward the plane) and

b. parallel to the plane is $W \sin \theta$ (down the plane). You will use this result many times in the study of engineering mechanics.

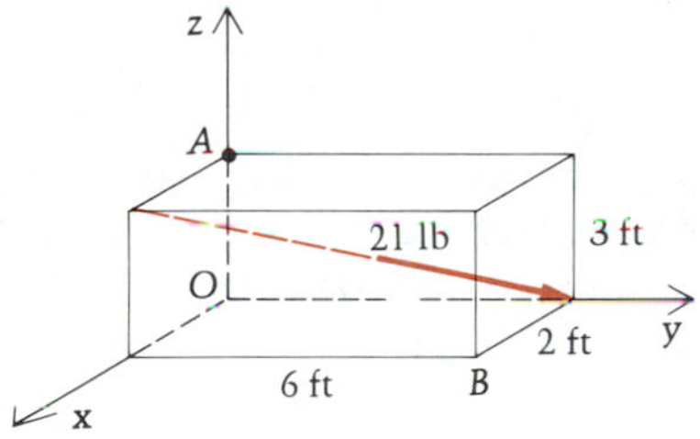

Figure P2.27

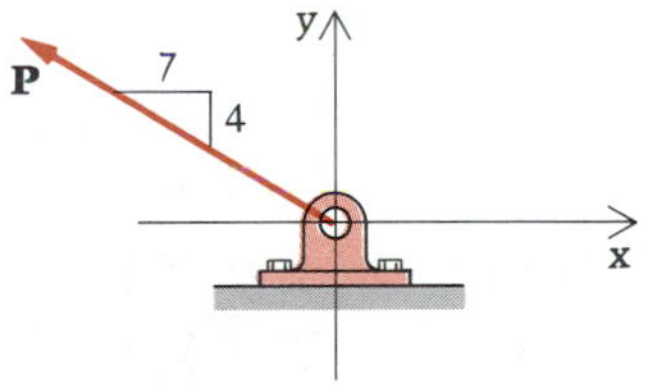

Figure P2.28

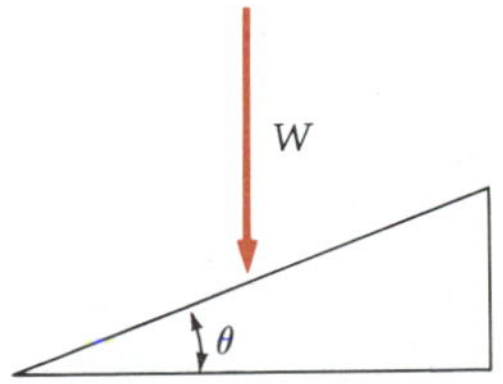

Figure P2.29

2.30 The girl in Figure P2.30 slowly pushes the lawnmower up the incline by exerting a 30-lb force **F** parallel to the handle as shown. Find the component of **F** (a) parallel to the incline; (b) normal to the incline; (c) parallel to the direction of gravity; (d) perpendicular to the direction of gravity.

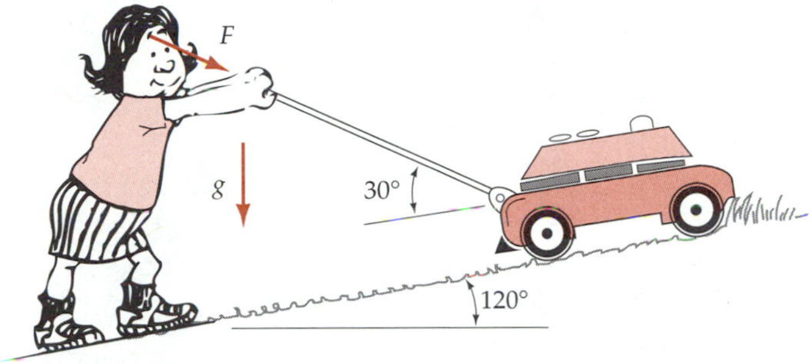

Figure P2.30

2.31 Resolve the 250-N force shown in Figure P2.31 into parts acting along the members PR and QR.

2.32 The x and z components of the force **F** in Figure P2.32 are known to be 100 N and −30 N, respectively. What is the force **F**, and what are its direction cosines?

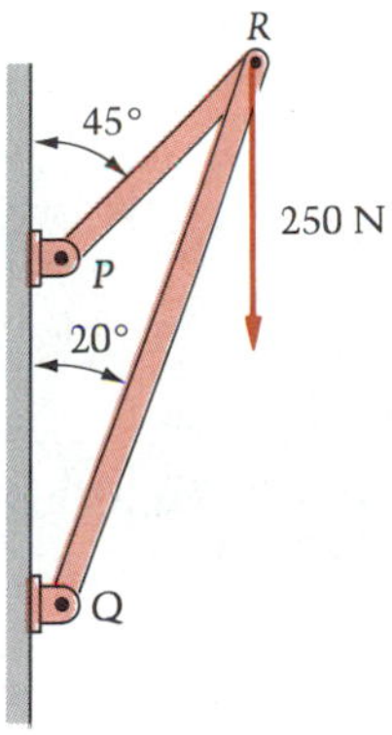

Figure P2.31

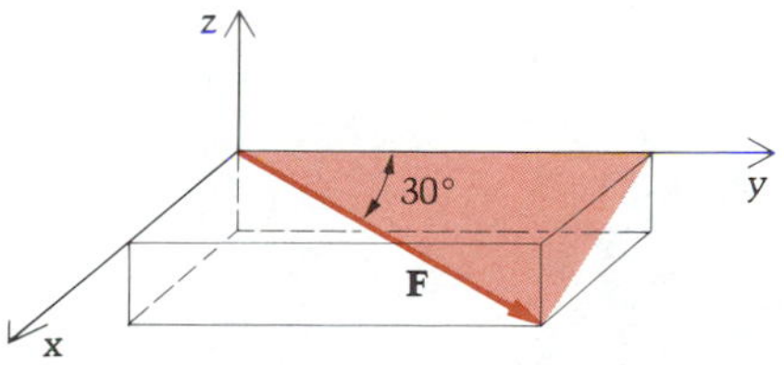

Figure P2.32

2.33 Resolve the 170-lb force, **F**, in Figure P2.33 into three parts—one of which is parallel to OQ, another parallel to OP, and the third parallel to the y axis. Are these the components of **F** in these directions?

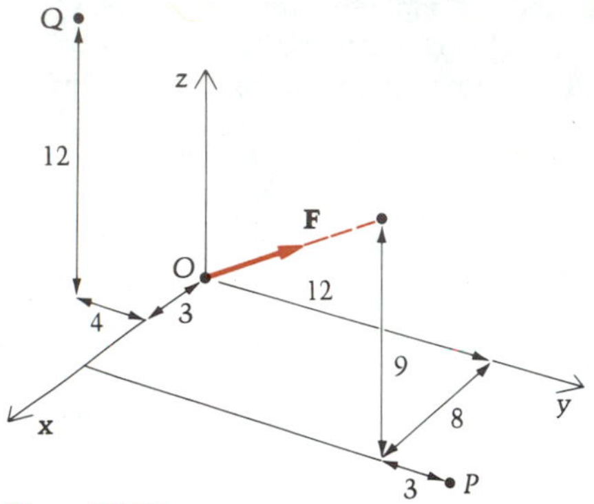

Figure P2.33

2.34 Two forces **P** and **Q** with respective magnitudes 100 and 200 N are applied to the upper corner of the crate in Figure P2.34. The sum of the two forces is a horizontal force to the right of magnitude 250 N. Find the angles that **P** and **Q** each make with their sum — that is, with the horizontal line through A.

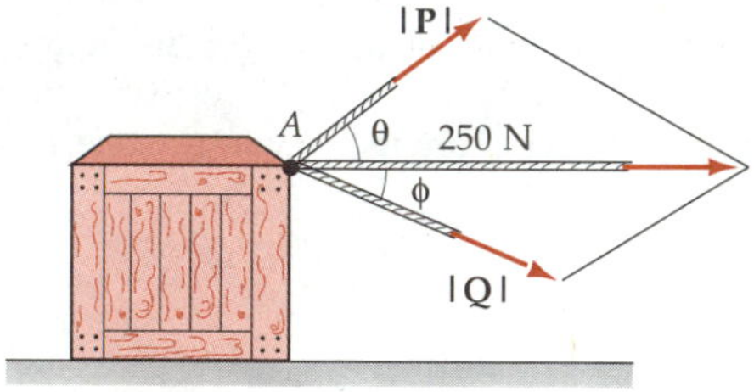

Figure P2.34

2.35 Find two 80-lb forces whose sum is the force $40\hat{\mathbf{i}}$ lb.

2.36 Given that $\mathbf{F}_1 = \hat{\mathbf{i}}$, $\mathbf{F}_2 = \hat{\mathbf{j}}$, $\mathbf{F}_3 = 3\hat{\mathbf{i}} - 4\hat{\mathbf{j}} + 5\hat{\mathbf{k}}$, $\mathbf{F}_4 = 6\hat{\mathbf{i}} - 4\hat{\mathbf{j}}$ lb, find a vector that is simultaneously in the plane of $\mathbf{F}_1$ and $\mathbf{F}_2$ and in the plane of $\mathbf{F}_3$ and $\mathbf{F}_4$.

2.37 Find all unit vectors that are perpendicular to each of the forces $\mathbf{F}_1 = \hat{\mathbf{i}} + 2\hat{\mathbf{j}} + 3\hat{\mathbf{k}}$ N and $\mathbf{F}_2 = 8\hat{\mathbf{i}} - 9\hat{\mathbf{j}} - 12\hat{\mathbf{k}}$ N.

2.38 Determine a unit vector in the plane of the forces $\hat{\mathbf{i}} + \hat{\mathbf{j}}$ lb and $\hat{\mathbf{j}} + \hat{\mathbf{k}}$ lb and simultaneously perpendicular to the force $\hat{\mathbf{i}} + \hat{\mathbf{j}} + \hat{\mathbf{k}}$ lb.

2.39 (a) The x-component of each of the four forces in Figure P2.39(a) is 10 lb. Which of the forces has the largest magnitude?

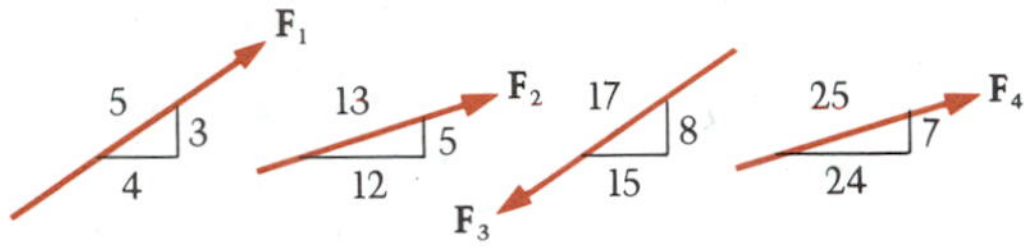

Figure P2.39(a)

(b) The vertical component of each of the six forces in Figure P2.39(b) is 1 N. Which force has the smallest magnitude?

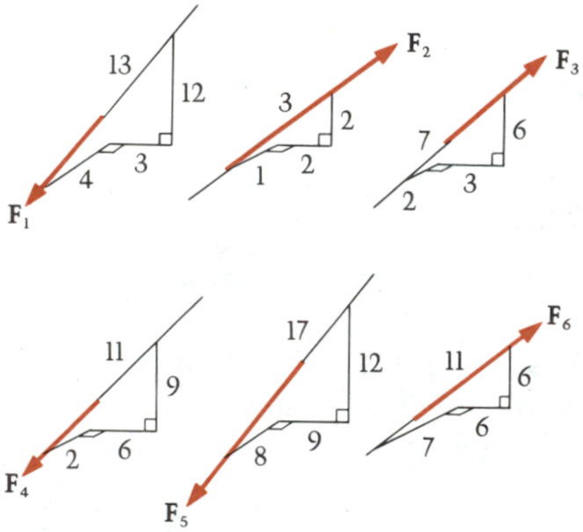

Figure P2.39(b)

2.40 Express the force of Examples 2.2 and 2.4, $\mathbf{F} = -5\hat{\mathbf{i}} + 8.66\hat{\mathbf{j}}$ N, as the sum of one force making equal angles with x, y, and $-z$, and a second force whose direction is in the yz plane.

2.3 Equilibrium of a Particle

When none of the particles (bits of material) making up a body is accelerating relative to an inertial frame of reference, Euler's laws (Section 1.3) take on particularly simple forms involving only the external forces acting on the body. The first of these two laws of **equilibrium** requires that the external forces exert no net thrust on the body. In equation form this is

$$\Sigma \mathbf{F} = \mathbf{0} \tag{2.6}$$

The second law is a statement that there be no net turning effect exerted by the external forces. To delve more deeply into this we need the concept of moment of a force, which is to come in the next chapter. For the remainder of this chapter we shall concern ourselves with simpler problems for which only Equation (2.6) is needed. These are problems for which the lines of action of the different external forces all intersect at a single point so that there can be no net turning effect and thus the actual size of the body is irrelevant. We often say that these are problems of *particle equilibrium,* a particle being a piece of material so small that we need not distinguish its different points.

In most engineering problems some of the external forces are known (or prescribed) before any analysis is carried out; we usually refer to these as **loads.** The external forces exerted by attached or supporting bodies are called **reactions;** usually we can think of these as forces that constrain the body against motion that the loads tend to produce.

The Free-Body Diagram The **free-body diagram** (FBD) is an extremely important and useful concept for the analysis of problems in mechanics. It is a figure, usually sketched, depicting and hence identifying precisely the body under consideration. On the FBD (Figure 2.6) we show, by arrows, all of the external forces which act on that body. Thus we have a catalog, graphically displayed, of all the forces that contribute to the equilibrium equations. Of particular importance is the fact that the free-body diagram provides us a way to express what we know about reactions (for example, that a certain reaction force has a known line of action) before applying the equations of equilibrium.

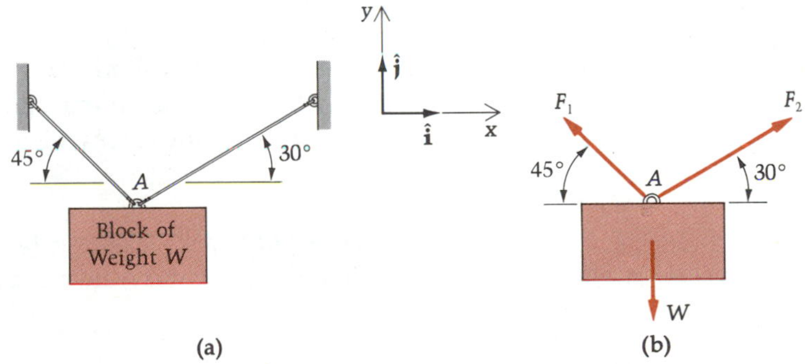

Figure 2.6

To illustrate the concept of a free-body diagram, consider the block shown in Figure 2.6a. It is supported (held in equilibrium) by two ropes or cables that counter the effect of gravity. The free-body diagram, Figure 2.6b, shows that three external forces act on the block. One of these is the weight W, which we show to have a line of action downward through the center of gravity of the block. That this single force can represent the cumulative effect of the distributed action of gravity will be shown in the next chapter. The other two forces, labeled F_1 and F_2, along with associated arrows represent the forces exerted *on* the block by the two cables. The directions of these forces are perceived to be along the (assumed) straight cables. The *arrow code* tells us that we are describing the force exerted on the block by the left cable by the expression

$$F_1(-\cos 45°\hat{\mathbf{i}} + \sin 45°\hat{\mathbf{j}})$$

Similarly the free-body diagram is conveying that the analytical description of the force exerted on the block by the right cable is to be

$$F_2\,(\cos 30°\hat{\mathbf{i}} + \sin 30°\hat{\mathbf{j}})$$

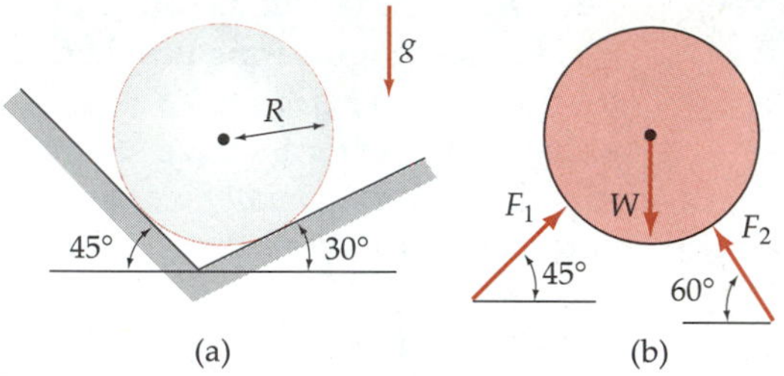

Figure 2.7

As a second illustration we consider the uniform sphere of weight W, supported by smooth (frictionless) surfaces as shown in Figure 2.7a. Because the contacting surfaces are smooth, the forces exerted on the sphere by the planes must be perpendicular to the surface. Thus we reason that the representations of those forces by F_1 and F_2 and the arrow codes are as depicted on the free-body diagram, Figure 2.7b.

Later on, in Chapter 4, when preparing to analyze general equilibrium problems involving the moment equation mentioned in Section 1.3 and non-concurrent force systems, we will dedicate an entire section of the book (4.2) to the free-body diagram. For now, we simply "sketch the particle" with all its forces intersecting at a point. Since all we are doing in this chapter is summing forces, displaying distances on the FBD is unimportant here.

Analysis of Equilibrium Let us now return to the block of Figure 2.6 and substitute into the equilibrium equation, $\Sigma\mathbf{F} = \mathbf{0}$, to obtain

$$F_1\,(-\cos 45°\hat{\mathbf{i}} + \sin 45°\hat{\mathbf{j}}) + F_2\,(\cos 30°\hat{\mathbf{i}} + \sin 30°\hat{\mathbf{j}}) + W(-\hat{\mathbf{j}}) = 0$$

or, because the $\hat{\mathbf{i}}$ and $\hat{\mathbf{j}}$ coefficients must separately add to zero, we get two equations in two unknowns:

$$-F_1 \cos 45° + F_2 \cos 30° = 0$$

and

$$F_1 \sin 45° + F_2 \sin 30° - W = 0$$

from which

$$F_1 = 0.897W$$

and

$$F_2 = 0.732W$$

We observe that there were only two independent scalar equations embodied in $\Sigma\mathbf{F} = \mathbf{0}$ because all the external forces have lines of action in a single plane and thus none of the forces has a component in the direction (z) perpendicular to that plane. In the language of Chapter 3, our force system here is coplanar and concurrent. Our two scalar equations express the vanishing of the sums of components in the x and y directions, respectively. We shall often write them using the notation

$$\Sigma F_x = 0 \quad \text{and} \quad \Sigma F_y = 0.$$

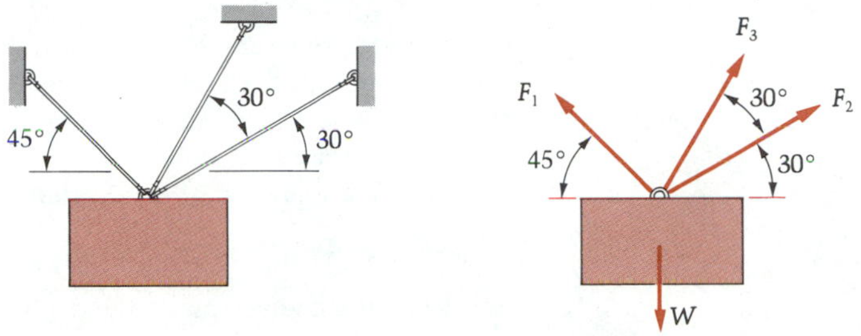

Figure 2.8

Before leaving this illustrative problem let us see what would happen if there were a third cable helping to support the block as shown in Figure 2.8. The force system is still coplanar so that there still will be only two independent equations, now relating F_1, F_2, and F_3. (The symbol $\xrightarrow{+}$ is a reminder that the unit vector in that direction has been suppressed from each term in the equation to follow. Some call it the "positive direction" for the force summation.)

$$\xrightarrow{+} \qquad \Sigma F_x = 0$$

$$-F_1 \cos 45° + F_2 \cos 30° + F_3 \cos 60° = 0$$

$$+\uparrow \qquad \Sigma F_y = 0$$

$$F_1 \sin 45° + F_2 \sin 30° + F_3 \sin 60° - W = 0$$

With three unknowns and only two equations the problem is now **statically indeterminate.** We need more information to determine F_1, F_2, and F_3. Problems such as this are solved in engineering courses variously titled Mechanics of Solids, Mechanics of Materials, Mechanics of Deformable Bodies, or Strength of Materials, where the equilibrium equations are supplemented by information describing the manner in which the cables stretch. Most of the problems in this book are statically determinate, but it is important to realize that the equations of equilibrium are necessary ingredients in the analysis of statically indeterminate problems as well. It's just that statics alone is insufficient to solve such problems.

There follow two more examples of problems of equilibrium. The student should note the central role played by the free-body diagram in each of these analyses. Throughout the book, whenever the weights of bodies subjected to other loads are not given as data in examples and in problems, it is to be understood that these weights may be neglected in comparison with the other loads.

EXAMPLE 2.6

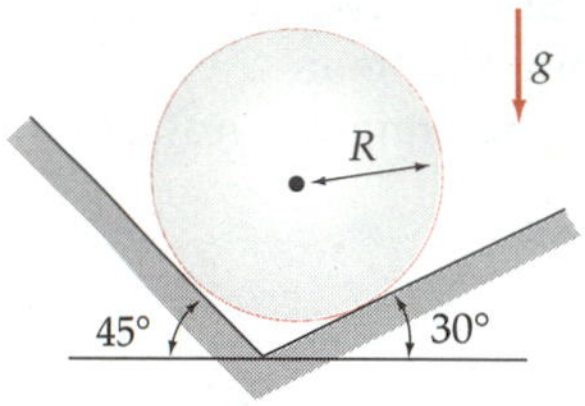

Figure E2.6a

A uniform sphere of weight W is supported by smooth (frictionless) plane surfaces as shown in Figure E2.6a. The plane of the page is vertical. Find the forces exerted by the supporting surfaces on the sphere.

Solution

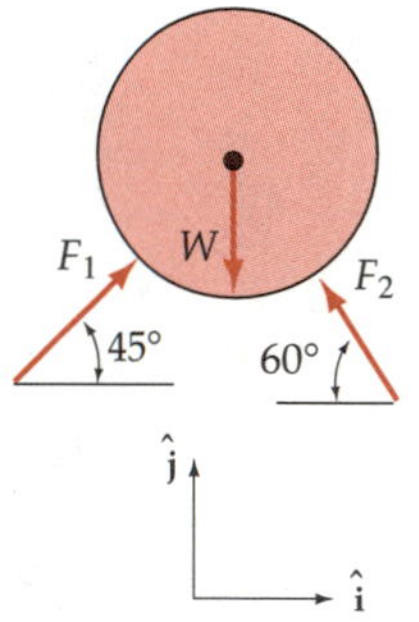

Figure E2.6b

The free-body diagram (Figure E2.6b) shows that each unknown force has a line of action through the sphere's center, which is also its mass center. This is an example of a body held in equilibrium by three forces (see Problem 4.141 for a general discussion).

The force-equation of equilibrium, $\Sigma\mathbf{F} = \mathbf{0}$, yields

$$F_1(\cos 45°\hat{\mathbf{i}} + \sin 45°\hat{\mathbf{j}}) + F_2(-\cos 60°\hat{\mathbf{i}} + \sin 60°\hat{\mathbf{j}}) + W(-\hat{\mathbf{j}}) = \mathbf{0}$$

Separating the coefficients of $\hat{\mathbf{i}}$ and $\hat{\mathbf{j}}$, we get

$$\hat{\mathbf{i}} \text{ coefficients:} \quad \frac{\sqrt{2}}{2}F_1 - \frac{1}{2}F_2 = 0 \tag{1}$$

which we could obtain directly by $\Sigma F_x = 0$, and

$$\hat{\mathbf{j}} \text{ coefficients:} \quad \frac{\sqrt{2}}{2}F_1 + \frac{\sqrt{3}}{2}F_2 = W \tag{2}$$

which we could get from $\Sigma F_y = 0$. (To do this, we would just sum up the vertical components of all forces and omit the unit vector $\hat{\mathbf{j}}$ as we go.) The solution to Equations (1) and (2) is

$$F_1 = 0.518W$$
$$F_2 = 0.732W$$

EXAMPLE 2.7

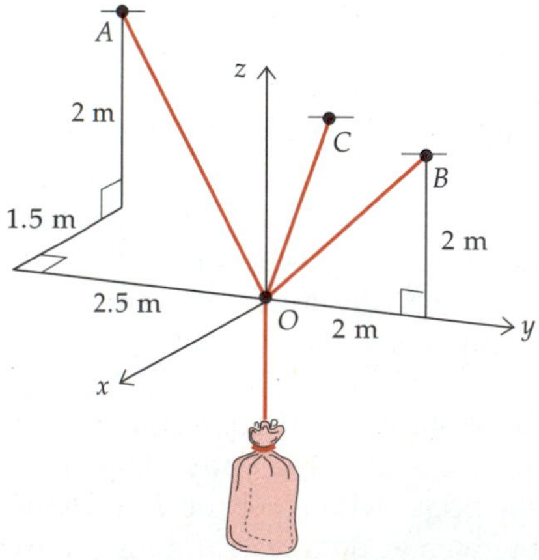

Figure E2.7a

The sandbag weighing 2500 N is held up by three cables as shown in Figure E2.7a. Each of the cables is tied to the ceiling 2 m above point O. The cable OC makes equal angles with the positive coordinate directions shown. Find the tension in each cable.

Solution

Let $\hat{\mathbf{i}}$, $\hat{\mathbf{j}}$, and $\hat{\mathbf{k}}$ be unit vectors along x, y, and z. To express the three forces exerted on the connection by the cables vertically, we first construct unit vectors along OA, OB, and OC. So

$$\hat{\mathbf{e}}_A = \frac{(-1.5\hat{\mathbf{i}} - 2.5\hat{\mathbf{j}} + 2\hat{\mathbf{k}})}{\sqrt{(1.5)^2 + (2.5)^2 + (2)^2}} = -0.424\hat{\mathbf{i}} - 0.707\hat{\mathbf{j}} + 0.566\hat{\mathbf{k}}$$

$$\hat{\mathbf{e}}_B = \frac{2\hat{\mathbf{j}} + 2\hat{\mathbf{k}}}{\sqrt{(2)^2 + (2)^2}} = 0.707\hat{\mathbf{j}} + 0.707\hat{\mathbf{k}}$$

$$\hat{\mathbf{e}}_C = \frac{2\hat{\mathbf{i}} + 2\hat{\mathbf{j}} + 2\hat{\mathbf{k}}}{\sqrt{(2)^2 + (2)^2 + (2)^2}} = 0.577\hat{\mathbf{i}} + 0.577\hat{\mathbf{j}} + 0.577\hat{\mathbf{k}}$$

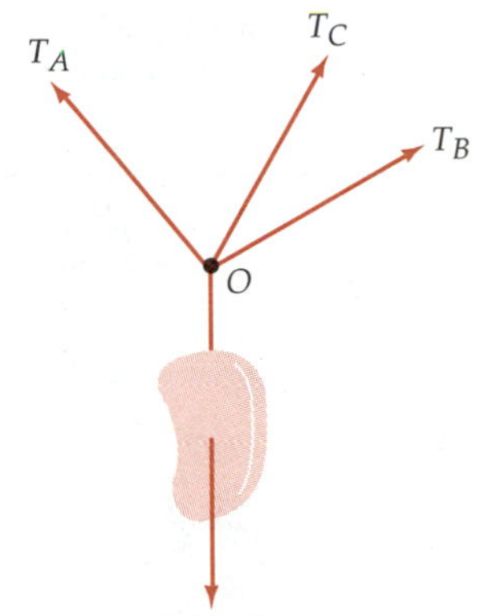

Figure E2.7b

Using these unit vectors, along with the scalars T_A, T_B, and T_C (see the FBD, Figure E2.7b), the equilibrium equation becomes

$$\Sigma \mathbf{F} = \mathbf{0}$$

$$T_A\hat{\mathbf{e}}_A + T_B\hat{\mathbf{e}}_B + T_C\hat{\mathbf{e}}_C + 2500\,(-\hat{\mathbf{k}}) = \mathbf{0}$$

Lifting off the coefficients of $\hat{\mathbf{i}}$, or equivalently writing $\Sigma F_x = 0$, we obtain

$$-0.424\ T_A + 0.577\ T_C = 0$$

and from $\Sigma F_y = 0$,

$$-0.707\ T_A + 0.707\ T_B + 0.577\ T_C = 0$$

and from $\Sigma F_z = 0$,

$$0.566\ T_A + 0.707\ T_B + 0.577\ T_C - 2500 = 0$$

The result of solving these three equations simultaneously is

$$T_A = 1970\text{ N}$$

$$T_B = 787\text{ N}$$

and

$$T_C = 1450\text{ N}$$

PROBLEMS ▶ Section 2.3

2.41 A large block of wood with an equilateral triangle cross section of side s as shown in Figure P2.41 is to be lifted by a sling of length $4s$. Without writing any equations, explain why the sling tension will be greater in configuration (b) than in (a).

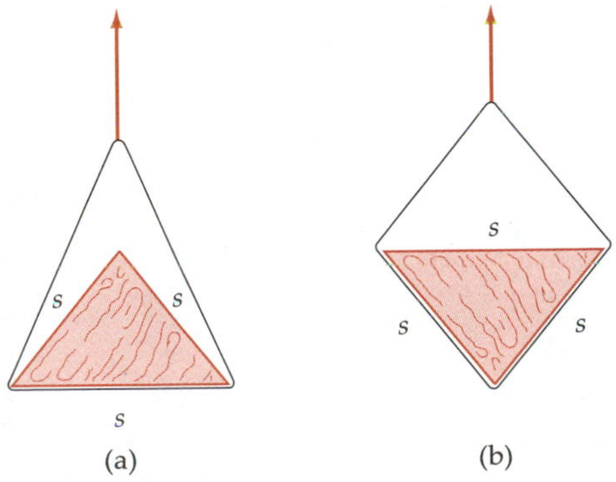

Figure P2.41

2.42 A boat is in the middle of a stream whose current flows from right to left (see Figure P2.42). If the forces F_1 and F_2, exerted on the ropes shown, are holding the boat in equilibrium against a force due to the current of 80 lb, what are the values of F_1 and F_2?

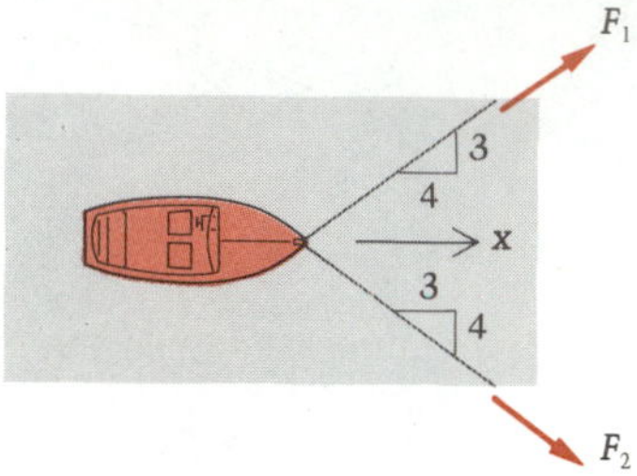

Figure P2.42

2.43 A 50-lb traffic light sags 1 ft in the center of a cable as shown in Figure P2.43. Determine the tension in the cable to which it is clamped.

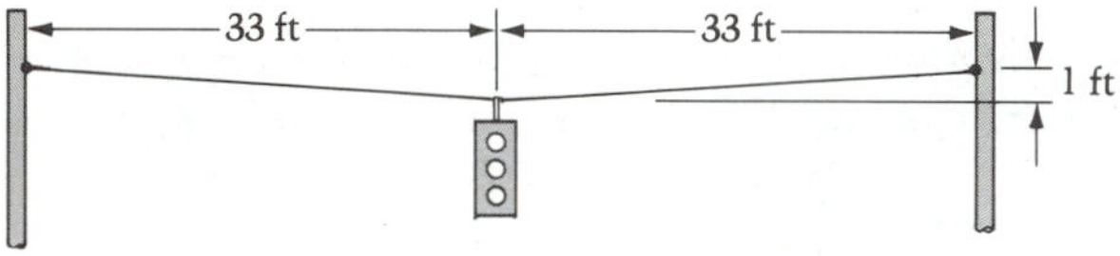

Figure P2.43

2.44 (a) A cable is being used to lift the 2000-N beam in Figure P2.44. Find the force in the lengths AB and BC of cable which are tied at B, in terms of the angle θ. (b) If the cable breaks when the tension exceeds 5000 N, what is the smallest angle θ that can be used?

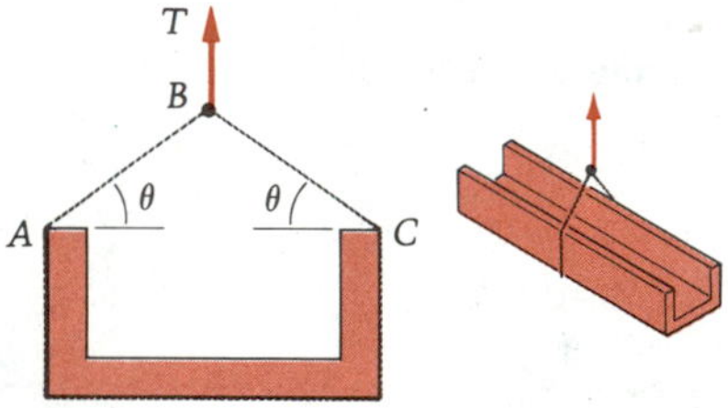

Figure P2.44

* **2.45** Find α such that the tension in cable AC is a minimum. (See Figure P2.45.)

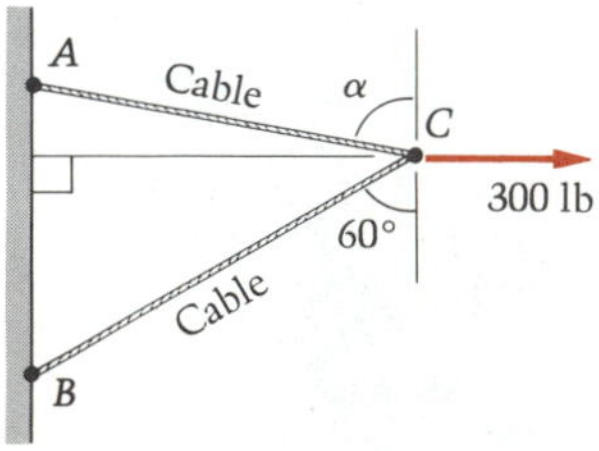

Figure P2.45

2.46 A smooth ball weighing 10 lb is supported by a cable and rests against a wall as shown in Figure P2.46. (a) Find the tension T in the cable and the normal force N exerted on the ball by the wall, as functions of the distance H. (b) Explain the limiting case results for T and N as H gets very large.

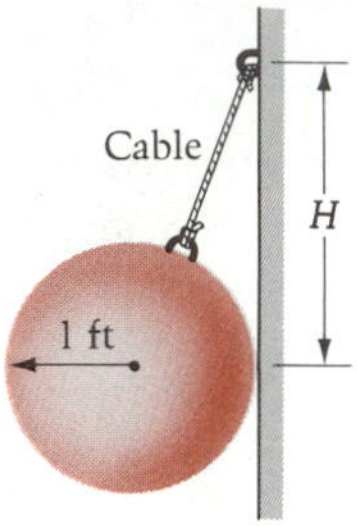

Figure P2.46

2.47 The 7-lb lamp in Figure P2.47 is suspended as indicated from a wall and a ceiling. Find the tensions in the two chains.

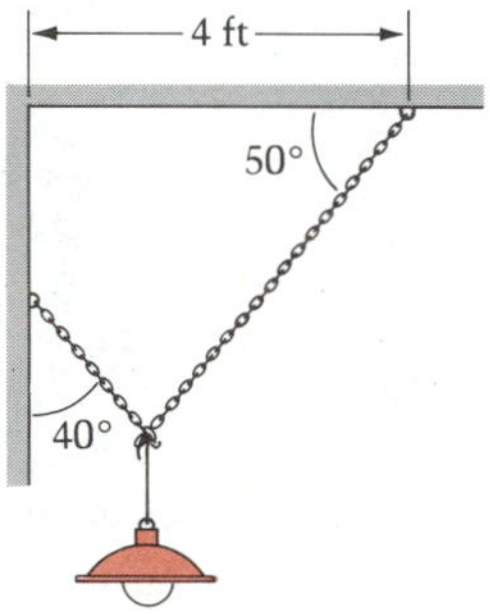

Figure P2.47

2.48 In the preceding problem, suppose the chains are replaced by a continuous 7-ft cord. (See Figure P2.48.) If it supports the lamp by passing through a smooth eye-hook so that the tension is the same on both sides, find this tension.

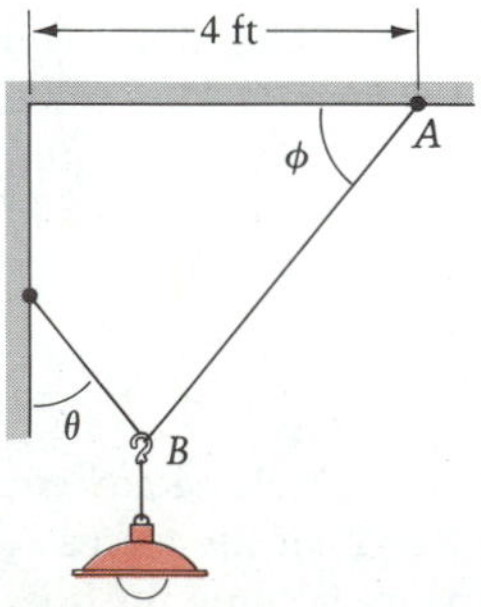

Figure P2.48

2.49 In Figure P2.49 force P is applied to a small wheel that is free to move on cable ACB. For a cable tension of 500 lb, find P and α.

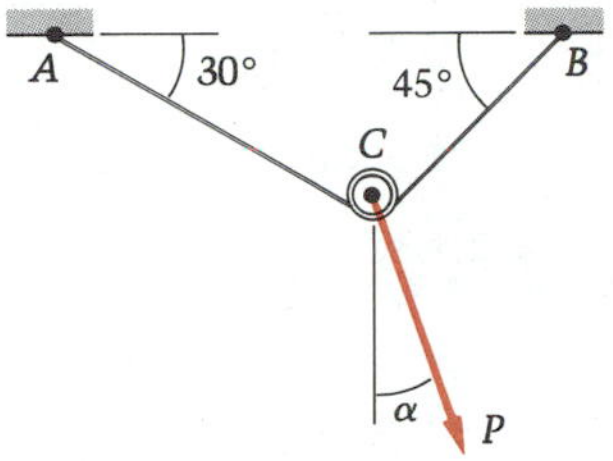

Figure P2.49

2.50 In Figure P2.50 a 1500-N weight is attached to a small, light pulley that can roll on the cable ABC. The pulley and weight are held in the position shown by a second cable DE, which is parallel to the portion BC of the main cable. Find the tension in cable ABC and the tension in cable DE.

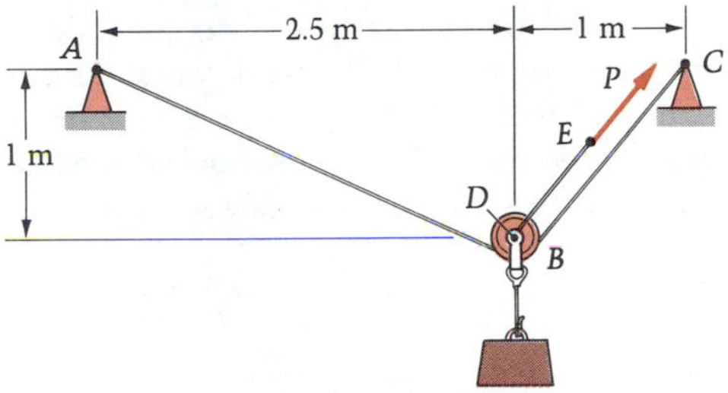

Figure P2.50

2.51 In Figure P2.51 the cable ABC is 10 feet long and flexible. A small pulley rides on the cable and supports a weight $W = 50$ lb. Find the tension T in the cable.

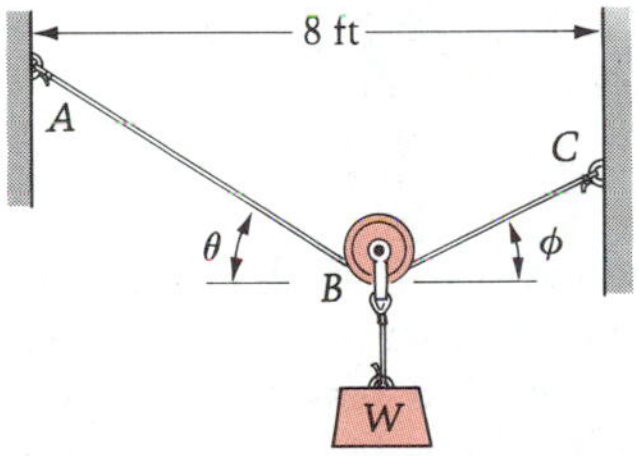

Figure P2.51

2.52 The man in Figure P2.52 is slowly pulling a drum over a circular hill. The drum weighs 60 N, and the hill is smooth. In the given position, find the tension in the rope (which does not vary along the rope if the hill is smooth).

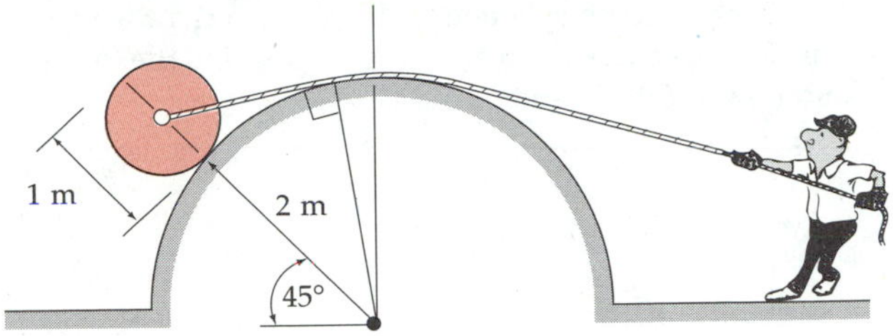

Figure P2.52

2.53 Find the forces exerted by the smooth planes on the 100-kg cylinder $\mathcal{C}$ shown in Figure P2.53. (The dotted cylinder $\mathcal{D}$ is absent from this problem.)

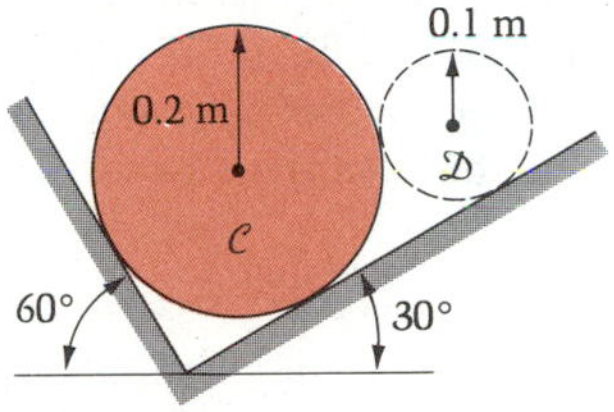

Figure P2.53

2.54 The man in Figure P2.54 is holding the 200-N drum in equilibrium at the point $(x, y) = (2, 1)$ m, with a force parallel to the plane at the point of contact. If friction is negligible, find the force exerted by the man.

2.55 The cylinder of weight W is in equilibrium between the two smooth planes. (See Figure P2.55.) Find the reactions N_1 and N_2 of the planes on the cylinder. Check your results by showing that $N_1 \to W$ and $N_2 \to 0$ as $\theta_1 \to 0$.

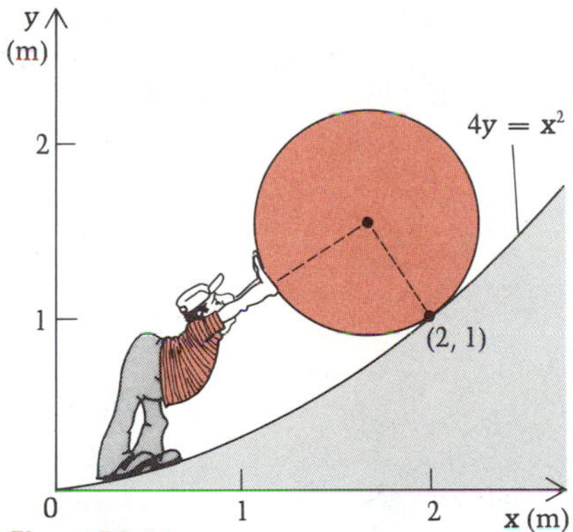

Figure P2.54

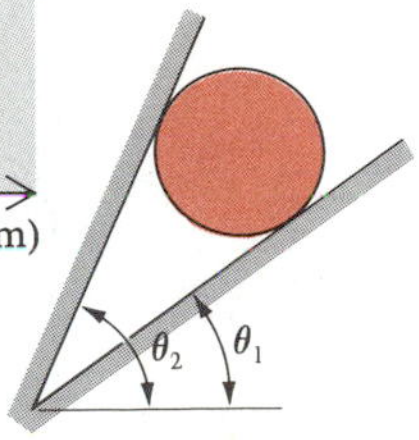

Figure P2.55

2.56 Three smooth cylinders *A*, *B*, and *C*, each of weight *W*, are arranged as shown in Figure P2.56. Find the forces exerted onto *C* by *A* and *B*.

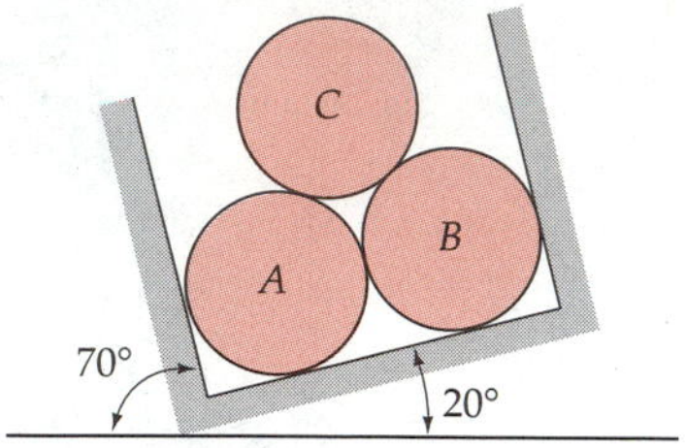

Figure P2.56

2.57 A shaft that carries a thrust of 1000 N terminates in a conical bearing as shown in Figure P2.57. If the angle θ is 25°, find the normal force that each of four equally spaced ball bearings exerts on the conical surface. Assume symmetry.

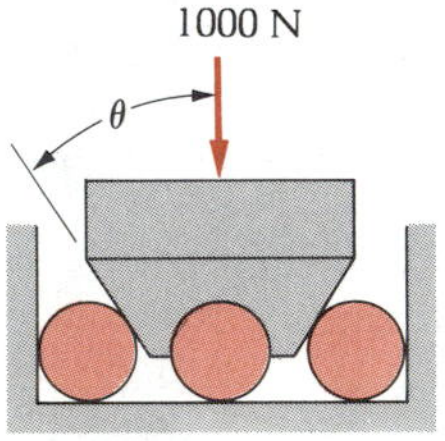

Figure P2.57

2.58 The 500-N reinforced concrete slab in Figure P2.58 is being slowly lowered by a winch at the end of the cable $\mathcal{C}$. The cables $\mathcal{A}$, $\mathcal{B}$, and $\mathcal{D}$ are each attached to the slab and to the hook. Find the forces in each of the four cables if the distance from the upper surface of the slab to the hook is 2 m.

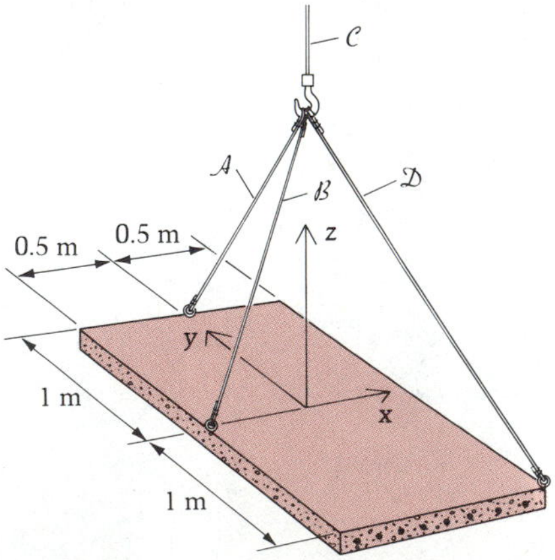

Figure P2.58

2.59 In Figure P2.59 the weight *W* is 600 N, and it is supported by cables *AD*, *BD*, and *CD*. Find the tension in each cable. (Points *A*, *B*, and *C* are all in the *xz* plane.)

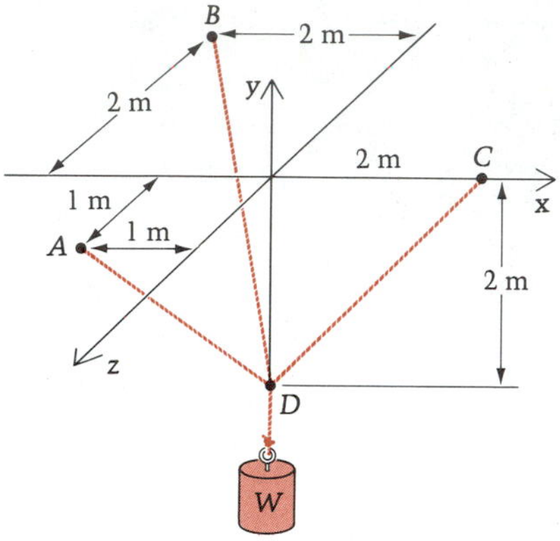

Figure P2.59

2.60 An equilateral triangular plate weighs 500 N. Three equal-length ropes (which break at a tension of 1500 N) are tied to the corners of the plate and to each other as shown in Figure P2.60, and hold up the horizontal plate. Find the shortest value of ℓ such that the ropes don't break.

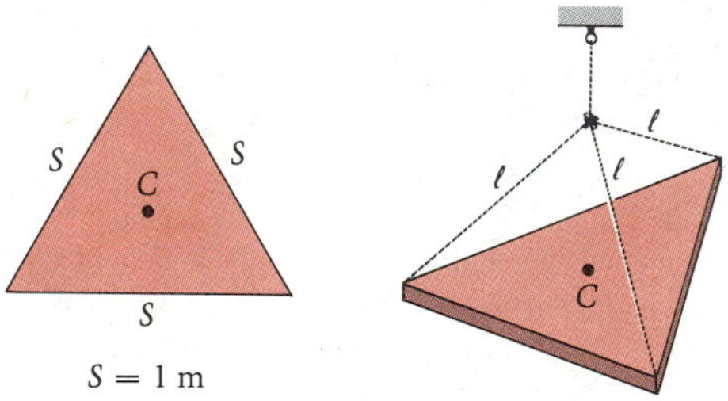

Figure P2.60

2.61 The weight in Figure P2.61 is to be suspended by 60 ft of cable in any possible number of equally spaced lengths (one 60-ft length, two 30-ft lengths, etc.). The cables are to be symmetrically attached to the 20-ft diameter ring as suggested in the figure for three lengths. Show that the load in the cables is less for four lengths than for one, two, three, or five, and that for six, the load in each cable is theoretically infinite. In each case, assume all the cable forces to be equal by symmetry, noting that for four or more cables, the problem is actually statically indeterminate.

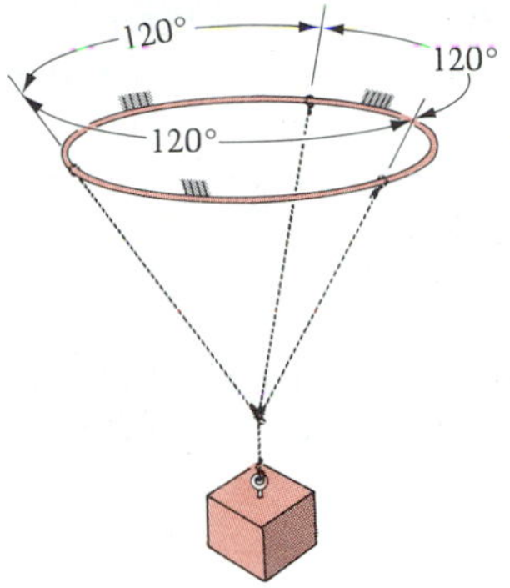

Figure P2.61

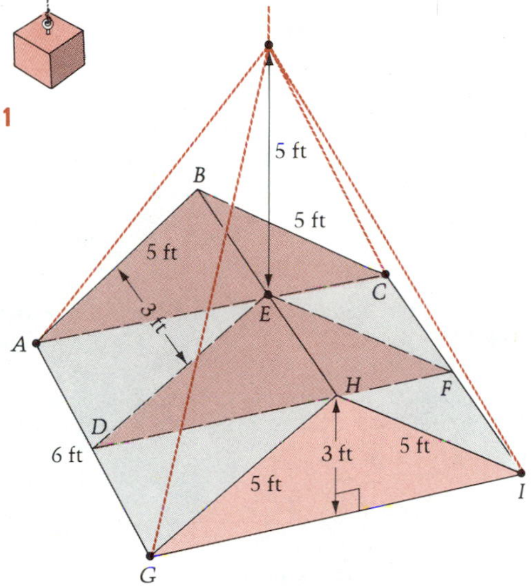

Figure P2.62

2.62 A small roof made of two identical rectangular plates (*ABHG* and *BCIH*), each weighing 60 lb, is supported symmetrically as shown in Figure P2.62. The plates are braced by three triangular plates (*ABC, DEF, GHI*) of the same density and thickness as the rectangular plates. Calculate the tension in each of the four cables, assuming them to be equal by symmetry. Could you work the problem without this assumption?

2.63 A particle located at the origin [$(x, y, z) = (0, 0, 0)$ m] is suddenly acted on by the three forces $\mathbf{F}_1$, $\mathbf{F}_2$, and $\mathbf{F}_3$, of magnitudes 14, 6, and 10 newtons, respectively. (See Figure P2.63.) $\mathbf{F}_1$ acts on the line from (0, 0, 0) to (3, 6, 2) m; $\mathbf{F}_2$ acts on the line from (0, 0, 0) to (3, 6, −6) m; and $\mathbf{F}_3$ acts on the line from (4, 3, 0) to (0, 0, 0) m. Is the particle still in equilibrium after the forces are applied? If not, what additional force $\mathbf{F}_4$ will keep it at the origin?

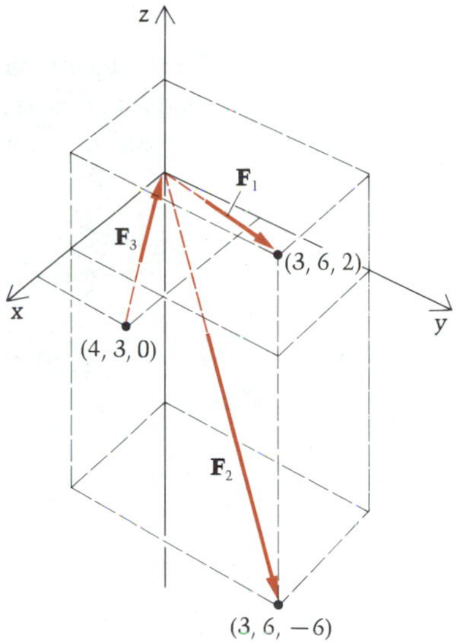

Figure P2.63

2.4 Equilibrium of a System of Particles

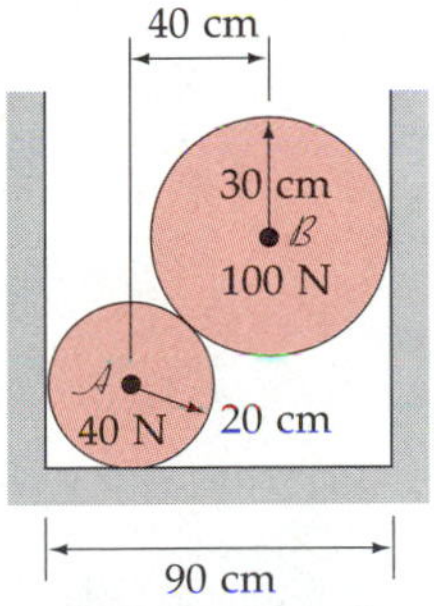

Figure 2.9

Sometimes it is useful in an analysis to separate a "body" into constituent parts, and it is necessary to do this if we wish to determine the forces of interaction between those parts. The key concept in such an analysis is the **action-reaction principle** which states that the force exerted on a first body by a second body is equal in magnitude but opposite in direction to the force exerted on the second by the first.

To illustrate these concepts consider the two cylinders of Figure 2.9. They are supported by the smooth floor and the two vertical smooth walls. In Figure 2.10 are free-body diagrams of each of the cylinders and

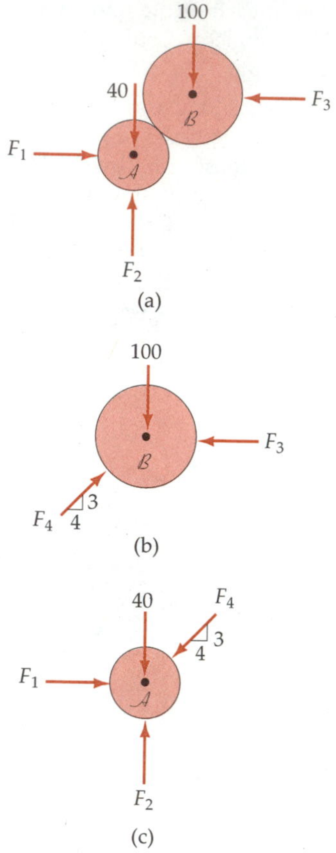

Figure 2.10

of the two-cylinder combination. The free-body diagram of the larger cylinder (Figure 2.10(b)) shows that the force exerted on it by the smaller one is given by

$$F_4\left(\frac{4}{5}\hat{\imath} + \frac{3}{5}\hat{\jmath}\right)$$

Execution of the action-reaction principle is seen in the free-body diagram, Figure 2.10(c), of the small cylinder where F_4 is associated with an oppositely directed unit vector. That is, the force exerted by the larger cylinder on the smaller must be

$$F_4\left(-\frac{4}{5}\hat{\imath} - \frac{3}{5}\hat{\jmath}\right)$$

The student should note that, by using the same scalar F_4 along with the arrow codes in the two free-body diagrams, we easily force satisfaction of the action-reaction principle.

Question 2.2 Why does F_4 not appear on the free-body diagram of the two-cylinder composite?

Turning to the free-body diagram of the larger cylinder and using the equilibrium equations, we obtain

$$+\uparrow \quad \Sigma F_y = 0$$

$$\frac{3}{5}F_4 - 100 = 0 \Rightarrow F_4 = 167 \text{ N}$$

and

$$\xrightarrow{+} \quad \Sigma F_x = 0$$

$$\frac{4}{5}F_4 - F_3 = 0$$

$$F_3 = \frac{4}{5}(167) = 133 \text{ N}$$

And then from the free-body diagram of the smaller cylinder, the equilibrium equations yield

$$\xrightarrow{+} \quad \Sigma F_x = 0$$

$$F_1 - \frac{4}{5}F_4 = 0 \Rightarrow F_1 = 133 \text{ N}$$

and

$$+\uparrow \quad \Sigma F_y = 0$$

$$F_2 - 40 - \frac{3}{5}F_4 = 0$$

$$F_2 = 140 \text{ N}$$

Answer 2.2 For the two-cylinder composite body, F_4 is an *internal* force. Only *external* forces are drawn on the free-body diagram of a body, because only they affect its equilibrium equations.

The student should note carefully that these forces, as they relate to the free-body diagram of the two-cylinder composite body, cause the equation $\Sigma \mathbf{F} = \mathbf{0}$ to be satisfied for that body.

An important class of multibody problems in which we may ignore the moments (turning effects) of forces is that of systems of pulleys. The following example illustrates application of equilibrium analysis to such a system. We must use in the analysis that, for a pulley in equilibrium with frictionless bearings, the tension in a rope or belt going around the pulley is everywhere the same. That fact will be rigorously proved in Chapter 4; for now, suffice it to say that if the tensions differed, a frictionless pulley would be turning.

Example 2.8

Determine the force the man in Figure E2.8a must exert to hold the blocks in equilibrium.

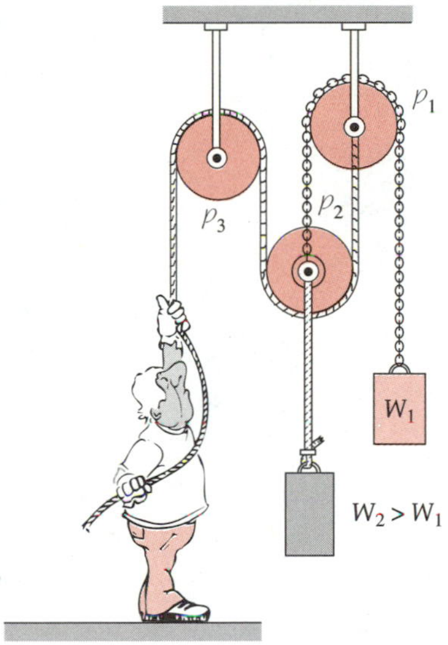

Figure E2.8a

Solution

In a problem involving frictionless (at the axle) pulleys in equilibrium, first remember that the tension in the rope is the same going on as it is coming off. The tensions would differ, of course, if either (a) the pulley were accelerating angularly, or (b) there were friction between the pulley and its axle.

Let us now start with the load W_1 and proceed to draw free-body diagrams of the various bodies. Normally the pulleys' weights are neglected, and we shall do so here.

The bodies are sketched below in proximity to one another to facilitate glancing from one to a neighboring one. The dashed lines are merely reminders of how the ropes or chain are connected to the bodies.

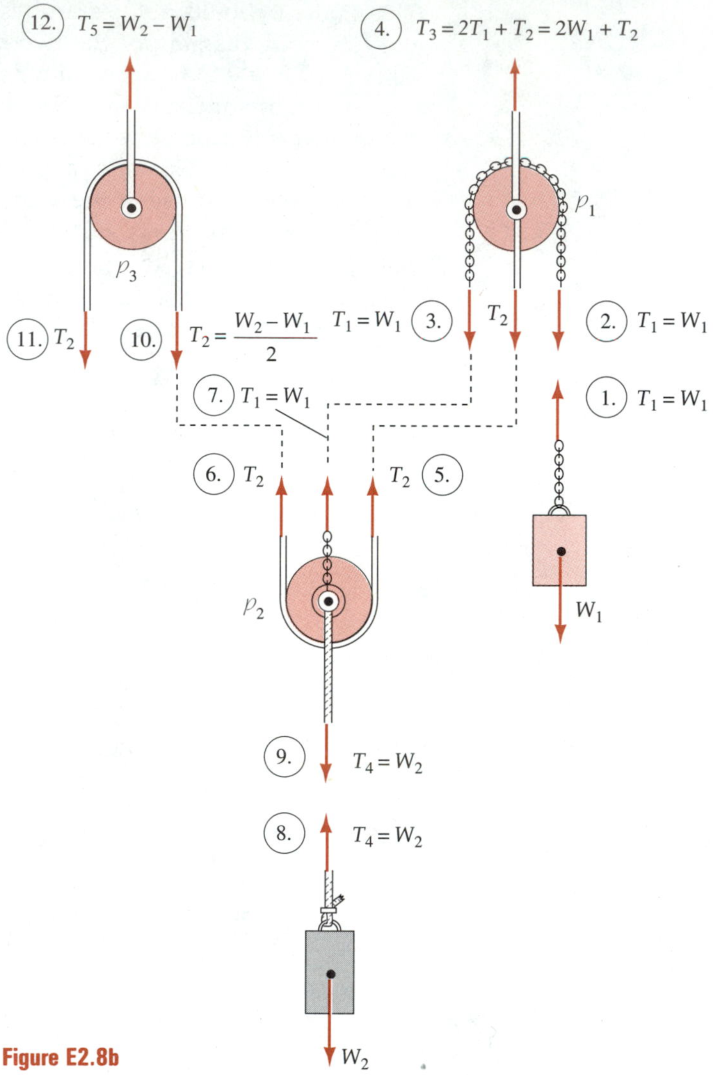

Figure E2.8b

Steps in the solution are correspondingly numbered in the various FBDs of Figure E2.8b; the reader should refer to these FBDs while carefully reading through the steps that follow this figure.

Step: Reason for result on figure:

1. $\Sigma F_y = 0$ on FBD of W_1.
2. Same straight portion of chain as in FBD of W_1, so same tension in it.
3. Same chain as on other side of this pulley.
4. $\Sigma F_y = 0$ on FBD of P_1; note T_2 remains unknown to this point.
5. Same rope as in FBD of P_1, so same tension in it.
6. Same rope as on other side of this pulley.
7. Same chain as in FBD of P_1, so same tension in it.
8. $\Sigma F_y = 0$ on FBD of W_2.
9. Same rope as in FBD of W_2, so same tension in it.

10. $\Sigma F_y = 0$ on FBD of P_2 yields

$$2T_2 + W_1 - W_2 = 0$$

so that

$$T_2 = \frac{W_2 - W_1}{2}$$

Then on FBD of P_3, same rope as in FBD of P_2.

11. Same rope as on other side of this pulley. Thus the man's pulling force on the rope is $(W_2 - W_1)/2$.

12. For completeness, $\Sigma F_y = 0$ on P_3 gives $T_5 - 2T_2 = 0$, or $T_5 = 2\left(\frac{W_2 - W_1}{2}\right) = W_2 - W_1$.

In the preceding example, it is interesting to show the results on the FBDs of the three separated pulleys (see Figure 2.11) . . . . .

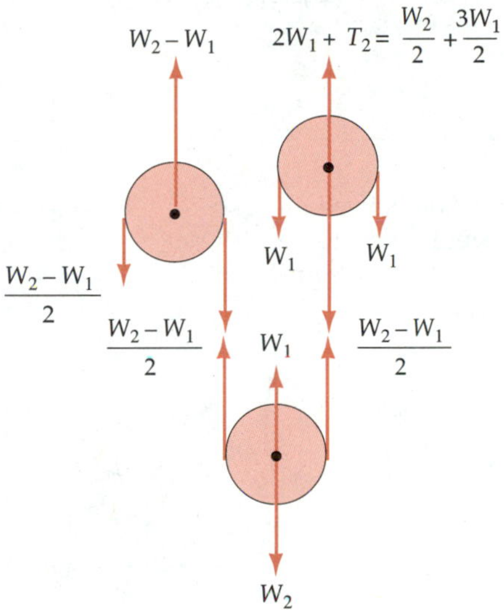

Figure 2.11

. . . . . and then to put the ropes and chains back together to form Figure 2.12. From this combined FBD we obtain a nice check on the solution:

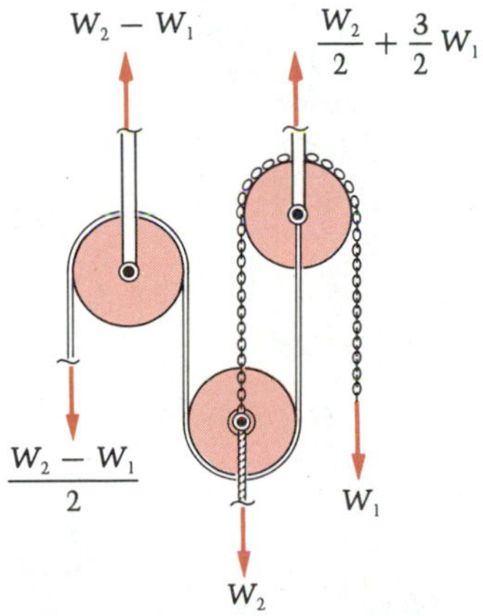

Figure 2.12

$$+\uparrow \quad \Sigma F_y = \underbrace{(W_2 - W_1)}_{\text{rope above } P_3} + \underbrace{\left(\frac{W_2}{2} + \frac{3}{2}W_1\right)}_{\text{rope above } P_1} - \underbrace{\left(\frac{W_2 - W_1}{2}\right)}_{\text{man's force}} - \underbrace{W_2}_{\text{weight of } W_2} - \underbrace{W_1}_{\text{weight of } W_1}$$

or

$$\Sigma F_y = 0 \checkmark$$

The reader should note that if $W_1 = W_2$, then no force is required to hold the blocks in equilibrium. This is because, in that case, the tensions on either side of $\mathcal{P}_1$ are equal to $W_1(= W_2)$, while the tensions on either side of $\mathcal{P}_3$ and $\mathcal{P}_2$ vanish.

Question 2.3 If the man begins to pull on the rope with a force slightly greater than the equilibrium value of $(W_2 - W_1)/2$, what happens to the blocks?

Question 2.4 How much does the man have to weigh to stay on the ground?

Answer 2.3 The center of $\mathcal{P}_2$ moves upward. W_2 moves up and W_1 moves down.
Answer 2.4 At least $(W_2 - W_1)/2$, or else he would need to pull down with more than his weight.

PROBLEMS ▸ Section 2.4

2.64 In Figure P2.64 the force of attraction between a pair of particles is 52 lb. What forces would have to be applied (if any), and where, for the system to be in equilibrium? (No other forces act on the particles.)

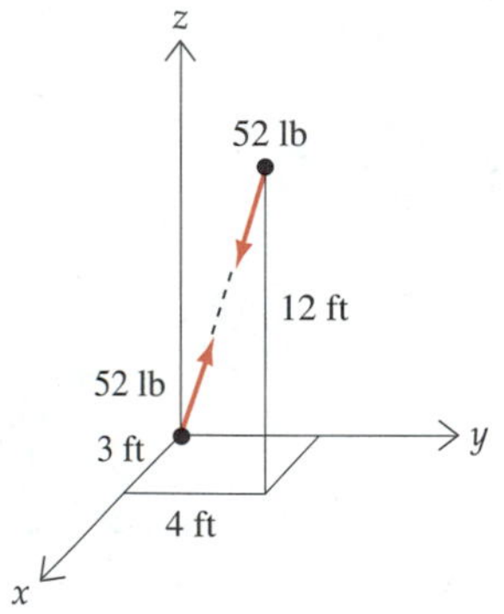

Figure P2.64

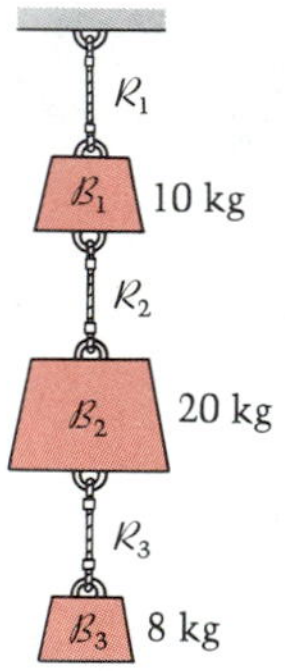

Figure P2.65

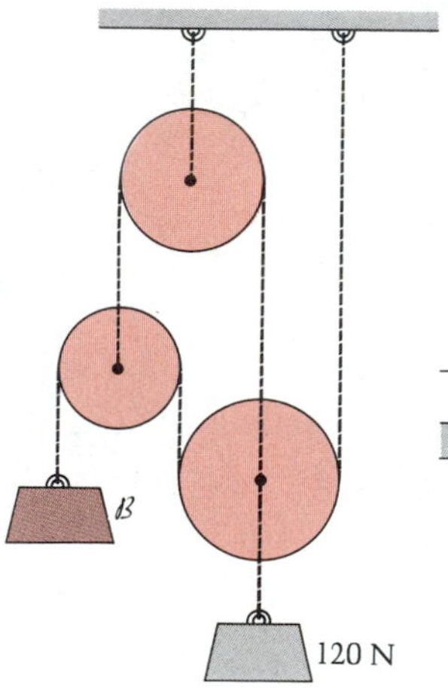

Figure P2.66

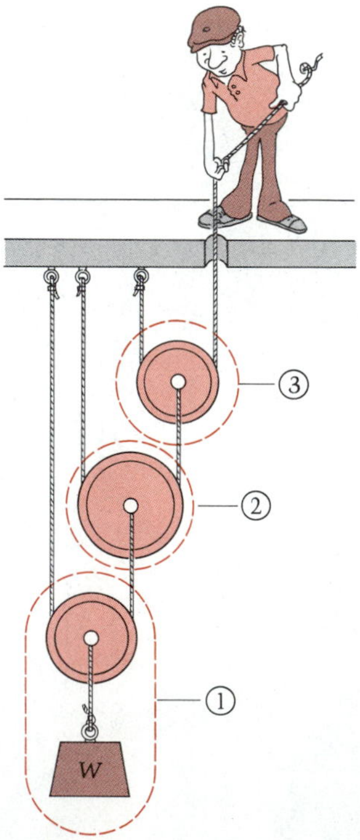

Figure P2.67

2.65 Using free-body diagrams, find the forces in the ropes $\mathcal{R}_1$, $\mathcal{R}_2$, and $\mathcal{R}_3$. (See Figure P2.65.)

2.66 Find the weight of $\mathcal{B}$ for equilibrium of the system shown in Figure P2.66.

2.67 Find, by successively drawing the free-body diagrams suggested by 1, 2, and 3 in Figure P2.67, the force that the man must exert to hold the weight in equilibrium.

2.68 The weight in Figure P2.68 is 1400 N. How much force P does the woman have to exert on the rope to lift the weight? How heavy does she have to be to stay on the ground?

2.69 (a) Show that the force F that the man must exert in order to lift the engine of W lb using the chain hoist is

$$F = \left(\frac{R - r}{2R}\right) W$$

(b) If $R = 10$ in., $r = 8.5$ in., and $W = 400$ lb, how many pounds are required? (See Figure P2.69.)

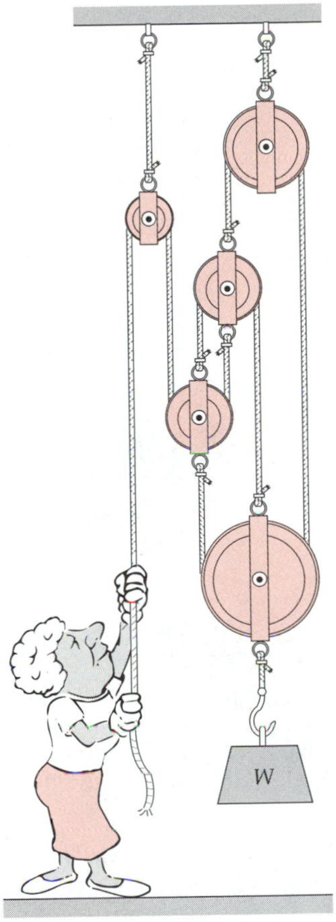

Figure P2.68

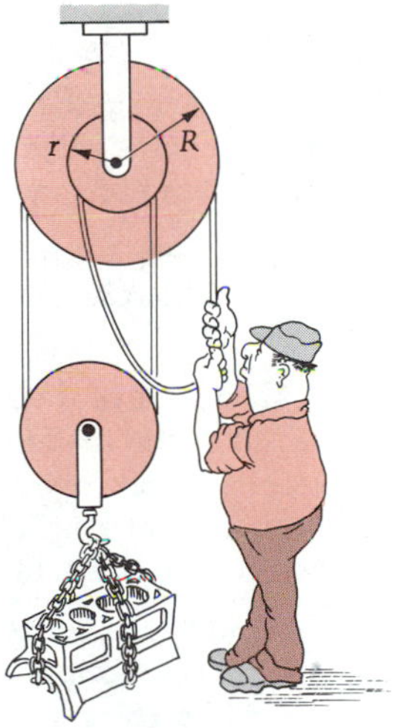

Figure P2.69

2.70 In the block and tackle shown in Figure P2.70, a single rope passes back and forth over pulleys that are free to rotate within the blocks about axes $\ell\ell$ and mm in the figure. The particular block and tackle shown has two pulleys in each block.

a. If the man in the preceding problem uses the block and tackle as indicated to raise the engine, how much force must he exert this time?

b. What is the ratio of r to R in the preceding problem for which the force he must exert is the same as it is with the block and tackle?

Assume all rope segments to be vertical.

2.71 In Figure P2.71, what is the force P needed to hold the weight W in equilibrium? Assume that all rope segments are vertical.

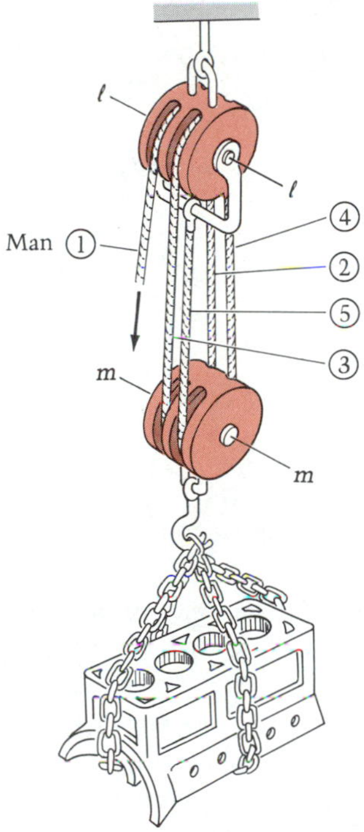

Figure P2.70

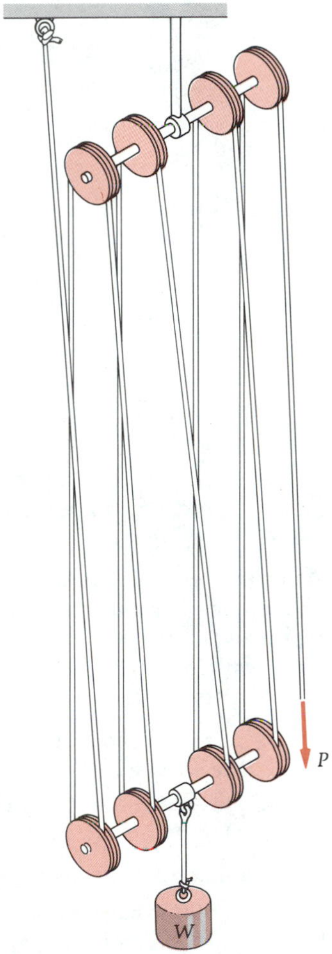

Figure P2.71

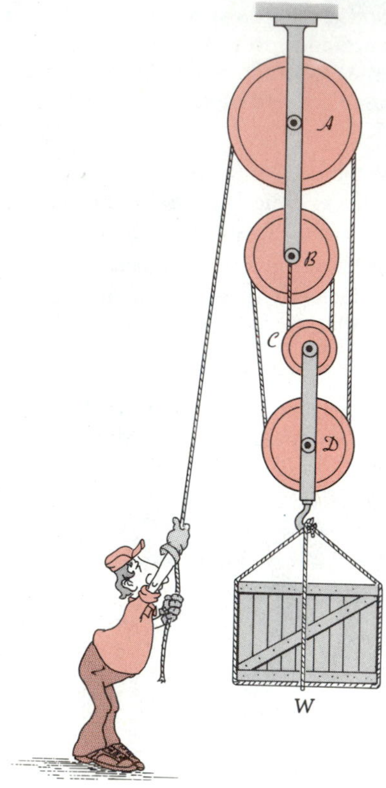

Figure P2.72

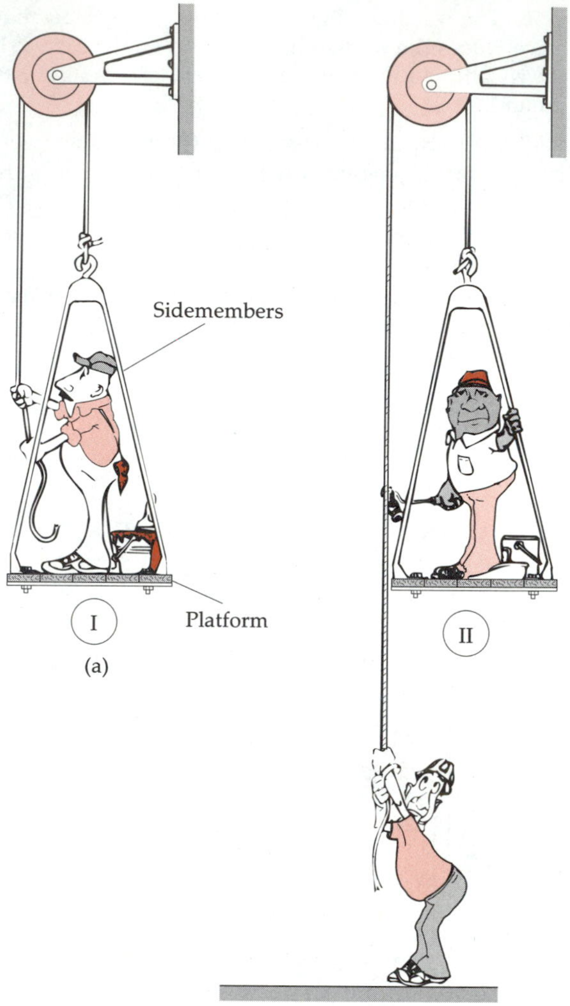

Figure P2.74

2.72 In Figure P2.72, what force must the man exert in order to raise the crate of weight W?

2.73 Repeat the preceding problem if the weights of the pulleys $\mathcal{A}$, $\mathcal{B}$, $\mathcal{C}$, and $\mathcal{D}$ are, respectively, $W/2$, $W/8$, $W/16$, and $W/4$.

2.74 The two painters in Figure P2.74 are slowly lifted on scaffolds. The first man lifts himself (case a). The second man is lifted by a colleague on the ground (case b). Each scaffold weighs 40 lb. Each painter weighs 180 lb. For each case:

- **a.** Draw free-body diagrams of the painter and the scaffold.
- **b.** Determine the magnitudes and the directions of all forces on the painter and on the scaffold.

Assume that the pulley is small and frictionless.

2.75 The 900-lb platforms in Figure P2.75 are supported by the light cable and pulley system as shown in the three configurations. Find the tension in the cable over pulley A and the tension in the cable over pulley B for each configuration. Assume mass center locations so that the platforms remain horizontal.

2.76 The five ropes in Figure P2.76 can each take 1500 N without breaking. How heavy can W be without breaking any?

2.77 The mass of the man in Figure P2.77 is 70 kg, and the mass of the scaffold on which he is sitting is 10 kg. The pulleys and ropes are light. Find the tension in the cable that the man is holding, and also the force he exerts directly on the scaffold.

2.78 Find the relationship between the load W and the force P for equilibrium of the differential winch shown in Figure P2.78. The rope is wrapped around the different-sized cylinders in opposite directions.

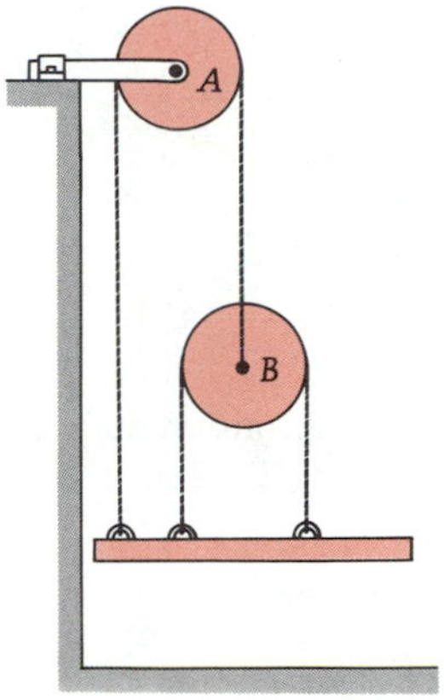

(a)

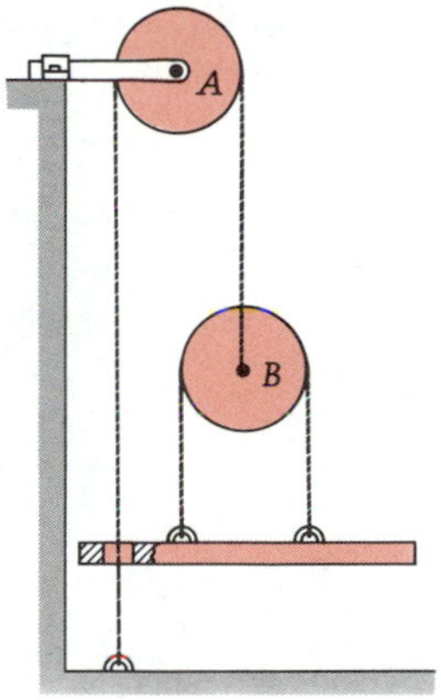

(b)

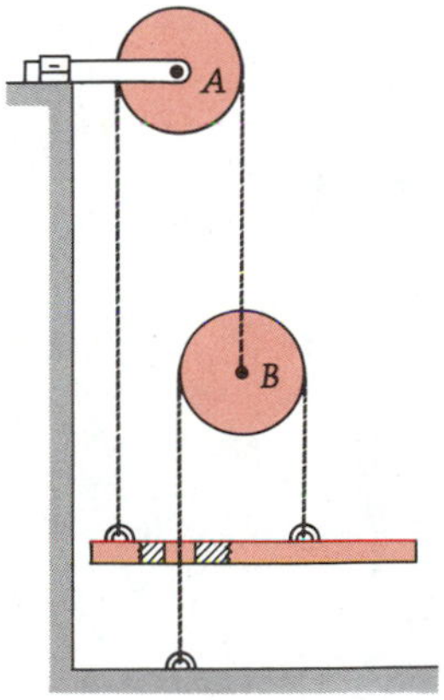

(c)

Figure P2.75

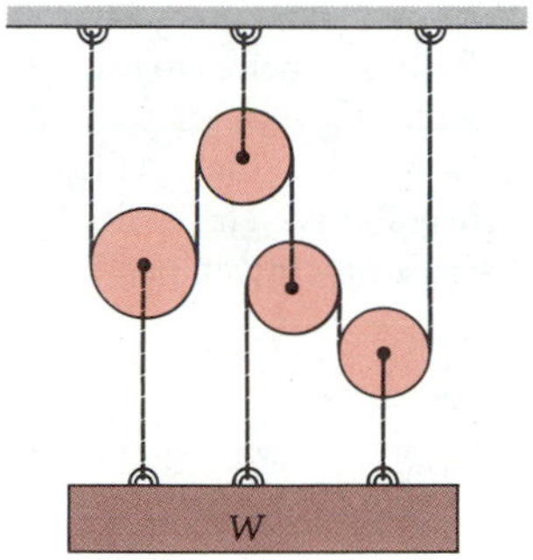

Figure P2.76

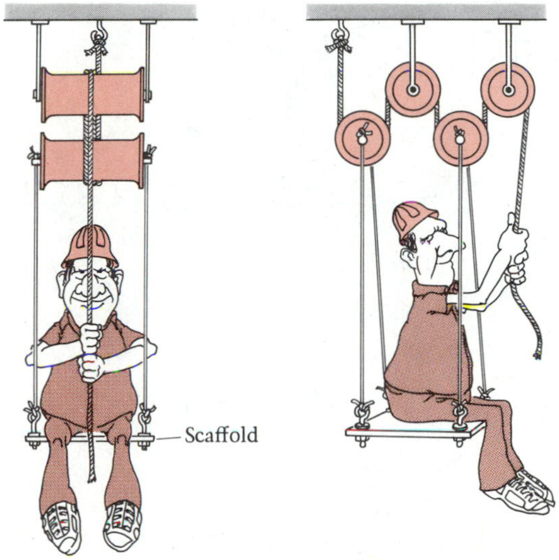

Figure P2.77

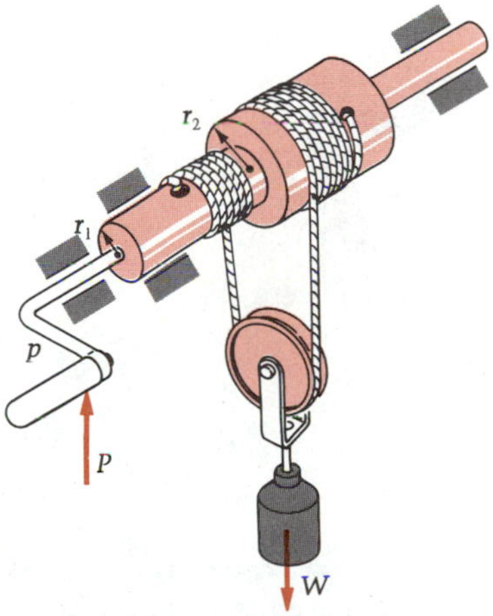

Figure P2.78

2.79 The collar in Figure P2.79 can slide along the rod without friction. The spring, which is attached to the collar and to the ceiling, exerts a force $k\delta$ proportional to its stretch δ, where k is the modulus of the spring. If $k = 50$ lb/in, $W = 50$ lb and the collar weighs 20 lb, find how much the spring is stretched in the given position.

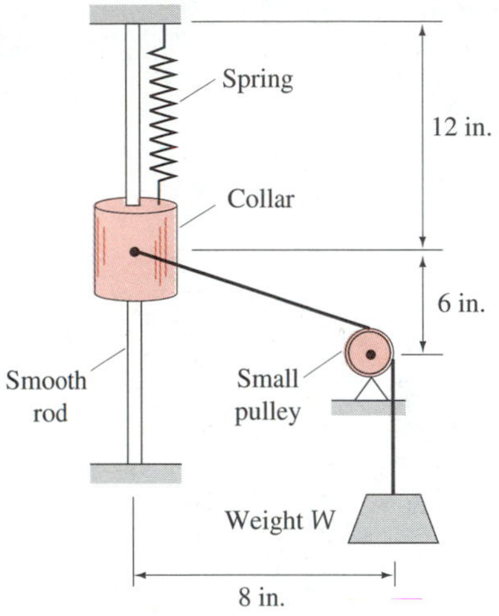

Figure P2.79

2.80 Two identical pieces of pipe rest against an incline and a vertical wall as shown in Figure P2.80. Each pipe has weight W, and all surfaces are smooth.

a. Draw free-body diagrams of the two pipes.
b. Determine the magnitude and show the direction of each force acting on each pipe.

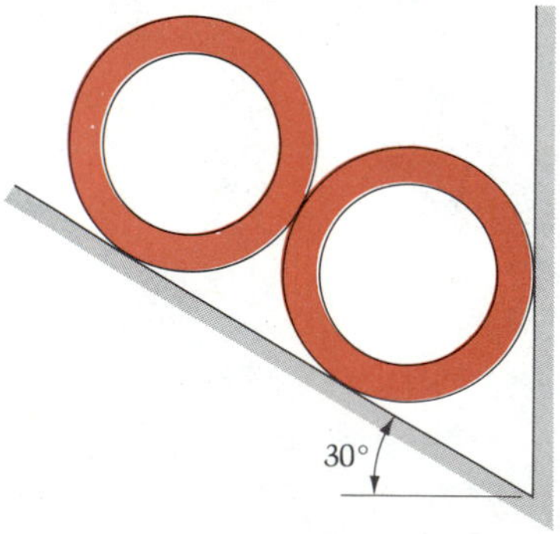

Figure P2.80

2.81 Two 50-lb traffic lights cause a cable sag of 8 in. as shown in Figure P2.81. Find the tensions in the three sections of the cable to which they are clamped.

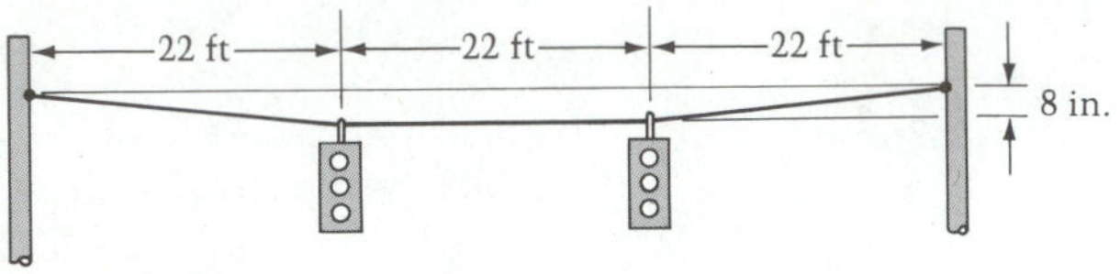

Figure P2.81

2.82 The two weights in Figure P2.82 are supported with six light, flexible, inextensible cords. Find the forces in the cords.

2.83 Find the weight of block $\mathcal{P}$ if the system is in equilibrium and $\mathcal{W}$ has a mass of 40 kg. (See Figure P2.83.)

2.84 The rope in Figure P2.84 has length ℓ. It is attached at one end to a pin at A and at the other to mass m after passing under the free, small pulley at D and over the fixed, small pulley at B. The mass M is suspended from the free pulley. Find the height H for equilibrium of the system.

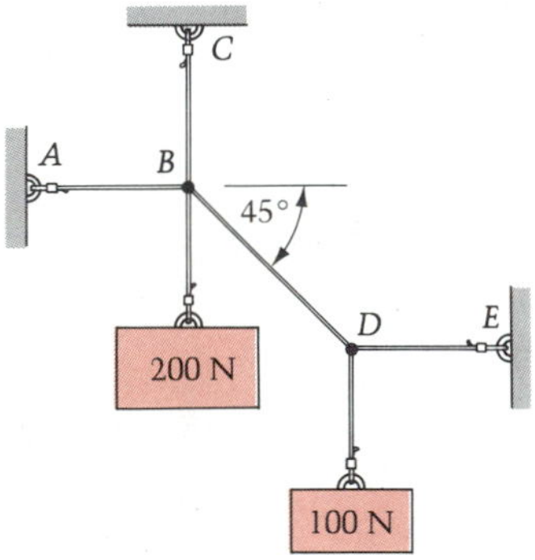

Figure P2.82

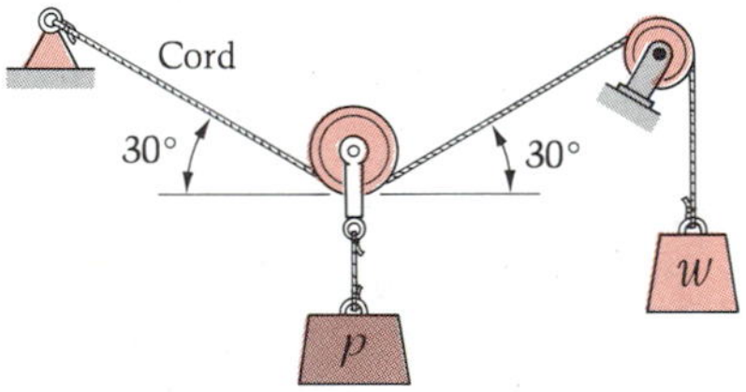

Figure P2.83

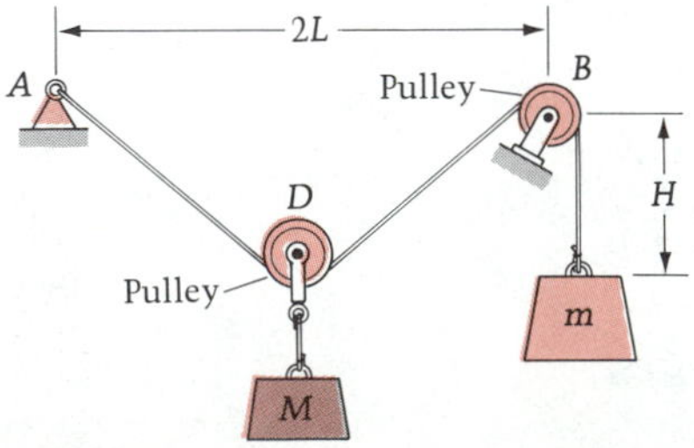

Figure P2.84

2.85 The man in Figure P2.85 weighs 800 N. He pulls down on the rope, raising the 250-N weight. He finds that the higher it goes, the more he must pull to raise it further. Explain this, and calculate and plot the rope tension T as a function of θ. What is the value of the tension, and the angle θ, when the man can lift it no further? Neglect the sizes and weights of the pulleys.

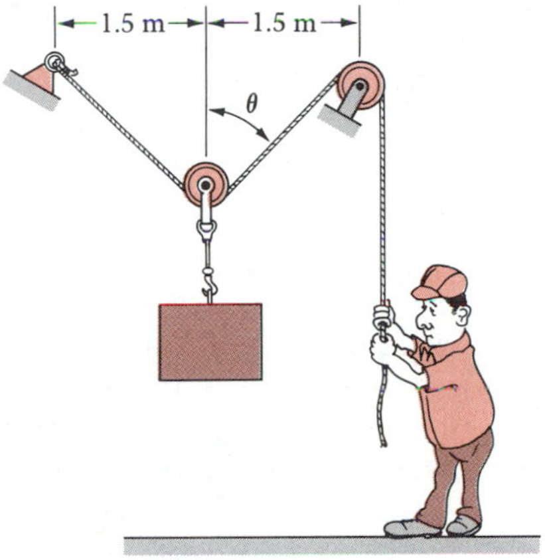

Figure P2.85

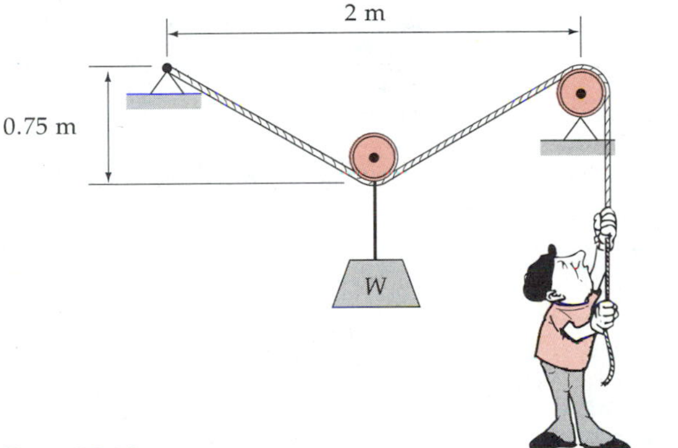

Figure P2.86

2.86 The 460-N man is holding the 360-N weight W in equilibrium as shown in Figure P2.86. (a) What is the tension in the rope? (b) How much higher can he raise the weight?

2.87 In the preceding problem, if in the figure there is a slack length of 0.5 m of rope below the man's hands and if he can reach 0.2 m higher than the position in the figure, what is the minimum rope tension possible?

2.88 Five identical smooth 5-kg cylinders are at rest on a 30° incline as shown in Figure P2.88. Find the normal force exerted by $\mathcal{B}$ on $\mathcal{C}$ at A. Repeat the problem if there are 100 cylinders instead of five.

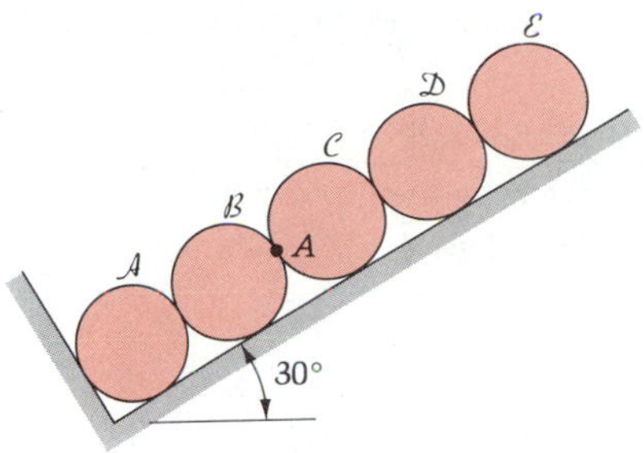

Figure P2.88

2.89 Cylinder $\mathcal{A}$ (mass 15 kg) rests on cylinder $\mathcal{B}$ (mass 20 kg) as shown in Figure P2.89. Find all forces acting on cylinder $\mathcal{B}$.

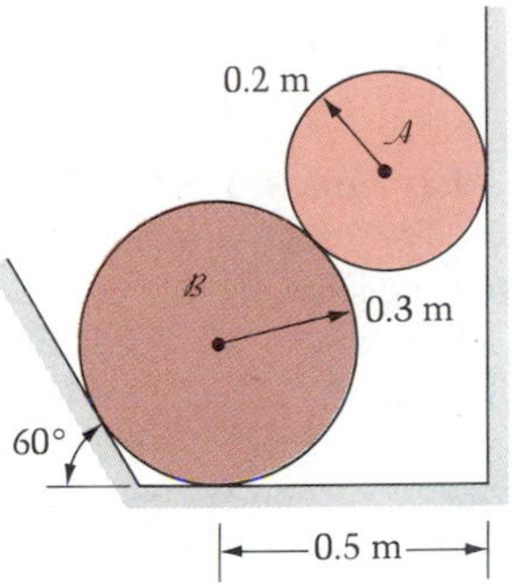

Figure P2.89

2.90 In Figure P2.90 the cylinders $\mathcal{A}$ (weight 50 N) and $\mathcal{B}$ (weight 150 N) are assumed to be smooth, and they rest on smooth planes oriented at right angles as shown. Find the angle Ψ between the horizontal line xx and the line joining the centers of the cylinders.

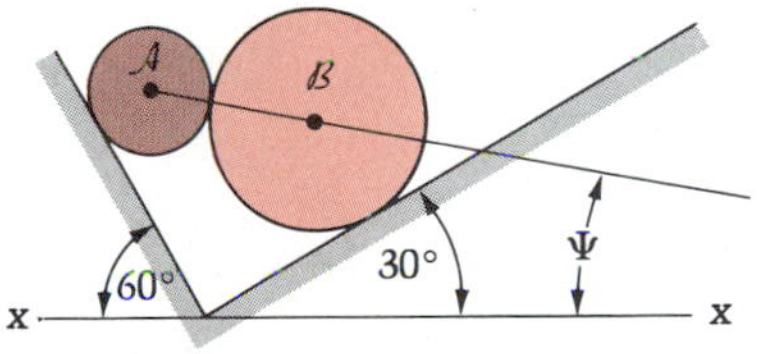

Figure P2.90

2.91 In Figure P2.91 two identical cylinders $\mathcal{C}_1$ and $\mathcal{C}_2$, each of radius r and weight P, are tied together by a cord. They support a third cylinder $\mathcal{C}_3$ of radius R and weight Q. There is no friction, and the tension in the cord is just sufficient to make the contact force between $\mathcal{C}_1$ and $\mathcal{C}_2$ zero. Find:

a. The tension in the cord

b. The force exerted by the ground on $\mathcal{C}_1$

c. The normal force between $\mathcal{C}_1$ and $\mathcal{C}_3$

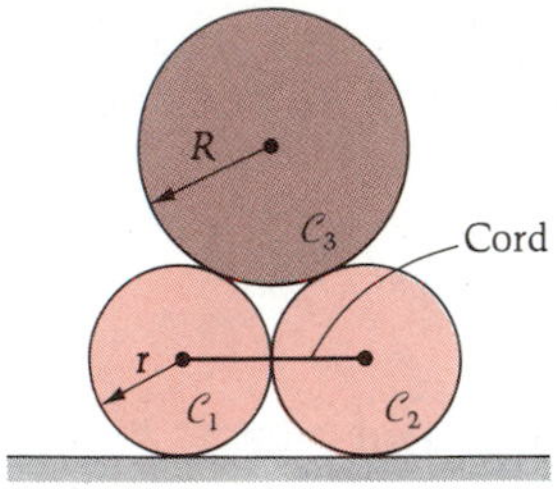

Figure P2.91

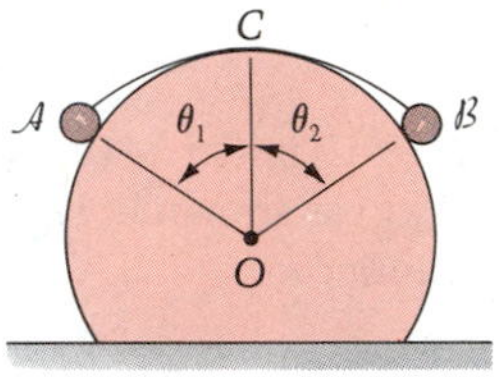

Figure P2.92

* **2.92** Two small balls $\mathcal{A}$ and $\mathcal{B}$ have masses m and $2m$, respectively. (See Figure P2.92.) They rest on a smooth circular cylinder with a horizontal axis and with radius R. They are connected by a thread of length $2R$. Find the angles θ_1 and θ_2 between the radii and the vertical line OC for equilibrium, as well as the tension in the thread and the forces exerted by $\mathcal{A}$ and $\mathcal{B}$ on the cylinder. Assume that the balls are very small and that the tension is constant.

* **2.93** Three identical spheres are at rest at the bottom of the spherical bowl shown in Figure P2.93. If a fourth sphere is placed on top, what is the largest ratio R/r for equilibrium if there is no friction?

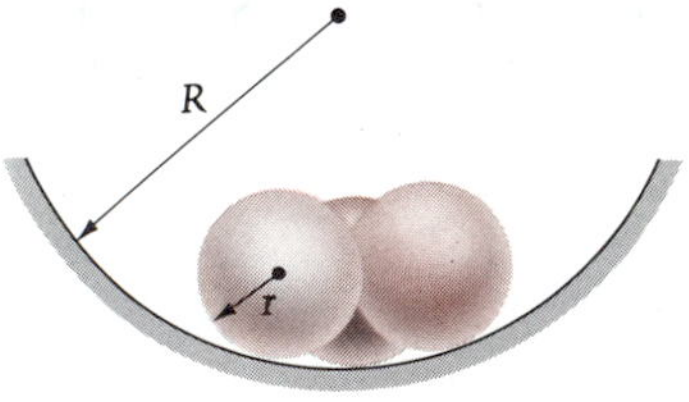

Figure P2.93

* **2.94** Repeat the previous problem if there are four spheres in the bowl and a fifth is placed on top.

2.95 Collars A and B in Figure P2.95 may slide smoothly on the rods OD and OE. Collar B weighs 20 pounds, and is connected to A by means of an inextensible cord which wraps around a pulley at C of negligible dimensions. The spring has a modulus (see Problem 2.79) of 50 lb/inch. How much is it stretched?

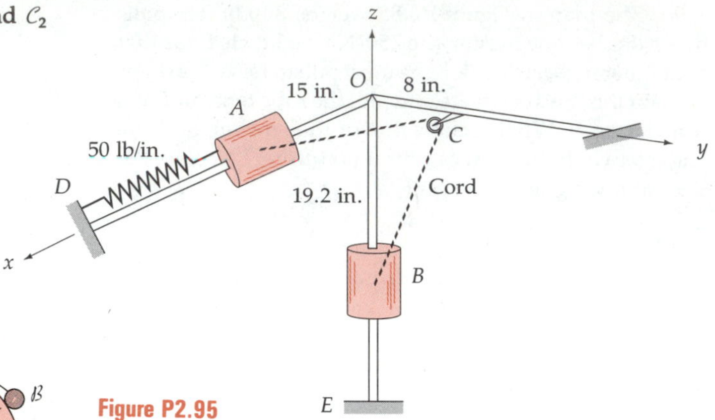

Figure P2.95

* **2.96** Three identical spheres that each weigh 10 N rest on a horizontal plane touching each other. They are tied together by a cord wrapped around their equatorial planes. A fourth 10-N sphere $\mathcal{D}$ is placed atop the others as shown in Figure P2.96(a). Neglecting friction, find the smallest cord tension needed to hold the spheres together. [Refer also to Figure P2.96(b).]

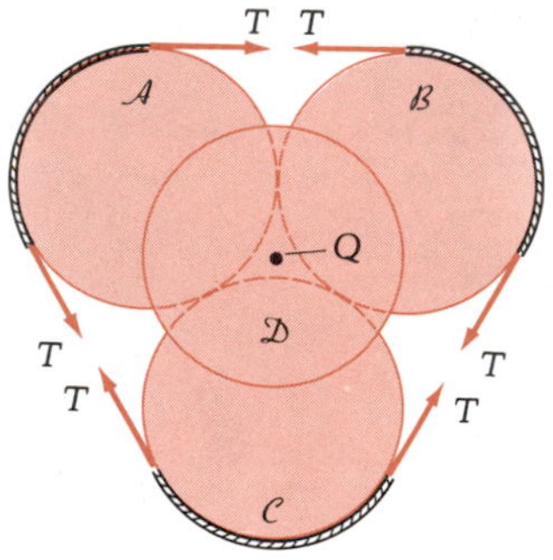

Figure P2.96(a)

Geometrical hints:

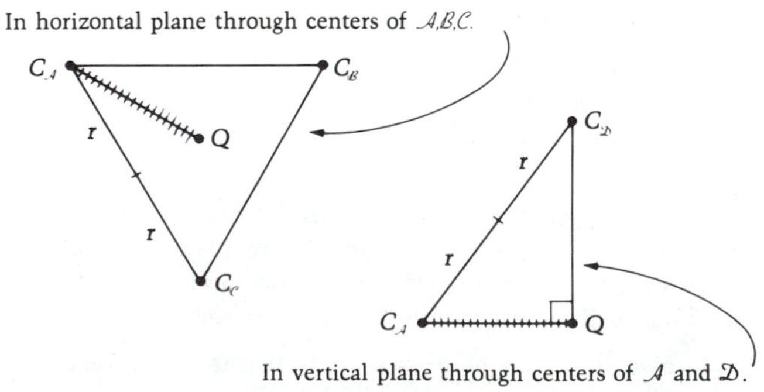

Q is in the plane of the three centers, equidistant from each.

Figure P2.96(b)

* **2.97** Four identical marbles are stacked as shown in Figure P2.97 with the three horizontal forces P maintaining equilibrium. There is no friction to be considered anywhere. Find the reactive forces between upper and lower marbles and give the minimum value of P for which equilibrium can exist.

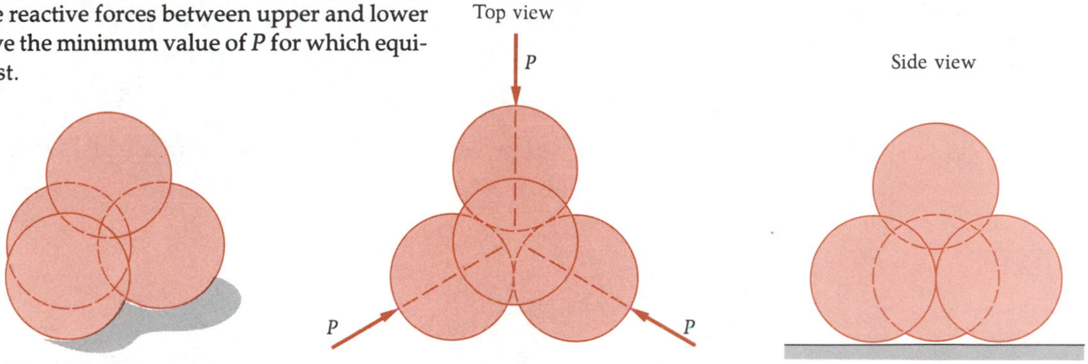

Figure P2.97

SUMMARY ▶ Chapter 2

In Chapter 2, we learned to express a force as a vector. Sometimes we do this in terms of the components (F_x, F_y, F_z) of the force:

$$\mathbf{F} = F_x\hat{\mathbf{i}} + F_y\hat{\mathbf{j}} + F_z\hat{\mathbf{k}},$$

while at other times we write the force as its magnitude $|\mathbf{F}|$ multiplied by a unit vector ($\hat{\mathbf{e}}$) in its direction:

$$\mathbf{F} = |\mathbf{F}|\hat{\mathbf{e}}$$

The two representations above are related by

$$\mathbf{F} = \underbrace{\sqrt{F_x^2 + F_y^2 + F_z^2}}_{|\mathbf{F}|} \quad \underbrace{\left(\frac{F_x\hat{\mathbf{i}} + F_y\hat{\mathbf{j}} + F_z\hat{\mathbf{k}}}{\sqrt{F_x^2 + F_y^2 + F_z^2}}\right)}_{\hat{\mathbf{e}} = e_x\hat{\mathbf{i}} + e_y\hat{\mathbf{j}} + e_z\hat{\mathbf{k}}}$$

where $F_z = 0$ if the force lies in the xy plane. The components (e_x, e_y, e_z) of the unit vector $\hat{\mathbf{e}}$ are also the cosines of the angles (θ_x, θ_y, θ_z) between the line of action of the force and the three coordinate axes (x, y, z).

To find the orthogonal component of a force $\mathbf{F}$ in the direction of a line L, we write a unit vector $\hat{\mathbf{u}}$ along the line, then form the dot product $\mathbf{F} \cdot \hat{\mathbf{u}}$. To write a unit vector along a line, we write *any* vector along the line, then divide by its magnitude. For example, if a 20-N force is directed from the point with coordinates (1, 3, 5) m to the point (4, 5, −6) m, then its vector representation is

$$\mathbf{F} = 20\,\frac{(4-1)\hat{\mathbf{i}} + (5-3)\hat{\mathbf{j}} + (-6-5)\hat{\mathbf{k}}}{\sqrt{3^2 + 2^2 + (-11)^2}}\ \text{N}$$

which may be simplified into component form as:

$$\mathbf{F} = 5.18\hat{\mathbf{i}} + 3.46\hat{\mathbf{j}} - 19.0\hat{\mathbf{k}}\ \text{N}$$

To analyze a particle in equilibrium, we first draw a free-body diagram of it, which is a sketch of the particle including all the external forces being exerted upon it. Then, with the help of the free-body diagram, the equilibrium equation $\Sigma \mathbf{F} = \mathbf{0}$ is written, which has the scalar component equations (if rectangular coordinates are used):

$$\Sigma F_x = 0 \qquad \Sigma F_y = 0 \qquad \Sigma F_z = 0$$

These equations may be written for each particle in a system if more than one constitute the system being analyzed.

REVIEW QUESTIONS ▶ Chapter 2

True or False?

1. A force on a body has to result from direct contact with another body.
2. Forces have magnitudes and directions but are not vectors because they do not obey the parallelogram law of addition.
3. It can be proved mathematically that forces are vectors.
4. Unit vectors have dimension of length.
5. The dot product of two vectors is a scalar.
6. The cross product of two vectors is a vector.
7. In particle equilibrium problems, only the equation $\Sigma \mathbf{F} = \mathbf{0}$ is needed.
8. Only the external forces acting on a body being analyzed for equilibrium are drawn on its free-body diagram.
9. If a frictionless pulley is in equilibrium, the tensions on either side, in a rope wrapped around it, are equal.

Answers: 1. F 2. F 3. F 4. F 5. T 6. T 7. T 8. T 9. T

3 The Moment of a Force; Resultants

3.1 Introduction

3.2 Moment of a Force About a Point

- Common Sense Definition
- Vector Representation
- Varignon's Theorem

3.3 Moment of a Force About a Line

- Definition of M_ℓ
- Physical Interpretation of M_ℓ
- Using "Force Times Perpendicular Distance" in Certain 3-D Applications to Find M_ℓ

3.4 The Couple

- Definition and Moment of a Couple
- The Most Important Property of a Couple

3.5 Laws of Equilibrium: Relationship Between Sums of Moments

- The Equilibrium Equations
- Relationship Between Sums of Moments

3.6 Equipollence of Force Systems

- The Meaning of Equipollence
- The Two Conditions for Equipollence

3.7 The Force-and-Couple Resultant of a System of Forces

- The Definition of a Resultant

3.8 The Simplest Resultant of a Force System

- The Single-Force Resultant
- Special Force Systems that are Equipollent to a Single Force
- The Simplest Resultant of a General Force-and-Couple System: A Collinear Force and Couple ("Screwdriver")

3.9 Distributed Force Systems

- Forces Distributed Along a Straight Line
- Dividing a Distributed Force System into Composite Parts
- Forces Distributed Over Surfaces
- Forces Distributed Throughout a Volume; Gravity

Summary

Review Questions

3.1 Introduction

In the preceding chapter, we learned that the external forces acting on a body in equilibrium sum to zero. For the relatively small subset of equilibrium problems examined in Chapter 2, that equation, $\Sigma \mathbf{F} = \mathbf{0}$, was all that was needed to complete the solution.

Most of the time, however, we will need a complementary, independent equation to complete the solution to statics problems. This second equation is that the moments, about an arbitrary point P, of all the forces acting on the body also add to zero. For this reason we shall spend a chapter learning a number of things about moments of forces.

In Section 3.2, we will begin by examining the moment of a force about a point, using three different definitions: a "common sense" formula, a vector representation, and a theorem which allows us to sum the moments of the *components* of a force and thereby obtain the moment of the entire force.

Sometimes we need to find the moment of a force about a line through a certain point, instead of about the point itself. We learn to do this in Section 3.3, where we also develop the physical interpretation that the moment of a force $\mathbf{F}$ about line ℓ is the turning effect, about ℓ, of the part of $\mathbf{F}$ that is perpendicular to ℓ.

Section 3.4 contains a study of the concept of the couple, which is a pair of non-collinear, equal-magnitude, oppositely directed forces. A couple will be seen to have a turning effect but no resultant force, and it has the same moment about every point of space. The couple is an important concept in the study of moments of forces.

In Section 3.5, we will present the other equilibrium equation (the "moment equation," $\Sigma \mathbf{M}_P = \mathbf{0}$) as a companion to the "force equation," $\Sigma \mathbf{F} = \mathbf{0}$, that we studied in Chapter 2. We then develop the relationship between the sum of the moments about two points ($\Sigma \mathbf{M}_P$ and $\Sigma \mathbf{M}_Q$) which leads in Section 3.6 to the concept of equipollent systems of forces, meaning they have equal power, or strength; more precisely, equipollent force systems make identical contributions to the equations of equilibrium (and also to the equations of motion in a later study of Dynamics).

Any system of forces and couples can be replaced at any point P by an equipollent system consisting of a force and couple there, which is called a resultant of the original system. This is proved and illustrated in Section 3.7, and followed in Section 3.8 by a further reduction to the simplest resultant. This resultant is just a simple force in the cases of concurrent, coplanar, and parallel force systems. For more complicated three-dimensional force systems, the simplest resultant is a collinear force and couple, which for obvious reasons is called a "screwdriver."

In the last section, 3.9, we examine distributed force systems. We shall find that the resultant of a continuously distributed system of parallel forces is the area beneath the loading curve, located at the centroid of this area.

3.2 Moment of a Force About a Point

Common Sense Definition

The moment of a force is a measure of the tendency of the force to turn a body to which the force is applied. The moment of a force *about a point* (or *with respect to a point*) is defined to be a vector whose magnitude is the product of (a) the magnitude of the force and (b) the perpendicular distance between the point and the line of action of the force. The vector is perpendicular to the plane defined by the point and the line of action of the force.

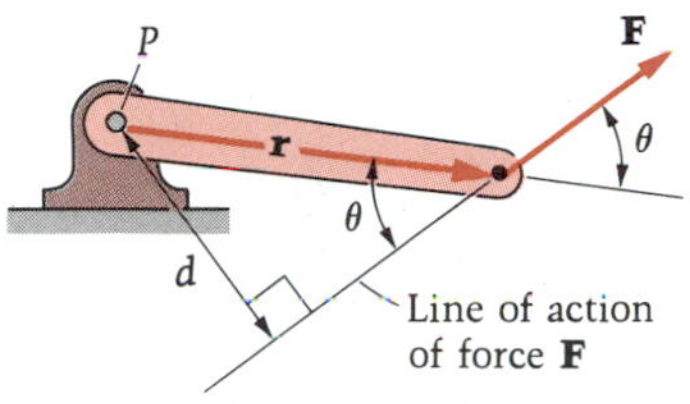

Figure 3.1

Question 3.1 Is there a case when the plane of P and $\mathbf{F}$ is not defined? If so, what is the moment?

The direction is assigned by the "right-hand rule": If the fingers of the right-hand curve in the direction of the perceived sense of the turning effect, the thumb will point in the direction of the moment. Thus, if we have the situation shown in Figure 3.1 and if we let $\mathbf{M}_P$ stand for the moment of $\mathbf{F}$ about P, then

$$\mathbf{M}_P = |\mathbf{F}|\,d\hat{\mathbf{n}} \tag{3.1}$$

In this case, the page is the plane of $\mathbf{r}$ and $\mathbf{F}$, and $\hat{\mathbf{n}}$ is a unit vector pointing out of the page toward the reader because we envision the turning effect to be counterclockwise in Figure 3.1. The magnitudes of the two sides of Equation (3.1) are of course equal:

$$|\mathbf{M}_P| = |\mathbf{F}|\,d \tag{3.2}$$

and this scalar equation allows us to find $|\mathbf{M}_P|$, $|\mathbf{F}|$, or d if we know the other two. Thus, for example, the perpendicular distance "d" from P to the line of action of the force is the magnitude of the moment $|\mathbf{M}_P|$ divided by the magnitude of the force $|\mathbf{F}|$.

Note that except for the case in Question 3.1, P, $\mathbf{r}$, and $\mathbf{F}$ always form a plane, so the above equations (3.1, 3.2) are always valid, whether the vectors are easy to depict (as in Figure 3.1) or not.

Vector Representation

Another way of representing $|\mathbf{M}_P|$ follows from the fact that, if $\mathbf{r}$ is the directed line segment from P to *any* point on the line of action of $\mathbf{F}$, then $d = |\mathbf{r}|\sin\theta$ as shown in Figure 3.1. Thus

$$\mathbf{M}_P = [|\mathbf{F}|\,|\mathbf{r}|\sin\theta]\hat{\mathbf{n}}$$

This result can be expressed in terms of the cross (or vector) product of the

Answer 3.1 If the point lies on the line of action of the force, then, of course, the plane isn't defined. But then the distance is zero, so the moment is zero.

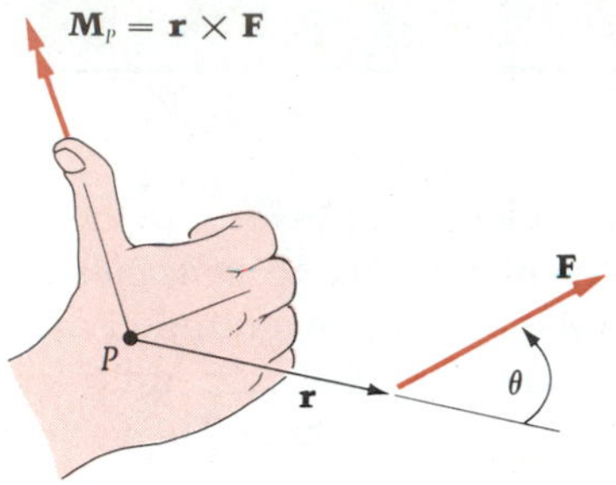

Figure 3.2

two vectors **r** and **F**, which is defined by

$$\mathbf{r} \times \mathbf{F} = [|\mathbf{r}||\mathbf{F}|\sin\theta]\hat{\mathbf{n}}$$

where $\hat{\mathbf{n}}$ is the unit vector determined by the right-hand rule as shown in Figure 3.2, that is, the right thumb points in the direction of $\hat{\mathbf{n}}$ (and hence of $\mathbf{r} \times \mathbf{F}$) if the fingers are turned from **r** into **F**.* Thus

$$\mathbf{M}_P = \mathbf{r} \times \mathbf{F} \tag{3.3}$$

Note from Figure 3.2 that we shall depict moment vectors by using double-headed arrows.

If we express **r** and **F** in component form, then

$$\begin{aligned}\mathbf{r} \times \mathbf{F} &= (r_x\hat{\mathbf{i}} + r_y\hat{\mathbf{j}} + r_z\hat{\mathbf{k}}) \times (F_x\hat{\mathbf{i}} + F_y\hat{\mathbf{j}} + F_z\hat{\mathbf{k}})\\ &= r_xF_x(\hat{\mathbf{i}} \times \hat{\mathbf{i}}) + r_xF_y(\hat{\mathbf{i}} \times \hat{\mathbf{j}}) + r_xF_z(\hat{\mathbf{i}} \times \hat{\mathbf{k}})\\ &\quad + r_yF_x(\hat{\mathbf{j}} \times \hat{\mathbf{i}}) + r_yF_y(\hat{\mathbf{j}} \times \hat{\mathbf{j}}) + r_yF_z(\hat{\mathbf{j}} \times \hat{\mathbf{k}})\\ &\quad + r_zF_x(\hat{\mathbf{k}} \times \hat{\mathbf{i}}) + r_zF_y(\hat{\mathbf{k}} \times \hat{\mathbf{j}}) + r_zF_z(\hat{\mathbf{k}} \times \hat{\mathbf{k}})\end{aligned}$$

But by the definition of the cross product,

$$\hat{\mathbf{i}} \times \hat{\mathbf{i}} = 1(1)\sin 0\ \hat{\mathbf{n}} = \mathbf{0}$$

and similarly

$$\hat{\mathbf{j}} \times \hat{\mathbf{j}} = \hat{\mathbf{k}} \times \hat{\mathbf{k}} = \mathbf{0}$$

If we specify $\hat{\mathbf{i}}$, $\hat{\mathbf{j}}$, and $\hat{\mathbf{k}}$ to constitute a right-handed system (as will be the case throughout this book), then

$$\begin{aligned}\hat{\mathbf{i}} \times \hat{\mathbf{j}} &= (1)(1)\sin 90^\circ\ \hat{\mathbf{k}} = \hat{\mathbf{k}}\\ \hat{\mathbf{j}} \times \hat{\mathbf{k}} &= \hat{\mathbf{i}}\\ \hat{\mathbf{k}} \times \hat{\mathbf{i}} &= \hat{\mathbf{j}}\end{aligned}$$

and similarly

$$\begin{aligned}\hat{\mathbf{j}} \times \hat{\mathbf{i}} &= -\hat{\mathbf{k}}\\ \hat{\mathbf{k}} \times \hat{\mathbf{j}} &= -\hat{\mathbf{i}}\\ \hat{\mathbf{i}} \times \hat{\mathbf{k}} &= -\hat{\mathbf{j}}\end{aligned}$$

Therefore

$$\begin{aligned}\mathbf{r} \times \mathbf{F} &= r_xF_y(\hat{\mathbf{k}}) + r_xF_z(-\hat{\mathbf{j}})\\ &\quad + r_yF_x(-\hat{\mathbf{k}}) + r_yF_z(\hat{\mathbf{i}})\\ &\quad + r_zF_x(\hat{\mathbf{j}}) + r_zF_y(-\hat{\mathbf{i}})\\ &= (r_yF_z - r_zF_y)\hat{\mathbf{i}} + (r_zF_x - r_xF_z)\hat{\mathbf{j}}\\ &\quad + (r_xF_y - r_yF_x)\hat{\mathbf{k}}\end{aligned}$$

* Through the smaller ($<180^\circ$) of the two angles between **r** and **F** in their plane.

The reader can easily verify that this can be put in the form of a determinant

$$\mathbf{r} \times \mathbf{F} = \begin{vmatrix} \hat{\mathbf{i}} & \hat{\mathbf{j}} & \hat{\mathbf{k}} \\ r_x & r_y & r_z \\ F_x & F_y & F_z \end{vmatrix}$$

The cross-product method of finding $\mathbf{M}_P$ is particularly useful in a situation in which the plane containing P and the line of action of $\mathbf{F}$ is not a natural reference plane for the problem under investigation. In such a circumstance the determination of d and $\hat{\mathbf{n}}$ by nonvector methods of analytic geometry becomes a difficult task. Equation (3.3) effectively reduces this task to a single straightforward operation.

Question 3.2 Why does $\mathbf{r} \times \mathbf{F}$ yield the same result for $\mathbf{M}_P$ regardless of which point on the line of action of $\mathbf{F}$ is intersected by $\mathbf{r}$?

Question 3.3 Does it matter whether $\mathbf{M}_P$ is computed as $\mathbf{r} \times \mathbf{F}$ or $\mathbf{F} \times \mathbf{r}$?

Varignon's Theorem

We now proceed to prove a very important theorem concerning the moments of forces. Suppose that we have the two forces of Figure 3.3 again acting on the crate at point A, and suppose that we are now interested in the moment about the lower right-hand corner point, P, of the sum, $\mathbf{F}$, of $\mathbf{F}_1$ and $\mathbf{F}_2$.

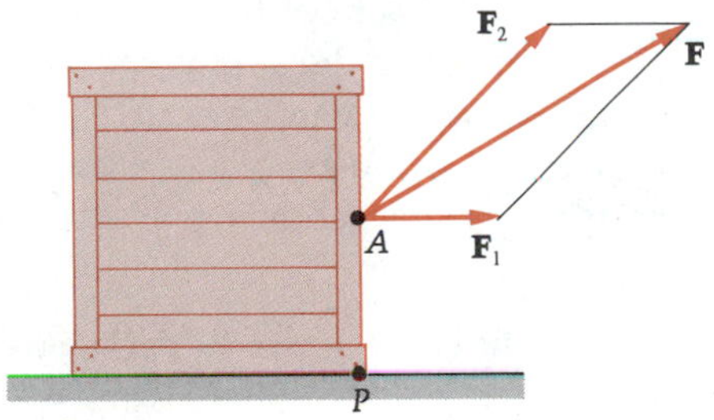

Figure 3.3

Then

$$\begin{aligned} \mathbf{M}_P &= \mathbf{r} \times \mathbf{F} \\ &= \mathbf{r} \times (\mathbf{F}_1 + \mathbf{F}_2) \end{aligned}$$

Answer 3.2 In Figure 3.1 we see that $|\mathbf{r}| \sin \theta$ is a constant, making $|\mathbf{M}_P|$ the same for all intersection points. This figure also shows by the right-hand rule that the *direction* of the cross product is also independent of the intersection point. Thus $\mathbf{M}_P$ is the same regardless of which point on $\mathbf{F}$ is intersected by $\mathbf{r}$ — that is, regardless of which $\mathbf{r}$ is used to form it.

Answer 3.3 Yes! $\mathbf{r} \times \mathbf{F} = -(\mathbf{F} \times \mathbf{r})$, so if $\mathbf{F} \times \mathbf{r}$ is used, the moment will be in the wrong direction.

and by the distributive property of the cross product,

$$\mathbf{M}_P = \mathbf{r} \times \mathbf{F}_1 + \mathbf{r} \times \mathbf{F}_2 = |\mathbf{F}_1| d_1 \hat{\mathbf{n}}_1 + |\mathbf{F}_2| d_2 \hat{\mathbf{n}}_2$$

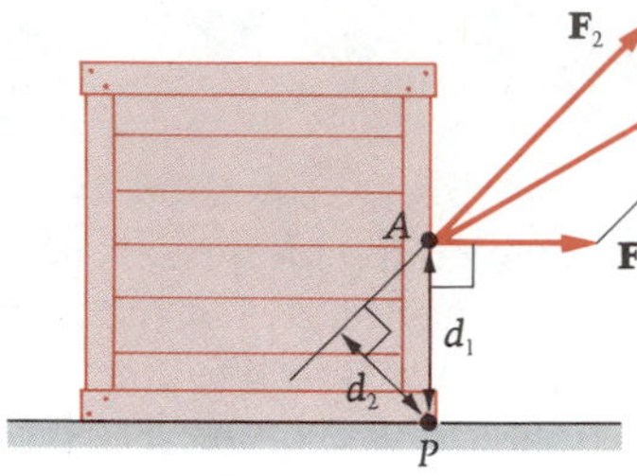

Figure 3.4

where d_1 and d_2 are shown in Figure 3.4 and where $\hat{\mathbf{n}}_1$ and $\hat{\mathbf{n}}_2$ are the same unit vector, directed into the paper. Thus we see that the moment of **F** about P is equal to the sum of the moments of $\mathbf{F}_1$ and $\mathbf{F}_2$ about P. By extension we can say in general that for any number n of forces acting at a common point A,

$$\begin{aligned}\mathbf{M}_P &= \mathbf{r} \times \mathbf{F}\\ &= \mathbf{r} \times (\mathbf{F}_1 + \mathbf{F}_2 + \cdots + \mathbf{F}_n)\\ &= \mathbf{r} \times \mathbf{F}_1 + \mathbf{r} \times \mathbf{F}_2 + \cdots + \mathbf{r} \times \mathbf{F}_n\\ \mathbf{M}_P &= |\mathbf{F}_1| d_1 \hat{\mathbf{n}}_1 + |\mathbf{F}_2| d_2 \hat{\mathbf{n}}_2 + \cdots + |\mathbf{F}_n| d_n \hat{\mathbf{n}}_n \qquad (3.4)\end{aligned}$$

Therefore the moment about a point P of the sum of n forces acting at a point A is the sum of the moments of the separate forces (more briefly, "the moment of the sum is the sum of the moments"). This statement, which perhaps seems obvious, is often known as *Varignon's Theorem.* It is of practical value especially when we can decompose a force like **F** into parts whose perpendicular distances from P are easily determined.

Question 3.4 Does the development of Equation 3.4 require that $\mathbf{F}_1$, $\mathbf{F}_2, \ldots, \mathbf{F}_n$ all lie in the same plane?

In the example to follow, we will use each of the above approaches [Equations (3.1), (3.3), and (3.4)] to calculate the moment of a force about a point.

Answer 3.4 No. This was never required in the derivation, and thus it isn't necessary.

EXAMPLE 3.1

Calculate the moment of the 10-N force with respect to point O, the origin of the rectangular coordinate system shown in Figure E3.1a. Use all three approaches suggested by Equation (3.1), (3.3), and (3.4).

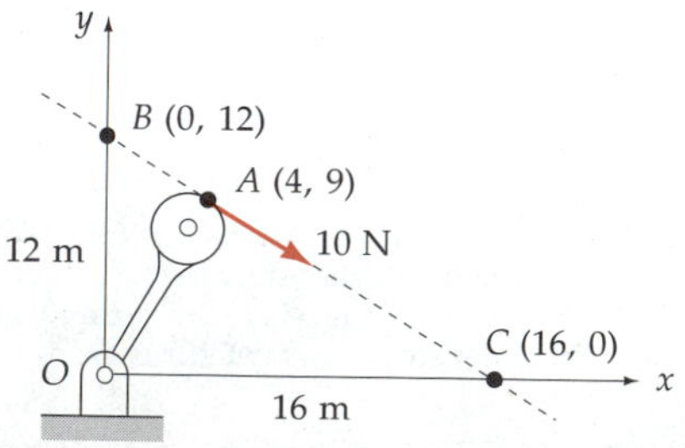

Figure E3.1a

Solution

To use Equation (3.1), we need the perpendicular distance d between O and the line of action of the force. It is calculated below to be 9.6 meters by using the similar (shaded) triangles, in Figure E3.1b.

$$\frac{16}{d} = \frac{5}{3} \Rightarrow d = \frac{48}{5} = 9.6 \text{ m}$$

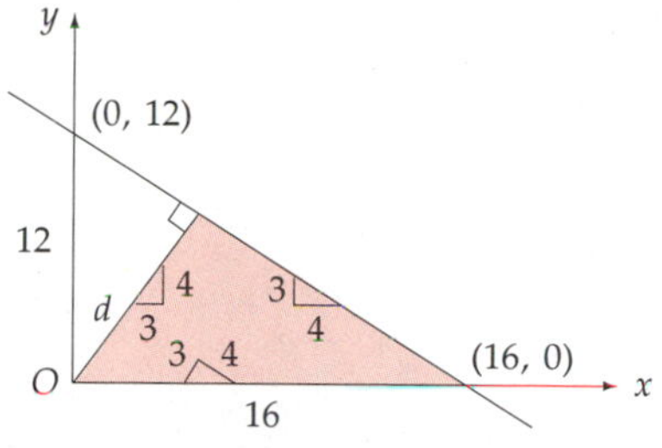

Figure E3.1b

Therefore, Equation (3.1) yields

$$\mathbf{M}_O = |\mathbf{F}|\, d\hat{\mathbf{n}} = (10)(9.6)(-\hat{\mathbf{k}}) = -96\hat{\mathbf{k}} \text{ N} \cdot \text{m or } 96 \circlearrowright \text{ N} \cdot \text{m}$$

We emphasize again that this approach is of little value in three-dimensional situations in which the perpendicular distance d is not easily calculated and the unit vector is not obvious.

To use Equation (3.3), we note that the force is $\mathbf{F} = 8\hat{\mathbf{i}} - 6\hat{\mathbf{j}}$ N and the directed line segment from O to point A on the line of action of $\mathbf{F}$ is $\mathbf{r}_{OA} = 4\hat{\mathbf{i}} + 9\hat{\mathbf{j}}$ m. So, by Equation (3.3),

$$\begin{aligned}\mathbf{M}_O = \mathbf{r}_{OA} \times \mathbf{F} &= (4\hat{\mathbf{i}} + 9\hat{\mathbf{j}}) \times (8\hat{\mathbf{i}} - 6\hat{\mathbf{j}}) \\ &= (4)(-6)(\hat{\mathbf{i}} \times \hat{\mathbf{j}}) + (9)(8)(\hat{\mathbf{j}} \times \hat{\mathbf{i}}) \\ &= -24\hat{\mathbf{k}} + 72(-\hat{\mathbf{k}}) = -96\hat{\mathbf{k}} \text{ N} \cdot \text{m} \qquad \text{(as above)}\end{aligned}$$

In arriving at this result, we have recalled that the cross-product of a vector with itself is zero, i.e.,

$$\hat{\mathbf{i}} \times \hat{\mathbf{i}} = \hat{\mathbf{j}} \times \hat{\mathbf{j}} = \hat{\mathbf{k}} \times \hat{\mathbf{k}} = \mathbf{0}.$$

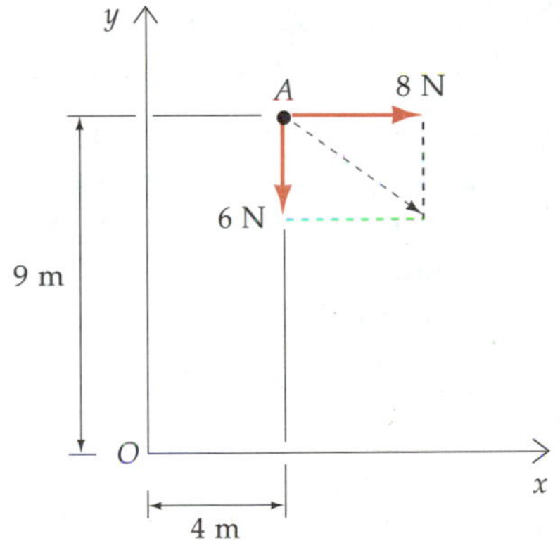

Figure E3.1c

To use Equation (3.4), we shall decompose the force $\mathbf{F}$ into the 8-N and 6-N forces at A as shown in Figure E3.1c. We then revert to the "force times perpendicular distance" definition to calculate separately the moments of these component forces with respect to O.* For the 8-N force we have

$$\mathbf{M}'_O = 8(9)(-\hat{\mathbf{k}}) = -72\hat{\mathbf{k}} \text{ N} \cdot \text{m}$$

and for the 6-N force we have

$$\mathbf{M}''_O = 6(4)(-\hat{\mathbf{k}}) = -24\hat{\mathbf{k}} \text{ N} \cdot \text{m}$$

Therefore the sum is $\mathbf{M}_O = -96\hat{\mathbf{k}}$ N · m, the same result we obtained in the first two parts of the example. Note how the unit vector is attached using the

* This is the practical importance of Varignon's Theorem.

right-hand rule. Also, it is important to observe the correspondence of $\mathbf{M}'_O$ and $\mathbf{M}''_O$ with the two terms in the cross-product calculation above. We have obtained the result without the formality of taking the cross product.

Before leaving this example, note that because point B is also on the line of action of $\mathbf{F}$, we could alternatively compute $\mathbf{M}_O$ by using $\mathbf{r}_{OB}$ instead of $\mathbf{r}_{OA}$:

$$\mathbf{M}_O = \mathbf{r}_{OB} \times \mathbf{F} = 12\hat{\mathbf{j}} \times (8\hat{\mathbf{i}} - 6\hat{\mathbf{j}}) = 12(8)(\hat{\mathbf{j}} \times \hat{\mathbf{i}})$$
$$= 96(-\hat{\mathbf{k}}) = -96\hat{\mathbf{k}} \text{ N} \cdot \text{m}$$

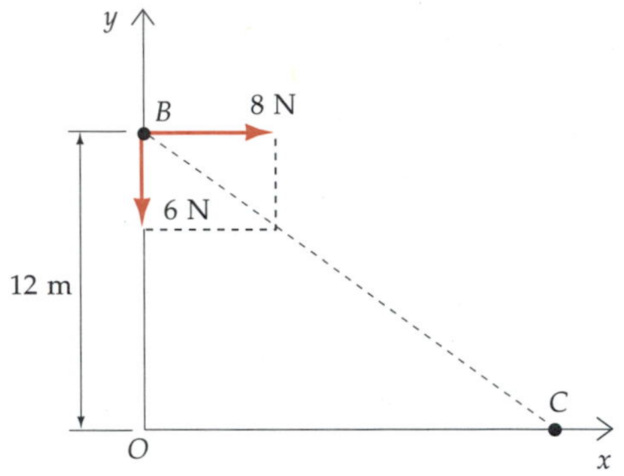

Figure E3.1d

The equivalent informal calculation follows from decomposing $\mathbf{F}$ into 8-N and 6-N components with lines of action intersecting at B (see Figure E3.1d). In this case the 6-N force produces no moment about O since its line of action passes through O, and the 8-N force produces a moment of $8(12)(-\hat{\mathbf{k}})$ so that (for the fifth time!):

$$\mathbf{M}_O = \mathbf{0} + (-96\hat{\mathbf{k}}) = -96\hat{\mathbf{k}} \text{ N} \cdot \text{m}$$

This illustrates the fact that it matters not where $\mathbf{F}$ is decomposed on its line of action in order to obtain a moment. In Problem 3.4, the student will be asked to recompute $\mathbf{M}_O$ by breaking the 10-N force into components at point C.

In the following example, the plane of $\mathbf{r}$ and $\mathbf{F}$ is skewed with respect to the xy, yz, and xz planes, and we shall use the cross product to advantage since finding the value of d needed in Equation (3.1) would actually require the same vector operation, the cross product.

EXAMPLE 3.2

A position vector from point P to point Q on the line of action of a force $\mathbf{F}$ is given by

$$\mathbf{r}_{PQ} = 2\hat{\mathbf{i}} + 3\hat{\mathbf{j}} - 4\hat{\mathbf{k}} \text{ ft}$$

The force $\mathbf{F}$ is

$$\mathbf{F} = \hat{\mathbf{i}} - 2\hat{\mathbf{j}} + 5\hat{\mathbf{k}} \text{ lb}$$

Find the moment of $\mathbf{F}$ about point P.

Solution

By Equation (3.3),

$$\begin{aligned}
\mathbf{M}_P = \mathbf{r}_{PQ} \times \mathbf{F} &= (2\hat{\mathbf{i}} + 3\hat{\mathbf{j}} - 4\hat{\mathbf{k}}) \times (\hat{\mathbf{i}} - 2\hat{\mathbf{j}} + 5\hat{\mathbf{k}}) \\
&= 2(1)(\hat{\mathbf{i}} \times \hat{\mathbf{i}}) + 2(-2)(\hat{\mathbf{i}} \times \hat{\mathbf{j}}) + 2(5)(\hat{\mathbf{i}} \times \hat{\mathbf{k}}) + 3(1)(\hat{\mathbf{j}} \times \hat{\mathbf{i}}) \\
&\quad + 3(-2)(\hat{\mathbf{j}} \times \hat{\mathbf{j}}) + 3(5)(\hat{\mathbf{j}} \times \hat{\mathbf{k}}) + (-4)(1)(\hat{\mathbf{k}} \times \hat{\mathbf{i}}) \\
&\quad + (-4)(-2)(\hat{\mathbf{k}} \times \hat{\mathbf{j}}) + (-4)(5)(\hat{\mathbf{k}} \times \hat{\mathbf{k}}) \\
&= (2)(0) + (-4)(\hat{\mathbf{k}}) + (10)(-\hat{\mathbf{j}}) + (3)(-\hat{\mathbf{k}}) + (-6)(0) \\
&\quad + (15)(\hat{\mathbf{i}}) + (-4)(\hat{\mathbf{j}}) + 8(-\hat{\mathbf{i}}) + (-20)(0) \\
&= 7\hat{\mathbf{i}} - 14\hat{\mathbf{j}} - 7\hat{\mathbf{k}} \text{ lb-ft}
\end{aligned}$$

Alternatively, we could use the determinant method of computing the cross product:

$$\begin{aligned}
\mathbf{M}_P = \mathbf{r}_{PQ} \times \mathbf{F} &= \begin{vmatrix} \hat{\mathbf{i}} & \hat{\mathbf{j}} & \hat{\mathbf{k}} \\ 2 & 3 & -4 \\ 1 & -2 & 5 \end{vmatrix} \\
&= \hat{\mathbf{i}} \begin{vmatrix} 3 & -4 \\ -2 & 5 \end{vmatrix} - \hat{\mathbf{j}} \begin{vmatrix} 2 & -4 \\ 1 & 5 \end{vmatrix} + \hat{\mathbf{k}} \begin{vmatrix} 2 & 3 \\ 1 & -2 \end{vmatrix} \\
&= \hat{\mathbf{i}}(15 - 8) - \hat{\mathbf{j}}(10 + 4) + \hat{\mathbf{k}}(-4 - 3) \\
&= 7\hat{\mathbf{i}} - 14\hat{\mathbf{j}} - 7\hat{\mathbf{k}} \text{ lb-ft}
\end{aligned}$$

(as before, but obtained much more quickly)

Sometimes we have to compute the position vector before we can take the moment, as in the next example:

EXAMPLE 3.3

A force, $\mathbf{F} = 3\hat{\mathbf{i}} - 5\hat{\mathbf{j}} + \hat{\mathbf{k}}$ lb, has a line of action through point A of Figure E3.3 with coordinates (0, 3, 4) ft. (a) Find the moment of $\mathbf{F}$ about point B whose coordinates are (4, 1, 2) ft. (b) Find the distance from point B to the line of action of $\mathbf{F}$.

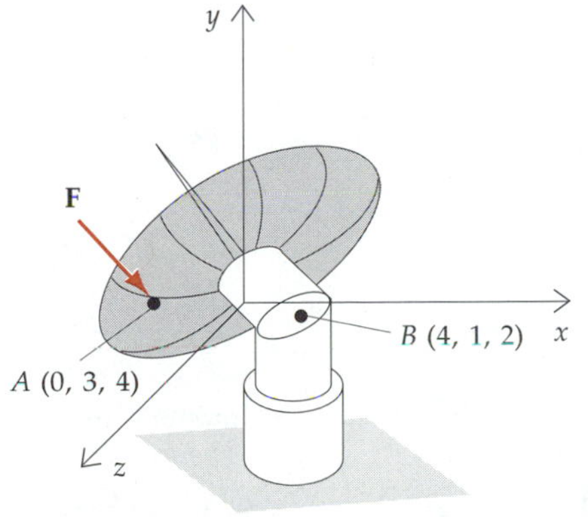

Figure E3.3

Solution

(a) The moment about B is given by

$$\mathbf{M}_B = \mathbf{r}_{BA} \times \mathbf{F}$$

where

$$\mathbf{r}_{BA} = (0 - 4)\hat{\mathbf{i}} + (3 - 1)\hat{\mathbf{j}} + (4 - 2)\hat{\mathbf{k}} = -4\hat{\mathbf{i}} + 2\hat{\mathbf{j}} + 2\hat{\mathbf{k}} \text{ ft}$$

Thus

$$\begin{aligned}\mathbf{M}_B &= (-4\hat{\mathbf{i}} + 2\hat{\mathbf{j}} + 2\hat{\mathbf{k}}) \times (3\hat{\mathbf{i}} - 5\hat{\mathbf{j}} + \hat{\mathbf{k}}) \\ &= (-4)(-5)\hat{\mathbf{k}} + (-4)(1)(-\hat{\mathbf{j}}) + (2)(3)(-\hat{\mathbf{k}}) + 2(1)\hat{\mathbf{i}} \\ &\quad + 2(3)\hat{\mathbf{j}} + 2(-5)(-\hat{\mathbf{i}}) \\ &= 12\hat{\mathbf{i}} + 10\hat{\mathbf{j}} + 14\hat{\mathbf{k}} \text{ lb-ft}\end{aligned}$$

(b) The distance from B to the line of action of $\mathbf{F}$ is found by Equation (3.2):

$$d = \frac{|\mathbf{M}_B|}{|\mathbf{F}|}$$

Computing the magnitudes of $\mathbf{M}_B$ and $\mathbf{F}$, we find:

$$|\mathbf{M}_B| = \sqrt{(12)^2 + (10)^2 + (14)^2} = \sqrt{440} = 21.0 \text{ lb-ft}$$
$$|\mathbf{F}| = \sqrt{(3)^2 + (-5)^2 + (1)^2} = \sqrt{35} = 5.92 \text{ lb}$$

Therefore

$$d = \frac{21.0}{5.92} = 3.55 \text{ ft}$$

PROBLEMS ▸ Section 3.2

3.1 Find the moment of the 12-N force about the origin in Figure P3.1.

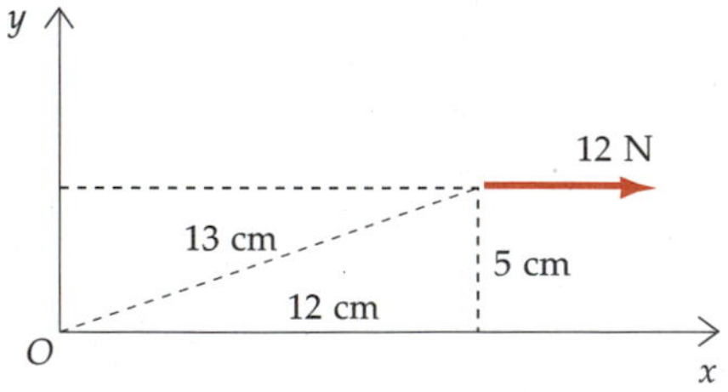

Figure P3.1

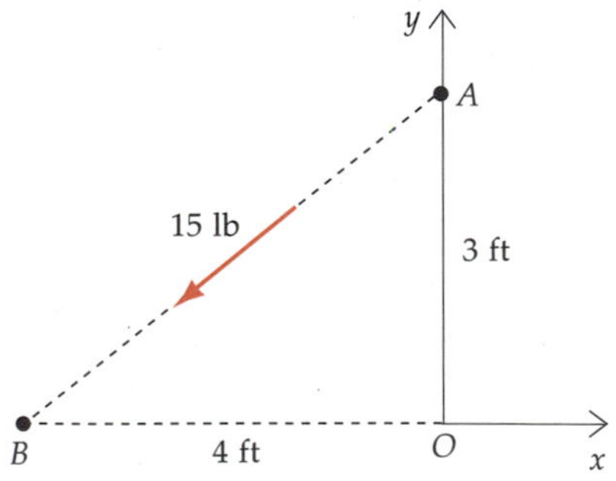

Figure P3.2

3.2 Find the moment of the force in Figure P3.2 about the origin using Equation (a) 3.1; (b) 3.3.

3.3 In the preceding problem, break the force into its components at (a) A and (b) B, and in each case find the moment of the force about the origin using Varignon's Theorem.

3.4 In Example 3.1, compute $\mathbf{M}_O$ by using the fact that C is also on the line of action of $\mathbf{F}$. Do this in two ways:

a. vectorially, with $\mathbf{r}_{OC} \times \mathbf{F}$;

b. by decomposing **F** into its x- and y-components at C and then summing the separate moments of the components.

3.5 A force **F** with magnitude 1000 lb is exerted on the tooth of the sector gear shown in Figure P3.5 by a tooth of another gear that is not shown. The force makes a 20° angle with the normal to the radius drawn to the tooth from point O, as shown in the small figure. Find the moment of **F** about O, and the perpendicular distance from O to the line of action of the force.

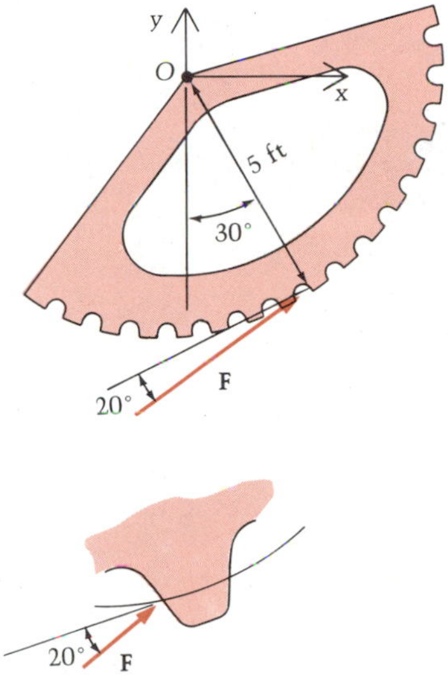

Figure P3.5

b. Calculate the moments about B of the three components of **F** acting at C by multiplying these forces by their perpendicular distances from B. Assign the correct unit vector to each using the right-hand rule and add them to form $\mathbf{M}_B$.

c. Calculate the moment of **F** about B using $\mathbf{r}_{BC} \times \mathbf{F}$, and compare it with $\mathbf{M}_B$ from part (b).

d. Calculate the moment of **F** about B using $\mathbf{r}_{BA} \times \mathbf{F}$, and note that the result is still the same because A also lies on the line of action of **F**.

3.7 Find the sum of the moments about point A of the five applied forces. (See Figure P3.7.)

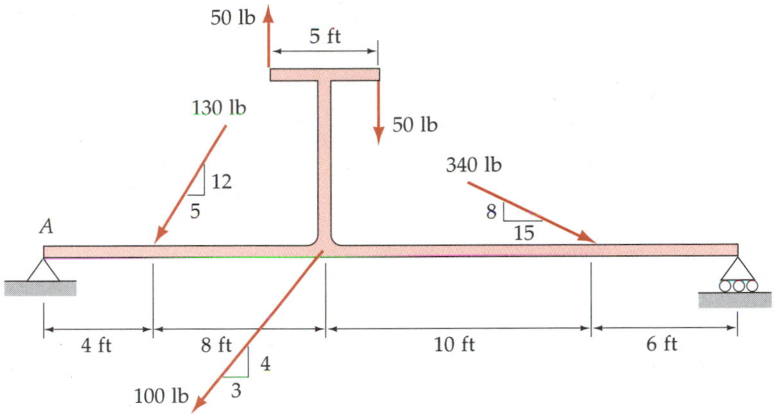

Figure P3.7

3.6 The force **F** in Figure P3.6 acts along the line CA and has a magnitude of 520 N.

a. Write the unit vector $\hat{\mathbf{u}}$ that has the direction CA, and use it to form the vector force **F**.

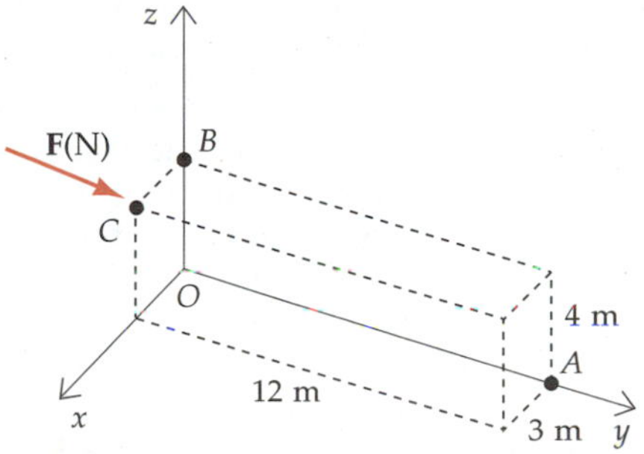

Figure P3.6

3.8 In Figure P3.8, find:

a. the moment of the 50-lb force about point P having the coordinates $(x, y, z) = (2, 3, 5)$ ft;

b. the perpendicular distance from P to the line of action of the force.

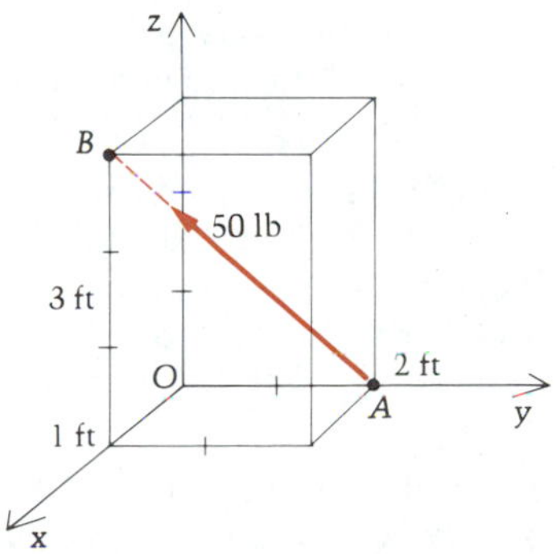

Figure P3.8

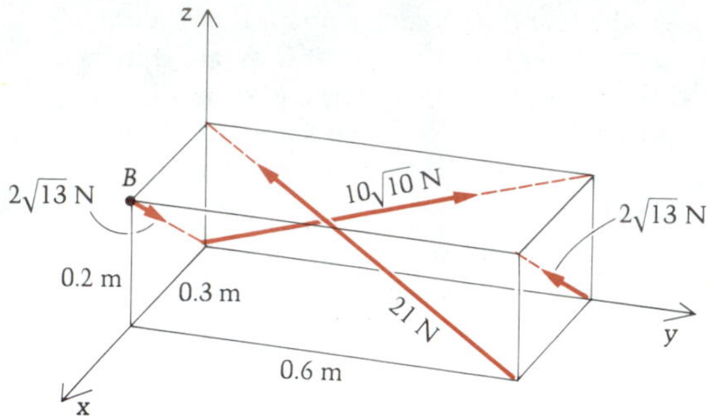

Figure P3.9

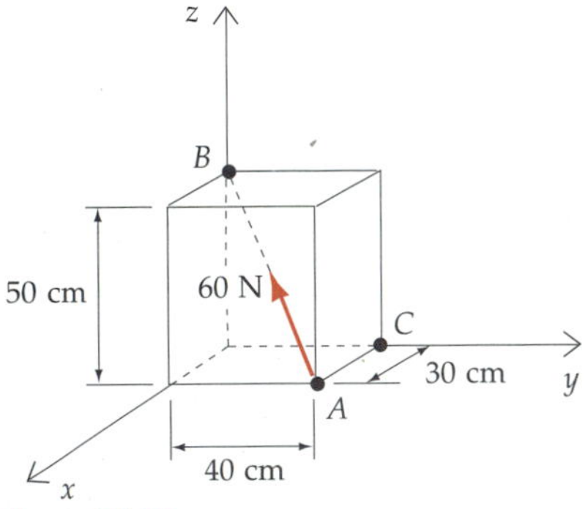

Figure P3.10

3.9 Find (see Figure P3.9):

a. The sum of the forces

b. The sum of their moments about point B.

3.10 Determine the moment of the 60-N force in Figure P3.10 with respect to corner C using

a. $\mathbf{r} \times \mathbf{F}$;

b. Varignon's Theorem after resolving the force into its components at A.

3.11 Find the moment about the origin due to the force $\mathbf{F} = -30\hat{\mathbf{i}} + 40\hat{\mathbf{k}}$ lb, which acts at the point $(x, y, z) = (20, 10, 0)$ ft.

3.12 In terms of angle θ, find the moment of the weight W about the base A of the mast in Figure P3.12 if the intersection line (dashed) of the deck plane (xy) with the plane of the mast, boom, and cable forms angles with x and y of 30° and 60°, respectively. The length of the boom is 10 m.

3.13 In raising the heavy mast $\mathcal{M}$ in Figure P3.13, the tension T in the cable is supplying a moment about O of magnitude 5000 N · m. Find the tension in the cable.

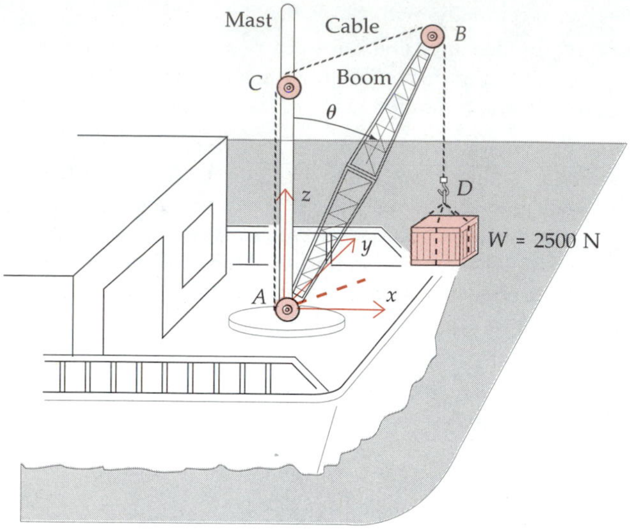

Figure P3.12

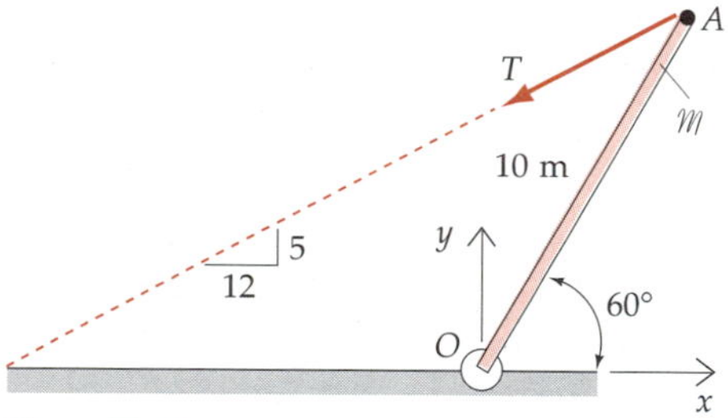

Figure P3.13

3.14 The force of 140 lb acts at A with a line of action directed toward B. (See Figure P3.14.) Find the moment of this force about O and the shortest distance from O to the line of action of the force.

* **3.15** A force of given magnitude P lb has a line of action through A, and is to act on the rod, as shown. Find (see Figure P3.15) the angle θ for which the moment of the force about B is a maximum.

* Asterisks identify the more difficult problems.

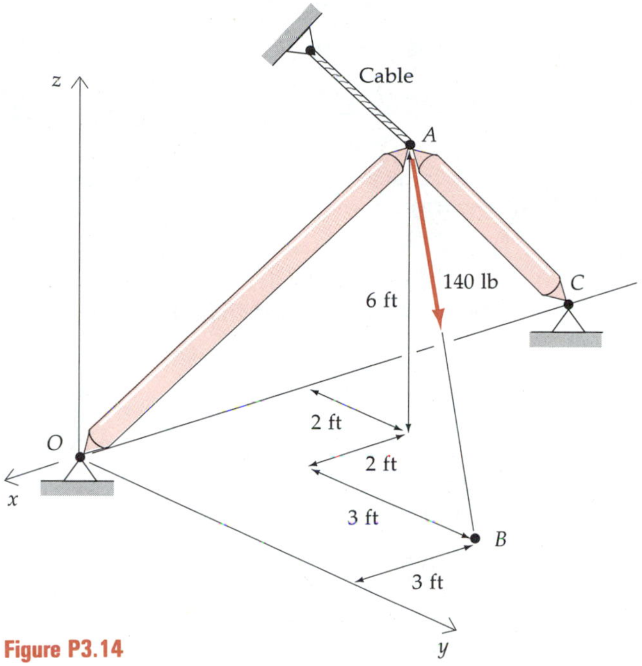

Figure P3.14

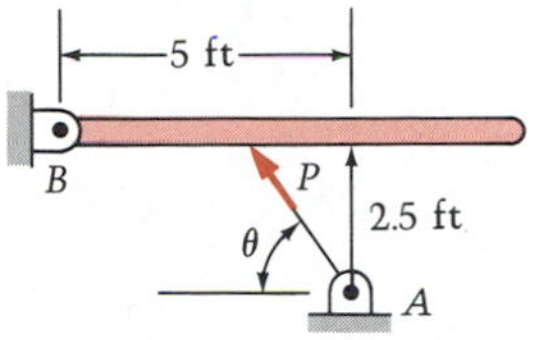

Figure P3.15

3.3 Moment of a Force About a Line

Definition of $\mathbf{M}_\ell$

In statics we are often interested in knowing the moment of a force about a line rather than about a point. As suggested by Figure 3.5, we shall define the moment of a force **F** about a line ℓ to be the projection along ℓ of the moment of **F** about any point P lying on ℓ. The moment $\mathbf{M}_\ell$ is thus defined to be

$$\mathbf{M}_\ell = (\mathbf{M}_P \cdot \hat{\mathbf{u}})\hat{\mathbf{u}} \tag{3.5}$$

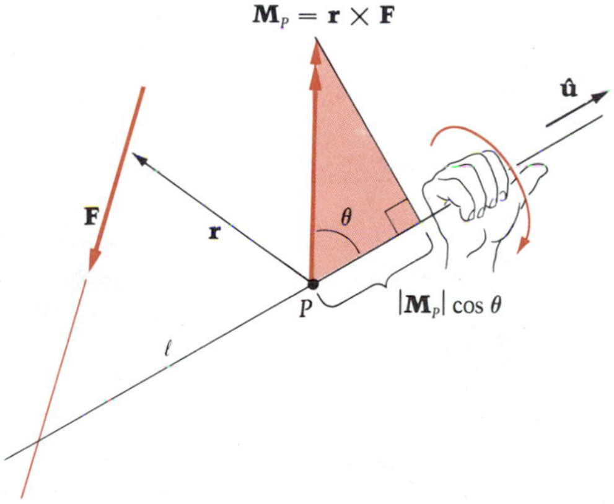

Figure 3.5

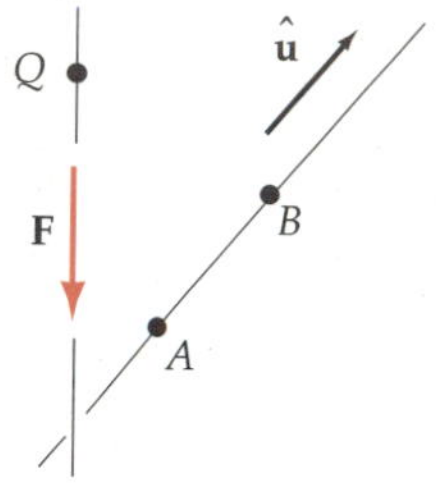

Figure 3.6

or

$$\mathbf{M}_\ell = [(\mathbf{r} \times \mathbf{F}) \cdot \hat{\mathbf{u}}]\hat{\mathbf{u}} \tag{3.6}$$

where $\hat{\mathbf{u}}$ is the unit vector in the direction of ℓ.

Question 3.5 How would the calculated $\mathbf{M}_\ell$ be different if $-\hat{\mathbf{u}}$ had been chosen as the reference unit vector instead of $\hat{\mathbf{u}}$ in Equation 3.6?

Our definition of $\mathbf{M}_\ell$ (Equation 3.5) suggests that the same value is obtained regardless of the location of P on the line ℓ. In fact we can prove, if A and B are points on ℓ (see Figure 3.6), that

$$\mathbf{M}_\ell = (\mathbf{M}_B \cdot \hat{\mathbf{u}})\hat{\mathbf{u}} = (\mathbf{M}_A \cdot \hat{\mathbf{u}})\hat{\mathbf{u}}$$

To do this, we use the fact that if Q is any point on the line of action of $\mathbf{F}$, then

$$\mathbf{M}_A = \mathbf{r}_{AQ} \times \mathbf{F}$$

But

$$\mathbf{r}_{AQ} = \mathbf{r}_{AB} + \mathbf{r}_{BQ}$$

so that

$$\mathbf{M}_A = \mathbf{r}_{AB} \times \mathbf{F} + \mathbf{r}_{BQ} \times \mathbf{F}$$

Thus

$$(\mathbf{M}_A \cdot \hat{\mathbf{u}})\hat{\mathbf{u}} = [(\mathbf{r}_{AB} \times \mathbf{F}) \cdot \hat{\mathbf{u}}]\hat{\mathbf{u}} + [(\mathbf{r}_{BQ} \times \mathbf{F}) \cdot \hat{\mathbf{u}}]\hat{\mathbf{u}}$$

But $\mathbf{r}_{AB} \times \mathbf{F}$ is perpendicular to ℓ, that is, to $\hat{\mathbf{u}}$, so that

$$(\mathbf{M}_A \cdot \hat{\mathbf{u}})\hat{\mathbf{u}} = 0 + [(\mathbf{r}_{BQ} \times \mathbf{F}) \cdot \hat{\mathbf{u}}]\hat{\mathbf{u}}$$

$$(\mathbf{M}_A \cdot \hat{\mathbf{u}})\hat{\mathbf{u}} = (\mathbf{M}_B \cdot \hat{\mathbf{u}})\hat{\mathbf{u}} \tag{3.7}$$

and thus $\mathbf{M}_\ell$ is independent of the point on ℓ which is chosen for the computation.

Physical Interpretation of $\mathbf{M}_\ell$

We now provide a physical interpretation of $\mathbf{M}_\ell$ as the turning effect, about ℓ, of the part of $\mathbf{F}$ that is perpendicular to ℓ. To do this we first identify, as in Figure 3.7, the plane defined by the line of action of $\mathbf{F}$ and any line intersecting it and parallel to ℓ. Let d be the distance between this plane and ℓ, as shown in Figure 3.7.

Question 3.6 If $\mathbf{F}$ is itself already parallel to ℓ, then this plane does not exist. Why? What is $\mathbf{M}_\ell$ in this case?

Answer 3.5 It would be the same.

Answer 3.6 If $\mathbf{F} \parallel \ell$, then the only line intersecting $\mathbf{F}$ and parallel to ℓ is coincident with $\mathbf{F}$; thus not a plane, but just a line, is defined. In that case, $\mathbf{r} \times \mathbf{F}$ is $\perp\hat{\mathbf{u}}$, so $\mathbf{M}_\ell = \mathbf{0}$ by Equation (3.6).

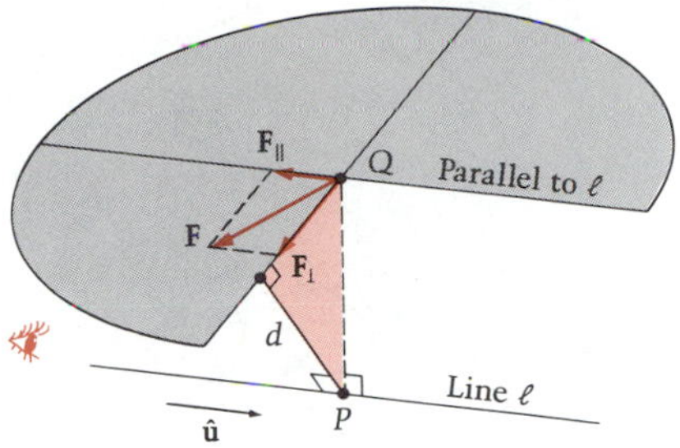

Figure 3.7

Next we construct a plane perpendicular to ℓ. (Again see Figure 3.7.) We shall let P be its intersection with ℓ, and Q its intersection with the line of action of $\mathbf{F}$. Furthermore, at Q we decompose $\mathbf{F}$ into components $\mathbf{F}_{\|}$ and $\mathbf{F}_{\perp}$, respectively, parallel and perpendicular to ℓ as shown.*

Taking the moment of $\mathbf{F}$ about point P, we obtain

$$\begin{aligned}\mathbf{M}_P &= \mathbf{r}_{PQ} \times \mathbf{F} = \mathbf{r}_{PQ} \times (\mathbf{F}_{\|} + \mathbf{F}_{\perp}) \\ &= \mathbf{r}_{PQ} \times \mathbf{F}_{\|} + \mathbf{r}_{PQ} \times \mathbf{F}_{\perp} \qquad (3.8)\end{aligned}$$

Our last step is to substitute $\mathbf{M}_P$ from Equation (3.8) into Equation (3.5) for $\mathbf{M}_\ell$:

$$\begin{aligned}\mathbf{M}_\ell &= (\mathbf{M}_P \cdot \hat{\mathbf{u}})\hat{\mathbf{u}} \\ &= [(\mathbf{r}_{PQ} \times \mathbf{F}_{\|} + \mathbf{r}_{PQ} \times \mathbf{F}_{\perp}) \cdot \hat{\mathbf{u}}]\hat{\mathbf{u}} \qquad (3.9)\end{aligned}$$

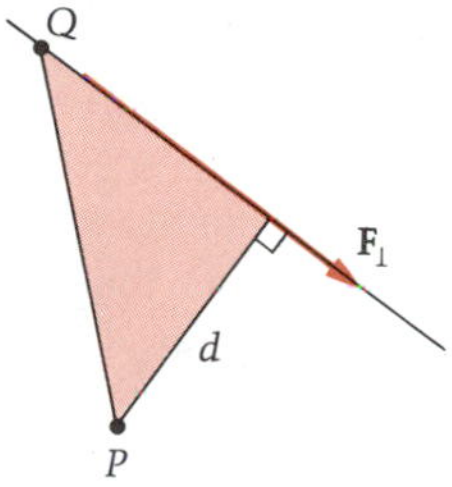

Figure 3.8 View along ℓ ($\hat{\mathbf{u}}$ is into the paper).

Because $\mathbf{r}_{PQ} \times \mathbf{F}_{\|}$ is perpendicular to $\mathbf{F}_{\|}$ and thus also to ℓ and $\hat{\mathbf{u}}$, the dot product $(\mathbf{r}_{PQ} \times \mathbf{F}_{\|}) \cdot \hat{\mathbf{u}}$ is zero. This means that, as expected, the component of $\mathbf{F}$ parallel to line ℓ produces no moment about ℓ.

Since $\mathbf{r}_{PQ}$ and $\mathbf{F}_{\perp}$ are each perpendicular to ℓ, then $\mathbf{r}_{PQ} \times \mathbf{F}_{\perp}$ is in the direction of ℓ, or, in other words, proportional to $\hat{\mathbf{u}}$. Referring to Figure 3.8, we see that

$$\mathbf{M}_\ell = \mathbf{r}_{PQ} \times \mathbf{F}_{\perp} = |\mathbf{F}_{\perp}|\, d\,\hat{\mathbf{u}} \qquad (3.10)$$

Finally, the moment $\mathbf{M}_\ell$ of a force $\mathbf{F}$ about a line ℓ can be summarized by Figure 3.9. Only the component $\mathbf{F}_{\perp}$ of $\mathbf{F}$ that is normal to ℓ has a turning effect about ℓ. The moment $\mathbf{M}_\ell$ has for its magnitude the product of $|\mathbf{F}_{\perp}|$ and the distance d between $\mathbf{F}_{\perp}$ and ℓ. The direction of $\mathbf{M}_\ell$ is given by the right thumb when the fingers of the right hand curl about ℓ in the direction of the turning effect of $\mathbf{F}_{\perp}$ about ℓ.†

* If $\mathbf{F}$ is itself *already* perpendicular to ℓ, i.e., if $\mathbf{F} = \mathbf{F}_{\perp}$, then Q is not a unique point. In this case we take the plane perpendicular to ℓ to be the one *containing* the entire line of action of $\mathbf{F}$, and Q to be any point on that line of action. For this case, in what follows, $\mathbf{F}_{\|}$ will of course be zero.

† We note that one point on the line of action of $\mathbf{F}$ in Figure 3.7 (nearer to the "eye" in the figure) will be closest to line ℓ; it will be the distance d from ℓ.

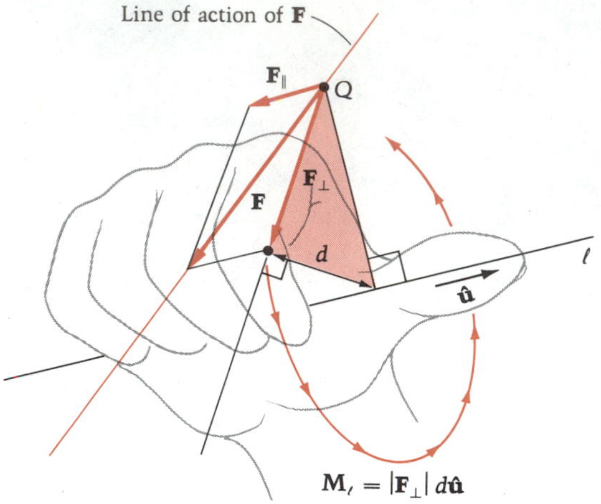

Figure 3.9

Question 3.7 Could the magnitude of the moment of a force about a line ever be greater than the magnitude of the moment of the force about any point on the line?

Question 3.8 Does the distance d in Equation (3.10) depend on the choice of point P on line ℓ?

Question 3.9 Suppose the moment of a force $\mathbf{F}$ about a point P has been computed using $\mathbf{M}_P = \mathbf{r} \times \mathbf{F}$, where $\mathbf{r}$ is a vector from P to some point on the line of action of $\mathbf{F}$. About some *line* through P, the moment of $\mathbf{F}$ will be the largest in magnitude. What is the unit vector in the direction of this line?

Answer 3.7 No! $\mathbf{M}_\ell$ is the same for all points on ℓ. But both $\mathbf{F}_\parallel$ and $\mathbf{F}_\perp$ make other contributions to the moment of $\mathbf{F}$ about points lying on ℓ— contributions perpendicular to $\mathbf{M}_\ell$— so that the magnitude of moments of $\mathbf{F}$ about points on ℓ is greater than or equal to $|\mathbf{M}_\ell|$.

Answer 3.8 No. Since both $\mathbf{M}_\ell$ and $\mathbf{F}_\perp$ are independent of the choice of P, Equation (3.10) shows that d is, too. The distance d is simply the distance indicated between the line ℓ and the plane containing $\mathbf{F}$ in Figure 3.7.

Answer 3.9 $(\mathbf{r} \times \mathbf{F})/|\mathbf{r} \times \mathbf{F}|$, for then the line is in the direction of $\mathbf{M}_P$ and none of it is "lost" in the dot product of Equation (3.6).

EXAMPLE 3.4

The chain in Figure E3.4 carries 20 lb of tension as it holds open the trap door. Find the moment of the chain force exerted at A, about the x-axis.

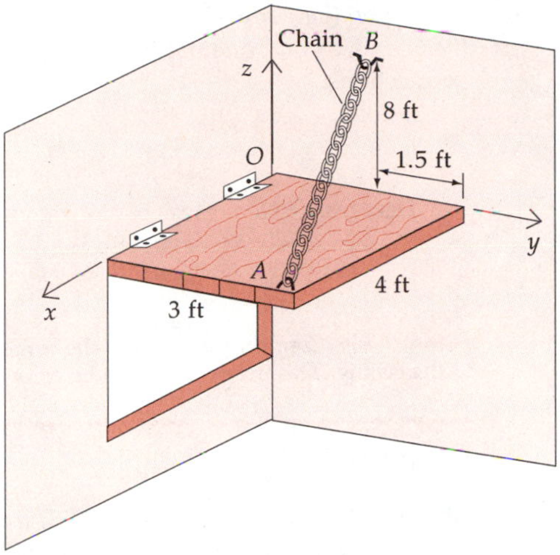

Figure E3.4

Solution

The force in the chain may be expressed as a vector (for use in Equation 3.6) by multiplying its magnitude by the unit vector in its direction:

$$\mathbf{F} = 20\left(\frac{\mathbf{r}_{AB}}{|\mathbf{r}_{AB}|}\right) \text{ lb}$$

The points A and B have the respective coordinates (4, 3, 0) and (0, 1.5, 8) ft; therefore,

$$\begin{aligned}\mathbf{F} &= 20\left(\frac{(0-4)\hat{\mathbf{i}} + (1.5-3)\hat{\mathbf{j}} + (8-0)\hat{\mathbf{k}}}{\sqrt{(-4)^2 + (-1.5)^2 + 8^2}}\right) \\ &= 20\left(\frac{-4\hat{\mathbf{i}} - 1.5\hat{\mathbf{j}} + 8\hat{\mathbf{k}}}{9.07}\right) \\ &= -8.82\hat{\mathbf{i}} - 3.31\hat{\mathbf{j}} + 17.6\hat{\mathbf{k}} \text{ lb} \qquad (1)\end{aligned}$$

The origin O is a convenient point on the line (x-axis) about which we desire the moment of $\mathbf{F}$. Therefore, using Equation 3.6,

$$\begin{aligned}\mathbf{M}_{x\text{-}axis} &= [\overbrace{(\mathbf{r}_{OA} \times \mathbf{F})}^{\mathbf{M}_O} \cdot \hat{\mathbf{i}}]\hat{\mathbf{i}} \\ &= [(4\hat{\mathbf{i}} + 3\hat{\mathbf{j}}) \times (-8.82\hat{\mathbf{i}} - 3.31\hat{\mathbf{j}} + 17.6\hat{\mathbf{k}}) \cdot \hat{\mathbf{i}}]\hat{\mathbf{i}}\end{aligned}$$

We note here that only the $\hat{\mathbf{i}}$-coefficient of the cross product will survive the subsequent dot product with $\hat{\mathbf{i}}$, so we only need to compute that component. Of the six vectors arising from the cross product, only $\hat{\mathbf{j}} \times \mathbf{k}$ results in an "$\hat{\mathbf{i}}$"; therefore,

$$\begin{aligned}\mathbf{M}_{x\text{-}axis} &= [(3\hat{\mathbf{j}} \times 17.6\hat{\mathbf{k}}) \cdot \hat{\mathbf{i}}]\hat{\mathbf{i}} \\ &= 52.8\hat{\mathbf{i}} \text{ lb-ft}\end{aligned}$$

Except possibly for a small amount of friction in the hinges, only the above (x) component of the total moment about O of the tension in the chain is holding the door open (i.e., opposing the moment of its weight). The rest of the moment $\mathbf{M}_O$ (its y and z components) is "wasted" and is only causing stress in the hinge connections.

Question 3.10 How could the chain be reattached so as to minimize the hinge reactions yet still hold the door open?

Answer 3.10 Connect the chain to the center of the free 4-ft edge and run it straight up to the ceiling. This may or may not be convenient, however.

EXAMPLE 3.5

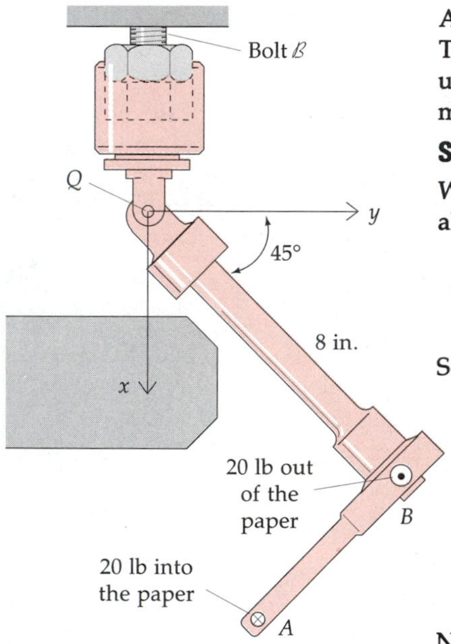

Figure E3.5

At the instant shown in Figure E3.5, the wrench lies in a horizontal (xy) plane. Two 20-lb forces are applied to a ratchet wrench, down (into the paper) at A and up (out of the paper) at B, in an effort to loosen the bolt $\mathcal{B}$. Find the sum of the moments of the two forces about the axis (x) of the bolt.

Solution

We shall find $\mathbf{M}_Q$ as the sum of the two $\mathbf{r} \times \mathbf{F}$'s each dotted with the unit vector along the line ($\hat{\mathbf{u}} = \hat{\mathbf{i}}$ here) as in Equation (3.5).

The vector from Q to B is seen to be

$$\mathbf{r}_{QB} = \frac{8}{\sqrt{2}}\hat{\mathbf{i}} + \frac{8}{\sqrt{2}}\hat{\mathbf{j}} \text{ in.}$$

So, for the force at B,

$$\mathbf{M}_x = [(\mathbf{r}_{QB} \times \mathbf{F}_B) \cdot \hat{\mathbf{i}}]\hat{\mathbf{i}}$$

$$= \left\{\left[\left(\frac{8}{\sqrt{2}}\hat{\mathbf{i}} + \frac{8}{\sqrt{2}}\hat{\mathbf{j}}\right) \times (20\hat{\mathbf{k}})\right] \cdot \hat{\mathbf{i}}\right\}\hat{\mathbf{i}}$$

$$= \left[\left(\frac{-160}{\sqrt{2}}\hat{\mathbf{j}} + \frac{160}{\sqrt{2}}\hat{\mathbf{i}}\right) \cdot \hat{\mathbf{i}}\right]\hat{\mathbf{i}} = \frac{160}{\sqrt{2}}\hat{\mathbf{i}} \text{ lb-in.}$$

Next, the vector from Q to A is:

$$\mathbf{r}_{QA} = \mathbf{r}_{QB} + \mathbf{r}_{BA} = \left(\frac{8}{\sqrt{2}}\hat{\mathbf{i}} + \frac{8}{\sqrt{2}}\hat{\mathbf{j}}\right) + \left(\frac{5}{\sqrt{2}}\hat{\mathbf{i}} - \frac{5}{\sqrt{2}}\hat{\mathbf{j}}\right)$$

$$= \frac{13}{\sqrt{2}}\hat{\mathbf{i}} + \frac{3}{\sqrt{2}}\hat{\mathbf{j}} \text{ in.}$$

so that, for the force at A,

$$\mathbf{M}_x = [(\mathbf{r}_{QA} \times \mathbf{F}_A) \cdot \hat{\mathbf{i}}]\hat{\mathbf{i}}$$

$$= \left\{\left[\left(\frac{13}{\sqrt{2}}\hat{\mathbf{i}} + \frac{3}{\sqrt{2}}\hat{\mathbf{j}}\right) \times (-20\hat{\mathbf{k}})\right] \cdot \hat{\mathbf{i}}\right\}\hat{\mathbf{i}}$$

$$= \left[\left(\frac{260}{\sqrt{2}}\hat{\mathbf{j}} - \frac{60}{\sqrt{2}}\hat{\mathbf{i}}\right) \cdot \hat{\mathbf{i}}\right]\hat{\mathbf{i}}$$

$$= \frac{-60}{\sqrt{2}}\hat{\mathbf{i}} \text{ lb-in.}$$

The net effect of the two forces toward loosening the bolt is the sum of their separate contributions to $\mathbf{M}_x$:

$$\mathbf{M}_x = \frac{160 - 60}{\sqrt{2}}\hat{\mathbf{i}} = 70.7\hat{\mathbf{i}} \text{ lb-in.}$$

The two noncollinear forces in the preceding example, being equal in magnitude but opposite in direction, form what is called a *couple*. We shall study the properties of a couple in the next section of this chapter and will then rework the example using the most powerful of those properties.

Example 3.6

Referring to Figure E3.6, find the moment of the 8-N force about the line *CD*.

Solution

To obtain $\mathbf{M}_{CD}$ we need the moment with respect to some point on the line such as *C* or *D*. Using *C*,

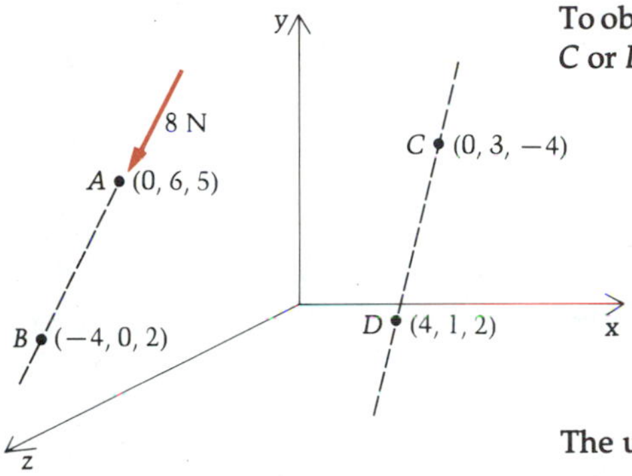

Figure E3.6

$$\begin{aligned}\mathbf{M}_C &= \mathbf{r}_{CA} \times \mathbf{F}\\ &= [(0-0)\hat{\mathbf{i}} + (6-3)\hat{\mathbf{j}} + (5-(-4))\hat{\mathbf{k}}] \times 8\hat{\mathbf{e}}_{AB}\\ &= (3\hat{\mathbf{j}} + 9\hat{\mathbf{k}}) \times 8\left[\frac{-4\hat{\mathbf{i}} - 6\hat{\mathbf{j}} - 3\hat{\mathbf{k}}}{\sqrt{4^2+6^2+3^2}}\right]\\ &= (3\hat{\mathbf{j}} + 9\hat{\mathbf{k}}) \times (-4.10\hat{\mathbf{i}} - 6.14\hat{\mathbf{j}} - 3.07\hat{\mathbf{k}})\\ &= 46.1\hat{\mathbf{i}} - 36.9\hat{\mathbf{j}} + 12.3\hat{\mathbf{k}} \text{ N}\cdot\text{m}\end{aligned}$$

The unit vector directed from *C* toward *D* is

$$\begin{aligned}\hat{\mathbf{e}}_{CD} &= \frac{\mathbf{r}_{CD}}{|\mathbf{r}_{CD}|} = \frac{1}{\sqrt{56}}(4\hat{\mathbf{i}} - 2\hat{\mathbf{j}} + 6\hat{\mathbf{k}})\\ &= 0.535\hat{\mathbf{i}} - 0.267\hat{\mathbf{j}} + 0.802\hat{\mathbf{k}}\end{aligned}$$

By Equation (3.5),

$$\begin{aligned}\mathbf{M}_{CD} &= (\mathbf{M}_C \cdot \hat{\mathbf{e}}_{CD})\hat{\mathbf{e}}_{CD}\\ &= [(46.1)(0.535) + (-36.9)(-0.267) + (12.3)(0.802)]\hat{\mathbf{e}}_{CD}\\ &= 44.4\hat{\mathbf{e}}_{CD} \text{ N}\cdot\text{m}\\ &= 23.8\hat{\mathbf{i}} + 11.9\hat{\mathbf{j}} + 35.6\hat{\mathbf{k}} \text{ N}\cdot\text{m}\end{aligned}$$

In the preceding example, the same result could have been obtained by first computing $\mathbf{M}_D$ instead of $\mathbf{M}_C$:

$$\begin{aligned}\mathbf{M}_D &= \mathbf{r}_{DA} \times \mathbf{F}\\ &= (-4\hat{\mathbf{i}} + 5\hat{\mathbf{j}} + 3\hat{\mathbf{k}}) \times (-4.10\hat{\mathbf{i}} - 6.14\hat{\mathbf{j}} - 3.07\hat{\mathbf{k}})\\ &= 3.07\hat{\mathbf{i}} - 24.6\hat{\mathbf{j}} + 45.1\hat{\mathbf{k}} \text{ N}\cdot\text{m}\end{aligned}$$

so that

$$\begin{aligned}\mathbf{M}_{CD} &= (\mathbf{M}_D \cdot \hat{\mathbf{e}}_{CD})\hat{\mathbf{e}}_{CD} \\ &= [(3.07)(0.535) + (-24.6)(-0.267) + (45.1)(0.802)]\hat{\mathbf{e}}_{CD} \\ &= 44.4\hat{\mathbf{e}}_{CD}\ \text{N} \cdot \text{m} \qquad \text{(as before)}\end{aligned}$$

One of the problems at the end of the section (Problem 3.17) is to rework this example using $\mathbf{r}_{CB}$ instead of $\mathbf{r}_{CA}$ (and $\mathbf{r}_{DB}$ instead of $\mathbf{r}_{DA}$).

Let us also give an illustration of the use of Equation 3.10, in conjunction with the preceding example, to find the shortest distance between the line of action of **F** and the line *CD*. First, the component of **F** parallel to *CD* is:

$$\begin{aligned}F_{\parallel} = \mathbf{F} \cdot \hat{\mathbf{e}}_{CD} &= (-4.10\hat{\mathbf{i}} - 6.14\hat{\mathbf{j}} - 3.07\hat{\mathbf{k}}) \\ &\quad \cdot (0.535\hat{\mathbf{i}} - 0.267\hat{\mathbf{j}} + 0.802\hat{\mathbf{k}}) \\ &= (-4.10)(0.535) + (-6.14)(-0.267) \\ &\quad + (-3.07)(0.802) \\ &= -3.02\ \text{N}\end{aligned}$$

so that

$$\begin{aligned}|\mathbf{F}_{\perp}| &= \sqrt{|\mathbf{F}|^2 - F_{\parallel}^2} \\ &= \sqrt{8^2 - 3.02^2} \\ &= 7.41\ \text{N}\end{aligned}$$

Now using Equation (3.10) and the result of Example 3.6, we can obtain the desired shortest distance between the line of action of **F**, and the line CD:

$$\mathbf{M}_{\ell} = \mathbf{M}_{CD} = |\mathbf{F}_{\perp}|\, d\hat{\mathbf{u}}$$

so that, equating the magnitudes of both sides,

$$\begin{aligned}44.4 &= 7.41\ d \\ d &= 5.99\ \text{m}\end{aligned}$$

EXAMPLE 3.7

Using Equation 3.10 and the result of Example 3.4, find the shortest distance between the x-axis and the line of action of the force in the chain.

Solution

Applying Equation (3.10),

$$|\mathbf{M}_{x\text{-axis}}| = |\mathbf{F}_{\perp}|\, d$$

or

$$d = \frac{52.8}{|\mathbf{F}_{\perp}|}$$

Now the part of **F** which is perpendicular to the line of interest (the x-axis) is in this case simply the vector sum of its y- and z-components (See Eq. (1) of Example 3.4):

$$\mathbf{F}_{\perp} = -3.31\hat{\mathbf{j}} + 17.6\hat{\mathbf{k}} \text{ lb}$$

or

$$|\mathbf{F}_{\perp}| = \sqrt{3.31^2 + 17.6^2} = 17.9 \text{ lb}$$

Therefore

$$d = \frac{52.8}{17.9} = 2.95 \text{ ft}$$

We can check the result of the preceding example by looking at Figure 3.10 below, which is a view looking from the positive x-axis:

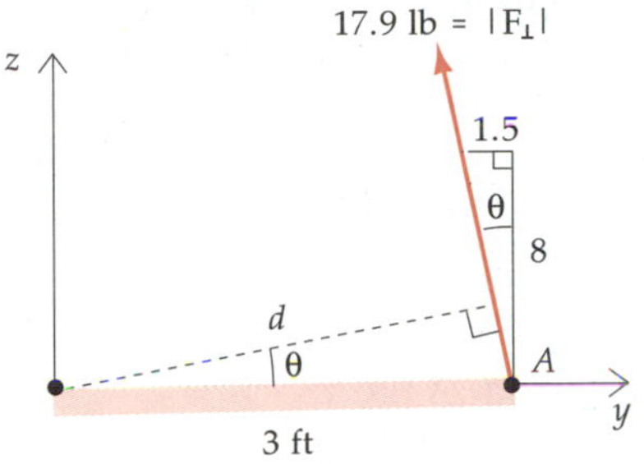

$$\begin{aligned} d &= 3\cos\theta \\ &= 3\left(\frac{8}{\sqrt{8^2 + 1.5^2}}\right) \\ &= 3\left(\frac{8}{8.14}\right) \\ &= 2.95 \text{ ft} \qquad \text{(as above)} \end{aligned}$$

Figure 3.10

Using "Force Times Perpendicular Distance" in Certain 3-D Applications to Find M_l

In finding the moment of a force **F** about a point P, the formal use of the cross product is not always necessary. To see an alternative approach that uses what we have learned about moments with respect to lines, we express the moment, $\mathbf{M}_P$, of **F** about P in component form. Let Q again be some point on the line of action of **F** as shown in Figure 3.11 and note that we have positioned the axes x, y, and z to intersect at P. Expressing both $\mathbf{r}_{PQ}$ and **F** in component form as

$$\begin{aligned} \mathbf{r}_{PQ} &= r_x\hat{\mathbf{i}} + r_y\hat{\mathbf{j}} + r_z\hat{\mathbf{k}} \\ \mathbf{F} &= F_x\hat{\mathbf{i}} + F_y\hat{\mathbf{j}} + F_z\hat{\mathbf{k}} \end{aligned}$$

then, as we noted in Section 3.2,

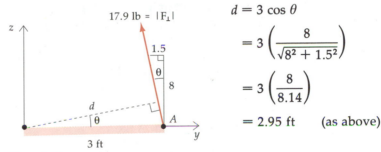

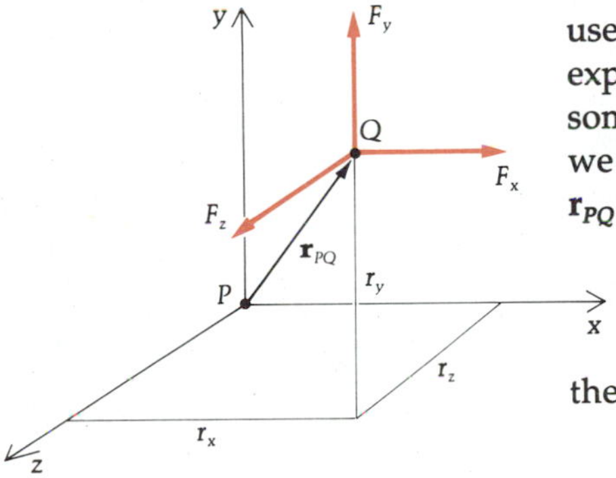

Figure 3.11

$$\begin{aligned} \mathbf{M}_P &= \mathbf{r}_{PQ} \times \mathbf{F} \\ &= (r_x\hat{\mathbf{i}} + r_y\hat{\mathbf{j}} + r_z\hat{\mathbf{k}}) \times (F_x\hat{\mathbf{i}} + F_y\hat{\mathbf{j}} + F_z\hat{\mathbf{k}}) \\ &= (r_yF_z - r_zF_y)\hat{\mathbf{i}} + (r_zF_x - r_xF_z)\hat{\mathbf{j}} + (r_xF_y - r_yF_x)\hat{\mathbf{k}} \qquad (3.11) \end{aligned}$$

Since the x component of this vector is $\mathbf{M}_P \cdot \hat{\mathbf{i}}$, then from the foregoing analysis and discussion we recognize $(r_yF_z - r_zF_y)\hat{\mathbf{i}}$ as the moment of $\mathbf{F}$ about the x axis through P. Similarly, the other components of $\mathbf{M}_P$ are the moments of $\mathbf{F}$ about the y and z axes. The term $r_yF_z\hat{\mathbf{i}}$ is clearly the moment of $F_z\hat{\mathbf{k}}$ about the x axis since r_y (or, more precisely, its magnitude) is the distance from the x axis to the line of action of $F_z\hat{\mathbf{k}}$. Likewise, $-r_zF_y\hat{\mathbf{i}}$ is the moment of $F_y\hat{\mathbf{j}}$ about the x axis. Obviously the force $F_x\hat{\mathbf{i}}$ produces no moment about the x axis, since its line of action is parallel to this axis.

Thus, we see from Equation (3.11) that the moment about a point may be constructed by finding the moments about three mutually perpendicular axes through the point, and each of these may be computed by decomposing the force into components parallel to the axes and summing the moments of the individual components about these axes.

The above process allows us to utilize the fundamental concept of "magnitude of force times perpendicular distance," and hence even in three dimensions we can avoid, if we wish, the formal (vector product) calculation. When we do this we must, of course, manually attach the correct unit vector (including the proper sign) with the help of the right-hand rule. We are not necessarily advocating the evasion of vector products here, but rather we are calling attention to the more physical interpretation that can be given to the terms in Equation (3.11). The following example illustrates these ideas.

EXAMPLE 3.8

Find the moment of the force $\mathbf{F}$ in Figure E3.8a about point Q, using the ideas expressed by Equation (3.11).

Solution

The force $\mathbf{F}$ has the components:

$$F_x = \frac{-4}{\sqrt{3^2 + 4^2 + 12^2}}\,260 = \left(\frac{-4}{13}\right)260 = -80 \text{ lb}$$

$$F_y = \frac{3}{13}\,260 = 60 \text{ lb}$$

$$F_z = \frac{12}{13}\,260 = 240 \text{ lb}$$

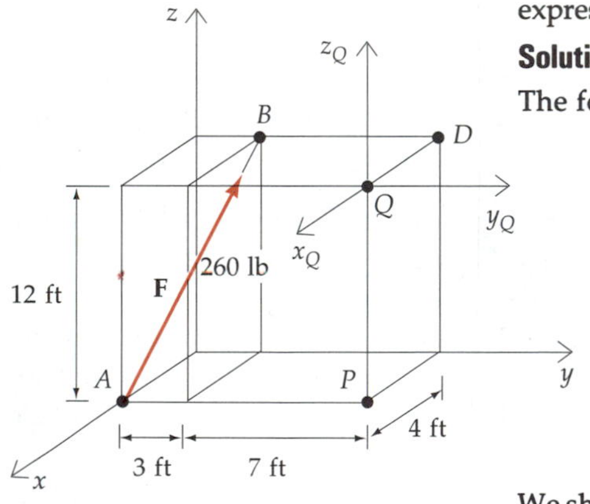

Figure E3.8a

We shall obtain $\mathbf{M}_Q$ by adding the separate moments of these components of $\mathbf{F}$ at A about the axes (x_Q, y_Q, z_Q) through Q.

The component $F_x\hat{\mathbf{i}}$, or $-80\hat{\mathbf{i}}$ lb, of $\mathbf{F}$ at A produces moments about lines through Q parallel to y and to z. These are 80(10) lb-ft about axis z_Q and 80(12) lb-ft about y_Q, as shown in Figures E3.8b,c on the next page. Thus the contribution of $F_x\hat{\mathbf{i}}$ to $\mathbf{M}_Q$ is $960\hat{\mathbf{j}} - 800\hat{\mathbf{k}}$ lb-ft. The reader should note how the unit vector is attached by using the right-hand rule. If the fingers curl in the direction of the turning effect about an axis through Q, the thumb will aim in the direction of the moment and the unit vector in this direction is then written down.

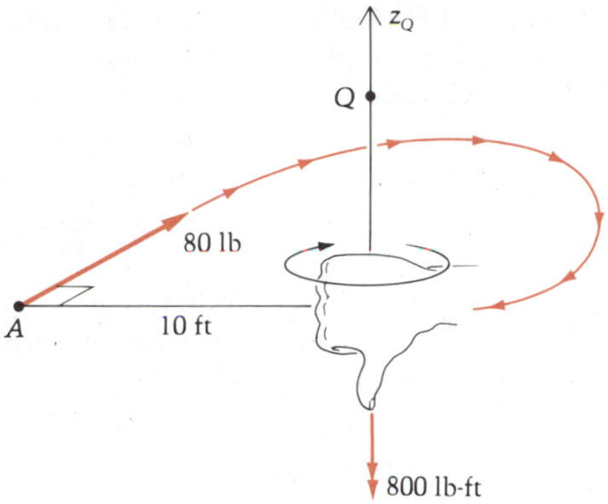

Figure E3.8b

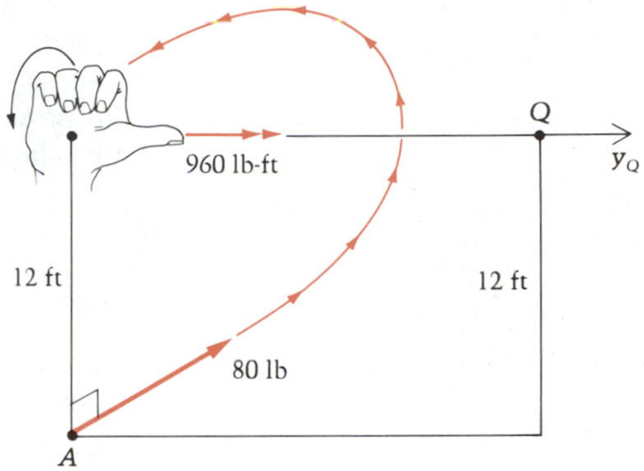

Figure E3.8c

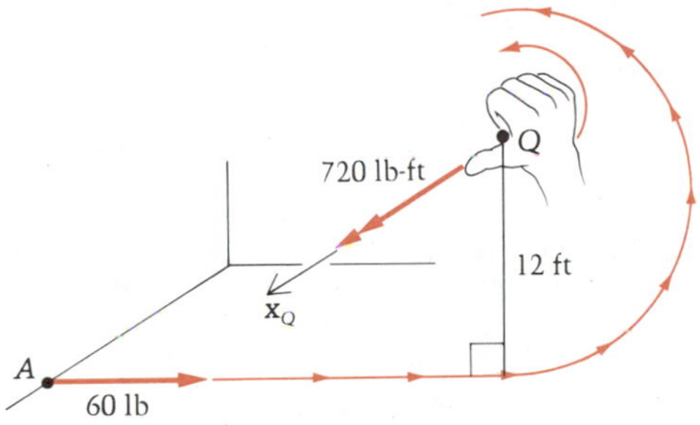

Figure E3.8d

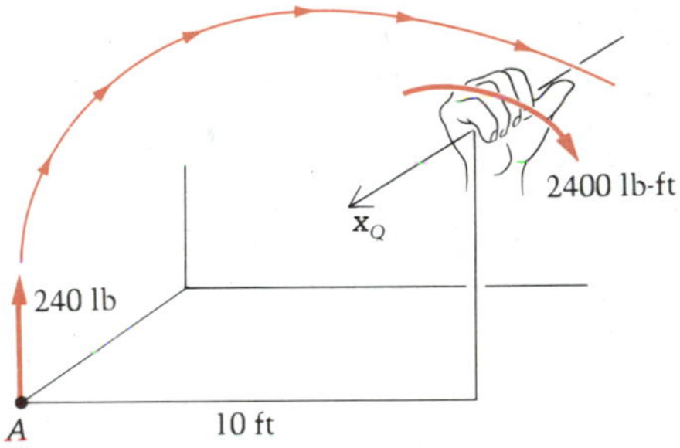

Figure E3.8e

Next, the component $F_y\hat{\mathbf{j}}$, or $60\hat{\mathbf{j}}$ lb, of **F** at A produces a moment about x_Q but not about y_Q or z_Q.

Question 3.11 Why not y_Q or z_Q?

The contribution is seen in Figure 3.8d to be 60(12) lb-ft about axis x_Q. Therefore, the contribution of $F_y\hat{\mathbf{j}}$ to $\mathbf{M}_Q$ is $720\hat{\mathbf{i}}$ lb-ft.

Finally, the third component $F_z\hat{\mathbf{k}}$, or $240\hat{\mathbf{k}}$ lb, of **F** at A also produces a moment about just one of the axes (x_Q, y_Q, z_Q) through point Q, and this axis is again x_Q. The moment contribution is seen in Figure E3.8e to be 240(10) lb-ft about x_Q, and so $F_z\hat{\mathbf{k}}$ adds $-2400\hat{\mathbf{i}}$ lb-ft to $\mathbf{M}_Q$. The total moment of **F** about Q is then

$$\mathbf{M}_Q = (960\hat{\mathbf{j}} - 800\hat{\mathbf{k}}) + (720\hat{\mathbf{i}}) + (-2400\hat{\mathbf{i}}) \text{ lb-ft}$$
$$= -1680\hat{\mathbf{i}} + 960\hat{\mathbf{j}} - 800\hat{\mathbf{k}} \text{ lb-ft}$$

Answer 3.11 $F_y\hat{\mathbf{j}}$ at A is parallel to y_Q and it intersects z_Q; thus, no moment about either of these lines!

The reader is encouraged to check this answer by finding and adding the moments about Q of F_x, F_y, and F_z placed at B instead of A and also by computing either $\mathbf{r}_{QA} \times \mathbf{F}$ or $\mathbf{r}_{QB} \times \mathbf{F}$. Comparing the two approaches, you will probably observe some loss of physical feeling for the moment when using the cross product.

The idea of "force times perpendicular distance" symbolized by Equation (3.11) and utilized in the preceding example can also be used to compute the moment about a *line* through a point. If in Example 3.8 we had been seeking the moment about axis y_Q, then only the second of the four figures would have been of interest to us because $\mathbf{F}_z$ passes through axis y_Q, and $\mathbf{F}_y$ is parallel to it; thus neither of these components contributes to the moment of $\mathbf{F}$ about axis y_Q. The answer would then have been simply the y-component of $\mathbf{M}_Q$, namely $\mathbf{M}_{y_Q} = 960\hat{\mathbf{j}}$ lb-ft.

Question 3.12 If in Example 3.8 we were using "force components (at A) times perpendicular distances" to find the moment about the line through point D parallel to the y-axis, how many products would be needed? Is there a better point than A to decompose the force in order to find this moment?

Answer 3.12 Two, because this time $\mathbf{F}_z$ doesn't pass through the line. Yes, point B, which lies *on* the line of interest so that by inspection $\mathbf{M}_D \cdot \hat{\mathbf{j}}$ vanishes.

EXAMPLE 3.9

Rework Example 3.4, using the ideas expressed by Equation (3.11) in conjunction with the preceding discussion. We are seeking the moment of the 20-lb chain force about the x-axis (see Figure E3.9a).

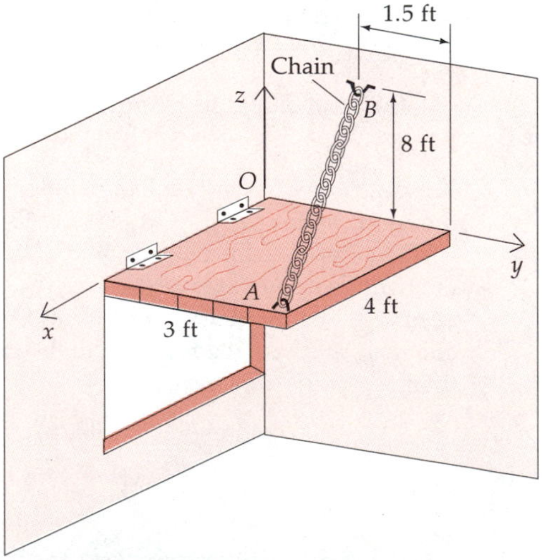

Figure E3.9a

Solution

We are again seeking the moment of the 20-lb chain force **F** at A, about the x-axis. The components of **F** at A were computed in Example 3.4 and are shown in Figure E3.9b. We see immediately that neither the x- nor y-components exert any moment (turning effect) about the x-axis. This is because the x-component of **F** is parallel to the line, and the y-component intersects it. Thus the moment about the x-axis is simply

$$\mathbf{M}_{x\text{-}axis} = (17.6)3\hat{\mathbf{i}} = 52.8\hat{\mathbf{i}} \text{ lb-ft},$$

as before.

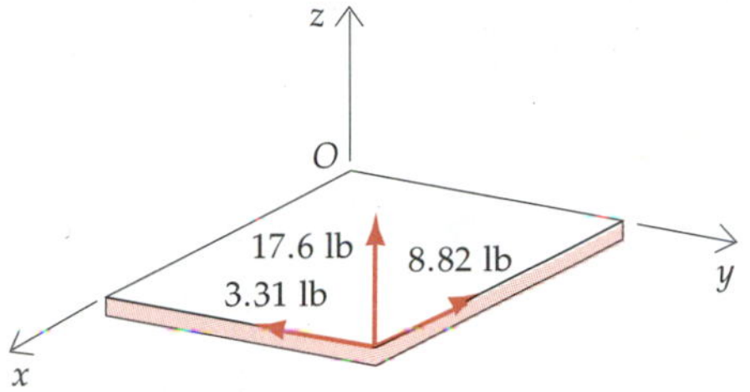

Figure E3.9b

PROBLEMS ▶ Section 3.3

3.16 The 200-N force in Figure P3.16 lies in a plane parallel to xz. Find the moment of this force about the axis of the vertical shaft (z).

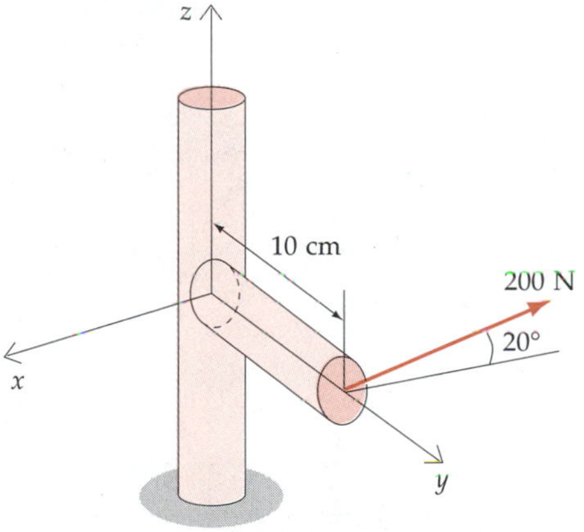

Figure P3.16

3.17 Recompute the moment $\mathbf{M}_{CD}$ in Example 3.6, using both of the suggestions following that example, namely, using $\mathbf{r}_{CB}$ instead of $\mathbf{r}_{CA}$ and $\mathbf{r}_{DB}$ instead of $\mathbf{r}_{DA}$.

3.18 Find the moment of the 21-lb force in Figure P3.18 about the line OD.

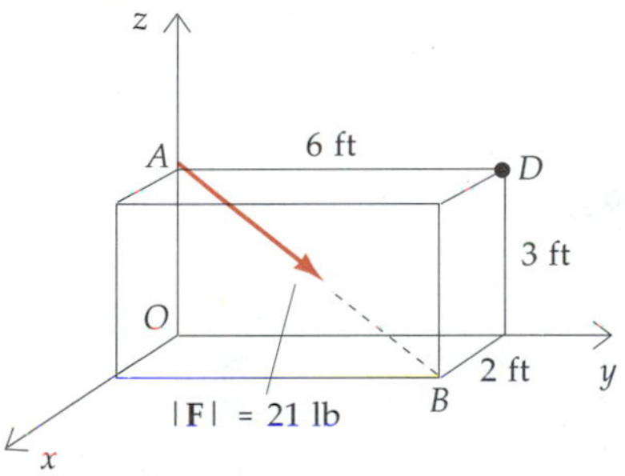

Figure P3.18

3.19 In Example 3.8, find the moment of **F** about line PQ using:

a. $[(\mathbf{r}_{PA} \times \mathbf{F}) \cdot \hat{\mathbf{k}}]\hat{\mathbf{k}}$
b. $[(\mathbf{r}_{PB} \times \mathbf{F}) \cdot \hat{\mathbf{k}}]\hat{\mathbf{k}}$
c. $[(\mathbf{r}_{QA} \times \mathbf{F}) \cdot \hat{\mathbf{k}}]\hat{\mathbf{k}}$
d. $[(\mathbf{r}_{QB} \times \mathbf{F}) \cdot \hat{\mathbf{k}}]\hat{\mathbf{k}}$

The answer should each time be $-800\hat{\mathbf{k}}$ lb-ft, which was the z-component of $\mathbf{M}_Q$ in Example 3.8.

3.20 Find the moment of the force in Problem 3.11 about the line that passes through points $y = 12$ ft on the y axis and $z = 5$ ft on the z axis.

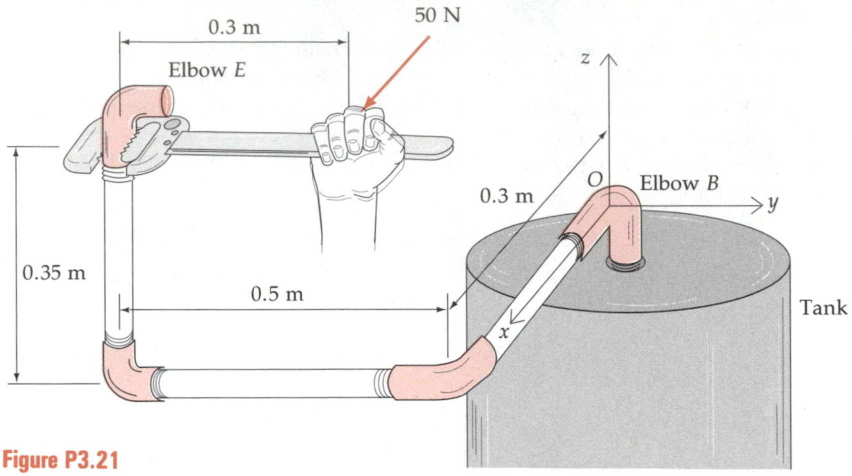

Figure P3.21

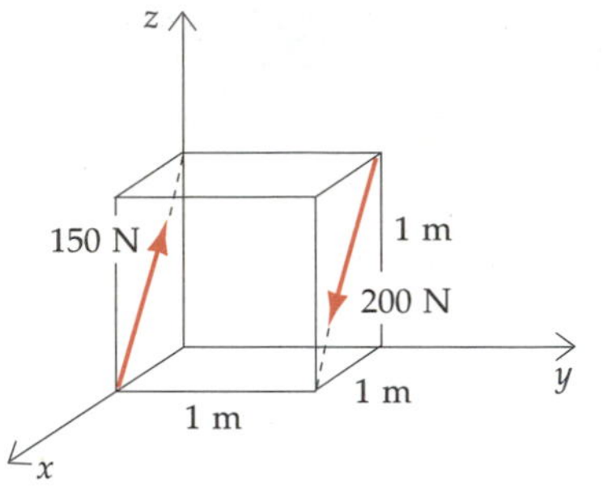

Figure P3.25

3.21 A do-it-yourselfer is (in a most unwise fashion) tightening an elbow onto a length of pipe. (See Figure P3.21.) Find the moment of the 50-N force about the z-axis through point O, and note from the answer that the "plumber" is actually exerting a force that is tending to unscrew the elbow B from the top of the tank.

3.22 Using the result of Problem 3.10, find the moment of the 60-N force about the y-axis (noting that it is a line through point C). (See Figure P3.10.)

3.23 In Problem 3.8, determine the shortest distance between the 50-lb force and the line through point P that makes equal angles with the coordinate axes.

3.24 With no writing, make a rough estimate of the moment of the 200-N force **F** about the line CE in Figure P3.24. Then calculate the moment and compare it with your estimate.

3.25 In Figure P3.25 find the moment of the two forces about:

a. The point $(x, y, z) = (1, 2, 3)$ m

b. The line defined by the intersection of the plane $x = 1$ m with the plane $z = 3$ m.

3.26 Find the moment of the 2000-N force in Figure P3.26 about the diagonal line AB.

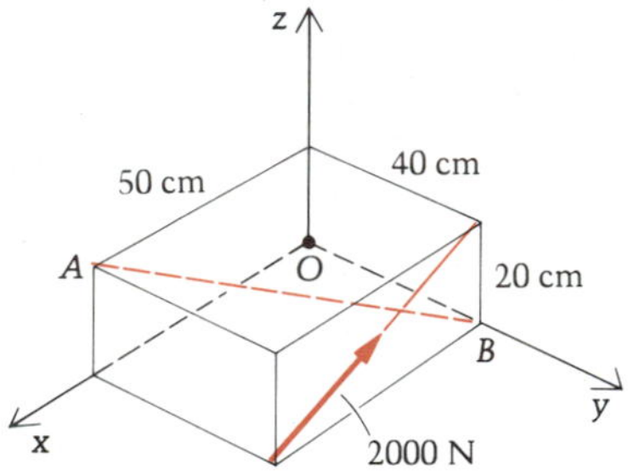

Figure P3.26

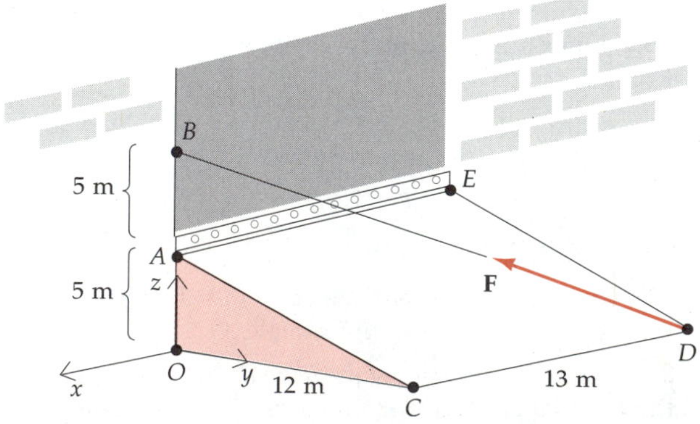

Figure P3.24

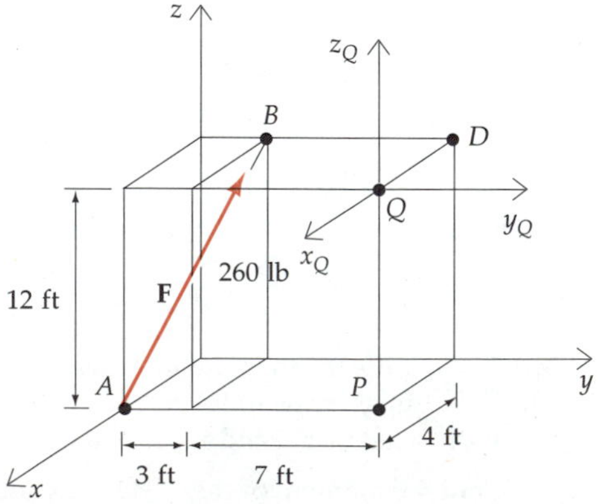

Figure P3.27

3.27 Find the moment of the force **F** in Example 3.8 about the line through point D parallel to the z-axis. (See Figure P3.27 on the preceding page.)

3.28 Using the result of Example 3.8, find the shortest distance between the force **F** and

a. point Q;

b. axis z_Q.

3.29 Determine the moment of the 180-lb force in Figure P3.29 about the line AC.

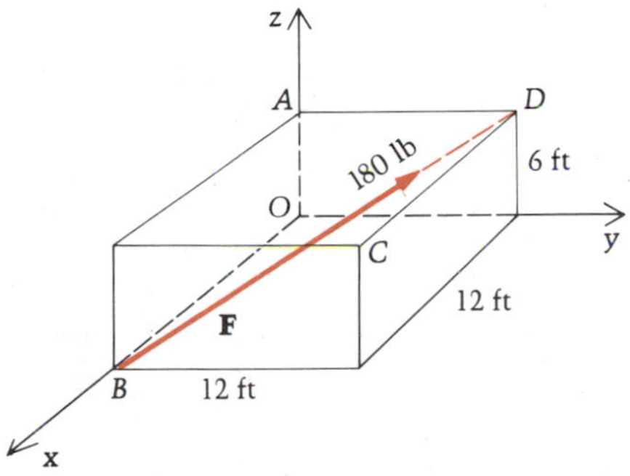

Figure P3.29

3.30 Find the moment of the force in Figure P3.30 about point C. Then find the moment about the line that passes through C and:

a. Point A

b. Point B

c. Point D.

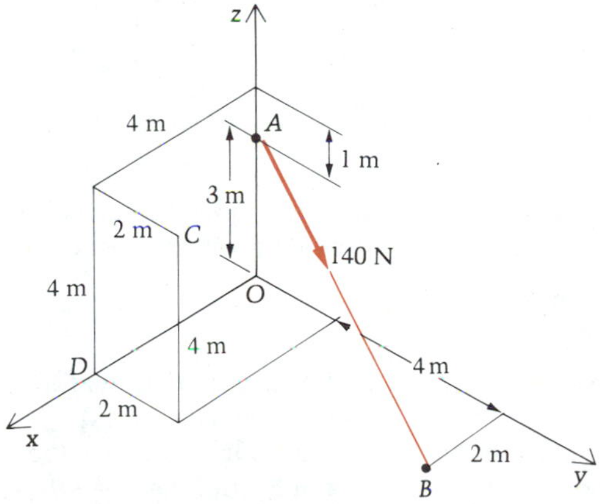

Figure P3.30

3.31 In Figure P3.31 determine the moment of the 280-lb force **F** with respect to:

a. Point A

b. Line OA

c. Point B

d. Line BR.

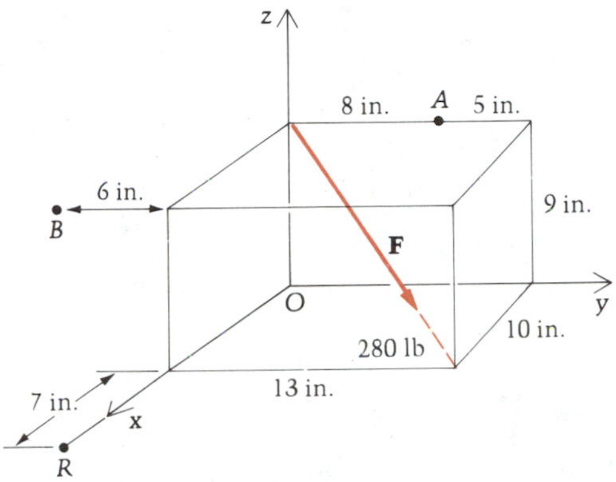

Figure P3.31

3.32 A particle located at $\mathbf{r}_{OP} = -\hat{\mathbf{i}} + 3\hat{\mathbf{j}} - 8\hat{\mathbf{k}}$ m is acted upon by the two forces

$$\mathbf{F}_1 = -2\hat{\mathbf{i}} + 3\hat{\mathbf{j}} - \hat{\mathbf{k}} \text{ N and } \mathbf{F}_2 = 7\hat{\mathbf{i}} + \hat{\mathbf{j}} - \hat{\mathbf{k}} \text{ N}$$

Find the moment of the sum of these forces about the z axis.

3.33 Find the moment of the 21-lb force **F** in Figure P3.33 about line AB by (a) using $(\mathbf{r}_{AP} \times \mathbf{F}) \cdot \hat{\mathbf{u}}_{AB}$, and (b) resolving **F** into its components at P and finding the moments of each about the line with the help of the right-hand rule.

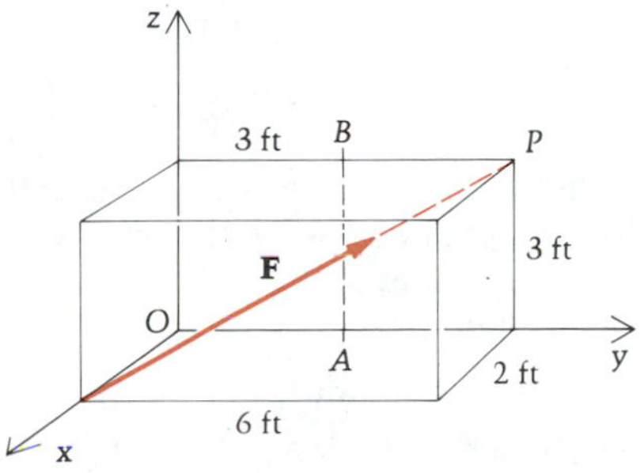

Figure P3.33

3.34 Find the moment of the force in Figure P3.34 with respect to:

- **a.** Line *BC*
- **b.** Point *A*
- **c.** Line *BA*.

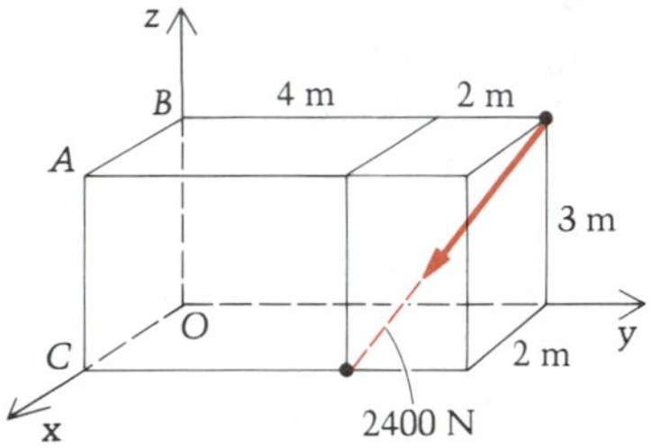

Figure P3.34

3.35 Find (in Figure P3.35):

- **a.** The direction cosines associated with the 39-lb force
- **b.** The 39-lb force expressed in terms of the unit vectors $\hat{\mathbf{i}}$, $\hat{\mathbf{j}}$, and $\hat{\mathbf{k}}$
- **c.** The moment of the 39-lb force about point *A*
- **d.** The moment of the 39-lb force about a line from *A* to *B*
- **e.** The moment of the 39-lb force about a line from *A* to *D*.

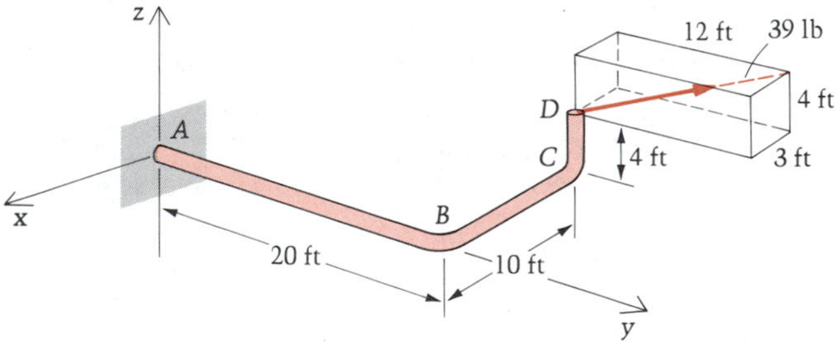

Figure P3.35

3.36 A bent bar is rigidly attached to a wall at the point (0, 0, 0). A force with magnitude $|\mathbf{F}| = 7$ lb acts at its free end with a line of action passing through the origin, as shown in Figure P3.36. Find:

- **a.** The moment of **F** about point *P*
- **b.** The moment about the line ℓ passing through *P* with a slope of 5 / 12 in the *yz* plane as shown.

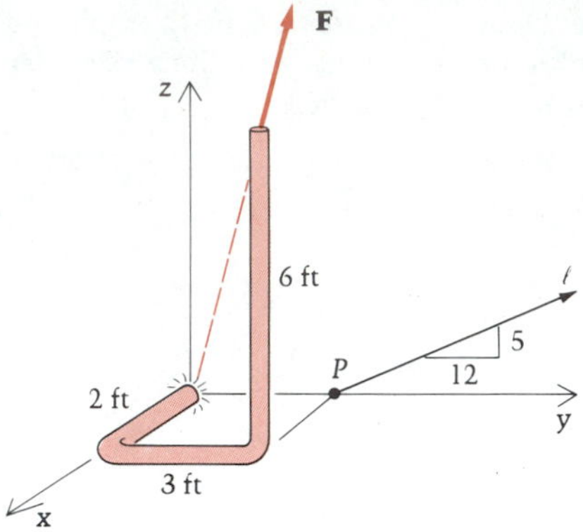

Figure P3.36

3.37 The force **F** of magnitude 10 lb in Figure P3.37 acts through *P* in the direction of the unit vector $\hat{\mathbf{e}}_F = 0.8\hat{\mathbf{i}} + 0.6\hat{\mathbf{j}}$. Determine:

- **a.** The moment of **F** about the origin *O*
- **b.** The moment of **F** about the *z* axis
- **c.** The direction cosines of the line through *O* about which **F** has the largest moment, and the value of this moment.

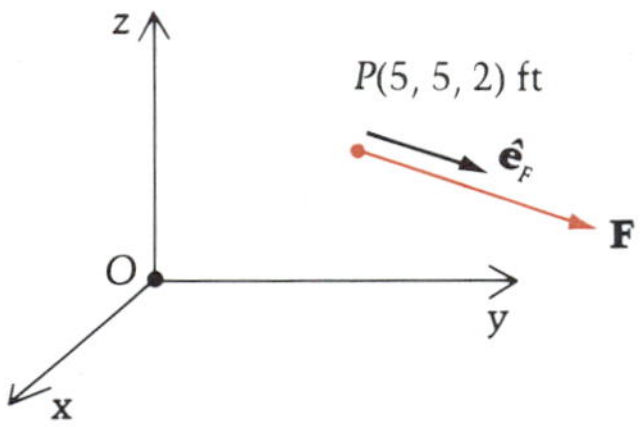

Figure P3.37

3.38 All coordinates in Figure P3.38 are given in inches. Find the moment of the 340-lb force with respect to:

- **a.** Point *A*
- **b.** Line *AB*.
- **c.** At any point *P* on the line of action of the force, it may be resolved into components $\mathbf{F}_{\parallel}$ and $\mathbf{F}_{\perp}$, parallel and perpendicular to line *AB*. Give the shortest distance between $\mathbf{F}_{\perp}$ and the line *AB* (which is independent of the choice of *P*).

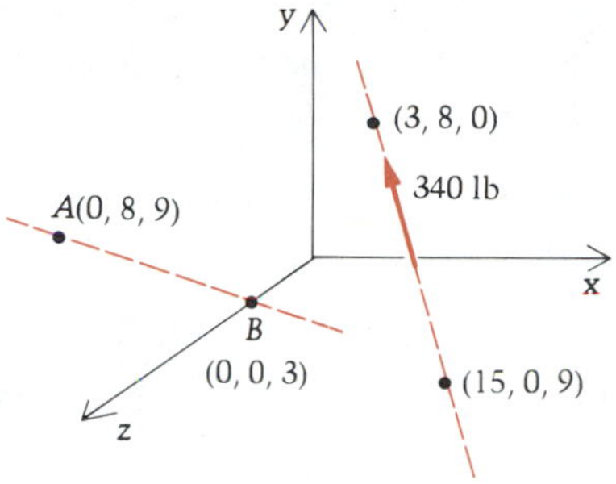

Figure P3.38

* **3.39** Find the moment of the force in Figure P3.39 about

a. the point A;

b. the line defined by:

$$\frac{x-1}{2} = \frac{y-1}{2} = \frac{z-0}{-4}$$

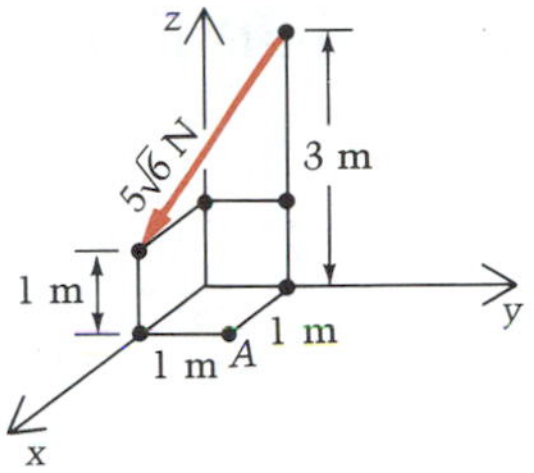

Figure P3.39

3.4 The Couple

Definition and Moment of a Couple

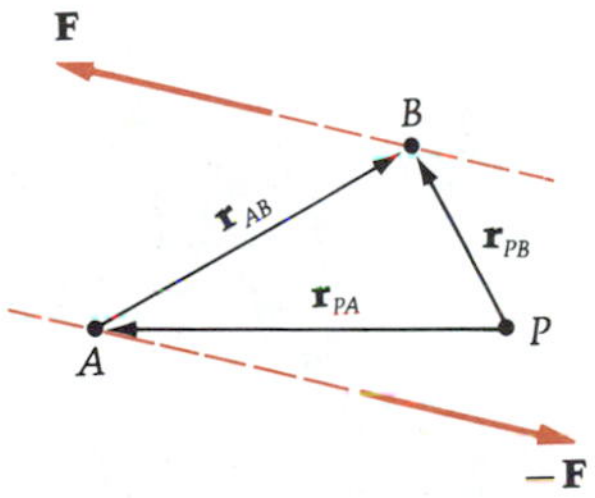

Figure 3.12

A pair of forces, equal in magnitude but having opposite directions and different (but, of course, parallel) lines of action, constitutes a **couple.** The pair of forces exerts no net thrust on a body, but there is obviously a turning effect. Hence the moment of the couple about a point is of great importance. Referring to Figure 3.12, we let A and B be points on the lines of action of a "couple" of forces $-\mathbf{F}$ and $\mathbf{F}$, respectively. By the moment of the couple we mean the sum of the moments of the two forces. Thus the moment $\mathbf{M}_P$ about point P is

$$\begin{aligned}\mathbf{M}_P &= \mathbf{r}_{PB} \times \mathbf{F} + \mathbf{r}_{PA} \times (-\mathbf{F}) \\ &= (\mathbf{r}_{PB} - \mathbf{r}_{PA}) \times \mathbf{F} = \mathbf{r}_{AB} \times \mathbf{F} \qquad (3.12)\end{aligned}$$

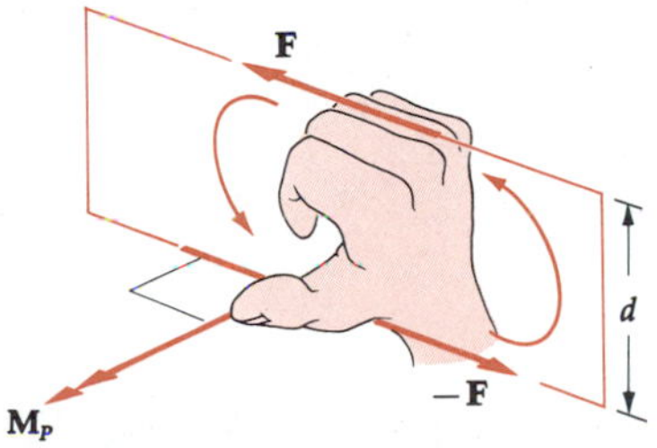

Figure 3.13

We see from Equation (3.12) that it makes no difference where A and B lie on the respective lines of action, since the cross product produces a vector whose magnitude is that of $\mathbf{F}$ multiplied by the distance d between the lines of action of the two forces. Moreover, recalling the direction associated with the cross product, we see from Figure 3.13 that $\mathbf{M}_P$ is perpendicular to the plane defined by the lines of action of the two forces and in the direction with which we would associate, by the right-hand rule, the sense of turning of the two forces. In Figure 3.13 the fingers show the turning effect of $\mathbf{F}$ and $-\mathbf{F}$ in their plane while the thumb points in the direction of the (axis of turning of the) couple.

Question 3.13 In the above derivation, is point P required to lie in the plane containing the lines of action of $\mathbf{F}$ and $-\mathbf{F}$?

The Most Important Property of a Couple

If we equate the magnitudes of the two sides of Equation (3.12), we obtain a very useful result:

$$|\mathbf{M}_P| = |\mathbf{F}|\,d \tag{3.13}$$

where, again, d is the distance between the lines of action of $\mathbf{F}$ and $-\mathbf{F}$ (see Figure 3.13). This simple scalar equation states that the magnitude of the moment of a couple (or its *strength*) is the magnitude of either force times the perpendicular distance between them.

In vector form,

$$\mathbf{M}_P = |\mathbf{F}|\,d\hat{\mathbf{u}} \tag{3.14}$$

where $\hat{\mathbf{u}}$ is the unit vector in the direction of $\mathbf{M}_P$ (normal to the plane of the two forces, along the "right thumb" of Figure 3.13).

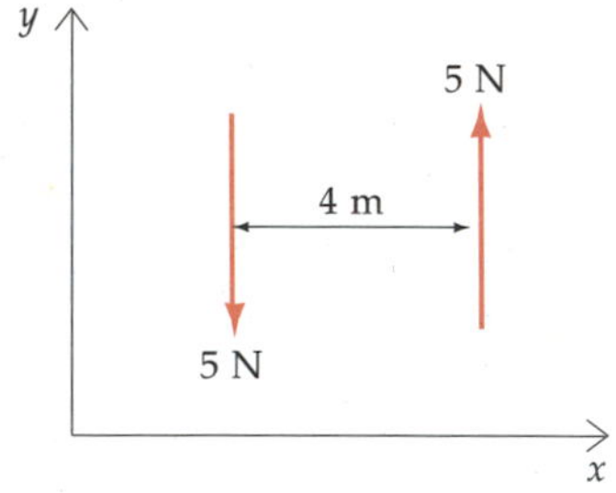

Figure 3.14

In Equation 3.12, all reference to point P has been lost; therefore, $\mathbf{M}_P$ does not depend in any way upon the location of point P. Thus, *the moment of a couple about every point is the same.* This is an extremely powerful property. For example, the moment of the couple in Figure 3.14 is 20 N · m counterclockwise or, in vector terms, directed out of the page. If the page is the xy-plane as shown, the moment of the couple is $20\hat{\mathbf{k}}$ N · m, which is its moment about the point (0, 0, 0), or the point (1, 2, 3), or *any other point of space!*

Furthermore, as far as the moment of a couple is concerned, its forces and the distance between them may be altered as desired, so long as the product $|\mathbf{F}|\,d$ and the couple's direction (right thumb in Figure 3.13) remain the same. The various couples in Figure 3.15 below thus have the same moment as did the couple of Figure 3.14:

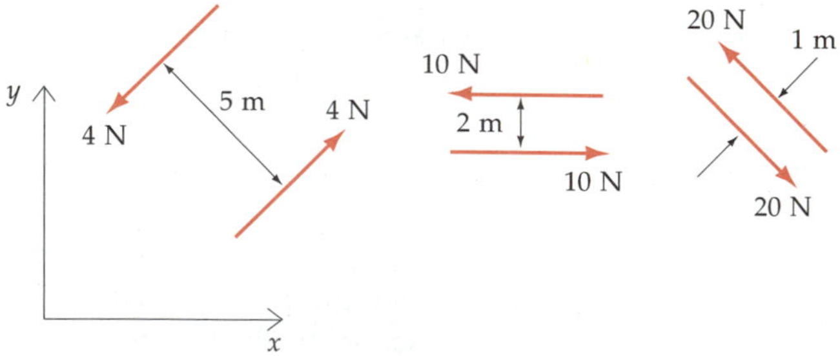

Figure 3.15

Answer 3.13 No. Nothing in the derivation of Equation (3.12) placed any restriction whatsoever on the location of point P.

The next four examples illustrate how to obtain the moment of a couple.

EXAMPLE 3.10

A mechanic applies two forces, each one of magnitude 20 lb as shown in Figure E3.10a, to a lug wrench in the process of changing a tire. Find the moment of the couple comprising the equal-magnitude, oppositely directed forces.

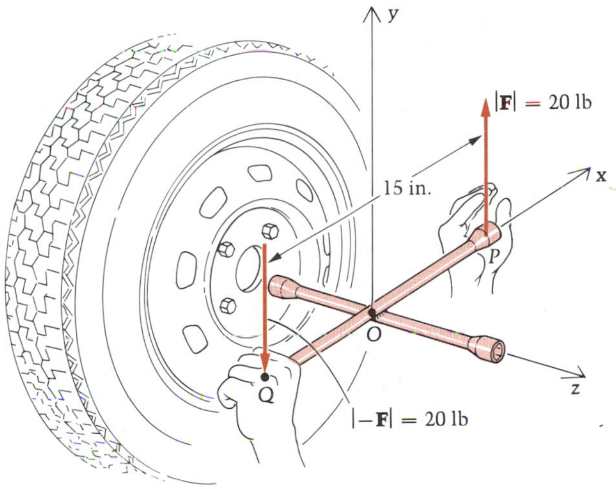

Figure E3.10a

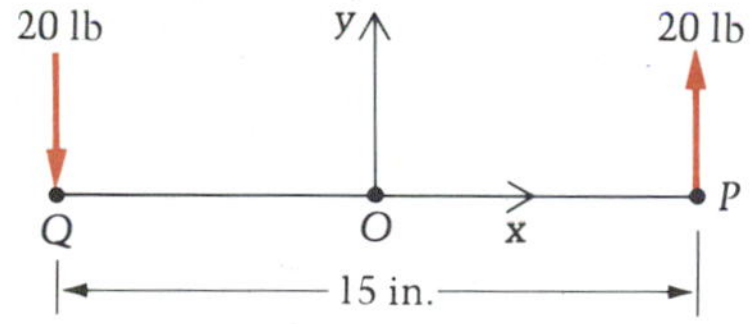

Figure E3.10b

Solution

The moment of the couple is calculated as the sum of the moments of **F** and −**F** about *any* point. Let us use point Q (at the mechanic's left hand (see Figure E3.10b):

$$\begin{aligned} \mathbf{C} &= \mathbf{r}_{QQ} \times (-\mathbf{F}) + \mathbf{r}_{QP} \times \mathbf{F} \\ &= \mathbf{0} + 15\hat{\mathbf{i}} \times 20\hat{\mathbf{j}} \\ &= 300\hat{\mathbf{k}} \text{ lb-in.} \end{aligned}$$

Alternatively, using point O,

$$\begin{aligned} \mathbf{C} &= -7.5\hat{\mathbf{i}} \times (-20\hat{\mathbf{j}}) + 7.5\hat{\mathbf{i}} \times 20\hat{\mathbf{j}} \\ &= +150\hat{\mathbf{k}} + 150\hat{\mathbf{k}} \\ &= 300\hat{\mathbf{k}} \text{ lb-in.} \qquad \text{(as before)} \end{aligned}$$

Regardless of the point chosen, **C**, as we have proved, always will be the same vector. We could also simply multiply either force times the perpendicular distance between them, obtaining:

$$\mathbf{C} = 20(15)\hat{\mathbf{k}} = 300\hat{\mathbf{k}} \text{ lb-in.} \qquad \text{(once again)}$$

where we attach the unit vector $\hat{\mathbf{k}}$ in accordance with the direction in which the forces are perceived to turn about an axis normal to their plane.

EXAMPLE 3.11

In Example 3.5, the two 20-lb forces perpendicular to the paper, down at A and up at B, are now seen to form a couple (see Figure E3.11a). Use the property that a couple has the same moment about any point of space to rework the previous example, i.e., to find the moment of the two forces about the axis (x) of the bolt.

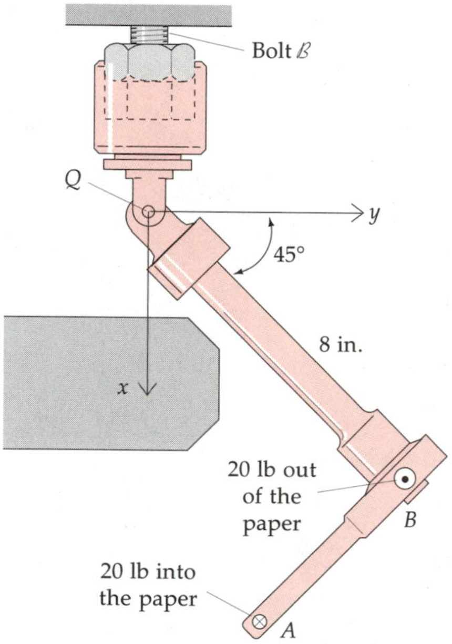

Figure E3.11a

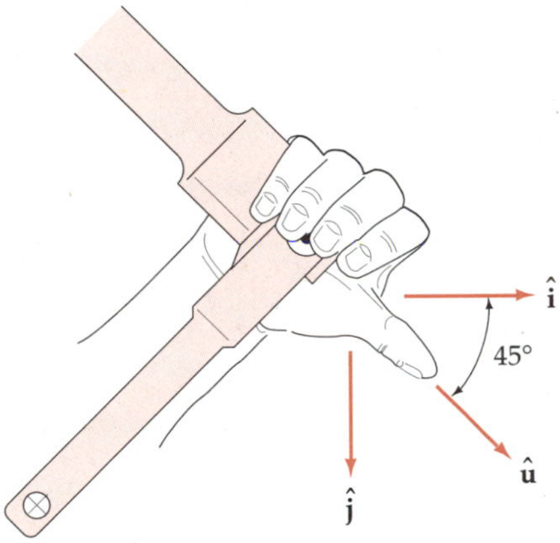

Figure E3.11b

Solution

Using Equation (3.14),

$$\mathbf{C} = |\mathbf{F}|\,d\hat{\mathbf{u}} = 20(5)\left(\frac{\hat{\mathbf{i}}+\hat{\mathbf{j}}}{\sqrt{2}}\right)$$

$$= \frac{100}{\sqrt{2}}(\hat{\mathbf{i}}+\hat{\mathbf{j}})$$

where Figure E3.11b gives the unit vector $\hat{\mathbf{u}}$ by the right-hand rule.

The vector $\mathbf{C}$ is the moment of the couple about any point, in particular about Q. Thus

$$\mathbf{M}_{x\text{-}axis} = (\mathbf{M}_Q \cdot \hat{\mathbf{i}})\,\hat{\mathbf{i}} = \left[\frac{100}{\sqrt{2}}(\hat{\mathbf{i}}+\hat{\mathbf{j}}) \cdot \hat{\mathbf{i}}\right]\hat{\mathbf{i}} = \frac{100}{\sqrt{2}}\,\hat{\mathbf{i}} = 70.7\hat{\mathbf{i}} \text{ lb-in},$$

a *much* simpler solution than was possible before!

Example 3.12

The two forces shown in Figure E3.12 have magnitudes of 50-N and are oppositely directed. Find the moment of the couple that they constitute.

Solution

As in the previous example, we compute the moment of the couple by adding the moments of the two forces. We can add them about any point, and we choose the origin:

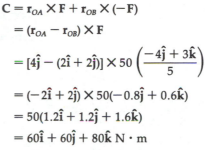

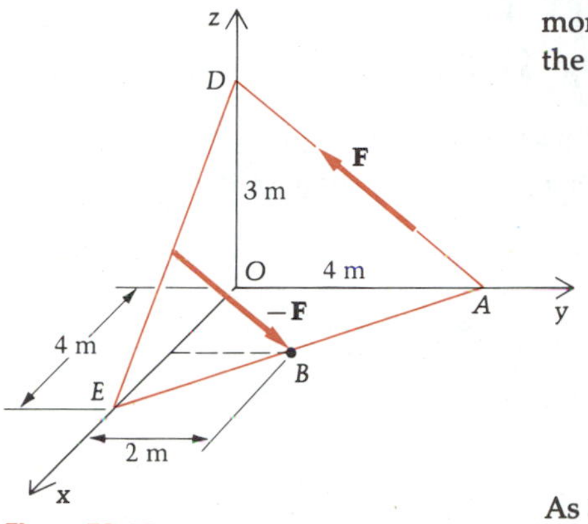

Figure E3.12

$$\begin{aligned}\mathbf{C} &= \mathbf{r}_{OA} \times \mathbf{F} + \mathbf{r}_{OB} \times (-\mathbf{F}) \\ &= (\mathbf{r}_{OA} - \mathbf{r}_{OB}) \times \mathbf{F} \\ &= [4\hat{\mathbf{j}} - (2\hat{\mathbf{i}} + 2\hat{\mathbf{j}})] \times 50\left(\frac{-4\hat{\mathbf{j}} + 3\hat{\mathbf{k}}}{5}\right) \\ &= (-2\hat{\mathbf{i}} + 2\hat{\mathbf{j}}) \times 50(-0.8\hat{\mathbf{j}} + 0.6\hat{\mathbf{k}}) \\ &= 50(1.2\hat{\mathbf{i}} + 1.2\hat{\mathbf{j}} + 1.6\hat{\mathbf{k}}) \\ &= 60\hat{\mathbf{i}} + 60\hat{\mathbf{j}} + 80\hat{\mathbf{k}}\ \text{N} \cdot \text{m}\end{aligned}$$

As a check, let us form the moment of the couple, **C**, by adding the moments of **F** and −**F** about *A*:

$$\begin{aligned}\mathbf{C} &= \mathbf{r}_{AA} \times \mathbf{F} + \mathbf{r}_{AB} \times (-\mathbf{F}) \\ &= \mathbf{0} + (2\hat{\mathbf{i}} - 2\hat{\mathbf{j}}) \times 50\left(\frac{4\hat{\mathbf{j}} - 3\hat{\mathbf{k}}}{5}\right) \\ &= \tfrac{50}{5}(8\hat{\mathbf{k}} + 6\hat{\mathbf{j}} + \mathbf{0} + 6\hat{\mathbf{i}}) \\ &= 60\hat{\mathbf{i}} + 60\hat{\mathbf{j}} + 80\hat{\mathbf{k}}\ \text{N} \cdot \text{m}\end{aligned}$$

as before. We note again that the couple has the same moment about *all* points.

Working the above example using $|\mathbf{F}|\, d\hat{\mathbf{u}}$ (Equation 3.14) is *much* harder (see Problem 3.56). Usually, when a couple's forces are not parallel to coordinate axes or do not lie in a coordinate plane (and the ones above are neither!), the use of vectors to compute its moment is easier.

Example 3.13

One of the two forces that constitute a couple is $\mathbf{F} = 3\hat{\mathbf{i}} - 4\hat{\mathbf{j}} + 5\hat{\mathbf{k}}$ lb, having a line of action that passes through the point *A* at (0, 6, 5) ft. The other force has a line of action that passes through point *B* at (−4, 0, 2) ft. Find the moment of the couple, and the distance between the lines of action of the forces.

Solution

The moment, **C**, of the couple is the sum of the moments of its two forces about any point. Choosing *B* as the point,

$$\begin{aligned}\mathbf{C} &= \mathbf{r}_{BA} \times \mathbf{F} + \mathbf{r}_{BB} \times (-\mathbf{F}) \\ &= \mathbf{r}_{BA} \times \mathbf{F}\end{aligned}$$

because A is on the line of action of $\mathbf{F}$ and B is on the line of action of the companion force $-\mathbf{F}$. The position vector $\mathbf{r}_{BA}$ is computed as

$$\mathbf{r}_{BA} = [0 - (-4)]\hat{\mathbf{i}} + (6 - 0)\hat{\mathbf{j}} + (5 - 2)\hat{\mathbf{k}}$$
$$= 4\hat{\mathbf{i}} + 6\hat{\mathbf{j}} + 3\hat{\mathbf{k}} \text{ ft}$$

Therefore

$$\mathbf{C} = (4\hat{\mathbf{i}} + 6\hat{\mathbf{j}} + 3\hat{\mathbf{k}}) \times (3\hat{\mathbf{i}} - 4\hat{\mathbf{j}} + 5\hat{\mathbf{k}})$$
$$= 4(-4)\hat{\mathbf{k}} + 4(5)(-\hat{\mathbf{j}}) + 6(3)(-\hat{\mathbf{k}}) + 6(5)\hat{\mathbf{i}} + 3(3)\hat{\mathbf{j}} + 3(-4)(-\hat{\mathbf{i}})$$
$$= 42\hat{\mathbf{i}} - 11\hat{\mathbf{j}} - 34\hat{\mathbf{k}} \text{ lb-ft}$$

The distance between the lines of action is, using Equation (3.13),

$$d = \frac{|\mathbf{C}|}{|\mathbf{F}|}$$
$$= \frac{\sqrt{(42)^2 + (11)^2 + (34)^2}}{\sqrt{(3)^2 + (4)^2 + (5)^2}} = \frac{\sqrt{3041}}{\sqrt{50}}$$
$$= 7.80 \text{ ft}$$

In the remainder of Chapter 3, we shall extend our study to include systems of forces and couples, in preparation for solving equilibrium problems in Chapters 4, 5, and 6 which will involve moments and thus be much more difficult than those of Chapter 2. We shall first present the full equations of equilibrium of a body, which will help us both motivationally and practically. Motivationally, the equilibrium equations show the importance of mastering force and moment relationships. Practically, the equilibrium equations will be used in our definition of equipollent force systems. Only after these ideas are mastered will we be fully prepared to solve problems involving the equilibrium of bodies.

PROBLEMS ▶ Section 3.4

Write the vector expressions for the moments of the couples in the following three problems (the forces are in the planes labeled in Figures P3.40–P3.42).

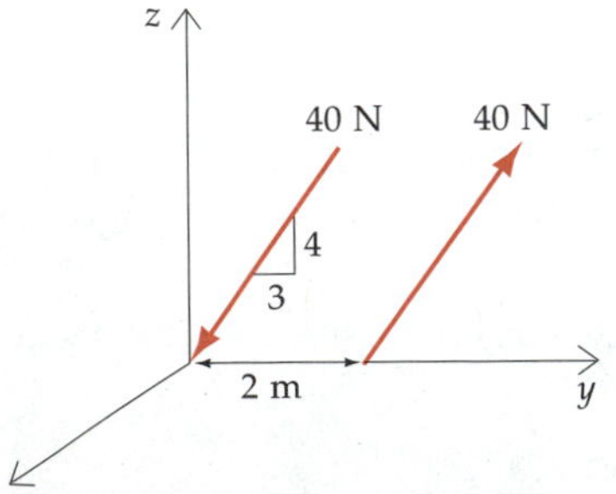

Figure P3.40

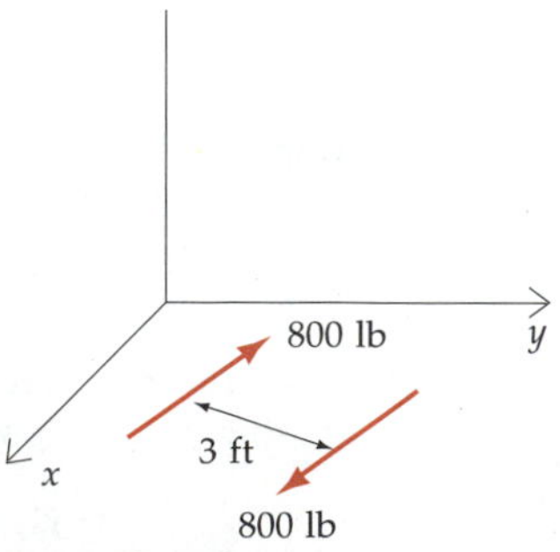

Figure P3.41

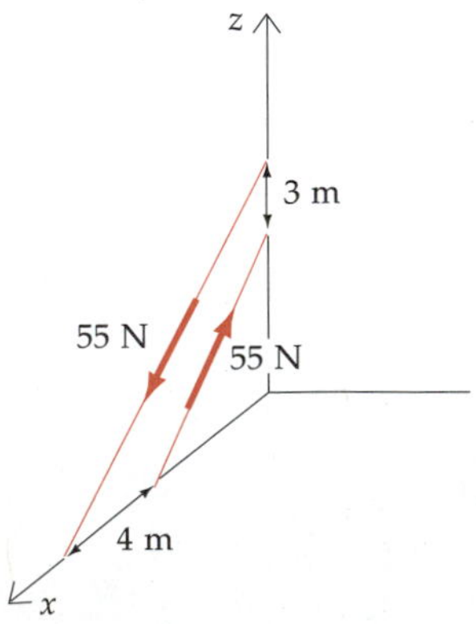

Figure P3.42

3.43 The radius of each of the two pulleys in Figure P3.43 is 1 ft. Determine the resultant moment of the two pulley tension forces about (a) point A and (b) any other point.

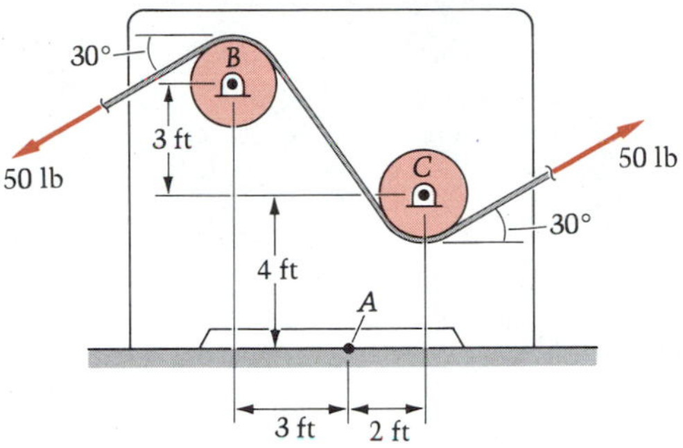

Figure P3.43

3.44 Friction is causing a set of uniformly distributed tangential forces of intensity 300 lb/in. to act on the circular ring of radius 16 in. shown in Figure P3.44. Determine the moment of these forces about the center, O, of the ring. What is the moment about A?

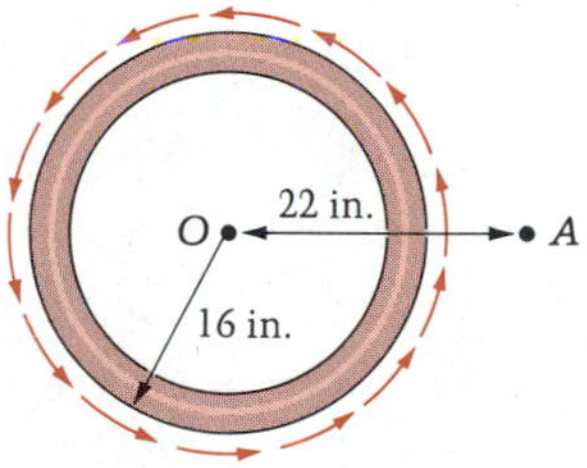

Figure P3.44

3.45 Determine the moment of the couple in Figure P3.45 about (a) point O, (b) point A, (c) the y axis, and (d) the line in the yz plane defined by $z = 4$ ft.

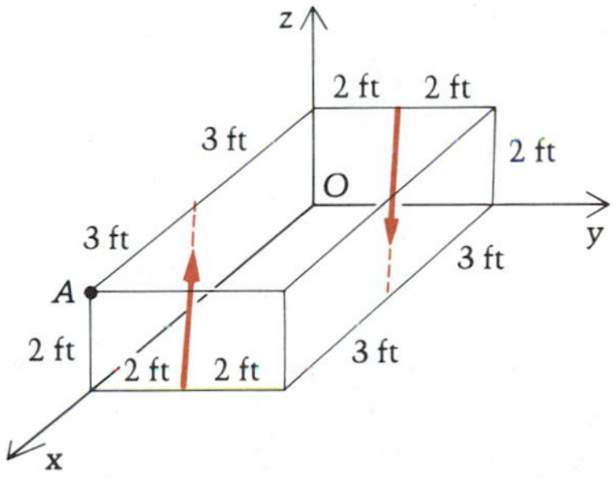

Figure P3.45

3.46 Find the moment of the couple about line AB in Figure P3.46. The lines of action of the forces are both in the yz plane.

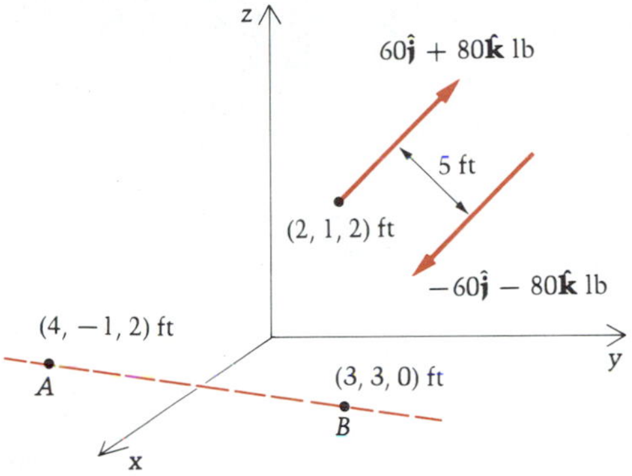

Figure P3.46

3.47 Write the moment of the couple formed by the two 30-lb forces shown in Figure P3.47. What is the moment of the couple about the point $(x, y, z) = (1, 5, -8)$?

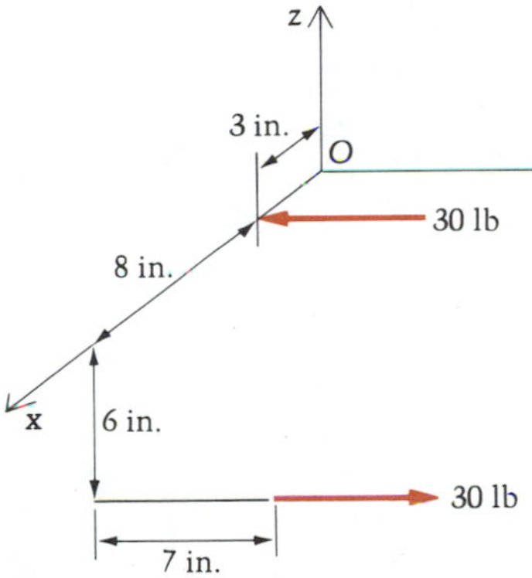

Figure P3.47

3.48 After the couple was defined and analyzed, it was found to have (1) the same moment about every point, and (2) no net thrust (or force). Show that the set of three forces shown in Figure P3.48 possesses these same two properties.

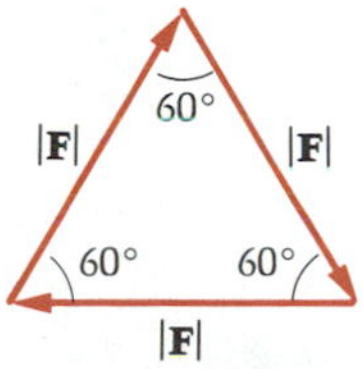

Figure P3.48

In each of the following four problems, show that the system of forces has no net thrust (or force). In Section 3.5 we shall see that whenever this happens, the moment of the forces about all points is the same. Find the moment about A and B in each problem and observe that they are equal.

3.49

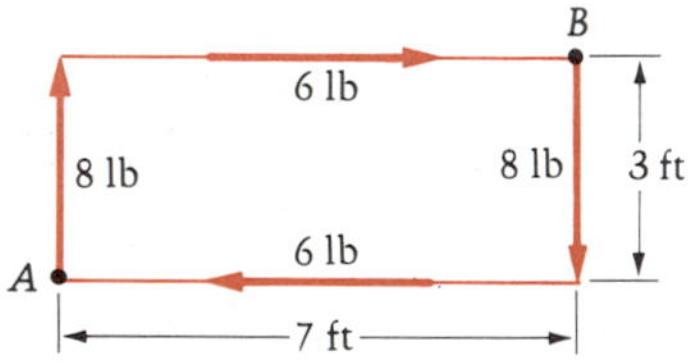

Figure P3.49

3.50

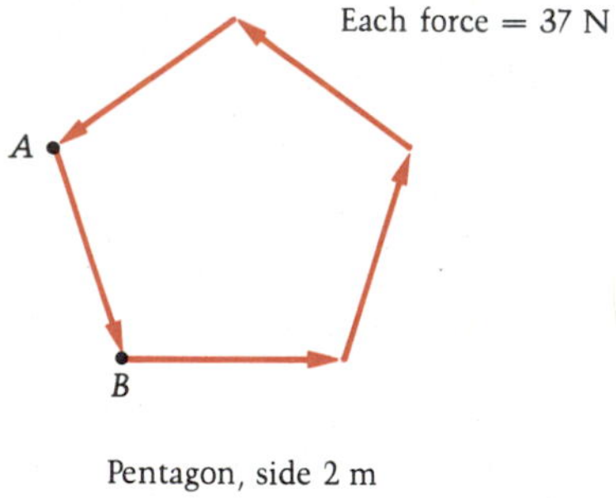

Figure P3.50

3.51

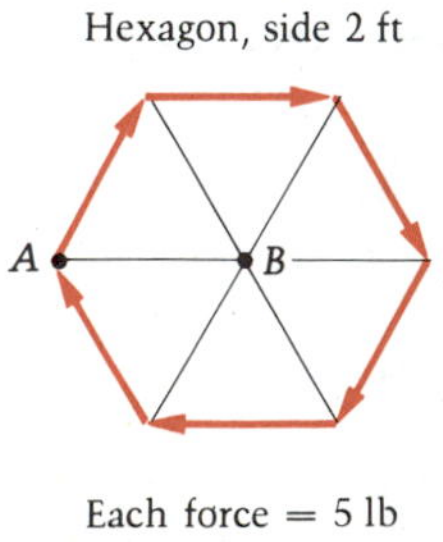

Figure P3.51

3.52 Sixteen forces of 14 lb each at 22.5° spacing on a 6-ft diameter circle. See Figure P3.52.

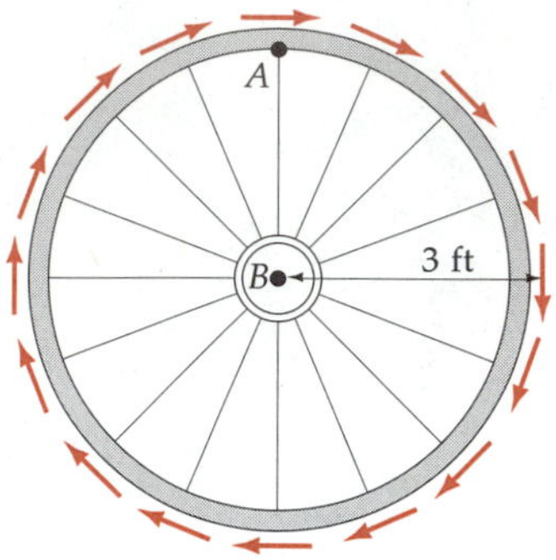

Figure P3.52

3.53 The two parallel forces in Figure P3.53 each have magnitude $10\sqrt{3}$ N. Find the (vector) moment of this system of forces about:

a. point A;

b. point B, which lies at (10, −6, 12) m;

c. line AB.

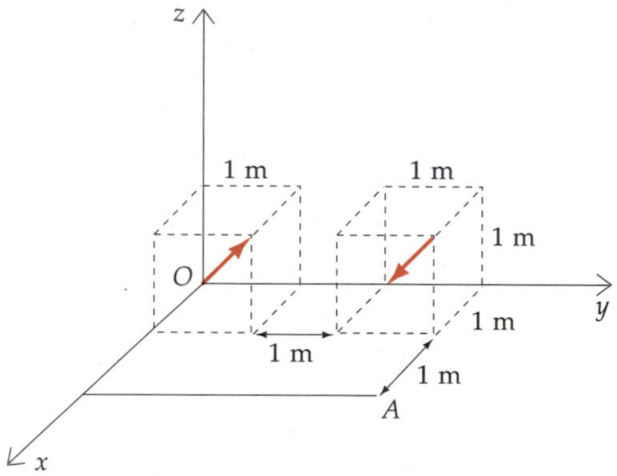

Figure P3.53

3.54 Force **F** in Figure P3.54 has magnitude $|\mathbf{F}| = 15$ lb, and couple **M** has magnitude $4\sqrt{41}$ lb-in. and is normal to plane ACD. Find the moment of the system of **F** and **M** with respect to point D. Then determine the moment of the system about line CD.

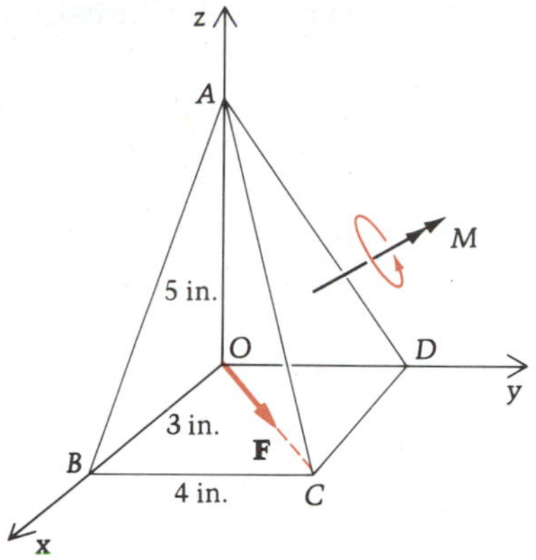

Figure P3.54

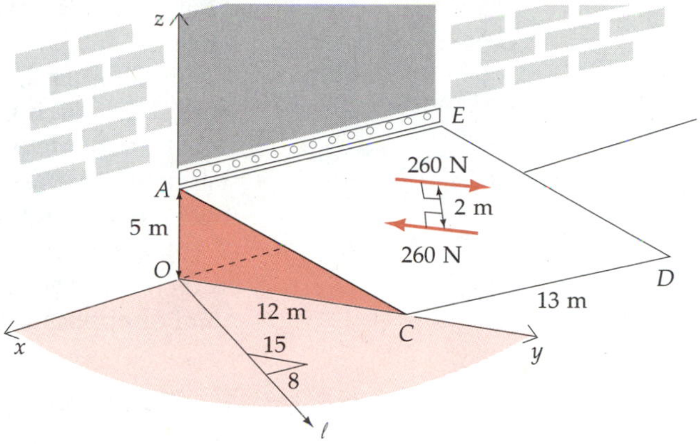

Figure P3.55

3.55 The two 260-N forces lie in the inclined plane *AEDC* shown in Figure P3.55. Find the moment of the couple, which they constitute, about the indicated line *ℓ* in the *xy* plane.

• **3.56** Find the moment of the couple in Example 3.12 by means of $\mathbf{M} = |\mathbf{F}|\,d\hat{\mathbf{u}}$, where $\hat{\mathbf{u}}$ is the unit vector in the direction of the couple (normal to the plane *ADE*). Hint: Crossing $\mathbf{r}_{EA}$ into $\mathbf{r}_{ED}$ gives a vector in that direction; then, dividing the result by its magnitude yields $\hat{\mathbf{u}}$. Finding the distance d is also tricky.

3.5 Laws of Equilibrium: Relationship Between Sums of Moments

The Equilibrium Equations

The purposes of this brief section are (a) to provide motivation for the topics covered in the remainder of the chapter, and (b) to establish a useful relationship having to do with the moments that a system of forces produce about different points in space. In particular, we need to clearly identify those characteristics of a force system that are essential to the analysis of problems in statics. In our view, the easiest way to do this is by displaying the equations of equilibrium; we are of course already familiar with the first of these from our studies in the preceding chapter.

When none of the particles making up a body is accelerating relative to an inertial frame of reference, Euler's laws (Section 1.3) are reduced to the statements (1) that the sum of the external forces vanishes, and (2) that the sum of their moments about any point also vanishes. A body that is stationary in an inertial frame clearly falls into this category. Letting the uppercase Greek letter sigma (Σ) indicate the process of summation, the two laws, called the **equilibrium equations**, are written symbolically as

$$\Sigma\mathbf{F} = \mathbf{0} \tag{3.15}$$

$$\Sigma\mathbf{M}_P = \mathbf{0} \tag{3.16}$$

where P denotes the point with respect to which the moments are calculated.*

In Chapter 2, we only needed Equation (3.15). That is because we were then solving only problems for which the "moment equation" (3.16) was not required. All we needed to know about forces in Chapter 2 was how to express them as vectors and sum them. A second characteristic of a force now comes into play with the second equilibrium equation (3.16), and that is the location of its line of action. As we know from our study of moments of forces earlier in this chapter, however, the specific point of application (along the line of action) is not important in statics.

Question 3.14 Why not?

Question 3.15 What about a couple acting on the body: How is it to be incorporated into the equations?

Relationship Between Sums of Moments

We shall not actually use the equilibrium Equations (3.15) and (3.16) together in problems until Chapter 4. But it is important to recognize the operations we shall then have to perform on a system of forces†; namely, we must sum the forces as we did in Chapter 2, and now we must in addition sum the moments of the forces. These very same operations are required in dynamics, where we analyze the motion of an accelerating body. Of immediate importance to us is the fact that, for a given system of forces, the sums of moments with respect to two different points are related in a particularly simple way. To establish this relationship, let the body (Figure 3.16) be acted upon by a number (N_F) of forces and a number (N_C) of couples. Let $\mathbf{F}_i$ be the ith force, for which the point of application is A_i (which actually could be *any* point on the line of action of $\mathbf{F}_i$). Also, let $\mathbf{C}_j$ be the moment of the jth couple. Recall from Section 3.4 that the moment of a couple about every point is the same. To distinguish, in figures such as Figure 3.16, couples from single forces we use

* Somewhat surprisingly the word "equilibrium," having a connotation of balance, does not have a universally accepted technical definition in the literature of mechanics. Some writers associate it with a force system, one that satisfies Equations (3.15) and (3.16); other writers give it the kinematic definition of stationarity of the body in an inertial frame. Still other writers define it by "stationarity" of the body in some, not necessarily inertial, frame of reference. The important point, as we study statics, is that the equilibrium equations are necessary conditions for a body to be stationary in an inertial frame of reference. In this text, whenever we refer to a "body in equilibrium" the reader should take that to mean that the body is stationary in an inertial frame.

† We use the phrases "system of forces" and "force system" to denote any collection of forces and/or couples of interest to us.

Answer 3.14 As we have seen, the moment of a force is the same regardless of its point of application on the line of action.

Answer 3.15 The couple may be ignored in $\Sigma\mathbf{F} = \mathbf{0}$ because its equal-magnitude, oppositely directed forces cancel. To the left-hand side of $\Sigma\mathbf{M}_P = \mathbf{0}$, we add the moments about $\mathbf{P}$ of the two forces constituting the couple (or add the moment of the couple directly, if we know it). Since the moment of a couple is the same about any point, any convenient point may be used to find the moment.

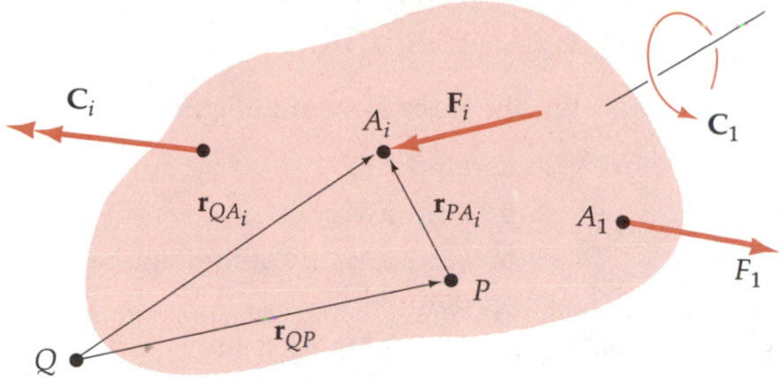

Figure 3.16

double-headed arrows, or sometimes curved arrows indicating sense of turning, to denote moments of couples.

Letting $\Sigma\mathbf{M}_Q$ denote the sum of the moments about Q of all the forces and couples in the system, we have

$$\Sigma\mathbf{M}_Q = \sum_{i=1}^{N_F} \mathbf{r}_{QA_i} \times \mathbf{F}_i + \sum_{j=1}^{N_C} \mathbf{C}_j \tag{3.17}$$

But

$$\mathbf{r}_{QA_i} = \mathbf{r}_{QP} + \mathbf{r}_{PA_i}$$

so that, making the substitution in Equation (3.17),

$$\begin{aligned}\Sigma\mathbf{M}_Q &= \sum_{i=1}^{N_F} (\mathbf{r}_{QP} + \mathbf{r}_{PA_i}) \times \mathbf{F}_i + \sum_{j=1}^{N_C} \mathbf{C}_j \\ &= \underbrace{\sum_{i=1}^{N_F} \mathbf{r}_{QP} \times \mathbf{F}_i}_{} + \underbrace{\sum_{i=1}^{N_F} \mathbf{r}_{PA_i} \times \mathbf{F}_i + \sum_{j=1}^{N_C} \mathbf{C}_j}_{}\end{aligned}$$

$$\Sigma\mathbf{M}_Q = \mathbf{r}_{QP} \times \Sigma\mathbf{F} \quad + \quad \Sigma\mathbf{M}_P \tag{3.18}$$

Equation (3.18) is very important because it tells us that the sum of the moments about one point is completely determined by the sum of the forces, the sum of the moments about a second point, and the relative locations of the two points. In particular we note that if Equations (3.15) and (3.16) both hold for a force system, then $\Sigma\mathbf{M}_Q$ is also zero for *any point* Q. We shall use Equation (3.18) a number of times in the remainder of this chapter.

Question 3.16 In view of Equation (3.18), would there be a system of forces for which the moment is (a) the same at all points? (b) zero at all points?

Question 3.17 Why is Equation (3.18) also valid when the body on which the forces and/or couples act is accelerating (i.e., in dynamics)?

Answer 3.16 (a) Yes, if and only if $\Sigma\mathbf{F} = \mathbf{0}$. (b) Yes, if and only if $\Sigma\mathbf{F} = \mathbf{0}$ *and* $\Sigma\mathbf{M}_P = \mathbf{0}$ at some one point P.

Answer 3.17 Because we didn't use the equilibrium equations (3.15, 3.16) in developing it. Equation (3.18) is just a useful result valid for any force system.

EXAMPLE 3.14

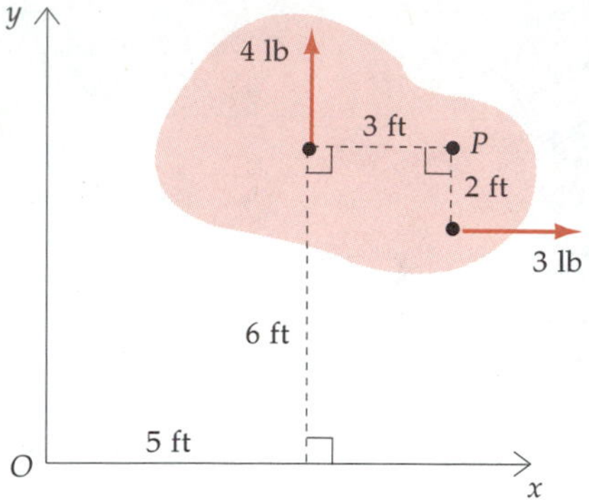

Figure E3.14

For the force system in Figure E3.14,

a. Find $\Sigma\mathbf{F}$;

b. Find $\Sigma\mathbf{M}_P$;

c. Find $\Sigma\mathbf{M}_O$ by summing the moments of the two forces with respect to point O;

d. Find $\Sigma\mathbf{M}_O$ by using Equation (3.18) together with the results from (a, b) above.

Solution

a. Adding the two forces,

$$\Sigma\mathbf{F} = 3\hat{\mathbf{i}} + 4\hat{\mathbf{j}}\text{ lb}$$

b.

$$\Sigma\mathbf{M}_P = (3\text{ lb})(2\text{ ft})\hat{\mathbf{k}} + (4\text{ lb})(3\text{ ft})(-\hat{\mathbf{k}})$$
$$= -6\hat{\mathbf{k}}\text{ lb-ft}$$

or, with vectors,

$$\Sigma\mathbf{M}_P = -2\hat{\mathbf{j}} \times 3\hat{\mathbf{i}} + (-3\hat{\mathbf{i}}) \times 4\hat{\mathbf{j}}$$
$$= 6\hat{\mathbf{k}} - 12\hat{\mathbf{k}} = -6\hat{\mathbf{k}}\text{ lb-ft}$$

c.

$$\Sigma\mathbf{M}_O = (4\text{ lb})(5\text{ ft})\hat{\mathbf{k}} + (3\text{ lb})(4\text{ ft})(-\hat{\mathbf{k}})$$
$$= 8\hat{\mathbf{k}}\text{ lb-ft}$$

d.

$$\Sigma\mathbf{M}_O = \Sigma\mathbf{M}_P + \mathbf{r}_{OP} \times \Sigma\mathbf{F}$$
$$= -6\hat{\mathbf{k}} + (8\hat{\mathbf{i}} + 6\hat{\mathbf{j}}) \times (3\hat{\mathbf{i}} + 4\hat{\mathbf{j}})$$
$$= -6\hat{\mathbf{k}} + 32\hat{\mathbf{k}} - 18\hat{\mathbf{k}}$$
$$= -8\hat{\mathbf{k}}\text{ lb-ft, as above.}$$

PROBLEMS ▶ Section 3.5

3.57 To the system of two forces in Example 3.14, add a third force of $5\hat{\mathbf{j}}$ lb along the line $x = 2$ ft. Answer the same four questions in the example for this three-force system.

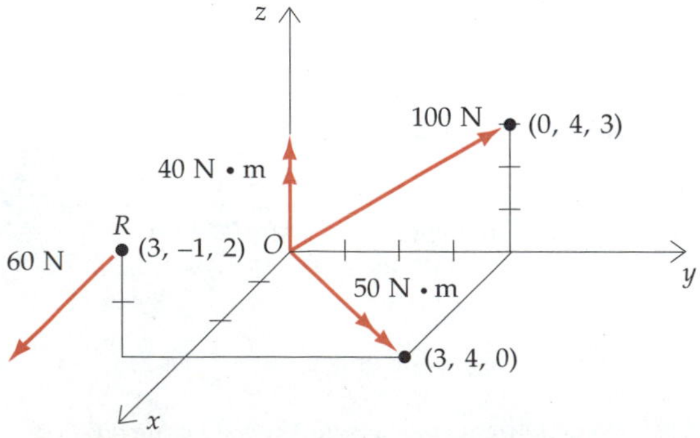

Figure P3.58

3.58 For the system shown in Figure P3.58, find:

a. $\Sigma\mathbf{F}$;

b. $\Sigma\mathbf{M}_R$ by summing the couples and the moments of forces about R;

c. $\Sigma\mathbf{M}_O$ by summing the couples and the moments of forces about O;

d. $\Sigma\mathbf{M}_O$ by using (a), (b), and Equation (3.18). The result must of course agree with that of (c).

3.59 A body $\mathcal{B}$ is subjected to five forces acting at points of an inertial reference frame given in the table on the next page.

a. Find $\Sigma\mathbf{F}$ and $\Sigma\mathbf{M}_P$, where P is the origin.

b. Show that $\mathcal{B}$ is not in equilibrium.

c. Add forces at (1, 0, 0) m and (4, 0, 0) m that would allow $\mathcal{B}$ to be in equilibrium.

Force (N)	Position Vector to Point of Application (m)
1. $2\hat{\mathbf{i}} + 3\hat{\mathbf{j}} - 5\hat{\mathbf{k}}$	$-\hat{\mathbf{i}} + \hat{\mathbf{j}}$
2. $7\hat{\mathbf{i}} - 2\hat{\mathbf{k}}$	$\hat{\mathbf{j}} - \hat{\mathbf{k}}$
3. $7\hat{\mathbf{i}} - 2\hat{\mathbf{k}}$	$\hat{\mathbf{i}} - \hat{\mathbf{j}}$
4. $-\hat{\mathbf{i}} - 3\hat{\mathbf{j}} + 5\hat{\mathbf{k}}$	$\hat{\mathbf{i}} - \hat{\mathbf{k}}$
5. $-14\hat{\mathbf{i}} + 4\hat{\mathbf{k}}$	$2\hat{\mathbf{j}} - \hat{\mathbf{k}}$

3.60 Find the moments of each of the five forces in the preceding problem about the point Q at $(x, y, z) = (1, 2, 3)$ m. Add them to obtain $\Sigma \mathbf{M}_Q$. Then illustrate Equation (3.18) by obtaining the same answer for $\Sigma \mathbf{M}_Q$ by using $\Sigma \mathbf{F}$ and $\Sigma \mathbf{M}_P$ (from the preceding problem) in this equation.

3.6 Equipollence of Force Systems

The Meaning of Equipollence

Two different systems of forces are said to be **equipollent** (meaning "of equal power, or strength") if they make the same contributions to the equations of equilibrium, (3.15) and (3.16). They will also make the same contributions to the corresponding equations of motion for a body that is *not* in equilibrium. Equipollence, then is a special kind of equivalence in which two force systems exert the same net push (or pull) on a body and also exert the same net turning action (moment).*

In general, a body will not respond in the same way to a force system S_1 as it will to an equipollent force system S_2, the exception being a rigid body whose responses to two such systems are indistinguishable. However, we are *not restricting our discussion to rigid bodies* because it is only the contributions to equilibrium equations that concern us here.

A simple example of equipollent forces is shown in Figure 3.17 where we can compare the processes of pushing and towing an automobile. The effects of the two 500-lb forces on the bumpers of the automobile are of course quite different. However, if the forces have the same line of action, they are equipollent because the equations of equilibrium (or motion, in the case of dynamics) relating the external forces acting on the automobile would not distinguish between them.

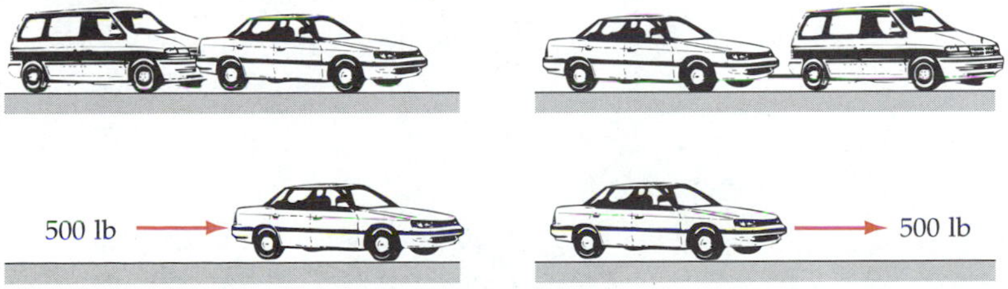

Figure 3.17

* Some authors use "statically equivalent," or "rigid-body equivalent," or simply "equivalent" to describe this relationship. We are using the less familiar word equipollent in order to emphasize that the relationship is restricted neither to statics nor to rigid bodies, and also to call attention to the special nature of the equivalence.

The Two Conditions for Equipollence

It follows from the definition that the two conditions for equipollence of systems S_1 and S_2 are:

1. *Force Condition:* $(\Sigma\mathbf{F})_1 = (\Sigma\mathbf{F})_2$ (3.19)
2. *Moment Condition:* $(\Sigma\mathbf{M}_P)_1 = (\Sigma\mathbf{M}_P)_2$ (3.20)

in which P is some common point in S_1 and S_2.

Question 3.18 Could a system consisting of a single force be equipollent to a system consisting only of a couple?

It is important to note that satisfaction of the two equipollence conditions, namely the force condition and the moment condition at a single point P, guarantees the satisfaction of the moment condition at *every* point. To show that this is the case, we let Q be any point and recall the important result, Equation (3.18):

$$(\Sigma\mathbf{M}_Q)_2 = \mathbf{r}_{QP} \times (\Sigma\mathbf{F})_2 + (\Sigma\mathbf{M}_P)_2$$

where we have applied the result to the force system of S_2. Next, by the equipollence of S_1 and S_2, we may replace $(\Sigma\mathbf{F})_2$ by $(\Sigma\mathbf{F})_1$ and $(\Sigma\mathbf{M}_P)_2$ by $(\Sigma\mathbf{M}_P)_1$ and get

$$(\Sigma\mathbf{M}_Q)_2 = \underbrace{\mathbf{r}_{QP} \times (\Sigma\mathbf{F})_1 + (\Sigma\mathbf{M}_P)_1}$$
$$= (\Sigma\mathbf{M}_Q)_1 \qquad \text{(by Equation (3.18) again)}$$

Therefore if the force summations are equal and the moment summations are equal about one point P, then they are also equal about *every other point.* Thus to determine whether two systems are equipollent, we need only to compare $\Sigma\mathbf{F}$, and then $\Sigma\mathbf{M}$ at any one point.

Question 3.19 Given three force systems S_1, S_2, and S_3, with S_1 equipollent to both S_2 and S_3, determine whether S_2 and S_3 are necessarily equipollent.

We now illustrate the equipollence of force systems by means of some examples. In the first one, a force and couple system equipollent to a system of two forces and a couple is found at a pre-assigned point.

Answer 3.18 No, because in this case $\Sigma\mathbf{F}$ could not possibly be the same. Recall that $\Sigma\mathbf{F} = \mathbf{0}$ for a couple! (This assumes the force and couple are not both zero, of course.)

Answer 3.19 Yes, they are.

EXAMPLE 3.15

Find a force and couple at point A which together form a system S_2 equipollent to the system S_1 of Figure E3.15a.

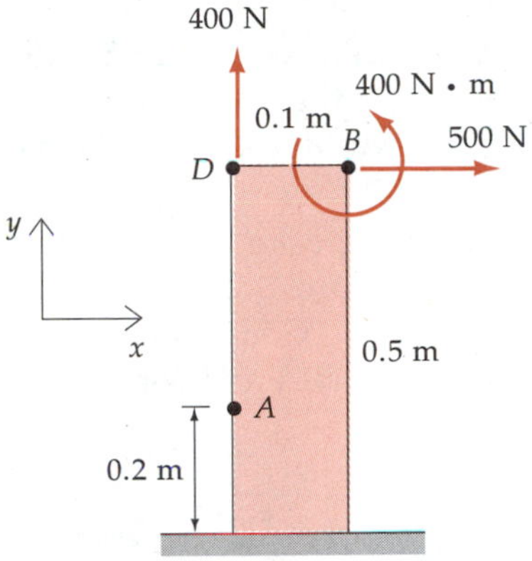

Figure E3.15a

Solution

We wish to determine **F** and **M** in Figure E3.15b such that our equipollence conditions 1 and 2 (Equations 3.19, 20) hold.

$$\text{Condition 1:} \qquad (\Sigma \mathbf{F})_1 = (\Sigma \mathbf{F})_2$$

$$500\hat{\mathbf{i}} + 400\hat{\mathbf{j}} = \mathbf{F}$$

Next we proceed to determine the couple **M**: we shall use A as the moment center (any point could be used):

$$\text{Condition 2:} \qquad (\Sigma \mathbf{M}_A)_1 = (\Sigma \mathbf{M}_A)_2$$

$$400\hat{\mathbf{k}} + 500(0.5 - 0.2)(-\hat{\mathbf{k}}) = \mathbf{M}$$

$$\mathbf{M} = 250\hat{\mathbf{k}} \text{ N} \cdot \text{m}$$

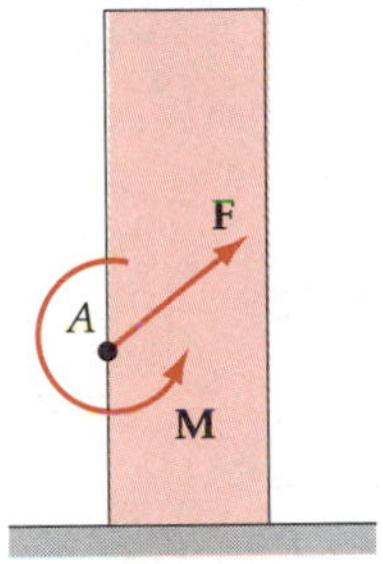

Figure E3.15b

Question 3.20 Why was A a good choice for the moment center here?

The equipollent system at A may therefore be drawn as shown in Figure E3.15c (with the force expressed in terms of its components), or in Figure E3.15d.

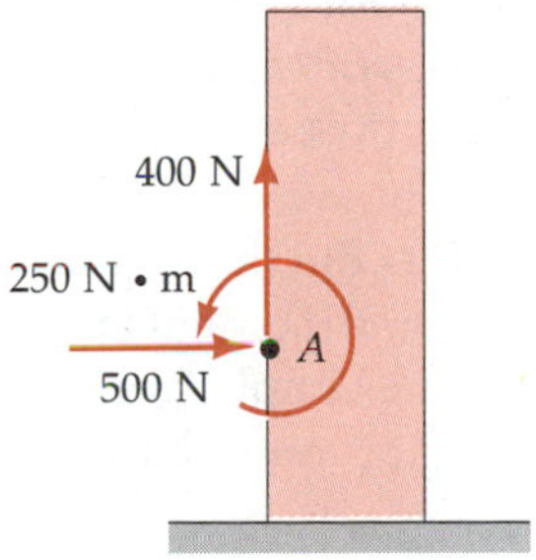

Figure E3.15c

or

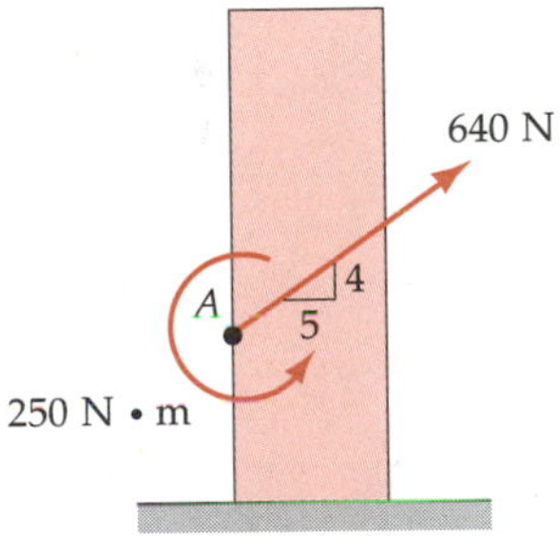

Figure E3.15d

It is interesting to note that the preceding example could alternately be worked in the following manner:

Answer 3.20 In system S_1, the 400-N force passes through A; and in system S_2, force **F** passes through A. These forces thus produce no moment about A, making for simpler moment calculations.

1. For any force (the 400-N force in this case) which passes through A, simply slide it along its line of action and place it at A. Also place and combine at A all couples in the system. See Figure 3.18.

2. "Add and subtract" copies of the remaining forces (just the 500-N force in this case) at the point A of interest. Note that this does not change $\Sigma\mathbf{F}$ or $\Sigma\mathbf{M}$ at any point since it's merely adding two cancelling, collinear forces. See Figure 3.19.

3. Recognize and replace each "equal-but-opposite" pair like the two 500-N forces enclosed by the dashed line in Figure 3.20 as a couple with moment = (500)(0.3) ↻ N · m = 150 ↻ N · m. Couples have the same moment about any point, so we may draw it at A and combine it with the couple already there.

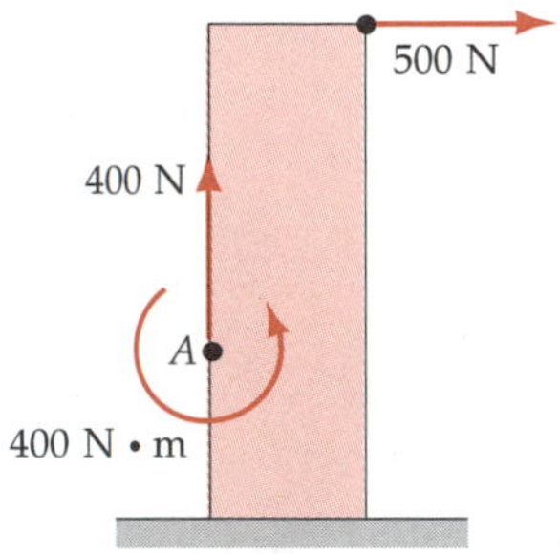

Figure 3.18

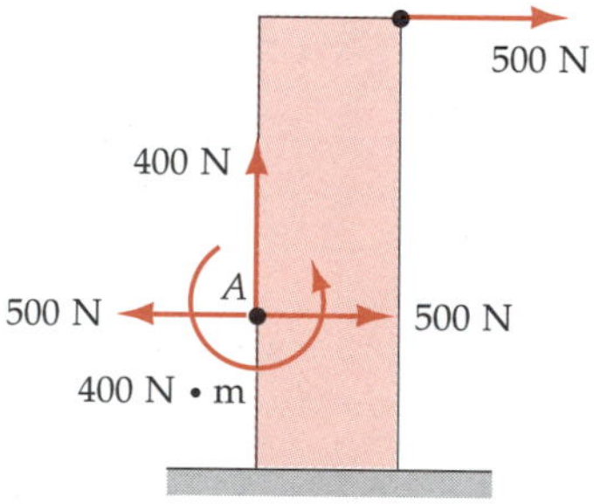

Figure 3.19

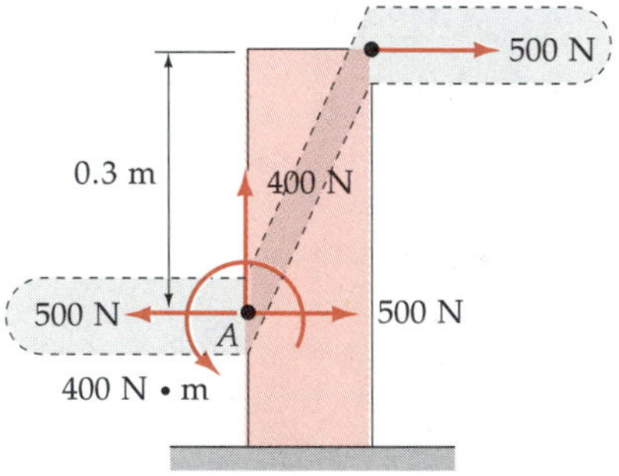

Figure 3.20

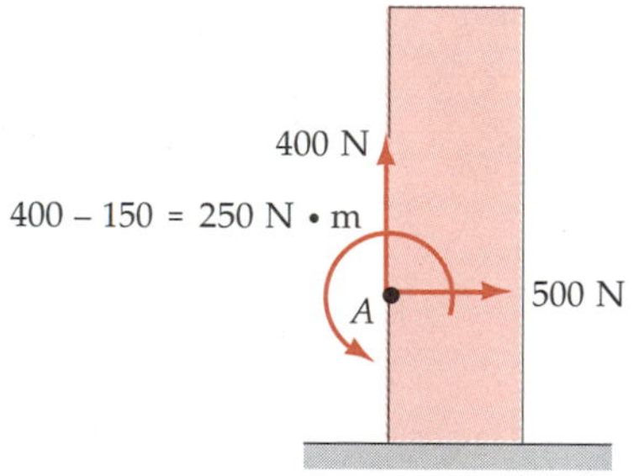

Figure 3.21

4. What now remains (see Figure 3.21) is of course the same result as we obtained more formally in the preceding example. The above approach, however, is often handy for simple two-dimensional systems as we shall see again in section 3.8.

The equipollent system S_2 will always have the same effect on the body, if it is rigid, as did the original system S_1: the same force tending to move it, and the same moment (about a common point) tending to turn it.

In our second example, we require the equipollent system to comprise forces acting at more than one pre-assigned point:

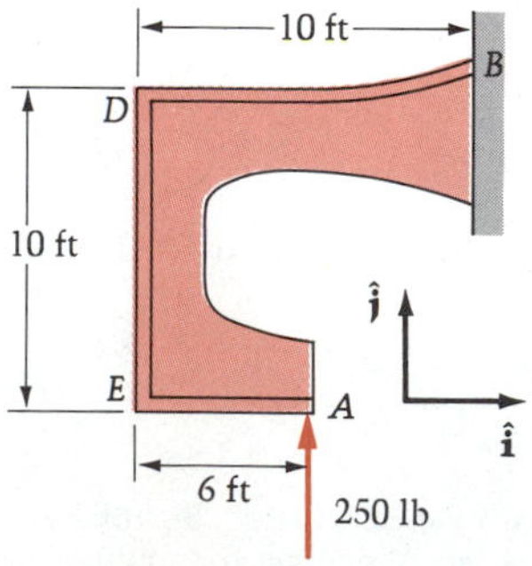

Figure E3.16a

EXAMPLE 3.16

Let the system in Figure E3.16a, consisting only of the 250-lb force at A, be called S_1. Determine a force system S_2 that is equipollent to S_1 and that consists of a vertical force acting at B and a pair of horizontal forces acting at D and E.

Solution

Since the original system S_1 contains no horizontal forces, the two horizontal forces at D and E must be equal in magnitude and opposite in direction; that is, they have to form a couple. But let us proceed as if we had not realized this, and

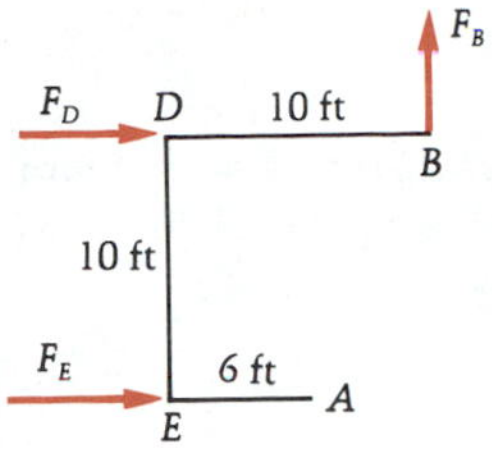

System S_2

Figure E3.16b

sketch the sought system S_2 shown in Figure E3.16b. Our force condition, which is necessary for equipollence of S_1 and S_2, gives

$$(\Sigma \mathbf{F})_1 = (\Sigma \mathbf{F})_2$$
$$250\hat{\mathbf{j}} = F_D\hat{\mathbf{i}} + F_E\hat{\mathbf{i}} + F_B\hat{\mathbf{j}}$$

Thus, from the $\hat{\mathbf{i}}$ coefficients, we find that the forces at D and E are equal in magnitude and opposite in direction:

$$F_D = -F_E \quad (1)$$

The $\hat{\mathbf{j}}$ coefficients give

$$250 = F_B \quad (2)$$

To find the value of F_D, we next ensure that our moment condition is also satisfied; we choose point A as our moment center (*any* point could be used!)

$$(\Sigma \mathbf{M}_A)_1 = (\Sigma \mathbf{M}_A)_2$$
$$\mathbf{0} = F_B(4)\hat{\mathbf{k}} + F_D(10)(-\hat{\mathbf{k}})$$

But $F_B = 250$ lb from Equation (2):

$$F_D = \frac{250(4)}{10} = 100 \text{ lb}$$

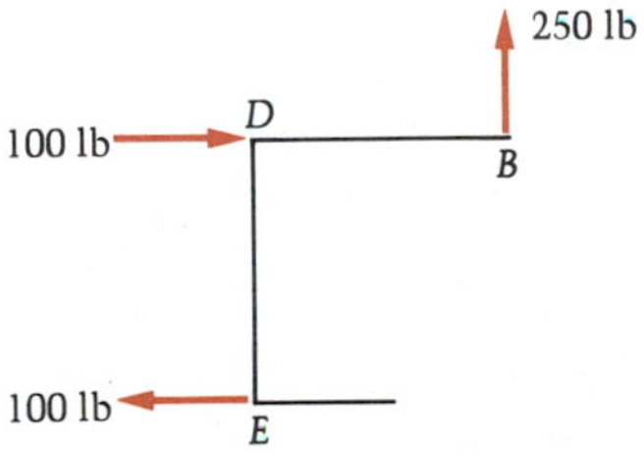

Figure E3.16c

Thus the final system S_2 appears as shown in Figure E3.16c.

As a check, we note that the moment condition is also satisfied at, say, B:

$$(\Sigma \mathbf{M}_B)_1 = (\Sigma \mathbf{M}_B)_2$$
$$250(10 - 6)(-\hat{\mathbf{k}}) = 100(10)(-\hat{\mathbf{k}})$$

or

$$-1000\hat{\mathbf{k}} = -1000\hat{\mathbf{k}} \text{ lb-ft}$$

In our third example, the original system S_1 consists of a simple couple:

EXAMPLE 3.17

Find a pair of vertical forces, $\mathbf{F}_1$ at A and $\mathbf{F}_2$ at B, that together form a force system S_2 equipollent to the pair of 0.2-lb forces acting on the spacecraft in Figure E3.17a.

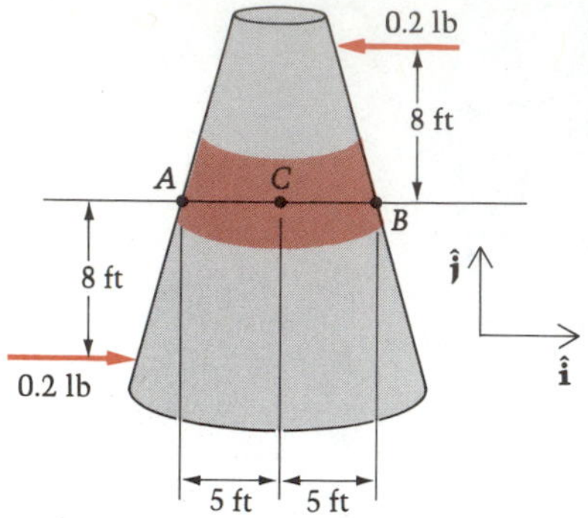

Figure E3.17a

Solution

If we call the original force system in the figure S_1, then using Condition 1 — that is, the force condition — we have

$$(\Sigma \mathbf{F})_1 = (\Sigma \mathbf{F})_2$$
$$-0.2\hat{\mathbf{i}} + 0.2\hat{\mathbf{i}} = \mathbf{F}_1 + \mathbf{F}_2$$
$$\mathbf{0} = \mathbf{F}_1 + \mathbf{F}_2$$

so that $\mathbf{F}_1 = -\mathbf{F}_2$. If $\mathbf{F}_2 = F_2\hat{\mathbf{j}}$, the system S_2 shown in Figure E3.17b results.

Next, we use Condition 2, or the moment condition, with point C selected as the moment center:

$$(\Sigma \mathbf{M}_C)_1 = (\Sigma \mathbf{M}_C)_2$$
$$8(0.2)\hat{\mathbf{k}} + 8(0.2)\hat{\mathbf{k}} = F_2(5)\hat{\mathbf{k}} + F_2(5)\hat{\mathbf{k}}$$
$$F_2 = \frac{3.2}{10} = 0.32 \text{ lb}$$

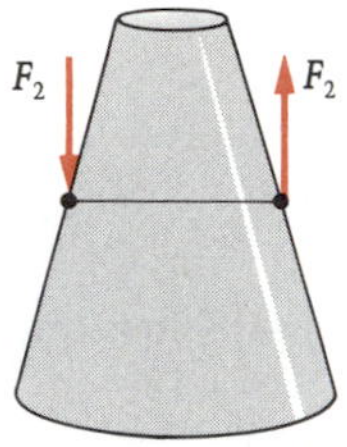

System S_2
Figure E3.17b

The reader has probably noticed that the original 0.2-lb forces form a couple; therefore, the two forces in S_2 must do the same, as we have found. Note that if the forces had been drawn as shown in Figure E3.17c, then the moment condition would have yielded

$$8(0.2)\hat{\mathbf{k}} + 8(0.2)\hat{\mathbf{k}} = F_1(5)(-\hat{\mathbf{k}}) + F_1(5)(-\hat{\mathbf{k}})$$
$$F_1 = -0.32 \text{ lb}$$

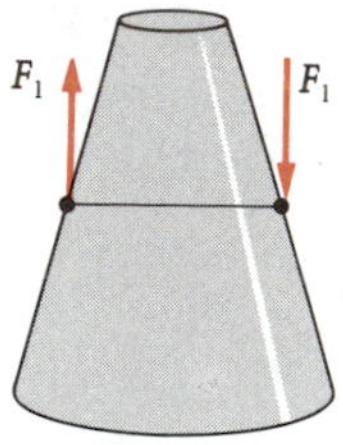

Figure E3.17c

which of course corresponds to the same pair of forces as the F_2's in the earlier figure (E3.17b) of system S_2.

Note also that if we had not required the forces at A and B to be vertical, then there would have been lots of correct answers, examples of which are shown in the figures below:

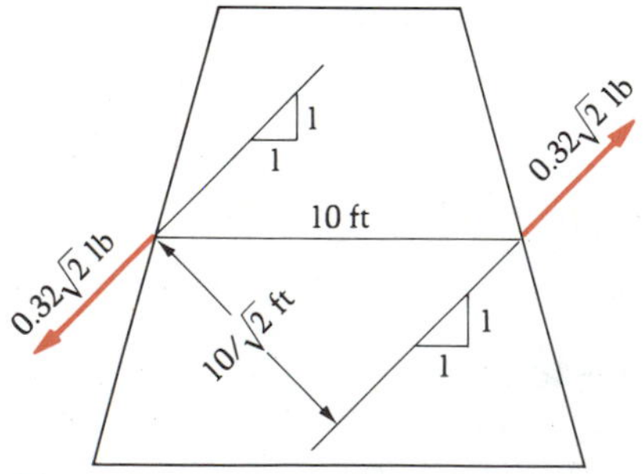

Figure E3.17d

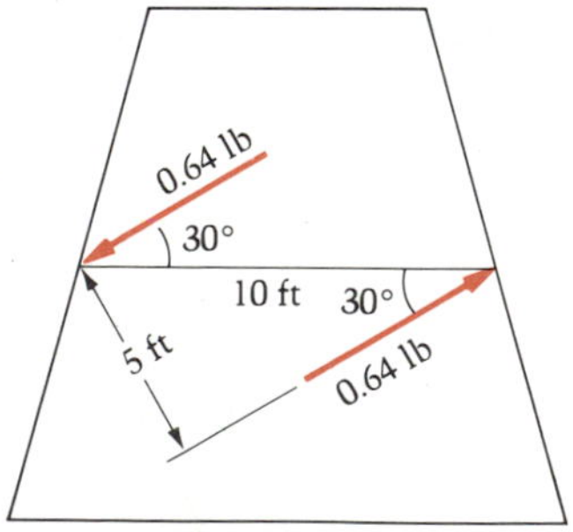

Figure E3.17e

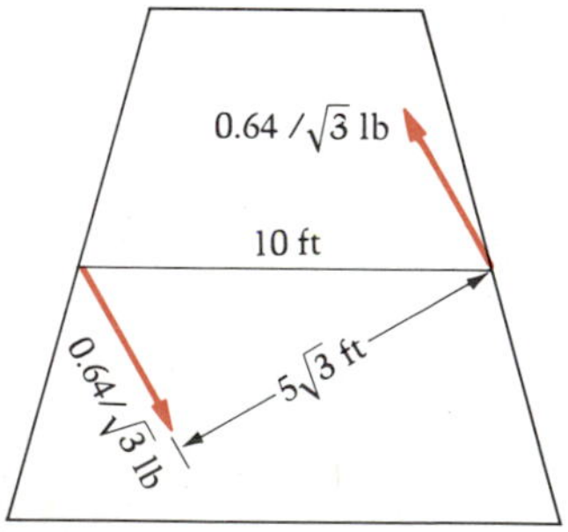

Figure E3.17f

The reader is encouraged to demonstrate that each of these systems is equipollent to S_1.

Question 3.21 Could the forces at A and B constituting system S_2 be (a) horizontal? (b) nonparallel?

In the next example in this section, we shall investigate a series of different force systems and determine which of them are equipollent.

Answer 3.21 (a) No, for they must form a couple. If they were horizontal, they would be collinear and their moment would be zero about points on their common line of action. (b) No, then they still couldn't form a couple. This time neither the force condition *nor* the moment condition can be satisfied.

Example 3.18

Consider a triangular plate under the six loading conditions shown in Figures E3.18a–f (SI units throughout). Determine which of the loadings, or force systems, are equipollent.

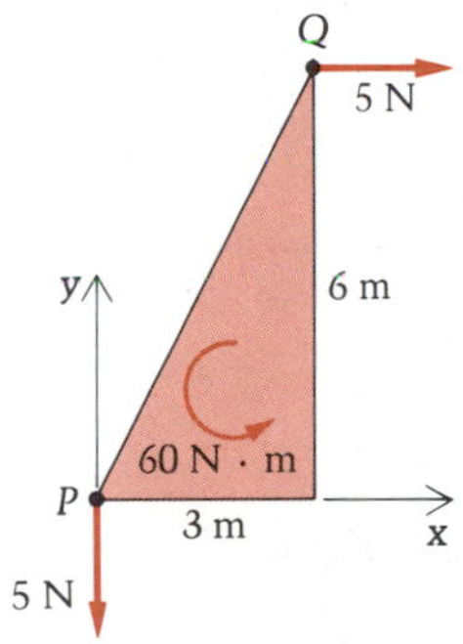

(1)

Figure E3.18a

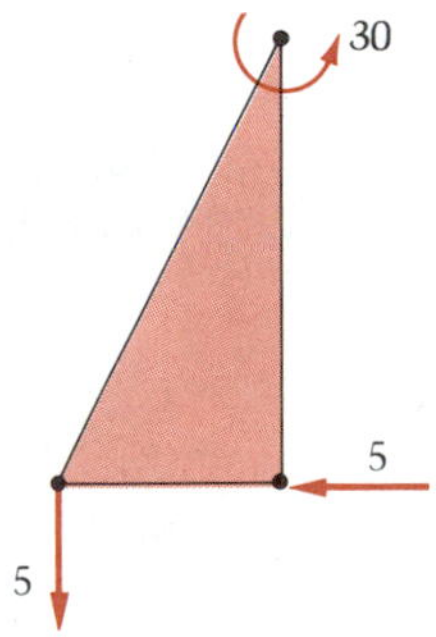

(2)

Figure E3.18b

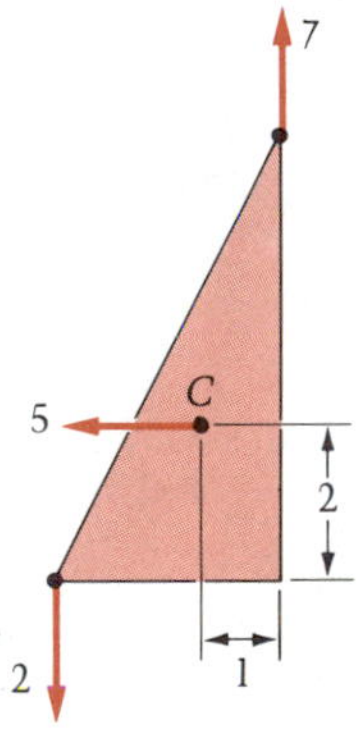

(3)

Figure E3.18c

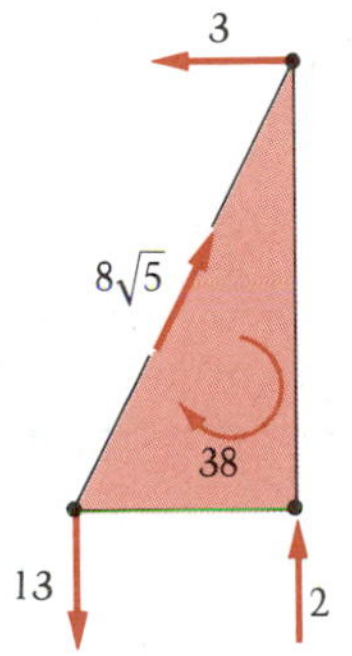

(4)

Figure E3.18d

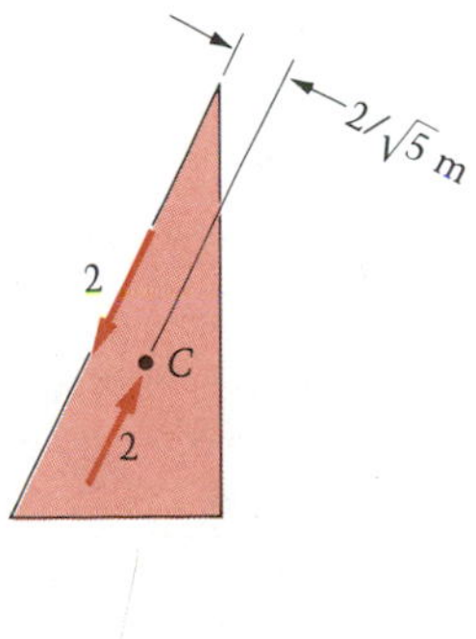

(5)

Figure E3.18e

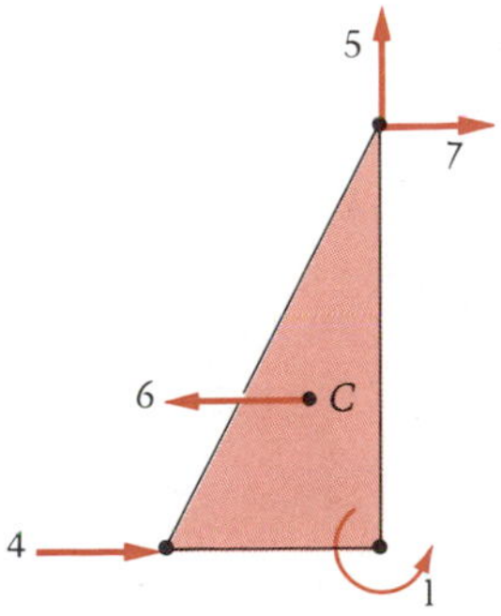

(6)

Figure E3.18f

Solution

Since we must compute $\Sigma\mathbf{F}$ and $\Sigma\mathbf{M}_P$ for each system and then compare these results, it would be useful to construct a table. We select the common moment center P to be the lower left-hand corner of the plate because forces pass through this point in each case, and thus their $\mathbf{r} \times \mathbf{F}$ is zero, making our calculations easier. The resulting table follows.

Force System	$\Sigma\mathbf{F}$ (N)	$\Sigma\mathbf{M}_P$ (N · m)
1	$5\hat{\mathbf{i}} - 5\hat{\mathbf{j}}$	$30\hat{\mathbf{k}}$
2	$-5\hat{\mathbf{i}} - 5\hat{\mathbf{j}}$	$30\hat{\mathbf{k}}$
3	$-5\hat{\mathbf{i}} + 5\hat{\mathbf{j}}$	$31\hat{\mathbf{k}}$
4	$5\hat{\mathbf{i}} + 5\hat{\mathbf{j}}$	$-14\hat{\mathbf{k}}$
5	$\mathbf{0}$	$1.79\hat{\mathbf{k}}$
6	$5\hat{\mathbf{i}} + 5\hat{\mathbf{j}}$	$-14\hat{\mathbf{k}}$

Notes on the entries of the six systems above are now listed:

1. Here, the value of $\Sigma\mathbf{F}$ is just the sum of the two forces, $5\hat{\mathbf{i}}$ and $-5\hat{\mathbf{j}}$. The moment about P is due to the $60\hat{\mathbf{k}}$ N · m couple plus the moment of the horizontal force, $\mathbf{r}_{PQ} \times 5\hat{\mathbf{i}} = (3\hat{\mathbf{i}} + 6\hat{\mathbf{j}}) \times 5\hat{\mathbf{i}} = 30\hat{\mathbf{k}}$ N · m. Note that the latter is merely 5 newtons times its perpendicular distance (6 m) from P, with the $-\hat{\mathbf{k}}$ attached since the force turns clockwise around P.

2. Both forces pass through P, leaving only the couple to contribute to $\Sigma\mathbf{M}_P$. Remember that couples have the same moment about all points! Note that although the moment condition is satisfied in comparing systems 1 and 2, the force condition is not, and it takes *both* for equipollence!

3. In the vertical direction, the force is the combination of 7 N up and 2 N down, or $5\hat{\mathbf{j}}$ N. Rotationally, the 7-N force and 5-N force each have moments around P, their "lever arms" being 3 m and 2 m, respectively.

4. In this case, the sum of the forces is calculated as

$$\Sigma\mathbf{F} = 8\sqrt{5}\left(\frac{\hat{\mathbf{i}} + 2\hat{\mathbf{j}}}{\sqrt{5}}\right) - 3\hat{\mathbf{i}} + 2\hat{\mathbf{j}} - 13\hat{\mathbf{j}} = (5\hat{\mathbf{i}} + 5\hat{\mathbf{j}})\text{ N}$$

The sum of moments is computed vectorially as follows:

$$\Sigma\mathbf{M}_P = (3\hat{\mathbf{i}} + 6\hat{\mathbf{j}}) \times (3\hat{\mathbf{i}}) + 3\hat{\mathbf{i}} \times 2\hat{\mathbf{j}} - 38\hat{\mathbf{k}} = -14\hat{\mathbf{k}}\text{ N} \cdot \text{m}$$

Alternatively, using forces times lever arms and attaching the unit vector with correct sign,

$$\Sigma\mathbf{M}_P \text{ also} = [3(6) + 2(3) - 38]\hat{\mathbf{k}} = -14\hat{\mathbf{k}}\text{ N} \cdot \text{m}$$

5. In this one, we have just a pure couple since the two forces cancel each other, making $\Sigma\mathbf{F} = \mathbf{0}$. The magnitude of the couple (whose unit vector and direction is seen to be $+\hat{\mathbf{k}}$) is given by either force times the distance d between them:

$$\Sigma\Sigma\mathbf{M}_P = 2\left(\frac{2}{\sqrt{5}}\right)\hat{\mathbf{k}} = 1.79\hat{\mathbf{k}}\text{ N} \cdot \text{m}$$

Note that by inspection of the table through force system (5), there are no equipollent loadings thus far.

6. The sum of forces is $\Sigma\mathbf{F} = (4 - 6 + 7)\hat{\mathbf{i}} + 5\hat{\mathbf{j}} = 5\hat{\mathbf{i}} + 5\hat{\mathbf{j}}$ N. And $\Sigma\mathbf{M}_P = 2\hat{\mathbf{j}} \times (-6\hat{\mathbf{i}}) + 6\hat{\mathbf{j}} \times 7\hat{\mathbf{i}} + 3\hat{\mathbf{i}} \times 5\hat{\mathbf{j}} + 1\hat{\mathbf{k}} = (12 - 42 + 15 + 1)\hat{\mathbf{k}} = -14\hat{\mathbf{k}}$ N · m, where we have used $2\hat{\mathbf{j}}$ and $6\hat{\mathbf{j}}$ m for the position vectors to points on the lines of action of the 6- and 7-N horizontal forces, respectively.

We see that $(\Sigma\mathbf{F})_4 = (\Sigma\mathbf{F})_6$ *and* $(\Sigma\mathbf{M}_P)_4 = (\Sigma\mathbf{M}_P)_6$, so that, in conclusion, the only equipollent systems among the six are numbers 4 and 6.

In our last example, we shall examine a three-dimensional force system; such a system is one in which (having expressed each couple as a pair of equal-but-opposite forces) not all the forces in the system can lie in a single plane. Note that according to this definition, all previous examples in this section have been *two*-dimensional.

Example 3.19

Find a system S_2 comprising a force at P and a couple that is equipollent to the 300-N force and 250-N · m couple in Figure E3.19. The couple's turning effect lies in the plane $PCBA$, so its unit vector is $\hat{\mathbf{k}}$.

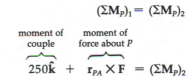

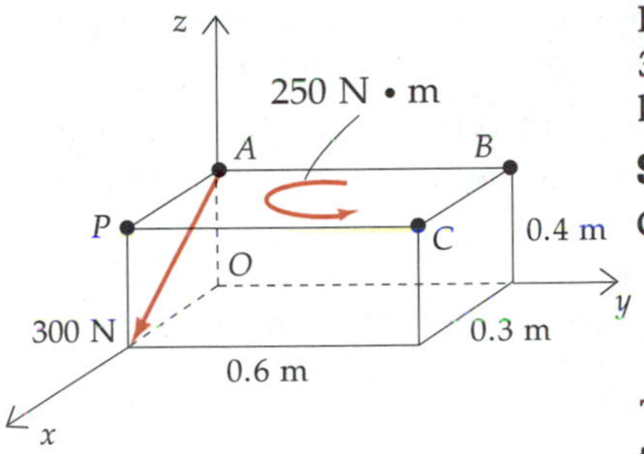

Figure E3.19

Solution

Condition (1) yields:

$$(\Sigma\mathbf{F})_2 = (\Sigma\mathbf{F})_1 = 300\left(\frac{3\hat{\mathbf{i}} - 4\hat{\mathbf{k}}}{5}\right) = 180\hat{\mathbf{i}} - 240\hat{\mathbf{k}}\text{ N}$$

To form system S_2, we place this force at P and then proceed to sum moments there to determine the accompanying couple. Condition (2) yields, with point P as the moment center:

$$(\Sigma\mathbf{M}_P)_1 = (\Sigma\mathbf{M}_P)_2$$

$$\underbrace{250\hat{\mathbf{k}}}_{\text{moment of couple}} + \underbrace{\mathbf{r}_{PA} \times \mathbf{F}}_{\text{moment of force about } P} = (\Sigma\mathbf{M}_P)_2$$

so that

$$(\Sigma\mathbf{M}_P)_2 = 250\hat{\mathbf{k}} + (-0.3\hat{\mathbf{i}}) \times (180\hat{\mathbf{i}} - 240\hat{\mathbf{k}})$$
$$= -72\hat{\mathbf{j}} + 250\hat{\mathbf{k}}\text{ N} \cdot \text{m},$$

which is the couple accompanying the force at P; together they form the system S_2 equipollent to the original system.

The reader should remember just two main things about equipollent systems in working the problems which follow: (1) Make the forces add up to the same vector in the two systems; (2) About some common point in the two systems, make the moments also add up to the same vector.

PROBLEMS ▶ Section 3.6

3.61 Replace the 2000-lb force in Figure P3.61 by an equipollent system consisting of a horizontal force through A, a vertical force along line CB, and a couple.

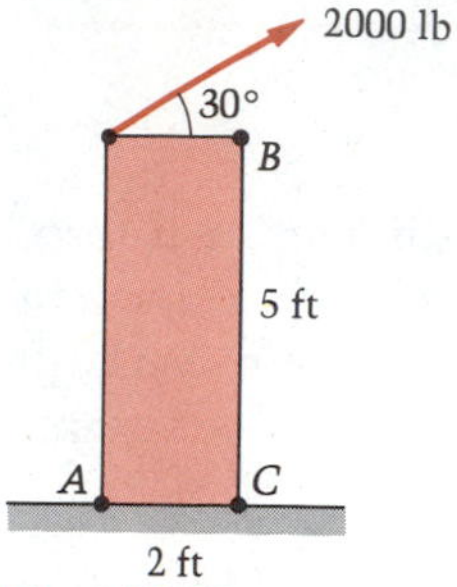

Figure P3.61

3.62 Find (and illustrate with a sketch) a pair of forces which (a) is equipollent to the couple in the figure, and (b) has one force equal to $50\hat{\mathbf{i}}$ N through the point $(x, y) = (10, 20)$ cm. (See Figure P3.62.)

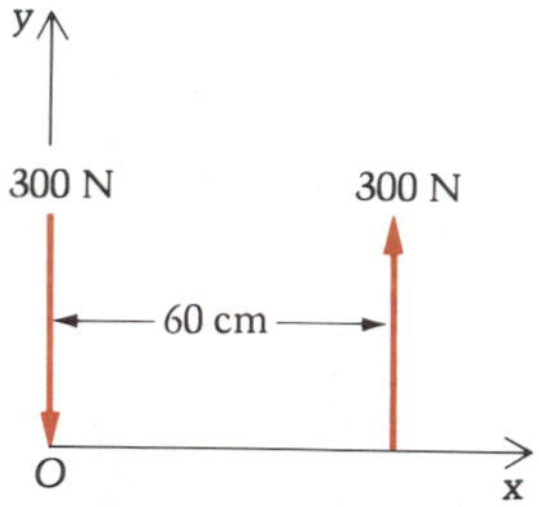

Figure P3.62

3.63 Repeat the preceding problem with condition (b) changed to a force of $180\hat{\mathbf{j}}$ N through $(x, y) = (30, 40)$ cm.

3.64 Replace the 250-lb force in Figure P3.64 by an equipollent system consisting of a force at Q and a couple. Then show that *both* systems have the same moment about point B.

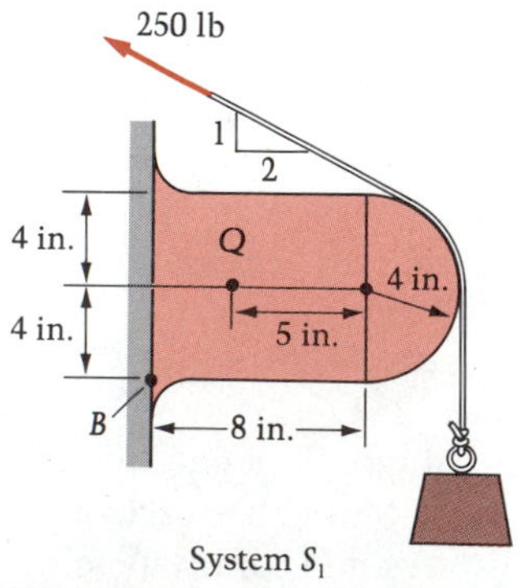

Figure P3.64

3.65 Find a system S_2 consisting of a force at A and a couple, which is equipollent to the system of two forces in Figure P3.65. Use the procedure of Examples 3.15–3.17.

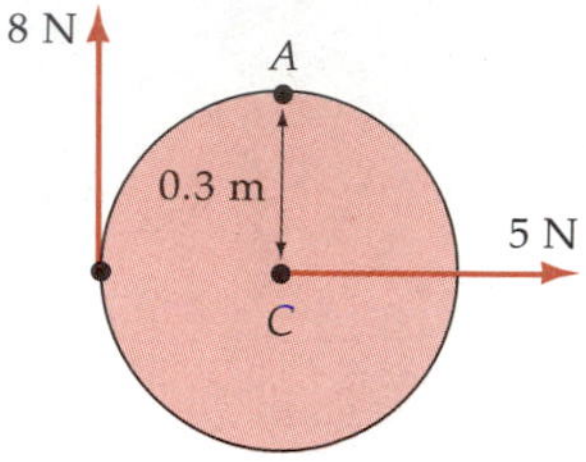

Figure P3.65

3.66 Repeat the preceding problem using the ideas which were illustrated following Example 3.15. Add and subtract two pairs of forces at A, both $\pm 5\hat{\mathbf{i}}$ N and $\pm 8\hat{\mathbf{j}}$ N. Then identify two couples and add them, leaving $5\hat{\mathbf{i}} + 8\hat{\mathbf{j}}$ N at point A, plus the combined couple.

3.67 By filling in the chart [Figure P3.67(b)], determine which of the six beam loadings in Figure P3.67(a) are equipollent.

3.68 Continue Example 3.18 by determining which of the force systems in Figure P3.68 are equipollent either to any of the loadings in that example or to one another. How many different comparisons of two systems at a time have now been made?

3.69 A bending moment of 20,000 lb-ft acts on a large plate bolted to the ground with eight equally spaced bolts as shown in Figure P3.69. Determine an equipollent system of eight vertical forces, each acting at one of the bolts and having a magnitude proportional to the distance of the bolt from line $\ell\ell$.

3.70 In the preceding problem, re-determine the eight forces if the bolt pattern is aligned with the moment vector as shown in Figure P3.70.

3.71 A body is acted on by two forces: one with components (10, 20, 30) N at a point having coordinates (3, 2, 1) meters and the other with components (30, 20, 10) N at a point (1, 2, 3) m. Find the equipollent system consisting of a force at the point (1, 1, 1) m and a couple.

3.72 In Figure P3.72 the magnitude of the moments of couples $\mathbf{M}_1$, $\mathbf{M}_2$, $\mathbf{M}_3$ are each $= \mathbf{M}_0 =$ constant. The forces act in planes ABC, ACD, and $BODC$, respectively, so that the moments of the couples are normal to these planes. Find a single couple that is equipollent to these three couples.

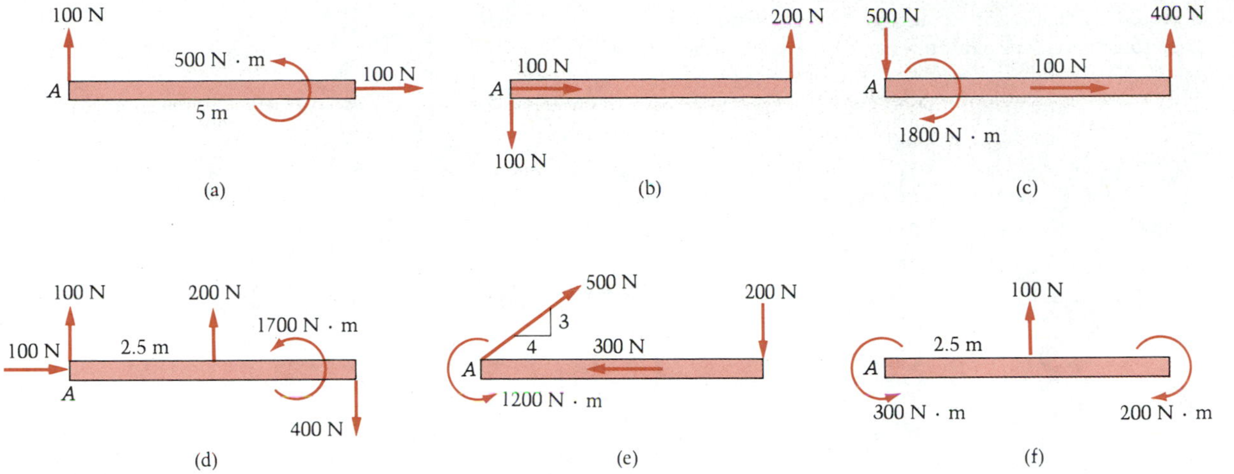

Figure P3.67(a)

Case	ΣF	$\Sigma\mathbf{M}_A$
(a)		
(b)		
(c)		
(d)		
(e)		
(f)		

Figure P3.67(b)

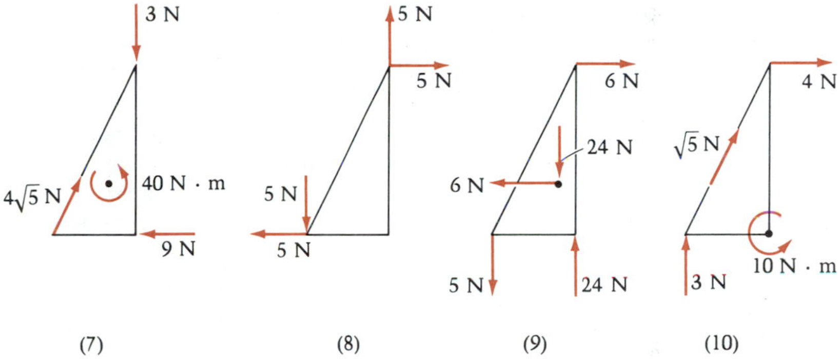

Figure P3.68

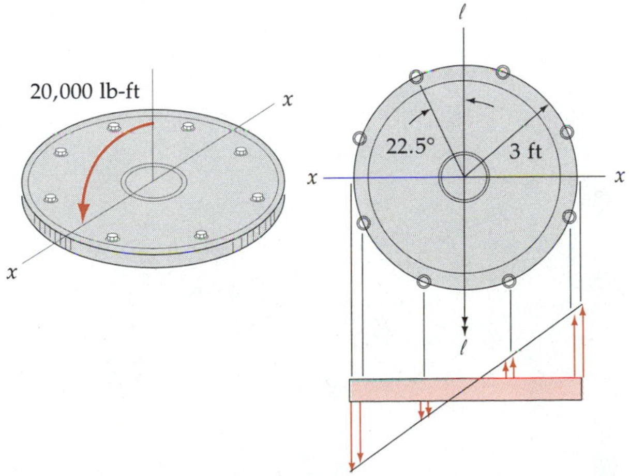

Figure P3.69

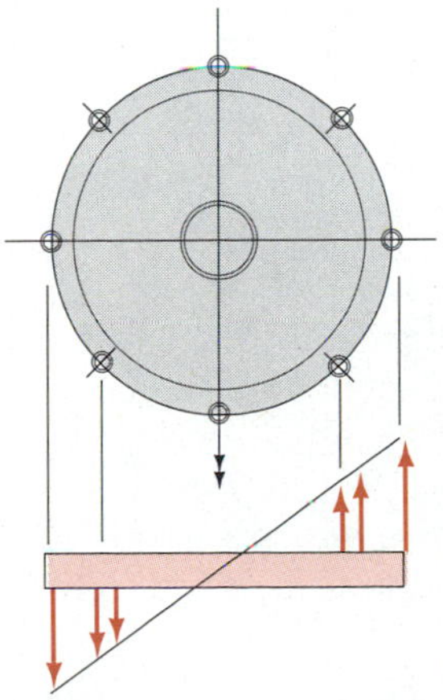

Figure P3.70

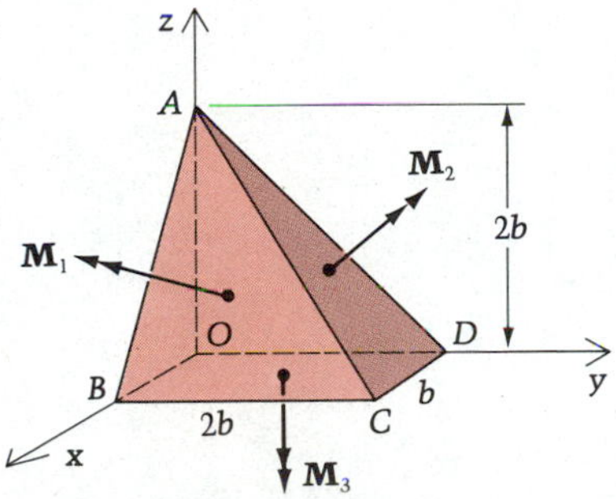

Figure P3.72

3.73 Find a force at E and an accompanying couple that together form a system equipollent to the one shown in Figure P3.73.

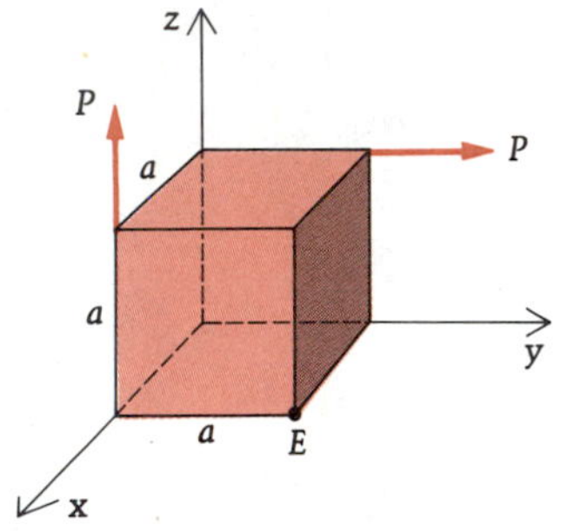

Figure P3.73

3.74 The couple's forces lie in the shaded plane, and there are two other applied forces in the system shown in Figure P3.74. Find an equipollent system consisting of a force at the origin O and a couple.

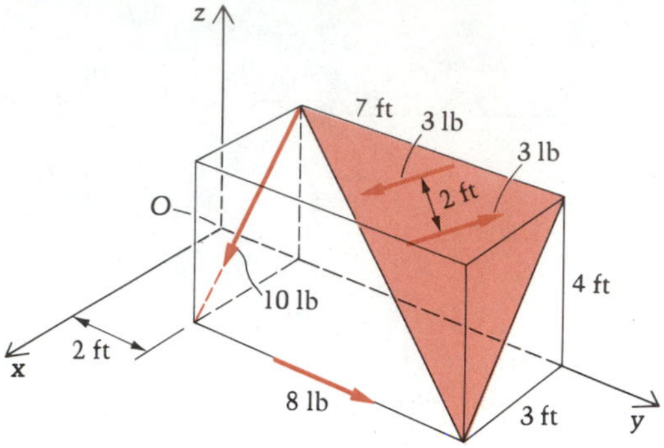

Figure P3.74

3.75 Find the force and couple at the origin equipollent to the five forces and the couple shown in Figure P3.75. Then repeat the problem considering the two vertical 30-N forces as a couple, and the two horizontal 40-N forces as another couple. That is, repeat the problem with the system treated as one force and three couples. Compare the results.

3.76 A couple of moment **C** lies in the shaded plane ABC and has a magnitude of $\sqrt{13}$ lb-ft and a direction indicated by the right-hand rule. (See Figure P3.76.) A force **F** of magnitude $2\sqrt{13}$ lb also acts as shown. Find the equipollent system consisting of a force at point C and a couple.

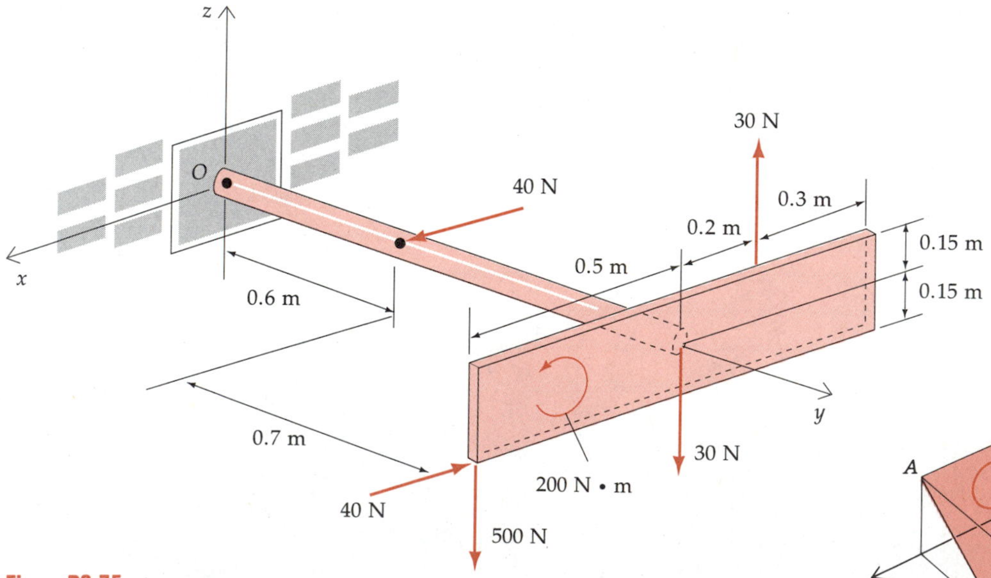

Figure P3.75

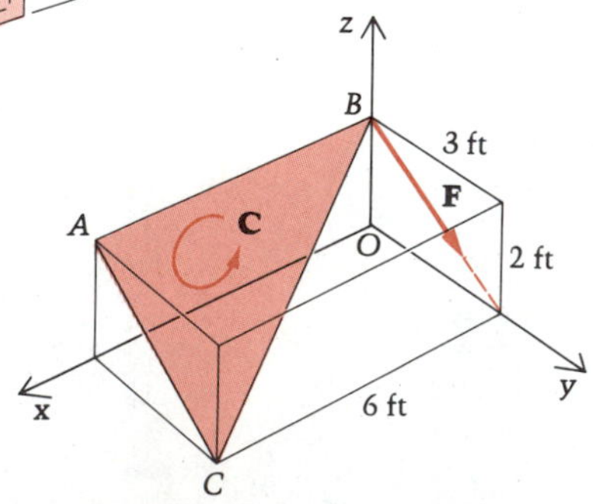

Figure P3.76

3.7 The Force-and-Couple Resultant of a System of Forces

The Definition of a Resultant

The definition of equipollent force systems correctly suggests that we may replace any system, no matter how complicated, by a force at any point P and a couple (see Figure 3.22).

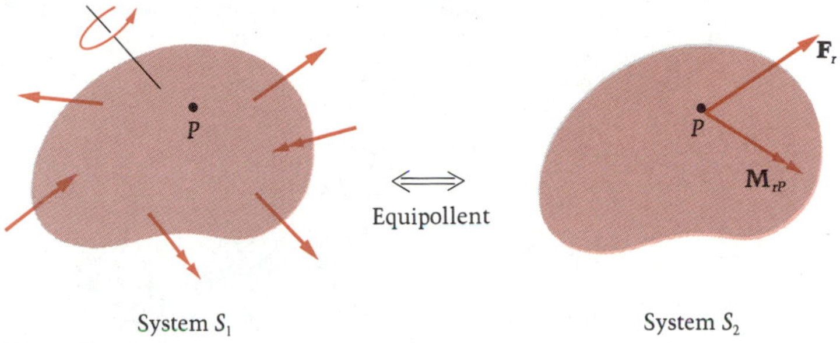

Figure 3.22

We have already seen that for the system S_2 to be equipollent to S_1 we need only to ensure that:

1. $\mathbf{F}_r$, the force at P in S_2, is the sum of all the forces acting in system S_1.

2. $\mathbf{M}_{rP}$, the moment of the couple in S_2, is the sum of the moments of all the couples in S_1 plus the sum of the moments about P of all the forces in S_1. Note that, in system S_2, $\mathbf{F}_r$ produces no moment about P.

The force-couple pair, $\mathbf{F}_r$ and $\mathbf{M}_{rP}$, is called a **resultant*** of the system S_1. For simplicity we are using the phrase "force and couple at point P" to describe the resultant, but it is important to realize that there is no reason to assign a location-point subscript to the couple since the moment of a couple about every point is the same. And while the force $\mathbf{F}_r$ does not depend upon the choice of reference point P, the couple $\mathbf{M}_{rP}$ depends upon the choice of the *line of action* of $\mathbf{F}_r$. Thus, by "force and couple at P" we mean *the* resultant (of S_1) when the line of action of $\mathbf{F}_r$ is chosen to pass through point P.

Question 3.22 The implication of the preceding paragraph is that for all other points on the line of action of $\mathbf{F}_r$ through P, the force and couple resultant is the same as it is at P. Why is this indeed so?

* Hence the subscript "r" for resultant, on each member of the pair.

Answer 3.22 Because (1) $\mathbf{F}_r$ is always the same from point to point, and (2) if Q is any other point on the line,

$$\mathbf{M}_{rQ} = \mathbf{M}_{rP} + \mathbf{r}_{QP} \times \mathbf{F}_r = \mathbf{M}_{rP}$$

since $\mathbf{r}_{QP}$ is coincident with the line of action of $\mathbf{F}_r$.

We consider now a number of examples of replacement of a system of forces and/or couples by an equipollent system of a force and a couple (that is, a resultant) at a preselected point.

EXAMPLE 3.20

Determine the resultant, at the pin P, of the two belt tensions shown in Figure E3.20a. The pulley has a radius $R = 0.6$ m.

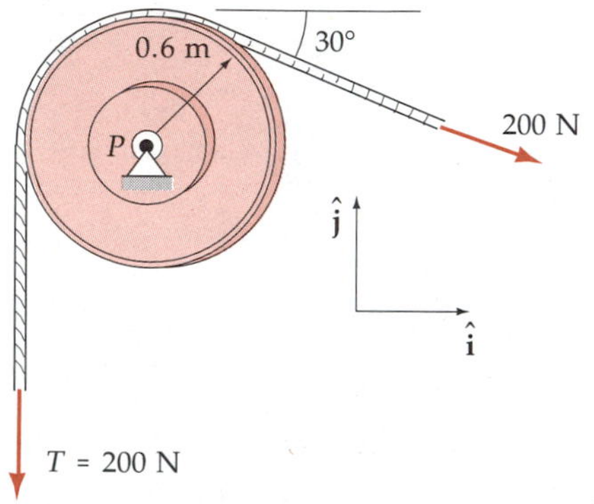

Figure E3.20a

Solution

Using the force condition, the force part of the resultant is calculated as follows:

$$(\Sigma \mathbf{F})_1 = (\Sigma \mathbf{F})_2$$

$$200(\cos 30°)\hat{\mathbf{i}} - [200 + 200(\sin 30°)]\hat{\mathbf{j}} = \mathbf{F}_r$$

or

$$\mathbf{F}_r = 173\hat{\mathbf{i}} - 300\hat{\mathbf{j}} \text{ N}$$

Next, the moment condition tells us that the couple part of the resultant at P vanishes this time:

$$(\Sigma \mathbf{M}_P)_1 = (\Sigma \mathbf{M}_P)_2$$

$$200(0.6)\hat{\mathbf{k}} + 200(0.6)(-\hat{\mathbf{k}}) = \mathbf{0} = \mathbf{M}_{rP}$$

Figure E3.20b shows what has happened in going from the original system (S_1) to the resultant at P (S_2).

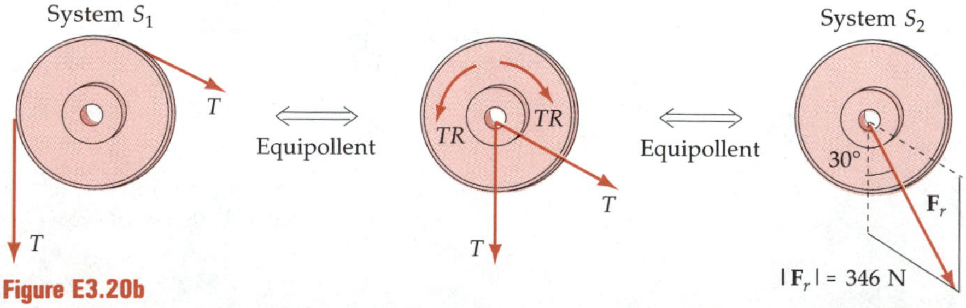

Figure E3.20b

In the next example, the original system consists only of couples.

EXAMPLE 3.21

Determine the resultant at point A of the four forces shown in Figure E3.21a.

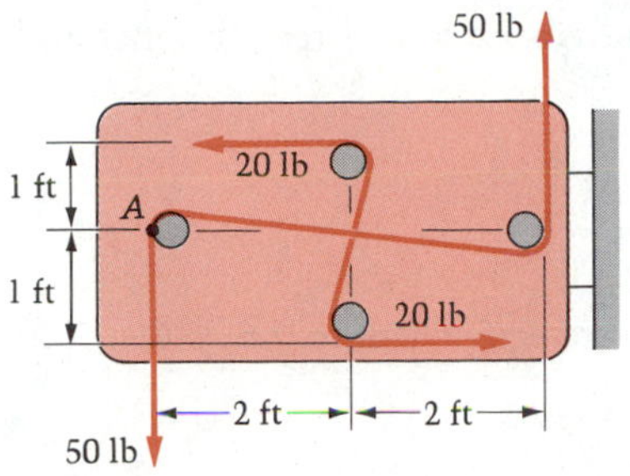

Figure E3.21a

Solution

Since the forces are seen to form a pair of couples, the force $\mathbf{F}_r$ of the resultant at A is zero. The moment of the couple is

$$\mathbf{M}_{rA} = \sum^{S_1} \mathbf{M}_A = 50(4)(\hat{\mathbf{k}}) + 20(2)\hat{\mathbf{k}} = 240\hat{\mathbf{k}} \text{ lb-ft}$$

The resultant is shown in Figure E3.21b.

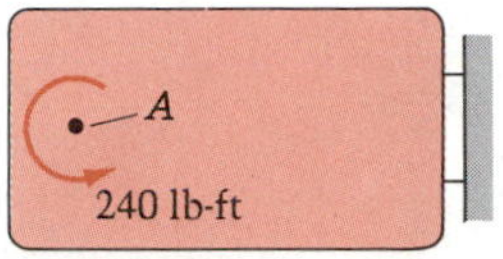

Figure E3.21b

Question 3.23 What is the resultant at the point 2 ft to the right of A?

In the next example, the original system comprises three forces in a plane and two couples with turning effects in that same plane:

Answer 3.23 At all points, the resultant in this problem is the same: a couple of $240\hat{\mathbf{k}}$ lb-ft, unaccompanied by a force.

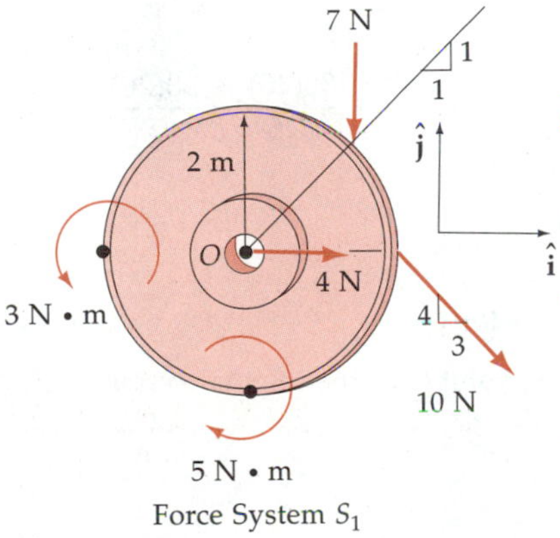

Figure E3.22a

EXAMPLE 3.22

Replace the force and couple system in Figure E3.22a by its resultant (an equipollent force and couple) at O.

Solution

The resultant force is

$$\mathbf{F}_r = [4 + \tfrac{3}{5}(10)]\hat{\mathbf{i}} + [-7 - \tfrac{4}{5}(10)]\hat{\mathbf{j}} = 10\hat{\mathbf{i}} - 15\hat{\mathbf{j}} \text{ N}$$

The resultant moment about O in S_1 is

$$\mathbf{M}_{rO} = \underbrace{(3\hat{\mathbf{k}} - 5\hat{\mathbf{k}})}_{\text{moments of couples}} + \underbrace{2\left(\frac{\hat{\mathbf{i}} + \hat{\mathbf{j}}}{\sqrt{2}}\right) \times (-7\hat{\mathbf{j}}) + 2\hat{\mathbf{i}} \times (6\hat{\mathbf{i}} - 8\hat{\mathbf{j}})}_{\text{moments of forces}}$$

$$= \left[-2 - \frac{14}{\sqrt{2}} - 16\right]\hat{\mathbf{k}} = -27.9\hat{\mathbf{k}} \text{ N} \cdot \text{m}$$

The equipollent system is shown in Figure E3.22b.

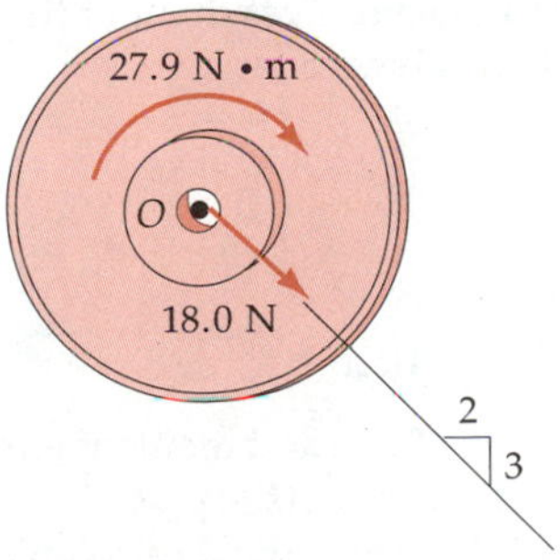

Figure E3.22b

In each of the next three examples, we have a three-dimensional system.

EXAMPLE 3.23

Find the force at the origin and couple that are equipollent to the system S_1 in Figure E3.23.

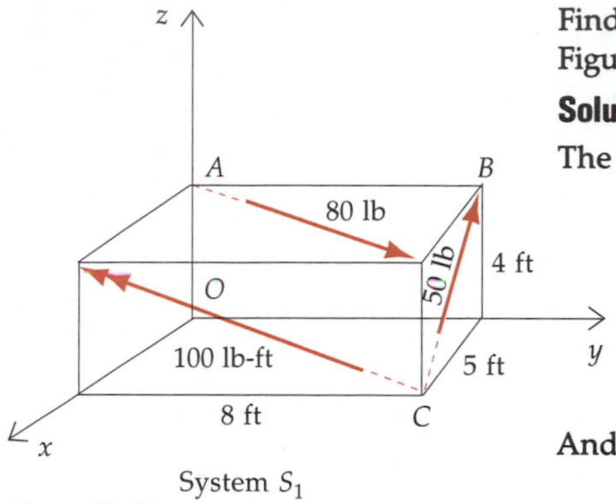

Figure E3.23

Solution

The force is

$$\mathbf{F}_r = \Sigma\mathbf{F} = 80\left(\frac{5\hat{\mathbf{i}} + 8\hat{\mathbf{j}}}{\sqrt{89}}\right) + 50\left(\frac{-5\hat{\mathbf{i}} + 4\hat{\mathbf{k}}}{\sqrt{41}}\right)\text{lb}$$
$$= (42.4 - 39.0)\hat{\mathbf{i}} + 67.8\hat{\mathbf{j}} + 31.2\hat{\mathbf{k}} \text{ lb}$$
$$= 3.4\hat{\mathbf{i}} + 67.8\hat{\mathbf{j}} + 31.2\hat{\mathbf{k}} \text{ lb}$$

And the moment of the couple is

$$\mathbf{M}_{rO} = \Sigma\mathbf{M}_O = 100\left(\frac{-8\hat{\mathbf{j}} + 4\hat{\mathbf{k}}}{\sqrt{80}}\right) + \overbrace{(4\hat{\mathbf{k}})}^{\mathbf{r}_{OA}} \times 80\left(\frac{5\hat{\mathbf{i}} + 8\hat{\mathbf{j}}}{\sqrt{89}}\right) + \overbrace{(8\hat{\mathbf{j}} + 4\hat{\mathbf{k}})}^{\mathbf{r}_{OB}}$$
$$\times 50\left(\frac{-5\hat{\mathbf{i}} + 4\hat{\mathbf{k}}}{\sqrt{41}}\right)$$
$$= \left[\frac{-80(32)}{\sqrt{89}} + \frac{50(32)}{\sqrt{41}}\right]\hat{\mathbf{i}} + \left[\frac{-800}{\sqrt{80}} + \frac{80(20)}{\sqrt{89}} - \frac{50(20)}{\sqrt{41}}\right]\hat{\mathbf{j}}$$
$$+ \left[\frac{400}{\sqrt{80}} + \frac{50(40)}{\sqrt{41}}\right]\hat{\mathbf{k}}$$
$$= -21.5\hat{\mathbf{i}} - 76.0\hat{\mathbf{j}} + 357\hat{\mathbf{k}} \text{ lb-ft}$$

Together, $\mathbf{F}_r$ and $\mathbf{M}_{rO}$ form the resultant at point O of the original system S_1.

EXAMPLE 3.24

In Figure E3.24a, a person exerts a force of 30 lb in the x direction to turn the elbow onto the threaded pipe. Determine the force and couple resultant at the origin O, where the pipe is screwed into another elbow above a tank.

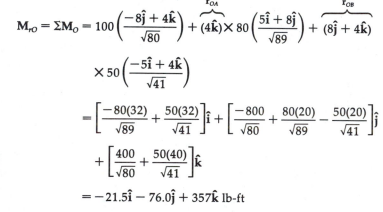

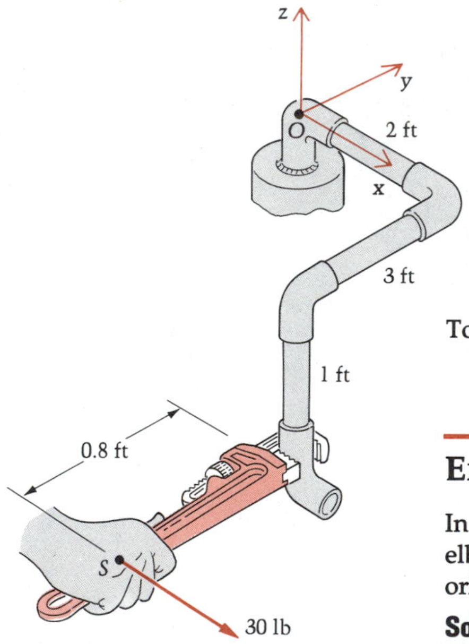

Figure E3.24a

Solution

The force is simply $\mathbf{F}_r = 30\hat{\mathbf{i}}$ lb. The moment at O is

$$\mathbf{M}_{rO} = \mathbf{r}_{OS} \times 30\hat{\mathbf{i}}$$
$$= (2\hat{\mathbf{i}} - 3.8\hat{\mathbf{j}} - \hat{\mathbf{k}}) \times 30\hat{\mathbf{i}} = -30\hat{\mathbf{j}} + 114\hat{\mathbf{k}} \text{ lb-ft}$$

Before moving to another example, we note that the force and moment are undesirable concerning stress and deflection (and maybe leaks!) in the pipes. This can be avoided by the wise use of a second pipe wrench (see Figure E3.24b). Now, we see that for the two forces, $\mathbf{F}_r = \mathbf{0}$. Furthermore, the z component of $\mathbf{M}_{rO}$ is

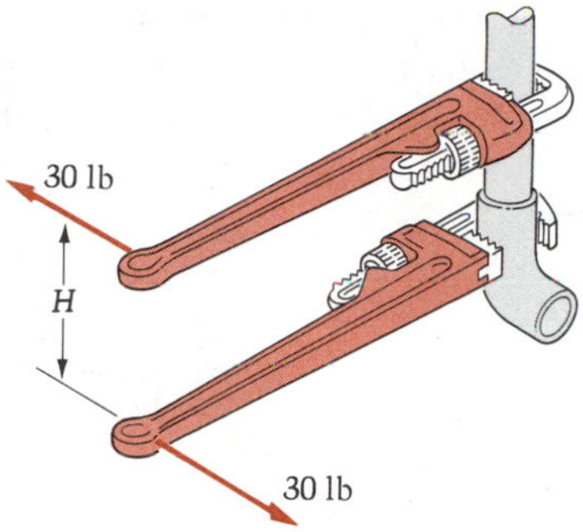

Figure E3.24b

eliminated. The couple at O formed by the two 30-lb forces is simply $30H(-\hat{\mathbf{j}})$, which, if H is, say, 2 in., amounts to only $-5\hat{\mathbf{j}}$ lb-ft. This illustrates the advantage that may accrue from using two forces to produce a desired moment, which here is $24\hat{\mathbf{k}}$ lb-ft on the elbow.

EXAMPLE 3.25

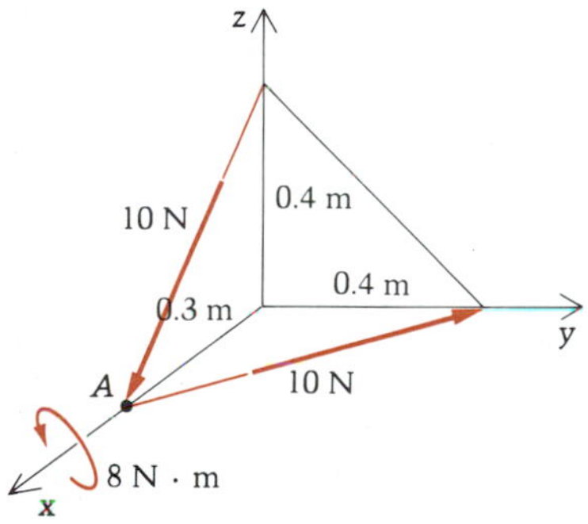

Figure E3.25

For the system shown in Figure E3.25, find the force and couple resultant at (a) the origin and (b) point A.

Solution

The force resultant is the same at all points, and is

$$\mathbf{F}_r = 10\left(\frac{3\hat{\mathbf{i}} - 4\hat{\mathbf{k}}}{5}\right) + 10\left(\frac{-3\hat{\mathbf{i}} + 4\hat{\mathbf{j}}}{5}\right)$$

$$= (6\hat{\mathbf{i}} - 8\hat{\mathbf{k}}) + (-6\hat{\mathbf{i}} + 8\hat{\mathbf{j}})$$

$$= 8\hat{\mathbf{j}} - 8\hat{\mathbf{k}}\text{ N}$$

Obtaining the resultant moment at the origin, and noting that $0.3\hat{\mathbf{i}}$ m is a vector from O to a point on the line of action of both forces,

$$\mathbf{M}_{rO} = 0.3\hat{\mathbf{i}} \times (6\hat{\mathbf{i}} - 8\hat{\mathbf{k}}) + 0.3\hat{\mathbf{i}} \times (-6\hat{\mathbf{i}} + 8\hat{\mathbf{j}}) + 8\hat{\mathbf{i}}$$

$$= 8\hat{\mathbf{i}} + 2.4\hat{\mathbf{j}} + 2.4\hat{\mathbf{k}}\text{ N}\cdot\text{m}$$

Thus at the origin, the original system is equipollent to

$$\mathbf{F}_r = 8\hat{\mathbf{j}} - 8\hat{\mathbf{k}}\text{ N}$$

$$\mathbf{M}_{rO} = 8\hat{\mathbf{i}} + 2.4\hat{\mathbf{j}} + 2.4\hat{\mathbf{k}}\text{ N}\cdot\text{m}$$

At point A, the moment is simply the couple of $8\hat{\mathbf{i}}$ N · m, since both forces pass through A. Therefore, at A the system equipollent to the given one is

$$\underbrace{\mathbf{F}_r = 8\hat{\mathbf{j}} - 8\hat{\mathbf{k}}\text{ N}}_{\text{(The force never changes from point to point!)}} \quad\text{and}\quad \mathbf{M}_{rA} = 8\hat{\mathbf{i}}\text{ N}\cdot\text{m}$$

We should note in this example that the two systems at A and O are of course not only each equipollent to the given system, but also to each other. To show this, note that their forces are each $8\hat{\mathbf{j}} - 8\hat{\mathbf{k}}$ N, and that the moment of the system at A, about O, is

$$\mathbf{M}_{rO} = \underbrace{\mathbf{M}_{rA}}_{\substack{\text{couple at } A \\ \text{(same moment} \\ \text{everywhere!)}}} + \underbrace{\mathbf{r}_{OA} \times \mathbf{F}_r}_{\substack{\text{moment about } O \text{ of} \\ \text{the force at } A}}$$

$$= 8\hat{\mathbf{i}} + 0.3\hat{\mathbf{i}} \times (8\hat{\mathbf{j}} - 8\hat{\mathbf{k}})$$

$$= 8\hat{\mathbf{i}} + 2.4\hat{\mathbf{j}} + 2.4\hat{\mathbf{k}} \text{ N} \cdot \text{m}$$

This is indeed what we had previously obtained for $\mathbf{M}_{rO}$.

Question 3.24 In terms of the $\mathbf{F}_r$ and $\mathbf{M}_{rP}$ "resultant" notation, what do the equilibrium equations (see Section 3.5) of a body look like?

Before closing this section, we remark that if the force ($\mathbf{F}_r$) and couple ($\mathbf{M}_{rP}$) resultant at point P has been computed for some system S_1 of forces and couples, then by Varignon's theorem the moment of these forces and couples about a line ℓ through P is simply

$$\mathbf{M}_\ell = (\mathbf{M}_{rP} \cdot \hat{\mathbf{u}}_\ell)\hat{\mathbf{u}}_\ell \tag{3.21}$$

where $\hat{\mathbf{u}}_\ell$ is a unit vector along ℓ.

As an illustration of Equation (3.21), the moment of the forces and couple of Example 3.25 about the y-axis (through O) is (with $\hat{\mathbf{u}}_\ell = \hat{\mathbf{j}}$):

$$\mathbf{M}_{y\text{-axis}} = (\mathbf{M}_{rO} \cdot \hat{\mathbf{j}})\hat{\mathbf{j}} = [(8\hat{\mathbf{i}} + 2.4\hat{\mathbf{j}} + 2.4\hat{\mathbf{k}}) \cdot \hat{\mathbf{j}}]\hat{\mathbf{j}}$$
$$= 2.4\hat{\mathbf{j}} \text{ N} \cdot \text{m}$$

Answer 3.24 $\Sigma\mathbf{F} = \mathbf{0}$ becomes $\mathbf{F}_r = \mathbf{0}$, and $\Sigma\mathbf{M}_P = \mathbf{0}$ becomes $\mathbf{M}_{rP} = \mathbf{0}$. Of course, in the context of equilibrium, the force-couple system we are considering is the set of *all* the external forces and couples exerted on the body.

PROBLEMS ▸ Section 3.7

3.77 Four truss members carrying the indicated forces have their center lines all intersecting at point P of the shaded gusset plate shown in Figure P3.77. Find the resultant of the four forces at P.

3.78 The resultant of the three-force system shown in Figure P3.78 is a single force of 300 lb pointing up along the y axis. Find the force $\mathbf{F}$ and the angle θ it forms with the x axis.

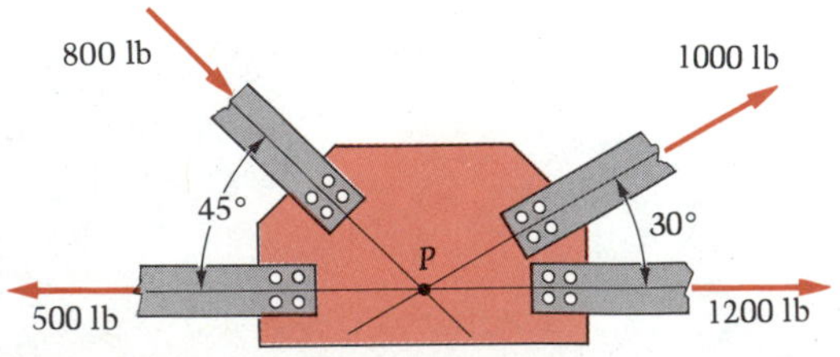

Figure P3.77

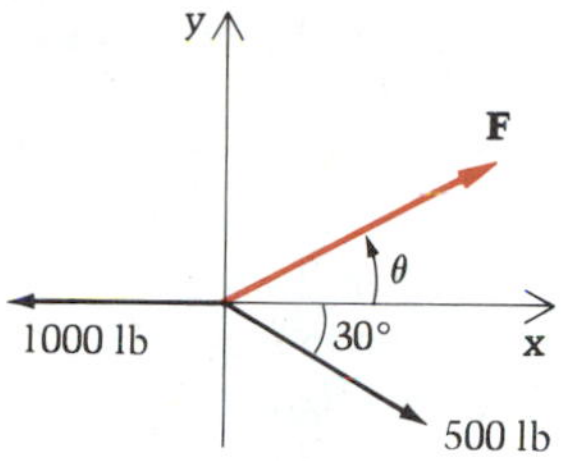

Figure P3.78

3.79 Find the resultant of the three concurrent forces at point C within the equilaterial triangle in Figure P3.79.

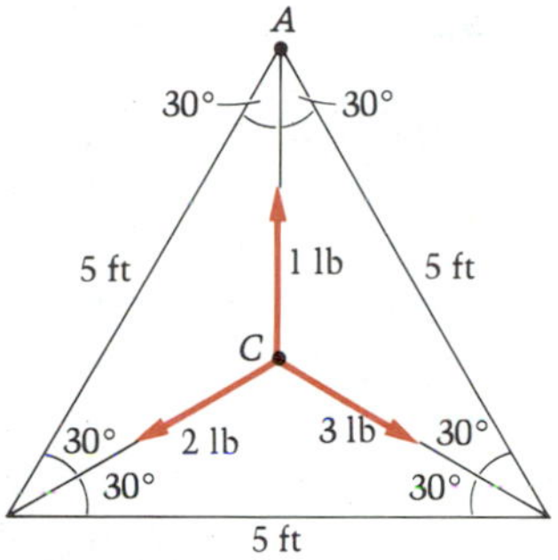

Figure P3.79

3.80 Referring to the preceding problem, find the force and couple at vertex A that are equipollent to the given system (that is, find the resultant at A).

3.81 Find the magnitude of the couple **C** (whose axis is in the xy plane), and its orientation angle θ, for which the three couples have a zero resultant (see Figure P3.81).

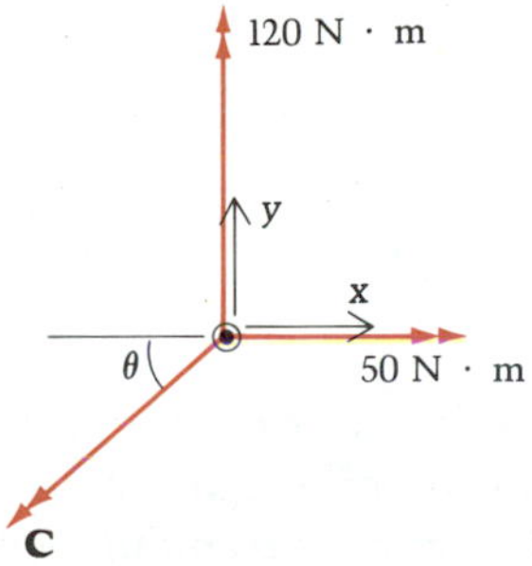

Figure P3.81

3.82 The 30-kN load is eccentrically applied to the column as shown in Figure P3.82. Determine the force and couple at C that are together equipollent to the 30-kN load.

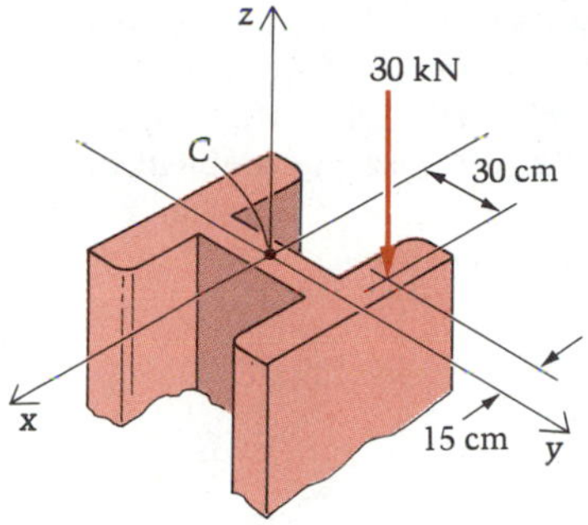

Figure P3.82

3.83 Determine the force and couple at O that constitute the resultant there of the three forces, the 40 lb-ft twisting couple, and the two bending couples acting on the end of the cantilever beam shown in Figure P3.83.

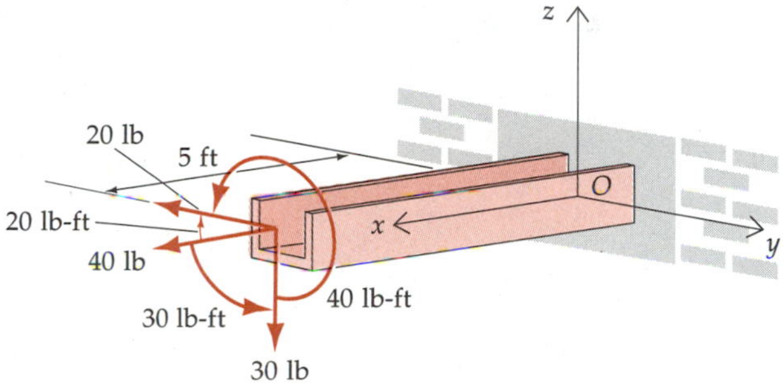

Figure P3.83

3.84 For the system shown in Figure P3.84, find the resultant at point B.

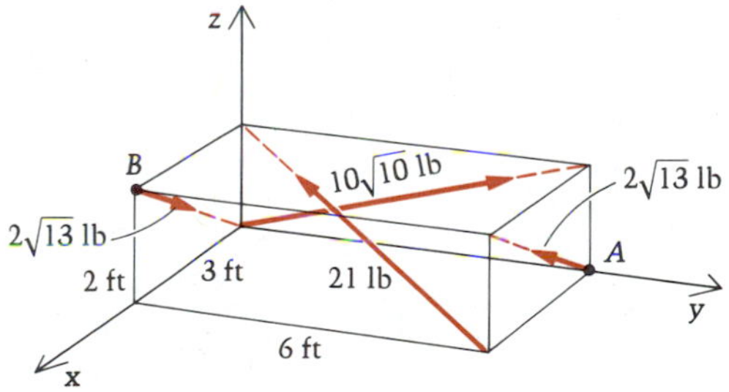

Figure P3.84

3.85 **a.** Repeat the preceding problem for A instead of B.

b. Demonstrate that $\mathbf{M}_{rA} = \mathbf{M}_{rB} + \mathbf{r}_{AB} \times \mathbf{F}_r$, which is, of course, generally true.

3.86 Determine the resultant at A of the forces applied to the bar shown in Figure P3.86. Then, find the moment at the origin and demonstrate that $\mathbf{M}_{rA} = \mathbf{M}_{rO} + \mathbf{r}_{AO} \times \mathbf{F}_r$. Finally, use Equation (3.21) to find the moment of the forces in the figure about the line through A that forms equal angles with the coordinate axes.

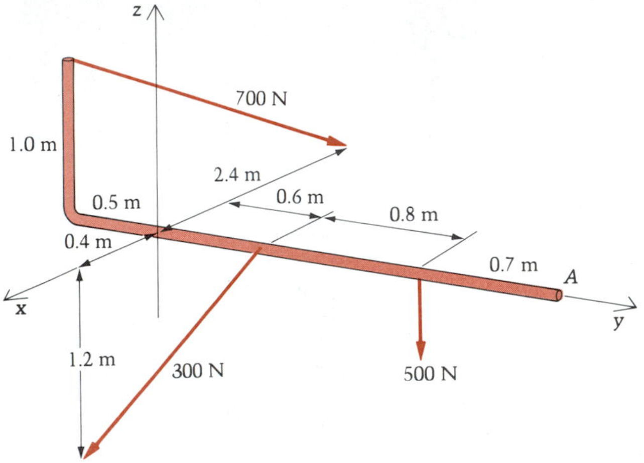

Figure P3.86

3.87 Check the answer to Example 3.23 by finding the resultant at C and then computing $\mathbf{M}_{rO} + \mathbf{r}_{CO} \times \mathbf{F}_r$ and comparing the result with $\mathbf{M}_{rC}$. The two results should of course be equal.

3.88 Determine the resultant of the four forces and the two couples that act on the shaft shown in Figure P3.88, expressed as a force and couple at the origin O.

3.89 In Figure 3.89, couple **C** lies in plane OBG, and has magnitude $2\sqrt{34}$ lb-ft. Find:

- **a.** the moment of $\mathbf{F}_1$ about O;
- **b.** the moment of $\mathbf{F}_2$ about O;
- **c.** the moment of **C** about O;
- **d.** the moment of the resultant (of $\mathbf{F}_1$, $\mathbf{F}_2$, and **C**) about line OG;
- **e.** the moment of the resultant (of $\mathbf{F}_1$, $\mathbf{F}_2$, and **C**) about line OE.
- **f.** Adding the answers to a, b, and c, then subtracting those of d and e, gives a vector normal to plane EOG. Without any calculations, explain why this must be so.

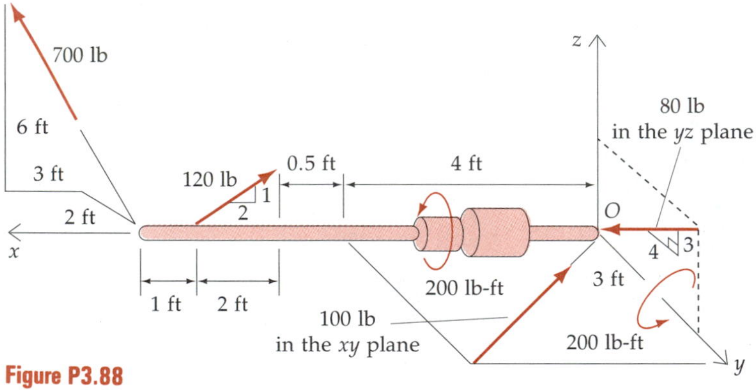

Figure P3.88

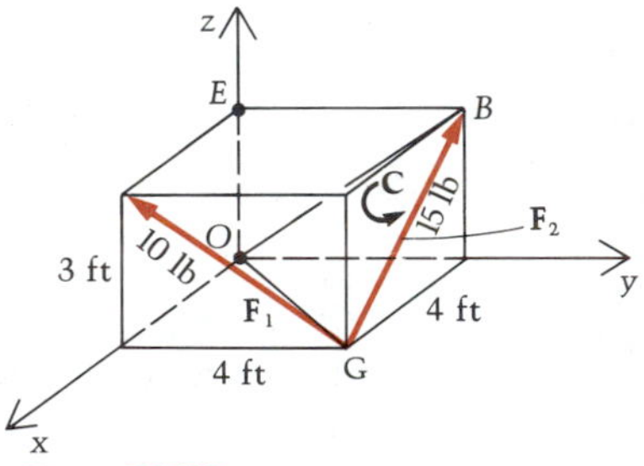

Figure P3.89

3.8 The Simplest Resultant of a Force System

The Single-Force Resultant

In Example 3.16, we constructed a force and couple that were equipollent to a single force. There are situations in which this procedure is desirable. Often, however, our motivation is to produce the simplest resultant — that is, where possible, simply a force or simply a couple.* Thus, in the

* Some authors use the term "resultant" to denote what we are calling "simplest resultant." Our choice of language is motivated by what the student will encounter in the analysis of stresses in deformable solids. There, the most useful form of resultant is often a force-couple pair with a preassigned reference point, and, even if further reduction to a single force is possible, it is not useful.

example mentioned above, the original force itself is already the simplest resultant. Similarly, the system of forces in Example 3.17 had $\mathbf{F}_r = \mathbf{0}$, and so the original *couple* was already the simplest resultant.

In this section we are going to think about the conditions under which we may reduce an arbitrary force system to an equipollent one consisting only of a force. Figure 3.23 depicts three systems; the first one being an "original system" S_1 comprising any number of forces and couples.

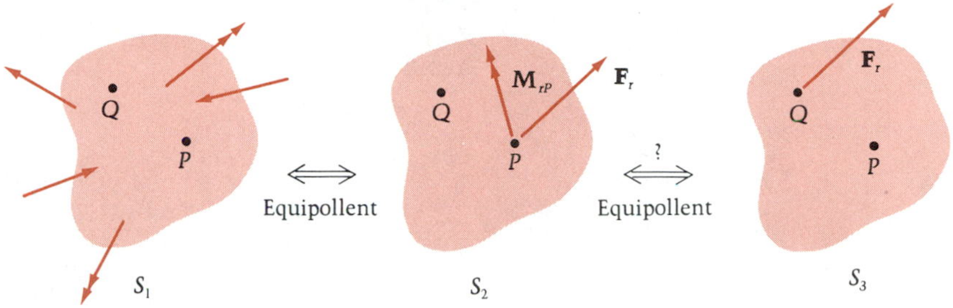

Figure 3.23

The second system, S_2, is equipollent to S_1 and consists of a force $\mathbf{F}_r$ and couple $\mathbf{M}_{rP}$ at some point P. This system can always be found, because, by the two conditions for equipollence,

$$\mathbf{F}_r = (\Sigma\mathbf{F})_1$$
$$\mathbf{M}_{rP} = (\Sigma\mathbf{M}_P)_1$$

The third system S_3, however, may or may not exist. It is required to be equipollent to S_2 (and hence also to S_1) and consist only of a "force alone" at some point Q. If this system is to exist, the force in S_3 must be the same as the sum of the forces in S_2 (and S_1):

$$\mathbf{F}_r = (\Sigma\mathbf{F})_1$$

which of course is already true.

The second condition that must be fulfilled for S_3 to exist is that the sum of the moments about P in S_3 be the same as that sum in S_2 (and in S_1); in S_2 it was called $\mathbf{M}_{rP}$:

$$(\Sigma\mathbf{M}_P)_3 = \mathbf{M}_{rP}$$

By using Equation (3.18),

$$(\Sigma\mathbf{M}_P)_3 = (\Sigma\mathbf{M}_Q)_3 + \mathbf{r}_{PQ} \times \mathbf{F}_r$$

But we want $(\Sigma\mathbf{M}_Q)_3$, or $\mathbf{M}_{rQ}$, to be zero. Thus, replacing $(\Sigma\mathbf{M}_P)_3$ with the equivalent notation $\mathbf{M}_{rP}$,

$$\mathbf{M}_{rP} = \mathbf{r}_{PQ} \times \mathbf{F}_r \tag{3.22}$$

Now, Equation (3.22) can be satisfied (or is meaningful) only if $\mathbf{M}_{rP}$ is orthogonal, or perpendicular, to $\mathbf{F}_r$. This is because a cross product is always perpendicular to each of the vectors making up the product, so that the cross product, $\mathbf{M}_{rP}$, is perpendicular both to $\mathbf{r}_{PQ}$ and to $\mathbf{F}_r$. Whenever this is the case, Equation (3.22) can be solved for vectors $\mathbf{r}_{PQ}$ identifying points on the line of action of the single force resultant $\mathbf{F}_r$ in system S_3.

We note that if in system S_1 we happen to have $\mathbf{F}_r = \mathbf{0}$ with $\mathbf{M}_{rP} \neq \mathbf{0}$, then by Equation (3.18) we have $\mathbf{M}_{rQ} = \mathbf{M}_{rP}$ and the resultant is a couple, period! And that if in system S_1 we have $\mathbf{F}_r \neq \mathbf{0}$ with $\mathbf{M}_{rP} = 0$, then we *already have* the location of the force-alone resultant — its line of action passes through P.*

We next explore three special force systems for which a single-force resultant exists.

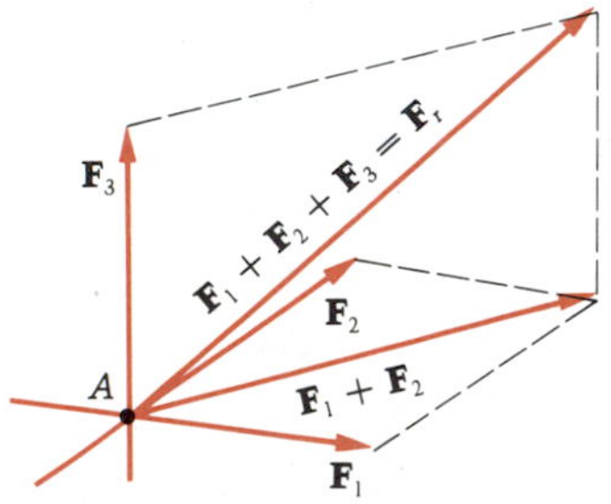

Figure 3.24

Special Force Systems that are Equipollent to a Single Force

1. Concurrent Force Systems A concurrent force system is one for which all of the lines of action intersect at a single point. If we call that point A as in Figure 3.24, then $\mathbf{M}_{rA} = \mathbf{0}$ because none of the forces produces a moment about A. Thus the resultant is simply $\mathbf{F}_r = \Sigma\mathbf{F}$ with its line of action passing through A.

EXAMPLE 3.26

Find the single-force resultant, or "force-alone" resultant, of the system of forces shown in Figure E3.26.

$$\mathbf{F}_1 = 2\hat{\mathbf{j}} \text{ N}$$
$$\mathbf{F}_2 = \hat{\mathbf{i}} + 2\hat{\mathbf{j}} + 5\hat{\mathbf{k}} \text{ N}$$
$$\mathbf{F}_3 = -4\hat{\mathbf{j}} + 7\hat{\mathbf{k}} \text{ N}$$
$$\mathbf{F}_4 = 6\hat{\mathbf{i}} \text{ N}$$

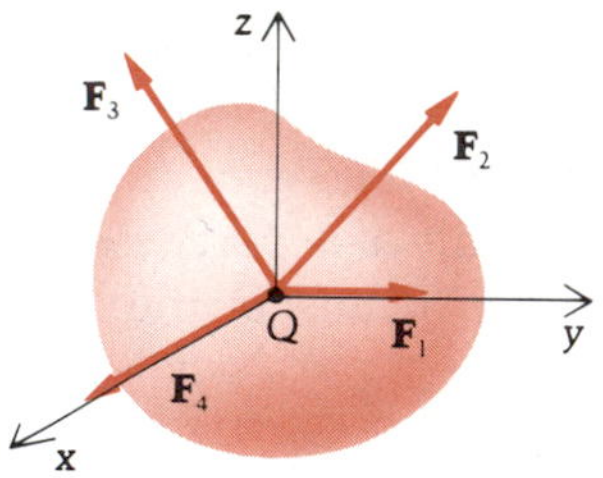

Figure E3.26

Solution

Because the four forces are concurrent at Q, there is a single force resultant that is simply their sum $\mathbf{F}_r$, acting at Q (or any other point on the line of $\mathbf{F}_r$ through Q):

$$\begin{aligned}\mathbf{F} &= \mathbf{F}_1 + \mathbf{F}_2 + \mathbf{F}_3 + \mathbf{F}_4 \\ &= 2\hat{\mathbf{j}} + (\hat{\mathbf{i}} + 2\hat{\mathbf{j}} + 5\hat{\mathbf{k}}) + (-4\hat{\mathbf{j}} + 7\hat{\mathbf{k}}) + 6\hat{\mathbf{i}} \\ &= 7\hat{\mathbf{i}} + 12\hat{\mathbf{k}} \text{ N}\end{aligned}$$

Note in Example 3.26 that at any point *off* the line of $\mathbf{F}_r$ through Q, the equipollent system will consist not only of the force $\mathbf{F}_r$, but also of a non-zero couple. For example, at the point B with coordinates (1, 1, 1),

* And if $\mathbf{F}_r$ and $\mathbf{M}_{rP}$ are *both* zero in S_1, then the sum of the forces and the sum of the moments about any point vanish for all systems equipollent to S_1. This in fact is what happens with the force system acting on a body in equilibrium.

we would still have

$$\mathbf{F}_r = 7\hat{\mathbf{i}} + 12\hat{\mathbf{k}}\ \text{N}$$

(the resultant force does not change from one equipollent system to another) accompanied by

$$\mathbf{M}_{rB} = \overset{0}{\cancel{\mathbf{M}_{rQ}}} + \mathbf{r}_{BQ} \times \mathbf{F}_r$$
$$= (-\hat{\mathbf{i}} - \hat{\mathbf{j}} - \hat{\mathbf{k}}) \times (7\hat{\mathbf{i}} + 12\hat{\mathbf{k}})$$
$$= -12\hat{\mathbf{i}} + 5\hat{\mathbf{j}} + 7\hat{\mathbf{k}}\ \text{N} \cdot \text{m}$$

which of course is not zero.

2. Coplanar Force Systems In this type of system, all of the forces have lines of action that lie in the same plane (say xy), while all the vectors representing moments of couples are normal to this plane (see Figure 3.25).

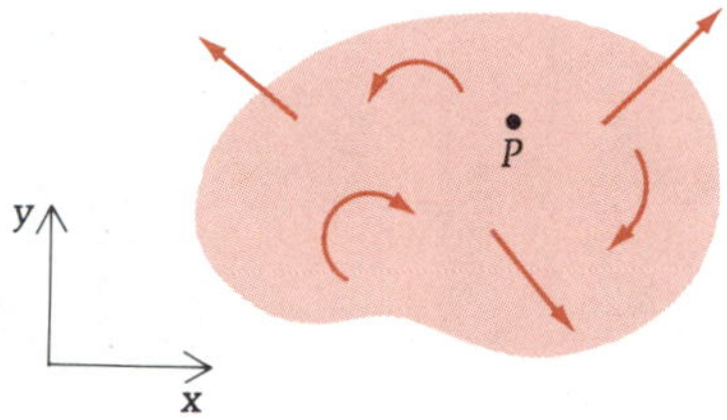

Figure 3.25

In this case if we sketch the equipollent system at P (that is, the resultant at P) in the xy plane (see Figure 3.26), we will necessarily have

$$\mathbf{F}_r = \mathbf{F}_{rx}\hat{\mathbf{i}} + \mathbf{F}_{ry}\hat{\mathbf{j}}$$
$$\mathbf{M}_{rP} = M_{rP}\hat{\mathbf{k}}$$

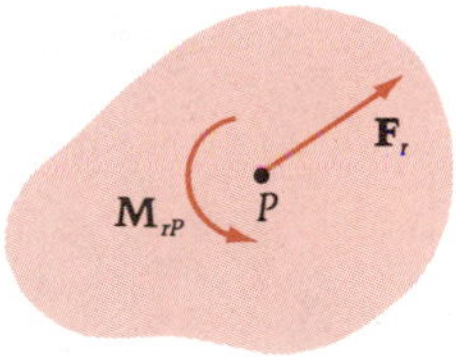

Figure 3.26

Question 3.25 (a) Why does $\mathbf{F}_r$ have no z component? (b) Why does $\mathbf{M}_{rP}$ have *only* a z component?

Now if we let P be the origin for rectangular coordinates and let x, y, and z be coordinates of a point through which the line of action of a "force-alone" resultant would pass, then by Equation (3.22),

$$M_{rP}\hat{\mathbf{k}} = (x\hat{\mathbf{i}} + y\hat{\mathbf{j}} + z\hat{\mathbf{k}}) \times (F_{rx}\hat{\mathbf{i}} + F_{ry}\hat{\mathbf{j}})$$

or

$$M_{rP}\hat{\mathbf{k}} = -zF_{ry}\hat{\mathbf{i}} + zF_{rx}\hat{\mathbf{j}} + (xF_{ry} - yF_{rx})\hat{\mathbf{k}}$$

From either the $\hat{\mathbf{i}}$- or $\hat{\mathbf{j}}$-coefficients of this equation, we see that $z = 0$. That is, as we would have anticipated, the line of action of the force-alone resultant lies in the xy plane (the plane of the forces in the original system). But also, from the $\hat{\mathbf{k}}$-coefficients, we find

$$xF_{ry} - yF_{rx} = M_{rP} \qquad \text{or} \qquad y = \frac{F_{ry}}{F_{rx}}x - \frac{M_{rP}}{F_{rx}} \tag{3.23}$$

which provides the equation of the line of action of the force-alone resultant.

Answer 3.25 (a) Because all forces in the system lie in the xy plane. (b) $\mathbf{M}_{rP}$ is made up of (a) couples of the form $C\hat{\mathbf{k}}$, and/or (b) moments of forces of form $(x\hat{\mathbf{i}} + y\hat{\mathbf{j}}) \times (F_x\hat{\mathbf{i}} + F_y\hat{\mathbf{j}})$. The latter is $(xF_y - yF_x)\hat{\mathbf{k}}$, thus only a z component.

By comparing Equation (3.23) with the familiar form of the equation of a straight line, $y = mx + b$, we see that the slope of the line is $m = F_{ry}/F_{rx}$, as it must be.

Question 3.26 Why must it?

Similarly, the y-intercept is $b = -M_{rP}/F_{rx}$. These results are depicted below in Figure 3.27:

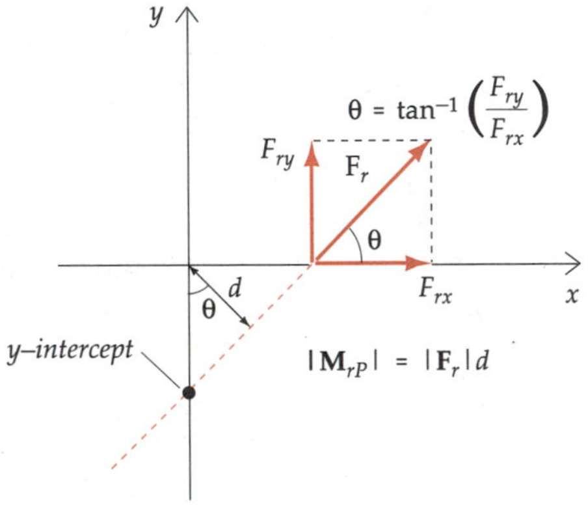

Figure 3.27

Note from the figure that:

$$m = \text{slope} = \tan\theta = \frac{F_{ry}}{F_{rx}}$$

and

$$b = y\text{-intercept} = \frac{-d}{\cos\theta} = -\frac{|\,\mathbf{M}_{rP}\,|/|\,\mathbf{F}_r\,|}{F_{rx}/|\,\mathbf{F}_r\,|} = -\frac{M_{rP}}{F_{rx}},$$

so that Equation (3.23) has been verified.

Before proceeding to examples of coplanar systems, we illustrate a slightly different view* of this process of finding the line of action of the force-alone resultant. Since $\mathbf{M}_{rP}$ and $\mathbf{F}_r$ are perpendicular, we may replace $\mathbf{M}_{rP}$ by an equipollent pair of forces $\mathbf{F}_r$ and $-\mathbf{F}_r$ as shown in the second frame of Figure 3.28, where we let the $-\mathbf{F}_r$ part of this pair have a line of action through our reference point P. Now the "canceling" of the $\mathbf{F}_r$ and $-\mathbf{F}_r$ pair through P leaves the third frame of the figure as the final result. Clearly this reasoning may be used any time $\mathbf{M}_{rP}$ and $\mathbf{F}_r$ are perpendicular.† This illustrates the fact that as long as $\mathbf{F}_r \perp \mathbf{M}_{rP}$ to start with, we are guaranteed a force-alone resultant for a coplanar force system.

Answer 3.26 Because the single force resultant is always $\mathbf{F}_r$.

* Which was previewed on page 90.

† See footnote on page 111.

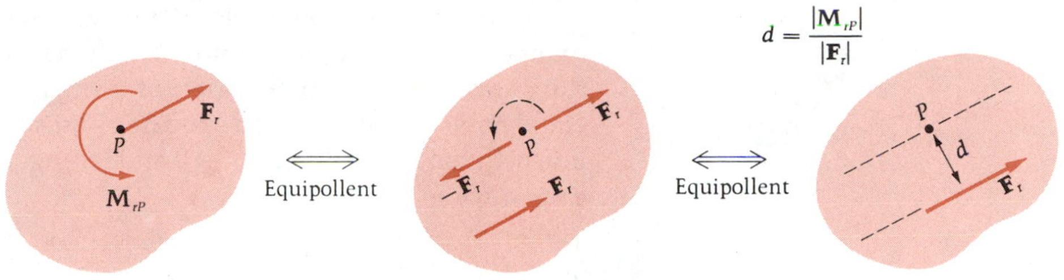

Figure 3.28

Example 3.27

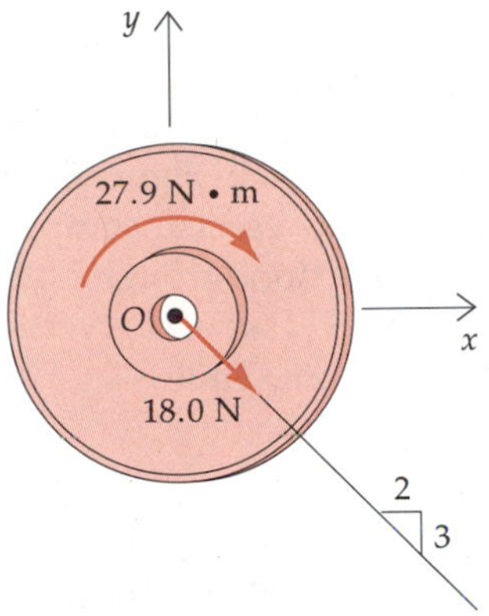

Figure E3.27a

In Example 3.22, the original system of forces and couples was shown to be equipollent to the single force and couple at O shown in Figure E3.27a. Now, further reduce this coplanar system to a single force.

Solution

We have $\mathbf{M}_{rO} \perp \mathbf{F}_r$. Therefore, the equation

$$\mathbf{r}_{OA} \times \mathbf{F}_r = \mathbf{M}_{rO}$$

will identify all points A on the line of action of the single force, $\mathbf{F}_r$, in an equipollent system consisting only of this force. Thus,

$$(x\hat{\mathbf{i}} + y\hat{\mathbf{j}}) \times (10\hat{\mathbf{i}} - 15\hat{\mathbf{j}}) = -27.9\hat{\mathbf{k}}$$

from which

$$-15x - 10y = -27.9$$

or

$$y = -1.5x + 2.79 \text{ m} \tag{1}$$

The solution, then, is the force $\mathbf{F}_r$ placed along the line defined by Equation (1), and shown below in Figure E3.27b:

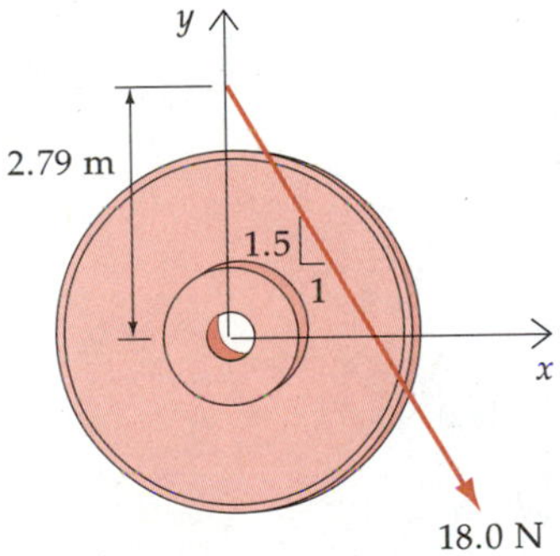

Figure E3.27b

† This is true whether or not the original system, to which $\mathbf{F}_r$ and $\mathbf{M}_{rP}$ are equipollent, is coplanar! We will see later in Example 3.31 an example of a non-coplanar system (which is also not concurrent or parallel) in which $\mathbf{F}_r \perp \mathbf{M}_{rP}$ so that a force-alone resultant exists.

In the previous example, we could have proceeded in a different manner: Realizing that we know $\mathbf{F}_r$, all that is really required to establish the line of action is the location of a single point on the line. Thus, we could have simply sought a specific point such as the intersection of the line with the x axis. Setting $y = 0$ and letting $x = a$ at that intersection, we can write

$$a\hat{\mathbf{i}} \times \mathbf{F}_r = \mathbf{M}_{rO}$$

or

$$a\hat{\mathbf{i}} \times (10\hat{\mathbf{i}} - 15\hat{\mathbf{j}}) = -27.9\,\hat{\mathbf{k}}\text{ N}\cdot\text{m}$$

Thus

$$15a = 27.9$$
$$a = 1.86\text{ m}$$

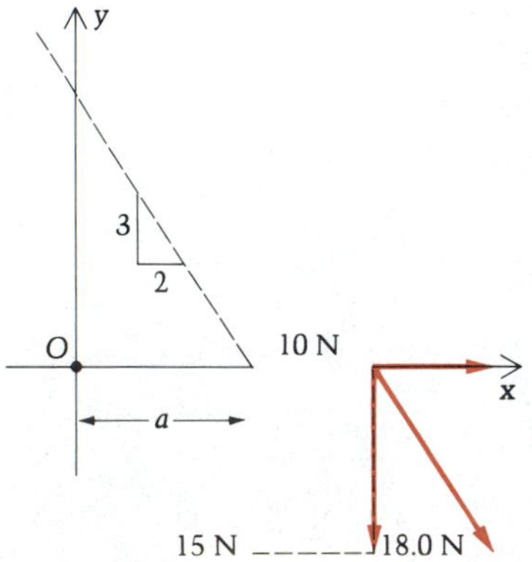

Figure 3.29

as seen in Figure 3.29.

The same result is obtained with less formality, and perhaps in a manner that facilitates physical insight, by referring again to Figure 3.29 where $\mathbf{F}_r$ has been decomposed into its horizontal and vertical components. Since the horizontal part (10 N) has a line of action through O, it produces no moment about O. The clockwise moment ($15a$) of the vertical part must then be equal to the clockwise 27.9 N · m. Thus

$$15a = 27.9$$
$$a = 1.86\text{ m}$$

(as before, and also obtainable from Equation (1) of Example 3.27 when y is set to zero).

Yet a fourth approach is to replace $\mathbf{M}_{rO}$ by two forces $\mathbf{F}_r$ and $-\mathbf{F}_r$ (as was suggested by Figure 3.28):

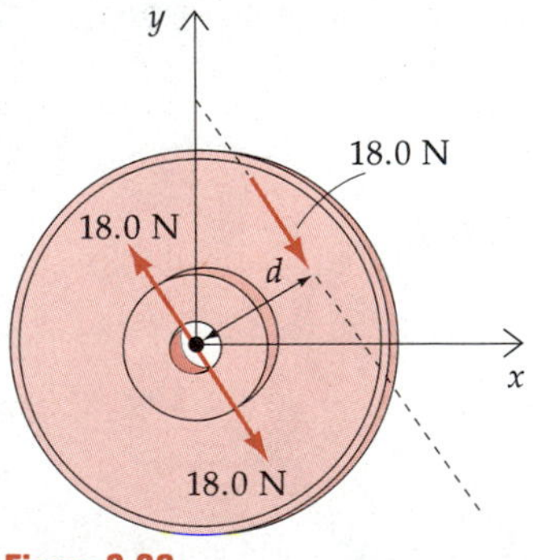

Figure 3.30

Since $\mathbf{M}_{rO}$ and $\mathbf{F}_r$ are given, the distance d in Figure 3.30 has to have the value

$$d = \frac{|\mathbf{M}_{rO}|}{|\mathbf{F}_r|} = \frac{27.9}{18.0} = 1.55\text{ m}$$

and the two 18.0 N forces at the origin cancel, leaving a force-alone resultant matching that of Figure E3.27b in the previous example. The

y-intercept of that figure is obtainable from Figure 3.30 and the similar triangles in Figure 3.31 below:

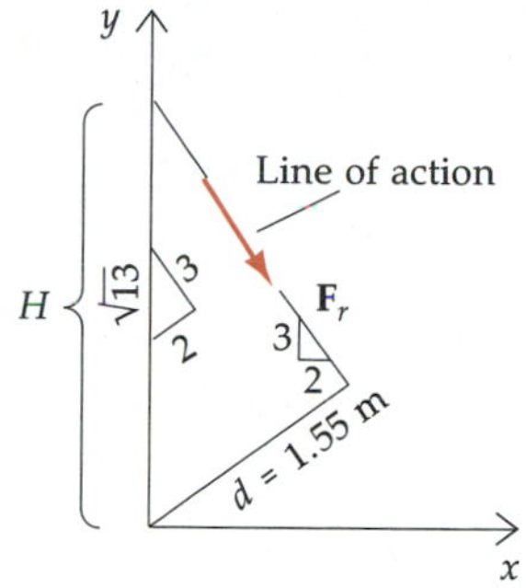

Figure 3.31

$$\frac{d}{H} = \frac{1.55}{H} = \frac{2}{\sqrt{13}}$$

or

$$H = 2.79 \text{ m, as before.}$$

Example 3.28

Find the single force that is equipollent to the system shown in Figure E3.28a.

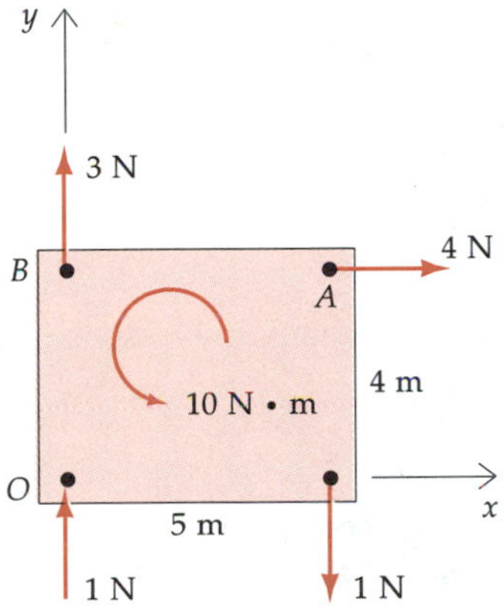

Figure E3.28a

Solution

The resultant force is

$$\mathbf{F}_r = 4\hat{\mathbf{i}} + (3 - 1 + 1)\hat{\mathbf{j}} = 4\hat{\mathbf{i}} + 3\hat{\mathbf{j}} \text{ N}$$

(Note that the 1-N forces form a couple and thus have no resultant force.) At the origin O, the resultant moment is

$$\mathbf{M}_{rO} = \underbrace{5\hat{\mathbf{i}} \times (-1\hat{\mathbf{j}}) + 10\hat{\mathbf{k}}}_{\text{couples}} + \underbrace{4\hat{\mathbf{j}} \times 4\hat{\mathbf{i}}}_{\text{moment of the 4-N force about } O}$$

$$= -11\hat{\mathbf{k}} \text{ N} \cdot \text{m}$$

In a case such as this where it is easy to identify the perpendicular distances from O to the lines of action of various forces, the moments can be calculated with less formality by using the "force times perpendicular distance" method in conjunction with the right-hand rule. With this approach

$$\mathbf{M}_{rO} = 1(5)(-\hat{\mathbf{k}}) + 10\hat{\mathbf{k}} + 4(4)(-\hat{\mathbf{k}})$$

$$= -11\hat{\mathbf{k}} \text{ N} \cdot \text{m} \quad \text{or} \quad 11 \circlearrowright \text{ N} \cdot \text{m}$$

Question 3.27 Why isn't the 3-N force included in the $\mathbf{M}_{rO}$ calculation?

Question 3.28 Why doesn't the vector $\mathbf{r}$, in the $\mathbf{r} \times \mathbf{F}$ calculation for the 4-N force, extend all the way to the application point A of the force?

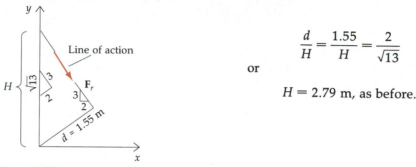

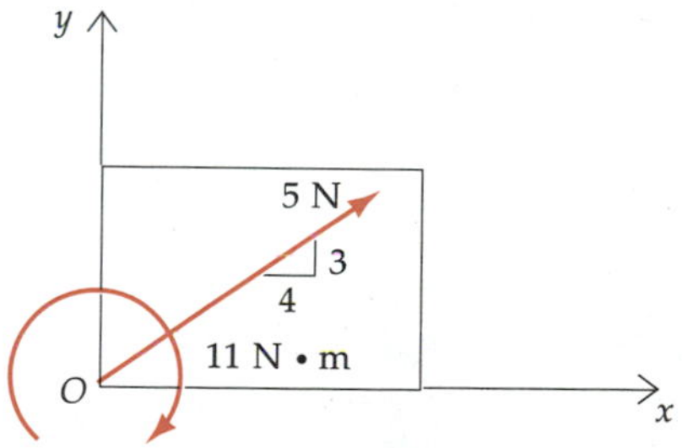

Figure E3.28b

The equipollent system at O thus appears as shown in Figure E3.28b.

Answer 3.27 It passes through O, and therefore has zero moment about O.

Answer 3.28 As we have seen in Section 3.2, the vector $\mathbf{r} \times \mathbf{F}$ may intersect $\mathbf{F}$ at any point on its line of action and the moment will be the same; in this case $\mathbf{r} = 4\hat{\mathbf{j}}$ intersects the line of action of the force at point B.

The force-alone resultant must now be positioned so that it produces a clockwise 11-N · m moment about point O. This means (see Figure E3.28c below) that this resultant intersects the y-axis *above* point O:

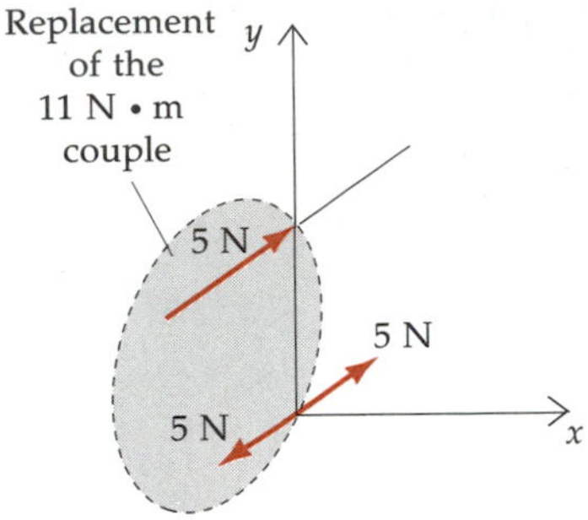

Figure E3.28c

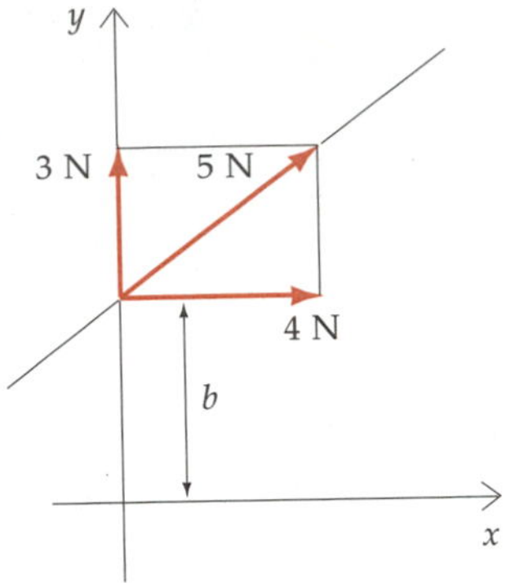

Figure E3.28d

By decomposing the 5-N resultant force into horizontal and vertical components where the line of action of $\mathbf{F}_r$ intersects the y axis (see Figure E3.28d above), we can easily determine that intersection by:

$$4b = 11\ \text{N} \cdot \text{m}$$

$$b = \tfrac{11}{4} = 2.75\ \text{m}$$

This pins down the line of action of the force-alone resultant and completes the example.

An alternative approach to the previous example would be to use the more formal vector algebra approach to determine the force-alone resultant and thereby obtain the equation of the line of action of the force, at the possible expense of a bit of physical understanding:

$$\mathbf{r}_{OA} \times \mathbf{F}_r = \mathbf{M}_{rO}$$

where A is any point on the line of action of the force. Therefore,

$$(x\hat{\mathbf{i}} + y\hat{\mathbf{j}}) \times (4\hat{\mathbf{i}} + 3\hat{\mathbf{j}}) = -11\hat{\mathbf{k}}$$

or

$$3x - 4y = -11$$

Question 3.29 Why hasn't a $z\hat{\mathbf{k}}$ been included as part of the vector $\mathbf{r}_{OA}$ above?

Thus

$$y = \tfrac{3}{4}x + \tfrac{11}{4}\ \text{m}$$

Answer 3.29 We have shown in general that the force-alone resultant for a coplanar force system lies in that same plane.

or

$$y = 0.75x + 2.75 \text{ m}$$

Note from Figure E3.28d that the point where the single-force resultant intercepts the y-axis, namely $(x, y) = (0, 2.75)$, obviously lies on this line.

3. Parallel Force Systems In this system all of the lines of action of the various forces are parallel; we can let the common direction of these forces be z (Figure 3.32).

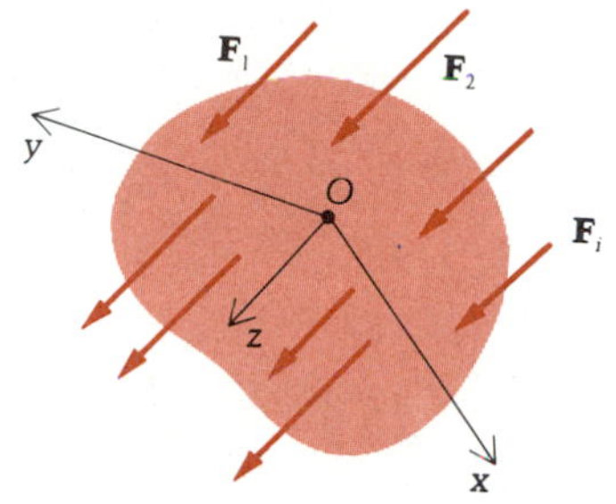

Figure 3.32

An intermediate equipollent system of one force and one couple at the origin has

$$\mathbf{F}_r = \Sigma\mathbf{F}_i = \Sigma F_i\hat{\mathbf{k}} = F_r\hat{\mathbf{k}}$$

where F_r could be negative here, together with the couple

$$\mathbf{M}_{rO} = \Sigma[(x_i\hat{\mathbf{i}} + y_i\hat{\mathbf{j}}) \times F_i\hat{\mathbf{k}}]$$

in which (x_i, y_i) locates the point where $\mathbf{F}_i$ pierces the xy plane. Performing the cross products,

$$\mathbf{M}_{rO} = \Sigma[F_i(-x_i\hat{\mathbf{j}} + y_i\hat{\mathbf{i}})]$$

and we see that $\mathbf{M}_{rO} \perp \mathbf{F}_r$. This means that the simplest resultant is once again a single force $\mathbf{F}_r$,* with no accompanying couple. To find where this resultant acts, we once again use Equation (3.22):

$$(x\hat{\mathbf{i}} + y\hat{\mathbf{j}} + z\hat{\mathbf{k}}) \times F_r\hat{\mathbf{k}} = \mathbf{M}_{rO} = \Sigma[F_i(y_i\hat{\mathbf{i}} - x_i\hat{\mathbf{j}})]$$

where (x, y, z) are the coordinates of a point on the line of action of $\mathbf{F}_r$. Matching the coefficients of $\hat{\mathbf{i}}$, $\hat{\mathbf{j}}$, and $\hat{\mathbf{k}}$, we obtain

$$\hat{\mathbf{i}} \text{ coefficients:} \quad y = \frac{\Sigma(F_i y_i)}{F_r} \tag{3.24a}$$

$$\hat{\mathbf{j}} \text{ coefficients:} \quad x = \frac{\Sigma(F_i x_i)}{F_r} \tag{3.24b}$$

$$\hat{\mathbf{k}} \text{ coefficients:} \quad 0 = 0 \quad \text{(means } z \text{ can have any value)}$$

Thus the line parallel to the z axis with x and y given by Equations (3.24a,b) is the line of action. At any point of this line, a system equipollent to the original one of Figure 3.32 is given quite simply by Figure 3.33.

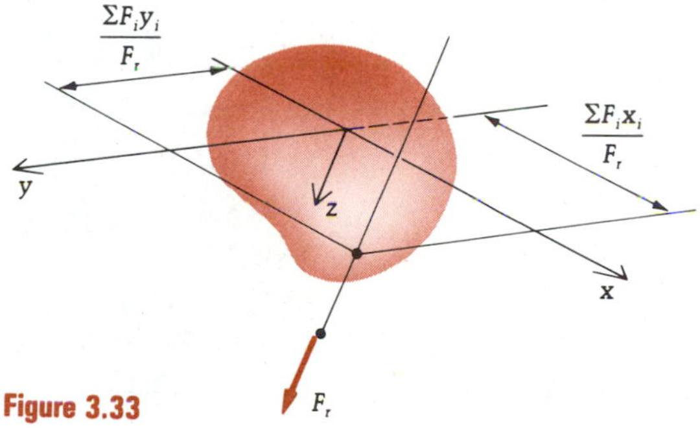

Figure 3.33

* Unless, of course, $\mathbf{F}_r = \mathbf{0}$ in which case the simplest resultant is the couple $\mathbf{M}_{rO}$.

We shall now consider two examples of parallel force systems:

EXAMPLE 3.29

Find the single-force resultant for the system of parallel forces acting on the plate in Figure E3.29.

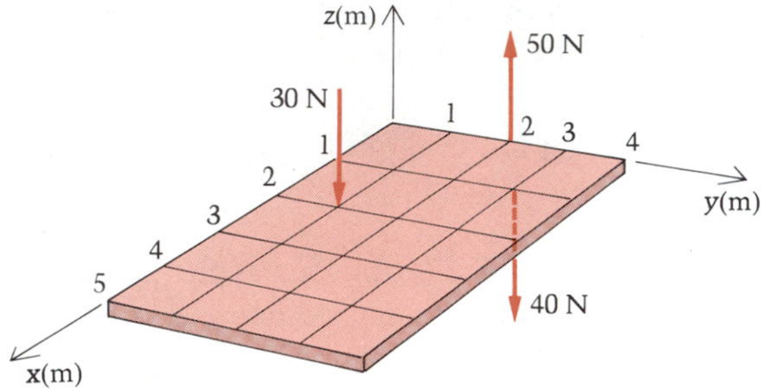

Figure E3.29

Solution

Let us call the given system S_1, and the system we are seeking, S_2. The force in S_2 is found to be

$$\mathbf{F}_r = \sum^{S_1} \mathbf{F} = -30\hat{\mathbf{k}} - 40\hat{\mathbf{k}} + 50\hat{\mathbf{k}}$$
$$= -20\hat{\mathbf{k}} \text{ N}$$

If we let x and y be the coordinates locating the intersection of the line of action of the force-alone resultant with the surface of the plate, then

$$(x\hat{\mathbf{i}} + y\hat{\mathbf{j}}) \times \mathbf{F}_r = \mathbf{M}_{rO}$$

or

$$(x\hat{\mathbf{i}} + y\hat{\mathbf{j}}) \times (-20\hat{\mathbf{k}}) = 2\hat{\mathbf{j}} \times 50\hat{\mathbf{k}} + (\hat{\mathbf{i}} + 3\hat{\mathbf{j}}) \times (-40\hat{\mathbf{k}}) + (2\hat{\mathbf{i}} + \hat{\mathbf{j}}) \times (-30\hat{\mathbf{k}})$$

or

$$20x\hat{\mathbf{j}} - 20y\hat{\mathbf{i}} = 100\hat{\mathbf{i}} + (40\hat{\mathbf{j}} - 120\hat{\mathbf{i}}) + (60\hat{\mathbf{j}} - 30\hat{\mathbf{i}})$$

Thus, collecting like terms,

$$\hat{\mathbf{i}}: \quad -20y = 100 - 120 - 30 = -50$$
$$y = 2.5 \text{ m}$$

and

$$\hat{\mathbf{j}}: \quad 20x = 40 + 60 = 100$$
$$x = 5 \text{ m}$$

These, of course, are the x and y coordinates of *every point* on the line of action of the force $\mathbf{F}_r$, since in this case the line of action parallels the z axis.

It is important to realize that the $\hat{\mathbf{i}}$ and $\hat{\mathbf{j}}$ parts of $\mathbf{M}_{rO}$ in the preceding example are the moments about the x and y axes, respectively. This fact may be used to locate the line of action of the force-alone resultant without recourse to the formalities of vector algebra. The force-alone resultant must produce the same moments about these axes as do the forces in the original system. And since all of the forces are perpendicular to the axes in question, the "force times perpendicular distance" method easily may be used to calculate the moment. Referring to Figure 3.34, then, with the 20-N resultant force shown dashed,

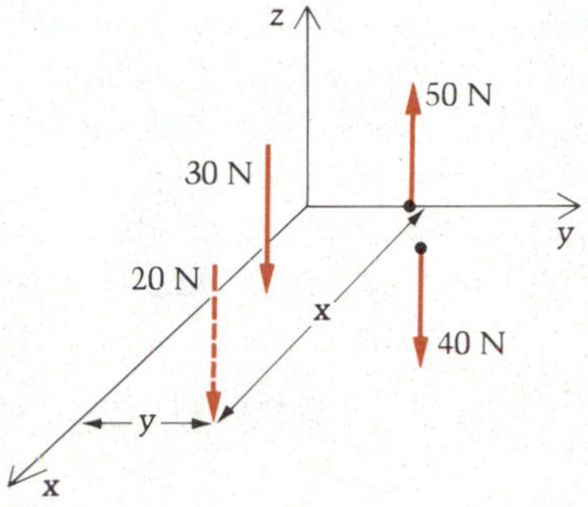

Figure 3.34

$$\text{Moment about } x = -20y = -30(1) + 50(2) - 40(3) = -50$$
$$y = 2.5 \text{ m} \qquad \text{(as before)}$$

and

$$\text{Moment about } y = 20x = 30(2) + 40(1) = 100$$
$$x = 5 \text{ m} \qquad \text{(as before)}$$

Example 3.30

Determine the force-alone resultant for the system of six parallel forces acting on the beam as shown in Figure E3.30.

Solution

The required resultant consists of a single force computed as

$$\mathbf{F}_r = \Sigma\mathbf{F} = (-10 - 20 - 30 - 40 - 50 - 60)\hat{\mathbf{k}}$$
$$= -210\hat{\mathbf{k}} \text{ lb}$$

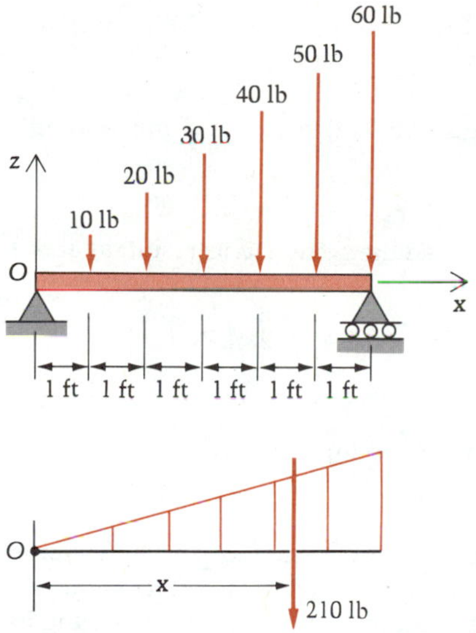

Figure E3.30

Its location is given by the value of x for which the moment of $\mathbf{F}_r$ about, say, O is the same as the moment of the six forces about O:

$$-210x = -10(1) - 20(2) - 30(3) - 40(4) - 50(5) - 60(6)$$
$$x = \frac{910}{210}$$
$$= 4.33 \text{ ft}$$

The y coordinate of the force-alone resultant vanishes this time because all the forces lie in the xz plane.

In this example, we have a set of forces, finite in number, that happen to increase proportionately. In Section 3.9 we shall consider what happens when a loading becomes *continuously distributed* across a line or an area, instead of acting at a few discrete points as above.

The Simplest Resultant of a General Force-and-Couple System: A Collinear Force and Couple ("Screwdriver")*

We close this section with two examples. In the first, the force and couple parts of the resultant at P, $\mathbf{F}_r$ and $\mathbf{M}_{rP}$, happen to be perpendicular. Thus, even though the original system is neither concurrent, coplanar, nor parallel, a force-alone resultant can be found.

In the second example, $\mathbf{F}_r$ and $\mathbf{M}_{rP}$ are not perpendicular, so a force-alone resultant does not exist. However, this example will show how we can always reduce such a system to *a force together with a parallel couple.* Since this is the mechanical action required to advance a screw with a screwdriver (see Figure 3.35), we call this simplest resultant the "equipollent screwdriver" for the system.† As the reader will see, the reduction is accomplished by applying Equation (3.22) to *that part of* $\mathbf{M}_{rP}$ *that is perpendicular to* $\mathbf{F}_r$.

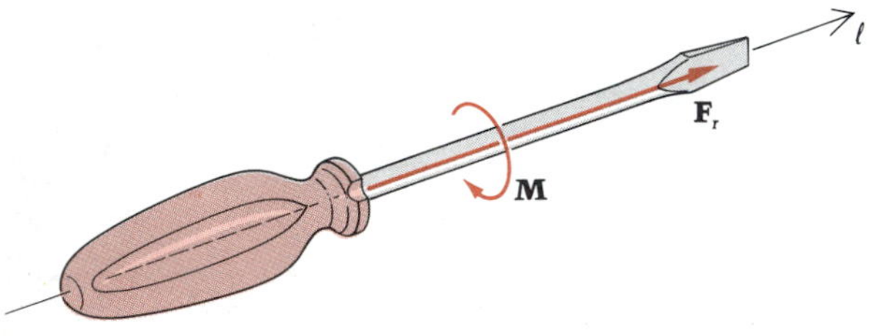

Figure 3.35

EXAMPLE 3.31

Reduce the system of Example 3.25 (see Figure E3.31a) to a force-alone resultant, if possible.

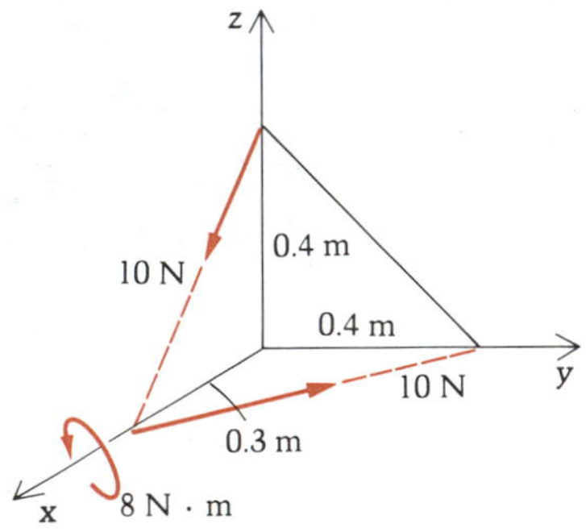

Figure E3.31a

Solution

We have in the earlier example already reduced the system to its resultant at the origin O:

$$\mathbf{F}_r = 8\hat{\mathbf{j}} - 8\hat{\mathbf{k}}\ \text{N} \qquad \text{and} \qquad \mathbf{M}_{rO} = 8\hat{\mathbf{i}} + 2.4\hat{\mathbf{j}} + 2.4\hat{\mathbf{k}}\ \text{N} \cdot \text{m}$$

We note that:

$$\mathbf{M}_{rO} \cdot \mathbf{F}_r = (8\hat{\mathbf{i}} + 2.4\hat{\mathbf{j}} + 2.4\hat{\mathbf{k}}) \cdot (8\hat{\mathbf{j}} - 8\hat{\mathbf{k}})$$
$$= 2.4(8) + 2.4(-8) = 0$$

This zero result means that the resultant moment at O is in fact perpendicular to the resultant force. Therefore, three-dimensional though it is, a force-alone resultant can be found in this example. We proceed then toward determining its

* This material has been included for completeness, but nothing essential to the study of statics is missed if this part of the section is omitted.

† In most texts this is called a "wrench," but the action of a screwdriver is more descriptive.

line of action. If A is a point on this line, then we may write

$$\mathbf{r}_{OA} = x\hat{\mathbf{i}} + y\hat{\mathbf{j}} + z\hat{\mathbf{k}}$$

and then from Equation (3.22),

$$\mathbf{r}_{OA} \times \mathbf{F}_r = \mathbf{M}_{rO}$$

we get

$$(x\hat{\mathbf{i}} + y\hat{\mathbf{j}} + z\hat{\mathbf{k}}) \times (8\hat{\mathbf{j}} - 8\hat{\mathbf{k}}) = 8\hat{\mathbf{i}} + 2.4\hat{\mathbf{j}} + 2.4\hat{\mathbf{k}}$$
$$(-8y - 8z)\hat{\mathbf{i}} + (8x)\hat{\mathbf{j}} + (8x)\hat{\mathbf{k}} = 8\hat{\mathbf{i}} + 2.4\hat{\mathbf{j}} + 2.4\hat{\mathbf{k}}$$

so that

$$\hat{\mathbf{i}} \text{ coefficients:} \quad -8y - 8z = 8 \qquad (1)$$
$$\hat{\mathbf{j}} \text{ coefficients:} \quad 8x = 2.4 \qquad (2)$$
$$\hat{\mathbf{k}} \text{ coefficients:} \quad 8x = 2.4 \qquad (3)$$

Equations (2) and (3) each give the result that $x = 0.3$; this means that the line lies in the plane parallel to yz at this value of x. In this plane, its equation is given by (1):

$$y + z = -1$$

The simplest resultant is shown in Figure E3.31b.

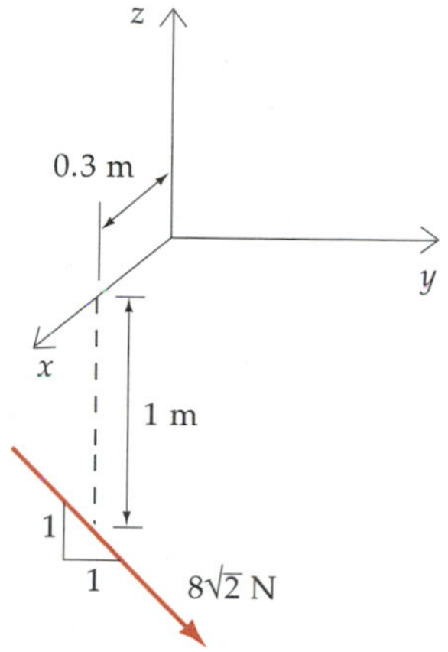

Figure E3.31b

EXAMPLE 3.32

The three forces in Figure E3.32a each have a magnitude of 10 lb. Find the screwdriver equipollent to this system.

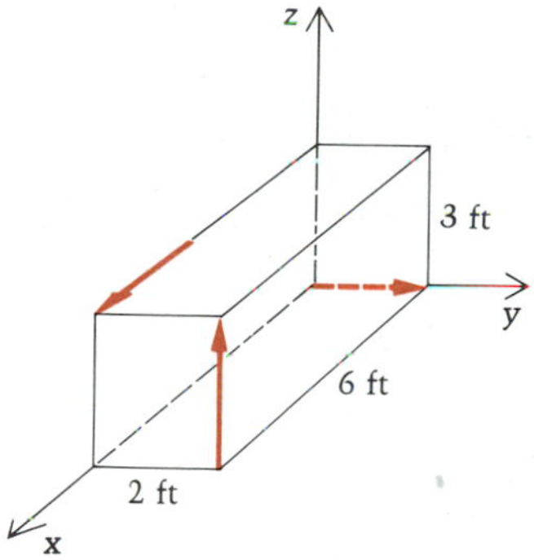

Figure E3.32a

Solution

The force resultant is

$$\mathbf{F}_r = 10\hat{\mathbf{i}} + 10\hat{\mathbf{j}} + 10\hat{\mathbf{k}} = 10\sqrt{3}\left(\frac{\hat{\mathbf{i}} + \hat{\mathbf{j}} + \hat{\mathbf{k}}}{\sqrt{3}}\right) \text{ lb}$$

Already, then, we know that (a) the force of the "screwdriver" is $10\sqrt{3}$ lb and (b) the orientation of its line in space is given by the unit vector $(\hat{\mathbf{i}} + \hat{\mathbf{j}} + \hat{\mathbf{k}})/\sqrt{3}$, which means in this case that the axis of the screwdriver makes equal angles (each $= \cos^{-1}(1/\sqrt{3}) = 54.7°$) with the coordinate axes.

At the origin, an equipollent system to the three given forces is $\mathbf{F}_r$ accompanied by the resultant couple there:

$$\mathbf{M}_{rO} = 3\hat{\mathbf{k}} \times 10\hat{\mathbf{i}} + (6\hat{\mathbf{i}} + 2\hat{\mathbf{j}}) \times 10\hat{\mathbf{k}}$$
$$= 20\hat{\mathbf{i}} - 30\hat{\mathbf{j}} \text{ lb-ft}$$

Here we pause to note that since

$$\mathbf{M}_{rO} \cdot \mathbf{F}_r = 20(10) - 30(10) + 0(10) \neq 0,$$

there is no possibility of a force-alone resultant this time. The non-zero dot product means there is a component of $\mathbf{M}_{rO}$ lying *along* the line of action of $\mathbf{F}_r$ so that the reduction illustrated by Figure 3.23 is not possible.

The "parallel component" of $\mathbf{M}_{rO}$ (the part parallel to $\mathbf{F}_r$) is

$$M_{rO\|} = \mathbf{M}_{rO} \cdot \left(\frac{\mathbf{F}_r}{|\mathbf{F}_r|}\right) = (20\hat{\mathbf{i}} - 30\hat{\mathbf{j}}) \cdot \left(\frac{\hat{\mathbf{i}} + \hat{\mathbf{j}} + \hat{\mathbf{k}}}{\sqrt{3}}\right)$$

$$= \frac{20 - 30}{\sqrt{3}} = \frac{-10}{\sqrt{3}} \text{ lb-ft}$$

Therefore,

$$\mathbf{M}_{rO\|} = \frac{-10}{\sqrt{3}} \frac{\mathbf{F}_r}{|\mathbf{F}_r|} = \frac{-10}{\sqrt{3}}\left(\frac{\hat{\mathbf{i}} + \hat{\mathbf{j}} + \hat{\mathbf{k}}}{\sqrt{3}}\right) = \frac{-10}{3}(\hat{\mathbf{i}} + \hat{\mathbf{j}} + \hat{\mathbf{k}}) \text{ lb-ft}$$

This moment, $\mathbf{M}_{rO\|}$, is the couple of the screwdriver. Note that this time it is *opposite* in direction to $\mathbf{F}_r$, as if a wood screw were being *un*screwed instead of advanced.

The "perpendicular component" of $\mathbf{M}_{rO}$ is the other part of $\mathbf{M}_{rO}$, this time *normal* to $\mathbf{F}_r$:

$$\mathbf{M}_{rO_\perp} = \mathbf{M}_{rO} - \mathbf{M}_{rO\|}$$

$$= (20\hat{\mathbf{i}} - 30\hat{\mathbf{j}}) - \left[\frac{-10}{3}(\hat{\mathbf{i}} + \hat{\mathbf{j}} + \hat{\mathbf{k}})\right]$$

$$= \tfrac{70}{3}\hat{\mathbf{i}} - \tfrac{80}{3}\hat{\mathbf{j}} + \tfrac{10}{3}\hat{\mathbf{k}} \text{ lb-ft}$$

Note as a check that, by inspection, $\mathbf{F}_r \cdot \mathbf{M}_{rO_\perp} = 0$.

Finally, if A is any point on the line of action of the screwdriver, located by the position vector

$$\mathbf{r}_{OA} = x\hat{\mathbf{i}} + y\hat{\mathbf{j}} + z\hat{\mathbf{k}}$$

then the condition

$$\mathbf{r}_{OA} \times \mathbf{F}_r = \mathbf{M}_{rO_\perp}$$

gives the equation of the axis of the equipollent screwdriver:

$$(x\hat{\mathbf{i}} + y\hat{\mathbf{j}} + z\hat{\mathbf{k}}) \times (10\hat{\mathbf{i}} + 10\hat{\mathbf{j}} + 10\hat{\mathbf{k}}) = \tfrac{70}{3}\hat{\mathbf{i}} - \tfrac{80}{3}\hat{\mathbf{j}} + \tfrac{10}{3}\hat{\mathbf{k}}$$

from which we obtain

$\hat{\mathbf{i}}$ coefficients: $10y - 10z = \frac{70}{3}$ (1)

$\hat{\mathbf{j}}$ coefficients: $10z - 10x = -\frac{80}{3}$ (2)

$\hat{\mathbf{k}}$ coefficients: $10x - 10y = \frac{10}{3}$ (3)

Note that by adding Equations (1) and (2), and multiplying the result by -1, we obtain Equation (3). This means that only two of the equations are independent and necessary to define the axis of the screwdriver. Indeed, a pair of equations such as (1) and (2) constitute the general form of a line in three-dimensional space. Before leaving this example, let us find two points on the line, and sketch the final answer:

(1) $y - z = \frac{7}{3}$

(3) $x - y = \frac{1}{3}$

Point 1: If $z = 0$, $y = \frac{7}{3}$ If $y = \frac{7}{3}$, $x = \frac{8}{3}$

Point 2: If $y = 0$, $x = \frac{1}{3}$ If $y = 0$, $z = -\frac{7}{3}$

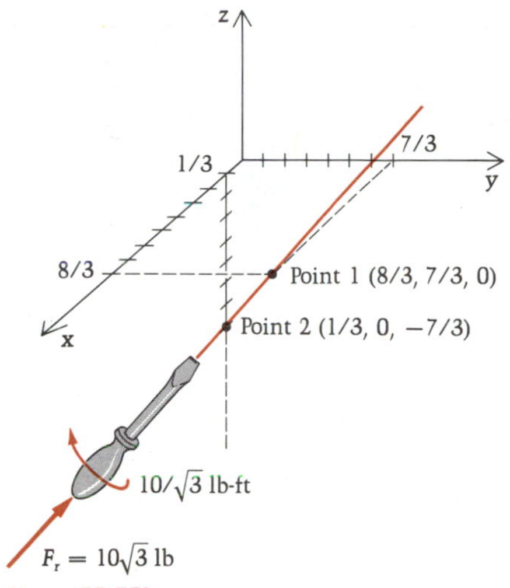

Figure E3.32b

The screwdriver may thus be sketched as shown in Figure E3.32b.

Question 3.30 Are the direction numbers of the screwdriver axis, obtained from the coordinates of points 1 and 2 in Example 3.32, in agreement with direction numbers obtained from $\mathbf{F}_r$ itself?

Answer 3.30 Yes, as they must be. Those of the lines are $[\frac{8}{3} - \frac{1}{3}, \frac{7}{3} - 0, 0 - (-\frac{7}{3})] = (\frac{7}{3}, \frac{7}{3}, \frac{7}{3})$, and these are proportional to the direction numbers of the line of action of $\mathbf{F}_r$, which are $(1/\sqrt{3}, 1/\sqrt{3}, 1/\sqrt{3})$.

PROBLEMS ► Section 3.8

3.90 Five members of a truss are exerting the indicated loads on the pin at point O as shown in Figure P3.90. Find the single-force resultant of the five forces.

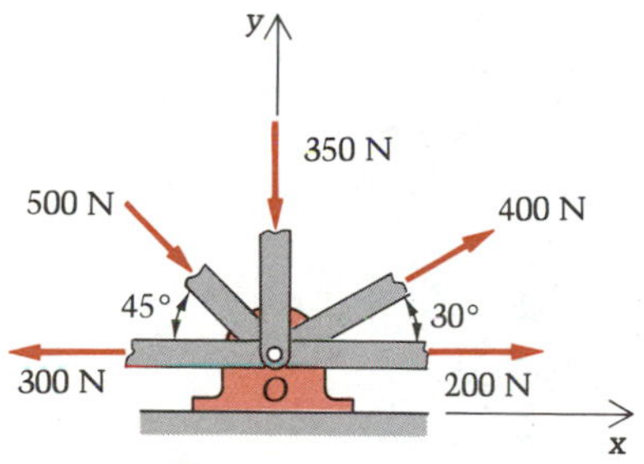

Figure P3.90

Find the force-alone resultants of the concurrent force systems in Figures P3.91–P3.96.

3.91

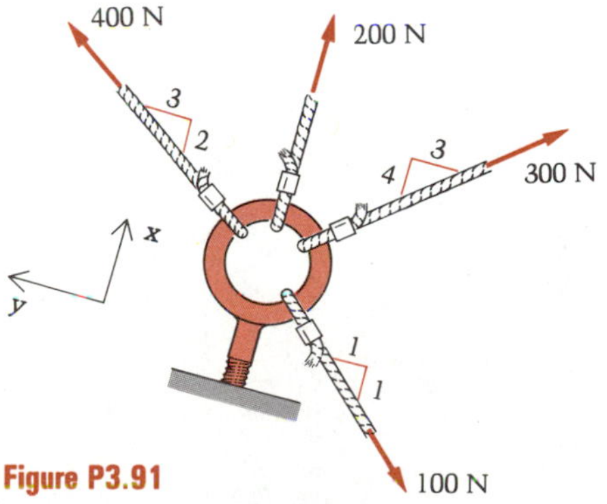

Figure P3.91

3.92

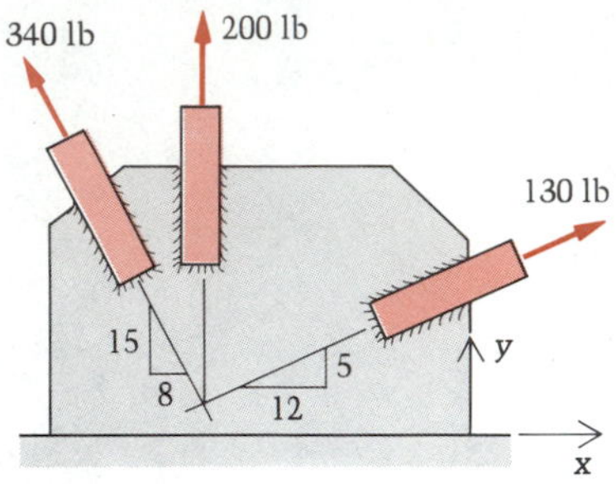

Figure P3.92

3.93

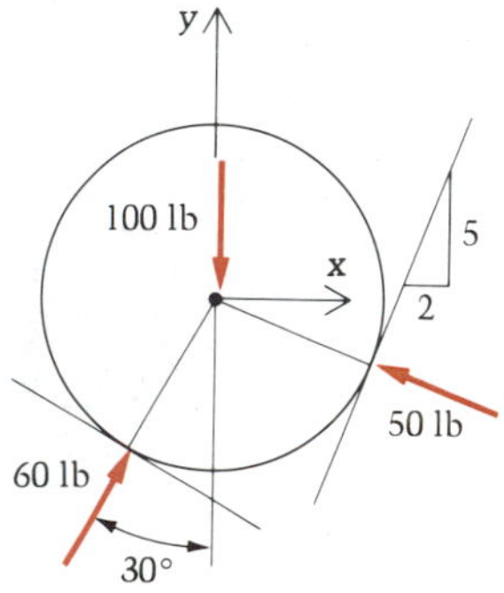

Figure P3.93

3.94

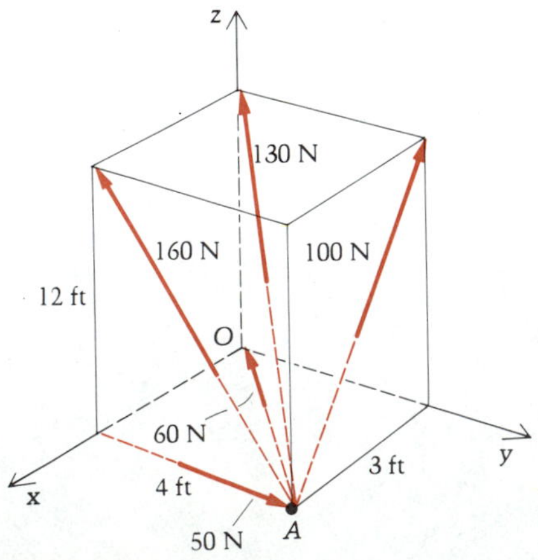

Figure P3.94

3.95

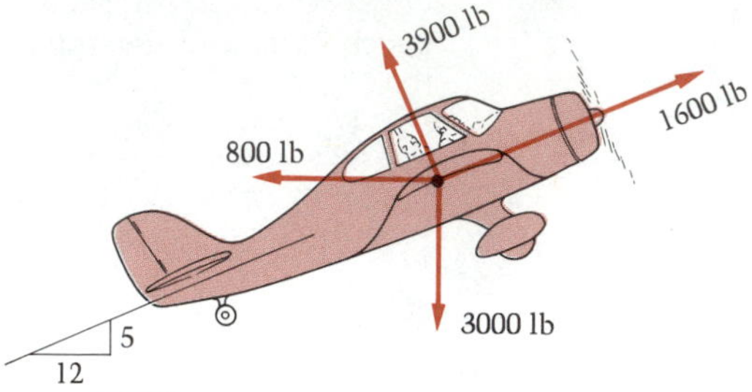

Figure P3.95

3.96

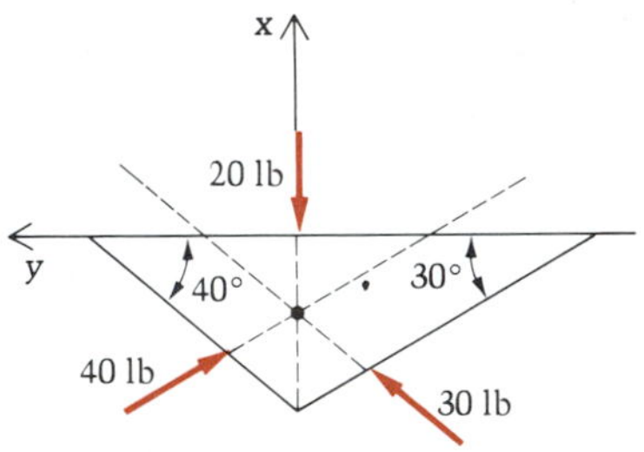

Figure P3.96

3.97 Find the force-alone resultant of the five forces shown in Figure P3.97. Locate the intersection of the line of action of the resultant with line BC.

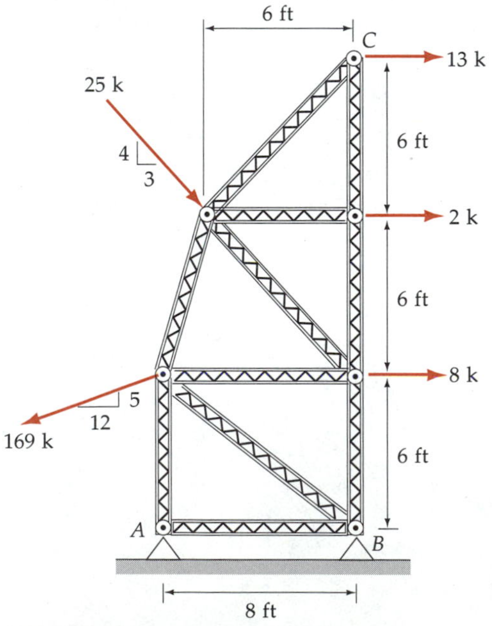

Figure P3.97

Find the simplest resultant for each of the coplanar force systems shown in Figures P3.98–P3.103. Give the equation of the line of action in each case, and determine where it crosses the x axis.

3.98

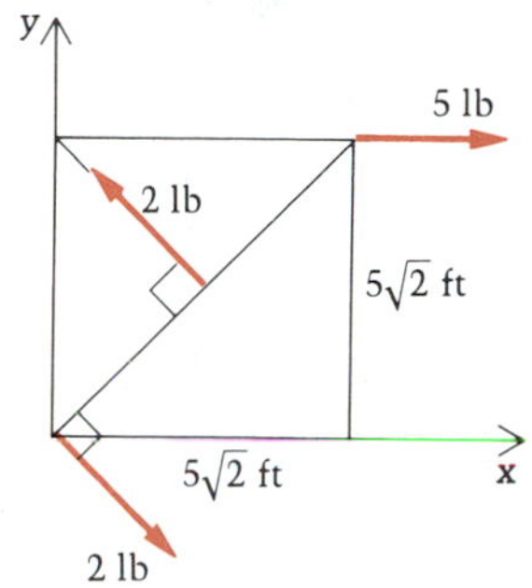

Figure P3.98

3.99

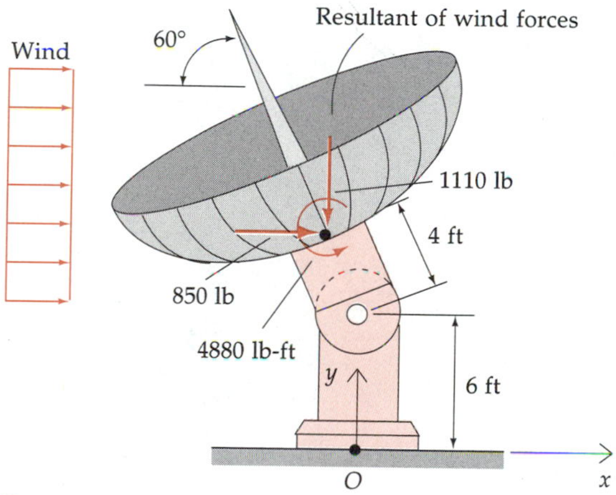

Figure P3.99

3.100

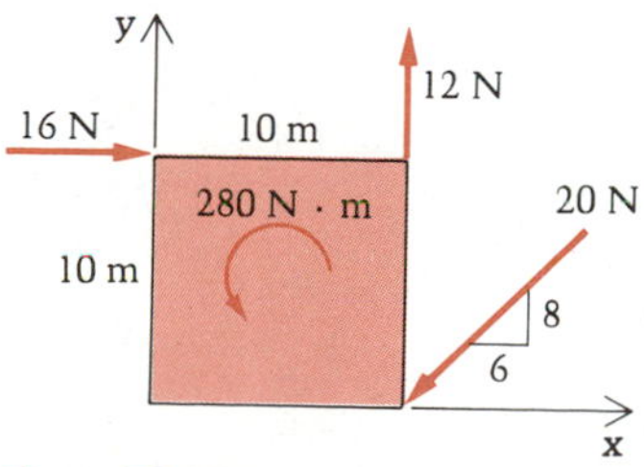

Figure P3.100

3.101

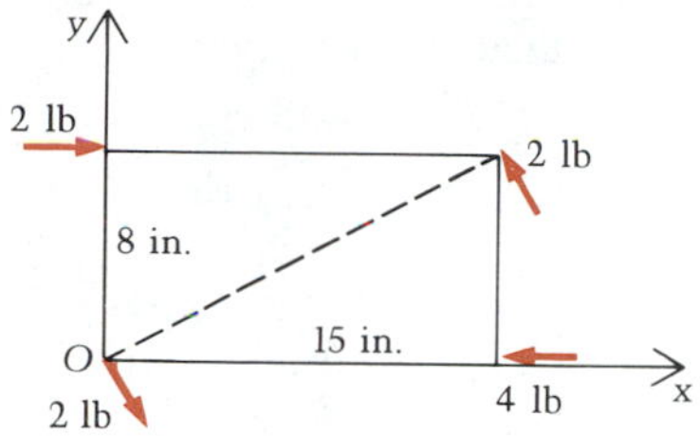

Figure P3.101

3.102

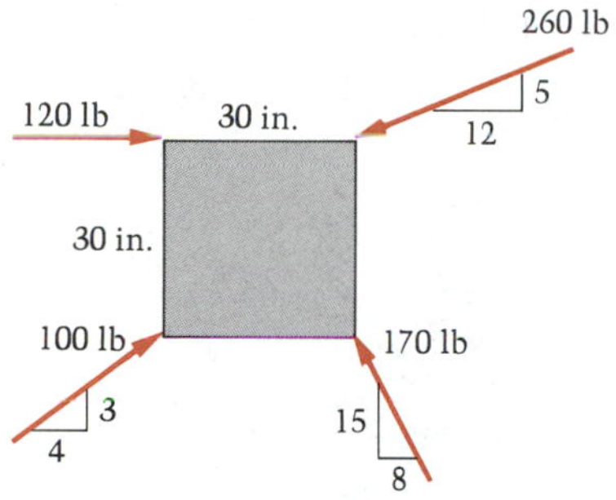

Figure P3.102

3.103

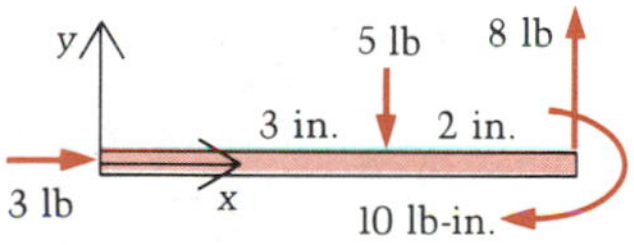

Figure P3.103

3.104 The forces have vertical lines of action and act through a horizontal unit grid as shown in Figure P3.104. Find and locate the single force that is equipollent to the given force system.

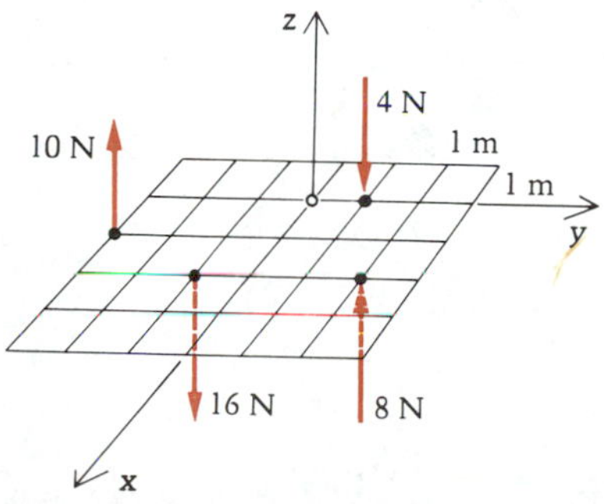

Figure P3.104

Find the simplest resultant for each of the parallel force systems shown in Figures P3.105–P3.109. Locate the line of action of each in the coordinate system given.

3.105

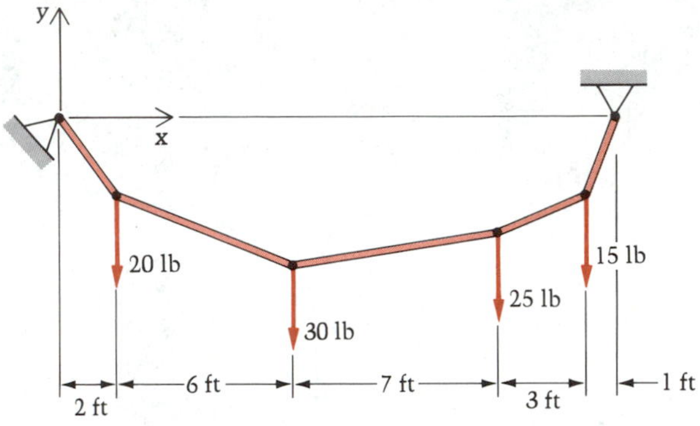

Figure P3.105

3.106

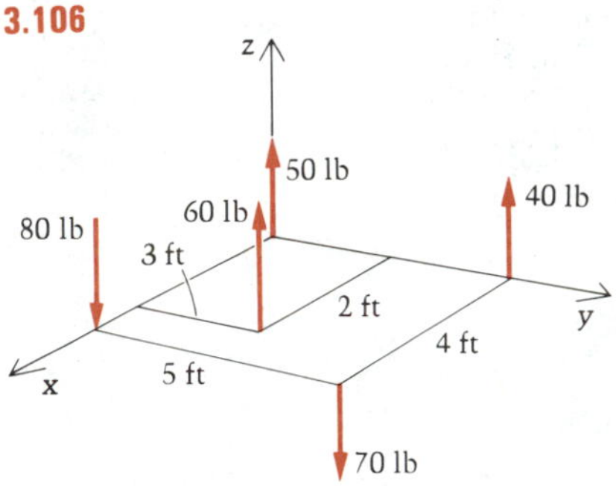

Figure P3.106

3.107

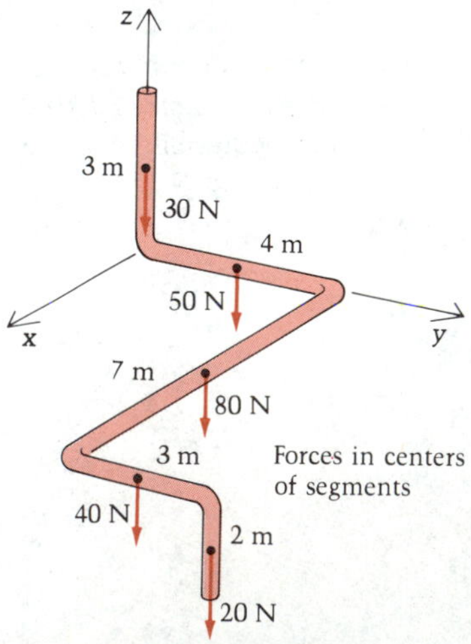

Figure P3.107

3.108

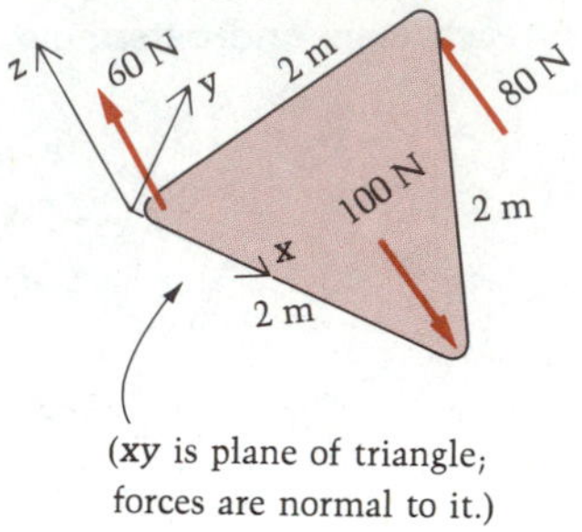

Figure P3.108

3.109

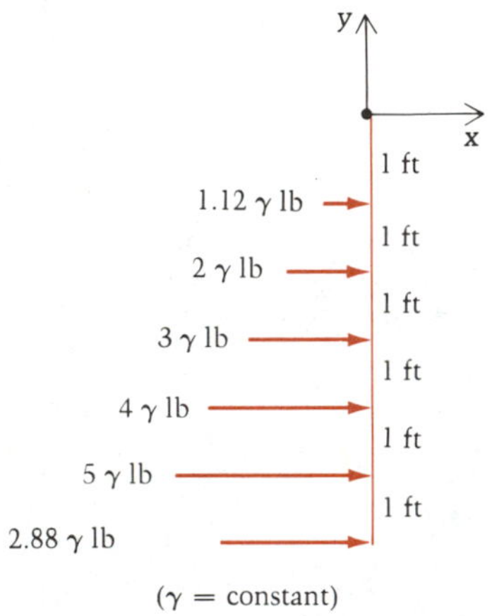

Figure P3.109

3.110 Replace the force system shown in Figure P3.110 by:

a. A force through O and a couple

b. A single force

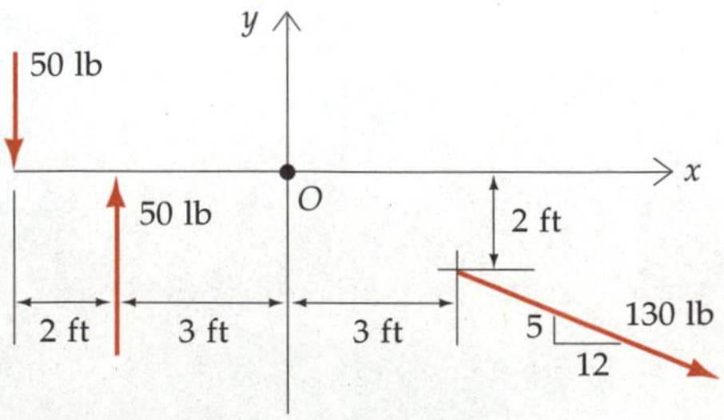

Figure P3.110

3.111 Is it possible to reduce the force system shown in Figure P3.111 to a single force? If so, what is the shortest distance from P to the line of action of this single force?

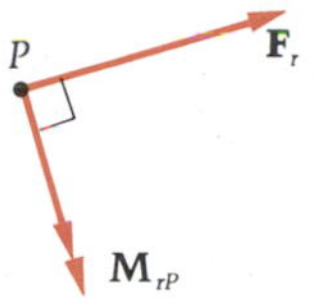

Figure P3.111

3.112 Suppose we take a complicated system of forces and couples and reduce them to an equipollent system, at P, of force $\mathbf{F}_r$ and moment $\mathbf{M}_{rP}$. When can the system be further reduced to (a) a single force? (b) a single couple? Hint: Think about the figure of the preceding problem.

3.113 Replace the two forces acting on the bar in Figure P3.113 by a single force. Give its magnitude, direction and line of action.

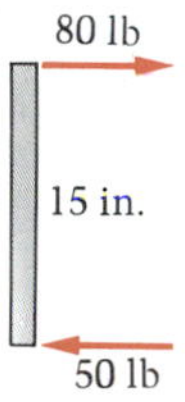

Figure P3.113

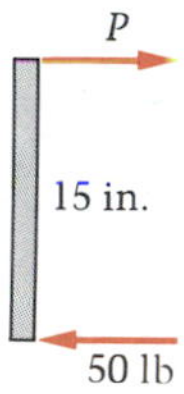

Figure P3.114

3.114 Find the force P so that the force-alone resultant of the two forces acts through the center of the bar in Figure P3.114.

3.115 With reference to Figure P3.115, find the following:

a. The force-alone resultant of the lift and drag forces and moment $\mathbf{M}_C$, which represent the resultant at C of the wind forces on the parabolic antenna dish.

b. The point where the resultant from part (a) crosses the line ℓ in the figure. (This line is the intersection of the rim plane and the xy plane.)

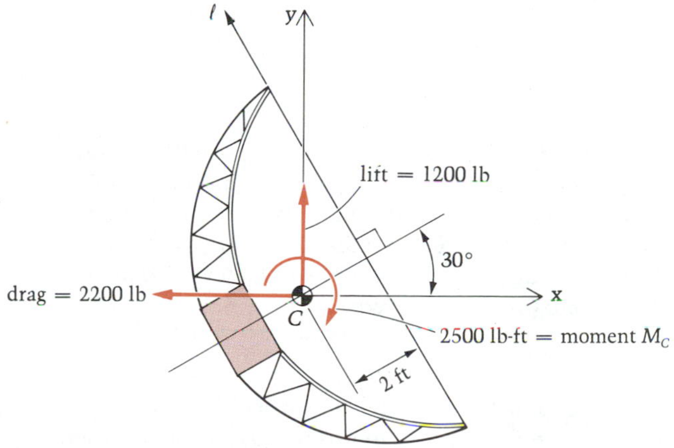

Figure P3.115

3.116 Figure P3.116 shows a force-couple system. The magnitudes are:

$$|\mathbf{F}_1| = 3\sqrt{13}\text{ lb}$$
$$|\mathbf{F}_2| = 2\sqrt{40}\text{ lb}$$
$$|\mathbf{C}_1| = 21\text{ lb-ft}$$
$$|\mathbf{C}_2| = 9\text{ lb-ft}$$

Give an equipollent system consisting of a single force and couple through point A (2, 0, 0). Can your system be further reduced to a single force or to a single couple?

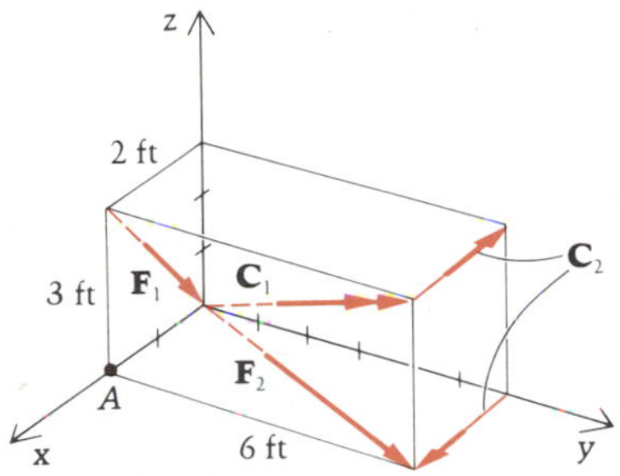

Figure P3.116

3.117 Replace the force system of Figure P3.117 by an equipollent system consisting of (a) a force through O and a couple; (b) a single force.

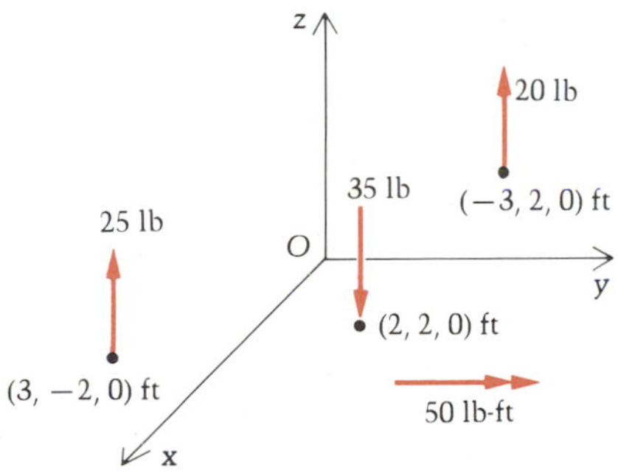

Figure P3.117

3.118 Show that if the elements of a force-couple resultant at one point are perpendicular, then the elements at any other point are also perpendicular.

3.119 Three forces of magnitude P, $2P$, and $3P$ act on a block as shown in Figure P3.119.

a. Find a equipollent system consisting of a force and a couple at point A.
b. Give the relation between a, b, and c so that the system may be reduced to a single force.
c. Can the system be reduced to a single couple? Why or why not?

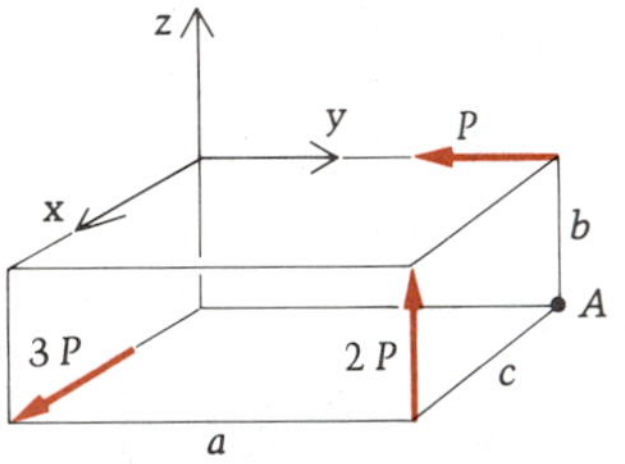

Figure P3.119

3.120 If possible, replace the system of forces in Figure P3.120 by a single equipollent force. Otherwise replace the system by a force through O and a couple.

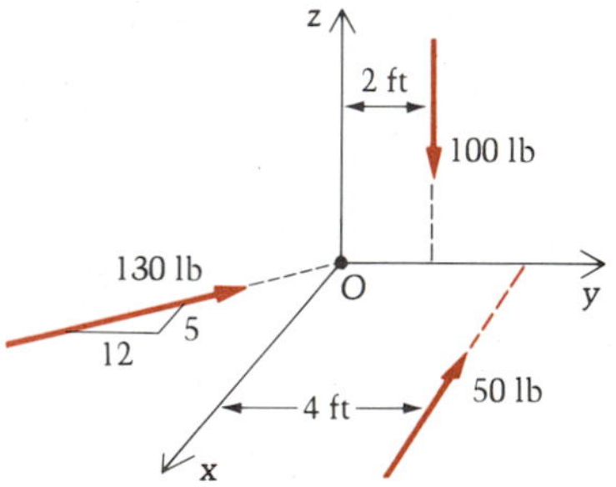

Figure P3.120

3.121 With reference to Figure P3.121, and in terms of ℓ, find the following:

a. The force and couple at the origin O that are equipollent to the three forces and three couples shown acting on the bent bar.
b. The value of ℓ for which the system can be further reduced to a single force.

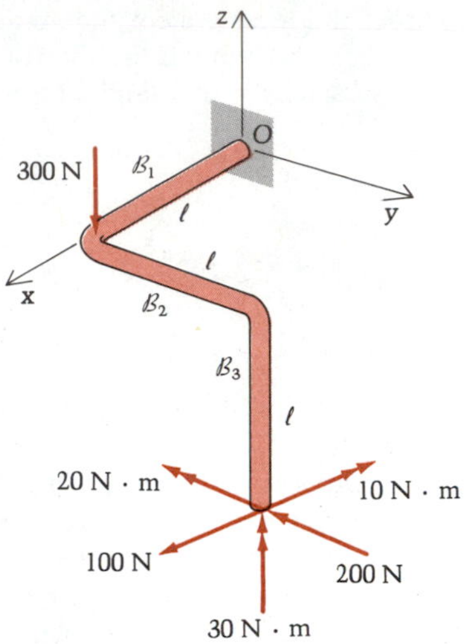

Figure P3.121

3.122 In Figure P3.122 a system of four forces and a couple is shown. Replace this system by an equipollent system consisting of a single couple and a single force whose line of action passes through point A, located at coordinates (2, 6, 0). Further reduce this new system to a single force, if possible.

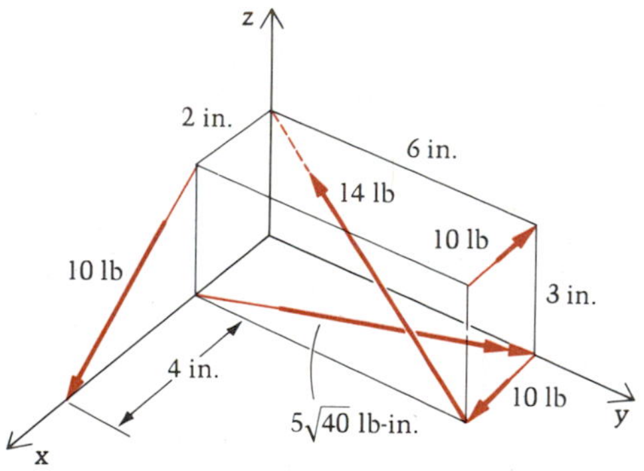

Figure P3.122

* **3.123** A circular plate of radius R supports three vertical loads as shown in Figure P3.123 on the next page. Determine the magnitude and point of application of the smallest additional vertical force that must be applied onto the surface of the plate if the four loads are to be equipollent to: (a) zero (that is, to a system with $\mathbf{F}_r = \mathbf{0}$ and $\mathbf{M}_{rO} = \mathbf{0}$); (b) a force through the center of the plate. What is this force?

3.124 Equation (3.23) in the text must obviously be rederived if $F_{rx} = 0$. Obtain the correct equation of the line for this special case.

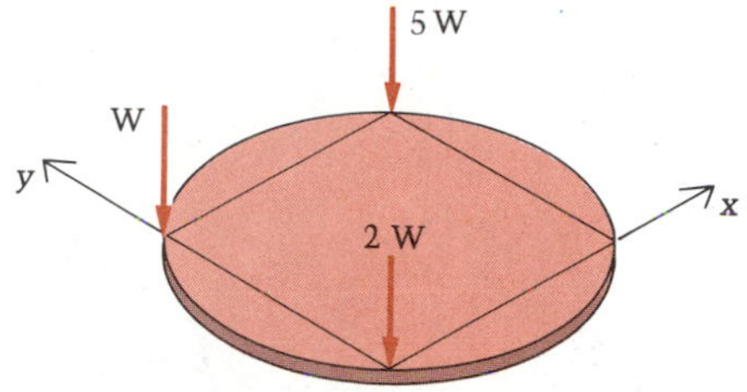

Figure P3.123

3.125 Return to Problem 3.74 and find the screwdriver equipollent to the forces and couples shown again in Figure P3.125. At what point does it pierce the ground (xy plane)?

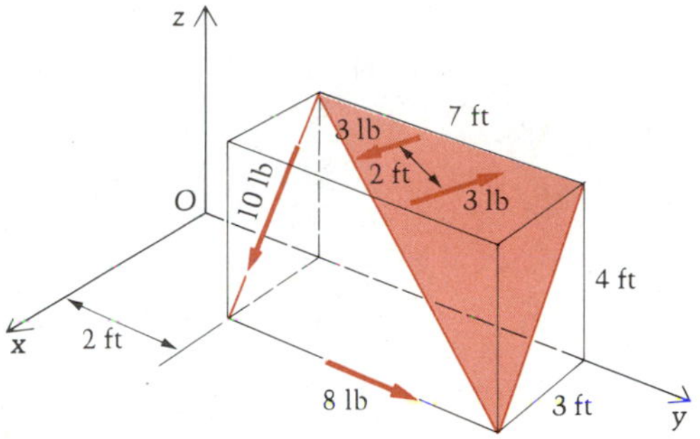

Figure P3.125

3.126 Return to Example 3.18 and find the screwdriver equipollent for each of the six force systems. Why can none of them be a single force through P?

3.127 Find the force and couple that must be added at point A to the force system shown at the left in Figure P3.127 so that it will be equipollent to the force system at the right. What is the screwdriver equipollent to the original system on the left?

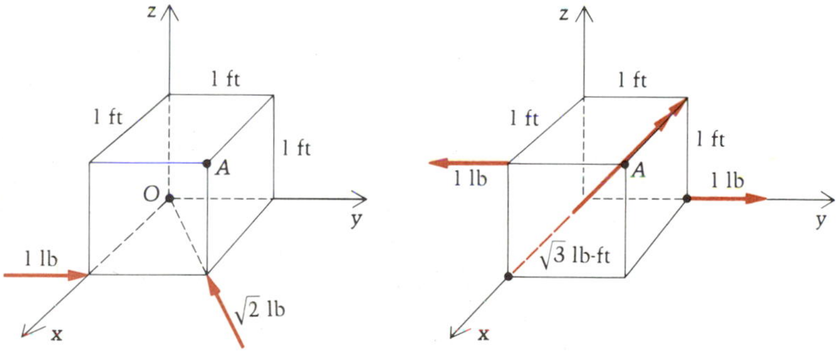

Figure P3.127

In Problems 3.128–3.139, the procedure outlined in the text is to be followed in establishing the "equipollent screwdriver" for the given system — that is, the resultant consisting of a collinear force and couple. Give, in addition to the force and couple, a complete description of the line of action of the screwdriver.

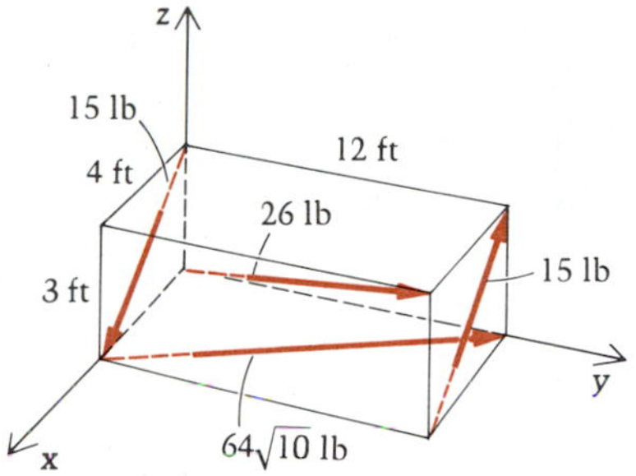

Figure P3.128

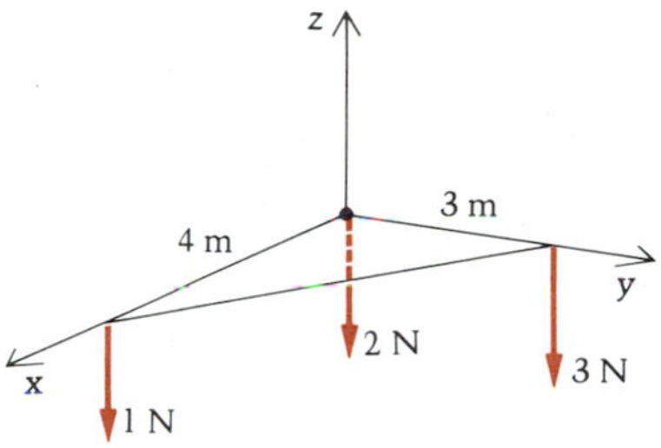

Figure P3.129

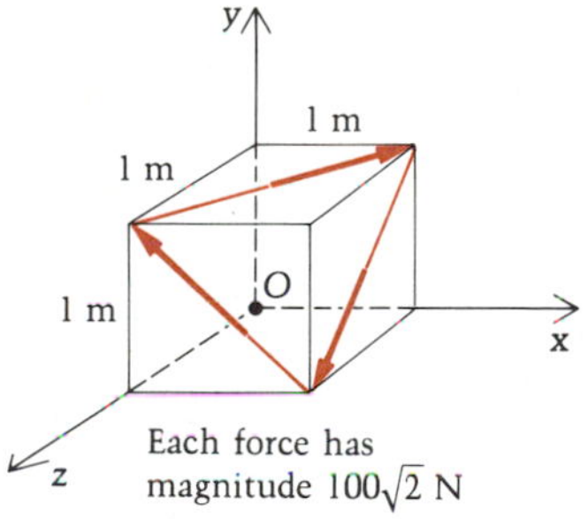

Figure P3.130

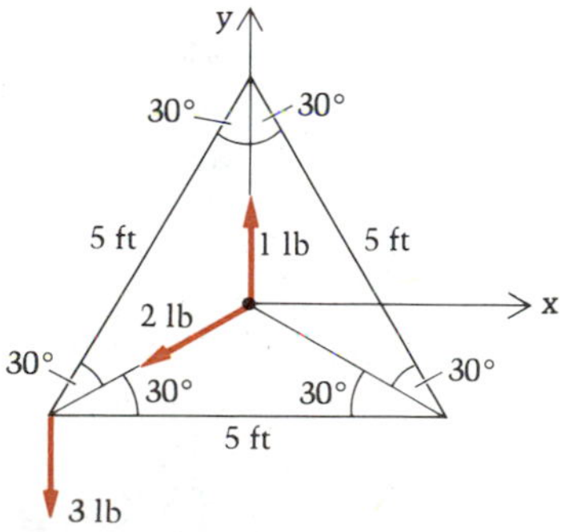

Figure P3.131

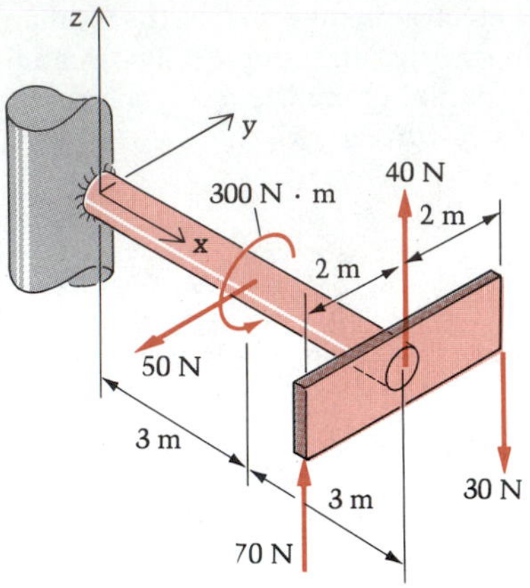

Figure P3.132

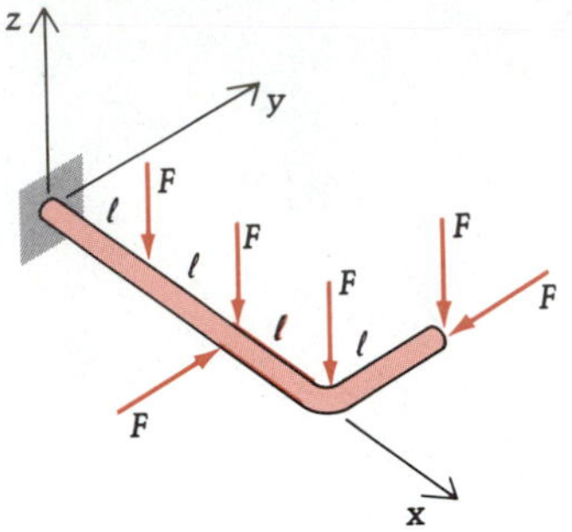

Figure P3.136

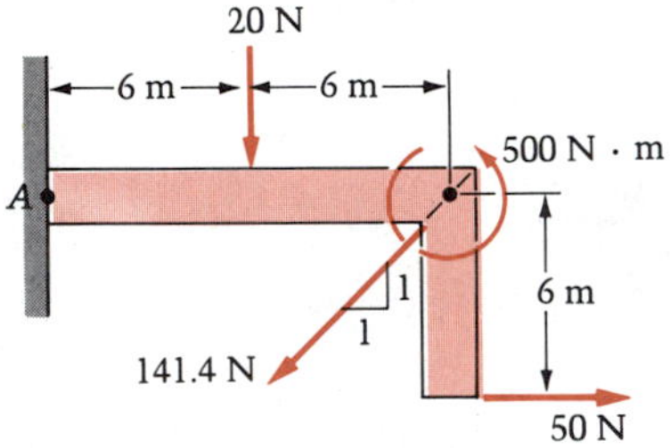

Figure P3.137

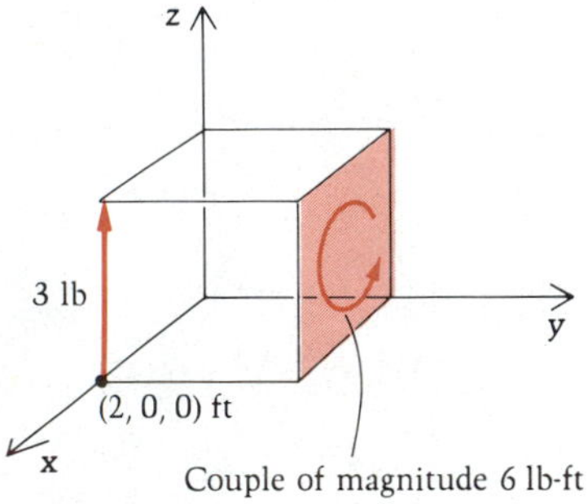

Figure P3.133

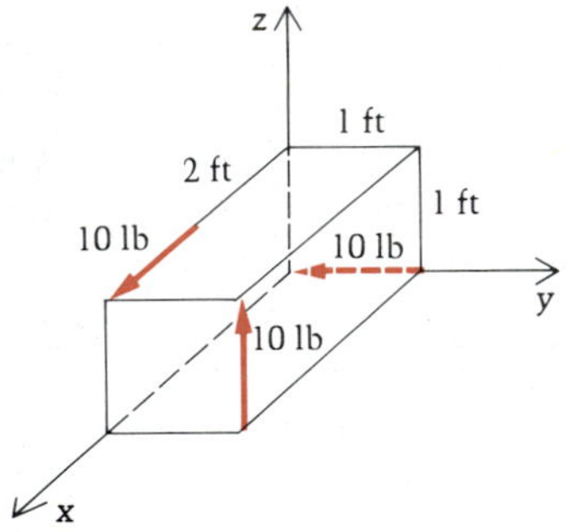

Figure P3.138

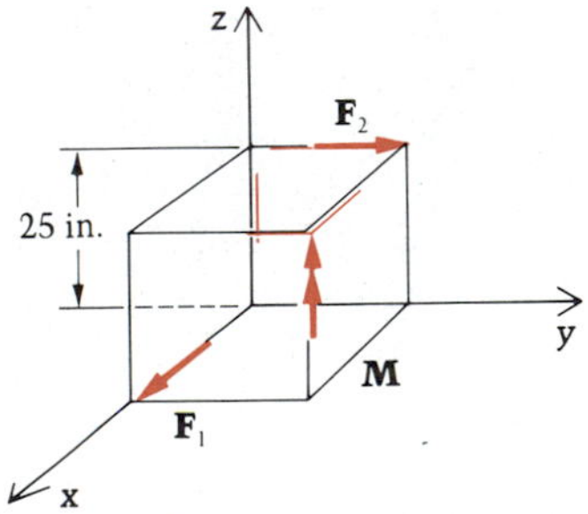

Figure P3.134 $F_1 = 3$ lb, $F_2 = 4$ lb, and $M = 50$ lb-in.

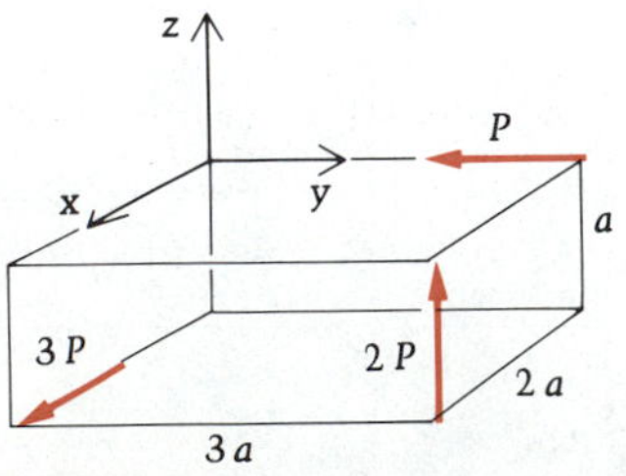

Figure P3.135

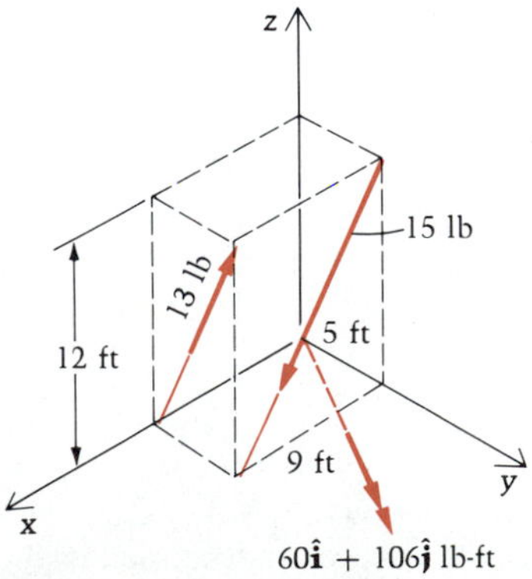

Figure P3.139

3.140 Even though the force system of Examples 3.25 and 3.31 was not concurrent, coplanar, or parallel, there *was* a force-alone resultant. From the force and couple resultant found in Example 3.25 at the point of intersection A of the two 10-N forces, note that $\mathbf{F}_r \cdot \mathbf{M}_{rA} = 0$. Now state why *any* change in the magnitude of just *one* of the two forces would make a single-force resultant impossible.

3.141 Explain the general process of going from the force and couple resultant of Figure P3.141(a) to the equipollent screwdriver of Figure P3.141(d). Tell what was done in each step.

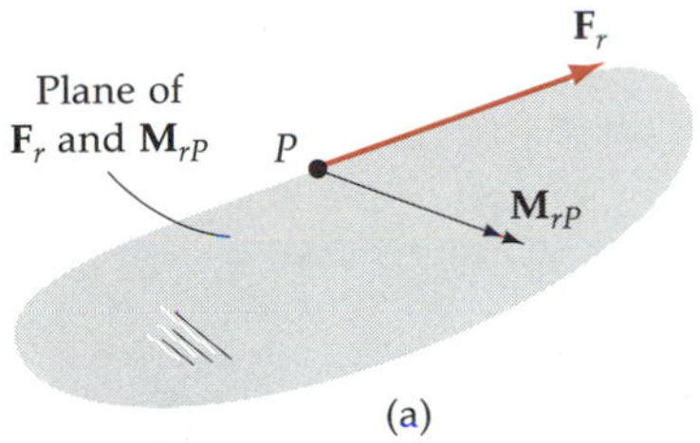

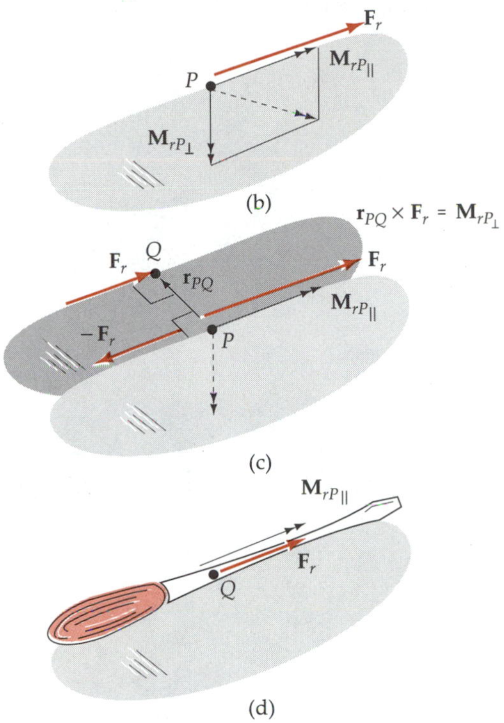

Figure P3.141

3.9 Distributed Force Systems

Probably the most important reason for studying the resultants of force systems is to have at our fingertips the resultants of frequently occurring distributed force systems. In engineering mechanics we usually are dealing with bodies on a large scale (macroscopic), where material is perceived (or modeled) to be continuously distributed in space. This is in contrast to a microscopic view where we might be distinguishing individual atoms and the spaces between them. We perceive mechanical actions to be exerted on bodies either by direct contact or by the action of a "field" such as gravity or electromagnetism. In the first instance (for example, pressing a finger against this book), it is natural to view the force exerted as the net effect of something distributed over the surface area of contact. In the second case (for example, the gravitational force exerted by the earth on the book), it is natural to view the force as the net effect of "weights" of individual particles, or elements of mass, which are distributed through the volume of the body. Thus, the mechanical actions that naturally arise in engineering mechanics may be classified as either **surface forces** or **body forces.**

Forces Distributed Along a Straight Line

We begin our calculations of resultants of distributed force systems by considering the simplest case, in which the distribution is over a *line.* This, of course, doesn't fit the classifications of the preceding paragraph, but it commonly arises as a system that is equipollent to one of those classifications. In particular we consider the frequently occurring case, illustrated in Figure 3.36, in which the force system is distributed over a straight line and the mechanical actions are all parallel to one another and perpendicular to the line.

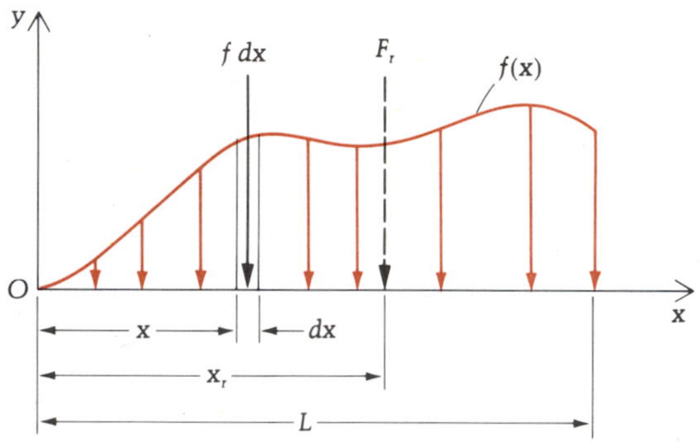

Figure 3.36

Letting $\mathbf{f} = -f\hat{\mathbf{j}}$ be the distributed force intensity — that is, force per unit of length along the line — then an elemental, or infinitesimal, force in the system at location x is $(-f\,dx)\hat{\mathbf{j}}$. To find the resultant force $\mathbf{F}_r$ we, as always, must simply add up all the forces in the system. Here the process of summation becomes integration so that

$$\mathbf{F}_r = \int_0^L \mathbf{f}\,dx = \int_0^L (-f\hat{\mathbf{j}})\,dx$$

$$= -\left(\int_0^L f\,dx\right)\hat{\mathbf{j}}$$

where the unit vector $-\hat{\mathbf{j}}$ is a constant, both in magnitude *and* direction. The couple part of the resultant at O is obtained by adding up (integrating) the moments of the elemental forces in the distributed system so that

$$\mathbf{M}_{rO} = \int_0^L x\hat{\mathbf{i}} \times (-f\hat{\mathbf{j}}\,dx) = -\hat{\mathbf{k}} \int_0^L xf\,dx$$

Thus if $\int_0^L f\,dx \neq 0$, there is a force-alone resultant (as we may have anticipated since our distributed system is both parallel *and* coplanar). Letting x_r locate the line of action of this force-alone resultant, we can find x_r by ensuring that the moment condition for equipollence is satisfied:

$$x_r\hat{\mathbf{i}} \times \mathbf{F}_r = \mathbf{M}_{rO}$$

or

$$x_r \int_0^L f\,dx = \int_0^L xf\,dx$$

or

$$x_r = \frac{\int_0^L xf\,dx}{\int_0^L f\,dx} \tag{3.25}$$

It is important to observe the similarity of Equation (3.25) to the second of Equations (3.24), where parallel discrete forces were being studied. It is also of interest to observe that $\int_0^L f\,dx$ may be interpreted as the "area" under the force intensity curve or loading curve, when $f(x)$ is graphed.

We shall see in Chapter 7 that x_r locates the x coordinate of the "centroid" of the area beneath the loading curve of Figure 3.31. The denominator of Equation (3.25) measures the "total" (or magnitude of the resultant) of the distributed force.

In summary, then, the force-alone resultant of a parallel distributed line loading has (a) a magnitude and sense given by the signed area beneath the loading curve, (b) a direction parallel to that of the parallel distributed forces, and (c) a line of action given by Equation (3.25). We shall now consider a series of examples in which we compute resultants of distributed line loadings.

Example 3.33

Find the resultant of the uniformly distributed weight of the sacks of cement stacked on the dock. There are 12 piles as shown, with 8 bags per pile, and each 94-lb bag is 20 in. wide by 2.5 ft long as shown in Figures E3.33a, b.

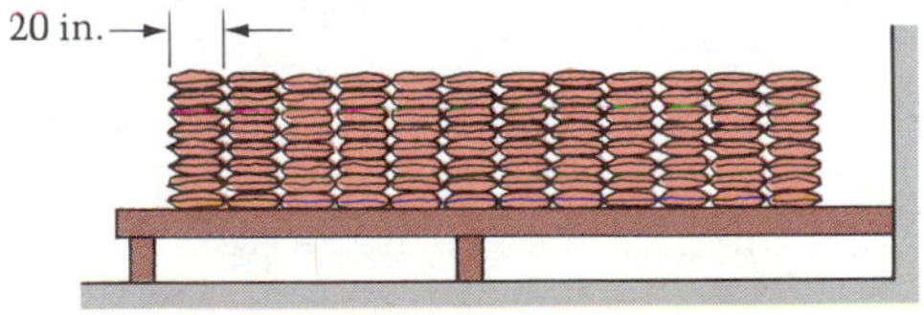

Figure 3.33a

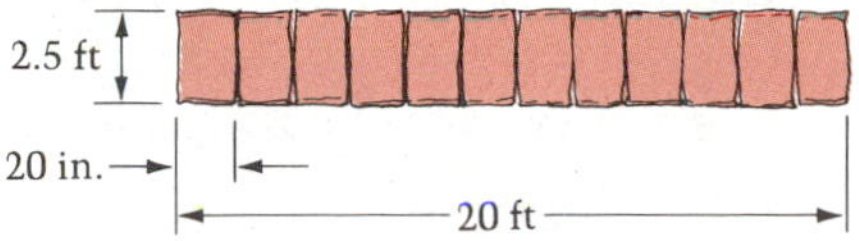

Figure 3.33b

Solution

The total weight of the cement is

$$94 \frac{\text{lb}}{\text{bag}} (8 \times 12) \text{ bags} = 9020 \text{ lb} \qquad \text{(to three digits)}$$

This of course is the magnitude of the resultant $\mathbf{F}_r$. By intuition, its line of action is downward, 10 feet from either end of the cement. But let us obtain the result by ensuring that the moments of the two systems shown in Figure E3.33c are the same about the left end A of the loading.

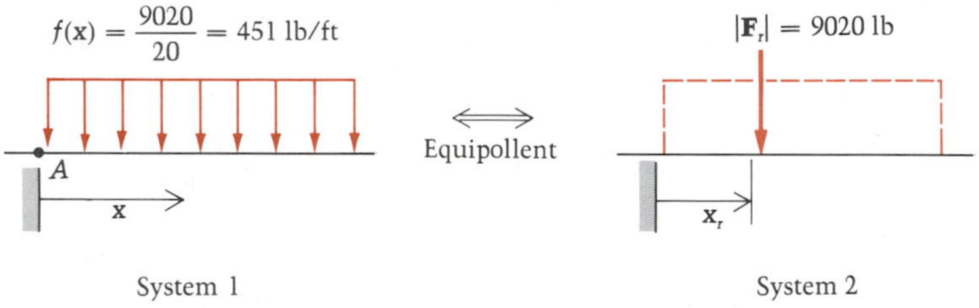

Figure 3.33c

$$(M_{rA})_1 = \int_0^L xf(x)\,dx = \int_0^{20} \frac{9020}{20} x\,dx = 451 \left.\frac{x^2}{2}\right|_0^{20} = 90{,}200 \text{ lb-ft}$$

This must equal the moment $(M_{rA})_2$ of the force-alone resultant:

$$9020\, x_r = 90{,}200$$

$$x_r = 10 \text{ ft (as expected)}$$

Note that in the solution we did not need the 2.5-ft dimension. Since the loading distribution does not vary in the direction perpendicular to the page, we have idealized the weight as a line load acting in the central plane of the bags. More will be said about this later in the section.

EXAMPLE 3.34

A concrete ramp has the weight distribution shown in Figure E3.34a, where p is the weight per unit length at the right end of the ramp. Find the resultant of the triangularly distributed weight.

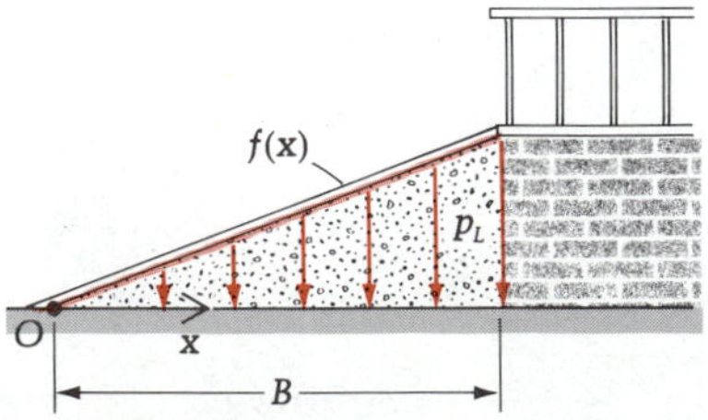

Figure 3.34a

Solution

The distributed loading has the equation (see Figure E3.34b):

$$f(x) = \frac{p_L}{B} x$$

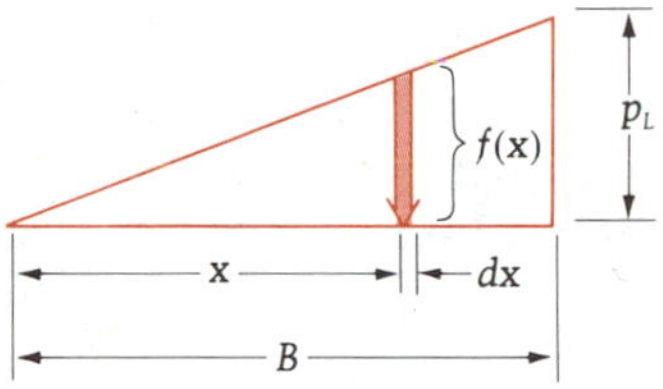

Figure 3.34b

The differential force dF at x is given as suggested in the figure by

$$df = f(x)\,dx = \frac{p_L}{B} x\,dx$$

so that the resultant $\mathbf{F}_r$ has magnitude

$$\int_0^B \frac{p_L}{B} x\,dx = \frac{p_L B}{2}$$

This is 1/2 the base times the height of the triangle, or the area beneath the loading curve, as we have previously shown. We next locate the line of action of the resultant by finding the value of x_r in Figure E3.34c for which the moments of the two systems are the same about some common point, say, O:

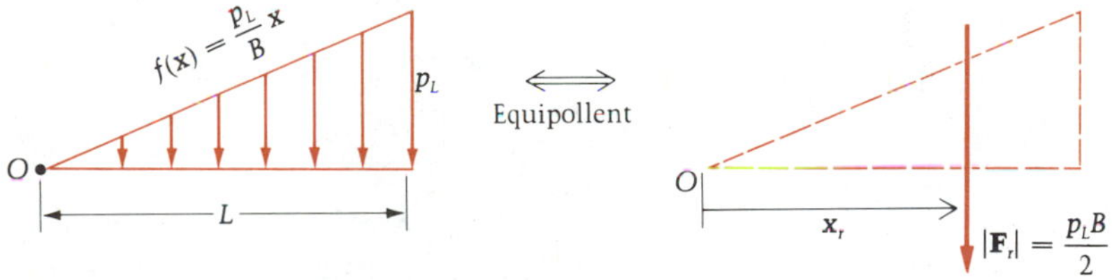

Figure 3.34c

We obtain

$$|\mathbf{F}_r|\,x_r = \int_0^B x f(x)\,dx = \int_0^B x \frac{p_L}{B} x\,dx$$

or

$$\frac{p_L B}{2} x_r = \frac{p_L}{B} \frac{x^3}{3}\bigg|_0^B = \frac{p_L B^2}{3}$$

$$x_r = \frac{2}{3} B$$

Therefore, whenever we encounter a triangularly distributed loading curve, as in the example to follow, we will know the magnitude of its resultant is its area, and that its line of action is 2/3 the distance from the vertex to the opposite side.

Question 3.31 How would you handle a case in which some of the distributed loading was upward and some downward?

We now address the fact that loadings such as we have just seen in Examples 3.33 and 3.34 are actually (that is, physically) distributed over *surfaces.* Force systems distributed over surfaces occur with great frequency in engineering mechanics. We shall treat here the special case in which each elemental force is perpendicular to the surface — that is, simple pressure. To begin, let us suppose that the surface is plane (flat) and rectangular in shape and that the pressure $p(x)$ varies only with the coordinate x, measured along one edge of the surface (Figure 3.37(a)).

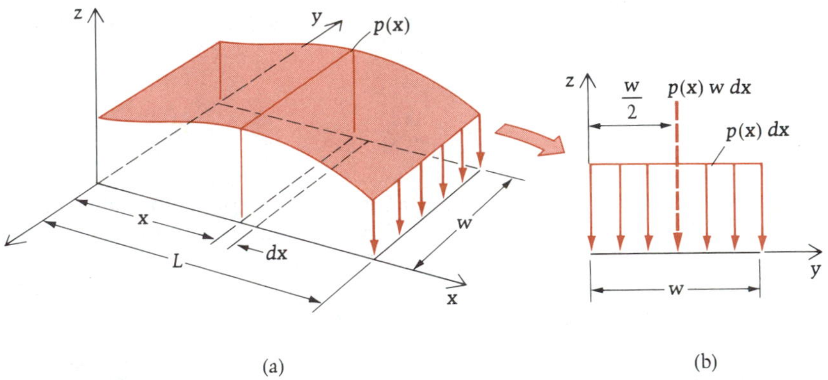

Figure 3.37

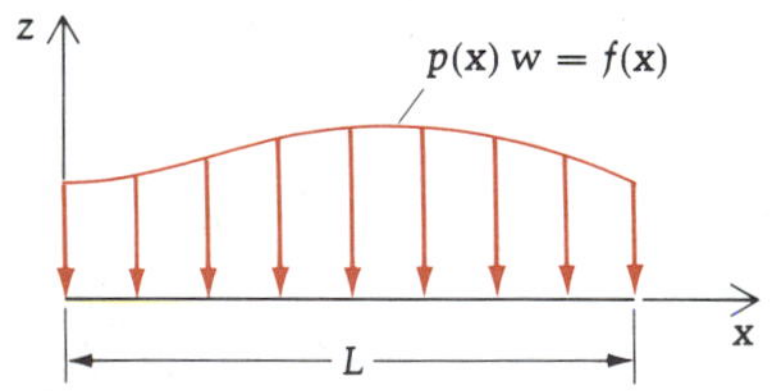

Figure 3.38

Along a strip of infinitesimal width dx and length w we have a set of infinitesimal forces $p(x)\,dx\,dy$, which can be viewed as a line-distributed force system, the intensity at each point being $p(x)\,dx$. The resultant of the strip of infinitesimal width is then $p(x)w\,dx$ [see Figure 3.37(b)] with line of action at $y = w/2$. Since this is true in *each* such strip, the pressure loading is equipollent to a line-distributed system along the line $y = (w/2)$ with $p(x)w = f(x)$ being the intensity as shown in Figure 3.38.* It is in this fashion that line-distributed force systems

Answer 3.31 In Equation (3.25), if f is negative over a part (or parts) of the interval $(0, L)$, it simply contributes a negative result for that portion of the integrals. Note that if $\int_0^L f\,dx = 0$, then the resultant will be (at most) simply a couple.

* The same type of equipollent line loading along $y = (w/2)$ occurs when $p = p(x, y)$ if the pressure is symmetric about $y = (w/2)$ — that is, if $p(x, y)$ is an even function of $(y - w/2)$. In that more general case, $f(x) = \int_0^w p(x, y)\,dy$.

arise in mechanics. When they are encountered in examples, the reader should always bear in mind that they are the result of the process we have just been through. This was the case in the two previous examples and will be again in the one to follow, in which we shall discuss its resolution into a line load.

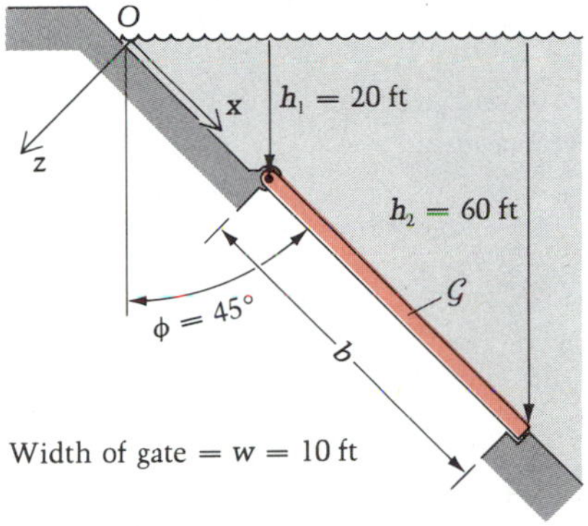

Figure E3.35a

Example 3.35

Water is held back by the submerged rectangular gate $\mathcal{G}$ in Figure E3.35a. Find the value of the single force (and its location) that is equipollent to the fluid pressure forces acting on $\mathcal{G}$.

Solution

As the reader may recall from previous studies, the pressure p in a fluid at rest:

1. is equal in all directions at a point (Pascal's Law)
2. is constant through the fluid in each horizontal plane
3. causes a force that is normal to every differential area of surface on which it acts
4. is equal to ρgh, where ρ is the mass density of the fluid (taken to be constant here), and h is the depth below the free surface

For a flat gate, the distributed loading caused by the water pressure therefore forms a parallel force system. Multiplying the pressure by the constant width w of the gate, we obtain the distributed line load as discussed in the preceding text and as illustrated in Figure E3.35b.

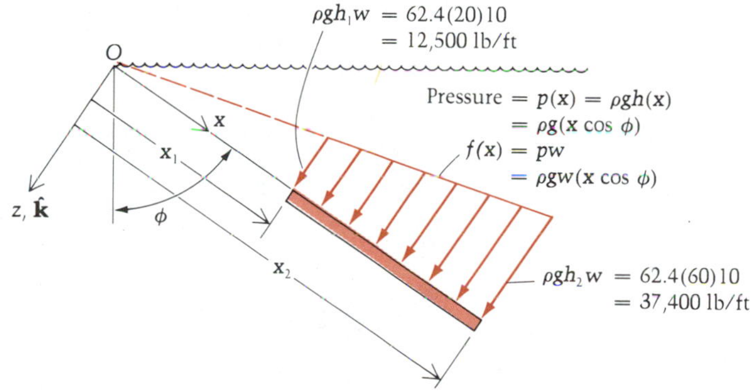

Figure E3.35b

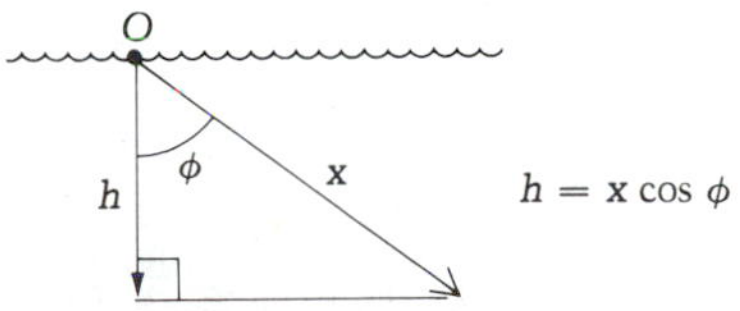

Figure E3.35c

In Figure E3.35b we have used the fact that h and the "slanted coordinate" x are related quite simply, as shown in Figure E3.35c. The single force that is equipollent to the system of parallel distributed forces caused by the water pressure is calculable from

$$F_r = \int f(x)\,dx = \int_{20\sqrt{2}}^{60\sqrt{2}} \underbrace{62.4(10)\frac{1}{\sqrt{2}}}_{\rho g w \cos\phi} x\,dx$$

$$= 441 \left.\frac{x^2}{2}\right|_{20\sqrt{2}}^{60\sqrt{2}}$$

$$= 1.41 \times 10^6 \text{ lb*}$$

The location of this resultant follows from equating the moment about O of the single-force resultant to that of the distributed loading:

$$\underbrace{F_r}_{1.41(10^6)\text{ lb}} \; x_r = \int x f(x)\,dx = \int_{20\sqrt{2}}^{60\sqrt{2}} 441x^2\,dx = 441 \left.\frac{x^3}{3}\right|_{20\sqrt{2}}^{60\sqrt{2}}$$

$$x_t = \frac{86.5 \times 10^6}{1.41 \times 10^6}$$

$$= 61.3 \text{ ft}$$

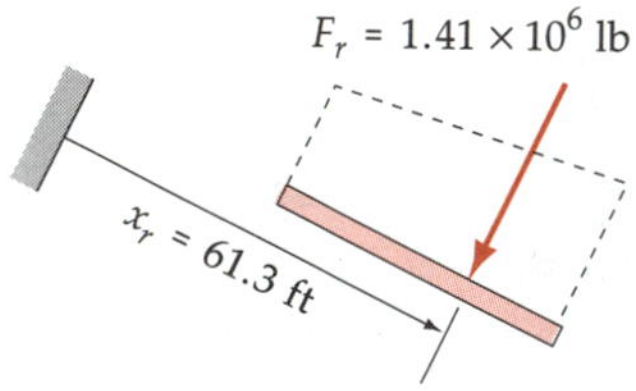

Figure E3.35d

The single-force resultant is shown in Figure E3.35d at the left.

Dividing a Distributed Force System into Composite Parts

We next illustrate an approach which can be used to avoid integrations whenever the loading can be divided into several individually familiar "composite parts":

EXAMPLE 3.36

Find the single-force resultant of the distributed loading acting on the beam shown in Figure E3.36a.

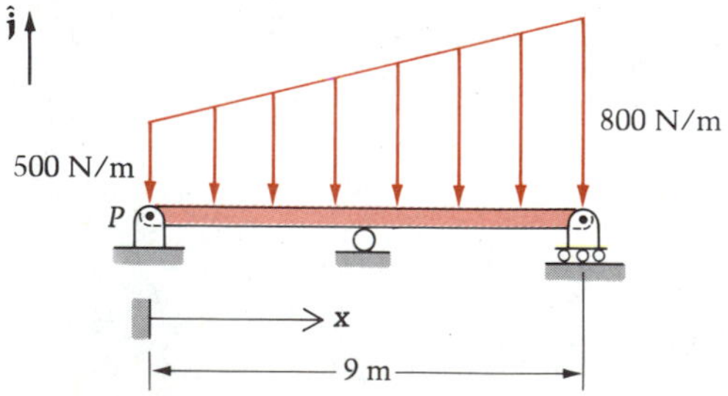

Figure E3.36a

* We shall show in Chapter 9 that this resultant equals the product of the pressure at the centroid of the gate and its area. Thus, as a check, $F_r = [(62.4)40][(60 - 20)\sqrt{2}(10)] = 1.41 \times 10^6$ lb.

Solution

We shall consider the loading as the sum of the two distributions illustrated in Figures E3.36b,c. Note that we are familiar with each of the loadings (1) and (2)

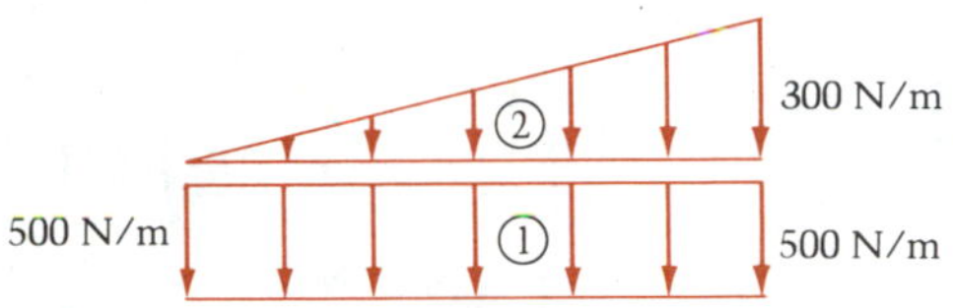

Figure E3.36b

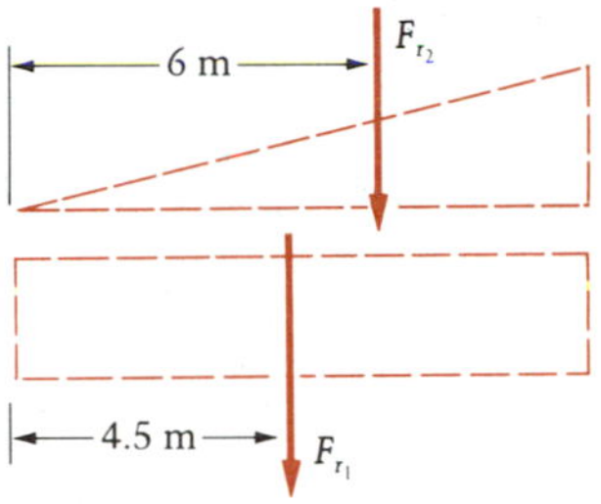

Figure E3.36c

from Examples (3.33) and (3.34). Therefore, the given loading is equipollent to the following pair of single forces:

$$F_{r2} = \tfrac{1}{2}(9)(300) = 1350 \text{ N} \qquad \text{(6m from left end of beam as shown)}$$

$$F_{r1} = (9)(500) = 4500 \text{ N} \qquad \text{(4.5m from left end of beam as shown)}$$

The final step is to find the single force $\mathbf{F}_r$ that is equipollent to this two-force system. Its value is obtained from the force condition:

$$\mathbf{F}_r = -1350\hat{\mathbf{j}} - 4500\hat{\mathbf{j}} = -5850\hat{\mathbf{j}} \text{ N}$$

And if the x coordinate of its vertical line of action is called x_r, we find the value of x_r from the moment condition

$$(5850)x_r = 4500(4.5) + 1350(6)$$

$$x_r = 4.85 \text{ m}$$

Therefore, our solution is shown in Figure E3.36d below.

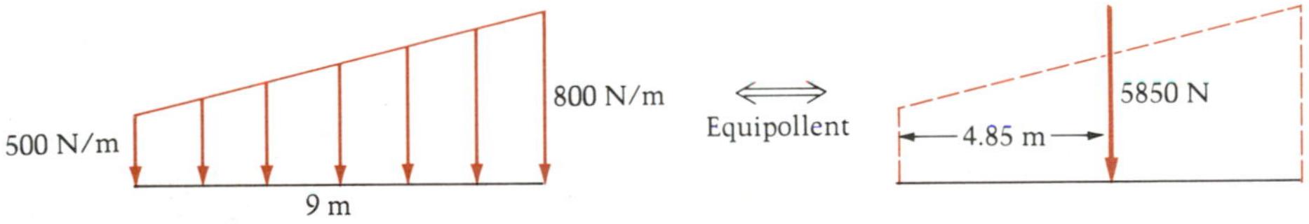

Figure E3.36d

We will see the method of composite parts again in Chapter 7 when we study a concept called centroids.

Let us now check the above results for $\mathbf{F}_r$ and its location by integrating the distributed load. Equating moments about the pin P we have

$$\underbrace{\left(\int_0^9 f(x)\,dx\right)}_{\substack{\text{magnitude}\\ \text{of force-alone}\\ \text{resultant}}} \quad \underbrace{(x_r)}_{\substack{\text{location}\\ \text{of force-alone}\\ \text{resultant}}} = \underbrace{\int_0^9 xf(x)\,dx}_{\substack{\text{moment of}\\ \text{distributed load}\\ \text{about } P}}$$

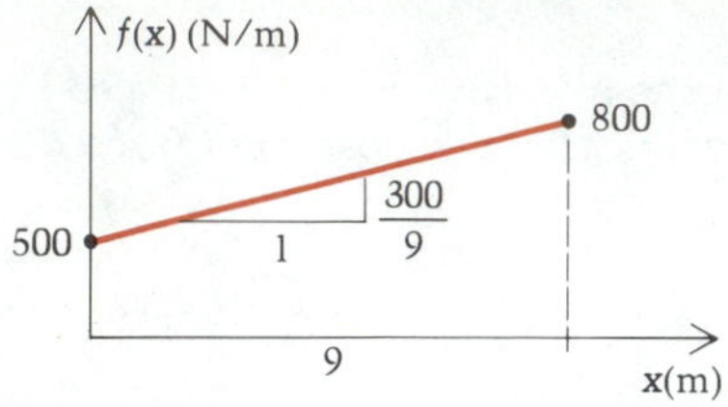

Figure 3.39

The loading curve in Figure 3.39 has the equation

$$f(x) = \frac{300}{9}x + 500$$

so that

$$\left[\int_0^9 \left(\frac{300x}{9} + 500\right) dx\right] x_r = \int_0^9 \left(\frac{300}{9}x^2 + 500x\right) dx$$

$$\underbrace{\left[\frac{300}{9}\frac{x^2}{2}\Big|_0^9 + 500x\Big|_0^9\right]}_{= |\mathbf{F}_r| = 5850 \text{ N}} x_r = \underbrace{\frac{300}{9}\frac{x^3}{3}\Big|_0^9 + 500\frac{x^2}{2}\Big|_0^9}_{28400 \text{ N} \cdot \text{m}}$$

or

$$x_r = 4.85 \text{ m} \qquad \text{(as before)}$$

EXAMPLE 3.37

Find the value of the single-force resultant (and its location) in Example 3.35 using the idea of composite parts.

Solution

The loading comprises the rectangular and triangular parts of Figure E3.37 below:

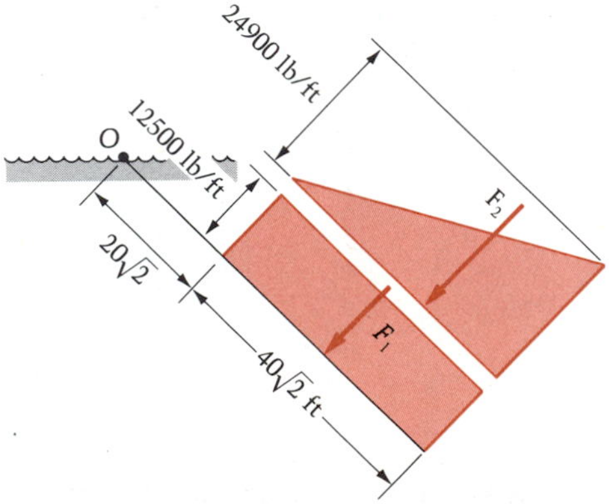

Figure E3.37

We have for the individual resultants:

$$F_1 = (40\sqrt{2})\ 12{,}500 = 707{,}000 \text{ lb}$$
$$F_2 = \tfrac{1}{2}(40\sqrt{2})\ 24{,}900 = 704{,}000 \text{ lb}$$

The single force that is equipollent to these two forces (and thus also to the original system of forces caused by the water) is obtained by the force condition for equipollence:

$$F_r = 707{,}000 + 704{,}000$$
$$= 1.41(10^6) \text{ lb} \qquad \text{(as before)}$$

The location of F_r is obtained by satisfying the moment condition for equipollence:

$$M_{rO} = F_r x_r = F_1\left(20\sqrt{2} + \frac{40\sqrt{2}}{2}\right) + F_2[20\sqrt{2} + \tfrac{2}{3}(40\sqrt{2})]$$
$$1.41(10^6)x_r = 40.0(10^6) + 46.5(10^6)$$
$$x_r = 61.3 \text{ ft} \qquad \text{(as before)}$$

As a practical matter, in a situation such as that of Example 3.36 or 3.37 we very seldom have an interest in determining a single-force resultant. What we really want to do is avoid, if possible, explicitly evaluating the integrals that arise when we evaluate the contributions of the distributed forces to the equilibrium equations for the beam. This was achieved in the two examples when we decomposed the distributed loading into two parts and were able to recognize the force-alone equipollent of each part. Our objective is then satisfied as suggested by Figure 3.40 below for Example 3.36. The next example will further illustrate this point.

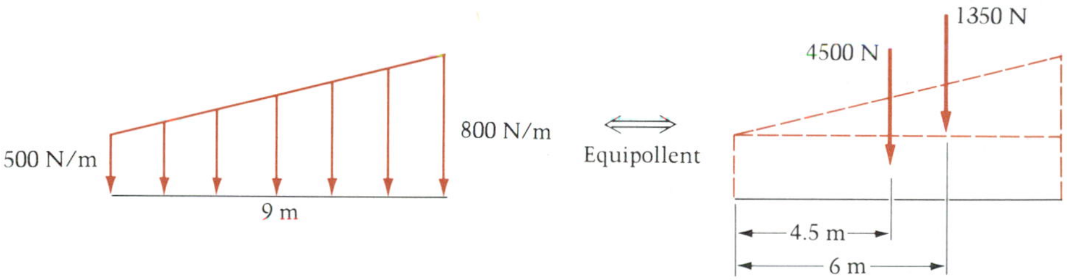

Figure 3.40

Example 3.38

Replace the distributed loading on the cantilever beam in Figure E3.38a by an equipollent set of forces.

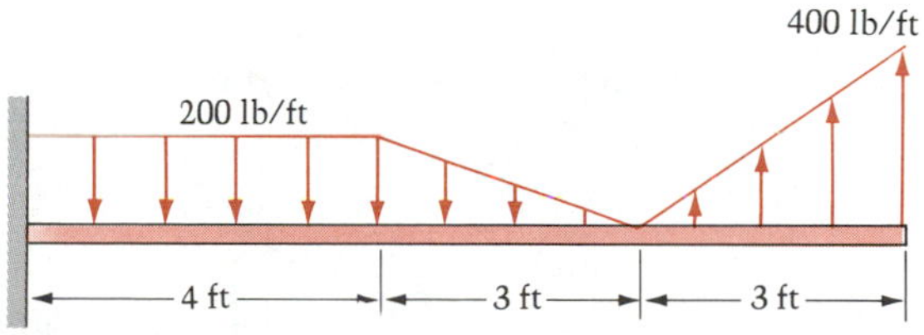

Figure E3.38a

Solution

By viewing the distributed loading as a uniform part plus two linearly varying parts, and using the results of Examples 3.33 and 3.34, we obtain the results shown in Figure E3.38b. These three forces will collectively make the same contributions to the equilibrium equations for the whole beam as will the original distributed loading.

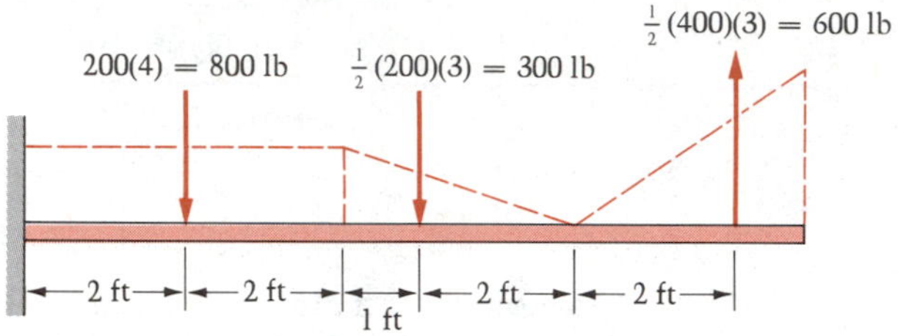

Figure E3.38b

Forces Distributed Over Surfaces

We now broaden the scope of our study by again considering pressure on a flat surface, S, but letting the surface boundary have any shape and allowing the pressure to vary arbitrarily with both of the coordinates x and y in the plane of the surface (Figure 3.41).

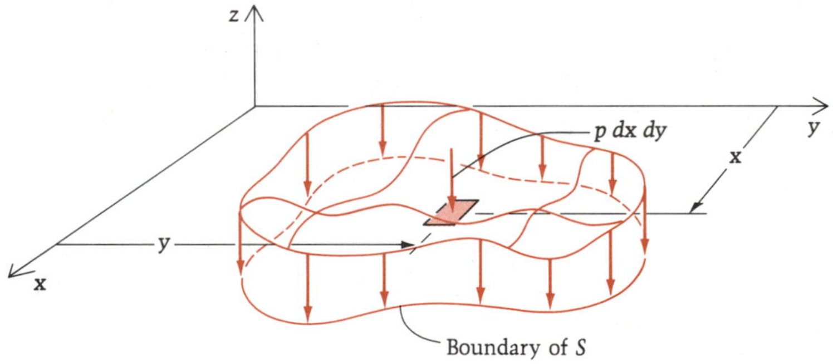

Figure 3.41

The resultant force $\mathbf{F}_r$ is given by

$$\mathbf{F}_r = (-\hat{\mathbf{k}}) \int_A p(x, y)\, dA = -P\hat{\mathbf{k}}$$

where A is the area of the surface S. The resultant couple at the origin is

$$\mathbf{M}_{rO} = \int_A (x\hat{\mathbf{i}} + y\hat{\mathbf{j}}) \times [p(x, y)\, dA(-\hat{\mathbf{k}})]$$

$$= -\hat{\mathbf{i}} \int_A yp(x, y)\, dA + \hat{\mathbf{j}} \int_A xp(x, y)\, dA$$

We pause at this point to note that sometimes there is negative pressure over a portion of the surface so that $P = \int\int p\,dx\,dy$ might vanish and *the* resultant is then the couple $\mathbf{M}_{rO}$. This occurs particularly in the mechanics of solids where we are concerned with an internal surface separating two portions of a body. In that application, "negative pressure" is called **tensile normal stress** and ordinary or positive pressure is called **compressive normal stress.** The resultant couple is usually then referred to as a **bending moment.**

When $P \neq 0$ there is, of course, a force-alone resultant, and if x_r and y_r locate its line of action,

$$x_r = \frac{\int xp\,dA}{\int p\,dA}$$

$$y_r = \frac{\int yp\,dA}{\int p\,dA}$$

In the special case of $p =$ constant,

$$x_r = \frac{\int x\,dA}{\int dA}$$

$$y_r = \frac{\int y\,dA}{\int dA} \tag{3.26}$$

where A is the area of surface S. We shall see in Chapter 7 that x_r and y_r locate the *centroid* of the area of the plane surface in this case.

With regard to pressure on curved surfaces, there are a few special properties of resultants we shall have occasion to use:

1. For uniform pressure the component of the resultant in a given direction is the pressure times the projection of the surface area onto the plane perpendicular to that direction. To show that this is the case, we refer to Figure 3.42, in which dA is an element of area on the surface and $\hat{\mathbf{n}}$ is a unit vector normal to the surface. The elemental force is $(p\,dA)\hat{\mathbf{n}}$ and the component in the x direction is $(p\,dA)\hat{\mathbf{n}} \cdot \hat{\mathbf{i}}$, so that

$$F_{rx} = \int_S (\hat{\mathbf{n}} \cdot \hat{\mathbf{i}})p\,dA$$

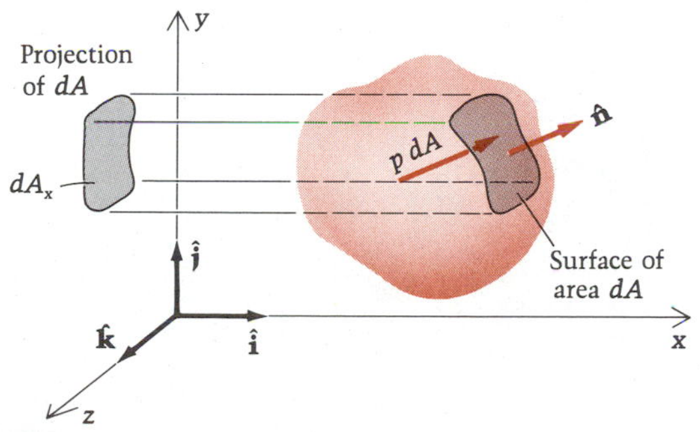

Figure 3.42

In the integration, $\hat{\mathbf{n}}$ varies since its direction is always normal to the surface. However, $\hat{\mathbf{n}} \cdot \hat{\mathbf{i}}$ is the cosine of the angle between the normal to the surface and x (which is perpendicular to the yz plane). Thus $(\hat{\mathbf{n}} \cdot \hat{\mathbf{i}})\, dA$ is the projection, dA_x, of that element of the surface onto the yz plane. Since p is uniform,

$$\begin{aligned} F_{rx} &= p \int \hat{\mathbf{n}} \cdot \hat{\mathbf{i}}\, dA = p \int dA_x \\ &= pA_x \end{aligned}$$

where A_x is the projection of the surface area A onto the yz plane.

2. The resultant of pressure on a spherical surface has a line of action through the center of the sphere (see Figure 3.43). This follows from the fact that each elemental force has a line of action through that point. That is, we have here a concurrent distributed force system.

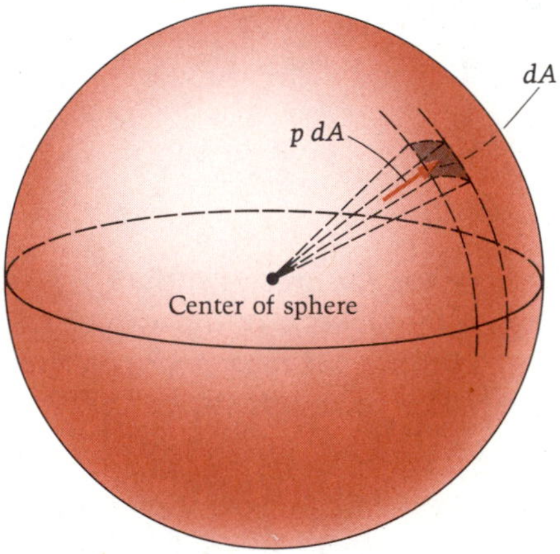

Figure 3.43

3. Pressure on a circular cylindrical surface produces no resultant moment about the cylinder axis (see Figure 3.44). This follows from the fact that each elemental force has a line of action that intersects that axis.

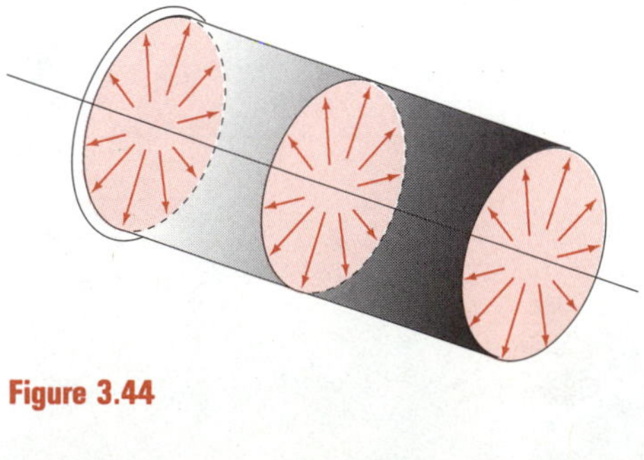
Figure 3.44

Question 3.32 Do the above results (2) and (3) concerning pressure on spherical and cylindrical surfaces require that the spheres and cylinders be complete (that is, whole)?

Question 3.33 Does the pressure in (1), (2), and (3) above have to be constant for the results to be valid?

Question 3.34 Would the moment about the axis of a cone, due to pressure on a portion of a cone's surface, be zero?

Forces Distributed Throughout a Volume; Gravity

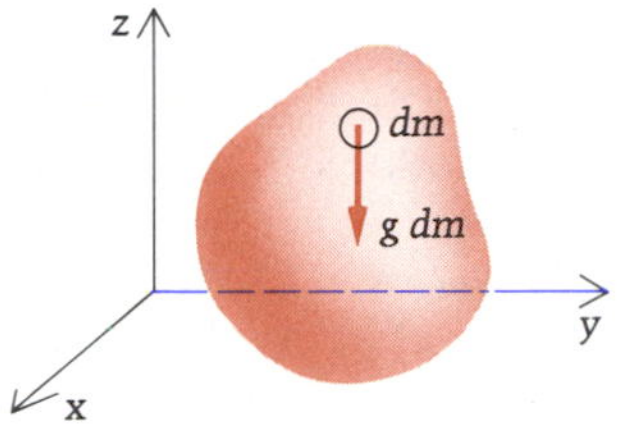

Figure 3.45

We now turn to the most common example of a parallel force system distributed over a volume. Suppose a body $\mathcal{B}$ can be treated as if the gravitational attractions of the earth on all of the particles of $\mathcal{B}$ are parallel (vertical). The magnitude of the force exerted by the earth on an element of mass in the body is $g(dm)$, where g is the strength of the gravitational field (or the acceleration of gravity), assumed constant over the body, and $dm = \rho\, dV$ is the mass of an elemental volume dV, where the density (mass per unit of volume) is ρ (see Figure 3.45).

For this (assumed) parallel force system, the resultant force (weight) is vertical (downward) with magnitude

$$F_r = \int_{\mathcal{B}} g\, dm = g \int_{\mathcal{B}} dm = mg$$

where m is the mass of the body. In order to produce the same moments about the x and y axes as does the distributed system, we must have

$$\begin{aligned} x_r F_r &= \int xg\, dm \\ &= g \int x\, dm \end{aligned} \tag{3.27}$$

and similarly,

$$\begin{aligned} y_r F_r &= \int yg\, dm \\ &= g \int y\, dm \end{aligned} \tag{3.28}$$

where x_r and y_r are coordinates of points on the line of action of the force-alone resultant.

Since $F_r = mg$, Equations (3.27) and (3.28) simplify to

$$x_r = \frac{\int x\, dm}{m}$$

$$y_r = \frac{\int y\, dm}{m} \tag{3.29}$$

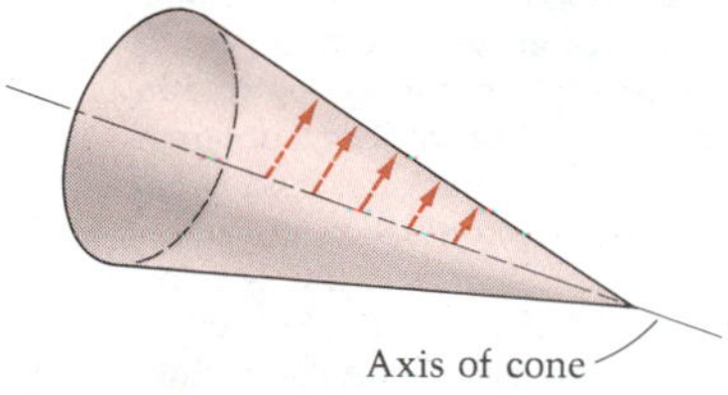

Figure 3.46

Answer 3.32 No. It is the fact that all the forces intersect the center of the sphere (or the axis of the cylinder) that makes the moment zero.

Answer 3.33 1: Yes; 2: No; 3: No.

Answer 3.34 Yes. Even though the forces caused by the pressure intersect the cone's axis nonperpendicularly, they still produce no moment about this line (see Figure 3.46).

The coordinates given by Equations (3.29) define a line on which the **center of mass** C of the body is located. A third equation,

$$z_r = \frac{\int z\,dm}{m}$$

pins down the actual point C. We usually write the coordinates of C as $(\bar{x}, \bar{y}, \bar{z})$ rather than (x_r, y_r, z_r), incidentally.

Thus we see that, when it is practical to assume that the gravity forces acting on a body are parallel and g is constant, the earth exerts a force (called the weight) with its line of action passing through the center of mass of the body. This will be the case in all the exercises in this book. In this case, the center of mass is sometimes called the **center of gravity.***

We note that if in Equations (3.29) we replace the differential mass dm by $\rho\,dV$, where ρ is the mass density and dV is the differential volume enclosing dm, then we obtain three coordinate equations, the first of which is

$$x_r = \frac{\int x\rho\,dV}{\int \rho\,dV} \tag{3.30}$$

in which the denominator is the mass m of the body:

$$m = \int dm = \int \rho\,dV$$

We see from Equation (2.30) that if the density ρ of the body is constant, it cancels, leaving

$$x_r = \frac{\int x\,dV}{\int dV} = \frac{\int x\,dV}{V} \tag{3.31}$$

and similarly

$$y_r = \frac{\int y\,dV}{V} \quad \text{and} \quad z_r = \frac{\int z\,dV}{V} \tag{3.32}$$

* The terms "center of mass" and "center of gravity," of course, refer to different physical concepts — the former having to do with the mass distribution of a body and the latter having to do with the resultant of distributed gravitational attractions. In the literature of mechanics "center of gravity" is used in two ways. One usage refers to the location of the equivalent particle (same mass as the body) that would cause this particle to be subjected to the same force by an attracting particle as would be the actual body in question; this center of gravity has a location which depends upon, among other things, the orientation of the body. The second and more common usage refers to the location of the point through which the resultant weight passes when the gravity field is uniform and parallel. This center of gravity is independent of the orientation of the body, and, as we have seen, it has the same location as the mass center of the body. Thus, in this case there is no reason to distinguish center of gravity from the center of mass except, perhaps, to remind the reader of the physical phenomenon motivating location of the point. However, it is important to realize that the solutions of some engineering problems require recognition of the fact that a gravitational field is not uniform and parallel. An example of this arises in the attitude control of an earth satellite. The resultant of the earth's gravitational attraction on the satellite is a force through the mass center and a couple whose small moment is called the "gravity gradient torque," and this is a very important factor in establishing control of the satellite.

The coordinates given by Equations (3.31) and (3.32) define the **centroid of volume** of the body. We see that, for a body having constant density, this centroid and the center of mass coincide.

When positions of non-obvious mass centers are needed in the equilibrium problems in Chapters 4–6, these mass centers will be indicated in the various figures.

We have presented in this section a number of special cases of distributed force systems, particularly those for which the force-alone resultant, or perhaps only its line of action, is easily recognized. It is for these cases that the concept of a resultant is most useful. For the sake of completeness, however, we set down here the method of handling the general case. Suppose $\mathbf{f}$ is the intensity of a distributed force system (force per unit of length or area or volume), and suppose we let dQ be an element of length or area or volume as the case may be. Let $\mathbf{r}$ be the directed line segment from point P to the point of application of an elemental force $\mathbf{f}\,dQ$. Then the resultant at P is

$$\mathbf{F}_r = \int \mathbf{f}\,dQ \qquad \text{and} \qquad \mathbf{M}_{rP} = \int \mathbf{r} \times \mathbf{f}\,dQ$$

This is nothing more than the obvious extension of our work with discrete systems; here, though, we are integrating instead of summing. Once $\mathbf{F}_r$ and $\mathbf{M}_{rP}$ are calculated, any further reduction follows in the same manner as for a system of discrete forces.

PROBLEMS ► Section 3.9

3.142 Specify and locate the single force that is equipollent to the concentrated and distributed loads acting on the shaded plate in Figure P3.142.

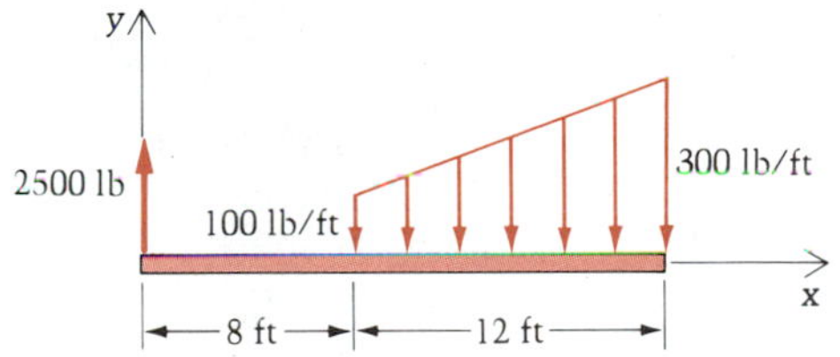

Figure P3.142

3.143 A cantilever beam is loaded as shown in Figure P3.143. Replace the distributed line load by an equipollent system at the wall.

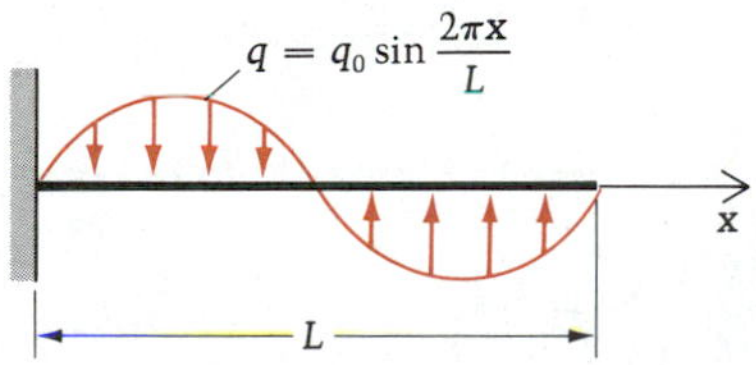

Figure P3.143

3.144 Pressure acts on a rectangular solid as shown in Figure P3.144. Find the single force that is equipollent to the given loading (magnitude *and* line of action).

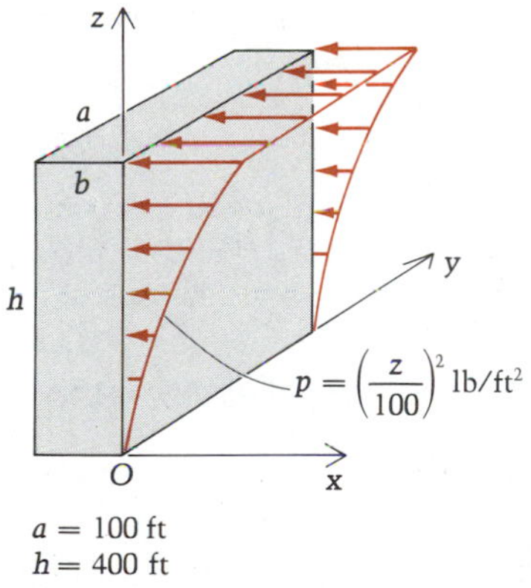

Figure P3.144

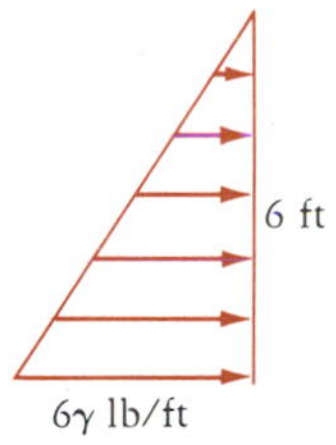

Figure P3.145

3.145 Compare the resultant of Problem 3.109 with that of the distributed load shown in Figure P3.145. Tell why the force is the same but the line of action is slightly different.

3.146 Wind velocity varies with height due to a number of factors, among which are the earth's angular velocity, wind path curvature pressure gradient, air density, latitude, and viscosity. In coastal areas, if the wind speed at 30 feet (called v_{30}) is under 60 mph, then it is assumed that at height z ft (for $z \le 600$ ft), the wind velocity is

$$v_z = v_{30}\left(\frac{z}{30}\right)^{0.3}$$

Use this equation to compute the total wind force on the windward sides of two buildings of equal areas facing the wind, one with height $h = 100$ ft and width $w = 50$ ft, the other with $h = 50$ ft and $w = 100$ ft. Take the value of v_{30} to be 20 ft/sec and the dynamic pressure to be $(1/2)\,\rho\, v_z^2\, C_D$, where ρ = wind density (use 0.0024 slug/ft^3) and C_D = drag coefficient (use 1.4).

3.147 A concrete ramp leading up to a hospital emergency room door is 1 m wide and has the length and height shown in Figure P3.147. If the concrete weighs 22,600 N/m^3, determine the ramp's weight and its line of action.

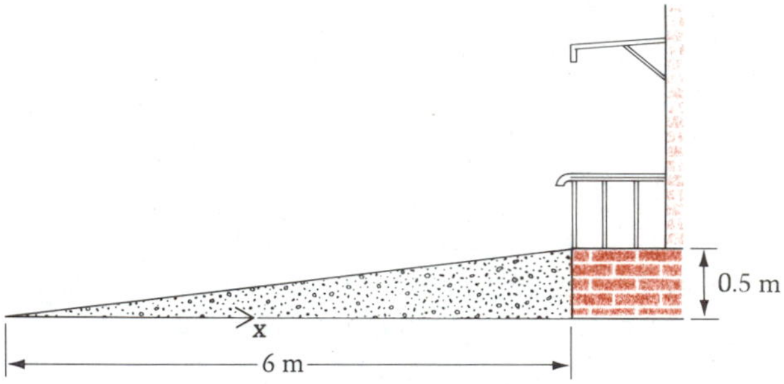

Figure P3.147

3.148 With reference to Figure P3.148,

a. Show that a uniform pressure p within a conically shaped tank with a closed, flat base produces no resultant moment about the axis of the cone.

b. Show further that the resultant of this distributed force system is zero.

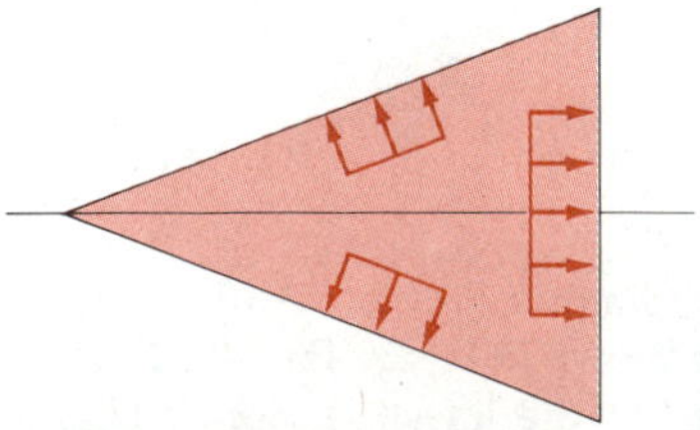

Figure P3.148

3.149 Replace the distributed loading shown in Figure P3.149 by a single force.

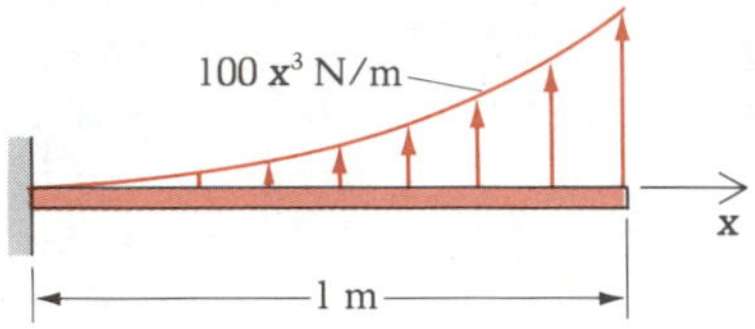

Figure P3.149

3.150 The distributed load on the arch shown in Figure P3.150 varies according to $q = 300 \cos\theta$ N/m. Find the resultant of this load, expressed as a force at Q and a couple.

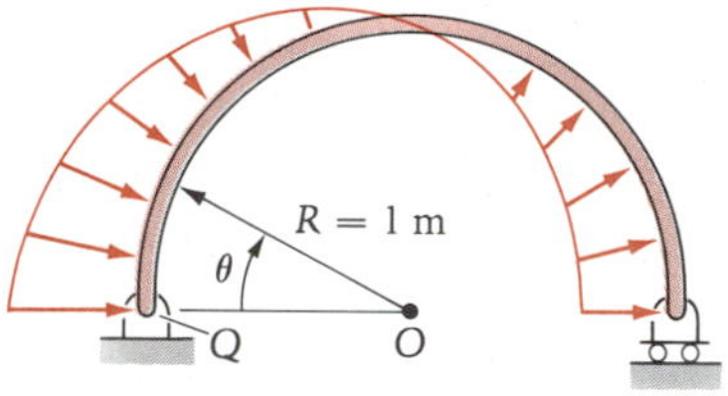

Figure P3.150

3.151 Repeat the preceding problem if $q = 300 \cos 2\theta$ (see Figure P3.151).

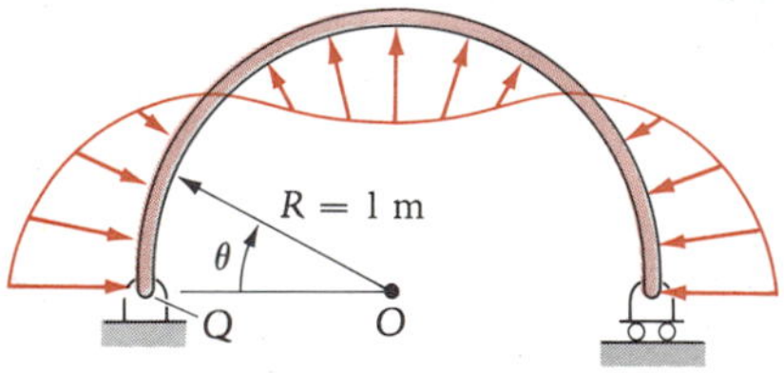

Figure P3.151

* **3.152** Repeat the preceding problem if $q = 300 \cos 3\theta$ (see Figure P3.152).

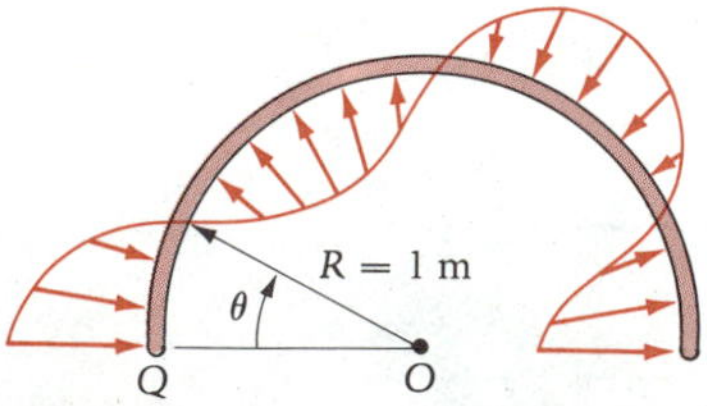

Figure P3.152

For each of the following five problems, replace the distributed loading on the beam by an equipollent set of forces.

3.153

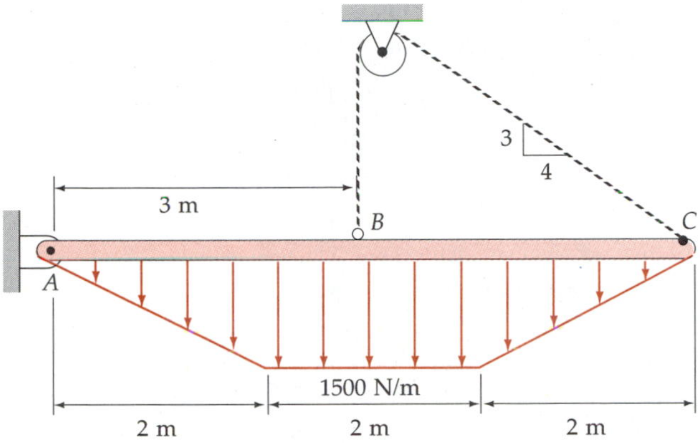

Figure P3.153

3.154

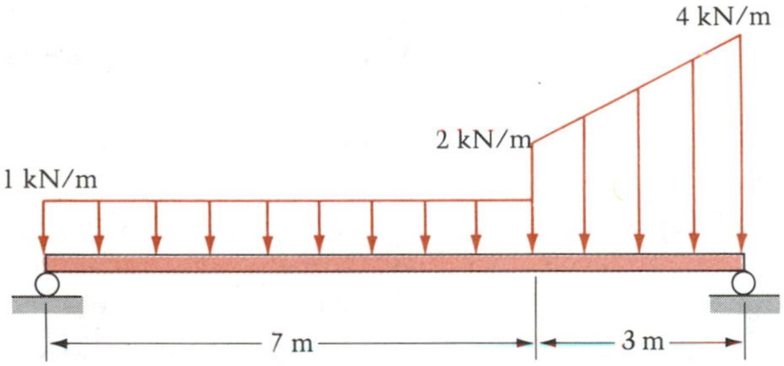

Figure P3.154

3.155

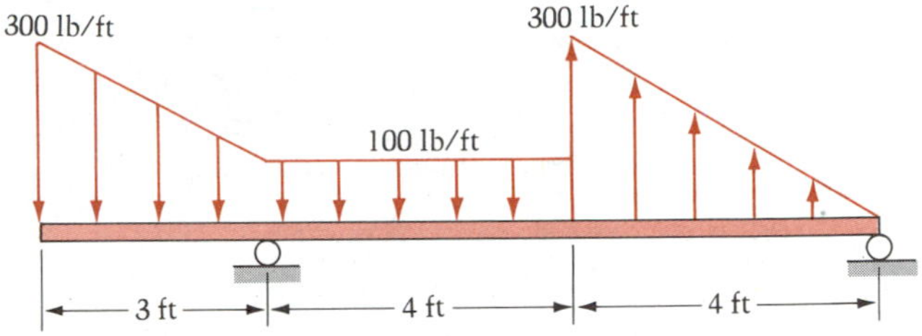

Figure P3.155

3.156

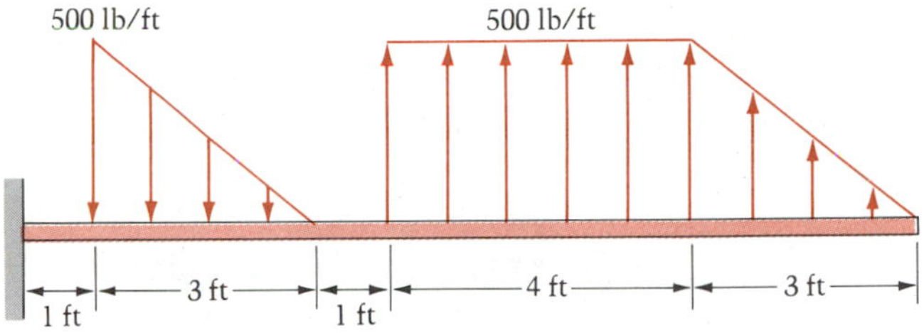

Figure P3.156

***3.157**

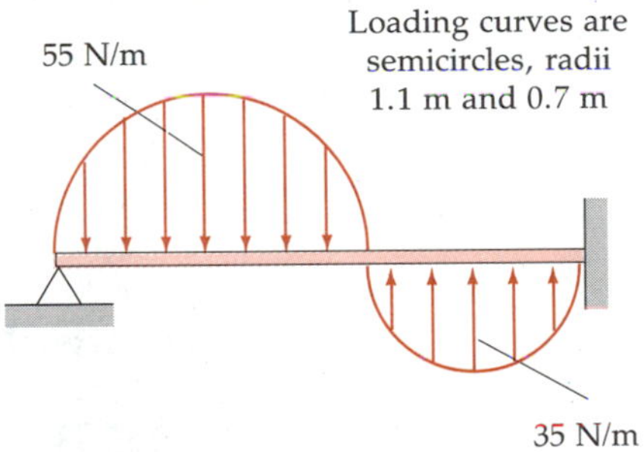

Figure P3.157

Computer Problems ▶ Chapter 3

3.158 Write a computer program that will read the components of the three-dimensional vectors $\mathbf{r}_{AB}$, $\mathbf{F}$ and $\mathbf{L}$, where B is a point on the line of action of force $\mathbf{F}$, $\mathbf{r}_{AB}$ is a position vector from point A to B, and $\mathbf{L}$ has as components a set of direction numbers of a line ℓ through A (see Figure P3.158). The program is to calculate and print (a) the moment of $\mathbf{F}$ about A; (b) the moment of $\mathbf{F}$ about line ℓ; and (c) the distance from A to the line of action of $\mathbf{F}$. Run the program for these data: $\mathbf{r}_{AB} = 2\hat{\mathbf{i}} + 3\hat{\mathbf{j}} - 6\hat{\mathbf{k}}$ m; $\mathbf{F} = \hat{\mathbf{i}} - 2\hat{\mathbf{j}} + 2\hat{\mathbf{k}}$ N; and $\mathbf{L} = 3\hat{\mathbf{i}} + 4\hat{\mathbf{j}} - 12\hat{\mathbf{k}}$.

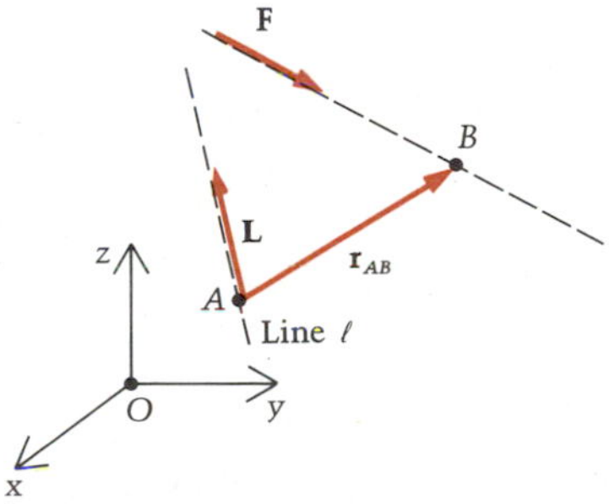

Figure P3.158

3.159 Write a program that will read any number of forces and couples in a coplanar force system, along with a point on the line of action of each of the forces. The program is then to compute and print the single-force resultant and the equation of its line of action. Run the program for these forces and couples:

$\mathbf{F}_1 = 100\hat{\mathbf{i}} - 200\hat{\mathbf{j}}$ lb, passing through $(x, y) = (1, 3)$ ft;

$\mathbf{F}_2 = 50\hat{\mathbf{i}} - 340\hat{\mathbf{j}}$ lb, passing through $(x, y) = (2, -5)$ ft;

$\mathbf{F}_3 = 250\hat{\mathbf{j}}$ lb, lying along the line $x = -7$ ft;

$\mathbf{F}_4 = 300$ lb, lying along the positive x axis;

$\mathbf{C}_1 = 400\hat{\mathbf{k}}$ lb-ft;

$\mathbf{C}_2 = -730\ \hat{\mathbf{k}}$ lb-ft.

3.160 Write a program that will read any number of forces parallel to the z axis (with $+\hat{\mathbf{k}}$ suppressed), along with the (x, y) coordinates of the points where each force pierces the plane $z = 0$. The program is then to compute and print the single-force resultant and *its* piercing point in the plane $z = 0$. Run the program for this data:

$\mathbf{F}_1 = 200$ N through the origin;

$\mathbf{F}_2 = 1000$ N through $(x, y) = (10, -20)$ m;

$\mathbf{F}_3 = -580$ N through $(x, y) = (-5, 15)$ m;

$\mathbf{F}_4 = -900$ N through $(x, y) = (-10, -2)$ m.

SUMMARY ▶ Chapter 3

In Section 3.2, we learned that the moment $\mathbf{M}_P$ of a force $\mathbf{F}$ about a point P has a magnitude equal to that of the force multiplied by the distance from the point to the line of action of the force. The direction of the moment vector is that of the right thumb when the right hand's fingers turn in the direction of the turning effect of $\mathbf{F}$ about P.

We also learned that a vector representation of $\mathbf{M}_P$ is $\mathbf{r} \times \mathbf{F}$, where $\mathbf{r}$ is a vector from P to *any point* on the line of action of $\mathbf{F}$. This cross product makes moment calculations much easier in cases of complicated geometry, particularly in three dimensions.

The final important topic in Section 3.2 was Varignon's theorem, which states that the moment about P, of the sum of a set of forces acting at a point A, is the same as the sum of the moments of the separate forces about P. This allows us to decompose a force into its components at A when the perpendicular distances along coordinate directions from A to P are easily seen or determined.

Section 3.3 was concerned with computing the moment of a force $\mathbf{F}$ about a line ℓ. This moment, $\mathbf{M}_\ell$, was the projection along ℓ of the moment of $\mathbf{F}$ about any point P lying on ℓ, i.e.,

$$\mathbf{M}_\ell = (\mathbf{M}_P \cdot \hat{\mathbf{u}})\, \hat{\mathbf{u}}$$

where $\hat{\mathbf{u}}$ is the unit vector in the direction of ℓ. We proved that $\mathbf{M}_\ell$ does not depend on the point P along ℓ used to compute it.

We proved and then displayed (in Figure 3.9) a physical interpretation of $\mathbf{M}_\ell$ as the turning effect, about ℓ, of the component of $\mathbf{F}$ that is perpendicular to ℓ. And finally, we illustrated the power of Varignon's theorem in finding the moment of a force about a line.

An interesting concept called a couple was studied in Section 3.4. Defined as a pair of equal-magnitude, oppositely directed, and non-collinear forces, the couple has no resultant force, but a moment *which is the same about all points of space.* If A lies on $-\mathbf{F}$ and B lies on $\mathbf{F}$, where $\mathbf{F}$ and $-\mathbf{F}$ constitute a couple, then this moment is simply $\mathbf{r}_{AB} \times \mathbf{F}$. But it is also $|\mathbf{F}|d$ in magnitude, where d is the distance between the lines of action of $\mathbf{F}$ and $-\mathbf{F}$, and directed according to the right-hand rule as illustrated in Figure 3.13.

The equations of equilibrium, $\Sigma\mathbf{F} = \mathbf{0}$ and $\Sigma\mathbf{M}_P = \mathbf{0}$, were presented in Section 3.5 for motivational purposes. We noted that there is more to the equilibrium of bodies than merely setting the summation of external forces equal to zero as we did in Chapter 2. For problems in which all the forces acting on a body are not concurrent, we must also set the sum of moments about a point P equal to zero.

In Section 3.5 we also proved that if Q is some other point besides P, then it is always true that

$$\Sigma\mathbf{M}_Q = \mathbf{r}_{QP} \times (\Sigma\mathbf{F}) + \Sigma\mathbf{M}_P$$

and we noted from this equation that if $\Sigma\mathbf{F} = \mathbf{0}$ and $\Sigma\mathbf{M}_P = \mathbf{0}$, then $\Sigma\mathbf{M}_Q$ is automatically also zero. This means that nothing is to be gained by a second moment equation once $\Sigma\mathbf{F} = \mathbf{0}$ and $\Sigma\mathbf{M}_P = \mathbf{0}$ have been used.

The equipollence (equal power) of force systems was defined and studied in Section 3.6. Two equipollent systems will make the same contributions to the equations of equilibrium (and also to the equations of motion in later studies in dynamics). Two systems are called equipollent if two things happen:

1. $(\Sigma\mathbf{F})$ in system 1 = $(\Sigma\mathbf{F})$ in system 2

and

2. $(\Sigma\mathbf{M}_P)$ in system 1 = $(\Sigma\mathbf{M}_P)$ in system 2 (where P is some common point in systems 1 and 2)

We showed that if conditions 1 and 2 above hold for two force systems, then the moment sums are also equal for *any other* point Q. Thus in checking for equipollence, the sum of the moments need only be compared at one point.

For any system of forces (and/or couples), we learned in Section 3.7 how to replace it by a force at any pre-selected point P and a couple. This force and couple pair, $\mathbf{F}_r$ and $\mathbf{M}_{rP}$, is called a resultant of the original system. The force, $\mathbf{F}_r$, is the summation $(\Sigma\mathbf{F})$ of all the forces in the original system, placed at P. The couple, $\mathbf{M}_{rP}$, is the sum of the moments about P of the forces in the original system, plus the sum of the moment vectors of the couples. Thus, regardless of how complicated a force-couple system may be, it can always be reduced to an equipollent force at any desired point and a couple.

We learned in Section 3.8 that in several cases of interest, the force and couple resultant, with $\mathbf{F}_r \neq \mathbf{0}$, can be further reduced to an equipollent system consisting of a single force. These cases are (a) concurrent force systems; (b) coplanar force systems; and (c) parallel force systems. Any other systems for which $\mathbf{F}_r$ is perpendicular to $\mathbf{M}_{rP}$ will also have a "force-alone" resultant. When $\mathbf{F}_r$ is *not* normal to $\mathbf{M}_{rP}$, the simplest resultant is a collinear force and couple, called a "screwdriver."

The final section in the chapter (3.9) dealt with an important subject: distributed force systems. When the distributed forces are all parallel and act on a body along a line normal to the forces, then the magnitude and sense of the force-alone resultant are given by the signed area beneath the loading curve. Its direction is that of the parallel distributed forces, and its line of action is given by Equation 3.25. (It will be seen in Chapter 7 that this resultant passes through a point called the centroid of the signed area beneath the loading curve.) These results were seen to facilitate problems in which distributed loadings due, for example, to weight or water pressure were involved.

Pressure varying in two dimensions over a flat surface was also considered, as were uniform pressure on curved surfaces and variable pressure on spherical and cylindrical surfaces.

Finally, the force of gravity was considered, and the meanings of, and differences between, the center of mass and the center of gravity were discussed.

REVIEW QUESTIONS ▶ Chapter 3

True or False?

1. The magnitude of the moment of a force about a point is the magnitude of the force multiplied by the distance from its line of action to the point.
2. A force has zero moment about any point on its line of action.
3. A force has no moment about a line parallel to its line of action.
4. The magnitude of the moment of a force about a line equals the product of the component of the force perpendicular to the line and the distance between this component and the line.
5. The moment about a point P of the sum of a set of concurrent forces, placed at the point of concurrency, equals the sum of the moments about P of the separate forces.
6. A couple has the same moment about any point.
7. If we know the moment of a couple about a point A, then we can find the moment of the couple about a specified line BC even if A does not lie on BC.
8. The sum of the moments, about any point, of the external forces acting on a body at rest in a noninertial frame is always zero.
9. Two force and couple systems are equipollent if they have either the same resultant force or if the moments about some point P are the same for both systems.
10. It is possible for a system consisting of a single couple to be equipollent to a system of one force and 17 couples.
11. Given any system of forces and couples, and any point P, the system may be reduced to an equipollent system comprising a force at P and a couple (where either or both might be zero).

12. The couple part of the resultant of a concurrent force system will vanish at all points.

13. If a system of forces is coplanar, there is a point Q in the plane where the resultant moment vanishes, and the moment also vanishes for any point on the line through Q normal to the plane of forces.

14. If a system of forces which don't sum to zero is parallel, then there is a line ℓ at every point of which the resultant moment is zero, and both the line ℓ and the resultant force are parallel to each of the forces.

15. If a system S_1 of forces and couples has a "force-alone" equipollent system S_2, then S_1 is either a concurrent, coplanar, or parallel force system.

16. The resultant of the distributed loading shown in Figure 3.47 is a zero force and a zero couple.

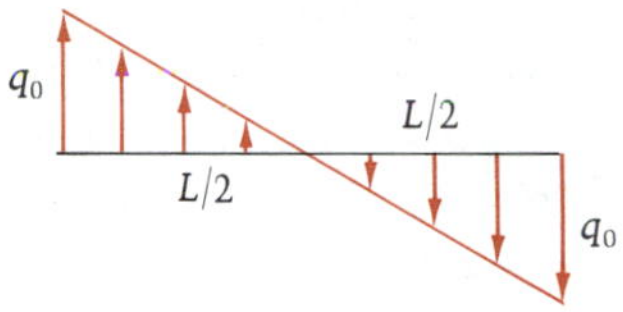

Figure 3.47

17. The simplest resultant of a general system of forces and couples is a "screwdriver," that is, a collinear force and couple along a specific axis in space.

18. Let a general system of forces and couples be resolved into a resultant force at point P and an accompanying couple. If the couple is normal to the force, then the equipollent screwdriver consists of a single force.

Answers: 1. T 2. T 3. T 4. T 5. T 6. T 7. T 8. F 9. F 10. F 11. T 12. F 13. F 14. T 15. F 16. F 17. T 18. T

4 Analysis of General Equilibrium Problems

4.1 Introduction

4.2 The Free-Body Diagram

What The Free-Body Diagram Is

Separating Two Bodies; The Action-Reaction Principle

Identifying Resultants Transmitted by Common Types of Connections

4.3 Fundamental Applications of the Equilibrium Equations

Number of Independent Equations

Static Indeterminacy

Types of Problems in This Section: Bodies Left Intact

Scalar Vs. Vector Approach

Special Results for Pulleys and Two-Force Members

Three-Dimensional Examples

4.4 Applications of the Equations of Equilibrium to Interacting Bodies or Parts of a Structure

Summary

Review Questions

4.1 Introduction

We are now in position to solve some general equilibrium problems. By "general," we mean problems which require us to use *both* of Euler's laws to effect a solution; i.e., we must invoke not only the "force equation" from Chapter 2:

$$\Sigma \mathbf{F} = 0 \tag{4.1}$$

but also the "moment equation" mentioned briefly back in Chapter 1:

$$\Sigma \mathbf{M}_P = 0 \tag{4.2}$$

where P is any point.

With Equation (4.2) now in our box of tools, we are no longer restricted to examining the relatively small number of equilibrium problems in which the forces acting on the body are concurrent. For a point-mass, of course, they must be so,* but most everyday, finite-sized bodies in the state of equilibrium are acted on by forces that do not all meet at one point. Examples are parked cars, desks, buildings — the list is endless.

So, when the external forces acting on a body are not concurrent, we must sum and set to zero the moments of forces as well as the forces themselves. This of course is the reason we spent Chapter 3 learning all about the moment of a force. Without such knowledge, we could not form Equation (4.2) from free-body diagrams of the bodies we wish to analyze.

This chapter is actually the heart of the book; the three chapters that preceded it are preparation *for* it, while the two that immediately follow will be special applications *of* it. By the end of this chapter the reader should be able to write and solve the equations relating the external forces acting on any body in equilibrium. Now the "body" might be a single indentifiable physical object, such as a table. But it can also be a collection of objects taken *together* to form a "combined body," such as a diver standing still on a diving board, or a table with a pitcher of water on it.

The equations of equilibrium [Equations (4.1) and (4.2)] relate *all* of the *external* forces acting on a body at rest in an inertial frame of reference. In most engineering problems some of these external forces are known (or prescribed) before any analysis is carried out; we usually refer to these as **loads.** The external forces exerted by attached or supporting bodies are called **reactions;** usually we can think of these as forces that constrain the body against motion the loads tend to produce. It is by the equations of equilibrium that we try to find these reactions. We emphasize, however, that a force "by any other name" is still a force as far as the equations of equilibrium are concerned; whether we think of a force as a cause (load) or as an effect (reaction) makes no difference in the equa-

*Because for a "point-mass," there is only one point on which they may act!

tions of equilibrium. If a body could be equipped with sensors to measure all of the forces acting on it, the sensors would not be able to distinguish applied loads from constraint reactions.

To illustrate one of the difficulties encountered in statics, we consider the problem of determining the forces (reactions) exerted by the supports of a diving board when a diver (whose weight is the load) stands on the end of the board as in Figure 4.1.

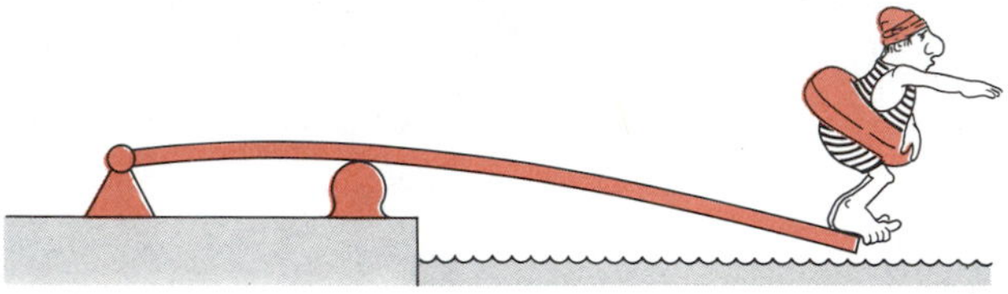

Figure 4.1

The configuration shown is that to which the equations of equilibrium must be brought to bear; the distances from the supports to the diver are important because of the moment equations. However, the diver's distance from each of the supports depends upon how much the board is bent — the greater the bending, or sag, of the board, the smaller the distance of the diver from each of the supports. Clearly, the bending of the board depends upon the weight of the diver and the stiffness of the board. Thus we might be tempted to conclude that we can do nothing useful until we study the geometry changes that occur when a body deforms. Fortunately, for many engineering problems the picture is not quite as bleak as the one we have painted. Frequently the deformations arising from the application of loads are small enough that gross changes in geometry can be ignored. A rigid body is the idealization in which no deformation at all occurs, a rigid body being one in which the distances between all possible pairs of points are unchanged when the body undergoes a change in configuration. That is, no portion of the body can change in shape or size. No real body is rigid, but it may be near-rigid in the sense just described — that is, for small deformations. If that can be assumed to be the case with the diving board, we have the situation shown in Figure 4.2:

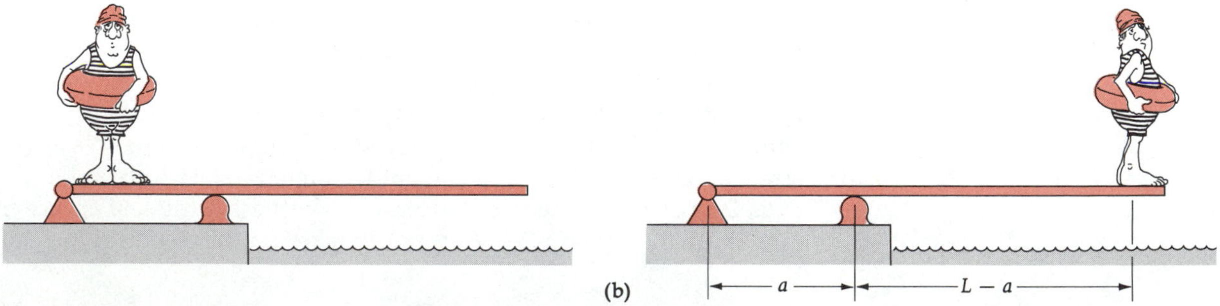

Figure 4.2 Near-rigid diving board.

Here the geometry with which we have to deal is, for all practical purposes, independent of the weight of the diver. The assumption of near-rigidity is to be understood throughout this book with few exceptions that will be obvious. However, as students proceed into further studies of mechanics (mechanics of deformable solids, mechanics of fluids), it is important for them to realize that the equations of equilibrium are valid for *any* body at rest in an inertial frame, regardless of the degree of deformability and regardless of the phase (solid, liquid, or gas).

In engineering mechanics, as in other areas of science and applied mathematics, we must be alert to the possibility of posing a meaningless, or silly, problem. For example, suppose we remove the interior support from the diving board and inquire as to the hinge reaction when the board is in equilibrium in the configuration of Figure 4.3. The difficulty here is that this configuration cannot be an equilibrium configuration. The equations of equilibrium will tell us that, but, of course, we don't need them here. The impossibility should be obvious from the fact that the board is free to rotate as a rigid body about the hinge. We shall usually be immune to the possibility of this situation if the body is constrained in such a way that a rigid-body change in configuration is prohibited. Such would be the case, for example, if one point of the body were fixed and rotation about each of three distinct nonplanar axes through the point were prohibited.

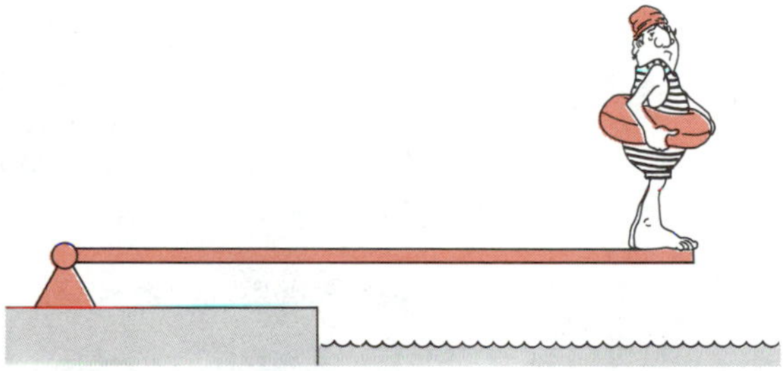

Figure 4.3 Equilibrium impossible.

Section 4.2 will be concerned with a much more complete treatment of the free-body diagram (FBD) than was possible in Chapter 2. We will learn, for example, to remove the supports in figures such as 4.1 and 4.2, thereby isolating for analysis bodies such as the diving board and/or diver. *We cannot overemphasize the importance of mastering the FBD concept.*

As we have said, we applied Equation (4.1) to a number of simple, particle-like problems in Chapter 2. The next step in analyzing equilibrium problems, now that we are knowledgeable about moments from Chapter 3, is to apply *both* Equations (4.1) and (4.2) to a single finite-sized* body (or to two or more physical objects left intact and considered as a single body). This will be the subject of Section 4.3.

Finally, in Section 4.4, we learn to separate bodies joined by simple connections such as pins, rollers, cables and the like, and to draw FBD's and write equilibrium equations for the separate bodies as well as the "combined body." In that process, we shall discover that not all the resulting equations are independent.

We will defer to Chapter 5 a last step in this graduated approach to studying equilibrium problems: that of imagining bodies to be sliced completely in two, so as to expose and then solve for the forces with*in* the body known as *internal forces.*

4.2 The Free-Body Diagram

In Chapter 2, we spoke briefly of the free-body diagram (FBD) when studying the equilibrium of a particle. In that context, the FBD actually needed be nothing more than a sketch of the forces acting on the particle, all interacting at the one point of concurrency.

As important as it was in equilibrium problems of particles, the FBD will be seen to be even *more* important in general equilibrium problems of finite-sized bodies, on which the forces need not be concurrent and on which couples may also act. This is because the FBD now gives not just a sketch of all the external forces and their directions for use in $\Sigma \mathbf{F} = \mathbf{0}$, but also illustrates the various points of application of these forces along with appropriate distances for use in $\Sigma \mathbf{M}_P = \mathbf{0}$. For this reason, it is now *vital* that the student sketch the actual body as more than just a point-mass.

Because there is this additional importance now attached to the free-body diagram, we devote this short but all-important section to its use. And to make the section self-contained, we shall not refer further to the brief FBD coverage in Section 2.2.

What the Free-Body Diagram Is

The **free-body diagram** is an extremely important and useful concept for the analysis of problems in mechanics. It is a figure, usually sketched, depicting (and hence identifying precisely) the body under consideration. On the figure we show, by arrows, all of the external forces, and moments of couples, that act on the body. Thus, we have a catalog, graphically displayed, of all the forces that contribute to the equilibrium equations (or the equations of motion if the problem is one of dynamics). Of particular importance is the fact that the free-body diagram provides us a way to express what we know about reactions (for example, that a

* "Finite-sized" means that the body cannot be analyzed as a particle unless all of the external forces are concurrent. The dimensions of the body and various points of application of the forces now become important because moment equations must also be written.

certain reaction force has a known line of action) before applying the equations of equilibrium. This is best illustrated by example.

Let us return to the diver and the diving board of the preceding section. We want to emphasize that a **body** is whatever collection of material we chose to focus on; we shall choose here to let the body be the diver and the diving board, taken together. The free-body diagram is

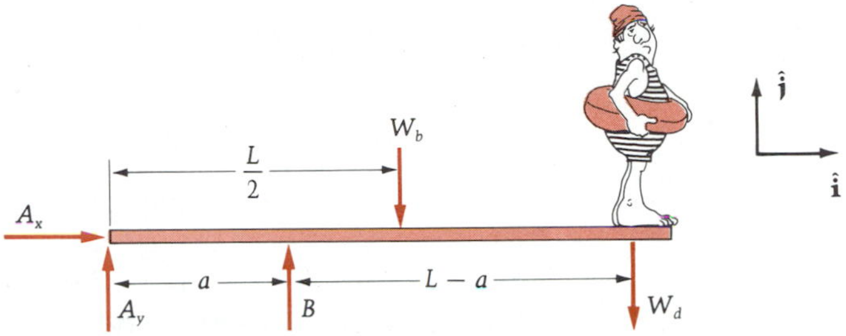

Figure 4.4 Free-body diagram of diver and board.

then shown in Figure 4.4.* The meaning of, and reasoning behind, the symbols appearing on the diagram are as follows:

1. W_d is the weight of the diver; thus the resultant force exerted by the earth on the diver by way of gravity is W_d times a unit vector downward. The line of action of this resultant is through the mass center of the diver. Similarly, W_b is the weight of the board, and the line of action of its resultant passes through the mass center of the board. In each case the letter and the arrow identify the vector description of the force, $W_d(-\hat{\mathbf{j}})$ or $W_b(-\hat{\mathbf{j}})$, and the location of its line of action. It is this information that will be needed in the equations of equilibrium.

2. The board is connected at its left end to the supporting structure by a hinge; basically, a cylindrical bar or pin fits into holes in brackets at the end of the board. In the absence of any significant friction we perceive that the pin exerts only pressure on the cylindrical surface of the hole. We saw in Section 2.9 that the resultant of this pressure will be a force with a line of action through the center of the hole. Because we don't know in advance the direction of this resultant, we express it by its unknown components A_x and A_y. That is, the force exerted on the left end is $A_x\hat{\mathbf{i}} + A_y\hat{\mathbf{j}}$. The letters and arrows at the left end constitute a code for how we have chosen to express the force, each arrow denoting a unit vector in the direction of the arrow.

3. The interior support is perceived to exert pressure perpendicular to the board. While we don't know precisely the location of the line of action of the resultant, if this pressure is distributed over a small region then we know the line of action close enough for engineering purposes.

* One need not be an artist to draw good-free body diagrams. The roughest of sketches will suffice as long as the body is clearly identified.

Because the pressure acts perpendicular to the board here, we know the direction of the force-alone resultant (it is vertical). Thus the force may be expressed in terms of a single unknown scalar B; that is, the diagram is communicating that the force may be expressed as $B\hat{\mathbf{j}}$ without any loss of generality.

Question 4.1 Why does the force exerted on the board by the diver's feet not appear on this free-body diagram?

It is important to realize that each of the forces appearing on the free-body diagram (Figure 4.4) is in fact the resultant of a distributed force system. We recall that the resultant of a force system embodies all of the characteristics of that system that show up in the equations of equilibrium. Thus, for example, when we write the equations of equilibrium for the diver-plus-board body, the only information we can hope to determine about the hinge pressure is its resultant, and that is precisely what is depicted on the free-body diagram. Shortly we shall look at several other kinds of mechanical connections between bodies and identify the nature of the resultants at these connections.

Separating Two Bodies; The Action-Reaction Principle

The hinge reaction in the diver-plus-board problem alternatively could be determined by an analysis in which the board alone is the body under consideration. The free-body diagram of the board is shown in Figure 4.5, where we see that the external forces are the same as before except that the weight of the diver does not appear. This is because the free-body diagram does not include him, and thus we have a new external force, P, that is the resultant of the pressure exerted on the board by the feet of the diver. The free-body diagram of the diver communicates (by the arrow) that the same scalar P multiplies an upward unit vector to express the force exerted *on* him *by* the board; that is, the force (vector) exerted by the board on the diver is the negative of the force (vector)

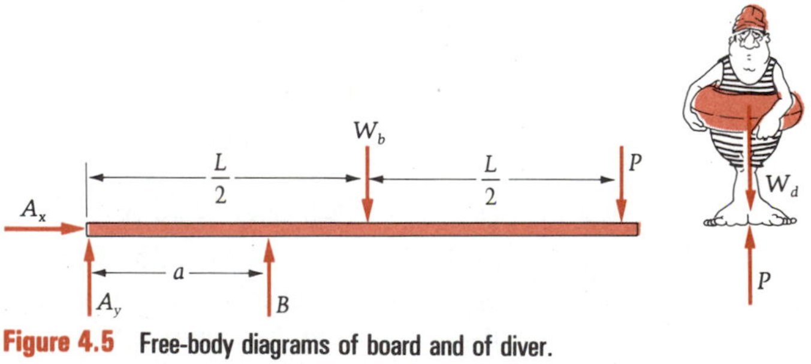

Figure 4.5 Free-body diagrams of board and of diver.

Answer 4.1 The body is the composite of the diver and board. This force is not external to that body; it is an internal force (interaction between parts of the body).

exerted by the diver on the board. This is an example of the **action-reaction principle,** which states that, when two bodies interact mechanically, the resultant exerted by the first body on the second body is the (vector) negative of that exerted by the second body on the first. This is often loosely referred to as Newton's Third Law. The principle is almost self-evident when we consider two bodies in intimate contact so that the interaction between the bodies is that of force systems distributed over the surface of contact. At each point on the common surface the mechanical action (force per unit area) of the first body on the second is the negative of what is exerted on the first body by the second. Because the force intensities are opposites, the resultants likewise must be the negatives of one another.

Identifying Resultants Transmitted by Common Types of Connections

Before we take up a variety of examples of free-body diagrams, it will be helpful to identify the resultants that are transmitted at different types of connections between bodies. To do this we follow the procedure we used on the diving-board hinge; that is, we study the particular "distributed-over-a-surface" force system involved and deduce what is the most general form of the resultant of that distributed system. This process could be tedious, but there is a shortcut that fortunately is in harmony with the intuition of most engineering students. The reasoning that leads to the correct conclusions is based upon the rigid-body motion that would be inhibited, or impeded, if each body attached to the one under consideration were to be fixed in the frame of reference. If there are several attachments, then to determine the resultant exerted by one of them we first imagine the others to be removed. We then determine how the single attachment hinders a rigid motion of the body. In particular, if we desire the resultant at a point A, then the force part of the resultant will have a component in any direction in which a displacement of A is inhibited. That is, if the connection resists a movement of A in a given direction, then the resultant force at A must have a component in that direction. The couple part of the resultant at A will have a component in the direction of any line through A about which the rotation of the body is inhibited. Results of applying this reasoning process are presented in Table 4.1 for several types of connections.

The first item (I) in Table 4.1 is a reminder that, when there is no special freedom associated with a connection, the resultant of the distributed forces of interaction is a general force and couple. In this case we know nothing about the direction of either, and hence each must be described by three unknown components. Under "Plane Counterpart" are displayed the symbols used to denote the connections when we have reason to be concerned only with forces having lines of action in a given plane together with moments perpendicular to that plane. (In our discussion of the diver and the diving board we tacitly assumed that to be the case.) We shall discuss this assumption in more detail later in one of the example problems. Unless otherwise indicated, items II-IX in Table 4.1

Table 4.1 Selected Connections and the Corresponding Unknown Components of Resultants on Body $\mathcal{B}$ (shaded)

Type of Connection	Reaction on $\mathcal{B}$ (Shaded)	Plane Counterpart
I. General interaction (no freedom) 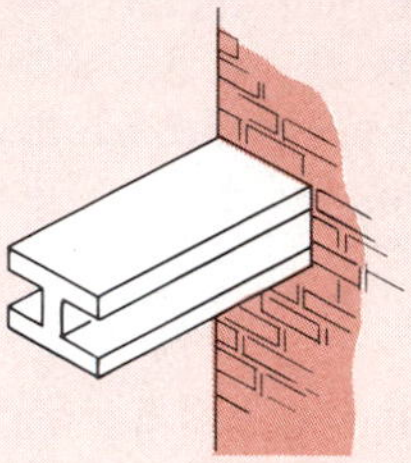 (a)	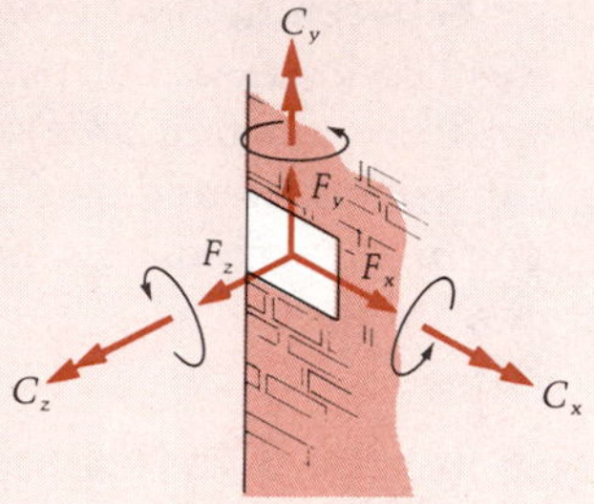 (b)	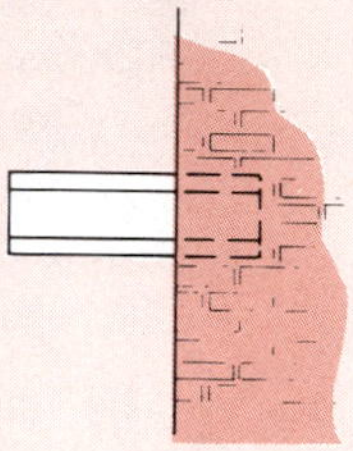 (c) 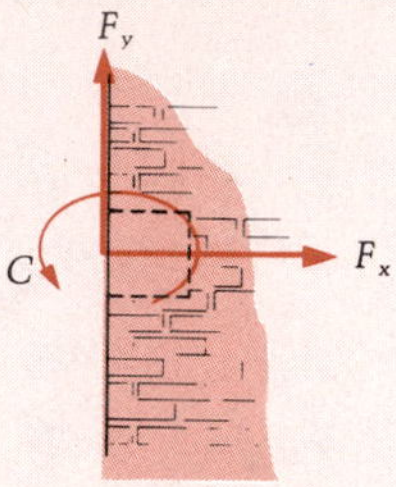 (d)
II. Hinge 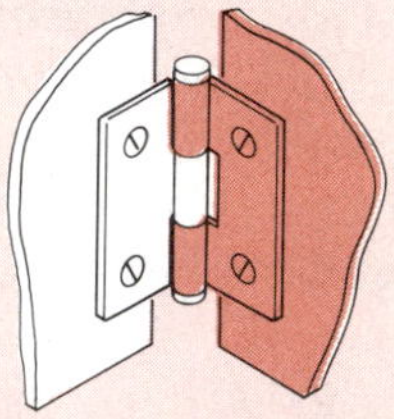 or clevis pin 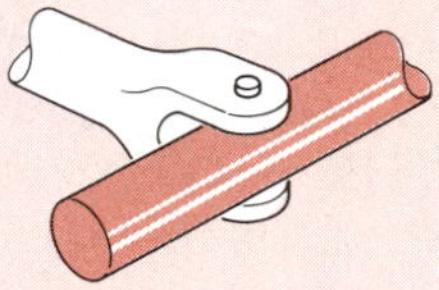	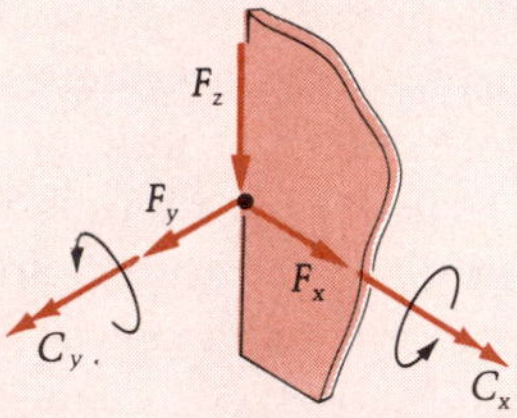 ↑ Same	Pin 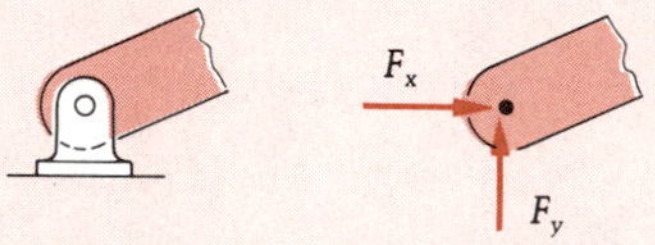 ↑ Same
III. Rod in sleeve (hinge without thrust support) 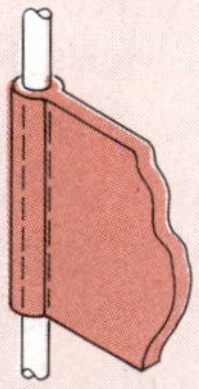	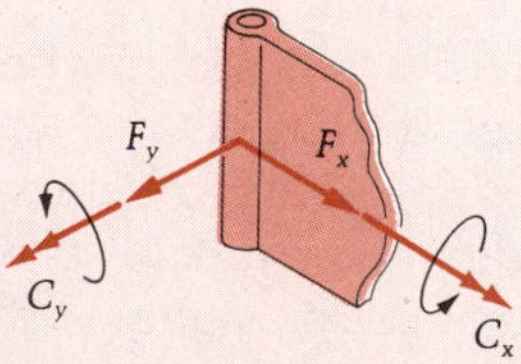	Same as II

Table 4.1 Continued

Type of Connection	Reaction on $\mathcal{B}$ (Shaded)	Plane Counterpart	
IV. Pin in slot, or roller, or line contact along smooth surface 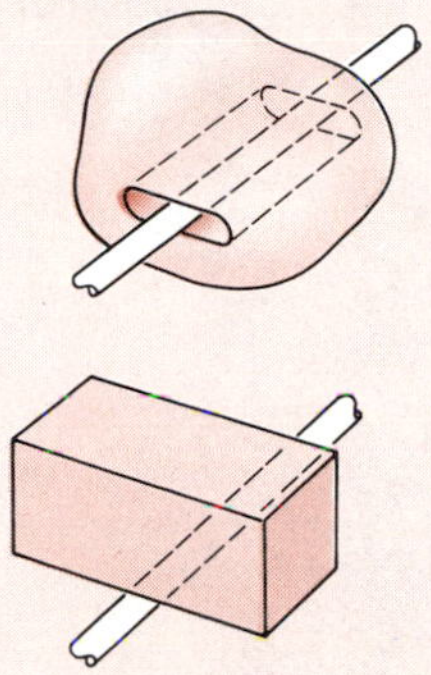	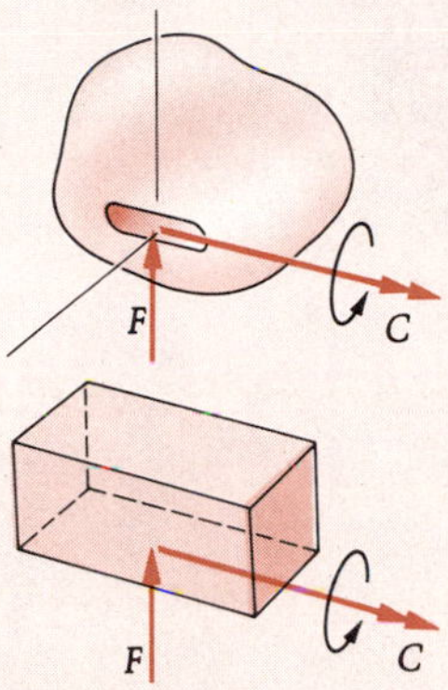	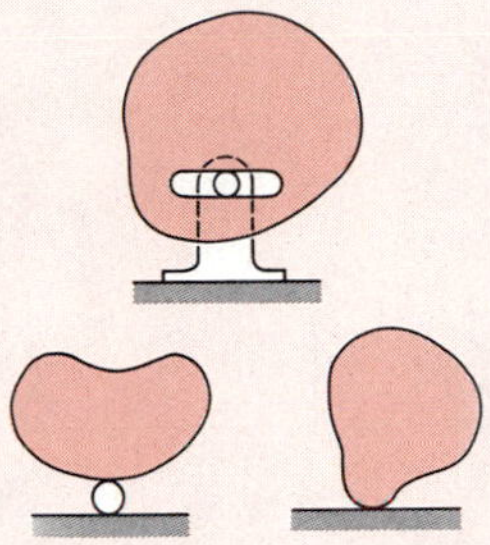	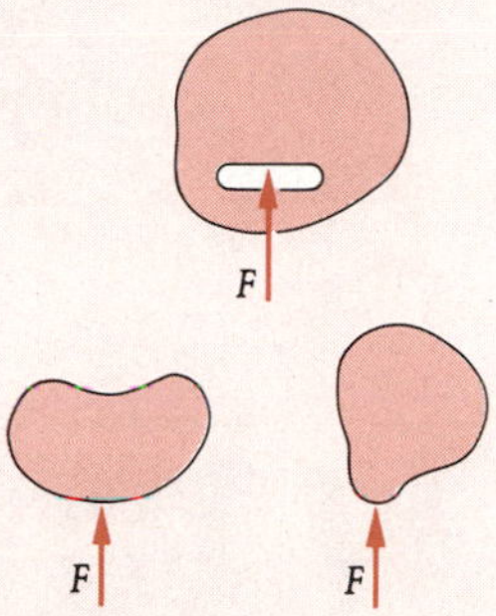
V. Ball-and-socket, or self-aligning bearing, or local contact with rough surface 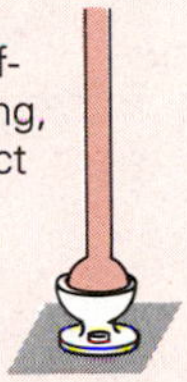 	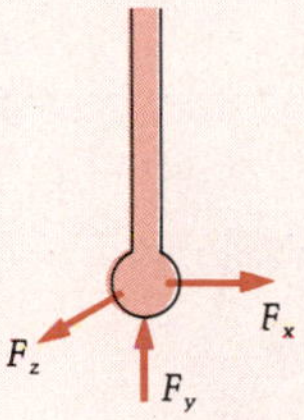 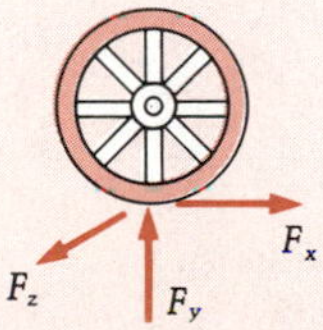	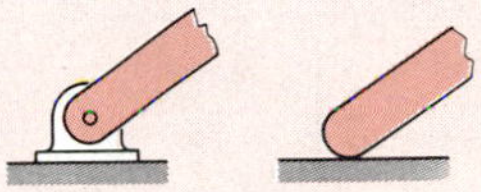	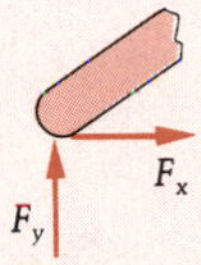
VI. Ball bearing 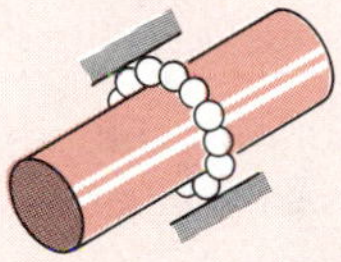	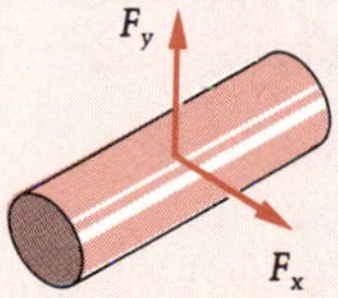	Same as II	

Table 4.1 Continued

Type of Connection	Reaction on $\mathcal{B}$ (Shaded)	Plane Counterpart
VII. Roller bearing 	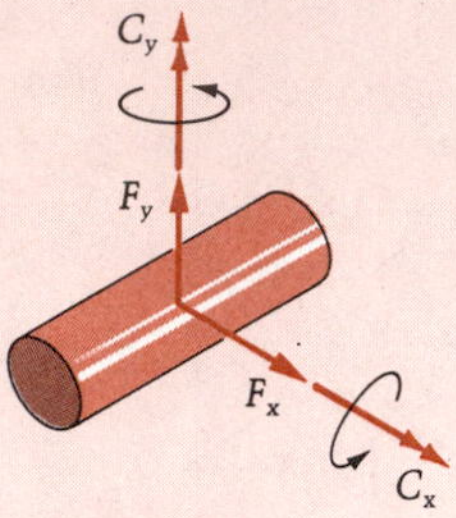	Same as II
VIII. Cable (rope, wire) 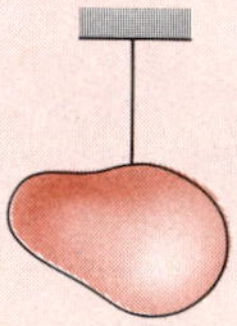	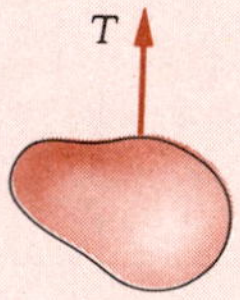	Same as in the figure at the left
IX. Clevis pinned to collar supporting smooth bar 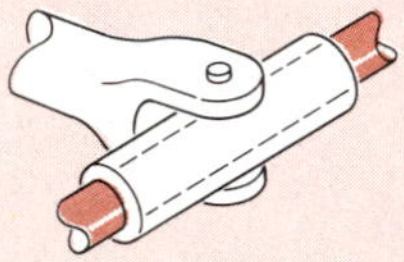	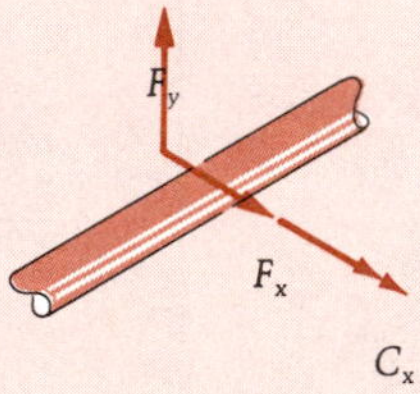	Same as II

are based on the assumption of negligible friction at contact surfaces. *The student should study them* to gain familiarity with the reasoning process and with the symbols that often will be used in the figures that depict exercise problems. A brief description of each follows:

II. An ordinary door hinge is the most common example of this connection. In the absence of friction, rotation is free to take place about the axis of the hinge; thus the only vanishing component of the resultant at a point on the axis is the component of the couple along the axis. Of course if the hinge is "rusty," friction will produce a component of couple in that direction, too, and the resultant will revert to that of I.

III. If there is no resistance to sliding along the hinge axis, the component of force in that direction also vanishes. An example of this connection is a cylindrical pin attached to one body and snugly fit into a cylindrical cavity in a second body if the pin is free to slide without friction along the cavity.

IV. If the pin of item III is inserted in a slot, there is only one component of force and one component of couple. That is, the pin is free to slide in two directions (in the plane of the slot) and to turn about an axis that is perpendicular to the plane of the slot. If a cylindrical roller is inserted between two (necessarily parallel) surfaces, the same kind of resultant is generated provided the contact is along a region sufficiently narrow to be approximated as line contact. In that case we could reason that the resultant should be a force alone. However, not knowing in advance where along the contact line the resultant will act, we express the resultant as a force with preassigned location on this contact line and a companion couple.

V. A ball-and-socket connection is a spherical ball on one body snugly fit into a spherical cavity in the other. No rotation is inhibited, but the attachment point (center of ball or socket) cannot move in any direction. In self-aligning bearings, the bearing housing is supported in this manner. The same resultant is also transmitted when one body is in local ("point") contact with a surface of a second body, there being friction at the interface.

VI. The balls in a ball bearing will exert what are essentially radial point loads on a circumferential line on the surface of a shaft. These produce no moments about diameters of the shaft that intersect that circumferential line. Consequently, the resultant is a force (with components F_x and F_y) having a line of action through the center of the shaft. All of this assumes that the bearing provides no thrust support — that is, resistance to motion of the shaft along its axis. The figure is of course simplified in that the bearing races are not shown.

VII. The rollers in a roller bearing exert loads essentially distributed over longitudinal (axial) lines on the surface of the shaft. Consequently, there is resistance to turning about diameters. Thus the possibility of a couple perpendicular to the axis of the shaft exists in addition to the resultant of F_x and F_y.

VIII. A flexible cable or wire exerts a tensile force in the direction of tangency to the cable at the attachment point. Unless otherwise indicated we shall assume that cables are sufficiently taut that the centerline will be a straight line joining its ends.

IX. The last item is a composite of the clevis pin (II) and rod-in sleeve (III). Note that freedom for rotation about the pin axis eliminates one component of the couple shown in III. That vanishing component is along the axis of the pin.

We now illustrate the construction of free-body diagrams through several examples. As we do this, we keep in mind that the equilibrium equations are

$$\Sigma \mathbf{F} = 0 \qquad \text{or} \qquad \mathbf{F}_r = 0$$

$$\Sigma \mathbf{M}_P = 0 \qquad \text{or} \qquad \mathbf{M}_{rP} = 0$$

where P is any point, and the subscript "r" denotes resultant (of the external force system). The free-body diagram will provide:

1. A catalog of all the external forces and/or couples acting on the body.
2. A graphical display of how much we know (directions, locations) about unknown reactions.
3. The geometric dimensions needed for establishing moments of the forces.

Thus, we shall find that the free-body diagram will include all of the information to be incorporated in the equations of equilibrium.

In many of the examples and subsequent exercise problems, information is given only for two spatial dimensions. It is reasonable for the student to be uneasy about the fact that we have not set down any criteria by which to decide when we can ignore considerations of the third dimension. Universally applicable criteria are not easy to establish, and this, like other issues of mathematical modeling in mechanics, requires experience. Working a large number of three-dimensional problems will provide students some of the experience with which to supplement their raw intuition.

EXAMPLE 4.1

Draw the free body diagram of the homogeneous block of Figure 2.6 if the supporting cables are vertical and attached to its upper corners as shown in Figure E4.1a, instead of meeting at point A.

Solution

We replace the cables by forces that act along their length, as shown in Figure E4.1b. The force of gravity is drawn as a weight W, acting downward through the center of gravity. Note that when the forces are not concurrent, it will become important to label the dimensions of bodies and precise locations of forces.

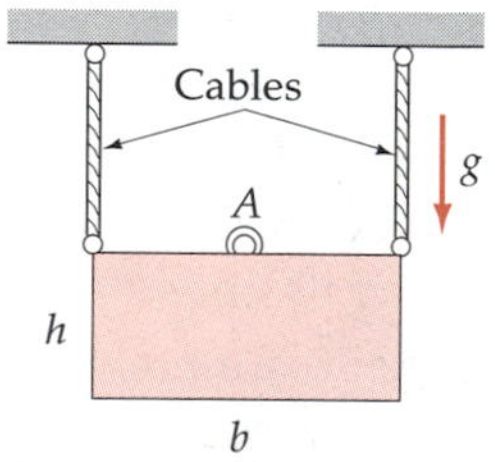

Figure E4.1a

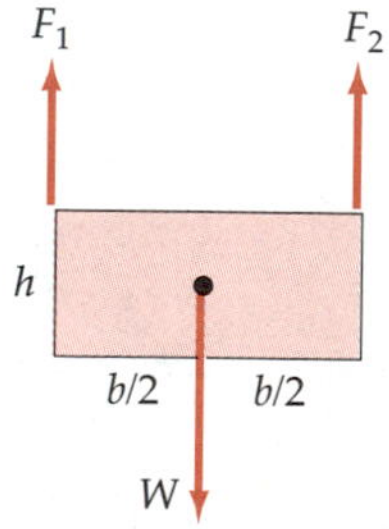

Figure E4.1b

In the extended Problem 4.141, we will outline a proof of the fact that if a body is in equilibrium under the action of three forces, then these forces are necessarily coplanar and either (a) concurrent or (b) parallel. The block B supported as in Figure 2.6 is an example of (a), while the same body, supported as in the preceding example, is an illustration of (b). It is interesting that any body in equilibrium under the action of only three forces is never more complicated than one of these two types!

Example 4.2

Sketch the free-body diagram of the small advertising sign and supporting post if there is a steady wind load producing the resultant shown in Figure E4.2a. The sign itself is sheet metal with a weight of 200 lb, and the post weighs 500 lb.

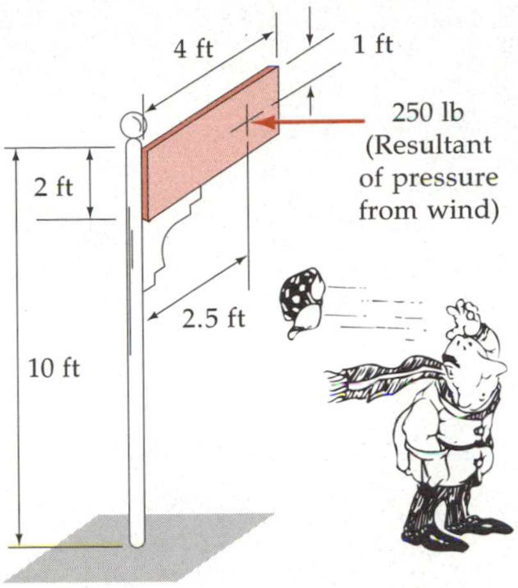

Figure E4.2a

Solution

When we separate the sign and post from the ground we see that the external forces on the body are:

a. The resultant of the pressure from the wind.

b. The weights of the sign and the post.

c. The reaction of the ground on the base of the post.

Thus the free-body diagram is as shown in Figure E4.2b.

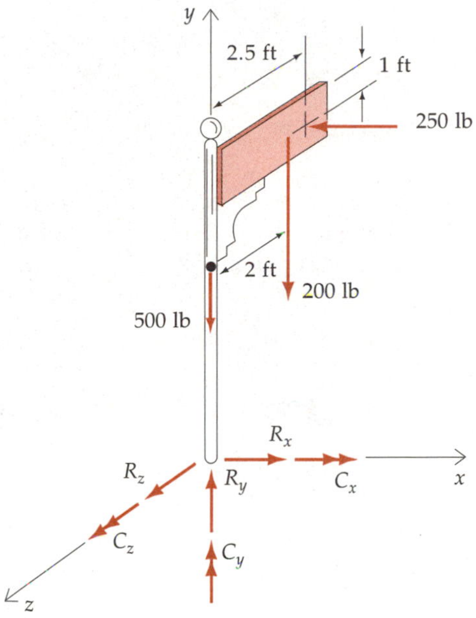

Figure E4.2b

Note that we have tacitly assumed that the mass center of the post is on its centerline and that the sign itself is of constant thickness and constant density so that its mass center is "in the middle." It is important to realize that the connection of the post to the ground provides resistance to displacement in every direction and resistance to rotation about any axis through the base. Thus the force and couple there will *each* have three unknown components.

Question 4.2 Why has it been unnecessary to specify the elevation of the mass center of the post on the free-body diagram?

Answer 4.2 It is only the *line of action* of the force that is important.

EXAMPLE 4.3

Sketch the free-body diagram for the pliers of Figure E4.3a and for each of its parts.

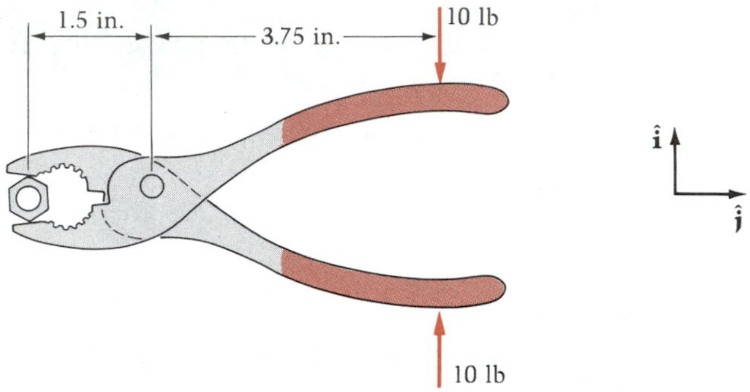

Figure E4.3a

Solution

The only external forces acting on the pliers are the hand-applied 10-pound forces and the reactions of the nut being gripped. The free-body diagram is shown in Figure E.4.3b.

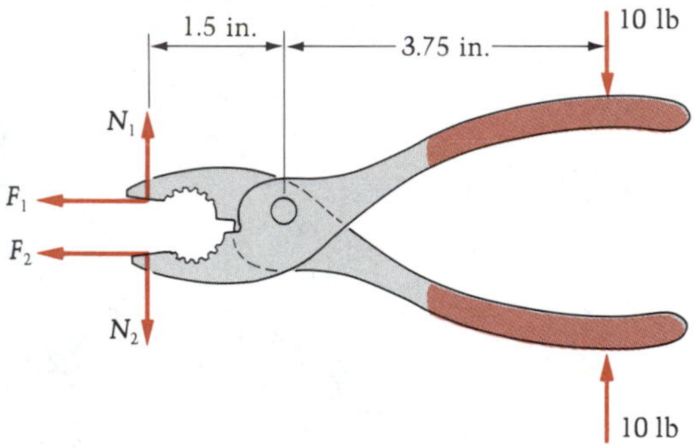

Figure E4.3b

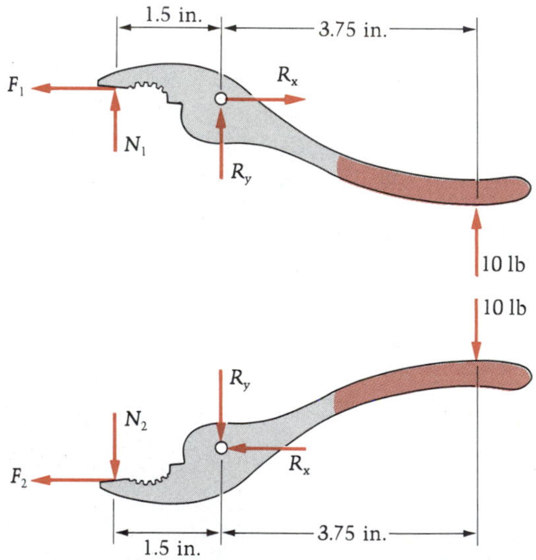

Figure E4.3c

Removing the pin that holds the two parts together, the free-body diagrams of the individual parts are shown in Figure E4.3c. Note that the action transmitted at the pin is just a force since the pin provides no resistance to relative rotation of the parts. Note further that the action-reaction principle has been satisfied automatically in depicting the pin forces.

Question 4.3 How have we guaranteed satisfaction of the action-reaction principle here?

Answer 4.3 The arrow code on the upper free-body diagram communicates the decision to express the force exerted by the lower part on the upper part as $R_x\hat{\mathbf{i}} + R_y\hat{\mathbf{j}}$. Similarly the arrow code on the lower free-body diagram communicates that the force exerted by the upper part on the lower part is $R_x(-\hat{\mathbf{i}}) + R_y(-\hat{\mathbf{j}}) = -(R_x\hat{\mathbf{i}} + R_y\hat{\mathbf{j}})$.

PROBLEMS ▶ Section 4.2

Draw a free-body diagram of body $\mathcal{A}$ in each of the following eight problems.

4.1 Body $\mathcal{A}$ in Figure P4.1 is a uniform 40-lb rod, 6 ft long.

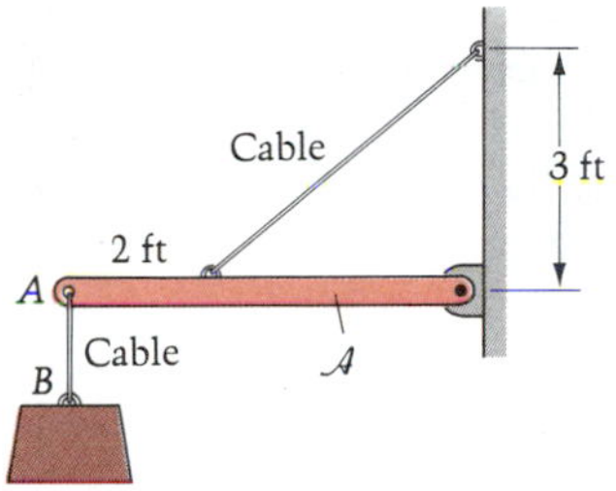

Figure P4.1

4.2 Body $\mathcal{A}$ in Figure P4.2 is a crowbar of negligible weight. Assume the force of the man's hand is directed along his arm, and that the force of the nail is along the axis of the exposed part of the nail.

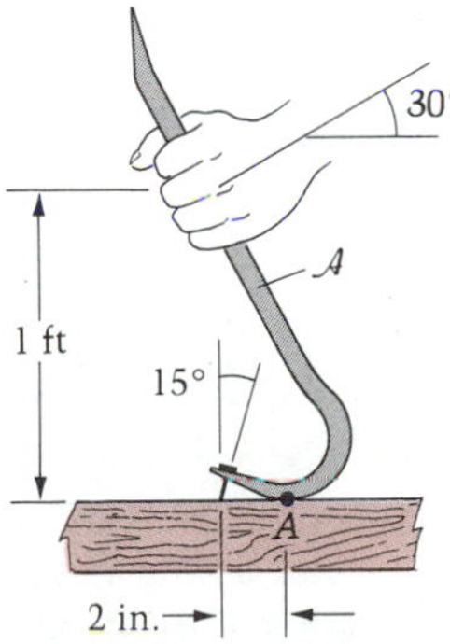

Figure P4.2

4.3 Body $\mathcal{A}$ in Figure P4.3 is the 200-N uniform cylinder in equilibrium on the rough plane. The cable is parallel to the plane.

4.4 Body $\mathcal{A}$ in Figure P4.4 is the 30-kg ladder, together with a 70-kg painter.

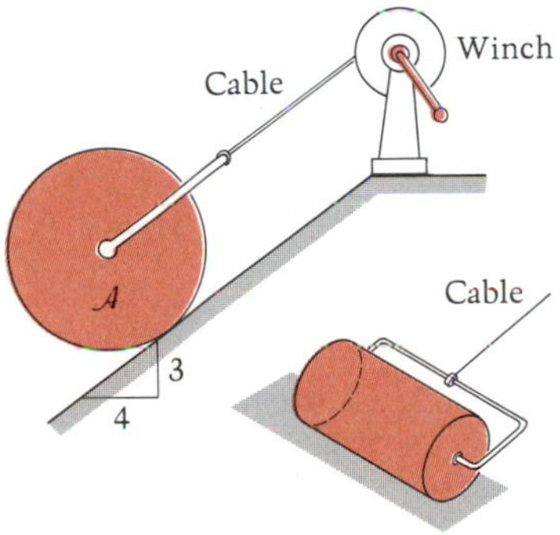

Figure P4.3

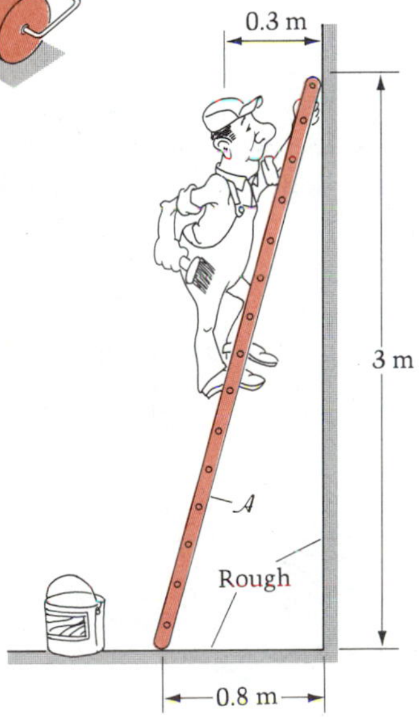

Figure P4.4

4.5 Body $\mathscr{A}$ in Figure P4.5 is the boom of the crane, weighing 1000 lb.

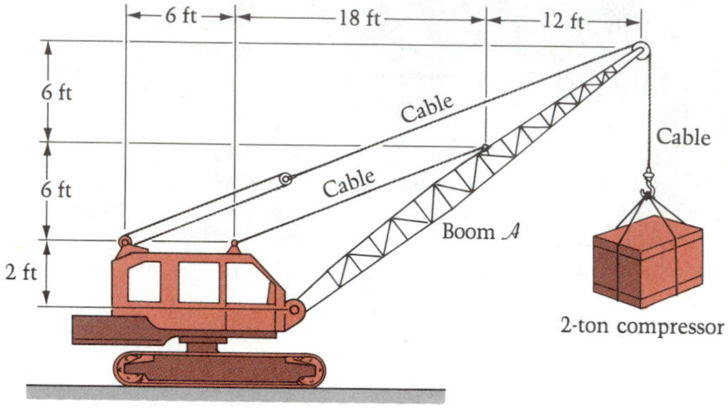

Figure P4.5

4.6 Body $\mathscr{A}$ in Figure P4.6 is a 50-lb door, supported by two hinges, each capable of exerting thrust in the direction of the hinge axis as well as a lateral force and couple.

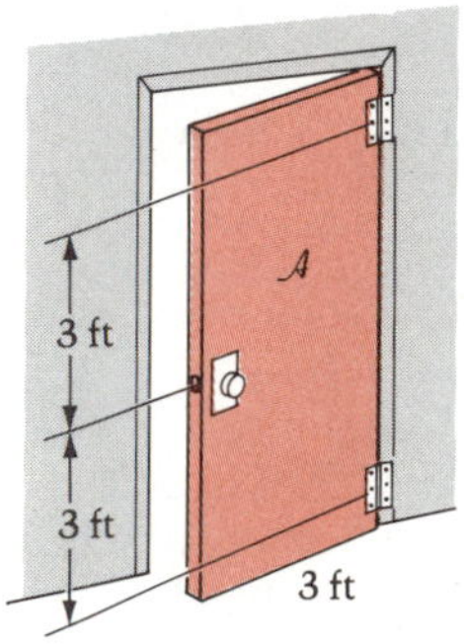

Figure P4.6

4.7 Body $\mathscr{A}$ in Figure P4.7 is a 100-N solid door. Each hinge can exert forces in all three directions and couples about both lateral axes (parallel to y and z, in this case).

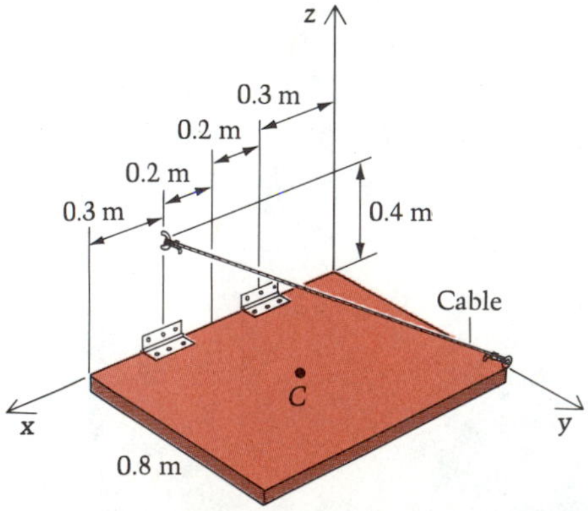

Figure P4.7

4.8 Body $\mathscr{A}$ in Figure P4.8 is the 180-N "corner bar," supported by three smooth eyebolts.

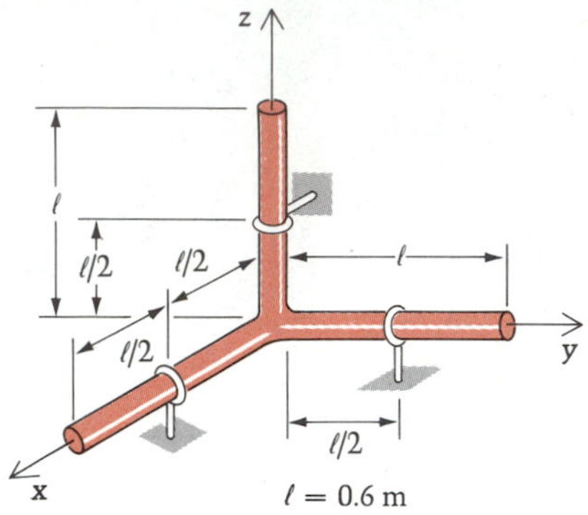

Figure P4.8

4.9 The block W in Figure P4.9 weighs 200 N and the uniform bar ABD weighs 300 N. Draw free-body diagrams of (a) W and (b) ABD including the roller at A as shown. The turning effect of the 1000-N couple applied to ABD is in the plane of the page.

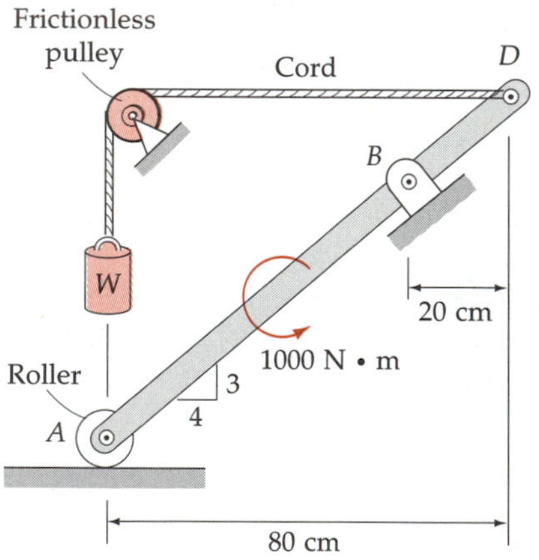

Figure P4.9

4.10 Draw a free-body diagram of the slotted bar ABC of Figure P4.10. The smooth pin bearing against the slot is fixed to the other bar, DBE, and the weights of the members may be neglected.

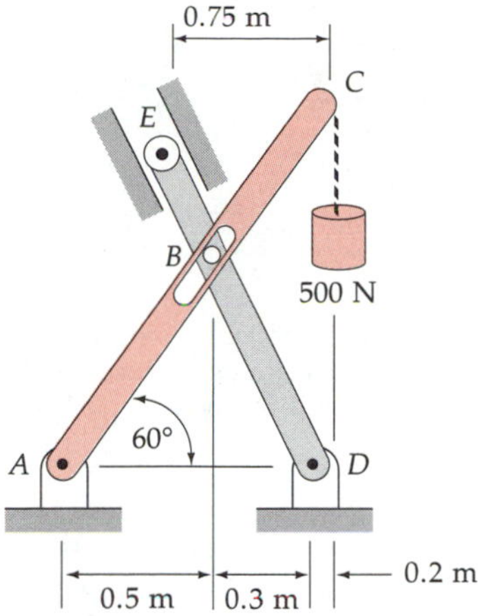

Figure P4.10

4.11 Draw free-body diagrams of the three bars *CE*, *BEF*, and *ABCD* in Figure P4.11. Their weights may be neglected in comparison with the 600-lb load applied.

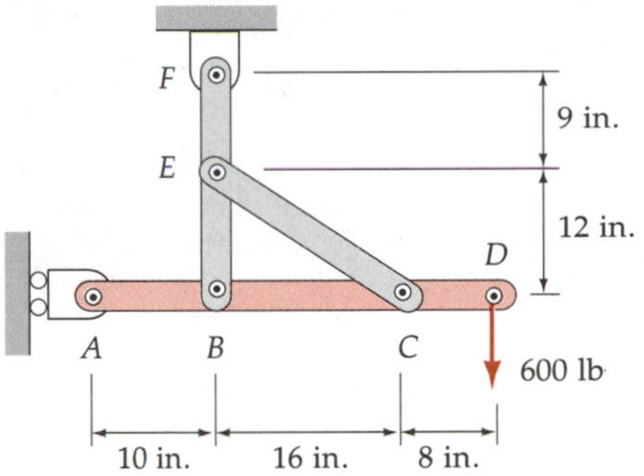

Figure P4.11

4.3 Fundamental Applications of the Equilibrium Equations

Number of Independent Equations

The most common problem in statics is the determination of unknown forces and couples which together with prescribed forces and couples hold a body in equilibrium. As we have seen in our discussion of free-body diagrams, the unknown forces and couples will be expressed in terms of the least number of unknown scalars (components) consistent with what is known about directions. There arises then the question: will the equilibrium equations provide a sufficient number of algebraic equations relating these scalars so that they may be found? At first glance we might be tempted to answer with an unqualified yes, since $\Sigma \mathbf{M}_P = \mathbf{0}$ for every point P so that there is no limit to the number of moment equations of equilibrium.

There is, however, a fallacy in this reasoning, because these additional moment equations are not independent. Let us first illustrate this for a planar equilibrium problem, in which all the forces (and turning effects of couples, if any) lie in a plane as in Figure 4.6(a). We recall that any system of forces and couples may be replaced by an equipollent system consisting of a force at a preselected point plus a couple. Thus the system at Point A in Figure 4.6(b), with the force broken into its x- and y-components, makes the same contribution to the equilibrium equations as does the original, actual system (a):

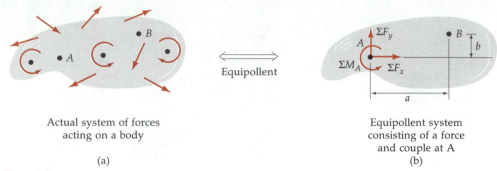

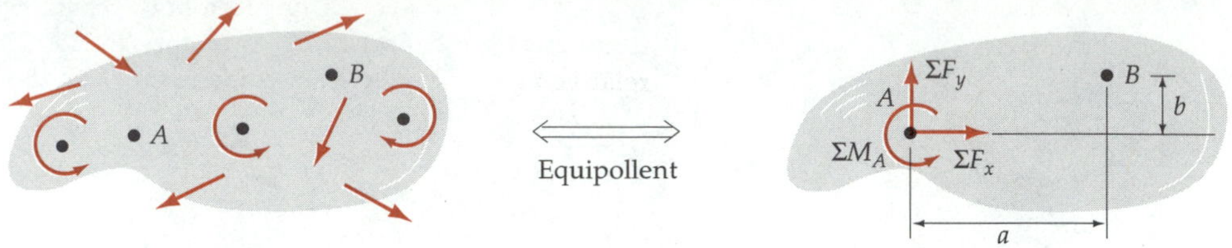

Actual system of forces acting on a body

(a)

Equipollent system consisting of a force and couple at A

(b)

Figure 4.6

Suppose now that we have written the three equilibrium equations $\Sigma F_x = 0$, $\Sigma F_y = 0$, and $\Sigma M_A = 0$, for the body. If we now additionally write $\Sigma M_B = 0$, *this will not be an independent equation,* because it is seen below to be *already satisfied:*

$$\overbrace{\underbrace{\Sigma M_A}_{0} + \underbrace{(\Sigma F_x)}_{0} b + \underbrace{(\Sigma F_y)}_{0}(-a)}^{\Sigma M_B = 0} = 0$$

$$0 + 0 + 0 = 0$$

$$0 = 0$$

Therefore, no new information is gained by writing additional moment equations once three independent equilibrium equations have been expressed. Problems 4.24–26 at the end of the section will deal with the question of when two (or three) moment equations could be used *if* we do not use one (or either) of the force equations.

A more formal vector proof of the above argument goes as follows for general (including three-dimensional) equilibrium problems: In Section 3.5 we found that for any system of forces:

$$\Sigma \mathbf{M}_Q = \Sigma \mathbf{M}_P + \mathbf{r}_{QP} \times (\Sigma \mathbf{F})$$

Thus, if $\Sigma \mathbf{M}_P = \mathbf{0}$ for some point P, and $\Sigma \mathbf{F} = \mathbf{0}$, then $\Sigma \mathbf{M}_Q$ *automatically* vanishes for *every* point Q. Therefore, the equations of equilibrium are the *two* independent vector equations

$$\Sigma \mathbf{F} = \mathbf{0}$$
$$\Sigma \mathbf{M}_P = \mathbf{0}$$

or in the notation of resultants

$$\mathbf{F}_r = \mathbf{0}$$
$$\mathbf{M}_{rP} = \mathbf{0}$$

Question 4.4 These equations pertain when the body is in equilibrium relative to what frames of reference?

Static Indeterminacy

The vanishing of a vector is guaranteed by the vanishing of three distinct (usually taken to be mutually perpendicular) components. Thus the two vector equations of equilibrium are equivalent to *six* component (scalar) equations. If we have more than six unknown scalars, then we cannot have enough independent equations, and the problem may fall into a category known as **static indeterminacy,** mentioned previously in Section 2.3. Just because we have twice as many vector equations of equilibrium (two instead of one) as we did for the particle (in Chapter 2) does not mean that we cannot still run into the problem of indeterminacy. For example, suppose we alter the problem of Example 2.6 and say that the surfaces of contact are *not* smooth, which means there is the possibility of friction forces as well as normal forces exerted by the planes on the sphere. The free-body diagram then becomes Figure 4.7 below:

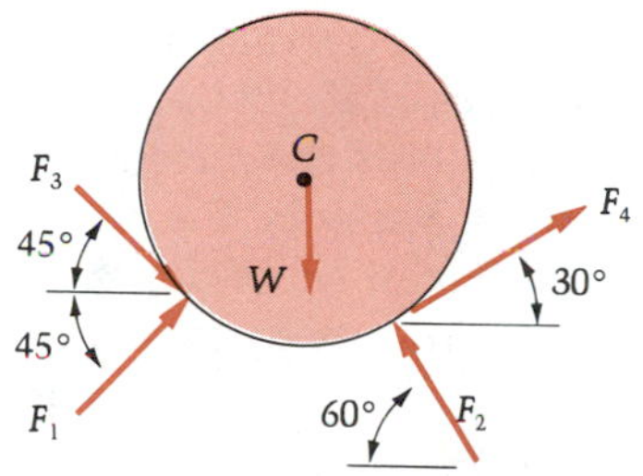

Figure 4.7

where F_3 and F_4 represent the friction forces.

The moment-equation of equilibrium $\Sigma \mathbf{M}_C = \mathbf{0}$ now yields

$$\circlearrowleft\!+ \quad F_4 R + F_3 R = 0$$

or

$$F_4 + F_3 = 0$$

From $\Sigma \mathbf{F} = \mathbf{0}$ we have, as before, two scalar equations, but now in the four unknowns F_1, F_2, F_3, and F_4. These two equations plus the above result of balancing moments, $F_4 + F_3 = 0$, constitute three equations in four unknowns and hence the problem is **statically indeterminate.**

We note that if just one surface is smooth, this condition renders the problem statically determinate because the remaining friction force vanishes (from the moment equation), and F_1 and F_2 revert to the same answers we found back in Example 2.6.

Answer 4.4 Inertial frames of reference.

Whenever we encounter indeterminacy, additional information, usually having to do with the deformability of the body, is needed to find the unknown forces.

Now in many situations the equilibrium equations provide *fewer* than six independent scalar equations, yet the problems need not be indeterminate. We have seen examples of this in Chapter 2. One of these was the block of Figure 2.6, supported by two taut cables and reproduced below as Figure 4.8:

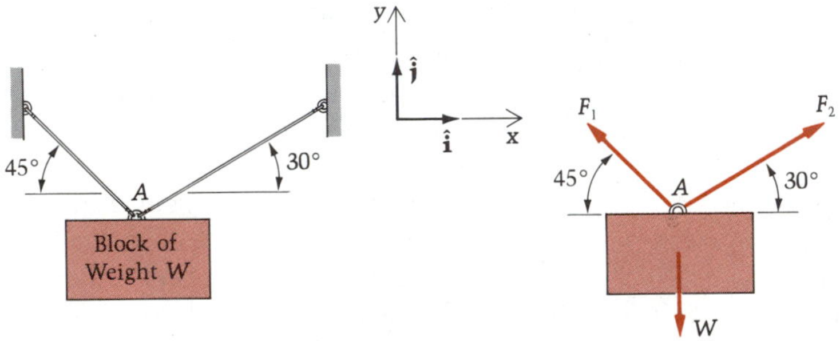

Figure 4.8

First we observe from the free-body diagram that all the external forces have lines of action passing through the same point (A);* in Chapter 3 we called this a concurrent force system. Since $\Sigma \mathbf{M}_A = \mathbf{0}$ regardless of the values of F_1 and F_2, the moment equation provides no information about these forces. More generally, we may state that whenever the external forces on a body constitute a concurrent system, there is a moment equation of equilibrium that is identically satisfied, and the greatest number of scalar equations relating the forces will be three.

In the above problem, the z-component of $\Sigma \mathbf{F} = \mathbf{0}$ also yields $0 = 0$, i.e., no information, so in fact only two equations were available. We found in Section 2.3 that they were sufficient to solve for the cable tensions, however.

Generally, then, a body in equilibrium will have between one and six useful equations available for determining forces, depending on the physical configuration in which the body is placed. We shall not attempt to catalog all the various circumstances in which one or more of the component-equations of equilibrium is identically satisfied, but we shall point out some commonly occurring situations as they arise in the examples.

Types of Problems in This Section: Bodies Left Intact

The examples and problems in this section will feature finite-sized bodies which are either (a) a single physical object, or (b) two or more physical objects in contact and not needing to be separated in order to effect a

---

* The block must be "hanging" in such a way that the mass center is directly below A; otherwise, $\Sigma \mathbf{M}_A \neq \mathbf{0}$.

solution to the equilibrium equations. In other words, this section comprises problems involving objects or groups of objects that are *left intact.* In Section 4.4 to follow, we will find it necessary to go a step further and separate the contacting objects in order to obtain solutions. We will then draw FBDs and write equilibrium equations for the separated objects, treating each in turn as the body being analyzed.

Finally, while we intentionally have not made a clear separation of discussions of two- and three-dimensional problems, the examples that follow are ordered so that the two-dimensional ones come first. In these, where we are dealing with a planar force system, it is important to realize that three of the component equations are satisfied identically. If the xy plane is the plane of the force system, $\Sigma F_z \equiv 0$ and each force produces, with respect to a point in the plane, a moment perpendicular to the plane. Thus, if P is a point in this plane, then $(\Sigma M_P)_x \equiv 0$ and $(\Sigma M_P)_y \equiv \mathbf{0}$. For this restricted class of problems, then, the component equations not automatically satisfied are

$$\Sigma F_x = 0$$
$$\Sigma F_y = 0$$
$$(\Sigma M_P)_z = 0$$

In the future, we shall shorten $(\Sigma M_A)_z$ to simply ΣM_A when the problem is a "plane" one; when this is the case, only "z moments" are normally written.

There follows a series of examples of equilibrium problems that can be solved by leaving the body intact. In each one, as in Chapter 2, the student should note carefully the great importance of the free-body diagram in formulating the equations of equilibrium. The first example is concerned with the equilibrium of a single physical object, a beam. Note that throughout the book, where weights of bodies subjected to other loads are not given as data in examples and in problems, it is to be understood that these weights may be neglected in comparison with the other loads. That is the case with the beam of this example, which also deals with a distributed load.

Example 4.4

Find the reactions on the ends of the simply supported beam shown in Figure E4.4a.

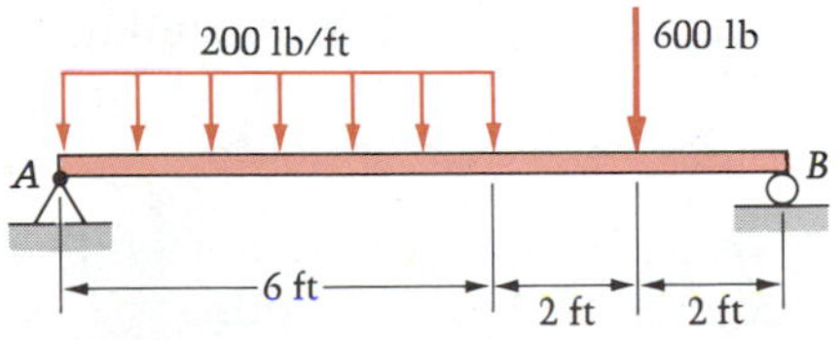

Figure E4.4a

Solution

From Section 3.9, we recognize the resultant of the distributed load to be the area beneath the loading curve, or 1200 lb. Its line of action is at the center of the

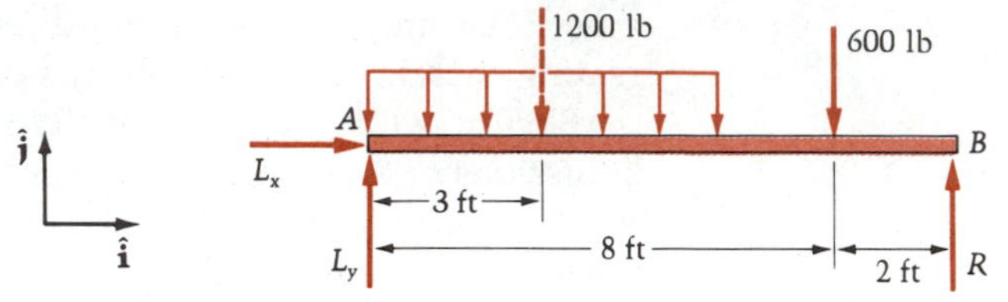

Figure E4.4b

loaded segment as indicated on the free-body diagram (Figure E4.4b) of the beam. Using the equilibrium equations,

$$\Sigma \mathbf{F} = \mathbf{0}$$

yields

$$L_x\hat{\mathbf{i}} + L_y\hat{\mathbf{j}} + 1200(-\hat{\mathbf{j}}) + 600(-\hat{\mathbf{j}}) + R\hat{\mathbf{j}} = \mathbf{0}$$

which has the following component equations:

$$L_x = 0 \tag{1}$$

and

$$L_y - 1200 - 600 + R = 0 \tag{2}$$

Also,

$$\Sigma \mathbf{M}_A = \mathbf{0}$$

yields

$$3(1200)(-\hat{\mathbf{k}}) + 8(600)(-\hat{\mathbf{k}}) + 10R\hat{\mathbf{k}} = \mathbf{0}$$

from which

$$-3600 - 4800 + 10R = 0$$

$$R = 840 \text{ lb} \tag{3}$$

Substituting R into Equation (1),

$$L_y = 1800 - R = 1800 - 840$$

$$L_y = 960 \text{ lb}$$

In the preceding example, note that the roller reaction at B was found with a single equation, by summing moments about the point of intersection (A) of the other two unknowns. We will often find this idea useful.

Scalar Vs. Vector Approach

We wish now to note a slightly different approach to the preceding example. Instead of writing the vector equations $\Sigma \mathbf{F} = \mathbf{0}$ and $\Sigma \mathbf{M}_A = \mathbf{0}$ and then picking off the coefficients of the unit vectors $\hat{\mathbf{i}}$, $\hat{\mathbf{j}}$, and $\hat{\mathbf{k}}$ to form the scalar equilibrium equations (1, 2, 3), we could have written, a bit more quickly,

$$\overset{+}{\rightarrow} \qquad \Sigma F_x = 0 = L_x$$

$$+\uparrow \qquad \Sigma F_y = 0 = L_y - 1200 - 600 + R$$

$$\circlearrowleft + \qquad \Sigma M_A = 0 = -3(1200) - 8(600) + 10R$$

This scalar approach has the advantage of omitting the unit vectors and proceeding immediately to the algebraic equations. One must be especially careful here to affix the correct sign on the various terms; it is also helpful to display a symbol such as $\overset{+}{\rightarrow}$ to the left of an equation as a reminder of the direction of the unit vector that is being suppressed.

One more point is worth mentioning in conjunction with the preceding example: It is often worthwhile to use another moment equation of equilibrium to check the numerical results. Let's see if $\Sigma M_B = 0$ is satisfied by the values we have calculated for L_x, L_y, and R:

$$\begin{aligned} \Sigma M_B &= 0 \\ 2(600) + 7(1200) - 10L_y &= 0 \\ 1200 + 8400 - 10(960) &= 0 \\ 9600 - 9600 &= 0 \end{aligned}$$

This check correctly suggests that we could have solved this problem using the three scalar equations: $\Sigma F_x = 0$, $\Sigma M_A = 0$, and $\Sigma M_B = 0$. As mentioned earlier, Problems 4.24–4.26 are concerned with the possibilities of using two and three moment equations when the force system (loads and reactions) is coplanar.

For an example of the equilibrium of a body comprising more than one physical object in contact, let us revisit the diver and diving board of Sections 4.1 and 4.2:

EXAMPLE 4.5

If $a = 5$ ft, $L = 14$ ft, and the respective weights of the diver and the board are 200 lb and 90 lb, find the reactions onto the uniform board in Figure E4.5a at the pin A and the roller B.

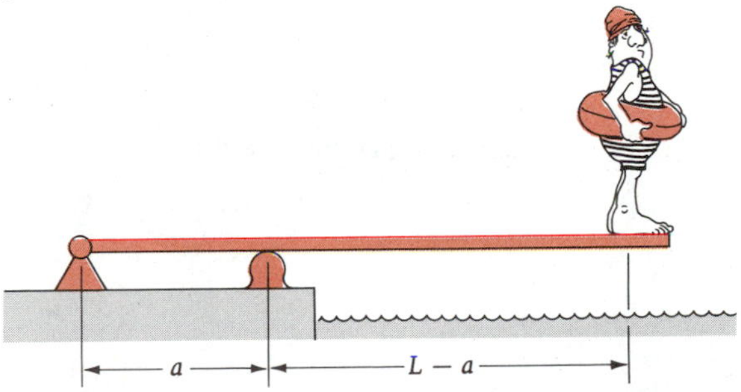

Figure E4.5a

Solution

The free-body diagram of the body (diver plus board) is shown in Figure E4.5b. Note that we must allow for the possibility of two force components (A_x and A_y) from the pin, but only a vertical component (B) from the roller:

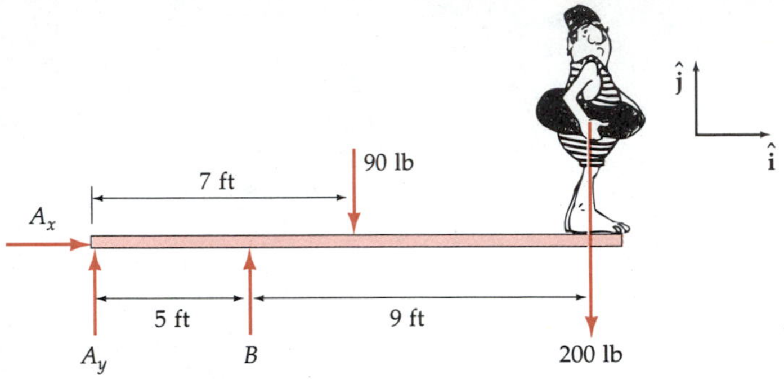

Figure E4.5b

We see from the FBD that we have three unknowns (A_x, A_y, B), and we know we can write three independent equations and solve for them. The first equation expresses the fact that the x-components of the external forces must sum to zero:

$$\overset{+}{\longrightarrow} \qquad \Sigma F_x = 0 = A_x$$

Thus A_x, being the only force in the x-direction acting on the body, must in this case vanish.

> **Question 4.5** Will the force A_x remain zero while the diver dives off the board?

Next we use the FBD to write the "y-equation" of equilibrium:

$$+\uparrow \qquad \Sigma F_y = 0 = A_y + B - 90 - 200$$

or

$$A_y + B = 290 \tag{1}$$

It is clear that we cannot solve for A_y and B without invoking the remaining equation — the "moment equation" of equilibrium:

$$\circlearrowleft\!+ \qquad \Sigma M_A = 0 = B(5 \text{ ft}) - (90 \text{ lb})(7 \text{ ft}) - (200 \text{ lb})(14 \text{ ft})$$

$$B = \frac{630 + 2800}{5} = \frac{3430}{5} = 686 \text{ lb}$$

Equation (1) then yields:

$$\begin{aligned} A_y &= 290 - B \\ &= 290 - 686 \\ &= -396 \text{ lb} \end{aligned}$$

Answer 4.5 No.

The minus sign tells us that the pin at A is pushing *down* on the board. Although we wrote the "y-equation" in scalar form, we were actually representing the force as $A_y\,\hat{\mathbf{j}}$, and we obtained $A_y = -396$. Thus the force in vector form is $-396\hat{\mathbf{j}}$, or $396\,(-\hat{\mathbf{j}})$, or $396 \downarrow$ lb.

As an overall check on our answers, let us sum the moments at C:

$$\begin{aligned}\Sigma M_C &= 396(7) - 686(7-5) - 200(7)\\ &= 2772 - 1372 - 1400\\ &= 0 \;\checkmark\end{aligned}$$

Just because the sum of the moments about C is zero does not *guarantee* that we haven't made a mistake, but it does make it highly unlikely.

In the next example, three bars are pinned together to form a body known as a *frame*.

Example 4.6

The "A-frame" is subjected to the 150-N load as shown in Figure E4.6a. Find the pin reaction at A and the force exerted by the roller at C.

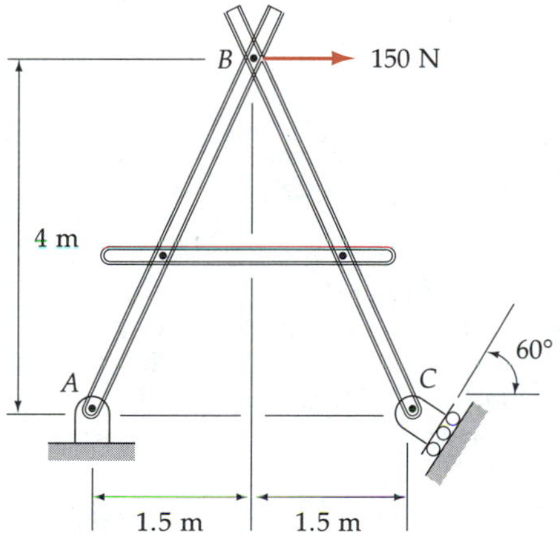

Figure E4.6a

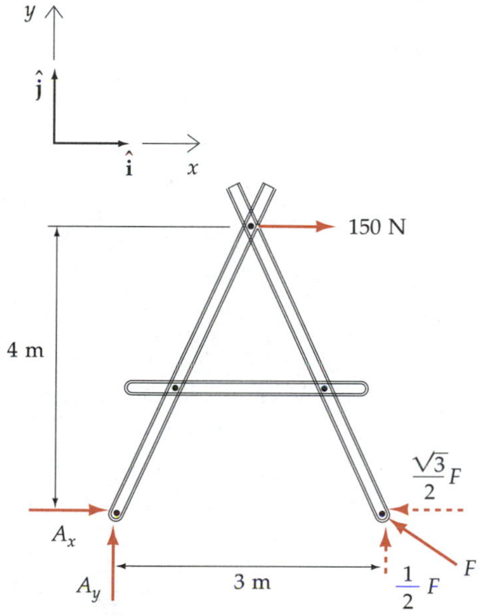

Figure E4.6b

Solution

Referring to the free-body diagram in Figure E4.6b, we see that we may obtain the reaction at C by summing moments at A. Since the force system is coplanar we may easily use the "scalar" form of the moment equation. Thus, counting counterclockwise moments as positive,

$$\circlearrowleft\!+ \quad \Sigma M_A = 0$$

$$3(F/2) - 4(150) = 0$$

$$F = 400 \text{ N}$$

The student should realize (or confirm) that this scalar equation is precisely what results when we write the vector equation $\Sigma \mathbf{M}_A = 0$ and then pick off coefficients of $\hat{\mathbf{k}}$. Were we instead to take *clockwise* moments as positive, then

$$+\!\circlearrowright \quad \Sigma M_A = 0$$

yields

$$-3(F/2) + 4(150) = 0$$

which is nothing more than the result of writing $\Sigma \mathbf{M}_A = \mathbf{0}$ and picking off coefficients of $(-\hat{\mathbf{k}})$.

The component forms of $\Sigma \mathbf{F} = \mathbf{0}$ are

$$\Sigma F_x = 0$$

and

$$\Sigma F_y = 0$$

From the first, we find

$$A_x + 150 - \frac{\sqrt{3}}{2}(400) = 0$$

so that

$$A_x = 196 \text{ N}$$

From the second,

$$A_y + F/2 = 0$$

so that

$$A_y = -200 \text{ N}$$

These are the same equations we would obtain by writing $\Sigma \mathbf{F} = \mathbf{0}$ and then picking off coefficients of $\hat{\mathbf{i}}$ and $\hat{\mathbf{j}}$, respectively.

In conclusion, the magnitude of the pin reaction at A is $\sqrt{A_x^2 + A_y^2} = \sqrt{(196)^2 + (200)^2} = 280$ N, and vectorially the force exerted by the pin on the frame at A is $196\hat{\mathbf{i}} - 200\hat{\mathbf{j}}$ N. The force exerted on the frame at C is

$$400\left(-\frac{\sqrt{3}}{2}\hat{\mathbf{i}} + \frac{1}{2}\hat{\mathbf{j}}\right) = -346\hat{\mathbf{i}} + 200\hat{\mathbf{j}} \text{ N}$$

Before moving on, we mention the concept of *stability*. Let the sphere of Example 2.6 (see Figure 4.9(a)) be nonuniform so that the mass center C does not coincide with the geometric center O. Taking the surfaces to be smooth, let us find all possible equilibrium configurations of the sphere.

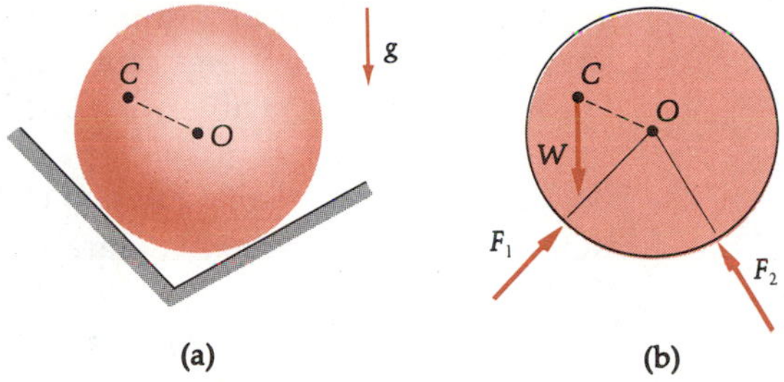

Figure 4.9

From the free-body diagram (Figure 4.9(b)) we see, as before, that the lines of action of the two reactions, F_1 and F_2, pass through the geometric center (O) of the sphere. Thus in order for $\Sigma \mathbf{M}_O = \mathbf{0}$ to be satisfied, the mass center C must lie either directly above or below O, so that the always-vertical weight can also pass through O.

The reader may recognize that this problem is similar to that of finding the equilibrium positions of a body supported in the manner of a pendulum, as shown (see Figure 4.10) in the next illustration. The equations of equilibrium tell us that there are two such positions: one where the mass center is directly below the support and one where the mass center is directly above. Our experience tells us that the body will not remain (without additional restraint) in this second position. This configuration is said to be an *unstable* equilibrium configuration and the first is said to be a *stable* equilibrium configuration. An important point to recognize is that the equilibrium equations do not themselves distinguish these two types of equilibrium states.

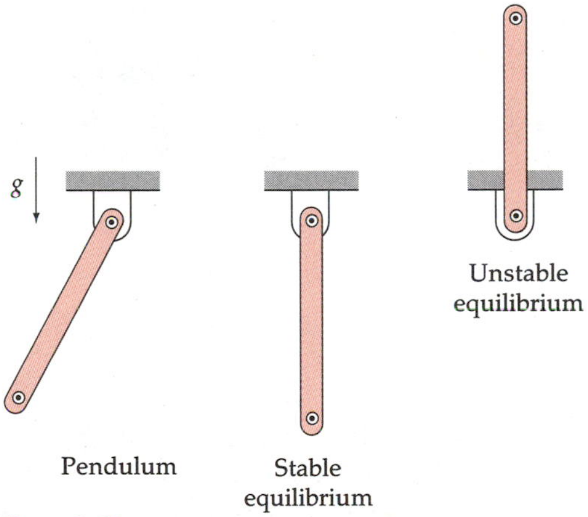

Figure 4.10

Special Results for Pulleys and Two-Force Members

We now wish to derive two extremely useful results before presenting another series of examples. One of these is concerned with the tensions in a belt on either side of a pulley; the other deals with the situation in which a body is held in equilibrium by the action of just two forces. These situations will arise many times in both examples and problems, and knowing the results will always facilitate the solutions.

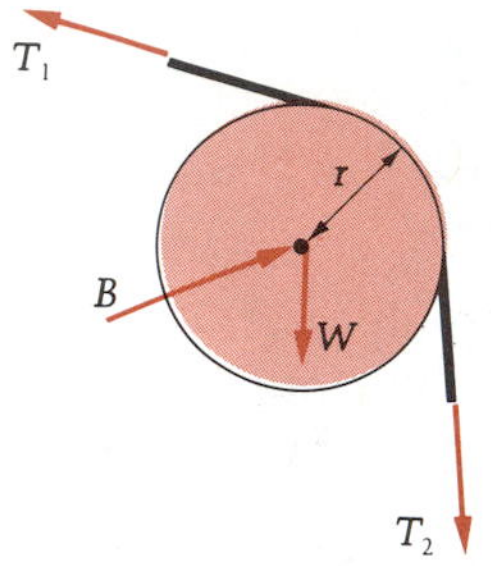

Figure 4.11

Belt, Rope, Cord, or Cable Passing Over a Pulley on Frictionless Bearings Referring to the free-body diagram (Figure 4.11), the equilibrium equation requiring that $\Sigma M_{\text{bearing axis}} = 0$ yields (provided the mass center of the pulley is on the bearing axis):

$$\circlearrowleft\!+ \qquad T_1 r - T_2 r = 0$$

(The symbol $\circlearrowleft\!+$ means that the unit vector out of the page has been suppressed from each term in the equation.) Thus we obtain

$$T_1 = T_2$$

That is, the belt tension is the same on both sides of a pulley supported in equilibrium by frictionless bearings. We used this result in Example 2.8, deferring the proof to the current section.

Two-force Body (or Member) Suppose that a light body is held in equilibrium by two (and only two) external forces. We let the plane of the page contain the points of application of these forces as shown below in Figure 4.12(a). In order that $\Sigma\mathbf{F} = \mathbf{0}$, we must have $\mathbf{F}_1 = -\mathbf{F}_2$ so that the free-body diagram might now appear as shown in Figure 4.12(b). But wait! These two forces, as drawn, constitute a couple and so $\Sigma\mathbf{M} \neq \mathbf{0}$ unless the lines of action of the two forces coincide. Thus the proper free-body diagram must be Figure 4.12(c) below:

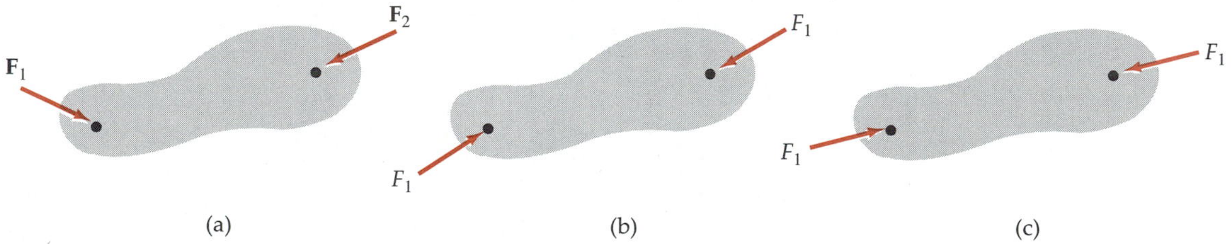

Figure 4.12

We now see that the two forces acting on the body must have a common line of action — *the line joining the two points of application* of the forces. Note in Figure 4.12(c) above that the forces are drawn acting *toward* each other, tending to compress the material between them. Alternately, they might in a given problem be directed *away from* each other, tending to stretch the material between them. These are the only two possibilities for the directions of the pair of forces acting on a two-force member.

In each of the next three examples a part of the body to be analyzed is either a pulley on bearings of negligible friction, or a two-force body.

EXAMPLE 4.7

The light pulley (radius 0.1 m) is supported by frictionless bearings at the end of the uniform 640-N beam *AB*, as shown in Figure E4.7a. A cable is wrapped around the pulley and connected to a ceiling at points *C* and *D*. Find the force in the cable and the reactions of the pin at *A* onto the beam.

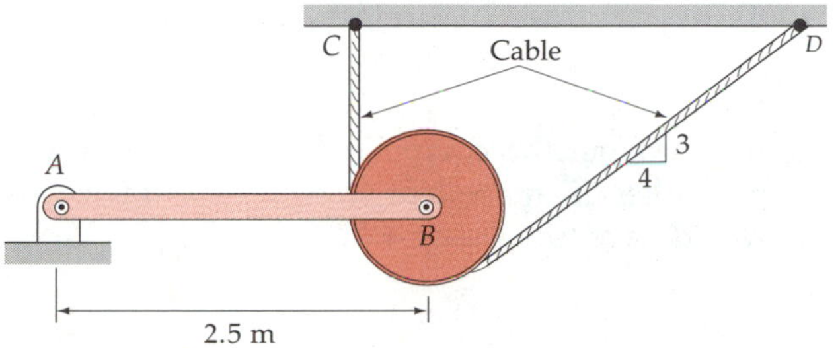

Figure E4.7a

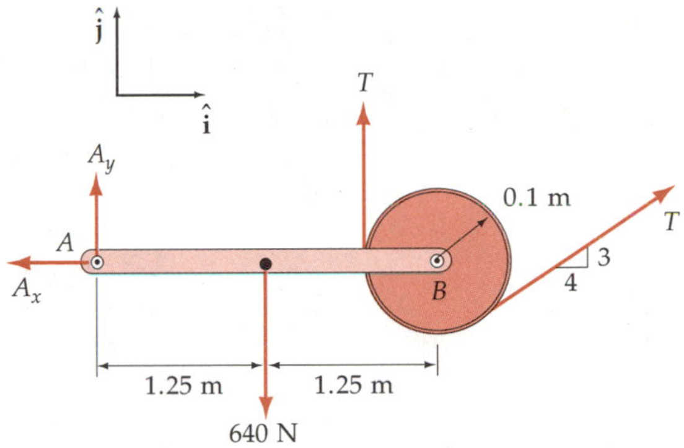

Figure E4.7b

Solution

In drawing the free-body diagram (see Figure E4.7b) we use the fact that the cable tension on either side of the pulley is the same, and we call it T. The pin reaction at A is represented by force components in the horizontal and vertical directions. We drew A_x acting to the left because in this case we are certain of its direction — it has to oppose the horizontal component of the right-most cable tension, which acts to the right. Observations like this are not essential, but they do cut down on the number of minus signs in the solution, and they encourage good thinking about such things as directions of forces.

If we now sum the moments about the pin A and equate the result to zero, the only unknown in the equation will be T, and we can thus find it with a single equation (the "moment equation"):

$$\circlearrowleft\!+ \qquad \Sigma M_A = 0 = -(640 \text{ N})(1.25 \text{ m}) + T(2.4 \text{ m}) + 0.6T(2.5 \text{ m}) + 0.8T(0) + T(0.1 \text{ m})$$

so that

$$T = \frac{800}{4.00} = 200 \text{ N}$$

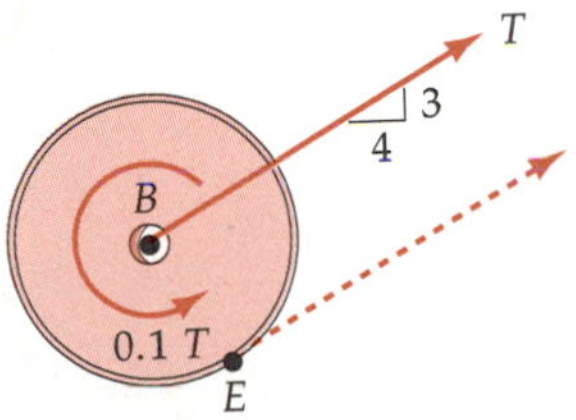

Figure E4.7c

In the above equation, we have for convenience replaced the rightmost tension by an equipollent system at B as shown in Figure E4.7c at the left. This is easier than resolving T into its components at E and then computing the moments of these components about A.* Recall from Section 3.6 that the equipollent system at B is the same force at the new point (B) accompanied by the moment ($0.1T$) that the force at its original point (E) exerts about B.

* See problem 4.14.

To obtain the pin reaction at A, we use the "force equations":

$$\Sigma F_x = 0 = -A_x + 0.8\,\cancelto{200}{T}$$

so

$$A_x = 160 \text{ N}$$

$$+\uparrow \qquad \Sigma F_y = 0 = A_y - 640 + \cancelto{200}{T} + 0.6\,\cancelto{200}{T}$$

$$A_y = 640 - 200 - 120 = 320 \text{ N}$$

Thus the pin reaction at A onto the beam is the vector force $-160\hat{\mathbf{i}} + 320\hat{\mathbf{j}}$ N.

Let us carry the discussion about the equipollent system at B in the preceding example a bit further. If equipollent systems to *both* tensions were drawn at B, we would have (see Figure 4.13):

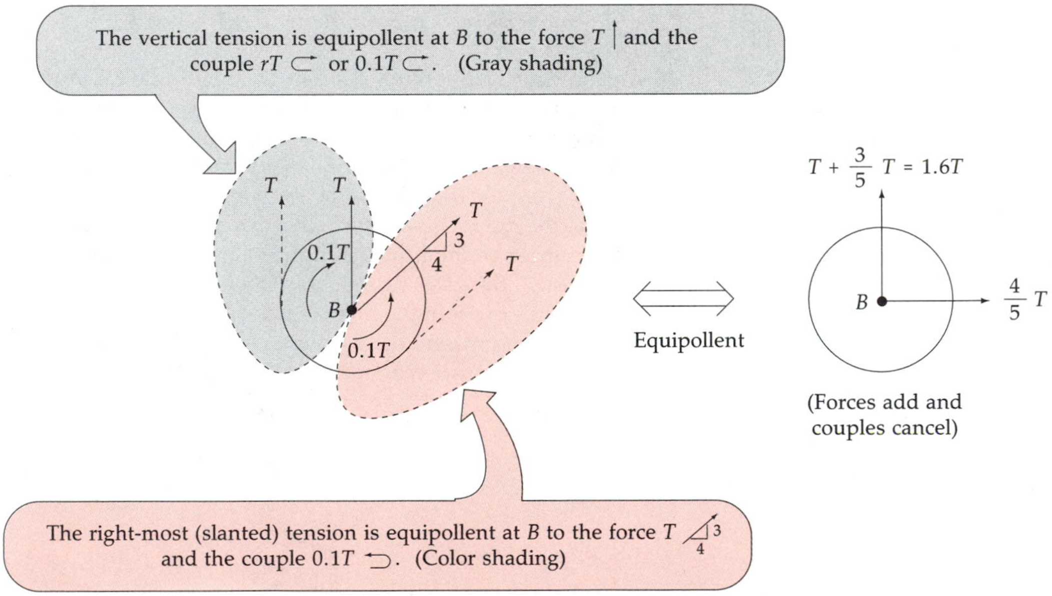
The vertical tension is equipollent at B to the force T ↑ and the couple rT or 0.1T. (Gray shading)
T
T
0.1T
B
0.1T
T
3
4
T
Equipollent
T + 3/5 T = 1.6T
B
4/5 T
(Forces add and couples cancel)
The right-most (slanted) tension is equipollent at B to the force T and the couple 0.1T. (Color shading)

Figure 4.13

This procedure would have made the solution for T even simpler; now we would have:

$$\circlearrowleft\!+ \qquad \Sigma M_A = 0 = (-640 \text{ N})(1.25 \text{ m}) + (1.6T)(2.5 \text{ m})$$

$$T = 200 \text{ N} \qquad \text{(as before)}$$

EXAMPLE 4.8

A dumping mechanism is shown in Figure E4.8a. The weight of the bed plus contents is 1200 lb. It is in equilibrium in the given position with mass center at C. Find the force in the strut AB, which contains a hydraulic cylinder for raising and lowering the bed.

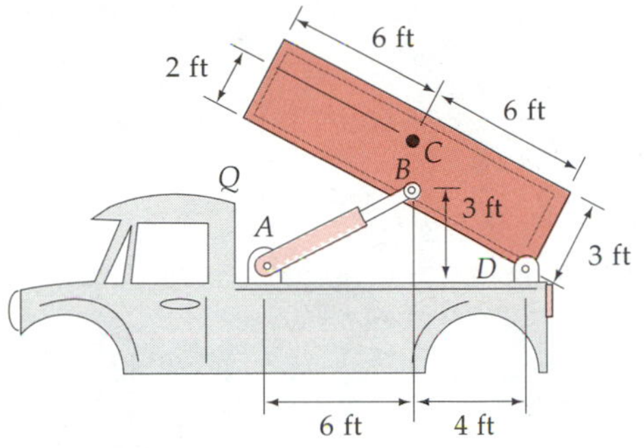

Figure E4.8a

Solution

Recognizing that the strut AB is a two-force member, we know that the force exerted on it at A by the pin lies along the line AB. Thus, we draw the free-body diagram as shown in Figure E4.8b. Note that the body we are analyzing consists of the bed, its contents, and the strut.

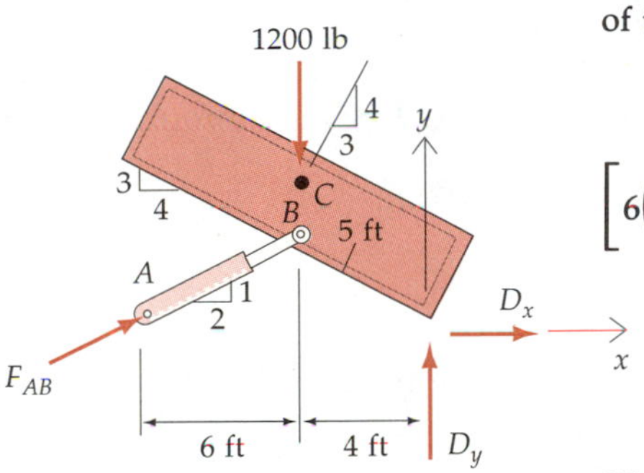

Figure E4.8b

Summing moments about the pin at D will give the force in AB:

$$\Sigma \mathbf{M}_D = \mathbf{0} = \mathbf{r}_{DC} \times (-1200\hat{\mathbf{j}}) + \mathbf{r}_{DA} \times \mathbf{F}_{AB}$$

$$\left[6\left(-\frac{4}{5}\hat{\mathbf{i}} + \frac{3}{5}\hat{\mathbf{j}}\right) + 2\left(\frac{3}{5}\hat{\mathbf{i}} + \frac{4}{5}\hat{\mathbf{j}}\right)\right] \times (-1200\hat{\mathbf{j}}) + (-10\hat{\mathbf{i}}) \times F_{AB}\left(\frac{2}{\sqrt{5}}\hat{\mathbf{i}} + \frac{1}{\sqrt{5}}\hat{\mathbf{j}}\right) = \mathbf{0}$$

$$(-3.60\hat{\mathbf{i}} + 5.20\hat{\mathbf{j}}) \times (-1200\hat{\mathbf{j}}) + F_{AB}(-4.47\hat{\mathbf{k}}) = \mathbf{0}$$

$$F_{AB} = 966 \text{ lb}$$

We say that the force "in" the two-force member AB is **compressive** since the strut is subjected to a pair of 966-pound forces that tend to cause it to shorten. Note that we defined (arbitrarily) the scalar F_{AB} in such a way that a positive value would indicate compression of the strut. Indeed it turned out that way as our intuition would suggest. Had F_{AB} turned out to be negative, we would have called the force **tensile** since the strut would then be subjected to a pair of forces tending to stretch it.

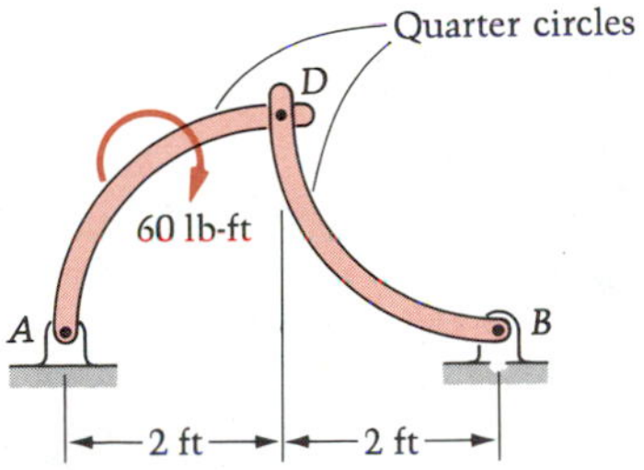

Figure E4.9a

EXAMPLE 4.9

Find the reactions exerted on the frame at A and B by the pins. See Figure E4.9a.

Solution

The member BD is a two-force member because, even though it isn't a straight bar, it is loaded by forces at only two points. Therefore, we know the direction of the reaction at B to be along BD. The overall free-body diagram (Figure E4.9b) will help us determine this reaction, and those at A as well:

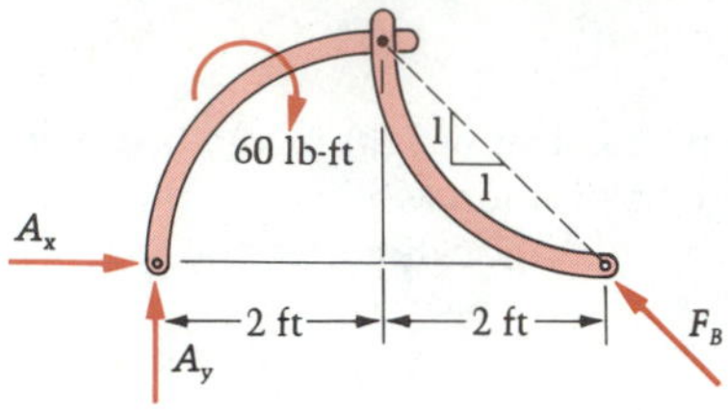

Figure E4.9b

$$\circlearrowleft\!+ \quad \Sigma M_A = 0 = -60 \text{ lb-ft} + \left(\frac{F_B}{\sqrt{2}}\right)(4 \text{ ft})$$

$$F_B = 15\sqrt{2} \text{ lb}$$

$$\overset{+}{\rightarrow} \quad \Sigma F_x = 0 = A_x - \frac{F_B}{\sqrt{2}} = A_x - 15$$

$$A_x = 15 \text{ lb}$$

and

$$+\!\uparrow \quad \Sigma F_y = 0 = A_y + \frac{F_B}{\sqrt{2}} = A_y + 15$$

$$A_y = -15 \text{ lb}$$

Therefore, the reactions onto the frame are

at A, $15\sqrt{2}$ 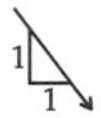 lb

at B, $15\sqrt{2}$  lb

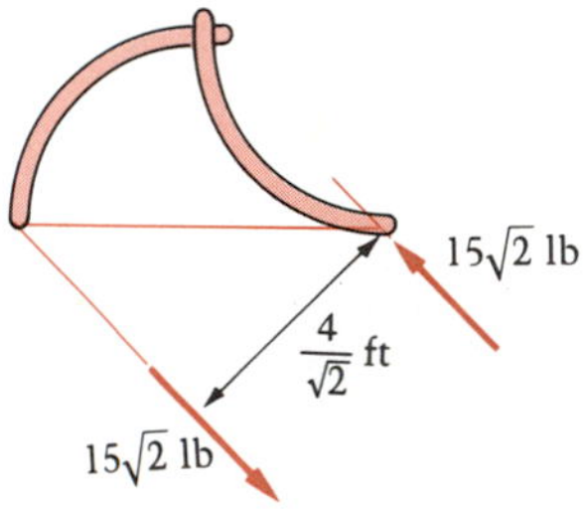

Figure E4.9c

Note from Figure E4.9c that these two reactions necessarily form a couple of 60 ↺ lb-ft.

Three-Dimensional Examples

We close the section now with a series of examples of equilibrium in three dimensions:

EXAMPLE 4.10

Find the force and couple reaction at the base of the advertising sign of Example 4.2, shown in Figure E4.10a.

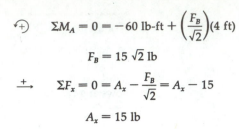

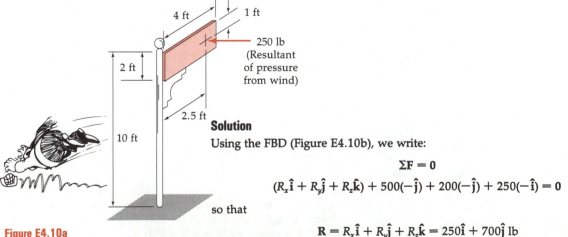

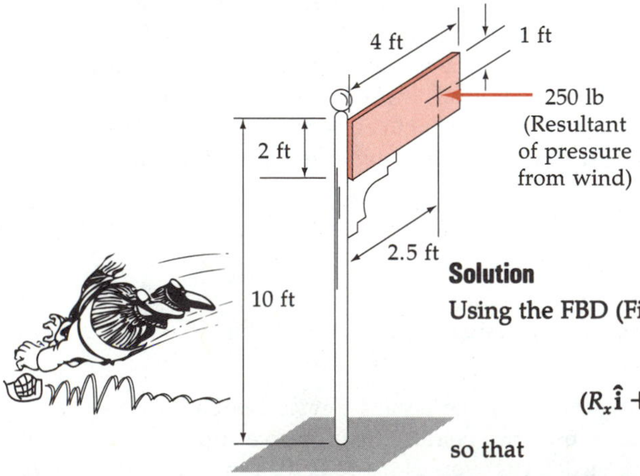

Figure E4.10a

Solution

Using the FBD (Figure E4.10b), we write:

$$\Sigma \mathbf{F} = \mathbf{0}$$

$$(R_x\hat{\mathbf{i}} + R_y\hat{\mathbf{j}} + R_z\hat{\mathbf{k}}) + 500(-\hat{\mathbf{j}}) + 200(-\hat{\mathbf{j}}) + 250(-\hat{\mathbf{i}}) = \mathbf{0}$$

so that

$$\mathbf{R} = R_x\hat{\mathbf{i}} + R_y\hat{\mathbf{j}} + R_z\hat{\mathbf{k}} = 250\hat{\mathbf{i}} + 700\hat{\mathbf{j}} \text{ lb}$$

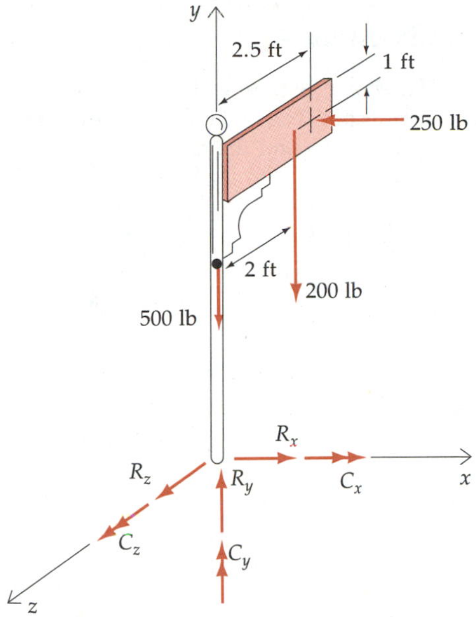

Figure E4.10b

is the resultant force exerted on the base of the post by the foundation. Thus

$$R_x = 250 \text{ lb} \qquad R_y = 700 \text{ lb} \qquad R_z = 0$$

Next, we sum the moments of all forces and couples about the base A of the sign:

$$\Sigma \mathbf{M}_A = \mathbf{0}$$

so that

$$(C_x\hat{\mathbf{i}} + C_y\hat{\mathbf{j}} + C_z\hat{\mathbf{k}}) + (9\hat{\mathbf{j}} - 2\hat{\mathbf{k}}) \times [200(-\hat{\mathbf{j}})] + (9\hat{\mathbf{j}} - 2.5\hat{\mathbf{k}}) \times [250(-\hat{\mathbf{i}})] = \mathbf{0}$$

Therefore,

$$(C_x\hat{\mathbf{i}} + C_y\hat{\mathbf{j}} + C_z\hat{\mathbf{k}}) + 9(200)(\mathbf{0}) + 2(200)(-\hat{\mathbf{i}}) + 9(250)(\hat{\mathbf{k}}) + 2.5(250)\hat{\mathbf{j}} = \mathbf{0}$$

so that

$$\mathbf{C} = C_x\hat{\mathbf{i}} + C_y\hat{\mathbf{j}} + C_z\hat{\mathbf{k}} = 400\hat{\mathbf{i}} - 625\hat{\mathbf{j}} - 2250\hat{\mathbf{k}} \text{ lb-ft}$$

is the moment of the resultant couple on the base of the post. Therefore

$$C_x = 400 \text{ lb-ft} \qquad C_y = -625 \text{ lb-ft} \qquad C_z = -2250 \text{ lb-ft}$$

Question 4.6 Why doesn't the 500-lb force appear explicitly in the moment equilibrium equation, $\Sigma \mathbf{M}_A = \mathbf{0}$?

Answer 4.6 Its line of action passes through A.

EXAMPLE 4.11

The boom, whose weight may be neglected, is supported by a ball-and-socket connection at A and two taut wires as shown in Figure E4.11a. Find the tensions in the wires and the reaction at A.

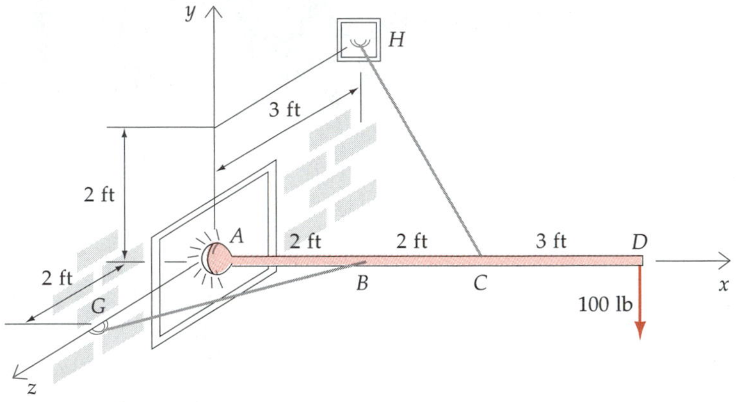

Figure E4.11a

Solution

From the free-body diagram (Figure E4.11b) we see that there are only five unknown scalars describing the reactions. For a three-dimensional problem such as this, we should then be concerned that there is not enough constraint provided against a rigid-body motion — that is, that we might not be able to satisfy the equations of equilibrium. However, in this case, one of the equations [the balance of moments about the axis of the boom (x)] is satisfied identically since each of the loads and reactions has a line of action intersecting that axis. The body would not be adequately restrained were there to be a loading that tended to turn the boom about its axis.

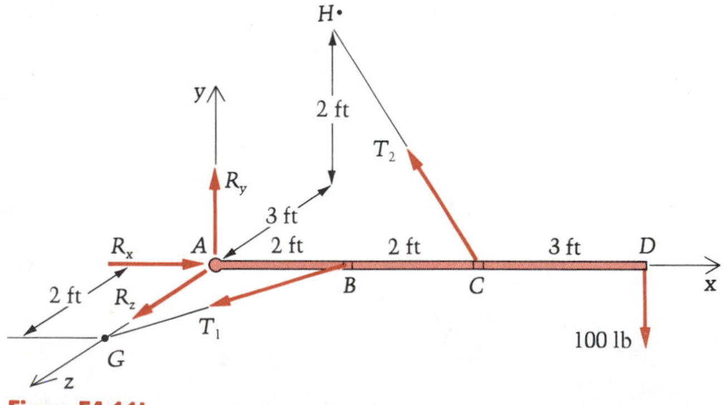

Figure E4.11b

The vector descriptions of the forces exerted by the wires on the boom are $T_1\hat{\mathbf{e}}_1$ and $T_2\hat{\mathbf{e}}_2$, where

$$\hat{\mathbf{e}}_1 = \frac{\mathbf{r}_{BG}}{|\mathbf{r}_{BG}|} = \frac{-2\hat{\mathbf{i}} + 2\hat{\mathbf{k}}}{\sqrt{(2)^2 + (2)^2}} = -0.707\hat{\mathbf{i}} + 0.707\hat{\mathbf{k}}$$

and

$$\hat{\mathbf{e}}_2 = \frac{\mathbf{r}_{CH}}{|\mathbf{r}_{CH}|} = \frac{-4\hat{\mathbf{i}} + 2\hat{\mathbf{j}} - 3\hat{\mathbf{k}}}{\sqrt{(4)^2 + (2)^2 + (3)^2}} = -0.743\hat{\mathbf{i}} + 0.371\hat{\mathbf{j}} - 0.557\hat{\mathbf{k}}$$

We first use the moment equation $\Sigma \mathbf{M}_A = \mathbf{0}$, since it will not involve the unknowns R_x, R_y, and R_z:

$$\Sigma \mathbf{M}_A = 0$$

$$\mathbf{r}_{AB} \times T_1\hat{\mathbf{e}}_1 + \mathbf{r}_{AC} \times T_2\hat{\mathbf{e}}_2 + \mathbf{r}_{AD} \times (-100\hat{\mathbf{j}}) = \mathbf{0}$$

$$2\hat{\mathbf{i}} \times T_1(-0.707\hat{\mathbf{i}} + 0.707\hat{\mathbf{k}}) + 4\hat{\mathbf{i}} \times T_2(-0.743\hat{\mathbf{i}} + 0.371\hat{\mathbf{j}} - 0.557\hat{\mathbf{k}}) + 7\hat{\mathbf{i}} \times (-100\hat{\mathbf{j}}) = \mathbf{0}$$

$$-2(0.707)T_1\hat{\mathbf{j}} + 4(0.371)T_2\hat{\mathbf{k}} + 4(0.557)T_2\hat{\mathbf{j}} - 7(100)\hat{\mathbf{k}} = \mathbf{0}$$

Note the absense of $\hat{\mathbf{i}}$ terms! Thus we have

$$\hat{\mathbf{k}}: \quad 4(0.371)T_2 - 7(100) = 0$$

$$T_2 = 472 \text{ lb}$$

and

$$\hat{\mathbf{j}}: \quad -2(0.707)T_1 + 4(0.577)T_2 = 0$$

$$T_1 = 744 \text{ lb}$$

Thus

$$T_1\hat{\mathbf{e}}_1 = -526\hat{\mathbf{i}} + 526\hat{\mathbf{k}} \text{ lb}$$

and

$$T_2\hat{\mathbf{e}}_2 = -351\hat{\mathbf{i}} + 175\hat{\mathbf{j}} - 263\hat{\mathbf{k}} \text{ lb}$$

Next we equilibrate the forces:

$$\Sigma \mathbf{F} = \mathbf{0}$$

$$(R_x\hat{\mathbf{i}} + R_y\hat{\mathbf{j}} + R_z\hat{\mathbf{k}}) + T_1\hat{\mathbf{e}}_1 + T_2\hat{\mathbf{e}}_2 + 100(-\hat{\mathbf{j}}) = \mathbf{0}$$

$$(R_x\hat{\mathbf{i}} + R_y\hat{\mathbf{j}} + R_z\hat{\mathbf{k}}) + (-526\hat{\mathbf{i}} + 526\hat{\mathbf{k}}) + (-351\hat{\mathbf{i}} + 175\hat{\mathbf{j}} - 263\hat{\mathbf{k}}) - 100\hat{\mathbf{j}} = \mathbf{0}$$

Thus, collecting like terms,

$$\hat{\mathbf{i}}: \quad R_x - 526 - 351 = 0$$

$$R_x = 877 \text{ lb}$$

$$\hat{\mathbf{j}}: \quad R_y + 175 - 100 = 0$$

$$R_y = -75 \text{ lb}$$

$$\hat{\mathbf{k}}: \quad R_z + 526 - 263 = 0$$

$$R_z = -263 \text{ lb}$$

Therefore, the force exerted on the boom by the ball-and-socket connection is $877\hat{\mathbf{i}} - 75\hat{\mathbf{j}} - 263\hat{\mathbf{k}}$ lb.

EXAMPLE 4.12

The uniform door weighing 1200 N is held in equilibrium in the horizontal position by the cable and by the smooth hinges at *A* and *B* (see Figure E4.12a). Find the force exerted on the door by the cable.

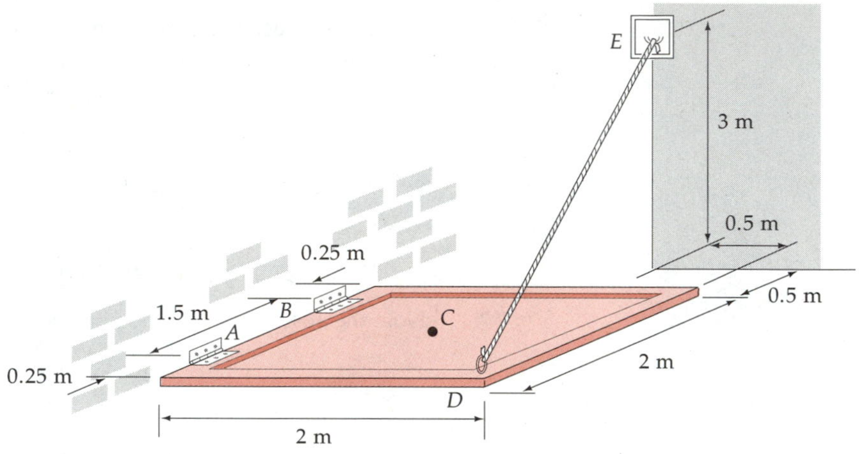

Figure E4.12a

Solution

As usual we express the unknown reactions in terms of the minimum number of unknown scalars (components). As indicated on the free-body diagram (Figure E4.12b), the hinge reactions are expressed as follows:

Force at *A*:

$$F_1\hat{\mathbf{i}} + F_2\hat{\mathbf{j}} + F_3\hat{\mathbf{k}}$$

Couple at *A*:

$$M_2\hat{\mathbf{j}} + M_3\hat{\mathbf{k}}$$

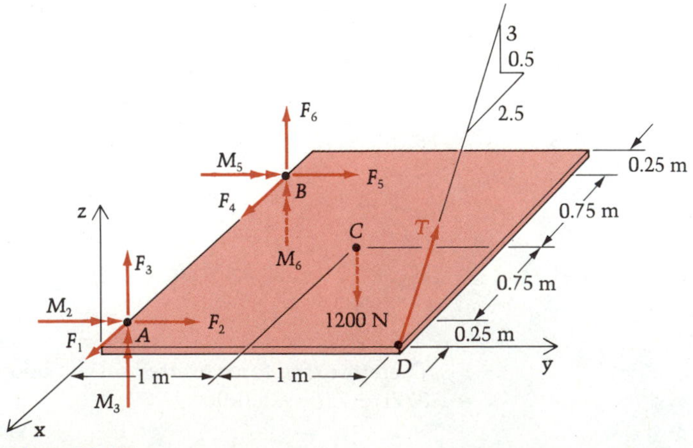

Figure E4.12b

Force at B:

$$F_4\hat{\mathbf{i}} + F_5\hat{\mathbf{j}} + F_6\hat{\mathbf{k}}$$

Couple at B:

$$M_5\hat{\mathbf{j}} + M_6\hat{\mathbf{k}}$$

The force exerted by the cable is

$$T\left(\frac{-2.5\hat{\mathbf{i}} - 0.5\hat{\mathbf{j}} + 3\hat{\mathbf{k}}}{\sqrt{(2.5)^2 + (0.5)^2 + (3)^2}}\right) = T(-0.635\hat{\mathbf{i}} - 0.127\hat{\mathbf{j}} + 0.762\hat{\mathbf{k}})$$

because we know that force to have the line joining points D and E as its line of action.

We see that ten unknown scalars are required to represent the hinge reactions. Including the unknown cable tension T, we have eleven unknown scalars, but we only have six independent equations available from the equations of equilibrium. Thus, the problem of finding all of the reactions is statically indeterminate. However, none of the hinge reactions produces a moment about the common axis of the hinges, but the cable force T does, and so it may be determined from the equilibrium equation requiring that the moments about the hinge axis sum to zero. This equation may be written as

$$\Sigma \mathbf{M}_A \cdot \hat{\mathbf{i}} = 0$$

Substituting the moments of the various forces and couples,

$$\begin{aligned}
&\hat{\mathbf{i}} \cdot [\mathbf{0} \times (F_1\hat{\mathbf{i}} + F_2\hat{\mathbf{j}} + F_3\hat{\mathbf{k}}) + M_2\hat{\mathbf{j}} + M_3\hat{\mathbf{k}}] \\
&\quad + \hat{\mathbf{i}} \cdot [(-1.5\hat{\mathbf{i}}) \times (F_4\hat{\mathbf{i}} + F_5\hat{\mathbf{j}} + F_6\hat{\mathbf{k}}) + M_5\hat{\mathbf{j}} + M_6\hat{\mathbf{k}}] \\
&\quad + \hat{\mathbf{i}} \cdot [(-0.75\hat{\mathbf{i}} + 1\hat{\mathbf{j}}) \times (-1200\hat{\mathbf{k}})] \\
&\quad + \hat{\mathbf{i}} \cdot [(0.25\hat{\mathbf{i}} + 2\hat{\mathbf{j}}) \times T(-0.635\hat{\mathbf{i}} - 0.127\hat{\mathbf{j}} + 0.762\hat{\mathbf{k}})] = 0 \\
&0 + 0 - (1)(1200) + 2(0.762)T = 0 \\
&\qquad\qquad T = 787 \text{ N}
\end{aligned}$$

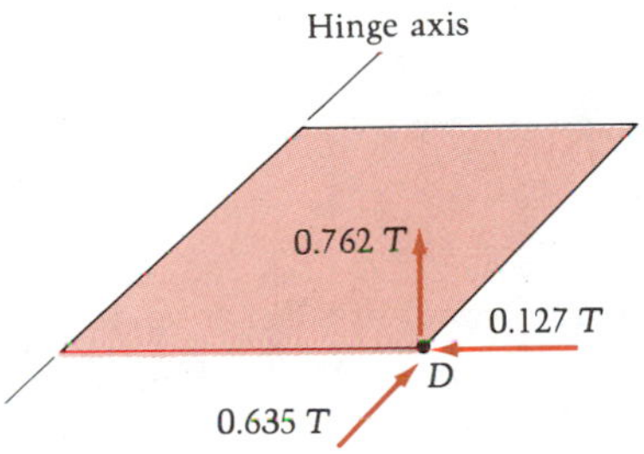

Figure E4.12c

We could have obtained the same result with a little less formality. Suppose we first decompose the cable tension into three parts at D as shown in Figure E4.12c. We see that only the $0.762T$ component produces a moment about the hinge axis and that moment (using the "perpendicular distance" method) is $2(0.762T)\hat{\mathbf{i}}$. Similarly the moment of the 1200-N weight is seen to be $(1)(1200)(-\hat{\mathbf{i}})$. These two moments must sum to zero since we can see that the hinge reactions produce no moments about the hinge axis. Thus

$$2(0.762T)\hat{\mathbf{i}} + (1)(1200)(-\hat{\mathbf{i}}) = \mathbf{0}$$

or

$$2(0.762T) - (1)(1200) = 0$$

$$T = 787 \text{ N}$$

as before.

Let us now see what information the other equations of equilibrium yield.

$$\Sigma \mathbf{F} = \mathbf{0}$$

$$(F_1\hat{\mathbf{i}} + F_2\hat{\mathbf{j}} + F_3\hat{\mathbf{k}}) + (F_4\hat{\mathbf{i}} + F_5\hat{\mathbf{j}} + F_6\hat{\mathbf{k}}) + 1200(-\hat{\mathbf{k}})$$
$$+ 787(-0.635\hat{\mathbf{i}} - 0.127\hat{\mathbf{j}} + 0.762\hat{\mathbf{k}}) = \mathbf{0}$$
$$(F_1\hat{\mathbf{i}} + F_2\hat{\mathbf{j}} + F_3\hat{\mathbf{k}}) + (F_4\hat{\mathbf{i}} + F_5\hat{\mathbf{j}} + F_6\hat{\mathbf{k}}) = 1200\hat{\mathbf{k}} + 500\hat{\mathbf{i}} + 100\hat{\mathbf{j}} - 600\hat{\mathbf{k}}$$

or

$$(F_1\hat{\mathbf{i}} + F_2\hat{\mathbf{j}} + F_3\hat{\mathbf{k}}) + (F_4\hat{\mathbf{i}} + F_5\hat{\mathbf{j}} + F_6\hat{\mathbf{k}}) = 500\hat{\mathbf{i}} + 100\hat{\mathbf{j}} + 600\hat{\mathbf{k}} \text{ N}$$

Notice that the left-hand side is the sum of the force parts of the hinge reactions at A and B; in other words, we have found the force part of the *resultant* of the hinge reactions.

Turning to the moment equation

$$\Sigma \mathbf{M}_A = \mathbf{0}$$

$$(M_2\hat{\mathbf{j}} + M_3\hat{\mathbf{k}}) + (-1.5\hat{\mathbf{i}}) \times (F_4\hat{\mathbf{i}} + F_5\hat{\mathbf{j}} + F_6\hat{\mathbf{k}}) + (M_5\hat{\mathbf{j}} + M_6\hat{\mathbf{k}})$$
$$+ (-0.75\hat{\mathbf{i}} + 1\hat{\mathbf{j}}) \times (-1200\hat{\mathbf{k}})$$
$$+ (0.25\hat{\mathbf{i}} + 2\hat{\mathbf{j}}) \times (-500\hat{\mathbf{i}} - 100\hat{\mathbf{j}} + 600\hat{\mathbf{k}}) = \mathbf{0}$$

We can recognize the first three terms (all of the unknowns) as the couple part of the resultant at A of the hinge reactions.

We see then that the reactions at the individual hinges cannot be found here, but their resultant can be. Unfortunately this resultant doesn't tell us much about conditions at the hinges in a problem such as this one where it is reasonable to expect the reactions at the two hinges to be substantially different. In the next example we explore a situation in which the two-hinge resultant might be expected to provide valuable information.

Question 4.7 If in the preceding example the hinge at B broke, could the body remain in equilibrium? If so, would the problem of finding the hinge reaction at A then be statically determinate?

Answer 4.7 Yes, there is still enough constraint to prohibit rigid-body motion. Yes, the six scalar equations from $\Sigma \mathbf{F} = \mathbf{0}$ and $\Sigma \mathbf{M}_A = \mathbf{0}$ can now be solved for the six unknowns F_1, F_2, F_3, M_2, M_3, and T.

EXAMPLE 4.13

The hinged door of Example 4.12 is supported by the (different) cable as shown in Figure E4.13a. Find the tension in the cable and the resultant (at O) of the hinge reactions.

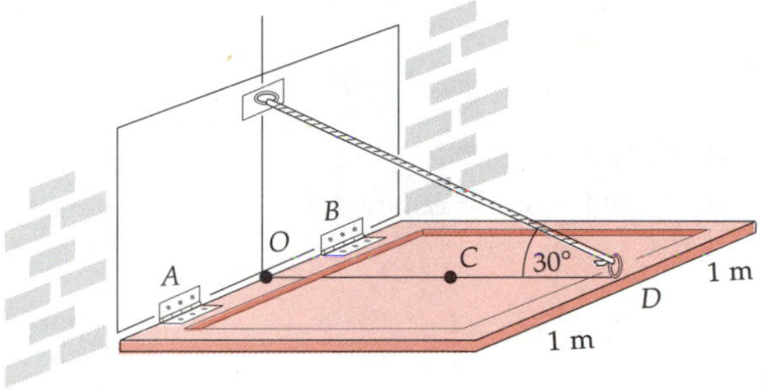

Figure E4.13a

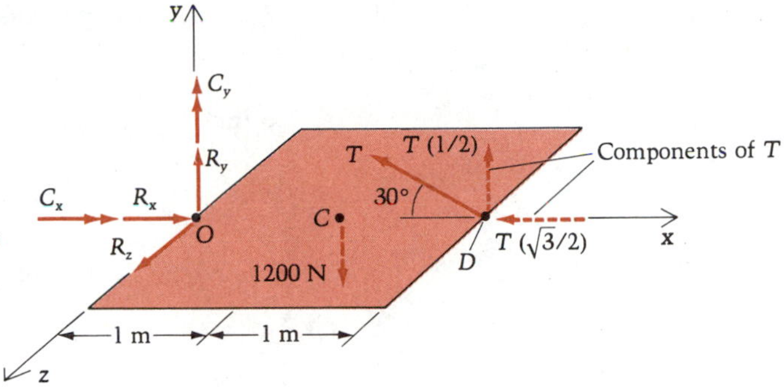

Figure E4.13b

Solution

Example 4.12 illustrated the futility of attempting to find the individual hinge reactions; thus here we are seeking their resultant, drawn in the free-body diagram as a force and couple at O. Because in this example the hinge axis is the z axis, the component of the couple in the z direction is zero as suggested by the free-body diagram (Figure 4.13b). We use here the component forms of the equilibrium equations and encourage the reader to write out the vector equations $\Sigma \mathbf{F} = \mathbf{0}$ and $\Sigma \mathbf{M}_o = \mathbf{0}$ in terms of unit vectors so as to compare with the results below. First, we sum moments about the hinge axis:

$$(\Sigma M_O)_z = 0$$

so that

$$2(T/2) - (1)(1200) = 0$$
$$T = 1200 \text{ N}$$

Next, we sum moments about the x axis (through O):

$$(\Sigma M_O)_x = 0$$

This yields

$$C_x = 0$$

A third moment equation is written about the y axis:

$$(\Sigma M_O)_y = 0$$

It gives

$$C_y = 0$$

Next, we sum the forces; first, the x components:

$$\xrightarrow{+} \quad \Sigma F_x = 0$$

giving

$$R_x - T\left(\frac{\sqrt{3}}{2}\right) = 0$$

so that

$$R_x = \frac{\sqrt{3}}{2}T = \frac{\sqrt{3}}{2}(1200) = 1040 \text{ N}$$

Next, the y components:

$$+\uparrow \qquad \Sigma F_y = 0$$

so that

$$R_y + T(\tfrac{1}{2}) - 1200 = 0$$
$$R_y = 1200 - 600 = 600 \text{ N}$$

Finally, the z components:

$$\overset{+}{\swarrow} \qquad \Sigma F_z = 0$$

giving

$$R_z = 0$$

Thus the resultant of the hinge reactions is $1040\hat{\mathbf{i}} + 600\hat{\mathbf{j}}$ N, with a line of action through O since $C_x = C_y = C_z = 0$.

As we mentioned before, the equilibrium equations won't tell us the individual hinge reactions. However, it would seem reasonable to assume with identical hinges symmetrically placed about the xy plane (as is the case here) that the components of the hinge reactions parallel to this plane will be identical. These components would each equal $520\hat{\mathbf{i}} + 300\hat{\mathbf{j}}$ N. At the hinge A, the z-directional force and the x- and y-direction couples will each be equal in magnitude but opposite in direction to the corresponding force or couple at B. These reactions, however, result from the tendency of the fairly rigid door to deform and are typically small in a problem such as this one.

The fact that there is a plane (xy) of symmetry for loads and supports suggests that this problem could have been treated as two-dimensional as shown

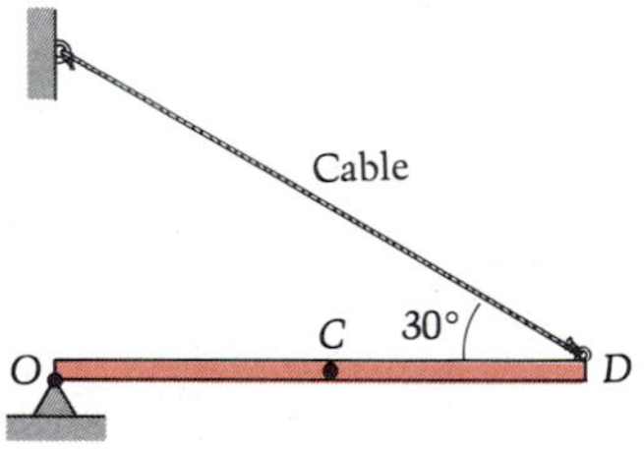

Figure E4.13c

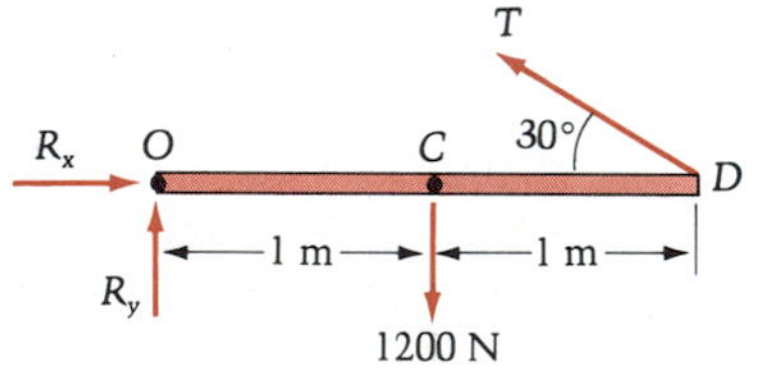

Figure E4.13d

in Figure E4.13c and its free-body diagram (Figure E4.13d). Invoking the equilibrium equations in the form appropriate to a coplanar system of loads and reactions, we obtain

$$\circlearrowleft + \qquad (\Sigma M_O)_z = 0$$
$$2(T \sin 30°) - 1(1200) = 0$$
$$T = 1200 \text{ N} \qquad \text{(as before)}$$
$$\xrightarrow{+} \qquad \Sigma F_x = 0$$
$$R_x - T \cos 30° = 0$$
$$R_x = T(\sqrt{3}/2) = 600\sqrt{3} = 1040 \text{ N} \qquad \text{(as before)}$$

and

$$+\!\uparrow \quad \Sigma F_y = 0$$

$$R_y - 1200 + T\sin 30^\circ = 0$$

$$R_y = 1200 - 1200(\tfrac{1}{2}) = 600 \text{ N} \qquad \text{(as before)}$$

EXAMPLE 4.14

The uniform thin triangular plate of Figure E4.14a is supported by a slider-on-smooth-guide welded to the plate at A and a similar slider attached at B by a ball-and-socket connection. The plate weighs 10 lb per square foot of plan area and the mass center of the plate is at C. Find the reactions at A and B when the plate is subjected to the 400-lb force shown.

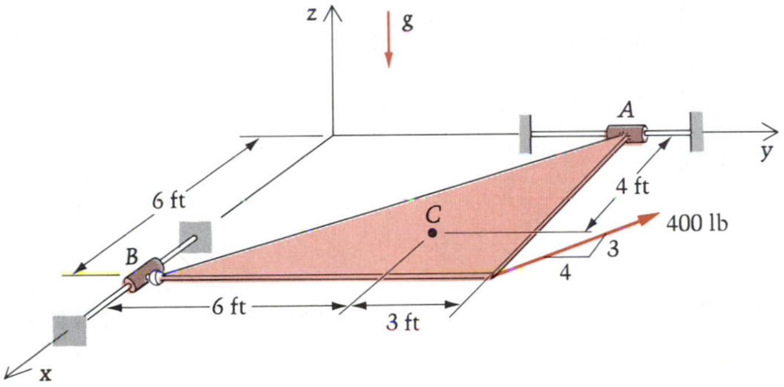

Figure E4.14a

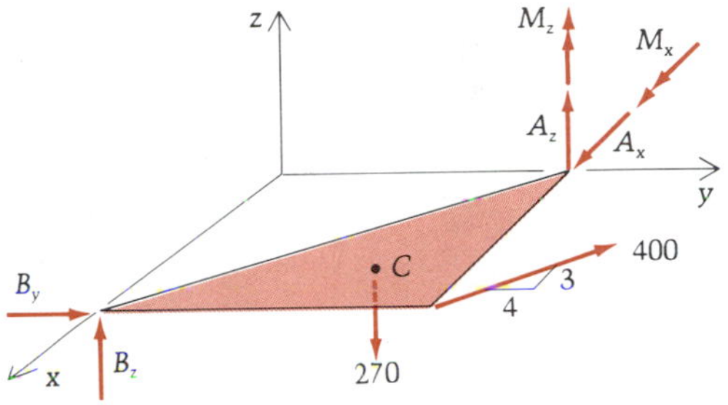

Figure E4.14b

Solution

The weight of the plate is 10[(9)(6)/2] = 270 lb and, of course, the line of action of this resultant of distributed gravitational attractions is downward through the mass center C as shown in the FBD of Figure E4.14b.

The ball-and-socket at B eliminates the possibility of a couple there, and the slider eliminates the possibility of an x component of force there. At A the slider eliminates the possibilities of a y component of force or of couple. Thus, we see from the free-body diagram that there are six unknown components of reactions. Since, in addition, the plate is adequately restrained against rigid motion, we should anticipate that this is a well-posed, statically determinate, three-dimensional problem; and thus that the six independent component-equations of equilibrium will provide a solution for the reactions.

Component equations guaranteeing that $\Sigma \mathbf{F} = \mathbf{0}$ are

$$\overset{+}{\swarrow} \quad \Sigma F_x = 0$$

$$A_x - \frac{3}{5}(400) = 0 \Rightarrow A_x = 240 \text{ lb} \tag{1}$$

$$\xrightarrow{+} \quad \Sigma F_y = 0$$

$$B_y + \frac{4}{5}(400) = 0 \Rightarrow B_y = -320 \text{ lb} \tag{2}$$

and

$$+\uparrow \qquad \Sigma F_z = 0$$

$$A_z + B_z - 270 = 0 \Rightarrow A_z + B_z = 270 \tag{3}$$

Component equations corresponding to $\Sigma \mathbf{M}_A = \mathbf{0}$ are obtained by balancing moments about axes through A that are respectively parallel to x, y, and z:

$$(\Sigma M_A)_x = 0$$

$$M_x + 3(270) - 9B_z = 0 \Rightarrow M_x - 9B_z = -810 \tag{4}$$

$$(\Sigma M_A)_y = 0$$

$$4(270) - 6B_z = 0 \Rightarrow B_z = 180 \text{ lb} \tag{5}$$

and

$$(\Sigma M_A)_z = 0$$

$$M_z + 6\left[\frac{4}{5}(400)\right] + 6B_y = 0 \tag{6}$$

or

$$M_z + 6(320) + 6(-320) = 0$$

and so

$$M_z = 0$$

Substituting the result $B_z = 180$ lb into Equations (3) and (4) we obtain

$$A_z + 180 = 270 \Rightarrow A_z = 90 \text{ lb}$$

and

$$M_x - 9(180) = -810 \Rightarrow M_x = 810 \text{ lb-ft}$$

We could have used other sets of component equations. For example, referring to the free-body diagram, if we had chosen to balance moments about the line AB, we would have been able to solve for M_x in one step. To show this, let $\hat{\mathbf{e}}_{AB}$ be a unit vector along AB:

$$\hat{\mathbf{e}}_{AB} = \frac{6\hat{\mathbf{i}} - 9\hat{\mathbf{j}}}{\sqrt{(6)^2 + (9)^2}} = \frac{2\hat{\mathbf{i}} - 3\hat{\mathbf{j}}}{\sqrt{13}}$$

The moment of the couple at A about line AB is

$$[(M_x\hat{\mathbf{i}} + M_z\hat{\mathbf{k}}) \cdot \hat{\mathbf{e}}_{AB}]\hat{\mathbf{e}}_{AB} = \frac{2}{\sqrt{13}} M_x \hat{\mathbf{e}}_{AB}$$

and the moment of the weight is, using $(\mathbf{M}_B \cdot \hat{\mathbf{e}}_{AB})\hat{\mathbf{e}}_{AB}$:

$$\{[(6\hat{\mathbf{j}} - 2\hat{\mathbf{i}}) \times (-270\hat{\mathbf{k}})] \cdot \hat{\mathbf{e}}_{AB}\}\hat{\mathbf{e}}_{AB} = (-1620\hat{\mathbf{i}} - 540\hat{\mathbf{j}}) \cdot \left(\frac{2\hat{\mathbf{i}} - 3\hat{\mathbf{j}}}{\sqrt{13}}\right)\hat{\mathbf{e}}_{AB}$$

$$= -\frac{1620}{\sqrt{13}}\hat{\mathbf{e}}_{AB}$$

Thus for $\Sigma \mathbf{M}_{AB} = \mathbf{0}$,

$$\frac{2}{\sqrt{13}} M_x - \frac{1620}{\sqrt{13}} = 0$$

or

$$M_x = 810 \text{ lb-ft} \qquad \text{(as before)}$$

The reader is encouraged to show that this equation expressing the balance of moments about AB is a linear combination of Equations (4) and (5). In this example we have chosen, out of convenience, to balance moments about axes that all intersect at the same point (A). Thus, the equations are all components of $\Sigma \mathbf{M}_A = \mathbf{0}$. However, because the sum of the moments about *any* line must vanish, we could have generated component equations by balancing moments about nonintersecting lines. The important thing is that we obtain three independent component equations that, together with $\Sigma \mathbf{F} = \mathbf{0}$, assure $\Sigma \mathbf{M}_P = \mathbf{0}$ for any point P, provided the three lines are nonparallel and nonplanar.

By this time the reader perhaps has developed some feel for the delicacy associated with providing adequate support for a body and at the same time having static determinacy. In the preceding example, if we were to fix the slider at B, a new component of reaction B_x would be introduced as shown on the free-body diagram in Figure 4.14. This problem is statically indeterminate, which we can clearly ascertain because we now have seven unknowns (B_x, B_y, B_z, A_x, A_z, M_x, and M_z) and only six independent scalar equations of equilibrium.

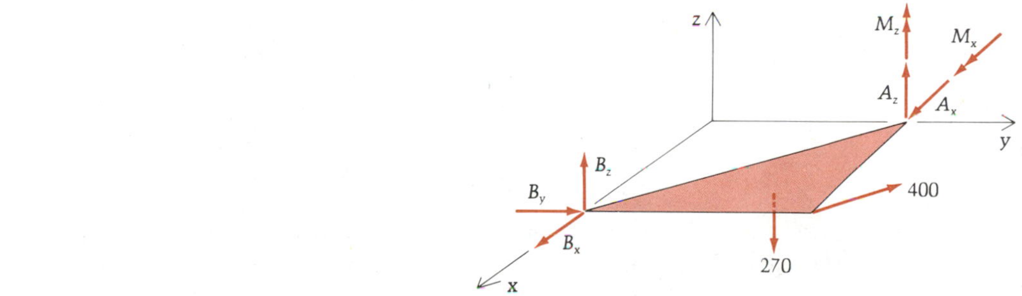

Figure 4.14

On the other hand, were we to provide only a "ball" support at B giving resistance only normal to the xz plane, so that $B_x = B_z = 0$, we would have only five unknowns. Then we would not be able to satisfy the equations of equilibrium unless, because of some special nature of the loading, one of the equations was identically satisfied. That would be the case, for example, were we here to neglect the weight (270 lb) of the plate. Without neglect of this force, we can see from the free-body diagram in Figure 4.15 that we cannot satisfy $\Sigma M_y = 0$! The inability to satisfy one (or more) equation(s) of equilibrium, except for special loadings, is characteristic of inadequate support (or constraint). When that is the case there usually will be one (or more) rigid motion(s) not prohibited by the support system. For the case at hand the supports provide no resistance to rotation of the plate about the y axis.

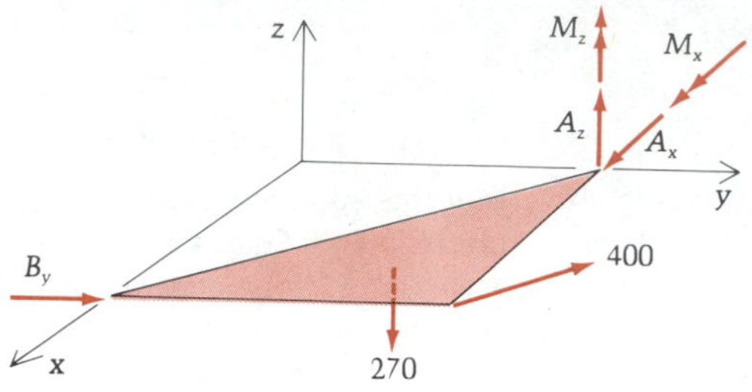

Figure 4.15

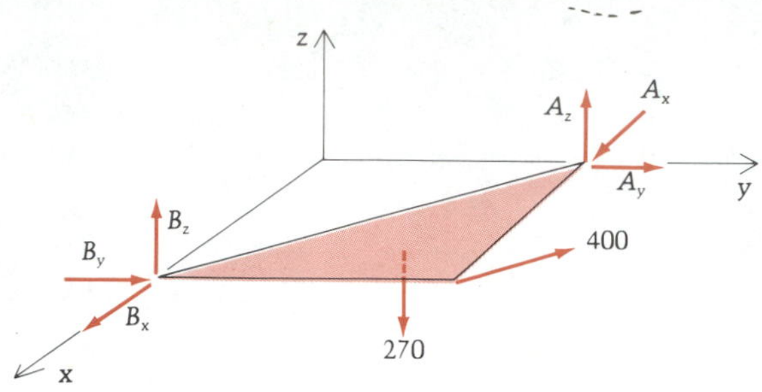

Figure 4.16

The preceding discussion might tempt us to conclude that if, in the three-dimensional situation, we have exactly six unknown reaction components, all is well. Unfortunately this is not the case. To illustrate this difficulty suppose that our plate is supported by ball-and-socket connections (no sliders) at both A and B as shown in Figure 4.16. The plate again is not adequately supported. We cannot satisfy the balance of moments about the line AB, and this correlates with the freedom to rotate about the line, which these (inadequate) supports do not curtail.

PROBLEMS ▸ Section 4.3

4.12 Is there a force **F** at P for which the system of three forces is in equilibrium? (See Figure P4.12.) If so, find it; if not, why not?

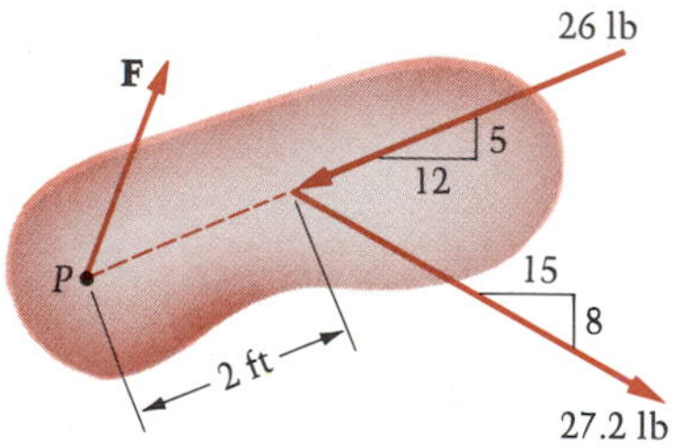

Figure P4.12

4.13 In Figure P4.13(a), the nail is on the verge of coming out of the board; that is, any larger force than 40 lb will pull it out. (a) Find the resultant of all the vertical differential friction forces acting on the nail, assuming the nail exerts no horizontal force on the claw of the hammer. (b) Suppose now that the nail turns out to be 3 in. long, and the claw hammer only succeeds in pulling it up 1 in. at first effort [see Figure P4.13(b)]. A board is then used as shown in Figure P4.13(c) to get enough leverage to finish extracting the nail. Assuming 150 lb of vertical friction

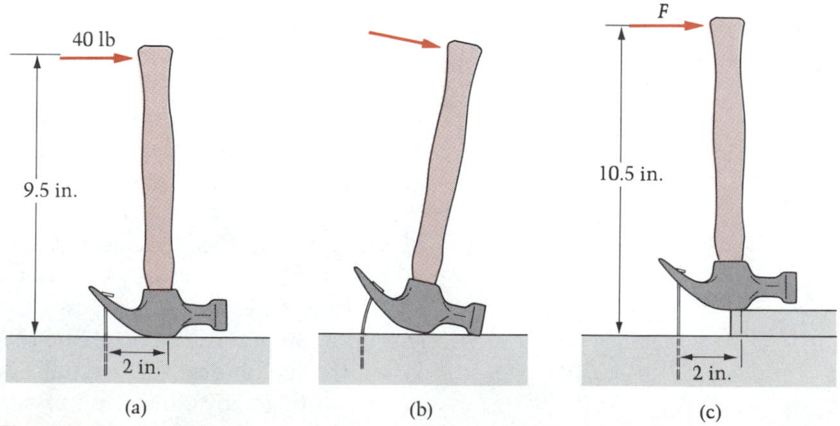

Figure P4.13

force must now be overcome, find the force **F** needed to complete the job. Again assume the nail exerts no horizontal force on the claw; that is, the nail has been straightened.

4.14 In Example 4.7, the moments about point A were summed and equated to zero in order to determine the tension T in the cable (see Figure P4.14). In that solution the right-most tension was replaced by an equipollent system of a force (itself) and couple $0.1T$ ↺ at point B. Without making this replacement, find the tension T (i.e., resolve T into its components at the point E where the cable leaves the pulley).

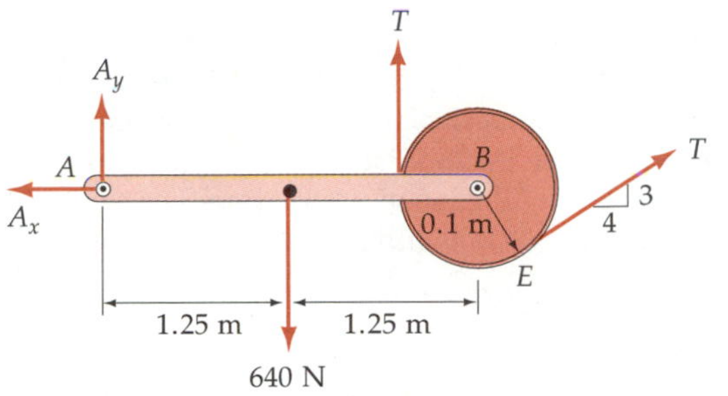

Figure P4.14

4.15 In Chapter 2, we found the wall and floor forces, and the internal force between the two cylinders (see Figures 2.9, 2.10), to be as shown in Figure P4.15. Show that when A and B are put back together as in Figure 2.10(a), the equilibrium equations $\Sigma F_x = 0$, $\Sigma F_y = 0$, and $\Sigma M_{\substack{\text{any}\\\text{point}}} = 0$ are satisfied, as they must be.

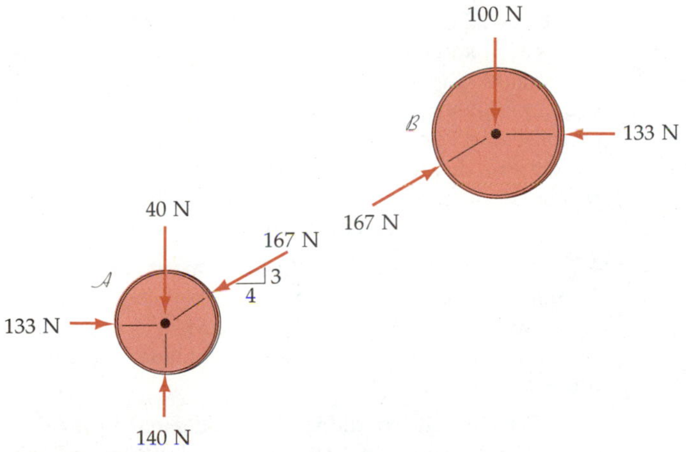

Figure P4.15

4.16 Find the pin reaction at A and the roller reaction at B onto the triangular block of weight 200 lb shown in Figure P4.16. The center of gravity of the block is as shown.

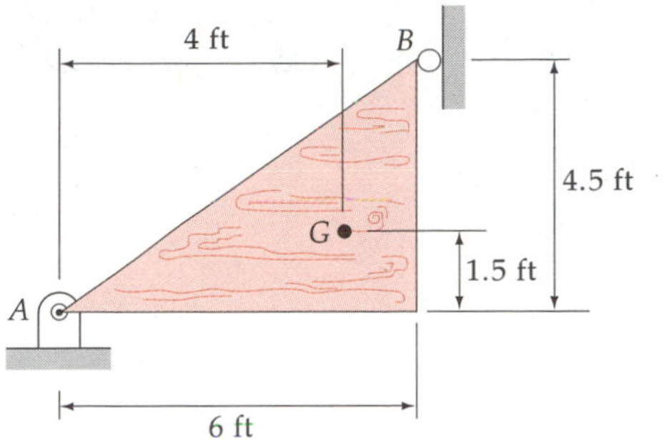

Figure P4.16

4.17 The uniform rod of mass m and length L is in equilibrium in the position shown in Figure P4.17. Find the tension in the string.

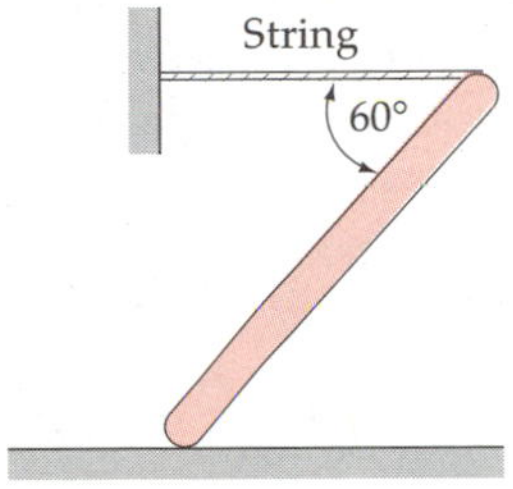

Figure P4.17

4.18 For the frame in Figure P4.18, find the reactions exerted by the ground onto member ABD at A.

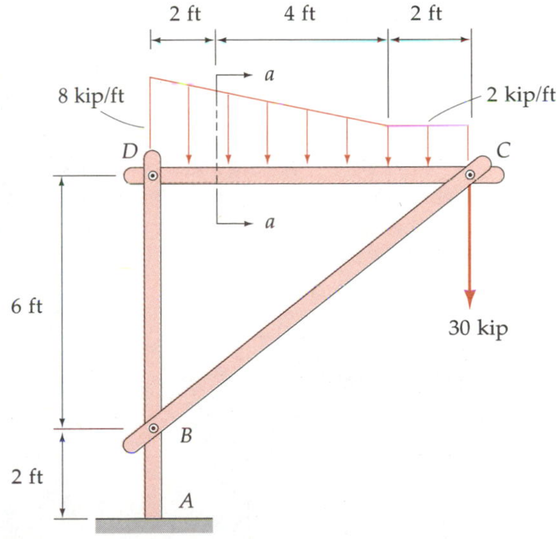

Figure P4.18

4.19 Find the force exerted by the slot on the pin at B (see Figure P4.19) which is fixed to the uniform bar $ABCD$. The weight of the bar is 100 N.

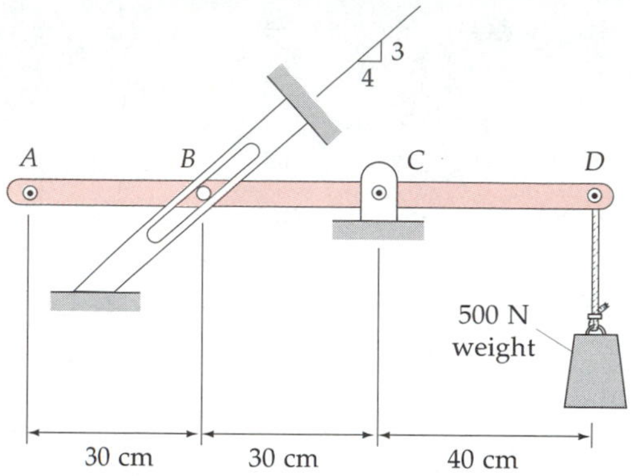

Figure P4.19

4.20 In Figure P4.20 joints A, B, and C are pinned, and the slender rods AC and BC are light in comparison to the applied forces. Determine the supporting force at B acting on member BC. Does this force put BC in tension or compression?

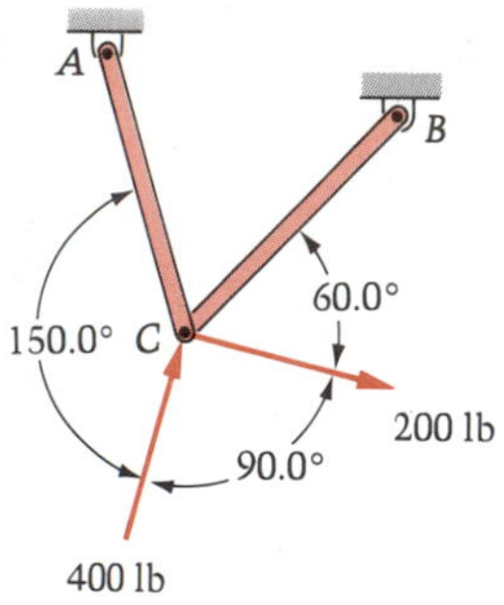

Figure P4.20

4.21 A straight uniform bar weighing 60 lb rests in a horizontal position against two frictionless slopes as shown in Figure P4.21. A concentrated vertical load of 200 lb acts at a distance x from the right end of the bar as shown. Find the distance x for equilibrium and determine the reactions at A and B.

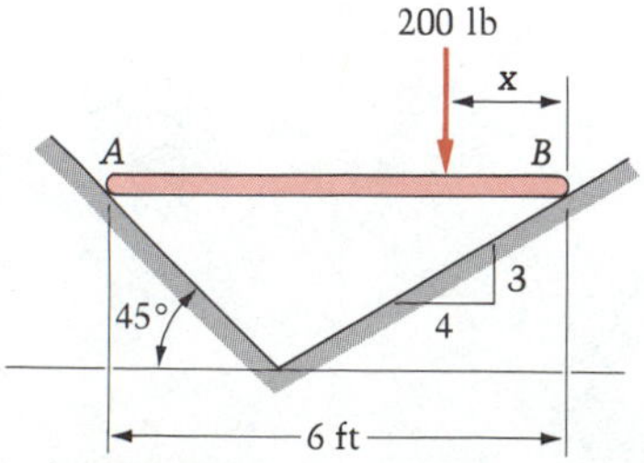

Figure P4.21

4.22 To problem 2.53, add the dotted cylinder $\mathcal{D}$ of Figure P4.22, which though smaller is more dense and has the same mass. The contact between $\mathcal{C}$ and $\mathcal{D}$ is frictionless. Find the forces exerted by the planes on $\mathcal{C}$ and $\mathcal{D}$.

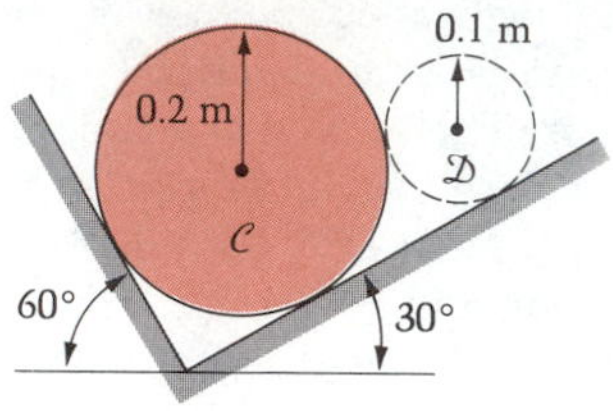

Figure P4.22

4.23 It is possible for the 20-N half-cylinder to be in equilibrium on the smooth plane for only one value of the angle ϕ. (See Figure P4.23.) For that angle, find the tension in the cord as a function of θ, and check your answer in the limiting cases $\theta = 0$ and $\theta \to \pi/2$.

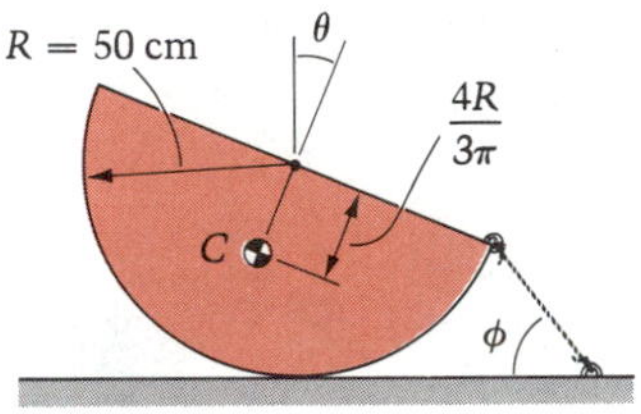

Figure P4.23

* **4.24** Show that the three scalar equations of equilibrium for a coplanar system of forces on a body ($\Sigma F_x = 0$, $\Sigma F_y = 0$, $\Sigma M_A = 0$) may be replaced by three moment equations $\Sigma M_A = 0$, $\Sigma M_B = 0$, $\Sigma M_C = 0$, provided the points A, B, and C (in the plane of the forces) are not collinear. *Hint:* Show that satisfaction of the three moment equations ensures a zero resultant. Use $\Sigma M_A = 0$ to establish the resultant as, at most, a force through A. Decompose this force into components parallel and perpendicular to the line joining A and B. Then apply the remaining moment equations. A sketch will help.

4.25 Find the conditions on the locations of points A and B so that the equations $\Sigma F_x = 0$, $\Sigma M_A = 0$, and $\Sigma M_B = 0$ are equivalent to $\Sigma F_x = 0$, $\Sigma F_y = 0$, $\Sigma M_P = 0$ as equilibrium equations of a body under a coplanar system of forces and couples.

4.26 What are the conditions on A and B in the previous exercise if the equivalent conditions include $\Sigma F_y = 0$ instead of $\Sigma F_x = 0$?

4.27 For equilibrium of the rectangular plate shown in Figure P4.27, what are the reactions at A and B?

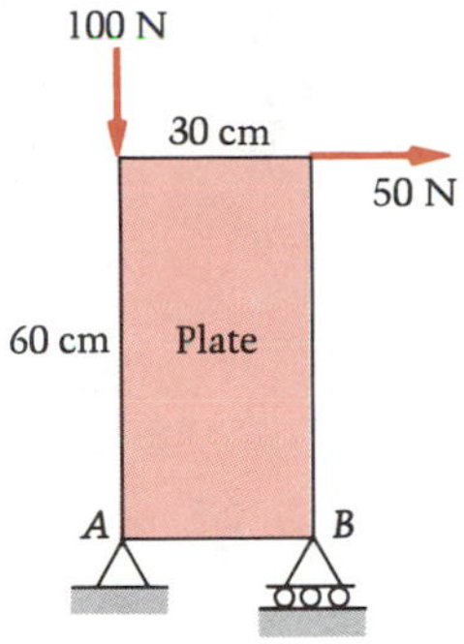

Figure P4.27

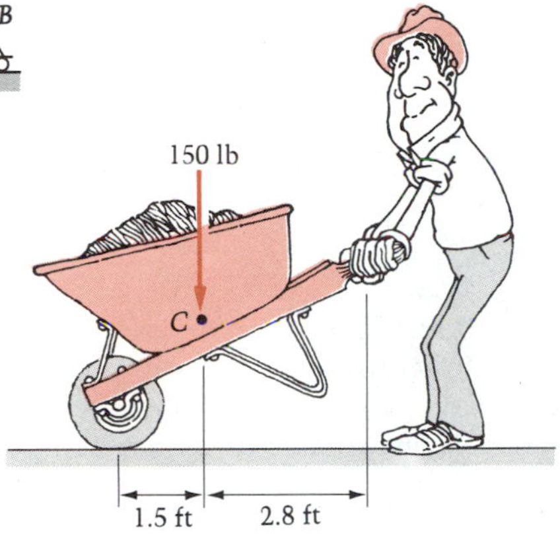

Figure P4.28

4.28 A wheelbarrow plus its load has the weight and center of mass location shown in Figure P4.28. Find the vertical force H exerted on the handle by the man and the force exerted on the tire by the ground. Assume the system is not moving.

4.29 In the preceding problem, determine the resultant N of the forces exerted on the man's shoes by the ground in the configuration shown in Figure P4.29. Also find the location of N; that is, find the distance d.

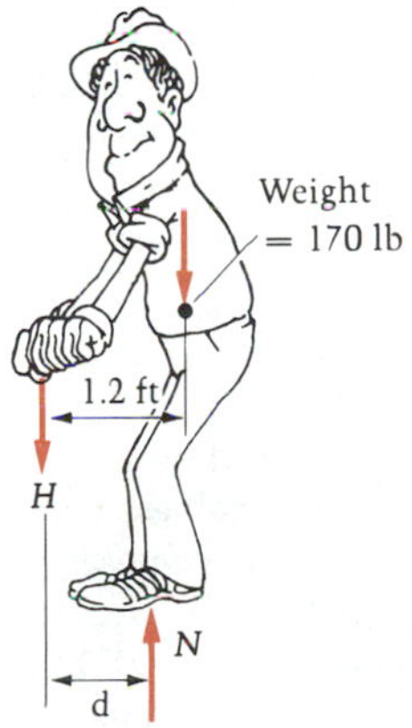

Figure P4.29

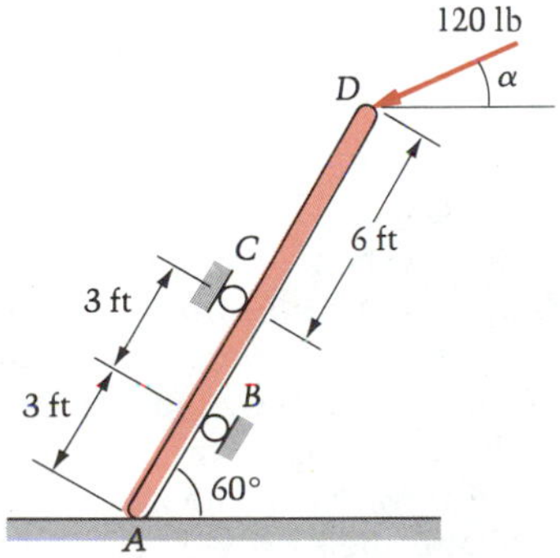

Figure P4.30

4.30 Rod $ABCD$ in Figure P4.30 rests against rollers at B and C, and against a smooth surface at A. (a) Find the angle α so the roller at B may be removed. (b) Find the resulting forces at A and C for this angle α.

4.31 The uniform bar BA in Figure P4.31 weighs 300 N and the block W weighs 500 N. The block rests on the bar but is secured by a taut wire wrapped around a smooth pulley and connected to the right end of the bar at A. Determine the tension in the wire and the reaction onto BA at B.

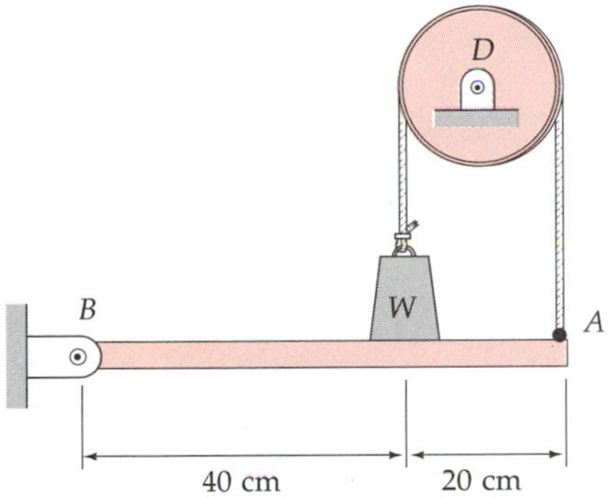

Figure P4.31

4.32 The bent bar $BAQRC$ of Figure P4.32 weighs 20 lb/ft. The cord is attached to the bar at A, and passes over pulleys P_1, P_2, P_3, and P_4, reattaching to the bar at Q. Find the tension in the cord using a single moment equation.

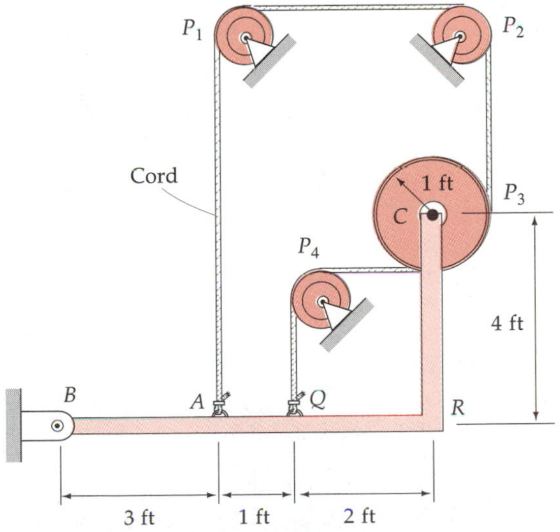

Figure P4.32

4.33 The pulley in Figure P4.33 is supported by frictionless bearings at the end of the uniform 750-N cantilever beam. The 1000-newton block is supported by the cable that passes over the pulley as shown. Neglect the weight of the pulley and find the reaction of the wall on the beam.

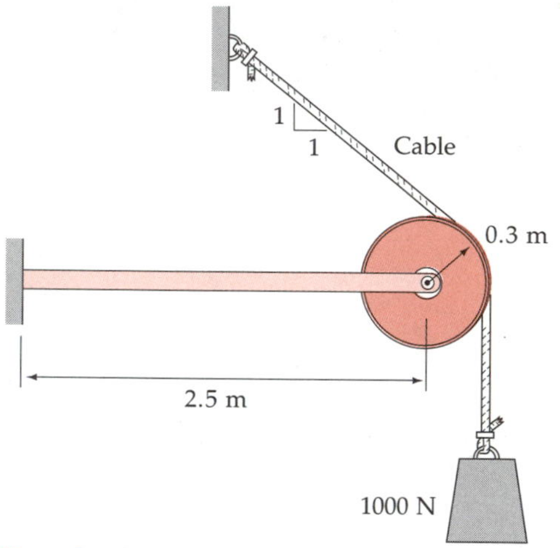

Figure P.4.33

4.34 The wrench in Figure P4.34 is applied to a hex-head bolt in an effort to loosen it. Determine the forces F_1 and F_2 on the bolt head if there is a slight clearance so that the contact is only at A and B.

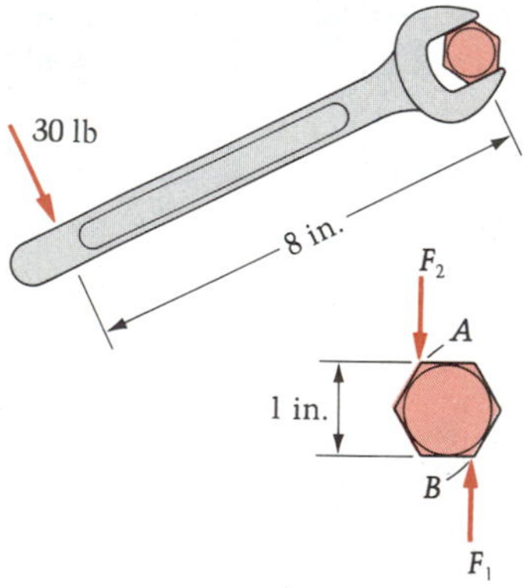

Figure P4.34

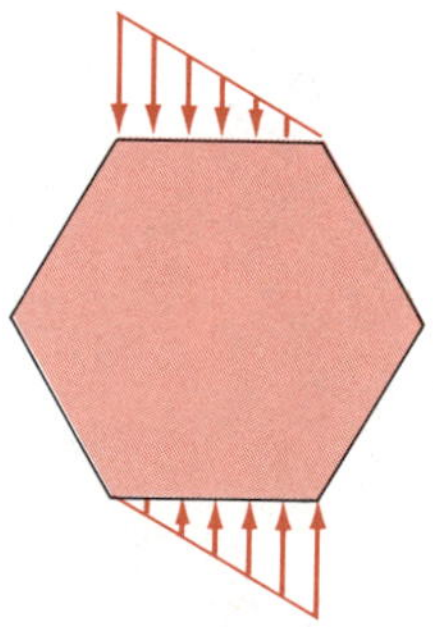

Figure P4.35

4.35 In the preceding problem, assume a tight fit between the bolt and the wrench and that the load is distributed linearly on the two faces as shown in Figure P4.35. Determine the force-alone equipollents of each of the two loadings. Explain the increases in these two values over the concentrated forces in the preceding problem.

4.36 Assume that the reaction of the wall onto the cantilever beam is the pair of linearly distributed forces shown in Figure P4.36. Find the intensities q_T and q_B in terms of F, L, and ℓ.

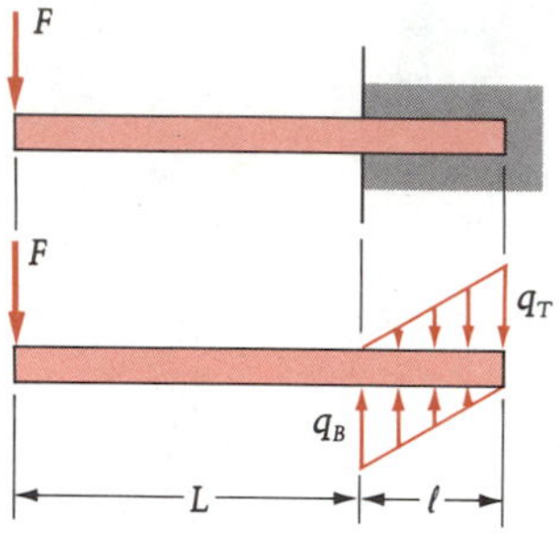

Figure P4.36

4.37 A slender uniform bar of length L and weight W is slung from the two cables shown in Figure P4.37. Find the tension in the cable on the right as a function of the ratio ℓ/L. Show that equilibrium is impossible if $\ell < L/2$.

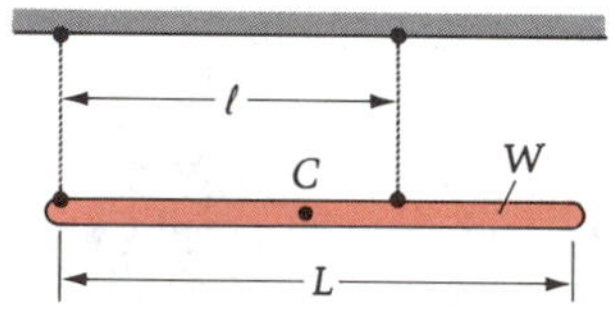

Figure P4.37

Figure P4.38

4.38 A man is painting a wall using a crude scaffold that consists of a pinned board and two cables. (See Figure P4.38.) (a) Find the tension T in each cable as a function of the man's weight and position (W and x), the length and weight of the board (ℓ, w), and the angle θ. Investigate the following limiting cases: (b) the board is light and the man is at $x = 0$; (c) the board is light and the man is at $x = \ell$; (d) the man is light. Assume symmetry.

4.39 In the "one size fits all" wrench of Figure P4.39, the handle $\mathcal{H}$ and the member $\mathcal{B}$ are free to turn relative to each other about the pin P. Neglecting friction between the wrench and nut, find:

- **a.** The force exerted on the nut by the handle if the nut is already tightened and nothing moves
- **b.** The magnitude of the force carried by the pin P
- **c.** The tightening moment (about the center line of the nut) that the wrench exerts on the nut.

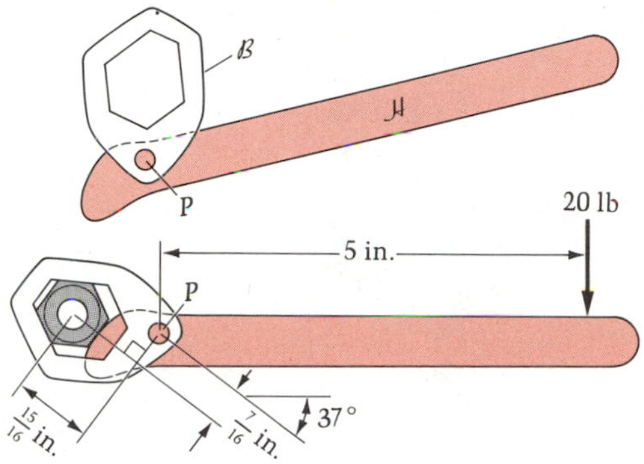

Figure P4.39

4.40 In Figure P4.40, find the reactions onto the beam at the pin (A) and roller (B).

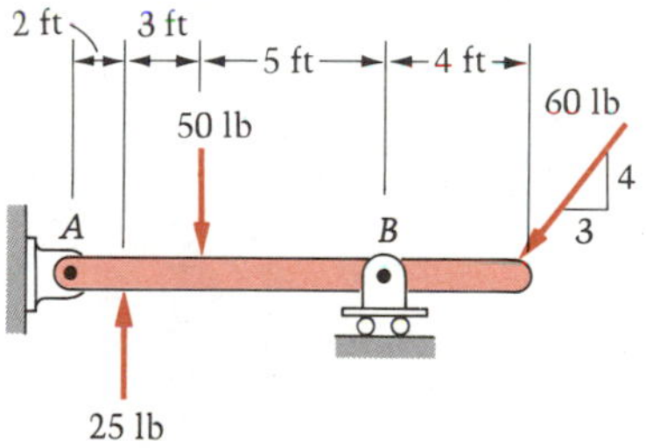

Figure P4.40

4.41 In Figure P4.41, find the reactions onto the bent bar at A, B, and C.

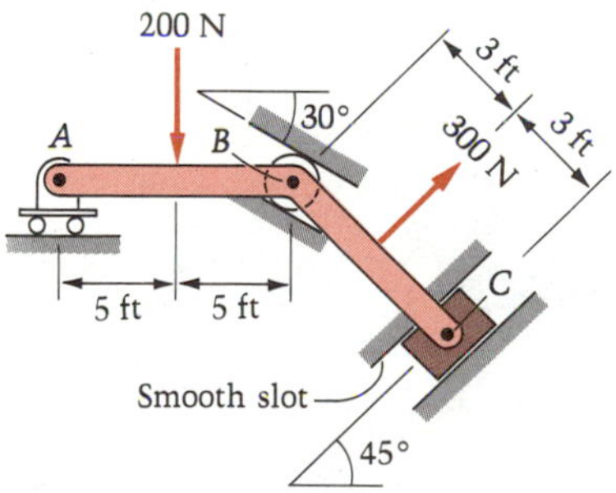

Figure P4.41

4.42 In Figure P4.42, find the reactions exerted by the wall on the beam.

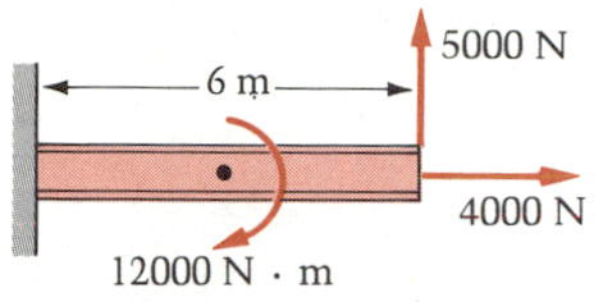

Figure P4.42

In Problems 4.43–4.54 find the reactions exerted on the members by the supports.

4.43

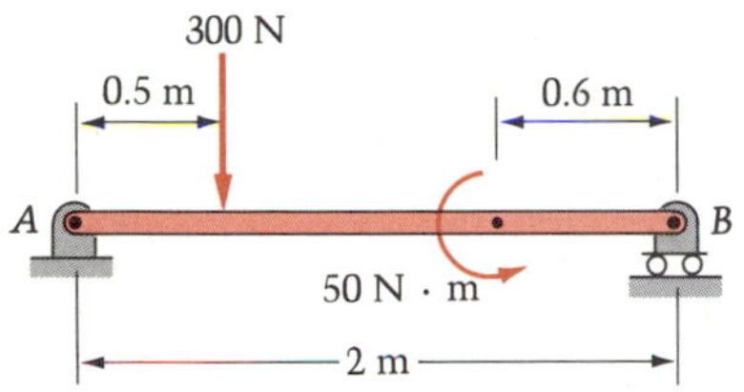

Figure P4.43

4.44

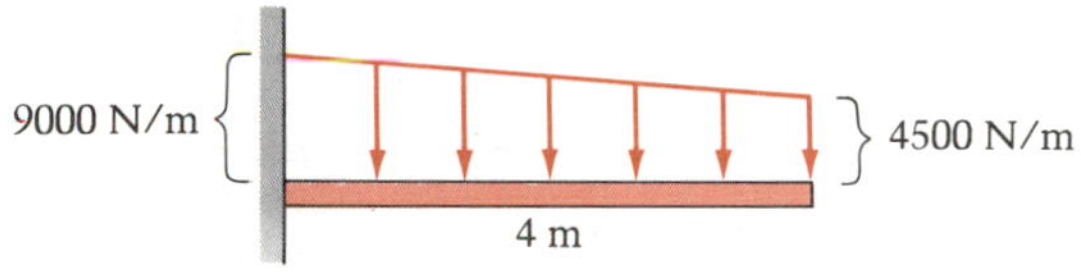

Figure P4.44

4.45

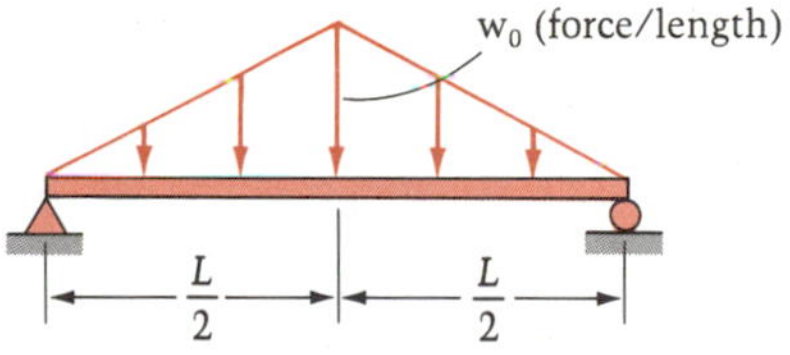

Figure P4.45

4.46

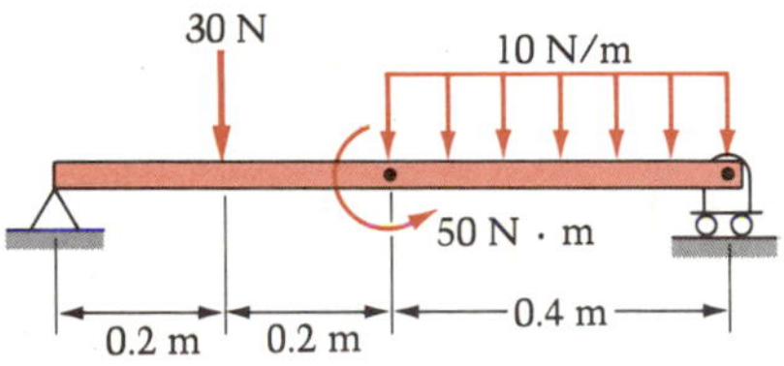

Figure P4.46

4.47

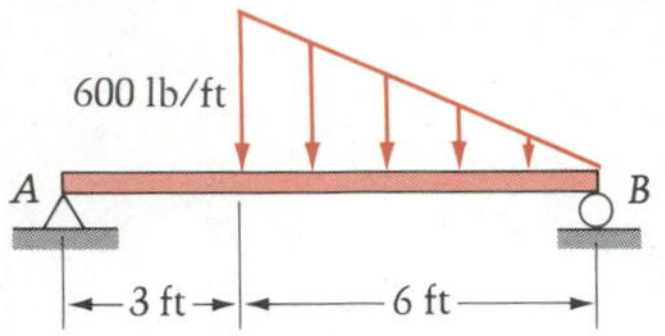

Figure P4.47

4.48

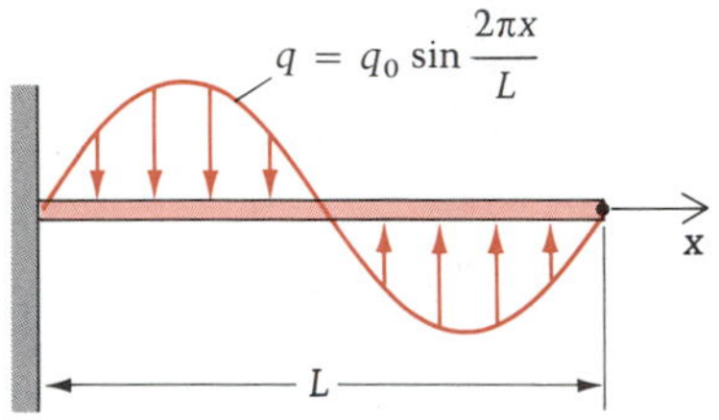

Figure P4.48

4.49

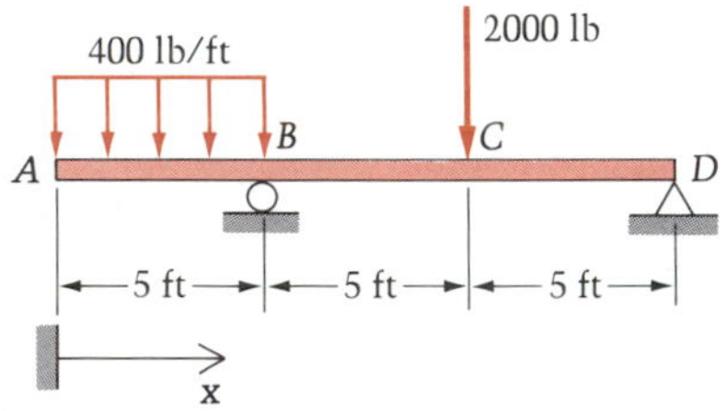

Figure P4.49

4.50

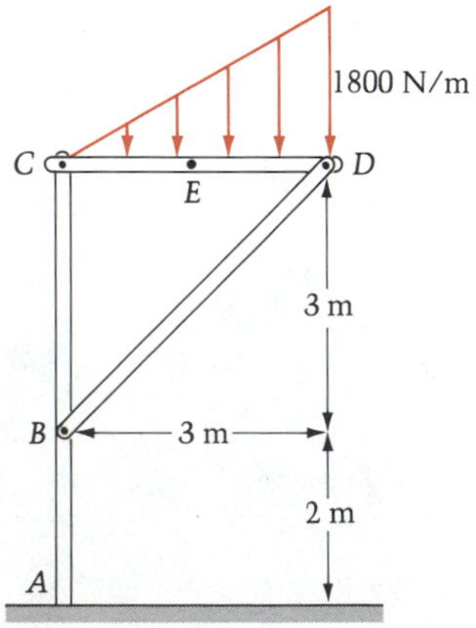

Figure P4.50

4.51

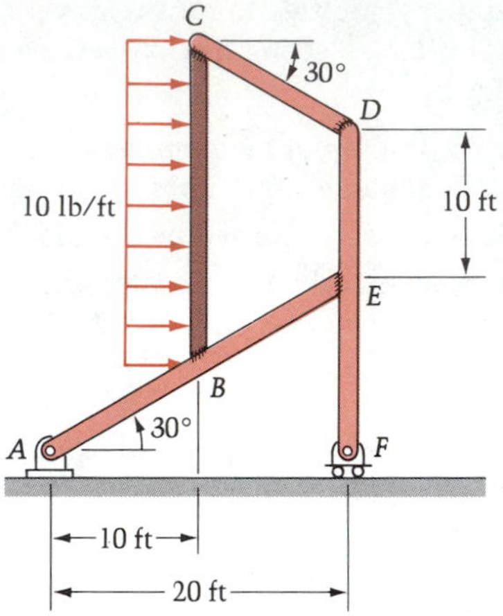

Figure P4.51

4.52

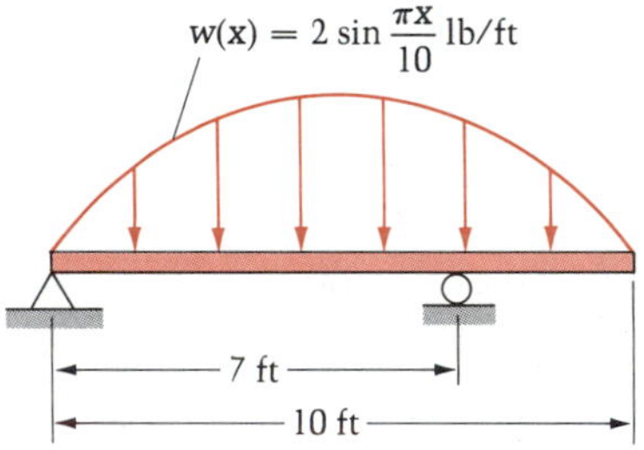

Figure P4.52

4.53

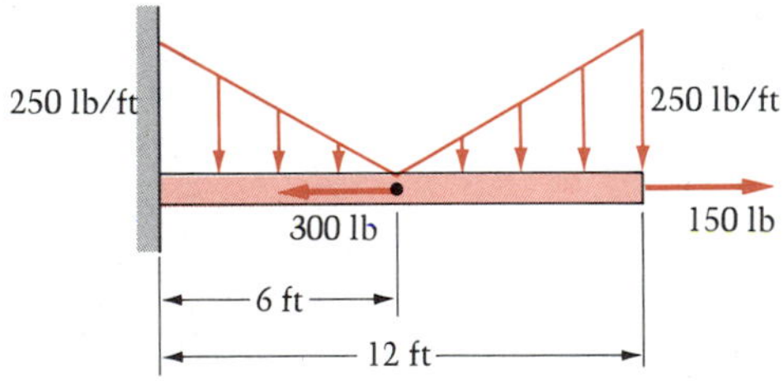

Figure P4.53

4.54

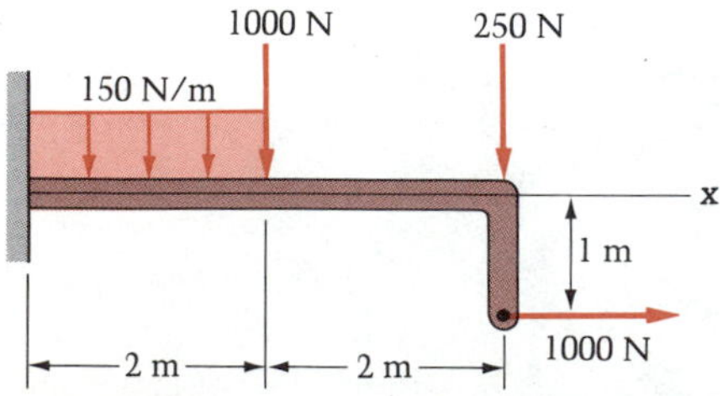

Figure P4.54

4.55 The crane in Figure P4.55 is pinned to the ground at O and to a screw-jack at A. The truss $\mathcal{T}$ and load $\mathcal{L}$ weigh 800 and 1200 pounds, respectively. In the given position, neglecting the distance between pin P and the nut, determine the tension in the screw and the pin reactions exerted onto $\mathcal{T}$ at O.

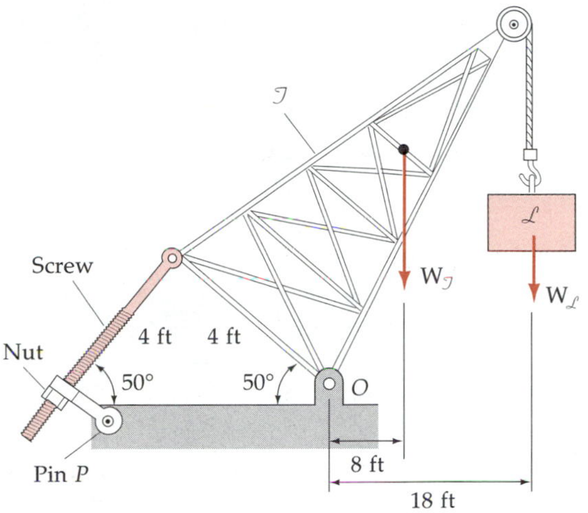

Figure P4.55

4.56 The crane of the preceding problem is lowered so that angle POA increases to 60°. (See Figure P4.56.) Again find the tension in the screw-jack and the reactions at O.

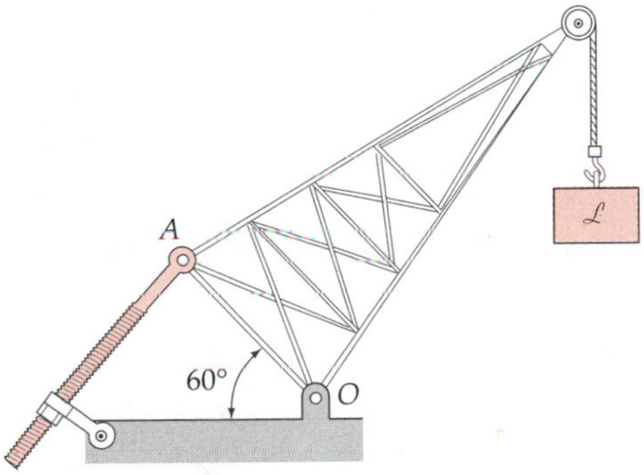

Figure P4.56

4.57 In Figure P4.57, the equilateral triangular plate has mass 80 kg and is supported by the light members $\mathcal{L}$ and $\mathcal{B}$, the latter of which is free to slide in a smooth slot. If the plate is in equilibrium, find the value of the couple M that is acting on it.

4.58 Repeat the preceding problem if M is replaced by a vertical force P at D whose magnitude is to be found.

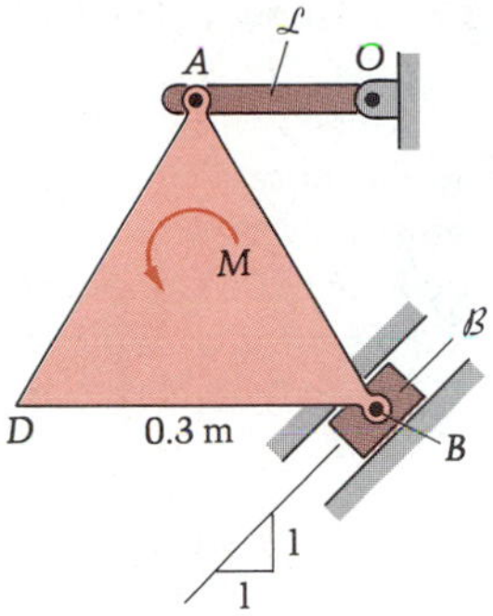

Figure P4.57

4.59 The weight of $\mathcal{W}$ is 300 lb, and the weight of wheel $\mathcal{C}$ is 200 lb (see Figure P4.59). Find the height (y coordinate) of the center of C of $\mathcal{C}$. Assume sufficient friction to prevent slipping.

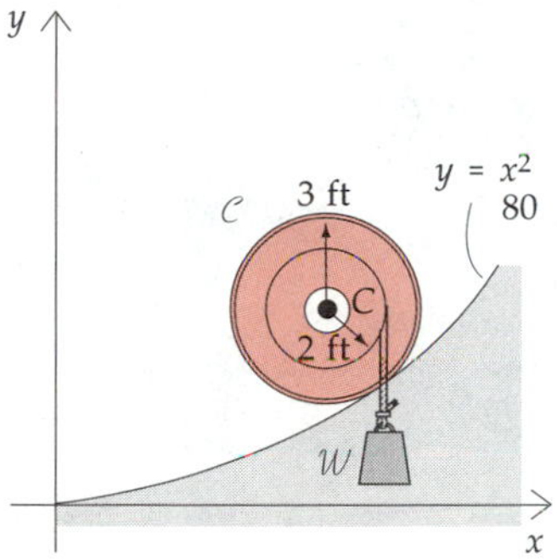

Figure P4.59

4.60 Repeat the preceding problem if the contact of the parabolic plane with the wheel is on its *inner* radius and if the weight is attached to a cord running over the *outer* radius. (See Figure P4.60)

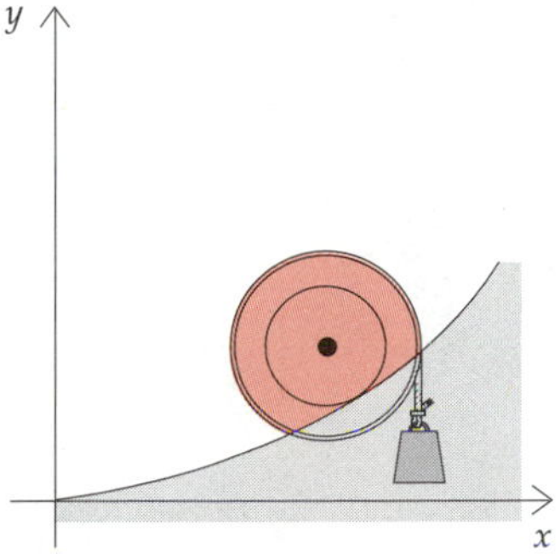

Figure P4.60

4.61 The 33-ft diameter antenna shown in Figure P4.61 (a) is supported at three pins. Pin A connects the "dish" to the jack AB, which extends to raise (and contracts to lower) the antenna in elevation. This rotation takes place about a horizontal line through two pins, shown as one at D in Figure P4.61(b). The entire structure suspended above A and D weighs 3500 lb, with mass center at C. Find the force in the elevation jack AB.

Figure P4.61(a)

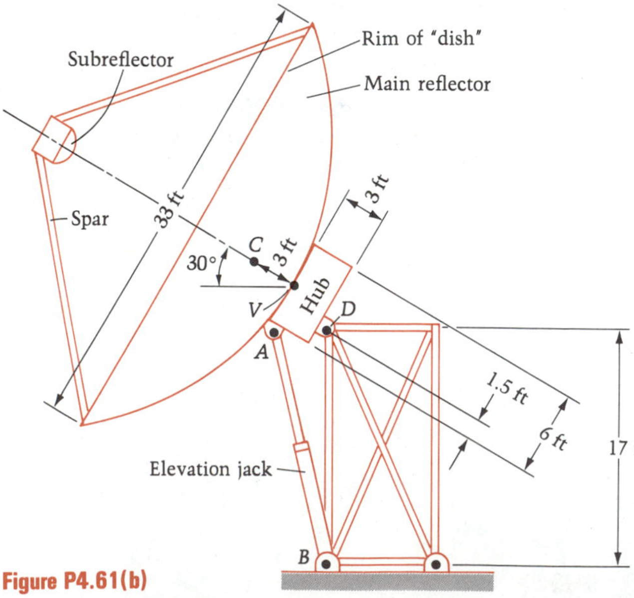

Figure P4.61(b)

4.62 To the preceding problem, we add the effect of the wind. Large antennas like the one shown in Figure P4.61(a) are designed to survive severe hurricanes and tornados of up to 125 mph wind velocities. At this wind speed, with the wind blowing horizontally into the dish (to the right), the wind forces on the dish have a resultant at the vertex V of: (1) Drag Force = 44,200 → lb; (2) "Lift" Force = 27,100 ↓ lb; and (3) Moment = 41,300 ↺ lb-ft. Find the force in the elevation jack and the combined reaction on the hub pin at D. [See Figure P4.61(b).]

* **4.63** Repeat the preceding problem, with the antenna pointing at 80° elevation. This time the 125 mph wind loads are: (1) Drag = 8790 → lb; (2) "Lift" = 19,100 ↓ lb; and (3) Moment = 106,000 ↻ lb-ft.

4.64 The fork-lift truck weighs 9500 lb. (a) If it carries a load of 5000 lb in the position shown in Figure P4.64, find the forces exerted by the ground on the tires. (b) How much greater load could be carried in the same position without the truck tipping forward?

4.65 The 50-kg uniform rod is pinned to the 80-kg wheel at A and B. (See Figure P4.65.) At C, the rod rests on a smooth floor, as does the wheel at D. Find the reactions at C and D.

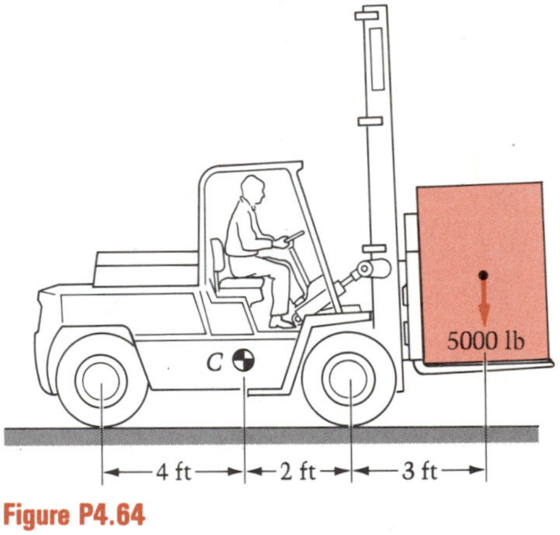

Figure P4.64

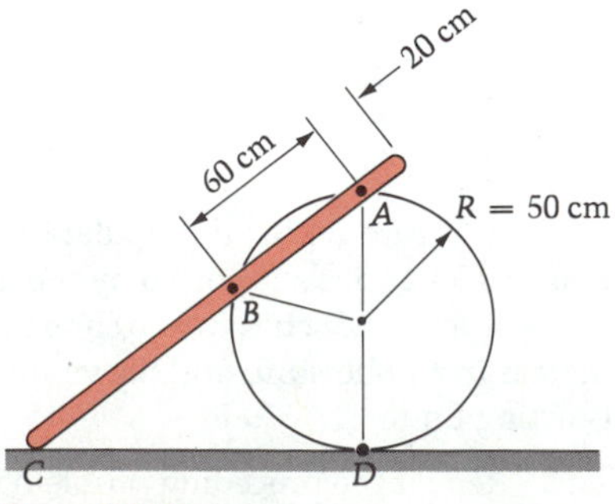

Figure P4.65

4.66 The rod weighs 64.4 lb and is held up by three cables as shown in Figure P4.66. Find the tension in each cable.

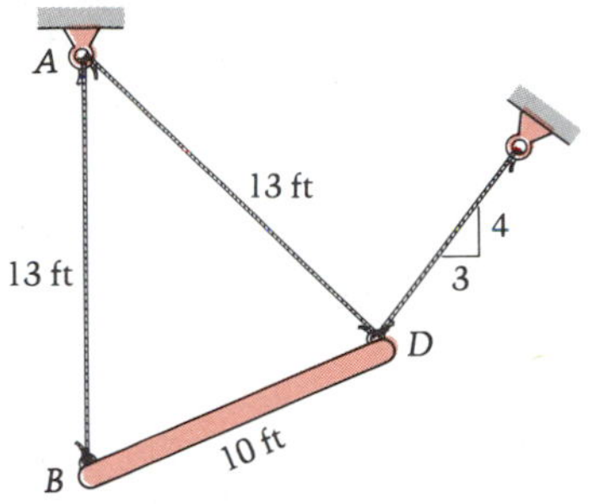

Figure P4.66

4.67 In Figure P4.67 the rod AB of length L and weight W is connected to smooth hinges at F and D by the light members AF and BD, each of length L. A cable CE completes the support of the rod. Find the force in the cable.

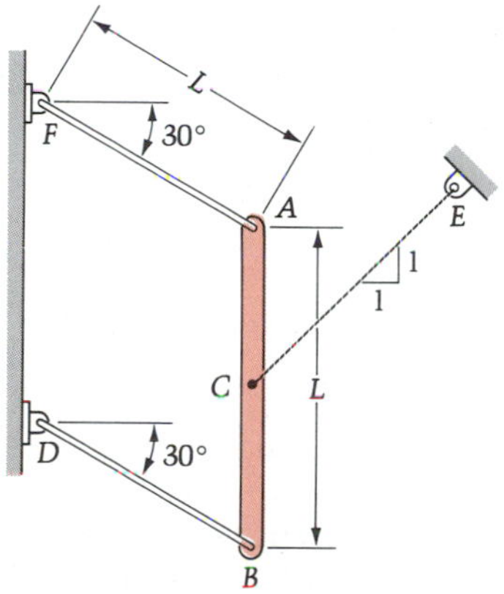

Figure P4.67

4.68 The man is holding the wheel of weight W in equilibrium on the rough inclined plane. (See Figure P4.68.) Find the tension in the rope, in terms of W. Assume sufficient friction to prevent slip.

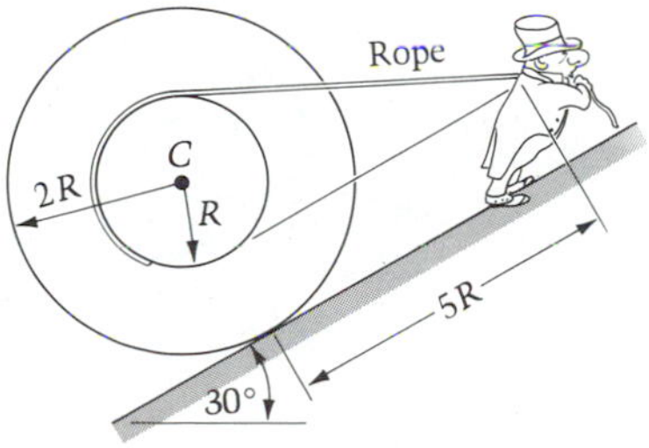

Figure P4.68

4.69 In Figure P4.69 disc $\mathscr{A}$ is pinned to the bent bar $\mathscr{B}$ at A; the bodies weigh 20 N ($\mathscr{A}$) and 30 N ($\mathscr{B}$). Body $\mathscr{B}$ has mass center at C and rests on the smooth plane. Find the force P required for equilibrium of the system of $\mathscr{A}$ and $\mathscr{B}$.

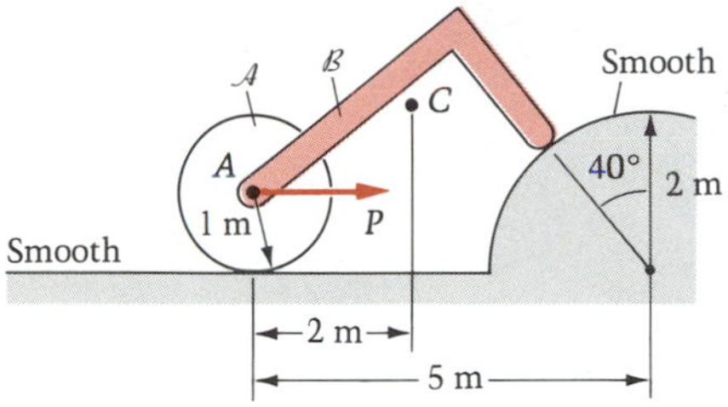

Figure P4.69

4.70 In Figure P4.70 the 150-kg cylinder is pinned at A to the 200-kg uniform bar. Find the force P required for equilibrium of the bodies.

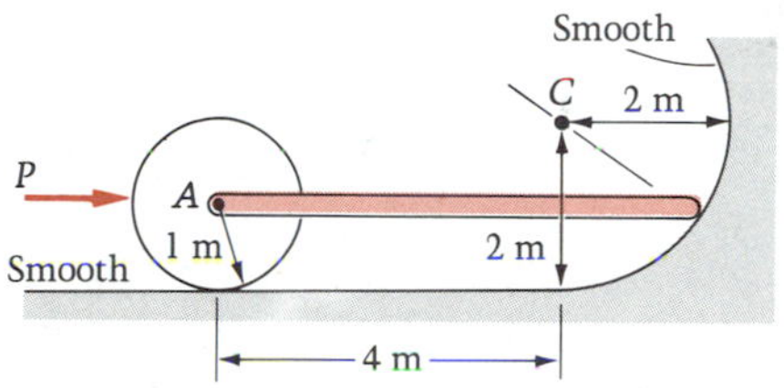

Figure P4.70

4.71 Repeat the preceding problem if a clockwise couple of 50 N · m acts upon the bar.

4.72 A beam with mass M and length ℓ is supported by a smooth wall and floor, and a cable, as shown in Figure P4.72. Find the tension T in the cable as a function of θ (different cables for different θ's). Show from $T(\theta)$ that as $\theta \rightarrow 0$, $T \rightarrow 0$, and that T approaches ∞ as $\theta \rightarrow 90°$. Verify that these results make sense using free-body diagrams at $\theta = 0$ and $\theta =$ almost $90°$.

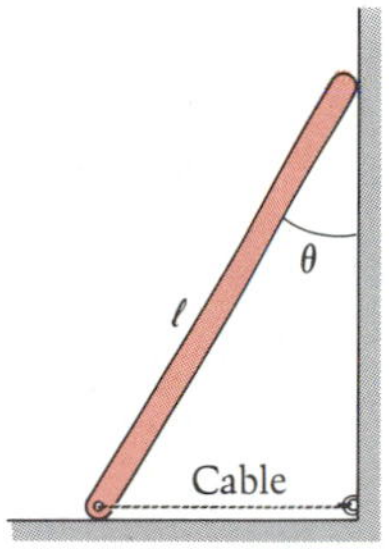

Figure P4.72

4.73 Show that if the cable in the preceding problem is instead attached to the top of the beam (see Figure P4.73), its tension is constant (i.e., this time it doesn't depend on θ).

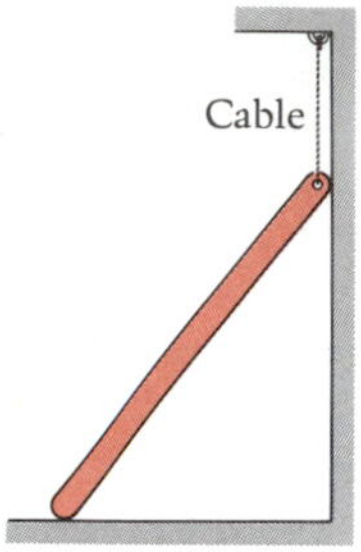

Figure P4.73

4.74 In Figure P4.74 the uniform circular disc of weight W and radius r has a uniform bar of length L and weight W welded to it at A such that the bar is perpendicular to AO. Find the angle θ for equilibrium of the combined body.

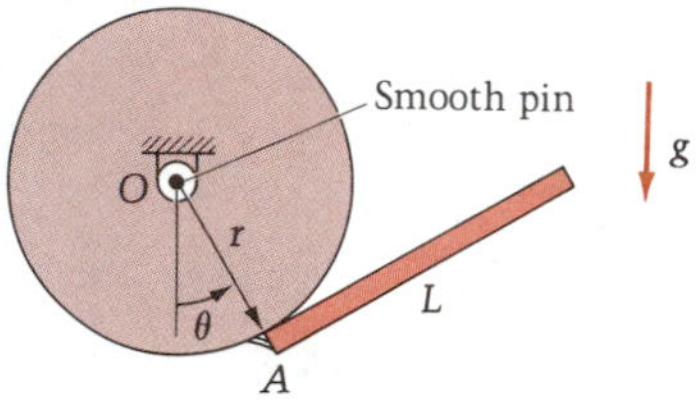

Figure P4.74

4.75 In Figure P4.75 the homogeneous, uniform bar weighs 10 lb/ft. Find the reaction the pin at A exerts on the bar.

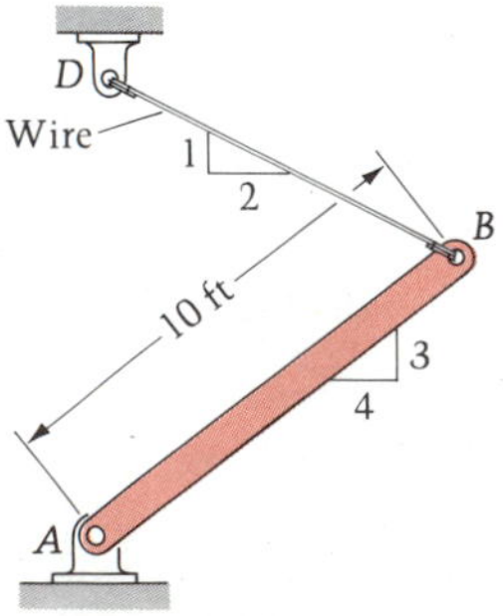

Figure P4.75

4.76 The uniform bar weighs 1000 N and is in equilibrium in the horizontal position shown in Figure P4.76. Find the tension in the cable and the reactions exerted by the pin at A onto the bar.

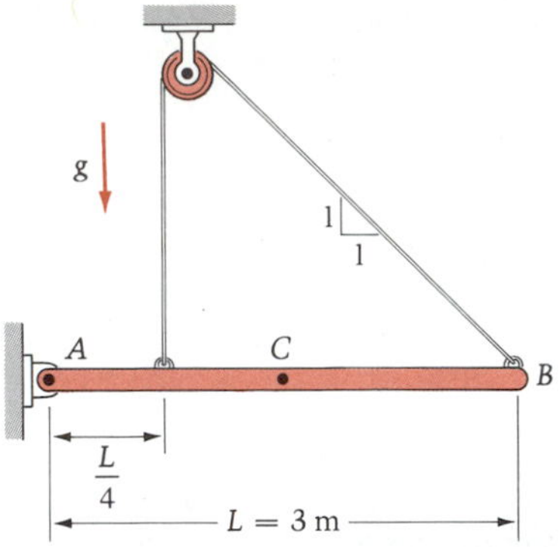

Figure P4.76

4.77 The device in Figure P4.77 is called a crusher. With a hand-force H applied as shown, a large force P can be developed onto material within the enclosure resisting the block moving to the right. Find the force exerted by the pin at C onto the bar ABC.

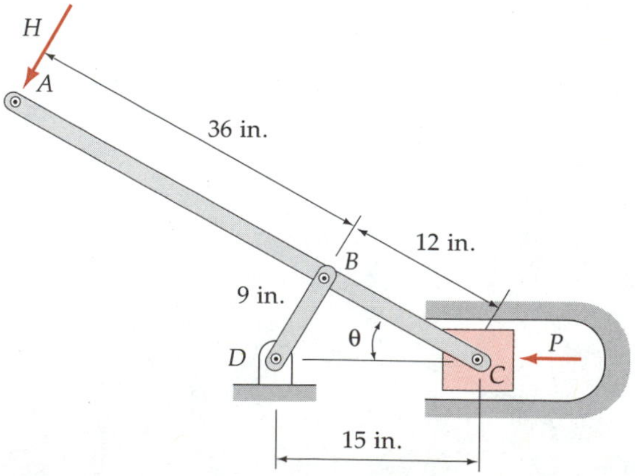

Figure P4.77

4.78 What horizontal force F will pull the 100-kg lawn roller over the step? (See Figure P4.78.) What is the value of F if it is directed normal to AC as shown dotted?

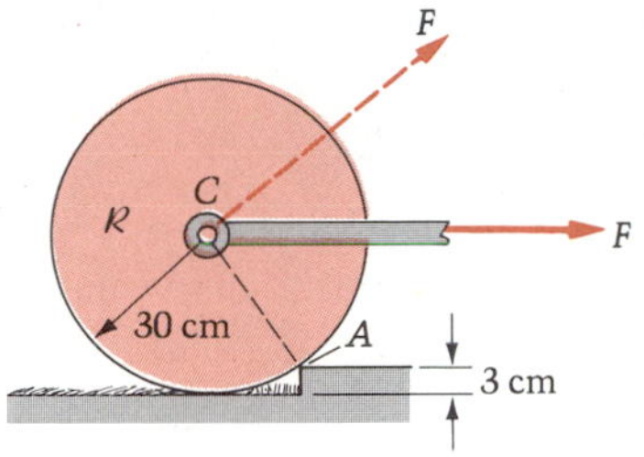

Figure P4.78

* **4.79** The 10-ton moving crane in Figure P4.79 has a mass center at C. It carries a maximum load of 18 tons. Find the smallest weight of counterweight $\mathcal{C}$, and also the largest distance d, so that: (a) the crane doesn't tip clockwise when the maximum load is lifted; and (b) the crane doesn't tip counterclockwise when there is no load present. Also find, for $d = 2.5$ ft, the *range* of weights of $\mathcal{C}$ for which (a) and (b) can be satisfied.

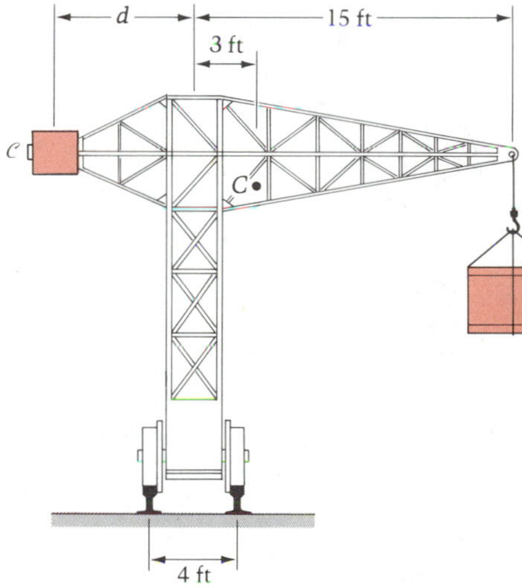

Figure P4.79

* **4.80** In the preceding problem, illustrate on a graph of $W_{\mathcal{C}}$ versus d the safe region of points $(d, W_{\mathcal{C}})$ for which *both* conditions (a) and (b) will be satisfied. Consider only values of $d > 2$ ft for practical reasons.

4.81 Repeat Problem 4.68 for the different man and rope arrangement shown in Figure P4.81.

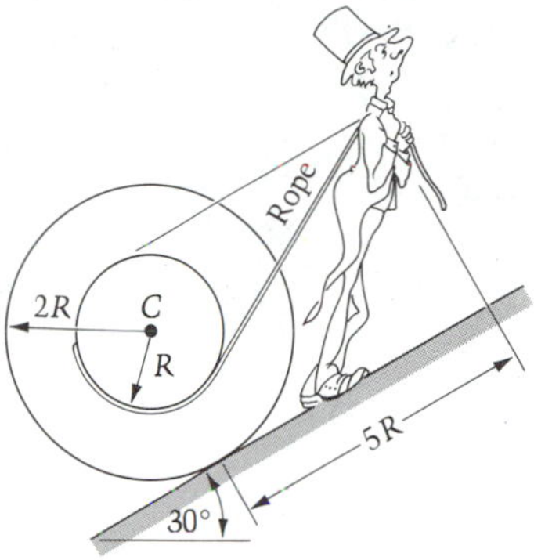

Figure P4.81

4.82 The car in Figure P4.82 has weight W. What is the resultant of the normal forces on the passenger-side tires if the car is parked on the indicated incline?

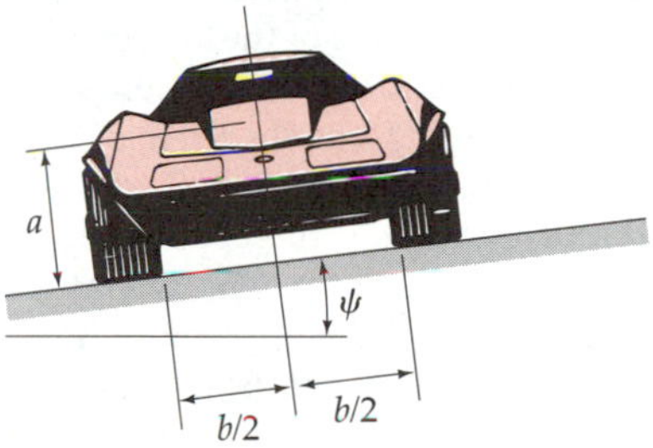

Figure P4.82

4.83 The 100-lb sign in Figure P4.83 is supported by a pin at A and a cable from B to C. If the cable breaks at 400-lb tension and the pin fails at 600 lb of force, find the safe values of θ.

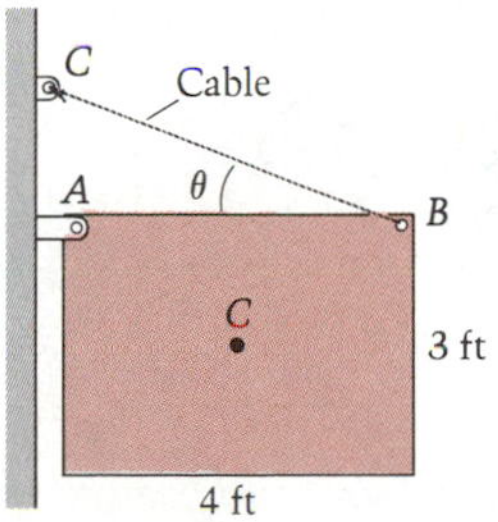

Figure P4.83

4.84 In Figure P4.84 the member weighs 50 N and has its mass center at C. At B, a pin, fixed to the ground, bears against a slot. The spring carries a tensile load of 100 N. Find the vertical component of the reaction at A.

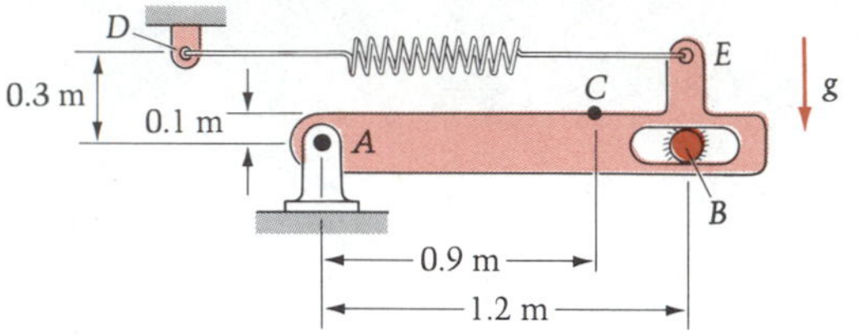

Figure P4.84

4.85 Zeke the moonshiner has built a "water" tower in which to store his liquid refreshment. The tank at the top is 5 ft high and 4.8 ft in diameter, and is mounted on four symmetrically located legs inclined to the horizontal. (See Figure P4.85.) The weight of the whole tower (legs plus tank) is 3800 lb. Wind force is to be computed on the basis of the pressure times the projected area of the tank on the plane perpendicular to the direction of the wind. (a) If this dynamic pressure is 62.5 lb/ft² (blowing left to right), find whether the tower will blow over or not. (Zeke sees the big storm coming and is running to get something with which to fill the tank.) (b) If Zeke begins filling the tank with moonshine having the density of water, how close to full must it get in order to prevent the tower from tipping over?

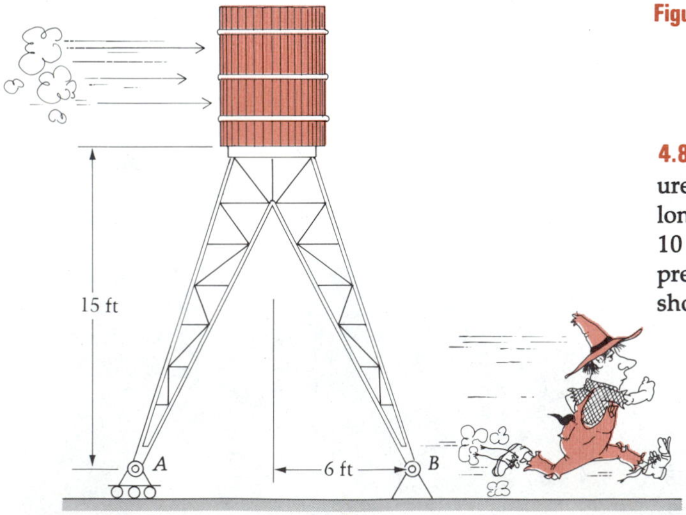

Figure P4.85

4.86 Bar AB in Figure P4.86 is supported by a roller at C, and by a smooth wall at A. The bar is uniform, has a mass of 20 kg and is 0.6 m long. What vertical force at B is necessary for equilibrium?

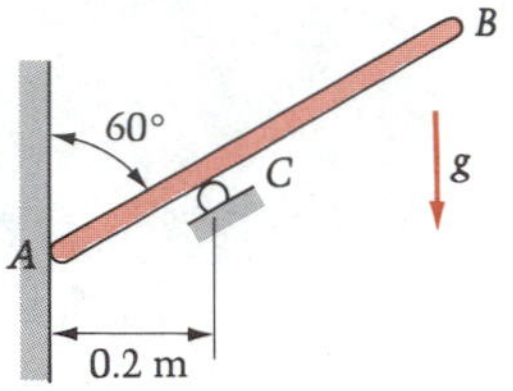

Figure P4.86

4.87 Find the reaction of the ground onto the pole in Problem 2.24, expressed as a force and couple at the point D, if the tension in the power line BE is 300 lb and the pole weighs 4000 lb.

4.88 In Figure P4.88 the beam AB and the pulley each weigh 30 lb. Find the tension in the rope, and the pin reaction at A.

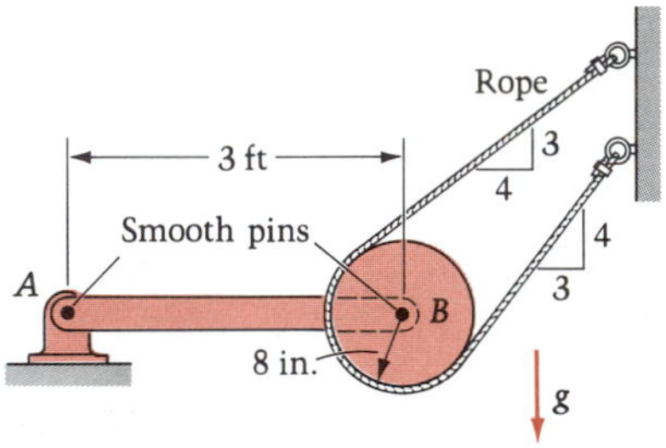

Figure P4.88

4.89 Five weights are suspended as shown in Figure P4.89 at equal distances along a uniform bar 40 cm long weighing 60 N. The first weight at the left end is 10 N, and each successive weight is 10 N heavier than the preceding one. At what distance x from the left end should the bar be suspended so as to remain horizontal?

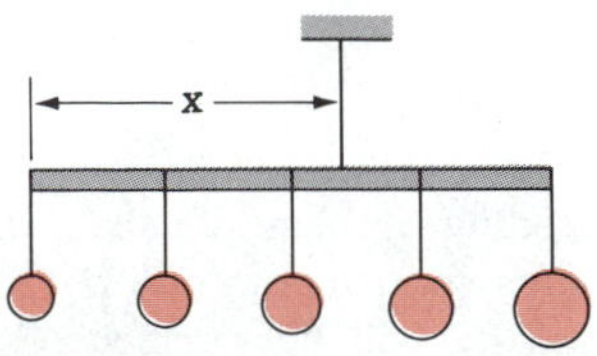

Figure P4.89

4.90 In Figure P4.90 the body *ADB* is a bell crank pinned to the floor at point *D*. Find the force *F* in the link $\mathcal{L}$, pinned to the bell crank at *B*, required for equilibrium.

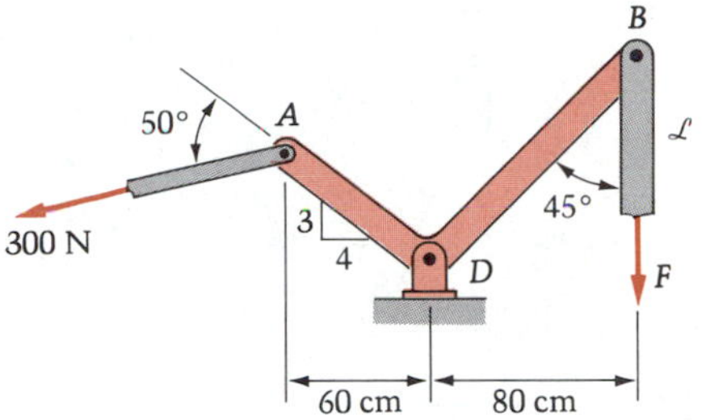

Figure P4.90

4.91 In Figure P4.91 the motor weighs 200 lb and is fixed to the 50-lb frame, which is pinned to the ground at *O*. The tension in the motor's belt prevents it from turning and falling. When the motor is turned off, the tensions T_a and T_b are equal if friction is neglected. In this case, find the tension in the belt.

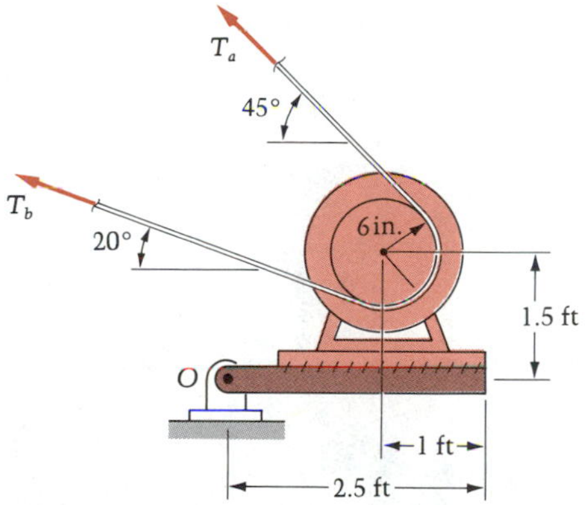

Figure P4.91

4.92 The frame in Figure P4.92 is supported by the pin at *A* and the cable. Find the pin reaction at *A* and the tension in the cable.

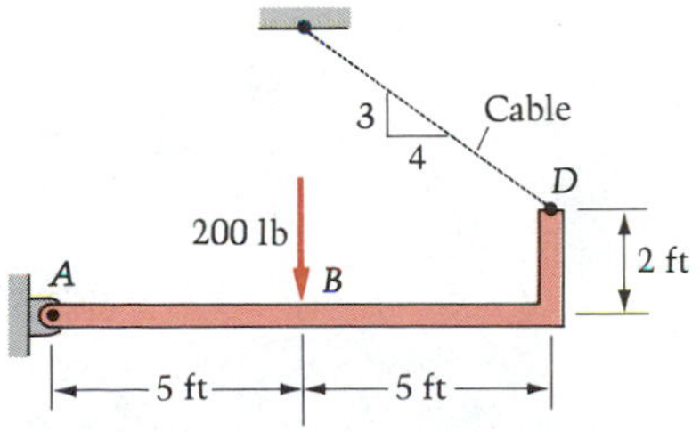

Figure P4.92

4.93 If the corner *Q* in Example 4.8 helps to support the 1200 lb weight with no force in *AB*, what is now the force in *AB* required to start the bed pivoting about the pin *D*? (See Figure P4.93.)

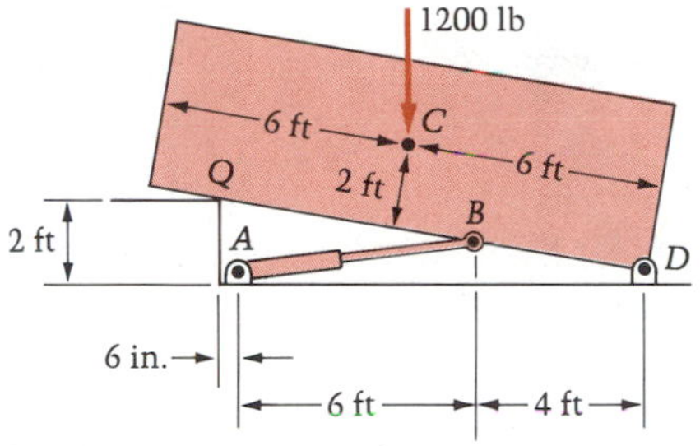

Figure P4.93

4.94 Bar *ABD* in Figure P4.94 is supported by a hinge at *B* and by a cable at *A*. Find (a) the tension in the cable, and (b) the components of the force at *B* on *ABD*.

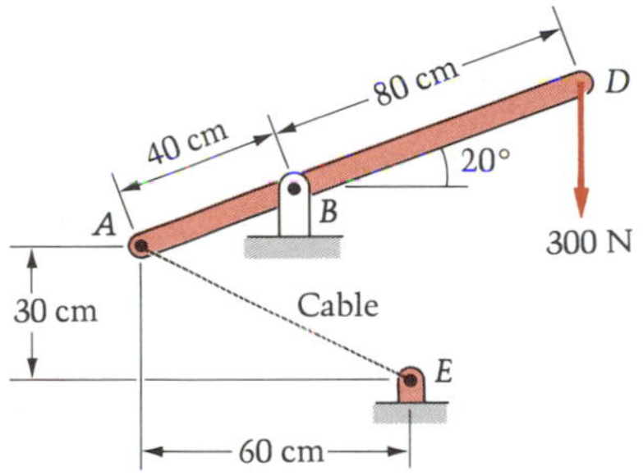

Figure P4.94

4.95 Find the pin reaction at *A* and the roller reaction at *B* for the bent bar situated and loaded as shown in Figure P4.95.

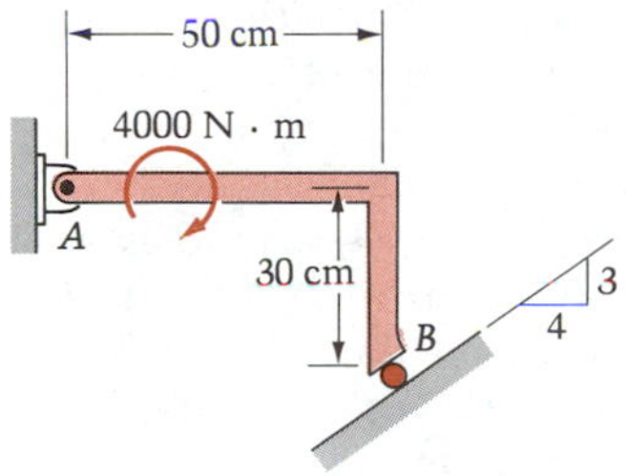

Figure P4.95

4.96 Find the force that the pin at A exerts on the rectangular plate shown in Figure P4.96.

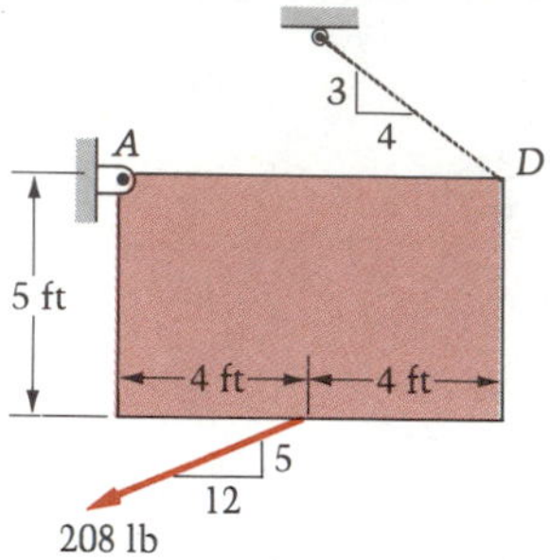

Figure P4.96

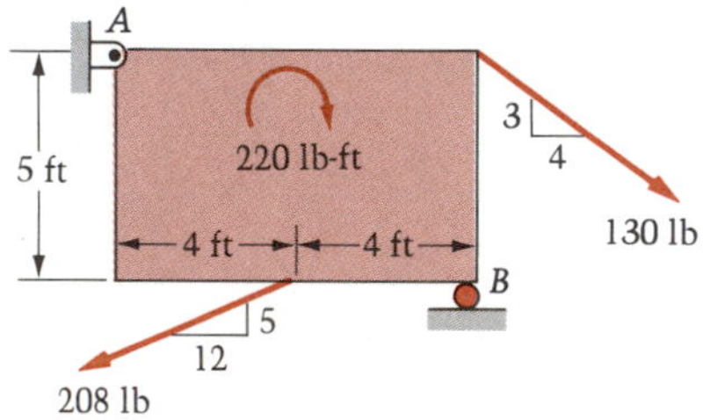

Figure P4.97

4.97 Repeat the preceding problem if a couple is applied as in Figure P4.97 and if the cable is replaced by an applied force of 130 lb. Also, the plate now rests on a roller at B as shown.

• **4.98** A davit is a crane on a ship for hoisting lifeboats, anchors, or cargo. A "gravity davit" uses gravity to help with the lowering. A boat is shown being held in the stow position by two gravity davits, one of which is shown in Figure P4.98(a). Assume that a removable pin is located at roller $\mathcal{A}$. The boat weighs 1500 lb, and each of the three 8-ft sections of the davit weighs 350 lb. Find the force in the cable (two per davit, i.e., four per boat) and also the reactions exerted on the smooth rollers $\mathcal{A}$ and $\mathcal{B}$. The rollers are pinned to the davit, as are the free pulleys $\mathcal{C}$, $\mathcal{D}$, and $\mathcal{E}$. (There are also pulleys on the other side of the davit at C, D, and E.) Neglect friction and the width of the davit, and note the photographs in Figures P4.98(b, c, d).

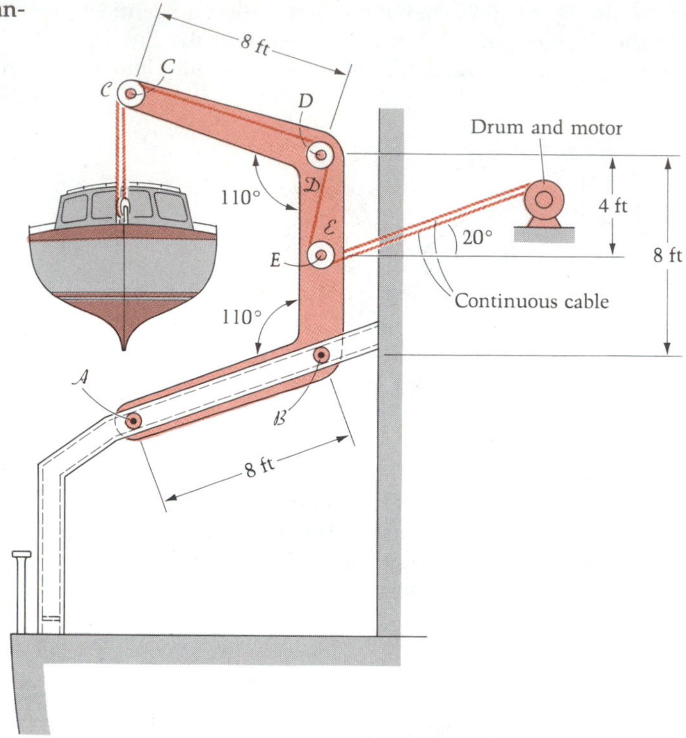

Figure P4.98(a)

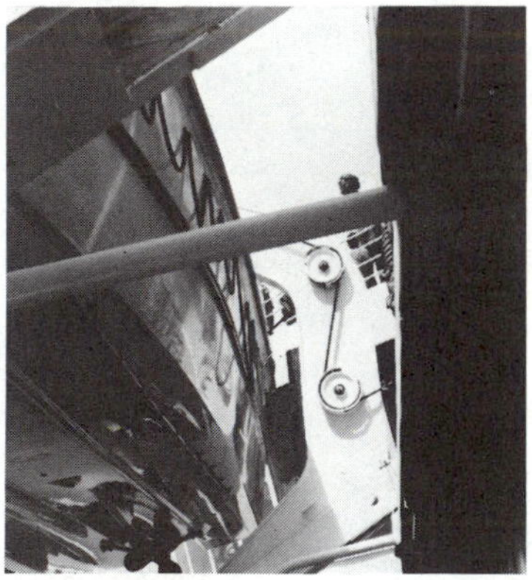
Figure P4.98(b)

Figure P4.98(c)

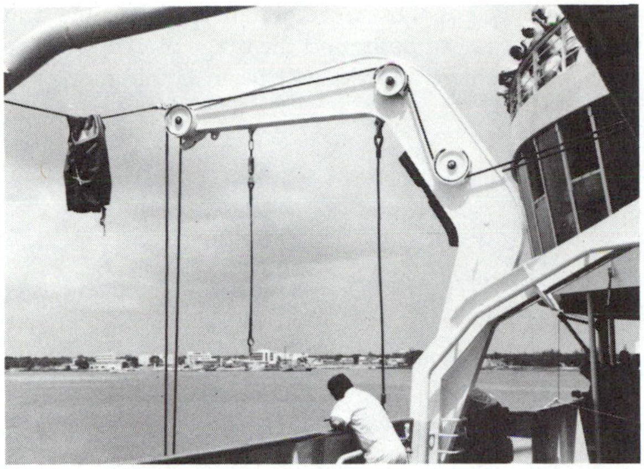

Figure P4.98(d)

4.100 (a) If the child and swing together weigh 200 N, find the force in each of the two ropes. (See Figure P4.100(a).) (b) If the child's father uses a horizontal force to pull the swing back 30° from the vertical plane of the ropes (see Figure P4.100(b)), what is this force and what is now the force in each rope?

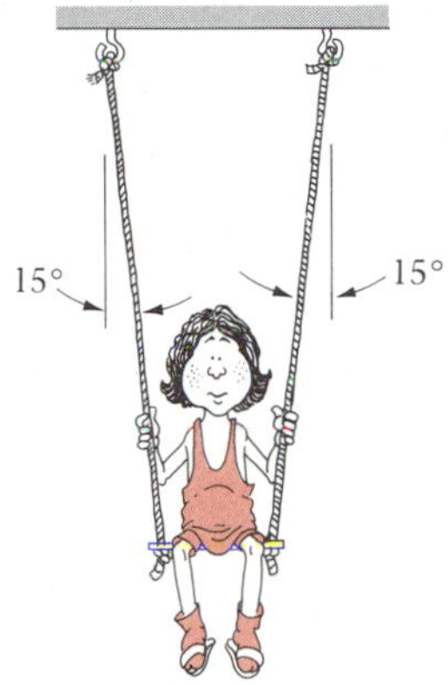

Figure P4.100(a)

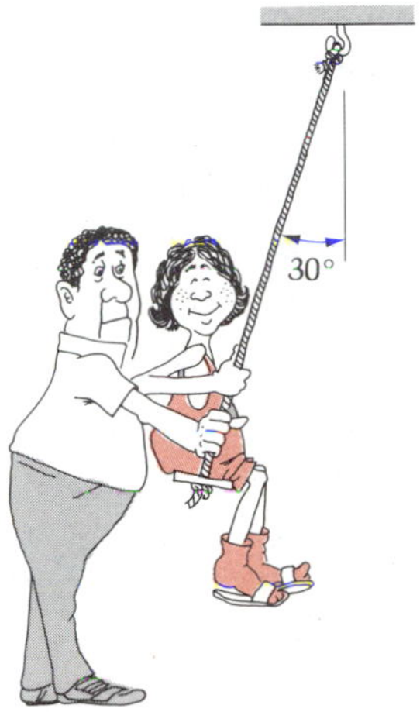

Figure P4.100(b)

* **4.99** When the cable is let out in the preceding problem, gravity causes the davit to move downward to the left in the slot. Eventually roller $\mathcal{A}$ reaches the bottom of the slot as shown in Figure P4.99, and the boat then lowers into the sea. After the cable is released from the boat, find the forces on the rollers $\mathcal{A}$ and $\mathcal{B}$ in the position shown.

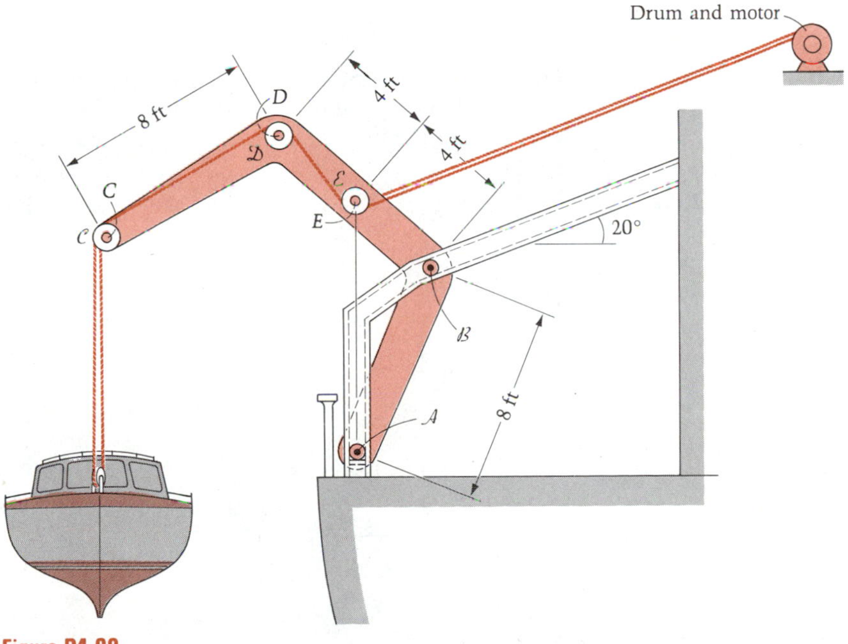

Figure P4.99

4.101 Find the reaction of the floor onto the roller in Figure P4.101. The block W weighs 200 N and the uniform bar ABD weighs 300 N.

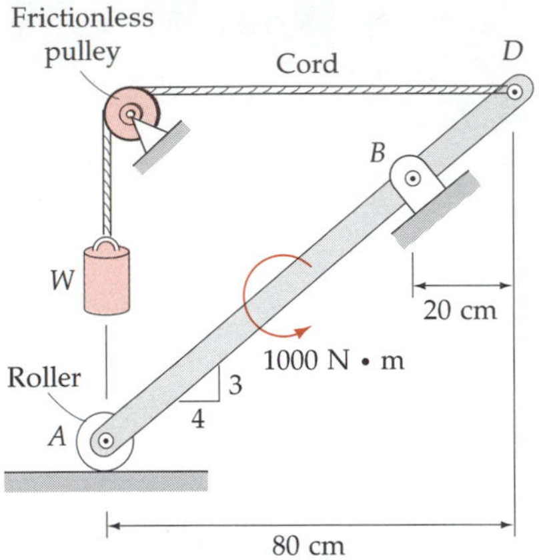

Figure P4.101

* **4.102** In Figure P4.102 determine the angle θ for which the ladder is in equilibrium if planes AB and BC are both smooth.

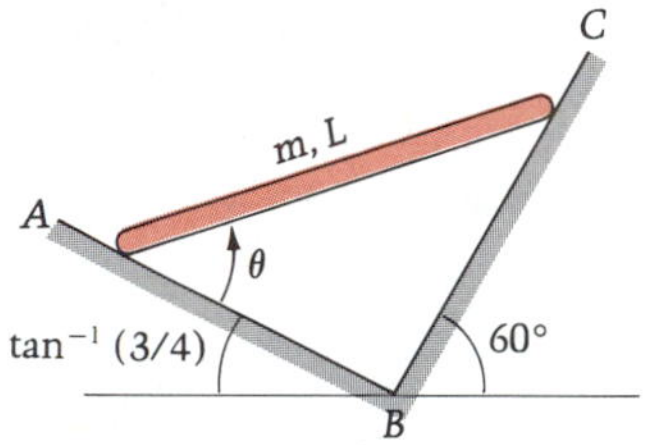

Figure P4.102

* **4.103** In Figure P4.103 the bar of mass m and length ℓ is pinned at B to a smooth collar, which is free to slide on a fixed vertical rod. The other end rests on the smooth parabolic surface shown. Find the x coordinate of the contact point A.

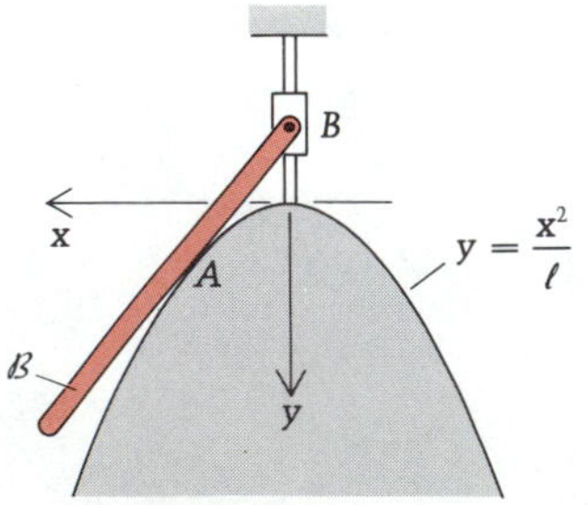

Figure P4.103

* **4.104** If the weight of the rod of length L is negligible, and if all surfaces are smooth, find the range of values of the angle β for which the rod will be in equilibrium. See Figure P4.104.

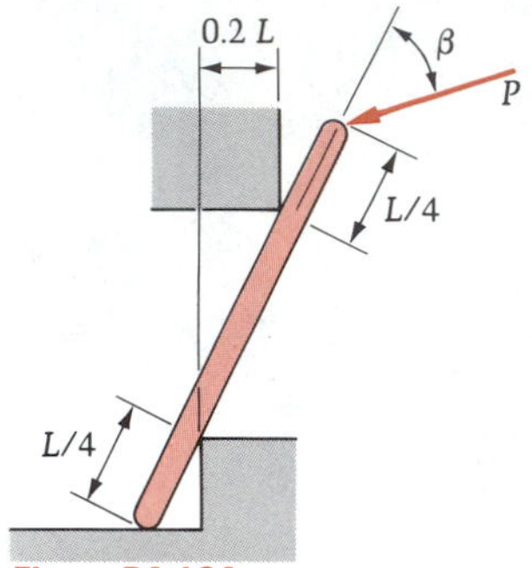

Figure P4.104

* **4.105** Find the angle θ assumed by the stirrer of length L, in equilibrium in a smooth hemispherical cup, for $L = 3R$. (See Figure P4.105.)

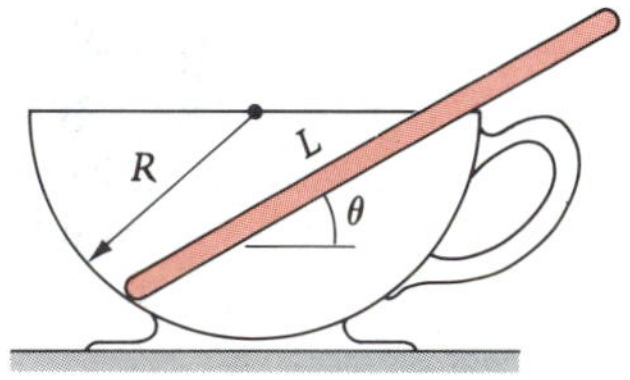

Figure P4.105

4.106 The boom in Figure P4.106, consisting of the identical struts $\mathcal{S}_1$ and $\mathcal{S}_2$ and the cable $\mathcal{C}$, holds up a compressor weighing 1300 lb. Find the forces in the struts and in the cable.

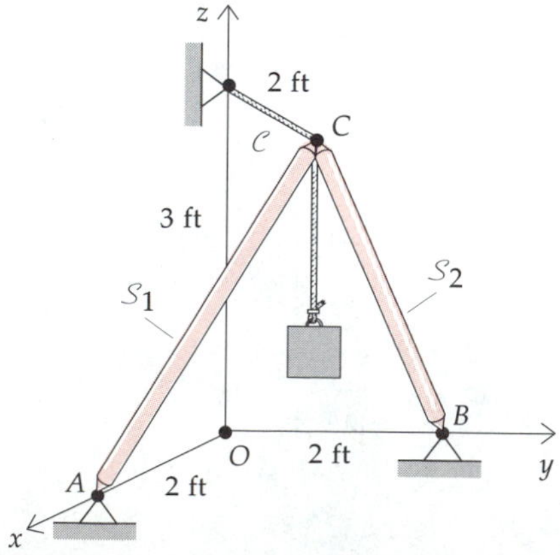

Figure P4.106

4.107 The plate in Figure P4.107 weighs 2 kN and is supported by the hinge at corner *A* and the cable *BC*. Find the tension in the cable, and the force and couple exerted by the hinge onto the plate.

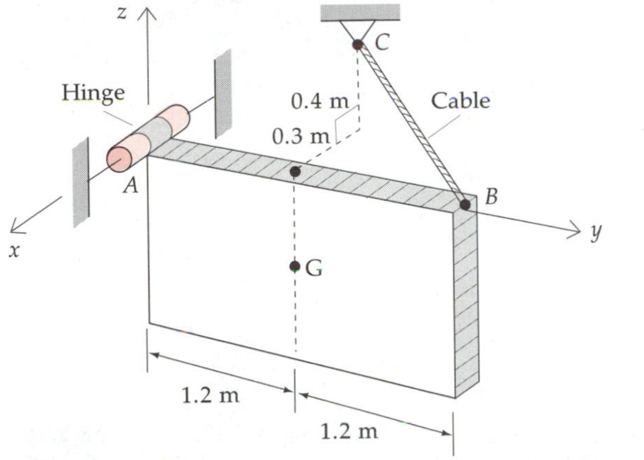

Figure P4.107

4.108 The 10-ft mast *OD* in Figure P4.108 can carry 10,000 lb without failing. The cables can each carry 4000 lb without breaking. Find the radius of the base circle, on which the cables are attached, for which both post and cables will reach their maximum loads simultaneously if the cables are gradually tightened.

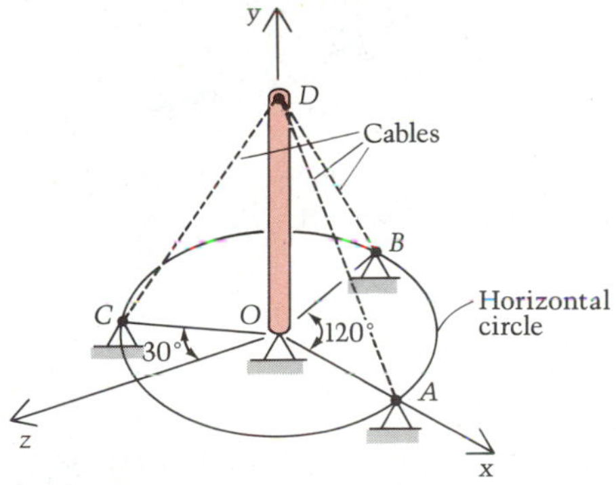

Figure P4.108

4.109 For the structure *ABCD* loaded and supported as shown in Figure P4.109, find the reactions at *A* and the cable tensions.

4.110 The uniform block in Figure P4.110 weighs 500 N. It is supported by ball joints at *O* and *A*, and a cable from *B* to *C*. Find the tension in the cable.

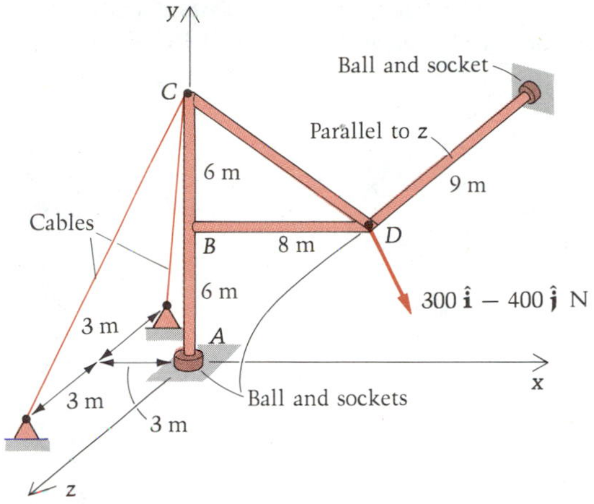

Figure P4.109

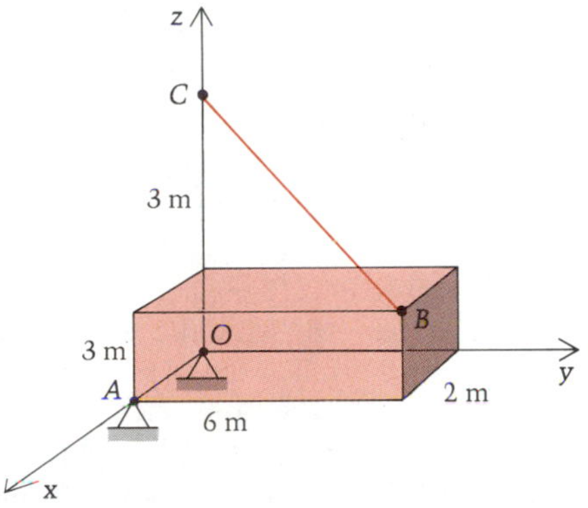

Figure P4.110

4.111 In Figure P4.111 determine the tensions in cables *BC* and *BE*. Neglect the weights of all members and assume that the support at *A* is a ball-and-socket joint. The 5200-N force has no *x* component.

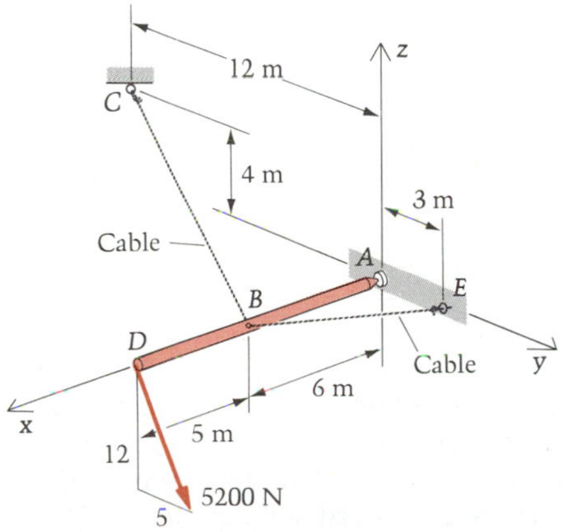

Figure P4.111

4.112 A 40-lb cellar door is propped open with a light stick, as shown in Figure P4.112. Find the force in the stick.

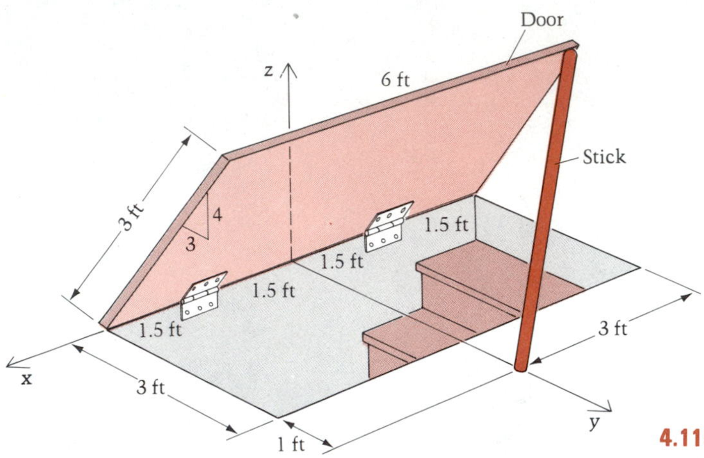

Figure P4.112

4.113 In Figure P4.113 the bar *CD* is welded to the center of another bar *AB*, with the end *D* resting against a smooth vertical wall in the *yz* plane. The bars aren't perpendicular, and *AB* is connected at *A* to a ball joint; at *B* it passes through a smooth eyebolt. The bars weigh 12 lb/ft. Find the force exerted on the bar at point *D*.

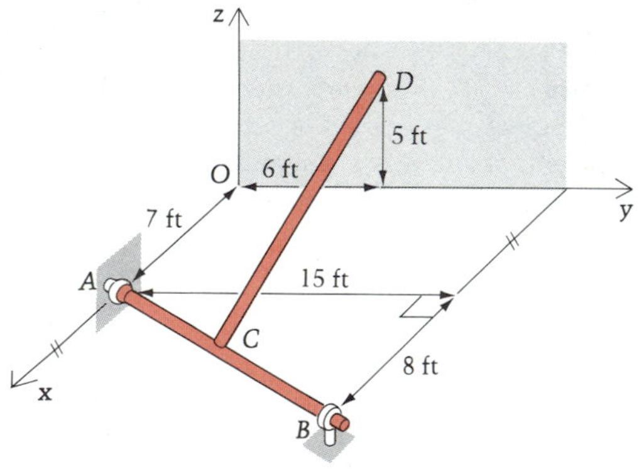

Figure P4.113

4.114 A rigid frame whose base is in the *xy* plane is shown in Figure P4.114. Calculate the forces in the cables, and the reactions on the frame at *A*, if the frame weighs 150 N/m. Neglect the cross-sectional dimensions of the beams constituting the frame.

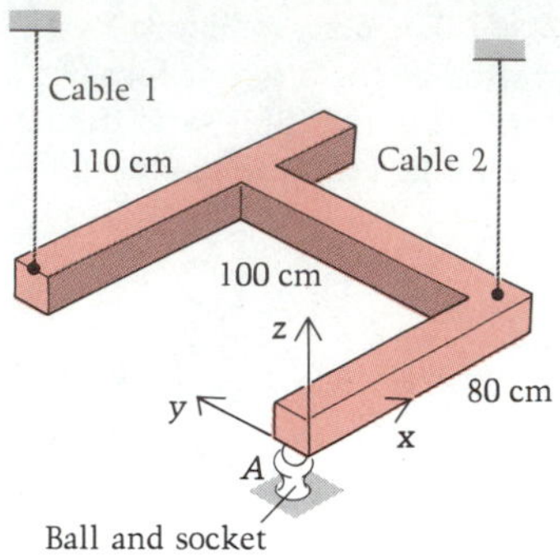

Figure P4.114

4.115 The triangular plate weighs 1000 N and is supported by the roller and hinge as shown in Figure P4.115. The plane of the plate is horizontal. Find the forces exerted by the roller, and the force- and couple-components exerted by the hinge (which cannot exert a couple in the *y*-direction).

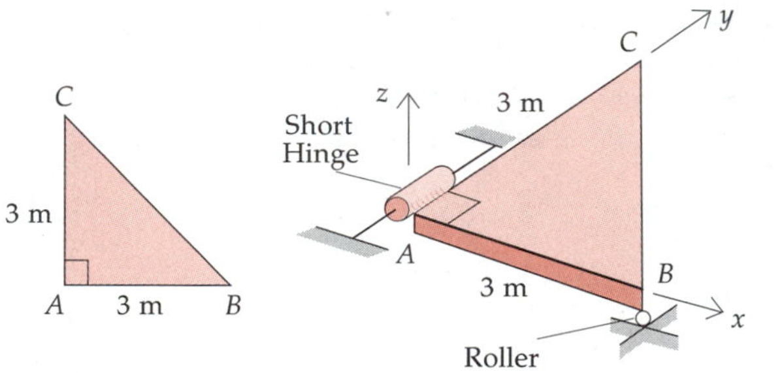

Figure P4.115

4.116 In Figure P4.116 the door weighs 120 lb. If there is no friction between the rope and the tree limb, what must the children collectively weigh to start the door swinging open about its hinges on the *x* axis?

4.117 The smooth collars (or sleeves) in Figure P4.117 are attached at C_1 and C_2 to the rod of mass *m* by ball and socket joints. In terms of mg, find the force **P** that, when applied parallel to the *x*-axis onto the lower collar, will result in equilibrium. Upon completing the solution, comment on why one of the six scalar component equations was redundant (i.e., yielded no new information).

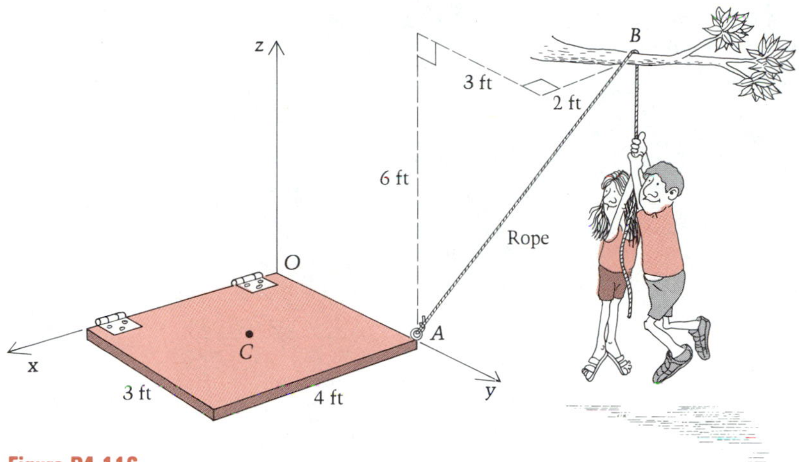

Figure P4.116

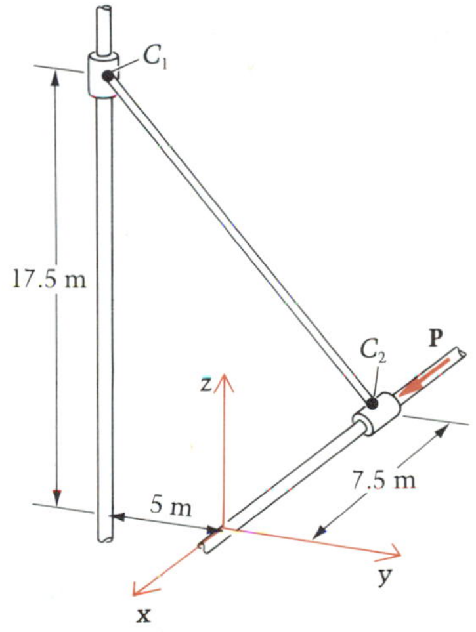

Figure P4.117

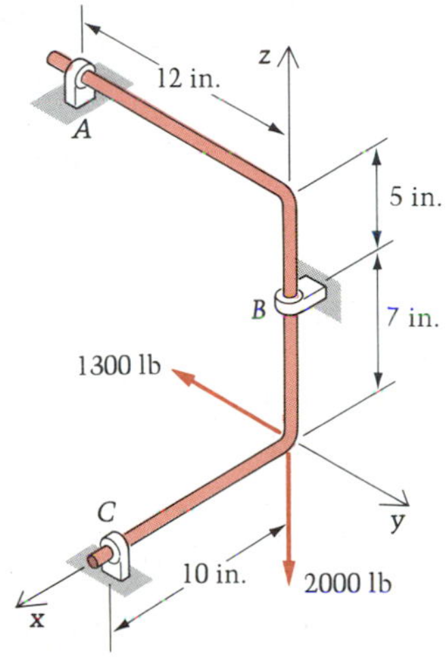

Figure P4.118

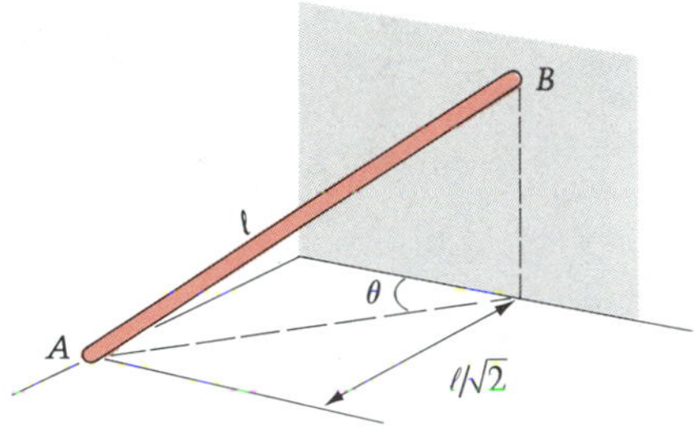

Figure P4.119

4.118 A shaft with two 90° bends is supported by three ball bearings as shown at A, B, and C in Figure P4.118. Find the reaction of bearing C on the shaft.

** **4.119** In Figure P4.119 a heavy uniform rod of length ℓ rests with one end, A, on the ground and the other, B, against a vertical wall. The vertical plane through the rod makes an angle θ with the wall. End A is $\ell/\sqrt{2}$ from the wall. Letting ρ and σ be the ratios of the tangential to normal reactions at the ground and wall, respectively, show that

$$\sqrt{\sigma^2 \sin^2\theta - \cos^2\theta} = \frac{1}{\rho} - 2\sqrt{2\sin^2\theta - 1}$$

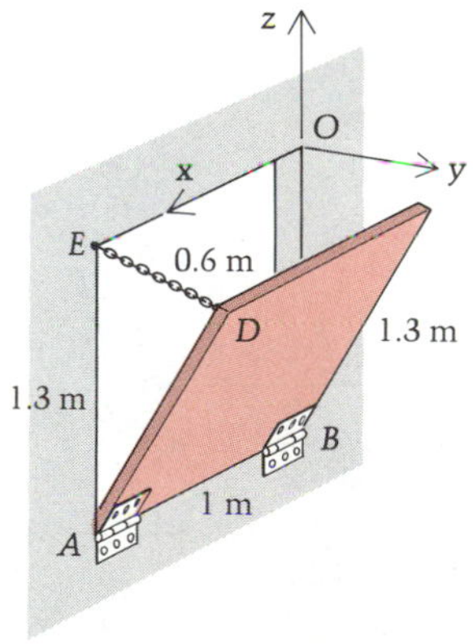

Figure P4.120

4.120 A 1 m × 1.3 m plate weighs 325 N and is supported by hinges at A and B. (See Figure P4.120.) It is held in the position shown by the 0.6-m chain ED. Find the tension in the chain.

4.121 Member AD in Figure P4.121 is supported by cables CE, BG, and BF, and by a ball-and-socket at A. Find the tensions in CE and BF if the tension in BG is 3000 N.

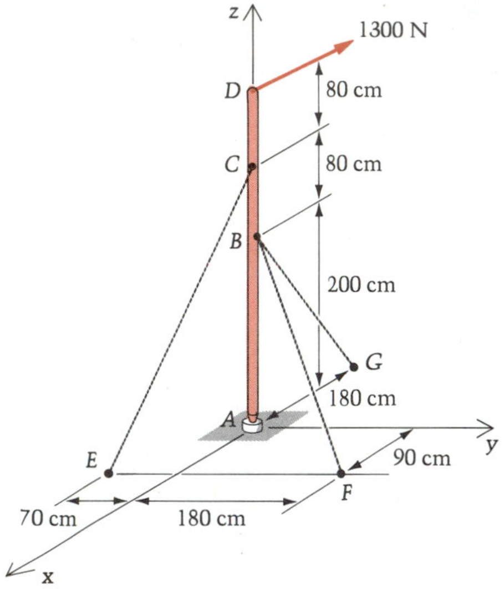

Figure P4.121

4.122 Find the tension T in the cable DE. This cable gives the same moment about line AC as does the weight of the uniform segment BD, which is 400 lb. (See Figure P4.122.)

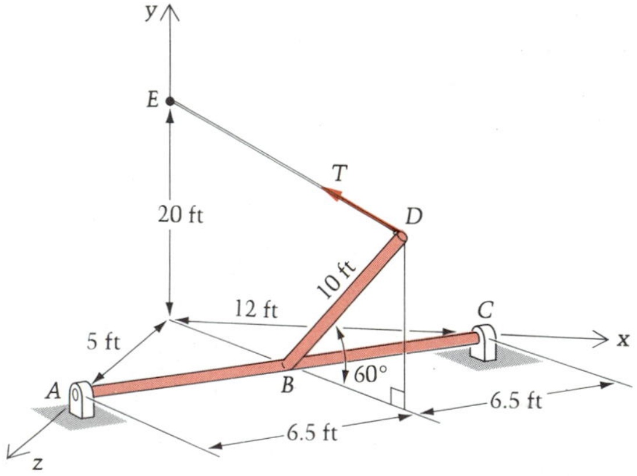

Figure P4.122

4.123 In Figure P4.123 find the force and couple exerted on the 200-lb plate by the hinge at O.

4.124 The bar in Figure P4.124 is supported by cables BD and CE, and by a ball-and-socket at A. Points D and E lie in the xz plane. Find the tension in each cable.

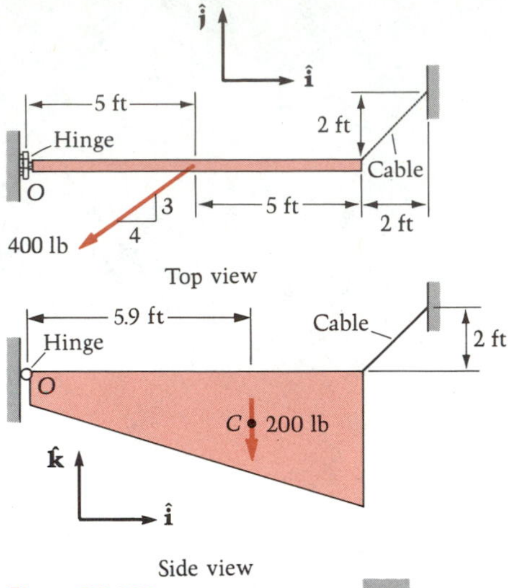

Figure P4.123

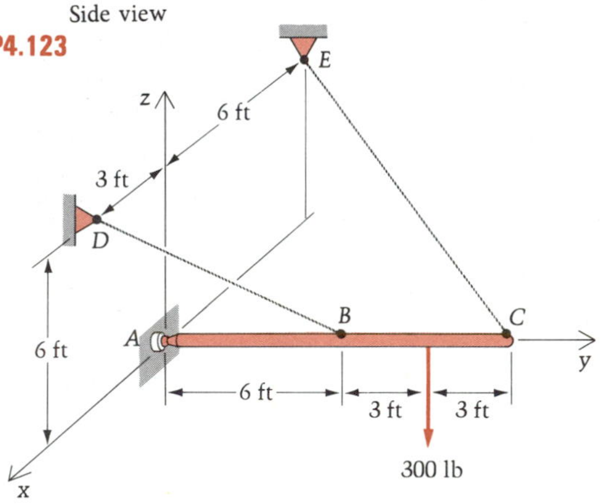

Figure P4.124

4.125 In Figure P4.125 the hinge at A has broken off. Find the tension in the cable. Then find the couples C_{Bz} and C_{By} that the remaining hinge must exert if the door remains in equilibrium.

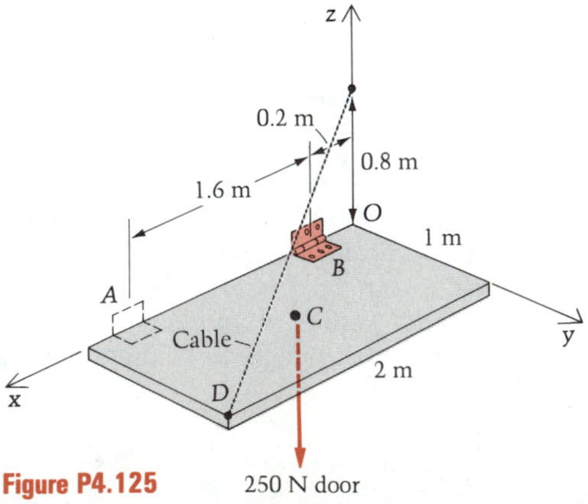

Figure P4.125

4.126 Find the force F and the bearing reactions. (See Figure P4.126.) Assume that one of the bearings cannot exert force in the z direction and that neither can exert a couple.

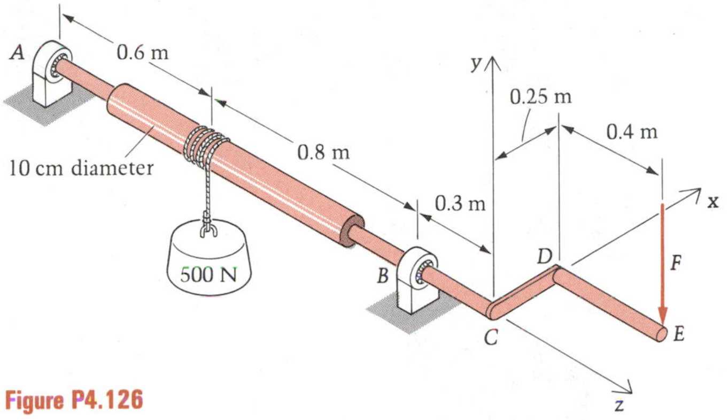

Figure P4.126

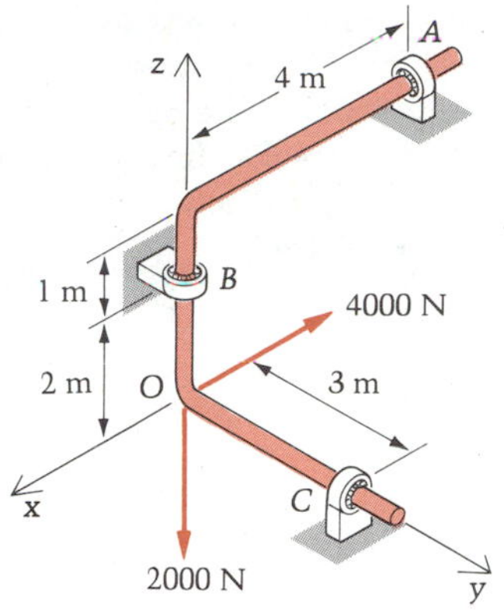

Figure P4.127

4.127 A shaft with two 90° bends is supported by three ball bearings at A, B, and C as shown in Figure P4.127. Find the reactions of the bearings on the shaft.

4.128 The semicircular bar of radius 30 cm is clamped at O. The 2000-N force acts downward through the bar's highest point, and the other force lies in the horizontal plane through its ends. Find the reactions $\mathbf{F}_O$ and $\mathbf{M}_O$ at the clamp. (See Figure P4.128.)

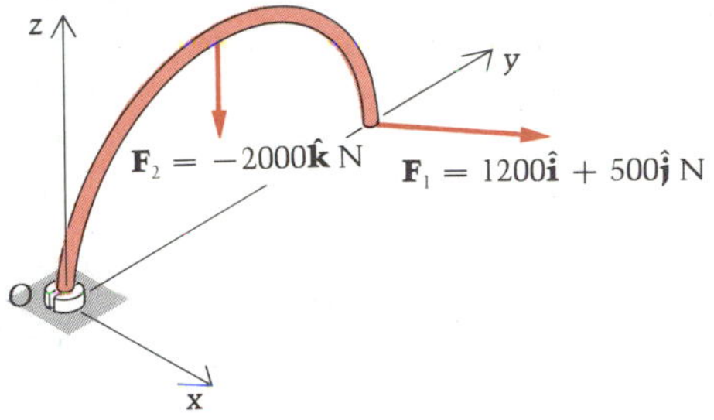

Figure P4.128

4.129 The rear door of the station wagon in Figure P4.129 is held up when open by the two gas-filled struts attached to the car by ball-joints. The door weighs 90 lb — 30 lb in part $\mathcal{A}$ and 60 lb in part $\mathcal{B}$. Find the forces in the two struts.

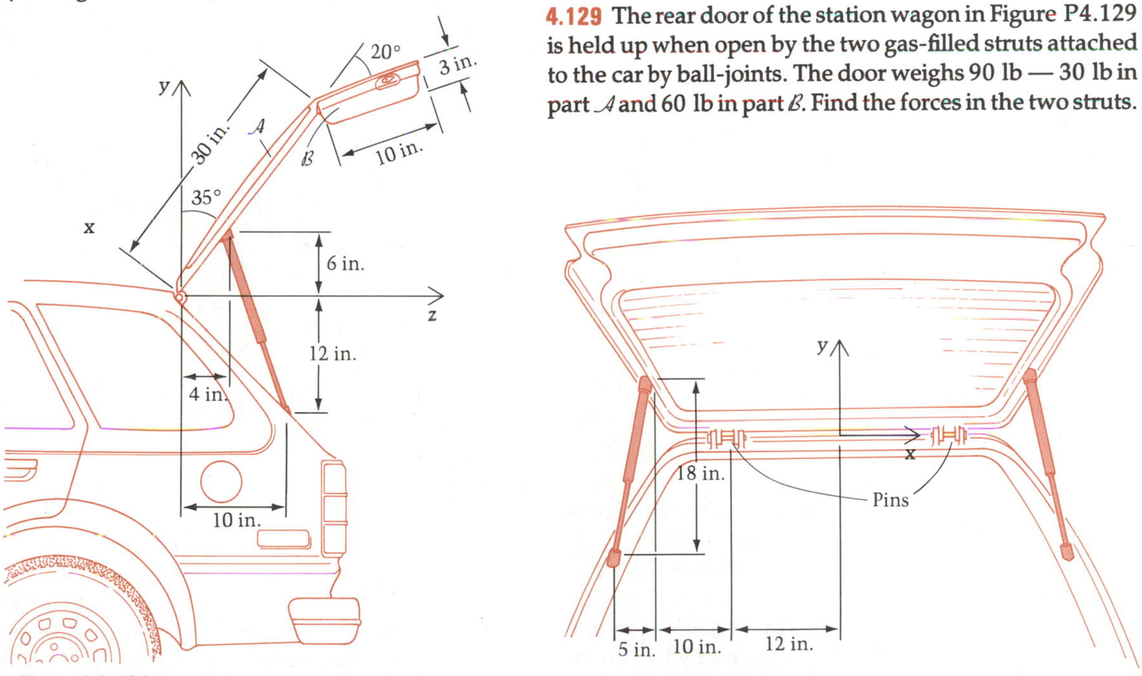

Figure P4.129

4.130 When the gas leaks out of one of the struts of Problem P4.130, it will no longer exert a force to hold up the door. If the right strut has become useless and the left pin breaks, find the resultant forces and couples exerted by the left strut and the right pin (hinge). (See Figure P4.130.)

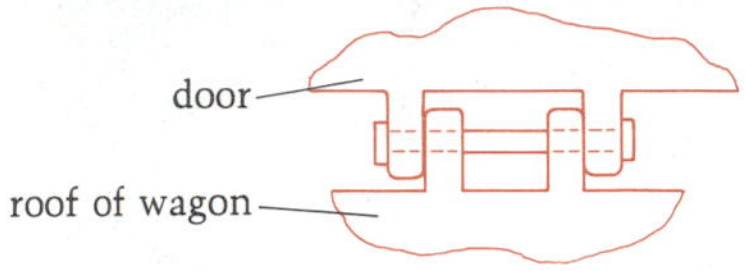

Figure P4.130

4.131 The horizontal homogeneous trap door in Figure P4.131 weighs 72 pounds. It is supported by the cable AB, a ball-and-socket at O, and a hinge at D that provides no support in the x direction. Find the force in the cable.

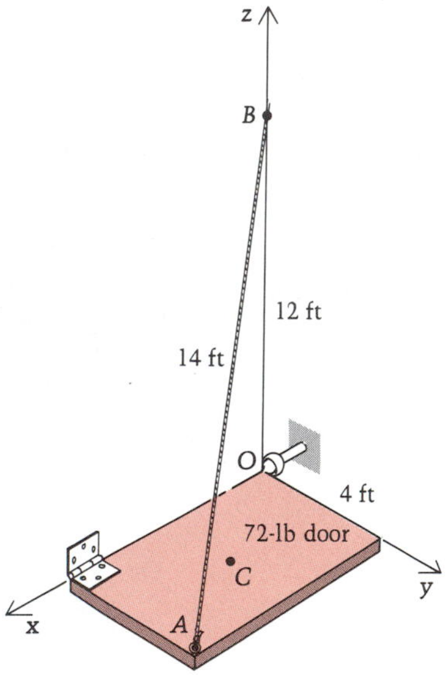

Figure P4.131

* **4.132** A ring is welded to a rod at a point A as shown in Figure P4.132. The cross-sectional area A and mass density of the rod are the same for the ring. The combined body is in equilibrium in the (precarious) position shown. Find the vertical reaction components exerted by the ball and socket onto the rod at B and by the smooth floor onto the ring at Q.

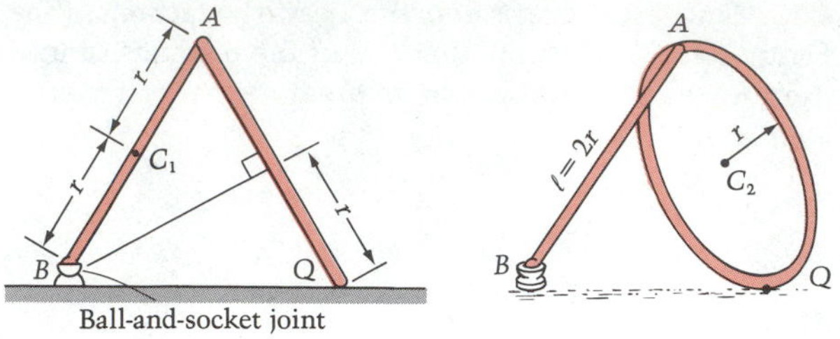

Figure P4.132

4.133 Find the supporting force system at A on the bent bar. (See Figures P4.133.)

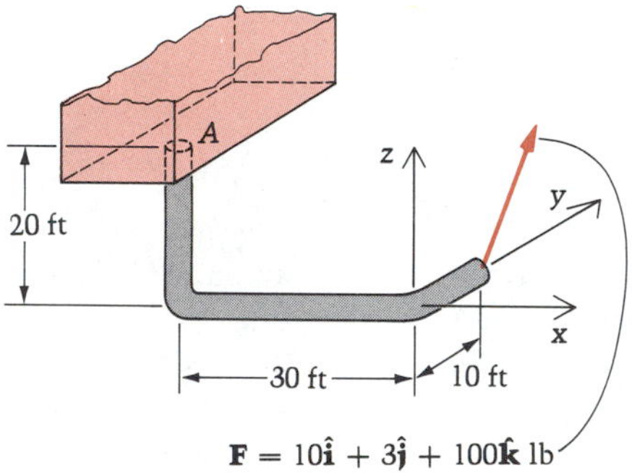

Figure P4.133

* **4.134** The plate weighs 100 lb and is supported as shown in Figure P4.134. Find the reactions at A, B, and D.

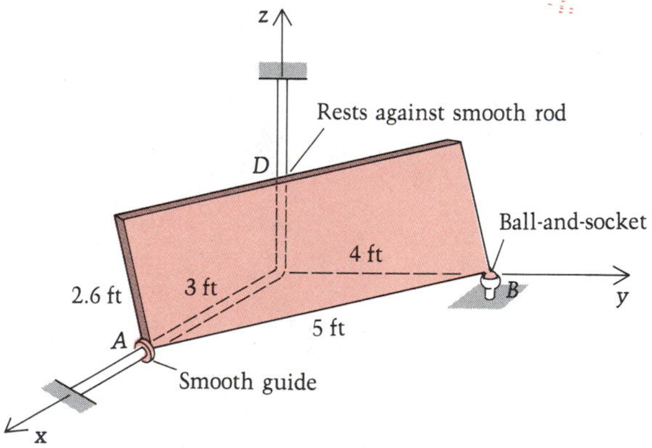

Figure P4.134

The spring is one of the most useful of all mechanical devices. It gives a resistive force proportional to its stretch if it is "linear," and the constant of proportionality is called the spring *modulus*. Problems 4.135–4.140 contain springs, and the last four are challenging.

4.135 The cart in Figure P4.135 weighs 500 N and is held in equilibrium on the inclined plane by a spring of modulus 5000 N/m. Find the force in the spring and its stretch.

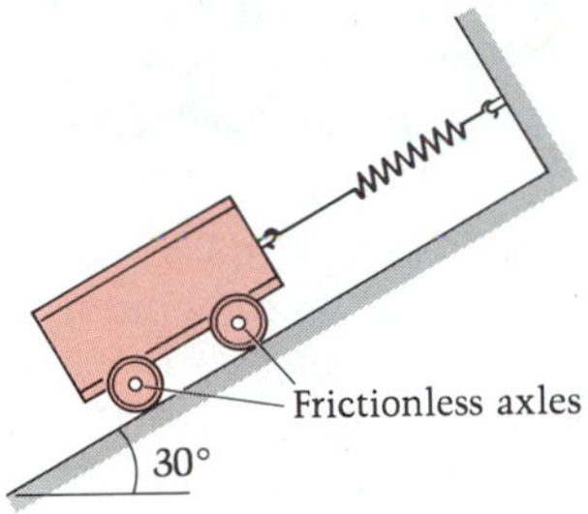

Figure P4.135

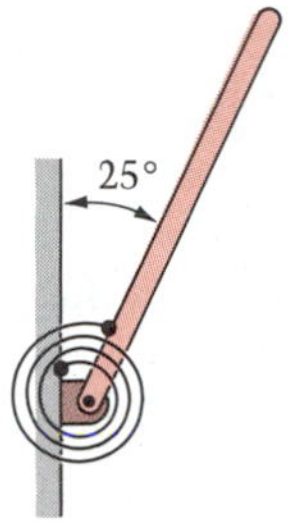

Figure P4.136

4.136 In Figure P4.136 the slender homogeneous rod 12-ft long and weighing 5 lb is connected by a pin and a moment spring to the vertical wall. The moment (or rotational or torsional) spring exerts a moment when the angle is changed between the two bodies to which it is fixed; an example is the springs in most flexible, metal-link watch bands. In a moment spring the modulus has units such as lb-ft per radian. If the rod is in equilibrium in the given position, and the natural (zero moment) position of the spring is when the bar is vertical, find the modulus of the moment spring.

* **4.137** The light bar in Figure P4.137 is fixed to the ground at O by means of a smooth pin, and is subjected to the vertical force P at its other end. The spring of modulus k is constrained to remain horizontal, and is unstretched when $\theta = 0$. If $P\ell < kh^2$, there is an equilibrium position for $0 < \theta < \pi/2$. Find this angle θ. Why must $P\ell < kh^2$ hold for this configuration to exist?

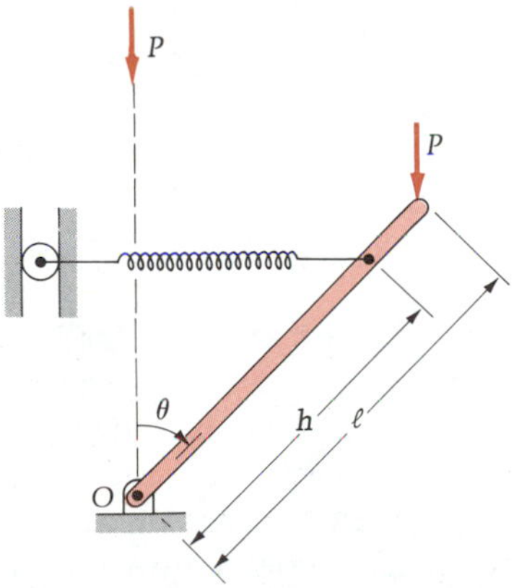

Figure P4.137

* **4.138** The spring in Figure P4.138 is unstretched with length ℓ when the uniform bar of mass m is vertical. (a) Write the moment equilibrium equation for the bar at an angle θ. (b) Show that $\theta = 0$ satisfies the equation for any k, ℓ, m, and g. (c) Let $k\ell = mg$ and show that there is another equilibrium angle θ between 0° and 180°. Find this angle numerically, using a calculator or computer.

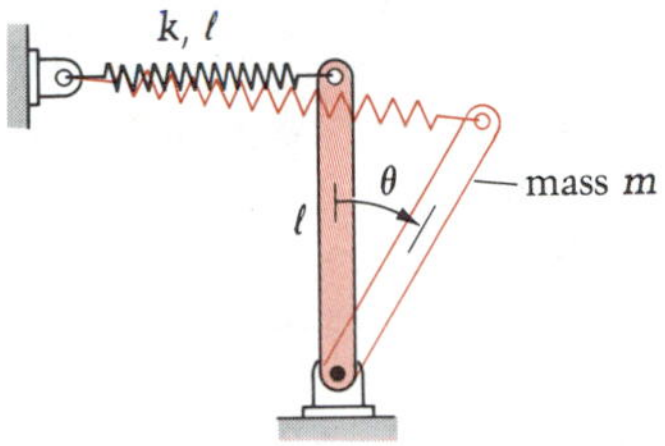

Figure P4.138

* **4.139** The bead of mass m can slide on the smooth semicircular hoop, and the spring connects the bead to the top A of the hoop (of radius R). (See Figure P4.139.) The spring has modulus $2mg/R$ and natural (unstretched) length $R/4$. Find the angle(s) θ at which the bead is in equilibrium. *Hint:* Prove and then use the fact that the three forces acting on the bead form a force triangle that is similar to ABC.

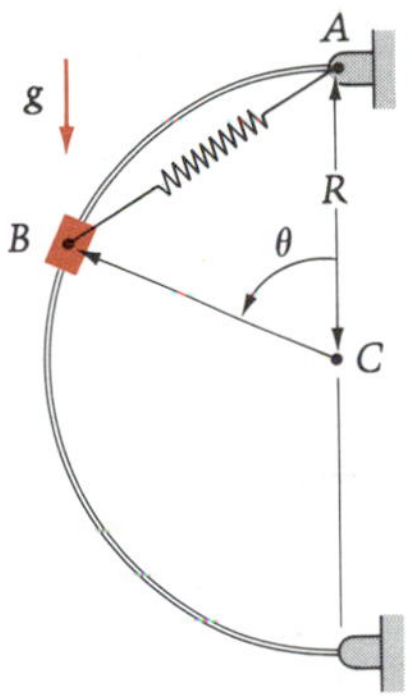

Figure P4.139

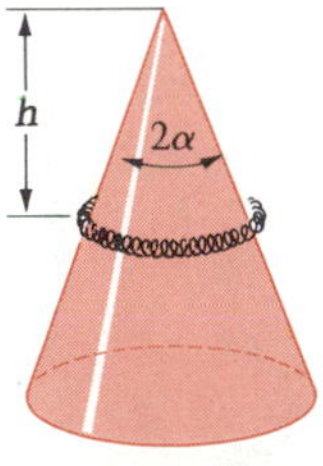

Figure P4.140

* **4.140** A linearly elastic endless spring with modulus k is placed around a smooth cone of vertex angle 2α. (See Figure P4.140.) The natural length of the spring is L, and it weighs W. Find the value of h for equilibrium.

* **4.141** Show that, if a body is held in equilibrium by three forces, the forces must have coplanar lines of action that are either parallel *or* concurrent. Note that we only need to consider the case in which none of the forces vanishes, for otherwise we have a two-force body that has already been discussed. *Hint:* We outline below a set of steps by which the result may be obtained; the student is encouraged also to think about alternative approaches.

1. Let $\mathbf{F}_1$, $\mathbf{F}_2$, and $\mathbf{F}_3$ be the three forces. Consider the two possibilities: (a) two lines of action are parallel and (b) no two lines of action are parallel.

2. For case (a), apply $\Sigma\mathbf{F} = \mathbf{0}$ to conclude that all three lines of action are parallel. Then consider the plane defined by the lines of action of $\mathbf{F}_1$ and $\mathbf{F}_2$ (if they are collinear then they, of course, must coincide with that of $\mathbf{F}_3$). Requiring that the sum of moments about any line in that plane vanish, conclude that all three of the (parallel) lines of action lie in the same plane.

3. For case (b), conclude from $\Sigma\mathbf{F} = \mathbf{0}$ that the three lines of action lie in parallel planes. Let a line, ℓ, be parallel to the line of action of $\mathbf{F}_2$ and intersect the line of action of $\mathbf{F}_1$. From $\Sigma M_\ell = 0$ conclude that the line of action of $\mathbf{F}_3$ intersects ℓ, and hence lies in the same plane as do ℓ and the line of action of $\mathbf{F}_1$. Now apply the moment equation of equilibrium for moments about the point of intersection of the lines of action of $\mathbf{F}_1$ and $\mathbf{F}_3$ to conclude that the three lines of action are concurrent and coplanar.

4.142 Illustrate the result of the preceding problem by showing that (a) the three forces at *A*, *B*, and *G* of Problem 4.16 have lines of action that all meet at one point, and (b) the three forces of Problem 4.23, for the correct angle ϕ, are parallel.

4.143 (a) If the light bar *ABCD* of Figure P4.143 is held in equilibrium by the two rollers at *B* and *C* and the smooth wall at *A*, show using the result of Problem 4.141 that none of the three support forces can vanish if $\theta = 30°$ as shown. (b) For what angle θ will the support force at *B* vanish?

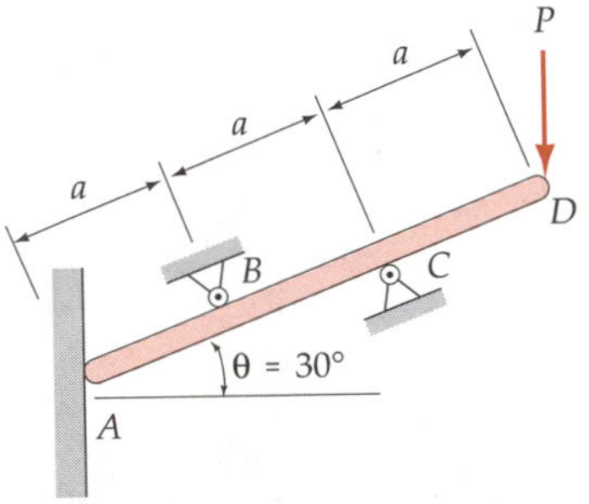

Figure P4.143

4.4 Applications of the Equations of Equilibrium to Interacting Bodies or Parts of a Structure

In Section 4.3, we studied the equilibrium of a body which was either a single physical object or else two or more objects which were left intact. These objects were connected in various ways, including pressing contact, pins, rollers, or cables. What was common about all those multi-object problems was that we could solve for the desired unknowns without separating the objects constituting the overall body. In this section, we shall see that sometimes we *must* separate the objects, or parts, of the body in order to obtain a solution.

As an example, suppose we wish to know the forces of interaction *between* the objects making up a body. These forces will never appear on free-body diagrams or in equations of the overall body because they are internal to it. But if we draw FBD's of the *separate* objects, treat each as a body in itself, and write equilibrium equations for each, these equations will now include the desired forces of interaction. This procedure makes use of the fact that *if a body is in equilibrium, each of its parts — being a body in its own right — is also in equilibrium.**

Let us re-examine Example 4.7 in the light of these ideas. Figure 4.17 (a) shows the problem, and (b) repeats the free-body diagram of the body consisting of the beam AB and the pulley supported on frictionless bearings at B:

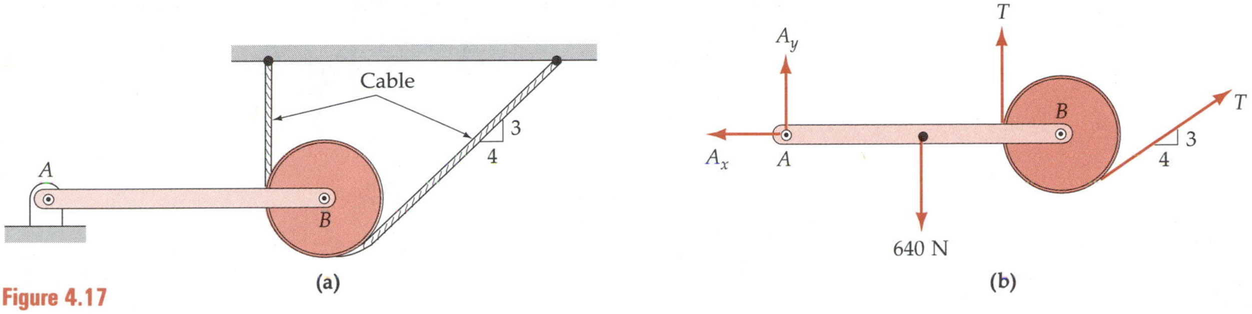

Figure 4.17

In the earlier example, we simply set the moments about A of the external forces acting on the intact body equal to zero, and thereby found the tension T in the cable from that single equation to be 200 N. We then used the force equations to find the components of the pin reaction at A: $A_x = 160$ N and $A_y = 320$ N.

But now suppose we desire to know the pin force exerted on the beam at B. That force can *only* be exposed, and subsequently solved for, by separating the beam and pulley, as shown below:

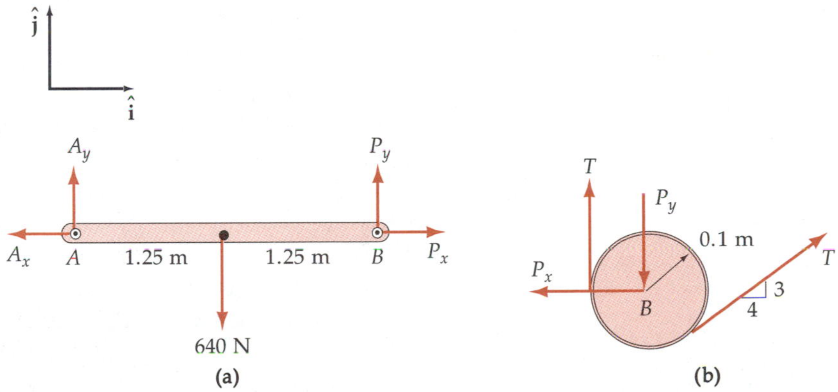

Figure 4.18

* More rigorously, when we say that a body is at rest (equilibrium) we mean that each of its material points (or particles) is at rest. Consequently, when a body is in equilibrium each of its parts, or subdivisions, is in equilibrium. Thus the equations of equilibrium apply to any part of a body as well as to the whole.

Now the desired pin force (expressed in terms of its x- and y-components P_x and P_y) appears as an *external force* on the separated bodies shown in Figure 4.18(a, b).

Note that the action-reaction principle (Newton's third law) has to be *rigorously enforced* when separating the objects or parts constituting a body. Thus P_x is shown acting to the left on the pulley but to the right on the bar, and P_y is depicted down on the pulley but up on the bar. Violation of the equal magnitude/opposite direction characteristic when dealing with interactive forces dooms a solution to failure.

On the FBD of the pulley, we have, using $T = 200$ N from Example 4.7:

$$\xrightarrow{+} \quad \Sigma F_x = 0 = \tfrac{4}{5}(200) - P_x$$

$$P_x = 160 \text{ N}$$

$$+\uparrow \quad \Sigma F_y = 0 = 200 + \tfrac{3}{5}(200) - P_y$$

$$P_y = 320 \text{ N}$$

Since the values of P_x and P_y came out positive, the directions of the pin forces onto the pulley are *as shown* in Figure 4.18(b), i.e., the force is $-160\hat{\mathbf{i}} - 320\hat{\mathbf{j}}$ N onto the pulley. Likewise, the equal magnitude/oppositely directed force exerted onto the bar is $160\,\hat{\mathbf{i}} + 320\hat{\mathbf{j}}$ N.

We have now reached a curious point: We still have an unused FBD [that of the beam, Figure 4.18(a)], but there are no more forces to compute! If we write the equilibrium equations from that free-body and substitute our previously obtained results, all we will find is a check on those answers:

$$\xrightarrow{+} \quad \Sigma F_x = -A_x + P_x = -160 + 160 = 0 \quad \checkmark$$

$$+\uparrow \quad \Sigma F_y = A_y + P_y - 640 = 320 + 320 - 640 = 0 \quad \checkmark$$

$$\circlearrowleft\!+ \quad \Sigma M_A = 2.5P_y - 640(1.25) = 800 - 800 = 0 \quad \checkmark$$

So these three equations are satisfied, but they give us no new information. This is because if we (a) divide a body into n parts; (b) consider separate FBD's of the n + 1 bodies consisting of the n parts plus the overall body; and (c) write a set of equilibrium equations for each of the n + 1 bodies, only n of the sets can be independent equations.

The proof of this statement is simple. If we picture all the parts and put them back together to form the overall original body, all of the interactions (like the two P_x's and the two P_y's in the foregoing example) between respective parts will disappear, leaving only the (external) loads and reactions on the overall body. In the same way, if we add all the separate bodies' $\Sigma F_x = 0$ equations,* all the x-forces of internal interaction will cancel, leaving only the x-components of loads and reactions on the overall body to add to zero. But that is precisely the $\Sigma F_x = 0$ equation for the overall body. Hence this equation is redundant. The same argu-

* Using the same sign convention for each!

ment is valid for the $\Sigma F_y = 0$ equation and for the moment equation (about a common point for all bodies). Therefore only n of the n + 1 possible sets of equations are independent.

Of course, instead of the equations of the n parts, we could use those of the overall body and n − 1 of the n parts. The point is that one set of equilibrium equations will always be redundant. In the foregoing example, that set was the equations of equilibrium for the bar, since we had already used the equations of the overall body and of the pulley.

Let us illustrate these general ideas for the problem of Figures 4.17,18 above. Note that if we start over and separately write the equations of equilibrium for the beam and the pulley using the two free-body diagrams above, we obtain:

Beam:

$$\Sigma F_x = 0 = P_x - A_x$$
$$\Sigma F_y = 0 = A_y + P_y - 640$$
$$\Sigma M_A = 0 = 2.5P_y - 1.25(640)$$

Pulley:

$$\Sigma F_x = 0 = -P_x + \tfrac{4}{5}T$$
$$\Sigma F_y = 0 = T + \tfrac{3}{5}T - P_y$$
$$\Sigma M_A = 0 = -2.5P_y + 2.4T + 2.5(\tfrac{3}{5})T + 0.1T$$

If we now add the respective pairs of equations, we have:

$$x\text{-equation:} \quad \Sigma F_x = 0 = (\tfrac{4}{5})T - A_x$$
$$y\text{-equation:} \quad \Sigma F_y = 0 = A_y + T + \tfrac{3}{5}T - 640$$
$$\text{moment-equation:} \quad \Sigma M_A = 0 = -640(1.25) + T(2.4) + \tfrac{3}{5}T(2.5) + 0.1T$$

These equations are identical to those that were written for the combined (intact) body in Example 4.7 consisting of beam plus pulley.

Thus when the separate FBDs are used to form the equilibrium equations, and those equations are then summed, the equations of equilibrium for the combined (or "overall") body will be reproduced. Therefore we can use the two parts independently, or the overall body and one of the parts, but not all three. The third set of equilibrium equations will be dependent on the other two. And if a body is separated into more than two parts, say seven parts, and we write equilibrium equations for the seven parts plus the overall (intact) body, only seven of the eight sets of equations will be independent.

The ideas expressed above are illustrated more rigorously for the general case of an arbitrary body $\mathcal{B}$ divided into parts $\mathcal{B}_1$ and $\mathcal{B}_2$ in Appendix E.

Sometimes, as previously mentioned, we find that the equilibrium equations for a combined body such as $\mathcal{B}$ contain more scalar unknowns than there are independent equations. This suggests that the problem might be statically indeterminate, but that is not always the case. The special nature of the connection between two parts ($\mathcal{B}_1$ and $\mathcal{B}_2$) of $\mathcal{B}$ may render one or more components of the interaction force (**R**) and couple (**C**) between the parts to vanish. If this happens, the equations of equilibrium for $\mathcal{B}$ together with those of $\mathcal{B}_1$ may yield an equal number of equations and unknowns, rendering the problem determinate. Some of the examples that follow illustrate such situations. We note, however, that when we know *nothing* in advance about **R** and **C** (for example, when $\mathcal{B}_1$ and $\mathcal{B}_2$ are "welded"), then the equilibrium equations for $\mathcal{B}_1$ will be of no assistance in solving for the external forces (reactions) on $\mathcal{B}$.

Question 4.8 Why won't they?

In most of the examples that follow, the body or device under consideration is an assembly of rigid (or near-rigid) parts.

When the assembly is intended to be a stationary structure for supporting loads, it is often called a **frame,** unless it is composed exclusively of straight two-force bodies or members. In that case it is called a **truss** (about which more will be said in Chapter 5). When the function of the device depends upon the freedom of the parts to move relative to one another, particularly for the purpose of doing mechanical work, it is often called a **machine.** The pliers of Example 4.15 will be seen to constitute such a device.

Answer 4.8 In this case, separating the body into two bodies will expose six new unknowns — three new force components and three new couple components. Thus the creation of six new equilibrium equations will gain us nothing in this case.

EXAMPLE 4.15

Find the forces transmitted through the pin in the pliers of Example 4.3 (see Figure E4.15a).

Solution

We first shall consider a free-body diagram of one of the parts of the pliers (Figure E4.15b) because the force we seek will be *external* to that body.

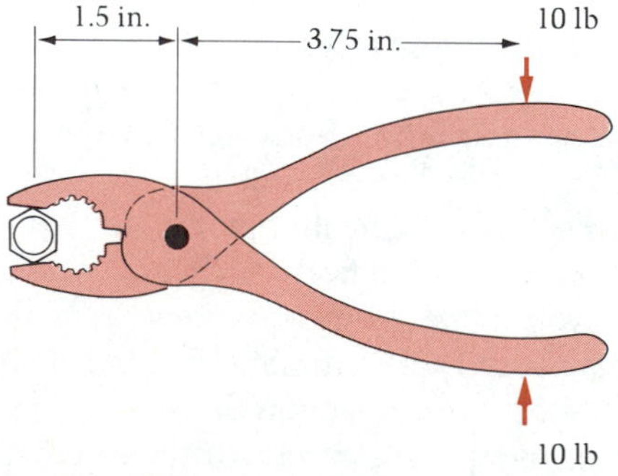

Figure E4.15a

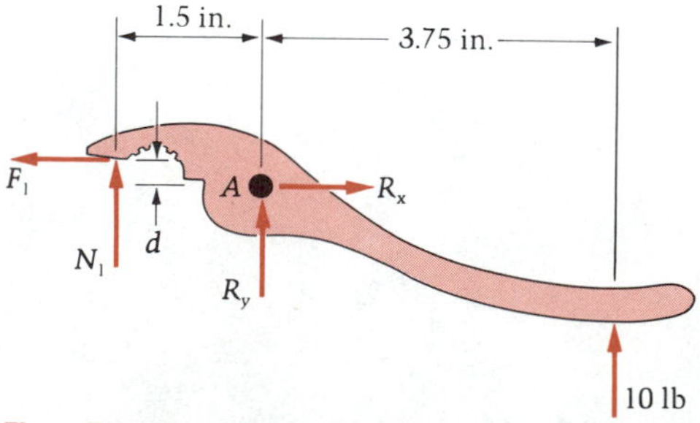

Figure E4.15b

We can see immediately that there are more (four) unknown forces appearing than independent component-equations (three) of equilibrium for this two-dimensional problem. Writing these equations, we have

$$\xrightarrow{+} \qquad \Sigma F_x = 0$$

$$R_x - F_1 = 0 \Rightarrow R_x = F_1 \tag{1}$$

and

$$+\uparrow \qquad \Sigma F_y = 0$$

$$N_1 + R_y + 10 = 0 \tag{2}$$

and

$$\circlearrowleft + \qquad \Sigma M_A = 0$$

$$-1.5N_1 + dF_1 + 3.75(10) = 0 \tag{3}$$

Turning to the free-body diagram (Figure E4.15c) of the whole pair of pliers, and summing moments at B, we obtain

$$\circlearrowleft + \qquad \Sigma M_B = 0$$

$$F_1 b = 0 \Rightarrow F_1 = 0$$

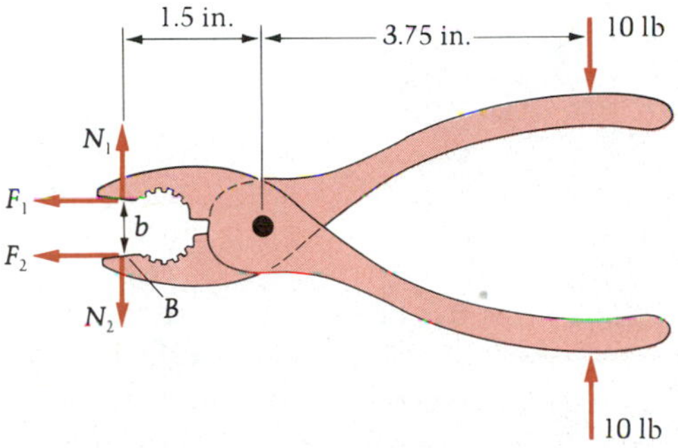

Figure E4.15c

The balance of forces yields

$$\xleftarrow{+} \qquad \Sigma F_x = F_1 + F_2 = 0 \Rightarrow F_2 = -F_1 = 0$$

and

$$+\uparrow \qquad \Sigma F_y = N_1 - N_2 + 10 - 10 = 0 \Rightarrow N_1 = N_2$$

Note that $F_1 = 0$ means from Equation (1) that R_x is also zero. Substituting $F_1 = 0$ into Equation (3),

$$1.5N_1 - 0 - 37.5 = 0$$

$$N_1 = 25 \text{ lb}$$

Then from Equation (2),

$$25 + R_y + 10 = 0$$

$$R_y = -35 \text{ lb}$$

The results displayed as free-body diagrams of the two parts are shown in Figure E4.15d.

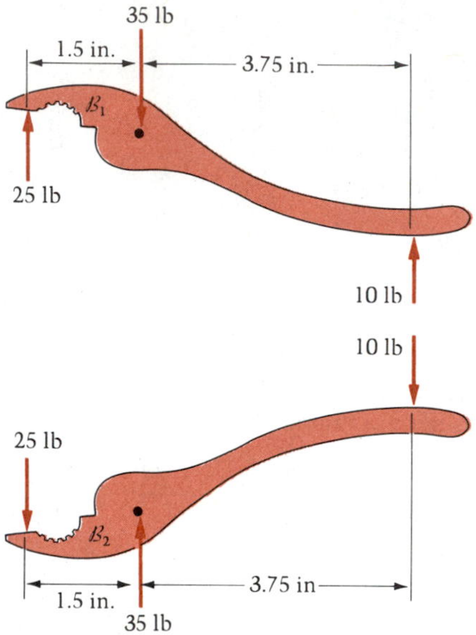

Figure E4.15d

In the preceding example, note that we have not actually specified which of the parts of the pliers (if either) would have the pin at A included in the free-body diagram. Which choice is made doesn't affect the analysis because the free-body diagram of the pin is as shown in Figure 4.19.

Figure 4.19

Note in the three figures that follow in Figure 4.20, the net result in each case is the pair of free-body diagrams of $\mathcal{B}_1$ and $\mathcal{B}_2$ just shown.

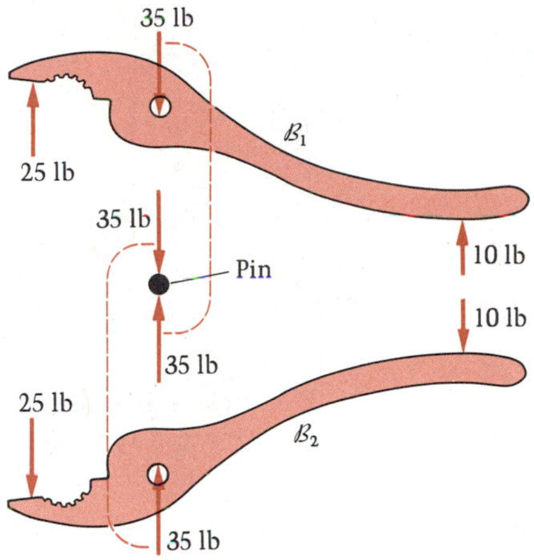

(a) The pin is a part of neither $\mathscr{B}_1$ nor $\mathscr{B}_2$.

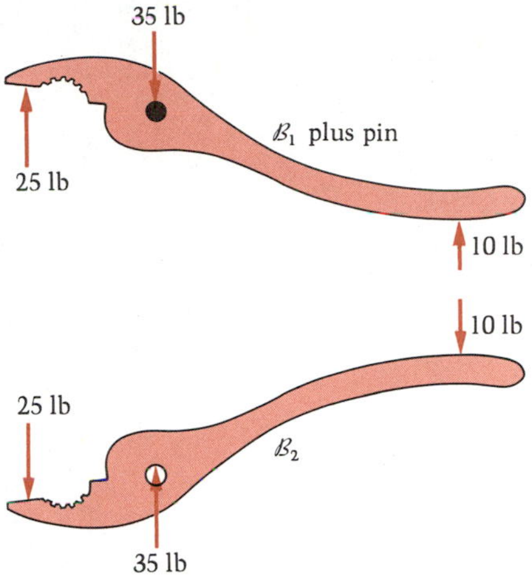

(b) The pin is included with $\mathscr{B}_1$.

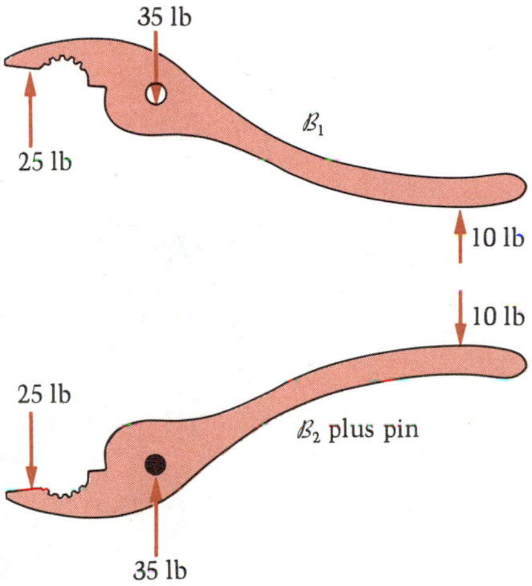

(c) The pin is included with $\mathscr{B}_2$.

Figure 4.20

We note also for completeness that the free-body diagram of the nut being gripped is Figure 4.21:

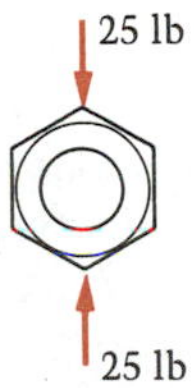

Figure 4.21

Question 4.9 If the hand force were different, would the gripping force still be 2.5 times it?

Answer 4.9 Yes.

EXAMPLE 4.16

For the frame of Example 4.6, shown in Figure E4.16a, find the force exerted by the pin at B onto member ADB. The 150 N force is applied to the pin at B.

Solution

In the earlier example, we found the reactions onto the frame at A to be $196\hat{\mathbf{i}} - 200\hat{\mathbf{j}}$ N and onto the frame at C to be $-346\hat{\mathbf{i}} + 200\hat{\mathbf{j}}$ N. Isolating member ADB and drawing its free-body diagram (see Figure E4.16b), we expose the components B_x and B_y of the force we seek. Note also that DE is a two-force member, so the force Q that it exerts at D lies along DE. Also, unlike Example 4.6, we now need to know the distance of DE above the baseline AC.

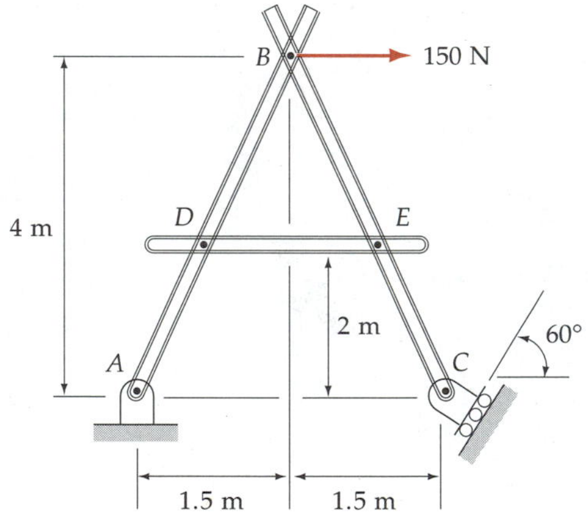

Figure E4.16a

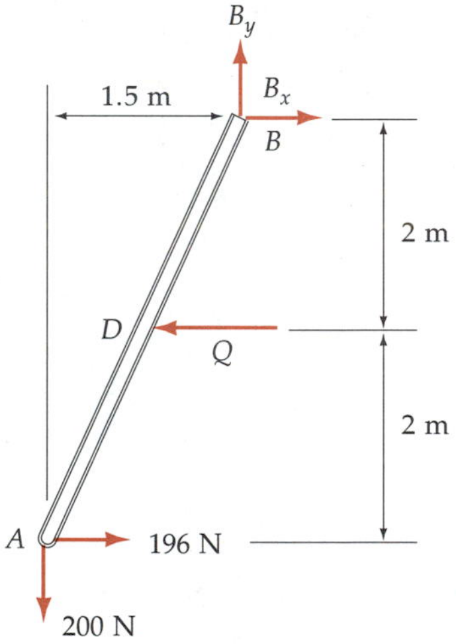

Figure E4.16b

Question 4.10 Why wasn't the position of DE needed in the earlier example (4.6)?

We now write the equilibrium equations of the separated member ADB. Note that taking moments about B to first determine the force Q in the two-force member

Answer 4.10 In Example 4.6, force Q was internal, thus not exposed; hence its distance from A was not needed.

allows us to determine B_x and B_y in a series of "one equation in a single unknown" steps:

$$\circlearrowleft + \quad \Sigma M_B = 0$$

$$(200\text{ N})(1.5\text{ m}) + (196\text{ N})(4\text{ m}) - Q(2\text{ m}) = 0$$

$$Q = 542\text{ N}$$

$$\xrightarrow{+} \quad \Sigma F_x = 0$$

$$B_x - Q + 196 = 0$$

$$B_x = 542 - 196 = 346\text{ N}$$

$$+\uparrow \quad \Sigma F_y = 0$$

$$B_y - 200 = 0$$

$$B_y = 200\text{ N}$$

Thus the force exerted by the pin at B onto ADB is

$$B_x\hat{\mathbf{i}} + B_y\hat{\mathbf{j}} = 346\hat{\mathbf{i}} + 200\hat{\mathbf{j}}\text{ N}.$$

It is interesting to examine a FBD of the pin at B from the preceding example (see Figure 4.22):

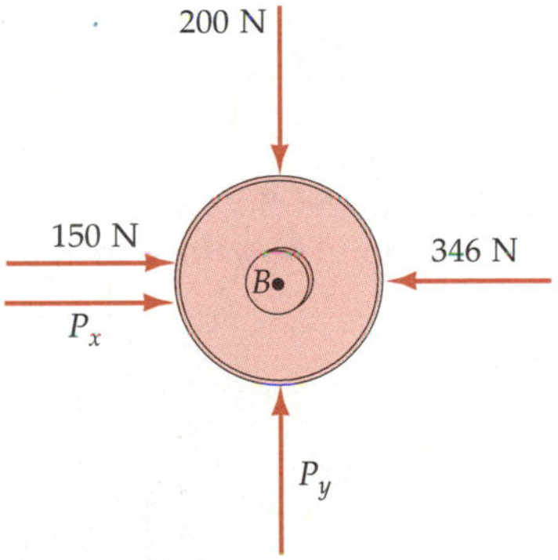

Figure 4.22

Acting on this pin are the external force (150 → N), the (equal magnitude, opposite direction) reaction from ADB, and the reaction from bar CEB, expressed on the FBD as $P_x\hat{\mathbf{i}} + P_y\hat{\mathbf{j}}$. Summing forces gives:

$$\xrightarrow{+} \quad \Sigma F_x = 0 \qquad\qquad +\uparrow \quad \Sigma F_y = 0$$

$$150 + P_x - 346 = 0 \qquad\qquad P_y - 200 = 0$$

$$P_x = 196\text{ N} \qquad\qquad P_y = 200\text{ N}$$

Let us check these results by seeing if the (now completely known) force system on CEB places it in equilibrium. Note in the FBD below (Figure 4.23) how we must reverse the forces P_x, P_y, and Q so as not to violate the principle of action and reaction (Newton's Third Law):

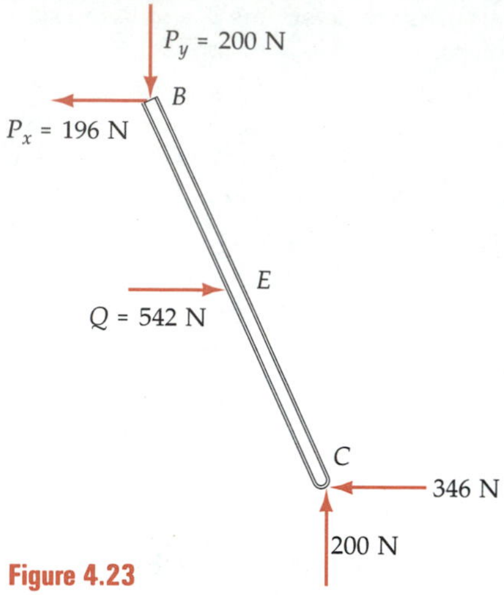

Figure 4.23

Checking,

$$\xrightarrow{+} \quad \Sigma F_x = 542 - 196 - 346 = 0 \quad \checkmark$$

$$+\uparrow \quad \Sigma F_y = 200 - 200 = 0 \quad \checkmark$$

$$\circlearrowleft + \quad \Sigma M_C = 196(4) + 200(1.5) - 542(2) = 0 \quad \checkmark$$

EXAMPLE 4.17

In Figure E4.17a, find the reactions at the supports *A*, *B*, and *C*, and the force between the beams transmitted by roller *D*.

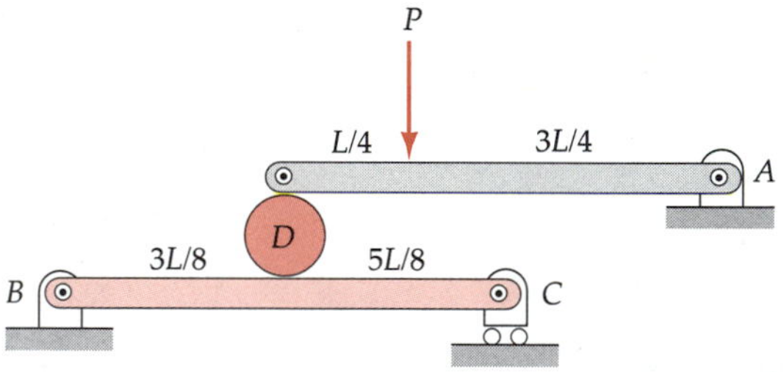

Figure E4.17a

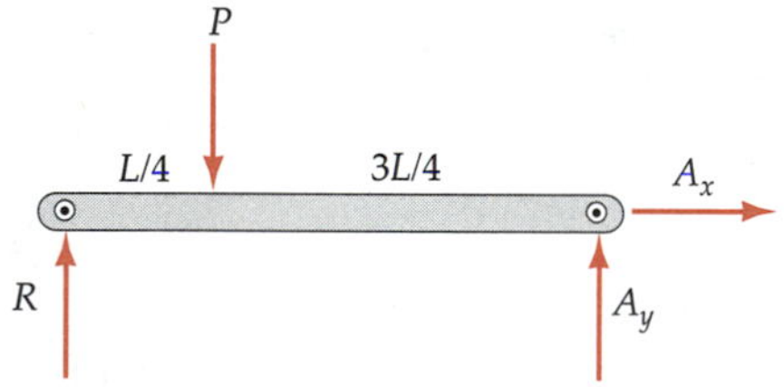

Figure E4.17b

Solution

The free-body diagrams of the separate beams are shown in Figures E4.17b and E4.17c. We see from these FBD's that A_x and B_x must respectively vanish in order that $\Sigma F_x = 0$ for each beam. Using Figure E4.17b,

$$\circlearrowleft + \quad \Sigma M_A = 0 = (3L/4)P - LR$$

$$R = (3/4)P$$

$$+\uparrow \quad \Sigma F_y = 0 = R + A_y - P$$

$$A_y = P - (3/4)P = P/4$$

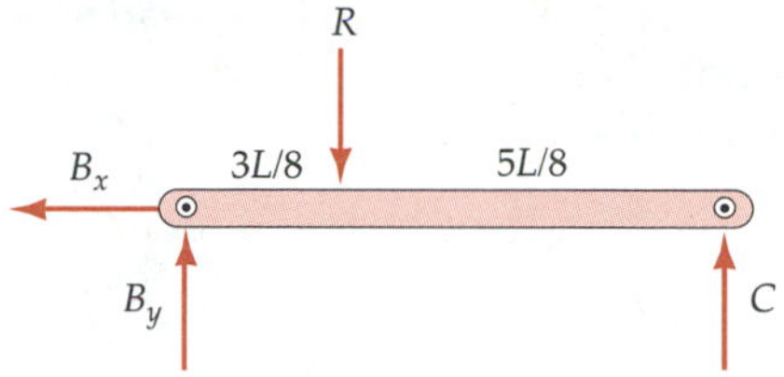

Figure E4.17c

And using Figure E4.17c,

$$\circlearrowleft\!+ \quad \Sigma M_B = \ 0 = LC - (3/8)LR$$

$$C = (3/8)(3/4)P = (9/32)P$$

$$+\uparrow \quad \Sigma F_y = \ 0 = B_y + C - R$$

$$B_y = (3/4)P - (9/32)P = (15/32)P$$

Putting the beams and roller back together in Figure E4.17e after displaying our results in Figure E4.17d, we see clearly how the roller forces, being internal, now cancel:

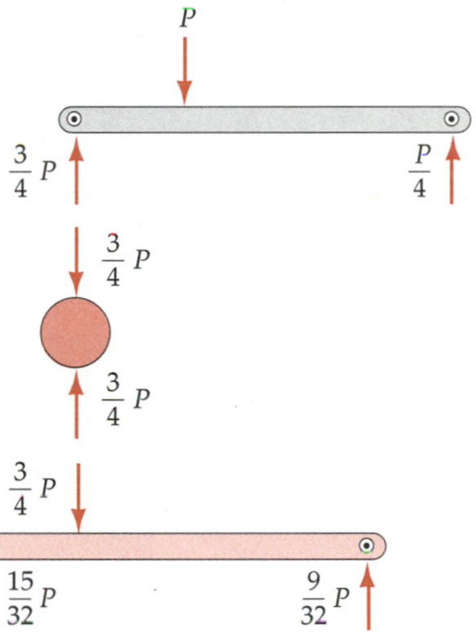

Figure E4.17d

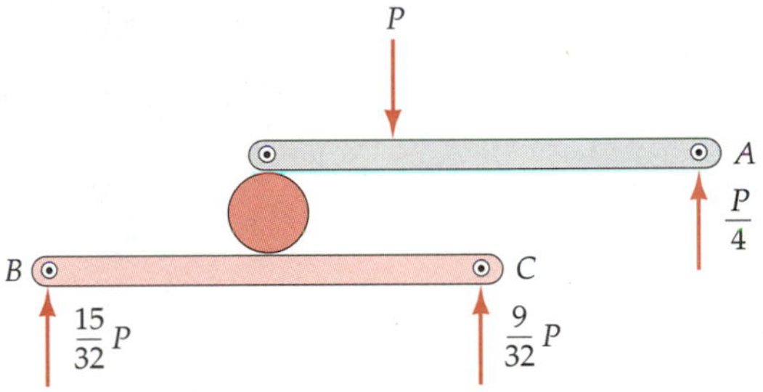

Figure E4.17e

Checking the equilibrium of the overall body with the help of Figure E4.17e, we find three verifications of our solution:

$$\overset{+}{\rightarrow} \quad \Sigma F_x = 0 \ \checkmark$$

$$+\uparrow \quad \Sigma F_y = (15/32 + 9/32 - 32/32 + 8/32)P = 0 \ \checkmark$$

$$\circlearrowleft\!+ \quad \Sigma M_B = -P(3/8 + 1/4)L + (9/32)PL + (P/4)(3/8 + 1)L$$

$$= PL[-20/32 + 9/32 + 11/32] = 0 \ \checkmark$$

It is important that the reader note that we cannot determine all the desired forces in the preceding example without disassembling the structure. First of all, the overall FBD, shown below in Figure 4.24, does not

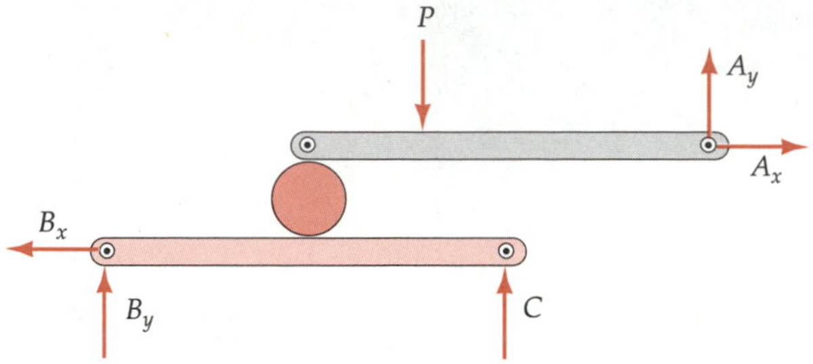

Figure 4.24

expose the roller force R which we seek. Secondly, all we know about the horizontal forces from this FBD is that $A_x = B_x$. Finally, even if we knew (from, say, mental FBD's of the separated beams) that $A_x = B_x = 0$, we would still be unable to find the three reactions A_y, B_y, and C from the two remaining equations $\Sigma F_y = 0$ and $\Sigma M_{\substack{any\\point}} = 0$.

EXAMPLE 4.18

In Figure E4.18a, the pin P, which bears against the slot in bar BA, is fixed to rod RT. The weight of W is 125 lb. Find the force the pin exerts onto BA, and the reaction onto BA at B exerted by the ground.

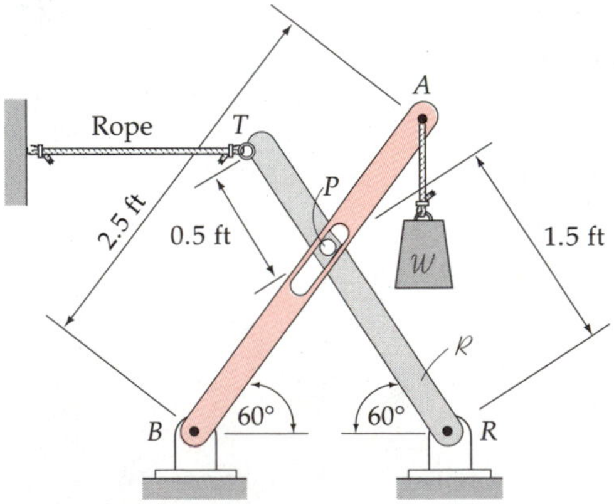

Figure E4.18a

Solution

The free-body diagrams of the structure and its two separate parts are shown in Figure E4.18b,c,d, although the only one we'll need in this problem is Figure E4.18c. Note that the line of action of the pin force P is known:

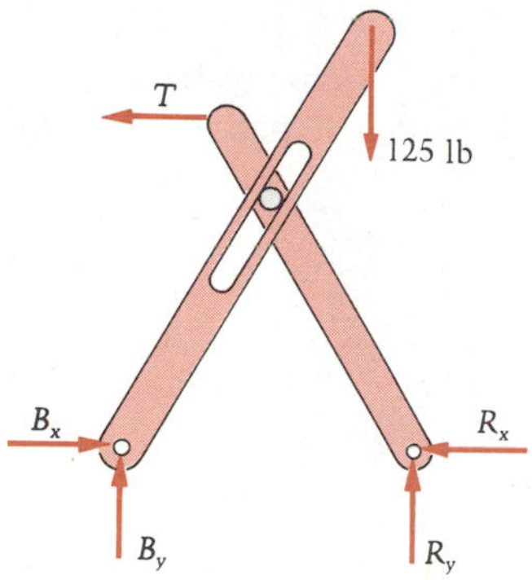

Figure E4.18b

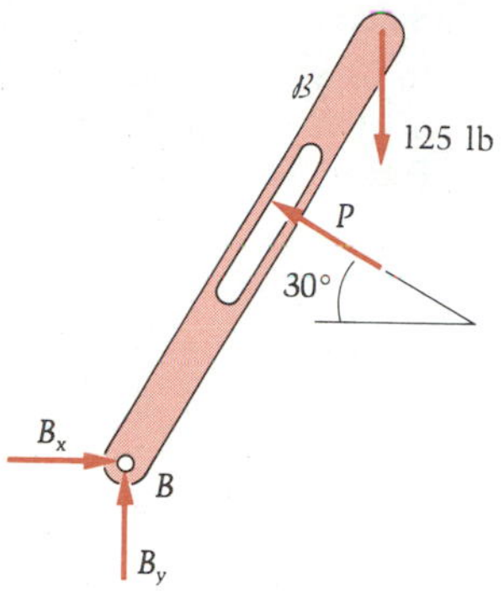

Figure E4.18c

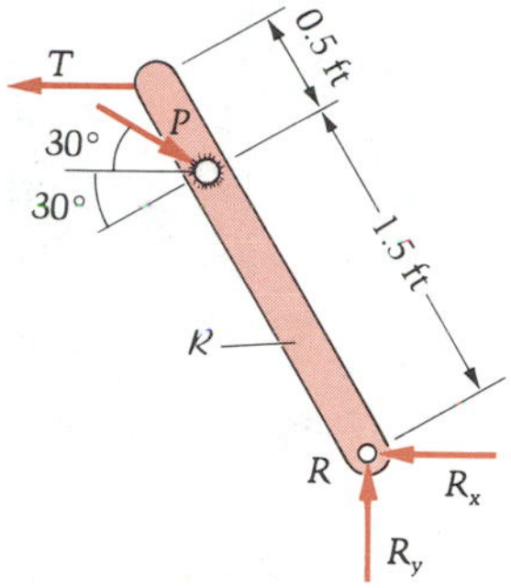

Figure E4.18d

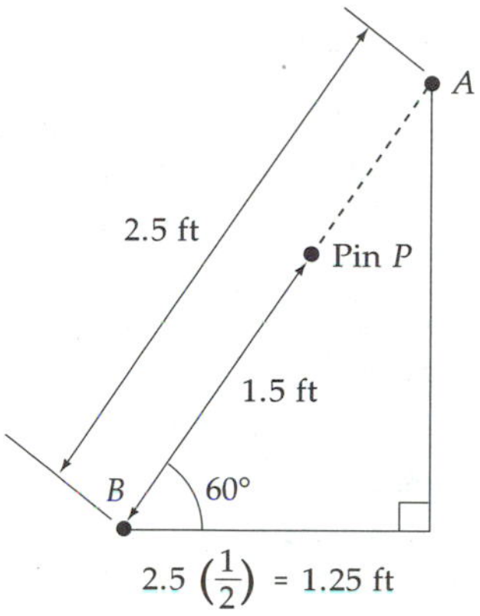

Figure E4.18e

Using Figures E4.18c and E4.18e, we may find the pin force immediately by summing moments about B:

$$\circlearrowleft + \qquad \Sigma M_B = 0 = (1.5 \text{ ft})(P) - (125 \text{ lb})(1.25 \text{ ft})$$

$$P = 104 \text{ lb}$$

Summing forces in the x direction on body BA then yields

$$\xrightarrow{+} \qquad \Sigma F_x = 0 = B_x - P \cos 30°$$

from which

$$B_x = (104)\sqrt{3}/2) = 90.1 \text{ lb}$$

Also, in the y direction,

$$+\uparrow \qquad \Sigma F_y = 0 = B_y - 125 + P \sin 30°$$

so that

$$B_y = 125 - 104(0.5) = 73 \text{ lb}$$

In vector form, with the convention $\hat{\mathbf{j}}\uparrow\!\!\rightarrow\hat{\mathbf{i}}$, the answers are: $\mathbf{P} = 104(-0.866\hat{\mathbf{i}} + 0.500\hat{\mathbf{j}})$ lb and $\mathbf{B} = 90.1\hat{\mathbf{i}} + 73.0\hat{\mathbf{j}}$ lb.

In contrast to Example 4.6, the preceding problem is an example of a **nonrigid frame.** Separation of the frame from its external supports by removal of the rope and removal of the pins at B and R leaves the two bars free to move relative to one another without deformation of the individual elements. Since T, R_x and R_y may be found with the help of the FBD of Figure E4.18d, the problem is statically determinate *in spite of the fact that the external reactions cannot be determined solely from equilibrium equations written for the assembled structure.*

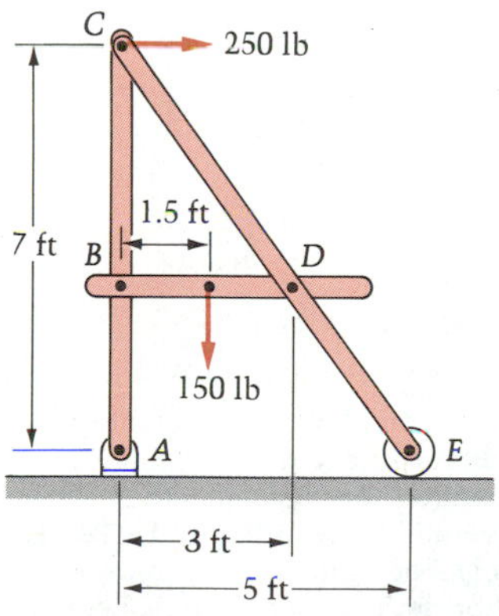

Figure E4.19a

Example 4.19

The 250 lb force in Figure E4.19a is applied to the pin at C. Find the x and y components of the force exerted on frame member CDE at D by the member BD.

Solution

In frame problems, it is wise to develop the habit of drawing free-body diagrams of the overall frame and its constituent parts.* See the five FBD's of Figure E4.19b.

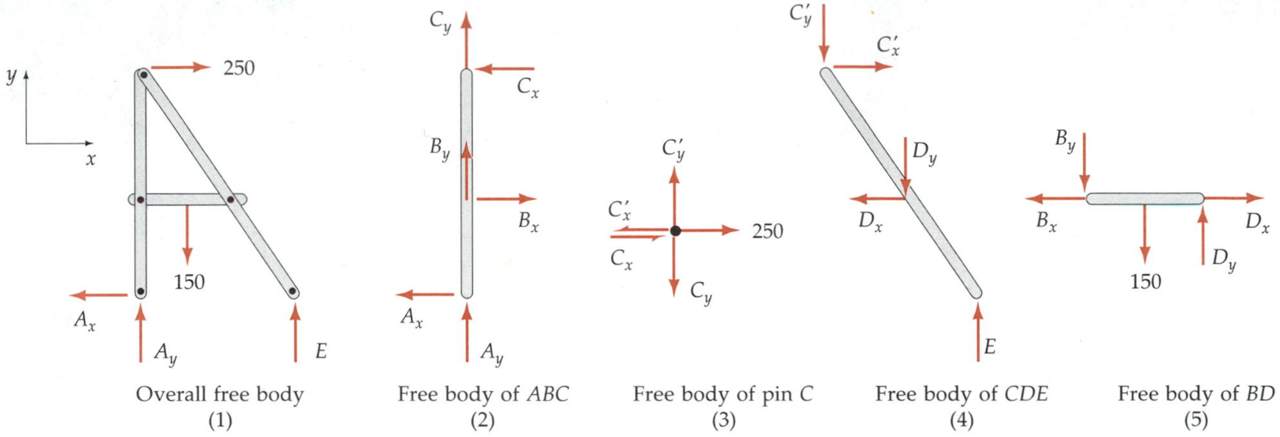

Figure E4.19b

What we are seeking in this problem are the forces D_x and D_y (see Figure E4.19b(4). Here are the steps we will employ to obtain them:

1. Sum the moments about A in FBD (1) to obtain the roller force E;
2. Sum the moments about point B in FBD (5) to obtain D_y;
3. Sum the moments about point C in FBD (4) to get an equation in E, D_x, and D_y. Solve for D_x, using D_y from Step 2 and E from Step 1.

Step (1): On FBD (1),

$$\circlearrowleft\!+ \qquad \Sigma M_A = 0 = (5\text{ ft})E - (150\text{ lb})(1.5\text{ ft}) - (250\text{ lb})(7\text{ ft})$$

$$E = \frac{1975}{5} = 395\text{ lb}$$

Step (2): On FBD (5),

$$\circlearrowleft\!+ \qquad \Sigma M_B = 0 = (3\text{ ft})D_y - (1.5\text{ ft})(150\text{ lb})$$

$$D_y = 75\text{ lb}$$

Step (3): On FBD (4),

$$\circlearrowleft\!+ \qquad \Sigma M_C = 0 = (5\text{ ft})\overset{395\text{ lb}}{\cancel{E}} - (3\text{ ft})\overset{75\text{ lb}}{\cancel{D_y}} - (4.2\text{ ft})D_x$$

$$D_x = 417\text{ lb}$$

* As we have seen in general at the start of this section, the complete set of equilibrium equations that we could write using each of the free-body diagrams in Figure E4.19b won't all be independent. This is because, when combined (or "put back together"), the bodies of Figures E4.19b(2–5) form or constitute the "overall" body of Figure E4.19b(1). Thus if we draw a "redundant" free-body diagram [any *one* of Figures 1–5 is redundant], we must be aware that some of the possible equilibrium equations (three, here) will be redundant also, and will serve us only as checks on our solutions.

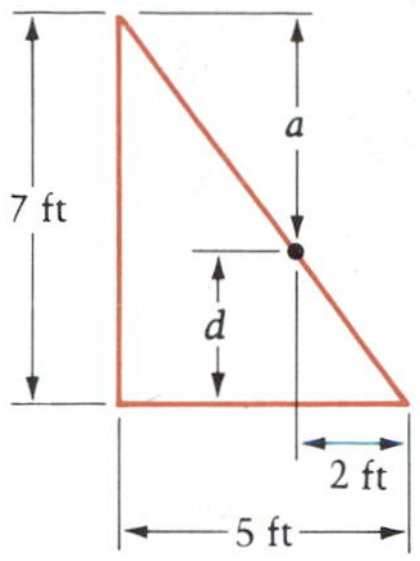

Figure E4.19c

where the dimension 4.2 ft for the moment arm of D_x was found as follows, using the similar triangles in Figure E4.19c:

$$\frac{2}{d} = \frac{5}{7} \Rightarrow d = \frac{14}{5} = 2.8 \text{ ft}$$

Thus

$$a = 7 - d = 4.2 \text{ ft}$$

Question 4.11 Draw a free-body diagram of the roller at E in the preceding example, and from it argue that the force exerted by the pin on CDE is the same as the normal force exerted by the ground onto the roller.

Answer 4.11

$$\left.\begin{aligned} \Sigma M_E &= 0 \Rightarrow F = 0 \\ \Sigma F_x &= 0 \Rightarrow E_x = 0 \\ \Sigma F_y &= 0 \Rightarrow N = E_y \end{aligned}\right\} \text{From the FBD:}$$

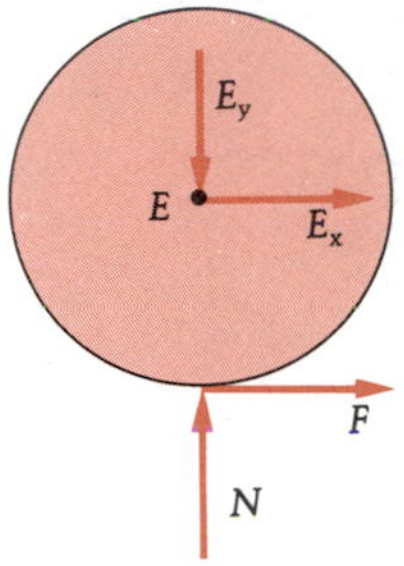

Example 4.20

In Figure E4.20a find the forces exerted by the pin at B on (a) the pulley, (b) the bar ABE, and (c) the bar BC.

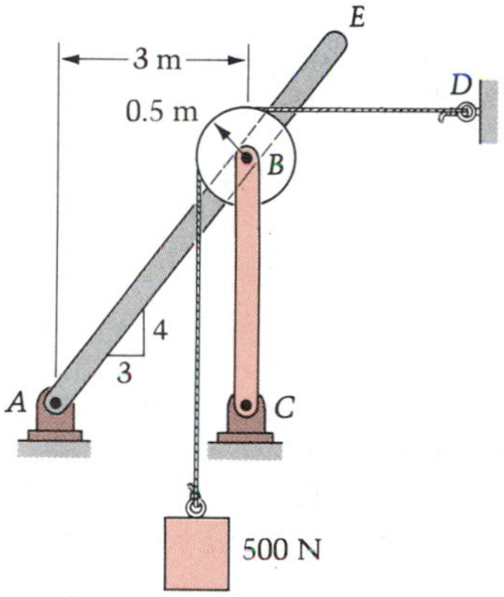

Figure E4.20a

Solution

We note by inspection that BC is a two-force member. The importance of such an observation is in realizing the reduction in unknowns from four (two pin reactions at C and at B) down to just one — the tensile or compressive force in the member. Thus we need no free-body diagram of that member.

The overall free-body diagram, which will prove helpful in analyzing the frame in this problem, is sketched in Figure E4.20b. Equilibrium requires

$$\circlearrowleft\!+ \quad \Sigma M_A = 0 = C_y(3 \text{ m}) - (500 \text{ N})(2.5 \text{ m}) - (500 \text{ N})(4.5 \text{ m})$$

$$C_y = \frac{3500}{3}$$

$$= 1167 \text{ N}$$

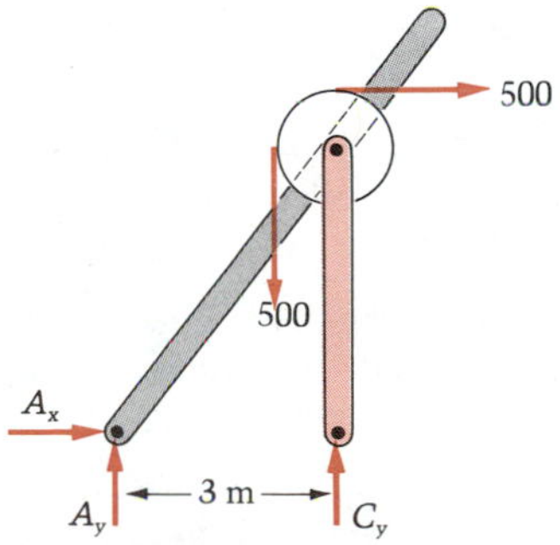

Figure E4.20b

where we start with four digits in this example. Also,

$$+\uparrow \quad \Sigma F_y = 0 = A_y + C_y - 500$$

$$A_y = 500 - 1167$$

$$= -667 \text{ N}$$

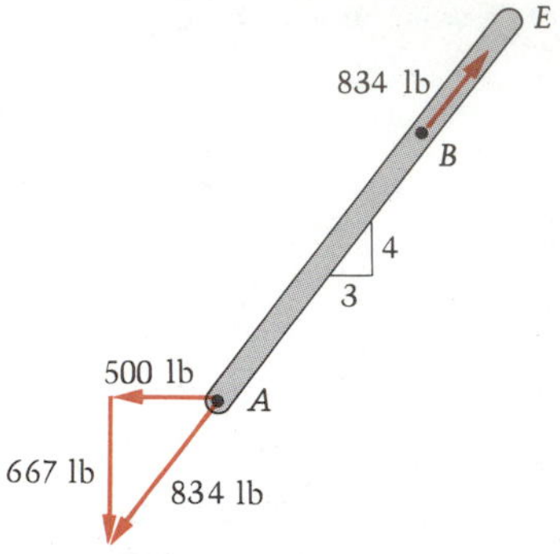

Figure E4.20c

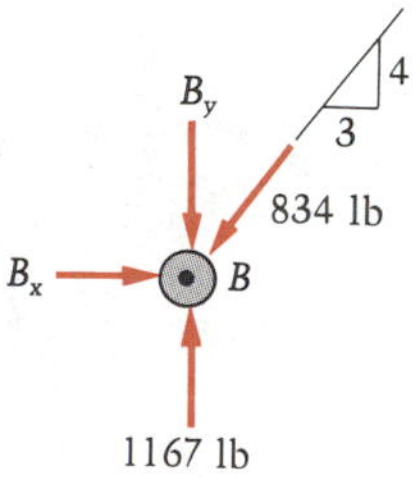

Figure E4.20d

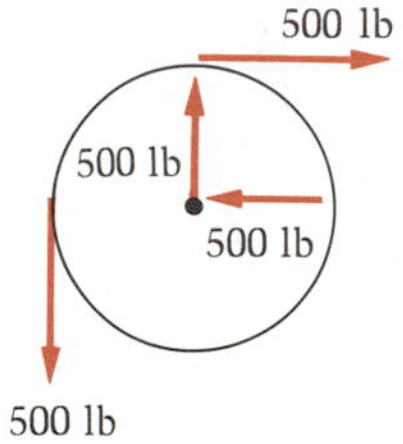

Figure E4.20e

Furthermore, we obtain:

$$\xrightarrow{+} \quad \Sigma F_x = 0 = A_x + 500$$

or

$$A_x = -500 \text{ N}$$

We note that the vector sum of $A_x\hat{\mathbf{i}}$ and $A_y\hat{\mathbf{j}}$ lies along BA because ABE, loaded by forces at just two points, is also a two-force member. Its free-body diagram is shown in Figure E4.20c. We see from this simple free-body diagram that the pin at B exerts the 834 lb force shown on the bar.

The free-body diagram of the pin at B is shown in Figure E4.20d, with B_x and B_y representing the components of force extended on the pin by the pulley. The 1167-lb and 834-lb forces are the forces from the pair of two-force members. Equilibrium of the pin requires

$$\xrightarrow{+} \quad \Sigma F_x = 0 = B_x - 834\left(\frac{3}{5}\right)$$

$$B_x = 500 \text{ lb}$$

$$+\!\uparrow \quad \Sigma F_y = 0 = 1167 - B_y - 834\left(\frac{4}{5}\right)$$

$$B_y = 500 \text{ lb}$$

Putting the reverses of these two forces onto the pulley [see its free-body diagram (Figure E4.20e)] provides an immediate check on our solution by inspection.

In conclusion, the force exerted by the pin at B onto:

a. the pulley is $-500\hat{\mathbf{i}} + 500\hat{\mathbf{j}}$ lb
b. the bar ABE is $500\hat{\mathbf{i}} + 667\hat{\mathbf{j}}$ lb
c. the bar BC is $-1167\hat{\mathbf{j}}$ lb

Note that in this problem, the forces exerted onto ABE and BC at B are not equal and opposite because of the presence of a third body (the pulley) there. What *is* true is that all three of these resultants add to zero because their negatives (Figure E4.20d) form the totality of external forces acting on the pin at B.

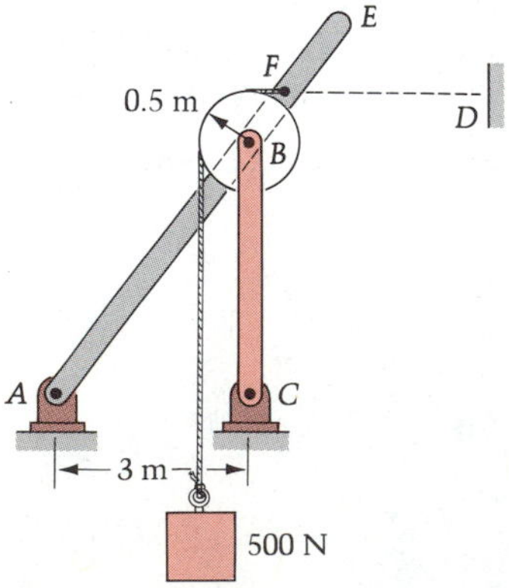

Figure E4.21a

EXAMPLE 4.21

Rework Example 4.20 if the horizontal portion of the cable is tied to ABE at F as shown in Figure E4.21a, instead of extending past it to the wall at D.

Solution

Recognizing again that BC is a two-force member, we have the overall free-body diagram shown in Figure E4.21b. Note that in order to isolate the frame, in this example we have to cut the rope only once. Thus $A_x = 0$ because

$$\xrightarrow{+} \quad \Sigma F_x = 0 = A_x$$

Note that ABE is not a two-force member in this example.

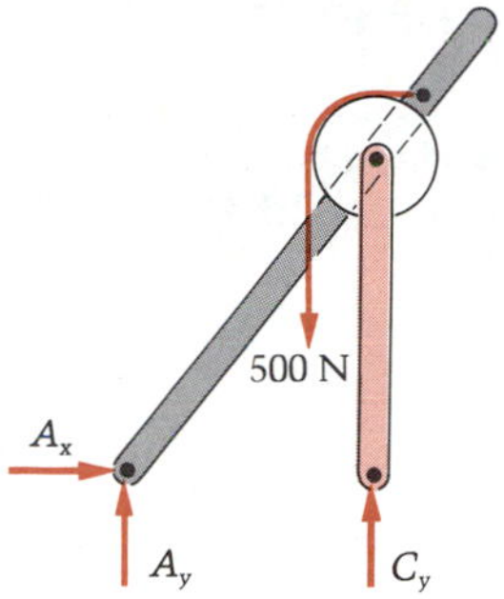

Figure E4.21b

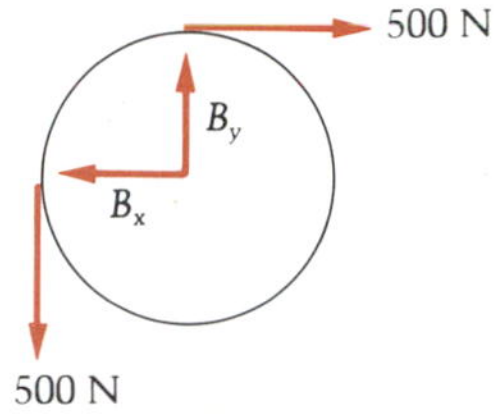

Figure E4.21c

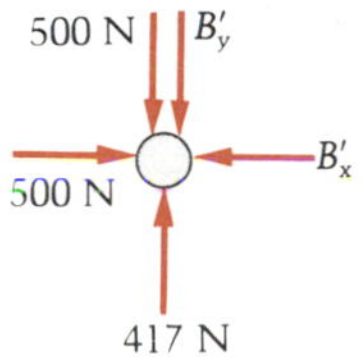

Figure E4.21d

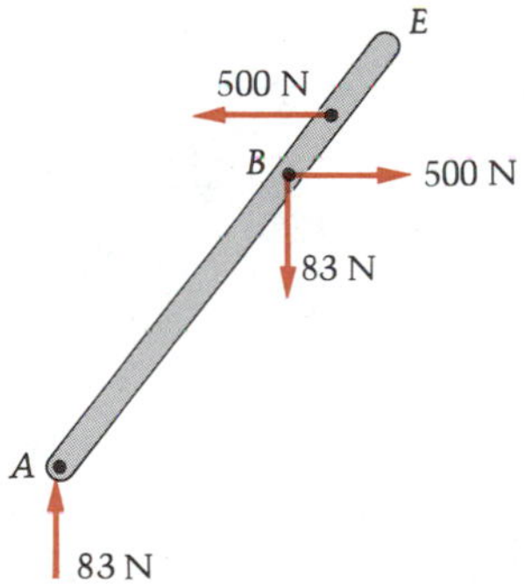

Figure E4.21e

Question 4.12 Why not?

Continuing,

$$\circlearrowleft\!+ \quad \Sigma M_A = 0 = C_y(3\text{ m}) - (500\text{ N})(2.5\text{ m})$$

$$C_y = \frac{1250}{3} = 417\text{ N}$$

from which

$$+\uparrow \quad \Sigma F_y = 0 = A_y + C_y - 500$$

$$A_y = 83\text{ N}$$

The free-body diagram of the pulley shown in Figure E4.21c is the same as in Example 4.20, and we obtain

$$\left.\begin{array}{l}\Sigma F_x = 0 \Rightarrow B_x = 500\text{ N}\\ \Sigma F_y = 0 \Rightarrow B_y = 500\text{ N}\end{array}\right\}\ \text{Both as shown. These are the forces exerted on the pulley by its pin, at } B.$$

The free-body diagram of the pin, with (B'_x, B'_y) representing the forces exerted on it by bar *ABE*, is shown in Figure E4.21d. Note that the force exerted on it by the bar *BC* is 417 $\uparrow$ N. It in turn exerts a force of 417 $\downarrow$ N on *BC* at *B*. Thus

$$\xrightarrow{+} \quad \Sigma F_x = 0 = 500 - B'_x$$

$$B'_x = 500\text{ N}$$

and

$$+\uparrow \quad \Sigma F_y = 0 = 417 - 500 - B'_y$$

$$B'_y = -83\text{ N or } 83 \uparrow \text{N} \qquad \text{(on the pin)}$$

Therefore the free-body diagram of member *ABE* is as shown in Figure E4.21e.

It is obvious that $\Sigma F_x = 0 = \Sigma F_y$. Checking for moment equilibrium,

$$\circlearrowleft\!+ \quad \Sigma M_A = -(83\text{ N})(3\text{ m}) - (500\text{ N})(4\text{ m}) + (500\text{ N})(4.5\text{ m})$$

$$= 1\text{ N}\cdot\text{m} \qquad \text{(differing from zero due to roundoff error)}$$

We next take up an example of equilibrium of a body in three dimensions. Recall that for such a body, there are six scalar equations, these being the components of $\Sigma\mathbf{F} = \mathbf{0}$ and $\Sigma\mathbf{M}_P = \mathbf{0}$, where P is any one point. Consequently, 3D-problems are generally more difficult than their planar counterparts, especially when the body must be taken apart.

Answer 4.12 It is loaded at more than two points.

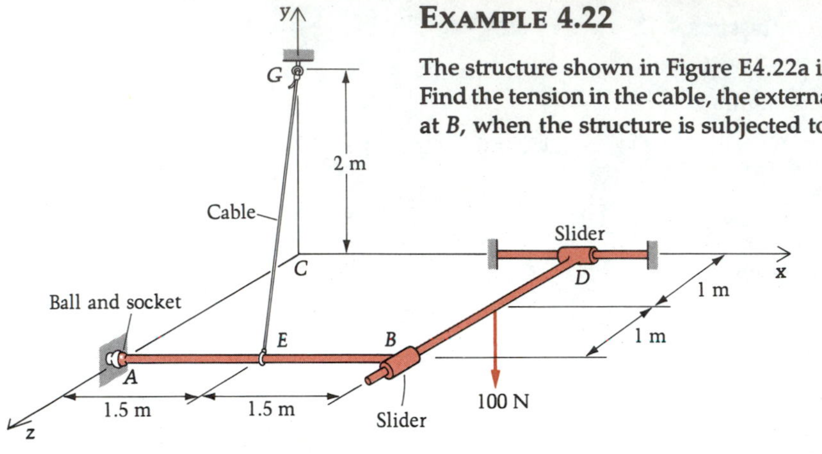

Figure E4.22a

EXAMPLE 4.22

The structure shown in Figure E4.22a is composed of two light bars and a cable. Find the tension in the cable, the external reactions at A and D, and the interaction at B, when the structure is subjected to the 100-N load.

Solution

First we shall consider the free-body diagram of the two bars taken together as shown in Figure E4.22b. We observe that, in this free-body diagram, eight unknown components of reaction appear. Thus we cannot determine all of the external reactions from the six corresponding equations of equilibrium. However, these equations taken together with those appropriate to *one* of the bars will turn out to be sufficient. Writing the equations of equilibrium for the two-bar system,

$$\Sigma \mathbf{F} = \mathbf{0}$$

$$(A_x\hat{\mathbf{i}} + A_y\hat{\mathbf{j}} + A_z\hat{\mathbf{k}}) + T\hat{\mathbf{e}}_{EG} + (D_y\hat{\mathbf{j}} + D_z\hat{\mathbf{k}}) + 100(-\hat{\mathbf{j}}) = \mathbf{0}$$

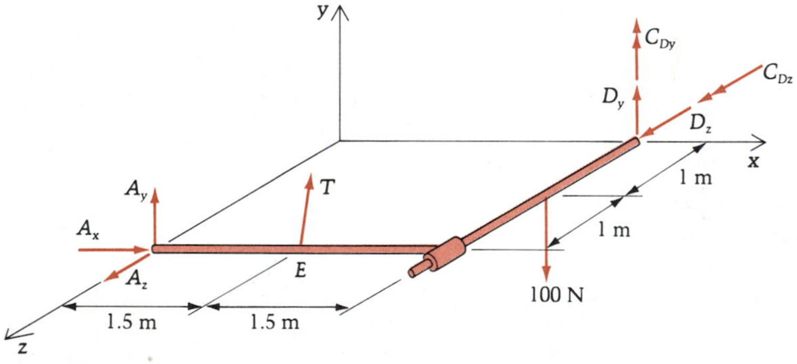

Figure E4.22b

Now $\hat{\mathbf{e}}_{EG}$, the unit vector along the line EG, is

$$\hat{\mathbf{e}}_{EG} = \frac{-1.5\hat{\mathbf{i}} + 2\hat{\mathbf{j}} - 2\hat{\mathbf{k}}}{\sqrt{(1.5)^2 + (2)^2 + (2)^2}}$$

$$= -0.469\hat{\mathbf{i}} + 0.625\hat{\mathbf{j}} - 0.625\hat{\mathbf{k}}$$

Thus from the coefficients of $\hat{\mathbf{i}}$, $\hat{\mathbf{j}}$, and $\hat{\mathbf{k}}$, respectively, in $\Sigma\mathbf{F} = \mathbf{0}$,

$$A_x - 0.469T = 0 \tag{1}$$

$$A_y + 0.625T + D_y - 100 = 0 \tag{2}$$

$$A_z - 0.625T - D_z = 0 \tag{3}$$

Summing moments at A,

$$\Sigma\mathbf{M}_A = \mathbf{0}$$

$$1.5\hat{\mathbf{i}} \times T(-0.469\hat{\mathbf{i}} + 0.625\hat{\mathbf{j}} - 0.625\hat{\mathbf{k}})$$
$$+ (3\hat{\mathbf{i}} - \hat{\mathbf{k}}) \times (-100\hat{\mathbf{j}}) + (3\hat{\mathbf{i}} - 2\hat{\mathbf{k}}) \times (D_y\hat{\mathbf{j}} + D_z\hat{\mathbf{k}})$$
$$+ (C_{Dy}\hat{\mathbf{j}} + C_{Dz}\hat{\mathbf{k}}) = \mathbf{0}$$

or

$$(1.5)(0.625T)(\hat{\mathbf{k}}) + (1.5)(-0.625T)(-\hat{\mathbf{j}}) + 3(-100)(\hat{\mathbf{k}})$$
$$+ (-1)(-100)(-\hat{\mathbf{i}}) + 3D_y(\hat{\mathbf{k}}) + 3D_z(-\hat{\mathbf{j}})$$
$$+ (-2D_y)(-\hat{\mathbf{i}}) + C_{Dy}\hat{\mathbf{j}} + C_{Dz}\hat{\mathbf{k}} = \mathbf{0}$$

From requiring that the coefficients of $\hat{\mathbf{i}}$, $\hat{\mathbf{j}}$, and $\hat{\mathbf{k}}$ vanish, respectively, we obtain:

$$-100 + 2D_y = 0 \Rightarrow D_y = 50 \text{ N} \tag{4}$$

$$+0.938T - 3D_z + 3D_z + C_{Dy} = 0 \tag{5}$$

$$0.938T - 300 + 3D_y + C_{Dz} = 0 \tag{6}$$

Holding Equations (1)–(6) in reserve, we now turn to the free-body diagram (Figure E4.22c) of bar BD. Note that our six equations of equilibrium will involve only four new unknowns, because of the special nature of the connection at B. Thus after writing six more equilibrium equations, we shall have $6 + 6 = 12$ equations in $8 + 4 = 12$ unknowns.

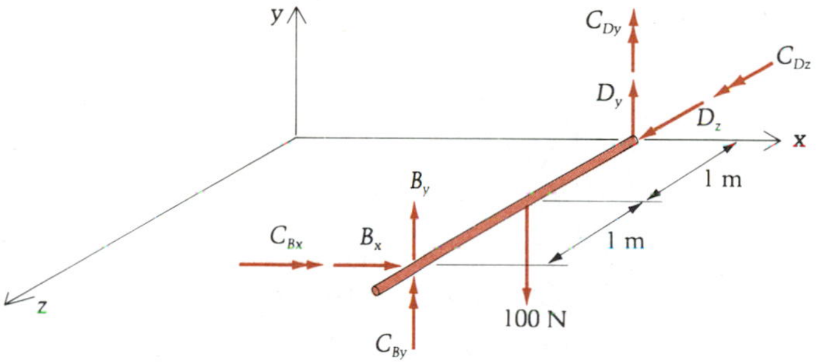

Figure E4.22c

The geometry here is so simple that it is relatively easy to write out the equations of equilibrium directly in component form:

$$\overset{+}{\rightarrow} \qquad \Sigma F_x = 0 \Rightarrow B_x = 0 \tag{7}$$

$$+\uparrow \qquad \Sigma F_y = 0 \Rightarrow B_y + D_y - 100 = 0$$

which, together with (4), yields

$$B_y + 50 - 100 = 0 \qquad \text{or} \qquad B_y = 50 \text{ N} \tag{8}$$

$$\overset{+}{\swarrow} \qquad \Sigma F_z = 0 \Rightarrow D_z = 0 \tag{9}$$

From $\Sigma \mathbf{M}_B = \mathbf{0}$, we obtain our final three equations:

$$\text{(i)} \qquad (\Sigma M_B)_x = 0$$

$$C_{Bx} + 2D_y - (1)(100) = 0$$

or, using (4)

$$C_{Bx} + 2(50) - 100 = 0 \Rightarrow C_{Bx} = 0 \tag{10}$$

$$\text{(ii)} \qquad (\Sigma M_B)_y = 0 \Rightarrow C_{By} + C_{Dy} = 0 \tag{11}$$

$$\text{(iii)} \qquad (\Sigma M_B)_z = 0 \Rightarrow C_{Dz} = 0 \tag{12}$$

Substituting (4) and (12) into (6), we obtain the cable tension:

$$0.938T - 300 + 3(50) + 0 = 0$$

$$T = \frac{150}{0.938} = 160 \text{ N}$$

This, together with (9), yields for (5) an equation we can solve for C_{Dy}:

$$0.938\left(\frac{150}{0.938}\right) - 3(0) + C_{Dy} = 0$$

$$C_{Dy} = -150 \text{ N} \cdot \text{m}$$

and hence from (11),

$$C_{By} = 150 \text{ N} \cdot \text{m}$$

Note that now we have determined the cable tension and the forces and moments associated with the connections at B and D. We now may return to Equations (1)–(3) to obtain the components of reaction at the ball-and-socket joint:

$$A_x - 0.469(160) = 0$$

$$A_x = 75.0 \text{ N}$$

$$A_y + 0.625(160) + 50 - 100 = 0$$

$$A_y = -50 \text{ N}$$

$$A_z - 0.625(160) + 0 = 0$$

$$A_z = 100 \text{ N}$$

Our final results, displayed on free-body diagrams of the individual bars, are shown in Figure E4.22d:

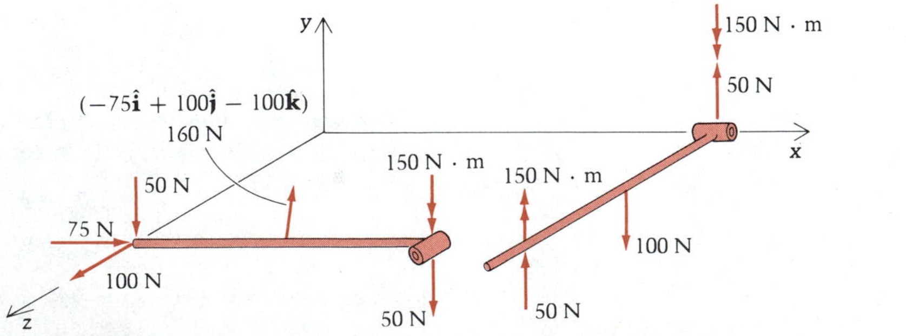

Figure E4.22d

The reader may wish to show that bar *AB*, which was not analyzed alone, is indeed in equilibrium under the forces acting on it in the final figure of the preceding example.

In the last two examples of this section, we shall have a preview of what is to come in Chapter 5. In the examples thus far, we have only removed supports, pins, ropes, and so on, leaving the various constituent members of the body intact (or whole). We have also made use of the special features of two-force bodies and pulleys. But there is nothing to prevent us from actually slicing through a member in order to expose, on a free-body diagram of only part of it, the forces and couples it transmits to the *other* part. This will be a distinguishing feature of Chapter 5. The following two examples will introduce the idea, although it is important to realize that there is nothing really new here in concept.* We again are using the fact that a body in equilibrium has each of its parts in equilibrium. If we wish to determine a certain force (and/or couple), we must choose a free-body diagram that exhibits the force as *external* so that it will appear in the corresponding equations of equilibrium.

EXAMPLE 4.23

For the beam of Example 4.4 (shown in Figure E4.23a), find the force-couple resultant transmitted at a cross section 3 feet from the left end.

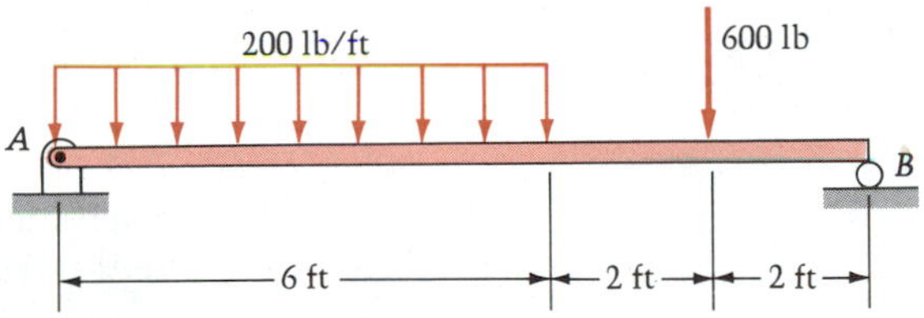

Figure E4.23a

Solution

We first separate (cut) the beam at the cross section of interest (point C of Figure E4.23b) and then sketch the free-body diagram of the material either to the left or to the right of the cut. Here we choose the material on the left. The free-body diagram is then shown with the dashed 600-lb force representing the resultant of the distributed load over the 3 feet. The 960-lb force at *A* is the reaction already determined in Example 4.4. Only a portion (3 feet) of the 200-lb / ft distributed load is external to the body we have chosen to analyze. The resultant of that distributed loading is 600 pounds as shown on the free-body diagram. For this two-dimensional problem the arrow code indicates that we have chosen to represent the force part of the resultant exerted on the material to the left of the cut by the material to the right of the cut as

* These distinctions—a single physical object, two or more objects left intact, physical objects separated, and a physical object imagined to be sliced into two parts—all obey the same simple laws of equilibrium. We have separated these topics because it is our experience that students master them more easily in graduated steps.

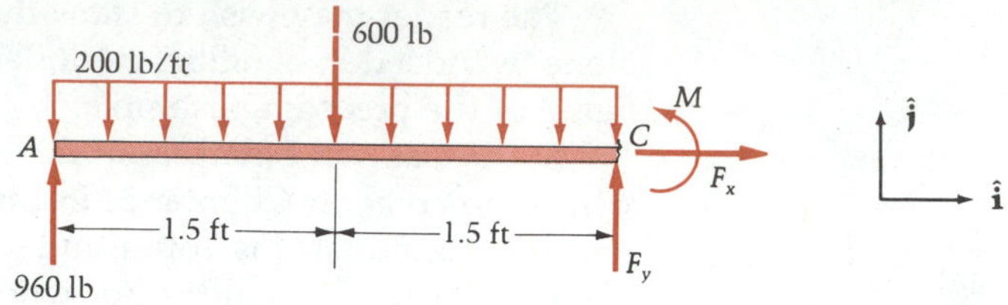

Figure E4.23b

by the material to the right of the cut as

$$F_x\hat{\mathbf{i}} + F_y\hat{\mathbf{j}}$$

The couple part of the resultant is expressed (vectorially) as $M\hat{\mathbf{k}}$.

Applying the equations of equilibrium in component form,

$$\xrightarrow{+} \qquad \Sigma F_x = 0$$

so that

$$F_x = 0$$

and

$$+\uparrow \qquad \Sigma F_y = 0$$

or

$$F_y + 960 - 600 = 0$$

so that

$$F_y = -360 \text{ lb}$$

The minus sign means that the vertical component of force acting on the cross section at C in Figure E4.23b is $360\downarrow$ lb. Finally, we sum moments to determine M:

$$\circlearrowleft\!+ \qquad \Sigma M_c = 0$$

or

$$M + 1.5(600) - 3(960) = 0$$

so that

$$M = 1980 \text{ lb-ft}$$

It is instructive to see what would have happened in the preceding example had we chosen instead to apply the equations of equilibrium to the material to the *right* of the cut. The appropriate free-body diagram is shown in Figure 4.25. Note the 840-lb right-end reaction previously found in Example 4.4. Note further that the arrow code and letters (F_x, F_y, M) represent automatic satisfaction of the action-reaction principle because the forces and couple they represent are equal in magnitude but opposite in direction to those of Figure E4.23b.

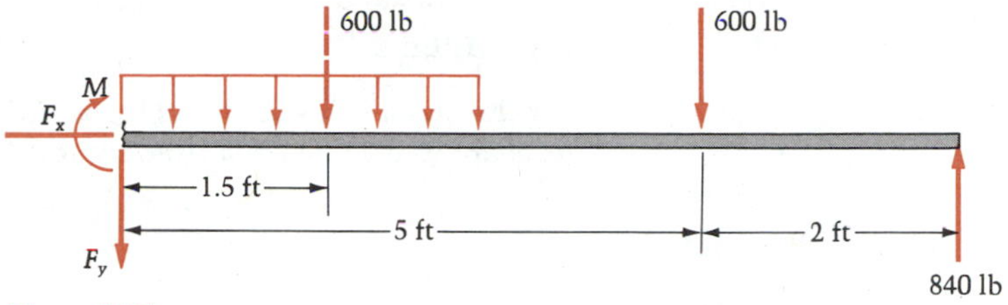

Figure 4.25

The equilibrium equations yield

$$\xrightarrow{+} \qquad \Sigma F_x = 0 = -F_x$$

$$F_x = 0 \qquad \text{(as before)}$$

and

$$+\uparrow \qquad \Sigma F_y = 0$$

$$-F_y - 600 - 600 + 840 = 0$$

$$F_y = -360 \text{ lb} \qquad \text{(as before)}$$

and

$$\circlearrowleft\!+ \qquad \Sigma M_C = 0$$

$$-M - 1.5(600) - 5(600) + 7(840) = 0$$

$$M = 1980 \text{ lb-ft} \qquad \text{(as before)}$$

Thus the answers are independent of which part of the cut body we use to obtain them.

Question 4.13 In the preceding example, why is there no force component F_z, nor couple components M_x or M_y, acting on the cut section in addition to the forces F_x and F_y and the couple M parallel to $\hat{\mathbf{k}}$?

Answer 4.13 The equations $\Sigma F_z = 0$, $\Sigma M_{Cx} = 0$ and $\Sigma M_{Cy} = 0$ would have respectively given $F_z = 0$, $M_x = 0$ and $M_y = 0$ if these components had been included on the FBD.

EXAMPLE 4.24

For the boom of Example 4.11, shown in Figure E4.24a, find the force-couple resultant transmitted at a cross section 4 feet from the right end of the boom.

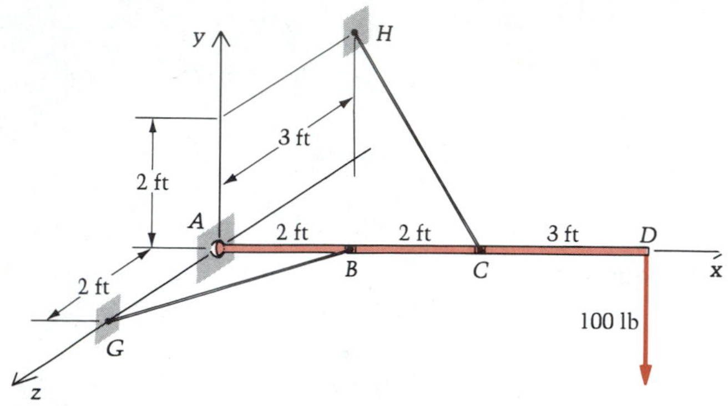

Figure E4.24a

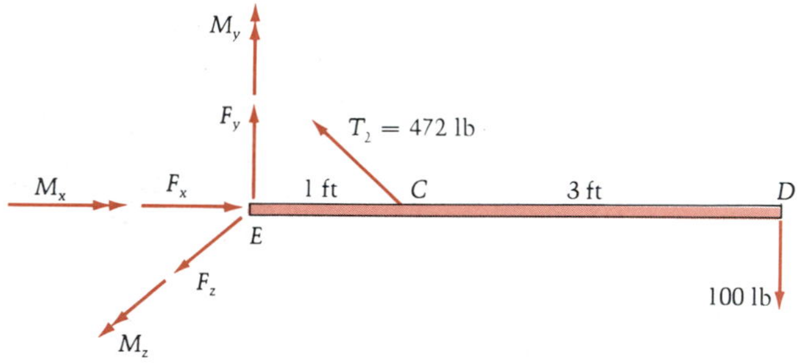

Figure E4.24b

Solution

Isolating the part of the boom to the right of the cross section in question, we obtain the free-body diagram shown in Figure E4.24b. We found in Example 4.11 that the force exerted by the wire is

$$T_2\hat{\mathbf{e}}_2 = T_2\frac{\mathbf{r}_{CH}}{|\mathbf{r}_{CH}|}$$

$$= 472(-0.743\hat{\mathbf{i}} + 0.371\hat{\mathbf{j}} - 0.577\hat{\mathbf{k}})$$

$$= -351\hat{\mathbf{i}} + 175\hat{\mathbf{j}} - 263\hat{\mathbf{k}} \text{ lb}$$

Applying the equations of equilibrium,

$$\Sigma\mathbf{F} = \mathbf{0}$$

$$(F_x\hat{\mathbf{i}} + F_y\hat{\mathbf{j}} + F_z\hat{\mathbf{k}}) + (-351\hat{\mathbf{i}} + 175\hat{\mathbf{j}} - 263\hat{\mathbf{k}}) + 100(-\hat{\mathbf{j}}) = \mathbf{0}$$

Thus

$$\hat{\mathbf{i}}: \quad F_x - 351 = 0$$

$$F_x = 351 \text{ lb}$$

$$\hat{\mathbf{j}}: \quad F_y + 175 - 100 = 0$$

$$F_y = -75 \text{ lb}$$

$$\hat{\mathbf{k}}: \quad F_z - 263 = 0$$

$$F_z = 263 \text{ lb}$$

Taking moments about point E will yield the couples there:

$$\Sigma \mathbf{M}_E = \mathbf{0}$$

$$(M_x\hat{\mathbf{i}} + M_y\hat{\mathbf{j}} + M_z\hat{\mathbf{k}}) + \mathbf{r}_{EC} \times T_2\hat{\mathbf{e}}_2 + \mathbf{r}_{ED} \times (-100\hat{\mathbf{j}}) = \mathbf{0}$$

$$M_x\hat{\mathbf{i}} + M_y\hat{\mathbf{j}} + M_z\hat{\mathbf{k}} + 1\hat{\mathbf{i}} \times (-351\hat{\mathbf{i}} + 175\hat{\mathbf{j}} - 263\hat{\mathbf{k}}) + 4\hat{\mathbf{i}} \times (-100\hat{\mathbf{j}}) = \mathbf{0}$$

$$M_x\hat{\mathbf{i}} + M_y\hat{\mathbf{j}} + M_z\hat{\mathbf{k}} + (175\hat{\mathbf{k}} + 263\hat{\mathbf{j}}) - 400\hat{\mathbf{k}} = \mathbf{0}$$

Thus

$\hat{\mathbf{i}}$: $M_x = 0$

$\hat{\mathbf{j}}$: $M_y + 263 = 0$

$M_y = -263$ lb-ft

$\hat{\mathbf{k}}$: $M_z + 175 - 400 = 0$

$M_z = 225$ lb-ft

Question 4.14 Why did the force exerted by the wire BG onto the boom appear neither on the free-body diagram nor in the equilibrium equations?

Answer 4.14 That force is not acting on the material that has been isolated here (free-body diagram) for analysis.

PROBLEMS ► Section 4.4

4.144 The uniform bar in Figure P4.144 weighs 100 lb and the man 140 lb. Find the tension in the cable for equilibrium. (Can the man exert this much force?)

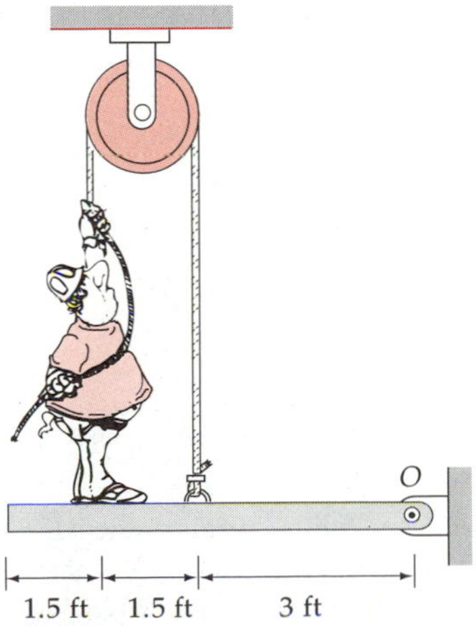

Figure P4.144

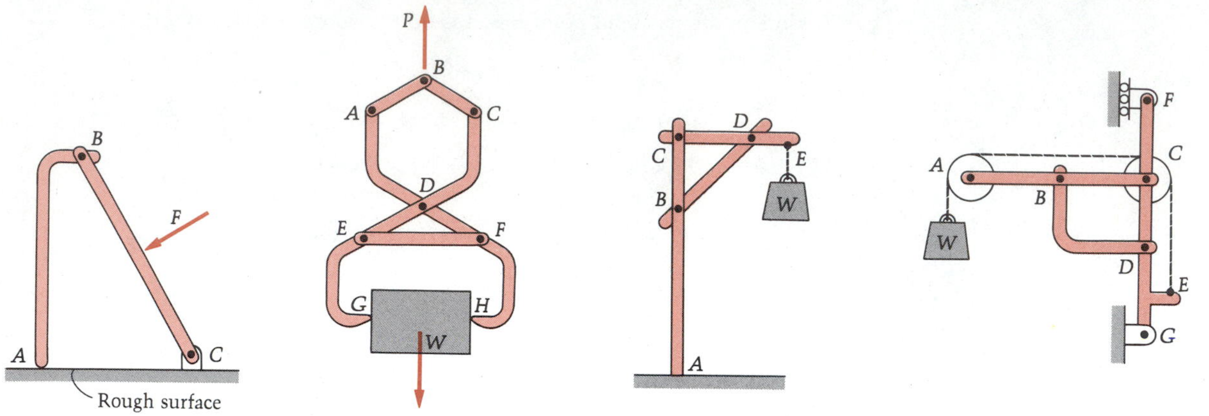

Figure P4.145

4.145 Name the two-force members (not counting cables) in the four structures shown in Figure P4.145 (three have one and one has three).

4.146 If b/a is 5, show that the compound lever system in Figure P4.146 will hold up a weight W that is 125 times the magnitude of the force F.

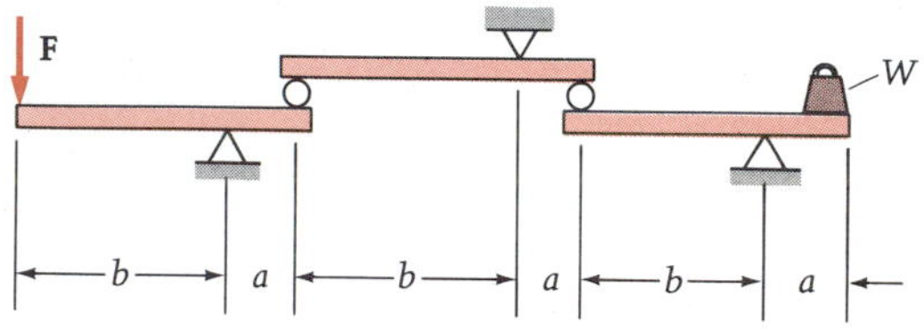

Figure P4.146

4.147 The uniform 1-ton (2000 lb) beam BD rests on a roller pinned to the truss structure at point B in Figure P4.147. Find the reactions onto the structure at A and C.

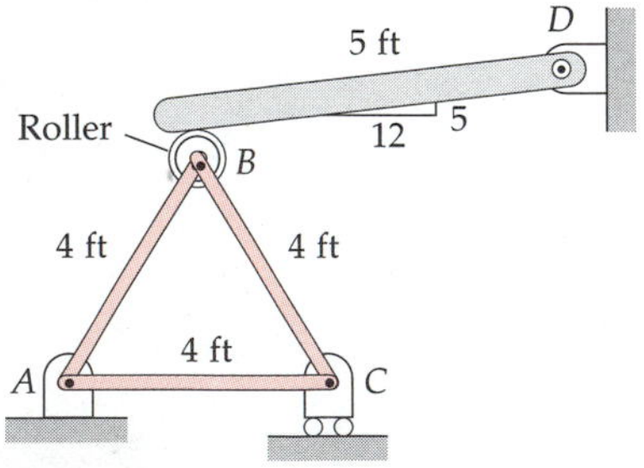

Figure P4.147

4.148 The object $\mathcal{B}$ is compressed by the toggle device. If a person pulls with 20 lb as shown in Figure P4.148, what is the compressive force exerted on $\mathcal{B}$ by block $\mathcal{A}$? Neglect friction between $\mathcal{A}$ and frame $\mathcal{F}$.

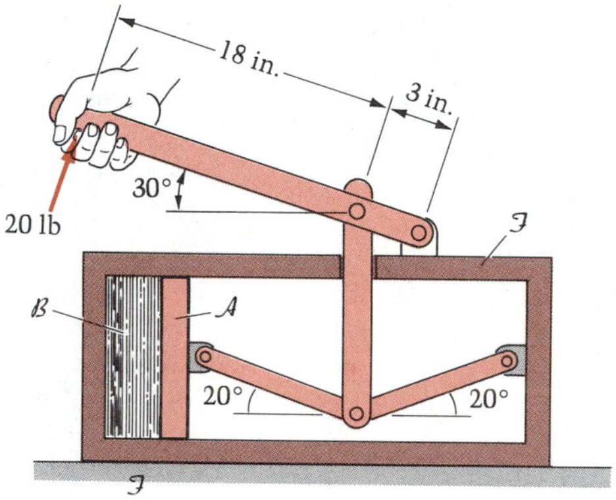

Figure P4.148

4.149 If the 3000-lb spherical boulder is in equilibrium, and if friction is negligible except at the fulcrum F, find the reaction of the wall on the boulder, and the angle θ. (See Figure P4.149.)

4.150 In Figure P4.150 the cord passes over a pulley at C and supports the block (mass 15 kg), in contact with the uniform bar AB (mass 10 kg). Find the horizontal and vertical components of the force at A on AB, and the force exerted on the bar by the block.

4.151 If the archer's left hand is pushing against the bow handle with a force $P = 50$ lb, what is the tension in the string (as a function of α)? What are the vertical forces exerted on the ends of the bow by the string? What is the horizontal component of the resultant force exerted on the archer's feet by the ground? (See Figure P4.151.)

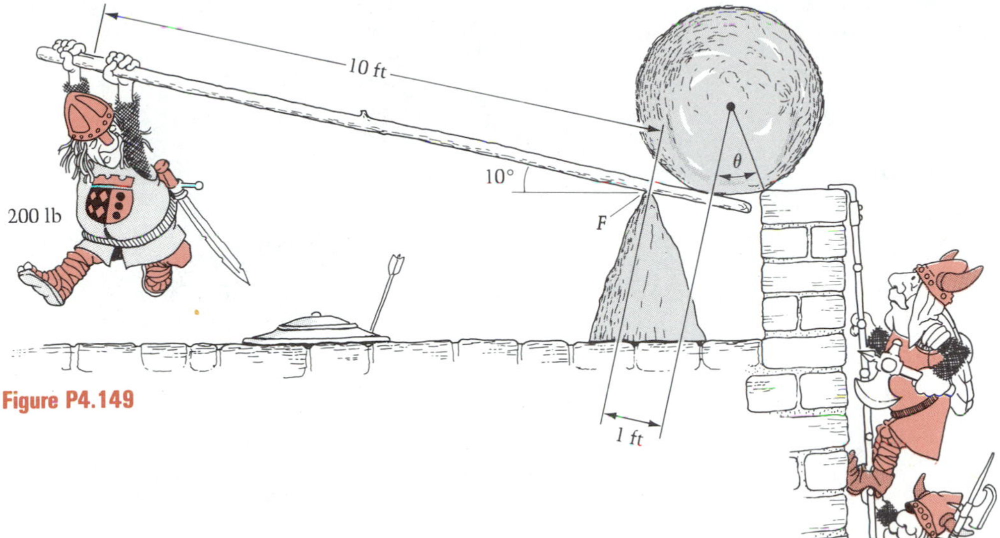

Figure P4.149

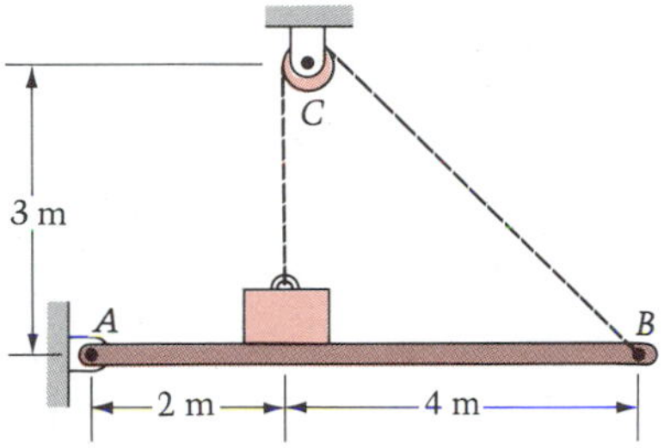

Figure P4.150

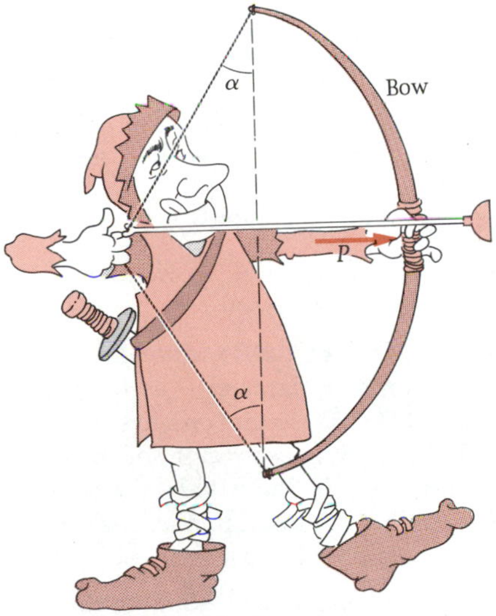

Figure P4.151

4.152 In Figure P4.152, the blocks $\mathscr{A}$ and $\mathscr{B}$ respectively weigh 223 N and 133 N, and the planes are smooth. The connecting rod is light. (a) Find the force P that will hold the system in equilibrium. (b) Repeat the problem with P applied instead horizontally to block $\mathscr{A}$.

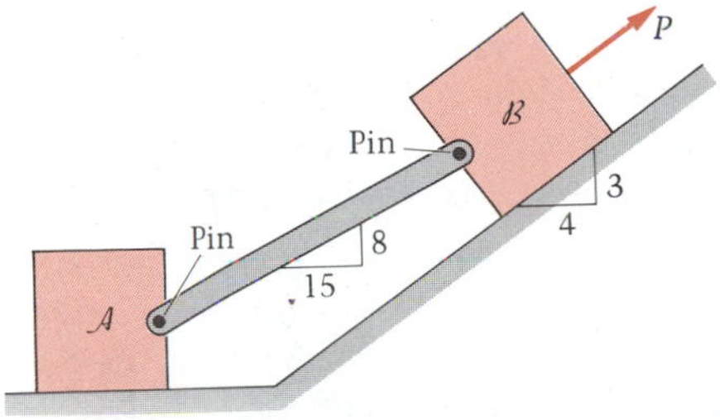

Figure P4.152

4.153 A block is lifted by the tongs shown in Figure P4.153. Find the force exerted on member $\mathcal{L}$, at C, by member $\mathcal{R}$.

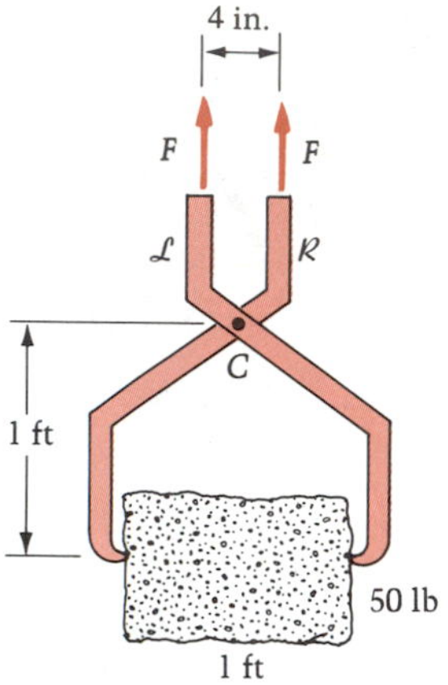

Figure P4.153

4.154 The identical wheels in Figure P4.154 each have mass 80 kg, and the rod has mass 40 kg and is pinned to the wheels as shown at A and B. The plane is smooth. Using free-body diagrams, show that the three bodies cannot be in equilibrium in the given position.

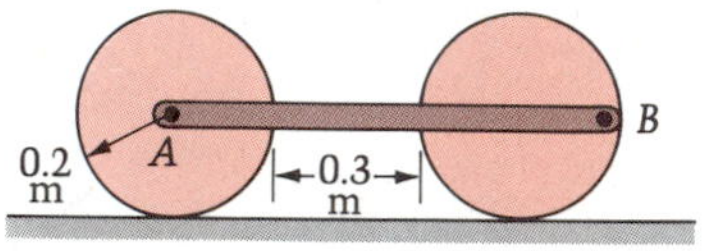

Figure P4.154

4.155 In the preceding problem, compute the moment of a couple that, when applied to the wheel on the right in the given position, results in equilibrium.

4.156 The ring gear $\mathcal{A}$ is fixed in a reference frame to which the centers of the sun gear $\mathcal{C}$ and spider arm $\mathcal{S}$ are pinned at C. (See Figure P4.156.) A clockwise couple M_o is applied to $\mathcal{S}$. Find the couple that must be applied to $\mathcal{C}$ in order that all bodies be in equilibrium.

4.157 In Figure P4.157 the pulley weighs 15 lb, the beam weighs 60 lb, the man weighs 160 lb, and the system is in equilibrium. Find the force in one of the two ropes at A, assuming the force is the same in each of the two ropes at an end.

4.158 Determine the gripping forces on (a) the nail in the most closed position of the pliers if $F = 20$ lb; (b) the pipe in the most open position of the pliers if $F = 20$ lb. (See Figure P4.158.)

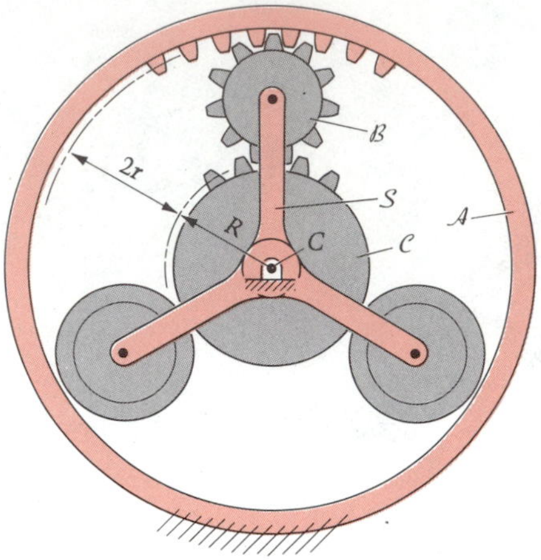

Figure P4.156

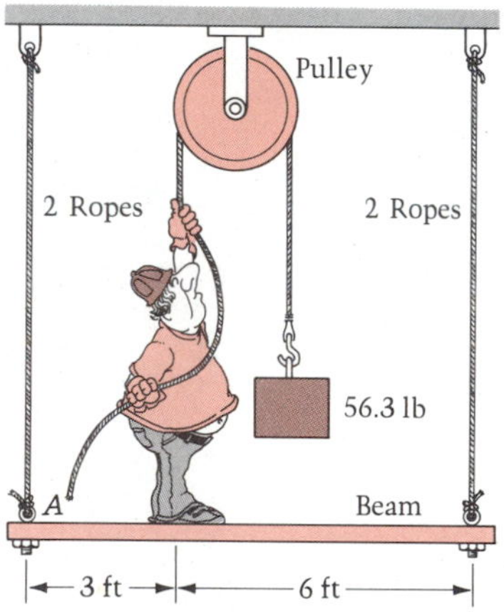

Figure P4.157

4.159 The rod AB in Figure P4.159 weighs 30 lb, and the block weighs 50 lb. Friction is negligible. What force P will hold the system of two bodies in equilibrium?

4.160 Find the forces in whichever of $\mathcal{B}_1$, $\mathcal{B}_2$, $\mathcal{B}_3$, and $\mathcal{B}_4$ are two-force members. (See Figure P4.160.)

* **4.161** (a) Two marbles, each of radius R and weight W, are placed inside a hollow tube of diameter D as shown in Figure P4.161. Note that $D < 4R$, so that only one marble touches the floor. Find the minimum weight of the tube such that it will not turn over. (b) Show that if the bottom of the tube is capped, then it will not turn over regardless of the weights or dimensions. All surfaces are smooth.

4.162 Find the force exerted on member $ABCD$ by the pin at B in Figure P4.162.

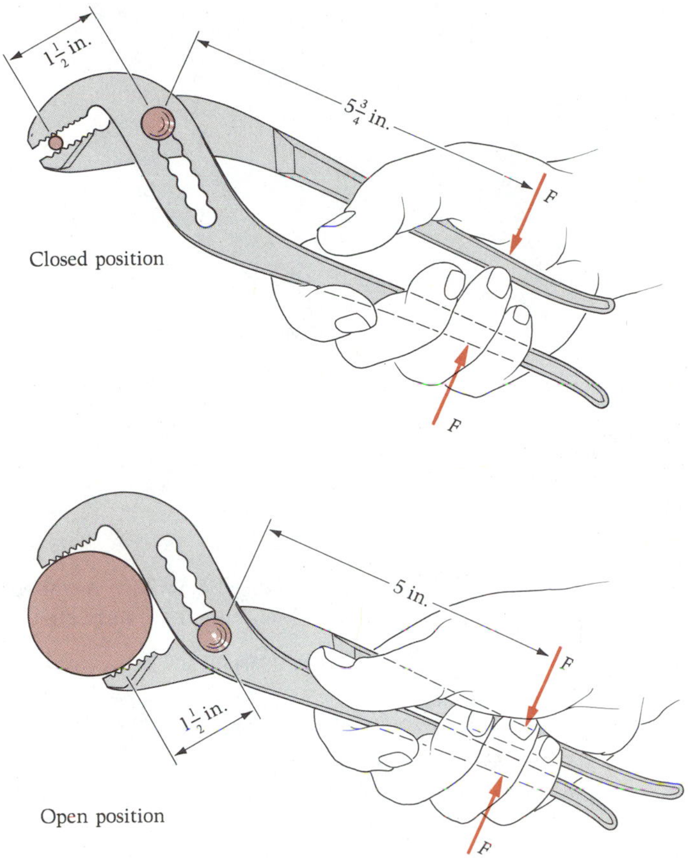

Figure P4.158

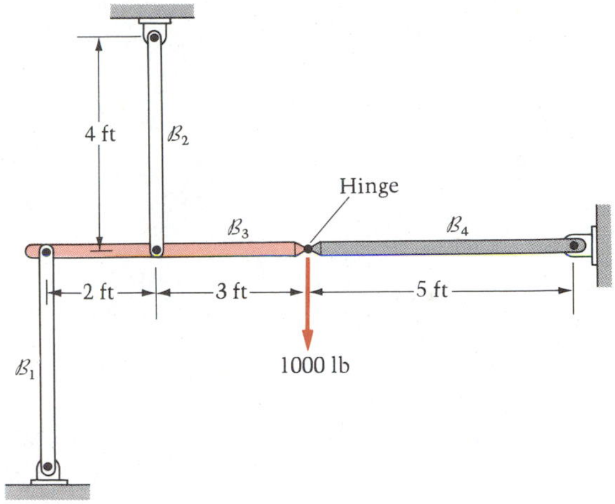

Figure P4.160

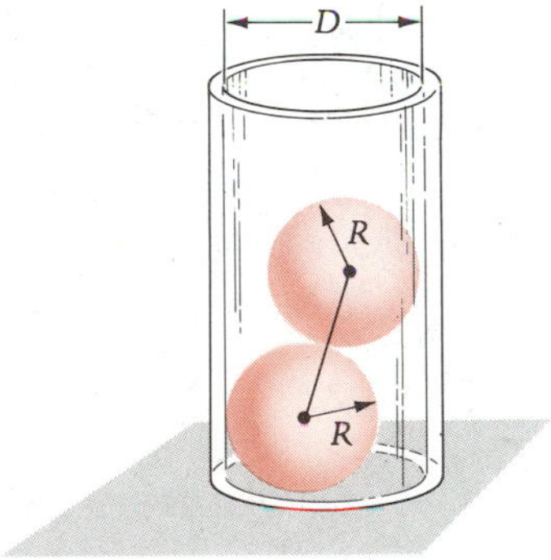

Figure P4.161

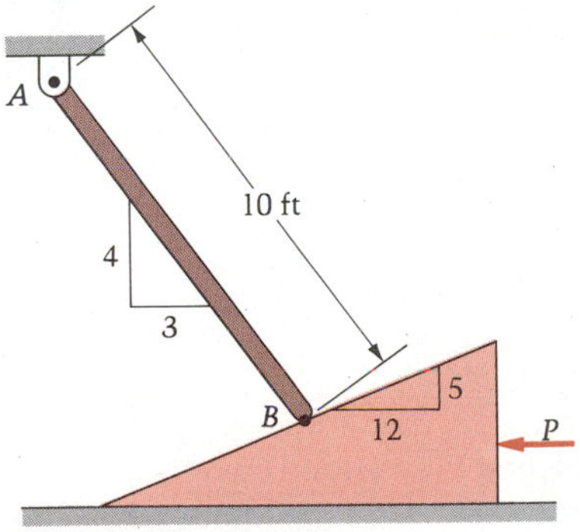

Figure P4.159

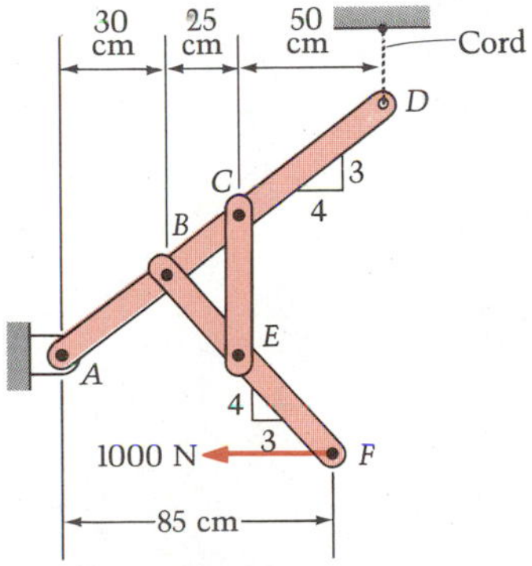

Figure P4.162

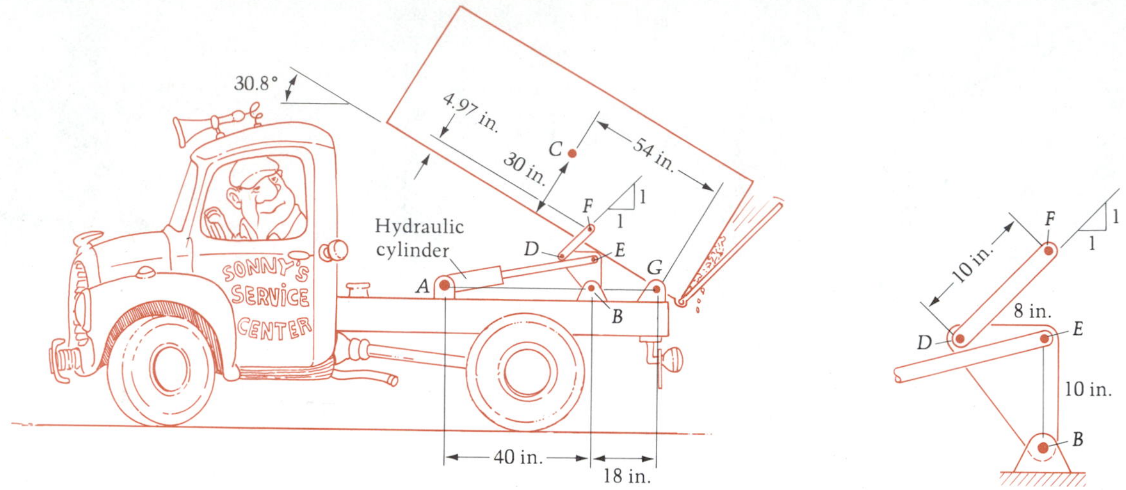

Figure P4.163

4.163 The dumptruck in Figure P4.163 is ready to release a load of gravel. The total weight being hoisted is 15,000 lb with mass center at C. Treating the cylinder rod AE as a two-force member, find the force in it at the indicated position.

4.164 Find the force that the bar CE exerts on the bar AF at C. (See Figure P4.164.)

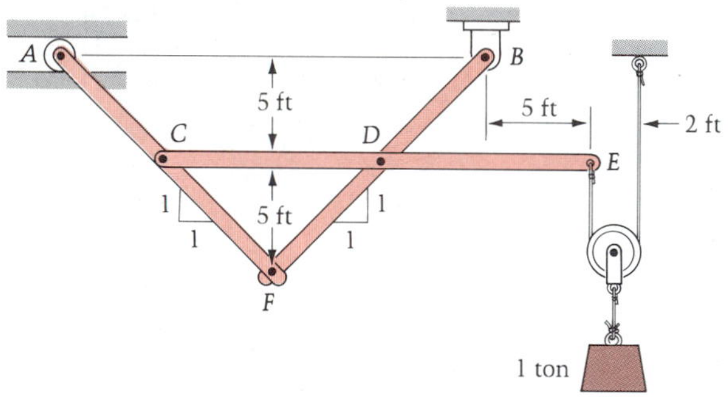

Figure P4.164

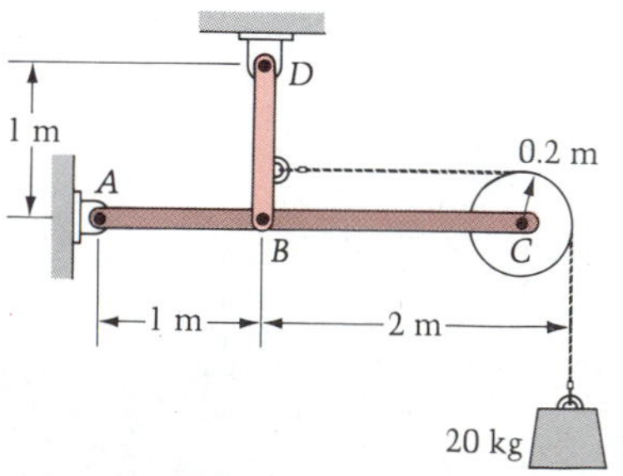

Figure P4.165

4.165 In the frame shown in Figure P4.165, the members are pin-connected and their weights can be neglected.

a. Find the external reactions on the frame at A and D.

b. Find the forces at B and C on member ABC.

4.166 In the frame shown in Figure P4.166, the members are pin-connected and their weights can be neglected. The 42 kN force is applied to the pin at C. Find:

a. The reactions on the frame at A and E

b. The components of the forces exerted by the pins at B and C on member ABC.

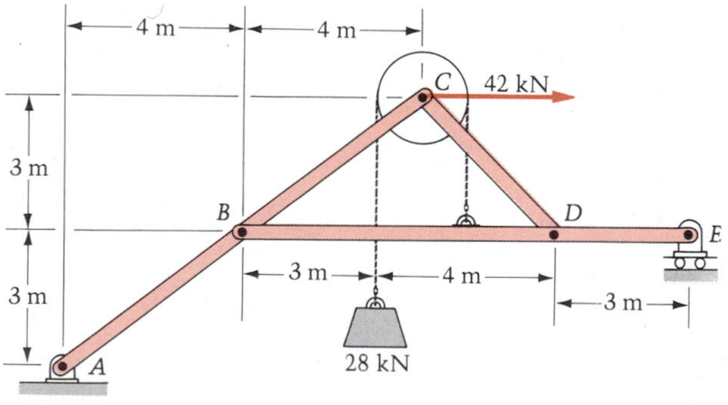

Figure P4.166

4.167 In Figure P4.167 find the reaction onto the frame at point F, and the force exerted on the pin at D by the member CF.

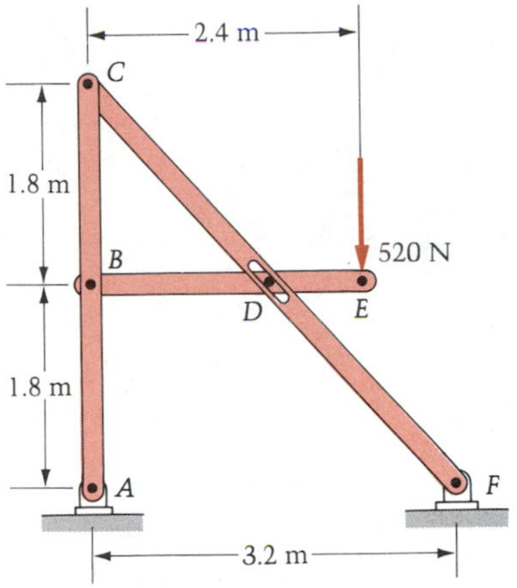

Figure P4.167

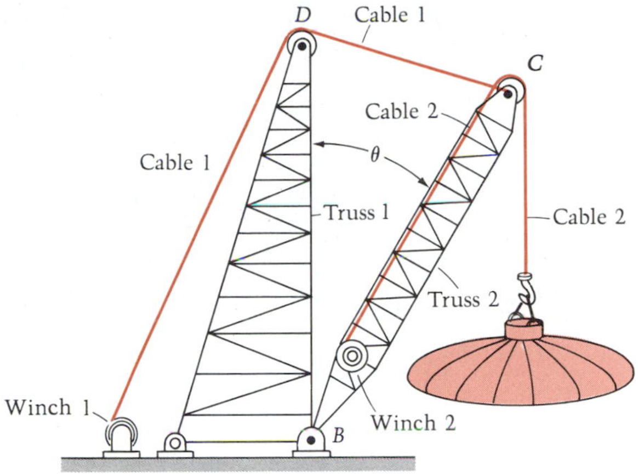

Figure P4.169

4.168 A slender, homogeneous, 20-ft rod weighing 64.4 lb is supported as shown in Figure P4.168. The bars *AB* and *DE* are of negligible mass. In terms of θ, find the force *P* that must be applied to the right end for equilibrium.

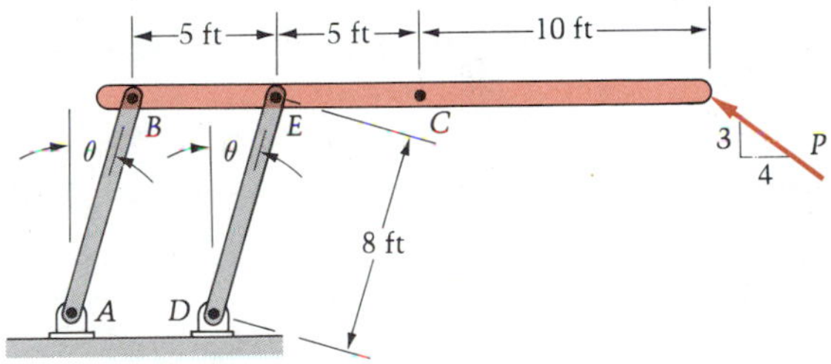

Figure P4.168

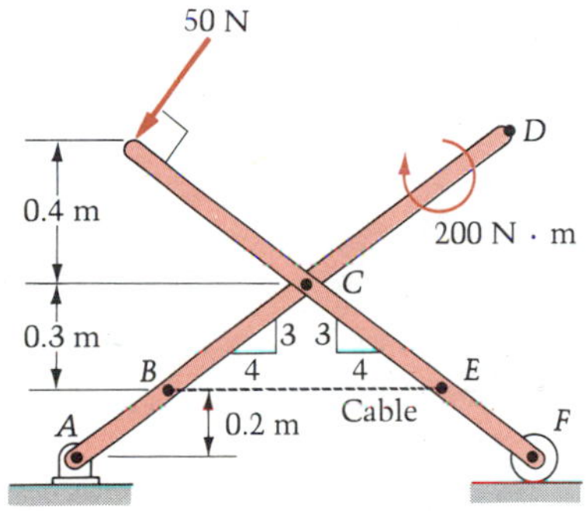

Figure P4.170

4.169 The crane in Figure P4.169 consists of two trusses, two cables, and two winches. Cable 1, let out by winch 1 to lower truss 2, passes over the small pulley at *D* and is attached to truss 2 at the center of the small pulley at *C*. Cable 2, let out by winch 2 attached to truss 2, lowers the 2000-lb antenna reflector after passing over the pulley at *C*. The lengths *BD* and *BC* are equal, and the weights of the trusses are to be neglected.

a. Find the tension in cable *DC* as a function of θ.

b. Find the force exerted by pin *B* on truss 2, and note that it is independent of θ.

4.170 In Figure P4.170 find the force exerted on *ABCD* by the pin at *C*.

4.171 Find the magnitude of the force of interaction between the two bars of Figure P4.171.

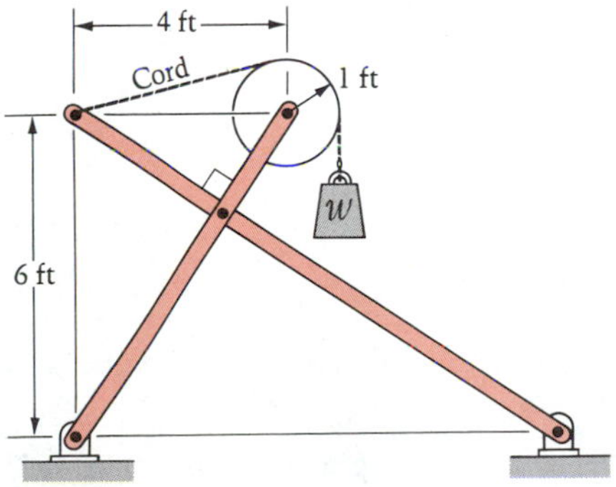

Figure P4.171

4.172 Find the pin reactions at B on member AB in Figure P4.172.

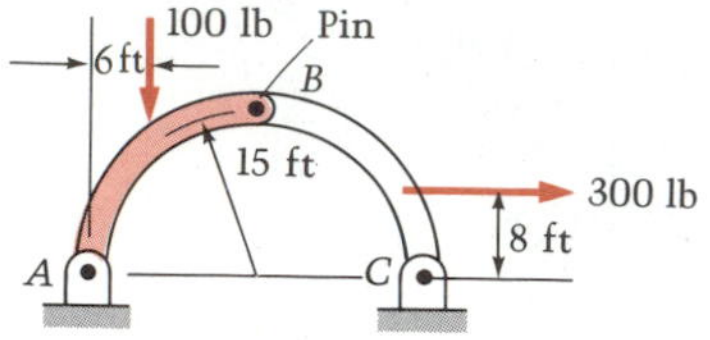

Figure P4.172

4.173 The blocks in Figure P4.173 each weigh 1250 N, with centers of mass at C_1 and C_2. The (shaded) platform on which they rest weighs 800 N with mass center at C_3, and is supported by two pairs of crossbars (one pair shown). Neglecting the weights of the crossbars, find the magnitude of the force transmitted by the pin that connects these two members at F. Assume that half the load is carried by each pair of crossbars.

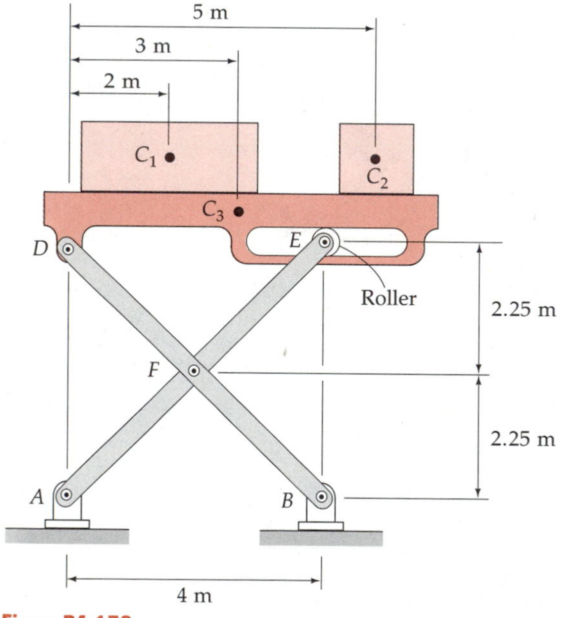

Figure P4.173

4.174 Draw the free-body diagrams of all members in Figure P4.174 and compute:

a. The force exerted by the pin at C on member ACD
b. The reactions at A and B.

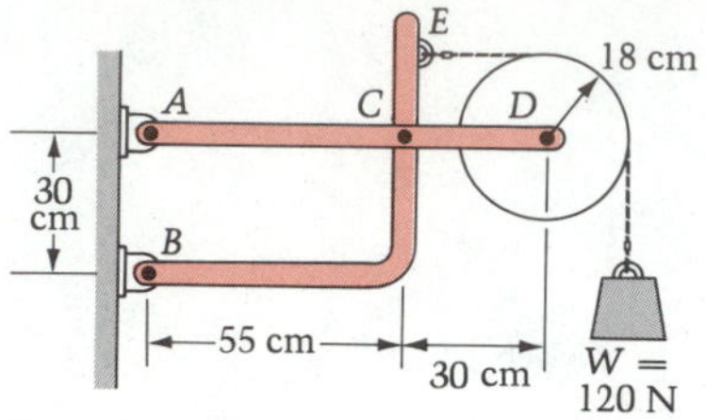

Figure P4.174

4.175 For the frame shown in Figure P4.175 find:

a. The reactions onto the frame members at A and E
b. The force exerted on BDF at D by the pin that joins the two members BDF and CDE to the pulley there.

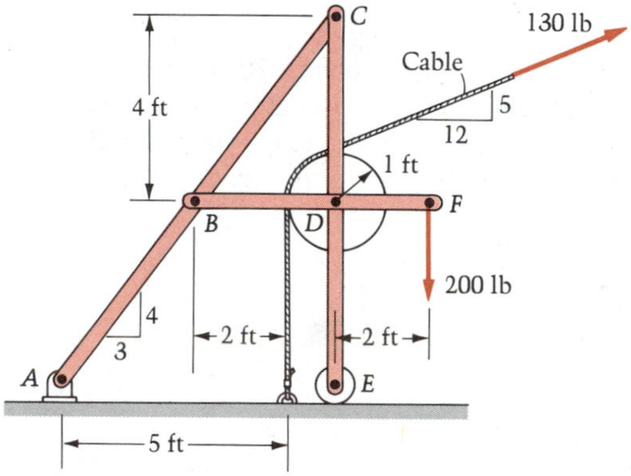

Figure P4.175

4.176 Find the reactions at A and B in Figure P4.176 when the horizontal force P is applied to the three-hinged arch. Neglect the weight of the arch.

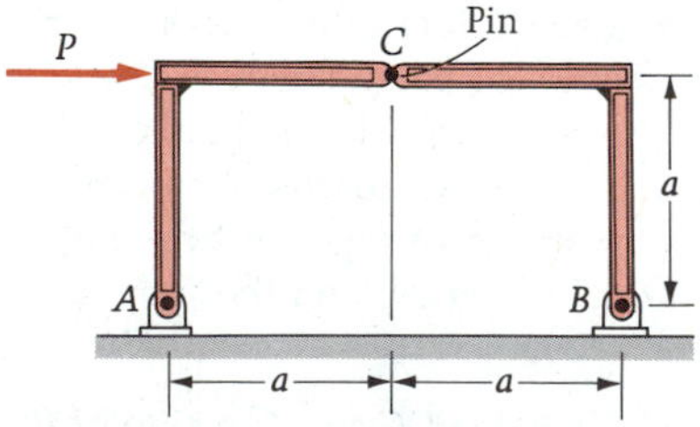

Figure P4.176

4.177 In Figure P4.177, the pin at D is a part of member EB. Find:

a. The reactions at A and B on the frame members
b. The force exerted by CD on EB at D.

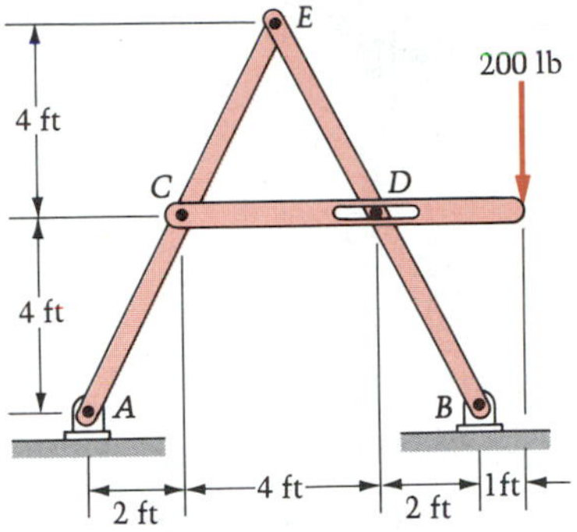

Figure P4.177

4.178 The length of the connecting rod $\mathcal{R}$ of the steam engine is 5 ft, and its crank $\mathcal{C}$, to which it is pinned at B, has length 10 in. (See Figure P4.178.) The front and back pressures on either side of the piston are indicated. Find the force in member $\mathcal{R}$, neglecting friction and assuming all bodies to be in equilibrium. *Hint:* Note that $\mathcal{C}$ carries a moment and is not a two-force member!

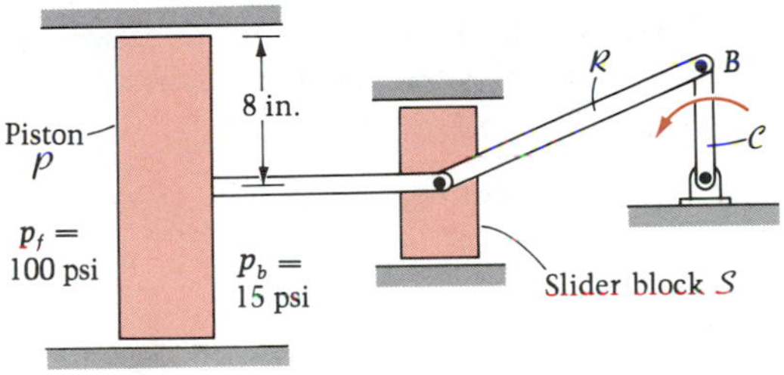

Figure P4.178

4.179 In Figure P4.179 determine the components of the pin force at C on member $\mathcal{A}$.

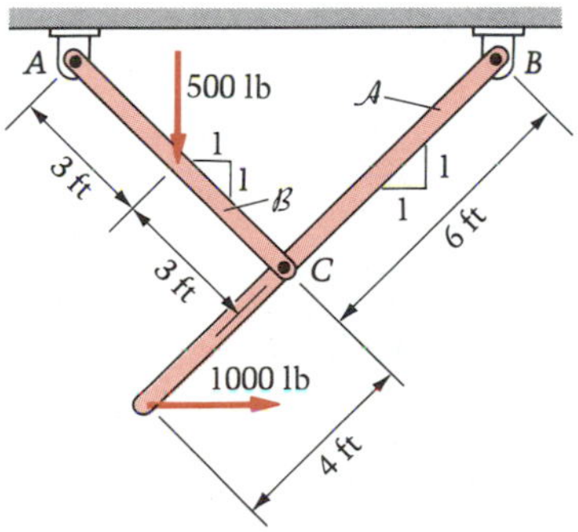

Figure P4.179

4.180 Find the reactions exerted on the bent bar AB at A and B. (See Figure P4.180.)

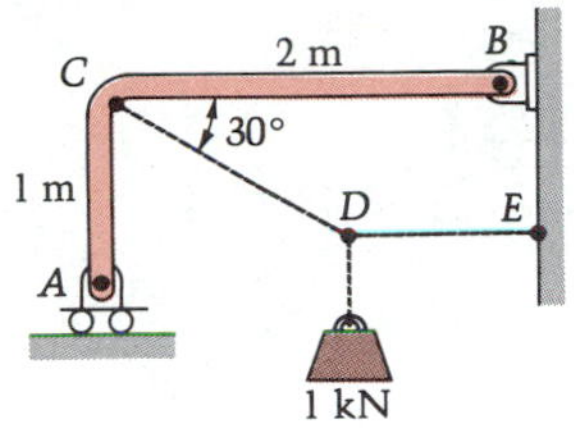

Figure P4.180

4.181 In Figure P4.181 what is the compressive force exerted by the nutcracker on the pecan? What is the force in the link AB?

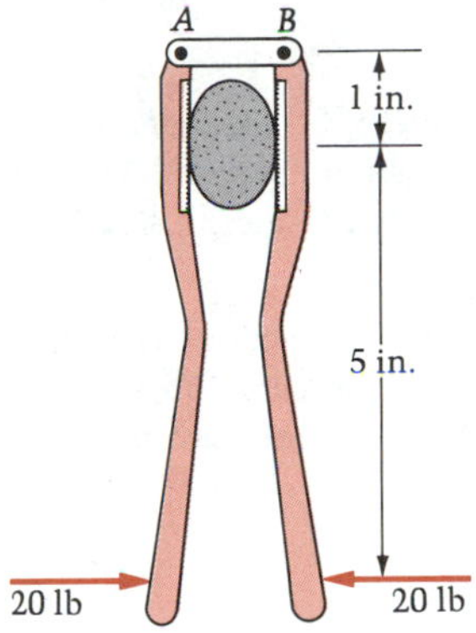

Figure P4.181

4.182 Find the force exerted by the pin at G onto member BEG. (See Figure P4.182.) Neglect the weights of the bars, but consider the weight of the 200-lb drum $\mathcal{D}$.

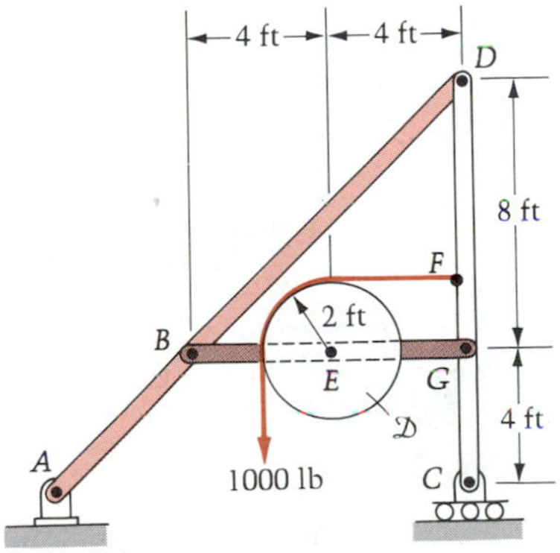

Figure P4.182

4.183 Find the force exerted by the pin at C onto member ABC. (See Figure P4.183.)

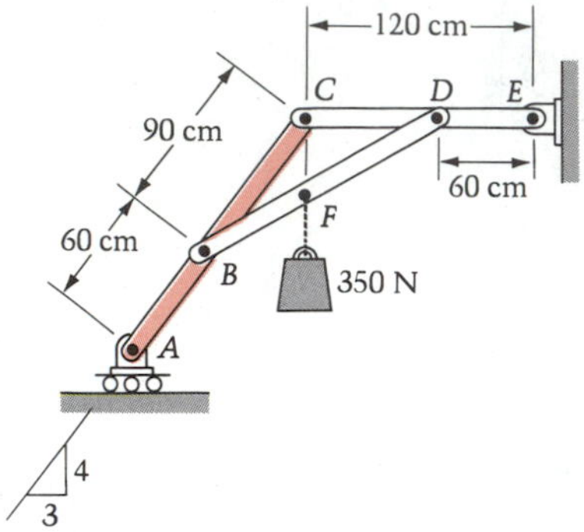

Figure P4.183

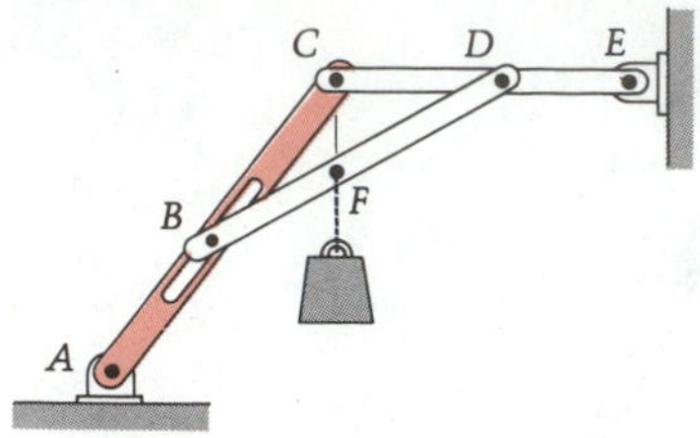

Figure P4.184

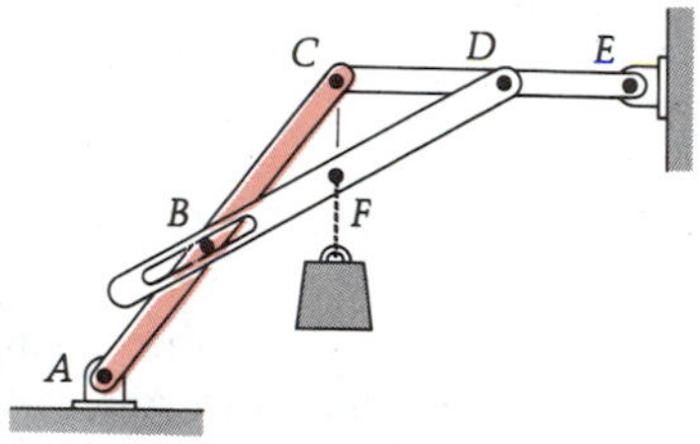

Figure P4.185

4.184 Repeat the preceding problem if the connections at A and B are changed to those shown in Figure P4.184. Pin B is attached to BD and bears against a smooth slot cut in ABC.

4.185 Repeat Problem 4.183 if the connections at A and B are changed to those shown in Figure P4.185. Pin B is attached to ABC and bears against a smooth slot cut in BD.

4.186 A worker in a "cherry picker" is installing cable TV equipment. (See Figure P4.186.) If the man plus bucket weigh 400 lb and the extendable member CE weighs 800 lb, find the force in the hydraulic cylinder BD and the pin reactions at C onto the extendable member CE. Neglect the weight of BD.

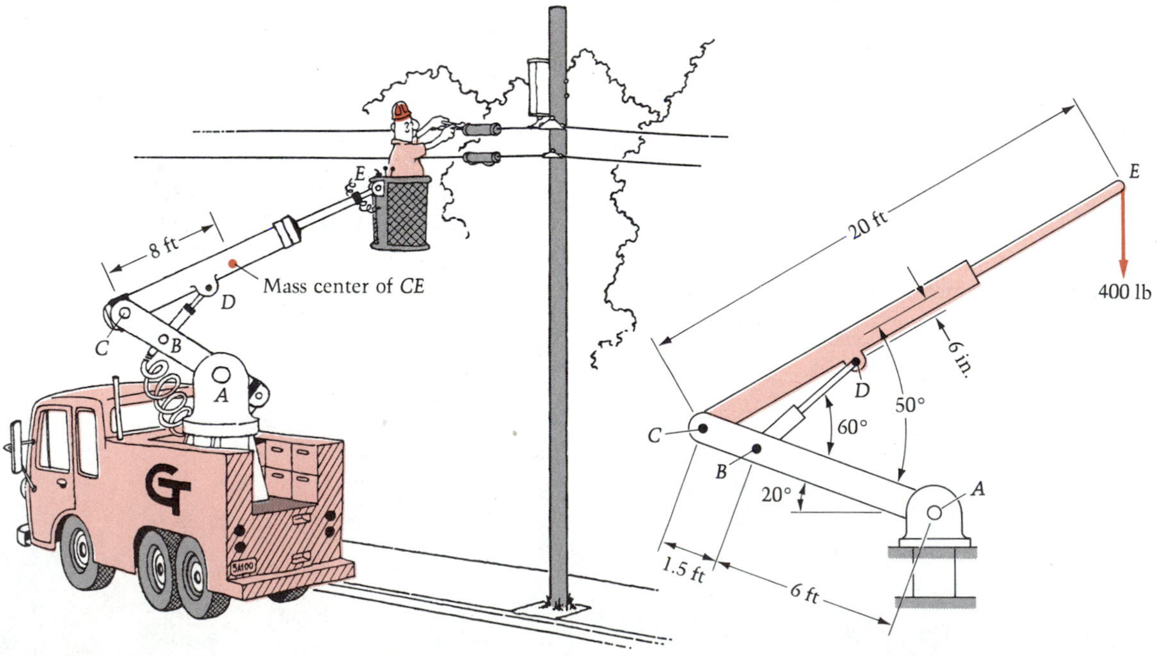

Figure P4.186

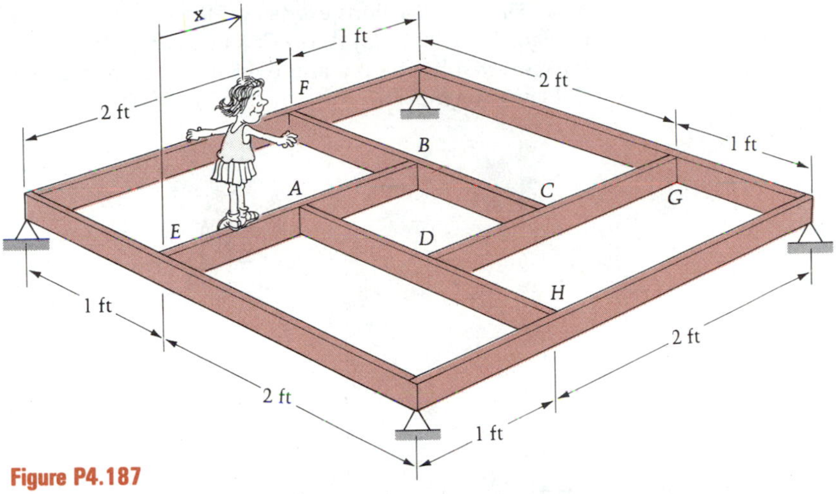

Figure P4.187

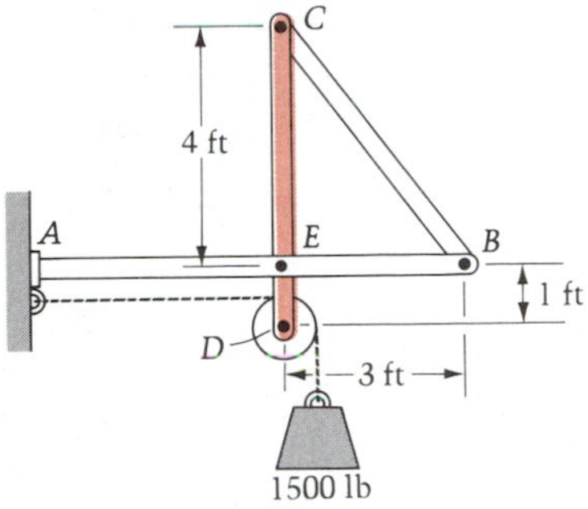

Figure P4.189

• **4.187** A carpenter has built a form for making concrete patio slabs. A little girl of weight W walks on the inside boards of the form as shown in Figure P4.187. Assuming no moments are exerted at the eight connections A, B, . . . , H, find the largest force ever exerted by any connection, and note that it is larger than the weight of the child. *Hint:* You need only consider all positions (use x) on board EAB, by symmetry.

4.188 A frame is loaded as shown in Figure P4.188. Find the forces exerted on member DG at D and at B.

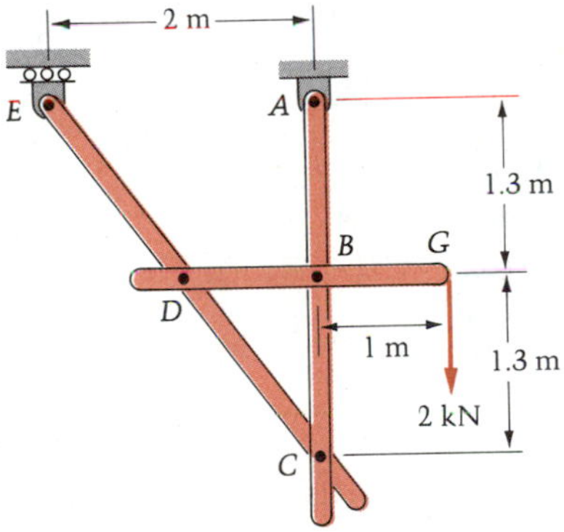

Figure P4.188

4.189 In Figure P4.189 find the force exerted on member AB at E.

4.190 In the preceding problem, find the reactions exerted on member AB at A if the distance AE is 6 ft.

4.191 Find: (a) the force that AB exerts on CD at B and (b) the force that the wall exerts on AB at A. (See Figure P4.191.)

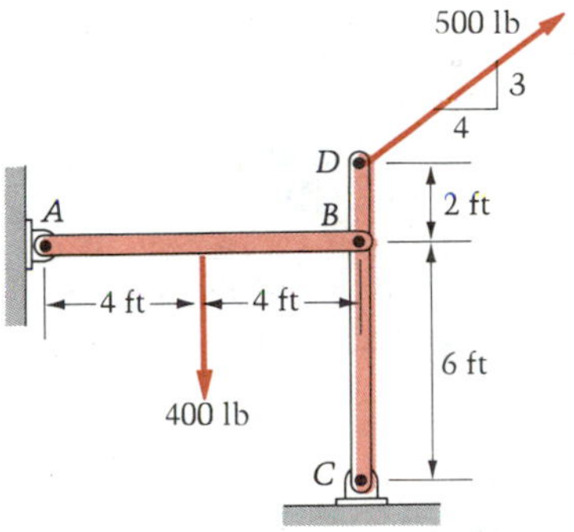

Figure P4.191

4.192 In Figure P4.192 find the vertical component of the reaction at C. Can the horizontal component be found by separating the members of the frame? If so, find it; if not, why not?

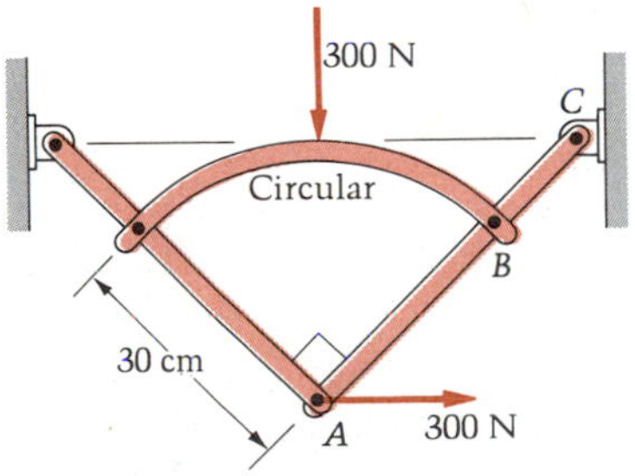

Figure P4.192

4.193 For the frame in Figure P4.193, find the forces exerted onto the member *ACE* at (a) *A*; (b) *C*; (c) *E*.

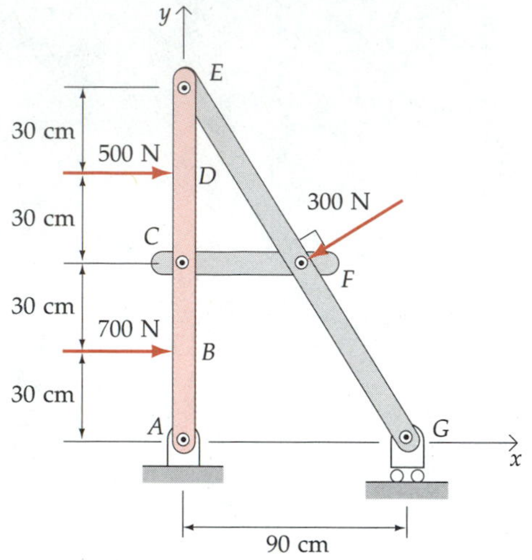

Figure P4.193

4.194 The three members *ABC*, *CDE*, and *BDF* are connected to form the frame in Figure P4.194. The cable over the pulley is fastened to the pin at *C*. The pin at *B* is fastened to member *ABC* and is free to slide in the horizontal slot in member *BDF*. Determine the force exerted on member *ABC* at *C*.

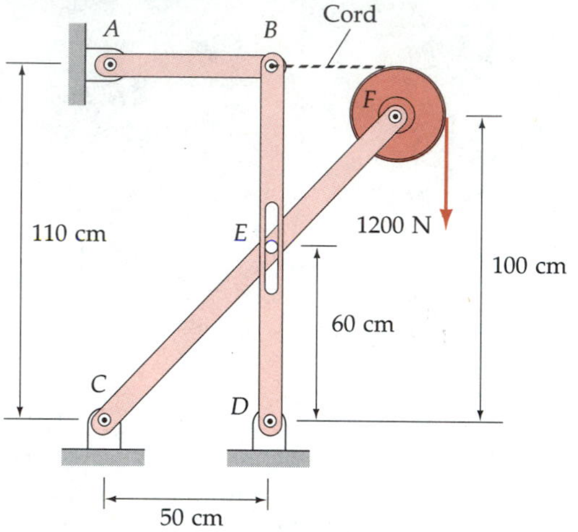

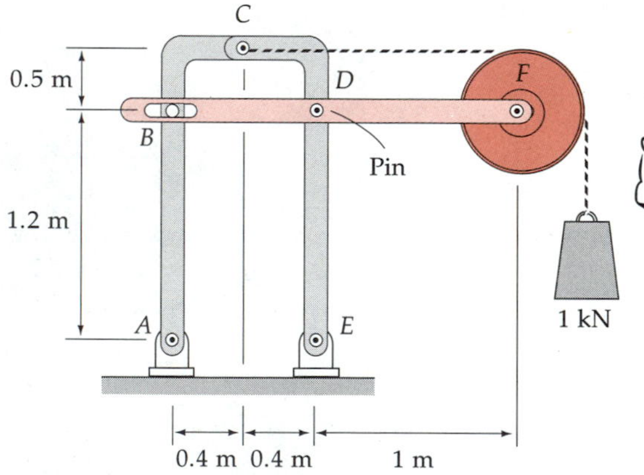

Figure P4.194

4.195 Find the reactions exerted on the structure shown in Figure P4.195 at (a) *A*; (b) *C*; (c) *D*. Also, (d) find the force exerted by the pin at *E* (which is fixed to member *CEF*) onto member *BD*.

Figure P4.195

4.196 A woman weighing 120 lbs has climbed 60% of the way up a lightweight folding ladder as shown in Figure P4.196. The ladder rests on a frictionless surface. Assume that the woman's weight acts through her body's center of gravity as shown in the figure. Find the force in the two identical symmetrical cross-braces (one on each side).

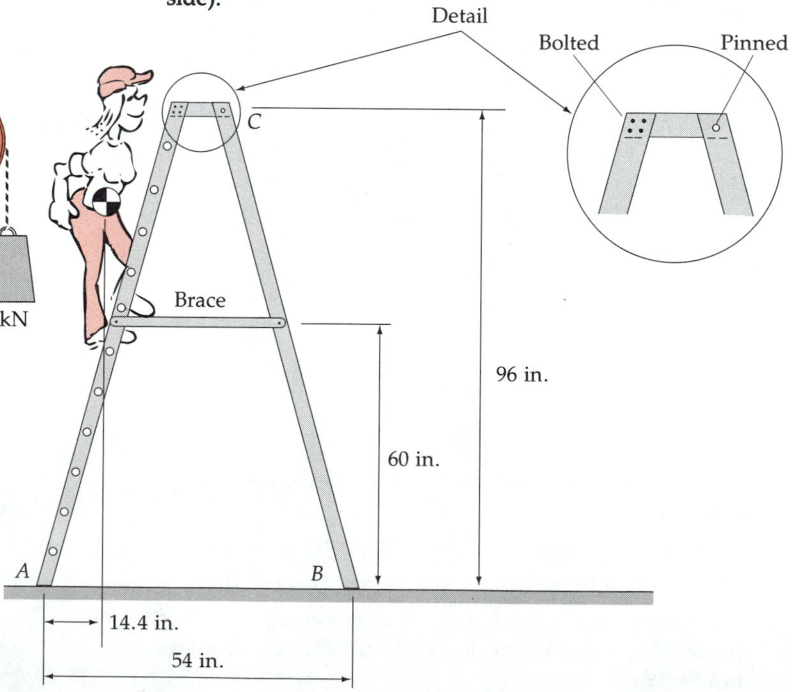

Figure P4.196

4.197 Two quarter-rings, each of mass m, are pinned smoothly together at Q and held in place by the two forces of magnitude P shown in Figure P4.197. The plane is smooth. Find the value of P for equilibrium.

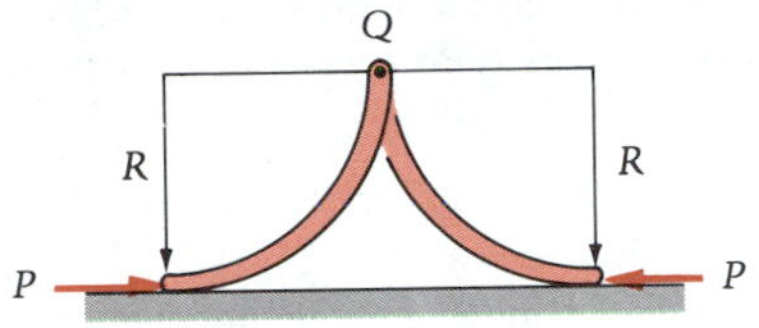

Figure P4.197

4.198 Two cylinders $\mathcal{A}$ and $\mathcal{B}$ are joined as shown by a stiff, light rod $\mathcal{R}$ and rest in equilibrium on two smooth planes (Figure P4.198). What is the angle between $\mathcal{R}$ and the horizontal?

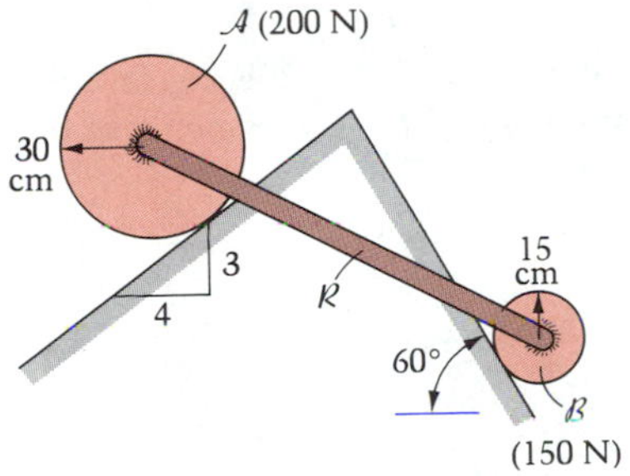

Figure P4.198

4.199 Repeat the preceding problem if the cylinders are turned around ($\mathcal{A}$ on the right, $\mathcal{B}$ on the left).

4.200 Show that if the dashed member $\mathcal{A}$ is pinned to $\mathcal{B}$ at A to form a simple pair of pliers [Figure P4.200(a)], the clamping force is $2P$. Then for the actual locking pliers comprising members $\mathcal{B}$, $\mathcal{C}$, $\mathcal{D}$, and the link $\mathcal{E}$ [Figure P4.200(b)], find and compare the new clamping force to the simple-pliers answer.

4.201 In the preceding problem, let the length of $\mathcal{E}$ be variable, with its upper end allowed to pin to $\mathcal{B}$ anywhere from directly above its fixed-position lower end to the point of application of P. find the largest possible clamping force.

4.202 Find the compressive clamping force on the object at C in Figure P4.202. (Note that the pin at B joins $\mathcal{B}$ and $\mathcal{D}$, but not $\mathcal{C}$!) Then rework the (simpler) problem if member $\mathcal{A}$ is pinned directly to $\mathcal{B}$ as shown in the lower figure. Compare the mechanical advantage of these simple snips with the compound snips.

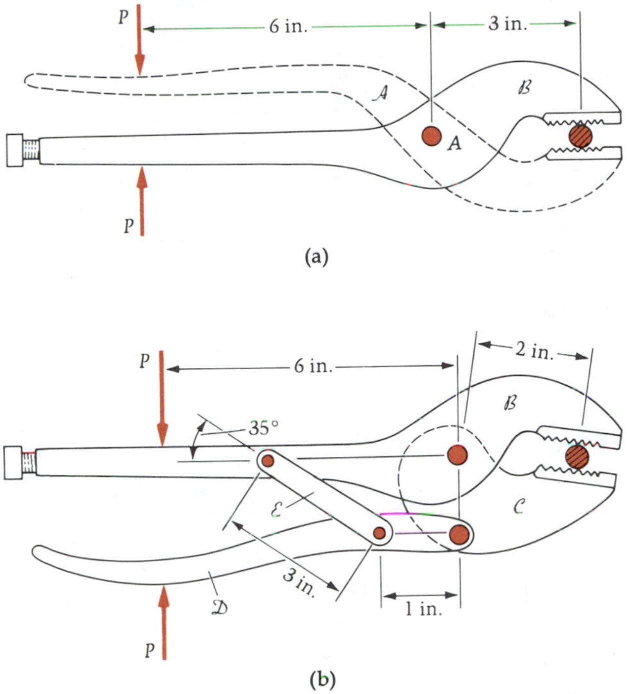

Figure P4.200

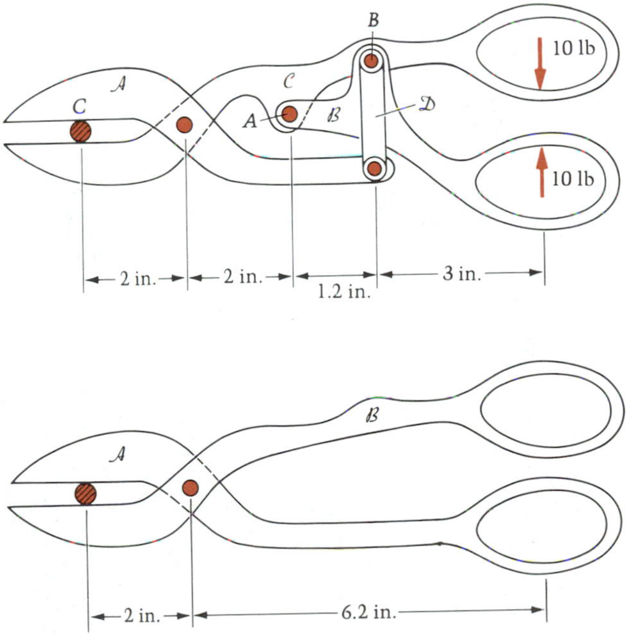

Figure P4.202

4.203 For the rivet squeezer shown in Figure P4.203, find the ratio of the clamping force at the point A to the gripping force H.

4.204 Repeat the preceding problem if the handle is member BCE instead of BCD. (Refer to Figure P4.204.) For the same dimensions, does this give a greater or lesser clamping force?

4.205 The toggle device in Figure P4.205 is being used to crush rocks. If the pressure in the chamber, p, is 70 psi and the radius of the piston is 6 in., find the force that the rock crusher exerts upon the rock. Members AB, BC, and BD are pinned at their ends.

4.206 For the frame shown in Figure P4.206, find the magnitude of the force exerted on each of the connecting pins at B, C, and D.

4.207 The uniform slender bars in Figure P4.207 are identical and each weighs 20 N. Find the angles α and β for equilibrium.

4.208 In the frame shown in Figure P4.208, (a) find the reactions at A and E and (b) find the components of the forces at B and C on member ABC.

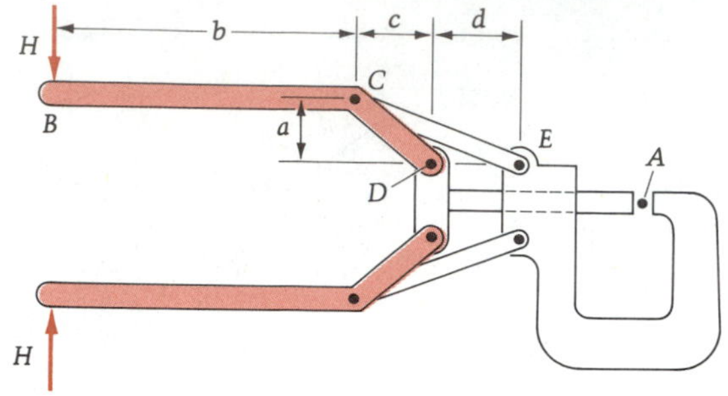

Figure P4.203

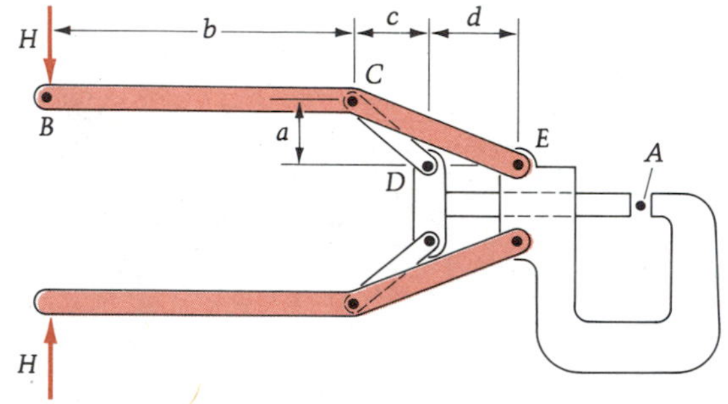

Figure P4.204

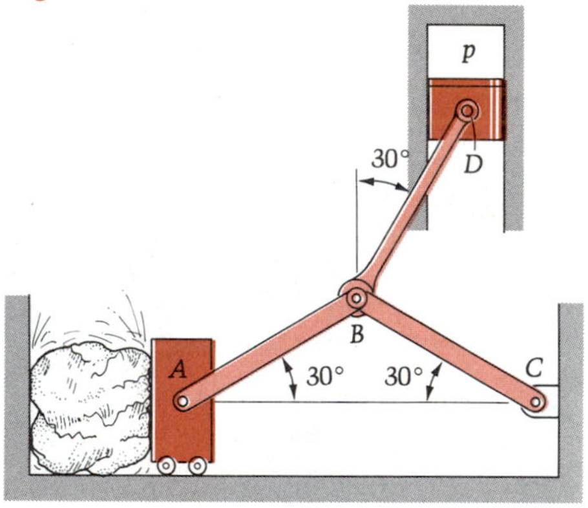

Figure P4.205

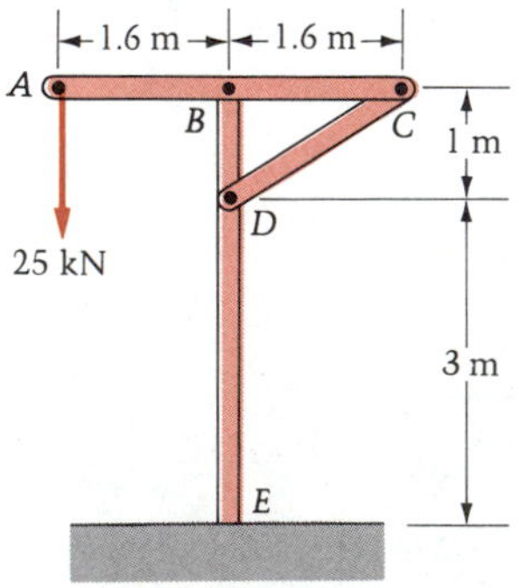

Figure P4.206

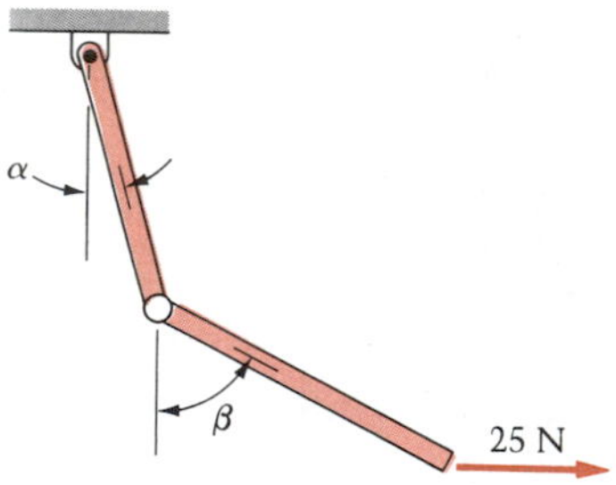

Figure P4.207

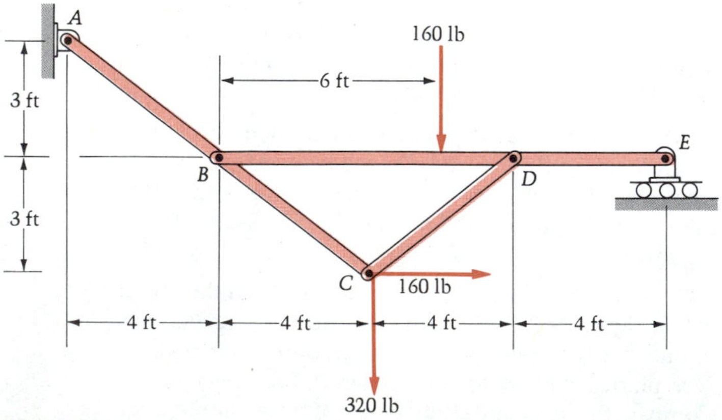

Figure P4.208

4.209 Find the reactions beneath each wheel, assuming symmetry, and the reaction on the ball (attached to the truck) of the ball-and-socket joint. (See Figure P4.209.)

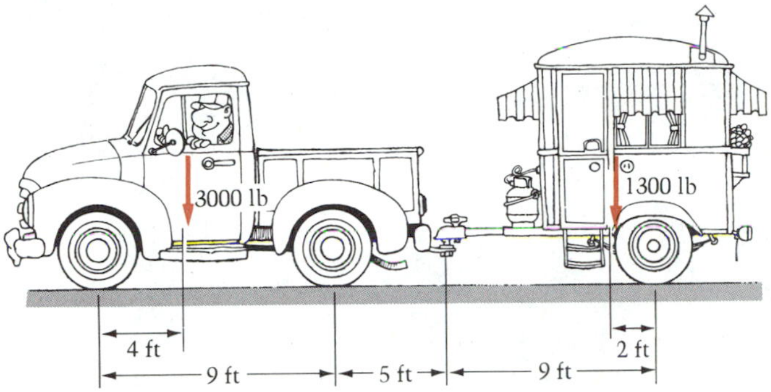

Figure P4.209

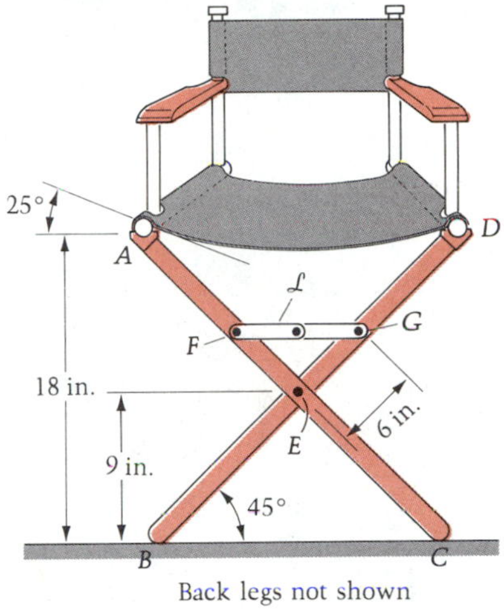

Figure P4.211

• **4.210** In the preceding problem the answers for the forces under the front tires are 753 lb (each of two), and under the back tires, 891 lb (each of two). A "load-leveler," or "equalizer," hitch can be used to more evenly distribute these forces. Each of two angle bars $\mathscr{A}$ fits and "bottoms" into a socket on the trailer side (see Figure P4.210), and its chain is pretensioned to 400 lb. Assume the ball and socket to be in the same position relative to truck and trailer as in the preceding problem. Find the distances a and b in the figure (which add to 2 ft) for which the reactions of the road on the four truck tires will be equal. For these values, show that the force between ball and socket (which was previously 289 lb) is greatly increased whereas the reactions of the road on the trailer tires (which were 506 lb each) are slightly increased.

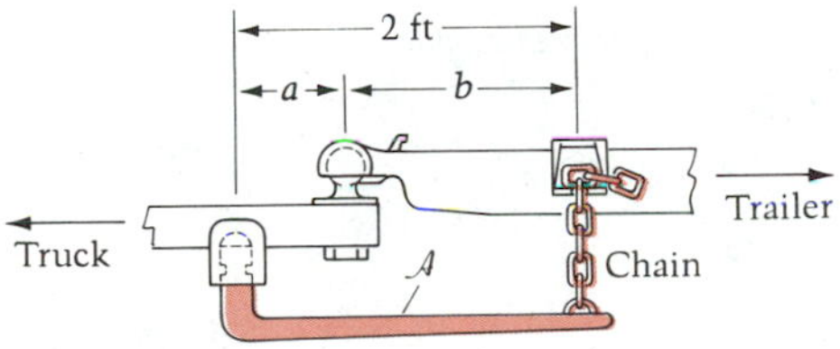

Figure P4.210

4.211 Shown in Figure P4.211 is a sketch of a director's chair. If the director is well-fed at 260 lb and if he sits with each of 4 legs supporting 20% of his weight, find:

a. The force exerted by the floor onto a leg, neglecting friction there

b. The force in the link member $\mathscr{L}$

c. The force exerted by the pin at E onto member BED.

4.212 In Figure P4.212 the sleeve is pinned to bar $\mathscr{B}$ and can slide smoothly on the rod $\mathscr{R}$. Find the force in the cord if the system is in equilibrium. Then repeat the problem if the sleeve is pinned to $\mathscr{R}$ and free to slide on $\mathscr{B}$.

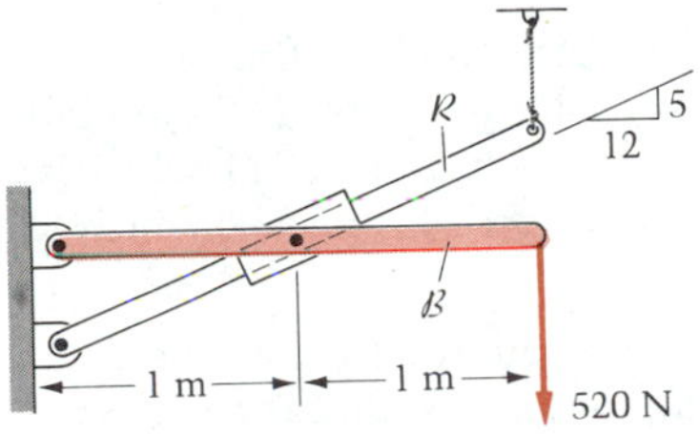

Figure P4.212

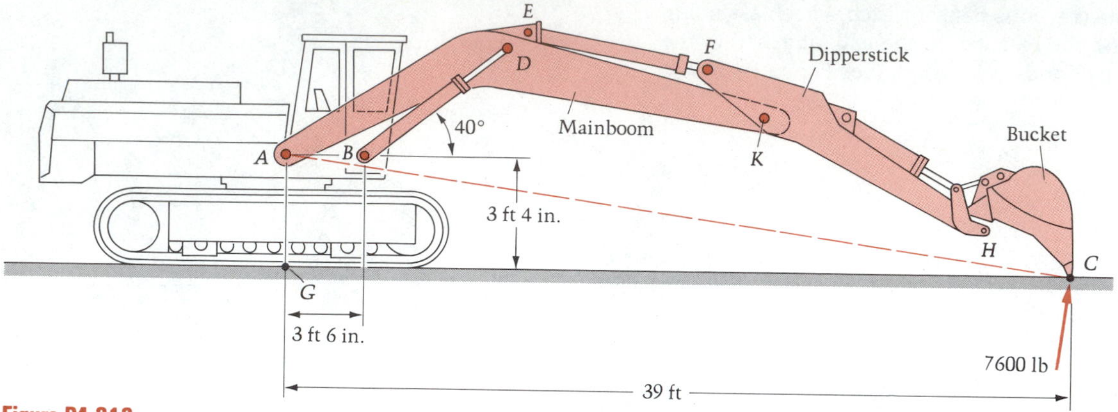

Figure P4.213

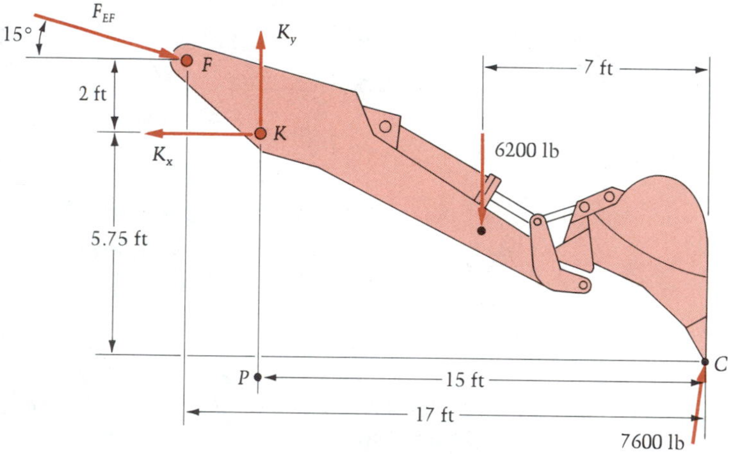

Figure P4.214

4.213 In Figure P4.213 the excavator is beginning to remove a section of earth. If the force from the ground is 7600 lb, approximately normal to the line AC as shown, find the force in the hydraulic cylinder strut BD, and the pin reactions onto the mainboom at A. The weight of the mainboom, dipperstick, bucket, and lift cylinders is 15,000 lb; assume the horizontal mass center location of the weight to be halfway between G and C.

4.214 In the preceding problem, find the force F_{EF} in the hydraulic cylinder between the mainboom (ADK) and the dipperstick (FH). *Hint:* Consider Figure P4.214, which is a free-body diagram of dipperstick plus bucket, which together weigh 6200 lb.

4.215 In Figure P4.215 compute the force in the cross-member $\mathcal{L}$ of the lifting tongs. Also find the horizontal component of the force at A acting on the 200-N block.

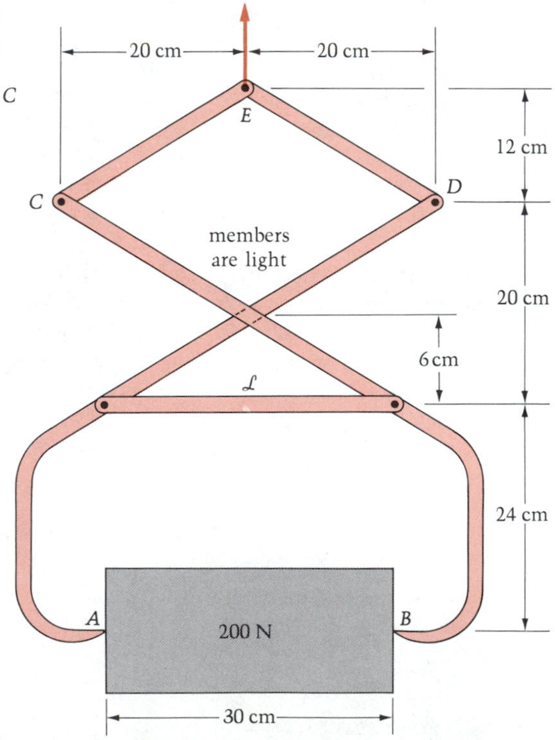

Figure P4.215

4.216 In the preceding exercise, let the link $\mathcal{L}$ be removed and let the members DA and CB be pinned where they cross. Find (a) the force exerted by this pin on DA, and (b) the horizontal component of the force at A acting on the block.

4.217 The utility loader (Figure P4.217) has raised 1600 lb of earth and rocks to its highest possible position. Exclusive of the bucket, each of two loader arms weighs 150 lb (including the dumping strut JH) with mass center at C, while the bucket weighs 280 lb with the mass center of it plus its contents at E. Determine the force in one of the hydraulic lifting struts AB, and the pin reactions at D, in the given position.

4.218 In the preceding problem, find the force in the hydraulic dumping strut JH and the reactions at pin P onto the bucket. (Refer to Figure P4.218.) Recall that the bucket plus contents weighs 1880 lb and that there are two of each of the struts in the figure.

4.219 A counterclockwise couple of 1000 lb-in. is applied to disk $\mathcal{G}$, which is free to turn about a pin at O', as shown in Figure P4.219. Pin P, attached to $\mathcal{G}$, bears against the slot in body $\mathcal{A}$. If the mass center of $\mathcal{A}$ is at C and the system is in equilibrium, find the weight of $\mathcal{A}$.

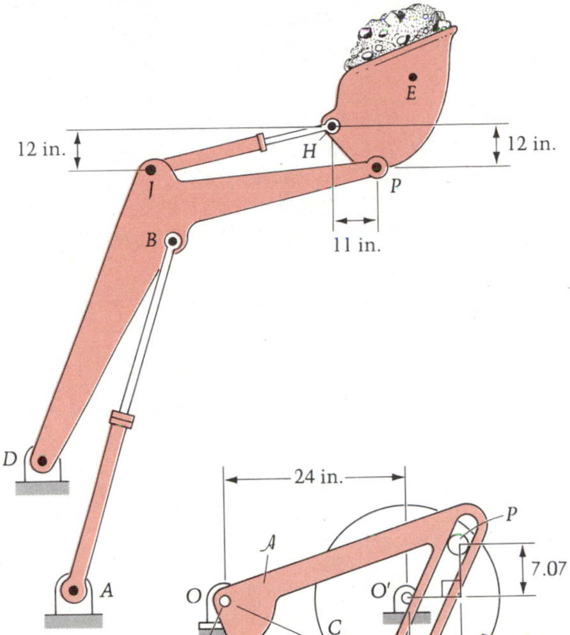

Figure P4.218

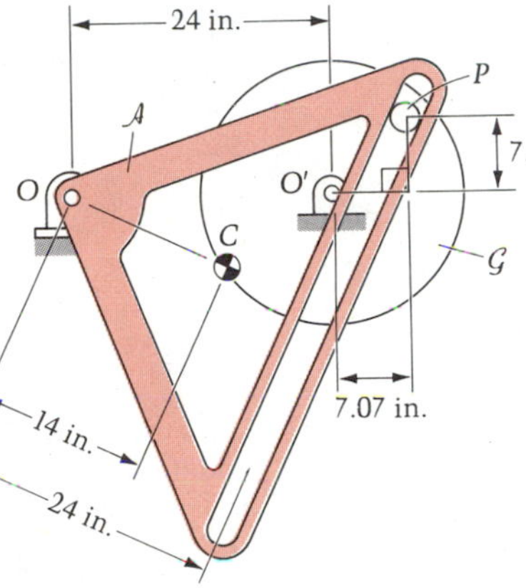

Figure P4.219

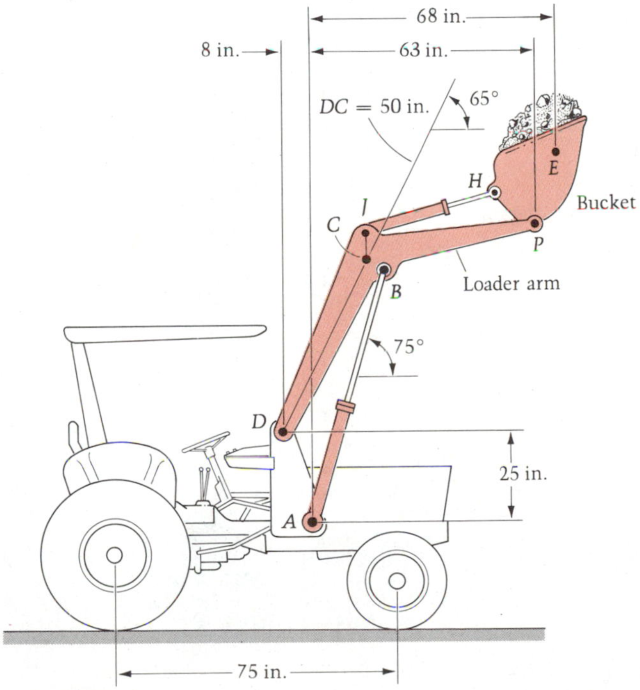

Figure P4.217

4.220 In Figure P4.220, AB, AC, and BC are light, slender bars, joined at their ends and supported by a hinge at A. Find (a) the angle θ for equilibrium and (b) the force in each bar.

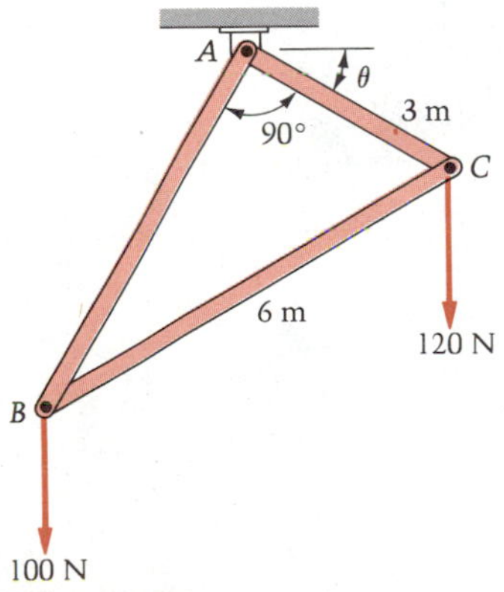

Figure P4.220

4.221 In Figure P4.221 find the reactions of (a) roller A onto bar $\mathcal{B}_1$ and (b) roller B onto bar $\mathcal{B}_2$.

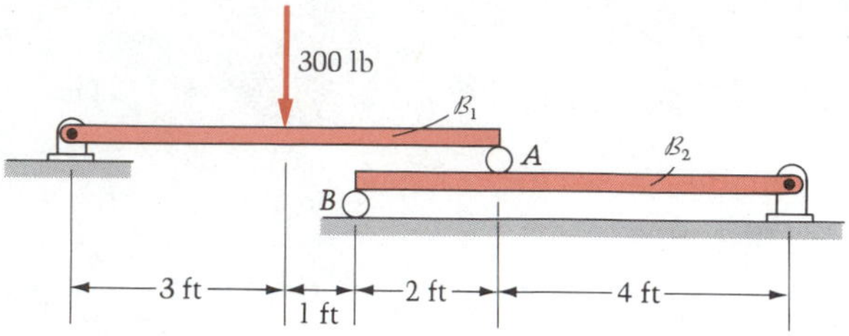

Figure P4.221

4.222 Find the compressive force in the spreader bar shown in Figure P4.222.

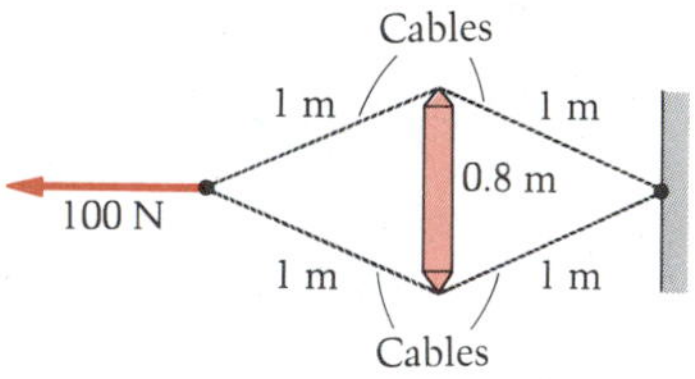

Figure P4.222

4.223 The shaded structure in Figure P4.223 holds up body $\mathcal{B}$, which weighs $2w$ with half the weight supported by each roller. Each half of the structure weighs W; the centers of gravity are at G_1 and G_2. Find the force exerted by the ground at E, and the magnitude of the roller reaction at C, in terms of b, L, e, d, w and W.

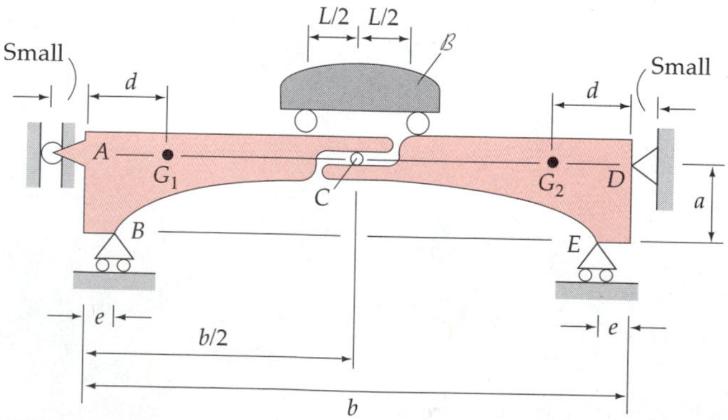

Figure P4.223

4.224 When running, the clothes dryer drum $\mathcal{D}$ is turned by means of a belt that passes around a motor pulley $\mathcal{M}$ as indicated in Figure P4.224. The belt also passes under an idler pulley $\mathcal{I}$ that is pinned to the bracket $\mathcal{B}$. The bracket is supported by the floor $\mathcal{F}$ of the dryer, which bears against extensions of $\mathcal{B}$ that fit through slots in $\mathcal{F}$. If the force at A is vertical, and if that at B has both x and y components, find these reaction forces when the dryer is turned off, if the belt tension then is 1 lb. Neglect the weights of $\mathcal{B}$ and $\mathcal{I}$, and note that this means you are actually finding the *differences* between the reactions with and without the belt.

4.225 The suspension shown in Figure P4.225 supports one-half of the front of a car. Find the force in the spring and the force exerted on the frame by the members AB and CD. The wheel and its brakes and support weigh 100 lb with mass center at G.

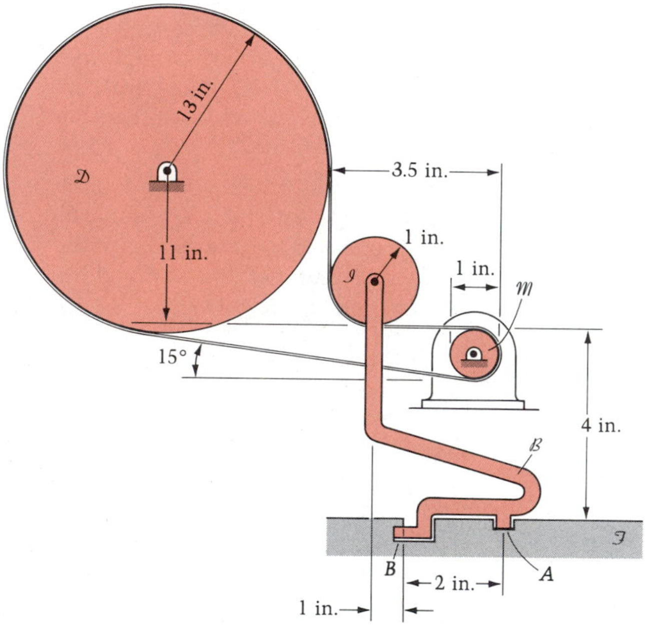

Figure P4.224

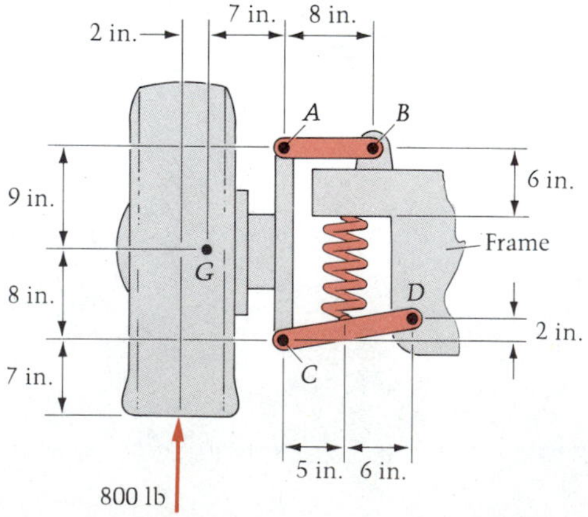

Figure P4.225

4.226 In Figure P4.226 winch 1 is used to raise and lower the boom of the derrick. After the desired angle θ is reached, winch 2 is then used to raise and lower the load. If $\theta = 50°$, find the forces in cables 1 and 2, and the compressive force in the boom.

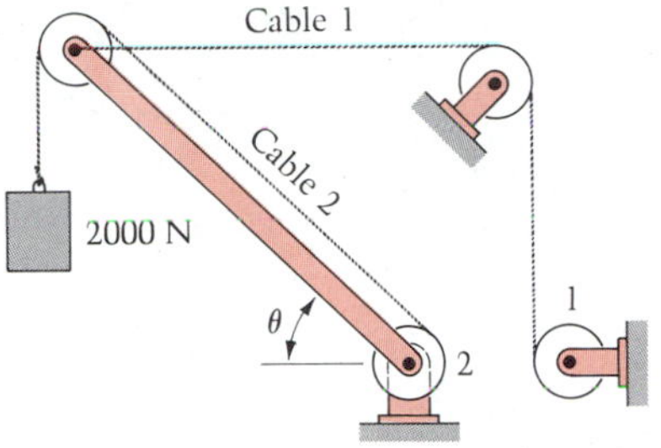

Figure P4.226

4.227 In Figure P4.227 find the force P that will hold the system of two 200-N cylinders and two light bars in equilibrium. (The other bar is behind the one shown.)

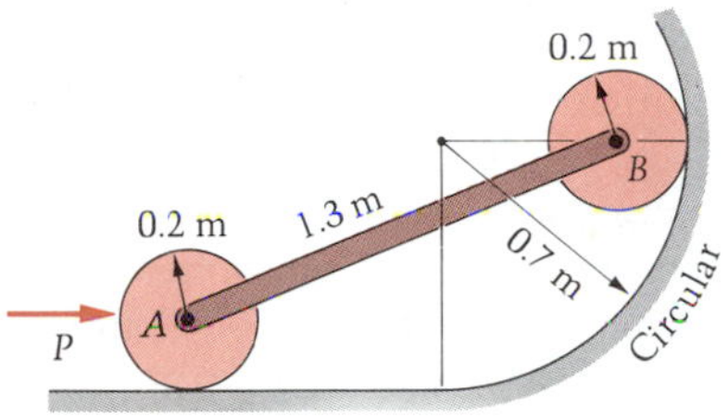

Figure P4.227

4.228 Determine the components of the pin forces onto the bars at A, B, and C. (See Figure P4.228.)

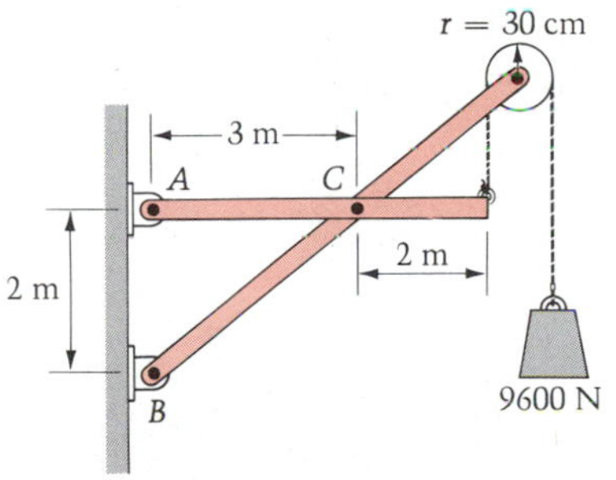

Figure P4.228

4.229 In Figure P4.229 find the force F needed for equilibrium of the system (called "Roberval's Balance"), and show that the value of F is independent of its position (i.e., doesn't depend on x). Neglect the weights of the members.

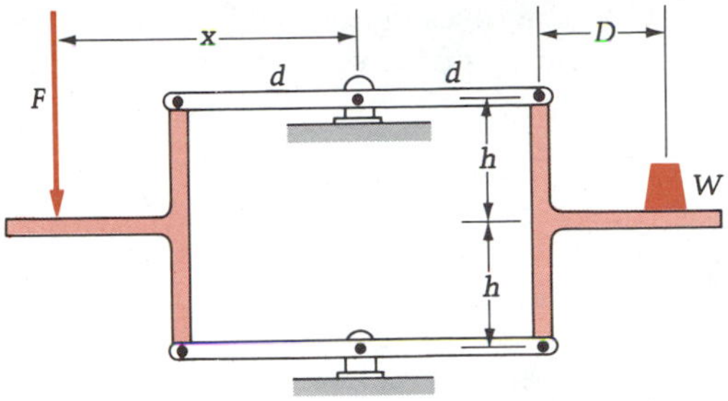

Figure P4.229

4.230 The linear spring exerts a force at each of its ends that is proportional to the amount of stretch it undergoes. In Figure P4.230, the spring modulus (proportionality constant) is 2 N/cm and its natural (unstretched) length is 1.5 m. Find the normal and friction forces (a) between cylinders $\mathcal{A}$ and $\mathcal{B}$ and (b) between $\mathcal{B}$ and the ground, if the weight of $\mathcal{A}$ is 500 N and that of each of $\mathcal{B}$ and $\mathcal{C}$ is 200 N.

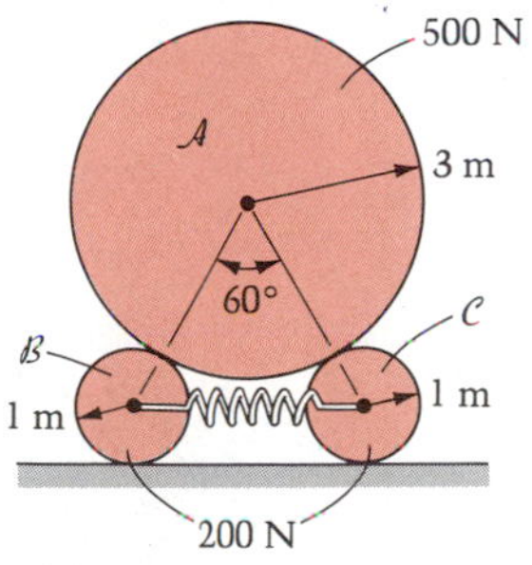

Figure P4.230

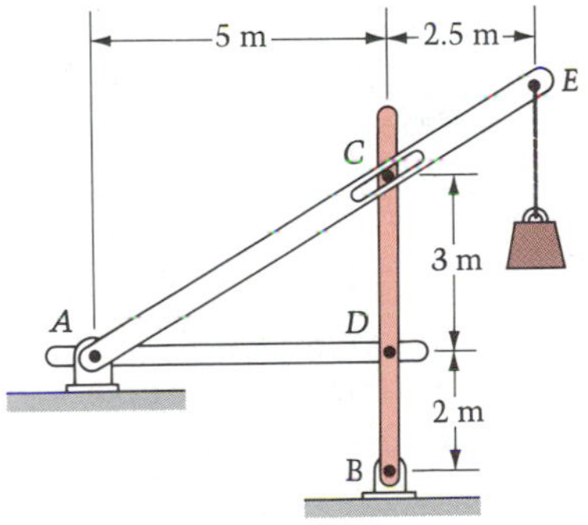

Figure P4.231

4.231 The weight (500 N) is suspended from point E by a cord. Find the force exerted at point D onto the shaded member BDC by the pin. (See Figure P4.231.)

4.232 Repeat the preceding problem if the slot is cut in BDC and the pin that slides in this slot is fixed to ACE.

4.233 Find the torque (twisting moment) carried by the sections $\mathcal{S}_1$, $\mathcal{S}_2$, and $\mathcal{S}_3$ of the stepped shaft shown in Figure P4.233.

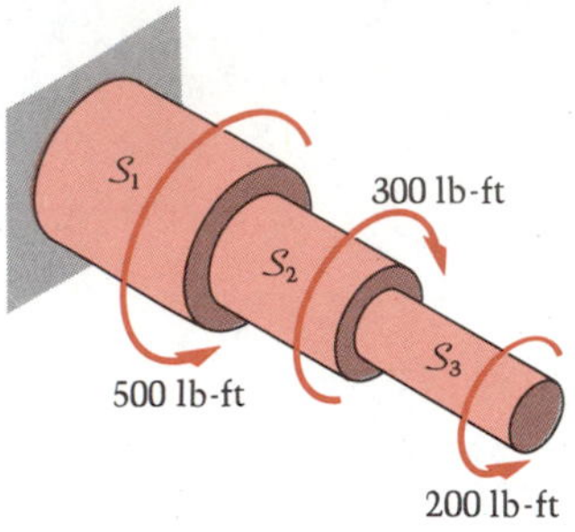

Figure P4.233

4.234 A cylinder weighing 2000 N is symmetrically lodged between two pairs of cross pieces of negligible weight. (See Figure P4.234.) Find the tension in the rope *AB*. (*AD* and *BC* are each continuous bars.)

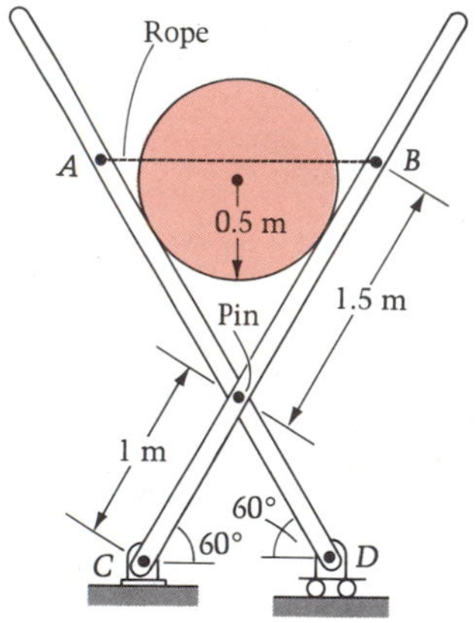

Figure P4.234

4.235 The 100-lb bar in Figure P4.235 rests in equilibrium against the 200-lb cubical block. The contact is smooth (frictionless) between the two bodies. Find the reaction of the plane onto the block.

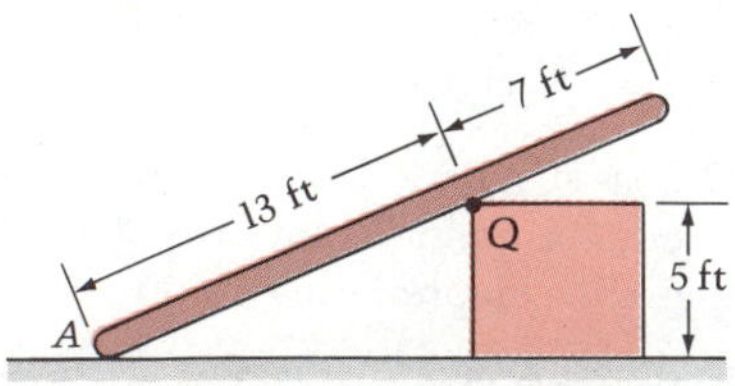

Figure P4.235

4.236 The two rods $\mathcal{B}_1$ and $\mathcal{B}_2$ are pinned as shown, and $\mathcal{B}_2$ is fit (with friction) through a sleeve in body $\mathcal{B}_3$, which is pinned smoothly to the ground. The 12 N · cm couple is applied to $\mathcal{B}_1$ as shown in Figure P4.236. Find the resultant of the force system exerted by $\mathcal{B}_3$ on $\mathcal{B}_2$, expressed at Q as a normal force, friction force, and a couple.

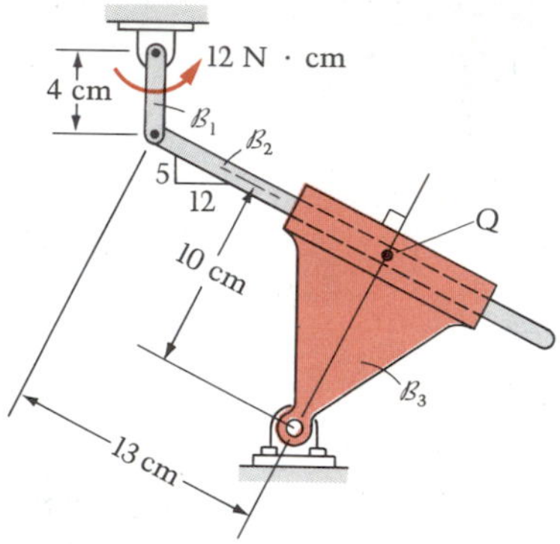

Figure P4.236

4.237 In Figure P4.237, the positioner supports a large paraboloidal antenna that is not shown. The antenna exerts the forces and couple (caused by wind and weight) shown at Q onto the positioner. [For information's sake, the antenna is "positioned" by (a) turning about the vertical around the azimuth bearing and (b) turning around a horizontal (elevation) axis normal to the page through O

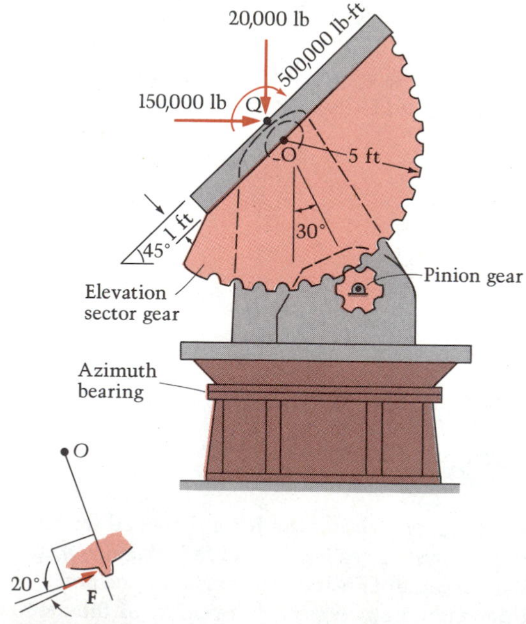

Figure P4.237

about *two* elevation bearings. This is called an elevation over azimuth positioner. Sometimes there is yet another azimuth rotation (for polarization) at the top.] If the tooth force **F** from the pinion onto one of *two* elevation sector gears is as shown, find the magnitude of **F**. Neglect the weight of the elevation assembly, which is in equilibrium, and assume just one tooth on each side is in contact.

4.238 The slender rod *AC* in Figure P4.238 is pinned to the small block at *C*. Friction prevents the block from sliding within the slotted body $\mathcal{B}$. Find the reactions onto the bar at *A* and onto $\mathcal{B}$ at *B*.

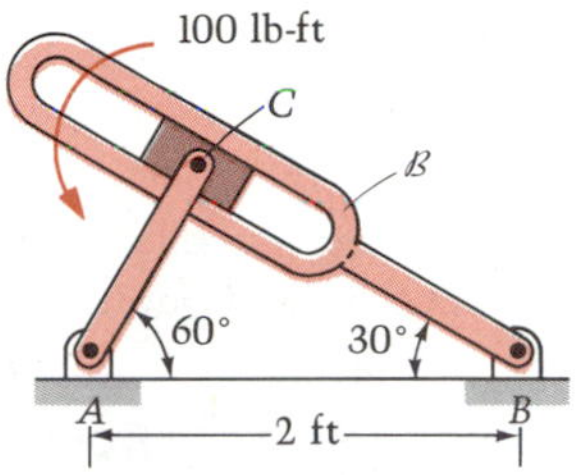

Figure P4.238

4.239 Repeat the preceding problem if the 60° angle is changed to 90°.

4.240 Figure P4.240 illustrates a jib crane. Its beam weighs 600 lb and is 10 ft in length. The weight of the suspended object is 400 lb. Plot the tension in the upper cable as a function of distance *d*, and find the pin reaction at *B* when $d = 6$ ft.

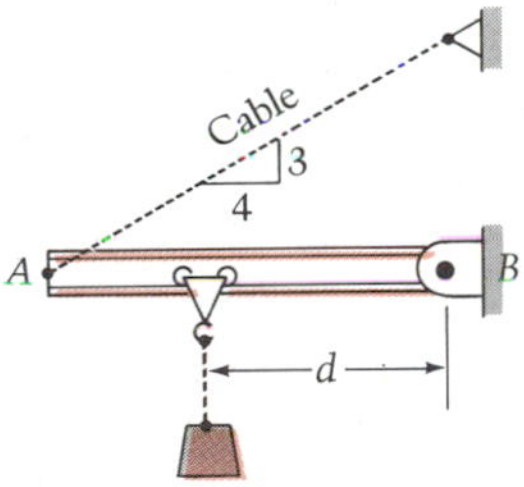

Figure P4.240

* **4.241** Four identical light rods are pinned together to form a square as shown in Figure P4.241, and supported by the four smooth platforms at the corners *A*, *B*, *C*, and *D*. A smooth sphere of radius *R* ($2R > a$) is then placed on the square. Show that the horizontal reaction between two adjacent rods has magnitude $Wa/(8\sqrt{2}\,h)$, where *h* is the height of the center of the sphere above the plane of the square, and *W* is the weight of the sphere.

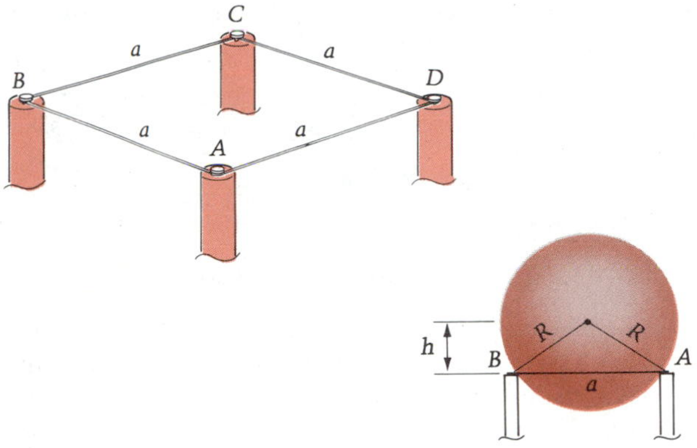

Figure P4.241

4.242 The two identical sticks shown in Figure P4.242 are pinned together at *A* and placed as shown onto the smooth block, the width of which is $\ell/2$. What is the angle ϕ for equilibrium?

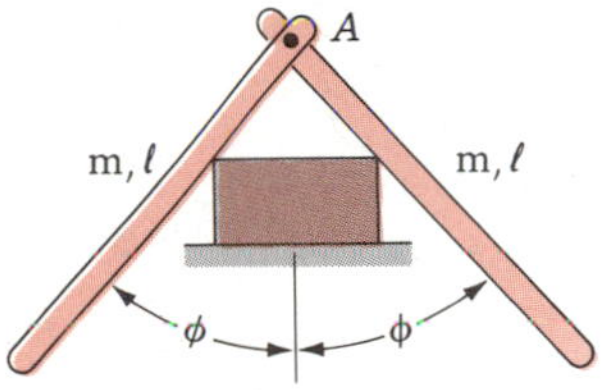

Figure P4.242

* **4.243** What is the maximum overhang for each identical slab shown in Figure P4.243 so that they are in equilibrium? There can be any number of slabs, each of length *b*. *Hint*: Start at the top instead of the bottom.

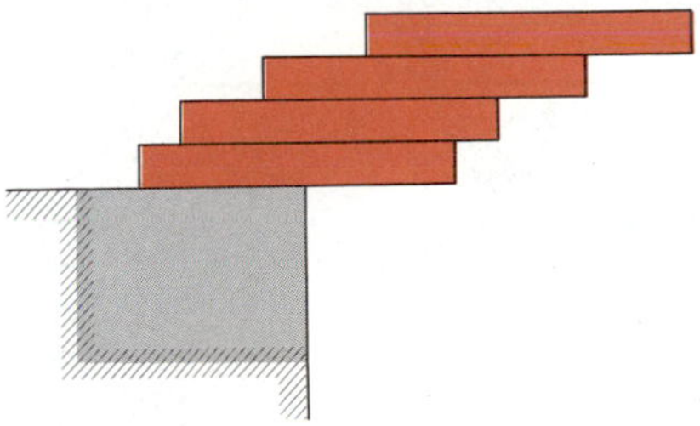

Figure P4.243

4.244 On a ten-speed bicycle, measure the following lengths (shown in Figure P4.244):

R_1, R_2 = large gear radii (measure to the middle of the teeth);

r_1, r_2, r_3, r_4, r_5 = small gear radii;

R = wheel radius;

R_f = radius to the middle of the pedal.

Let the bike and rider be traveling at constant speed, so that the force f is balanced by air resistance and we may consider the problem to be one of statics. Let F_f be a foot force, taken here to be constant for all gear combinations, with the pedal in the same position (shown) for each. Show, using appropriate free-body diagrams, that:

1. $TR_i = F_f R_f$ (where R_i is R_1 or R_2, depending on the gear being used);

and

2. $Tr_j = fR$ (where r_j is r_1, r_2, r_3, r_4 or r_5, depending on the gear being used);

so that

3. $$\frac{F_f R_f}{R_i} = \frac{fR}{r_j}$$

or

$$f = \text{friction force that moves the bike} = \frac{F_f R_f r_j}{R R_i}$$

Thus the driving force f is largest when r_j / R_i is largest, i.e., the "easiest" gear ratio is r_5 / R_1. Make a table ordering the ratios from the easiest to the "hardest," r_1 / R_2. Then re-do the chart by using ratios of numbers of teeth and compare. Ideally, the answers should be the same. Why should they?

4.245 In Figure P4.18, write one moment equation and find the force in the two-force member BC, without using the result of Problem 4.18.

4.246 Find the force exerted by the smooth pin at B onto bar ABC in Figure P4.246. Then find the force exerted by the smooth roller at E onto bar DBE.

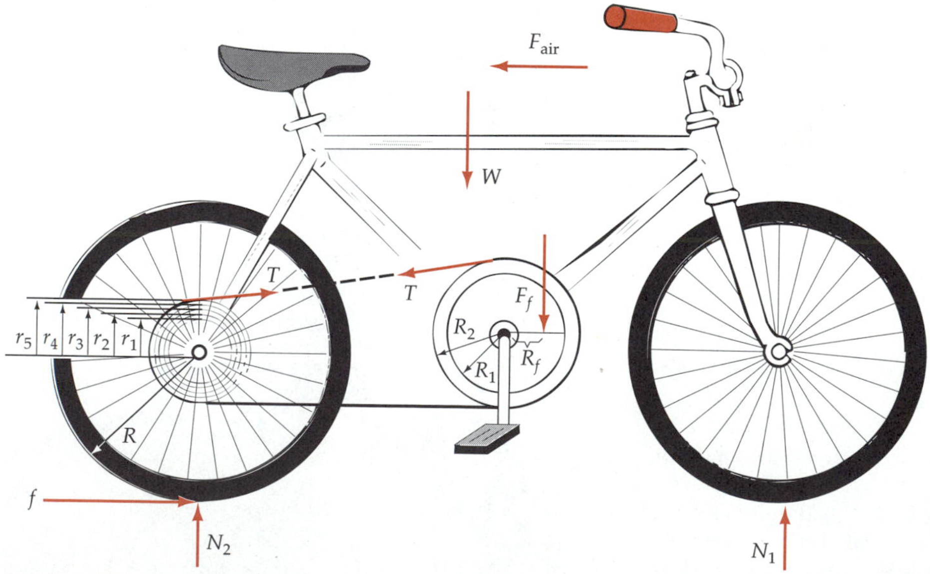

Figure P4.244

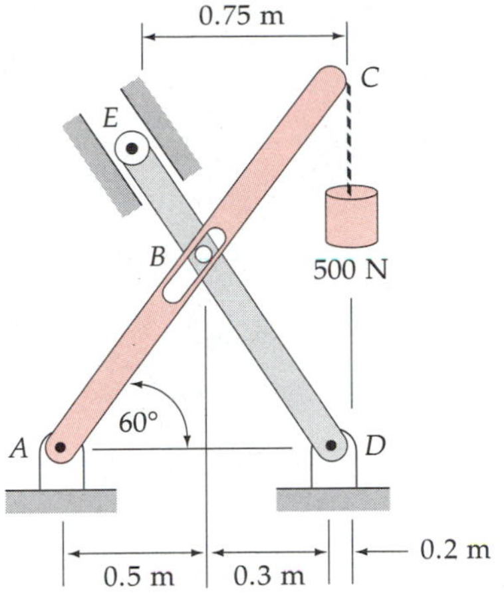

Figure P4.246

4.247 Find the force in the 2-force member of the frame in Figure P4.247.

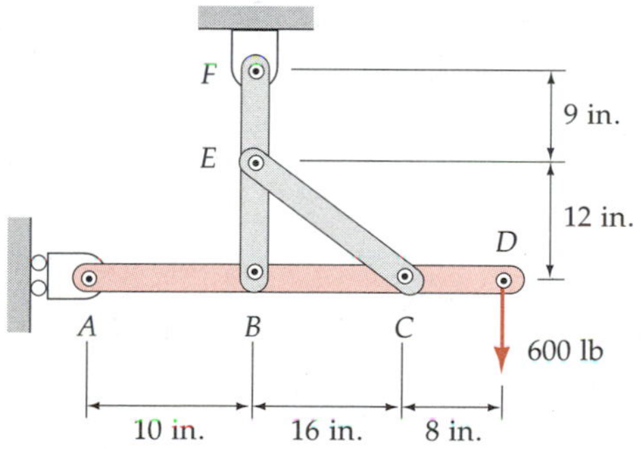

Figure P4.247

4.248 In Example 4.18, find the tension in the rope and the reactions exerted by the ground onto the structure at R. Then enter values for B_x, B_y, R_x, R_y, and T onto the overall FBD (Figure E4.18b) in the example, and verify its equilibrium by summing forces in the x and y directions, and moments about any point you choose.

4.249 Show that if in Examples 4.6 and 4.16 the roller at C is removed and the end C of bar BEC is instead pinned to the ground, the reactions onto ADB at B cannot be found. Show this by separating the members of the frame, writing their equilibrium equations, and attempting to solve them. The reason no solution is forthcoming from the equations of statics alone is that the structure contains more members than are needed for it to be stable. Note that DE *was* needed when there was a roller at C, but not now.

4.250 In the preceding problem, remove member DE and solve for the reactions B_x and B_y exerted by the pin at B onto member ADB. Do this by making use of the fact that AB and CB are now two-force members. Note the large differences in the x-components of the reactions when compared to the Examples 4.6 and 4.16.

4.251 In Example 4.19, find the forces exerted on members ABC and EDC at point C, assuming again that the 250 lb force is applied to the *pin* at C. Then draw the FBD of the pin at C to explain why the x-components of the forces found above are necessarily *not* equal and opposite.

4.252 The three light bars AB, CD, and EF of Figure P4.252 are connected together and to the ground by ball-and-socket joints. The 10-kN and 20-kN loads are applied parallel to the y- and x-axes, respectively. Find the force exerted onto member AB at C by member CD.

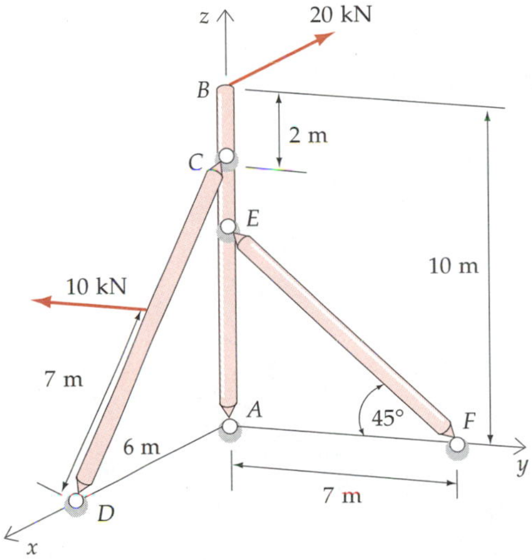

Figure P4.252

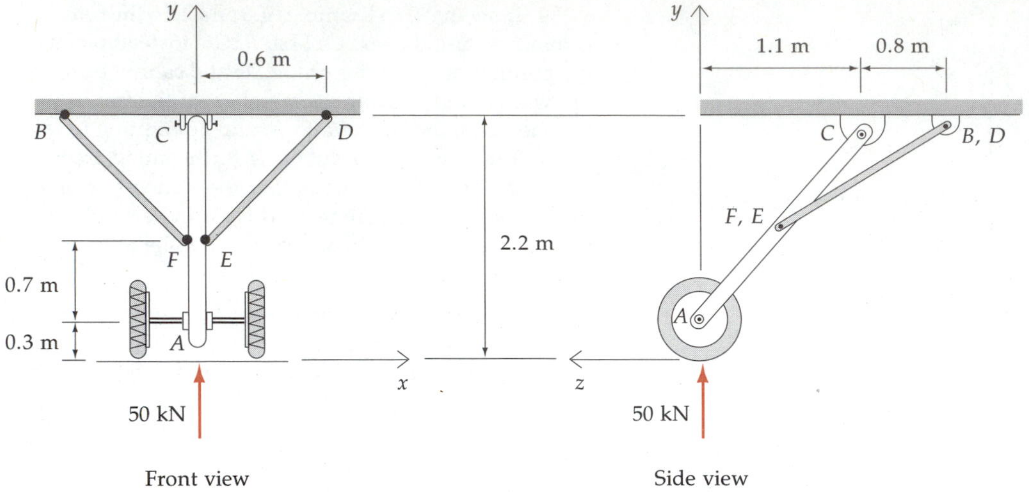

Figure P4.253

4.253 The struts *BF* and *ED* in Figure P4.253 are connected to the continuous member *AC* by ball-and-socket joints. The resultant force on the two wheels is 50↑ kN. Find the force in either strut (their magnitudes being equal by symmetry).

4.254 Bars *AB* and *DE* in Figure P4.254 are connected to *BDC* by ball-and-socket joints at *B* and *D*, respectively. Find the forces in *DE* and *AB*, the latter by mental inspection of a free-body diagram of bar *BDC*.

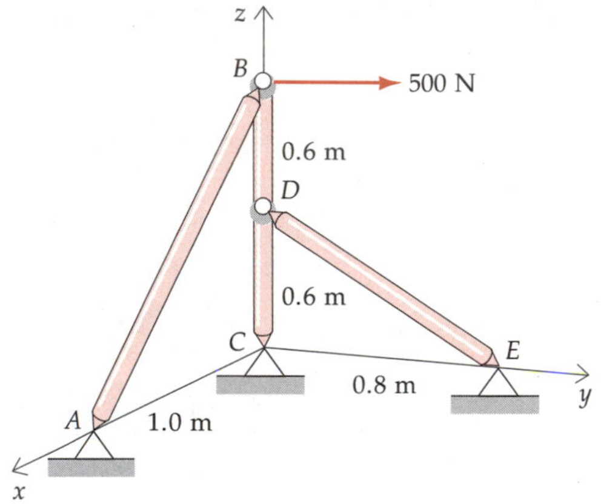

Figure P4.254

4.255 The three rods *AB*, *DB*, and *CB* (Figure P4.255) are joined together at *B*, and are also joined by the three horizontal bracing members connected to them at *F*, *G*, and *E*. Find the forces in the bracing members if a weight of 1200 lb is hung from *B*. Treat the joints as ball-and-socket connections. The floor is smooth.

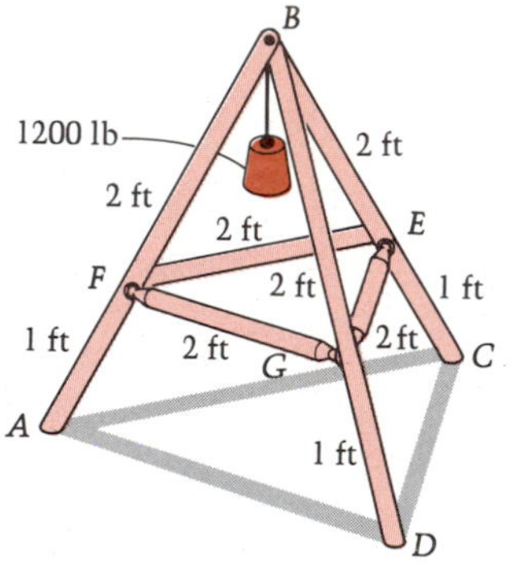

Figure P4.255

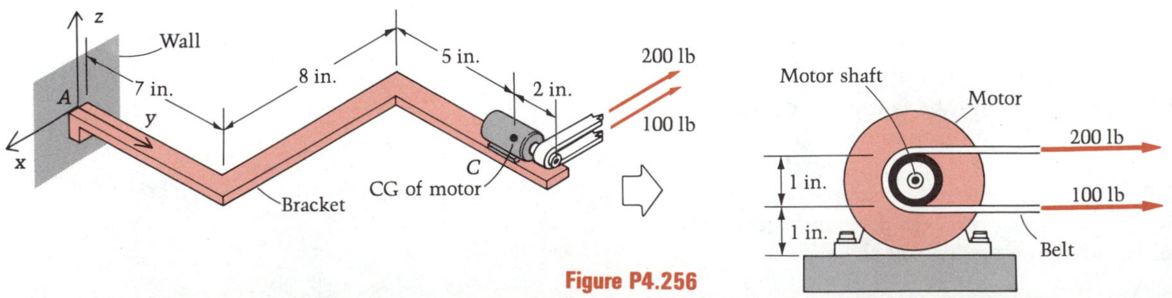

Figure P4.256

4.256 A steel bracket is bolted to a wall at A, and supports a 10-lb motor at B. The motor is delivering torque, with the forces in the belt shown in Figure P4.256 on the preceding page. The bracket weighs 5 lb. Find:

- **a.** The torque exerted by the two belt forces about the centerline of the motor shaft
- **b.** The force exerted on the bracket by the wall at A
- **c.** The moment exerted on the bracket by the wall at A.

4.257 So as not to interfere with other bodies, an antenna was designed and built with an offset axis as shown in Figure P4.257. The antenna is composed of a 12-ft, 1200-lb parabolic reflector $\mathcal{B}_1$, a counterweight $\mathcal{B}_2$, a reflector support structure $\mathcal{B}_3$ and a positioner. The positioner consists of (1) a pedestal $\mathcal{B}_4$ that is fixed to the ground; (2) an azimuth bearing at O and ring gear by means of which the housing $\mathcal{B}_5$ is made to rotate about the vertical; and (3) an elevation torque motor at E that rotates the support structure $\mathcal{B}_3$ with respect to $\mathcal{B}_5$. The purpose of the counterweight is to place the center of gravity of the combined body $\mathcal{B}_6(\mathcal{B}_1 + \mathcal{B}_2 + \mathcal{B}_3)$ on its elevation axis (x). If the reflector is modeled as indicated by a simple disk, determine the reactions onto $\mathcal{B}_6$ at E, and onto $\mathcal{B}_5$ at O, if the system is in equilibrium in the given position. Neglect the weights of $\mathcal{B}_3$ and $\mathcal{B}_5$.

* **4.258** In Figure P4.258 a quarter-ring is formed by two sections (AB and BC) of a circular bar being connected by a ball joint at B. The other ends are fixed to the reference frame, also by ball joints, at A and C. The three cables then hold the bar in the xz plane as shown. The radius R = 2 m, and the ring weighs 10 N/m. Find the cable tensions.

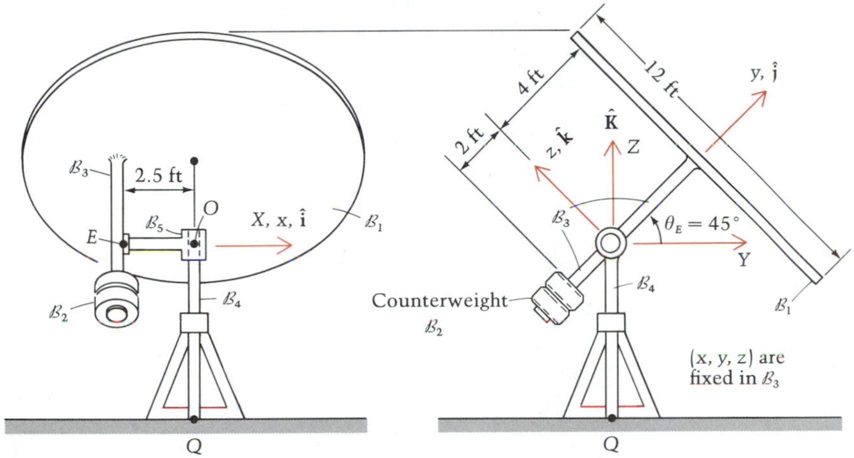

Figure P4.257

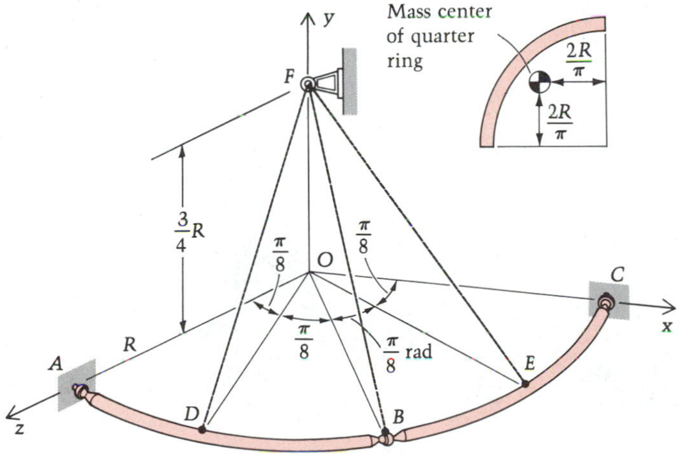

Figure P4.258

COMPUTER PROBLEMS ▶ Chapter 4

4.259 There is a range of values of L/R for which the uniform stick in Figure P4.259 can rest in equilibrium in the smooth hemispherical bowl. Show, using the equations of equilibrium, that this range is $\frac{2}{3}\sqrt{6} < \frac{L}{R} < 4$. Then write a computer program that will divide this range into 101 equally spaced values of L/R and, for each such value, to calculate and print the value of θ for equilibrium.

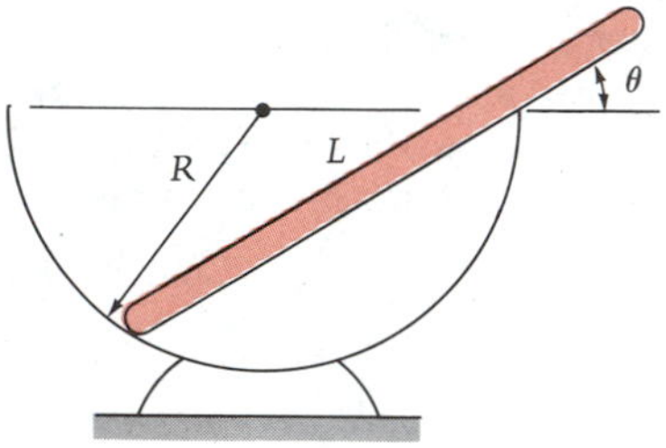

Figure P4.259

4.260 The uniform stick of mass m and length L rests in equilibrium between two smooth walls as shown in Figure P4.260. Write a computer program to help you construct a plot of the equilibrium angle θ of the rod versus the angle α of the right-hand plane.

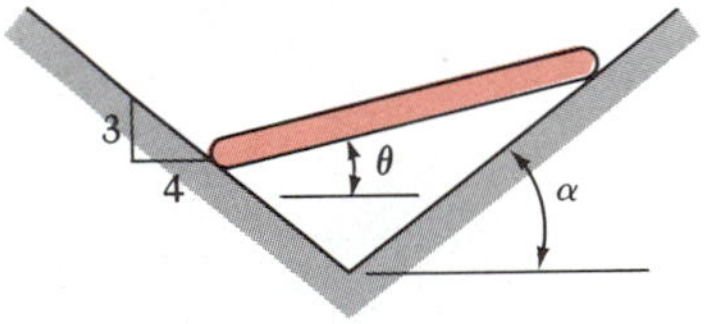

Figure P4.260

4.261 The distributed load shown in Figure P4.261 has a resultant of 300 ↓ lb acting 4 ft from the left end. The roller reaction is 200 ↑ lb. Note that two concentrated loads increasing proportionately and adding to 300 ↓ lb result in a roller reaction of 250 ↑ lb; three loads in 233 ↑ lb; four loads in 225 ↑ lb; etc. Write a computer program that will compute the value of this reaction for *any number* of concentrated loads. Use the program to compute the smallest number of concentrated loads needed to make the roller reaction within 1% of the continuously distributed limiting case.

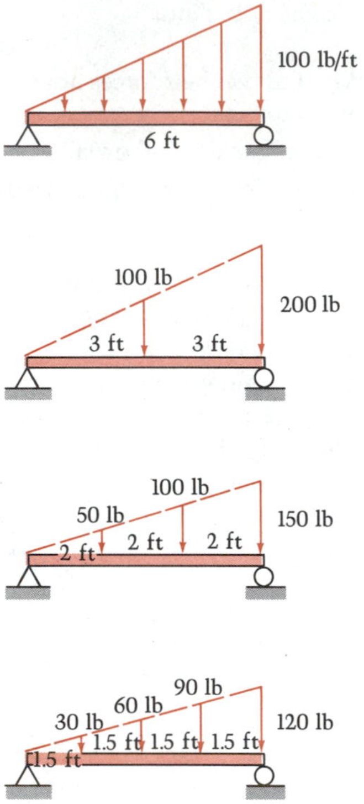

Figure P4.261

4.262 Clearly, when the boom angle θ in Figure P4.262(a) is either O or π, no force in the winch cable is needed for equilibrium. Somewhere between these values, the winch cable tension T is a maximum. We wish to find the

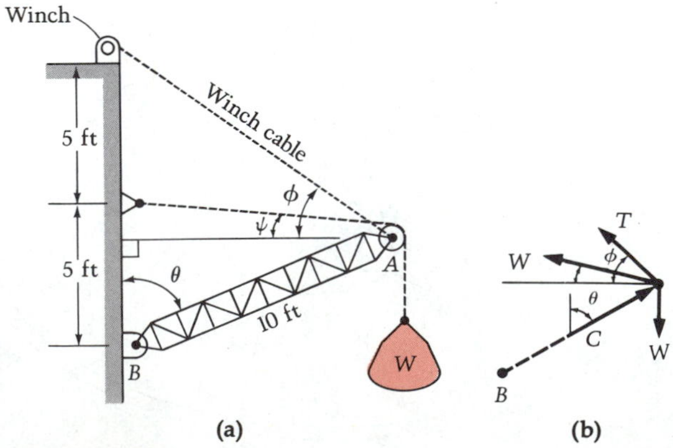

Figure P4.262

value of θ at which this occurs. First, use the free-body diagram (Figure 4.262(b)) of the pin at A, sum moments about B, and obtain an equilibrium equation, free of the boom compression C. Solve the equation for the ratio T/W as a function only of the angles θ, ϕ, and ψ in the figure. Prove that $\phi = \theta/2$ and that

$$\sin\psi = \frac{5(1 - 2\cos\theta)}{\sqrt{125 - 100\cos\theta}} \quad \text{and}$$

$$\cos\psi = \frac{10\sin\theta}{\sqrt{125 - 100\cos\theta}}$$

Use the computer to generate data for a plot of T/W versus θ for the range $0 < \theta < \pi$.

4.263 In Problem 4.77, DBC was a right triangle with the angle θ (see Figure P4.77) equalling $\tan^{-1}(3/4)$. Consider now other values of θ obtained by moving pin D to different points on the horizontal line through C as suggested by Figure P4.263.

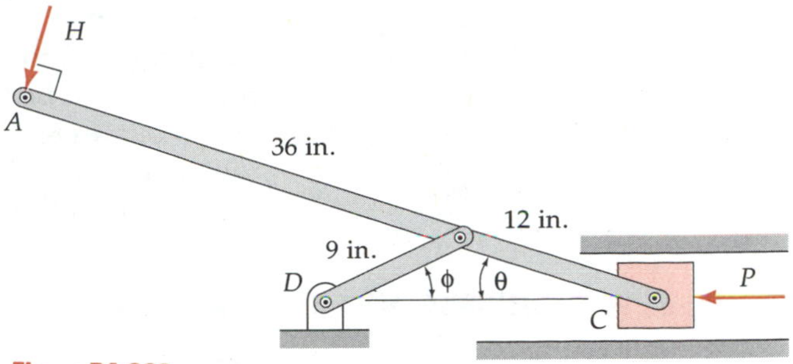

Figure P4.263

Show that: (a) $\phi = \sin^{-1}(\frac{4}{3}\sin\theta)$; (b) the force B in the 2-force member is $4H/\sin(\theta + \phi)$, compression; (c) the crushing force $P = B\cos\phi - H\sin\theta$. (d) Write a computer program and generate data for a plot of P/H versus θ, for θ in the range from 2° to 47°. (e) Show that the largest angle θ for which crushing is possible is, to five significant digits, 47.929°.

SUMMARY ▶ Chapter 4

This Chapter has introduced the student to a large number of realistic equilibrium problems, each one solvable through the use of only two equations: the sum of the external forces acting on a body is zero ($\Sigma\mathbf{F} = \mathbf{0}$), and the sum of the moments of the external forces about any common point is zero ($\Sigma\mathbf{M_P} = \mathbf{0}$). This process is facilitated immensely by the use of the free-body diagram (FBD), covered in minute detail in Section 4.2.

We had used $\Sigma\mathbf{F} = \mathbf{0}$ alone to solve some particle equilibrium problems in Chapter 2. In Section 4.3, we expanded our knowledge to the solution of equilibrium problems of finite-sized bodies requiring the moment equation as well as the force equation. In Section 4.3, however, we restricted ourselves to problems involving only a single body (which includes the idea of two or more physical objects left intact and thus considered as a single body).

In Section 4.4, we took yet another step and learned to separate bodies joined by simple connections by removing the pins, rollers, cables, etc., and drawing FBD's of the separate bodies. With the help of these FBD's, we then wrote equilibrium equations of the various separated bodies as well as of the overall (unseparated) "combined body." We noted that this process contains both good news and bad news: It is bad news that if the equilibrium equations are written for the combined body and all its separated bodies, not all these equations will be independent. But it is good news that the separation process often renders a problem solvable which was indeterminate on the basis of the combined body's equations alone.

In Chapter 5, we will take a final step in the "particle / single finite-sized body / finite-sized body separated into two or more parts" sequence. It will be to imagine slices through bodies which expose, on the cut-through cross sections of the separated parts, the internal forces within the body. These forces are of paramount importance in determining the stresses existing in bodies under load.

REVIEW QUESTIONS ▶ Chapter 4

True or False?

1. Free-body diagrams help us considerably in writing correct equations of equilibrium in statics.
2. If a free-body diagram of body $\mathcal{B}$ is to be useful, then $\mathcal{B}$ must be in equilibrium.
3. On a two-force member, the two forces are equal in magnitude, opposite in direction, and each acts along the line joining their two points of application.
4. If a body is in equilibrium under the action of three forces, and two of these intersect at a point P, then the line of action of the third *also* passes through P.
5. If a body is in equilibrium under the action of three forces, and two of these are parallel, then the third need not be parallel to the first two.
6. If a body is in equilibrium under four forces — two of which form a couple — then the other two also form a couple.
7. If a body is in equilibrium under the action of three forces, the forces need not be coplanar.
8. A body acted on only by a single couple cannot be in equilibrium.
9. Let the external forces on a body be such that $\Sigma \mathbf{F} = \mathbf{0}$ and $\Sigma \mathbf{M} = \mathbf{0}$; then the body must be at rest.
10. On a three-force member in equilibrium, the forces are either (a) coplanar and concurrent or (b) coplanar and parallel.
11. The tensions in a cable passing over a pulley in equilibrium are always equal.
12. It is possible for a body $\mathcal{B}$ to be in equilibrium with two separate parts comprising $\mathcal{B}$ not being in equilibrium separately.
13. One of the most important things to keep in mind when drawing free-body diagrams of various parts of a body is the action-reaction principle.

Answers: 1. T 2. F 3. T 4. T 5. F 6. T 7. F 8. T 9. F 10. T 11. F 12. F 13. T

The study of classical mechanics is a profound experience. The deeper one delves into it, the more he appreciates the contributions of the great masters. Y. C. FUNG

INTRODUCTION

COMPREHENDING MECHANICS GOES beyond reading the textbook and working problems. Knowing statics means that you understand the physics embodied in the laws of mechanics, recognize their limitations and assumptions, and can correctly apply them to situations you encounter in practicing engineering. Knowing mechanics requires that you also develop a reasonable sense of the physical consequences of the fundamental principles along with the mathematical consequences.

GREAT MASTERS OF MECHANICS such as Galileo, Leonardo da Vinci, Hooke, Kepler, and Newton formulated the laws of statics and dynamics from the results of numerous observations and experiments. They devised simple experiments to test and clarify their ideas. Using empirical findings, they developed theories for predicting the behavior of mechanical systems. The principles they discovered and the mathematical expressions that describe these principles are the cornerstones of engineering mechanics.

STUDENTS (AND TEACHERS) of statics often overemphasize analysis and pay too little attention to the relationship between theory and the actual physical behavior of mechanical systems. Understanding both aspects of mechanics is essential. Engineers cannot successfully model the behavior of a mechanical system if they are unsure of the physics of the system. And the ability to predict successfully the behavior of physical systems is fundamental to the process of engineering design.

OVERVIEW

THIS SECTION PRESENTS experiments not unlike those used in early empirical studies of mechanics. These experiments demonstrate actual behaviors of simple mechanical systems and are intended to strengthen your understanding of the basic laws of statics. These exercises emphasize physical reality to help you develop qualitative intuitive skills that are essential in the practice of engineering. In addition, the demonstrations provide a way to check the soundness of certain mathematical models that are used to describe real-world problems.

THESE EXPERIMENTS PROVIDE only a starting point for your explorations and will raise additional questions as you conduct them. Do not leave those questions unanswered. To answer them you may need to modify a demonstration, design new experiments, or simply concentrate on interpreting a mathematical model. The important point is that you should pursue the answers. Along the way you will develop new insights into statics, you will become more proficient, and your ability to explain and predict the physical world will improve.

THE EXERCISES IN statics are keyed to specific sections and problems in the text. Be sure to review the text material before attempting these exercises. Each demonstration requires that you compare your observations with behavior predicted from a mathematical model. In most cases, experimental and theoretical results should be reasonably close. Remember, however, that models are only approximate and that experiments are never perfect. So, if your results disagree, find out why. To do so, verify your measurements and, if necessary, repeat or redesign the demonstration. Review the assumptions and limitations of the theory and check your analysis or computer program for mistakes. If you still find a disparity between the results, you may be applying the wrong principles or using incorrect equations.

MATERIALS

THE EXPERIMENTAL setups are simple and easy to construct. To conduct them, however, you will need some materials, all of which are readily available and can be obtained at little or no cost. These materials can be found at hardware stores, hobby shops, toy stores, and in the engineering shop at your college. We encourage the use of scrap materials and creative scrounging!

FOR MECHANICAL PARTS you should collect an assortment of cylinders, tubes, spheres, wheels, and rectangular blocks. The only requirement is that the parts be homogeneous and reasonably uniform. For example, if you need a cylindrical tube, select one that is straight and has a constant diameter and thickness. Manufactured tubes such as the following are excellent:

- Wood dowel
- PVC pipe, copper pipe, steel pipe
- Conduit tubing
- Aluminum rod, steel rod
- Empty coffee can with ends removed, tennis ball can
- Cardboard tube from roll of paper towels or toilet tissue
- Cardboard mailing tube
- Hockey puck
- Thread spool, metal adhesive tape spool

Be creative and resourceful when selecting materials.

IN ADDITION, you will need some laboratory supplies. They include string, duct tape, protractor, graph paper, stopwatch, tape measure or ruler, scissors, inexpensive calipers, and a scale or access to a scale for weighing parts.

STATIC EQUILIBRIUM OF A SYSTEM OF BODIES

The external forces acting on a body at rest satisfy the equilibrium equations $\sum \boldsymbol{F} = 0$ and $\sum \boldsymbol{M} = 0$. This demonstration examines static equilibrium for a system of bodies in contact. Use a cylindrical cardboard tube from a roll of paper towels or toilet tissue and two identical balls with a diameter of 60% to 80% of the tube diameter. Hold the hollow tube vertically on a flat surface and drop the two balls into the tube as shown in the figure. Slowly release the tube and observe its behavior. If the tube remains upright, reduce its length by 10% and repeat the process. At some point the system will not be able to remain in static equilibrium and the tube will fall over when you release it. Can you predict when this will occur? Show that the relative diameters and weights of the balls and tube are the parameters that determine whether the system remains at rest. Cap the bottom end of the tube with stiff paper and tape and repeat the experiment. Explain why the tube does not tip over when the bottom is capped.

You can perform this demonstration with different tubes and balls of various weights and sizes. In this case the balls need not be identical; however, the sum of their diameters must be greater than the inside diameter of the tube. Use the equations of static equilibrium to verify or predict your observations and to study the relationships among the system parameters. Be sure to include the wall thickness of the tube and the order of placement of the balls if their sizes or weights are different.

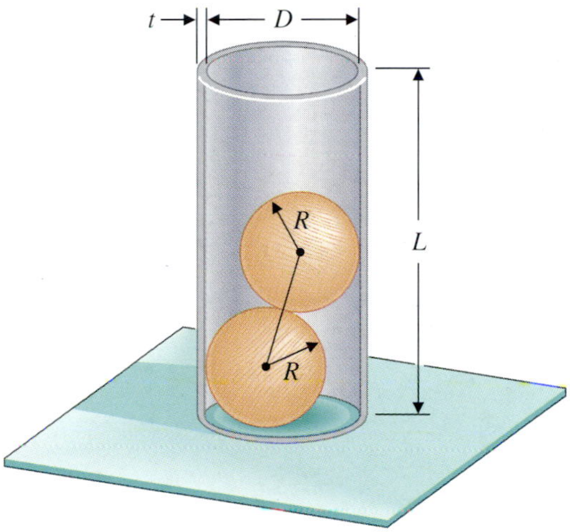

REFERENCE: *Statics* Sections 4.2-4.4

MEASURING COEFFICIENTS OF STATIC FRICTION

Recall from statics that the coefficient of static friction, μ_s, between two materials is the tangent of the angle θ_s at which a sample of one material will start to slide down a plane made of the other. (See the figure.) Use this relationship to determine μ_s for different combinations of materials. Compare your values with some of those given in the text. Explain how the following affect μ_s: the contact area between samples, the roughness of their surfaces, and the weight and geometry of the sample that slides. Devise a method to measure μ_s if one of the samples is a length of pipe.

REFERENCE: *Statics* Section 6.2

Static Equilibrium of an Unbalanced Cylinder

Build an unbalanced but round cylinder. One approach shown in the figure is to tape or glue a rod or small cylinder to the inside of a cylindrical tube. (Be sure that the two axes are parallel.) Equal lengths of 3 1/2″ diameter PVC pipe and 1″ diameter steel conduit tubing work well for this demonstration. Measure and calculate the eccentricity e of the mass center. Place the cylinder on a slightly inclined plane, remove the support provided by your finger as shown below, and observe the cylinder's motion. If the cylinder continues to roll, reduce the slope of the plane until the cylinder remains at rest. Slowly increase the slope and find the maximum angle of inclination β_{max} for which the cylinder will not roll down the incline. Measure β_{max} and the corresponding resting angle ϕ which is defined in the bottom figure. Using the equations of static equilibrium, derive an expression that relates the angles ϕ and β in terms of the ratio e/R. Calculate the maximum inclination angle β_{max} and the corresponding resting angle from that equation. Note that other values of ϕ and β are possible if β is less than β_{max}. Plot these predicted values of ϕ vs. β for several values of e/R between 0 and 1. Be sure to include a curve corresponding to the value of e/R for the unbalanced cylinder. Measure ϕ vs. β for the cylinder and plot those values over your predictions. Do the results agree? Explain.

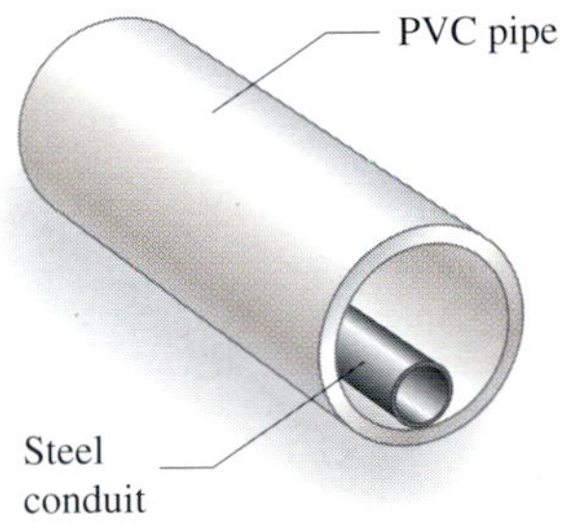

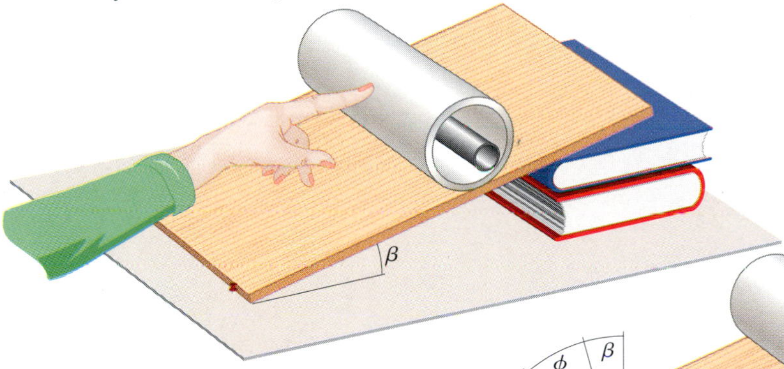

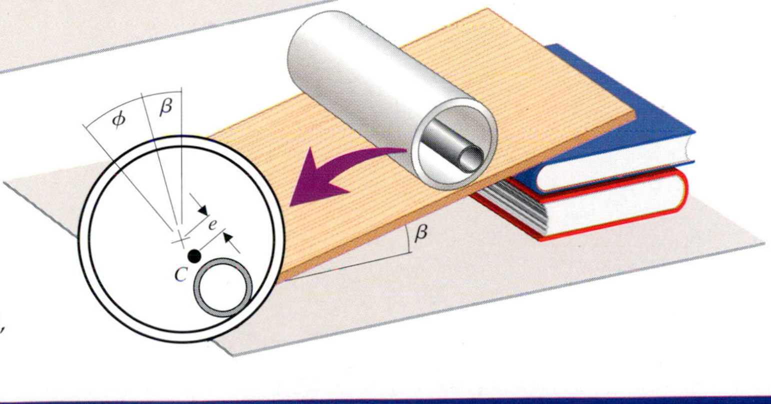

REFERENCE: *Statics* Sections 4.2-4.4, 7.3, and 7.4

HOLDING A SPOOL ON AN INCLINE

Assemble a spool using two identical disks or cylinders and a single axle as shown in the figure. A metal adhesive tape spool works well for this demonstration. Tape a string to the center of the axle. Bring the string over the top of the axle and hold it parallel to the surface of an inclined board as in the top figure. Make sure the string pulls in the midplane of the axle. Notice that for small values of slope θ the spool remains in static equilibrium. Confirm this observation and predict the angle θ_{max} at which the spool will begin to move down the incline. You will need to know the coefficient of friction between the spool and the board. Measure θ_{max} and compare it to your prediction. Repeat the above procedures when the string is pulled from the bottom of the axle. (See the bottom figure.) Explain why these two cases are markedly different. Plot the theoretical relationships between $\tan\theta_{max}$ and the axle/spool radius ratio r/R for the two cases. Can you anticipate the initial motion of the spool when θ exceeds θ_{max}? Could you use this experimental procedure to determine μ_s?

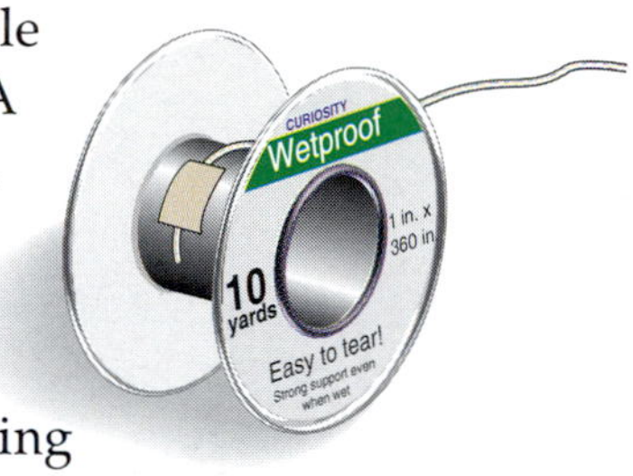

REFERENCE: *Statics* Section 4.3 and 6.2

Balancing a Cylinder on an Incline

This demonstration explores the static equilibrium of a cylinder resting against a step on an inclined plane. Select a cylinder of radius R and install a step of height $H \cong 0.1R$ on a flat board. Place the cylinder against the step as shown in the figure below. Slowly increase the inclination of the board and measure the angle θ_{max} at which the cylinder just rolls over the step. Repeat the experiment for several values of step heights $H \leq R$. One easy way to vary the height is to build the steps from layers of stiff cardboard cut from the backing of a note pad. Use the equations of static equilibrium to obtain an expression for $\tan \theta_{max}$ in terms of the ratio H/R. Graph the measured and calculated values of $\tan \theta_{max}$ vs. H/R and compare the results. Show that your results are independent of the coefficient of friction and weight of the cylinder.

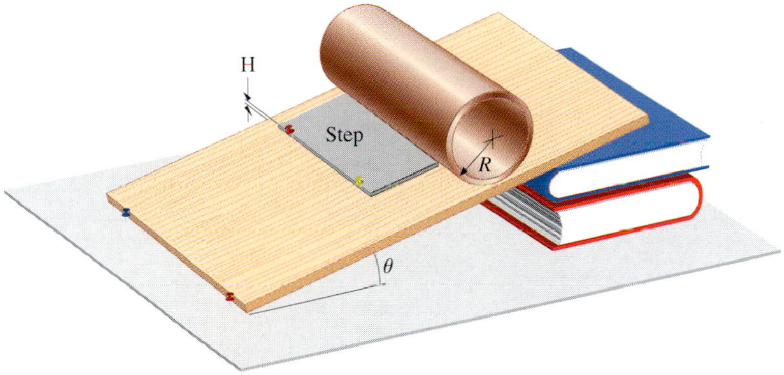

REFERENCE: *Statics* Sections 4.2-4.4, and 6.2

Balancing a Rectangular Block on an Incline

Install a small lip or bump on an inclined flat board and place a rectangular block against the lip, as shown in the figure below. Slowly increase the inclination of the board and measure the angle θ_{max} at which the block just tips over. Use the equations of static equilibrium to calculate a theoretical value for θ_{max} in terms of the length-to-height ratio L/H. Compare the two values and explain any difference. Why should the bump be very small? Why do you not need to know the coefficient of friction? Repeat the demonstration for rectangular blocks of various dimensions. Graph the measured and predicted values of tan θ_{max} vs. L/H and compare the results.

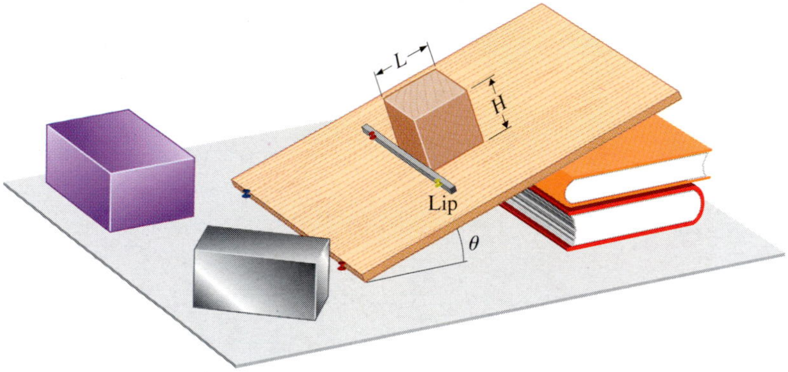

REFERENCE: *Statics* Section 4.3 and 6.2

5 STRUCTURAL APPLICATIONS

5.1 Introduction

I TRUSSES

5.2 Definition of a Truss; Examples of Trusses
Forces in Truss Members

5.3 The Method of Joints (or Pins)

5.4 Shortcuts and Rigidity/Determinacy Results
Zero-Force Members
Other Shortcuts
Numbers of Members and Pins in a Truss

5.5 The Method of Sections
Cutting the Truss into Two Distinct Sections

5.6 Space Trusses

5.7 A Brief Introduction to the Mechanics of Deformable Bodies
Normal Stress
Extensional Strain
Hooke's Law
Elongation of a Truss Element
Statically Indeterminate Problems
Thermal Effects

II SYSTEMS CONTAINING MULTIFORCE MEMBERS

5.8 Axial and Shear Forces and Bending Moments
The Multiforce Member

5.9 Beams/Shear and Moment Distributions
Sign Conventions for Internal Forces in Beams

5.10 Differential Relationships Between $q(x)$, $V(x)$, and $M(x)$ in a Beam/Shear and Bending Moment Diagrams
Relationship Between Shear Force and Distributed Lateral Loading
Relationship Between Bending Moment and Shear Force
The Other Sign Convention for Shear Force

III CABLES

5.11 Parabolic and Catenary Cables
Vertical Loadings Depending upon the Horizontal Coordinate
Vertical Loadings Depending upon Arclength Along the Cable

5.12 Cables Under Concentrated Loads

SUMMARY
REVIEW QUESTIONS

5.1 Introduction

In building a structure, we must always ensure that its members are designed with sufficient strength to carry the loads intended (and then some, to account for a factor of safety). This "strength" of the materials, covered in detail in later courses in the mechanics of deformable bodies, is measured by a quantity called stress, which has units of force per unit of area.

To be properly prepared for such later studies, a student must exit statics knowing how to find the resultants of forces *within* elements of structures such as trusses, beams, and frames. In other words, if a beam is imagined sliced into two parts, we must be able to determine the resultant forces (and/or couples) exerted on each "half" by the other. These forces are called internal forces for obvious reasons.

We will again use the result that if a body is in equilibrium, then every part of it is also in equilibrium. Thus for the beam in the preceding paragraph, we will learn to write the equilibrium equations for one of the parts, and to then solve them for the forces and moments exerted on the "face" that was exposed by the cut. These forces, while *external* to the part of the structure that is being analyzed, are *internal* to the complete (uncut) structure, and are resultants of the stresses on the face.

This chapter differs from Chapter 4 only in the level of sophistication in separating a body into parts: the bodies here will be cut completely through to expose internal forces, whereas in Chapter 4 we were simply separating bodies by removing simple connection devices such as pins, rollers and cables. We emphasize, however, that the *equations are the same* ($\Sigma\mathbf{F} = \mathbf{0}$ and $\Sigma\mathbf{M}_P = \mathbf{0}$ for each part analyzed).

The chapter is divided into three parts. In Part I, we deal with the truss. Each member of a truss is a straight bar in which the internal force resultant is a tensile or compressive force parallel to the member's axis. In Sections 5.2–5.6, we will examine the various means of determining these forces in truss members, or elements. Then in Section 5.7, we present a brief introduction to the mechanics of deformable bodies, treating the uniaxial stress and strain found in truss members.

The second part of Chapter 5 deals with multiforce members. Frames are structures comprising a number of members (or elements) connected together for some structural purpose, at least one member of which is not a straight, two-force, truss-like element. Such members, which in general can carry shear forces and bending and twisting moments in addition to axial forces, are called multiforce members. How to find these more complicated internal forces in frames is the subject of Section 5.8, and in Sections 5.9 and 5.10 this study is continued for another very common structural element known as a beam.

Part III deals with cables, in Sections 5.11 and 5.12. The first of these two sections looks at cables under distributed loads; in the second we examine cables acted on by a finite number of concentrated forces.

I / TRUSSES

5.2 Definition of a Truss; Examples of Trusses

We define a **truss** to be an *idealized* structure consisting of straight and slender bars, each of which is pinned to the rest of the structure and/or to the ground at its two endpoints by frictionless pins (or, in three-dimensional trusses, by ball-and-socket joints). In addition, the structure is loaded only by forces at such pins. Thus trusses are composed entirely of two-force members. See the examples in Figure 5.1.

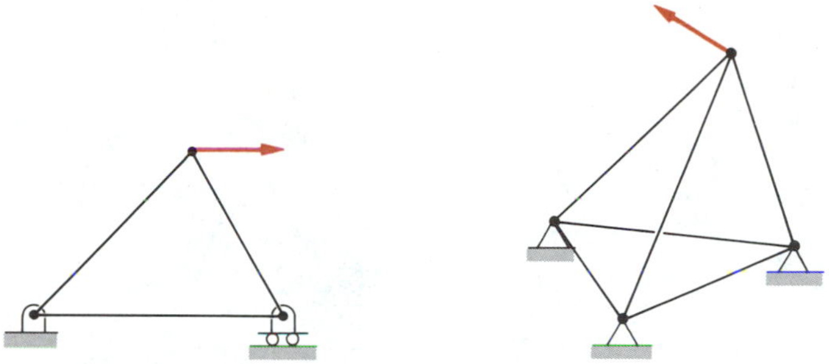

(a) Plane truss.

(b) Space truss.

Figure 5.1

We hasten to mention that the above idealization is a "mathematical truss," and the connections are, in reality, rarely smooth pins or ball joints. However, this idealization of a physical truss gives good results if the centerlines of the members at each connection all intersect at a single point; more will be said about this later. For now, the reader is invited to note, in the set of photographs on the next two pages, just a few of the many uses of trusses. These photographs indicate that there are many practical applications of the truss. Furthermore, even within a given application, there are often a large number of different types of trusses. For example, Figure 5.2 (see page 278) shows a number of common roof truss configurations. It should be mentioned that the names of the trusses may vary from one manufacturer to another. Each type has its own special use and span capability.

If the members of the truss all lie in a plane [as in Figure 5.1(a)], then we have a "plane truss"; if not [Figure 5.1(b)], the truss is called a "space truss." We shall introduce the methods of truss analysis with plane trusses, then consider the more complex space trusses later in the chapter.

Water towers.

Signs.

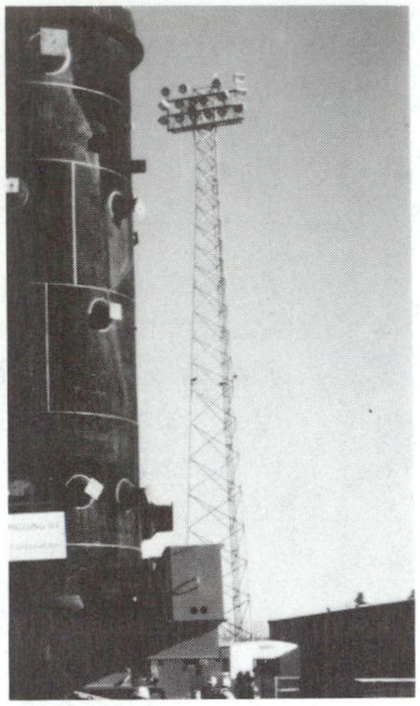

Lighting.

Backing and mounting structures for antennas.

Crane supports in steel mills.

Roof structures.

Highway signs.

Temporary support for new highways.

Conveyors.

Electric power transmission towers.

Construction cranes.

Radio and TV towers.

Supports for amusement park rides.
(*Courtesy of Six Flags Over Georgia*)

Derricks.

Bridges.

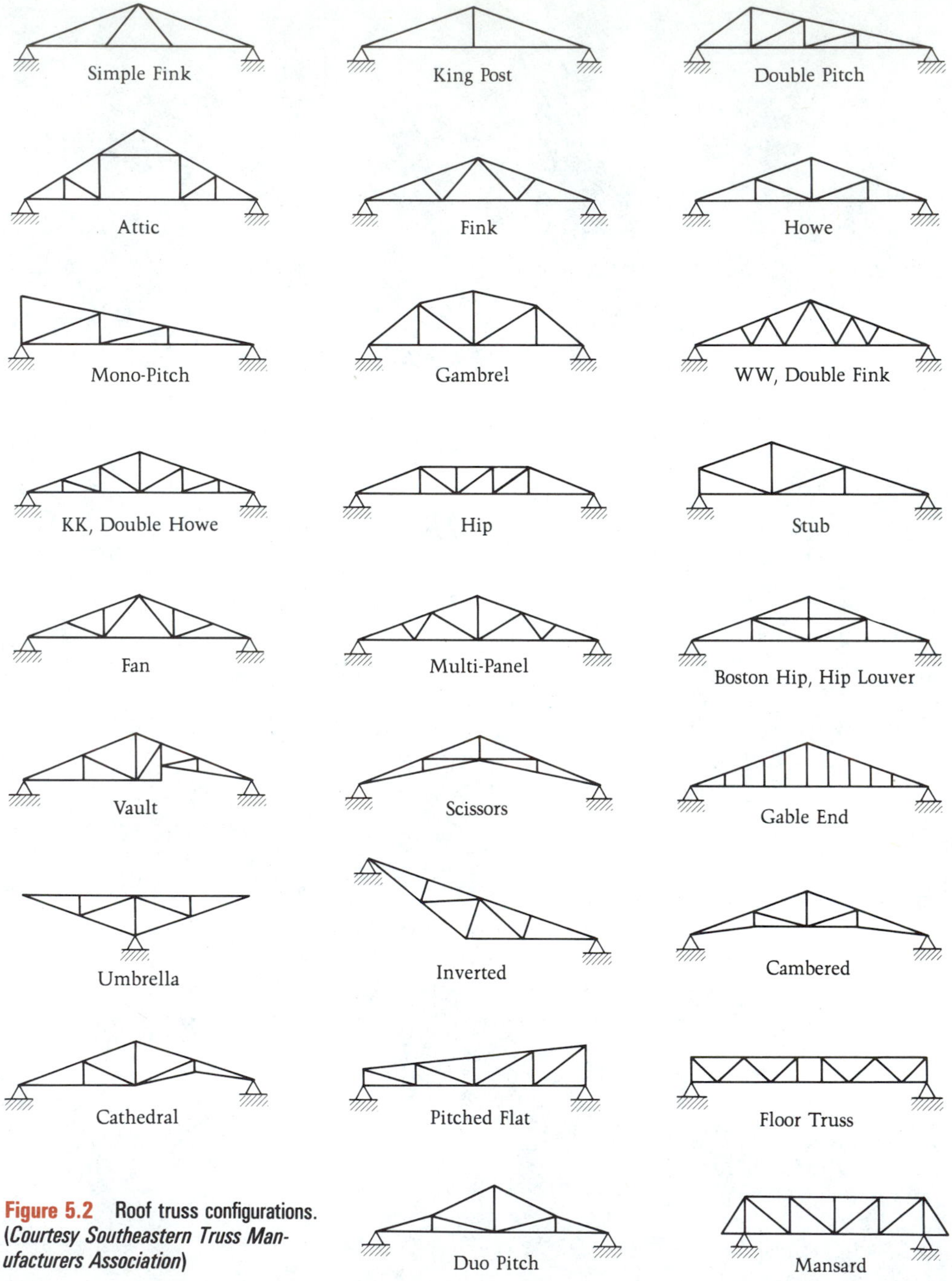

Figure 5.2 Roof truss configurations. (*Courtesy Southeastern Truss Manufacturers Association*)

Forces in Truss Members

By its definition, a truss is made up only of two-force members, and the force distribution across any cross section of a member has a very simple resultant. To determine it, we note, recalling the discussion of two-force members in Chapter 4, that a truss member carries only a pair of equal magnitude, oppositely directed, forces along its length (see Figure 5.3).

Therefore, if we cut a section through the respective members shown in Figure 5.3, we obtain the free-body diagrams shown in Figure 5.4.

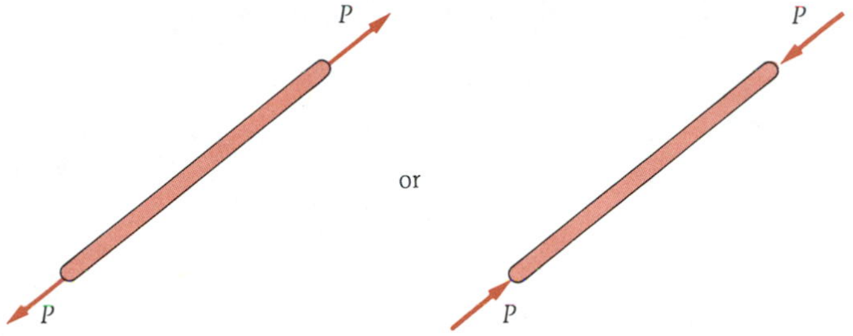

(a) Tension member of a truss. (b) Compression member of a truss.

Figure 5.3

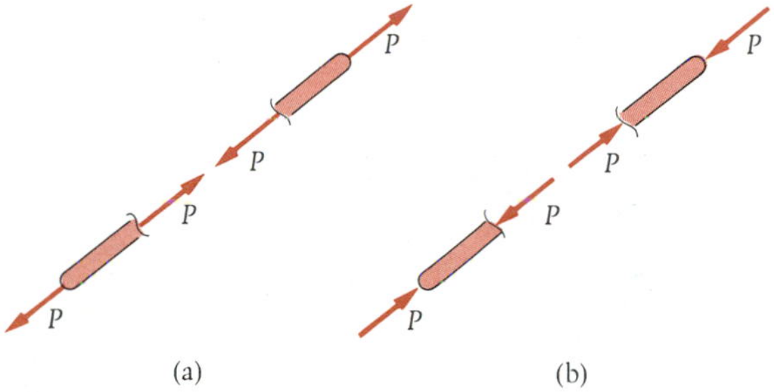

Figure 5.4

Figure 5.5

In both parts of Figure 5.4, the resultant is simply an axial force. This is because, if we were to assume the more general distribution of internal forces across the lower section of Figure 5.4(a) to be as shown in Figure 5.5, equilibrium would immediately require that the shear force (V) and the bending moment (M) vanish and that the axial force (A) be equal to P.*

A truss member in the condition of Figure 5.4(a) is being stretched and is said to be in tension; the member of Figure 5.4(b), however, is being compressed and is said to be in compression. (If $P = 0$, of course, the member is not loaded.) These are the only possibilities for truss member forces, and thus the answers are easy to present. For example, if we say that the force "in" a member extending from A to B (member AB) is 647 N Ⓣ, we mean that a free-body diagram of AB looks like Figure 5.3(a) with $P = 647$ N. Similarly, if we say the force in a member

* Figure 5.5 refers to a member of a plane truss. If the member is from a space truss, then equilibrium would require that *two* shear forces vanish, that *two* bending moments *and* a twisting couple also vanish, and that, again, $A = P$.

DE is 212 lb ©, we mean that a free-body diagram of *DE* looks like Figure 5.3(b) with $P = 212$ lb.

For example, the truss shown in Figure 5.6 has seven members that, if the truss is solved,* carry the forces indicated in Figure 5.7 (all in kips, or "kilo-pounds"; 1 kip = 1 k = 1000 lb). The forces exerted on the various pins are also shown.

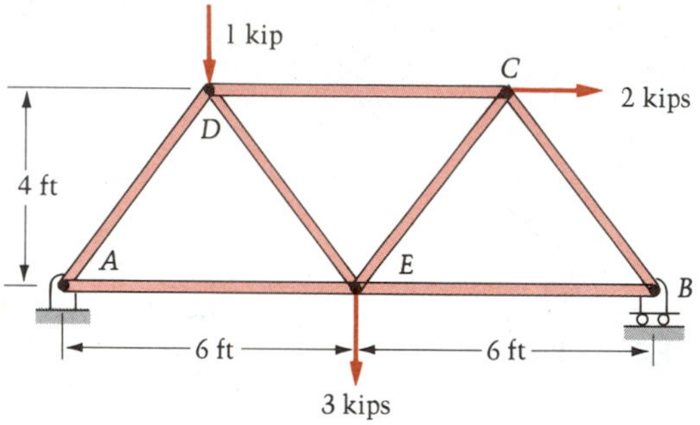

Figure 5.6

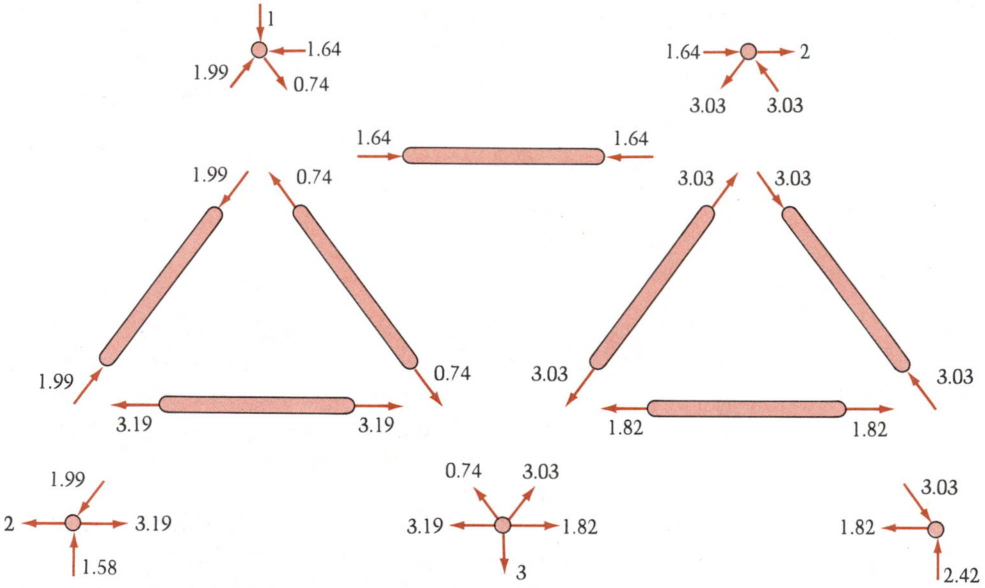

Figure 5.7 (All forces in kips)

We shall return to that truss in Example 5.1 and actually compute the indicated member forces after more preliminary discussion.

Though truss members are always assumed to be pinned at their ends, in reality this is seldom the case. The members of roof trusses and bridges, for example, are normally connected by means of a plate to

* Meaning all the forces in its members have been found.

which the members are joined by nails, rivets, pins, welds, or bolts at a number of points, as suggested in Figure 5.8.

Although it may seem like a bad assumption to replace a plate and a large number of bolts with a single pin, this is not the case. If the centerlines of the members intersect approximately at a point as shown in Figure 5.8, then the structure turns out to behave very much as though it were an ideal truss; that is, the transverse (shear) forces and the bending moments in the members will be small. Such structures are thus usually analyzed as ideal trusses.

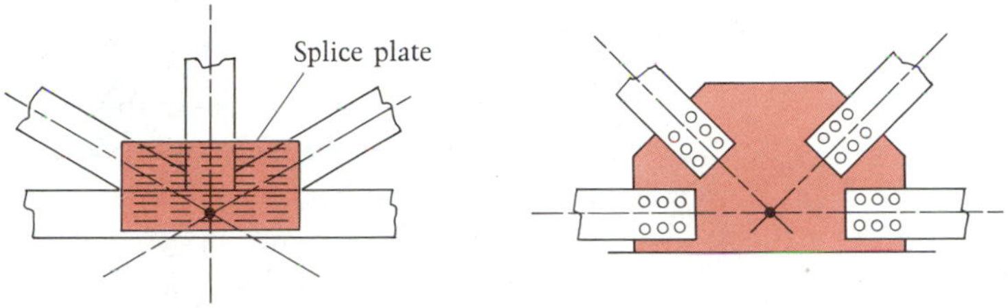

(a) Roof truss connection.

(b) Gusset plate connection.

Figure 5.8

There are two main methods commonly used in truss analysis: the method of joints and the method of sections. We shall begin our study with the method of joints (or pins).

5.3 The Method of Joints (or Pins)

In this section, we will be studying the use of the method of joints in truss analysis. This method is simply to isolate one pin at a time (with a free-body diagram) and to write the equilibrium equations for it. Oftentimes, we can at the outset find at least one pin on which only two unknown member forces act; when this is the case,* both these forces may be found from $\Sigma F_x = 0$ and $\Sigma F_y = 0$. After doing so, we repeat the procedure at another joint and thus work our way into and through the truss.

Before moving into some examples, we wish to first discuss the directions of unknown member forces acting on pins. Consider the free-body diagram on the next page of the pin at point *B* [Figures 5.9(a,b)], where the 2.42-kip force is an already-determined roller reaction. Now, when we sketch a force such as F_{CB}† pushing on the pin of a joint, we are assuming the member which is doing the pushing (*CB* in this case) to be in compression (Ⓒ). This is because if the member pushes on the pin, then by action and reaction, the pin pushes *back*, compressing the member. Thus the four forces associated with member *CB* (the two it exerts, and

* This always happens with "simple plane trusses," as we shall see later.

† We attach no meaning to the order of the subscripts on the force; thus $F_{CB} = F_{BC}$.

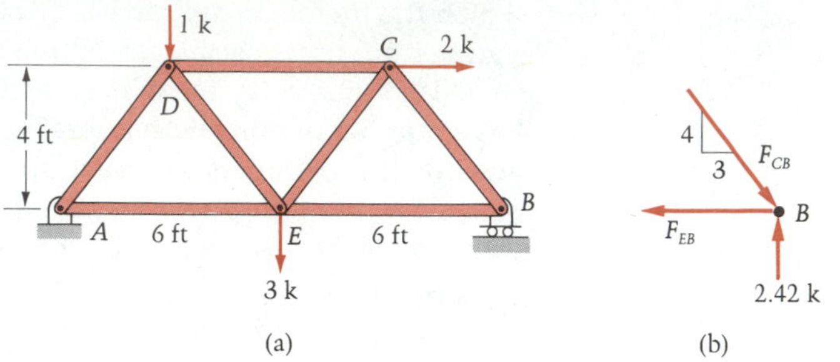

Figure 5.9

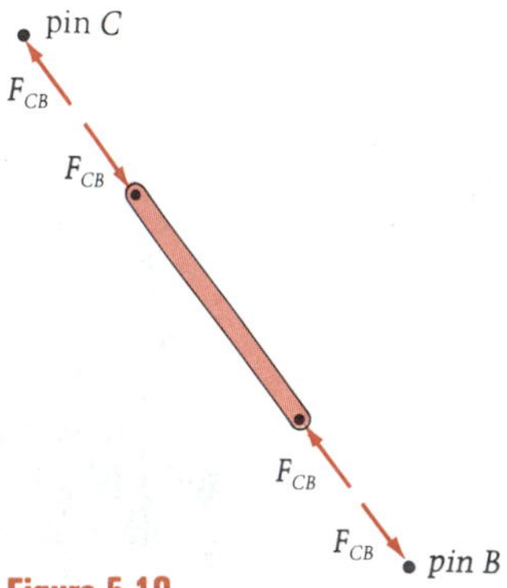

Figure 5.10

the two exerted on it) are in the directions shown in Figure 5.10 if member CB turns out to be in compression. [In this particular case, we are actually *sure* that CB is in compression because in the free-body diagram of joint B, only member CB can have a (vertical) component to balance the upward roller reaction of 2.42 kips, and the vertical component of F_{CB} onto the pin at B can be downward only if CB is in compression.]

On the other hand, if we draw a force such as F_{EB} *pulling* on a pin as in Figure 5.9(b), then we are assuming the member (EB here) to be in tension (Ⓣ). Again the reason is action and reaction. If the member pulls on the pin, the pin likewise pulls back on the member. (See Figure 5.11.) We are also sure here that EB is in tension, once we have seen that CB pushes down and to the right onto pin B. In the free-body diagram of pin B [Figure 5.9(b)], only F_{EB} can balance the horizontal component of F_{CB}, which is to the right.

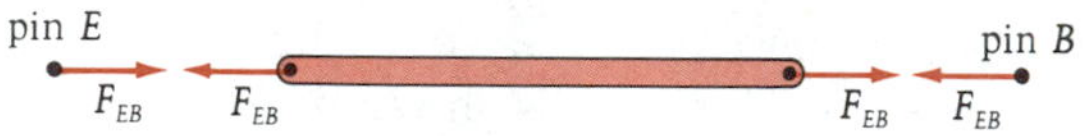

Figure 5.11

Sometimes, however, we are not certain of the direction of a force. (For example, does it push, or pull, on a pin?) In such a case, we simply guess one or the other, and if the solution for the force F turns out in the algebraic solution to be negative, this means the bar is in tension if we "assumed" compression, and vice-versa.* For example, suppose we assumed that a member PQ was in compression and later found $F_{PQ} = -300$ lb. It is common to then communicate that the force in PQ is

* Usually in an engineering analysis, if we make an assumption and then find it to be false, we must revise our assumption and repeat the analysis. That is not the case here because the "assumption" isn't really an assumption in the usual sense. Rather, it is nothing more than a statement of the physical significance of a positive value of the scalar used (together with a unit vector) to represent a certain force vector. There is actually no prejudgment about the sign of the scalar since the laws of mechanics and mathematics will dictate its sign. Because the practice is widespread, however, and because it provides such a concise means of communication, we use this weak form of the word "assume."

tensile by writing $F_{PQ} = 300$ lb Ⓣ. This *second* use of the symbol F_{PQ} is not an algebraic statement, but merely a shorthand means of reporting that the force in PQ has been found to be tensile with magnitude 300 lb.

We are now ready for our first example, in which we shall solve (for all the forces in the members of) the truss of Figure 5.9(a).

EXAMPLE 5.1

Find the forces in each member of the truss in Figure E5.1a by the method of joints.

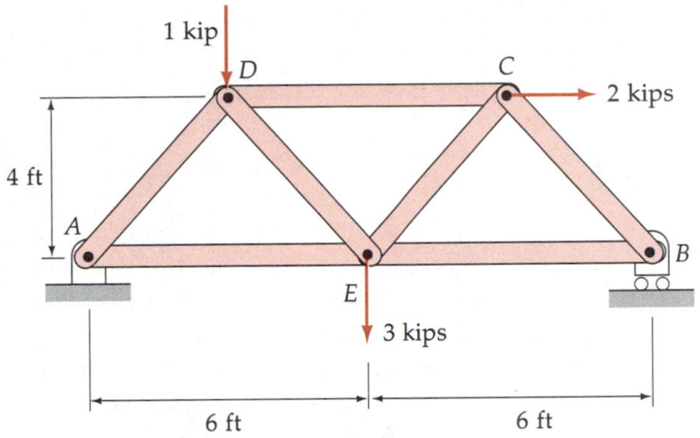

Figure E5.1a

Solution

We begin by finding the reactions at A and B, which are the forces exerted on the truss by its supports. Using the free-body diagram of the overall structure, we obtain

$$\circlearrowleft\!+ \qquad \Sigma M_A = 0 = (12\text{ ft})F_B - (6\text{ ft})(3\text{ k}) - (3\text{ ft})(1\text{ k}) - (4\text{ ft})(2\text{ k})$$

Thus the reaction at the roller is

$$F_B = \frac{29}{12} = 2.42\text{ kips}$$

Then,

$$+\!\uparrow \qquad \Sigma F_y = 0 = F_{Ay} + \overset{2.42\text{ k}}{F_B} - 1\text{ k} - 3\text{ k}$$

so that the vertical component of the pin reaction is

$$F_{Ay} = 1.58\text{ k}$$

(The reader is encouraged to check this value of F_{A_y} by using $\Sigma M_B = 0$.) Continuing,

$$\overset{+}{\rightarrow} \qquad \Sigma F_x = 0 = 2\text{ k} - F_{Ax} \Rightarrow F_{Ax} = 2\text{ k}$$

We are now ready to use the method of joints. We use the free-body diagram [Figure 4.9(b)] discussed earlier and repeated in Figure E5.1b. We obtain the forces F_{CB} and F_{EB} by satisfying the equilibrium equations of the pin:

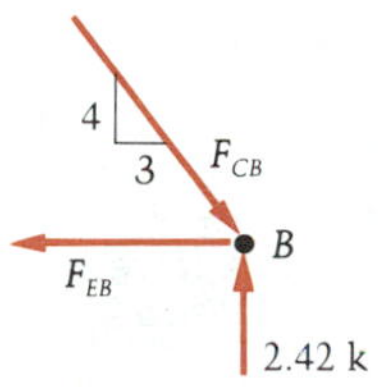

Figure E5.1b

$$+\uparrow \qquad \Sigma F_y = 0 = 2.42 - \frac{4}{5} F_{CB}$$

$$F_{CB} = \frac{5}{4}(2.42) = 3.03 \text{ k} \qquad (\text{or } F_{CB} = 3.03 \text{ k Ⓒ})$$

Because F_{CB} came out positive, the bar CB is in compression as assumed. Then,

$$\xrightarrow{+} \qquad \Sigma F_x = 0 = \frac{3}{5} F_{CB} - F_{EB}$$

$$F_{EB} = 0.6(3.03) = 1.82 \text{ k} \qquad (\text{or } F_{EB} = 1.82 \text{ k Ⓣ}).$$

Again the answer came out positive; this time we had assumed FB to be in tension, so it actually is.

As we have mentioned, letters Ⓒ and Ⓣ beside the force in a truss respectively indicate to us whether the member is in compression or tension. In the case of a truss member, these letters tell much more than a direction arrow or even than a vector representation. For example, if we were to say

$$\mathbf{F}_{CB} = 3.03 \; {}^{4}\!\searrow_{3} \text{ kips} \qquad \text{or} \qquad 3.03\left(\frac{3\hat{\mathbf{i}} - 4\hat{\mathbf{j}}}{5}\right) \text{ kips,}$$

then this is OK if what is meant is the force exerted by CB onto the pin at B, *and* if such is stated clearly. But if this force vector was used as the force exerted *by* the pin onto CB at B, *or* as the force exerted onto the pin at C by CB, then the result would be 180° away from the correct direction. Therefore, a Ⓒ beside the answer "$F_{CB} = 3.03$ kips" removes all this uncertainty.

Next we analyze the pin at C. Its free-body diagram is shown in Figure E5.1c. Enforcing the equilibrium of joint C,

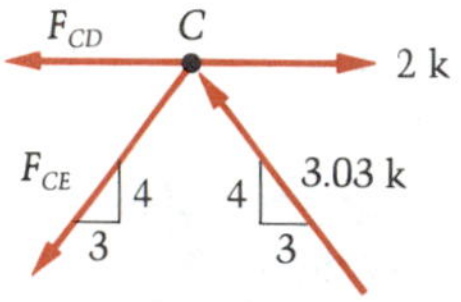

Figure E5.1c

$$+\uparrow \qquad \Sigma F_y = 0 = -F_{CE}\left(\frac{4}{5}\right) + 3.03\left(\frac{4}{5}\right)$$

$$F_{CE} = 3.03 \text{ k} \qquad (\text{or } F_{CE} = 3.03 \text{ k Ⓣ})$$

and

$$\xrightarrow{+} \qquad \Sigma F_x = 0 = 2 - F_{CD} - 3.03\left(\frac{3}{5}\right) - \overset{3.03\text{ k}}{\not{F}_{CE}}\left(\frac{3}{5}\right)$$

$$F_{CD} = -1.64 \text{ k} \qquad (\text{or } F_{CD} = 1.64 \text{ k Ⓒ})$$

Question 5.1 Why do we use pin C prior to E at this stage of the solution?

Question 5.2 Why is the 3.03 kip force acting upward and to the left on pin C in Figure E5.1c?

Question 5.3 Why is the direction associated with F_{CE} in the diagram bound to be correct here, and not just a guess?

Answer 5.1 Pin C will at this stage have two unknown forces acting on it; pin E has four.

Answer 5.2 Because we found it to be compressive when we examined pin B. If we were to now draw it tensile on pin C, we would be in violation of the action-reaction principle!

Answer 5.3 Because unless F_{CE} is tensile, ΣF_y cannot vanish for pin C.

This time, we have encountered an incorrect guess for the first time. By the way we drew the force F_{CD} in Figure 5.1c, we had "assumed" member CD to be in tension. We see now that it is in compression, as evidenced by the solution $F_{CD} = -1.64$ k. Thus, the force in CD is expressed as 1.64 k Ⓒ.

We next examine pin D. Assuming members DA and DE to be in compression yields the free-body diagram in Figure E5.1d.

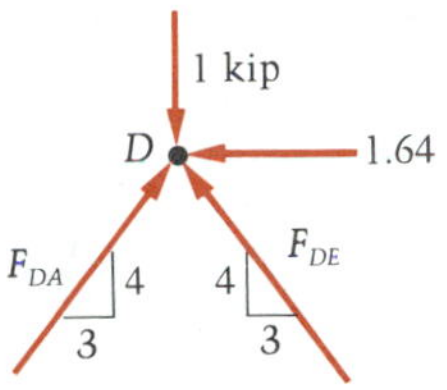

Figure E5.1d

Question 5.4 Explain why these two sets of assumed directions:

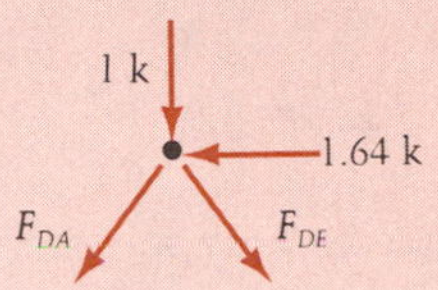

and

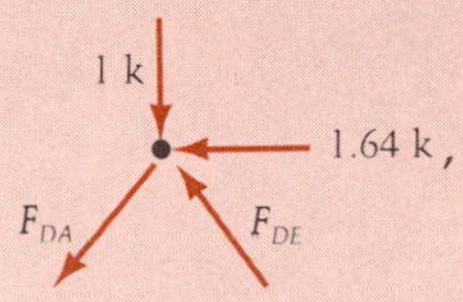

, have no chance of being correct.

Writing the equilibrium equations,

$$\overset{+}{\rightarrow} \qquad \Sigma F_x = 0 = F_{DA}\left(\frac{3}{5}\right) - F_{DE}\left(\frac{3}{5}\right) - 1.64$$

$$+\uparrow \qquad \Sigma F_y = 0 = F_{DA}\left(\frac{4}{5}\right) + F_{DE}\left(\frac{4}{5}\right) - 1$$

or

$$F_{DA} - F_{DE} = +\frac{5}{3}(1.64) = +2.73 \text{ k}$$

$$F_{DA} + F_{DE} = \frac{5}{4}(1) = 1.25 \text{ k}$$

Adding, we eliminate F_{DE} and find F_{DA}:

$$2F_{DA} = +3.98 \Rightarrow F_{DA} = 1.99 \text{ k}$$

Subtracting,

$$2F_{DE} = -1.48 \Rightarrow F_{DE} = -0.74 \text{ k}$$

Thus we have guessed the wrong direction for the force in DE. Because we assumed compression, it is actually in tension:

$$F_{DE} = 0.74 \text{ k Ⓣ} \qquad F_{DA} = 1.99 \text{ k Ⓒ}$$

At the pin at A in Figure E5.1e there is now but one unknown, which is the force in AE. Checking the equilibrium in the vertical direction, we see that

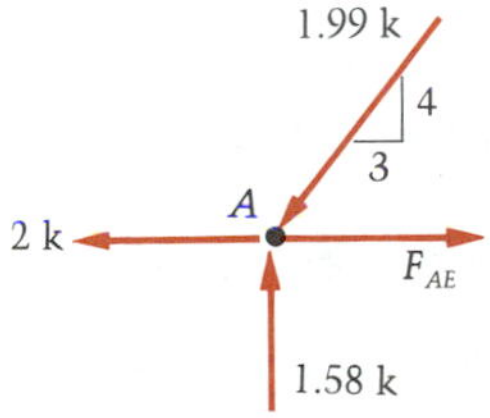

Figure E5.1e

$$+\uparrow \qquad \Sigma F_y = 1.58 - 1.99\left(\frac{4}{5}\right) = -0.01 \approx 0$$

the difference being due to numerical roundoff. In the x direction,

$$\overset{+}{\rightarrow} \qquad \Sigma F_x = 0 = F_{AE} - 2 - 1.99\left(\frac{3}{5}\right)$$

$$F_{AE} = 3.19 \text{ kips} \qquad (\text{or } F_{AE} = 3.19 \text{ kips Ⓣ})$$

Answer 5.4 If both F_{DA} and F_{DE} were to turn out positive as indicated, then: (a) the first sketch leaves $\Sigma F_y \neq 0$; (b) the second sketch leaves $\Sigma F_x \neq 0$.

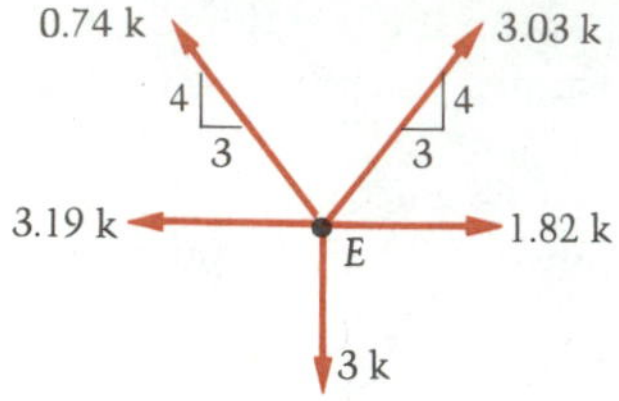

Figure E5.1f

Finally, we may use pin E as a check on the results. We come into it with calculated forces from five different directions (see Figure E5.1f).

The four bar forces in the free-body diagram were all computed to be in tension; thus, they each pull on the pin at E. Checking the equilibrium of the pin, we see that

$$\overset{+}{\longrightarrow} \qquad \Sigma F_x = 1.82 + 3.03\left(\frac{3}{5}\right) - 0.74\left(\frac{3}{5}\right) - 3.19 = 0.00$$

and

$$+\!\uparrow \qquad \Sigma F_y = 3.03\left(\frac{4}{5}\right) + 0.74\left(\frac{4}{5}\right) - 3 = 0.02 \approx 0$$

Thus we have successfully solved the truss.

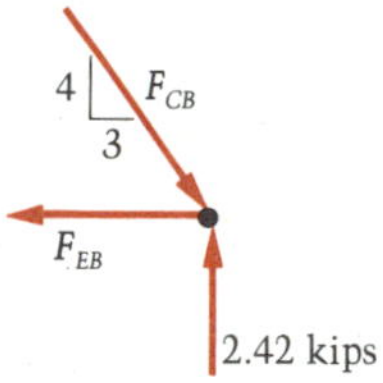

Figure 5.12

In the preceding example, the first free-body diagram was of the pin at B. (See Figure 5.12.) It could equally well have been drawn as shown in Figure 5.13, which is a free-body diagram of the connection, plus short, cut lengths of members CB and EB. With this approach, one sees more clearly the tension or compression in the bars.

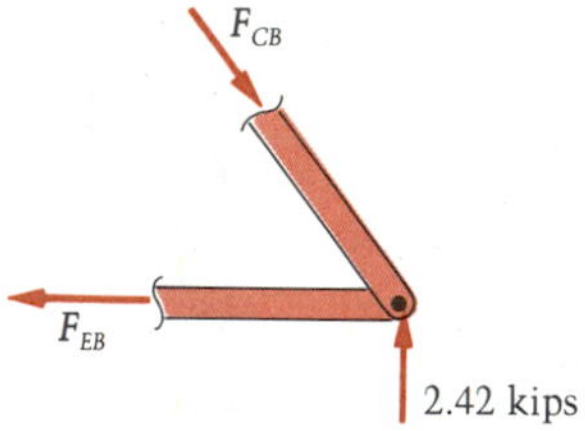

Figure 5.13

Question 5.5 Why does

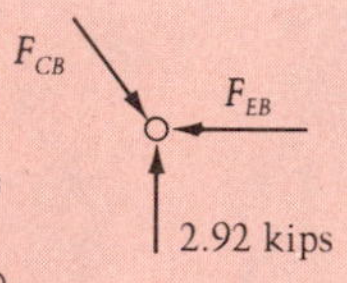

appear more *phys-ically* realistic than

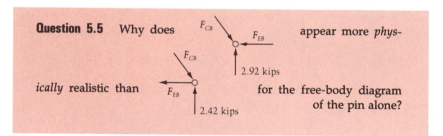

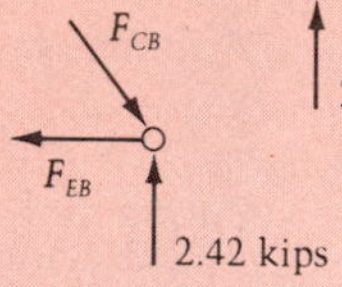

for the free-body diagram of the pin alone?

In Example 5.1 the reader may have noticed that no member weights were considered. If a structure is to be analyzed as a truss, then, by definition, its members can be loaded only by forces at their pinned ends. Weights of truss members will be neglected in the truss analyses in this book.

In our second example, we shall not explain each step in quite as much detail.

Answer 5.5 The force exerted by the inside of the hole (in member EB) onto the pin comes from the right side of the hole; that is, it pushes to the left instead of pulling to the left. Nonetheless, because forces may be translated along their lines of action, the free-body diagram of the pin is often drawn as in the second sketch.

Example 5.2

Find the forces in the members of the truss in Figure E5.2a.

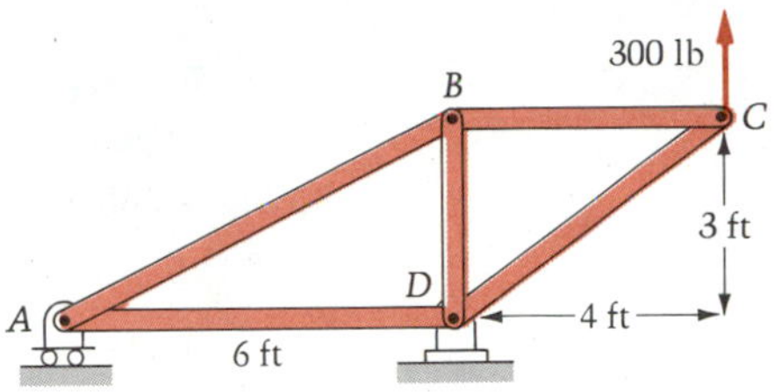

Figure E5.2a

Solution

We calculate the reactions, but only for use at the end in checking our solution. We use the overall free-body diagram; Figure E5.2b, and obtain:

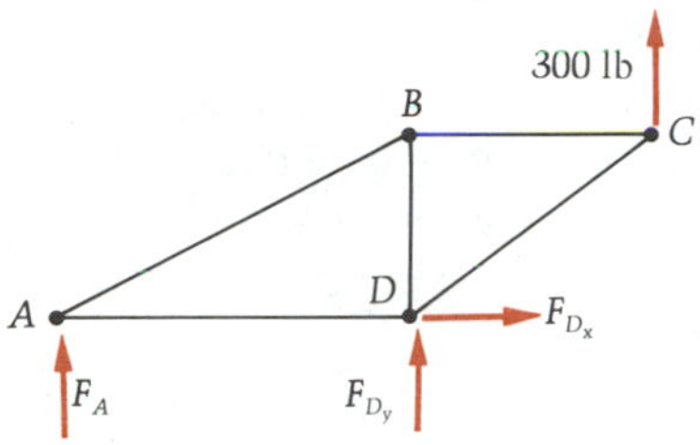

Figure E5.2b

$$\circlearrowleft + \quad \Sigma M_D = 0 = (300 \text{ lb})(4 \text{ ft}) - F_A\,(6 \text{ ft})$$

$$F_A = 200 \text{ lb} \qquad \text{(the roller reaction)}$$

$$\xrightarrow{+} \quad \Sigma F_x = 0 = F_{Dx}$$

$$+\uparrow \quad \Sigma F_y = 0 = F_{Dy} + F_A + 300$$

$$F_{Dy} = -300 - 200 \quad = -500 \text{ lb}$$

Thus the pin reaction is 500 lb ↓. We now use the method of joints to determine the forces in the members. From a free-body diagram of pin C (see Figure E5.2c), we have

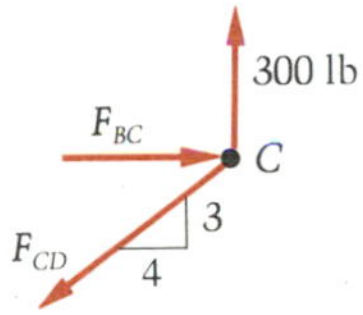

Figure E5.2c

$$+\uparrow \quad \Sigma F_y = 0 = 300 - F_{CD}\left(\frac{3}{5}\right) \Rightarrow F_{CD} = 500 \text{ lb} \qquad (\text{or } 500 \text{ lb} \text{ Ⓣ})$$

$$\xrightarrow{+} \quad \Sigma F_x = 0 = F_{BC} - F_{CD}\left(\frac{4}{5}\right) = F_{BC} - (500)\left(\frac{4}{5}\right)$$

$$F_{BC} = 400 \text{ lb} \qquad (\text{or } F_{BC} = 400 \text{ lb Ⓒ})$$

(Note that in finding F_{CD} or F_{BC}, we did not need the reactions.) Next, from a free-body diagram of pin B (Figure E5.2d), we obtain:

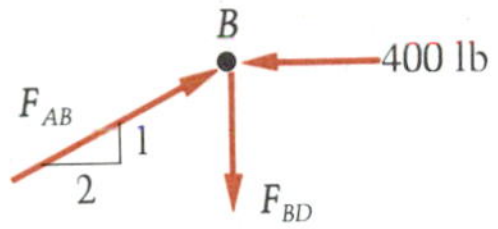

Figure E5.2d

$$\xrightarrow{+} \quad \Sigma F_x = 0 = F_{AB}\frac{2}{\sqrt{5}} - 400 \Rightarrow F_{AB} = 200\sqrt{5} \text{ lb } (\text{or } 200\sqrt{5} \text{ lb Ⓒ})$$

$$+\uparrow \quad \Sigma F_y = 0 = F_{AB}\left(\frac{1}{\sqrt{5}}\right) - F_{BD} = 200 - F_{BD}$$

or

$$F_{BD} = 200 \text{ lb} \qquad (\text{or } 200 \text{ lb Ⓣ})$$

And from a free-body diagram of A (Figure E5.2e), we may write

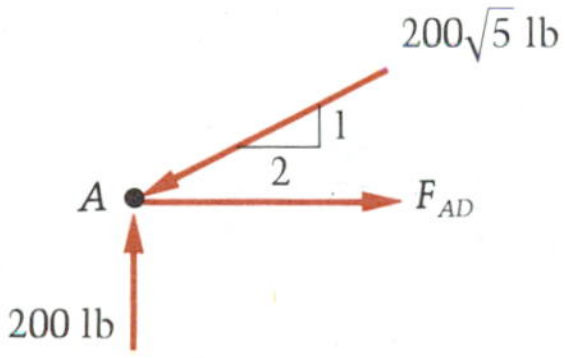

Figure E5.2e

$$\xrightarrow{+} \quad \Sigma F_x = 0 = -200\sqrt{5}\,\frac{2}{\sqrt{5}} + F_{AD} \Rightarrow F_{AD} = 400 \text{ lb} \qquad (\text{or } F_{AD} = 400 \text{ lb Ⓣ})$$

and for a check, using the precalculated 200-lb roller reaction,

$$+\uparrow \quad \Sigma F_y = 200 - 200\sqrt{5}\,\frac{1}{\sqrt{5}} = 0 \quad ✓$$

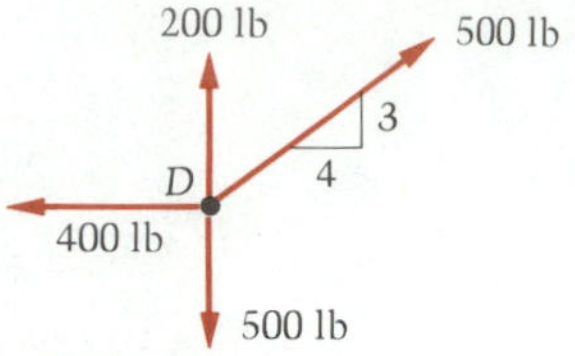

Figure E5.2f

This completes the solution of the truss. As in our other example, though, we can check our results at the remaining joint D. Applying the now-known member forces and reactions to the pin, we get the free-body diagram in Figure E5.2f and the following equations:

$$\xrightarrow{+} \quad \Sigma F_x = 500\left(\frac{4}{5}\right) - 400 = 0 \quad \checkmark$$

$$+\uparrow \quad \Sigma F_y = 200 + 500\left(\frac{3}{5}\right) - 500 = 0 \quad \checkmark$$

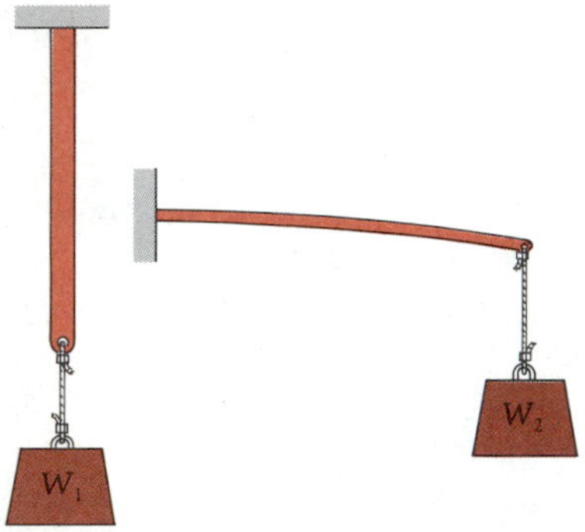

Figure 5.14

Because of the way its members are loaded, a truss is an extremely efficient, lightweight structure. To emphasize this idea, we encourage the reader to think about the difference between the weights W_1 and W_2 that could be safely supported by a yardstick in the two manners shown in Figure 5.14. In the diagram at the left, the member is loaded as it would be in a truss; at the right, the member is a beam and is not a two-force member because the reaction at the wall includes a couple and a force component normal to the member's axis.

We shall work through one more plane truss example with the method of joints in the next section, after first discussing some shortcuts that sometimes make truss analysis easier.

PROBLEMS ▶ Section 5.3

Find the forces in each member of the trusses in Problems 5.1–5.11.

5.1

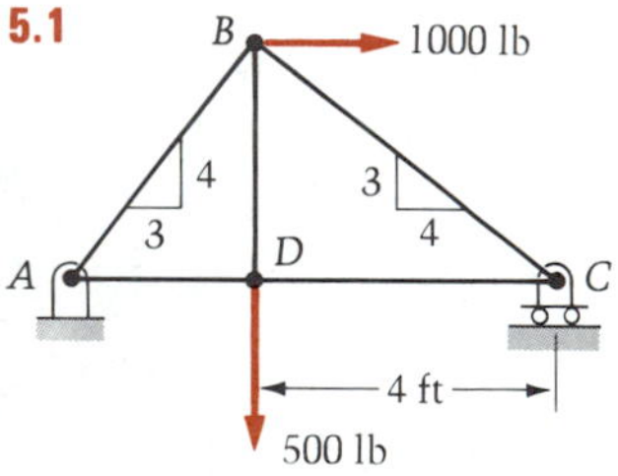

Figure P5.1

5.2

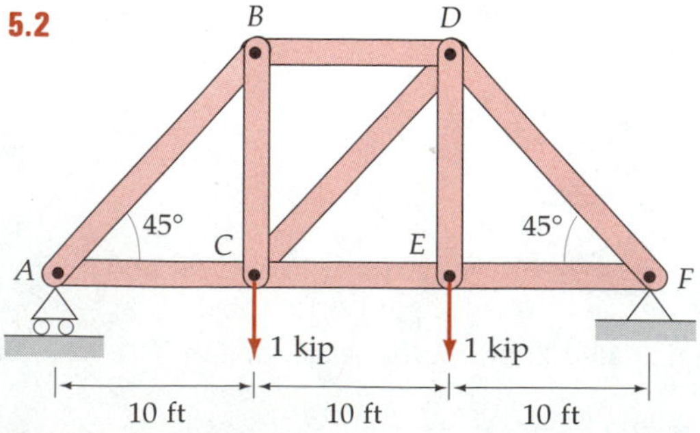

Figure P5.2

5.3

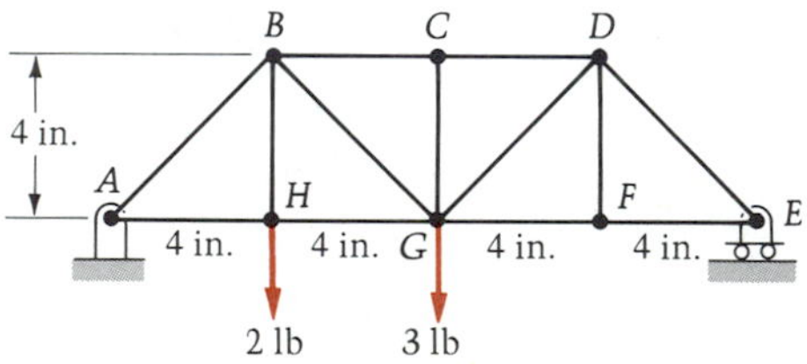

Figure P5.3

5.4

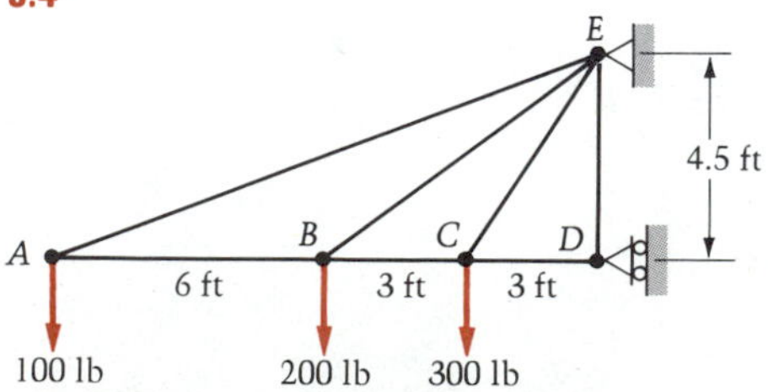

Figure P5.4

5.5

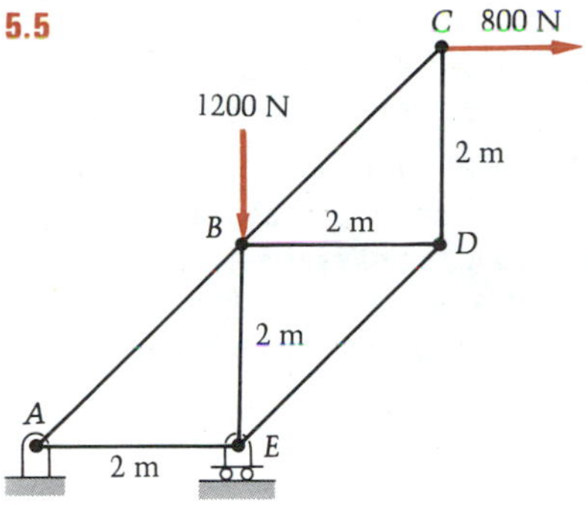

Figure P5.5

*** 5.9**

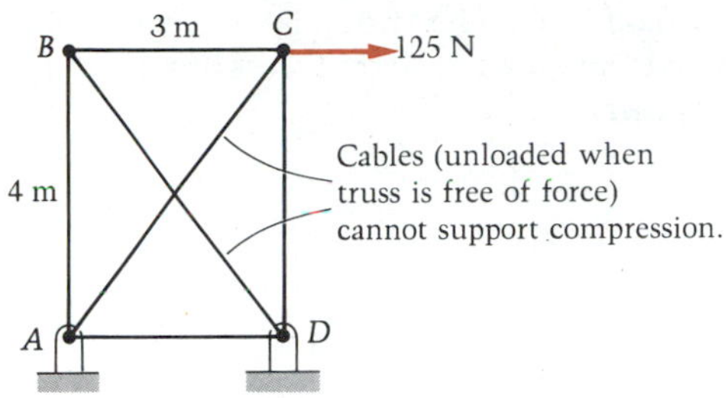

Figure P5.9

5.6

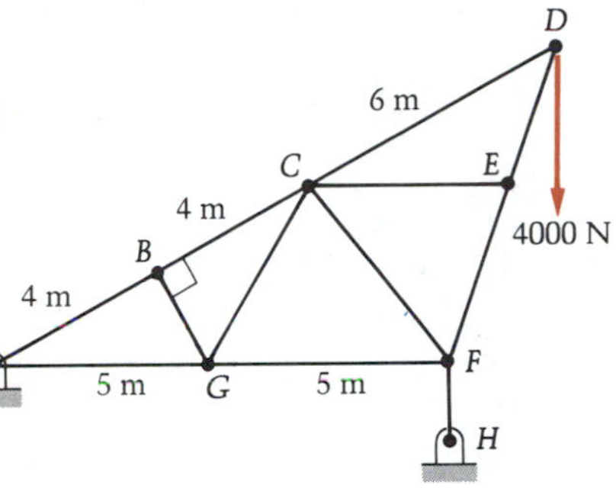

Figure P5.6

5.10

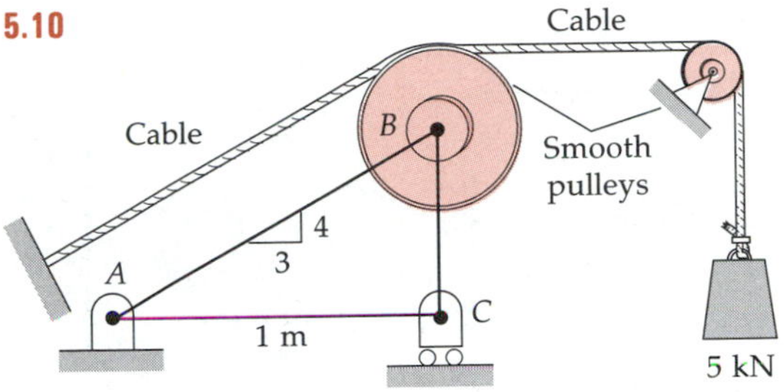

Figure P5.10

5.7

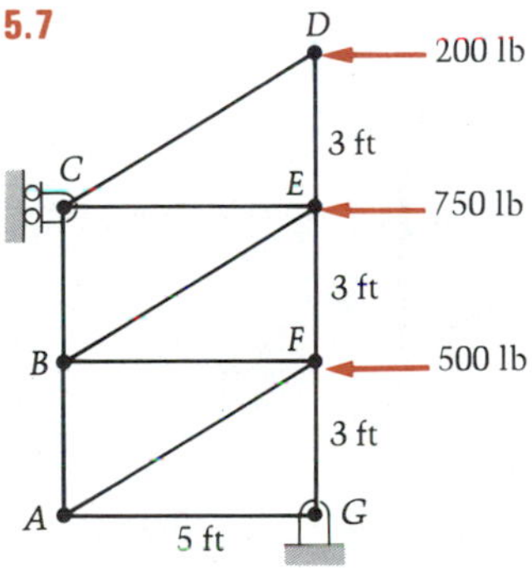

Figure P5.7

5.11

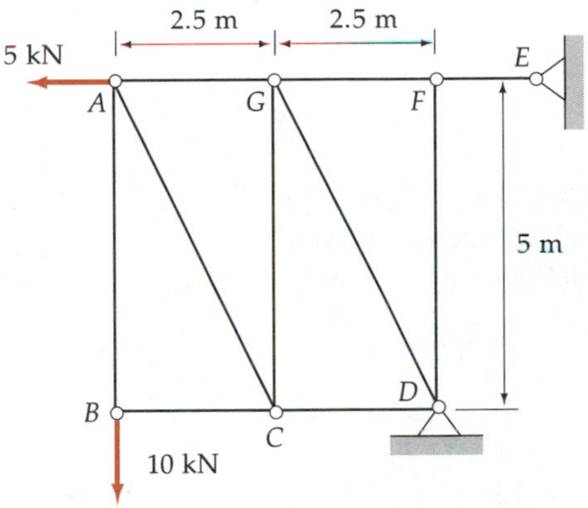

Figure P5.11

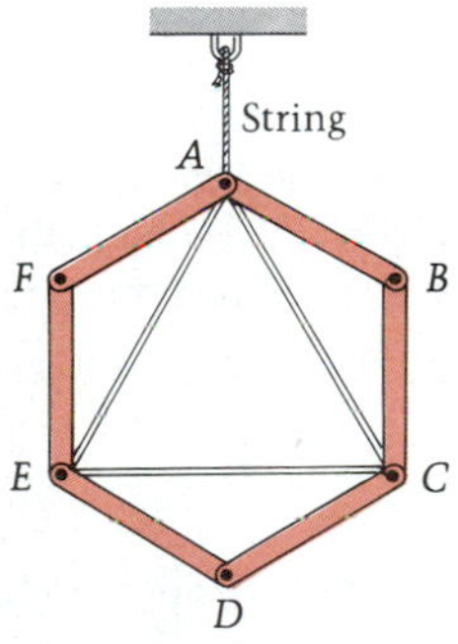

Figure P5.12

5.8

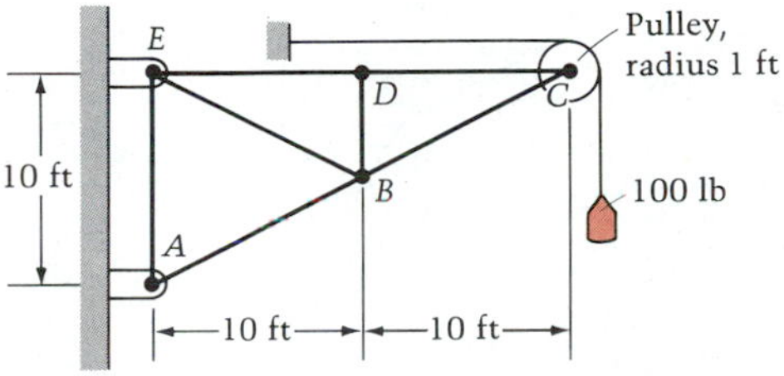

Figure P5.8

*** 5.12** In Figure P5.12 the hexagon *ABCDEF* of six uniform pinned rods *AB*, *BC*, *CD*, *DE*, *EF*, and *FA* of equal lengths and weights *W*, is stiffened by light rods *AC*, *CE*, and *EA* and suspended by the string at *A*. Find the tension in *AC*, and then the upward force that must be applied at *D* to reduce the force in *AC* to zero.

* Asterisks identify the more difficult problems.

5.13 Members *AE* and *EQ* are pinned together as shown in Figure P5.13 to form a billboard. It is subjected to the given distributed wind load. Find the forces in each of the two-force members.

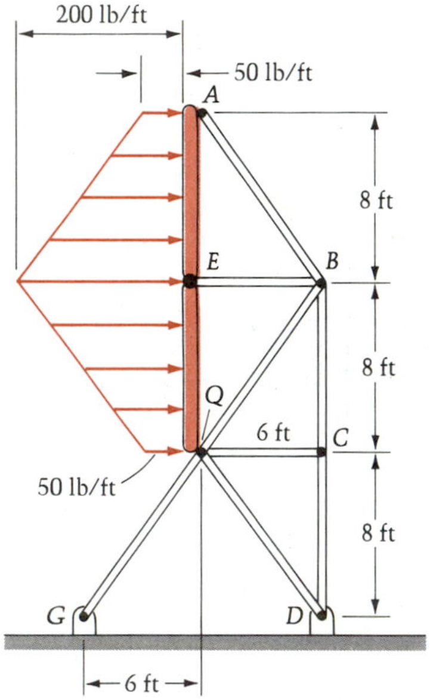

Figure P5.13

5.14 Find the force in member *BD* of the truss shown in Figure P5.14. Note that the external reactions are indeterminate and cannot be found by statics alone.

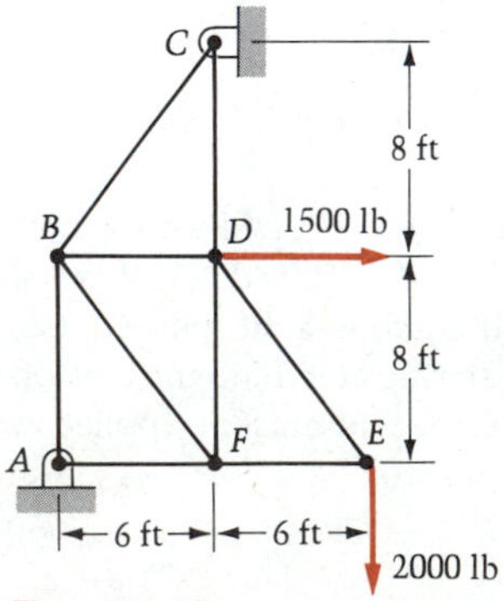

Figure P5.14

5.15 Find the forces in truss members *AB*, *BH*, and *OH* shown in Figure P5.15. Note that the external reactions are indeterminate and cannot be found by statics alone.

5.16 The truss in Figure P5.16 is pin-connected and is supported by the pin at *A*, and by the cable attached at *B* and *C*. The cable passes over a pulley that is connected to the reference frame by a smooth pin. Neglecting the weights of the truss members, find the force in *BD*.

* **5.17** In Figure P5.17 find the forces in all the members of the truss, in terms of *W*.

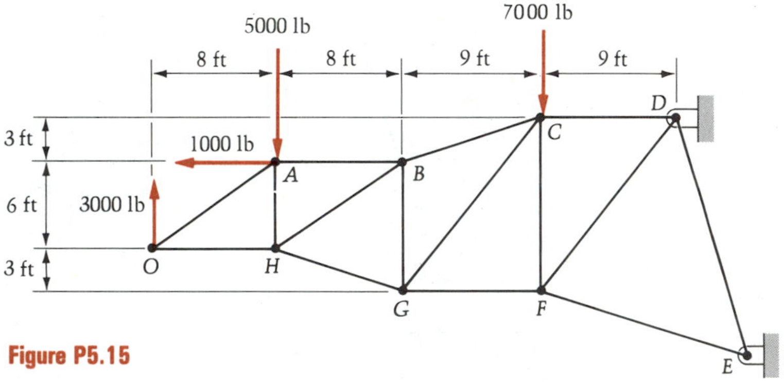

Figure P5.15

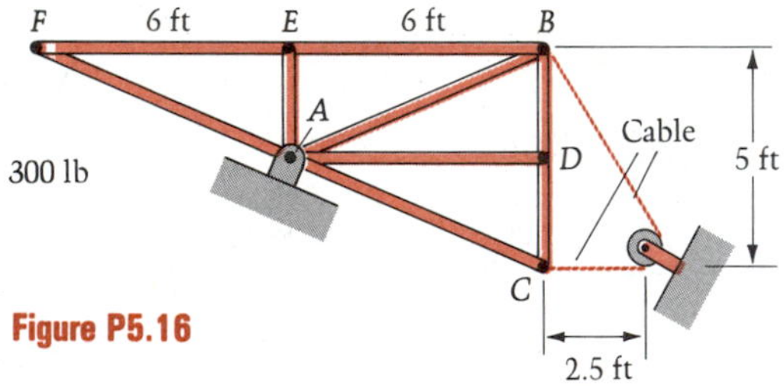

Figure P5.16

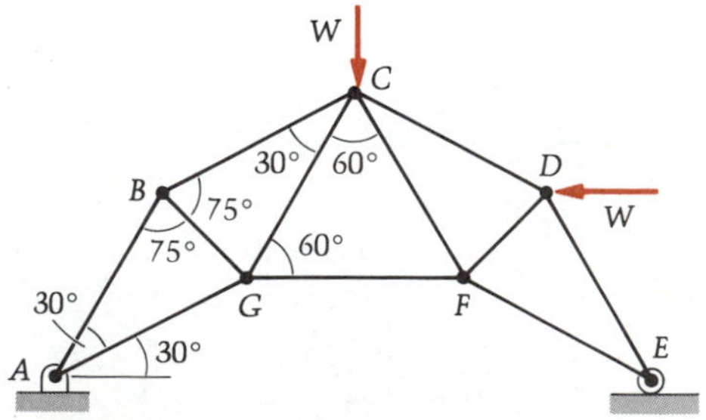

Figure P5.17

5.18 Determine the force in each member of the truss shown in Figure P5.18 if $a = 4$ ft, $b = 8$ ft, and $c = d = 3$ ft.

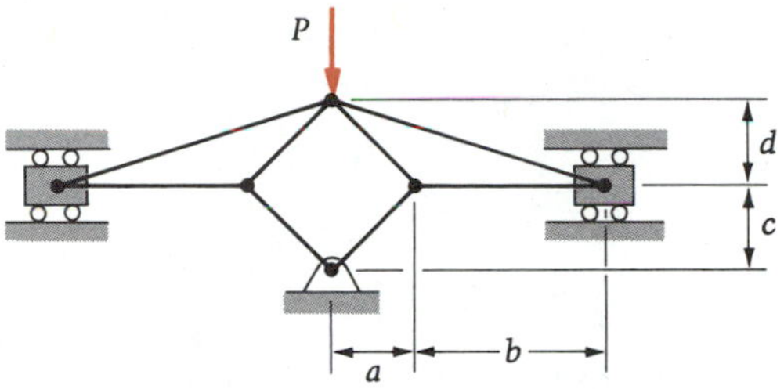

Figure P5.18

5.19 Find the forces in all members of the truss shown in Figure P5.19.

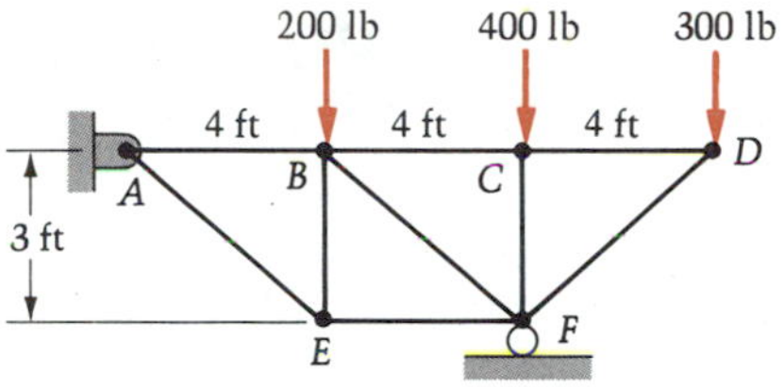

Figure P5.19

5.20 Find the force in member *CG* of the truss in Figure P5.20.

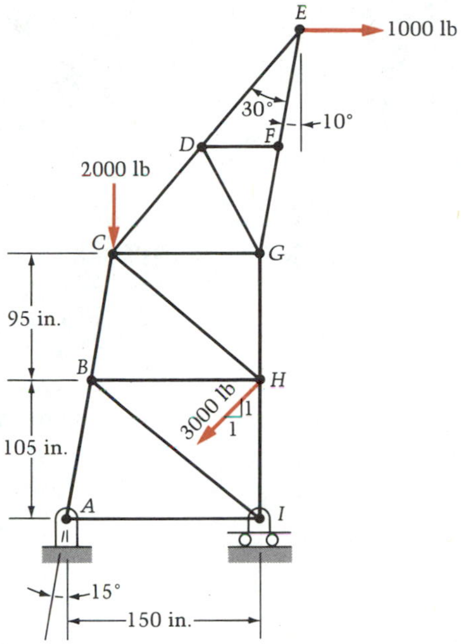

Figure P5.20

5.21 Find the forces in members *AB*, *AF*, *AG*, and *CD* of the truss shown in Figure P5.21.

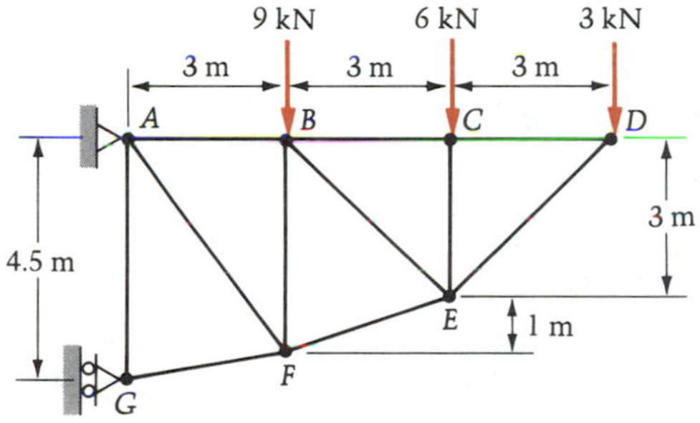

Figure P5.21

5.4 Shortcuts and Rigidity/Determinacy Results

Zero-force Members

There are several common shortcuts to watch for in analyzing a truss. The first involves what are called "zero-force members." Sometimes in a truss we will find at a joint J that only one member could carry a component of force in a certain direction and that there are no external loads at J in that direction. For example, at joint A in Figure 5.15, only member AE can have a force component normal to line CAD. If we call this normal direction u, then from the free-body diagram of pin A (see Figure 5.16 on the next page), it is seen that F_{AE} must be zero:

$$\overset{+}{\nwarrow} \qquad \Sigma F_u = 0 = F_{AE} \cos \theta$$

and so

$$F_{AE} = 0 \qquad (\text{note } \theta \neq 90°)$$

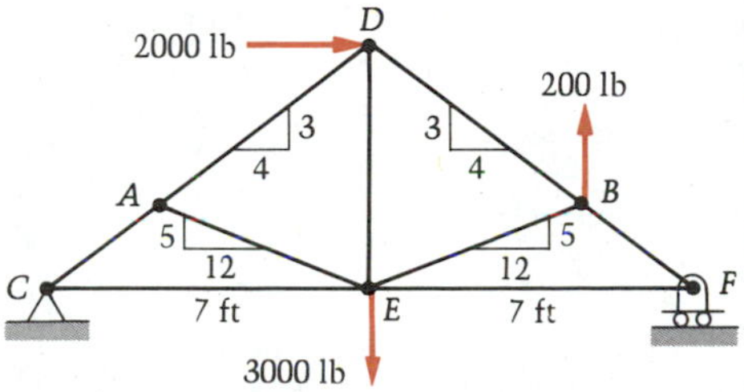

Figure 5.15

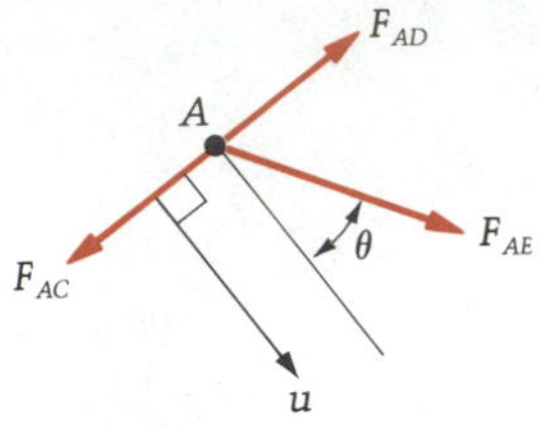

Figure 5.16

Thus AE is a "zero-force member" of the truss. In general, this only happens when there is no external loading at the pin with a component along the direction of the bar (such as AE) that is being examined for a possible zero value. For example, at B (Figure 5.17) we have only BE able

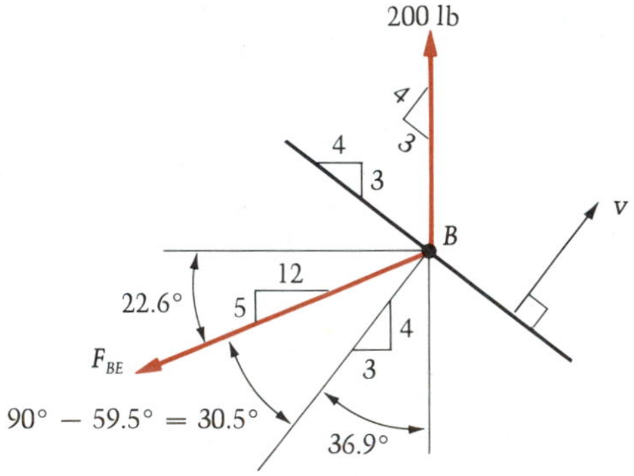

Figure 5.17

to carry a force normal to DBF. This time, however, this normal component balances the component of the 200-lb external force in that direction:

$$\nearrow + \qquad \Sigma F_v = 0 = 200\left(\frac{4}{5}\right) - F_{BE}\cos 30.5°$$

$$F_{BE} = 186 \text{ lb}$$

Even though $F_{BE} \neq 0$ in this case, its value is nonetheless determined from a single equation, which is still a help.

Question 5.6 Give reasons why a zero-force member like AE in Figure 5.16 is still an important part of the truss even if it isn't carrying any load.

Answer 5.6 For *other* external loads, the member may carry a force. Also, the member really does carry load due to the (neglected) weights of the members. It also adds stability to the structure.

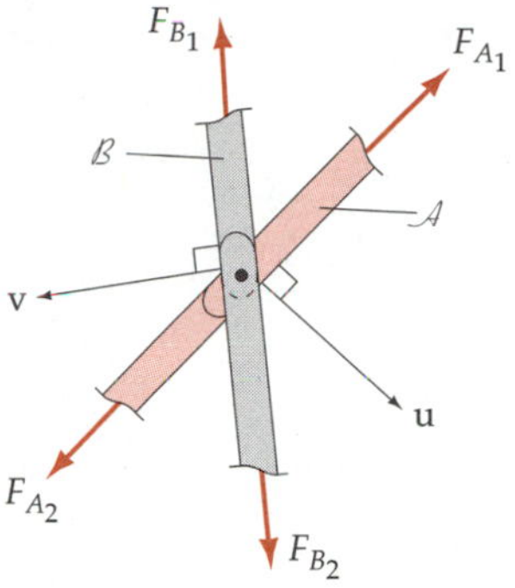

Figure 5.18

Other Shortcuts

A second shortcut in truss analysis arises in a situation where four members that are lined up in pairs as indicated in Figure 5.18 meet at a pin. If there are no other members or external forces at that joint, then we see from Figure 5.18 that the forces F_{A_1} and F_{A_2} are equal, from the equation $\Sigma F_v = 0$ applied to the free body shown. In the same way, $\Sigma F_u = 0$ gives $F_{B_1} = F_{B_2}$.

A third shortcut arises when two non-collinear members are joined at a pin where no other bar or external force appears (see Figure 5.19(a)). At pin B the equilibrium equations, written with the help of the free-body diagram, show that both F_{BA} and F_{BC} are zero (see Figure 5.19(b)):

$$\Sigma F_u = 0 = -F_{BC} \cos \theta$$

$$F_{BC} = 0$$

$$\Sigma F_v = 0 = -F_{BA} \cos \beta$$

$$F_{BA} = 0$$

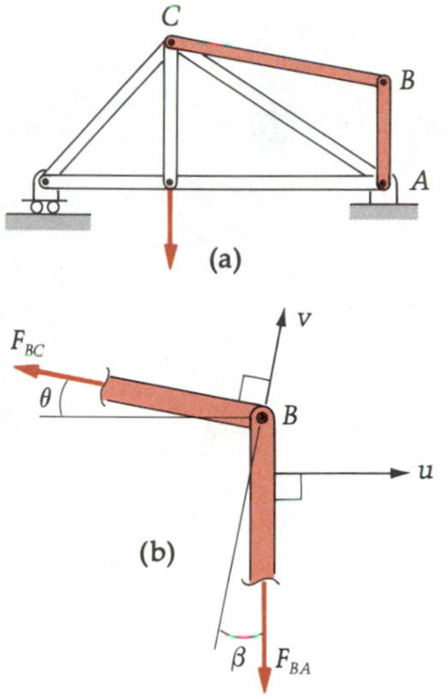

Figure 5.19

EXAMPLE 5.3

Find the forces in the members of the truss in Figure E5.3a.

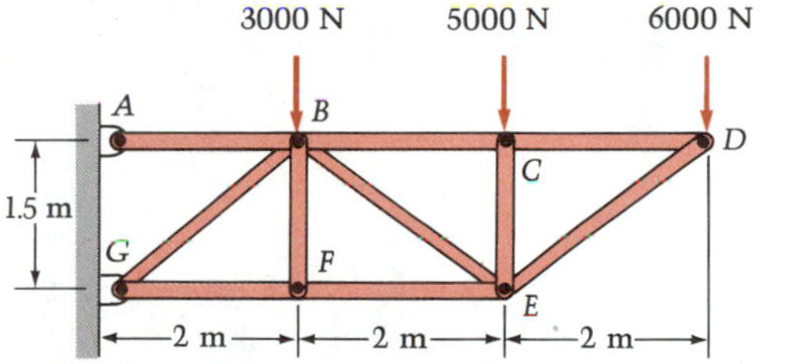

Figure E5.3a

Solution

We shall retain four digits in this example for numerical checking purposes. Two of the answers are known by inspection:

1. $F_{BF} = 0$, by considering $\Sigma F_y = 0$ at joint F.
2. $F_{CE} = 5000$ N Ⓒ, by considering $\Sigma F_y = 0$ at joint C.

In a simple truss such as this one, as we shall see later in general, we can always find a joint where only two unknown bar forces act. Here, it is joint D shown in Figure E5.3b:

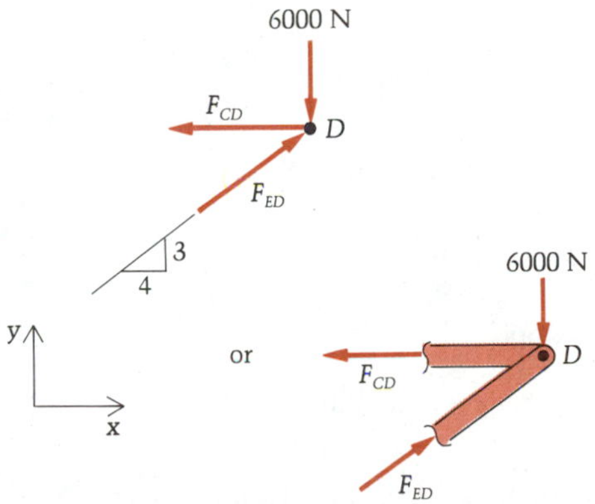

Figure E5.3b

$$+\uparrow \qquad \Sigma F_y = 0 = F_{ED}\left(\frac{3}{5}\right) - 6000 \Rightarrow F_{ED} = 10{,}000 \text{ N Ⓒ}$$

$$\xrightarrow{+} \qquad \Sigma F_x = 0 = F_{ED}\left(\frac{4}{5}\right) - F_{CD} = 10{,}000\left(\frac{4}{5}\right) - F_{CD}$$

or

$$F_{CD} = 8000 \text{ N} \qquad (\text{or } F_{CD} = 8000 \text{ N Ⓣ})$$

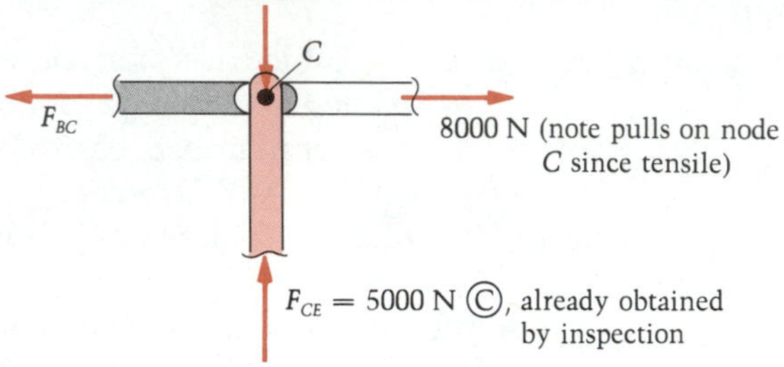

Figure E5.3c

Next, we isolate joint C (see Figure E5.3c), and obtain:

$$\overset{+}{\longrightarrow} \qquad \Sigma F_x = 0 = 8000 - F_{BC}$$

$$F_{BC} = 8000 \text{ N} \qquad (\text{or } 8000 \text{ N Ⓣ})$$

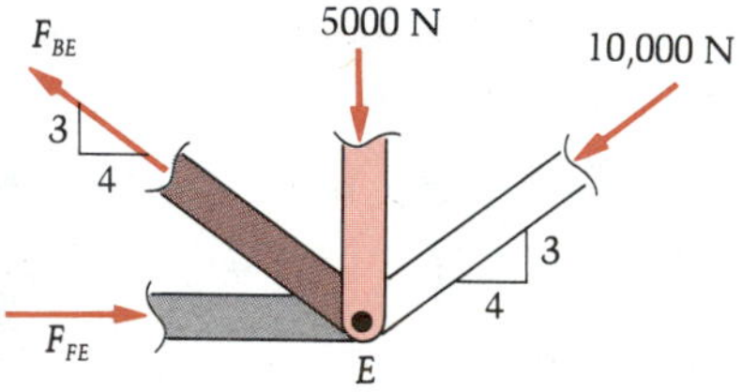

Figure E5.3d

We next use joint E (see Figure E5.3d), because at present B has three unknown forces acting on it.

Question 5.7 In this case, we know ahead of time that F_{BE} is Ⓣ and F_{FE} is Ⓒ. Why?

The equilibrium equations are:

$$+\!\uparrow \qquad \Sigma F_y = 0 = F_{BE}\left(\frac{3}{5}\right) - 10{,}000\left(\frac{3}{5}\right) - 5000$$

$$F_{BE} = 18{,}330 \text{ N} \qquad (\text{or } F_{BE} = 18{,}330 \text{ N Ⓣ})$$

$$\overset{+}{\longrightarrow} \qquad \Sigma F_x = 0 = F_{FE} - F_{BE}\left(\frac{4}{5}\right) - 10{,}000\left(\frac{4}{5}\right)$$

$$F_{FE} = 18{,}330(0.8) + 8000$$

$$F_{FE} = 22{,}660 \text{ N} \qquad (\text{or } F_{FE} = 22{,}660 \text{ N Ⓒ})$$

Answer 5.7 First, F_{BE} has to be tensile so that ΣF_y can be zero. Then F_{FE} has to be compressive in order that ΣF_x can vanish.

By inspection of joint F, the force in GF is the same as FE. This is because no other horizontal forces act on the pin at F, so that GF and FE have to equilibrate each other there. Thus,

$$F_{GF} = 22{,}660 \text{ N} \qquad (\text{or } F_{GF} = 22{,}660 \text{ N Ⓒ})$$

At B, the pin has the forces acting as shown in Figure E5.3e. We obtain:

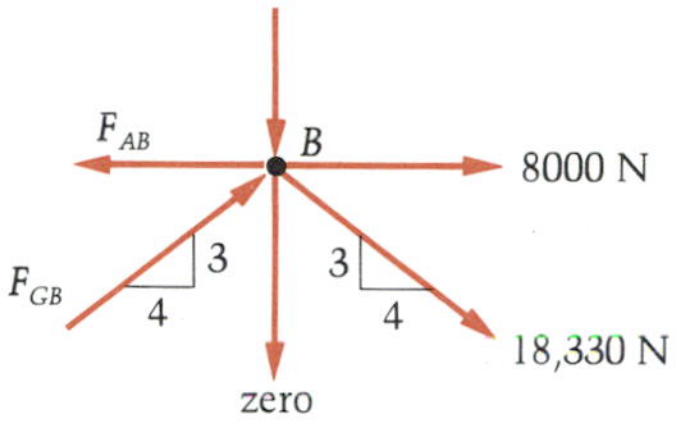

Figure E5.3e

$$+\uparrow \qquad \Sigma F_y = 0 = F_{GB}\left(\frac{3}{5}\right) - 18{,}330\left(\frac{3}{5}\right) - 3000$$

$$F_{GB} = 23{,}330 \text{ N} \qquad (\text{or } F_{GB} = 23{,}330 \text{ N Ⓒ})$$

$$\overset{+}{\rightarrow} \qquad \Sigma F_x = 0 = F_{GB}\left(\frac{4}{5}\right) + 18{,}330\left(\frac{4}{5}\right) + 8000 - F_{AB}$$

$$F_{AB} = 41{,}330 \text{ N} \qquad (\text{or } F_{AB} = 41{,}330 \text{ N Ⓣ})$$

In practice, truss members such as AB that carry much more load than others for typical expected loadings will be made larger in cross section.

Finally, free-body diagrams of the pins at A and G (see Figures E5.3f and E5.3g) allow us to compute the reactions there (onto the pins from the clevis attached to the wall).

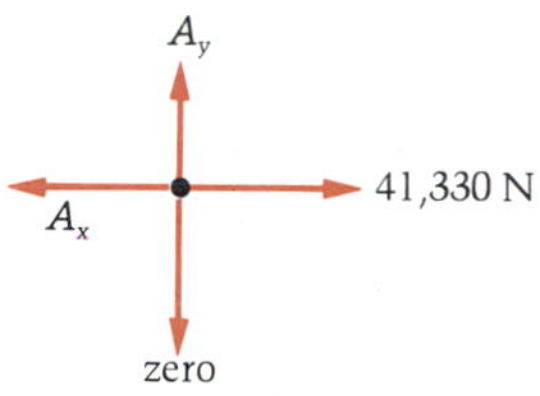

Figure E5.3f

$$+\uparrow \qquad \Sigma F_y = 0 \Rightarrow A_y = 0$$

$$\overset{+}{\rightarrow} \qquad \Sigma F_x = 0 \Rightarrow A_x = 41{,}330 \text{ N}$$

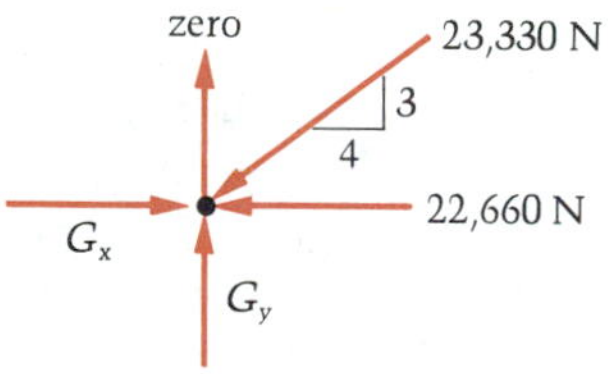

Figure E5.3g

$$\overset{+}{\rightarrow} \qquad \Sigma F_x = 0 = G_x - 22{,}660 - 23{,}330\left(\frac{4}{5}\right)$$

$$G_x = 41{,}320 \text{ N}$$

$$+\uparrow \qquad \Sigma F_y = 0 = G_y - 23{,}330\left(\frac{3}{5}\right)$$

$$G_y = 14{,}000 \text{ N}$$

Note that the pin at G actually *feels* the vector sum of G_x and G_y in shear; that is, the clevis exerts on it the force

$$\mathbf{G} = 41{,}320\hat{\mathbf{i}} + 14{,}000\hat{\mathbf{j}} \text{ N}$$ (or 43,630 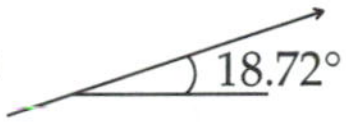 18.72° N)

Back-checking with the overall free-body diagram shown in Figure E5.3h, we see that our calculations look correct:

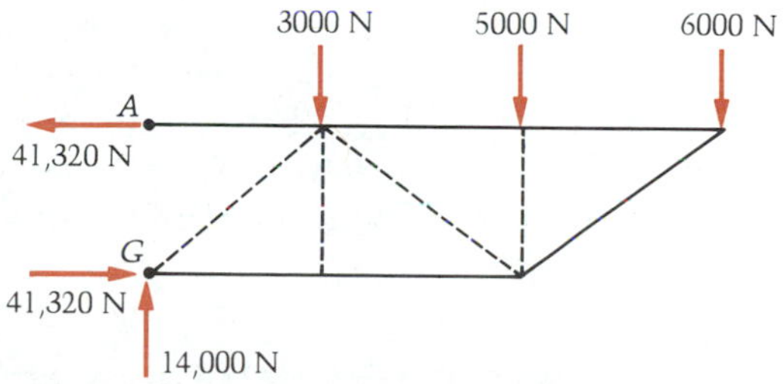

Figure E5.3h

It is obvious by inspection that ΣF_x and ΣF_y each vanish. Checking moments,

$$\circlearrowleft + \quad \Sigma M_G = (41{,}320 \text{ N})(1.5 \text{ m}) - (3000 \text{ N})(2 \text{ m}) - (5000 \text{ N})(4 \text{ m}) - (6000 \text{ N})(6 \text{ m})$$

$$= -20 \text{ N} \cdot \text{m} \quad \text{(differing from zero due to rounding to four digits)}$$

Note from the final figure that when considering the entire "overall" truss as a free-body diagram, one need not take time to draw in all the dashed internal members; the outside profile is sufficient unless there are forces applied at "internal" joints. We must, of course, remember that all the elements of the truss are actually being included.

Numbers of Members and Pins in a Truss

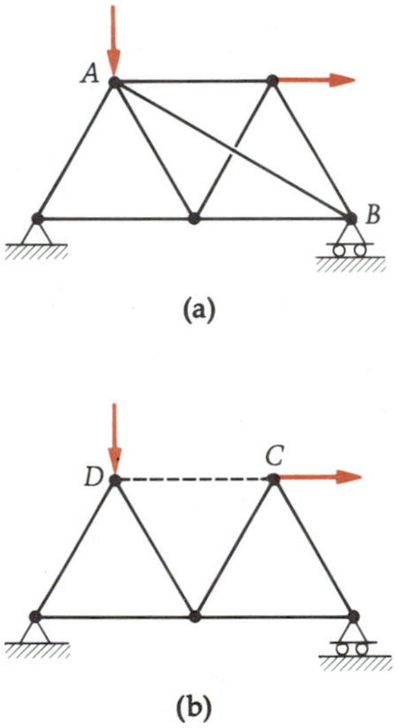

Figure 5.20

In Example 5.1 we recall that there were three equations used as checks (one at A and two at E) because they involved no new unknowns. These three redundant equations resulted because there are obviously $2 \times$ (number of pins) $= 2p$ independent equilibrium equations, and we used up three independent equations in finding the external reactions (using the overall free-body diagram). If a plane truss with three statically determinate reactions has more than $m = 2p - 3$ members, then we cannot solve for the forces in them all from the equilibrium equations alone, and the truss is then appropriately deemed statically indeterminate.* Though we may find the reactions, we cannot solve the truss if $m > 2p - 3$. On the other hand, if there are *fewer* than $2p - 3$ members with three statically determinate reactions, we then do not have enough member forces to satisfy all the equilibrium equations at the pins. In this case, the truss is not rigid. Figures 5.20(a) and 5.20(b) illustrate these two ideas. In Figure 5.20(a), adding member AB to the truss of Figure 5.6 makes the truss statically indeterminate; p still $= 5$, but m now $= 8$ and $2p - 3 = 7$, which is now $<m$. In Figure 5.20(b), on the other hand, removing member DC gives $m = 6$ while p still $= 5$, so that $2p - 3 = 7$, which is now $>m$. This leaves a non-rigid structure, which is unstable and will collapse as the reader may visualize.

Of course, the word *rigid* does not mean that a truss will not deform at all under loading. It will undergo very small deformations, very nearly retaining its original shape, as we shall see in Section 5.7.

We have seen that in a two-dimensional (or plane) truss, the pin connections are called joints (or nodes). They are the points where the members, or bars, that constitute the truss are joined together. If, for example, we begin with a single triangle (Figure 5.21(a)), it has three pins (p) (or joints) and three members (m). Thus $p = 3$ and $m = 3$. If we add,

* In Section 5.7, we will see how an indeterminate structure *can* be solved if *deformations* are included in the analysis.

from two distinct joints A and B of this triangle, two new members and pin them at a new, common joint C not on line AB, then we have (see Figure 5.21(b)):

$$p = 3 + 1$$
$$m = 3 + 2$$

Repeating this procedure once more, we obtain a truss similar to that of Figure 5.6 (see Figure 5.21(c)):

$$p = 3 + 1 + 1 = 5$$
$$m = 3 + 2 + 2 = 7$$

For a plane truss constructed in this way (and they often are), the number of pins (joints) and the number of members are related by

$$m = 3 + 2(p - 3) = 2p - 3$$

a relation we have seen before. For the last truss, $m = 2(5) - 3 = 7$. A plane truss constructed in the above manner is called a **simple truss.** If a simple truss is supported with a pin and roller (or their equivalent) so as to satisfy overall equilibrium for any loading, it will be a rigid, stable, determinate structure. Note in Figure 5.21(d) that the simple plane truss need not be made up of a series of connected triangles (though it often is). Note also that in a simple plane truss, we can always find at least one joint where there are only two unknown member forces.

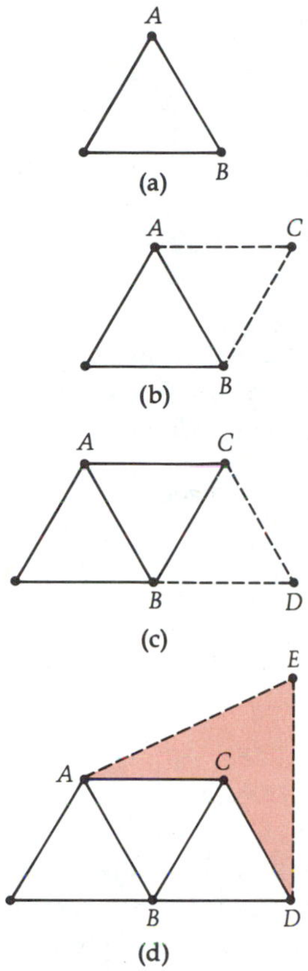

Figure 5.21

Question 5.8 How?

As we have already seen, this is a natural starting place for solving such a truss by the method of joints.

It is important to note that the condition $m = 2p - 3$ is not generally either sufficient or necessary for a *non*-simple plane truss to be rigid and statically determinate. For example, the (silly) truss shown in Figure 5.22 has $m = 2p - 3$ but is not rigid. The truss shown in Figure 5.23, however, has $m \neq 2p - 3$ but is, while not rigid, both stable *and* statically determinate. (Note the *two* support pins!)

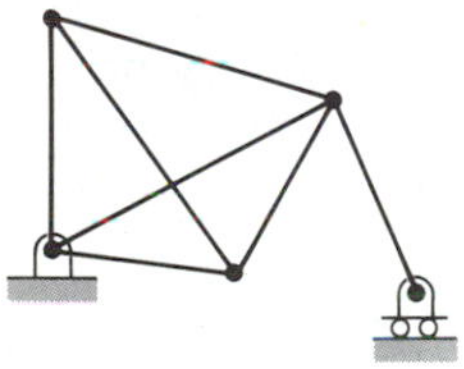
Figure 5.22 $p = 5; m = 7 = 2p - 3$

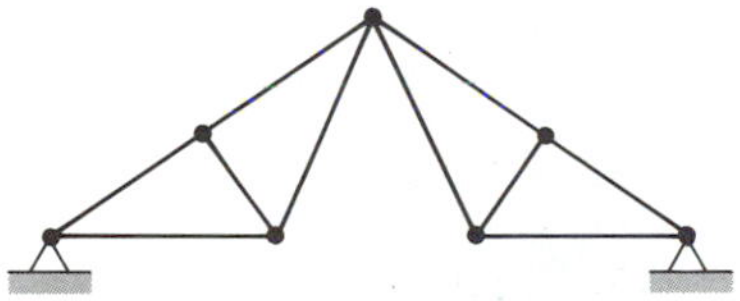
Figure 5.23 $p = 7; m = 10 \neq 2p - 3$

There are also well-known and commonly used plane trusses that with three determinate support reactions are both rigid and statically determinate but not simple (buildable from a triangle by successively adding two new members and one new pin at a time). One such truss is shown in Figure 5.24. Note again that the equation $m = 2p - 3$ does not guarantee a truss to be simple.

The truss of Figure 5.24 is called a compound truss, which is a truss comprising two or more simple trusses connected together so as to leave a rigid, determinate truss as the result. If the connection is made as

Answer 5.8 The last vertex drawn has but two members terminating at it.

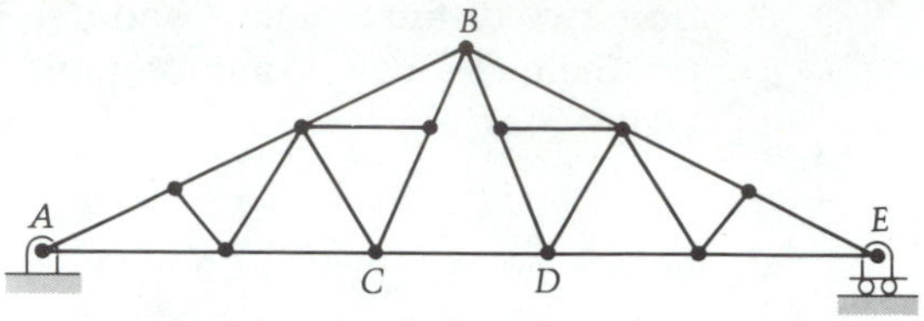

Figure 5.24 A Fink truss. $p = 15$; $m = 27 = 2p - 3$

in Figure 5.24—that is, by joining the simple trusses ABC and BED at B and adding one member (CD)—then, in general, if there are p_{ABC} pins and m_{ABC} members in ABC, and p_{BDE} pins and m_{BDE} members in BDE, we have

$$2p_{ABC} - 3 = m_{ABC} \qquad \text{and} \qquad 2p_{BDE} - 3 = m_{BDE}$$

and we see that for the combined truss,

$$\begin{aligned} 2p - 3 &= 2(p_{ABC} + p_{BDE} - 1) - 3 \\ &= (3 + m_{ABC}) + (3 + m_{BDE}) - 2 - 3 \\ &= m_{ABC} + m_{BDE} + 1 \end{aligned}$$

or

$$2p - 3 = m$$

Thus $m = 2p - 3$ even though (as the reader may wish to show) the compound truss is itself not simple. With proper support reactions such as a pin and a roller, a compound truss, like the simple trusses of which it is made, will be rigid (non-collapsible upon release from its supports) and statically determinate.

Question 5.9 Three of the entries in Figure 5.2, if made only of pinned-together two-force members, would not be able to maintain their shapes if detached from their supports. Which ones are they?

Answer 5.9 Attic, gable end, floor truss.

PROBLEMS ▶ Section 5.4

5.22 Find the forces in members GB and DF of the truss shown in Figure P5.22.

5.23 Find the forces in all members of the truss shown in Figure P5.23.

5.24 By inspection, identify six zero-force members in Figure P5.24 and explain why each vanishes.

5.25 Find the forces in members 1, 2, 3, 4, and 5 of the truss shown in Figure P5.25.

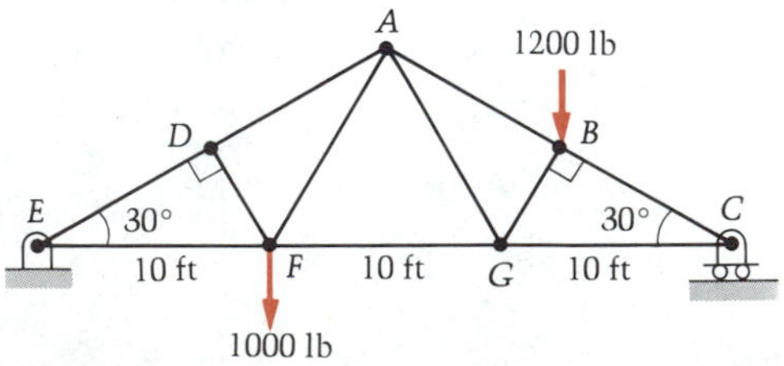

Figure P5.22

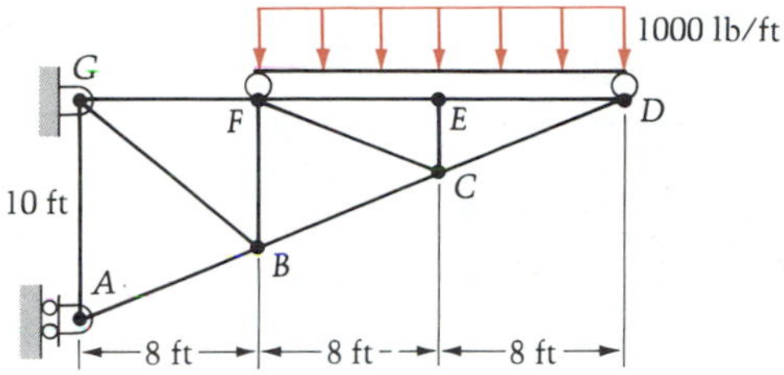

Figure P5.23

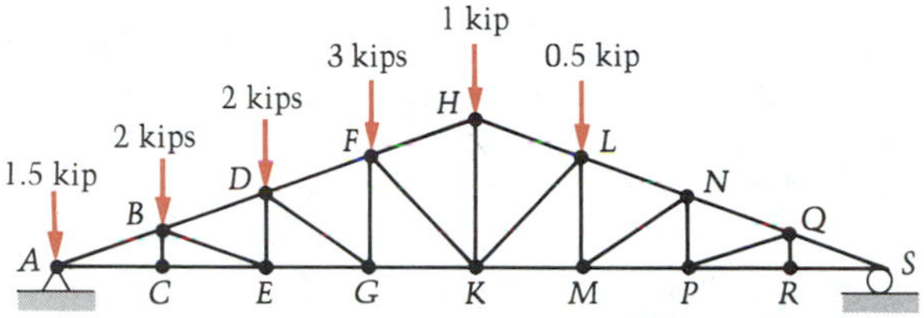

Figure P5.24

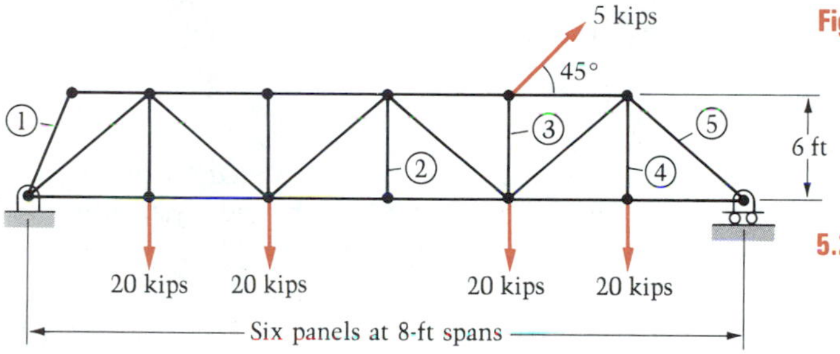

Figure P5.25

5.26 Find the forces in members *AB*, *BH*, *BC*, and *DF* of the truss shown in Figure P5.26.

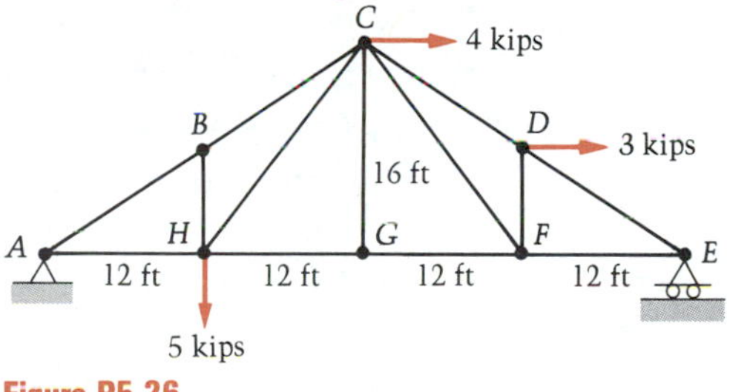

Figure P5.26

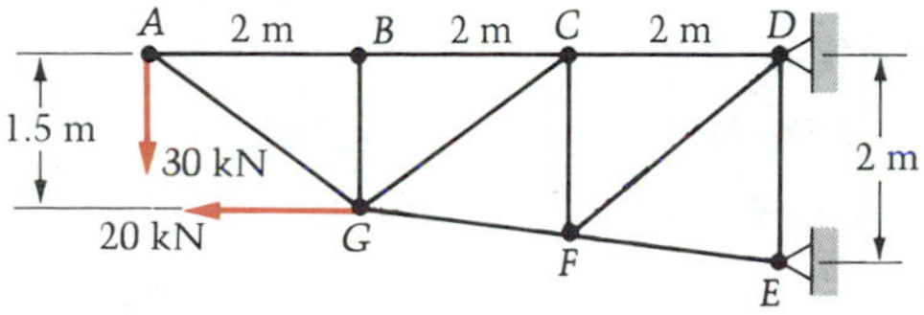

Figure P5.27

5.27 Find the forces in members *AG*, *BG*, and *CG* of the truss shown in Figure P5.27.

5.28 Find the force in each of the members of the truss in Figure P5.28.

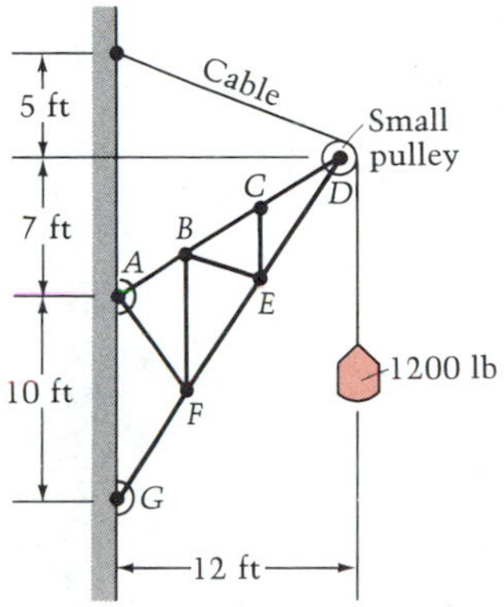

Figure P5.28

5.29

a. How many zero-force members can be found by inspection for the truss shown in Figure P5.29?

b. Determine the force in member *HF*.

c. The roller at *G* is replaced by a pin, and the pin at *A* is replaced by a roller. With no calculations, argue by means of a free-body diagram of pin *A* that the force in *AJ* will have changed.

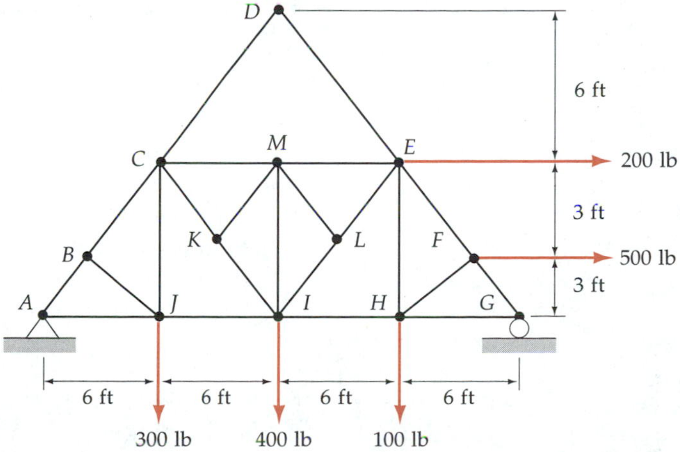

Figure P5.29

5.30 For the truss in Figure P5.30.

a. Show that half the members are zero-force members for the given loading;

b. find the reaction at G;

c. find the force in member DF.

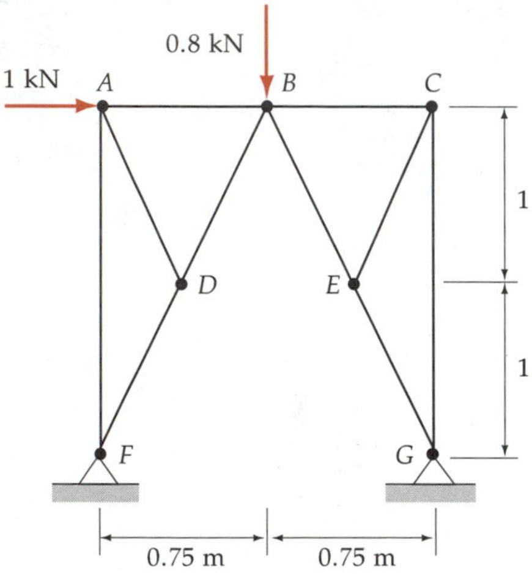

Figure P5.30

5.31 Show with a series of sketches that the K-truss shown in Figure P5.31 is a simple truss.

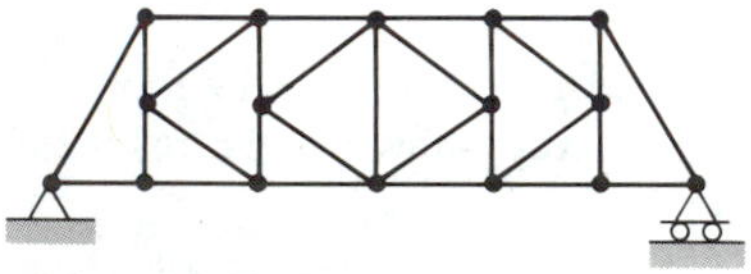

Figure P5.31 K-truss. $p = 16$; $m = 29 = 2p - 3$

5.32 Show that the Baltimore truss in Figure P5.32 is not a simple truss, even though $m = 2p - 3$.

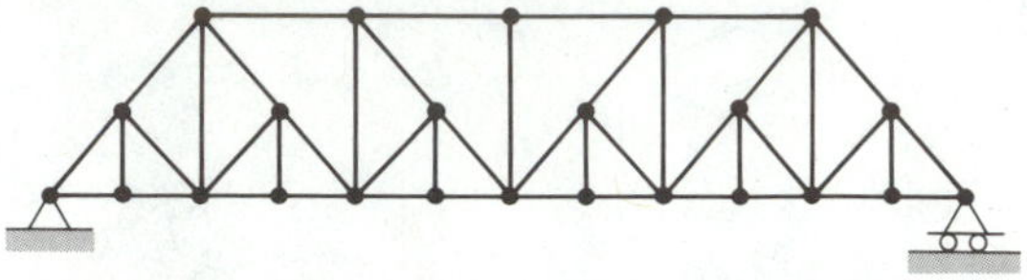

Figure P5.32 Baltimore truss. $p = 24$; $m = 45 = 2p - 3$

5.33 Show that if two simple trusses are joined to form a compound truss by simply bringing two joints together as shown in Figure P5.33 and omitting bar $B'C'$, the resulting truss will have $m = 2p - 3$.

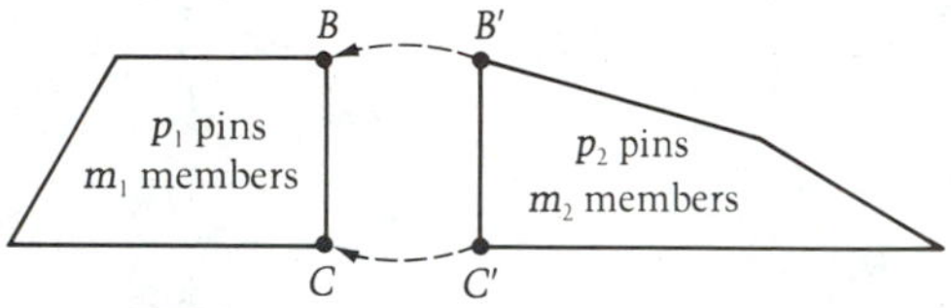

Figure P5.33

5.34 Two simple trusses are connected by the dashed bars in Figure P5.34 to form a compound truss. Prove that $m = 2p - 3$ for this compound truss.

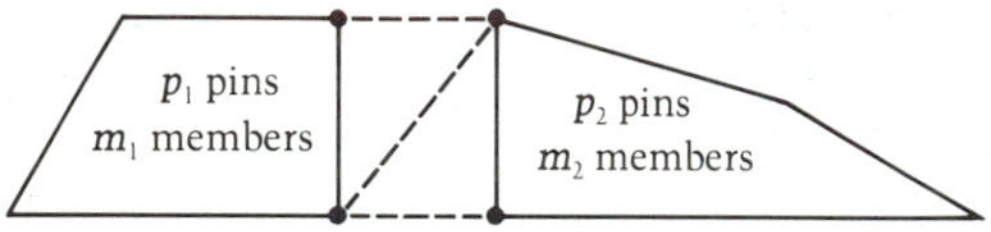

Figure P5.34

5.5 The Method of Sections

Another method commonly used to determine the forces in the members of a truss is called the method of sections. This name comes from the fact that in using this method, the truss is divided into two sections, which are both in equilibrium. *(If a body is in equilibrium, any and all parts of it are.)*

The advantage in using the method of sections is that member forces of interest may be found very quickly without solving the entire truss. For example, suppose we wished to know only the force in member GH of the (symmetrical) truss shown in Figure 5.25. By the method of joints, we would probably arrive at F_{GH} by:

1. Finding the roller reaction at F by $\Sigma M_A = 0$ on the overall truss.
2. Finding F_{EF} and F_{GF} by enforcing equilibrium of pin F.
3. Finding F_{GE} by enforcing equilibrium of pin E.
4. Finding F_{GH} by enforcing equilibrium of pin G.

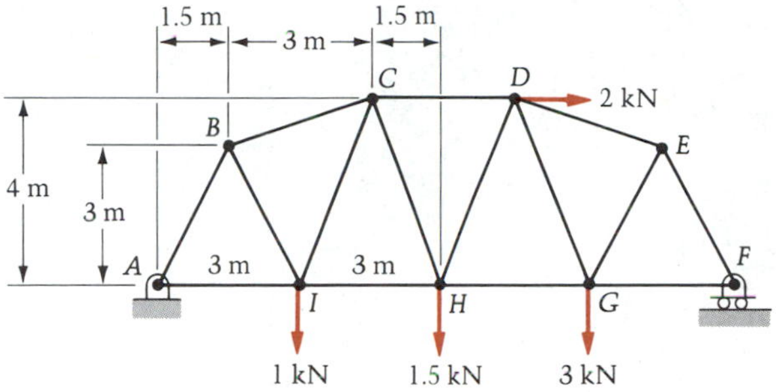

Figure 5.25

Cutting the Truss into Two Distinct Sections

A much quicker way to determine F_{GH} is to use the method of sections, which will be described in the example to follow. In this method, we cut the truss into two separate sections. One of the cut members is the one whose force we seek. This force thus appears as an *external* force on each of the two "halves" of the cut truss. If only three members have been cut and the external reactions are known, then the three equilibrium equations for either "half" will yield the desired member force. Sometimes we can find it by just summing moments about the point of intersection (if there is one) of the other unknown forces; this is the case in the following example.

Example 5.4

Determine the force in member *GH* of the truss shown in Figure E5.4a.

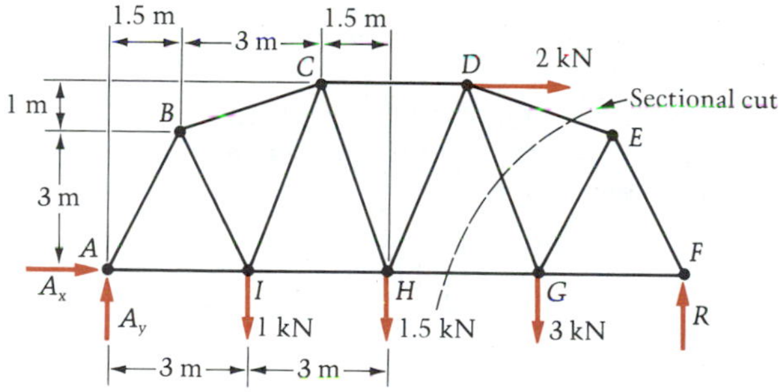

Figure E5.4a

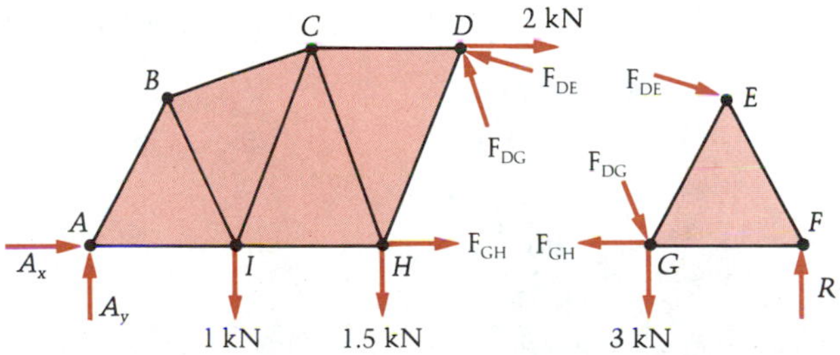

Figure E5.4b

Solution

In Figure E5.4b, we have divided the truss into two sections. In each section one of the unknown forces exposed by the cut is the desired F_{GH}.

As we mentioned (following Example 5.1), we can think of the forces of the cut members in either of the ways shown in Figure E5.4c at the top of the next page.

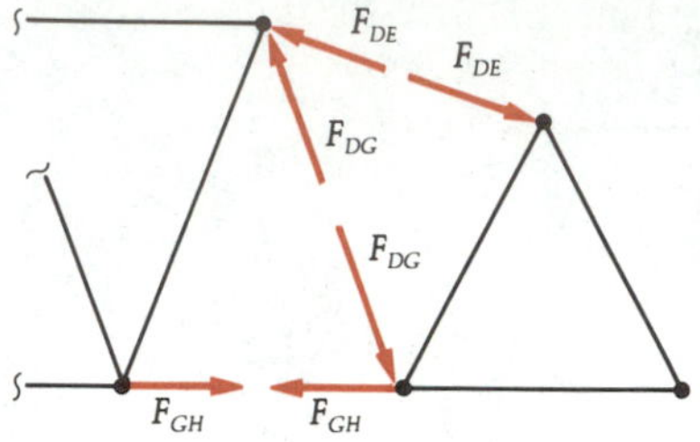

Figure E5.4c

The three bars DE, DG, and GH have been removed and replaced by the forces they exert on their respective pins.

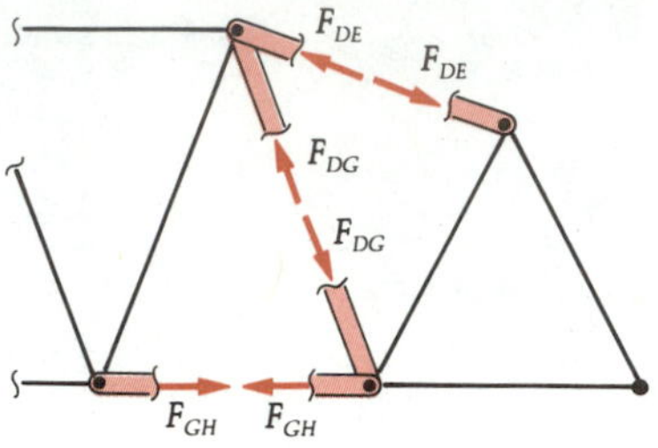

The bars are still there, but they are each cut in two. Thus their internal (axial) forces become exposed and are *external* to the separate sections, each of which is in equilibrium alone.

Summing moments about A, we find the roller reaction on the overall (uncut!) free-body diagram (Figure E5.4a):

$$\circlearrowleft\!+\quad \Sigma M_A = 0 = R(12 \text{ m}) - (1 \text{ kN})(3 \text{ m}) - (1.5 \text{ kN})(6 \text{ m}) - (3 \text{ kN})(9 \text{ m}) - (2 \text{ kN})(4 \text{ m})$$

$$R = \frac{47}{12} = 3.92 \text{ kN}$$

We are ready to use one of the cut sections of Figure E5.4b to find F_{GH}. We see that if we use the section in Figure E5.4d and simply equate the sum of the moments about D to zero, we shall find the desired force in a single step. This is because the other two unknown forces F_{DG} and F_{DE} both pass through point D, and thus the only unknown appearing in the moment equation will be the desired F_{GH}:

$$\circlearrowleft\!+\quad \Sigma M_D = 0 = -F_{GH}(4 \text{ m}) - (3 \text{ kN})(1.5 \text{ m}) + (3.92 \text{ kN})(4.5 \text{ m})$$

$$F_{GH} = 3.29 \text{ kN}$$

(The force is tensile because it was drawn that way and the scalar F_{GH} turned out to be positive in the solution.)

Note that if we were looking for F_{DE}, the same cut section could be used, and the summation of moments about G would give that force, again in one step (i.e., one equation in the single desired unknown)*.

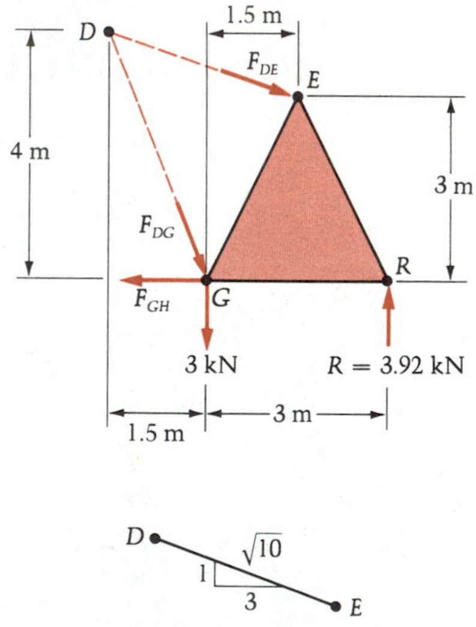

Figure E5.4d

$$\circlearrowleft\!+\quad \Sigma M_G = 0 = (3.92 \text{ kN})(3\text{m}) - \left(F_{DE}\frac{3}{\sqrt{10}}\right)(3 \text{ m}) - \left(F_{DE}\frac{1}{\sqrt{10}}\right)(1.5 \text{ m})$$

$$F_{DE} = 3.54 \text{ kN}$$

(The force is compressive because it was drawn that way and the scalar F_{DE} turned out positive in the solution.)

* We could, once F_{GH} has been found, solve for F_{DE} and F_{DG} by $\Sigma F_x = 0$ together with $\Sigma F_y = 0$. But $\Sigma M_G = 0$ gives F_{DE} in one equation, even if F_{GH} has not been previously determined. Thus errors we may have made in F_{GH} will not propagate into our solution for F_{DE}.

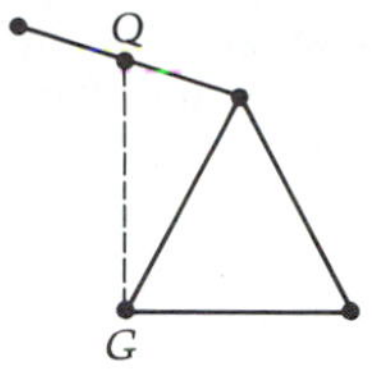

Figure E5.4e

In the $\Sigma M_G = 0$ equation, note that the force in DE was resolved into its two components at E. The horizontal component, $F_{DE}(3/\sqrt{10})$, has a moment arm (or perpendicular distance) from G of 3 m. The vertical component, $F_{DE}(1/\sqrt{10})$, has a moment arm of 1.5 m. The reader should note that if the point Q in Figure E5.4e is used, then the vertical component of F_{DE} has no moment about G and the computation is shortened to

$$\circlearrowleft\!+ \qquad \Sigma M_G = (3.92 \text{ kN})(3 \text{ m}) - \left(F_{DE}\frac{3}{\sqrt{10}}\right)(3.5 \text{ m})$$

$$F_{DE} = 3.54 \text{ kN} \qquad (\text{or } 3.54 \text{ kN } ©), \text{ again}$$

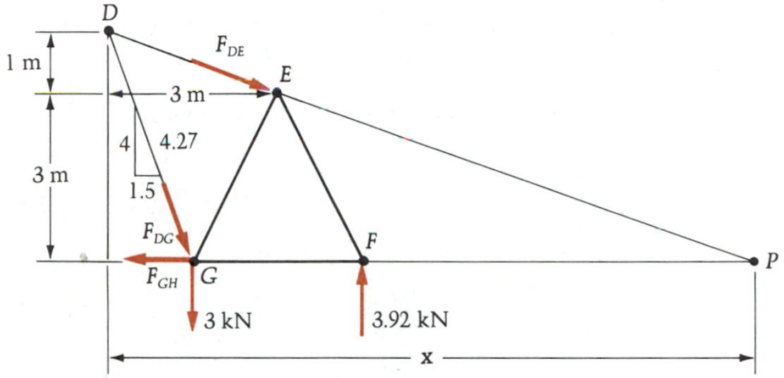

Figure E5.4f

If we wanted F_{DG}, we could now solve for it from $\Sigma F_x = 0$ or $\Sigma F_y = 0$. But, to further emphasize the idea of summing moments about the point of intersection P of other unknown forces, we find the intersection of F_{DE} and F_{GH} as shown in Figure E5.4f above. By similar triangles,

$$\frac{1}{3} = \frac{4}{x} \qquad \text{or} \qquad x = 12 \text{ m}$$

Therefore,

$$\circlearrowleft\!+ \qquad \Sigma M_P = 0 = (3 \text{ kN})(10.5 \text{ m}) - (3.92 \text{ kN})(7.5 \text{ m}) + F_{DG}\left(\frac{4}{4.27}\right)(10.5 \text{ m})$$

$$F_{DG} = -0.214 \text{ kN} \qquad \text{or} \qquad 0.214 \text{ kN } Ⓣ$$

Let us now check our three results by using the force equilibrium equations on the cut section:

$$\overset{+}{\rightarrow} \qquad \Sigma F_x = F_{DE}\frac{3}{\sqrt{10}} + F_{DG}\frac{1.5}{4.27} - F_{GH}$$

$$= 3.54(0.949) + (-0.214)(0.351) - 3.29$$

$$= -0.006 \approx 0 \qquad \text{(roundoff!)}$$

and

$$+\uparrow \qquad \Sigma F_y = 3.92 - 3 - F_{DE}\frac{1}{\sqrt{10}} - F_{DG}\frac{4}{4.27}$$

$$= 0.92 - 3.54(0.316) + 0.214(0.937)$$

$$= 0.002 \approx 0$$

The reader is encouraged to find the three forces F_{DE}, F_{DG}, and F_{GH} using the *left* section of Figure E5.4b, after determining A_x and A_y from the equilibrium equations for the overall free-body diagram. The answers for the three forces should, of course, agree with those found above.

Question 5.10 Why would a sectional cut that only went partly through a truss (i.e., didn't divide it into two parts) be useless?

EXAMPLE 5.5

Find the force in member *HI* of the truss shown in Figure E5.5a.

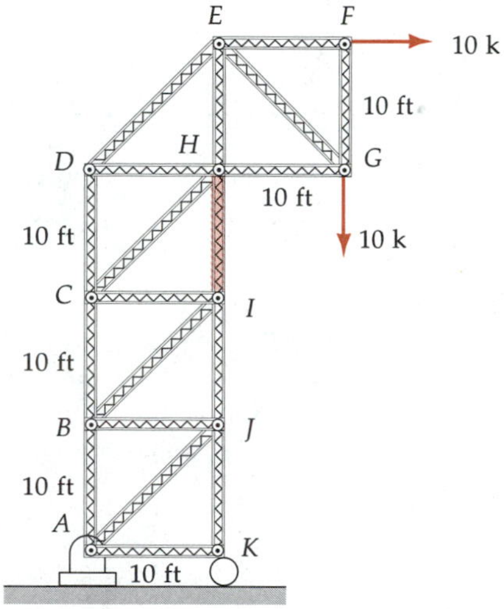

Figure E5.5a

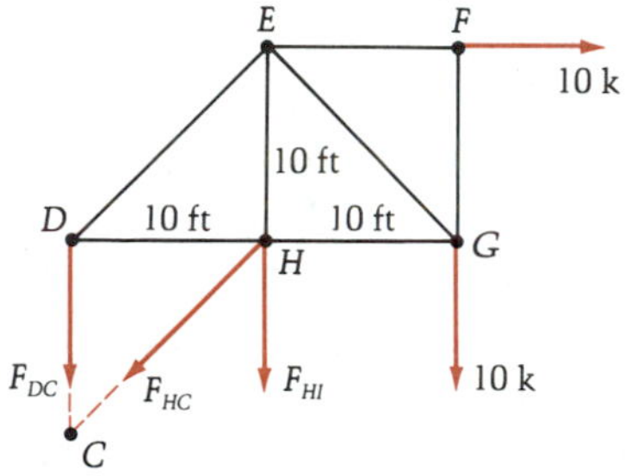

Figure E5.5b

Solution

We make a horizontal cut through the truss that exposes the force in the member *HI*, as suggested by the free-body diagram of Figure E5.5b. In this problem, we use the "upper half" of the truss because by so doing we avoid having to find the reactions. We see that summing moments at point *C* will eliminate the unknown forces F_{DC} and F_{HC} and allow an immediate solution for F_{HI}:

$$\circlearrowleft + \quad \Sigma M_C = 0 = -(10\text{ k})(20\text{ ft}) - (10\text{ k})(20\text{ ft}) - F_{HI}(10\text{ ft})$$

$$F_{HI} = -40\text{ k}$$

Because the force in member *HI* was drawn as if the member were in tension, and F_{HI} turned out negative, the force in *HI* is 40 kips Ⓒ.

Answer 5.10 Depicted as external forces on any legitimate free-body diagram thus produced would be a cancelling pair of forces associated with each cut member. Only the whole truss can be used as a free-body unless a part of it is completely cut away.

EXAMPLE 5.6

Find the forces in members *EF*, *RF*, *RN*, and *ON* of the *K*-truss shown in Figure E5.6a.

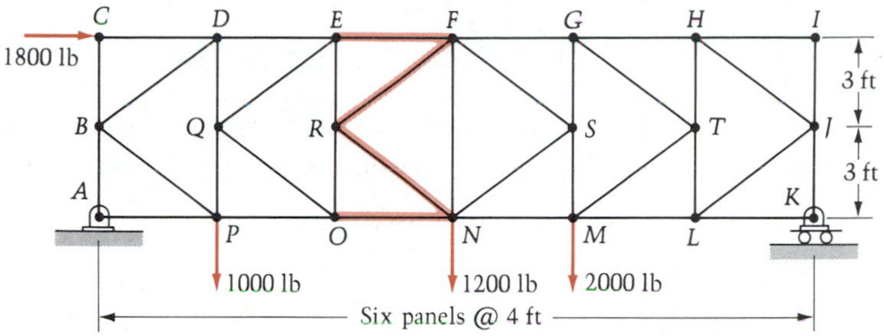

Figure E5.6a

Solution

In this truss we will not be able to cut sections and find all the desired forces one by one. For instance, a mental vertical cut through the four bars of interest shows that on each resulting part we will have three equations and four unknowns. Furthermore, there is no point where three of the four bars intersect. However, we shall use another section to find two of the four member forces, then employ force equilibrium on a second section to complete the solution.

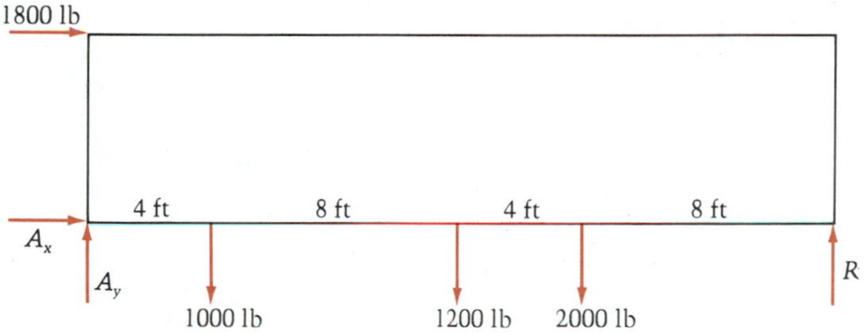

Figure E5.6b

First, though, we find the reactions; we use the overall *FBD* in Figure E5.6b:

$$\circlearrowleft\!+\quad \Sigma M_A = 0 = R(24 \text{ ft}) - (1000 \text{ lb})(4 \text{ ft}) - (1200 \text{ lb})(12 \text{ ft}) - (2000 \text{ lb})(16 \text{ ft}) - (1800 \text{ lb})(6 \text{ ft})$$

$$R = 2550 \text{ lb}$$

$$+\uparrow\quad \Sigma F_y = 0 = A_y + 2550 - 1000 - 1200 - 2000$$

$$A_y = 1650 \text{ lb}$$

$$\xrightarrow{+}\quad \Sigma F_x = 0 = A_x + 1800$$

$$A_x = -1800 \text{ lb}$$

Next, we consider the sectioning of the truss shown in Figure E5.6c on the following page.

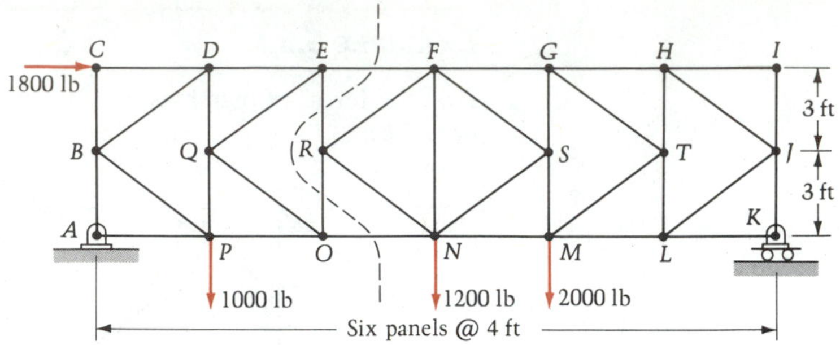

Figure E5.6c

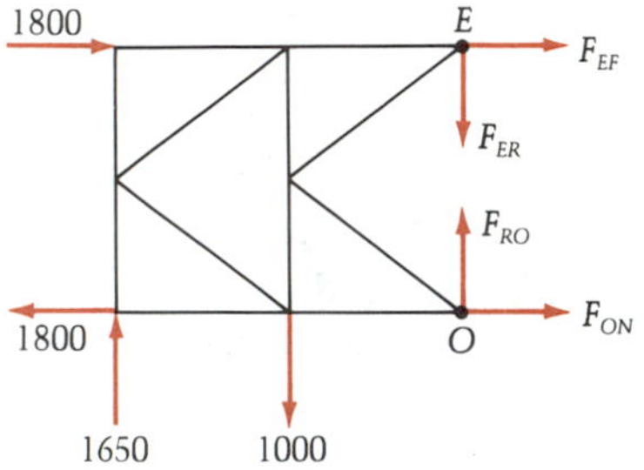

Figure E5.6d

Using the left section (see Figure E5.6d), we can determine two of the desired member forces: $\Sigma M_E = 0$ will give us F_{ON}; $\Sigma F_x = 0$ will then give us F_{EF}. Thus,

$$\circlearrowleft + \quad \Sigma M_E = 0 = F_{ON}(6 \text{ ft}) + (1000 \text{ lb})(4 \text{ ft}) - (1650 \text{ lb})(8 \text{ ft}) - (1800 \text{ lb})(6 \text{ ft})$$

$$F_{ON} = 3330 \text{ lb} \qquad (\text{or } F_{ON} = 3330 \text{ lb } Ⓣ)$$

and

$$\xrightarrow{+} \quad \Sigma F_x = 0 = F_{EF} + F_{ON} + 1800 - 1800$$

$$F_{EF} = -F_{ON} = -3330 \text{ lb} \qquad (\text{or } F_{EF} = 3330 \text{ lb } Ⓒ)$$

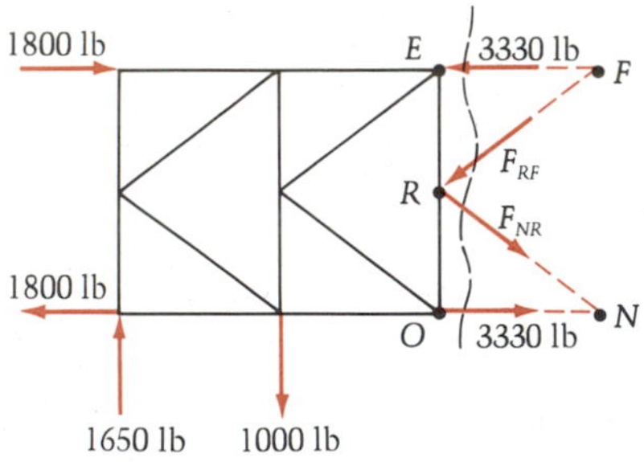

Figure E5.6e

Now we examine the previously mentioned vertical cut shown in Figure E5.5e. By inspection, the forces F_{RF} and F_{NR} have equal magnitudes. This is because all the other external forces with x components are in balance; thus the horizontal components of F_{RF} and F_{NR} must cancel. Because these bars make the same angle $[\cos^{-1}(\frac{4}{5})]$ with the horizontal through R, then F_{RF} must be equal to F_{NR}. As to whether they are

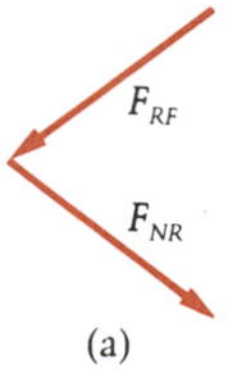

or

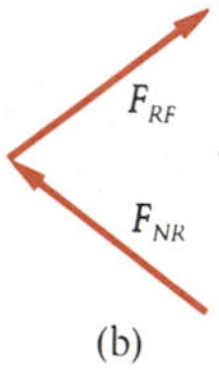

we can again answer by inspection that part (a) in the figure is the only possibility. This is because the vertical components of these two forces must add to 1650 − 1000 = 650 lb downward. Therefore,

$$+\uparrow \quad \Sigma F_y = 0 = 1650 - 1000 - F_{RF}\left(\frac{3}{5}\right)2$$

or

$$F_{RF} = 542 \text{ lb} \qquad (\text{or } 542 \text{ lb } Ⓒ)$$

so that

$$F_{NR} = 542 \text{ lb} \qquad (\text{or } 542 \text{ lb } Ⓣ)$$

It is unnecessary (as we have seen a number of times) to try to determine in advance whether an unknown load is Ⓣ or Ⓒ as we are drawing the free-body diagram. However, it is very instructive and improves the student's feel for the equilibrium analysis.

The other section can always be used as a check on our solutions for the bar forces; using Figure E5.6f, we see:

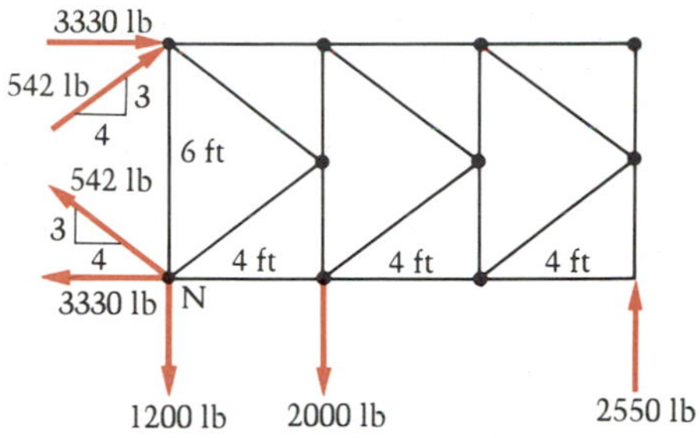

Figure E5.6f

$$\xrightarrow{+} \qquad \Sigma F_x = 0 \qquad \text{(by inspection)}$$

$$+\uparrow \qquad \Sigma F_y = 542\left(\frac{3}{5}\right)2 - 1200 - 2000 + 2550$$

$$= 0.40 \text{ lb} \approx 0$$

$$\circlearrowleft\!+ \qquad \Sigma M_N = -(3330 \text{ lb})(6 \text{ ft}) - (542 \text{ lb})\left(\frac{4}{5}\right)(6 \text{ ft})$$

$$-(2000 \text{ lb})(4 \text{ ft}) + (2550 \text{ lb})(12 \text{ ft})$$

$$= 18.4 \text{ lb-ft}$$

These values differ from zero due to roundoff error. (If four digits are retained instead of three, then ΣM_N becomes less than 0.5 lb-ft.)

Example 5.7

Find the force in member *EL*, the only member of the truss in Figure E5.7a that has a length other than 1 m.

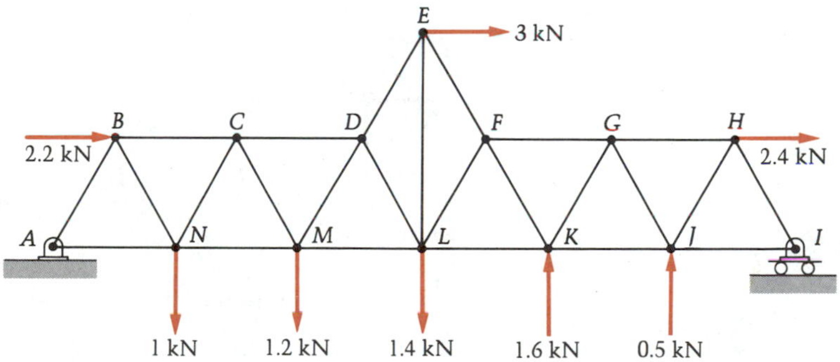

Figure E5.7a

Solution

Except for isolating joint *E* where there are three unknown bar forces acting, any complete section through *EL* will have to cut at least three other members that are not all concurrent. Thus the solution cannot be obtained with just one moment equation using the method of sections; nor could we obtain it by a combination of equilibrium equations from one sectioning. We shall have to use a combination of the methods of sections and joints before arriving at F_{EL}.

First, we find the roller reaction from equilibrium of the overall truss shown in Figure E5.7b on the following page.

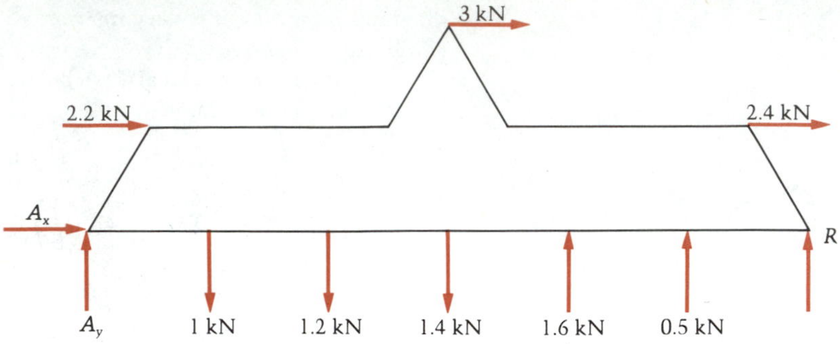

Figure E5.7b

$$\circlearrowleft + \quad \Sigma M_A = 0 = R(6 \text{ m}) + (0.5 \text{ kN})(5 \text{ m}) + (1.6 \text{ kN})(4 \text{ m})$$
$$- (1.4 \text{ kN})(3 \text{ m}) - (1.2 \text{ kN})(2 \text{ m}) - (1 \text{ kN})(1 \text{ m})$$
$$- (2.4 + 2.2) \text{ kN } (1 \sin 60° \text{ m})$$
$$- (3 \text{ kN})(2 \sin 60° \text{ m})$$
$$R = 1.31 \text{ kN}$$

Next we cut a section that will allow us to find F_{DE} (see Figure E5.7c). Then we can isolate pin E and find F_{EL}.

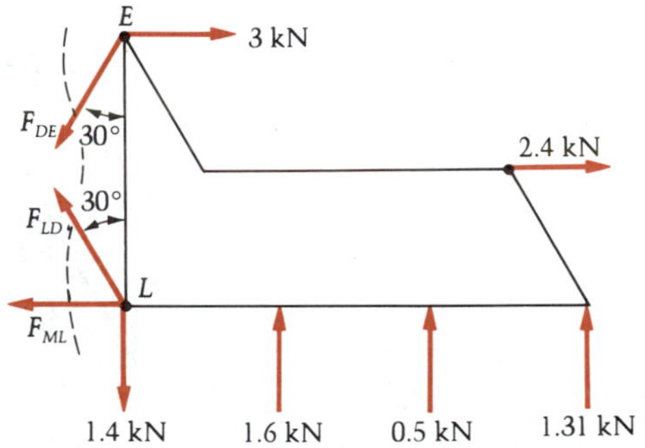

Figure E5.7c

$$\circlearrowleft + \quad \Sigma M_L = 0 = (1.6 \text{ kN})(1 \text{ m}) + (0.5 \text{ kN})(2 \text{ m}) + (1.31 \text{ kN})(3 \text{ m})$$
$$- (2.4 \text{ kN})(1 \sin 60° \text{ m})$$
$$- (3 \text{ kN})(2 \sin 60° \text{ m}) + (F_{DE} \sin 30°)(2 \sin 60° \text{ m})$$
$$F_{DE} = 0.860 \text{ kN} \qquad (\text{or } 0.860 \text{ kN } \text{Ⓣ})$$

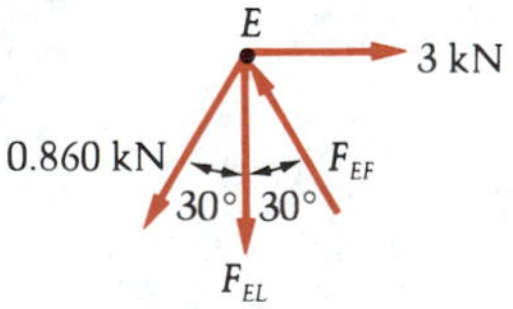

Figure E5.7d

Now we shift our attention to the pin at E shown in Figure E5.7d, using the method of joints to find F_{EL}.

$$\xrightarrow{+} \quad \Sigma F_x = 0 = 3 - 0.860 \sin 30° - F_{EF} \sin 30°$$
$$F_{EF} = 5.14 \text{ kN} \qquad (\text{or } 5.14 \text{ kN } \text{Ⓒ})$$

5.14

$$+\uparrow \quad \Sigma F_y = 0 = -0.860 \cos 30° + F_{EF} \cos 30° - F_{EL}$$
$$F_{EL} = 3.71 \text{ kN} \qquad (\text{or } 3.71 \text{ kN } \text{Ⓣ})$$

As a check, consider the free-body diagram* shown below in Figure E5.7e:

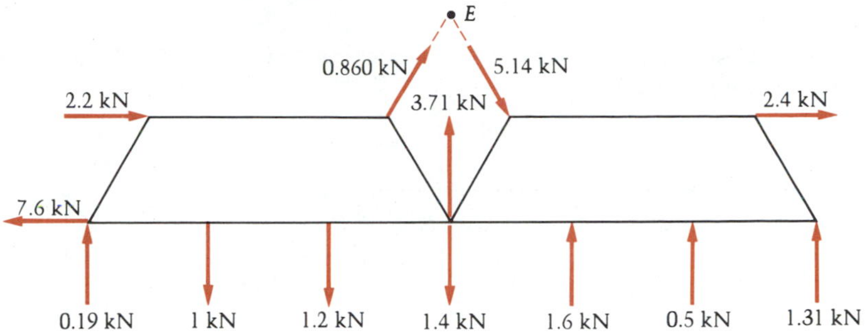

Figure E5.7e

$\overset{+}{\longrightarrow}$ $\Sigma F_x = (2.2 + 2.4 - 7.6 + 5.14 \cos 60° + 0.860 \cos 60°)$ kN $= 0$ ✓

$+\uparrow$ $\Sigma F_y = (0.19 - 1 - 1.2 - 1.4 + 1.6 + 0.5 + 1.31 + 3.71 + 0.860 \sin 60° - 5.14 \sin 60°)$ kN $= 0.003$ kN ≈ 0 ✓

$$\Sigma M_E = (2.2 + 2.4) \text{ kN } (1 \sin 60° \text{ m}) + (1.31 \text{ kN})(3 \text{ m}) - (0.19 \text{ kN})(3 \text{ m}) - (7.6 \text{ kN})(2 \sin 60° \text{ m}) + (1.6 + 1.2) \text{ kN } (1 \text{ m}) + (0.5 + 1) \text{ kN } (2 \text{ m}) = -0.02 \text{ kN} \cdot \text{m} \approx 0 \checkmark$$

We note that these checks will *usually* indicate an error. However, they are only necessary conditions for equilibrium and are *no guarantee* that an error has not been made. For instance, if the 0.860 kN Ⓣ force F_{DE} had been incorrectly drawn as compressive on Figure E5.7d, erroneous results would have resulted for F_{EF} and F_{EL}. Applying these (with $F_{DE} = 0.860$ kN ©) onto the free-body diagram of Figure E5.7e would interestingly *and erroneously* have given the same three "checks" in this case!

Question 5.11 Return to the place in the solution where the cut was made that resulted in the solution for F_{DE}. Describe a different sectioning that would have allowed us to find a *different* force at E by means of a simpler free-body diagram (with one less contribution to ΣM_L).

* Where, as the reader may wish to show from Figure E5.7b with $R = 1.31$ kN, $A_x = -7.6$ kN and $A_y = 0.19$ kN.

Answer 5.11 Use a vertical cut through EF, FL, and LK. $\Sigma M_L = 0$ now gives F_{EF}, and the 3-kN force is not used.

PROBLEMS ▸ Section 5.5

5.35 Each truss member in Figure P5.35 has length 2 m. What is the force in *CD*?

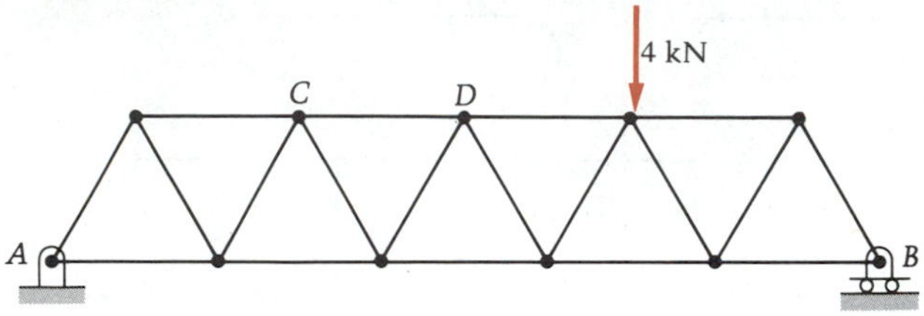

Figure P5.35

5.36 Find the force in members *BD* and *CD* of the truss shown in Figure P5.36.

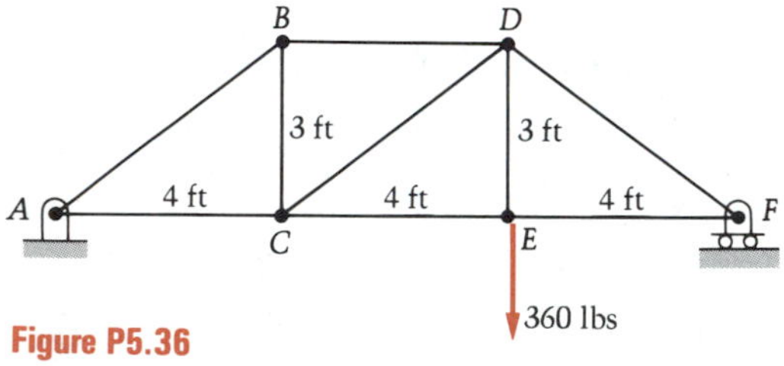

Figure P5.36

5.37 Determine the force in member *GH* of the truss shown in Figure P5.37.

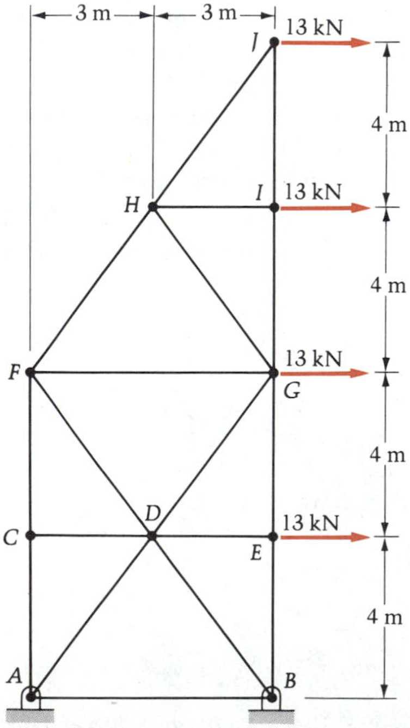

Figure P5.37

5.38 Find the forces in members *BC*, *BH*, and *AH* of the truss, loaded as shown in Figure P5.38.

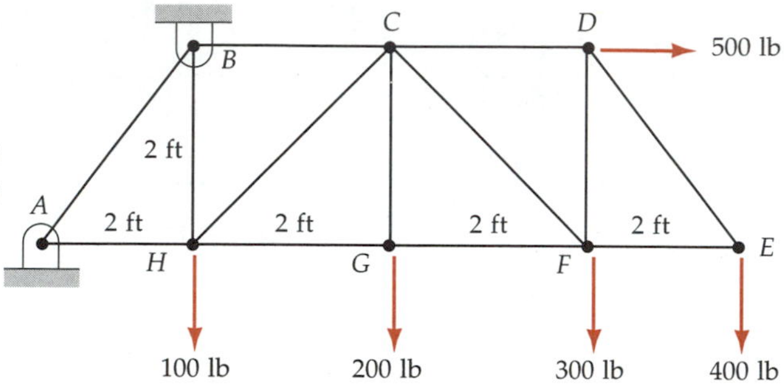

Figure P5.38

5.39 Find the forces in the darkened members of the trusses in Problems 5.39–5.43.

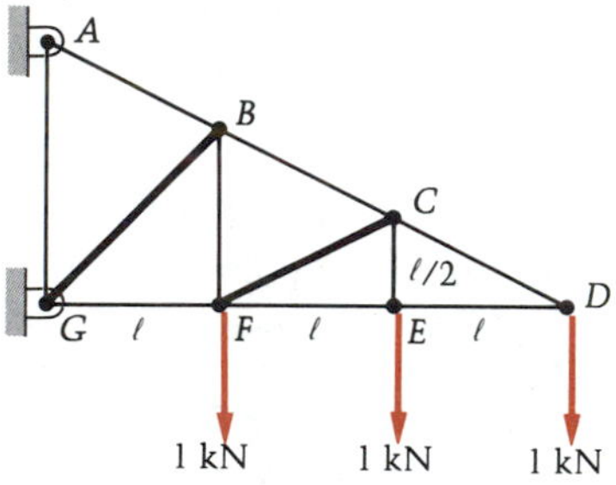

Figure P5.39

* **5.40**

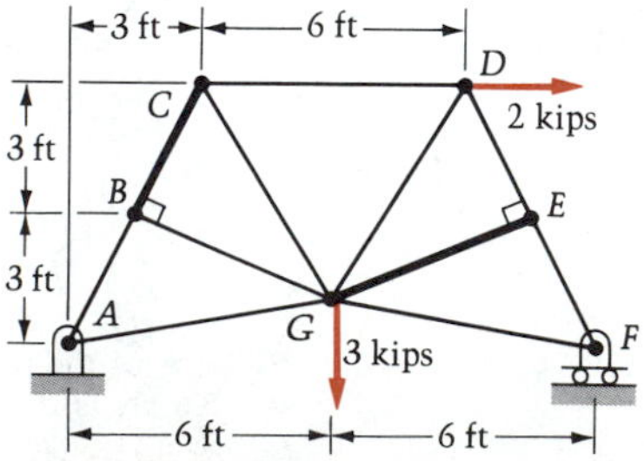

Figure P5.40

5.41

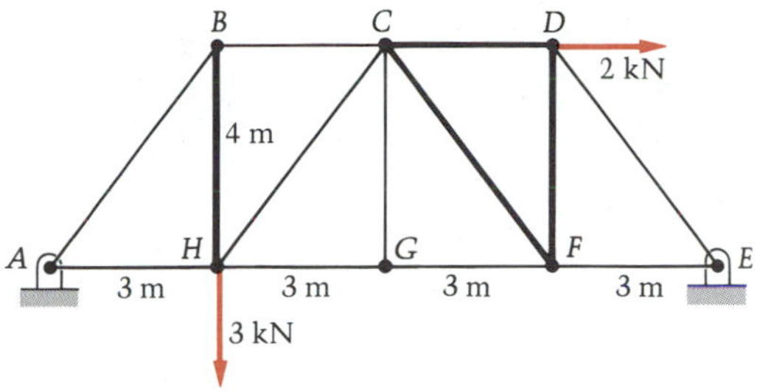

Figure P5.41

5.42

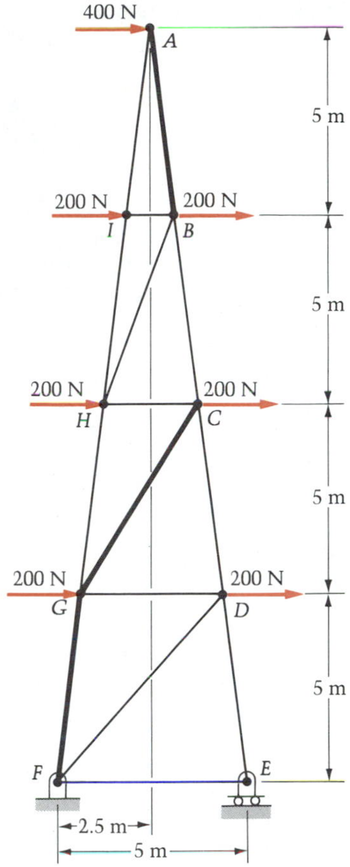

Figure P5.42

5.43

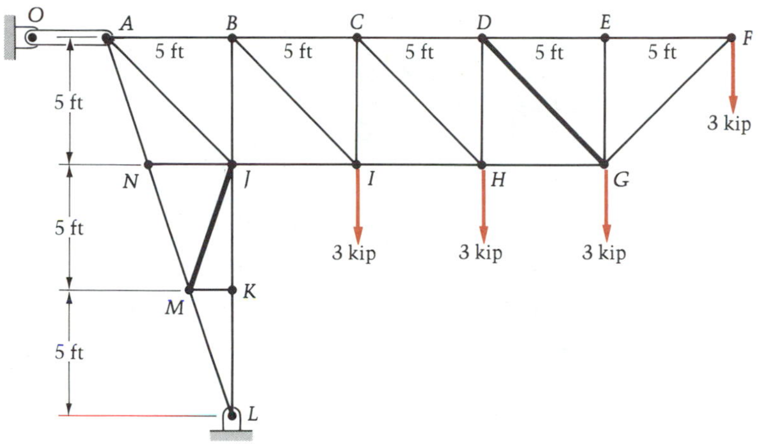

Figure P5.43

5.44 Find the force in members *FH*, *IH*, and *JK* of the truss shown in Figure P5.44.

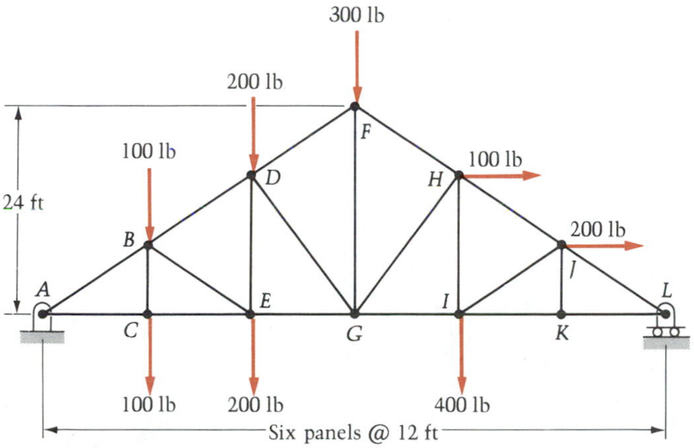

Figure P5.44

5.45 For the pin connected truss shown in Figure P5.45:

a. Using the method of joints, find the force in member *BC*.

b. Using the method of sections, find the force in member *CE*.

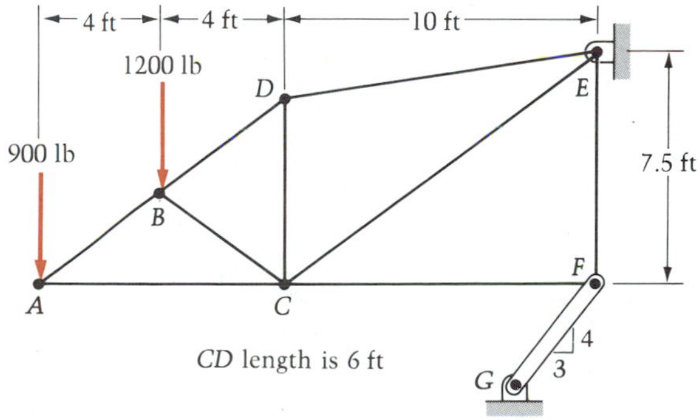

Figure P5.45

5.46 Find the forces in members *IE*, *JD*, *KJ*, and *CJ* of the pin-connected truss shown in Figure P5.46.

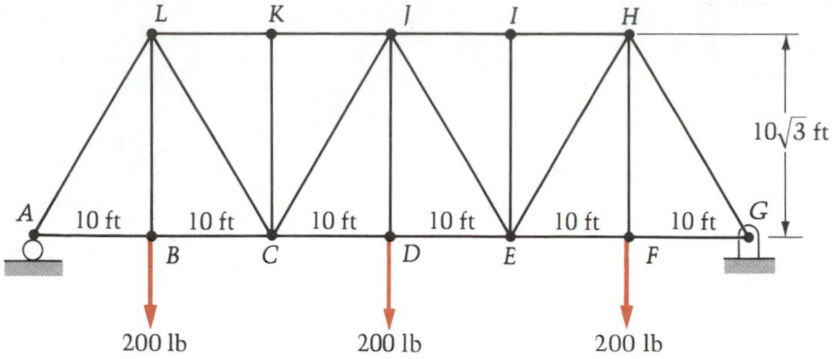

Figure P5.46

5.47 Find the forces in members *CF* and *AF* shown in Figure P5.47.

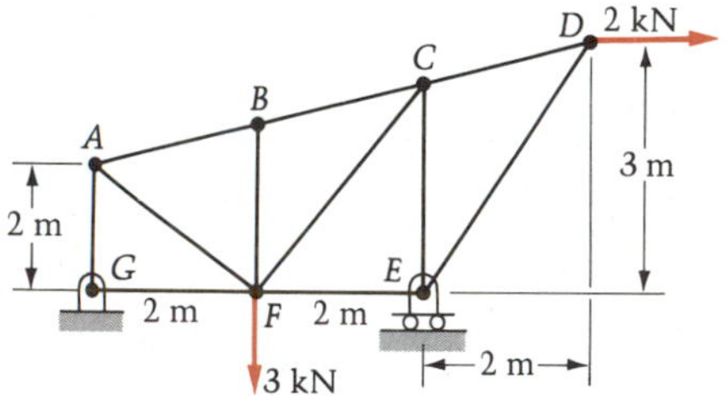

Figure P5.47

5.48 In Figure P5.48 find the forces in members *AD*, *ED*, and *AB*. Can the forces in all members of this truss be found from statics equations alone?

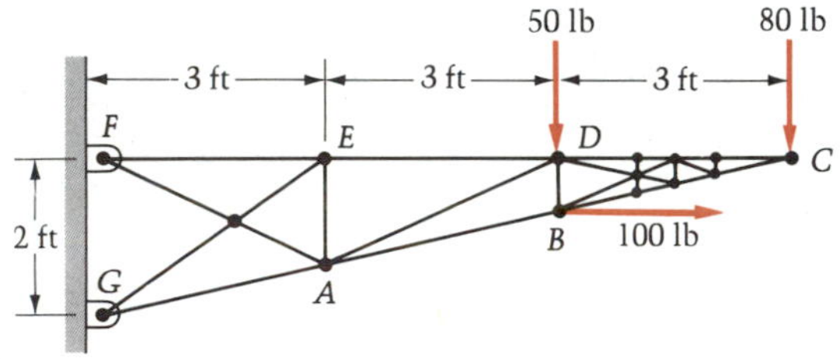

Figure P5.48

5.49 Find the forces in members *IE*, *JC*, *KC*, and *DE* of the truss of Problem 5.46, shown again in Figure P5.49 with a new loading.

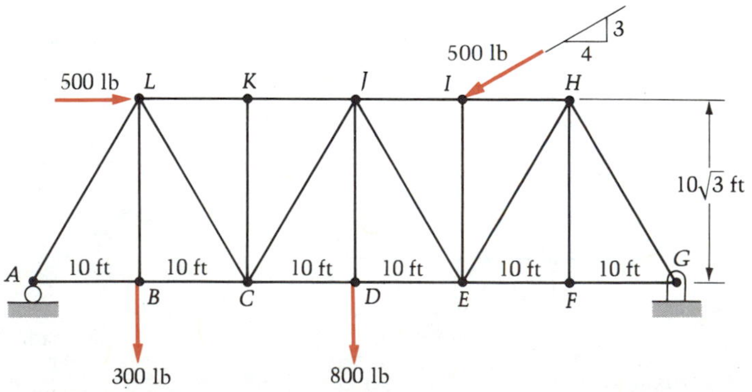

Figure P5.49

5.50 An antenna on a testing range, together with its supporting structure, weighs 300 lb with mass center at *C*. Assume that $\frac{1}{4}$ of the weight is supported by each of the pins at *A* and *B* that connect the structure to the truss tower. See Figure P5.50. (The other half is carried by the pins on an identical truss behind the one in the figure.)

a. Compute the horizontal components of the pin reactions at *A* and *B* on the truss.

b. Find the forces in members *DF*, *DG*, *AH*, and *AB*. Which of these four forces is dependent on the assumption about "$\frac{1}{4}$ of the weight"?

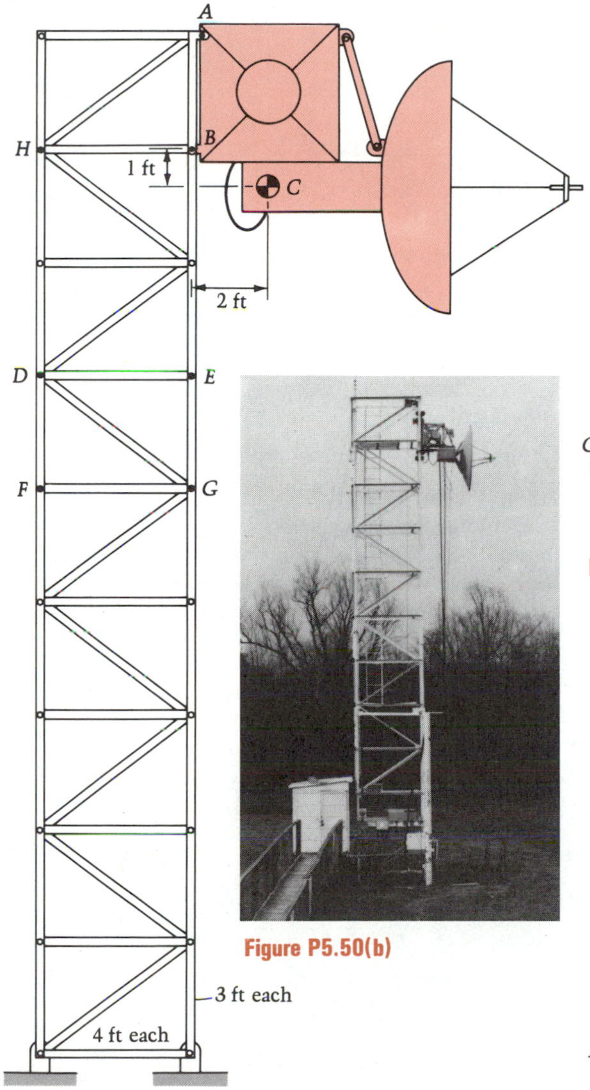

Figure P5.50(a)

Figure P5.50(b)

5.51 For the truss in Figure P5.51 calculate the forces in members *BF* and *BC*.

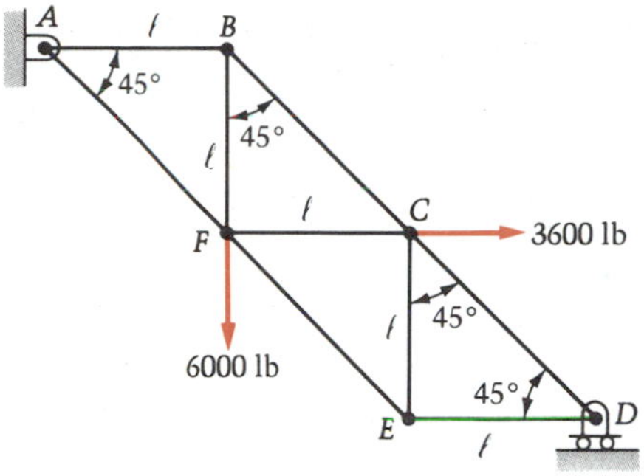

Figure P5.51

5.52 Given: $GK = KE = EB = BF = FL = LH = 10$ ft. Find the force in member *AB* for the plane truss shown in Figure P5.52.

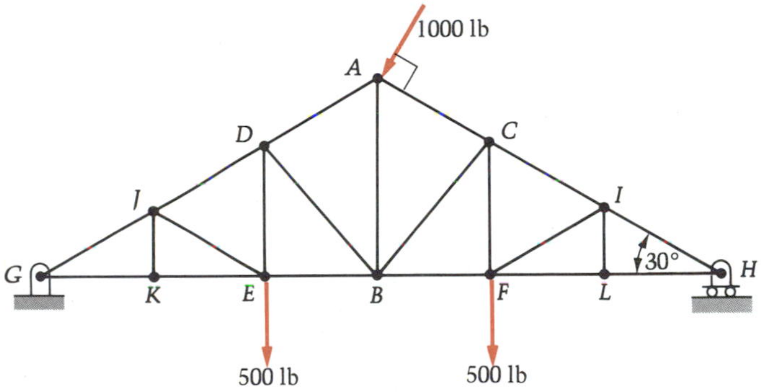

Figure P5.52

5.53 Find the forces in members *GH*, *CH*, and *BC* of the roof truss shown in Figure P5.53.

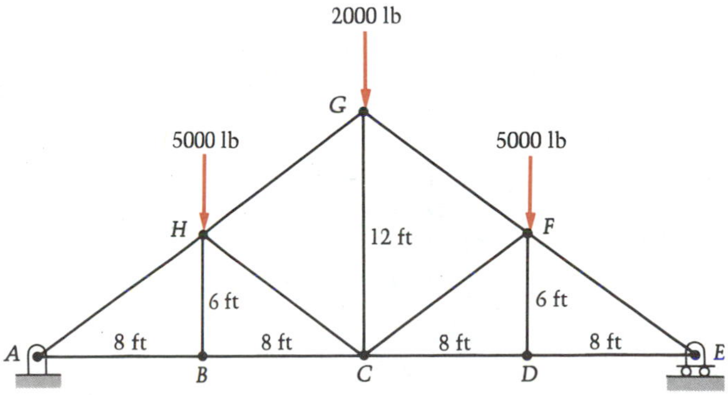

Figure P5.53

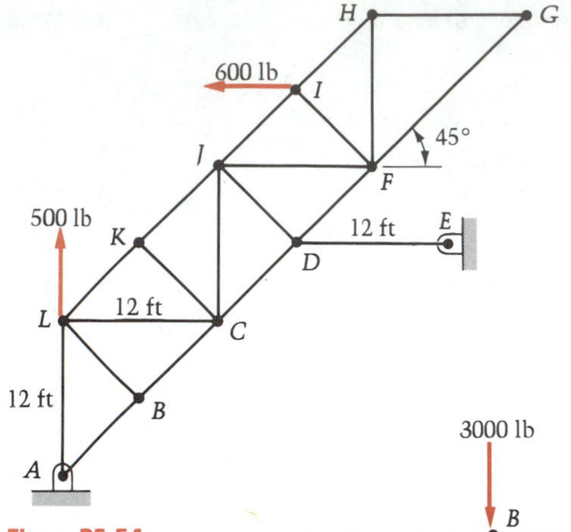

Figure P5.54

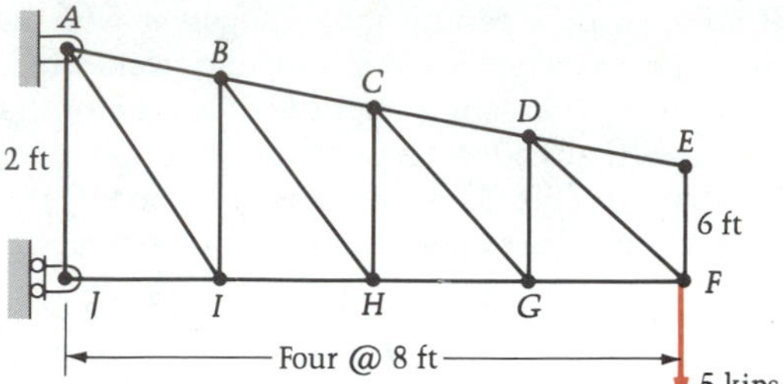

Figure P5.55

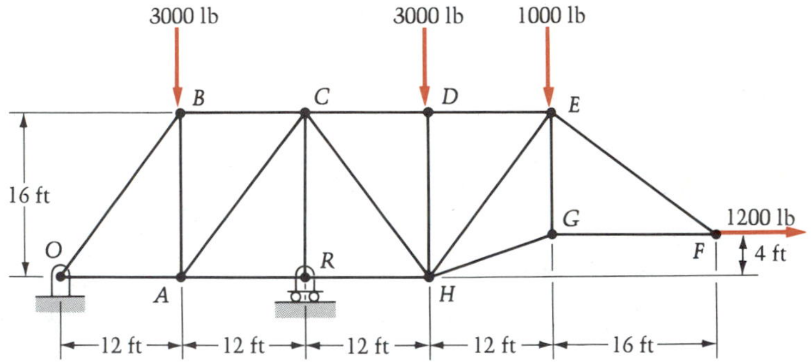

Figure P5.56

5.54 Find the forces in members CJ, BL, FI, and HG shown in Figure P5.54. All members have length 12 ft or $12/\sqrt{2}$ ft, and all angles between members are 0°, 45°, 90°, 135° or 180°.

5.55 Find the force in member BH of the truss in Figure P5.55. Identify three zero-force members.

5.56 Find the forces in members DE, CD, and CH of the truss in Figure P5.56.

5.57 Find the force in member EF of the truss shown in Figure P5.57.

5.58 Demonstrate with a series of sketches that the truss in Figure P5.58 is simple. Find the force in BE.

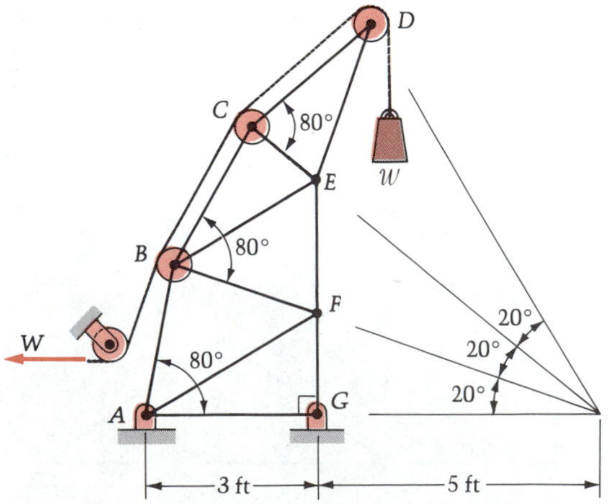

Figure P5.57

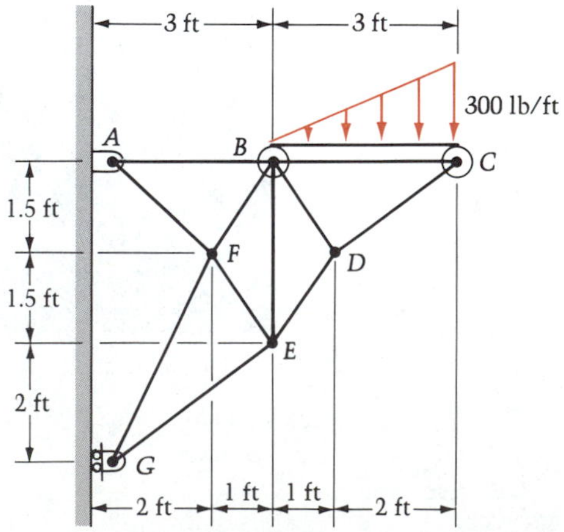

Figure P5.58

5.59 Find the forces in members *DC*, *DG*, and *DF* of the truss in Figure P5.59.

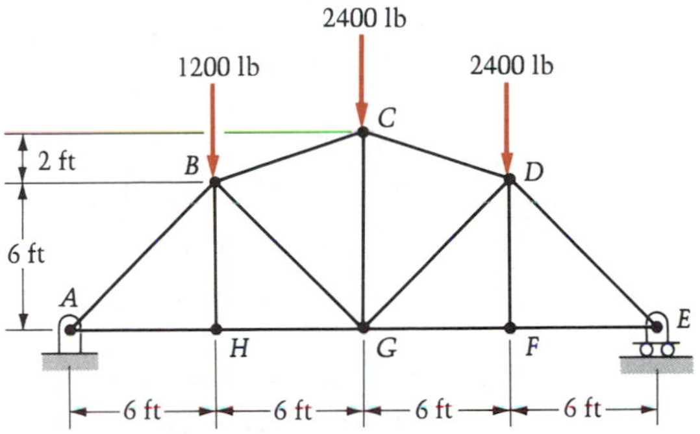

Figure P5.59

5.60 Find the force in member *BD* of the truss in Figure P5.60 if the weight of $\mathcal{W}$ is 1000 newtons.

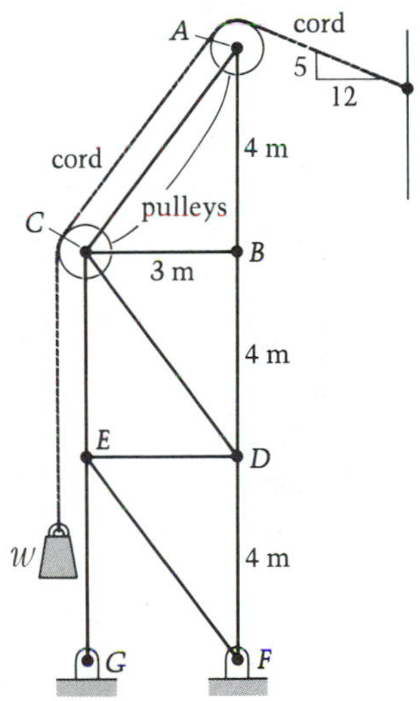

Figure P5.60

5.61 Find the forces in members *CD*, *DH*, and *CH* of the truss in Figure P5.61.

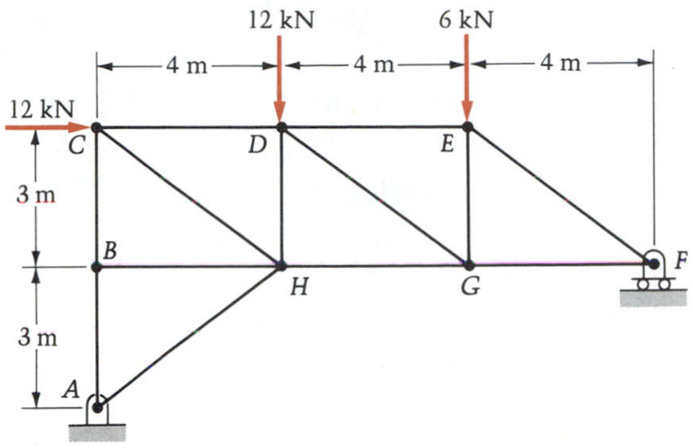

Figure P5.61

5.62 Find the forces in members *AB* and *BC* of the truss in Figure P5.62.

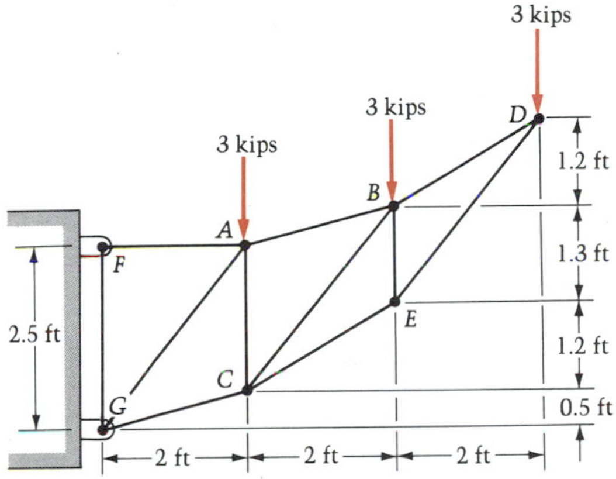

Figure P5.62

5.63 Find the forces in members *DC* and *DE* of the truss in Figure P5.63.

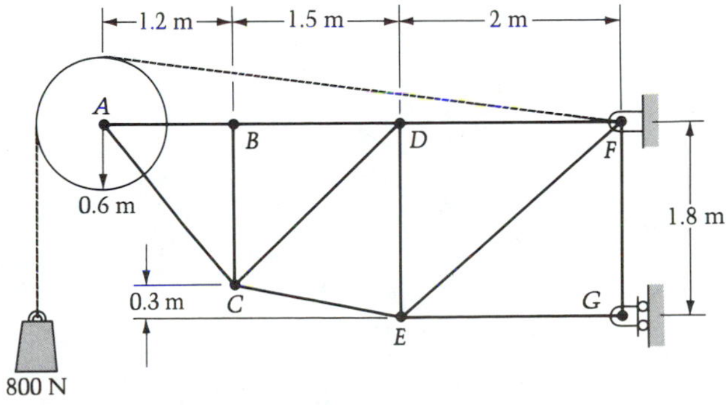

Figure P5.63

5.64 Find the force in member *CF* of the truss in Figure P5.64.

5.65 Find the forces in members *CD*, *KJ*, and *LJ* of the truss in Figure P5.65.

5.66 Find the forces in *FE*, *BE*, and *BC* of the truss shown in Figure P5.66.

5.67 Find the forces in members *CD* and *DJ* of the truss shown in Figure P5.67.

5.68 Find the forces in members *DE*, *QE*, and *OP* of the truss of Example 5.6 shown in Figure P5.68. Use the reactions from that example.

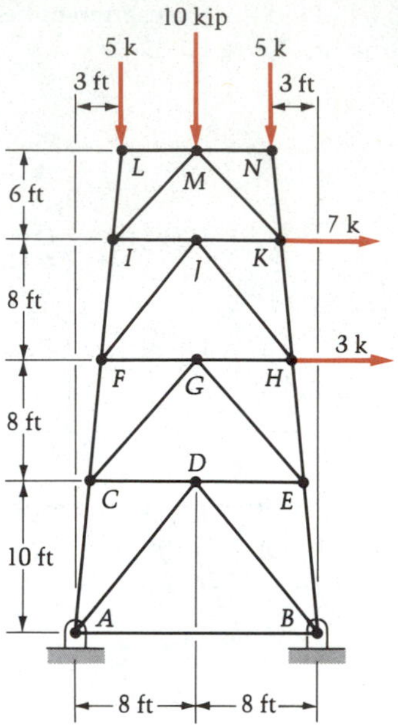

Figure P5.64

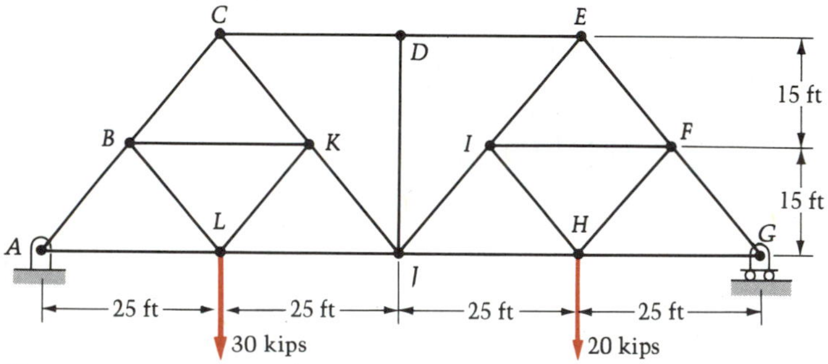

Figure P5.65

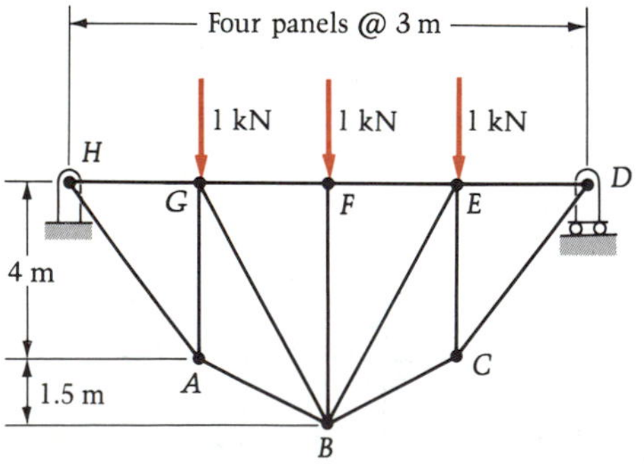

Figure P5.66

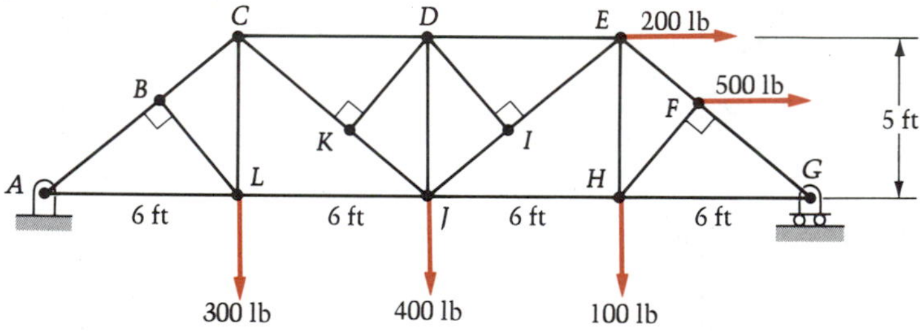

Figure P5.67

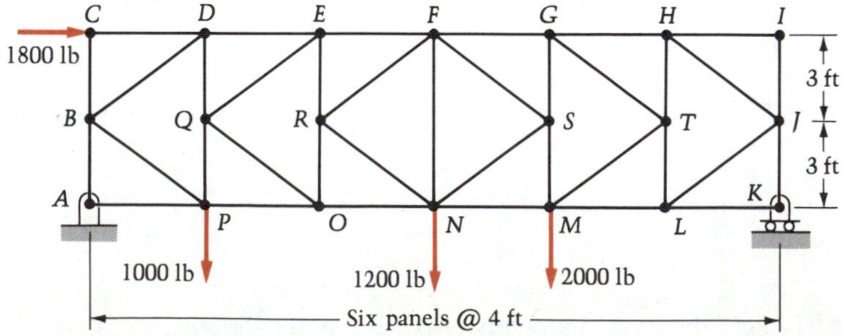

Figure P5.68

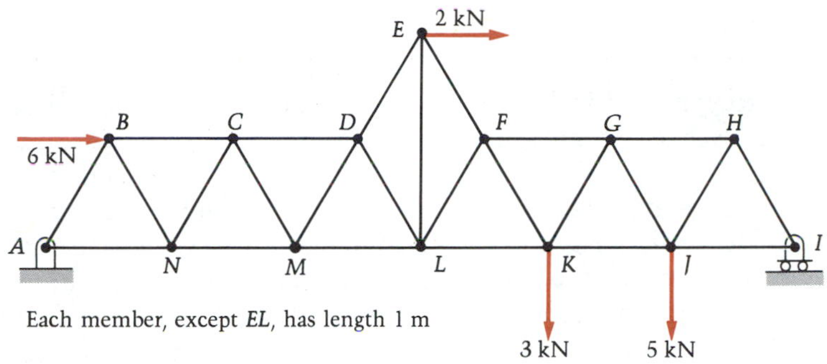

Figure P5.69

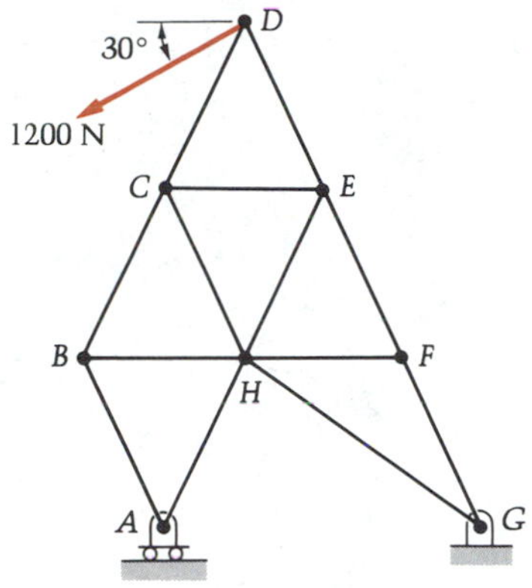

Figure P5.70

5.69 Find the forces in members EL and GH of the truss of Example 5.7 under the new loading shown in Figure P5.69. (Do not use the reactions of that example!)

5.70 Determine the force in member HE for the truss in Figure P5.70. All members are of length 0.8 m except GH.

* **5.71** In Figure P5.71(a), the two pulleys at A are independently mounted.

a. Show that three members of the truss are zero-force members.

b. Find the forces in members FG, FD, FE, and CD.

Hint:

$$\tan\theta = \frac{1 + 0.9\cos\theta}{6 - 0.9\sin\theta} \qquad \text{[See Figure P5.71(b).]}$$

which can be solved by trial-and-error for θ.

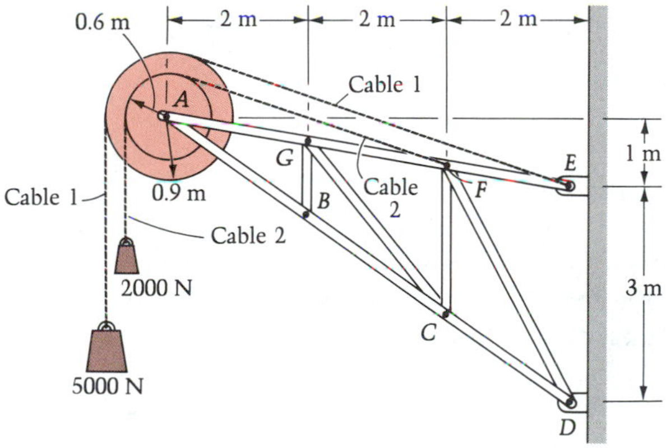

Figure P5.71(a)

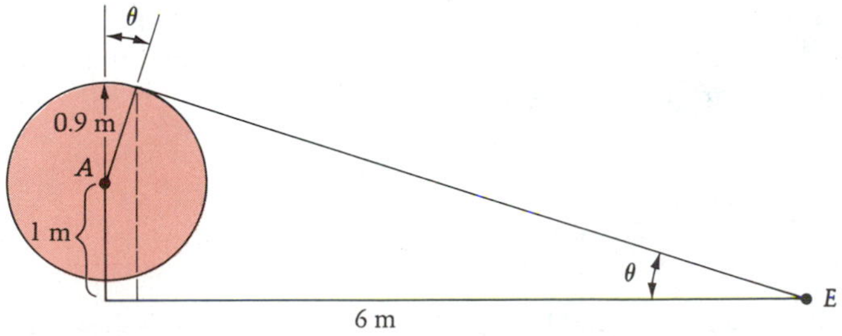

Figure P5.71(b)

5.72 The cable in Figure P5.72 extends from a winch at A around two pulleys $\mathcal{P}_1$ and $\mathcal{P}_2$ to the weight W. Find the forces in truss members CJ and IJ. The pulleys each have a radius of 0.2 m.

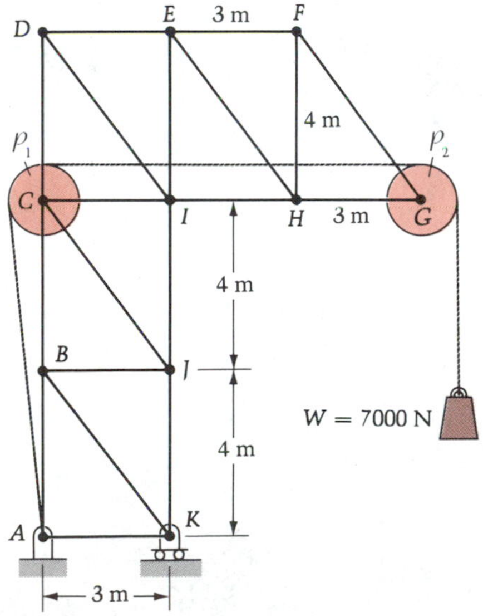

Figure P5.72

5.6 Space Trusses

A nonplanar truss is called a "space truss." Its analysis is complicated by the fact that at each joint there are not two, but three nontrivial equations of equilibrium.

The basic space truss is a tetrahedron, formed by six members pinned at their ends as shown in Figure 5.26. By adding three members at a time and connecting them to create a new joint, a gradually larger structure is formed, called (as in the plane case) a **simple space truss**. This time, instead of $2p - 3$ members as in the plane case, the space truss has $3p - 6$ members, as the reader may wish to show. Instead of the three constraint reactions required in the plane case, in three dimensions we need six to completely constrain the body.

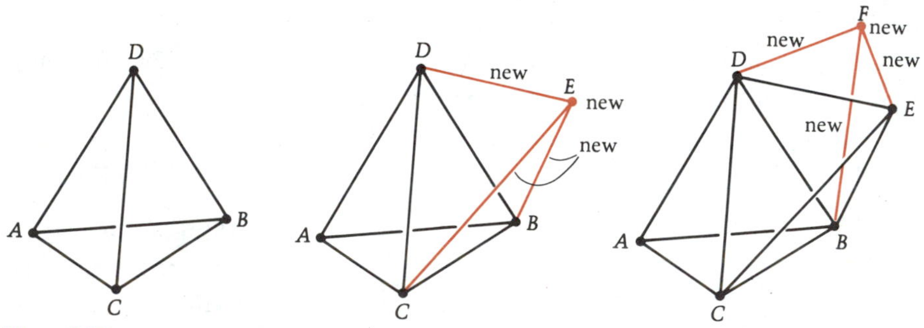

Figure 5.26

The methods of joints and sections apply as well to space trusses as they did to plane trusses. The problem with the space truss, however, is the much larger number of unknown member forces. In a simple space truss, a joint (the last one added in the above procedure) with only three member forces acting on it can always be found. This joint is a good place to start if the truss is to be completely solved. On the other hand, the method of sections, as in the plane case, can sometimes be used to great advantage in finding an isolated member force somewhere in the middle of a space truss. This time, if we are to find this force with a single equation, we have to look for a *line* (instead of a point) through which all the undesired unknown forces pass. If we can find one, then equating the moments about the line to zero will yield the desired member force. We now examine some examples of space trusses.

EXAMPLE 5.8

Find the forces in the three members *AB*, *AC*, and *AD* of the space truss shown in Figure E5.8a.

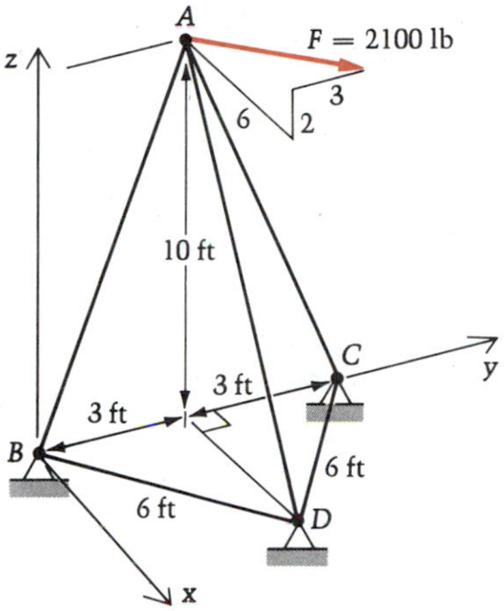

Figure E5.8a

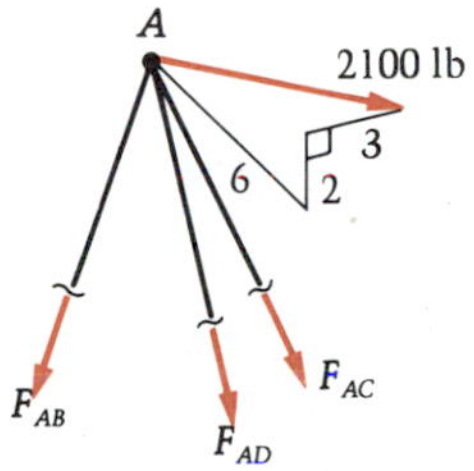

Figure E5.8b

Solution

Just as in a plane truss, the bars of a space truss are all two-force members. Therefore, isolating pin A, we have the free-body diagram shown in Figure E5.8b. The three bar forces are expressible as:

$$\mathbf{F}_{AB} = F_{AB}\left(\frac{-3\hat{\mathbf{j}} - 10\hat{\mathbf{k}}}{\sqrt{109}}\right) = F_{AB}(-0.287\hat{\mathbf{j}} - 0.958\hat{\mathbf{k}})$$

$$\mathbf{F}_{AC} = F_{AC}\left(\frac{3\hat{\mathbf{j}} - 10\hat{\mathbf{k}}}{\sqrt{109}}\right) = F_{AC}(0.287\hat{\mathbf{j}} - 0.958\hat{\mathbf{k}})$$

$$\mathbf{F}_{AD} = F_{AD}\left(\frac{3\sqrt{3}\hat{\mathbf{i}} - 10\hat{\mathbf{k}}}{\sqrt{127}}\right) = F_{AD}(0.461\hat{\mathbf{i}} - 0.887\hat{\mathbf{k}})$$

The applied force can be written as

$$\mathbf{F} = 2100\left(\frac{6\hat{\mathbf{i}} + 3\hat{\mathbf{j}} + 2\hat{\mathbf{k}}}{7}\right) = 1800\hat{\mathbf{i}} + 900\hat{\mathbf{j}} + 600\hat{\mathbf{k}}$$

Equilibrium of joint A requires that

$$\Sigma\mathbf{F} = \mathbf{F}_{AB} + \mathbf{F}_{AC} + \mathbf{F}_{AD} + \mathbf{F} = \mathbf{0}$$

Therefore, equating the coefficients of $\hat{\mathbf{i}}$, $\hat{\mathbf{j}}$, and $\hat{\mathbf{k}}$ to zero,

$$\hat{\mathbf{i}}\text{-coefficients} \Rightarrow 0 + 0 + 0.461\,F_{AD} + 1800 = 0$$

from which

$$F_{AD} = -3900 \text{ lb}$$

Therefore, member AD is in compression (because we assumed tension and obtained a negative answer), and it carries 3900 lb of compressive force.

$$\hat{\mathbf{j}}\text{-coefficients} \Rightarrow -0.287F_{AB} + 0.287F_{AC} + 900 = 0$$

$$\hat{\mathbf{k}}\text{-coefficients} \Rightarrow -0.958F_{AB} - 0.958F_{AC} - 0.887(-3900) + 600 = 0$$

The solution to these two equations is

$$F_{AB} = 3690 \text{ lb}$$

$$F_{AC} = 554 \text{ lb}$$

Both these members are in tension, because we assumed they were and positive answers were then obtained for F_{AB} and F_{AC}.

Example 5.9

The space truss shown in Figure E5.9a supports a parabolic antenna. The antenna is positioned in azimuth* by altering the lengths of AQ and BQ, and in elevation by the screwjack member OR. The wind and gravity loads for a certain orientation of the antenna are equipollent to the forces at P, Q, and R given in the

* "Azimuth" is the rotation around the z axis (local vertical).

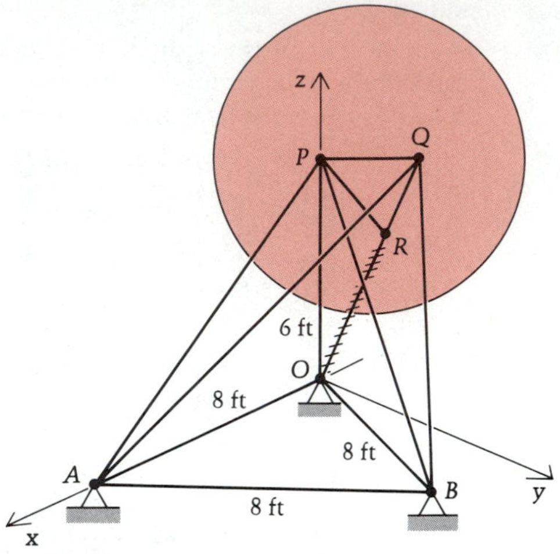

Figure E5.9a

following equations. The locations of Q and R are also indicated. Find the force in the screwjack.

$$\text{Coordinates of } Q\text{:} \quad (-1.5, 2, 6) \text{ ft}$$

$$\text{Coordinates of } R\text{:} \quad (-2.75, -0.25, 3) \text{ ft}$$

$$\mathbf{F}_P = 2000\hat{\mathbf{i}} + 500\hat{\mathbf{j}} - 800\hat{\mathbf{k}} \text{ lb}$$

$$\mathbf{F}_Q = -1400\hat{\mathbf{i}} - 300\hat{\mathbf{j}} - 800\hat{\mathbf{k}} \text{ lb}$$

$$\mathbf{F}_R = 700\hat{\mathbf{i}} + 500\hat{\mathbf{j}} - 400\hat{\mathbf{k}} \text{ lb}$$

Solution

The free-body diagram of the antenna dish is shown in Figure E5.9b.

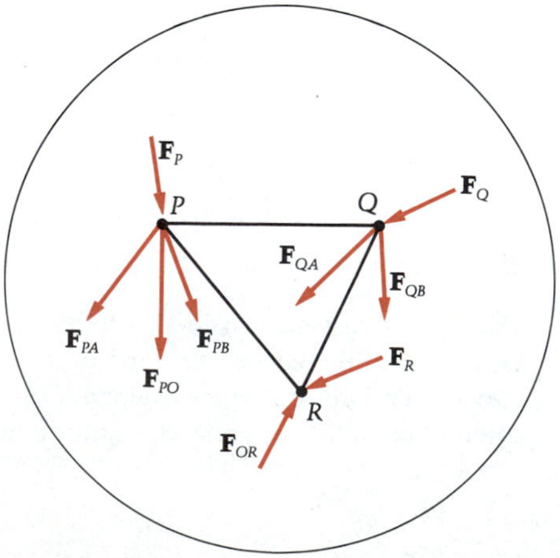

Figure E5.9b

A moment summation about the line PQ will yield the screwjack force $\mathbf{F}_{OR}$:

$$\Sigma \mathbf{M}_P \cdot \mathbf{u}_{PQ} = \Sigma M_{\text{line}\,PQ} = 0$$

The only forces contributing to this equation are $\mathbf{F}_{OR}$ and $\mathbf{F}_R$. Therefore,

$$\mathbf{r}_{PR} \times (\mathbf{F}_R + \mathbf{F}_{OR}) \cdot \hat{\mathbf{u}}_{PQ} = 0$$

$$[(-2.75 - 0)\hat{\mathbf{i}} + (-0.25 - 0)\hat{\mathbf{j}} + (3 - 6)\hat{\mathbf{k}}]$$

$$\times \left[(700\hat{\mathbf{i}} + 500\hat{\mathbf{j}} - 400\hat{\mathbf{k}}) + F_{OR}\left(\frac{-2.75\hat{\mathbf{i}} - 0.25\hat{\mathbf{j}} + 3\hat{\mathbf{k}}}{4.08}\right)\right]$$

$$\left[\frac{(-1.5 - 0)\hat{\mathbf{i}} + (2 - 0)\hat{\mathbf{j}} + (6 - 6)\hat{\mathbf{k}}}{2.50}\right] = 0$$

The scalar triple product can be expressed as a determinant:

$$\begin{matrix} \hat{\mathbf{u}}_{PQ} \text{ components} \rightarrow \\ \mathbf{r}_{PR} \text{ components} \rightarrow \\ \mathbf{F}_R + \mathbf{F}_{RO} \text{ components} \rightarrow \end{matrix} \begin{vmatrix} -0.6 & 0.8 & 0 \\ -2.75 & -0.25 & -3 \\ (700 - 0.674F_{OP}) & (500 - 0.0613F_{OP}) & (-400 + 0.735F_{OR}) \end{vmatrix} = 0$$

Adding $\frac{4}{3}$ times the first column to the second column will simplify the evaluation of the determinant:

$$\begin{vmatrix} -0.6 & 0 & 0 \\ -2.75 & -3.92 & -3 \\ (700 - 0.674F_{OR}) & (1430 - 0.960F_{OR}) & (-400 + 0.735F_{OR}) \end{vmatrix} = 0$$

Thus,

$$-0.6[-3.92(-400 + 0.735F_{OR}) + 3(1430 - 0.960F_{OR})] = 0$$

$$1570 - 2.88F_{OR} + 4290 - 2.88F_{OR} = 0$$

$$F_{OR}\,\frac{5860}{5.76} = 1020 \text{ lb} \qquad \text{(compressive as assumed)}$$

Question 5.12 At the outset of the preceding example, is there a line about which the moments, summed and equated to zero, will yield by this single equation the force in BP? OP? AP? AQ? BQ?

Answer 5.12 No to all five questions.

Example 5.10

The space truss in Figure E5.10a (see on the next page) was used to support a large "cross log periodic antenna" (which is shown in Figure P7.93(a) of Chapter 7). A certain loading under heavy wind, self-weight, and ice weight is transmitted to the truss as indicated in the figure. Find the forces in the truss members.*

* The actual truss had a seventh member from point B to D. Because this makes the truss statically indeterminate, we omit it in this example.

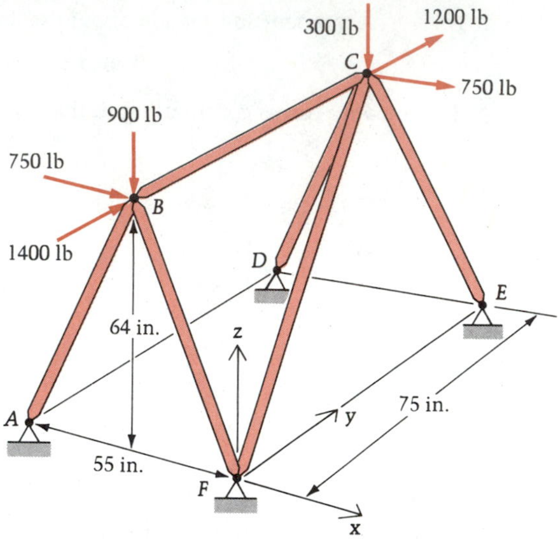

Figure E5.10a

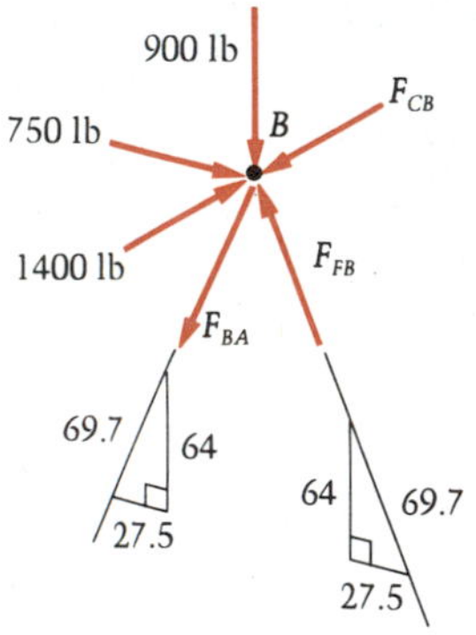

Figure E5.10b

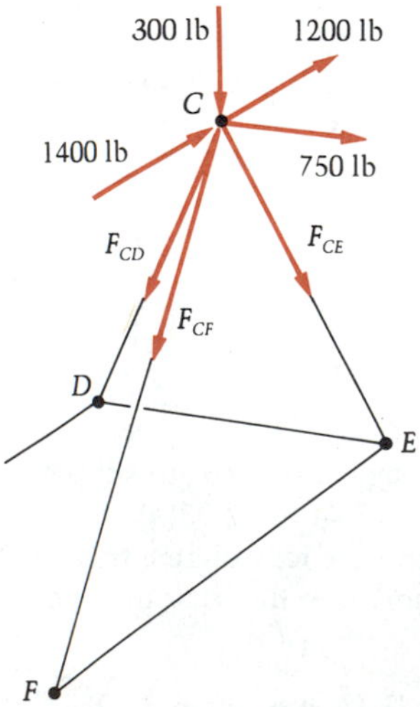

Figure E5.10c

Solution

We consider first a free-body diagram of joint B (see Figure E5.10b). Writing the equilibrium equations,

$$\Sigma F_y = 0 = 1400 - F_{CB} \Rightarrow F_{CB} = 1400 \text{ lb} \text{ Ⓒ}$$

$$\Sigma F_x = 0 = 750 - F_{BA}\left(\frac{27.5}{69.7}\right) - F_{FB}\left(\frac{27.5}{69.7}\right) \quad (1)$$

$$\Sigma F_z = 0 = -900 - F_{BA}\left(\frac{64}{69.7}\right) + F_{FB}\left(\frac{64}{69.7}\right) \quad (2)$$

Solving Equations (1) and (2),

$$F_{FB} = 1440 \text{ lb Ⓒ} \qquad \text{and} \qquad F_{BA} = 460 \text{ lb Ⓣ}$$

We note that F_{FB} could have been found without the need to solve two equations simultaneously:

$$+\circlearrowleft \quad \Sigma M_{AD} = 0 = -\left(F_{BP}\frac{64}{69.7}\right)(55 \text{ in.}) + (750 \text{ lb})(64 \text{ in.})$$

$$+(900 \text{ lb})(27.5 \text{ in.})$$

$$F_{BF} = +1440 \text{ lb Ⓒ}$$

Then F_{BA} would follow from either force equation.

Moving next to a free-body diagram of joint C (see Figure E5.10c), we see that we can avoid three equations in three unknowns (which we would get from $\Sigma \mathbf{F} = \mathbf{0}$) by summing moments about line FE:

$$\circlearrowleft + \quad \Sigma M_{FE} = 0 = (750 \text{ lb})(64 \text{ in.}) - (300 \text{ lb})(27.5 \text{ in.})$$

$$-F_{CD}\left(\frac{64}{69.7}\right)(55 \text{ in.})$$

$$F_{CD} = 787 \text{ lb Ⓣ}$$

Similarly, moments about line DE will yield F_{CF}:

$$\circlearrowleft^{+}\ \Sigma M_{DE} = [(1400 + 1200)\text{lb}](64 \text{ in.}) + F_{CF}\left(\frac{75}{\sqrt{64^2 + 27.5^2 + 75^2}}\right)(64 \text{ in.}) = 0 \qquad (3)$$

$$F_{CF} = 3550 \text{ lb (T)}$$

Note that the force F_{CF} was broken up into components at joint C. If we had broken it up at F, the last term in Equation (3) would have been

$$F_{CF}\left(\frac{64}{\sqrt{64^2 + 27.5^2 + 75^2}}\right)(75 \text{ in.})$$

which is the same result but with a different lever arm. Finally,

$$\searrow^{+}\quad \Sigma F_x = 0 = F_{CE}\left(\frac{27.5}{69.7}\right) - 787\left(\frac{27.5}{69.7}\right) + 3550\left(\frac{27.5}{102}\right) + 750$$

$$F_{CE} = -3540 \text{ lb} \qquad (\text{or } 3540 \text{ lb (C)})$$

As checks,

$$\nearrow^{+}\quad \Sigma F_y = 1200 + 1400 - 3550\left(\frac{75}{102}\right)$$

$$= 2600 - 2610 = -10 \text{ lb} \ \checkmark$$

$$+\uparrow\quad \Sigma F_z = -300 + 3540\left(\frac{64}{69.7}\right) - 3550\left(\frac{64}{102}\right) - 787\left(\frac{64}{69.7}\right)$$

$$= -300 + 3250 - 2230 - 720^* = 0 \ \checkmark$$

The −10-lb result for ΣF_y differs from zero due to rounding to three digits. Had we kept more digits, ΣF_y would have been smaller, as the reader may wish to show, but we must always remember that an answer is no more accurate than the least accurate number in the data.

* To three digits, 787(64 / 69.7) is 723; however, the units digit in the second and third terms (3250 and −2230) is insignificant.

PROBLEMS ▶ Section 5.6

5.73 A set of three forces in coordinate directions is applied at point C in Figure P5.73. Find the forces in the three bars, which are pinned together at C and rest on ball joints at A, O, and B.

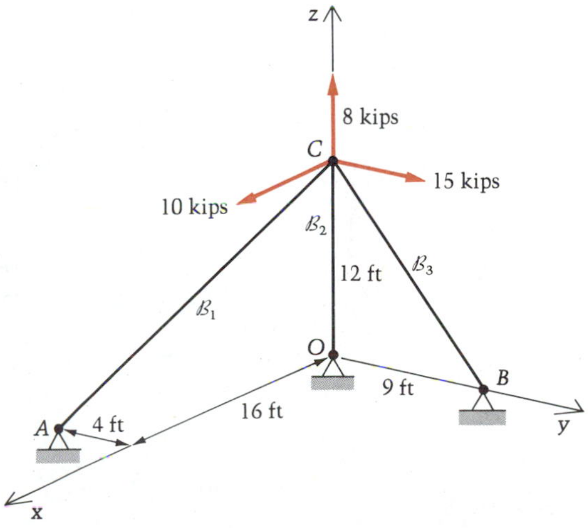

Figure P5.73

5.74 Two uniform light bars BD and DC are pinned together at D, where a cable is also attached as shown in Figure P5.74. The bars are connected to the reference frame at B and C by smooth ball joints. The system supports the 300-lb force pulling at D in the negative x direction. Find the force in the cable, and the reactions at B and C. The bars are in the yz plane.

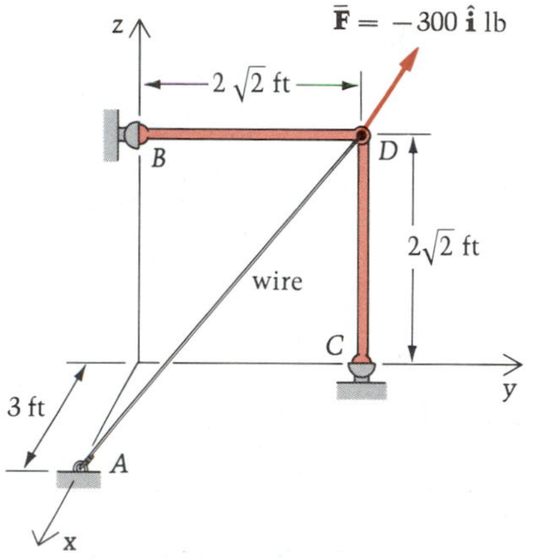

Figure P5.74

5.75 The portable jack stand in Figure P5.75 carries a downward load of 600 lb. How much compressive force is there in each of the four symmetrical legs OA, OB, OC, and OD? Assume that all connections are ball joints, and that the only reactions from the ground are equal vertical forces (using symmetry) at A, B, C, and D. Also, neglect the dimensions of the top plate.

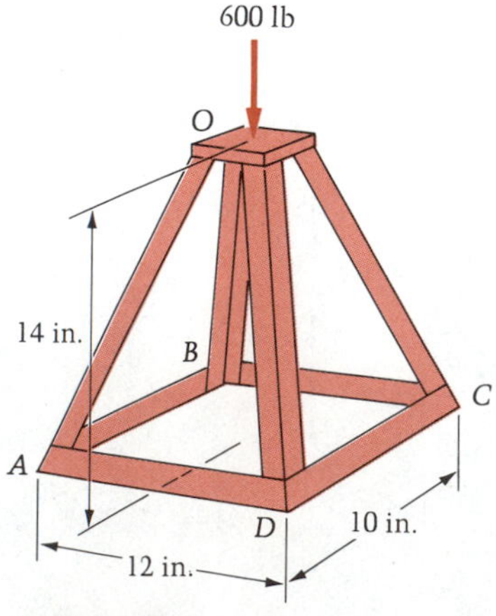

Figure P5.75

5.76 Consider all connections of the weightless struts in Figure P5.76 as being ball-and-socket. The points A, C, and D all lie in the xz plane. A vertical load of 960 lb is applied downward at B. Determine the force in member AB.

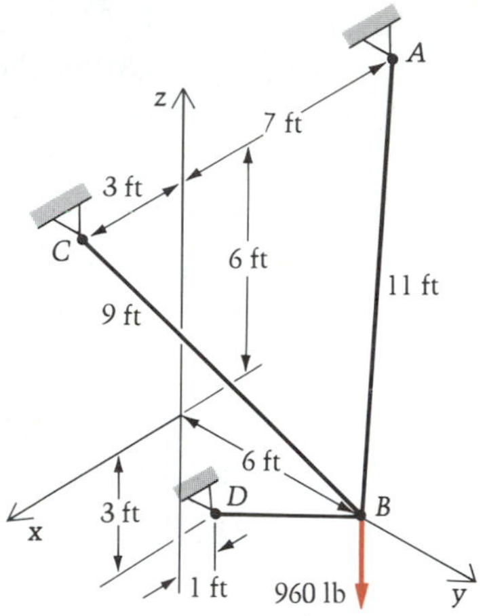

Figure P5.76

5.77 The light tripod in Figure P5.77 supports a weight of 144 lb. If the legs $\mathcal{A}$, $\mathcal{C}$, and $\mathcal{D}$ are pinned at B and if the ground is rough, find the force in leg $\mathcal{C}$.

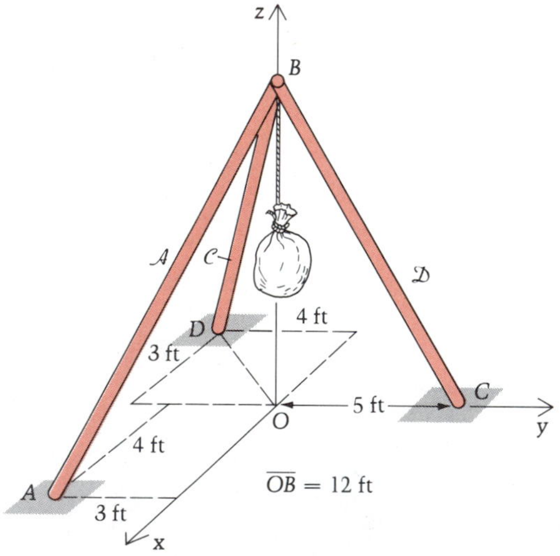

Figure P5.77

5.78 By inspection (it takes thinking, but not writing!), find the force in member AB of the truss in Figure P5.78. Then find the force in DB.

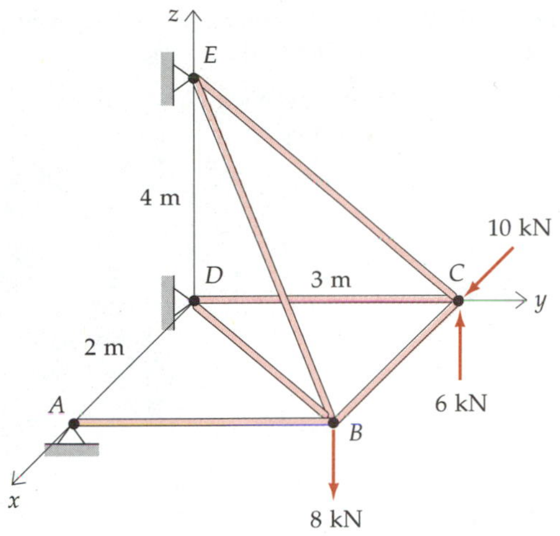

Figure P5.78

5.79 For the space truss shown in Figure P5.79:

a. Find the reactions at *A*, *B*, and *C*.

b. Using the method of sections, with a section cut by a plane parallel to *xz*, find the force in members *AD* and *CD*.

c. Find the forces in members *AB*, *BC*, and *BD* using the method of joints.

d. Find the force in *AC* by inspection.

(F_{A_x}, F_{A_y}, and F_{B_x} represent the only nonzero reaction components at *A* and *B*.)

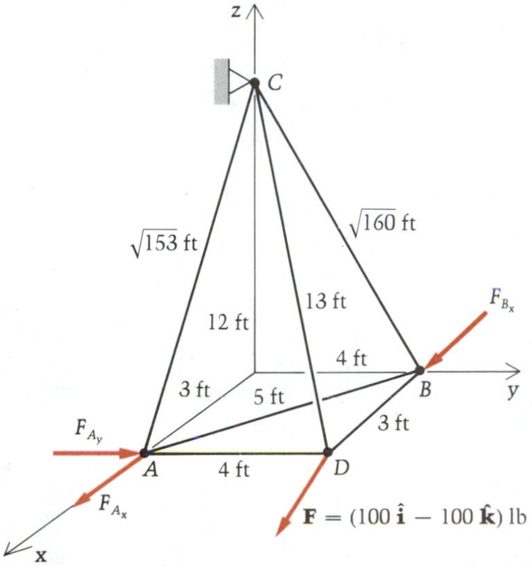

Figure P5.79

* **5.80** In Example 5.9 find the forces in the adjustable legs *AQ* and *BQ*.

* **5.81** In Figure P5.81 the nine members all have the same length ℓ. Find the forces in the members, noting that, by symmetry, (a) the forces in *AB*, *AC*, *AD*, *EB*, *EC*, and *ED* are equal; and (b) the forces in *BC*, *CD*, and *DB* are equal.

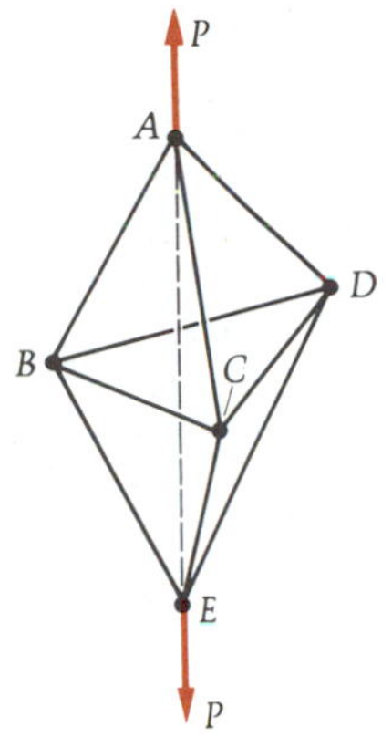

Figure P5.81

5.82 The 2000-N force is in the negative *x* direction; the 2400-N force is in the *yz* plane, and the 2100-N force has direction cosines $(-\frac{3}{7}, \frac{2}{7}, -\frac{6}{7})$. Find all the members' forces that are not indeterminate in the space truss shown in Figure P5.82.

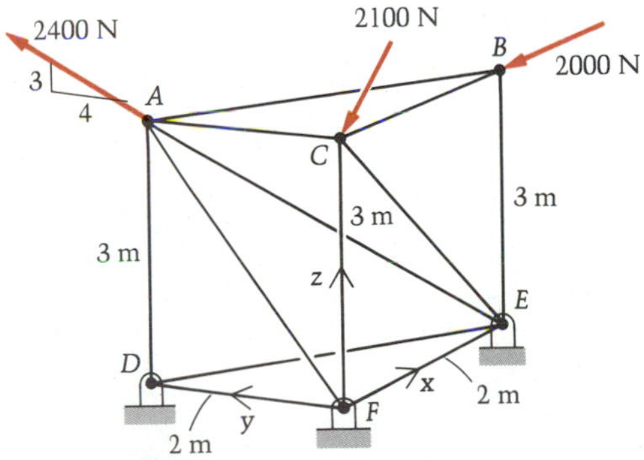

Figure P5.82

Problems 5.83–5.87 are based upon Figure P5.83 and the following text: Bars $\mathcal{B}_1$, $\mathcal{B}_2$, and $\mathcal{B}_3$ are pinned at A to form a space truss. They are likewise pinned (ball-jointed) to the ground at O_1, O_2, and O_3, respectively. Find the forces in the three members if the force **F** applied at A is as given.

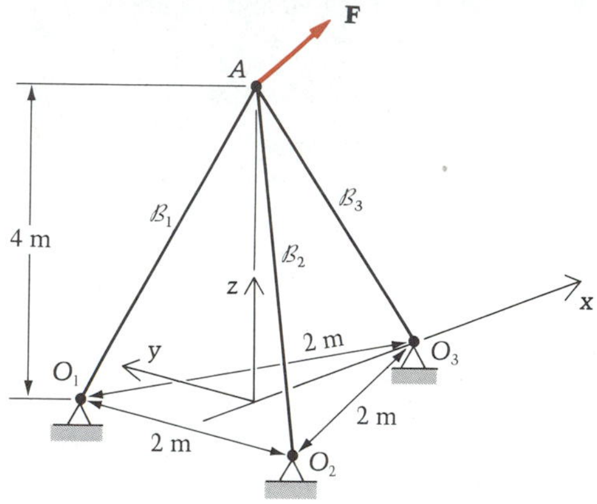

Figure P5.83

5.83 $\mathbf{F} = -1000\hat{\mathbf{k}}$ N

5.84 $\mathbf{F} = 1000\hat{\mathbf{i}}$ N

5.85 $\mathbf{F} = 1000\hat{\mathbf{j}}$ N

5.86 $\mathbf{F} = 707\hat{\mathbf{i}} + 707\hat{\mathbf{j}}$ N

5.87 $\mathbf{F} = 577\hat{\mathbf{i}} + 577\hat{\mathbf{j}} + 577\hat{\mathbf{k}}$ N

5.88 The space truss in Figure P5.88 carries the loads

$$\mathbf{F}_1 = -1000\hat{\mathbf{i}} + 1200\hat{\mathbf{j}} + 2000\hat{\mathbf{k}} \text{ lb}$$

and

$$\mathbf{F}_2 = 500\hat{\mathbf{i}} - 1000\hat{\mathbf{j}} + 1500\hat{\mathbf{k}} \text{ lb}$$

a. By inspection, find a zero-force member.

b. Find the force in member AD.

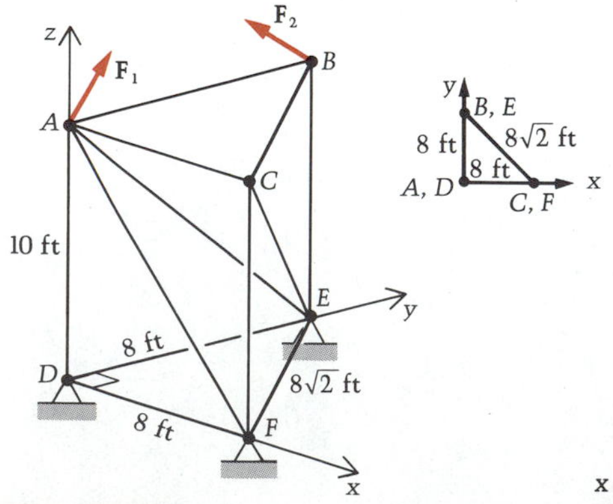

Figure P5.88

5.89 In Figure P5.89 line AB of the space truss is vertical, and the 2000-lb force is parallel to the x axis.

a. Show that the structure is a simple space truss.

b. Find the force in member EF.

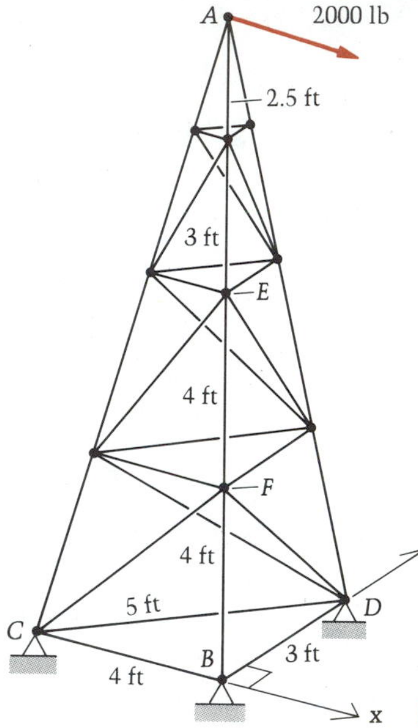

Figure P5.89

5.90 For the space truss in Figure P5.90, find:

a. The reactions at A, B, C, and D

b. The force in member BE using the method of sections with the section shown and a single equation

c. The force in member BE by the method of joints.

Note: The support 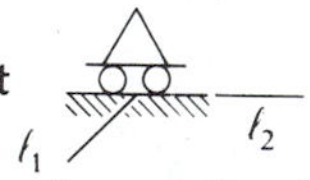 transmits force only normal to the plane defined by the two perpendicular lines ℓ_1 and ℓ_2.

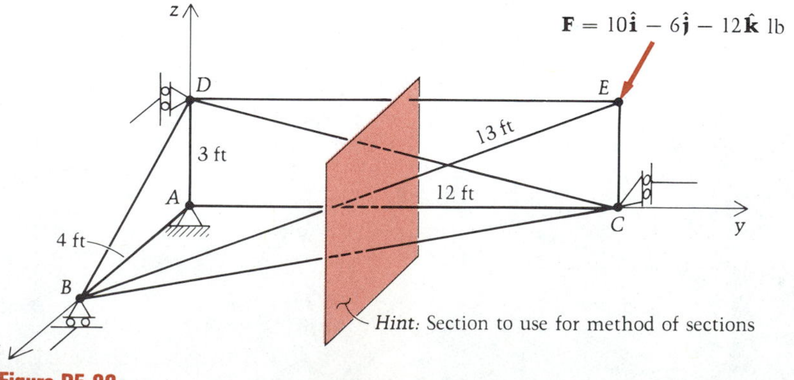

Figure P5.90

5.7 A Brief Introduction to the Mechanics of Deformable Bodies

We will now use the truss to give the reader a brief introduction to the mechanics of deformable bodies, also known as strength of materials, or mechanics of materials.

In this section, we shall introduce — in a one dimensional analysis appropriate to the truss — the concepts of stress, strain, and Hooke's law. We will then demonstrate how a consideration of deformations allows the solution of certain types of problems previously deemed statically indeterminate. Lastly, we will look at the effect of temperature.

Normal Stress

The major reason why we have learned in Sections 5.2–5.6 to find the forces in a truss is so that we can later determine the *stresses* within the members. Stresses, which have units of force per unit of area, are important because materials fail when stresses exceed certain limits; thus in design we are interested in knowing the maximum stress within each of the members of a structure. It turns out that this is very easy to do in a truss, for two reasons:

1. In each member of a truss, the resultant force F, acting on every cross section (of area A) normal to the member's axis, is the same.

2. The normal stress at every point in the above cross section, in the direction of the member's axis, is simply F/A, as we shall shortly see.

Suppose now that the member AB in Figure 5.27(a) is a truss member carrying the 10 kN tensile force as shown.

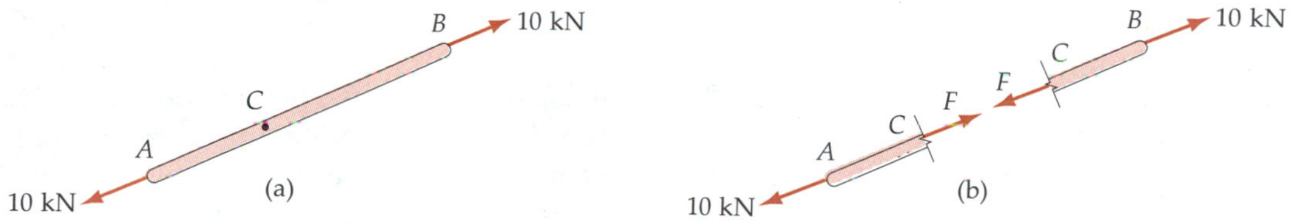

Figure 5.27

We know from previous sections of the chapter that if we were to mentally slice through the member at point C as in Figure 5.27(b), then using either of the two free-body diagrams in that figure, we would find F to be a 10-kN tensile force. But let us now look more carefully at the cross-sectional cut on the FBD of CB.

We see from Figure 5.28(a), on the next page, that F is merely the *resultant* of infinitely many small forces ΔF, each acting on a small area ΔA. At a point P within one of these ΔA's (see Figure 5.28(b)), we define the **normal stress** σ to be the limit of the resultant force ΔF acting on ΔA, divided by ΔA, as ΔA shrinks to zero while always including point P:

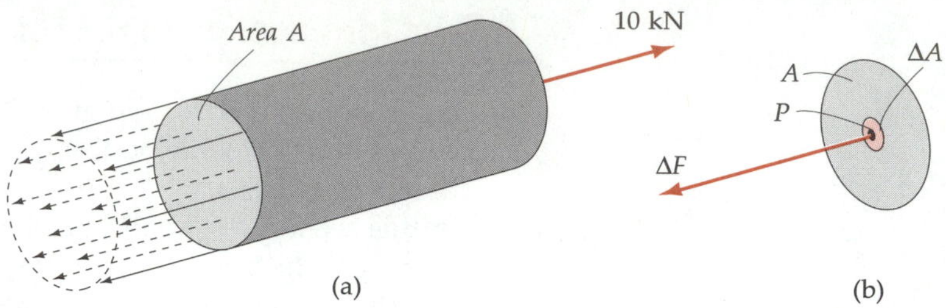

Figure 5.28

$$\sigma = \lim_{\Delta A \to 0} \frac{\Delta F}{\Delta A}$$

A small distance away from the ends of the truss member (where the load typically is applied through a pin), the force distribution across the cross section becomes uniform, so that

$$\sigma = \frac{F}{A} \tag{5.1}$$

and thus the normal stress at every point of a truss member in the direction of its axis is the force in the member divided by its cross-sectional area. Furthermore, because σ is the same for all points within the member, we may simply speak of the stress in the truss member as F/A. Thus if the member in Figure 5.27 had a cross-sectional area of 0.01 m^2, the stress in the member would be

$$\sigma = \frac{10{,}000 \text{ N}}{0.01 \text{ m}^2} = 1 \times 10^6 \text{ N/m}^2$$

This is a tensile stress, because (see Figure 5.28) it "pulls" on the area on which it acts. If the two 10 kN forces were compressing the bar AB instead of stretching it, the force F on the section at C would be "pushing" on the area, and the stress in that case would be called a compressive stress. All normal stresses σ are either tensile or compressive, and the sign convention is that they are positive if tensile, and thus negative if compressive.

In the SI system of units, one N/m^2 is called a Pascal (Pa); thus the above stress is $\sigma = 1 \times 10^6$ Pa or 1MPa (megapascal). In the U.S. system of units, stress is usually written in pounds per square inch, or psi. In an inflated tire, for example, 32 psi is a typical recommended air pressure.

Before moving to an example of stress calculation in a truss, we mention that in structures other than trusses, stress analysis is in general *much* more complicated. For one thing, as forces vary from point to point, so does the resulting stress. Secondly, we may have normal stress in two and even three dimensions to be concerned about. And thirdly, there is yet another kind of stress, known as shear stress and which we shall not treat in this brief section, which complicates the picture still further. While normal stress is force per unit area perpendicular to an area, shear stress is force per unit area *in the plane* of an area. Thus, for example, the intensity of friction force is shear stress because it resists the tendency of one body to slide past another in their plane of contact.

EXAMPLE 5.11

In the truss of Figure 5.6, shown in Figure E5.11a, the members are bars with cross-sectional areas 3 in.^{2} (*AE* and *EB*), 2 in.^{2} (*AD*, *DE*, *CE*, and *CB*), and 1.5 in.^{2} (*DC*). Find which member is being subjected to the largest stress for the given loading.

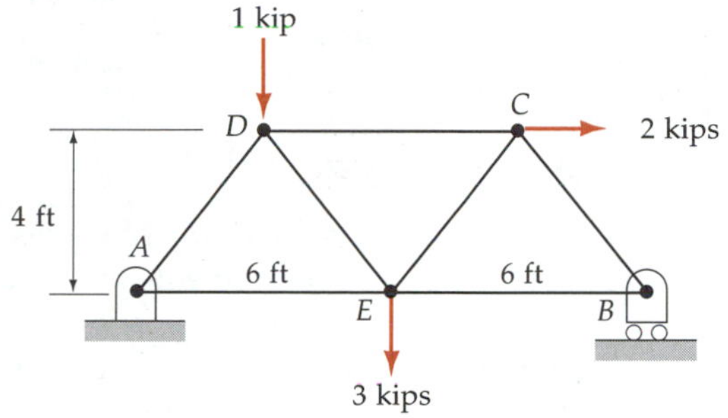

Figure E5.11a

Solution

The forces in the members were given in kips near the end of Section 5.2, and they are shown pictorially in Figure E5.11b:

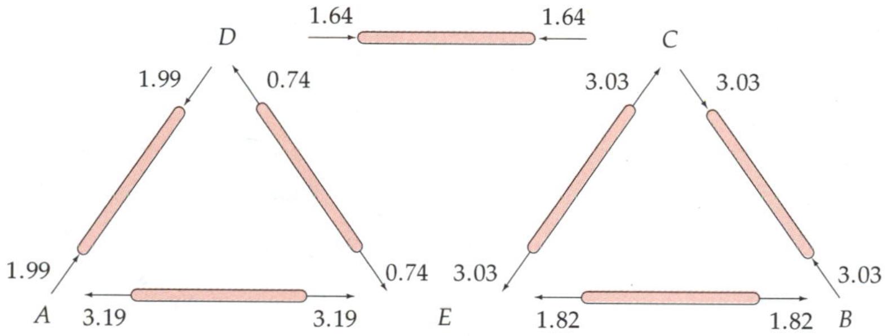

Figure E5.11b

Thus the stresses carried by the members are:

$$\sigma_{AE} = P_{AE}/A_{AE} = \frac{3190}{3} = 1063 \text{ psi, tensile}$$

$$\sigma_{EB} = \frac{1820}{3} = 607 \text{ psi, tensile}$$

$$\sigma_{AD} = \frac{1990}{2} = 995 \text{ psi, compressive}$$

$$\sigma_{DE} = \frac{740}{2} = 370 \text{ psi, tensile}$$

$$\sigma_{CE} = \frac{3030}{2} = 1515 \text{ psi, tensile}$$

$$\sigma_{CB} = \frac{3030}{2} = 1515 \text{ psi, compressive}$$

$$\sigma_{DC} = \frac{1640}{1.5} = 1093 \text{ psi, compressive}$$

Members *CE* and *CB* are seen to carry the largest stress, and it is worth noting that these two members do not carry the largest force, nor do they have the smallest area. It is the *combination* of force and area that counts in computing the maximum stress.

Extensional Strain

Normal stress, which we have discussed above in conjunction with trusses, causes line elements in a body to extend or contract, depending on whether σ is positive (tensile) or negative (compressive). These extensions and contractions strain a truss element (for example, the one depicted previously in Figure 5.27), and in fact we may give a formal definition for **extensional strain** ϵ in the direction of the member axis at a point such as C (see Figure 5.29(a)) located at the coordinate x along the axis:

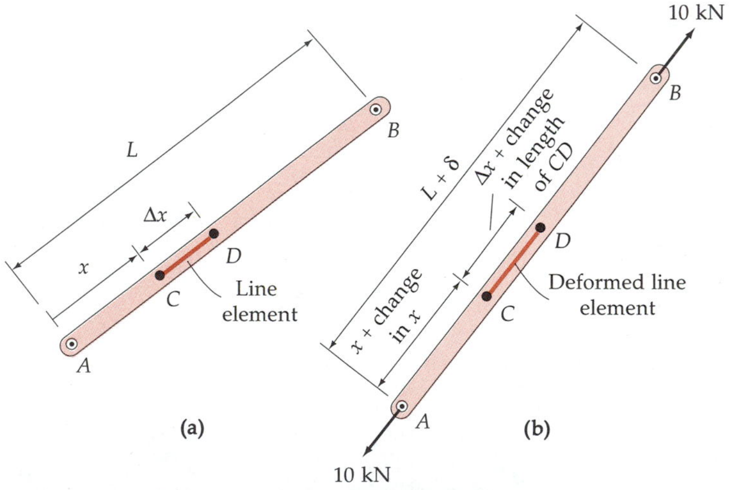

Figure 5.29 (Line element length and deformations highly exaggerated)

The extensional strain ϵ is the limit of the change in length of a line element at C divided by the original length of that line element, as the length of the element approaches zero. In Figure 5.29(a), the short line element CD at point C has a length Δx before the truss (of which AB is a member) is loaded. Therefore,

$$\begin{matrix}\text{Strain} \\ \text{at } C\end{matrix} = \epsilon = \lim_{\Delta x \to 0} \left(\frac{\text{Change in Length of } CD}{\Delta x} \right)$$

After the truss is loaded, member AB carries its 10 kN load, and member AB has moved slightly in space and has elongated. (It would have shortened had the 10 kN force been compressive.) As will be seen in the next section, stress and strain are proportional for linearly elastic materials. Thus the strain in a truss element such as AB is also constant, meaning that the change in length per original length is the same for every line element from one end of the member to the other. If the *total* elongation of member AB is δ, then we may write for a truss element:

$$\epsilon = \frac{\delta}{L} \tag{5.2}$$

i.e., the constant strain in the member is its change in length divided by its original length. If member AB elongates 0.0020 m under the 10 kN load, then if its original length was 1.5 m, the strain in the member is

$$\epsilon = \frac{0.0020}{1.5} = 0.0013 \text{ m}/\text{m}$$

Note that strain is a dimensionless quantity, although if we expressed the above answer as 1.3 mm / m, the units would need to be displayed. Also, sometimes strain is written as "0.13% strain," which simply means $\epsilon = 0.0013$ (or 0.0013 m / m, or 0.0013 in. / in., etc.)

We will do more with Equation (5.2) after we study the relationship between stress and strain in the following sub-section.

Hooke's Law

For elastic materials, stress and strain are proportional — up to a point. The linear relationship, discovered by Robert Hooke in the 17th century and known as **Hooke's Law,** is written for uniaxial stress and strain, such as we have in a truss element, as

$$\sigma = E\epsilon \tag{5.3}$$

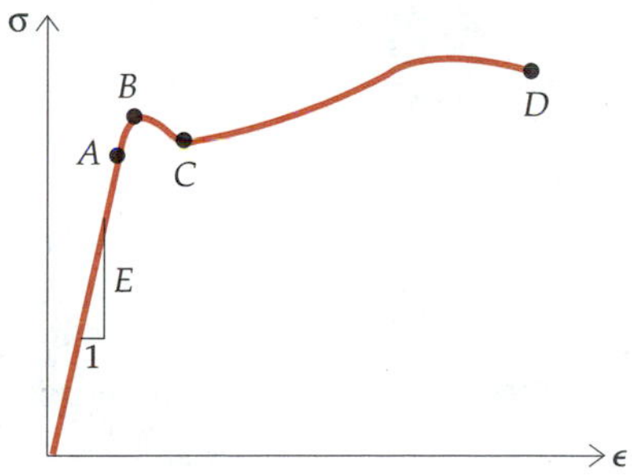

Figure 5.30

where the constant of proportionality, E, is called Young's modulus of elasticity. As strain is dimensionless, E has the same units as stress, namely pascals (N/m^2) or psi ($lb/in.^2$). Figure 5.30 shows the relationship pictorially, in a drawing called a stress-strain diagram. A figure like this will be produced if a mild steel specimen is mounted in grips in a testing machine (see Figure 5.31 on the next page) and subjected to a tensile test. The machine stretches the specimen with a gradually increasing load and measures its increasing elongation until it fails. From the force and elongation data, σ and ϵ are calculated and plotted. With respect to the stress-strain diagram, we note the following:

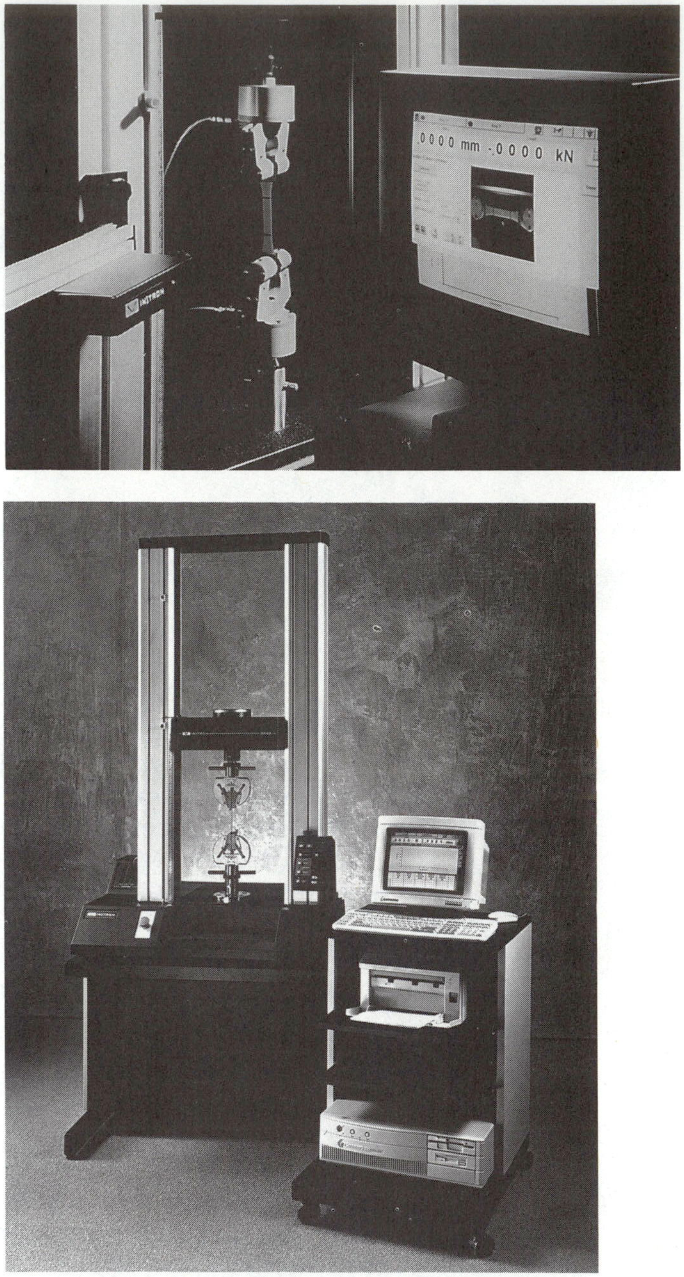

Figure 5.31 Tensile Tests (Courtesy of Instron Corporation.)

1. The slope of the "straight-line portion" is $\sigma / \epsilon = E$ at any point.

2. Beyond a point A called the proportional limit, stress and strain are no longer proportional.

3. Beyond a point B, called the elastic limit, the curve will not retrace its path if the load is then reduced back to zero.

4. Beyond a point C, called the yield point, the material begins to give, or yield, with large increase in strain for very little increase in stress.

5. At a point D, the material finally breaks (or fractures, or fails) at a value of stress called the ultimate strength.

Some materials such as aluminum behave elastically up to an elastic limit but do not map out a straight line portion like that of Figure 5.30 if stress and strain are plotted from a tensile test. The curve for such a material, which might look more like Figure 5.32, may be used to obtain a yield stress by the offset method. It turns out that if a line is drawn from the point A of 0.002 strain on the ϵ-axis in the figure, with the same slope as the curve has at the origin, this dotted line will intersect the stress-strain curve at a value of σ very close to the yield stress σ_y of the material. Furthermore, the slope of this line, i.e., the slope of the curve at the origin, serves well for Young's modulus E in calculations.

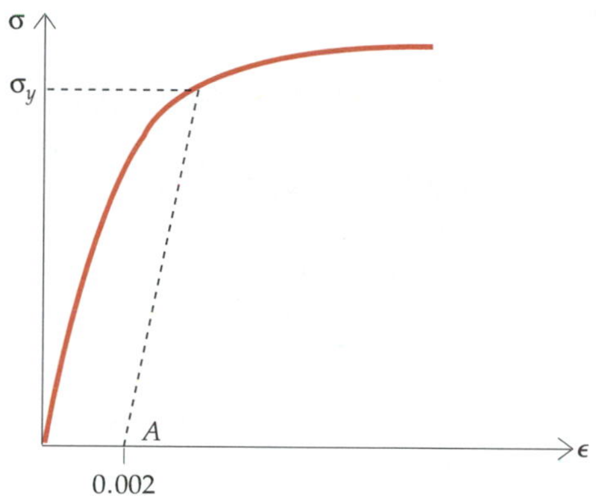

Figure 5.32

Table 5.1 below gives values of Young's modulus of elasticity for several commonly used elastic materials:

Table 5.1

Material	Young's Modulus of Elasticity E		Yield Stress σ_y	
	ksi	MPa	ksi	MPa
Aluminum Alloy 6061-T6	10,000	69,000	40	280
Structural Steel ASTM-A36	30,000	210,000	36	250
Titanium Alloy	16,000	110,000	120	830
Brass	15,000	104,000	70	480
Bronze	15,000	104,000	65	450
Hard-drawn Copper	17,000	117,000	48	330
Medium-Strength Concrete in Compression	3,600	25,000	—	—

Elongation of a Truss Element

If we now assume that our truss members are made of elastic material, we may use Hooke's law to link the results (Equations 5.1,2,3) of the previous three subsections, as follows:

$$\underbrace{\frac{\delta}{L} = \epsilon}_{\substack{\text{strain-deformation} \\ \text{relation, Eq. (5.2)}}} = \overbrace{\frac{\sigma}{E}}^{\substack{\text{stress-strain relation} \\ \text{(Hooke's law), Eq. (5.3)}}} \underbrace{= \frac{F}{AE}}_{\substack{\text{Eq. (5.1) for stress in} \\ \text{terms of force and area}}} \tag{5.4}$$

Therefore, we obtain the following relation between the elongation δ and the force F in the truss element:

$$\delta = \frac{FL}{AE} \tag{5.5}$$

We note that all of the equations (5.1–5.5) of this section are equally applicable to any straight two-force member that we may encounter within a frame. Such a member, like a truss member, is also in a state of uniaxial stress and strain. Other members of frames, however, may carry forces in lateral directions as well as bending and twisting moments; their analysis is beyond the scope of this brief "one-dimensional" treatment and must await a course in the mechanics of deformable bodies.

As an example of the use of Equation (5.5), the steel truss members DE and CB of Example 5.11 respectively extend and contract the amounts:

$$\delta_{DE} = \frac{P_{DE}\,L_{DE}}{A_{DE}\,E_{DE}} = \frac{740\,(5 \times 12)}{2\,(30 \times 10^6)} = 0.000740 \text{ in.}$$

$$\delta_{CB} = \frac{P_{CB}\,L_{CB}}{A_{CB}\,E_{CB}} = \frac{3030\,(5 \times 12)}{2\,(30 \times 10^6)} = 0.00303 \text{ in.}$$

which are tiny in comparison to the 60-inch original lengths of DE and CB, but are by no means unimportant.

In modern times, structures large and small may be analyzed by means of computer programs that utilize advanced techniques known as finite elements to quickly solve for stresses in the members, and deflections of the joints, of a truss. Just the same, for the insight it will afford us, let us use Equation (5.5) to compute the deflection of a joint in a simple truss.

EXAMPLE 5.12

Find the location of joint A of the truss after the 300 N load is applied as shown in Figure E5.12a. The members are each made of steel with $E = 207{,}000$ MPa and have cross-sectional area $= 0.004$ m^2.

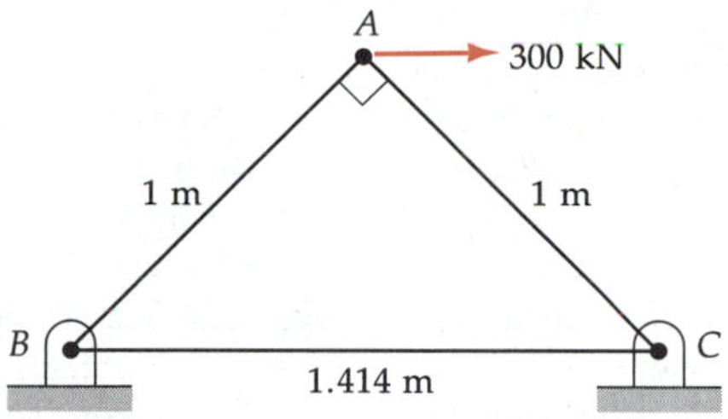

Figure E5.12a

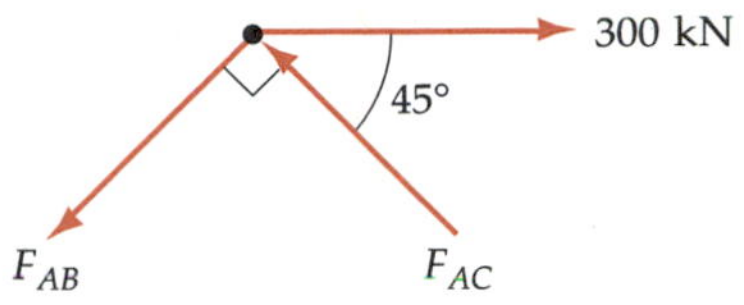

Figure E5.12b

Solution

At joint A, we have the FBD shown in Figure E5.12b. If we sum forces along the CA-direction (to get rid of one of the forces), we obtain:

$$\overset{+}{\nwarrow} \qquad \Sigma F_{CA} = 0 = F_{AC} - 300{,}000\,\frac{\sqrt{2}}{2}$$

so that

$$F_{AC} = 212{,}000 \text{ N} \quad Ⓒ$$

Then,

$$+\!\uparrow \qquad \Sigma F_y = 0 = 212{,}000\left(\frac{\sqrt{2}}{2}\right) - F_{AB}\left(\frac{\sqrt{2}}{2}\right)$$

so

$$F_{AB} = 212{,}000 \text{ N} \quad Ⓣ$$

From these results, we find

$$\delta_{AB} = \frac{212{,}000(1)}{(0.004)207 \times 10^9} = 0.256 \times 10^{-3} \text{ m (extension)} \qquad \text{and} \qquad \delta_{AC} = \frac{-212{,}000(1)}{(0.004)207 \times 10^9} = -0.256 \times 10^{-3} \text{ m (contraction)}$$

Separating the members for clarity in Figure E5.12c, we see that the extended end A of AC, and the contracted end A of AC, must lie on the circles shown in the figure in order that they have the respective proper final lengths

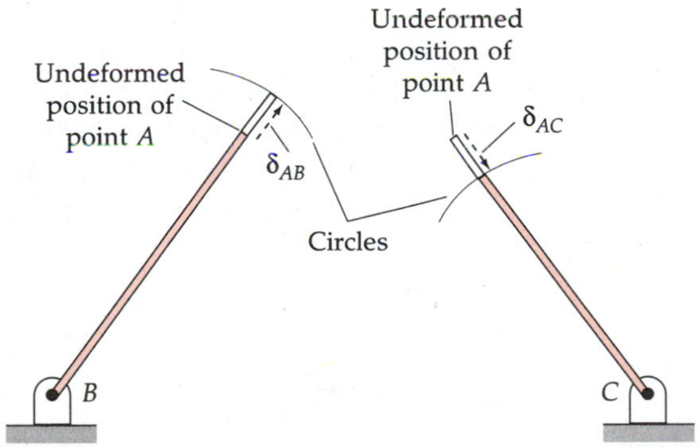

Figure E5.12c

$1 + 0.256 \times 10^{-3} = 1.0003$ m and $1 - 0.256 \times 10^{-3} = 0.9997$ m. But these deflections are so small with respect to the original lengths of AB and AC that we may treat these curves as straight lines; when we then put the members back together (Figure E5.12d), we find that the final position of point A has to lie on the intersection of the two straight lines shown. Thus point A experiences a deflection of $.000256\sqrt{2} = 0.000362$ m$\rightarrow$. The fact that the deflection is in the exact direction as the applied force is a peculiarity here of the symmetry and the equal "AE's"; the reader should note, for example, that if the "AE" of bar AB were different from that of AC, the δ's would differ and the resulting displacement of A would contain a vertical component (see Problem 5.118).

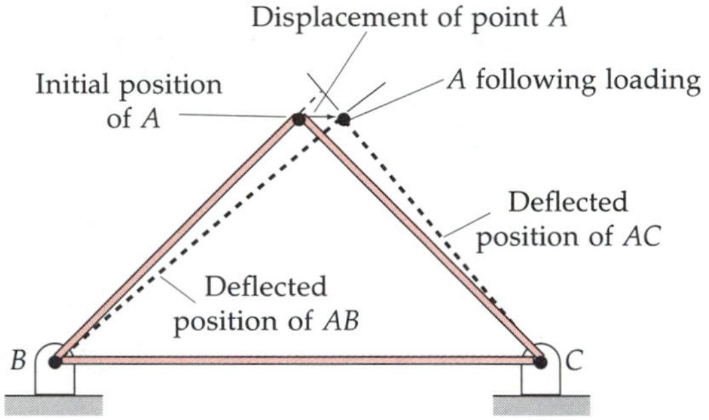

Figure E5.12d

Note in the preceding example that the stresses in the members AB and AC of the truss are:

$$\sigma_{AB} = \frac{F}{A} = \frac{212{,}000}{0.004} = 53 \times 10^6 \text{ Pa} = 53 \text{ MPa} \quad \text{(tensile)}$$

and

$$\sigma_{AC} = \frac{F}{A} = \frac{-212{,}000}{0.004} = -53 \text{ MPa, or } 53 \text{ MPa} \quad \text{(compressive)}$$

These values are less than one-fourth the yield stress for steel.

Statically Indeterminate Problems

In Chapters 2 (just below Figure 2.8) and 4 (surrounding Figure 4.7) we learned that not all equilibrium problems can be solved by means of the equations of statics alone. Those that cannot, because there are more unknowns than equations, were called statically indeterminate, for obvious reasons. For example, for the light bar in Figure 5.33,

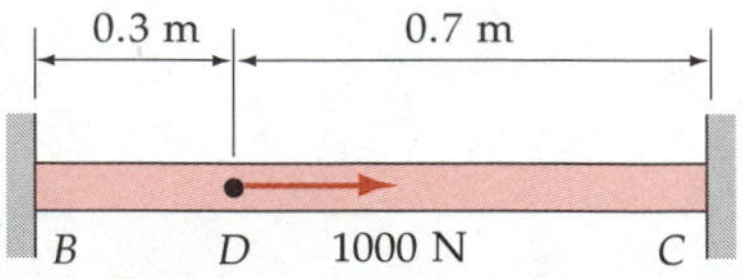

Figure 5.33

the free-body diagram is as shown in Figure 5.34:

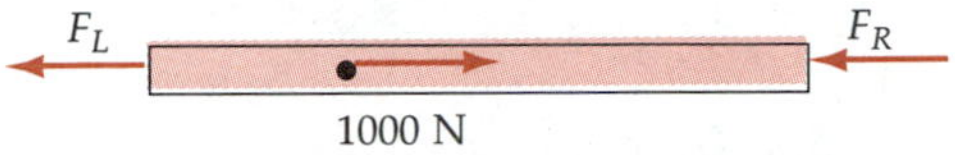

Figure 5.34

Clearly, we know that

$$\overset{+}{\longrightarrow} \quad \Sigma F_x = 0 = 1000 - F_L - F_R$$

$$F_L + F_R = 1000 \tag{5.6}$$

but that is *all* we know from statics, because any moment equation will contain the sum of F_L and F_R and thus F_L and F_R cannot be separately found.

However, our brief introduction to mechanics of materials will allow us to determine F_L and F_R by considering the *deformation* of the bar. Note from the sectionings in Figure 5.35 that:

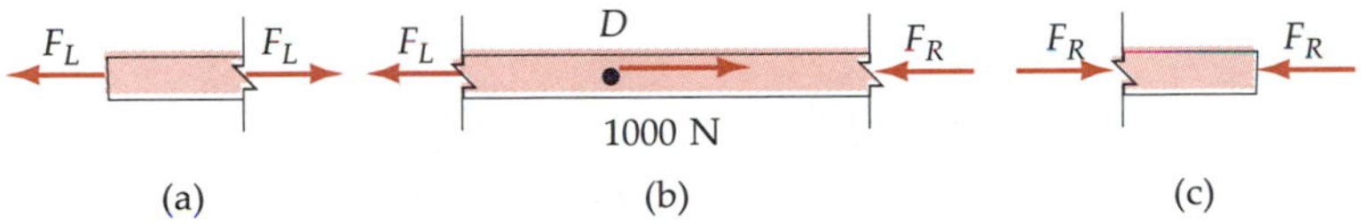

Figure 5.35

a. on any cross-section to the left of D (see Figure 5.35(a or b)) the force is F_L (tension); and

b. on any cross-section to the right of D (see Figure 5.35(b or c)), the force is F_R (compression).

Therefore, using Equation (5.5), the elongations of the two portions of the bar are:

$$\delta_{BD} = \frac{F_L\, 0.3}{AE} \quad \text{and} \quad \delta_{DC} = \frac{-F_R\, 0.7}{AE}$$

where F_R is positive because it is already drawn as compressive.

Now we are ready for the big step: The total elongation of BC has to be zero, because it is built into the (assumed-to-be) rigid walls. Thus:

$$\delta_{BC} = 0 = \delta_{BD} + \delta_{DC} \tag{5.7}$$

or

$$\frac{F_L\, 0.3}{AE} - \frac{F_R\, 0.7}{AE} = 0$$

Therefore,

$$F_L = (7\,/\,3)\, F_R \tag{5.8}$$

and we have found a second equation in the two unknown forces.

Substituting Equation (5.8) into (5.6), we arrive at:

$$(7/3)F_R + F_R = 1000$$
$$F_R = 300 \text{ N}$$

and from (3),

$$F_L = 700 \text{ N}$$

The deformations of a typical elastic structural member are nearly always very, very, small. Yet we have used them above to solve a problem which was hopeless using the equations of Statics alone. Now let us look at a second example, this time with bars of different materials and areas constituting a composite member between the walls:

EXAMPLE 5.13

Find the forces exerted by the rigid walls on the ends of the composite member *BDC* in Figure E5.13a.

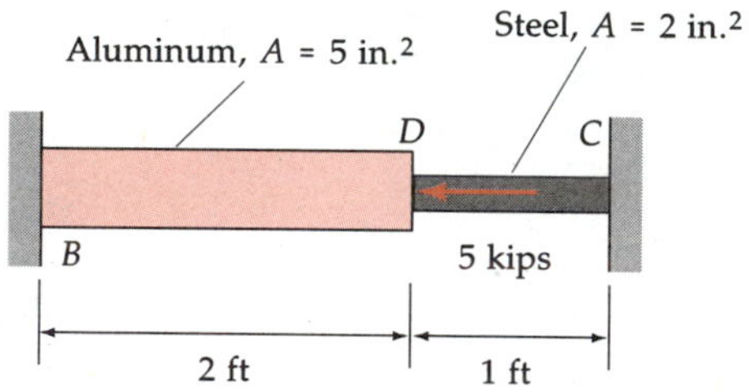

Figure E5.13a

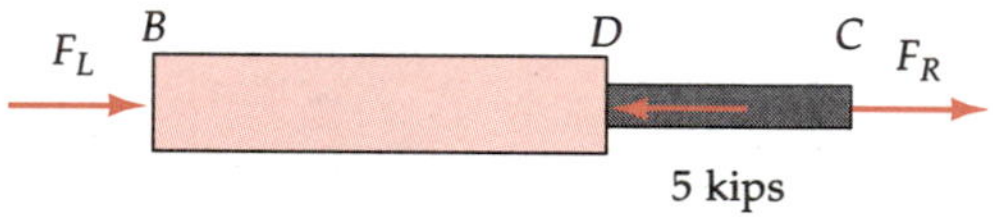

Figure E5.13b

Solution

From the FBD of the entire composite bar in Figure E5.13b, we have

$$\xrightarrow{+} \quad \Sigma F_x = 0 = F_L + F_R - 5$$

or

$$F_L + F_R = 5 \tag{1}$$

And from the FBD's in Figure E5.13c, we see that everywhere to the right of *D*, the force on a cross-section is F_R (tensile); and on all cross-sections to the left of *D*, the force is F_L (compressive).

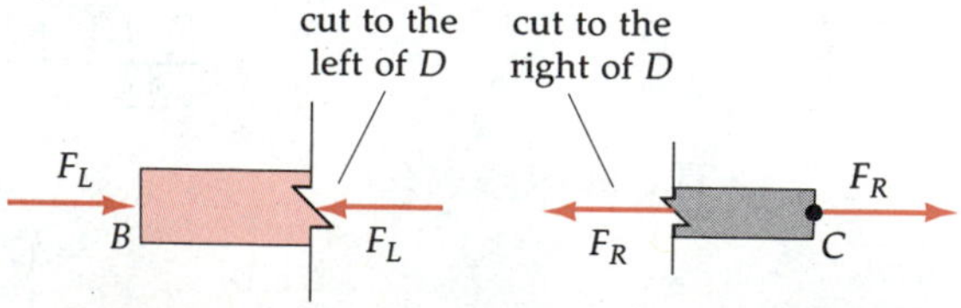

Figure E5.13c

Enforcing now the fact that the total elongation δ between the walls must again vanish, we obtain (using Table 5.1 for the elastic moduli):

$$\delta_{BC} = 0 = \delta_{BD} + \delta_{DC}$$

$$\frac{-F_L\,24}{5(10 \times 10^6)} + \frac{F_R\,12}{2(30 \times 10^6)} = 0$$

from which $F_L = (5/12)F_R$, and substituting this result into Equation (1) yields

$$\left(\frac{5}{12} + 1\right) F_R = 5$$

or

$$F_R = \frac{60}{17} = 3.53 \text{ kips}$$

and thus

$$F_L = \frac{5}{12}(3.53) = 1.47 \text{ kips}$$

In the preceding example, the steel carries more load because even though it has only 40% the area of the aluminum, it has three times the modulus and only half the length; these last two factors combine to make the steel "spring" (see Figure 5.36) $3 \times 2 = 6$ times as stiff, while the area makes it 40% as stiff. All things considered, then, the steel is $6 \times 0.4 = 2.4$ times stiffer and indeed the force in it, 3.53 kips, is 2.4 times that (1.47 kips) in the aluminum. A straight, two-force structure such as a truss element is perfectly analogous to a spring with modulus $k = AE/L$.

Spring modulus $= k_{alum} = \dfrac{AE}{L} = \dfrac{5(10 \times 10^6)}{24} = 2.08 \times 10^6$ lb/in.

Spring modulus $= k_{steel} = \dfrac{AE}{L} = \dfrac{2(30 \times 10^6)}{12} = 5 \times 10^6$ lb/in.

Figure 5.36

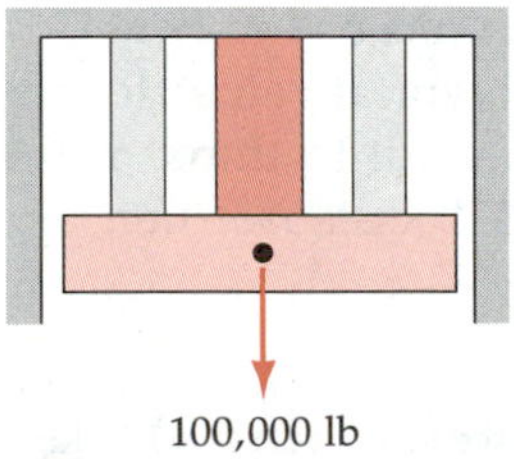

Figure E5.14a

Example 5.14

The three bars in Figure E5.14a support the 100,000-lb weight. The two outer bars, symmetrically placed, are steel, with areas 6 in.^{2} each; the central bar is copper with area 8 in.^{2} Find the forces in the bars.

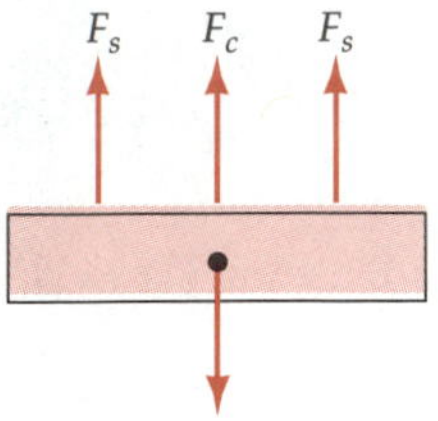

Figure E5.14b

Solution

The vertical walls force the weight to deflect straight down (which if all is symmetrical it would do anyway), so that the deflections δ_{steel} and δ_{copper} will be equal. We have from Figure E5.14b:

$$+\uparrow \qquad \Sigma F_y = 0 = 2F_S + F_C - 100{,}000$$

$$2F_S + F_C = 100{,}000 \tag{1}$$

The fact that the δ's are the same here yields (note the lengths are the same):

$$\delta_S = \delta_C$$

$$\frac{F_S L}{A_S E_S} = \frac{F_C L}{A_C E_C}$$

$$F_S = \frac{A_S E_S F_C}{A_C E_C} = \frac{(6)30 \times 10^6}{(8)17 \times 10^6} F_C = 1.32\ F_C \tag{2}$$

Question 5.13 Why did we use the area of just one steel bar in the preceding equation?

Substituting Equation (2) into (1),

$$2(1.32\ F_C) + F_C = 100{,}000$$

$$F_C = 27{,}500 \text{ lb}$$

$$F_S = 1.32\ F_C = 36{,}300$$

As a check, $2F_S + F_C = 100{,}100$ lb, off by a tenth of a percent due to roundoff error.

Now that we have seen that previously indeterminate problems can be solved by considering deformations in addition to equilibrium equations; we also see the importance of realizing that these equations are not restricted to rigid bodies.

Thermal Effects

The strain due to a temperature change ΔT in a truss member free to expand is given by

$$\epsilon = \alpha\, \Delta T \tag{5.9}$$

where α is the coefficient of thermal expansion, in units of 1 / °F (in U.S. units) or 1 / °C (in SI units). Table 5.2 gives some typical values for the coefficient of thermal expansion. It is perhaps helpful to think of the units of α as, for example, "inches / inch per °F," because then one

Table 5.2

Material	$\alpha \times 10^6$	
	1/°F	1/°C
Structural Steel	6.5	11.7
Aluminum	13	23.4
Copper	9.5	17.1
Brass	11	19.8
Bronze	10.5	18.9
Titanium Alloy	5	9
Concrete	6	10.8

Answer 5.13 While *both* steel bar forces add to the copper force to support the 100 kip weight, the elongation of *each* steel bar equals that of the copper bar.

realizes that once α is multiplied by the ΔT, what remains is in./in., i.e., the change in length per original length, or dimensionless strain.

The change in length of a free-to-expand member is, since the strain due to temperature is uniform, ϵL or $\alpha L(\Delta T)$. If, however, the member is in any way constrained from expanding (or contracting if ΔT is negative), then there is also strain present from the tensile or compressive load carried by the member. Consider the following example:

Example 5.15

The brass bar in Figure E5.15a has area 2 in.^{2} If subjected to a temperature rise of 30°F, what are the forces exerted by the rigid walls that squeeze the ends together, preventing the expansion?

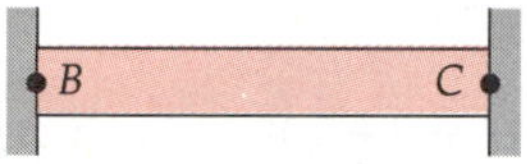

Figure E5.15a

Figure E5.15b

Solution

It is obvious from Figure E5.15b that this problem is statically indeterminate, since $\Sigma F_x = 0$ yields $0 = 0$. Thus we must again consider the deformation of the member. We mentally free the right end C, allow the expansion to occur, then apply the precise wall force which will bring end C back to its given distance from B, as suggested by Figure E5.15c:

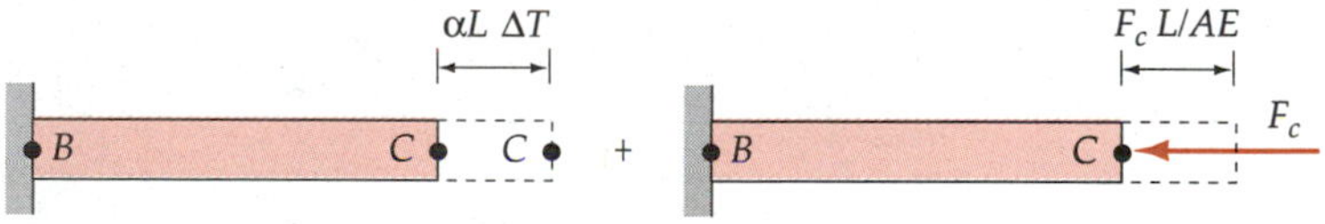

Figure E5.15c

As in the preceding sub-section, we have $\delta_{BC} = 0$:

$$\delta_{BC} = 0 = \delta_{BC_{\substack{\text{due to}\\ \text{temp.}}}} + \delta_{BC_{\substack{\text{due to}\\ F_C}}} = \alpha L(\Delta T) - F_C\, L/(AE)$$

or

$$\alpha L(\Delta T) = F_C L/(AE)$$

Therefore

$$\begin{aligned} F_C &= AE\,\alpha\,\Delta T \\ &= 2(15 \times 10^6)(11 \times 10^{-6})(30) \\ &= 9900 \text{ lb}, \end{aligned}$$

a surprisingly large force.

EXAMPLE 5.16

The truss (see Figure E5.16a) of Example 5.12, unloaded, experiences a temperature drop of 20°C. Find the deflection of joint A.

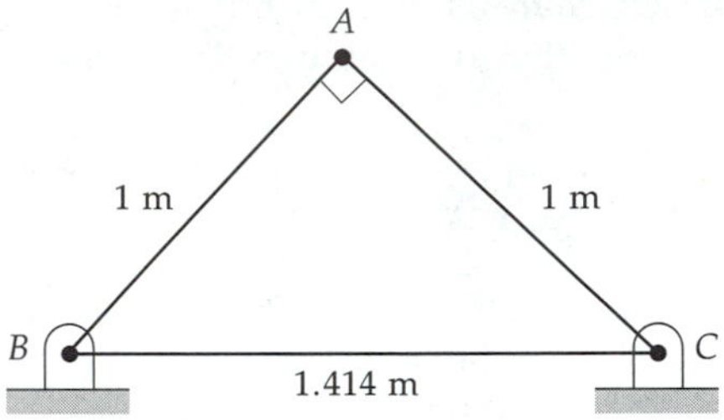

Figure E5.16a

Solution

The members AB and AC, if the pin at A were removed, would each contract the amount

$$\begin{aligned}\delta &= \alpha L\, \Delta T \\ &= 11.7 \times 10^{-6}(1)20 = 234 \times 10^{-6}\ \text{m} \quad (\text{or } 234\ \mu\text{m}^*)\end{aligned}$$

The final position of A lies on the intersection of the two lines drawn perpendicular to the bars after their shortenings, as seen in Figure E5.16b. Therefore, the displacement of A is $(234 \times 10^{-6})\dfrac{\sqrt{2}}{2}(2)$, or 331 μm↓.

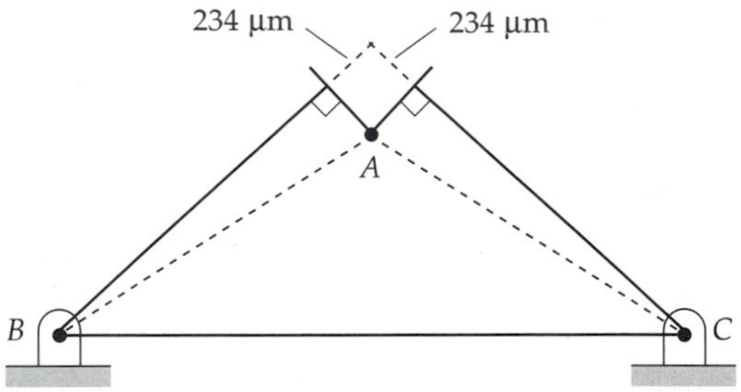

Figure E5.16b

In the preceding example, the bar BC, constrained by the pins at its ends, cannot relieve its thermal stress by straining, as did AB and AC. Instead it is left with the tensile stress (see Example 5.15):

$$\begin{aligned}\sigma_{BC} = \frac{F_{BC}}{A_{BC}} &= \frac{\not{A} E\, \alpha\, \Delta T}{\not{A}} = 207(10^9)(11.7 \times 10^{-6})(20)\ \text{Pa} \\ &= 48.4\ \text{MPa}\end{aligned}$$

* One μm ("micro-meter") is 10^{-6} m.

PROBLEMS ▶ Section 5.7

Find the stresses in each member of the trusses in Problems 5.91–93.

5.91

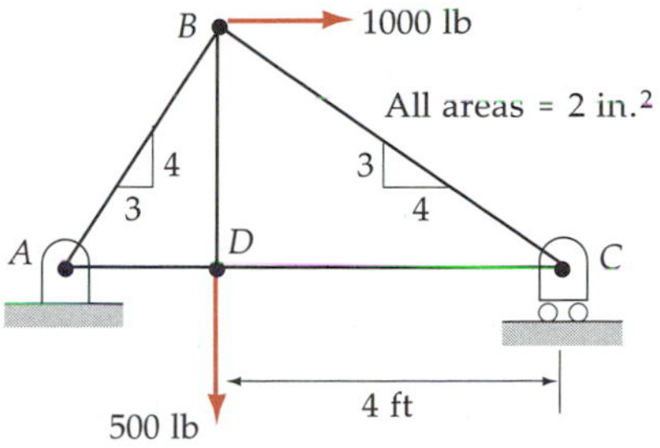

Figure P5.91

5.92

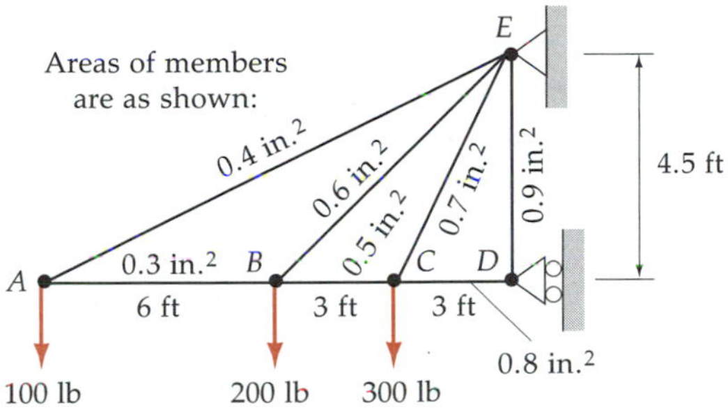

Figure P5.92

5.93

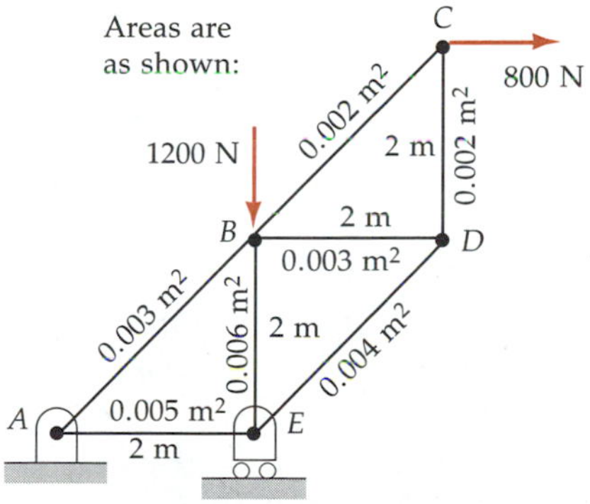

Figure P5.93

5.94 A wooden box-beam has the cross-section shown in Figure P5.94(a). Its modulus of elasticity is 1800 ksi, and it is centrally loaded through a plate as shown in Figure P5.94(b). Find the stress in the member.

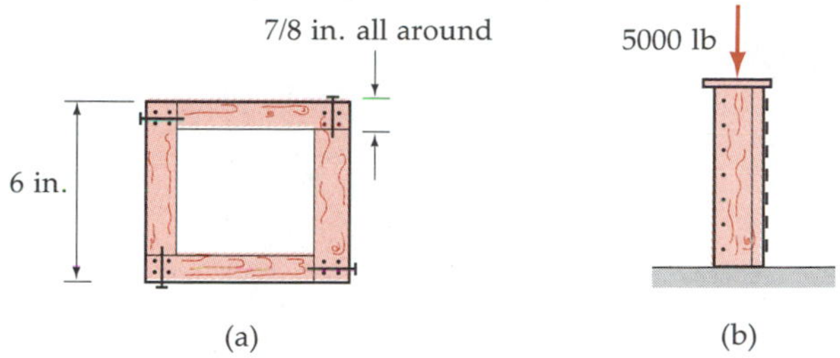

Figure P5.94

5.95 Find the stresses in members BG and CF of the truss in Figure P5.95 if their respective areas are 0.002 m² and 0.0025 m².

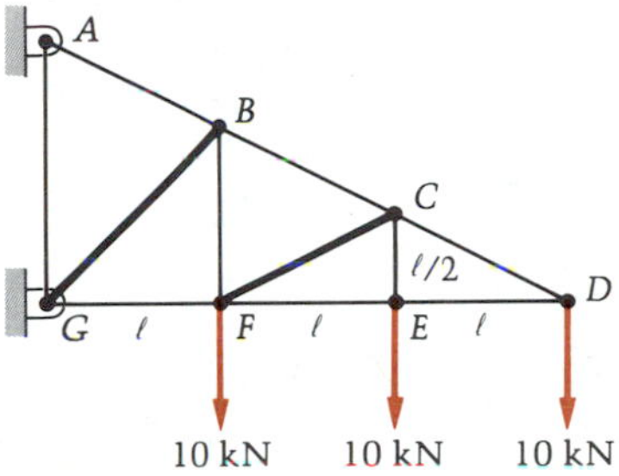

Figure P5.95

5.96 Find the stress in member DF of the truss in Figure P5.96 if the area of each member is 3 in.² and it is made of a material having $E = 24 \times 10^6$ psi.

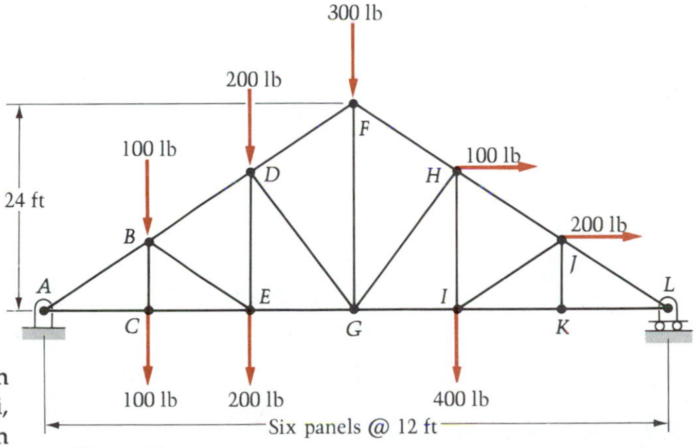

Figure P5.96

5.97 The 500-N force is applied to a bar made of two connected parts as shown in Figure P5.97. Find the normal stresses in the sections LC and CR.

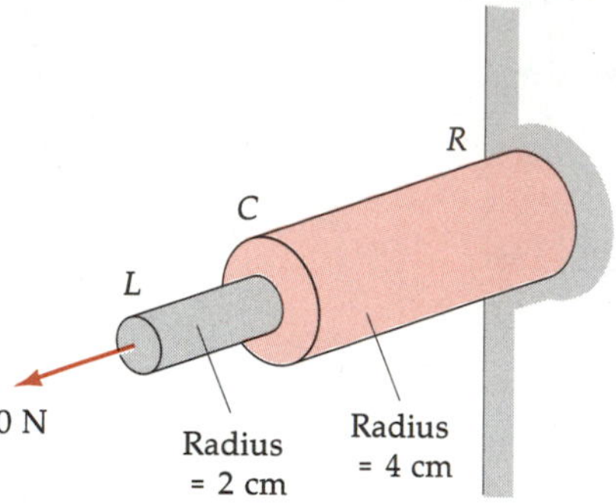

Figure P5.97

5.98 In Figure P5.98, the nylon cords each have area 0.1 in.^{2} If the breaking stress of the cord is 8 ksi, determine how much weight W can be suspended.

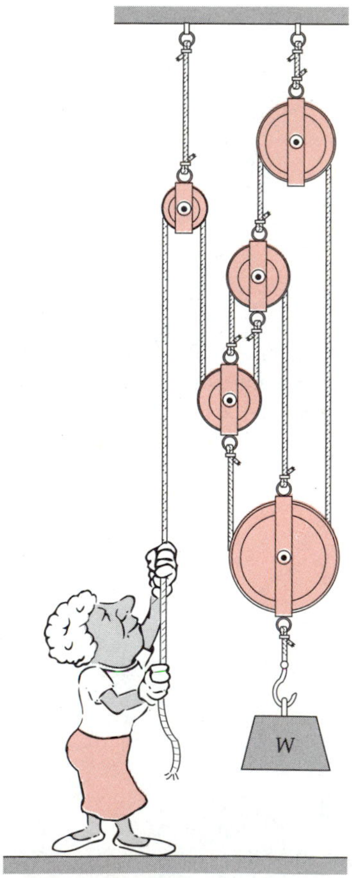

Figure P5.98

5.99 Members DE, DG, and HG of the truss in Figure P5.99 have respective areas 0.0038 m^2, 0.0030 m^2, and 0.0060 m^2. Which of the three members is being subjected to the largest stress?

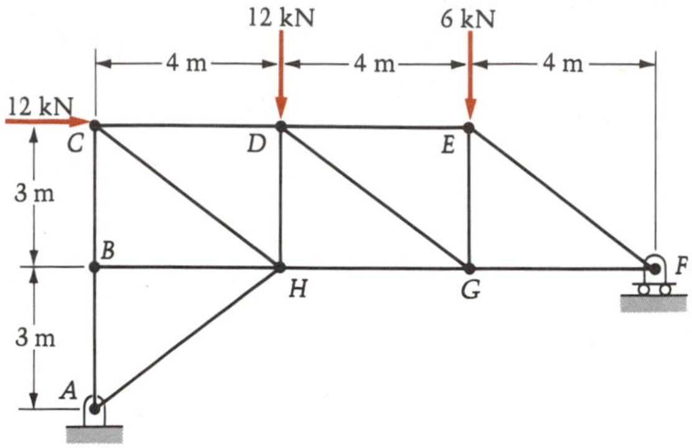

Figure P5.99

5.100 Find the stress in the two-force member AB of the frame in Figure P5.100, if AB is a 1-inch diameter steel rod.

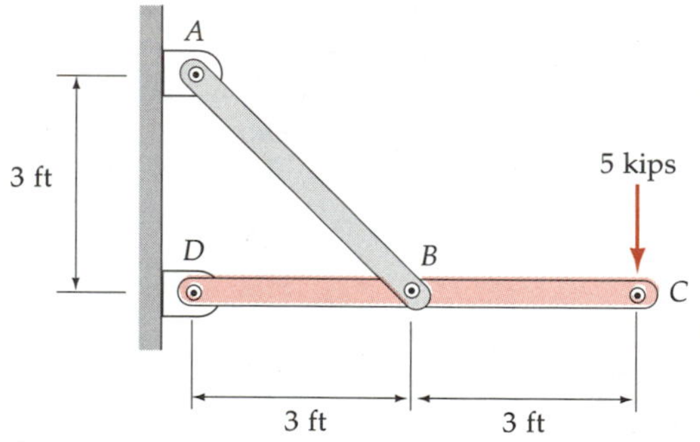

Figure P5.100

5.101 Find the stress in the two-force member BD of the frame in Figure P5.101, if BD is a 2-cm diameter aluminum rod.

5.102 Find the stress in member AB of the truss of Figure P5.102. The area of AB is 3 in.^{2}.

5.103 The toggle device in Figure P5.103 is being used to crush rocks. If the pressure in the chamber, p, is 70 psi and the radius of the piston is 6 in., find the stresses in the three two-force members AB, BC, and BD. The piston rod BD has diameter 1 in., and AB and BC each have diameter 1.5 in.

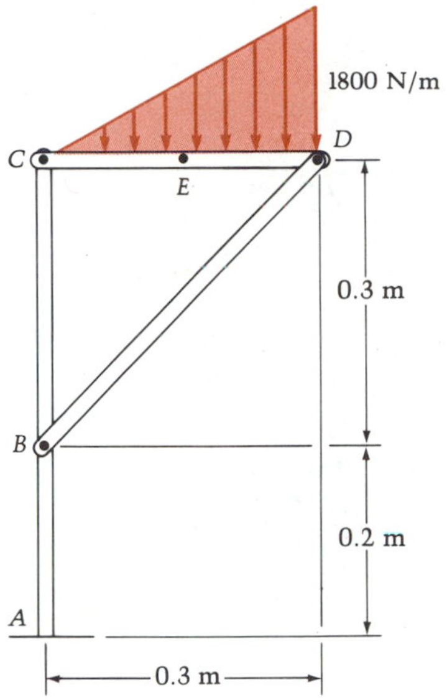

Figure P5.101

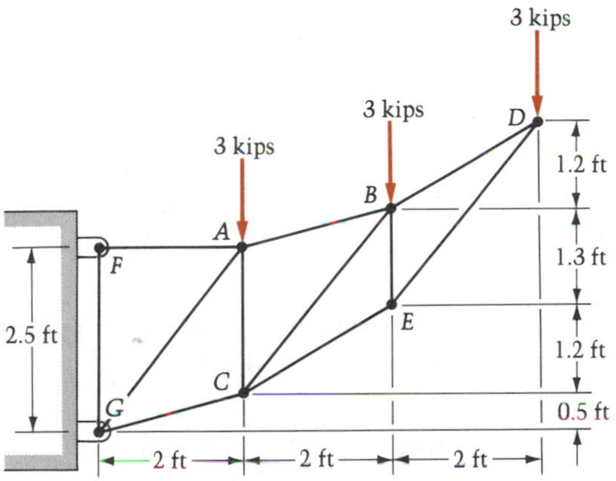

Figure P5.102

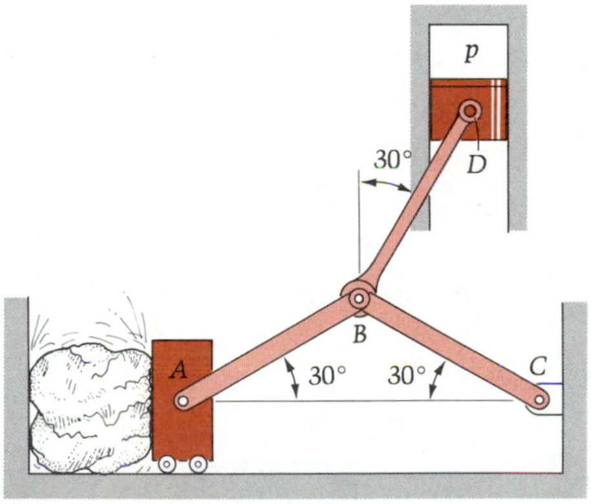

Figure P5.103

5.104 Find the maximum stress in the hollow cylinder in Figure P5.104 due to its own weight.

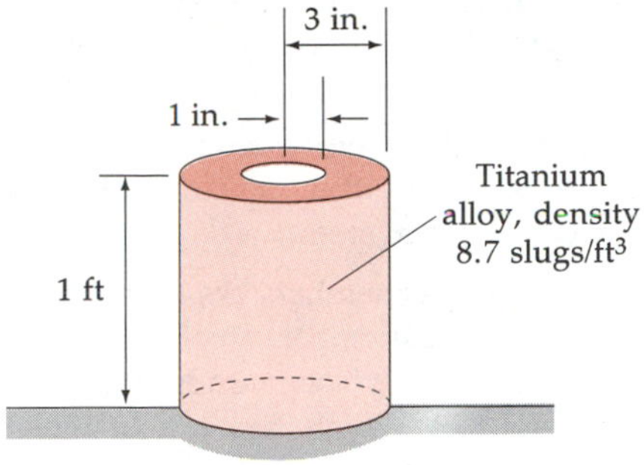

Figure P5.104

5.105 The solid steel bar (density 7850 kg/m^3) in Figure P5.105 is suspended from a ceiling. Find the maximum stress in the bar caused by its own weight.

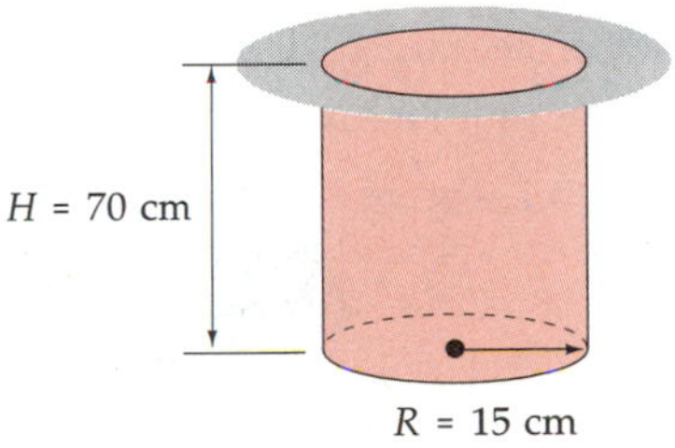

Figure P5.105

* **5.106** Where on line *xx* in Figure P5.106 should the pin *D* be placed (with *B* staying beneath it) so that the stress in *DC* is minimized? The volume of *DC*, i.e. $A_{DC}L_{DC}$, is to remain equal to that of *DB* which is $A_{DB}L_{DB} = (4 \text{ in.}^2)(48 \text{ in.}) = 192 \text{ in.}^3$.

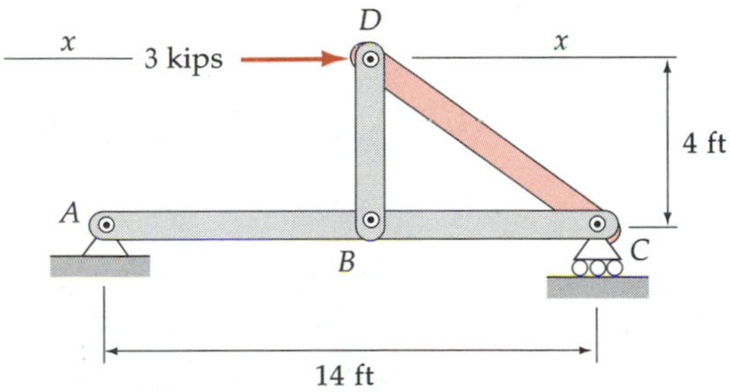

Figure P5.106

5.107 In Problem 5.100, find the elongation of the two-force member *AB*.

5.108 In Problem 5.101, find the elongation of member *BD*.

5.109 In Problem 5.102, find the elongation of member *AB* if the truss is constructed of aluminum.

5.110 Find the changes in length of the steel members *AB*, *BC*, and *BD* in Problem 5.103 if they each have length 2 ft.

5.111 In Problem 5.95, find the elongations of the two members *BG* and *CF* caused by the applied loads. The members are made of structural steel.

5.112 In Problem 5.99, if the members are all made of aluminum, which of the three members *DE*, *DG*, and *HG* has the largest elongation or contraction under the given loading? (See Figure P5.99.) Which has the smallest? Would the order be the same if the members were made of steel?

5.113 If the box-beam in Problem 5.94 is 6 feet long, how much does it shorten under the applied 5-kip loading?

5.114 In Problem 5.97 (see Figure P5.114), find the total elongation of the bar (made of two parts) if:

a. section *LC* is made of steel and *CR* is of aluminum;

b. section *LC* is made of aluminum and *CR* is of steel.

5.115 The truss member of Figure P5.115 increases in length by 1 cm after the 10 kN load is applied. From this information, find the modulus of elasticity of the member.

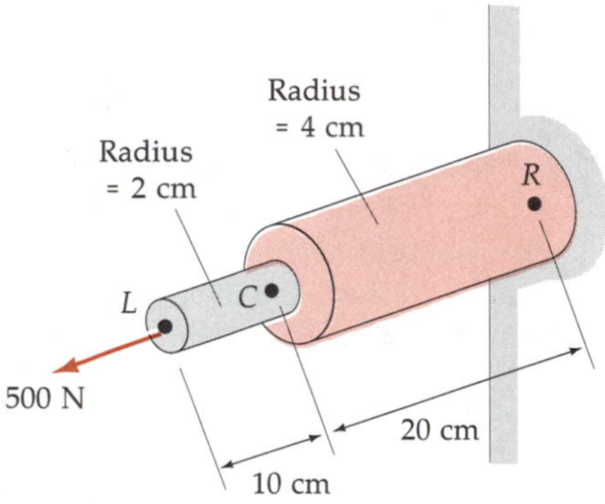

Figure P5.114

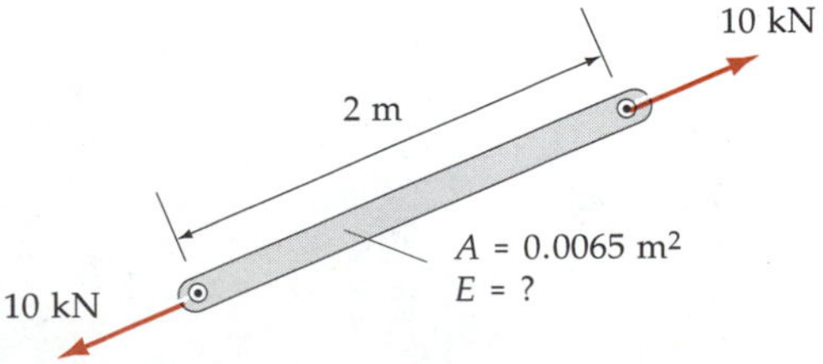

Figure P5.115

5.116 The bar *AB* in Figure P5.116 is a brass rod of diameter 1.5 in. Find the total change in length of *AB* after the four forces are applied as shown.

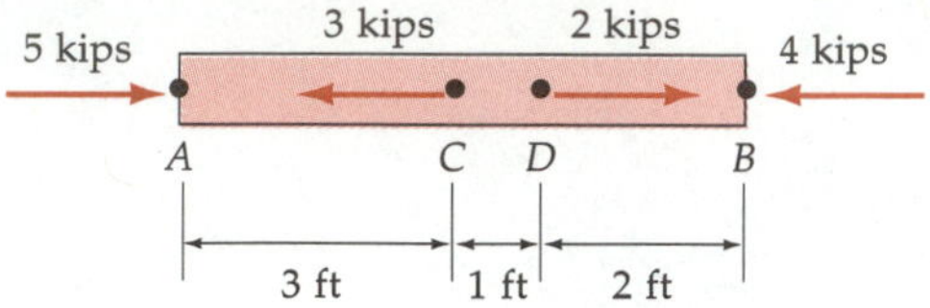

Figure P5.116

5.117 The beam *BD* in Figure P5.117 is supported by the wires *AB* and *CD*, which have equal areas and elastic moduli. Find the length of *CD*, in terms of *L*, for which *BD* stays horizontal after the force *P* is applied as shown.

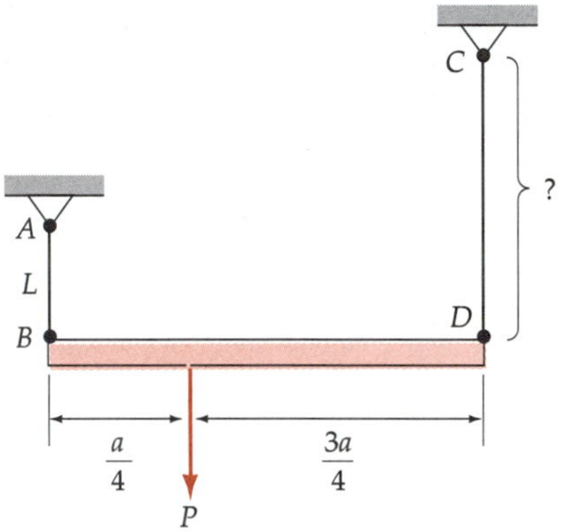

Figure P5.117

5.118 Find the displacement of joint *A* in Example 5.12 if the areas are changed to 0.003 m² for member *AB* and 0.005 m² for *AC*.

* **5.119** Find the displacement of joint *C* following application of the 2000 lb force shown in Figure P5.119. Data on the members is:

BC: aluminum, $L = 3$ ft, $A = 3$ in.^{2}

AC: steel, $L = 4$ ft, $A = 2$ in.^{2}

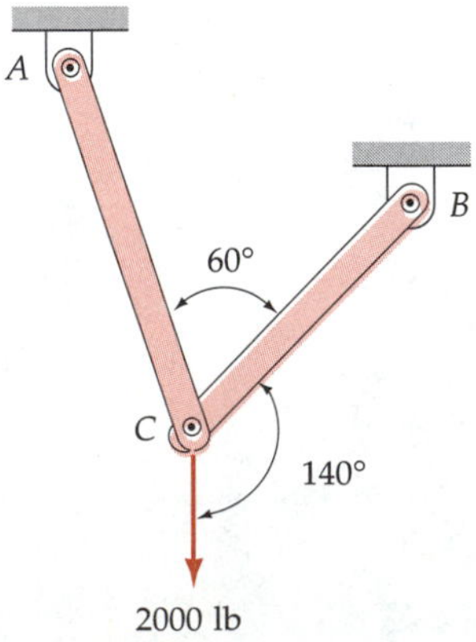

Figure P5.119

• **5.120** Find the final position of joint C in Figure P5.120 if the rods are constituted as follows:

a. AC: bronze, Area 0.5 in.^{2}

b. BC: copper, Area 0.4 in.^{2}

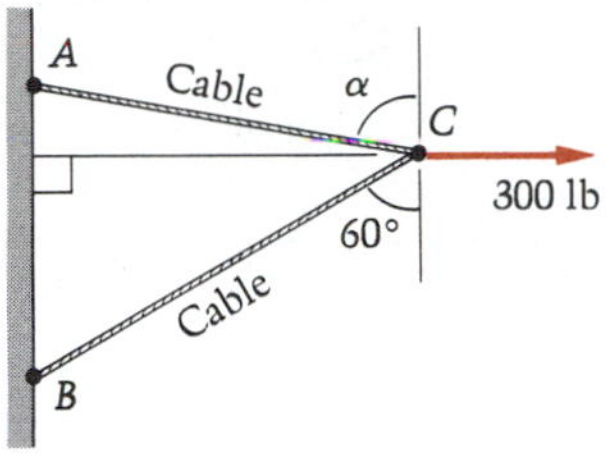

Figure P5.120

5.121 Show that the vertical component δ_y of the deflection of joint Q caused by the force P acting on the simple plane truss in Figure P5.121 is given by the equation

$$\delta_y = \frac{PL}{AE}\left(\frac{1}{\sin^2\theta\cos\theta} + \frac{1}{\tan^2\theta}\right)$$

where P, L, and θ are shown in the figure, A is the area of members BQ and CQ, and E is a constant.

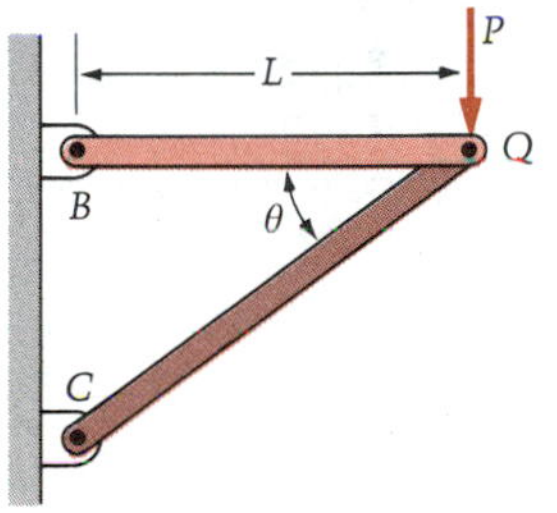

Figure P5.121

5.122 Find the displacements of joints B and C of the truss of Problem 5.10 (see Figure P5.122). The bars are made of steel, each with an area 0.004 m^2.

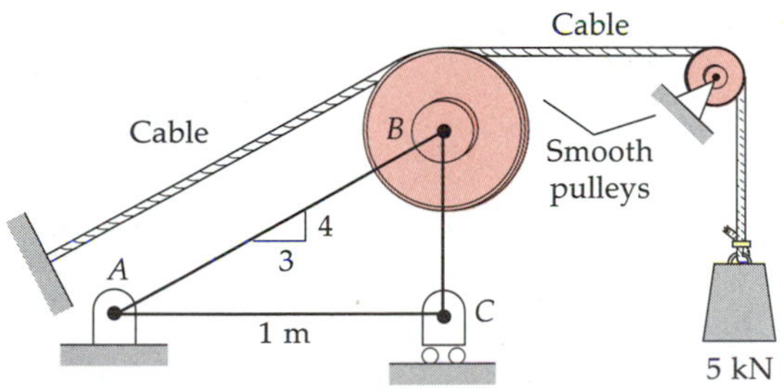

Figure P5.122

• **5.123** Find the displacement of joint A in Example 5.12 if there is a roller at C as shown in Figure P5.123, instead of the joint being pinned to the ground.

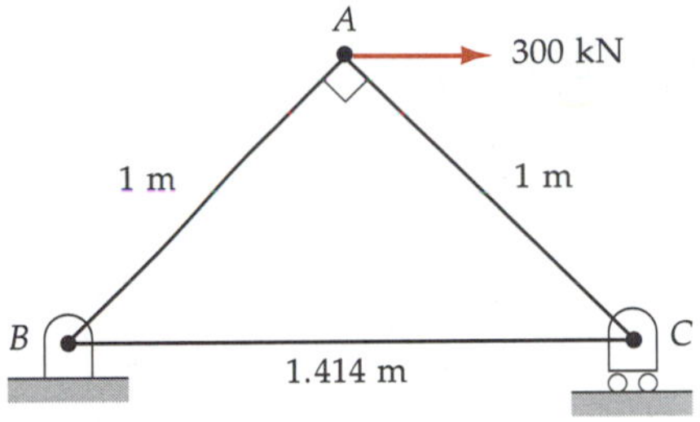

Figure P5.123

5.124 Repeat Example 5.13 if the 5 kN force is applied at the mid-point of BD (in the same direction), instead of at the juncture of the two bars.

5.125 Find the downward displacement of the 100-kip rigid block W in Figure P5.125. Also find the force in each bar.

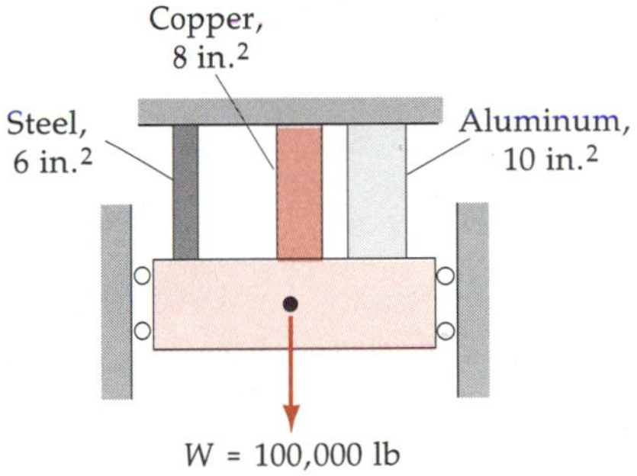

Figure P5.125

5.126 The rigid bar AB is supported by the pin at A and the two members CE and DF in Figure P5.126. Find the small angular clockwise rotation of AB about A when the 5-kN force is applied as shown.

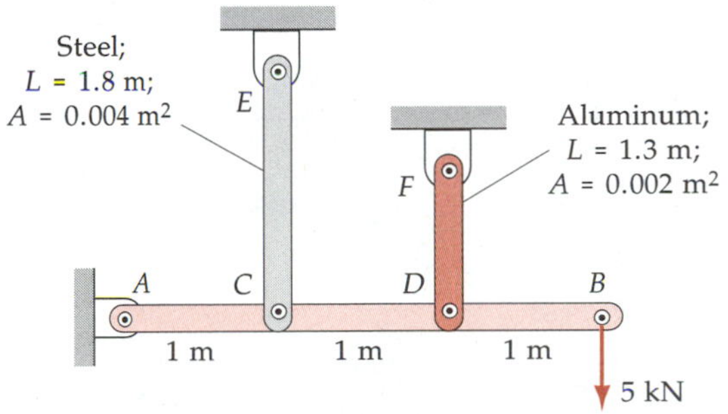

Figure P5.126

5.127 Find the temperature increase in °C that will expand the bar so as to barely close the gap in Figure P5.127.

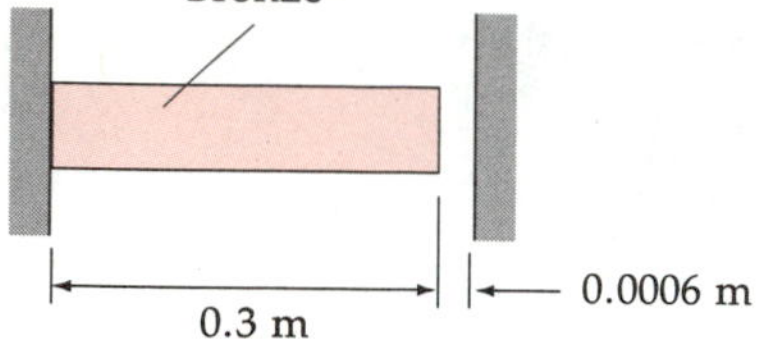

Figure P5.127

5.128 In the preceding problem, if the bar has area 0.005 m^2 and if the temperature change from the position shown in the figure is 110°C, what are the reaction forces from the walls onto the bar?

* **5.129** In Problem 5.125, find the forces in the bars if an 80°F temperature rise is imposed on the system.

* **5.130** In Problem 5.126, determine the angular rotation change if the structure is cooled by the amount $\Delta T = -20°C$ with the 5kN force acting as shown.

II / SYSTEMS CONTAINING MULTIFORCE MEMBERS

5.8 Axial and Shear Forces and Bending Moments

The Multiforce Member

In this section we extend the work we have done on trusses to structures containing "multiforce members." Such a member is not a straight two-force bar, as were all the members of a truss. Consequently, on cross-sectional cuts, a member will be subjected to more than just an axial force.

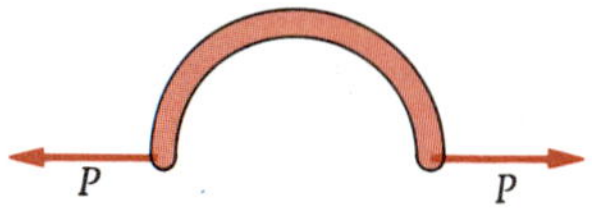

Figure 5.37

Consider Figure 5.37. The half-ring is pulled at the ends by forces P as shown. Though the half-ring is a two-force member, it is not a *straight* two-force member. Hence its various cross sections are subjected to more than just the now-familiar axial forces of a truss. Consider the free-body diagrams shown in Figure 5.38. For equilibrium, we need equal magnitude, oppositely directed forces P as shown, together with the indicated moments M. Considering the shaded section, equilibrium requires

$$\circlearrowleft\!+ \quad \Sigma M_A = 0 = M - P(R \sin \theta)$$

or

$$M = PR \sin \theta$$

This moment (called a bending moment because it tends to bend the ring) is seen to vanish only at $\theta = 0$ and 180°.

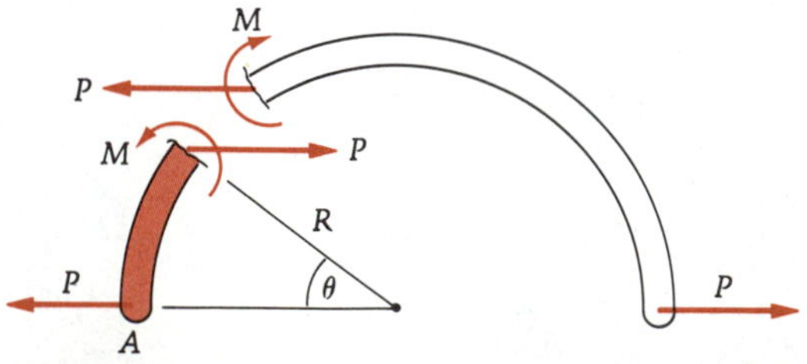

Figure 5.38

Question 5.14 Looking at Figure 5.38, why is it clear that M must vanish at $\theta = 0$ and 180°?

Our member above thus differs from a truss member in that it carries a bending moment. But also, the force acting on the cross sections has become complicated.

We see in Figure 5.39 that if we resolve the force at the cut into its two components (a) tangent to the center line of the curved ring, and (b) normal to the center line, we obtain, respectively, in addition to the familiar axial force (N) studied in trusses, a shear force (V). Note from the figure that N and V, as was the case with M, vary with θ. This is another distinct difference from a truss, in which the (axial) member forces were constant from end to end.

The body need not be curved for V and M to be present. Consider a second example, in which we seek N, V, and M at isolated points.

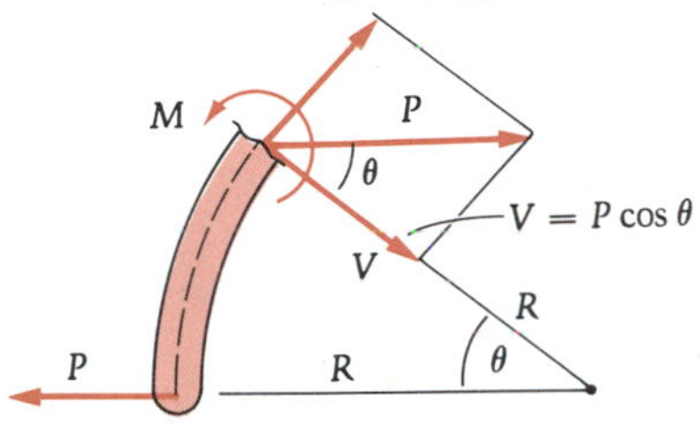

Figure 5.39

Answer 5.14 Because the distance between the pair of forces P goes to zero at those two points.

Example 5.17

Find the axial force, shear force, and bending moment at points P and Q of the frame in Figure E5.17a.

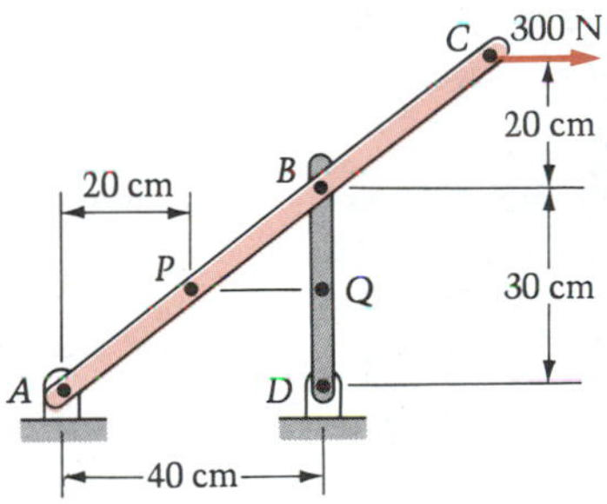

Figure E5.17a

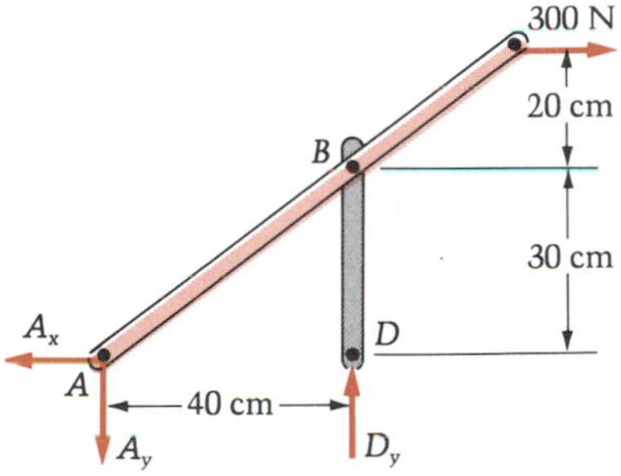

Figure E5.17b

Solution

The free-body diagram of the complete frame is shown in Figure E5.17b. Note that BD is a two-force member, which will allow us to determine A_x and A_y, and also the force D_y in BD, from this single free-body diagram.

$$\circlearrowleft + \qquad \Sigma M_A = 0 = D_y(40 \text{ cm}) - (300 \text{ N})(50 \text{ cm})$$

$$D_y = 375 \text{ N}$$

Thus

$$+\uparrow \qquad \Sigma F_y = 0 = D_y - A_y = 375 - A_y$$

$$A_y = 375 \text{ N}$$

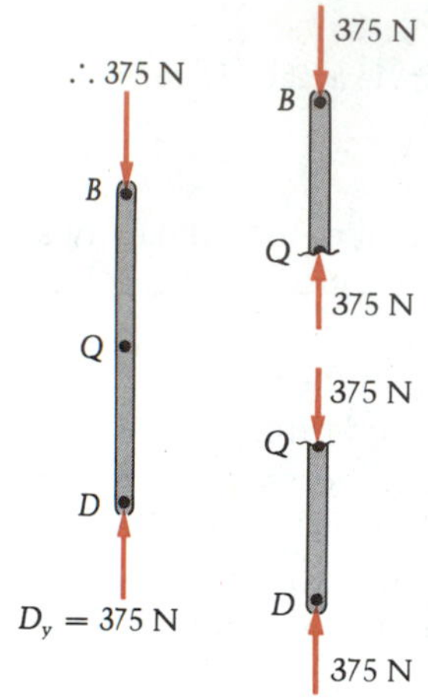

Figure E5.17c

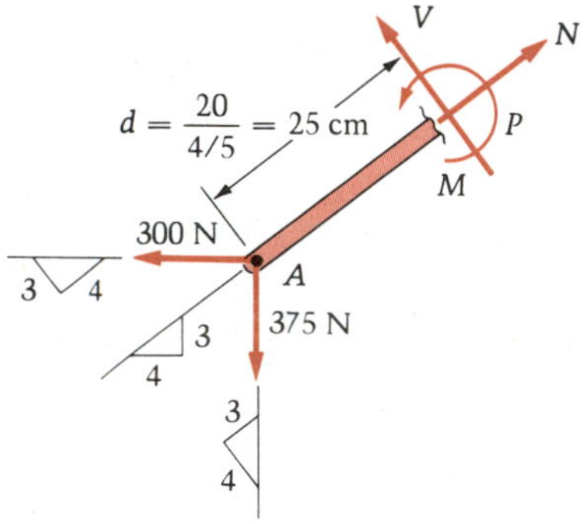

Figure E5.17d

Also,

$$\xrightarrow{+} \qquad \Sigma F_x = 0 = 300 - A_x \Rightarrow A_x = 300 \text{ N}$$

The force at Q is therefore 375 N Ⓒ, as shown on both the upper and lower cut sections of BD (Figure E5.17c).

Sectioning member ABC at P will expose the forces N, V, and M that we seek (see Figure E5.17d, the free-body diagram of AP):

$$\Sigma F_x = 0 = N - 0.8(300) - 0.6(375)$$

$$N = 465 \text{ N}$$

$$\Sigma F_y = 0 = V + 0.6(300) - 0.8(375)$$

$$V = 120 \text{ N}$$

$$\Sigma M_P = 0 = M + 0.8(375)25 - 0.6(300)25$$

$$M = -3000 \text{ N} \cdot \text{m}$$

Therefore the forces are in the directions sketched in Figure E5.17d, while the bending moment is in the opposite (or clockwise) direction to that shown on the free-body diagram.

The next example is similar to the last one, but it is a bit longer and requires more free-body diagrams.

EXAMPLE 5.18

In the frame of Example 4.19, find the axial force, shear force, and bending moment at the points P, Q, and R in Figure E5.18a.

Solution

We had found the results in Example 4.19 by dismantling the frame, as shown in the three figures E5.18b, E5.18c and E5.18d below. Making a cut at P in member

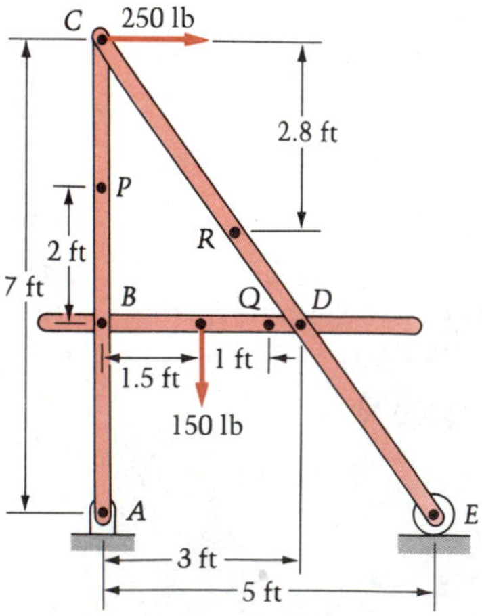

Figure E5.18a

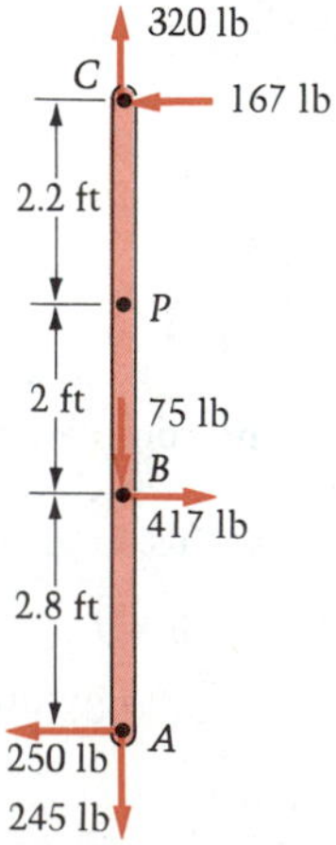

Member ABC

Figure E5.18b

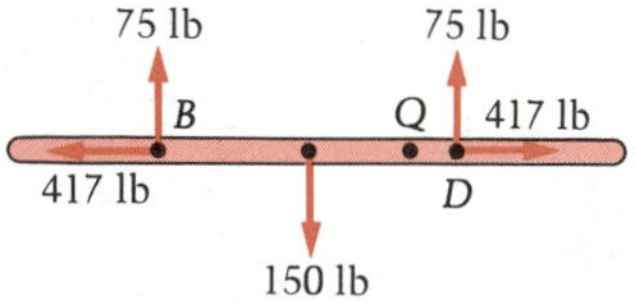

Member BD

Figure E5.18c

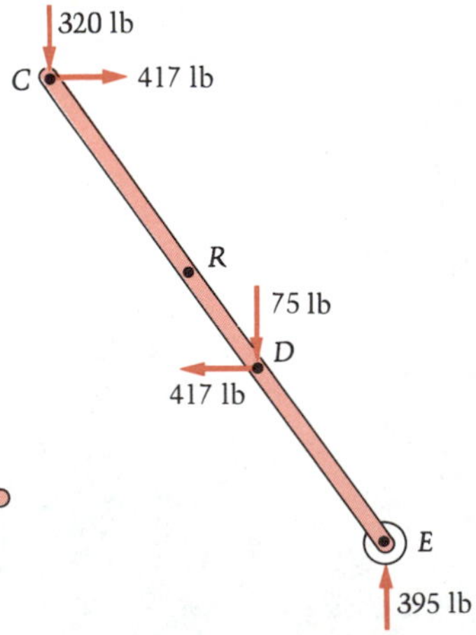

Member CDE

Figure E5.18d

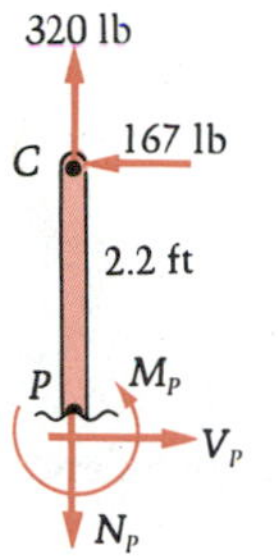

Figure E5.18e

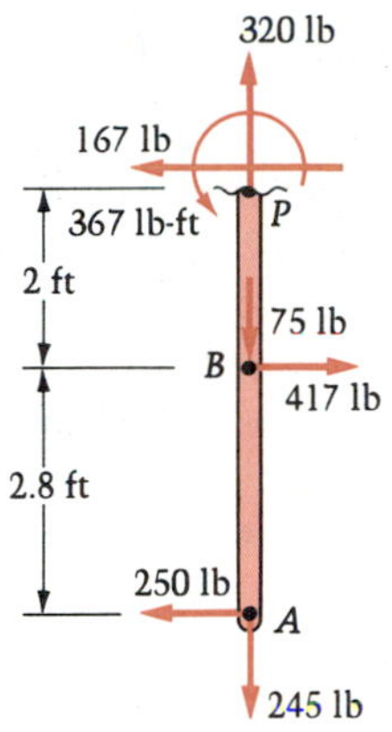

Figure E5.18f

ABC of Figure E5.18b, we have, with the help of the resulting free-body diagram in Figure E5.18e:

$$\overset{+}{\rightarrow} \quad \Sigma F_x = 0 = V_P - 167 \Rightarrow V_P = 167 \text{ lb}$$

$$+\uparrow \quad \Sigma F_y = 0 = 320 - N_P \Rightarrow N_P = 320 \text{ lb}$$

$$\circlearrowleft\!+ \quad \Sigma M_P = 0 = M_P + (167 \text{ lb})(2.2 \text{ ft}) \Rightarrow M_P = -367 \text{ lb-ft}$$

Note that if we had chosen to use the lower section of *ABC*, shown in Figure E5.18f, we would have had more forces to deal with but we would have obtained consistent results — the opposites* of those on the upper section, in agreement with action and reaction. Either set of answers (shown on the proper section) is correct.

For the internal force system at *Q*, we cut the bar *BD* of Figure E5.18c there. We can use either side of the cut:

Using the left section, Figure E5.18g below,

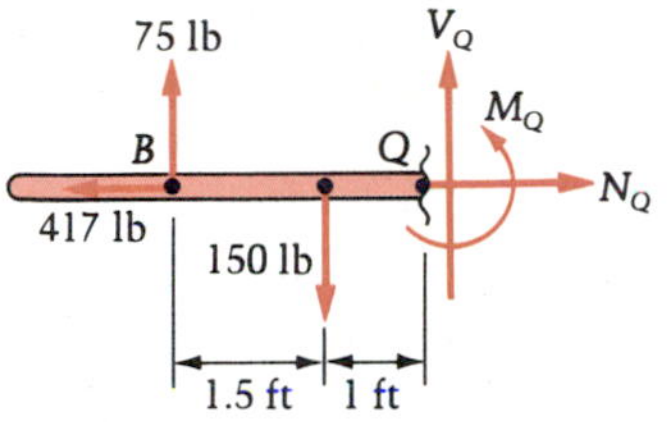

Figure E5.18g

$$\overset{+}{\rightarrow} \quad \Sigma F_x = 0 = N_Q - 417 \Rightarrow N_Q = 417 \text{ lb}$$

$$+\uparrow \quad \Sigma F_y = 0 = 75 - 150 + V_Q \Rightarrow V_Q = 75 \text{ lb}$$

$$\circlearrowleft\!+ \quad \Sigma M_Q = 0 = M_Q + (150 \text{ lb})(1 \text{ ft}) - (75 \text{ lb})(2.5 \text{ ft})$$

$$M_Q = 37.5 \text{ lb-ft}$$

Alternatively, using the right section, Figure E5.18h below,

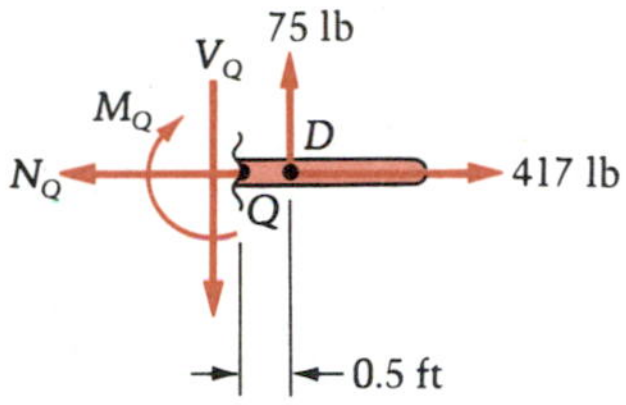

Figure E5.18h

$$\overset{+}{\rightarrow} \quad \Sigma F_x = 0 = 417 - N_Q \Rightarrow N_Q = 417 \text{ lb}$$

$$+\uparrow \quad \Sigma F_y = 0 = 75 - V_Q \Rightarrow V_Q = 75 \text{ lb}$$

$$\circlearrowleft\!+ \quad \Sigma M_Q = 0 = -M_Q + (75 \text{ lb})(0.5 \text{ ft})$$

$$M_Q = 37.5 \text{ lb-ft}$$

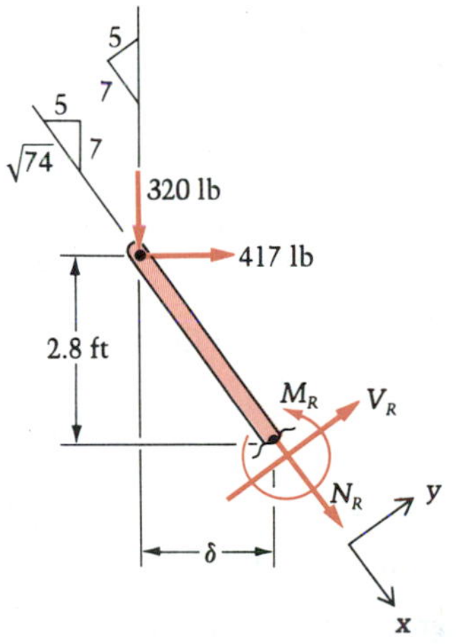

Figure E5.18i

The results agree, as they must, for each of the axial force (N_Q), shear force (V_Q), and bending moment (M_Q). Note that we took care of "Newton's Third Law" when we assigned the directions on the free-body diagrams.

For the internal force system at *R*, we use a cut section of *CDE* from Figure E5.18d, as Figure E5.18i. From similar triangles, $\frac{2.8}{\delta} = \frac{7}{5} \Rightarrow \delta = 2$ ft. We can avoid having to solve two equations in V_R and N_R by summing forces in the slanted coordinate directions *x* and *y* in the figure.

$$\overset{+}{\searrow} \quad \Sigma F_x = 0 = N_R + 320\left(\frac{7}{\sqrt{74}}\right) + 417\left(\frac{5}{\sqrt{74}}\right)$$

$$N_R = -503 \text{ lb}$$

* The moment is off by 1 lb-ft due to rounding to three digits in the original example.

$$\Sigma F_y = 0 = V_R - 320\left(\frac{5}{\sqrt{74}}\right) + 417\left(\frac{7}{\sqrt{74}}\right)$$

$$V_R = -153 \text{ lb}$$

$$\Sigma M_R = 0 = M_R + (320 \text{ lb})(2 \text{ ft}) - (417 \text{ lb})(2.8 \text{ ft})$$

$$M_R = 528 \text{ lb-ft}$$

(Note that for the moment calculation we have used the original horizontal and vertical force and moment arm components.)

The reader is encouraged to check the above results using the lower cut section of member *CDE*.

In the last example in this section, we take a detailed look at the variation in the internal force system elements (*N*, *V*, and *M*) along the axial direction of a member. This example is in preparation for our study of beams, to follow in the next section.

EXAMPLE 5.19

Determine the distribution of internal forces in the bent member *ABC* shown in Figure E5.19a.

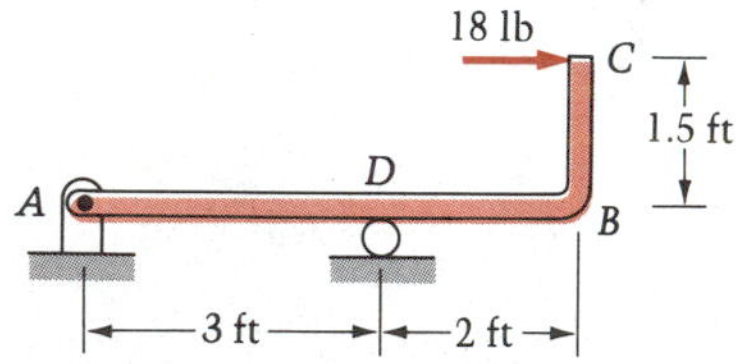

Figure E5.19a

Solution

In contrast to the first two examples, we are looking for *N*, *V*, and *M* at *all* points instead of just two or three. In other words, we wish to know how *N*, *V*, and *M* vary along the member.

The three reactions are found first. Using the overall free-body diagram in Figure E5.19b,

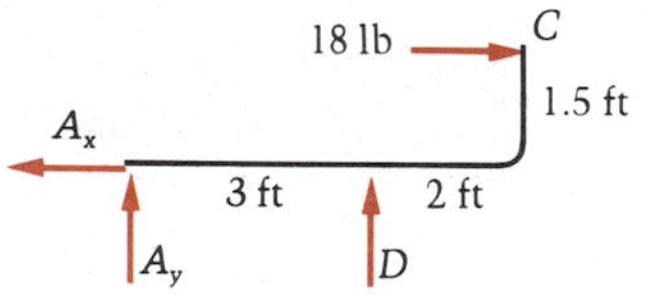

Figure E5.19b

$$\Sigma M_A = 0 = D(3 \text{ ft}) - (18 \text{ lb})(1.5 \text{ ft})$$

$$D = 9 \text{ lb}$$

$$\Sigma F_x = 0 = 18 \text{ lb} - A_x \Rightarrow A_x = 18 \text{ lb}$$

$$\Sigma F_y = 0 = A_y + D = A_y + 9$$

or

$$A_y = -9 \text{ lb}$$

Next we consider a free-body diagram of a portion of the horizontal section *AB* of the bar, extending from the left end *A* to a cut section with $x < 3$ ft (see Figure E5.19c).

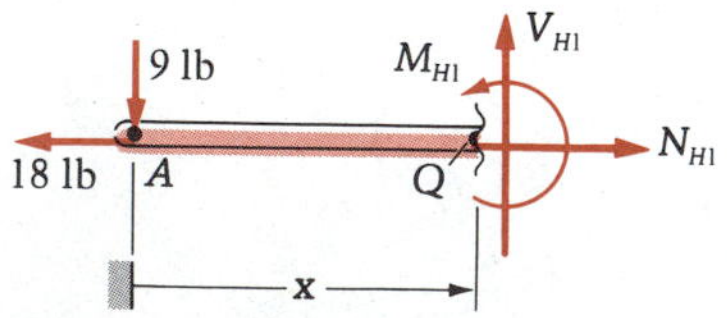

Figure E5.19c

For equilibrium, we see that not only do we need the normal force N_{H1} (as with trusses), but also a shear force V_{H1} and bending moment M_{H1}:

$$\Sigma F_x = 0 = N_{H1} - 18 \text{ lb} \Rightarrow N_{H1} = 18 \text{ lb}$$

$$\Sigma F_y = 0 = -9 \text{ lb} + V_{H1} \Rightarrow V_{H1} = 9 \text{ lb}$$

$$\Sigma M_Q = 0 = M_{H1} + (9 \text{ lb})(x \text{ ft}) \Rightarrow M_{H1} = -9x \text{ lb-ft}$$

After *x* becomes greater than 3 ft, the free-body diagram changes. At a cut section such that $3 \text{ ft} < x < 5 \text{ ft}$, we now have (see Figure E5.19d):

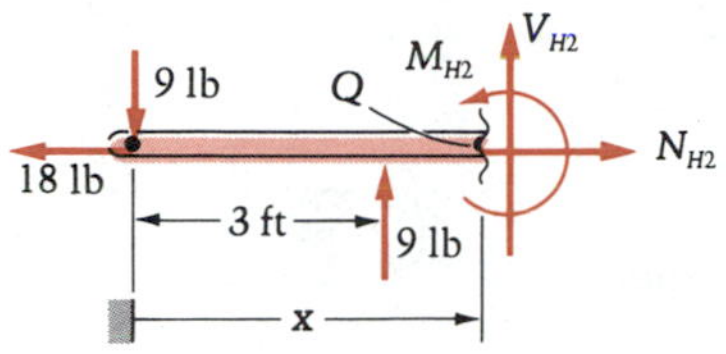

Figure E5.19d

$$\xrightarrow{+} \quad \Sigma F_x = 0 = N_{H2} - 18 \text{ lb} \Rightarrow N_{H2} = 18 \text{ lb} \qquad \text{(still the same)}$$

but

$$+\uparrow \quad \Sigma F_y = 0 = 9 \text{ lb} - 9 \text{ lb} + V_{H2} \Rightarrow V_{H2} = 0$$

and

$$\circlearrowleft + \quad \Sigma M_Q = 0 = M_{H2} + \underbrace{(9 \text{ lb})(x \text{ ft}) - (9 \text{ lb})(x - 3) \text{ ft}}$$

Note that this equals the constant 27 lb-ft, the strength of the couple formed by the 9-lb forces!

$$M_{H2} = -27 \text{ lb-ft}$$

Note that N_{H2} and M_{H2} are continuous across the cut at $x = 3$ ft, but that there has occurred a jump (or discontinuity) in the shear force V_{H2} caused by the concentrated reaction at the roller.

The above expressions for N_{H2}, V_{H2}, and M_{H2} are valid until $x = 5$ ft when the bend occurs. On a cut past the bend at P, we may use the upper section from P to C to most easily determine what is happening in the vertical part (BC) of the bar. (See Figure E5.19e.) Enforcing equilibrium gives

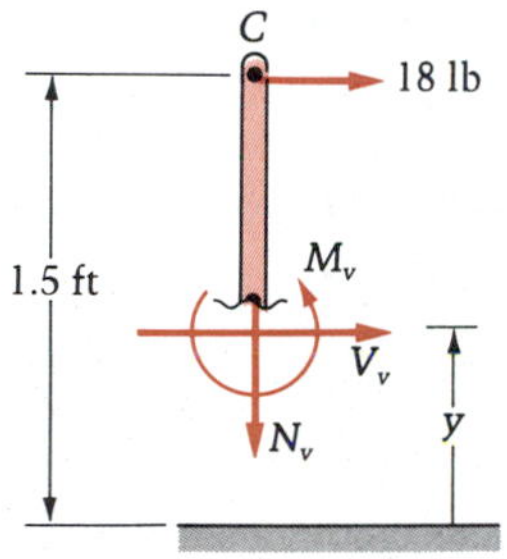

Figure E5.19e

$$\xrightarrow{+} \quad \Sigma F_x = 0 = 18 \text{ lb} + V_v \Rightarrow V_v = -18 \text{ lb}$$

$$+\uparrow \quad \Sigma F_y = 0 = -N_v \Rightarrow N_v = 0$$

$$\circlearrowleft + \quad \Sigma M_P = 0 = M_v - (18 \text{ lb})(1.5 - y) \text{ ft}$$

$$M_v = (27 - 18y) \text{ lb-ft}$$

If we now look at a free-body diagram of the bend alone, we see that it, too, is of course in equilibrium. (See Figure E5.19f.) We note that the discontinuity of V and N in "rounding the bend" is somewhat artificial. It is due to their exchange of roles caused by the discontinuity at the corner in the orientation of the center line.

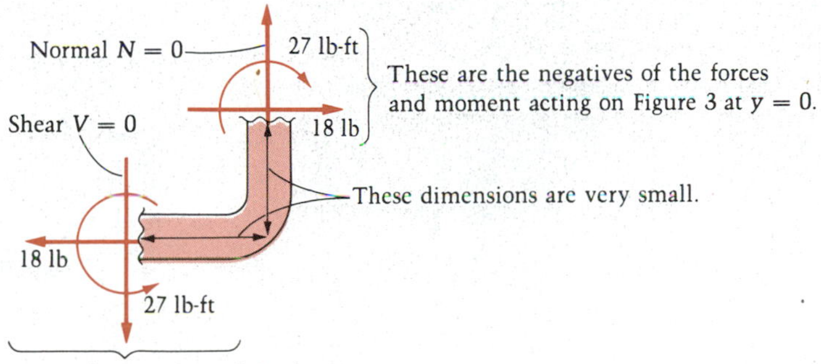

Figure E5.19f

We mention again that when we wish to know the internal force resultant (N, V, and M) at a point, we may use either of the two free-body diagrams formed by the cut section. For instance, if the section to the right of the first cut is used (as in Figure E5.19g) instead of the section in Figure E5.19c, then equilibrium requires:

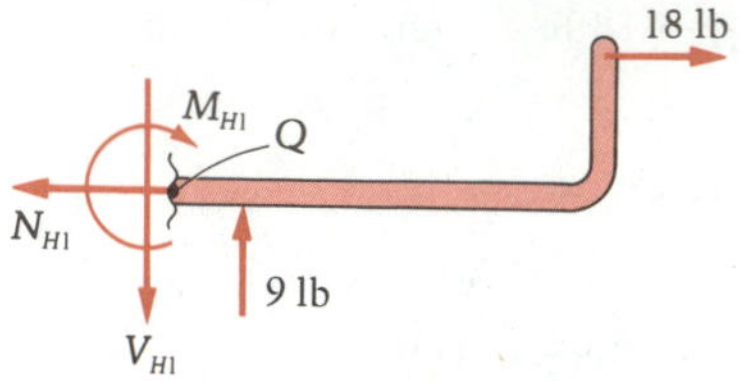

Figure E5.19g

$$\xrightarrow{+} \quad \Sigma F_x = 0 = 18 - N_{H1} \Rightarrow N_{H1} = 18 \text{ lb} \quad \text{(as before)}$$

$$+\uparrow \quad \Sigma F_y = 0 = -V_{H1} + 9 \Rightarrow V_{H1} = 9 \text{ lb} \quad \text{(as before)}$$

$$\circlearrowleft\!+ \quad \Sigma M_Q = 0 = -M_{H1} - (18 \text{ lb})(1.5 \text{ ft}) + (9 \text{ lb})(3 - x) \text{ ft}$$

$$M_{H1} = -9x \text{ lb-ft} \quad \text{(as before)}$$

Thus all three results are in agreement with those we obtained using the material on the other side of the cut.

PROBLEMS ▸ Section 5.8

5.131 Find the internal forces at section *a*-*a* on bar $\mathcal{B}$ shown in Figure P5.131.

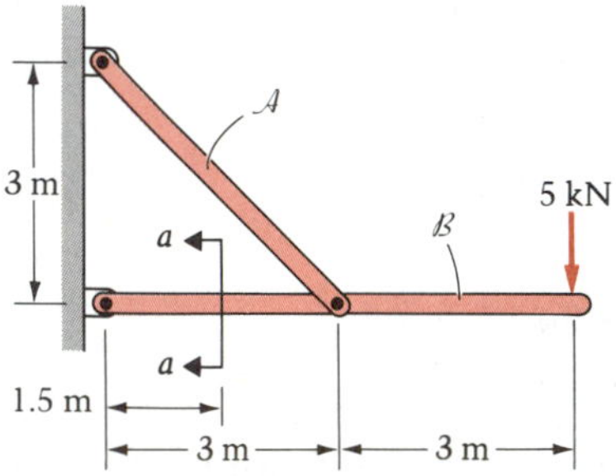

Figure P5.131

5.132 Find the internal forces at section *a*-*a* on bar *BCD* shown in Figure P5.132.

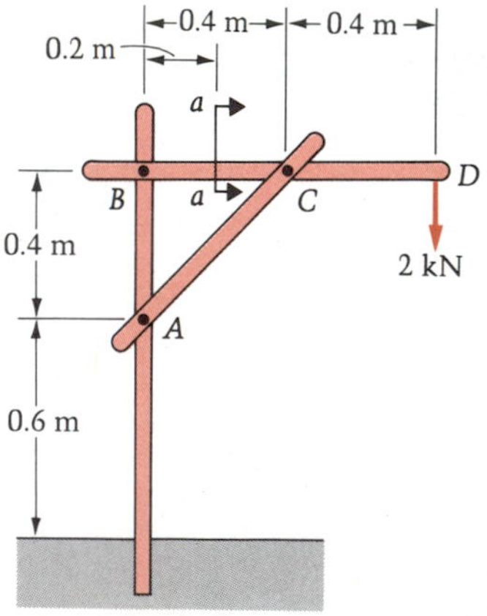

Figure P5.132

5.133 In Figure P5.133 a weight of 3000 N is suspended from beam $\mathcal{B}$ at point C. The beam is connected to a vertical wall by a frictionless pin and a cable as shown. If the beam (of negligible weight) has length 3 m and the length *AC* is 1 m, find the internal forces and moment transmitted at section *E*-*E*.

5.134 In Figure P5.134:

a. Find the force exerted on the shaded member in Figure P5.134 by pin *A*.

b. Find the internal forces (axial, shear, and bending moment) at the midpoint *M* of the horizontal part of the shaded member.

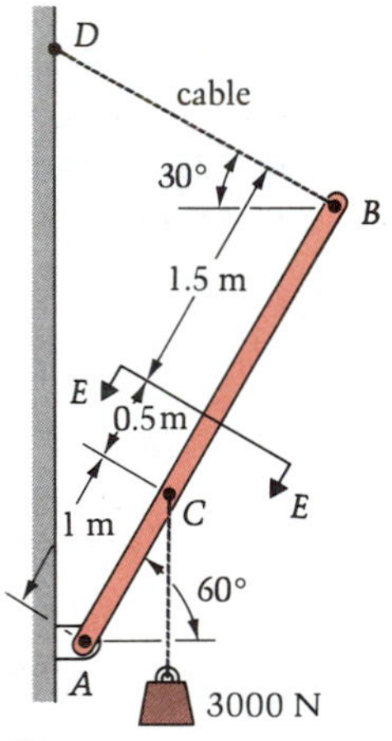

Figure P5.133

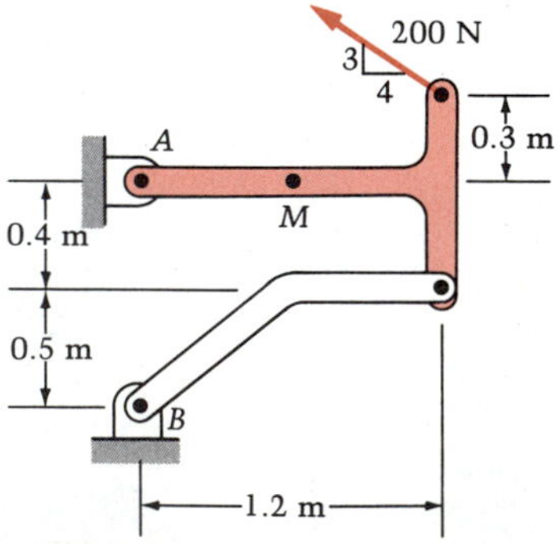

Figure P5.134

5.135 The two boards in Figure P5.135 have been glued together and are being held as they dry by a C-clamp. If the compressive force holdingthe boards together is 20 lb, find the internal force system at sections *A-A*, *B-B*, and *C-C*.

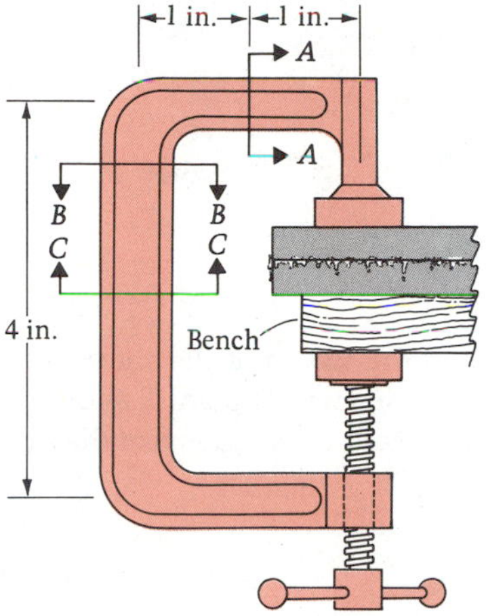

Figure P5.135

5.136 In Figure P5.136 a beam of length L is fastened into a wall at angle α with respect to the horizontal. An oil drum of weight W is slung under the beam using a cable as shown fastened at $L/4$ and $3L/4$ to the beam. If $\alpha = 30°$ and $\beta = 30°$, find the axial force, shearing force, and bending moment at the midpoint of the beam. Assume that the beam is weightless.

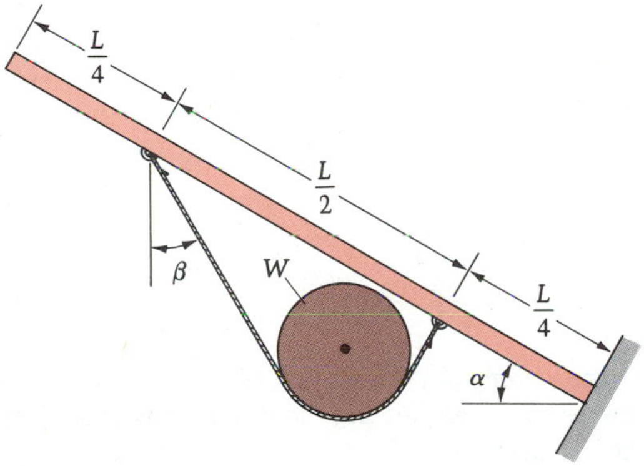

Figure P5.136

5.137 Find the internal force resultants (axial, shear, and bending moment) at sections *A-A*, *B-B*, and *C-C* in Figure P5.137.

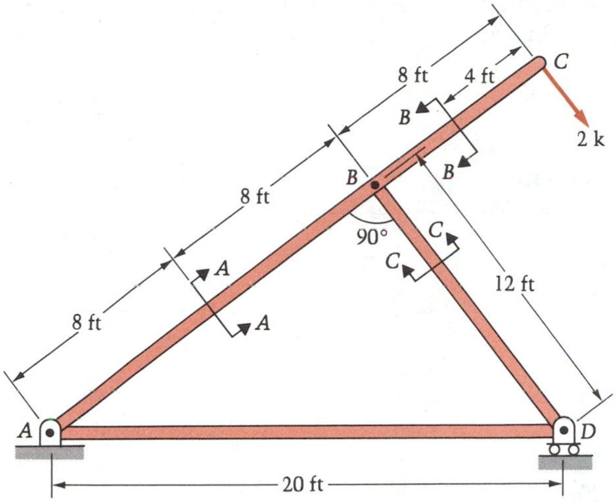

Figure P5.137

5.138 In Figure P5.138 a weight of 1000 lb is suspended from the beam at point *C*. The beam is connected to a vertical wall by a frictionless pin at *A* and by a cable between *B* and *D*. Neglecting the weight of the beam, determine the internal forces and moment transmitted across section *E-E*.

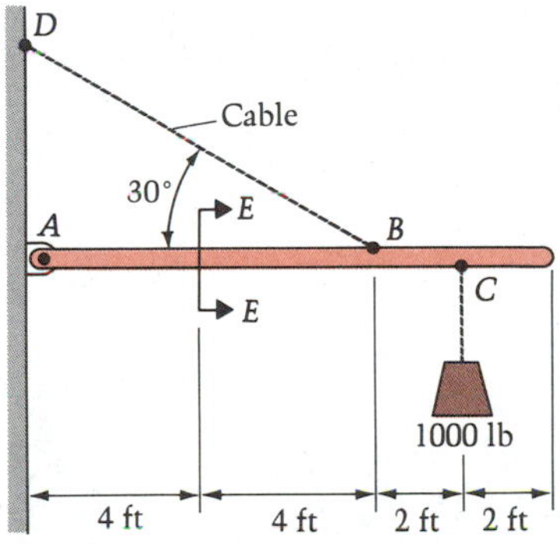

Figure P5.138

In Problems 5.139–5.144 find the shear force and bending moment in the beam at point *P*. Show the result on a free-body diagram of the part of the beam between *P* and *B*.

5.139

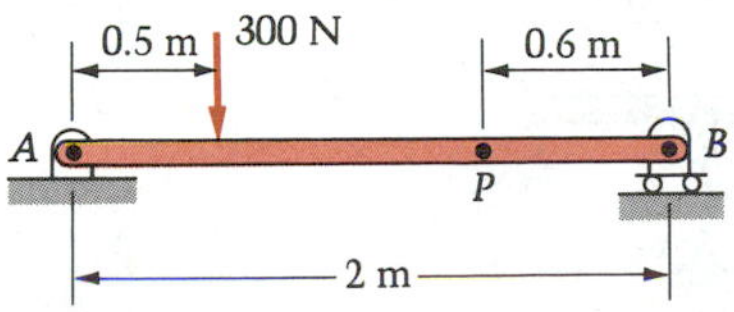

Figure P5.139

(For Problems 5.140–144, see the instructions preceding problem 5.139.)

5.140

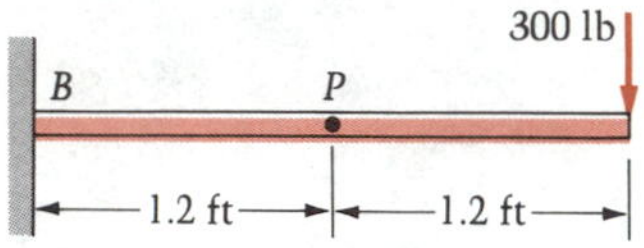

Figure P5.140

5.141

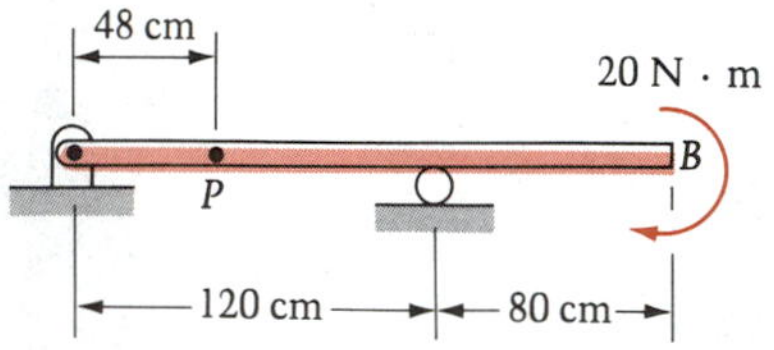

Figure P5.141

5.142

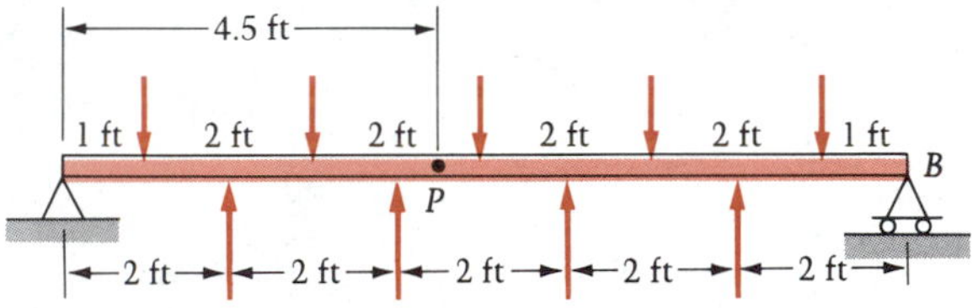

Figure P5.142

5.143

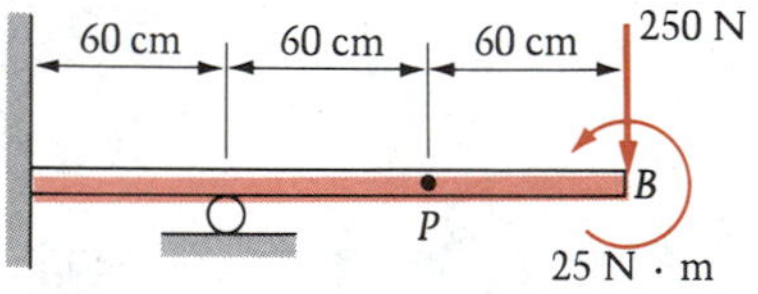

Figure P5.143

5.144

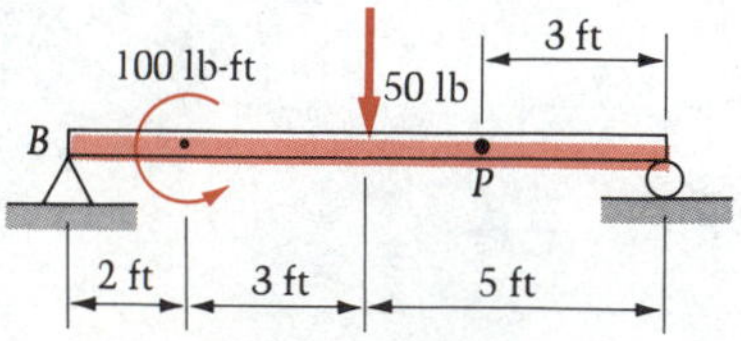

Figure P5.144

5.145 Find the shear force and bending moment on section *A-A* of the beam in Figure P5.145.

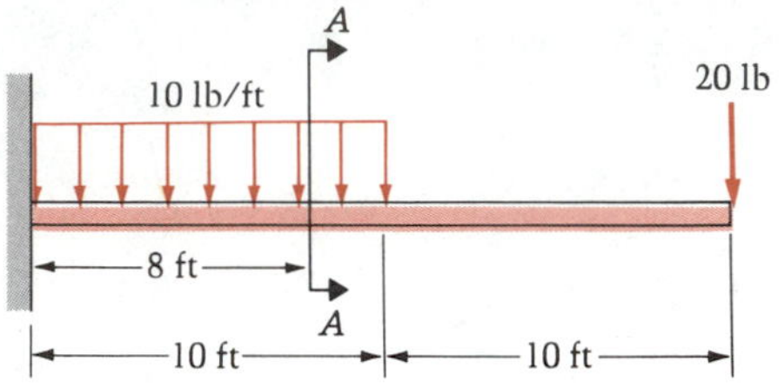

Figure P5.145

5.146 In Figure P5.146 find the values of the shear and axial forces and the bending moment at point *E* midway between *C* and *D*. Show your results on a sketch of the cut section at *E*.

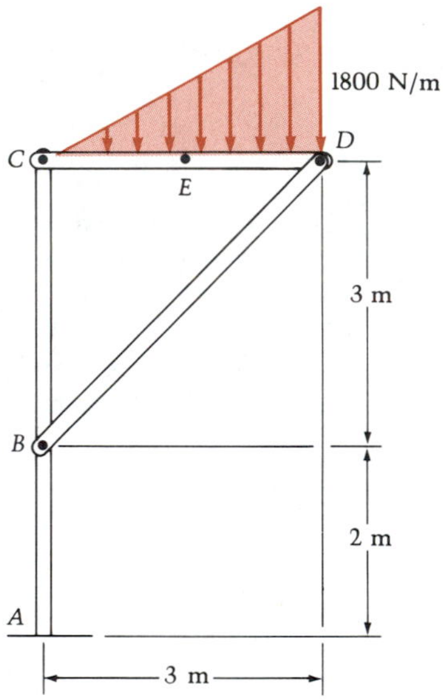

Figure P5.146

5.147 For the frame in Figure P5.147, compute the internal forces and bending moment acting within the member *DC* on section *a-a*.

5.148 Find the internal forces and moments at points *E* and *F* of the light frame shown in Figure P5.148.

5.149 Calculate the axial force, shear force, and bending moment at the midpoint *D* of the portion *AB* of the bent bar *ABC* in Figure P5.149. Show the results on a complete free-body diagram of *AD*.

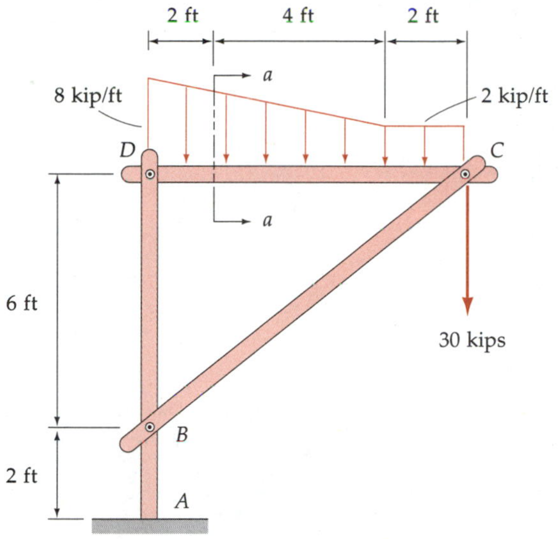

Figure P5.147

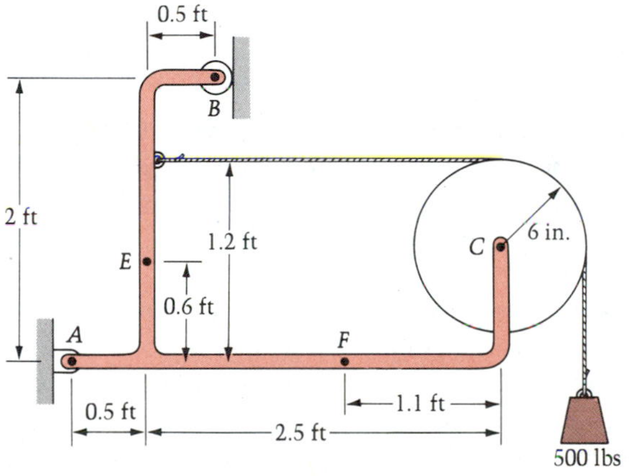

Figure P5.148

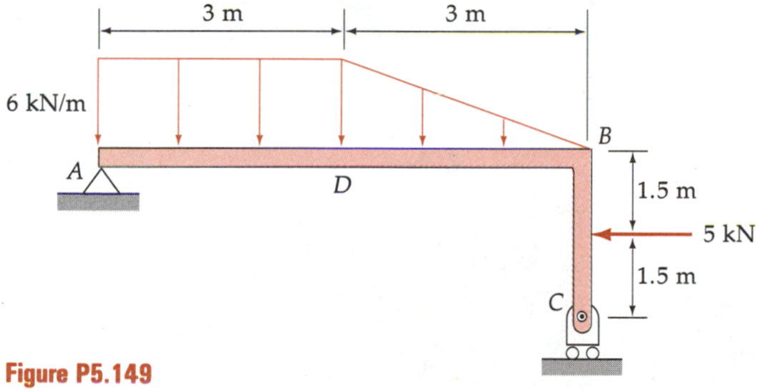

Figure P5.149

5.150 In Problem 4.195, find the internal forces and moments acting on a cross-section of member *BED* 30 cm above point *D* (see Figure P5.150).

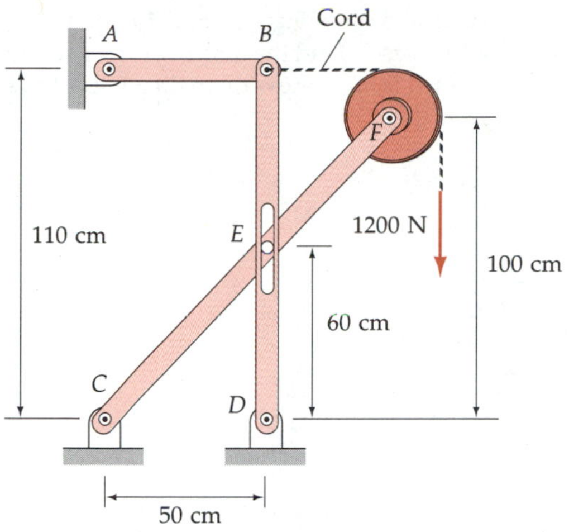

Figure P5.150

5.151 In Problem 4.195, find the axial force, shear force, and bending moment acting on a cross-section of member *CEF* 40 cm from point *C* (see Figure P5.150).

5.152 In Examples 4.11 and 4.24, draw a free-body diagram of the part of the boom between 1 ft and 3 ft from the left end. Use results from the two examples to determine all forces acting on the ends of the free-body, then verify that the equilibrium equations are satisfied.

5.153 In Problem 4.32, find the internal forces (V, N, M) within the bar at a point 5 ft to the right of pin *B*. Show the results on a sketch which includes the part of the bar to the left of the cut section (i.e., which includes points *B*, *A*, and *Q*). See Figure P5.153.

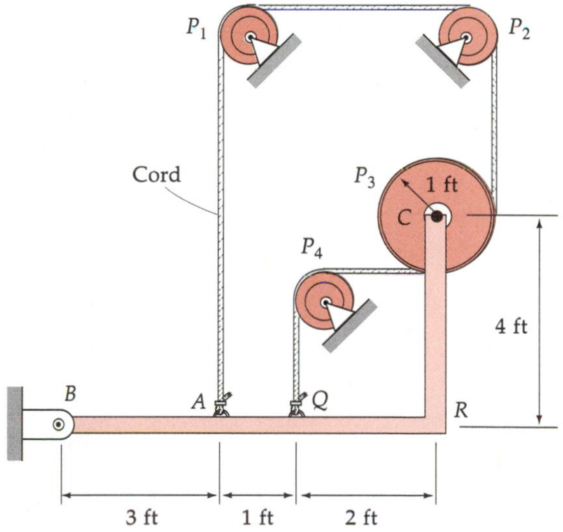

Figure P5.153

5.154 The semicircular arch in Figure P5.154 is loaded as shown by the two forces P. Find the internal forces acting on the cross-section at D, and show them on a FBD of the half DB of the arch.

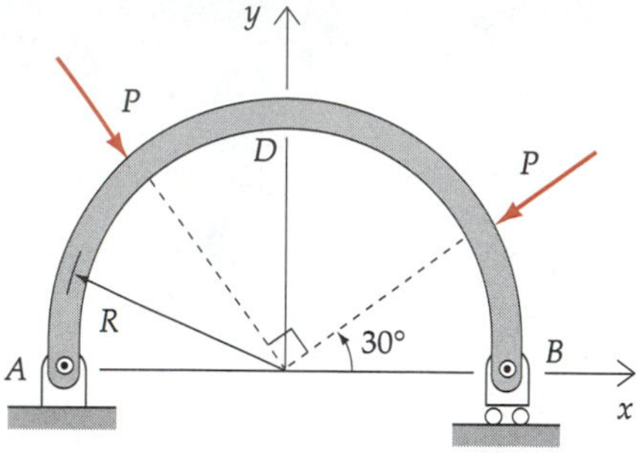

Figure P5.154

5.9 Beams/Shear and Moment Distributions

In the preceding section, we studied the internal force system (N, V, M) acting within a cross section of a planar structure. More precisely, N, V, and M represent a force and couple system equipollent to the infinitely many differential forces exerted on the material on one side of the cut by the material on the other side.

To describe the resultant internal force system under a more general loading than the planar case, we need not just three, but *six* components at a selected point in the section: three of force and three of moment. To see this, let us slice a plane section through a body $\mathcal{B}$, producing two parts $\mathcal{B}_1$ and $\mathcal{B}_2$ as shown in Figure 5.40.

The forces exerted upon $\mathcal{B}_1$ by $\mathcal{B}_2$ are equipollent to a force and a couple at any selected point (such as A). In general, these vectors each have three components, as shown in Figure 5.41 at the point A in the plane of the cut.

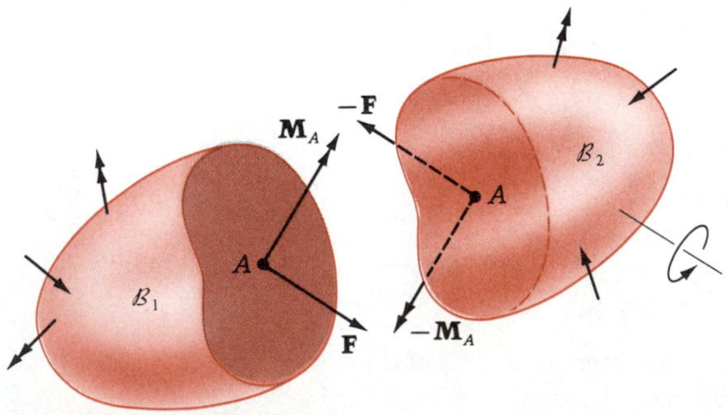

Figure 5.40

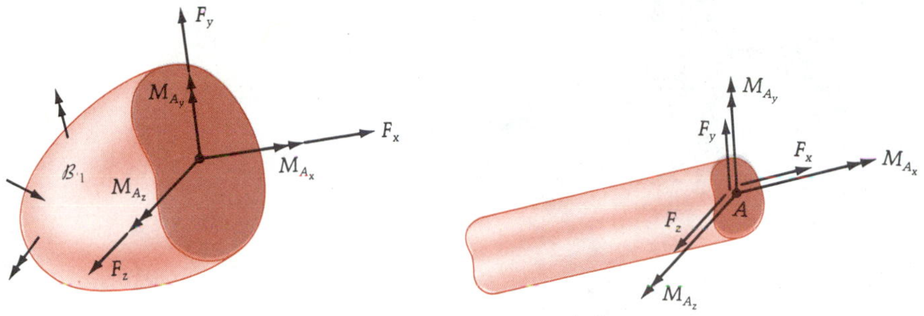

(a) (b)

Figure 5.41

The two force components lying *in* the plane of the cut, F_y and F_z in this case (see Figure 5.41), are called shear forces. The force perpendicular to the plane of the cut (here, F_x) is known as a normal force. When $\mathcal{B}_1$ is a slender member, as shown in Figure 5.41(b), the two moment components whose vector directions (thumb of right hand) lie *in* the plane of the cut are called bending moments (M_{Ay}, M_{Az} here), while the other one, M_{Ax}, with vector *normal* to the plane of the cut and turning effect in the plane, is called a twisting moment.* The same six components acting at A on $\mathcal{B}_2$ would each be respectively opposite in direction, by the principle of action and reaction.

When $\mathcal{B}$ is a slender member that carries shear force(s), bending moment(s), and / or a twisting moment, in addition to the axial force of a truss member, it is called a **beam.**

A beam can be loaded in two planes such that it carries a perpendicular pair of shearing force and bending moment components. It may also be loaded so that the twisting moment component is present. In this elementary look at beams, we shall, however, restrict our attention to the case in which the beam is loaded in just one (xy) plane. That is, all the forces will lie in a plane and all the couples will be normal to that plane (in the $\pm z$ direction).

Thus, only one shear force (F_y, which we shall call V) and one bending moment (M_{Az}, which we shall call M) will be produced, in addition to the axial load (F_x, called N). The twisting moment M_{Ax} will necessarily be absent, as will the other component of shear force (F_z) and of bending moment (M_{Ay}). In Figure 5.42, on the following page, we illustrate these ideas. With all the forces in the plane of the paper, and with the couples normal to this plane, the only resultant of all the internal forces at a cut section at x is expressible as the shear force V, axial force N, and bending moment M.

* The student can readily appreciate this by successively applying moment components to a slender body such as a yardstick.

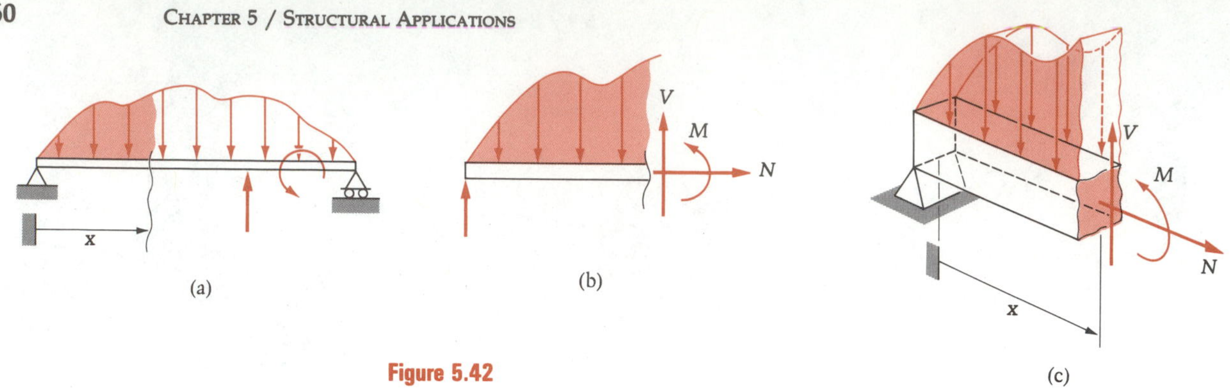

Figure 5.42

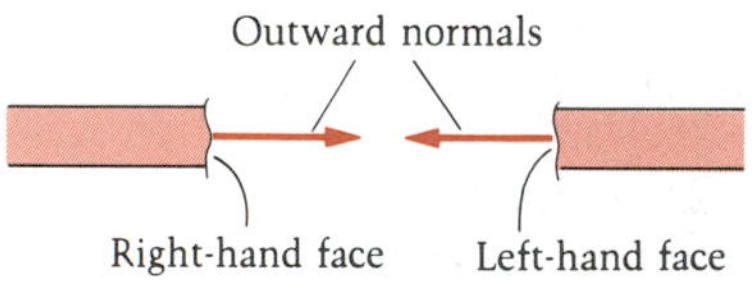

Figure 5.43

Sign Conventions for Internal Forces in Beams

We next set down our sign conventions for N, V, and M.* Suppose a beam has been cut, or "sectioned," into two parts. If we are considering the section on the left, then the exposed face is called a "right-hand face" (with outward normal to the right). If we are considering the section on the right, then the exposed face is on the left (outward normal to the left), and is called a "left-hand face" (see Figure 5.43).

We are now in a position to define our sign conventions for N, V, and M (see Figure 5.44).

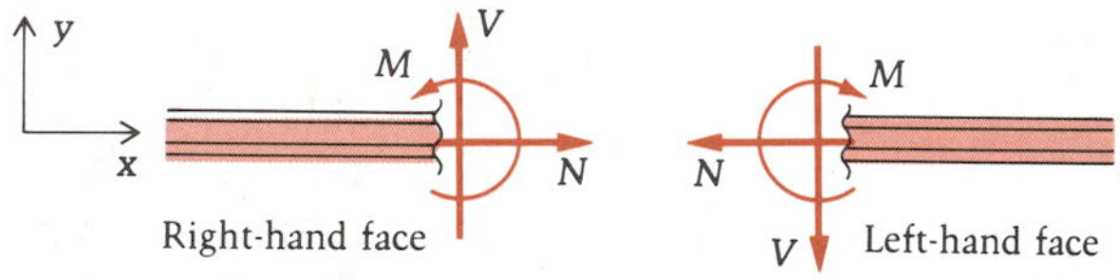

Figure 5.44 Sign convention for axial force, shear force, and bending moment in beams.

As seen in Figure 5.44, the normal (or axial) force N is defined as positive on either type of face if it is in the direction of the outward normal — that is, to the right for a right-hand face and to the left for a left-hand face. In other words, in both cases N is positive if it tends to produce tension in (or to stretch) the axial fibers aligned with x. Saying that "N is in the direction of the outward normal" is shorthand for saying that the scalar N multiplies a unit vector in the direction of the outward normal to form the axial force.

As for the shear force V, it is defined to be positive if directed upward on a right-hand face and positive if downward on a left-hand face. The bending moment is positive if counterclockwise on a right-hand face and if clockwise on a left-hand face. Thus M is positive in both instances if it

* Unfortunately, these conventions vary from book to book, especially for the shear force V, and one must be aware of this when referring to other texts.

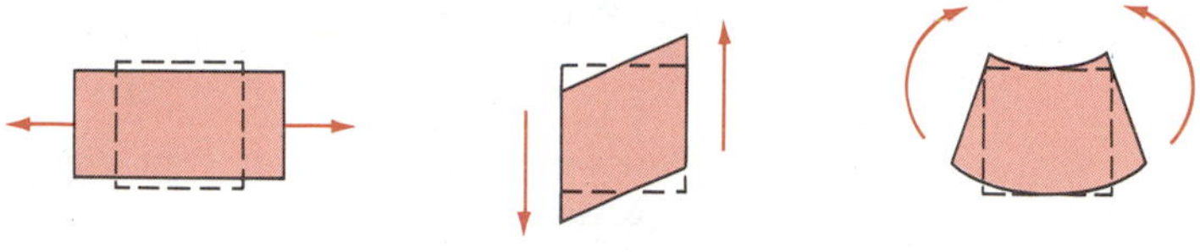

(a) Positive N (b) Positive V (c) Positive M

Figure 5.45

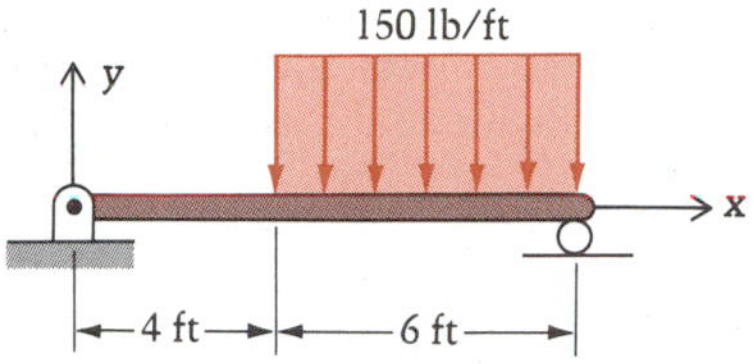

Figure 5.46

bends the section upward (i.e., toward a concave upward configuration). Perhaps the summary sketches shown in Figure 5.45 will be helpful.

Consider next the typical beam shown in Figure 5.46. Such a beam, supported on one end by a pin and on the other by a roller, is said to be "simply supported."

In presenting beam problems, the pin is usually drawn as either

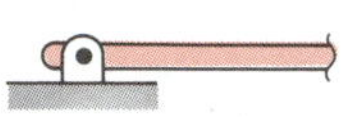 or  while the roller appears

as either 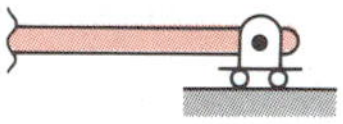 or 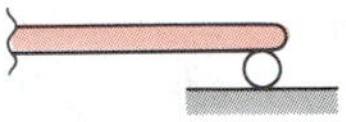.

As we have already seen in Chapter 4, the pin is able to exert two components of force (for example, horizontally and vertically) on the beam, while the roller exerts only a component normal to its contact plane with the beam. (This is a vertical force for the roller in Figure 5.46.) The purpose of the roller is to allow the beam to slightly move so as to prevent or relieve large axial stretching stresses when it is loaded, and/or the potentially large axial stretching or compressing stresses when it is cooled or heated. Thus for a simply supported beam without applied axial (x direction) loads, the horizontal component of the pin reaction will be zero. Let us proceed to find the distribution of shear force and bending moment in the beam of Figure 5.46.

Example 5.20

Find the distribution (as functions of x) of the normal (N) and shear (V) forces and bending moment (M) for the beam shown in Figure E5.20a.

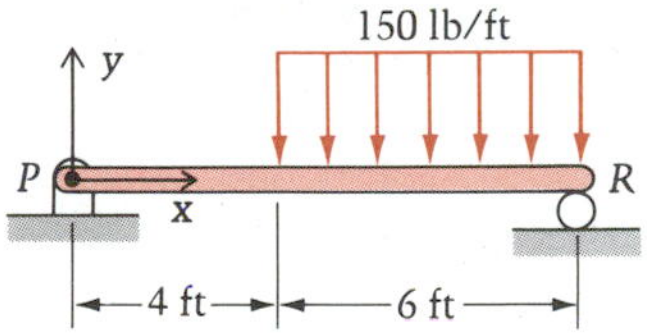

Figure E5.20a

Solution

For finding the reactions, we replace the uniformly distributed load by the equipollent single force shown in the free-body diagram (Figure E5.20b). Note that $P_x = 0$ because it is the only external force in the x direction. Summing moments about the pin gives the roller reaction:

$$\circlearrowleft + \quad \Sigma M_P = 0 = R(10 \text{ ft}) - (900 \text{ lb})(7 \text{ ft})$$

$$R = 630 \text{ lb}$$

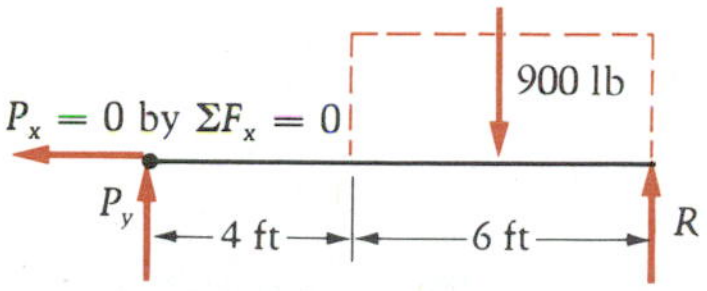

Figure E5.20b

Then

$$+\uparrow \quad \Sigma F_y = 0 = P_y + R - 900 \Rightarrow P_y = 270 \text{ lb}$$

As a check,

$$\circlearrowleft + \quad \Sigma M_R = 0 = (900 \text{ lb})(3 \text{ ft}) - (270 \text{ lb})(10 \text{ ft}) = 0$$

Now we proceed to the determination of $V(x)$ and $M(x)$; note that $N(x) \equiv 0$ in this beam.

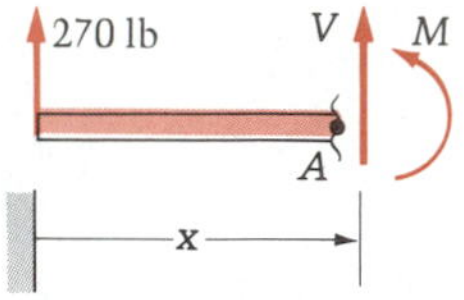

Figure E5.20c

Question 5.15 Why?

If we make a sectional cut between $x = 0$ and $x = 4$ ft, we can obtain expressions for V and M valid throughout that interval (see Figure E5.20c).

We always place V and M (and N, in beams where it is present) on the cut sections in positive directions according to our sign conventions. Hence in Figure E5.20c, V is drawn upward and M counterclockwise on the exposed right-hand face. Equilibrium then requires

$$+\uparrow \quad \Sigma F_y = 0 = 270 + V \Rightarrow V = -270 \text{ lb} \tag{1}$$

$$\circlearrowleft + \quad \Sigma M_A = 0 = M - 270x \Rightarrow M = 270x \text{ lb-ft} \tag{2}$$

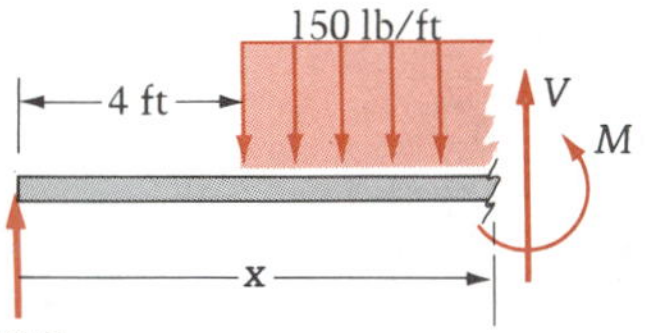

Figure E5.20d

The reader should note that at $x = 0$, these expressions give the correct V and M—namely, a negative shear force of 270 lb (up on a left-hand face is negative V) and a zero moment.

From $x = 4$ ft to $x = 10$ ft, Equations (1) and (2) will not be valid because the distributed load "starts" at $x = 4$ ft and was not included in the free-body diagram of Figure E5.20c. Thus we must make a new cut whenever something changes, such as a distributed load starting or ending, or a concentrated load or couple appearing. The new free-body diagram (Figure E5.20d) will be valid for the rest of the beam (4 ft $< x <$ 10 ft).

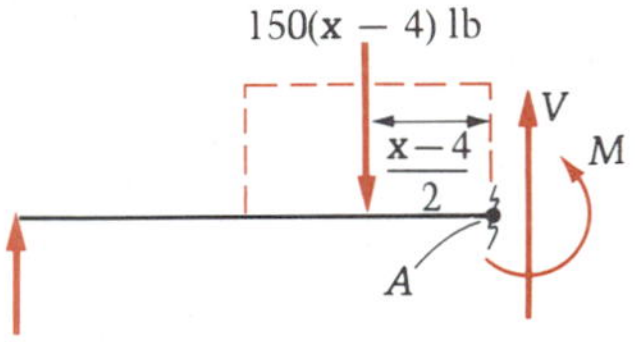

Figure E5.20e

For purposes of finding V and M from the equilibrium equations, we may replace the (part of the) distributed load by its equipollent "force-alone resultant," shown in Figure E5.20e. Then we obtain, from the equilibrium equations for the material to the left of the cut section,

$$+\uparrow \quad \Sigma F_y = 0 = 270 - 150(x - 4) + V$$

$$V = 150(x - 4) - 270 \text{ lb}$$

and

$$\circlearrowleft + \quad \Sigma M_A = 0 = -270 \text{ lb}(x \text{ ft}) + [150(x - 4) \text{ lb}]\left[\left(\frac{x - 4}{2}\right) \text{ft}\right] + M$$

$$M = 270x - 75(x - 4)^2 \text{ lb-ft}$$

Note as a check that at $x = 10$ ft,

$$V|_{x=10\,\text{ft}} = 150(10 - 4) - 270 = 630 \text{ lb}$$

which is the shear (value and sign) produced by the roller reaction R on the right-hand face at $x = 10$ ft. Also,

$$M|_{x=10\,\text{ft}} = 270(10) - 75(10 - 4)^2 = 0$$

Answer 5.15 Because on any free-body diagram such as Figure E5.20c, N would be the only force in the x direction so that $\Sigma F_x = 0 = N$.

This result also agrees with what is actually going on at the right end of the beam, for there is no moment there. These "self-checks" should always be made. They are not foolproof, but the chance of making errors in each of $V(x)$ and $M(x)$ that still give the correct values at the right end of the beam is highly unlikely.

The reader is asked to note for future reference that, in both segments of the beam in Example 5.20, we have $dM/dx = -V$. Also, if we define $q(x)$ to be the distributed load intensity, positive if upward, then $dV/dx = -q$:

$0 < x < 4$ ft	4 ft $< x < 10$ ft
$\dfrac{dV}{dx} = \dfrac{d(-270)}{dx}$	$\dfrac{dV}{dx} = \dfrac{d[150(x-4) - 270]}{dx}$
$= 0$	$= 150$ lb/ft
This is $-q(x)$ because $q(x) = 0$ here.	This is $-q(x)$ because $q(x) = -150$ lb/ft here.
$\dfrac{dM}{dx} = \dfrac{d(270x)}{dx}$	$\dfrac{dM}{dx} = \dfrac{d[270x - 75(x-4)^2]}{dx}$
$= 270$ lb	$= 270 - 150(x-4)$ lb
This is $-V(x)$ because $V(x) = -270$ lb here.	This is $-V(x)$ because $V(x) = -270 + 150(x-4)$ lb here.

These relationships turn out to be true in general; we shall derive and study them in the next section of this chapter.

Example 5.21

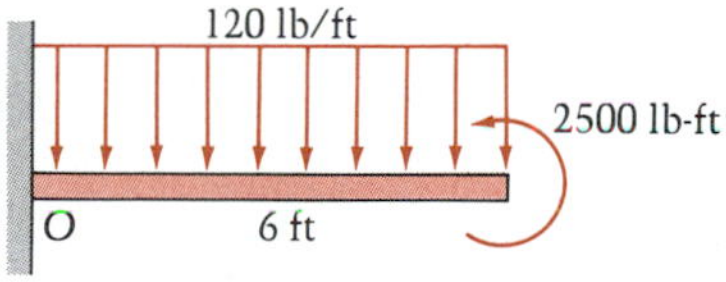

Figure E5.21a

Obtain expressions for the shear force $V(x)$ and bending moment $M(x)$ for the cantilever beam loaded as shown in Figure E5.21a.

Solution

A cantilever beam is one which is fixed at one end into a structure (such as a wall) and juts out with its other end unsupported, or "free." The free-body diagram in Figure E5.21b will help us compute the reactions at the wall. [Note that $N(x)$ is again identically zero here.]

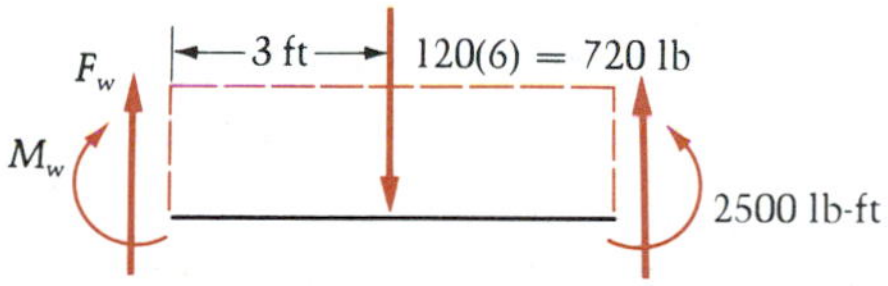

Figure E5.21b

$$+\uparrow \quad \Sigma F_y = 0 = -720 + F_w \Rightarrow F_w = 720 \text{ lb} \tag{1}$$

$$\circlearrowleft + \quad \Sigma M_O = 0 = 2500 \text{ lb-ft} - (720 \text{ lb})(3 \text{ ft}) - M_w \tag{2}$$

$$M_w = 340 \text{ lb-ft}$$

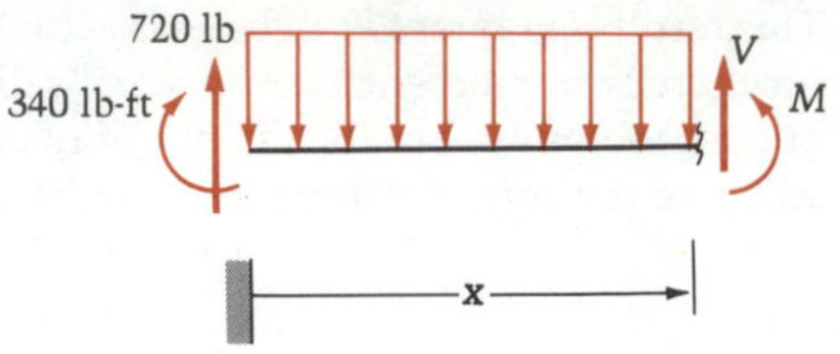

Figure E5.21c

In this problem we are lucky because a single cut section (see Figure E5.21c) will yield $V(x)$ and $M(x)$ expressions valid for the entire beam.

To find V and M, we replace the distributed load by the force $120x$ lb located at $x/2$ as shown in Figure E5.21d below:

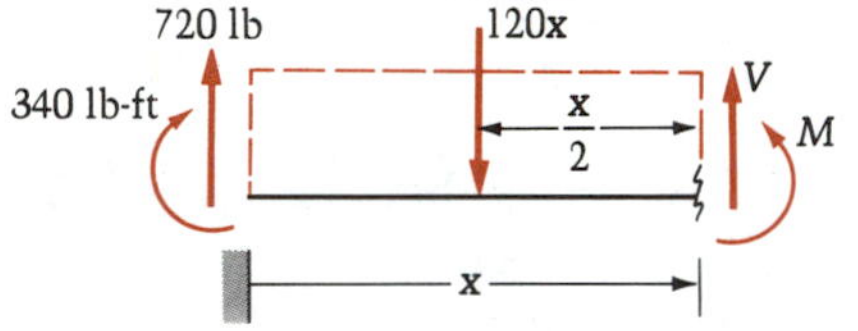

Figure E5.21d

Equilibrium of the cut section requires

$$+\uparrow \qquad \Sigma F_y = 0 = V + 720 - 120x$$

$$V = 120x - 720 \text{ lb} \tag{3}$$

$$\circlearrowleft + \qquad \Sigma M_A = 0 = M + (120x \text{ lb})\left(\frac{x}{2}\text{ ft}\right) - 340 \text{ lb-ft} - (720 \text{ lb})(x \text{ ft})$$

$$M = -60x^2 + 720x + 340 \text{ lb-ft} \tag{4}$$

Note that:

a. At $x = 0$, $V = -720$ lb, which checks.
b. At $x = 0$, $M = +340$ lb-ft, which checks.
c. At $x = 6$ ft, $V = 120(6) - 720 = 0$, which checks.
d. At $x = 6$ ft, $M = -60(6)^2 + 720(6) + 340 = 2500$ lb-ft, which checks.

It is our experience that students often have trouble, at least at first, in distinguishing between the meanings of:

a. The sign beside an equilibrium equation, such as the $+\uparrow$ in Equation (1) of the above example; and
b. The sign given to the scalar V (or to the scalar M) representing a shear force (or a bending moment).

The difference between these signs is explained as follows:

The sign $+\uparrow$ is a temporary convenience, where we are agreeing on which unit vector is factored when the forces (or moments) are summed and equated to zero. In the preceding example, Equation (1) could be written more formally as the y component of $\Sigma\mathbf{F} = \mathbf{0}$:

$$720(-\hat{\mathbf{j}}) + F_w(\hat{\mathbf{j}}) = 0$$

We see that if we suppress the "$\hat{\mathbf{j}}$," Equation (1) of the example remains. Thus the sign $+\uparrow$ in front of that equation is merely saying it is "plus $\hat{\mathbf{j}}$" that is omitted, and has nothing to do with a shear force sign convention. If we, however, want to know the shear force on the left end of that same beam, it comes from an upward force (of $F_w = 720$ lb) on a left-hand face, which is negative shear force by the sign convention for V; that is, $V = -720$ lb there at $x = 0$.

In our final example of this section, we shall consider a simply supported beam loaded in four commonly occurring ways: by concentrated axial and lateral forces, by a concentrated couple, and by a distributed loading.

Example 5.22

Find the axial force, shear force, and bending moment distributions throughout the beam shown in Figure E5.22a.

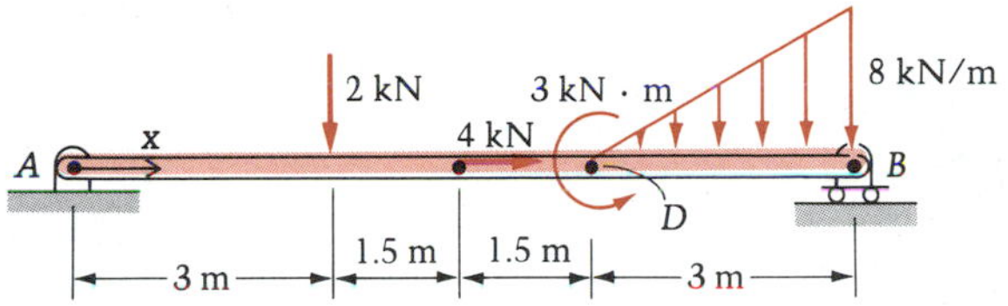

Figure E5.22a

Solution

For the reactions, we shall use the free-body diagram in Figure E5.22b, in which the distributed load (see Section 3.9) has been replaced, solely for purposes of using the equilibrium equations, by the area under the loading curve acting along the indicated line.

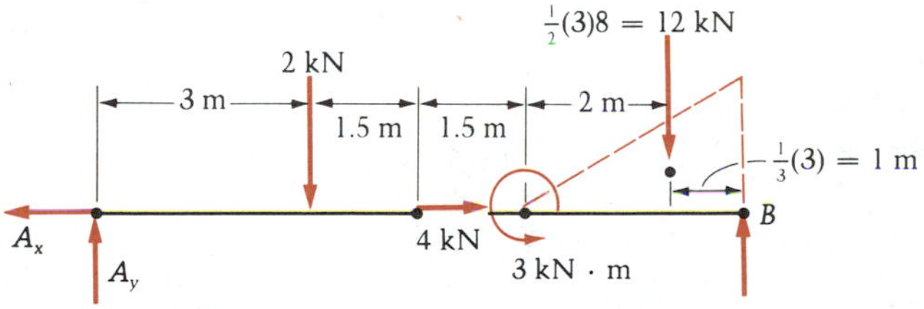

Figure E5.22b

Because the force-alone resultant of 12 kN ↓ acting 1 meter from the right end of the beam produces the same contribution to the equilibrium equations as does the actual distributed loading, it can be used to advantage in computing the three reactions A_x, A_y, and B:

$$\xrightarrow{+} \quad \Sigma F_x = 0 = 4\text{ kN} - A_x \Rightarrow A_x = 4\text{ kN} \qquad \text{(to the left as assumed)}$$

$$+\uparrow \quad \Sigma F_y = 0 = A_y - 2\text{ kN} - 12\text{ kN} + B$$

$$\circlearrowleft\!+ \quad \Sigma M_A = 0 = -2\text{ kN}(3\text{ m}) + 3\text{ kN}\cdot\text{m} - 12\text{ kN}(8\text{ m}) + B(9\text{ m})$$

$$B = \frac{(6 - 3 + 96)\text{ kN}\cdot\text{m}}{9\text{ m}} = 11\text{ kN}$$

Therefore, from the "y equation,"

$$A_y = (2 + 12 - 11)\text{ kN}$$

$$A_y = 3\text{ kN}$$

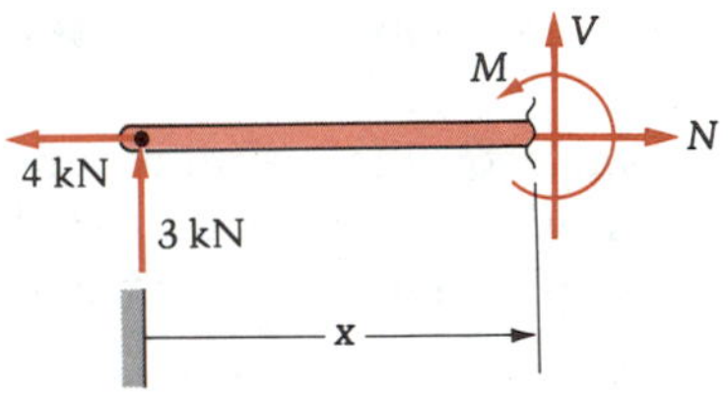

Figure E5.22c

and the reactions have been found. Now we proceed to find $N(x)$, $V(x)$, and $M(x)$ in the various segments of the beam. There will be four such segments: (0, 3), (3, 4.5), (4.5, 6), and (6, 9) because at $x = 0$, 3, 4.5, and 6 m, we have concentrated forces, couples, and/or the beginning of the distributed load.

We first cut the beam at a value of x between 0 and 3 meters, $0 < x < 3$ m. We consider the section to the left of the cut shown in Figure E5.22c.

Note that the sign convention is always in agreement with Figure 5.44 for the "right-hand face" above. To determine N, V, and M, we simply ensure that the section is in equilibrium:

$$\xrightarrow{+} \quad \Sigma F_x = 0 = -4\text{ kN} + N \Rightarrow N = 4\text{ kN}$$

$$+\uparrow \quad \Sigma F_y = 0 = 3\text{ kN} + V \Rightarrow V = -3\text{ kN}$$

$$\circlearrowleft\!+ \quad \Sigma M_A = 0 = M - (3\text{ kN})(x\text{ m})$$

$$M = 3x\text{ kN}\cdot\text{m}$$

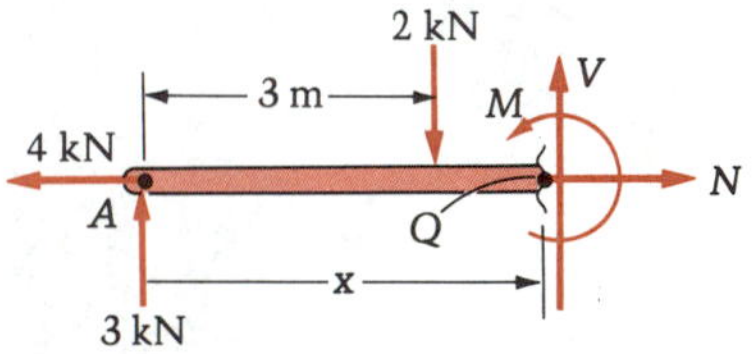

Figure E5.22d

Therefore, over the first 3 meters of the beam, the axial and shear forces are constant while the bending moment grows linearly with x. Note too that the derivative of M with respect to x equals the value of $(-V)$. As we have mentioned, this will later be seen to be true in general.

We next cut the beam between the 2-kN and 4-kN forces. From the equilibrium equations, we obtain (see Figure E5.22d):

$$\xrightarrow{+} \quad \Sigma F_x = 0 = N - 4\text{ kN} \Rightarrow N = 4\text{ kN}$$

$$+\uparrow \quad \Sigma F_y = 0 = (3 - 2)\text{ kN} + V$$

$$V = -1\text{ kN}$$

$$\circlearrowleft\!+ \quad \Sigma M_Q = 0 = M + (2\text{ kN})(x - 3)\text{ m} - (3\text{ kN})(x\text{ m})$$

$$M = (6 + x)\text{ kN}\cdot\text{m}$$

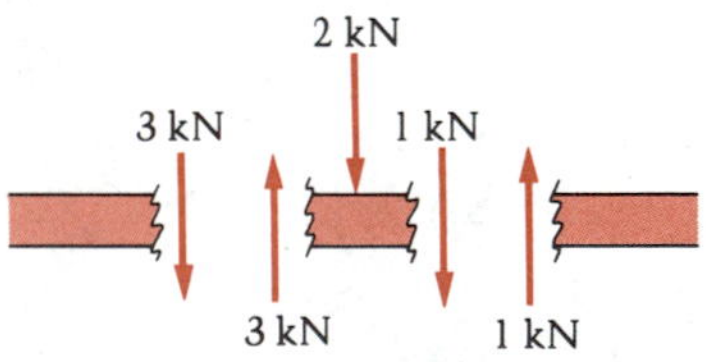

(showing only shear forces)

Figure E5.22e

We note that the values of N are the same on either side of the point $x = 3$ m. This is also true for M; approaching $x = 3$ m from the left, $M \to 3(3) = 9$ kN · m, and from the right, $M \to 6 + 3 = 9$ kN · m. Thus N and M are continuous through the point, but not so for V.* A free body surrounding the point shows the situation, and why there is no single value of V at the point $x = 3$ (see Figure E5.22e).

* This is to be expected because we have no concentrated longitudinal force or couple applied at the point, but we do have a concentrated transverse external force there. Concentrated forces or moments will always and only give rise to discontinuities in the respective related quantity N, V, or M.

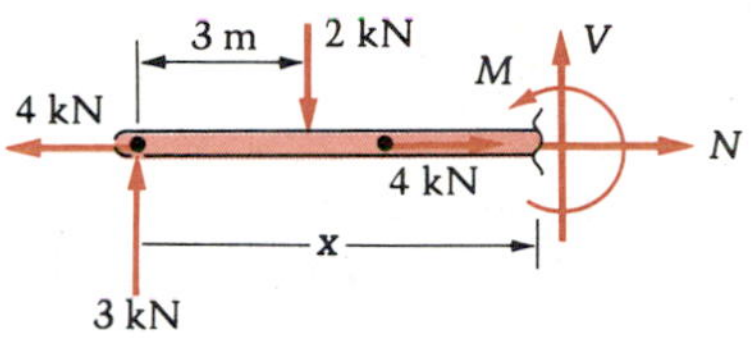

Figure E5.22f

Passing next over the concentrated axial force of 4 kN at $x = 4.5$ m, we see (Figure E5.22f) that to the right of this force, there is no axial force in the beam:

$$\xrightarrow{+} \quad \Sigma F_x = 0 = N - 4\text{ kN} + 4\text{ kN}$$

$$N = 0$$

V and M this time retain their earlier expressions, as the reader may note:

$$V = -1\text{ kN}$$

$$M = (6 + x)\text{ kN}\cdot\text{m}$$

When we reach point D, two things happen:

1. There is a concentrated couple there that causes a jump in the expression for $M(x)$.
2. A distributed load begins that will affect both V and M.

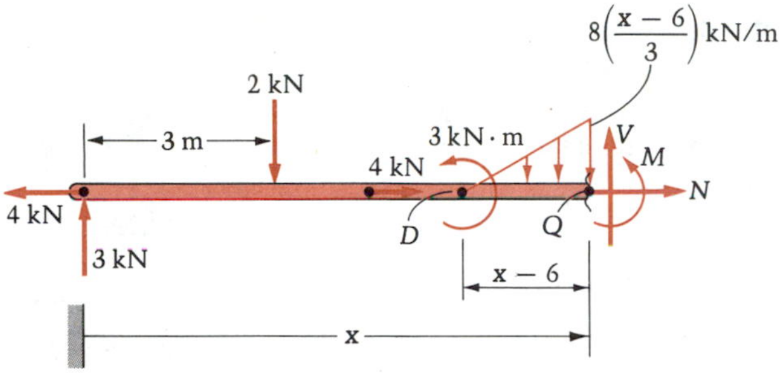

Figure E5.22g

Shown in Figure E5.22g is a free-body diagram of a section of the beam to the left of a cut at a point with $x > 6$ m. Equilibrium requires

$$\xrightarrow{+} \quad \Sigma F_x = 0 = -4 + 4 + N \Rightarrow N = 0$$

(Only the portion of the beam being stretched carries any axial force.)

$$+\uparrow \quad \Sigma F_y = 0 = 3 - 2 + V - \underbrace{\frac{1}{2}(x-6)\left[8\left(\frac{x-6}{3}\right)\right]}_{\text{area of triangular loading curve (see Example 3.34)}}$$

$$V = -1 + \frac{4}{3}(x-6)^2\text{ kN}$$

(Note that as x grows from 6 m, the shear must become more and more positive.)

$$\circlearrowleft\!+ \quad \Sigma M_Q = 0 = -3x + 2(x-3) + 3 + M$$

$$+\underbrace{\frac{4}{3}(x-6)^2}_{\substack{\text{downward force}\\\text{from distributed}\\\text{load from}\\\text{preceding equation}}} \bullet \underbrace{\left(\frac{x-6}{3}\right)}_{\substack{\text{location of}\\\text{resultant of}\\\text{distributed load}\\\text{(see Example 3.34)}}}$$

$$M = 3 + x - \frac{4}{9}(x-6)^3\text{ kN}\cdot\text{m}$$

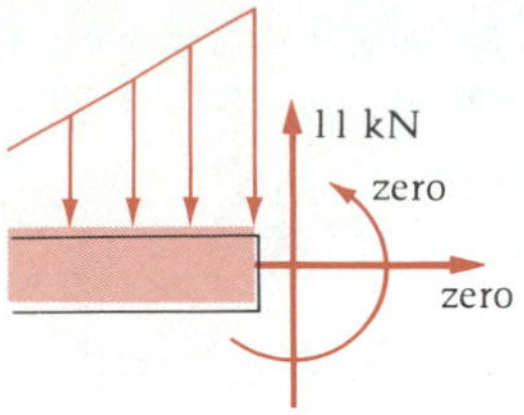

Figure E5.22h

As checks on the solution, note that at $x = 9$ m:

$$N = 0$$

$$V = -1 + \frac{4}{3}(3^2) = 11 \text{ kN}$$

$$M = 3 + 9 - \frac{4}{9}(3)^3 = 0$$

This check is at the right end of the beam shown in Figure E5.22h. We have:

N should be zero. ✓
V should be 11 kN. ✓
M should be zero. ✓

PROBLEMS ▶ Section 5.9

5.155 At what point(s) on the beam shown in Figure P5.155 do the following occur?

a. Zero shear
b. Maximum shear (give the value)
c. Zero moment
d. Maximum moment (give the value)

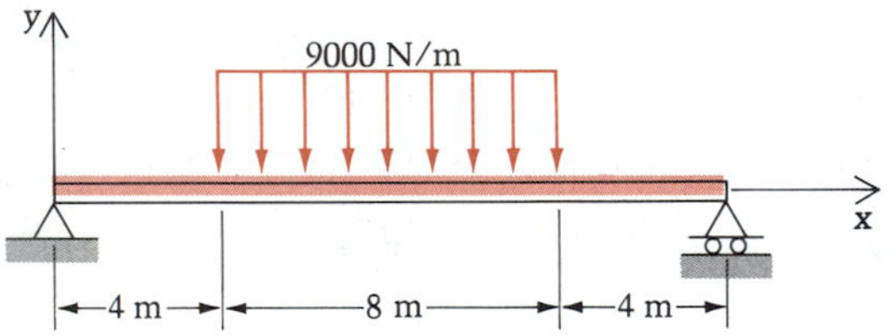

Figure P5.155

5.156 Repeat the preceding problem for the beam of Figure P5.156. Assume the reaction of the ground onto the beam is uniformly distributed.

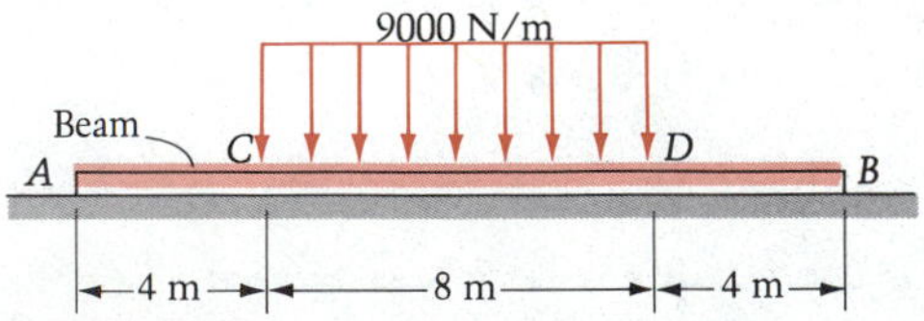

Figure P5.156

5.157 For the beam in Figure P5.157, what is the bending moment of largest magnitude and where does it occur?

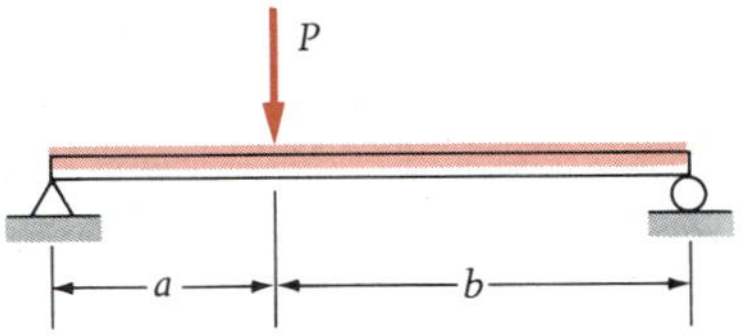

Figure P5.157

5.158 How do the shear force and bending moment vary in the central 3-ft segment of the beam in Figure P5.158?

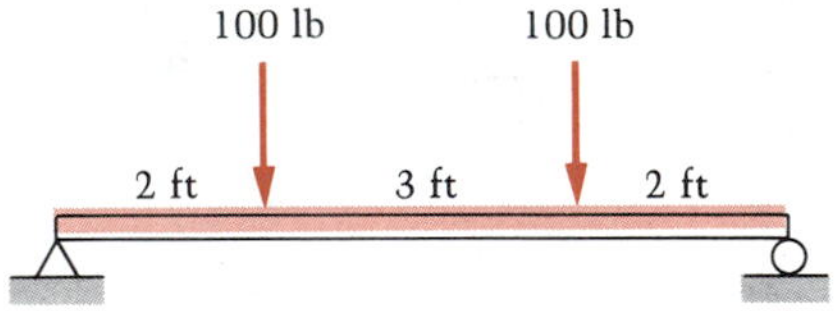

Figure P5.158

5.159 Find the maximum shear and bending moment in the beam shown in Figure P5.159.

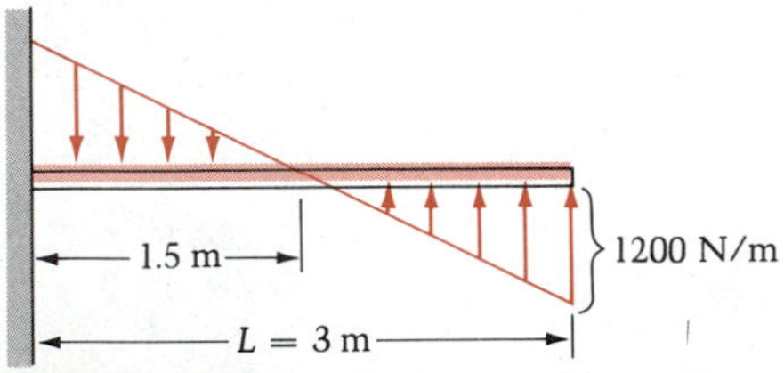

Figure P5.159

5.160 Given the beam shown in Figure P5.160:

a. Find the axial force (N), shear force (V), and bending moment (M) in the portion of the beam along x, as functions of x.

b. Sketch the graph of the functions in part (a) and label the values at $x = 0$, 2 m, and 4 m.

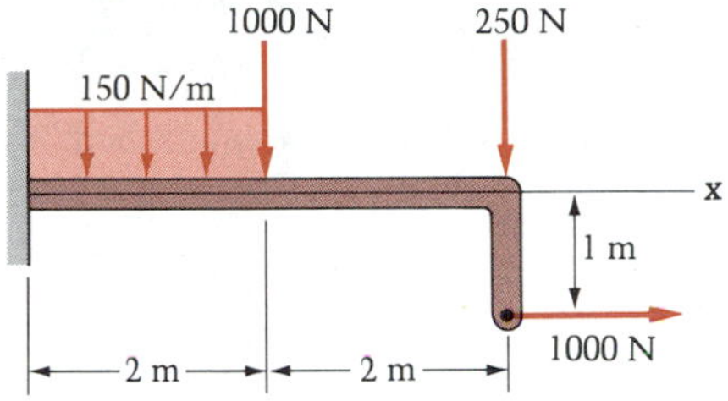

Figure P5.160

5.161 In Figure P5.161:

a. Find the reactions at the wall for the cantilever beam shown.

b. Find the shear force, bending moment, and axial force in the beam as functions of x.

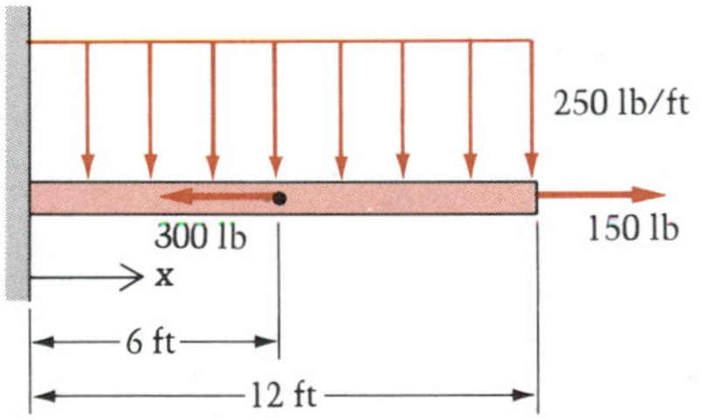

Figure P5.161

5.162 A signboard, BC, is to withstand a windload of $w = 80$ lb/ft. It is supported by the structure shown in Figure P5.162.

a. Determine the magnitude of the windload on the signboard, BC.

b. Identify each of the two-force members(s), if any.

c. Determine the force in each two-force member.

d. Determine the horizontal and vertical components of the reaction forces at A and F.

e. Show the axial and shear forces and the bending moment at point H of member DEF, and show the results on a free-body diagram that includes the cut section there.

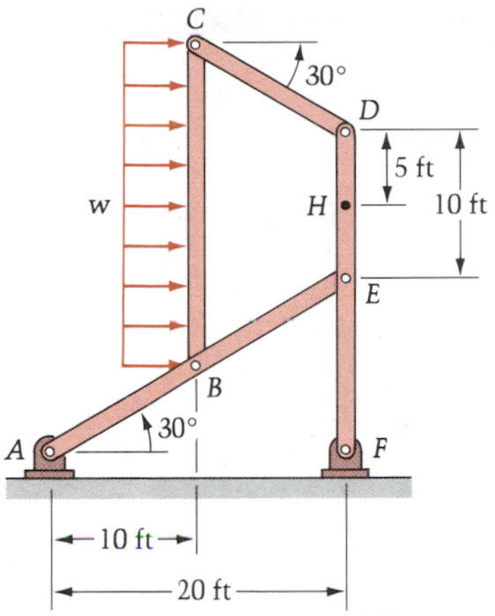

Figure P5.162

5.163 In Figure P5.163:

a. Find the reactions at the wall for the cantilever beam.

b. Find the shear force, bending moment, and axial force as functions of x.

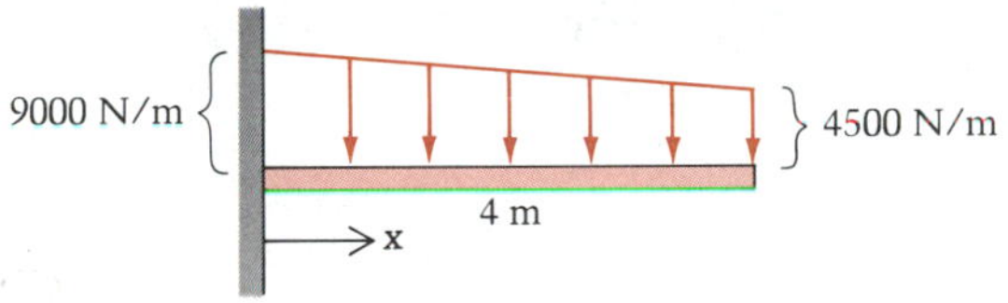

Figure P5.163

In terms of P and L, find the extreme values of the shear force V and bending moment M in the beams in Problems 5.164–5.169. Also give the locations along the beam where these extreme values occur. (It will help to graph $V(x)$ and $M(x)$.)

5.164

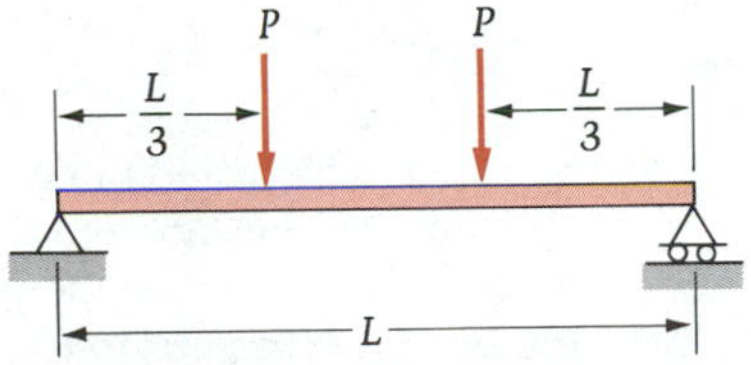

Figure P5.164

(For Problems 5.165–169, see the instructions preceding problem 5.164.)

5.165

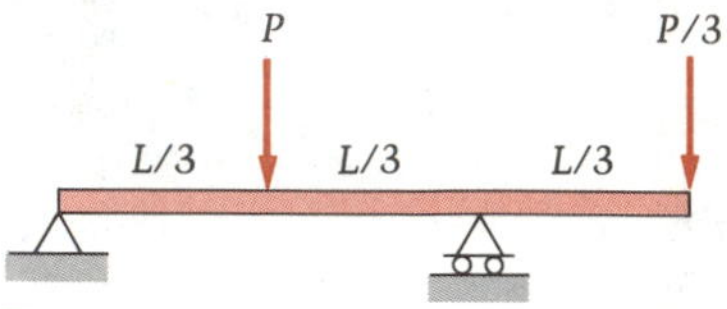

Figure P5.165

5.166

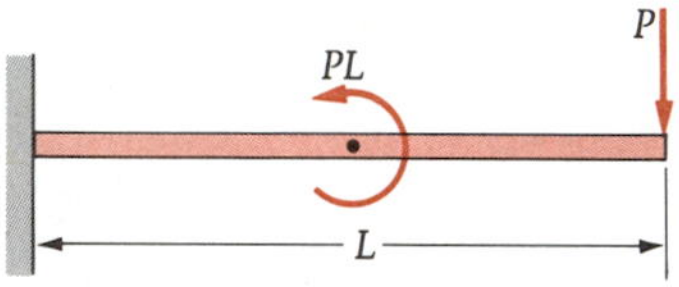

Figure P5.166

5.167

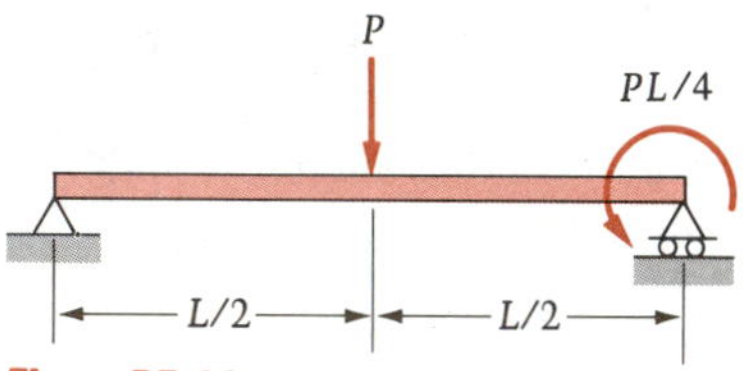

Figure P5.167

5.168

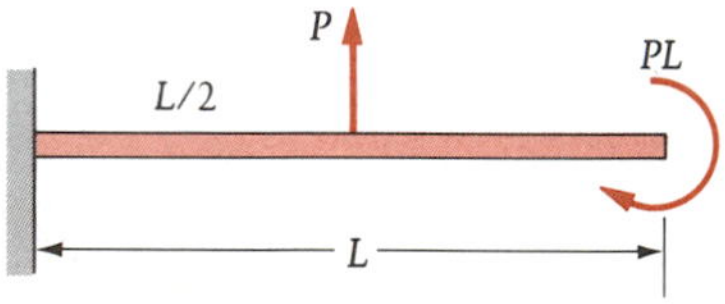

Figure P5.168

5.169

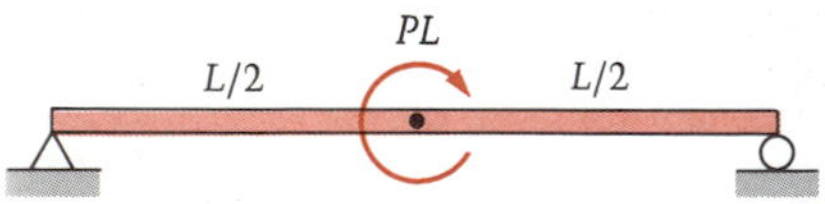

Figure P5.169

5.170 For the cantilever beam loaded as indicated in Figure P5.170, find the shear force V and bending moment M as functions of x. Then find the shear force and bending moment at the point where the bending moment in the beam is the largest in absolute value.

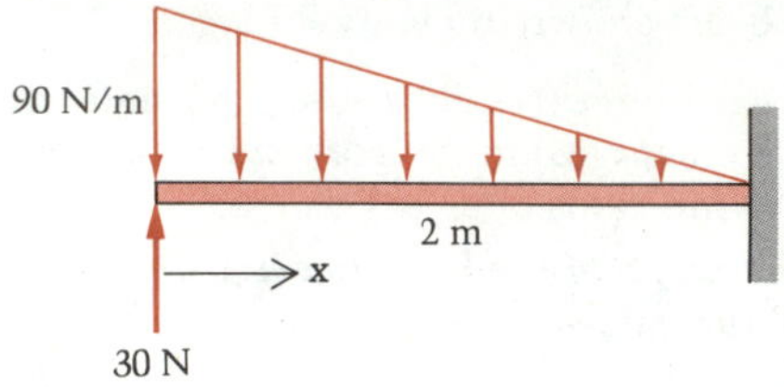

Figure P5.170

Write expressions for the shear force $V(x)$ and bending moment $M(x)$ in the beams shown in Problems 5.171–5.174.

5.171

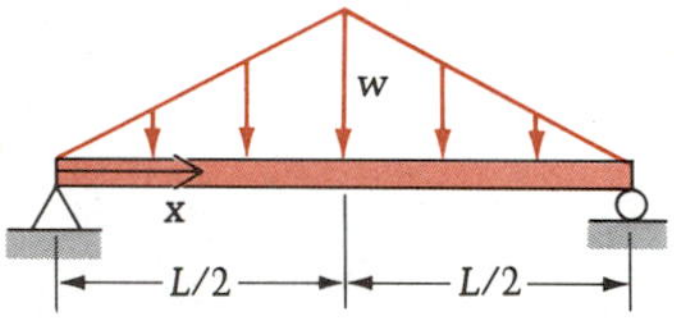

Figure P5.171

5.172

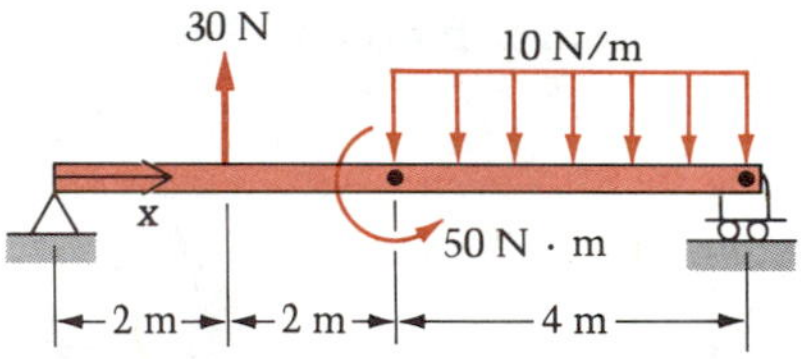

Figure P5.172

5.173

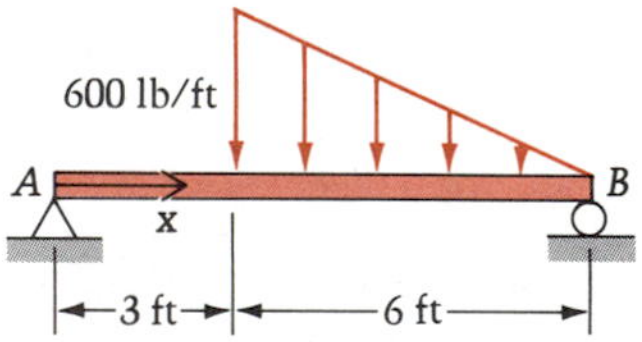

Figure P5.173

5.174

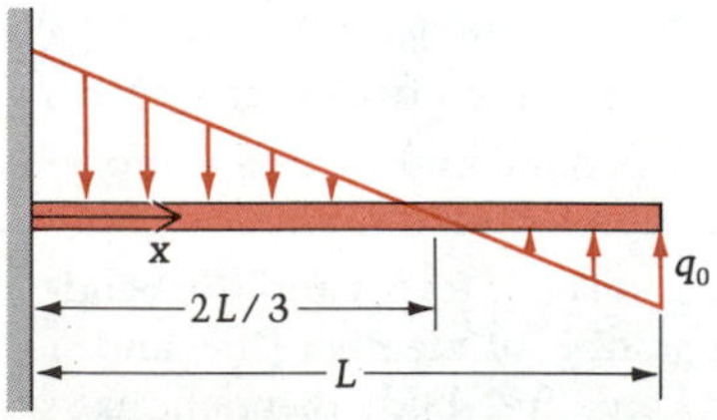

Figure P5.174

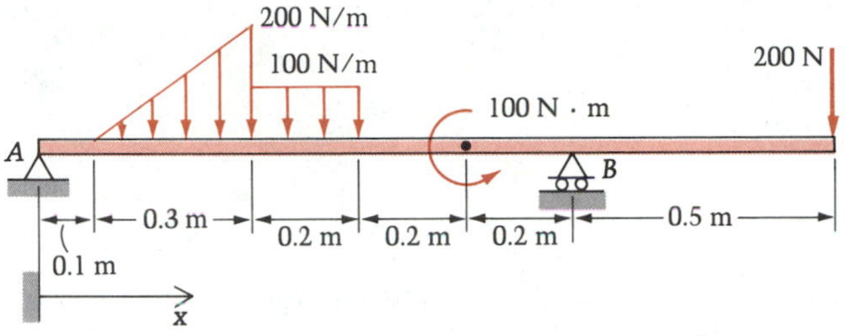

Figure P5.175

5.175 For the beam shown in Figure P5.175, write equations for the shear force V and bending moment M as functions of x, in the intervals $0 < x < 0.1$ m, $0.1 < x < 0.4$ m, $0.4 < x < 0.6$ m, $0.6 < x < 0.8$ m, $0.8 < x < 1.0$ m, and $1.0 < x < 1.5$ m.

5.176 For the beam shown in Figure P5.176, write equations for the shear force V and bending moment M as functions of x, in the intervals $0 < x < 5$ ft, $5 < x < 25$ ft, and $25 < x < 31$ ft.

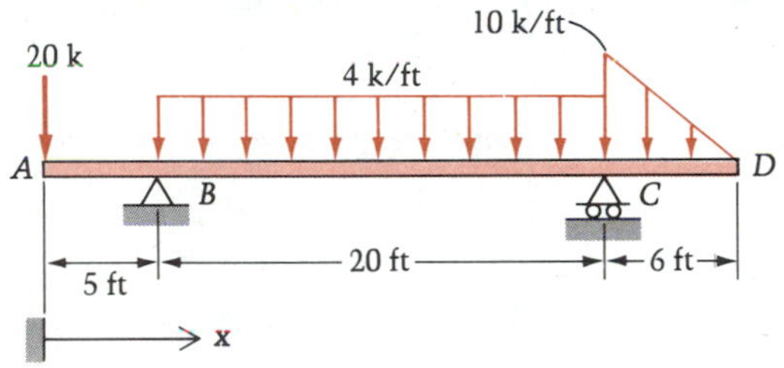

Figure P5.176

5.177 In Figure P5.177:

a. Write algebraic expressions for the shear force and bending moment in the beam.

b. Find the shear force of largest magnitude in the beam.

c. Find the bending moment of largest magnitude in the beam.

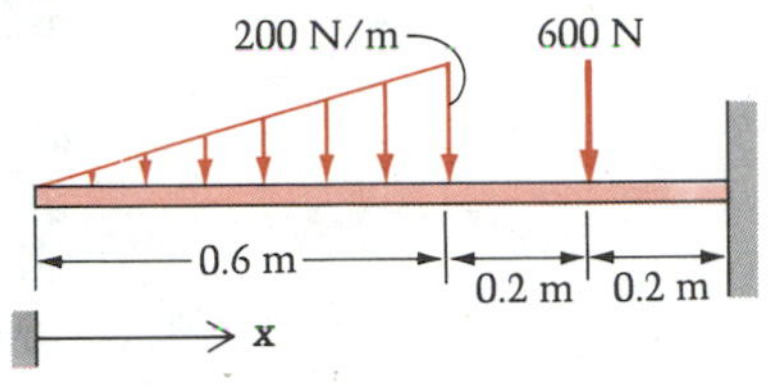

Figure P5.177

5.178 Find expressions for the shear force and bending moment in the beam shown in Figure P5.178. What is the bending moment of largest magnitude and where does it occur?

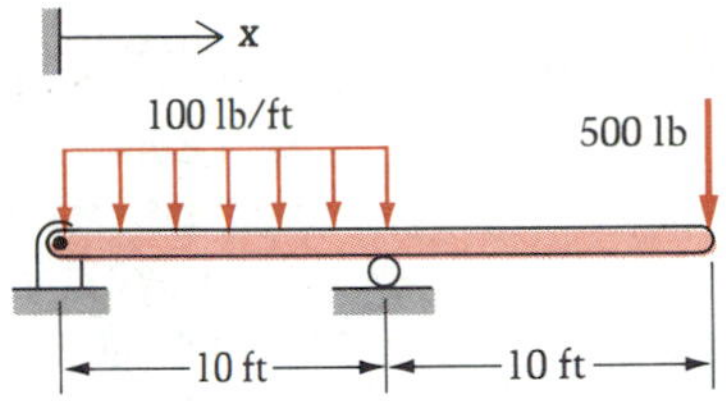

Figure P5.178

5.179 In Figure P5.179 find the shear force $V(x)$ in the segment of the beam $0 < x < 5$ ft and the bending moment $M(x)$ in the segment $5 < x < 10$ ft.

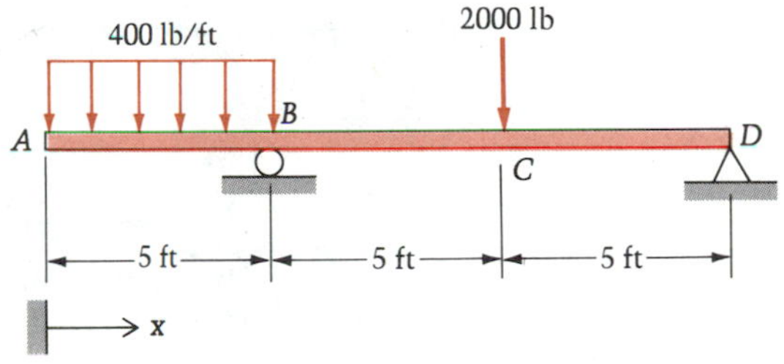

Figure P5.179

5.10 Differential Relationships Between $q(x)$, $V(x)$, and $M(x)$ in a Beam / Shear and Bending Moment Diagrams

We noted just after Example 5.20 that in both of the analyzed segments of the beam, the derived expressions for $q(x)$, $V(x)$, and $M(x)$ satisfied the equations

$$\frac{dV}{dx} = -q \quad \text{and} \quad \frac{dM}{dx} = -V$$

The same is true in Example 5.21, where

$$\frac{dV}{dx} = \frac{d(120x - 720)}{dx} = 120 = -q \qquad \text{(since } q = -120\text{)}$$

and

$$\frac{dM}{dx} = \frac{d(-60x^2 + 720x + 340)}{dx} = -120x + 720 = -V$$

$$\text{(since } V = 120x - 720\text{)}$$

The reader is encouraged to observe that the same two equations hold in Example 5.22 in each of the four analyzed segments of the beam.

Relationship Between Shear Force and Distributed Lateral Loading

The previous observations are *no coincidence.* There is this definite relationship between the shear force and bending moment in a beam. Let us assume we don't know the result and derive it "from scratch." We begin by isolating an elemental segment of a uniformly loaded beam as shown in Figure 5.47:

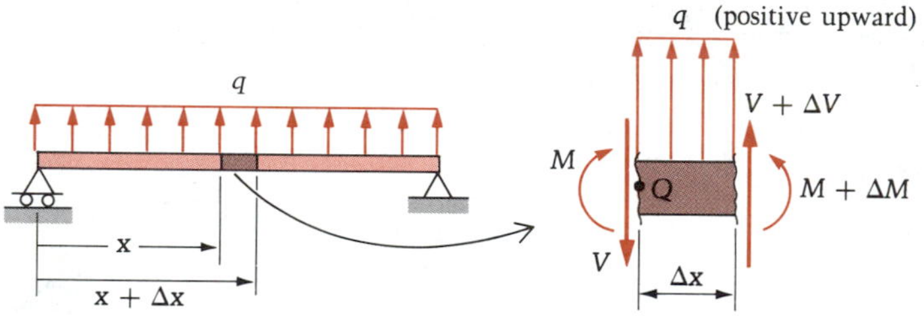

Figure 5.47

Note that, as we have seen, the values of V and M are generally functions of x, so that from the left-hand face (at x) to the right-hand face (at $x + \Delta x$), V and M respectively change to $V + \Delta V$ and $M + \Delta M$.

We now establish the equilibrium of the element:

$$\uparrow + \qquad \Sigma F_y = 0 = -V + (V + \Delta V) + q\Delta x \tag{5.10}$$

Dividing Equation (5.10) by Δx and simplifying,

$$\frac{\Delta V}{\Delta x} = -q$$

Taking the limit of both sides,

$$\lim_{\Delta x \to 0} \frac{\Delta V}{\Delta x} = \frac{dV}{dx} = -q \tag{5.11}$$

When q is $q(x)$, the value of the distributed load intensity at x, the result is the same.*

Thus the slope of the shear diagram (a plot of V versus x) at any point equals the negative of the distributed load value at the point. Let us see by two examples how we might use this result in obtaining the distribution of V throughout a beam.

Example 5.23

Draw the shear diagram (V versus x) for the beam loaded as shown in Figure E5.23a.

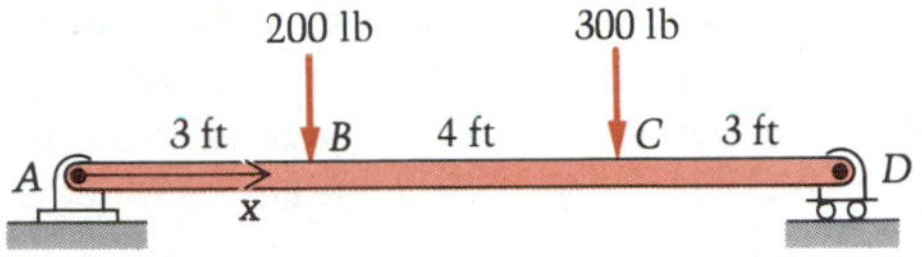

Figure E5.23a

Solution

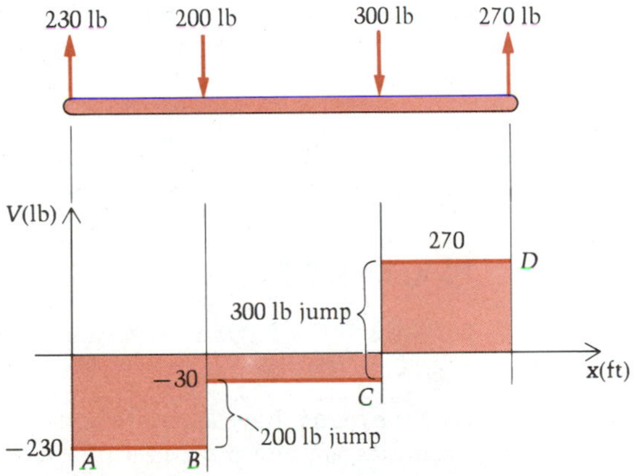

Figure E5.23b

In this problem, the reactions are 230 lb ↑ at the pin A and 270 lb ↑ at the roller B. At the left end of the beam, which is a left-hand face, an upward force corresponds to negative shear. Thus the shear diagram (Figure E5.23b) starts at the point $V = -230$ at $x = 0$. In this beam, there is no $q(x)$ anywhere, so the slope of the V diagram — that is, dV/dx or V' — is zero throughout. Hence we draw a horizontal line from point A of the diagram to the point where $x = 3$ ft, point B.

At B, there will be a discontinuity, or jump, in V because of the 200-lb downward load there. As soon as we pass the point B, the shear force on a

* Note that if the distributed load $q(x)$ is continuous *but not uniform* (i.e., not constant), the argument is slightly more complicated. By the mean-value theorem for integrals, there is a number X such that $x \le X \le x + \Delta x$ and $\int_{\xi=x}^{x+\Delta x} q(\xi)\, d\xi = \Delta x\, q(X)$. Then, as the limit is taken, $\lim_{\Delta x \to 0} X \to x$ so that $q(X) \to q(x)$, and the result is the same.

right-hand face will have to increase by 200 lb to maintain equilibrium (see Figure E5.23c).

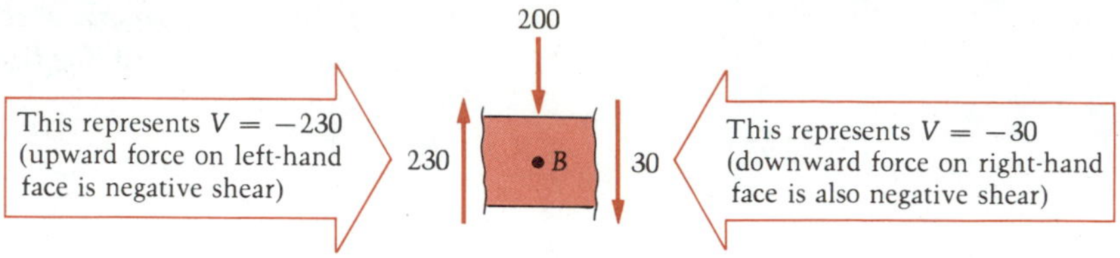

Figure E5.23c

Therefore the value of V becomes -30 lb (or $-230 + 200$) just past B, and stays at that value until we reach point C. There, we get a second jump, this time of 300 lb in V by the same reasoning as at B. This puts us at $V = -30 + 300 = 270$ lb. Again, because $q(x) = 0$ in this beam, we extend the line straight across to D, and end at $V = 270$ lb.

It is important to note that we have an automatic check on our solution because we know the answer for V at the right end of the beam, a right-hand face. In this problem, we have an upward reaction of 270 lb there. Thus (up on the right is positive) V has to be $+270$ lb at $x = 10$ ft; otherwise we have made a mistake. This automatic check should always be made because if you end at the correct value of V, the chances are very good that the diagram was drawn correctly.

EXAMPLE 5.24

Draw the shear diagram (V versus x) for the beam of Example 5.22.

Solution

We have found the reactions to be as indicated in Figure E.5.24a. We present the shear diagram and proceed in the series of steps below to explain how it was constructed.

1. The reaction at the left end gives the starting value of V. It is 3 kN upward on a left-hand face, which is negative shear. Thus we begin with a dot at $V = -3$ on the V axis (see Figure E5.24b).

2. There is no change in V between $x = 0$ and $x = 3$ m. This is because in that interval, $q = 0$ (no distributed load), so that $dV/dx = 0$. Thus the slope of V vs. x is constant at zero, and we draw a horizontal line across from the starting point to B.

3. As we cross B, there is a sudden jump in V of 2 kN because of the downward 2 kN load there.

4. Again, there is no $q(x)$ between $x = 3$ m and 6 m, so $dV/dx = 0$ again. We draw another horizontal line from C to D on the shear diagram.

5. At D, the couple has no effect on V. But starting there, the value of $q(x)$ begins to get more and more negative, linearly. Because $dV/dx = -q$, the *slope* of

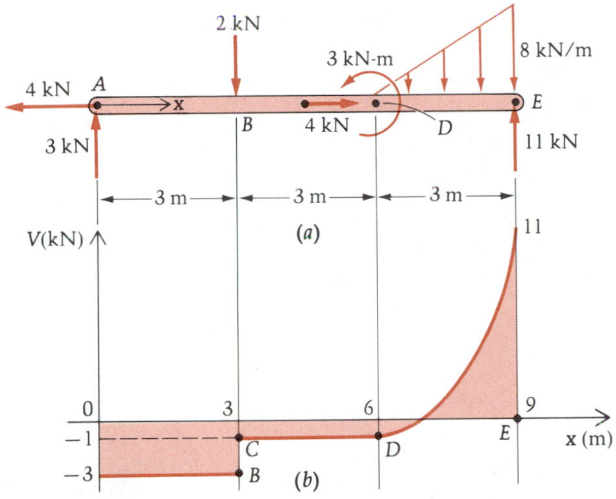

Figure E5.24a,b

the V diagram must get more and more *positive*, linearly. The *change* in V between two points is seen to be the negative of the area beneath the q diagram between those points:

$$\frac{dV}{dx} = -q \Rightarrow \int_{V_1}^{V_2} dV = -\int_{x_1}^{x_2} q\,dx$$

$$V_2 - V_1 = -\int_{x_1}^{x_2} q\,dx$$

From $x = 6$ m to $x = 9$ m, the area beneath "q versus x" is $\frac{1}{2}(3\text{ m})(-8\text{ kN/m}) = -12$ kN. Thus V climbs by 12 kN from -1 kN, ending at 11 kN.

6. We have an automatic check on our diagram because the reaction at the right end is applying an upward force of 11 kN on a right-hand face, which corresponds to $V = 11$ kN, a built-in check.

Relationship Between Bending Moment and Shear Force

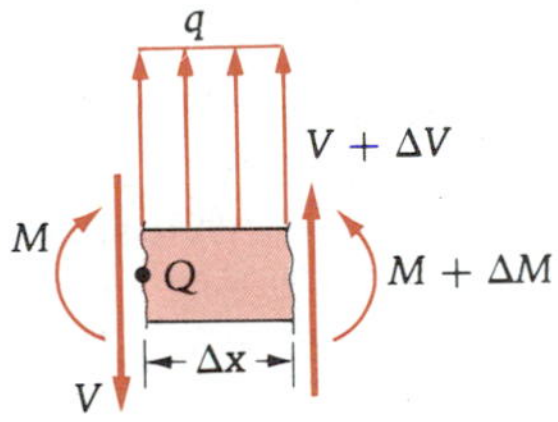

Figure 5.48

Now that we have the shear diagram in hand, we are ready to study its companion, the (bending) moment diagram. We begin by writing the moment equilibrium equation for the segment of Figure 5.47, reproduced at the left as Figure 5.48.

$$\Sigma M_Q = 0 = -M + (M + \Delta M) + (V + \Delta V)\,\Delta x + [q\,\Delta x]\left(\frac{\Delta x}{2}\right)$$

Therefore,

$$\frac{\Delta M}{\Delta x} = -V - \Delta V - \frac{q(\Delta x)}{2}$$

As the limit is taken ($\Delta x \to 0$), the last two terms in this equation vanish,* leaving

$$\frac{dM}{dx} = -V \tag{5.12}$$

Thus at each point where M' is defined, the slope of the bending moment diagram is equal to the negative of the value of the shear force at that point.

Also, we have

$$\int_{x_1}^{x_2} \frac{dM}{dx}\, dx = -\int_{x_1}^{x_2} V\, dx$$

or

$$M_2 - M_1 = -\int_{x_1}^{x_2} V\, dx$$

so that the *change* in the bending moment between two points x_1 and x_2 equals the negative of the area beneath the shear diagram between those two points. Let us use these results to extend the last two examples to include moment diagrams.

EXAMPLE 5.25

Draw the moment diagram for the beam of Example 5.23, which carried the loading shown in Figure E5.25a and had the shear diagram in Figure E5.25b.

Solution

It is helpful to draw the various diagrams beneath one another because we can use the relationships such as $V' = -q$ and $M' = -V$ more easily. Also, we can for our convenience extend lines downward at places of importance, such as points where loads appear.

The moment diagram starts at the origin (see Figure E5.25c) because there is no moment applied to the beam by the pin and no external moment is being exerted at the end $x = 0$. The slope of M versus x, for $0 < x < 3$ ft, is $-V$, or $+230$ lb. Over 3 ft, this slope will result in a buildup of M from zero to 690 lb-ft. Note that the *change* in M from its value of zero at $x = 0$ is the negative of the area beneath the V diagram, or $-(3\text{ ft})(-230\text{ lb})$, which is, again, 690 lb-ft.

The jump in V at $x = 3$ ft means to us that there is a discontinuity in the *slope* of the moment diagram. The M versus x curve will come out of the point B with a different slope [namely $-(-30)$ or 30] than it had when it went in. Over the next four feet, the change in M, then, will be the negative of the area under the shear diagram between B and C, or $-(4\text{ ft})(-30\text{ lb}) = 120$ lb-ft. Hence the new (and the maximum) value of M is the old plus the change, or $690 + 120 = 810$ lb-ft.

* If $q(x)$ is continuous but not constant, then again (as in the footnote on page 373) using the mean-value theorem, there is a number X such that $x \le X \le x + \Delta x$ and $\int_{\xi=x}^{x+\Delta x} (\xi - x)q(\xi)\, d\xi = \Delta x(X - x)q(X)$. Then, as the $\lim_{\Delta x \to 0}$ is taken, $X \to x$ and the moment of the distributed load $q(x)$ about Q again vanishes.

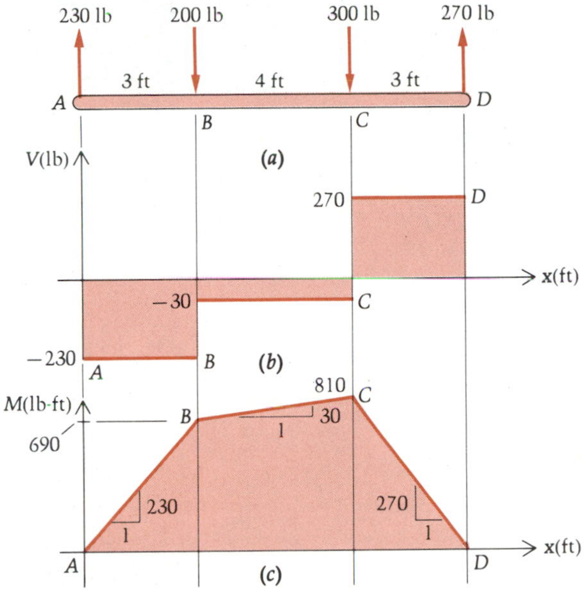

Figure E5.25a,b,c

Another change in slope occurs in the M diagram, this time at point C. To the right of this point, the slope is $-(270) = -270$ lb. The change in M between C and D is $-(270)(3\text{ ft}) = -810$ lb-ft. Thus the value of M at the right end of the beam is $810 + (-810) = 0$. This result serves as a check on our diagram (just as we had with the shear diagram examples) because the value of M at $x = 10$ ft is indeed zero at a roller with no external couple applied on the end of the beam.

Example 5.26

Draw the moment diagram (M versus x) for the beam of Example 5.24 (see Figure E5.26a), which had the shear diagram depicted in Figure E5.26b.

Solution

We now detail the steps we underwent in sketching the moment diagram (M versus x) in Figure E5.26c.

1. There is no moment applied to the left end of the beam, and the pin cannot transmit a moment to the beam there. Thus the moment diagram starts at zero.

2. The shear force is -3 kN between $x = 0$ and $x = 3$ m. Thus the slope of the M diagram equals a constant $+3$ kN/m (it is minus the value of V) for $3 > x \geq 0$. Thus at $x = 3$, $M = 3(3) = 9$ kN · m.

3. At $x = 3$, the concentrated 2-kN force does not produce a discontinuity in the M diagram. However, as soon as we pass over $x = 3$ m, the negative of the value of V (and hence the slope of M vs. x) changes to 1 kN. Over three more meters, this positive slope causes a further rise in M of $1(3) = 3$ kN · m, which

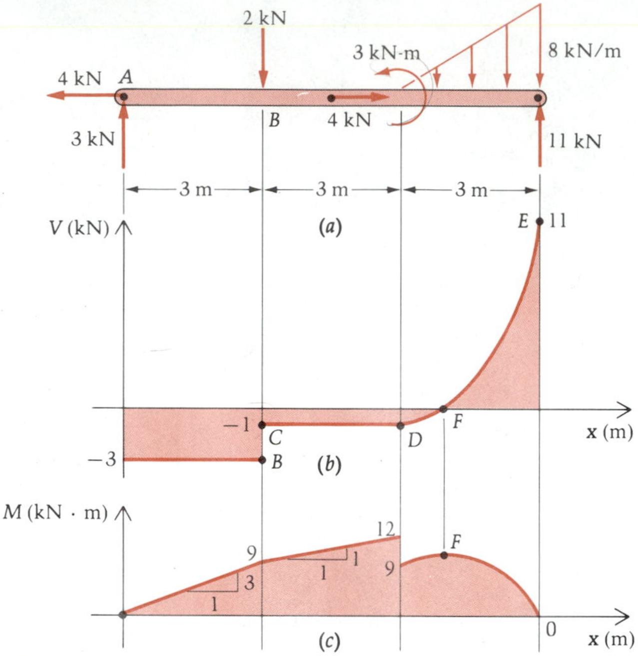

Figure E5.26a,b,c

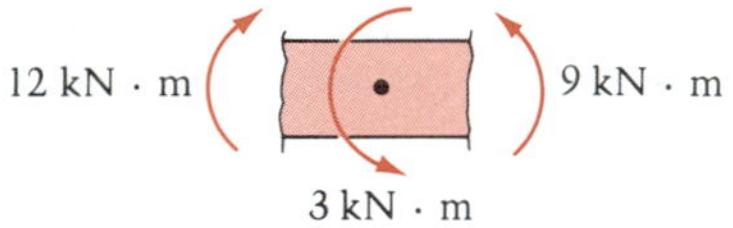

Figure E5.26d

equals the negative of the area under V vs. x between $x = 3$ and $x = 6$. Thus M, at $x = 6$, is $9 + 3 = 12$ kN · m.

4. As we pass over $x = 6$ m, the moment diagram must experience a discontinuity because of the concentrated couple applied there. The moment on the right-hand face just past $x = 6$ must drop by 3 kN · m, as shown in Figure E5.26d. Thus the moment just to the right of $x = 6$ m is $12 - 3 = 9$ kN · m.

5. Between D and F on the shear diagram of Figure E5.26b the values of the shear V get less and less negative, ending at zero. Thus the slope of M must get less and less positive ending at zero because $dM/dx = -V$; the M diagram is thus sketched with a horizontal tangent at F.

6. Between F and E, the shear gets more and more positive, ending at 11; thus dM/dx, the slope of the moment diagram, gets more and more negative, ending with a slope value of -11 at the right end of the beam.

Just as with the shear diagram, there is an automatic check available for the moment diagram. We simply see if the computed value of M at the right end of the beam is the actual (known) value there. In this case, we know that M must be zero at $x = 9$ m. To see if it calculates out to be zero, we need the area beneath V vs. x in the interval $6 < x < 9$. In that interval,

$$q = -\frac{8}{3}(x - 6)$$

so that, integrating,

$$V = -\int q\,dx = \frac{4(x-6)^2}{3} + C_1 = \frac{4(x-6)^2}{3} - 1$$

where C_1 is found by the condition $V = -1$ at $x = 6$. Integrating a second time,

$$M = -\int V\,dx = -\frac{4(x-6)^3}{9} + x + C_2$$

$$= -\frac{4(x-6)^3}{9} + x + 3,$$

because $M = 9$ at $x = 6^+$ (just to the right of $x = 6$). Thus the change in M that we need is

$$\left[-\frac{4(x-6)^3}{9} + x + 3\right]_6^9 = -12 + 9 + 3 + 0 - 6 - 3 = -9$$

This result gives a change in M of -9 from its $+9$ value, and it ends at zero at $x = 9$ m correctly. Alternatively, we could simply substitute into $M(x)$ to get $M = 0$ at $x = 9$ m:

$$M = -\frac{4(9-6)^3}{9} + 9 + 3 = 0$$

Or finally, if we are just interested in the change in M between $x = 6$ and 9 m, we could integrate the function $-V$ with definite integral limits:

$$\Delta M = -\int_{x=6}^{9} V\,dx = -\int_{x=6}^{9}\left[\frac{4(x-6)^2}{3} - 1\right]dx$$

$$= -\left[\frac{4(x-6)^3}{9} - x\right]_6^9 = -[12 - 9 - (0 - 6)] = -9$$

as before.

It is also very important to know the value of M at a point such as F where M is a local maximum. Maxima of moment will often correspond to locations of highest stress in sections of the beam. At F, the shear is zero:

$$V = \frac{4(x-6)^2}{3} - 1 = 0$$

so that

$$(x_F - 6)^2 = 0.75$$

$$x_F = 6.87 \text{ m}$$

Now that F has been located, we can find the bending moment there:

$$M_F = \frac{-4(6.87-6)^3}{9} + 6.87 + 3 = 9.58 \text{ N} \cdot \text{m}$$

which is less than the 12 N · m at $x = 6$ m, but which is still a local maximum.

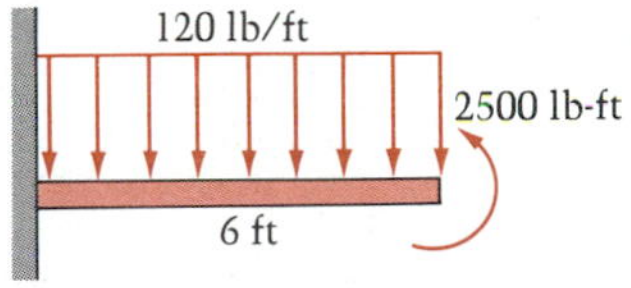

Figure E5.27a

Example 5.27

For the beam of Example 5.21 (see Figure E5.27a), draw the shear and moment diagrams.

Solution

We could, of course, plot the curves $V = 120x - 720$ and $M = -60x^2 + 720x + 340$ from Example 5.21. However, we shall work this problem as though we did not know these relationships, and then use them as a check.

The free-body diagram will help us find the reactions at the wall. Note that for "statics purposes" (i.e., use of equilibrium equations for the whole beam) we may replace the distributed load by the "force-alone resultant" of 720 lb as indicated in Figure E5.27b. The equations are:

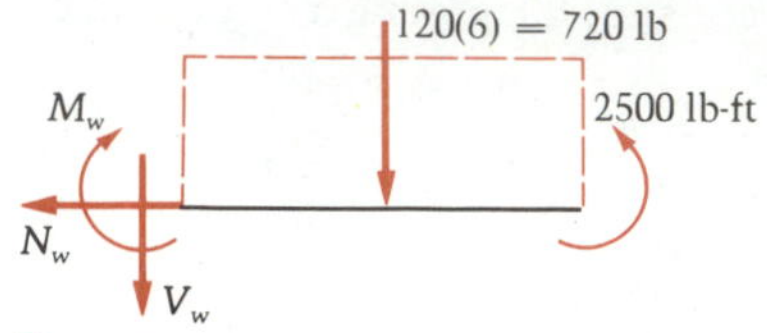

Figure E5.27b

$$\overset{+}{\rightarrow} \quad \Sigma F_x = 0 = N_w$$

$$+\uparrow \quad \Sigma F_y = 0 = -V_w - 720 \Rightarrow V_w = -720 \text{ lb}$$

$$\circlearrowleft + \quad \Sigma M_A = 0 = 2500 - 720(3) - M_w \Rightarrow M_w = 340 \text{ lb-ft}$$

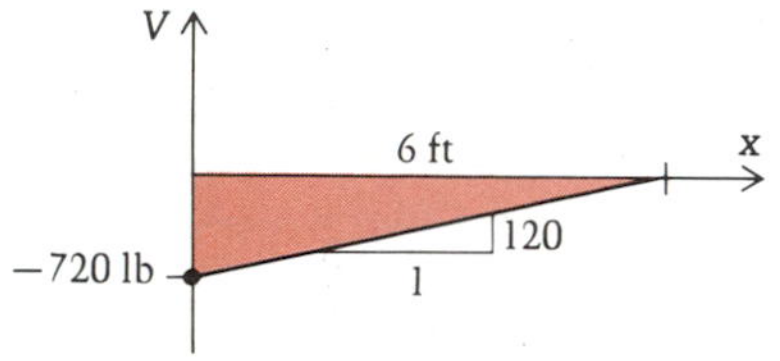

Figure E5.27c

Thus the shear diagram (Figure E5.27c) starts at a value of -720 lb when $x = 0$. The distributed load $q(x)$ is a constant (-120 lb / ft), hence the slope of V will be constant too ($V' = -q = 120$). Thus the value reached by the curve at the free end is $V = -720 + 120(6) = 0$, which checks because there is no shear force on the cantilever beam at $x = 6$ ft. Note also that the change in V from $x = 0$ to 6 ft is the negative of the area beneath the $q(x)$ diagram — that is, $-(-120)6 = +720$.

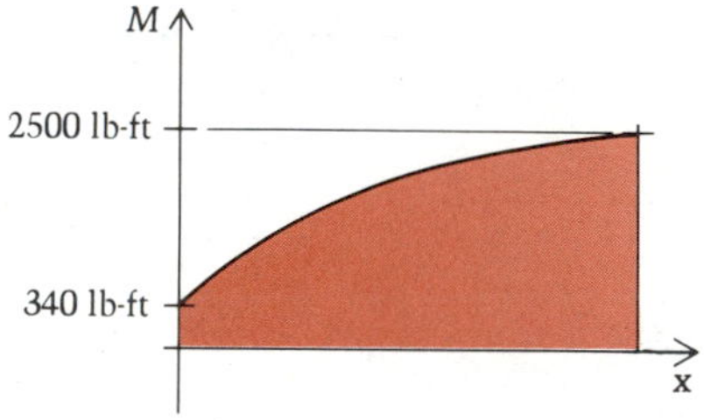

Figure E5.27d

For the moment diagram, we begin with the wall moment of $+340$ lb-ft (see Figure E5.27d). As x increases, the values of V become less and less negative. Thus the slopes of M versus x become less and less positive. The *value* of M changes over the beam by the negative of the area beneath the V diagram, which is $-[\frac{1}{2}(6)(-720)] = 2160$ lb-ft. Therefore $M|_{x=6\text{ft}} = 340 + 2160 = 2500$ lb-ft, which checks the applied moment at the free end.

Note that the V vs. x curve has the equation $V = 120x - 720$, which checks the result in Example 5.21. The M vs. x curve is a parabola (it is second order in x because it is the integral of a straight line equation) with vertex at $x = 6$. Thus $M - 2500 = -k(x - 6)^2$, and the constant is evaluated using $M = 340$ when $x = 0$:

$$340 - 2500 = -k(-6)^2 \Rightarrow k = 60 \text{ lb / ft}$$

Thus

$$M = 2500 - 60(x - 6)^2$$
$$= -60x^2 + 720x + 340 \text{ lb-ft}$$

This result also agrees with Example 5.21.

EXAMPLE 5.28

Draw the shear and moment diagrams for the beam loaded as shown in Figure E5.28a.

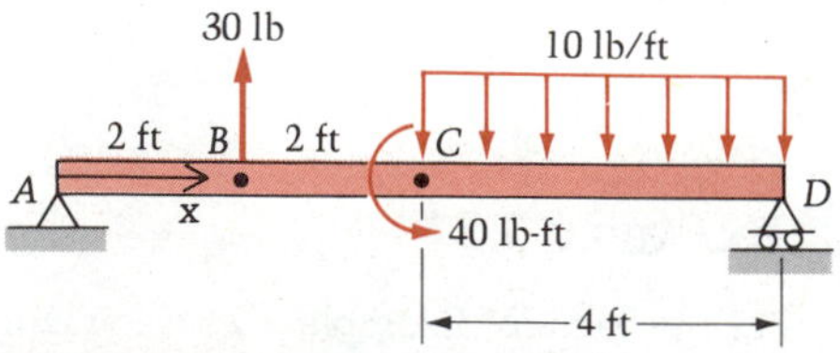

Figure E5.28a

Solution

First we find the reactions. The uniformly distributed load is replaced by its force-alone resultant, which has:

a. Magnitude = area beneath loading curve
= (10 lb / ft)(4 ft) = 40 lb

b. Location 2 ft to the left of D.

Thus we have the following free-body diagram (Figure E5.28b) and equilibrium equations:

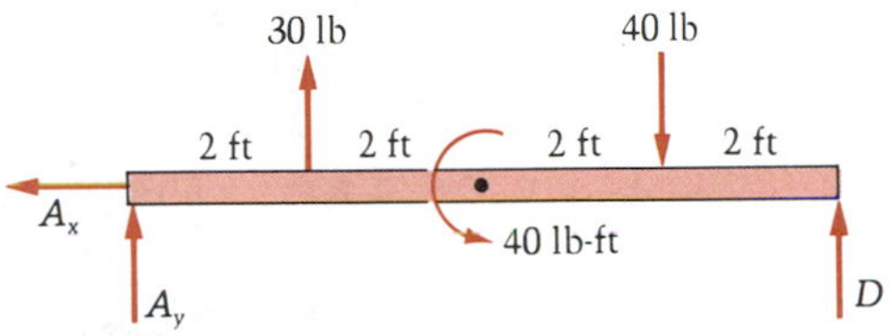

Figure E5.28b

$$\overset{+}{\leftarrow} \quad \Sigma F_x = 0 \Rightarrow A_x = 0$$

$$+\uparrow \quad \Sigma F_y = 0 = A_y + 30 + D - 40$$

$$\circlearrowleft + \quad \Sigma M_A = 0 = (30 \text{ lb})(2 \text{ ft}) + 40 \text{ lb-ft} - (40 \text{ lb})(6 \text{ ft}) + (8 \text{ ft})D$$

$$D = 17.5 \text{ lb} \tag{1}$$

Thus from Equation (1), $A_y = -7.5$ lb.

Question 5.16 How might the reactions be checked before proceeding?

Here are the steps we have followed in constructing the V and M diagrams on the next page:

1. Redraw the original loading diagram, being sure to return to the distributed load — i.e., do *not* use the force-alone resultant (see Figure E5.28c).

Question 5.17 Why not stick with the force-alone resultant?

2. Draw the axes beneath one another.

Question 5.18 What is the advantage of this?

Answer 5.16 By $\Sigma M_D = 0$, we get the same answer for A_y.

Answer 5.17 The force-alone resultant makes the same contribution to the equilibrium equations, hence giving the correct external reactions. But it does not create the same internal force distributions within the beam!

Answer 5.18 Whenever $q = 0$, V will have a zero slope; whenever $V = 0$, M will have a zero slope. Thus if the curves are drawn beneath one another, these locations can be projected down to advantage.

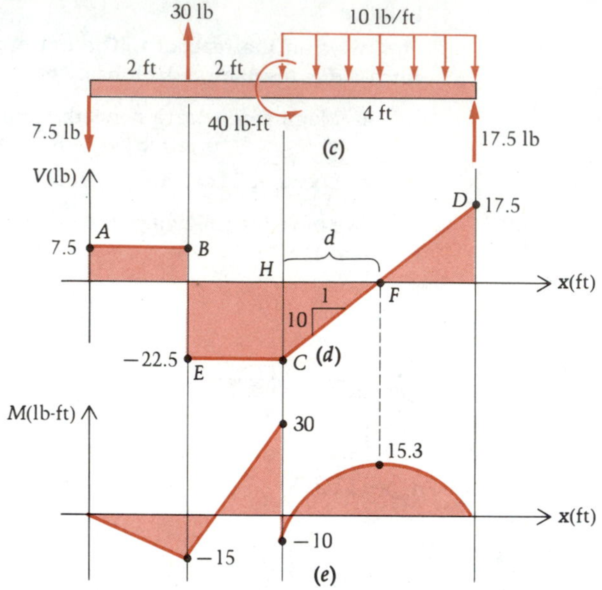

Figure E5.28c,d,e

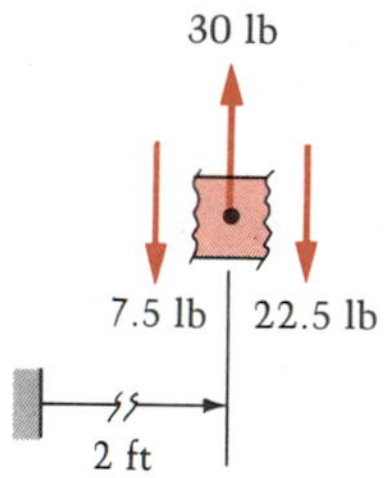

Figure E5.28f

3. Start (Figure E5.28d) at $V = +7.5$. (Recall that a downward force on a left-hand face is positive!)

4. Since $q = 0$ between $x = 0$ and 2 ft, draw the line AB with $dV/dx = 0$ as shown.

5. The shear must decrease by 30 as we cross $x = 2$ ft to equilibrate the 30-lb concentrated force there. Thus put a dot at E, which has a V coordinate of $7.5 - 30 = -22.5$ lb; see Figure E5.28f at the left which shows a small element at $x = 2$ ft.

6. Again draw a horizontal line, EC this time, because $dV/dx = 0$ in the interval $2 < x < 4$ ft.

7. Note that the couple has no effect on V at C (zero resultant force!).

8. Because $dV/dx = -q$, and because $q(x) = -10$ for $4 < x < 8$ ft, the slope of the V diagram between $x = 4$ and 8 ft is $+10$ lb/ft. Thus it ends at $V = -22.5 + 10(4) = +17.5$ lb.

9. The check is that V really is $+17.5$ lb at $x = 8$ ft because the reaction is 17.5 lb ↑, and an upward force on a right-hand face is positive shear.

10. Preparatory to discussing the moment diagram, let us find the location of the point F where V is zero. This will correspond to a horizontal tangent in the M diagram. By the similar triangles CDG and CFH, we have

$$\frac{22.5}{d} = \frac{10}{1}$$

so that

$$d = 2.25 \text{ ft}$$

11. The moment M (see Figure E5.28e) is zero at $x = 0$ (the pin cannot transmit a moment, and none is externally applied directly to the beam).

12. $dM/dx = -V = -7.5$ lb between $x = 0$ and 2 ft, so that M reaches a value of -15 lb-ft at $x = 2$ ft.

13. Between $x = 2$ and 4 ft, the slope of M changes to the negative of the new value of V, or $M' = +22.5$. The change in M from $x = 2$ ft to $x = 4$ ft is the negative of the area beneath V vs. x in that interval, or (22.5 lb)(2 ft) = 45 lb-ft. Thus M at $x = 4$ ft is $-15 + 45 = 30$ lb-ft.

14. At $x = 4$ ft, the moment diagram drops by the amount of the 40 lb-ft couple; see Figure E5.28g below of a small element at $x = 4$ ft:

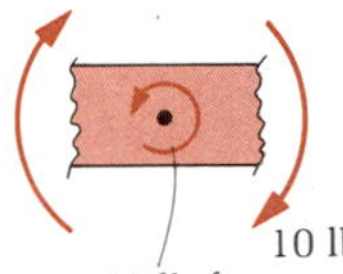

Figure E5.28g

Therefore, at $x = 4^+$ ft, M is $30 - 40 = -10$ lb-ft.

15. As we move from H to F(see the V diagram), the values of V are negative numbers heading toward zero. The slopes of M vs. x, being equal to $-V$, are thus positive numbers heading toward zero. The change in M is

$$-\begin{pmatrix}\text{Area of } V \text{ diagram}\\ \text{between } H \text{ and } F\end{pmatrix} = \frac{1}{2}(2.25 \text{ ft})(22.5 \text{ lb})$$

$$= 25.3 \text{ lb-ft}$$

Thus M at F is $-10 + 25.3 = 15.3$ lb-ft.

16. From F to D, V is now getting more and more positive, ending at $+17.5$ lb. The slope of M thus gets more and more negative, and the change in M is, this time,

$$-\begin{pmatrix}\text{Area of } V \text{ diagram}\\ \text{between } F \text{ and } D\end{pmatrix} = -\frac{1}{2}(4 - 2.25) \text{ ft } (17.5 \text{ lb})$$

$$= -15.3 \text{ lb-ft}$$

Thus M at $x = 8$ ft is $15.3 - 15.3 = 0$, as it should be because no moment is applied to the right end of the beam externally or by the roller.

We note that the maximum magnitude of moment M is 30 lb-ft just to the left of $x = 4$ ft; the maximum magnitude of shear force V is 22.5 lb throughout the interval $2 < x < 4$ ft.

The Other Sign Convention for Shear Force

In closing, we note that some prefer to define the shear force as positive if downward on a right-hand face. This convention, although not in agreement with definitions used in the theory of elasticity, has the advantage that the Equations (5.11) and (5.12) become $V' = +q$ and $M' = +V$. This makes it easier to construct loading, shear, and moment diagrams using the other curves, because one needn't keep up with the minus signs. The only difference between the diagrams drawn with V positive according to ↑□↓ and our convention ↓□↑ is that the shear diagram is "flipped," as shown in Figure 5.49:

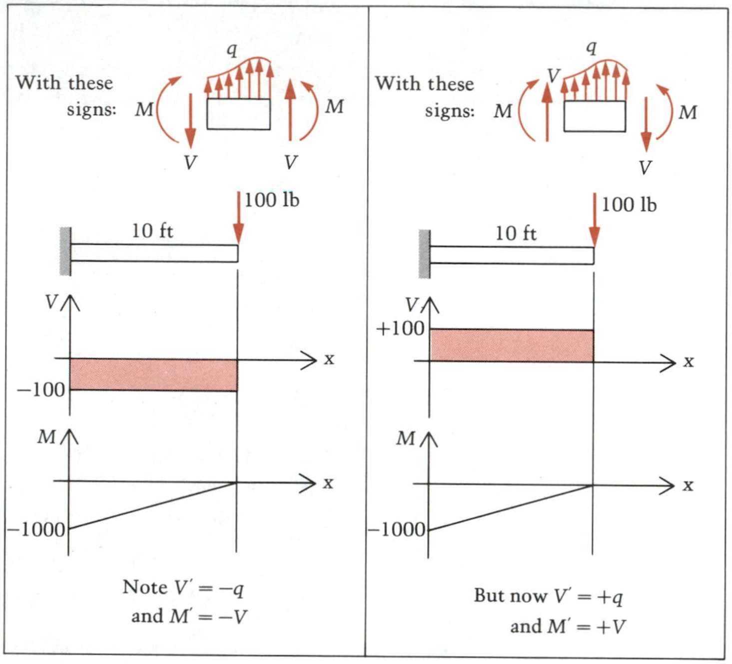

Figure 5.49

Note that the moment diagram is unaffected by the change in the sign of V.

PROBLEMS ▶ Section 5.10

5.180 A distributed load in a beam varies linearly from zero on each end to a maximum value of 8000 N / m in the center, as shown in Figure P5.180. Draw the shear and moment diagrams for this beam.

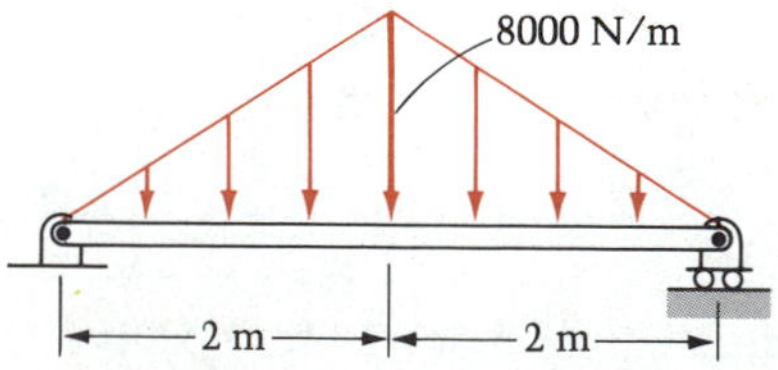

Figure P5.180

In Problems 5.181–5.185 draw the shear and bending moment diagrams for the beams loaded as shown.

5.181

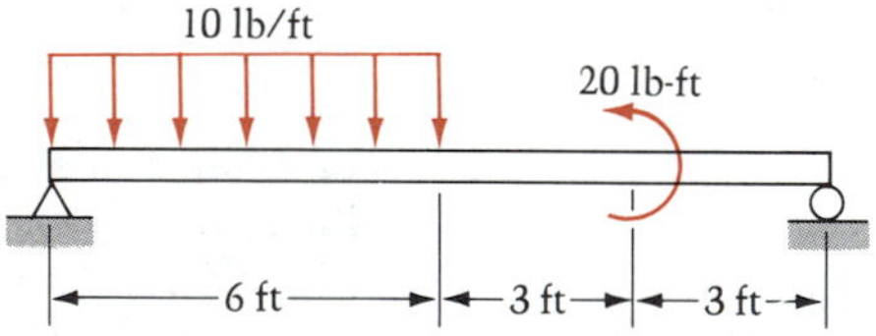

Figure P5.181

5.182

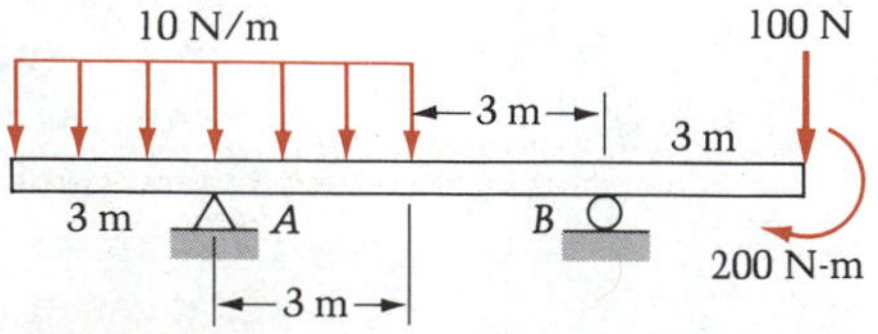

Figure P5.182

5.183

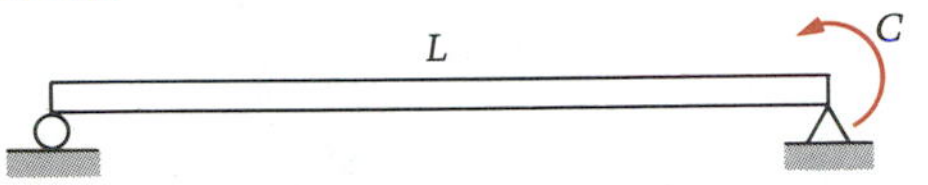

Figure P5.183

5.184

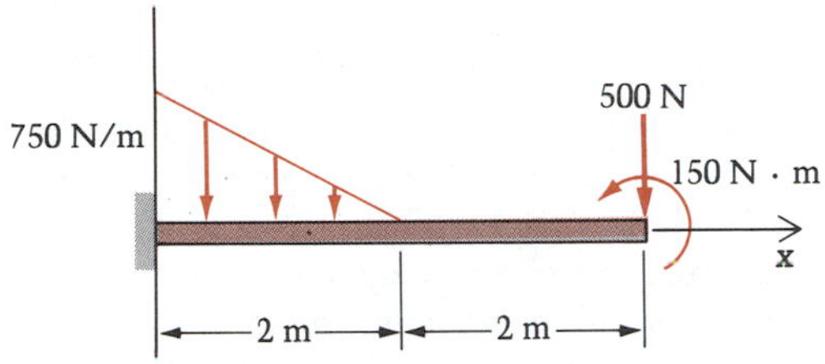

Figure P5.184

5.185

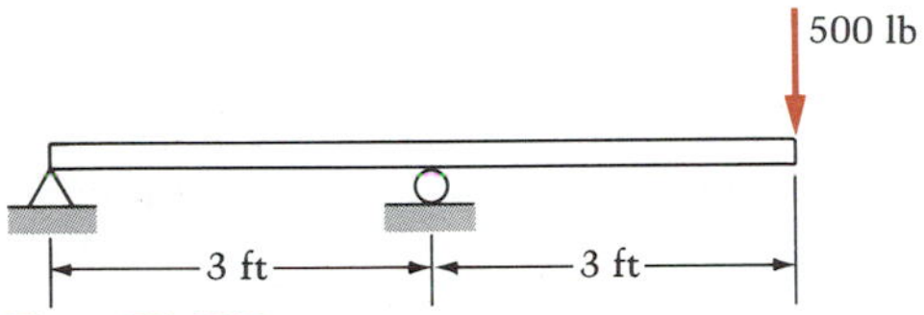

Figure P5.185

5.186 Draw the shear and moment diagram for the beam shown in Figure P5.186.

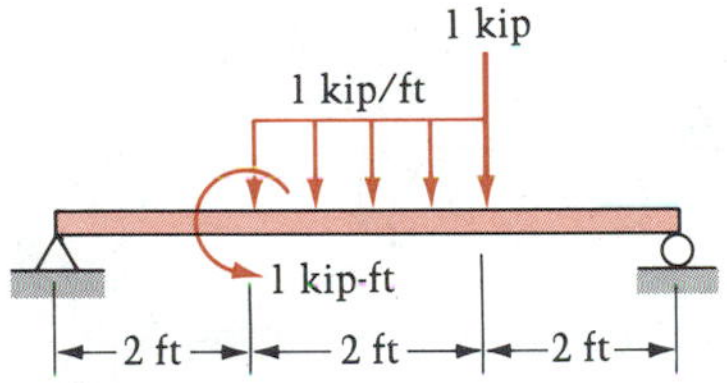

Figure P5.186

5.187 Draw the shear and bending moment diagrams for the beam shown in Figure P5.187. Select an appropriate free-body and write the equations for shear and bending moment for $2 < x < 6$ m.

* **5.188** In Figure P5.188:

a. Draw shear and moment diagrams for the beam shown when $a = 6$ ft. Indicate the values of all pertinent ordinates.

b. Where should the roller be located ($a = ?$) in order that the greatest magnitude of bending moment in the beam be as small as possible?

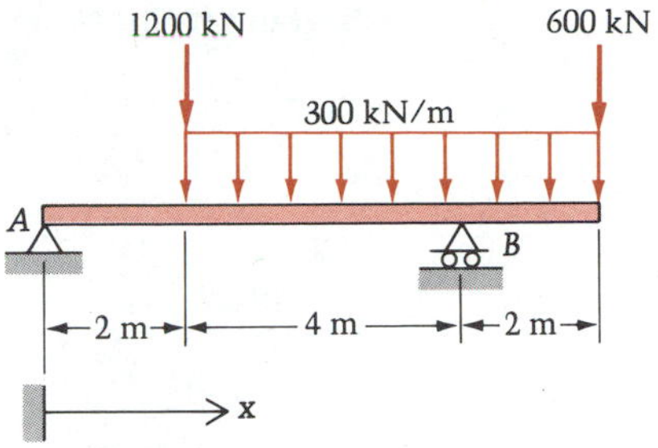

Figure P5.187

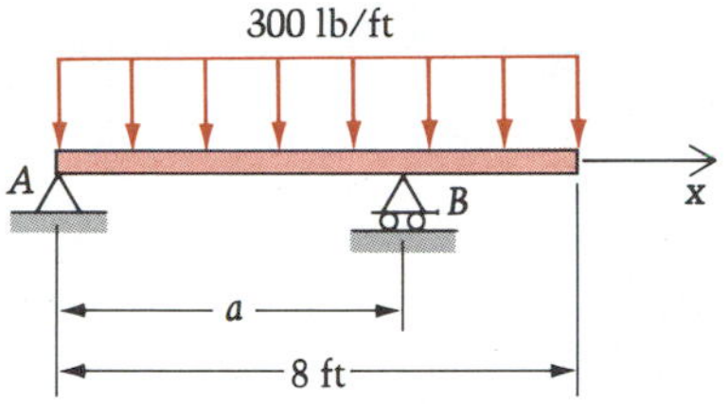

Figure P5.188

5.189 Draw the shear and bending moment diagrams for the beam shown in Figure P5.189.

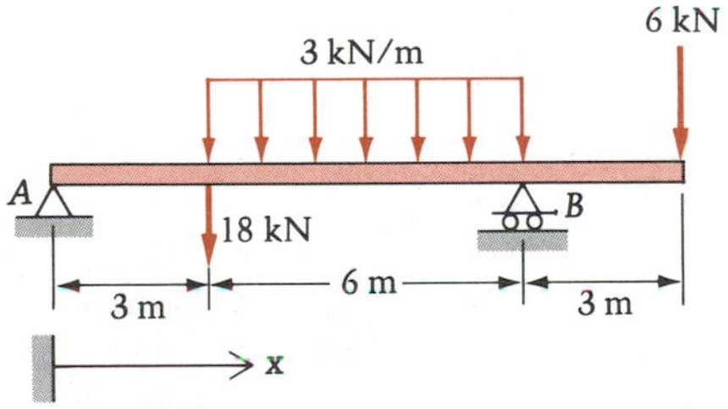

Figure P5.189

5.190 Assuming the upward reaction of the ground to be uniformly distributed, draw the shear and moment diagrams for the beam shown in Figure P5.190.

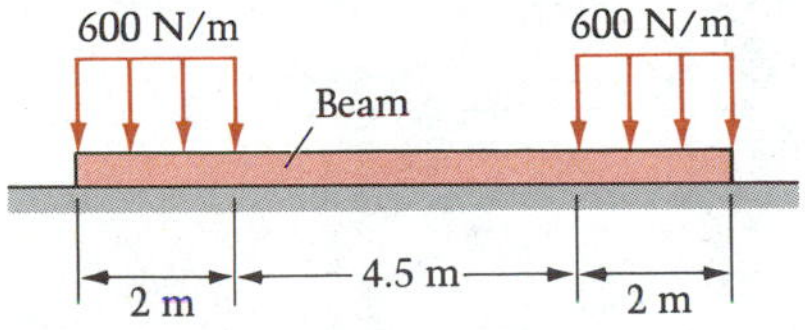

Figure P5.190

5.191 Draw the shear and moment diagrams for the beam in Figure P5.191.

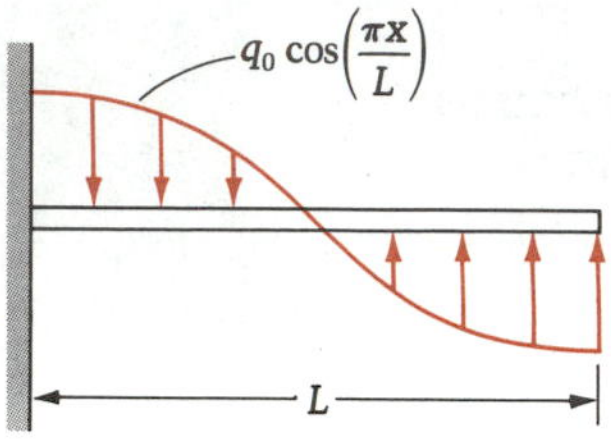

Figure P5.191

Draw shear and moment diagrams for the beams shown in Problems 5.192–5.195.

5.192

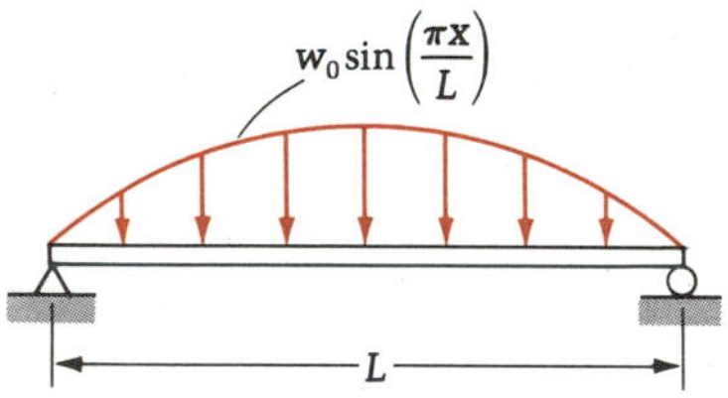

Figure P5.192

5.193

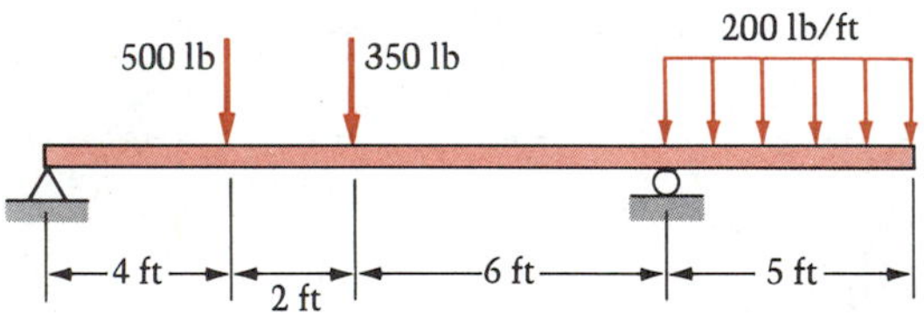

Figure P5.193

5.194

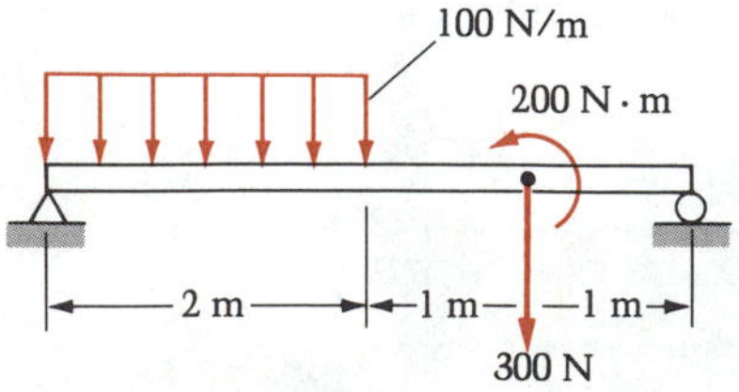

Figure P5.194

5.195

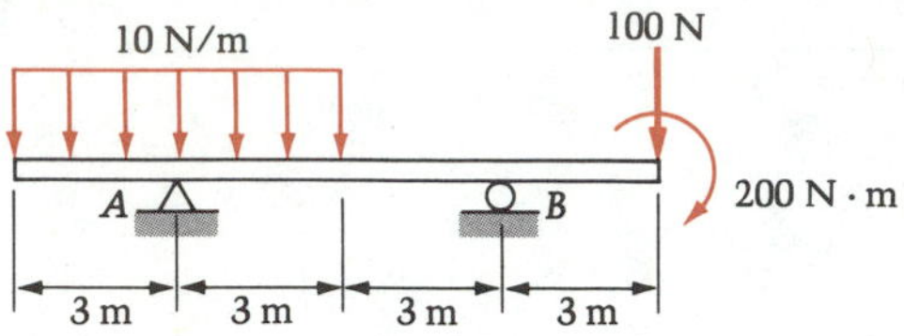

Figure P5.195

* **5.196** The simply supported beam of length L in Figure P5.196 is loaded as indicated with a uniformly distributed moment of intensity μ per unit length. Draw the shear and bending moment diagrams. *Hint:* $dM/dx = -V$ does not apply since distributed moments were not included in its derivation.

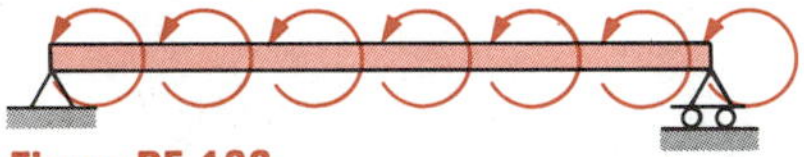

Figure P5.196

5.197 *ABE* in Figure P5.197 is a single continuous member. Draw the shear and moment diagrams for the portion *AB* of the member.

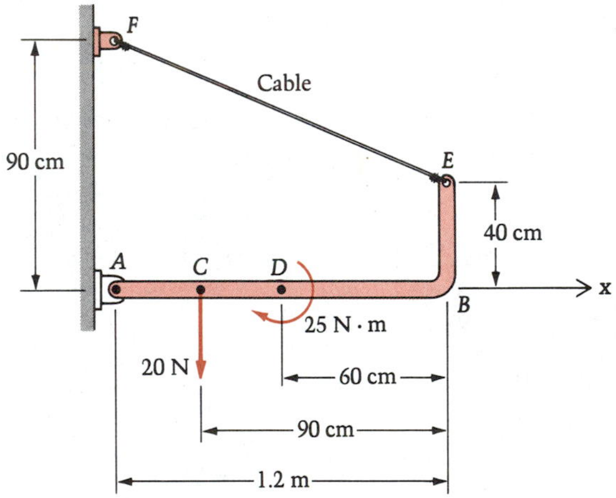

Figure P5.197

* **5.198** In Figure P5.198 the uniform rod has weight W and rests on two props as shown, with $b > a$. Draw the shear and bending moment diagrams and show that the largest bending moment is at B if $a\sqrt{2} > b$.

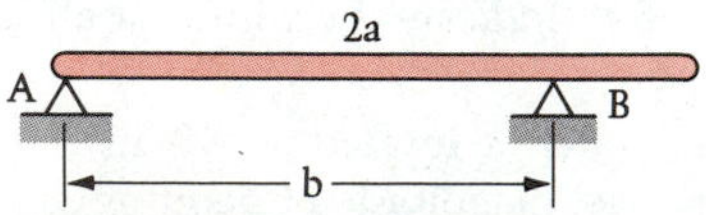

Figure P5.198

III / CABLES

5.11 Parabolic and Catenary Cables

Vertical Loadings Depending upon the Horizontal Coordinate

The flexible cable is yet another common method of supporting loads. For example, the suspension bridge has been used for many centuries and is perhaps the best example of the engineering use of cables. We shall first consider a cable suspended in a plane between the two points A and B, as indicated in Figure 5.50. Let us investigate the shape of the curve in which the cable will hang, under the action of a loading $q(x)$ that is everywhere vertical. In our analysis of cables, we shall take $q(x)$ to be positive *downward.*

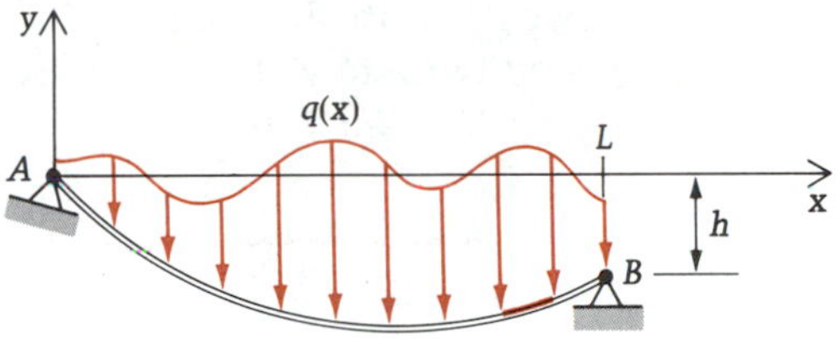

Figure 5.50

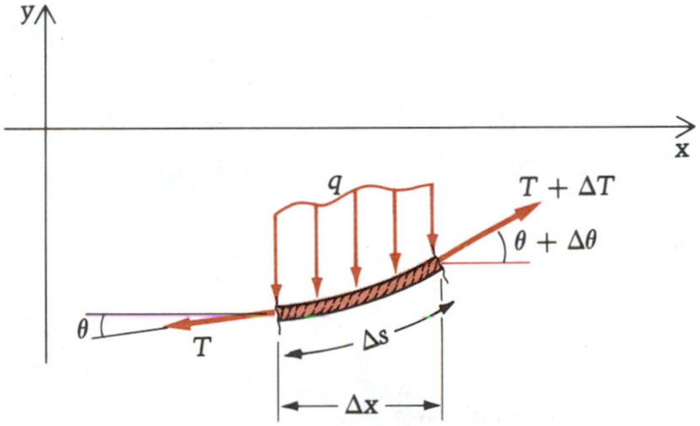

Figure 5.51 Free-body diagram of cable element.

We consider the equilibrium of a small element of such a cable, as shown in Figure 5.51. Summing forces,

$$\xrightarrow{+} \qquad \Sigma F_x = 0 = (T + \Delta T)\cos(\theta + \Delta\theta) - T\cos\theta = 0$$

Dividing by Δx and taking the limit as $\Delta x \rightarrow 0$, we see that

$$\frac{d(T\cos\theta)}{dx} = 0$$

which implies that

$$T\cos\theta = \text{constant} = T_H \tag{5.13}$$

Thus the horizontal component of cable tension does not change from one end to the other. Continuing, if $q(x)$ is constant,*

$$+\uparrow \qquad \Sigma F_y = 0 = (T + \Delta T)\sin(\theta + \Delta\theta) - T\sin\theta - q\,\Delta x$$

Dividing again by Δx and letting Δx approach zero, we get this time

$$\frac{d(T\sin\theta)}{dx} = q \tag{5.14}$$

and substituting $T = T_H/\cos\theta$ from Equation (5.13),

$$T_H\frac{d(\tan\theta)}{dx} = q \tag{5.15}$$

or because $\tan\theta = dy/dx = y'$,

$$y'' = \frac{q}{T_H} \tag{5.16}$$

If we know the loading as a function of x, this equation may then be integrated twice to yield the cable deflection, as shown in the following example.

EXAMPLE 5.29

In the foregoing theory, let $q(x) = \text{constant} = q_0$ and let h (Figure 5.50) be zero; that is, let the suspension points lie on the same horizontal level. Compute the deflection of the cable and the tension in it as functions of x. Also find the length of the cable for a given sag H and span L. (See Figure E5.29b.)

Solution

This problem is similar to that of a suspension bridge held up by many vertical cables connected to the main cable as shown in Figure E5.29a. The more vertical

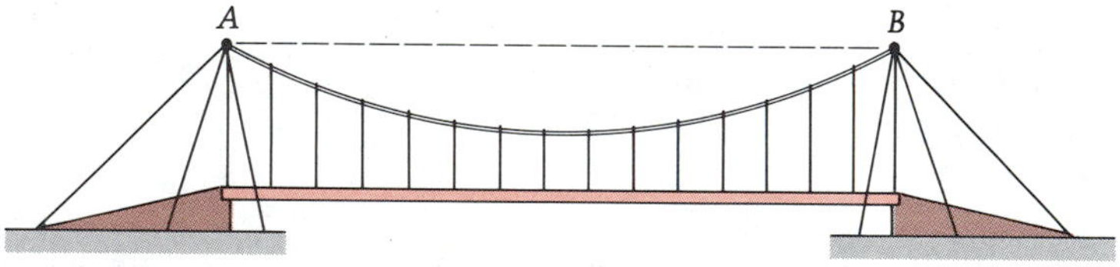

Figure E5.29a

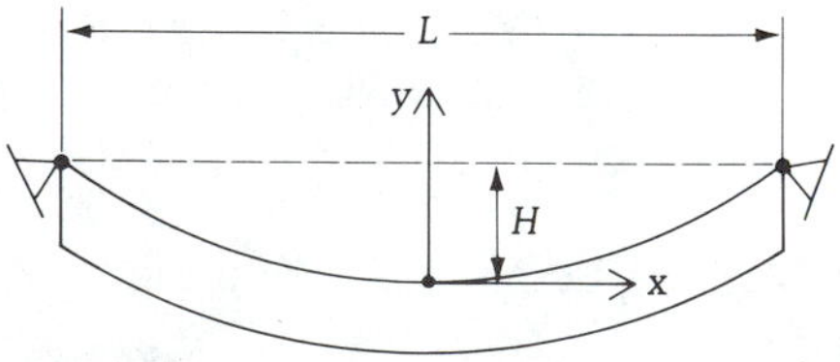

Figure E5.29b

* If $q(x)$ is not constant, the resulting Equations (5.14)–(5.16) are the same. The proof uses the mean-value theorem as described in the footnote on page 373.

cables there are, the closer we are to the distributed loading $q(x)$ that was acting on the main cable in the preceding text.

The change to new coordinates shown in Figure E5.29b will allow us to take advantage of symmetry when we evaluate our integration constants. Integrating Equation (5.16),

$$y' = \frac{q_0 x}{T_H} + \cancel{C_1} \quad \text{(0, because } y' = 0 \text{ at } x = 0) \tag{1}$$

And integrating again,

$$y = \frac{q_0 x^2}{2T_H} + \cancel{C_2} \quad \text{(0, because } y = 0 \text{ at } x = 0) \tag{2}$$

Thus this cable loading results in a parabolic deflection shape.

To find the tension, Equations (5.13) and (5.14) give [with $q = q_0$ in Equation (5.14)]

$$T \cos\theta = T_H = \text{constant} \tag{3}$$

$$T \sin\theta = q_0 x + \cancel{C_3} \quad \text{(0, because } \theta = 0 \text{ at } x = 0) \tag{4}$$

Squaring and adding,

$$T = \sqrt{T_H^2 + q_0^2 x^2} \tag{5}$$

The tension is seen to be minimum at the lowest point, where $x = 0$:

$$T_{\text{MIN}} = T_H \tag{6}$$

and maximum at the support points A and B, where $x = L/2$:

$$T_{\text{MAX}} = \sqrt{T_H^2 + q_0^2 L^2/4}^{*} \tag{7}$$

If we wish, we can express T_H in terms of the sag H, at the lowest point of the cable, by Equation (2):

$$H = \frac{q_0 L^2}{8T_H} \Rightarrow T_H = \frac{q_0 L^2}{8H} \tag{8}$$

Question 5.19 Does Equation (8) make sense in the limiting case of large values of q_0, L, and T_H? How about small values?

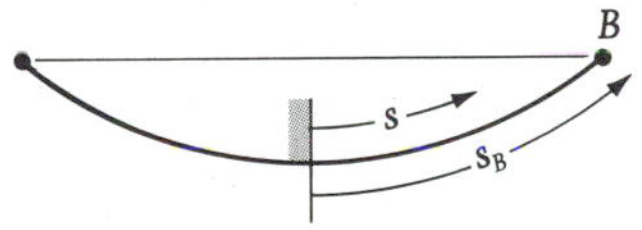

s = arclength

Figure E5.29c

Thus, alternatively,

$$T_{\text{MIN}} = \frac{q_0 L^2}{8H} \quad \text{and} \quad T_{\text{MAX}} = \frac{q_0 L}{2}\sqrt{1 + \frac{L^2}{16H^2}} \tag{9}$$

The required length of the cable for given endpoints and sag is also of interest. It is (see Figure E5.29c):

$$\ell = 2\int_{s=0}^{s_B} ds = 2\int_{s=0}^{s_B} \sqrt{dx^2 + dy^2} = 2\int_{x=0}^{L/2} \sqrt{1 + y'^2}\, dx$$

* Note that $q_0 L$ is the total load supported by the cable.

Answer 5.19 Yes. The larger the values of loading q_0 and length L, the larger the deflection. And the larger the tension, the smaller the deflection. The reverses also make sense, with each producing the opposite effect on deflection.

Using Equations (1) and (8),

$$\ell = 2\int_0^{L/2} \sqrt{1 + \left(\frac{8Hx}{L^2}\right)^2}\, dx \tag{10}$$

Using integral tables,*

$$\ell = \left[\frac{L^2}{8H}\sinh^{-1}\frac{8Hx}{L^2} + x\sqrt{1 + \left(\frac{8Hx}{L^2}\right)^2}\right]\Bigg|_{x=0}^{L/2}$$

$$\ell = \frac{L^2}{8H}\sinh^{-1}\frac{4H}{L} + \frac{L}{2}\sqrt{1 + \left(\frac{4H}{L}\right)^2} \tag{11}$$

Defining $\mathcal{H}$ to be the ratio H/L (sag-to-span),

$$\ell = \frac{L}{2}\left[\frac{1}{4\mathcal{H}}\sinh^{-1}(4\mathcal{H}) + \sqrt{1 + (4\mathcal{H})^2}\right] \tag{12}$$

For example, if $\mathcal{H} = H/L = 0.2$, then $\sinh^{-1} 0.8 \approx 0.733$ and $\ell = 1.10L$. Thus the cable is about 10% longer than the span in this case.

Vertical Loadings Depending upon Arclength Along the Cable

Even if the cable in the preceding example is not supported with A and B (Figure 5.50) at the same level, the solution is still a parabolic curve as long as the distributed load, viewed as a function of x, is uniform. As just mentioned, this is the case for a suspension bridge.† But what if the cable is an electric power transmission line, not supporting anything but its own weight and perhaps that of adhered ice in a winter storm? In this instance, the important difference is that q, while still a vertical loading, is now force per unit of *arclength* s along the cable, and not of its horizontal projection x. The cable now hangs in a different shape than a parabola, though the two curves are close together if the maximum of y is small in relation to the span L. Let us, then, determine the shape of the cable when $q = q(s)$.

If $q = q(s)$, then appyling $\Sigma F_x = 0$ to the element of Figure 5.51 still yields

$$T\cos\theta = \text{constant} = T_H \tag{5.17}$$

The y-equilibrium equation, however, changes:

$$+\!\uparrow \qquad \Sigma F_y = 0 = (T + \Delta T)\sin(\theta + \Delta\theta) - T\sin\theta - q(s)\,\Delta s$$

so that

$$\frac{(T\sin\theta)|_{x+\Delta x} - T\sin\theta|_x}{\Delta x} = q(s)\frac{\Delta s}{\Delta x} \tag{5.18}$$

* An approximate answer for ℓ may be obtained by expanding the radical using the binomial theorem, integrating each term, and truncating the series (which converges for $|8Hx/L^2| < 1$, that is, for $|y'| < 1$).

† Provided the suspended load far outweighs the cable itself.

Taking the limit,*

$$\frac{d(T \sin \theta)}{dx} = q(s) \frac{ds}{dx} \tag{5.19}$$

$$= q(s) \sqrt{1 + y^2} \frac{dx}{dx}$$

$$= q(s) \sqrt{1 + y'^2} \tag{5.20}$$

Substituting for T from Equation (5.17) as before,

$$T_H \frac{d(\tan \theta)}{dx} = q(s) \sqrt{1 + y'^2} \tag{5.21}$$

or

$$T_H y'' = q(s) \sqrt{1 + y'^2} \tag{5.22}$$

Therefore, we find that we must solve a different differential equation this time, which is

$$y'' = \frac{q}{T_H} \sqrt{1 + y'^2} \tag{5.23}$$

The solution to (5.23) is usually accomplished via a change of variable. We let y' be renamed

$$y' = p \tag{5.24}$$

so that

$$y'' = p' \tag{5.25}$$

Then Equation (5.23) becomes

$$\frac{p'}{\sqrt{1 + p^2}} = \frac{q}{T_H} \tag{5.26}$$

or

$$\frac{dp}{\sqrt{1 + p^2}} = \frac{q}{T_H} dx \tag{5.27}$$

Integrating (with integral tables), for the case when $q = q_0 =$ constant,†

$$\ln(p + \sqrt{1 + p^2}) = \frac{q_0}{T_H} x + K_1 \tag{5.28}$$

from which

$$\sqrt{1 + p^2} = e^{\left(\frac{q_0}{T_H} x + K_1\right)} - p \tag{5.29}$$

---

* See the footnote on page 373.

† This would be the case if the cable were both uniform prior to suspension and also practically inextensible.

Squaring,

$$1 + \cancel{p^2} = e^{\left(\frac{2q_0x}{T_H}+2K_1\right)} - 2pe^{\left(\frac{q_0x}{T_H}+K_1\right)} + \cancel{p^2} \tag{5.30}$$

Replacing p by y' and solving for this quantity,

$$y' = \frac{e^{\left(\frac{2q_0x}{T_H}+2K_1\right)} - 1}{2e^{\left(\frac{q_0x}{T_H}+K_1\right)}}$$

$$y' = \frac{e^{\left(\frac{q_0x}{T_H}+K_1\right)} - e^{-\left(\frac{q_0x}{T_H}+K_1\right)}}{2} \tag{5.31}$$

the right-hand side of which can be recognized as the hyperbolic sine. Therefore,

$$y' = \sinh\left(\frac{q_0x}{T_H} + K_1\right) \tag{5.32}$$

Integrating one last time,

$$y = \frac{T_H}{q_0}\cosh\left(\frac{q_0x}{T_H} + K_1\right) + K_2 \tag{5.33}$$

If we now again choose the origin to be the lowest point of the curve (see Figure 5.52), then $y = 0$ and $y' = 0$ at $x = 0$, so that

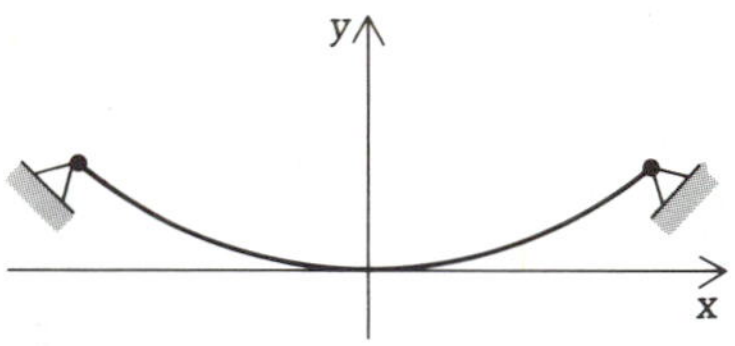

Figure 5.52

$$(1) \quad 0 = \frac{T_H}{q_0}\cosh K_1 + K_2 \tag{5.34}$$

and

$$(2) \quad 0 = \sinh K_1 \tag{5.35}$$

Equation (5.35) gives $K_1 = 0$, after which it follows from Equation (5.34) that $K_2 = -T_H/q_0$. Therefore,

$$y = \frac{T_H}{q_0}\left(\cosh\frac{q_0x}{T_H} - 1\right)^* \tag{5.36}$$

Nondimensionalizing the deflection, we obtain

$$\frac{q_0y}{T_H} = \cosh\frac{q_0x}{T_H} - 1 \tag{5.37}$$

Note that because

$$\cosh\left(\frac{q_0x}{T_H}\right) = 1 + \left(\frac{q_0x}{T_H}\right)^2 \Big/ 2! + \left(\frac{q_0x}{T_H}\right)^4 \Big/ 4! + \cdots \tag{5.38}$$

* The shape represented by this answer is often called the "catenary curve."

we have, if $q_0L/T_H \ll 1$ (where L is the span),

$$\frac{q_0 y}{T_H} \approx \left(1 + \frac{q_0^2 x^2}{2T_H^2}\right) - 1$$

or

$$y = \frac{q_0 x^2}{2T_H} \tag{5.39}$$

which was the solution to Example 5.29!

Example 5.30

The cable in Figure E5.30 holds the balloon in equilibrium with a tension at the ground attachment of 100 lb. If the cable weighs 0.25 lb/ft and is 150 ft long, what is the height H of the balloon?

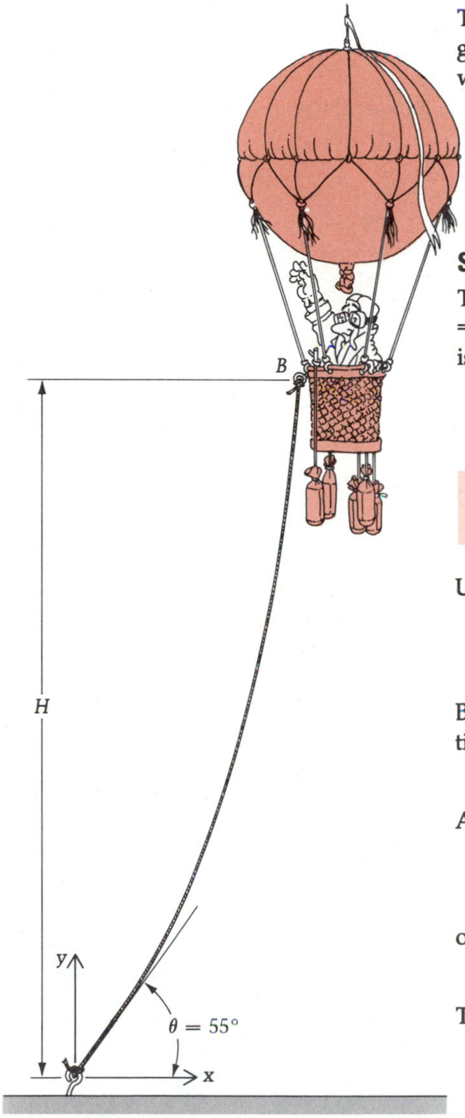

Figure E5.30

Solution

This is a "catenary cable" because it is loaded by its weight — that is, $q = q(s) = q_0$. Therefore we begin with the general solution for this class of cables, which is Equation (5.33):

$$y = \frac{T_H}{q_0} \cosh\left(\frac{q_0 x}{T_H} + K_1\right) + K_2 \tag{1}$$

Question 5.20 Why not start with Equation (5.36)?

Using the coordinates shown in the figure, we have $y = 0$ at $x = 0$, so that

$$0 = \frac{T_H}{q_0} \cosh K_1 + K_2 \tag{2}$$

Because $q_0 = 0.25$ lb/ft, and $T_H = \text{constant} = 100 \cos 55° = 57.4$ lb, Equation (2) becomes

$$K_2 = -230 \cosh K_1 \tag{3}$$

Also, $y' = \tan\theta = \tan 55°$ at $x = 0$, so that

$$\tan 55° = \left[\sinh\left(\frac{q_0 x}{T_H} + K_1\right)\right]\Bigg|_{x=0} \tag{4}$$

or

$$1.43 = \sinh K_1$$

Therefore, after evaluating the inverse hyperbolic sine,

$$K_1 = 1.16$$

Answer 5.20 Equation (5.36) is based on the condition $y = y' = 0$ at $x = 0$.

Thus from Equation (3),

$$K_2 = -403$$

And so we get

$$y = 230 \cosh\left(\frac{x}{230} + 1.16\right) - 403$$

The length ℓ of the cable is given by

$$\ell = 150 = \int_{s=0}^{150} ds = \int_{s=0}^{150} \sqrt{dx^2 + dy^2} = \int_{x=0}^{x_B} \sqrt{1 + y'^2}\, dx$$

where x_B is the x coordinate of the attachment (at B) to the balloon. Because

$$y' = \sinh\left(\frac{x}{230} + 1.16\right)$$

and because $1 + \sinh^2\theta = \cosh^2\theta$, the integral is

$$150 = \int_{x=0}^{x_B} \cosh\left(\frac{x}{230} + 1.16\right) dx = 230 \sinh\left(\frac{x}{230} + 1.16\right)\Big|_0^{x_B}$$

or

$$0.652 = \sinh\left(\frac{x_B}{230} + 116\right) - 1.44$$

or

$$\sinh^{-1} 2.09 = \frac{x_B}{230} + 1.16$$

Thus

$$1.48 = \frac{x_B}{230} + 1.16$$

or

$$x_B = 73.6 \text{ ft}$$

Therefore, the height of the balloon is

$$H = y\Big|_{x=73.6} = 230 \cosh\left(\frac{73.6}{230} + 1.16\right) - 403$$

$$H = 531 - 403 = 128 \text{ ft}$$

Question 5.21 Can we summarize the cable study by saying that we get the parabolic shape for small deflections and the catenary for large deflections?

Question 5.22 Would the analyses of this section apply to suspended chains?

Answer 5.21 No! The parabolic curve comes from a load uniformly distributed with respect to x; the catenary from a load uniformly distributed with respect to arclength s. In fact, transmission cables have *small* deflections but hang in the catenary shape, while suspension bridge cables have *large* deflections with $q(x) = q_0$.

Answer 5.22 Yes. The difference between cables and chains is that there is a small bending stiffness in a cable, and none in a chain; but because this stiffness was neglected, the results apply to chains. Of course, the lengths of (near-rigid) links must be very small compared to the span of the chain.

PROBLEMS ▶ Section 5.11

5.199 The cable in Figure P5.199 is to support a uniform load of 50 lb / ft between P and Q, with the lowest point B being 10 ft below the level of Q. Find the horizontal location of B and then the maximum cable tension.

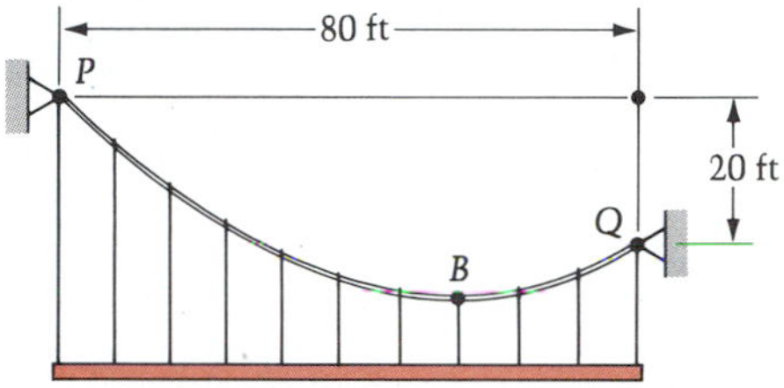

Figure P5.199

5.200 Two cables of the suspension bridge shown in Figure P5.200 symmetrically support a total load of 80×10^6 lb. The sag in the cables at mid-span is 500 ft. Find the mid-span tension and the angle θ the cables make with the tower. Neglect the cable weight.

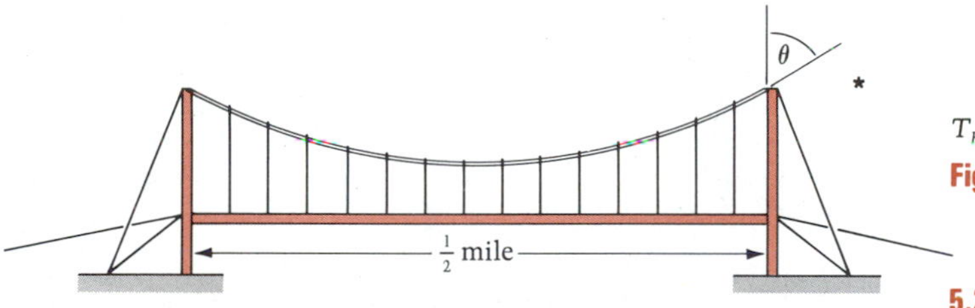

Figure P5.200

5.201 A clothesline is deflected in an ice storm as suggested by Figure P5.201. Assume that the ice load varies cosinusoidally with x, with intensity q_0 at $x = 0$. Neglect the weight of the clothesline and determine its deflection curve.

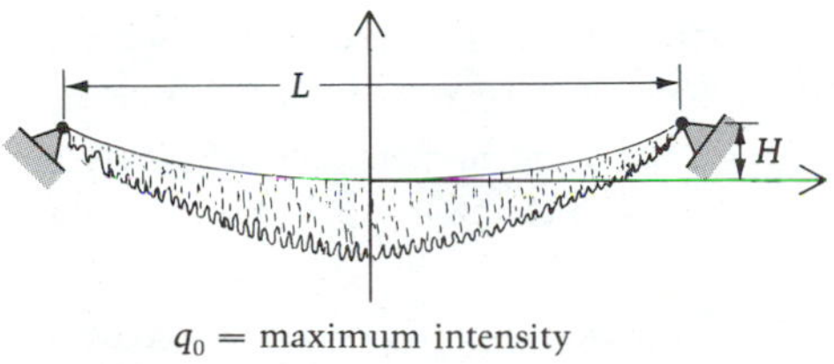

Figure P5.201

5.202 For the cable in Figure P5.202 carrying a uniform load $q(x) = q_0$, prove that at the supports, the tensions are

$$T_A = q_0 L \sqrt{1 + \frac{L^2}{4H_L^2}}$$

and

$$T_B = q_0 R \sqrt{1 + \frac{R^2}{4H_R^2}}$$

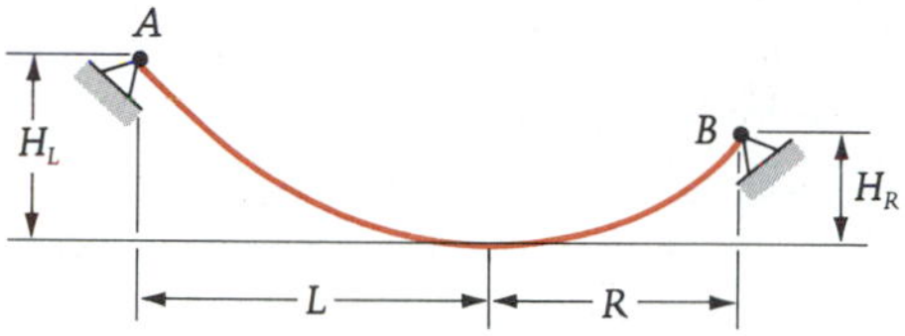

Figure P5.202

5.203 Obtain the result $T = \sqrt{T_H^2 + q_0^2 x^2}$ (from Example 5.29) by using the free-body diagram shown in Figure P5.203.

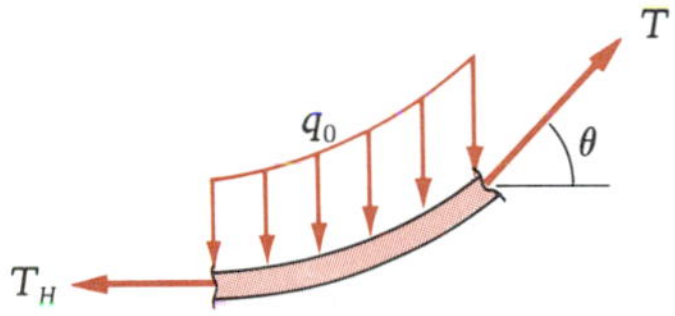

Figure P5.203

5.204 Show that summing moments about the left end of the element of Figure 5.51 adds nothing to the results of the analysis.

* **5.205** Note that if the ends of the cable are not at the same level, as shown in Figure P5.205, then we don't know the location of the lowest point ahead of time, and cannot

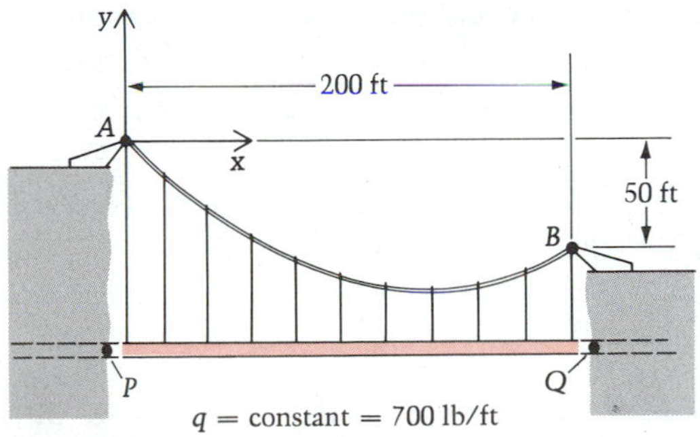

Figure P5.205

take advantage of symmetry. For the parabolic cable, begin with

$$y = \frac{q_0 x^2}{2T_H} + C_1 x + C_2$$

and solve the following problem: The pipeline weighs 700 lb / ft. It is to be assumed that over the gorge it is supported entirely by cables (i.e., neglect any forces from the rest of the pipe entering the earth at P and Q). Find the required length of cable if the lowest point of the parabola has an x coordinate of 150 ft. Also find the tensions at the cable suspension points A and B.

5.206 In the preceding problem, find the length of cable from the lowest point to the support at B.

* **5.207** A light cable shown in Figure P5.207 carries a load that grows with x according to $q = kx^n$, where k is a constant and n is a positive integer. If the cable crosses the x axis at $x = L/2$, find the deflection $y(x)$.

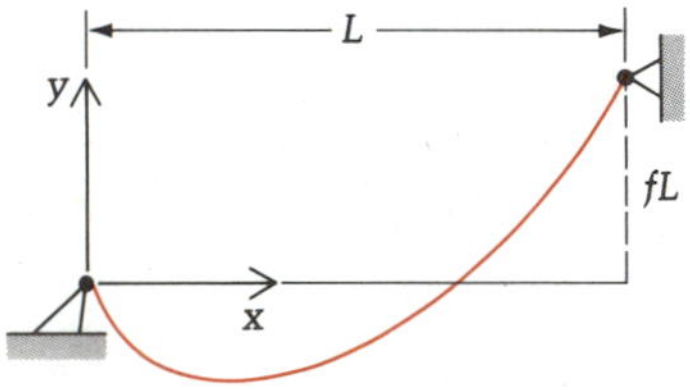

Figure P5.207

5.208 Expand the integrand of Equation (10) of Example 5.29 into the following power series:

$$\sqrt{1 + \left(\frac{8Hx}{L^2}\right)^2} = 1 + \frac{1}{2}\left(\frac{8Hx}{L^2}\right)^2 - \frac{1}{8}\left(\frac{8Hx}{L^2}\right)^4 (\pm) \cdots$$

Integrate term by term. For $H/L = 0.2$ compare ℓ/L from the example with this series result for: (a) one term; (b) two terms; (c) three terms.

5.209 Given a cable with maximum allowable tension T_0, length ℓ, and weight per unit length q_0, find the maximum horizontal span that can exist for this cable without exceeding T_0.

5.210 In the preceding problem, find the maximum span if $T_0 = q_0\ell$.

* **5.211** In Figure P5.211 determine the vertical distance from B (the center of the pulley) to A.

5.212 If a rope 22 ft long is to be suspended from a tree limb to make a swing, and the attachment points are 3 ft apart, what is the sag distance H? Assume that 1 ft is used on each side for the knotting. Compare the answer with the straight line result of 9.89 ft shown in Figure P5.212.

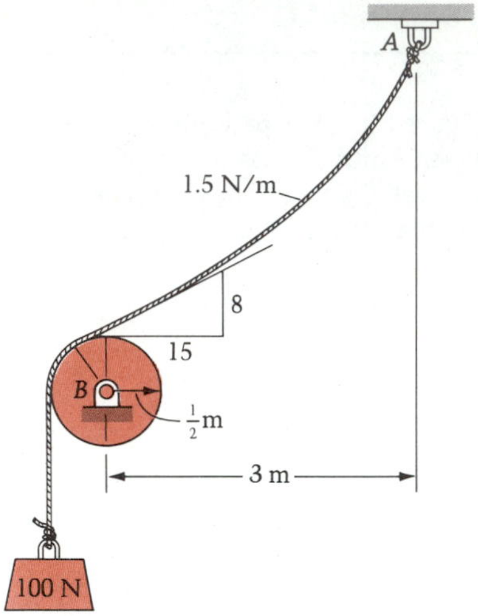

Figure P5.211

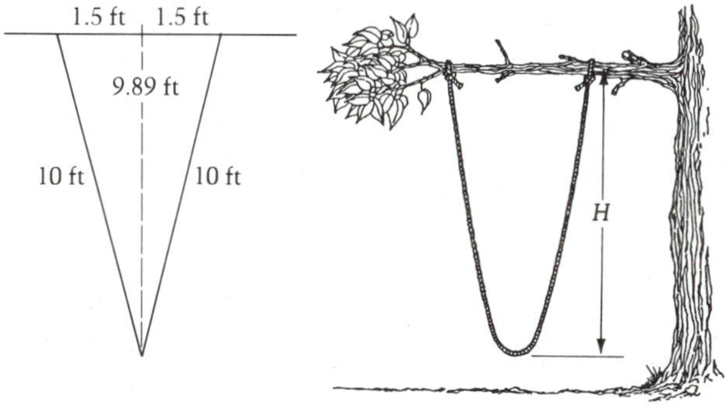

Figure P5.212

* **5.213** Find the error incurred in using the parabolic solution [Equation (5.39)] to get the center sag of a symmetrical catenary cable if the sag ÷ span is: (a) $\frac{1}{10}$; (b) $\frac{1}{4}$; (c) 1.

5.214 A cable of length 80 m hangs between the tops of two identical towers that are 66 m apart. Find the maximum sag in the cable.

5.215 A cable of length 80 m hangs between the tops of two towers. If the maximum sag is 5 m and the maximum tension is 480 N, what is the weight per unit length of the cable?

5.216 Repeat Problem 5.199 if there is no suspended load, but rather the cable *itself* weighs 50 lb / ft.

5.217 In Example 5.30, if the diameter of the base (where the cable is attached) is 8 ft, and the wind is horizontal, find:

a. The difference between the vertical force of the air and the weight

b. The resultant horizontal force caused by the wind and the height of its line of action above B.

* **5.218** In Figure P5.218 the balloon of Example 4.24 has come loose from the ground and has begun very slowly moving horizontally across an open field. The balloonist lets out some air, and the balloon lowers to a height of 30 ft and begins to drag its cable and its spike. If the net (horizontal) wind force is 50 lb, find the length of cable on the ground and the maximum cable tension.

* **5.219** The cable in Figure P5.219 has a weight per unit length of q_0. A length L hangs over the pulley as shown. Find the maximum tension in the cable, the total length of the cable, and the center sag.

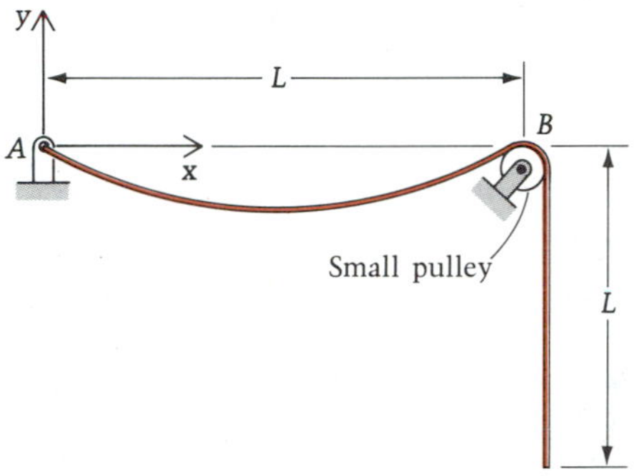

Figure P5.219

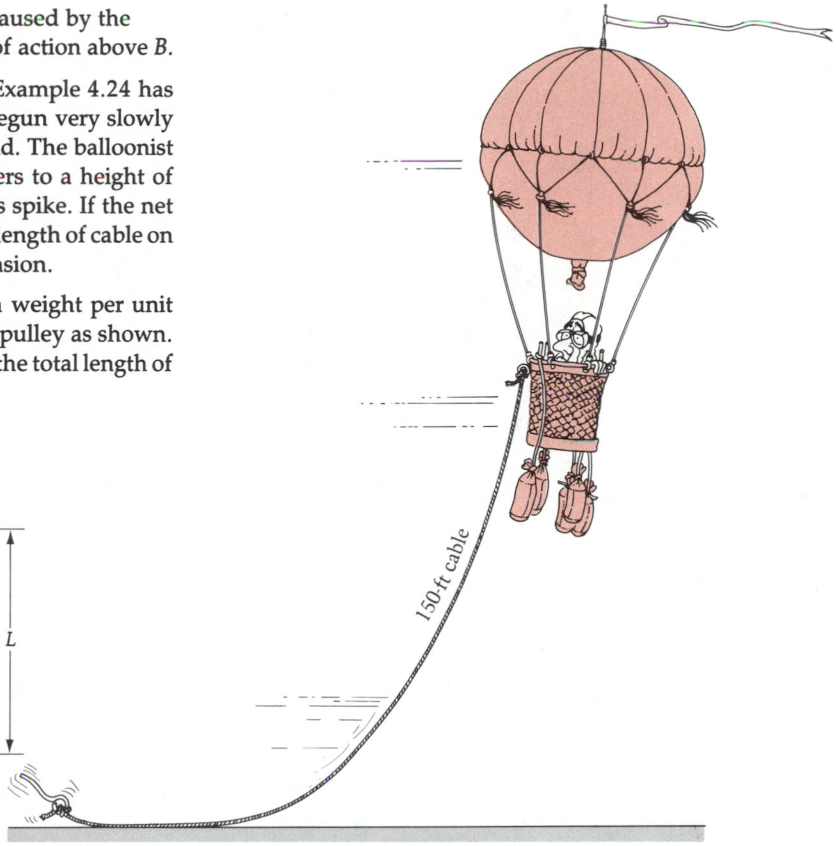

Figure P5.218

5.12 Cables Under Concentrated Loads

Sometimes we are faced with the necessity of analyzing a cable of negligible weight from which a number of forces ($W_1, W_2, \ldots, W_n$) are suspended at a series of load points $P_1, P_2 \ldots, P_n$. (See Figure 5.53.) Suppose we are given the lengths of the segments ($L_1, L_2 \ldots, L_{n+1}$ in Figure 5.53) and coordinates (D, H) of the right end with respect to the left. Then if we wish to know the shape in which the cable will hang (given by $\theta_1, \theta_2, \ldots, \theta_{n+1}$) and the tensions in the segments ($T_1, T_2, \ldots, T_{n+1}$), we have an extremely difficult set of equations to solve, consisting of $2n$ equilibrium equations, two for each load point, like (see Figure 5.54)

$$\xrightarrow{+} \qquad \Sigma F_x = 0 = -T_1 \cos\theta_1 + T_2 \cos\theta_2 \tag{5.40}$$

$$+\uparrow \qquad \Sigma F_y = 0 = T_1 \sin\theta_1 - T_2 \sin\theta_2 - W_1 \tag{5.41}$$

plus two geometric equations relating the L_i's and θ_i's to D and H:

$$D = \sum_{i=1}^{n+1} L_i \cos\theta_i \qquad H = -\sum_{i=1}^{n+1} L_i \sin\theta_i \tag{5.42}$$

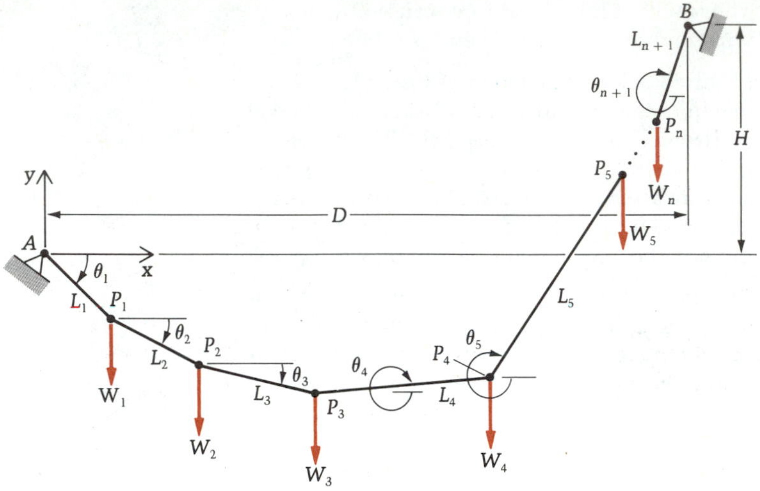

Figure 5.53

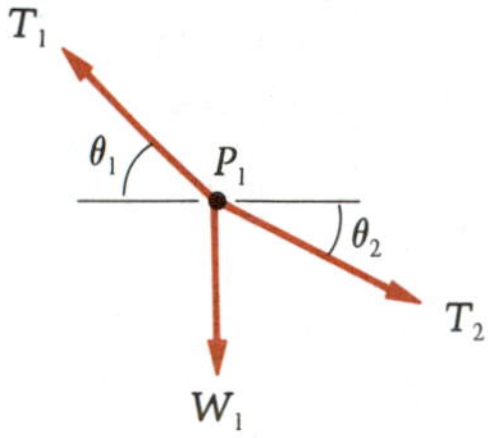

Figure 5.54

These $2n + 2$ equations are very hard to solve because of the nonalgebraic, trigonometric functions of half the unknowns (the θ_i's), *even for just two segments!*

Another approach to the problem is to seek the tensions and "sags" when we know the x coordinates of the n load attachment points $P_1, P_2, \ldots, P_n$, as suggested by Figure 5.55.

First we note that the horizontal components of all the tensions in the $n + 1$ segments are the same. From Equation (5.40) we have

$$T_1 \cos \theta_1 = T_2 \cos \theta_2$$

Because this can be repeated for each load point, we have

$$T_1 \cos \theta_1 = T_2 \cos \theta_2 = \cdots = T_{n+1} \cos \theta_{n+1} = T_H \qquad (5.43)$$

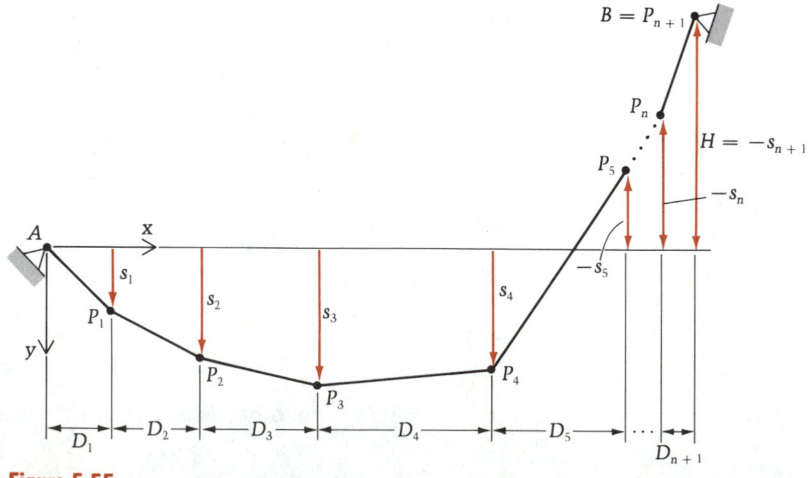

Figure 5.55

which is the same result we obtained in Section 5.11 for continuously loaded cables. We can use this "common horizontal component of tension" result to write the vertical components of each tension in terms of the unknown sags $s_1, s_2, \ldots, s_{n+1}$. Consider segment 2 in Figure 5.56 (from P_1 to P_2) and the force that its tension exerts on P_1:

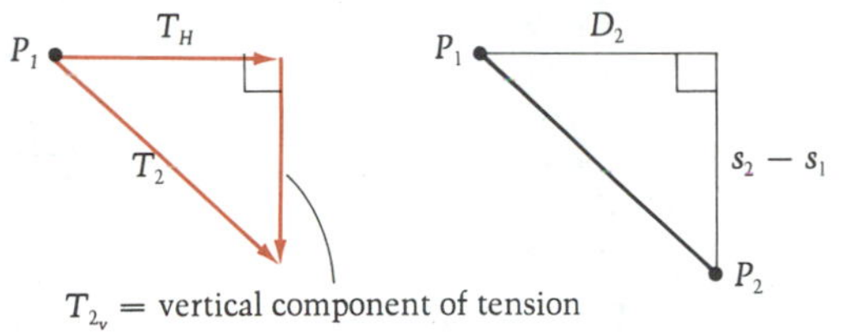

Figure 5.56

These triangles are similar, so that

$$\frac{T_{2v}}{T_H} = \frac{s_2 - s_1}{D_2} \Rightarrow T_{2v} = T_H\left(\frac{s_2 - s_1}{D_2}\right)$$

and in general*

$$T_{iv} = T_H\left(\frac{s_i - s_{i-1}}{D_i}\right) \tag{5.44}$$

Thus if we know T_H and the n sags $s_1, s_2, \ldots, s_n$, then the problem is solved because the tensions may be found as follows:

$$T_i = \sqrt{T_H^2 + T_{iv}^2} = T_H\sqrt{1 + \left(\frac{s_i - s_{i-1}}{D_i}\right)^2} \tag{5.45}$$

It is clear that the largest tension will occur in either the first or last segment.

Note now that in using Equation (5.43) we automatically satisfy $\Sigma F_x = 0$ at all n load points. If we now set the summation of forces in the y direction at each load point equal to zero, n equations will be generated. But all except the first and last will involve three of the sags. For example, at P_2, we have (see Figure 5.57):

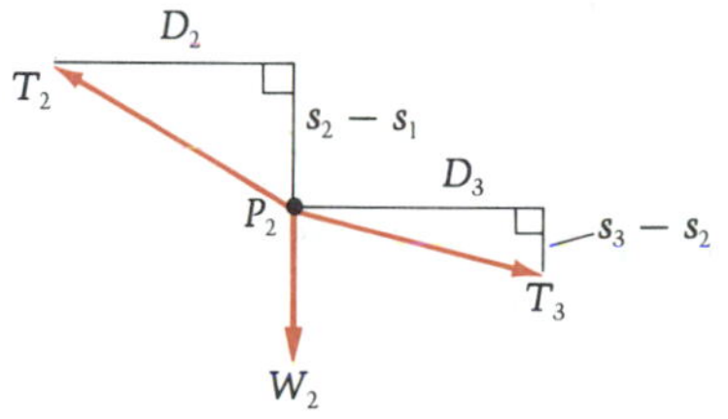

Figure 5.57

$$+\uparrow \quad \Sigma F_y = 0 = T_{2v} - T_{3v} - W_2$$

Using Equation (5.44),

$$T_H\left(\frac{s_2 - s_1}{D_2}\right) - T_H\left(\frac{s_3 - s_2}{D_3}\right) = W_2$$

or

$$s_2(D_2 + D_3) - s_1 D_3 - s_3 D_2 = \frac{W_2 D_2 D_3}{T_H}$$

* Note that $s_i - s_{i-1} < 0$ for the segments with negative slopes.

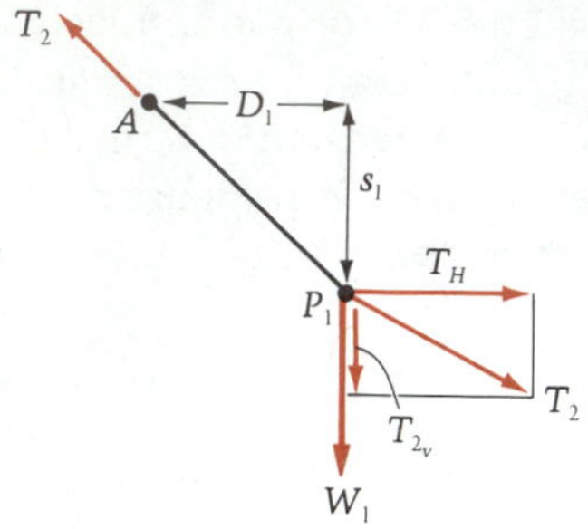

Figure 5.58

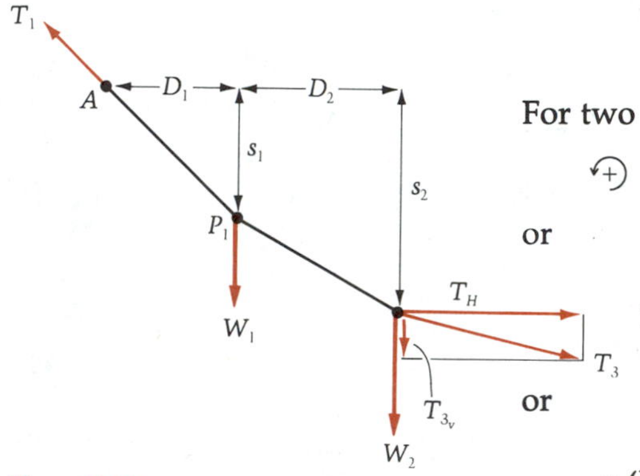

Figure 5.59

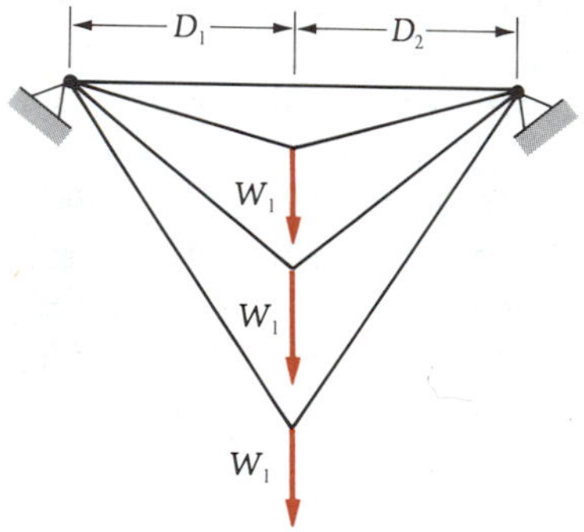

Figure 5.60

which involves s_1, s_2, and s_3 as well as T_H. If instead of summing forces we sum moments on the following series of free-body diagrams,* only two sags at most are involved in each equation. For one full segment (see Figure 5.58 at the left),

$$\circlearrowleft + \quad \Sigma M_A = 0 = T_H s_1 - T_{2v} D_1 - W_1 D_1$$

or

$$T_H s_1 - T_H \left(\frac{s_2 - s_1}{D_2} \right) D_1 = W_1 D_1$$

or

$$s_1(D_2 + D_1) - s_2 D_1 = \frac{W_1}{T_H} D_1 D_2$$

For two full segments (see Figure 5.59),

$$\circlearrowleft + \quad \Sigma M_A = 0 = T_H s_2 - T_{3v}(D_1 + D_2) - W_1 D_1 - W_2(D_1 + D_2)$$

or

$$T_H s_2 - T_H \left(\frac{s_3 - s_2}{D_3} \right) (D_1 + D_2) = W_1 D_1 + W_2(D_1 + D_2)$$

or

$$s_2(D_3 + D_2 + D_1) - s_3(D_1 + D_2) = \frac{[W_1 D_1 + W_2(D_1 + D_2)]}{T_H} D_3$$

This procedure will give us n equations, as we sum moments on a series of free-body diagrams that involve successive cuts of segments $2, 3, \ldots, n + 1$. Our problem, however, is that we have $n + 1$ unknowns ($s_1, s_2, \ldots, s_n$, and T_H), and until we specify one more condition on the geometry, there is no unique solution; that is, there are lots of cables that could satisfy the conditions. For example, each of the cables in Figure 5.60 has the same set values of D_1, D_2, and W_1, yet clearly the sought-after sags and tensions are different for each. If for our $(n + 1)$st equation we specify the overall length of the cable,

$$L = \sqrt{D_1^2 + s_1^2} + \sqrt{D_2^2 + (s_2 - s_1)^2} + \cdots + \sqrt{D_{n+1}^2 + (H - s_n)^2}$$

then we are back to formidable analytical difficulties. A better condition is to specify one of the sags. Then we can use the moment equations above and solve for the other sags in terms of T_H, one at a time, never having over one unknown sag per equation. Another more realistic possibility for the $(n + 1)$st equation is to be given a maximum allowable tension. This tension will occur in the first or last segment, as we have noted. We now give examples of these ideas.

* Note that the reactions at A and B necessarily equal the tensions T_1 and T_{n+1} and act along the lines of the first and last segments.

Example 5.31

The endpoints of a cable are tied to the points A and B in Figure E5.31a. The cable carries the indicated loads at $x = 2$ ft, 4 ft, and 6 ft. The maximum tension in the cable is 1000 lb. Find the tensions in all four segments of the cable, and the sags of the load points P_1, P_2, and P_3.

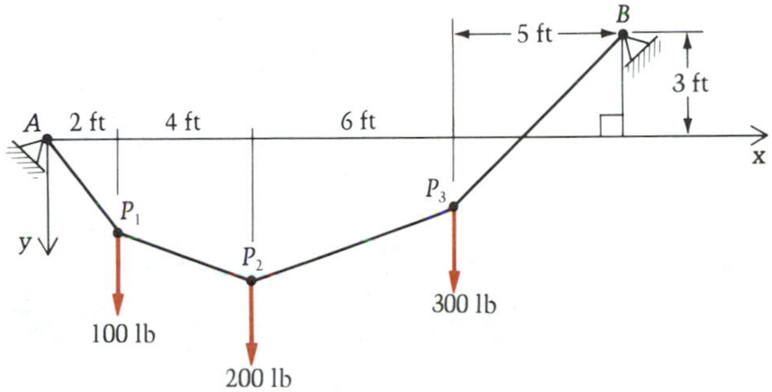

Figure E5.31a

Solution

We shall work to four significant digits in this example. We write the moment equations described in the foregoing discussion (see Figure E5.31b):

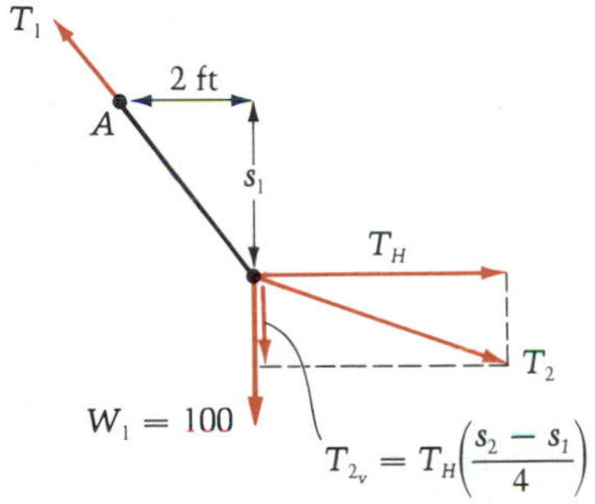

Figure E5.31b Free-body diagram of segment 1 and part of 2.

$$\circlearrowleft + \quad \Sigma M_A = T_H s_1 - \left[T_H\left(\frac{s_2 - s_1}{4}\right)\right]2 - 100(2) = 0$$

$$s_1(4+2) - s_2 2 = \frac{200(4)}{T_H}$$

$$6s_1 - 2s_2 = \frac{800}{T_H} \tag{1}$$

and using Figure E5.31c,

$$\circlearrowleft + \quad \Sigma M_A = T_H s_2 - \left[T_H\left(\frac{s_3 - s_2}{6}\right)\right]6 - 100(2) - 200(6) = 0$$

$$s_2(6+6) - s_3(6) = \frac{1400}{T_H}(6)$$

$$12s_2 - 6s_3 = \frac{8400}{T_H} \tag{2}$$

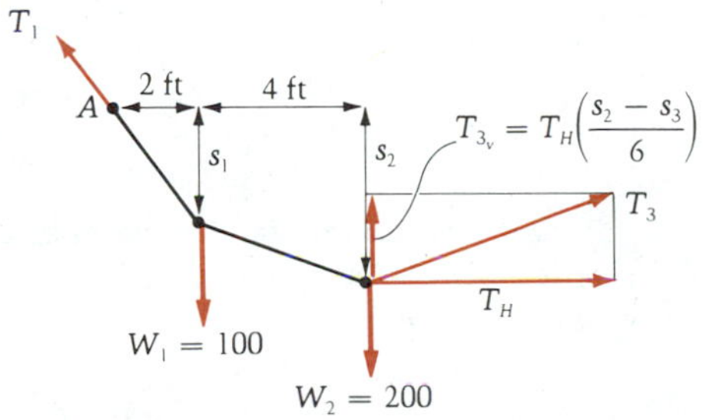

Figure E5.31c Free-body diagram of segments 1, 2, and part of 3.

Next, using Figure E5.31d,

$$\circlearrowleft\!+\quad \Sigma M_A = T_H s_3 - \left[T_H\left(\frac{-3 - s_3}{5}\right)\right] 12 - 200 - 1200 - 3600 = 0$$

$$17s_3 + 12\,(3) = \frac{5000}{T_H}\,(5) \tag{3}$$

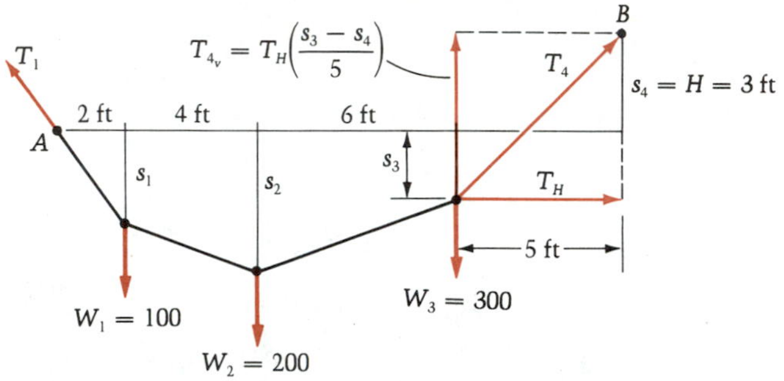

Figure E5.31d Free-body diagram of segments 1, 2, and 3, and part of 4.

We now have four equations [(1), (2), (3), and (5.45)] in four unknowns. First we find the sags in terms of T_H:
From (3),

$$s_3 = \frac{1471}{T_H} - 2.118 \tag{4}$$

From (2),

$$s_2 = \frac{700.0}{T_H} + \frac{1}{2}\overbrace{\left(\frac{1471}{T_H} - 2.118\right)}^{s_3}$$

or

$$s_2 = \frac{1436}{T_H} - 1.059 \tag{5}$$

From (1),

$$s_1 = \frac{133.3}{T_H} + \frac{1}{3}\overbrace{\left(\frac{1436}{T_H} - 1.059\right)}^{s_2}$$

$$s_1 = \frac{612.0}{T_H} - 0.3530 \tag{6}$$

We now assume (and must later check) that the maximum tension is in segment AP_1. (The other possibility is P_3B.) Then $T_1 = 1000$ lb, and Equation (5.45) gives

$$1000 = T_H\sqrt{1 + \left(\frac{s_1}{2}\right)^2} \tag{7}$$

or

$$\frac{10^6}{T_H^2} = 1 + \frac{s_1^2}{4} \Rightarrow s_1 = \sqrt{\frac{4 \times 10^6}{T_H^2} - 4} \qquad (8)$$

Equating the two expressions (6) and (8) for s_1 and squaring,

$$\frac{4 \times 10^6}{T_H^2} - 4 = \frac{374{,}500}{T_H^2} - \frac{432.1}{T_H} + 0.1246$$

Rearranging, $4.125T_H^2 - 432.1\ T_H - 3{,}626{,}000$. Solving with the quadratic formula gives

$$T_H = 991.4 \text{ lb}$$

Back-substituting,

$$s_1 = 0.2643 \text{ ft}$$

$$s_2 = 0.3895 \text{ ft}$$

$$s_3 = -0.6342 \text{ ft}$$

(meaning the third load point is *above* the x axis through A)

The other tensions are given by Equation (5.45):

$$T_2 = T_H \sqrt{1 + \left(\frac{s_2 - s_1}{D_2}\right)^2}$$

$$= 991.9 \text{ lb}$$

$$T_3 = T_H \sqrt{1 + \left(\frac{s_3 - s_2}{D_3}\right)^2}$$

$$= 1006 \text{ lb}$$

$$T_4 = T_H \sqrt{1 + \left(\frac{s_4 - s_3}{D_4}\right)^2} \qquad (s_4 = -H = -3 \text{ ft})$$

$$= 1097 \text{ lb}$$

We therefore see that we have guessed wrong and that the maximum tension is instead in the *right-most* cable. Rewriting Equation (7) (the three moment equations are still valid):

$$1000 = T_H \sqrt{1 + \left(\frac{s_4 - s_3}{5}\right)^2} \qquad (s_4 = -3 \text{ ft})$$

$$\frac{10^6}{T_H^2} = 1 + \left(\frac{-3 - s_3}{5}\right)^2$$

$$(3 + s_3)^2 = 25\frac{10^6}{T_H^2} - 25$$

$$s_3 = 5\sqrt{\frac{10^6}{T_H^2} - 1} - 3$$

Equating expressions for s_3 gives

$$5\sqrt{\frac{10^6}{T_H^2} - 1} = \frac{1471}{T_H} + 0.8820$$

Squaring,

$$25\left(\frac{10^6}{T_H^2}-1\right)=\frac{2{,}164{,}000}{T_H^2}+\frac{2595}{T_H}+0.7779$$

$$1.031T_H^2+103.8T_H-913{,}400=0$$

$$T_H=892.2\text{ lb}$$

so that, using Equations (6), (5), and (4) in order,

$$s_1=0.3329\text{ ft}$$

$$s_2=0.5505\text{ ft}$$

$$s_3=-0.4693\text{ ft}$$

and

$$T_1=T_H\sqrt{1+\left(\frac{s_1}{D_1}\right)^2}=904.5\text{ lb}$$

$$T_2=T_H\sqrt{1+\left(\frac{s_2-s_1}{D_2}\right)^2}=893.5\text{ lb}$$

$$T_3=T_H\sqrt{1+\left(\frac{s_3-s_2}{D_3}\right)^2}=905.0\text{ lb}$$

This time all the tensions are ≤ 1000 lb, and we have also solved for the sags.

To illustrate the procedure when a sag is specified instead of a maximum tension, we shall specify that $s_2 = 0.5505$ ft, which came out of the preceding solution, and solve for all the tensions in the following example.

EXAMPLE 5.32

If the sag s_2 of load point P_2 is 0.5505 ft, find the tensions in the four cable segments and the sags of P_1 and P_3 (see Figure E5.32a).

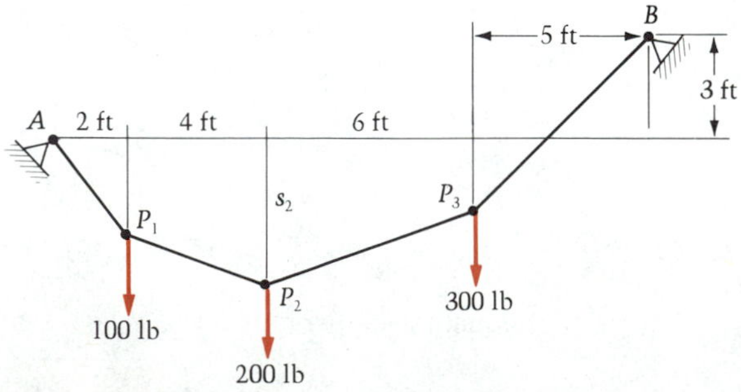

Figure E5.32a

Solution

For the overall free-body diagram shown in Figure E5.32b, we have:

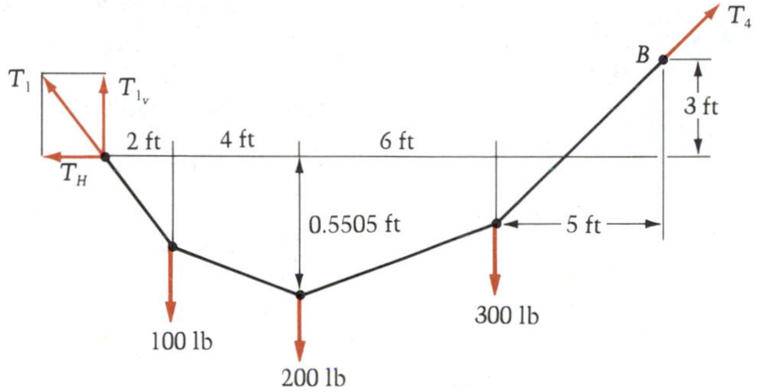

Figure E5.32b

$$\Sigma M_B = 0 = (300 \text{ lb})(5 \text{ ft}) + (200 \text{ lb})(11 \text{ ft}) + (100 \text{ lb})(15 \text{ ft}) - T_H(3 \text{ ft}) - T_{1v}(17 \text{ ft})$$

or

$$3T_H + 17T_{1v} = 5200$$

And using the free-body diagram of AP_1P_2 in Figure E5.32c,

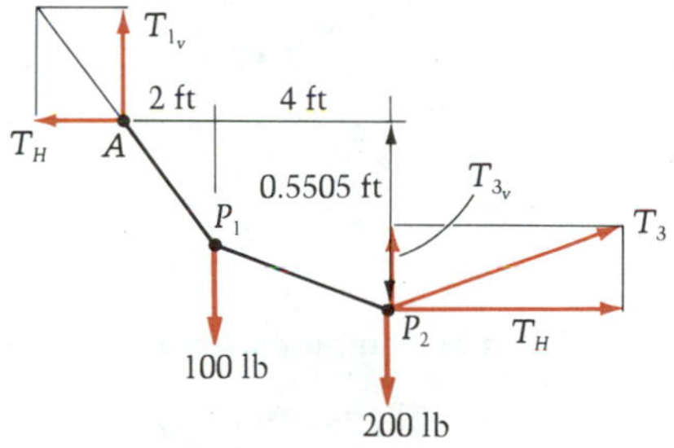

Figure E5.32c

$$\circlearrowleft + \quad \Sigma M_{P_2} = 0 = (100 \text{ lb})(4 \text{ ft}) + T_H(0.5505 \text{ ft}) - T_{1v}(6 \text{ ft})$$

or

$$0.5505T_H - 6T_{1v} = -400$$

Solving,

$$\left.\begin{aligned} T_H &= 891.9 \text{ lb} \\ T_{1v} &= 148.5 \text{ lb} \end{aligned}\right\} T_1 = 904.2 \text{ lb}$$

From the same free-body, remembering that $T_{3x} = T_H = 891.9$ lb,

$$\circlearrowleft + \quad \Sigma M_A = 0 = (891.9 \text{ lb})(0.5505 \text{ ft}) + T_{3v}(6 \text{ ft}) - (100 \text{ lb})(2 \text{ ft}) - (200 \text{ lb})(6 \text{ ft})$$

$$T_{3v} = 151.5 \text{ lb}$$

$$T_3 = \sqrt{891.9^2 + 151.5^2} = 904.7 \text{ lb}$$

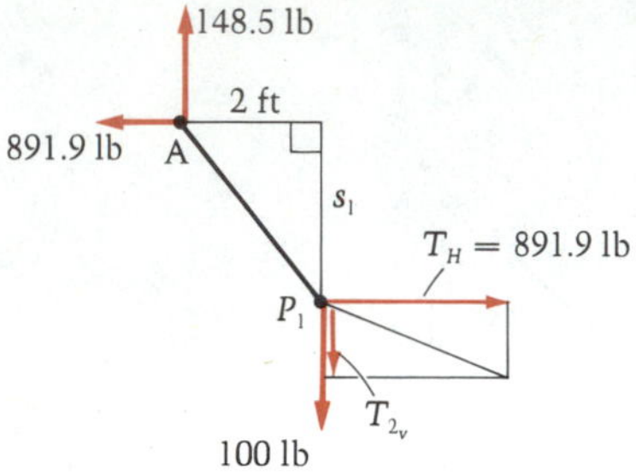

Figure E5.32d

Then we use other free-body diagrams to get the unknown sags; from Figure E5.32d,

$$\circlearrowleft\!+ \quad \Sigma M_{P_1} = 0 = (891.9 \text{ lb})(s_1) - (148.5 \text{ lb})(2 \text{ ft})$$

$$s_1 = 0.3330 \text{ ft}$$

For T_2, note that its horizontal component is $T_H = 891.9$ lb and

$$+\uparrow \quad \Sigma F_y = 0 = 148.5 - 100 - T_{2v}$$

$$T_{2v} = 48.5 \text{ lb}$$

Thus

$$T_2 = \sqrt{891.9^2 + 48.5^2} = 893 \text{ lb}$$

Then, using the free-body diagram in Figure E5.32e,

$$\circlearrowleft\!+ \quad \Sigma M_{P_3} = 0 = (100 \text{ lb})(10 \text{ ft}) + (200 \text{ lb})(6 \text{ ft}) + (891.9 \text{ lb})(s_3) - (148.5 \text{ lb})(12 \text{ ft})$$

$$s_3 = -0.4687 \text{ ft} \qquad \text{(as before)}$$

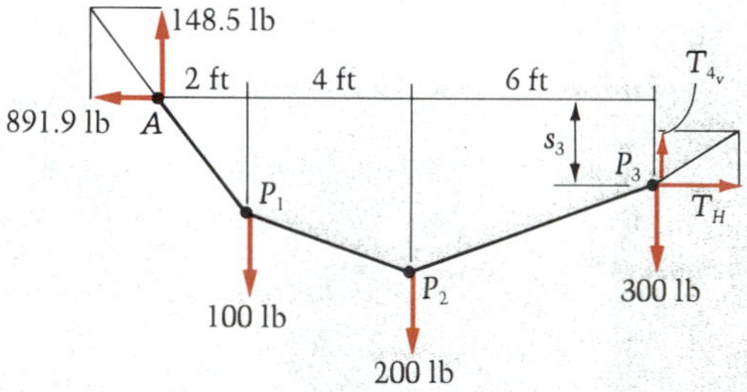

Figure E5.32e

Also, T_{4v} can be computed from

$$+\uparrow \quad \Sigma F_y = 0 = 148.5 - 100 - 200 - 300 + T_{4v}$$

$$T_{4v} = 451.5 \text{ lb}$$

so that

$$T_4 = \sqrt{891.9^2 + 451.5^2} = 999.7 \text{ lb}$$

differing from 1000 lb due to rounding to four digits.

Partial checks of the two solutions obtained in Example 5.31 are afforded by writing the overall vertical equations of equilibrium (see Figure 5.61):

$$+\uparrow \quad \Sigma F_y = -600 + 1000\left(\frac{0.2643}{\sqrt{0.2643^2 + 2^2}}\right) + 1097\left(\frac{2.366}{\sqrt{2.366^2 + 5^2}}\right)$$

$$= -600 + 131.0 + 469.2 = 0.200 \text{ lb} \quad ✓$$

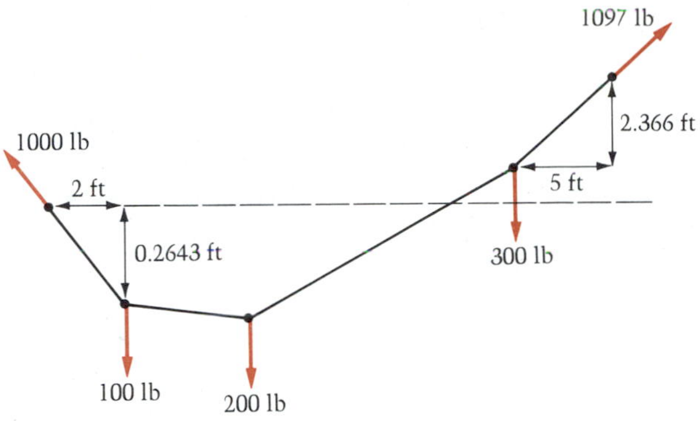

Figure 5.61

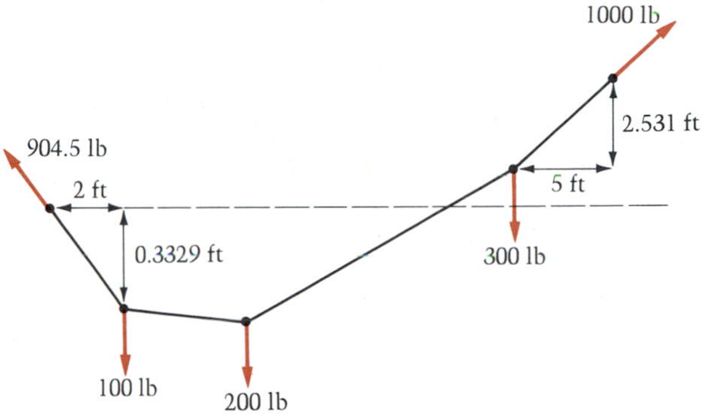

Figure 5.62

and (see Figure 5.62):

$$+\uparrow \quad \Sigma F_y = -600 + 904.5\left(\frac{0.3329}{\sqrt{0.3329^2 + 2^2}}\right) + 1000\left(\frac{2.531}{\sqrt{2.531^2 + 5^2}}\right)$$

$$= -600 + 148.5 + 451.6 = 0.100 \text{ lb} \quad ✓$$

Problems ► Section 5.12

5.220 The cable supports two vertical loads at P_1 and P_2 as shown in Figure P5.220. Find the elevation of the load point P_1 with respect to the horizontal line through A, and the tensions in the three segments (AP_1, P_1P_2, and P_2B) of the cable.

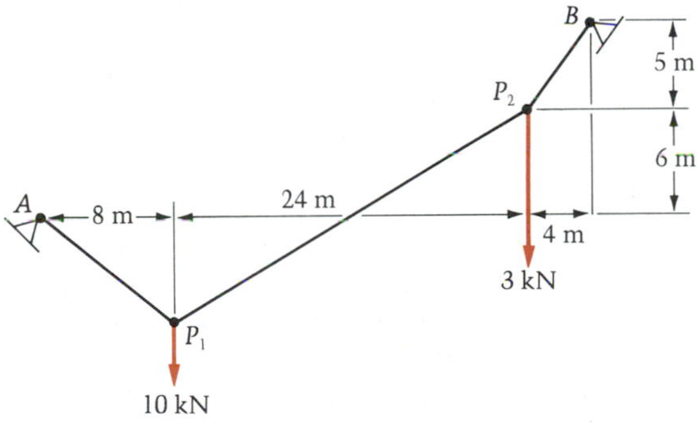

Figure P5.220

5.221 In Figure P5.221 find the sags of P_1 and P_2, and the tensions in the three cable segments, if the maximum tension is known to be 1400 lb.

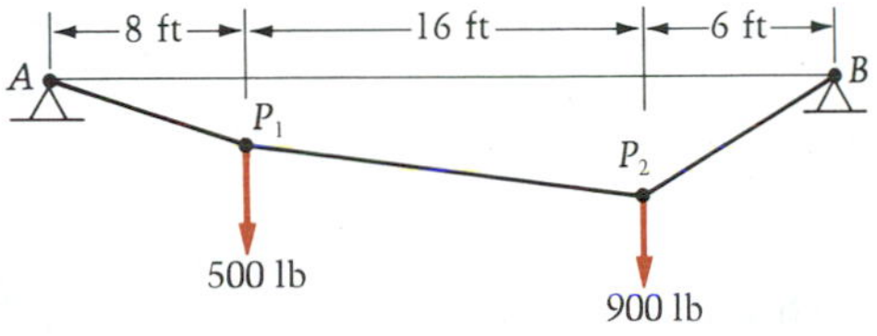

Figure P5.221

5.222 In Figure P5.222:

a. If the horizontal force P is holding the cable in equilibrium, what is the value of P?

b. Find the distances d_1 and d_2.

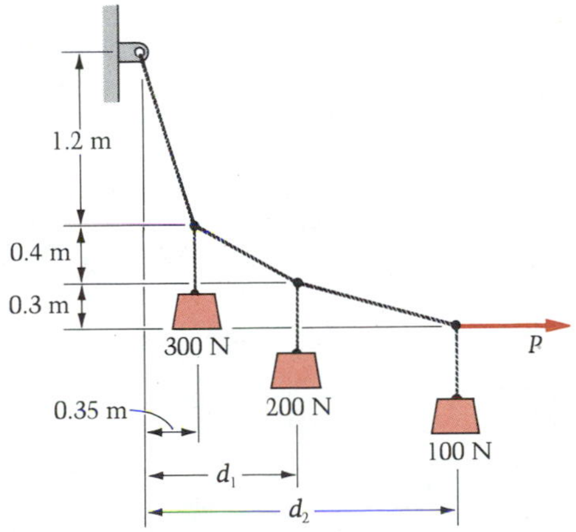

Figure P5.222

5.223 The beam weighs 200 lb and is to be supported in a horizontal position by two cables as shown in Figure P5.223. Find:

a. The angles ϕ_1 and ϕ_2
b. The tensions in segments AP_1 and P_2B
c. The elevation H of B.

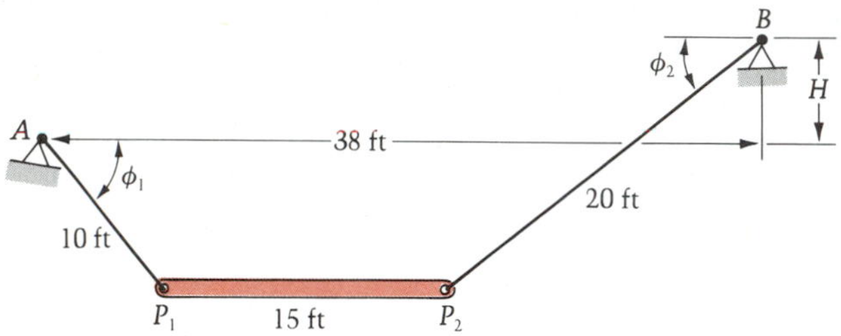

Figure P5.223

5.224 The light cable is loaded by the two downward forces as indicated in Figure P5.224. Find:

a. The elevation of the load application point P_1 with respect to A.
b. The tensions in the three cable segments.

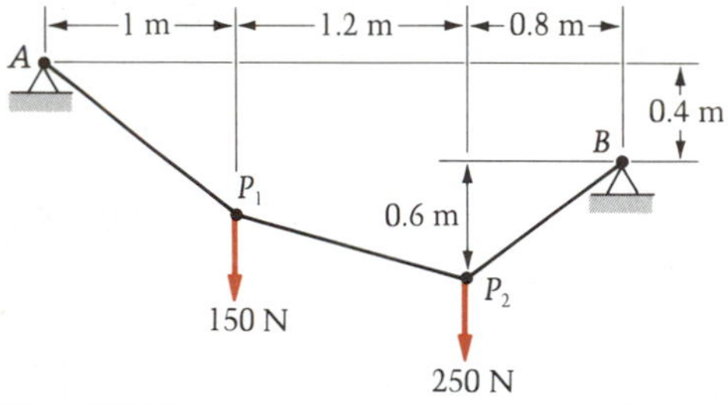

Figure P5.224

5.225 Repeat the preceding problem if the 0.6-m dimension is deleted and replaced by the condition that the maximum cable tension is to be 600 N.

5.226 Find the angles θ and ϕ and the tensions in the two segments of the cable loaded by the 200-lb force shown in Figure P5.226(a). Check your answers for θ and ϕ by making the construction suggested in Figure P5.226(b) and measuring the angles with a protractor.

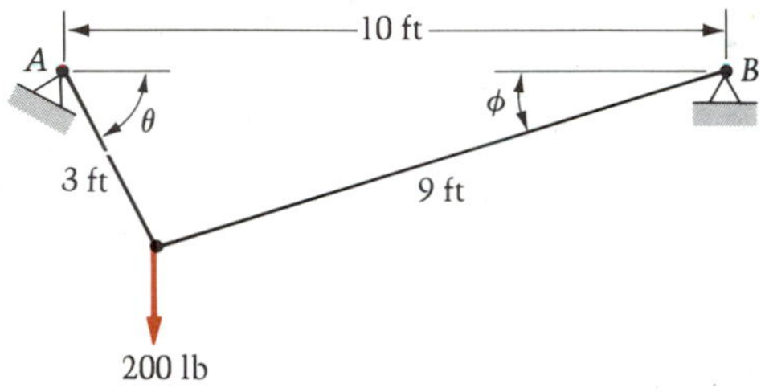

Figure P5.226(a)

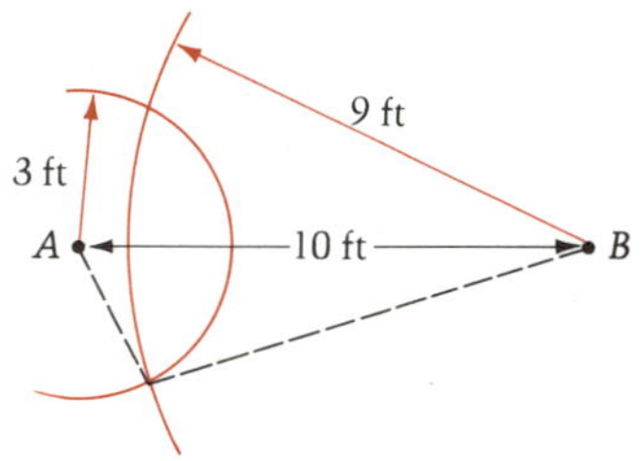

Figure P5.226(b)

COMPUTER PROBLEMS ▶ Chapter 5

5.227 Use a computer to find the angle θ for which δ_y is a minimum in Problem 5.121. You may either list values of the parenthesized function of θ and watch for the minimum, or else use calculus and then solve a resulting cubic polynomial for the cosine of θ.

5.228 Solve Problem 5.213 using a computer.

SUMMARY ▶ Chapter 5

Chapter 5 has been concerned with finding the resultants of internal force distributions within structural elements. In Part I (Sections 5.2–5.7) we studied the truss, and learned in Section 5.2 that all its elements are by definition straight, two-force members. We found in Sections 5.3–5.5 that there are two basic means of determining the forces in a truss: (1) the method of joints (or pins), in which we isolate one pin at a time, writing and solving equilibrium equations until we have found the member force(s) of interest to us; and (2) the method of sections, in which we imagine the truss cut completely in two, with the slice exposing the forces in the cut members. Equilibrium equations on one or the other "half" of the sliced-through truss are then written and solved to determine this force.

Space (3D) trusses are examined in Section 5.6, and in Section 5.7. We introduced the reader to the mechanics of deformable bodies by studying the uniaxial stress and strain found in a truss element.

Part II of the chapter dealt with the internal forces and moments in frames (Section 5.8) and beams (Sections 5.9, 5.10). Here we again slice through members, this time at locations where we wish to know the axial force, shear force and bending moments. (Unlike trusses, where there is but one constant axial force in each member from end to end, the internal forces and moments in frames and beams generally vary from point to point.) For the beam, we also studied in Section 5.10 the differential relationships between the bending moment, the transverse shearing force, and the externally applied lateral distributed loading. This led to the concepts of shear and bending moment diagrams. Once completed, these diagrams tell us at a glance how these two stress-related quantities vary over the beam's length.

We examined the cable in Part III of the chapter. We found in Section 5.11 that when the distributed load on a cable is a uniform function of the x-coordinate, the cable hangs in a parabolic arc, while when the load varies uniformly with arclength along the cable, it hangs in a catenary (hyperbolic cosine) curve. In Section 5.12 we studied the light cable under a set of concentrated parallel forces, and found that, surprisingly, such problems are quite complicated. The reason is that the equations are non-algebraic with variables occurring as arguments of sines and cosines.

REVIEW QUESTIONS ▶ Chapter 5

True or False?

1. Any truss can be solved (i.e., the forces in all its members can be found) by enough applications of the method of joints.
2. In a simple plane truss, a joint can always be found with only two unknown member forces acting on it.
3. For the method of sections to be helpful, the truss has to be sectioned into two distinct, unconnected parts.
4. It is always possible to cut a section that allows the force in any truss member to be found by a single moment equation.
5. No cross section of a two-force member ever has to resist a bending moment or a shear force.

6. No member of a truss ever has to resist a bending moment or a shear force.
7. If a truss member pulls on a pin, then it is in tension; if it pushes on a pin, it is in compression.
8. If a truss member pulls on a pin at one of its ends, then it must push on the pin at its other end.
9. The normal stress in a truss member is the force in the member divided by its cross-sectional area.
10. In a tensile test, if stress versus strain is plotted, we will get a straight line for any material.
11. Statically indeterminate problems may be solved by including consideration of the deformations of the structure.
12. The rate of change of bending moment in a beam with respect to x equals the axial force.
13. The area beneath the loading curve between two points P_1 and P_2 of a beam equals the negative of the change in the shear force between P_1 and P_2.
14. If shear and moment diagrams are sketched and correct values at the right end of the beam are obtained, then the diagrams have probably been drawn correctly.
15. If a cable load is uniformly distributed with respect to x, then the shape of the cable is parabolic. If the load is uniformly distributed with respect to arclength s, the shape of the cable is a catenary.
16. For a cable under any vertical loading, the horizontal component of tension is constant across the cable.

Answers: 1. F 2. T 3. T 4. F 5. F 6. T 7. T 8. F 9. T 10. F 11. T 12. F 13. T 14. T 15. T 16. T

6 FRICTION

6.1 Introduction

6.2 Laws, Coefficients, and Basic Applications of Coulomb Friction

Normal and Friction Forces

Maximum Value of Friction; Coefficient of Friction

Self-locking Mechanisms

6.3 Special Applications of Coulomb Friction

Wedges

Flexible Flat Belts and V-Belts

Screws

Disk Friction

SUMMARY

REVIEW QUESTIONS

6.1 Introduction

This chapter is concerned with one and only one topic: Coulomb friction. Friction is the force resisting one body's tendency to slide past another. If the two bodies are dry, then the friction is often known as Coulomb friction.

We have actually already drawn friction forces on free body diagrams in Chapter 4. Each was the "tangent-plane-of-contact component" of the interactive force of one body onto another. But there, we didn't know those forces as friction forces, nor did we study the principle of Coulomb friction (postponing that until now), which is that a Coulomb friction force f is limited to a maximum value f_{MAX} which is proportional to the normal (pressing) force N between the bodies, according to:

$$0 \le f \le f_{MAX} = \mu N$$

In this inequality, μ is the "coefficient of static friction," which depends on the types of materials in contact and on the roughnesses of their surfaces. A low value of μ means not much of a friction force can be developed; the limiting value of $\mu = 0$ corresponds to "smooth" surfaces on which friction is absent. A high value of μ (say about 0.5 and above), conversely, indicates that a relatively large friction force may be developed between the two surfaces.

It is only when a body is on the verge of slipping past another (or is actually doing so) that $f = \mu N$. Usually this is not the case, and the magnitude of the friction force f lies somewhere between zero and the maximum possible value: $f_{MAX} = \mu N$. Beneath a wheel, for example, the friction force is rarely ever equal to μN.

In actual *slipping* situations, the friction force is written as $f = \mu_k N$ where μ_k is the coefficient of kinetic friction, generally less than the static coefficient μ; the latter is written μ_s when a distinction between it and μ_k needs to be made.

Section 6.2 is concerned with basic applications of Coulomb friction. The examples and problems of this section are like those of Chapter 4, *except that now there is the possibility of slipping (sliding).* In the other section, 6.3, we introduce the reader to a few special engineering applications of Coulomb friction: wedges, flexible flat belts and V-belts, square-threaded screws, and disk friction.

The reader should already appreciate the fact that friction is of enormous importance in engineering.

6.2 Laws, Coefficients, and Basic Applications of Coulomb Friction

The world would continue to go round and round without friction, but the lives of just about everything living on it would change drastically if friction were suddenly to disappear. Friction is therefore a very important topic, worthy of much study. It serves both good and bad purposes.

Without friction, automobiles, trains, and bicycles would not work at all because their wheels need friction to produce acceleration of the vehicles to which they are attached. And without friction on the soles of your shoes, you could not walk out of the room in which you are sitting or lying. We wish to maximize the friction force in situations where it is being used to our advantage, such as in the use of tires, brakes, wedges, screws, belts, clutches, and in simple but important applications such as at the foot of a ladder. Conversely, friction has many deleterious effects: it causes mechanical parts to wear out, and much energy is expended in overcoming frictional resistance to motions of machines and vehicles.

We shall now examine the law of dry friction between a pair of surfaces; this law was first set forth by Coulomb in the 18th century and is called **Coulomb's Law of Dry Friction.**

Normal and Friction Forces

For two bodies $\mathcal{B}_1$ and $\mathcal{B}_2$ in contact, the **friction force** on, say, $\mathcal{B}_1$, is the component of the resultant force (exerted on $\mathcal{B}_1$ by $\mathcal{B}_2$) that lies in the tangent plane of contact (see Figure 6.1). The other component, perpendicular to this tangent plane, is called the **normal force** exerted on $\mathcal{B}_1$ by $\mathcal{B}_2$.

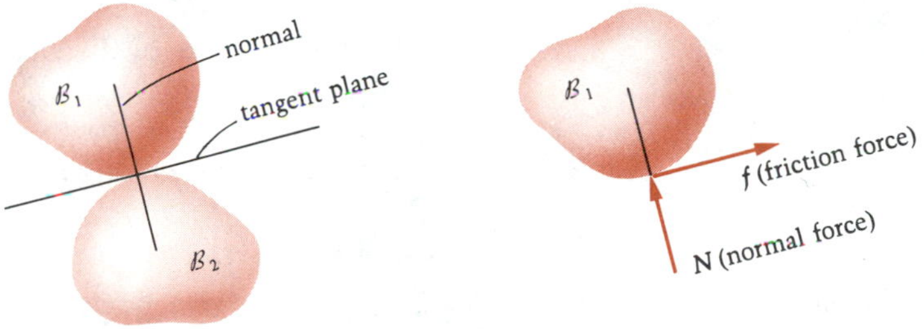

Figure 6.1

Let us now imagine a crate resting on a horizontal surface. The free-body diagram is shown in Figure 6.2(a). The resultant of all the "pressures times areas" at each of the points of contact beneath the box is depicted as a single resultant force N, which equals the weight mg of the crate and which acts upward through the mass center C. In this state, there is no friction at all. Friction resists the tendency to slide and if there is no such tendency, it will not be present.

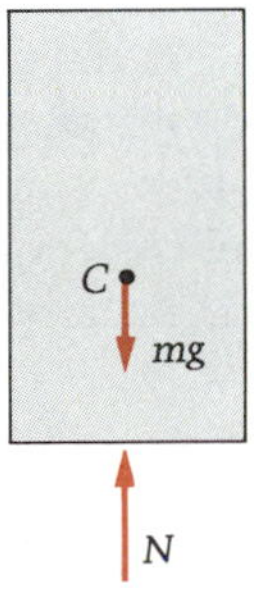

Figure 6.2(a)

Now suppose a child exerts a slowly and linearly increasing (with time t) horizontal force P in an effort to move the crate to the right as shown in Figure 6.2(b) (see the following page). Two things of importance to our study then happen at once. They are:

1. A friction force f develops simultaneously that opposes the tendency of the crate to slide to the right.
2. The normal force N is still equal to mg because there is no vertical acceleration and the y forces are still in balance. But the *position* of

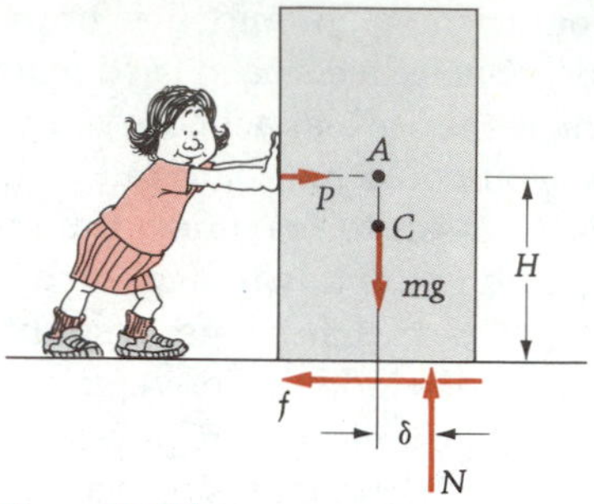

Figure 6.2(b)

N — that is, of the resultant of the many pressing forces under the crate — shifts to the right. The *way* this happens is by the pressure increasing on the right and decreasing on the left. The *reason* it happens is that N must act so as to overcome the tendency of the crate to tip forward. Said another way, the moments of f and N about A* must add to zero to prevent rotation. Thus N moves to the right so that the moment $fH\circlearrowright$ can be balanced by the moment $N\delta\circlearrowleft$, where δ is (see the figure) the shift, to the right, of the force N from the centerline of the crate.

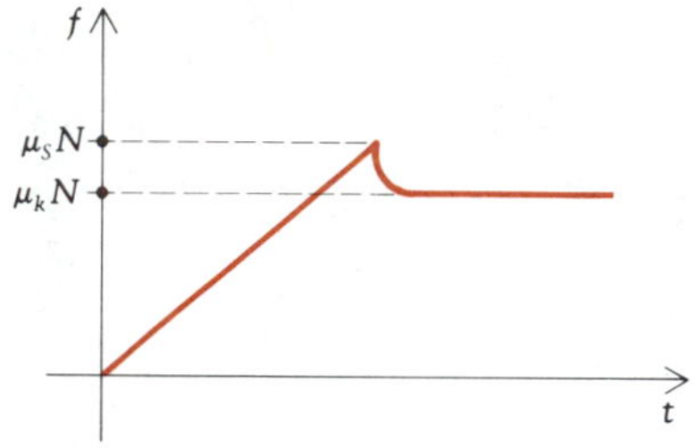

Figure 6.3

As the applied force of the child grows, so do f and δ. If the crate is wide enough and/or short enough, it will not tip before sliding. Let us assume this to be the case for now. The friction force f will continue to balance the force P for as long as it is able.

Maximum Value of Friction; Coefficient of Friction

At some point in the above process, a maximum value is reached for f, as suggested by Figure 6.3, and the crate will start to slide. This value was observed by Coulomb to be proportional to the normal force:

$$f_{\text{MAX}} = \mu_s N \tag{6.1}$$

in which the proportionality constant μ_s is the **coefficient of static friction,** which depends upon the types of materials in contact and upon the roughnesses of their surfaces. The value of μ_s varies from near zero for very smooth surfaces to 1 or even more than 1 for very rough surfaces.

Now let us return to the child and the crate. As most of us have experienced, it is harder to "get something moving than to keep it moving." When P reaches a value equal to the static maximum of f, motion is said to be impending. Any further increase in P will cause the crate to slide, and the friction force normally drops off to a second, usually slightly lower value.

* Note that P and mg produce no moment about A.

Question 6.1 Sketch a possible plot of P vs. t for the case when the block accelerates over Δt to a constant velocity.

This reduced value of f is assumed to be the normal force N times a different coefficient of friction, this time called the coefficient of kinetic friction, μ_k:

$$f_k = \mu_k N \tag{6.2}$$

where f_k is the value of friction to be used if sliding has started and is actually ongoing. The force f_k is thus used in dynamics. There, if no subscript is placed on the μ, it is to be assumed that $\mu_s \approx \mu_k$. But actually, μ_k is normally dependent upon the speed of sliding, and, as the above discussion suggests, it is less than μ_s. A simple model that is often used is to take μ_k to be a constant (independent of sliding speed) less than μ_s.

In statics, however, we are usually only concerned with μ_s, so that we may drop the subscript on μ in most cases. The student should always remember that

$$0 \le f \le f_{\text{MAX}} = \mu N \tag{6.3}$$

and that in the absence of relative motion, $f = \mu N$ *if and only if* the bodies at rest and in contact are at the point (on the verge) of slipping relative to each other. This is called "impending slipping." When surfaces are called "smooth," it means μ is very small, approaching zero. In this case,

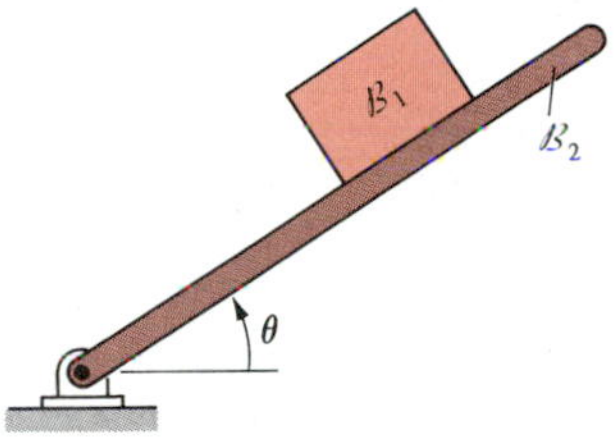

Figure 6.4(a)

$$0 \le f \le \mu N \approx 0 \tag{6.4}$$

so that no friction is available and f is to be assumed as zero.

The static coefficient of friction may be determined for two materials by constructing a plane of one of the materials and a block of the other (see Figure 6.4):

As the angle θ of the plane is increased from zero, the friction force beneath $\mathcal{B}_1$ will increase (see the free-body diagram of $\mathcal{B}_1$) because it has to equilibrate an increasing component of the weight (mg) down the plane. Friction always acts on each body in the direction that opposes sliding of the two contacting bodies past each other.

Answer 6.1 A plot of P versus time will be identical to f vs. t *except* during the short interval Δt during which the block accelerates up to constant speed. (See the figure.) During that time P is larger than f, and their difference, $P - f$, produces the acceleration of the block.

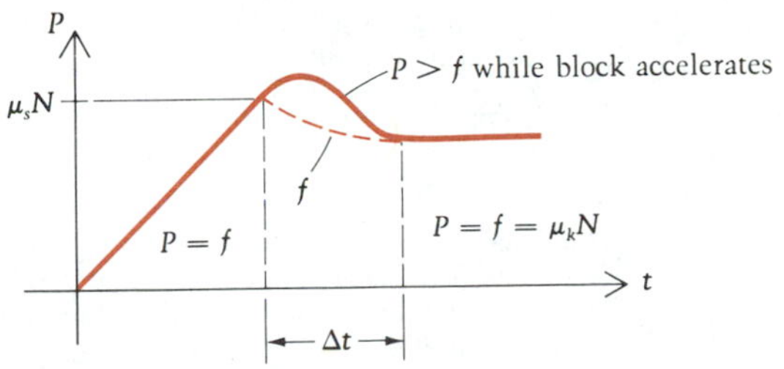

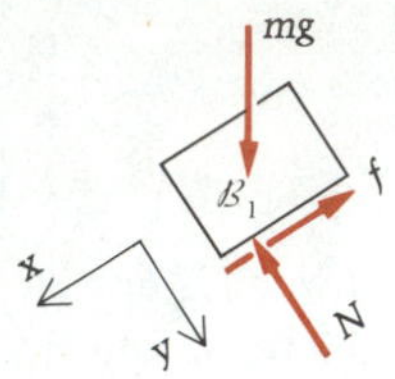

Figure 6.4(b)

Using the indicated directions for x and y, the equilibrium equations of $\mathcal{B}_1$ are seen from Figure 6.4(b) to be

$$\overset{+}{\swarrow} \qquad \Sigma F_x = 0 = mg \sin\theta - f \Rightarrow f = mg\sin\theta$$

$$\overset{+}{\searrow} \qquad \Sigma F_y = 0 = mg\cos\theta - N \Rightarrow N = mg\cos\theta$$

(Note that the moment equation would give us the location of the normal force N.)

If we slowly increase θ until the block slides, then at this critical angle (call it θ_S), we have $f = \mu N$. Therefore, at $\theta = \theta_S$,

$$f = mg\sin\theta_S = \mu N = \mu\, mg\cos\theta_S$$

or

$$\mu = \tan\theta_S \tag{6.5}$$

so that the tangent of the angle at which slipping begins is the coefficient of friction for the two materials of interest. The angle θ_S is called the **angle of friction,** or angle of repose.

In the following table, some typical values of μ_s are presented for various selected pairs of materials.

Table 6.1

Approximate* Values of the Coefficient of Static Friction for Clean, Dry Surfaces

Surface Materials	Coefficient of Friction
Steel on steel	0.75
Rubber on concrete	0.8
Rubber on asphalt	0.85
Rubber on ice	0.1
Oak on oak	0.6
Wood on metal	0.4
Cast iron on cast iron	1.1
Cast iron on copper	1.05
Teflon on teflon or steel	0.04
Copper on steel	0.5
Aluminum on aluminum	1.1
Steel on lead	0.95
Aluminum on steel	0.6
Brass on brass	0.9
Hemp rope on wood	0.7
Metal on stone	0.5
Metal on leather	0.45
Metal on ice	0.04
Oak on leather	0.6
Bonded carbide on iron	0.8
Glass on glass	0.9
Copper on copper	1.2

* Values of μ may vary widely for a pair of materials, depending on such things as whether the steel is mild or hard, whether the friction on the wood is with or across the grain, and so forth. The tabulated values are intended just to give the reader a general idea of values of μ.

We now proceed to some examples in which we have friction acting between various flat surfaces. In the first one, we shall make use of Equation 6.5 at the outset.

Example 6.1

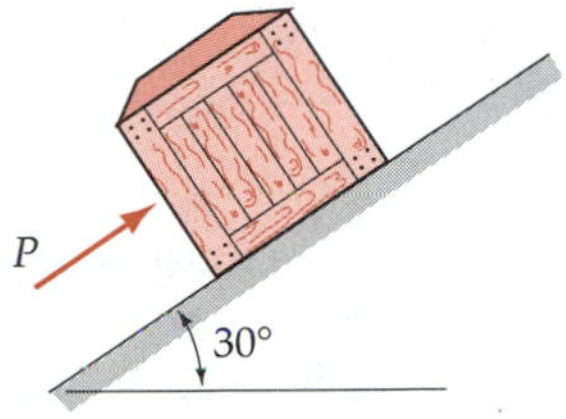

Figure E6.1a

In Figure E6.1a find the smallest force P that will prevent the crate from sliding* down the plane. The crate weighs 200 lb and the friction coefficient between it and the plane is 0.4.

Solution

We note first from Equation 6.5 that the crate will not remain in equilibrium without force P because $\tan 30° = 0.577 > \mu = 0.4$. For the smallest force P, the friction force will be at its maximum, acting up the plane as shown in the free-body diagram of Figure E6.1b. Equilibrium requires:

$$\overset{+}{\nearrow} \qquad \Sigma F_x = 0 = 200 \sin 30° - 0.4N - P \qquad (1)$$

and

$$\overset{+}{\searrow} \qquad \Sigma F_y = 0 = 200 \cos 30° - N \qquad (2)$$

From Equation (2), $N = 173$ lb. Substituting this result into Equation (1) yields

$$P = 31 \text{ lb}$$

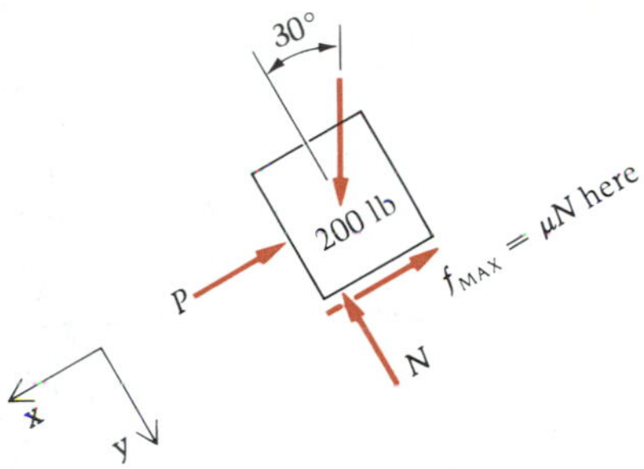

Figure E6.1b

Question 6.2 What change in Equation (1) would result in the *maximum* value of P for equilibrium of the crate?

In the preceding example, we did not concern ourselves with the possibility of the crate tipping over. In the next example, we examine both cases: slipping *and* tipping.

* Slipping and sliding mean the same thing here, and the two verbs are used interchangeably throughout the chapter.

Answer 6.2 If $0 = 200 \sin 30° \oplus 0.4N - P$, then the crate would be on the verge of slipping up the plane. This would result in the maximum P, for then P would be opposing friction as well as gravity, instead of being aided by friction as in the example.

Example 6.2

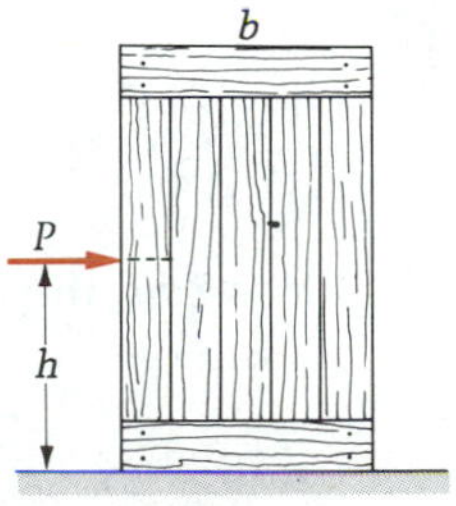

Figure E6.2a

Find the condition for which the uniform crate of Figure E6.2a pushed to the right by a slowly increasing force P, will slide before it tips over.

Solution

The equations of equilibrium are seen to be (note the free-body diagram, Figure E6.2b):

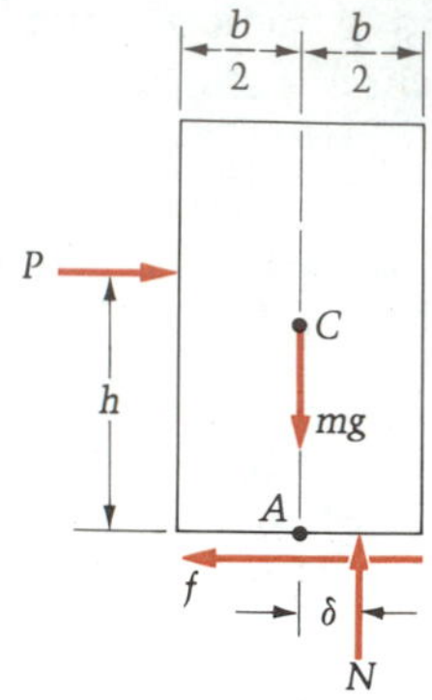

Figure E6.2b

$$\xrightarrow{+} \quad \Sigma F_x = 0 = P - f \Rightarrow P = f$$

$$+\uparrow \quad \Sigma F_y = 0 = N - mg \Rightarrow N = mg$$

$$\circlearrowleft + \quad \Sigma M_A = 0 = N\delta - Ph \Rightarrow N\delta = Ph$$

If we assume that the block slips before tipping, then at the point of slipping (or sliding), $f = \mu N$. The equations then give

$$N\delta = Ph = fh = \mu Nh$$

or

$$\delta = \mu h \qquad \text{and} \qquad P = \mu mg$$

To check whether our assumption of slipping was correct, we must determine whether or not N acts between the centerline and the lower right corner of the block; that is, δ must lie between zero and $b/2$ for slipping to occur first.

$$\delta = \mu h < \frac{b}{2}$$

or

$$b > 2\mu h$$

Thus if the base dimension b is greater than $2\mu h$, the block will slip. Equality of b and $2\mu h$ means slipping and tipping occur simultaneously, and $2\mu h > b$ implies that tipping happens first.

Let us check the results of the preceding example by starting over with a different assumption: that tipping occurs first. If so, both N and f act at the lower right-hand corner of the block as shown in Figure 6.5. We obtain, from the moment equation,

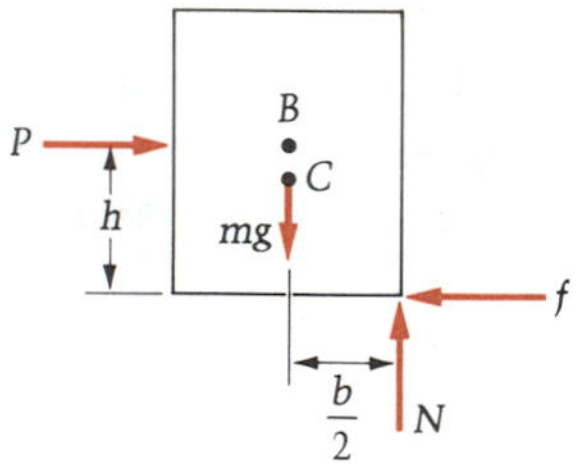

Figure 6.5

$$\circlearrowleft + \quad \Sigma M_B = 0 = N\frac{b}{2} - fh$$

or

$$f = \frac{Nb}{2h}$$

To check our assumption and see whether slipping has already occurred, we must find out if $f < \mu N$. If so, our tipping assumption was correct:

$$f = \frac{Nb}{2h} < \mu N$$

or $b < 2\mu h$ for tipping, as we had previously obtained. If $b = 2\mu h$, the crate slips and tips together, and if $b > 2\mu h$, the assumption was incorrect and it slides first, again as previously determined.

Sometimes we have to decide between more than one sliding possibility, as in the next example:

Example 6.3

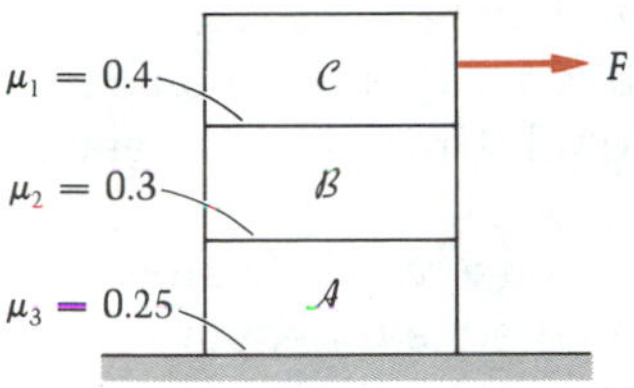

Figure E6.3a

The masses of blocks $\mathcal{C}$, $\mathcal{B}$, and $\mathcal{A}$ of Figure E6.3a are 30 kg, 5 kg, and 15 kg, respectively. Find the largest value of the force F for which no sliding will take place between any pair of surfaces. Assume the blocks to be wide enough so that tipping will not occur.

Solution

There are three cases to consider:

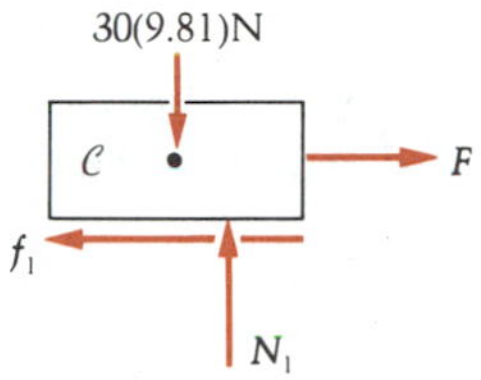

Figure E6.3b

1. Impending sliding of $\mathcal{C}$ on $\mathcal{B}$, with $\mathcal{A}$ and $\mathcal{B}$ stationary (see Figure E6.3b):

$$+\uparrow \quad \Sigma F_y = 0 = N_1 - 30(9.81) \Rightarrow N_1 = 294 \text{ N}$$

$$f_1 = f_{1\,\text{MAX}} = \mu_1 N_1 = 0.4(294) = 118 \text{ N}$$

$$\overset{+}{\rightarrow} \quad \Sigma F_x = 0 = F - f_1 \Rightarrow F = 118 \text{ N}$$

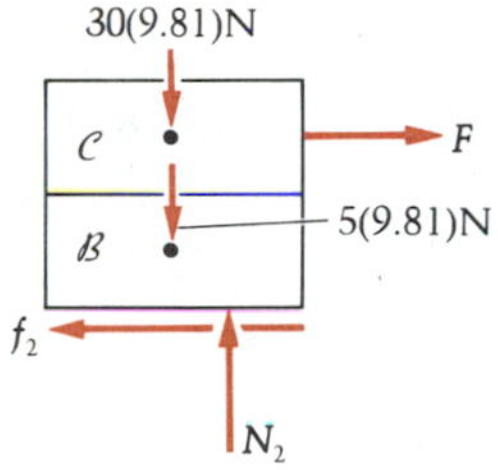

Figure E6.3c

2. Impending motion of $\mathcal{C}$ and $\mathcal{B}$ together as one body, with impending sliding of $\mathcal{B}$ on a stationary block $\mathcal{A}$ (see Figure E6.3c):

$$+\uparrow \quad \Sigma F_y = 0 = N_2 - 35(9.81) \Rightarrow N_2 = 343 \text{ N}$$

$$f_2 = f_{2\,\text{MAX}} = \mu_2 N_2 = 0.3(343) = 103 \text{ N}$$

$$\overset{+}{\rightarrow} \quad \Sigma F_x = 0 = F - f_2 \Rightarrow F = 103 \text{ N}$$

which is less than the F value to move $\mathcal{C}$ alone.

> **Question 6.3** The results of cases 1 and 2 prove that $\mathcal{C}$ will not be on the verge of slipping on $\mathcal{B}$ with $\mathcal{B}$ simultaneously on the verge of slipping on $\mathcal{A}$. Why?

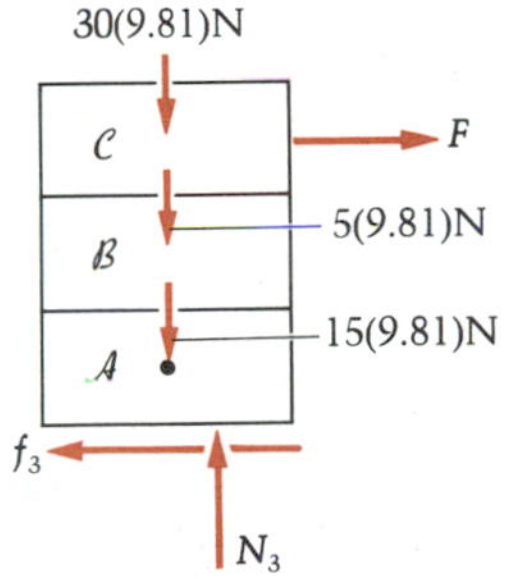

Figure E6.3d

3. All three blocks are about to move together, with impending slipping of $\mathcal{A}$ on the plane (see Figure E6.3d):

$$+\uparrow \quad \Sigma F_y = 0 = N_3 - 50(9.81) \Rightarrow N_3 = 491 \text{ N}$$

$$f_3 = f_{3\,\text{MAX}} = \mu_3 N_3 = 0.25(491) = 123 \text{ N}$$

$$\overset{+}{\rightarrow} \quad \Sigma F_x = 0 = F - f_3 \Rightarrow F = 123 \text{ N},$$

the highest of the three values of F. Therefore the maximum force F for which equilibrium of the stack of three blocks can exist is 103 newtons.

In the preceding example, intuition tells us that the case of $\mathcal{A}$ and $\mathcal{C}$ sliding to the right with $\mathcal{B}$ remaining stationary is impossible. It is interesting to study the reason why.

The free-body diagrams in Figure 6.6 show the directions of the various friction forces f_1, f_2, and f_3 upon application of F. Note that if $\mathcal{B}$ remains stationary, then f_1 balances f_2 in the figure. But for $\mathcal{A}$ to slide to

Answer 6.3 Because when F reaches 103 N, we have impending slippage of $\mathcal{B}$ on $\mathcal{A}$; however, 118 N is required to simultaneously cause impending slippage of $\mathcal{C}$ on $\mathcal{B}$.

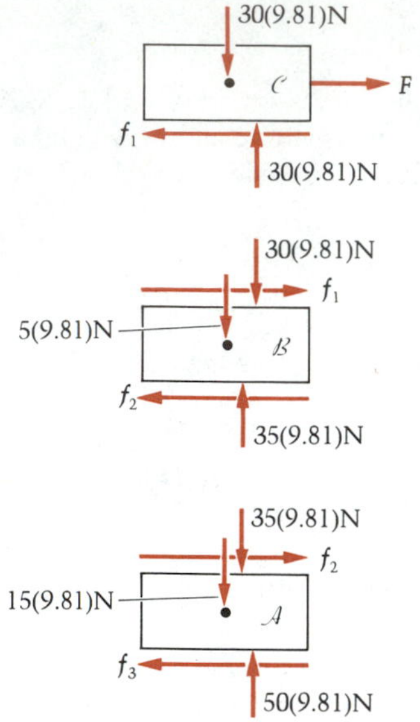

Figure 6.6

the right relative to $\mathcal{B}$ would require f_2 to be *opposite* to the directions indicated on both $\mathcal{A}$ and $\mathcal{B}$ opposing the relative sliding. Two contradictions would then result: (a) $\mathcal{A}$ would no longer have an applied force to the right to tend to make it move that way; (b) on $\mathcal{B}$, f_1 and f_2 would add and thus not balance; $\mathcal{B}$ could not then remain still. Thus $\mathcal{B}$ cannot remain stationary unless $\mathcal{A}$ does also.

The next example illustrates the fact that often a *range* of values of a quantity (here, of weights of body $\mathcal{A}$) will be possible for equilibrium:

EXAMPLE 6.4

From the discussion surrounding Equation 6.5, we know that without the rope and body $\mathcal{A}$ present in Figure E6.4a block $\mathcal{B}$ will slide down the incline if $\tan\phi > \mu$. Let $\phi = 60°$ and $\mu = 0.6$, so that

$$\tan\phi = \sqrt{3} > \mu = 0.6$$

Determine the range of values of the weight of $\mathcal{A}$ that will result in equilibrium of the system.

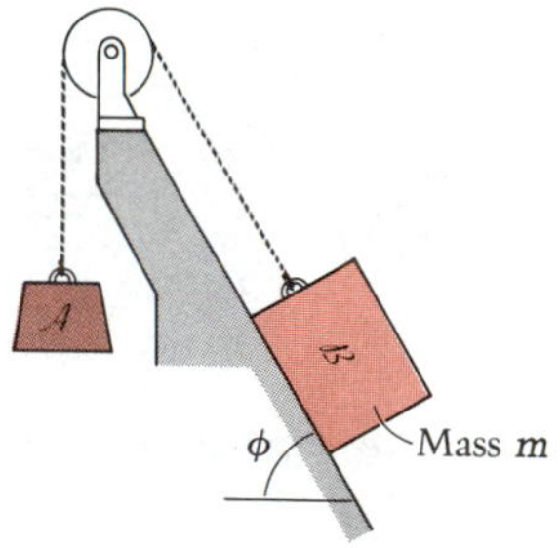

Figure E6.4a

Solution

There is a minimum value of $\mathcal{A}$ that will prevent $\mathcal{B}$ from sliding down the plane; similarly, there is also a *maximum* weight of $\mathcal{A}$ above which $\mathcal{A}$ will pull $\mathcal{B}$ *up* the plane. Between these two weights, the bodies will be in equilibrium with the friction force f beneath $\mathcal{B}$ lying between its maxima up and down the plane, respectively. In the first case, the motion of $\mathcal{B}$ impends downward (see the FBDs in Figure E6.4b). Equilibrium requires

$$\Sigma F_x = 0 = mg \sin 60° - T - f_{MAX} \tag{1}$$

$$\Sigma F_y = 0 = mg \cos 60° - N \Rightarrow N = \frac{mg}{2} \tag{2}$$

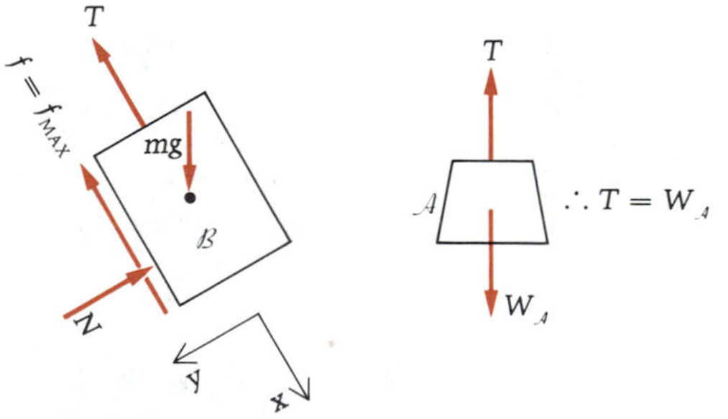

Figure E6.4b

But $f_{MAX} = \mu N = 0.6mg/2 = 0.3mg$, so that, from Equation (1),

$$T = mg\frac{\sqrt{3}}{2} - f_{MAX} = mg\left(\frac{\sqrt{3}}{2} - 0.3\right) = 0.566mg$$

and

$$W_{\mathcal{A}MIN} = T = 0.566mg$$

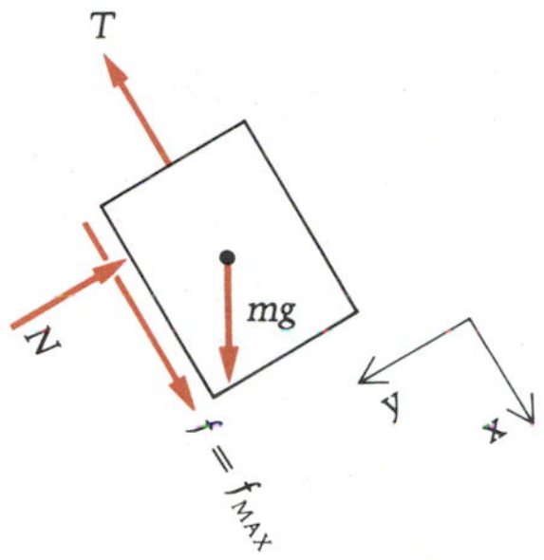

Figure E6.4c

In the second case, the motion of $\mathcal{B}$ impends upward, with friction resisting this tendency, acting downward (see Figure E6.4c and note the change in direction of f). The equations become:

$$\searrow^{+} \qquad \Sigma F_x = 0 = mg \sin 60° - T + f_{\text{MAX}} \tag{3}$$

$$\nearrow^{+} \qquad \Sigma F_y = 0 = mg \cos 60° - N \Rightarrow N = \frac{mg}{2} \quad \text{as before.} \tag{4}$$

Again, $f_{\text{MAX}} = \mu N = 0.3mg$. This time, the first equation (3) gives

$$T = mg\frac{\sqrt{3}}{2} + f_{\text{MAX}} = mg\left(\frac{\sqrt{3}}{2} + 0.3\right) = 1.17mg$$

and

$$W_{\mathcal{A}\text{MAX}} = T = 1.17mg$$

Any weight of $\mathcal{A}$ between $0.566mg$ and $1.17mg$ will result in equilibrium of the two bodies. Note that one of the values of the friction force, for $W_{\mathcal{A}}$ in this range, is zero. This happens when $T = mg \sin 60°$, or for $W_{\mathcal{A}} = mg\sqrt{3}/2 = 0.866mg$.

Our first four examples have been concerned with impending sliding of flat surfaces in contact. Friction also develops in rolling situations, as we shall see in the next few examples. In the first one, we have to determine whether the two lower cylinders are about to roll on the ground (while slipping on $\mathcal{A}$), or about to roll on $\mathcal{A}$ (while slipping on the ground):

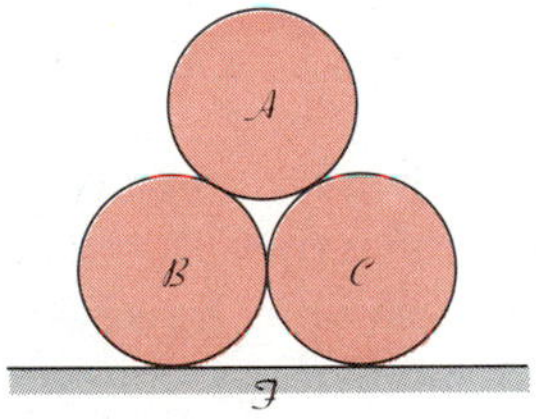

Figure E6.5a

Example 6.5

Three identical cylinders are stacked as shown in Figure E6.5a. Assuming that the friction coefficient μ is the same for all pairs of contacting surfaces, determine the minimum value of μ for which the cylinders will remain stacked in equilibrium.

Solution

From the free-body diagram of the lower-left cylinder $\mathcal{B}$ in Figure E6.5b, we can sum moments about B and immediately find a relation between the forces f and N:

$$\Sigma M_B = 0 = N\frac{r}{2} - fr(1 + \sqrt{3}/2)$$

from which

$$f = 0.268N \tag{1}$$

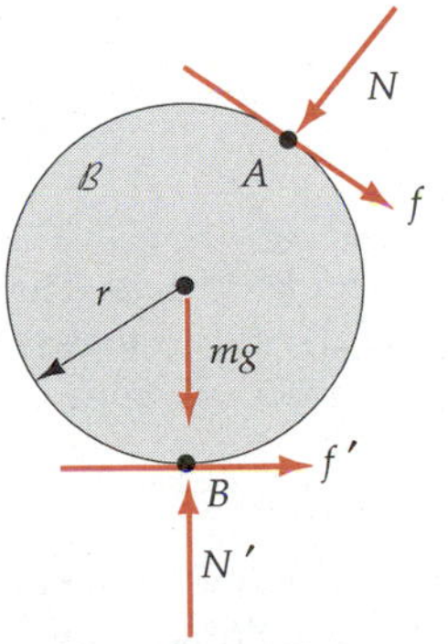

Figure E6.5b

Since f is limited by μN, the above result means that the critical (smallest) value of μ that will prevent $\mathcal{B}$ and $\mathcal{C}$ from rolling outward on the ground (while each slips on $\mathcal{A}$) is 0.268.

Question 6.4 In the free-body diagram of cylinder $\mathcal{B}$, why is there no force exerted by $\mathcal{C}$?

We must also investigate the possibility of $\mathcal{B}$ and $\mathcal{C}$ *skidding* outward on the plane while maintaining rolling (i.e., no slip) contact with the upper cylinder $\mathcal{A}$. To this end, we sum moments about A on the same free-body (Figure E6.5b):

$$\Sigma M_A = 0 = (mg - N')\frac{r}{2} + f'r(1 + \sqrt{3}/2) \tag{2}$$

If we knew the relation between mg and N', we could obtain the relation between f' and N' that we need. From the "overall" free-body in Figure E6.5c, we see that

$$\Sigma F_y = 0 = 2N' - 3mg$$

so that

$$mg = \frac{2}{3}N'$$

Substituting $\frac{2}{3}N'$ for mg in Equation (2) gives:

$$f' = \frac{0.268}{3}N' = 0.089N' \tag{3}$$

Therefore, to prevent "skidding" of $\mathcal{B}$ and $\mathcal{C}$ on the plane, the friction coefficient μ needs to be only one-third that of the other case. Therefore the value needed to prevent *both* motions is the larger value, 0.268. Note that Equations (1) and (3) tell us the manner in which equilibrium can be lost.

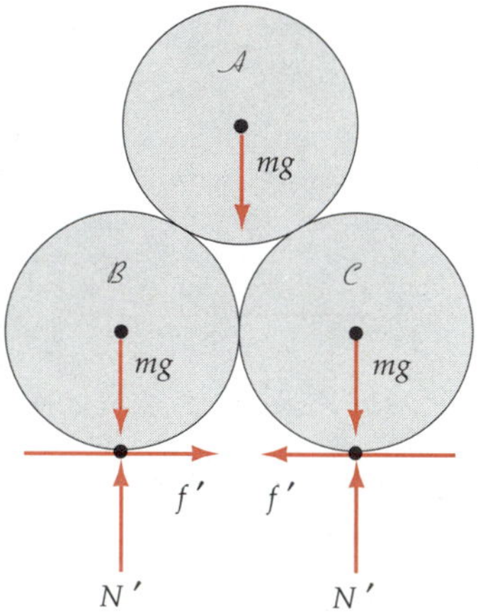

Figure E6.5c

We note from the preceding example that it is not always necessary to use all of the available equilibrium equations (we did not write $\Sigma F_x = 0$ or $\Sigma F_y = 0$ in the first part of the solution). In the next example, however, we will need all three equations for the body being studied:

Answer 6.4 For impending motion of $\mathcal{B}$ to the left and $\mathcal{C}$ to the right, the force of $\mathcal{C}$ on $\mathcal{B}$ has gone to zero.

EXAMPLE 6.6

The truck in Figure E6.6a is preparing to push a drum up an 8° incline. The coefficient of friction between the drum and the incline is 0.3. Find the range of positive values of the friction coefficient μ between the drum and truck bumper so that the drum rolls (does not slip) on the ground.

Figure E6.6a

Solution

Let us assume that the drum is in equilibrium — that is, that its motion has not quite yet begun. Making use of the FBD in Figure E6.6b, the equilibrium equations are:

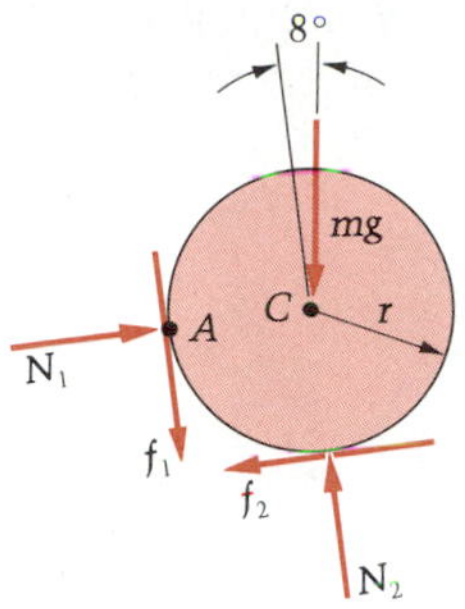

Figure E6.6b

$$\Sigma F_x = 0 = N_1 - f_2 - mg \sin 8° \tag{1}$$

$$\Sigma F_y = 0 = N_2 - f_1 - mg \cos 8° \tag{2}$$

$$\Sigma M_C = 0 = f_1 r - f_2 r \Rightarrow f_1 = f_2 \tag{3}$$

Thus we have three equations in the four unknowns N_1, N_2, f_1,and f_2. Assuming that slipping impends at A (which it will if the drum is about to roll on the ground) means

$$f_1 = \mu N_1 \tag{4}$$

Thus (1) and (3) give

$$N_1(1 - \mu) = mg \sin 8°$$

$$N_1 = \frac{mg \sin 8°}{1 - \mu} \tag{5}$$

Equation (2) gives

$$N_2 = \mu N_1 + mg \cos 8°$$

$$= \frac{\mu}{1-\mu} mg \sin 8° + mg \cos 8°$$

$$N_2 = mg\left(\cos 8° + \frac{\mu}{1-\mu} \sin 8°\right) \tag{6}$$

And from Equations (3), (4), and (5) we have

$$f_2 = f_1 = \frac{\mu mg \sin 8°}{1 - \mu} \tag{7}$$

If the drum is not to slip on the incline, then

$$f_2 \le 0.3\, N_2 \tag{8}$$

Substituting for f_2 from Equation (7) and N_2 from (6), and canceling mg,

$$\frac{\mu}{1-\mu} \sin 8° \le 0.3\left[\cos 8° + \frac{\mu}{1-\mu} \sin 8°\right]$$

$$0.7 \frac{\mu}{1-\mu} \sin 8° \le 0.3 \cos 8°$$

$$\mu 0.0974 \le (1 - \mu)(0.297)$$

$$\mu \le \frac{0.297}{0.394} = 0.754$$

In the preceding example, note that if the "$f \le \mu N$" *in*equality is carried forward (as was done by substituting into (8)), then the final result is an entire *range* of values of μ ("anything less than or equal to 0.754") instead of just a number. The inequality thus gives us a more complete result than if we were to set $f = \mu N$ and solve for one number for μ, for then we would have to think about whether the desired range consisted of all values above this μ, or all values below it.

In the next example, the cylinder is prevented from rolling on the plane; it slips there instead, for a large enough θ, because of the constraint imposed by the cord.

EXAMPLE 6.7

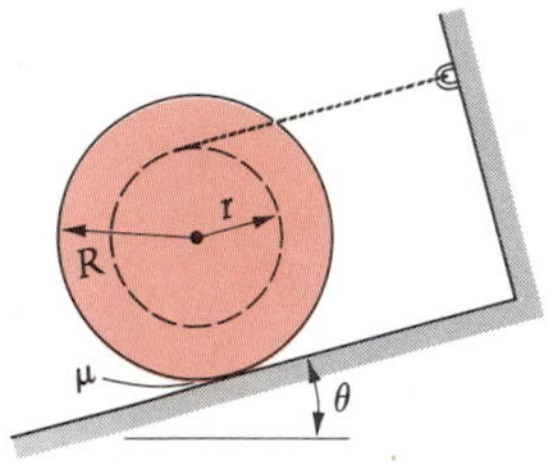

Figure E6.7a

The angle θ in Figure E6.7a is slowly increased from zero. Find the value of θ at which the slotted cylinder of mass m will begin to slip.

Solution

The cylinder cannot roll on the plane because it is restrained by the cord, which is assumed to be wrapped tightly and not slipping in the slot. At a certain value of θ, the cylinder will slip downward on the plane while pivoting about the point A shown on the free-body diagram. It will do this when the value of f needed for equilibrium exceeds the maximum value possible for f, which is μN. Referring to the FBD, Figure E6.7b, we write the equilibrium equations:

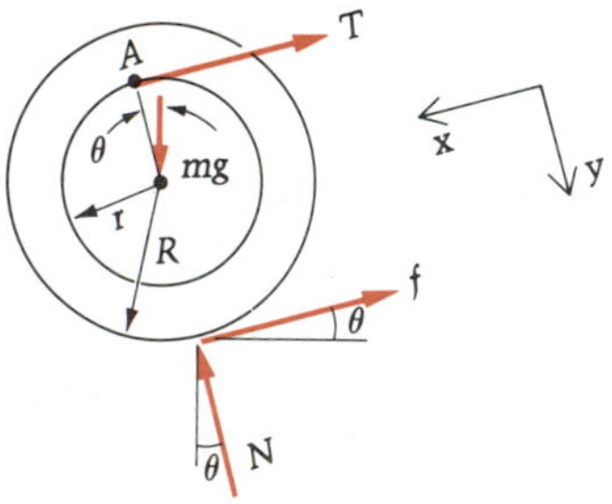

Figure E6.7b

$$\Sigma F_x = 0 = mg \sin\theta - T - f \quad (1)$$

$$\Sigma F_y = 0 = mg \cos\theta - N \Rightarrow N = mg\cos\theta \quad (2)$$

$$\Sigma M_A = 0 = -rmg\sin\theta + f(R + r) \Rightarrow f = \frac{rmg\sin\theta}{R + r} \quad (3)$$

Enforcing the no-slipping condition,

$$f \le \mu N$$

$$\frac{rmg\sin\theta}{R + r} \le \mu mg\cos\theta$$

$$\tan\theta \le \frac{\mu(R + r)}{r}$$

$$\theta_{\text{slip}} = \tan^{-1}\left(\frac{\mu(R + r)}{r}\right)$$

This angle could be substituted if desired into Equations (3) and (1) to yield the values of the friction force f and the tension T at the slipping angle θ_{slip}.

In the next example, because a moment equation is involved, we must locate the center of mass before we can proceed to the friction part of the problem.

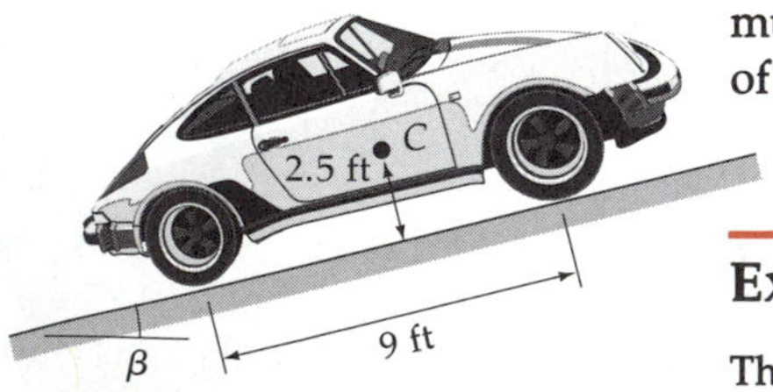

Figure E6.8a

EXAMPLE 6.8

The front wheels of a certain car (see Figure E6.8a) support 55% of its weight W on level ground. The coefficient of friction between the tires and an inclined road surface is 0.3. What is the largest incline angle β for which the car will not slip when parked? (Only the rear wheels are locked.)

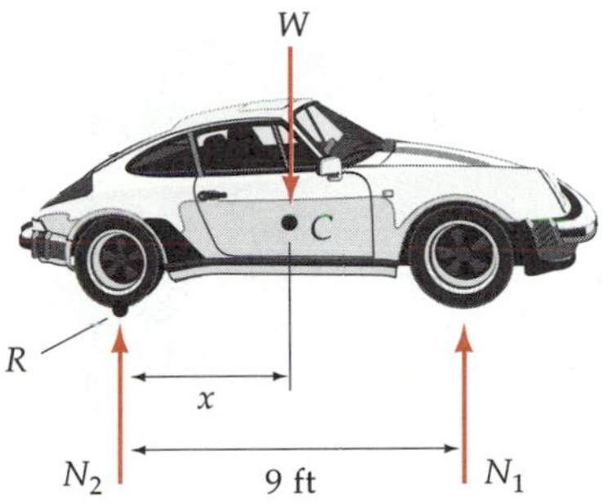

Figure E6.8b

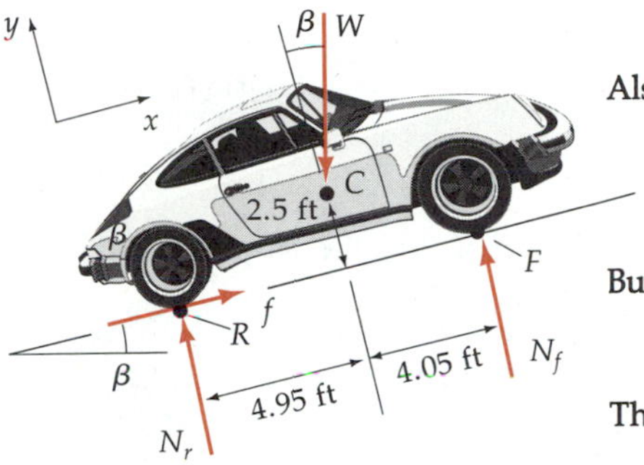

Figure E6.8c

Solution

In equilibrium on level ground, by summing moments about R we find that the mass center C of the car is 4.95 ft forward of the rear wheels (see the free-body diagram in Figure E6.8b):

$$\circlearrowleft\!+ \quad \Sigma M_R = 0 = 9N_1 - W$$
$$= 9(0.55\ W) - xW$$
$$x = 4.95 \text{ ft}$$

Putting the car on the incline, we sum moments about the bottom (F) of the front wheel and determine N_r (see the free-body diagram in Figure E6.8c):

$$\circlearrowleft\!+ \quad \Sigma M_F = 0 = W \cos \beta(4.05\ \text{ ft}) + W \sin \beta(2.5 \text{ ft}) - N_r(9 \text{ ft}) \tag{1}$$
$$N_r = W[0.45 \cos \beta + 0.278 \sin \beta] \tag{2}$$

Also,

$$\Sigma F_x = 0 = f - W \sin \beta$$
$$f = W \sin \beta \tag{3}$$

But for the car not to slip,

$$f \le \mu N_r \tag{4}$$

Thus, substituting N_r from Equation (2) and f from (3) into (4),

$$W \sin \beta \le 0.3W[0.45 \cos \beta + 0.278 \sin \beta]$$

or

$$0.917 \sin \beta \le 0.3(0.45) \cos \beta$$
$$\tan \beta \le 0.147$$
$$\beta \le 8.36°$$

In the preceding example we note that, all other parameters remaining equal, raising the mass center (increasing the 2.5-ft dimension) increases the "$W \sin \beta$" coefficient in Equation (1). This then has the effect of raising both N_r and the angle β that comes out of the inequality above. Physically, a larger N_r with the same μ gives more maximum frictional resistance to slipping.

In the next example, the beam and the attached shoe act as a brake on the cylinder:

EXAMPLE 6.9

The uniform beam $\mathcal{B}$ in Figure E6.9a weighs 1000 N, and is attached to the light shoe $\mathcal{S}$, which rests on the cylinder $\mathcal{C}$. How much weight can be suspended in equilibrium by a cord wrapped tightly around the hub of $\mathcal{C}$ as shown?

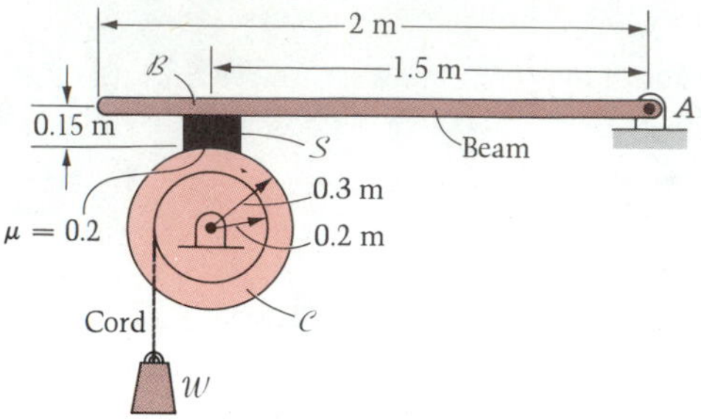

Figure E6.9a

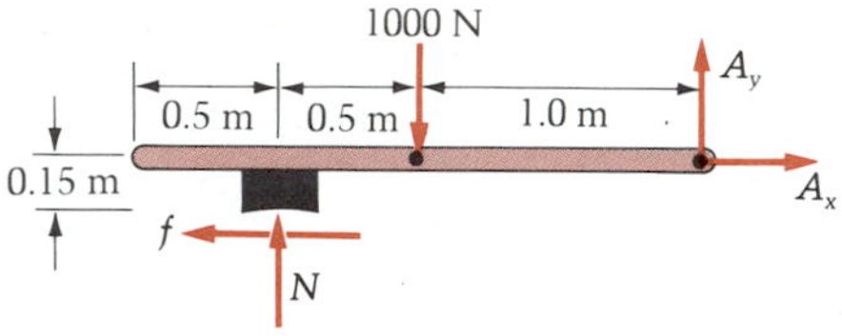

Figure E6.9b

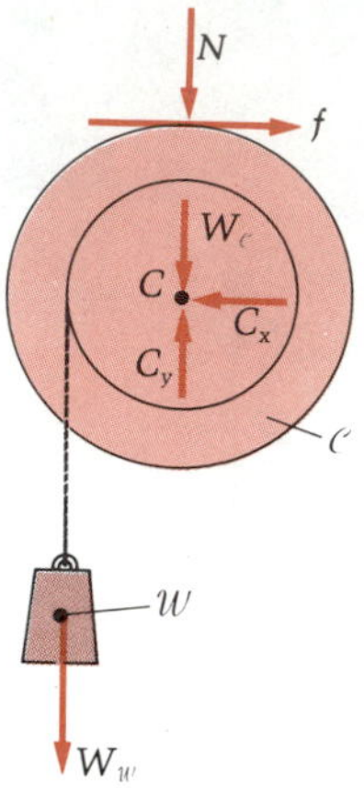

Figure E6.9c

Solution

Free-body diagrams of (a) beam plus shoe, and of (b) cylinder plus weight are shown in Figure E6.9b and E6.9c.

If we imagine the weight W to be slowly increased, it will eventually reach a critical value at which cylinder $\mathcal{C}$ will begin to turn counterclockwise. This has to happen because the free-body diagram of ($\mathcal{C} + \mathcal{W}$) shows that the only moment about C available to equilibrate that of W_w is the moment of the friction force f. That moment is bounded by μN times the radius 0.3 m. Thus we first need N. From the free-body diagram of ($\mathcal{B} + \mathcal{S}$), Figure E6.9b, we see that:

1. Writing $\Sigma F_x = 0$ and $\Sigma F_y = 0$ will not be of help because they will include A_x and A_y, which are undesired unknowns.
2. Moments about A will allow us to determine N without finding A_x and A_y:

$$\circlearrowleft\!+ \quad \Sigma M_A = 0 = (1000 \text{ newtons})(1 \text{ m}) - N(1.5 \text{ m}) - f(0.15 \text{ m}) \quad (1)$$

At the critical value of W, which we seek, f becomes f_{MAX}, or μN, so that

$$N = \frac{1000}{1.53}$$

$$= 654 \text{ newtons}$$

We return to the free-body diagram of ($\mathcal{C} + \mathcal{W}$), Figure E6.9c. The same remark (1) above again holds, this time with the pin reactions C_x and C_y unknown and unwanted. Thus again we proceed directly to the moment equation that eliminates these reactions from consideration:*

$$\circlearrowleft\!+ \quad \Sigma M_C = 0 = W\,(0.2 \text{ m}) - (0.2N)(0.3 \text{ m}) \quad (2)$$

so that

$$W = \frac{(0.2)(654)(0.3)}{0.2} = 196 \text{ newtons}$$

* If we were seeking G_x and G_y in addition to W, we would *still* write $\Sigma M_C = 0$ first to get W. Then, writing $\Sigma F_x = 0$ and $\Sigma F_y = 0$ would yield the pin reactions.

It is instructive to consider what changes might occur in the solution if the weight comes off the *right* side of the hub of the cylinder instead of the left side.

Question 6.5 Will the answer then be the same?

Self-locking Mechanisms

In the next example, we consider the phenomenon called **self-locking.** This refers to a mechanism in which there will be no slipping regardless of increases in the size of the applied force. The reason for self-locking, as you will see, is that in such situations the normal force and friction force increase proportionately so that f can never exceed μN if μ is large enough to begin with.

Answer 6.5 No, it won't. The friction will reverse direction on both $\mathcal{C}$ and $\mathcal{S}$, and Equation (1) will then give a different (larger) N. Thus f will be larger and so will W.

EXAMPLE 6.10

A cam consisting of a cylinder mounted off-center (see Figure E6.10a) is to be used as a self-locking mechanism. Whenever the force P acts downward on the plate, the cam tends to rotate clockwise and grip the plate, holding it fast. Neglecting the weights of cam and plate in comparison with the large force P, and assuming the wall that bears against the plate is smooth, find the smallest value of the coefficient of friction between cam and plate for which the self-locking will occur.

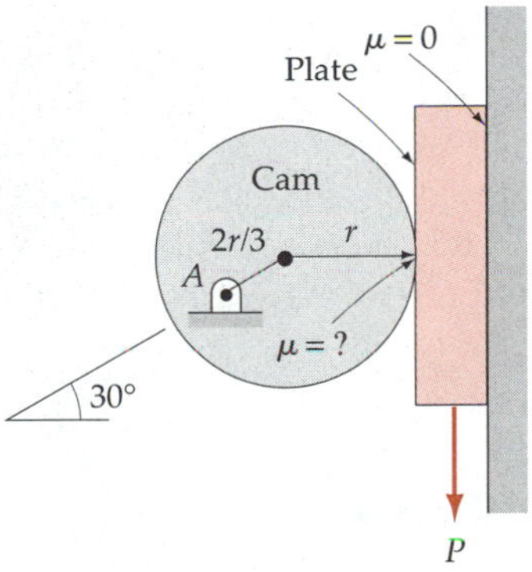

Figure E6.10a

Solution

Free-body diagrams of the cam and plate are shown in Figures E6.10b and E6.10c:

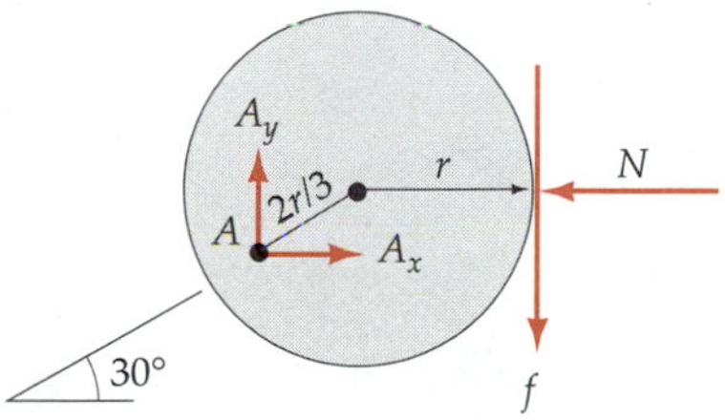

Figure E6.10b

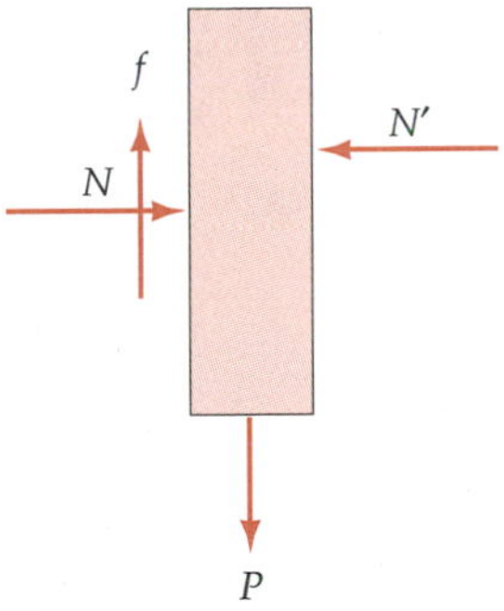

Figure E6.10c

We see by inspection of Figure E6.10c that $f = P$ for equilibrium of the plate.*

* Also, $N' = N$ and, for a balance of moments, f and P form a clockwise couple which is the negative of another formed by N and N'.

Thus the magnitude of f cannot be limited if equilibrium is to exist regardless of the size of P. We move next to the FBD of the cam, Figure E6.10b.

As the pin reactions A_x and A_y are not of interest to us here, we sum moments about A in Figure E6.10b and thereby eliminate them:

$$\circlearrowleft\!+ \quad \Sigma M_A = 0 = -f\left(r + \frac{2r}{3}\frac{\sqrt{3}}{2}\right) + N\frac{2r}{3}\frac{1}{2}$$

or

$$f = 0.211N \qquad (1)$$

But

$$f \leq f_{\text{MAX}} = \mu N$$

or, substituting for f using Equation (1),

$$0.211N \leq \mu N$$

so that if $\mu > 0.211$, the plate will never slip downward past the cam regardless of the size of P. As P gets larger, (a) so does f since $f = P$, and (b) so does N by Equation (1).

Question 6.6 In the preceding example, could the self-locking occur if the cam were pinned at the center? to the left of the center?

The last example in the section is longer than the others. It involves a body (tool) that must be taken apart to complete the solution. The reader should note that working such a problem in terms of symbols gives a designer much more flexibility when it comes time to analyze the effects of different dimensions on the answers.

Answer 6.6 No, it would develop no friction. No, it would lose contact.

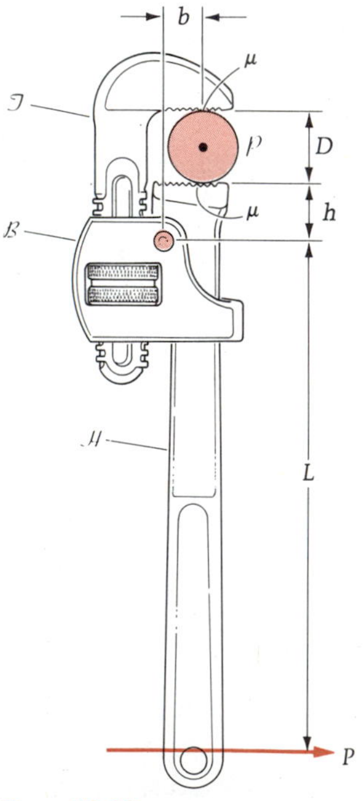

Figure E6.11a

EXAMPLE 6.11

The pipe wrench (or Stillson wrench) in Figure E6.11a is gripping a pipe $\mathcal{P}$ that is about to be unscrewed from a fitting by the force P as shown in the figure. The member $\mathcal{B}$ is pinned to $\mathcal{H}$ and is loose-fitting on $\mathcal{J}$. The friction coefficients μ between $\mathcal{J}$ and $\mathcal{P}$ and between $\mathcal{H}$ and $\mathcal{P}$ are given to be equal. Find the relationship between μ, D, h, b, and L for which the pipe wrench will be "self-locking," meaning that it will not slip on the pipe *regardless of the size of P*. Neglect all weights, and for the purposes of this example, make the assumption that the teeth of the wrench do not significantly dig into the surface of the pipe.

Solution

Separate free-body diagrams of the wrench as well as two of the parts of the wrench are shown in Figures E6.11b–d. We see immediately from Figure E6.11b that

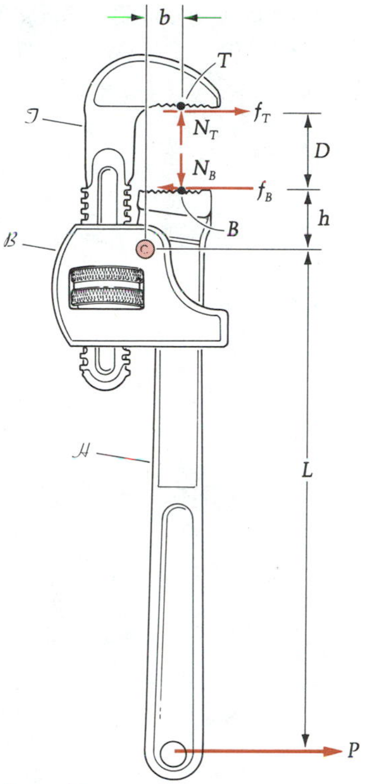

Figure E6.11b

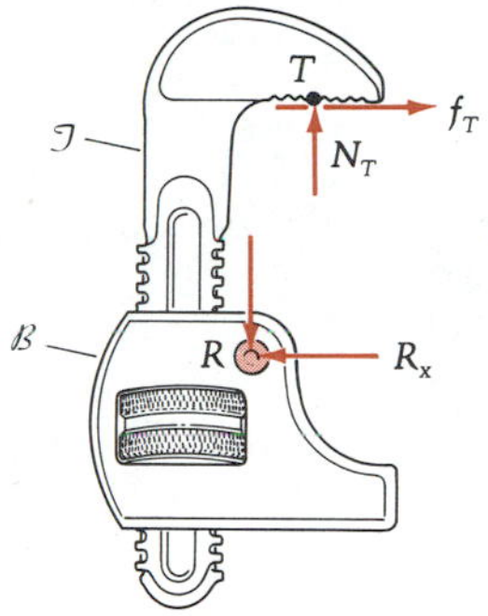

Figure E6.11c

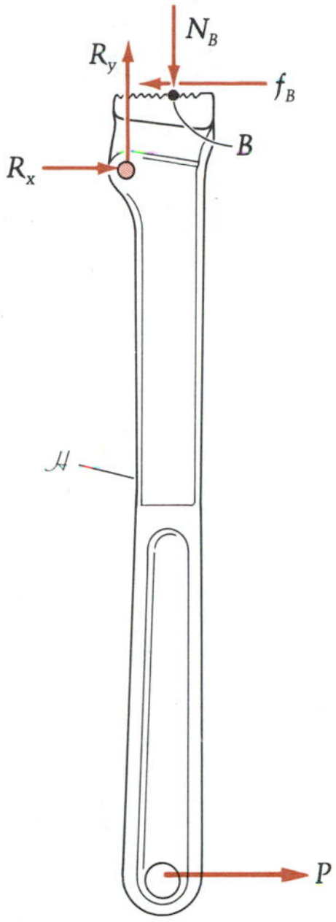

Figure E6.11d

$$\xrightarrow{+} \quad \Sigma F_x = 0 = f_T - f_B + P \Rightarrow P = f_B - f_T \qquad (1)$$

$$+\uparrow \quad \Sigma F_y = 0 = N_T - N_B \Rightarrow N_T = N_B \qquad (2)$$

$$\circlearrowleft\!+ \quad \Sigma M_T = 0 = (D + h + L)P - f_B D \qquad (3)$$

Notice that we are not yet stating that either of the friction forces f is equal to μN. We shall first get the equilibrium equations written, then investigate the slipping possibilities.

Using the free-body diagram (Figure E6.11c) of the combined body ($\mathscr{B} + \mathscr{T}$), we get

$$\xrightarrow{+} \quad \Sigma F_x = 0 = f_T - R_x \Rightarrow f_T = R_x \qquad (4)$$

$$+\uparrow \quad \Sigma F_y = 0 = N_T - R_y \Rightarrow N_T = R_y \qquad (5)$$

$$\circlearrowright\!+ \quad \Sigma M_R = 0 = N_T b - f_T(D + h) \qquad (6)$$

Using Equation (6) we can immediately obtain the minimum value of the coefficient of friction μ needed to prevent slipping at the top of the pipe (point T):

$$f_T = N_T \frac{b}{D + h} \le f_{T\text{MAX}} = \mu N_T \qquad (7)$$

or

$$\mu \ge \frac{b}{D + h} \qquad (8)$$

We proceed to obtain a similar condition on μ for which slipping will not occur at the bottom point (B) of the pipe. Equations (1) and (3) give, eliminating P,

$$f_B - f_T = \frac{f_B D}{D + h + L} \tag{9}$$

or, using Equation (7),

$$f_B\left(\frac{h + L}{D + h + L}\right) = f_T = N_T \frac{b}{D + h} \tag{10}$$

But $N_T = N_B$ by Equation (2); thus

$$f_B\left(\frac{h + L}{D + h + L}\right) = \frac{b}{D + h} N_B \tag{11}$$

or

$$f_B = \frac{b(D + h + L)}{(h + L)(D + h)} N_B \leq f_{B_{MAX}} = \mu N_B \tag{12}$$

so that, this time,

$$\mu \geq \frac{b(D + h + L)}{(h + L)(D + h)} \tag{13}$$

This coefficient is slightly higher than the one given in inequality (8), so there will be no slipping on either contact point (T or B) with the pipe if $\mu \geq b(D + h + L)/[(h + L)(D + h)]$.

For the larger wrench in the photo (see Figure E6.11e), for example, we see that D, h, b, and L are, respectively, 1 in., 1.25 in., 0.75 in., and 10.3 in. Therefore, the inequalities (8) and (13) give

$$\mu \geq \frac{0.75}{1 + 1.25} = 0.33 \quad \text{needed to prevent slip at } T$$

and

$$\mu \geq \frac{0.75(1 + 1.25 + 10.3)}{(1.25 + 10.3)(1 + 1.25)} = 0.36 \quad \text{needed to prevent slip at } B.$$

With the teeth that are actually built into the jaws of a pipe wrench, 0.36 is easily obtained.

Note from the photos that the jaws are not exactly parallel; resistance to slip is further increased because of this angle. Note also that this result, $\mu = 0.36$, is for a 1-inch pipe ($D = 1$ in.); for other diameters, different answers for μ will be obtained.

Question 6.7 In designing a pipe wrench, which should be used in determining the minimum coefficient of friction needed: the largest or the smallest pipe expected to be turned by the wrench?

The reader may wish to show that for the smaller wrench in the figure, less friction is needed for self-locking against a 1-inch pipe. Another follow-up exer-

Answer 6.7 The smallest D, using Equation (13), will give us the minimum μ needed.

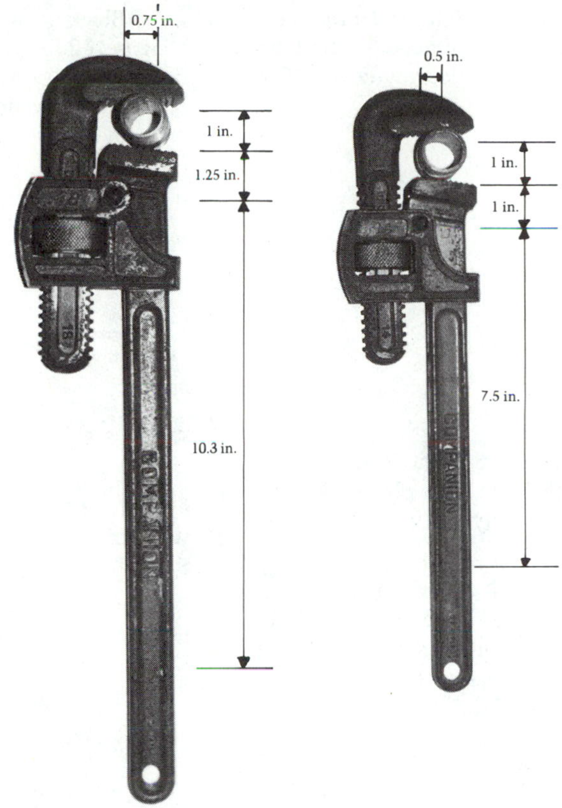

Figure E6.11e

cise is to solve Equations (1)-(6) for f_T, N_T, f_B, N_B, R_x, and R_y in terms of P, and then to show that the equilibrium equations for body $\mathcal{H}$ (Figure E6.11d) are satisfied by these results.

PROBLEMS ▶ Section 6.2

6.1 The electronics cabinet in Figure P6.1 is 7 ft high, weighs 200 lb and has its mass center at C. The coefficient of friction between the cabinet and the floor is 0.3. Without disturbing equilibrium, how large a force P can be applied

a. to the right?

b. to the left?

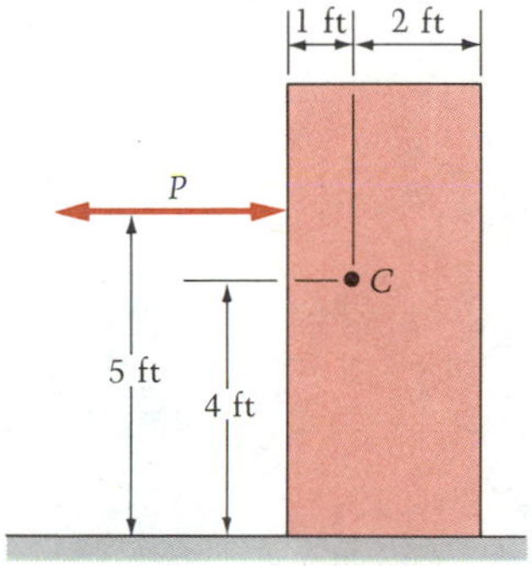

Figure P6.1

6.2 The friction coefficients for the bodies in Figure P6.2 are: 0.4 between $\mathcal{B}_1$ and $\mathcal{B}_2$ and 0.2 between the floor and $\mathcal{B}_2$. The mass of $\mathcal{B}_1$ is 10 kg. Find the minimum force P needed to disturb equilibrium if (a) mass of $\mathcal{B}_2 = 8$ kg; (b) mass of $\mathcal{B}_2 = 12$ kg.

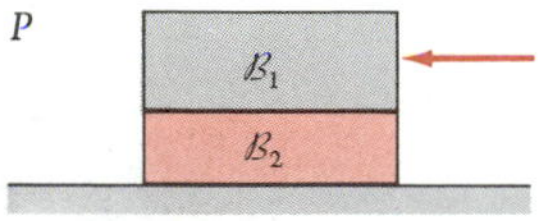

Figure P6.2

6.3 If each of $\mathcal{W}_1$ and $\mathcal{W}_2$ weighs 100 lb, and if the coefficient of friction for all surfaces of contact is 0.2, find the angle α for which $\mathcal{W}_1$ begins to slide downward. (See Figure P6.3.)

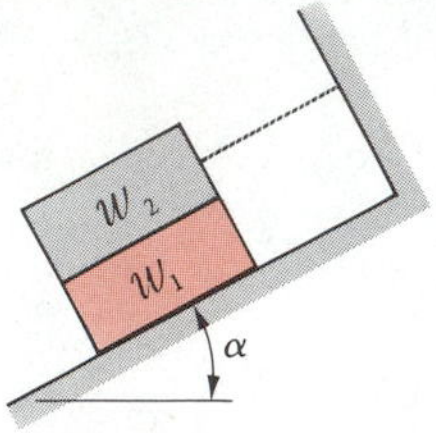

Figure P6.3

6.4 Return to Example 6.1 and find the *largest* force P for which the block will not slide *up* the plane. What do you conclude will happen for any value of P between 31 lb and your answer?

6.5 The block of weight $W = 100$ newtons is in equilibrium on the inclined plane (see Figure P6.5). The force T is applied as shown, and slowly increases from zero. At what value of T will equilibrium cease to exist? How is the equilibrium lost at this value of T?

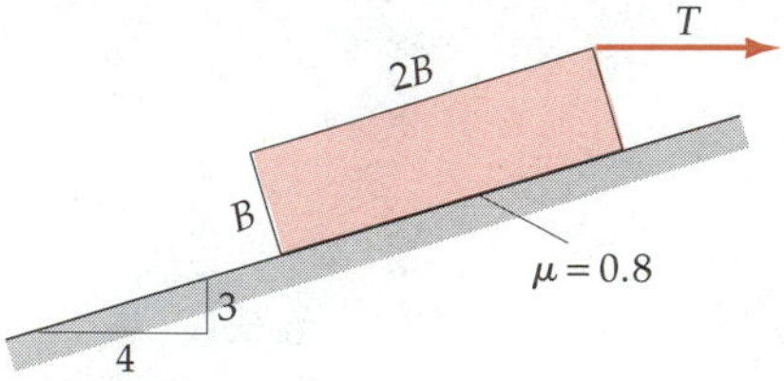

Figure P6.5

6.6 In Figure P6.6 blocks $\mathcal{A}$ and $\mathcal{B}$ weigh 50 lb and 100 lb, respectively. The coefficients of friction are: $\mu_1 = 0.2$ between $\mathcal{A}$ and $\mathcal{B}$, and $\mu_2 = 0.35$ between $\mathcal{B}$ and the horizontal plane. Find the smallest force P that will cause $\mathcal{A}$ to move to the left.

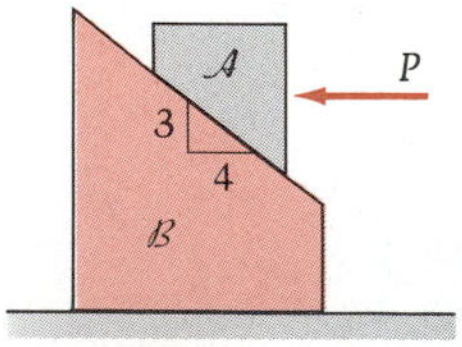

Figure P6.6

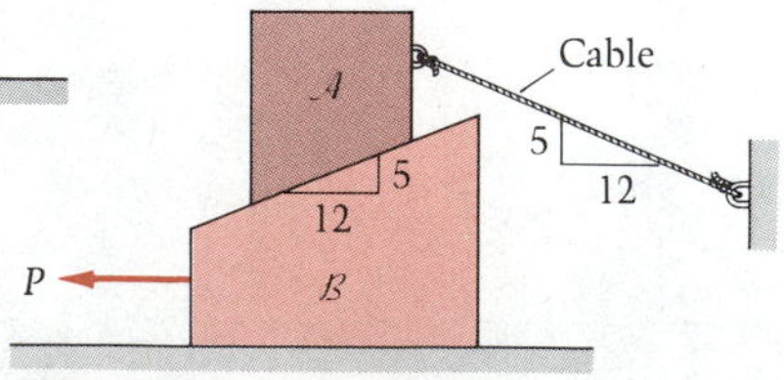

Figure P6.7

6.7 Block $\mathcal{A}$, weighing 50 lb, rests on block $\mathcal{B}$, which weighs 100 lb. (see Figure P6.7.) The coefficient of friction between $\mathcal{A}$ and $\mathcal{B}$ is 0.4. Neglecting friction between $\mathcal{B}$ and the floor, find the largest force P for which $\mathcal{B}$ does not move.

6.8 In Figure P6.8 find the smallest force P for which the blocks will slide if angle ϕ is 20°.

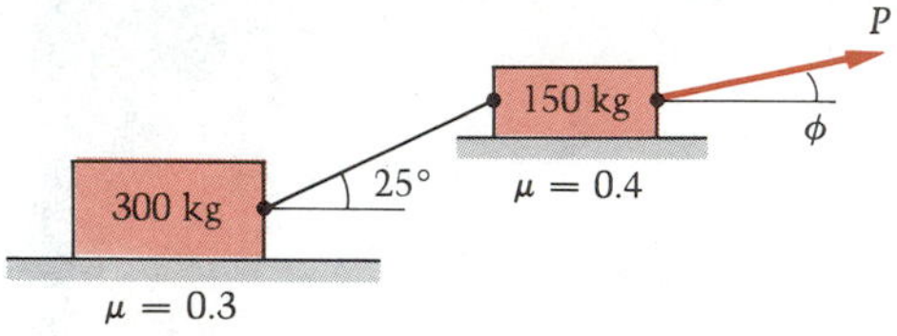

Figure P6.8

6.9 Find the maximum force P for which no slipping will occur anywhere for the system of blocks in Figure P6.9.

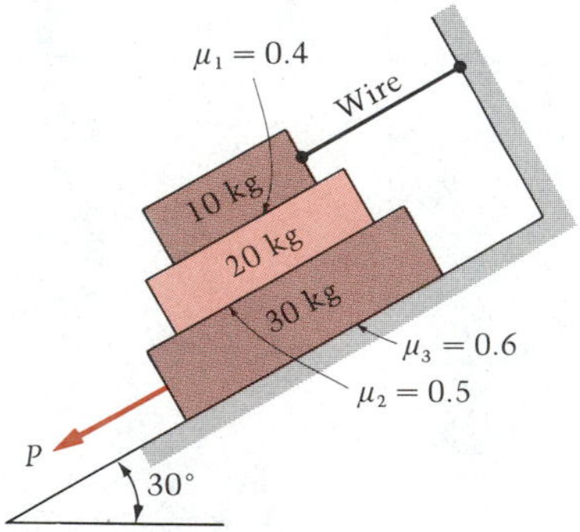

Figure P6.9

6.10 Repeat the preceding problem, if the values of μ_1 and μ_3 are reversed.

6.11 Two identical triangular blocks, each of mass m, are put together to form a rectangular solid, and they are being held in equilibrium by the force P (see Figure P6.11). If P is then increased until motion impends, at what value of P will this occur? Consider *all* possibilities.

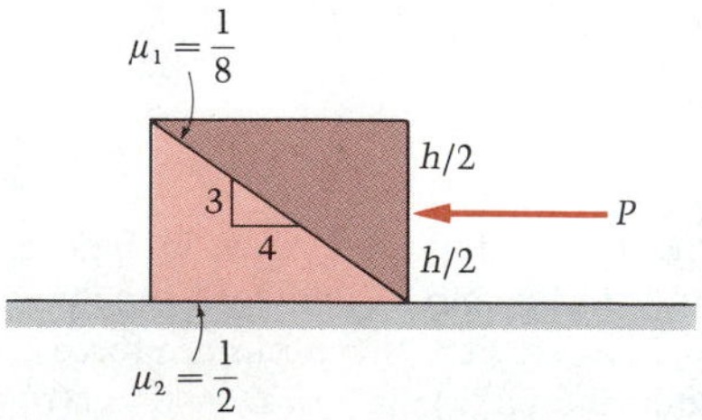

Figure P6.11

6.12 In Figure P6.12 a small box rests on the parabolic incline at the point $(1, \frac{1}{2})$. If the box is on the verge of slipping, what is the coefficient of friction between it and the incline? Also, find the force P that must be exerted on the box, tangent to the incline, to start it moving *upward.*

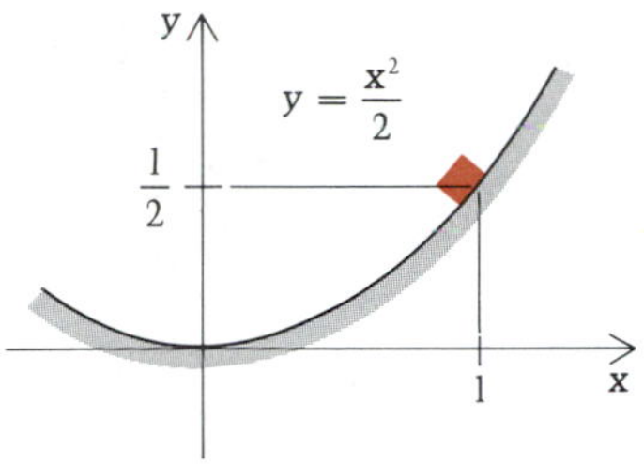

Figure P6.12

6.13 In Figure P6.13 a bug falls into a bowl that has the shape of an inverted spherical cap of radius R. The coefficient of friction between the bowl and the bug's feet is $\mu = \frac{1}{4}$. The bug is barely able to crawl out without slipping. Find the depth D of the bowl.

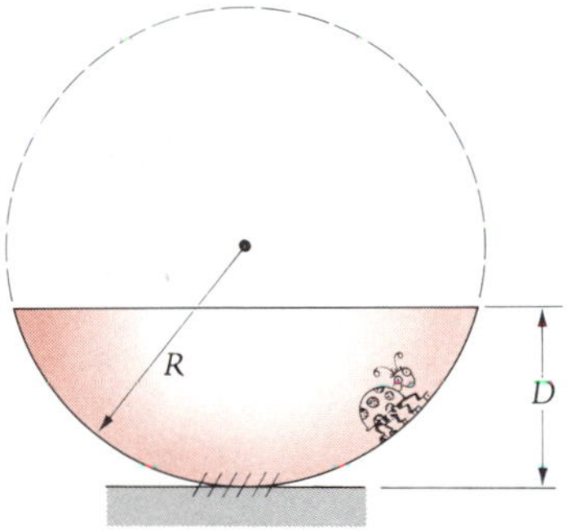

Figure P6.13

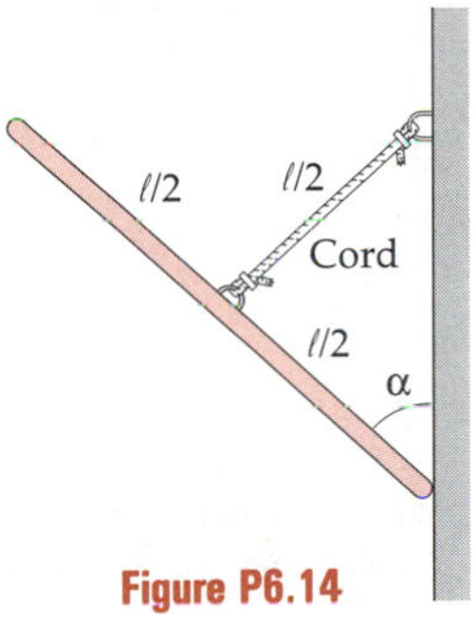

Figure P6.14

6.14 The uniform rod of mass m and length ℓ rests in the position shown in Figure P6.14. Find the minimum coefficient of friction μ between the rod and the wall for which the equilibrium can occur, as a function of angle α.

6.15 In a strength test of fiber optic cable, the nut in Figure P6.15(a) was attached to the jacket of the cable, which was then slipped through a slot in the channel (Figure P6.15(b)). The coefficient of friction between nut and channel was 0.1. The load T in Figure P6.15(c) was then slowly increased, and at 200 lb it was observed that a bend induced in the channel caused the nut to slip off the channel. At what angle θ did this occur?

6.16 The 140-N triangular block in Figure P6.16 is pulled at the top by a gradually increasing horizontal force P. The coefficient of friction between the block and the floor is $\mu = 0.4$. Find the smallest base dimension B for which the uniform block will slide instead of tipping.

6.17 The cylinder in Figure P6.17 weighs 200 lb and is in equilibrium on the inclined plane. What is the friction force at its point of contact? If it is on the verge of slipping on the plane, what is the coefficient of friction?

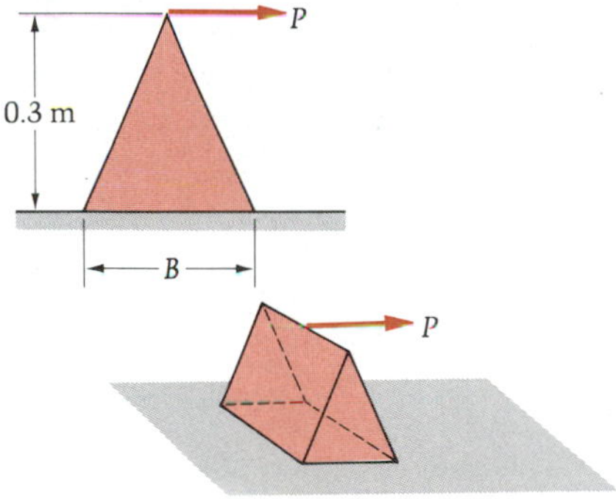

Figure P6.16

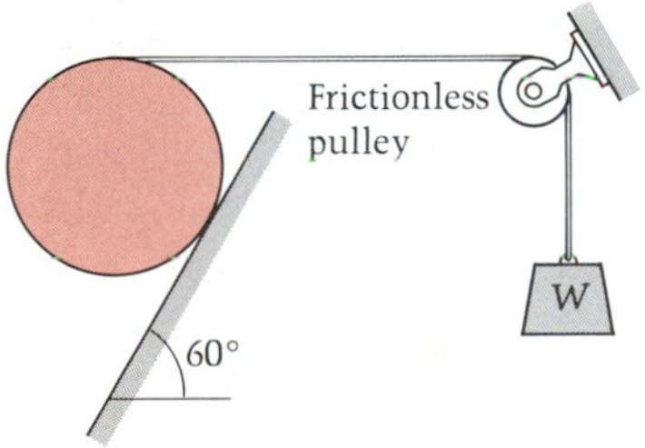

Figure P6.17

(a) (b) (c)

Figure P6.15

6.18 In Figure P6.18 block $\mathscr{B}$ weighs 50 lb. Determine the range of values of the weight of $\mathscr{A}$ that will maintain equilibrium of the system. (Be sure to check all cases!)

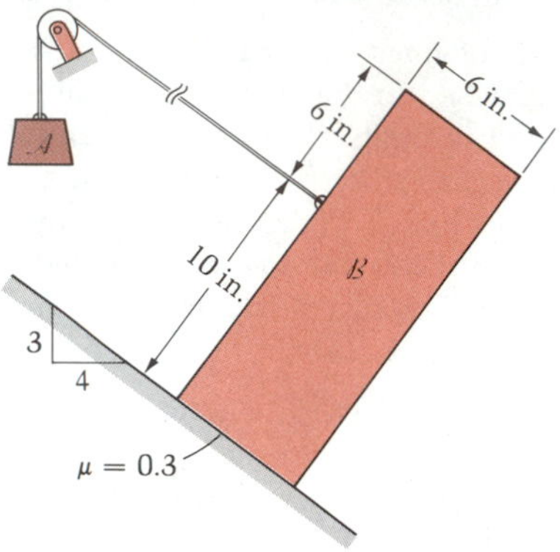

Figure P6.18

6.19 In Figure P6.19 body $\mathscr{A}$ weighs 270 N and body $\mathscr{B}$ weighs 180 N. The coefficient of static friction between body $\mathscr{A}$ and the plane is 0.30. Determine the range of values of the force P for which body $\mathscr{A}$ will be in equilibrium.

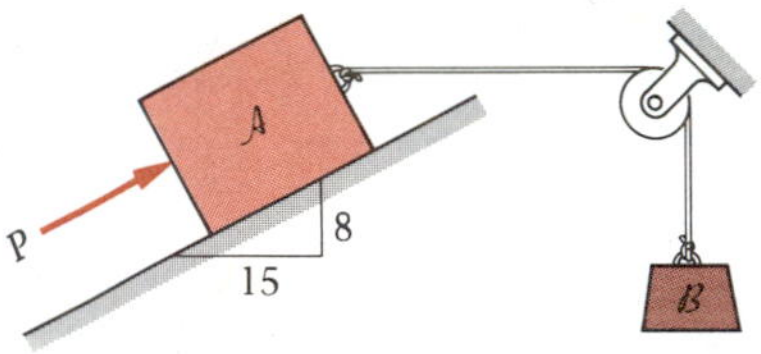

Figure P6.19

6.20 In Figure P6.20 what is the largest angle ϕ for which the slotted cylinder will remain in equilibrium when released from rest on the inclined plane? The coefficient of friction between the cylinder and the plane is $\mu = \frac{1}{6}$.

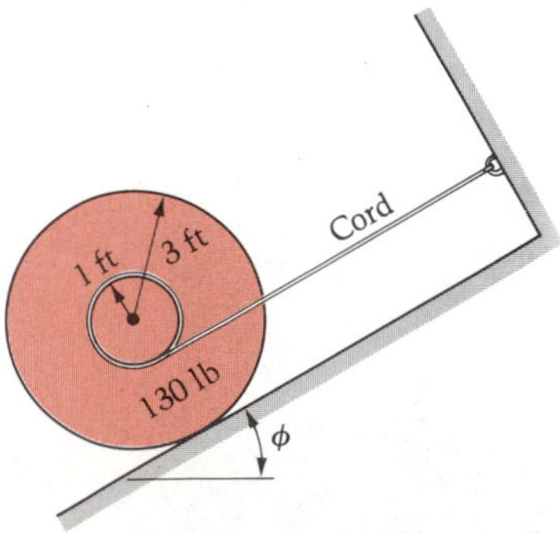

Figure P6.20

6.21 The slotted cylinder is in equilibrium on the inclined plane in Figure P6.21. If the spring is stretched 4 in., what is its modulus? What is the minimum friction coefficient between the cylinder and the plane for which the equilibrium can exist?

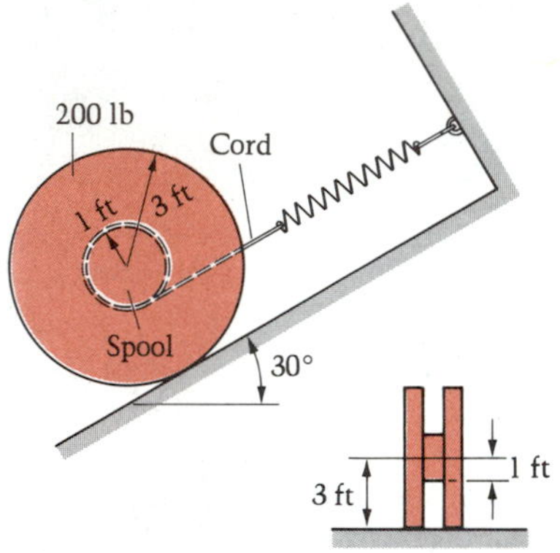

Figure P6.21

6.22 What force P will cause motion of the 100-lb cylinder in Figure P6.22 to impend?

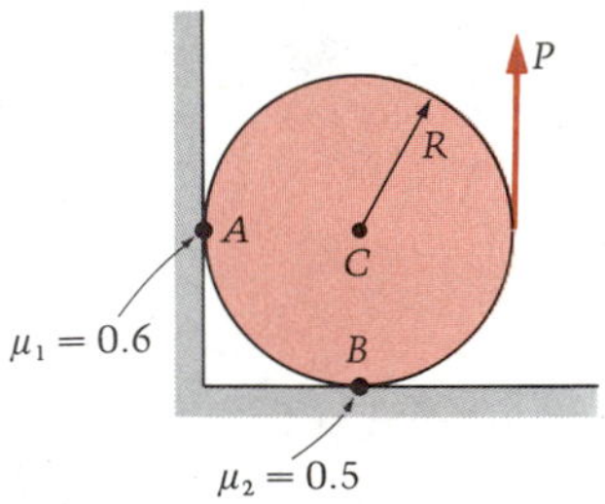

Figure P6.22

6.23 The cylinder in Figure P6.23 has a mass of 30 kg and a radius of 0.6 meter. The coefficients of static friction are 0.2 between the cylinder and the inclined plane, and 0.3 between the cylinder and the floor. Find the value of the counterclockwise couple M_0 that will cause the cylinder to begin to rotate.

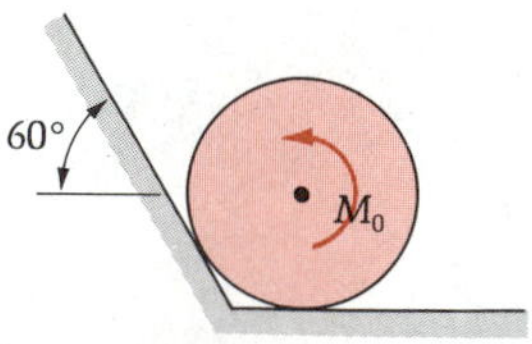

Figure P6.23

6.24 In Figure P6.24 a force P is applied to the rope by a man who erroneously thinks that if he pulls to the left, the 200-lb wheel should roll to the left. If the friction coefficient between the wheel and both the wall and floor is 0.35, determine the force P that will cause the wheel to move. *How* will it move?

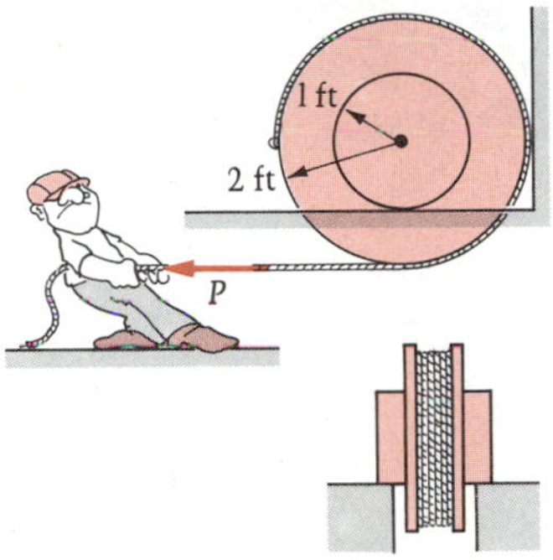

Figure P6.24

* **6.25** The spool is in equilibrium as shown in Figure P6.25. Find the minimum value of the coefficient of friction for which this is possible. *Hint:* Angle θ can be found from the dimensions given. Also, assume that the rope is tightly wrapped.

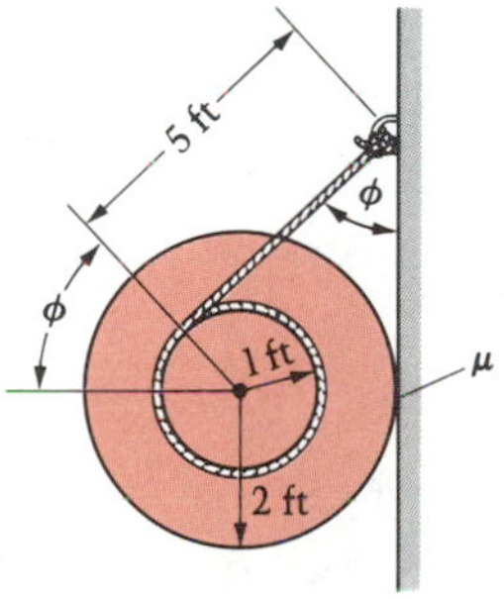

Figure P6.25

6.26 In Figure P6.26 find the largest force that can be developed in the tow cable if the coefficient of friction between the back tires of the truck and the ground is 0.25. The truck weighs 5000 lb and is rear-wheel driven.

6.27 In Figure P6.27 find the minimum coefficient of friction between the bar and the floor for which the bar can be in equilibrium in the given position.

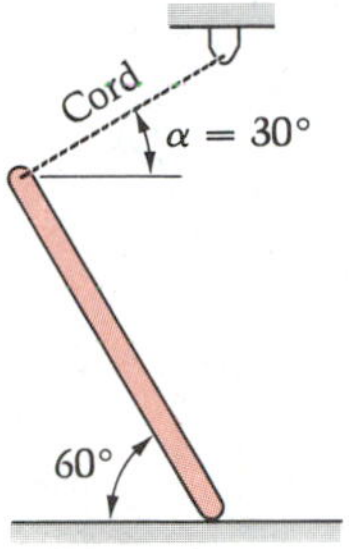

Figure P6.27

6.28 Repeat the preceding problem if the upper angle α is 60°.

6.29 The stick of mass m was originally in equilibrium with the string force S acting vertically ($\theta = 90°$). The force S is now slowly turned clockwise as indicated in Figure P6.29, with the stick remaining in equilibrium in the same position.

a. Find the force S as a function of θ (and mg and ϕ), and show that it must increase in magnitude.
b. Let $\phi = 30°$ and $mg = 5$ lb. If the stick is observed to slip at $\theta = 30°$, find the coefficient of friction between the stick and the floor.

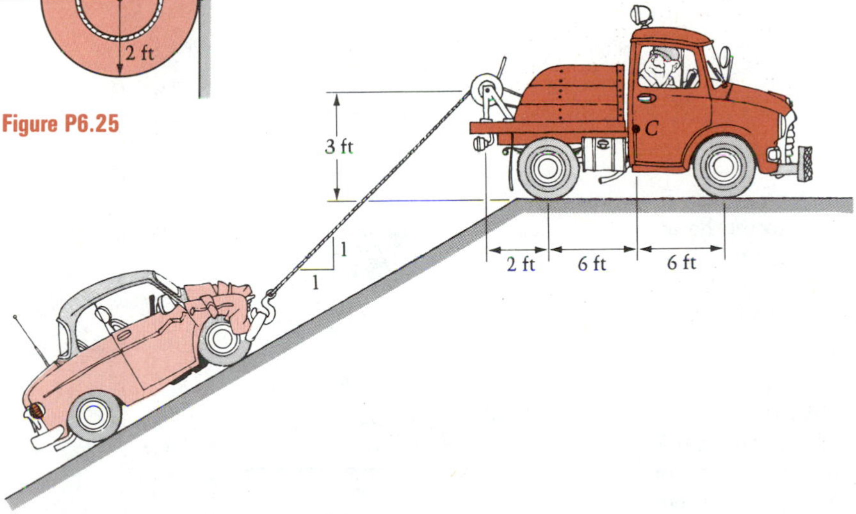

Figure P6.26

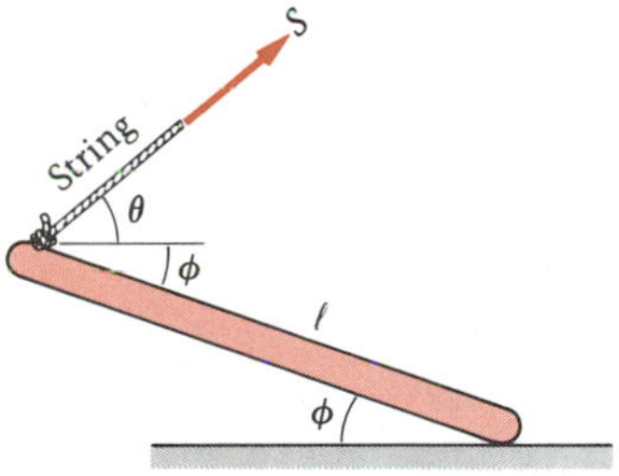

Figure P6.29

6.30 Find the weight of the block that will prevent slipping of the 300-N ladder if the coefficients of friction are: 0.3 at A, 0.1 at B, zero at C, and 0.4 between the block and the floor. (See Figure P6.30).

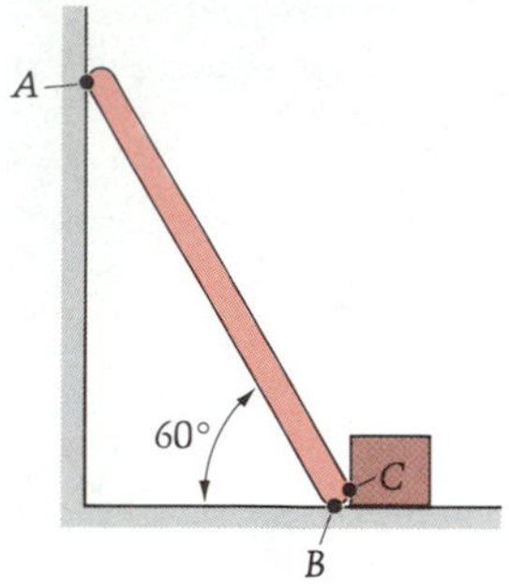

Figure P6.30

6.31 Given a homogeneous block on an inclined plane as shown in Figure P6.31, find the angle θ for which the block is on the verge of tipping. Then find the coefficient of friction μ for which the block is also on the verge of sliding. Note that C_1 and C_2 are the centers of gravity of the triangular and rectangular parts of the block, respectively.

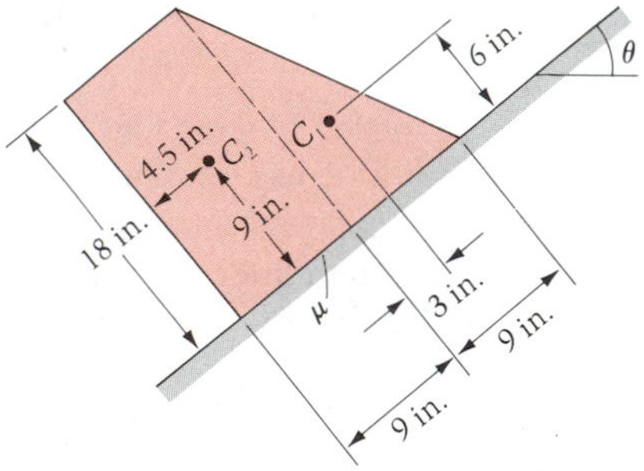

Figure P6.31

6.32 The girl in Figure P6.32 is learning to use a wheelchair. What force must she exert tangent to the handwheel (radius r) to make the wheelchair begin to roll up the incline (angle ϕ)? The radius of the large wheels is R; the masses of patient and wheelchair are M and m, respectively.

6.33 In Figure P6.33 bodies $\mathcal{A}$ and $\mathcal{B}$ respectively weigh 2000 N and 1500 N. The various coefficients of friction are: between $\mathcal{A}$ and the plane, 0.2; between $\mathcal{A}$ and $\mathcal{B}$, 0.3; between $\mathcal{B}$ and the plane, 0.4. Find the magnitude of couple $\mathcal{C}$ that will cause body $\mathcal{B}$ to have impending motion.

Figure P6.32

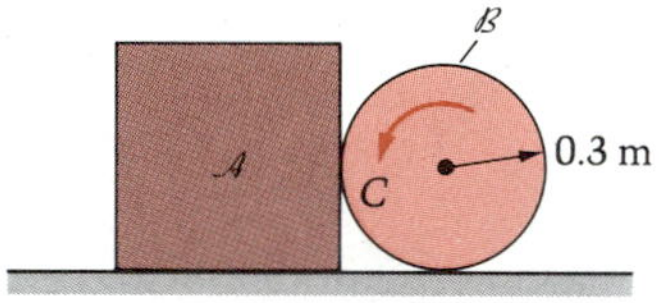

Figure P6.33

6.34 A 100-kg man climbs a 30-kg ladder that is 11 m in length and makes a 30° angle with the vertical. (See Figure P6.34.) The man has placed a solid block at the foot of the ladder to prevent slipping.

a. Is the block necessary?
b. If so, what is the minimum block mass to keep the man and ladder from slipping?

Figure P6.34

6.35 In Figure P6.35 a solid circular disk of weight 65 lb is connected to a 130-lb block by a light bar AB. Find the smallest coefficient of friction between the block and the plane for which the system will be in equilibrium. Assume that the block is wide enough so that it will not tip.

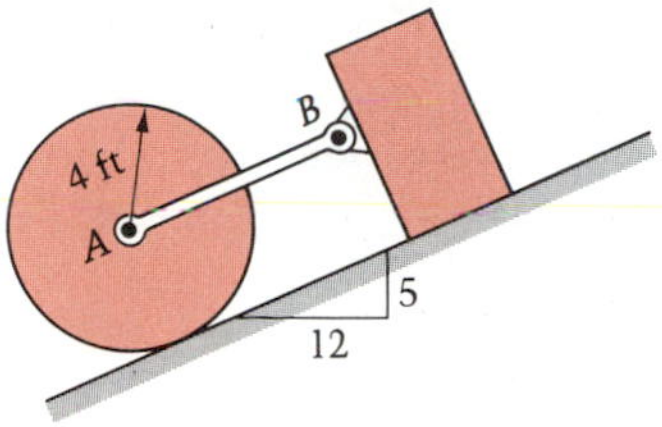

Figure P6.35

6.36 In Figure P6.36, the light rod is connected to the drum and the block at pins. The block weighs 26 lb and the coefficient of static friction for all surfaces is 0.6. If the system is to be in equilibrium, what is the maximum allowable value for the weight of the drum?

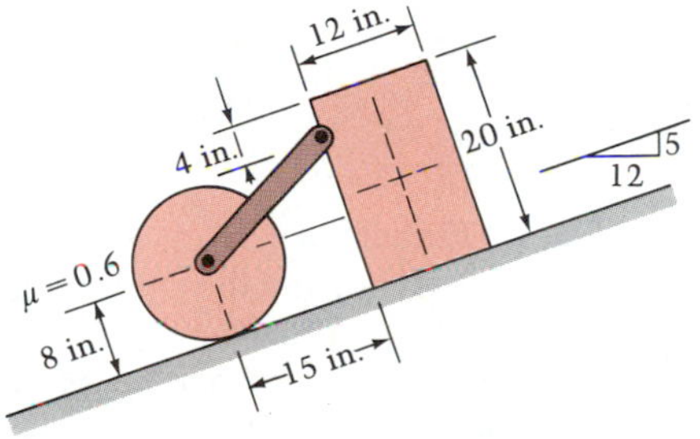

Figure P6.36

6.37 The coefficient of friction between the homogeneous beam and the corner at A is $\mu = 0.3$. (See Figure P6.37.) Find the coordinates of the mass center of the beam if it is in equilibrium but on the verge of slipping at A.

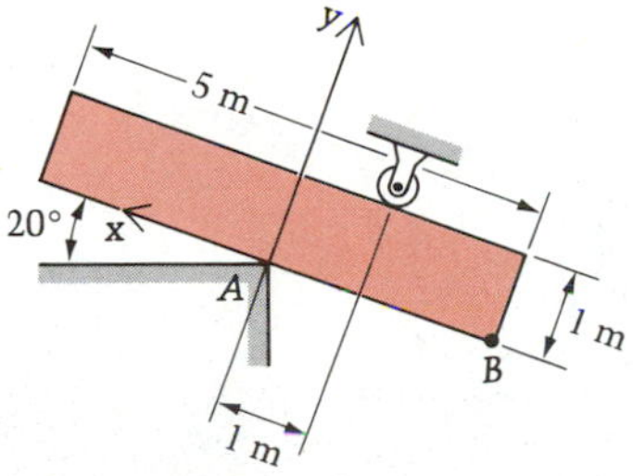

Figure P6.37

6.38 Using Figure P6.37, suppose that the distance from A to B is 1.6 m and that the beam is now uniform. Find the minimum coefficient of friction between the beam and the corner support that will prevent slipping of the beam.

6.39 The cylinder $\mathcal{C}$ and ring $\mathcal{R}$ in Figure P6.39 are of the same radius r and W, and are connected by the light rod AB.

- **a.** Find the couple M, which when applied to $\mathcal{C}$ results in equilibrium. Assume no slipping.
- **b.** Argue from new free-body diagrams that if M is applied instead to R, the same answer is obtained for the required couple.

6.40 In Figure P6.40 the coefficient of friction μ is the same between $\mathcal{A}$ and $\mathcal{B}$ as it is between $\mathcal{B}$ and the plane. Find the minimum value of μ for which the system will remain in equilibrium, and the tension in the cord for this value of μ.

6.41 Repeat the preceding problem with the altered pulley arrangement shown in Figure P6.41.

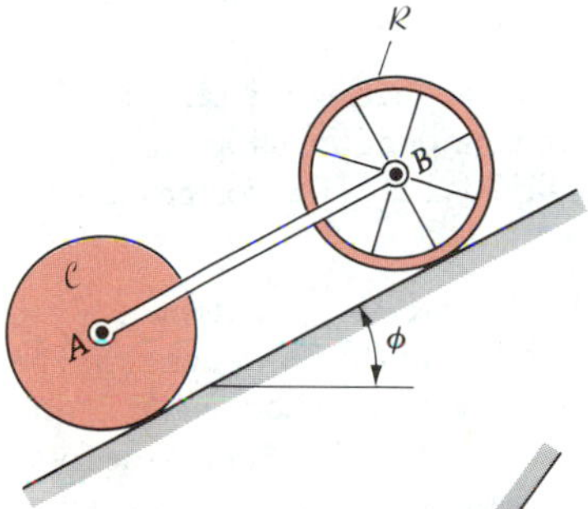

Figure P6.39

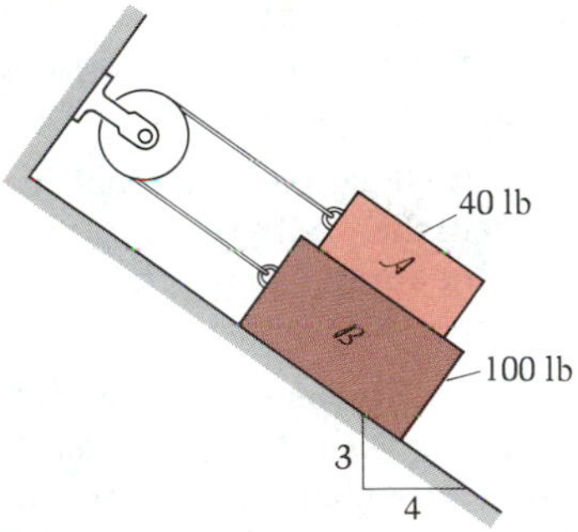

Figure P6.40

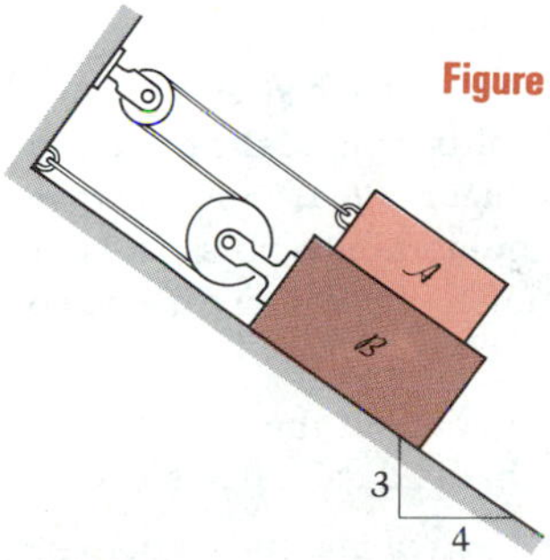

Figure P6.41

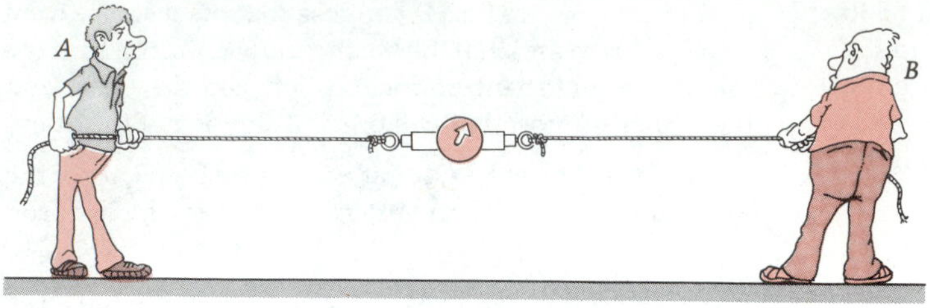

Figure P6.43

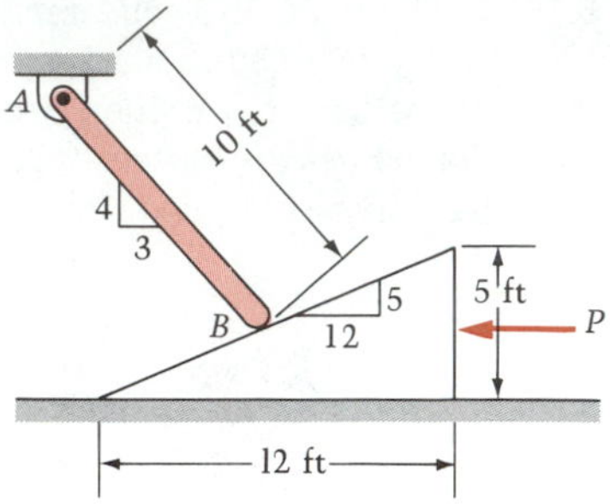

Figure P6.44

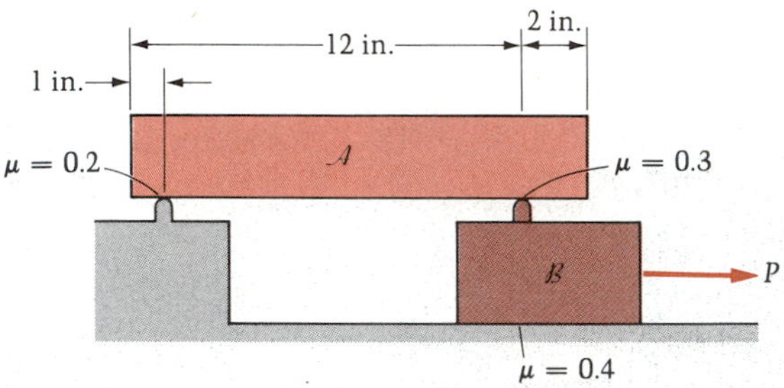

Figure P6.47

6.42 In Problem 6.40, suppose the common friction coefficient is $\mu = 0.2$, and the angle of the plane is now variable. Find the largest angle of the plane for which the blocks will remain in equilibrium.

6.43 In Figure P6.43 Mr. A weighs 160 lb and Mr. B weighs 200 lb. Each man pulls on the rope with a force of 20 lb.

- **a.** What force does the fish scale register?
- **b.** What are the magnitudes of the friction forces on Mr. A and Mr. B?
- **c.** If Messrs. A and B are wearing the same kind of shoes and if there is impending slipping, which man is about to slip? What is the coefficient of static friction μ_s?

6.44 In Figure P6.44, the 50-lb rod is pinned smoothly to the ceiling at A, and it rests on the 100-lb wedge at B. The coefficient of friction between the two bodies is 0.2, and that between the wedge and the floor is 0.1. Find the force P for which motion of the wedge to the left will be impending.

6.45 In the preceding problem, is any force P needed to prevent the wedge sliding to the right? If so, find it; if not, find the force P *to the right* that will start motion of the wedge to the right.

6.46 In Problems 4.238 and 4.239, find the minimum coefficient of friction between the block and the slotted body for which the equilibrium can exist.

6.47 The uniform blocks $\mathscr{A}$ and $\mathscr{B}$ weigh 90 lb and 60 lb, respectively. The coefficients of friction at the three contacting surfaces are shown in Figure P6.47. Find the maximum value of P for which equilibrium is possible. Is the answer the same if P is pushing to the left on $\mathscr{B}$ instead of pulling to the right?

6.48 Repeat the preceding problem if the coefficient of friction between $\mathscr{A}$ and $\mathscr{B}$ is changed from 0.3 to 0.1.

6.49 In Figure P6.49 a heavy man of mass m has placed a light ladder of length L at a dangerous angle, and has begun to climb it. If the coefficients of friction are 0.25 between both the ladder and ground, and the ladder and wall, find how far up the ladder (fL in the figure) the man can climb before the ladder slips. Show further that prior to the slipping position, the ground and wall reactions cannot be found (i.e., that the problem is statically indeterminate).

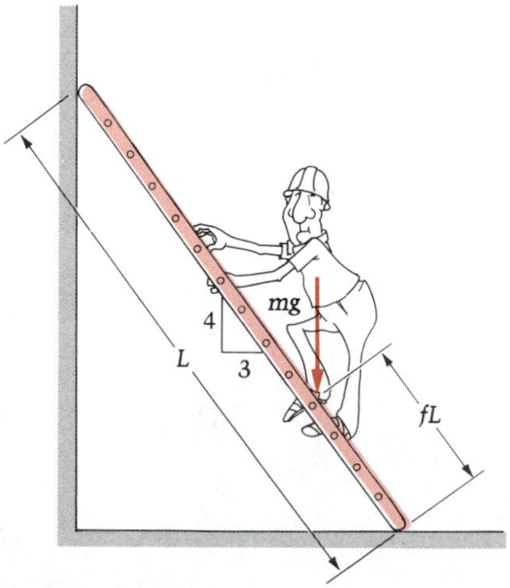

Figure P6.49

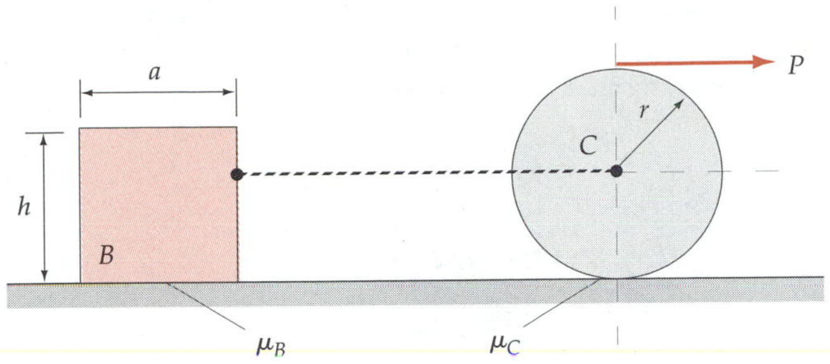

Figure P6.50

6.50 Block B and the homogeneous cylinder C in Figure P6.50 each weigh W. A light cable connects the center of the cylinder to the block B. The coefficients of static friction are μ_B and μ_C, as indicated.

a. If $\mu B = 0.5$ and $\mu_C = 0.2$, what is the largest force P (in terms of W) that can be applied without disturbing equilibrium? (Assume that dimension "a" is large enough to prevent the block from tipping.) Describe in a sentence the impending motion at this value of P.

b. Repeat the problem if the friction coefficients are reversed, i.e., $\mu_B = 0.2$ and $\mu_C = 0.5$ (again assume no tipping).

6.51 What ratio of the coefficients of friction, μ_B/μ_C, would be necessary for *simultaneous* impending motion of B and C in the preceding problem?

6.52 In Problem 6.50, let $\mu_C = 0.5$ and $\mu_B = 0.2$. In terms of r, find the smallest value of the width "a" of the block so that it will not tip over for any value of P less than $0.1W$.

6.53 In Figure P6.53 the block and cylinder each weigh 300 N.

a. Show that the block will be on the verge of slipping when P is increased to the value $120/\cos\theta$ N.

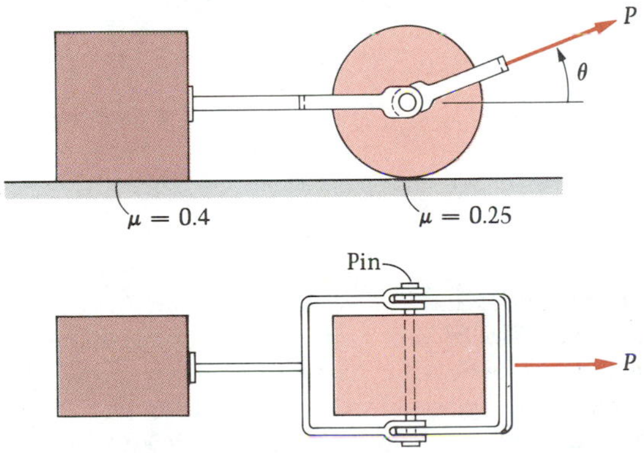

Figure P6.53

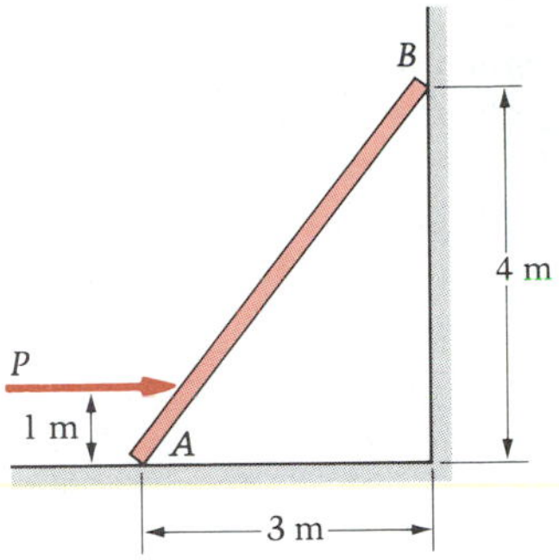

Figure P6.54

b. At this value of P, what is the cylinder on the verge of doing?

6.54 A uniform ladder, mass 12 kg, rests against a wall and against a horizontal surface, as shown in Figure P6.54. If the coefficient of friction at A and B is 0.3, find the smallest force P that will prevent motion of the ladder to the left at A.

6.55 The bodies $\mathcal{A}$, $\mathcal{B}$ and $\mathcal{C}$ in Figure P6.55 each weigh 20 lb and are in equilibrium. $\mathcal{A}$ and $\mathcal{C}$ are pinned to $\mathcal{B}$ at its ends. Find the force P (magnitude and sense) and the forces exerted by the slot on $\mathcal{A}$ and by the circular track on $\mathcal{C}$.

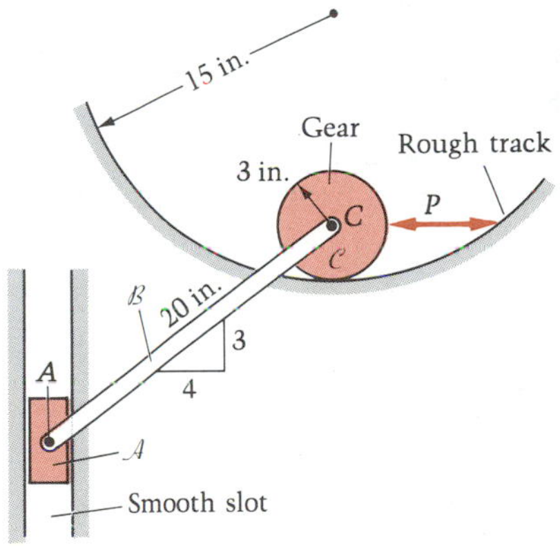

Figure P6.55

6.56 In the preceding problem, let P be replaced by a couple M applied to $\mathcal{C}$. Find M. Also find the minimum coefficient of friction between $\mathcal{C}$ and the rough track for equilibrium.

6.57 The uniform 10-lb rod and 100-lb cylinder are pinned smoothly at C. If the 2-lb force is applied as shown in Figure P6.57 and the system remains in equilibrium, then the coefficient of friction between rod and floor has to be larger than a certain value. What is this value?

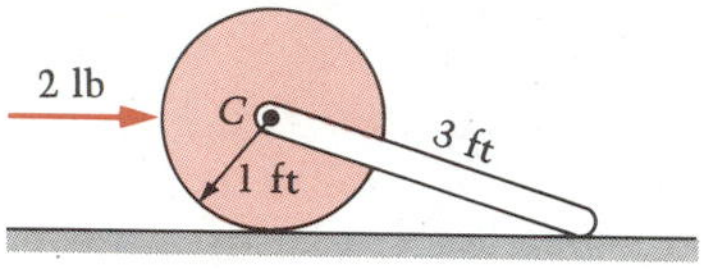

Figure P6.57

6.58 In Figure P6.58 the collar $\mathcal{C}$ is very light compared with the force P. It is fitted over the rough vertical shaft $\mathcal{S}$ with a slight amount of clearance. Show that if the distance d is greater than $H/(2\mu) - r$, then $\mathcal{C}$ will not slip down the shaft regardless of the size of P. *Hint:* There will be contact at two points of $\mathcal{S}$.

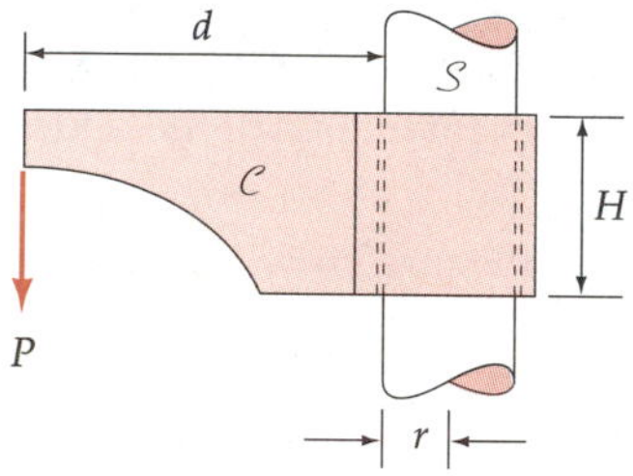

Figure P6.58

6.59 In Example 6.8 suppose the car is to be parked on a 15° incline. (See Figure P6.59.) What must be the coefficient of friction between tires and street in order to prevent slipping of the car if only the rear wheels are locked?

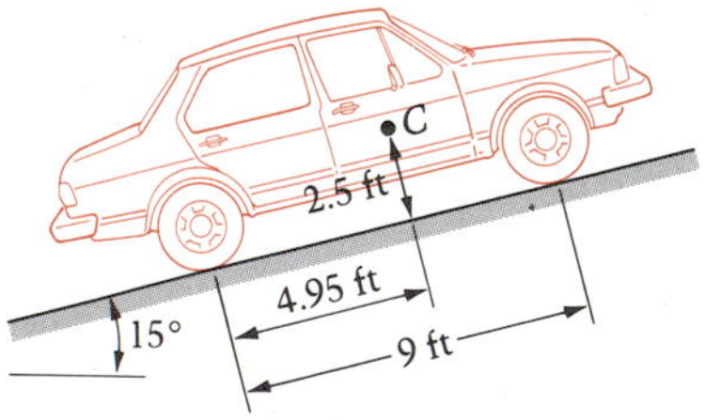

Figure P6.59

6.60 Repeat the preceding problem if all four wheels are locked.

6.61 The homogeneous cylinder in Figure P6.61 weighs 425 lb and the block weighs 300 lb. The coefficient of friction for all surfaces of contact is 0.25. Moment T_0 is slowly increased from zero. Determine the value of T_0 that will cause the cylinder to have impending motion, assuming that the block does not tip.

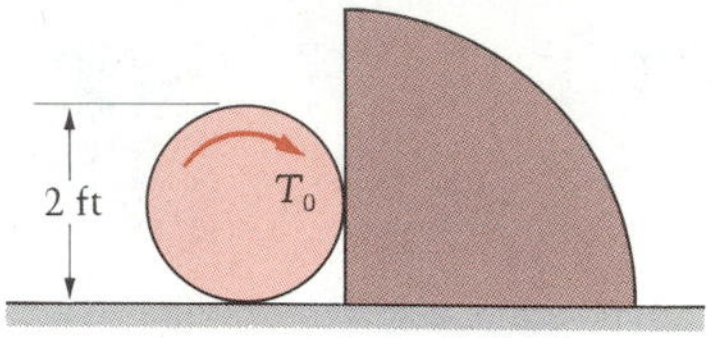

Figure P6.61

6.62 In Figure P6.62 if $\mu_1 = \mu_2$, then as P increases from its lowest possible equilibrium value, will the cylinder $\mathcal{C}$ first roll, or slide, on the plane for (a) $\theta = 30°$. (b) $\theta = 45°$, (c) $\theta = 60°$? *Hint:* Show that for rolling,

$$P_1 = \frac{W \sin\theta}{1 - \mu_1}$$

and for sliding,

$$P_2 = W\left[\sin\theta + \frac{\mu_2 \cos\theta}{1 - \mu_2}\right]$$

and compare the numerical values.

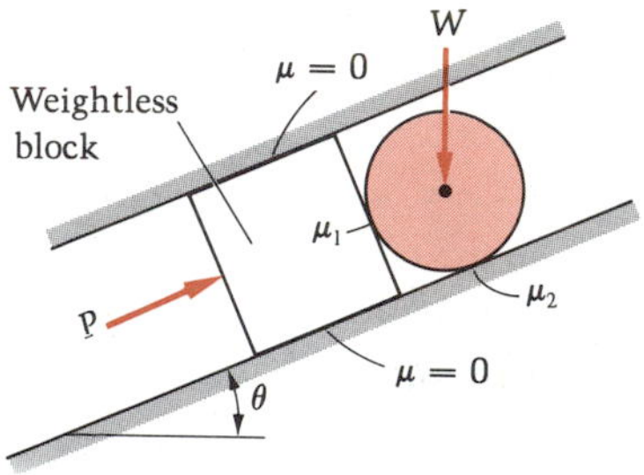

Figure P6.62

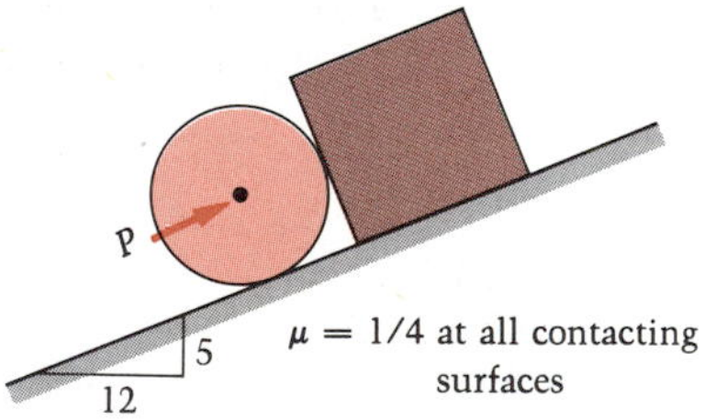

Figure P6.63

6.63 The cylinder and the block each weigh 26 N. (See Figure P6.63.) The friction coefficient is $\frac{1}{4}$ between block and cylinder, $\frac{1}{4}$ between cylinder and plane, and $\frac{1}{4}$ between block and plane. Find the smallest force P for which motion will begin up the plane if it is known that the block won't tip.

6.64 In the preceding problem, find the value of the common friction coefficient below which the cylinder will roll on the plane and above which it will slip on the plane, provided $\mu \neq 0$.

6.65 Without block $\mathcal{B}$ present, the center of the wheel $\mathcal{C}$ will move to the right on the plane. If $\mathcal{A}$, $\mathcal{B}$, and $\mathcal{C}$ each weigh W, and if the friction coefficients at the three contacting surfaces are all equal (μ), find the minimum value of μ for which equilibrium will exist. (See Figure P6.65.) Assume that $\mathcal{B}$ is wide enough so that it will not tip.

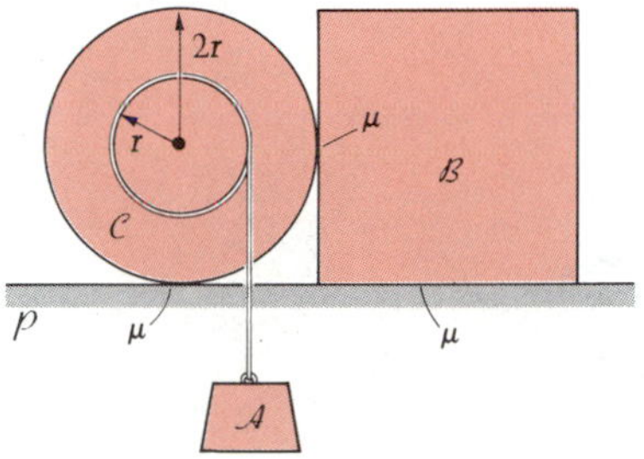

Figure P6.65

6.66 Two identical cylinders are placed together on an incline as shown in Figure P6.66. Show that the cylinders cannot remain in equilibrium. *Hint:* Write the three equilibrium equations for each cylinder. Show that they require the normal force between them to be zero. Thus conclude that: (a) the friction between them is also zero; (b) there can be no friction beneath the cylinders if there is equilibrium; and finally (c) the component of the weights down the plane is unbalanced, which is a contradiction.

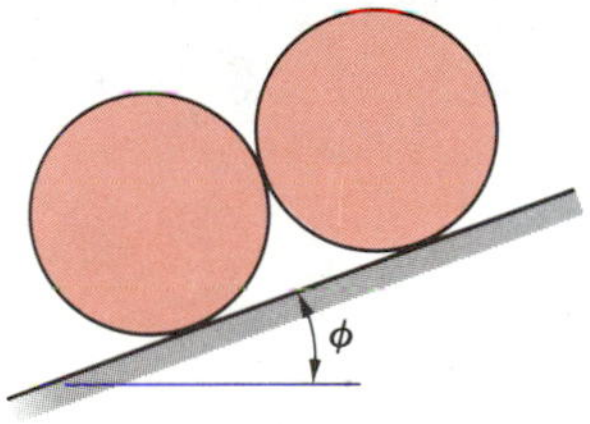

Figure P6.66

6.67 In the preceding problem, suppose that a normal force between the cylinders is created by a pair of springs (see Figure P6.67), one at each end of the cylinders. If $m = 20$ kg, $\mu = 0.6$ at all surfaces, and $\phi = 20°$, find the spring tension required for equilibrium of the two cylinders.

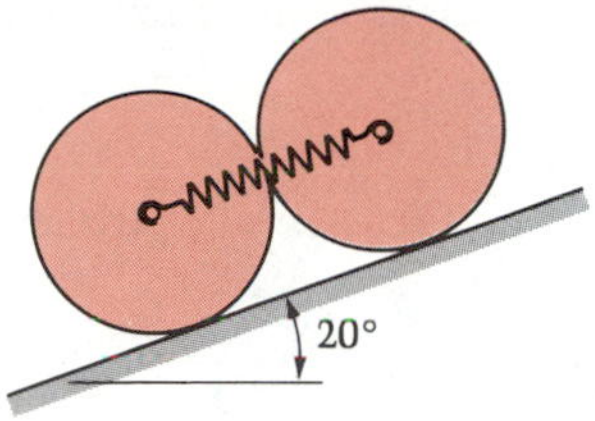

Figure P6.67

6.68 A vehicle of mass m in Figure P6.68 is driven by a gear and pinion as follows: A motor exerts a constant moment M_0 onto the pinion $\mathcal{P}$. It then drives the gear $\mathcal{G}$ that is attached to a rear wheel, thereby turning the wheel counterclockwise and moving the vehicle up the plane. What value of M_0 is needed to barely move the vehicle if:

a. Axle friction is negligible?

b. Axle friction gives a constant resistive moment of M_f opposing the rolling?

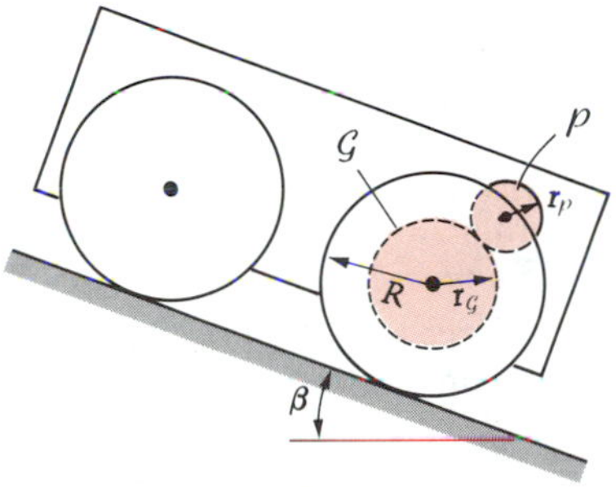

Figure P6.68

Figure P6.69

6.69 In Example 6.6 let the two friction coefficients be μ_P between drum and plane and μ_T between drum and truck. Find the relationship between μ_P and μ_T for which the drum will roll up the plane (i.e., will not slip on the ground). (See Figure P6.69.) The answer should be of the form $\mu_T \leq f(\mu_P)$.

6.70 In Example 6.6 let both friction coefficients be 0.3. Find the maximum incline angle ϕ for which the drum will roll (i.e., not slip on the inclined plane).

6.71 In Figure P6.71 the cylinder is in equilibrium. Determine:

a. The stretch in the spring.
b. The minimum coefficient of friction μ for which the equilibrium is possible.

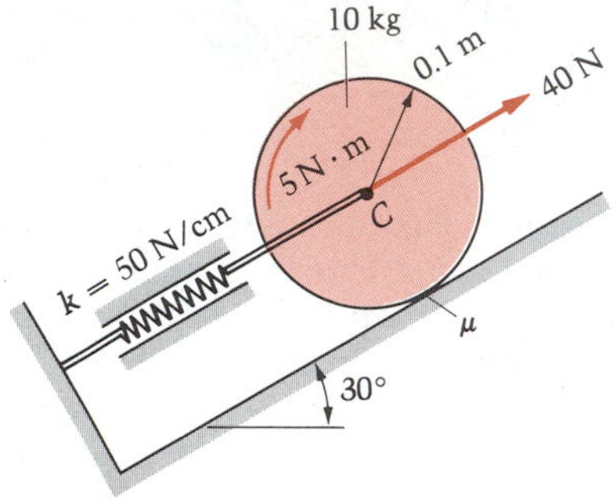

Figure P6.71

6.72 The 130-lb homogeneous plank of length 6 ft is placed on the two homogeneous cylindrical rollers, each of weight 30 lb, and held there by the force P. (See Figure P6.72.)

a. Find P for equilibrium of the system, assuming no slipping occurs anywhere.
b. If the coefficient of friction μ at all surfaces of contact is the same, find the smallest μ needed for equilibrium.

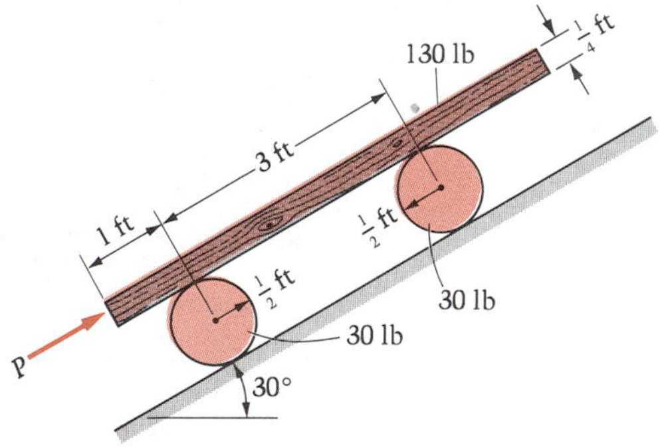

Figure P6.72

In Problems 6.73–6.76 the left support is 1.2 m from the left end of the beam, and the force P is holding the beam in equilibrium. Find the force P required to start the beam moving to the left. The thin beam in each figure weighs 2600 N.

6.73

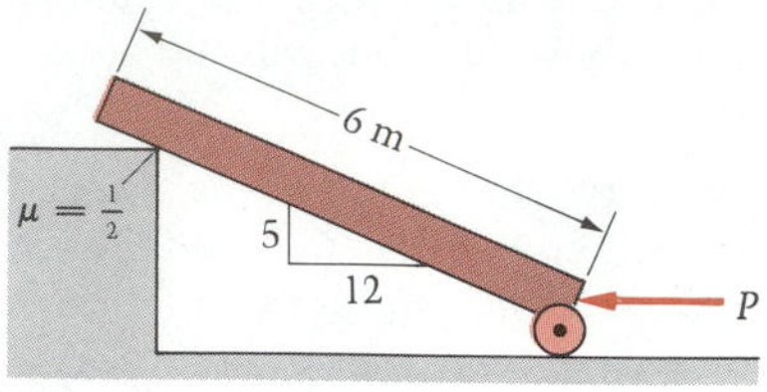

Figure P6.73

6.74

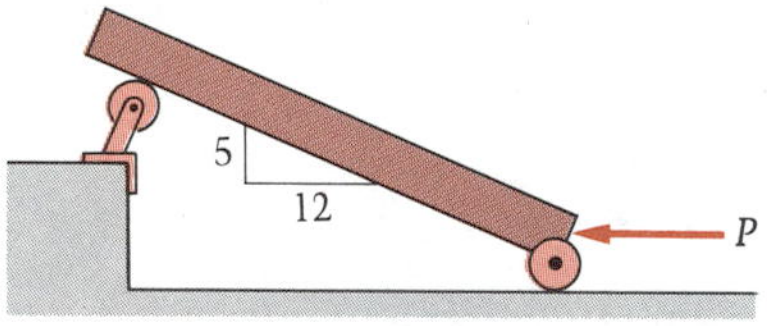

Figure P6.74

6.75

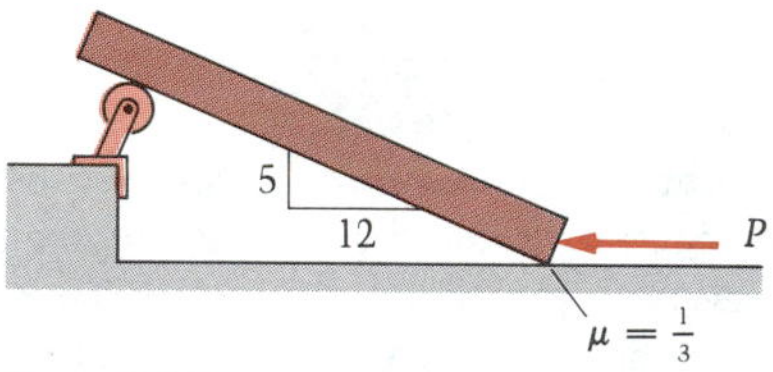

Figure P6.75

6.76

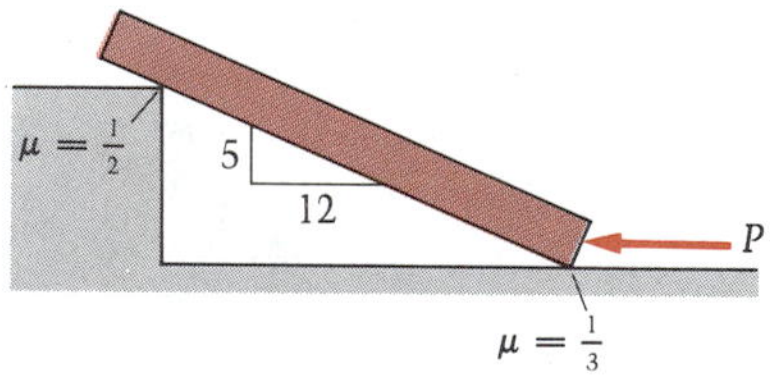

Figure P6.76

6.77–6.80 Reverse the direction of P in Problems 6.73–6.76, respectively, and find the value of P required to start the beam moving to the right.

6.81 The two identical sticks are pinned together at A and are in equilibrium as shown in Figure P6.81 (on the following page). What is the lowest coefficient of friction (as a function of α) for which this can take place?

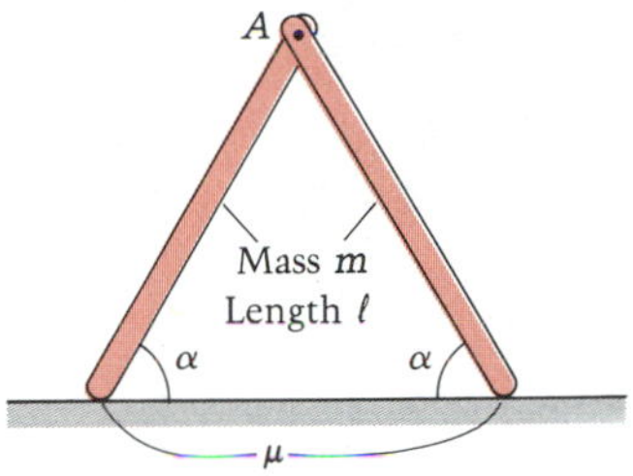

Figure P6.81

6.82 The cylinder has mass M, including an attached hub around which a cord is wrapped. The cord, after passing over a pulley, is attached to a mass m. (See Figure P6.82.)

- **a.** Show that if the plane is smooth, the cylinder cannot be in equilibrium, regardless of the values of the parameters.
- **b.** Find the minimum coefficient of friction μ between the cylinder and the plane for which the system is in equilibrium. Find the relationship between M, m, ϕ, R, and r for equilibrium.

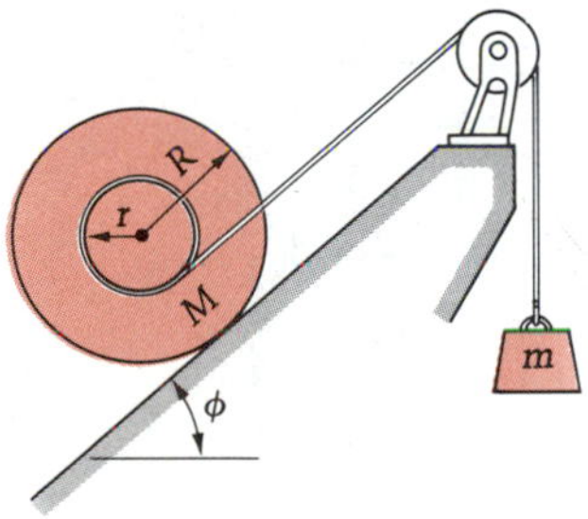

Figure P6.82

6.83 In Figure P6.83 the cord is wrapped around the light hub attached to the 130-lb cylinder. Find the minimum coefficient of friction between the cylinder and the plane for which the cylinder will be in equilibrium.

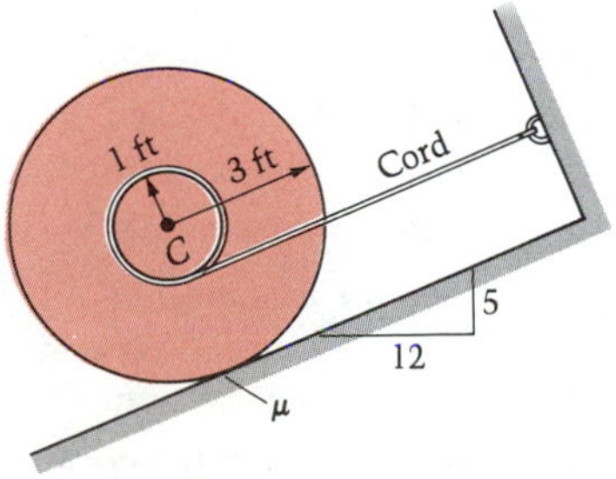

Figure P6.83

6.84 Find the braking force B if motion impends between the brake and drum at A, with friction coefficient 0.4. Neglect the weight of the brake and its thickness. (See Figure P6.84.) The couple T has magnitude 80 N · m.

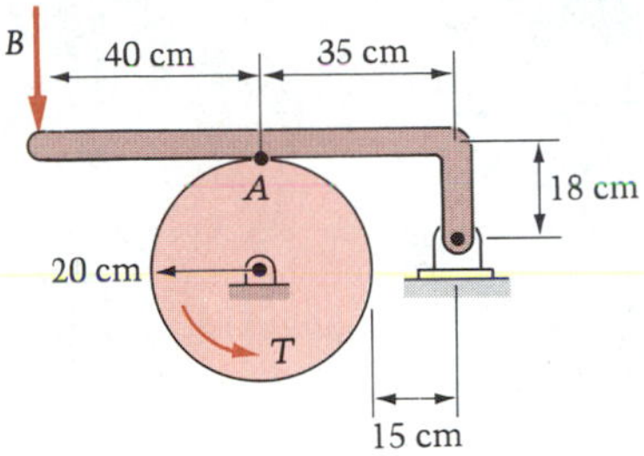

Figure P6.84

6.85 Repeat the preceding problem if the torque T is applied clockwise.

6.86 In Problem 6.84, if $T = 100$ N · m and if the slender brake arm weighs 3 N/cm, find the force B if motion is impending.

6.87 The cable exerts 320 lb on the 400-lb weight as shown in Figure P6.87. The truck then moves slowly to the right, while rotating the arm $\mathcal{A}$ counterclockwise so as to keep the tension approximately constant. At what angle ϕ will the weight begin to move? Will it slide, or will it tip?

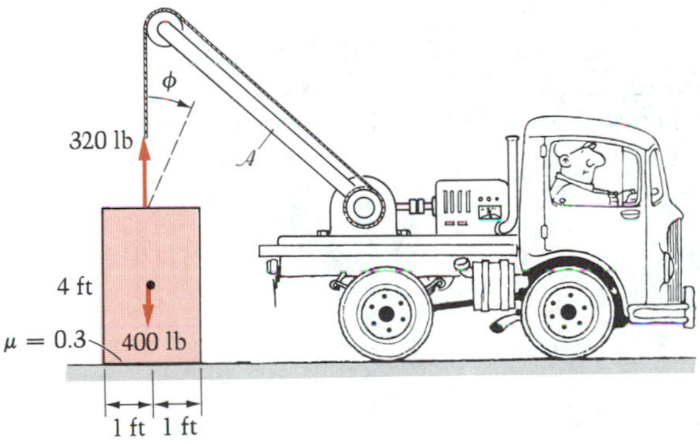

Figure P6.87

6.88 The coefficient of limiting static friction between the 100-lb block and the plane is 0.5. The force P is applied horizontally as shown in Figure P6.88 (see the following page). Determine:

- **a.** The magnitude of the force P for which the friction force vanishes.
- **b.** The force P required to start the block moving up the plane.

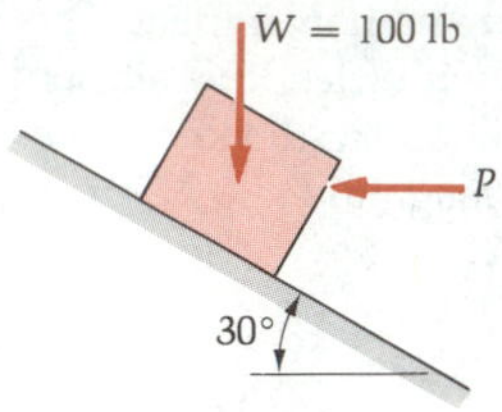

Figure P6.88

6.89 A 200-lb cylinder is in equilibrium in the position shown in Figure P6.89. Find:

a. The friction force at A.

b. The coefficient of friction if slipping impends at A in the position shown.

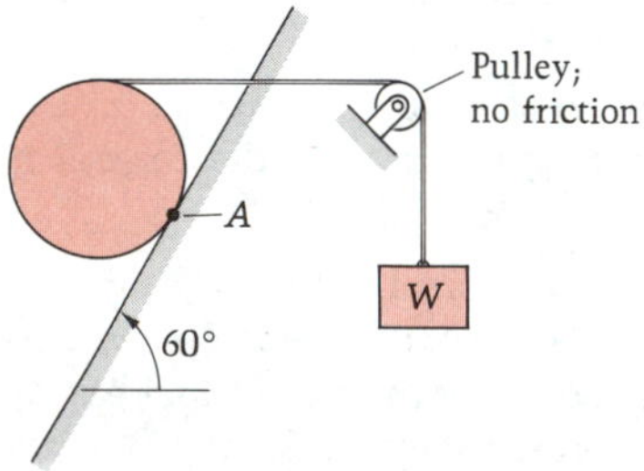

Figure P6.89

6.90 Determine the force P that will move the two identical 10-kg cylindrical rollers up the inclined plane shown in Figure P6.90. The friction coefficient is 0.3 at A, B, and C.

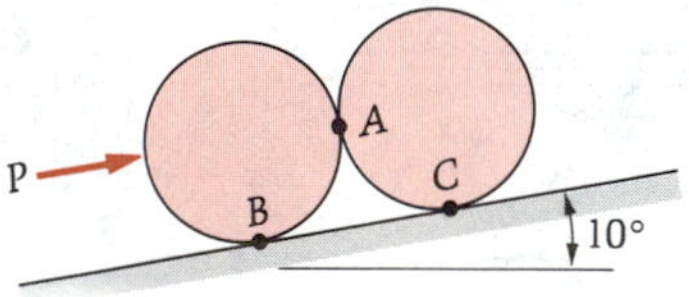

Figure P6.90

6.91 The hydraulic cylinder in Figure P6.91 exerts a force to the right on the brake arm at point B. This brings the shoe into contact with a turning wheel, in order to stop it by friction. While the wheel is not in equilibrium (indeed, it is being decelerated), the arm is. Find the ratio of the force exerted by the hydraulic cylinder to the normal force between the shoe and the wheel if the wheel rotates (a) clockwise; (b) counterclockwise. Neglect the weights of arm and shoe.

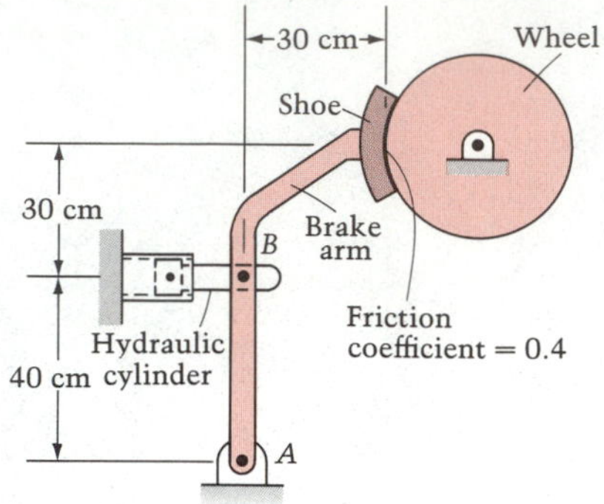

Figure P6.91

6.92 In Figure P6.92 find the smallest couple M that will cause the cylinder of weight W and radius R to have impending motion. The coefficient of friction for all surfaces is μ.

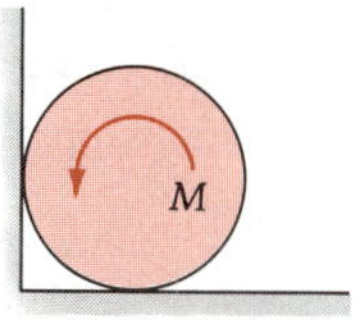

Figure P6.92

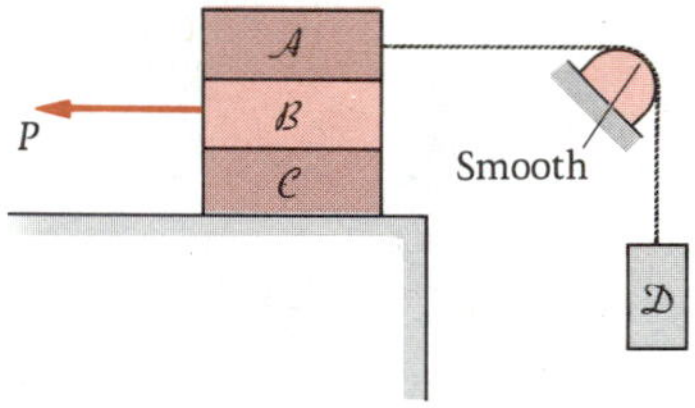

Figure P6.93

6.93 The blocks $\mathcal{A}$, $\mathcal{B}$, $\mathcal{C}$, and $\mathcal{D}$ in Figure P6.93 weigh 100 lb, 50 lb, 75 lb, and 40 lb, respectively. The friction coefficients are 0.5 between $\mathcal{A}$ and $\mathcal{B}$, 0.3 between $\mathcal{B}$ and $\mathcal{C}$, and 0.25 between $\mathcal{C}$ and the plane. Because the peg is smooth, the tension in the cord is the same on both sides. Force P is slowly increased from zero. At what value of P will the equilibrium be disturbed?

6.94 A slender uniform rod of length 17 in. and weight 50 lb is smoothly hinged to a fixed support at A and rests on a half-cylinder block at B, where the contact is smooth. (See Figure P6.94.)

a. Find the smallest coefficient of friction μ between the 70-lb block and the plane for which the system can be in equilibrium as shown.

b. For this value of μ, find the location of the normal force resultant between the block and the plane.

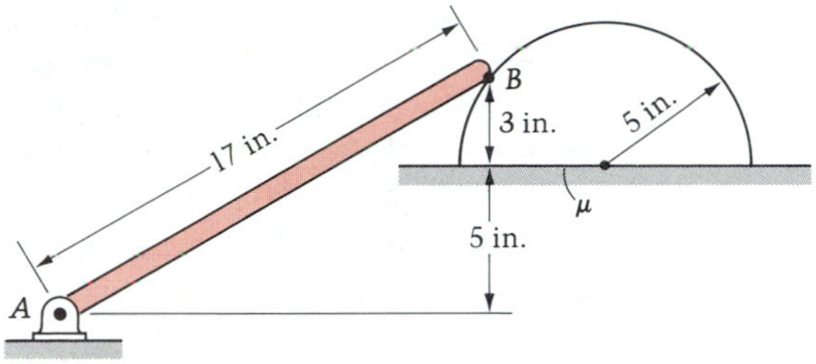

Figure P6.94

6.95 The homogenous rod AB in Figure P6.95 weighs 50 pounds and is supported by the half-cylinder C and by the pin at A. If $D = 6$ ft, $R = 3$ ft, and $L = 8$ ft and if the contact between C and the floor is smooth, find

a. the normal and friction forces acting on C at E;

b. the minimum coefficient of friction between the rod and half-cylinder for which the equilibrium can occur.

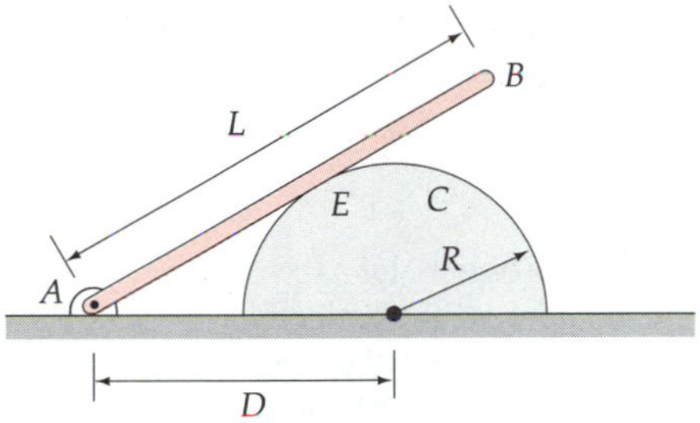

Figure P6.95

6.96 The rod in Figure P6.96 has a mass of 3 slugs. It rests on a rough floor at A and a smooth, fixed half-cylinder at B. Find:

a. The reactions exerted on the rod at A and B.

b. The minimum coefficient of friction for which the rod will not slip at A.

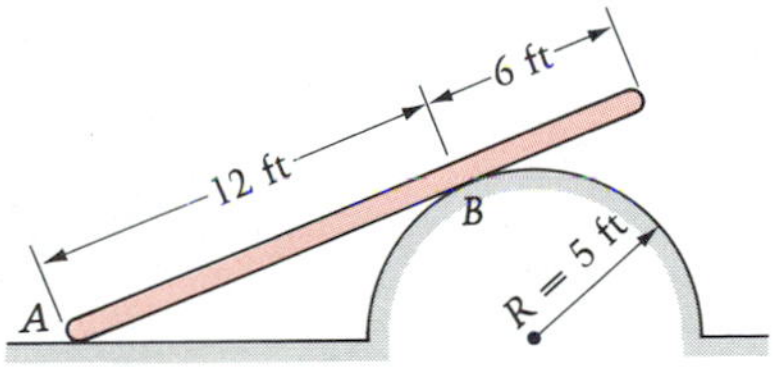

Figure P6.96

6.97 The bent bar in Figure P6.97 is pinned to a disc at one end (A) and rests on a semicircular fixed support at the other end (B). The bar weighs 20 N and its center of gravity is indicated in the figure. There is also a force P acting on the pin at the center of the disc. The coefficient of friction μ between the bar and the support is 0.35, and the pin is smooth.

a. Find the normal force exerted by the semicircular support on the bar as a function of the applied force P.

b. Likewise, find the frictional force exerted by the semicircular support on the bar in terms of P.

c. Using the results of (a) and (b), find the least value of P for which the bar-disc assembly will remain in equilibrium.

d. Using the results of (a) and (b), find the greatest value of P for which the bar-disc assembly will remain in equilibrium.

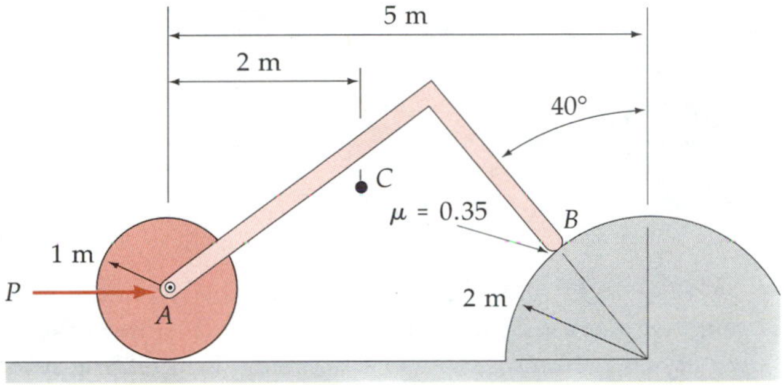

Figure P6.97

6.98 The weightless members in Figure P6.98 are supported by pins C and B and by a rough surface at A. Determine:

a. The minimum coefficient of friction at A to assure equilibrium of the structure.

b. The forces exerted on member BC by the pins at B and C.

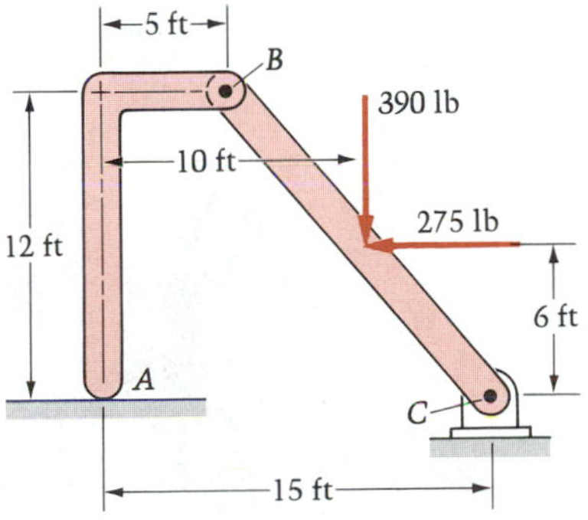

Figure P6.98

6.99 A person plugs in an electric appliance, pushing the plug into the receptacle with a horizontal force F. (See Figure P6.99.) The prongs initially touch the contacts as shown, and when the force F reaches 3 lb, the prongs slide past them, completing the contact. When the sliding starts, what is the component of force on each contact perpendicular to the prong? Assume a friction coefficient of 0.25.

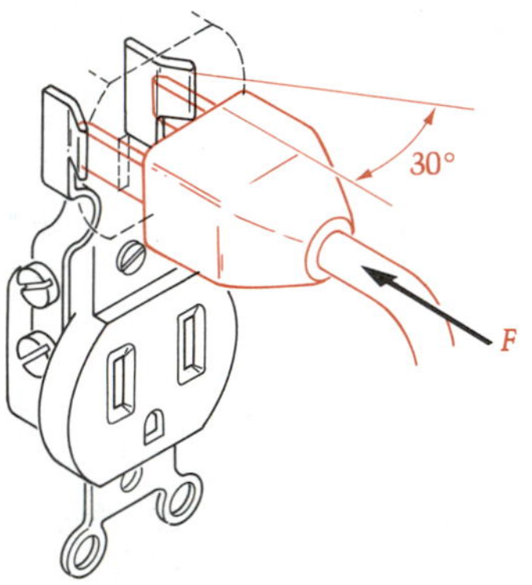

Figure P6.99

6.100 The coefficient of friction between the rod and the triangular block is μ_0, and the plane is smooth. Show that the only condition for the block to remain in equilibrium is that $\tan\theta \le \mu_0$. (See Figure P6.100.)

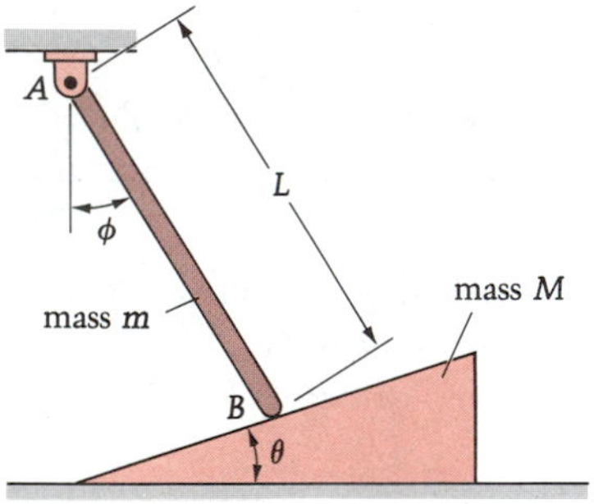

Figure P6.100

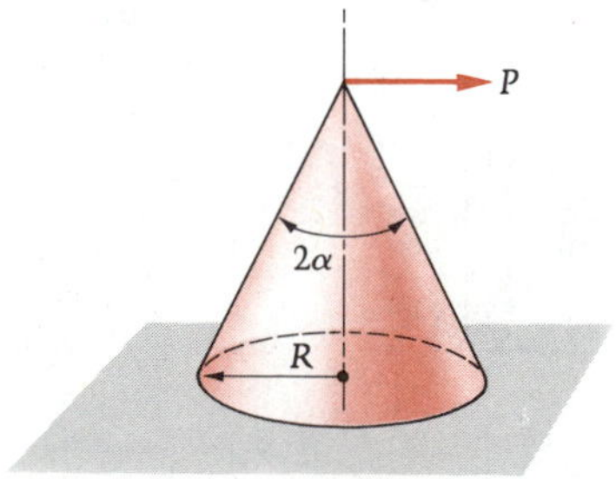

Figure P6.101

6.101 The force P is applied to the cone of mass m as shown in Figure P6.101, and gradually increased in magnitude from zero. If the coefficient of friction between the cone and the plane is μ_0, find the value of P at which motion of the cone impends. Is it tipping, or slipping?

6.102 The edge AB of block ABC in Figure P6.102 is a parabola whose vertex is at A. Its equation is $x^2 = \frac{64}{3}y$. If the coefficient of friction between the shaft and the block ABC is μ_0, find the largest value of x for which the system can be in equilibrium. Neglect friction beneath the rollers and let m be the mass of the shaft plus the plate attached above it.

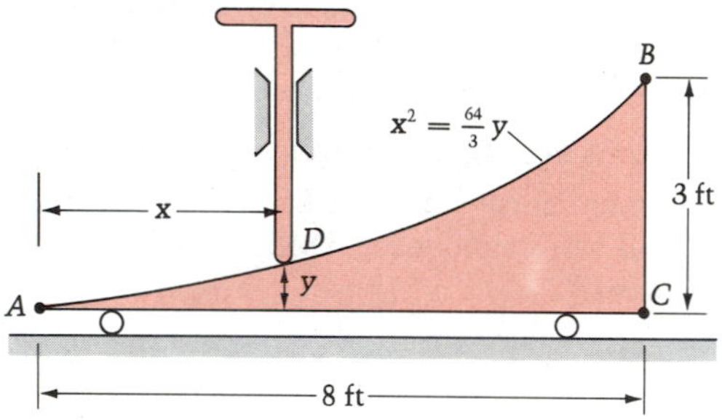

Figure P6.102

6.103 In the preceding problem, let a force P be applied to the left on face BC of block ABC. Neglecting friction in the vertical slot in which the shaft slides, find the value of P, in terms of x, for which motion of the shaft will be impending to the left.

6.104 Show that the bar $\mathcal{B}$ in Figure P6.104 will not slip to the right, regardless of the size of F, if the friction coefficient μ is greater than or equal to $\cot\theta$. The weight of $\mathcal{R}$ is W.

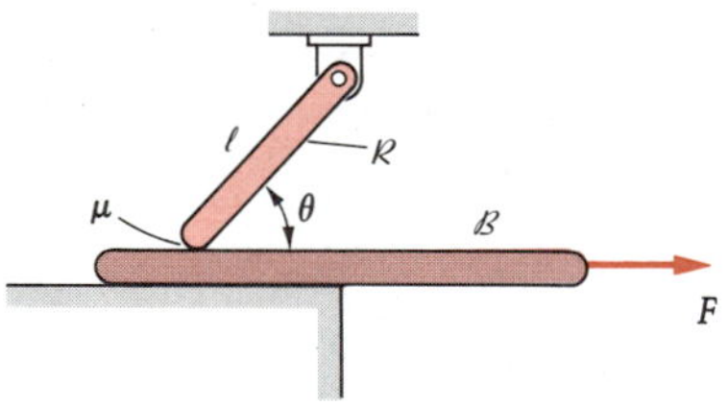

Figure P6.104

6.105 The uniform rod rests atop a uniform circular cylinder of the same weight W. The coefficients of friction at the contact points A and B are each equal to μ. Calculate the smallest value of μ for which the cylinder can stay in equilibrium with the thin rod maintaining its horizontal position. (See Figure P6.105 on the following page.)

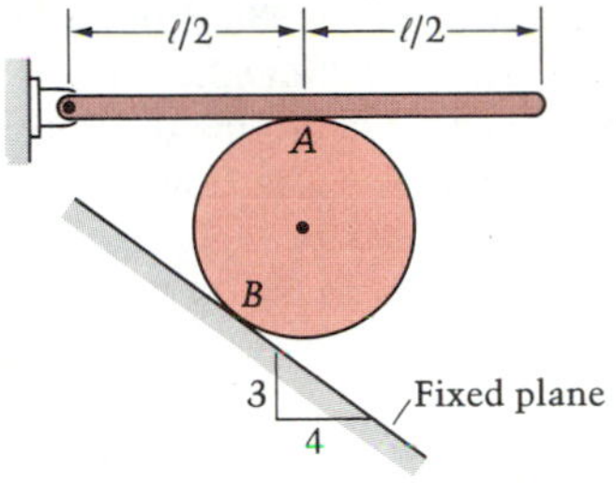

Figure P6.105

6.106 In Figure P6.106 find the minimum coefficient of friction between the small block and the wall for which equilibrium of the system is possible.

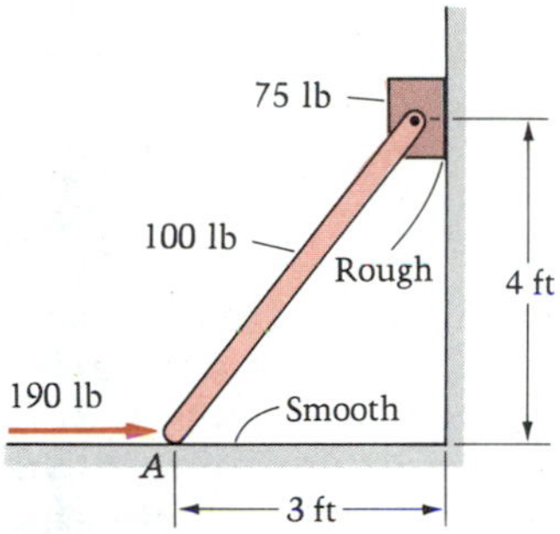

Figure P6.106

6.107 The 200-lb block rests between the smooth vertical guides as shown in Figure P6.107. It is supported by the 50-lb rod. The coefficients of static friction between the rod and the block at A, and the rod and the ground at B, are each 0.5. Determine the maximum angle θ for equilibrium.

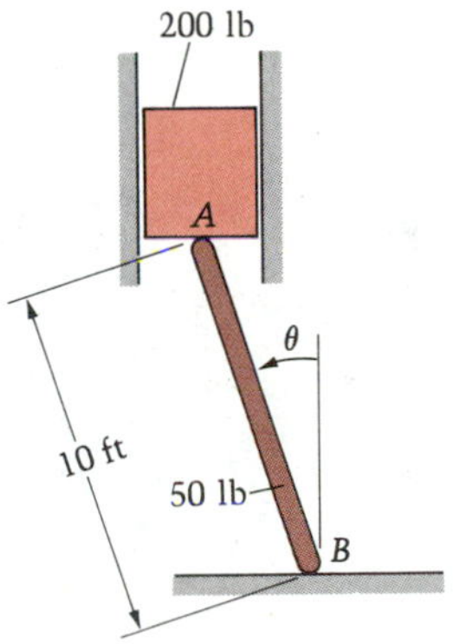

Figure P6.107

6.108 The uniform bar of mass 50 kg is in equilibrium. The wall is smooth and the floor is rough. (See Figure P6.108.) Find the reactions at A and B, and the smallest coefficient of friction between the bar and the floor for which the equilibrium is possible.

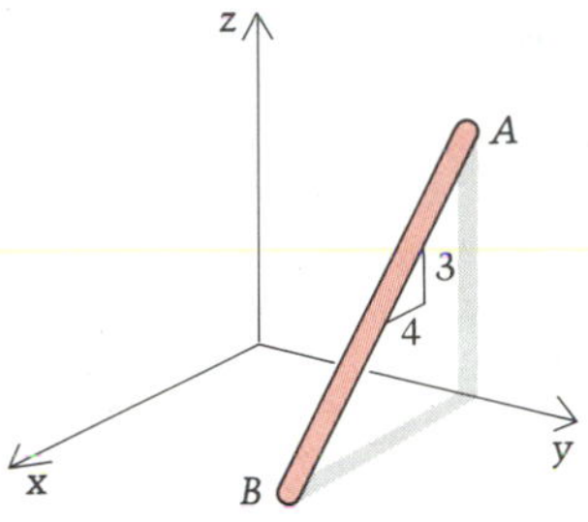

Figure P6.108

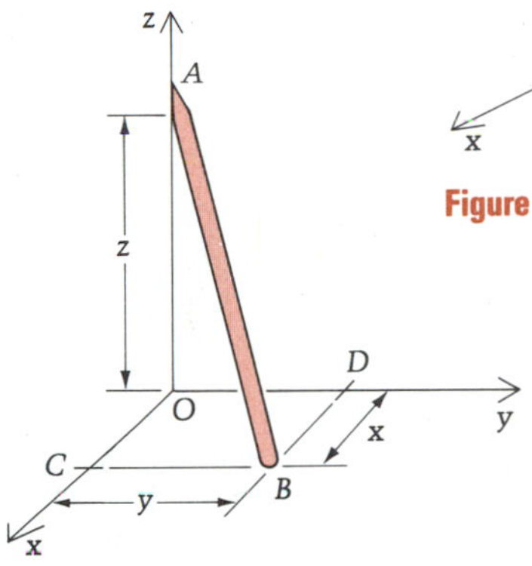

Figure P6.109

6.109 In the preceding problem, suppose the stick is sharpened at A and that this upper end is in the corner (touching the z axis). (See Figure P6.109).

- **a.** Show that the reaction from the corner is in the plane containing the triangle AOB.
- **b.** Find, in terms of x, y, and z, the minimum coefficient of friction between the bar and the floor for which equilibrium is possible.

6.110 The block weighs 200 lb and rests on the 300-lb wheel with a friction coefficient between them of 0.4. (See Figure P6.110.) If the friction coefficient between the wheel and the plane is 0.3, find the lowest value of T for which the wheel will move. *How* will it start to move?

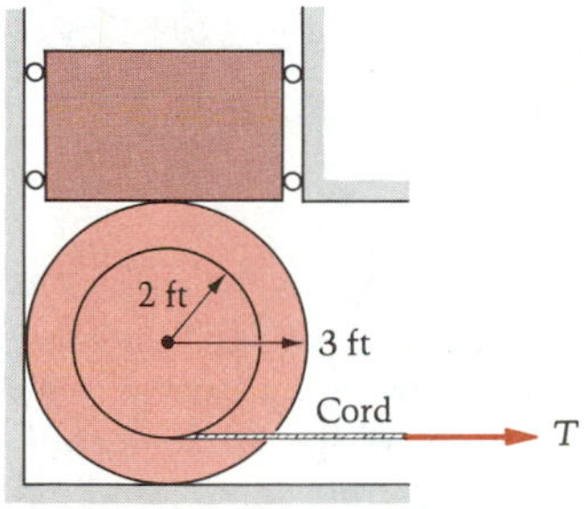

Figure P6.110

6.111 Repeat the preceding problem if the coefficient of friction between the wheel and the plane is (a) 0.4; (b) 0.9.

6.112 Repeat Problem 6.110 if the cord comes off the *top* of the inner radius instead of the bottom.

6.113 The frame in Figure P6.113 consists of the bar AD, the curved bar DC (quarter circle), and the pulley. Determine whether the frame is in equilibrium without the box. If not, find the minimum weight of the box for equilibrium. Assume that all the forces exerted on DC, by the floor and by the box, act at the point C. Also find the force exerted on AD by the pin at D, for your equilibrium case.

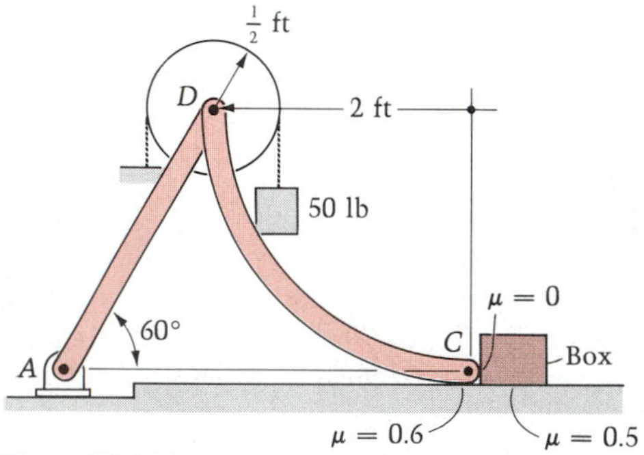

Figure P6.113

6.114 The ladder AB in Figure P6.114 is 30 ft long and weighs 60 lb. If the coefficient of friction is 0.5 between the ladder and both the floor and the wall, find the force P required to start the ladder moving up at A.

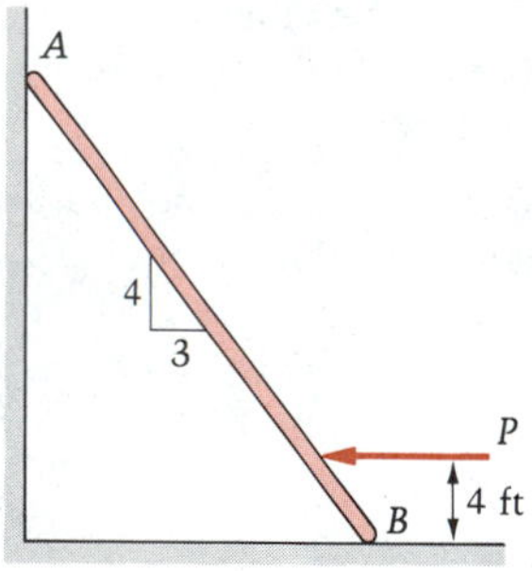

Figure P6.114

6.115 In the preceding problem, suppose the force P is applied at height y instead of 4 ft. Find the smallest value of y for which the ladder will not slide upward at A no matter how large the force P. *Hint:* Write the three equilibrium equations, eliminate the normal forces, and examine carefully the remaining equation.

6.116 The coefficient of friction between the drum and brake shown in Figure P6.116 is 0.5. (a) Show that there is adequate friction for equilibrium. (b) Determine the horizontal and vertical components of the pin reaction at O on the brake. The drum weighs 100 lb and the weight of the brake is 10 lb/ft. The brake is a uniform bar.

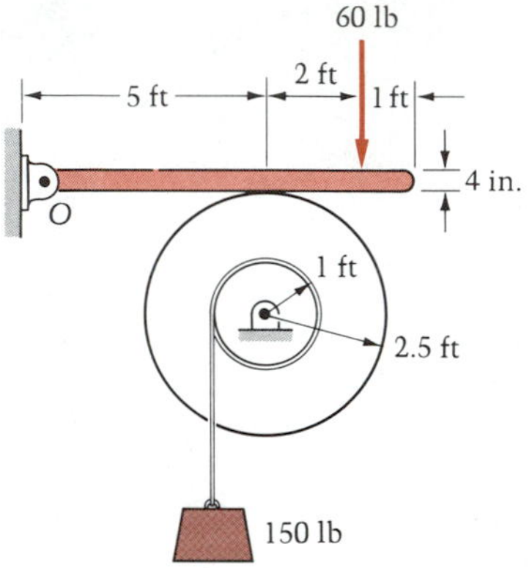

Figure P6.116

6.117 The man in Figure P6.117 is trying to ride a one-speed bicycle up a steep incline. He is tiring, and the bike is moving *very* slowly. Find the normal forces beneath each tire, and the friction force exerted up the plane by the pavement on the rear tire. Neglect the friction beneath the non-driven front wheel, and take the total weight of bike and man to be 190 lb with mass center at C. Also find the minimum coefficient of friction between tire and incline for which the wheel will not slip.

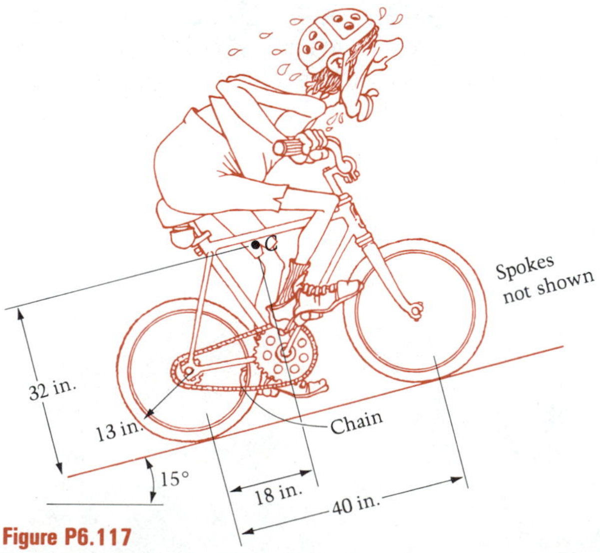

Figure P6.117

6.118 Extend the preceding problem. Knowing the friction force, use the free-body diagram of the rear wheel to determine the force T in the upper segment of the chain. Then use the free-body diagram of the pedal sprocket to find the force F that the man's left foot is exerting at the instant shown in Figure P6.118.

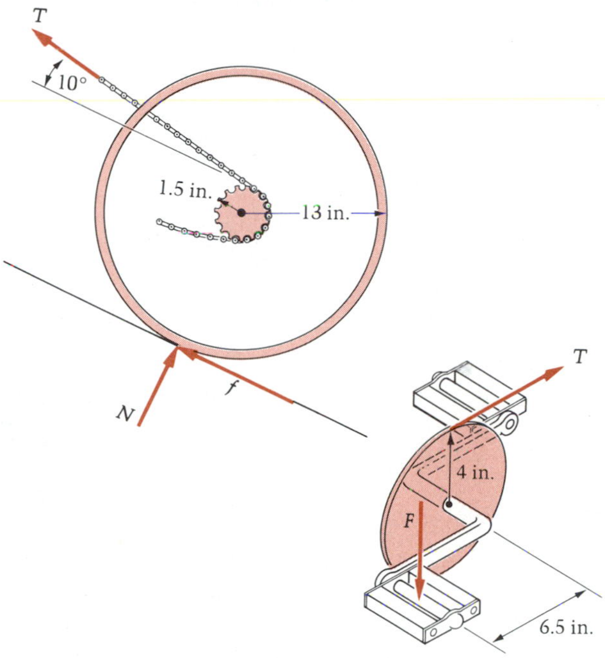

Figure P6.118

6.119 As P increases from an equilibrium value, where does the cylinder slip first, and at what value of P does this occur? (See Figure P6.119.)

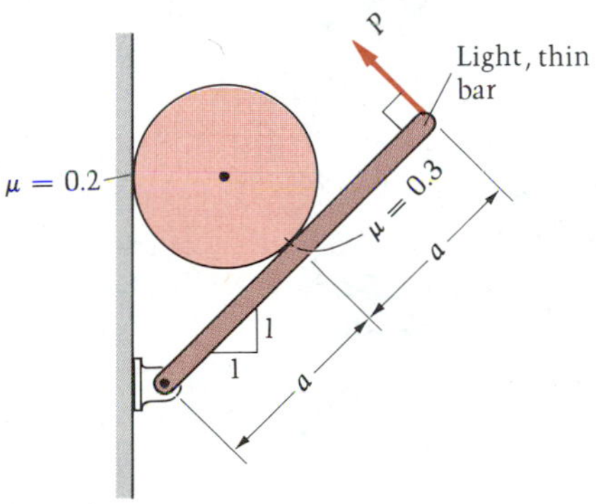

Figure P6.119

6.120 Repeat the preceding problem with the friction coefficients reversed.

6.121 The 240-N · m couple is applied to the bar DB. Bar AC rests on DB in equilibrium as shown in Figure P6.121. Find the minimum coefficient of friction between the two bars for which the equilibrium can exist.

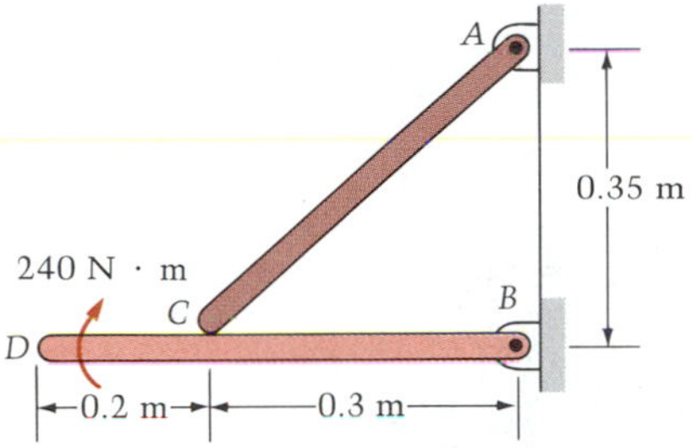

Figure P6.121

6.122 Repeat the preceding problem if the 240-N · m couple is replaced by an upward force of 480 N at D. Tell why the answer for μ_{min} is the same in the two problems, but not the reaction at B.

6.123 In Problem 4.235, suppose the bar and block are made of the same material and their coefficients of friction μ with the plane are the same. What is the minimum value of μ for which the equilibrium can exist?

6.124 The rope is held securely by the self-clamping device in Figure P6.124.

- **a.** Explain how both the actual and maximum friction forces increase with P.
- **b.** Find the minimum coefficient of friction between device and rope to prevent slipping.
- **c.** Find the resultant force exerted by the pin at A onto the support if $P = 500$ lb.

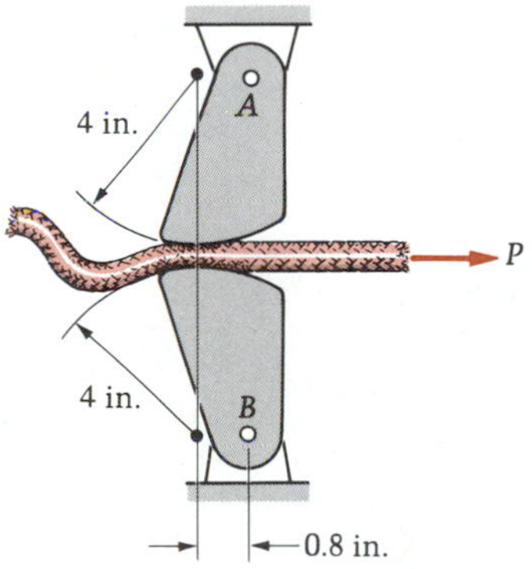

Figure P6.124

6.125 The light rod AB in Figure P6.125 is pinned smoothly to the ground at A and to the cylinder at B. The cylinder rests against the block, which in turn rests against the wall. The cylinder and block each weigh W. Find the minimum coefficient of friction between the block and the wall so that the block will not slip downward regardless of the value of W.

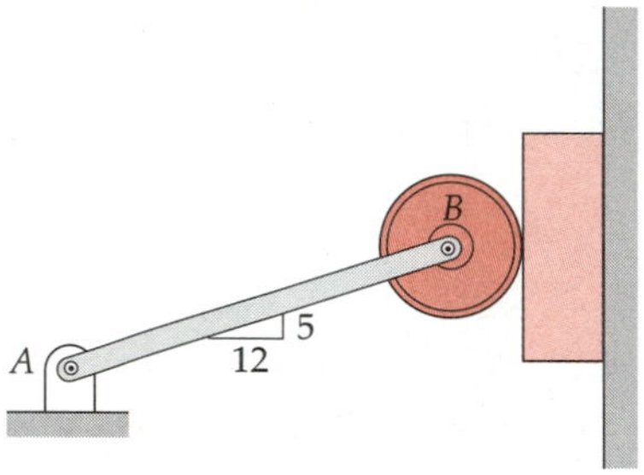

Figure P6.125

6.126 Repeat the preceding problem if the cylinder weighs W_C and the block weighs W_B.

6.127 Repeat the preceding problem if in addition the rod weighs W_R and has length ℓ.

6.128 Show that the identical, light cylindrical rollers in Figure P6.128 will hold the heavy bar of weight W in equilibrium in a self-locking manner (i.e., independent of the value of W) if the friction coefficients μ_1 and μ_2 between roller and bar, and roller and wall, respectively, are greater than a certain number. Find this common value of the two μ's, which depends on the angle α. Note that the smaller the angle α, the better the self-locking, because then the smaller μ needs to be to prevent slip regardless of the size of W.

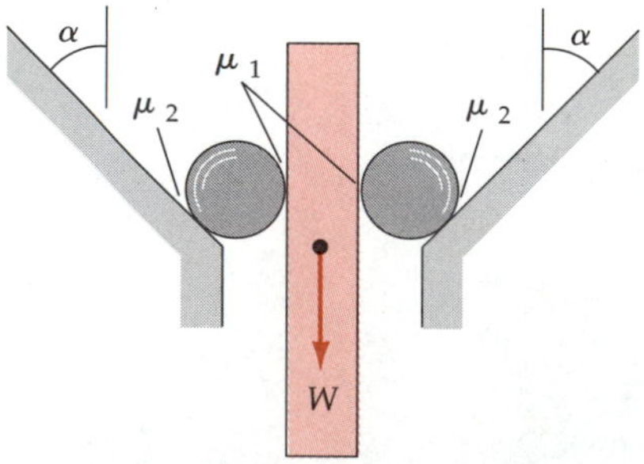

Figure P6.128

6.129 The light scaffolding hook in Figure P6.129 rests against a rough ceiling at A and a smooth wall at B. (a) What must be the friction coefficient between the hook and the ceiling to prevent the hook from slipping at A regardless of the size of P? (b) Change the 0.4m dimension (i.e., alter the shape of the hook) so as to make the answer become $\mu = 0.2$, for a safer result.

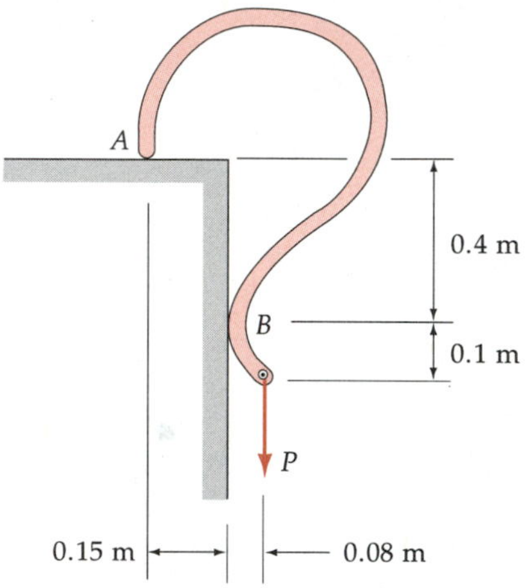

Figure P6.129

* **6.130** The plank rests on the floor at A and on the smooth, fixed block at C. (See Figure P6.130.) If the coefficient of friction between the plank and the floor at A is 0.3, find the range of heights H of the block for which the plank won't slip on the floor.

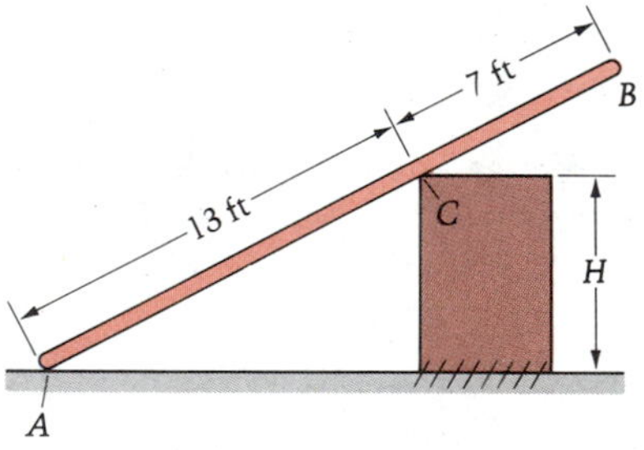

Figure P6.130

* **6.131** In Problem 6.130, let the coefficient of friction be 0.3 at *both* surfaces of contact. Find the lowest height of the block for which the plank will slip.

6.132 Solve Example 6.9 if the cord to which $\mathcal{W}$ is attached comes off the *right* side of the hub of $\mathcal{C}$.

* Asterisks identify the more difficult problems.

6.133 The tongs in Figure P6.133 are holding the light cylinder in equilibrium. What is the minimum coefficient of friction between the cylinder and the tongs such that the cylinder can't slip upward regardless of the value of P?

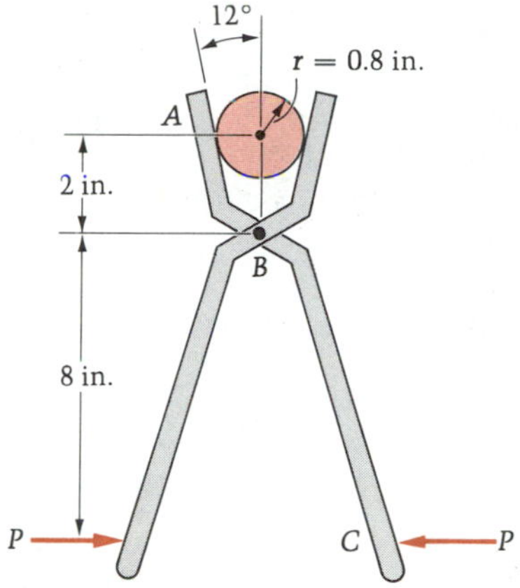

Figure P6.133

6.134 In the preceding problem, find the force exerted by the pin at B onto member ABC when $P = 15$ lb. Assume a friction coefficient large enough to prevent slip.

6.135 The 20-N half-cylinder is in equilibrium on the plane, with a cord between the center of the right edge and the plane as shown in Figure P6.135. If the cylinder is on the verge of slipping with $\theta = 30°$ and $\phi = 20°$, find the tension in the cord and the coefficient of friction between the cylinder and the plane.

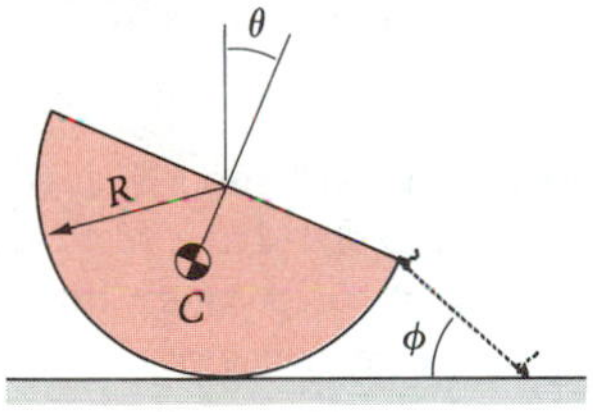

Figure P6.135

6.136 Find the angle ϕ (in terms of T, W, and R), for which the cylinder of weight W can be in equilibrium in the cylindrical cavity under the action of the couple of moment T (see Figure P6.136). Also find the required minimum coefficient of friction μ for which this equilibrium can exist. It is given that $WR > T$.

* **6.137** In Figure P6.137 find the smallest coefficient of friction for which the thin rod ABC will not slip on the cylindrical surface.

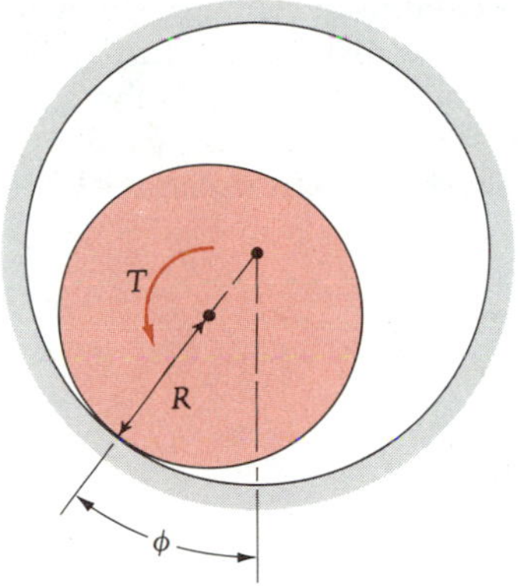

Figure P6.136

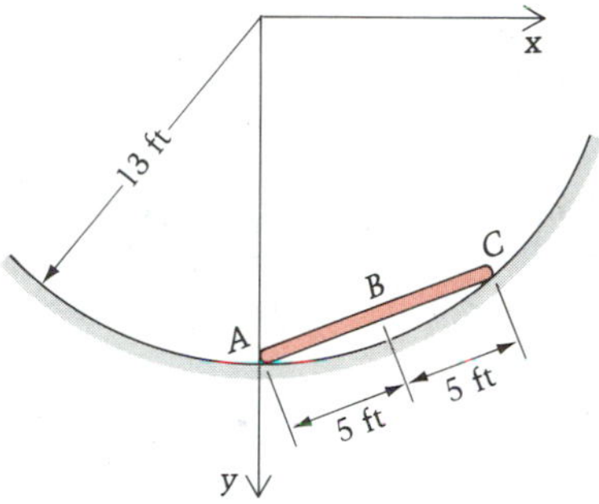

Figure P6.137

** **6.138** The friction coefficient between the stick and the wall is μ. The stick (mass m), connected to the floor at A with a ball-joint, can clearly rest on the wall at P in equilibrium. Find the largest angle between the stick and line AP for which the stick can be in equilibrium. (See Figure P6.138.) Assume that the friction force at Q acts in a direction so as to oppose movement of that end of the stick.

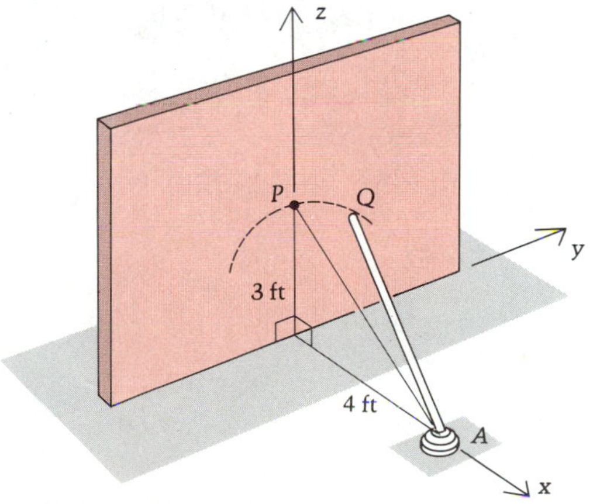

Figure P6.138

6.139 The block of mass m is at rest on an inclined plane as shown in Figure P6.139. The coefficient of friction between the block and the plane is μ, which is therefore $\geq \tan \phi$. If a horizontal force H is applied parallel to the incline, find the largest value of H for which the block will not slip.

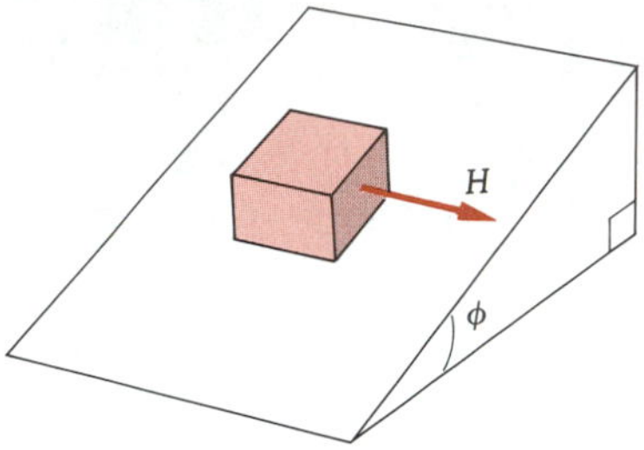

Figure P6.139

* **6.140** Repeat the preceding problem if H is applied at a 45° angle downward from its previous horizontal position (but still parallel to the incline). (See Figure P6.140.)

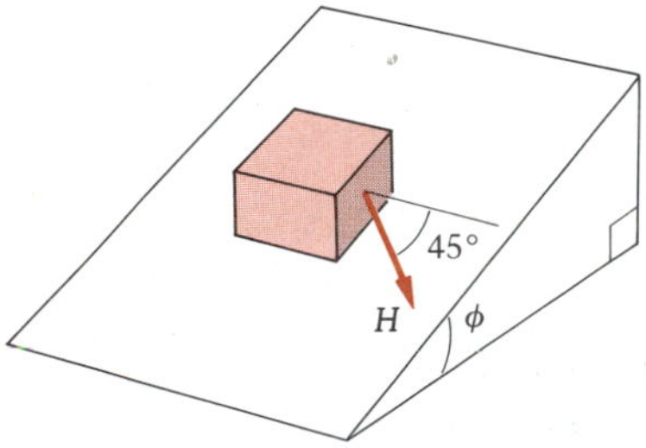

Figure P6.140

Figure P6.141

* **6.141** In Figure P6.141 the coefficient of friction μ between any two of the four identical spheres is the same as between any sphere and the ground. The three lower spheres are as shown in the lower figure, and the upper sphere rests on the top. What is the minimum value of μ for which the spheres can remain in equilibrium? *Hint:* Use symmetry.

* **6.142** Two small balls of equal masses m are in equilibrium on a cylinder as shown in Figure P6.142, connected by a light string of length $1.5R$. The balls are on the verge of slipping clockwise; their friction coefficient with the cylinder is 0.3 for each. Find the angles θ_1 and θ_2, the (constant) tension in the string, and the normal forces N_1 and N_2, respectively, exerted on the left and right masses.

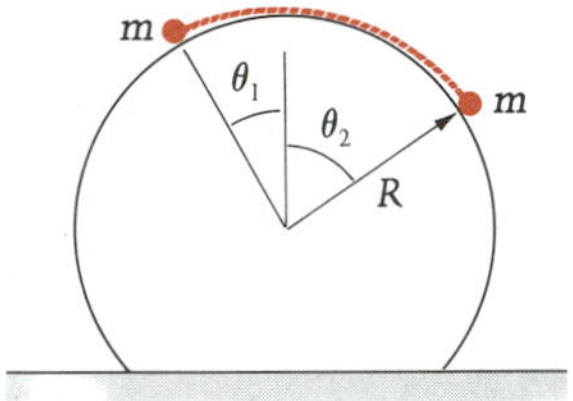

Figure P6.142

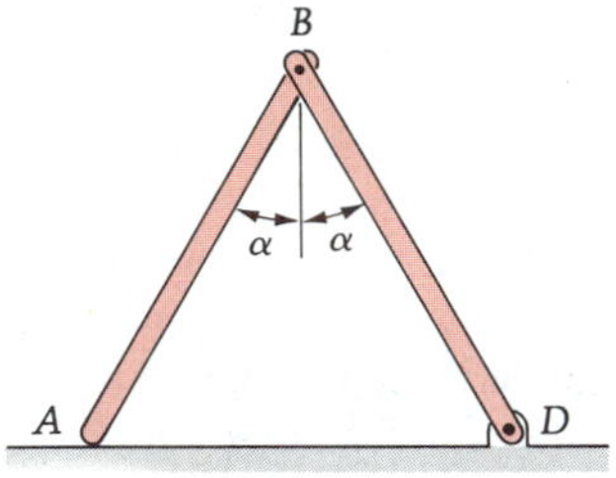

Figure P6.143

6.143 Two identical uniform heavy rods AB and BD, each of weight W, are freely pivoted at B and stand in a vertical plane. (See Figure P6.143). End A rests on a rough horizontal ground, to which D is pinned. In terms of α, find the minimum friction coefficient between AB and the ground for which equilibrium can exist.

* **6.144** In the preceding problem, let $\alpha = 30°$ and $\mu = 0.4$. Let a force P be applied downward at B and turned downward slowly toward BD (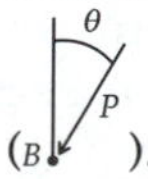
). Find the value of P if slipping occurs at A when $\theta = 30°$.

* **6.145** The pinned, identical bars in Figure P6.145 rest symmetrically on the cylinder. What is the range of angles between the bars for equilibrium if $L = 2R$ and the coefficient of friction is $\mu = 0.2$?

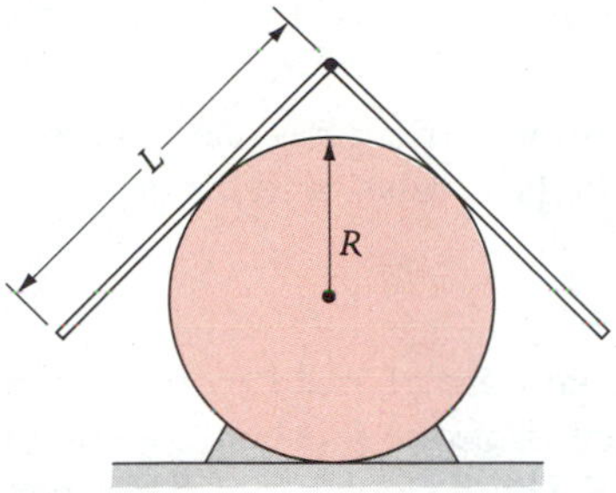

Figure P6.145

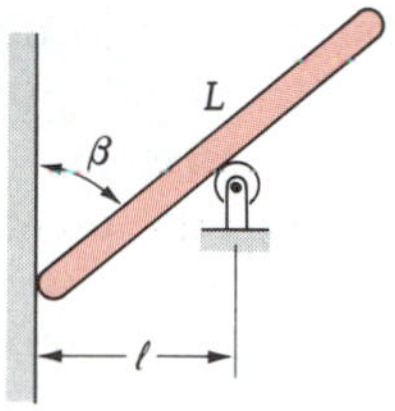

Figure P6.146

* **6.146** The friction coefficient between the bar and the wall is 0.5. The uniform bar has length $L = 2\ell$, where ℓ is the distance of the small, fixed pulley from the wall as shown in Figure P6.146. determine the smallest angle β for which the bar can be in equilibrium.

6.147 In the preceding problem, find the *largest* angle β for which the bar can be in equilibrium.

* **6.148** In Figure P6.148:

  a. Show that if the system in the figure is in equilibrium, then $m_1 \geq m_2/2$.

  b. If the friction coefficient between the bars and the plane is μ, find the minimum value of μ for equilibrium of the system, as a function of θ, m_1, and m_2.

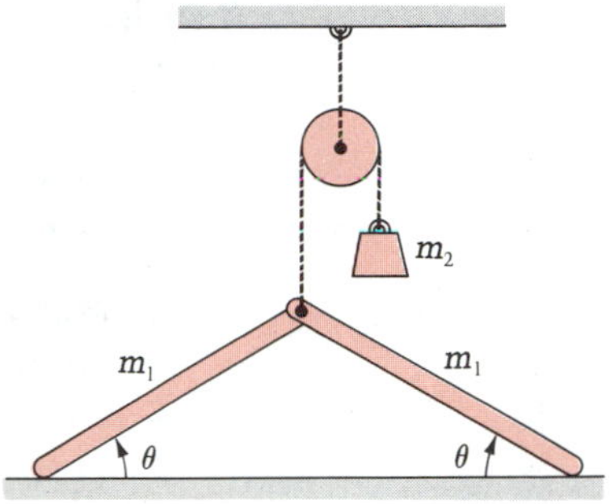

Figure P6.148

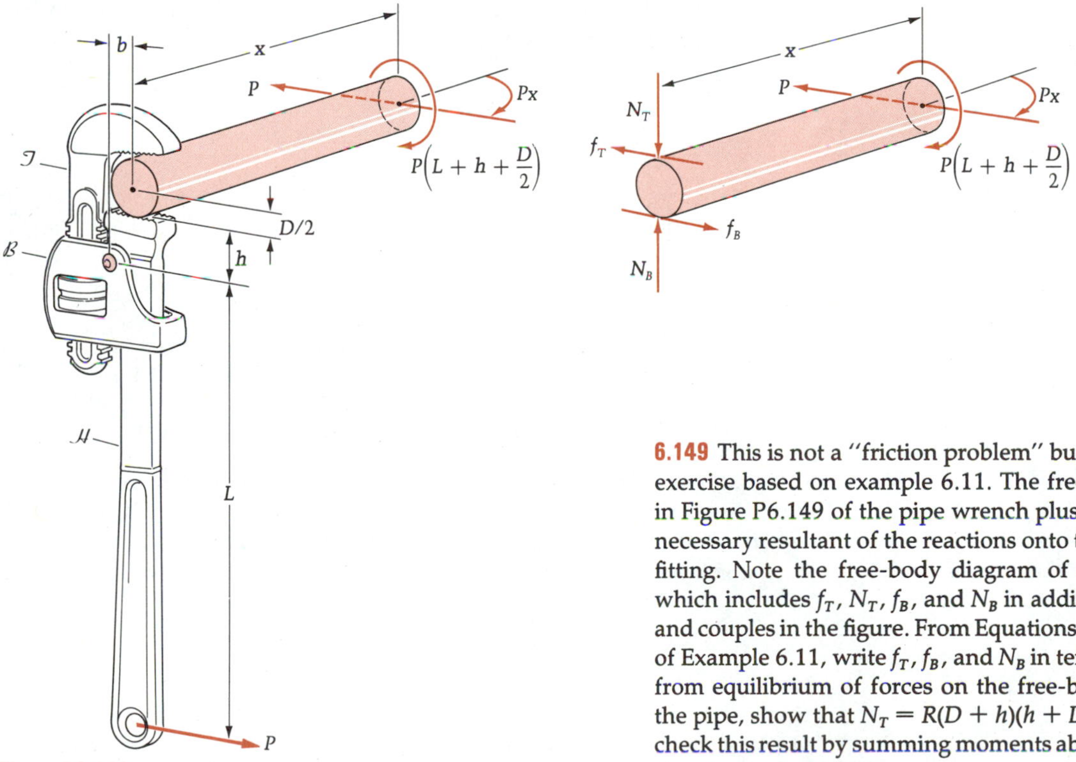

Figure P6.149

6.149 This is not a "friction problem" but an equilibrium exercise based on example 6.11. The free-body diagram in Figure P6.149 of the pipe wrench plus pipe shows the necessary resultant of the reactions onto the pipe from its fitting. Note the free-body diagram of the pipe alone, which includes f_T, N_T, f_B, and N_B in addition to the force and couples in the figure. From Equations (2), (6), and (11) of Example 6.11, write f_T, f_B, and N_B in terms of N_T. Then from equilibrium of forces on the free-body diagram of the pipe, show that $N_T = R(D + h)(h + L)/(bD)$. Finally, check this result by summing moments about any point of the pipe and obtaining zero.

6.3 Special Applications of Coulomb Friction

In this section, we are going to see how Coulomb friction is used to great advantage in four special ways: wedges, belts, screws, and disks.

Wedges

Wedges are simple machines in which large normal and friction forces are developed between two bodies by means of one ($\mathcal{B}_1$) being wedged between the other ($\mathcal{B}_2$) and a fixed surface $\mathcal{I}$. This is done in order to effect a (usually) small change in the position of $\mathcal{B}_2$, as in the following two examples.

Example 6.12

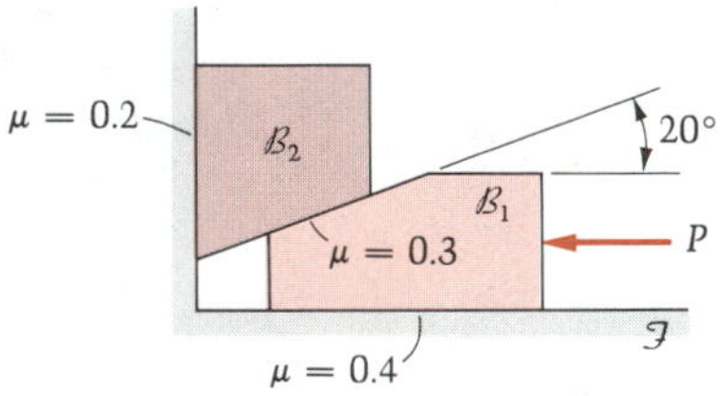

Figure E6.12a

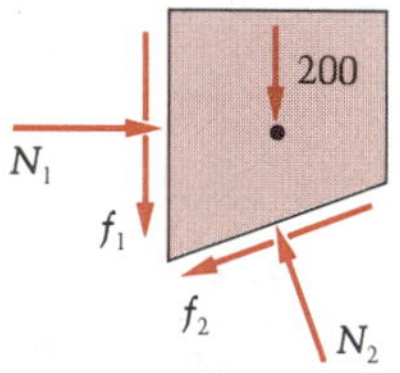

Figure E6.12b

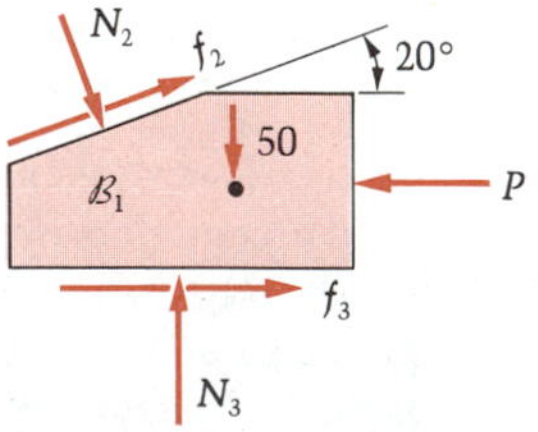

Figure E6.12c

Find the force P for which the 200-lb block $\mathcal{B}_2$ in Figure E6.12a will be on the verge of sliding upward. The wedge $\mathcal{B}_1$ weighs 50 lb.

Solution

In order for block $\mathcal{B}_2$ to be at the point of impending slipping, block $\mathcal{B}_1$ must also be at the point of slipping at both of its surfaces of contact with $\mathcal{B}_2$ and $\mathcal{I}$. Sometimes, as was seen in previous examples of Section 6.2, two or more cases of slipping are possible in a problem. Normally in wedge problems, slip occurs at all surfaces if it occurs at all.

The free-body diagrams and resulting equilibrium equations, with each friction force being maximum, are as follows:

For $\mathcal{B}_2$ (see Figure E6.12b):

$$\overset{+}{\longrightarrow} \qquad \Sigma F_x = N_1 - \overset{0.3\,N_2}{\cancel{f_2}} \cos 20° - N_2 \sin 20° = 0$$

$$+\!\uparrow \qquad \Sigma F_y = -200 - \overset{0.2\,N_1}{\cancel{f_1}} - \overset{0.3\,N_2}{\cancel{f_2}} \sin 20° + N_2 \cos 20° = 0$$

For $\mathcal{B}_1$ (see Figure E6.12c):

$$\overset{+}{\longrightarrow} \qquad \Sigma F_x = N_2 \sin 20° + \overset{0.3\,N_2}{\cancel{f_2}} \cos 20° + \overset{0.4\,N_3}{\cancel{f_3}} - P = 0$$

$$+\!\uparrow \qquad \Sigma F_y = N_3 - N_2 \cos 20° + \overset{0.3\,N_2}{\cancel{f_2}} \sin 20° - 50 = 0$$

(Note that here we have not used any moment equations, as they would merely serve to locate the lines of action of the resultants of the normal forces; we assume the blocks to be large enough that tipping will not occur.)

The solution to the above four equations in the unknowns N_1, N_2, N_3, and P is*

* Note that if P is given and motion is not impending or occurring, the problem of finding the three f's and the three N's is statically indeterminate.

$$N_1 = 175 \text{ lb}$$
$$N_2 = 281 \text{ lb}$$
$$N_3 = 285 \text{ lb}$$
$$P = 289 \text{ lb}$$

Note that all three normal forces have to be positive in a problem such as this, which serves as a partial check on the algebra. Also, on the system of $\mathcal{B}_1$ *plus* $\mathcal{B}_2$, note that (see the free-body diagram below, Figure E6.12d):

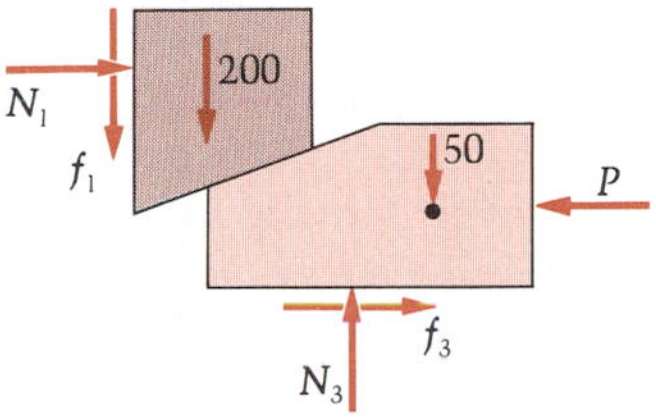

Figure E6.12d

$$\xrightarrow{+} \quad \Sigma F_x = N_1 + f_3 - P$$
$$= 175 + 0.4(285) - 289$$
$$= 0 \quad \checkmark$$

$$+\uparrow \quad \Sigma Fy = 0 = N_3 - f_1 - 200 - 50$$
$$= 285 - 0.2(175) - 250$$
$$= 0 \quad \checkmark$$

which are further checks. In the above two equations, note that f_2 and N_2 are internal forces within the system that do not therefore appear in the "overall" equilibrium equations. Only when $\mathcal{B}_1$ and $\mathcal{B}_2$ are separated do f_2 and N_2 become *external* forces, and thus appear on the free-body diagrams.

Example 6.13

A wedge is being forced into a log by a maul (see Figure E6.13a) in an effort to split the log into firewood. What is the minimum coefficient of friction between the log and the wedge for which the wedge will not pop out of the log when it is not in contact with the maul?

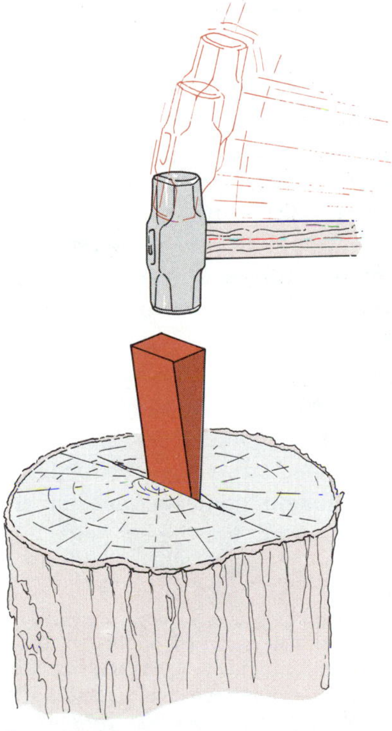
Figure E6.13a

Solution

Note that very large normal forces N are developed on the faces of the wedge. Without friction, when $P = 0$ the vertical components of the N's could pop the wedge upward because these components would ordinarily be considerably greater than the weight (w) of the wedge. The friction opposes the entry of the wedge during the striking of it by the maul. (See Figure E6.13b.) But then, when the force P is absent, the forces f change direction as shown and try to keep the wedge within the log. (Friction is a peacemaker; it always opposes the relative motion, or tendency for relative motion, of the contacting points of the bodies.)

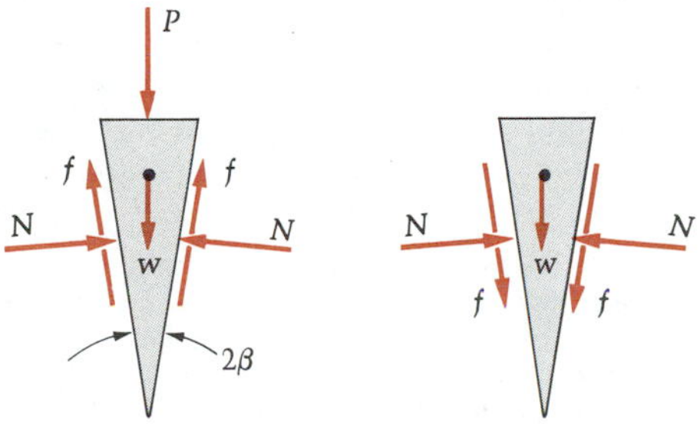

Figure E6.13b

Only the vertical equilibrium equation is of value in this problem. It is:

$$+\uparrow \quad \Sigma F_y = 0 = 2N \sin \beta - 2f \cos \beta - w \qquad (1)$$

If the wedge is on the verge of popping out, then $f = \mu N$:

$$N(2 \sin \beta - 2\mu \cos \beta) - w = 0$$

If, now, the weight is neglected (it is typically much smaller than the wedging forces in this application), we obtain

$$\mu = \tan \beta \qquad (2)$$

Thus any friction coefficient $\geq \tan \beta$ will insure that the wedge is "self-locking"—that is, that it will not move in the absence of the driving force P.*

Another slightly different approach to this problem, which makes use of the friction inequality, is to use Equation (1) when w is small to obtain

$$f = N \tan \beta$$

and then from $0 \leq f \leq f_{\text{MAX}} = \mu N$, we get

$$N \tan \beta \leq \mu N \Rightarrow \mu \geq \tan \beta$$

for equilibrium.

Question 6.8 Using equation (1), and noting that the weight of the wedge was neglected, determine whether or not $\mu = \tan \beta$ is a conservative answer for the minimum friction coefficient for self-locking.

Flexible Flat Belts and V-Belts

Belts are used to turn pulleys, sheaves, and drums in countless applications. For example, a fan belt turns the alternator shaft beneath the hood of an automobile. A string on pulleys once allowed us to tune to different stations on an old radio. A flat belt surrounds the rotating parts of a clothes dryer and makes it turn when a motor shaft drives the belt.

In this section we shall derive the relationship between the tensions in the two sides of a belt wrapped around a circular drum or restrained pulley when slipping is impending.

As shown in Figure 6.6(a), the angle between the entering and leaving points P_1 and P_2 is traditionally called β. We isolate an element of the belt at the angle θ, somewhere between 0 and β, and show the forces acting on it in Figure 6.6(b). The belt is assumed to be perfectly flexible, which means it has no bending stiffness. Thus no moments (and in turn no shear forces) appear on the free-body diagram. For the element, the

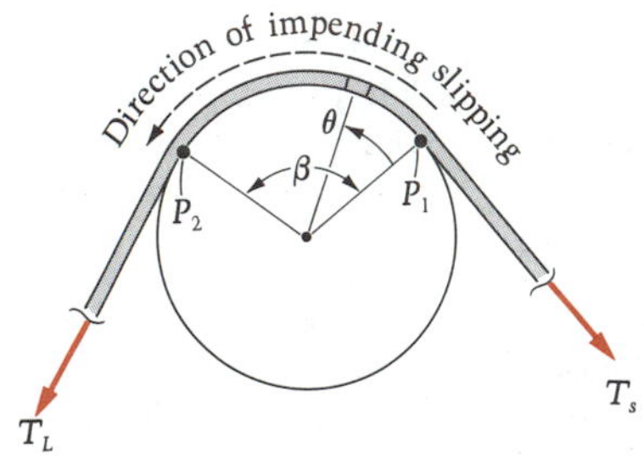

(a)

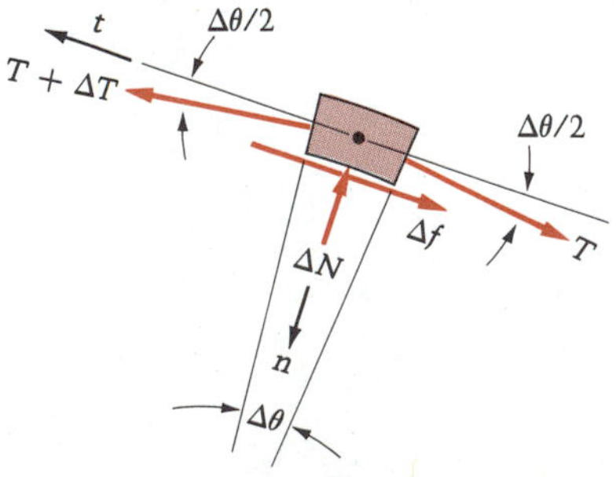

(b)

Figure 6.6

---

* One of the authors has split hundreds of logs and has never seen a wedge pop out of one, though he has heard of it happening. Wedges have half angles (β) of about 6°, so that μ between the wood and wedge need only be about 0.1 for the wedge to stay in place.

Answer 6.8 If w is larger than zero, then f must decrease for the equation to remain valid. Therefore, the maximum f need not be as great, and so $\tan \beta$ was a conservative answer for μ.

equilibrium equations in the normal (n) and tangential (t) directions are

$$\overset{+}{\swarrow} \quad \Sigma F_n = T \sin\frac{\Delta\theta}{2} + (T + \Delta T)\sin\frac{\Delta\theta}{2} - \Delta N = 0 \tag{6.6}$$

$$\overset{\nwarrow}{+} \quad \Sigma F_t = (T + \Delta T)\cos\frac{\Delta\theta}{2} - T\cos\frac{\Delta\theta}{2} - \Delta f = 0 \tag{6.7}$$

Dividing Equations (6.6) and (6.7) by $\Delta\theta$ yields

$$T\frac{\sin\dfrac{\Delta\theta}{2}}{\left(\dfrac{\Delta\theta}{2}\right)} + \frac{\Delta T}{2}\frac{\sin\dfrac{\Delta\theta}{2}}{\left(\dfrac{\Delta\theta}{2}\right)} - \frac{\Delta N}{\Delta\theta} = 0 \tag{6.8}$$

$$\frac{\Delta T}{\Delta\theta}\cos\frac{\Delta\theta}{2} - \frac{\Delta f}{\Delta\theta} = 0 \tag{6.9}$$

Taking the limits of these two equations as $\Delta\theta \to 0$ gives

$$T = \frac{dN}{d\theta} \tag{6.10}$$

and

$$\frac{dT}{d\theta} = \frac{df}{d\theta} \tag{6.11}$$

where we have used:

$$\lim_{\Delta\theta\to 0}\left[\frac{\sin\left(\dfrac{\Delta\theta}{2}\right)}{\left(\dfrac{\Delta\theta}{2}\right)}\right] = 1; \quad \lim_{\Delta\theta\to 0}\left[\cos\left(\frac{\Delta\theta}{2}\right)\right] = 1; \quad \lim_{\Delta\theta\to 0}(\Delta T) = 0$$

We are interested in the case when the belt is at the point of slipping, for which $f = \mu N$. Thus from Equation (6.11):

$$\frac{1}{\mu}\frac{dT}{d\theta} = \frac{dN}{d\theta} \tag{6.12}$$

and Equations (6.10) and (6.12) give

$$\frac{1}{T}\frac{dT}{d\theta} = \mu \tag{6.13}$$

from which, by integrating, we find

$$\ln T = \mu\theta + C_1 \tag{6.14}$$

The constant can be evaluated to be $\ln T_s$ because the tension is T_s when $\theta = 0$. Therefore,

$$\ln\frac{T}{T_s} = \mu\theta \tag{6.15}$$

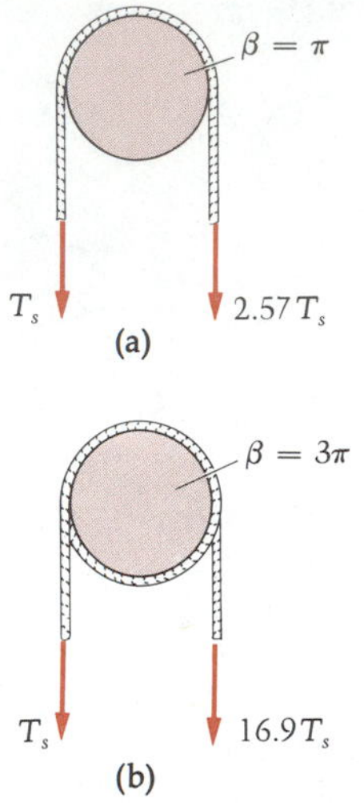

Figure 6.7

And the tension at any angle θ (necessarily in radians!) is given by

$$T = T_s e^{\mu\theta} \tag{6.16}$$

The maximum T (called T_L) is seen to be, when $\theta = \beta$,

$$T_L = T_s e^{\mu\beta} \tag{6.17}$$

The ratio of large to small tensions (leaving to entering the contact surface) is therefore an exponential function of the product of the coefficient of friction and the angle of wrap β. This can build up dramatically. For example, if a rope passes over a tree limb (Figure 6.7(a)) and $\mu = 0.3$ between the two materials, then for $\beta = \pi$,

$$T_L = T_s e^{0.3\pi} = 2.57T_s$$

But with a wrap and a half (Figure 6.7(b)),

$$T_L = T_s e^{0.3(3\pi)} = 16.9T_s$$

an increase of about 5.6 times!

As another example, just a few wraps of a guitar or banjo string around its peg enables it to be tightened to large tensions without the string being securely tied or clamped at all. (See Figure 6.8.) The friction buildup, even on such a relatively smooth shaft, is sufficiently great to prevent slippage.

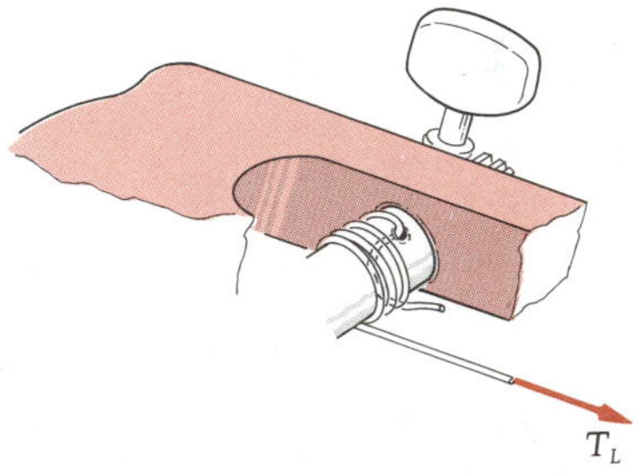

Figure 6.8

Question 6.9 What stiffness of the belt was neglected in the development of Equation (6.17)?

We now examine two formal examples of the use of Equation (6.17).

Answer 6.9 Its bending stiffness, which has a slight effect on the angle of wrap. If the belt is moving, its inertia also becomes a factor.

EXAMPLE 6.14

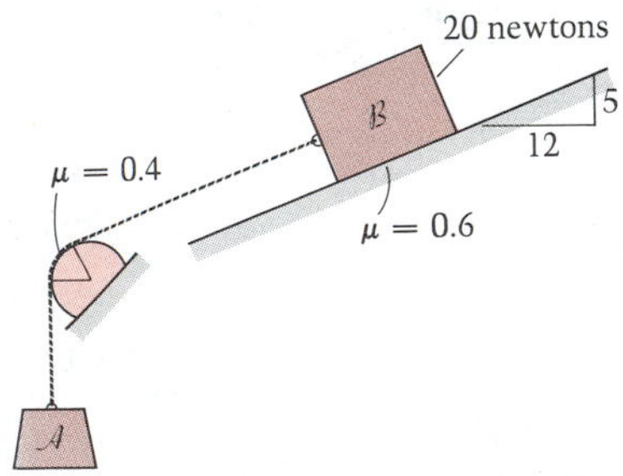

Figure E6.14a

Determine the minimum weight of block $\mathcal{A}$ for which block $\mathcal{B}$ will have impending motion down the plane in Figure E6.14a.

Solution

We note that $\tan\phi = \frac{5}{12} = 0.417 < 0.6 = \mu$, so that $\mathcal{B}$ will not begin to slide on its own. The angle of wrap of the rope is (see Figure E6.14b):

$$\beta = 90° - \tan^{-1}(5/12) = 67.4° = 1.18 \text{ rad}$$

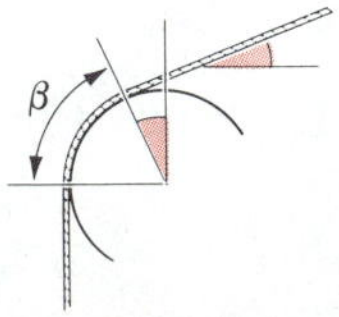

Figure E6.14b

Thus the tension T_s in the portion of rope between the fixed drum and $\mathcal{B}$ is obtained from Equation 6.17, with T_L equalling the weight of $\mathcal{A}$:

$$W_{\mathcal{A}} = T_s e^{0.4(1.18)} = 1.60T_s$$

or

$$T_s = W_{\mathcal{A}}/1.60$$

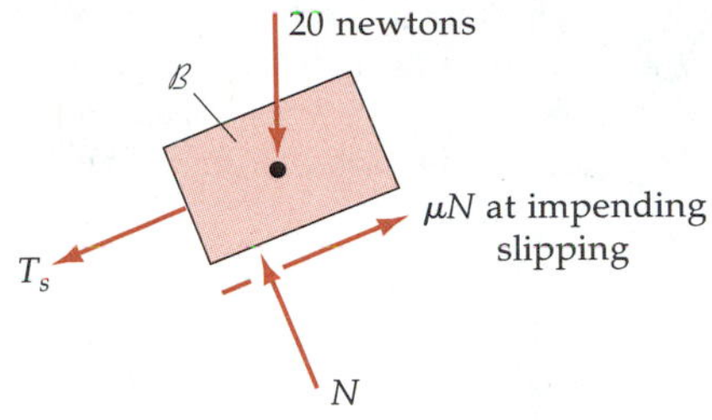

Figure E6.14c

Question 6.10 How do we know that $W_{\mathcal{A}}$ is the *large* tension and not the small one?

The free-body diagram of ℬ is shown in Figure E6.14c. The equilibrium equations give:

$$+\nwarrow \quad \Sigma F_y = 0 = N - 20\left(\frac{12}{13}\right) \Rightarrow N = 18.5 \text{ newtons}$$

$$\overset{+}{\swarrow} \quad \Sigma F_x = 0 = T_s + 20\left(\frac{5}{13}\right) - \underbrace{0.6(18.5)}_{\mu N}$$

from which

$$T_s = 3.41 \text{ newtons} = \frac{W_{\mathcal{A}}}{1.60}$$

$$W_{\mathcal{A}} = 5.46 \text{ newtons}$$

This is the weight of 𝒜 for which slipping of ℬ down the plane will be impending.

Answer 6.10 Because if ℬ is to slide downward, then the rope must be at the point of slipping counterclockwise on the drum.

Example 6.15

The oil-filter wrench in Figure E6.15a consists of a handle and a thin, flexible band that fits over the cylindrical filter. To remove the filter, force P is applied as shown in the figure. The tensions in the band produce friction forces around the filter whose resultant is a force together with a counterclockwise couple which

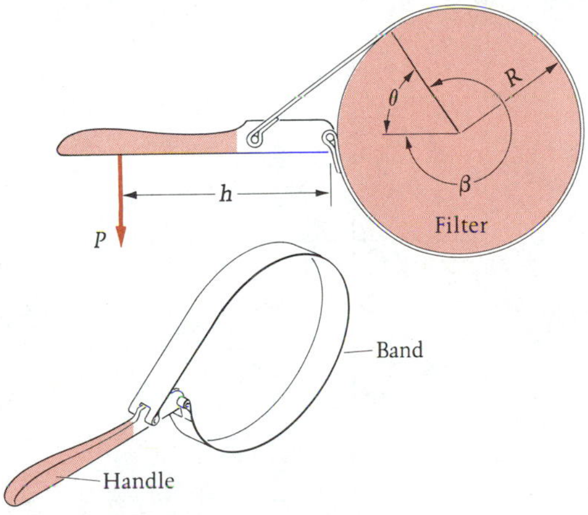

Figure E6.15a

loosens it. For a given filter radius R, friction coefficient μ, and wrap angle $\beta(2\pi - \theta)$, there is a minimum handle length h above which the band will not slip on the filter. Find $h = h(R, \mu, \beta)$. Then for $R = 2$ in. and $\mu = 0.3$, find $h_{\min}$ for $\theta = 60°, 45°$, and $30°$, and note that practical wrench handles are all longer than this.

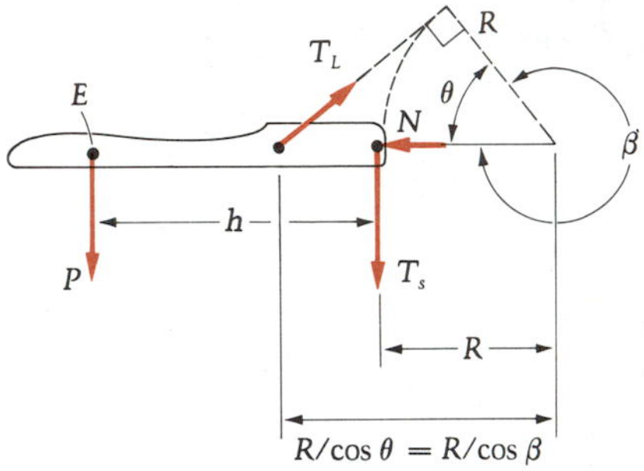

Figure E6.15b

Solution

Taking moments about the point E of the handle (using its free-body diagram, Figure E6.15b), we eliminate P and therefore can relate T_L and T_s immediately:

$$\circlearrowleft\!+\quad \Sigma M_E = 0 = T_L \cos\theta \left[h - \left(\frac{R}{\cos\theta} - R \right) \right] - T_s h \tag{1}$$

$$T_L = \frac{T_s h}{(h + R)\cos\theta - R}$$

This gives the relationship between the large and small tensions at the ends of the band. Now we know that if T_L never reaches $T_s e^{\mu\beta}$, the band will never slip on the filter. Therefore,

$$T_L = \frac{T_s h}{(h + R)\cos\theta - R} < T_s e^{\mu\beta} \tag{2}$$

or

$$h < e^{\mu\beta}[(h + R)\cos\theta - R] \tag{3}$$

The denominator in Equation (2) is positive for a practical wrench so that the

θ (deg)	β (rad)	h (in.)	Sketch of Wrench
60°	5.24	3.42	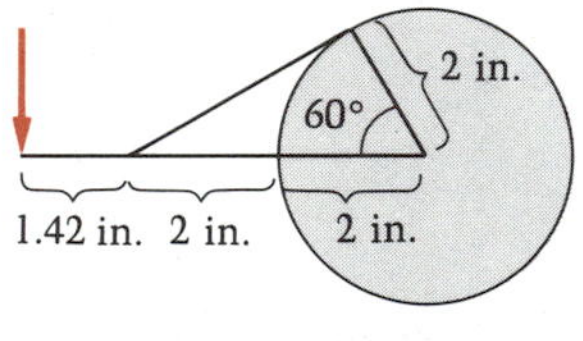
45°	5.50	1.14	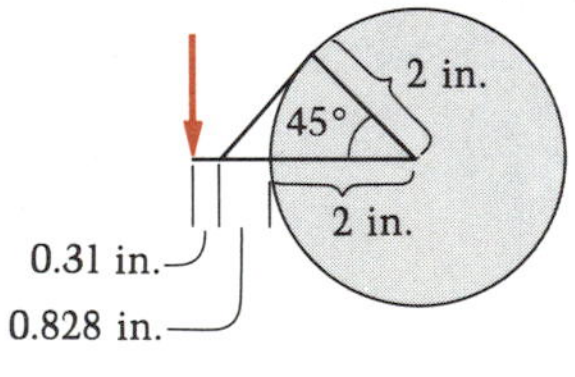
30°	5.76	0.389	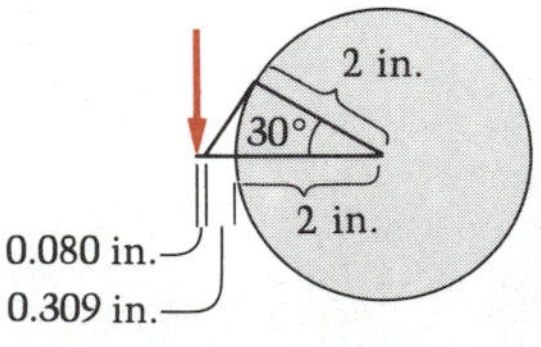

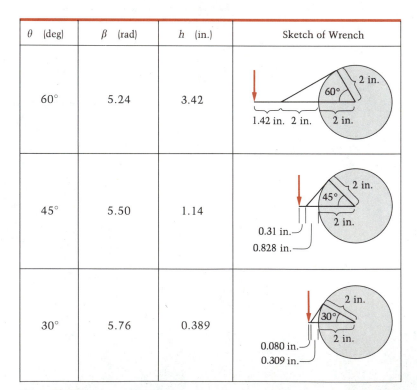

Figure E6.15c

sense in inequality (3) is unchanged. Also, we note that the cosines of θ and β are identical because $\beta = 2\pi - \theta$ so that, rearranging inequality (3),

$$e^{\mu\beta}(R - R\cos\beta) < h(e^{\mu\beta}\cos\beta - 1)$$

If $e^{\mu\beta}\cos\beta > 1$,* we obtain

$$\frac{h}{R} > \frac{e^{\mu\beta}(1 - \cos\beta)}{e^{\mu\beta}\cos\beta - 1}$$

as our final equation. For $\mu = 0.3$ and $R = 2$ in., we then compute h for the three required values of θ. The results are shown in Figure E6.15c on previous page.

We see that even at $\theta = 60°$, the average handle is longer than the minimum h calculated for no slip. This is to accommodate the hand width of an average person; note that h is the dimension to the *resultant* of the force of the hand.

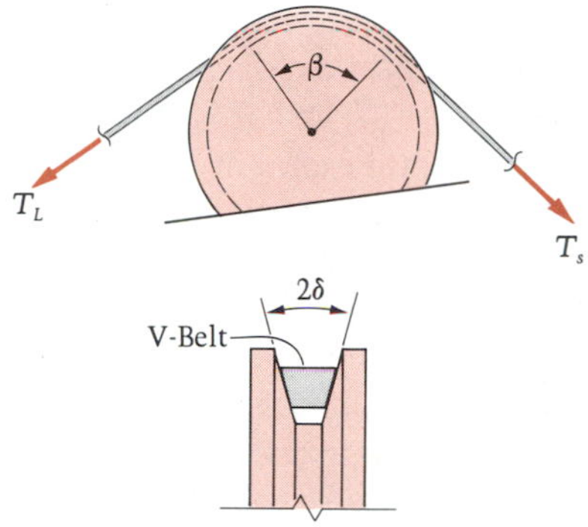

Figure 6.9

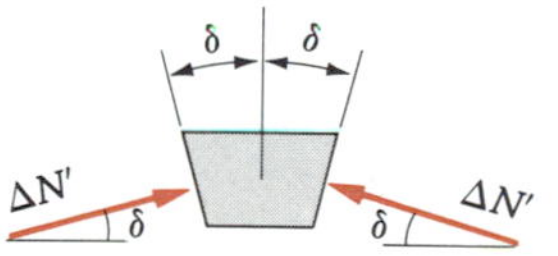

Figure 6.10

If the belt in our derivation is a V-belt instead of a flat belt, then the relationship between T_L and T_s changes and much larger tensions are possible. Suppose the angle of the "V" is 2δ as shown in Figure 6.9. It is seen that in contrast to our previous derivation, there is now no normal force *beneath* the belt at all. Instead, all the normal forces are exerted on the slanted faces on the *sides* of the belt as seen in Figure 6.10.

The outward component of the total increment ΔN of normal force acting on the element of Figure 6.6(b) is seen now in Figure 6.10 to be replaced, for a V-belt, by ($2\,\Delta N'\sin\delta$). Furthermore, it follows that the friction increment Δf will be $2\mu\,\Delta N'$, which is the total friction developed on *both* faces of the element of the V-belt. Substituting these new values of ΔN and Δf into Equations (6.6) and (6.7) leads to the following revised relation between the large and small tensions for a V-belt:

$$T_L = T_s e^{[\mu\beta/\sin\delta]} \tag{6.18}$$

Typical groove angle (2δ) values range from 30° to 38°. For a leather belt on a fixed iron pulley, $\mu \approx 0.26$. With an angle of wrap of 180° and with $2\delta = 30°$, the value of T_L/T_s is

$$e^{(0.26)(\pi)/\sin 15°} = 23.5$$

as compared to a value for the flat belt that is less than one-tenth this amount:

$$e^{(0.26)\pi} = 2.26$$

Screws

We shall consider here only the square-threaded screw, which is used with more efficiency than the V-thread in transmitting power or in causing motion.

Suppose that a load F is to be raised by applying a moment M to a square-threaded jackscrew as shown in Figure 6.11.

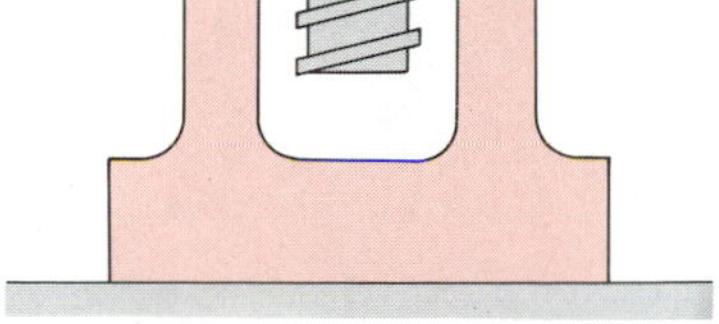

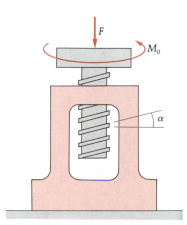

Figure 6.11

* Which is usually the case. For example, if $\theta < 45°$ ($\beta > 315°$), it is true for any $\mu > 0.063$.

We consider the forces acting on the bottom surfaces of an elemental segment S of the screw thread in Figure 6.12.

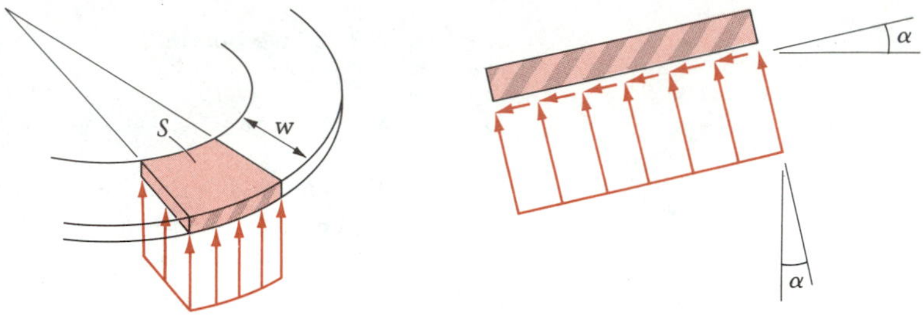

Figure 6.12

We emphasize that Figure 6.12 is not a free-body diagram of the segment S because it only shows the forces on one "face" of S. However, the accumulation of these forces, and the load F and moment M_0, are the only external forces acting on the overall screw. Thus they are the only forces that are important in expressing its equilibrium equations.

In what follows we shall assume that the pressure p and the shearing (frictional) stress s are constants. These stresses are exerted by the jack threads onto the screw threads, over whatever number of threads are in contact. On the elemental segment, the resultant normal and friction forces respectively caused by the stresses p and s are simply the stresses times the area beneath s (see Figure 6.13):

$$dN = p\,dA \qquad df = s\,dA \tag{6.19}$$

If we exert the minimum moment that will turn the screw, then we have impending slipping and

$$df = \mu\,dN$$

$$dN = p \times \text{Differential area}$$
$$= p\left(\frac{r\,d\theta}{\cos\alpha}\right)w$$

$$df = s \times \text{Differential area}$$
$$= s\left(\frac{r\,d\theta}{\cos\alpha}\right)w$$

Centerline of screw

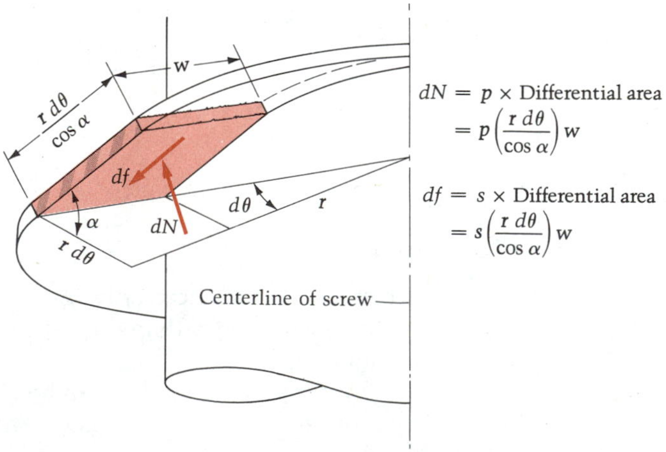

Figure 6.13

Summing the vertical components of the external forces on the *entire screw* then gives

$$+\uparrow \quad \Sigma F_y = 0 = -F + \int_{\substack{\text{contacting}\\ \text{threads}}} \cos\alpha \, dN - \int_{\substack{\text{contacting}\\ \text{threads}}} \sin\alpha \, df$$

or

$$F = \int_{\substack{\text{contacting}\\ \text{threads}}} (\cos\alpha - \mu \sin\alpha)\, dN$$

$$= \int_0^{\theta_c} (\cos\alpha - \mu\sin\alpha)\frac{prw}{\cos\alpha}\, d\theta$$

$$F = rw(1 - \mu \tan\alpha)\int_0^{\theta_c} p\, d\theta \tag{6.20}$$

where θ_c is the total angle of contact between the threads of the screw and jack. For example, if $5\frac{1}{2}$ threads are in contact, then $\theta = 11\pi$ rad.

Equation (6.20) then gives the average pressure p in terms of the load:

$$p_{ave} = \frac{F}{rw\theta_c(1 - \mu\tan\alpha)} \tag{6.21}$$

We also have to ensure that the total of all the differential moments caused by df and dN equilibrate the applied moment M_0. Summing moments about the axis, z, of the screw,

$$(\Sigma M)_z = 0 = M_0 - \int_{\substack{\text{contacting}\\ \text{threads}}} r\, dN \sin\alpha - \int_{\substack{\text{contacting}\\ \text{threads}}} r\, df \cos\alpha$$

or

$$M_0 = \int_0^{\theta_c} r\sin\alpha\left(\frac{prw\, d\theta}{\cos\alpha}\right) + \int_0^{\theta_c} r\cos\alpha\left(\frac{\mu prw\, d\theta}{\cos\alpha}\right)$$

$$M_0 = r^2 w(\tan\alpha + \mu)\int_0^{\theta_c} p\, d\theta \tag{6.22}$$

Using Equation (6.20) to eliminate the pressure,

$$M_0 = \frac{Fr(\mu + \tan\alpha)}{1 - \mu\tan\alpha} \tag{6.23}$$

This is the moment required to begin to move the load upward.

If M_0 is now removed, let us determine the condition under which the screw will unwind itself. In this case, the motion is impending *downward*, and in Figure 6.13 the friction direction would have to be reversed. This reversal can be accomplished in Equation (6.20) by changing the sign of the "friction force term" (the term with μ). Therefore, for impending motion downward without M_0, we get

$$F = rw(1 + \mu \tan \alpha) \int_0^{\theta_c} p \, d\theta \quad (6.24)$$

and

$$\tan \alpha = \mu$$

Note that this time, the friction *helps* the normal force to support the load. Therefore, if

$$\tan \alpha > \mu$$

then df will exceed $\mu \, dN$ and the screw will "unwind itself," moving downward without any applied moment M_0. Such a screw is called an "overhauling" screw.

If we have $\tan \alpha < \mu$—that is, $\alpha < \tan^{-1}\mu$—then, regardless of the size of the load F, the screw will not move downward. Such a screw is said to be "self-locking." In this case, a moment opposite in direction to that of Figure 6.11 is needed to lower the load and the screw. Equation (6.24) applies, together with the moment equation of equilibrium:

$$-M_0 - r(\sin \alpha - \mu \cos \alpha)\left[\left(\int_0^{\theta_c} p \, d\theta\right) rw / \cos \alpha\right] = 0 \quad (6.25)$$

Thus if we use Equation (6.24) to eliminate p, we obtain

$$M_0 = Fr\left(\frac{\mu - \tan \alpha}{1 + \mu \tan \alpha}\right) \quad (6.26)$$

which has both a smaller numerator and a larger denominator than Equation (6.23), indicating that not nearly as much moment is needed to lower the load as to raise it.

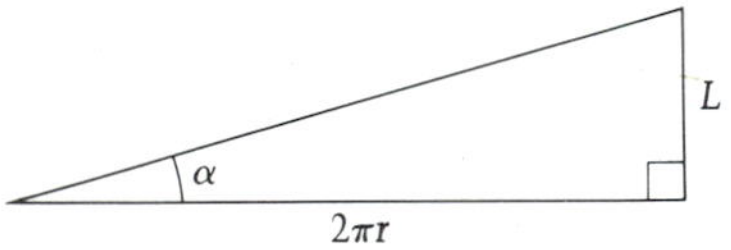

Figure 6.14

We note that the path of a screw is a helix. We can think of it as an inclined plane spiraling around a cylinder. A property of the helix is that the *lead* L (the amount the screw will advance or retract in one turn) is a constant. The pitch P of the screw is the axial distance between similar points on successive threads of the screw. Thus, if we imagine the inclined plane to be "unwound," we have (see Figure 6.14) $\tan \alpha = L/(2\pi r)$. For a single-threaded screw, we have $L = P$, but for a double* or triple or n-threaded screw, we would have $L = nP$. Equations (6.23) and (6.26) may be rewritten in terms of the pitch P:

$$M_0 \text{ (advancing against the load)} = Fr\left(\frac{2\pi r\mu + nP}{2\pi r - nP\mu}\right) \quad (6.27)$$

$$M_0 \text{ (retracting away from the load)} = Fr\left(\frac{2\pi r\mu - nP}{2\pi r + nP\mu}\right) \quad (6.28)$$

In these formulas, we would use the static coefficient of friction μ_s to compute the moments required to begin the turning. For very slow rota-

* A double-threaded screw is one that has two separate threads intertwined.

tions so that inertial effects are negligible, the kinetic coefficient μ_k must be used, but the equilibrium equations still apply.

A final remark prior to two examples is that screws, like wedges, are statically indeterminate when motion is not impending. We can see this from our analysis, because without the relationship $df = \mu\, dN$, we would not be able to relate M_0 and F.

Example 6.16

The scissors jack is supporting an automobile while a flat tire is being changed. The screw of the jack has a single square thread with 5-mm radius and 2-mm pitch, with a friction coefficient of 0.3. The owner of the car notices that he has jacked it up high enough to remove the flat tire, but not high enough to put on the inflated spare. What force B will begin to raise the car higher? (See Figure E6.16a.)

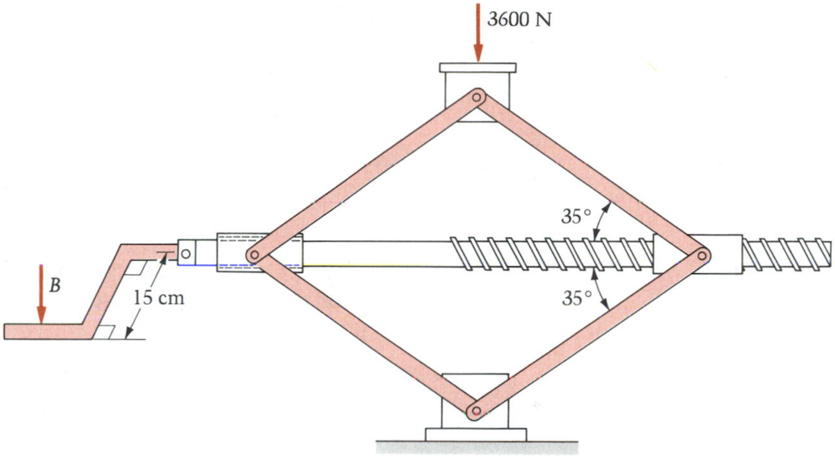

Figure E6.16a

Solution

The two free-body diagrams (see Figures E6.16b and E6.16c) and the accompanying equilibrium equations show that the force F developed in the jackscrew is 5140 N:

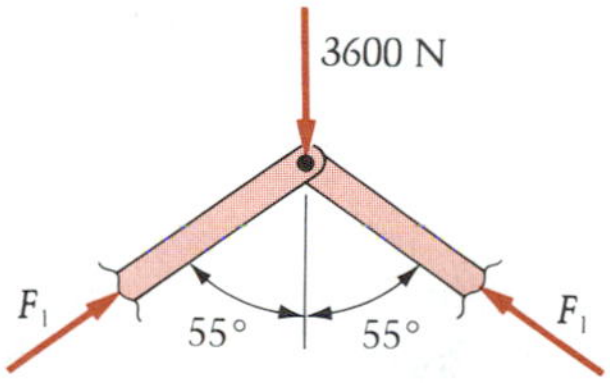

Figure E6.16b

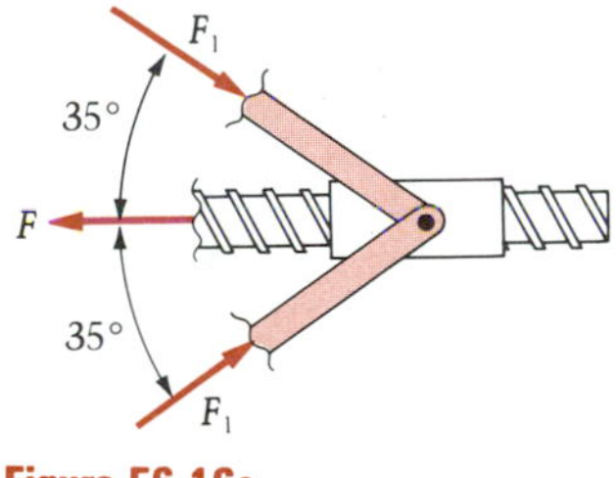

Figure E6.16c

$$+\uparrow \quad \Sigma F_y = 0 = 2F_1 \cos 55° - 3600 \qquad \overset{+}{\rightarrow} \quad \Sigma F_x = 0 = 2F_1 \cos 35° - F$$

$$F_1 = 3140 \text{ N} \qquad\qquad F = 5140 \text{ N}$$

In other words, *more* than the 3600 N supported weight of part of the car must be exerted by the jackscrew because of the angles involved.

Next, the moment that turns the jack is seen from the original figure to be $0.15B$ N · m, with B in newtons. To raise the car higher, Equation (5.27) gives

$$0.15B = 5140(0.005)\left(\frac{2\pi(0.005)(0.3) + 1(0.002)}{2\pi(0.005) - 1(0.002)(0.3)}\right)$$

$$B = 63.5 \text{ N}$$

When the tire has been changed and the owner is ready to lower the car, the force B required then is obtained from equation (5.28):

$$0.15B = 5140(0.005)\left(\frac{2\pi(0.005)(0.3) - 1(0.002)}{2\pi(0.005) + 1(0.002)(0.3)}\right)$$

$$B = 39.7 \text{ N}$$

about 63% of the load that was required to lift the car.

EXAMPLE 6.17

A single-threaded (or single-pitch) bolt has square threads* with 0.5-inch mean radius and $\frac{1}{16}$-inch pitch. The bolt is being used to securely fasten two plates which have kinetic coefficients of friction with the bolt of $\mu_k = 0.3$. (a) Find the axial force in the bolt if it is slowly turning under a 12 lb-ft torque. (b) If the rotation stops, what torque is required to start the bolt turning again if the static coefficient of friction is $\mu_s = 0.4$?

Solution

(a) Equation (6.27) gives

$$F = \frac{M_0}{r}\left(\frac{2\pi r - nP\mu}{2\pi r\mu + nP}\right) = \frac{12(12)}{0.5}\left(\frac{2\pi(0.5) - 1(\frac{1}{16})0.3}{2\pi(0.5)0.3 + 1(\frac{1}{16})}\right)$$

$$F = 895 \text{ lb}$$

(b) To begin again, we would use μ_s instead of μ_k. The same equation, but this time solved for M_0, gives

$$M_0 = Fr\left(\frac{2\pi r\mu + nP}{2\pi r - nP\mu}\right) = (895)\left(\frac{0.5}{12}\right)\left(\frac{2\pi(0.5)0.4 + 1(\frac{1}{16})}{2\pi(0.5) - 1(\frac{1}{16})0.4}\right)$$

$$= 15.8 \text{ lb-ft, almost a 32\% increase!}$$

Disk Friction

Disk friction is developed when two flat circular or annular surfaces come into contact, with each either turning or tending to turn relative to the other, about a central axis normal to their contact plane. Common examples are disk brakes and clutch plates.

* Note that most fastenings are made with V-threaded screws, which provide slightly more friction. See, for example, *Statics* by Goodman and Warner, Wadsworth, Belmont, CA, 1963, p. 325.

Suppose we wish to know how much moment resistance to the relative motion (or tendency toward it) is developed over the area of contact by the friction. We shall assume that:

a. the contact is over an annular area having inner and outer radii R_i and R_0;

b. the friction coefficient and pressure p are constant over the area;

c. there is no lubrication between the surfaces.

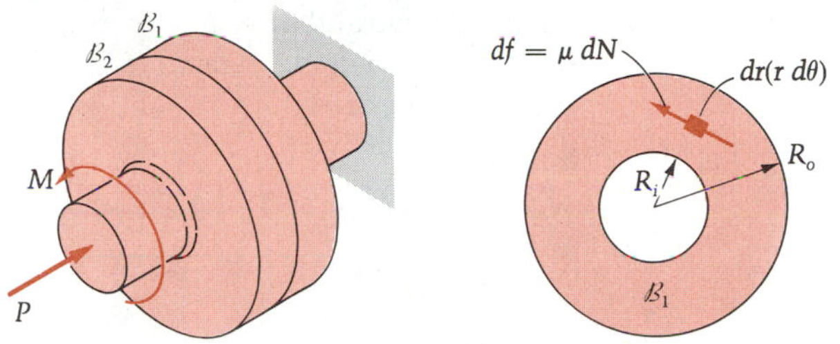

Figure 6.15

With these assumptions, the normal force on a differential area $r\,dr\,d\theta$ of one of the surfaces, say S, of body $\mathcal{B}_1$, is (see Figure 6.15)

$$dN = p\, dA$$

so that the moment developed on $\mathcal{B}_1$ by all the elemental friction forces at the point of slipping is

$$M = \int \underbrace{(\mu\, dN)}_{df} r = \int \mu p r\, dA = \int_{\theta=0}^{2\pi} \int_{r=R_i}^{R_0} \mu p r^2\, dr\, d\theta \tag{6.29}$$

Therefore, after integrating we find

$$M = \frac{\mu p (R_0^3 - R_i^3) 2\pi}{3} \tag{6.30}$$

Or, in terms of the compressing force P between the bodies, which is

$$P = pA = p\pi(R_0^2 - R_i^2), \tag{6.31}$$

we may eliminate the pressure p from Equation (6.30) and get

$$M = \frac{2\mu P}{3} \frac{R_0^3 - R_i^3}{R_0^2 - R_i^2} = \frac{2\mu P}{3} \frac{(R_0^2 + R_0 R_i + R_i^2)}{R_0 + R_i} \tag{6.32}$$

where it is seen that if $R_i = 0$ (solid circular area of contact),

$$M = \frac{2\mu P R_0}{3} \tag{6.33}$$

which is the moment that can be transmitted by friction across two circular disks. It decreases somewhat as the surfaces wear, as one might expect. Initially the material at larger r will wear faster because it has greater speed. This will alter the pressure distribution. If instead of constant pressure we reach a condition of constant rate of wear, it can be shown that the moment is reduced by 25% to $\frac{1}{2}\mu P R_0$.

Here is an example of the use of Equation (6.32):

EXAMPLE 6.18

A clutch is being tested. Its contacting surfaces are annular disks of inner and outer radii two and three inches, respectively. If with an axial force of 500 lb the clutch slips when the transmitted moment reaches 400 lb-in., what is the coefficient of friction?

Solution

We use Equation (6.32), which when solved for μ is

$$\mu = \frac{3M}{2P}\frac{R_0 + R_i}{R_0^2 + R_0 R_i + R_i^2}$$

Therefore,

$$\mu = \frac{3(400)}{2(500)}\frac{3 + 2}{3^2 + 3(2) + 2^2} = 0.316$$

PROBLEMS ► Section 6.3

6.150 The bodies $\mathcal{A}$ and $\mathcal{B}$ in Figure P6.150 have masses 50 kg and 75 kg, respectively. The various coefficients of friction between pairs of surfaces are shown in the figure. Find the force F that will cause $\mathcal{A}$ to begin to move up the wall.

6.151 In the preceding problem, let the force F be instead applied downward on the top of $\mathcal{A}$. Show that regardless of the value of F, block $\mathcal{B}$ will not move to the right.

6.152 In Figure P6.152 block $\mathcal{A}$, mass 20 kg, rests on block $\mathcal{B}$, mass 30 kg. If the coefficient of friction is 0.25 for all surfaces, find the force P necessary to raise block $\mathcal{A}$.

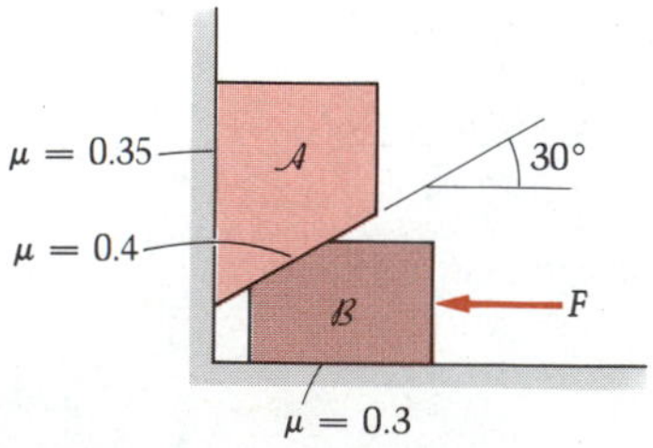

Figure P6.150

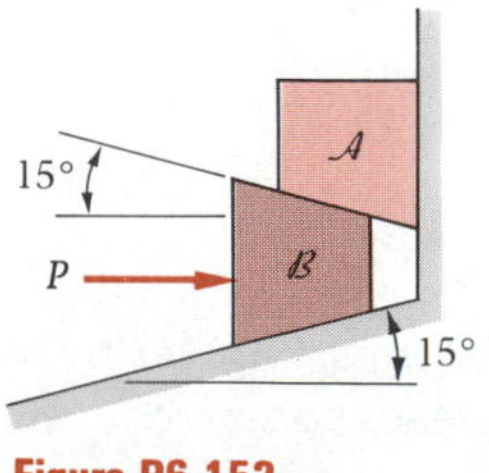

Figure P6.152

6.153 In Figure P6.153 find the smallest force F that will raise the body $\mathcal{W}$, which weighs 1000 lb. The friction coefficients are 0.2 between $\mathcal{A}$ and $\mathcal{B}$, and 0.15 between $\mathcal{A}$ and $\mathcal{W}$. Neglect the weight of $\mathcal{A}$.

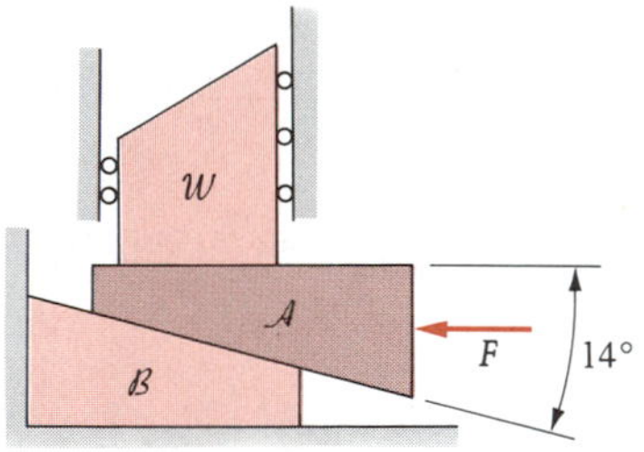

Figure P6.153

6.154 In the preceding problem, will the system remain in equilibrium if force F is removed? If not, how much force is needed?

6.155 In Figure P6.155 find the smallest force F that will cause the center C of the cylinder to move up the inclined plane. The cylinder has a mass of 300 kg, and the wedge is light.

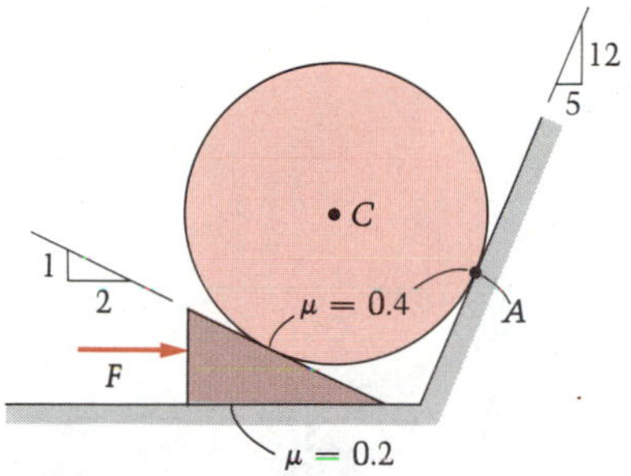

Figure P6.155

6.156 In the preceding problem, determine whether equilibrium is possible if force F is removed. If not, how much force is needed?

6.157 Find the range of values of the weight of $\mathscr{A}$ that will hold $\mathscr{B}$ in equilibrium. The friction coefficient is 0.45 between the cable and the fixed drum $\mathscr{D}$. (See Figure P6.157.)

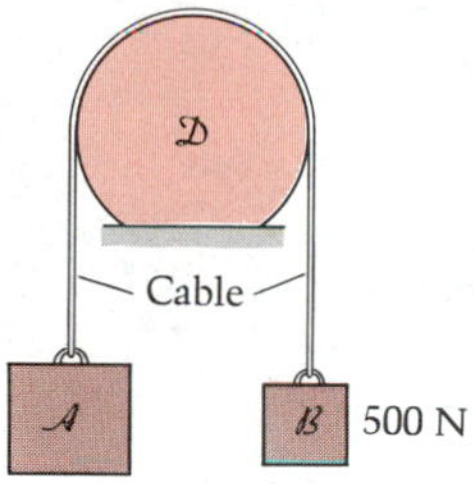

Figure P6.157

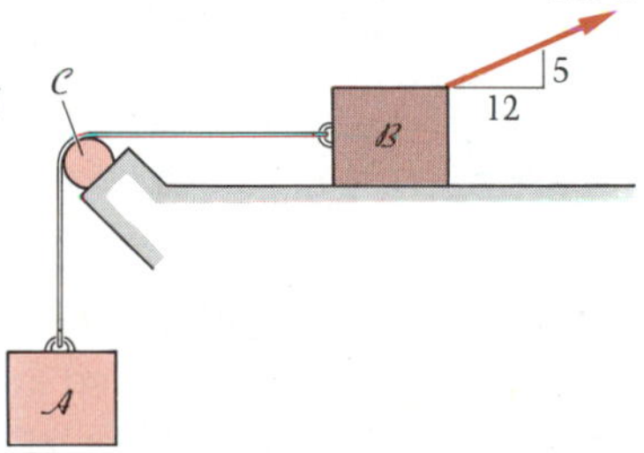

Figure P6.158

6.158 Homogeneous blocks $\mathscr{A}$ and $\mathscr{B}$ are connected by a belt that passes over the fixed drum $\mathscr{C}$. (See Figure P6.158.) The force of 130 lb acts upon block $\mathscr{B}$ as shown. The coefficients of friction are 0.4 between drum and belt, and 0.2 between $\mathscr{B}$ and the plane. Block $\mathscr{B}$ weighs 150 lb and is wide enough so that it will not tip. Find the range of weights that $\mathscr{A}$ can have for equilibrium of the system.

6.159 A 150-lb man stands on a 60-lb bar, to which a cable is fixed at B. (See Figure P6.159.) The cable passes over two fixed pegs with coefficients of friction as indicated. Assume that the normal force exerted by the man on the bar acts downward through A and find the smallest force the man can exert on the cable if the bar is to remain horizontal.

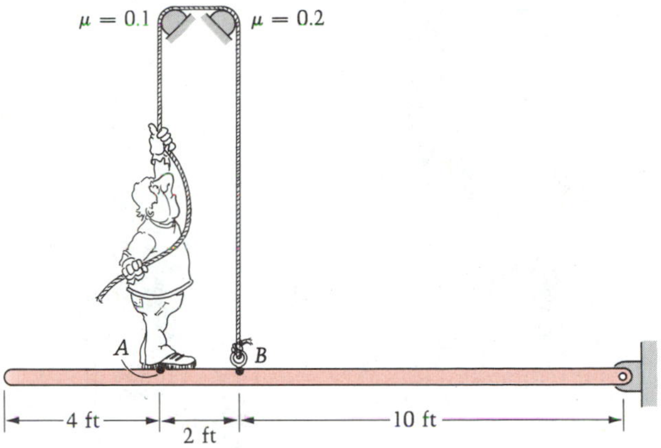

Figure P6.159

6.160 In the preceding problem, find the *largest* force the man can exert on the cable if the bar is to remain horizontal.

6.161 In Figure P6.161 a rope is wrapped around the fixed drum $1\frac{1}{4}$ times. If the girl can pull with 40 lb, what maximum weight $\mathscr{W}$ can be held in equilibrium if $\mu = 0.3$ between rope and drum?

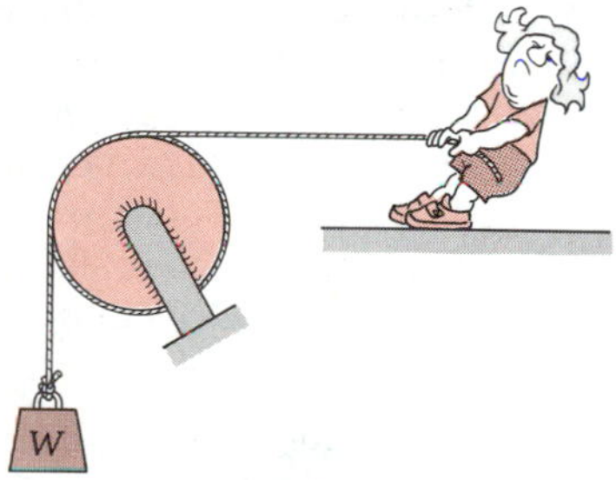

Figure P6.161

6.162 In Figure P6.162 the block weighs 40 lb. The friction coefficients are 0.5 between the rope and drum, and 0.35 between the block and the plane. Find the lowest force T that will cause the block to move.

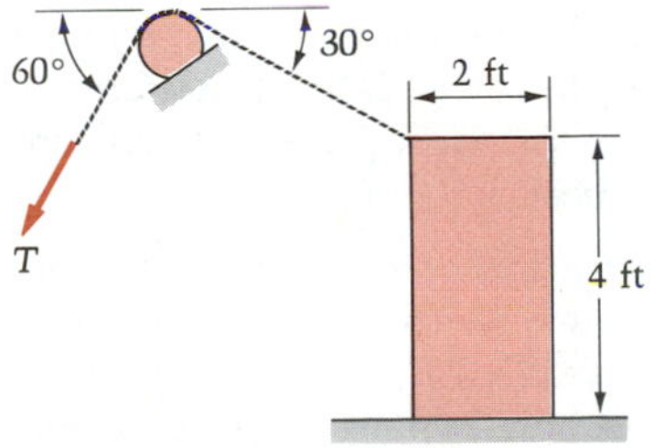

Figure P6.162

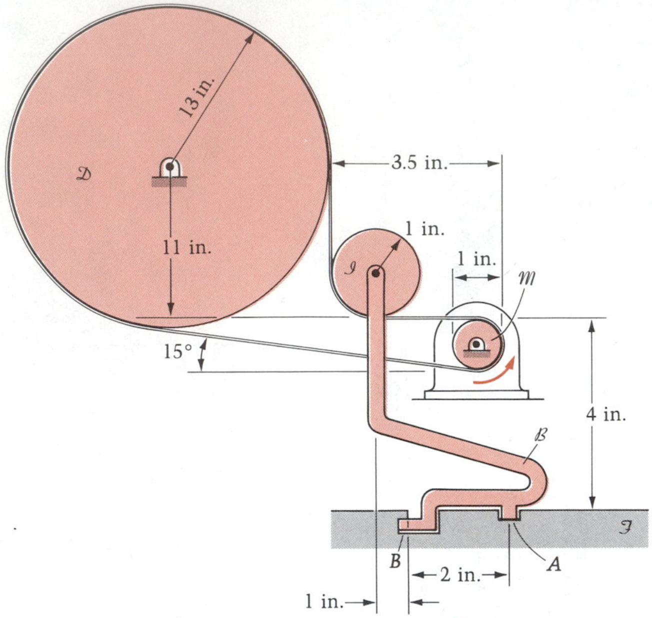

Figure P6.163

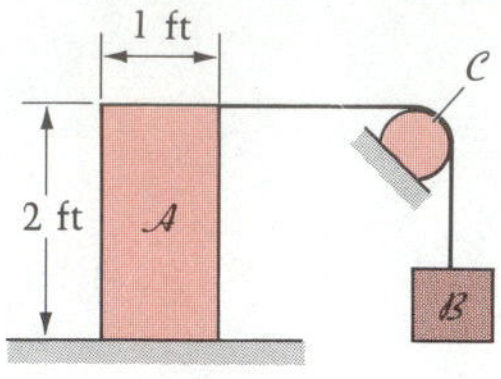

Figure P6.164

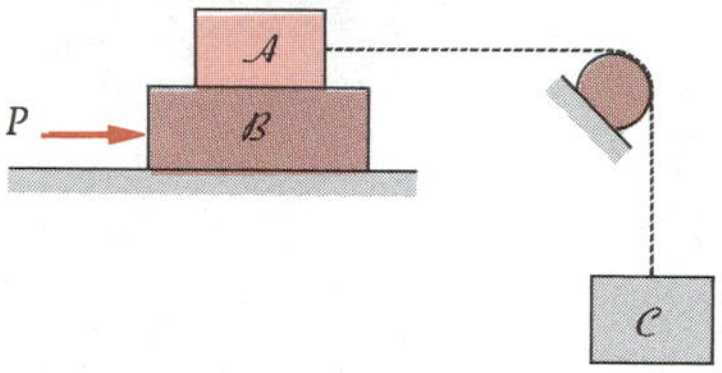

Figure P6.165

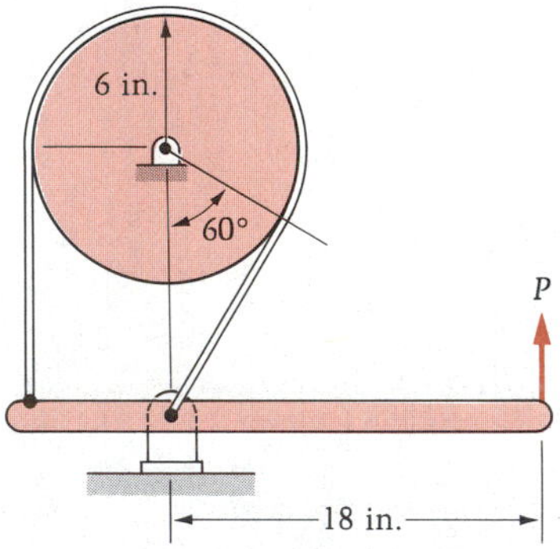

Figure P6.166

6.163 In the clothes dryer, the pulley of motor $\mathcal{M}$ turns counterclockwise as shown in Figure P6.163. If the dryer drum $\mathcal{D}$ is turning at constant angular velocity, then the moments of the large (lower) and the small (upper) tensions are in balance with the moments due to friction at the support wheels (not shown). At the point of impending slipping of the belt on the drum, find the ratio of large to small tensions with and without the idler pulley $\mathcal{I}$ present. Let $\mu = 0.3$.

6.164 The homogeneous block $\mathcal{A}$ in Figure P6.164 is connected by a belt to another homogeneous block, $\mathcal{B}$. Block $\mathcal{A}$ weighs 350 lb, and the coefficients of friction are 0.3 between $\mathcal{A}$ and the horizontal plane and 0.2 between the belt and the fixed drum $\mathcal{C}$. Find the maximum weight $\mathcal{B}$ can have without disturbing the equilibrium of the two bodies.

6.165 In Figure P6.165 body $\mathcal{B}$ weighs 1200 lb and body $\mathcal{C}$ weighs 680 lb. The coefficients of friction μ are $2/\pi$ between rope and drum, 0.4 between $\mathcal{A}$ and $\mathcal{B}$, and 0.3 between $\mathcal{B}$ and the plane. The force P is 350 lb. Determine the minimum weight of $\mathcal{A}$ that will prevent downward motion of $\mathcal{C}$. Neglect the possibility of tipping.

6.166 The coefficient of friction between the drum and the band is 0.4. (See Figure P6.166.) Find the friction moment on the drum developed by the band brake, in terms of force P, for impending slipping. The drum is held fixed by a couple that is not shown.

6.167 Can the 100-lb painter in Figure P6.167 stand on the left end of the scaffold and drink his lemonade without the rope slipping and dumping him to the ground below?

6.168 In Figure P6.168, given that $\mathcal{A}$ weighs 1000 lb, $\mathcal{B}$ weighs 750 lb, and the coefficients of friction μ are 0.3 between $\mathcal{A}$ and $\mathcal{B}$, 0.3 between $\mathcal{B}$ and the plane, and $1/\pi$ between rope and drum, find the minimum force P that will cause $\mathcal{A}$ to move.

6.169 Repeat Problem 6.93 if the peg is not smooth but has a friction coefficient with the rope of 0.4. (See Figure P6.169.) The blocks $\mathcal{A}$, $\mathcal{B}$, $\mathcal{C}$, and $\mathcal{D}$ weigh 100 lb, 50 lb, 75 lb, and 40 lb, respectively. Force P is slowly increased from zero. At what value of P will the equilibrium be disturbed?

6.170 The flat belt in Figure P6.170 is transmitting power from the small pulley $\mathcal{P}_1$ to the large pulley $\mathcal{P}_2$. The applied torque T_a turns $\mathcal{P}_1$ clockwise, and the belt then turns $\mathcal{P}_2$ in the same direction, with the load (being rotated by $\mathcal{P}_2$ and

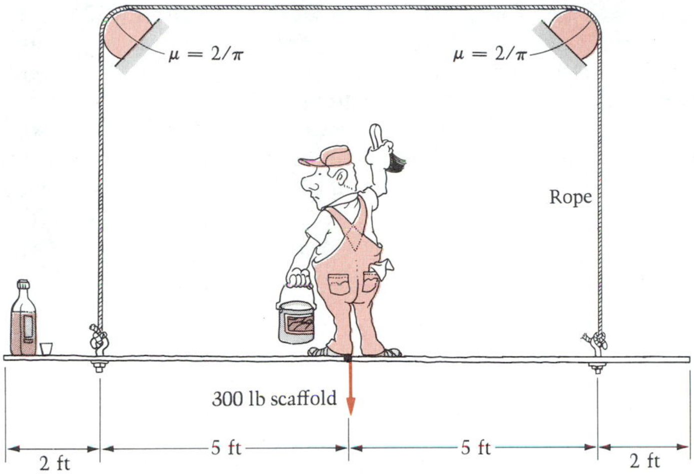

Figure P6.167

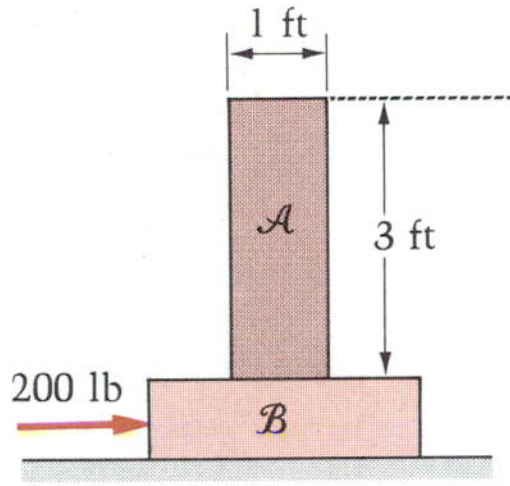

Figure P6.168

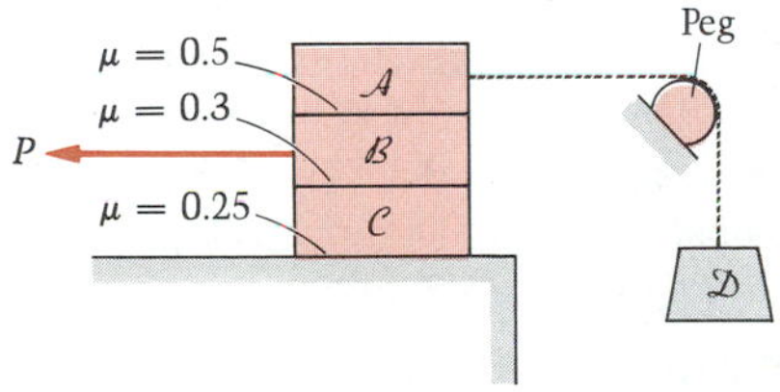

Figure P6.169

not shown) resisting the rotation with torque T_R. The friction coefficient between the belt and each pulley is 0.25. If the largest safe belt tension is 200 N, find:

a. The maximum torque T_a that can be applied without slipping the belt on $\mathscr{P}_1$.

b. The maximum load torque T_R to which the pulley $\mathscr{P}_2$ should be connected.

Note: Strictly speaking, this is a dynamics problem. However, for the pulleys turning at constant rates, the equilibrium equations hold for each pulley. Conditions in the belt are as developed in this section provided the inertia of

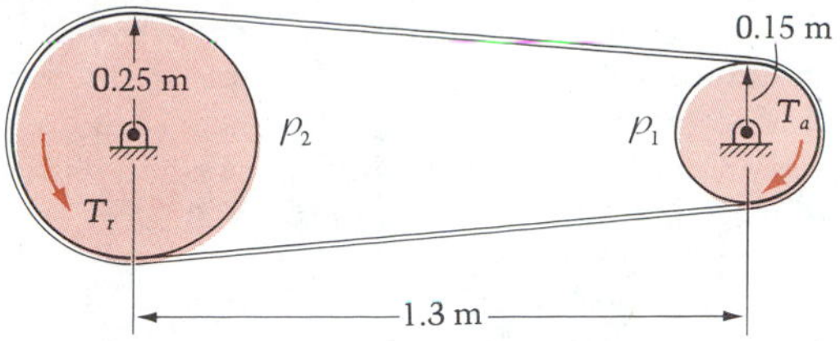

Figure P6.170

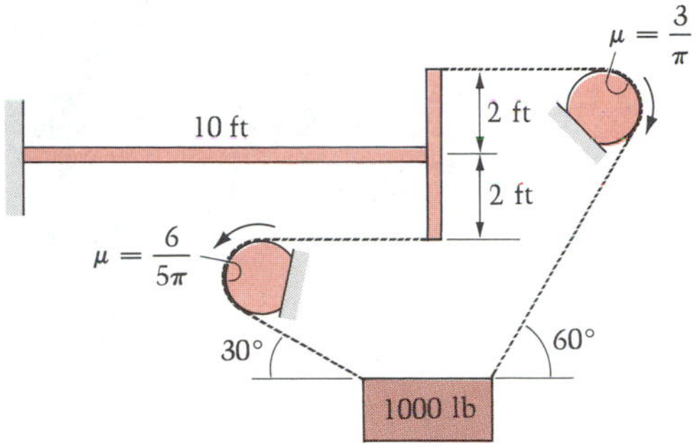

Figure P6.171

the belt can be neglected, which really means that the pulley speeds are not too large.

6.171 The two cables are on the verge of slipping in the indicated directions in Figure P6.171. Find the reactions exerted by the wall onto the left end of the flexible beam, whose weight is to be neglected.

6.172 The sandbag in Figure P6.172 weighs 50 lb. The coefficient of friction between the rope and the tree branch is 0.4. What range of force can the man exert and maintain equilibrium if he weighs 150 lb and if the coefficient of friction is 0.3 between his shoes and the ground?

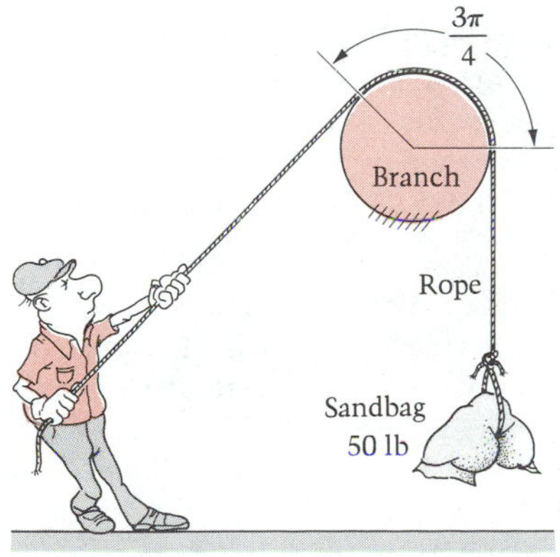

Figure P6.172

6.173 The ferry boat in Figure P6.173 is being docked, and exerts 1000 lb on each of the ropes. How many wraps around the posts (mooring bitts) must each dock worker make in order to hold the boat in position with less than 40 lb? The friction coefficient is $\mu = 0.3$.

Figure P6.173

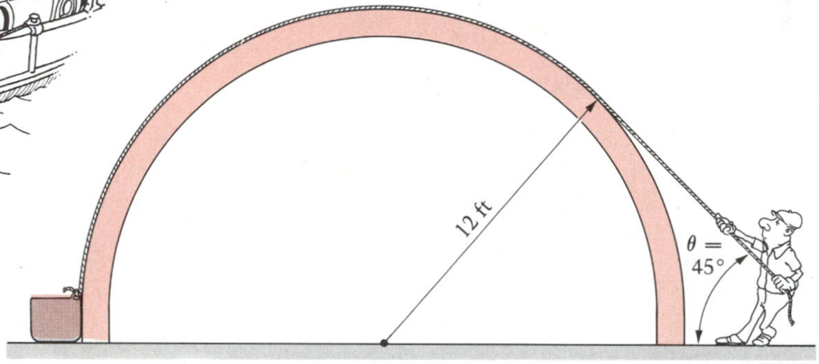

Figure P6.175

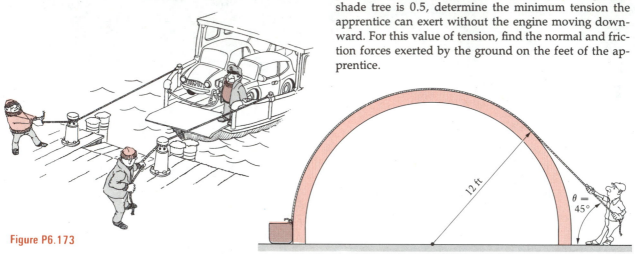

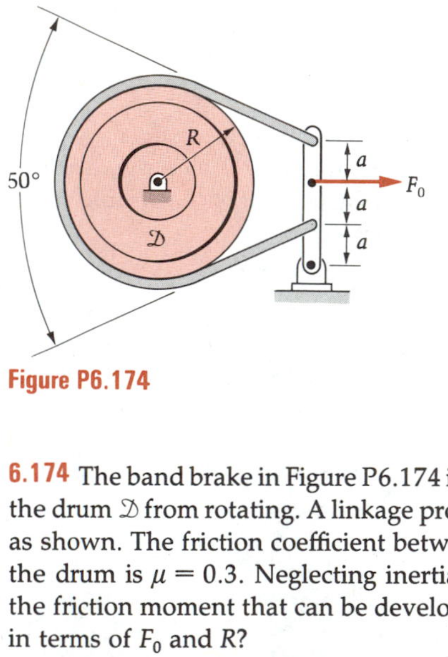

Figure P6.174

6.174 The band brake in Figure P6.174 is designed to stop the drum $\mathcal{D}$ from rotating. A linkage provides the force F_0 as shown. The friction coefficient between the band and the drum is $\mu = 0.3$. Neglecting inertial effects, what is the friction moment that can be developed by the brake, in terms of F_0 and R?

6.175 The man in Figure P6.175 is trying to pull the block over the round hut. He is able to exert 50 lb of force on the rope; in the position shown he is unable to lift the block. He backs up until the angle θ becomes 35°, and finds he is then barely able, with his 50 lb, to raise the block. The friction coefficient between rope and hut is 0.1. Neglecting friction between the block and the hut, find the weight of the block.

6.176 The shade tree mechanic in Figure P6.176 is removing some concrete blocks from beneath an engine, preparatory to reinstalling it in his car. The engine weighs 700 lb, and is being held in equilibrium by a 150-lb apprentice. If the coefficient of friction between rope and shade tree is 0.5, determine the minimum tension the apprentice can exert without the engine moving downward. For this value of tension, find the normal and friction forces exerted by the ground on the feet of the apprentice.

6.177 An old bench vise has a square-threaded screw $\mathcal{S}$ with a mean diameter of $\frac{1}{2}$ inch and a pitch of $\frac{1}{8}$ inch. (See Figure P6.177.) The screw turns within the guide $\mathcal{G}$, which is fixed to one jaw ($\mathcal{J}_1$) of the vise. The other jaw, $\mathcal{J}_2$, is part of body $\mathcal{A}$, which is clamped to the bench as shown. Body $\mathcal{A}$ contains the threaded sleeve $\mathcal{C}$ that engages the screw with coefficients of friction 0.25 (kinetic) and 0.3 (static).

When the vise is clamping something between its jaws, the guide $\mathcal{G}$ contacts $\mathcal{A}$ at B and at D, and the head of the screw bears against $\mathcal{G}$ at A. If a board is being held in the vise with a clamping force of 200 lb:

a. What are the forces exerted at B and D onto guide $\mathcal{G}$, neglecting friction at B and D and the weight of the vise?

b. What was the force at H on handle $\mathcal{H}$ just before the tightening was completed?

6.178 In the preceding problem, after the handle was released (clamping force = 200 lb), it was decided to increase the clamping action.

a. What force must be applied at H to initiate further compression of the board?

b. What force must be applied to initiate a reduction (from 200 lb) of the clamping force?

6.179 The power input while raising the load of Figure 6.11 at a constant angular speed ω_0 rad/sec of the

Figure P6.176

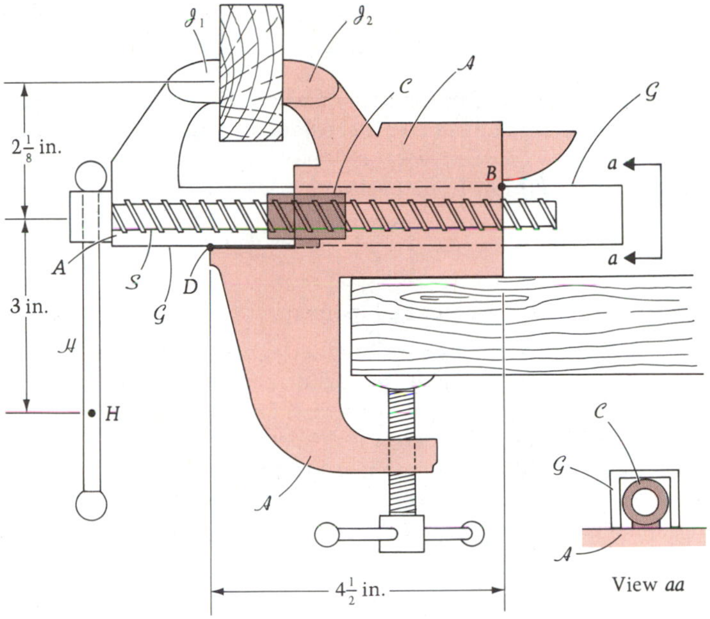

Figure P6.177

screw is $M\omega_0$. The "power output" is the product of F and the (vertical) velocity of the load, or $F[(r\omega_0)\tan\alpha]$. If the efficiency of the jackscrew is the ratio of the power output to the power input, find the efficiency in terms of α and μ.

6.180 In the preceding problem, if $\mu = 0.3$, find the pitch angle giving maximum efficiency to the jackscrew. What is this maximum efficiency?

6.181 In Example 6.17 find the torque required to retract the bolt when it has stopped with the 895-lb tensile force in it.

6.182 The single-threaded turnbuckle in Figure P6.182 is being used to move the heavy crates slightly closer. When the tension in the turnbuckle is at the point of causing impending slipping of the crates, what moment M_0 applied to the turnbuckle will start them slipping? The turnbuckle data are: pitches = 0.06 in., radii = 0.25 in., and static friction coefficients = 0.35. Assume symmetry.

6.183 In the preceding problem, let the respective weights of $\mathcal{A}$ and $\mathcal{B}$ be 1200 lb and 700 lb, with friction coefficients of 0.2 between $\mathcal{A}$ and the floor and 0.3 between $\mathcal{B}$ and the floor. Find the couple being applied to the turnbuckle at the point when one of the crates moves.

6.184 Show that for the square-threaded jackscrew, the moment M to turn the screw against the load F may be expressed in terms of the angle of friction ϕ as $M = Fr\tan(\alpha + \phi)$.

6.185 Prove the statement, made in the text just prior to Example 6.18, that the moment transferred once the condition of constant rate of wear is reached is reduced by 25% to $\mu PR_0/2$. *Hint*: Constant rate of wear means that the rate of work done per unit area by friction is constant,

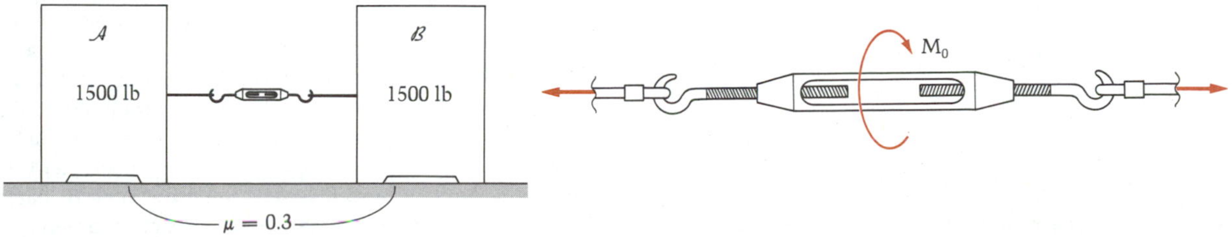

Figure P6.182

so that

$$\mu p v = \text{constant} = k_1$$

where v = velocity of the point in question. This velocity magnitude is the radius times a constant angular speed. Therefore, pr = constant = k_2, and the constant k_2 can be found by $P = \int_A p\, dA$. Then the moment M is balanced by

$$M = \int_A r(p\, dA)\mu = \int_{\theta=0}^{2\pi} \int_{r=0}^{R} r\frac{k_2}{r}\mu\, r\, dr\, d\theta$$

Integrate and find the required moment.

6.186 If a person pushes a disk sander against a surface with a force of 10 lb, and if the friction coefficient is 0.4 between the sandpaper and the surface, find the moment the motor must exert to overcome the developed friction. (See Figure P6.186.) Assume the pressure to be constant over the surface.

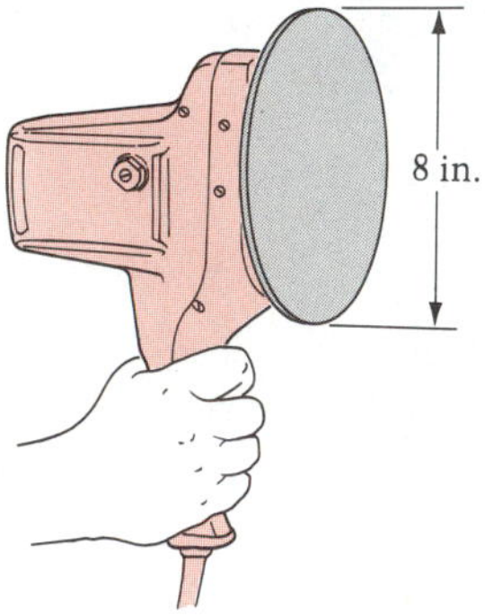

Figure P6.186

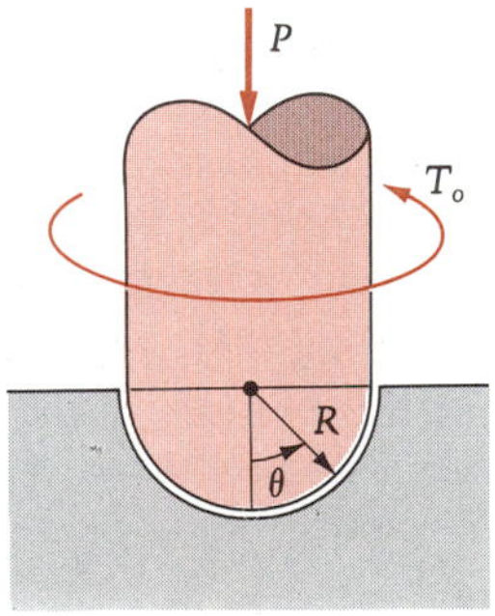

Figure P6.188

6.187 In the preceding problem, assume the pressure drops off linearly with radial distance from the center to a value at the rim that is 60% of that at the center. Find the new value of the moment, and compare.

* **6.188** Find the largest torque that may be applied to the shaft in Figure P6.188 without it slipping in the hemispherical seat. Assume the pressure varies as $k\left(\frac{\pi - 2\theta}{\pi}\right)$ and the coefficient of friction is μ.

6.189 Repeat the preceding problem if the pressure distribution is cosinusoidal: $p = k\cos\theta$.

6.190 Find the maximum torque T_0 that may be applied to the shaft in Figure P6.190 without it slipping in the conical seat. Assume a uniform pressure distribution. Check your answer against Equation (6.30) for the case when $d = 0$ and $\beta = 90°$.

6.191 The collar bearing in Figure P6.191 supports the 200-lb thrust in the shaft. What is the value of the torque T at which the shaft will begin to turn?

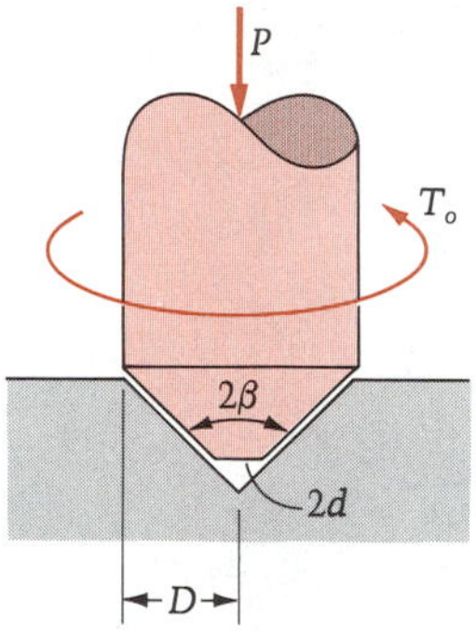

Figure P6.190

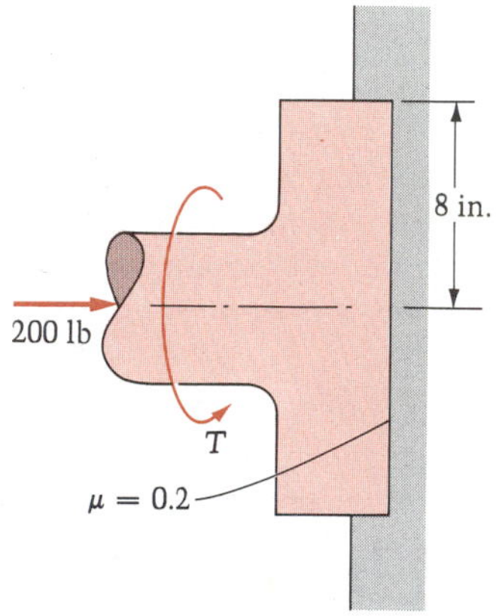

Figure P6.191

6.192 Repeat the preceding problem for the new support conditions on the flange shown in Figure P6.192.

6.193 In Figure P6.193 the shaft $\mathcal{A}$ and disk $\mathcal{B}$ are welded together to form one body that is free to rotate around its horizontal axis in bearings not shown in the figure. Disk $\mathcal{C}$, larger than $\mathcal{B}$, is free to slide on $\mathcal{A}$ but is normally held apart from $\mathcal{B}$. When the separating forces are removed, the tension in four equally spaced springs (two are shown) cause $\mathcal{C}$ to contact $\mathcal{B}$. The friction then stops the rotation. The lugs attached to $\mathcal{C}$ slide in fixed slots and prevent rotation of $\mathcal{C}$ during contact. If the coefficient of friction between $\mathcal{B}$ and $\mathcal{C}$ is 0.3, find the force in each spring required to produce a disk friction moment of 5 N · m.

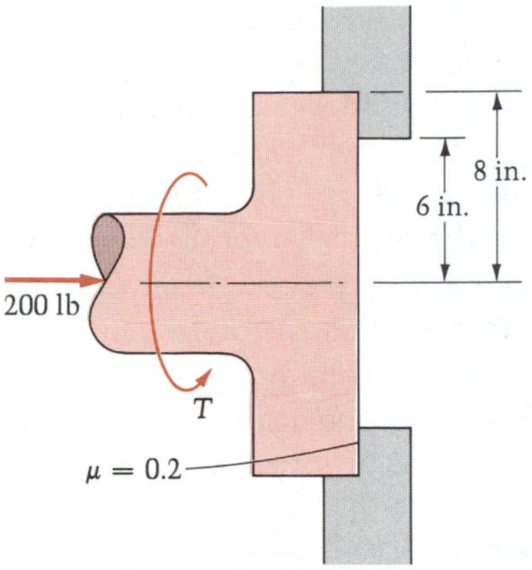

Figure P6.192

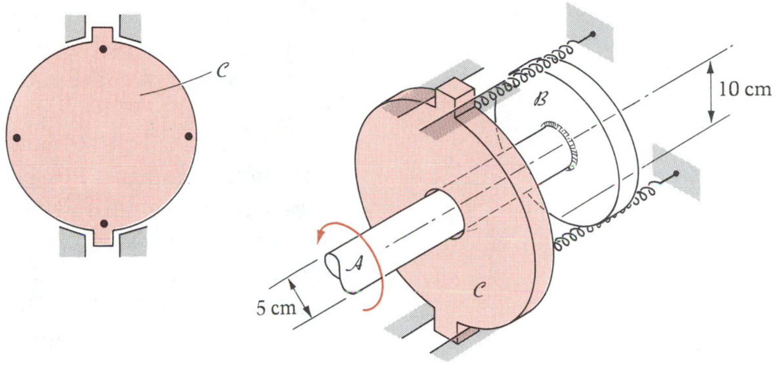

Figure P6.193

COMPUTER PROBLEMS ▶ Chapter 6

6.194 Work Problem 6.142 with the help of a computer. *Hint*: There will be two equilibrium equations for each mass. The fifth equation in the five unknowns is simply that $1.5R = R(\theta_1 + \theta_2)$. After eliminating the tension, the two normal forces, and the angle θ_2, you should be left with the following equation in the remaining unknown θ_1:

$$0.3 \cos \theta_1 + 0.3 \cos(1.5 - \theta_1) + \sin(\theta_1 - 1.5) + \sin \theta_1 = 0.$$

Use the computer to solve for θ_1, then backtrack through your equations to obtain θ_2, T, N_1, and N_2.

6.195 Read Problem 4.119. Note that if the bar is on the verge of slipping at both surfaces of contact, then the friction coefficients at these surfaces are ρ and σ. Assume that these coefficients are equal, and plot the value of ρ required for equilibrium of the stick as a function of θ. Note from the equation that for values of θ smaller than 45°, *no* amount of friction can result in equilibrium.

6.196 Solve Problem 6.145 with the help of a computer. *Hint*: In each case (impending slipping up and down), after eliminating the two normal forces from your equilibrium equations for one of the sticks, you will obtain a cubic equation in $\sin \alpha$ to solve, where α is half the angle between the sticks for equilibrium.

6.197 Solve Problem 6.146 with the help of a computer.

SUMMARY ▶ Chapter 6

In this chapter we learned that resistance to the tendency of two bodies to slide past each other is called friction, and if the surfaces are dry, it is often called Coulomb friction. This resistance is in the form of a friction force f that obeys the inequality

$$0 \le f \le f_{MAX} = \mu N$$

In other words, with Coulomb friction the friction force is limited to a maximum value proportional to the normal, or pressing force between the two bodies. The proportionality constant μ is called the coefficient of friction, and it is dependent on the types of materials in contact and the roughnesses of their surfaces. The table on page 416 gives approximate values of μ for a number of different pairs of materials. We learned that if a block of one material is placed on a plane made of the other material and the angle of the plane is slowly increased, the angle at which the block will slip on the plane is $\tan^{-1}\mu$; thus this is one way of determining μ.

In Section 6.2 we studied a number of examples of equilibrium problems involving impending slipping. As we did so, we saw that the study of friction often entails decisions—whether a body is about to slip or not, whether a body slips or tips first, whether it slips on one surface or another (or yet another, etc.), whether a wheel rolls or slips against a surface, and so on. It is for the most part these various cases which must be considered that make this chapter's problems different from those of Chapter 4.

Some special applications of Coulomb friction were taken up in Section 6.3. These were wedges, flexible flat belts and V-belts, square-threaded screws, and disk friction. Wedges are used to create large normal and friction forces that effect a small change in the position of one body when another is wedged between it and a second, usually fixed, surface.

When a flat belt is wrapped around a fixed circular surface and then placed under tension, the large tension T_L was seen to be related to the small tension T_s by the formula

$$T_L = T_s\, e^{\mu\beta}$$

where e is the base of natural logarithms and β is the angle (in radians) of wrap. For a V-belt with angle 2δ, the corresponding formula is

$$T_L = T_s\, e^{\mu\beta/\sin\delta}$$

The above two formulae show the great effect of the buildup of friction around wrapped belts.

We developed a number of useful formula for the square-threaded screw, including the following, which gives the ratio of M_o to (Fr), where F is a weight to be raised, M_o is the moment applied to the screw to raise the weight, and r is the radius of the screw:

$$\frac{M_o}{Fr} = \frac{\mu + \tan\alpha}{1 - \mu\tan\alpha}$$

where μ is the coefficient of friction and α is the angle of the thread.

Disk friction occurs when two flat circular or annular surfaces (such as disk brakes or clutch plates) come into contact, with each turning or tending to turn relative to the other about a central axis normal to their contact plane. If μ is the coefficient of friction as usual, R_i and R_o are the inner and outer radii of contact, and P is the pressing force between the bodies, then the resisting moment that can be developed by the friction between the bodies at the point of impending slipping is

$$M = \frac{2\,\mu\,P}{3}\,\frac{R_o^2 + R_o R_i + R_i^2}{R_o + R_i}$$

which if $R_i = 0$ (solid circular area of contact) reduces to

$$M = \frac{2\,\mu\,PR_o}{3}$$

Review Questions ▶ Chapter 6

True or False?

1. The Coulomb law is appropriate for surfaces lubricated by grease.
2. The coefficient of friction for two surfaces is independent of the materials comprising the surfaces.
3. The friction force always has magnitude μN, unless it is zero.
4. Friction is never helpful.
5. Friction is a force that resists the tendency of two bodies to slide past each other.
6. If the normal force between two bodies is zero at the point of contact, there can be no friction force developed there according to the Coulomb law.
7. It is not possible for an object to start to slide and tip at the same time.
8. If body A is rolling on body B, the friction force between A and B has to be zero.
9. In working friction problems, we always know in advance whether or not there is slipping between two surfaces.
10. Suppose we assume a certain state of impending slipping exists between two bodies, and we then draw the force as μN on the free-body diagram of one of the bodies. It is OK if we get the direction of the force wrong because the sign of the answer will indicate it is really the other way.
11. The formula $T_L = T_s e^{\mu\beta}$ relates the small and large end tensions in a flat belt wrapped on a drum, at the point of slipping of the belt.
12. The formula in the preceding question applies only for angles β between 0 and 2π radians.
13. For the same coefficient of friction and angle of wrap, the quotient T_L/T_s of the large and small tensions, at the point of slipping, is smaller for a V-belt than for a flat belt.
14. It is harder (takes more moment) to advance a double-threaded screw against a load than a single-threaded screw, but it is easier (less moment needed) to retract the double-threaded screw away from the load than the single-threaded screw.
15. The moment that can be transmitted by disk friction is independent of the radii of the contacting surfaces.

Answers: 1. F 2. F 3. F 4. F 5. T 6. T 7. F 8. F 9. F 10. F 11. T 12. F 13. F 14. T 15. F

7 CENTROIDS AND MASS CENTERS

7.1 Introduction
7.2 Centroids of Lines, Areas, and Volumes
The Recurring Integral
Use of Centroids in Expressing Resultants of Distributed Forces
7.3 The Method of Composite Parts
7.4 Center of Mass
7.5 The Theorems of Pappus
SUMMARY
REVIEW QUESTIONS

7.1 Introduction

In this chapter we will learn about the center of arclength, center of area, center of volume, and center of mass. The first three of these, involved with spatial quantities, are commonly called centroids (of arclength, area, and volume).

All four of the above quantities may be defined by the same formula, in which the quantity (be it arclength, area, volume, or mass) appears each time in an integral which we therefore call the recurring integral. Section 7.2 is concerned with finding centroids using the recurring integral. Also in Section 7.2 we shall revisit some results from Section 3.9 about distributed forces because we are able now to understand their relationship to centroids.

It happens that when an entity is made up of several known parts (e.g., an area made up of several smaller, known areas), and the centroid (or center of mass, if the entity is mass) of each of the parts is known, we can find the centroid of the "overall" entity without integration. This procedure, called the Method of Composite Parts, is presented and studied in detail in Section 7.3. This interesting method not only works in combining parts to form the total entity whose centroid is of interest, but also in *subtracting* one part from another to yield the entity of interest.

The center of mass, or mass center, covered in Section 7.4, is of monumental importance in engineering mechanics. In Statics, for uniform gravity fields it becomes indistinguishable from the center of gravity — which of course is the point of application* of the equipollent single force representing a body's weight. In a later course in dynamics, the mass center is the point of a body whose acceleration equals the external force exerted on the body divided by its mass. Clearly, then, we could get nowhere without the mass center concept.

When the density of a body is constant (or "uniform"), its center of mass will be seen in Section 7.4 to be the very same point as the centroid of volume, but if the density varies with x, y, or z, this will ordinarily not be the case. The reader will recall that density is "mass per unit volume," so that if a density function varies over the body's volume, the integrand of the recurring integral will generally be more complicated and, as a result, mass center computation will be more difficult.

A pair of very old theorems, known as The Theorems of Pappus, are at once so elegant and useful that we have included them in their own section, 7.5. These theorems allow us to find surface areas and volumes which are respectively formed by rotating plane curves and plane areas around an axis; alternately, if we know the areas and volumes of revolution, we may use the theorems to find the centroids of the generating curves and areas.

* Meaning, for a body B of mass m attracted by a particle P of mass M, the point Q on the line of action of the weight force $\mathbf{F}$ such that $r_{PQ} = \left(\frac{GmM}{|\mathbf{F}|}\right)^{1/2}$.

7.2 Centroids of Lines, Areas, and Volumes

The Recurring Integral

In this section we shall learn the meaning of the word centroid, or, more generally, the "center of Q," where the letter Q denotes any scalar quantity associated with a region of space. The most commonly occurring quantities Q are arclength, area, volume, and mass; we shall examine the first three of these in this section, using s, A, and V as the respective symbols to represent them. The center of mass will be examined in Section 7.3, even though it occurs naturally in the development of dynamics.

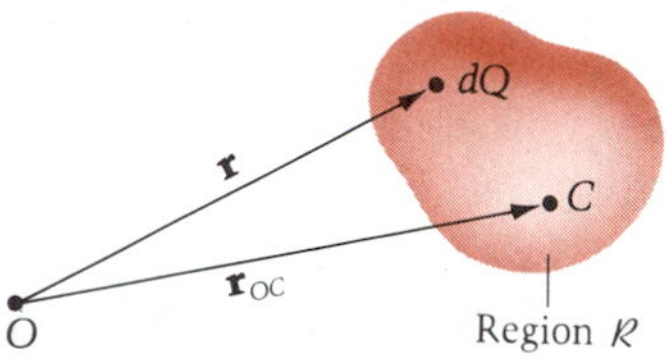

Figure 7.1

The "center of Q" is a point C whose location (see Figure 7.1) is defined by the equation

$$Q\mathbf{r}_{OC} = \int_{\mathcal{R}} \mathbf{r}\, dQ \tag{7.1}$$

where Q itself is given by

$$Q = \int_{\mathcal{R}} dQ \tag{7.2}$$

and $\mathcal{R}$ denotes the region in space over which the integration is carried out; it may be a one-, two-, or three-dimensional region. If, for example, we are seeking a "center of volume," the region of integration is three-dimensional and Equation (7.1) becomes

$$V\mathbf{r}_{OC} = \int_{\mathcal{R}} \mathbf{r}\, dV$$

where $\mathcal{R}$ is the spatial region whose volume is of interest. Similarly, when Q denotes an area the region of integration is two-dimensional, and when Q denotes length of a line (either curved or straight), the region of integration is one-dimensional. When, as in each of these cases, the quantity Q is geometric — that is, arclength, area, or volume — then the "center of Q" is more commonly known as the **centroid** (of arclength, area, or volume). An important fact is that the centroid is unique; that is, point C located by Equation (7.1) is independent of the choice of origin O. The proof of this is left to the student as Problem 7.1.

To locate the centroid, we use the scalar orthogonal components of the vectors in Equation (7.1). If we set up a rectangular coordinate system with origin at O, then $\mathbf{r} = x\hat{\mathbf{i}} + y\hat{\mathbf{j}} + z\hat{\mathbf{k}}$ and $\mathbf{r}_{OC} = \bar{x}\hat{\mathbf{i}} + \bar{y}\hat{\mathbf{j}} + \bar{z}\hat{\mathbf{k}}$ where $\bar{x}$, $\bar{y}$, and $\bar{z}$ are the coordinates of the centroid C in the rectangular coordinate system. Thus

$$\begin{aligned} Q\bar{x} &= \int x\, dQ \\ Q\bar{y} &= \int y\, dQ \\ Q\bar{z} &= \int z\, dQ \end{aligned} \tag{7.3}$$

Scalar integrals* of the type in Equation (7.3) (and occasionally the vector form $\int \mathbf{r}\, dQ$) naturally arise again and again in different areas of mechanics as well as in other branches of mathematical physics and engineering. The recurrence of these integrals is the principal reason for identifying and studying centroids. Suppose, for example, that we encounter and must evaluate the integral $\int y\, dA$, which naturally arises in the analysis of stresses in solids. If we know the location of the centroid of the area in question, we need only multiply the total area by the y coordinate of the centroid to obtain the value of the integral.

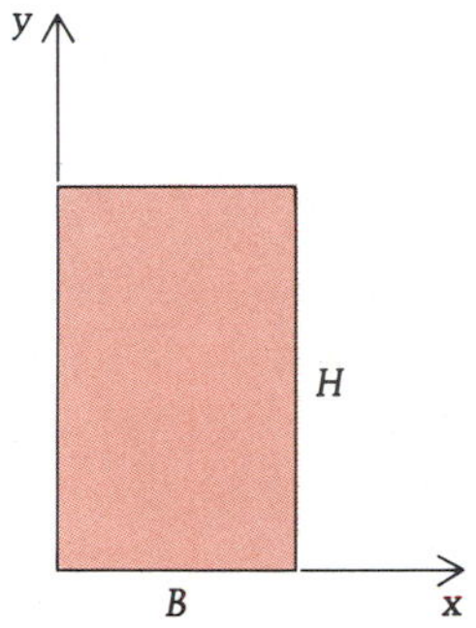

Figure E7.1a

Question 7.1 What then is $\int y\, dA$ if the origin of the coordinate system is *at* the centroid C?

We now proceed to a series of examples in each of which we shall locate the centroid of a line, an area, or a volume. It will be seen that sometimes one or more of $(\bar{x}, \bar{y}, \bar{z})$ will be zero by symmetry, and we will not have to perform as many integrations in order to find the centroid.

* The integrals are often called "first moments" of Q because of their relationship to certain properties of distributed force systems. See Section 3.9.

Answer 7.1 Zero, because $\bar{y} = 0$.

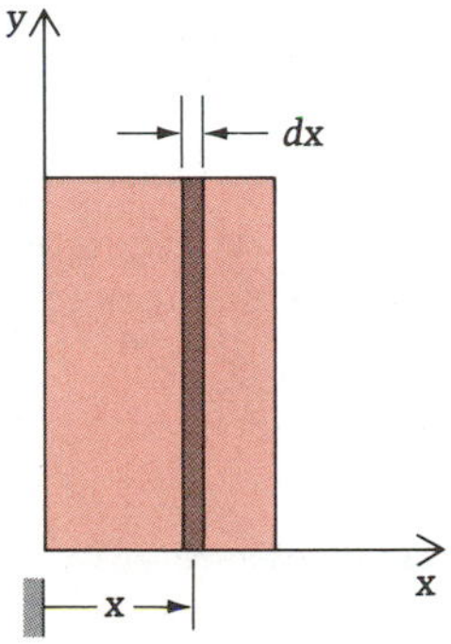

Figure E7.1b

Example 7.1

Locate the centroid of the area of a rectangle of base B and height H (see Figure E7.1a).

Solution

Our use of the term "center" suggests that the centroid will be at $x = B/2$, $y = H/2$. We are simply going to confirm that this is so.

To find $\bar{x}$, the x-coordinate of the centroid we seek, we use the first of Equations (7.3), with Q replaced by area (A):

$$A\bar{x} = \int x\, dA$$

Using the vertical strip of width dx and height H for our differential area dA (see Figure E7.1b), and substituting $A = BH$,

$$\bar{x} = \frac{\displaystyle\int_0^B xH\, dx}{BH} = \frac{H \left.\dfrac{x^2}{2}\right|_0^B}{BH} = \frac{B}{2}$$

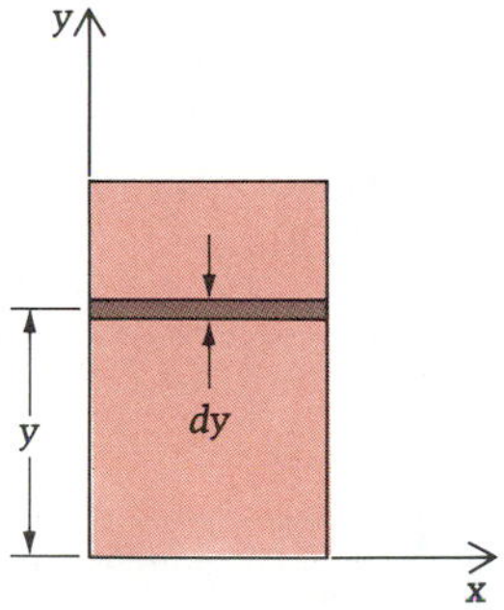

Figure E7.1c

We see that the centroid is, as expected, on a vertical line halfway across the rectangle.

To obtain $\bar{y}$, we shall use the horizontal strip shown in Figure E7.1c so that $dA = B\, dy$ this time:

$$A\bar{y} = \int y\,dA$$

$$BH\bar{y} = \int_0^H yB\,dy$$

$$\bar{y} = \frac{B\left.\frac{y^2}{2}\right|_0^H}{BH} = \frac{H}{2}$$

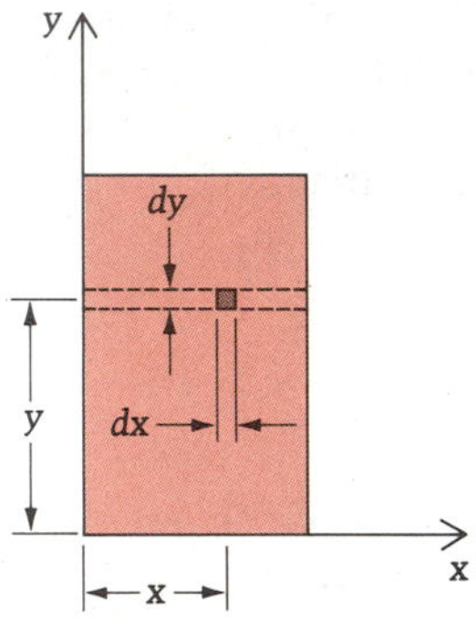

Figure E7.1d

and so the centroid is indeed in the middle of the rectangle.

Question 7.2 What could be the problem in using the vertical strip as the dA in calculating $\bar{y}$?

For readers familiar with double integrals at this point of their study, then with $dA = dx\,dy$ we have (see Figure E7.1d):

$$A\bar{y} = (BH)\bar{y} = \int_0^H \int_0^B y\,dx\,dy$$

$$= \int_0^H y\left(x\Big|_0^B\right) dy$$

$$= \int yB\,dy = B\left.\frac{y^2}{2}\right|_0^H$$

Thus

$$(BH)\bar{y} = BH^2/2$$

so that $\bar{y} = H/2$ as expected. Note that the first integration (on x) generates coverage of the horizontal strip, B by dy, that was used above in the single integration solution.

Answer 7.2 The coordinate y would not be the same for all parts of the element.

EXAMPLE 7.2

Find the centroid of the semicircle shown in Figure E7.2a.

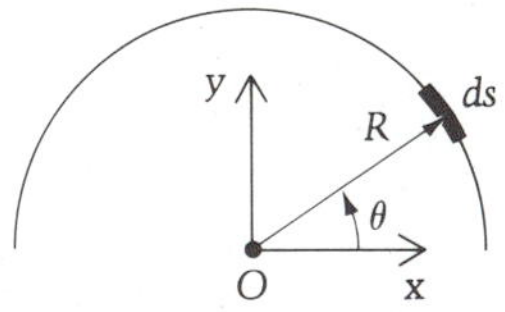

Figure E7.2a

Solution

This time the quantity Q is arclength s, and we want to find the point C for which

$$s\mathbf{r}_{OC} = \int \mathbf{r}\,ds$$

or, for which

$$s\bar{x} = \int x\,ds \qquad s\bar{y} = \int y\,ds \qquad s\bar{z} = \int z\,ds$$

where the total arclength s is πR.

In this problem, by symmetry, $\bar{x} = 0$. This is because for every element ds to the right of the y axis with x coordinate "x," there is an element to the *left* of the y axis with x coordinate "$-x$" (and vice-versa). Thus the entire integral vanishes. We never *have* to use symmetry, however. Letting $ds = R\,d\theta$, and noting that $x = R\cos\theta$,

$$\pi R\bar{x} = \int_{\theta=0}^{\pi} (R\cos\theta)R\,d\theta$$

$$= R^2 \sin\theta \Big|_0^{\pi}$$

$$= 0$$

or

$$\bar{x} = 0$$

The z coordinate of C is also zero, this time because the line is in the xy plane, so that the z coordinate of each "ds" is zero. Thus $\int z\,ds = \int(0)\,ds = 0$.

Thus the only coordinate not immediately obvious is $\bar{y}$. Noting that the y coordinate of ds is $R\sin\theta$, we find, from $s\bar{y} = \pi R\bar{y} = \int y\,ds$,

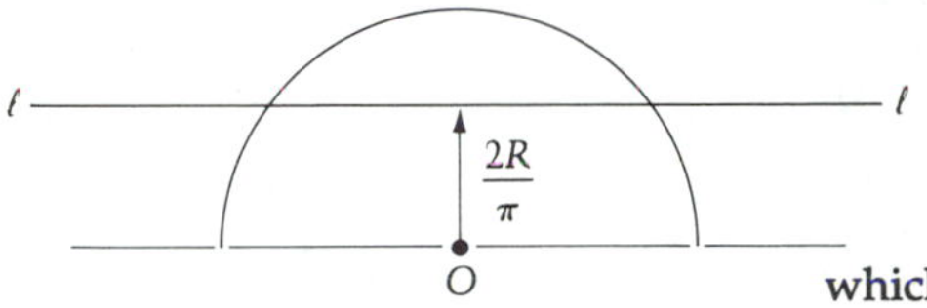

Figure E7.2b

$$\bar{y} = \frac{\int_{\theta=0}^{\theta=\pi} yR\,d\theta}{\pi R} = \frac{\int_0^{\pi} R^2\sin\theta\,d\theta}{\pi R} = \frac{R^2(-\cos\theta)\Big|_0^{\pi}}{\pi R} = \frac{2R}{\pi}$$

which is slightly less than two-thirds of the way from O to the top of the semicircle, as shown in Figure E7.2b.

Question 7.3 Does the arclength above the line $\ell\ell$ equal that below?

Answer 7.3 No. The *integrals* of arclength *times* y-distance are what are balanced; i.e., the "first moments" are balanced above and below the line $\ell\ell$.

Example 7.3

Find the centroid of the triangular area shown in Figure E7.3a.

Solution

The differential area dA may be taken to be $Y\,dx$, as suggested by the shaded strip in the figure.

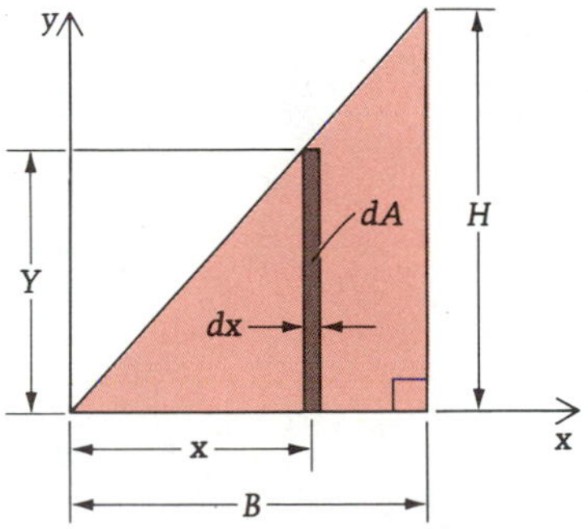

Figure E7.3a

$$A\bar{x} = \int x\,dA$$

$$\frac{BH}{2}\bar{x} = \int_0^B x(Y\,dx)$$

But Y is a function of x, given by $Y = \frac{H}{B}x$; therefore,

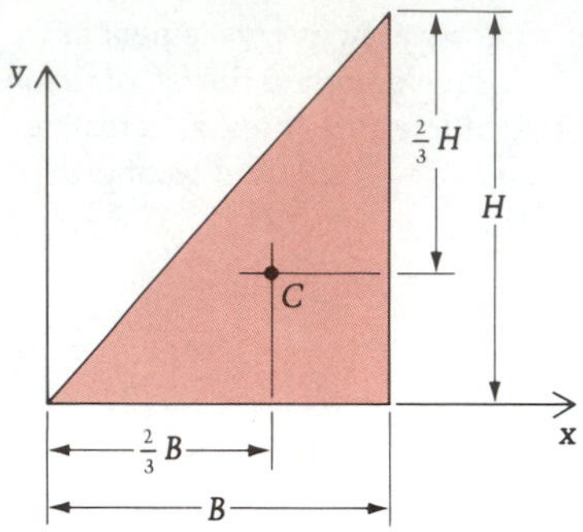

Figure E7.3b

$$\left(\frac{BH}{2}\right)\bar{x} = \int_0^B x\left(\frac{H}{B}x\right)dx$$
$$= \frac{HB^2}{3}$$

or

$$\bar{x} = \frac{2}{3}B$$

Therefore the centroid of a right triangle is $\frac{2}{3}$ of the distance from a vertex on either end of the hypotenuse to the opposite side, as shown in Figure E7.3b:

Question 7.4 How do we know the y coordinate of C without integrating?

We shall later see in Example 7.12 that the "$\frac{2}{3}$ rule" stated above is in fact true for *any* triangle.

Answer 7.4 Relabeling dimensions, $\bar{x} = 2H/3$ in the figure.

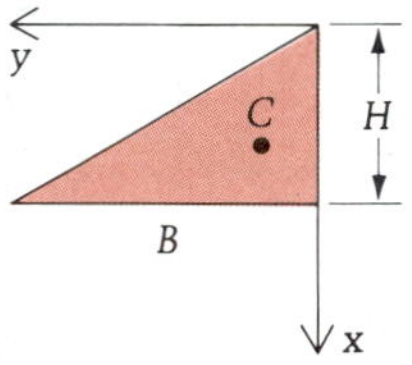

EXAMPLE 7.4

Find the centroid of the shaded area beneath the parabola shown in Figure E7.4a.

Solution

Using the vertical strip as shown in Figure E7.4b, we have:

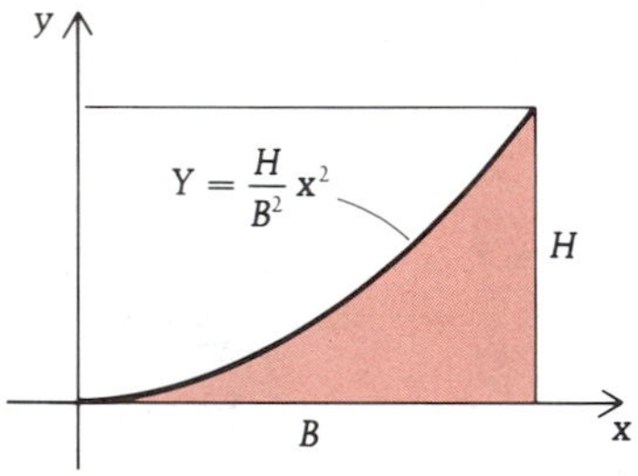

Figure E7.4a

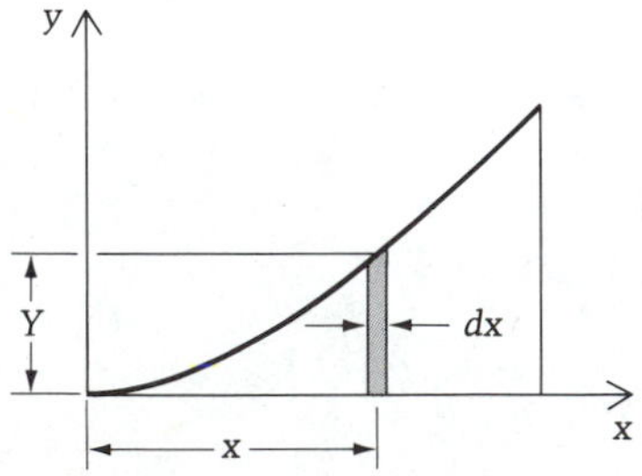

Figure E7.4b

$$A\bar{x} = \int x\,dA$$
$$= \int_0^B x(Y\,dx)$$
$$= \int_0^B x\left(\frac{H}{B^2}x^2\right)dx$$

or

$$A\bar{x} = \frac{HB^2}{4}$$

But

$$A = \int_0^B Y\,dx$$

$$= \int_0^B \frac{H}{B^2} x^2\,dx$$

$$A = \frac{BH}{3}$$

so that $\bar{x} = \frac{3}{4}B$.

For the y coordinate, we may still use the above area element $Y\,dx$, provided we recognize that *its own centroid* is at $Y/2$ (see Example 7.1). We shall "add up" the areas times the y coordinates of the centroids of the vertical strips:

$$A\bar{y} = \int \frac{Y}{2}\,dA \qquad \text{(again, } Y/2 \text{ is the centroid distance for "}dA\text{")}$$

Thus

$$\bar{y} = \frac{\displaystyle\int \frac{Y}{2} Y\,dx}{BH/3}$$

$$= \frac{3}{2BH} \int_{x=0}^{B} \frac{H^2}{B^4} x^4\,dx$$

So

$$\bar{y} = \frac{3}{2BH}\frac{H^2}{B^4}\frac{B^5}{5} = \frac{3}{10}H$$

The student is encouraged to confirm this result either by double integration or by single integration using a horizontal strip dA.

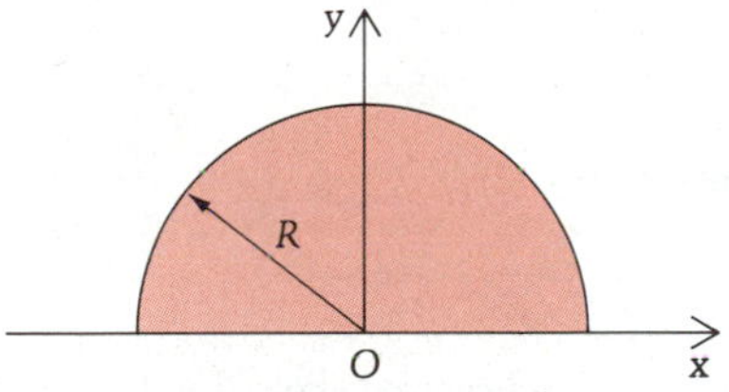

Figure E7.5a

Example 7.5

Find the location of the centroid of the area enclosed by the semicircle and the (diametral) x axis as shown in Figure E7.5a.

Solution

For the same reasons as in Example 7.2, $\bar{x} = \bar{z} = 0$. To find $\bar{y}$, we again use Equation (7.1) where this time the quantity Q is area; and we shall also use the horizontal strip, shown in Figure E7.5b, for our "dA." We get, then, with $dA = 2\sqrt{R^2 - y^2}\,dy$,

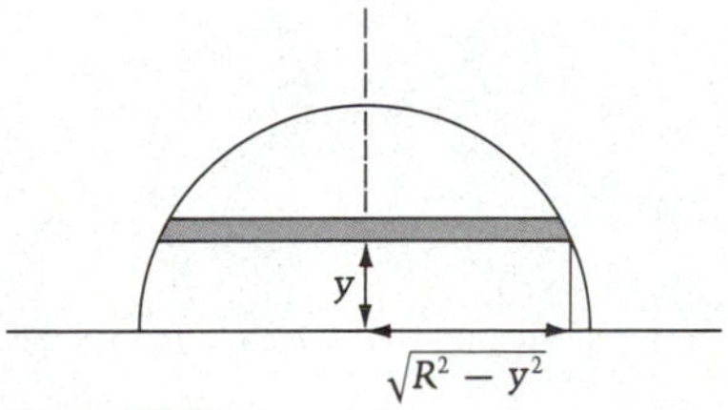

Figure E7.5b

$$A\bar{y} = \int y\,dA$$

$$\frac{\pi R^2}{2}\bar{y} = \int_0^R y(2\sqrt{R^2 - y^2}\,dy)$$

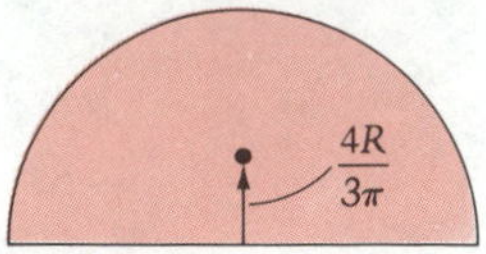

Figure E7.5c

Thus

$$\frac{\pi R^2}{2}\,\bar{y} = \int_0^R 2y\sqrt{R^2 - y^2}\,dy = \left.\frac{(R^2 - y^2)^{3/2}}{-\frac{3}{2}}\right|_0^R$$

$$= \frac{2}{3}R^3$$

And therefore

$$\bar{y} = \frac{4R}{3\pi}$$

This centroid (see Figure E7.5c) is lower than the centroid of the semicircle (see Example 7.2).

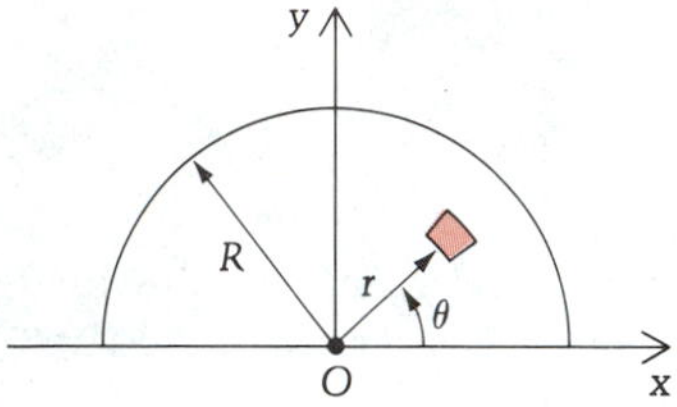

Figure E7.5d

Question 7.5 Explain this, from the standpoint of where the quantities Q (arclength, area) are located in the two problems.

For the reader familiar with double integration, polar coordinates (see Figure E7.5d) offer a nice alternative to the solution above. This time, the differential area is seen in Figure E7.5e to be $dA = r\,dr\,d\theta$. We obtain

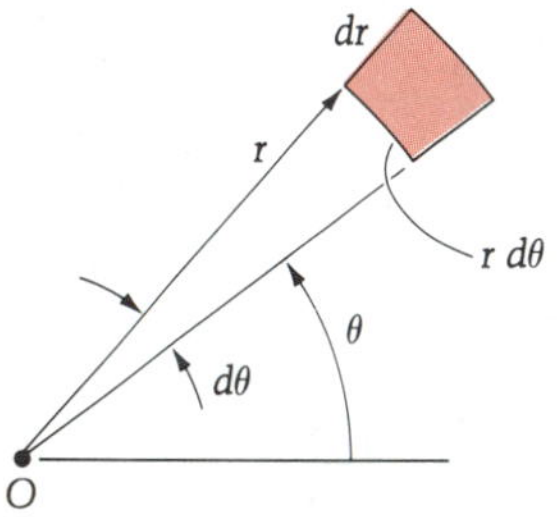

Figure E7.5e

$$A\bar{y} = \int y\,dA$$

$$\left(\frac{\pi R^2}{2}\right)\bar{y} = \int_0^\pi \int_0^R (r\sin\theta) r\,dr\,d\theta$$

$$= \int_0^\pi \left.\frac{r^3}{3}\right|_0^R \sin\theta\,d\theta$$

$$= \frac{R^3}{3}(-\cos\theta)\Big|_0^\pi$$

$$= \frac{2R^3}{3}$$

$$\bar{y} = \frac{4R}{3\pi} \qquad \text{(as before)}$$

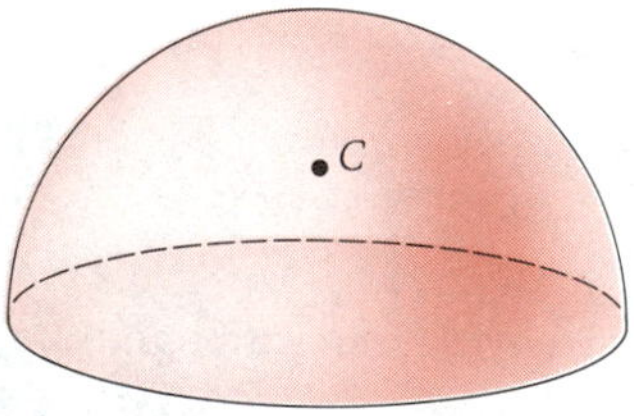

Figure E7.6a

Answer 7.5 This time it is "y-distance times *area*," which is balanced above and below the horizontal line through the centroid. The semicircle is now "filled-in" and there is much more quantity Q (area) lower than there was when Q was arclength. This time, the "first moments of *area*" are balanced.

EXAMPLE 7.6

Find the centroid of the volume of a hemisphere of radius R (see Figure E7.6a).

Solution

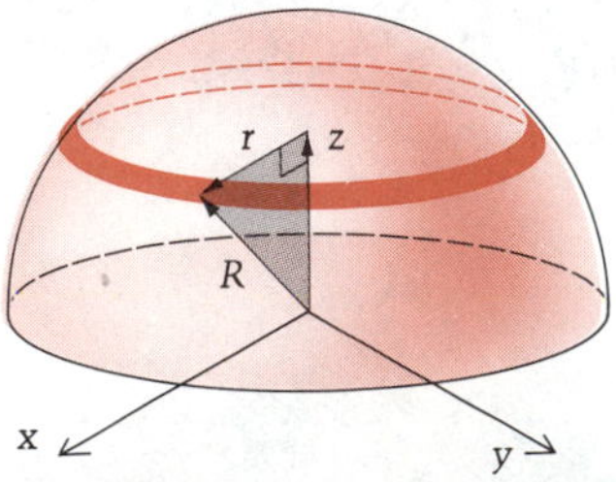

Figure E7.6b

By symmetry it is clear that $\bar{x} = \bar{y} = 0$. To find $\bar{z}$, we shall use the volume element shown in Figure E7.6b, which is a disk having the differential volume dV

$= \pi r^2\,dz$. But we see from the shaded triangle that

$$R^2 = r^2 + z^2$$

so that as a function of z, the radius is expressible as

$$r = \sqrt{R^2 - z^2}$$

Therefore, because all points of the disk have equal z coordinates,

$$\begin{aligned} V\bar{z} &= \int z\,dV = \int z\pi r^2\,dz \\ &= \pi \int_0^R z(R^2 - z^2)\,dz = \pi \int_0^R (R^2 z - z^3)\,dz \\ &= \pi \left(\frac{R^2 z^2}{2} - \frac{z^4}{4} \right) \Bigg|_0^R \\ &= \frac{\pi R^4}{4} \end{aligned}$$

Because the volume V is $\frac{1}{2}(\frac{4}{3}\pi R^3)$, we obtain

$$\bar{z} = \frac{\pi R^4/4}{2\pi R^3/3} = \frac{3}{8} R$$

Question 7.6 Why should this answer of $\frac{3}{8}R$ be less than the $4R/3\pi$ for the semicircular area?

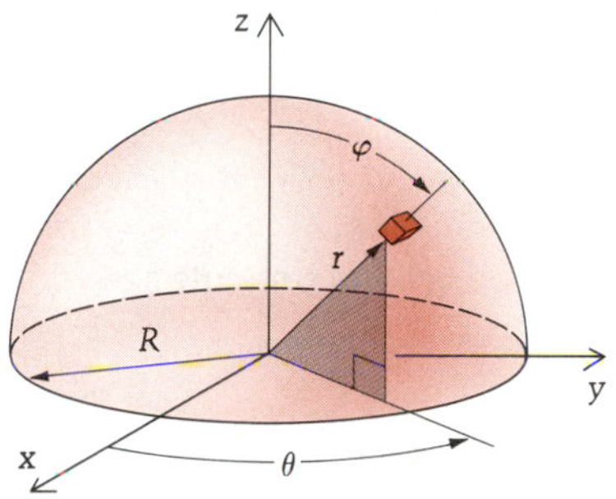

Figure E7.6c

The centroid of the hemisphere may also be computed using the spherical coordinates (r, φ, θ) (see Figure E7.6c). For those familiar with multiple integration, this alternative approach is presented next:

$$\begin{aligned} V\bar{z} &= \int z\,dV = \int_0^{2\pi} \int_0^{\pi/2} \int_0^R \overbrace{(r\cos\varphi)}^{z}\overbrace{(r^2 \sin\varphi\,dr\,d\varphi\,d\theta)}^{dV} \\ &= \frac{R^4}{4} \int_0^{2\pi} \int_0^{\pi/2} \sin\varphi \cos\varphi\,d\varphi\,d\theta \\ &= \frac{R^4}{4} \int_0^{2\pi} \left[\frac{\sin^2\varphi}{2} \right]_0^{\pi/2} d\theta \\ &= \frac{R^4}{8} \int_0^{2\pi} d\theta \\ &= \frac{\pi R^4}{4} \end{aligned}$$

Answer 7.6 This time it is the first moments of *volume* that are balanced above and below the plane $z = \frac{3}{8}R$. There is proportionately much more volume near the plane $z = 0$ than there was area near the line $y = 0$ in Example 7.5.

so that

$$\bar{z} = \frac{\pi R^4/4}{2\pi R^3/3} = \frac{3}{8}R \qquad \text{(as before)}$$

EXAMPLE 7.7

Find the centroid of the conical volume shown in Figure E7.7a.

Solution

As in the previous example we shall use a disk as the element of volume for integration. The radius of the disk is $(R/H)\,z$ so that the element of volume is $\pi[(R/H)\,z]^2\,dz$ (see Figure E7.7b). We take this opportunity to derive the cone's volume:

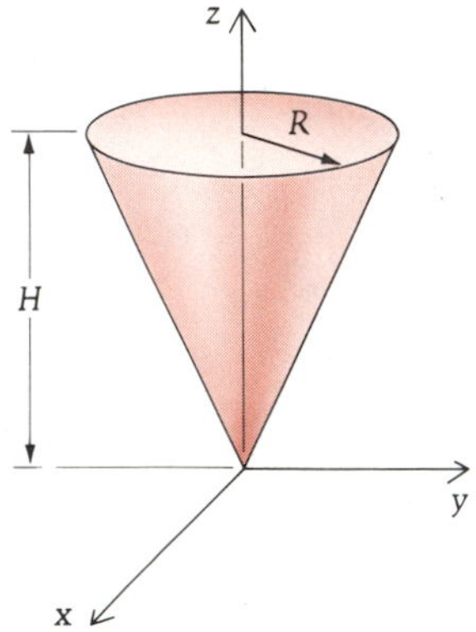

Figure E7.7a

$$V = \int_0^H \pi\left(\frac{R}{H}z\right)^2 dz$$

$$= \frac{\pi R^2}{H^2}\int_0^H z^2\,dz$$

$$= \frac{\pi R^2 H}{3}$$

Noting next that $\bar{x} = \bar{y} = 0$ by symmetry, we proceed to compute $\bar{z}$:

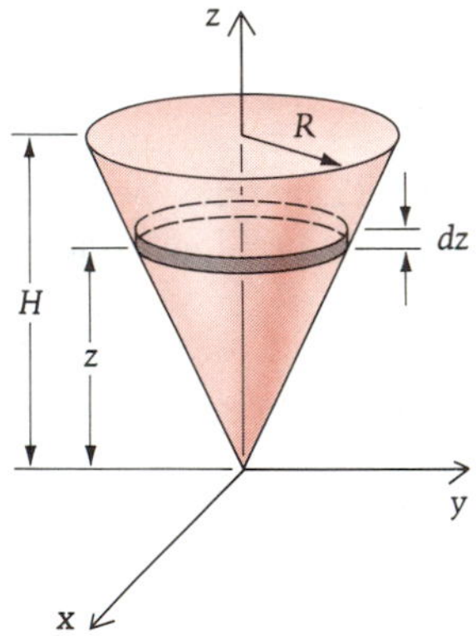

Figure E7.7b

$$V\bar{z} = \int z\,dV$$

$$= \int_0^H z\left(\frac{\pi R^2}{H^2}z^2\right) dz$$

$$= \frac{\pi R^2 H^2}{4}$$

Thus

$$\bar{z} = \frac{\dfrac{\pi R^2 H^2}{4}}{\dfrac{\pi R^2 H}{3}} = \frac{3}{4}H$$

Therefore the centroid of a conical volume is on its axis, three-fourths of the distance from the vertex to the base.

The reader familiar with multiple integration is encouraged to reproduce this result using cylindrical coordinates.

Use of Centroids in Expressing Resultants of Distributed Forces

We close this section by reviewing some results from Section 3.9 about distributed forces. In that section, we saw that for a distributed line loading of parallel forces, the resultant force F_r equals the "area" beneath

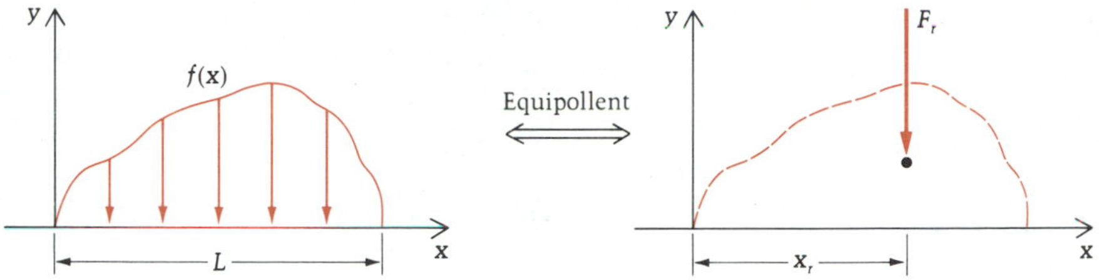

Figure 7.2

the loading diagram (see Figure 7.2), with a line of action given by Equation (3.25):

$$x_r = \frac{\int_0^L xf(x)\,dx}{\int_0^L f(x)\,dx} \tag{7.4}$$

By comparing Equations (7.4) and the first of (7.3), we now see clearly what we were only able to mention in Chapter 3 — namely, that the force-alone resultant acts through the centroid of the "area" beneath the loading curve.

For pressure acting on a plane surface we found, in Section 3.9, a force-alone resultant as indicated in Figure 7.3:

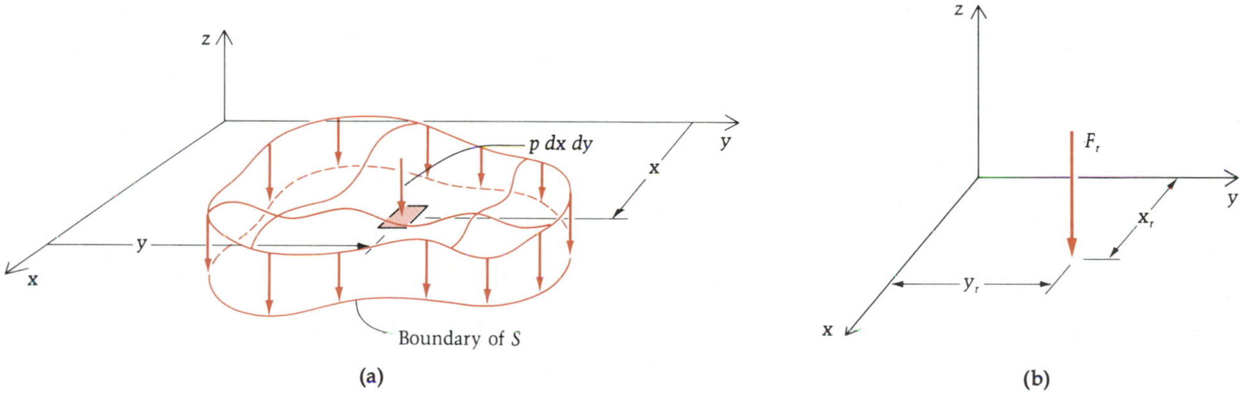

Figure 7.3

which was

$$F_r = \int p(x, y)\,dA$$

with a line of action given by (x_r, y_r):

$$F_r x_r = \int xp(x, y)\,dA \qquad \text{and} \qquad F_r y_r = \int yp(x, y)\,dA$$

Thus we see that the resultant force is, this time, the "volume" of the region for which the loaded plane surface is the base and the altitude is the varying pressure $p(x, y)$. Furthermore, the resultant has a line of action through the centroid of this volume.

A special case of the above with great practical importance arises when the pressure is uniform. Then, we obtain

$$F_r = p \int dA = pA$$

and

$$Ax_r = \int x \, dA$$

$$Ay_r = \int y \, dA$$

so that the resultant force has a line of action through the centroid of the plane area on which the pressure acts.

EXAMPLE 7.8

Find the force-alone resultant of the distributed line loading shown in Figure E7.8a, and its line of action.

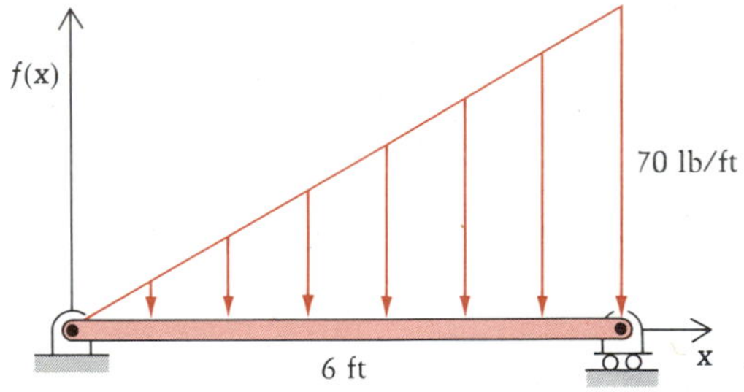

Figure E7.8a

Solution

The resultant is:

$$F_r = \underbrace{\begin{array}{c}\text{Area beneath}\\ \text{loading curve}\end{array}} \qquad \left(\text{or} \quad \int_0^6 \underbrace{f(x)}_{\frac{70}{6}x} \, dx\right)$$

$$= \frac{1}{2}(6)70 \qquad \left(\text{or} \quad \frac{70}{6}\frac{x^2}{2}\bigg|_0^6\right)$$

$$= 210 \text{ lb}$$

This force acts downward (parallel to the given loading) at the centroid of the area beneath the loading curve.

$$x_r = \frac{2}{3}(6) \qquad \left(\text{or } \frac{\displaystyle\int_0^6 \overbrace{xf(x)}^{\frac{70}{6}x}\,dx}{210} = \frac{\dfrac{70}{6}\dfrac{x^3}{3}\Big|_0^6}{210}\right)$$

$$= 4 \text{ ft}$$

Therefore, we have found the results shown in Figure E7.8b with the aid of centroids:

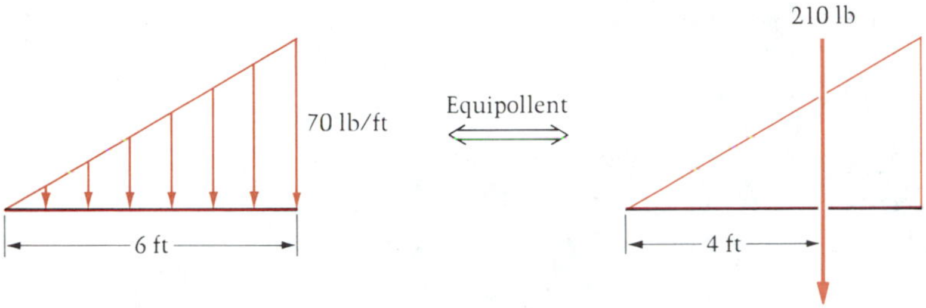

Figure E7.8b

It is interesting to contrast this problem with Example 3.30, which consisted of the six forces shown in Figure E7.8c, and which was shown to be equipollent to the single force shown in Figure E7.8d. We see that even though the six forces sum to the same 210 pounds, and even though the forces' magnitudes are proportional to their distances from O, the single force resultants of the two systems are *not the same because the lines of action differ.*

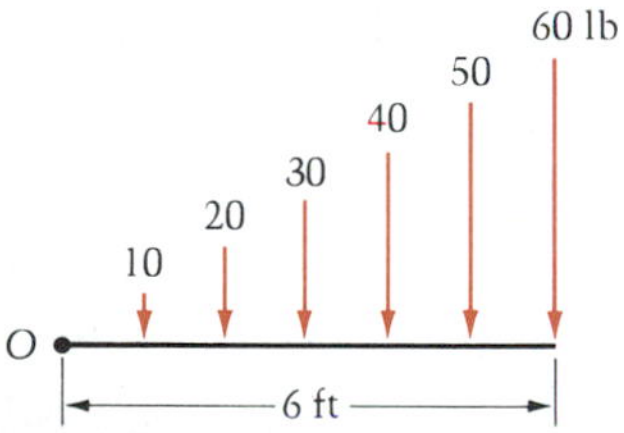

Figure E7.8c

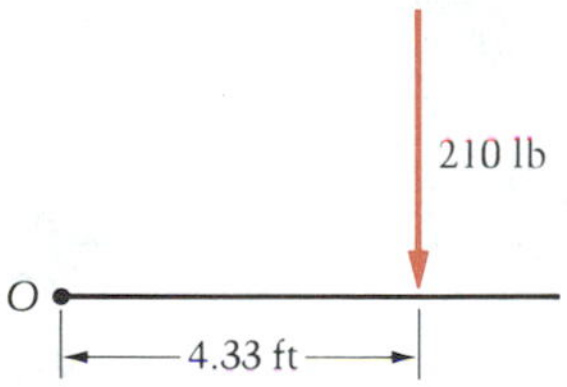

Figure E7.8d

PROBLEMS ▸ Section 7.2

7.1 Show that the centroid C of the area in Figure P7.1 does not depend upon the choice of origin in the reference frame. *Hint:* Consider the two centroids resulting from Equation (7.1) with two separate origins O_1 and O_2

$$\mathbf{r}_{O_1C_1} = \frac{1}{A}\int \mathbf{r}_1\, dA, \qquad \mathbf{r}_{O_2C_2} = \frac{1}{A}\int \mathbf{r}_2\, dA$$

and relate $\mathbf{r}_1$ to $\mathbf{r}_2$. Show then that $\mathbf{r}_{O_1C_1} = \mathbf{r}_{O_2C_2}$, which means that C_1 and C_2 are the same point.

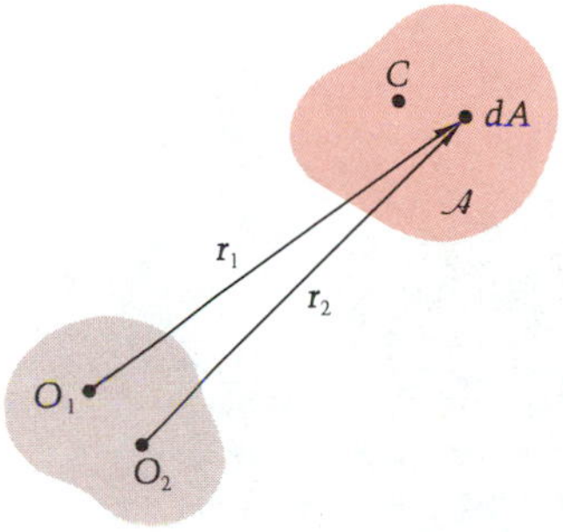

Figure P7.1

7.2 Find the centroid of the area bounded by the parabola and the line $x = a$ in Figure P7.2.

7.3 Find the centroid of the circular arc shown in Figure P7.3.

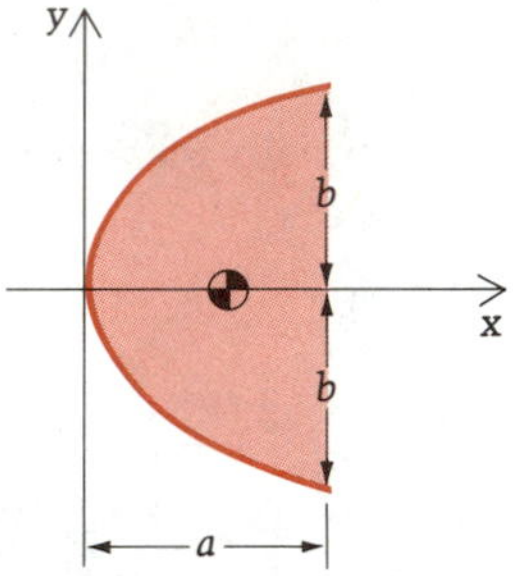

Figure P7.2

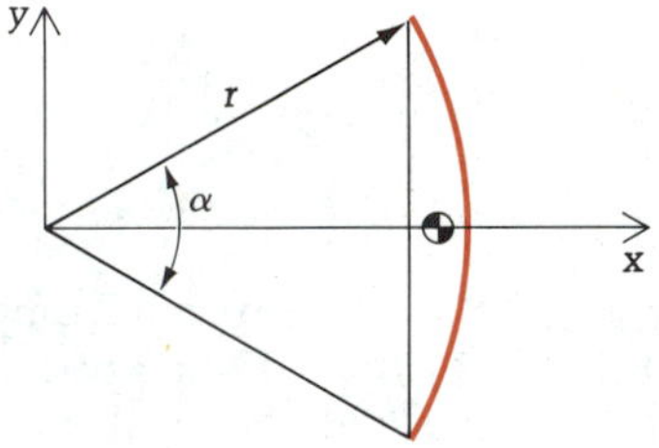

Figure P7.3

7.4 Which of the following expressions may be used to find the centroid location $\bar{y}$ of the area under the curve $y = Y(x)$? (See Figure P7.4.)

(a) $\dfrac{\int_0^{x_1} xY\,dx}{\int_0^{x_1} Y\,dx}$ (b) $\dfrac{\frac{1}{2}\int_0^{x_1} xY\,dx}{\int_0^{x_1} Y\,dx}$

(c) $\dfrac{\int_0^{x_1} xY\,dx}{\int_0^{x_1} Y^2\,dx}$ (d) $\dfrac{\frac{1}{2}\int_0^{x_1} Y^2\,dx}{\int_0^{x_1} Y\,dx}$

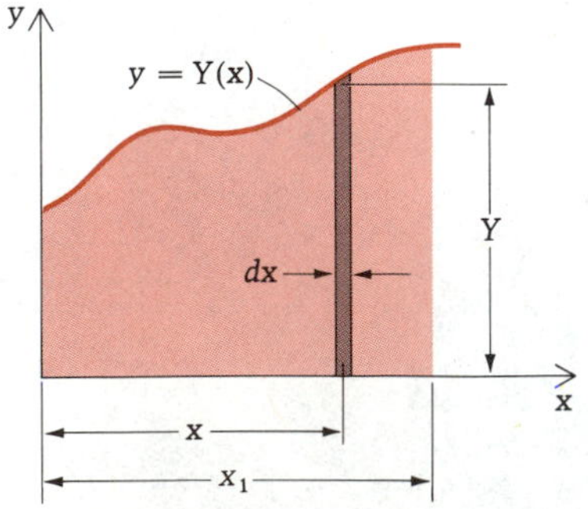

Figure P7.4

7.5 In Problem 7.4, find $\bar{y}$ if $Y(x) = e^x$ and $x_1 = 2$.

In Problems 7.6–7.9 find the centroids of the shaded areas.

7.6

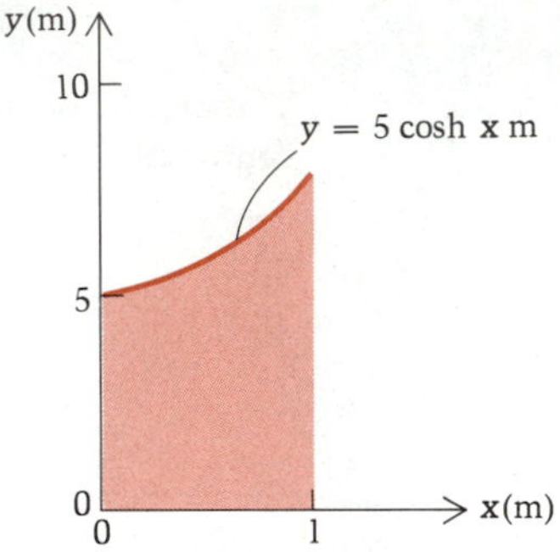

Figure P7.6

* **7.7**

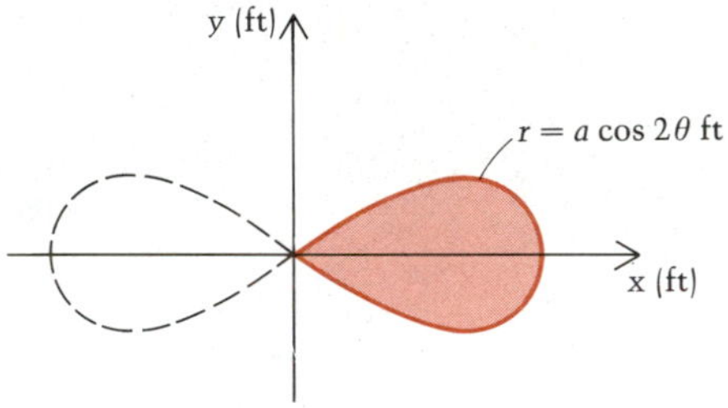

Figure P7.7

7.8

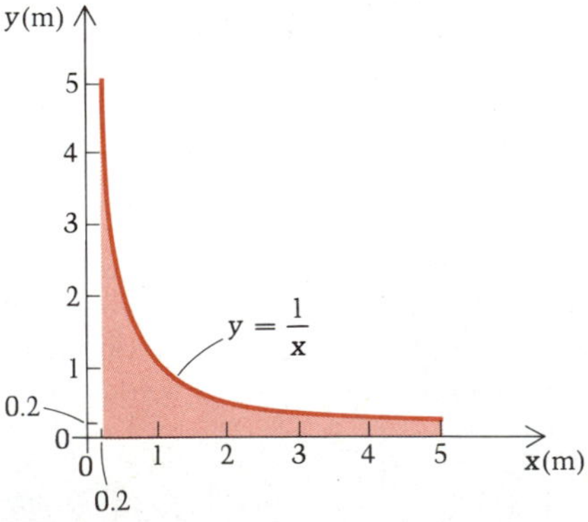

Figure P7.8

* Asterisks identify the more difficult problems.

7.9

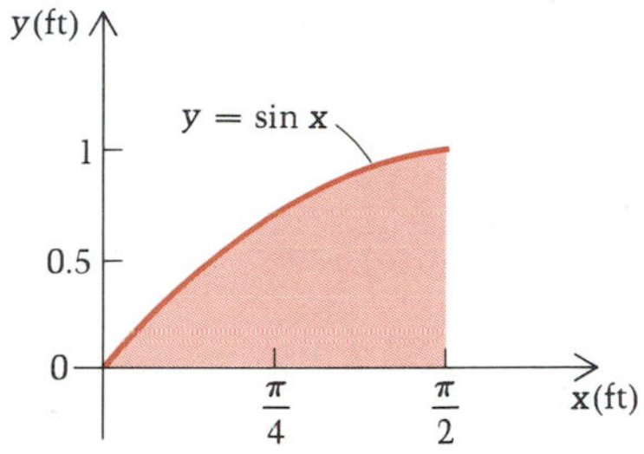

Figure P7.9

7.10 Find the y coordinate of the centroid of the shaded area shown in Figure P7.10.

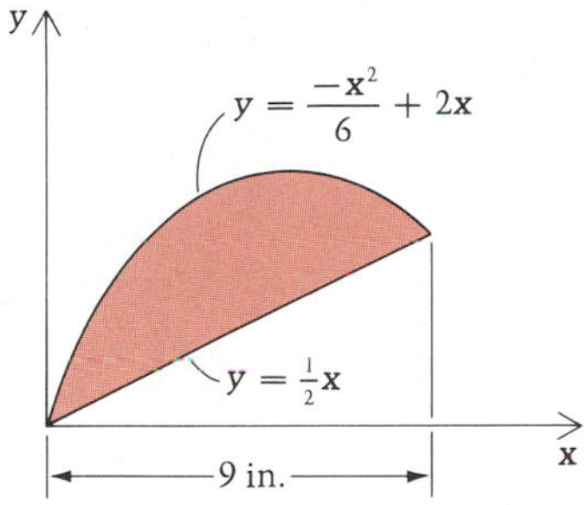

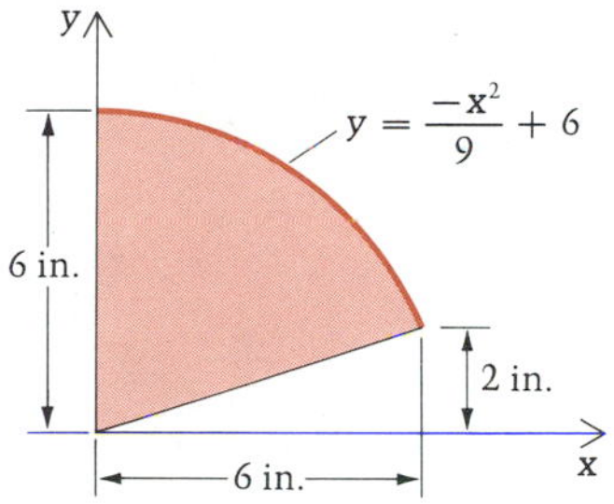

Figure P7.10

7.11 In Problem 7.10, find the x coordinate of the centroid.

In Problems 7.12 and 7.13 find the x coordinates of the centroids of the shaded areas.

7.12

Figure P7.12

7.13

Figure P7.13

7.14 Find the y coordinate of the centroid of the shaded area in Problem 7.12.

7.15 Find the y coordinate of the centroid of the shaded area in Problem 7.13.

In Problems 7.16–7.20 find the x coordinates of the centroids of the *lines* by integration. *Hint* for 7.17–7.19: $ds = \sqrt{dx^2 + dy^2} = \sqrt{1 + (y')^2}\, dx.$

7.16

Figure P7.16

7.17

Figure P7.17

7.18

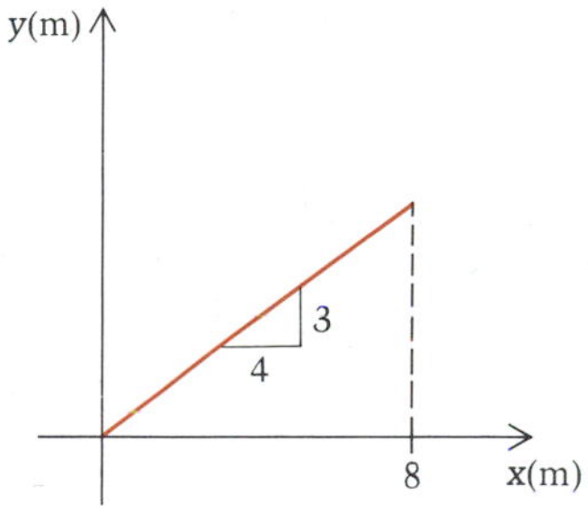

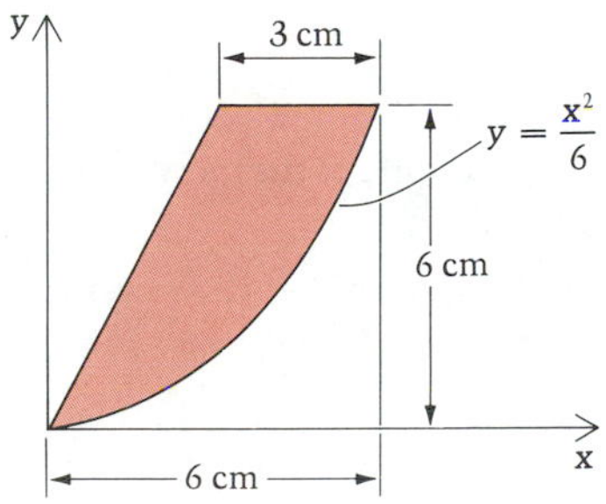

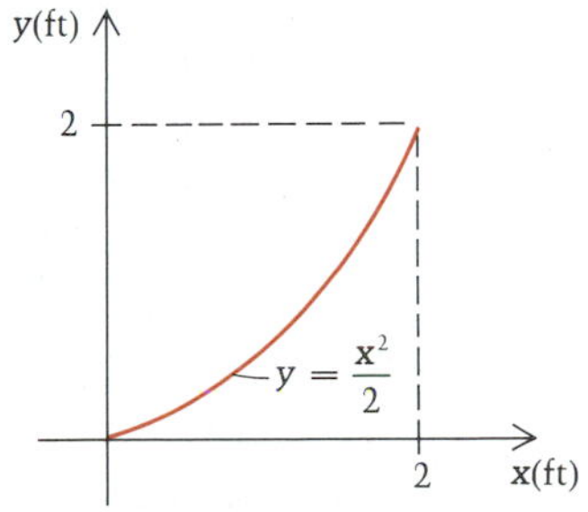

Figure P7.18

For Problems 7.19 and 7.20, see instructions above Problem 7.16.

7.19

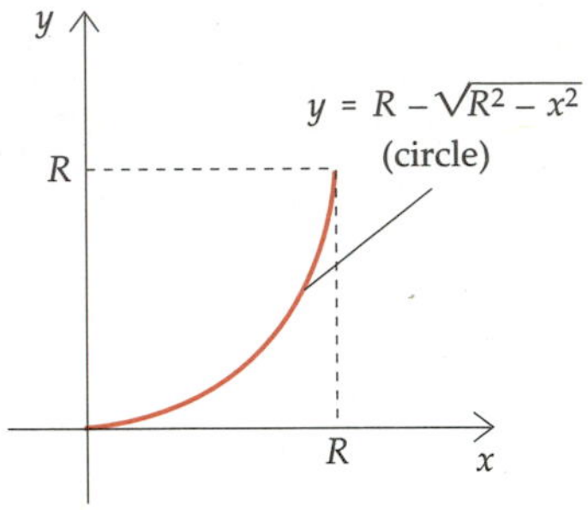

Figure P7.19

7.20

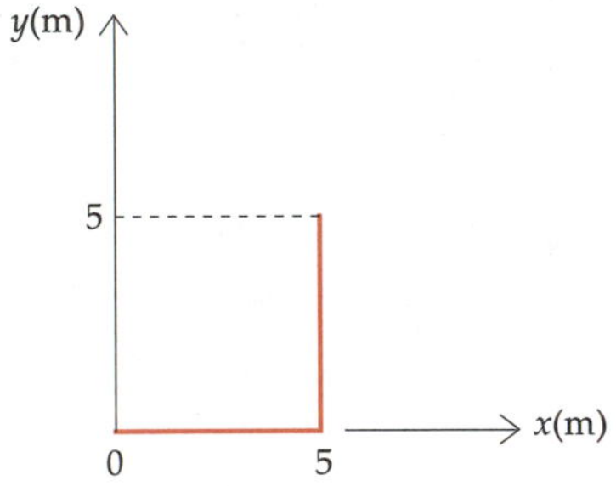

Figure P7.20

7.21 Find by integration the y-coordinate of the centroid of the straight line in Problem 7.17.

7.22 Find by integration the y-coordinate of the centroid of the bent line in Problem 7.20.

7.23 Find the centroid of the shaded area in Figure P7.23.

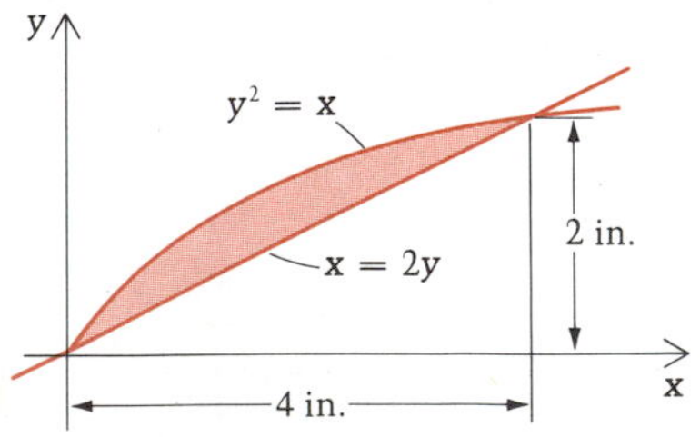

Figure P7.23

7.24 Find the centroid of the shaded area shown in Figure P7.24.

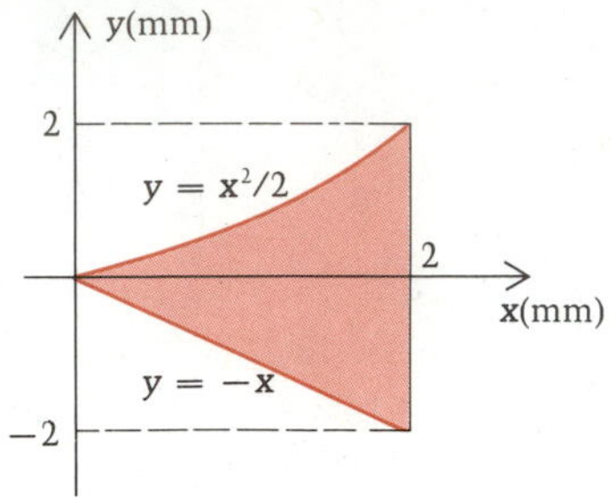

Figure P7.24

7.25 Determine the x coordinate of the centroid of the shaded area in Figure P7.25. The equation of the curve is $y^2 = x$.

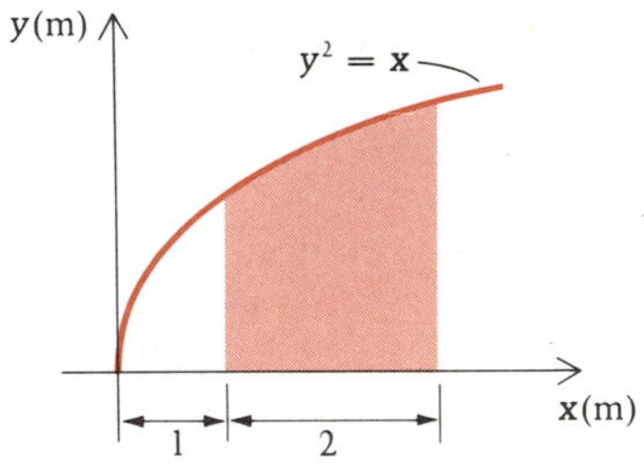

Figure P7.25

7.26 Find the centroid of the solid of revolution shown in Figure P7.26.

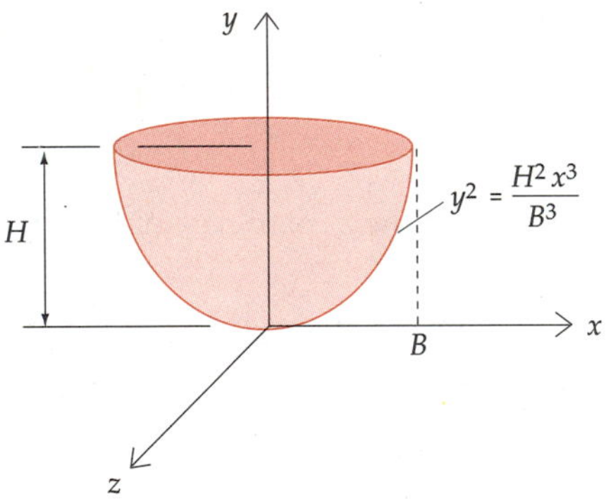

Figure P7.26

7.27 Find the centroid of the volume of the paraboloid of revolution shown in Figure P7.27.

* **7.28** In Problem 7.27 find the centroid of the outer surface area of the paraboloid. *Hint:* You'll need integral tables!

* Asterisks identify the more difficult problems.

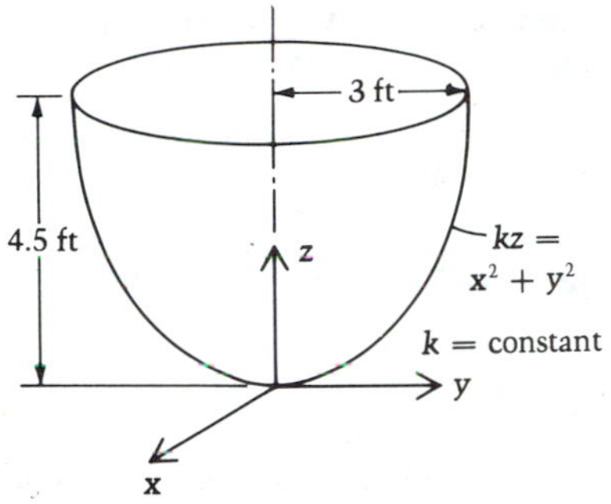

Figure P7.27

7.29 Show that the centroid of the area under the "nth degree parabola" in Figure P7.29 is at

$$(\bar{x}, \bar{y}) = \left(\frac{n+1}{2n+1}b, \frac{n+1}{2(n+2)}h\right),$$

and that the area is

$$\frac{n}{n+1}bh.$$

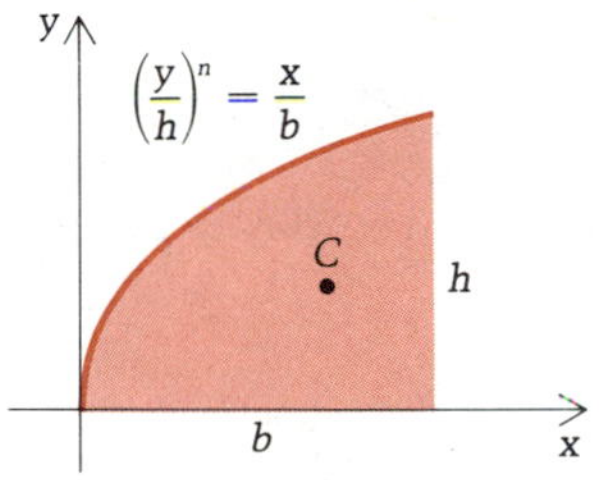

Figure P7.29

7.30 Find the centroid of the volume of the pyramid shown in Figure P7.30.

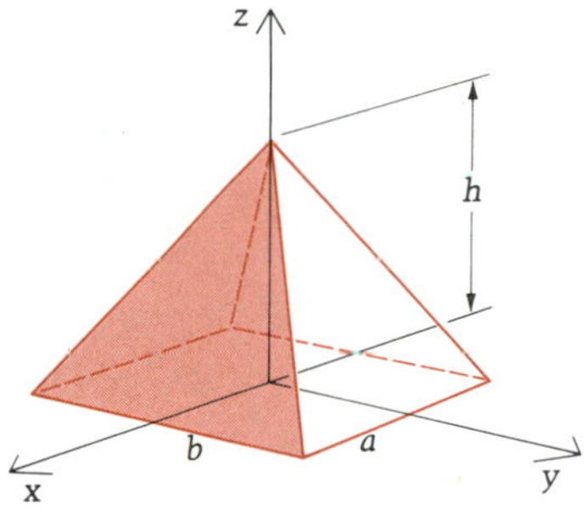

Figure P7.30

7.31 Find the centroid of the volume of the tetrahedron shown in Figure P7.31.

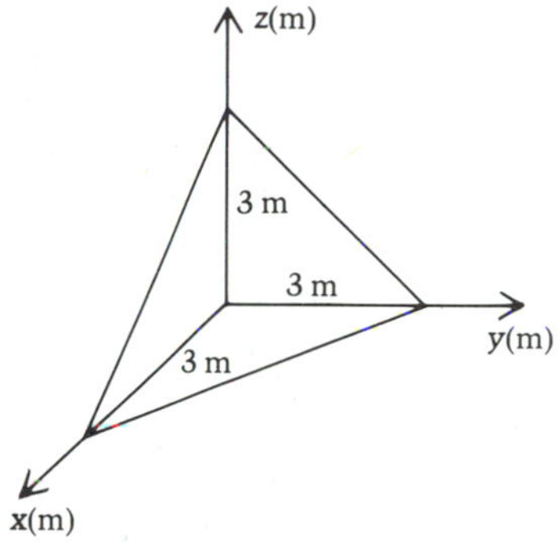

Figure P7.31

7.3 The Method of Composite Parts

When two (or more) familiar areas* A_1 and A_2 are to be considered together as a single area, we do not have to integrate over this combined area in order to find the centroid. Fortunately, an approach called the method of composite parts allows us to use the known areas and centroid locations of the separate parts to establish the centroid of the combined area ($A_1 + A_2 = A$).

The theory goes as follows, with areas A_1 and A_2 having respective centroidal x coordinates $\bar{x}_1$ and $\bar{x}_2$:

$$A\bar{x} = \int_R x\,dA$$

* Or volumes or lines, as the case may be.

But the integral over region $\mathcal{R}$ may be taken separately over subregions $\mathcal{R}_1$ and $\mathcal{R}_2$ for which the areas are, respectively, A_1 and A_2:

$$(A_1 + A_2)\bar{x} = \int_{\mathcal{R}_1} x\,dA + \int_{\mathcal{R}_2} x\,dA$$

Now separately invoking the basic centroid equation (7.3) over each of A_1 and A_2, we obtain

$$(A_1 + A_2)\bar{x} = A_1\bar{x}_1 + A_2\bar{x}_2 \tag{7.5}$$

and we see that all need for integration has been circumvented if we know A_1, A_2, $\bar{x}_1$, and $\bar{x}_2$. Any number of areas may be combined in this way.

In the more general form (where the quantity whose centroid we seek is any scalar Q), the steps to the similar vector expression are

$$\begin{aligned} Q\mathbf{r}_{OC} &= \int \mathbf{r}\,dQ \\ &= \int_{\mathcal{R}_1} \mathbf{r}\,dQ + \int_{\mathcal{R}_2} \mathbf{r}\,dQ \\ &= Q_1\mathbf{r}_{OC_1} + Q_2\mathbf{r}_{OC_2} \end{aligned}$$

where

$$Q = Q_1 + Q_2$$

We now consider a number of examples of the use of "composite parts."

EXAMPLE 7.9

Find the centroid of the three line segments connected as shown in Figure E7.9a.

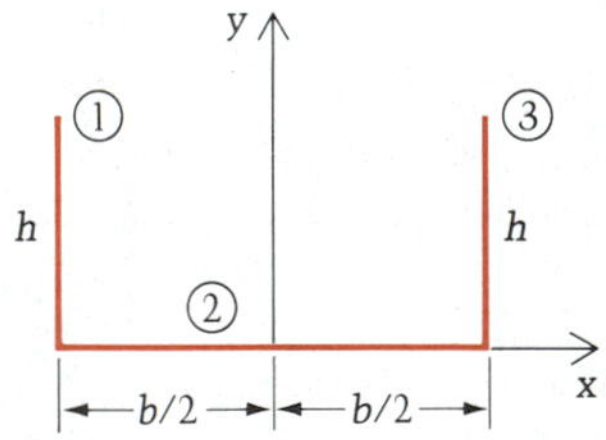

Figure E7.9a

Solution

Labeling the segments as shown, we have, using composite parts,

$$\begin{aligned} (h + b + h)\bar{y} &= L_1\bar{y}_1 + L_2\bar{y}_2 + L_3\bar{y}_3 \\ &= h\left(\frac{h}{2}\right) + b(0) + h\left(\frac{h}{2}\right) \end{aligned}$$

where the middle term vanishes because the y coordinate of the centroid of element ② is zero. Therefore,

$$(2h + b)\bar{y} = \frac{h^2}{2} + \frac{h^2}{2} = h^2$$

so that

$$\bar{y} = \frac{h^2}{2h + b}$$

To illustrate the vanishing of "first moments" when measurements are made from the centroid, consider the special case in this example for which $b = h$;

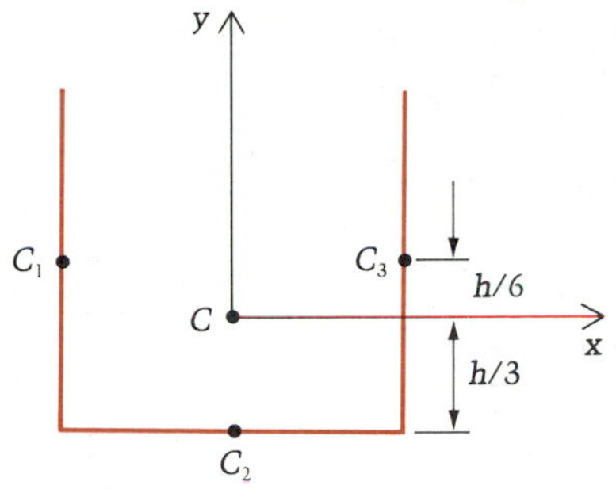

Figure E7.9b

then $\bar{y} = h/3$. Now let us set up *new* axes with origin at the centroid as shown in Figure E7.9b.

Relative to these new axes, we see that the "first moment" is

$$\int y\, ds = h(h/6) + h(-h/3) + h(h/6)$$
$$= 0$$

as, of course, must be the case since the y coordinate of the centroid in this system is zero.

Example 7.10

The circles in Figure E7.10 represent cross sections of bolts, with the bolts holding down a gear-box cover whose shape is the curve shown. It is important to locate the centroid C of the bolt pattern because, for example, the distance from C to each bolt is used both to compute the stiffness of the overall pattern against twisting and also to calculate the shear stress in each of the bolts. Letting each bolt's cross-sectional area be A, find the centroid of the pattern of ten identical bolts.

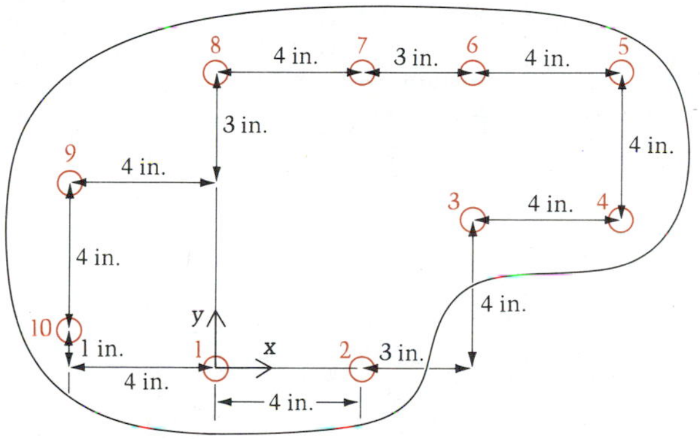

Figure E7.10

Solution

Using composite areas,

$$\left(\sum_{i=1}^{10} A_i\right)\bar{x} = \sum_{i=1}^{10} A_i \bar{x}_i$$

$$10A\bar{x} = A\Sigma\bar{x}_i$$

$$\bar{x} = \frac{\Sigma\bar{x}_i}{10}$$

and similarly,

$$\bar{y} = \frac{\Sigma\bar{y}_i}{10}$$

while $\bar{z}$ is zero because all of the areas lie in the x-y plane.

Substituting (using the bolt-numbering system shown in the figure), and moving counterclockwise from bolt ① at the origin,

$$\bar{x} = (0 + 4 + 7 + 11 + 11 + 7 + 4 + 0 - 4 - 4)/10$$

$$\bar{x} = 3.60 \text{ in.}$$

Question 7.7 Why do the contributions of bolts 2, 7, 9, and 10 cancel in the $\bar{x}$ computation?

Continuing,

$$\bar{y} = (0 + 0 + 4 + 4 + 8 + 8 + 8 + 8 + 5 + 1)/10$$

$$\bar{y} = 4.60 \text{ in.}$$

Therefore $(\bar{x}, \bar{y}, \bar{z}) = (3.60, 4.60, 0)$ in.

Answer 7.7 Because, with the four areas equal, two +4 coordinates cancel two −4 coordinates.

EXAMPLE 7.11

Find the centroid of the cross-sectional area (of a channel beam) shown in Figure E7.11a.

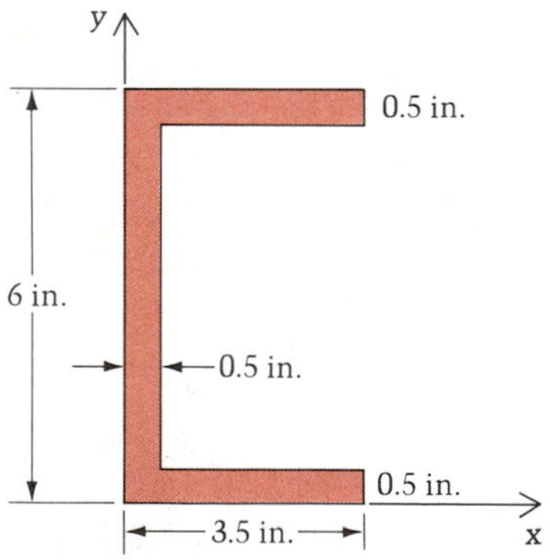

Figure E7.11a

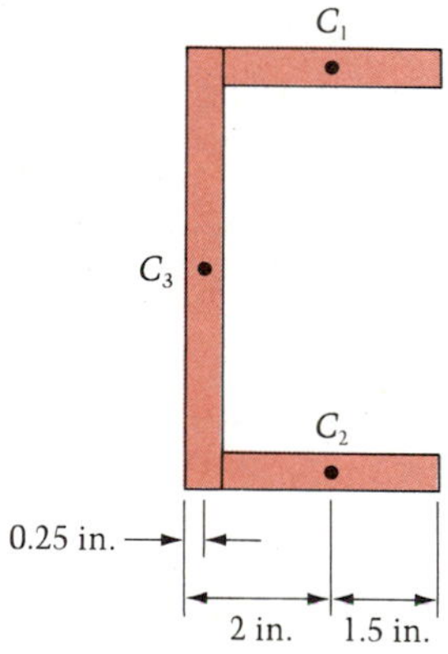

Figure E7.11b

Solution

The axis of symmetry ($y = 3$ in.) allows us to conclude by inspection that $\bar{y} = 3$ in. We shall determine the x coordinate of C in two ways. First we note that the area in question is the composite of three rectangles as shown in Figure E7.11b, and we obtain:

$$A\bar{x} = A_1\bar{x}_1 + A_2\bar{x}_2 + A_3\bar{x}_3$$

$$\bar{x} = \frac{[3(0.5)](2) + [3(0.5)](2) + [6(0.5)](0.25)}{3(0.5) + 3(0.5) + 6(0.5)}$$

$$= \frac{6.75}{6} = 1.12 \text{ in.}$$

An alternative approach is to note that the area of interest is obtained by removing a 3 in. × 5 in. rectangle from a 3.5 in. × 6 in. rectangle as shown in Figure E7.11c below:

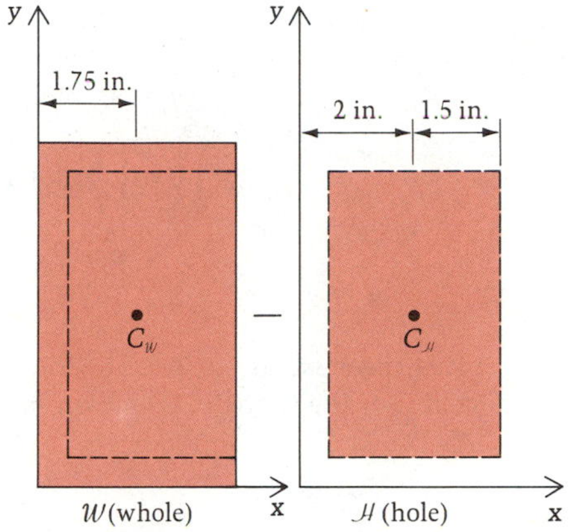

Figure E7.11c

Applying the method of composite parts to the whole (3.5 in. × 6 in.) rectangle, we have

$$A_W\bar{x}_W = A\bar{x} + A_H\bar{x}_H$$

where A and $\bar{x}$ refer to the (net) area of interest. Rearranging,

$$A\bar{x} = A_W\bar{x}_W - A_H\bar{x}_H{}^*$$

$$[(6)(3.5) - 5(3)]\bar{x} = [6(3.5)](1.75) - [5(3)](2)$$

or

$$\bar{x} = (3.5)(1.75) - 5$$

$$= 6.12 - 5$$

$$= 1.12 \text{ in.} \qquad \text{(as before)}$$

* Some like to see this as a direct application of the method of composite parts by thinking of the "hole" as a negative area; that is,

$$A\bar{x} = A_W\bar{x}_W + (-A_H)\bar{x}_H$$

EXAMPLE 7.12

Show that the centroid of *any* triangle is on the line that is two-thirds the distance from any vertex V to the opposite side and parallel to that side.

Solution

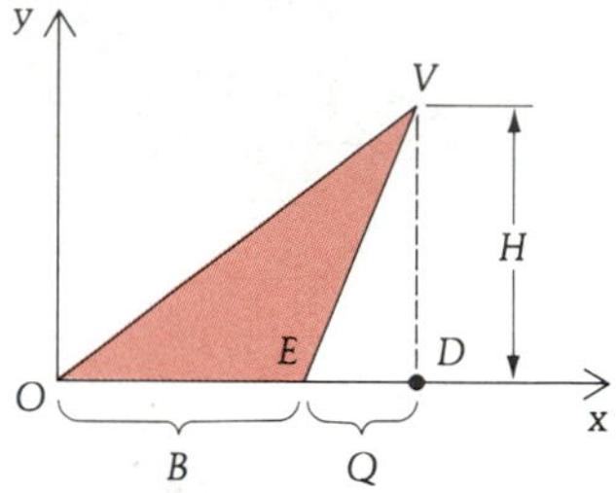

Figure E7.12a

This example provides an opportunity to obtain a useful, general result without using integration. Consider the actual shaded, scalene triangle of Figure E7.12a to be made up of the right triangle *OVD* less the smaller right triangle *EVD*. Then, using what we have learned about the centroids of right triangles,

$$A_{OVD}\bar{y}_{OVD} = A\bar{y} + A_{EVD}\bar{y}_{EVD}$$

or

$$A\bar{y} = A_{OVD}\bar{y}_{OVD} - A_{EVD}\bar{y}_{EVD}$$

$$\left(\frac{1}{2}BH\right)\bar{y} = \left[\frac{1}{2}(B+Q)H\right]\frac{H}{3} - \left(\frac{1}{2}QH\right)\frac{H}{3}$$

$$\bar{y} = \frac{BH^2/6}{BH/2} = \frac{H}{3} \qquad \text{(again)}$$

This means that all three of the lines ℓ_1, ℓ_2, ℓ_3 below will intersect at the centroid C of the triangle, as shown in Figure E7.12b.

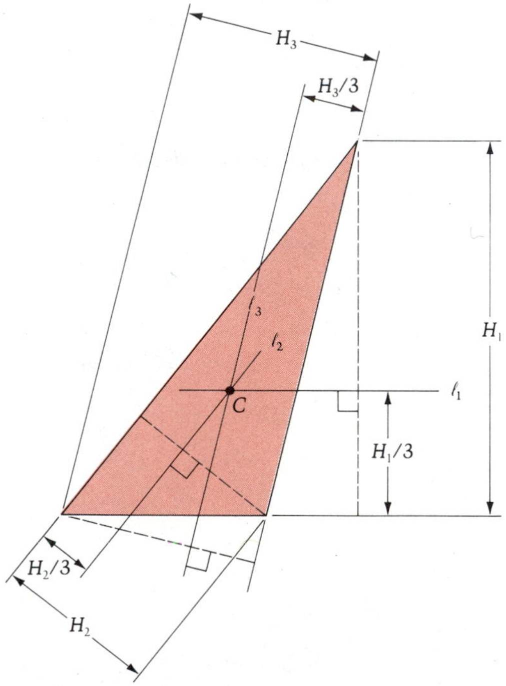

Figure E7.12b

EXAMPLE 7.13

Use the method of composite areas and the result of Example 7.4 to find the centroid of the shaded area in Figure E7.13a.

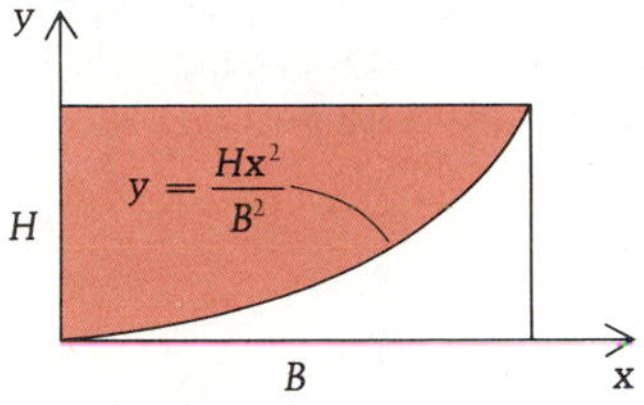

Figure E7.13a

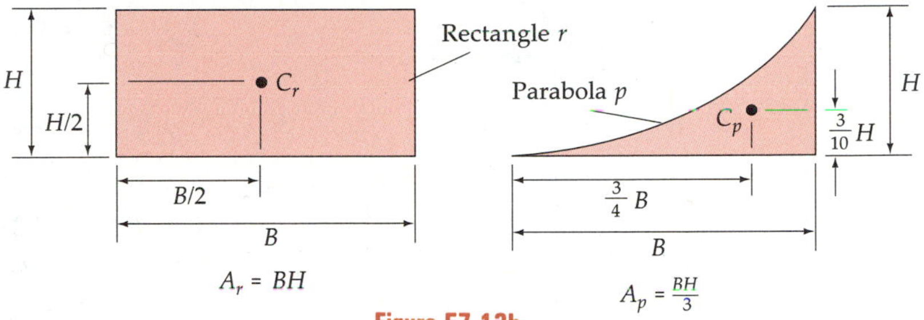

Figure E7.13b

Solution

The shaded area is obtained by subtracting the area beneath the parabola (p) in Figure E7.13b from the rectangle (r). Their areas and centroid locations are shown in the figures. Therefore,

$$A = A_r - A_p$$

$$= BH - \frac{BH}{3} = \frac{2}{3}BH$$

and

$$A_r\bar{x}_r = A\bar{x} + A_p\bar{x}_p$$

$$(BH)\left(\frac{B}{2}\right) = \left(\frac{2}{3}BH\right)\bar{x} + \left(\frac{BH}{3}\right)\left(\frac{3}{4}B\right)$$

$$\bar{x} = \frac{3}{8}B$$

The same procedure will yield the y coordinate of the centroid:

$$A_r\bar{y}_r = A\bar{y} + A_p\bar{y}_p$$

$$(BH)\left(\frac{H}{2}\right) = \left(\frac{2}{3}BH\right)\bar{y} + \left(\frac{BH}{3}\right)\left(\frac{3}{10}H\right)$$

$$\bar{y} = \frac{3}{5}H$$

EXAMPLE 7.14

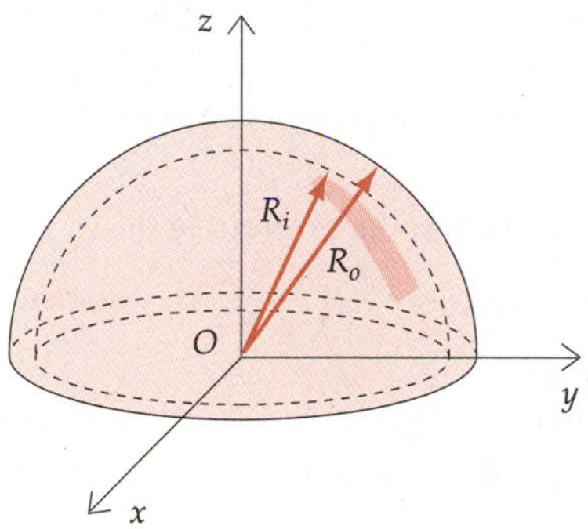

Figure E7.14

Locate the centroid of the volume of a hemispherical shell of inner radius R_i and outer radius R_o (see Figure E7.14).

Solution

From symmetry we recognize that $\bar{x} = \bar{y} = 0$. The volume in question may be viewed as a hemisphere of radius R_o out of which a hemisphere of radius R_i has been removed. From Example 7.6, we know that the z coordinates of centroids for these are $\bar{z}_o = \frac{3}{8}R_o$ and $\bar{z}_i = \frac{3}{8}R_i$. Using the method of composite parts,

$$V_o\bar{z}_o = V\bar{z} + V_i\bar{z}_i$$

or

$$V\bar{z} = V_o\bar{z}_o - V_i\bar{z}_i$$

$$\left(\frac{2}{3}\pi R_o^3 - \frac{2}{3}\pi R_i^3\right)\bar{z} = \left(\frac{2}{3}\pi R_o^3\right)\left(\frac{3}{8}R_o\right) - \left(\frac{2}{3}\pi R_i^3\right)\left(\frac{3}{8}R_i\right)$$

$$\bar{z} = \frac{3(R_o^4 - R_i^4)}{8(R_o^3 - R_i^3)} = \frac{3(R_o^2 + R_i^2)(R_o + R_i)\cancel{(R_o - R_i)}}{8(R_o^2 + R_oR_i + R_i^2)\cancel{(R_o - R_i)}} \qquad (1)$$

Note that if we set $R_i = 0$, we recover $\bar{z} = \frac{3}{8}R_o$.

A limiting case sometimes of interest is that for which $R_i \approx R_o$, where the volume of interest is a thin shell. In that circumstance we obtain from Equation (1)

$$\bar{z} \approx \frac{3(2)R_o^2)(2R_o)}{8(3R_o^2)} = \frac{R_o}{2}$$

which is precisely the z coordinate of the centroid of the (curved) surface *area* of the hemisphere. (The student is encouraged to verify this by direct calculation.)

Question 7.8 Is this a coincidence?

Answer 7.8 No. As the thickness gets smaller and smaller (let $R_i \rightarrow R_o$), the volume elements get closer and closer to the surface of the outer hemisphere.

PROBLEMS ▶ Section 7.3

7.32 Find the centroid of the shaded area in Figure P7.32.

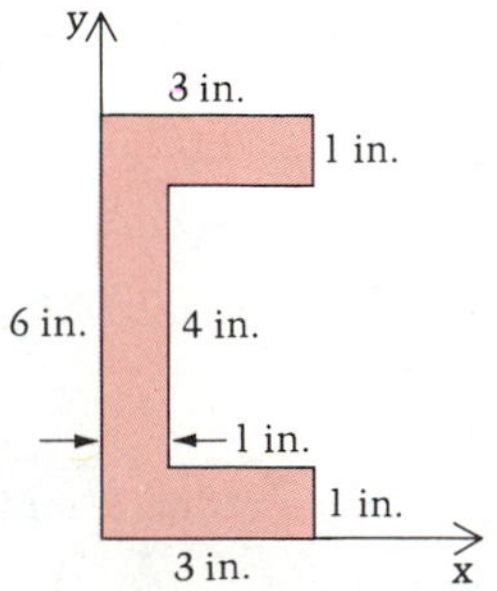

Figure P7.32

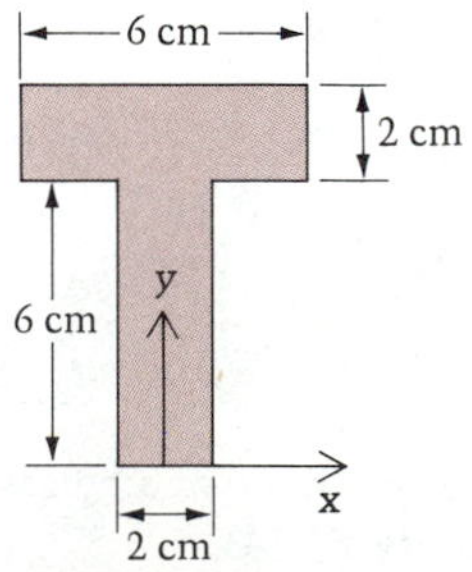

Figure P7.33

7.33 Find the y coordinate of the centroid of the T-shaped area shown in Figure P7.33.

7.34 In Figure P7.34 locate the centroid of the shaded area.

7.35 In Figure P7.35 show that the centroid of the trapezoidal area has a y coordinate of $\dfrac{H}{3}\left(\dfrac{a + 2b}{a + b}\right)$, independent of the angles α_1 and α_2.

7.36 Find the centroid of the area shown in Figure P7.36.

7.37 Find the centroid of the area shown in Figure P7.37.

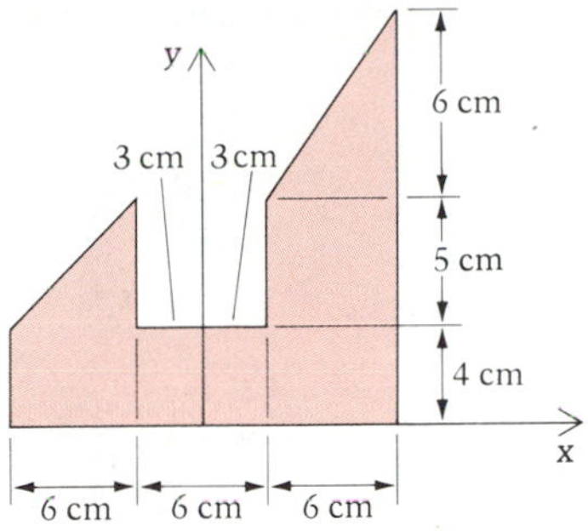

Figure P7.34

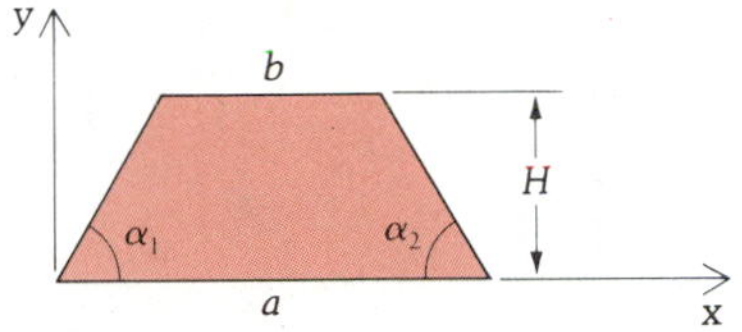

Figure P7.35

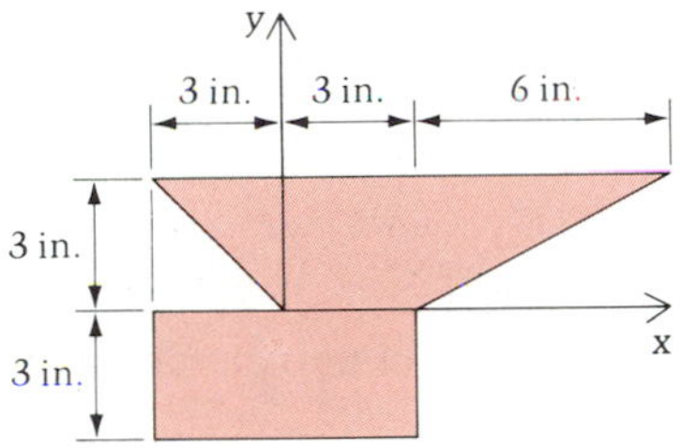

Figure P7.36

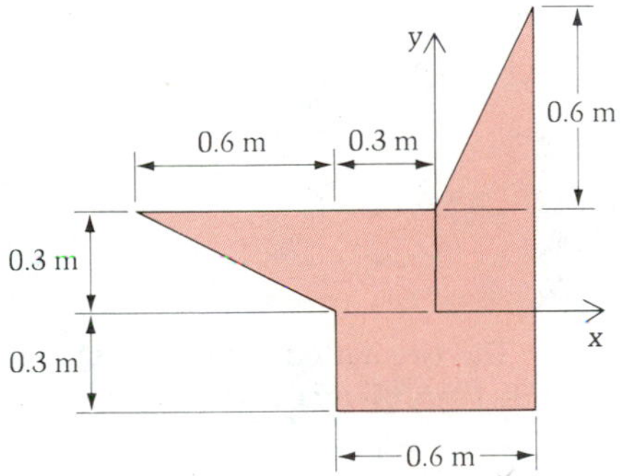

Figure P7.37

7.38 In Figure P7.38, A_1 and A_2 are two areas with centroids C_1 and C_2 separated by the distance ℓ as shown. Show that the centroid of the composite area (A_1 plus A_2) lies on line C_1C_2, and is a distance $\dfrac{A_2}{A_1 + A_2}\,\ell$ from C_1.

7.39 Find the centroid of the three *lines* in Figure P7.39, and compare its location with the centroid of the enclosed *area.*

7.40 Find the centroid of the tee rail cross section shown in Figure P7.40. (We are neglecting the fact that the web sides and the top of the head are actually segments of large circles for this particular rail.)

7.41 Locate the centroid of the shaded area in Figure P7.41.

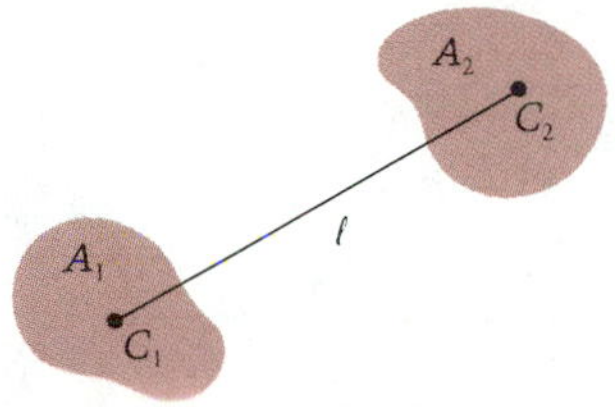

Figure P7.38

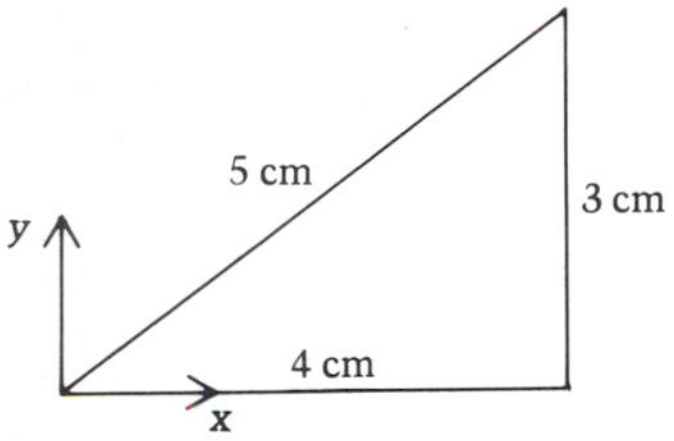

Figure P7.39

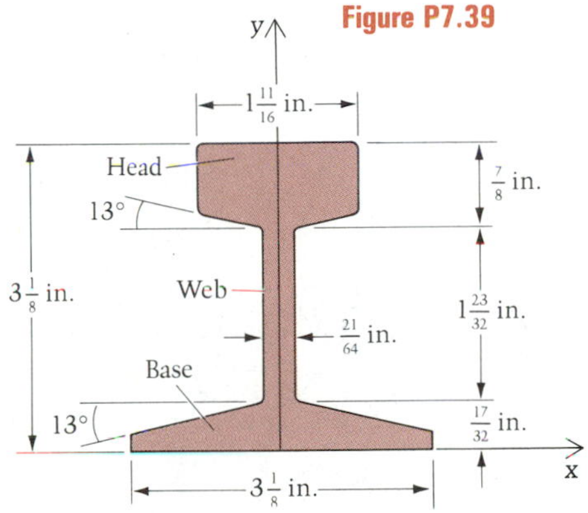

Figure P7.40

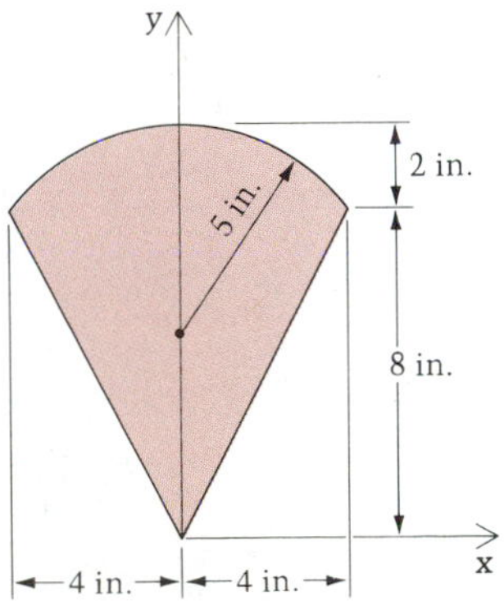

Figure P7.41

7.42 In Figure P7.42 locate the centroid of the homogeneous square plate with a triangular cut-out.

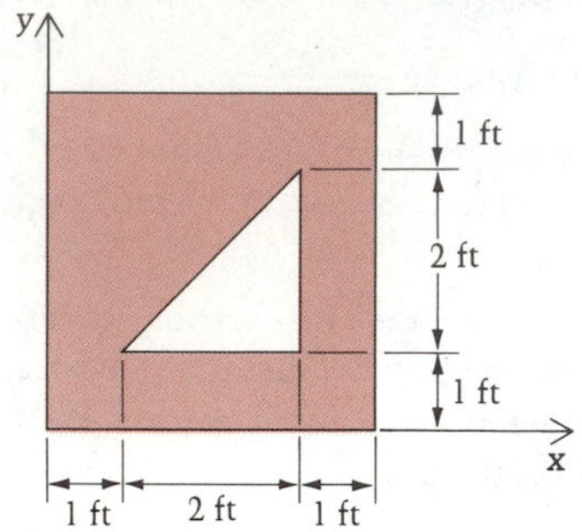

Figure P7.42

7.43 Find the centroid of the shaded area in Figure P7.43.

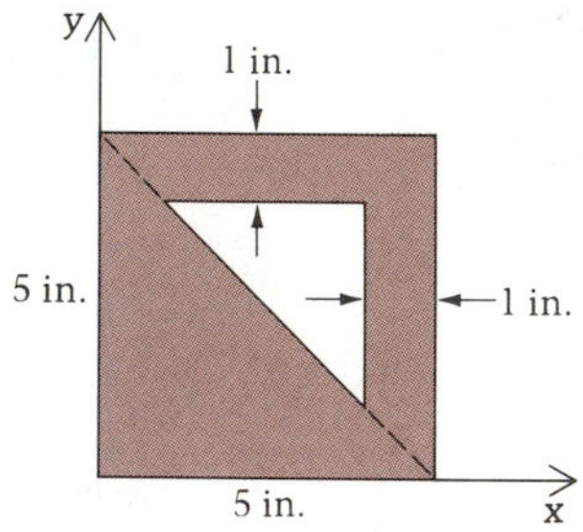

Figure P7.43

7.44 Find the centroid of the shaded area in Figure P7.44.

7.45 Find the centroid of the shaded area in Figure P7.45.

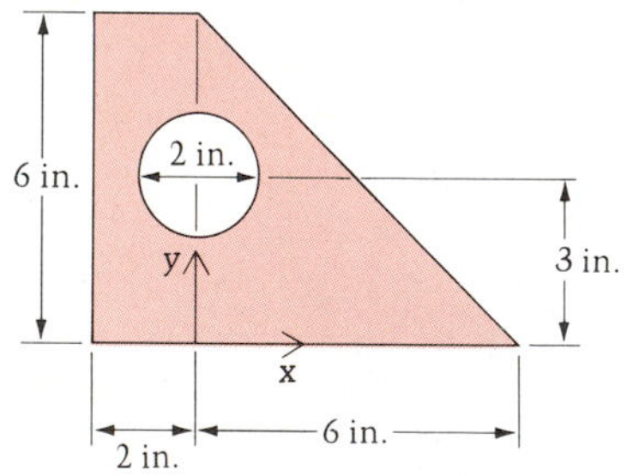

Figure P7.44

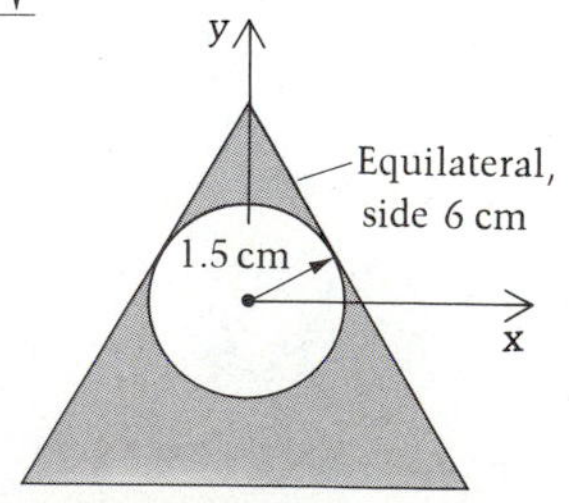

Figure P7.45

7.46 Find the centroid of the shaded area in Figure P7.46.

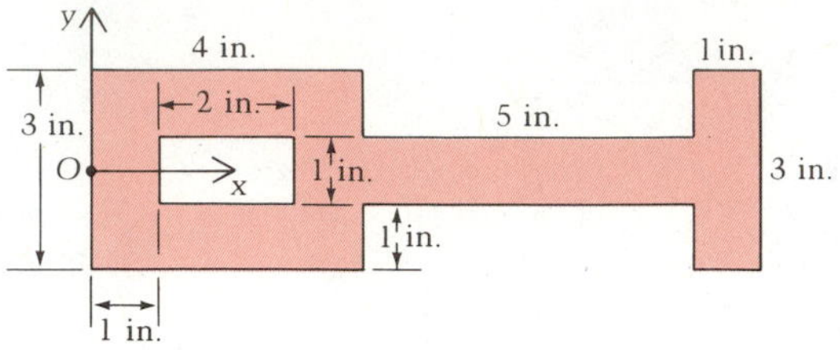

Figure P7.46

7.47 Find the centroid of the shaded area in Figure P7.47.

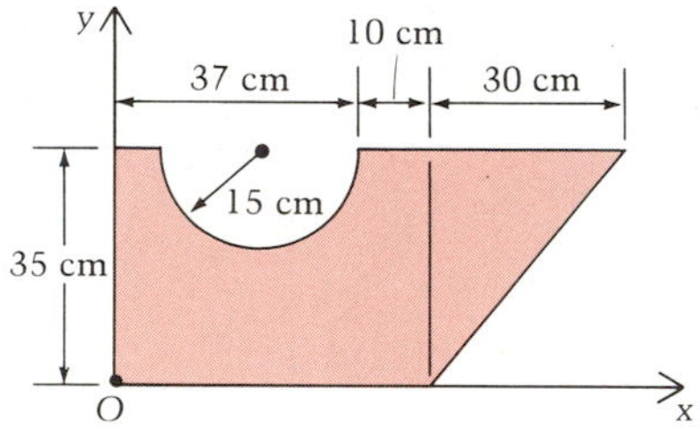

Figure P7.47

7.48 Find the x and y coordinates of the centroid of the shaded area in Figure P7.48.

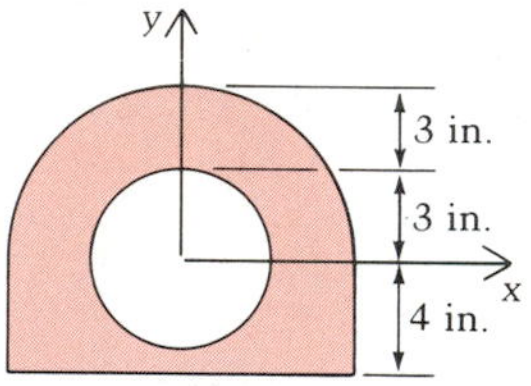

Figure P7.48

7.49 The circular cutout shown in Figure P7.49 has an area of 12 in.^{2} Find the centroid of the shaded area.

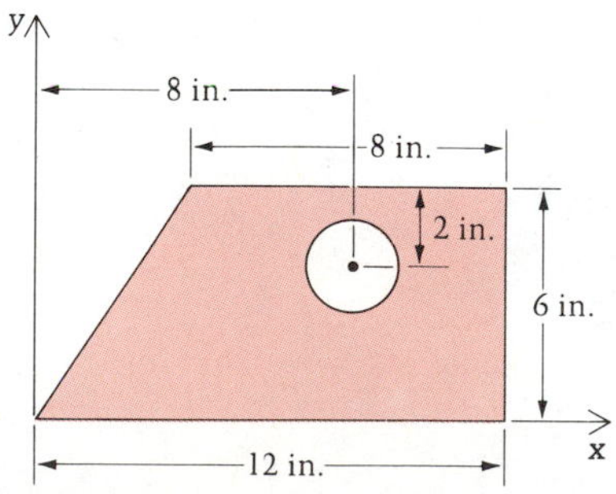

Figure P7.49

7.50 Find the centroid of the shaded area in Figure P7.50.

7.51 In Figure P7.51 find the centroid of the shaded square, which has a circular and triangular cutout as shown.

7.52 Find the centroid of the shaded area in Figure P7.52.

7.53 Find the centroid of the shaded area in Figure P7.53.

7.54 Find the y coordinate of the centroid of the shaded area in Figure P7.54.

7.55 In Figure P7.55 find the centroid of the shaded triangular area with the semicircular cutout shown.

7.56 In Figure P7.56 the holes in the trapezoidal area have 1-in. radii. Find the x coordinate of the centroid of the area.

7.57 Find the centroid of the shaded area in Figure P7.57.

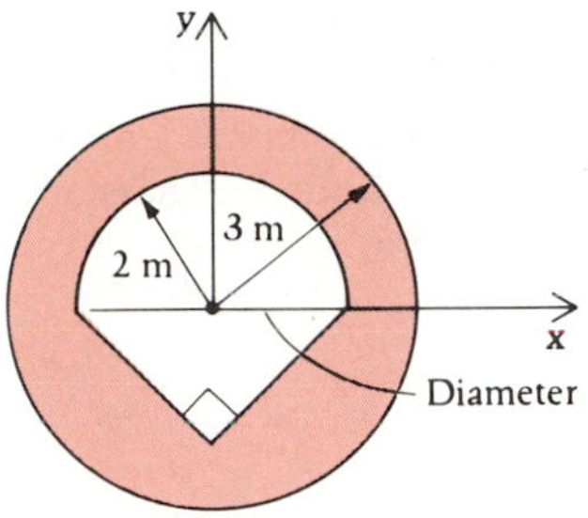

Figure P7.50

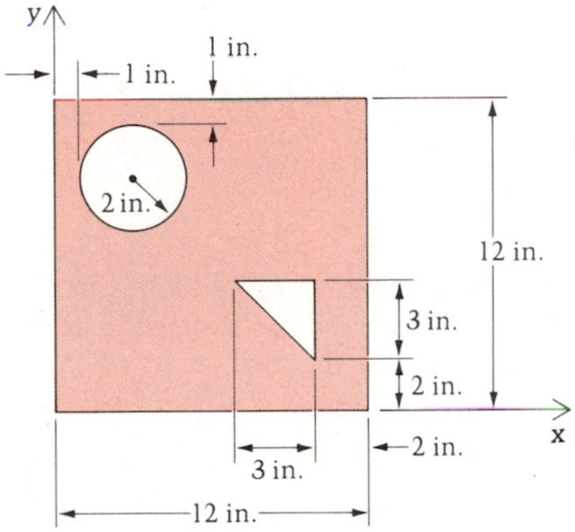

Figure P7.51

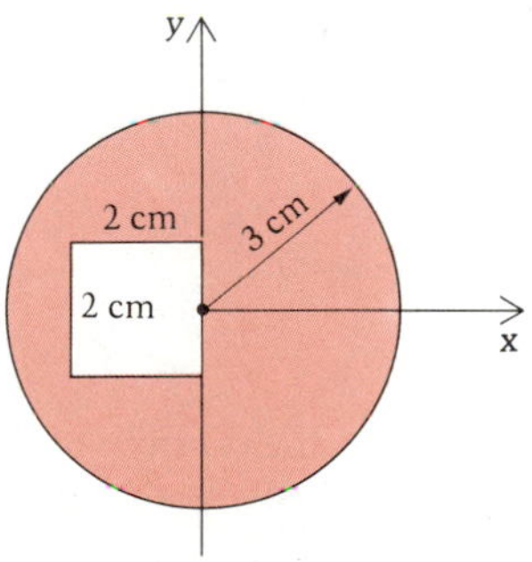

Figure P7.52

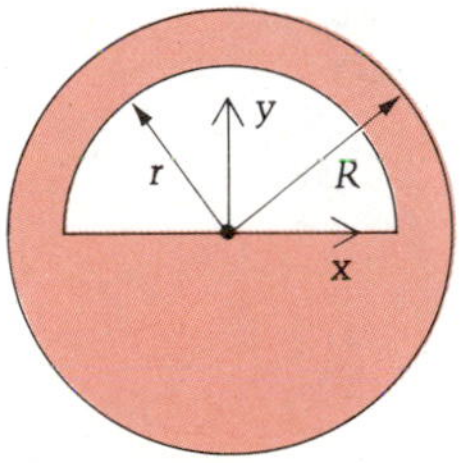

Figure P7.53

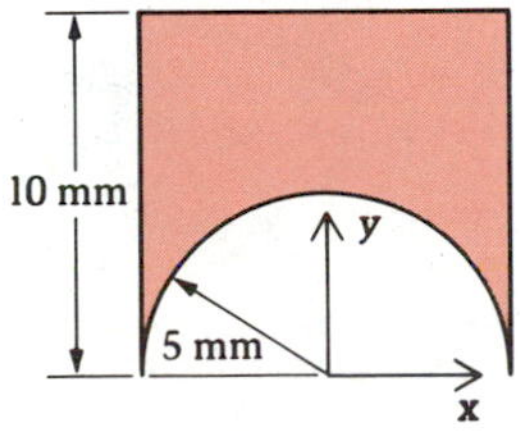

Figure P7.54

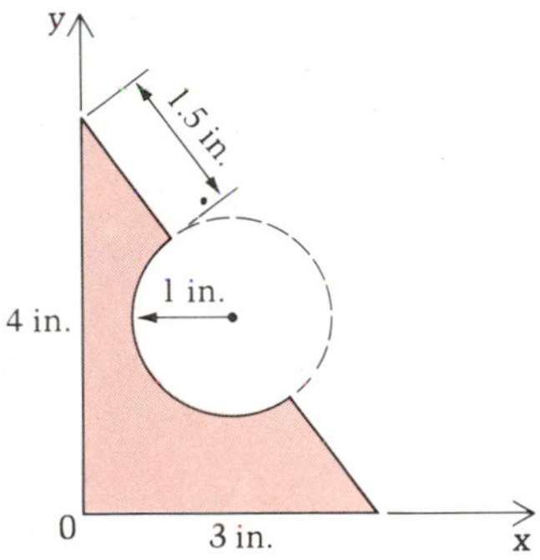

Figure P7.55

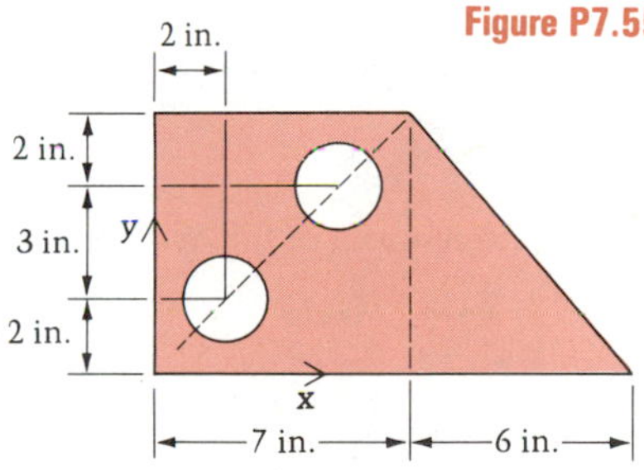

Figure P7.56

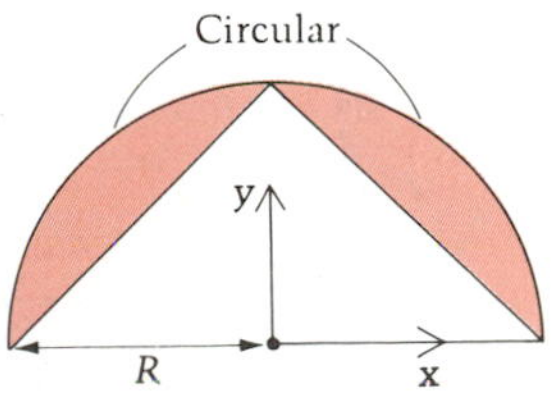

Figure P7.57

7.58 For the angle section shown in Figure P7.58, find the coordinates of the centroid.

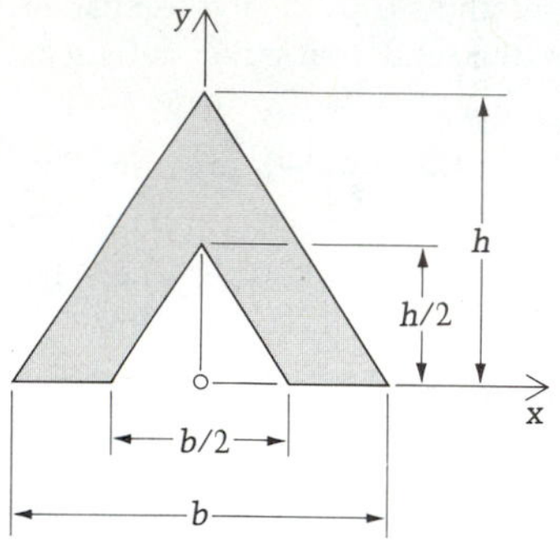

Figure P7.58

7.59 Find the centroid of the composite area shown in Figure P7.59.

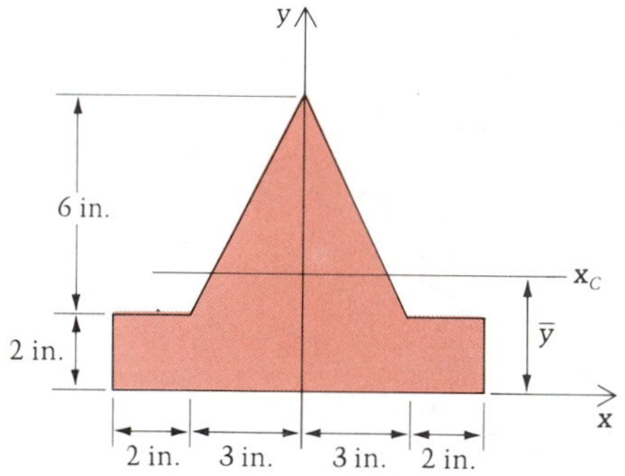

Figure P7.59

7.60 In Figure P7.60 find the centroid of the five circular areas. Each area is 0.75 in.²

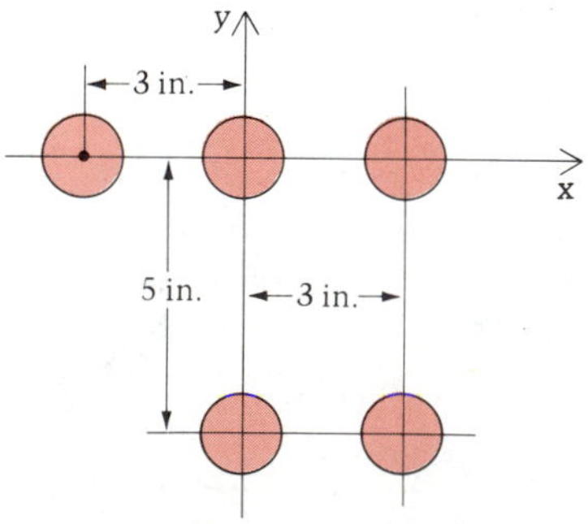

Figure P7.60

7.61 Find the centroid of the cross-sectional area of the structural steel angle beam shown in Figure P7.61. Assume a constant thickness of $\frac{5}{8}$ in., and compare your answers with the actual (considering rounded corners and fillets) values of $(\bar{x}, \bar{y}) = (1.03, 2.03)$ in.

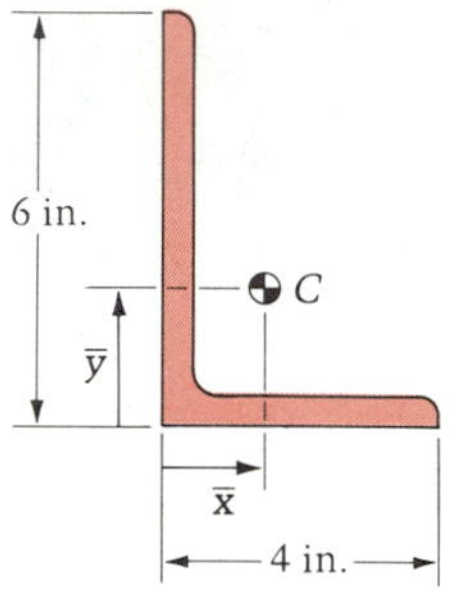

Figure P7.61

7.62 In Figure P7.62 determine the height h of the "T" (formed by two perpendicular lines as shown) in order that the centroid be 2 cm below the top.

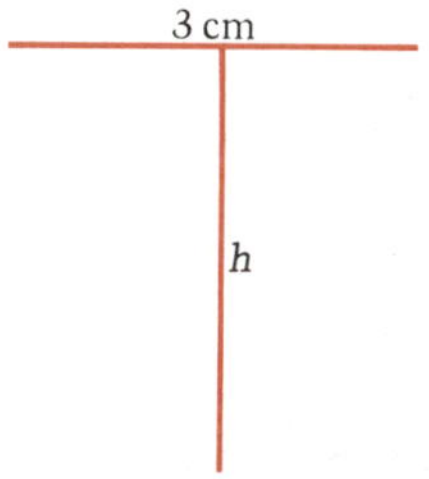

Figure P7.62

7.63 Find the centroid of the center line of the bar consisting of three straight sections parallel to the (x, y, z) axes as shown in Figure P7.63.

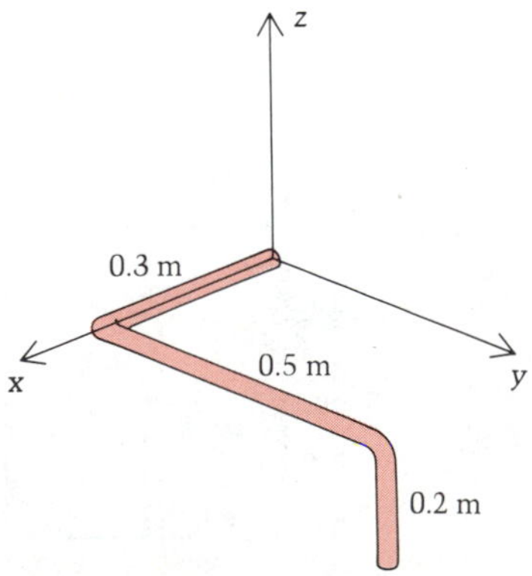

Figure P7.63

7.64 Rework Problem 7.20 using the method of composite parts.

7.65 Find the centroid of the "bent line" *ABCD* in Figure P7.65.

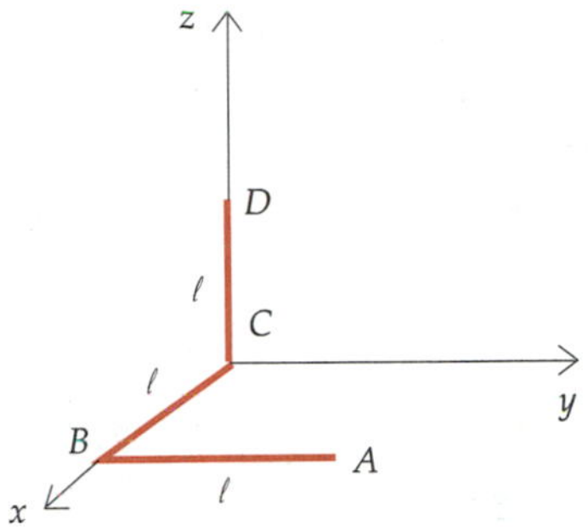

Figure P7.65

7.66 Find the centroid of the set of five lines shown in Figure P7.66.

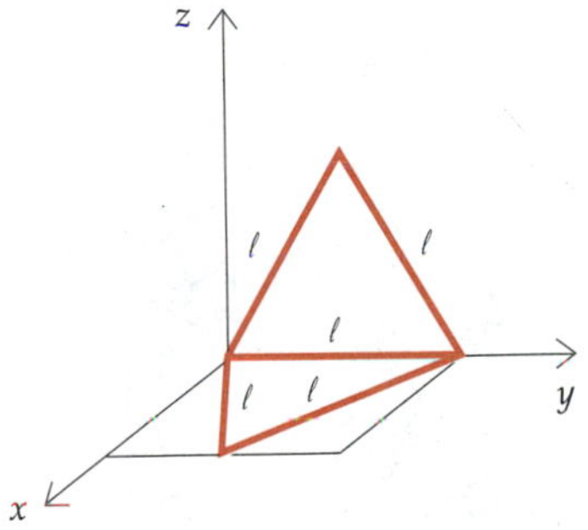

Figure P7.66

7.67 Find the centroid of the set of nine lines shown in Figure P7.67. (Which one of $\bar{x}, \bar{y}, \bar{z}$ can actually be seen by inspection?)

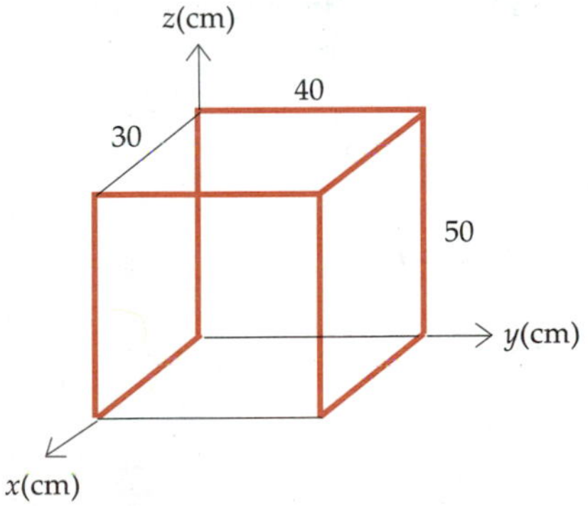

Figure P7.67

7.68 Find the centroid of the bent line *ABCDA* in Figure P7.68.

7.69 Find the centroid of the set of three areas shown in Figure P7.69. The triangle lies in the *yz*-plane, the rectangle in the *xy*-plane, and the semicircle in the *xz*-plane.

7.70 Locate the centroid of the enclosed area *ABCD* shown in Figure P7.70.

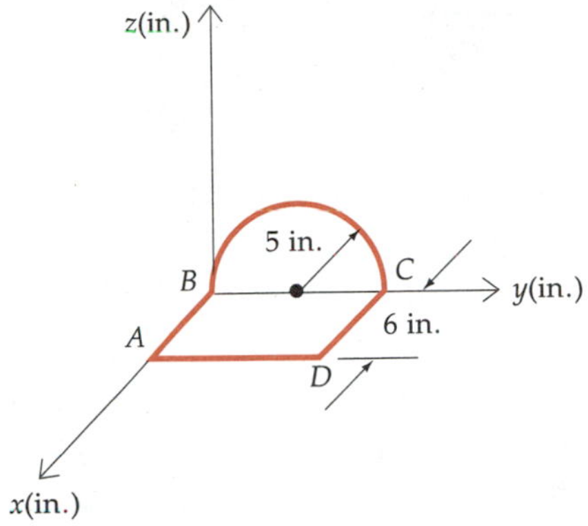

Figure P7.68

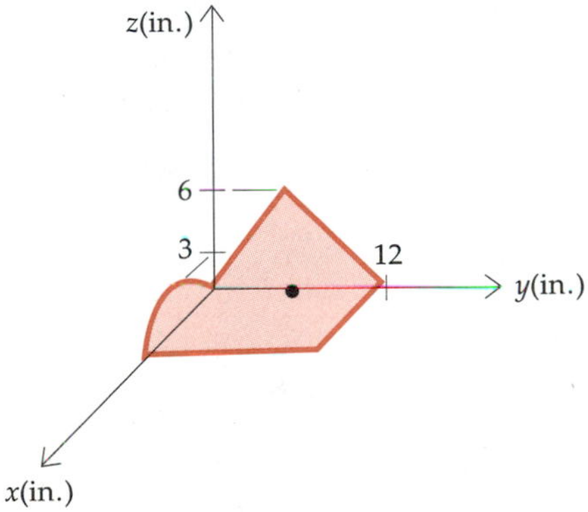

Figure P7.69

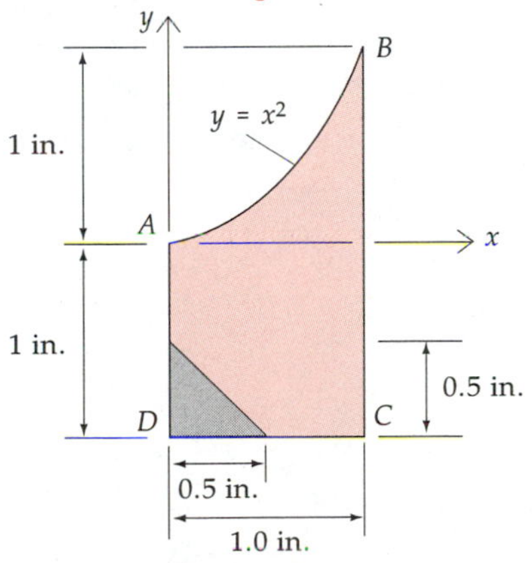

Figure P7.70

7.71 If the shaded triangle is removed from the area in the preceding problem, find the centroid of the remaining area.

7.72 Locate the centroid of the enclosed area shown in Figure P7.72.

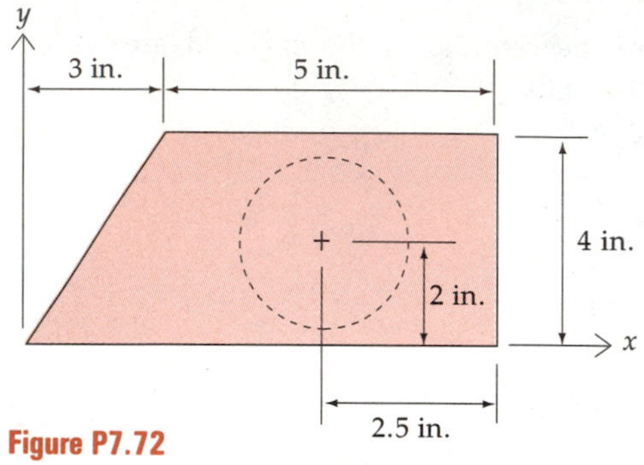

Figure P7.72

7.73 In the preceding problem, a circular hole of radius 1.5 inches is cut from the rectangular part of the plate at the location indicated by the dashed circle. Determine the new location of the centroid.

7.74 In Figure P7.74 determine the height h of the cutout rectangle that will place the centroid C of the *shaded* area at the position shown.

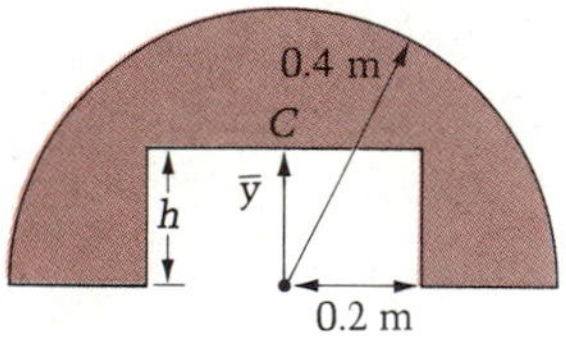

Figure P7.74

7.75 In Problem 7.33, to what value should the 6-cm vertical dimension be changed in order that the new centroid be on the line separating the two rectangular areas?

7.76 Locate the centroid of the shaded area shown in Figure P7.76.

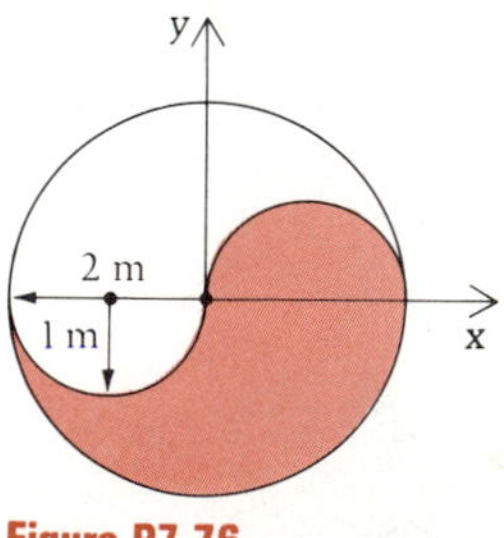

Figure P7.76

7.77 In Figure P7.77 find the centroid of the shaded area.

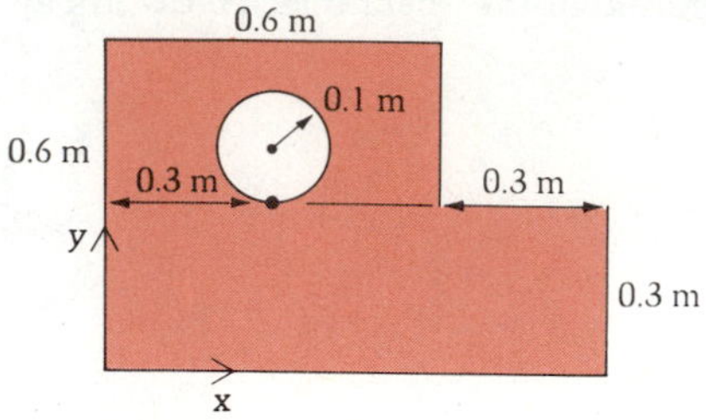

Figure P7.77

7.78 In Figure P7.78 find $\bar{x}$ for the shaded area, cut from a circle.

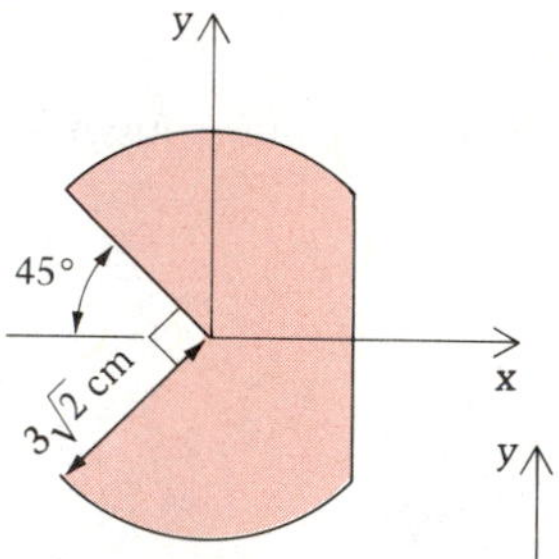

Figure P7.78

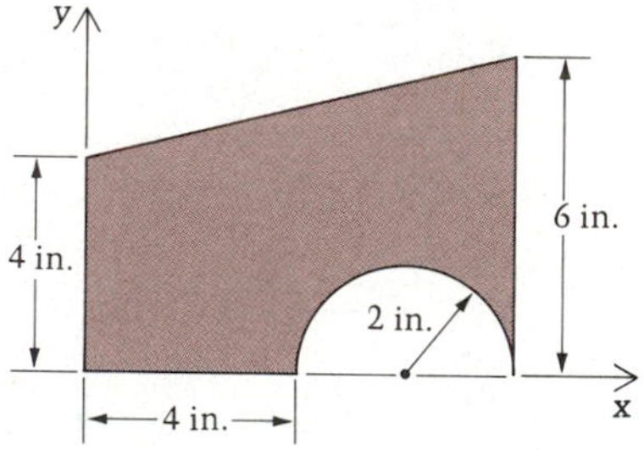

Figure P7.79

7.79 In Figure P7.79 find the centroid of the shaded area.

7.80 Find the area and the centroid of the *dotted* area in Figure P7.80.

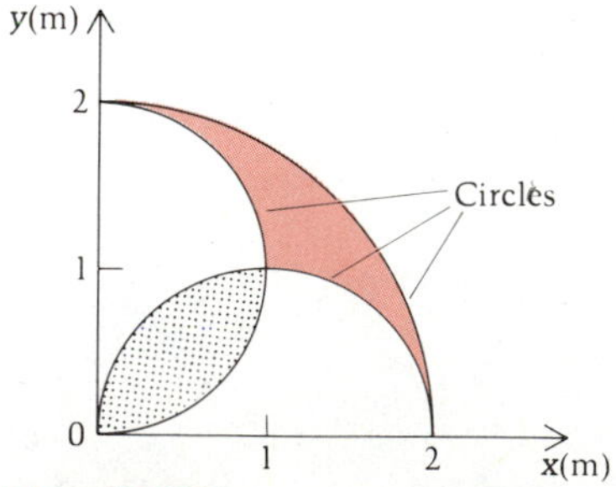

Figure P7.80

7.81 Find the centroid of the *shaded* area in Figure P7.80.

7.82 A circular plate with radius r_1 and unit thickness has five holes drilled in it as shown in Figure P7.82. The centers of the holes lie on an arc of radius r_2. The spacing between the centers is 60°. Locate the centroid of the volume of the plate if $r_1 = 0.5$ m, $r_2 = 0.3$ m, and $r_3 = 0.1$ m.

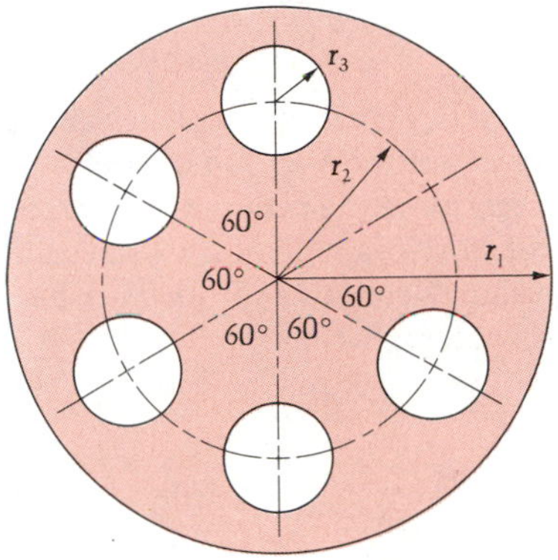

Figure P7.82

7.83 Show by (a) redrawing Figure P7.29 with the y-axis horizontal and x vertical; then (b) changing the names of x, y, b, and h; and finally (c) letting n in the answer be 2, that the centroid of the shaded area in Figure P7.83 is as shown, and that its area is $(2/3)bh$. Use the results of Problem 7.29, and note that the solution agrees with that of Example 7.13.

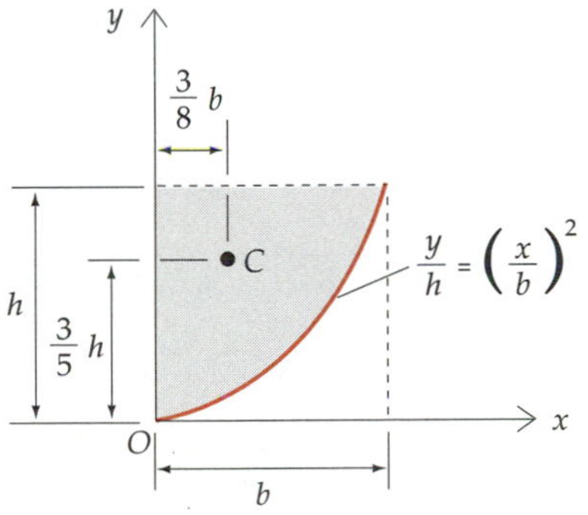

Figure P7.83

7.84 Use the results of the preceding problem together with the method of composite parts to show that the centroid of the dotted area beneath the parabolic segment in Figure P7.84 is $(\bar{x}, \bar{y}) = (3b/4, 3h/10)$ and that its area is $bh/3$, all in agreement with Example 7.4.

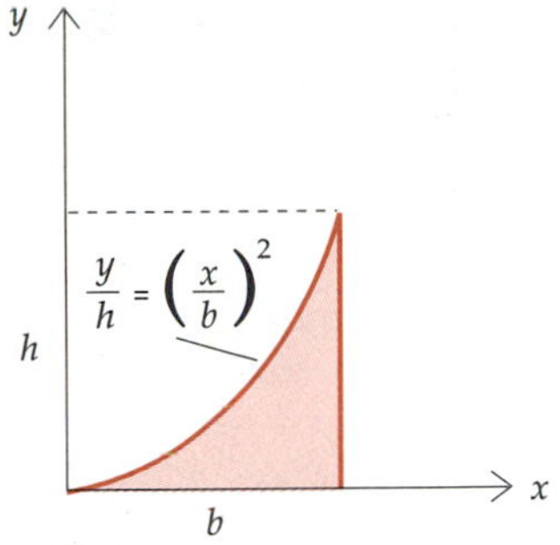

Figure P7.84

7.85 Using the results of the preceding problem and the method of composite parts, find the x-coordinate of the centroid of the shaded area in Figure P7.85.

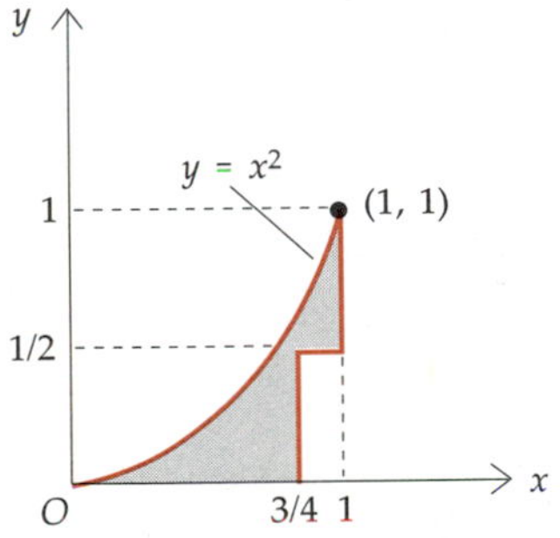

Figure P7.85

7.86 Using the results of Problem 7.84 and the method of composite parts, find the y-coordinate of the centroid of the same shaded area as in the preceding problem.

7.87 Find the ratio of R_i to R_o for which the centroid of the semicircular annulus in Figure P7.87 is at the highest point of the inner bounding circle.

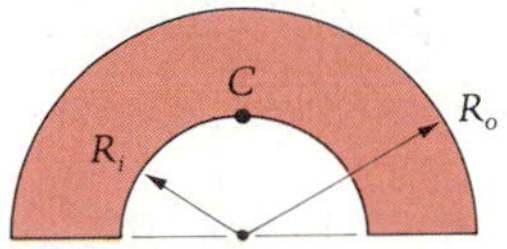

Figure P7.87

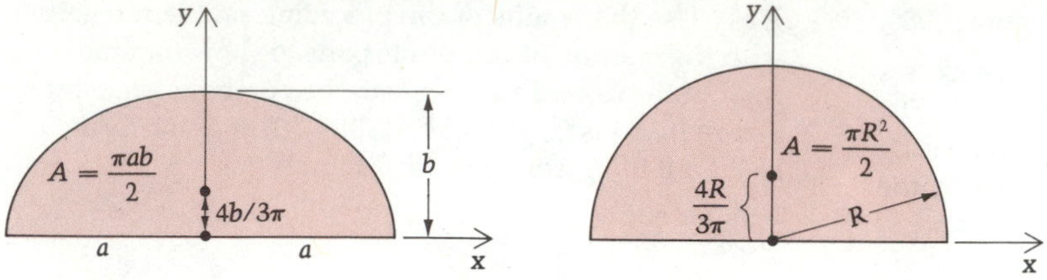

Figure P7.88a

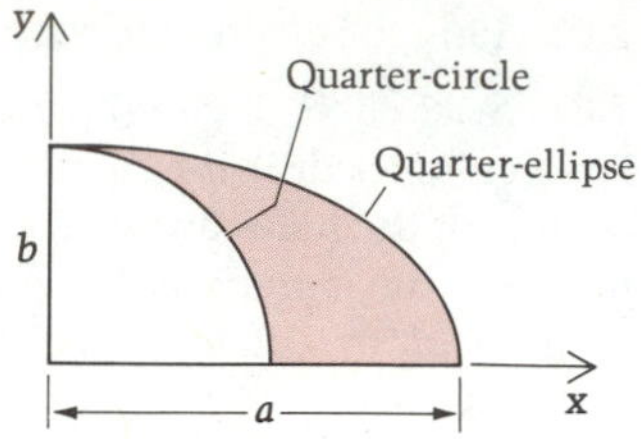

Figure P7.88b

7.88 Given the two results in Figure P7.88a for the y coordinates of the centroids, use the method of composite areas to find the y coordinate of the centroid of the shaded area in Figure P7.88b. Do you really need the areas in this problem?

7.89 Find the x coordinate of the centroid in Problem 7.88. Are the areas needed now?

7.90 In Figure P7.90 show by integration that the centroid of the circular sector is given by

$$\bar{y} = \frac{2r \sin \alpha}{3\alpha}$$

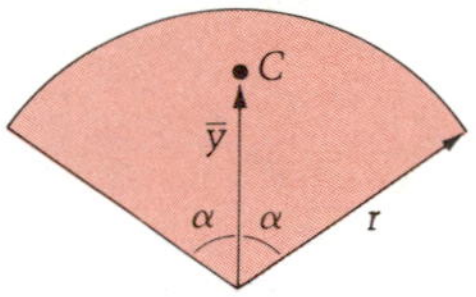

Figure P7.90

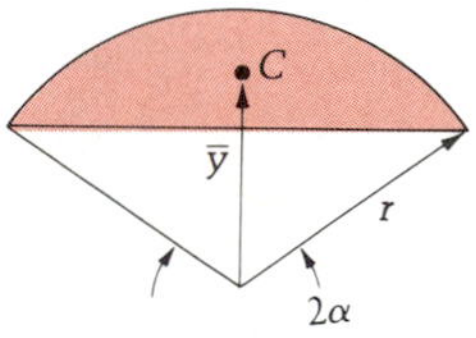

Figure P7.92

7.91 In Problem 7.90 use L'Hospital's rule to show that as the sector becomes a sliver ($\alpha \to 0$), the result for $\bar{y}$ approaches that for a triangle, as it should.

7.92 Using the result of Problem 7.90, show by composite areas (sector minus triangle) that the centroid of the shaded circular segment shown in Figure P7.92 is given by

$$\bar{y} = \frac{2r \sin^3 \alpha}{3(\alpha - \sin \alpha \cos \alpha)}$$

* Show that your answer approaches r as $\alpha \to 0$, as it must.

* **7.93** Find the centroid of the shaded area in Figure P7.93.

7.94 Find the centroid of the shaded area in Figure P7.94.

7.95 The flat circular cam in Figure P7.95 rotates about an axis $\frac{3}{8}$ in. from the center of the 6-in. radius circle. To balance the cam — that is, place the centroid of its cross section at the axis of rotation — a hole of 2-in. radius is machined through the cam. Determine the distance b to the hole center that will accomplish this.

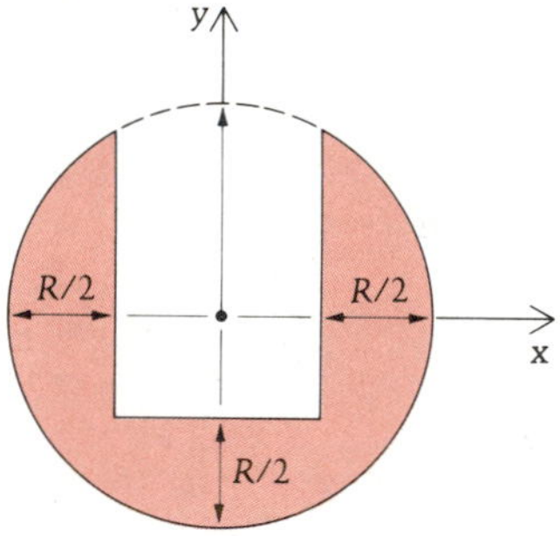

Figure P7.93

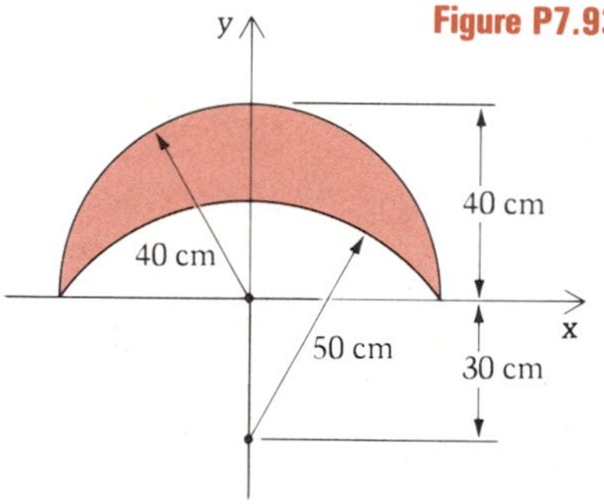

Figure P7.94

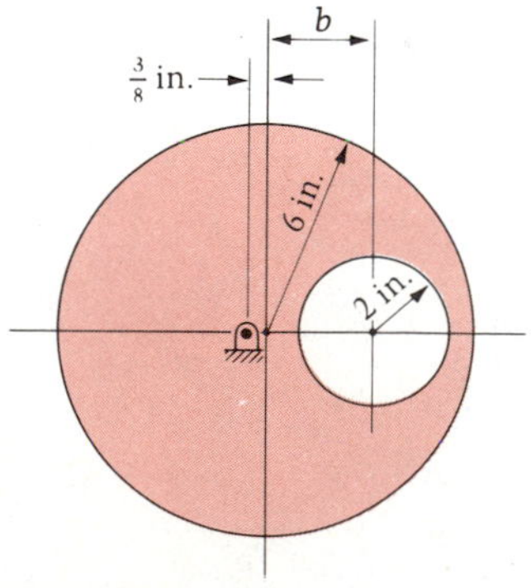

Figure P7.95

• **7.96** Approximate the y coordinate of the centroid of the (shaded) cross-sectional area of the cam, which is symmetrical about the y axis. (See Figure P7.96.)

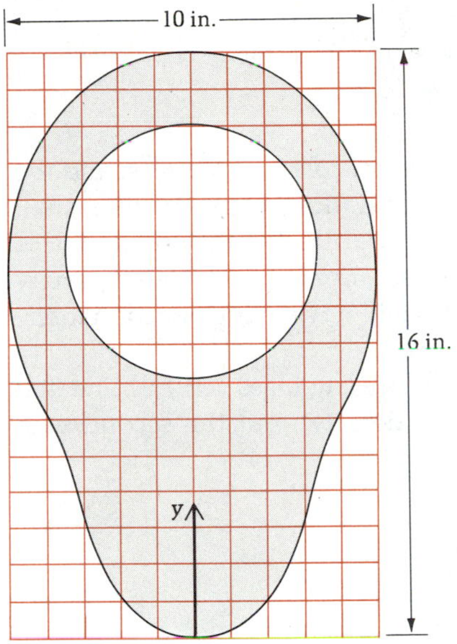

Figure P7.96

7.97 Find the centroid of the volume of the very thin piece of sheet metal in Figure P7.97.

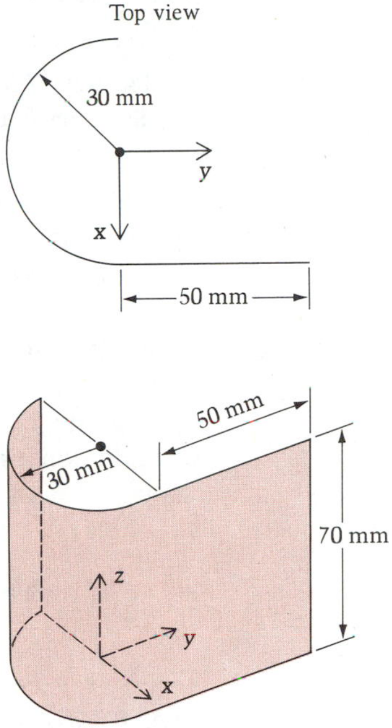

Figure P7.97

7.4 Center of Mass

The center of mass, or mass center, of a body $\mathcal{B}$ is the point C defined by Equation (7.1) when the quantity Q becomes mass:

$$m\mathbf{r}_{OC} = \int \mathbf{r}\, dm \tag{7.6}$$

where $m = \int_{\mathcal{B}} dm$ is the mass of the body and the differential mass dm is related to its volume dV through the mass density ρ according to $dm = \rho\, dV$ (see Figure 7.4). Thus

$$m\mathbf{r}_{OC} = \int \mathbf{r}\rho\, dV \tag{7.7}$$

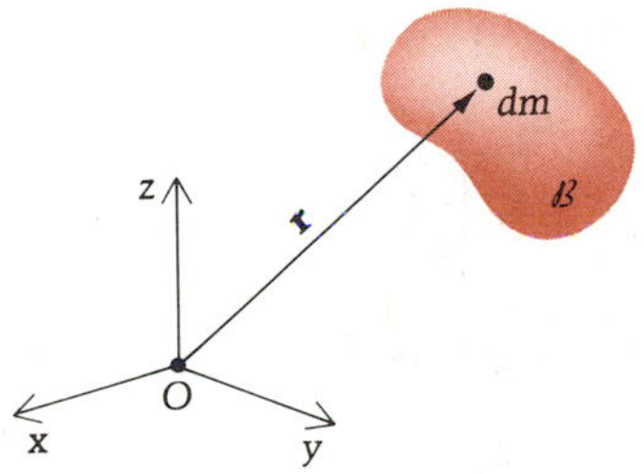

Figure 7.4

If the density ρ is constant, then $m = \rho V$, where V is the volume of body $\mathcal{B}$ and

$$\rho V\mathbf{r}_{OC} = \rho \int \mathbf{r}\, dV$$

or

$$V\mathbf{r}_{OC} = \int \mathbf{r}\, dV \tag{7.8}$$

which is seen to be the equation defining the location of the centroid of the volume of $\mathcal{B}$. *Thus the mass center always coincides with the centroid of volume for a body of constant density.*

Question 7.9 Can you think of a case in which the density is *not* constant and yet the two points coincide?

The mass center of a body plays a central role in dynamics, and the student will encounter it naturally there. In statics its significance is that, assuming a uniform field of gravity, then regardless of the orientation of a body $\mathcal{B}$ on the earth's surface, the resultant of the distributed gravity forces (weight) exerted on $\mathcal{B}$ by the earth will have a line of action intersecting the center of mass of $\mathcal{B}$.

We now consider some examples of locating mass centers. The first example contains a body with variable density, and the second uses composite parts.

Answer 7.9 A sphere whose density depends only upon the radius.

EXAMPLE 7.15

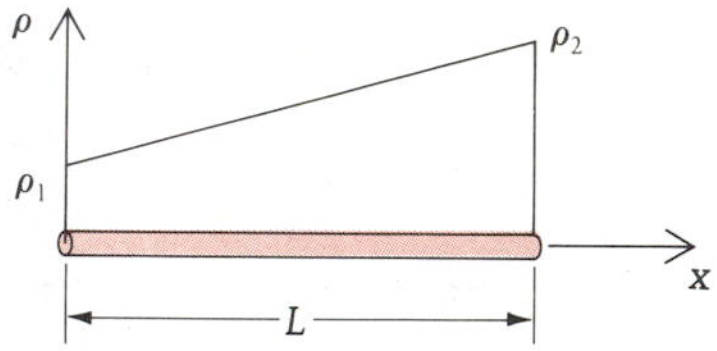

Figure E7.15

The mass density of a uniform, circular cylindrical rod of constant cross-sectional area A varies linearly with x as indicated in Figure E7.15. Find the center of mass.

Solution

The equation for the spatial variation of the density is

$$\rho = \frac{\rho_2 - \rho_1}{L} x + \rho_1$$

Therefore, with $\bar{y} = \bar{z} = 0$ by symmetry,

$$\underbrace{(\int \rho \, dV)}_{m}\bar{x} = \int \rho x \, dV$$

$$\left\{\int \left[\left(\frac{\rho_2 - \rho_1}{L}\right) x + \rho_1\right] A \, dx\right\}\bar{x} = \int \left[\left(\frac{\rho_2 - \rho_1}{L}\right) x + \rho_1\right] xA \, dx$$

$$\left[\frac{\rho_2 - \rho_1}{L} \frac{x^2}{2}\Big|_0^L + \rho_1 x\Big|_0^L\right]\bar{x} = \left(\frac{\rho_2 - \rho_1}{L}\right)\frac{x^3}{3}\Big|_0^L + \rho_1 \frac{x^2}{2}\Big|_0^L$$

$$\left[(\rho_2 - \rho_1)\frac{L}{2} + \rho_1 L\right]\bar{x} = \left(\frac{\rho_2 - \rho_1}{L}\right)\frac{L^3}{3} + \rho_1 \frac{L^2}{2}$$

Therefore,

$$\bar{x} = \left(\frac{\rho_1 + 2\rho_2}{3(\rho_1 + \rho_2)}\right) L$$

As a check, if $\rho_1 = \rho_2$, then $\bar{x} = L/2$, as it should be for a uniform rod.

Question 7.10 Does the solution check for the cases $\rho_1 = 0$ and $\rho_2 = 0$?

Answer 7.10 Yes; the answers are $\frac{2}{3}L$ and $\frac{1}{3}L$. These solutions are the same as for the thin plates  and 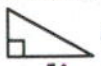 of uniform density. The solutions should match, for with these plates we have a linear variation of height instead of density.

Example 7.16

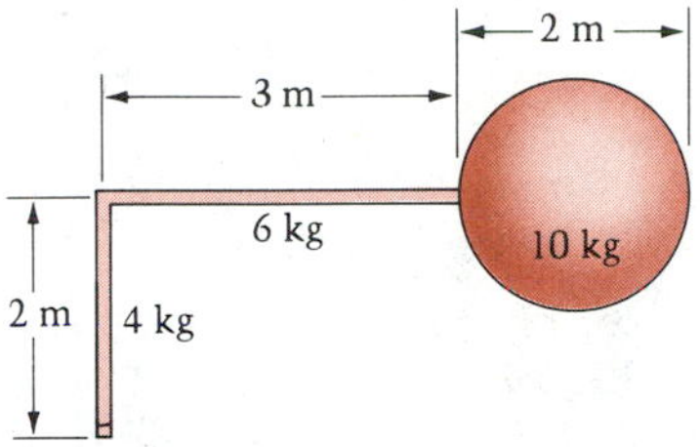

Figure E7.16a

Find the center of mass of the body shown in Figure E7.16a, composed of two uniform slender bars and a uniform sphere.

Solution

We shall use the method of composite parts. The masses and mass-center coordinates (see Figure E7.16b) are:

Body	Mass	Mass-Center Coordinates (x, y)
1	10 kg	(0, 0) m
2	6 kg	(−2.5, 0) m
3	4 kg	(−4, −1) m

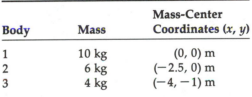

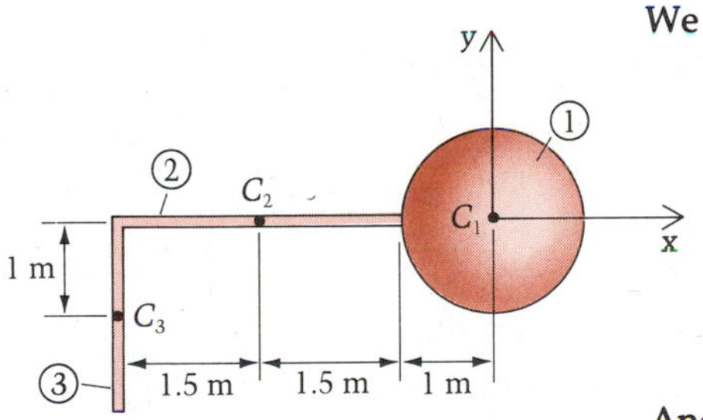

Figure E7.16b

We have, for the x coordinate of the mass center,

$$m\bar{x} = m_1\bar{x}_1 + m_2\bar{x}_2 + m_3\bar{x}_3$$

$$(10 + 6 + 4)\bar{x} = 10(0) + 6(-2.5) + 4(-4)$$

$$20\bar{x} = -31$$

$$\bar{x} = -1.55 \text{ m}$$

And for the y coordinate,

$$m\bar{y} = m_1\bar{y}_1 + m_2\bar{y}_2 = m_3\bar{y}_3$$

$$20\bar{y} = 10(0) + 6(0) + 4(-1)$$

$$\bar{y} = -0.200 \text{ m}$$

And, of course, $\bar{z} = 0$ by symmetry.

Question 7.11 In the preceding example, it turns out that the mass center lies inside the sphere. Does the mass center of a body have to be a material point of the body? If not, give an example.

Question 7.12 Give an example to illustrate the fact that the mass center need not always have the same location relative to material points of the body. *Hint*: Clearly for this to be true, the body has to be deformable!

Answer 7.11 No, it does not. A length of straight pipe has a mass center on its axis, in the space inside it.

Answer 7.12 Bend a pipe cleaner. The mass center before bending is on the axis of the pipe cleaner, while afterwards it is not.

PROBLEMS ▶ Section 7.4

7.98 Find the mass center of the body in Figure P7.98, which is a hemisphere glued to a solid cylinder of the same density, if $L = 2R$.

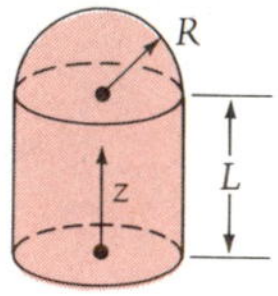

Figure P7.98

7.99 In the preceding problem, for what ratio of L to R is the mass center in the interface between the sphere and the cylinder?

7.100 In Figure P7.100 find the distance d such that the center of mass of the uniform thin wire is at the center Q of the semicircle.

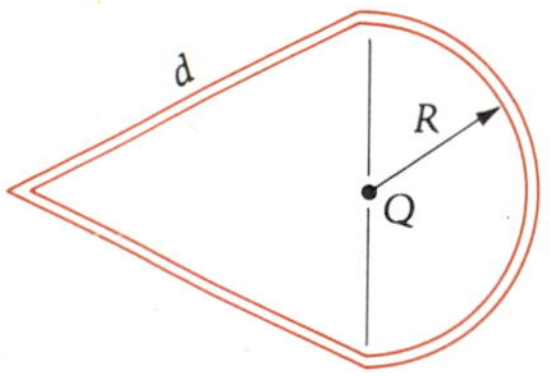

Figure P7.100

7.101 Repeat Problem 7.100 if instead of a wire, the body is a uniform thin plate with the same periphery.

7.102 Determine the center of mass of the composite body shown in Figure P7.102. The density of $\mathcal{A}$ is 2000 kg/m³, the density of $\mathcal{B}$ is 3000 kg/m³, and the density of $\mathcal{C}$ is 4000 kg/m³.

7.103 In Figure P7.103 find the center of mass of the rigid frame, comprising seven identical rigid rods, each of length ℓ.

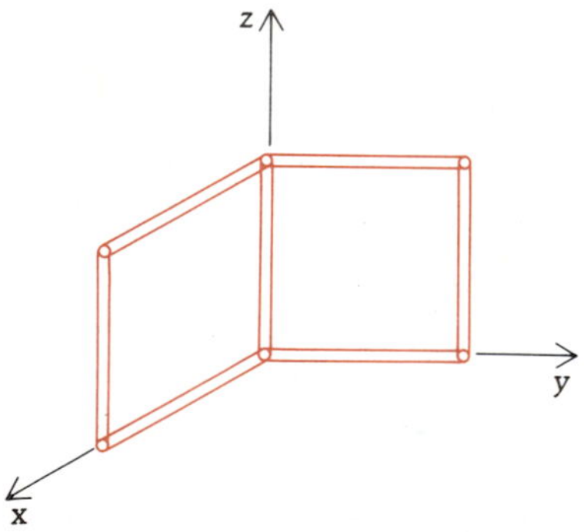

Figure P7.103

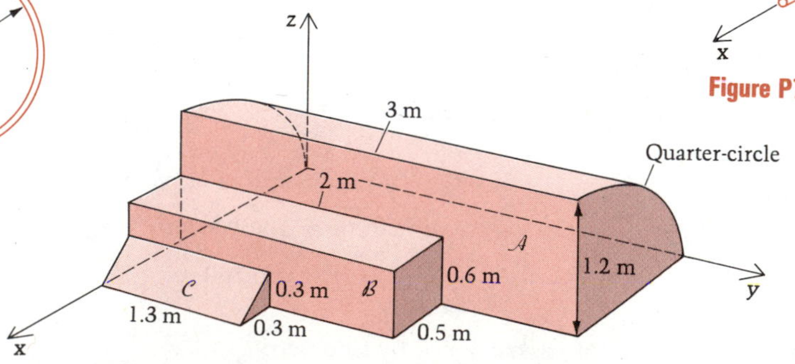

Figure P7.102

7.104 Find the center of mass of the bent bar, each leg of which is parallel to a coordinate axis and has uniform density and mass m. (See Figure P7.104).

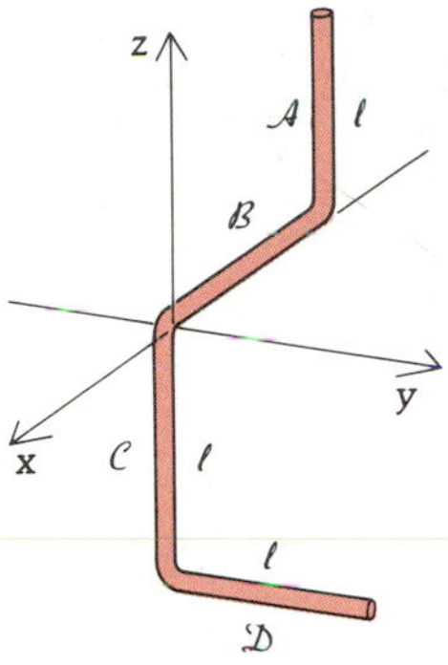

Figure P7.104

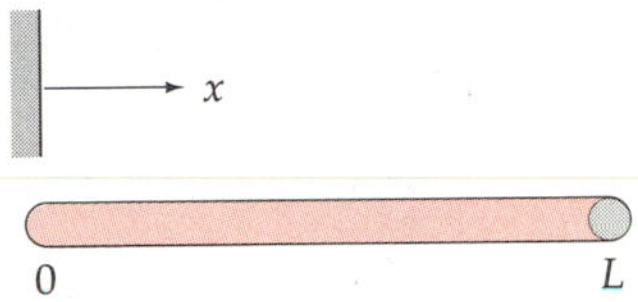

Figure P7.105

7.105 Find the mass center of the wire in Figure P7.105 if its density varies according to $\rho = \rho_o x/L$, where ρ_o is a constant, and so is the cross-sectional area A.

7.106 The density of the thin plate of thickness t in Figure P7.106 is given by $\rho = \rho_0 xy$, where ρ_0 is a constant. Find the mass center of the plate. Note the differences between your $(\bar{x}, \bar{y})$ and the answer $(\frac{2}{3}, \frac{1}{3})$ for a uniform density.

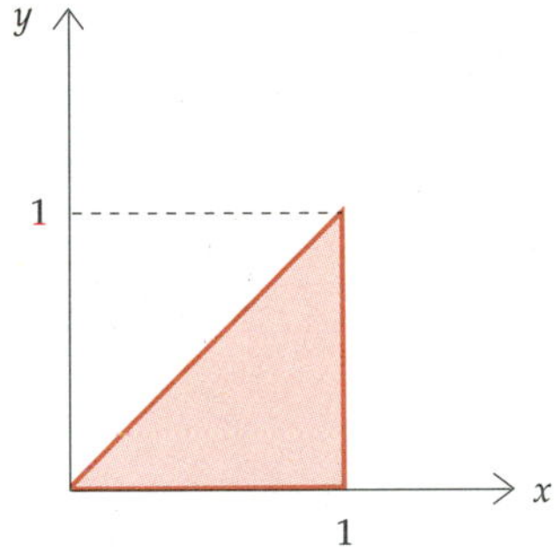

Figure P7.106

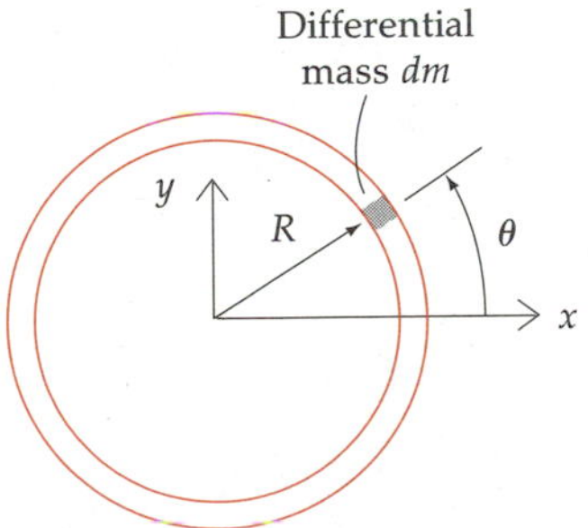

Figure P7.107

7.107 Find the mass center of the thin circular wire in Figure P7.107 if its area A is constant and its density varies according to $\rho = \rho_o\,\theta/(2\pi)$. In the figure, θ measures to a differential element of mass $dm = \rho A R\,d\theta$.

7.108 Repeat Problem 7.104 if the four legs have uniform, but different, densities, so that the masses of $\mathcal{A}$, $\mathcal{B}$, $\mathcal{C}$, and $\mathcal{D}$ are respectively, m, $2m$, $3m$, and $4m$.

7.109 See Figure P7.109.

a. How far over the edge can the can extend without falling over? It is open on the left and closed on the right, and the thickness is the same throughout.

b. How far can it extend if the closed end goes first?

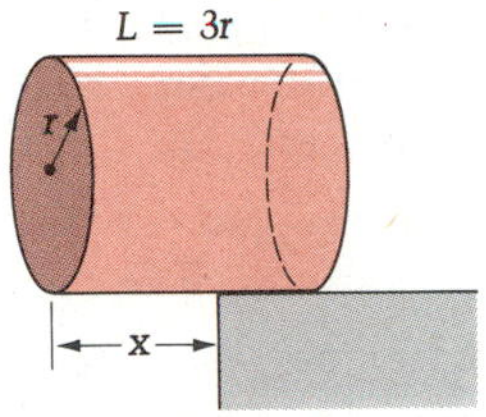

Figure P7.109

7.110 Use spherical coordinates, as suggested in Figure P7.110, to show that the center of mass of the uniform body formed by the intersection of the sphere of radius R, and the cone having vertex angle 2α, is at

$$\bar{z} = \frac{3R}{8}(1 + \cos\alpha)$$

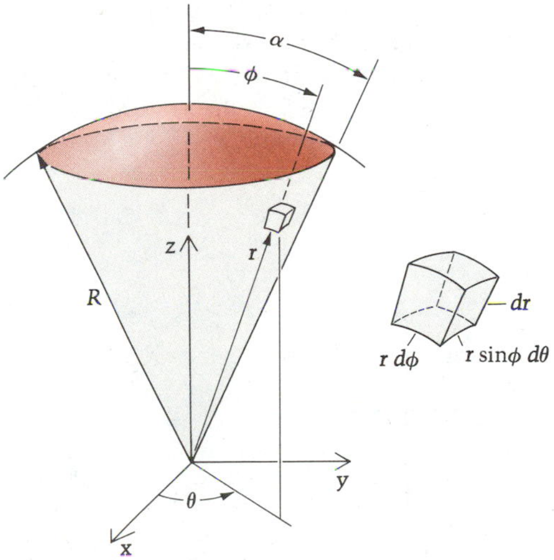

Figure P7.110

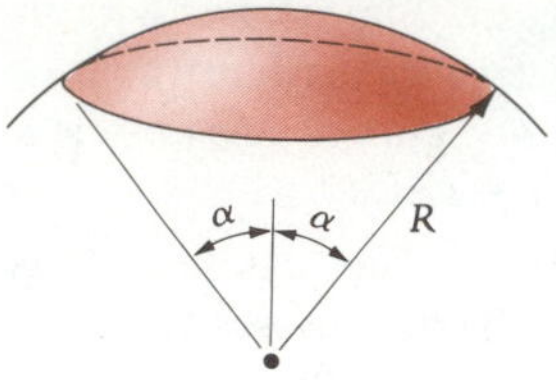

Figure P7.111

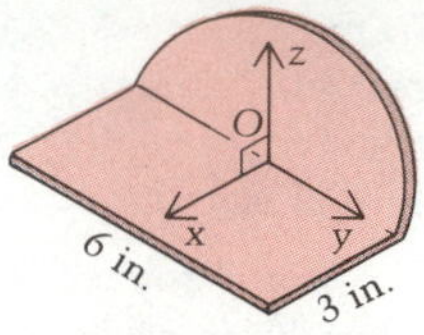

Figure P7.112

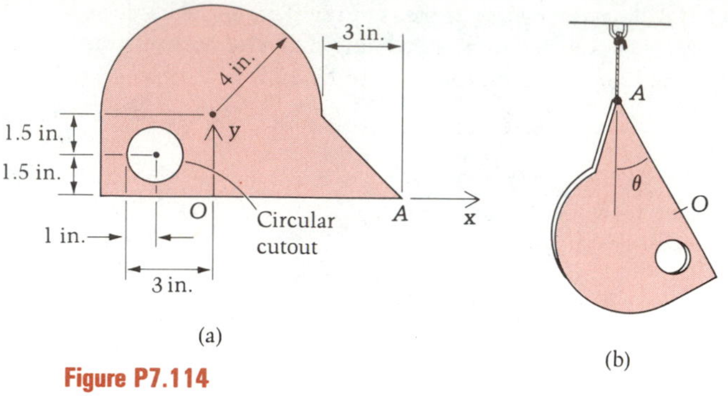

Figure P7.114

7.111 Use the results of Problem 7.110, together with the knowledge that the centroid of a cone is three-fourths the distance from vertex to base, to find the mass center of a uniform spherical cap without further integration. (See Figure P7.111.)

7.112 The body in Figure P7.112 is made of a single piece of sheet metal, cut into a rectangle and semicircle and bent at 90° as shown. Find the mass center.

7.113 A trailer carries a load of specific weight γ lb/ft^3, which varies with y (but not with z) as shown in Figure P7.113.

- **a.** Find the weight and the position of the mass center of the load.
- **b.** Find the reactions F_A, F_B, and F_C due to the load only (do not include the trailer weight in this exercise).

7.114 Find the angle between line OA and the vertical if the thin plate in Figure P7.114(a) is suspended at A by a string, as shown in the sketch (b).

7.115 The object in Figure P7.115(a) is constructed of a thin but rigid uniform semicircular ring and a right triangular plate. The ring and the plate are made of the same material. Locate, with the angle θ, the point on the ring that would contact a smooth horizontal cylinder if the object was hung on the cylinder as shown in Figure P7.115(b). The mass center C of the semicircular ring is at a distance of $2r/\pi$ from point Q.

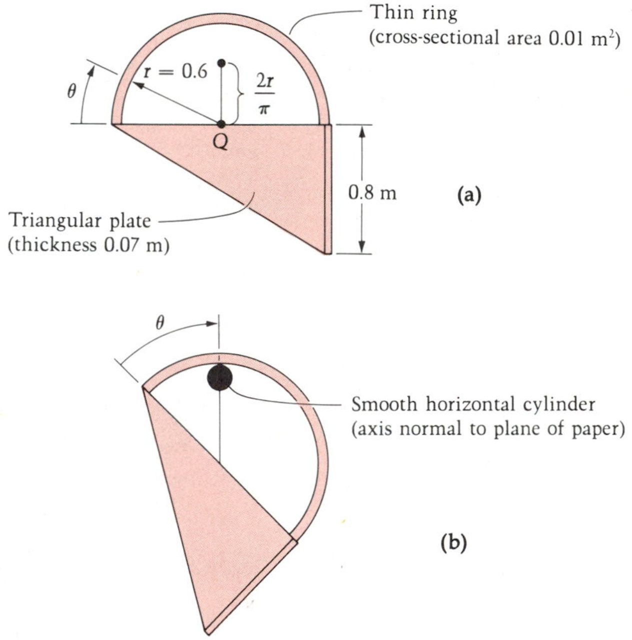

Figure P7.115

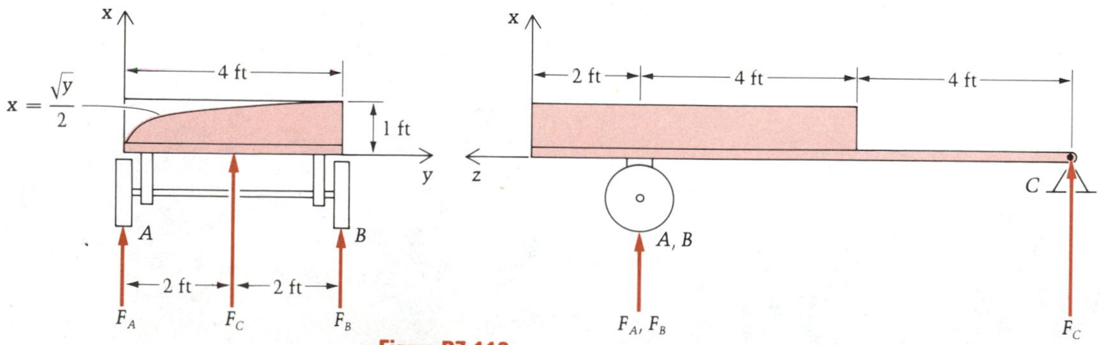

Figure P7.113

• **7.116** Shown in Figure P7.116 is a scale model of a thin parabolic shell structure. The shell's outer surface is a curve formed by revolving the indicated parabola about the y axis. Assume a constant mass density ρ and that the constant thickness t of the model is very small. Find the center of mass of the shell model. Note that because of the two assumptions you are actually seeking the centroid of the outer (curved) surface area.

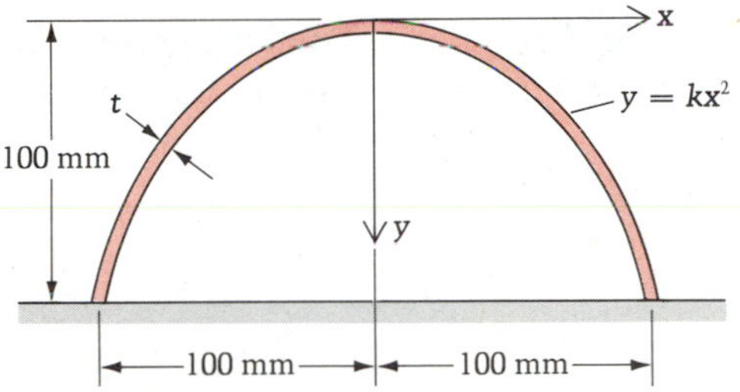

Figure P7.116

7.117 The half cylinder (weight W_C) and the stick (W_S) are attached and in equilibrium as shown in Figure P7.117. Find the x and y coordinates of the combined mass center, and use them to determine the "lean angle" ϕ.

7.118 When a structure is very thin (thickness much smaller than other dimensions) and flat, it is called a plate. If it is very thin but curved, it is called a shell. These are very important structural elements. Give an argument proving that if a plate or a shell is homogeneous, its center of mass is at its centroid of volume, which in turn is approximately located at the centroid of its (flat or curved) surface area.

• **7.119** Referring to Problem 7.118, find the center of mass of the thin conical shell of constant density shown in Figure P7.119.

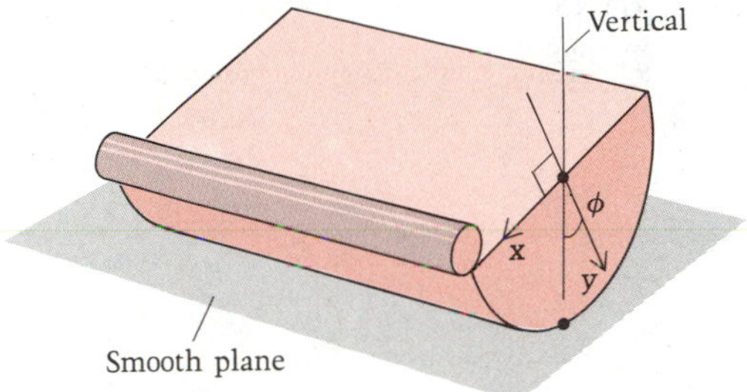

Figure P7.117

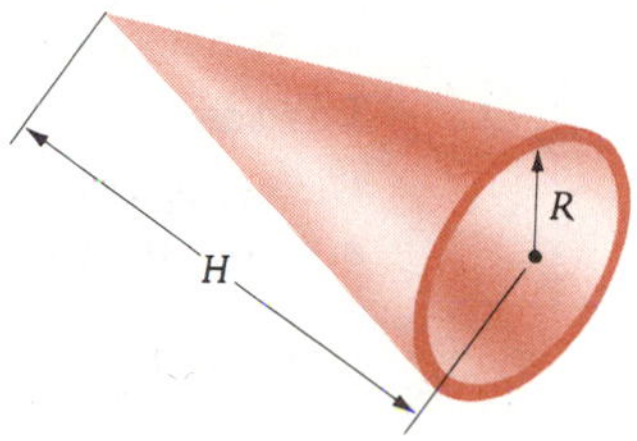

Figure P7.119

7.5 The Theorems of Pappus

There are two very old theorems of value when we are calculating either:

a. The surface area generated by revolving a plane curve around an axis (which doesn't cross the curve) in the plane, or

b. The volume generated by revolving a plane area around an axis (which doesn't cross the area) in the plane.

The two theorems are due to a Greek geometer of the third century, Pappus of Alexandria, and they are developed below.

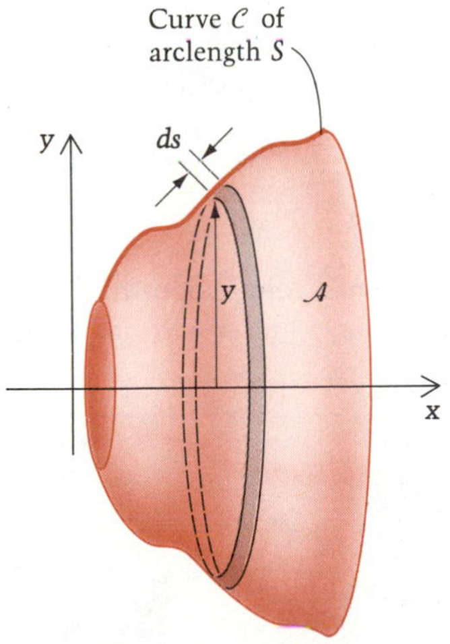

Figure 7.5

(I) The area of A of the surface $\mathcal{A}$ generated by revolving curve $\mathcal{C}$ about the x axis (see Figure 7.5) is equal to $\int 2\pi y\, ds$, or

$$A = 2\pi \int y\, ds$$

So from our knowledge of centroids,

$$A = 2\pi \bar{y} s$$

where s is the arclength of the generating curve $\mathcal{C}$ and $\bar{y}$ is the y coordinate of the centroid of $\mathcal{C}$. Thus the generated area is the arclength (s) of the

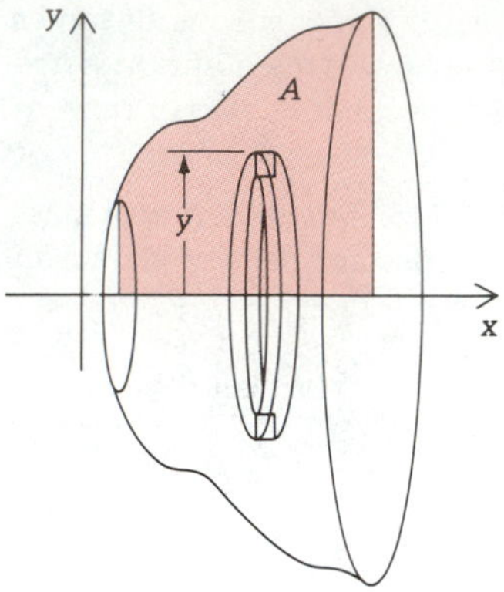

Figure 7.6

curve times the distance ($2\pi\bar{y}$) traveled during the revolving by the centroid of the curve. If the surface is formed by less than a full revolution, say by angle ϕ between 0 and 2π radians, then, of course, $A = \phi\bar{y}s$, still the arclength times the distance traveled by C.

(II) The volume V generated by revolving the shaded area A shown in Figure 7.6 about the x axis is

$$V = \int 2\pi y \, dA = 2\pi \int y \, dA = 2\pi\bar{y}A$$

where $\bar{y}$ is this time the centroid of the shaded *area*. Thus the generated *volume* is this area multiplied by the distance traveled by its centroid during the revolving. Again, for less than a full revolution, we would simply have $V = \phi\bar{y}A$ with $0 < \phi < 2\pi$.

We now consider five examples of the use of the theorems of Pappus.

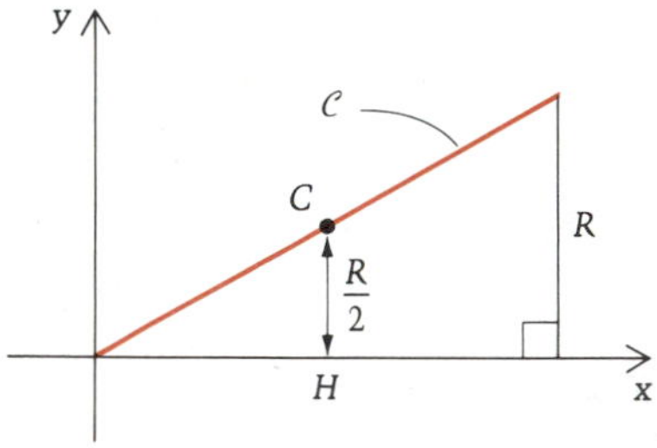

Figure E7.17

EXAMPLE 7.17

Determine the surface area of a right circular cone of radius R and height H.

Solution

Using the first theorem of Pappus, we avoid integration completely. The desired area will be generated by revolving the line $\mathcal{C}$ in Figure E7.17 about the x axis:

$$A = 2\pi\bar{y}s = 2\pi \frac{R}{2} \sqrt{R^2 + H^2}$$

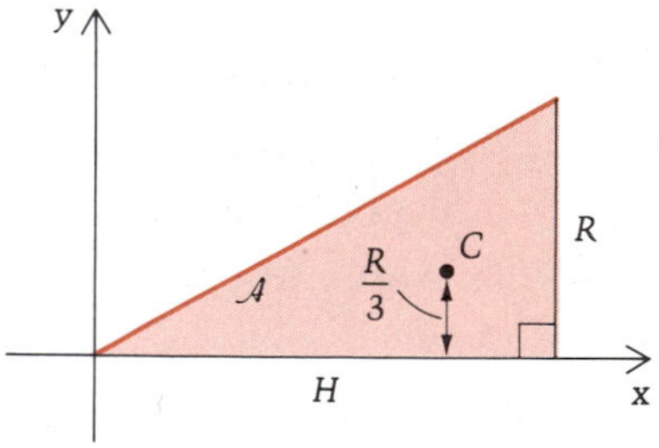

Figure E7.18

EXAMPLE 7.18

Find the volume of a right circular cone of radius R and height H.

Solution

The volume will be generated if we revolve the shaded area $\mathcal{A}$ of Figure E7.18 around the x axis. Then, by the second theorem of Pappus,

$$V = 2\pi\bar{y}A = 2\pi \frac{R}{3} \left(\frac{1}{2} HR\right) = \frac{\pi R^2 H}{3}$$

as we have seen in Example 7.7. Note that the volume of a cone is one-third that of the smallest cylinder that will enclose it — that is, the cylinder with the same height and base radius.

EXAMPLE 7.19

Find the surface area of a complete torus (doughnut).

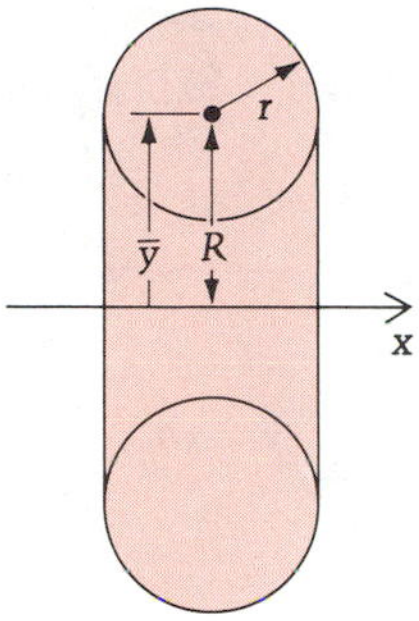

Figure E7.19

Solution

We shall generate the surface area by revolving the circle about the x axis (see Figure E7.19) and using the first theorem of Pappus:

$$A = 2\pi\bar{y}s$$

$$= 2\pi R(2\pi r)$$

$$A = 4\pi^2 Rr$$

Example 7.20

Find the volume of a hollowed torus.

Solution

We shall revolve the darker shaded area in Figure E7.20 about the y axis to generate the required volume. By the second theorem of Pappus,

$$V = 2\pi\bar{x}A = 2\pi R\pi(R_o^2 - R_i^2)$$

$$V = 2\pi^2 R(R_o^2 - R_i^2)$$

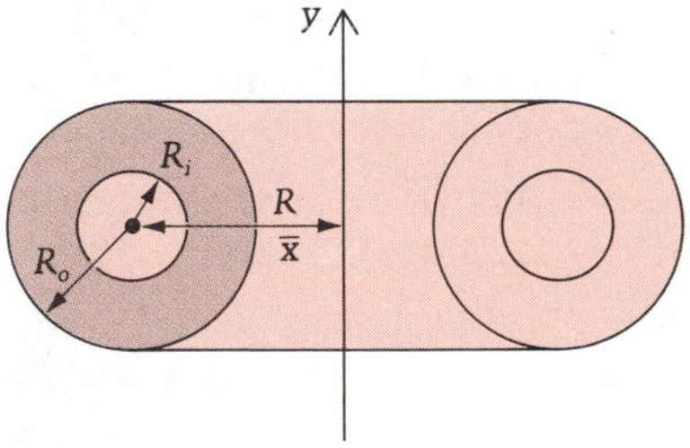

Figure E7.20

The theorems of Pappus are also very useful in finding centroids when volumes and areas (or areas and arcs) are known. We illustrate this with a final example.

Example 7.21

Find the centroid of a semicircular annulus.

Solution

Revolving the area in Figure E7.21a about x, we obtain a hollow sphere (with the cross-section shown in Figure E7.21b), and from the second theorem of Pappus,

$$V = 2\pi\bar{y}A$$

$$\underbrace{\frac{4}{3}\pi(R_o^3 - R_i^3)}_{\text{Volume of generated hollow sphere}} = 2\pi\bar{y}\underbrace{\frac{\pi(R_o^2 - R_i^2)}{2}}_{\text{Generating area}}$$

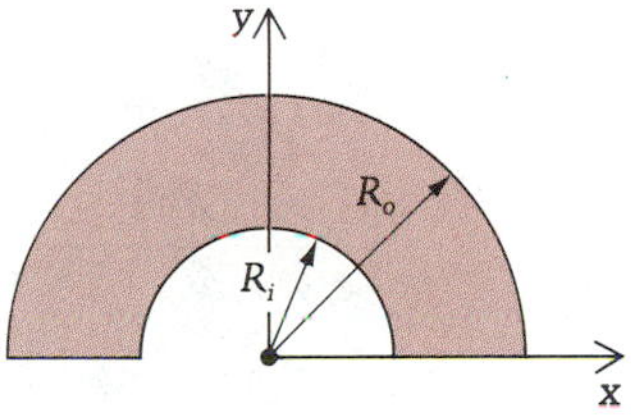

Figure E7.21a

Thus,

$$\bar{y} = \frac{4\cancel{(R_o - R_i)}(R_o^2 + R_oR_i + R_i^2)}{3\pi(R_o + R_i)\cancel{(R_o - R_i)}}$$

or

$$\bar{y} = \frac{4(R_o^2 + R_oR_i + R_i^2)}{3\pi(R_o + R_i)}$$

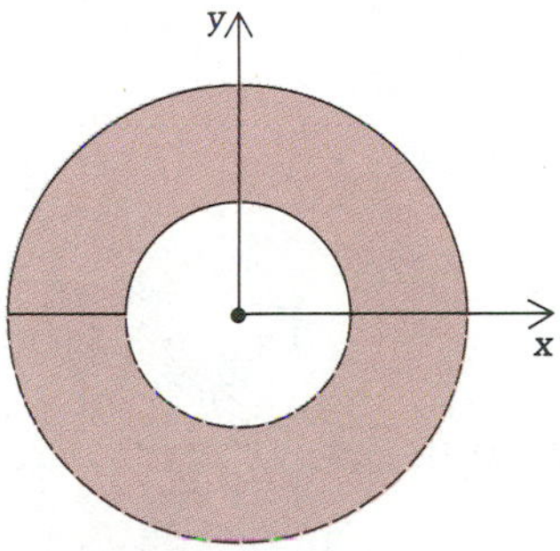

Figure E7.21b

Question 7.13 Does the answer check out in the two limiting cases $R_i \rightarrow 0$ and $R_i \rightarrow R_o$?

Answer 7.13 Yes. As $R_i \rightarrow 0$, $\bar{y} \rightarrow 4R_o/3\pi$ and as $R_i \rightarrow R_o$, $\bar{y} \rightarrow 4(3R_o^2)/[3\pi(2R_o)] = 2R_o/\pi$. We obtained these results in Examples 7.5 and 7.2, respectively.

PROBLEMS ▶ Section 7.5

7.120 Find the surface area (exterior plus interior) of the object of revolution shown in Figure P7.120. It has a rectangular cross section with a circle removed.

7.121 Find the volume of the object in Problem 7.120.

7.122 In Figure P7.122 find the surface area of the body of revolution.

7.123 Find the volume of the body in Problem 7.122.

7.124 Find the surface area (exterior plus interior) of the tube, which spans 100° of a circle as shown in Figure P7.124.

7.125 Find the volume of the tube in Problem 7.124.

7.126 The 10-m diameter parabolic antenna in Figure P7.126 has an f/D ratio (focal distance to diameter) of 0.28. Find the surface area of one side of the thin shell.

7.127 Find the volume of water that the parabolic shell of revolution in Problem 7.126 could hold.

7.128 The shaft in Figure P7.128 ends in a conical thrust bearing. Find the surface area of the truncated cone if its upper and lower radii are 2 in. and 1 in. and its height is 1.5 in.

7.129 Find the volume of the truncated cone in Problem 7.128.

7.130 Find the centroid of the semicircular area in Figure P7.130 by revolving it about axis x to generate a spherical volume, using the second theorem of Pappus.

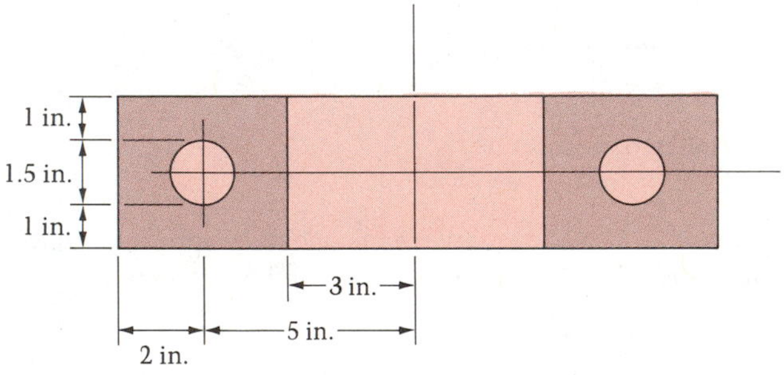

Figure P7.120

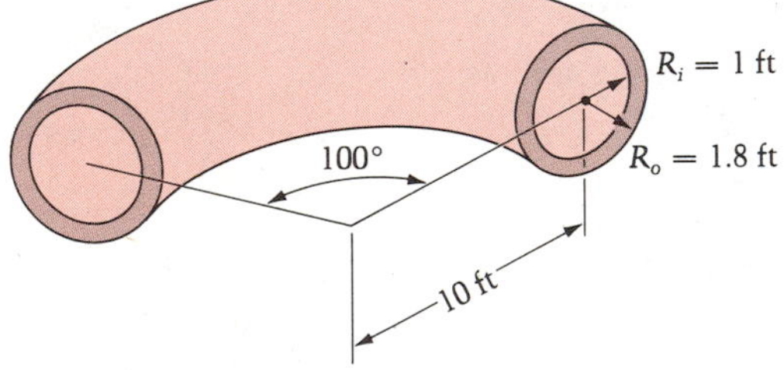

Figure P7.124

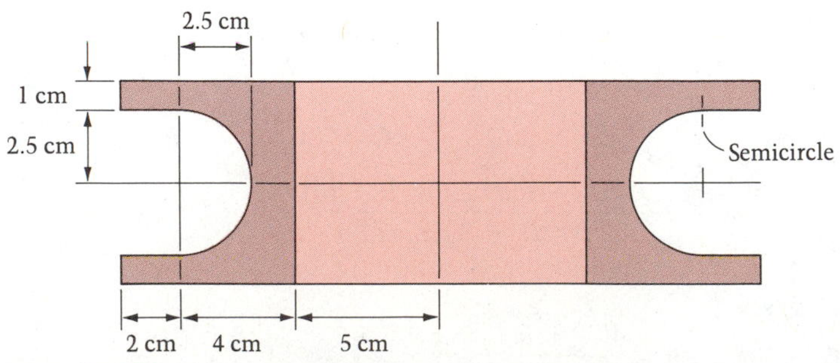

Figure P7.122

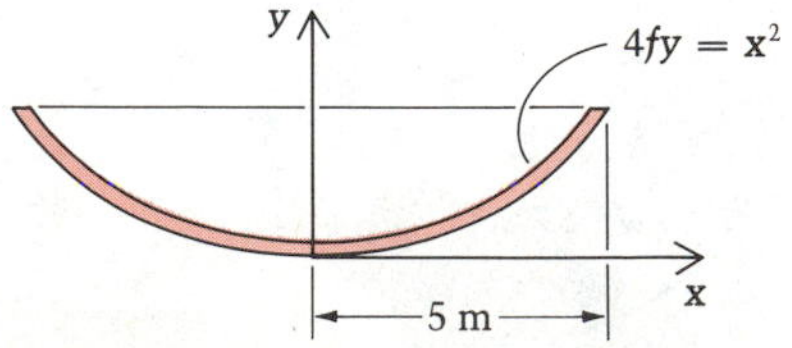

Figure P7.126

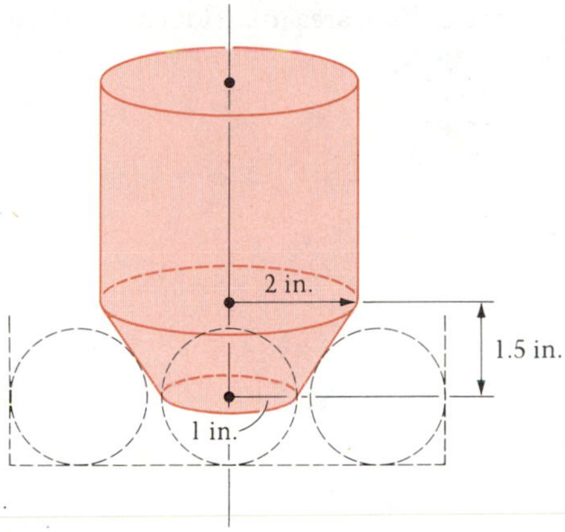

Figure P7.128

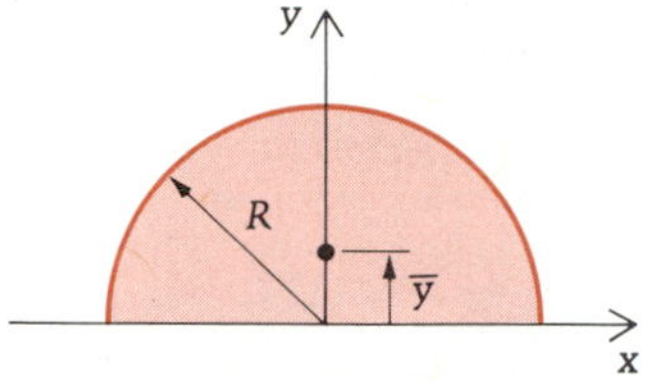

Figure P7.130

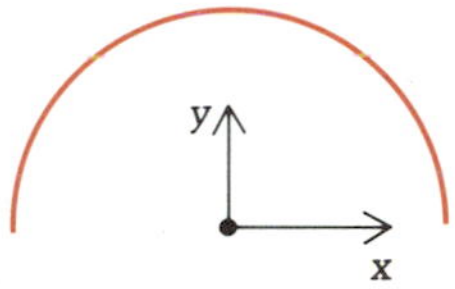

Figure P7.131

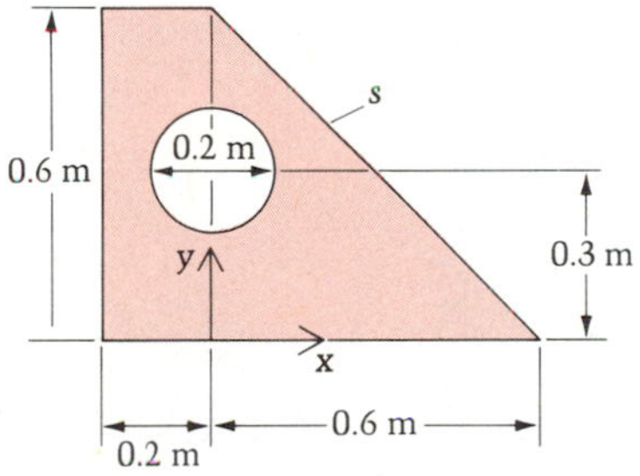

Figure P7.134

7.131 Find the centroid of the semicircle (arc) in Figure P7.131 by revolving it about axis x to generate the surface area of a sphere.

7.132 Find the area of the surface formed by revolving an equilateral triangle of side s about one of its sides.

7.133 Find the volume generated by the revolution described in Problem 7.132.

7.134 Find the volume of the three-dimensional figure generated by revolving the area about the line $x = 0.8$ m. See Figure P7.134.

7.135 Repeat Problem 7.134 but with the revolution about the line $y = 0$.

7.136 Find the area generated by the slanted line segment s during the revolution of Problem 7.134.

7.137 Find the area generated by the slanted line segment s during the revolution of Problem 7.135.

7.138 Find the approximate volume of the thin, half-toroidal shell shown in Figure P7.138.

7.139 Refer to Problem 7.116. Find $s\bar{x}$ for the parabolic segment between the points (0, 0) and (100, 100) mm. Then use the first theorem of Pappus to determine the outer surface area of the shell.

7.140 In the preceding problem, find $A\bar{x}$ for the area bounded by $x = 0$, $y = 100$ mm, and the segment of the curve between (0, 0) and (100, 100) mm. Use this with the second theorem of Pappus to determine the volume enclosed within the inner surface of the shell.

7.141 Find the shaded surface area of the hollow axisymmetric shell section shown in Figure P7.141.

7.142 In Problem 7.141 find the volume bounded by the inside of the shell and the planes $x = 0$ and $x = 3$ in.

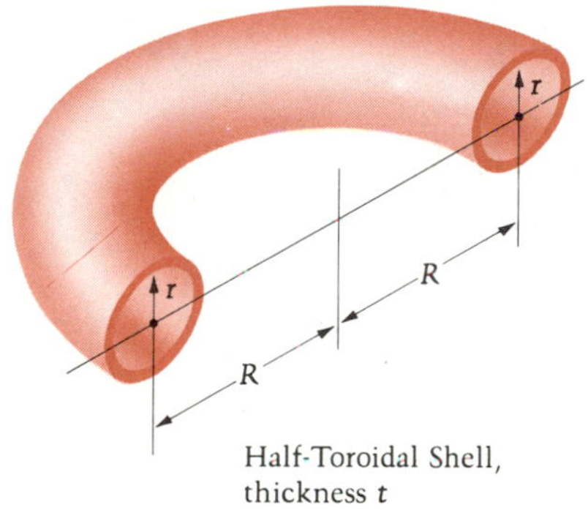

Figure P7.138

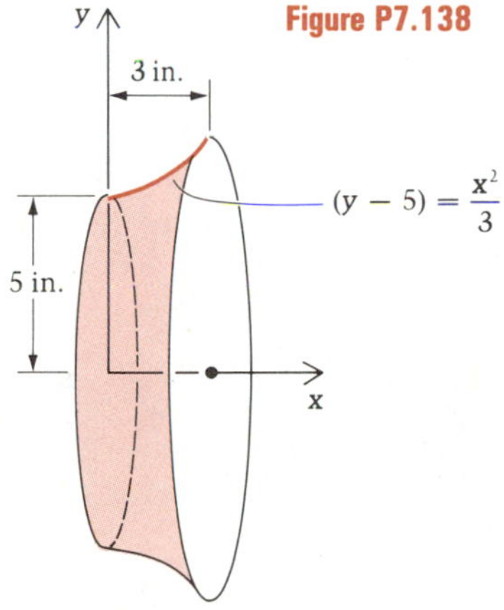

Figure P7.141

Problems 7.143–7.147 deal with various parts of a thin half-toroidal shell.

7.143 Find the shaded surface area (one side) of the shell shown in Figure P7.143.

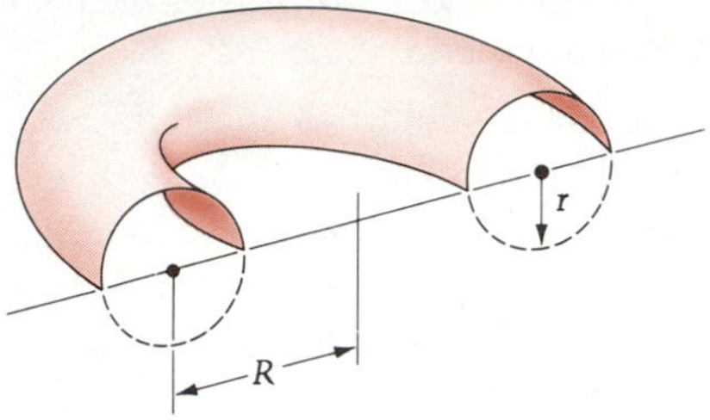

Figure P7.143

7.144 Find the shaded surface area (one side) of the shell shown in Figure P7.144.

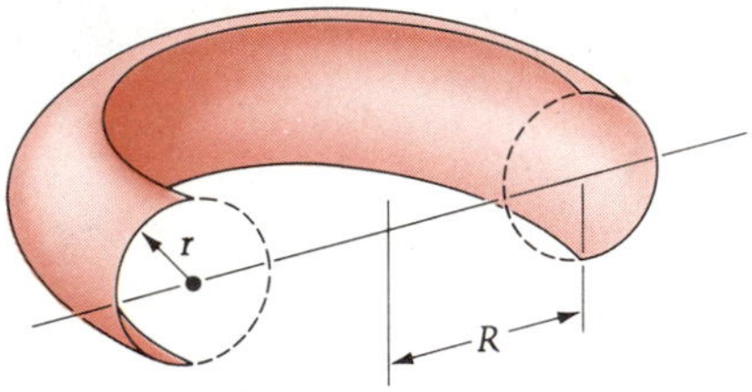

Figure P7.144

7.145 Find the shaded surface area (one side) of the shell shown in Figure P7.145.

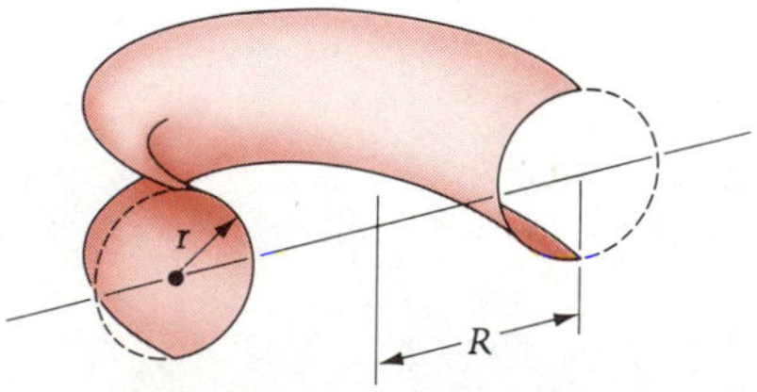

Figure P7.145

7.146 Find the shaded surface area (one side) of the shell shown in Figure P7.146.

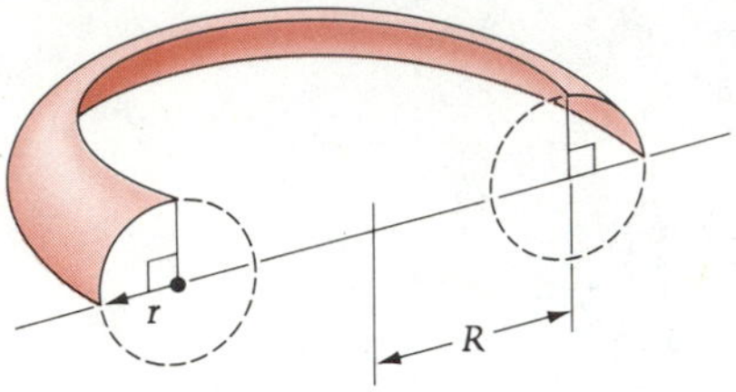

Figure P7.146

7.147 Find the shaded surface area (one side) of the shell shown in Figure P7.147.

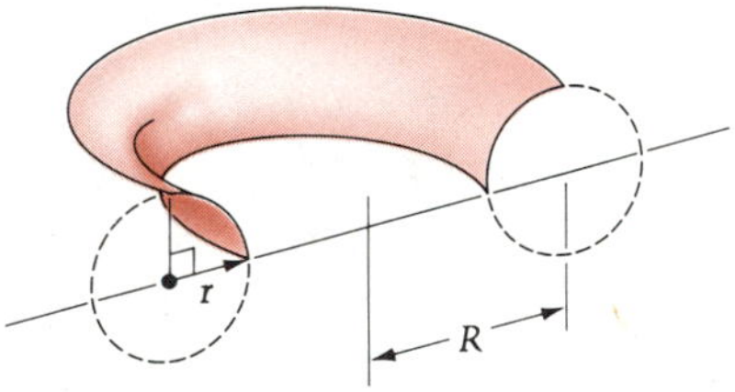

Figure P7.147

7.148–7.152 Find the volumes enclosed by the shells of Figures P7.148–P7.152, respectively.

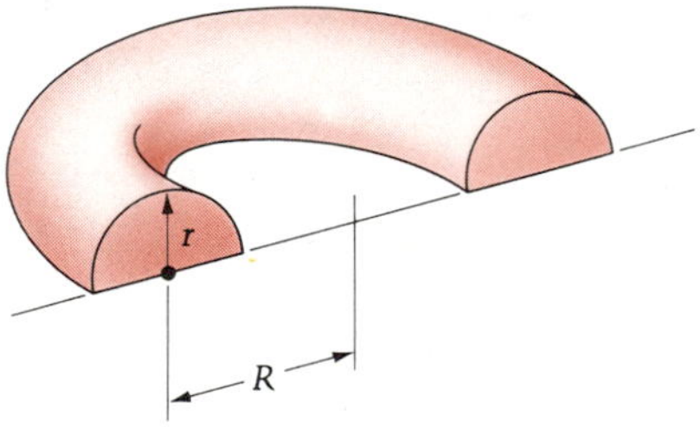

Figure P7.148

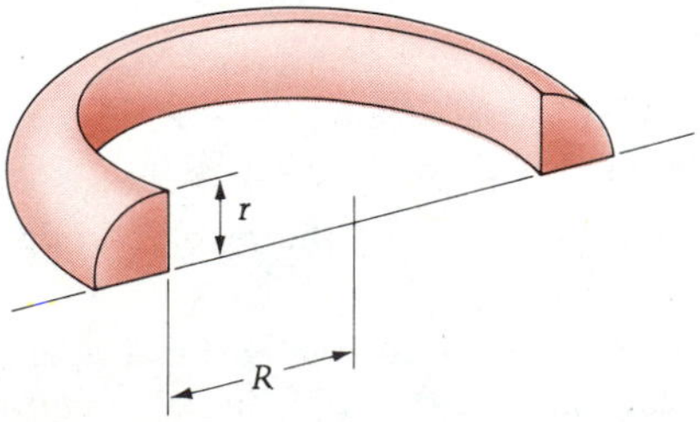

Figure P7.149

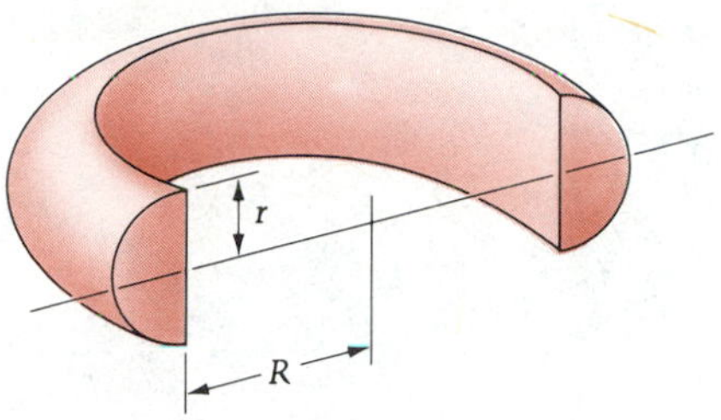

Figure P7.150

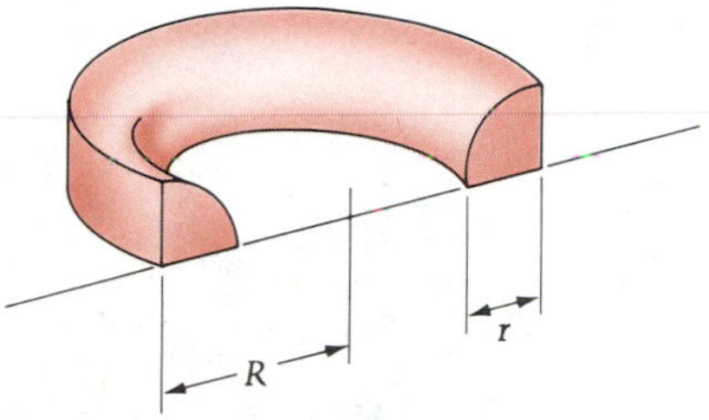

Figure P7.151

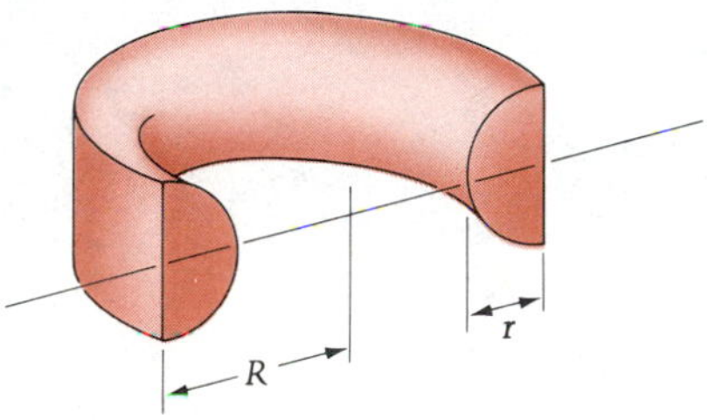

Figure P7.152

7.153 The dam spans 90°, as shown in Figure P7.153. Find the volume of concrete that was required to build it.

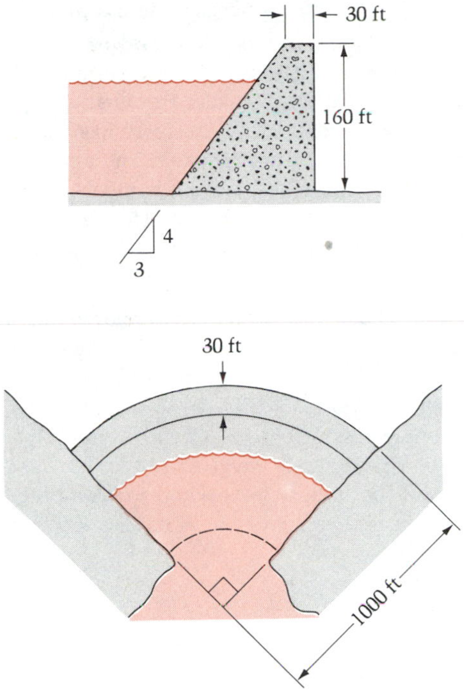

Figure P7.153

COMPUTER PROBLEMS ▸ Chapter 7

7.154 Write a computer program that will accept any number of triplets of numbers $A_i, \bar{x}_i, \bar{y}_i$, where A_i is a plane area and $(\bar{x}_i, \bar{y}_i)$ are the x and y coordinates of its centroid. The program is to read N (the number of areas making up a composite area), then read N triplets $(A_i, \bar{x}_i, \bar{y}_i)$, then compute and print the total area A and the coordinates of its centroid. Use the program to find the centroid of the area in Figure P7.154.

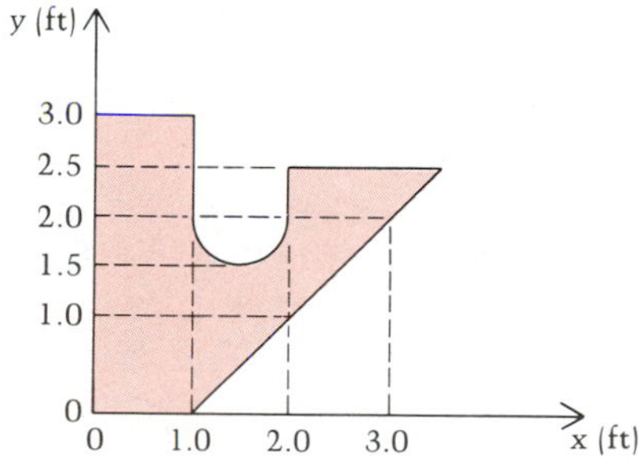

Figure P7.154

* **7.155** A crossed log-periodic dipole array (antenna) looks like a Christmas tree (see Figure P7.155a) with tubes extending in four directions from a mast. One of the four sets of tubes is shown in Figure P7.155b. Not shown are identical sets downward and both into and out of the paper. The figure shows a simplified model of such an antenna; the tubes actually get smaller from left to right and the mast is in sections that also decrease in size.

  Find the mass center of the antenna. The x coordinates of the elements (in inches) are the numbers below the mast. Assume the elements are attached to the outside of the mast. Would the mass center be the same point if the entire antenna was covered with an inch of ice?

* **7.156** A man built a basketball goal for his children out of a 15-ft steel pipe weighing 3.03 lb/ft, and having an inside diameter of 3.33 in. The goal vibrated too much when the basketball hit the backboard, so the man filled the pipe with concrete, which increased the bending stiffness of the pole. The concrete weighed 144 lb/ft³. How full (percent) was the hollow pipe when the combined mass center of pipe and concrete was at its lowest point?

Figure P7.155a

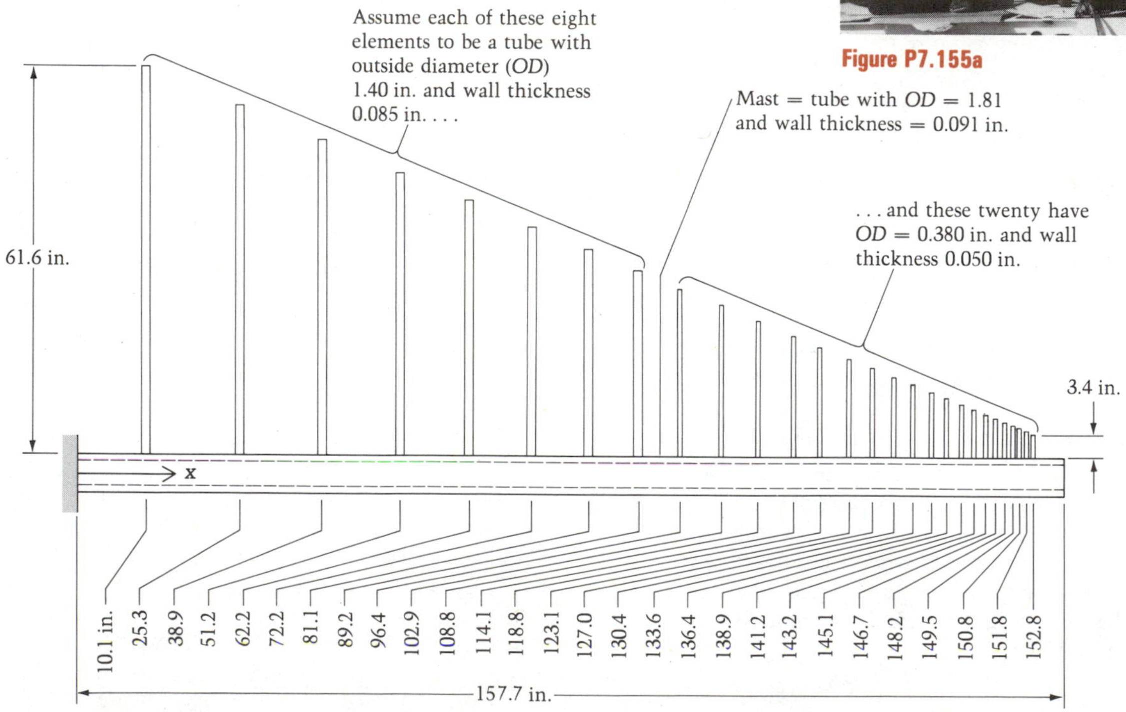

Figure P7.155b

(See Figure P7.156.) *Hint:* Write a program that will list or plot the mass center height versus concrete height, and pick off the answer from the list or plot. Check your answer by using calculus to obtain the exact solution.

* **7.157** The density of the earth varies greatly through its depth. It can be fairly well approximated by a graph consisting of straight lines through the following points:

Depth from Surface of Earth (miles)	Density (g / cc)
0	3.50
400	4.30
1000	5.20
1800	5.70
1800 (jump)	9.60
2500	10.8
3100	11.7
3300	14.0
3500	15.0
3960	16.3

Assuming spherical symmetry, write a computer program to calculate the mass of the earth, and use that result to calculate the average density of the earth.

Figure P7.156

SUMMARY ▶ Chapter 7

In this chapter we learned (in Section 7.2) that the centroid of arclength s, area A, or volume V, as well as the center of mass m, when defined over a region $\mathcal{R}$ can all be given by the same formula:

$$\mathbf{r}_{OC} = \frac{\int \mathbf{r} dQ}{\int dQ} = \frac{\int \mathbf{r} dQ}{Q}$$

in which Q represents s, A, V, or m; $\mathbf{r}$ is a vector from an origin O to a differential element of Q, and C is the "center of Q," meaning the centroid of s, A, or V or the mass center, as the case may be.

We usually work with scalar components of the above equation, such as $A\bar{x} = \int x dA$, where $\bar{x}$ is the x-component of the vector $\mathbf{r}_{OC}$ from the origin O to the centroid (of area in this case) C.

We studied the Method of Composite Parts in Section 7.3. This extremely useful procedure allows us to find the centroid of a composite area* when we know the areas and centroids of two or more parts that constitute the entire (or "overall") area. For example, the x-component of the centroid of an area made up of two areas A_1 and A_2 is given by

* or arclength or volume or mass, as the case may be.

$$\bar{x} = \frac{A_1\bar{x}_1 + A_2\bar{x}_2}{A_1 + A_2},$$

where $\bar{x}_1$ and $\bar{x}_2$ are the respective centroids of the areas A_1 and A_2.

The center of mass, covered in Section 7.4, differs in complexity from centers (centroids) of arclength, area, and volume because of the presence of the density function $\rho = \rho(x, y, z)$. The mass center is defined by

$$\mathbf{r}_{OC} = \frac{\int \mathbf{r}dm}{\int dm} = \frac{\int \mathbf{r}\rho dV}{\int \rho dv}$$

so that if ρ is constant ("uniform density"), it cancels and the center of mass is the same point as the centroid of volume. If ρ is not constant, however, we must include it in the integrand of the numerator and denominator, and the calculation is ordinarily harder.

The Theorems of Pappus were studied in Section 7.5. They are very useful when calculating either (a) the surface area generated by revolving a plane curve around an axis (which doesn't cross the curve) in the plane, or (b) the volume generated by revolving a plane area around an axis (which doesn't cross the area) in the plane. The formulae which correspond to (a) and (b) are, respectively:

$$\text{(a)} \qquad A = 2\,\pi \int y\; ds = 2\,\pi\bar{y}s$$

and

$$\text{(b)} \qquad V = 2\,\pi \int y\; dA = 2\,\pi\bar{y}A$$

These formulae allow us to use known centroidal locations to compute surface areas and volumes of objects of (full or partial) revolution about an axis. If the revolution is less than complete, say through ϕ radians, then we simply replace the 2π in the above formulae (Theorems of Pappus) by ϕ.

REVIEW QUESTIONS ▶ Chapter 7

True or False?

1. The centroid of an area depends upon which reference point is used to locate it in the defining equation $A\mathbf{r}_{OC} = \int \mathbf{r}\; dA$.
2. The centroid of a quantity Q is the point where all of Q could be *concentrated* with the same resulting first moment as has the actual distribution of Q.
3. All centroid calculations for areas require two integrations; centroid calculations for volumes require three integrations.
4. Any line through the centroid of an area divides the area into two equal parts.
5. The centroid of a semicircle is farther from the circle center than is the centroid of the semicircular area contained within it.

6. Suppose that in an arbitrary triangle we drop a perpendicular from a vertex V to the opposite base (line ℓ, extended if necessary) and find that the distance from V to ℓ is 9 in. The centroid of the area of the triangle is then 3 in. from ℓ.
7. If an area A is made up of two parts A_1 and A_2 such that $A_1 + A_2 = A$, then the x coordinate of its centroid is given by $\bar{x} = (A_1\bar{x}_1 + A_2\bar{x}_2)/(A_1 + A_2)$.
8. Finding centroids of areas with cutouts is often simplified by the method of composite parts.
9. The center of mass of a body has to be a material (physical) point of the body.
10. The mass center of a rigid body is always the same point of the body (or of a rigid extension of the body) regardless of the position of the body.
11. The centroid of the volume of a body coincides with its center of mass only if its density is constant.
12. The integral $\int_{\mathcal{B}} \mathbf{r}\, dm$ equals the mass of $\mathcal{B}$ multiplied by the vector from the origin of $\mathbf{r}$ to the mass center of $\mathcal{B}$.
13. The use of the Theorem of Pappus always involves the revolution of a curve or an area about a given axis in the same plane.

Answers: 1. F 2. T 3. F 4. F 5. T 6. T 7. T 8. T 9. F 10. T 11. F 12. T 13. T

8 Inertia Properties of Plane Areas

8.1 Introduction
8.2 Moments of Inertia of a Plane Area
8.3 The Polar Moment of Inertia of a Plane Area
8.4 The Parallel-Axis Theorem (or Transfer Theorem) for Moments of Inertia/The Radius of Gyration
Development of the Parallel Axis Theorem
Radius of Gyration
8.5 The Method of Composite Areas
8.6 Products of Inertia of Plane Areas
8.7 The Parallel-Axis Theorem for Products of Inertia
8.8 Moments and Products of Inertia with Respect to Rotated Axes Through a Point/Mohr's Circle
Developing the Equation of Mohr's Circle
Using Mohr's Circle
Summary
Review Questions

8.1 Introduction

In this chapter we study the inertia properties of plane areas. One reason for studying this topic in statics is that these properties arise in the formulas for locating the resultant of hydrostatic pressure forces on a submerged body (which we shall examine in Section 9.2). A more important reason for this study is that it is sometimes considered a prerequisite for courses in strength of materials (or deformable bodies), which follow statics. In these later courses the student will find that stresses in a transversely loaded beam are, under special but important circumstances, inversely proportional to a moment of inertia of the cross-sectional area of the beam. Deflections of the beam will likewise be inversely proportional to this moment of inertia, which forms part of the resistance to bending of the beam. Similarly, the "polar moment of inertia" is a factor in the resistance of a shaft to torsion, or twisting.

The four sections 8.2–8.5 of this chapter can be read by a student who is familiar with no more than single integration. These are the sections normally needed in the first course in the mechanics of deformable solids. The last three sections, however, utilize double integrals when dealing with products of inertia.

Moments and products of inertia of *mass* are needed in dynamics; we shall cover this related topic in our second volume at the point where the subject arises naturally.

8.2 Moments of Inertia of a Plane Area

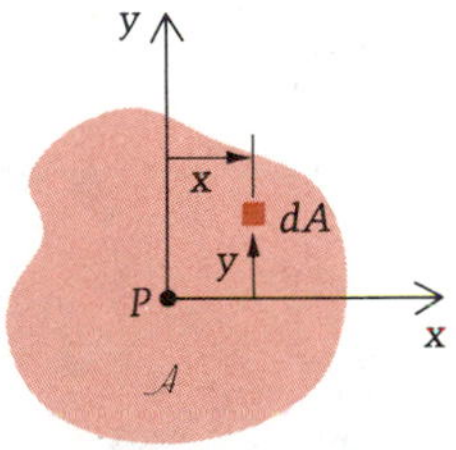

Figure 8.1

For the plane area shown in Figure 8.1, the moments of inertia* with respect to the x and y axes are defined to be

$$I_x = \int_{\mathcal{A}} y^2 \, dA \tag{8.1}$$

and

$$I_y = \int_{\mathcal{A}} x^2 \, dA \tag{8.2}$$

These definitions illustrate why a moment of inertia is sometimes called a "*second* moment" — because of the square of the distance from the x axis for I_x (and from the y axis for I_y). We have seen "first moments" in Chapter 7 in relation to the concept of centroid. Because a moment of inertia is made up of areas multiplied by squares of distances, it has the dimension (length)^4.

* Many prefer to use the term "second moments of area," feeling that the word "inertia" suggests mass and should be reserved for similar integrals that reflect the mass distribution of a body. Nevertheless, the terms "area moment of inertia" and "moment of inertia of area" are widely used in texts on the mechanics of deformable solids.

Equations (8.1) and (8.2) also tell us that a moment of inertia is always positive and is a measure of "how much area is located how far" from a line. If we wish to be specific about the origin of the x and y coordinates, we may write, for example, I_{x_C} if the origin is the centroid, or I_{x_P} if the origin is some other point P.

We now proceed to use the above definitions to find moments of inertia of several common shapes in the following examples.

EXAMPLE 8.1

Find the moments of inertia of the rectangular area of Figure E8.1a about the centroidal x and y axes.

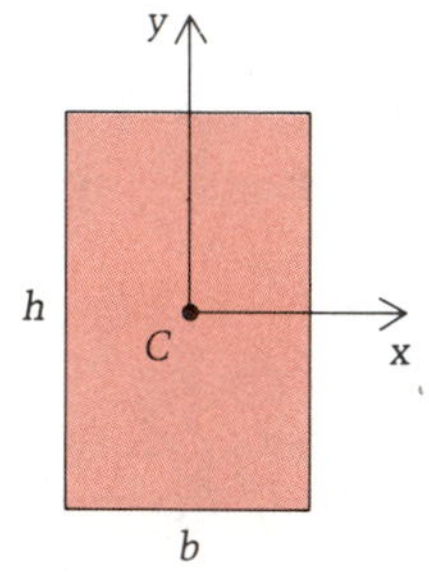

Figure E8.1a

Solution

To find I_{x_C}, we need the integral $\int y^2\,dA$. Using the horizontal strip shown in Figure E8.1b for our differential area dA, and noting that the y coordinate is the same for all parts of the strip, we get

$$I_{x_C} = \int y^2\,dA = \int_{-h/2}^{h/2} y^2(b\,dy)$$

$$= \frac{by^3}{3}\bigg|_{-h/2}^{h/2}$$

$$= \frac{bh^3}{12}$$

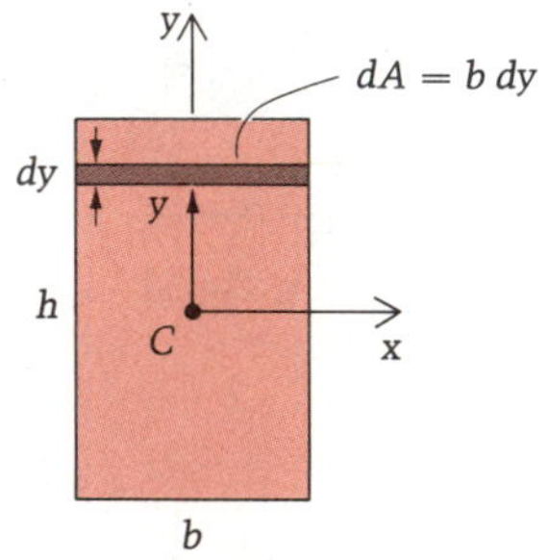

Figure E8.1b

A similar integration with $dA = h\,dx$ as shown in Figure E8.1c gives I_{y_C}:

$$I_{y_C} = \int_{x=-b/2}^{b/2} \underbrace{x^2(h\,dx)}_{dA} = h\,\frac{x^3}{3}\bigg|_{-b/2}^{b/2}$$

$$= \frac{hb^3}{12}$$

Question 8.1 Was the I_{y_C} calculation really necessary? That is, could the answer for I_{y_C} have been deduced from the result for I_{x_C} obtained first?

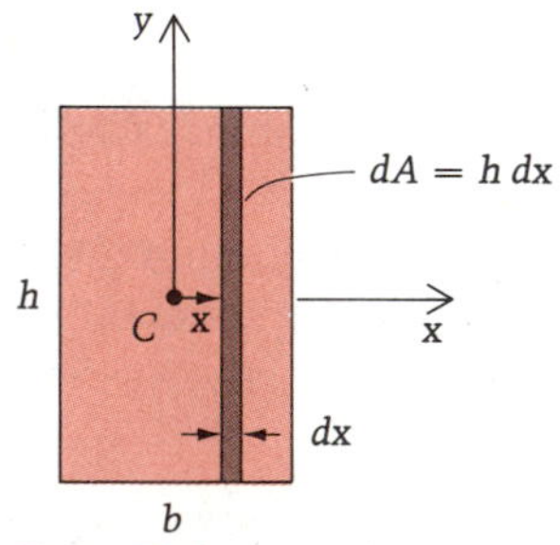

Figure E8.1c

For those familiar with double integrals, we use them below to reproduce the result for I_{x_C}:

$$I_{x_C} = \int_{y=-h/2}^{h/2}\int_{x=-b/2}^{b/2} y^2 \underbrace{dx\,dy}_{dA}$$

Answer 8.1 No, it was not necessary, because if we change the names of x and y, and of b and h, then the same integral as before yields $I_{y_C} = hb^3/12$ without having to integrate again.

$$= \int_{y=-h/2}^{h/2} x \Big|_{-b/2}^{b/2} y^2\,dy = \int_{-h/2}^{h/2} y^2\,b\,dy = b\,\frac{y^3}{3}\Big|_{-h/2}^{h/2} = \frac{bh^3}{12}$$

Note that the first integration (on x) *produces* the strip $b\,dy$ used earlier.

Question 8.2 Could the strip $h\,dx$ have been used for "dA" in the I_{x_C} calculation?

Answer 8.2 No, because then y is not the same for every element of the differential strip.

EXAMPLE 8.2

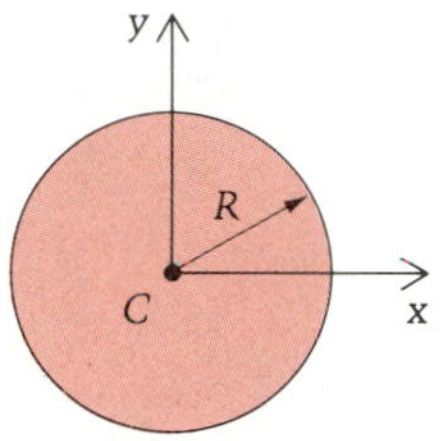

Figure E8.2a

Find the moment of inertia of a circular area (see Figure E8.2a) about any diameter.

Solution

Since $x^2 + y^2 = R^2$ on the boundary, our dA in Figure E8.2b is given by

$$dA = 2\sqrt{R^2 - y^2}\,dy$$

so that

$$I_{x_C} = \int y^2\,dA = \int_{-R}^{R} 2y^2\sqrt{R^2 - y^2}\,dy$$

$$= 2\int_0^R 2y^2\sqrt{R^2 - y^2}\,dy$$

$$= 4R^4 \int_0^1 \frac{y^2}{R^2}\sqrt{1 - \left(\frac{y}{R}\right)^2}\,d\left(\frac{y}{R}\right)$$

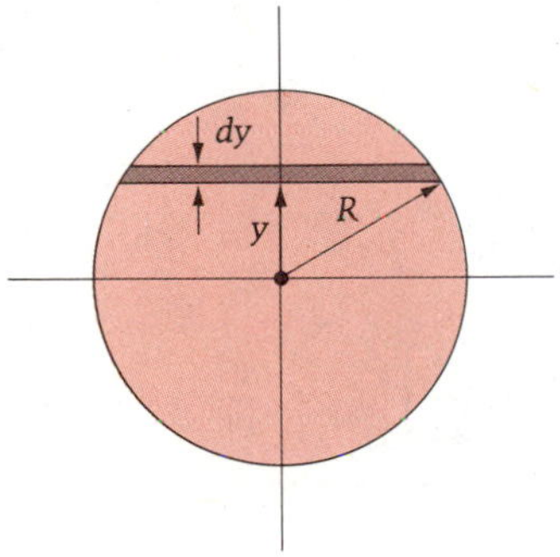

Figure E8.2b

Substituting $\sin\theta$ for y/R, and noting that $d(y/R) = \cos\theta\,d\theta$,

$$I_{x_C} = 4R^4 \int_0^{\pi/2} \sin^2\theta\cos\theta(\cos\theta\,d\theta)$$

where for the integral limits, $y/R = 0$ when $\theta = 0$, and $y/R = 1$ when $\theta = \pi/2$. Continuing,

$$I_{x_C} = 4R^4 \int_0^{\pi/2} \frac{\sin^2 2\theta}{4}\,d\theta$$

$$= R^2 \int_0^{\pi/2} \frac{1 - \cos 4\theta}{2}\,d\theta$$

$$= \frac{R^4}{2}\left(\theta - \frac{\sin 4\theta}{4}\right)\Big|_0^{\pi/2} = \frac{\pi R^4}{4}$$

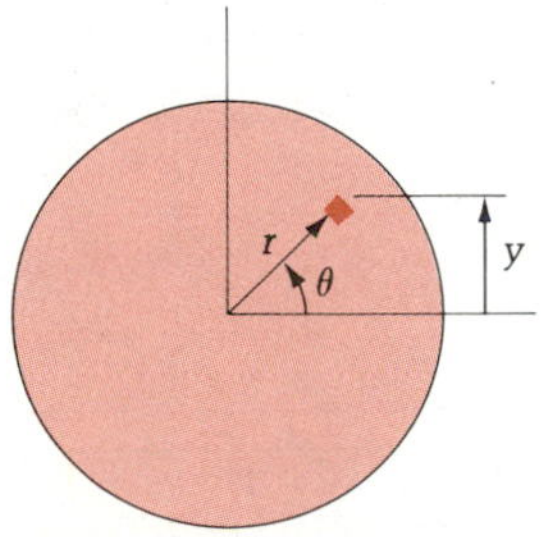

Figure E8.2c

For the reader acquainted with double integrals, we can obtain the above result with somewhat less effort using polar coordinates (see Figure E8.2c) as follows:

$$I_{x_C} = \int y^2\,dA = \int_{\theta=0}^{2\pi}\int_{r=0}^{R} (r\sin\theta)^2 \underbrace{r\,dr\,d\theta}_{dA}$$

$$= \int_0^{2\pi} \frac{r^4}{4}\bigg|_0^R \sin^2\theta\,d\theta$$

$$= \frac{R^4}{4}\int_0^{2\pi}\left(\frac{1-\cos 2\theta}{2}\right)d\theta = \frac{R^4}{8}\left[\theta - \frac{\sin 2\theta}{2}\right]\bigg|_0^{2\pi}$$

$$I_{x_C} = \frac{\pi R^4}{4}$$

This is, of course, also I_{y_C} or I about any other diameter of the circle.

EXAMPLE 8.3

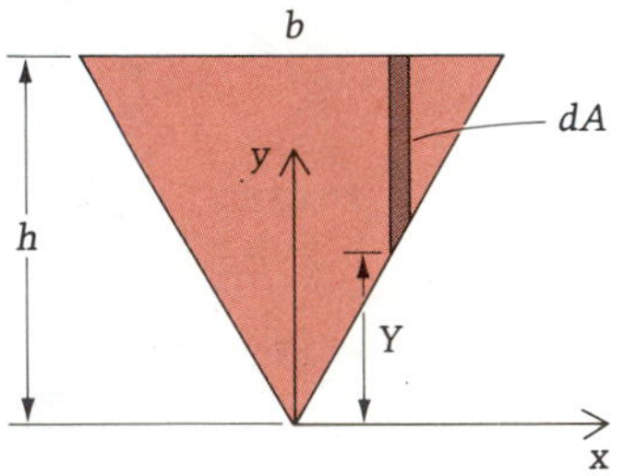

Figure E8.3

Find the moment of inertia of the triangular area in Figure E8.3 about the y axis.

Solution

For our differential area, we shall use the shaded strip in the figure; thus, using Y to locate the lowest (boundary) point of the strip,

$$dA = (h - Y)\,dx$$

But $Y = \frac{2h}{b}x$ for the side of the triangle in the first quadrant, so that

$$dA = \left(h - \frac{2hx}{b}\right)dx$$

Therefore,

$$I_y = \int x^2\,dA = 2\int_0^{b/2} x^2\left(h - \frac{2h}{b}x\right)dx$$

$$= 2h\int_0^{b/2}\left(x^2 - \frac{2x^3}{b}\right)dx = 2h\left[\frac{x^3}{3} - \frac{x^4}{2b}\right]\bigg|_0^{b/2} = \frac{hb^3}{48}$$

PROBLEMS ▶ Section 8.2

8.1 In Figure P8.1:

a. Find the moment of inertia of the shaded area about the x axis.

b. Tell why I_y is identical to I_{y_C} of Example 8.1.

8.2 Find I_x for the shaded area in Figure P8.2.

8.3 In Problem 8.2 find I_y for the same area.

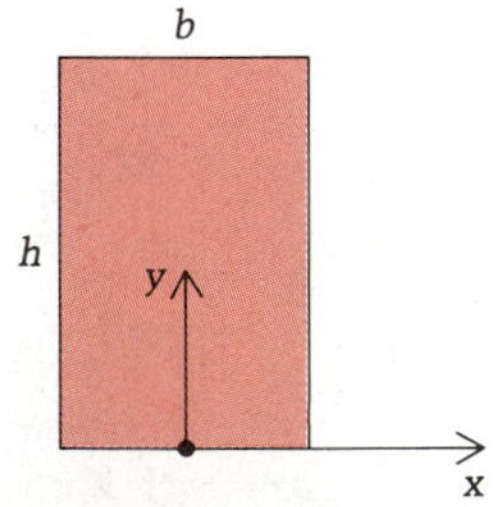

Figure P8.1

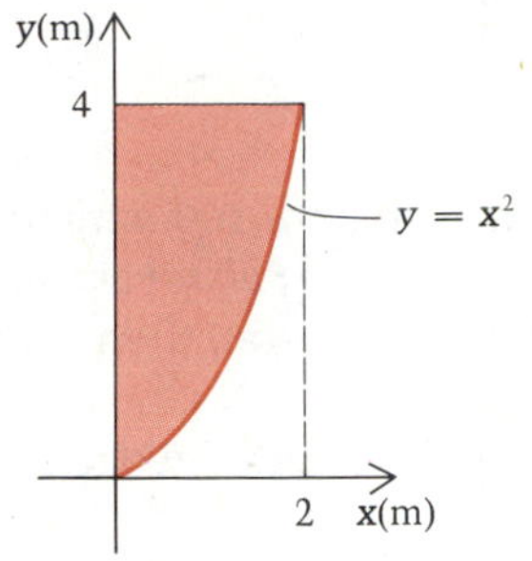

Figure P8.2

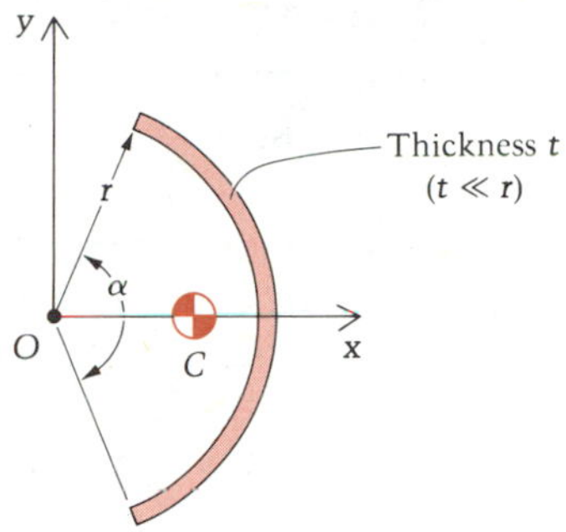

Figure P8.7

8.4 Review Examples 8.1 and 8.3.

a. Find the moment of inertia of the shaded area in Figure P8.4 about the y axis.

b. Add the result to that of Example 8.3, and note that the answer is, as it should be, that for the combined, rectangular area.

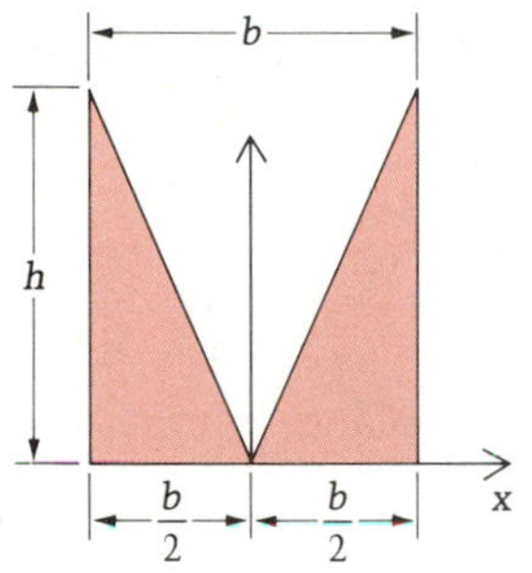

Figure P8.4

8.5 Determine the moment of inertia of the shaded area in Figure P8.5 with respect to the x axis.

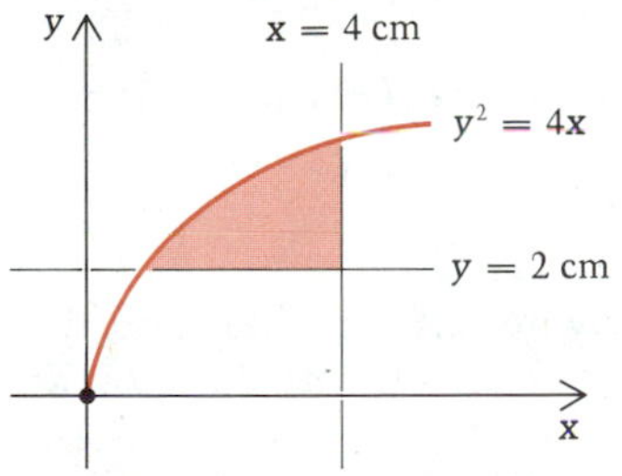

Figure P8.5

8.6 Repeat Problem 8.5 for the moment of inertia about the y axis.

8.7 Find I_x for the thin circular section shown in Figure P8.7. *Hint*: Use polar coordinates, with $dA = tr\, d\theta$.

8.8 In Problem 8.7, find I_y.

8.9 Find the moment of inertia about the x axis of the area under the nth-degree curve shown in Figure P8.9. (n is an integer, $n \geq 1$.)

8.10 In Problem 8.9, find I_y.

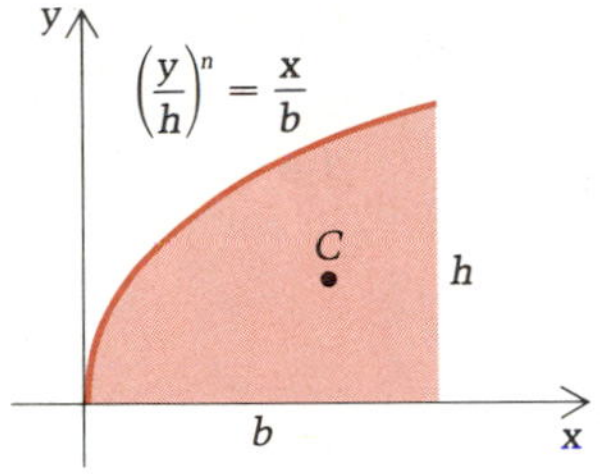

Figure P8.9

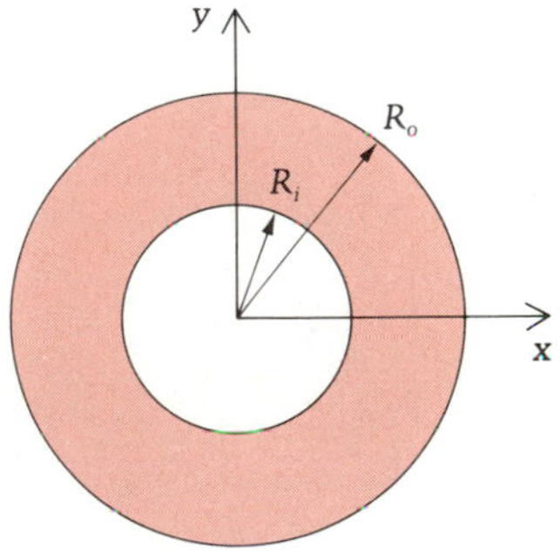

Figure P8.11

8.11 Show that the moment of inertia I_x of the shaded, annular area in Figure P8.11 is $\pi(R_o^4 - R_i^4)/4$ by:

a. Integrating as in Example 8.2 but with limits of $\int_{R_i}^{R_o}$ on the integration with respect to r.

b. Subtracting the moments of inertia of the two circles. Why does this work?

8.12 In Problem 8.11, let the figure depict the cross section of a round tube. If the tube is very thin ($R_i \to R_o$), show that I_x is approximately $\pi R^3 t$, where $R = R_i \approx R_o$ and $t =$ thickness of tube $= R_o - R_i$.

8.3 The Polar Moment of Inertia of a Plane Area

In the study of deformable solids, the "torsion problem" refers to what happens to a shaft when it is twisted. In the same way that the moment of inertia forms part of the resistance of a beam to bending, the *polar moment of inertia* forms part of its resistance to twisting. For this reason, then, we shall discuss the polar moment of inertia in this section.

The polar moment of inertia of an area about a point P is defined to be (see Figure 8.2).

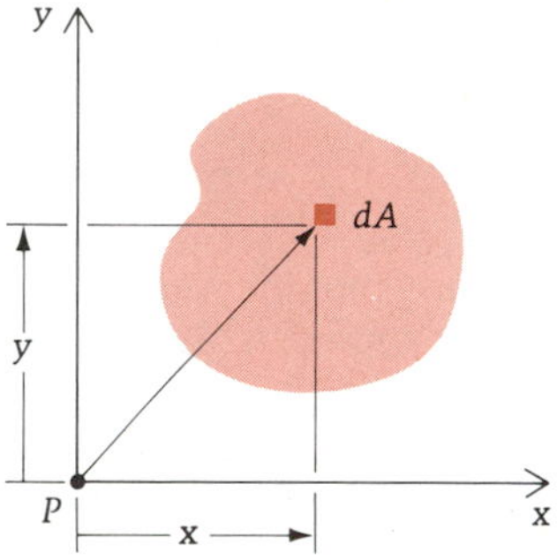

Figure 8.2

$$J_P = \int (x^2 + y^2)\, dA$$

where the axes (x, y) have origin at P. Since the polar coordinate r is given by* $x^2 + y^2 = r^2$, this simplifies to

$$J_P = \int r^2\, dA$$

If we recall that both I_x and I_y are the sums (in integration) of (differential areas) times (squares of their distances from the axis), then we see that J_P is the same type of quantity, this time with respect to the z axis (*normal* to the area) through a point P. Also,

$$J_P = \int y^2\, dA + \int x^2\, dA$$

$$= I_{x_P} + I_{y_P} \tag{8.3}$$

so that for an area moment of inertia, the x and y (in-plane) inertias add up to the z (out-of-plane) inertia.

EXAMPLE 8.4

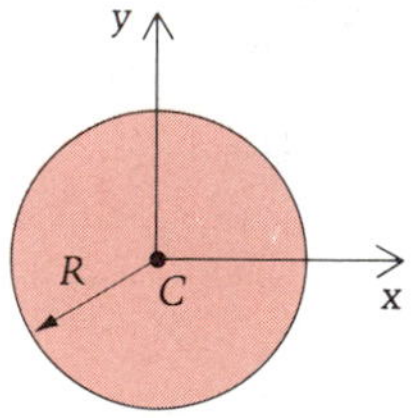

Figure E8.4a

Find the polar moment of inertia of the circular cross section of Figure E8.4a with respect to point C.

Solution

By $J_C = I_{x_C} + I_{y_C}$, we obtain immediately from the results of Example 8.2:

$$J_C = \frac{\pi R^4}{4} + \frac{\pi R^4}{4} = \frac{\pi R^4}{2}$$

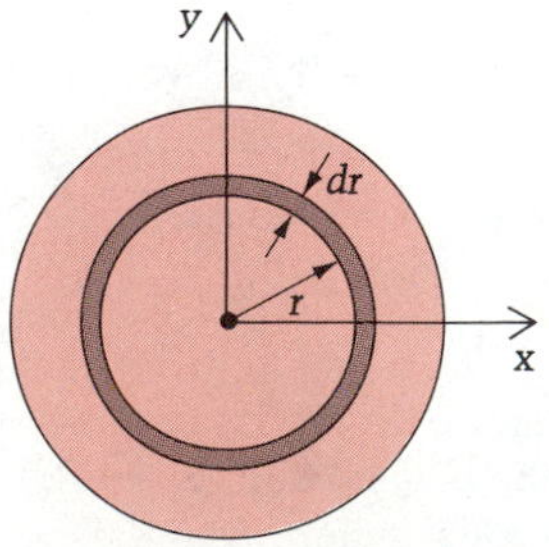

Figure E8.4b

Alternatively, the shaded "circular strip" in Figure E8.4b has the feature that all its elements are the same distance r from the z axis. Thus with $dA = 2\pi r\, dr$,

$$J_C = \int r^2\, dA = \int_0^R r^2(2\pi r\, dr)$$

$$= 2\pi \left.\frac{r^4}{4}\right|_0^R = \frac{\pi R^4}{2}$$

And if double integrals are familiar to the reader, we obtain the same result a

* Hence the name "polar moment of inertia."

third time using polar coordinates:

$$J_C = \int r^2\,dA = \int_{\theta=0}^{2\pi} \int_{r=0}^{R} r^2(r\,dr\,d\theta)$$

$$= \int_{\theta=0}^{2\pi} \frac{r^4}{4}\bigg|_0^R d\theta = \frac{R^4}{4}\,\theta\bigg|_0^{2\pi} = \frac{\pi R^4}{2}$$

EXAMPLE 8.5

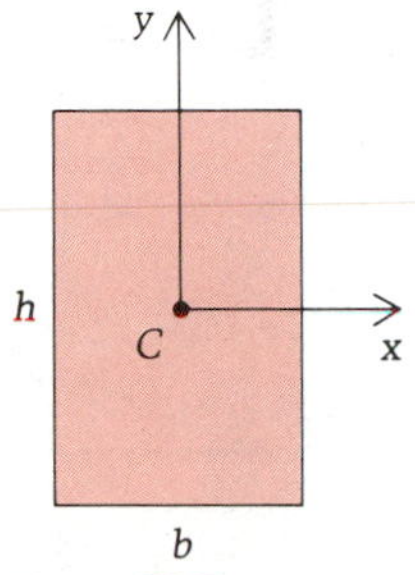

Figure E8.5

Find the polar moment of inertia of the rectangular cross section of Figure E8.5, at point C.

Solution

Because $J_C = I_{x_C} + I_{y_C}$, we use previous results and get

$$J_C = \frac{bh^3}{12} + \frac{hb^3}{12} = \frac{bh(b^2 + h^2)}{12}$$

or

$$J_C = \frac{A(b^2 + h^2)}{12}$$

PROBLEMS ▶ Section 8.3

In Problems 8.13–8.19 find the polar moment of inertia with respect to the origin for each of the shaded areas.

8.13

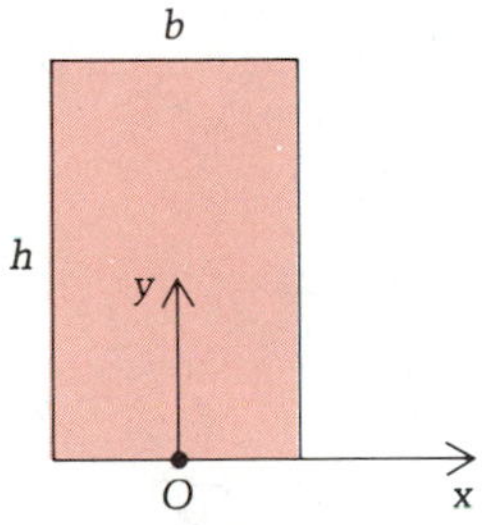

Figure P8.13

8.14

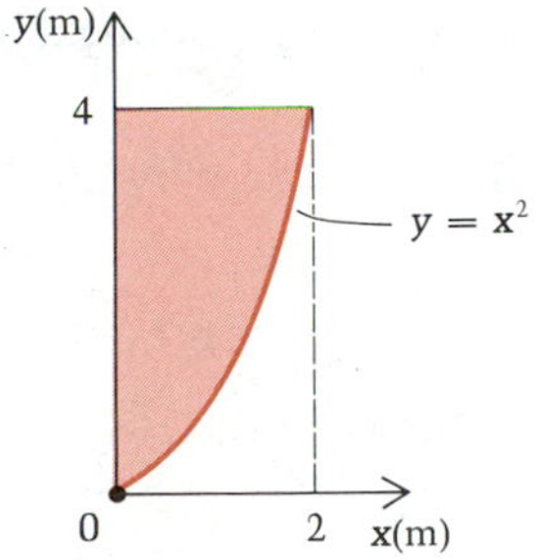

Figure P8.14

8.15

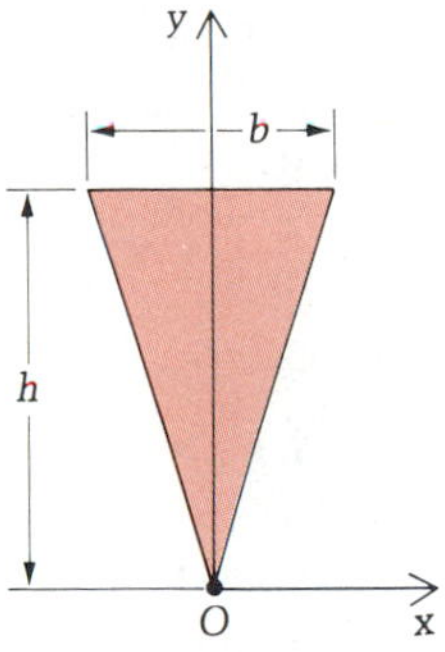

Figure P8.15

8.16

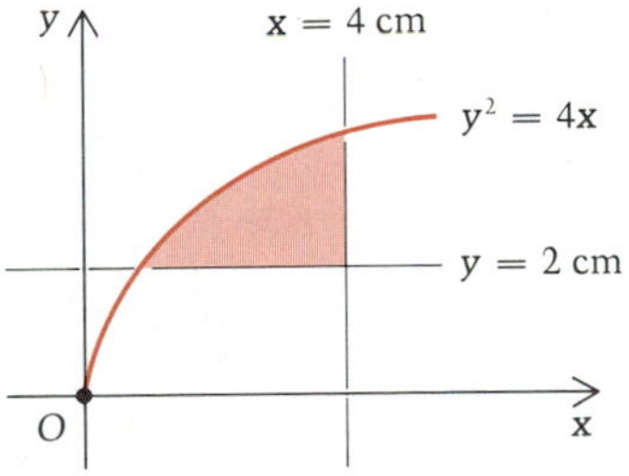

Figure P8.16

8.17

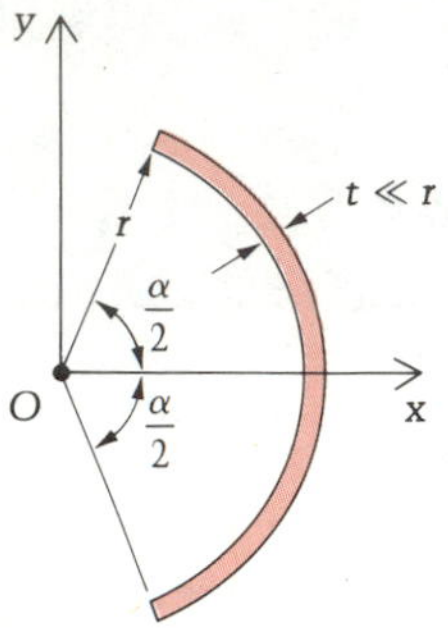

Figure P8.17

8.18

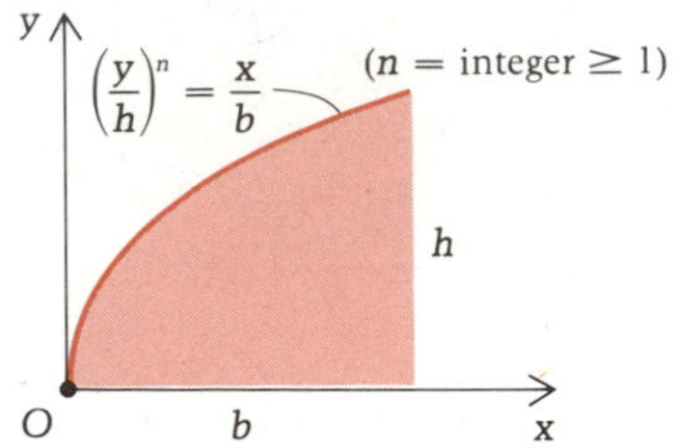

Figure P8.18

8.19

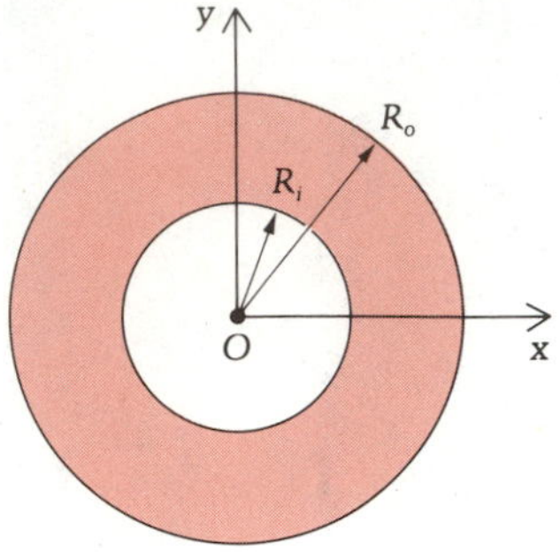

Figure P8.19

8.4 The Parallel-Axis Theorem (or Transfer Theorem) for Moments of Inertia/The Radius of Gyration

Development of the Parallel Axis Theorem

Fortunately, if we wish to determine the moment of inertia I_{x_P} about an axis through a point P other than the centroid, we don't have to integrate if we already know I_{x_C}. All we need do is "transfer" this result, adding the area A times the square of the distance d between the parallel x axes through C and P (see Figure 8.3):

$$I_{x_P} = I_{x_C} + Ad^2 \tag{8.4}$$

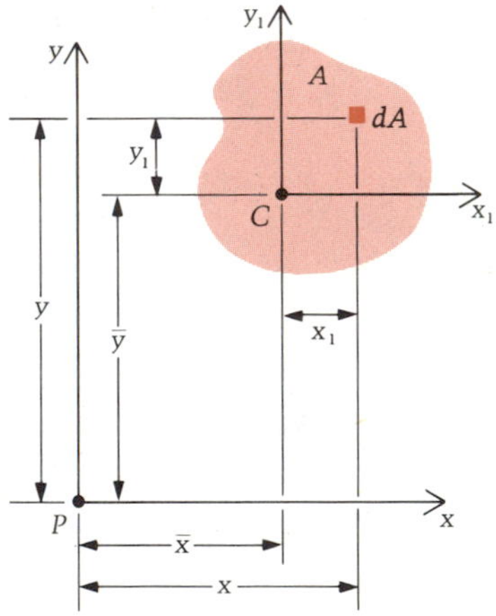

Figure 8.3

We now proceed to prove this very helpful theorem. We have

$$I_{x_P} = \int_A y^2\, dA = \int_A (\bar{y} + y_1)^2\, dA$$

$$I_{x_P} = \underbrace{\int_A y_1^2\, dA}_{I_{x_C}} + \bar{y}^2 \int_A dA + 2\bar{y} \int_A y_1\, dA$$

where from Figure 8.1 we see that the y coordinate of dA is $\bar{y} + y_1$, where $\bar{y}$ is the y coordinate of C.

Question 8.3 Why may the $\bar{y}$ be brought outside the integrals above?

The last integral, $\int y_1\, dA$, is zero by the definition of the centroid, and therefore

$$I_{x_P} = I_{x_C} + A\bar{y}^2 = I_{x_C} + Ad^2$$

where d is the distance between the two parallel x axes through P and C.

Answer 8.3 It's constant.

This is the parallel-axis theorem that was stated in Equation (8.4). Note from Equation (8.4) that the smallest value for the moment of inertia of an area is with respect to an axis through the body's centroid. For y axes, we have

$$I_{y_P} = I_{y_C} + A\bar{x}^2 = I_{y_C} + Ad^2 \tag{8.5}$$

where *this* time d is the absolute value of $\bar{x}$, and is again the distance between the parallel axes through P and C.

For the polar moment of inertia, we also have a parallel-axis theorem:

$$\begin{aligned} J_P \text{ (or } I_{z_P}) &= \int (x^2 + y^2)\, dA = \int [(\bar{x} + x_1)^2 + (\bar{y} + y_1)^2]\, dA \\ &= \int (x_1^2 + y_1^2)\, dA + \bar{y} \int x_1\, dA + \bar{x} \int y_1\, dA + \int (\bar{x}^2 + \bar{y}^2)\, dA \\ &= J_C + 0 + 0 + Ad^2 \end{aligned}$$

$$J_P = J_C + Ad^2 \tag{8.6}$$

where once again d is the distance between the parallel axes, this time z axes through P and C, of interest. It is also the distance between the points P and C in the plane of the area.

Question 8.4 Give an alternative proof of Equation (8.6) using Equation (8.3).

Answer 8.4 $J_P = I_{x_P} + I_{y_P} = I_{x_C} + A\bar{y}^2 + I_{y_C} + A\bar{x}^2 = (I_{x_C} + I_{y_C}) + A(\bar{x}^2 + \bar{y}^2) = J_C + Ad^2$

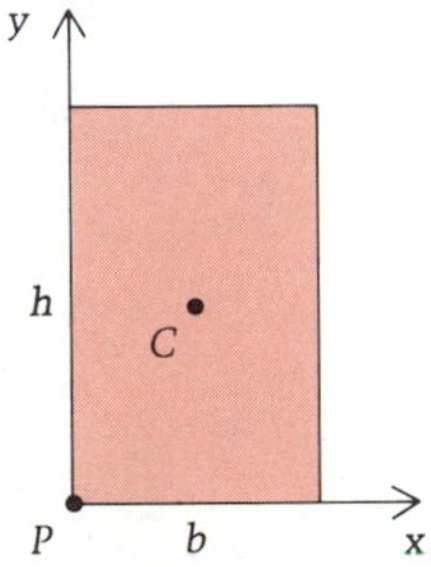

Figure E8.6

Example 8.6

Find the moments of inertia of the rectangular area about the indicated axes through the lower left corner P in Figure E8.6.

Solution

Using the parallel-axis theorem together with the results of Example 7.1,

$$I_x \text{ (or } I_{x_P}) = I_{x_C} + Ad^2 = \frac{bh^3}{12} + bh\left(\frac{h}{2}\right)^2 = \frac{bh^3}{3}$$

$$I_y \text{ (or } I_{y_P}) = I_{y_C} + Ad^2 = \frac{hb^3}{12} + hb\left(\frac{b}{2}\right)^2 = \frac{hb^3}{3}$$

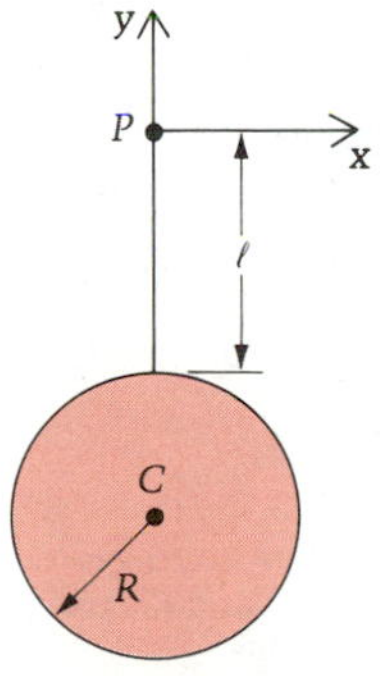

Figure E8.7

Example 8.7

Find the moments of inertia of the circular area about the axes x and y shown in Figure E8.7.

Solution

Using the parallel-axis theorem, and the results of Example 8.2,

$$I_x = I_{x_c} + Ad^2 = \frac{\pi R^4}{4} + (\pi R^2)(\ell + R)^2 = \frac{\pi R^4}{4}\left(5 + \frac{8\ell}{R} + \frac{4\ell^2}{R^2}\right)$$

$$I_y = I_{y_c} + Ad^2 = \frac{\pi R^4}{4} + 0$$

[because the distance between the y axes through the two origins vanishes. (They are the same line!)]

EXAMPLE 8.8

(a) Find the moment of inertia of the triangular area in Figure E8.8a about the x axis through O. (b) Then use the transfer theorem to determine I_{x_c}.

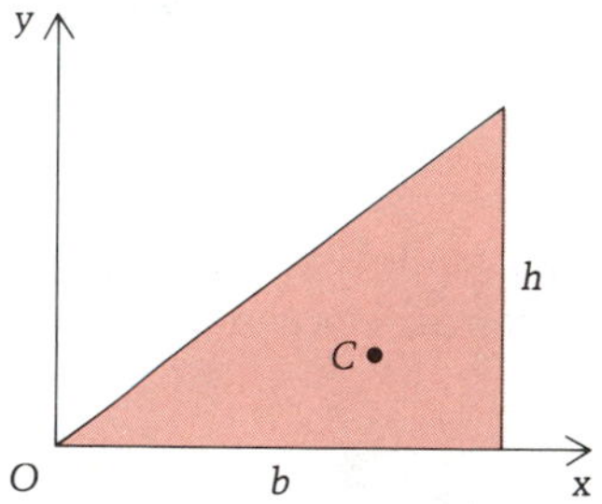

Figure E8.8a

Solution

(a) By the definition

$$I_{x_O} = \int_A y^2\, dA$$

Noting that the hypotenuse has the equation $y = \left(\frac{h}{b}\right) x$, we have (see Figure E8.8b):

$$I_{x_O} = \int_0^h y^2 \left(b - \frac{b}{h} y\right) dy = \frac{bh^3}{3} - \frac{bh^3}{4} = \frac{bh^3}{12}$$

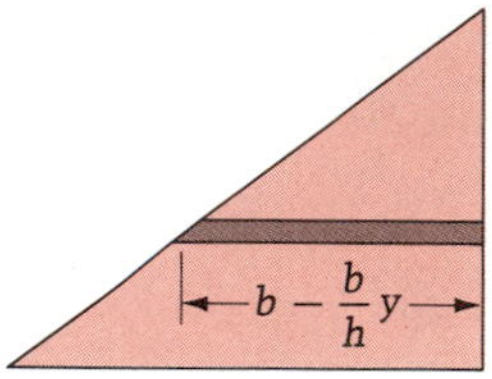

Figure E8.8b

Question 8.5 Show how to obtain this answer without integrating by using Example 8.3.

(b) The transfer theorem gives

$$I_{x_O} = I_{x_c} + Ad^2$$

$$\frac{bh^3}{12} = I_{x_c} + \frac{bh}{2}\left(\frac{h}{3}\right)^2 = I_{x_c} + \frac{bh^3}{18}$$

so that

$$I_{x_c} = \frac{bh^3}{36}$$

Notice that we don't always transfer *from* C to other points, in this case we knew the answer for I_x at O, and we used it "in reverse" to *find* I_{x_c}.

Answer 8.5 Working from the answer to Example 8.3, to make the notations and geometry agree, we must do three things: (1) swap b and h; (2) double the old h; and (3) use half the answer. Thus I_x in Example 8.8 is $\left[\frac{b(2h)^3}{48}\right]\Big/ 2 = \frac{bh^3}{12}$, which checks.

Question 8.6 Suppose we know the moment of inertia of an area about a line through point P. May we circumvent the centroid and transfer directly from P to another point Q? That is, is $I_{x_Q} = I_{x_P} + Ad^2$?

Answer 8.6 No! We can only transfer to or from an axis through the centroid.

Example 8.9

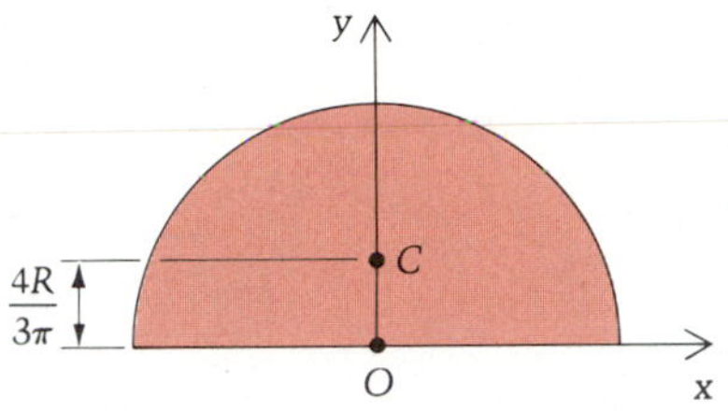

Figure E8.9

Find the polar moment of inertia of the semicircular area of Figure E8.9 with respect to its centroid C.

Solution

At point O we have

$$J_O = \int r^2\, dA = \frac{\pi R^4}{4}$$

because (see Example 8.4) we must get half of the answer for the full circular area. Using the parallel-axis theorem,

$$J_O = J_C + Ad^2$$

$$\frac{\pi R^4}{4} = J_C + \frac{\pi R^2}{2}\left(\frac{4R}{3\pi}\right)^2$$

$$J_C = \pi R^4\left(\frac{1}{4} - \frac{8}{9\pi^2}\right)$$

$$= 0.160\pi R^4 = 0.503R^4$$

Example 8.10

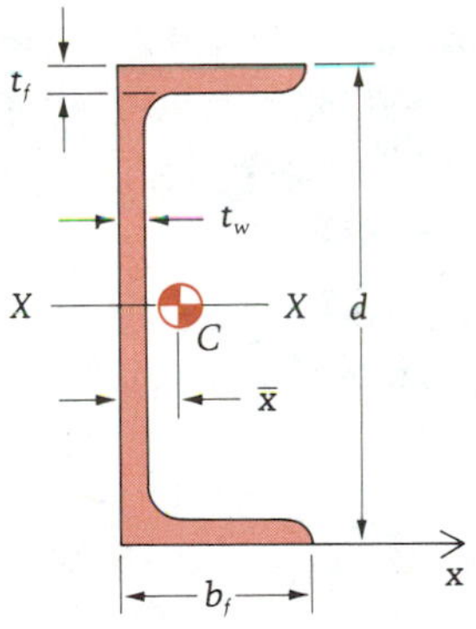

d = depth
b_f = width of flange
t_f = average thickness of flange
t_w = web thickness
$\bar{x}$ = distance locating centroid of section

Figure E8.10

For the C15 × 50 channel section,* (see Figure E8.10) the manual of the American Institute of Steel Construction lists the area of the cross section as 14.7 in.^{2} and the centroidal moment of inertia about line X-X as 404 in.^{4} If $d = 15.0$ in., find the moment of inertia about the baseline axis x.

Solution

We simply use the parallel-axis theorem, and obtain, without integration,

$$I_x = I_{x_C} + Ad^2$$

$$= 404 + 14.7\,(7.5)^2$$

$$= 1230 \text{ in.}^4$$

* Meaning "C" for channel and that the nominal depth is 15 in. and the weight per lineal foot is 50 lb / ft. See the AISC Manual of Steel Construction, American Institute of Steel Construction, Inc., 101 Park Avenue, New York, N.Y. 10017.

Radius of Gyration

This is a good place to introduce the reader to a distance called the *radius of gyration.** The radius of gyration is a concept associated with the moment of inertia of an area about an axis. The radius of gyration is simply the length that, when squared and multiplied by the area, gives the moment of inertia about the axis. Thus, for example,

$$I_x = Ak_x^2$$

For a physical interpretation of the radius of gyration, consider the area A in Figure 8.4(a) having the moment of inertia I_x about the x axis. If

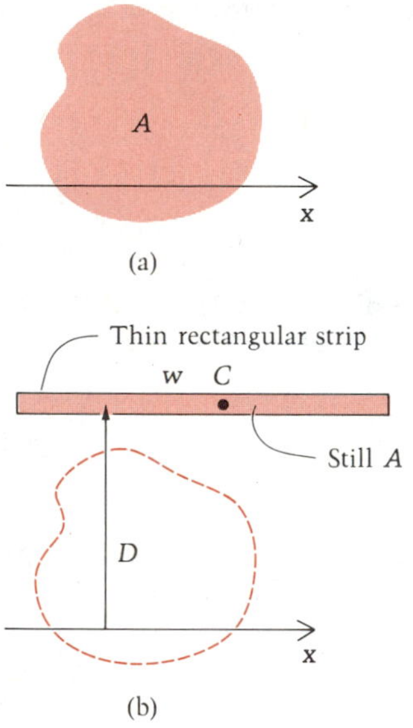

Figure 8.4

we squeeze the same area into a very thin strip and place it a distance D from the x axis (and parallel to it as shown in Figure 8.4(b), then the moment of inertia of the strip about the x axis is

$$I_x^{\text{STRIP}} = I_{x_C}^{\text{STRIP}} + AD^2 \approx 0 + AD^2$$

Clearly, then, there is a distance D for which I_x^{STRIP} equals I_x of the actual area. This value of D is the radius of gyration k_x and is therefore $\sqrt{I_x/A}$, as in the definition above.

Question 8.7 Why is $I_{x_C}^{\text{STRIP}} \ll AD^2$ in the above equation?

* This parameter arises naturally in the analysis of buckling of columns.

Answer 8.7 Because $I_{x_C}^{STRIP} = \dfrac{wt^3}{12} = \dfrac{At^2}{12}$, much smaller than AD^2 because the strip is given to be very thin.

As examples, the radii of gyration of the areas in Figure 8.5 about the indicated lines are shown beneath the various figures:

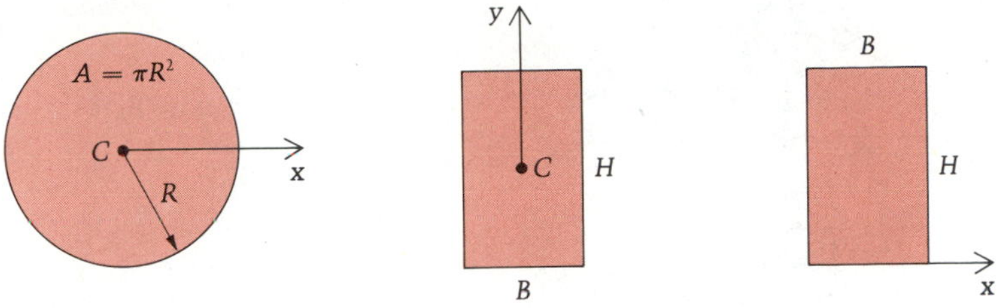

(a) $k_x = \dfrac{R}{2}$ since $I_x = \dfrac{\pi R^4}{4}$

(b) $k_y = \dfrac{\sqrt{3}B}{6}$ since $I_y = \dfrac{HB^3}{12}$

(c) $k_x = \dfrac{\sqrt{3}H}{3}$ since $I_x = \dfrac{BH^3}{3}$

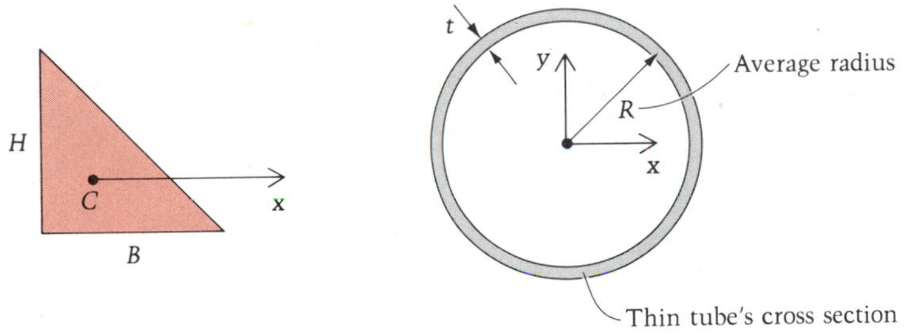

(d) $k_x = \dfrac{\sqrt{2}H}{6}$ since $I_x = \dfrac{BH^3}{36}$

(e) $k_z \approx R$ since $J \approx 2\pi R^3 t$ (and $A = 2\pi Rt$)

Figure 8.5

PROBLEMS ▶ Section 8.4

For each of the areas shown in Problems 8.20–8.23 find the moment of inertia about a line through the centroid parallel to x.

8.20 (See Problem 8.2.)

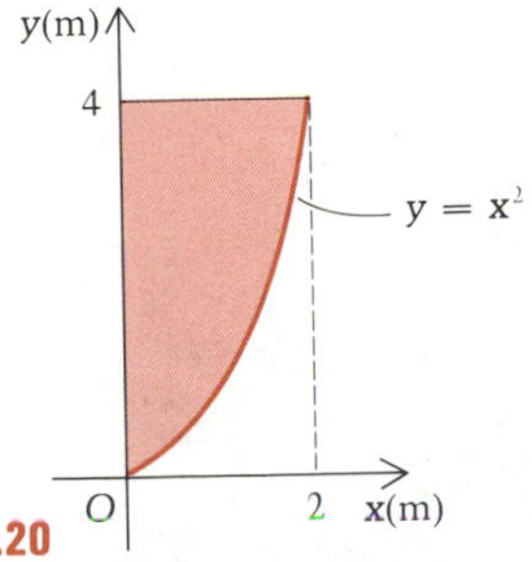

Figure P8.20

8.21

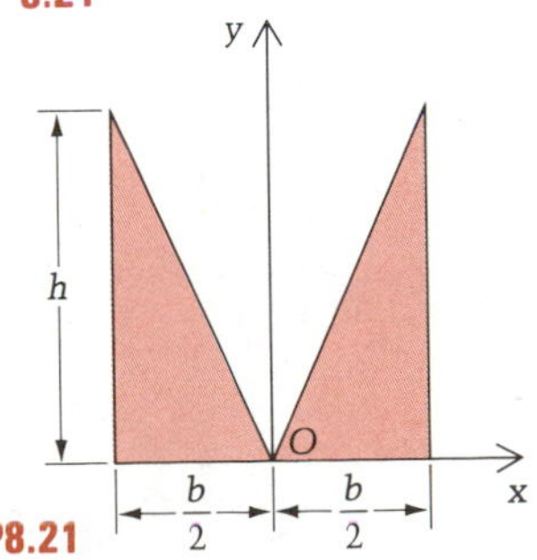

Figure P8.21

8.22 (See Problem 8.5.)

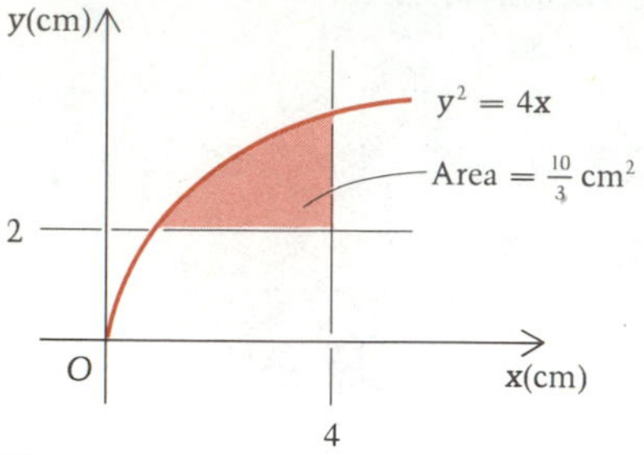

Figure P8.22

8.23 (See Problems 8.9 and 7.23)

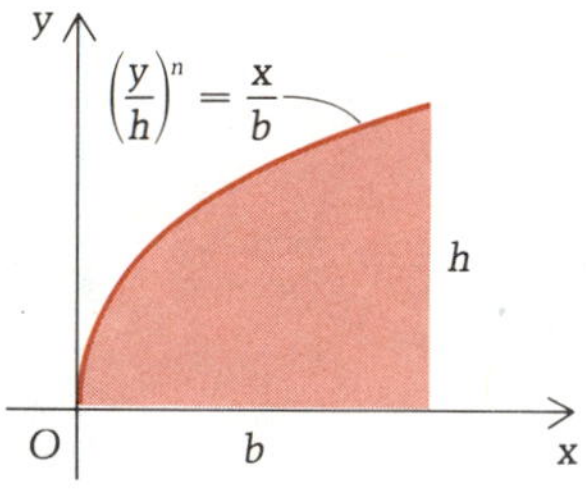

Figure P8.23

In Problems 8.24–8.26, for the problem referred to, find the moment of inertia about a line through the centroid parallel to y.

8.24 Problem 8.20.

8.25 Problem 8.22.

8.26 Problem 8.23.

8.27 In Problem 8.8, the distance from the origin to the centroid of the very thin section is approximately $[r \sin(\alpha/2)]/(\alpha/2)$. Use the parallel-axis theorem to find I_{y_C}.

8.28 Use the parallel-axis theorem to find the moment of inertia of the semicircular area about an axis through C parallel to x. (See Figure P8.28.)

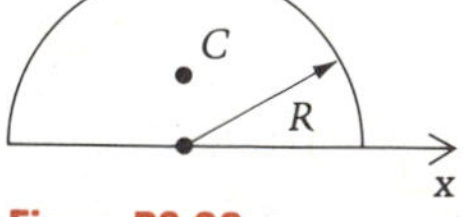

Figure P8.28

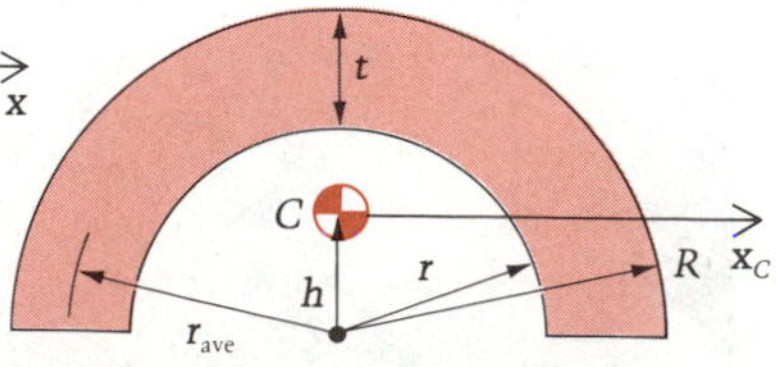

Figure P8.29

8.29 Show that the centroidal moment of inertia I_{x_C} of the annular area shown in Figure P8.29 is

$$0.110(R^4 - r^4) - \frac{0.283R^2r^2(R - r)}{R + r}$$

Show that this result becomes $I_{x_C} \approx 0.3tr^3$ if $r \approx R$. Note that

$$h = \frac{4}{3\pi}\left(\frac{R^2 + Rr + r^2}{R + r}\right)$$

8.30 The shaded region $\mathcal{A}$ in Figure P8.30 has an area of 10 ft². If the moment of inertia of the area about the y axis is known to be 600 ft⁴, what is the moment of inertia of the area about the η axis (i.e., about the line $x = 3$ ft)? The centroid C of this area is located at coordinates (7, 3) ft.

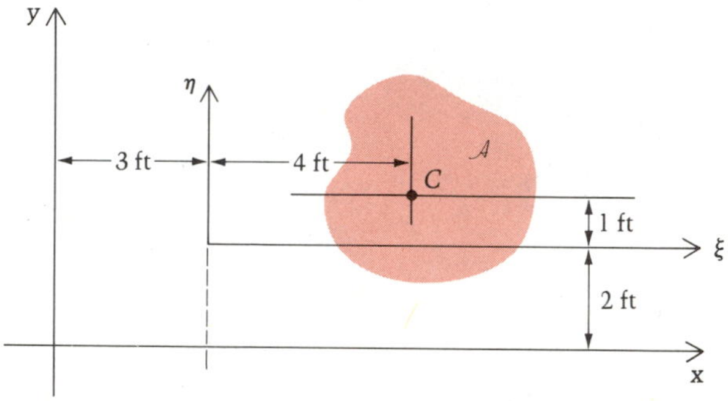

Figure P8.30

8.31 The same shaded region in Problem 8.30 has a radius of gyration with respect to the ξ axis of 2 ft. What is its moment of inertia with respect to the x axis?

8.32 In Figure P8.32:

a. By integration, determine the moment of inertia of the elliptical area about the x axis.

b. What is the radius of gyration of this area with respect to the x axis?

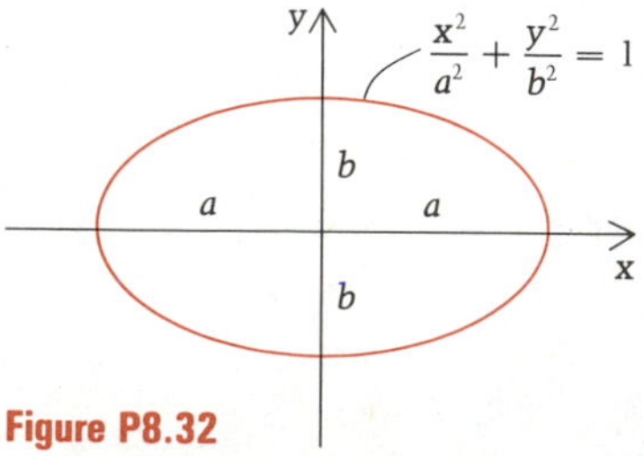

Figure P8.32

8.33 In Figure P8.33:

- **a.** Find the polar moment of inertia J_C of the equilateral triangular area with respect to its centroid C.
- **b.** Compare the result of part (a) with J_C of a square cross section having the same area. Think about which should be larger before making the calculation.

8.34 Find the polar moment of inertia of the triangular area of Problem 8.33 about one of its vertices.

For Problems 8.35–8.37 find the polar moments of inertia of the areas about their respective centroids in the problems referred to.

8.35 Problem 8.20.

8.36 Problem 8.22.

8.37 Problem 8.23.

8.38 In Problem 8.20, find the radius of gyration k_x of the shaded area.

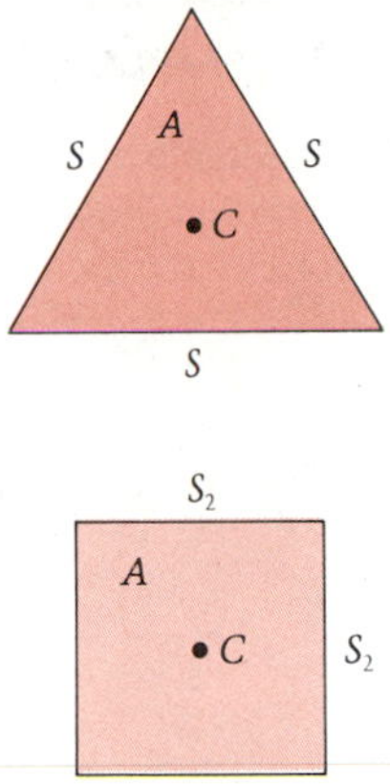

Figure P8.33

8.39 In Problem 8.20, find the radius of gyration k_y of the shaded area.

8.40 Find the radius of gyration k_x for the shaded area of Problem 8.21.

8.5 The Method of Composite Areas

We learned in Chapter 7 that in finding the centroid of a composite area A, the integration could be divided into separate integrals over the various areas comprising A. Thus, for example, if $A_1 + A_2 + A_3 = A$, then

$$A\bar{y} = \int_A y\,dA = \int_{A_1} y\,dA + \int_{A_2} y\,dA + \int_{A_3} y\,dA$$
$$= A_1\bar{y}_1 + A_2\bar{y}_2 + A_3\bar{y}_3$$

or

$$\bar{y} = \frac{A_1\bar{y}_1 + A_2\bar{y}_2 + A_3\bar{y}_3}{A_1 + A_2 + A_3}$$

This same idea, or method of "composite areas," may be used in calculating moments and polar moments of inertia:

$$I_x = \int_A y^2\,dA = \int_{A_1} y^2\,dA + \int_{A_2} y^2\,dA + \int_{A_3} y^2\,dA$$
$$= I_x^{A_1} + I_x^{A_2} + I_x^{A_3}$$

where $I_x^{A_1}$ is the moment of inertia of the area A_1 about the x axis, and similarly for A_2 and A_3.

We consider some examples made simpler by the use of composite areas.

EXAMPLE 8.11

Calculate the polar moment of inertia of the ten identical small bolt cross sections (studied previously in Example 7.10 and shown in Figure E8.11) about both the origin and point C.

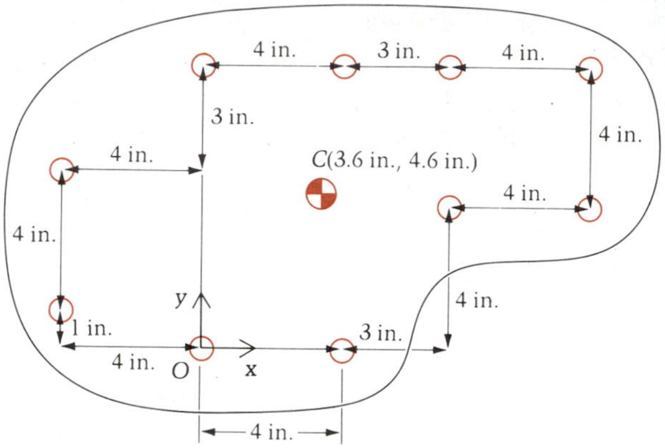

Figure E8.11

Solution

If, as the figure suggests, each bolt diameter is small in comparison with the spacings between bolts, then all points in one of the areas are approximately the same distance from the point with respect to which we desire to calculate the polar moment of inertia. Then

$$J_O = \int r^2\,dA = \int_{A_1} r^2\,dA + \cdots + \int_{A_{10}} r^2\,dA$$

$$\approx r_1^2 \int_{A_1} dA + \cdots + r_{10}^2 \int_{A_{10}} dA = \sum_{i=1}^{10} A_i r_i^2$$

where r_i is the distance from O to the center of the ith area. The expression above corresponds to thinking of each bolt's area as "concentrated" at its center.

With identical areas ($A_i = A$), we obtain

$$J_O = A \sum_{i=1}^{10} r_i^2$$

Starting at O and proceeding counterclockwise we have

$$J_O = A\Big[0 + 4^2 + \sqrt{7^2+4^2}^{\,2} + \sqrt{11^2+4^2}^{\,2} + \sqrt{11^2+8^2}^{\,2}$$

$$+ \sqrt{7^2+8^2}^{\,2} + \sqrt{4^2+8^2}^{\,2} + 8^2 + \sqrt{(-4)^2+5^2}^{\,2} + \sqrt{(-4)^2+1^2}^{\,2}\Big]$$

$$= A[4^2 + 7^2 + 4^2 + 11^2 + 4^2 + 11^2 + 8^2 + 7^2 + 8^2 + 4^2 + 8^2$$

$$+ 8^2 + 4^2 + 5^2 + 4^2 + 1^2]$$

$$= 718A \text{ in.}^4$$

where A is in in.^{2} Note, in the above steps, that it is a waste of time to take the square root to get r_i because we are going to then immediately square it to get r_i^2.

Now let us back up and put this analysis on a more rigorous basis. Let J_{C_i} be the polar moment of inertia of the ith area with respect to its own centroid. Using the fact that polar moments of inertia of composite parts add to yield the polar moment of inertia of the whole area, and also using the parallel-axis theorem for each of the parts,

$$J_O = \sum_{i=1}^{10} (J_{C_i} + A_i r_i^2)$$

where r_i is the distance from O to C_i. Thus

$$J_O = \sum_{i=1}^{10} J_{C_i} + \sum_{i=1}^{10} A_i r_i^2$$

and the second sum is precisely what was calculated above. For circular areas (see Example 7.4)

$$J_{C_i} = \frac{A_i R_i^2}{2}$$

where R_i is the radius of the ith bolt's area. With identical areas, we have $R_i = R$, $A_i = A$, and:

$$\begin{aligned} J_O &= \frac{10AR^2}{2} + A\sum_{i=1}^{10} r_i^2 \\ &= 5R^2A + 718A \\ &= (5R^2 + 718)A \text{ in.}^4 \end{aligned}$$

and the smaller is R the better is the concentrated-area approximation. For example, if $R = \frac{1}{4}$ in., then $5R^2A = \frac{5}{16}A$ and the error associated with the concentrated-area approximation is about 0.04%.

To get J_C (the polar moment of inertia about the centroid), we shall use the transfer (or parallel-axis) theorem. The centroid C was computed in Example 7.10 to be at $(\bar{x}, \bar{y}) = (3.6, 4.6)$ in. Therefore,

$$J_O = J_C + A_{\text{Total}}\, d^2$$

where $d = |\mathbf{r}_{OC}|$. Substituting,

$$\begin{aligned} J_C &= 718A - (10A)(3.6^2 + 4.6^2) \\ &= 718A - 341A \\ &= 377\,A \text{ in.}^4 \end{aligned}$$

Note that the transfer theorem can be used to *find* J_C if J is known with respect to some other point such as O in this example.

Example 8.12

Find the moment of inertia I_{x_C} for the cross-sectional area of the channel beam shown in Figure E8.12a.

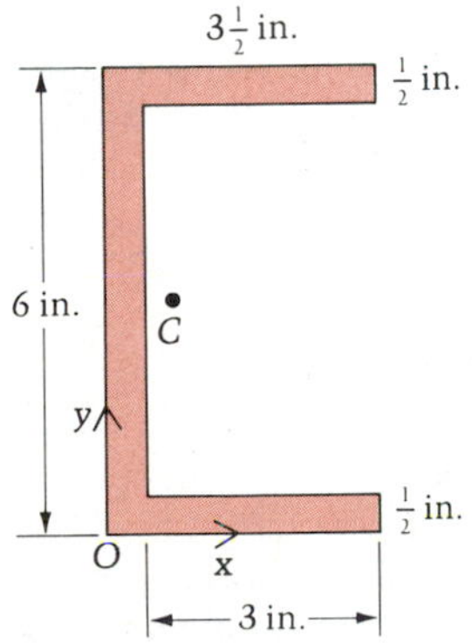

Figure E8.12a

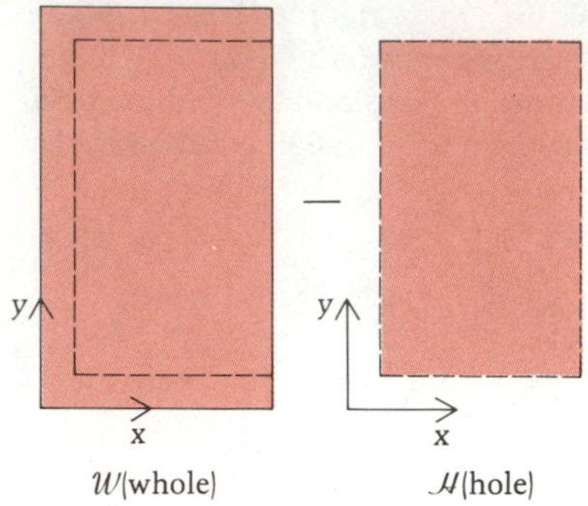

Figure E8.12b

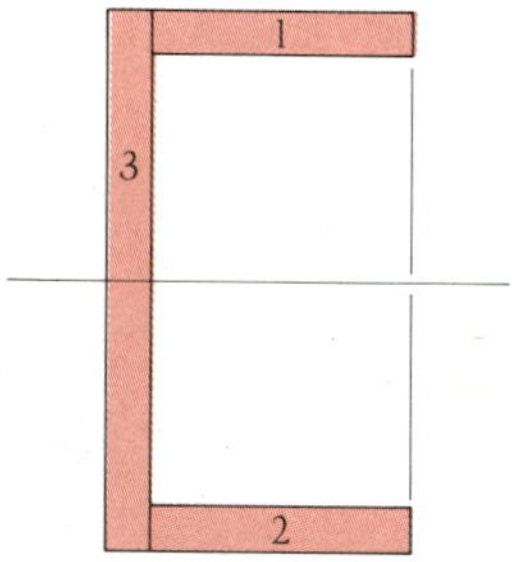

Figure E8.12c

Solution

We shall solve the problem by finding I_x and then using the transfer theorem. We shall find I_x in two different ways. First, using Figure E8.12b,

$$I_x = \underbrace{I_x^{\mathcal{W}}} - \underbrace{I_x^{\mathcal{H}}}$$

$$= \frac{3.5(6^3)}{3} - \left[\frac{3 \times 5^3}{12} + (3 \times 5)(2.5 + 0.5)^2\right]$$

$$= 252 - 166$$

$$= 86 \text{ in.}^4$$

Alternatively, using Figure E8.12c,

$$I_x = I_x^{A_1} + I_x^{A_2} + I_x^{A_3}$$

$$= \left[\frac{3(0.5^3)}{12} + (3 \times 0.5)(5.75^2)\right] + \left[\frac{3(0.5^3)}{12} + (3 \times 0.5)(0.25^2)\right]$$

$$+ \left[\frac{0.5(6^3)}{12} + (0.5 \times 6)(3^2)\right]$$

$$= 49.6 + 0.125 + 36.0 = 85.7 \approx 86 \text{ in.}^4 \qquad \text{(as before)}$$

Using the parallel-axis theorem,

$$I_x = I_{x_C} + Ad^2$$

$$85.7 = I_{x_C} + [6 \times 3.5 - 5 \times 3](3)^2$$

$$I_{x_C} = 85.7 - 6(3^2) = 31.7 \text{ in.}^4$$

The calculation of I_{y_C} will be given as an exercise. (See Problem 8.73.)

EXAMPLE 8.13

Show that even if a triangle, such as the one shown in Figure E8.13, has no right angle, its moment of inertia about a centroidal axis parallel to a base B is still $BH^3/36$, where H is the height of the triangle in the direction normal to that base.

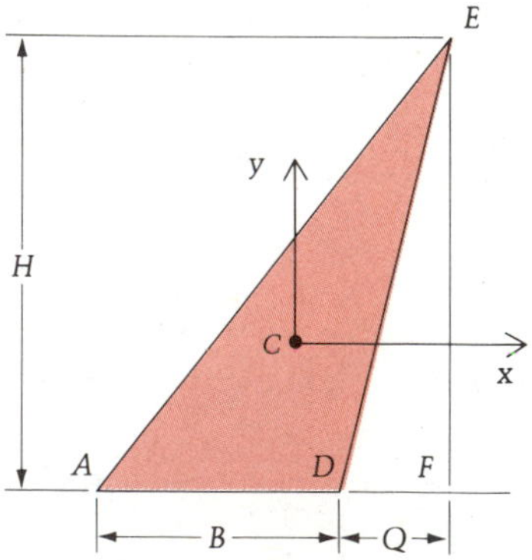

Figure E8.13

Solution

Let us call the shaded triangle AED of interest $\mathcal{J}$, and denote the two right triangles AEF and DEF by $\mathcal{W}$ and $\mathcal{H}$, respectively. Note that $A_{\mathcal{J}} = A_{\mathcal{W}} - A_{\mathcal{H}}$, so that

$$I_{x_C}^{\mathcal{J}} = \int_{\mathcal{J}} y^2\, dA = \int_{\mathcal{W}} y^2\, dA - \int_{\mathcal{H}} y^2\, dA$$

$$= \frac{(B+Q)H^3}{36} + \frac{(B+Q)H}{2}(d^2_{\mathcal{W}\mathcal{J}}) - \left[\frac{QH^3}{36} + \frac{QH}{2}(d^2_{\mathcal{H}\mathcal{J}})\right]$$

where $d_{\mathcal{W}\mathcal{J}}$ and $d_{\mathcal{H}\mathcal{J}}$ are the respective distances between the x axis in the figure and the x axes through the centroids of $\mathcal{W}$ and $\mathcal{H}$. But all three of these x axes are coincident, a distance $H/3$ above the base line ADF, because of earlier general results for centroids of triangles. Therefore, $d_{\mathcal{W}\mathcal{J}} = d_{\mathcal{H}\mathcal{J}} = 0$, and we obtain

$$I_{x_C}^{\mathcal{T}} = \frac{BH^3}{36}$$

the same result that we obtained for the right triangle in Example 8.8.

Example 8.14

From previous examples, we have the moments of inertia shown in Figure E8.14. Show that I_x for the rectangle follows from the indicated moments of inertia for the two right triangles, using the transfer theorem and the method of composite areas.

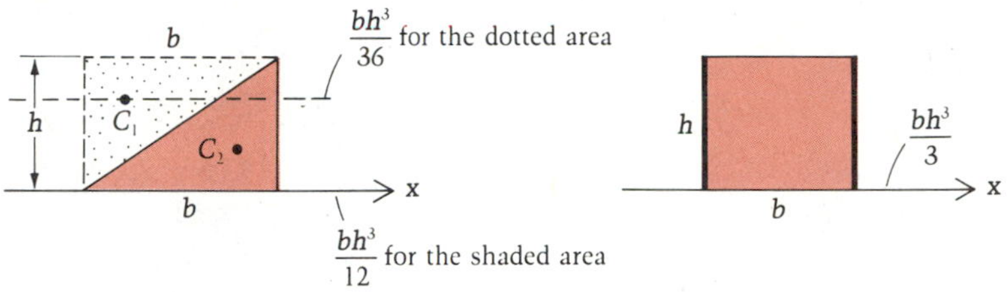

Figure E8.14

Solution

Transferring the moment of inertia of the dotted area to the x axis gives

$$I_x^{\text{dotted}} = \underbrace{\frac{bh^3}{36}}_{I_{x_{C_1}}^{\text{dotted}}} + \underbrace{\frac{1}{2}bh}_{A}\underbrace{\left(\frac{2}{3}h\right)^2}_{d^2} = \frac{bh^3}{4}$$

Then,

$$I_x^{\text{rectangle}} = I_x^{\text{dotted}} + I_x^{\text{shaded}}$$

$$= \frac{bh^3}{4} + \frac{bh^3}{12} = \frac{bh^3}{3}$$

which agrees with the previously obtained answer for the rectangle.

Example 8.15

Find the moments of inertia about the x and y axes through the centroid of the cross-sectional area shown in Figure E8.15. This cross section of a wide-flange beam is called a W14 × 84 because its nominal depth is 14 in. and it weighs 84 lb per ft.

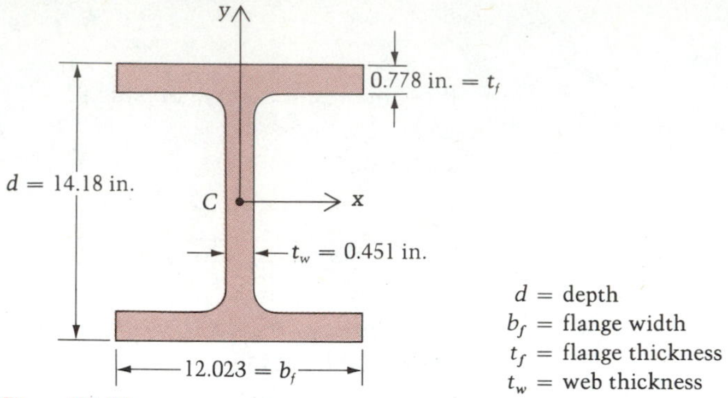

Figure E8.15

Solution

$$I_{x_C} = I_{x_C}^{\text{web}} + I_{x_C}^{\text{flange}}$$

$$= \underbrace{\frac{0.451[14.18 - 2(0.778)]^3}{12}}_{\text{web}}$$

$$+ \underbrace{2\left[\frac{12.023(0.778)^3}{12} + 12.023(0.778)\left(\frac{14.18}{2} - \frac{0.778}{2}\right)^2\right]}_{\text{flanges}}$$

Thus

$$I_{x_C} = 75.6 + 2[0.472 + 420] = 916 \text{ in.}^4$$

Notice that the flanges contribute most of the inertia (91.7%!) because of the large "transfer term." Continuing,

$$I_{y_C} = \underbrace{2\left[\frac{0.778(12.023)^3}{12}\right]}_{\text{flanges}} + \underbrace{\frac{[14.18 - 2(0.778)](0.451)^3}{12}}_{\text{web}}$$

$$= 225 + 0.0965 = 225 \text{ in.}^4$$

Notice how tiny the web contribution is here. (Its area is all very close to the y axis!)

The American Institute of Steel Construction manual lists I_{x_C} and I_{y_C} to be 928 in.^{4} and 225 in.^{4} The table values will generally be higher than the results of calculations such as we have made above because of the additional material due to the rounded fillets where the web and flanges are joined. To further illustrate this point, note that we would compute the area of the section to be

$$A = 2(12.023)(0.778) + [14.18 - 2(0.778)](0.451)$$

$$= 18.7 + 5.69 = 24.4 \text{ in.}^2$$

whereas the true are (again from the steel construction manual) is 24.7 in.^{2}

Question 8.8 Why do you suppose I_{x_C} is higher in the manual than we have calculated it to be, while the I_{y_C} values are the same?

Answer 8.8 Because the fillets are so close to the y axis and so relatively far away from the x axis, they will contribute, relatively, a lot to I_{x_C} and very little to I_{y_C}.

PROBLEMS ▸ Section 8.5

8.41 Compute I_x for the I section shown in Figure P8.41.

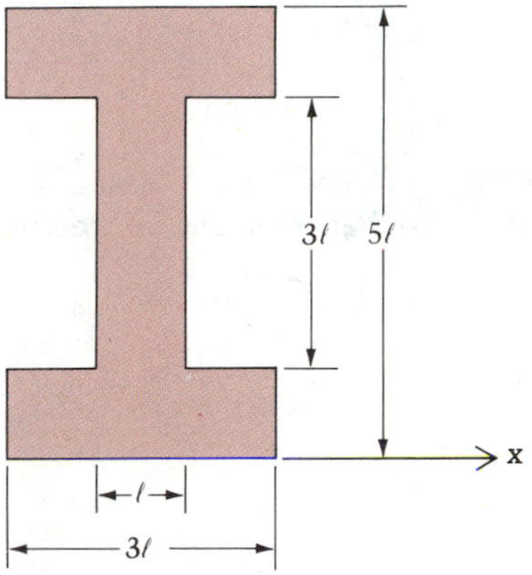

Figure P8.41

8.42 Find the moment of inertia I_{x_C} of the hollow rectangular area shown in Figure P8.42.

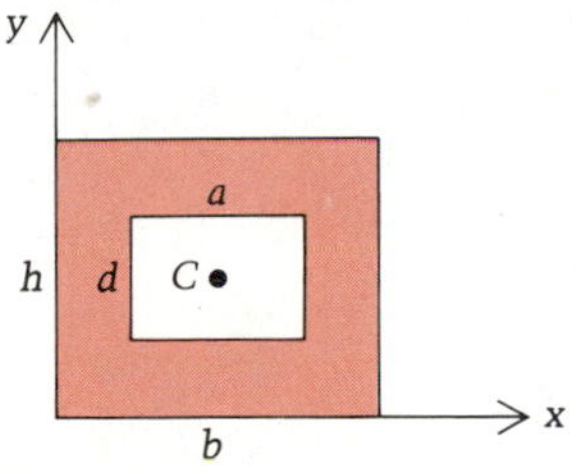

Figure P8.42

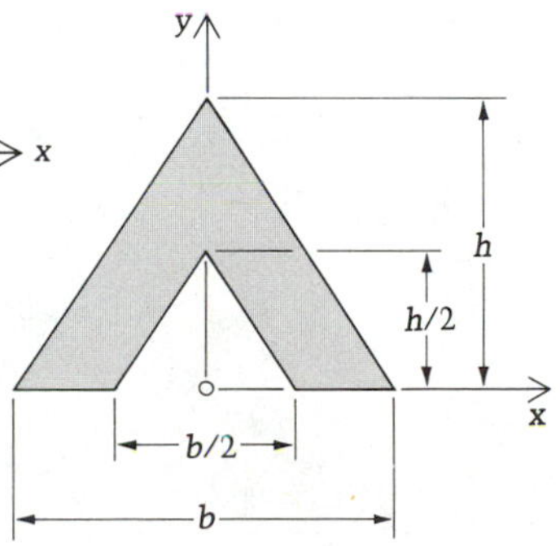

Figure P8.43

8.43 Find I_x for the area shown in Figure P8.43.

8.44 In Problem 8.43 find I_y.

In Problems 8.45–8.68 find the moment of inertia about the x axis (odd numbers) and y axis (even numbers) for the areas.

8.45, 8.46

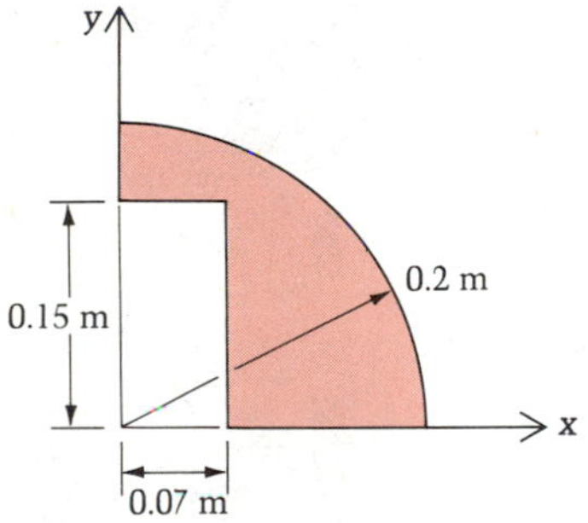

Figure P8.45, P8.46

8.47, 8.48

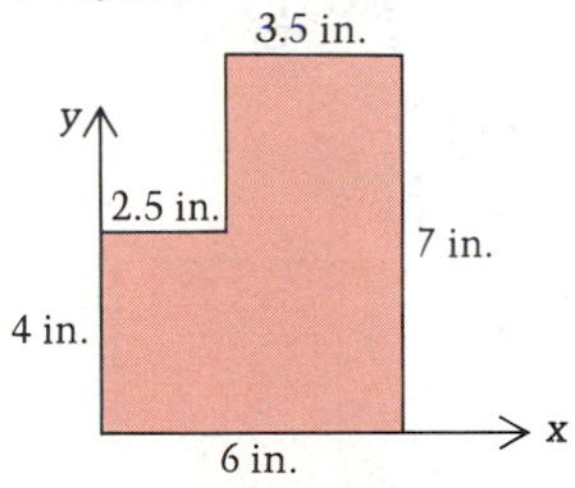

Figure P8.47, P8.48

8.49, 8.50

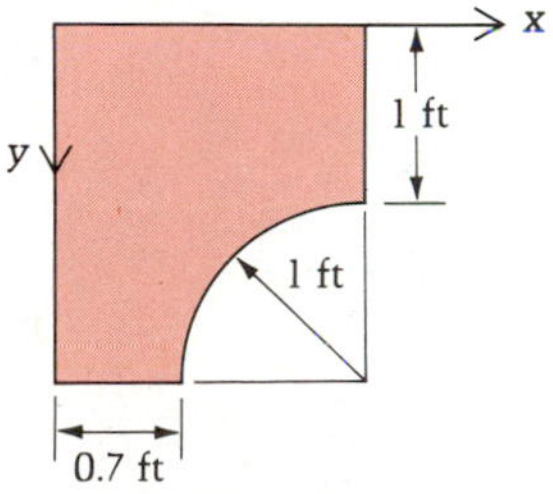

Figure P8.49, P8.50

For Problems 8.51–8.68, see the instructions above Problem 8.45.

8.51, 8.52

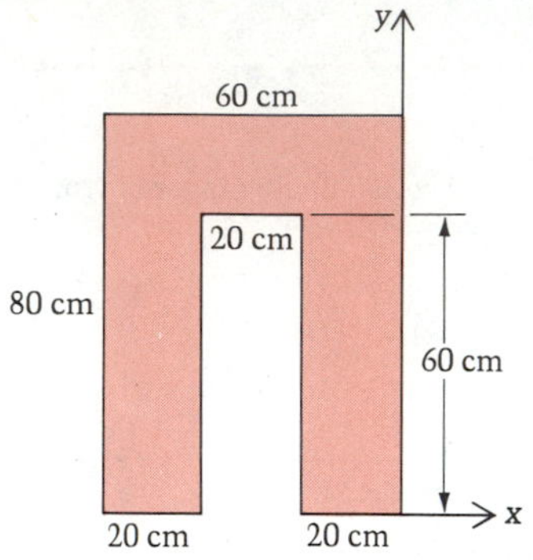

Figure P8.51, P8.52

8.53, 8.54

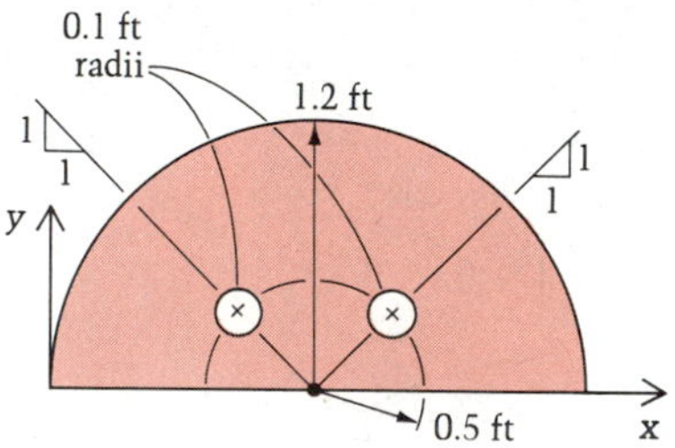

Figure P8.53, P8.54

8.55, 8.56

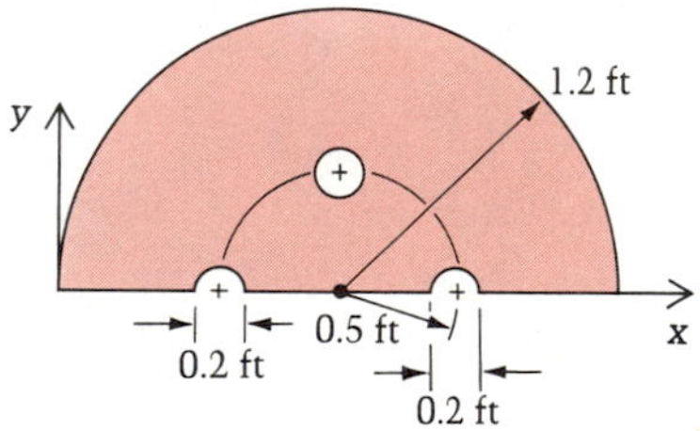

Figure P8.55, P8.56

8.57, 8.58

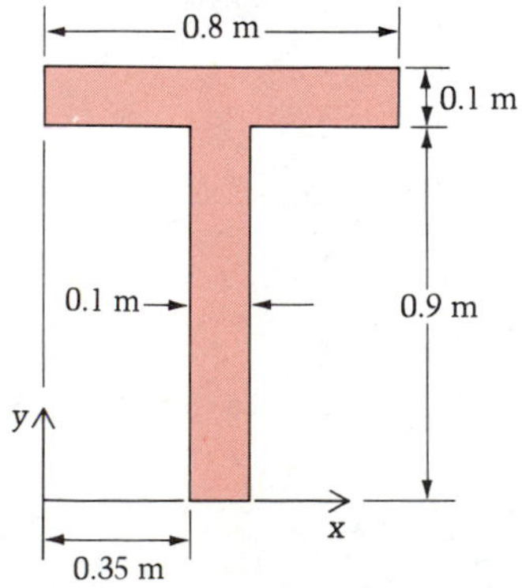

Figure P8.57, P8.58

8.59, 8.60

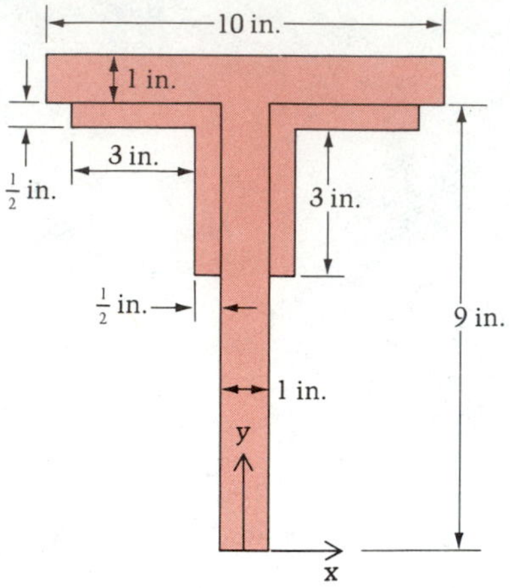

Figure P8.59, P8.60

Problem 8.61 is actually an S24 × 120 I beam, for which the published centroidal moment of inertia is $I_x = 3030\text{ in.}^4$

8.61, 8.62

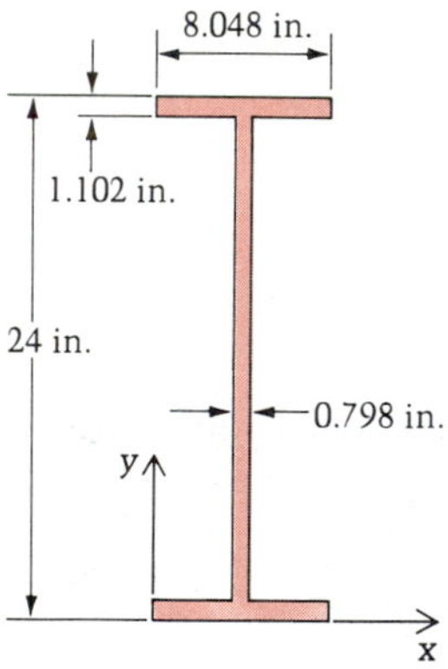

Figure P8.61, P8.62

8.63, 8.64

Hexagon, side length *s*

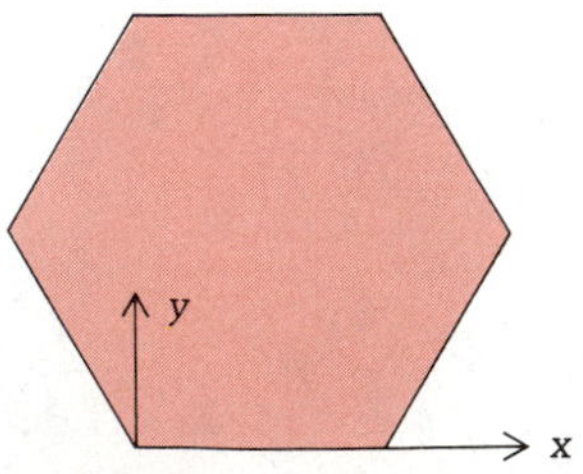

Figure P8.63, P8.64

8.65, 8.66

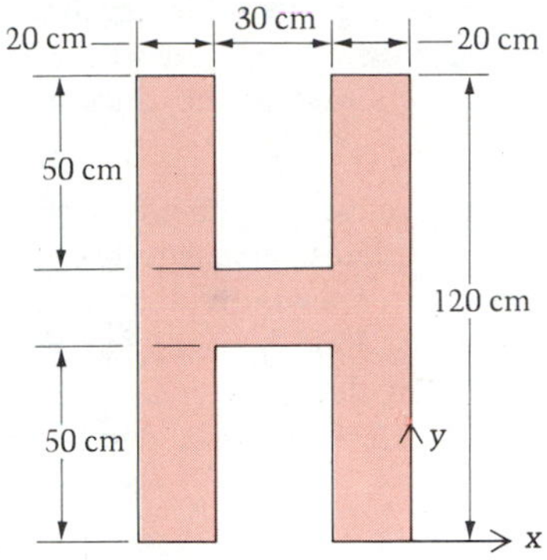

Figure P8.65, P8.66

*** 8.67, 8.68**

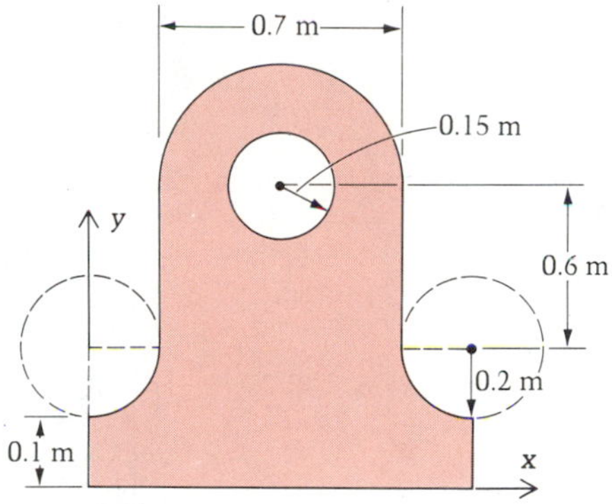

Figure P8.67, P8.68

8.69 Show that the moment of inertia I_{x_C} of the hexagonal area in Figure P8.69 has the value indicated.

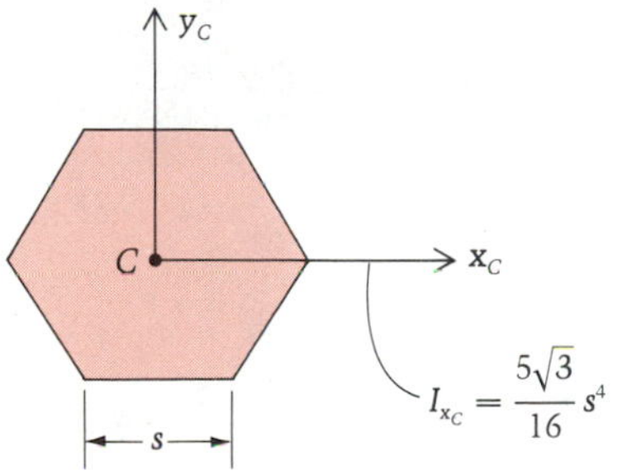

Figure P8.69

8.70 In Problem 8.69 find I_{y_C} and then note that it is the same as I_{x_C}.

* Asterisks identify the more difficult problems.

8.71 Find the moment of inertia I_{x_C} of the cross-sectional area of a steel angle beam. Assume a constant thickness of $\frac{5}{8}$ in., and compare your answer with the actual (considering rounded corners and fillets) value of 21.1 in.[4] The x and y coordinates of C in Figure P8.71 are (1.03, 2.03) in.

8.72 Find I_{y_C} in Problem 8.71. Compare with the table value of 7.5 in.[4]

8.73 Find I_{y_C} for the cross-sectional area of the channel beam of Example 8.12. Do it in two ways as was done in the example in finding I_{x_C}. (See Figure P8.73.)

8.74 Find I_{x_C} for the shaded Z section in Figure P8.74.

8.75 Find I_{y_C} for the area of Problem 8.74.

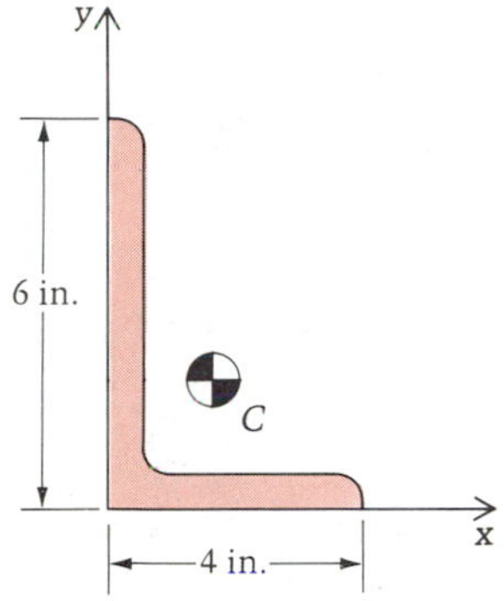

Figure P8.71

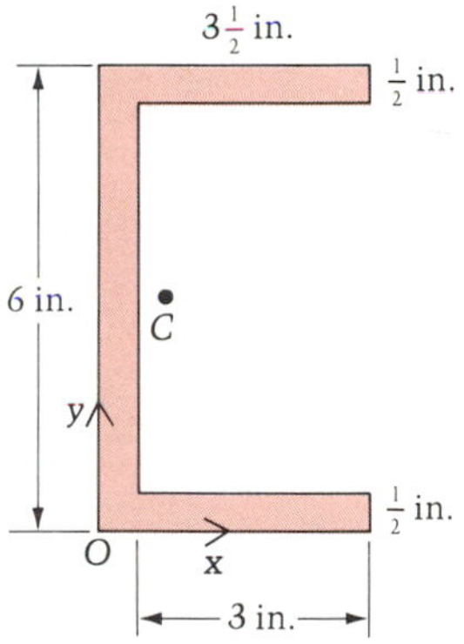

Figure P8.73

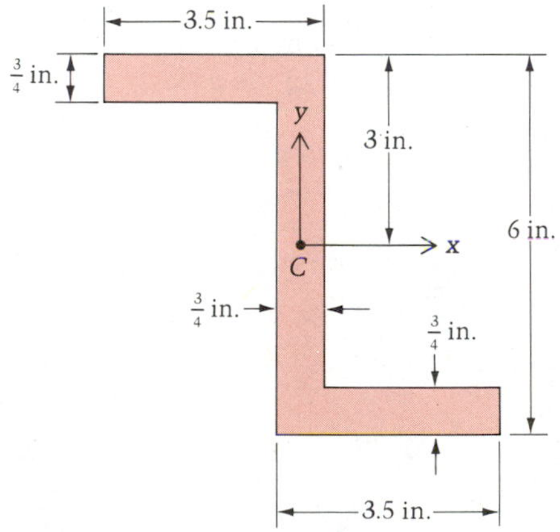

Figure P8.74

8.76 Find the moment of inertia of the composite area about the x axis through the centroid.

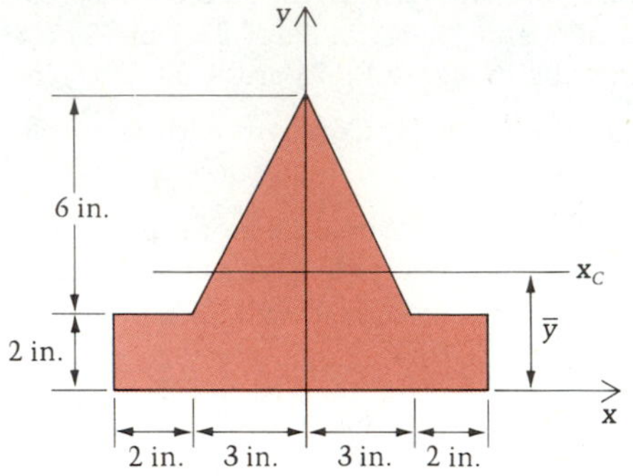

Figure P8.76

8.77 Find the moment of inertia of the composite area in Problem 8.76 about the y axis.

8.78 Prove that the moments of inertia of the triangular area about the two indicated lines have the values shown in Figure P8.78.

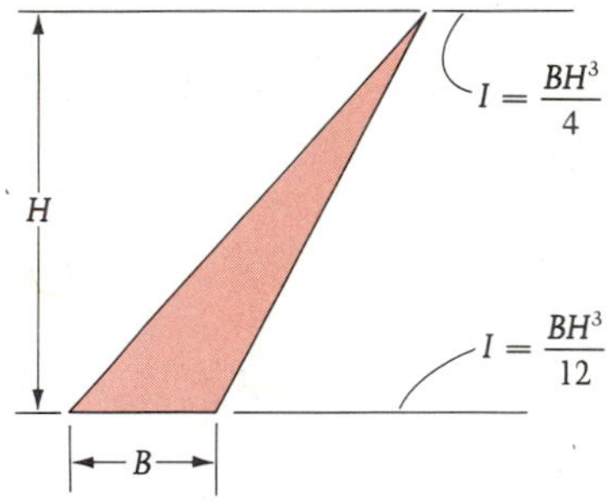

Figure P8.78

8.79 In Figure P8.79 find the moment of inertia of the shaded area about (a) the x axis; (b) the y axis.

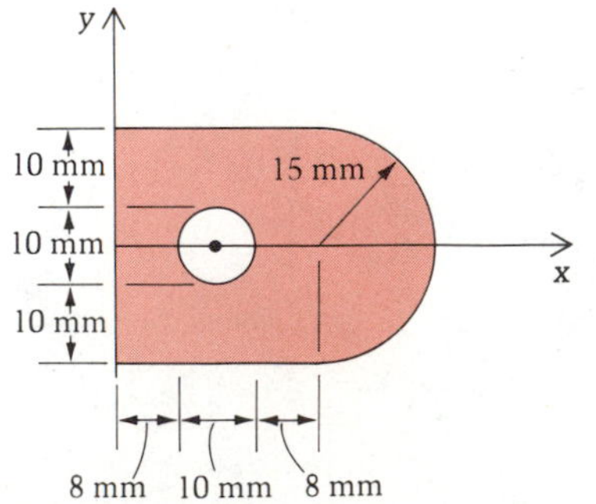

Figure P8.79

8.80 In Figure P8.80 find the polar moment of inertia with respect to C of the annular area by (a) integration; (b) composite areas. (c) Show that your answer approaches $2\pi R_o^3 t$ as $R_i \rightarrow R_o$, where t is the thickness $(R_o - R_i)$.

8.81 A box beam is formed by welding together two steel C15 × 50 channels and two plates as suggested by Figure P8.81(a). The properties of an individual channel are shown in Figure P8.81(b). Find I_{x_C} and I_{y_C} for the area of the box beam. (*Note*: The designation C15 × 50 means: C = channel section; 15 = nominal depth in inches; 50 = weight in lb/ft.)

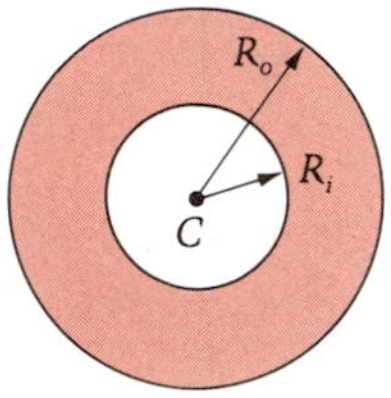

Figure P8.80

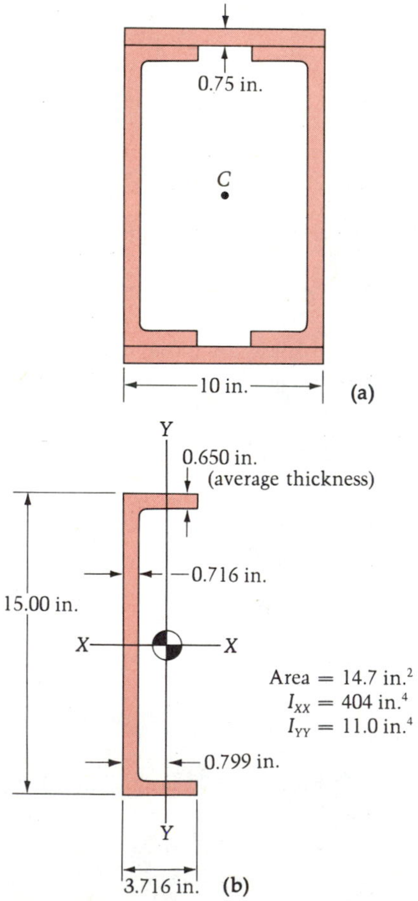

Figure P8.81

8.82 The W12 × 65 steel beam cross section shown in Figure P8.82 has the centroidal moments of inertia $I_{XX} = 533$ in.^{4} and $I_{YY} = 175$ in.^{4} Find the centroidal moments of inertia if two of these beams are welded together to form the combined cross section.

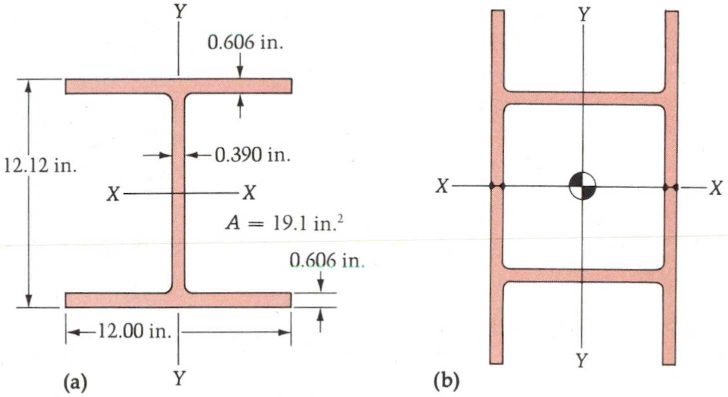

Figure P8.82

8.83 (a) Find I_x and I_y for the section shown in Figure P8.83 in which $t \ll b, h$. (Neglect all powers of t higher than the first.) (b) Find the ratio b/h for which $I_x = I_y$. *Hint*: You'll have to solve a cubic polynomial equation! Use trial and error with a calculator, and note that there is only one positive root.

8.84 The two 6 × 4 × $\frac{7}{8}$ angles are welded to the shaded steel plate as shown in Figure P8.84. Find the moment of inertia and radius of gyration of the combined area about the x axis.

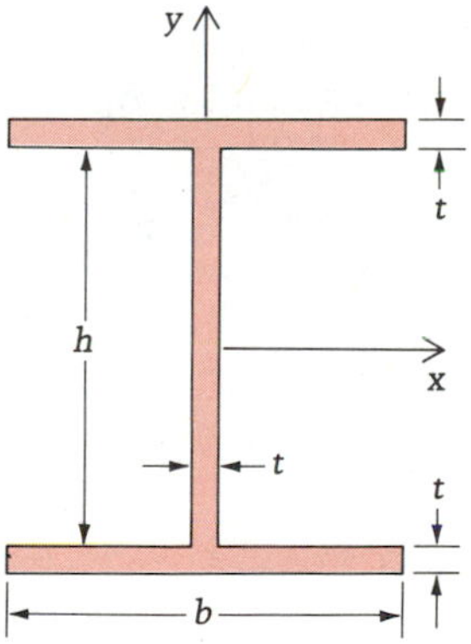

Figure P8.83

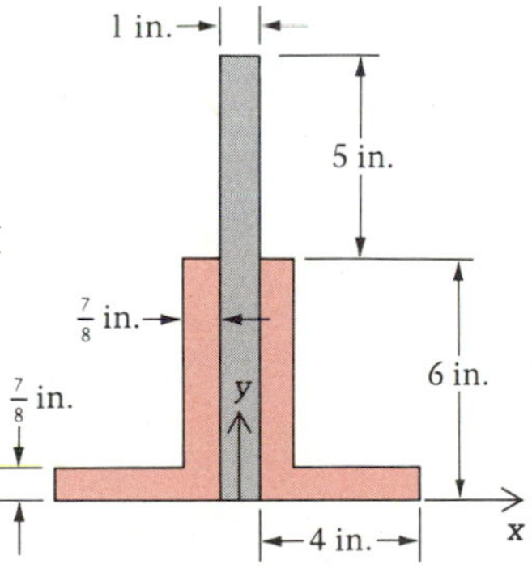

Figure P8.84

8.85 Repeat Problem 8.84 for the moment of inertia about the y axis.

8.86 In Figure P8.86 show that the moment of inertia of the trapezoidal area about the x axis is $(3b + a)H^3/12$, independent of the angles α_1 and α_2.

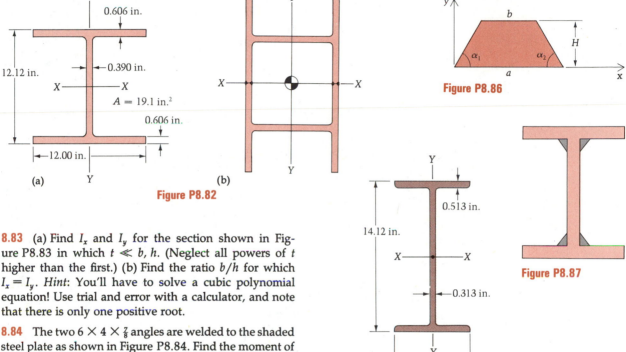

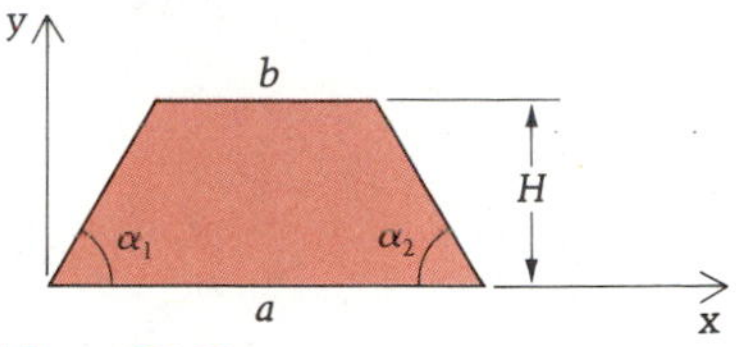

Figure P8.86

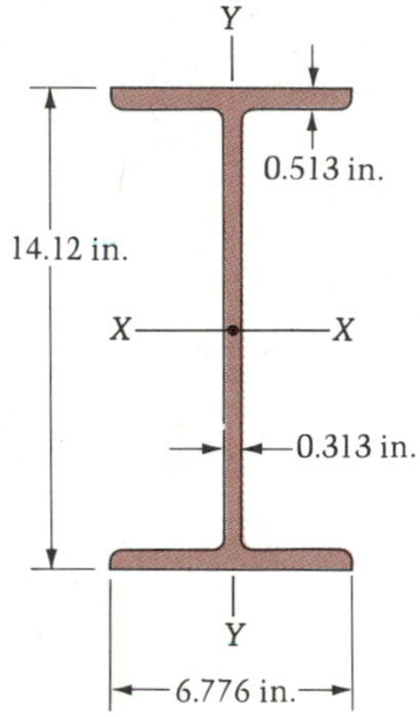

Figure P8.88

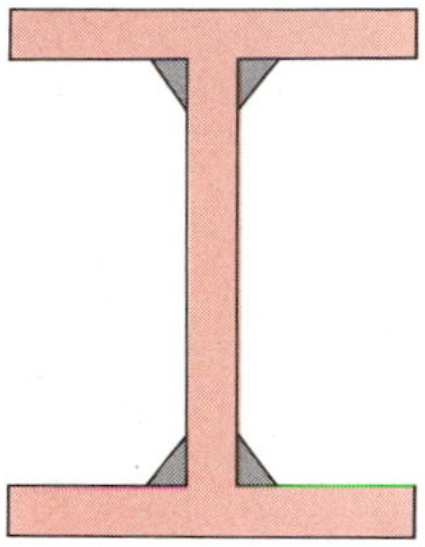

Figure P8.87

8.87 In Example 8.15 assume that four isosceles triangles form the fillets that give the difference between the true area, 24.7 in.^{2}, and the 24.4-in.^{2} area of the web and flanges. (See Figure P8.87.) Show that these triangles (a) contribute the difference between I_{x_c} in the AISC manual (928 in.^{4}) and in the example (916 in.^{4}) and that (b) the triangles contribute negligibly to I_{y_c} so that the manual and example values (225 in.^{4}) should in fact agree as they do.

8.88 A W14 × 38 (meaning the depth is about 14 in. and the weight per ft is 38 lb) wide-flange beam has the cross section shown in Figure P8.88. Calculate the moment of inertia about the centroidal axis X-X and the radius of gyration about this axis. Compare your answers with the table (American Institute of Steel Construction Manual) values of 386 in.^{4} and 5.88 in. (They will differ slightly due to fillets.)

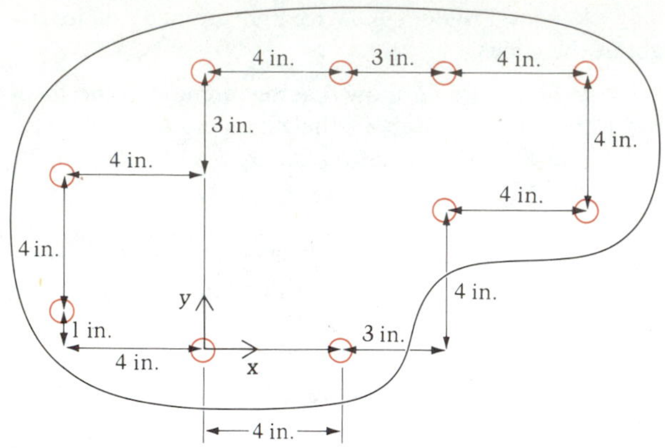

Figure P8.91

8.89 Repeat Problem 8.88 for axis Y-Y. The table results this time are 26.6 in.^{4} and 1.54 in.

8.90 Find the polar moment of inertia of the area of Problem 8.41 about its centroid.

8.91 Find the centroidal polar moment of mertia of the bolt pattern in Example 8.11 by direct calculation using $\Sigma A_i r_i^2$, where r_i = distance from A_i to C. Your answer should, of course, agree with J_C from that example. See Figure P8.91.

8.92 Find the polar moment of inertia of the area of Problem 8.79 about the center of the hole.

8.6 Products of Inertia of Plane Areas

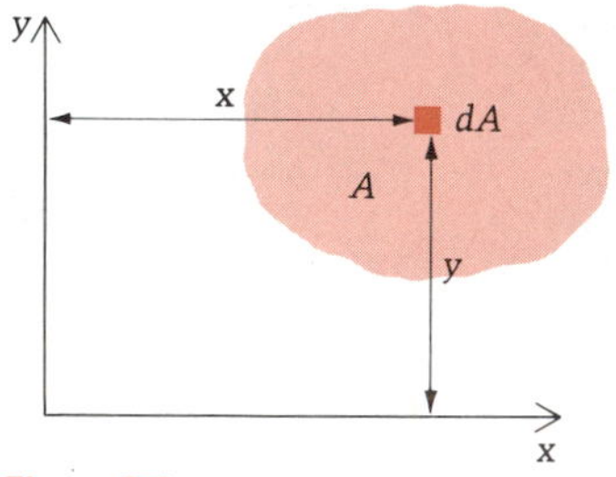

Figure 8.6

Suppose our area A of interest is again in the xy plane, as suggested by Figure 8.6. Then the product of inertia with respect to the axes x and y is defined to be*

$$I_{xy} = -\int_A xy\, dA$$

Thus the product of inertia is a measure of the imbalance of the area with respect to the two axes. Two important special cases in which the product of inertia vanishes are:

a. If y is an axis of symmetry, then $I_{xy} = 0$ (see Figure 8.7(a)). This is because *each dA* has a mirror image on the other side of the y axis so that $xy\, dA$ and $(-x)y\, dA$ add to zero. Over the entire region, then, the integral vanishes.

b. If x is an axis of symmetry, then I_{xy} again $= 0$ (see Figure 8.7(b)). This time, *each dA* has a mirror image on the other

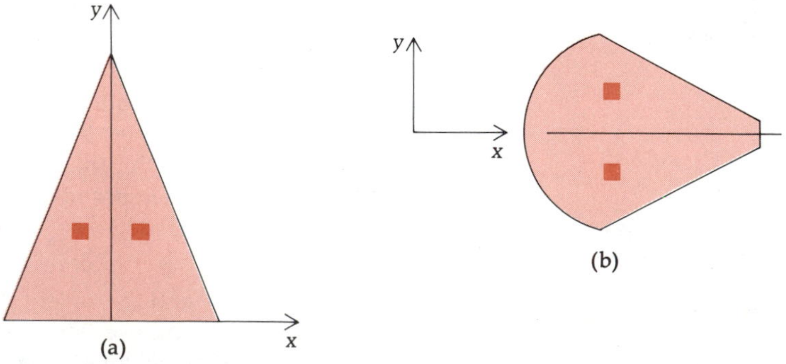

Figure 8.7

* Some authors define I_{xy} without the minus sign. The reason for our choice here is noted at the end of Section 8.7.

side of the x axis, and $xy\,dA$ plus $x(-y)\,dA$ add to zero. I_{xy} thus again equals zero.

Frequently, however, we have to deal with *un*symmetrical areas for which I_{xy} might not be zero. Four such examples follow.

Example 8.16

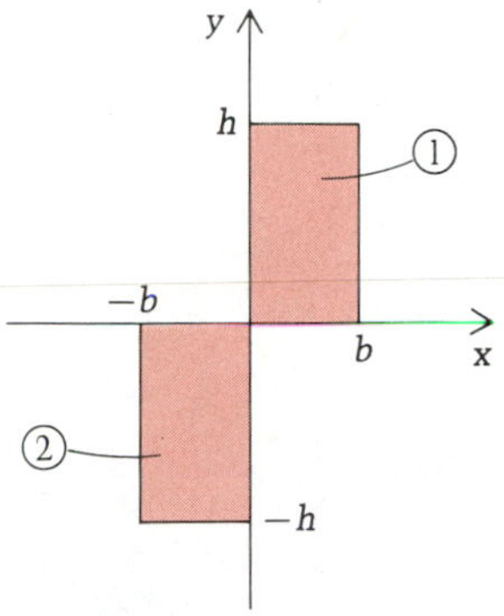

Figure E8.16

Find I_{xy} for the shaded area in Figure E8.16.

Solution

Letting b and h be the base and height of each rectangle, we have

$$I_{xy} = -\int_{A_1} xy\,dA - \int_{A_2} xy\,dA$$

$$= -\int_{y=0}^{h}\int_{x=0}^{b} xy\,dx\,dy - \int_{y=-h}^{0}\int_{x=-b}^{0} xy\,dx\,dy$$

$$= \int_{y=0}^{h} -\frac{x^2}{2}\bigg|_0^b y\,dy - \int_{y=-h}^{0} \frac{x^2}{2}\bigg|_{-b}^{0} y\,dy$$

$$= \frac{-b^2}{2}\frac{y^2}{2}\bigg|_0^h + \frac{b^2}{2}\frac{y^2}{2}\bigg|_{-h}^{0} = \frac{-b^2h^2}{4} - \frac{b^2h^2}{4}$$

or

$$I_{xy} = \frac{-b^2h^2}{2}$$

Example 8.17

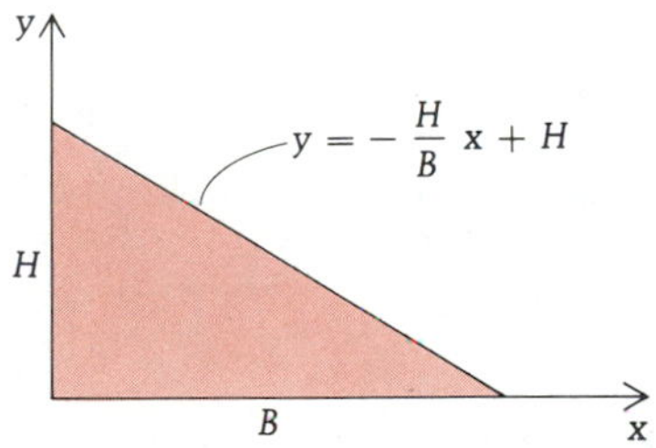

Figure E8.17

Find I_{xy} for the shaded area in Figure E8.17.

Solution

$$I_{xy} = -\int xy\,dA$$

$$I_{xy} = -\int_{y=0}^{H}\int_{x=0}^{\frac{B}{H}(H-y)} xy\,dx\,dy = -\int_{y=0}^{H} y\frac{x^2}{2}\bigg|_0^{\frac{B}{H}(H-y)} dy$$

$$= -\int_0^H \frac{B^2y}{2H^2}(H-y)^2\,dy = \frac{-B^2}{2H^2}\int_0^H (H^2y - 2Hy^2 + y^3)\,dy$$

$$= \frac{-B^2}{2H^2}H^4\left(\frac{1}{2} - \frac{2}{3} + \frac{1}{4}\right) = \frac{-B^2H^2}{24}$$

Example 8.18

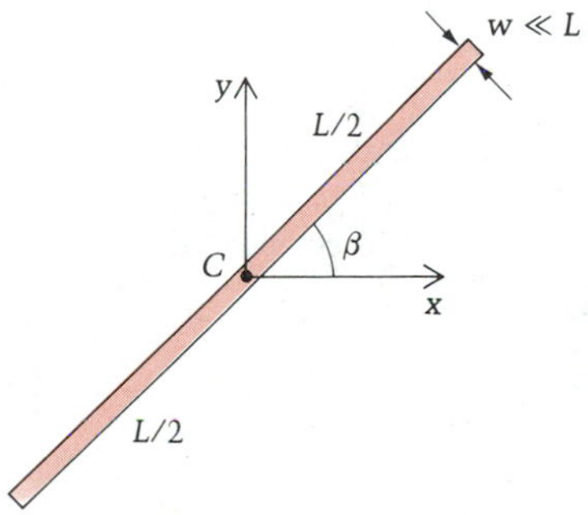

Figure E8.18a

Find the centroidal product of inertia I_{xy_C} for the *thin* strip shown in Figure E8.18a. The strip lies in the xy plane.

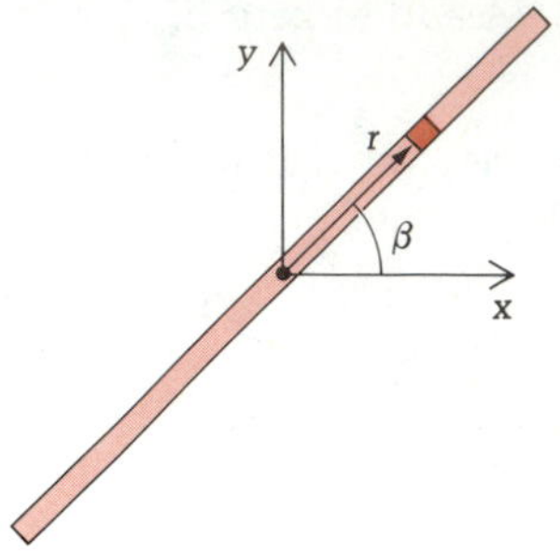

Figure E8.18b

Solution

We shall use the polar coordinate r as shown in Figure E8.18b. The differential area will be $w\,dr$:

$$I_{xy_C} = -\int xy\,dA = -\int_{r=-\frac{L}{2}}^{L/2} (r\cos\beta)(r\sin\beta)w\,dr$$

$$= -w\sin\beta\cos\beta \int_{-L/2}^{L/2} r^2\,dr$$

$$= \frac{-wL^3\sin\beta\cos\beta}{12}$$

$$= \frac{-wL^3\sin 2\beta}{24}$$

Question 8.9 Why is this result only approximately correct?

Answer 8.9 Because $r\cos\beta$ and $r\sin\beta$ are not the (x, y) coordinates of all the points in the "dA." The smaller is w relative to L, the better the approximation.

EXAMPLE 8.19

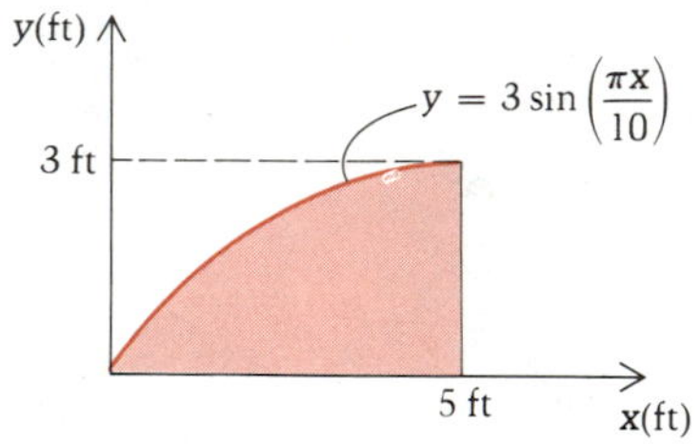

Figure E8.19

Find I_{xy} for the shaded area shown in Figure E8.19.

Solution

$$I_{xy} = -\int xy\,dA = -\int_{x=0}^{x=5}\int_{y=0}^{y=3\sin\frac{\pi x}{10}} xy\,dy\,dx$$

$$= -\int_{x=0}^{5} \left.\frac{xy^2}{2}\right|_{y=0}^{3\sin\frac{\pi x}{10}} dx$$

$$= -\int_{x=0}^{5} \frac{x}{2}\,9\sin^2\frac{\pi x}{10}\,dx$$

$$= -4.5\int_{x=0}^{5} x\left(\frac{1-\cos(\pi x/5)}{2}\right)dx$$

and using integration by parts,

$$I_{xy} = -4.5\left.\frac{x^2}{4}\right|_0^5 + 2.25\left.\left(x\frac{\sin(\pi x/5)}{\pi/5}\right)\right|_0^5 - 2.25\int_{x=0}^{5}\frac{\sin(\pi x/5)}{\pi/5}\,dx$$

$$= -28.1 + 0 + 2.25\left(\frac{5}{\pi}\right)^2 \left.\cos(\pi x/5)\right|_0^5$$

$$= -28.1 - 11.4 = -39.5\ \text{ft}^4$$

PROBLEMS ► Section 8.6

8.93 Find the product of inertia I_{xy} of the shaded area in Figure P8.93.

8.94 In Figure P8.94:

a. Find the product of inertia I_{xy} of the thin strip bent into two quarter circles as shown.

b. Compare your result with that of Example 8.18 for the case in which $\beta = \pi/4$ and when $L/2 = 2\pi R/4$ — that is, when the lengths of the strips in the two problems are the same.

8.95 Find the moments and product of inertia I_x, I_y, and I_{xy} of the shaded area shown in Figure P8.95.

8.96 In Figure P8.96 show that for the shaded triangular area,

$$I_{xy} = -\frac{bH^2(2d + b)}{24}$$

8.97 Show that $I_x = I_y = 2b^4/3$ and $I_{xy} = b^4/2$ for the shaded area in Figure P8.97.

8.98 Find I_{xy} for the shaded area in Figure P8.98.

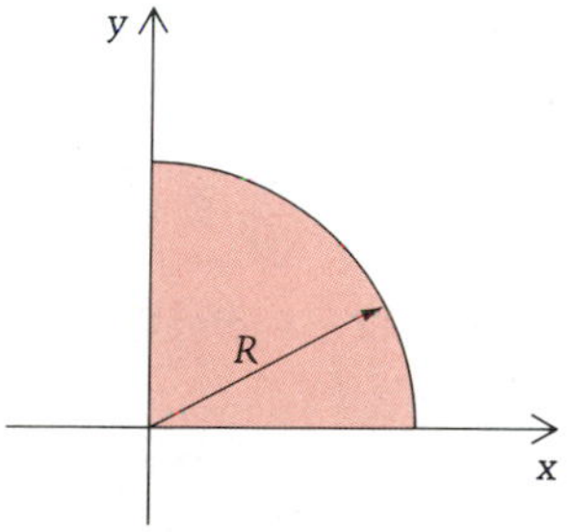

Figure P8.93

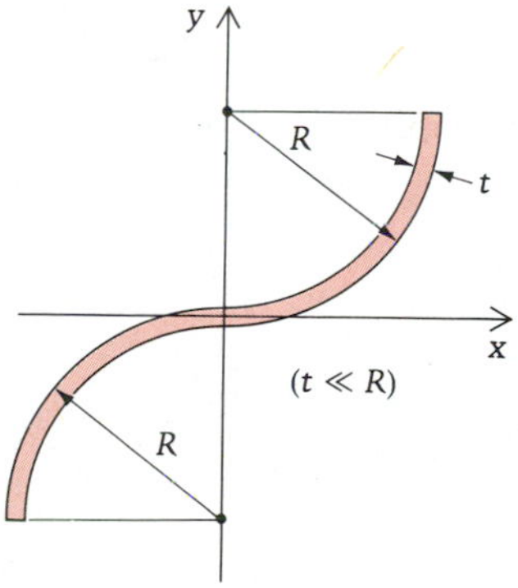

Figure P8.94

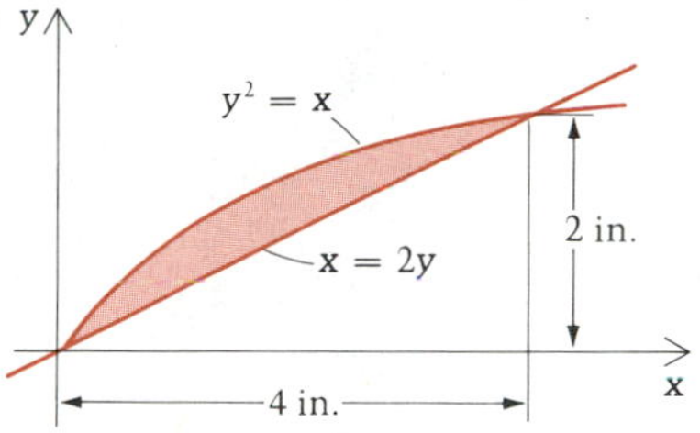

Figure P8.95

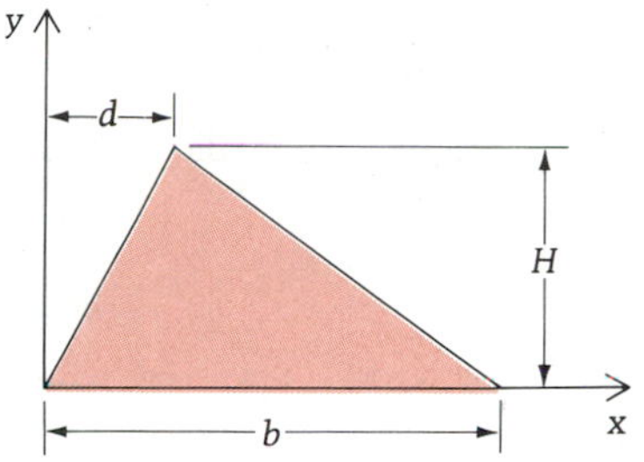

Figure P8.96

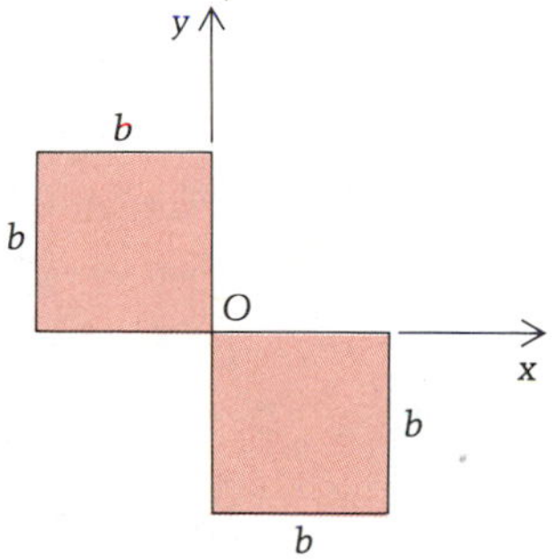

Figure P8.97

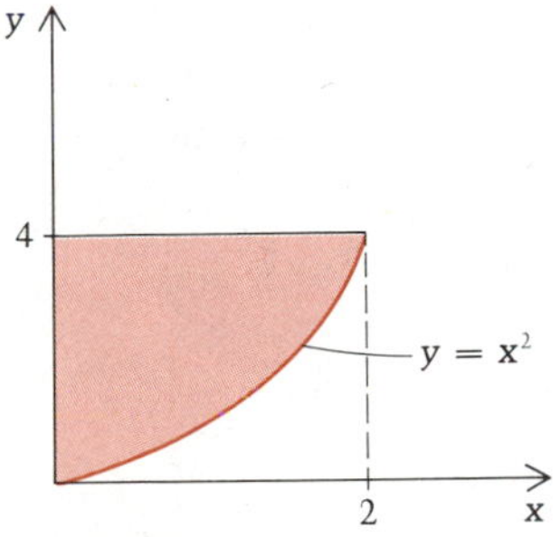

Figure P8.98

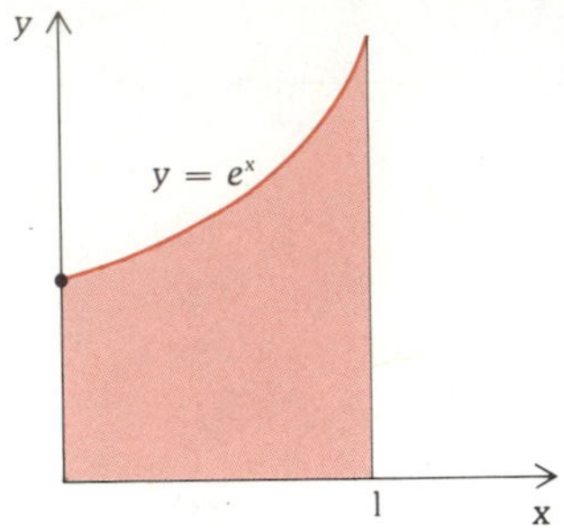

Figure P8.99

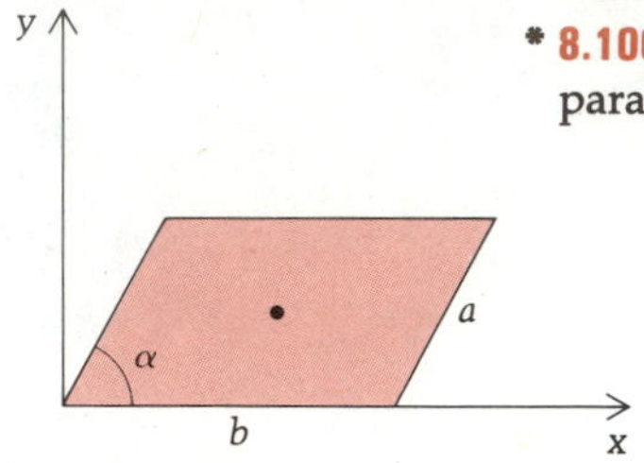

Figure P8.100

8.99 Find the product of inertia I_{xy} of the area shown in Figure P8.99.

* **8.100** By integration, find the product of inertia I_{xy} of the parallelogram shown in Figure P8.100.

8.7 The Parallel-Axis Theorem for Products of Inertia

There is a parallel-axis theorem, or transfer theorem, for products of inertia just as there is for moments of inertia. It is

$$I_{xy_P} = I_{xy_C} - A\bar{x}\bar{y}$$

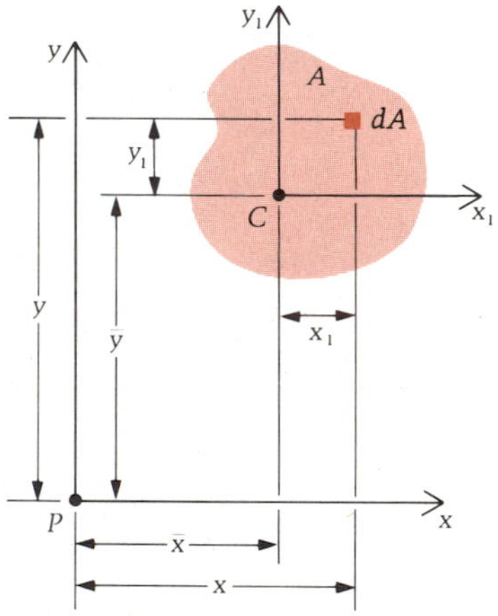

Figure 8.8

The proof goes as follows: Referring to Figure 8.8,

$$I_{xy_P} = -\int x\, y\; dA = -\int (\bar{x} + x_1)(\bar{y} + y_1) dA$$

$$I_{xy_P} = -\bar{x} \int y_1\, dA - \bar{y} \int x_1\, dA - \int x_1 y_1\, dA - \bar{x}\bar{y} \int dA$$

The first two integrals vanish by the definition of the centroid, leaving the desired theorem:

$$I_{xy_P} = I_{xy_C} - A\bar{x}\bar{y}$$

Question 8.10 Do we need to memorize whether the coordinates $(\bar{x}, \bar{y})$ locate "C with respect to P" as compared to "P with respect to C"?

Answer 8.10 No, because *both* signs change if we go the other way!

EXAMPLE 8.20

Obtain the product of inertia I_{xy} for the area of Example 8.16, shown here as Figure E8.20, by using composite areas and the transfer theorem.

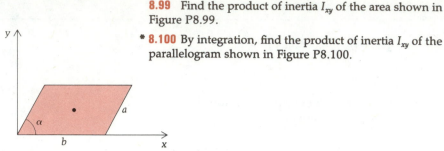

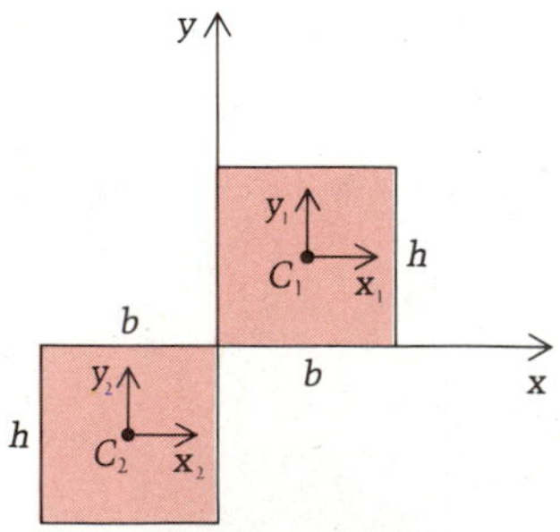

Figure E8.20

Solution

$$I_{xy} = (I^{A_1}_{x_1y_1} - A_1\bar{x}_1\bar{y}_1) + (I^{A_2}_{x_2y_2} - A_2\bar{x}_2\bar{y}_2)$$

$$= \left[0 - bh\left(\frac{b}{2}\right)\left(\frac{h}{2}\right)\right] + \left[0 - bh\left(-\frac{b}{2}\right)\left(-\frac{h}{2}\right)\right]$$

$$= -\frac{b^2h^2}{4} - \frac{b^2h^2}{4} = -\frac{b^2h^2}{2} \qquad \text{(as before)}$$

Question 8.11 Do we have to have one of the symmetries described at the beginning of Section 8.5 in order for I_{xy} to be zero? If not, think of a case in which $I_{xy} = 0$ even though the area is not symmetric with respect to the x or y axes.

Answer 8.11 No. I_{xy} is zero for this area:

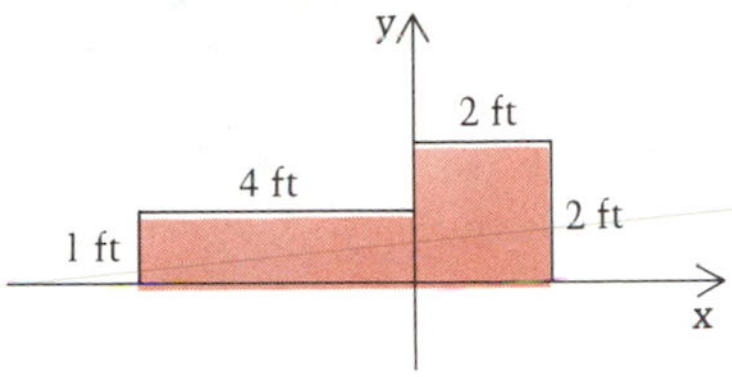

EXAMPLE 8.21

Find I_{xy_C} for the shaded right triangular area shown in Figure E8.21.

Solution

From Example 8.17 we had $I_{xy} = \dfrac{-B^2H^2}{24}$. Using the parallel-axis theorem,

$$I_{xy} = I_{xy_C} - A\bar{x}\bar{y}$$

$$I_{xy_C} = \frac{-B^2H^2}{24} + \frac{BH}{2}\frac{H}{3}\frac{B}{3} = \frac{B^2H^2}{72}$$

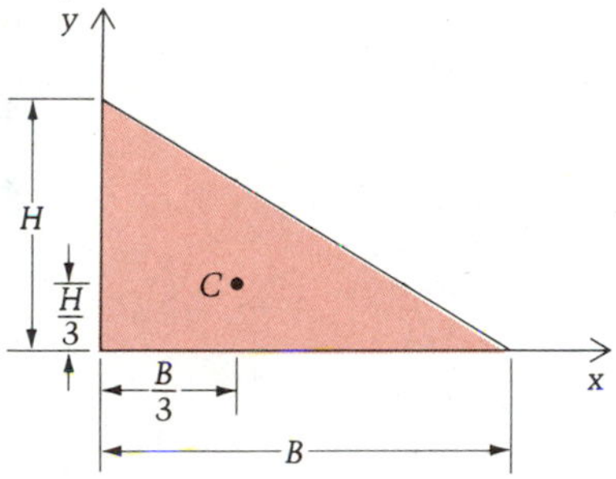

Figure E8.21

EXAMPLE 8.22

Determine the centroidal product of inertia I_{xy} for the "zee" section shown in Figure E8.22.

Solution

We shall label the three parts of the "zee" as indicated. Using composite areas together with the parallel-axis theorem,

$$I_{xy} = I_{xy}^{①} + I_{xy}^{②} + I_{xy}^{③}$$

$$= \Big[0 - \underbrace{0.5(3.5)}_{A_①}\underbrace{\left(\frac{3.5 + 0.5}{2}\right)}_{\bar{x}_①}\underbrace{(4)}_{\bar{y}_①}\Big] + 0 + \Big[0 - \underbrace{0.5(3.5)}_{A_③}\underbrace{\left(\frac{-3.5 - 0.5}{2}\right)}_{\bar{x}_③}\underbrace{(-4)}_{\bar{y}_③}\Big]$$

$$= -14.0 - 14.0$$

$$I_{xy} = -28.0 \text{ in.}^4$$

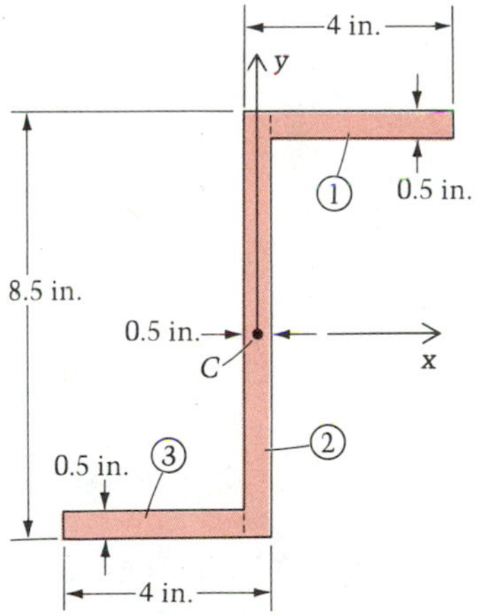

Figure E8.22

Note that the centroidal axes of ①, ②, and ③, which are parallel to x and y through C are axes of symmetry for the respective parts. Thus each centroidal product of inertia is zero, leaving only the transfer terms. Furthermore, even the transfer term vanishes for ② because its centroid is the same as C so that $\bar{x}_2$ and $\bar{y}_2$ are zero. As expected, then, the only contributions to I_{xy_C} are due to the areas ① and ③ being in the "opposite quadrants" (first and third) so that their transfer terms have the same sign.

PROBLEMS ▶ Section 8.7

8.101 What is the product of inertia I_{xy_C} of the Z section shown in Figure P8.101?

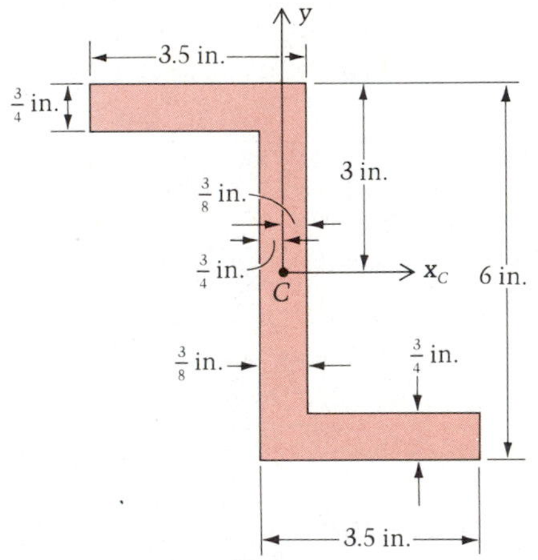

Figure P8.101

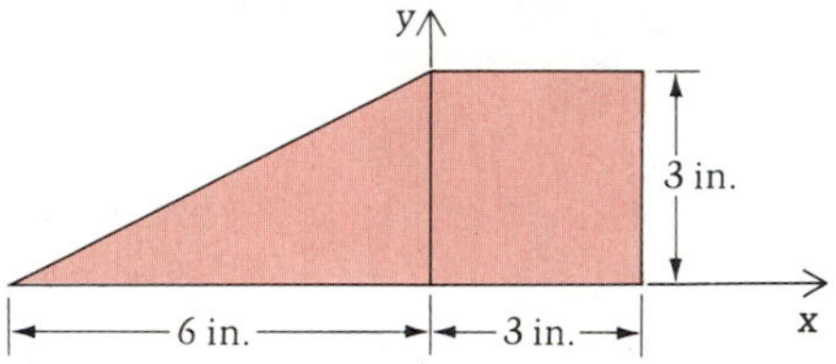

Figure P8.115

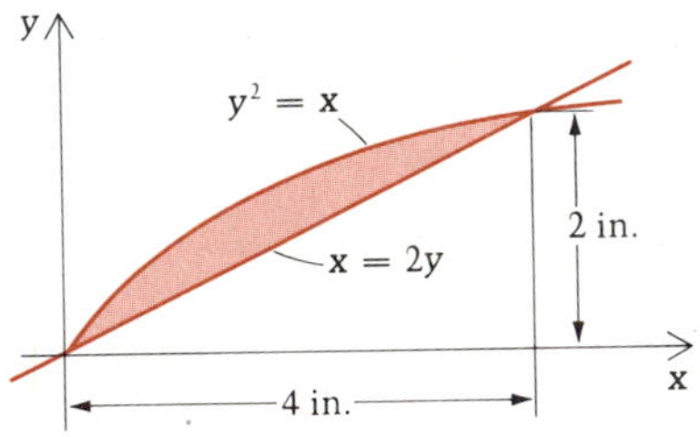

Figure P8.117

8.102 Find the product of inertia I_{xy} for the hollow rectangular area of Problem 8.42.

8.103–8.114 Find the products of inertia I_{xy} for the 12 areas in Problems 8.45–8.68, respectively.

8.115 Find the centroidal moments of inertia I_{x_C} and I_{y_C} and the product of inertia I_{xy_C} of the composite area shown in Figure P8.115.

8.116 Repeat Problem 8.100, this time using the transfer theorem and the fact that $I_{xy_C} = -\dfrac{a^3 b}{14} \sin^2 \alpha \cos \alpha$.

8.117 Use the results of Problems 7.23 and 8.95 to find the moments and product of inertia I_{x_C}, I_{y_C}, and I_{xy_C} for the shaded area in Figure P8.117.

8.118 Find I_{xy} for the shaded area in Figure P8.118. (Make use of the parallel-axis theorem!)

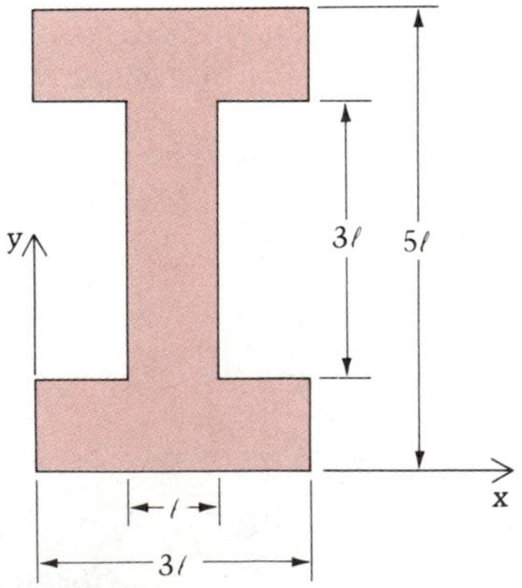

Figure P8.118

8.8 Moments and Products of Inertia with Respect to Rotated Axes Through a Point / Mohr's Circle

Developing the Equation of Mohr's Circle

Suppose we know the moments of inertia I_x and I_y with respect to the orthogonal axes x and y in the plane of an area A, and also the product of inertia I_{xy}. It is possible to derive a pair of equations that will allow us to then find the moments and products of inertia associated with *arbitrarily oriented* orthogonal lines through the point. We shall find these equations in this section, and we shall also see that they may be combined to form the equation of a circle (called **Mohr's circle**), which is a handy tool in finding inertia properties with respect to rotated axes.

We begin by using Figure 8.9 to derive the formula for $I_{x'}$ (the moment of inertia about the arbitrary axis x'). It is assumed that I_x, I_y, and I_{xy} are known for the x and y axes at P.

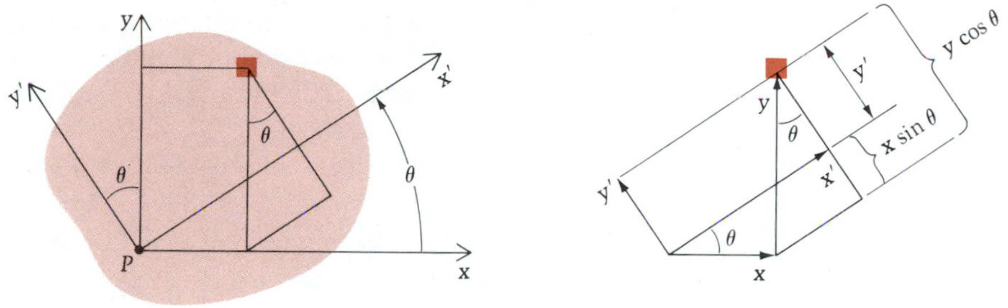

Figure 8.9

Using the definition of the moment of inertia, we obtain

$$I_{x'} = \int y'^2\, dA = \int (y \cos\theta - x \sin\theta)^2\, dA$$

in which y' is seen in Figure 8.9 to be $(y \cos\theta - x \sin\theta)$. Expanding,

$$I_{x'} = \cos^2\theta \int y^2\, dA - 2 \sin\theta \cos\theta \int xy\, dA + \sin^2\theta \int x^2\, dA$$

or, recognizing the definitions of I_x, I_y, and I_{xy},

$$I_{x'} = I_x \cos^2\theta + I_y \sin^2\theta + 2I_{xy} \sin\theta \cos\theta \tag{8.7}$$

Using the following trigonometric identities in Equation (8.7), $I_{x'}$ can be expressed as a function of the angle 2θ:

$$\cos^2\theta = \frac{1 + \cos 2\theta}{2}$$

$$\sin^2\theta = \frac{1 - \cos 2\theta}{2}$$

$$\sin\theta \cos\theta = \frac{\sin 2\theta}{2}$$

We thus obtain

$$I_{x'} - \frac{I_x + I_y}{2} = \frac{I_x - I_y}{2} \cos 2\theta + I_{xy} \sin 2\theta \tag{8.8}$$

We will also need $I_{x'y'}$ in terms of I_x, I_y, and I_{xy}:

$$\begin{aligned} I_{x'y'} &= -\int x'y' \, dA \\ &= -\int (y \sin \theta + x \cos \theta)(y \cos \theta - x \sin \theta) \, dA \\ &= \sin \theta \cos \theta \left[-\int (y^2 - x^2) \, dA\right] + \underbrace{(\cos^2 \theta - \sin^2 \theta)}_{\cos 2\theta} \left[-\int xy \, dA\right] \end{aligned}$$

or

$$I_{x'y'} = -\left(\frac{I_x - I_y}{2}\right) \sin 2\theta + I_{xy} \cos 2\theta \tag{8.9}$$

Equations (8.8) and (8.9) are the transformation equations that give the inertia properties for rotated axes. They can be manipulated into the equation of a circle by squaring both sides of each and adding:

$$\left[I_{x'} - \left(\frac{I_x + I_y}{2}\right)\right]^2 + I_{x'y'}^2 = \left(\frac{I_x - I_y}{2}\right)^2 + I_{xy}^2 \tag{8.10}$$

This is the equation of a circle, which the reader may recognize more easily in the form

$$(x - x_0)^2 + y^2 = R^2$$

The abscissa (x) is seen to be the "moments of inertia axis," while the ordinate is the "products of inertia axis." Thus a point on the circle has the coordinates ($I_{x'}$, $I_{x'y'}$). The center C of the circle is located at

$$\left(\frac{I_x + I_y}{2}, 0\right)$$

while the radius R is given by

$$R = \sqrt{\left(\frac{I_x - I_y}{2}\right)^2 + I_{xy}^2}$$

Therefore, Mohr's circle appears as shown in Figure 8.10.

Using Mohr's Circle

Now that we know how to *draw* Mohr's circle, let us proceed to demonstrate how to *use* it. Suppose we wish to know (a) the moment of inertia of an area about an axis x', located angle θ ahead of x as in Figure 8.9, and / or (b) the product of inertia $I_{x'y'}$ with respect to x' and y' (again refer to Figure 8.9). We could, of course, use Equations (8.8) and (8.9) to find these inertia properties directly. But some find it easier to remember, and to make use of, the circle that was derived from these two equations.

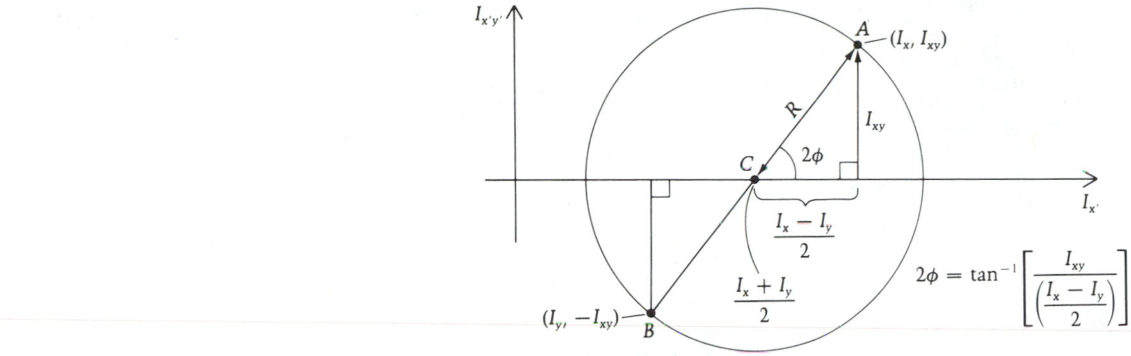

Figure 8.10 Mohr's circle.

Point A in Figure 8.10 has coordinates (I_x, I_{xy}), which correspond to x' lining up with x, and to y' with $y(\theta = 0)$. Point B has the coordinates $(I_y, -I_{xy})$, which corresponds to x' lining up with y, and hence y' with $-x$ ($\theta = 90°$). These points are 180° apart on the circle because of the "2θ" in the equations — *all the angles on the circle are double what they are on the actual area.* To find $I_{x'}$ for an arbitrary axis x', we rotate, from CA, through 2θ in the opposite direction (i.e., ⤼) on the circle.* We then reach the point P (see Figure 8.11) whose coordinates are $(I_{x'}, I_{x'y'})$. This is all there is to it, but before using this simple procedure, we need to *prove* that the point so identified gives the correct answers for $I_{x'}$ and $I_{x'y'}$ — that is, *that the coordinates of the point P, thus identified, agree with Equations (8.8) and (8.9).*

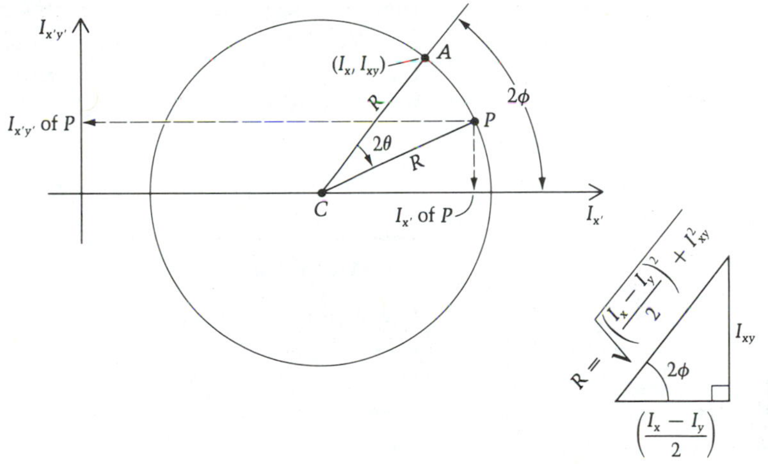

Figure 8.11

* If the product of inertia axis is plotted downward instead of upward, the rotation directions on the circle and the actual area will then be the same.

To do this, we first write the $I_{x'}$ coordinate of point P, using Figure 8.11:

$$I_{x'} = \text{center} + (\text{radius}) \cdot [\cos(2\phi - 2\theta)]$$

$$I_{x'} = \left(\frac{I_x + I_y}{2}\right) + R(\cos 2\phi \cos 2\theta + \sin 2\phi \sin 2\theta)$$

$$I_{x'} = \frac{I_x + I_y}{2} + R\frac{\left(\dfrac{I_x - I_y}{2}\right)}{R}\cos 2\theta + R\frac{I_{xy}}{R}\sin 2\theta$$

$$I_{x'} = \frac{I_x + I_y}{2} + \frac{I_x - I_y}{2}\cos 2\theta + I_{xy}\sin 2\theta$$

This is the same equation as (8.8); therefore, the procedure has yielded the correct answer, in general, for $I_{x'}$. Next we check $I_{x'y'}$. From Figure 8.11 again,

$$I_{x'y'} = (\text{radius})[\sin(2\phi - 2\theta)] = R(\sin 2\phi \cos 2\theta - \cos 2\phi \sin 2\theta)$$

$$= R\frac{I_{xy}}{R}\cos 2\theta - R\frac{\left(\dfrac{I_x - I_y}{2}\right)}{R}\sin 2\theta$$

$$I_{x'y'} = I_{xy}\cos 2\theta - \left(\frac{I_x - I_y}{2}\right)\sin 2\theta$$

which agrees with Equation (8.9). Therefore, reiterating, once the circle is drawn we can find the values of $I_{x'}$ and $I_{x'y'}$ at any angle $\angle\theta$ with the x axis by rotating on the circle through 2θ ↻ from the reference line CA and then writing down the coordinates of the point thus located.

We now work an example problem using Mohr's circle, which was named for a German engineer, Otto Mohr (1835–1918), who discovered its relationship to Equations (8.8) and (8.9).

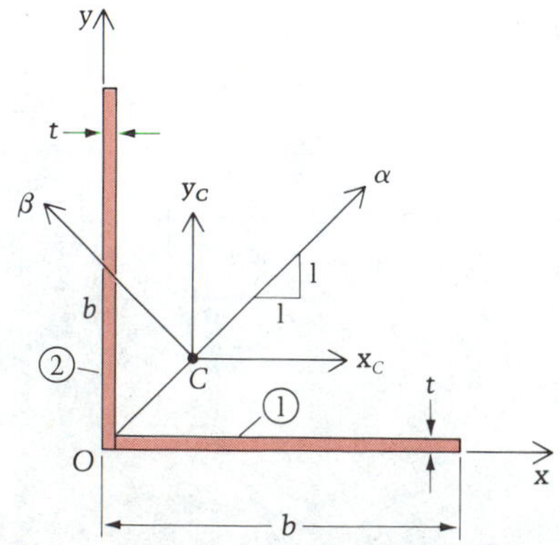

Figure E8.23a

EXAMPLE 8.23

Use Mohr's circle to find the moments of inertia of the cross-sectional area shown in Figure E8.23a about the α and β axes through its centroid C. The thickness t is very small compared to b.

Solution

The steps in this solution will be:

1. Find the centroid C.
2. Find I_x, I_y, and I_{xy}.
3. Use the transfer theorems to get I_{x_C}, I_{y_C}, and I_{xy_C}.
4. Draw Mohr's circle and use it to obtain I_α and I_β.

(1) The centroid: Let the horizontal rectangle have area A_1 and the vertical one A_2:

$$A\bar{x} = A_1\bar{x}_1 + A_2\bar{x}_2 = (b-t)t\left(\frac{b-t}{2}+t\right) + bt\,\frac{t}{2}$$

$$\bar{x} = \frac{b^2t/2 + bt^2/2 - t^3/2}{2bt - t^2} \approx \frac{b}{4} \qquad (\text{since } t \ll b)$$

and

$$\bar{y} \approx \frac{b}{4} \qquad (\text{by symmetry})$$

(2) The inertia properties at O:

$$I_x = I_x^{①} + I_x^{②} = \underbrace{\frac{(b-t)t^3}{12} + (b-t)t\left(\frac{t}{2}\right)^2}_{①} + \underbrace{\frac{tb^3}{12} + tb\left(\frac{b}{2}\right)^2}_{②}$$

$$= \frac{(b-t)t^3}{3} + \frac{tb^3}{3}$$

$$\approx \frac{b^3t}{3}$$

And

$$I_y \approx \frac{b^3t}{3} \qquad (\text{by symmetry})$$

Also,

$$I_{xy} = \underbrace{0 - (b-t)t\left(\frac{b-t}{2}+t\right)\frac{t}{2}}_{①} + \underbrace{0 - (bt)\frac{t}{2}\frac{b}{2}}_{②}$$

which we shall neglect since $t \ll b$ and the largest term is a "t^2 term."

(3) The inertia properties at C:

$$I_x = I_{x_C} + Ad^2$$

so that

$$I_{x_C} = I_x - Ad^2 \approx \frac{b^3t}{3} - (2bt)\left(\frac{b}{4}\right)^2 = \frac{5}{24}b^3t \Rightarrow I_{x_C} \approx \frac{5}{24}b^3t$$

$$I_{y_C} \approx \frac{5}{24}b^3t \qquad (\text{by symmetry})$$

$$I_{xy} = I_{xy_C} - A\bar{x}\bar{y}$$

$$I_{xy_C} \approx 0 + (2bt)\frac{b}{4}\frac{b}{4} = \frac{b^3t}{8} \qquad \text{or} \qquad \frac{3b^3t}{24}$$

The reader may wish to check these values by transferring the inertia properties of ① and ② *directly* to C.

(4) We now use Mohr's circle to get the inertias about the α and β axes at C. Factoring $\frac{b^3t}{24}$, the circle is drawn as shown in Figure E8.23b in which:

a. Center $= \frac{I_{xc} + I_{yc}}{2} = 5\left(\frac{b^3t}{24}\right)$

b. Radius $= R = \sqrt{\left(\frac{I_{xc} - I_{yc}}{2}\right)^2 + I_{xyc}^2} = 3\left(\frac{b^3t}{24}\right)$

Point A represents the values (I_{xc}, I_{xyc}), while point B represents $(I_{yc}, -I_{xyc})$.

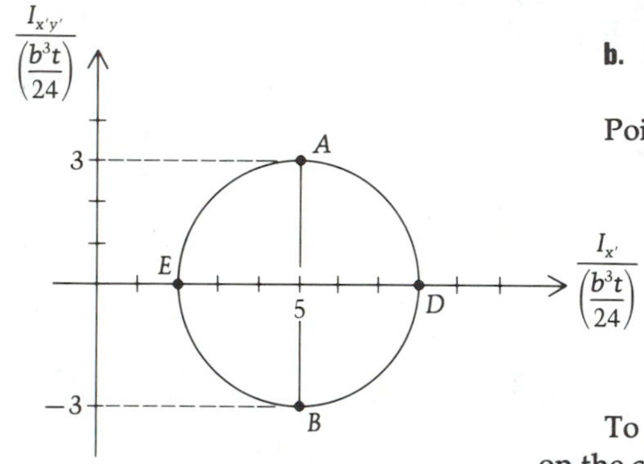

Figure E8.23b

To get from x to α on the actual area, we rotate 45° ↺. Thus we go 90° ↻ on the circle, and reach point D of Figure E8.23b, where the moment of inertia is maximum, equaling $I_{x'} = I_\alpha =$ center + radius $= 8\frac{b^3t}{24} = \frac{b^3t}{3}$, and where $I_{x'y'} = I_{\alpha\beta} = 0$.

Question 8.12 Does it make sense that α should be the axis of maximum moment of inertia? That $I_{\alpha\beta}$ should $= 0$? Explain.

Next, to get to β, we continue 90° ↺ *more* on the actual area, therefore 180° more ↻ on the circle. This puts us at point E, where $I_{x'} = I_\beta =$ center − radius $= 2\frac{b^3t}{24} = \frac{b^3t}{12}$, and where again $I_{x'y'} = 0$. Note that the area is distributed fairly close to the β axis through point C, so that it is not surprising that $I_{x'}$ is minimum (see the circle) for this line.

In general, I_x will not equal I_y, so that calculation of the desired $I_{x'}$'s will be slightly more complicated than in the preceding example. However, that example has served to ease us into the use of the circle, and has given us a feel for the axes of maximum and minimum moments of inertia through a point. For these two axes, called the **principal axes,** the product of inertia always vanishes. The associated points on the $I_{x'}$ axis (abscissa) that are farthest to the left and right on the circle give us the **principal moments of inertia.** The circle clearly shows that they are the minimum and maximum moments of inertia at the point and that, for their associated axes, the products of inertia vanish. This has very important implications in the mechanics of deformable solids.

Answer 8.12 Yes; α appears to be the axis for which the area is distributed farthest away, in an overall "$r^2\,dA$" sense. And by symmetry, $I_{\alpha\beta}$ should vanish.

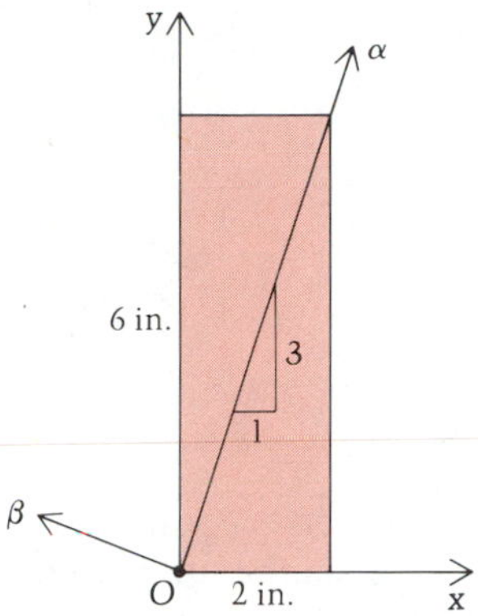

Figure E8.24a

EXAMPLE 8.24

For the rectangular area shown in Figure E8.24a, find the values of I_α, I_β, and $I_{\alpha\beta}$ at the origin (corner O).*

Solution

First we find I_x, I_y, and I_{xy} while reviewing the use of the transfer theorem. We recall that

$$I_{y_C} = \frac{hb^3}{12} = \frac{6(2^3)}{12} = 4 \text{ in.}^4$$

$$I_{x_C} = \frac{bh^3}{12} = \frac{2(6^3)}{12} = 36 \text{ in.}^4$$

$$I_{xy_C} = 0 \qquad \text{(by symmetry)}$$

Transferring these centroidal inertia properties to the corner O, we obtain

$$I_{x_O} = I_{x_C} + Ad^2 = 36 + (12)3^2 = 144 \text{ in.}^4$$

$$I_{y_O} = I_{y_C} + Ad^2 = 4 + (12)1^2 = 16 \text{ in.}^4$$

$$I_{xy_O} = I_{xy_C} - A\bar{x}\bar{y} = 0 - (12)(+1)(+3) = -36 \text{ in.}^4$$

Now we are ready to construct Mohr's circle. We need:

1. The center; it is at

$$\left(\frac{I_x + I_y}{2}, 0\right) = \left(\frac{144 + 16}{2}, 0\right) = (80, 0) \text{ in.}^4$$

2. The reference point A, corresponding to $\theta = 0$, is at $(I_x, I_{xy}) = (144, -36)$ in.^{4}

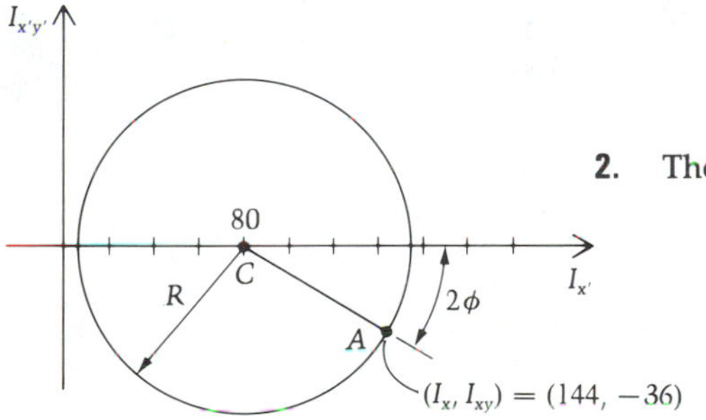

Figure E8.24b

Therefore the circle is constructed as shown in Figure E8.24b. We see from the circle that

$$R = \sqrt{(144 - 80)^2 + (36)^2}$$
$$= \sqrt{(64)^2 + (36)^2} = 73.430 \text{ in.}^4$$

and

$$2\phi = \tan^{-1}\left(\frac{36}{64}\right) = 29.358°$$

Next, the line α is seen in Figure E8.24a to make an angle with the x axis of:

$$\theta = \tan^{-1}\left(\frac{6}{2}\right) = 71.565°$$

* We shall start with five-digit accuracy in this example so as not to obscure certain results with roundoff error.

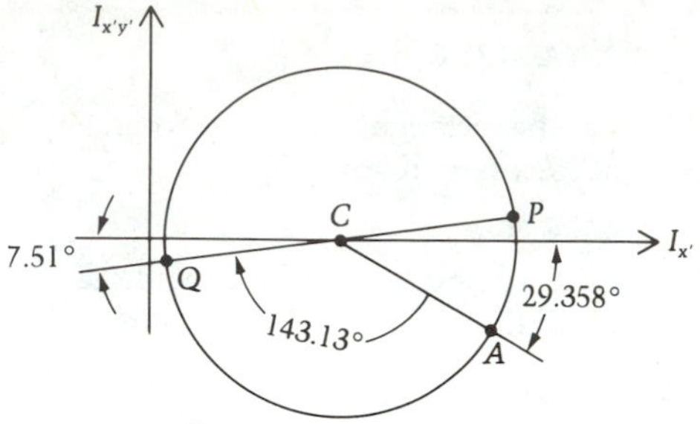

Figure E8.24c

Thus $2\theta = 143.13°$, and we turn on the circle through 2θ ↻ (opposite to the θ direction (↺) on the actual area). We reach the point Q on the circle whose $(I_{x'}, I_{x'y'})$ coordinates correspond to I_α and $I_{\alpha\beta}$ (see Figure E8.24c). The answers for I_α and $I_{\alpha\beta}$ are readily and simply written down with the help of Mohr's circle:

$$I_{x'} = I_\alpha = \text{center} - R \cos 7.51° = 80 - 73.4(0.991) = 7.26 \text{ in.}^4$$

$$I_{x'y'} = I_{\alpha\beta} = -R \sin 7.51° = -73.4(0.131) = -9.59 \text{ in.}^4$$

The value of I_β corresponds to the abscissa of the point P on the circle 180° away (or diametrically opposite) from Q:

$$I_{x'} = I_\beta = \text{center} + R \cos 7.51° = 80 + 73.4(0.991) = 153 \text{ in.}^4$$

It is interesting to note that α (the diagonal of the rectangle) is not the line of minimum moment of inertia. That line is located $143.13 + 7.51 = 151°$ clockwise from CA on the circle; hence 75.5° counterclockwise from x on the actual area (see Figure E8.24d).

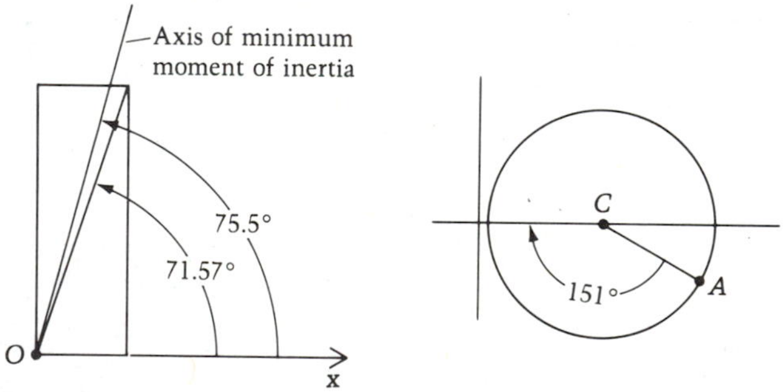

Figure E8.24d

Similarly, to get to the axis of maximum moment of inertia, we turn either 29.36° ↺ from CA on the circle or $151 + 180 = 331°$ ↻ from CA on the circle. Both these rotations (which, rounded, add to 360°) will reach the point farthest to the right on Mohr's circle — clearly the point with the largest $I_{x'}$, as we have seen before. It has the value of $I_{x'} = \text{center} + \text{radius} = 80 + 73.43 = 153.4 \text{ in.}^4$, and is located 90° away from the previously found axis of minimum moment of inertia. (See Figure E8.24e.)

Note that θ_1 and θ_2 in Figure E8.24e add to 180°, and thus they locate the same line.

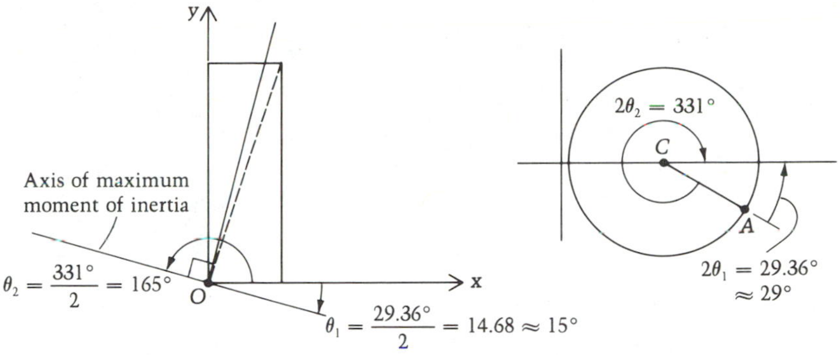

Figure E8.24e

The reader may also come across Mohr's circle again, but in a different context, when studying the mechanics of deformable solids. In that subject, two quantities (stress and strain) each have transformation formulas mathematically identical to Equations (8.8) and (8.9). Therefore, they too have Mohr's circles, by the same derivation as we have presented for inertia properties. Actually, the equations are the transformation formulas for a mathematical entity called a second-order tensor; inertia properties, stress, and strain are thus three examples of second-order tensors. It is a fact that without the minus sign in the definition of the product of inertia ($I_{xy} = -\int xy\, dA$), the inertia properties do not obey the tensor transformation equations, and this is why we have included it in the definition.

PROBLEMS ▶ Section 8.8

8.119 In Figure P8.119 find the principal moments and axes of inertia through the centroid of the shaded area by using Mohr's circle and the results of Problem 8.117.

8.120 Shown in Figure P8.120 is the cross section of a length of $3 \times 3 \times \frac{1}{4}$ "angle iron." Find the exact values of I_α and I_β and compare the results with those of Example 8.23 in which the thickness was neglected.

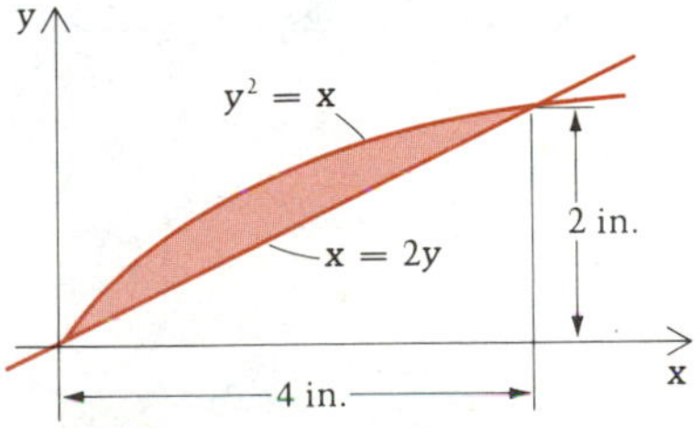

Figure P8.119

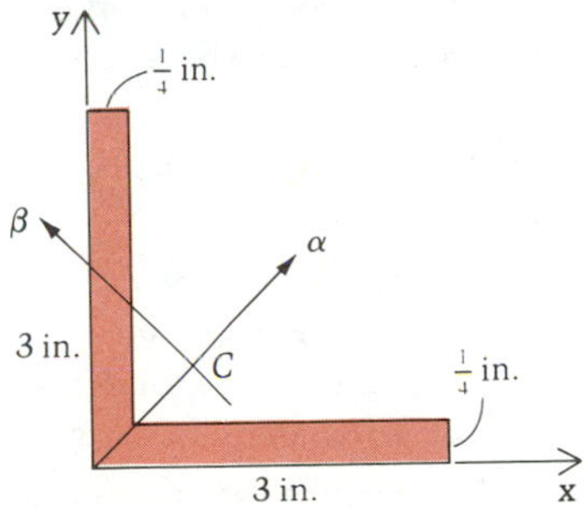

Figure P8.120

8.121 From Examples 8.8 and 8.21 we have these results for the isosceles right triangle shown in Figure P8.121(a):

$$I_{x_C} = I_{y_C} = \frac{h^4}{36} \quad \text{and} \quad I_{xy_C} = \frac{h^4}{72}$$

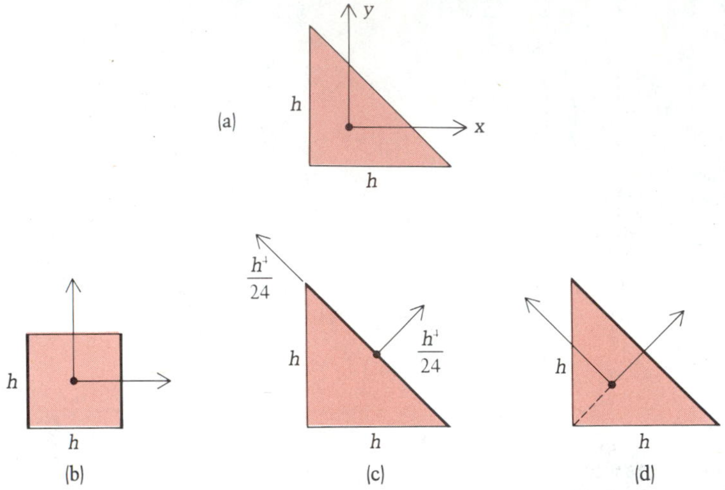

Figure P8.121

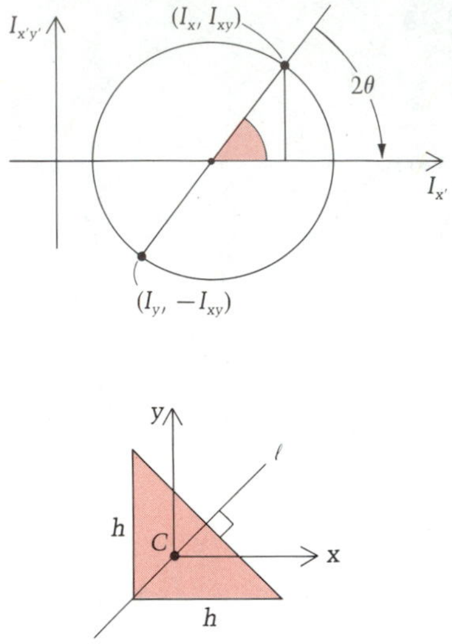

Figure P8.122

a. Use Mohr's circle to compute the principal moments and axes of inertia (i.e., the maximum and minimum moments of inertia and their associated axes) at C.

b. Give an argument that at the center C of the square in Figure P8.121(b), the Mohr's circle is just a point, and that therefore the moment of inertia about all lines through C is $h^4/12$, with the product of inertia vanishing for all pairs of axes there 90° apart.

c. Using Part (b), argue that the moments of inertia of the isosceles right triangle shown in Figure P8.121(c) about the indicated axes are as given in the figure, with the product of inertia associated with these two axes vanishing.

d. Transfer the results of Part (c) to the centroid C of the triangular area [Figure P8.121(d)], and show that the answers agree with those of part (a).

8.122 In Figure P8.122, using Mohr's circle, derive a formula for the angle θ through which we must turn from the x axis in order to reach the axis of maximum moment of inertia. The angle should be in terms of I_x, I_y, and I_{xy}. Test your equation on the bottom sketch shown in Figure P8.122 for which $I_{x_C} = I_{y_C} = h^4/36$ and $I_{xy_C} = h^4/72$, and for which line ℓ is the axis of maximum moment of inertia.

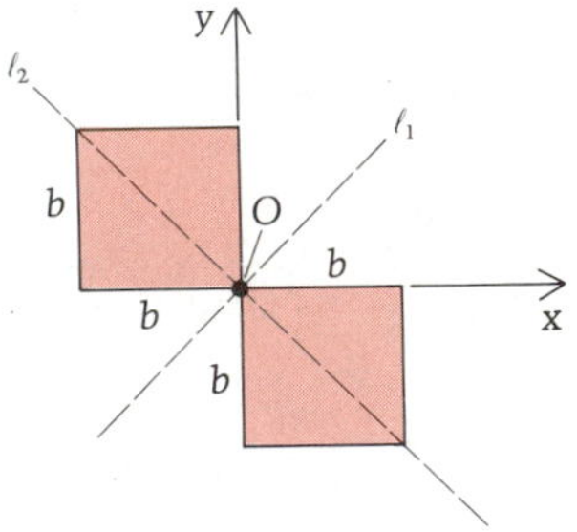

Figure P8.123

8.123 In an earlier problem, it was shown that $I_x = I_y = \frac{2}{3}b^4$ and $I_{xy} = b^4/2$ for the shaded area shown in Figure P8.123. Now use Mohr's circle to find the maximum and minimum moments of inertia at O and their associated axes. You should find that these axes are ℓ_1 and ℓ_2 in the figure. Explain why these lines make sense. Also note that your results are larger and smaller than $\frac{2}{3}b^4$ as they must be.

8.124 Prove with the help of Mohr's circle that $I_x + I_y = I_{x'} + I_{y'}$ for the plane area shown in Figure P8.124.

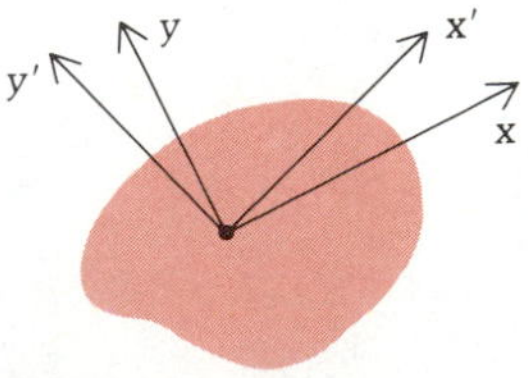

Figure P8.124

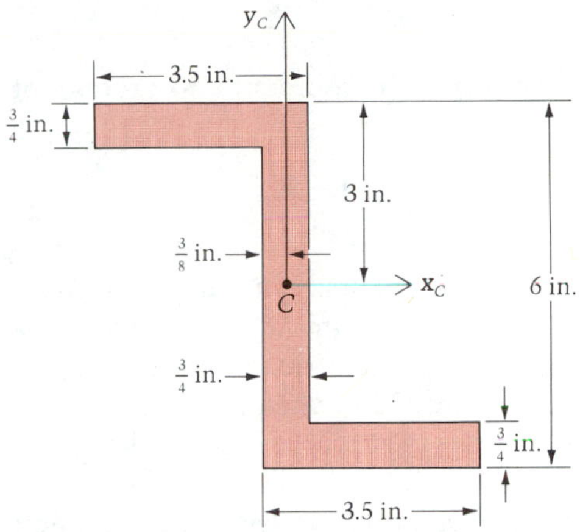

Figure P8.125

8.125 Using Mohr's circle and the results of Problems 8.74, 8.75, and 8.101, find the principal moments and axes of inertia at C for the Z section shown in Figure P8.125.

8.126 Find the smallest moment of inertia for any line drawn through C in Problems 8.71 and 8.72. Compare with the table value of 4.37 in.^{4}

Computer Problems ▶ Chapter 8

8.127 Add to the computer program written for Problem 7.131 the following feature: The program is to also read the centroidal moments of inertia $I_{x_{i_c}}$ and $I_{y_{i_c}}$ of each area A_i. It is then to use these two moments of inertia to *directly* calculate and print (a) the moments of inertia I_x and I_y of the total area about the x and y axes; and (b) the moments of inertia I_{x_C} and I_{y_C} of the total area about axes parallel to x and y through the (calculated) centroid. Also to be printed are the results $I_{x_C} + A\bar{y}^2$ and $I_{y_C} + A\bar{x}^2$, where $\bar{x}$ and $\bar{y}$ are the coordinates of the centroid C. These results should match I_x and I_y if everything is done right. Test the program with the example from Problem 7.131.

8.128 The first three examples in Chapter 8 resulted in centroidal moments of inertia for rectangular, circular, and triangular areas as depicted in Figure P8.128.

Write a program that will use these results, plus the parallel-axis theorem, to place these three areas side by side in the six possible ways shown in the figure and compute I_{y_o} about the left vertical edge. Take $h = 2R = b = 10$ in.

The program should contain an algorithm that does all six problems; i.e., you must teach the computer to permute the bodies and do the transferring of moments of inertia for each permutation.

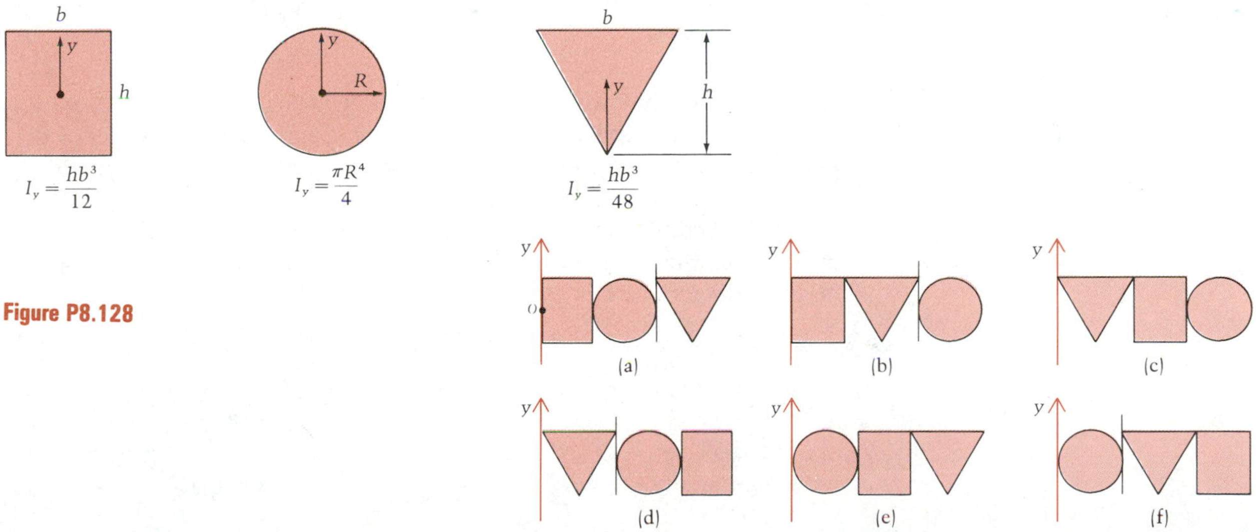

Figure P8.128

SUMMARY ▶ Chapter 8

In this chapter we learned the meaning of the moment and product of inertia of a plane area. With the basic formulas

$$I_x = \int_A y^2\, dA \qquad \text{and} \qquad I_y = \int_A x^2\, dA,$$

we derived in Section 8.2 values for the moments of inertia of a number of common shapes, including rectangular, circular, and triangular areas (a number of others are found in Appendix C). The quantities I_x and I_y are important concepts in determining bending and shearing stresses, and also deflections, of laterally loaded beams in studies of the mechanics of deformable bodies.

The polar moment of inertia of a plane area was defined in Section 8.3 to be

$$J_p = \int_A (x^2 + y^2)\, dA \qquad \text{or} \qquad \int_A r^2\, dA$$

The quantity J_p is crucial in deformable-bodies calculations of shear stress and of the angle of twist in shafts subjected to torsion, or twisting.

In Section 8.4 we discovered the Parallel-Axis Theorem, or Transfer Theorem, for moments of inertia; it is

$$I_{x_p} = I_{x_C} + Ad^2$$

The Transfer Theorem allows us to find I_x about an x-axis through P by adding, to I_x for an x-axis through C, the area multiplied by the square of the distance between the two axes. This means that we don't need to re-integrate to find I_x every time we need the moment of inertia about a different parallel line. The theorem applies as well for y-axes and I_y, and also for the polar moment of inertia.

We also learned in Section 8.4 the meaning of the radius of gyration k_x associated with an x-axis and an area A. It is defined by

$$I_x = Ak_x^2,$$

which shows that the radius of gyration is a number which, when squared and multiplied by the area, gives the moment of inertia of the area about the x-axis. There are also radii of gyration associated with I_y and J_p for an area, defined in the same way.

In Section 8.5 we learned that the idea of composite areas (introduced in conjunction with centroids in Section 7.3) may also be used in inertia calculations by

$$I_x = I_x^{A_1} + I_x^{A_1} + \cdots + I_x^{A_n}$$

where there can be any number of divisions of an area A into $A = A_1 + A_2 + \cdots + A_n$, and where $I_x^{A_i}$ is the moment of inertia of area A_i about the x-axis. It is here that the transfer theorem has its greatest use, for we generally know $I_{x_i}^{A_i}$, the moment of inertia of A_i about the x_i-axis through its own centroid C_i and parallel to x. Thus we merely add $A_i d_i^2$ to $I_{x_i}^{A_i}$ to form $I_x^{A_i}$ and then easily sum all such terms on the right-hand side to form I_x. This idea of course also applies to I_y and to J_p.

Sections 8.6 and 8.7 deal respectively with the product of inertia

$$I_{xy} = -\int_A xy\, dA$$

of a plane area, and with the parallel-axis theorem for products of inertia:

$$I_{xy_p} = I_{xy_C} - A\overline{xy}$$

These concepts become important in deformable bodies and fluid statics whenever x or y are not axes of symmetry for the area in question.

In Section 8.8 we developed formulas (8.8, 8.9) which allow us to find the moments and products of inertia associated with arbitrarily oriented lines through any point at which we already know I_x, I_y and I_{xy}. We then presented the simplifying graphical interpretation of the use of these formulas known as Mohr's circle.

REVIEW QUESTIONS ▶ Chapter 8

True or False?

1. The moment of inertia of a plane area A, about any axis, is always a positive number.
2. The dimension of a moment of inertia of an area is L^3, where L denotes length.
3. The polar moment of inertia of an area A about any axis (z) normal to the plane (xy) of A through a point P in this plane is larger than each of I_{x_P} and I_{y_P}, and in fact equals their sum.
4. Let the moment of inertia of an area A about the x axis through its centroid be called I_{x_C}. Consider the moments of inertia of A about all other axes parallel to x. All these moments of inertia are less than I_{x_C}.
5. The moment of inertia of a circular area about any line in its plane tangent to the circle is $\frac{5}{4}\pi R^4$.
6. If we know the moment of inertia of a plane area A about an axis x in the plane, then the moment of inertia about the parallel axis through the centroid of A is $I_x + Ad^2$, where d is the distance between the axes.
7. The utility of the method of composite areas is significantly enhanced by the parallel-axis theorem.
8. The product of inertia of an area can be positive or negative, but never zero.

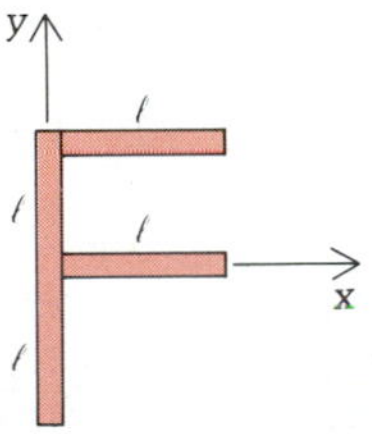

9. The product of inertia I_{xy} of the given letter "F" in the drawing at the left is $\approx -\dfrac{\ell^3 t}{2}$ where t is the thickness of the letter and $t \ll \ell$.

10. Mohr's circle is useful in finding the axes and values of the largest and smallest moments of inertia of a plane area with respect to axes through any point.
11. It is possible for Mohr's circle (for area moments of inertia) to cross the ordinate (product of inertia axis).

Answers: 1. T 2. F 3. T 4. F 5. T 6. F 7. T 8. F 9. T 10. T 11. F

9 Special Topics

9.1 Introduction

9.2 The Principle of Virtual Work

Kinematics of a Rigid Body in Two Dimensions

Work of a Force and Work of a Couple for Infinitesimal Displacements

Virtual Displacement and Virtual Work

Principle of Virtual Work

General Equivalence of the Principle of Virtual Work and the Equilibrium Equations

Virtual Work of Various Common Types of Connections

9.3 Hydrostatic Pressure on Submerged Bodies

Hydrostatic Pressure on Plane Surfaces

Hydrostatic Pressure on Curved Surfaces

Buoyant Force

Review Questions

9.1 Introduction

In this chapter, we introduce the interested reader to two additional subjects in Statics. The first of these (in Section 9.2) is the very old and very interesting principle of virtual work. An alternative to Newton's and Euler's laws, this principle is at the heart of more advanced energy principles of mechanics. But, of course, it can also be used to solve simple statics problems, and, more importantly, to render certain very complicated statics problems simple.

The second special topic in the chapter, examined in Section 9.3, is that of hydrostatic pressure on submerged bodies. In this section, principles of fluids such as Pascal's law and Archimedes' principle are coupled with what we have learned about equilibrium in Chapters 2–4 to solve problems in which some of the forces on a body are caused by pressure from a fluid at rest.

9.2 The Principle of Virtual Work

In Section 4.4 we encountered bodies or structures composed of a number of rigid (or near-rigid) parts with special connections such as pins. When there is a large number of such elements, many free-body diagrams and associated sets of equilibrium equations may be required in the analysis. Among the unknowns appearing in these equations will be all the forces of interaction between adjacent elements or members. It sometimes occurs that most, if not all, of these interactions are not of interest, and we simply have to contend with them in order to get to the results that are important. In such situations there is an equivalent and economical form of analysis that can be used, and it is based upon what is called the *Principle of Virtual Work.* Before stating the principle we must establish some preliminary concepts. For simplicity we restrict our attention here to two-dimensional (plane) problems or situations.

Kinematics of a Rigid Body in Two Dimensions

In two dimensions the general change in configuration possible for a rigid body is shown in Figure 9.1. The vector $\mathbf{r}_{OA}$ is called a position vector for material point A if point O is a fixed point in the frame of reference; $\mathbf{r}_{OA_i}$ denotes that position vector for the initial location of A and $\mathbf{r}_{OA_f}$ the position vector for the final location of A. The vector $\mathbf{u}_A$ is called the displacement that A undergoes when A moves from its initial to final position. So, using the laws of vector addition, we see that

$$\mathbf{u}_A = \mathbf{r}_{OA_f} - \mathbf{r}_{OA_i} \tag{9.1}$$

The rigid-body change in configuration may be accomplished by first moving point A to its final position without changing the orientation of the body (this is called translation), and then rotating the body about an axis (perpendicular to the plane) through A. The angle of rotation is φ.

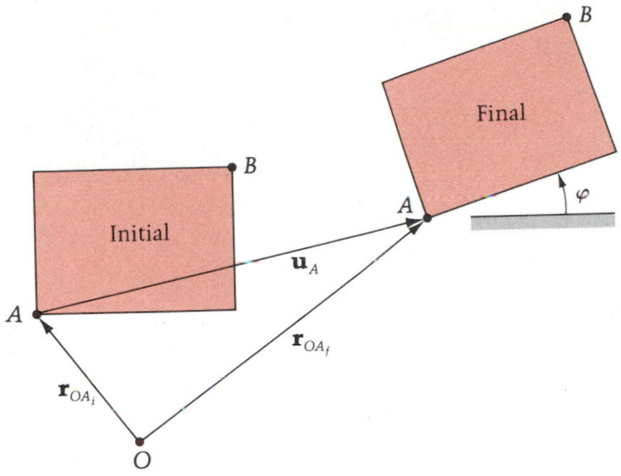

Figure 9.1

The student should use sketches to verify that the same displacement $\mathbf{u}_A$ and angle φ yield the final configuration if the order of the operations is reversed; that is, the body is first rotated about A and then translated to the final configuration. Furthermore, the student should verify that the same change in configuration could be accomplished by a translation taking B to its final position followed by a rotation about B of the *same* angle φ. Thus the most general change in configuration is characterized by the displacement of some point plus a rotation about an axis through that point.

Referring still to Figure 9.1, we note that

$$\mathbf{r}_{OB_f} = \mathbf{r}_{OA_f} + \mathbf{r}_{AB_f}$$

and

$$\mathbf{r}_{OB_i} = \mathbf{r}_{OA_i} + \mathbf{r}_{AB_i}$$

Subtracting, and using Equation (9.1),

$$\mathbf{u}_B = \mathbf{u}_A + (\mathbf{r}_{AB_f} - \mathbf{r}_{AB_i}) \tag{9.2}$$

From the preceding discussion we see that the term $\mathbf{r}_{AB_f} - \mathbf{r}_{AB_i}$ is the displacement that B undergoes when the body is rotated through angle φ about the axis through A. This vector can be put into a particularly useful form if the angle of rotation is infinitesimal; in that case $\mathbf{r}_{AB_f} - \mathbf{r}_{AB_i}$ is infinitesimal and may be written as the differential $d\mathbf{r}_{AB}$. To evaluate it we refer to Figure 9.2 (where $\hat{\mathbf{i}}$ and $\hat{\mathbf{j}}$ are fixed in the frame of reference). The angle θ describes the orientation of the rigid body in which A and B reside, and a change in θ thus represents a rotation of the body. We see from Figure 9.2 that

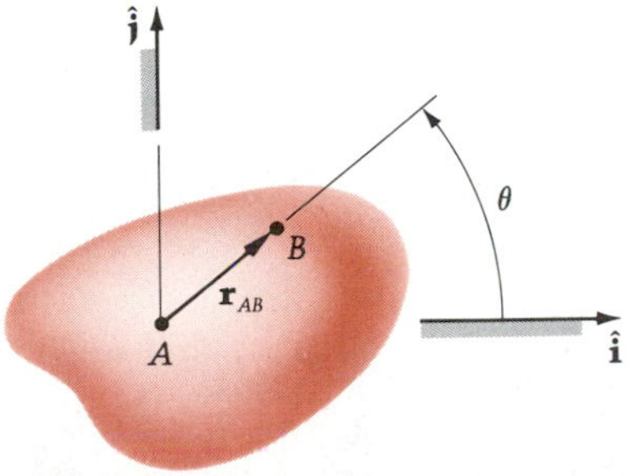

Figure 9.2

$$\mathbf{r}_{AB} = |\mathbf{r}_{AB}|(\cos\theta\hat{\mathbf{i}} + \sin\theta\hat{\mathbf{j}}) \tag{9.3}$$

where we note that $|\mathbf{r}_{AB}|$ can't change since the body is rigid. Thus with a differential change in θ we find

$$d\mathbf{r}_{AB} = |\mathbf{r}_{AB}|(-\sin\theta\hat{\mathbf{i}} + \cos\theta\hat{\mathbf{j}})\,d\theta$$

which may be expressed as*

$$= |\mathbf{r}_{AB}|(d\theta)\hat{\mathbf{k}} \times (\cos\theta\hat{\mathbf{i}} + \sin\theta\hat{\mathbf{j}})$$
$$= (d\theta)\hat{\mathbf{k}} \times \mathbf{r}_{AB} \tag{9.4}$$

The vector $(d\theta)\hat{\mathbf{k}}$ is called the infinitesimal rotation vector and it can be seen to conform to a right-hand rule. That is, if the fingers curl in the direction of rotation, the thumb points in the direction of the vector. It is important to recognize that $d\mathbf{r}_{AB}$ is perpendicular to $\mathbf{r}_{AB}$.

We may now return to Equation (9.2) to write

$$\mathbf{u}_B = \mathbf{u}_A + (d\theta)\hat{\mathbf{k}} \times \mathbf{r}_{AB} \tag{9.5}$$

when the body undergoes an infinitesimal rotation $d\theta$. If $\mathbf{u}_B$ and $\mathbf{u}_A$ are infinitesimal, then they are differentials of $\mathbf{r}_{OB}$ and $\mathbf{r}_{OA}$ and

$$d\mathbf{r}_{OB} = d\mathbf{r}_{OA} + (d\theta)\hat{\mathbf{k}} \times \mathbf{r}_{AB} \tag{9.6}$$

This equation will be of considerable importance to us shortly, but first we must examine the concept of work.

Work of a Force and Work of a Couple for Infinitesimal Displacements

The increment of work done by a force **F** acting on point P of body $\mathcal{B}$ during the infinitesimal displacement $d\mathbf{r}_{OP}$ (see Figure 9.3) is defined to be the scalar

$$dW = \mathbf{F} \cdot d\mathbf{r}_{OP} \tag{9.7}$$

in which $\mathbf{r}_{OP}$ is a position vector of the point P in the inertial reference frame $\mathcal{I}$, and where it must be stipulated that **F** is constant (or at most changes infinitesimally) during the displacement. Note that the definition [Equation (9.7)] tells us that only the component of **F** that lies along the displacement $d\mathbf{r}_{OP}$ will contribute to dW; thus this mechanical definition of work is different from our ordinary, everyday perception.†

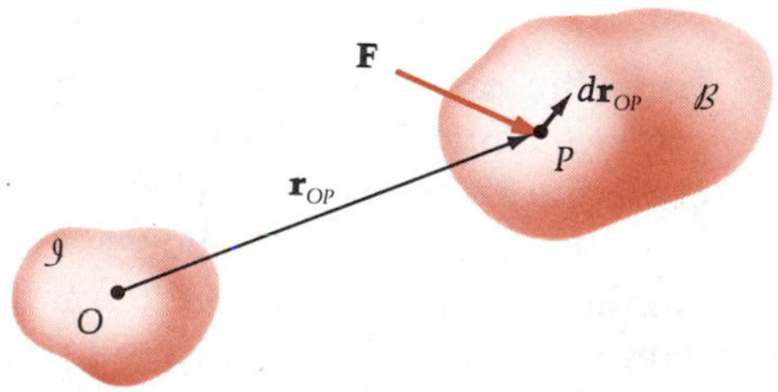

Figure 9.3

* An expression of similar form holds in three dimensions, but it is more difficult to derive and to visualize. It is for this reason that the development of this section is restricted to two dimensions.

† Try telling a farmer who has carried a 100-lb sack of feed across a field that he hasn't done any work!

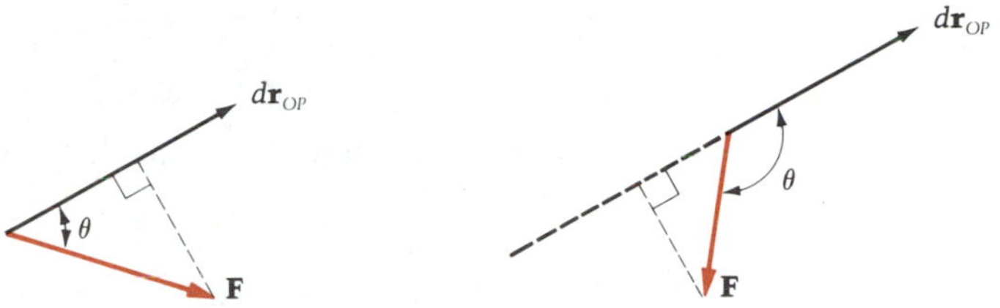

$\theta < \pi/2$
F does positive work
(a)

$\theta > \pi/2$
F does negative work
(b)

Figure 9.4

Also, note that if the smaller of the two angles between **F** and $d\mathbf{r}_{OP}$ is less than 90°, **F** does positive work during the displacement $d\mathbf{r}_{OP}$ [see Figure 9.4(a)]. If the smaller angle is greater than 90°, however [see Figure 9.4(b)], then the work of **F** during the displacement $d\mathbf{r}_{OP}$ will be negative. We see then that, speaking loosely, we may say that a force does positive work when it "gets to move in the direction it wants to."

Question 9.1 What work is done during $d\mathbf{r}_{OP}$ if the angle between **F** and $d\mathbf{r}_{OP}$ is 90°?

Note too that because, if $F = |\mathbf{F}|$,

$$dW = \mathbf{F} \cdot d\mathbf{r}_{OP} = F(|d\mathbf{r}_{OP}| \cos \theta) = (F \cos \theta)|d\mathbf{r}_{OP}| \tag{9.8}$$

we can view the work increment as either (a) the magnitude of the force times the component of the displacement of P in the direction of the force; or as (b) the magnitude of the displacement of P times the component of **F** in the direction of the displacement.

If we write the two vectors in Equation (9.7) in terms of rectangular Cartesian components parallel to directions fixed in $\mathcal{I}$, then

$$\mathbf{F} = F_x\hat{\mathbf{i}} + F_y\hat{\mathbf{j}}$$
$$\mathbf{r}_{OP} = x\hat{\mathbf{i}} + y\hat{\mathbf{j}}$$

so that

$$d\mathbf{r}_{OP} = dx\hat{\mathbf{i}} + dy\hat{\mathbf{j}}$$

The following equation is another way of writing the work, this time in terms of the rectangular components of the force (see equation 9.7):

$$dW = F_x dx + F_y dy \tag{9.9}$$

As an example, the work of the force of gravity (or weight) is, with $\hat{\mathbf{j}}$ and positive y downward,

Answer 9.1 Zero work.

$$dW = mg\hat{\mathbf{j}} \cdot d\mathbf{r}_{OC} = mg\hat{\mathbf{j}} \cdot (dx\hat{\mathbf{i}} + dy\hat{\mathbf{j}})$$
$$dW = mg\,dy \tag{9.10}$$

which is the weight times the downward component of the displacement of the mass center.

The units of work or energy are seen to be the same as are the moments of forces. In the SI system the N · m is used for moment, and the joule (J), which is 1 N · m, is used for work so as to identify the nature of the quantity. Often this distinction is made in the U.S. system by using lb-ft for moment and ft-lb for work.

The increment of work done by a constant couple (of moment $C\hat{\mathbf{k}}$) acting on a body during an infinitesimal rigid change in configuration is defined by

$$dW = C\hat{\mathbf{k}} \cdot [(d\theta)\hat{\mathbf{k}}]$$
$$= C\,d\theta \tag{9.11}$$

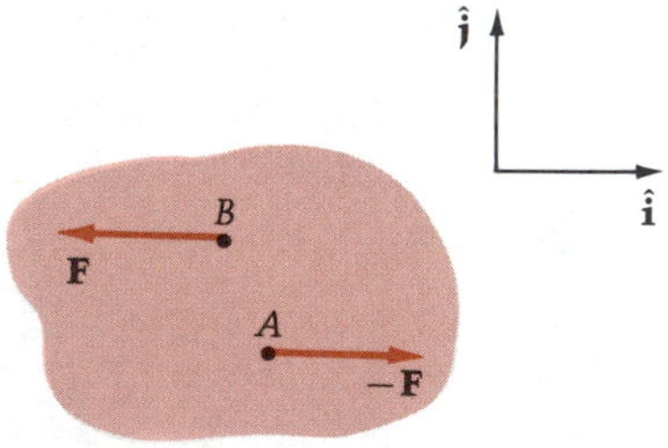

Figure 9.5

We see that the work is positive if the sense of turning of the moment of the couple is the same as the direction of the infinitesimal rotation, and negative otherwise.

For this definition to be useful, we surely expect the sum of the increments of work of a pair of equal-in-magnitude but opposite forces to equal the increment of work of the corresponding moment of the couple. To establish this connection, consider the situation depicted in Figure 9.5. The net increment of work done by the forces $\mathbf{F}$ and $-\mathbf{F}$ is

$$dW = \mathbf{F} \cdot d\mathbf{r}_{OB} + (-\mathbf{F}) \cdot d\mathbf{r}_{OA}$$
$$= \mathbf{F} \cdot (d\mathbf{r}_{OB} - d\mathbf{r}_{OA})$$

and using Equation (9.6),

$$dW = \mathbf{F} \cdot [(d\theta)\hat{\mathbf{k}} \times \mathbf{r}_{AB}]$$
$$= (d\theta)\hat{\mathbf{k}} \cdot (\mathbf{r}_{AB} \times \mathbf{F})$$
$$= (d\theta)\hat{\mathbf{k}} \cdot C\hat{\mathbf{k}} = C\,d\theta$$

where we have recognized $C\hat{\mathbf{k}} = \mathbf{r}_{AB} \times \mathbf{F}$ as the moment of the couple composed of the pair of forces.

Virtual Displacement and Virtual Work

A virtual displacement is an infinitesimal displacement that a material point might be imagined to undergo. The word *virtual* is used to convey that this displacement is imaginary or fictitious; the point in question might be "tied down" (fixed) so that no actual change in configuration of the body would cause the point to move. However, a virtual displacement is a displacement that could take place if the constraint were removed. The symbol δ is used to communicate "virtual" (or imaginary) increments, so that $\delta\mathbf{r}_{OP}$ stands for a virtual displacement and $\delta\theta$ stands for a virtual rotation. Thus, in two dimensions, the most general virtual change of configuration for a rigid body is characterized by a virtual displacement of a point together with a virtual rotation. The virtual displacements of two points A and B are related by

$$\delta \mathbf{r}_{OB} = \delta \mathbf{r}_{OA} + (\delta\theta)\hat{\mathbf{k}} \times \mathbf{r}_{AB} \tag{9.12}$$

where we have simply substituted δ's for d's in Equation (9.6).

The virtual work of a force is the work that would be done if the point of application of the force were to undergo a virtual displacement *and* the force were not to change during that displacement. Similarly the virtual work of a couple acting on a rigid body is the work that would be done if the body were to undergo a virtual rotation and the couple were not to change during that rotation.

Principle of Virtual Work

The Principle of Virtual Work may be stated as follows. *If a body in equilibrium in an inertial frame is given a virtual rigid change in configuration, starting from the equilibrium configuration, the net virtual work of all of the external forces and couples acting on the body in its equilibrium state vanishes.** We use the symbol δW to stand for the sum of the virtual works of all of the external (equilibrium) forces and couples. Thus the principle is

$$\delta W = 0$$

Before proceeding to a proof of the equivalence of the principle of virtual work and the equilibrium equations, we illustrate this equivalence by a simple example. A coffee cup $\mathcal{B}$ of mass m is in equilibrium on a horizontal table top as shown in Figure 9.6. In the free-body diagram (two-dimensionality has been assumed) we show the weight of the cup and the reaction of the table; this reaction is assumed to have a line of action through B whose distance from the line of action of mg is the unknown a. The problem, of course, is to find the unknown scalars f, N, and a.

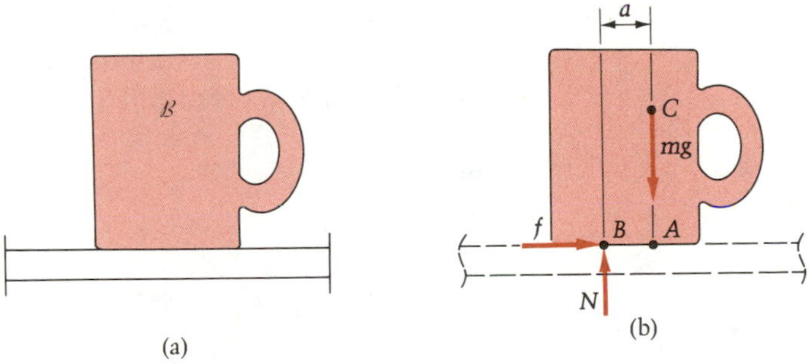

Figure 9.6

* When the body is caused to deform by a virtual change in configuration, the appropriate generalization of this statement is that the sum of the virtual work of the external forces and the virtual work of the internal forces vanishes. Nontrivial applications of this form are beyond the scope of this book.

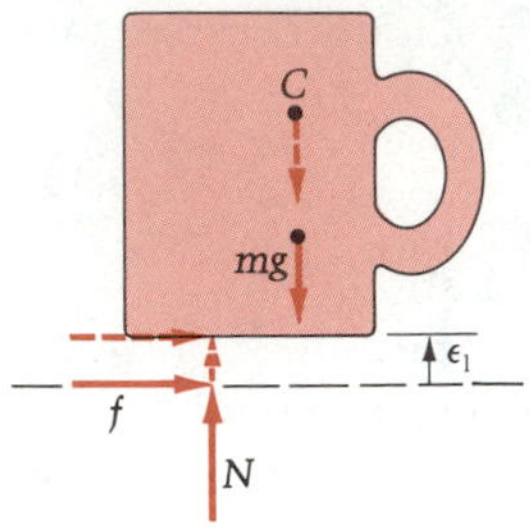

Figure 9.7

First let us give the cup an upward virtual translation, each point in the cup having an upward virtual displacement ϵ_1 (Figure 9.7). We note the virtual work of f is zero since that friction component of reaction is perpendicular to the virtual displacement ϵ_1 of $\mathscr{B}$. The virtual work of N is $N\epsilon_1$, and the virtual work of the weight is $-mg\epsilon_1$ because the mass center C, like every other point, moves upward. Thus

$$\delta W = N\epsilon_1 - mg\epsilon_1 = 0$$
$$(N - mg)\epsilon_1 = 0$$

Because ϵ_1 is arbitrary, we obtain

$$N - mg = 0$$
$$N = mg$$

This, of course, is what we alternatively obtain from the equilibrium equation $\Sigma F_y = 0$. This exercise is instructive with regard to the concept of virtual work. The force N will vanish as soon as the cup breaks contact with the table top, and so an *actual* upward displacement of the cup would be accompanied by zero *actual* work done by N.

Question 9.2 In what way, if at all, does the virtual work calculation above depend upon ϵ_1 being infinitesimal?

As an alternative to the above let us give the cup a downward virtual displacement ϵ_2; in this case (see Figure 9.8 at the left)

$$\delta W = -N\epsilon_2 + mg\epsilon_2 = 0$$

or

$$N = mg \qquad \text{(as before)}$$

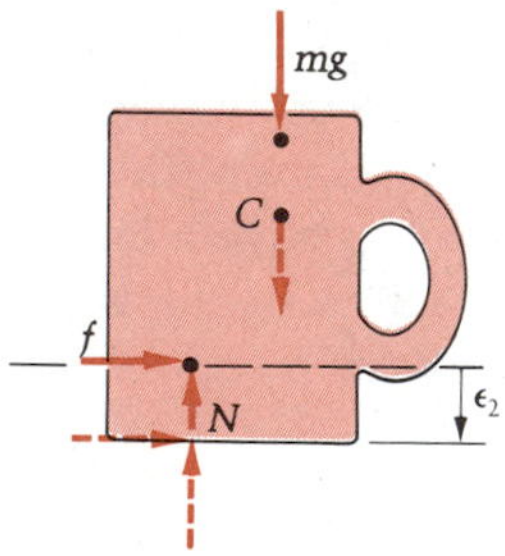

Figure 9.8

This instructive aspect of this part of the exercise is that the virtual displacement ϵ_2 represents a change in configuration that is impossible for the cup to undergo in actuality because it requires the cup to penetrate the table top. Thus in forming virtual changes in configurations, it is not necessary that we conform to actual constraints imposed upon the body.

A horizontal translation of the cup by a displacement ϵ_3 to the right yields

$$\delta W = f\epsilon_3 = 0$$

or

$$f = 0$$

which we recognize as the result of applying the equilibrium equation $\Sigma F_x = 0$.

Not at all.

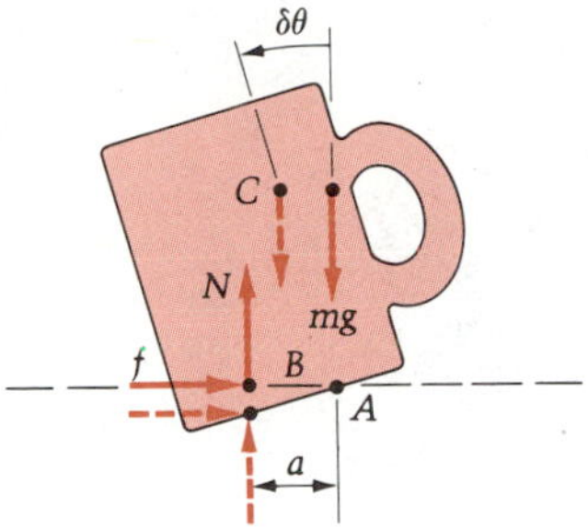

Figure 9.9

Finally let us give the cup a virtual rotation ($\delta\theta$ counterclockwise) about A (see Figure 9.9). We note from the figure that point C will be displaced horizontally, so no virtual work will be done by the weight mg. Point B will move downward $a\ \delta\theta$, so no virtual work would be done by f even if f were not already known to vanish. Therefore,

$$\delta W = -N(a\ \delta\theta) = 0$$

or

$$a = 0$$

which we recognize as the result that would have been obtained by applying the equilibrium equation $\Sigma M_A = 0$.

Question 9.3 In what way, if at all, does this virtual work-calculation depend upon $\delta\theta$ being infinitesimal?

This example has illustrated the way in which virtual displacements differ from real displacements and the way in which virtual work differs from real work. And we see that the application of the principle for three independent rigid changes in configuration has produced the same results that would be obtained from the three independent equations of equilibrium:

$$\Sigma F_x = 0 \qquad (\text{gives } f = 0)$$
$$\Sigma F_y = 0 \qquad (\text{gives } N = mg)$$
$$\Sigma M_A = 0 \qquad (\text{gives } a = 0)$$

General Equivalence of the Principle of Virtual Work and the Equilibrium Equations

We are now in a position to demonstrate the generality of this equivalence.

Consider the body $\mathcal{B}$ to be acted upon by the external forces $\mathbf{F}_1$, $\mathbf{F}_2, \ldots$ and the couples $C_1\hat{\mathbf{k}}$, $C_2\hat{\mathbf{k}}, \ldots$, as seen in Figure 9.10.

Now let the body undergo a virtual rigid change in configuration producing changes in the position vectors ($\mathbf{r}_{OP_1}, \mathbf{r}_{OP_2}, \ldots$) of the points of application of the forces (call these virtual changes $\delta\mathbf{r}_{OP_1}, \delta\mathbf{r}_{OP_2}, \ldots$) and a change in the orientation of the body $(\delta\theta)\hat{\mathbf{k}}$. Then the virtual work of the $\mathbf{F}_i$'s and $\mathbf{C}$'s is

$$\begin{aligned} \delta W &= \text{Virtual Work} \\ &= \mathbf{F}_1 \cdot \delta\mathbf{r}_{OP_1} + \mathbf{F}_2 \cdot \delta\mathbf{r}_{OP_2} + \cdots \\ &\quad + C_1\hat{\mathbf{k}} \cdot (\delta\theta)\hat{\mathbf{k}} + C_2\hat{\mathbf{k}} \cdot (\delta\theta)\hat{\mathbf{k}} + \cdots \end{aligned} \qquad (9.13)$$

Answer 9.3 Only for infinitesimal $\delta\theta$ does C move horizontally and does the point of application of N and f move vertically.

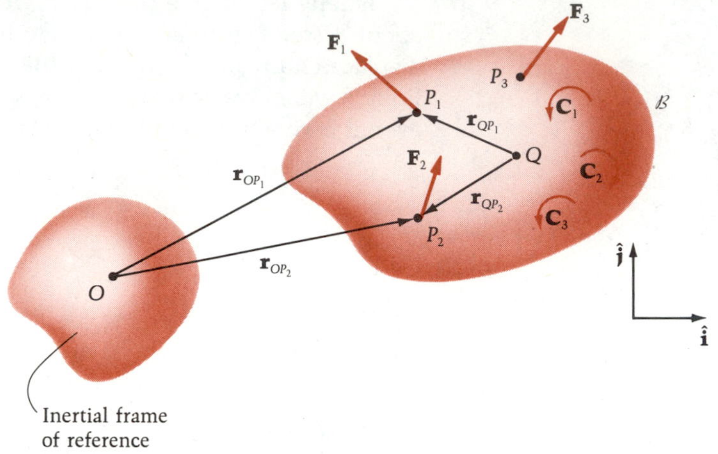

Figure 9.10

Relating the infinitesimal virtual displacements of $P_1, P_2, \ldots$ to that of an arbitrary point of the body, say Q,

$$\delta W = \mathbf{F}_1 \cdot (\delta\mathbf{r}_{OQ} + \delta\theta\hat{\mathbf{k}} \times \mathbf{r}_{QP_1}) + \mathbf{F}_2 \cdot (\delta\mathbf{r}_{OQ} + \delta\theta\hat{\mathbf{k}} \times \mathbf{r}_{QP_2}) + \cdots + (C_1\hat{\mathbf{k}} + C_2\hat{\mathbf{k}} + \cdots) \cdot \delta\theta\hat{\mathbf{k}}$$

but because

$$\mathbf{F}_1 \cdot (\delta\theta\hat{\mathbf{k}} \times \mathbf{r}_{QP_1}) = (\delta\theta\hat{\mathbf{k}} \times \mathbf{r}_{QP_1}) \cdot \mathbf{F}_1 = \delta\theta\hat{\mathbf{k}} \cdot (\mathbf{r}_{QP_1} \times \mathbf{F}_1)$$

we obtain

$$\delta W = (\mathbf{F}_1 + \mathbf{F}_2 + \cdots) \cdot \delta\mathbf{r}_{OQ} + \delta\theta\hat{\mathbf{k}} \cdot (\mathbf{r}_{QP_1} \times \mathbf{F}_1 + \mathbf{r}_{QP_2} \times \mathbf{F}_2 + \cdots + C_1\hat{\mathbf{k}} + C_2\hat{\mathbf{k}} + \cdots)$$

$$\delta W = \Sigma\mathbf{F} \cdot \delta\mathbf{r}_{OQ} + \delta\theta\hat{\mathbf{k}} \cdot \Sigma\mathbf{M}_Q \tag{9.14}$$

Since the virtual displacement $\delta\mathbf{r}_{OQ}$ of point Q is independent of the virtual angular displacement (rotation) $\delta\theta\hat{\mathbf{k}}$, each can be taken zero while the other is not.* Thus $\delta W = 0$ implies

$$\Sigma\mathbf{F} = \mathbf{0} \qquad \text{and} \qquad \Sigma\mathbf{M}_Q = \mathbf{0} \tag{9.15}$$

which are the equilibrium equations. Conversely, if $\Sigma\mathbf{F} = \mathbf{0}$ and $\Sigma\mathbf{M}_Q = \mathbf{0}$, it follows from Equation (9.14) that $\delta W = 0$. This establishes (for two dimensions) the equivalence of the principle of virtual work and the equations of equilibrium. This equivalence also holds in three dimensions, but we have not attempted to establish it here because of the more complicated kinematics of rigid bodies in three dimensions.

We now consider an example of the use of the principle of virtual work to establish an equilibrium configuration of a (near-) rigid body.

* The virtual work vanishes for *any and all* virtual displacements.

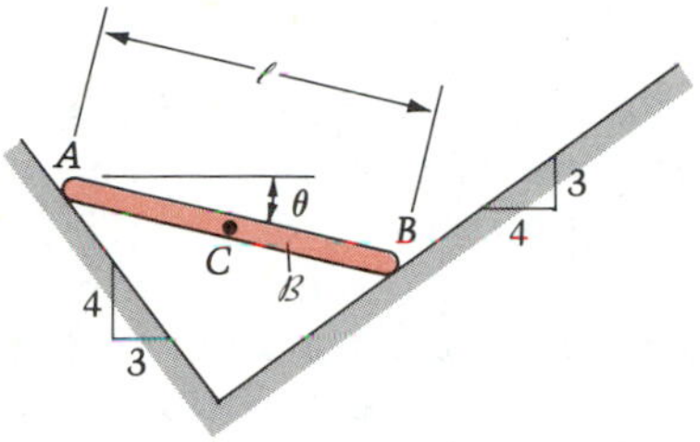

Figure E9.1a

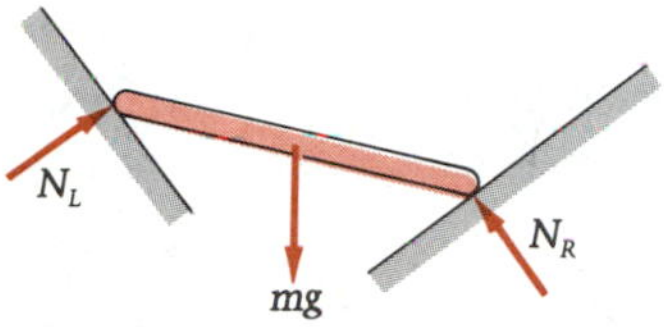

Figure E9.1b

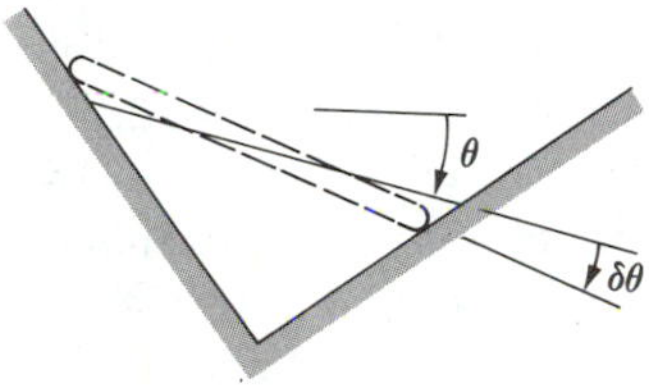

Figure E9.1c

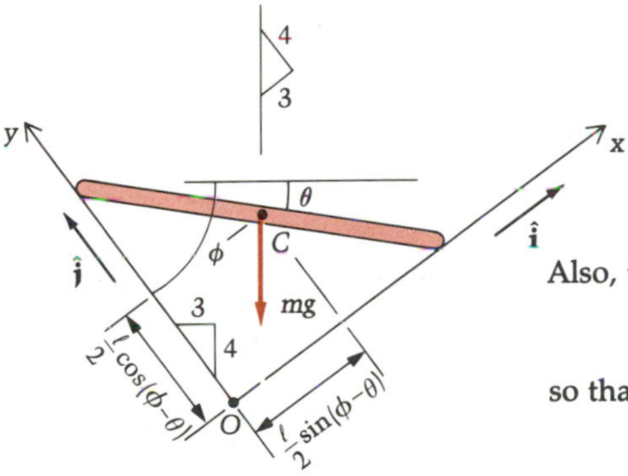

Figure E9.1d

Example 9.1

Find the angle θ for equilibrium of the uniform bar $\mathcal{B}$ shown in Figure E9.1a. Both planes are smooth.

Solution

This problem illustrates the way in which we commonly use the principle of virtual work. We can frequently give the body a virtual displacement in which one or more of the unwanted unknowns in the problem do no work and thus do not appear in the equation $\delta W = 0$. We do this here with a virtual displacement in which A and B move along their respective planes.

Question 9.4 Why then must B move down its plane if A moves up its plane?

Our sought-after angle increases by $\delta\theta$, as seen in Figure E9.1c. The normal forces N_L and N_R do no virtual work because each is normal to the virtual displacement of the point on which it acts (see Figures E9.1b and E9.1c). Hence only gravity will contribute to the virtual work, which vanishes:

$$\delta W = \mathbf{F}_g \cdot \delta \mathbf{r}_{OC} = 0$$

With the help of Figure E9.1d, we find $\mathbf{r}_{OC}$ and then $\delta\mathbf{r}_{OC}$

$$\mathbf{r}_{OC} = \frac{\ell}{2}[\sin(\phi - \theta)\hat{\mathbf{i}} + \cos(\phi - \theta)\hat{\mathbf{j}}]$$

$$\delta\mathbf{r}_{OC} = \frac{\ell}{2}[\cos(\phi - \theta)\hat{\mathbf{i}} - \sin(\phi - \theta)\hat{\mathbf{j}}]\,\underbrace{\delta(\phi - \theta)}_{-\delta\theta}$$

$$= \frac{\ell}{2}[\cos\phi\cos\theta + \sin\phi\sin\theta)\hat{\mathbf{i}} - (\sin\phi\cos\theta - \cos\phi\sin\theta)\hat{\mathbf{j}}](-\delta\theta)$$

$$= \frac{-\ell}{2}\delta\theta[(\tfrac{3}{5}\cos\theta + \tfrac{4}{5}\sin\theta)\hat{\mathbf{i}} - (\tfrac{4}{5}\cos\theta - \tfrac{3}{5}\sin\theta)\hat{\mathbf{j}}]$$

Also, the gravity force, or weight, is

$$\mathbf{F}_g = mg(-\tfrac{3}{5}\hat{\mathbf{i}} - \tfrac{4}{5}\hat{\mathbf{j}})$$

so that, substituting into Equation (1),

$$\delta W = mg\frac{\ell\,\delta\theta}{2}[\tfrac{3}{5}(\tfrac{3}{5}\cos\theta + \tfrac{4}{5}\sin\theta) - \tfrac{4}{5}(\tfrac{4}{5}\cos\theta - \tfrac{3}{5}\sin\theta)] = 0$$

and since $\delta\theta$ is arbitrary, the bracketed expression vanishes:

$$-\tfrac{7}{25}\cos\theta + \tfrac{24}{25}\sin\theta = 0$$

$$\theta = \tan^{-1}(\tfrac{7}{24}) = 16.3^\circ$$

Answer 9.4 The distance, ℓ, from A to B is constant during the rigid change in configuration.

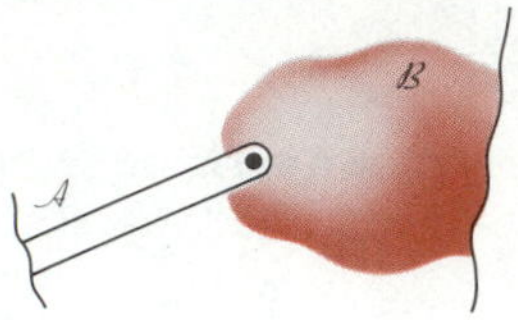

Virtual Work of Various Common Types of Connections

The real advantage of virtual work is not in analyzing a single rigid body,* but rather in studying problems with systems of rigid bodies, joined together in frictionless ways such as by (a) smooth pins or ball joints, (b) inextensible, taut cables, or (c) linear springs. With (a) and (b), we now proceed to show that in a virtual displacement that is consistent with the connections, no net work is done on the two connected bodies by the pins, ball points, or cables.

If $\mathcal{A}$ and $\mathcal{B}$ are joined by a smooth pin or ball joint, no net work will be done by the forces **F** (exerted on $\mathcal{A}$ at A) and $-$**F** (exerted on $\mathcal{B}$ at B), as shown in Figure 9.11. This is because the virtual displacements of the points A and B are the same (they are pinned together) while the forces are equal in magnitude but opposite in direction (in accordance with the principle of action and reaction). Thus,

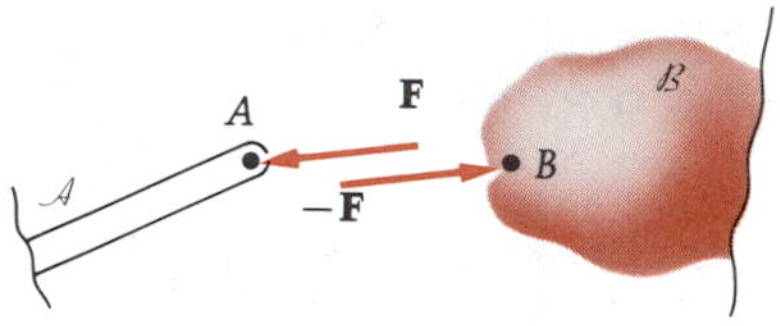

Figure 9.11

$$\delta W = \underbrace{\mathbf{F}\cdot\delta\mathbf{r}_{OA}}_{\text{Work of }\mathbf{F}\text{ on }\mathcal{A}} + \underbrace{-\mathbf{F}\cdot\delta\mathbf{r}_{OB}}_{\text{Work of }-\mathbf{F}\text{ on }\mathcal{B}} = \mathbf{F}\cdot\underbrace{(\delta\mathbf{r}_{OA}-\delta\mathbf{r}_{OB})}_{\text{Zero}} = 0$$

We next let bodies $\mathcal{A}$ and $\mathcal{B}$ be joined by a flexible but inextensible, taut cable, with or without the pulley in between. (See Figure 9.12.) Let the unit vectors $\hat{\mathbf{u}}_{\parallel}$ and $\hat{\mathbf{v}}_{\parallel}$ be along the cable† as shown, with $\hat{\mathbf{u}}_{\perp}$ and $\hat{\mathbf{v}}_{\perp}$ being normal to the cable, in the respective directions of the normal (to the cable) components of the virtual displacements $\delta\mathbf{r}_{OA}$ of A and $\delta\mathbf{r}_{OB}$ of B. Then the virtual work done by the cable on the two bodies is

$$\begin{aligned}\delta W &= -T\hat{\mathbf{u}}_{\parallel}\cdot(\delta\mathbf{r}_{OA}\,\hat{\mathbf{u}}_{\parallel}+\delta\mathbf{r}_{OA_{\perp}}\hat{\mathbf{u}}_{\perp}) + T\hat{\mathbf{v}}_{\parallel}\cdot(\delta\mathbf{r}_{OB_{\parallel}}\hat{\mathbf{v}}_{\parallel}+\delta\mathbf{r}_{OB_{\perp}}\hat{\mathbf{v}}_{\perp})\\ &= -T\delta\mathbf{r}_{OA_{\parallel}} + T\delta\mathbf{r}_{OB_{\parallel}} \qquad (9.16)\\ &= T(\delta\mathbf{r}_{OB_{\parallel}} - \delta\mathbf{r}_{OA_{\parallel}}) = 0\end{aligned}$$

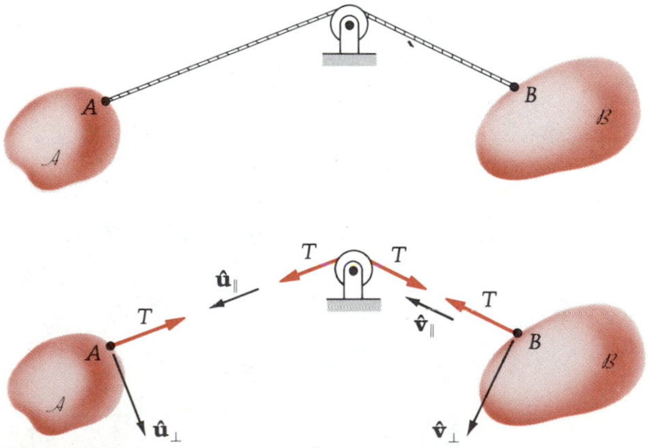

Figure 9.12

* For the remainder of this section we shall use the term "rigid body" as a shorthand expression to specify that the body is near rigid *and* that it is to be subjected only to rigid virtual changes in configuration.

† Note that $\hat{\mathbf{u}}_{\parallel} = \hat{\mathbf{v}}_{\parallel}$ and $\hat{\mathbf{u}}_{\perp} = \hat{\mathbf{v}}_{\perp}$ if there is no pulley.

because the components of virtual displacement of A and B *along the cable* must be equal if it is inextensible (and, of course, taut).

Question 9.5 Does this result, $\delta W = 0$, depend upon the pulley being frictionless?

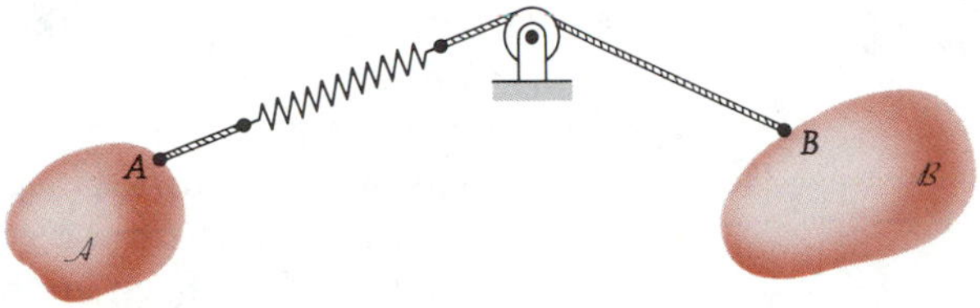

Figure 9.13

If we place a spring in the cable as shown above in Figure 9.13, then

$$\delta W = T(\delta \mathbf{r}_{OB_\parallel} - \delta \mathbf{r}_{OA_\parallel})$$

need not vanish, because the spring can stretch. In fact, if we let S be the stretch of the spring, we obtain

$$\delta \mathbf{r}_{OA_\parallel} - \delta \mathbf{r}_{OB_\parallel} = \delta S,$$

the "virtual stretch" of the spring. For a linear spring, the tension in the spring (and cable) is kS where k is the spring modulus, or stiffness, and Equation (9.16) yields

$$\delta W = -kS\,\delta S \tag{9.17}$$

This means that:

(a) Assuming $S > 0$ (spring is stretched), then if

(i) $\delta S > 0, \qquad \delta W < 0$

(ii) $\delta S = 0, \qquad \delta W = 0$

(iii) $\delta S < 0, \qquad \delta W > 0$

(b) Assuming $S < 0$ (spring is compressed),* then if

(i) $\delta S > 0, \qquad \delta W > 0$

(ii) $\delta S = 0, \qquad \delta W = 0$

(iii) $\delta S < 0, \qquad \delta W < 0$

Of course, if $S = 0$—that is, the spring is unstretched—then it will do *no*

Answer 9.5 Yes, otherwise the cable tensions on the two sides of the pulley would not have to be the same.

* Of course, a transversely flexible cable (or cord) won't support compression. However, a special case of the above analysis is that where the cable and pulley are removed and the ends of the spring are attached directly to bodies $\mathcal{A}$ and $\mathcal{B}$. In that circumstance we can have a compression-supporting spring connection.

virtual work. In the various cases above, the reader is urged to note carefully that the spring does positive virtual work during the virtual displacement if it gets to return *toward* its unstretched length, and negative virtual work if it is forced *farther away* from it.

When we consider two or more rigid bodies connected by frictionless pins or ball joints, and / or by flexible, inextensible cables, we see that the virtual work done by the external forces and couples on the *combined* system of bodies is zero (as was the case for a single rigid body). This is because $\delta W = 0$ on each body and thus it also adds to zero on the combination. Moreover, the virtual work done by a pin (or ball joint or cable) on one body of the connection cancels its work on the other provided the virtual displacements are consistent with the connections, as we have shown. Thus such an internal interaction need not even be considered in the equation $\delta W = 0$! This is the huge advantage in using virtual work: *the non-working forces can be ignored.*

For a linear spring performing non-zero virtual work on a system of rigid bodies, we have seen that its virtual work is $-kS\ \delta S$. If the spring connects the rigid bodies $\mathcal{A}$ and $\mathcal{B}$, then we get the result shown in the following equation:

$$\underbrace{\underset{\substack{\text{(on }\mathcal{A}\text{ except}\\ \text{for spring)}}}{\delta W} + \underset{(\text{of spring on }\mathcal{A})}{\delta W}}_{0} + \underbrace{\underset{\substack{\text{(of spring}\\ \text{on }\mathcal{B})}}{\delta W} + \underset{\substack{\text{(on }\mathcal{B}\text{ except}\\ \text{for spring)}}}{\delta W}}_{0} = 0$$

$$\underbrace{\underset{(\text{of spring on }\mathcal{A})}{\delta W} + \underset{\substack{\text{(of spring}\\ \text{on }\mathcal{B})}}{\delta W}}_{-kS\ \delta S}$$

This way of computing the virtual work done by the spring, as well as the ignoring of non-working internal and reactive forces in computing δW, will now be illustrated by some examples. The reader is urged to mentally consider the difficulty of solving these problems with a set of equilibrium equations. *And* to remember that, if *any* of the forces acting on a body was omitted from the free-body diagram and thence from the equations of equilibrium in Chapters 2, 4, 5, and 6, our solution was doomed!

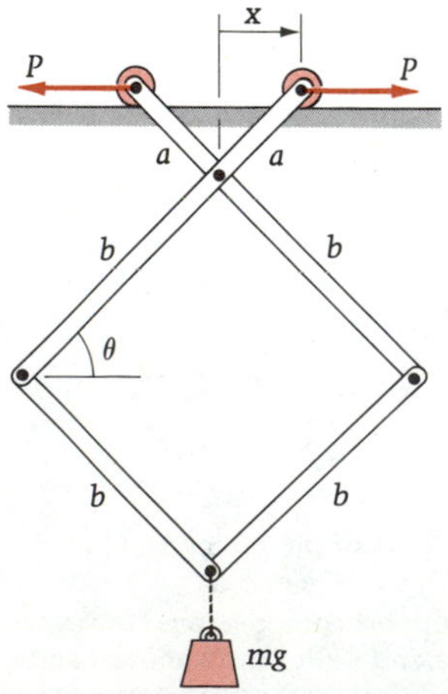

Figure E9.2a

EXAMPLE 9.2

Find the relationship between the forces P and mg and the angle θ, for equilibrium. Neglect the weights of the four bars and the sizes of the two rollers. (See Figure E9.2a.)

Solution

For a virtual displacement (which will be described below), the virtual work will vanish:

$$\delta W = 0 \tag{1}$$

From the equilibrium configuration, we imagine a virtual displacement δx, a slight movement outward of the two points where the forces P are applied. Then the virtual work of these forces is $2[P\ \delta x]$.

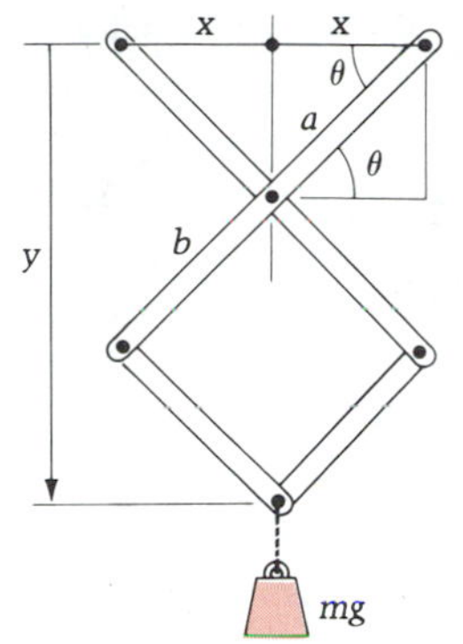

Figure E9.2b

The only other force that will perform virtual work is the weight mg. We must determine how much mg "moves" in response to the imagined δx. From Figure E9.2b, we see that $x = a \cos\theta$ so that

$$\delta x = -a \sin\theta\, \delta\theta \tag{2}$$

which relates δx and the corresponding virtual change $\delta\theta$ in θ according to the rules for differentiation and the formation of differentials. Similarly,

$$y = (a + b)\sin\theta + b\sin\theta = (a + 2b)\sin\theta$$

so that

$$\delta y = (a + 2b)\cos\theta\, \delta\theta$$

which is the virtual displacement of the mass center of the weight. The virtual work of gravity is then $mg\, \delta y$. Thus Equation (1) becomes

$$2P\delta x + mg(a + 2b)\cos\theta\, \delta\theta = 0$$

and using equation (2),

$$2P(-a\sin\theta\, \delta\theta) + mg(a + 2b)\cos\theta\, \delta\theta = 0$$

or

$$[-2Pa\sin\theta + mg(a + 2b)\cos\theta]\, \delta\theta = 0$$

Since $\delta\theta$ is arbitrary, the term in brackets must vanish, giving

$$P = \frac{mg(a + 2b)\cot\theta}{2a}$$

We see from this answer that as $\theta \to 90°$, $P \to 0$; this is because if the rods are vertical, there is no need for any force P to maintain equilibrium.

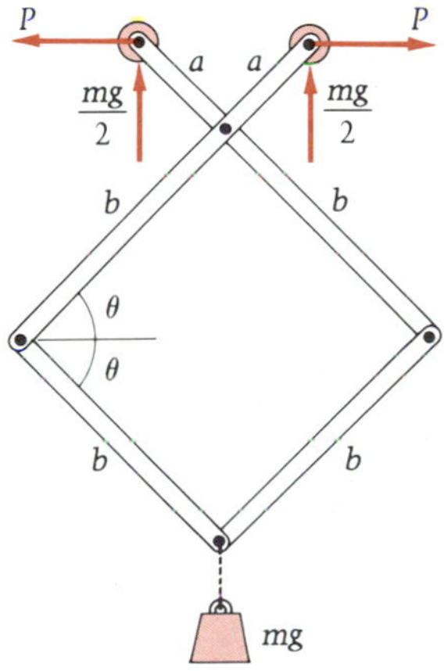

Figure 9.14

We now work the problem of the preceding example by the usual method of free-body diagrams accompanied by equilibrium equations of various parts of the structure (see Figure 9.14). From the overall free-body diagram, the roller reactions are each $mg/2$.

The forces exerted on the two lower (two-force!) members shown in Figure 9.15 below are obtained from

$$+\uparrow \qquad \Sigma F_y = 0 = 2F_1 \sin\theta - mg$$

$$F_1 = \frac{mg}{2\sin\theta}$$

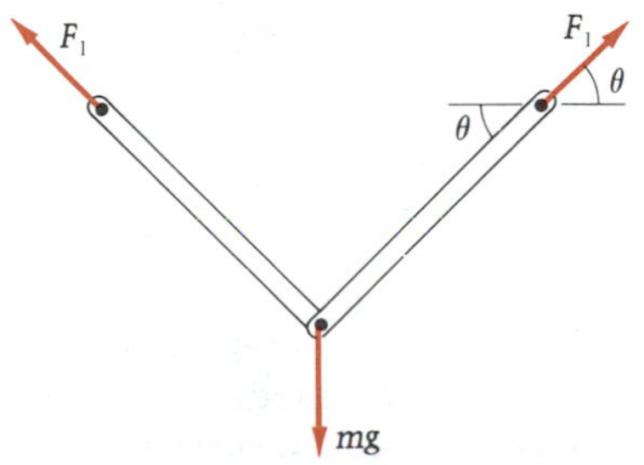

Figure 9.15

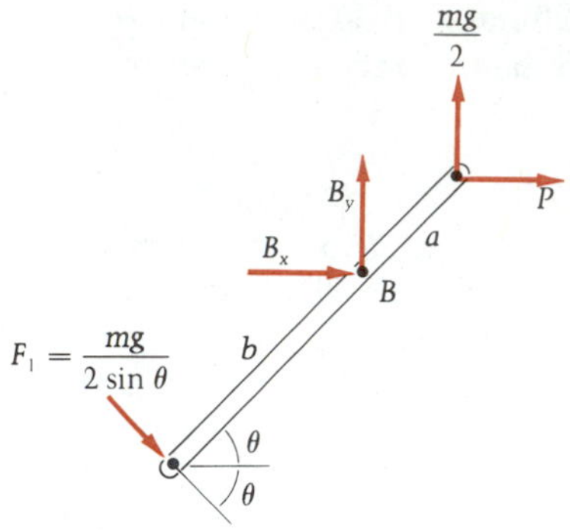

Figure 9.16

Moments about B will now give the value of P for equilibrium. Using the free-body diagram in Figure 9.16,

$$\circlearrowleft\!+\quad \Sigma M_B = 0$$

$$= \frac{mg}{2}\, a \cos\theta + \frac{mg}{2\sin\theta}\sin(\pi - 2\theta)b - Pa\sin\theta$$

$$P = \left[\frac{mga}{2}\cos\theta + \frac{mgb}{2\sin\theta}\overbrace{(2\sin\theta\cos\theta)}^{\sin 2\theta}\right]\Big/(a\sin\theta)$$

or

$$P = \frac{mg(a + 2b)\cot\theta}{2a}$$

as we obtained earlier using virtual work. Note now that as $\theta \to 0$, $\cot\theta$ (and therefore P) get very large. This can be seen most easily from the figure, where if $\theta \to 0$ only a small fraction ($\sin\theta$) of each F_1 is available to hold up half of mg.

Comparing the two approaches, we see that virtual work is in a sense a "cleaner" approach because forces (such as F_1 above) that do not do any net virtual work on the overall structure do not have to be considered at all in the solution.

EXAMPLE 9.3

Six bars, four of length 2ℓ and two of length ℓ, are pinned together to form a gate as shown in Figure E9.3a. The springs are attached at the midpoints of the indicated members, and the modulus of each is k. The springs can be compressed, and they are unstretched when the gate is fully retracted. Find the angle θ for equilibrium.

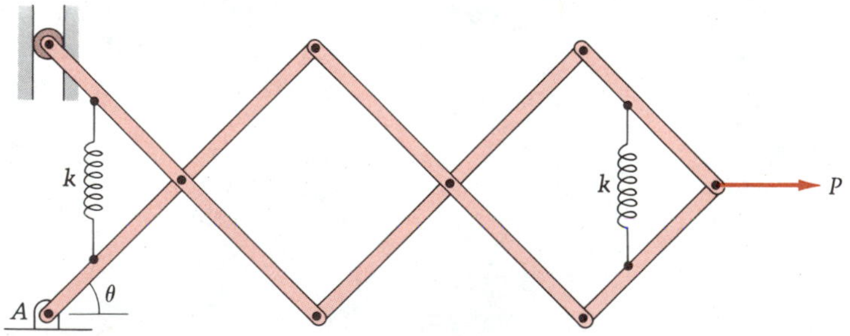

Figure E9.3a

Solution

Provided the actual constraints are not violated, the horizontal reaction of the slot onto the roller does no work during a virtual displacement $\delta\theta$ from the equilibrium position, nor do the pin reactions at A. Only the springs and P will contribute to the virtual work δW, which vanishes.

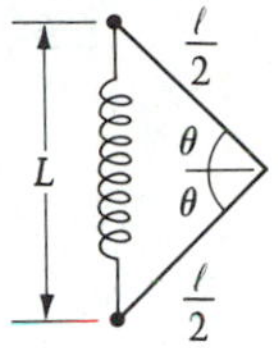

Figure E9.3b

The length L of each spring in the equilibrium configuration is seen in Figure E9.3b to be

$$L = 2\left(\frac{\ell}{2}\sin\theta\right)$$
$$= \ell\sin\theta$$

Therefore, the stretch in the spring is given by

$$S = \ell\sin\theta - \underbrace{2\left(\frac{\ell}{2}\right)}_{\text{Unstretched length}} = \ell(\sin\theta - 1)$$

which is negative because the spring is compressed. The virtual change in S, or "virtual stretch," is

$$\delta S = \ell\cos\theta\,\delta\theta$$

and the virtual work of two of them (with equal S and δS) is

$$\delta W_{\text{springs}} = -2kS\,\delta S = -2k\ell(\sin\theta - 1)\ell\cos\theta\,\delta\theta$$
$$= 2k\ell^2\cos\theta(1 - \sin\theta)\,\delta\theta$$

As seen in Figure E9.3c, the force P is located at $x = 5\ell\cos\theta$ so that

$$\delta x = -5\ell\sin\theta\,\delta\theta$$

and thus the virtual work of P is

$$\delta W_{\text{by } P} = P\,\delta x = -5P\ell\sin\theta\,\delta\theta$$

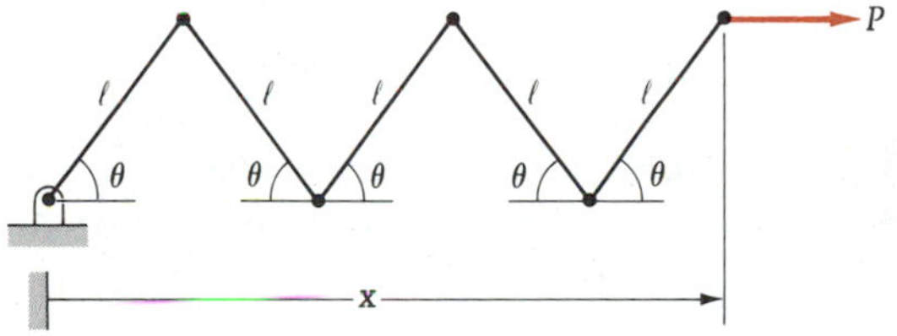

Figure E9.3c

Using $\delta W = 0$, we find

$$\delta W = 2k\ell^2\cos\theta(1 - \sin\theta)\delta\theta - 5P\ell\sin\theta\,\delta\theta = 0$$

or, since $\delta\theta$ is arbitrary,

$$1 - \sin\theta - 2.5\left(\frac{P}{k\ell}\right)\tan\theta = 0 \qquad (1)$$

The solution to this equation may be obtained by trial and error using a calculator or a computer. If the calculator is not programmable, we could proceed as follows for the case $P/k\ell = 1$.

A rough sketch of the functions $1 - \sin\theta$ and $2.5\tan\theta$ is shown in Figure E9.3d. It appears that they are equal [thus satisfying Equation (1)] at

$\theta \approx 0.4$. Using this as an initial guess and proceeding to compute the left-hand side of Equation (1), call it $f(\theta)$ (with the calculator computing in radians!), we obtain:

θ	$f(\theta)$
0.4	−0.446401
0.3	−0.068861
0.29	−0.031984
0.28	0.004759
0.281	0.001090
0.282	−0.002580

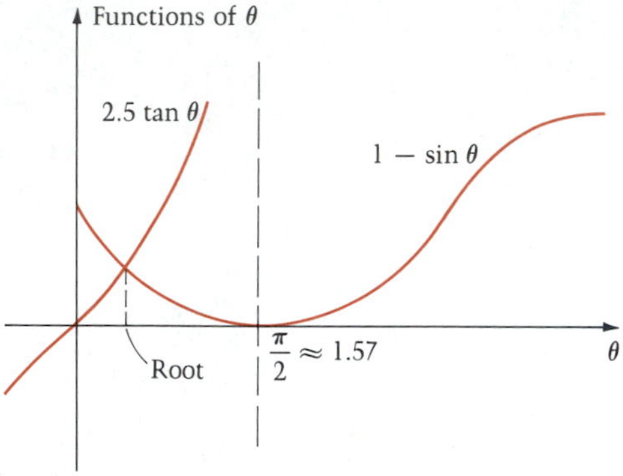

Figure E9.3d

Therefore θ for equilibrium is about 0.281 rad or about 16.1°.

EXAMPLE 9.4

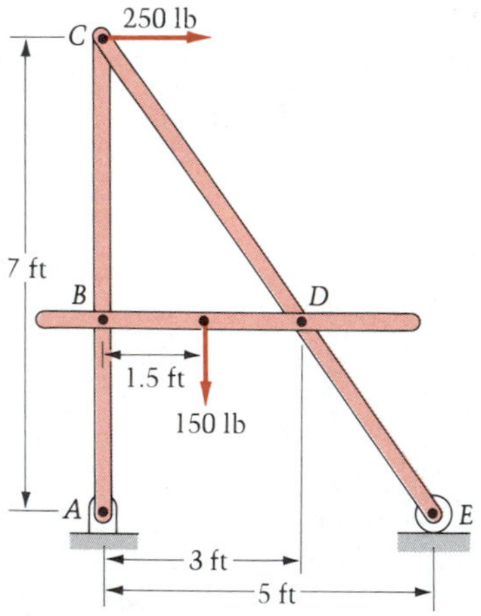

Figure E9.4a

In Figure E9.4a find the force exerted by the pin at D on the member BD in the rigid frame of Example 4.19.

Solution

Referring to the free-body diagram of BD, Figure E9.4b, we see that a virtual rotation with B fixed would cause D_y to be the only unknown component to do virtual work. Remember that the angle of rotation is to be infinitesimal so the displacement of D is essentially vertical. Letting $\delta\theta$ be the virtual rotation, then (see Figures E9.4b,c)

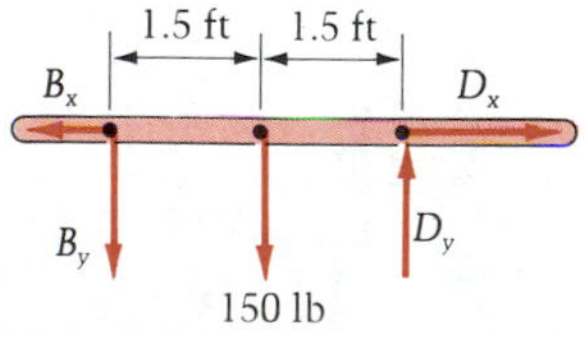

Figure E9.4b

$$\delta W = 0$$

$$D_y(3\ \delta\theta) - 150(1.5\ \delta\theta) = 0$$

$$D_y = 75 \text{ lb}$$

Of course, this is the same result obtained from applying $\Sigma M_B = 0$.

Now giving the bar a virtual horizontal translation, ϵ, as shown in Figure E9.4d, then

$$\delta W = 0$$

$$D_x\epsilon - B_x\epsilon = 0$$

$$D_x = B_x$$

which, of course, is nothing more than the equivalent of applying $\Sigma F_x = 0$.

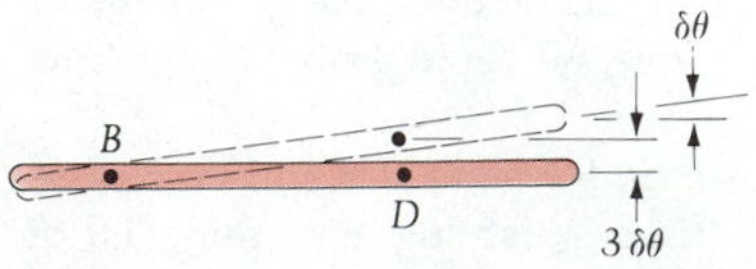

Figure E9.4c

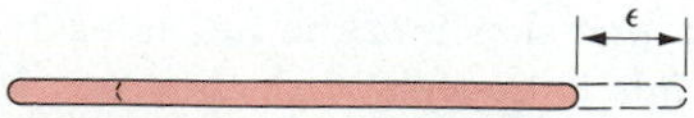

Figure E9.4d

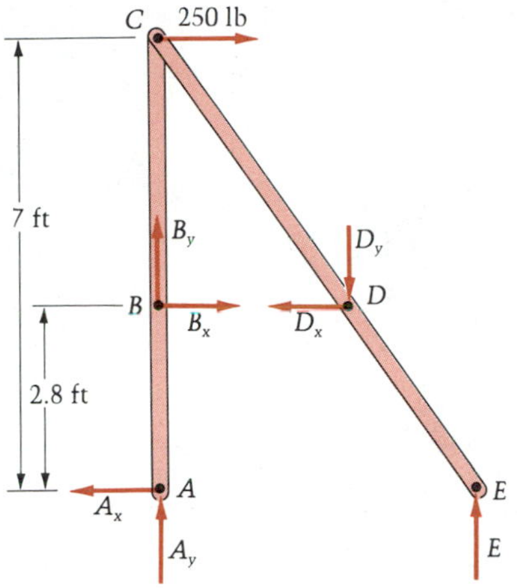

Figure E9.4e

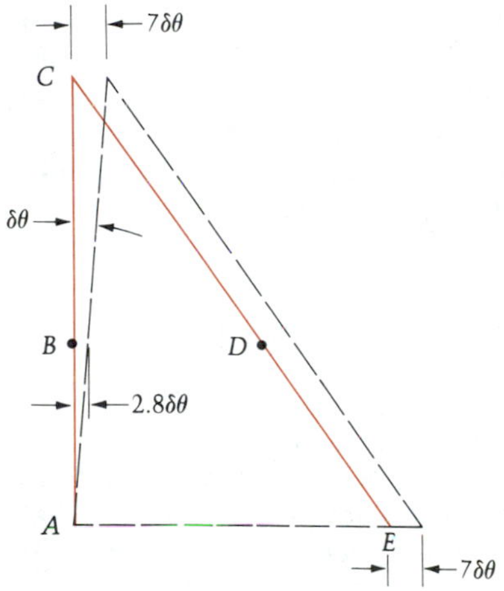

Figure E9.4f

To complete the solution of the problem, we now apply the method of virtual work in a manner that does not provide such an easily recognizable correspondence with the equilibrium equations. We focus our attention on the composite of the bars *ABC* and *CDE* for which we have the free-body diagram shown in Figure E9.4e. Now we choose the virtual change in configuration to be given by an infinitesimal rotation, $\delta\theta$, of *ABC* about *A* and a horizontal translation, $7\ \delta\theta$, of *CDE* so that the bars remain connected at *C*, and *E* moves horizontally as shown in Figure E9.4f. Thus, point *B* moves $2.8\ \delta\theta$ to the right and points *C* and *D* each move $7\ \delta\theta$ to the right. Applying the principle of virtual work to this assembly and noting that there is no net work done by the forces of interaction at *C*,

$$\delta W = 0$$

$$B_x(2.8\ \delta\theta) + 250(7\ \delta\theta) - D_x(7\ \delta\theta) = 0$$

or

$$2.8B_x + 7(250) - 7D_x = 0$$

and since

$$B_x = D_x$$

$$4.2D_x = 7(250)$$

or

$$D_x = 417 \text{ lb}$$

as was found before in Example 4.19.

Example 9.5

Twelve light rods of length *R* are connected to six equally spaced points *A*, *B*, *C*, *D*, *E*, and *F* on a circular base as shown in Figure E9.5a. They are joined by ball and socket connections to the base and to each other. The six bar-to-bar connections

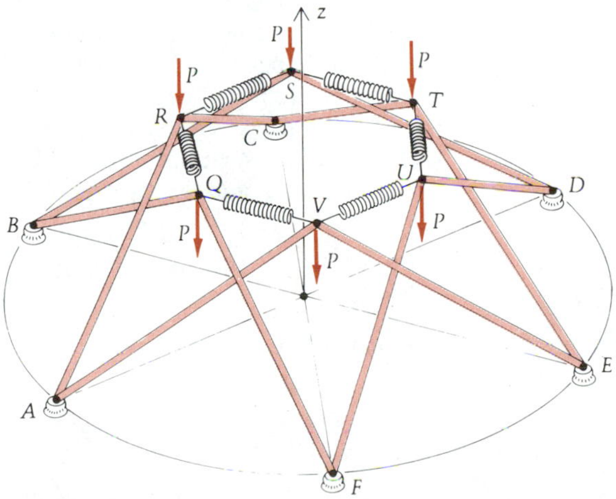

Figure E9.5a

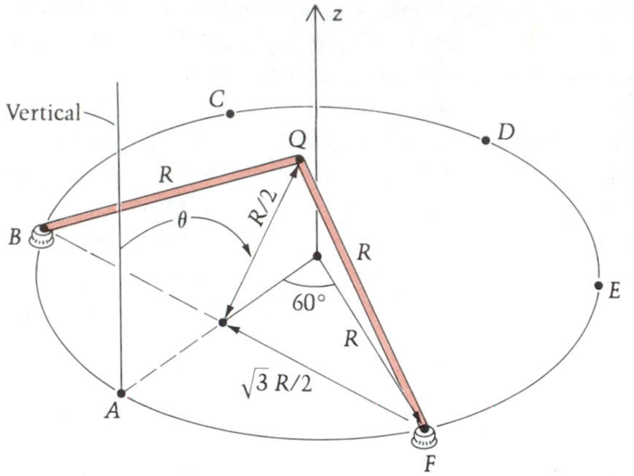

Figure E9.5b

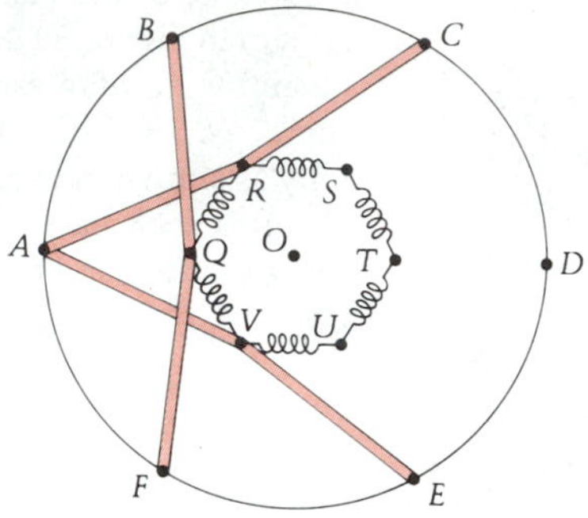

Figure E9.5c

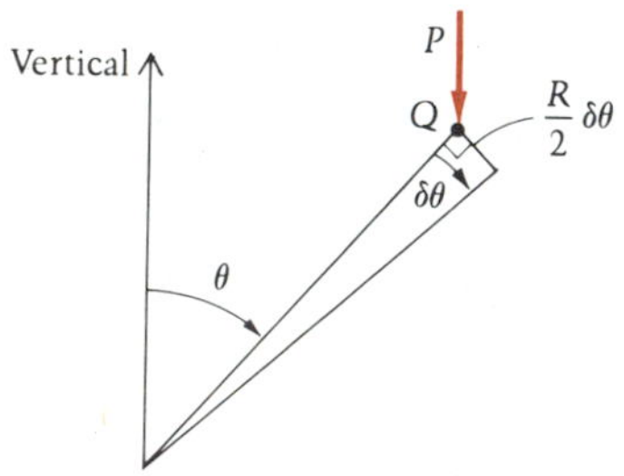

Figure E9.5d

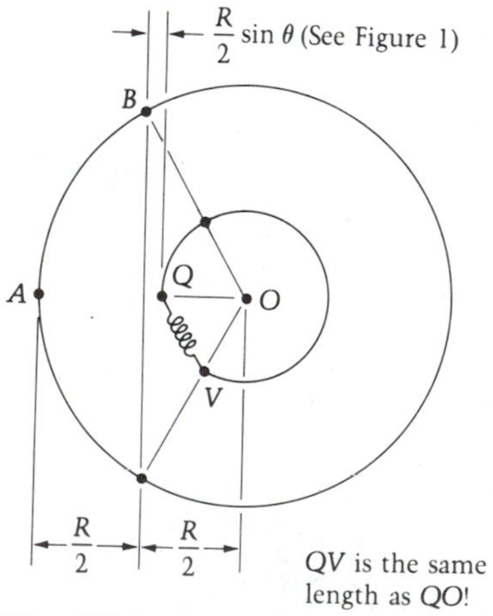

Figure E9.5e

at the top are joined by six stiff springs of unstretched lengths $0.4R$. The springs, which can carry compression, form a hexagon in a horizontal plane. If the spring moduli are each equal to k, find the angle(s) θ, between the vertical through A and the plane BQF (see Figure E9.5b), for which the system is in equilibrium under six downward loads P applied at Q, R, S, T, U, and V. Consider only symmetrical displacements—that is, those in which the six top connections Q, R, S, T, U, and V (Figures E9.5a and E9.5c) each move inward and downward identical amounts.

Solution

The sum of the virtual work done by the forces P and the springs will add to zero:

$$\delta W = 0$$

From symmetrical equilibrium positions, the forces P will each move inward toward the z axis (see Figure E9.5a) and downward, during a virtual increase in θ of amount $\delta\theta$. The virtual work done by P is seen in Figure E9.5d to be equal to $P(R/2)\ \delta\theta \sin\theta$, and there are six of these.

Next we consider the virtual work done by the six springs. During the virtual rotation $\delta\theta$ of triangle BQF (Figure E9.5b) about the line BF, the length of each spring will change slightly. In the equilibrium position of the structure at θ, the length of each spring (see Figure E9.5e and consider spring QV) is seen to equal $R - (R/2) - (R/2)\sin\theta$. Thus the stretch in each spring is $0.1R - (R/2)\sin\theta$. This means the "virtual stretch" occurring during the virtual rotation is $-(R/2)\cos\theta\ \delta\theta$, and the virtual work of each spring is

$$-kS\ \delta S = -k\left(0.1R - \frac{R}{2}\sin\theta\right)\left(-\frac{R}{2}\cos\theta\ \delta\theta\right)$$

The total virtual work may now be equated to zero:

$$\delta W = 0 = 6\left[P\frac{R}{2}\delta\theta\sin\theta - kS\ \delta S\right]$$

$$0 = \left[\frac{PR}{2}\sin\theta - k\left(0.1R - \frac{R}{2}\sin\theta\right)\left(-\frac{R}{2}\cos\theta\right)\right]\delta\theta$$

or

$$\frac{4P}{kR}\sin\theta + (0.4 - 2\sin\theta)\cos\theta = 0$$

For $P = kR/4$, the only roots to this equation for $0 \le \theta \le \pi/2$ are $\theta = 27.2°$ and $46.3°$. It is interesting to note that some equilibrium positions are unstable, such as a marble balanced on an inverted spherical bowl. It can be shown using more advanced mechanics that the 27.2° equilibrium position that we have found above is stable, while the 46.3° position is unstable.

PROBLEMS ▶ Section 9.1

Use the principle of virtual work to solve:

9.1 Problem 4.72

9.2 Problem 4.92

9.3 Problem 4.93

9.4 Problem 4.95

9.5 Problem 4.222

9.6 Problem 4.168

9.7 Problem 4.188

9.8 Problem 4.177

9.9 Problem 6.142

9.10 Problem 6.114

9.11 Find the relationship between P and F for equilibrium of the linkage shown in Figure P9.11.

9.12 Find the force P that must be exerted for equilibrium of the double lever system shown in Figure P9.12.

9.13 In Figure P9.13 find the force F that will hold the weight W in equilibrium.

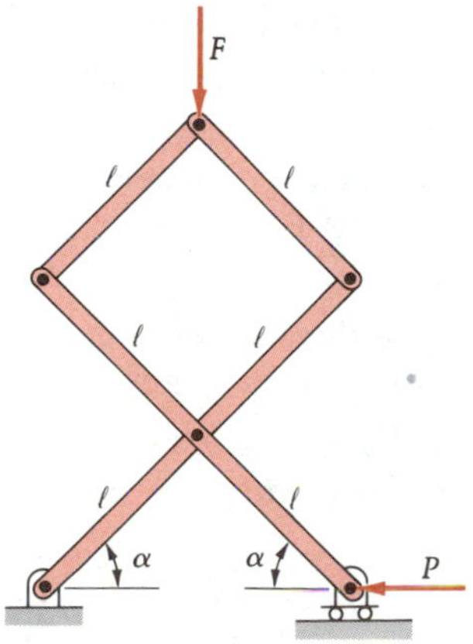

Figure P9.11

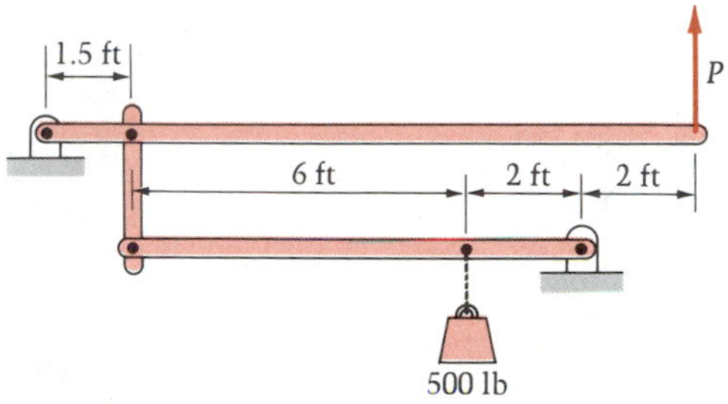

Figure P9.12

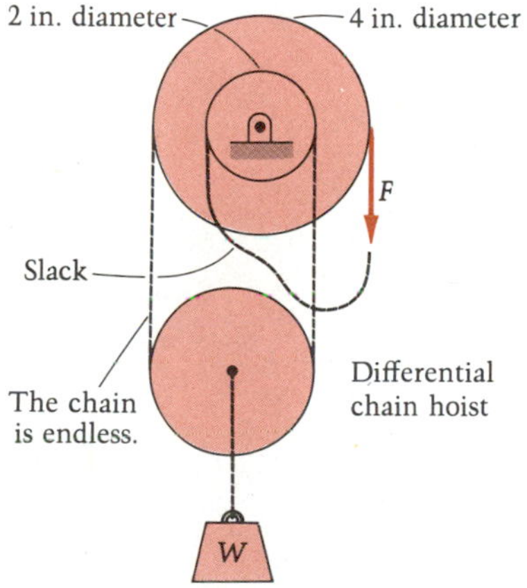

Figure P9.13

9.14 In Figure P9.14 the light, homogeneous, 1-m bars $\mathcal{B}_1$ and $\mathcal{B}_2$ are pinned together at D and are also connected by the spring AB of modulus 200 N/m; the unstretched length of the spring is 0.15 m. What is force P if the system is in equilibrium when triangle DEF is equilateral?

9.15 Repeat Problem 9.14 if $\mathcal{B}_1$ and $\mathcal{B}_2$ each have a mass of 2 kg with respective mass centers at C_1 and C_2.

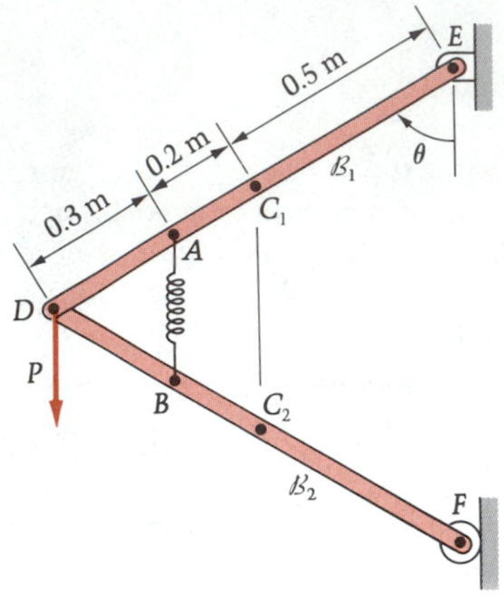

Figure P9.14

9.16 For the data of Problem 9.14, find the angle θ for equilibrium if the force P has a magnitude of 20 newtons.

9.17 Find P as a function of a, b, c, W, and θ for equilibrium of the system shown in Figure P9.17.

9.18 Find the angles θ and φ for equilibrium of the system of two light bars and two heavy weights shown in Figure P9.18.

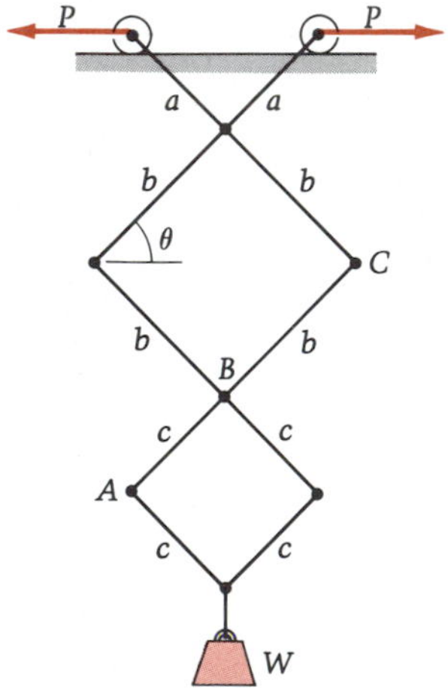

(Members such as ABC are continuous bars.)

Figure P9.17

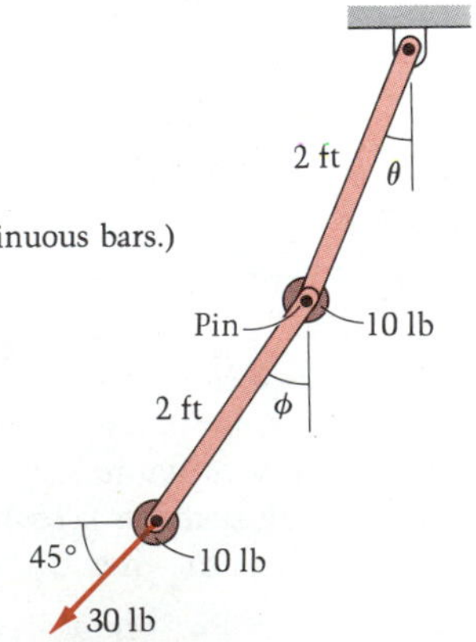

Figure P9.18

***9.19** In Figure P9.19 find the force exerted on the cargo lift by the hydraulic cylinder $\mathcal{C}$ if it is pinned to the lift at (a) point A; (b) point B. (c) Use the result of (a) to compute the reaction of pin C onto the member CDE at C. Is this reaction *along* CD?

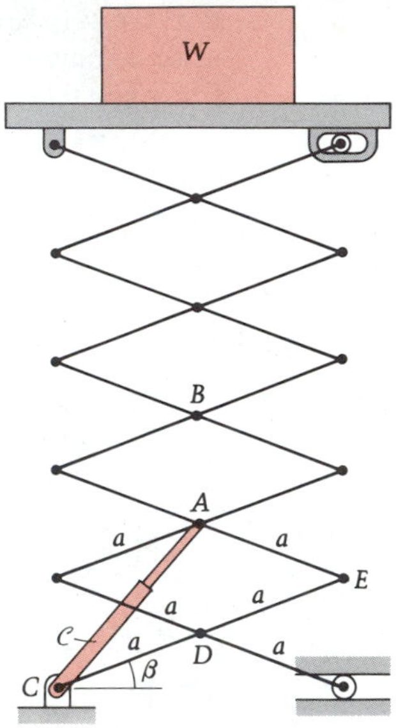

Figure P9.19

9.20 If the system in Figure P9.20 is in equilibrium, find the couple M being applied to crank $\mathcal{A}$ as a function of the angle θ and the force P on the piston. The crank, piston, and connecting rod $\mathcal{R}$ are light, and friction is negligible.

9.21 The linkage is made up of 10 light bars of length 2 ft (such as AB) and two bars of length 1 ft (on the extreme right), pinned together as indicated in Figure P9.21. The 300-lb force is to be applied as shown, and it is desired to have the system in equilibrium at $\theta = 60°$. Find the upward force at C on bar CD that will accomplish this.

9.22 In Problem 9.21, suppose that instead of the force at C, the equilibrium is to be maintained by applying a counterclockwise couple onto bar AB. Find the magnitude of the couple.

9.23 In a problem similar to 9.22, there are just four bars, with all other data remaining the same. See Figure P9.23.

a. Determine the couple M acting on AB for equilibrium using the principle of virtual work.

* Asterisks identify the more difficult problems.

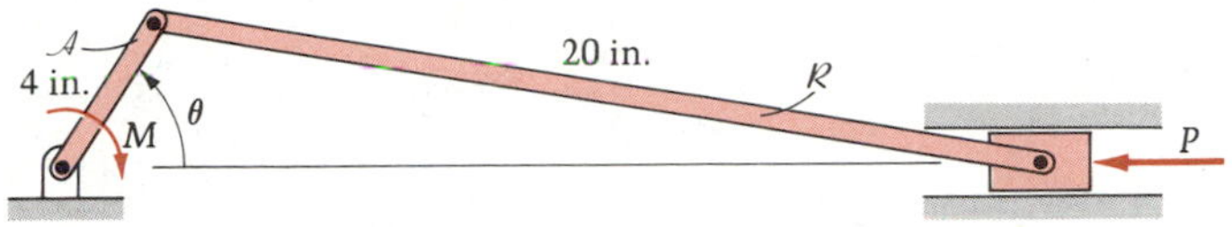

Figure P9.20

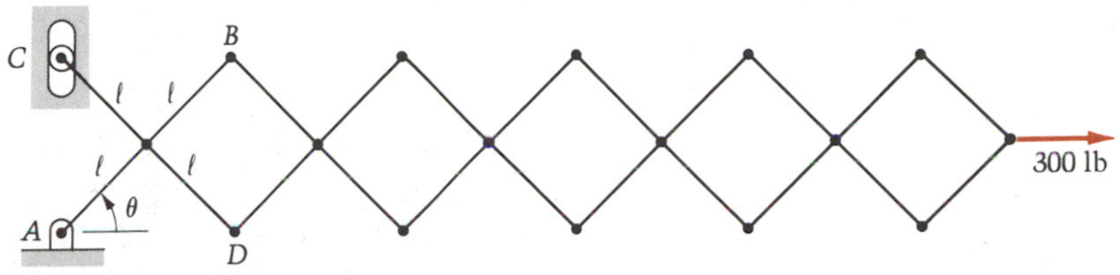

Figure P9.21

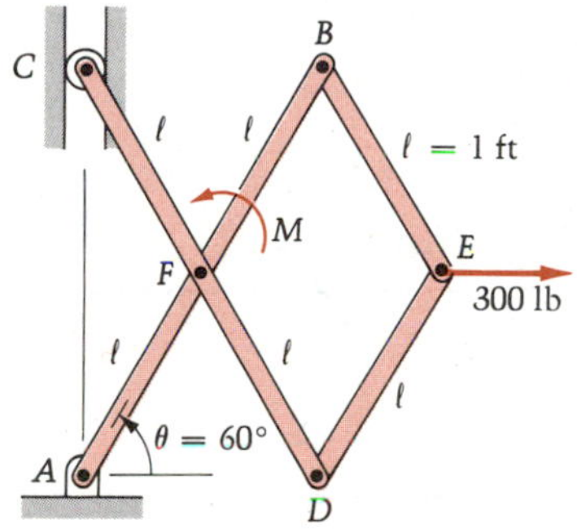

Figure P9.23

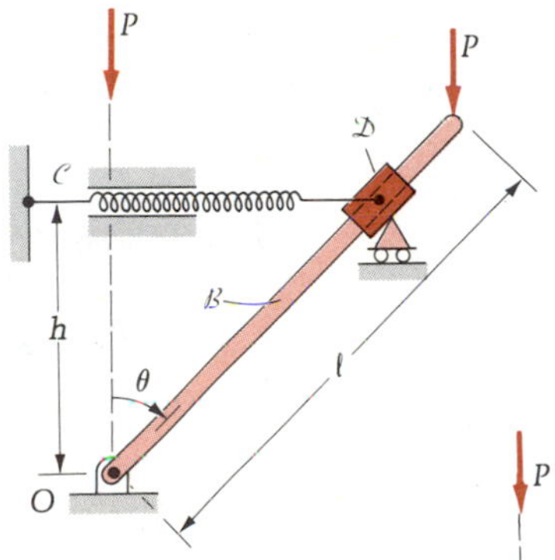

Figure P9.25

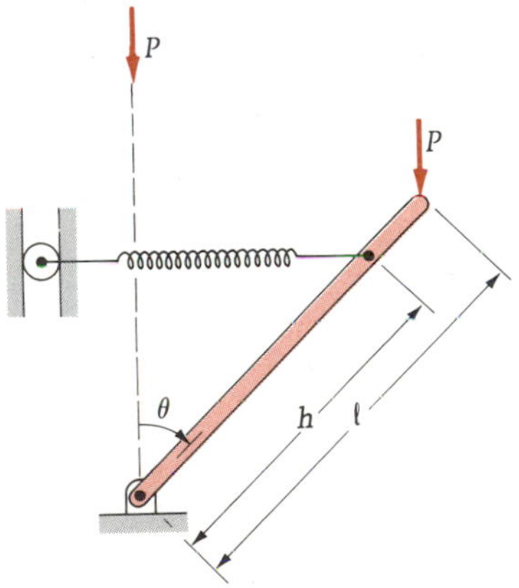

Figure P9.26

b. Find the force exerted on the roller at C by the slot, using equilibrium equations. Then take the four bars apart, and, on free-body diagrams of each, show all forces acting on the separate bars.

c. Show (with both virtual work and with equilibrium equations) that with the couple omitted the linkage cannot be in equilibrium.

9.24 In the preceding problem, delete the couple M and add a vertically attached stiff spring joining the midpoints of segments CF and AF with natural length 1 ft. Find the spring modulus if equilibrium exists at $\theta = 60°$.

9.25 The smooth collars $\mathscr{C}$ and $\mathscr{D}$ in Figure P9.25 ensure that the spring remains on the same horizontal level. The spring is unstretched at $\theta = 0$. Show that there is an equilibrium position of bar $\mathscr{B}$ (besides the obvious one at $\theta = 0$) given by

$$\theta = \cos^{-1}\left[\left(\frac{kh^2}{P\ell}\right)^{1/3}\right]$$

9.26 This problem is similar to Problem 9.25 except that now the spring changes its level while remaining horizontal. Find the "not-obvious" equilibrium angle θ. (See Figure P9.26, where the spring is unstretched at $\theta = 0$.)

9.27 Use virtual work to solve Problem 4.205.

9.28 Verify the result of Example 9.5 by using the equilibrium equations.

In Problems 9.29–9.32 find the relationship between P and W using the principle of virtual work.

9.29

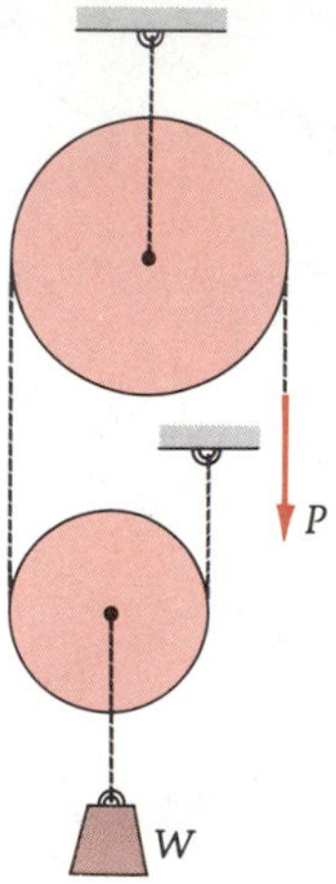

Figure P9.29

9.30

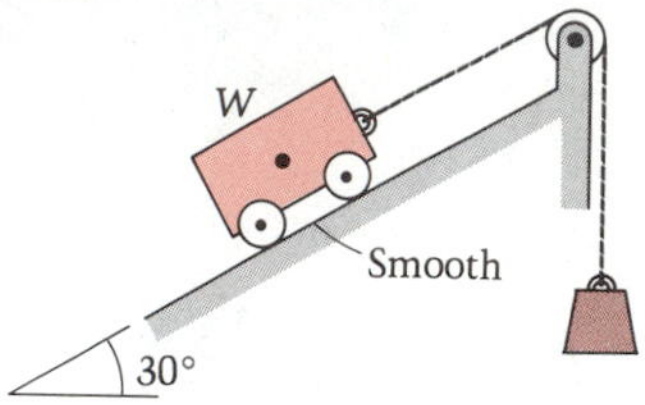

Figure P9.30

9.31

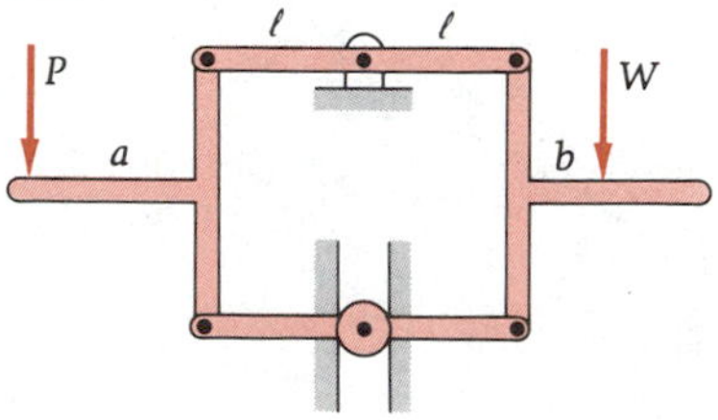

Figure P9.31

9.32

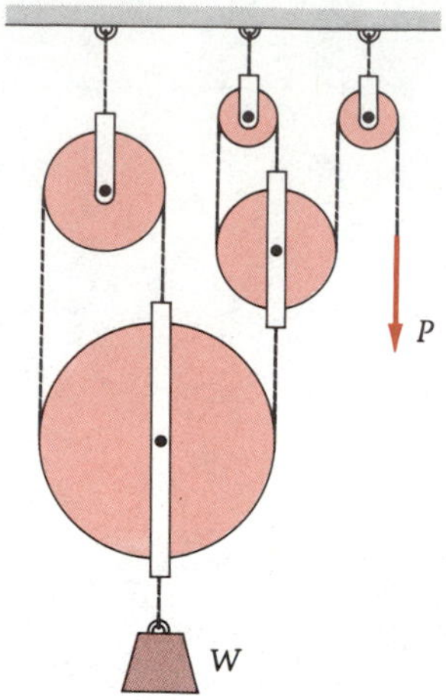

Figure P9.32

9.3 Hydrostatic Pressure on Submerged Bodies

In Example 3.35 we recalled that in a fluid at rest:

a. The pressure is equal in all directions at a point (Pascal's law).

b. The pressure is constant through the fluid in each horizontal plane.

c. The pressure causes a force that is normal to every differential area of surface on which it acts.

d. The pressure is equal to γh, where γ is the specific weight of the fluid* (taken constant here) and h is the depth below the free surface.

Also, in Section 7.2 we found that when pressure acts on a plane surface, the resultant of the pressure forces has:

a. A magnitude F_r equal to the volume beneath the loading surface.

b. A line of action normal to the plane and directed toward it, and passing through the centroid of this volume.

In this section we shall first broaden our treatment to allow a *hydrostatic* pressure to act on *non-rectangular* plane areas. It will turn out that the formulas we obtain for $\mathbf{F}_r$ and its line of action will be intimately tied to the very same properties of areas we have been studying in detail in Chapters 7 and 8: the amount of area, its centroid, and its inertia properties.

---

* γ is its weight per unit volume, related to the mass density ρ by $\gamma = \rho g$.

We shall look at two results that allow us to obtain the resultant of fluid pressures on *curved* surfaces, often without the need for integration. Finally, we shall complete our look at fluid statics by briefly discussing Archimedes' principle of buoyancy.

Hydrostatic Pressure on Plane Surfaces

We begin by developing some results for hydrostatic pressure on *arbitrary* plane areas. At the depth $x \cos\theta$ (see Figure 9.17), the pressure* on all elements of the line $\ell\ell$ is given by

$$p = \gamma x \cos\theta \tag{9.18}$$

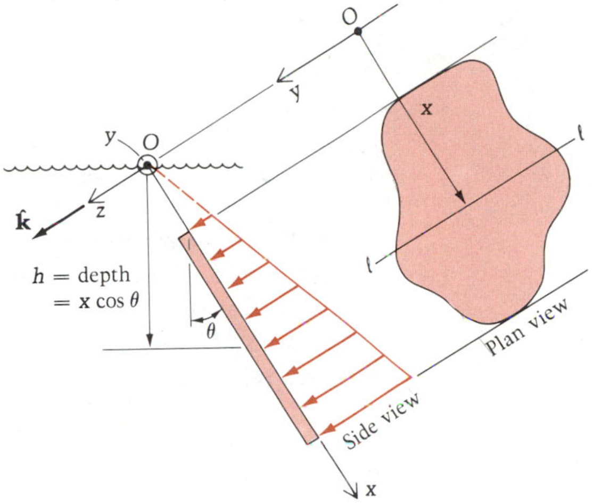

Figure 9.17

where θ is the inclination of the plane area A with the vertical. Noting that all the differential forces on area A due to the pressure ($dF = p\,dA$) are parallel, we know that there will be a "force-alone resultant" in this direction, and we may write

$$d\mathbf{F} = p\,dA\hat{\mathbf{k}}$$

Integrating, we then obtain the resultant:

$$\mathbf{F}_r = \hat{\mathbf{k}} \int \underbrace{\gamma x \cos\theta}_{p}\, dA$$

$$\mathbf{F}_r = \hat{\mathbf{k}}\gamma \cos\theta \int x\,dA \tag{9.19}$$

* Throughout this section, we are using gauge pressure — that is, the pressure above atmospheric. If the pressure *behind* one of the submerged surfaces we shall be studying is atmospheric, then the actual front-to-back pressure difference is indeed the gauge pressure we are using. If this is not the case, however, then both should be expressed as absolute pressures or both as gauge pressures so that correct differences will be used in determining resultant fluid forces on the surfaces.

But from our knowledge of centroids,

$$\bar{x}A = \int x\, dA \tag{9.20}$$

so that

$$\mathbf{F}_r = \hat{\mathbf{k}}\, \underbrace{\gamma\underbrace{(\bar{x} \cos \theta)}_{\text{Depth of centroid of area}}}_{\text{Pressure at centroid } = p_c} A = p_c A \hat{\mathbf{k}}$$

Thus the resultant force, normal to the plane of the area, has a magnitude equal to the *pressure at the centroid times the area.* We now seek the *location* of this force-alone resultant which is equipollent to the fluid pressure forces. It will *not* be at the centroid of the plane area unless $\theta = 90°$.

We are looking for the x and y coordinates of a point P where $\mathbf{F}_r$ (above) may be placed and have the same moment about any point (we use the origin O) as does the original system of fluid forces. This point is called the center of pressure.

Thus, we wish to find the vector $\mathbf{r}_{OP}$ in the equation

$$\mathbf{r}_{OP} \times \mathbf{F}_r = \int \mathbf{r} \times d\mathbf{F} \tag{9.21}$$

We represent the sought vector $\mathbf{r}_{OP}$ as

$$\mathbf{r}_{OP} = x_P \hat{\mathbf{i}} + y_P \hat{\mathbf{j}} \tag{9.22}$$

and we write

$$(x_P \hat{\mathbf{i}} + y_P \hat{\mathbf{j}}) \times (p_c A \hat{\mathbf{k}}) = \int (x \hat{\mathbf{i}} + y \hat{\mathbf{j}}) \times (p\, dA \hat{\mathbf{k}}) \tag{9.23}$$

$$-x_P p_c A \hat{\mathbf{j}} + y_P p_c A \hat{\mathbf{i}} = -\hat{\mathbf{j}} \int xp\, dA + \hat{\mathbf{i}} \int yp\, dA \tag{9.24}$$

Equating the $\hat{\mathbf{j}}$ coefficients of Equation (9.24) will give us the x coordinate of the center of pressure:

$$x_p = \frac{\int xp\, dA}{p_c A} = \frac{\int x(\gamma x \cos \theta)\, dA}{\gamma \cos \theta\, \bar{x} A}$$

$$x_p = \frac{\cancel{\gamma} \cancel{\cos \theta} \int x^2\, dA}{\cancel{\gamma} \cancel{\cos \theta}\, \bar{x} A}$$

and recognizing the moment of inertia (see Section 8.1),

$$x_P = \frac{I_{yO}}{\bar{x}A} \tag{9.25}$$

Now, using the parallel-axis (transfer) theorem (see Section 8.3),

$$x_P = \frac{I_{y_c} + A\bar{x}^2}{\bar{x}A} = \bar{x} + \frac{I_{y_c}}{\bar{x}A} \tag{9.26}$$

We see that, since I_{y_c} is always positive, the center of pressure is always lower than the centroid of the area unless $\theta = 90°$, in which case the pressure is the same at all points of A, and then P and C are at the same depth.

Question 9.6 Why doesn't Equation (9.26) apply for this case?

We next equate the $\hat{\mathbf{i}}$ coefficients of Equation (9.24) and find the lateral position (y_P) of the center of pressure:

$$y_P = \frac{\int yp\, dA}{p_C A} = \frac{\int y\gamma x \cos\theta\, dA}{(\gamma \cos\theta\, \bar{x})A}$$

$$y_P = \frac{\int xy\, dA}{\bar{x}A} = \frac{-I_{xyO}}{\bar{x}A} \tag{9.27}$$

where $I_{xyO} = -\int xy\, dA$ is the product of inertia of the area with respect to the axes x and at y at point O. Using the parallel-axis (or transfer) theorem (see Section 7.6),

$$y_P = \frac{-(I_{xy_c} - A\bar{y}\bar{x})}{\bar{x}A} = \bar{y} - \frac{I_{xy_c}}{\bar{x}A} \tag{9.28}$$

If either of the axes through C, parallel to x or y, is an axis of symmetry of the plane area, then $I_{xy_c} = 0$ and then $y_P = \bar{y}$. This means that the lateral distance from the x axis to the centroid in this case is the same as the distance from the x axis to the center of pressure* (see Figure 9.18).

If, however, $I_{xy_c} \neq 0$, then the center of pressure will be displaced laterally from C; y_P will not equal $\bar{y}$. We shall now consider three examples. In the first, the area is rectangular; in the others, the width varies with depth.

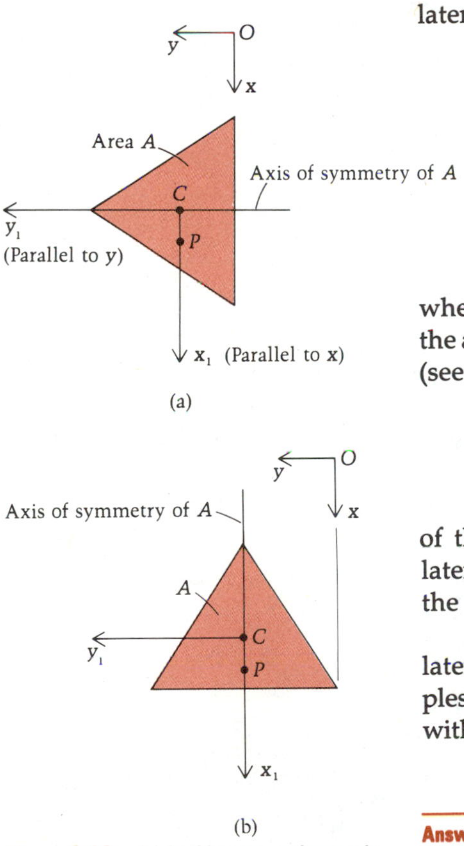

Figure 9.18 (a) In this case, y_1 is an axis of symmetry of area A, so that $I_{xy_c} = 0$ and thus $y_P = \bar{y}$. (b) This time, x_1 is an axis of symmetry of A, so again $I_{xy_c} = 0$ and $y_P = \bar{y}$.

Answer 9.6 If $\theta = 90°$, we have divided by zero in deriving Equation (9.26).

* For the case when the axis through C parallel to x is an axis of symmetry, the result $y_P = \bar{y}$ becomes obvious when we recall that the line of action of $\mathbf{F}_r$ passes through the centroid of the volume beneath the loading (pressure) surface; the other case's (same) result does not, however, follow from this knowledge about $\mathbf{F}_r$'s line of action.

EXAMPLE 9.6

Show that the results of Example 3.35 follow from Equations (9.20) and (9.26) of this section.

Solution

We have seen that the resultant force on any plane area A equals the centroidal pressure times A; therefore, using Figure E9.6a,

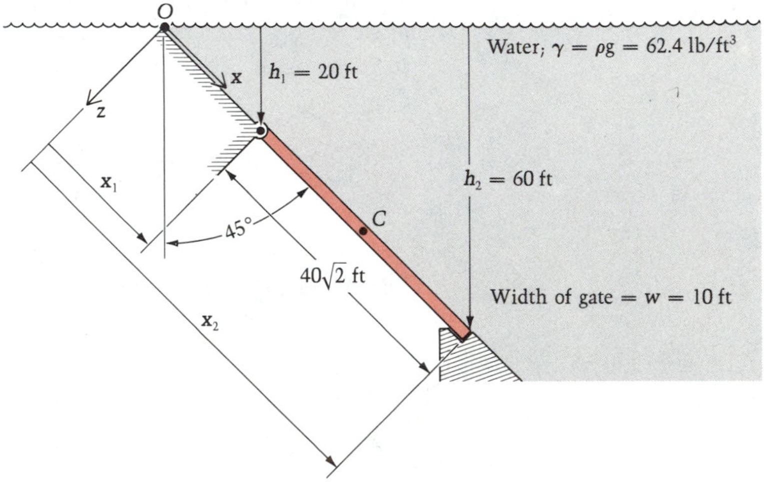

Figure E9.6a

$$|\mathbf{F}_r| = \underbrace{\gamma\left(\frac{x_1+x_2}{2}\right)(\cos\phi)}_{\text{Pressure at } C}\underbrace{(x_2-x_1)w}_{\text{Area of gate}} = 62.4(x_2^2 - x_1^2)\,\frac{1/\sqrt{2}}{2}\,(10)$$

$$= 624[(60\sqrt{2})^2 - (20\sqrt{2})^2]/(2\sqrt{2}) = 1.41 \times 10^6 \text{ lb}$$

which is the same answer as we obtained by integration in Example 3.35. For the distance d to the center of pressure P, we use Equation (9.26):

$$x_P = \bar{x} + \frac{I_{yc}}{\bar{x}A}$$

Using Figure E9.6b, the terms become

$$x_P = \frac{x_1+x_2}{2} + \frac{\overbrace{\dfrac{w(x_2-x_1)^3}{12}}^{I_{yc}}}{\underbrace{\dfrac{(x_1+x_2)}{2}}_{\bar{x}}\underbrace{[(x_2-x_1)w]}_{A}}$$

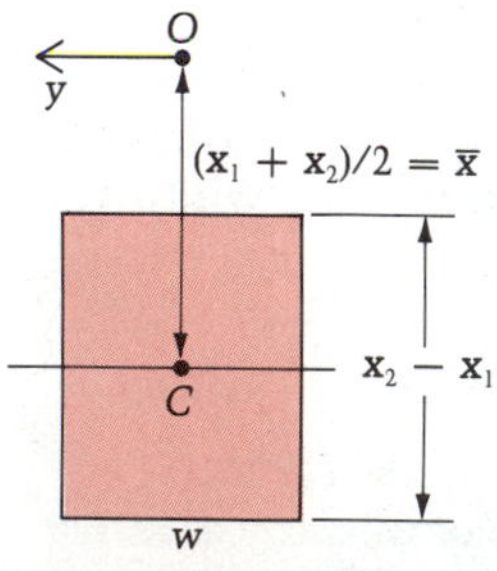

$I_{y_C} = \dfrac{(\text{base})(\text{height})^3}{12}$

Figure E9.6b

After simplifying, this results in

$$d = \frac{2}{3}\frac{x_2^2 + x_1x_2 + x_1^2}{x_1 + x_2} = 61.3 \text{ ft}$$

which is again the same answer we obtained in Example 3.35.

Example 9.7

Find the center of pressure of the triangular fluid gate, shaped and situated as shown in Figure E9.7

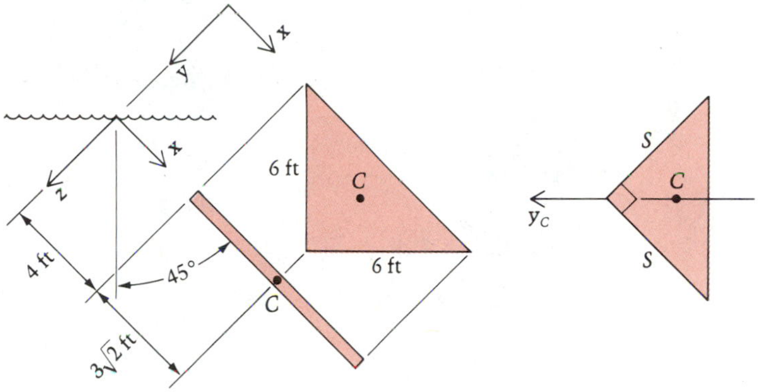

Figure E9.7

Solution

Using Example 8.3, we can obtain

$$I_{y_c} = \frac{S^4}{24}$$

We then obtain, for our triangle,

$$I_{y_c} = \frac{6^4}{24} = 54 \text{ ft}^4$$

Therefore,

$$x_p = \bar{x} + \frac{I_{y_c}}{\bar{x}A}$$

$$= (4 + 3\sqrt{2}) + \frac{54}{(4 + 3\sqrt{2})\underbrace{\tfrac{1}{2}(6)6}_{A}}$$

$$= 8.61 \text{ ft}$$

For the y coordinate of P,

$$y_p = \bar{y} - \frac{\overset{0 \text{ by symmetry}}{\cancel{I_{xy_c}}}}{\bar{x}A}$$

$$y_p = \bar{y} - 0 = \tfrac{1}{3}(3\sqrt{2})$$

$$= \sqrt{2} = 1.41 \text{ ft}$$

EXAMPLE 9.8

If the same gate as in Example 9.7 is reoriented as shown in Figure E9.8, find the new position of the center of pressure.

Solution

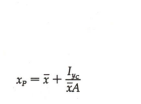

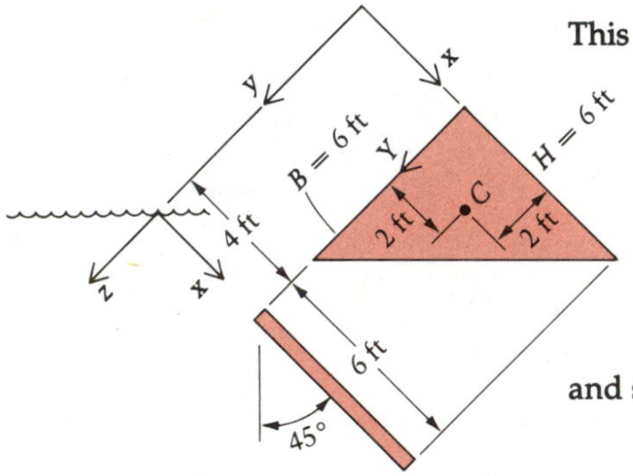

Figure E9.8

This time, I_Y (see the figure) $= \dfrac{BH^3}{12} = \dfrac{6(6^3)}{12} = 108 \text{ ft}^4$ so that

$$I_{y_c} = 108 - Ad^2$$

$$= 108 - \tfrac{1}{2}(6)6(2^2) = 36 \text{ ft}^{4*}$$

and so

$$x_p = \bar{x} + \frac{I_{y_c}}{\bar{x}A}$$

$$= (4 + 2) + \frac{36}{(4 + 2)18}$$

$$= 6 + (\tfrac{1}{3}) = 6.33 \text{ ft}$$

The product of inertia I_{xy_c} is needed for y_P. From Example 8.21, it is

$$I_{xy_c} = \frac{(\text{base} \times \text{height})^2}{72}$$

$$= \frac{(6 \times 6)^2}{72} = 18 \text{ ft}^4$$

Therefore,

$$y_P = \bar{y} - \frac{I_{xy_c}}{\bar{x}A}$$

$$= 2 - \frac{18}{(6)(18)} = 1.83 \text{ ft}$$

* Or we could alternatively use $I_{y_c} = \dfrac{BH^3}{36} = \dfrac{6(6^3)}{36} = 36 \text{ ft}^4$.

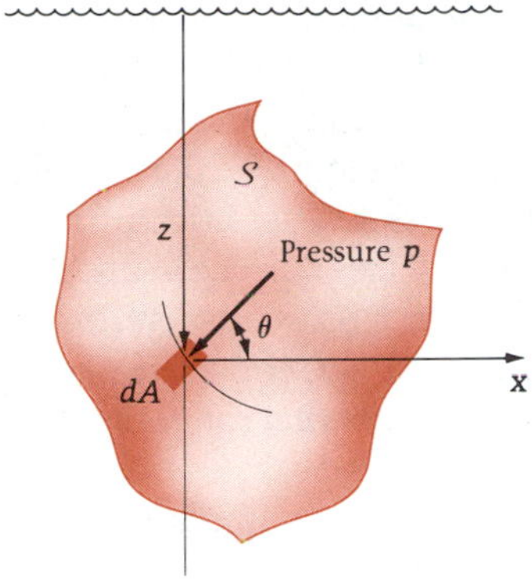

Figure 9.19

Hydrostatic Pressure on Curved Surfaces

If the submerged surface $\mathcal{S}$ is not necessarily flat, then formulas such as (9.26) and (9.28) do not apply. Instead we may use the following two results: Let the resultant force due to hydrostatic fluid pressure on one side of a curved surface* be called $\mathbf{F}_r$. Then:

Result (I) The component of $\mathbf{F}_r$ in a horizontal direction x is the force caused by the same pressure acting on the vertical projection of the area onto a plane normal to x. The proof is as follows, with θ being the angle between the normal to the differential area dA and the x direction (see Figure 9.19):

$$F_{r_x} = \underbrace{\int \underbrace{(p\,dA)}_{dF} \cos\theta}_{\substack{\text{Differential force} \\ \text{in } x \text{ direction}}} = \underbrace{\int p \underbrace{(dA \cos\theta)}_{\substack{\text{Vertical protection} \\ \text{of } dA\text{, normal to } x}}}_{\substack{\text{Differential force on} \\ \text{projected area normal to } x}}$$

> **Question 9.7** What is the difference between this result and the first of the three special properties of resultants on page 141 in Section 3.9?

Result (II) The vertical component of $\mathbf{F}_r$ is the weight of the column of fluid above $\mathcal{S}$.

> **Question 9.8** Why can't the liquid surrounding the column help hold it up?

In this second result, we see that there is no confusion if the fluid above the surface all extends vertically up to the free surface, as shown on the left in Figure 9.20(a). A surface $\mathcal{S}$ made up of the curve AB and a

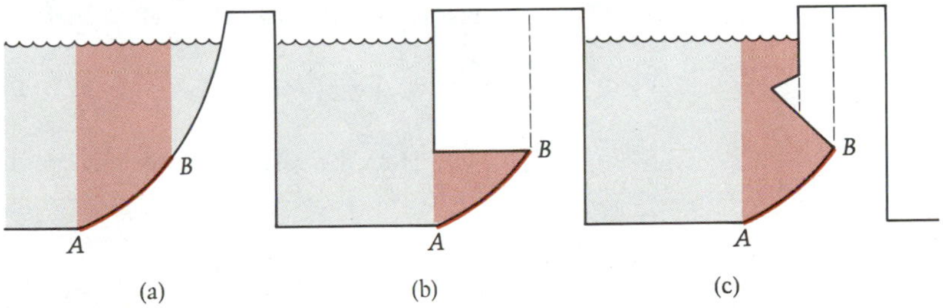

Figure 9.20

* With no vertical outward normals except possibly along part of its edge.

Answer 9.7 In the prior result (in Section 3.9) the pressure was constant, and so *any* projection could be used. In this chapter, the (hydrostatic) pressure varies linearly with depth. Thus the vertical projection of the area is the only one it makes sense to use in the theorem.

Answer 9.8 It is only able to exert normal (horizontal, here) forces on the column.

uniform width w normal to the page is being considered, and the vertical component of the fluid force resultant on $\mathcal{S}$ is obviously the weight of the shaded fluid. In Figure 9.20(b), however, the fluid above the same surface $\mathcal{S}$ does not extend up to the free surface. And in Figure 9.20(c), it does so over some of the area elements, on others it does so after interruption, and on still others it doesn't extend upward to any free surface at all. The point is that none of this matters. The vertical force on $\mathcal{S}$ acts as though the fluid existed in a column up to the free surface (extended, if necessary) in all cases because *it is the pressure on the surface that counts and the pressure in the fluid is the same across any horizontal level.*

We now present two related examples. In the second, we shall consider the pressure on a curved surface; in both, we shall use results I and II.

EXAMPLE 9.9

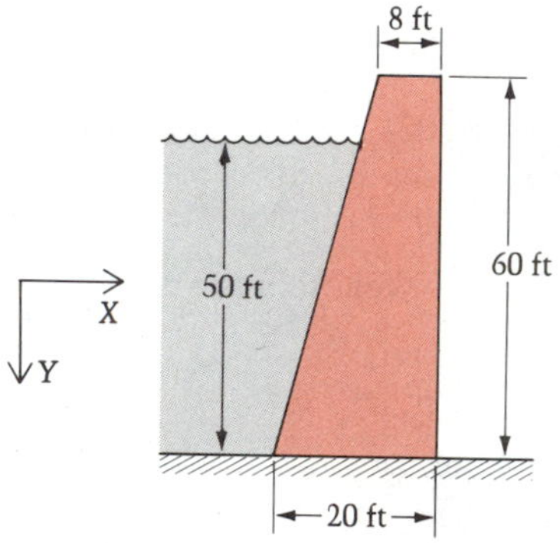

Figure E9.9a

If the dam in Figure E9.9a is 200 ft wide, determine and locate the resultant force exerted on it by the water pressure.

Solution

We note first that in this case, all differential forces exerted on the dam by the water are parallel; therefore, there is a "force-alone" resultant that is normal to the slanted surface.

The horizontal component of the resultant force equals the pressure at the centroid of the projected (shaded) vertical area (see Figure E9.9b) multiplied by this area:

$$F_{rX} = \underbrace{(\gamma h_C)}_{\text{Pressure at the centroid}} \underbrace{(A_P)}_{\text{Projected area}}$$

$$= \left(62.4\ \frac{\text{lb}}{\text{ft}^3}\right)(25\text{ ft})(10{,}000\text{ ft}^2) = 15.6 \times 10^6\text{ lb}$$

The vertical component of the resultant force equals the weight of water (see Figure E9.9c) above the wetted surface:

$$F_{rY} = \gamma(\text{Vol}) = 62.4\ \frac{\text{lb}}{\text{ft}^3}\left(\tfrac{1}{2} \times 10\text{ ft} \times 50\text{ ft} \times 200\text{ ft}\right) = 3.12 \times 10^6\text{ lb}$$

Thus the resultant force has the magnitude

$$F_r = \sqrt{15.6^2 + 3.12^2} \times 10^6 = 15.9(10^6)\text{ lb}$$

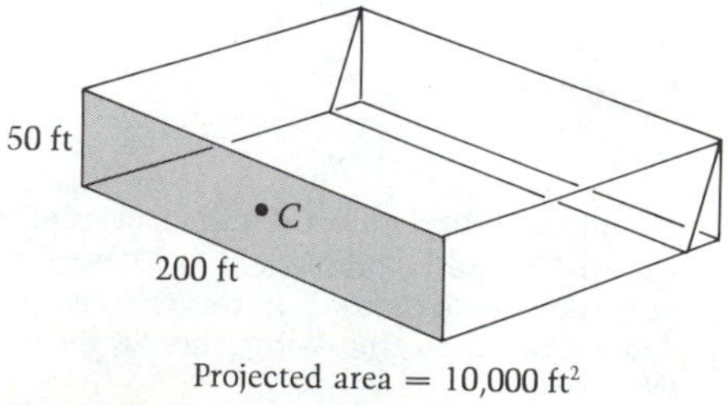

Figure E9.9b

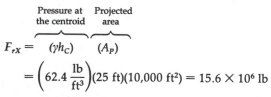

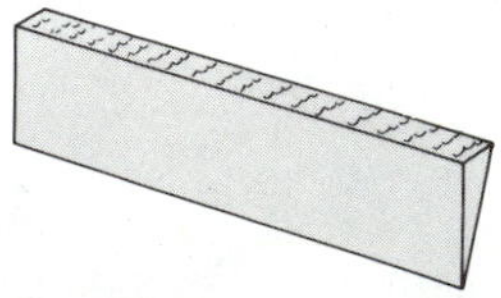

Figure E9.9c

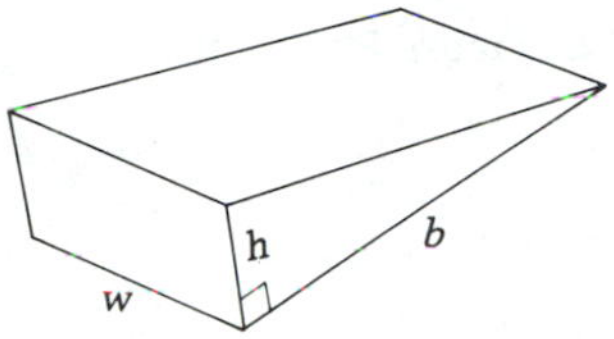

Figure E9.9d

This magnitude of $\mathbf{F}_r$ may be checked. As we have seen, it is also the volume (of the triangular prism in Figure E9.9d) under the loading (or pressure) surface:

$$\begin{aligned} F_r &= \tfrac{1}{2}bhw \\ &= \tfrac{1}{2}(51)(\gamma 50)(200) \text{ lb} \\ &= 25.5(62.4)50(200) \text{ lb} \\ &= 15.9(10^6) \text{ lb} \qquad \text{(as above)} \end{aligned}$$

As for the *direction* of $\mathbf{F}_r$, it is, as previously mentioned, normal to the surface because it comprises of infinitely many parallel differential forces, *each one* normal to the surface with the sense

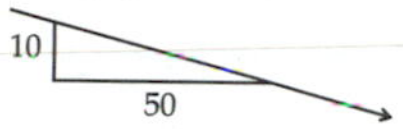

We note that the theorems used to obtain F_{rX} and F_{rY} also yield the correct direction for the resultant:

$$\theta = \tan^{-1}\left(\frac{15.6}{3.12}\right) = \tan^{-1} 5 \qquad \text{(as above and as seen in Figure E9.9e)}$$

where

$$\phi = \tan^{-1}\frac{50}{10} = \tan^{-1} 5 \qquad \text{(also)}$$

$\therefore$ $\mathbf{F}_r$ is seen again to be normal to the slanted surface of the dam.

The place on the dam where $\mathbf{F}_r$ acts (known as the center of pressure) is at the centroid of the distributed loading, which is in this case a triangular prism. Thus the load acts normal to the dam a distance $\frac{2}{3}(51) = 34$ ft along the slant from the water surface (see Figure E9.9f), and, of course, halfway across the dam (100 feet from each end).

We may check this result in two ways: First, we use the components F_{rX} and F_{rY} obtained earlier, and let M_{rO} be the resultant moment about O of the forces due to hydrostatic fluid pressure. Then (see Figure E9.9g),

$$\begin{aligned} M_{rO} &= F_{rX}[\tfrac{2}{3}(50)\text{ ft}] + F_{rY}[\tfrac{2}{3}(10)\text{ ft}] \\ &= 15.6(10^6)\left[\frac{100}{3}\right] + 3.12(10^6)\left[\frac{20}{3}\right] \text{ lb-ft} \\ &= 520(10^6) + 20.8(10^6) = 541(10^6) \text{ lb-ft} \end{aligned}$$

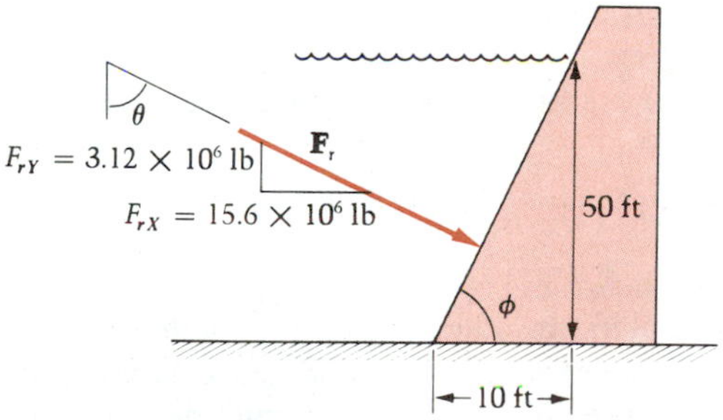

Figure E9.9e

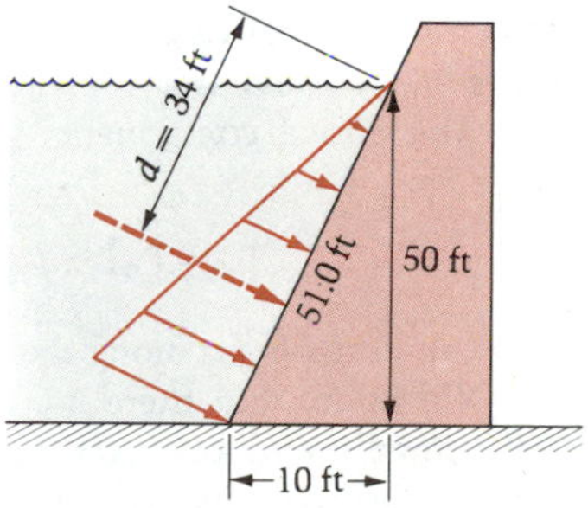

Figure E9.9f

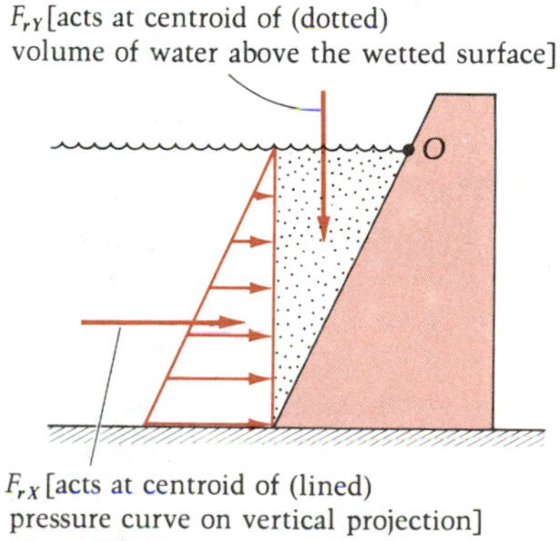

Figure E9.9g

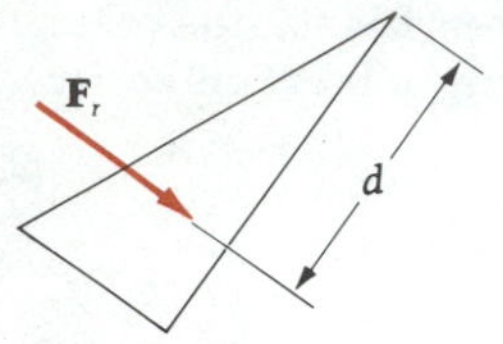

Figure E9.9h

If this is to be equipollent to $\mathbf{F}_r$ at distance d (see Figure E9.9h), then

$$15.9(10^6)d = 541(10^6)$$

$$d = 34.0 \text{ ft} \qquad \text{(as before)}$$

Alternatively, we may determine the distance d by using Equation (9.26):

$$d \text{ or } x_p = \frac{I_{yc}}{\bar{x}A} + \bar{x}$$

$$d = \frac{\left[\dfrac{200(51^3)}{12}\right]}{25.5[200 \times 51]} + 25.5 = 8.5 + 25.5$$

$$d = 34 \text{ ft}$$

EXAMPLE 9.10

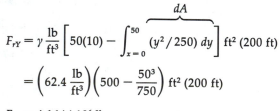

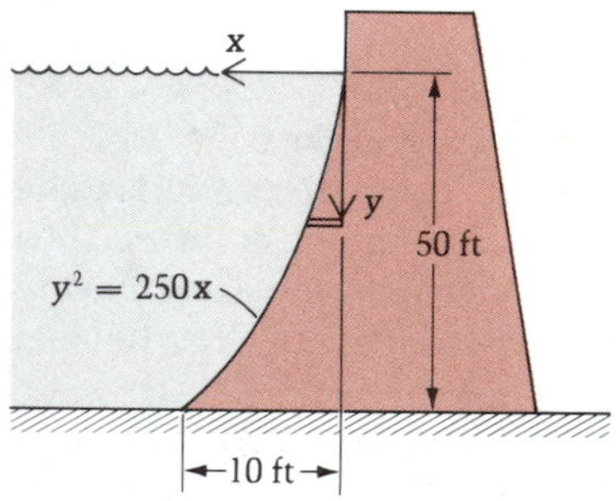

Figure E9.10a

Repeat Example 9.9 if the dam is parabolic instead of linear. (See Figure E9.10a, noting the different coordinates.)

Solution

The horizontal component of $\mathbf{F}_r$ is the same because the projected area is unchanged:

$$F_{rX} = -15.6 \times 10^6 \text{ lb}$$

For F_{rY}, we need the specific weight times the volume of the water above the wetted surface of the dam:

$$F_{rY} = \gamma \frac{\text{lb}}{\text{ft}^3}\left[50(10) - \int_{x=0}^{50} \overbrace{(y^2/250)\,dy}^{dA}\right] \text{ft}^2\,(200 \text{ ft})$$

$$= \left(62.4 \frac{\text{lb}}{\text{ft}^3}\right)\left(500 - \frac{50^3}{750}\right) \text{ft}^2\,(200 \text{ ft})$$

$$F_{rY} = 4.16 \times 10^6 \text{ lb}$$

which is a bigger force than we found for the triangular dam because there is more water above this one.

The above components of course add vectorially to form the resultant force $\mathbf{F}_r$:

$$\mathbf{F}_r = F_{rX}\hat{\mathbf{i}} + F_{rY}\hat{\mathbf{j}} = -15.6(10^6)\hat{\mathbf{i}} + 4.16(10^6)\hat{\mathbf{j}} \text{ lb}$$

The components act along the lines shown in Figure E9.10b.

We know that there is a force-alone resultant of all the infinitely many differential forces acting on the dam due to water pressure because these forces form a coplanar distributed system once the pressures are multiplied by the 200-ft width. The location of the line along which this force-alone resultant acts is shown in Figure E9.10c.

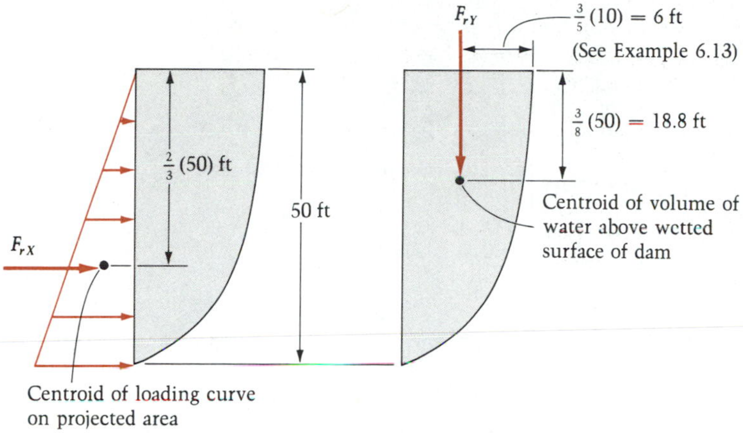

Figure E9.10b

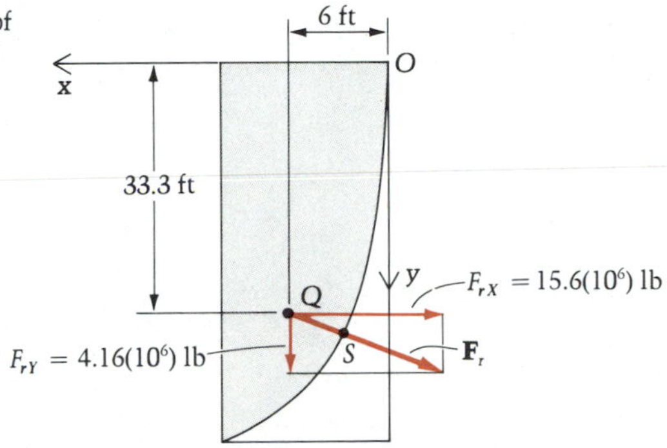

Figure E9.10c

Question 9.9 Why is there no couple accompanying $\mathbf{F}_r$ at Q?

An exercise problem will be to locate the spot S where $\mathbf{F}_r$ intersects the surface of the dam, and to show that $\mathbf{F}_r$ is not in this case normal to the surface, as would *always* be the case for a *plane* area as we have seen in the previous example.

To check the location of the resultant, we shall integrate to get $\mathbf{M}_{rO}$:

$$\mathbf{M}_{rO} = \int_{\substack{\text{Wetted}\\ \text{surface}}} \mathbf{r} \times d\mathbf{F}$$

$$\mathbf{M}_{rO} = \int \underbrace{[(x\hat{\mathbf{i}} + y\hat{\mathbf{j}})}_{\mathbf{r}} \times \underbrace{\underbrace{\gamma y}_{\text{Pressure}} \; \underbrace{w\,ds}_{dA} \; \underbrace{(-\cos\theta\hat{\mathbf{i}} + \sin\theta\hat{\mathbf{j}})]^*}_{\substack{\text{Unit vector in}\\ \text{direction of}\\ \text{differential force}}}}_{\text{Differential force } d\mathbf{F}}$$

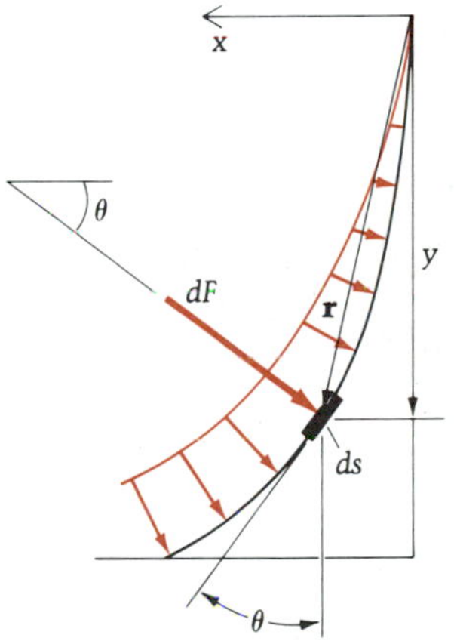

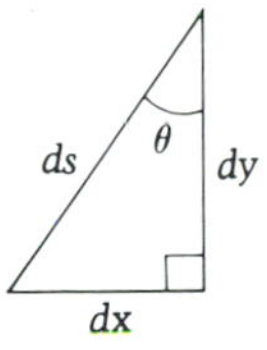

Figure E9.10d

$$\mathbf{M}_{rO} = \gamma w\hat{\mathbf{k}} \int (xy \sin\theta + y^2 \cos\theta)\, ds$$

But as can be seen in Figure E9.10d,

$$\sin\theta = \frac{dx}{ds} \qquad \text{and} \qquad \cos\theta = \frac{dy}{ds}$$

Answer 9.9 Q is the point of intersection of the components constituting $\mathbf{F}_r$.

* Note that $\theta = \tan^{-1}\left(\dfrac{dx}{dy}\right) = \tan^{-1}\left(\dfrac{y}{125}\right)$, which is not needed in this example.

so that

$$\mathbf{M}_{rO} = 62.4(200)\hat{\mathbf{k}}\left[\int_{x=0}^{10} x\underset{\sqrt{250x}}{y}\,dx + \int_{y=0}^{50} y^2\,dy\right]$$

$$\mathbf{M}_{rO} = 12500\hat{\mathbf{k}}\left\{\left[\sqrt{250}\,\frac{x^{5/2}}{5/2}\right]\Bigg|_0^{10} + \left[\frac{y^3}{3}\right]\Bigg|_0^{50}\right\}$$

$$= 12500\hat{\mathbf{k}}[2000 + 41700]$$

$$= 546(10^6)\hat{\mathbf{k}} \text{ lb-ft}$$

The resultant $\mathbf{F}_r$ has the same moment about O as do the above distributed forces:

$$\mathbf{M}_{rO} = \underset{15.6 \times 10^6 \text{ lb}}{F_{rx}}\left(\tfrac{2}{3} \times 50\right) + \underset{4.16 \times 10^6 \text{ lb}}{F_{ry}}\left(\tfrac{3}{5} \times 10\right)$$

$$= 545(10^6) \text{ lb-ft,}$$

off by about $\frac{1}{5}$ of 1% due to roundoff.

Sometimes we need to compute fluid pressure forces *beneath* a surface. The force due to such pressure is the negative of the force produced by an imaginary column of fluid *above* the same surface and extending up to the free surface level. Thus we may use our previous two results *(with all forces reversed)* for such problems, if we wish. We illustrate this idea with an example worked two ways: first with straightforward integration, and then using the idea of "imaginary fluid *above* the surface."

EXAMPLE 9.11

Find the resultant of the fluid pressure forces acting on the sluice gate AB in the left diagram of Figure E9.11a, expressed as a force at the pin A and a couple. The gate is a thick, quarter-cylindrical surface with radius R and length L perpendicular to the paper.

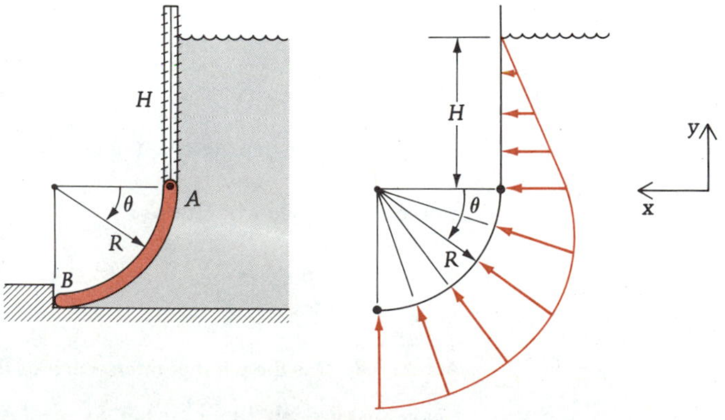

Figure E911.a

Solution

Since the pressure varies linearly with distance from the free surface, as suggested in the right diagram of Figure E9.11a, and since this distance is $H + R\sin\theta$, the resultant force $\mathbf{F}_r$ has the components:

$$F_{rx} = \int_{\theta=0}^{\pi/2} \underbrace{\underbrace{\underbrace{\gamma(H + R\sin\theta)}_{\text{Pressure}}\underbrace{LR\,d\theta}_{dA}}_{\text{Differential force } dF}\cos\theta}_{dF_x} \tag{1}$$

and

$$F_{ry} = \int_{\theta=0}^{\pi/2} \underbrace{\underbrace{\underbrace{\gamma(H + R\sin\theta)}_{\text{Pressure}}\underbrace{LR\,d\theta}_{dA}}_{\text{Differential force } dF}\sin\theta}_{dF_y} \tag{2}$$

Carrying out the integrations leads to

$$\begin{aligned}\mathbf{F}_r &= F_{rx}\hat{\mathbf{i}} + F_{ry}\hat{\mathbf{j}} \\ &= \gamma RL\left(H + \frac{R}{2}\right)\hat{\mathbf{i}} + \gamma RL\left(H + \frac{\pi}{4}R\right)\hat{\mathbf{j}}\end{aligned} \tag{3}$$

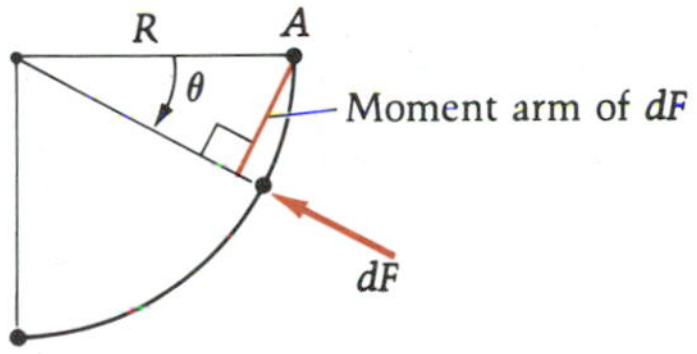

Figure E9.11b

For the moment about the pin A, which together with $\mathbf{F}_r$ completes the resultant at A of the fluid forces on the sluice gate, we integrate one more time (see Figure E9.11b):

$$\begin{aligned}M_{rA} &= \int_{\theta=0}^{\pi/2} \underbrace{\gamma(H + R\sin\theta)LR\,d\theta}_{dF \text{ (as above)}}\underbrace{R\sin\theta}_{\text{Moment arm}} \\ &= \gamma R^2 L\left(H + \frac{\pi}{4}R\right)\end{aligned} \tag{4}$$

Thus the resultant at A is the force $\mathbf{F}_r$ together with a clockwise couple of strength M_{rA}.

Consider now the forces resulting from water *imagined* to be *above* the sluice gate (see Figure E9.11c on following page).

For the horizontal component of the resultant force, we add the forces on the vertical projection CB of the gate:

$$\underbrace{\gamma LHR}_{\substack{\text{Volume of rectangular}\\ \text{part of load surface}\\ \text{(shaded, Figure E9.11c)}\\ \text{(acts at } z = H + R/2,\\ \text{the centroid of}\\ \text{this volume)}}} + \underbrace{\tfrac{1}{2}(\gamma LR)R}_{\substack{\text{Volume of triangular}\\ \text{part of load surface}\\ \text{(dotted, Figure E9.11c) (acts}\\ \text{at } z = H + (2/3)R,\\ \text{the centroid of}\\ \text{this volume)}}}$$

It is seen that this force, to the right, agrees with the actual F_{rx} of Equation (3).

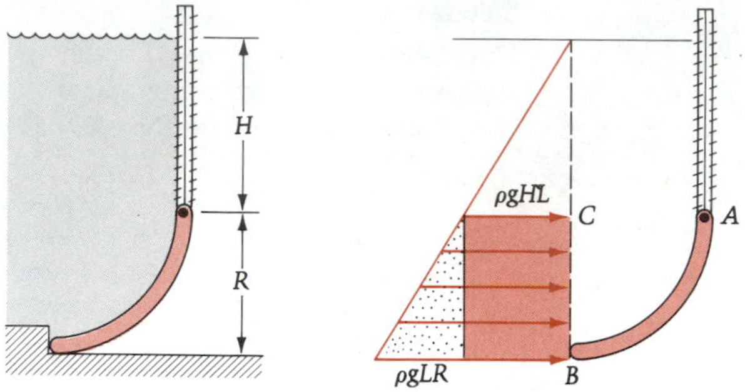

Figure E9.11c

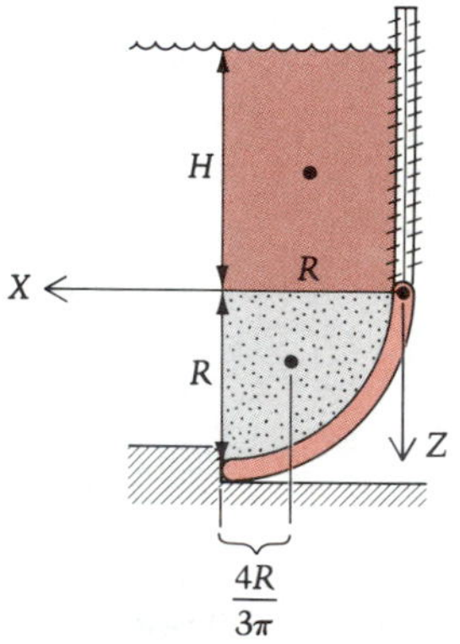

Figure E9.11d

For the y component, we see in Figure E9.11d that the weight of the water shown shaded is $\gamma(HRL)$, acting at the center of the block, and the weight of the water shown dotted is $\gamma\left(\frac{\pi R^2}{4}L\right)$, acting at $X = Z = R - \frac{4R}{3\pi}$ and $Y = \frac{L}{2}$.

It is also seen that the (downward) total y force, the sum of the weights of the two volumes of water in Figure E9.11d, is the same as F_{ry} in Equation (3).

Finally, the above four contributors to $-\mathbf{F}_r$ may also be used, each in reverse, to check M_{rA}:

$$\circlearrowleft\!\!+\quad M_{rA} = \overbrace{(\gamma HRL)\frac{R}{2}}^{\text{Rectangle}} + \overbrace{\left(\gamma\frac{\pi R^2}{4}L\right)\left(R - \frac{4R}{3\pi}\right)}^{\text{Quarter-circle}} + \overbrace{(\gamma LHR)\left(\frac{R}{2}\right)}^{\text{Rectangle}} + \overbrace{\left(\frac{\gamma LR^2}{2}\right)\left(\frac{2}{3}R\right)}^{\text{Triangle}}$$

Moments of y forces (first two terms); Moments of x forces (last two terms)

$$= \gamma LHR^2 + \frac{\pi}{4}\gamma R^3 L$$

This is the same result that we obtained (Equation 4) much more directly by integration.

Buoyant Force

Archimedes' principle (220 B.C.!) is that the buoyant force on a submerged body equals the weight of the fluid displaced by the submerged portion of the body. The buoyant force acts through the center of mass of the displaced fluid, a point known as the center of buoyancy.

For the proof of the principle, let the *submerged portion* of the body be replaced by an identical volume V_F, in amount and shape, of the fluid itself as suggested by Figure 9.21. Such a volume is obviously in equilibrium, so that the buoyant force B on the fluid is the weight of the fluid volume, acting upward through the mass center C_F of this fluid volume. In Figure 9.22, ρ_F is the mass density of the fluid and B is the single-force resultant of all the differential forces caused by fluid pressure on the fluid volume.

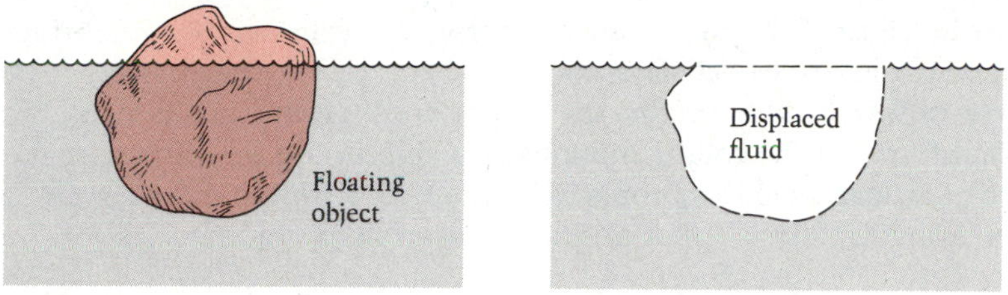

Figure 9.21

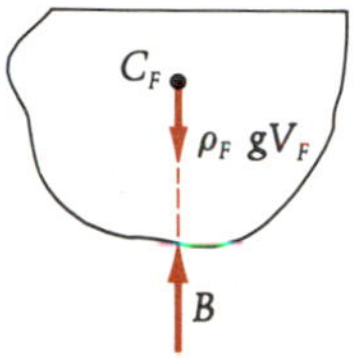

Figure 9.22

Question 9.10 How do we know that there is a single-force resultant in this case?

Now, returning to the actual body, the fluid pressure on its submerged portion is everywhere the same as we had on the substituted volume of fluid because this pressure, a function of depth and fluid density, is unaffected by what really comprises the volume V_F. Hence the resultant B of the forces caused by this pressure is the same. The free-body diagram of the actual floating body $\mathcal{B}$ is seen in Figure 9.23. The body is thus buoyed upward by a force equal to the weight of displaced fluid. In Figure 9.23(a), ρ_T and ρ_S are the average mass densities of the parts of $\mathcal{B}$ above and below the free surface of the fluid. Note that the mass center of the part of $\mathcal{B}$ occupying V_F need not be at C_F, or even on the same vertical line as C_F. But the mass center C of *all* of $\mathcal{B}$ is on this line as

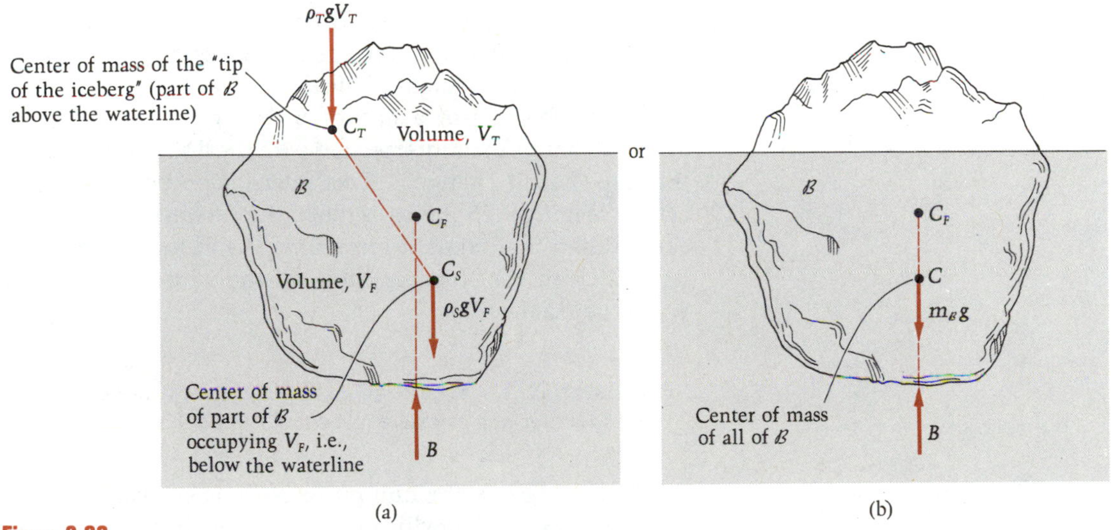

Figure 9.23

Answer 9.10 The resultant of the forces caused by water pressure has to be the negative of the weight of the displaced fluid ($\rho_F gV_F$). Thus if the resultant of these water pressure forces at C_F contains a moment, then ΣM_{C_F} cannot vanish there.

shown in Figure 9.23(b), because otherwise the moment equilibrium equation could not be satisfied.

We now consider briefly the importance of the position of the buoyant force on the rolling motions of a ship in the sea. Consider the sketch of a floating ship's profile in Figure 9.24 with the mass center C_F representing the mass center of the displaced fluid volume V_F.

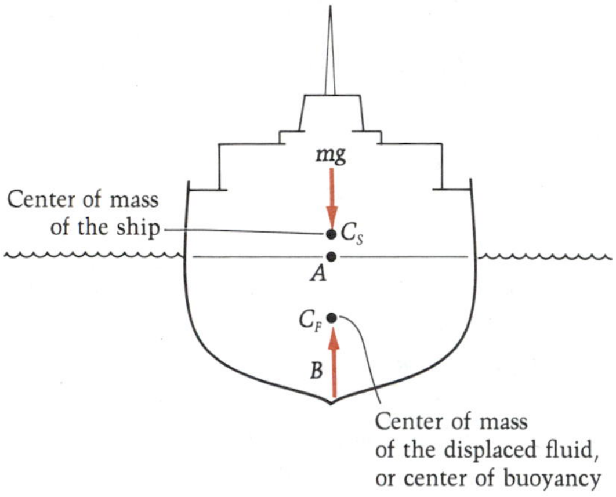

Figure 9.24

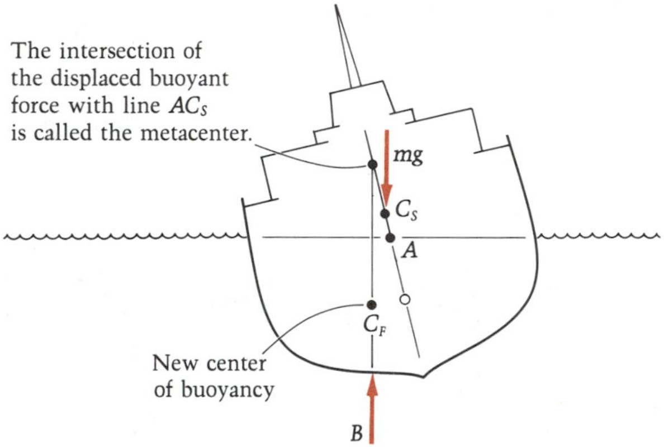

Figure 9.25

Now let the ship roll about the z axis normal to the paper through, say, point A in a motion caused by wind and water forces (see Figure 9.25). The buoyant force will, in general, change. There is now more fluid displaced on the left and less on the right. Depending upon the profile of the ship's hull, the weight of this displaced fluid may or may not even equal the weight of the ship. In any case, there will be a definite change in the line of action of B drawn on the ship's profile, and a possible change in its intersection with the line AC_s (which line, relative to the ship, hasn't changed). This intersection point is called the "metacenter." Figure 9.25 shows us that if the "metacenter" lies above the mass center, then the moment formed by B and mg is "restoring" — that is, tends to turn the ship back toward the vertical. In this case, the ship would be stable.

Question 9.11 What would happen if the metacenter were *below* the mass center at a certain angle of roll?

In ship design, the hull profile and mass distributions are carefully computed so as to keep the metacenter above the mass center and also to keep it close to the same spot for fairly large roll angles of the ship, up to 20° and more in some cases.

Answer 9.11 Disaster! The moment of B and mg would then be in a direction that would overturn the ship further.

EXAMPLE 9.12

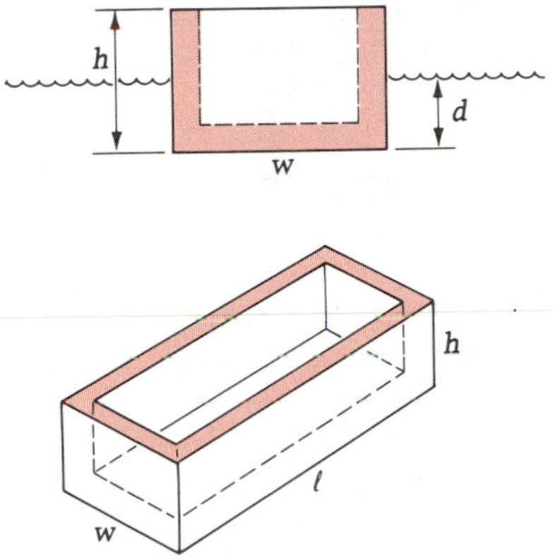

Figure E9.12

Compute the largest (uniform) thickness t so that a tub hollowed out of a steel block will float (see Figure E9.12). Let $\ell = 7$ ft, $w = 3$ ft, $h = 2$ ft, and $\gamma_W = 62.4 \text{ lb/ft}^3$, $\gamma_s = 490 \text{ lb/ft}^3$.

Solution

Using Archimedes' principle, we shall equate the weight of the tub (in terms of t) with the buoyant force (weight of water displaced) when the water line is up to the top of the tub:

$$\text{Weight of tub} = \gamma_s\,(\text{Volume of steel in tub}).$$

$$W_T = 490[w\ell h - (h - t)(w - 2t)(\ell - 2t)]$$

Substituting and simplifying,

$$W_T = 490(4t^3 - 28t^2 + 61t)$$

The buoyant force is

$$B = \gamma_W(\text{Volume of displaced water})$$
$$= 62.4(w\ell h) = 62.4(42)$$
$$= 2620 \text{ lb}$$

For equilibrium,

$$+\downarrow \qquad \Sigma F_y = 0 = W_T - B$$

which gives the following equation in the thickness t:

$$f(t) = t^3 - 7t^2 + 15.3t - 1.34 = 0 \tag{1}$$

This equation has but one real root, which is found below by trial and error to be $t = 0.0913$ ft or about 1.1 inch:

t	$f(t)$
0	−1.34
1	7.96
0.1	0.121
0.09	−0.0190
0.091	−0.0049
0.0912	−0.0021
0.0913	−0.0007
0.0914	+0.0007

PROBLEMS ▶ Section 9.3

9.33 Show that the pressure in a fluid at rest (a) is constant through the fluid in each horizontal plane; (b) obeys the equation $dp = \gamma\, dz$, where z is measured downward. Thus if z is zero at the free surface of a constant density fluid, then $p = \gamma z$.

9.34 Explain why the fluid force is normal to every differential area on which it acts.

9.35 Prove Pascal's law, that the pressure in a fluid at rest is constant in all directions. (See Figure P9.35.) *Hint:* Sum forces on the infinitesimal element of fluid in the x

and z directions, obtaining $p_s = p_x$ and $p_s = p_z$. Then note that θ and the orientation of the slanted surface around the vertical are arbitrary.

9.36–9.40 Find the force-alone resultant of the water pressure forces on the shaded side of the submerged surfaces in Figures P9.36–P9.40.

9.41 For this third orientation of the gate of Examples 9.7 and 9.8, find the center of pressure. See Figure P9.41.

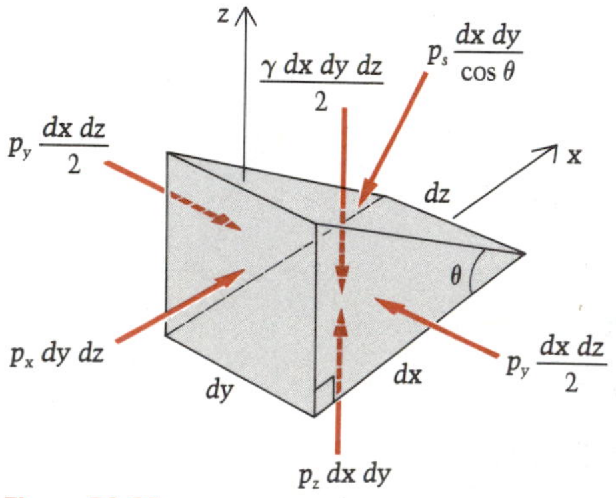

Figure P9.35

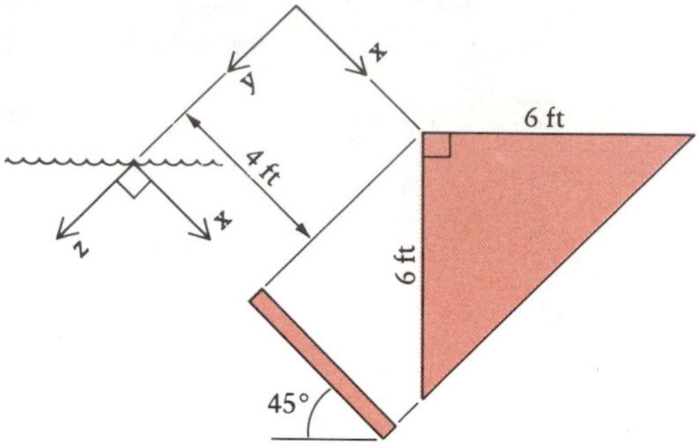

Figure P9.41

9.42 Show that the x coordinate of the center of pressure of fluid force acting on the equilateral triangular area $\mathcal{A}$ in Figure P9.42 is unchanged by any rotation of $\mathcal{A}$ about an axis through the centroid normal to the area. What is this force if the fluid is water with specific weight 62.4 lb/ft^3 and if $x_A = 30$ ft and $s = 3$ ft?

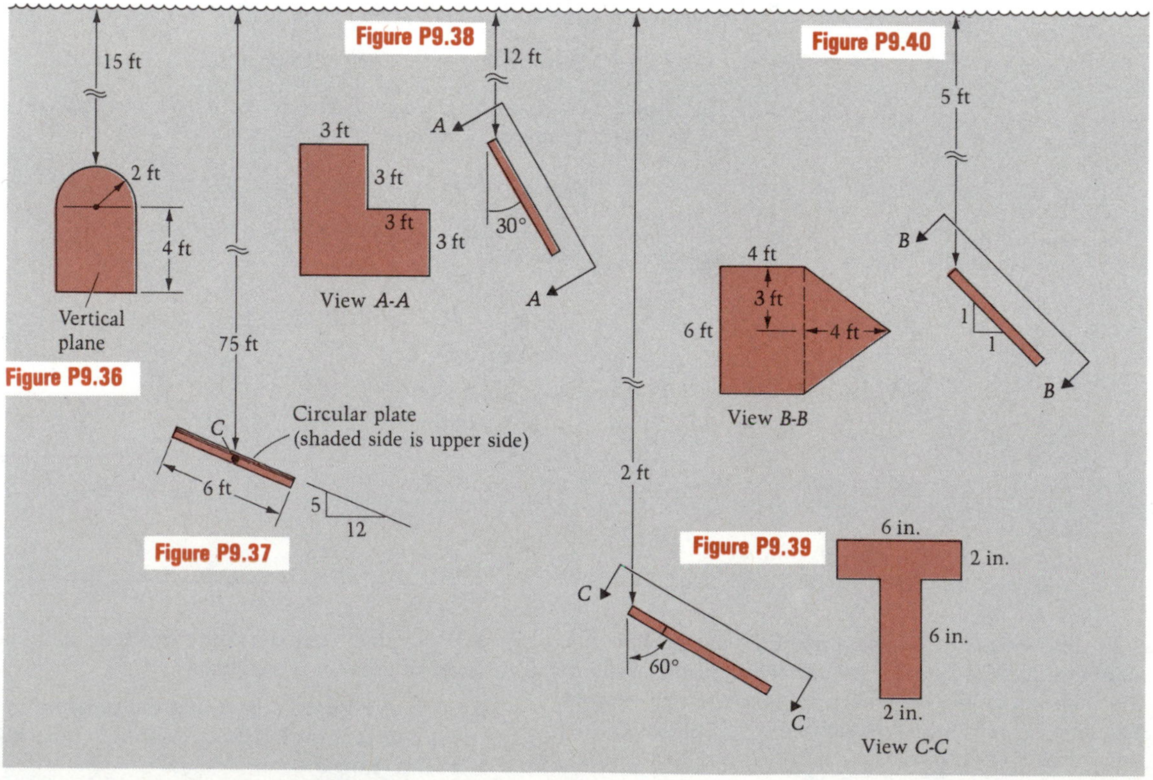

Figure P9.36

Figure P9.37

Figure P9.38

Figure P9.39

Figure P9.40

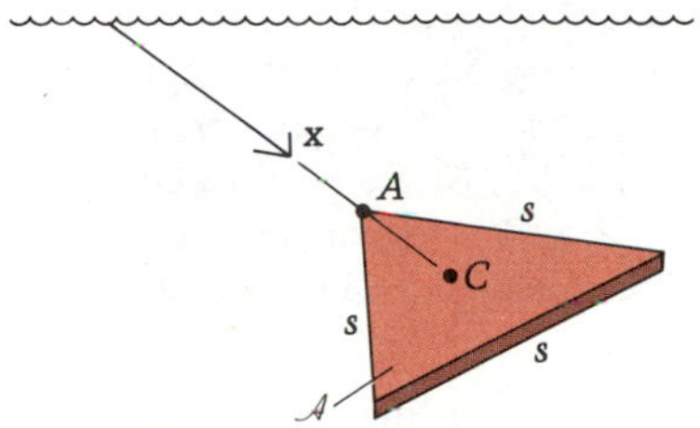

Figure P9.42

9.43 The gate in Figure P9.43 holds back a reservoir of water. The pressure at any depth z is γz, where γ = specific weight of the water. The gate is 5 ft wide perpendicular to the paper. Thus the pressure load may be considered as a distributed line load of intensity $5\gamma z$ lb / ft acting on the "beam" shown. Draw the shear and moment diagrams as suggested by the axes to the left of the figure.

9.44 Find the resultant force caused by fluid pressure on the shaded door in Figure P9.44 if the door is rectangular with a width of w and a height of h.

9.45 Work Problem 9.44 if the door is circular with diameter h.

9.46 Work Problem 9.44 if the door's shape is an equilateral triangle as shown in Figure P9.46.

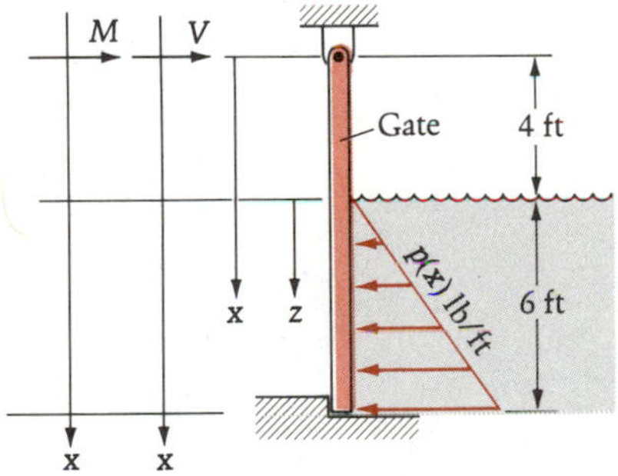

Figure P9.43

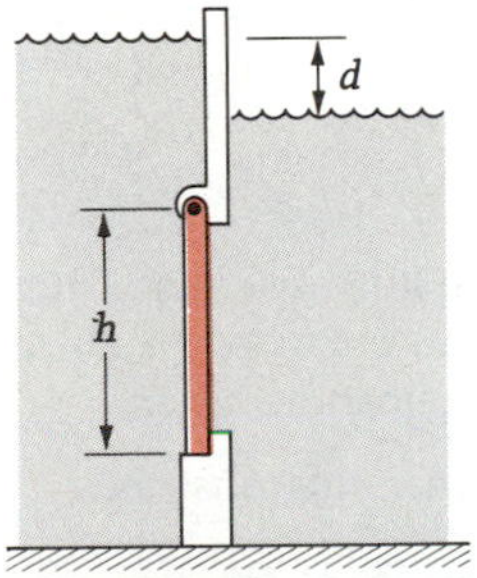

Figure P9.44

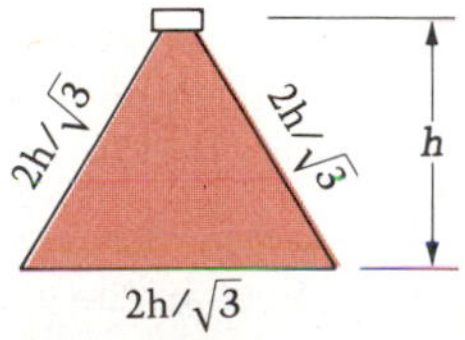

Figure P9.46

9.47 Find the force exerted on the rectangular gate at A by the stop. (See Figure P9.47.)

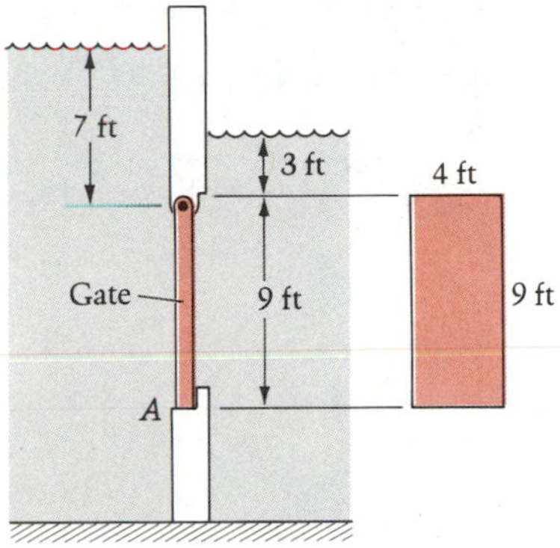

Figure P9.47

9.48 The circular gate in Figure P9.48 has a diameter of 6 ft. It swings about a horizontal pin located 4 in. below its center. At what depth h of water will the gate be in equilibrium?

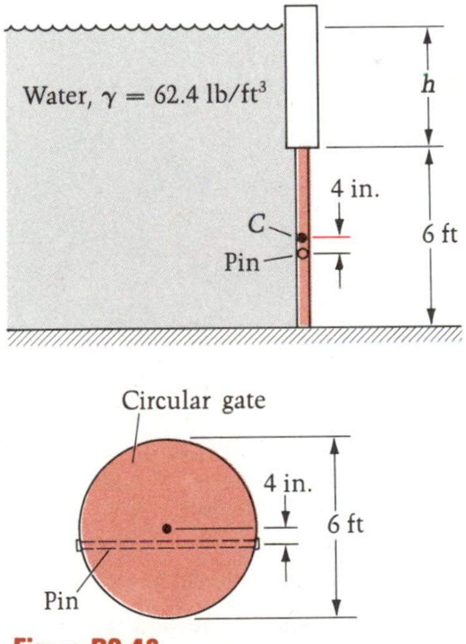

Figure P9.48

9.49 Find the force due to fluid pressure acting on the gate in Example 9.7.

9.50 Find the force due to fluid pressure acting on the gate in Example 9.8.

9.51 In Example 9.10 verify the resultant force by integration. See Figure P9.51. *Hint:* Use:

$$d\mathbf{F} = (\text{pressure})(dA)(-\cos\theta\hat{\mathbf{i}} + \sin\theta\hat{\mathbf{j}})$$

with pressure $= \gamma y$, and $dA = (\text{width})\, ds = 200\, ds$. Then use:

$$ds = \sqrt{dx^2 + dy^2} = dy\sqrt{1 + \left(\frac{dx}{dy}\right)^2}$$

or

$$ds = \sqrt{1 + \left(\frac{y}{125}\right)^2}\, dy$$

Further,

$$\cos\theta = \frac{dy}{ds} \quad \text{and} \quad \sin\theta = \frac{dx}{ds}$$

and substitution gives

$$d\mathbf{F} = 200\gamma y[-dy\hat{\mathbf{i}} + \underbrace{(y\, dy/125)}_{dx}\hat{\mathbf{j}}]$$

which can be integrated to give $\mathbf{F}_r$.

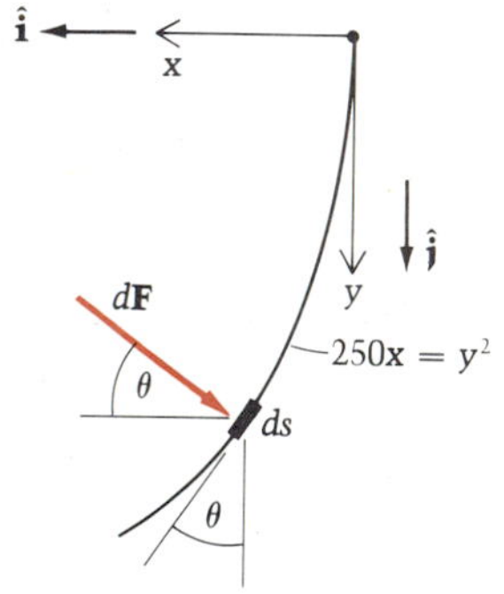

Figure P9.51

*** 9.52** In Example 9.10, find the coordinates of the point S where $\mathbf{F}_r$ intersects the parabolic surface of the dam. Show that $\mathbf{F}_r$ is not normal to the surface at S.

9.53 The dam in Figure P9.53, similar to that of Example 9.9, is holding back water, this time against a vertical (instead of slanted) plane. The dam is made of concrete with average specific weight 155 lb/ft³. There are forces beneath the dam due to two sources: (a) the reaction of the earth ("intergranular," i.e., earth particles onto concrete), and (b) the "uplift" force due to water pressure from fluid that has seeped underneath the dam. The first of these will have a horizontal and a vertical component and act somewhat between B and D; the second is the vertical force-alone resultant of the upward fluid pressure forces (assumed to act over approximately the entire bottom surface of the dam). Assume the uplift pressure to vary linearly from the full 50γ at B to zero at D, and determine the magnitude and direction of the earth's (force-alone resultant) reaction, and the point where it intersects line BD.

9.54 In Problem 9.53 find the factor of safety against overturning of the dam about D, defined as the ratio of the counterclockwise moment about D due to the weight of the dam, to the clockwise moment caused by the forces from the water pressure on AB and BD. The numerator resists overturning and the denominator promotes it.

9.55 Let $H = 2R$ and find the magnitude, direction, and line of action of the single-force resultant of the fluid forces on the gate of Example 9.11.

9.56 Find the force at B in Figure P9.56 required to hold the gate AB of Example 9.11 in equilibrium.

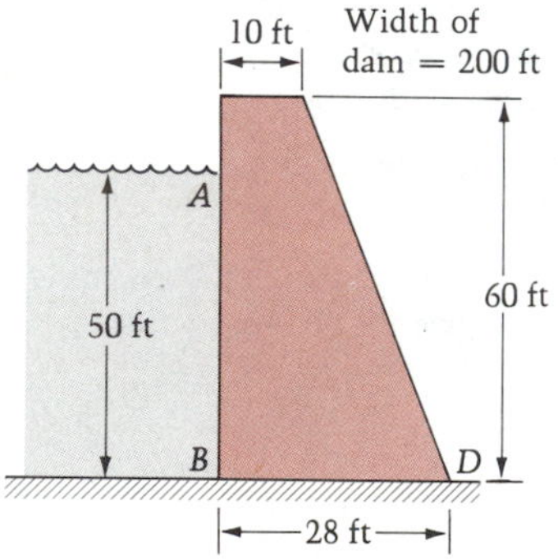

Figure P9.53

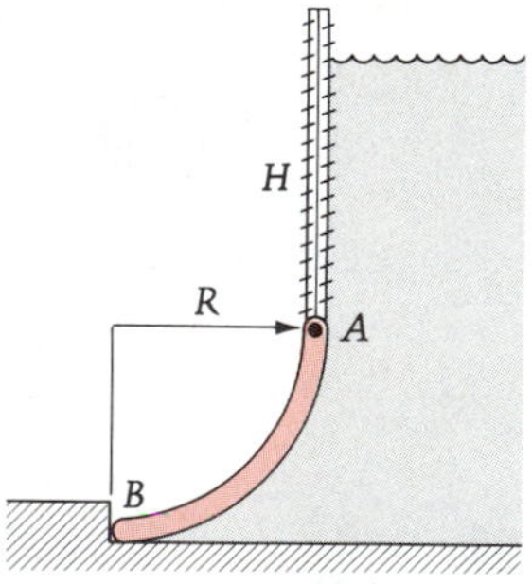

Figure P9.56

9.57 In Example 9.11 compute the line of action of the horizontal component F_{rx} of the fluid force on the sluice gate in two ways, and check one against the other:

a. Using the rectangular and triangular distributions in the example

b. Using Equation (9.26).

9.58 The steel gate has the cross section shown in Figure P9.58. It is 10 ft long, and has a specific weight of 490 lb/ft³. Determine the depth H of water that will cause the gate to open (i.e., pivot clockwise, letting water flow under).

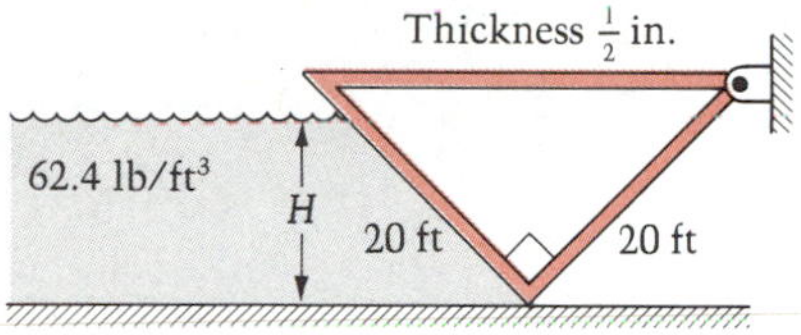

Figure P9.58

9.59 In Figure P9.59 find the water level H at which the hinged gate will open and let water spill out along the line PQ. Neglect the weight of the gate.

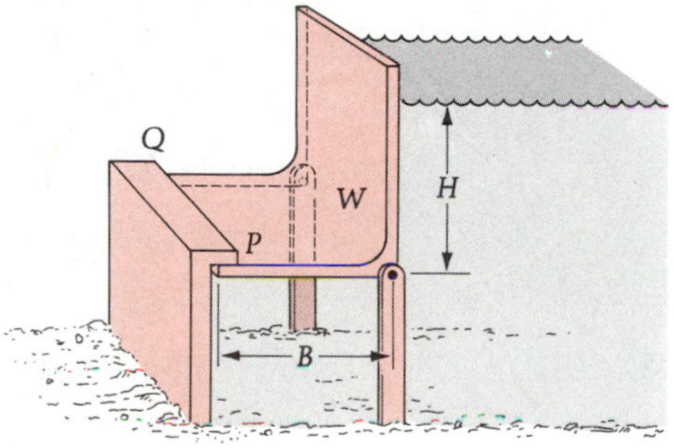

Figure P9.59

9.60 A bowl with the dimensions shown in Figure P9.60 is filled with a liquid of density ρ. By integration, find the resultant force exerted by the liquid on the bowl.

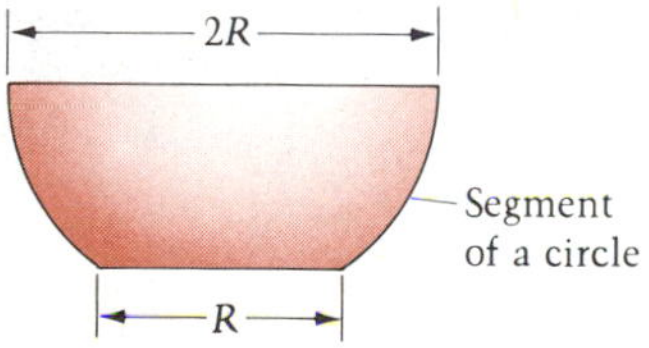

Figure P9.60

9.61 In Figure P9.61:

a. Find the single-force resultant acting on the gate $\mathcal{G}$ caused by hydrostatic pressure, for the case $\theta = 0$. The width of the gate is 5 m and the mass density of the water = 1 g/cm³.

b. Find the reactions at the pin A and floor B.

9.62 Repeat Problem 9.61 when $\theta = 20°$.

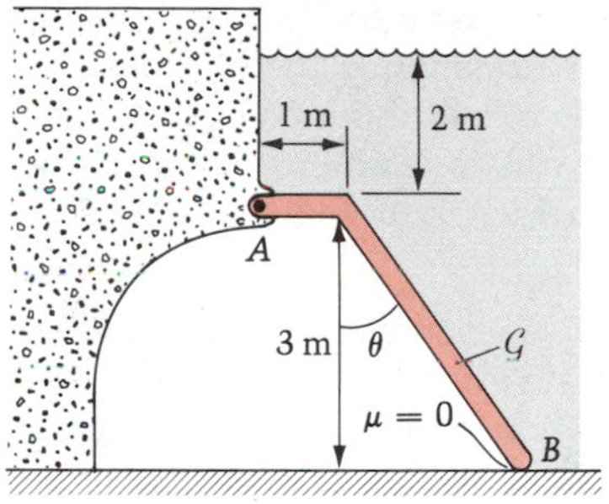

Figure P9.61

9.63 In Figure P9.63 find the magnitude, direction, and line of action of the force-alone resultant caused by hydrostatic pressure acting on a 1-m width of the approximately parabolic surface.

9.64 Consider the law of buoyancy as it relates to ships floating in the sea. For the simplified "ship" in Figure P9.64, assume that the center of mass is at C on the waterline in calm sea. Assume that the ship then rolls as indicated. Locate the centroid of the displaced fluid (which used to be at F) and show that the same buoyant force has now shifted to the right sufficiently to cause the ship to begin to roll back counterclockwise.

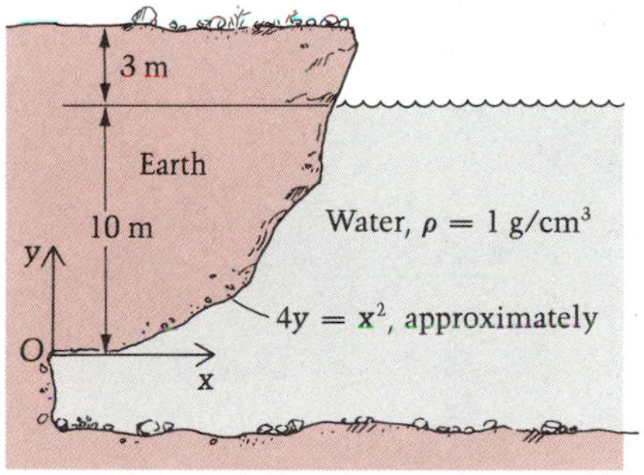

Figure P9.63

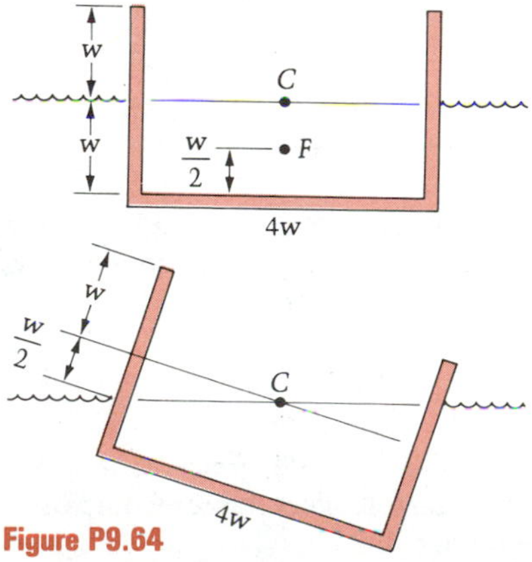

Figure P9.64

9.65 In Problem 9.64 show that if the simplified ship is only 2w wide, then it will continue to roll clockwise toward capsizing if it has rolled to the point where the same $3w/2$ is above the waterline on the left side. (This angle of roll is 26.6°, which is *very* large.) See Figure P9.65.

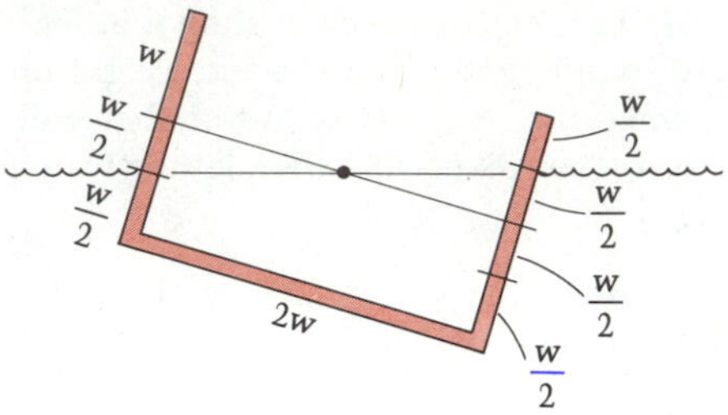

Figure P9.65

9.66 The hollowed sphere in Figure P9.66 is sealed at its base and is being used as a tank. How full is it $(H - z)$ at the instant it lifts off the ground? (Assume the seal exerts no downward force on the tank.)

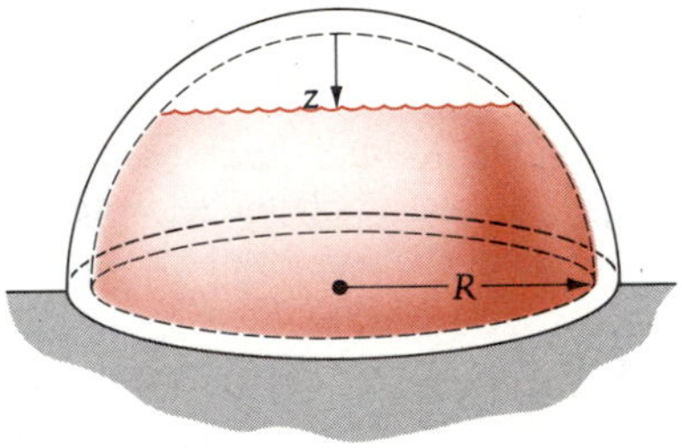

Figure P9.66

9.67 Huck Finn has gone to sleep with one end of his fishing pole resting between his toes. If the density of the pole is half that of water, find the fraction of the pole that is under water. See Figure P9.67.

Figure P9.67

Computer Problems ▶ Chapter 9

9.68 Write a Newton-Raphson iterative algorithm (see Appendix D) for solving equations of the form $f(x) = 0$. Use the algorithm to solve the equation

$$f(\theta) = 1 - \sin\theta - \frac{2.5P}{k\ell}\tan\theta = 0$$

from Example 9.3, for a wide range of values of $P/k\ell$ from zero to very large. Compare with the text when $P/(k\ell) = 1$.

9.69 Use the algorithm in the preceding problem to solve the equation

$$\begin{aligned} f(t) &= t^3 - 7t^2 + 15.3t - 1.34 \\ &= 0 \end{aligned}$$

for the critical tub thickness in Example 9.12; compare with the result found in the text.

* **9.70** The hollowed cone in Figure P9.70 is sealed at its base and is being used as a tank. How full is it $(H - z)$ at the instant it lifts off the ground? (Assume the seal cannot exert a downward force on the tank.)

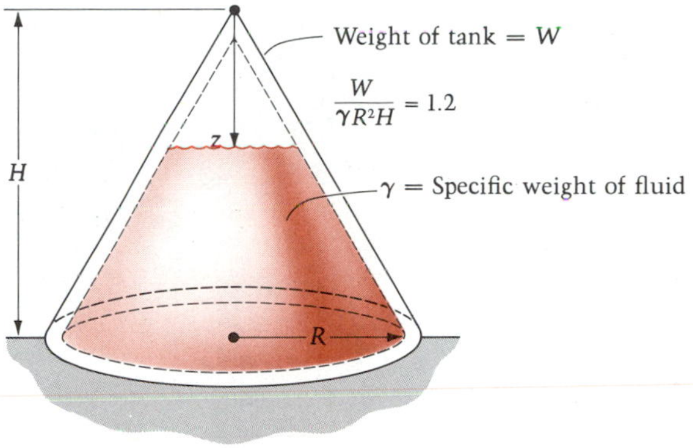

Figure P9.70

REVIEW QUESTIONS ▶ Chapter 9

True or False?

1. A virtual displacement is an imagined one that constraints on the body may prevent from occurring in reality.
2. An advantage of virtual work is that forces not of interest may be suppressed in the analysis by appropriate choices of virtual displacements.
3. In using virtual work, we need not worry about the body being in equilibrium in an *inertial* frame; any old frame will do.
4. Virtual displacements really need not be restricted to infinitesimals provided rotations are infinitesimal or zero.
5. The method of virtual work cannot be used to find the force of interaction where two parts of a body are pinned together.
6. The Principle of Virtual Work is restricted to situations where there is no friction.
7. The resultant force due to hydrostatic pressure on a submerged plane surface always equals the product of the projected area on a vertical plane and the pressure at the centroid of this projection.
8. For hydrostatic pressure on a submerged plane area, the center of pressure is closer to the free surface of the liquid than is the centroid of the area.
9. Consider the vertical plane that (a) contains the centroid of a plane area submerged in a liquid; and (b) is perpendicular to the plane area. The center of pressure has to lie in this vertical plane.
10. The magnitude of the resultant force on a curved surface equals the pressure at the centroid of the curved area multiplied by this area.
11. The buoyant force is always vertical.

Answers: 1. T 2. T 3. F 4. T 5. F 6. F 7. F 8. F 9. F 10. F 11. T

Appendix Contents

Appendix A VECTORS

- **A.1** Vectors: Addition, Subtraction, and Multiplication by a Scalar
- **A.2** Unit Vectors and Orthogonal Components
- **A.3** Scalar (Dot) Product
- **A.4** Vector (Cross) Product
- **A.5** Scalar and Vector Triple Products

Appendix B TABLES RELATING TO UNITS

Appendix C MOMENTS AND PRODUCTS OF INERTIA OF AREAS

Appendix D EXAMPLES OF NUMERICAL ANALYSIS / THE NEWTON-RAPHSON METHOD

Appendix E EQUILIBRIUM: A BODY AND ITS PARTS

Appendix F ANSWERS TO ODD-NUMBERED PROBLEMS

A VECTORS

A.1 Vectors: Addition, Subtraction, and Multiplication by a Scalar*

Vectors are mathematical entities possessing the qualities of magnitude and direction and obeying certain algebraic rules. Although the concepts involved may be extended to spaces of higher dimensions, our concern shall be with vectors in ordinary three-dimensional space. In this circumstance it is possible to represent a vector by an "arrow" — that is, a line segment whose length is proportional to the magnitude of the vector, with directionality along the line segment indicated by the arrowhead.

Many of the features of the algebra of vectors may be either deduced or readily seen from the *parallelogram law* by which vector addition is defined (Figure A.1). Using boldfaced type to denote vectors, we write

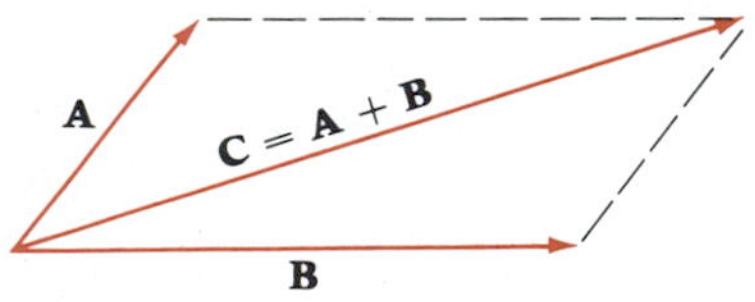

Figure A.1 Parallelogram Law of Addition

$$\mathbf{C} = \mathbf{A} + \mathbf{B}$$

Clearly from the parallelogram law there is no significance attributable to order, so that also

$$\mathbf{C} = \mathbf{B} + \mathbf{A}$$

which is to say that vector addition is commutative. We also note that the parallelogram law suggests a slightly different way of depicting vector addition; that is, the same result is obtained by a head-to-tail arrangement as shown in Figure A.2.

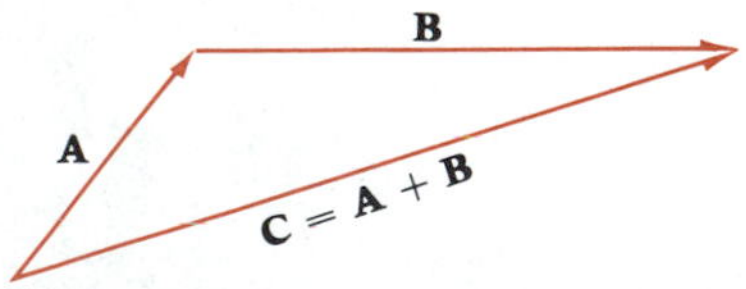

Figure A.2

We are now in a position to illustrate (at least for two dimensions) the fact that vector addition is associative. In Figure A.3 we see that

$$\mathbf{A} + (\mathbf{B} + \mathbf{C}) = (\mathbf{A} + \mathbf{B}) + \mathbf{C}$$

Thus, in vector addition, order is irrelevant.

The negative of a vector **A** is the vector having the same magnitude as **A** but the opposite direction. Writing $(-\mathbf{A})$ for the negative of **A**, the parallelogram law dictates that

$$\mathbf{A} + (-\mathbf{A}) = \mathbf{0}$$

* Examples pertinent to Sections A.1–A.3 may be found in Section 2.2 of the text.

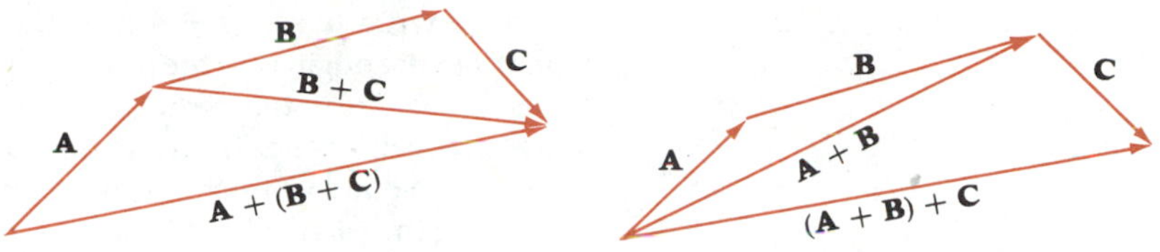

Figure A.3

Furthermore, letting **D** be the sum of a vector **B** and (−**A**) we have

$$\mathbf{D} = \mathbf{B} + (-\mathbf{A})$$

or more simply

$$\mathbf{D} = \mathbf{B} - \mathbf{A}$$

Thus we have the process of *subtraction.*

The simplest of the processes of multiplication involving a vector is multiplication by a scalar. A scalar is an entity whose measure is a real number. If α is a scalar, then by $\alpha\mathbf{A}$ we mean the vector whose magnitude is the product of the absolute value of α and the magnitude of **A**, and $\alpha\mathbf{A}$ has the same direction as **A** if α is positive and the opposite direction if α is negative. The parallelogram law shows that multiplication by a scalar is distributive — that is

$$\alpha(\mathbf{A} + \mathbf{B}) = \alpha\mathbf{A} + \alpha\mathbf{B}$$

A.2 Unit Vectors and Orthogonal Components

The vector $\hat{\mathbf{e}} = \mathbf{A}/|\mathbf{A}|$, where $|\mathbf{A}|$ is the magnitude of **A**, is dimensionless, has a magnitude of unity, and has the same direction as **A**. Thus $\hat{\mathbf{e}}$ is called the *unit vector* in the direction of **A**.

Unit vectors of preassigned directions provide the mechanism by which vectors are usually expressed. Suppose we let (as in Figure A.4) x, y, and z be mutually perpendicular axes, or reference directions, and we let $\hat{\mathbf{i}}$, $\hat{\mathbf{j}}$, and $\hat{\mathbf{k}}$ be unit vectors in those directions. The parallelogram law allows us to decompose a vector **A** into three mutually perpendicular parts written $A_x\hat{\mathbf{i}}$, $A_y\hat{\mathbf{j}}$, and $A_z\hat{\mathbf{k}}$ so that

$$\mathbf{A} = A_x\hat{\mathbf{i}} + A_y\hat{\mathbf{j}} + A_z\hat{\mathbf{k}}$$

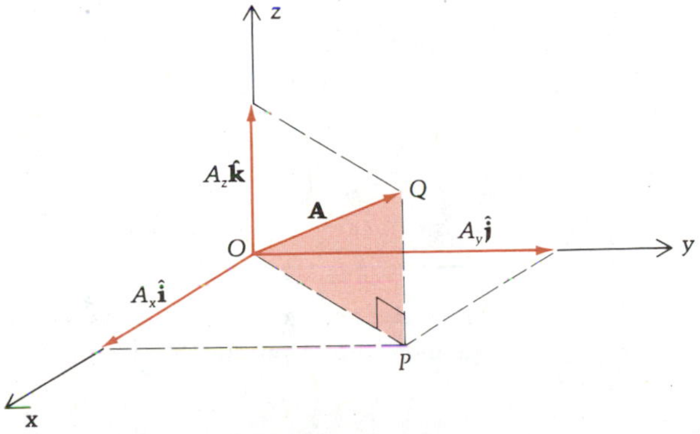

Figure A.4 Components of a Vector

If we write $\mathbf{A} = (A_x\hat{\mathbf{i}} + A_y\hat{\mathbf{j}}) + A_z\hat{\mathbf{k}}$, then the vector in the parentheses extends from the origin O to the projection point P of $\mathbf{A}$ in the xy plane. This vector, by the Pythagorean theorem, has magnitude $\sqrt{A_x^2 + A_y^2}$. If to this vector we then add $A_z\hat{\mathbf{k}}$ (which extends from P to the tip Q of the arrowhead of $\mathbf{A}$), then we have succeeded in building up the vector $\mathbf{A}$ from its three components. The magnitude of $\mathbf{A}$, using the (shaded) right triangle OPQ, is then given in terms of the components of $\mathbf{A}$ as

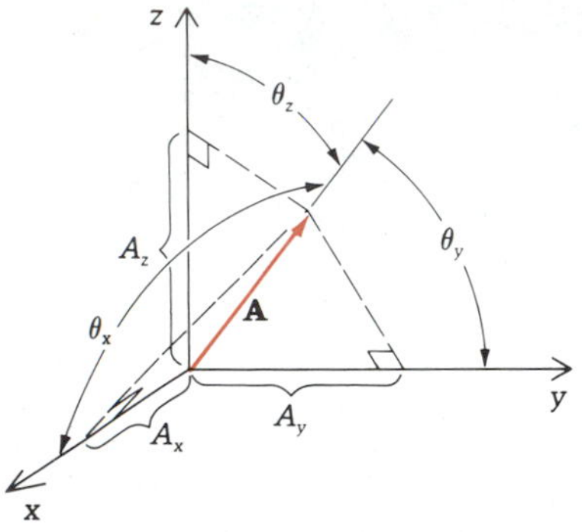

Figure A.5 Angles Whose Cosines Are the Direction Cosines of **A**

$$
\begin{aligned}
|\mathbf{A}| &= \sqrt{\sqrt{A_x^2 + A_y^2}^{\,2} + A_z^2} \\
&= \sqrt{A_x^2 + A_y^2 + A_z^2} \qquad \text{(A.1)}
\end{aligned}
$$

We can see that the vectors $A_x\hat{\mathbf{i}}$, $A_y\hat{\mathbf{j}}$, and $A_z\hat{\mathbf{k}}$ are the (orthogonal) projections of $\mathbf{A}$ onto the x, y, and z directions. They are often called orthogonal components of $\mathbf{A}$; the same terminology is also used for the scalars A_x, A_y, and A_z. When context does not make clear which components are intended, we distinguish them by calling the former "vector components" and the latter "scalar components." If θ_x, θ_y, and θ_z are the angles between vector $\mathbf{A}$ and x, y, and z, respectively, then from Figure A.5 we see that

$$
\begin{aligned}
A_x &= |\mathbf{A}| \cos \theta_x \\
A_y &= |\mathbf{A}| \cos \theta_y \\
A_z &= |\mathbf{A}| \cos \theta_z \qquad \text{(A.2)}
\end{aligned}
$$

These cosines are called the direction cosines of $\mathbf{A}$. We note that since a unit vector has a magnitude of unity, Equations (A.2) show that the direction cosines of a unit vector *are* its components.

Squaring both sides in each of Equations (A.2), adding, and using Equation (A.1), we confirm an important property of the direction cosines of a line — that is,

$$\cos^2 \theta_x + \cos^2 \theta_y + \cos^2 \theta_z = 1 \qquad \text{(A.3)}$$

Finally, it is of great practical importance to realize that, from the properties of addition and multiplication by a scalar enumerated in Section A.1,

$$\alpha\mathbf{A} = \alpha A_x\hat{\mathbf{i}} + \alpha A_y\hat{\mathbf{j}} + \alpha A_z\hat{\mathbf{k}} \qquad \text{(A.4)}$$

α being any scalar. Moreover, if

$$\mathbf{C} = \mathbf{A} + \mathbf{B}$$

then

$$
\begin{aligned}
C_x &= A_x + B_x \\
C_y &= A_y + B_y \\
C_z &= A_z + B_z \qquad \text{(A.5)}
\end{aligned}
$$

A.3 Scalar (Dot) Product

The scalar product of two vectors $\mathbf{A}$ and $\mathbf{B}$ is denoted $\mathbf{A} \cdot \mathbf{B}$ (from whence comes the name "dot" product) and is defined by

$$\mathbf{A} \cdot \mathbf{B} = |\mathbf{A}||\mathbf{B}| \cos \theta$$

where $\cos \theta$ is the cosine of the angle between the two vectors. This product, as its name communicates, is a scalar, and from the definition it is clear that scalar multiplication is commutative — that is,

$$\mathbf{A} \cdot \mathbf{B} = \mathbf{B} \cdot \mathbf{A}$$

For a unit vector $\hat{\mathbf{e}}$,

$$\mathbf{A} \cdot \hat{\mathbf{e}} = |\mathbf{A}|(1) \cos \theta$$

so that $\mathbf{A} \cdot \hat{\mathbf{e}}$ is the scalar component of $\mathbf{A}$ associated with the direction of $\hat{\mathbf{e}}$. Thus if, as in the preceding section, we express $\mathbf{A}$ by

$$\mathbf{A} = A_x\hat{\mathbf{i}} + A_y\hat{\mathbf{j}} + A_z\hat{\mathbf{k}}$$

then

$$A_x = \mathbf{A} \cdot \hat{\mathbf{i}}$$
$$A_y = \mathbf{A} \cdot \hat{\mathbf{j}}$$
$$A_z = \mathbf{A} \cdot \hat{\mathbf{k}}$$

A very important property is that the scalar product is distributive over addition — that is,

$$\mathbf{A} \cdot (\mathbf{B} + \mathbf{C}) = \mathbf{A} \cdot \mathbf{B} + \mathbf{A} \cdot \mathbf{C}$$

We may obtain this property easily if we now define the unit vector $\hat{\mathbf{e}}$ to be $\mathbf{A}/|\mathbf{A}|$, so that

$$\mathbf{A} \cdot (\mathbf{B} + \mathbf{C}) = |\mathbf{A}|\hat{\mathbf{e}} \cdot (\mathbf{B} + \mathbf{C})$$

From Figure A.6 it is clear that

$$\hat{\mathbf{e}} \cdot (\mathbf{B} + \mathbf{C}) = \hat{\mathbf{e}} \cdot \mathbf{B} + \hat{\mathbf{e}} \cdot \mathbf{C})$$

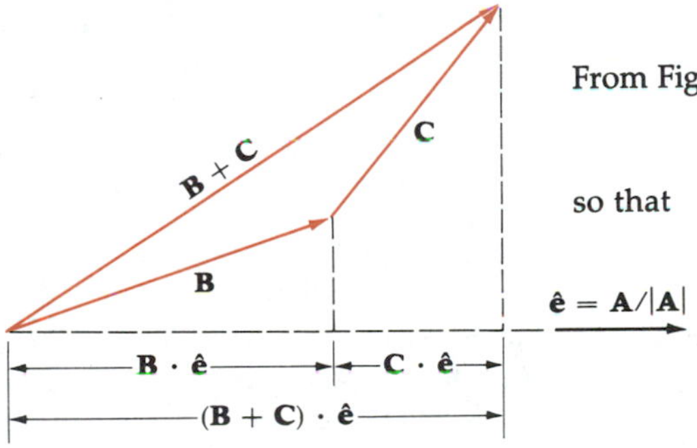

Figure A.6 Distributive Property of the Scalar Product

so that

$$\begin{aligned}\mathbf{A} \cdot (\mathbf{B} + \mathbf{C}) &= |\mathbf{A}|\hat{\mathbf{e}} \cdot (\mathbf{B} + \mathbf{C}) \\ &= |\mathbf{A}|(\hat{\mathbf{e}} \cdot \mathbf{B} + \hat{\mathbf{e}} \cdot \mathbf{C}) \\ &= |\mathbf{A}|\hat{\mathbf{e}} \cdot \mathbf{B} + |\mathbf{A}|\hat{\mathbf{e}} \cdot \mathbf{C} \\ &= \mathbf{A} \cdot \mathbf{B} + \mathbf{A} \cdot \mathbf{C}\end{aligned}$$

The commutative and distributive properties of the scalar product allows us to obtain a result of great practical importance. Suppose we express the vectors $\mathbf{A}$ and $\mathbf{B}$ as

$$\mathbf{A} = A_x\hat{\mathbf{i}} + A_y\hat{\mathbf{j}} + A_z\hat{\mathbf{k}}$$
$$\mathbf{B} = B_x\hat{\mathbf{i}} + B_y\hat{\mathbf{j}} + B_z\hat{\mathbf{k}}$$

Then

$$\begin{aligned}\mathbf{A} \cdot \mathbf{B} &= (A_x\hat{\mathbf{i}} + A_y\hat{\mathbf{j}} + A_z\hat{\mathbf{k}}) \cdot (B_x\hat{\mathbf{i}} + B_y\hat{\mathbf{j}} + B_z\hat{\mathbf{k}}) \\ &= A_xB_x(\hat{\mathbf{i}} \cdot \hat{\mathbf{i}}) + A_xB_y(\hat{\mathbf{i}} \cdot \hat{\mathbf{j}}) + A_xB_z(\hat{\mathbf{i}} \cdot \hat{\mathbf{k}}) \\ &\quad + A_yB_x(\hat{\mathbf{j}} \cdot \hat{\mathbf{i}}) + A_yB_y(\hat{\mathbf{j}} \cdot \hat{\mathbf{j}}) + A_yB_z(\hat{\mathbf{j}} \cdot \hat{\mathbf{k}}) \\ &\quad + A_zB_x(\hat{\mathbf{k}} \cdot \hat{\mathbf{i}}) + A_zB_y(\hat{\mathbf{k}} \cdot \hat{\mathbf{j}}) + A_zB_z(\hat{\mathbf{k}} \cdot \hat{\mathbf{k}})\end{aligned}$$

But, since $\hat{\mathbf{i}}$, $\hat{\mathbf{j}}$, and $\hat{\mathbf{k}}$ are mutually perpendicular unit vectors, then by the definition of the scalar product we obtain

$$\hat{\mathbf{i}} \cdot \hat{\mathbf{i}} = \hat{\mathbf{j}} \cdot \hat{\mathbf{j}} = \hat{\mathbf{k}} \cdot \hat{\mathbf{k}} = 1$$
$$\hat{\mathbf{i}} \cdot \hat{\mathbf{j}} = \hat{\mathbf{i}} \cdot \hat{\mathbf{k}} = \hat{\mathbf{j}} \cdot \hat{\mathbf{k}} = 0$$

Substituting above, then, we obtain

$$\mathbf{A} \cdot \mathbf{B} = A_x B_x + A_y B_y + A_z B_z \tag{A.6}$$

This important result tells us how to compute the scalar product of two vectors in terms of its scalar components. When we apply it to $\mathbf{B} = \mathbf{A}$ we obtain

$$\begin{aligned}\mathbf{A} \cdot \mathbf{A} &= A_x^2 + A_y^2 + A_z^2 \\ &= |\mathbf{A}|^2\end{aligned}$$

which was already available by the definition — that is,

$$\begin{aligned}\mathbf{A} \cdot \mathbf{A} &= |\mathbf{A}||\mathbf{A}| \cos(0) \\ &= |\mathbf{A}|^2\end{aligned}$$

Finally, before leaving this section we point out that the scalar product gives us an elegant and practical way to express the angle between any two vectors, and, of particular importance in mechanics, it gives us a compact way to compute the orthogonal component of a vector in any given direction.

A.4 Vector (Cross) Product

A second multiplicative operation between two vectors is called the vector product and is usually denoted $\mathbf{A} \times \mathbf{B}$ from whence comes the term "cross" product. The vector product is defined to be a vector so that, if $\mathbf{C} = \mathbf{A} \times \mathbf{B}$, then $|\mathbf{C}| = |\mathbf{A}||\mathbf{B}| \sin \theta$, where θ is the angle (not greater than 180°) between $\mathbf{A}$ and $\mathbf{B}$, and $\mathbf{C}$ is perpendicular to both $\mathbf{A}$ and $\mathbf{B}$ with its direction determined by the "right hand rule" as indicated in Figure A.7. The assignment of the direction of $\mathbf{C}$ may be visualized as in the figure by first giving $\mathbf{A}$ and $\mathbf{B}$ a common point of origin or "tail," and then by rotating $\mathbf{A}$ about its tail so as to line it up with $\mathbf{B}$. If the sense of turning of $\mathbf{A}$ (through angle θ) to line it up with $\mathbf{B}$ is mimicked by curling the fingers of the right hand, then the extended thumb points in the direction of $\mathbf{C} = \mathbf{A} \times \mathbf{B}$. This is also the correspondence between the turning and advance of a screw or bolt with the customary right-hand threads. A sometimes useful geometric interpretation of the cross product is that its magnitude is the area of a parallelogram whose sides have as their lengths the magnitudes of $\mathbf{A}$ and $\mathbf{B}$. (See Problem A.13 at the end of this appendix.)

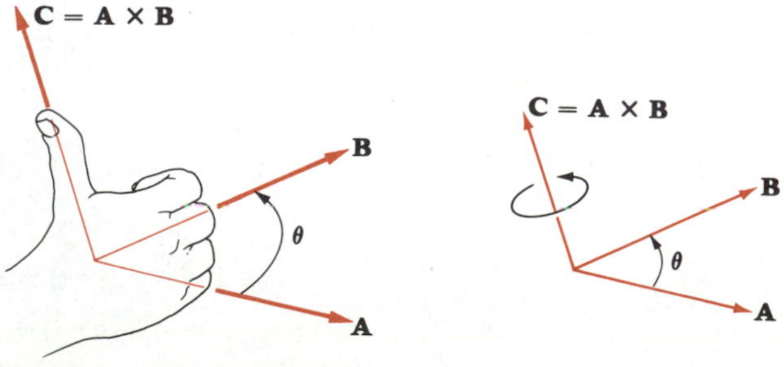

Figure A.7 Direction of the Vector Product

From its definition, vector multiplication is not commutative, and, in fact,

$$\mathbf{B} \times \mathbf{A} = -\mathbf{A} \times \mathbf{B}$$

Vector multiplication is, however, distributive, and this is an important property of which we shall take advantage many times. The property is that

$$\mathbf{A} \times (\mathbf{B} + \mathbf{C}) = \mathbf{A} \times \mathbf{B} + \mathbf{A} \times \mathbf{C} \tag{A.7}$$

To establish this property we begin with a special case. Let **B** and **C** be vectors, each perpendicular to **A** as shown in Figure A.8. Because **A** is perpendicular to **B**, to **C** and to **B** + **C**, then the vector product of **A** with each of these is obtained by rotating the original vector through 90° about **A** as shown, and multiplying its magnitude by $|\mathbf{A}|$. Thus by the parallelogram law of addition,

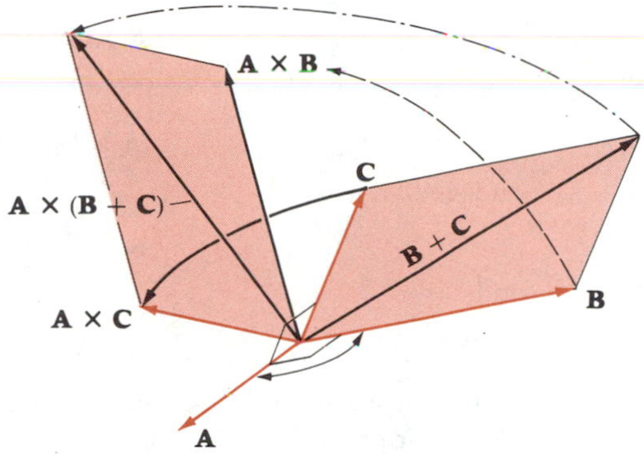

Figure A.8 Distributive Property of the Vector Product

$$\mathbf{A} \times (\mathbf{B} + \mathbf{C}) = \mathbf{A} \times \mathbf{B} + \mathbf{A} \times \mathbf{C}$$

We now need to relax the restriction that **B** and **C** are perpendicular to **A**. To that end we decompose the vector **B** into two parts: one part, $\mathbf{B}_1$, in the direction of **A** and the remainder, $\mathbf{B}_2$, necessarily perpendicular to **A**. Thus

$$\mathbf{B}_1 + \mathbf{B}_2 = \mathbf{B}$$

We now need the result that

$$\mathbf{A} \times \mathbf{B} = \mathbf{A} \times \mathbf{B}_2$$

Figure A.9 shows us that the direction of $\mathbf{A} \times \mathbf{B}_2$ will be the same as that of $\mathbf{A} \times \mathbf{B}$. As for the magnitudes, we have

$$|\mathbf{A} \times \mathbf{B}| = |\mathbf{A}||\mathbf{B}| \sin \theta$$

and this equals $|\mathbf{A} \times \mathbf{B}_2|$, since

$$|\mathbf{A} \times \mathbf{B}_2| = |\mathbf{A}||\mathbf{B}_2| \sin 90°$$
$$= |\mathbf{A}|(|\mathbf{B}| \sin \theta)1$$

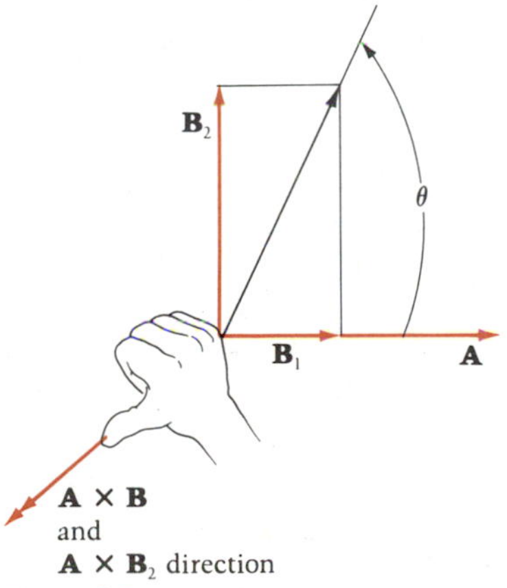

Figure A.9

Now let us call **D** the sum of **B** and **C** and similarly decompose **C** and **D**. Thus

$$\mathbf{D} = \mathbf{B} + \mathbf{C}$$

and

$$\mathbf{D}_1 = \mathbf{B}_1 + \mathbf{C}_1$$

and thus

$$\mathbf{D}_2 = \mathbf{B}_2 + \mathbf{C}_2$$

Because $\mathbf{B}_2$ and $\mathbf{C}_2$ (and hence $\mathbf{D}_2$) are each perpendicular to $\mathbf{A}$, we use our restricted distributive law, which is that if $\mathbf{B}_2$ and $\mathbf{C}_2$ are normal to $\mathbf{A}$,

$$\mathbf{A} \times (\mathbf{B}_2 + \mathbf{C}_2) = \mathbf{A} \times \mathbf{B}_2 + \mathbf{A} \times \mathbf{C}_2$$

or

$$\mathbf{A} \times \mathbf{D}_2 = \mathbf{A} \times \mathbf{B}_2 + \mathbf{A} \times \mathbf{C}_2$$

But

$$\mathbf{A} \times \mathbf{B}_2 = \mathbf{A} \times \mathbf{B}$$

and similarly,

$$\mathbf{A} \times \mathbf{C}_2 = \mathbf{A} \times \mathbf{C}$$

and

$$\mathbf{A} \times \mathbf{D}_2 = \mathbf{A} \times \mathbf{D}$$

so that $\mathbf{A} \times (\mathbf{B} + \mathbf{C}) = \mathbf{A} \times \mathbf{B} + \mathbf{A} \times \mathbf{C}$ without any restriction, which completes our proof.

We close this section with a formula for the cross product in terms of components of the multiplying vectors. As usual let $\hat{\mathbf{i}}$, $\hat{\mathbf{j}}$, and $\hat{\mathbf{k}}$ be mutually perpendicular unit vectors in the x, y, and z directions, but let us also specify that this is a *right-handed system;* that is

$$\hat{\mathbf{i}} \times \hat{\mathbf{j}} = \hat{\mathbf{k}}$$
$$\hat{\mathbf{k}} \times \hat{\mathbf{i}} = \hat{\mathbf{j}}$$

and

$$\hat{\mathbf{j}} \times \hat{\mathbf{k}} = \hat{\mathbf{i}}$$

It then follows that

$$\hat{\mathbf{j}} \times \hat{\mathbf{i}} = -\hat{\mathbf{k}}$$
$$\hat{\mathbf{i}} \times \hat{\mathbf{k}} = -\hat{\mathbf{j}}$$

and

$$\hat{\mathbf{k}} \times \hat{\mathbf{j}} = -\hat{\mathbf{i}}$$

and of course

$$\begin{aligned} \hat{\mathbf{i}} \times \hat{\mathbf{i}} &= \hat{\mathbf{j}} \times \hat{\mathbf{j}} \\ &= \hat{\mathbf{k}} \times \hat{\mathbf{k}} = 0 \end{aligned}$$

Expressing $\mathbf{A}$ and $\mathbf{B}$ in terms of components along x, y, and z,

$$\begin{aligned} \mathbf{A} \times \mathbf{B} &= (A_x\hat{\mathbf{i}} + A_y\hat{\mathbf{j}} + A_z\hat{\mathbf{k}}) \times (B_x\hat{\mathbf{i}} + B_y\hat{\mathbf{j}} + B_z\hat{\mathbf{k}}) \\ &= A_xB_y(\hat{\mathbf{i}} \times \hat{\mathbf{j}}) + A_xB_z(\hat{\mathbf{i}} \times \hat{\mathbf{k}}) + A_yB_x(\hat{\mathbf{j}} \times \hat{\mathbf{i}}) \\ &\quad + A_yB_z(\hat{\mathbf{j}} \times \hat{\mathbf{k}}) + A_zB_x(\hat{\mathbf{k}} \times \hat{\mathbf{i}}) + A_zB_y(\hat{\mathbf{k}} \times \hat{\mathbf{j}}) \\ &= (A_yB_z - A_zB_y)\hat{\mathbf{i}} + (A_zB_x - A_xB_z)\hat{\mathbf{j}} + (A_xB_y - A_yB_x)\hat{\mathbf{k}} \end{aligned} \tag{A.8}$$

Sometimes it is useful to express this result in the form of a determinant,

$$\begin{vmatrix} \hat{\mathbf{i}} & \hat{\mathbf{j}} & \hat{\mathbf{k}} \\ A_x & A_y & A_z \\ B_x & B_y & B_z \end{vmatrix}$$

which will be recognized, when expanded, as the right-hand side of Equation (A.8).

It is necessary for the reader to become thoroughly familiar with the vector, or cross, product, because it is this operation that lies at the heart of the concept of the *moment of a force* (see Section 3.2).

A.5 Scalar and Vector Triple Products

Two different multiplicative operations, each involving three vectors, naturally arise in mechanics. The first of these is called the *scalar triple product.* It is, as the name implies, a scalar, and it results from the scalar (dot) product of a vector with the vector (cross) product of two others, as in $\mathbf{C} \cdot (\mathbf{A} \times \mathbf{B})$. A sometimes useful geometric interpretation of this product is that its magnitude is the volume of a parallelepiped whose sides have as their lengths the magnitudes of **A**, **B**, and **C**. Of greater importance to us here is that the value of the scalar triple product is not changed by certain exchanges of elements; in particular;

$$\mathbf{C} \cdot (\mathbf{A} \times \mathbf{B}) = \mathbf{A} \cdot (\mathbf{B} \times \mathbf{C}) = \mathbf{B} \cdot (\mathbf{C} \times \mathbf{A})^* \tag{A.9}$$

Equation (A.9) may be established by returning to the expression for $\mathbf{A} \times \mathbf{B}$ given by Equation (A.8):

$$\mathbf{A} \times \mathbf{B} = (A_yB_z - A_zB_y)\hat{\mathbf{i}} + (A_zB_x - A_xB_z)\hat{\mathbf{j}} + (A_xB_y - A_yB_x)\hat{\mathbf{k}}$$

If we similarly express **C** by

$$\mathbf{C} = C_x\hat{\mathbf{i}} + C_y\hat{\mathbf{j}} + C_z\hat{\mathbf{k}}$$

then

$$\mathbf{C} \cdot (\mathbf{A} \times \mathbf{B}) = (A_yB_z - A_zB_y)C_x + (A_zB_x - A_xB_z)C_y + (A_xB_y - A_yB_x)C_z \tag{A.10}$$

The right-hand side of Equation (A.10) may be expressed as a determinant so that

$$\mathbf{C} \cdot (\mathbf{A} \times \mathbf{B}) = \begin{vmatrix} C_x & C_y & C_z \\ A_x & A_y & A_z \\ B_x & B_y & B_z \end{vmatrix} \tag{A.11}$$

Because the value of a determinant is not altered by two interchanges of its rows, the right-hand side of Equation (A.11) has the same value as

$$\begin{vmatrix} A_x & A_y & A_z \\ B_x & B_y & B_z \\ C_x & C_y & C_z \end{vmatrix} \quad \text{or} \quad \begin{vmatrix} B_x & B_y & B_z \\ C_x & C_y & C_z \\ A_x & A_y & A_z \end{vmatrix} \tag{A.12}$$

* This third expression may also be written $(\mathbf{C} \times \mathbf{A}) \cdot \mathbf{B}$ so that Equation (A.9) expresses that an identity follows from "interchanging" the "dot" and "cross." It is of course understood that the parentheses, indicating the order of the operations, always group the elements of the cross product.

The first of these may now be recognized as $\mathbf{A}\cdot(\mathbf{B}\times\mathbf{C})$ and the second as $\mathbf{B}\cdot(\mathbf{C}\times\mathbf{A})$, which establishes the identity (A.9). We note that, because $\mathbf{B}\times\mathbf{A} = -\mathbf{A}\times\mathbf{B}$, then $\mathbf{C}\cdot(\mathbf{B}\times\mathbf{A}) = -\mathbf{C}\cdot(\mathbf{A}\times\mathbf{B})$. This result is also immediately available from the determinant form because one interchange of rows changes the sign of a determinant.

Finally, we point out a feature of the scalar triple product that the reader may already have deduced. That is, if any two of the three vectors are proportional,* the scalar triple product vanishes. This follows from the fact that the cross product vanishes if the two vectors are proportional, since one of the forms of Equation (A.9) will have the two proportional vectors in the cross product. The result may also be deduced from the determinant form since a determinant vanishes if two rows are proportional. It is also seen from the fact that, in this case, the parallelepiped mentioned in the opening paragraph has zero volume.

EXAMPLE A.1

For the vectors:

$$\mathbf{A} = 2\hat{\imath} - 3\hat{\jmath} + \hat{k}$$
$$\mathbf{B} = 5\hat{\imath} + \hat{\jmath} - 4\hat{k}$$
$$\mathbf{C} = 3\hat{\imath} + \hat{k}$$

find the scalar triple product $\mathbf{C}\cdot(\mathbf{A}\times\mathbf{B})$.

Solution

$$\mathbf{C}\cdot(\mathbf{A}\times\mathbf{B}) = \mathbf{C}\cdot[(2\hat{\imath} - 3\hat{\jmath} + \hat{k})\times(5\hat{\imath} + \hat{\jmath} - 4\hat{k})]$$

and omitting the zero cross products such as $\hat{\imath}\times\hat{\imath}$,

$$\begin{aligned}
&= \mathbf{C}\cdot[2(1)(\hat{\imath}\times\hat{\jmath}) + 2(-4)(\hat{\imath}\times\hat{k}) + (-3)5(\hat{\jmath}\times\hat{\imath}) + (-3)(-4)(\hat{\jmath}\times\hat{k}) \\
&\quad + 1(5)(\hat{k}\times\hat{\imath}) + 1(1)(\hat{k}\times\hat{\jmath})] \\
&= (3\hat{\imath} + \hat{k})\cdot[2\hat{k} + 8\hat{\jmath} + 15\hat{k} + 12\hat{\imath} + 5\hat{\jmath} - \hat{\imath}] \\
&= (3\hat{\imath} + \hat{k})\cdot[11\hat{\imath} + 13\hat{\jmath} + 17\hat{k}] \\
&= 3(11) + 0(13) + 1(17) = 50
\end{aligned}$$

Interchanging the dot and cross and recomputing as a check,

$$\begin{aligned}
(\mathbf{C}\times\mathbf{A})\cdot\mathbf{B} &= [3(-3)(\hat{\imath}\times\hat{\jmath}) + 3(1)(\hat{\imath}\times\hat{k}) + 1(2)(\hat{k}\times\hat{\imath}) \\
&\qquad + 1(-3)(\hat{k}\times\hat{\jmath})]\cdot\mathbf{B} \\
&= [-9\hat{k} - 3\hat{\jmath} + 2\hat{\jmath} + 3\hat{\imath}]\cdot\mathbf{B} \\
&= (3\hat{\imath} - \hat{\jmath} - 9\hat{k})\cdot(5\hat{\imath} + \hat{\jmath} - 4\hat{k}) \\
&= 3(5) + (-1)(1) + (-9)(-4) = 50
\end{aligned}$$

The reader is encouraged to confirm this result by using the determinant form, Equation (A.11).

* For example, $\mathbf{A} = \alpha\mathbf{B}$, where α is a scalar.

A *vector triple product* is the result of the cross product of a vector with the result of a preceding cross product as in $\mathbf{A} \times (\mathbf{B} \times \mathbf{C})$. The parentheses, which denote the order of multiplication, were really unnecessary in the case of the scalar triple product because the result of a dot product is a scalar and the cross product of a scalar and a vector is without meaning. Here, however, the parentheses *are* needed because $(\mathbf{A} \times \mathbf{B}) \times \mathbf{C}$ is itself a legitimate vector triple product. Moreover, in general

$$\mathbf{A} \times (\mathbf{B} \times \mathbf{C}) \neq (\mathbf{A} \times \mathbf{B}) \times \mathbf{C}$$

We may easily illustrate this fact by an example: let $\hat{\mathbf{i}}$, $\hat{\mathbf{j}}$, and $\hat{\mathbf{k}}$ be the usual right-handed system of mutually perpendicular unit vectors; then

$$\hat{\mathbf{i}} \times (\hat{\mathbf{i}} \times \hat{\mathbf{j}}) = \hat{\mathbf{i}} \times \hat{\mathbf{k}} = -\hat{\mathbf{j}}$$

But

$$(\hat{\mathbf{i}} \times \hat{\mathbf{i}}) \times \hat{\mathbf{j}} = \mathbf{0} \times \hat{\mathbf{j}} = \mathbf{0}$$

which establishes the importance of the order of multiplication.

Identities involving vector triple products that will be particularly useful in dynamics are

$$\mathbf{A} \times (\mathbf{B} \times \mathbf{C}) = (\mathbf{A} \cdot \mathbf{C})\mathbf{B} - (\mathbf{A} \cdot \mathbf{B})\mathbf{C} \tag{A.13}$$

and

$$(\mathbf{A} \times \mathbf{B}) \times \mathbf{C} = (\mathbf{A} \cdot \mathbf{C})\mathbf{B} - (\mathbf{B} \cdot \mathbf{C})\mathbf{A} \tag{A.14}$$

Both (A.13) and (A.14) can perhaps be remembered by noting that in both formulas, the answer is "the vector in the middle times the dot product of the other two, *minus* the other one in parentheses times the dot product of the other two."

To prove the identity (A.13), we apply the component form (A.7) of the cross product to $\mathbf{B} \times \mathbf{C}$:

$$\mathbf{B} \times \mathbf{C} = (B_yC_z - B_zC_y)\hat{\mathbf{i}} + (B_zC_x - B_xC_z)\hat{\mathbf{j}} + (B_xC_y - B_yC_x)\hat{\mathbf{k}}$$

Now letting $\mathbf{B} \times \mathbf{C}$ play the role of $\mathbf{B}$ in Equation (A.7),

$$\begin{aligned}
\mathbf{A} \times (\mathbf{B} \times \mathbf{C}) &= [A_y(B_xC_y - B_yC_x) - A_z(B_zC_x - B_xC_z)]\hat{\mathbf{i}} \\
&\quad + [A_z(B_yC_z - B_zC_y) - A_x(B_xC_y - B_yC_x)]\hat{\mathbf{j}} \\
&\quad + [A_x(B_zC_x - B_xC_z) - A_y(B_yC_z - B_zC_y)]\hat{\mathbf{k}} \\
&= (A_yC_y + A_zC_z)B_x\hat{\mathbf{i}} + (A_zC_z + A_xC_x)B_y\hat{\mathbf{j}} \\
&\quad + (A_xC_x + A_yC_y)B_z\hat{\mathbf{k}} - [(A_yB_y + A_zB_z)C_x\hat{\mathbf{i}} \\
&\quad + (A_zB_z + A_xB_x)C_y\hat{\mathbf{j}} + (A_xB_x + A_yB_y)C_z\hat{\mathbf{k}}]
\end{aligned}$$

If on the right-hand side we add and subtract the vector

$$A_xB_xC_x\hat{\mathbf{i}} + A_yB_yC_y\hat{\mathbf{j}} + A_zB_zC_z\hat{\mathbf{k}}$$

then we have, after regrouping,

$$\begin{aligned}
\mathbf{A} \times (\mathbf{B} \times \mathbf{C}) &= (A_xC_x + A_yC_y + A_zC_z)(B_x\hat{\mathbf{i}} + B_y\hat{\mathbf{j}} + B_z\hat{\mathbf{k}}) \\
&\quad - (A_xB_x + A_yB_y + A_zB_z)(C_x\hat{\mathbf{i}} + C_y\hat{\mathbf{j}} + C_z\hat{\mathbf{k}})
\end{aligned}$$

But by Equation (A.6),

$$A_xC_x + A_yC_y + A_zC_z = \mathbf{A} \cdot \mathbf{C}$$

and

$$A_xB_x + A_yB_y + A_zB_z = \mathbf{A} \cdot \mathbf{B}$$

Therefore,

$$\mathbf{A} \times (\mathbf{B} \times \mathbf{C}) = (\mathbf{A} \cdot \mathbf{C})\mathbf{B} - (\mathbf{A} \cdot \mathbf{B})\mathbf{C}$$

which is the desired result. Identity (A.14) may be similarly established.

EXAMPLE A.2

Determine $\mathbf{A} \times (\mathbf{B} \times \mathbf{C})$ for the vectors

$$\mathbf{A} = 2\hat{\imath} - 3\hat{\jmath} + \hat{k}$$
$$\mathbf{B} = 5\hat{\imath} + \hat{\jmath} - 4\hat{k}$$
$$\mathbf{C} = 3\hat{\imath} + \hat{k}$$

Solution

$$\begin{aligned}\mathbf{A} \times (\mathbf{B} \times \mathbf{C}) &= (\mathbf{A} \cdot \mathbf{C})\mathbf{B} - (\mathbf{A} \cdot \mathbf{B})\mathbf{C} \\ &= [2(3) + (-3)0 + 1(1)](5\hat{\imath} + \hat{\jmath} - 4\hat{k}) \\ &\quad - [2(5) + (-3)1 + 1(-4)](3\hat{\imath} + \hat{k}) \\ &= 7(5\hat{\imath} + \hat{\jmath} - 4\hat{k}) - 3(3\hat{\imath} + \hat{k}) \\ &= 26\hat{\imath} + 7\hat{\jmath} - 31\hat{k}\end{aligned}$$

The reader is encouraged to confirm this result by first computing $\mathbf{B} \times \mathbf{C}$ and then "crossing" $\mathbf{A}$ into it.

PROBLEMS ▶ Appendix A

A.1 Find $\mathbf{A} + \mathbf{B}$ and $\mathbf{A} - \mathbf{B}$ if $\mathbf{A} = 2\hat{\imath} + 6\hat{\jmath} - 4\hat{k}$ and $\mathbf{B} = -3\hat{\imath} - \hat{\jmath} + 10\hat{k}$.

A.2 What is the unit vector in the direction of the vector $(3\hat{\imath} + \hat{\jmath} - \hat{k})$?

A.3 Find the dot product of the vectors:

$$\mathbf{F} = 2\hat{\imath} + 3\hat{\jmath} + 6\hat{k}$$
$$\mathbf{Q} = -\hat{\imath} - 6\hat{\jmath} - 3\hat{k}$$

A.4 Make up a rule about the direction of $\mathbf{A} \times \mathbf{B}$ that relates to a right-handed screw turning from $\mathbf{A}$ into $\mathbf{B}$. (See Figure PA.4.)

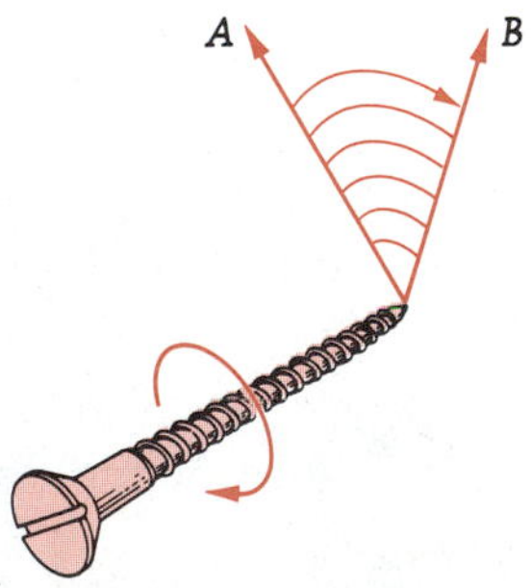

Figure PA.4

A.5 Given:

$$\mathbf{A} = 6\hat{\imath} + 2\hat{\jmath} + 9\hat{k}, \qquad \mathbf{B} = 10\hat{\imath} + 6\hat{k},$$
$$\mathbf{C} = 2\hat{\imath} + 4\hat{\jmath} + 6\hat{k}$$

Find:

$$(\mathbf{A} \cdot \mathbf{B})\mathbf{C}, \qquad \mathbf{A} \cdot \mathbf{B} \times \mathbf{C},$$
$$(\mathbf{A} \times \mathbf{B}) \times \mathbf{C}, \qquad \mathbf{A} \times (\mathbf{B} \times \mathbf{C})$$

For the following sets of vectors in Problems A.6–A.9, find: (a) $\mathbf{A}+\mathbf{B}$; (b) $\mathbf{B}-\mathbf{C}$; (c) $\mathbf{A}\cdot\mathbf{B}$; (d) $\mathbf{B}\times\mathbf{C}$; (e) $\mathbf{A}\cdot(\mathbf{B}\times\mathbf{C})$; (f) $\mathbf{A}\times(\mathbf{B}\times\mathbf{C})$; (g) the unit vector in the direction of $\mathbf{C}$; (h) the direction cosines of $\mathbf{B}$.

A.6 $\mathbf{A}=\hat{\imath}+\hat{\jmath}$; $\mathbf{B}=2\hat{\imath}-3\hat{\jmath}+4\hat{k}$; $\mathbf{C}=-5\hat{\imath}+4\hat{\jmath}-2\hat{k}$

A.7 $\mathbf{A}=2.4\hat{\imath}-6.3\hat{k}$; $\mathbf{B}=4.1\hat{\imath}-20.5\hat{\jmath}+6.0\hat{k}$; $\mathbf{C}=5.1\hat{\imath}+8.7\hat{\jmath}$

A.8 $\mathbf{A}=3\hat{\imath}$; $\mathbf{B}=-7\hat{\jmath}+\hat{k}$; $\mathbf{C}=\hat{\imath}+\hat{\jmath}+\hat{k}$

A.9 $\mathbf{A}=2\hat{\imath}-\hat{\jmath}+\hat{k}$; $\mathbf{B}=15\hat{\imath}-20\hat{\jmath}+18\hat{k}$; $\mathbf{C}=\hat{\imath}+7\hat{k}$

A.10 Find the cross product of the vectors

$$\mathbf{P}=2\hat{\imath}+3\hat{\jmath}+6\hat{k} \quad \text{and} \quad \mathbf{Q}=-\hat{\imath}-6\hat{\jmath}-3\hat{k}$$

A.11 Find the dot and cross products of

$$\mathbf{A}=2\hat{\imath}-\hat{\jmath}+3\hat{k} \quad \text{and} \quad \mathbf{B}=3\hat{\imath}+2\hat{\jmath}-5\hat{k}$$

A.12 Find the angle between $\hat{k}$ and the sum of the vectors in Problem A.11.

A.13 Show that the magnitude of $\mathbf{A}\times\mathbf{B}$ equals the "area" of a parallelogram having side lengths equal to $|\mathbf{A}|$ and $|\mathbf{B}|$.

A.14 Find the angle between the forces $\mathbf{F}_1$ and $\mathbf{F}_2$ in Figure PA.14.

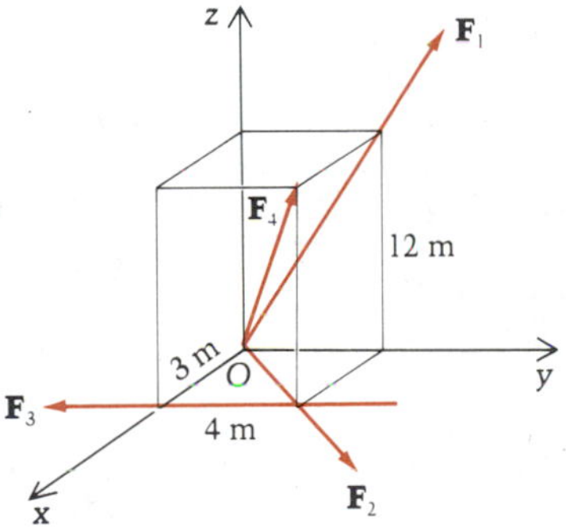

Figure PA.14

* **A.15** In Problem A.14, find the distance between forces $\mathbf{F}_4$ and $\mathbf{F}_3$ (i.e., the length of the line segment intersecting the two and perpendicular to them both).

A.16 Show that $\mathbf{A}\cdot(\mathbf{B}\times\mathbf{C})$ is the volume of the parallelepiped in Figure PA.16.

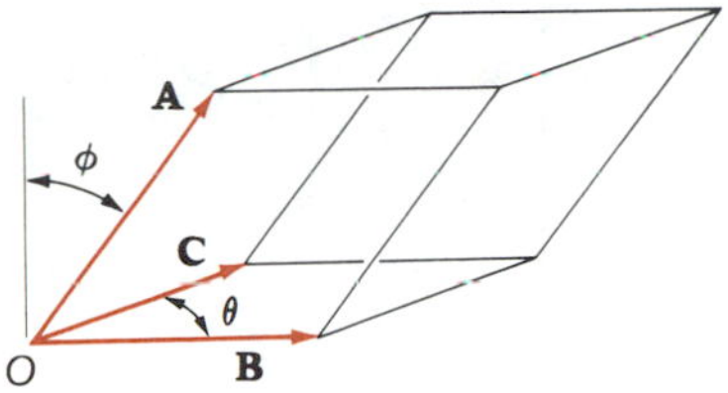

Figure PA.16

A.17 The three vectors $\mathbf{A}=2\hat{\imath}+3\hat{\jmath}$, $\mathbf{B}=5\hat{\jmath}+\hat{k}$, and $\mathbf{C}=\hat{\imath}+\hat{\jmath}+\hat{k}$ define a parallelepiped with the origin O as one vertex. Find the length of the diagonal from O to the opposite corner.

A.18 Show that any vector $\mathbf{C}$ perpendicular to both $\mathbf{A}$ and $\mathbf{B}$ has the form

$$\mathbf{C}=\pm|\mathbf{C}|\frac{\mathbf{A}\times\mathbf{B}}{|\mathbf{A}\times\mathbf{B}|}$$

A.19 Given $\mathbf{A}=3\hat{\imath}+4\hat{\jmath}$ and $\mathbf{B}=4\hat{\jmath}+\hat{k}$, write unit vectors (a) having the same direction as $\mathbf{A}$; (b) perpendicular to $\mathbf{B}$ and lying in the xy plane; (c) perpendicular to both $\mathbf{A}$ and $\mathbf{B}$.

A.20 The vector $\mathbf{D}$ in Figure PA.20 has a magnitude of 60. Resolve the vector into two nonorthogonal components, one along AB and another along BC.

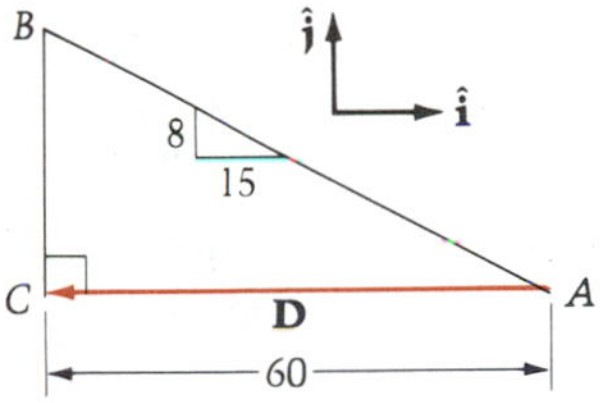

Figure PA.20

* Asterisks identify the more difficult problems.

B TABLES RELATING TO UNITS

The tables in this appendix are useful in identifying and converting units between the SI and U.S. systems.

Table B.1 Units Commonly Used in Engineering Mechanics

Quantity	SI (Standard International or "Metric") Unit	U.S. Unit
force	newton (N)	pound (lb)
mass	kilogram (kg)	slug
length	meter (m)	foot (ft)
time	second (s)	second (sec)
moment of force	N · m	lb-ft
work or energy	joule (J)(= N · m)	ft-lb
pressure or stress	pascal (Pa)(= N/m^2) –	lb/ft^2
velocity	m/s	ft/sec
angular velocity	rad/s	rad/sec
acceleration	m/s^2	ft/sec^2
angular acceleration	rad/s^2	rad/sec^2
mass moment of inertia	kg · m^2	slug-ft^2
moment of inertia of area	m^4	ft^4
momentum	kg · m/s	slug-ft/sec
moment of momentum	kg · m^2/s	slug-ft^2/sec
impulse	N · s	lb-sec
angular impulse	N · m · s	lb-ft-sec
mass density	kg/m^3	slug/ft^3
specific weight	N/m^3	lb/ft^3
power	watt (W)(= J/s)	ft-lb/sec
frequency	hertz (Hz)(= 1 cycle/s)	Hz (same)

Table B.2 Standard Prefixes Used in the SI System of Units

tera	T	10^{12}	centi	c	10^{-2}
giga	G	10^{9}	milli	m	10^{-3}
mega	M	10^{6}	micro	μ	10^{-6}
kilo	k	10^{3}	nano	n	10^{-9}
hecto	h	10^{2}	pico	p	10^{-12}
deka	da	10^{1}	femto	f	10^{-15}
deci	d	10^{-1}	atto	a	10^{-18}

Table B.3 Conversion Factors for SI and U.S. Units*

To Convert From	To	Multiply By	Reciprocal (to Get from SI to U.S. Units)
Length, area, volume			
foot (ft)	meter (m)	0.30480	3.2808
inch (in.)	m	0.025400	39.370
statute mile (mi)	m	1609.3	6.2137×10^{-4}
foot² (ft²)	meter² (m²)	0.092903	10.764
inch² (in.²)	m²	6.4516×10^{-4}	1550.0
foot³ (ft³)	meter³ (m³)	0.028317	35.315
inch³ (in.³)	m³	1.6387×10^{-5}	61024
Velocity			
feet/second (ft/sec)	meter/second (m/s)	0.30480	3.2808
feet/minute (ft/min)	m/s	0.0050800	196.85
knot (nautical mi/hr)	m/s	0.51444	1.9438
mile/hour (mi/hr)	m/s	0.44704	2.2369
mile/hour (mi/hr)	kilometer/hour (km/h)	1.6093	0.62137
Acceleration			
feet/second² (ft/sec²)	meter/second² (m/s²)	0.30480	3.2808
inch/second² (in./sec²)	m/s²	0.025400	39.370
Mass			
slug (lb-sec²/ft)	kg	14.594	0.068522
Force			
pound (lb) or pound-force (lbf)	newton (N)	4.4482	0.22481
Density			
slug/foot³ (slug/ft³)	kg/m³	515.38	0.0019403
Energy, work, or moment of force			
foot-pound or pound-foot (ft-lb) (lb-ft)	joule (J) or newton · meter (N · m)	1.3558	0.73757

(Continued)

* Rounded to the five digits cited. Note, for example, that 1 ft = 0.30480 m, so that

$$(\text{Number of feet}) \times \left(\frac{0.30480\text{ m}}{1\text{ ft}}\right) = \text{Number of meters}$$

Table B.3 continued

To Convert From	To	Multiply By	Reciprocal (to Get from SI to U.S. Units)
Power			
foot-pound/second (ft-lb/sec)	watt (W)	1.3558	0.73756
horsepower (hp) (550 ft-lb/sec)	W	745.70	0.0013410
Stress, pressure			
pound/inch^2 (lb/in.^{2} or psi)	N/m^2 (or Pa)	6894.8	1.4504×10^{-4}
pound/foot^2 (lb/ft^2)	N/m^2 (or Pa)	47.880	0.020886
Mass moment of inertia			
slug-foot^2 (slug-ft^2 or lb-ft-sec^2)	$kg \cdot m^2$	1.3558	0.73756
Momentum (or linear momentum)			
slug-foot/second (slug-ft/sec)	$kg \cdot m/s$	4.4482	0.22481
Impulse (or linear impulse)			
pound-second (lb-sec)	$N \cdot s$ (or $kg \cdot m/s$)	4.4482	0.22481
Moment of momentum (or angular momentum)			
slug-foot^2/second (slug-ft^2/sec)	$kg \cdot m^2/s$	1.3558	0.73756
Angular impulse			
pound-foot-second (lb-ft-sec)	$N \cdot m \cdot s$ (or $kg \cdot m^2/s$)	1.3558	0.73756

C Moments and Products of Inertia of Areas

	Shape	Area	Moment of Inertia
Parallelogram	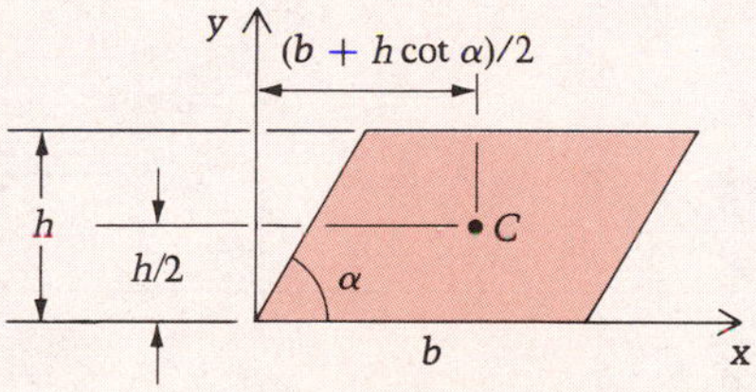	bh	$I_x = \frac{bh^3}{3}$ $I_y = \frac{bh}{3}(b + h\cot\alpha)^2 - \frac{b^2h^2}{6}\cot\alpha$ $I_{xy} = -\frac{bh^2}{12}(3b + 4h\cot\alpha)$
Rectangle	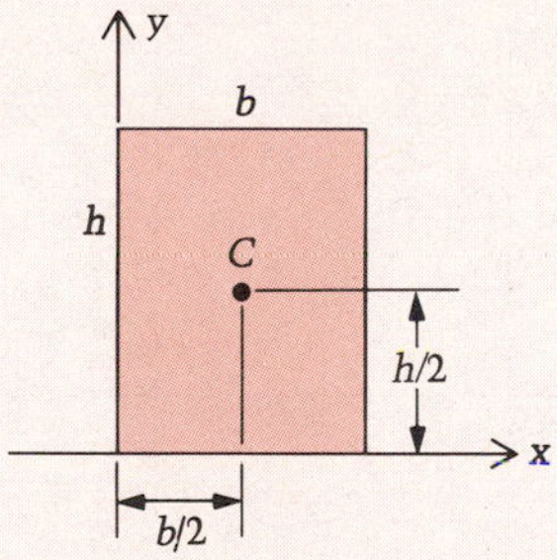	bh	$I_x = \frac{bh^3}{3}$ $I_y = \frac{bh^3}{3}$ $I_{xy} = -\frac{b^2h^2}{4}$ $I_{x_C} = \frac{bh^3}{12}$ $I_{y_C} = \frac{hb^3}{12}$ $I_{xy_C} = 0$

	Shape	Area	Moment of Inertia
Triangle	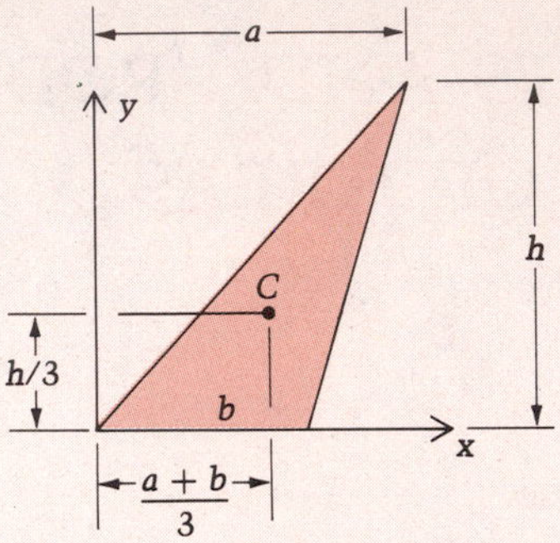	$\frac{1}{2}bh$	$I_x = \frac{bh^3}{12}$ $I_y = \frac{bh}{12}(a^2 + ab + b^2)$ $I_{xy} = -\frac{bh^2}{24}(2a + b)$
Right Triangle	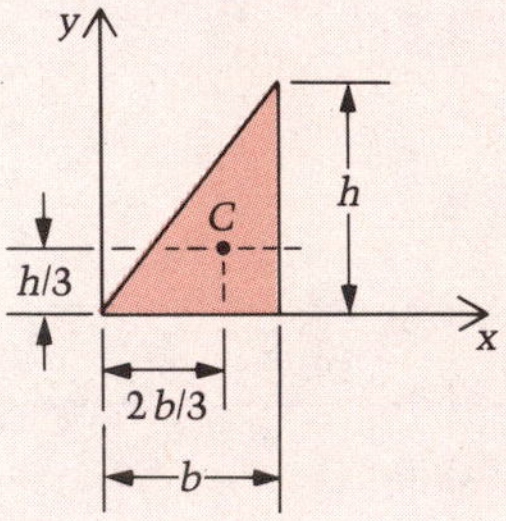	$\frac{1}{2}bh$	$I_x = \frac{bh^3}{12}$ $I_y = \frac{hb^3}{12}$ $I_{xy} = -\frac{b^2h^2}{8}$ $I_{x_C} = \frac{bh^3}{36}$ $I_{y_C} = \frac{hb^3}{36}$ $I_{xy_C} = -\frac{b^2h^2}{72}$
Circle	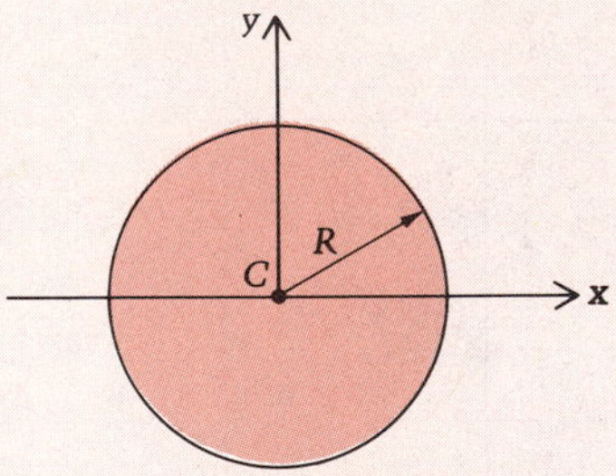	πR^2	$I_{x_C} = I_{y_C} = \frac{\pi R^4}{4}$ $I_{xy_C} = 0$
Semicircular Area	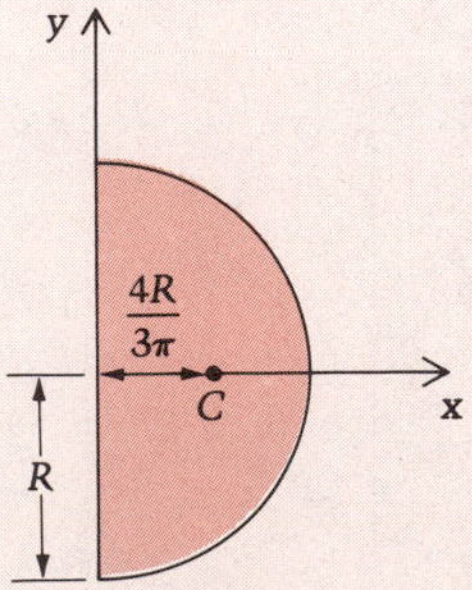	$\frac{\pi R^2}{2}$	$I_x = I_y = \frac{\pi R^4}{8}$ $I_{xy} = 0$

	Shape	Area	Moment of Inertia
Quarter Circle	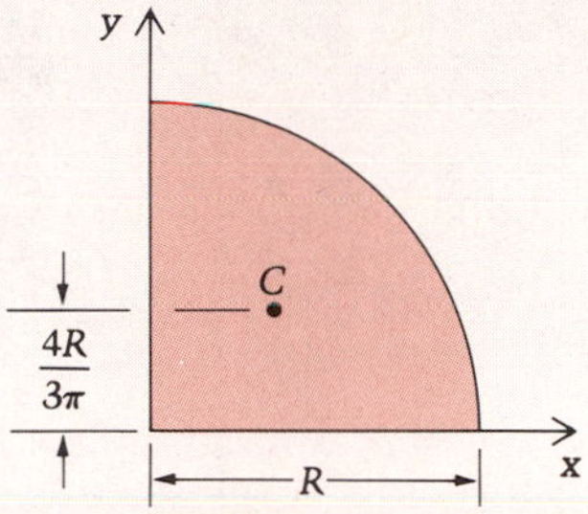	$\dfrac{\pi R^2}{4}$	$I_x = I_y = \dfrac{\pi R^4}{16}$ $I_{xy} = -\dfrac{R^4}{8}$
Annulus	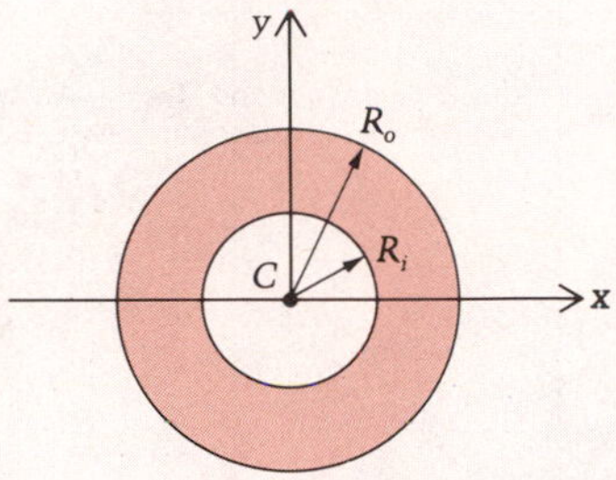	$\pi(R_o^2 - R_i^2)$	$I_{x_C} = I_{y_C} = \dfrac{\pi(R_o^4 - R_i^4)}{4}$ $I_{xy_C} = 0$
Thin Annulus	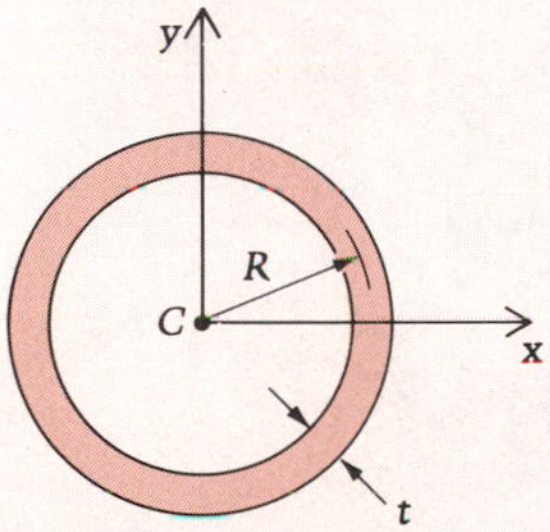	$2\pi Rt \quad (t \ll R)$	$I_{x_C} = I_{y_C} \approx \pi R^3 t$ $I_{xy_C} = 0$
Thin Annular Sector	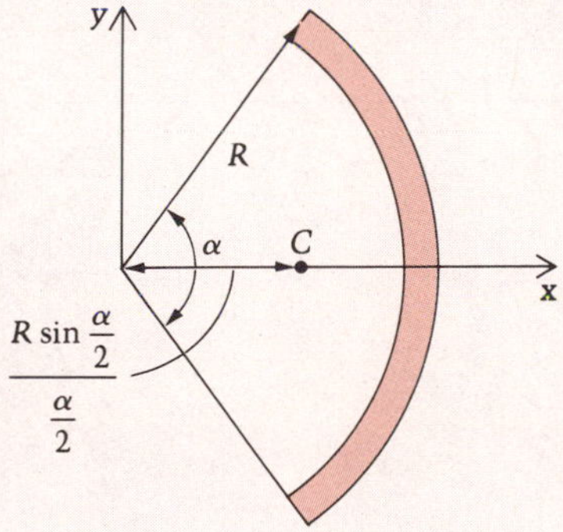	$tR\alpha \quad (t \ll R)$	$I_x \approx \dfrac{tR^3}{2}(\alpha - \sin\alpha)$ $I_y \approx \dfrac{tR^3}{2}(\alpha + \sin\alpha)$ $I_{xy} = 0$

	Shape	Area	Moment of Inertia
Circular Sector	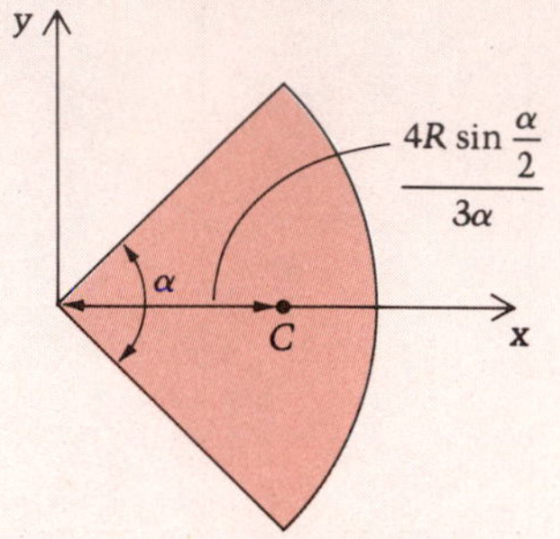	$\frac{\alpha R^2}{2}$	$I_x = \frac{\alpha R^4}{8}\left(1 - \frac{\sin\alpha}{\alpha}\right)$ $I_y = \frac{\alpha R^4}{8}\left(1 + \frac{\sin\alpha}{\alpha}\right)$ $I_{xy} = 0$
Segment of a Circle	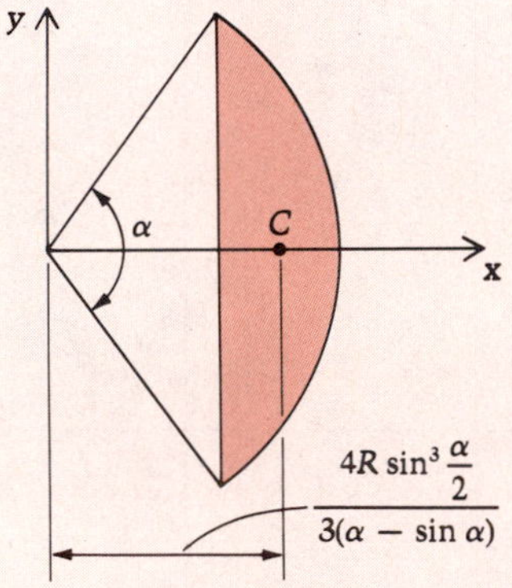	$\frac{R^2}{2}(\alpha - \sin\alpha)$	$I_x = \frac{R^4}{24}(3\alpha - 4\sin\alpha + \sin\alpha\cos\alpha)$ $I_y = \frac{R^4}{8}(\alpha - \sin\alpha\cos\alpha)$ $I_{xy} = 0$
Ellipse	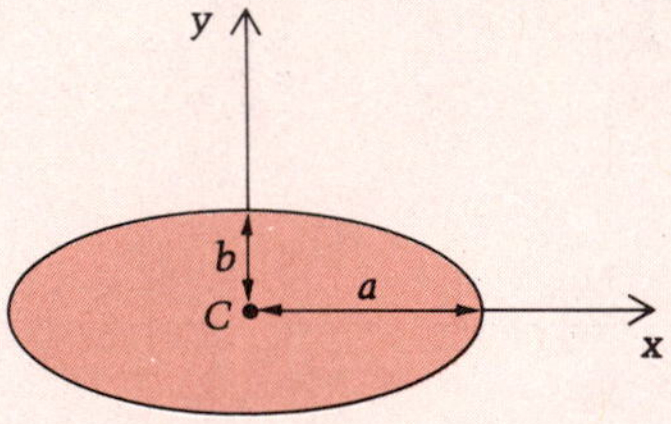	πab	$I_{x_C} = \frac{\pi ab^3}{4}$ $I_{y_C} = \frac{\pi ba^3}{4}$ $I_{xy_C} = 0$
Semiellipse	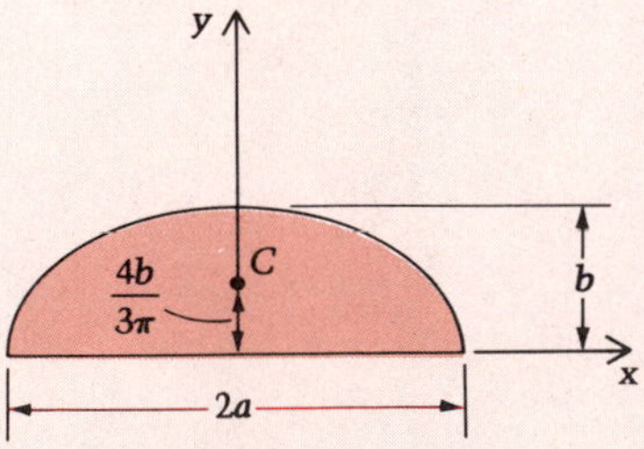	$\frac{\pi ab}{2}$	$I_x = \frac{\pi ab^3}{8}$ $I_y = \frac{\pi a^3 b}{8}$ $I_{xy} = 0$

	Shape	Area	Moment of Inertia
Segment of a Parabola	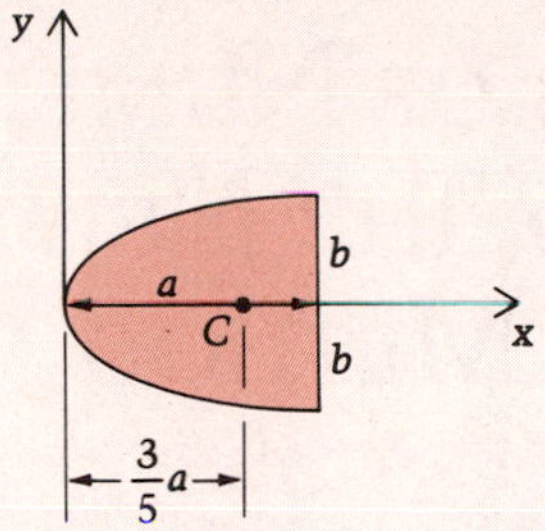	$\frac{4}{3}ab$	$I_x = \frac{4}{15}ab^3$ $I_y = \frac{4}{7}a^3b$ $I_{xy} = 0$
Bent Strip	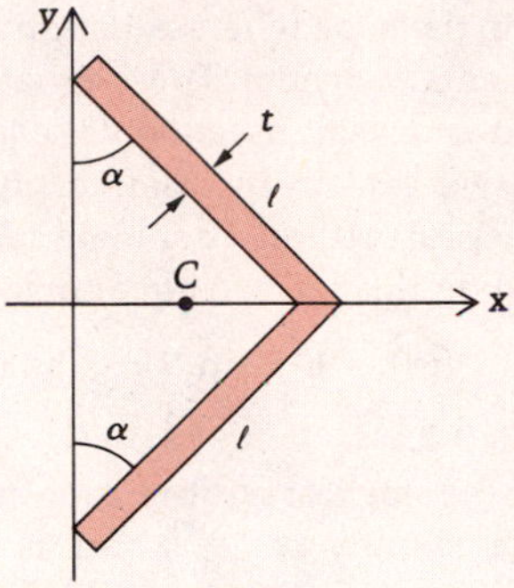	$2\ell t \quad (t \ll \ell)$	$I_x \approx \frac{2\ell^3 t \cos^2 \alpha}{3}$ $I_y \approx \frac{2\ell^3 t \sin^2 \alpha}{3}$ $I_{xy} = 0$

D ▸ Examples of Numerical Analysis/The Newton-Raphson Method

There are a few places in the book where equations arise whose solutions cannot easily be found by elementary algebra. These equations are either polynomials of degree higher than two, or else transcendental equations.

In this appendix, we explain in brief the fundamental idea behind the Newton-Raphson numerical method for solving such equations and others of the form $f(\theta) = 0$. We shall do this for two equations. The first is

$$f(\theta) = 1 - \sin\theta - 2.5\tan\theta \tag{D.1}$$

which occurs in Section 9.2.

We are looking for the smallest positive root θ of Equation (D.1). We see that the function has the value unity at $\theta = 0$ and its slope, $-\cos\theta - 2.5\sec^2\theta$, is -3.5 there. Furthermore (again by inspection), $f(\theta)$ is negative at $\theta = \pi/4$. Thus (see Figure D.1) we expect a root between 0 and $\pi/4$. The Newton-Raphson algorithm, found in more detail in any book on numerical analysis, works as follows: If θ_0 is an initial estimate of a root θ, then a better approximation is

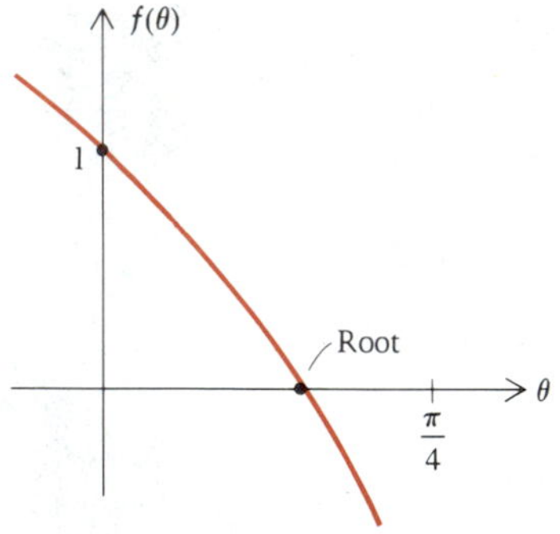

Figure D.1

$$\theta_1 = \theta_0 - \frac{f(\theta_0)}{f'(\theta_0)}$$

Figure D.2 shows what is happening. The quantity $f(\theta_0)/f'(\theta_0)$ causes a backup, from the initial θ_0 approximation, toward the actual root we seek.*

Let us use the initial guess $\theta_0 = \pi/4$ (see Example 9.3). Then $f(\pi/4) = -2.207106781$ and $f'(\pi/4) = -5.707106786$ so that the improved estimate is

$$\theta_1 = 0.785398164 - \frac{-2.207106781}{-5.707106786} = 0.398668624$$

Repeating the algorithm, $f(\theta_1) = -0.441253558$, $f'(\theta_1) = -3.865154335$, and so

$$\theta_2 = 0.398668624 - \frac{-0.441253558}{-3.865154335} = 0.284506673$$

Further applications of the Newton-Raphson algorithm yield

$$\theta_3 = 0.281298920$$

$$\left.\begin{aligned}\theta_4 &= 0.281297097\\ \theta_5 &= 0.281297097\\ \theta_6 &= 0.281297097\end{aligned}\right\} \text{convergence!}$$

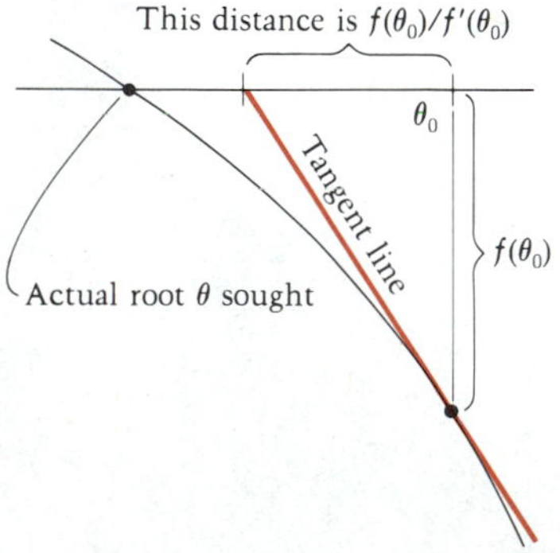

Figure D.2

* The reader may wish to show that the idea works as well for the other three possible sign combinations of f and f' (i.e., $++$, $+-$, and $-+$).

We have nine good digits after just four iterations, and the approximation $\theta \approx 0.281$ agrees with the solution in the text.

As a second example, we consider the cubic polynomial equation from Example 9.12:

$$f(t) = t^3 - 7t^2 + 15.3t - 1.34 = 0 \tag{D.2}$$

We begin with the estimate $t_0 = 1$ in. and quickly obtain, using $f'(t) = 3t^2 - 14t + 15.3$,

$$t_1 = t_0 - \frac{f(t_0)}{f'(t_0)} = 1 - \frac{7.96000000}{4.30000000} = -0.851162791$$

Continuing, the results are:

$$\begin{aligned}
t_1 &= 1.000000000 \\
t_2 &= -0.851162791 \\
t_3 &= -0.168924556 \\
t_4 &= 0.063694406 \\
t_5 &= 0.090990090 \\
t_6 &= 0.091349686 \\
&\left.\begin{aligned} t_7 &= 0.091349748 \\ t_8 &= 0.091349748 \\ t_9 &= 0.091349748 \end{aligned}\right\} \quad \text{convergence!}
\end{aligned}$$

This time we have eight significant digits after only seven iterations. Note that we have in fact found the *only* root of Equation (D.2), as the reader may wish to show by examining the values of f at the two roots of $f' = 0$. (Note that f is negative at $t = 0$ and positive for large t, and that by Descartes' rule of signs, the *maximum number* of positive real roots is three, and of negative ones is zero.)

E ▶ EQUILIBRIUM: A BODY AND ITS PARTS

In this appendix, we present a general proof of the ideas given by example in Section 4.4.

Suppose we decompose a body $\mathcal{B}$ into two parts $\mathcal{B}_1$ and $\mathcal{B}_2$ as shown in Figure E.1:

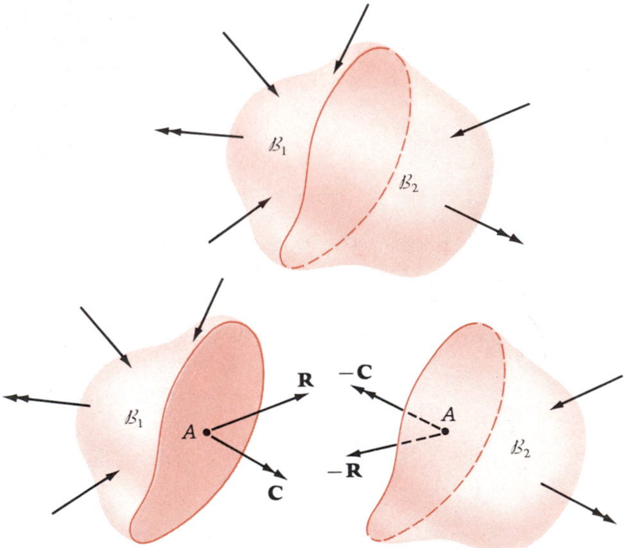

Figure E.1

Furthermore:

1. Let $\mathbf{F}_r$ be the sum of all the external forces on $\mathcal{B}$, the combined body.
2. Let $\mathbf{M}_{rP}$ be the sum of the moments, about an arbitrary point P, of all the external forces and couples on $\mathcal{B}$.
3. Let $\mathbf{F}'_r$ be the sum of all the forces external to $\mathcal{B}$ that act on $\mathcal{B}_1$, and $\mathbf{M}'_{rP}$ be the sum of the moments of those forces and couples about P.
4. Let $\mathbf{F}''_r$ be the sum of all the forces external to $\mathcal{B}$ that act on $\mathcal{B}_2$, and $\mathbf{M}''_{rP}$ be the sum of the moments of those forces and couples about P.
5. Let force $\mathbf{R}$ and couple $\mathbf{C}$ constitute the resultant at A of all the forces that $\mathcal{B}_2$ exerts on $\mathcal{B}_1$; these are *internal* to $\mathcal{B}$.

From the definitions just given we see that

$$\mathbf{F}_r = \mathbf{F}'_r + \mathbf{F}''_r$$

and

$$\mathbf{M}_{rP} = \mathbf{M}'_{rP} + \mathbf{M}''_{rP}$$

We now apply the equilibrium equations to $\mathcal{B}_1$ and obtain

$$\mathbf{F}'_r + \mathbf{R} = \mathbf{0} \tag{E.1}$$

$$\mathbf{M}'_{rP} + \mathbf{r}_{PA} \times \mathbf{R} + \mathbf{C} = \mathbf{0} \tag{E.2}$$

Applying the equilibrium equations to $\mathcal{B}_2$ and using the action-reaction principle as indicated on the free-body diagram,

$$\mathbf{F}''_r + (-\mathbf{R}) = \mathbf{0} \tag{E.3}$$

$$\mathbf{M}''_{rP} + \mathbf{r}_{PA} \times (-\mathbf{R}) + (-\mathbf{C}) = \mathbf{0} \tag{E.4}$$

Adding Equations (E.1) and (E.3),

$$\mathbf{F}'_r + \mathbf{F}''_r = \mathbf{0} \qquad \text{or} \qquad \mathbf{F}_r = \mathbf{0} \tag{E.5}$$

which is of course the force-equation of equilibrium for the combined body $\mathcal{B}$. Adding Equations (E.2) and (E.4),

$$\mathbf{M}'_{rP} + \mathbf{M}''_{rP} = \mathbf{0}$$

or

$$\mathbf{M}_{rP} = \mathbf{0} \tag{E.6}$$

which is the moment equation of equilibrium for $\mathcal{B}$.

Two important conclusions can be drawn from this analysis. First, we note that the equilibrium equations for body $\mathcal{B}$ are the sums of the equilibrium equations for its constituents $\mathcal{B}_1$ and $\mathcal{B}_2$. Therefore, any two of the sets will be independent, but not all three. Thus while there may be profit to be gained from writing the equilibrium equations for two, say $\mathcal{B}$ and $\mathcal{B}_1$, nothing additional can be obtained from the equilibrium equations for $\mathcal{B}_2$.

Second, if we know, or can determine, all of the external forces on $\mathcal{B}$, then the resultant interaction between $\mathcal{B}_1$ and $\mathcal{B}_2$ can be determined from the equations of equilibrium for $\mathcal{B}_1$ (or $\mathcal{B}_2$). This is the basis for much of stress analysis in the mechanics of deformable solids [for example, the determination of shear and bending moments in beams (Chapter 5)].

F

Answers to Odd-Numbered Problems

Unless suggested otherwise by the problem statement or figure, $\hat{\mathbf{i}}$, $\hat{\mathbf{j}}$, and $\hat{\mathbf{k}}$ are unit vectors in the respective directions $\rightarrow$, $\uparrow$, and out of the page.

Chapter 1

1.1 An automobile is parked on a hill. Find the forces exerted on the tires by the road.
1.3 Velocity is defined from the concepts of space and time. The dimension of velocity is L/T. Energy is defined from the concepts of force and space; the dimension of energy is $L \cdot F$. **1.5** 1.609 km
1.7 1060 J **1.9** Yes; both sides have dimension L/T.
1.11 3.44×10^{-8} ft^3/(slug-sec^2) [or ft^4/(lb-sec^4)]
1.13 2.07×10^{11} Pa; 2.07×10^{8} kPa; 2.07×10^{5} MPa; 207 GPa **1.15** 1% **1.17** 0.006%; 0.000006%

Chapter 2

2.1 $\mathbf{F}_2$, $|\mathbf{F}_2| = 9.00$ N
2.3 Result is apparent from Figure 2.3.
2.5 (a) 87.8 N (b) 0.456, 0.569, -0.684
2.7 $0.545\hat{\mathbf{i}} - 0.545\hat{\mathbf{j}} + 0.636\hat{\mathbf{k}}$
2.9 From 2.6, $\mathbf{F} = 70\hat{\mathbf{i}} - 20\hat{\mathbf{j}} \pm 68.6\hat{\mathbf{k}}$ lb
For +, $(\mathbf{F} \cdot \hat{\mathbf{e}})\hat{\mathbf{e}} = 42.1(-0.3\hat{\mathbf{i}} + 0.1\hat{\mathbf{j}} + 0.949\hat{\mathbf{k}})$ lb
For −, $(\mathbf{F} \cdot \hat{\mathbf{e}})\hat{\mathbf{e}} = -88.1(-0.3\hat{\mathbf{i}} + 0.1\hat{\mathbf{j}} + 0.949\hat{\mathbf{k}})$ lb
2.11 $18\hat{\mathbf{i}} + 3\hat{\mathbf{j}} - 48\hat{\mathbf{k}}$ lb **2.13** $0.8\hat{\mathbf{i}} + 0.6\hat{\mathbf{j}}$ **2.15** 33 N
2.17 (a) $25.7\hat{\mathbf{i}} - 10.7\hat{\mathbf{j}} + 16\hat{\mathbf{k}}$ lb
(b) $55.1\hat{\mathbf{i}} - 37.1\hat{\mathbf{j}} + 16\hat{\mathbf{k}}$ lb
2.19 12.9 kN; $-18.8\hat{\mathbf{j}}$ kN **2.21** $11.5\hat{\mathbf{i}} - 16.4\hat{\mathbf{j}}$ kN
2.23 $(\mathbf{A} \cdot \mathbf{B})^2 = |\mathbf{A}|^2 |\mathbf{B}|^2 \cos^2\theta \le |\mathbf{A}|^2 |\mathbf{B}|^2$

2.25 50.2 ↑ N and 3.98 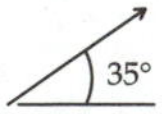

N
2.27 (a) $21\left(\dfrac{-2\hat{\mathbf{i}} + 6\hat{\mathbf{j}} - 3\hat{\mathbf{k}}}{7}\right)$ lb
(b) $-6\hat{\mathbf{i}} + 18\hat{\mathbf{j}} - 9\hat{\mathbf{k}}$ lb
2.29 Answer given in problem.

2.31 $\mathbf{F}_{PR} = 202$ 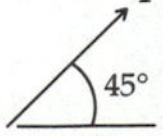

N; $\mathbf{F}_{QR} = 418$ 

N

2.33 $97.5\left(\dfrac{3}{13}\hat{\mathbf{i}} - \dfrac{4}{13}\hat{\mathbf{j}} + \dfrac{12}{13}\hat{\mathbf{k}}\right)$ lb;
$122\left(\dfrac{8}{17}\hat{\mathbf{i}} + \dfrac{15}{17}\hat{\mathbf{j}}\right)$ lb; $42.2\hat{\mathbf{j}}$ lb; no
2.35 $20\hat{\mathbf{i}} + 77.5\hat{\mathbf{j}}$ lb and $20\hat{\mathbf{i}} - 77.5\hat{\mathbf{j}}$ lb
2.37 $\pm(0.0683\hat{\mathbf{i}} + 0.820\hat{\mathbf{j}} - 0.569\hat{\mathbf{k}})$
2.39 (a) $|\mathbf{F}_1| = 12.5$ lb; (b) $|\mathbf{F}_1| = 1.08$ N
2.41 The angle is greater in (b) between the sections of the sling above the block and the vertical. The vertical components of force in these sections add to the weight, so the larger angle means the tension must be larger in (b). **2.43** 825 lb **2.45** 30°
2.47 4.57 lb each **2.49** 609 lb; 7.5° **2.51** 41.7 lb

2.53 491 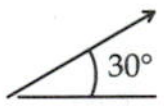

N; 850 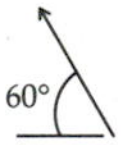

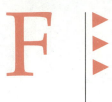

N
2.55 $N_1 = W \sin\theta_2 / \sin(\theta_2 - \theta_1)$;
$N_2 = W \sin\theta_1 / \sin(\theta_2 - \theta_1)$;
As $\theta_1 \rightarrow 0$, $N_1 \rightarrow W$ and $N_2 \rightarrow 0$ ✓. **2.57** 592 N
2.59 $F_{AD} = 294$ N; $F_{BD} = 208$ N; $F_{CD} = 339$ N
2.61 $F_1 = W$; $F_2 = 0.530\ W$; $F_3 = 0.385\ W$;
$F_4 = 0.335\ W$; $F_5 = 0.362\ W$; $F_6 \rightarrow \infty$ (They are horizontal, so cannot support the vertical load!)
2.63 $\Sigma\mathbf{F} = 10\hat{\mathbf{j}}$ N $\neq \mathbf{0}$ so *no*! Need $\mathbf{F}_4 = -10\hat{\mathbf{j}}$ N
2.65 In $\mathcal{R}_1$, 373 N; $\mathcal{R}_2$, 275 N; $\mathcal{R}_3$, 78.5 N **2.67** $W/8 \uparrow$
2.69 (a) Answer given in problem; (b) 30 lb
2.71 $W/8$ **2.73** $\dfrac{21}{64}W$
2.75 (a) A: 450 lb, B: 225 lb; (b) A: 900 lb, B: 450 lb; (c) A: 600 lb, B: 300 lb
2.77 Tension = 157 N; His force onto scaffold = 530 ↓ N **2.79** 1 in.
2.81 Outside sections: 1651 lb; center section: 1650 lb
2.83 392 N
2.85 From a FBD of the pulley, as θ goes up so does

the tension; $T = 125/\cos\theta$ N, which is 125 N @ $\theta = 0$ and 483 N @ $\theta = 75°$. When T = man's weight, $\theta = 81.0°$. **2.87** 230 N

2.89 From ground, 280 ↑ N;

from incline, 127 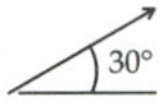

N;

from gravity, 196 ↓ N; from $\mathcal{A}$, 184 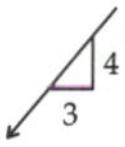

N

2.91 (a) $\dfrac{Qr}{2\sqrt{R^2 + 2Rr}}$ (b) $P + Q/2$ (c) $\dfrac{Q(r + R)}{2\sqrt{R^2 + 2Rr}}$

2.93 8 **2.95** 0.383 in. **2.97** $W/\sqrt{6}$; $W\sqrt{2}/6$

CHAPTER 3

3.1 $-60\hat{\mathbf{k}}$ N · cm **3.3** $36\hat{\mathbf{k}}$ lb-ft
3.5 $4700\hat{\mathbf{k}}$ lb-ft; 4.7 ft **3.7** 5210 ↻ lb-ft
3.9 (a) $-9\hat{\mathbf{i}} + 12\hat{\mathbf{j}} + 16\hat{\mathbf{k}}$ N
(b) $8.4\hat{\mathbf{i}} + 4.8\hat{\mathbf{j}} - 7.2\hat{\mathbf{k}}$ N · m
3.11 $400\hat{\mathbf{i}} - 800\hat{\mathbf{j}} + 300\hat{\mathbf{k}}$ lb-ft **3.13** 824 N
3.15 117° **3.17** $44.4\hat{\mathbf{e}}_{CD}$ N · m **3.19** $-800\hat{\mathbf{k}}$ lb-ft
3.21 $10\hat{\mathbf{k}}$ N · m **3.23** 1.63 ft
3.25 (a) $-71\hat{\mathbf{i}} - 105\hat{\mathbf{j}} - 71\hat{\mathbf{k}}$ N · m (b) $-105\hat{\mathbf{j}}$ N · m
3.27 $-560\hat{\mathbf{k}}$ lb-ft **3.29** $360\hat{\mathbf{i}} + 360\hat{\mathbf{j}}$ lb-ft
3.31 (a) $1080\hat{\mathbf{i}} + 1200\hat{\mathbf{k}}$ lb-in.
(b) $900(0.667\hat{\mathbf{j}} + 0.750\hat{\mathbf{k}})$ lb-in.
(c) $-810\hat{\mathbf{i}} - 1350\hat{\mathbf{j}} - 2850\hat{\mathbf{k}}$ lb-in.
(d) $923(0.543\hat{\mathbf{i}} + 0.466\hat{\mathbf{j}} - 0.699\hat{\mathbf{k}})$ lb-in.
3.33 $18\hat{\mathbf{k}}$ lb-ft
3.35 (a) -0.231, 0.923, 0.308 (b) $-9\hat{\mathbf{i}} + 36\hat{\mathbf{j}} + 12\hat{\mathbf{k}}$ lb
(c) $96\hat{\mathbf{i}} + 84\hat{\mathbf{j}} - 180\hat{\mathbf{k}}$ lb-ft (d) $84\hat{\mathbf{j}}$ lb-ft (e) **0**
3.37 (a) $-12\hat{\mathbf{i}} + 16\hat{\mathbf{j}} - 10\hat{\mathbf{k}}$ lb-ft (b) $-10\hat{\mathbf{k}}$ lb-ft
(c) $(-0.537, 0.716, -0.447)$; $-12\hat{\mathbf{i}} + 16\hat{\mathbf{j}} - 10\hat{\mathbf{k}}$ lb-ft
3.39 (a) $15\hat{\mathbf{i}} + 5\hat{\mathbf{j}} + 5\hat{\mathbf{k}}$ N · m
(b) $\frac{5}{3}\hat{\mathbf{i}} + \frac{5}{3}\hat{\mathbf{j}} - \frac{10}{3}\hat{\mathbf{k}}$ N · m **3.41** $-2400\hat{\mathbf{k}}$ lb-ft
3.43 (a) $355\hat{\mathbf{k}}$ lb-ft (b) same about any point!
3.45 (a) $-258\hat{\mathbf{i}} - 388\hat{\mathbf{j}} - 774\hat{\mathbf{k}}$ lb-ft
(b) same as (a) (c) $-388\hat{\mathbf{j}}$ lb-ft (d) $-388\hat{\mathbf{j}}$ lb-ft
3.47 $180\hat{\mathbf{i}} + 240\hat{\mathbf{k}}$ lb-in., same about all points.
3.49 $-74\hat{\mathbf{k}}$ lb-ft **3.51** $-52.0\hat{\mathbf{k}}$ lb-ft
3.53 (a) $-20\hat{\mathbf{i}} + 20\hat{\mathbf{k}}$ N · m (b) same as (a);
(c) $\frac{80}{17}\left(\frac{8\hat{\mathbf{i}} - 9\hat{\mathbf{j}} + 12\hat{\mathbf{k}}}{17}\right)$ N · m **3.55** $-176\hat{\mathbf{u}}_t$ N · m
3.57 (a) $3\hat{\mathbf{i}} + 9\hat{\mathbf{j}}$ lb (b) $-36\hat{\mathbf{k}}$ lb-ft (c) $18\hat{\mathbf{k}}$ lb-ft
(d) $18\hat{\mathbf{k}}$ lb-ft
3.59 (a) $\hat{\mathbf{i}}$ N; $20\hat{\mathbf{k}}$ N · m
(b) Both answers to (a) must be zero for equilibrium
(c) $a\hat{\mathbf{i}} + \frac{20}{3}\hat{\mathbf{j}}$ N and $(-a - 1)\hat{\mathbf{i}} - \frac{20}{3}\hat{\mathbf{j}}$ N,
respectively, where "a" is arbitrary.

3.61 $1{,}730\hat{\mathbf{i}}$ lb; $1{,}000\hat{\mathbf{j}}$ lb; $-10{,}700\hat{\mathbf{k}}$ lb-ft
3.63 Second force is $-180\hat{\mathbf{j}}$ N, along $x = -70$ cm.
3.65 $5\hat{\mathbf{i}} + 8\hat{\mathbf{j}}$ N; $-0.9\hat{\mathbf{k}}$ N · m **3.67** (c) and (d)
3.69 Forces with largest x are 1540 lb;
others are 639 lb. **3.71** $40\hat{\mathbf{i}} + 40\hat{\mathbf{j}} + 40\hat{\mathbf{k}}$ N, **0** N · m
3.73 $P\hat{\mathbf{j}} + P\hat{\mathbf{k}}$; $-Pa(2\hat{\mathbf{i}} + \hat{\mathbf{k}})$
3.75 $-500\hat{\mathbf{k}}$ N; $-650\hat{\mathbf{i}} + 462\hat{\mathbf{j}} + 28\hat{\mathbf{k}}$ N · m
3.77 $2130\hat{\mathbf{i}} - 66\hat{\mathbf{j}}$ lb
3.79 $\mathbf{F}_r = 0.866\hat{\mathbf{i}} - 1.50\hat{\mathbf{j}}$ lb; $\mathbf{M}_{rC} = \mathbf{0}$
3.81 130 N · m; $\theta = 67.4°$
3.83 $\mathbf{F} = 40\hat{\mathbf{i}} - 20\hat{\mathbf{j}} - 30\hat{\mathbf{k}}$ lb;
$\mathbf{M}_{rO} = 40\hat{\mathbf{i}} + 180\hat{\mathbf{j}} - 120\hat{\mathbf{k}}$ lb-ft
3.85 $\mathbf{F}_r = -9\hat{\mathbf{i}} + 12\hat{\mathbf{j}} + 16\hat{\mathbf{k}}$ lb;
$\mathbf{M}_{rA} = -36\hat{\mathbf{i}} - 18\hat{\mathbf{j}} - 90\hat{\mathbf{k}}$ lb-ft
3.87 Resultant at C: $3.4\hat{\mathbf{i}} + 67.8\hat{\mathbf{j}} + 31.2\hat{\mathbf{k}}$ lb and
$-271\hat{\mathbf{i}} + 80\hat{\mathbf{j}} + 45\hat{\mathbf{k}}$ lb-ft
3.89 (a) $24\hat{\mathbf{i}} - 24\hat{\mathbf{j}} - 32\hat{\mathbf{k}}$ lb-ft
(b) $36\hat{\mathbf{i}} - 36\hat{\mathbf{j}} + 48\hat{\mathbf{k}}$ lb-ft
(c) $6\hat{\mathbf{i}} - 6\hat{\mathbf{j}} + 8\hat{\mathbf{k}}$ lb-ft (d) **0** (e) $24\hat{\mathbf{k}}$ lb-ft
(f) $\Sigma\mathbf{M}_0$, less its 2 ⊥ components *in* the plane *EOG*, leaves its component ⊥ *EOG*.

3.91 $591\hat{\mathbf{i}} + 82.1\hat{\mathbf{j}}$ N **3.93** 33.7 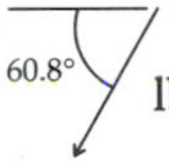

lb

3.95 $-823\hat{\mathbf{i}} + 1215\hat{\mathbf{j}}$ lb

3.97 145.4 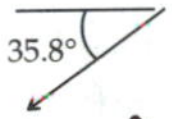

kips; 9.24 ft above B.
3.99 $850\hat{\mathbf{i}} - 1110\hat{\mathbf{j}}$ lb; $y + 1.35x = 1.11$; at $x = 0.82$ ft
3.101 $\mathbf{F}_r = -2\hat{\mathbf{i}}$ lb, along the line $y = 9$ in.
(doesn't cross x axis!)
3.103 $\mathbf{F}_r = 3\hat{\mathbf{i}} + 3\hat{\mathbf{j}}$ lb, crossing at $x = 5$ in. Equation is $y = x - 5$ **3.105** $-90\hat{\mathbf{j}}$ lb, 10.3 ft to the right of O.
3.107 $-220\hat{\mathbf{k}}$ N, passing through (x, y) = (3.18, 3.55) m **3.109** $18\gamma\hat{\mathbf{i}}$ lb; $y = -4.02$ ft
3.111 Yes; $|\mathbf{M}_{rP}|/|\mathbf{F}_r|$
3.113 $30\hat{\mathbf{i}}$ lb; 25 in. above 80-lb force
3.115 (a) $-2200\hat{\mathbf{i}} + 1200\hat{\mathbf{j}}$ lb,
with intercepts $x = -2.08$ ft, $y = -1.14$ ft
(b) $(x, y) = (4.33, -3.50)$ ft
3.117 (a) $10\hat{\mathbf{k}}$ lb, $-80\hat{\mathbf{i}} + 105\hat{\mathbf{j}}$ lb-ft
(b) $10\hat{\mathbf{k}}$ lb at $(-10.5, -8)$ ft
3.119 (a) $3P\hat{\mathbf{i}} - P\hat{\mathbf{j}} + 2P\hat{\mathbf{k}}$, $Pb\hat{\mathbf{i}} - 2Pc\hat{\mathbf{j}} + 3Pa\hat{\mathbf{k}}$
(b) $6a + 3b + 2c = 0$ (c) No, $\Sigma\mathbf{F} \neq \mathbf{0}$
3.121 (a) $100\hat{\mathbf{i}} - 200\hat{\mathbf{j}} - 300\hat{\mathbf{k}}$ N;
$(-200\ell - 10)\hat{\mathbf{i}} + (200\ell - 20)\hat{\mathbf{j}} + (-300\ell + 30)\hat{\mathbf{k}}$ N · m
(b) $\ell = 0.200$ m
3.123 (a) $8W\hat{\mathbf{k}}$ at $(x, y) = (0.884R, 1.06R)$
(b) $3.16W\hat{\mathbf{k}}$ at $(1.15R, 1.60R)$ producing $-4.84W\hat{\mathbf{k}}$ at middle of plate; or $-3.16W\hat{\mathbf{k}}$ at $(0.260R, -0.187R)$ producing $-11.2W\hat{\mathbf{k}}$ at middle of plate.
3.125 $6\hat{\mathbf{i}} + 8\hat{\mathbf{j}} - 8\hat{\mathbf{k}}$ lb, piercing xy plane at (3, 1.4) ft

3.127 Add $1\hat{\mathbf{i}}$ lb at A; original's screwdriver: $-1\hat{\mathbf{i}}$ lb, 1ft below A
3.129 $-6\hat{\mathbf{k}}$ N along $x = 0.667$ m, $y = 1.5$ m
3.131 $-1.73\hat{\mathbf{i}} - 3\hat{\mathbf{j}}$ lb through $x = -2.5$ ft, $y = 0$
3.133 $3\hat{\mathbf{k}}$ lb through the origin
3.135 $\mathbf{F}_r = 3P\hat{\mathbf{i}} - P\hat{\mathbf{j}} + 2P\hat{\mathbf{k}}$ along $28y + 14z = 9a$, $28x - 42z = 73a$; $\mathbf{C} = Pa\left(\frac{75}{14}\hat{\mathbf{i}} - \frac{25}{14}\hat{\mathbf{j}} + \frac{50}{14}\hat{\mathbf{k}}\right)$
3.137 $-50\hat{\mathbf{i}} - 120\hat{\mathbf{j}}$ N intersecting a horizontal line through A 4.33 m to the right of A
3.139 $\mathbf{F}_r = 9\hat{\mathbf{i}} + 5\hat{\mathbf{j}}$ lb along $z = 9$ ft, $5x - 9y = 0$; $\mathbf{C} = 45\hat{\mathbf{i}} + 25\hat{\mathbf{j}}$ lb-ft
3.141 (a) Replace system by $\mathbf{F}_r$ and $\mathbf{M}_{rP}$ at P; (b) Replace $\mathbf{M}_{rP}$ by its components; (c) Replace $\mathbf{M}_{rP_\perp}$ by $\mathbf{F}_r$ and $-\mathbf{F}_r$; (d) Cancel $\mathbf{F}_r$ and $-\mathbf{F}_r$ at P; slide $\mathbf{M}_{rP_\parallel}$ over to Q to form, with $\mathbf{F}_r$, the screwdriver there. **3.143** $\dfrac{q_0L^2}{2\pi}\hat{\mathbf{k}}$
3.145 $F_r = \frac{1}{2}(6y)6 = 18y$ lb at $z_r = 2$ ft. In 3.109, lines of action of concentrated loads are not properly placed for equipollence. **3.147** 33,900 N at $x_G = 4$ m
3.149 25 ↑ N at $x_r = 0.8$ m
3.151 $200\hat{\mathbf{j}}$ N and $200\hat{\mathbf{k}}$ N · m
Let x be measured from the left end of the beam in the next three solutions:
3.153 1500 N down at $x = 1.33$ m, 3000 N down at $x = 3$ m, and 1500 N down at $x = 4.67$ m
3.155 Letting x be measured from left end of beam, 300 lb down at $x = 1$ ft, 700 lb down at $x = 3.5$ ft, and 600 lb up at $x = 8.33$ ft.
3.157 95.0 N down at $x = 1.1$ m and 38.5 N up at $x = 2.9$ m
3.159 $450\hat{\mathbf{i}} - 290\hat{\mathbf{j}}$ lb; $y = -0.644x + 6.69$

CHAPTER 4

4.1

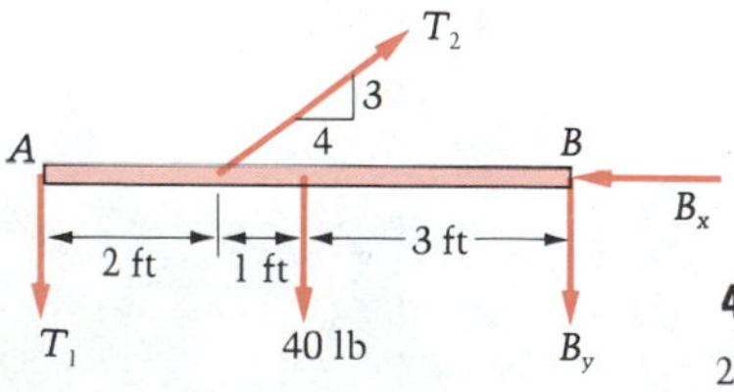

4.3

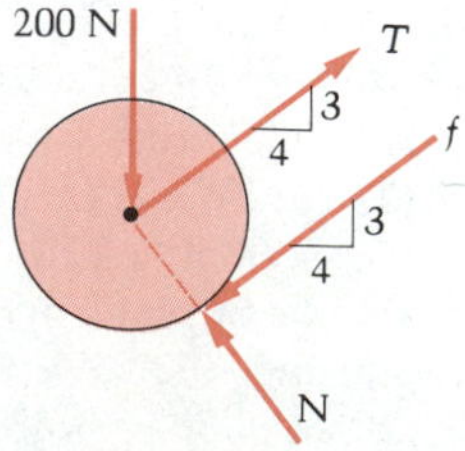

4.5

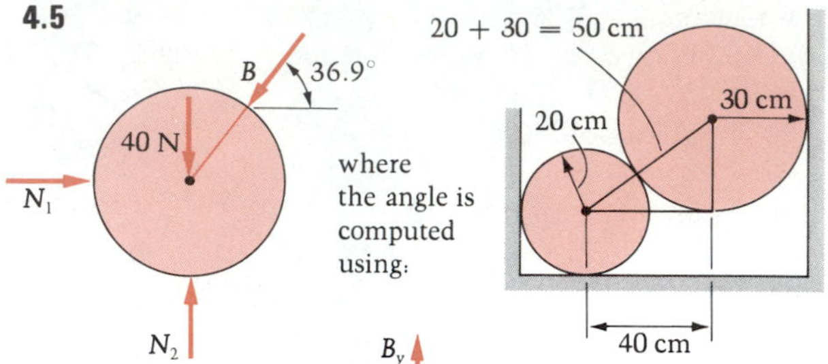

4.7

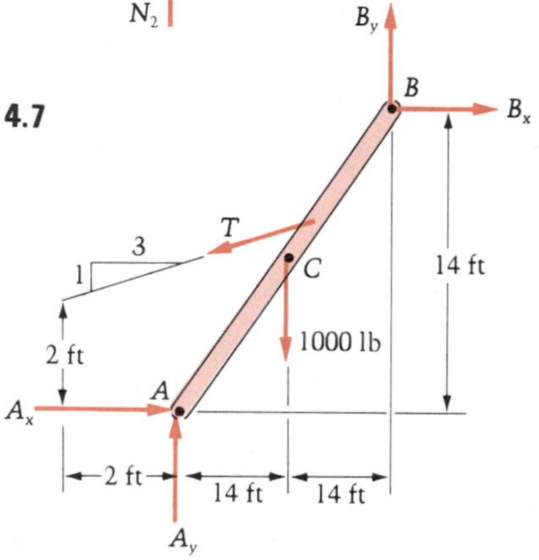

4.9

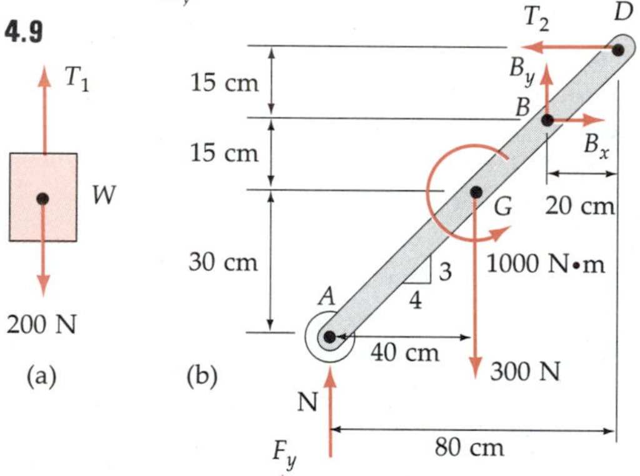

4.11

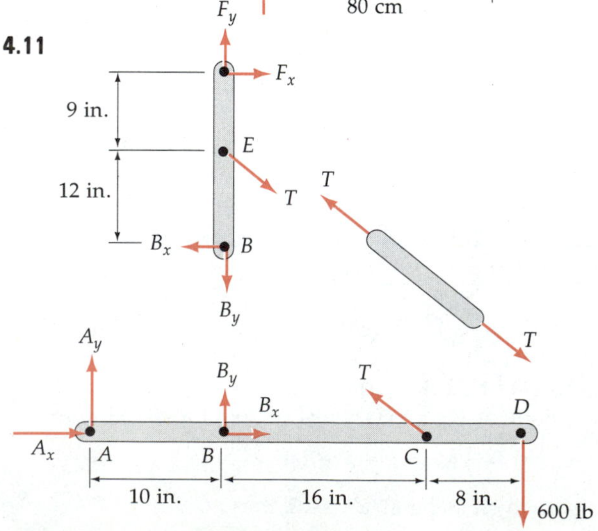

4.13 190 ↓ lb; 31.6 → lb
4.15 Answer given in problem

4.17 0.289 mg **4.19** 792 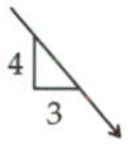

N

4.21 2.45 ft; 158 lb and 186 lb, respectively, normal to the planes
4.23 $(4 \tan \theta)/(3\pi)$; $T = 0$ at $\theta = 0$, $T \to \infty$ as $\theta \to \pi/2$
4.25 B is not on a line parallel to y drawn through A
4.27 At A, 50 ← N; at B, 100 ↑ N
4.29 222 ↑ lb; 0.919 ft **4.31** 290 N; 220 ↑ N
4.33 $707\hat{\mathbf{i}} + 1040\hat{\mathbf{j}}$, 1670 ↺ N · m
4.35 On top face, 1263 lb ↓, one-third of distance from left end; on bottom face, 1233 lb ↑, one-third of distance from right end; smaller distance between resultant forces. **4.37** $WL/(2\ell)$; the other tension is <0 if $\ell < L/2$, and "you can't push a rope."

4.39 (a) 229 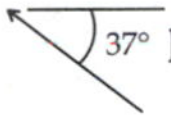

lb; 242 lb; 120 ⇄ lb-in.

4.41 At A, 190 ↑ N; at B, 303 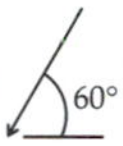

N; at C, 85.6 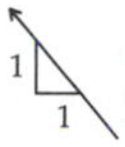

N

4.43 At A, 250 ↑ N; at B, 50 ↑ N
4.45 $wL/4$ ↑ at each end
4.47 At A, 800 ↑ lb; at B, 1000 ↑ lb
4.49 At B, 3500 ↑ lb; at D, 500 ↑ lb
4.51 At A, $-216\hat{\mathbf{i}} - 178\hat{\mathbf{j}}$ lb; at F, 178 ↑ lb
4.53 $150\hat{\mathbf{i}} + 1500\hat{\mathbf{j}}$ lb and 9000 ↻ lb-ft
4.55 7110 lb; $4570\hat{\mathbf{i}} + 7450\hat{\mathbf{j}}$ lb
4.57 86.2 ↻ N · m **4.59** 6.58 ft
4.61 4790 lb (compression)
4.63 89,600 lb (tension); $-17{,}800\hat{\mathbf{i}} + 112{,}000\hat{\mathbf{j}}$ lb
4.65 At C, 220 ↑ N; at D, 1060 ↑ N **4.67** 0.897 W
4.69 12.1 → N **4.71** 1710 → N
4.73 Tension is $mg/2$
4.75 $40\hat{\mathbf{i}} + 80\hat{\mathbf{j}}$ lb **4.77** $-1.80H\hat{\mathbf{i}} - 2.40H\hat{\mathbf{j}}$
4.79 3.03 ft and 48.5 tons; $54.2 \leqq W_C \leqq 100$ tons
4.81 $13W/11$
4.83 $\theta \geqq 7.18°$ (considered only θ's between 0 and 90°)
4.85 (a) It blows over; the moment imbalance about B is 3450 lb-ft. (b) 0.51 ft deep
4.87 $300\hat{\mathbf{i}} + 4800\hat{\mathbf{j}}$ lb; 6000 ↻ lb-ft **4.89** 24.8 cm
4.91 89.6 lb **4.93** 5260 lb compression
4.95 At A, $10{,}900\hat{\mathbf{i}} - 14{,}500\hat{\mathbf{j}}$ N; at B,

18,200 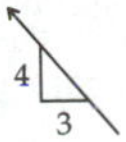

N

4.97 $88\hat{\mathbf{i}} - 108\hat{\mathbf{j}}$ lb

4.99 On $\mathscr{A}$, $-112\hat{\mathbf{i}} + 1360\hat{\mathbf{j}}$ lb; on $\mathscr{B}$, 328 

lb

4.101 167 ↑ N **4.103** 0.309ℓ **4.105** 23.2°
4.107 3250 N; $750\hat{\mathbf{i}} + 3000\hat{\mathbf{j}} + 1000\hat{\mathbf{k}}$ N; $1800\hat{\mathbf{k}}$ N · m **4.109** $117\hat{\mathbf{i}} + 2070\hat{\mathbf{j}}$ N; 884 N
4.111 BC: 30800 N, BE: 50800 N
4.113 $149\hat{\mathbf{i}}$ lb **4.115** 333 ↑ N; $667\hat{\mathbf{k}}$ N, $1000\hat{\mathbf{i}}$ N · m
4.117 $0.214\hat{\mathbf{i}}$ mg; all forces pass through the bar, so that moments about its centerline add to zero automatically.
4.119 Answer given in problem.
4.121 $F_{BF} = 3840$ N; $F_{CE} = 7510$ N
4.123 $202\hat{\mathbf{i}} + 122\hat{\mathbf{j}} + 82\hat{\mathbf{k}}$ lb; $20\hat{\mathbf{k}}$ lb-ft
4.125 371 N; $C_{B_y} = 24.9$ N · m; $C_{B_z} = -31.2$ N · m
4.127 At A, $-12{,}000\hat{\mathbf{j}} + 6{,}000\hat{\mathbf{k}}$ N; at B, $-12{,}000\hat{\mathbf{i}} + 12{,}000\hat{\mathbf{j}}$ N; at C, $16{,}000\hat{\mathbf{i}} - 4{,}000\hat{\mathbf{k}}$ N
4.129 144 lb compression each
4.131 42 lb **4.133** $-10\hat{\mathbf{i}} - 3\hat{\mathbf{j}} - 100\hat{\mathbf{k}}$ lb; $-1060\hat{\mathbf{i}} + 3200\hat{\mathbf{j}} + 10\hat{\mathbf{k}}$ lb-ft
4.135 250 N tension; 0.0500 m
4.137 $\cos^{-1}\left(\dfrac{P\ell}{kh^2}\right)$; cosines can't be greater than 1
4.139 29.0°
4.141 Solution outlined in problem.
4.143 (a) If any one vanished, would have a 3-force, coplanar system with the forces neither concurrent nor parallel; If two vanished, the remaining one isn't "equal and opposite" to P; If all three vanished, $\Sigma F_y = -P \neq 0$. (b) 35.3°
4.145 From left to right: (a) AB (b) AB; BC; EF (c) BD (d) BD **4.147** $-355\hat{\mathbf{i}} + 119\hat{\mathbf{j}}$ lb; $733\hat{\mathbf{j}}$ lb

4.149 1110 

lb; 27.9°

4.151 $25/\sin\alpha$ lb; $25 \cot\alpha$ lb (↓ on top; ↑ on bottom); zero **4.153** 16.7 ← lb **4.155** 39.2 ↺ N · m
4.157 49.7 lb **4.159** 14.1 ← lb
4.161 (a) $2W(D - 2R)/D$ (b) Answer given in problem.
4.163 24,000 lb compression
4.165 (a) At A, $39.0\hat{\mathbf{i}} - 353\hat{\mathbf{j}}$ N; at D, $-39.0\hat{\mathbf{i}} + 549\hat{\mathbf{j}}$ N; (b) at B, $157\hat{\mathbf{i}} + 549\hat{\mathbf{j}}$ N; at C, $-196\hat{\mathbf{i}} - 196\hat{\mathbf{j}}$ N

4.167 At F, $436\hat{\mathbf{i}} + 390\hat{\mathbf{j}}$ N; 1170 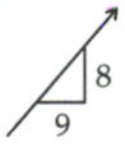

N onto BDE at D

4.169 (a) $4000 \sin(\theta/2)$ lb (b) 2000 lb directed from B toward C **4.171** 1.09 W
4.173 1860 N

4.175 (a) At A, $-120\hat{\mathbf{i}} - 147\hat{\mathbf{j}}$ lb; at E, $427\hat{\mathbf{j}}$ lb (b) $120\hat{\mathbf{i}} + 333\hat{\mathbf{j}}$ lb
4.177 (a) At A, $25\hat{\mathbf{i}} - 25\hat{\mathbf{j}}$ lb; at B, $-25\hat{\mathbf{i}} + 225\hat{\mathbf{j}}$ lb (b) on EB at D, 350 ↓ lb **4.179** $-958\hat{\mathbf{i}} + 708\hat{\mathbf{j}}$ lb
4.181 120 lb of compression on pecan; 100 lb of tension in link AB
4.183 261 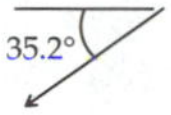

N **4.185** 189 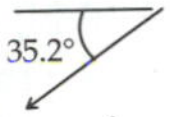

N
4.187 $16\,W/15$ **4.189** $-1880\hat{\mathbf{i}} - 1000\hat{\mathbf{j}}$ lb
4.191 (a) $-533\hat{\mathbf{i}} - 200\hat{\mathbf{j}}$ lb (b) $-533\hat{\mathbf{i}} + 200\hat{\mathbf{j}}$ lb
4.193 (a) $-960\hat{\mathbf{i}} - 483\hat{\mathbf{j}}$ N (b) $620\hat{\mathbf{i}}$ N (c) $-860\hat{\mathbf{i}} + 483\hat{\mathbf{j}}$ N
4.195 (a) $-1020\hat{\mathbf{i}}$ N (b) $870\hat{\mathbf{i}} + 1200\hat{\mathbf{j}}$ N (c) $150\hat{\mathbf{i}}$ N **4.197** $2\,mg/\pi$
4.199 27.2° **4.201** 15.0 P
4.203 $2(b+c)(c+d)/(ad)$ **4.205** 9140 ← lb
4.207 $\alpha = 39.8°$, $\beta = 68.2°$
4.209 Trailer, 506 lb each; truck rear, 891 lb each; truck front, 753 lb each; 289 lb downward on ball
4.211 (a) 52 ↑ lb (b) 14.9 lb Ⓒ (needs a sleeve over pin to prevent buckling) (c) 96 → lb
4.213 2210 lb Ⓣ; $1040\hat{\mathbf{i}} + 8850\hat{\mathbf{j}}$ lb
4.215 451 N Ⓣ; 284 → N
4.217 5250 lb Ⓒ; $-1360\hat{\mathbf{i}} - 3980\hat{\mathbf{j}}$ lb **4.219** 177 lb
4.221 (a) 150 ↑ lb (b) 100 ↑ lb
4.223 At E, $\dfrac{W}{e}\left[d + \dfrac{b(d-e)}{b-2e}\right] + \dfrac{w}{e}\left[\dfrac{b-L}{2} + \dfrac{b(b-L-2e)}{2(b-2e)}\right]$;
At C, $\dfrac{2W(d-e)}{b-2e} + w\left(\dfrac{b-L-2e}{b-2e}\right)$
4.225 1410 lb compression; 382 → lb by AB; $-382\hat{\mathbf{i}} - 710\hat{\mathbf{j}}$ lb by CD **4.227** 480 → N
4.229 $F = W$ for any x **4.231** 828 ← N
4.233 With $\hat{\mathbf{i}}$ toward the wall: $\mathbf{T}_3 = 200\hat{\mathbf{i}}$, $\mathbf{T}_2 = -100\hat{\mathbf{i}}$, $\mathbf{T}_1 = 400\hat{\mathbf{i}}$ lb-ft, each on a FBD including the part of the shaft farther from the wall. **4.235** $-27.3\hat{\mathbf{i}} + 266\hat{\mathbf{j}}$ lb
4.237 63,000 lb **4.239** At A, $50\hat{\mathbf{j}}$ lb; at B, $-50\hat{\mathbf{j}}$ lb
4.241 Answer given in problem. **4.243** $25\,L/24$
4.245 70.8 kips Ⓒ **4.247** 1500 lb Ⓣ
4.249 Solution outlined in problem.
4.251 ABC: $-167\hat{\mathbf{i}} + 320\hat{\mathbf{j}}$ lb; EDC: $417\hat{\mathbf{i}} - 320\hat{\mathbf{j}}$ lb; The applied load is the difference in the x-components acting on the pin. **4.253** 57.6 kN Ⓣ
4.255 245 lb Ⓣ each
4.257 At E: 3600 $\hat{\mathbf{K}}$ lb with $-3000\,\hat{\mathbf{J}}$ lb-ft; at O: 3600 $\hat{\mathbf{K}}$ lb with the couple 6000 $\hat{\mathbf{J}}$ lb
4.259 Answer given in problem.
4.261 50 **4.263** Answers given in problem.

CHAPTER 5

5.1 BC: 1100 lb Ⓒ; CD: 880 lb Ⓣ; BD: 500 lb Ⓣ; AD: 880 lb Ⓣ; AB: 200 lb Ⓣ
5.3 AB: 4.24 lb Ⓒ; AH and HG: 3 lb Ⓣ; BH, EF, and FG: 2 lb Ⓣ; BC and CD: 4 lb Ⓒ; CG and DF: 0; BG: 1.41 lb Ⓣ; DG: 2.83 lb Ⓣ; DE: 2.83 lb Ⓒ
5.5 CD: 800 N Ⓒ; BD: 800 N Ⓣ; AE: 800 N Ⓒ; BE: 2000 N Ⓒ; AB: 2260 N Ⓣ; BC: 1130 N Ⓣ; DE: 1130 N Ⓒ
5.7 AG: 150 lb Ⓣ; AF: 175 lb Ⓒ; AB: 90 lb Ⓣ; EF: 90 lb Ⓒ; BF: 350 lb Ⓒ; CD: 233 lb Ⓒ; DE: 120 lb Ⓣ; BE: 408 lb Ⓣ; CE: 1100 lb Ⓒ; BC: 120 lb Ⓒ; $FG = 0$
5.9 CD: 167 N Ⓒ; AC: 208 N Ⓣ; others: 0
5.11 AB: 10 kN Ⓣ; BC: 0; FE: 15 kN Ⓣ; FD: 0; GF: 15 kN Ⓣ; GD: 11.2 kN Ⓒ; CD: 5 kN Ⓒ; AC: 11.2 kN Ⓒ; AG: 10 kN Ⓣ; GC: 10 kN Ⓣ
5.13 AB: 667 lb Ⓒ; BE: 1200 lb Ⓒ; BC: 2670 lb Ⓒ; CD: 2670 lb Ⓒ; BQ: 2670 lb Ⓣ; GQ: 3330 lb Ⓣ; QC: 0; QD: 0
5.15 OH: 4 k Ⓣ; AB: 3 k Ⓒ; BH: 3.89 k Ⓣ
5.17 AB: 0.415W Ⓒ; AG: 0.915W Ⓒ; BG: 0.215W Ⓣ; BC: 0.414W Ⓒ; CG: 0.704W Ⓒ; FG: 0.289W Ⓒ; CD: 1.05W Ⓒ; CF: 0.395W Ⓣ; DF: 0.354W Ⓒ; DE: 0.317WⒸ; EF: 0.183W Ⓣ
5.19 AB: 67 lb Ⓣ; AE: 83 lb Ⓒ; BC: 400 lb Ⓣ; BE: 50 lb Ⓣ; BF: 417 lb Ⓒ; CD: 400 lb Ⓣ; CF: 400 lb Ⓒ; DF: 500 lb Ⓒ; EF: 67 lb Ⓒ
5.21 AG: 3.33 kN Ⓣ; AF: 18.3 kN Ⓣ; AB: 9.0 kN Ⓣ; CD: 3 kN Ⓣ
5.23 CE and CF: 0; DE, EF, and FG: 19.2 k Ⓣ; BF: 8 k Ⓒ; BC and CD: 20.8 k Ⓒ; AB: 27.7 k Ⓒ; AG: 10.7 k Ⓣ; BG: 8.33 k Ⓣ
5.25 $F_1 = F_2 = 0$; $F_3 = 3.54$ k Ⓣ; $F_4 = 20$ k Ⓣ; $F_5 = 63.5$ k Ⓒ
5.27 BG: 0; AG: 50 kN Ⓒ; CG: 39.3 kN Ⓣ
5.29 (a) 6 (b) 354 lb Ⓣ (c) A_y is same either way, but A_x is zero for the roller. F_{AB} is same either way, so F_{AJ} must change with a roller at A.
5.31 Start with a triangle in the middle and work outward.
5.33 $m = m_1 + m_2 - 1$, $p = p_1 + p_2 - 2$, $m_1 = 2p_1 - 3$, and $m_2 = 2p_2 - 3$ yields $m = 2p - 3$
5.35 2.77 kN Ⓒ **5.37** 10.8 kN Ⓒ
5.39 CF: 1.12 kN Ⓒ; BG: 1.41 kN Ⓒ
5.41 BH: 1.58 kN Ⓣ; DF: 1.42 kN Ⓣ; CF: 1.78 kN Ⓒ; CD: 0.935 kN Ⓣ **5.43** JM: 0; DG: 8.49 kips Ⓣ
5.45 (a) BC: 1000 lb Ⓒ (b) CE: 3000 lb Ⓣ
5.47 AF: 0; CF: 3.75 kN Ⓣ
5.49 IE: 300 lb Ⓒ; JC: 486 lb Ⓒ; KC: 0; DE: 901 lb Ⓣ
5.51 BF: 1600 lb Ⓣ; BC: 2260 lb Ⓒ

5.53 *GH*: 5830 lb Ⓒ; *CH*: 4170 lb Ⓒ; *BC*: 8000 lb Ⓣ
5.55 *BH*: 4.19 kips Ⓣ; *FE*, *DE*, *AJ* are zero-force members. **5.57** 1.40*W* Ⓒ
5.59 *DC*: 3320 lb Ⓒ; *DG*: 212 lb Ⓒ; *DF*: 0
5.61 *CD*: 29.3 kN Ⓒ; *DH*: 4 kN Ⓒ; *CH*: 21.6 kN Ⓣ
5.63 *DC*: 893 N Ⓣ; *DE*: 631 N Ⓒ
5.65 *CD*: 20.8 kips Ⓒ; *KJ*: 3.25 kips Ⓒ; *LJ*: 22.9 kips Ⓣ **5.67** *CD*: 473 lb Ⓒ; *DJ*: 0
5.69 *EL*: 16.3 kN Ⓣ; *GH*: 8.79 kN Ⓒ
5.71 (a) *BG*, *CG*, *CF* are zero-force members
(b) *FG*: 9600 N Ⓣ; *FD*: 389 N Ⓣ; *FE*: 11300 N Ⓣ; *CD*: 19400 N Ⓒ
5.73 *AC*: 12.8 kN Ⓒ; *OC*: 38.8 kN Ⓣ; *BC*: 29.2 kN Ⓒ
5.75 172 lb Ⓒ **5.77** 52 lb Ⓒ **5.79** (a) at *A*: $33.3\hat{\mathbf{j}}$ lb; at *B*: $-75\hat{\mathbf{i}}$ lb; at *C*: $-25\hat{\mathbf{i}} - 33.3\hat{\mathbf{j}} + 100\hat{\mathbf{k}}$ lb
(b) *AD*: 33.3 lb Ⓒ; *CD*: 108 lb Ⓣ
(c) *BC*: 0; *AB*: 0; *BD*: 75 lb Ⓣ (d) 0
5.81 *AB*: 0.410*P* Ⓣ; *BC*: 0.273*P* Ⓒ
5.83 347 N Ⓒ each
5.85 $\mathscr{B}_1$: 2080 N Ⓒ; $\mathscr{B}_2$: 2080 N Ⓣ; $\mathscr{B}_3$: 0
5.87 $\mathscr{B}_1$: 307 N Ⓒ; $\mathscr{B}_2$: 2100 N Ⓣ; $\mathscr{B}_3$: 1190 N Ⓒ
5.89 (a) Start with tetrahedron *CBDF*; (b) 6750 lb Ⓒ
5.91 In psi: *AB*: 100, tensile; *BD*: 250, tensile; *BC*: 550, compressive; *DC* and *AD*: 440, tensile
5.93 In MPa: *AB*: 0.753 Ⓣ; *BE*: 0.333 Ⓒ; *BC*: 0.565 Ⓣ; *BD*: 0.267 Ⓣ; *AE*: 0.160 Ⓒ; *ED*: 0.283 Ⓒ; *CD*: 0.400 Ⓒ where Ⓣ means tensile stress and Ⓒ means compressive.
5.95 $\sigma_{BG} = 7.05$ MPa; $\sigma_{CF} = 4.48$ MPa, both compressive
5.97 *LC*: 398 kPa, *CR*: 99.5 kPa, both tensile
5.99 *DE* (stress is 4.92 MPa, compressive)
5.101 0.812 MPa, compressive
5.103 *AB*: 6000 psi, *BC*: 2990 psi, and *BD*: 11,600 psi, all compressive **5.105** 53.9 kPa
5.107 0.0305 in. elongation
5.109 0.00742 in. elongation
5.111 $\delta_{BG} = 0.0000475\ell$ m, $\delta_{CF} = 0.0000239\ell$ m, both shortened **5.113** 0.0112 in. **5.115** 308 MPa
5.117 3*L* **5.119** $0.111 \times 10^{-4}\hat{\mathbf{i}} - 12.5 \times 10^{-4}\hat{\mathbf{j}}$ in.
5.121 Answer given in problem.
5.123 $(0.490\hat{\mathbf{i}} - 0.128\hat{\mathbf{j}}) \times 10^{-3}$ m
5.125 $2.41 \times 10^{-4}L$ in. (with *L* in inches); $F_A = 24{,}100$ lb; $F_s = 43{,}300$ lb; $F_C = 32{,}700$ lb **5.127** 105.8°C
5.129 $F_A = 7600$ lb Ⓒ; $F_S = 79{,}900$ lb Ⓣ; $F_C = 27{,}700$ lb Ⓣ
5.131 $-10\hat{\mathbf{i}} + 5\hat{\mathbf{j}}$ kN with the couple $-7.5\hat{\mathbf{k}}$ kN · m
5.133

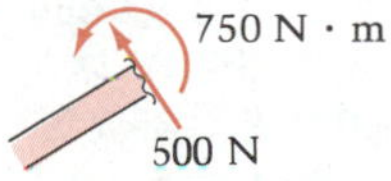

5.135 On *A-A*, 20 ↓ lb with 20 ↻ lb-in.; on *B-B*, 20 ↑ lb with 40 ↺ lb-in.; on *C-C*, 20 ↓ lb with 40 ↻ lb-in.
5.137 At *AA*,

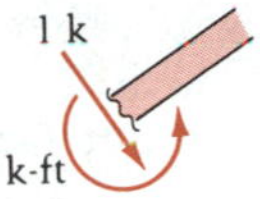

; at *BB*,

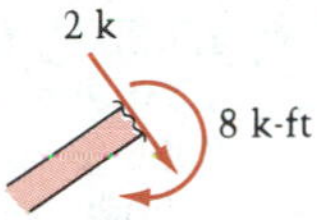

at *CC*,

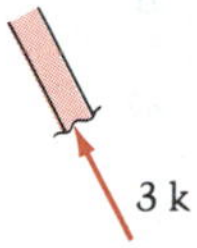

5.139 75 ↓ N with 45 ↻ N · m (or $V = 75$ N; $M = 45$ N · m)
5.141 16.7 ↓ N with 8 ↺ N · m (or $V = 16.7$ N; $M = -8$ N · m)
5.143 250 ↑ N with 125 ↺ N · m (or $V = -250$ N; $M = -125$ N · m)
5.145 40 ↑ lb and 260 ↺ lb-ft
5.147 56.6 ← kips with 7.52 ↑ kips with 28.2 ↻ kip-ft
5.149

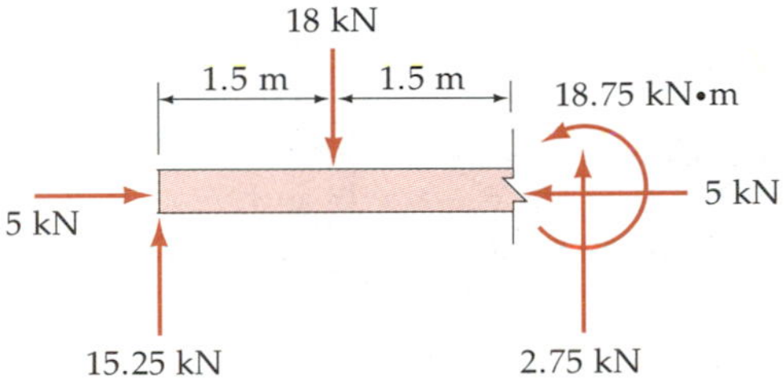

5.151

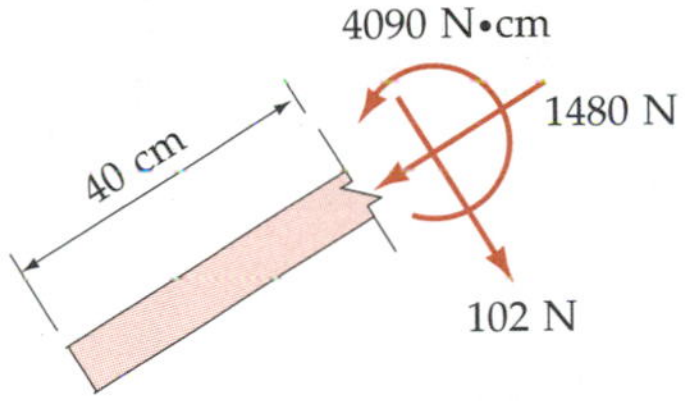

5.153

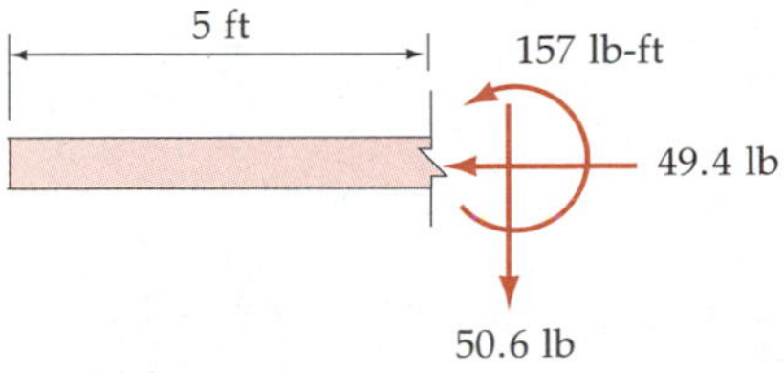

5.155 (a) $x = 8$ m
(b) 36 kN for $0 \le x \le 4$ and $12 \le x \le 16$ m
(c) $x = 0$, 16 m (d) 216 kN · m at $x = 8$ ft

5.157 Pab/L, at $x = a$
5.159 900 N at center; 1800 N · m at the wall
5.161 (a) $150\hat{\imath} + 3000\hat{\jmath}$ lb and $18000\hat{k}$ lb-ft
(b) $N = 150$ lb Ⓒ for $0 \le x < 6$ ft; $N = 150$ lb Ⓣ for $6 < x \le 12$ ft; $V = 250x - 3000$ lb, $0 \le x \le 12$ ft; $M = -18000 + 3000x - 125x^2$ lb-ft, $0 \le x \le 12$ ft
5.163 (a) 27000 ↑ N with 48000 ⟲ N · m
(b) $N = 0$; $V = -27000 + 9000x - 563x^2$; $M = -48000 + 27000x - 4500x^2 + 188x^3$
5.165 $V = \dfrac{2P}{3}$ for $\dfrac{L}{3} < x < \dfrac{2L}{3}$; $M = \dfrac{PL}{9}$ at $x = \dfrac{L}{3}$ and $\dfrac{-PL}{9}$ at $x = \dfrac{2}{3}L$
5.167 $V = \dfrac{-3P}{4}$ for $0 < x < \dfrac{L}{2}$; $M = \dfrac{3PL}{8}$ at $x = \dfrac{L}{2}$
5.169 $V = P$ for $0 \le x \le L$; $M = \dfrac{PL}{2}$ at $x = \left(\dfrac{L}{2}\right)^+$; $M = -\dfrac{PL}{2}$ at $x = \left(\dfrac{L}{2}\right)^-$
5.171 $V = \dfrac{wx^2}{L} - \dfrac{wL}{4}$ and $M = \dfrac{wLx}{4} - \dfrac{wx^3}{3L}$, for $0 \leqq x \leqq \dfrac{L}{2}$. For $\dfrac{L}{2} \leqq x \leqq L$, obtain V and M by symmetry.
5.173 For $0 \leqq x \leqq 3$ ft, $V = -800$ lb and $M = 800x$ lb-ft; for $3 \leqq x \leqq 9$ ft, $V = -50x^2 + 900x - 3050$ lb and $M = \dfrac{50}{3}x^3 - 450x^2 + 3050x - 3150$ lb-ft.
5.175 In the six intervals, from left to right, $(V, M) = (-31, 31x)$; $\left[\dfrac{2000}{6}(x - 0.1)^2 - 31, -\dfrac{1000}{9}(x - 0.1)^3 + 31x\right]$; $(100x - 41, -50x^2 + 41x + 1)$; $(19, -19x + 19)$; $(19, -19x - 81)$: $(-200, 200x - 300)$ with units of (N, N · m).
5.177 (a) For $0 < x < 0.6$ m, $V = \dfrac{500}{3}x^2$ and $M = \dfrac{-500}{9}x^3$; for $0.6 < x < 0.8$, $V = 60$ and $M = -60x + 24$; for $0.8 < x < 1$, $V = 660$ and $M = -660x + 504$. Units are N for V, and N · m for M.
(b) 660 N for $0.8 < x \le 1$ (c) -156 N · m at the wall
5.179 $V = 400x$ lb; $M = 1500x - 12500$ lb-ft
5.181 $V_{\text{extremes}} = 13.3$ lb for $6 < x < 12$ ft, and -46.7 lb at $x = 0$; $M_{\text{extremes}} = 109$ lb-ft at $x = 4.67$ ft, and 0 at $x = 0$ and 12 ft
5.183 $V = \text{constant} = -\dfrac{C}{L}$ throughout; $M_{\text{extremes}} = 0$ at $x = 0$ and C at $x = L$
5.185 $V_{\text{extremes}} = 500$ lb for $0 \le x < 3$ ft and -500 lb for $3 < x \le 6$ ft, $M_{\text{extremes}} = 0$ at $x = 0$, 6 ft and -1500 lb-ft at $x = 3$ ft
5.187 $V_{\text{extremes}} = 1500$ kN at $x = 6^-$ m, and -1200 kN at $x = 6^+$ m; $M_{\text{extremes}} = \pm 1800$ kN · m at $x = 2$ and 6 m, respectively. For $2 < x < 6$ m, $V = 300(x - 1)$ kN and $M = -150x^2 + 300x + 1800$ kN · m.
5.189 $V_{\text{extremes}} = 20$ kN at $x = 9^-$ m, and -16 kN for $0 \leqq x < 3$ m; $M_{\text{extremes}} = 48$ kN · m at $x = 3$ m, and -18 kN · m at $x = 9$ m
5.191 $V_{\text{extremes}} = \dfrac{q_0 L}{\pi}$ at $x = \dfrac{L}{2}$, and 0 at $x = 0$ and L; $M_{\text{extremes}} = \dfrac{2q_0 L^2}{\pi^2}$ at $x = 0$, and 0 at $x = L$
5.193 $V_{\text{extremes}} = 550$ lb for $6 < x < 12$ ft, and -1000 lb at $x = 12^+$ ft; $M_{\text{extremes}} = 1200$ lb-ft at $x = 4$ ft, and -2500 lb-ft at $x = 12$ ft
5.195 $V_{\text{extremes}} = 83.3$ N for $6 \leqq x < 9$ m, and -100 N for $9 < x \leqq 12$ m; $M_{\text{extremes}} = 0$ at $x = 0$, and -500 N · m at $x = 9$ m
5.197 $V_{\text{extremes}} = 14.4$ N for $0.3 < x \leqq 1.2$ m, and -5.6 N for $0 \leqq x < 0.3$ m; $M_{\text{extremes}} = 22.4$ N · m at $x = 0.6^+$ m, and -2.64 N · m at $x = 0.6^-$ m
5.199 50.7 ft from P; 3310 lb at P
5.201 $y = H\left(1 - \cos\dfrac{\pi x}{L}\right)$
5.203 Answer given in problem.
5.205 214 ft; $T_A = 175{,}000$ lb, $T_B = 144{,}000$ lb
5.207 $fL\left[\dfrac{2^{n+2}(x/L)^{n+2} - 2(x/L)}{2^{n+2} - 2}\right]$
5.209 $\ell\sqrt{K - 1}\,\cosh^{-1}\sqrt{K/(K - 1)}$, where $K = \left(\dfrac{2T_0}{q_0\ell}\right)^2$
5.211 2.3 m **5.213** (a) 1.3% (b) 6.8% (c) 38.3%
5.215 2.95 N/m
5.217 (a) 120 ↑ lb; (b) 57.4 → lb with line of action 8.3 ft above B **5.219** $q_0 L$; $2.06L$; $0.152L$
5.221 $S_1 = 3.94$ ft; $S_2 = 4.61$ ft; $T_1 = 1240$ lb; $T_2 = 1110$ lb; $T_3 = 1400$ lb
5.223 Assuming a uniform beam,
(a) $\phi_1 = \phi_2 = 39.9°$ (b) Each = 156 lb
(c) 6.42 ft above the horizontal line through A
5.225 (a) 0.436 m (b) $T_1 = 600$ N; $T_2 = 557$ N; $T_3 = 572$ N **5.227** 60°

CHAPTER 6

6.1 (a) 60 lb; (b) 40 lb **6.3** 31.0°
6.5 $1.38W$, by tipping **6.7** 82.9 lb **6.9** 94.6 N
6.11 $28mg/29$ **6.13** $0.030R$ **6.15** 5.71°
6.17 115 lb; 0.577 **6.19** $-129.0 \le P \le 65.1$ N
6.21 0.289 **6.23** 52.9 N · m **6.25** 0.885

6.27 0.247 **6.29** (a) $mg \cos \phi / [2 \sin (\theta + \phi)]$ (b) 0.577 **6.31** 41.2°; 0.875 **6.33** 166 N · m
6.35 0.625 **6.37** $(x, y) = (0.395, 0.500)$ m
6.39 (a) $2Wr \sin \phi$ ↻ (b)Answer given in problem.
6.41 0.0577; 25.8 lb
6.43 (a) 20 lb; (b) 20 lb each; (c) Mr. A; 0.125
6.45 No; 6.56 → lb
6.47 51.8 lb; answer is same if P pushes to the left.
6.49 0.373L; before slip occurs have 5 unknowns but only three independent equations. **6.51** 2
6.53 (a) Answer given in problem; (b) rolling to the right.
6.55 $P = 40$ lb to the right; 40 ← lb by slot on $\mathcal{A}$; 60 ↑ lb by track on $\mathcal{C}$. **6.57** 0.350 **6.59** 0.511
6.61 100 lb-ft **6.63** 31.3 N **6.65** 0.333
6.67 112 N **6.69** $\mu_T \leqq \mu_P / [(1 - \mu_P) \tan 8° + \mu_P]$
6.71 (a) 0.819 cm (in tension) (b) 0.588
6.73 1270 N **6.75** 982 N **6.77** 115 N **6.79** 0
6.81 $(\cot \alpha)/2$ **6.83** 0.208 **6.85** 563 N
6.87 3.61°, tips **6.89** (a) 115 lb (b) 0.577
6.91 (a) 2.05; (b) 1.45 **6.93** 85 lb
6.95 (a) $N = 33.3$ lb; $f = 19.2$ lb directed along AB (b) 0.577

6.97 (a) $8.250 + 0.533P$ (b) $6.925 - 0.858P$ 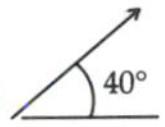

(c) 3.87 N (d) 14.6 N (by reversing the direction of f)
6.99 1.55 lb
6.101 If $\tan \alpha > \mu_o$, slips first at $P = \mu_o mg$. If $\tan \alpha < \mu_o$, tips first at $P = mg \tan \alpha$. If $\tan \alpha = \mu_o$, slips and tips simultaneously at $P = \mu_o mg$.
6.103 $mg(32\mu_o + 3x)/(32 - 3\mu_o x)$
6.105 2/3 **6.107** 24.0°
6.109 (a) $\Sigma M_{\text{line } OB} = 0$ shows that N is in plane AOB; (b) $\sqrt{x^2 + y^2}/(2z)$
6.111 (a) 240 lb, slips only on plane and to the right as $\mathcal{A}$ turns ↺ (b) 480 lb, slips only on $\mathcal{B}$ and to the right as $\mathcal{A}$ turns ↻
6.113 $\mathcal{B}$ is needed for equilibrium; $W_{\mathcal{B}_{\min}} = 29.2$ lb;
73.2 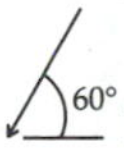

lb **6.115** 12 ft
6.117 $N_{\text{rear}} = 140$ lb; $N_{\text{front}} = 43.2$ lb; $f = 49.2$ lb; $\mu_{\min} = 0.351$ **6.119** at the wall; 1.17mg **6.121** 0.857
6.123 0.791 **6.125** 5/12
6.127 $5W_B/(12W_C + 6W_R)$
6.129 (a) 0.575 (b)1.15 m **6.131** 5.04 ft
6.133 0.213 **6.135** 5.53 N; 0.237 **6.137** 0.348
6.139 $W\sqrt{\mu^2 \cos^2 \phi - \sin^2 \phi}$ **6.141** 0.318
6.143 $(\tan \alpha)/2$
6.145 $107° < 2\alpha < 116°$, where 2α = angle between bars **6.147** 90° **6.149** Answer given in problem.
6.151 Answer given in problem. **6.153** 623 lb
6.155 3070 N **6.157** $122 \text{ N} \leq W_{\mathcal{A}} \leq 2060$ N
6.159 81.4 lb **6.161** 422 lb **6.163** 3.17
6.165 800 lb **6.167** Yes. (He could go 8.2 ft further out before it slipped.) **6.169** 95 lb
6.171 67 → lb and 602 ↺ lb-ft
6.173 2 **6.175** 40.2 lb
6.177 (a) B: 94.4 ↓ lb, D: 94.4 ↑ lb (b) 5.60 lb
6.179 $(1 - \mu \tan \alpha) \tan \alpha / (\mu + \tan \alpha)$
6.181 14.1 lb-ft **6.183** 41.2 lb-in.
6.185 The integration yields $M = P\mu R_o/2$, which is 25% less than $2P\mu R_o/3$.
6.187 10.2 lb-in. (95% of the moment in 6.186)
6.189 μPR **6.191** 213 lb-in. **6.193** 107 N
6.195 Some results: $\theta = 90° \Rightarrow \rho = 0.414$; $\theta = 45° \Rightarrow \rho = 1.414$
6.197 72.0° (Equation to solve is $\sin^3 \beta + \frac{1}{2} \sin^2 \beta \cos \beta - 1 = 0$)

CHAPTER 7

7.1 Answer given in problem.
7.3 $\bar{x} = [2r \sin(\alpha/2)]/\alpha$ **7.5** 2.10
7.7 (0.549a, 0) ft **7.9** (1.00, 0.393) ft **7.11** 2.73 in.
7.13 4.50 in. **7.15** 3.60 in. **7.17** 4 m **7.19** 0.637R
7.21 3 m **7.23** 1 in. **7.25** 2.087 m **7.27** $\bar{z} = 3$ ft
7.29 Answer given in problem. **7.31** $(\frac{3}{4}, \frac{3}{4}, \frac{3}{4})$ m
7.33 5 cm **7.35** Answer given in problem.
7.37 (−0.05, 0.117) m
7.39 lines: (2.5, 1.0) cm; triangle: (2.67, 1.0) cm
7.41 (0, 6.24) in. **7.43** (2.39, 2.39) in.
7.45 (0, −0.849) cm **7.47** (33.5, 17.0) cm
7.49 (6.67, 2.50) in. **7.51** (6.19, 5.77) in.
7.53 $(0, -4r^3/[3\pi(2R^2 - r^2)])$ **7.55** (0.943, 1.19) in.
7.57 (0, 0.584R) **7.59** (0, 2.42) in.
7.61 (1.03, 2.03) in.; no difference to 3 digits
7.63 (0.255, 0.225, −0.0200) m **7.65** $\left(\frac{\ell}{2}, \frac{\ell}{6}, \frac{\ell}{6}\right)$
7.67 (15, 21.6, 28.4) cm; $\bar{x}$
7.69 (2.21, 4.77, 0.794) in. **7.71** (0.602, −0.244) in.
7.73 (4.40, 1.79) in. **7.75** 3.46 cm
7.77 (0.397, 0.260) m **7.79** (3.94, 2.85) in.
7.81 (1.25, 1.25) m **7.83** Answer given in problem.
7.85 0.675 units of length
7.87 0.494 **7.89** $\frac{4}{3\pi}(a + b)$; yes
7.91 $\lim_{\alpha \to 0} \left(\frac{2r \sin \alpha}{3\alpha}\right) = \frac{2r}{3} = \bar{y}$ for the triangle
7.93 (0, −0.198R) **7.95** 3.00 in.
7.97 (10.4, −3.81, 35) mm **7.99** 0.707
7.101 1.73R **7.103** $\left(\frac{2\ell}{7}, \frac{2\ell}{7}, \frac{\ell}{2}\right)$ **7.105** $\frac{2}{3}L$

7.107 $(0, -R/\pi)$ **7.109** (a) $1.71r$ (b) $1.29r$
7.111 $\bar{z} = 3R(1 + \cos\alpha)^2/[4(2 + \cos\alpha)]$
7.113 (a) 16γ; $(\frac{3}{8}, \frac{12}{5}, -3)$ (b) 5.4γ, 8.6γ, 2γ, respectively **7.115** 14.7° **7.117** $\tan^{-1}\left(\dfrac{3\pi W_s}{4W_c}\right)$
7.119 On the axis, $H/3$ from the base toward the vertex **7.121** 384 in.^{3} **7.123** 993 cm^3
7.125 123 ft^3 **7.127** 87.5 m^3 **7.129** 11.0 in.^{3}
7.131 $(0, 2R/\pi)$ **7.133** 0.785 s^3 **7.135** 0.393 m^3
7.137 1.60 m^2 **7.139** 8,484 mm^2; 53,300 mm^2
7.141 174 in.^{2} **7.143** $\pi^2 Rr$ **7.145** $\pi r(\pi R - 2r)$
7.147 $\pi r\left(\dfrac{\pi R}{2} - r\right)$ **7.149** $\dfrac{\pi r^2}{12}(3\pi R + 4r)$
7.151 $\dfrac{\pi r^2}{12}(3\pi R - 4r)$ **7.153** 21,500,000 ft^3
7.155 (57.5, 0, 0); Because the mast's diameter is much larger, $\bar{x}$ would increase with an inch of ice covering everything.
7.157 6.10×10^{24} kg (Handbook actual mass is 5.98×10^{24} kg, so not bad!); 5.63 g/cc

CHAPTER 8

8.1 (a) $bh^3/3$; (b) y and y_C are the same line!
8.3 4.27 m^4 **8.5** 25.1 cm^4 **8.7** $tr^3(\alpha - \sin\alpha)/2$
8.9 $bh^3n/[3(n+3)]$ **8.11** Answer given in problem.
8.13 $bh(4h^2 + b^2)/12$ **8.15** $bh(12h^2 + b^2)/48$
8.17 αtr^3 **8.19** $\pi(R_0^4 - R_i^4)/2$ **8.21** $bh^3/36$
8.23 $bh^3n\left[\dfrac{n^2 + 4n + 7}{12(n+3)(n+2)^2}\right]$ **8.25** 1.82 cm^4
8.27 $r^3t\left(\dfrac{\alpha}{2} + \dfrac{\sin\alpha}{2} - \dfrac{4}{\alpha}\sin^2\dfrac{\alpha}{2}\right)$
8.29 Answers given in problem. **8.31** 120 ft^4
8.33 0.0361 s^4; 0.0313 s^4 = 0.867 of J_C for the triangle, which has some areas farther from the axis.
8.35 7.15 m^4
8.37
$$bh^3n\left[\frac{n^2 + 4n + 7}{12(n+3)(n+2)^2}\right] + hb^3n\left[\frac{n^2}{(3n+1)(2n+1)^2}\right]$$
8.39 0.894 m **8.41** 83.0 ℓ^4 **8.43** $\frac{5}{64}bh^3$
8.45 0.000235 m^4 **8.47** 453 in.^{4} **8.49** 2.53 ft^4
8.51 8,800,000 cm^4 **8.53** 0.806 ft^4
8.55 0.806 ft^4 **8.57** 0.0966 m^4 **8.59** 1560 in.^{4}
8.61 8070 in.^{4} **8.63** 2.49 s^4 **8.65** 0.252 m^4
8.67 0.326 m^4 **8.69** $5\sqrt{3}\, s^4/16$
8.71 21.1 in.^{4}, the same as table value **8.73** 6.91 in.^{4}
8.75 15.4 in.^{4} **8.77** 194 in.^{4}
8.79 (a) 77,900 mm^4 (b) 538,000 mm^4
8.81 $I_{x_C} = 1740$ in.^{4}; $I_{y_C} = 666$ in.^{4}
8.83 (a) $I_x = \dfrac{th^3}{12} + \dfrac{tbh^2}{2}$; $I_y = \dfrac{tb^3}{6}$ (b) 1.81
8.85 62.5 in.^{4} **8.87** Answers given in problem.
8.89 $I_{YY} = 26.6$ in.^{4}, right on; $k_Y = 1.55$ in., off by less than 1% **8.91** 377 A **8.93** $-R^4/8$
8.95 $\frac{8}{5}$ in.^{4}; $\frac{32}{7}$ in.^{4}; $\frac{-8}{3}$ in.^{4}
8.97 Answers given in problems. **8.99** −1.05
8.101 19.0 in.^{4} **8.103** −0.000172 m^4
8.105 −1.33 ft^4 **8.107** −1.36 ft^4
8.109 −0.0466 m^4 **8.111** −1700 in.^{4}
8.113 11,300,000 cm^4
8.115 12.4 in.^{4}; 79.9 in.^{4}; −12.4 in.^{4}
8.117 0.267 in.^{4}; 1.16 in.^{4}; −0.533 in.^{4}
8.119 1.41 in.^{4}, 64.9° clockwise from x_C; 0.0174 in.^{4}, 25.1° counterclockwise from x_C
8.121 (a) $h^4/24$, $h^4/72$ (respectively normal and parallel to the hypotenuse) (b) If $I_x = I_y$ and $I_{xy} = 0$, the circle is a point. (c) The triangle contributes half the inertia of the square in all directions. (d) Answer given in problem. **8.123** $7b^4/6$; $b^4/6$
8.125 $I_{max} = 52.0$ in.^{4} at 27.5° ↻ from x_C; $I_{min} = 5.60$ in.^{4} at 62.5° ↺ from x_C.
8.127 (a) $I_x = 14.96$ ft^4, $I_y = 10.51$ ft^4; (b) $I_{x_C} = 3.091$ ft^4, $I_{y_C} = 3.883$ ft^4; the results check.

CHAPTER 9

9.1 $\dfrac{mg}{2}\tan\theta$; the limits are as stated. **9.3** 5250 lb
9.5 43.7 N **9.7** At D, 2 ↓ kN; at B, 4 ↑ kN
9.9 $\theta_1 = 26.27°$; $\theta_2 = 59.67°$; $T = 0.7116mg$; $N_{left} = 0.8967mg$; $N_{right} = 0.5050mg$
9.11 $P = (1.5\cot\alpha)F$ **9.13** $\dfrac{W}{2}\left(1 - \dfrac{r}{R}\right) = \dfrac{W}{4}$
9.15 21.3 ↑ N **9.17** $P = \dfrac{(a + 2b + 2c)\cot\theta}{2a}W$
9.19 (a) $\dfrac{5W\sqrt{9\sin^2\beta + \cos^2\beta}}{4\sin\beta}$
(b) $\dfrac{5W\sqrt{25\sin^2\beta + \cos^2\beta}}{12\sin\beta}$
(c) $-\dfrac{5}{4}W(\cot\beta)\hat{\mathbf{i}} - \dfrac{13}{4}W\hat{\mathbf{j}}$; no **9.21** 2860 ↑ lb
9.23 (a) 779 ↻ lb-ft (b) 300 → lb, (c) Answer given in problem. (See FBDs at top of next page.)
9.25 Answer given in problem.
9.27 9140 lb **9.29** $P = W/2$
9.31 $P = W$ (independent of a and b!)
9.33 Answers given in problem. (Consider thin horizontal and vertical cylinders of fluid, respectively.)
9.35 Answer given in problem.
9.37 132,000 lb, normal to the area at depth 75.004 ft. (Note from the fifth decimal digit that,

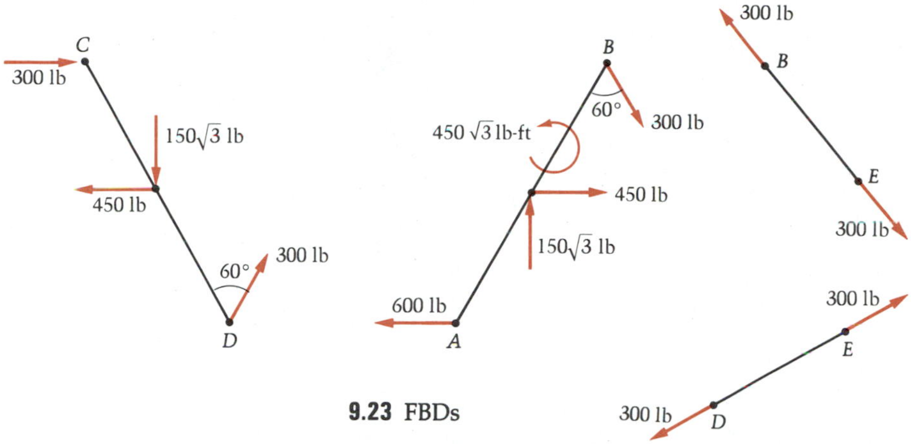

9.23 FBDs

as depth increases, distance between centroid and center of pressure decreases!)
9.39 22.1 lb normal to the area at depth 2.130 ft
9.41 $(x_P, y_P, z_P) = (6.98, 0, 0)$ ft
9.43

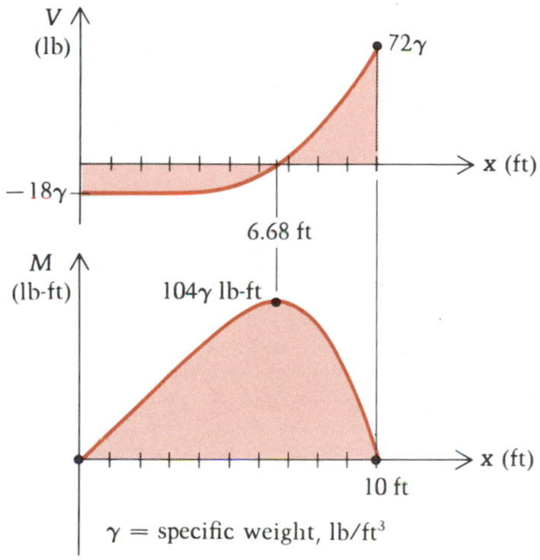

9.45 $\dfrac{\pi\gamma d h^2}{4} \rightarrow$ at center of door **9.47** $72\gamma \leftarrow$

9.49 105γ 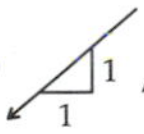 , at the center of pressure of Example 9.7
9.51 Answer given in problem.
9.53 $(-15.6\hat{\mathbf{i}} + 26.6\hat{\mathbf{j}}) \times 10^6$ lb, 7.71 ft to the left of D

9.55 $3.74\gamma R^2 L$ 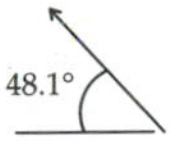 through center of quarter-circle

9.57 $\dfrac{2(3H^2 + 3HR + R^2)}{3(2H + R)}$ below the water line
9.59 $\sqrt{3}B$

9.61 (a) $-515{,}000\hat{\mathbf{i}} - 98{,}100\hat{\mathbf{j}}$ N on the line

$y = 0.190x - 1.81$ with 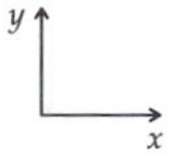 at A (b) at A,

$515{,}000\hat{\mathbf{i}} - 834{,}000\hat{\mathbf{j}}$ N; at B, 932,000 ↑ N

9.63 642,000 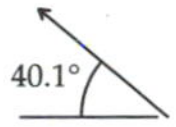 N, along the line $y = -0.843x + 5.33$ **9.65** Answer given in problem.
9.67 0.293 **9.69** 0.091349748

APPENDIX A

A.1 $-\hat{\mathbf{i}} + 5\hat{\mathbf{j}} + 6\hat{\mathbf{k}}$; $5\hat{\mathbf{i}} + 7\hat{\mathbf{j}} - 14\hat{\mathbf{k}}$ **A.3** -38
A.5 $228\hat{\mathbf{i}} + 456\hat{\mathbf{j}} + 684\hat{\mathbf{k}}$; 120; $404\hat{\mathbf{i}} - 112\hat{\mathbf{j}} - 60\hat{\mathbf{k}}$; $512\hat{\mathbf{i}} - 456\hat{\mathbf{j}} - 240\hat{\mathbf{k}}$
A.7 (a) $6.5\hat{\mathbf{i}} - 20.5\hat{\mathbf{j}} - 0.3\hat{\mathbf{k}}$ (b) $-\hat{\mathbf{i}} - 29.2\hat{\mathbf{j}} + 6\hat{\mathbf{k}}$ (c) -28.0 (d) $-52.2\hat{\mathbf{i}} + 30.6\hat{\mathbf{j}} + 140\hat{\mathbf{k}}$ (e) -1010 (f) $193\hat{\mathbf{i}} - 7.14\hat{\mathbf{j}} + 73.4\hat{\mathbf{k}}$ (g) $0.506\hat{\mathbf{i}} + 0.863\hat{\mathbf{j}}$ (h) $(0.189, -0.943, 0.276)$
A.9 (a) $17\hat{\mathbf{i}} - 21\hat{\mathbf{j}} + 19\hat{\mathbf{k}}$ (b) $14\hat{\mathbf{i}} - 20\hat{\mathbf{j}} + 11\hat{\mathbf{k}}$ (c) 68 (d) $-140\hat{\mathbf{i}} - 87\hat{\mathbf{j}} + 20\hat{\mathbf{k}}$ (e) -173 (f) $67\hat{\mathbf{i}} - 180\hat{\mathbf{j}} - 314\hat{\mathbf{k}}$ (g) $0.141\hat{\mathbf{i}} + 0.990\hat{\mathbf{k}}$ (h) $(0.487, -0.649, 0.584)$
A.11 -11; $\mathbf{A} \times \mathbf{B} = -\hat{\mathbf{i}} + 19\hat{\mathbf{j}} + 7\hat{\mathbf{k}}$
A.13 Area of shaded parallelogram = Area of dashed rectangle = (Base)(Height) = $|\mathbf{A}|(|\mathbf{B}|\sin\theta)$ and $|\mathbf{A} \times \mathbf{B}| = |\mathbf{A}||\mathbf{B}|\sin\theta$

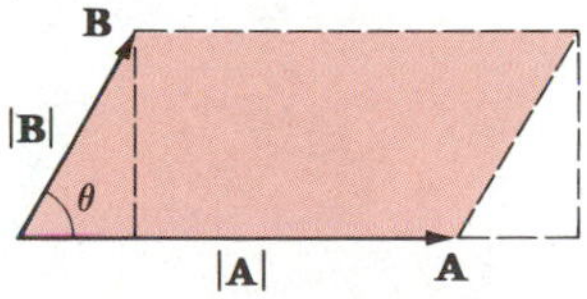

A.15 2.91 m **A.17** 9.70
A.19 (a) $\frac{3}{5}\hat{\mathbf{i}} + \frac{4}{5}\hat{\mathbf{j}}$ (b) $\hat{\mathbf{i}}$ (c) $\frac{4}{13}\hat{\mathbf{i}} - \frac{3}{13}\hat{\mathbf{j}} + \frac{12}{13}\hat{\mathbf{k}}$

INDEX

Acceleration of gravity, 5, 143
Accuracy of solutions, 8–9
Action
  lines of, 109
  of a force, 12
Action-reaction principle, 4, 33, 158–159, 222, 646–647
Angle of friction, 416
Archimedes' principle, 576, 599
Area(s)
  centroid of, 481, 489
  first moment of, 481
  moments of inertia of, 529–530, 639–643
  polar moments of inertia of, 534
  products of inertia of, 554–555, 558, 639–643
  second moments of, 529
Associative property for addition of vectors, 624
Axes
  moments and products of inertia and, 561
  principal, 566
Axial force, 279
  sign convention for, 360

Beams
  axial force, 360
  bending moment, 359
  cantilever, 363
  shear force, 359
  sign conventions for internal forces in, 360–361
  simply supported, 361, 365
  twisting moment, 359
Belts, 456, 461
Bending moment, 141, 348, 359
  and shear force, 375
  sign convention for, 360
Body(ies), 157
  centroid of volume of, 145
  interacting, 220–221
  deformable, mechanics of, 327
  rigid, kinematics of, 576–578
  submerged, hydrostatic pressure on, 598, 605
  two-force, 179–180
Body forces, 129
Buoyancy, 599, 612
Buoyant force, 612–614

Cables, 390
  catenary, 387
  under concentrated loads, 397–400
  parabolic, 387
Catenary cables, 387
Center
  of buoyancy, 612
  of gravity, 144
  of mass, 144, 511–512
  of pressure, 602
  of Q, 480
  of volume, 480
Centroids, 480–481, 488
  of area, 481, 489
  by composite parts, 495
  of lines, 481
  of volume, 145, 481, 490, 511
Coefficient
  of friction, 414, 416
  of thermal expansion, 340
Commutative property(ies)
  for addition of vectors, 624
  of scalar product, 627
Components
  dot product to find, 19–20
  orthogonal, 15–16, 19
  scalar, 15–16
  vector, 15, 16

Composite areas, method of, 543
Composite parts, centroid, 495
Compressive normal stress, 141
Concurrent force systems, 108
Connections, virtual work of, 586
Consistency, dimensional, 6–8
Coplanar force systems, 109
Coulomb's law of dry friction, 413, 454
Couple, 65
  moment of, 75–76
  work of, 580
Cross (vector) product, 49, 628–631

Deformable bodies, mechanics of, 327–329
Dimension, 6
Dimensional consistency, 6–8
Direction cosines, 626
Disk friction, 466
Displacement, 577
  infinitesimal, 578–579
  virtual, 580
Distributed forces, use of centroids in expressing resultants of, 488
Distributed force systems, 129–130, 136, 140
Distributive properties
  of scalar product, 627
  for vector product, 625
Dot product, 19, 626–628
Dry friction, 413

Elasticity, Young's modulus of, 331, 333
Equilibrium
  analysis of, 26–28
  of a particle, 24–28
  of a system of particles, 33–34
Equilibrium equations, 83–84, 169, 583
  applications of, 220–221
  three-dimensional examples of, 184
Equipollence, 87, 88, 118
Euler's laws, 4, 24, 83, 153
Extensional strain, 330–331
External forces, 24, 25, 34, 87

Face, 360
FBD (free-body diagram), 25–26, 33, 34, 156
First moments, 481
Flat belt, 456, 461
Force(s), 3
  application of, 12
  axial, 279
  body, 129
  buoyant, 612–614
  dimensions of, 5
  distributed, 488
    throughout a volume, 143–145
  equipollent, 87
  external, 24, 25, 34, 87
  friction, 413
  gravitational, 4
  internal, 34
  magnitude of
    times perpendicular distance, 67–68
  moment of, 35
    about a line, 59, 61
    about a point, 49, 67, 68
  newton as a unit of, 5
  normal, 413
  point of application of, 12
  pound as a unit of, 5, 6
  shear, 359, 372, 375, 383–384
  surface, 129
  system. *See* Force system
  as vectors, 12–14
  work of, 578–580
Force-and-couple resultant, 99
Force system
  concurrent, 108
  coplanar, 109
  distributed, 129–130, 136, 140
  equipollent, 87–88
  force-and-couple resultant of, 99
  internal, 352
  parallel, 115, 143
  single-force resultant of, 106–107
Frame(s)
  nonrigid, 233
  of reference, 2, 3–4
  rigid, 224
Free-body diagram (FBD), 25–26, 33, 34, 156
Friction, 412–413
  angle of, 416
  belt, 456
  coefficient of, 414, 416
  Coulomb, 413, 454
  disk, 466
  dry, 413
  kinetic, 415
  maximum value of, 414
  screw, 461
  static, 414

Gravitation
  Newton's law of, 4–5
  universal constant, 4
Gravity
  acceleration of, 5, 143
  center of, 144
Gyration, radius of, 540

Hooke's law, 331, 334
Hydrostatic pressure
  on curved surfaces, 605
  on plane surfaces, 599
  on submerged bodies, 598, 605

Impending slipping, 415
Indeterminacy, static, 171
Inertia
  area moments of, 529–530
  area products of, 554–555, 558
  parallel-axis theorems of, 536, 540, 558
  polar moments of, 534, 537
  principle moments of, 566
Inertial frame of reference, 2, 3–4, 83
Infinitesimal displacement, 578–579
Integral
  recurring, 480
  scalar, 481
Interacting bodies, 220–221
Internal force, 34, 352
International System (SI) of Units, 5, 637–638

Joints, method of, 281–282

Kilogram, 5, 6
Kilo-pound, 280
Kinematics of rigid body, 576–578
Kinetic friction, 415
Kip, 280

Lead of a screw, 464
"Left-hand face," 360
Length, 5
Line(s)
  of action, 99, 109
  centroid of, 481
  of force, 12
Loads, 25, 153

Machine, 224
Magnitude of a vector, 624
Mass, 3
  center of, 144, 511–512
  kilogram as a unit of, 5, 6
  units of, 5
Mechanics, 2, 576
  basic laws of, 3–5
Megapascal, 328
Metacenter, 614
Meter, 5
Mohr's circle, 569
  developing equations of, 561–562
  use of, 562–563
Moment(s)
  bending. *See* Bending moment
  of a couple, 75–76
  first, 481
  of a force, 35
    about a line, 59, 61
    about a point, 49, 67, 68
  of inertia of areas. *See* Moments of inertia of areas
  relationship between sums of, 84–85
  second of area, 529
  twisting, 359
Moments of inertia of areas, 529–530, 639–643
  minimum and maximum, 566
  using Mohr's circle, 561, 562–563, 569
  and parallel-axis theorem, 536
  polar, 534, 537
  principal, 566
  and radius of gyration, 540
  and rotated axes through a point, 561
Motion, Newton's laws of, 3
Multiforce member, 348–349
Mutually perpendicular components, 15–16

Newton, as a unit of force, 5
Newton's law(s)
  of gravitation, 4–5
  of motion, 3
  third, 222
Newton-Raphson method, 644–645
Newtonian frame of reference, 2, 3–4
Nonrigid frame, 233
Normal force, 413
Normal stress, 327

Orthogonal components, 15–16, 19

Pappus, theorems of, 517–518
Parabolic cables, 387
Parallel force systems, 115, 143
Parallel-axis theorem
  for area moments of inertia, 536
  for area products of inertia, 558
Parallelogram law of addition, 12–13, 14, 624
Particles, 24–28, 33–34
Pascal, 328
Pascal's law, 576, 598
Pins, 281–282
Pitch of a screw, 464
Plane truss, 279
Point of application of a force, 12
Polar moments of inertia, 534, 537
Pound, 5, 6
Pressure
  center of, 602
  hydrostatic, 598, 599, 605

resultant, 134, 142
Primitives, 2–3
Principal axes of inertia, 566
Problem solving, 8–9
Product
of inertia of area, 554–555, 558, 639–643
scalar, 19, 626–627, 631–634
vector, 631–634
Pulleys, special results for, 179–180

Q, center of, 480

Radius of gyration, 540
Reactions, 25, 153
Recurring integral, 480
Resultant
of distributed force systems, 130
equipollent "screwdriver," 118
force-and-couple, 99
of parallel force system, 143
of pressure
on a flat surface, 134
on a spherical surface, 142
single-force, 106–107
identifying, 159
use of centroids in expressing, 488
Revolution
surface of, 517–518
volume of, 518
"Right-hand face," 360, 383
Right-hand rule, 49
Rigid body, kinematics of, 576–578
Rigid-body equivalent, 87
Rolling, 421

Scalar components, 15–16
Scalar integrals, 481
Scalar product, 19, 626–627, 631–634
Screw, 461, 464
"Screwdriver" resultant, equipollent, 118
Second, 5
Second moment, 529
Sections, method of, 300–301
Self-locking mechanisms, 427
Shear force, 359
and bending moment, 375
and distributed lateral loading, 372
sign convention for, 383–384
Shear stress, 328–329
SI of units, 5, 637–638
Sign conventions
for internal forces in beams, 360–361
for shear force, 383–384
Simple space truss, 318
Simple truss, 297
Single-force resultant, 106–107
Slipping, impending, 415
Slug, 5
Solutions, accuracy of, 8–9
Space, 2–3
Space truss, 279, 318
simple, 318
Specific weight, 598
Spring, 339
Static equivalent, 87
Static friction, 414, 416
Static indeterminacy, 171
Statically indeterminate problems, 336–337
Statics, study of, 2
Strain, extensional, 330–331
Stress
normal, 327
shear, 328–329
tensile, 141, 328
Stress-strain diagram, 331
Submerged bodies, hydrostatic pressure on, 598, 605
Surface forces, 129
Symbols, importance of using, 8

Tensile stress, 141, 328
Tension, 456
Thermal expansion, coefficient of, 340
Time, 3, 5
Transfer theorem (parallel-axis)
for area moments of inertia, 536
for area products of inertia, 558
Truss
analysis of, 291–293
definition of, 275
forces in, 278–279
number of members and pins in, 296
plane, 279
simple, 297
space, 279, 318
thermal effects and, 340–341
Truss element, elongation of, 334
Twisting moment, 359
Two-force members, 278
special results for, 179–180

Unit(s), 5, 636
Unit conversion, 5–6
Unit vectors, 15
orthogonal components of, 625–626

V-belt, 456, 461
Varignon's theorem, 51–52, 104
Vector (cross) product, 49, 628–631
Vectors
addition of, 624
components, 15, 16

dot product of, 19
forces as, 12–14
magnitude of, 624
representation, 49–50
scalar product of, 19
triple product, 631–634
unit, 15, 625–626
Virtual displacement, 580
Virtual work
of connections, 586
general equivalence of, 583
principle of, 576, 581
Volume
center of, 480
centroid of, 145, 480, 481, 490
forces distributed throughout, 143–145

Wedge, 454
Weight, 26
Work
of a couple, 580
of a force, 578–580
virtual, 576, 581, 583, 586
Wrench ("screwdriver") form of resultant, 118

Young's modulus of elasticity, 331
values of, 333

Zero-force members, 291–292

Engineering Mechanics

An Introduction to

Dynamics

PWS SERIES IN ENGINEERING

Anderson, *Thermodynamics*
Askeland, *The Science and Engineering of Materials, Third Edition*
Borse, *FORTRAN 77 and Numerical Methods for Engineers, Second Edition*
Bolluyt/Stewart/Oladipupo, *Modeling for Design Using SilverScreen*
Clements, *68000 Family Assembly Language*
Clements, *Microprocessor Systems Design, Second Edition*
Clements, *Principles of Computer Hardware, Second Edition*
Das, *Principles of Foundation Engineering, Third Edition*
Das, *Principles of Geotechnical Engineering, Third Edition*
Das, *Principles of Soil Dynamics*
Duff/Ross, *Freehand Sketching for Engineering Design*
El-Wakil/Askeland, *Materials Science and Engineering Lab Manual*
Fleischer, *Introduction to Engineering Economy*
Gere/Timoshenko, *Mechanics of Materials, Third Edition*
Glover/Sarma, *Power System Analysis and Design, Second Edition*
Janna, *Design of Fluid Thermal Systems*
Janna, *Introduction to Fluid Mechanics, Third Edition*
Kassimali, *Structural Analysis*
Keedy, *An Introduction to CAD Using CADKEY 5 and 6, Third Edition*
Keedy/Teske, *Engineering Design Using CADKEY 5 and 6*
Knight, *The Finite Element Method in Mechanical Design*
Knight, *A Finite Element Method Primer for Mechanical Design*
Logan, *A First Course in the Finite Element Method, Second Edition*
McDonald, *Continuum Mechanics*
McGill/King, *Engineering Mechanics: Statics, Third Edition*
McGill/King, *Engineering Mechanics: An Introduction to Dynamics, Third Edition*
McGill/King, *Engineering Mechanics: Statics and An Introduction to Dynamics, Third Edition*
Meissner, *Fortran 90*
Raines, *Software for Mechanics of Materials*
Ray, *Environmental Engineering*
Reed-Hil/Abbaschian, *Physical Metallurgy Principles, Third Edition*
Reynolds, *Unit Operations and Processes in Environmental Engineering*
Russ, *CD-ROM for Materials Science*
Schmidt/Wong, *Fundamentals of Surveying, Third Edition*
Segui, *Fundamentals of Structural Steel Design*
Segui, *LRFD Steel Design*
Shen/Kong, *Applied Electromagnetism, Second Edition*
Sule, *Manufacturing Facilities, Second Edition*
Vardeman, *Statistics for Engineering Problem Solving*
Weinman, *VAX FORTRAN, Second Edition*
Weinman, *FORTRAN for Scientists and Engineers*
Wempner, *Mechanics of Solids*
Wolff, *Spreadsheet Applications in Geotechnical Engineering*
Zirkel/Berlinger, *Understanding FORTRAN 77 and 90*

Third Edition

ENGINEERING MECHANICS
AN INTRODUCTION TO
DYNAMICS

DAVID J. MCGILL AND WILTON W. KING
Georgia Institute of Technology

PWS Publishing Company
Boston

I(T)P An International Thomson Publishing Company

Boston • Albany • Bonn • Cincinnati • Detroit • London • Madrid
Melbourne • Mexico City • New York • Paris • San Francisco
Singapore • Tokyo • Toronto • Washington

PWS PUBLISHING COMPANY
20 Park Plaza, Boston, Massachusetts 02116-4324

This book is printed on recycled, acid-free paper

I(T)P™
International Thomson Publishing
The trademark ITP is used under license.

For more information, contact:

PWS Publishing Company
20 Park Plaza
Boston, MA 02116

International Thomson Publishing Europe
Berkshire House I68-I73
High Holborn
London WC1V 7AA
England

Thomas Nelson Australia
102 Dodds Street
South Melbourne, 3205
Victoria, Australia

Nelson Canada
1120 Birchmount Road
Scarborough, Ontario
Canada M1K 5G4

International Thomson Editores
Campos Eliseos 385, Piso 7
Col. Polanco
11560 Mexico D.F., Mexico

International Thomson Publishing GmbH
Konigswinterer Strasse 418
53227 Bonn, Germany

International Thomson Publishing Asia
221 Henderson Road
#05-10 Henderson Building
Singapore 0315

International Thomson Publishing Japan
Hirakawacho Kyowa Building, 31
2-2-1 Hirakawacho
Chiyoda-ku, Tokyo 102
Japan

Library of Congress Cataloging-in-Publication Data

McGill, David J., 1939–
Engineering Mechanics, an introduction to dynamics / David J. McGill and Wilton W. King.
— 3rd ed.
p. cm.
At head of title: Engineering mechanics.
Includes index.
ISBN 0-534-93399-8
1. Dynamics. I. King, Wilton W., . II. Title.
III. Title: Engineering mechanics.
TA352.M385 1994
620.1'04 — dc20 94-33846
CIP

Printed and bound in the United States of America
95 96 97 98 99 — 10 9 8 7 6 5 4 3 2 1

Sponsoring Editor: Jonathan Plant
Editorial Assistant: Cynthia Harris
Developmental Editor: Mary Thomas
Production Coordinator: Kirby Lozyniak
Marketing Manager: Nathan Wilbur
Manufacturing Coordinator: Marcia Locke
Production: York Production Services

Cover Designer: Julie Gecha
Interior Designer: York Production Services
Compositor: Progressive Information Technologies
Cover Photo: © by Walter Bibikow, courtesy of Image Bank
Cover Printer: John P. Pow Company, Inc.
Text Printer: Quebecor Printing, Hawkins

To our wives, Carolyn and Kay

Contents

Preface xi

1 Kinematics of Material Points or Particles 1

1.1 Introduction, 2
1.2 Reference Frames and Vector Derivatives, 3
1.3 Position, Velocity, and Acceleration, 6
1.4 Kinematics of a Point in Rectilinear Motion, 8
1.5 Rectangular Cartesian Coordinates, 24
1.6 Cylindrical Coordinates, 31
1.7 Tangential and Normal Components, 43
Summary, 53
Review Questions, 54

2 Kinetics of Particles and of Mass Centers of Bodies 55

2.1 Introduction, 56
2.2 Newton's Laws and Euler's First Law, 56
2.3 Motions of Particles and of Mass Centers of Bodies, 62
2.4 Work and Kinetic Energy for Particles, 87
2.5 Momentum Form of Euler's First Law, 101
2.6 Euler's Second Law (The Moment Equation), 117
Summary, 125
Review Questions, 128

3 Kinematics of Plane Motion of a Rigid Body 129

3.1 Introduction, 130
3.2 Velocity and Angular Velocity Relationship for Two Points of the Same Rigid Body, 134
3.3 Translation, 147
3.4 Instantaneous Center of Zero Velocity, 149
3.5 Acceleration and Angular Acceleration Relationship for Two Points of the Same Rigid Body, 163

3.6 Rolling, 170
3.7 Relationship Between the Velocities of a Point with Respect to Two Different Frames of Reference, 198
3.8 Relationship Between the Accelerations of a Point with Respect to Two Different Frames of Reference, 207
Summary, 215
Review Questions, 216

4 KINETICS OF A RIGID BODY IN PLANE MOTION/ DEVELOPMENT AND SOLUTION OF THE DIFFERENTIAL EQUATIONS GOVERNING THE MOTION 217
4.1 Introduction, 218
4.2 Rigid Bodies in Translation, 219
4.3 Moment of Momentum (Angular Momentum), 227
4.4 Moments and Products of Inertia/The Parallel-Axis Theorems, 229
4.5 The Mass-Center Form of the Moment Equation of Motion, 246
4.6 Other Useful Forms of the Moment Equation, 272
4.7 Rotation of Unbalanced Bodies, 291
Summary, 301
Review Questions, 303

5 SPECIAL INTEGRALS OF THE EQUATIONS OF PLANE MOTION OF RIGID BODIES: WORK-ENERGY AND IMPULSE-MOMENTUM METHODS 304
5.1 Introduction, 305
5.2 The Principle(s) of Work and Kinetic Energy, 305
5.3 The Principles of Impulse and Momentum, 343
Summary, 375
Review Questions, 377

6 KINEMATICS OF A RIGID BODY IN THREE-DIMENSIONAL MOTION 379
6.1 Introduction, 380
6.2 Relation Between Derivatives/The Angular Velocity Vector, 380
6.3 Properties of Angular Velocity, 384
6.4 The Angular Acceleration Vector, 398
6.5 Velocity and Acceleration in Moving Frames of Reference, 401
6.6 The Earth as a Moving Frame, 410
6.7 Velocity and Acceleration Equations for Two Points of the Same Rigid Body, 414
6.8 Describing the Orientation of a Rigid Body, 428
6.9 Rotation Matrices, 434
Summary, 441
Review Questions, 442

7 KINETICS OF A RIGID BODY IN GENERAL MOTION 444

7.1 Introduction, 445
7.2 Moment of Momentum (Angular Momentum) in Three Dimensions, 446
7.3 Transformations of Inertia Properties, 448
7.4 Principal Axes and Principal Moments of Inertia, 455
7.5 The Moment Equation Governing Rotational Motion, 472
7.6 Gyroscopes, 492
7.7 Impulse and Momentum, 499
7.8 Work and Kinetic Energy, 504
Summary, 513
Review Questions, 514

8 SPECIAL TOPICS 516

8.1 Introduction, 517
8.2 Introduction to Vibrations, 517
8.3 Euler's Law for a Control Volume, 535
8.4 Central Force Motion, 543
Review Questions, 554

APPENDICES CONTENTS 557

Appendix A UNITS, 558
Appendix B EXAMPLES OF NUMERICAL ANALYSIS/THE NEWTON-RAPHSON METHOD, 564
Appendix C MOMENTS OF INERTIA OF MASSES, 567
Appendix D ANSWERS TO ODD-NUMBERED PROBLEMS, 573
Appendix E ADDITIONAL MODEL-BASED PROBLEMS FOR ENGINEERING MECHANICS, 583
INDEX, 587

Preface

An Introduction to Dynamics is the second of two volumes covering basic topics of mechanics. The first two-thirds of the book contains most of the topics traditionally taught in a first course in dynamics at most colleges of engineering.

In the writing of this text we have followed one basic guideline — to write the book the same way we teach the course. To this end, we have written many explanatory footnotes and included frequent questions interspersed throughout the chapters. These questions are the same kind as the ones we ask in class; to make the most of them, treat them as serious homework as you read, and look at the answers only after you have your own answer in mind. The questions are intended to encourage thinking about tricky points and to emphasize the basic principles of the subject.

In addition to the text questions, a set of approximately one dozen review questions and answers are included at the end of each chapter. These true-false questions are designed for both classroom discussion and for student review. Homework problems of varying degrees of difficulty appear at the end of every major section. There are over 1,100 of these exercises, and the answers to the odd-numbered ones constitute Appendix D in the back of the book.

There are a number of reasons (besides carelessness) why it may be difficult to get the correct answer to a homework problem on the first try. The problem may require an unusual amount of thinking and insight; it may contain tedious calculations; or it may challenge the student's advanced mathematics skills. We have placed an asterisk beside especially difficult problems falling into one or more of these categories.

Some examples and problems are presented in SI (Système International) metric units, whereas others use traditional United States engineering system units. Whereas the United States is slowly and painfully converting to SI units, our consulting activities make it clear that much engineering work is still being performed using traditional units. Most United States engineers still tend to think in pounds instead of newtons and in feet instead of meters. We believe students will become much better engineers, scientists, and scholars if they are thoroughly familiar with both systems.

Dynamics is a subject rich in its varied applications; therefore, it is important that the student develop a feel for realistically modeling an engineering situation. Consequently, we have included a large number of actual engineering problems among the examples and exercises. Being aware of the assumptions and accompanying limitations of the model and the solution method is a valuable skill that can only be developed by sweating over many problems outside the classroom. Only in this way can a student develop the insight and creativity that must be brought to bear on engineering problems.

Kinematics of the particle, or of a material point of a body, is covered in Chapter 1. The associated kinetics of particles and mass centers of bodies follows logically in Chapter 2. Here it will be seen that we have not dwelt at length upon the "point-mass" model of a body. Since the engineering student will be dealing with bodies of finite dimensions, we believe that it is important to present equations of motion valid for such bodies as quickly as possible. Thus Euler's laws have been introduced relatively early; this provides for a compact presentation of general principles without, in our opinion and experience, any loss of understanding on the student's part. This is not meant to deprecate the point-mass model, which surely plays an important role in classical physics and can be utilized in a number of engineering problems. As we shall see in Chapter 2, however, these problems may be attacked directly through the equation of motion of the mass center of a body without detracting from the view that the body has finite dimensions. Trajectory problems, sometimes placed with particle kinematics, will be found in this kinetics chapter also, since a law of motion is essential to their formulation.

The rigid body in plane motion is treated in detail in the center of the book — the kinematics in Chapter 3 and the kinetics in Chapters 4 and 5. In Chapter 3 the topic of rolling is discussed only after both the velocity and acceleration equations relating two points of a rigid body have been covered. Further, we treat the equations of velocity and acceleration of a point moving relative to two frames of reference ("moving frames") in Chapters 3 (plane kinematics) and 6 (three-dimensional, or general, kinematics), after the student has been properly introduced to the angular velocity vector in these chapters.

Chapter 4 approaches plane kinetics from the equations of motion, written with the aid of a free-body diagram of the body being studied — that is, a sketch of the body depicting all the external forces and couples but excluding any vectors expressing acceleration. Thus the free-body diagram means the same thing in dynamics as it does in statics, facilitating the student's transition to the more difficult subject. Moments and products of inertia are covered right where they appear in the development of kinetics. This presentation gives students an appreciation of these concepts, as well as a sense of history, as they encounter them along the same paths that were traveled by the old masters.

Chapter 5 is dedicated to solving plane kinetics problems of rigid bodies with certain special yet general solutions (or integrals) of the equations of Chapters 2 and 4. These are known as the principles of work and kinetic energy, impulse and momentum, and angular impulse and angular momentum.

Chapters 6 and 7 deal comprehensively with the kinematics and kinetics, respectively, of rigid bodies in three dimensions. There is no natural linear extension from plane to general motion, and the culprit is the angular velocity vector $\boldsymbol{\omega}$, which depends in a much more complicated way than "$\dot{\theta}\hat{\mathbf{k}}$" on the angles used to orient the body in three dimensions. We have found that if students understand the angular velocity vector $\boldsymbol{\omega}$, they will have little trouble with the general motion of rigid bodies. Thus we begin Chapter 6 with a study of $\boldsymbol{\omega}$ and its properties. In three dimensions, the definition of angular velocity is motivationally developed through the relationship between derivatives of a vector in two different frames of reference. While this point of view is often associated with more advanced texts, we have found that college students at the junior level are quite capable of appreciating and exploiting the power of this approach. In particular, it allows the student to attack, in an orderly way, intimidating problems such as motions of gear systems and those of universal joints connecting noncollinear shafts.

Chapter 8 is an introduction to three special topics in the area of dynamics: vibrations, mass redistribution problems, and central force motion.

We have received a number of helpful suggestions from those who taught from earlier editions of the text, and we are especially grateful to Lawrence Malvern of the *University of Florida* and to our colleagues at *Georgia Institute of Technology*, in particular Don Berghaus, Mike Bernard, Al Ferri, Satya Hanagud, Dewey Hodges, Larry Jacobs, John Papastavridis, Jianmin Qu, George Rentzepis, Virgil Smith, Charles Ueng, Ray Vito, James Wang, Gerry Wempner and Wan-Lee Yin.

We also wish to thank our reviewers this time around:

William Bickford
Arizona State University

Donald E. Carlson
University of Illinois at Urbana

Robert L. Collins
University of Louisville

John Dickerson
University of South Carolina

John F. Ely
North Carolina State University

Laurence Jacobs
Georgia Institute of Technology

Seymour Lampert
University of Southern California

Vincent WoSang Lee
University of Southern California

Joseph Longuski
Purdue University

Robert G. Oakberg
Montana State University

Joseph E. Panarelli
University of Nebraska

Mario P. Rivera
Union College

Wallace S. Venable
West Virginia University

Carl Vilmann
Michigan Technological University

We are grateful to the following professors, who each responded to a questionnaire we personally sent out in 1991: Don Carlson, *University of Illinois*; Patrick MacDonald and John Ely, *North Carolina State University*; Vincent Lee, *University of Southern California*; Charles Krousgrill, *Purdue University*; Samuel Sutcliffe, *Tufts University*; Larry Malvern and Martin Eisenberg, *University of Florida*; John Dickerson, *University of South Carolina*; Bill Bickford, *Arizona State University*; James Wilson, *Duke University*; Mario Rivera, *Union College*; and Larry Jacobs, *Georgia Institute of Technology*. Their comments were also invaluable.

Special appreciation is expressed to our insightful editor, Jonathan Plant, our typist, Meghan Root, and to the following individuals involved in the smooth production of this third edition: Mary Thomas and Kirby Lozyniak of *PWS Publishing Company*, and Tamra Winters of *York Production Services*.

David J. McGill
Wilton W. King

We are pleased to introduce to this edition a new set of model-based problems. These problems, presented in a full-color insert bound into the book, introduce students to the process of building three-dimensional models from commonly found objects in order to observe as well as calculate mechanical behavior. Many students beginning their engineering education lack a hands-on, intuitive feel for this behavior, and these specially designed problems can help build confidence in their observational and analytical abilities.

We wish to acknowledge the following contributors to the model-based problems insert:

David J. McGill and Wilton W. King for their initial conception and presentation of the model-based problem idea in *Dynamics Model Problems*, written to accompany *Engineering Mechanics: An Introduction to Dynamics*, Third Edition.

David Barnett, *Stanford University*, Mario P. Rivera, *Union College*; Robert G. Oakberg, *Montana State University*; John F. Ely, *North Carolina State University*; Carl Vilmann, *Michigan Technological University*; Robert L. Collins, *University of Louisville*; Nicholas P. Jones, *Johns Hopkins University* and William B. Bickford, *Arizona State University* for their evaluations of McGill and King's *Dynamics Model Problems*.

We thank Mario P. Rivera, Robert G. Oakberg, and John F. Ely, for developing additional model problems for the insert. And a very special thanks to Michael K. Wells, *Montana State University*, for developing and editing the final text of the insert, and for providing an introduction and additional problems.

PWS Publishing Company

Engineering Mechanics

An Introduction to

Dynamics

1

Kinematics of Material Points or Particles

1.1 Introduction

1.2 Reference Frames and Vector Derivatives

1.3 Position, Velocity, and Acceleration

1.4 Kinematics of a Point in Rectilinear Motion

The v-t Diagram

1.5 Rectangular Cartesian Coordinates

1.6 Cylindrical Coordinates

1.7 Tangential and Normal Components

Summary

Review Questions

1.1 Introduction

Dynamics is the general name given to the study of the motions of bodies and the forces that accompany or cause those motions. The branch of the subject that deals only with considerations of space and time is called **kinematics.** The branch that deals with the relationships between forces and motions is called **kinetics,** but since the force-motion relationships involve kinematic considerations, it is necessary to study kinematics first.

In this chapter we present some fundamentals of the kinematics of a material point or, equivalently, an infinitesimal element of material. We shall use the term **particle** for such an element, but we shall also use this term in a broader sense to denote a piece of material sufficiently small that the locations of its different material points need not be distinguished. The vagueness of this definition correctly suggests that, for *some* purposes, a truck or a space vehicle or even a planet might be modeled adequately as a particle.

The key elements of the kinematics of a point (or particle) are its position, velocity and acceleration. Velocity is rate-of-change of position, and acceleration is rate-of-change of velocity. It is the acceleration in Newton's Second Law and the position-velocity-acceleration relationships that allow us to deduce two things: the forces that must act for a particle to achieve a certain motion; or the evolution of a particle's position under the action of a set of prescribed forces.

Position, velocity, and acceleration vectors are defined in Section 1.3. These definitions are independent of the choice of any particular coordinate system. However, solution of a practical problem almost always involves the use of some specific coordinate system, the most common being rectangular (Cartesian), cylindrical and spherical. The rectangular and the cylindrical coordinate systems are treated in great detail in this chapter (Sections 1.4–1.6) because we judge them to be of greatest practical use, particularly for problems of motion confined to a plane. These developments are sufficient for establishing the procedures to follow should the reader later find it desirable to develop counterpart relationships using some other system.

If we focus our attention on the path being traversed by a point (or particle), we find that the velocity of the point is tangent to the path and that the acceleration has parts normal and tangent to the path with special significances. These characteristics are developed and exploited in Section 1.7.

We now begin our study of particle kinematics with a preliminary section devoted to the calculus of vectors that depend upon scalars. In particular, we need to understand how to take derivatives of vectors with respect to time and to acknowledge the crucial role of the frame of reference.

1.2 Reference Frames and Vector Derivatives

In the next section and throughout the book, we are going to be differentiating vectors; the derivative of the position vector of a point will be its velocity, for example. Thus in this preliminary section it seems wise to examine the concept of the derivative of a vector **A**, which is a function of time t. The definition of $d\mathbf{A}/dt$, which is also commonly written as $\dot{\mathbf{A}}$, is deceptively simple:

$$\frac{d\mathbf{A}}{dt} \equiv \lim_{\Delta t \to 0} \left[\frac{\mathbf{A}(t + \Delta t) - \mathbf{A}(t)}{\Delta t} \right] \tag{1.1}$$

This definition closely parallels the definition of the derivative of a scalar, such as dy/dx, as found in any calculus text. But what we must realize about a vector is that it can change with time to *two* ways—in direction as well as in magnitude. This means that $\dot{\mathbf{A}}$ is intrinsically tied to the frame of reference in which the derivative is taken.

To illustrate this idea, consider the two points P and Q on the surface of the phonograph record in Figure 1.1. The record rests on a turntable that revolves in the indicated direction at, say, $33\frac{1}{3}$ rpm.

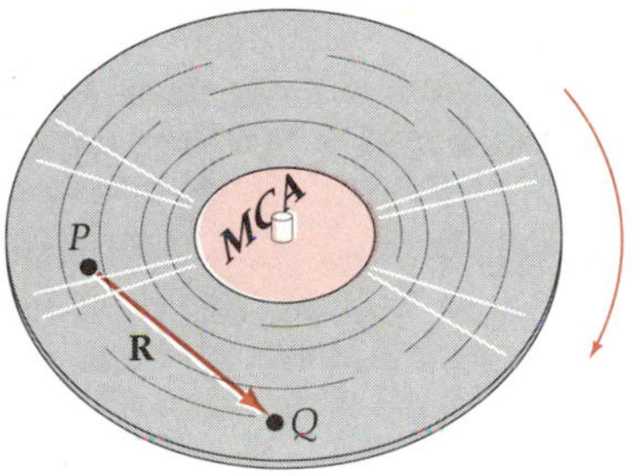

Figure 1.1

Suppose we call **R** the vector that is the directed line segment from P to Q, and inquire about the rate at which **R** changes with time as the turntable rotates. Even though we perceive the record and turntable to behave as a rigid body so that the distance between P and Q (that is, the magnitude of **R**) is constant, most of us would judge $d\mathbf{R}/dt$ to be nonzero, owing to the varying direction of **R**. This conclusion follows from our automatically having adopted the building (or earth) as our frame of reference. If we were to ride on the turntable and blind ourselves to the surroundings, however, our perception would be that **R** is a constant vector and consequently has a vanishing derivative. Thus **R**, relative to the turntable, is a constant vector and, relative to the building, is a constant-magnitude but varying-direction vector. It is therefore seen that $d\mathbf{R}/dt$ cannot be evaluated except by specific association with a **frame of reference,** which is nothing more nor less than a **rigid body.** We shall discuss the frame of reference concept further in the next section and again in Chapter 3.

We shall be needing several vector derivative relationships that have analogs in the calculus of scalars; these relationships follow directly from the definition (Equation 1.1). If α is a scalar and **A** and **B** are vector functions of t, then

$$\frac{d}{dt}(\alpha\mathbf{A}) = \left(\frac{d\alpha}{dt}\right)\mathbf{A} + \alpha\left(\frac{d\mathbf{A}}{dt}\right) \tag{1.2}$$

$$\frac{d}{dt}(\mathbf{A} + \mathbf{B}) = \frac{d\mathbf{A}}{dt} + \frac{d\mathbf{B}}{dt} \tag{1.3}$$

$$\frac{d}{dt}(\mathbf{A} \cdot \mathbf{B}) = \left(\frac{d\mathbf{A}}{dt}\right) \cdot \mathbf{B} + \mathbf{A} \cdot \left(\frac{d\mathbf{B}}{dt}\right) \tag{1.4}$$

$$\frac{d}{dt}(\mathbf{A} \times \mathbf{B}) = \left(\frac{d\mathbf{A}}{dt}\right) \times \mathbf{B} + \mathbf{A} \times \left(\frac{d\mathbf{B}}{dt}\right) \tag{1.5}$$

The first and second of these equations allow us to be more specific about the manner in which differentiation is linked to a frame of reference. Suppose that $\hat{\mathbf{e}}_1, \hat{\mathbf{e}}_2, \hat{\mathbf{e}}_3$ are mutually perpendicular unit vectors* and A_1, A_2, A_3 are the corresponding scalar components of a vector $\mathbf{A}$ so that

$$\mathbf{A} = A_1\hat{\mathbf{e}}_1 + A_2\hat{\mathbf{e}}_2 + A_3\hat{\mathbf{e}}_3 \tag{1.6}$$

If $\mathcal{I}$ is the frame of reference** and we denote the derivative of $\mathbf{A}$ relative to $\mathcal{I}$ by ${}^{\mathcal{I}}d\mathbf{A}/dt$, then

$$\begin{aligned}\frac{{}^{\mathcal{I}}d\mathbf{A}}{dt} = \left(\frac{dA_1}{dt}\right)^{\dagger}\hat{\mathbf{e}}_1 + A_1\,{}^{\mathcal{I}}\!\left(\frac{d\hat{\mathbf{e}}_1}{dt}\right) + \left(\frac{dA_2}{dt}\right)\hat{\mathbf{e}}_2 \\ + A_2\,{}^{\mathcal{I}}\!\left(\frac{d\hat{\mathbf{e}}_2}{dt}\right) + \left(\frac{dA_3}{dt}\right)\hat{\mathbf{e}}_3 + A_3\,{}^{\mathcal{I}}\!\left(\frac{d\hat{\mathbf{e}}_3}{dt}\right)\end{aligned} \tag{1.7}$$

Now if we choose $\hat{\mathbf{e}}_1$, $\hat{\mathbf{e}}_2$, $\hat{\mathbf{e}}_3$ to have fixed directions in $\mathcal{I}$, they are each constant there and

$$\frac{{}^{\mathcal{I}}d\mathbf{A}}{dt} = \frac{dA_1}{dt}\hat{\mathbf{e}}_1 + \frac{dA_2}{dt}\hat{\mathbf{e}}_2 + \frac{dA_3}{dt}\hat{\mathbf{e}}_3 \tag{1.8}$$

which is the most straightforward way to express the derivative of a vector and its intrinsic association with a frame of reference. We now give one example of the use of Equation (1.8) and, assuming the reader to be familiar with Equations (1.1) to (1.5), then move on to Section 1.3 and the task of describing the motion of a point (particle) P.

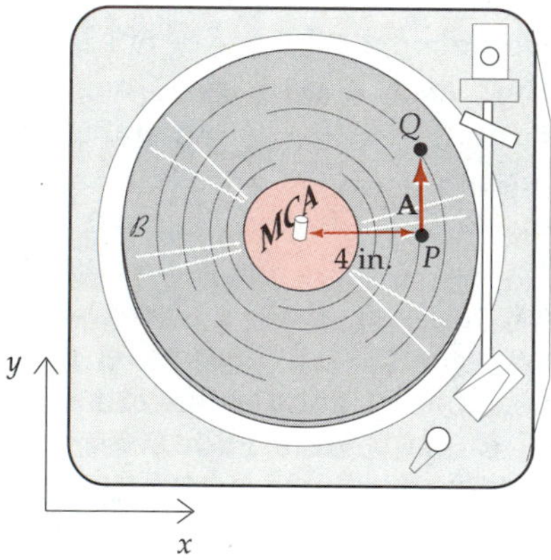

Figure E1.1a

EXAMPLE 1.1

If the distance from P to Q on the $33\frac{1}{3}$ rpm record in Figure E1.1a is 3 in. and if $\mathbf{A}$ is the vector from P to Q, find ${}^{\mathcal{I}}\dot{\mathbf{A}}$, where the frame $\mathcal{I}$ is the cabinet of the stereo in which the axes (x, y, z) are embedded. It is also given that the line PQ is in the indicated position (parallel to y) when $t = 0$.

Solution

At a later time t (in seconds), the vector $\mathbf{A}$ is seen in Figure E1.1b to make an angle $\theta(t)$ with y of

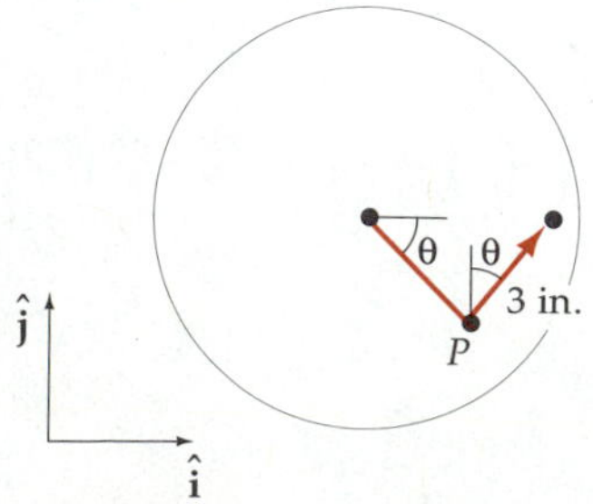

Figure E1.1b

* Note that we could use any set of base vectors (that is, linearly independent reference vectors) here, in which case A_1, A_2, A_3 are not necessarily orthogonal components of $\mathbf{A}$. Equation (1.6) simply illustrates the most common choice of scalars and base vectors. When this is the case, the magnitude of $\mathbf{A}$, written $|\mathbf{A}|$ or sometimes A, is $\sqrt{A_1^2 + A_2^2 + A_3^2}$.

** Throughout the book, frames (rigid bodies) are denoted by capital script letters. These are intended simply to be the capital cursive letters we use in writing; thus we *write* the names of bodies and *print* the names of points. We do this because, as we shall see in Chapter 3, points and rigid bodies have very different motion properties.

† Note that the derivatives of the scalar components of $\mathbf{A}$, such as dA_1/dt, need not be "tagged" since they are the same in any frame.

$$\theta = (33\tfrac{1}{3})\left(\frac{2\pi}{60}\right) t = 3.49t \text{ rad}$$

The vector **A**, expressed in terms of the unit vectors $\hat{\mathbf{i}}$ and $\hat{\mathbf{j}}$ in the respective directions of x and y, then has the following form:

$$\mathbf{A} = 3(\sin\theta\hat{\mathbf{i}} + \cos\theta\hat{\mathbf{j}}) \text{ in.}$$

Noting that the unit vectors do not change in direction in $\mathcal{I}$, we obtain, using Equation (1.8),

$$\frac{{}^{\mathcal{I}}d\mathbf{A}}{dt} = {}^{\mathcal{I}}\dot{\mathbf{A}} = 3\cos\theta\frac{d\theta}{dt}\hat{\mathbf{i}} - 3\sin\theta\frac{d\theta}{dt}\hat{\mathbf{j}}$$

$$= 3\cos(3.49t)(3.49)\hat{\mathbf{i}} - 3\sin(3.49t)(3.49)\hat{\mathbf{j}}$$

$$= 10.5(\cos\theta\hat{\mathbf{i}} - \sin\theta\hat{\mathbf{j}}) \text{ in./sec}$$

We see from this result, for example, that:

1. At $\theta = 0$, ${}^{\mathcal{I}}\dot{\mathbf{A}}$ is in the x direction.
2. At $\theta = \pi/2$, ${}^{\mathcal{I}}\dot{\mathbf{A}}$ is in the $-y$ direction.
3. At $\theta = \pi$, ${}^{\mathcal{I}}\dot{\mathbf{A}}$ is in the $-x$ direction.
4. At $\theta = 3\pi/2$, ${}^{\mathcal{I}}\dot{\mathbf{A}}$ is in the y direction.

In all four cases, and at all intermediate angles as well, the derivative of **A** in $\mathcal{I}$ is seen to be that of the cross product:

$$[\dot{\theta}(-\hat{\mathbf{k}})] \times \mathbf{A}$$

The bracketed vector represents what will come to be called *the angular velocity* of the record ($\mathcal{B}$) in the reference frame (stereo cabinet) $\mathcal{I}$. In later chapters we shall see that it is precisely this cross product that must be added to ${}^{\mathcal{B}}\dot{\mathbf{A}}$ to obtain ${}^{\mathcal{I}}\dot{\mathbf{A}}$. Here, of course, **A** is constant relative to the turntable $\mathcal{B}$ so that its derivative in $\mathcal{B}$ (that is, ${}^{\mathcal{B}}\dot{\mathbf{A}}$) vanishes.

PROBLEMS ▶ Section 1.2

In Problems 1.1–1.8, $\hat{\mathbf{i}}$, $\hat{\mathbf{j}}$, $\hat{\mathbf{k}}$ are mutually perpendicular unit vectors having directions fixed in the frame of reference. In each case t is time measured in seconds ("s" in SI units). Determine at $t = 3$ s the rate of change (with respect to time) of vector **L**.

1.1 $\mathbf{L} = 2\hat{\mathbf{i}} + 3t^2\hat{\mathbf{j}} - 8t\hat{\mathbf{k}}$ kg-m/s

1.2 $\mathbf{L} = 20\sin(\pi t/4)\hat{\mathbf{i}} - 20\cos(\pi t/4)\hat{\mathbf{k}}$ slug-ft/sec

1.3 $\mathbf{L} = -100e^{-t/2}\hat{\mathbf{i}} + 20t\hat{\mathbf{j}} - 5t^2\hat{\mathbf{k}}$ slug-ft/sec

1.4 $\mathbf{L} = 20\cosh(\pi t/4)\hat{\mathbf{i}} + 20\sinh(\pi t/4)\hat{\mathbf{k}}$ kg-m/s

1.5 $\mathbf{L} = 5t\hat{\mathbf{i}} - \dfrac{12}{t^2}\hat{\mathbf{j}} + \dfrac{6}{t^3}\hat{\mathbf{k}}$ slug-ft/sec

1.6 $\mathbf{L} = e^{-6t}(5\hat{\mathbf{i}} + 8t\hat{\mathbf{j}})$ slug-ft/sec

1.7 $\mathbf{L} = -2te^{-t^2}\hat{\mathbf{i}} + 3^t\hat{\mathbf{j}}$ kg-m/s

1.8 $\mathbf{L} = 50\hat{\mathbf{i}} + 60\ln t\hat{\mathbf{j}}$ kg-m/s

1.9–1.16 If the vectors enumerated in Problems 1.1–1.8 represent various forces **F**, find the integral of each force over the time interval from $t = 2$ through 5 sec. Let the metric units become newtons and the U.S. units become pounds.

1.3 Position, Velocity, and Acceleration

In this short but important section, we present the definitions of the position, velocity, and acceleration vectors of a material point P as it moves relative to a frame of reference $\mathcal{I}$. It is important to mention that while a frame of reference is usually identified by the material constituting the reference body (for example, the earth, the moon, or the body of a truck), the frame is actually composed of all those material points *plus* the points generated by a rigid extension of the body to all of space. Thus, for example, we refer to a point on the centerline of a straight pipe as a point in (or of) the pipe.

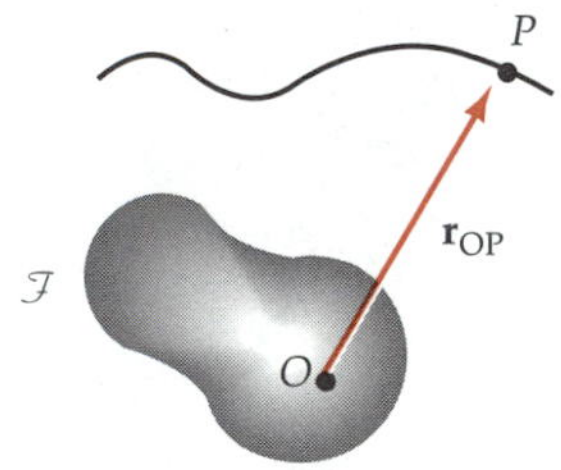

Figure 1.2 Position vector for P in $\mathcal{I}$.

We now consider a point P as it moves along a path as shown in Figure 1.2. The **path** is the locus of points of $\mathcal{I}$ that P occupies as time passes. If we select a point O of $\mathcal{I}$ to be our reference point (or origin), then the depicted vector from O to P is called a **position vector** for P in $\mathcal{I}$ and is written $\mathbf{r}_{OP}$.

The first and second derivatives (with respect to time) of the position vector are respectively called the **velocity** ($\mathbf{v}_P$) and **acceleration** ($\mathbf{a}_P$) of point P in $\mathcal{I}$:

$$\mathbf{v}_P = \frac{d\mathbf{r}_{OP}}{dt} = \dot{\mathbf{r}}_{OP} \qquad \text{(The } \textit{magnitude} \text{ of } \mathbf{v}_P \text{ is called the } \textbf{speed} \text{ of } P.) \tag{1.9}$$

$$\mathbf{a}_P = \frac{d^2\mathbf{r}_{OP}}{dt^2} = \ddot{\mathbf{r}}_{OP} = \frac{d\mathbf{v}_P}{dt} = \dot{\mathbf{v}}_P \tag{1.10}$$

The derivatives in Equations (1.9) and (1.10) are calculated in frame $\mathcal{I}$, the only frame under consideration here. Later, however, we shall sometimes find it necessary to specify the frame in which derivatives, velocities, and accelerations are to be computed; we shall then tag the derivatives as in Equation (1.8) and write

$$\mathbf{v}_{P/\mathcal{I}} = {}^{\mathcal{I}}\dot{\mathbf{r}}_{OP} \tag{1.11}$$

Whenever there is just one frame involved, we shall omit the $\mathcal{I}$ on both sides and write an equation such as (1.11) in the form of (1.9).

Throughout the text we have inserted questions for the reader to think about. (The answer is always on the same page as the question.) Here is the first question:

Question 1.1 Do the velocity and acceleration of a point P depend upon: (a) the choice of reference frame? (b) the origin selected for the position vector?

Answer 1.1 (a) Yes; we could simply define a frame in which P is fixed, and it would then have $\mathbf{v}_P = \mathbf{0} = \mathbf{a}_P$. (b) No; letting O' be a second origin in $\mathcal{I}$ and differentiating the relationship $\mathbf{r}_{OP} = \mathbf{r}_{OO'} + \mathbf{r}_{O'P}$ in $\mathcal{I}$ shows that: $\mathbf{v}_P$ (with origin O) $= \mathbf{v}_P$ (with origin O'). The derivative of $\mathbf{r}_{OO'}$ in $\mathcal{I}$ is, of course, zero! See Problem 1.17.

At this point it is reasonable to wonder why we have not chosen to introduce time derivatives of the position vector higher than the second. The reason is that the relationships between forces and motions do not involve those higher derivatives. As we shall see later when we study kinetics, if we know the accelerations of the particles making up a body, the force-motion laws will yield the external forces; conversely, for rigid bodies, if we know the external forces, we can calculate the accelerations and then, by integrating twice, the position vectors. The force-motion laws turn out to be valid only in certain frames of reference; for that reason writers sometimes refer to motion relative to such a frame as *absolute motion.* We have not used the word *absolute* here because we wish to emphasize that kinematics inherently expresses relationships of geometry and time, independent of any laws linking forces and motions. *Thus in kinematics all frames of reference are of the same importance.*

Finally, we note that positions (or locations) of points are normally established through the use of a coordinate system. The ways in which positions, velocities, and accelerations are expressed in two of the most common systems, rectangular and cylindrical, are presented in the next three sections.

PROBLEMS ▶ Section 1.3

1.17 Show that the velocity (and therefore the acceleration also) of a point P in a frame $\mathcal{I}$ does not depend on the choice of the origin. *Hint*: Differentiate the following relationship in $\mathcal{I}$ (see Figure P1.17):

$$\mathbf{r}_{OP} = \mathbf{r}_{OO'} + \mathbf{r}_{O'P}$$

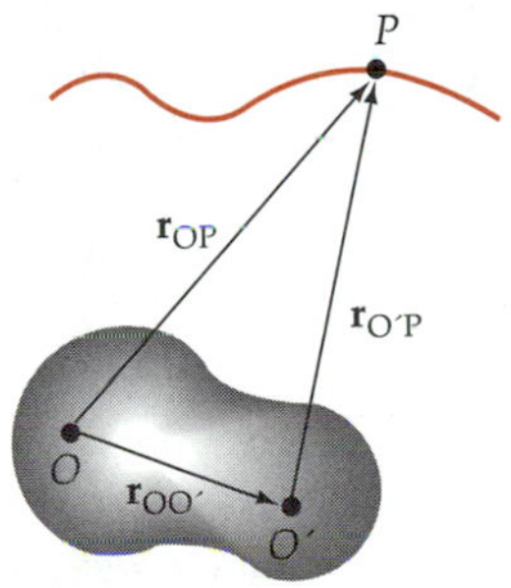

Figure P1.17

In Problems 1.18–1.22, $\hat{\mathbf{i}}, \hat{\mathbf{j}}, \hat{\mathbf{k}}$ are mutually perpendicular unit vectors having directions fixed in the frame of reference; $\mathbf{v}_P$ is the velocity of a point P moving in the frame; t is time measured in seconds. Determine at $t = 2$ s the acceleration of the point for the velocity given.

1.18 $\mathbf{v}_P = 12t\hat{\mathbf{i}} + \dfrac{36}{t^2}\hat{\mathbf{j}} - 3t^2\hat{\mathbf{k}}$ m/s

1.19 $\mathbf{v}_P = 20e^{-0.1t}\left(\sin\dfrac{\pi t}{4}\hat{\mathbf{i}} - \cos\dfrac{\pi t}{4}\hat{\mathbf{j}}\right)$ ft/sec

1.20 $\mathbf{v}_P = 20\sin\dfrac{\pi t}{4}\hat{\mathbf{i}} - 20\cos\dfrac{\pi t}{4}\hat{\mathbf{j}}$ m/s

1.21 $\mathbf{v}_P = t\left(\sin\dfrac{\pi t}{4}\hat{\mathbf{i}} + \cos\dfrac{\pi t}{4}\hat{\mathbf{j}}\right)$ ft/sec

1.22 $\mathbf{v}_P = 5e^{-0.1t}\hat{\mathbf{i}} - 4e^{-0.4t}\hat{\mathbf{k}}$ m/s

1.23–1.27 The **displacement** of a point over a time interval t_1 to t_2 is defined to be the difference of the position vectors—that is, $\mathbf{r}(t_2) - \mathbf{r}(t_1)$. For the cases enumerated in Problems 1.18–1.22, find the displacement and the magnitude of the displacement over the interval $t = 4$ s to $t = 6$ s.

1.4 Kinematics of a Point in Rectilinear Motion

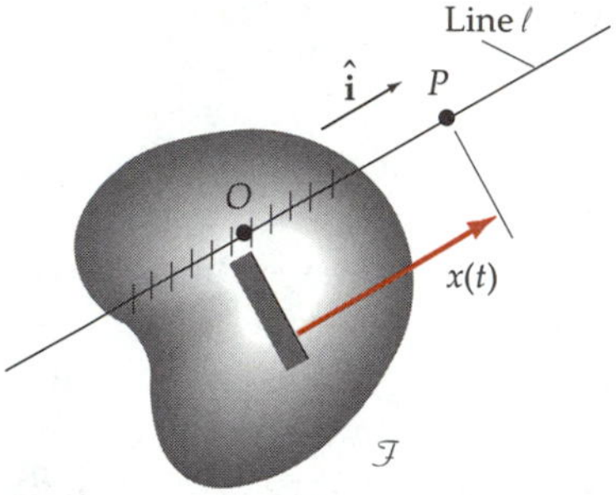

Figure 1.3

In this section we study problems in which point P moves along a straight line in the reference frame $\mathcal{I}$. This situation is called **rectilinear motion,** and the position of P may be expressed with a single coordinate x measured along the fixed line on which P moves (see Figure 1.3).

A position vector for P is simply

$$\mathbf{r}_{OP} = x\hat{\mathbf{i}} \tag{1.12}$$

in which the unit vector $\hat{\mathbf{i}}$ is parallel to the line as shown in Figure 1.3 and hence does not change in either magnitude or direction in $\mathcal{I}$. Therefore P has the following very simple velocity and acceleration expressions:

$$\mathbf{v}_P = \dot{\mathbf{r}}_{OP} = \dot{x}\hat{\mathbf{i}} \tag{1.13}$$

$$\mathbf{a}_P = \ddot{\mathbf{r}}_{OP} = \ddot{x}\hat{\mathbf{i}} \tag{1.14}$$

In rectilinear motion, there are three interesting cases worthy of special note:

1. Acceleration is a known function of time, $f(t)$.
2. Acceleration is a known function of velocity, $g(v)$, where $v = \dot{x}$.
3. Acceleration is a known function of position, $h(x)$.

In each case, we can go far with general integrations. We shall consider each case in turn and give an example.

First, if $\ddot{x} = f(t)$, then

$$\ddot{x} = f(t) \Rightarrow \dot{x} = \int f(t)\,dt + C_1 \Rightarrow x = \int\int f(t)\,dt + C_1 t + C_2 \tag{1.15}$$

in which C_1 and C_2 are to be determined by the initial conditions on velocity and position, respectively, once the problem (and thus $f(t)$) is stated and the indefinite integrals are performed. Alternatively, we might know the values of x at two times, instead of one position and one velocity. In any case, we need two constants.

EXAMPLE 1.2

The acceleration of a point P in rectilinear motion is given by the equation $\ddot{x} = 5t^2$ m/s², with initial conditions $\dot{x}(0) = 2$ m/s and $x(0) = -7$ m. Find $x(t)$.

Solution

We note that this is the problem of a point moving with a quadratically varying acceleration magnitude and with the initial conditions being the position and velocity of P at $t = 0$ as shown in Figure E1.2. Integrating as above,

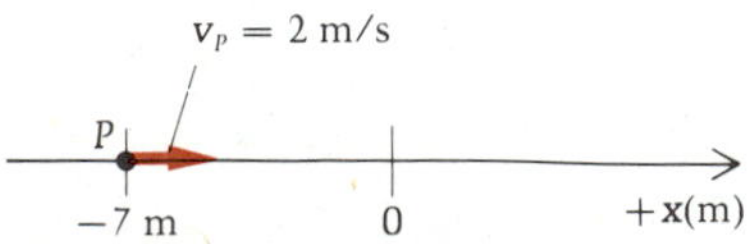

Figure E1.2

$$\dot{x} = \int 5t^2\,dt + C_1 = \tfrac{5}{3}t^3 + C_1 \text{ m/s}$$

And integrating once more,

$$x = \tfrac{5}{12}t^4 + C_1 t + C_2 \text{ m}$$

The constants are found from the initial conditions to be $C_1 = 2$ m/s and $C_2 = -7$ m, as follows:

$$\dot{x}(0) = 2 = (\tfrac{5}{3})(0)^3 + C_1 \Rightarrow C_1 = 2 \text{ m/s}$$

$$x(0) = -7 = (\tfrac{5}{12})(0)^4 + 2(0) + C_2 \Rightarrow C_2 = -7 \text{ m}$$

Thus the motion of the point P is given by the integrated function of time:

$$x = \tfrac{5}{12}t^4 + 2t - 7 \text{ m}$$

If $\ddot{x} = g(v)$, then

$$\ddot{x} = \frac{dv}{dt} = g(v)$$

Suppose $v(t)$ can be inverted to give $t(v)$; then

$$\frac{dt}{dv} = \frac{1}{g(v)}$$

and

$$t + C_3 = \int \frac{dv}{g(v)}$$

If the integral can be found as a function $p(v)$, then we may be able to solve the equation $p(v) = t + C_3$ for the velocity:

$$v = q(t)$$

If so, then

$$v = \frac{dx}{dt} = q(t)$$

so that

$$x = \int q(t)\, dt + C_4 \tag{1.16}$$

This procedure should become clearer with the following example.

Example 1.3

Suppose that the acceleration of a point P in one-dimensional motion is proportional to velocity according to $\ddot{x} = -2v$ m/s^2 with the same initial conditions as in the previous example. Solve for the motion $x(t)$.

Solution

$$\ddot{x} = \frac{dv}{dt} = -2v \Rightarrow \int \frac{dv}{-2v} = t + C_3$$

so that, integrating,* we get

$$t + C_3 = \frac{-\ln v}{2} \Rightarrow v = e^{-2t - 2C_3}$$

Since $v = 2$ when $t = 0$, then $C_3 = (-\ln 2)/2$ and

$$\frac{dx}{dt} = v = e^{-2t + \ln 2} = 2e^{-2t} \text{ m/s}$$

Therefore

$$x = \int 2e^{-2t}\, dt + C_4 = -e^{-2t} + C_4 \text{ m}$$

But $x = -7$ m when $t = 0$ s gives $C_4 = -6$ m, and so we obtain our solution:

$$x = -6 - e^{-2t} \text{ m}$$

When acceleration is a function of position, $a = \ddot{x} = h(x)$, we may combine $a = \dot{v}$ and $v = \dot{x}$ to obtain the useful relation

$$a \frac{dx}{dt} = v \frac{dv}{dt} \tag{1.17}$$

Then, if a function $r(x)$ exists such that $h(x) = \dfrac{d\, r(x)}{dx}$, we obtain, from Equation (1.17), the following:

$$\frac{dr}{dx} \frac{dx}{dt} = v \frac{dv}{dt}$$

$$\frac{dr}{dt} = v \frac{dv}{dt} \tag{1.18}$$

and integrating with respect to time,

$$r(x) = \frac{v^2}{2} + C_5 \tag{1.19}$$

Thus the square of the speed is

$$v^2 = 2r(x) - 2C_5$$

Equation (1.19) will be called an energy integral in Chapters 2 and 5.

EXAMPLE 1.4

Let $\ddot{x} = h(x) = -4x$ m/s^2. Find $v^2(x)$ if the initial conditions are the same as in Examples 1.2 and 1.3.

---

* This problem could also be solved by first integrating the linear differential equation $\dot{v} + 2v = 0$, observing that Ae^{-2t} is the general solution.

Solution

We are dealing with the equation

$$\ddot{x} + 4x = 0$$

Actually we know that the solution to this equation, by the theory of differential equations, is $x = A \sin 2t + B \cos 2t$—which, with $x(0) = -7$ m and $\dot{x}(0) = 2$ m/s, becomes $x = \sin 2t - 7 \cos 2t$ meters. But let us obtain the desired result by using the procedure described above, which applies even when $h(x)$ is *not* linear. Here $h(x) = -4x$, so with $r(x) = -2x^2$, Equation (1.19) gives

$$-2x^2 = \frac{v^2}{2} + C_5$$

or

$$v^2 = -4x^2 - 2C_5 = 200 - 4x^2$$

where C_5 has been found by using $v = 2$ m/s and $x = -7$ m at $t = 0$.

The *v-t* Diagram

In problems of rectilinear kinematics in which the acceleration is a known function of time (Case 1), we sometimes use what is called the v-t diagram. We shall give just one example of its use because it is a method somewhat limited in application. (We discuss this shortcoming at the end of the example.)

Example 1.5

A point P moves on a line, starting from rest at the origin with constant acceleration of 0.8 m/s² to the right. After 10 s, the acceleration of P is suddenly reversed to 0.2 m/s² to the left. Determine the total time elapsed when P is again passing through the origin.

Solution

If we graph the velocity versus time, the acceleration (dv/dt) will of course be the slope of the curve at every point. The v-t diagram for this problem is shown in Figure E1.5a.

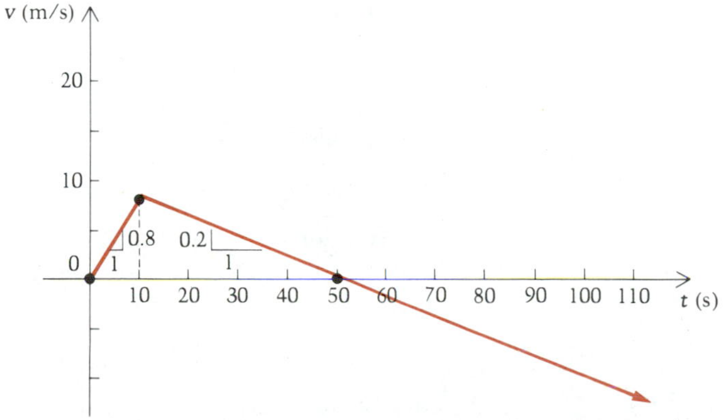

Figure E1.5a

We note not only that

$$a = \frac{dv}{dt} = \text{slope of diagram}$$

but also that

$$x = \int v\, dt + \text{constant}$$

Hence the change in the position x between any two times is nothing more than the area beneath the v-t diagram between those points. Thus four points, or times, are important in the diagram for this problem:

t_1 = starting time (in this case zero)

t_2 = time when acceleration changes (given to be 10 s)

t_3 = time when velocity has been reduced to zero (deceleration causes P to stop before moving in opposite direction)

t_4 = required total time elapsed before point P is again at origin

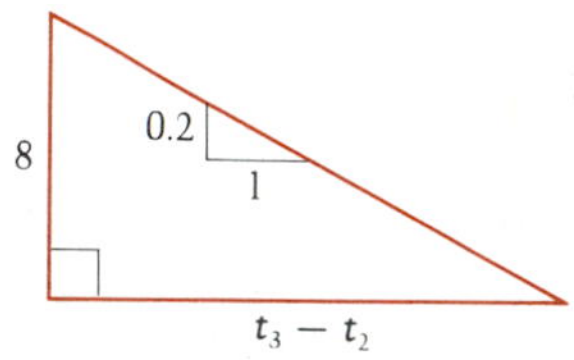

Figure E1.5b

The velocity at time t_2 is seen to be $0.8 \text{ m/s}^2 \times 10 \text{ s} = 8 \text{ m/s}$. To find the time interval $t_3 - t_2$, we use the similar triangles shown in Figure E1.5b.

$$\frac{0.2}{1} = \frac{8}{t_3 - t_2} \Rightarrow t_3 - t_2 = 40$$

$$t_3 = 50 \text{ s}$$

The total distance traveled before the point (momentarily) stops is thus

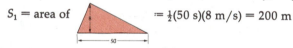

S_1 = area of 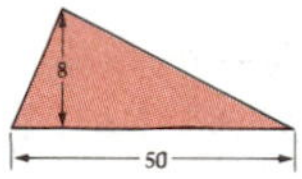 $= \frac{1}{2}(50 \text{ s})(8 \text{ m/s}) = 200 \text{ m}$

This is the distance traveled by the point in the positive direction (to the right).

The point will be back at $x = 0$ when the absolute value of the negative area *beneath* the t axis (the distance traveled back to the left) equals the 200 m traveled to the right (represented by the area *above* the axis):

$$\underbrace{\tfrac{1}{2}(t_4 - 50)}_{\text{base of triangle}}\underbrace{[0.2(t_4 - 50)]}_{\text{height of triangle}} = 200$$

which can be rewritten as

$$t_4^2 - 100t_4 + 500 = 0$$

The only root of this equation larger than 50 s is $t_4 = 94.7$ s, and this is the answer to the problem.

An alternative approach to the preceding v-t diagram solution is as follows. Integrating the acceleration during the interval $0 \le t < 10$ s, with x during this interval called x_1,

$$\ddot{x}_1 = 0.8 \Rightarrow \dot{x}_1 = 0.8t + C_1 = 0.8t \text{ m/s} \qquad (\text{since } \dot{x}_1 = 0 \text{ at } t = 0)$$

Integrating again (over the same interval), we get

$$x_1 = 0.4t^2 + C_2 = 0.4t^2 \text{ m} \qquad (\text{since } x_1 = 0 \text{ at } t = 0)$$

Thus at $t = 10$ s, by substitution,

$$x_1 = 40 \text{ m} \qquad \text{and} \qquad \dot{x}_1 = 8 \text{ m/s}$$

Next, after the deceleration starts, using x_2 in this interval,

$$\ddot{x}_2 = -0.2 \Rightarrow \dot{x}_2 = -0.2t + C_3 \qquad (\text{for } t \geq 10 \text{ s})$$

and since $\dot{x}_2 = 8$ m/s when $t = 10$ s, we obtain $C_3 = 10$. Therefore

$$\dot{x}_2 = -0.2t + 10 \text{ m/s}$$

Integrating again, we get

$$x_2 = -0.1t^2 + 10t + C_4 \text{ m}$$

And with $x_2 = 40$ m when $t = 10$ s, then $C_4 = -50$ m:

$$x_2 = -0.1t^2 + 10t - 50 \text{ m}$$

When $x_2 = 0$, we can solve for the time; the equation is the same as in the v-t diagram solution:

$$t_4^2 - 100t_4 + 500 = 0$$

Of the roots, $t_4 = 5.28$ and 94.7 s, only the latter is valid since 5.28 s occurs prior to the change of acceleration expressions.

Even though both approaches yield the correct answer of 94.7 s in the preceding example, we must recommend the latter approach of integrating the accelerations and matching velocities and positions between intervals. The reason is that when we are faced with *nonconstant* accelerations, the v-t diagram approach requires us to find areas under curves, the formulas for which are not ordinarily memorized.

It is interesting, in using the equations, to start a new time measurement t_2 at the beginning of the second interval:

$$\ddot{x}_2 = -0.2 \text{ m/s}^2 \Rightarrow \dot{x}_2 = -0.2t_2 + C_3 = -0.2t_2 + 8 \text{ m/s}$$

Integrating again, we get

$$x_2 = -0.1t_2^2 + 8t_2 + C_4 = -0.1t_2^2 + 8t_2 + 40 \text{ m}$$

Then $x_2 = 0$ yields the equation

$$t_2^2 - 80t_2 - 400 = 0$$

which has the positive root $t_2 = 84.7$—which, added to the 10-s duration of the first interval, gives again 94.7 s of total time elapsed. It is slightly easier to calculate the integration constants with this approach of "starting time over" than to use the same t throughout. The only price we pay for this convenience is that we must add the times at the end.

EXAMPLE 1.6

A point B starts from rest at the origin at $t = 0$ and accelerates at a constant rate k m/s² in rectilinear motion. After 6 s, the acceleration changes to the time-dependent function $0.006t_2^2$ m/s² in the *opposite* direction, where $t_2 = 0$ when

$t = 6$ s. If the point stops at $t = 26$ s (from the starting time) and reverses direction, find the acceleration k during the first interval and the distance traveled by B before it reverses direction. Then find the total time elapsed before B passes back through the origin.

Solution

We begin the solution by determining the motion ($x_1(t)$) during the first time interval; we integrate the acceleration to obtain the velocity and then again to get the position:

$$\ddot{x}_1 = k \text{ m/s}^2$$

$$\dot{x}_1 = kt_1 + c_1 = kt_1 \text{ m/s} \qquad (\text{since } \dot{x}_1 = 0 \text{ when } t_1 = 0)$$

$$x_1 = \frac{kt_1^2}{2} + c_2 = \frac{kt_1^2}{2} \text{ m} \qquad (\text{since } x_1 = 0 \text{ when } t_1 = 0)$$

At $t_1 = 6$ s, the acceleration changes to a negative value and point B "decelerates." At the beginning of this second interval, the speed and position of B are given by the "ending" values during the first interval. These values are $\dot{x}_1$ and x_1 at $t_1 = 6$:

$$x_2|_{t_2=0} = x_1|_{t_1=6} = \frac{k6^2}{2} = 18k \text{ m}$$

$$\dot{x}_2|_{t_2=0} = \dot{x}_1|_{t_1=6} = 6k \text{ m/s}$$

Note that we start time t_2 at the beginning of the second interval, during which we have

$$\ddot{x}_2 = -0.006t^2 \text{ m/s}^2$$

where we note that the minus sign is needed to express the *deceleration.* Integrating, we get

$$\begin{aligned}\dot{x}_2 &= -0.002t_2^3 + c_3 \\ &= -0.002t_2^3 + 6k \text{ m/s}\end{aligned}$$

since $\dot{x}_2 = 6k$ m/s when $t_2 = 0$. Integrating a second time, we get

$$\begin{aligned}x_2 &= -0.0005t_2^4 + 6kt_2 + c_4 \\ &= -0.0005t_2^4 + 6kt_2 + 18k \text{ m}\end{aligned}$$

where c_4 was computed by using the initial condition that $x_2 = 18k$ meters when $t_2 = 0$.

Now we use the fact that $\dot{x}_2$ is zero at time $t_2 = 26 - 6 = 20$ s; this strategy will allow us to determine k:

$$0 = -0.002(20^3) + 6k$$

$$k = 2.67 \text{ m/s}^2$$

Substituting k into the x_2 expression at $t_2 = 20$ s gives us the position of B at the "turnaround":

$$\begin{aligned}x_{2\text{STOP}} &= -0.0005(20^4) + 6(2.67)(20) + 18(2.67) \\ &= 288 \text{ m}\end{aligned}$$

Finally, to obtain the time $t_{2\text{END}}$ when B is passing back through the origin we set

$$x_2 = 0 = -0.0005t_{2\text{END}}^4 + 6(2.67)t_{2\text{END}} + 18(2.67)$$

Rewriting, we get

$$t_{2\text{END}}^4 - 32{,}000t_{2\text{END}} - 96{,}100 = 0 \tag{1}$$

On a calculator, the only positive root to this equation* is found in a matter of minutes to be (to three significant figures):

$$t_{2\text{END}} = 32.7 \text{ s}$$

The total time is $t_{2\text{END}}$ plus the duration of the first interval, or 38.7 s.

Before we leave this example we wish to note that during the first time interval, *while the acceleration is constant,*

$$x = \frac{kt^2}{2} + v_0 t + x_0 \text{ m} \tag{1.20}$$

where

$$x_0 = x(0) \text{ m}$$
$$v_0 = \dot{x}(0) \text{ m/s}$$

Letting $v = \dot{x}$, we note further that

$$v = kt + v_0 \text{ m/s} \tag{1.21}$$

and eliminating t we obtain

$$v^2 = v_0^2 + 2k(x - x_0) \text{ m}^2/\text{s}^2 \tag{1.22}$$

This expression gives us the magnitude of the velocity in terms of displacement. Most students have used this relationship in high school or perhaps elementary college physics. There is sometimes a tendency, however, to forget the conditions under which it is valid; it holds only for *rectilinear* motion with *constant acceleration.* Thus it could not be used during the second interval of the preceding example, nor could the equations for x and v from which it was derived.

Example 1.7

Two cars in a demolition derby are approaching a common point (the origin in Figure E1.7), each at 55 mph in a straight line as indicated. Car $\mathcal{C}_1$ does not speed up or slow down; the driver of $\mathcal{C}_2$ applies the brakes. Find the smallest rate of deceleration of $\mathcal{C}_2$ that will allow $\mathcal{C}_1$ to precede it through the intersection, if:

a. $d_2 = 200$ ft
b. $d_2 = 100$ ft

* Descartes' rule of signs tells us that the maximum number of positive real roots to Equation (1) is one (the number of changes in sign on the left-hand side). And there will be exactly one because the left side is negative at $t_{2\text{END}} = 0$ and positive for large values of $t_{2\text{END}}$.

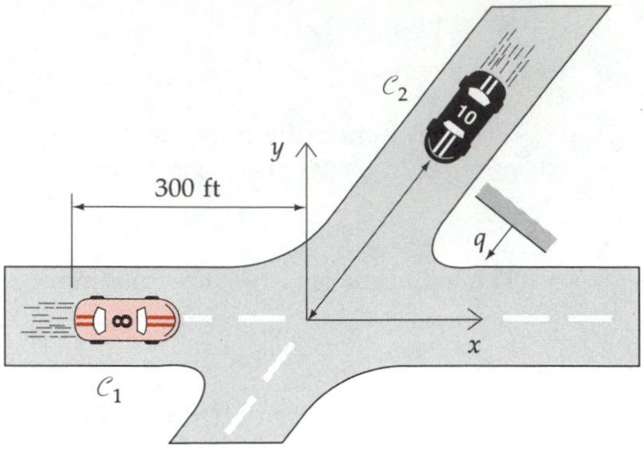

Figure E1.7

Solution

Placing the origin at the point of intersection of the two cars' paths, we have the following for $\mathcal{C}_1$:

$$\dot{x}_1 = 55 \text{ mph}\left(\frac{88 \text{ ft/sec}}{60 \text{ mph}}\right) = 80.7 \text{ ft/sec}$$

$$x_1 = 80.7t + C_1 \text{ ft}$$

Using the initial condition that $x_1 = -300$ ft when $t = 0$, we get

$$x_1 = 80.7t - 300 \text{ ft}$$

The back of car $\mathcal{C}_1$ will be at the origin (point of possible collision) when $x = 0$:

$$0 = 80.7t_0 - 300$$

$$t_0 = 3.72 \text{ sec}$$

Now let us study the motion of car $\mathcal{C}_2$. We use the coordinate q as shown for this car. Calling the unknown deceleration K, we obtain

$$\ddot{q} = -K \text{ ft/sec}^2$$

so that

$$\dot{q} = -Kt + C_2 = -Kt + 80.7 \text{ ft/sec}$$

and

$$q = \frac{-Kt^2}{2} + 80.7t + C_3 \text{ ft}$$

But $C_3 = 0$, since $q = 0$ at $t = 0$.

Next we see that at 3.72 sec the position of $\mathcal{C}_2$ is

$$q = \frac{-K(3.72^2)}{2} + 80.7(3.72)$$

$$= -6.92K + 300 \text{ ft}$$

Finally, car $\mathcal{C}_2$ just passes the rear of $\mathcal{C}_1$ if q is d_2 at this time:

$$d_2 = -6.92K + 300 \text{ ft}$$

Hence:

a. If $d_2 = 200$ ft, then $K = 14.5$ ft/sec^2.
b. If $d_2 = 100$ ft, then $K = 28.9$ ft/sec^2.

Note also that if $d_2 = 300$ ft, then $K = 0$; this is because *no* deceleration is needed for the same distances at the same speeds. Further, if $d_2 > 300$, then K is negative, meaning that car $\mathcal{C}_2$ would have to *accelerate* to arrive at the intersection at the same time as car $\mathcal{C}_1$.

Sometimes there are special conditions in a problem that require ingenuity in expressing the kinematics. If there is an inextensible rope, string, cable, or cord present, for example, we may have to express the constancy of length mathematically. This is the case in the following example.

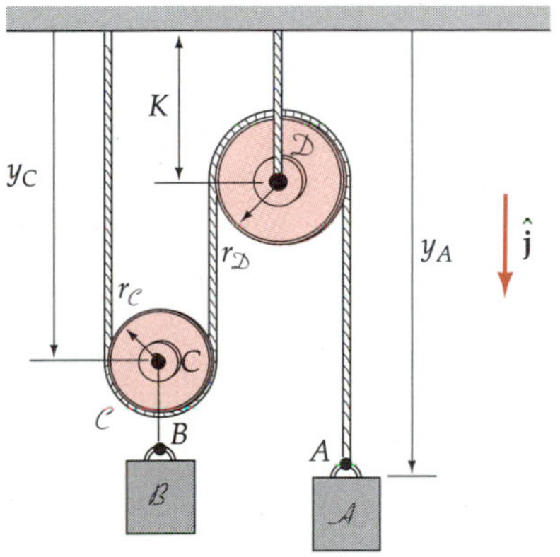

Figure E1.8

Example 1.8

Block $\mathcal{A}$ travels downward with $\mathbf{v}_A = 3t^2 \downarrow$ m/s. Find the velocity of block $\mathcal{B}$ when $t = 4$ s.

Solution

The length L of the rope that passes around both small pulleys is a constant. This is a constraint equation that must be used in the solution. The procedure is as follows (see Figure E1.8):

$$L = y_C + \pi r_{\mathcal{C}} + (y_C - K) + \pi r_{\mathcal{D}} + (y_A - K)\text{ m}$$

Differentiating and noting that L, π, $r_{\mathcal{C}}$, $r_{\mathcal{D}}$, and K are constants, we get

$$0 = 2\dot{y}_C + \dot{y}_A \Rightarrow \dot{y}_C = -\frac{\dot{y}_A}{2} = -\frac{2}{3}t^2$$

The velocities of C and B are equal since both points move on the same path with a constant length separating them. Hence

$$\mathbf{v}_B = -\frac{3}{2}t^2\hat{\mathbf{j}}\text{ m/s} \qquad \text{(Note that } C \text{ moves upward since } \hat{\mathbf{j}} \text{ is downward!)}$$

Therefore

$$\mathbf{v}_B|_{t=4} = -24\hat{\mathbf{j}}\text{ m/s} \qquad \text{or} \qquad 24 \uparrow \text{m/s}$$

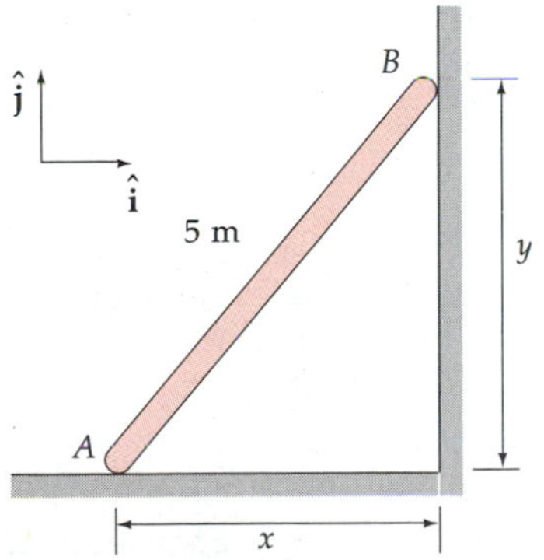

Figure E1.9

Example 1.9

The ends A and B of the rigid bar in Figure E1.9 are to move along the horizontal and vertical guides as shown. End A moves to the right at a constant speed of 8 m/s. Find the velocity and acceleration of B at the instant when A is 3 m from the corner.

Solution

In terms of the parameters and unit vectors shown in Figure E1.9,

$$\mathbf{v}_A = -\dot{x}\hat{\mathbf{i}} = 8\hat{\mathbf{i}} \text{ m/s}, \qquad \mathbf{v}_B = \dot{y}\hat{\mathbf{j}}$$
$$\mathbf{a}_A = -\ddot{x}\hat{\mathbf{i}} = 0 \qquad \mathbf{a}_B = \ddot{y}\hat{\mathbf{j}}$$

Using the fact that the distance from A to B is a constant 5 m,

$$x^2 + y^2 = 25$$

so that

$$2\,x\dot{x} + 2y\dot{y} = 0$$

or

$$x\dot{x} + y\dot{y} = 0$$

Thus when $x = 3$ m, $y = 4$ m and

$$3(-8) + 4\dot{y} = 0$$

or

$$\dot{y} = 6 \text{ m/s}$$

so that

$$\mathbf{v}_B = 6\hat{\mathbf{j}} \text{ m/s}$$

Differentiating again,

$$\dot{x}\dot{x} + x\ddot{x} + \dot{y}\dot{y} + y\ddot{y} = 0$$

Therefore at the instant of interest

$$(-8)(-8) + (3)(0) + (6)(6) + 4\ddot{y} = 0$$

or

$$\ddot{y} = -25 \text{ m/s}^2$$

and

$$\mathbf{a}_B = -25\hat{\mathbf{j}} \text{ m/s}^2$$

Our last example illustrates a different kind of constraint, that of a point on a body maintaining contact with a surface (or line) on another moving body.

EXAMPLE 1.10

The curve AB on block $\mathcal{B}$ (see Figure E1.10a) is a parabola whose vertex is at A. Its equation is $x^2 = (64/3)y$. The block $\mathcal{B}$ is pushed to the left with a constant velocity of 10 ft/sec. The rod $\mathcal{R}$ slides on the parabola so that the plate $\mathcal{P}$ is forced upward. Find the acceleration of the plate.

Solution

We first note that plate $\mathcal{P}$ and rod $\mathcal{R}$ together constitute a single body, each of whose points has one-dimensional (y) motion. The velocities and accelerations of

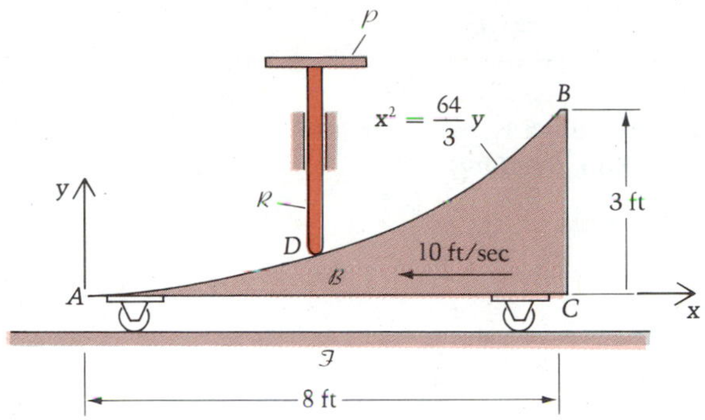

Figure E1.10a

all these points are therefore the same. We shall then focus on point D, the lowest point of $\mathcal{R}$, which is in contact with $\mathcal{B}$.

Defining the ground to be the reference frame $\mathcal{I}$, we establish its origin at O as shown in Figure E1.10b.

$$\mathbf{r}_{OA} = -x\hat{\mathbf{i}} \qquad \text{and} \qquad \mathbf{r}_{OD} = y\hat{\mathbf{j}} \text{ ft} \tag{1}$$

But because D always rests on the parabolic surface of $\mathcal{B}$, $y = (3/64)x^2$ so that

$$\mathbf{r}_{OD} = \tfrac{3}{64}x^2\hat{\mathbf{j}} \text{ ft} \tag{2}$$

To get the acceleration of D, we first find its velocity:

$$\mathbf{v}_D = \tfrac{3}{32}x\dot{x}\hat{\mathbf{j}} \text{ ft/sec} \tag{3}$$

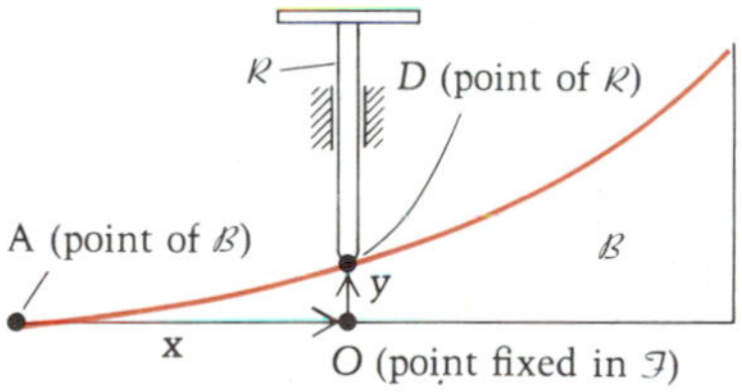

Figure E1.10b

To obtain $\dot{x}$, we differentiate $\mathbf{r}_{OA}$ from Equation (1):

$$\dot{\mathbf{r}}_{OA} = \mathbf{v}_A = -\dot{x}\hat{\mathbf{i}} = -10\hat{\mathbf{i}} \text{ ft/sec} \tag{4}$$

since it is given that all points of the body $\mathcal{B}$ have the constant velocity 10 ft/sec to the left.

Substitution of $\dot{x} = 10$ into Equation (3) then gives

$$\mathbf{v}_D = \tfrac{3}{32}x(10)\hat{\mathbf{j}} = \tfrac{15}{16}x\hat{\mathbf{j}} \text{ ft/sec} \tag{5}$$

and we see that the velocity of D depends upon x. Differentiating $\mathbf{v}_D$ will give us the acceleration of D:

$$\mathbf{a}_D = \frac{15}{16}\dot{x}\hat{\mathbf{j}} = \frac{150}{16}\hat{\mathbf{j}} = 9.38\hat{\mathbf{j}} \text{ ft/sec}^2 \tag{6}$$

Equation (6) gives the acceleration of all the points of the plate. Note that the acceleration of D is a constant.

Question 1.2 Would $\mathbf{a}_D$ be a constant if instead of being quadratically shaped, the inclined surface were (a) flat or (b) cubic?

Answer 1.2 If the surface is flat, then $\mathbf{a}_D$ vanishes. If it is cubic, then $\mathbf{a}_D$ is linear in x.

PROBLEMS ▶ Section 1.4

1.28 A slider block moves rectilinearly in a slot (see Figure P1.28) with an acceleration given by

$$\mathbf{a}_s = \ddot{x}\hat{\mathbf{i}} = -\pi^2 \sin \pi t \hat{\mathbf{i}} \text{ m/s}^2$$

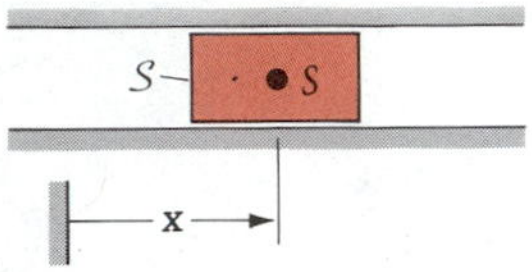

Figure P1.28

Find the motion $x(t)$ of the slider block if at $t = 0$:

a. It is passing through the origin, and

b. It has velocity $\dot{x}\hat{\mathbf{i}} = 2\pi\hat{\mathbf{i}}$ m/s.

1.29 Suppose an airplane touches down smoothly on a runway at 60 mph. If it then decelerates to a stop at the constant deceleration rate of 10 ft/sec², find the required length of runway.

1.30 A train is traveling at 60 km/hr. If its brakes give the train a constant deceleration of 0.5 m/s², find the distance from the station where the brakes should be applied so that the train will come to a stop at the station. How long will it take the train to stop?

1.31 A point P starts from rest and accelerates uniformly (meaning $\ddot{x}$ = constant) to a speed of 88 ft/sec after traveling 120 ft. Find the acceleration of P.

1.32 If in the preceding problem a braking deceleration of 2 ft/sec² is experienced beginning when P is at 120 ft, determine the time and distance required for stopping.

1.33 A car is traveling at 55 mph on a straight road. The driver applies her brakes for 6 sec, producing a constant deceleration of 5 ft/sec², and then immediately accelerates at 2 ft/sec². How long does it take for the car to return to its original velocity?

1.34 In the preceding problem, suppose the acceleration following the braking is not constant but is instead given by $\ddot{x} = 0.6t$ ft/sec². *Now* how long does it take to return to 55 mph?

In Problems 1.35–1.37, the graph depicts the velocity of a point P in rectilinear motion. Draw curves showing the position $x(t)$ and acceleration $a(t)$ of P if the point is at the indicated position x_0 at $t = 0$.

1.35 $x_0 = -1125$ m
Time interval: $0 \le t \le 20$ s (See Figure P1.35.)

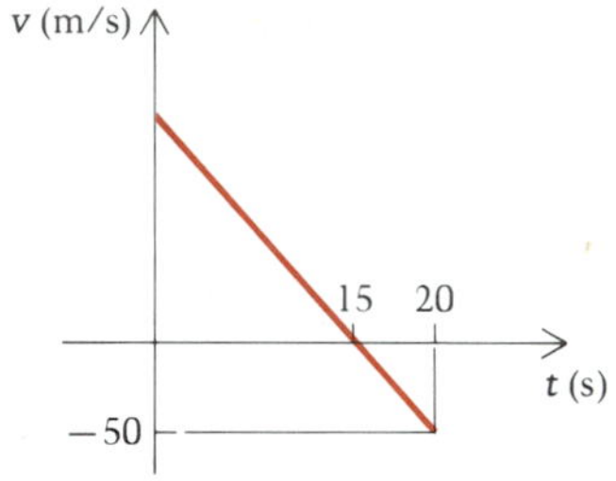

Figure P1.35

1.36 $x_0 = 10$ m
Time interval: $0 \le t \le 30$ s (See Figure P1.36.)

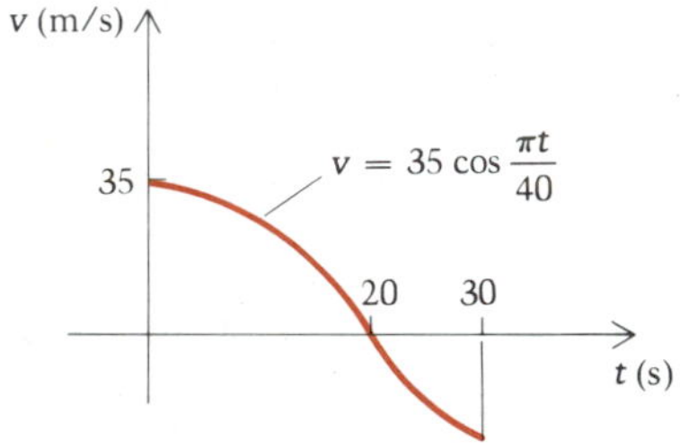

Figure P1.36

1.37 $x_0 = 10$ m
Time interval: $2 \le t \le 5$ s (See Figure P1.37.)

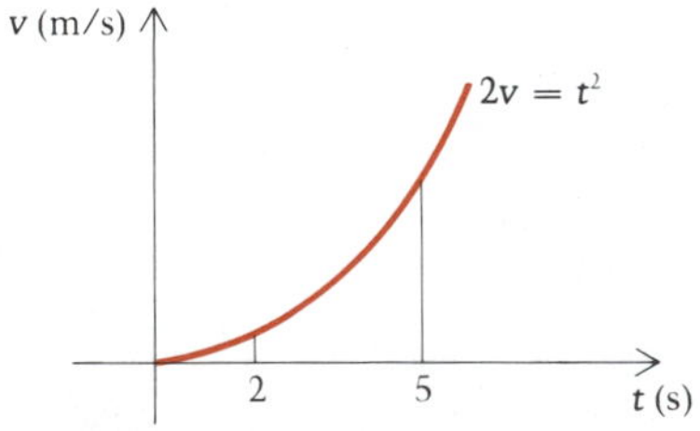

Figure P1.37

1.38 A hot rod enthusiast accelerates his dragster along a straight drag strip at a constant rate of acceleration from zero to 120 mph. Then he immediately decelerates at a constant rate to a stop. He finds that he has traveled a total distance of $\frac{1}{4}$ mi from start to stop. How much time passes from the instant he starts to the time he stops? *Hint*: Sketch a v-t diagram.

1.39 Ben Johnson set a world record of 9.83 seconds in the 100-meter dash on August 31, 1987. He had also set the record for the 60-meter dash of 6.40 seconds that same year. Assuming that in each race Johnson accelerated uniformly up to a certain speed v_o and then held that same maximum speed to the end of both races, find (a) the time t_o to reach v_o; (b) the value of v_o; and (c) the distance traveled while accelerating.

1.40 A train travels from one city to another which is 134 miles away. It accelerates from rest to a maximum speed of 100 mph in 4 min, averaging 65 mph during this time interval. It maintains maximum velocity until just before arrival, when it decelerates to rest at an average speed during the deceleration of 40 mph. If the total travel time was 110 min, find the deceleration interval.

1.41 A point Q in rectilinear motion passes through the origin at $t = 0$, and from then until 5 seconds have passed, the acceleration of Q is 6 ft/sec^2 to the right. Beginning at $t = 5$ seconds, the acceleration of Q is $12t$ ft/sec^2 to the left. If after 2 more seconds point Q is 13 feet to the right of the origin, what was the velocity of Q at $t = 0$?

1.42 A point begins at rest at $x = 0$ and experiences constant acceleration to the right for 10 s. It then continues at constant velocity for 8 more seconds. In the third phase of its motion, it decelerates at 5 m/s^2 and is observed to be passing again through the origin when the total time of travel equals 28 s. Determine the acceleration in the first 10 s.

1.43 An automobile passes a point P at a speed of 80 mph. At P it begins to decelerate at a rate proportional to time. If after 5 sec the car has slowed to 50 mph, what distance has it traveled?

1.44 Work the preceding problem, but suppose the deceleration is proportional to the square of time. The other information is the same.

1.45 A particle has a linearly varying rectilinear acceleration of $\mathbf{a} = \ddot{x}\hat{\mathbf{i}} = 12t\hat{\mathbf{i}}$ m/s^2. Two observations of the particle's motion are made: Its velocity at $t = 1$ s is $\dot{x}\hat{\mathbf{i}} = 2\hat{\mathbf{i}}$ m/s, and its position at $t = 2$ s is given by $x\hat{\mathbf{i}} = 3\hat{\mathbf{i}}$ meters.

a. Find the displacement of the particle at $t = 5$ s relative to where it was at $t = 0$.

b. Determine the distance traveled by the particle over the same time interval.

1.46 A point P moves on a line. The acceleration of P is given by $\mathbf{a}_P = \ddot{x}_P\hat{\mathbf{i}} = (3t^2 - 30t + 56)\hat{\mathbf{i}}$ m/s^2. The velocity of P at $t = 0$ is $-60\hat{\mathbf{i}}$ m/s, with the point at $x_P = 7$ m at that time. Find the distance traveled by P in the time interval $t = 0$ to $t = 13$ s.

* **1.47** The position of a point P on a line is given by the equation $x = t \sin(\pi t/2)$. The point starts moving at $t = 0$. Find the total distance traveled by P when it passes through the origin (counting the start as the first pass) for the third time.

1.48 A particle moving on a straight line is subject to an acceleration directly proportional to its distance from a fixed point P on the line and directed toward P. Initially the particle is 5 ft to the left of P and moving to the right with a velocity of 24 ft/sec. If the particle momentarily comes to rest 10 ft to the right of P, find its velocity as it passes through P.

1.49 A particle moving on the x axis has an acceleration always directed to the origin. The magnitude of the acceleration is nine times the distance from the origin. When the particle is 6 m to the left of the origin, it has a velocity of 3 m/s to the right. Find the time for the particle to get from this position to the origin.

1.50 A point P has an acceleration that is position-dependent according to the equation $\ddot{x} = -5x^2$ m/s^2. Determine the velocity of P as a function of its position x if P is at 0.3 m with $\dot{x} = 0.6$ m/s when $t = 0$.

1.51 Suppose initial conditions are the same as in the preceding problem but $\ddot{x} = -5\dot{x}^2$. Find $\dot{x}$ as a function of time.

1.52 The velocity of a particle moving along a horizontal path is proportional to its distance from a fixed point on the path. When $t = 0$, the particle is 1 ft to the right of the fixed point. When $v = 20$ ft/sec to the right, $a = 5$ ft/sec^2 to the right. Determine the position of the particle when $t = 4$ sec. (See Figure P1.52.)

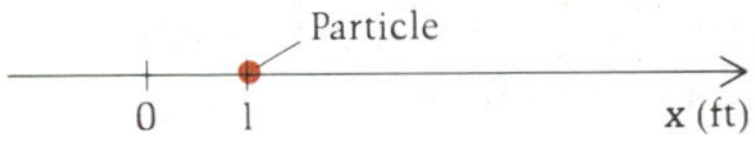

Figure P1.52

* Asterisks identify the more difficult problems.

1.53 A speeder zooms past a parked police car at a constant speed of 70 mph (Figure P1.53). Then, 3 sec later, the policewoman starts accelerating from rest at 10 ft/sec^2 until her velocity is 85 mph. How long does it take her to overtake the speeding car if it neither slows down nor speeds up?

Figure P1.53

1.54 In the preceding problem, suppose the speeder sees the policewoman 10 sec after she begins to move, and decelerates at 3 ft/sec^2. How long does it take the policewoman to *pass* the car if she is actually chasing a faster speeder ahead of it?

1.55 Two cars start from rest at the same location and at the same instant and race along a straight track. Car $\mathcal{A}$ accelerates at 6.6 ft/sec^2 to a speed of 90 mph and then runs at a constant speed. Car $\mathcal{B}$ accelerates at 4.4 ft/sec^2 to a speed of 96 mph and then runs at a constant speed.

- **a.** Which car will win the 3-mi race, and by what distance?
- **b.** What will be the maximum lead of $\mathcal{A}$ over $\mathcal{B}$?
- **c.** How far will the cars have traveled when $\mathcal{B}$ passes $\mathcal{A}$?

* **1.56** A car is 40 ft behind a truck; both are moving at 55 mph. (See Figure P1.56.) Suddenly the truck driver slams on his brakes after seeing an obstruction in the road ahead, and he decelerates at 10 ft/sec^2. Then, 2 sec later, the driver of the car reacts by slamming on her brakes, giving her car a deceleration a_C. Find the minimum value of a_C for which the car will not collide with the truck. *Hint:* Enforce $x_T > x_C$ for *all* time t before the vehicles are stopped.

Figure P1.56

1.57 Point B of block $\mathcal{B}$ has a constant acceleration of 10 m/s^2 upward. At the instant shown in Figure P1.57, it is 30 m below the level of point A of $\mathcal{A}$. At this time, $\mathbf{v}_A$ and $\mathbf{v}_B$ are zero. Determine the velocities of A and B as they pass each other.

1.58 The accelerations of the translating blocks $\mathcal{A}$ and $\mathcal{B}$ are 2 m/s^2 ↓ and 4 m/s^2 ↑, respectively. (See Figure P1.58.)

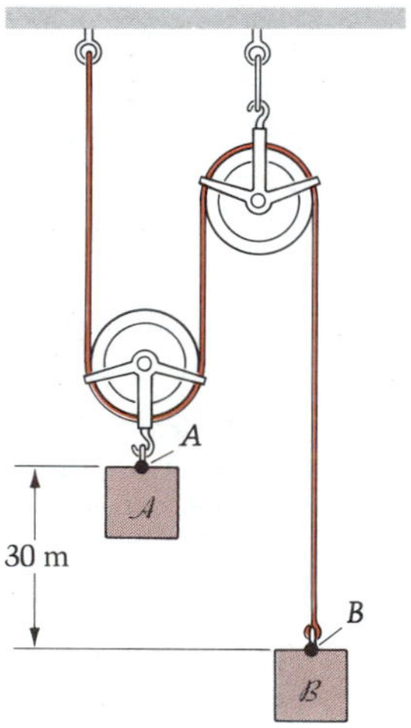

Figure P1.57

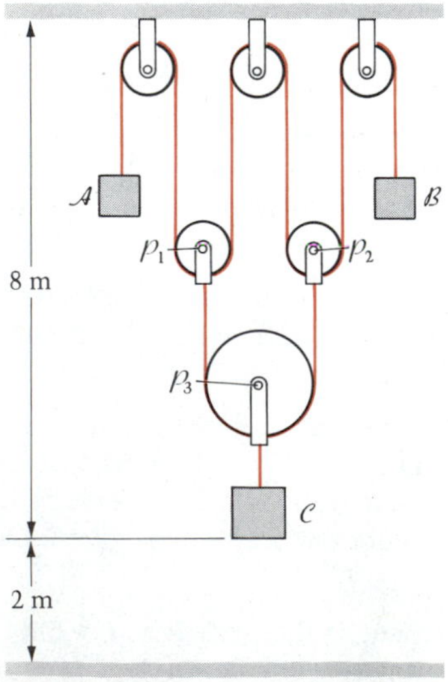

Figure P1.58

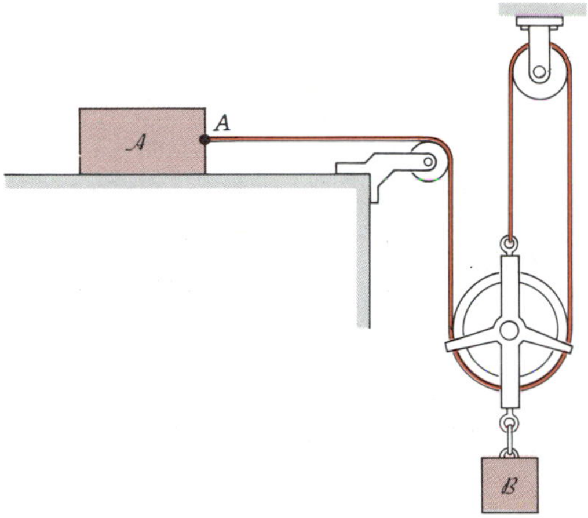

Figure P1.59

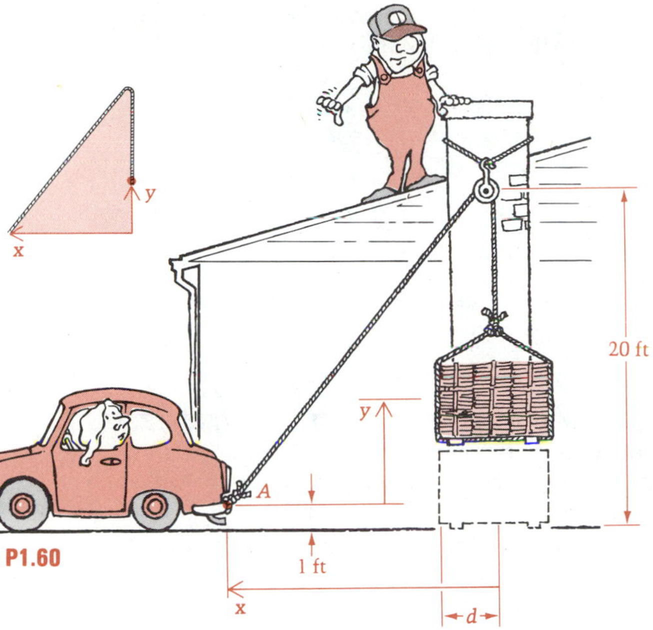

Figure P1.60

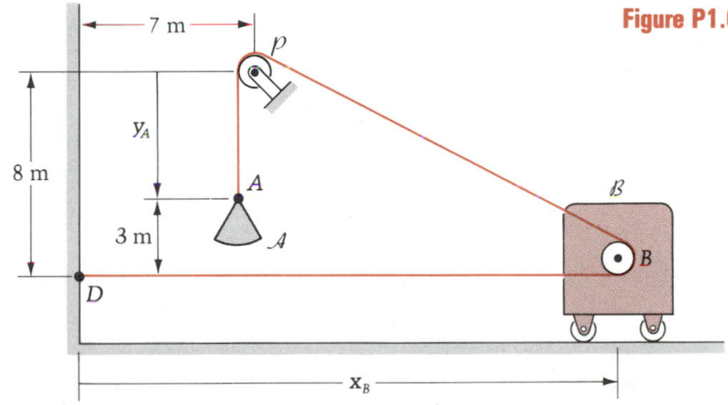

Figure P1.61

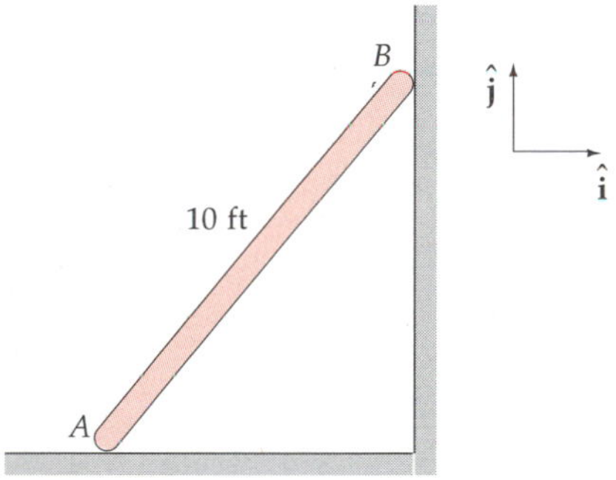

Figure P1.62

The entire system is at rest at the given instant. Find how long it will take for block $\mathcal{C}$ to hit the ground. (Do not assume that pulleys $\mathcal{P}_1$ and $\mathcal{P}_2$ remain at the same level!)

* **1.59** Block $\mathcal{A}$ has $\mathbf{v}_A = 10$ m/s to the right at $t = 0$ and a constant acceleration of 2 m/s² to the left. Find the distance traveled by block $\mathcal{B}$ during the interval $t = 0$ to 8 s. (See Figure P1.59.)

1.60 A man and his daughter have figured out an ingenious way to hoist 8000 lb of shingles onto their roof, several bundles at a time. They have rigged a pulley onto a frame around the chimney (Figure P1.60) and will use the car to raise the weights. When the bumper of the car is at $x = 0$ (neglect d), the pallet of shingles is on the ground with no slack in the rope. While the car is traveling to the left at a constant speed of $v_A = 2$ mph, find the velocity and acceleration of the shingles as a function of x. Do this by using the triangle to the left of the figure to express y as a function of x; then differentiate the result.

* **1.61** The cord shown in Figure P1.61, attached to the wall at D, passes around a small pulley fixed to $\mathcal{B}$ at B; it then passes around another small pulley $\mathcal{P}$ and ends at point A of body $\mathcal{A}$. The cord is 44 m long, and the system is being held at rest in the given position. Suddenly point B is forced to move to the right with constant acceleration $a_B = 2$ m/s². Determine the velocity of A just before it reaches the pulley.

1.62 The ends of the rigid bar in Figure P1.62 move while maintaining contact with floor and wall. End A moves toward the wall at the constant rate of 2 ft/sec. What is the acceleration of B at the instant when A is 6 feet from the wall?

1.63 The velocity of point A in Figure P1.63 is a constant 2 m/s to the right. Find the velocity of B when $x = 10$ m.

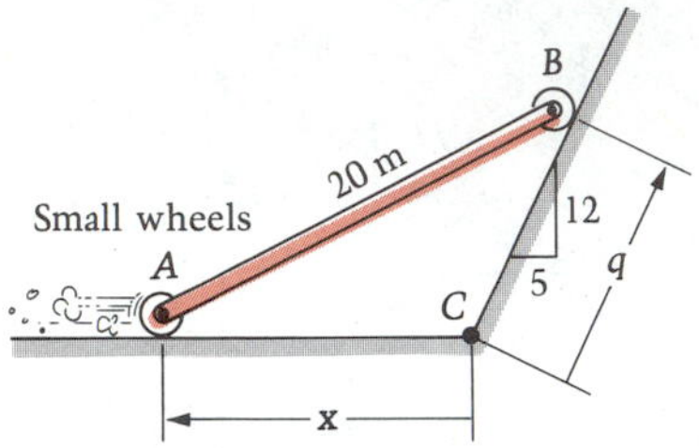

Figure P1.63

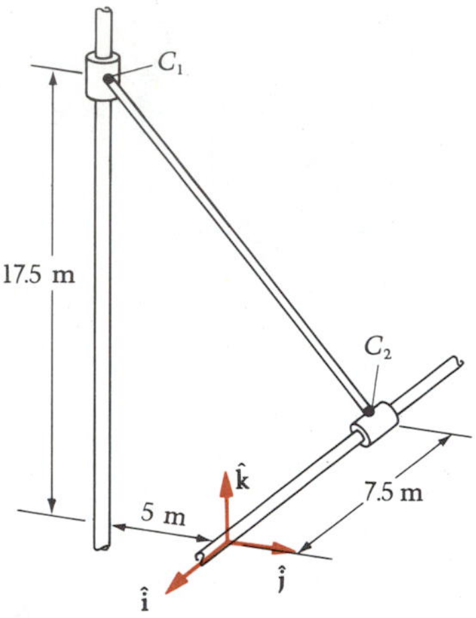

Figure P1.64

1.64 The collars in Figure P1.64 are attached at C_1 and C_2 to the rod by ball and socket joints. Point C_2 has a velocity of $-3\hat{\mathbf{i}}$ m/s and no acceleration at the instant shown. Find the velocity and acceleration of C_1 at this instant.

1.65 The wedge-shaped cam in Figure P1.65 is moving to the left with constant acceleration a_0. Find the acceleration of the follower $\mathcal{F}$.

1.66 In Example 1.10, let the equation of the incline be given by $x^3 = (512/3)y$. If the motion starts when $x = y = 0$, find the acceleration of the plate $\mathcal{P}$ when $y = 2$ ft.

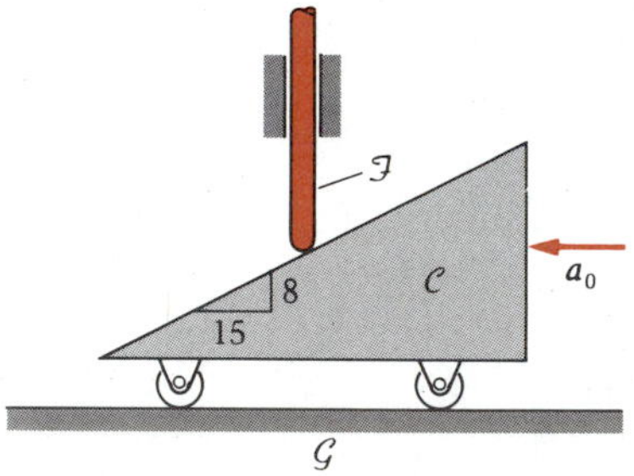

Figure P1.65

1.5 Rectangular Cartesian Coordinates

In this section we merely add the y and/or z components of position to the rectilinear (x) component studied in the preceding section. This step allows the point P to move on a curve in two- or three-dimensional space instead of being constrained to movement along a straight line in the reference frame $\mathcal{I}$.

Suppose that P is in a state of general (three-dimensional) motion in frame $\mathcal{I}$. We may study this motion by embedding a set of orthogonal axes in $\mathcal{I}$ as shown in Figure 1.4 on the next page. The position vector of point P may then be expressed as

$$\mathbf{r}_{OP} = x\hat{\mathbf{i}} + y\hat{\mathbf{j}} + z\hat{\mathbf{k}} \tag{1.23}$$

in which (x, y, z) are **rectangular Cartesian coordinates** of P measured along the embedded axes and $(\hat{\mathbf{i}}, \hat{\mathbf{j}}, \hat{\mathbf{k}})$ are unit vectors respectively parallel to these axes (Figure 1.4). Using the basic definitions (Equations 1.9 and 1.10), we may differentiate $\mathbf{r}_{OP}$ and obtain expressions for velocity and acceleration in rectangular Cartesian coordinates:

$$\mathbf{v}_P = \dot{x}\hat{\mathbf{i}} + \dot{y}\hat{\mathbf{j}} + \dot{z}\hat{\mathbf{k}} \tag{1.24}$$

$$\mathbf{a}_P = \ddot{x}\hat{\mathbf{i}} + \ddot{y}\hat{\mathbf{j}} + \ddot{z}\hat{\mathbf{k}} \tag{1.25}$$

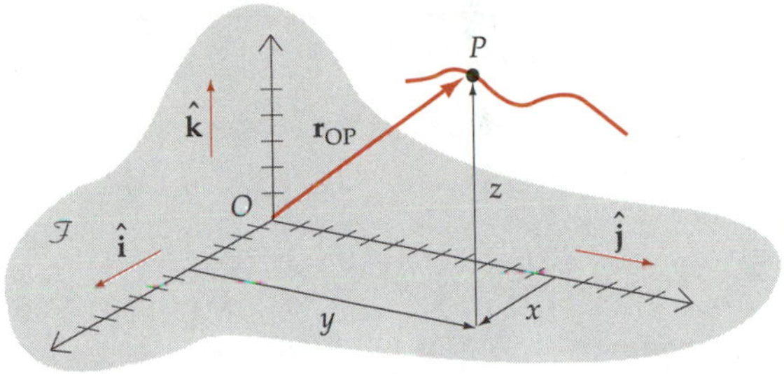

Figure 1.4 Rectangular Cartesian coordinates of P.

We shall now consider examples in which points move in two and three dimensions.

Example 1.11

The position vector of a point P is given as

$$\mathbf{r}_{OP} = 2t\hat{\mathbf{i}} + t^3\hat{\mathbf{j}} + 3t^2\hat{\mathbf{k}} \text{ ft}$$

Find the velocity and acceleration of P at $t = 1$ sec.

Solution

Differentiating the position vector, we obtain the velocity vector of P:

$$\mathbf{v}_P = 2\hat{\mathbf{i}} + 3t^2\hat{\mathbf{j}} + 6t\hat{\mathbf{k}} \text{ ft/sec}$$

Another derivative yields the acceleration of P:

$$\mathbf{a}_P = 6t\hat{\mathbf{j}} + 6\hat{\mathbf{k}} \text{ ft/sec}^2$$

Therefore, at $t = 1$ sec, the velocity and acceleration of P are

$$\mathbf{v}_P|_{t=1} = 2\hat{\mathbf{i}} + 3\hat{\mathbf{j}} + 6\hat{\mathbf{k}} = 7\left(\frac{2\hat{\mathbf{i}} + 3\hat{\mathbf{j}} + 6\hat{\mathbf{k}}}{\sqrt{2^2 + 3^2 + 6^2}}\right) \text{ ft/sec}$$

$$\mathbf{a}_P|_{t=1} = 6\hat{\mathbf{j}} + 6\hat{\mathbf{k}} = 6\sqrt{2}\left(\frac{\hat{\mathbf{j}} + \hat{\mathbf{k}}}{\sqrt{2}}\right) \text{ ft/sec}^2$$

Note that the speed (magnitude of velocity) of P at $t = 1$ is $|\mathbf{v}_P| = 7$ ft/sec and the magnitude of the acceleration at $t = 1$ is $6\sqrt{2}$ ft/sec^2. We shall return to this example in Section 1.7.

We see from the previous example that if the position vector of P is known as a function of time, it is a very simple matter to obtain the velocity and acceleration of the point. In the following example we are given the acceleration of P and asked for its *position.* Since this problem requires integration instead of differentiation, initial conditions enter the picture. These conditions allow us to compute the constants of integration, just as they did for rectilinear motion in the preceding section.

EXAMPLE 1.12

A point Q has the acceleration vector

$$\mathbf{a}_Q = 4\hat{\mathbf{i}} - 6t\hat{\mathbf{j}} + \sin 0.2t\hat{\mathbf{k}} \text{ m/s}^2$$

At $t = 0$, the point Q is located at $(x, y, z) = (1, 3, -5)$ m and has a velocity vector of $2\hat{\mathbf{i}} - 7\hat{\mathbf{j}} + 3.4\hat{\mathbf{k}}$ m/s. When $t = 3$ s, find the speed of Q and its distance from the starting point.

Solution

Integrating, we get

$$\mathbf{v}_Q = 4t\hat{\mathbf{i}} - 3t^2\hat{\mathbf{j}} - 5 \cos 0.2t\hat{\mathbf{k}} + \mathbf{c} \text{ m/s}$$

in which $\mathbf{c}$ is a vector constant. Using the initial condition for velocity at $t = 0$, we obtain

$$\mathbf{v}_Q|_{t=0} = 0\hat{\mathbf{i}} - 0\hat{\mathbf{j}} - 5\hat{\mathbf{k}} + \mathbf{c} = 2\hat{\mathbf{i}} - 7\hat{\mathbf{j}} + 3.4\hat{\mathbf{k}} \text{ m/s}$$

so that

$$\mathbf{c} = 2\hat{\mathbf{i}} - 7\hat{\mathbf{j}} + 8.4\hat{\mathbf{k}} \text{ m/s}$$

Therefore

$$\mathbf{v}_Q = (4t + 2)\hat{\mathbf{i}} - (3t^2 + 7)\hat{\mathbf{j}} + (8.4 - 5 \cos 0.2t)\hat{\mathbf{k}} \text{ m/s}$$

Integrating again, we get

$$\mathbf{r}_{OQ} = (2t^2 + 2t)\hat{\mathbf{i}} - (t^3 + 7t)\hat{\mathbf{j}} + (8.4t - 25 \sin 0.2t)\hat{\mathbf{k}} + \mathbf{c}' \text{ m}$$

where $\mathbf{c}'$ is another vector constant, evaluated below from the initial condition for the *position* of Q at $t = 0$:

$$\mathbf{r}_{OQ}|_{t=0} = 0\hat{\mathbf{i}} - 0\hat{\mathbf{j}} + 0\hat{\mathbf{k}} + \mathbf{c}' = \hat{\mathbf{i}} + 3\hat{\mathbf{j}} - 5\hat{\mathbf{k}} \text{ m}$$

so that

$$\mathbf{c}' = \hat{\mathbf{i}} + 3\hat{\mathbf{j}} - 5\hat{\mathbf{k}} \text{ m}$$

and thus

$$\mathbf{r}_{OQ} = (2t^2 + 2t + 1)\hat{\mathbf{i}} - (t^3 + 7t - 3)\hat{\mathbf{j}} + (8.4t - 25 \sin 0.2t - 5)\hat{\mathbf{k}} \text{ m}$$

Substituting the required time, $t = 3$ s, into the expressions for $\mathbf{v}_Q$ and $\mathbf{r}_{OQ}$ will give the answers:

$$\mathbf{v}_Q|_{t=3} = 14\hat{\mathbf{i}} - 34\hat{\mathbf{j}} + (8.4 - 5 \cos 0.6)\hat{\mathbf{k}} = 14\hat{\mathbf{i}} - 34\hat{\mathbf{j}} + 4.27\hat{\mathbf{k}} \text{ m/s}$$

Thus the speed of Q is given by

$$v_Q|_{t=3} = \sqrt{14^2 + (-34)^2 + 4.27^2} = 37.0 \text{ m/s}$$

Continuing, we have

$$\mathbf{r}_{OQ}|_{t=3} = 25\hat{\mathbf{i}} - 45\hat{\mathbf{j}} + (20.2 - 25 \sin 0.6)\hat{\mathbf{k}} = 25\hat{\mathbf{i}} - 45\hat{\mathbf{j}} + 6.08\hat{\mathbf{k}} \text{ m}$$

The distance d between Q and its starting point is therefore given by

$$d = |\mathbf{r}_{OQ}(3) - \mathbf{r}_{OQ}(0)|$$
$$= \sqrt{(25-1)^2 + (-45-3)^2 + [6.08 - (-5)]^2}$$
$$= 54.8 \text{ m}$$

Example 1.13

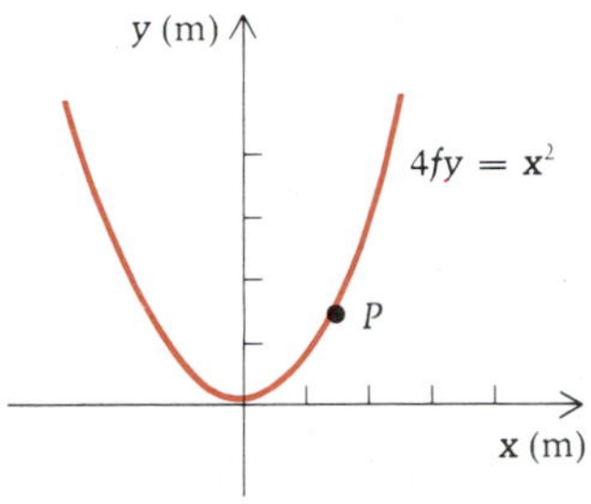

Figure E1.13

The point P in Figure E1.13 travels on the parabola (with focal distance $f = \frac{1}{2}$ m) at the constant speed of 0.2 m/s. Determine the acceleration of P: (a) as a function of x and (b) at $x = 2$ m.

Solution

We may obtain the velocity components by differentiating:

$$2y = x^2$$
$$2\dot{y} = 2x\dot{x} \Rightarrow \dot{y} = x\dot{x}$$

Thus

$$\mathbf{v}_P = \dot{x}\hat{\mathbf{i}} + \dot{y}\hat{\mathbf{j}} = \dot{x}\hat{\mathbf{i}} + x\dot{x}\hat{\mathbf{j}} \text{ m/s} \tag{1}$$

Similarly the acceleration of P is

$$\mathbf{a}_P = \ddot{x}\hat{\mathbf{i}} + \ddot{y}\hat{\mathbf{j}} = \ddot{x}\hat{\mathbf{i}} + (\dot{x}^2 + x\ddot{x})\hat{\mathbf{j}} \text{ m/s}^2 \tag{2}$$

Since $|\mathbf{v}_P|$, or v_P, is constant, we have

$$v_P = 0.2 = \sqrt{\dot{x}^2 + (x\dot{x})^2} \text{ m/s}$$

$$\dot{x} = \frac{0.2}{\sqrt{1+x^2}} \text{ m/s} \tag{3}$$

We also see from (2) that we need $\ddot{x}$; differentiating (3), we get

$$\ddot{x} = \frac{-0.2\, x\dot{x}}{(1+x^2)^{3/2}} = \frac{-0.2\left(\dfrac{0.2}{\sqrt{1+x^2}}\right)x}{(1+x^2)^{3/2}} \text{ m/s}^2$$

or

$$\ddot{x} = \frac{-0.04x}{(1+x^2)^2} \text{ m/s}^2 \tag{4}$$

Substituting (3) and (4) into (2), we get

$$\mathbf{a}_P = \frac{-0.04x}{(1+x^2)^2}\hat{\mathbf{i}} + \left[\frac{0.04}{1+x^2} - \frac{0.04x^2}{(1+x^2)^2}\right]\hat{\mathbf{j}} \text{ m/s}^2$$

When $x = 2$ m,

$$\mathbf{a}_P = \frac{-0.04(2)}{5^2}\hat{\mathbf{i}} + \left[\frac{0.04}{5} - \frac{0.04(2^2)}{5^2}\right]\hat{\mathbf{j}}$$
$$= -0.0032\hat{\mathbf{i}} + 0.0016\hat{\mathbf{j}} \text{ m/s}^2$$

In closing this section, we remark that the simple forms of Equations (1.24) and (1.25) are due to the fact that the unit vectors $\hat{\mathbf{i}}$, $\hat{\mathbf{j}}$, $\hat{\mathbf{k}}$ remain constant in both magnitude and direction when the axes are fixed in the frame of reference. For *planar* applications (Chapter 3), we shall set the z component of velocity identically to zero, obtaining the following for a point in plane motion (moving only in a plane parallel to the xy plane):

$$\mathbf{r}_{OP} = x\hat{\mathbf{i}} + y\hat{\mathbf{j}} + z\hat{\mathbf{k}} \quad \text{(where } z \text{ is constant)} \tag{1.26}$$

$$\mathbf{v}_P = \dot{x}\hat{\mathbf{i}} + \dot{y}\hat{\mathbf{j}} \tag{1.27}$$

$$\mathbf{a}_P = \ddot{x}\hat{\mathbf{i}} + \ddot{y}\hat{\mathbf{j}} \tag{1.28}$$

PROBLEMS ▶ Section 1.5

1.67 The moving pin P of a rotating crank has a location defined by

$$x = 20 \cos \pi t \text{ m}$$

$$y = 20 \sin \pi t \text{ m}$$

Find the velocity of P when $t = 0, \frac{1}{2}, \frac{3}{2}$, and 2 s.

1.68 A bar of length $2L$ moves with its ends in contact with the guides shown in Figure P1.68. Find the velocity and acceleration of point C in terms of θ and its derivatives.

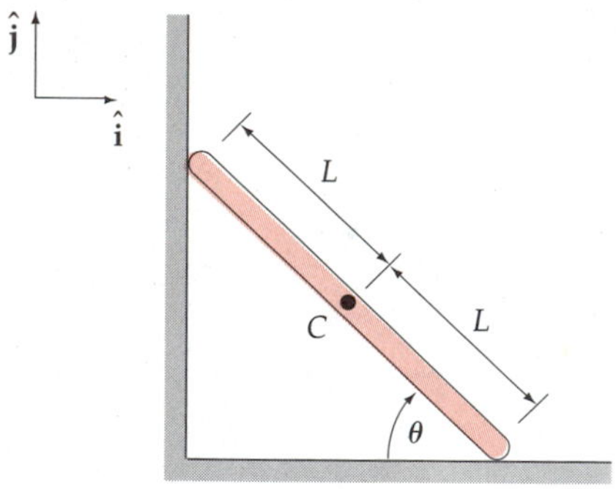

Figure P1.68

1.69 A point P moves on a circle in the direction shown in Figure P1.69. Express $\mathbf{r}_{OP}$ in (x, y, z) coordinates and differentiate to obtain $\mathbf{v}_P$ and $\mathbf{a}_P$. (Angle θ is in radians.)

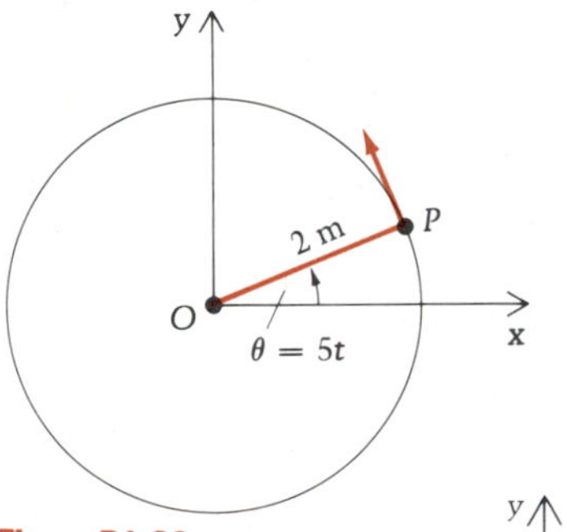

Figure P1.69

1.70 Repeat the preceding problem. In this case, however, the angle θ increases quadratically, instead of linearly, with time according to $\theta = 2t^2$ rad.

1.71 A point P starts at the origin and moves along the parabola shown in Figure P1.71 with a constant x-component of velocity, $\dot{x} = 3$ ft/sec. Find the velocity and acceleration of P at the point $(x, y) = (1, 1)$.

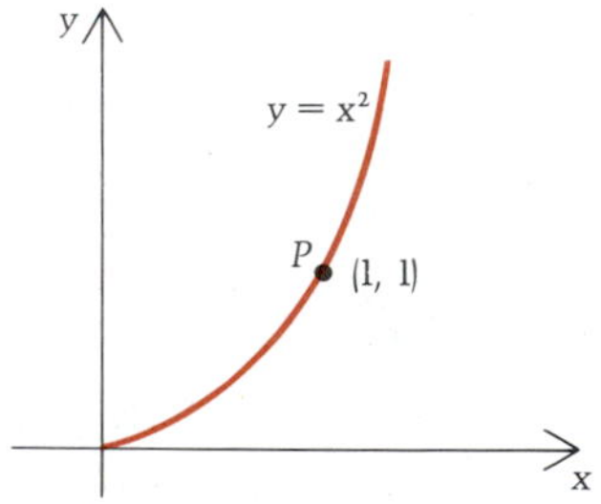

Figure P1.71

1.72 Point P is constrained to move in the two slots shown: one cut in the body $\mathcal{A}$, the other cut in the reference frame $\mathcal{R}$. The constant acceleration of $\mathcal{A}$ is given to be 4 cm/s² to the left. If point P reaches the bottom of the slot (in $\mathcal{A}$) 2 sec after the instant shown in Figure P1.72,

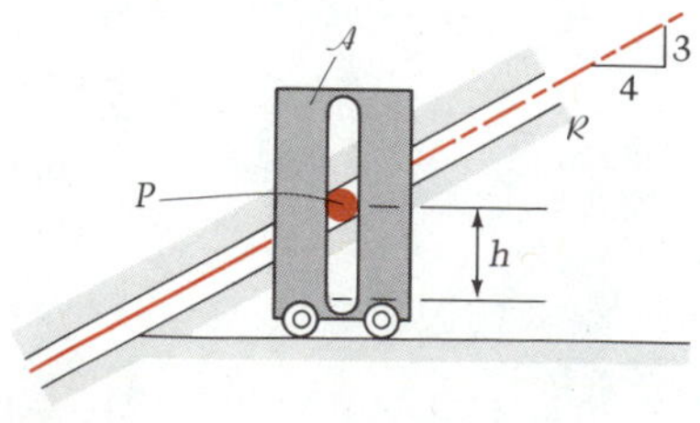

Figure P1.72

when $\mathcal{A}$ is at rest,

a. Through what height h did the marble move?

b. What distance did the marble travel?

1.73 The pin P shown in Figure P1.73 moves in a parabolic slot cut in the reference frame $\mathcal{I}$ and is guided by the vertical slot in body $\mathcal{B}$. For body $\mathcal{B}$, $x = 0.05t^3$ m locates the centerline of its slot.

a. Find the acceleration of P at $t = 5$ s.

b. Find the time(s) when the x and y components of $\mathbf{a}_P$ are equal.

1.74 A pin P moves in a slot that is cut in the shape of a hyperbolic sine as shown in Figure P1.74. It is guided along by the vertical slotted body $\mathcal{B}$, all the points of which have velocity 0.08 m/s to the right. Find $\mathbf{v}_P$ and $\mathbf{a}_P$ when $x = 0.2$ m.

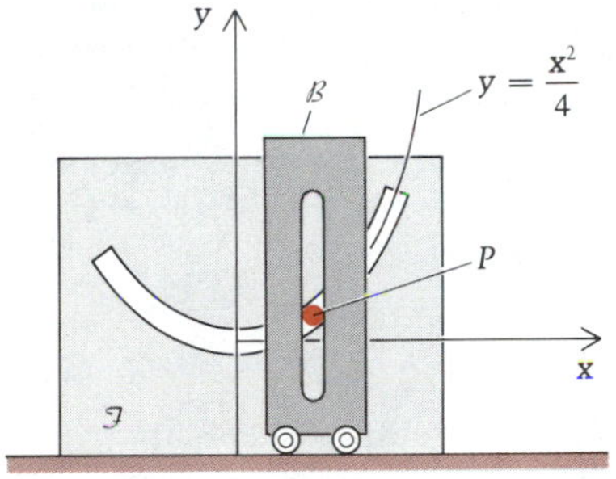

Figure P1.73

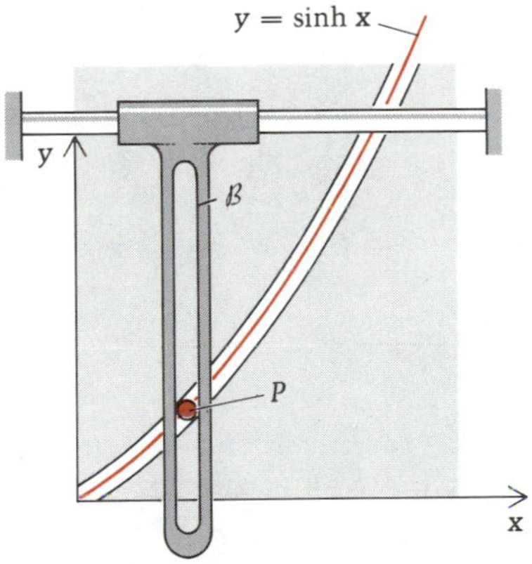

Figure P1.74

1.75 A point P travels on a path and has the following coordinates as functions of time t (in seconds):

$$x = 12 \cos \frac{\pi t}{2} \text{ m} \qquad y = 8 \sin \frac{\pi t}{2} \text{ m}$$

$$z = 0$$

a. Find the velocity $\mathbf{v}_P(t)$ and acceleration $\mathbf{a}_P(t)$ of P.

b. Find the position, velocity, and acceleration of P when $t = 4$ s.

c. Eliminate the time t from the x and y expressions and obtain the equation of the path of P.

1.76 The motion of a particle P is given by $x = C \cosh kt$ and $y = C \sinh kt$, where C and k are constants. Find the equation of the path of P by eliminating time t.

1.77 In the preceding problem, find the speed of P as a function of the distance $r (= \sqrt{x^2 + y^2})$ from the origin to P.

1.78 Describe precisely the path of a particle's motion if its xy coordinates are given by $(2.5t^2 + 7, 6t^2 + 9)$ meters when t is in seconds.

1.79 A particle P moves in the xy plane. The motion of P is given by

$$x = 30t + 6 \text{ ft}$$

$$y = 20t - 7 \text{ ft}$$

Find the equation of the path of P in the form $y = f(x)$.

1.80 Repeat Problem 1.79 if:

$$x = 5t \text{ m}$$

$$y = -250t^3 \text{ m}$$

1.81 Repeat Problem 1.79 if:

$$x = 2 + 3 \sin t \text{ ft}$$

$$y = 4 \cos t \text{ ft}$$

1.82 A cycloid is the curve traced out by a point (such as P) on the rim of a rolling wheel. In terms of the parameter θ (the angle in Figure P1.82), the equations of the cycloid are:

$$x = a(\theta - \sin \theta)$$

$$y = a(1 - \cos \theta)$$

Noting that θ changes with time, find the speed of P at $\theta = 0, \pi/2, \pi$, and $3\pi/2$ radians, in terms of a and $\dot{\theta}$.

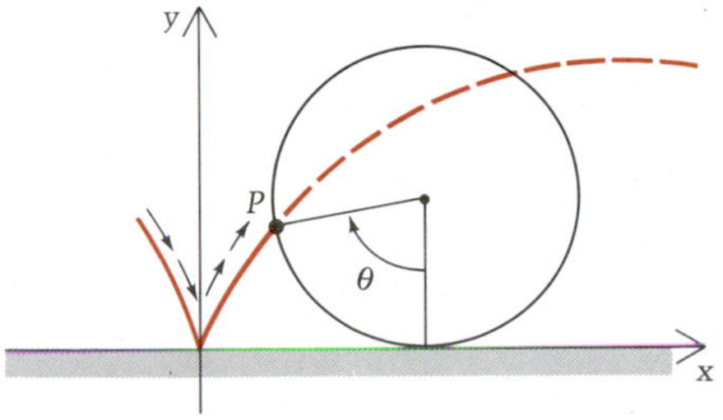

Figure P1.82

In Problems 1.83–1.86 (see Figures P1.83–P1.86), a point P travels on the curve with a constant x component of velocity, $\dot{x} = 3$ in./sec. Each starts on the curve at $x = 1$ when $t = 0$. Find the velocity vector of P when $t = 10$ sec in each case.

1.83 Logarithmic curve

1.84 Exponential curve

1.85 First-quadrant branch of rectangular hyperbola

1.86 First-quadrant branch of semicubical parabola

1.87–1.90 Find the respective acceleration vectors at $t = 10$ s of the points whose motions are described in Problems 1.83–1.86.

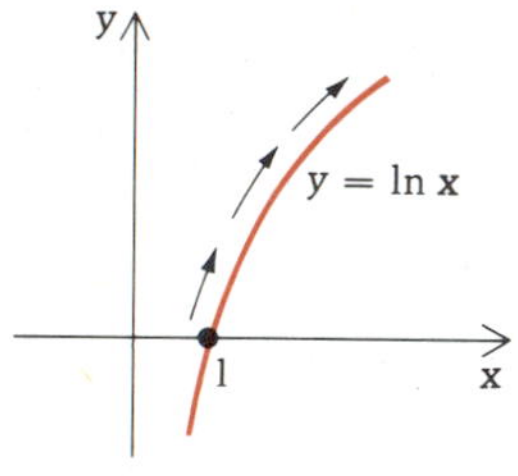

Figure P1.83

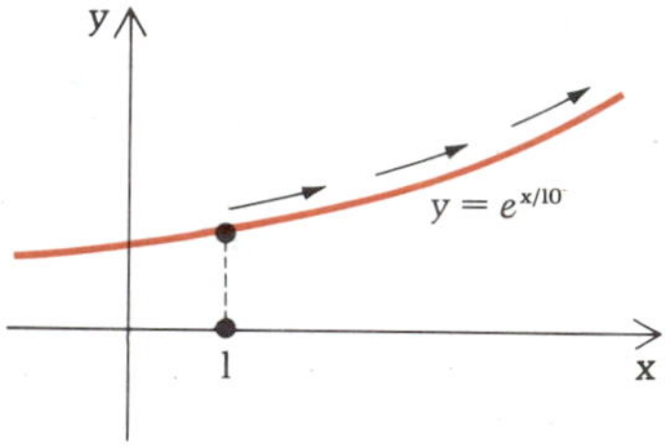

Figure P1.84

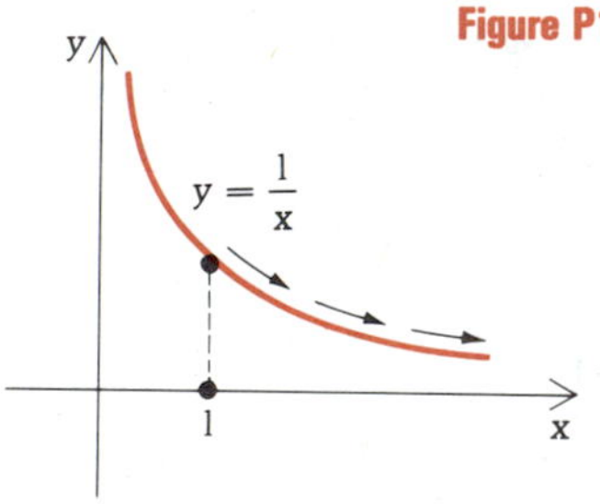

Figure P1.85

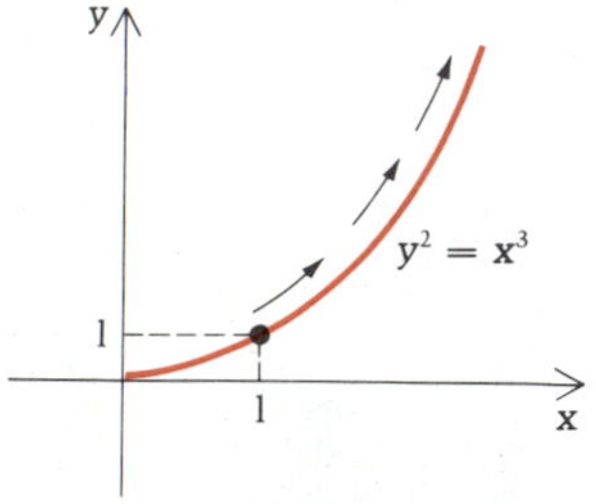

Figure P1.86

1.91 Two points P and Q have position vectors in a reference frame that are given by $\mathbf{r}_{OP} = 50t\hat{\mathbf{i}}$ meters and $\mathbf{r}_{OQ} = 40\hat{\mathbf{i}} - 20t\hat{\mathbf{j}}$ meters. Find the minimum distance between P and Q and the time at which this occurs.

1.92 Describe the path of a point P that has the following rectangular Cartesian coordinates as functions of time: $x = a \cos \omega t$, $y = a \sin \omega t$, and $z = bt$, where a, b, and ω are constants. Identify the meanings of the three constants.

1.93 For the following values of the constants, find the velocity of P at $t = 5$ s in the preceding problem: $a = 2$ m, $b = 0.5$ m/s, and $\omega = 1.2$ rad/s.

1.94 The acceleration of a point is given by

$$\mathbf{a}_P = 6t\hat{\mathbf{i}} + 12t^2\hat{\mathbf{j}} - 4\hat{\mathbf{k}} \text{ m/s}^2$$

At $t = 0$, the initial conditions are that $\mathbf{v}_P = 2\hat{\mathbf{i}}$ m/s and $\mathbf{r}_{OP} = \hat{\mathbf{i}} + 3\hat{\mathbf{j}} + 9\hat{\mathbf{k}}$ meters. Find the position vector of P at $t = 3$ s, and determine how far P then is from its position at $t = 0$.

1.95 A point moves on a path, with a position vector as a function of time given by $\mathbf{r}_{OP} = \sin 2t\hat{\mathbf{i}} + 3t\hat{\mathbf{j}} + e^{6t}\hat{\mathbf{k}}$, in units of meters when t is in seconds. Find:

a. The speed of the point at $t = 0$.

b. Its acceleration at $t = \pi/2$ s.

c. The component of the velocity vector, at $t = 0$, which is parallel to the line ℓ in the xy plane given by $y = \frac{5}{12}x - 6$ shown in Figure P1.95.

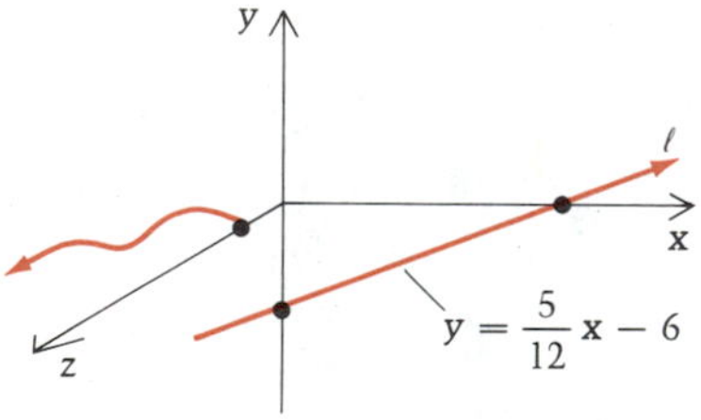

Figure P1.95

*** 1.96** A car travels on a section of highway that approximates the cosine curve in Figure P1.96. If the driver

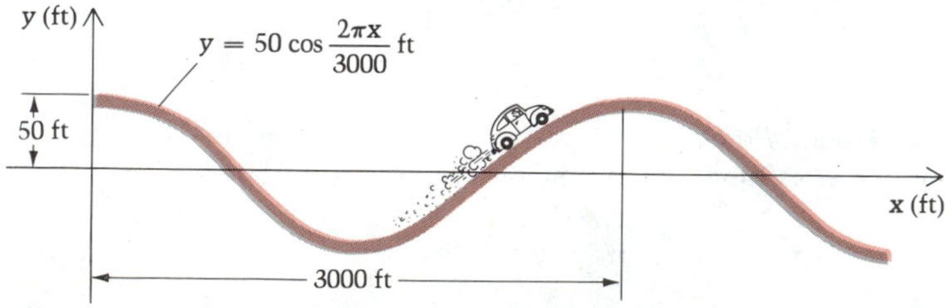

Figure P1.96

maintains a constant speed of 55 mph, determine his x and y components of velocity when $x = 2500$ ft.

* **1.97** A car travels along the highway of the preceding problem with a constant x-component of velocity of 54.9 mph. Over what sections of the highway does the driver exceed the speed limit of 55 mph?

1.98 Determine the minimum magnitude of acceleration of the car in Problem 1.96. Where on the curve is this acceleration experienced?

1.99 Find the maximum magnitude of acceleration of the car in Problem 1.97. Where does it occur on the curve?

1.6 Cylindrical Coordinates

If a point P is moving in such a way that its projection into the xy plane is more easily described with polar (r and θ) coordinates than with x and y, then we may use **cylindrical coordinates** to advantage. These coordinates are nothing more than the **polar coordinates** r and θ together with an "axial" coordinate z. Thus r and θ locate the projection point of P in a plane, while z gives the distance of P *from* the plane.

Embedding the same set of rectangular axes (x, y, z) in the reference frame $\mathcal{I}$ as we did in the preceding section, we now show the coordinates r and θ as well (Figure 1.5). Note that P' is the projection of P into the plane xy. From Figure 1.5 we see that the unit vectors $\hat{\mathbf{e}}_r$ and $\hat{\mathbf{e}}_\theta$ are drawn in the xy plane and that:

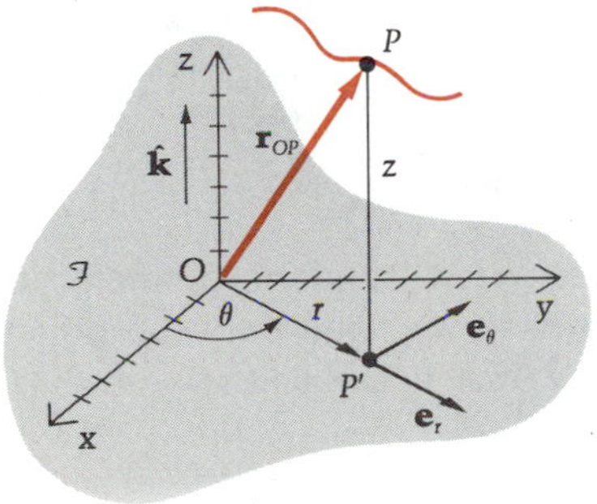

Figure 1.5 Cylindrical coordinates of P.

1. The direction of $\hat{\mathbf{e}}_r$ is that of OP'.
2. $\hat{\mathbf{e}}_\theta$ is normal to $\hat{\mathbf{e}}_r$ in the direction of increasing θ.

It will be helpful later in the section to note carefully at this point that $\hat{\mathbf{e}}_r$ and $\hat{\mathbf{e}}_\theta$ change (in direction) with changes in θ, but not with r or z. Thus if the point P moves along either a radial line or a vertical line, the two unit vectors remain the same. But if P moves in such a way as to alter θ, then the directions of $\hat{\mathbf{e}}_r$ and $\hat{\mathbf{e}}_\theta$ will vary.

The rectangular and cylindrical coordinates (both having z in common) are related through

$$x = r\cos\theta$$
$$y = r\sin\theta \tag{1.29}$$

which can be differentiated to produce, by Equations (1.24) and (1.25), formulas for velocity and acceleration in terms of the cylindrical coordinates (and their derivatives) and the unit vectors $\hat{\mathbf{i}}$, $\hat{\mathbf{j}}$, and $\hat{\mathbf{k}}$. It is usually more desirable, however, to express the velocity and acceleration in terms of the set of unit vectors ($\hat{\mathbf{e}}_r$, $\hat{\mathbf{e}}_\theta$, $\hat{\mathbf{k}}$), which are naturally associated with cylindrical coordinates. Thus it is useful to express a position vector $\mathbf{r}_{OP}$ as

$$\mathbf{r}_{OP} = r\hat{\mathbf{e}}_r + z\hat{\mathbf{k}} \tag{1.30}$$

Question 1.3 Why is there no $\hat{\mathbf{e}}_\theta$ term in Equation (1.30)?

Answer 1.3 From Figure 1.5, we see that $\hat{\mathbf{e}}_\theta$ is perpendicular to $\mathbf{r}_{OP}$. Note, however, that implicit in the writing and use of Equation (1.30) is the polar angle θ.

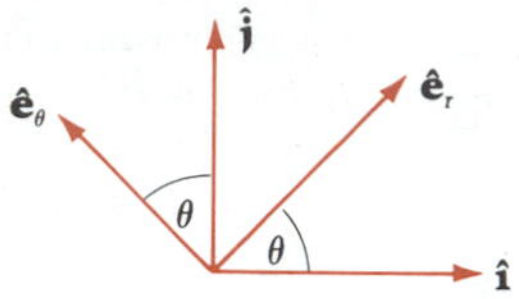

Figure 1.6

Differentiating Equation (1.30), we obtain the velocity of P:

$$\mathbf{v}_P = \dot{r}\hat{\mathbf{e}}_r + r\dot{\hat{\mathbf{e}}}_r + \dot{z}\hat{\mathbf{k}} \tag{1.31}$$

To evaluate $\dot{\hat{\mathbf{e}}}_r$ we note from Figure 1.6 that

$$\hat{\mathbf{e}}_r = \cos\theta\hat{\mathbf{i}} + \sin\theta\hat{\mathbf{j}} \tag{1.32a}$$

$$\hat{\mathbf{e}}_\theta = -\sin\theta\hat{\mathbf{i}} + \cos\theta\hat{\mathbf{j}} \tag{1.32b}$$

Hence

$$\dot{\hat{\mathbf{e}}}_r = \dot{\theta}(-\sin\theta\hat{\mathbf{i}} + \cos\theta\hat{\mathbf{j}}) = \dot{\theta}\hat{\mathbf{e}}_\theta \tag{1.33}$$

and thus the velocity in cylindrical coordinates takes the form

$$\mathbf{v}_P = \dot{r}\hat{\mathbf{e}}_r + r\dot{\theta}\hat{\mathbf{e}}_\theta + \dot{z}\hat{\mathbf{k}} \tag{1.34}$$

Differentiating again, we get

$$\mathbf{a}_P = \ddot{r}\hat{\mathbf{e}}_r + \dot{r}\dot{\hat{\mathbf{e}}}_r + \dot{r}\dot{\theta}\hat{\mathbf{e}}_\theta + r\ddot{\theta}\hat{\mathbf{e}}_\theta + r\dot{\theta}\dot{\hat{\mathbf{e}}}_\theta + \ddot{z}\hat{\mathbf{k}} \tag{1.35}$$

Using Equations (1.32), we find that

$$\dot{\hat{\mathbf{e}}}_\theta = \dot{\theta}(-\cos\theta\hat{\mathbf{i}} - \sin\theta\hat{\mathbf{j}}) = -\dot{\theta}\hat{\mathbf{e}}_r \tag{1.36}$$

Thus the acceleration expression in cylindrical coordinates is

$$\mathbf{a}_P = (\ddot{r} - r\dot{\theta}^2)\hat{\mathbf{e}}_r + (r\ddot{\theta} + 2\dot{r}\dot{\theta})\hat{\mathbf{e}}_\theta + \ddot{z}\hat{\mathbf{k}} \tag{1.37}$$

In the special case for which the motion is in a plane defined by $z =$ constant, we have $\dot{z} = \ddot{z} = 0$. In this case we need only the polar coordinates r and θ, and the directions of $\hat{\mathbf{e}}_r$ and $\hat{\mathbf{e}}_\theta$ are said to be **radial** and **transverse**, respectively.

Question 1.4 If a point P moves with $z \equiv 0$, then $\mathbf{r}_{OP} = r\hat{\mathbf{e}}_r$; thus $|\mathbf{r}_{OP}| = r$. Why then isn't $|\mathbf{v}_P| = \dot{r}$?

Before turning to the examples in this section, we return briefly to calculation of the derivatives of the unit vectors $\hat{\mathbf{e}}_r$ and $\hat{\mathbf{e}}_\theta$. We note that each derivative turns out to be perpendicular to the vector being differentiated. To understand why this is the case, we consider an alternative derivation of the formula for $\dot{\hat{\mathbf{e}}}_r$. The time dependence of $\hat{\mathbf{e}}_r$ is due to the time dependence of the coordinate θ on which $\hat{\mathbf{e}}_r$ depends explicitly; thus we can write

$$\dot{\hat{\mathbf{e}}}_r = \frac{d\hat{\mathbf{e}}_r}{d\theta}\dot{\theta}$$

Answer 1.4 For one thing, $\dot{r}$ can be negative. But $|\mathbf{v}_P| \neq |\dot{r}|$ either, because the magnitude of the derivative of a vector is not equal to the absolute value of the derivative of the magnitude of the vector. Differentiating $\mathbf{r}_{OP}$ produces a term $(r\dot{\theta}\hat{\mathbf{e}}_\theta)$ in $\mathbf{v}_P$ in addition to $\dot{r}\hat{\mathbf{e}}_r$.

Let us study the derivative $d\hat{\mathbf{e}}_r/d\theta$. By definition,

$$\frac{d\hat{\mathbf{e}}_r}{d\theta} = \lim_{\Delta\theta \to 0} \left[\frac{\hat{\mathbf{e}}_r(\theta + \Delta\theta) - \hat{\mathbf{e}}_r(\theta)}{\Delta\theta}\right] = \lim_{\Delta\theta \to 0} \left(\frac{\Delta\hat{\mathbf{e}}_r}{\Delta\theta}\right)$$

With the aid of Figure 1.7 we can see that:

1. The direction of $\Delta\hat{\mathbf{e}}_r$ (and hence $\Delta\hat{\mathbf{e}}_r/\Delta\theta$) approaches that of $\hat{\mathbf{e}}_\theta$ as $\Delta\theta$ approaches zero.
2. The magnitude of $\Delta\hat{\mathbf{e}}_r/\Delta\theta$ is $[2(1)\ \sin(\Delta\theta/2)]/\Delta\theta$, which approaches unity as $\Delta\theta$ approaches zero.

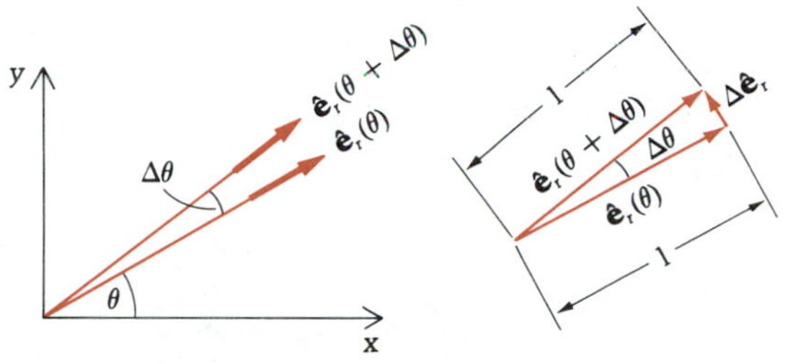

Figure 1.7 Change in $\hat{\mathbf{e}}_r$ as θ changes.

Thus $d\hat{\mathbf{e}}_r/d\theta = \hat{\mathbf{e}}_\theta$, and we obtain $\dot{\hat{\mathbf{e}}}_r = \dot{\theta}\hat{\mathbf{e}}_\theta$ in agreement with Equation (1.33). The reader may wish to sketch a similar geometric proof of Equation (1.36).

This mutual orthogonality of a vector and its derivative, incidentally, is not just restricted to unit vectors. It is in fact a property of all vectors of constant magnitude. We can show that this is the case by noting that if $\mathbf{A}$ is such a vector, then

$$\mathbf{A} \cdot \mathbf{A} = |\mathbf{A}|^2 = \text{constant}$$

and thus

$$\frac{d}{dt}(\mathbf{A} \cdot \mathbf{A}) = 0$$

or

$$\frac{d\mathbf{A}}{dt} \cdot \mathbf{A} + \mathbf{A} \cdot \frac{d\mathbf{A}}{dt} = 0$$

or

$$2\mathbf{A} \cdot \frac{d\mathbf{A}}{dt} = 0 \qquad (1.38)$$

Hence, provided that neither the vector nor its derivative vanishes, the vector and the derivative are mutually perpendicular. This is a result we shall make use of frequently throughout the book. We now proceed to some examples of velocity and acceleration in cylindrical coordinates.

EXAMPLE 1.14

The pin P in Figure E1.14a moves outward with respect to a horizontal circular disk, and its radial coordinate r is given as a function of time by $r = 3t^2/2$ meters. The disk $\mathcal{D}$ turns with the time-dependent angle $\theta = 4t^2/3$ rad. Find the velocity and acceleration of P at $t = 1$ s.

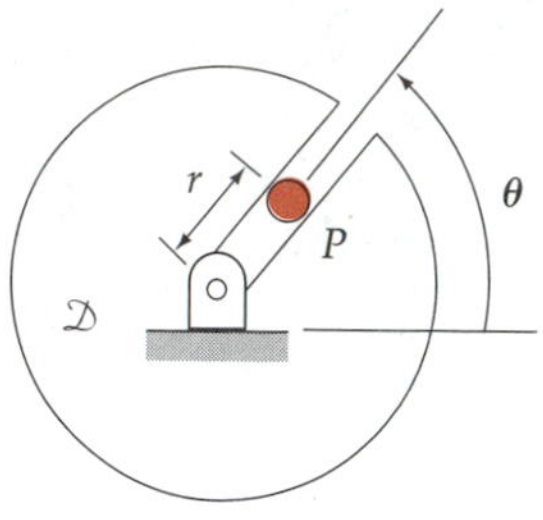

Figure E1.14a

Solution

From Equation (1.34) we have

$$\mathbf{v}_P = \dot{r}\hat{\mathbf{e}}_r + r\dot{\theta}\hat{\mathbf{e}}_\theta + \dot{z}\hat{\mathbf{k}}$$
$$= 3t\hat{\mathbf{e}}_r + \left(\frac{3}{2}t^2\right)\left(\frac{8}{3}t\right)\hat{\mathbf{e}}_\theta + 0$$
$$= 3t\hat{\mathbf{e}}_r + 4t^3\hat{\mathbf{e}}_\theta \text{ m/s}$$

Thus

$$\mathbf{v}_P|_{t=1} = 3\hat{\mathbf{e}}_r + 4\hat{\mathbf{e}}_\theta \text{ m/s}$$

and we note that the speed of P at $t = 1$ s is 5 m/s.

Continuing, from Equation (1.37) we get

$$\mathbf{a}_P = (\ddot{r} - r\dot{\theta}^2)\hat{\mathbf{e}}_r + (r\ddot{\theta} + 2\dot{r}\dot{\theta})\hat{\mathbf{e}}_\theta + \ddot{z}\hat{\mathbf{k}}$$
$$= \left[3 - \left(\frac{3}{2}t^2\right)\left(\frac{8}{3}t\right)^2\right]\hat{\mathbf{e}}_r + \left[\left(\frac{3}{2}t^2\right)\frac{8}{3} + 2(3t)\left(\frac{8}{3}t\right)\right]\hat{\mathbf{e}}_\theta + 0$$
$$= \left(3 - \frac{32}{3}t^4\right)\hat{\mathbf{e}}_r + 20t^2\hat{\mathbf{e}}_\theta$$

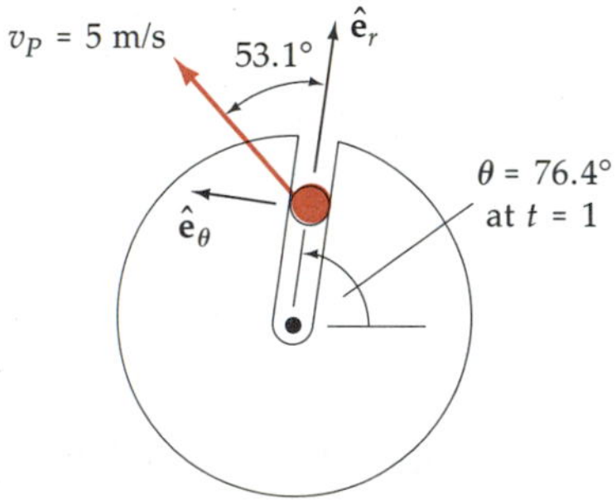

Velocity $\mathbf{v}_P$

Figure E1.14b

Thus

$$\mathbf{a}_P|_{t=1} = \frac{-23}{3}\hat{\mathbf{e}}_r + 20\hat{\mathbf{e}}_\theta \text{ m/s}^2$$

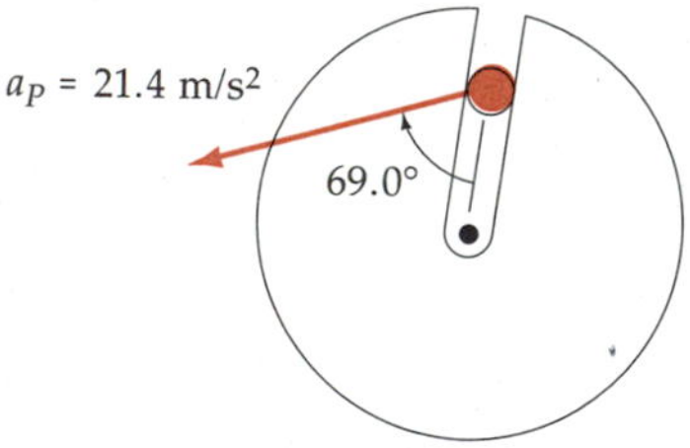

Acceleration $\mathbf{a}_P$

Figure E1.14c

Since at $t = 1$ we have $r = 3/2$ m and $\theta = 4/3$ rad, we show the preceding results pictorially in Figures E1.14b and c.

Note that there is a time, $t = \sqrt[4]{9/32}$ s, when the $\ddot{r}$ and $-r\dot{\theta}^2$ parts of the radial component of $\mathbf{a}_P$ cancel each other, making this component zero at that instant of time. The reader is urged to compute and sketch $\mathbf{v}_P$ and $\mathbf{a}_P$ at another time, say $t = 2$ s.

EXAMPLE 1.15

In the preceding example, discard the given r and θ. Suppose instead that $\dot{\theta}_{\mathcal{D}}$ = constant = 0.3 rad/s and that the pin slides not only in the slot of disk $\mathcal{D}$ (see Figure E1.15a), but also in the spiral slot cut in the reference frame and defined by $r = 0.1\theta$ meters, with θ in radians. Find the velocity and acceleration of the pin when $\theta = \pi$ rad.

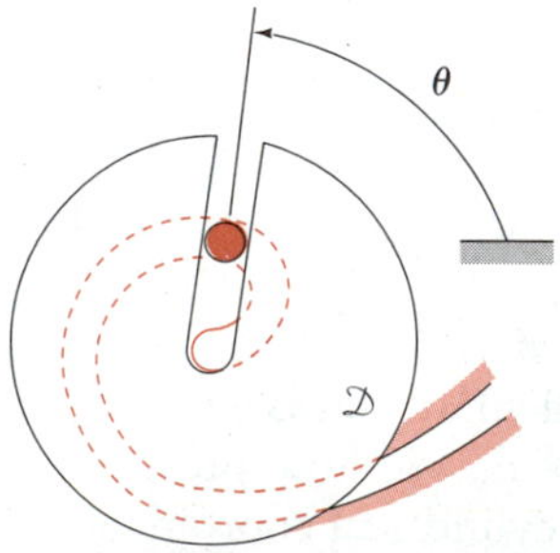

Figure E1.15a

Solution

From $r = 0.1\theta$, we have $\dot{r} = 0.1\dot{\theta}$ and $\ddot{r} = 0.1\ddot{\theta}$, which is zero since $\dot{\theta}$ = constant. Therefore

$$\mathbf{v}_P = \dot{r}\hat{\mathbf{e}}_r + r\dot{\theta}\hat{\mathbf{e}}_\theta = 0.1\dot{\theta}\hat{\mathbf{e}}_r + 0.1\theta\dot{\theta}\hat{\mathbf{e}}_\theta$$
$$= 0.0300\hat{\mathbf{e}}_r + 0.0942\hat{\mathbf{e}}_\theta \text{ m/s}$$

Now for the acceleration:

$$\mathbf{a}_P = (\ddot{r} - r\dot{\theta}^2)\hat{\mathbf{e}}_r + (r\ddot{\theta} + 2\dot{r}\dot{\theta})\hat{\mathbf{e}}_\theta$$
$$= (0 - 0.1\theta\dot{\theta}^2)\hat{\mathbf{e}}_r + [0 + 2(0.1\dot{\theta})\dot{\theta}]\hat{\mathbf{e}}_\theta$$
$$= -0.0283\hat{\mathbf{e}}_r + 0.0180\hat{\mathbf{e}}_\theta \text{ m/s}^2$$

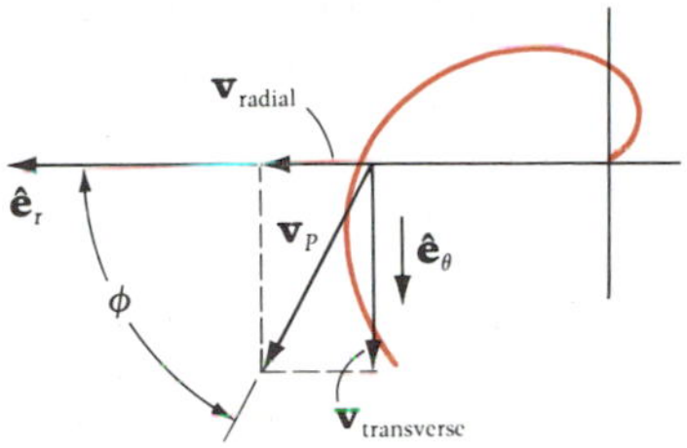

Figure E1.15b

We shall learn in the next section that the velocity is always tangent to the path of the point. Thus the angle ϕ between the path and the $-x$ axis can be found from the velocity components, as shown in Figure E1.15b, as follows:

$$\phi = \tan^{-1}\left(\frac{0.0942}{0.0300}\right) = 72.3°$$

In our next examples, there is motion in the z direction as well as the radial (r) and transverse (θ) of the previous examples.

Example 1.16

A point Q moves on a helix as shown in Figure E1.16a. The pitch, p, of the helix is 0.2 m, and the point travels at constant speed 20 m/s. Find the velocity of Q in terms of its cylindrical components.

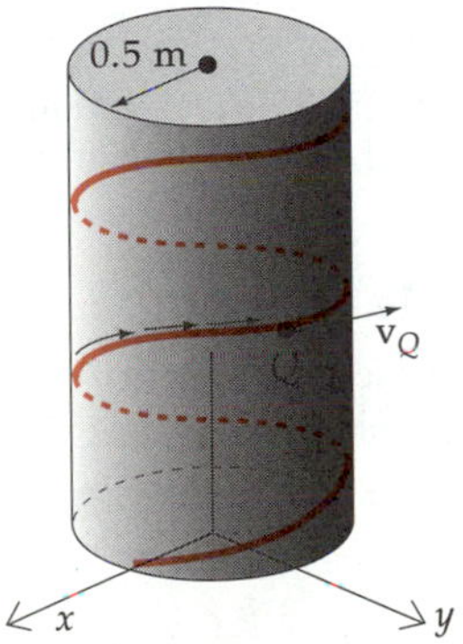

Figure E1.16a

Solution

The meaning of the *pitch* of a helix is the (constant) advance of Q in the z direction for each revolution in θ. Therefore

$$\frac{p\theta}{2\pi} = z \qquad (1)$$

so that

$$\dot{\theta} = \frac{2\pi\dot{z}}{p} \qquad (2)$$

or, for this problem,

$$\dot{\theta} = 31.42\dot{z} \qquad (3)$$

Noting that $\dot{r} = 0$ since Q travels on a cylinder (with r therefore constant), Equation (1.34) then gives the following for the point's velocity:

$$\mathbf{v}_Q = r\dot{\theta}\hat{\mathbf{e}}_\theta + \dot{z}\hat{\mathbf{k}} \qquad (4)$$
$$= 0.5\dot{\theta}\hat{\mathbf{e}}_\theta + \dot{z}\hat{\mathbf{k}} \qquad (5)$$

or, using (3),

$$\mathbf{v}_Q = 15.71\dot{z}\hat{\mathbf{e}}_\theta + \dot{z}\hat{\mathbf{k}} \qquad (6)$$

The speed of Q is constant at 20 m/s; thus

$$\sqrt{(15.71\dot{z})^2 + \dot{z}^2} = 20 \qquad (7)$$
$$\dot{z} = 1.271 \text{ m/s} \qquad (8)$$

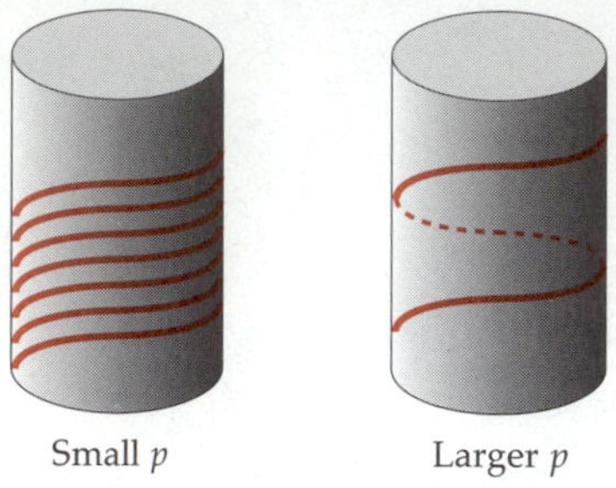

Figure E1.16b

From Equation (3), we then get

$$\dot{\theta} = 31.42(1.271) = 39.94 \text{ rad/s} \tag{9}$$

Hence the velocity vector of Q is (substituting (9) and (8) into (5))

$$\mathbf{v}_Q = 19.97\hat{\mathbf{e}}_\theta + 1.271\hat{\mathbf{k}} \text{ m/s} \tag{10}$$

Note that $|\mathbf{v}_Q| = 20.0$ m/s, as it must be. Note also that a larger pitch will spread out the helix (see Figure E1.16b). The equations of this example then show that the $\hat{\mathbf{k}}$ component will become larger in comparison to the $\hat{\mathbf{e}}_\theta$ component for larger p.

EXAMPLE 1.17

Find the acceleration of Q in the preceding example.

Solution

From Equation (1.37) we get

$$\mathbf{a}_Q = (\ddot{r} - r\dot{\theta}^2)\hat{\mathbf{e}}_r + (r\ddot{\theta} + 2\dot{r}\dot{\theta})\hat{\mathbf{e}}_\theta + \ddot{z}\hat{\mathbf{k}}$$

Because r is constant on the cylinder, this equation reduces to

$$\mathbf{a}_Q = -r\dot{\theta}^2\hat{\mathbf{e}}_r + r\ddot{\theta}\hat{\mathbf{e}}_\theta + \ddot{z}\hat{\mathbf{k}}$$

Furthermore, since $|\mathbf{v}_Q|$ is constant, Equations (8) and (3) of the previous example show that $\dot{z}$ and $\dot{\theta}$ are constants. Therefore there is only one nonvanishing acceleration component here:

$$\mathbf{a}_Q = -r\dot{\theta}^2\hat{\mathbf{e}}_r = -0.5(39.9^2)\hat{\mathbf{e}}_r = -796\hat{\mathbf{e}}_r \text{ m/s}^2$$

Note that even though point Q *never* has a radial component of velocity (see Figure E1.17), it has *only* a radial component of acceleration!

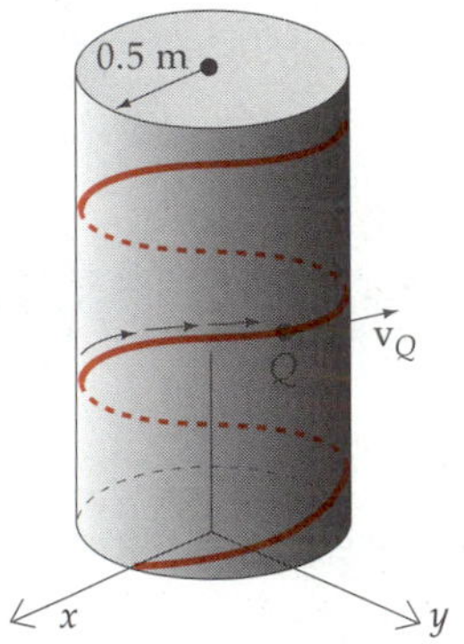

Figure E1.17

EXAMPLE 1.18

Find the velocity and acceleration vectors of point Q in Example 1.16 if, instead of the speed of Q being constant, we have its vertical position given as the function of time:

$$z = 0.08t^3 \text{ m}$$

Solution

Referring to Example 1.16 (see Figure E1.18), we find

$$\dot{\theta} = 31.4\dot{z} = 31.4(0.24t^2)$$
$$= 7.54t^2 \text{ rad/s}$$

Therefore

$$\mathbf{v}_Q = 0.5(7.54t^2)\hat{\mathbf{e}}_\theta + 0.24t^2\hat{\mathbf{k}}$$
$$= 3.77t^2\hat{\mathbf{e}}_\theta + 0.24t^2\hat{\mathbf{k}} \text{ m/s}$$

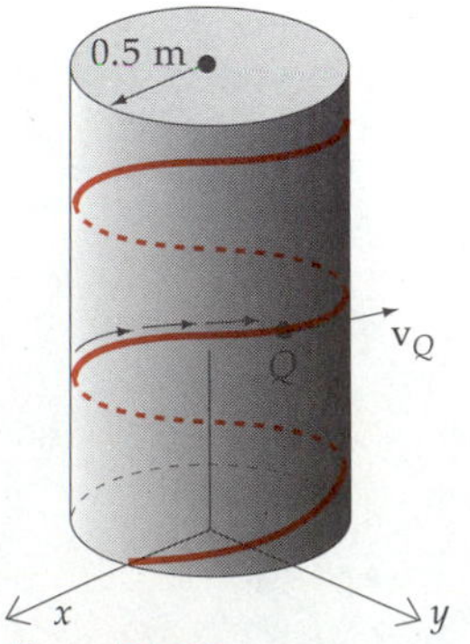

Figure E1.18

This time, however, the velocity is seen to depend on the time; for example, at $t = 10$ s,

$$\mathbf{v}_Q|_{t=10} = 377\hat{\mathbf{e}}_\theta + 24.0\hat{\mathbf{k}} \text{ m/s}$$

For the acceleration, we note that $\dot{r}$ is still zero, so that

$$\mathbf{a}_Q = -r\dot{\theta}^2\hat{\mathbf{e}}_r + r\ddot{\theta}\hat{\mathbf{e}}_\theta + \ddot{z}\hat{\mathbf{k}}$$

This time, all three terms are nonzero. We have

$$\dot{z} = 0.24t^2 \text{ m/s} \qquad \dot{\theta} = 7.54t^2 \text{ rad/s}$$
$$\ddot{z} = 0.48t \text{ m/s}^2 \qquad \ddot{\theta} = 15.1t \text{ rad/s}^2$$

Thus

$$\begin{aligned}\mathbf{a}_Q &= -0.5(7.54t^2)^2\hat{\mathbf{e}}_r + 0.5(15.1t)\hat{\mathbf{e}}_\theta + (0.48t)\hat{\mathbf{k}} \\ &= -28.4t^4\hat{\mathbf{e}}_r + 7.55t\hat{\mathbf{e}}_\theta + 0.48t\hat{\mathbf{k}} \text{ m/s}^2\end{aligned}$$

In the final example of this section, we consider the case in which, in addition to the changing θ and z of the preceding three examples, the radius varies.

Example 1.19

A point P moves on a spiraling path that winds around the paraboloid of revolution shown in Figure E1.19. The focal distance f is $\frac{1}{4}$ m, and the point P advances 4.0 m vertically with each revolution. If the speed of P is 0.7 m/s, a constant, determine the vertical component of the velocity vector of P as a function of r.

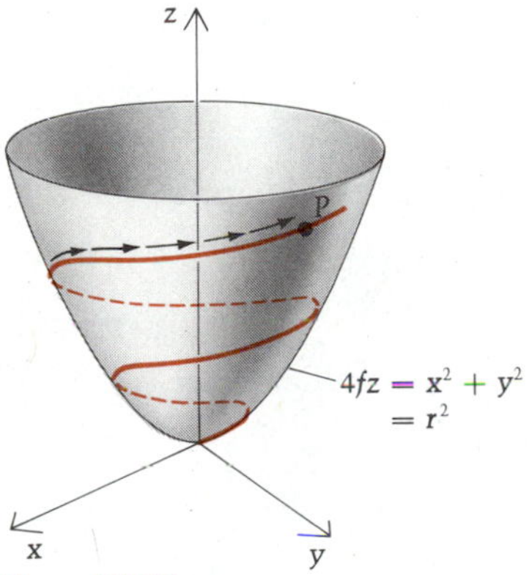

Figure E1.19

Solution

From $z = r^2$, we obtain

$$\dot{z} = 2r\dot{r} \Rightarrow \dot{r} = \frac{\dot{z}}{2r} \text{ m/s} \tag{1}$$

And from the pitch relationship $p\theta/2\pi = z$, we get

$$4.0\dot{\theta} = 2\pi\dot{z} \Rightarrow \dot{\theta} = \frac{\pi}{2}\dot{z} \text{ rad/s} \tag{2}$$

Therefore the speed of P may be expressed as

$$\begin{aligned}|\mathbf{v}_P| = 0.7 &= \sqrt{\dot{r}^2 + (r\dot{\theta})^2 + \dot{z}^2} \\ &= \sqrt{\left(\frac{\dot{z}}{2r}\right)^2 + \left(\frac{\pi}{2}r\dot{z}\right)^2 + \dot{z}^2} \text{ m/s}\end{aligned}$$

Thus the answer is

$$\dot{z} = \frac{1.4r}{\sqrt{1 + 4r^2 + \pi^2 r^4}} \text{ m/s}$$

Let us extend the preceding example slightly. We can see that $\dot{z}$ varies with the radius r (distance from the z axis to P) and that it is zero initially and approaches zero again for large r. Its maximum may be determined from calculus:

$$\frac{d\dot{z}}{dr} = 0$$

or

$$0 = \frac{\sqrt{1 + 4r^2 + \pi^2 r^4}\,\dfrac{d(1.4r)}{dr} - 1.4r\,\dfrac{d\sqrt{1 + 4r^2 + \pi^2 r^4}}{dr}}{(\sqrt{1 + 4r^2 + \pi^2 r^4})^2}$$

This yields the equation and result:

$$\pi^2 r^4 = 1 \Rightarrow r = 0.56 \text{ m}$$

at which

$$\dot{z}_{max} = 0.44 \text{ m/s}$$

Note from Equation (2) in the example that at this value of $\dot{z}$,

$$\dot{\theta} = 0.69 \text{ rad/s}$$

From Equation (1), we see that at the same time

$$\dot{r} = \frac{\dot{z}}{2r} = 0.39 \text{ m/s}$$

and therefore, when $\dot{z}$ is maximum, the speed is

$$|\mathbf{v}_P| = \sqrt{\dot{r}^2 + (r\dot{\theta})^2 + \dot{z}^2}$$
$$= \sqrt{0.39^2 + (0.56 \times 0.69)^2 + 0.44^2} = 0.70 \text{ m/s}$$

as it should be, since it does not change with time.

Question 1.5 By inspection (with little or no writing), what is the maximum magnitude of the radial component of $\mathbf{v}_P$?

Answer 1.5 When $r = 0$, we have $\dot{z} = 0 = \dot{\theta}$. Thus $\dot{r}$, the radial component of $\mathbf{v}_P$, is maximum there at the value 0.7, which is the constant speed. (Note that $\dot{r}$ decreases continuously toward zero from there.)

PROBLEMS ▶ Section 1.6

1.100 The airplane in Figure P1.100 travels at constant speed at a constant altitude. The radar tracks the plane and computes the distance D, the angle θ, and the rate of change of θ ($\dot{\theta}$) at all times. In terms of θ, $\dot{\theta}$, and D, find the speed of the airplane.

1.101 A ball bearing is moving radially outward in a slotted horizontal disk that is rotating about the vertical z axis. At the instant shown in Figure P1.101, the ball bearing is 3 in. from the center of the disk. It is traveling radially outward at a velocity of 4 in./sec relative to the disk

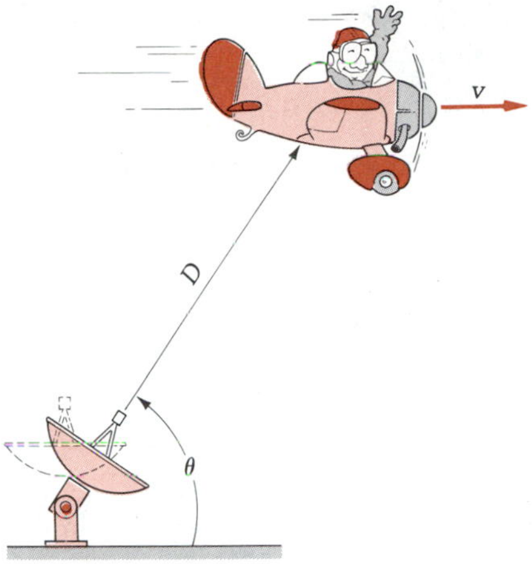

Figure P1.100

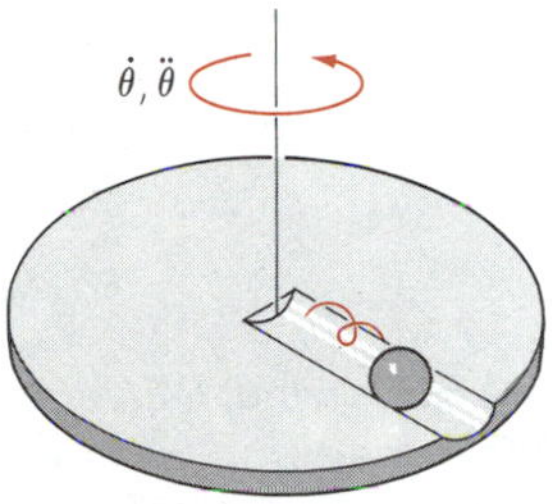

Figure P1.101

and has a radial acceleration with respect to the disk of 5 in./sec² outward. What would $\dot{\theta}$ and $\ddot{\theta}$ have to be at the instant shown for the ball bearing to have a total acceleration of zero?

1.102 The disk shown in Figure P1.102 is horizontal and turns so that $\dot{\theta} = ct$ about the vertical. Forces cause a marble to move in a slot such that its radial distance from the center equals kt^2. Note that c and k are constants.

a. Find the acceleration of the marble.

b. At what time does the radial acceleration vanish?

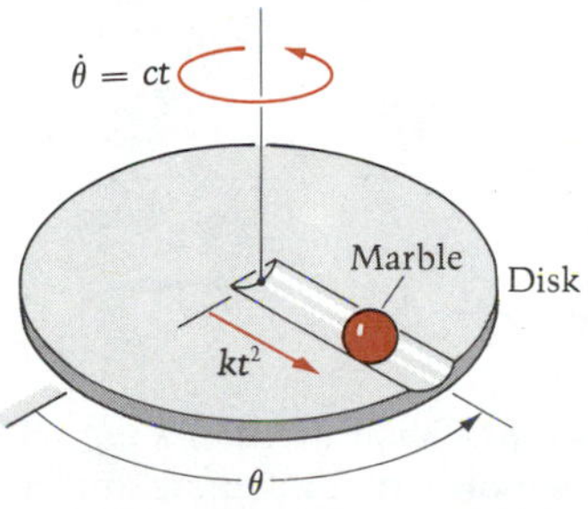

Figure P1.102

1.103 A particle moves on a curve called the "Lemniscate of Bernoulli," defined by $r^2 = 2 \cos 2\theta$ ft². It moves along the branch shown in Figure P1.103 with arrows, and passes through point P at $t = 0$. The angle θ increases with time according to $\theta = 3t^2 + 2t$ rad, with t measured in seconds. At the point P, find the velocity and acceleration of the particle.

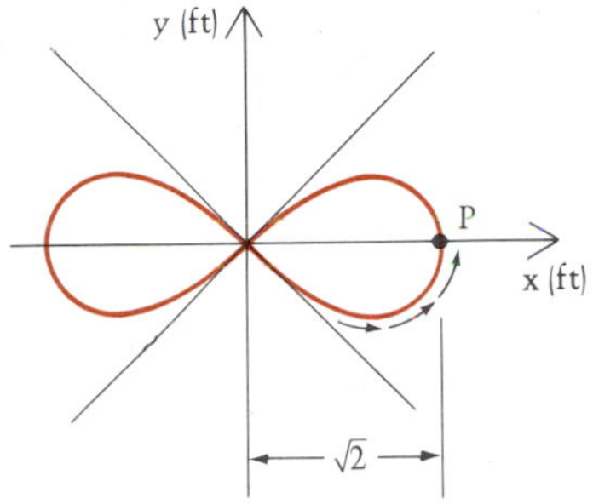

Figure P1.103

* **1.104** A point P moves on the "Spiral of Archimedes" at constant speed 2 m/s. (See Figure P1.104.) The equation of the spiral is $r = 3\theta$. Find the acceleration of P when $\theta = 180°$.

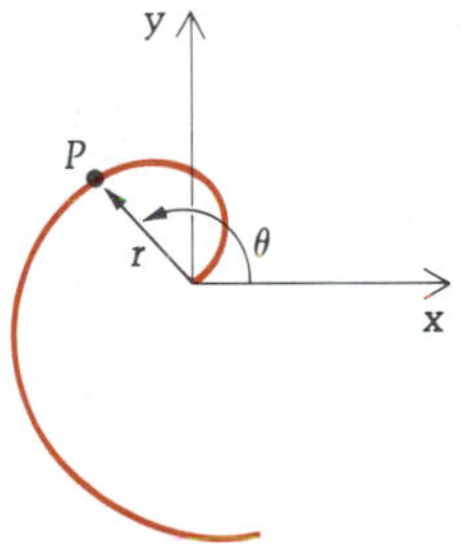

Figure P1.104

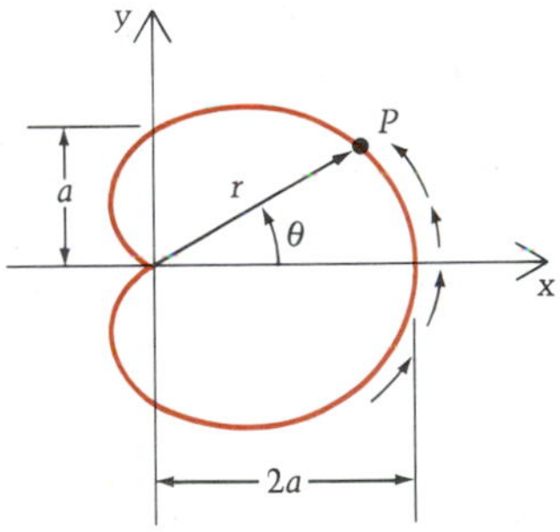

Figure P1.105

1.105 The cardioid in Figure P1.105 has the equation $r = a(1 + \cos \theta)$. Point P travels around this curve, in the direction indicated, in such a way that $\dot{\theta} = K =$ constant. In terms of K and the length a, find the velocity of P at the four points where its path intersects the coordinate axes. Express the result in terms of radial and transverse components, and then convert to rectangular components by expressing $\hat{\mathbf{e}}_r$ and $\hat{\mathbf{e}}_\theta$ in terms of $\hat{\mathbf{i}}$ and $\hat{\mathbf{j}}$ at each position.

1.106 In the preceding problem, find the acceleration of P at the same four points. Again, do the problem first in $(\hat{\mathbf{e}}_r, \hat{\mathbf{e}}_\theta)$ components and then convert the results to $(\hat{\mathbf{i}}, \hat{\mathbf{j}})$ components.

1.107 A point P starts at the origin and moves along the parabola shown in Figure P1.107 with a constant x-component of velocity, $\dot{x} = 3$ ft/sec. Using the following approach, find the radial and transverse components of the velocity and acceleration of P at the point $(x, y) = (1, 1)$: Find $\mathbf{v}_P$ and $\mathbf{a}_P$ in rectangular components (see Problem 1.71); then resolve these vectors along $\hat{\mathbf{e}}_r$ and $\hat{\mathbf{e}}_\theta$ to obtain their radial and transverse components.

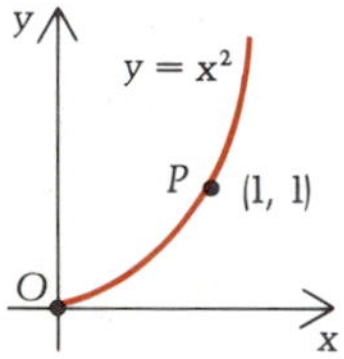

Figure P1.107

1.108 Solve the preceding problem by a different approach: Recall the polar coordinate relations $r = \sqrt{x^2 + y^2}$ and $\theta = \tan^{-1}(y/x)$, and differentiate to obtain $\dot{r}, \ddot{r}, \dot{\theta}$, and $\ddot{\theta}$ for entry into Equations (1.34) and (1.37).

1.109 The four-leaf rose in Figure P1.109 has the Equation $r = 3 \sin 2\theta$ ft. A particle P starts at the origin and travels on the indicated path with $\dot{\theta} = 1/6$ rad/sec = constant. When P is at the highest point in the first quadrant, find:

a. the speed of P
b. the acceleration of P

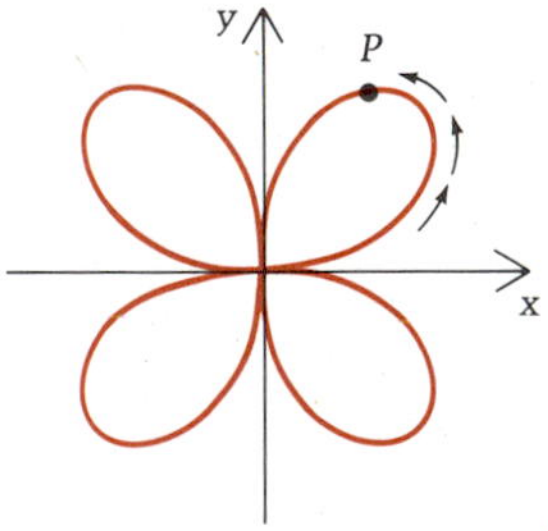

Figure P1.109

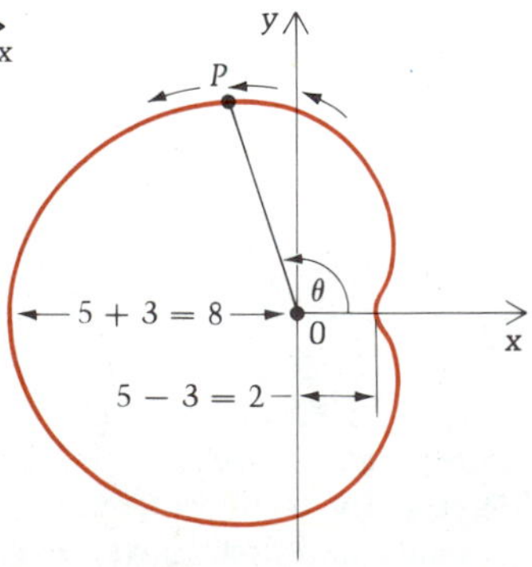

Figure P1.110

* **1.110** The point P in Figure P1.110 moves on the limacon defined in polar coordinates by

$$r = 5 - 3\cos\theta \text{ m}$$

If the polar angle is quadratic in time according to $\theta = 10t^2$ rad, find the velocity of P when it is at its highest point.

1.111 In the preceding problem, determine the acceleration of P at (a) the same highest point and (b) $\theta = \pi$ rad.

* **1.112** A point P moves on the figure eight in the indicated direction (Figure P1.112) at constant speed 2 m/s. Find the acceleration vector of P the next time its velocity is horizontal.

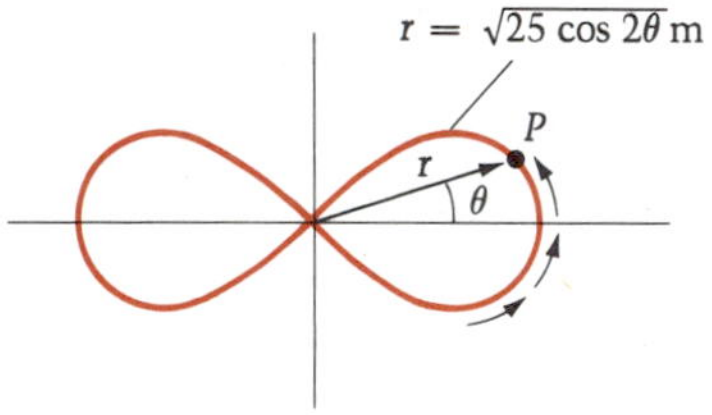

Figure P1.112

* **1.113** An insect is asleep on a $33\frac{1}{3}$ rpm record, 6 in. from the spindle. When the record is turned on, the insect wakes up and dizzily heads toward the center, in a straight line relative to the disk, at 1 in./sec (Figure P1.113). If the bug can withstand a maximum acceleration magnitude of 100 in./sec², does it make it to the spindle (a) if it starts after the record is up to speed? (b) if it starts as soon as the record is turned on? Assume that the turntable accelerates linearly (with time) up to speed in one revolution, and that $\ddot{r} = -\frac{1}{4}$ in./sec² until $\dot{r} = -1$ in./sec.

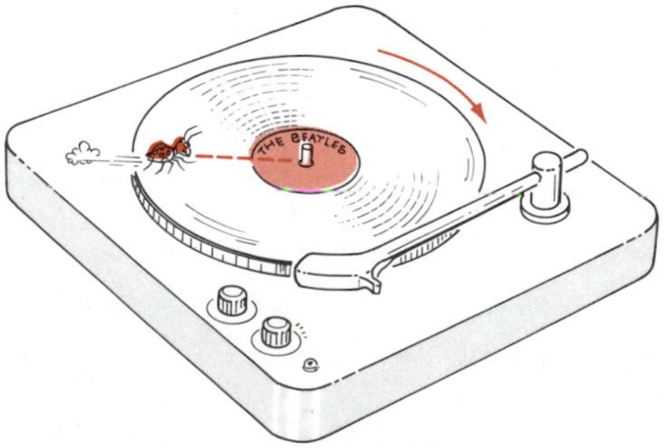

Figure P1.113

1.114 David throws a rock at Goliath with a sling. He whirls it around one revolution plus 135° more and releases it there, as shown in Figure P1.114, at 50 ft/sec.

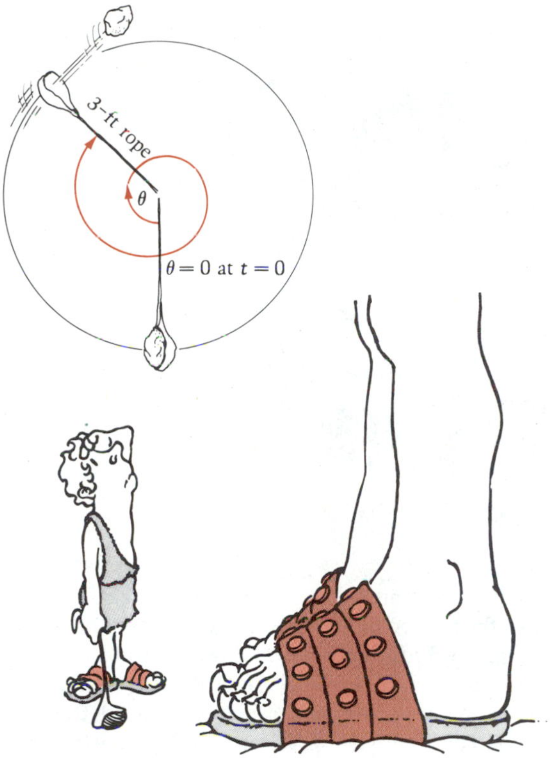

Figure P1.114

As he whirls the sling, the speed of the rock increases linearly with the time t; that is, $\dot{\theta} = kt$, where k is a constant. Find the acceleration of the rock just prior to release.

1.115 In Problem 1.60 show that the velocity of the shingles may also be obtained by simply taking the component of the velocity of the bumper attachment point A *along the rope.* Using the cylindrical coordinate expression for velocity, explain why this procedure works.

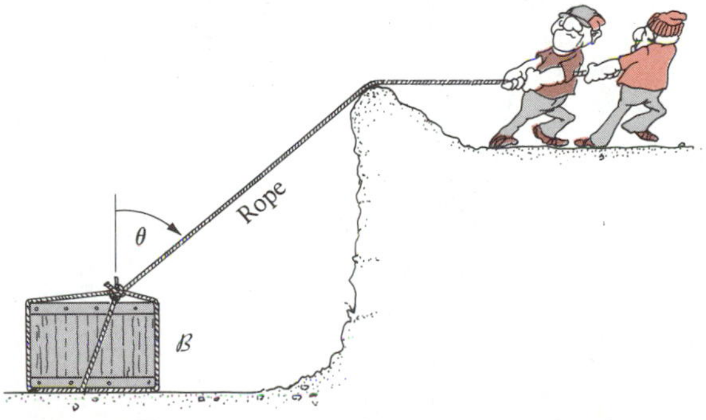

Figure P1.116

1.116 Two people moving at 2 ft/sec to the right are using a rope to drag the box $\mathcal{B}$ along the ground at the lower level (Figure P1.116). Determine the speed of $\mathcal{B}$ as a function of the angle θ between the rope and the vertical.

1.117 The rigid rod $\mathcal{R}$ in Figure P1.117 moves so that its ends, A and B, remain in contact with the surfaces. If, at the instant shown, the velocity of A is 0.5 ft/sec to the right, find the velocity of B.

1.118 In Problem 1.117, find the acceleration of B at the instant in question if the acceleration of A is 2 ft/sec² to the left at that time.

1.119 An ant travels up the banister of a spiral staircase (Figure P1.119) according to

$$\mathbf{r}_{OA} = 2\cos\frac{t}{50}\hat{\mathbf{i}} + 2\sin\frac{t}{50}\hat{\mathbf{j}} + \frac{t}{50}\hat{\mathbf{k}}\text{ m}$$

Find the position and velocity of the ant when $t = 30$ s.

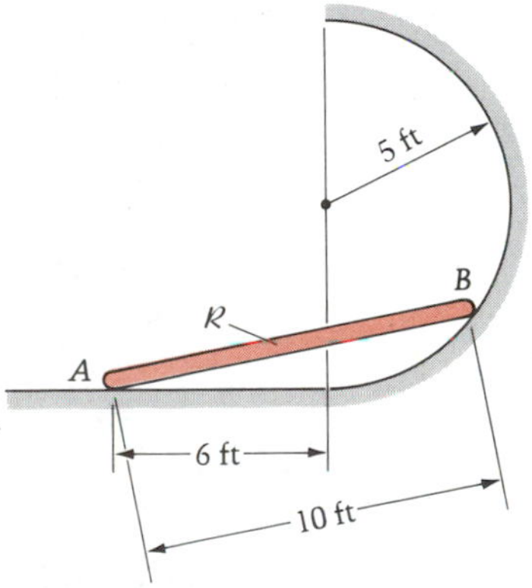

Figure P1.117

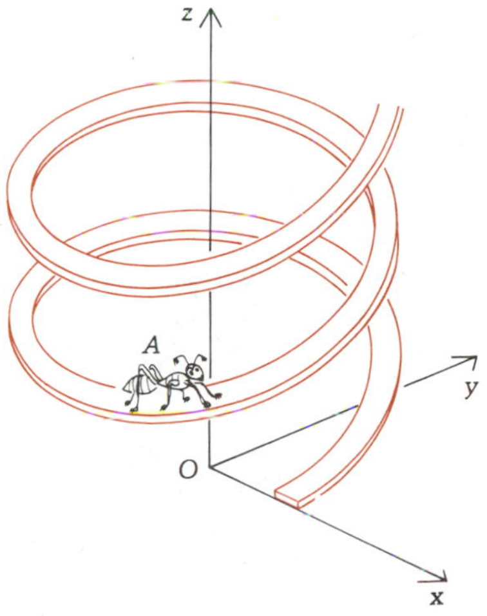

Figure P1.119

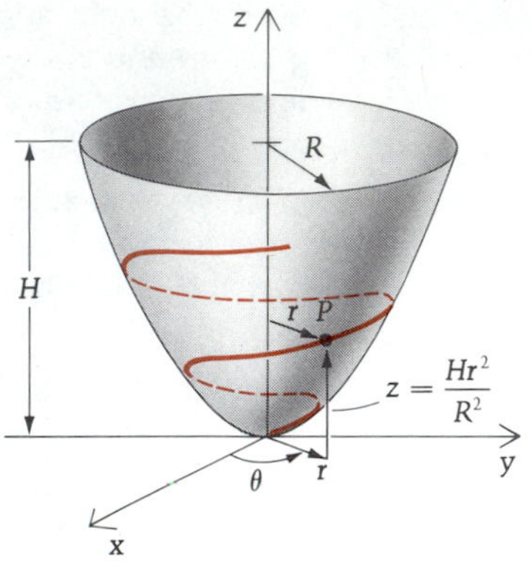

Figure P1.121

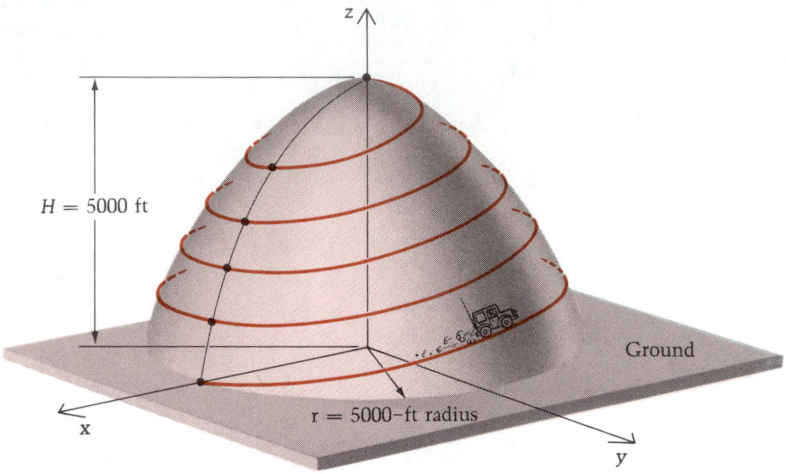

Figure P1.123

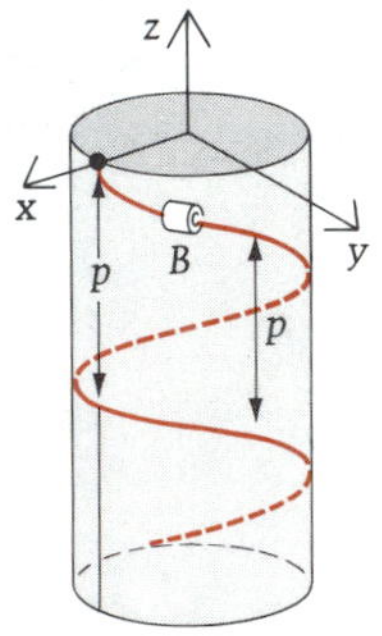

Figure P1.122

1.120 Find the acceleration of the ant (again at $t = 30$ s) in the preceding problem.

1.121 A point P starts at $t = 0$ at the origin and proceeds along a path on the paraboloid of revolution shown in Figure P1.121. The path is described (with time as a parameter) by

$$r = k_1 t$$

$$\theta = k_2 t^2$$

Find the position, velocity, and acceleration vectors of the point when it reaches the top edge of the paraboloid. (H, R, k_1, and k_2 are constants.)

1.122 A bead B slides down and around a cylindrical surface on a helical wire (Figure P1.122). The vertical drop of the bead as θ changes by 2π is called the pitch p of the helix; R is the radius of the helix.

a. Noting that θ (and therefore also z) is a function of time, write the equations for $\mathbf{r}_{OB}$, $\mathbf{v}_B$, and $\mathbf{a}_B$ in cylindrical coordinates.

b. For the values $R = 0.3$ m, $p = 0.2$ m, and $\theta = 0.6t$ rad/s, find and sketch the velocity and acceleration vectors of B when $t = 10$ s.

* **1.123** The mountain shown in Figure P1.123 is in the shape of the paraboloid of revolution $H - z = kr^2$, where H = height = 5000 ft, r is the radius at z, and k is a constant. The base radius is also 5000 ft. A car travels up the mountain on a spiraling path. Each time around, the car's altitude is 1000 ft higher. The car travels at the *constant speed* of 50 mph. Find the largest and smallest absolute values of the radial component of velocity on the journey, and tell where the car is at these two times.

* **1.124** In the preceding problem, find the locations of the car (r, θ, z) for which the following velocity components are equal.

a. Radial ($\dot{r}$) and transverse ($r\dot{\theta}$)

b. Radial and vertical ($\dot{z}$)

c. Transverse and vertical

1.125 Show that the velocity of a point P in spherical coordinates (r, θ, ϕ) is given by

$$\mathbf{v}_P = \dot{r}\hat{\mathbf{e}}_r + r\dot{\theta}\hat{\mathbf{e}}_\theta + r\dot{\phi}\sin\theta\hat{\mathbf{e}}_\phi$$

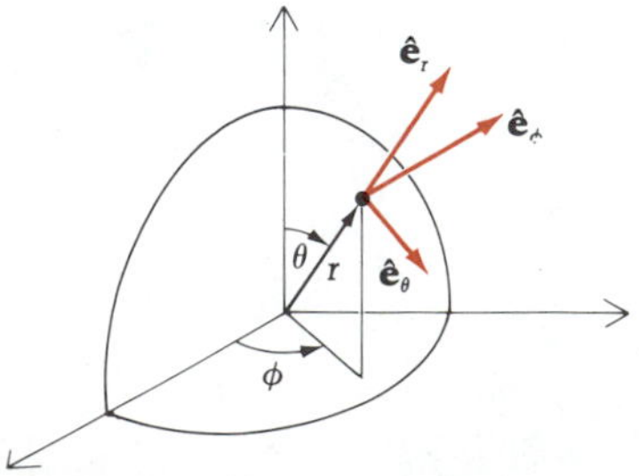

Figure P1.125

See Figure P1.125. *Hint*: As intermediate steps, obtain the results

$$\dot{\hat{\mathbf{e}}}_r = \dot{\theta}\hat{\mathbf{e}}_\theta + \dot{\phi}\sin\theta\hat{\mathbf{e}}_\phi$$
$$\dot{\hat{\mathbf{e}}}_\theta = -\dot{\theta}\hat{\mathbf{e}}_r + \dot{\phi}\cos\theta\hat{\mathbf{e}}_\phi$$
$$\dot{\hat{\mathbf{e}}}_\phi = -\dot{\phi}(\sin\theta\hat{\mathbf{e}}_r + \cos\theta\hat{\mathbf{e}}_\theta)$$

Then differentiate the simple position vector $\mathbf{r}_{OP} = r\hat{\mathbf{e}}_r$.

1.126 Show by differentiating $\mathbf{v}_P$ in the preceding problem that the corresponding expression for the acceleration in spherical coordinates is

$$\mathbf{a}_P = (\ddot{r} - r\dot{\theta}^2 - r\dot{\phi}^2\sin^2\theta)\hat{\mathbf{e}}_r$$
$$+(2\dot{r}\dot{\theta} + r\ddot{\theta} - r\dot{\phi}^2\sin\theta\cos\theta)\hat{\mathbf{e}}_\theta$$
$$+(2\dot{r}\dot{\phi}\sin\theta + 2r\dot{\theta}\dot{\phi}\cos\theta + r\ddot{\phi}\sin\theta)\hat{\mathbf{e}}_\phi$$

* **1.127** The velocity of a point P moving in a plane is the resultant of one part, $v\hat{\mathbf{e}}_r$, along the radius from a fixed point O to the point P, and another part, $u\hat{\mathbf{i}}$, which is always parallel to a fixed line. (See Figure P1.127.) Prove that the acceleration of P may be written as $a_v\hat{\mathbf{e}}_r + a_u\hat{\mathbf{i}}$, where

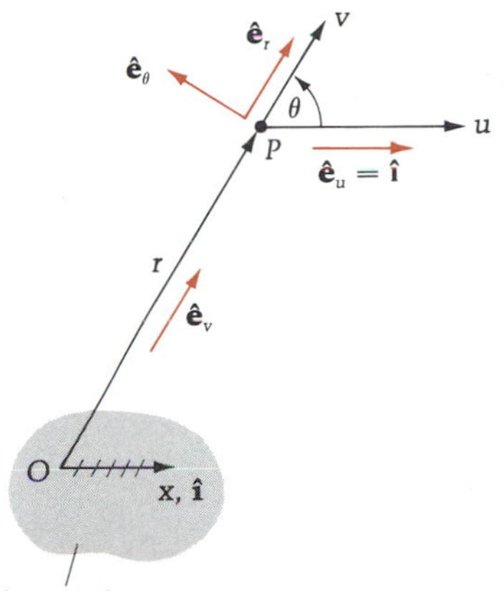

Figure P1.127

$$a_v = \frac{dv}{dt} - \frac{uv}{r}\cos\theta \qquad \text{and} \qquad a_u = \frac{du}{dt} + \frac{uv}{r}$$

where r is the length of the radius vector from O to P and θ is the angle it makes with the fixed direction.

1.7 Tangential and Normal Components

In this section we examine yet another means of expressing the velocity and acceleration of a point P. Instead of focusing on a specific coordinate system, this time we shall study the way in which the motion of P is related to its path. Consequently, the components of velocity and acceleration that result are sometimes called *intrinsic* or *natural.*

The path of point P, as mentioned in Section 1.3, is the locus of points of the reference frame $\mathcal{I}$ successively occupied by P as it moves. We begin, then, by defining some reference point on the path. From this arclength origin we then measure the arclength s along the path to the point P. Clearly, the arclength coordinate depends on the time; that is, $s = s(t)$.

In Figure 1.8 we see a position vector, $\mathbf{r}_{OP}$, for point P. This vector was seen in preceding sections to define the location of P, and thus it may be considered a function of the arclength s:

$$\mathbf{r}_{OP} = \mathbf{r}_{OP}(s) = \mathbf{r}_{OP}[s(t)] \tag{1.39}$$

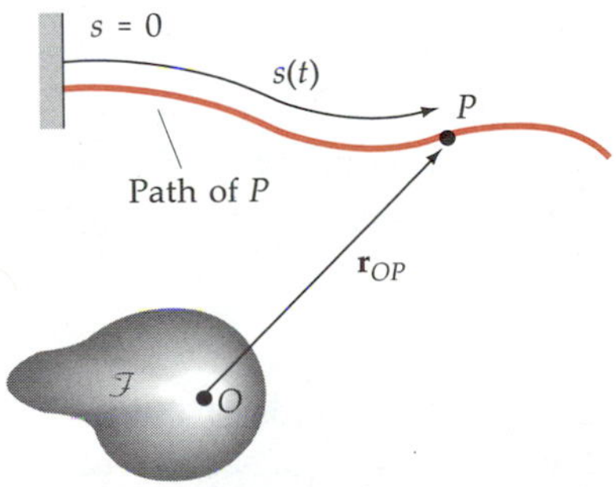

Figure 1.8 Arclength measurement of point *P* on its path.

Forming the velocity of P by differentiation (the definition is the same, regardless of how we choose to represent the vectors), we get

$$\mathbf{v}_P = \dot{\mathbf{r}}_{OP} = \frac{d\mathbf{r}_{OP}}{dt}$$

and, by the chain rule,

$$\begin{aligned}\mathbf{v}_P &= \frac{d\mathbf{r}_{OP}}{ds}\frac{ds}{dt} \\ &= \dot{s}\lim_{\Delta s\to 0}\left[\frac{\mathbf{r}_{OP}(s+\Delta s)-\mathbf{r}_{OP}(s)}{\Delta s}\right] \\ &= \dot{s}\lim_{\Delta s\to 0}\left(\frac{\Delta\mathbf{r}_{OP}}{\Delta s}\right)\end{aligned} \tag{1.40}$$

Figure 1.9 shows the quantities $\Delta\mathbf{r}_{OP}$ and Δs.

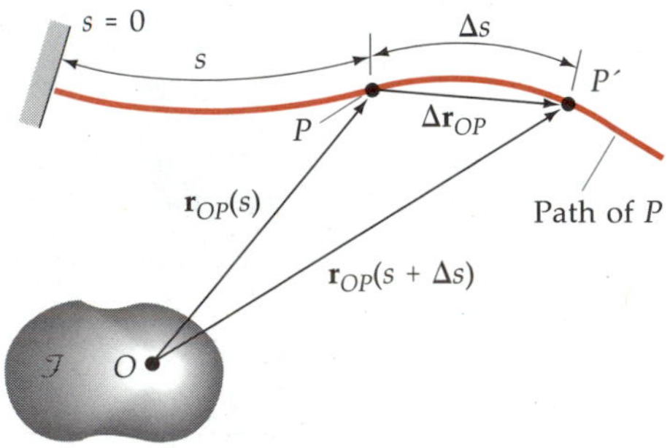

Figure 1.9 Changes in $\mathbf{r}_{OP}$ as s changes.

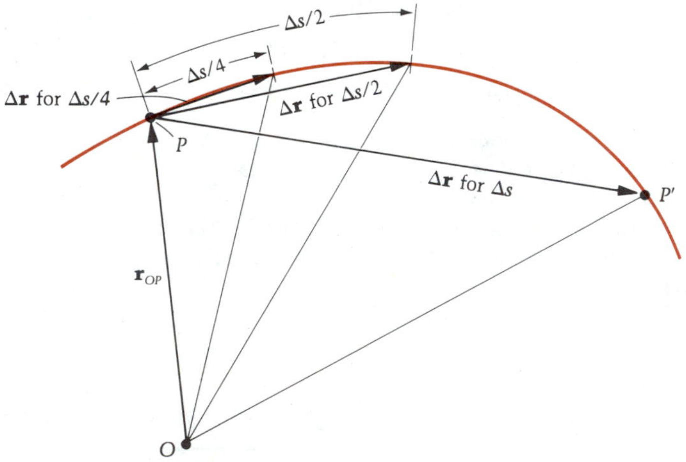

Figure 1.10 Shrinking Δs toward zero.

We suggest that the reader sketch an arc on a large sheet of paper and then use a straightedge to draw the triangle OPP' (Figure 1.10). Then the limit in Equation (1.40) can be taken by, let us say, dividing Δs in half each time. After just a few more divisions of Δs on the large sheet, it will become clear that as Δs approaches zero — that is, as P' backs up toward P — two interesting things happen:

1. $\Delta\mathbf{r}_{OP}$ becomes tangent to the path of P at arclength s.
2. The magnitude of $\Delta\mathbf{r}_{OP}/\Delta s$ approaches $\Delta s/\Delta s = 1$.

These two results, taken together, prove that $d\mathbf{r}_{OP}/ds$ is always a unit vector that is tangent to the path and pointing in the direction of increasing s. It is for these reasons that this vector is called $\hat{\mathbf{e}}_t$, the **unit tangent.** Equation (1.40) may then be rewritten as

$$\mathbf{v}_P = \dot{s}\hat{\mathbf{e}}_t \tag{1.41}$$

From Equation (1.41) we see that the velocity vector of point P is *always tangent to its path.* The absolute value $|\dot{s}|$ of the scalar part — which is the same as the magnitude $|\mathbf{v}_P|$ of the velocity vector $\mathbf{v}_P$ — is called the *speed* of P in $\mathcal{I}$, as we mentioned in Section 1.3.

Next we shall differentiate again in order to obtain the acceleration of P. Using Equation (1.41), we get

$$\mathbf{a}_P = \dot{\mathbf{v}}_P = \ddot{s}\hat{\mathbf{e}}_t + \dot{s}\frac{d\hat{\mathbf{e}}_t}{dt}$$

$$= \ddot{s}\hat{\mathbf{e}}_t + \dot{s}^2\frac{d\hat{\mathbf{e}}_t}{ds} \tag{1.42}$$

Since $\hat{\mathbf{e}}_t$ is a unit vector, $d\hat{\mathbf{e}}_t/ds$ is perpendicular to $\hat{\mathbf{e}}_t$ and hence perpendicular, or normal, to the path. Equation (1.42) shows an important separation of the acceleration into two parts, one tangent and the other normal to the path of P. The component tangent to the path, $\ddot{s}$, is (for $\dot{s} > 0$) the rate of change of the velocity *magnitude,* or speed, of P. The component normal to the path reflects the rate of change of the *direction* of the velocity vector.

Further examination of $d\hat{\mathbf{e}}_t/ds$ in Equation (1.42) is facilitated if we first restrict our attention to the case of a two-dimensional (plane) curve. To that end, let θ be the inclination of a tangent to the plane curve as shown in Figure 1.11. We can visualize that as s increases, $\hat{\mathbf{e}}_t$ turns in such a way that $d\hat{\mathbf{e}}_t/ds$ points toward the inside of the curve — that is, in the direction of $\hat{\mathbf{e}}_n$ shown in the figure. We can obtain this result analytically if we write

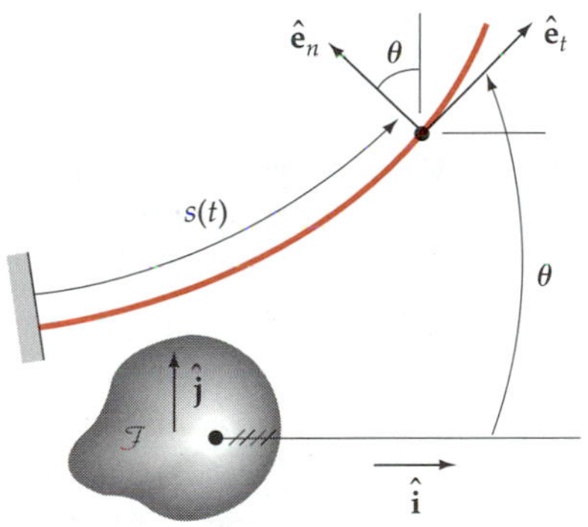

Figure 1.11 Tangent and normal to a plane curve.

$$\frac{d\hat{\mathbf{e}}_t}{ds} = \frac{d\theta}{ds}\frac{d\hat{\mathbf{e}}_t}{d\theta} \tag{1.43}$$

Noting from Figure 1.11 that

$$\hat{\mathbf{e}}_t = \cos\theta\hat{\mathbf{i}} + \sin\theta\hat{\mathbf{j}}$$

we may differentiate to obtain

$$\frac{d\hat{\mathbf{e}}_t}{d\theta} = -\sin\theta\hat{\mathbf{i}} + \cos\theta\hat{\mathbf{j}}$$

which is a unit vector normal to the curve. If $d\theta/ds$ is positive, as is illustrated in Figure 1.11, then $d\hat{\mathbf{e}}_t/d\theta = \hat{\mathbf{e}}_n$. If $d\theta/ds$ is negative (curve concave downward), then $d\hat{\mathbf{e}}_t/d\theta$ points toward the outside of the curve, as the reader may wish to confirm with a sketch, and $d\hat{\mathbf{e}}_t/ds$ again points toward the inside of the curve. Thus, in either case

$$\frac{d\hat{\mathbf{e}}_t}{ds} = \left|\frac{d\theta}{ds}\right|\hat{\mathbf{e}}_n \tag{1.44}$$

where $\hat{\mathbf{e}}_n$ is understood to point toward the inside of the curve.

From studies in calculus the reader probably recognizes $|d\theta/ds|$ as the **curvature** of a plane curve. The reciprocal of the curvature is the **radius of curvature** ρ. The radius of curvature is the radius of the circle that provides the best local approximation to an infinitesimal segment of the curve. Equation (1.43) may thus be written

$$\frac{d\hat{\mathbf{e}}_t}{ds} = \frac{1}{\rho}\hat{\mathbf{e}}_n \tag{1.45}$$

In three dimensions the situation is more difficult to visualize. We cannot use the preceding development because $\hat{\mathbf{e}}_t$ cannot be expressed as a function of a single angle such as θ. Consequently, in the general case we adopt a definition of curvature that, in two dimensions, reduces to what we have just established. That is, we simply define the curvature $1/\rho$ to be the magnitude of the vector $d\hat{\mathbf{e}}_t/ds$. Then the unit vector $\hat{\mathbf{e}}_n$ as defined by

$$\hat{\mathbf{e}}_n = \frac{1}{|d\hat{\mathbf{e}}_t/ds|}\frac{d\hat{\mathbf{e}}_t}{ds} = \rho\frac{d\hat{\mathbf{e}}_t}{ds} \tag{1.46}$$

is called the **principal unit normal** to the curve. Upon substituting into Equation (1.42), we then obtain

$$\mathbf{a}_P = \ddot{s}\hat{\mathbf{e}}_t + \frac{(\dot{s})^2}{\rho}\hat{\mathbf{e}}_n \tag{1.47}$$

An alternative form in which the arclength parameter s is not explicitly involved follows if we choose the measurement of s so that at the instant of interest $\dot{s} > 0$. Hence $\dot{s} = |\mathbf{v}_P|$ and

$$\mathbf{a}_P = \left(\frac{d}{dt}|\mathbf{v}_P|\right)\hat{\mathbf{e}}_t + \frac{|\mathbf{v}_P|^2}{\rho}\hat{\mathbf{e}}_n \tag{1.48}$$

This expression more vividly depicts the natural decomposition of acceleration into parts related to rate of change of magnitude of velocity and rate of change of direction of velocity. We now consider some examples of the use of tangential and normal components.

EXAMPLE 1.20

A car starts at rest at A and increases its speed around the track at 6 ft/sec², traveling counterclockwise (see Figure E1.20). Determine the position and the time at which the car's acceleration magnitude reaches 20 ft/sec².

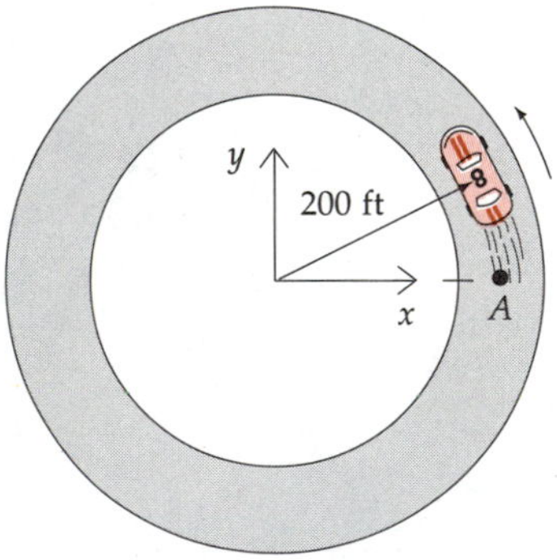

Figure E1.20

Solution

$$\ddot{s} = 6 \text{ ft/sec}^2$$

$$\dot{s} = 6t + C_1 \quad \text{(}C_1 = 0 \text{ ft/sec (The constant is zero since } \dot{s} = 0 \text{ at } t = 0.))$$

$$s = 3t^2 + C_2 \quad \text{(}C_2 = 0 \text{ ft (The constant is zero since } s = 0 \text{ at } t = 0.))$$

The acceleration magnitude of Q is

$$|\mathbf{a}_Q| = a_Q = \sqrt{\ddot{s}^2 + (\dot{s}^2/\rho)^2}$$
$$= \sqrt{36 + (36t^2/200)^2} \text{ ft/sec}^2$$

When $a_Q = 20$ ft/sec², we obtain the equation

$$20^2 = 36 + (0.18t^2)^2 = 36 + 0.0324t^4$$

$$t = \sqrt[4]{\frac{364}{0.0324}} = 10.3 \text{ sec}$$

At $t = 10.3$ sec, $s = 318$ ft, which represents $318/(2\pi r) = 0.253$ of a revolution, or 91.1° counterclockwise from the x axis.

Example 1.21

Verify the results of Example 1.13, at $x = 2$ m, by using $\hat{\mathbf{e}}_t$ and $\hat{\mathbf{e}}_n$ components.

Solution

We are given that $\dot{s} = |\mathbf{v}_P| = 0.2$ m/s. Since $\mathbf{v}_P$ is tangent to the path of P, we can calculate $\hat{\mathbf{e}}_t$:

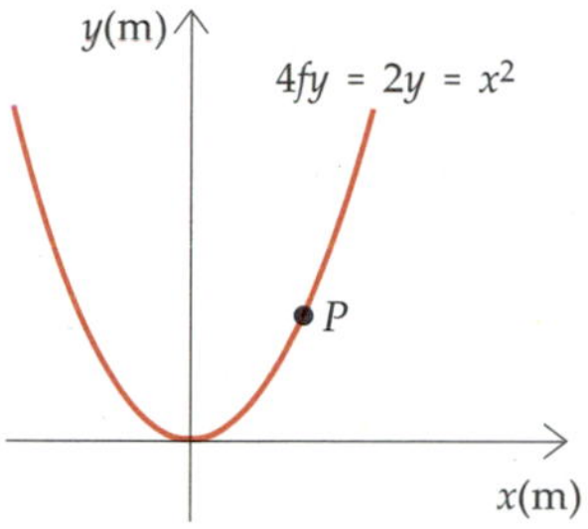

Figure E1.21a

$$y = \frac{x^2}{4f} = \frac{x^2}{2} \text{ for } f = \tfrac{1}{2} \qquad \text{(see Figure E1.21a)}$$

$$\tan\theta = \frac{dy}{dx} = x = 2 \qquad \text{(at the given point)}$$

$$\theta = \tan^{-1}(2) = 63.4°$$

$$\hat{\mathbf{e}}_t = \cos\theta\hat{\mathbf{i}} + \sin\theta\hat{\mathbf{j}} \qquad \text{(see Figure E1.21b)}$$

$$= 0.448\hat{\mathbf{i}} + 0.894\hat{\mathbf{j}}$$

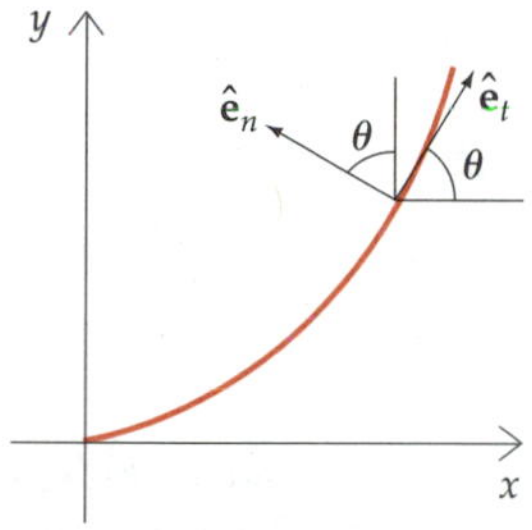

Figure E1.21b

The radius of curvature comes from calculus:

$$\frac{1}{\rho} = \left|\frac{y''}{(1 + y'^2)^{3/2}}\right| = \left|\frac{1}{(1 + x^2)^{3/2}}\right| = \frac{1}{5^{3/2}}$$

or

$$\rho = 11.2 \text{ m}$$

Thus with $\hat{\mathbf{e}}_n$ seen to be $-\sin\theta\hat{\mathbf{i}} + \cos\theta\hat{\mathbf{j}}$, we have

$$\mathbf{a}_P = \ddot{s}\hat{\mathbf{e}}_t + \frac{\dot{s}^2}{\rho}\hat{\mathbf{e}}_n$$

$$= 0 + \frac{\dot{s}^2}{\rho}\hat{\mathbf{e}}_n \qquad \text{(since } \dot{s} = \text{constant)}$$

$$= \frac{0.04}{11.2}(-0.894\hat{\mathbf{i}} + 0.448\hat{\mathbf{j}})$$

$$= -0.0032\hat{\mathbf{i}} + 0.0016\hat{\mathbf{j}} \text{ m/s}^2 \qquad \text{(as before)}$$

Example 1.22

In Example 1.11 find the following for point P at $t = 1$ sec: tangential and normal components of acceleration, radius of curvature, and the principal unit normal.

Solution

We obtained

$$\mathbf{r}_{OP} = 2t\hat{\mathbf{i}} + t^3\hat{\mathbf{j}} + 3t^2\hat{\mathbf{k}} \text{ ft}$$

$$\mathbf{v}_P = 2\hat{\mathbf{i}} + 3t^2\hat{\mathbf{j}} + 6t\hat{\mathbf{k}} \text{ ft/sec}$$

$$\mathbf{a}_P = 6t\hat{\mathbf{j}} + 6\hat{\mathbf{k}} \text{ ft/sec}^2$$

If we write the velocity $\mathbf{v}_P$ as a magnitude times a unit vector, we can determine $\dot{s}$ and $\hat{\mathbf{e}}_t$ for P:

$$\mathbf{v}_P = \sqrt{4 + 9t^4 + 36t^2}\left(\frac{2\hat{\mathbf{i}} + 3t^2\hat{\mathbf{j}} + 6t\hat{\mathbf{k}}}{\sqrt{4 + 9t^4 + 36t^2}}\right) \text{ ft/sec}$$

$$= \dot{s}\hat{\mathbf{e}}_t$$

where we note that $\dot{s} > 0$ since we are choosing the direction of increasing s to be that of the velocity.

Let us find the tangential and normal components of the acceleration of P at $t = 1$ sec:

$$\mathbf{v}_P|_{t=1} = 7\left(\frac{2\hat{\mathbf{i}} + 3\hat{\mathbf{j}} + 6\hat{\mathbf{k}}}{7}\right) = |\mathbf{v}_P|\hat{\mathbf{e}}_t \text{ ft/sec}$$

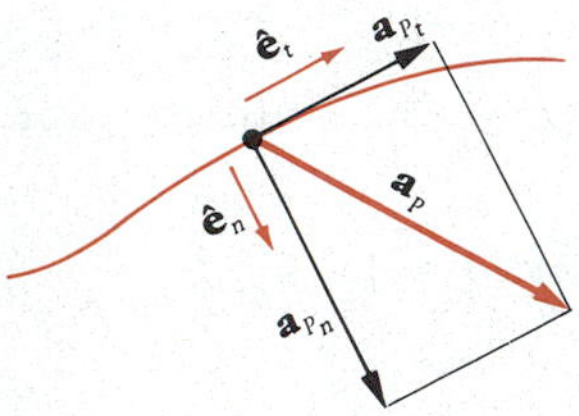

Figure E1.22

Now that we have $\hat{\mathbf{e}}_t$, we can use it to split the acceleration $\mathbf{a}_P = 6\hat{\mathbf{j}} + 6\hat{\mathbf{k}}$ ft/sec² (at $t = 1$) into its tangential and normal components (see Figure E1.22). The tangential component of $\mathbf{a}_P$ (that is, the component parallel to $\hat{\mathbf{e}}_t$) is seen from the figure to be the dot product of $\mathbf{a}_P$ with $\hat{\mathbf{e}}_t$:

$$a_{P_t}|_{t=1} = \mathbf{a}_P|_{t=1} \cdot \hat{\mathbf{e}}_t|_{t=1}$$

$$= 6\sqrt{2}\left(\frac{\hat{\mathbf{i}} + \hat{\mathbf{k}}}{\sqrt{2}}\right) \cdot \left(\frac{2\hat{\mathbf{i}} + 3\hat{\mathbf{j}} + 6\hat{\mathbf{k}}}{7}\right)$$

$$= \frac{6}{7}(3 + 6) = \frac{54}{7} \text{ ft/sec}^2$$

Next we obtain the normal acceleration component by vectorially subtracting the component $\mathbf{a}_{P_t}(=(54/7)\hat{\mathbf{e}}_t)$ from the total acceleration $\mathbf{a}_P$. That is, since

$$\mathbf{a}_P = \mathbf{a}_{P_t} + \mathbf{a}_{P_n}$$

we obtain

$$a_{P_n}|_{t=1} = |\mathbf{a}_{P_n}|\Big|_{t=1} = \left|\mathbf{a}_P|_{t=1} - \mathbf{a}_{P_t}|_{t=1}\right|$$

$$= \left|6\hat{\mathbf{j}} + 6\hat{\mathbf{k}} - \frac{54}{7}\left(\frac{2\hat{\mathbf{i}} + 3\hat{\mathbf{j}} + 6\hat{\mathbf{k}}}{7}\right)\right| \text{ ft/sec}^2$$

$$= \left|\frac{-108\hat{\mathbf{i}} + 132\hat{\mathbf{j}} - 30\hat{\mathbf{k}}}{49}\right| = 3.53 \text{ ft/sec}^2$$

And since $a_{P_n} = \dot{s}^2/\rho = |\mathbf{v}_P|^2/\rho$, we obtain the radius of curvature:

$$\rho = \frac{7^2}{3.53} = 13.9 \text{ ft}$$

The unit vector $\hat{\mathbf{e}}_n$ follows from

$$\hat{\mathbf{e}}_n = \frac{\mathbf{a}_{P_n}}{a_{P_n}} = \frac{\mathbf{a}_P - \mathbf{a}_t}{a_{P_n}} = \frac{-108\hat{\mathbf{i}} + 132\hat{\mathbf{j}} - 30\hat{\mathbf{k}}}{49(3.53)}$$

$$= -0.624\hat{\mathbf{i}} + 0.763\hat{\mathbf{j}} - 0.173\hat{\mathbf{k}}$$

It is instructive to make a direct calculation of $d|\mathbf{v}_P|/dt$ since we here know $|\mathbf{v}_P|$ as a function of time:

$$|\mathbf{v}_P| = \sqrt{4 + 9t^4 + 36t^2} \text{ ft/sec}$$

$$\frac{d|\mathbf{v}_P|}{dt} = \frac{\frac{1}{2}(36t^3 + 72t)}{\sqrt{4 + 9t^4 + 36t^2}} \text{ ft/sec}^2$$

Thus

$$\left.\frac{d|\mathbf{v}_P|}{dt}\right|_{t=1} = \frac{\frac{1}{2}(108)}{\sqrt{49}} = \frac{54}{7} \text{ ft/sec}^2$$

which is, of course, the result we have already obtained by investigating the components of the acceleration vector.

Question 1.6 How would you find the position vector from the origin O to the center of curvature at $t = 1$ sec?

In closing this section, we remark that tangential and normal components of velocity and acceleration will be very useful to us later when we happen to know the path of a point (the center C of a wheel rolling on a curved track, for instance). We can then use Equations (1.41) and (1.47) to express $\mathbf{v}_P$ and $\mathbf{a}_P$.

Answer 1.6 If we call the center of curvature C, then $\mathbf{r}_{OC} = \mathbf{r}_{OP} + \mathbf{r}_{PC} = \mathbf{r}_{OP} + \rho\hat{\mathbf{e}}_n$, with everything evaluated at the time of interest (in this case $t = 1$ sec).

PROBLEMS ▶ Section 1.7

1.128 Particle P moves on a circle (Figure P1.128) with an arclength given as a function of time as shown. Find the time(s) and the angle(s) θ when the tangential and normal acceleration components are equal.

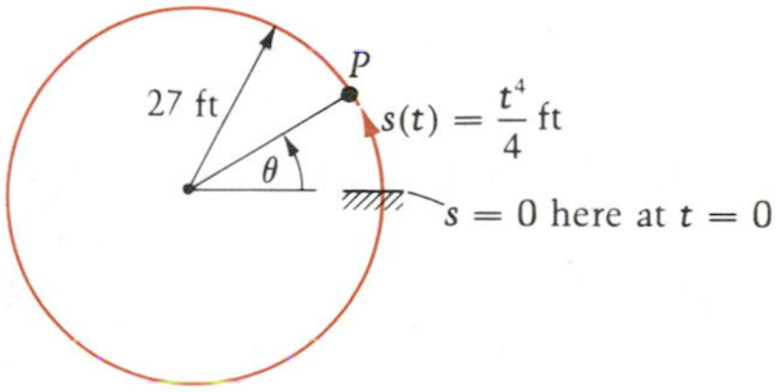

Figure P1.128

1.129 In Problem 1.78 find the arclength s as a function of time.

1.130 In Problem 1.67 determine the expression for $\dot{s}(t)$. Integrate, for a motion beginning at $t = 0$ at $(x, y) = (20, 0)$ m, and obtain $s(t)$. Evaluate the arclength at $t = 2$ s and show that the result, as it should be, is the circumference of the circle on which P travels.

1.131 A point P moves on a path with $s = ct^3$ where $c = \text{constant} = 1$ ft/sec^3. At $t = 2$ sec, the magnitude of the acceleration is 15 ft/sec^2. At that time, find the radius of curvature of the path of P.

1.132 A point D moves along a curve in space with a speed given by $\dot{s} = 6t$ m/s, where t is measured from zero when D is at the arclength origin $s = 0$. If at a certain time t' the acceleration magnitude of D is 12 m/s^2 and the radius of curvature is 3 m, determine t'.

1.133 At a certain instant the velocity and acceleration of a point are as shown in Figure P1.133. At this instant find

a. $\dfrac{d|\mathbf{v}|}{dt}$

b. the radius of curvature of the path

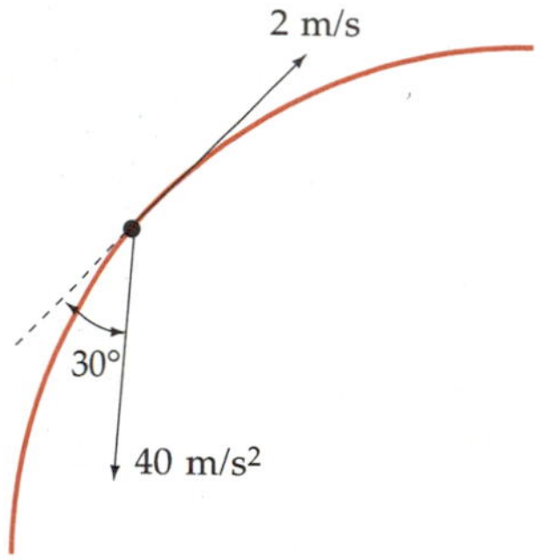

Figure P1.133

1.134 At a certain instant, the velocity and acceleration of a point are

$$\mathbf{v} = 3\hat{\mathbf{i}} + 4\hat{\mathbf{j}} \text{ m/s}$$
$$\mathbf{a} = -10\hat{\mathbf{k}} \text{ m/s}^2$$

At this instant find (a) $\dfrac{d|\mathbf{v}|}{dt}$, (b) the radius of curvature of the path, (c) the principal unit normal.

1.135 At an instant the velocity and acceleration of a point are

$$\mathbf{v} = -2\hat{\mathbf{i}} \text{ m/s}$$
$$\mathbf{a} = -4\hat{\mathbf{i}} + 3\hat{\mathbf{j}} - 2\hat{\mathbf{k}} \text{ m/s}^2$$

At this instant find:

a. $\dfrac{d}{dt}|\mathbf{v}|$

b. the radius of curvature of the path.

1.136 At a certain instant, the velocity and acceleration of a point are

$$\mathbf{v} = 4\hat{\mathbf{i}} - 3\hat{\mathbf{j}} \text{ m/s}$$
$$\mathbf{a} = -10\hat{\mathbf{i}} + 20\hat{\mathbf{j}} + 12\hat{\mathbf{k}} \text{ m/s}^2$$

Find:

a. $\dfrac{d}{dt}|\mathbf{v}|$

b. $\hat{\mathbf{e}}_n$

1.137 In Problem 1.103, find the radius of curvature of the path of P at the instant given. Note that $\hat{\mathbf{e}}_t = \hat{\mathbf{j}} = \hat{\mathbf{e}}_\theta$ and $\hat{\mathbf{e}}_n = -\hat{\mathbf{i}} = -\hat{\mathbf{e}}_r$.

1.138 Find the radius of curvature of the "Witch of Agnesi" curve at $x = 0$. (See Figure P1.138.)

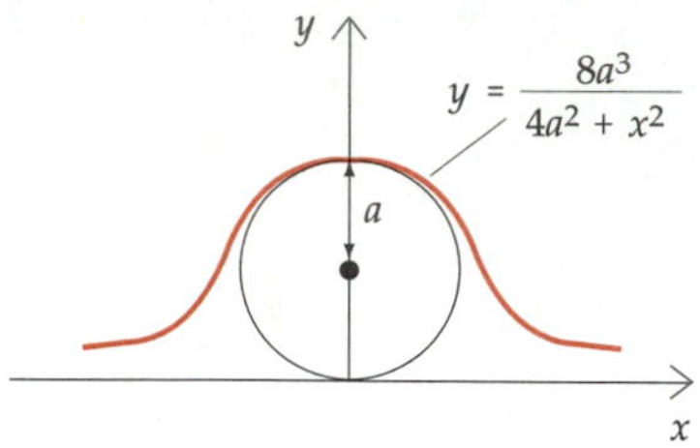

Figure P1.138

1.139 A point P moves from left to right along the curve defined in the preceding problem with a constant x component ($\dot{x}_0$) of velocity. Find the acceleration of P when it reaches the point $(x, y) = (0, 2a)$.

1.140 In Problem 1.105, at the same four points express $\mathbf{v}_P$ in terms of tangential and normal components.

1.141 In Problem 1.105, for the position $\theta = \pi/2$, express $\mathbf{a}_P$ in terms of tangential and normal components, and find the radius of curvature of the path of P at that point.

1.142 A point P starts at the origin and moves along the parabola shown in Figure P1.142 with a constant x-component of velocity, $\dot{x} = 3$ ft/sec. Find the tangential and normal components of the velocity and acceleration of P at the point $(x, y) = (1, 1)$.

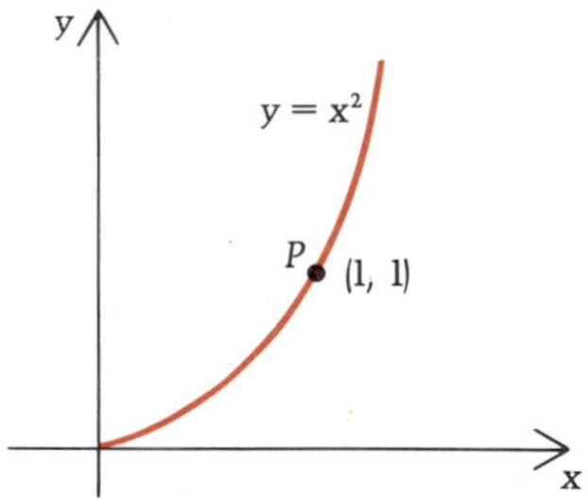

Figure P1.142

* **1.143** In Problem 1.104, find the center of curvature of the path of P when $\theta = 270°$.

1.144 At a particular instant a point has a velocity $3\hat{\mathbf{i}} + 4\hat{\mathbf{j}}$ in./sec and an acceleration $\hat{\mathbf{i}} + \hat{\mathbf{j}} - \hat{\mathbf{k}}$ in./sec². At this instant find: (a) the principal unit normal, (b) the curvature of the path, and (c) the time rate of change of the point's speed.

1.145 A point P has position vector $\mathbf{r}_{OP} = t^2\hat{\mathbf{i}} + t^3\hat{\mathbf{j}} - t^6\hat{\mathbf{k}}$ meters. Find the vector from the origin to the center of curvature of the path of P at $t = 1$ s. Find $d|\mathbf{v}_P|/dt$ at the same instant.

1.146 The position vector of a point is given as a function of time by $\mathbf{r}(t) = t^2\hat{\mathbf{i}} - t\hat{\mathbf{j}} + t^3\hat{\mathbf{k}}$ ft. Find the tangential and normal components of acceleration at $t = 1$ sec and determine the radius of curvature at that time.

1.147 A particle P has the x, y, and z coordinates $(3t, 0, 4 \ln t)$ meters as functions of time. What is the vector from the origin to the center of curvature of the path of P at $t = 1$ s?

1.148 Show by expressing the velocity and acceleration in tangential and normal components that

$$|\mathbf{v} \times \mathbf{a}| = va_n = \frac{|\dot{s}^3|}{\rho}$$

so that

$$\frac{1}{\rho} = \frac{|\mathbf{v} \times \mathbf{a}|}{|\mathbf{v}|^3} \qquad \text{or} \qquad \frac{|\mathbf{v} \times \mathbf{a}|}{v^3}$$

1.149 There is another formula for the radius of curvature ρ from the calculus; this one is in terms of a parameter such as time t, and for a plane curve:

$$\frac{1}{\rho} = \left|\frac{\dot{x}\ddot{y} - \dot{y}\ddot{x}}{(\dot{x}^2 + \dot{y}^2)^{3/2}}\right|$$

Derive this from the result of the preceding problem and use it to find the radius of curvature at 8 sec if

$$x = 6 \sinh 0.02t \text{ ft} \qquad y = t^3 - 10t^2 + 7.2 \text{ ft}$$

1.150 Find the difference between the velocities (and also the accelerations) of cars A and B in Figure P1.150 if, at the instant shown,

$$\dot{s}_A = 30 \text{ mph} \qquad \ddot{s}_A = 500 \text{ mph}^2$$
$$\dot{s}_B = 50 \text{ mph} \qquad \ddot{s}_B = -1000 \text{ mph}^2$$

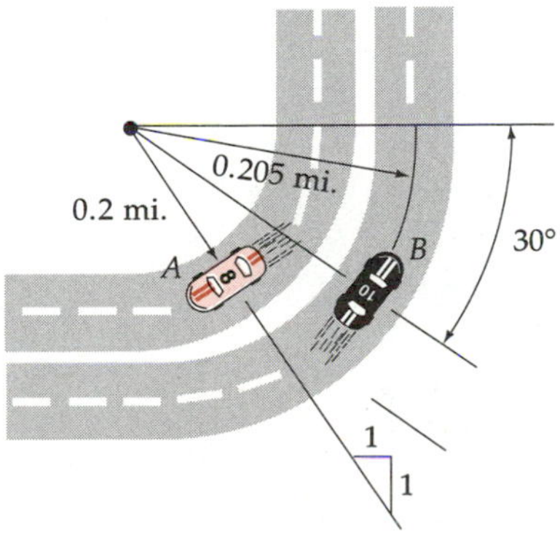

Figure P1.150

1.151 A particle moves on the curve $(x - a)^2 + y^2 = a^2$, where a is a constant distance in meters. The first and second time derivatives of the arclength s are related by

$$\dot{s} = Ka\ddot{s}$$

in which the constant K has the value 1 second/meter. The distance s is measured counterclockwise on the curve from the point $(2a, 0)$ meters. When $t = 0$, the speed of the particle is $\dot{s} = 1$ meter/second and $s = a$. Find the normal and tangential components of the acceleration at time $t = 0$. Show these components on a diagram.

1.152 Use Example 1.21 to show that the center of curvature does not have to be on the y axis for a curve symmetric about y. *Hint*: Use $f = 1$ and $x = 2$, find ρ, and compare with the distance from (2, 1) to the y axis along the normal to the tangent at this point.

• **1.153** In Problem 1.96 find the tangential and normal components of the car's acceleration when $x = 2500$ ft. Check your result by also computing $\ddot{x}$ and $\ddot{y}$ there and showing that $\sqrt{\ddot{x}^2 + \ddot{y}^2} = \sqrt{a_t^2 + a_n^2}$.

• **1.154** A particle P starts from rest at the origin and moves along the parabola shown in Figure P1.154. Its speed is given by $\dot{s} = 3s + 2$, where $\dot{s}$ is in meters per second when s is in meters. Determine the velocity of P when its x coordinate is 5 m. Also determine the elapsed time. *Hint*: $ds = \sqrt{dx^2 + dy^2} = \sqrt{1 + y'^2}\, dx$, so that $s = \int \sqrt{1 + y'^2}\, dx$. Substitute y' and use a table of integrals to get $s(x)$.

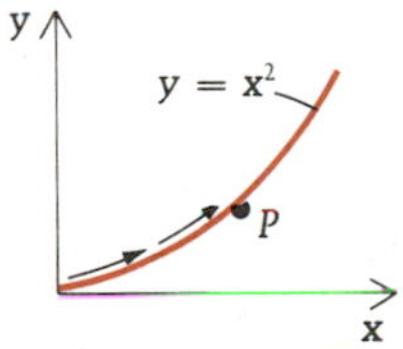

Figure P1.154

1.155 There is a third unit vector associated with the motion of a point on its path. It is called the binormal $\hat{\mathbf{e}}_b$ and forms an orthogonal moving trihedral with $\hat{\mathbf{e}}_t$ and $\hat{\mathbf{e}}_n$, defined by $\hat{\mathbf{e}}_b = \hat{\mathbf{e}}_t \times \hat{\mathbf{e}}_n$.

a. Differentiate $\hat{\mathbf{e}}_b \cdot \hat{\mathbf{e}}_t = 0$ with respect to s. Then, using $d\hat{\mathbf{e}}_t/ds = \hat{\mathbf{e}}_n/\rho$,† prove that $d\hat{\mathbf{e}}_b/ds \cdot \hat{\mathbf{e}}_t = 0$ and therefore that $d\hat{\mathbf{e}}_b/ds$ is parallel to $\hat{\mathbf{e}}_n$.

b. Using part (a), let $d\hat{\mathbf{e}}_b/ds = \tau\hat{\mathbf{e}}_n$.† ($\tau$ is called the torsion of the path or curve.) Then differentiate $\hat{\mathbf{e}}_n = \hat{\mathbf{e}}_b \times \hat{\mathbf{e}}_t$ with respect to s and prove that

$$\frac{d\hat{\mathbf{e}}_n}{ds} = -\left(\frac{1}{\rho}\hat{\mathbf{e}}_t + \tau\hat{\mathbf{e}}_b\right)^\dagger$$

The three equations marked with daggers give the derivatives of the three unit vectors associated with a space curve and are called the Serret-Frenet formulas.

1.156 The derivative of acceleration is called the *jerk* and is studied in the dynamics of vehicle impact and in the kinematics of mechanisms involving cams and followers. Show that the jerk of a point has the following form in terms of its intrinsic components:

$$\mathbf{J}_P = \dot{\mathbf{a}}_P = \left(\dddot{s} - \frac{\dot{s}^3}{\rho^2}\right)\hat{\mathbf{e}}_t + \left(\frac{3\dot{s}\ddot{s}}{\rho} - \frac{\dot{\rho}\dot{s}^2}{\rho^2}\right)\hat{\mathbf{e}}_n - \frac{\dot{s}^3}{\rho}\tau\hat{\mathbf{e}}_b$$

* **1.157** The following "pursuit" problem is very difficult, yet it illustrates exceptionally well the idea that the velocity vector is tangent to the path. Thus we include it along with a set of steps for the courageous student who wishes to "pursue" it. A dog begins at the point $(x, y) = (D, O)$ and runs toward his master at constant speed $2V_0$. (See Figure P1.157.) The dog's velocity direction is always toward his master, who starts at the same time at the origin and moves along the positive y direction at speed V_0. Find the man's position when his dog overtakes him, and determine how much time has elapsed. *Hints*: The man's y coordinate is y_M (which of course is $V_0 t$). Show that:

1. $\dfrac{-dy_D}{dx} = \dfrac{y_M - y_D}{x}$, where (x, y_D) represent the dog's coordinates at any time.

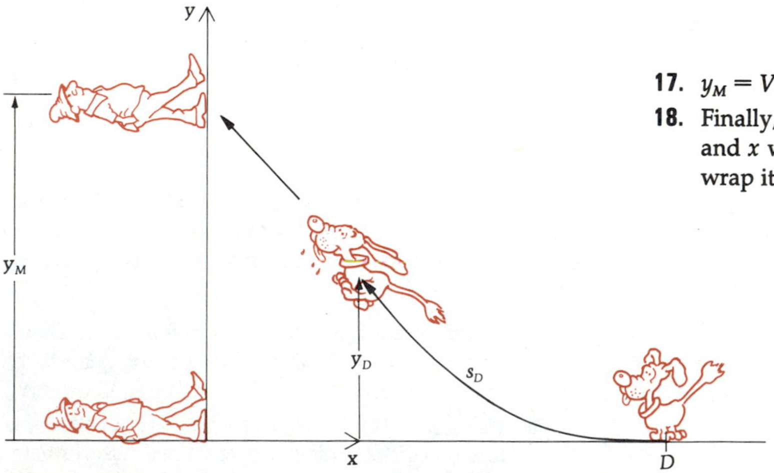

Figure P1.157

2. $2V_0 = \dfrac{ds_D}{dt} = \dfrac{\sqrt{dx^2 + dy_D^2}}{dt} = \dfrac{-dx\sqrt{1 + y'^2_D}}{dt}$.
3. $V_0 = dy_M/dt$.
4. From dividing and rearranging steps 2 and 3, we get $2y'_M = -\sqrt{1 + y'^2_D}$ $\quad (y'_M = dy_M/dx)$.
5. From step 1, we have $y'_M = -xy''_D$.
6. From steps 4 and 5, we have $2xy''_D = \sqrt{1 + y'^2_D}$.
7. Letting $y'_D = p$, from step 6 we get $dp/\sqrt{1 + p^2} = dx/2x$.
8. By integrating step 7 with a table of integrals, we get
$$\ln(p + \sqrt{1 + p^2}) = \tfrac{1}{2}\ln(x) - \tfrac{1}{2}\ln(C_1) = \ln\left(\frac{x}{C_1}\right)^{1/2}.$$
9. From step 8, we get $p + \sqrt{1 + p^2} = (x/C_1)^{1/2}$.
10. From step 9, we get $p - \sqrt{x/C_1} = -\sqrt{1 + p^2}$.
11. From step 10, squaring both sides and solving for p,
$$2p = \sqrt{\frac{x}{C_1}} - \sqrt{\frac{C_1}{x}} = 2y'_D$$
12. $y'_D = 0$ when $x = D$ (initial condition).
13. From steps 11 and 12 we have $C_1 = D$, so that $2y'_D = \sqrt{x/D} - \sqrt{D/x}$.
14. Integrating step 13, we get
$$y_D = x^{3/2}/(3\sqrt{D}) - \sqrt{xD} + C_2.$$
15. $y_D = 0$ when $x = D$ (initial condition).
16. From steps 14 and 15 we have $C_2 = 2D/3$, so that
$$y_D = \frac{x^{3/2}}{3\sqrt{D}} - \sqrt{xD} + \tfrac{2}{3}D$$
17. $y_M = V_0 t$.
18. Finally, write the conditions relating to y_M, y_D, and x when the dog overtakes his master, and wrap it up!

Computer Problem ▶ Chapter 1

• **1.158** A particle moves in the xy plane according to the equation $r = k\theta$, where k is a constant, and has the constant speed v_0. The particle passes through the origin with $r = \theta = 0$ at $t = 0$. (See Figure P1.158.)

a. Show that $k\dot{\theta}\sqrt{1 + \theta^2} = v_0$.

b. With the trigonometric substitution $\theta = \tan\phi$, and then consulting integral tables, integrate the equation and obtain:

$$\frac{\theta\sqrt{1+\theta^2}}{2} + \frac{1}{2}\ln\left[\frac{\sqrt{1+\theta^2}+1+\theta}{\sqrt{1+\theta^2}+1-\theta}\right] = \frac{v_0 t}{k}$$

c. For the case $v_0/k = 1$, use the computer to plot θ versus time until θ has increased from 0 to 2π radians.

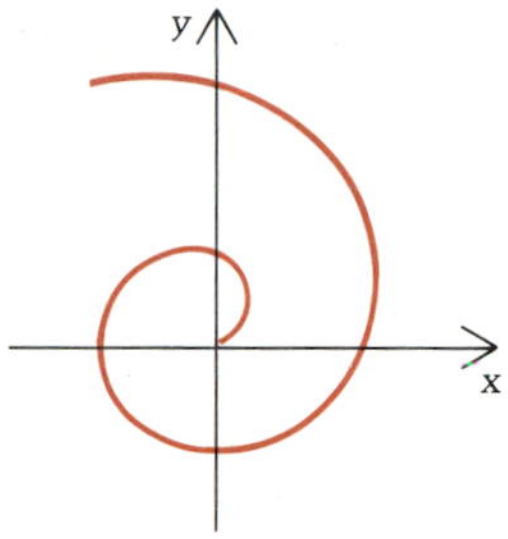

Figure P1.158

Summary ▶ Chapter 1

In this chapter we have studied the position, velocity and acceleration of a point (or particle). With O being a point fixed in the frame of reference and P denoting the moving point, then $\mathbf{r}_{OP}$ is a position vector for P and we defined

$$\text{Velocity:} \qquad \mathbf{v}_P = \frac{d\mathbf{r}_{OP}}{dt}$$

$$\text{Acceleration:} \qquad \mathbf{a}_P = \frac{d\mathbf{v}_P}{dt} = \frac{d^2\mathbf{r}_{OP}}{dt^2}$$

With rectangular coordinates and associated unit vectors, and with O chosen as the origin of the coordinate system,

$$\mathbf{r}_{OP} = x\hat{\mathbf{i}} + y\hat{\mathbf{j}} + z\hat{\mathbf{k}}$$
$$\mathbf{v}_P = \dot{x}\hat{\mathbf{i}} + \dot{y}\hat{\mathbf{j}} + \dot{z}\hat{\mathbf{k}}$$
$$\mathbf{a}_P = \ddot{x}\hat{\mathbf{i}} + \ddot{y}\hat{\mathbf{j}} + \ddot{z}\hat{\mathbf{k}}$$

In similar fashion for cylindrical coordinates,

$$\mathbf{r}_{OP} = r\hat{\mathbf{e}}_r + z\hat{\mathbf{k}}$$
$$\mathbf{v}_P = \dot{r}\hat{\mathbf{e}}_r + r\dot{\theta}\,\hat{\mathbf{e}}_\theta + \dot{z}\hat{\mathbf{k}}$$
$$\mathbf{a}_P = (\ddot{r} - r\dot{\theta}^2)\hat{\mathbf{e}}_r + (r\ddot{\theta} + 2\dot{r}\dot{\theta})\hat{\mathbf{e}}_\theta + \ddot{z}\hat{\mathbf{k}}$$

With a path-length parameter $s(t)$, describing the motion of P on a given curve (path) and with $\hat{\mathbf{e}}_t$ and $\hat{\mathbf{e}}_n$ being unit tangent and principal unit normal, respectively,

$$\mathbf{v}_P = \dot{s}\hat{\mathbf{e}}_t$$

$$\mathbf{a}_P = \ddot{s}\hat{\mathbf{e}}_t + \frac{(\dot{s})^2\hat{\mathbf{e}}_n}{\rho},$$

where ρ is the radius of curvature of the path at the point occupied by P at time t. Sometimes it is convenient to choose to measure $s(t)$ so that $\dot{s} > 0$ in an interval of time of interest. In this case, we have expressions that don't involve s explicitly:

$$\mathbf{v}_P = |\mathbf{v}_P|\,\hat{\mathbf{e}}_t$$

$$\mathbf{a}_P = \left(\frac{d}{dt}|\mathbf{v}_P|\right)\hat{\mathbf{e}}_t + \frac{|\mathbf{v}_P|^2}{\rho}\,\hat{\mathbf{e}}_n$$

REVIEW QUESTIONS ▶ Chapter 1

True or False?

1. The velocity $\mathbf{v}_P$ of a point P is always tangent to its path.
2. $\mathbf{v}_P$ depends on the reference frame chosen to express the position of P.
3. $\mathbf{v}_P$ depends on the origin chosen in the reference frame.
4. The magnitude and direction in space of $\mathbf{v}_P$ depend on the choice of coordinates used to locate the point relative to the reference frame.
5. $\mathbf{a}_P$ always has a nonvanishing component normal to the path.
6. For any point P, $|\mathbf{a}_P| = \sqrt{\ddot{s}^2 + \dot{s}^4/\rho^2}$.
7. A point can have $\ddot{r} = 0$ but still have a nonvanishing radial component of acceleration.
8. If a ball on a string is being whirled around in a horizontal circle at constant speed, the center of the ball has zero acceleration.
9. Studying the kinematics of a particle results in the same equations as studying the kinematics of a point.
10. In our study of the kinematics of a point, the following terms have not appeared in any of the equations: mass, force, moments, gravity, momentum, moment of momentum, inertia, or Newtonian (inertial) frames.
11. The acceleration vector of P, at the indicated point on the path shown in the figure, can lie in any of the four quadrants.

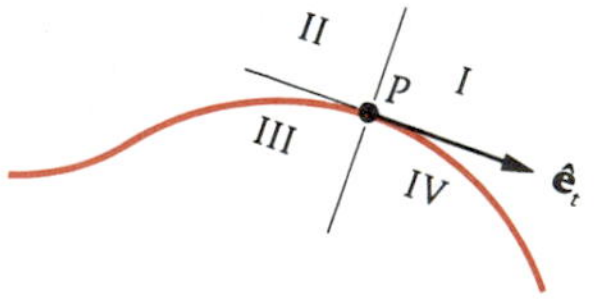

12. A particle moving in a plane, with constant values of $\dot{r}$ and $\dot{\theta}$, will at all times have zero acceleration.

Answers: 1. T 2. T 3. F 4. F 5. F 6. T 7. T 8. F 9. T 10. T 11. F 12. F

2 Kinetics of Particles and of Mass Centers of Bodies

2.1 Introduction

2.2 Newton's Laws and Euler's First Law

Motion of the Mass Center

2.3 Motions of Particles and of Mass Centers of Bodies

The Free-Body Diagram

2.4 Work and Kinetic Energy for Particles

Work and Kinetic Energy for a Particle
Work Done by a Constant Force
Work Done by a Central Force
Work Done by a Linear Spring
Work Done by Gravity
Conservative Forces
Conservation of Energy
Work and Kinetic Energy for a System of Particles

2.5 Momentum Form of Euler's First Law

Impulse and Momentum; Conservation of Momentum
Impact
Coefficient of Restitution

2.6 Euler's Second Law (The Moment Equation)

Moment of Momentum
Momentum Forms of Euler's Second Law
Conservation of Moment of Momentum

Summary
Review Questions

2.1 Introduction

In this chapter we begin to consider the manner in which the motion of a body is related to external mechanical actions (forces and couples). Our kinematics notions of space and time must now be augmented by those of mass and force, which, like space and time, are **primitives** of the subject of mechanics. We simply have to agree in advance that some measures of **quantity of matter (mass)** and **mechanical action (force)** are basic ingredients in any attempt to analyze the motion of a body. We assume that the reader has a working knowledge, probably from a study of statics, of the characteristics of forces and moments and their vector descriptions. We use the term **body** to denote some material of fixed identity; we could think of a specific set of atoms, although the model we shall employ is based upon viewing material on a spatial scale such that mass is perceived to be distributed continuously. A body need not be rigid or even a solid, but, since our subject is classical dynamics (no relativistic effects), a body necessarily has constant mass.

In Section 2.2 we use Newton's laws for a particle and for interacting particles to deduce that the sum of the external forces on a body of any size is equal to the sum of the $m\mathbf{a}$'s of the body, alternatively expressed as the total mass multiplied by the acceleration of the mass center. This result is usually called Euler's first law. Applications of this are developed in Section 2.3 along with a review of the critically important concept of the free-body diagram.

The Principle of Work and Kinetic Energy for a particle is developed in Section 2.4, wherein are found expressions for the work done by several special types of forces. The concept of a conservative force is introduced, and the condition for which Work and Kinetic Energy becomes Conservation of Mechanical Energy is established. Finally, the implications of Work and Kinetic Energy for a system of particles are explored.

In Section 2.5 the impulse-momentum form of Euler's first law is developed, and conditions for conservation of momentum are demonstrated. Applications are made to problems of impact.

Euler's second law is the subject of Section 2.6. We return to Newton's laws so as to derive this important result which states that the sum of the moments of the external forces on a body (or system of particles) equals the sum of the moments of the body's $m\mathbf{a}$'s. Momentum forms of this are developed, primarily for later applications in Chapters 4, 5, and 7.

2.2 Newton's Laws and Euler's First Law

The usual starting point for relating the external forces on a body to its motion is Newton's laws. These were proposed in the Englishman Isaac Newton's famous work the *Principia,* published in 1687, and are commonly expressed today as:

1. If the resultant force $\mathbf{F}$ on a particle is zero, then the particle has constant velocity.
2. If $\mathbf{F} \neq \mathbf{0}$, then $\mathbf{F}$ is proportional to the time derivative of the particle's momentum $m\mathbf{v}$ (product of mass and velocity).
3. The interaction of two particles is through a pair of self-equilibrating forces. That is, they have the same magnitude, opposite directions, and a common line of action.

Clearly the first law may be regarded a special case of the second, and one must add an assumption about the frame of reference, since a point may have its velocity constant in one frame of reference and varying in another. Frames of reference in which these laws are valid are variously called Newtonian, Galilean, or **inertial.** Furthermore, the constant of proportionality in the second law can be made unity by appropriate choices of units so that the law becomes

$$\mathbf{F} = \frac{d}{dt}(m\mathbf{v}) = m\mathbf{a}$$

where $\mathbf{a}$ is the acceleration of the particle.

As we mentioned briefly in Chapter 1, a particle is a piece of material sufficiently small that we need not make distinctions among its material points with respect to locations (or to velocities or accelerations). We also noted that this definition allows, for *some* purposes, a truck or a space vehicle or even a planet to be adequately modeled as a particle. In the *Principia,* Newton used heavenly bodies as the particles in his examples and treated them as moving points subject only to universal gravitation and their own inertia.* Newton did not extend his work to problems for which it is necessary to account for the actual sizes of the bodies and how their masses are distributed. It was to be over 50 years before the Swiss mathematician Leonhard Euler presented the first of the two principles that have come to be called **Euler's laws.**

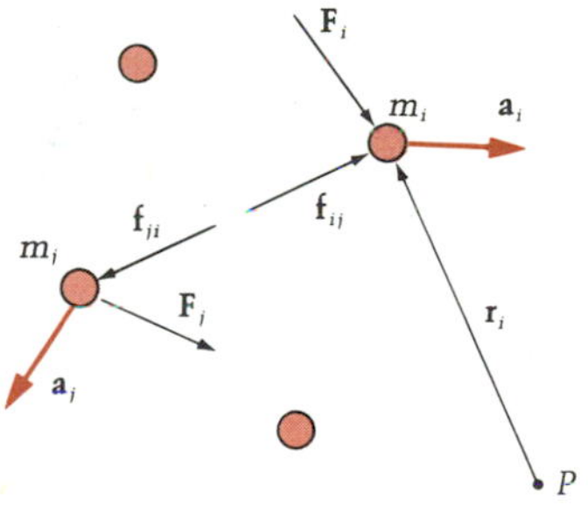

Figure 2.1

For a body composed of a set of N particles, we may deduce Euler's laws from Newton's laws. As suggested by Figure 2.1, we separate the forces acting on the ith particle into two groups: there are the N-1 forces exerted by other particles of the system, $\mathbf{f}_{ij}$ being that exerted by the jth particle; then there is $\mathbf{F}_i$, the net force exerted on the ith particle by things *external* to the system. Applying Newton's second law to the ith particle whose mass is m_i and whose acceleration is $\mathbf{a}_i$,

$$\mathbf{F}_i + \sum_{j=1}^{N} \mathbf{f}_{ij} = m_i\mathbf{a}_i \tag{2.1}$$

in which we understand

$$\mathbf{f}_{ii} = \mathbf{0}$$

* See C. Truesdell, *Essays in the History of Mechanics* (Berlin: Springer-Verlag, 1968).

Now we sum the N such equations to obtain

$$\sum_{i=1}^{N} \mathbf{F}_i + \sum_{i=1}^{N}\sum_{j=1}^{N} \mathbf{f}_{ij} = \sum_{i=1}^{N} m_i\mathbf{a}_i \tag{2.2}$$

But Newton's third law tells us that

$$\mathbf{f}_{ij} = -\mathbf{f}_{ji}$$

so that

$$\Sigma\Sigma\mathbf{f}_{ij} = 0$$

Thus we conclude that

$$\Sigma\mathbf{F}_i = \Sigma m_i\mathbf{a}_i \tag{2.3}$$

which is the particle-system form of Euler's first law and states that the sum of the *external forces* on the system equals the sum of the $m\mathbf{a}$'s of the particles making up the system.

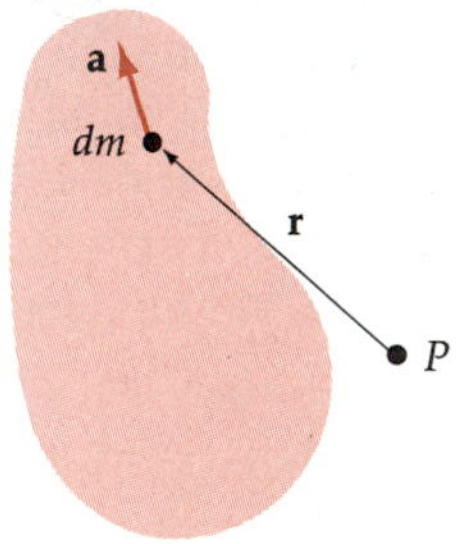

Figure 2.2

For a body whose mass is continuously distributed, as depicted in Figure 2.2, the counterpart to equation (2.3) is

$$\Sigma\mathbf{F} = \int \mathbf{a}\, dm \tag{2.4}$$

where dm* is a differential element of mass, $\mathbf{a}$ is its acceleration, and $\Sigma\mathbf{F}$ is the sum of the external forces acting on the body.

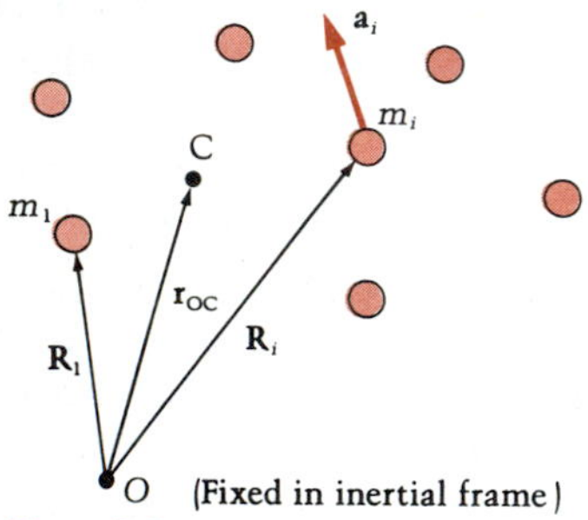

Figure 2.3a

Motion of the Mass Center

We close this section by developing the relationship between the external forces acting on a body and the motion of its mass center. To do this we first construct position vectors for the particles of a system as shown in Figure 2.3a. Thus the acceleration of the ith particle may be written

$$\mathbf{a}_i = \frac{d^2\mathbf{R}_i}{dt^2} \tag{2.5}$$

Applying this to Equation (2.3),

$$\begin{aligned}\Sigma\mathbf{F}_i &= \Sigma m_i \frac{d^2\mathbf{R}_i}{dt^2}\\ &= \frac{d^2}{dt^2}(\Sigma m_i\mathbf{R}_i)\end{aligned} \tag{2.6}$$

The location of the mass center, C, of a system of particles is defined by

$$\mathbf{r}_{OC} = \frac{\Sigma m_i\mathbf{R}_i}{m} \tag{2.7}$$

* $dm = \rho\, dV$ where ρ is mass density and dV is an infinitesimal element of volume. When using rectangular coordinates x, y, z, then $dV = dx\, dy\, dz$.

where m is the mass (Σm_i) of the system. Thus Euler's first law becomes

$$\Sigma \mathbf{F} = \frac{d^2}{dt^2}(m\mathbf{r}_{OC}) = m\frac{d^2\mathbf{r}_{OC}}{dt^2}$$

or

$$\Sigma \mathbf{F} = m\mathbf{a}_C \tag{2.8}$$

Question 2.1 What happened to the $\dot{m}$ and $\ddot{m}$ terms in going from $\frac{d^2}{dt^2}(m\mathbf{r}_{OC})$ to $m\frac{d^2\mathbf{r}_{OC}}{dt^2}$?

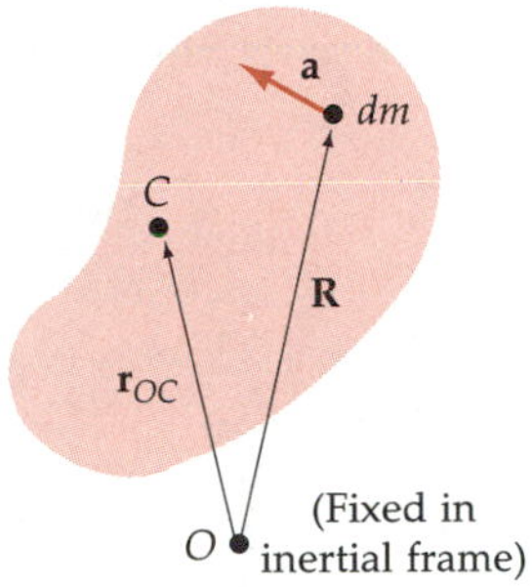

Figure 2.3b

For a continuous body (Figure 2.3b), the counterparts to Equations (2.5–8) are

$$\mathbf{a} = \frac{d^2\mathbf{R}}{dt^2}$$

$$\Sigma \mathbf{F} = \int \frac{d^2\mathbf{R}}{dt^2}\,dm = \frac{d^2}{dt^2}\int \mathbf{R}\,dm$$

and

$$m\mathbf{r}_{OC} = \int \mathbf{R}\,dm$$

resulting again in Equation (2.8). Thus we see that the resultant external force on the body is the product of the constant mass m of the body and the acceleration $\mathbf{a}_C$ of its mass center. Hence the motion of the mass center of a body is governed by an equation identical in form to Newton's second law for a particle. It is very important to realize that for a rigid body the mass center C coincides at every instant with a specific material point of the body or of its rigid extension (for example, the center of a hollow sphere). This is not the case for a deformable body.

Sometimes it is useful to subdivide a body into two parts, say of masses m_1 and m_2 with mass-center locations C_1 and C_2. Recalling a property of mass centers,

$$m\mathbf{r}_{OC} = m_1\mathbf{r}_{OC_1} + m_2\mathbf{r}_{OC_2}$$

so that after differentiating twice with respect to time,

$$m\mathbf{a}_C = m_1\mathbf{a}_{C_1} + m_2\mathbf{a}_{C_2}$$

They are zero since our definition of a body requires that its mass be constant.

We see that Equation (2.8) can also be written

$$\Sigma \mathbf{F} = m_1 \mathbf{a}_{C_1} + m_2 \mathbf{a}_{C_2} \tag{2.9}$$

The principal purpose of this section has been the derivation of Equation (2.8), and the next section is devoted wholly to applications of it. But the natural occurrence of the mass center between Equations (2.6) and (2.8) motivates a brief review of the calculation of its location. The example and problems that follow are designed to provide that review in instances for which the body comprises several parts, the mass centers of which are known.

EXAMPLE 2.1

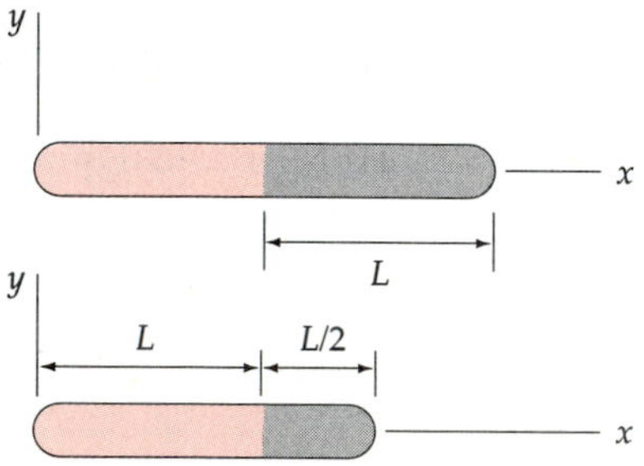

Figure E2.1

A uniform prismatic rod of density ρ and length $2L$ is deformed in such a way that the right half is uniformly compressed to length $L/2$ with no change in cross-sectional area A. (See Figure E2.1.) The left half of the rod is not altered. Letting the x axis be the locus of cross-sectional centroids, find the coordinates of the mass center in the deformed configuration.

Solution

In the first configuration the center-of-mass coordinates are $(L, 0, 0)$; that is, the center of mass is at the interface of the two segments. In the second configuration, however,

$$2\rho AL\bar{x} = \rho AL\left(\frac{L}{2}\right) + \rho AL\left(\frac{5}{4}L\right)$$

$$\bar{x} = \frac{L}{4} + \frac{5}{8}L$$

$$= \frac{7}{8}L$$

Thus the mass center no longer lies in the interface. This example illustrates that the mass center of a deformable body does not in general coincide with the same material point in the body at different times.

PROBLEMS ▶ Section 2.2

2.1 Show that the mass center C of a body $\mathcal{B}$ is unique. *Hint*: Consider the two mass centers C_1 and C_2, respectively:

$$\mathbf{r}_{O_1C_1} = \frac{1}{m}\int \mathbf{R}_1 \, dm$$

$$\mathbf{r}_{O_2C_2} = \frac{1}{m}\int \mathbf{R}_2 \, dm$$

and relate $\mathbf{R}_1$ to $\mathbf{R}_2$. (See Figure P2.1.) Using this relation,

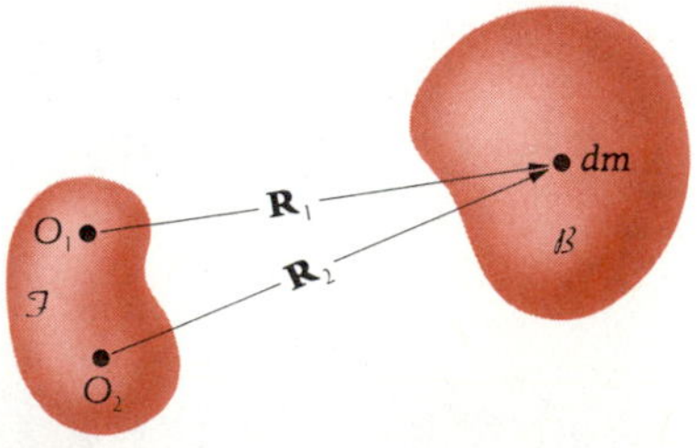

Figure P2.1

show that $\mathbf{r}_{O_1C_1} = \mathbf{r}_{O_1C_2}$, which means that C_1 and C_2 are the same point!

2.2 Find the mass center of the composite body shown in Figure P2.2. Note that the three parts are composed of different materials.

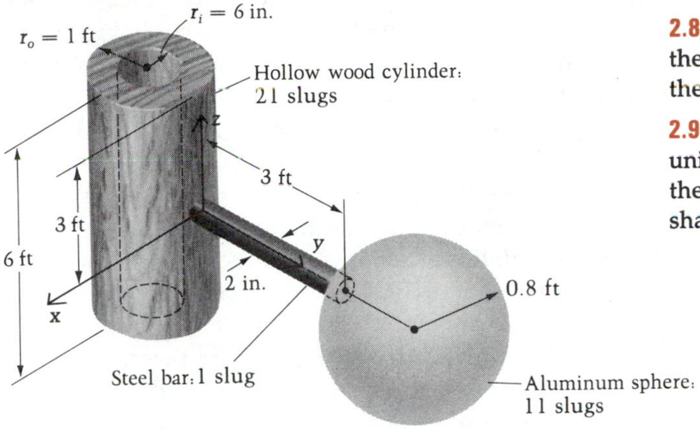

Figure P2.2

2.3 Find the center of mass of the body composed of two uniform slender bars and a uniform sphere in Figure 2.3.

2.4 Find the center of mass of the bent bar, each leg of which is parallel to a coordinate axis and as uniform density and mass m. (See Figure P2.4.)

2.5 Repeat Problem 2.4 if the four legs have uniform, but different, densities, so that the masses of $\mathcal{A}$, $\mathcal{B}$, $\mathcal{C}$, and $\mathcal{D}$ are, respectively, m, $2m$, $3m$, and $4m$.

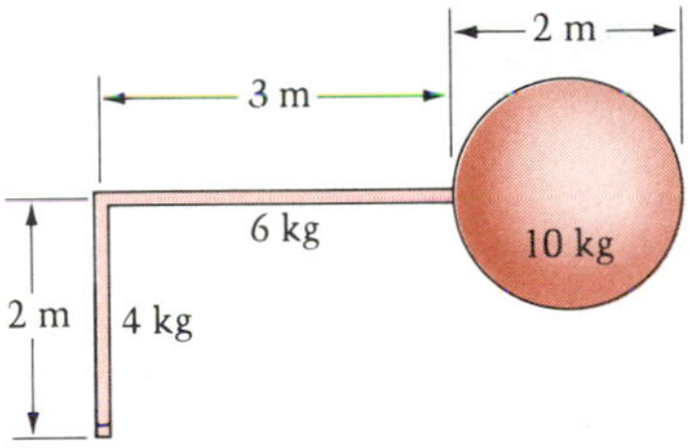

Figure P2.3

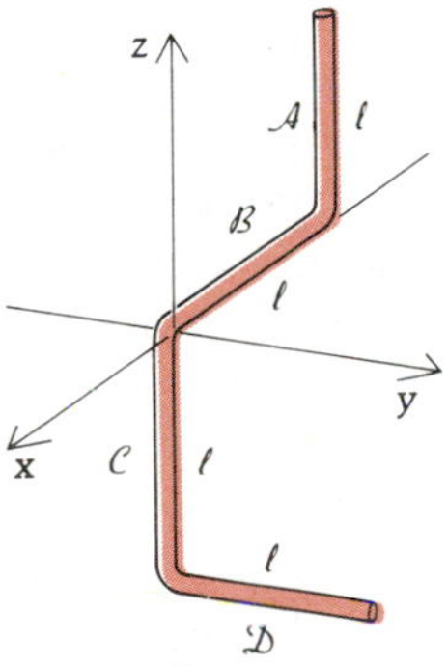

Figure P2.4

2.6 Consider a body that is a composite of a uniform sphere and a uniform cylinder, each of density ρ. Find the mass center of the body. (See Figure P2.6.)

2.7 Find the mass center of the body in Figure P2.7 which is a hemisphere glued to a solid cylinder of the same density, if $L = 2R$.

2.8 In the preceding problem, for what ratio of L to R is the mass center in the interface between the sphere and the cylinder?

2.9 In Figure P2.9 find the height H of the cone of uniform density (in terms of R) so that the mass center of the cone plus hemisphere is at the interface of the two shapes (i.e., $z = 0$).

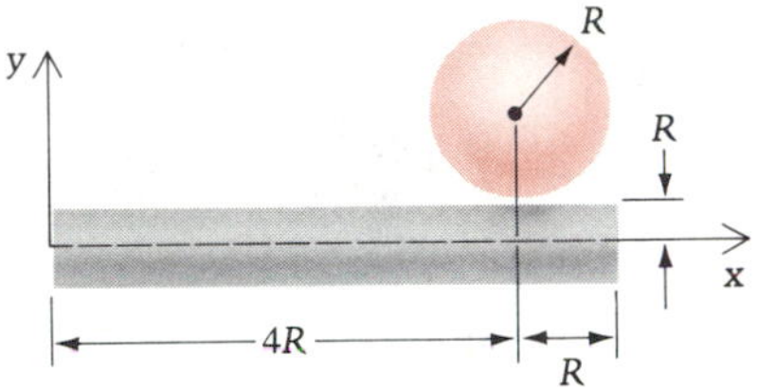

Figure P2.6

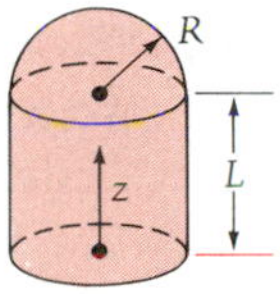

Figure P2.7

A thin wire is bent into the shape of an isosceles triangle (Figure P2.10). Find the mass center of the object, and show that it is at the same point as the mass center of a triangular plate of equal dimensions only if the triangle is equilateral. (Area of cross section $= A$ and mass density $= \rho$, both constant.)

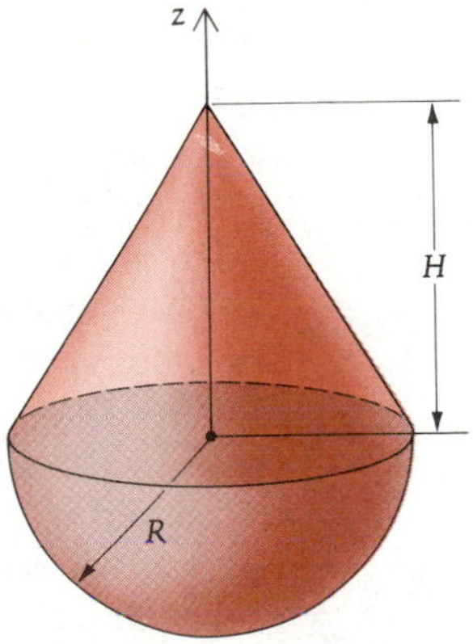

Figure P2.9

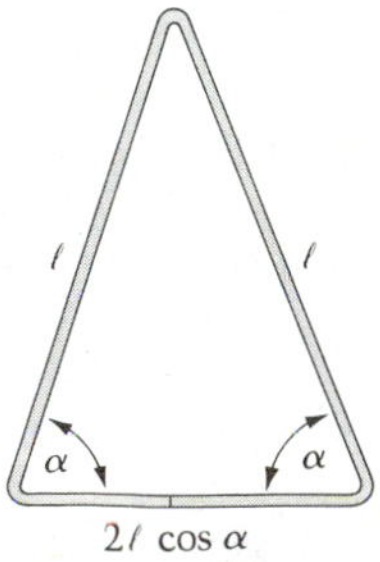

Figure P2.10

2.3 Motions of Particles and of Mass Centers of Bodies

Although the mass center of a body does not always coincide with a specific material point of the body, the mass center is nonetheless clearly an important point reflecting the distribution of the body's mass. Furthermore, there are a number of situations in which our objectives are satisfied if we can determine the motion of any material or characteristic point of the body. Clearly this is the case when we attempt to describe the orbits in which the planets move around the sun. Closer to home, a football coach is overjoyed if he finds a punter who can consistently kick the ball 60 yards in the air, regardless of whether the ball gets there end over end, spiraling, or floating like a "knuckleball." In such cases the material point upon which we focus our attention is unimportant. However, there is a strong computational advantage in focusing on the mass center: It is that the motion of that point is directly related to the external forces acting on the body.

We are more likely to think of the football as particle-like when exhibiting the knuckleball behavior than when it is rapidly spinning. Nonetheless, the *mass center's* motion in each case is governed by Euler's first law, although those motions might be quite different because of the different sets of external force induced by the differing interactions of the ball with the air.

If the external forces acting on the body are known functions of time, the mass center's motion can be calculated from Euler's first law:

$$\Sigma \mathbf{F} = m\mathbf{a}_C \tag{2.8}$$

or, alternatively,

$$\Sigma \mathbf{F} = m \frac{d^2\mathbf{r}_{OC}}{dt^2} \tag{2.10}$$

where $\mathbf{r}_{OC}$ is a position vector for the mass center. It is easily seen that two integrations of (2.10) with respect to time yield $\mathbf{r}_{OC}(t)$ provided that initial values of $\mathbf{r}_{OC}$ and $\dot{\mathbf{r}}_{OC}(= \mathbf{v}_C)$ are known.

The Free-Body Diagram

Only one thing remains to be done prior to studying several examples that make use of Euler's first law to analyze the motions of the mass centers of bodies. It is to review the concept of the **free-body diagram** which the reader should already have mastered in the study of statics. Without the ability to identify the external forces (and later the moments also), the student will not be able to write a correct set of equations of motion.

A free-body diagram is a sketch of a body in which all the external forces and couples acting upon it are carefully drawn with respect to location, direction, and magnitude. These forces might result from pushes or pulls, as the boy and girl are exerting on the crate and rope in Figure 2.4a. Or the forces might result from gravity, such as the weight of

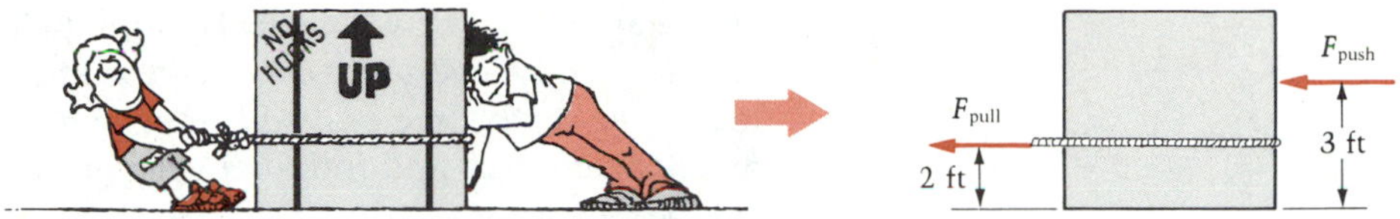

Figure 2.4a

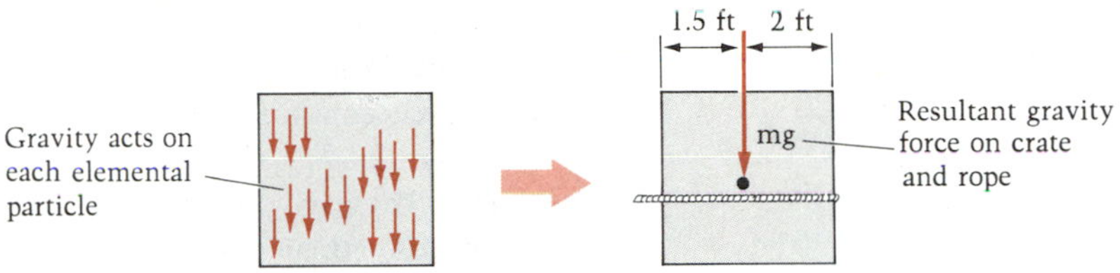

Figure 2.4b

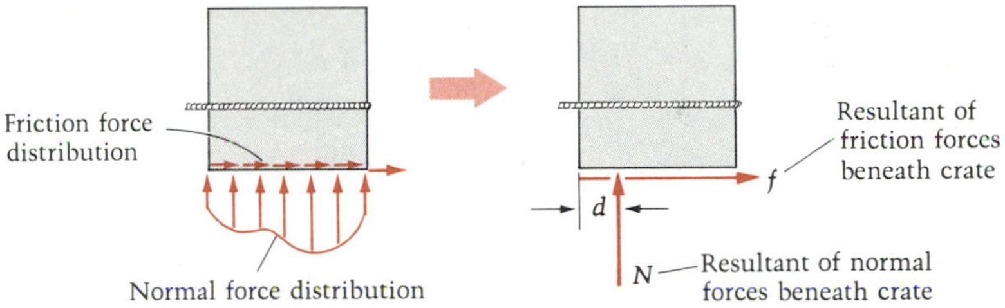

Figure 2.4c

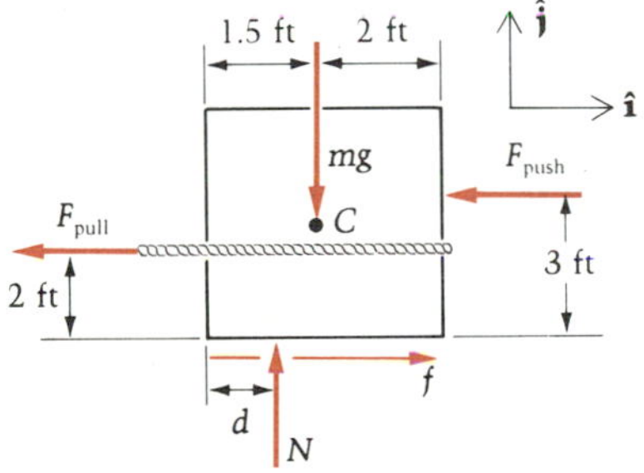

Figure 2.5

the crate in Figure 2.4b. (Note that the forces need not *touch* the body to be included in the free-body diagram; another such example is electromagnetic forces.) Or the forces might result from supports, such as the floor beneath the crate in Figure 2.4c. If the crate/rope body is acted upon simultaneously by all the forces in these figures, its complete free-body diagram is as shown in Figure 2.5.

It is important to recognize that the free-body diagram:

1. Clearly identifies the body whose motion is to be analyzed.
2. Provides a catalog of all the *external* forces (and couples) *on* the body.
3. Allows us to express, in a compact way, what we know or can easily conclude about the lines of action of known and unknown forces. For example, we know that the pressure (distributed normal force) exerted by the floor on the bottom of the box has a resultant that is a force with a vertical line of action. The symbol N along with the arrow is a code for communicating the fact that we have decided to express that unknown (vector) force as $N\hat{\mathbf{j}}$. The fact that we do not know the location of the line of action of that force is displayed by the presence of the unknown length d.

In dynamics, as in statics, the only characteristics of a force that are manifest in the equations of motion are the vector describing the force and the location of its line of action; that is, we must sum up all the

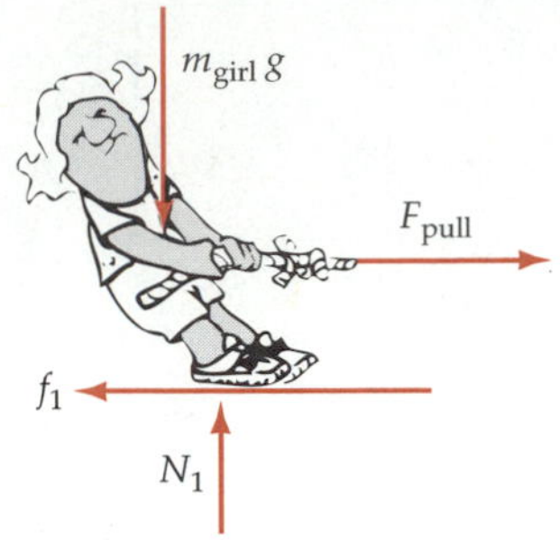

Figure 2.6

external forces, and we must also sum their moments about some point. Consequently, everything we need to know about the external forces is displayed on the free-body diagram, and we may readily check our work by glancing back and forth between our diagram and the equations we are writing.

When we focus individually on two or more interacting bodies, the free-body diagrams provide an economical way to satisfy—and show that we have satisfied—the principle of action and reaction. The free-body diagram of the girl in our example is shown in Figure 2.6. Since we have already established by Figure 2.5 that the force exerted by the girl on the rope will be $F_{\mathbf{pull}}(-\hat{\mathbf{i}})$, then, by the action-reaction principle, the force exerted by the rope on the girl must be $F_{\mathbf{pull}}(+\hat{\mathbf{i}})$ as shown in Figure 2.6. In other words, consistent forces of interaction are expressed through the single scalar $F_{\mathbf{pull}}$ and the arrow code.

EXAMPLE 2.2

Ignoring air resistance, find the trajectory of a golf ball hit off a tee at speed v_0 and angle θ with the horizontal.

Solution

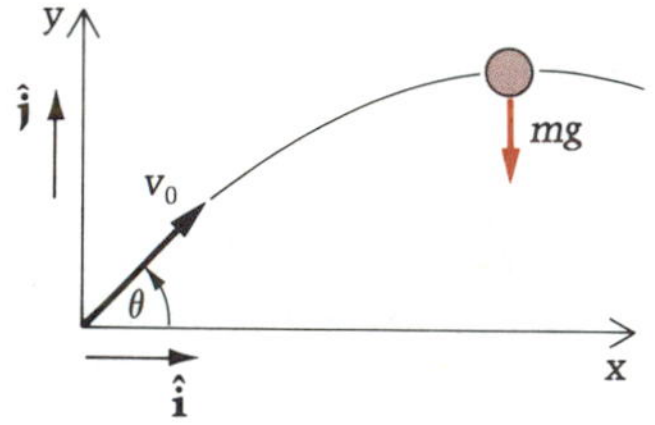

Figure E2.2

It is convenient here to set up a rectangular coordinate system as shown in Figure E2.2 and let time $t = 0$ be the instant at which the ball leaves the club. With x, y, and z as the coordinates of the mass center of the ball and since the only external force on the ball is its weight, $-mg\hat{\mathbf{j}}$, we have from Equation (2.10):

$$-mg\hat{\mathbf{j}} = m(\ddot{x}\hat{\mathbf{i}} + \ddot{y}\hat{\mathbf{j}} + \ddot{z}\hat{\mathbf{k}})$$

Thus, collecting the coefficients of $\hat{\mathbf{i}}$, $\hat{\mathbf{j}}$, and $\hat{\mathbf{k}}$, we obtain

$$\ddot{x} = 0 \qquad \ddot{y} = -g \qquad \ddot{z} = 0$$

Integrating, we get

$$\dot{x} = C_1 \qquad \dot{y} = -gt + C_2 \qquad \dot{z} = C_3$$

Because of the way we have aligned the x and z axes,

$$\dot{x}(0) = v_0 \cos\theta \qquad \dot{y}(0) = v_0 \sin\theta \qquad \dot{z}(0) = 0$$

Therefore

$$C_1 = v_0 \cos\theta \qquad C_2 = v_0 \sin\theta \qquad C_3 = 0$$

Integrating again, we get

$$x = v_0(\cos\theta)t + C_4$$

$$y = \frac{-gt^2}{2} + v_0(\sin\theta)t + C_5$$

$$z = C_6$$

Our location of the origin of the coordinate system at the "launch" site yields $x(0) = y(0) = z(0) = 0$, so that $C_4 = C_5 = C_6 = 0$ and the trajectory of the mass center of the ball is given by

$$x = v_0(\cos\theta)t$$

$$y = -\frac{gt^2}{2} + v_0(\sin\theta)t$$

$$z = 0$$

which describes a parabola in the xy plane—that is, in the vertical plane defined by the launch point and the direction of the launch velocity.

Letting the time of maximum elevation be t_1, we find that $\dot{y}(t_1) = 0$ yields

$$0 = -gt_1 + v_0 \sin\theta$$

so that $t_1 = (v_0/g)\sin\theta$ and the maximum elevation is

$$y(t_1) = -\frac{v_0^2}{2g}\sin^2\theta + \frac{v_0^2}{g}\sin^2\theta = \frac{v_0^2}{2g}\sin^2\theta$$

If t_2 is the time the ball strikes the fairway (assumed level), then

$$y(t_2) = 0 = -\frac{gt_2^2}{2} + v_0 t_2 \sin\theta$$

$$t_2 = \frac{2v_0}{g}\sin\theta$$

which is, not surprisingly, twice the time (t_1) to reach maximum elevation. The length of the drive is

$$x(t_2) = v_0(\cos\theta)\left(\frac{2v_0}{g}\sin\theta\right)$$

$$= \frac{2v_0^2}{g}\sin\theta\cos\theta$$

$$= \frac{v_0^2}{g}\sin 2\theta$$

which, with v_0 fixed, is maximized by $\theta = 45°$. That is, for a given launch speed we get maximum range when the launch angle is $45°$.

The results of this analysis apply to the unpowered flight of any projectile as long as the path is sufficiently limited that the gravitational force is constant (magnitude and direction) and we can ignore the medium (air) through which the body moves. Interaction with the air is responsible not only for the drag (retarding of motion) on a golf ball but also for the fact that its path is usually not planar (slice or hook!). On one of the Apollo moon landings in the early 1970s, astronaut Alan Shepard drove a golf ball a "country mile" on the moon because of the absence of air resistance and, more important, because the gravitational acceleration at the moon's surface is only about one-sixth that at the surface of the earth.

Example 2.3

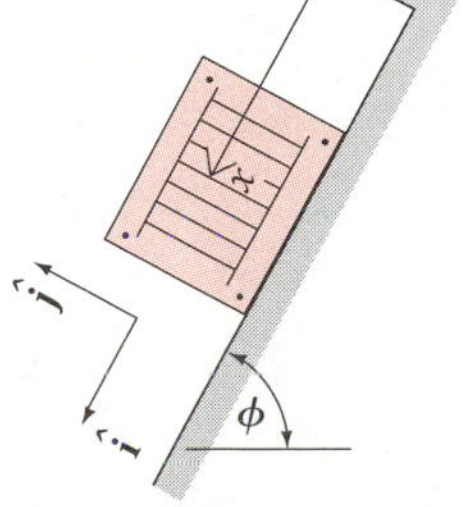

Figure E2.3a

If the 20-kg block shown in Figure E2.3a is released from rest, find its speed after it has descended a distance $d = 5$ m down the plane. The angle $\varphi = 60°$ and the (Coulomb) coefficients of friction are

$$\mu_s \text{ (static)} = 0.30$$
$$\mu_k \text{ (kinetic)} = 0.25$$

Solution

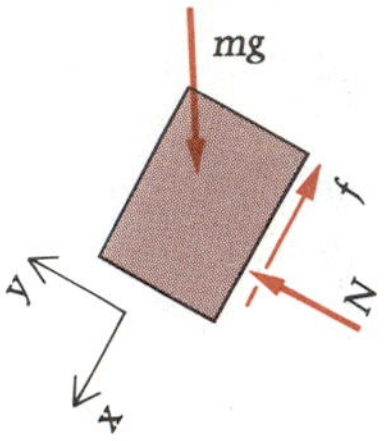

Figure E2.3b

In the statement of the problem we are using some loose but common terminology in referring to the speed of the block. In fact we may only speak of the speed of a *point,* but here we are tacitly assuming that the block is *rigid* and *translating* so that every point in the block has the same velocity and the same acceleration. In contrast to the preceding example, note that here we do not know all the external forces on the body before we carry out the analysis, because the surface touching the block constrains its motion. That constraint is acknowledged by expressing the velocity of (the mass center of) the block by $\dot{x}\hat{\mathbf{i}}$ and its acceleration by $\ddot{x}\hat{\mathbf{i}}$.

Referring to the free-body diagram shown in Figure E2.3b,

$$\Sigma \mathbf{F} = m\mathbf{a}_C$$
$$N\hat{\mathbf{j}} + mg(\sin\varphi\hat{\mathbf{i}} - \cos\varphi\hat{\mathbf{j}}) - f\hat{\mathbf{i}} = m\ddot{x}\hat{\mathbf{i}}$$

or

$$N = mg\cos\varphi$$

and

$$mg\sin\varphi - f = m\ddot{x}$$

First we must determine if in fact the block will move. For equilibrium, $\ddot{x} = 0$ and f is limited by $0 \le f \le f_{\max} = \mu_s N$. Hence

$$f = mg\sin\varphi \le \mu_s\, mg\cos\varphi = \mu_s N$$

or

$$\tan\varphi \le \mu_s$$

But

$$\tan 60° = 1.73 > 0.3 = \mu_s$$

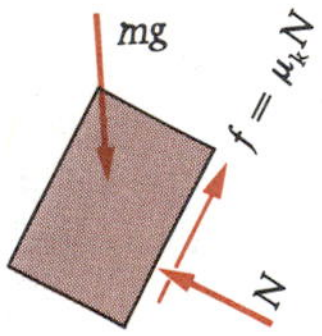

Figure E2.3c

Thus the block moves (and, as it does, is acted on by $\mu_k N$ up the plane as shown in Figure E2.3c). We note that $\tan^{-1}(\mu_s)$ is sometimes called the *angle of friction*. Here $\tan^{-1}(\mu_s)$ is 16.7°, and this of course is the angle for which $\tan\varphi = \mu_s$; it means that any angle $\varphi > 16.7°$ (like our 60°) will result in sliding, or a loss of equilibrium.

Having checked the statics and briefly reviewed friction, we now solve the equation of motion for $\ddot{x}$:

$$\begin{aligned} m\ddot{x} &= mg\sin\varphi - \mu_k N \\ &= mg\sin\varphi - \mu_k(mg\cos\varphi) \\ &= mg(\sin\varphi - \mu_k\cos\varphi) \end{aligned}$$

or

$$\ddot{x} = g(\sin\varphi - \mu_k\cos\varphi)$$

Thus

$$\dot{x} = g(\sin\varphi - \mu_k\cos\varphi)t + C_1$$

and $C_1 = 0$ since $\dot{x}(0) = 0$ if $t = 0$ is the instant at which the block is released. Hence

$$x = \frac{g}{2}(\sin\varphi - \mu_k \cos\varphi)t^2 + C_2$$

and $C_2 = 0$ if we choose the measurement of x so that $x(0) = 0$.

If we let t_1 be the time at which $x = d$, then

$$d = \frac{g}{2}(\sin\varphi - \mu_k \cos\varphi)t_1^2$$

For $\varphi = 60°$, $\mu_k = 0.25$, $d = 5$ m, and $g = 9.81$ m/s², we get

$$5 = \left(\frac{9.81}{2}\right)[0.866 - 0.25(0.5)]t_1^2$$

from which $t_1 = 1.17$ s.

Since the velocity is given by $\dot{x}\hat{\mathbf{i}}$, the speed at t_1 is merely the magnitude (or absolute value) of $\dot{x}(t_1)$ and

$$\begin{aligned}\dot{x}(t_1) &= (9.81)[0.866 - 0.25(0.5)](1.17)\\ &= 8.50 \text{ m/s}\end{aligned}$$

Finally we should note that the plausibility of our numerical results can be verified from the fact that, owing to the steep angle and moderate coefficient of friction, they should be of the same orders of magnitude as those arising from a free vertical drop (acceleration g) for which

$$t_1 = \sqrt{\frac{2d}{g}} = \sqrt{\frac{2(5)}{9.81}} = 1.01 \text{ s}$$

and

$$\text{Speed} = \sqrt{2gd} = \sqrt{2(9.81)(5)} = 9.90 \text{ m/s}$$

Example 2.4

A ball of mass m (see Figure E2.4) is released from rest with the cord taut and $\theta = 30°$. Find the tension in the cord during the ensuing motion.

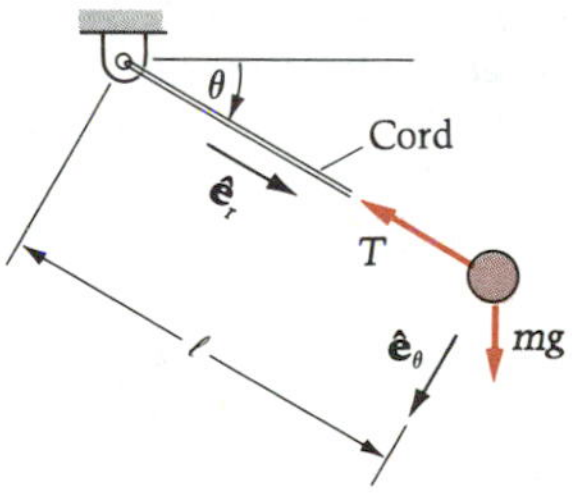

Figure E2.4

Solution

In this problem we make two basic assumptions:

1. The cord is inextensible.
2. The cord is attached to the ball at its mass center (or equivalently the ball is small enough to be treated as a particle). Either way the point whose motion is to be described has a path that is a circle. Thus the problem is similar to Example 2.3 in that the path of the mass center is known in advance (a circle here and a straight line there) and consequently among the external forces are unknowns caused by constraints (the tension in the cord here and the surface reaction in the preceding problem).

Using polar coordinates (Section 1.6), we may express the acceleration as

$$\mathbf{a} = (\ddot{r} - r\dot{\theta}^2)\hat{\mathbf{e}}_r + (r\ddot{\theta} + 2\dot{r}\dot{\theta})\hat{\mathbf{e}}_\theta$$

Since the polar coordinate r is the constant ℓ here, referring to the free-body diagram in Figure E2.4 we have

$$-T\hat{\mathbf{e}}_r + mg(\hat{\mathbf{e}}_\theta \cos\theta + \hat{\mathbf{e}}_r \sin\theta) = m(-\ell\dot{\theta}^2\hat{\mathbf{e}}_r + \ell\ddot{\theta}\hat{\mathbf{e}}_\theta)$$

so that

$$T = mg\sin\theta + m\ell\dot{\theta}^2 \tag{1}$$

and

$$m\ell\ddot{\theta} - mg\cos\theta = 0 \tag{2}$$

The first of these component equations (Equation (1)) yields the tension T if we know $\theta(t)$; the second (2) is the differential equation that we must integrate to obtain $\theta(t)$. In Example 2.3 the counterpart of Equation (2) is $\ddot{x}$ = constant, which of course was easily integrated.

Here not only do we have a nontrivial differential equation in that $\ddot{\theta}$ is a function of θ, but we have the substantial complication that Equation (2) is nonlinear because $\cos\theta$ is a nonlinear function of θ. However, a partial integration of Equation (2) can be accomplished; to this end we write the equation in the standard form

$$\ddot{\theta} - \frac{g}{\ell}\cos\theta = 0$$

and then multiply by $\dot{\theta}$ to obtain

$$\dot{\theta}\ddot{\theta} - \frac{g}{\ell}\dot{\theta}\cos\theta = 0$$

which we recognize to be

$$\frac{d}{dt}\left(\frac{\dot{\theta}^2}{2} - \frac{g}{\ell}\sin\theta\right) = 0$$

or

$$\frac{\dot{\theta}^2}{2} - \frac{g}{\ell}\sin\theta = C_1 \qquad \text{(a constant)} \tag{3}$$

Equation (3) is called an **energy integral** of Equation (2) and is closely related to the "work and kinetic energy" principle that is introduced in the next section.

For the problem at hand the constant C_1 may be obtained from the fact that when $\theta = 30°$, then $\dot{\theta} = 0$; thus

$$0 - \frac{g}{\ell}\sin 30° = C_1$$

or

$$C_1 = \frac{-g}{2\ell}$$

Thus from Equation (3) we get

$$\dot{\theta}^2 = \frac{g}{\ell}(2\sin\theta - 1) \tag{4}$$

which we may substitute in Equation (1) to obtain

$$T = mg \sin \theta + mg(2 \sin \theta - 1)$$

or

$$T = mg(3 \sin \theta - 1) \tag{5}$$

Even though we have not obtained the time dependence of the tension,* the energy integral has enabled us to find the way in which the tension depends on the position of the ball. As we would anticipate intuitively, the maximum tension occurs when $\theta = 90°$, at which time $T = [3(1) - 1]\, mg = 2\, mg$.

EXAMPLE 2.5

A planet P of mass m moves in a circular orbit around a star of mass M, far away from any other gravitational or other forces (see Figure E2.5). If the planet completes one orbit in τ units of time, find the orbit radius, using the fact that in a circular orbit the speed is constant.

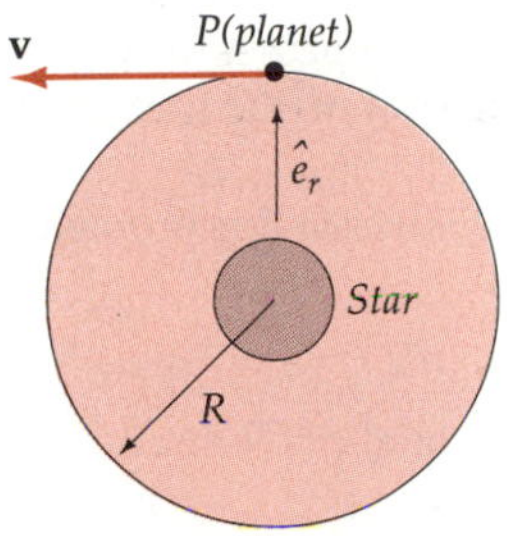

Figure E2.5

Solution

Writing the component of the equation of motion for the planet P in the radial direction,

$$\Sigma F_r = m\mathbf{a}_{Cr}$$

The only external force on P is gravity, in the $-\hat{\mathbf{e}}_r$ direction (towards the star). Letting G be the universal gravitational constant, we substitute and obtain:

$$\frac{-GMm}{R^2} = m[\ddot{r} - r\dot{\theta}^2]$$

where from Equation (1.37), the radial acceleration component is $\ddot{r} - r\dot{\theta}^2$. Since $r = R =$ constant, we have, with $\dot{\theta} = \omega =$ orbital rate,

$$\frac{GM}{R^2} = R\omega^2$$

But $\omega = \dfrac{2\pi}{\tau}$, so that

$$R^3 = \frac{GM}{\left(\dfrac{2\pi}{\tau}\right)^2} = \frac{GM\tau^2}{4\pi^2}$$

or

$$R = \left(\frac{GM\tau^2}{4\pi^2}\right)^{1/3}$$

* This would require solving the differential equation (4) for $\theta(t)$ and substituting into Equation (5).

EXAMPLE 2.6

A car accelerates from rest, increasing its speed at the constant rate of $K = 6$ ft/sec^2. (See Figure E2.6a.) It travels on a circular path starting at point A. Find the time and the position of the car when it first leaves the surface due to excessive speed.

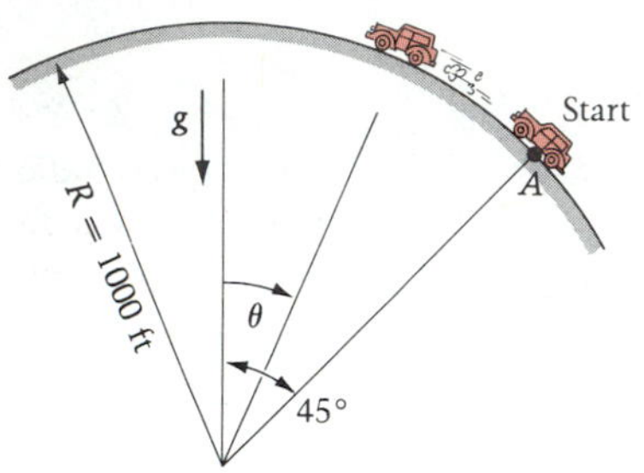

Figure E2.6a

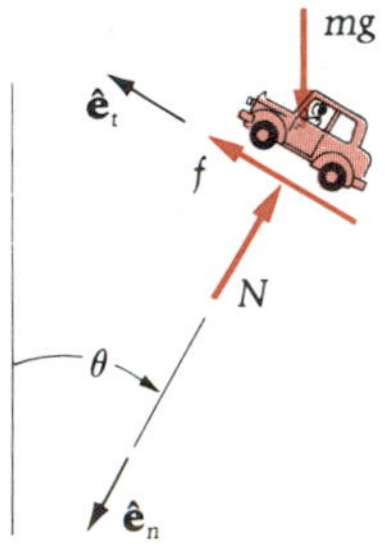

Figure E2.6b

Solution

Before the car (treated as a particle) leaves the surface, the free-body diagram is as shown in Figure E2.6b. We shall work this problem in general (without substituting numbers until the end). The purpose is to illustrate the concept of nondimensional parameters. The equation of motion in the tangential ($\hat{\mathbf{e}}_t$) direction is

$$\Sigma F_t = m\ddot{s}$$

$$f - mg \sin \theta = mK \tag{1}$$

Equation (1) shows that the friction exerted on the tires by the road is the external force which moves the car up the path. Note that after it passes the top of the circular hill, we have $\sin \theta < 0$ and then the gravity force *adds* to the friction in accelerating the car on the way down.

The following equation of motion is the one that will help us in this problem; it equates $\Sigma\mathbf{F}$ and $m\mathbf{a}_C$ in the normal direction.

Question 2.2 Are the components of $\Sigma\mathbf{F}$ and $m\mathbf{a}_C$ equal in *all* directions or just in coordinate directions?

$$\Sigma F_n = m\frac{\dot{s}^2}{\rho}$$

$$mg \cos \theta - N = \frac{m(Kt)^2}{R} \tag{2}$$

where $\dot{s} = \int \ddot{s}\, dt = Kt + C_1 = Kt$, since $\dot{s} = 0$ when $t = 0$.

We note that the car will lose contact with the road when N becomes zero. (The ground cannot pull down on the car for further increases of t, which would require $N < 0$!) Therefore, at the point of leaving the ground,

$$\cancel{m}g \cos \theta = \frac{\cancel{m}K^2t^2}{R} \tag{3}$$

Question 2.3 What is the *meaning* of the fact that m cancels in Equation (3)?

Now θ is related to s according to

$$s = R\left(\frac{\pi}{4} - \theta\right) \tag{4}$$

Answer 2.2 The components of the two sides of a vector equation are equal in *any* direction.
Answer 2.3 It means the answer does not depend on the mass of the car.

And from $\dot{s} = Kt$, we get another expression for s:

$$s = \frac{Kt^2}{2} + C_2 = \frac{Kt^2}{2} \qquad \text{(since } s = 0 \text{ when } t = 0) \tag{5}$$

Hence from equations (4) and (5) we get

$$\frac{Kt^2}{2} = R\left(\frac{\pi}{4} - \theta\right) \Rightarrow \theta = \frac{\pi}{4} - \frac{Kt^2}{2R} \tag{6}$$

Substituting for θ from (6) into (3), we have

$$g \cos\left(\frac{\pi}{4} - \frac{Kt^2}{2R}\right) = \frac{K^2t^2}{R}$$

or

$$\cos\left(\frac{\pi}{4} - \frac{Kt^2}{2R}\right) = \frac{Kt^2}{2R} \cdot \frac{2K}{g} \tag{7}$$

Equation (7) allows us to solve for the dimensionless parameter $q = (Kt^2/2R)$, once we have selected a value of the car's dimensionless acceleration K/g. In this problem, for example,

$$\cos\left(\frac{\pi}{4} - q\right) = 2q\left(\frac{6}{32.2}\right) = 0.373q \tag{8}$$

The following table shows how (with a calculator)* we can quickly arrive at the value of q that solves Equation (8):

q	$\cos\left(\frac{\pi}{4} - q\right)$	$0.373q$
0.1	0.7742	0.0373
0.5	0.9595	0.1865
07.854 (at the top)	1	0.2930
1.0	0.9771	0.3730
1.3	0.8705	0.4849
1.6	0.6862	0.5968
1.7	0.6101	0.6341
1.69	0.6180	0.6304
1.68	0.6258	0.6266

Thus at $q = Kt^2/2R \approx 1.68$, the car leaves the circular track due to excessive speed. Therefore for $K = 6$ ft/sec^2 and $R = 1000$ ft,

$$t = \sqrt{\frac{1.68 \times 2 \times 1000}{6}} = 23.7 \text{ sec}$$

* See Appendix B for a numerical solution to this problem using the Newton-Raphson method.

The angle at loss of contact is given by Equation (6):

$$\theta = \frac{\pi}{4} - \frac{Kt^2}{2R} = \frac{\pi}{4} - q$$

$$= \frac{\pi}{4} - 1.68$$

$$= -0.895 \text{ rad}$$

$$= -51.3°$$

The speed at the point of loss of contact is $Kt = 6(23.7) = 142$ ft/sec $= 97.0$ mph. Note that for $K/g = 6/32.2$, the angle $-51.3°$ is the angle of leaving for many combinations of t and R (so long as $Kt^2/2R = 1.68$).

EXAMPLE 2.7

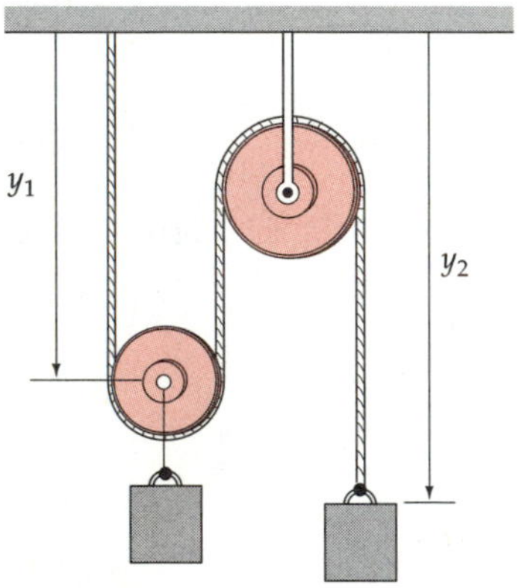

Figure E2.7a

In the system shown in Figure E2.7a, each of the blocks weighs 10 lb and the pulleys are very much lighter. Find the accelerations of the blocks, assuming the belt (or rope) to be inextensible and of negligible mass.

Solution

If there is negligible friction in the bearings of the pulley, and the pulley is much lighter than other elements of the system, then the belt tension won't change from one side of the pulley to the other. So, referring to Figure E2.7b,

$$T_1 = T_2$$

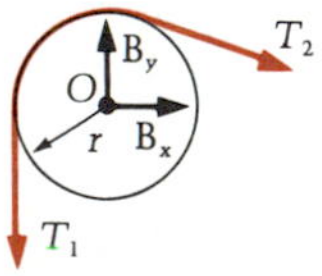

Figure E2.7b

For the block in Figure E2.7c:

$$10 - T = \frac{10}{32.2}\ddot{y}_2 \tag{1}$$

Figure E2.7c

And for the block and pulley in Figure E2.7d:

$$10 - 2T = \frac{10}{32.2}\ddot{y}_1 \tag{2}$$

We also have a kinematic constraint relationship between y_1 and y_2 because the belt is inextensible. For this problem it is that $\dot{y}_2 = -2\dot{y}_1$ (see Example 1.8) and consequently

$$\ddot{y}_2 = -2\ddot{y}_1 \tag{3}$$

Solving Equations (1), (2), and (3) simultaneously,

$$T = 6 \text{ lb}$$

$$\ddot{y}_1 = -6.44 \text{ ft/sec}^2$$

$$\ddot{y}_2 = 12.9 \text{ ft/sec}^2$$

so the left block accelerates upward at 6.44 ft/sec^2 and the right block accelerates downward at 12.9 ft/sec^2.

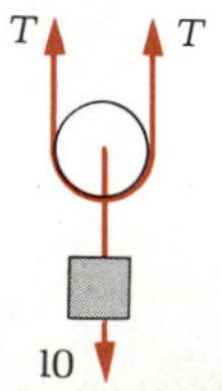

Figure E2.7d

Example 2.8

Find the accelerations of the blocks shown in Figure E2.8a when released from rest. Then repeat the problem with the friction coefficients reversed.

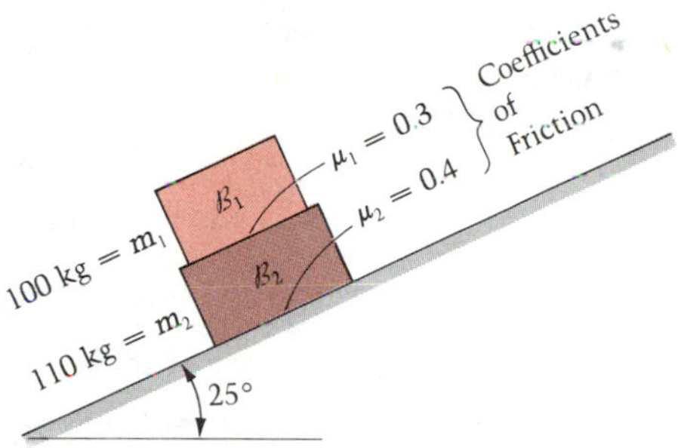

Figure E2.8a

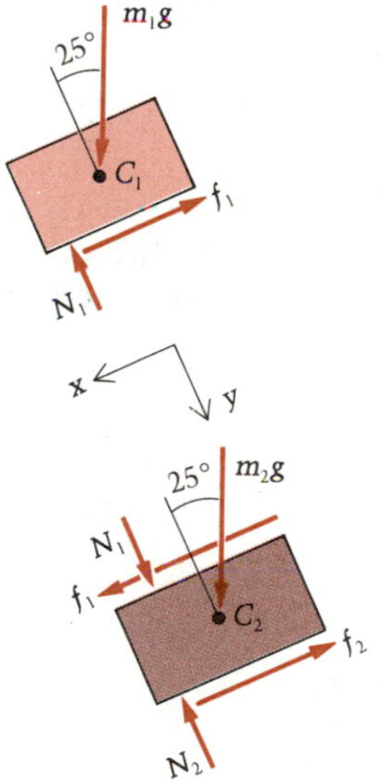

Figure E2.8b

Solution

We know from statics that *if* the two blocks move as a unit, their motion will occur when

$$\tan 25° > \mu_2$$

that is, when

$$0.466 > 0.4$$

which is the case here. But before our solution is complete we must determine whether either block moves without the other. We consider the free-body diagrams of each translating block (see Figure E2.8b) and write the equations of motion:

$$\Sigma F_x = m_1 \ddot{x}_{C_1} \Rightarrow 100(9.81) \sin 25° - f_1 = 100\ddot{x}_{C_1} \quad (1)$$

$$\Sigma F_y = m_1 \ddot{y}_{C_1} \Rightarrow 100(9.81) \cos 25° - N_1 = 100)\overset{0}{\cancel{\ddot{y}_{C_1}}} \quad (2)$$

$$\Sigma F_x = m_2 \ddot{x}_{C_2} \Rightarrow 110(9.81) \sin 25° + f_1 - f_2 = 110\ddot{x}_{C_2} \quad (3)$$

$$\Sigma F_y = m_2 \ddot{y}_{C_2} \Rightarrow 110(9.81) \cos 25° + N_1 - N_2 = 110)\overset{0}{\cancel{\ddot{y}_{C_2}}} \quad (4)$$

We mention that the sum of Equations (1) and (3) gives the "x equation" of the overall system; the sum of (2) and (4) yields the "y equation" (C is the mass center of the combined blocks):

$$\begin{aligned} \Sigma F_x &= (100 + 110)9.81 \sin 25° - f_2 \\ &= 100\ddot{x}_{C_1} + 110\ddot{x}_{C_2} \quad (5) \\ &= (100 + 110)\ddot{x}_C \end{aligned}$$

$$\begin{aligned} \Sigma F_y &= (100 + 110)9.81 \cos 25° - N_2 \\ &= (100 + 110)\overset{0}{\cancel{\ddot{y}_C}} = 0 \quad (6) \end{aligned}$$

Note that f_1 and N_1 disappear in (5) and (6), as they become internal forces on the combined system.

Equation (6) tells us that $N_2 = 1870$ N, regardless of which motion takes place. The equation for the x motion (Equation 5) shows again that if $f_{2_{max}} < mg(\sin 25°)$, then one or both of the blocks must slide:

$$f_{2_{max}} = \mu_2 N_2 = 0.4(1870) = 748 < 210(9.81)(0.423) = 871 \text{ N}$$

and so $\ddot{x}_C$ cannot be zero. Assuming first that the blocks *both* move, then f_2 is at its maximum:

$$f_2 = f_{2_{max}} = 748 \text{ N}$$

If they move *together* as one body, then Equation (5) gives us

$$\begin{aligned}\ddot{x}_C (= \ddot{x}_{C_1} = \ddot{x}_{C_2}) &= (mg \sin 25° - f_{2_{max}})/m \\ &= (871 - 748)/210 \\ &= 0.586 \text{ m/s}^2\end{aligned}$$

Substituting this acceleration into Equation (1), we can check to see if body $\mathcal{B}_1$ additionally slides relative to $\mathcal{B}_2$:

$$f_1 = -100(0.586) + 415 = 356 \text{ N}$$

But the maximum value that f_1 can have is given by

$$f_{1_{max}} = \mu_1 N_1 = 0.3(889) = 267 \text{ N}$$

Hence block $\mathcal{B}_1$ slides on $\mathcal{B}_2$ and the blocks do not move together; our assumption was incorrect. We then substitute $f_1 = \mu_1 N_1$ into Equation (1) and proceed:

$$\ddot{x}_{C_1} = (415 - 267) \div 100 = 1.48 \text{ m/s}^2$$

This is then the acceleration of the top block. Substituting f_1 into Equation (3) gives

$$\begin{aligned}456 + 267 - f_2 &= 110\ddot{x}_{C_2} \\ 723 - f_2 &= 110\ddot{x}_{C_2}\end{aligned}$$

For no motion of the bottom block, $f_{2_{max}}$ clearly needs to be at least 723 N. Since it is in fact 748 N, the bottom block does *not* move for this combination of parameters, and $\ddot{x}_{C_2} = 0$.

If the friction coefficients are now swapped, nothing changes until we begin to analyze the six equations. We have

$$f_{2_{max}} = \mu_2 N_2 = 0.3(1870) = 561 < mg \sin 25° = 871 \text{ N} \qquad \text{(as before)}$$

Again, then, $\ddot{x}_C$ cannot be zero. Assuming again that the blocks both move, f_2 is its maximum and Equation (5) gives

$$\ddot{x}_C = (871 - 561) \div 210 = 1.48 \text{ m/s}^2$$

Substituting this acceleration into Equation (1), we get

$$f_1 = 415 - 100(1.48) = 267 < \mu_1 N_1 = 0.4(889) = 356 \text{ N}$$

This time we have more friction than we need in order to prevent $\mathcal{B}_1$ from slipping on $\mathcal{B}_2$. Thus both $\ddot{x}_{C_1}$ and $\ddot{x}_{C_2}$ are 1.48 m/s².

PROBLEMS ▸ Section 2.3

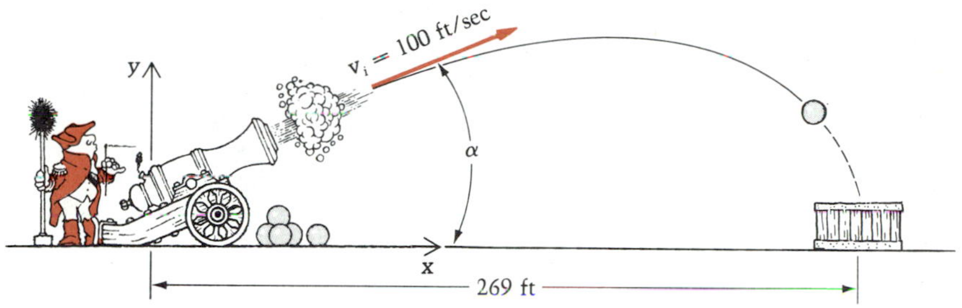

Figure P2.12

2.11 In Problem 1.114 what is the acceleration of the rock just *after* release?

2.12 A cannonball is fired as shown in Figure P2.12. Neglecting air resistance, find the angle α that will result in the cannonball landing in the box.

2.13 A baseball slugger connects with a pitch 4 ft above the ground. The ball heads toward the 10-ft-high center-field fence, 455 ft away. The ball leaves the bat with a velocity of 125 ft/sec and a slope of 3 vertical to 4 horizontal. Neglecting air resistance, determine whether the ball hits the fence (if it does, how high above the ground?) or whether it is a home run (if it is, by how much does it clear the fence?).

2.14 From a high vantage point in Yankee Stadium, a baseball fan observes a high-flying foul ball. Traveling vertically upward, the ball passes the level of the observer 1.5 sec after leaving the bat, and it passes this level again on its away down 4 sec after leaving the bat. Disregarding air friction, find the maximum height reached by the baseball and determine the ball's initial velocity as it leaves the bat (which is 3 ft above the ground at impact).

2.15 A soccer ball (Figure P2.15) is kicked toward the goal from 60 ft. It strikes the top of the goal at the highest point of its trajectory. Find the velocity and angle θ at which the ball was kicked, and determine the time of travel.

2.16 The motorcycle in Figure P2.16 is to be driven by a stunt man. Find the minimum takeoff velocity at A for which the motorcycle can clear the gap, and determine the corresponding angle θ for which the landing will be tangent to the road at B and hence smooth.

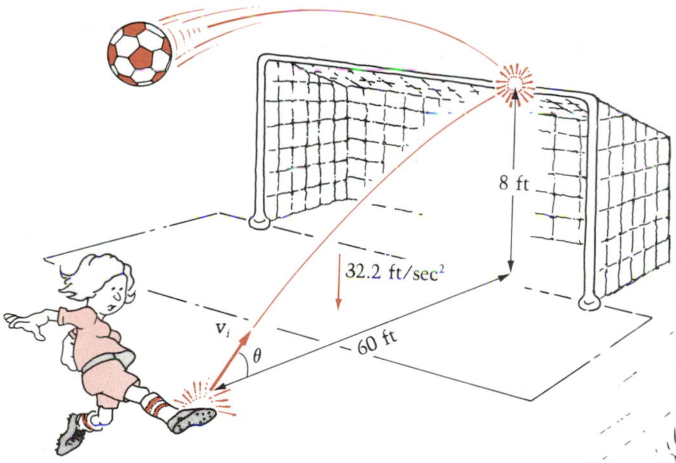

Figure P2.15

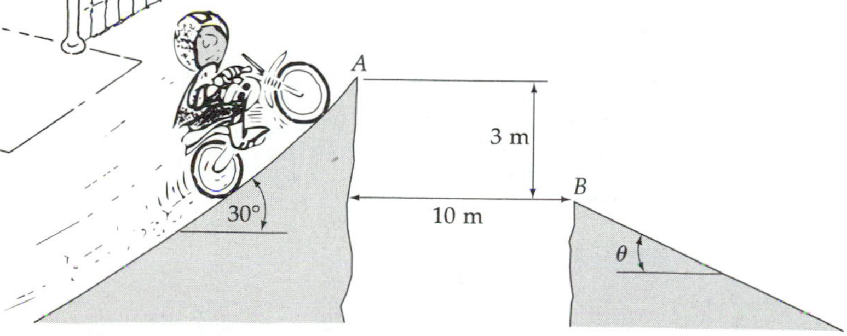

Figure P2.16

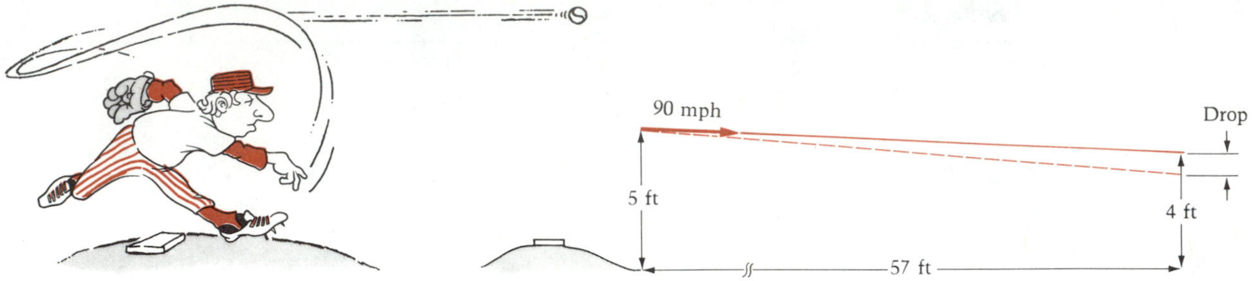

Figure P2.17

2.17 A baseball pitcher releases a 90-mph fastball 5 ft off the ground (Figure P2.17). If in the absence of gravity the ball would arrive at home plate 4 ft off the ground, find the drop in the actual path caused by gravity. Neglect air resistance.

2.18 In the preceding problem, find the radius of curvature of the path of the baseball's center at the instant it arrives at the plate.

2.19 In the preceding problem, the batter hits a pop-up that leaves the bat at a 45° angle with the ground. The shortstop loses the ball in the sun and it lands on second base, $90\sqrt{2}$ ft from home plate. What was the velocity of the baseball when it left the bat?

2.20 The pilot of an airplane flying at 300 km / hr wishes to release a package of mail at the right position so that it hits spot A. (See Figure P2.20.) What angle θ should his line of sight to the target make at the instant of release?

2.21 A darts player releases a dart at the position indicated in Figure P.2.21 with the initial velocity vector making a 10° angle with the horizontal. What must the dart's initial speed be if it scores a bull's-eye?

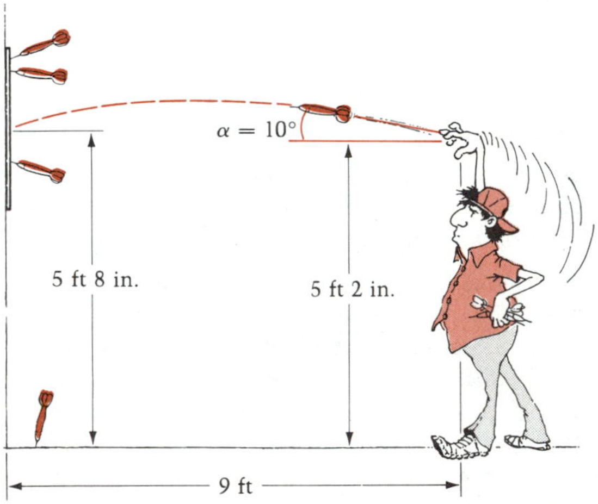

Figure P2.21

2.22 In the preceding problem, suppose the initial speed of the dart is 20 ft / sec. What must the angle α be if a bull's-eye is scored?

Figure P2.20

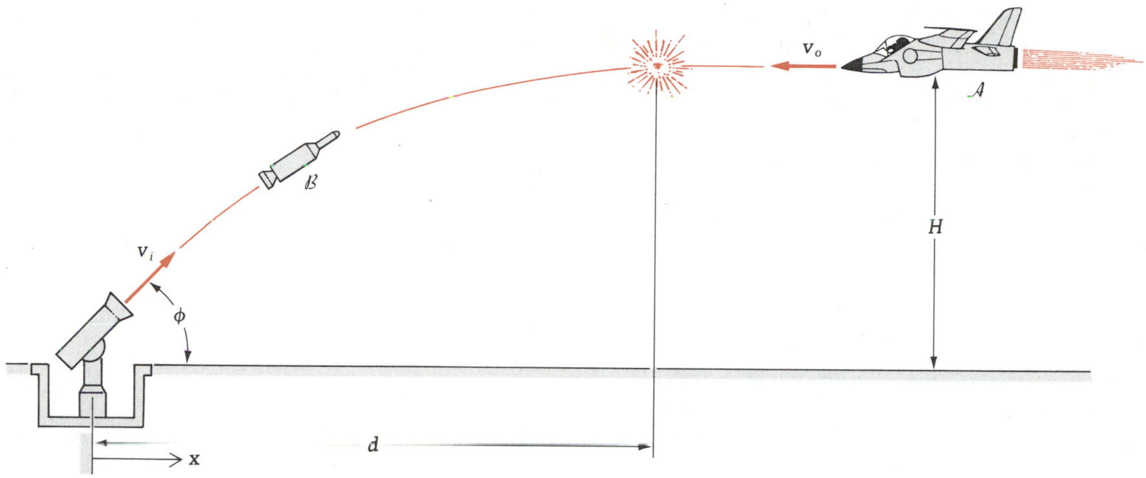

Figure P2.23

2.23 Find the angle ϕ, firing velocity v_i, and time t_f of intercept so that the ballistic missile shown in Figure P2.23 will intercept the bomber when $x = d$. The bomber, at $x = D$ when the missile is launched, travels horizontally at constant speed v_0 and altitude H. What has been neglected in your solution?

2.24 The garden hose shown in Figure P2.24 expels water at 13 m/s from a height of 1 m. Determine the maximum height H and horizontal distance D reached by the water.

* **2.25** In the preceding problem, use calculus to find the angle θ that will give maximum range D to the water.

* **2.26** Find the range R for a projectile fired onto the inclined plane shown in Figure P2.26. Determine the maximum value of R for a given muzzle velocity u. (Angle α = constant.)

2.27 If a baseball player can throw a ball 90 m on the fly on earth, how far can he throw it on the moon where the gravitational acceleration is about one-sixth that on earth? Neglect the height of the player and the air resistance on earth.

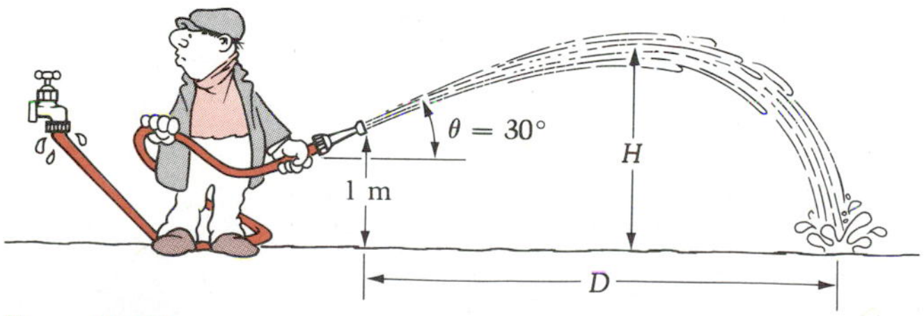

Figure P2.24

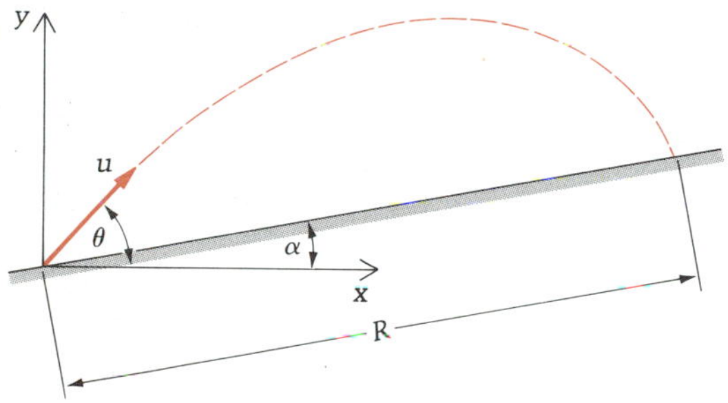

Figure P2.26

2.28 A child drops a rock into a well and hears it splash into the water at the bottom exactly 2 sec later. (See Figure P2.28.) If she is at a location where the speed of sound is v_s = 1100 ft/sec, determine the depth of the well with and without considering v_s. Compare the two results.

Figure P2.28

2.29 At liftoff the space shuttle is powered upward by two solid rocket boosters of 12.9×10^6 N each and by the three Orbiter main liquid-rocket engines with thrusts of 1.67×10^6 N each. At liftoff, the total weight of the shuttle (orbiter, tanks, payload, boosters) is about 19.8×10^6 N. Determine the acceleration experienced by the crew members at liftoff. (This differs from the initial acceleration on earlier manned flights; demonstrate this by comparing with the Apollo moon rocket, which weighed 6.26×10^6 lb at liftoff and was powered by five engines each with a thrust of 1.5×10^6 lb.) Neglect the change in mass between ignition and liftoff.

2.30 What is the apparent weight, as perceived through pressure on the feet, of a 200-lb passenger in an elevator accelerating at the rate of 10 ft / sec^2 upward (a) or downward (b)?

2.31 When a man stands on a scale at one of the poles of the earth, the scale indicates weight W. Assuming the earth to be spherical (4000-mile radius) and assuming the earth to be an inertial frame, what will the scale read when the man stands on it at the equator?

2.32 Assuming the earth's orbit around the sun to be circular and supposing that a frame containing the earth's center and poles and the center of the sun is inertial, repeat Problem 2.31. Neglect the earth's tilt.

2.33 In an emergency the driver of an automobile applies the brakes and locks all four wheels. Find the time and distance required to bring the car to rest in terms of the coefficient of sliding friction μ, the initial speed v, and the gravitational acceleration g.

2.34 A box is placed in the rear of a pickup truck. Find the maximum acceleration of the truck for which the block does not slide on the truck bed. The coefficient of friction between the box and truck bed is μ.

2.35 The truck in Figure P2.35 is traveling at 45 mph. Find the minimum stopping distance such that the 250-lb crate will not slide. Assume the crate cannot tip over.

Figure P2.35

2.36 The 200-lb block is at rest on the floor ($\mu = 0.2$) before the 50-lb force is applied as shown in Figure P2.36. What is the acceleration of the block immediately after application of the force? Assume the block is wide enough that it cannot tip over.

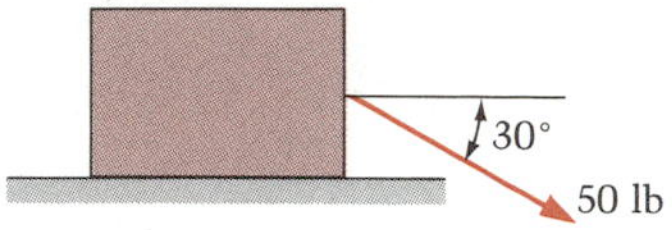

Figure P2.36

2.37 Repeat Problem 2.36 with $\mu = 0.1$.

2.38 Repeat Problem 2.36 for the case in which the 50-lb force is applied as shown in Figure P2.38.

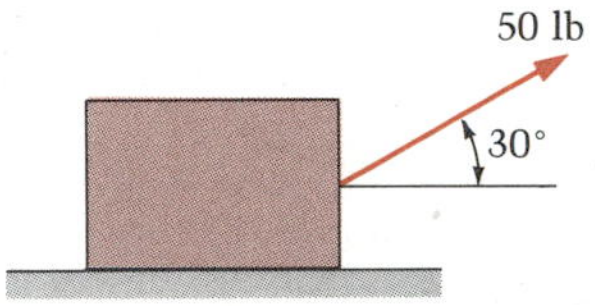

Figure P2.38

2.39 The two blocks in Figure P2.39 are at rest before the 100-Newton force is applied. If friction between $\mathcal{B}$ and the floor is negligible and if $\mu = 0.4$ between $\mathcal{A}$ and $\mathcal{B}$, find the magnitude and direction of the subsequent friction force exerted *on* $\mathcal{A}$ *by* $\mathcal{B}$.

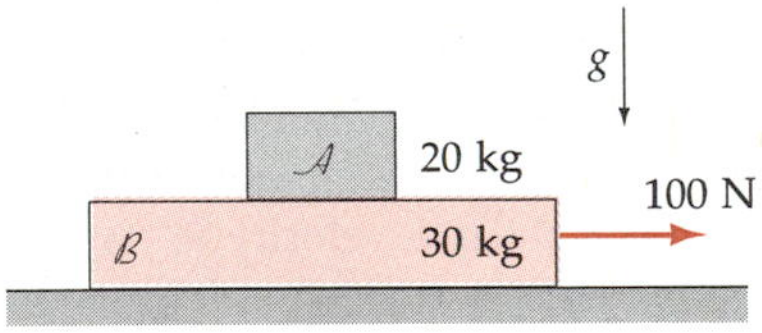

Figure P2.39

2.40 Find the largest force P for which $\mathcal{A}$ in Figure P2.40 will not slide on $\mathcal{B}$.

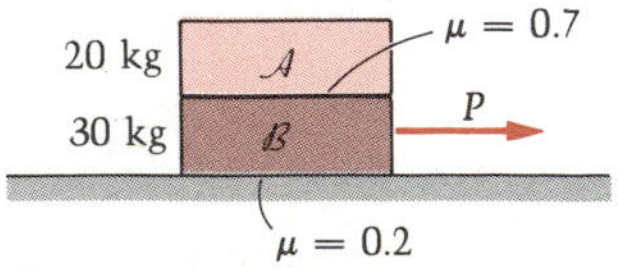

Figure P2.40

2.41 Work the preceding problem if P is applied to $\mathcal{A}$ instead of $\mathcal{B}$.

2.42 The blocks in Figure P2.42 are in contact as they slide down the inclined plane. The masses of the blocks are $m_{\mathcal{B}} = 25$ kg and $m_{\mathcal{A}} = 20$ kg, and the friction coefficients between the blocks and the plane are 0.5 for $\mathcal{A}$ and 0.1 for $\mathcal{B}$. Determine the force between the blocks and find their common acceleration.

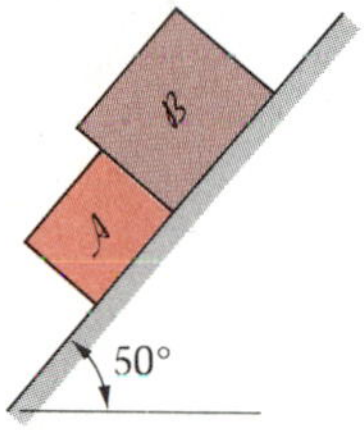

Figure P2.42

2.43 In the preceding problem, let μ be the coefficient of friction between $\mathcal{A}$ and the plane. Using the two motion equations of the blocks, find the range of values of μ for which the blocks will separate when released from rest.

2.44 If all surfaces are smooth for the setup of blocks and planes in Figure P2.44, find the force P that will give block $\mathcal{B}$ an acceleration of 4 ft/sec² up the incline.

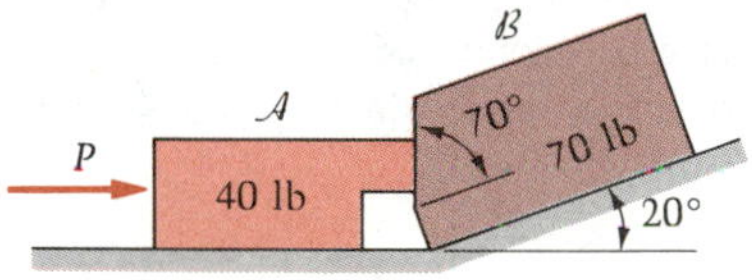

Figure P2.44

2.45 Work the preceding problem if the planes are still smooth but the friction coefficient between $\mathcal{A}$ and $\mathcal{B}$ is $\mu_s \approx \mu_k = 0.3$.

2.46 Work the preceding problem if the coefficient of friction is 0.3 for *all* contacting surfaces.

* **2.47** Generalizing Example 2.8, let the blocks, friction coefficients, and angle of the plane be as shown in Figure P2.47. Show that:

a. If $\tan\varphi > \mu_2$, motion will occur, and if so:

b. If $\mu_2 \le \mu_1$, the blocks move together

c. If $\mu_2 > \mu_1$, then $\mathcal{B}_1$ slides on $\mathcal{B}_2$. In this case, the lower block does not move if

$$\tan\varphi \le \mu_2 + (\mu_2 - \mu_1)\frac{m_1}{m_2}$$

d. If $\tan\varphi \le \mu_2$, then the lower block will not move. In this case, the upper block slides on it if and only if $\tan\varphi > \mu_1$.

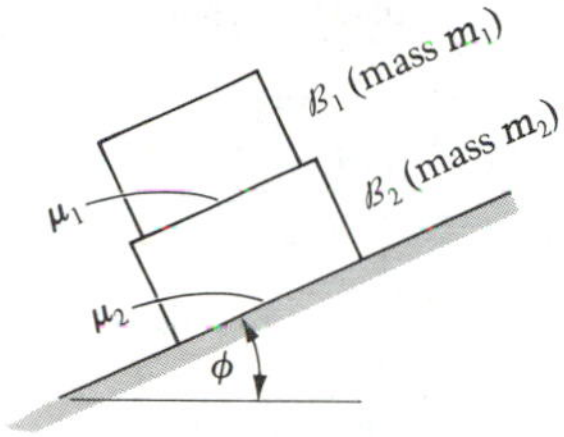

Figure P2.47

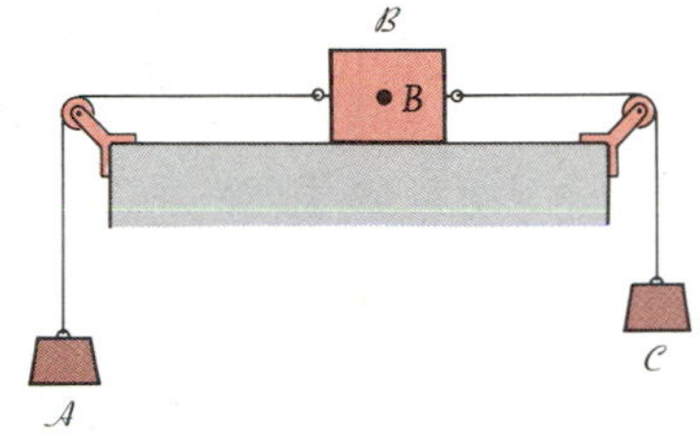

Figure P2.48

2.48 In Figure P2.48 the masses of $\mathcal{A}$, $\mathcal{B}$, and $\mathcal{C}$ are 10, 60, and 50 kg, respectively. The coefficient of friction between $\mathcal{B}$ and the plane is $\mu = 0.35$, and the pulleys have negligible mass and friction. Find the tensions in each cord, and the acceleration of B, upon release from rest.

2.49 If the system in Figure P2.49 is released from rest, how long does it take the 5-lb block to drop 2 ft? Neglect friction in the light pulley and assume the cord connecting the blocks to be inextensible.

2.50 The coefficient of friction $\mu = 0.1$ is the same between $\mathcal{A}$ and $\mathcal{B}$ as it is between $\mathcal{B}$ and the plane. (See Figure P2.50.) Find the tension in the cord at the instant the system is released from rest. Neglect friction in the light pulley.

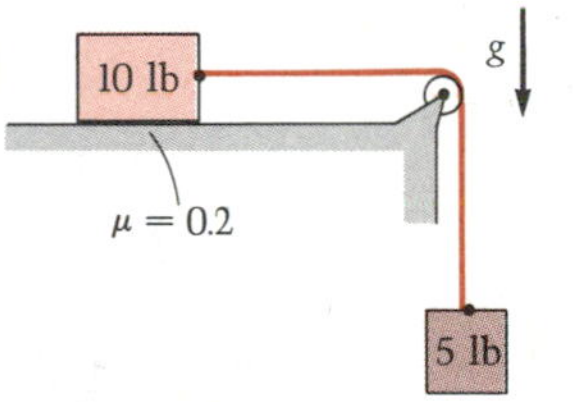

Figure P2.49

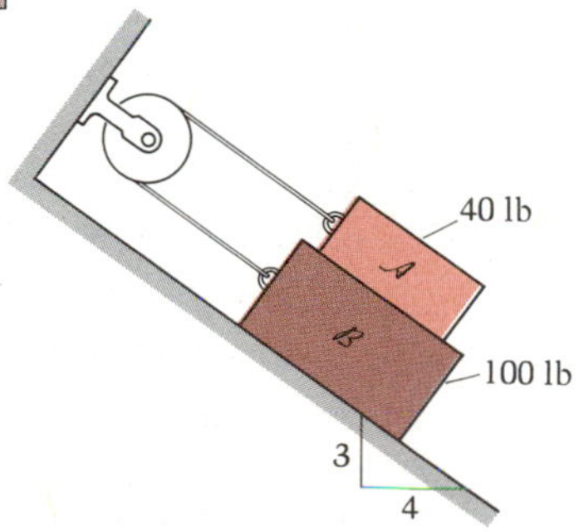

Figure P2.50

2.51 The system in Figure P2.51 is released from rest.

a. How far does block A move in 2 sec?

b. How would the solution be changed if the coefficient of friction between the floor and A were $\mu = 0.2$?

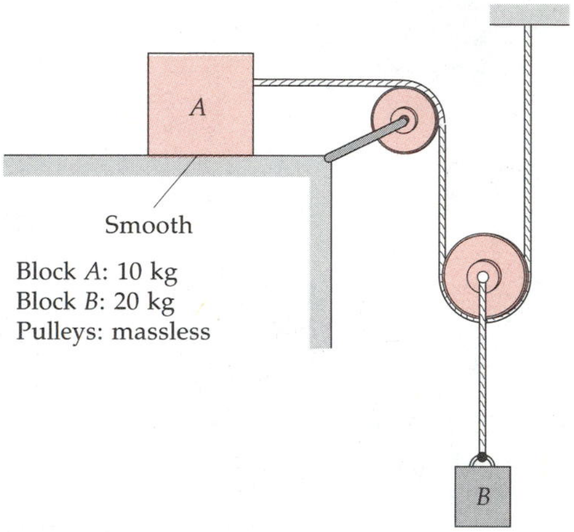

Figure P2.51

2.52 A child notices that sometimes the ball $\mathcal{M}$ does not slide down the inclined surface of toy $\mathcal{T}$ when she pushes it along the floor. (See Figure P2.52.) What is the minimum acceleration $a_{\min}$ of $\mathcal{T}$ to prevent this motion? Assume all surfaces are smooth.

Figure P2.52

2.53 In the preceding problem, suppose the acceleration of $\mathcal{T}$ is $2a_{\min}$. What is the normal force between the vertical surface of $\mathcal{T}$ and the ball? The ball's weight is 0.06 lb.

2.54 Let the mass of $\mathcal{B}$ in Problem 1.57 be 20 kg. What then must be the mass of $\mathcal{A}$ to produce the prescribed motion? Neglect the masses of the pulleys.

* **2.55** Find the tension in the cord in Problem 1.61 at the onset of the motion if the mass of $\mathcal{A}$ is 10 kg.

2.56 For the cam-follower system of Problem 1.65 find the force that must be applied to the cam to produce the motion. Let the masses of cam and follower be m_1 and m_2 and neglect friction.

2.57 In Figure P2.57 the masses of blocks $\mathcal{A}$, $\mathcal{B}$, and $\mathcal{C}$ are 50, 20, and 30 kg, respectively. Find the accelerations of each if the table is removed. Which block will hit the floor first? How long will it take?

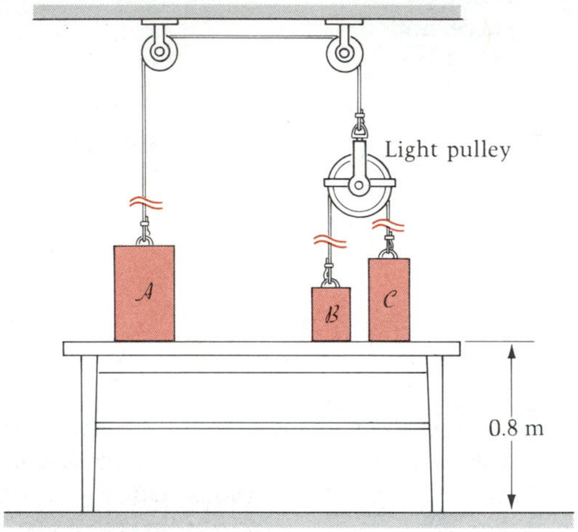

Figure P2.57

2.58 Body $\mathcal{A}$ in Figure P2.58 weighs 223 N and body $\mathcal{B}$ weighs 133 N. Neglect the weight of the rigid member connecting $\mathcal{A}$ and $\mathcal{B}$. The coefficient of friction is 0.3 between all surfaces. Determine the accelerations of $\mathcal{A}$ and $\mathcal{B}$ just after the cord is cut.

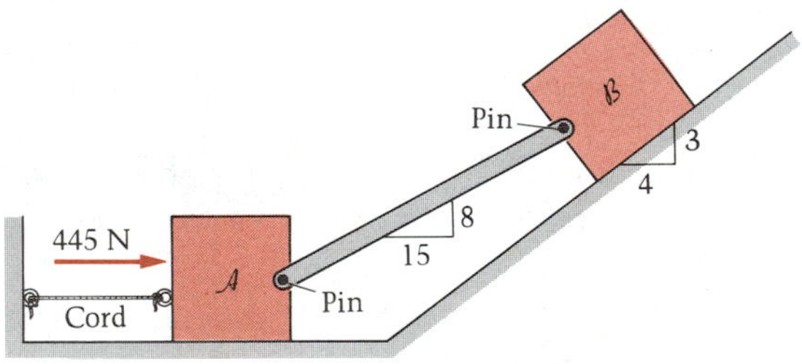

Figure P2.58

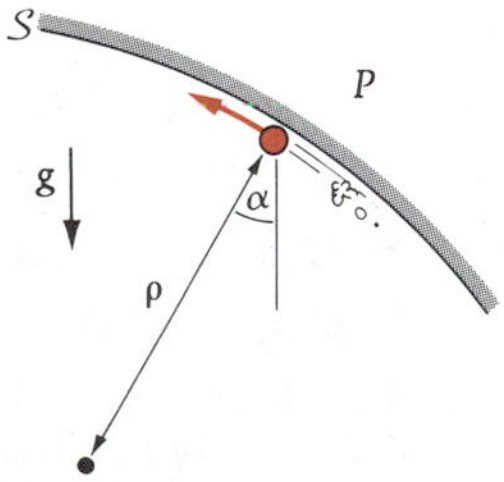

Figure P2.59

2.59 A particle P moves along a curved surface $\mathcal{S}$ as shown in Figure P2.59. Show that P will remain in contact with $\mathcal{S}$ provided that, at all times, $v \geq \sqrt{\rho g \cos \alpha}$.

2.60 Find the condition for retention of contact if P moves along the *outside* of a surface defined by the same curve as in the preceding problem. (See Figure P2.60.)

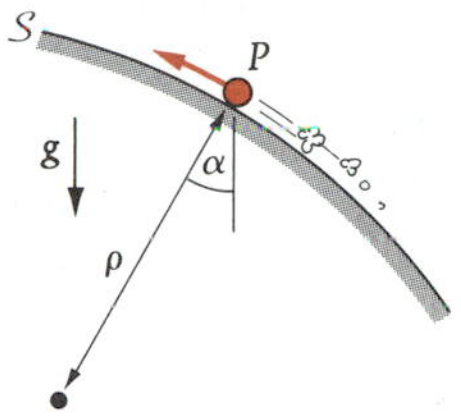

Figure P2.60

2.61 A ball of mass m on a string is swung at constant speed v_0 in a horizontal circle of radius R by a child. (See Figure P2.61.)

- **a.** What holds up the ball?
- **b.** What is the tension in the string?
- **c.** If the child increases the speed of the ball, what provides the force in the forward direction needed to produce the $\ddot{s}$? Explain.

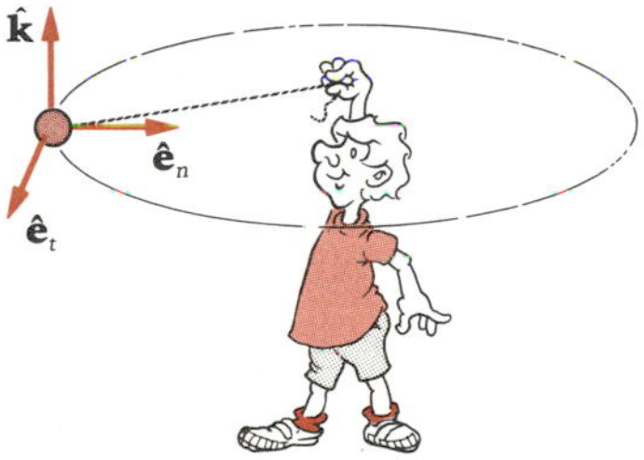

Figure P2.61

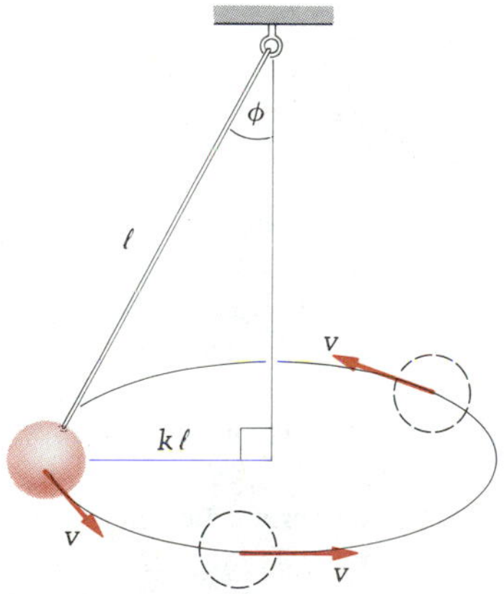

Figure P2.62

2.62 There is a speed, called the *conical speed*, at which a ball on a string, in the absence of all friction, moves on a specific horizontal circle (with the string sweeping out a conical surface) with no radial or vertical component of velocity (Figure P2.62). If ℓ is the length of the string and $k\ell$ is the radius of the circle on which the ball moves, find the conical speed in terms of k, ℓ, and the acceleration of gravity.

2.63 For an object at rest on the earth's surface, we can write $mg = (GMm)/R_e^2$, so that the unwieldy constant GM may be replaced by gR_e^2, which for the earth is approximately $32.2[3960(5280)]^2$ ft^3/sec^2. Use this, plus the result of Example 2.5, to solve for the distance above the earth of a satellite in a circular, 90-minute orbit. (Let the satellite replace the planet, and the earth replace the star, in the example.)

2.64 Communications satellites are placed in *geosynchronous orbit*, an orbit in which the satellites are always located in the same position in the sky (Figure P2.64).

- **a.** Give an argument why this orbit must lie in the equatorial plane. Why must it be circular?
- **b.** If the satellites are to remain in orbit without expending energy, find the important ratio of the orbit radius r_s to the earth's radius r_e. *Hint*: Use Newton's law of universal gravitation

$$F = \frac{Gm_s m_e}{r_s^2}$$

together with the law of motion in the radial direction, and note that if the satellite were sitting on the earth's surface, the force would be

$$F = m_s g = \frac{Gm_s m_e}{r_e^2}$$

so that the product Gm_e may be rewritten as gr_e^2, as in Problem 2.63. Use $r_e = 3960$ mi.

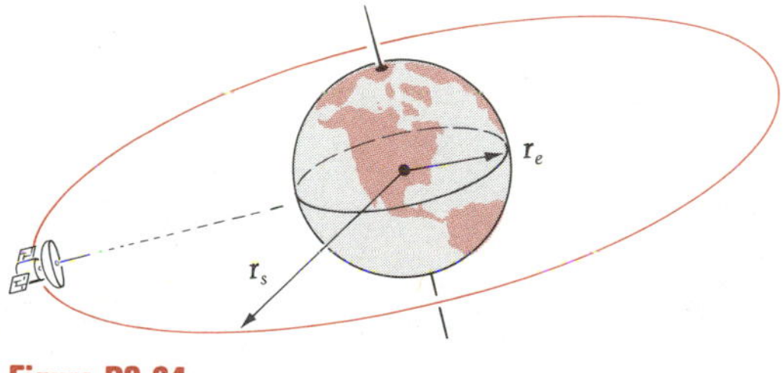

Figure P2.64

2.65 Using the result of the preceding problem, show that a minimum of three satellites in geosynchronous orbit are required for continuous communications coverage over the whole earth except for small regions near the poles.

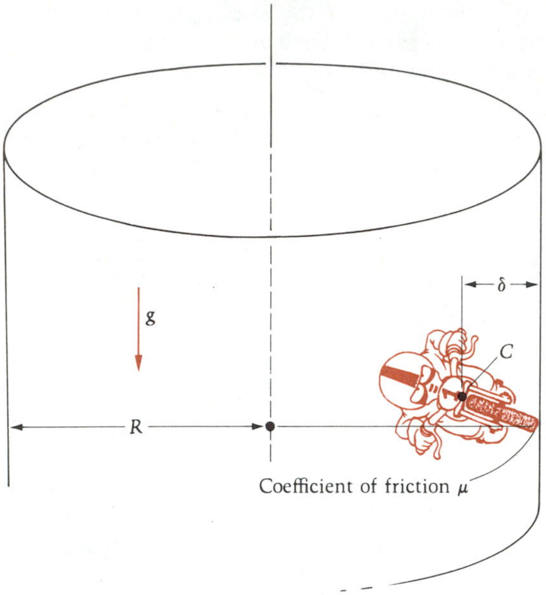

Figure P2.66

2.66 In terms of the parameters δ, R, μ, and g defined in Figure P2.66, find the minimum speed for which the motorcycle will not slip down the inside wall of the cylinder.

2.67 In the "spindle top" ride in an amusement park, people stand against a cylindrical wall and the cylinder is then spun up to a certain angular velocity ω_0. (See Figure P2.67.) The floor is then lowered, but the people remain against the wall at the same level. Use the equation $\Sigma F_n = ma_{C_n} = mv^2/R$ to explain the phenomenon. Noting that each person is "in equilibrium vertically," solve for the minimum ω_0 to prevent people from slipping if $R = 2$ m and the expected friction coefficient between the rough wall and the clothing is $\mu = 0.5$.

Figure P2.67

2.68 In preparation for Problem 2.69, for the ellipse shown in Figure P2.68, the equation is

$$\frac{x^2}{a^2} + \frac{y^2}{b^2} = 1$$

Show that the radius of curvature ρ of the ellipse, as a function of x, is

$$\rho = \frac{[a^2(a^2 - x^2) + b^2x^2]^{3/2}}{a^4 b}$$

Hint: Recall from calculus that if $y = y(x)$, then

$$\rho = \left| \frac{(1 + y'^2)^{3/2}}{y''} \right|$$

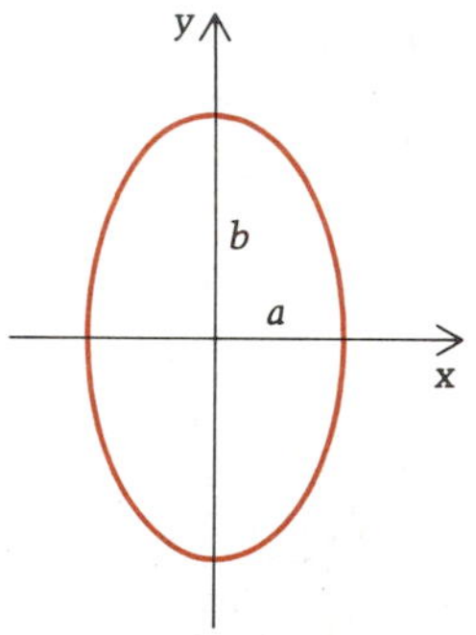

Figure P2.68

2.69 In a certain amusement park, the tallest loop in a somersaulting ride (Figure P2.69) is 100 ft high and shaped approximately like an ellipse with a width of 95 ft. The ride advertises "five times the earth's pull at over 50 mph." Use the result of the preceding exercise to compute the radius of curvature at the bottom of the loop. Assuming that the normal force resultant is 5 mg, determine whether or not the maximum speed is over 50 mph. Treat the cars as a single particle.

2.70 If bar $\mathcal{B}$ shown in Figure P2.70 were raised *slowly*, block $\mathcal{A}$ would start to slide at the angle $\theta = \tan^{-1}(\mu)$, which was seen in statics to be one way of determining the friction coefficient μ. Suppose now that the bar is suddenly rotated, starting from the position $\theta = 0$, at constant angular velocity ω_0 ↺. For $\mu = 0.5$ and $r\omega_0^2 = 0.1g$, compute the angle θ at which $\mathcal{A}$ slips downward on $\mathcal{B}_1$ and compare the result with $\tan^{-1}\mu = \tan^{-1}(0.5)$.

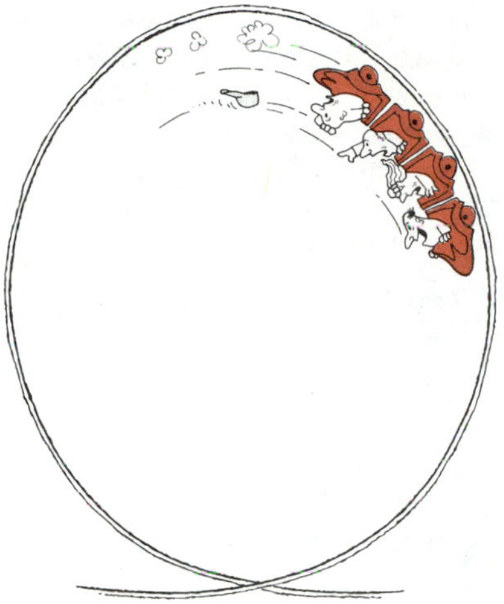

Figure P2.69

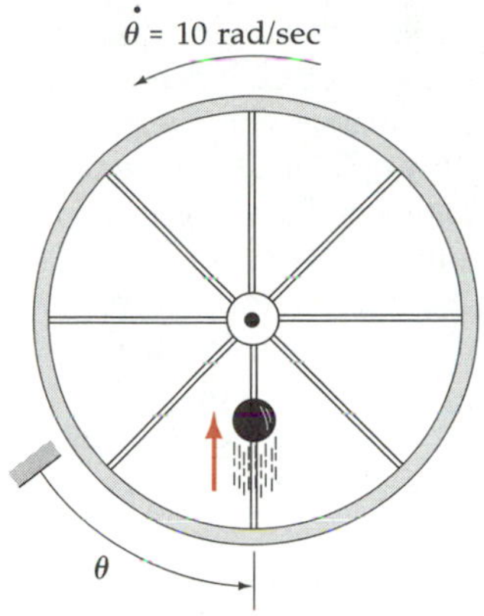

Figure P2.72

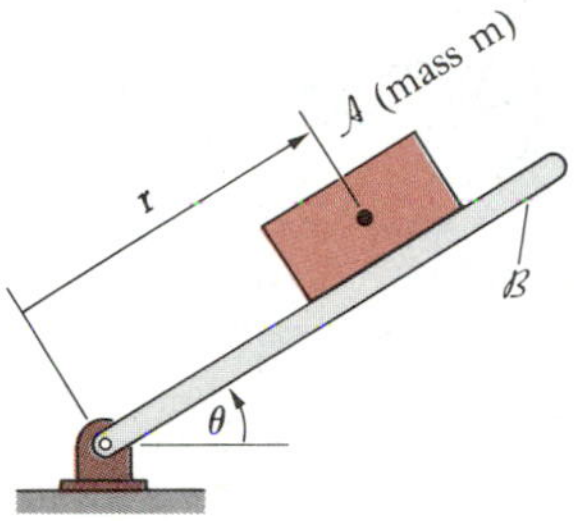

Figure P2.70

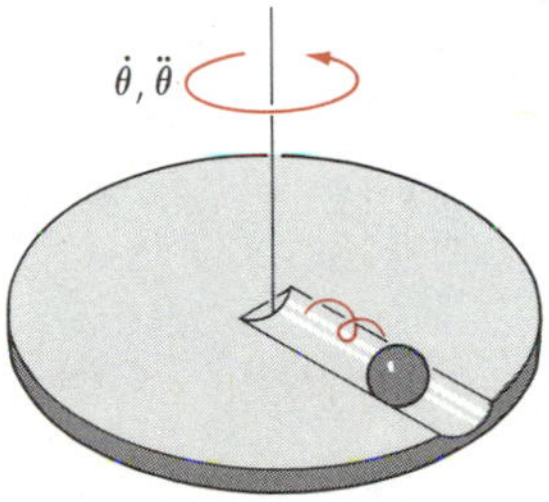

Figure P2.73

2.71 In the preceding problem, let μ remain at 0.5 but consider increasing the parameter $r\omega_0^2/g$. At what value of this parameter will $\mathcal{A}$ slide *outward* on $\mathcal{B}$? At what angle will this occur?

2.72 A horizontal wheel is rotating about its fixed axis at a rate of 10 rad/sec, and this angular speed is increasing at the given time at $\ddot{\theta} = 5$ rad/sec². (See Figure 2.72.) At this same instant, a bead is sliding inward relative to the spoke on which it moves at 5 ft/sec; this speed is slowing down at this time at 2 ft/sec². If the bead weighs 0.02 lb and is 1 ft from the center in the given configuration, find the external force exerted on the bead. Is it possible that this force can be exerted solely by the spoke and not in part by other external sources?

2.73 A ball bearing is moving radially outward in a slotted horizontal disk that is rotating about the vertical z axis. At the instant shown in Figure P2.73, the ball bearing is 3 in. from the center of the disk. It is traveling radially outward at a velocity of 4 in./sec relative to the disk. If $\dot{\theta} = 2$ rad/sec and is constant, find $\ddot{r}$ and the force exerted on the ball by the disk at this instant. Assume no friction and take the weight of the ball to be 0.05 lb.

2.74 A bead slides down a smooth circular hoop that, at a certain instant, has $\dot{\varphi} = 2$ rad/sec and $\ddot{\varphi} = 3$ rad/sec² in the direction shown in Figure P2.74. The angular speed of line OP at this time is $\dot{\theta} = 0.5$ rad/sec and $\theta = 135°$. Find the value of $\ddot{\theta}$ and the force exerted on the bead by the hoop at the given instant, if the mass of the bead is 0.1 kg and the radius of the hoop is 20 cm. *Hint*: Use spherical coordinates.

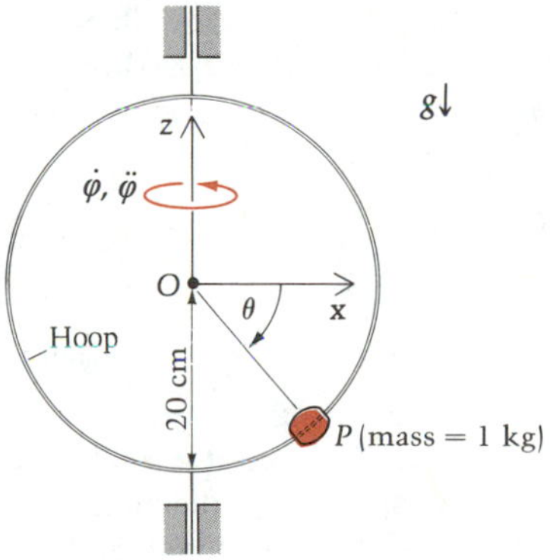

Figure P2.74

* **2.75** The four light rods are pinned at the origin and at each mass in such a way that as these seven bodies are spun up about the vertical, the masses m move outward and the mass M slides smoothly up along the vertical rod Oy. There is a relationship between ϕ, ω_0, ℓ, g, m, and M such that at the particular spin-speed ω_0, the bodies behave as one rigid body (meaning ϕ remains constant). Find the relationship. *Hint*: Use separate free-body diagrams of m and M, and write equations of motion for each. The unknowns are F_T (force in each top rod) and F_B (force in each bottom rod), ω_0 and ϕ. There will be three useful equations. See Figure P2.75.

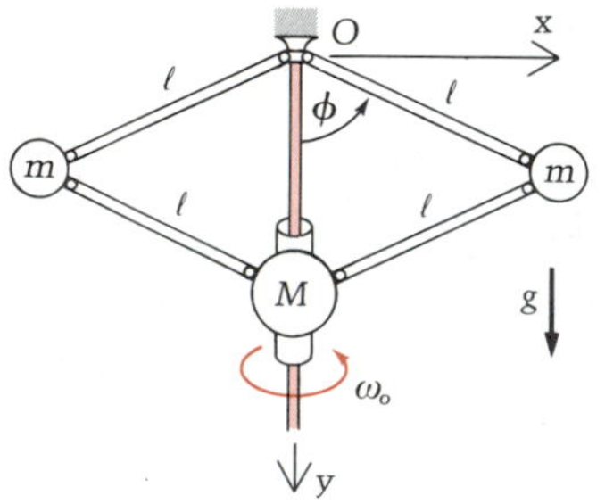

Figure P2.75

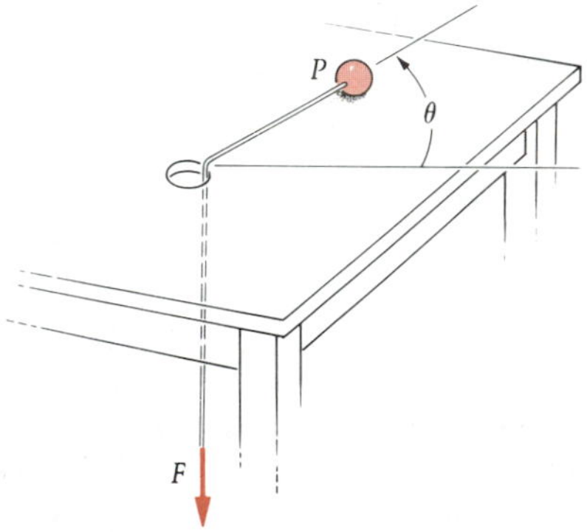

Figure P2.76

2.76 A particle P of mass m moves on a smooth, horizontal table and is attached to a light, inextensible cord that is being pulled downward by a force $F(t)$ as shown in Figure P2.76. Show that the differential equations of motion of P are

$$-F = m(\ddot{r} - r\dot{\theta}^2) \qquad (1)$$

$$0 = r\ddot{\theta} + 2\dot{r}\dot{\theta} \qquad (2)$$

Then show that Equation (2) implies that $r^2\dot{\theta} =$ constant.

2.77 In the preceding problem, let the particle be at $r = r_0$ at $t = 0$, and let the part of the cord beneath the table be descending at constant speed v_C. If the transverse component of velocity of P is $r\dot{\theta} = r_0\dot{\theta}_0$ at $t = 0$, find the tension in the cord as a function of time t.

2.78 A wintertime fisherman of mass 70 kg is in trouble—*he* is being reeled in by Jaws on a lake of frozen ice. At the instant shown in Figure P2.78, the man has a velocity component, perpendicular to the radius r, of $v_{1_\perp} = 0.3$ m/s at an instant when $r = R_1 = 5$ m. If Jaws pulls in the line with a force of 100 N, find the value of $v_{2_\perp}$ when the radius is $R_2 = 1$ m. *Hint*:

$$r\ddot{\theta} + 2\dot{r}\dot{\theta} = \frac{(d/dt)(r^2\dot{\theta})}{r} \quad \text{and} \quad \Sigma F_\theta = ?$$

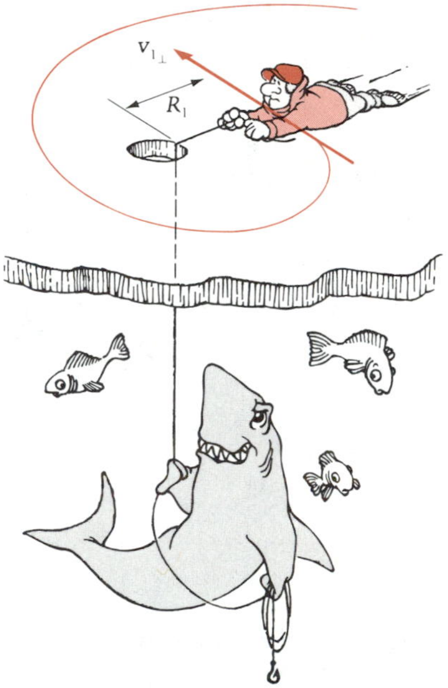

Figure P2.78

2.79 In the preceding problem, show that the differential equation of the man's radial motion is $\ddot{r} = (2.25/r^3) - (10/7)$. Use $\dot{r}\ddot{r} = d/dt\,(\dot{r}^2/2)$ to integrate this, and *if* $\dot{r} = 0$ when $r = 5$ m, show that the radial component ($\dot{r}$) of the man's velocity when $r = 1$ m is 3.04 m/s.

2.80 Particle P of mass m travels in a circle of radius a on the smooth table shown in Figure P2.80. Particle P is connected by an inextensible string to the stationary particle of mass M. Find the period of one revolution of P.

* **2.81** A weight of 100 lb hangs freely from a light rope (Figure P2.81). It is pulled up by a force that is 150 lb at $t = 0$ but diminishes uniformly in magnitude at 1 lb per foot pulled up. Find the time required to pull the weight up to the platform from rest, and determine its velocity upon reaching the top.

* **2.82** Rework Problem 2.81, but this time assume that the force *increases* by 1 lb per foot pulled up.

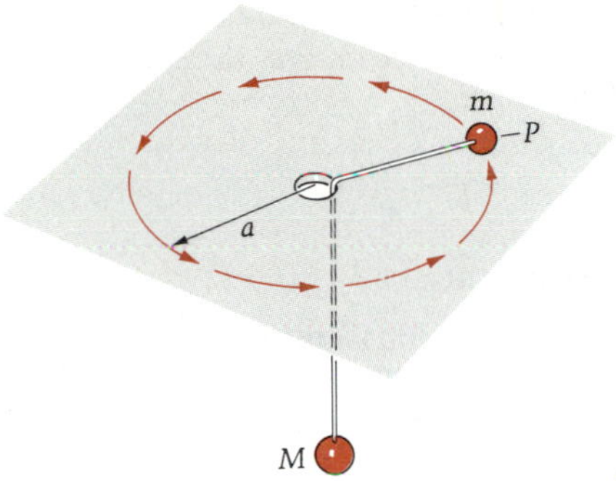

Figure P2.80

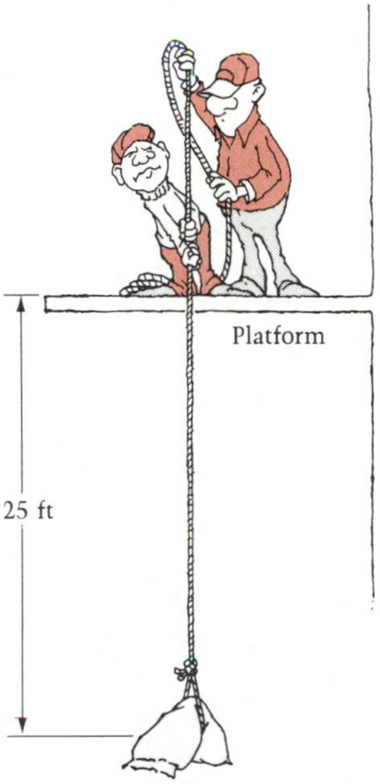

Figure P2.81

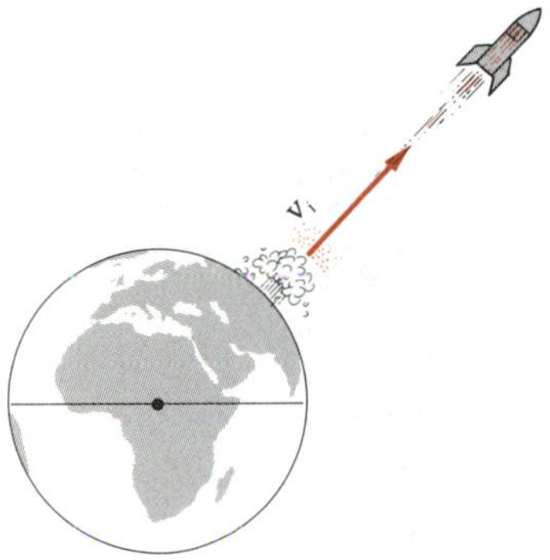

Figure P2.83

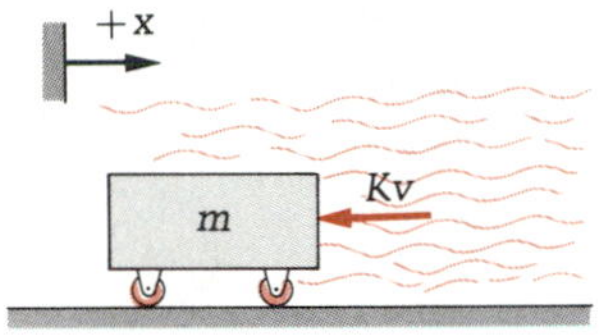

Figure P2.84

2.83 The acceleration of gravity varies with distance z above the earth's surface as

$$\ddot{z} = \frac{-gR^2}{(R+z)^2}$$

where g is the acceleration of gravity on the surface and R is the earth's radius. Find the minimum firing velocity v_i that a projectile must have in order to escape the earth if fired straight up (Figure P2.83). *Hint*: Not to return to earth requires the condition that $v \to 0$ as z gets large for the *minimum* possible v_i.

2.84 The mass m shown in Figure P2.84 is given an initial velocity of v_0 in the x direction. It moves in a medium that resists its motion with force proportional to its velocity, with proportionality constant K. By solving for $v(x)$, determine how far the mass travels before stopping. Then solve in a different manner for $v(t)$ if $v = v_0$ when $t = 0$.

2.85 Using the result of the preceding problem and expressing v as dx/dt, solve for $x(t)$ if $x = 0$ when $t = 0$.

2.86 The identical plastic scottie dogs shown in Figure P2.86 are glued onto magnets and attract each other with a force $F = K/(2x)^2$, where K is a constant related to the strength of the magnets. Find the speeds at which the dogs collide if the magnets are initially separated by the distance S.

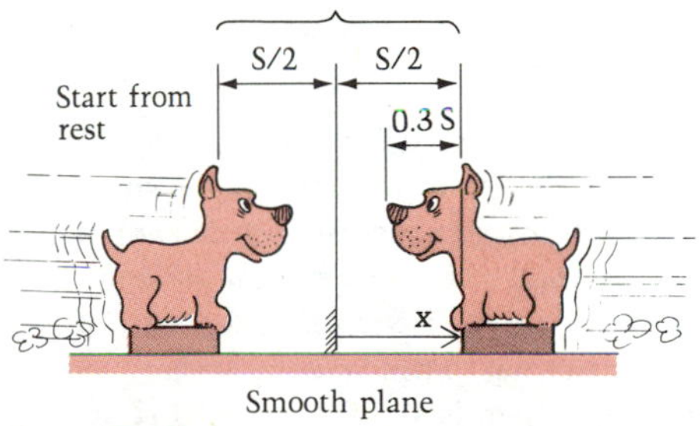

Figure P2.86

2.87 A ball is dropped from the top of a tall building. The motion is resisted by the air, which exerts a drag force given by Dv^2; D is a constant and v is the speed of the ball. Find the *terminal speed* (the limiting speed of fall) if there is no limit on the drop height. What is the drop height for which the ball will strike the ground at 95 percent of the terminal speed?

Figure P2.88

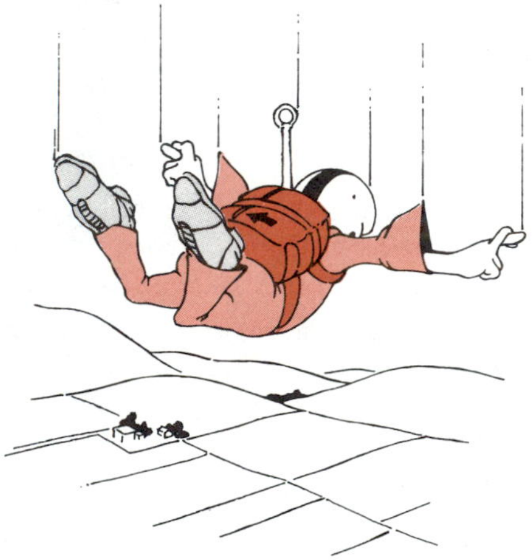

Figure P2.90

* **2.88** Over a certain range of velocities, the effect of air resistance on a projectile is proportional to the square of the object's speed. If the object can be regarded as a particle, the *drag force* is expressible as $F_D = \frac{1}{2}\rho A C_D v^2$, in which ρ is the density of the air, A is the projected area of the object onto a plane normal to the velocity vector, and C_D is a coefficient that depends on the object's shape. If $\rho A C_D = 0.0004$ lb-sec^2/ft^2 for the 76-lb cannonball of Figure P2.88, find the maximum height it reaches. Compare your result with the answer neglecting air resistance.

* **2.89** In the preceding problem, find the velocity of the cannonball just before it hits the ground; again compare with the case of no air resistance.

* **2.90** A 160-lb parachutist in the "free-fall spread-stable position" (Figure P2.90) reaches a velocity of 174 ft/sec in 12 sec after exiting a stationary blimp. Assuming velocity-squared air resistance, solve the differential equation of motion

$$\ddot{y} = g - \frac{k}{m}\dot{y}^2$$

and determine the constant k.

* **2.91** In the preceding problem, suppose the parachutist opens his chute at a height of 1000 ft. If the value of k then becomes 0.63 lb/(ft/sec)^2, find the velocity at which the parachutist strikes the ground, if $v_i = 174$ ft/sec.

2.92 The drag car of mass m shown in Figure P2.92, traveling at speed v_0, is to be initially slowed primarily by the deployment of a parachute. The parachute exerts a force F_d proportional to the square of the velocity of the car, $F_d = Cv^2$. Neglecting friction and the inertia of the wheels, determine the distance traveled by the car before its velocity is 40 percent of v_0. If the car and driver weigh 1000 lb and $C = 0.182$ lb-sec^2/ft^2, find the distance in feet.

Figure P2.92

2.93 In the preceding problem, suppose the drag car's speed at parachute release is 237 mph. Find the time it takes to reach 40 percent speed.

* **2.94** A 50-lb shell is fired from the cannon shown in Figure P2.94. The pressure of the expanding gases is inversely proportional to the volume behind the shell. Initially this pressure is 10 tons per square inch; just before exit, it is one-tenth this value. Find the exit velocity of the shell.

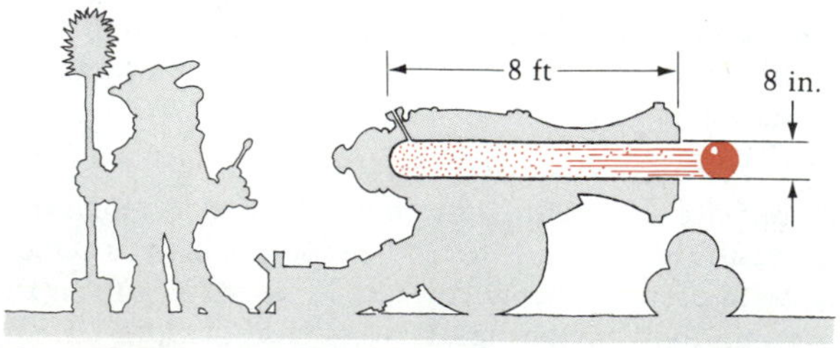

Figure P2.94

2.95 The block of mass m shown in Figure P2.95 is brought slowly down to the point of contact with the end of the spring, and then (at $t = 0$) the block is released. Write the differential equation governing the subsequent motion, clearly defining your choice of displacement parameter. What are the initial conditions? Find the maximum force induced in the spring and the first time at which it occurs.

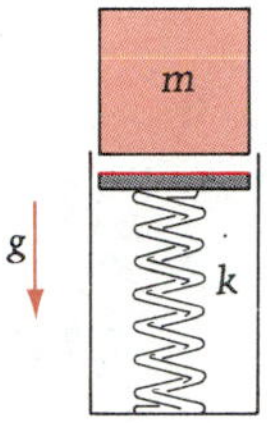

Figure P2.95

2.96 Show that for a particle P moving in a viscous medium in which the air resistance is proportional to velocity (Figure P2.96), the differential equations of motion are

$$m\ddot{x} = -k\dot{x}$$

$$m\ddot{y} = -k\dot{y} - mg$$

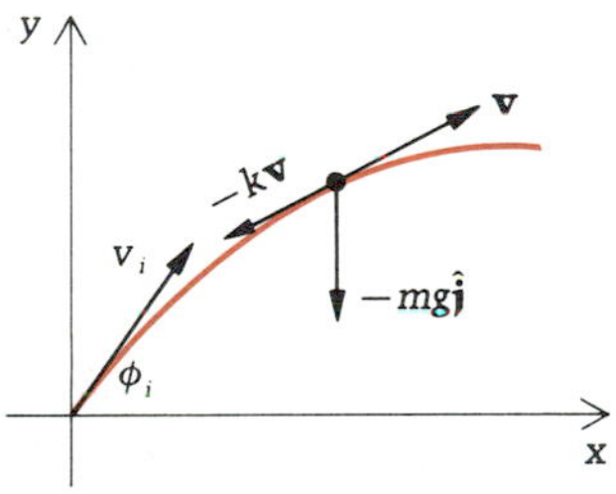

Figure P2.96

2.97 In the preceding problem, show by integration that the components of velocity of P are given by

$$\dot{x} = v_i \cos\phi_i\, e^{-kt/m}$$

$$\dot{y} = \frac{-mg}{k} + \left(v_i \sin\phi_i + \frac{mg}{k}\right) e^{-kt/m}$$

2.98 Continue the preceding exercise and show that the particle's position is given by the equations

$$x = \frac{mv_i \cos\phi_i}{k}(1 - e^{-kt/m})$$

$$y = \frac{m}{k}\left[\left(\frac{mg}{k} + v_i \sin\phi_i\right)(1 - e^{-kt/m}) - gt\right]$$

Show further that both x and $\dot{y}$ approach asymptotes as $t \to \infty$. (The limiting value of $\dot{y}$ is known as the terminal velocity of P, after which the weight is balanced by the viscous resistance so that the acceleration goes to zero.)

* **2.99** A particle moves on the inside of a fixed, smooth vertical hoop of radius a. It is projected from the lowest point A with velocity $\sqrt{7ga/2}$. Show that it will leave the hoop at a height $3a/2$ above A and meet the hoop again at A.

* **2.100** The two particles in Figure P2.100 are at rest on a smooth horizontal table and connected by an inextensible string that passes through a small, smooth ring fixed to the table. The lighter particle (mass m) is then projected at right angles to the string with velocity v_0. Prove that the other particle will strike the ring with velocity $v_0\sqrt{3}/(2\sqrt{n}+1)$. *Hint*: Use polar coordinates and note that $r^2\dot{\theta}$ is constant for each particle.

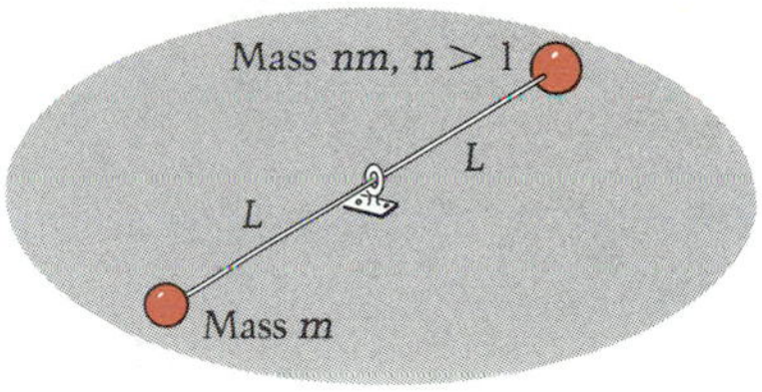

Figure P2.100

2.4 Work and Kinetic Energy for Particles

In Example 2.4 we were able to get useful information from an energy integral of the governing differential equation. The same result may be obtained in general by an integration of

$$\Sigma\mathbf{F} = m\mathbf{a}_C$$

Forming the dot product of each side with the velocity $\mathbf{v}_C$ of the mass center, we have

$$\begin{aligned}(\Sigma\mathbf{F}) \cdot \mathbf{v}_C &= m\mathbf{a}_C \cdot \mathbf{v}_C \\ &= m\frac{d\mathbf{v}_C}{dt} \cdot \mathbf{v}_C \\ &= \frac{m}{2}\frac{d}{dt}(\mathbf{v}_C \cdot \mathbf{v}_C) \\ &= \frac{d}{dt}\left(\frac{m}{2}\mathbf{v}_C \cdot \mathbf{v}_C\right) \\ &= \frac{d}{dt}\left(\frac{m}{2}|\mathbf{v}_C|^2\right)\end{aligned}$$

Integrating,* we get

$$\int_{t_1}^{t_2} (\Sigma\mathbf{F}) \cdot \mathbf{v}_C\, dt = \frac{m}{2}\left[|\mathbf{v}_C(t_2)|^2 - |\mathbf{v}_C(t_1)|^2\right] \tag{2.11}$$

or, for a particle,

$$\int_{t_1}^{t_2} (\Sigma\mathbf{F}) \cdot \mathbf{v}\, dt = \frac{m}{2}\left[|\mathbf{v}(t_2)|^2 - |\mathbf{v}(t_1)|^2\right] \tag{2.12}$$

Work and Kinetic Energy for a Particle

For a particle, $\int_{t_1}^{t_2}(\Sigma\mathbf{F}) \cdot \mathbf{v}\, dt$ is called the work done on the particle by the resultant of external forces.† We note that if there are N forces acting on the particle, then the resultant $\Sigma\mathbf{F}$ is given by $\mathbf{F}_1 + \mathbf{F}_2 + \cdots + \mathbf{F}_N$ and

$$\begin{aligned}(\Sigma\mathbf{F}) \cdot \mathbf{v} &= (\mathbf{F}_1 + \mathbf{F}_2 + \cdots + \mathbf{F}_N) \cdot \mathbf{v} \\ &= \mathbf{F}_1 \cdot \mathbf{v} + \mathbf{F}_2 \cdot \mathbf{v} + \cdots + \mathbf{F}_N \cdot \mathbf{v}\end{aligned}$$

Each term of this equation represents the rate of work of one of the forces. Thus the left side of Equation (2.12) may be read as the sum of the works of the individual forces acting on the particle. These statements are all consistent with the presentation to come in Chapter 5 in which we define the rate of work done by a force $\mathbf{F}$ to be $\mathbf{F} \cdot \mathbf{v}$, where $\mathbf{v}$ is the velocity of the point of the body at which the force is applied. The left side of Equation (2.11) may then be interpreted as the work that *would* be done by the external forces *were* each to have a line of action through the mass center.

For a *particle*, $(m/2)|\mathbf{v}|^2$ is called the kinetic energy, usually written T. Thus for the particle, Equation (2.12) is the work and kinetic energy principle:

Work done on the particle = Change in the particle's kinetic energy

* Sometimes (t_i, t_f), referring to "initial" and "final," are used instead of (t_1, t_2).

† Thus the appropriate unit of work and of energy in SI is the joule (J), the joule being 1 N · m; in U.S. units the ft-lb is the unit of work and energy. The N · m and lb-ft are usually reserved for the moment of a force. Note that work, energy, and moment of force all have the same dimension.

or

$$W = \Delta T \tag{2.13}$$

For a *body*, the kinetic energy is defined to be the sum of the kinetic energies of the particles constituting the body. If all the points in a body $\mathcal{B}$ have the same velocity (which is then $\mathbf{v}_C$), then $(m/2)|\mathbf{v}_C|^2$ is the *total* kinetic energy of $\mathcal{B}$. In general, however, the body is turning or deforming (or both) and this is not the case; the body then has *additional* kinetic energy due to its changes in orientation (that is, due to its angular motion) or due to the deformation. For a body $\mathcal{B}$, we shall also see in Chapter 5 that, in general, the left side of Equation (2.11) does not constitute the total work done on $\mathcal{B}$ by the external forces and couples. This is because, for a body, the forces do not have to be concurrent as they are for a particle. Equation (2.13) still turns out to be true for a rigid body, however, with the two sides of Equation (2.11) representing *parts* of W and ΔT.

Finally, with no restrictions on the size of the body, the energy integral (Equation 2.11) states that the work that would be done if the external forces acted at the mass center equals the change in what would be the kinetic energy if every point in the body had the velocity of the mass center. We could call this result the "mass center work and kinetic energy principle."

Work Done by a Constant Force

Before attempting to apply the work-energy principle to a specific problem, it is helpful to determine the work done by two classes of forces. First, suppose $\mathbf{F}$ is a constant force and suppose we let $\mathbf{r}$ be a position vector for the particle. Then

$$\begin{aligned}\int_{t_1}^{t_2} \mathbf{F} \cdot \mathbf{v}\, dt &= \mathbf{F} \cdot \int_{t_1}^{t_2} \mathbf{v}\, dt \\ &= \mathbf{F} \cdot \int_{t_1}^{t_2} \frac{d\mathbf{r}}{dt}\, dt \\ &= \mathbf{F} \cdot [\mathbf{r}(t_2) - \mathbf{r}(t_1)] \end{aligned} \tag{2.14}$$

which states that the work done is the dot product of the force with the displacement of the particle. We recall that this dot product can be expressed as the product of the force magnitude and the component of displacement in the direction of the force *or* as the product of the displacement magnitude and the component of force in the direction of the displacement.

Work Done by a Central Force

The second case to which we give special attention is that of a central force. Such a force is defined to have a line of action always passing through the same fixed point in the frame of reference and a magnitude

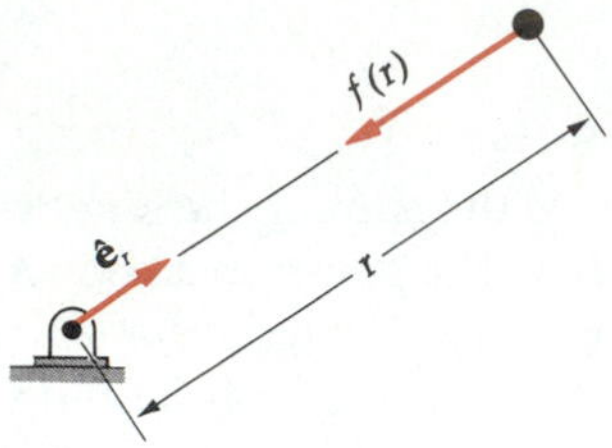

Figure 2.7

that depends only upon the distance r of the particle from that fixed point, as shown in Figure 2.7.

The velocity of the particle may be expressed as

$$\mathbf{v} = \frac{d}{dt}(r\hat{\mathbf{e}}_r) = \dot{r}\hat{\mathbf{e}}_r + r\dot{\hat{\mathbf{e}}}_r$$

$$\mathbf{F}\cdot\mathbf{v} = [-f(r)\hat{\mathbf{e}}_r]\cdot[\dot{r}\hat{\mathbf{e}}_r + r\dot{\hat{\mathbf{e}}}_r]$$
$$= -\dot{r}f(r)$$

since by Equation (1.38) we know that $\hat{\mathbf{e}}_r \cdot \dot{\hat{\mathbf{e}}}_r = 0$. Thus the work done by $\mathbf{F} = -f(r)\hat{\mathbf{e}}_r$ is

$$W = \int_{t_1}^{t_2} -f(r)\frac{dr}{dt}\,dt$$
$$= -\int_{r(t_1)}^{r(t_2)} f(r)\,dr \tag{2.15}$$

If φ is a function of r so that $f = d\varphi/dr$, then the work may be written as

$$W = -\int_{r(t_1)}^{r(t_2)} \frac{d\varphi}{dr}\,dr$$
$$= -\varphi[r(t_2)] + \varphi[r(t_1)] \tag{2.16}$$

Work Done by a Linear Spring

A special central force is that exerted by a spring on a particle when the other end of the spring is fixed. In the case of a linear spring of instantaneous length r, we note that $f = k(r - L_0)$, where L_0 is its natural, or unstretched, length and k is called the spring modulus or stiffness. In this case, $\varphi = (k/2)(r - L_0)^2$ or, more simply, $\varphi = (k/2)\delta^2$, where δ is the spring stretch. Thus by equation (2.16),

$$W = -\frac{k}{2}[\delta^2(t_2) - \delta^2(t_1)] \tag{2.17}$$

Question 2.4 What assumption about the mass of the spring is to be understood in the force-stretch relationship?

Answer 2.4 $(k)\cdot$(stretch) gives the equal-in-magnitude but opposite-in-direction force acting at the ends of a spring in equilibrium. If particles in the spring are accelerating, as is generally the case in dynamics problems, there is no simple force-stretch law. If the spring is very light, however, so that its mass may be neglected (compared to the masses of other bodies in the problem), $\int_{\text{spring}} \mathbf{a}\,dm \approx \mathbf{0}$ and the forces on the spring are instantaneously related just as if the spring were in equilibrium. The hidden assumption is that the mass of the spring may be neglected.

Work Done by Gravity

A second special case of a central force is the gravitational force exerted on a body by the earth. By Newton's law of universal gravitation,

$$f = \frac{mgr_e^2}{r^2} \tag{2.18}$$

where r_e is the radius of the earth, m is the mass of the attracted body, and g is the gravitational strength (or acceleration) at the surface of the earth.* By Equation (2.15) we have

$$W = -\int_{r(t_1)}^{r(t_2)} \frac{mgr_e^2}{r^2}\,dr = mgr_e^2\left[\frac{1}{r(t_2)} - \frac{1}{r(t_1)}\right]$$

We note that for this case the function φ is given by

$$\varphi = \frac{-mgr_e^2}{r} \tag{2.19}$$

If the motion is sufficiently near the surface of the earth,

$$\frac{r_e^2}{r^2} \approx 1$$

and so $f \approx mg$, a constant. In this case,

$$\begin{aligned} W &= \frac{mgr_e^2[r(t_1) - r(t_2)]}{r(t_1)r(t_2)} \\ &\approx mg[r(t_1) - r(t_2)] \\ &= \text{(weight)} \times \text{(decrease in altitude of mass center of body)} \end{aligned} \tag{2.20}$$

and the φ function becomes (if z_C is positive upward)

$$\varphi = mgz_C \tag{2.21}$$

Conservative Forces

In each case we have considered, the work has depended only on the initial and final positions of the point where the force is applied. Such a force whose work is independent of the path traveled by the point on

* The force of gravity in fact results in infinitely many differential forces, each tugging on one of the body's particles. For nearly all applications on the planet earth, these forces may be thought of as equivalent to a single force through the mass center of the body. For applications in astronomy or in space vehicle dynamics, however, the gravity moment that accompanies the force at the mass center becomes important. In Skylab, for example, three huge control-moment gyros were present to "take out" the angular momentum built up by a gravity moment of only a few lb-ft. And the gravity moment exerted on the earth by the sun and moon's gravitation causes the earth's axis to precess in the heavens once every 25,800 yr. The gravity moment vanishes if the body is a uniform sphere (which the earth is not, being bulged at the equator and having varying density). A further discussion of this *luni-solar precession* is presented in Chapter 7.

which it acts is called **conservative.** Furthermore, the work may be expressed as the change in a scalar function of position; we saw this to be the case for the central force, and we may make the same statement for the constant force by defining φ to be $-\mathbf{F} \cdot \mathbf{r}$.

Question 2.5 Why the minus sign in front of $\mathbf{F} \cdot \mathbf{r}$?

Conservation of Energy

If all forces acting are conservative and if φ is the sum of all their φ functions, then the work and kinetic energy equation (2.13) becomes

$$\varphi(t_1) - \varphi(t_2) = T(t_2) - T(t_1)$$

or

$$T(t_2) + \varphi(t_2) = T(t_1) + \varphi(t_1)$$

or

$$T + \varphi = \text{constant} \tag{2.22}$$

We call φ the **potential energy** and $T + \varphi$ the **total (mechanical) energy.** Thus Equation (2.22) is a statement of conservation of mechanical energy when all the forces are conservative (path-independent).

Question 2.6 How would Equation (2.22) read if instead of φ we had chosen to construct the scalar function ψ so that the work done by a force is the *increase* in its ψ?

In closing it is important to realize that not all forces are conservative. An example is the force of friction acting on a block sliding on a fixed surface. That force does negative work regardless of the direction of the motion, and a potential function φ cannot be found for it.

Answer 2.5 It is needed so that the work equals the decrease in φ; that is, $\varphi[\mathbf{r}(t_1)] - \varphi[\mathbf{r}(t_2)]$.
Answer 2.6 $T - \psi =$ constant; we use the φ simply so that we may say that mechanical energy is the sum of its two parts.

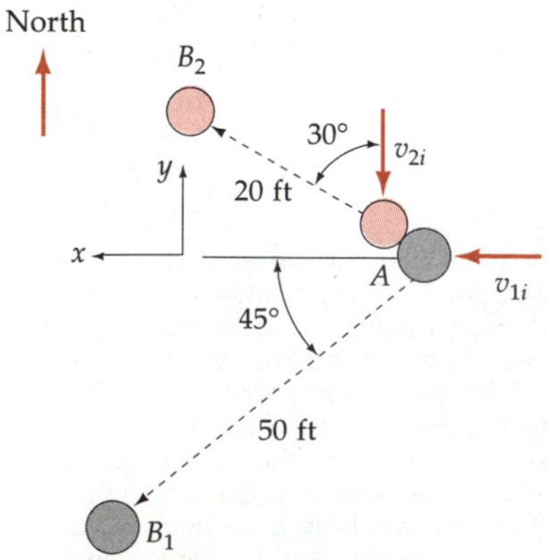

Figure E2.9

EXAMPLE 2.9

In an accident reconstruction, the following facts are known:

1. Identical cars 1 and 2, respectively headed west and south as indicated in Figure E2.9, collided at point A in an intersection.
2. With locked brakes indicated by skid marks, the cars skidded to the final positions B_1 and B_2 shown in the figure.

Assuming the cars are particles, determine their velocities immediately following the collision if the friction coefficient between the tires and road is 0.5.

Solution

After separation, each car is brought to rest by the friction force acting on its tires. For Car 1, we have

$$W = \Delta T = T_f - T_i$$

$$-\mu mg\,(50) = 0 - \frac{1}{2}mv_1^2$$

in which $T_f = 0$ since the cars end up at rest, and v_1 is the speed of Car 1 just after the cars separate. Thus

$$v_1 = \sqrt{2(50)0.5(32.2)} = 40.1 \text{ ft/sec}$$

Similarly for Car 2,

$$v_2 = \sqrt{2(50)0.5(32.2)} = 25.4 \text{ ft/sec}$$

In the next section, we will return to this example and use the principle of impulse and momentum to approximate the speeds of the cars *before* the impact.

Example 2.10

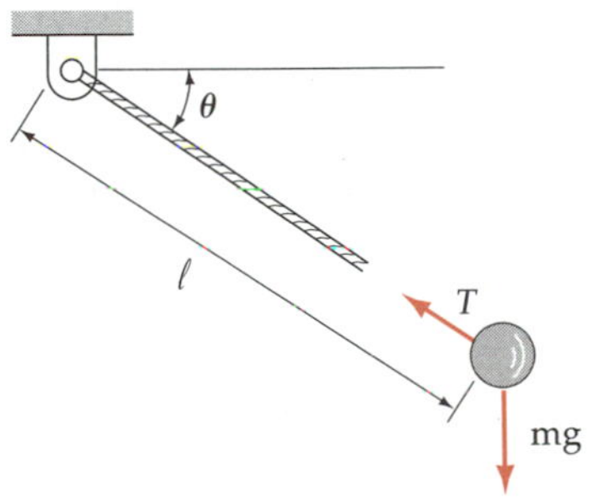

Figure E2.10

We repeat Example 2.4 (see Figure E2.10): For a ball of mass m released from rest with the cord taut and $\theta = 30°$, we wish to find the tension in the cord as a function of θ.

Solution

If, as before, we write the force-acceleration component equation in the radial direction, we have

$$T = mg\sin\theta + m\ell\dot{\theta}^2 \qquad (1)$$

Now we apply Equation (2.13) by letting t_1 be the initial time at which $\theta = 30°$, and letting t_2 be the time at which we are applying (1). We note that

$$|\mathbf{v}(t_1)| = 0$$

$$|\mathbf{v}(t_2)|^2 = \ell^2\dot{\theta}^2$$

and the work done by the cord tension T is zero since that force is always perpendicular to the velocity of (the center of mass of) the ball. By Equation (2.20) the work of the weight is $mg[\ell\sin\theta - \ell\sin 30°]$. Thus Equation (2.13) yields

$$mg[\ell\sin\theta - \ell\sin 30°] = \frac{1}{2}m(\ell^2\dot{\theta}^2) - 0$$

or

$$m\ell\dot{\theta}^2 = 2mg\left(\sin\theta - \frac{1}{2}\right)$$

Substituting in Equation (1) above, we get

$$T = mg\sin\theta + 2mg\sin\theta - mg$$

$$= mg(3\sin\theta - 1)$$

which is precisely the result obtained previously.

EXAMPLE 2.11

The block shown in Figure E2.11a slides on an inclined surface for which the coefficient of friction is $\mu = 0.3$. Find the maximum force induced in the spring if the motion begins under the conditions shown.

Solution

We assume that the block can be treated as rigid; thus the end of the spring, once it contacts the block, will undergo the same displacements as the mass center (or for that matter any other point) of the block. To apply the mass-center work and kinetic energy principle we let t_1 be the initial time shown above and let t_2 be the time of maximum compression Δ of the spring. To catalog the external forces that do work, we consider a free-body diagram at some arbitrary instant between t_1 and t_2. (See Figure E2.11b.) Since the mass center has a path parallel to the inclined plane, there is no component of acceleration perpendicular to it and

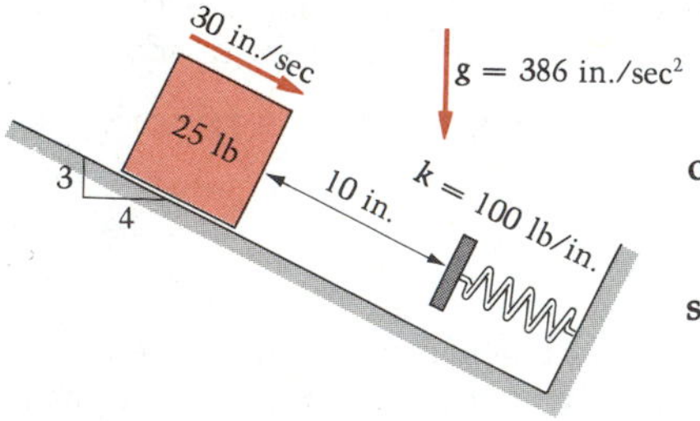

Figure E2.11a

$$N - \frac{4}{5}(25) = 0$$

or

$$N = 20 \text{ lb}$$

so that the friction force is $0.3(20) = 6$ lb.

Denoting the left side of Equation (2.11) by Work (t_1, t_2) we have

$$\text{Work}(t_1, t_2) = \frac{1}{2} m[|\mathbf{v}_C(t_2)|^2 - |\mathbf{v}_C(t_1)|^2]$$

where

$$|\mathbf{v}_C(t_2)| = 0$$

$$|\mathbf{v}_C(t_1)| = 30 \text{ in./sec}$$

and the various works are

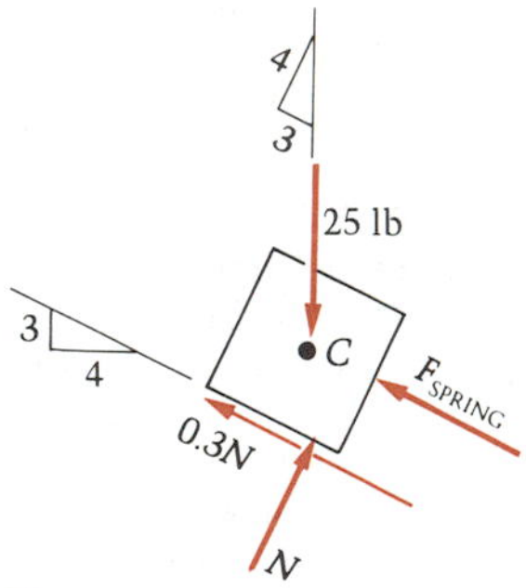

Figure E2.11b

1. For N, work$(t_1, t_2) = 0$ since the force is perpendicular to $\mathbf{v}_C$ at each instant.
2. For friction, work$(t_1, t_2) = -6(10 + \Delta)$ in.-lb
3. For the weight, work$(t_1, t_2) = [(\frac{3}{5})(25)](10 + \Delta)$ in.-lb
4. For the spring,

$$\text{Work}(t_1, t_2) = \text{work}(t_1, \text{contact}) + \text{work}(\text{contact}, t_2)$$

$$= 0 - \frac{100}{2}[(-\Delta)^2 - 0] \text{ in.-lb}$$

Thus, $W = \Delta T$ gives (with $g = 32.2 \times 12 = 386$ in./sec²)

$$-6(10 + \Delta) + 15(10 + \Delta) - 50\Delta^2 = 0 - \frac{1}{2}\left(\frac{25}{386}\right)(30)^2$$

or

$$50\Delta^2 - 9\Delta - 119 = 0$$

$$\Delta^2 - 0.180\Delta - 2.38 = 0$$

From the quadratic formula,

$$\Delta = 0.09 \pm \sqrt{0.0081 + 2.38} \text{ in.}$$

from which only the positive root is meaningful:

$$\Delta = 0.09 + 1.55 = 1.64 \text{ in.}$$

The corresponding force is 100(1.64) = 164 lb.

Example 2.12

In the preceding example, find the next position at which the block comes to rest.

Solution

At time t_2 the spring force (164 lb) exceeds the sum of the component of weight along the plane (15 lb) and the maximum frictional resistance (6 lb), so we know that the block is not in equilibrium and must then begin to move back up the plane, with the friction force now acting down the plane as shown in Figure E2.12. Suppose we let t_3 be the time at which the block next comes to rest and let d represent the corresponding compression of the spring. Then, since $|\mathbf{v}_C(t_2)| = |\mathbf{v}_C(t_3)| = 0$, we have

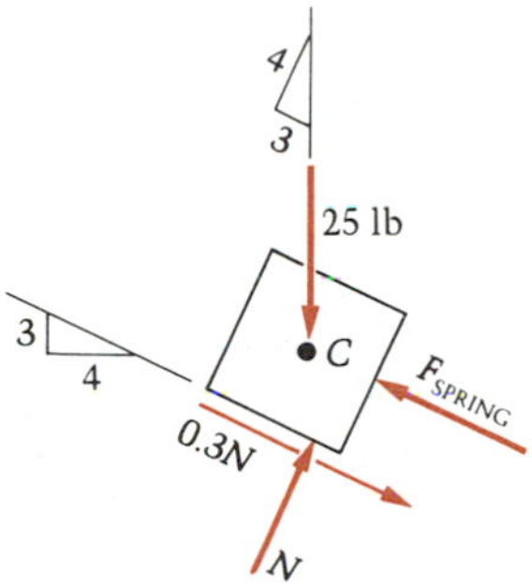

Figure E2.12

$$\text{Work}(t_2, t_3) = 0$$

For the spring, the work is $(-100/2)[d^2 - (-\Delta)^2]$; for the friction force, the work is $-6(\Delta - d)$; for the weight, the work is $-(\frac{3}{5})(25)(\Delta - d)$. Thus

$$(\Delta - d)\left[\frac{100}{2}(\Delta + d) - 6 - 15\right] = 0$$

or

$$d = \frac{21}{50} - 1.64 = -1.22 \text{ in.}$$

The negative sign here tells us that the spring must be stretched 1.22 in. when the block again comes to rest. If, as intended here, the spring does not become permanently attached to the block on first contact (that is, contact is maintained only in compression), our analysis only tells us that contact is broken before the block comes to rest. We therefore need to modify the expression for the work done by the spring, which we now see should have been $-(100/2)(0 - \Delta^2)$. It is convenient to let e be the distance from the end of the spring to the block (measured up the plane). Then

$$-\frac{100}{2}[0 - \Delta^2] - 6(\Delta + e) - \frac{3}{5}(25)(\Delta + e) = 0$$

$$21(\Delta + e) = 50\Delta^2$$

$$e = \frac{50}{21}\Delta^2 - \Delta$$

$$= \frac{50}{21}(1.64)^2 - 1.64$$

$$= 4.76 \text{ in.}$$

It is instructive to obtain this result by using the work and kinetic energy principle over the interval t_1 to t_3, for which the net work done by the spring is zero. Noting then that the mass center of the block drops $\frac{3}{5}(10 - e)$ in. vertically and that the distance traveled by the block on the plane is $(10 + 1.64 + 1.64 + e)$ in., we have

$$\text{Work}(t_1, t_3) = \frac{1}{2} m|\mathbf{v}_C(t_3)|^2 - \frac{1}{2} m|\mathbf{v}_C(t_1)|^2$$

$$25\left[\frac{3}{5}(10 - e)\right] - 6(13.3 + e) = 0 - \frac{1}{2}\left(\frac{25}{386}\right)(30)^2$$

$$150 - 15e - 79.8 - 6e = -29.2$$

$$e = 4.73 \text{ in.}$$

which is the same result we obtained before except for the third significant figure — a consequence of rounding off at an intermediate step.

EXAMPLE 2.13

A particle P of mass m rests atop a smooth spherical surface. (See Figure E2.13a.) A slight nudge starts it sliding downward in a vertical plane. Find the angle θ_L at which the particle leaves the surface.

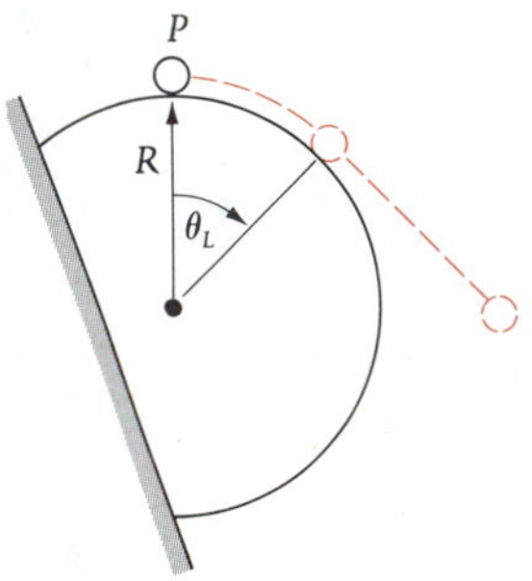

Figure E2.13a

Solution

Mechanical energy is conserved here because (1) there is no friction, (2) the normal force does not work since it is always normal to the velocity of P, and (3) the only other force is gravity. (See Figure E2.13b.) Therefore, using Equations (2.21) and (2.22),

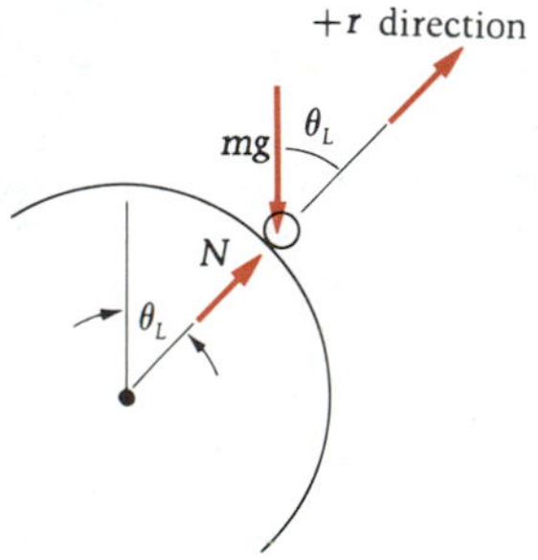

Figure E2.13b

$$T_1 + \varphi_1 = T_2 + \varphi_2$$

$$0 + mgR = \frac{1}{2} mv_2^2 + mgR \cos \theta_L$$

$$v_2^2 = 2gR(1 - \cos \theta_L) \qquad (1)$$

Equation (1) contains two unknowns; to eliminate the velocity v_2, we use the equation of motion in the radial direction:

$$\Sigma F_r = ma_r$$

$$-mg \cos \theta_L + N = m\left(-\frac{v_2^2}{R}\right) \qquad (2)$$

But N has just become zero when P is at the point of leaving. Therefore

$$v_2^2 = gR \cos \theta_L \qquad (3)$$

Equating the right sides of Equations (1) and (3), we get

$$gR \cos \theta_L = 2gR(1 - \cos \theta_L)$$

$$3 \cos \theta_L = 2$$

$$\theta_L = \cos^{-1}\left(\frac{2}{3}\right) = 48.2°$$

Work and Kinetic Energy for a System of Particles

As has been mentioned before, the kinetic energy of a system of particles is given by

$$T = \sum_{i=1}^{N} \frac{1}{2} m_i |\mathbf{v}_i|^2$$

or

$$T = \sum_{i=1}^{N} T_i$$

Now if we let W_i be the work of all the forces acting on the ith particle,

$$W_i = \Delta T_i$$

and, summing over all the particles,

$$\sum_{i=1}^{N} W_i = \Delta \sum_{i=1}^{N} T_i$$

or

$$W = \Delta T \tag{2.23}$$

for the system. The work, W, however, is the net work of all the forces external *and internal* that act on the particles. Sometimes Equation (2.23) can be used effectively because the net work of internal forces can be evaluated. For example, if two moving particles are joined by a linear spring, no simple formula can be written for the work of the spring force on *one* of the particles. However, the net work of the equal and opposite spring forces on the two particles is given by Equation (2.17) (see Problem 2.128). Particles that are rigidly connected interact through forces for which no net work is done. Thus, we shall find in Chapter 5 that when we use $W = \Delta T$ for a rigid body, the work only involves external forces.

PROBLEMS ▸ Section 2.4

2.101 A truck body weighing 4000 lb is carried by four light wheels that roll on the sloping surface. (See Figure P2.101.) The truck has a velocity of 5 ft/sec in the position shown. Determine the modulus of the spring if the truck is brought to rest by compressing the spring 6 in. *Note:* Light wheels with good bearings imply negligible friction.

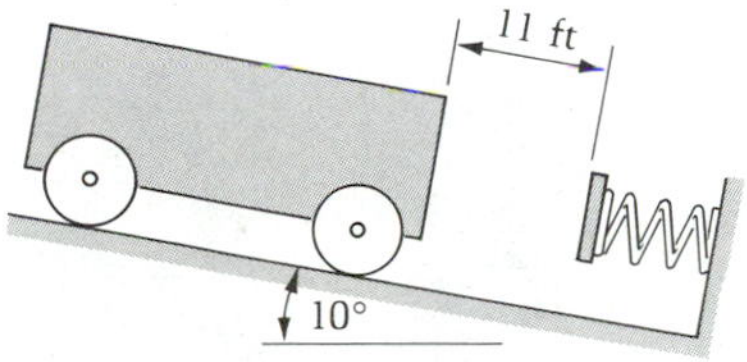

Figure P2.101

2.102 The block shown in Figure P2.102 weighs 100 lb and the spring's modulus is 10 lb/ft. The spring is unstretched when the block is released from rest. Find the minimum coefficient of friction μ such that the block will not start back up the plane after it stops.

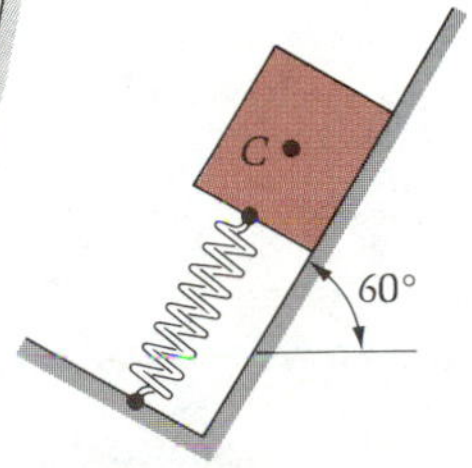

Figure P2.102

2.103 The block shown in Figure P2.103 is released from rest. What is its velocity when it first hits the spring?

2.104 How far does the block in the preceding problem rebound back up the plane after compressing the spring?

2.105 The 6-lb block shown in Figure P2.105 is released from rest when it just contacts the end of the unstretched spring. For the subsequent motion, find: (a) the maximum force in the spring; (b) the maximum speed of the block.

2.106 Block $\mathcal{A}$ weighs 16.1 lb and translates along a smooth horizontal plane with a speed of 36 ft/sec. (See Figure P2.106.) The coefficient of friction between $\mathcal{A}$ and the inclined surface is $\mu = 0.5$, and the spring constant is 100 lb/ft. Determine the distance that $\mathcal{A}$ moves up the incline before coming to rest.

2.107 The weight shown in Figure P2.107 is prevented from sliding down the inclined plane by a cable. An engineer wishes to lower the weight to the dashed position by inserting a spring and then cutting the cable. Find the modulus of a spring that will accomplish this task without allowing the block to move back up the incline after it stops. *Hint*: You are free to specify the initial stretch — try zero!

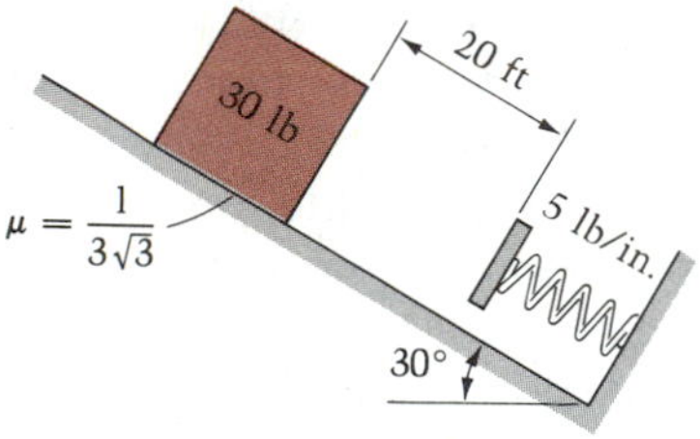

Figure P2.103

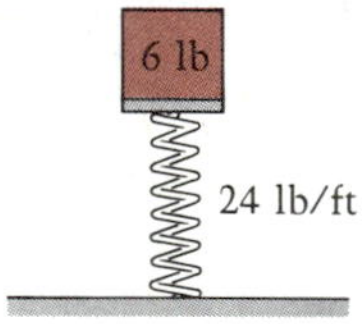

Figure P2.105

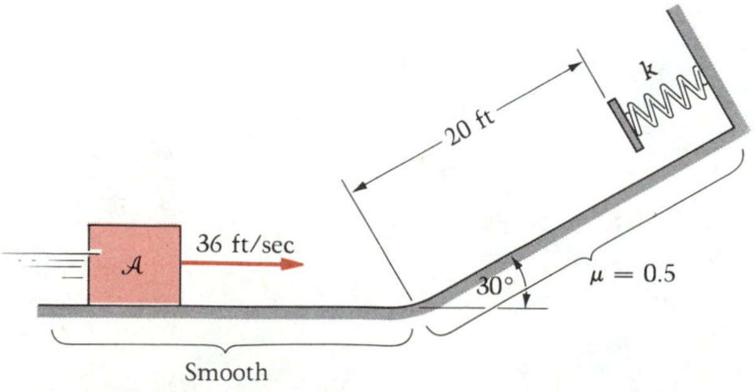

Figure P2.106

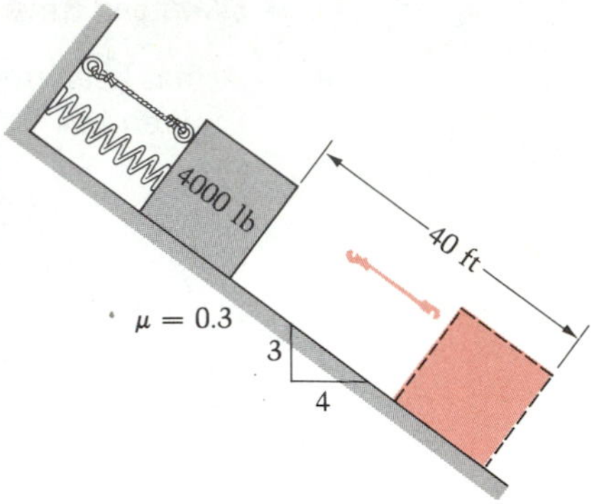

Figure P2.107

2.108 At the instant shown in Figure P2.108, the block is traveling to the left at 7 m/s and the spring is unstretched. Find the velocity of the block when it has moved 4 m to the left.

2.109 Repeat the preceding problem with $\mu = 0.2$. You will need

$$\int \frac{dx}{\sqrt{a^2 + x^2}} = \ln(x + \sqrt{a^2 + x^2}) + c$$

where a and c are constants.

2.110 The Bernoulli brothers posed and then solved the "brachistochrone problem." (See Figure P2.110.) The problem was to determine on which single-valued, continuous, smooth path a particle would arrive at B in minimum time under uniform gravity, after beginning at rest at a higher point A. Their solution, beyond the scope of this book, was that this path of "quickest descent" is a cycloid. Show that, regardless of the path, the *speed on arrival* is as if the particle had been dropped freely through the same height H.

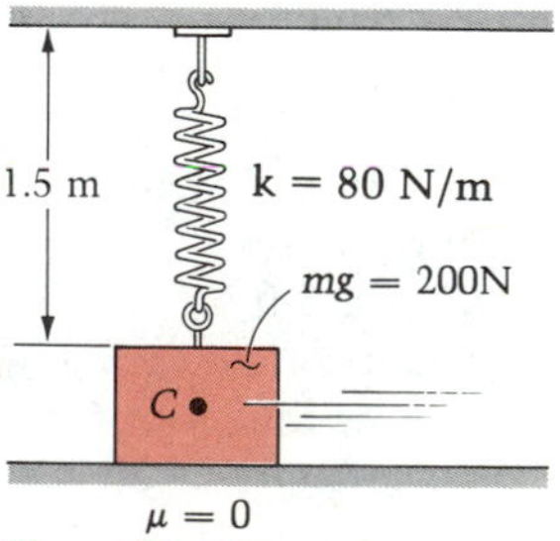

Figure P2.108

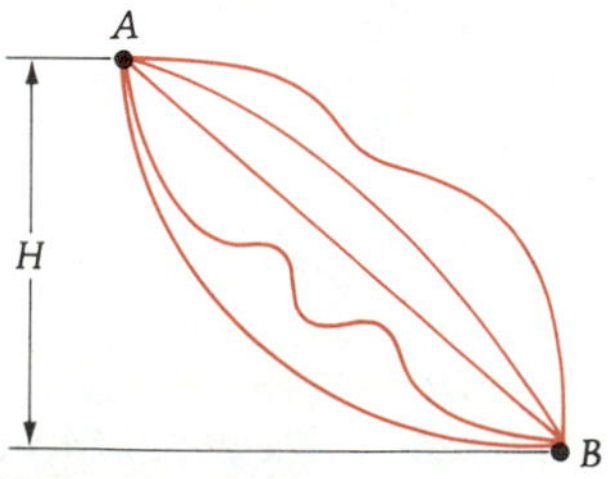

Figure P2.110

2.111 A small box $\mathcal{B}$ (see Figure P2.111) slides from rest down a rough inclined plane from A to B and then falls onto the loading dock. The coefficient of sliding friction between the box and plane is $\mu = 0.4$. Find the distance D to the point C where the box strikes the dock.

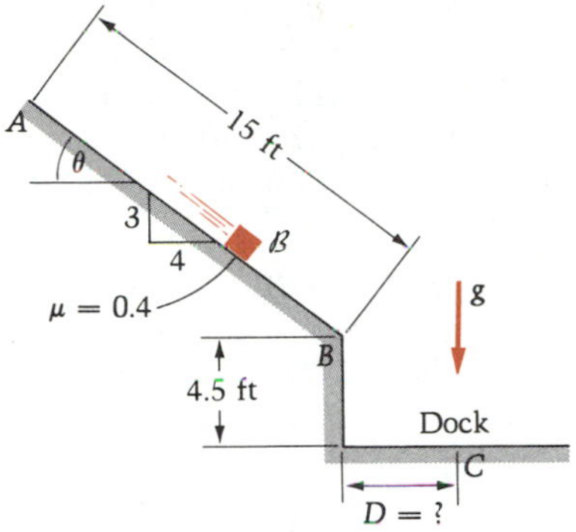

Figure P2.111

2.112 A particle is released at rest at A and slides on the smooth parabolic surface to B, where it flies off. (See Figure P2.112.) Find the total horizontal distance D that it travels before hitting the ground at C.

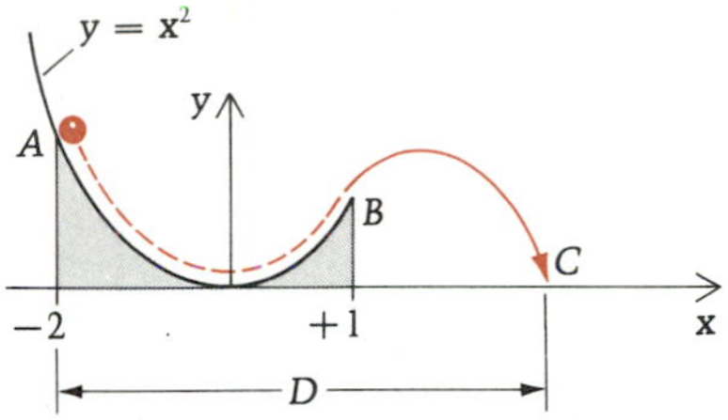

Figure P2.112

2.113 A particle of mass m slides down a frictionless chute and enters a circular loop of diameter d. (See Figure P2.113.) Find the minimum starting height h in order that the particle will make a complete circuit of the loop and exit normally (without having lost contact with the loop).

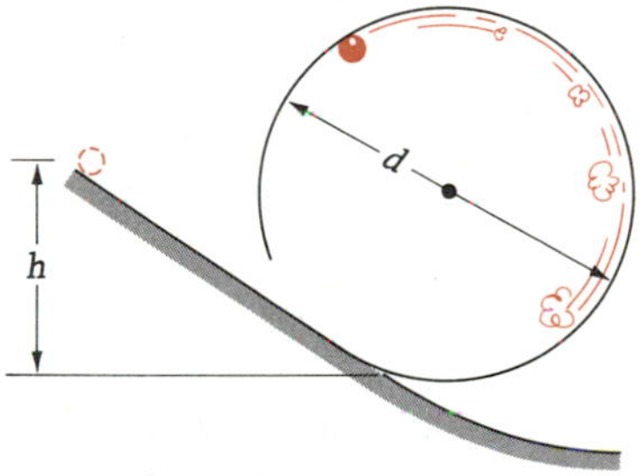

Figure P2.113

2.114 An 80-lb child rides a 10-lb wagon down an incline (Figure P2.114). Neglecting all frictional losses, find the "weight" of the child at A as indicated by a scale upon which she is sitting.

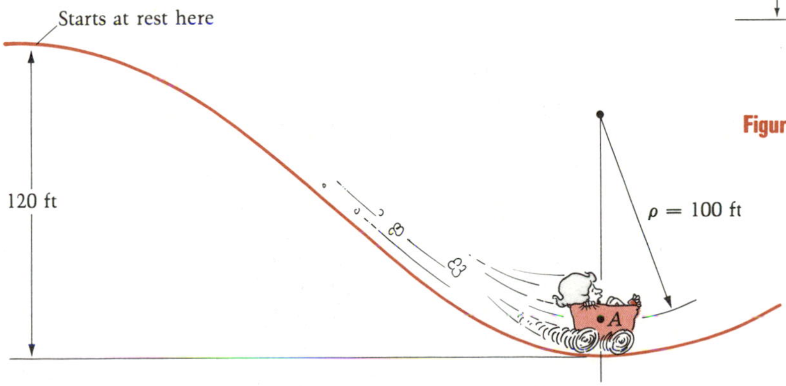

Figure P2.114

2.115 A skier descends the smooth slope, which may be approximated by the parabola $y = \frac{1}{20}x^2 - 5$ (see Figure P2.115). If she starts from rest at "A" and has a mass of 52 kg, determine the normal force she exerts on the ground the instant she arrives at "B", and her acceleration there. Note: treat the skier as a particle, and neglect friction.

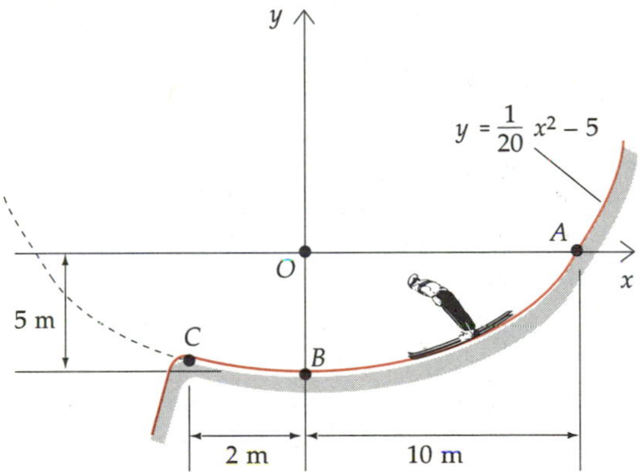

Figure P2.115

2.116 Find the speed sought in Problem 2.81 using work and kinetic energy.

2.117 Use work and kinetic energy to solve Problem 2.83.

2.118 Use work and kinetic energy to solve Problem 2.94.

2.119 Use work and kinetic energy to solve Problem 2.86.

2.120 Show that the equation $m\ell\dot{\theta}^2 = 2mg(\sin\theta - \frac{1}{2})$ can be obtained by conservation of mechanical energy ($T + \varphi =$ constant) in Example 2.10. Why can this principle not be used in Examples 2.11 and 2.12?

2.121 Check the solutions to Problems 2.78 and 2.79 by using the principle of work and kinetic energy.

2.122 Show that for central gravitational force $[\mathbf{F} = (-GMm/r^2)\hat{\mathbf{e}}_r]$, as distinct from the uniform gravity $(-mg\hat{\mathbf{k}})$ in the text, the potential is given by

$$\varphi = \frac{-GMm}{r}$$

where G is the universal gravitational constant and M and m are the masses of the two attracting bodies. Note that in view of Equation (2.19), one simply needs to show that $gr_e^2 = GM$.

2.123 Using the result of the preceding problem, calculate the work done by the earth's gravity on a satellite between the times of launch and insertion into a geosynchronous orbit with radius 6.61 times the radius of earth. (See Problem 2.64.)

2.124 For Problem 2.49, use $W = \Delta T$ to find the speeds of the blocks when the 5-lb block has droppped 2 ft.

2.125 Block $\mathcal{A}$ in Figure P2.125 is moving downward at 5 ft/sec at a certain time when the spring is compressed 6 in. The coeefficient of friction between block $\mathcal{B}$ and the plane is 0.2, the pulley is light, and the weights of $\mathcal{A}$ and $\mathcal{B}$ are 161 and 193 lb, respectively.

a. Find the distance that $\mathcal{A}$ falls from its initial position before coming to zero speed.

b. Determine whether or not body $\mathcal{A}$ will start to move back upward.

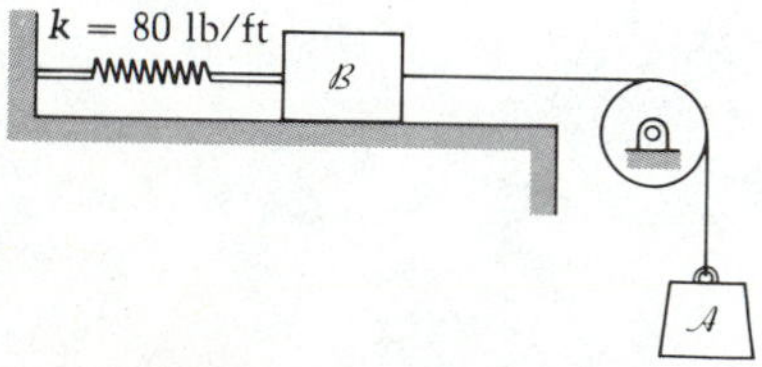

Figure P2.125

2.126 The system in Figure P2.126 consists of the 12-lb body $\mathcal{A}$, the light pulley $\mathcal{B}$, the 8-lb "rider" $\mathcal{C}$, and the 10-lb body $\mathcal{D}$. Everything is released from rest in the given position. Body $\mathcal{D}$ then falls through a hole in bracket $\mathcal{E}$, which stops body $\mathcal{C}$. Find how far $\mathcal{D}$ descends from its original position.

2.127 At the instant shown in Figure P2.127, block $\mathcal{B}$ is 30 m below the level of block $\mathcal{A}$. At this time, $\mathbf{v}_A$ and $\mathbf{v}_B$ are zero. Determine the velocities of $\mathcal{A}$ and $\mathcal{B}$ as they pass each other. $\mathcal{A}$ and $\mathcal{B}$ have masses of 15 kg and 5 kg, respectively. The pulleys are light.

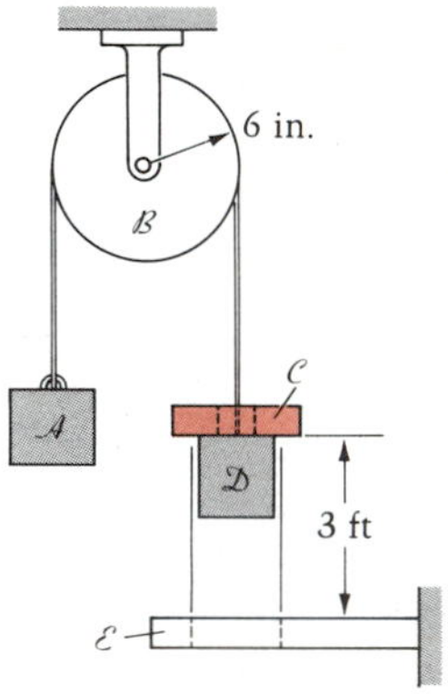

Figure P2.126

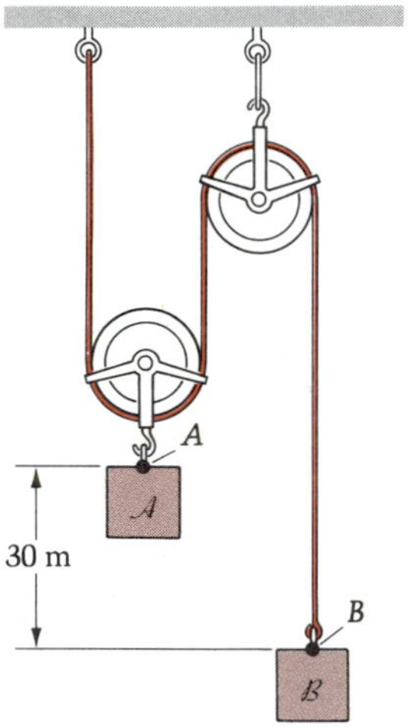

Figure P2.127

2.128 Suppose the ends of a spring are attached to "particles" of mass m_1 and m_2. Show that the sum of the works of the spring forces on the particles is given by Equation (2.17).

* **2.129** The blocks in Figure P2.129 are released from rest. Determine where they are when they stop permanently. What is the spring force then? *Hint:* Write the work-energy equation for each block, add the two equations, and use the result of Problem 2.125. Also, think about the motion of the mass center.

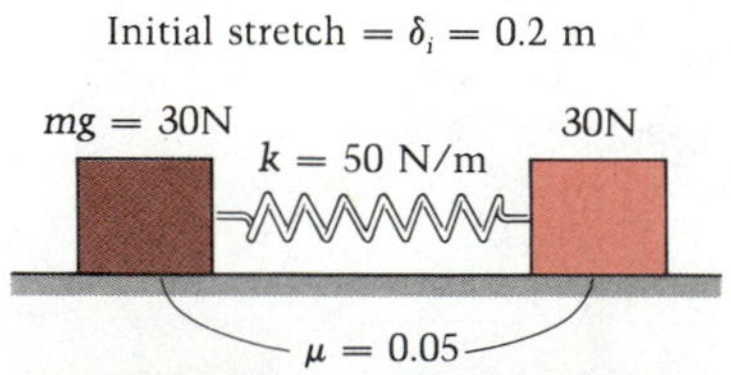

Figure P2.129

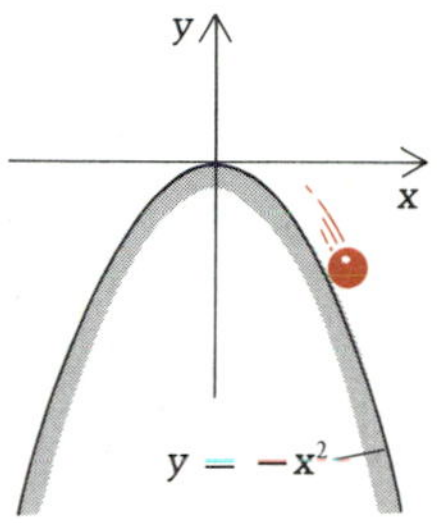

Figure P2.130

• **2.130** Show that if the surface in Example 2.13 is the parabola shown in Figure P2.130, the particle will *never* leave the surface. *Hint*: Show that

$$\rho = \left|\frac{(1 + y'^2)^{3/2}}{y''}\right| = \frac{(1 + 4x^2)^{3/2}}{2}$$

and use this in our equation:

$$\Sigma F_n = m\,\frac{\dot{s}^2}{\rho}$$

together with $W = \Delta T$.

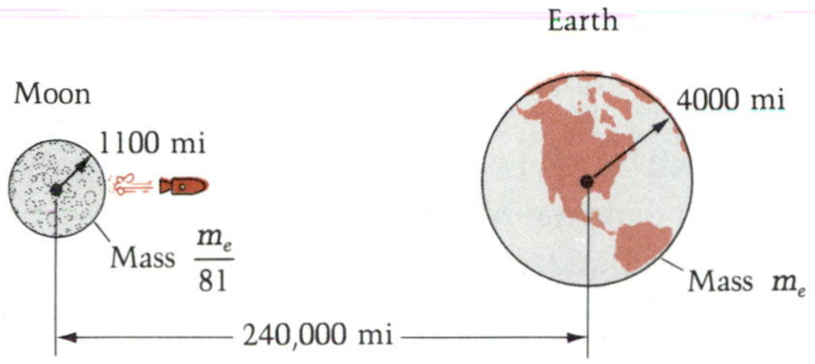

Figure P2.131

• **2.131** Find the least velocity with which a particle could be projected from the moon and reach the earth. (See Figure P2.131.) For this problem assume that the centers of the moon and earth are both fixed in an inertial frame.

2.5 Momentum Form of Euler's First Law

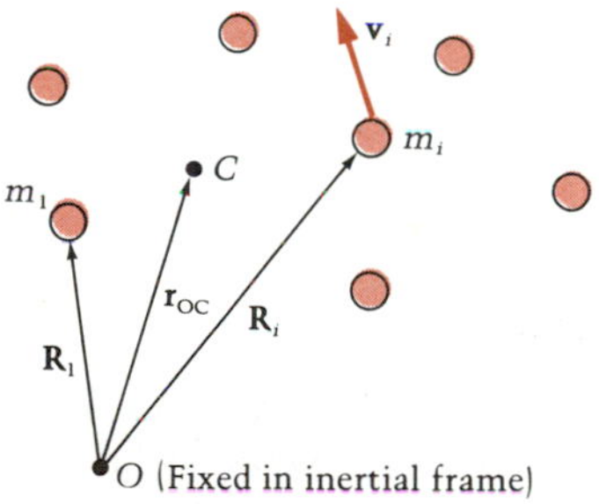

Figure 2.8a

The **momentum** of a particle is defined to be the product of its mass and its velocity. For a system of particles (Figure 2.8a) the momentum is defined to be the sum of the momenta of the particles in the system. Thus, if we denote the momentum of a system (or body) by **L**,* then

$$\mathbf{L} \equiv \sum_{i=1}^{N} m_i \mathbf{v}_i \tag{2.24}$$

or, for a body of continuously distributed mass (Figure 2.8b),

$$\mathbf{L} \equiv \int \mathbf{v}\, dm \tag{2.25}$$

If $\mathbf{R}_i$ is a position vector for the ith particle, then

$$\mathbf{v}_i = \frac{d\mathbf{R}_i}{dt}$$

and Equation (2.24) becomes

$$\mathbf{L} = \sum_{i=1}^{N} m_i \frac{d\mathbf{R}_i}{dt}$$

or

$$\mathbf{L} = \frac{d}{dt}\,(\Sigma m_i \mathbf{R}_i)$$

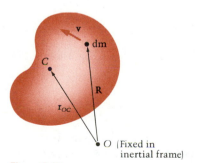

Figure 2.8b

* Momentum is sometimes called *linear momentum.*

Recalling [see Equation (2.7) and Figure 2.3a] that

$$\Sigma m_i \mathbf{R}_i = m\mathbf{r}_{OC}$$

then

$$\mathbf{L} = \frac{d}{dt}(m\mathbf{r}_{OC})$$

$$= m\frac{d\mathbf{r}_{OC}}{dt}$$

or

$$\mathbf{L} = m\mathbf{v}_C \tag{2.26}$$

where $\mathbf{v}_C$ is the velocity of the mass center of the system or body.

The connection between external forces and momentum now can be made easily by differentiating Equation (2.26) to obtain

$$\frac{d\mathbf{L}}{dt} = m\frac{d\mathbf{v}_C}{dt}$$

$$= m\mathbf{a}_C$$

But

$$\Sigma\mathbf{F} = m\mathbf{a}_C$$

so that

$$\Sigma\mathbf{F} = \frac{d\mathbf{L}}{dt} \tag{2.27}$$

which is the momentum form of Euler's first law.

Question 2.7 Should we expect Equation (2.27) to be valid for a system for which the mass is changing with time, such as a rocket with its varying-mass contents?

Impulse and Momentum; Conservation of Momentum

A straightforward integration of the first law of motion (Equation 2.27) yields

$$\int_{t_1}^{t_2} \Sigma\mathbf{F}\,dt = \mathbf{L}(t_2) - \mathbf{L}(t_1) = \text{change in momentum} = \Delta\mathbf{L} \tag{2.28}$$

where the integral is usually called the impulse imparted to the body by the external forces;* note that the impulse is intrinsically associated with

Answer 2.7 No, at several places in the development we have needed to require that the mass be constant.

* The impulse is sometimes called the *linear impulse.*

a specific time interval. If, during some time interval, the sum of the external forces vanishes, then $\dot{\mathbf{L}} = \mathbf{0}$ and hence the momentum is a constant, or is *conserved,* during that interval.

Since Equation (2.28) is a vector equation (unlike the scalar work and kinetic energy equation), we may use any or all of its component equations. For example:

$$\int_{t_1}^{t_2} \Sigma F_x \, dt = L_x(t_2) - L_x(t_1) = m\dot{x}_C(t_2) - m\dot{x}_C(t_1) \qquad (2.29)$$

and similarly for y and z. We note that we may have a planar situation, for example, in which $\Sigma F_x = 0$ but $\Sigma F_y \neq 0$ over an interval. If this is the case, momentum is conserved in the x direction but not in the y direction.*

Impact

Sometimes it is possible, by conservation of momentum, to obtain limited quantitative information about the motions of colliding bodies. As a rule this can be done when the bodies interact for a relatively brief interval — *before* and *after* which it is reasonable to treat their motions as rigid. While the analysis is best discussed with examples, we make the observation here that generally it makes little sense to treat the bodies as rigid during the collision. If we wish to describe the motion that ensues when a bullet is fired into a wooden block, for example, the block clearly cannot be regarded as rigid during the penetration process. On the other hand, it may be quite plausible to assume that rigid motion of the block and embedded bullet occurs subsequent to permanent reorientation of material.

A key feature in the analysis of collision (or impact) problems is the fact that the momentum of a body made up of two parts is the sum of the momenta of the individual parts. This feature follows directly from the definition of a body's momentum as the integral

$$\mathbf{L} = \int_{\mathcal{B}} \mathbf{v} \, dm = \int_{\mathcal{B}_1} \mathbf{v} \, dm + \int_{\mathcal{B}_2} \mathbf{v} \, dm$$
$$= \mathbf{L}_1 + \mathbf{L}_2 \qquad (2.30)$$

where the subscripts (1 and 2) identify the two constituent parts of the body.

Example 2.14

A wooden block of mass m_1 is at rest on a smooth horizontal surface when it is struck by a bullet of mass m_2 traveling at a speed v as shown in Figure E2.14a. After the bullet becomes embedded in the block, the block slides to the right at speed V. Find the relationship between v and V.

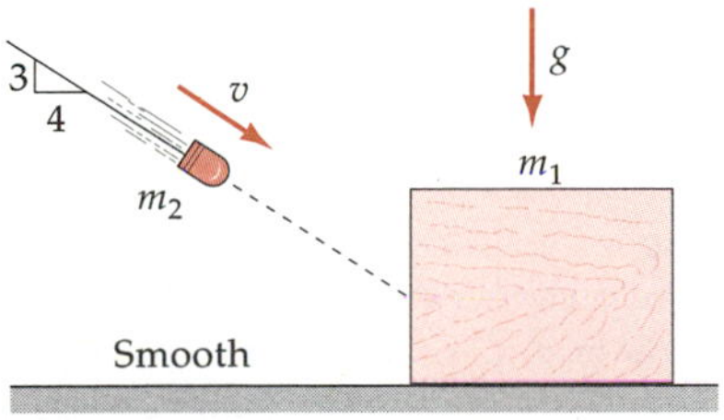

Figure E2.14a

* Ballistics problems are of this type if air resistance is neglected.

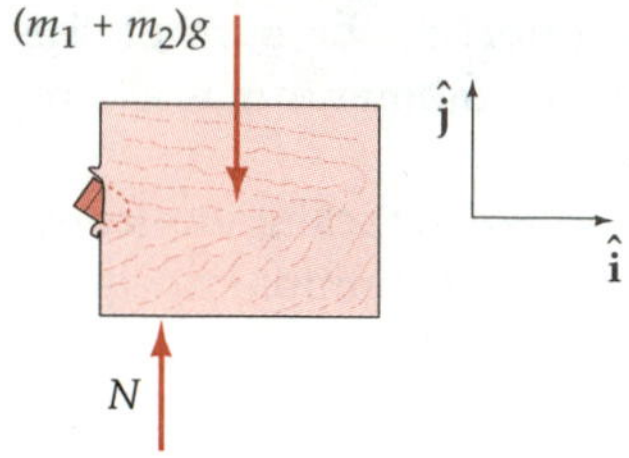

Figure E2.14b

Solution

Let t_1 be the time at which the bullet first contacts the block and let t_2 be the time after which the bullet/block composite behaves as a rigid body in translation. For $t_1 < t < t_2$, a complex process of deformation and redistribution of mass is occurring within the block. If we isolate the block/bullet system during this interval (see Figure E2.14b), Equation (2.28) yields

$$\int_{t_1}^{t_2} [N(t) - m_1 g - m_2 g]\hat{\mathbf{j}}\, dt = \mathbf{L}(t_2) - \mathbf{L}(t_1)$$

But

$$\mathbf{L}(t_2) = (m_1 + m_2)V\hat{\mathbf{i}}$$

and

$$\mathbf{L}(t_1) = m_2(0.8v\hat{\mathbf{i}} - 0.6v\hat{\mathbf{j}})$$

since v is the speed of the mass center of the bullet and the block has no momentum at t_1. Therefore, equating the $\hat{\mathbf{i}}$ coefficients, we get

$$(m_1 + m_2)V = 0.8m_2 v$$

or

$$V = \frac{0.8m_2 v}{m_1 + m_2}$$

We note that, in the absence of an external force with a horizontal component, the horizontal component of momentum is conserved.

While we cannot calculate the reaction N during the collision, we *can* calculate its impulse:

$$\int_{t_1}^{t_2} [N - (m_1 + m_2)g]\, dt = 0.6m_2 v$$

or

$$\int_{t_1}^{t_2} N\, dt = (m_1 + m_2)g(t_2 - t_1) + 0.6m_2 v$$

Similarly, the impulse of the force $\mathbf{F}$ exerted on the bullet by the block can be calculated if we apply Equation (2.28) to the bullet:

$$\int_{t_1}^{t_2} (\mathbf{F} - m_2 g\hat{\mathbf{j}})\, dt = m_2 V\hat{\mathbf{i}} - m_2(0.8v\hat{\mathbf{i}} - 0.6v\hat{\mathbf{j}})$$

$$\int_{t_1}^{t_2} \mathbf{F}\, dt = m_2(V - 0.8v)\hat{\mathbf{i}} + m_2[g(t_2 - t_1) + 0.6v]\hat{\mathbf{j}}$$

$$= m_2\left(\frac{0.8m_2 v}{m_1 + m_2} - 0.8v\right)\hat{\mathbf{i}} + m_2[g(t_2 - t_1) + 0.6v]\hat{\mathbf{j}}$$

$$= -\frac{0.8m_1 m_2}{m_1 + m_2} v\hat{\mathbf{i}} + m_2[g(t_2 - t_1) + 0.6v]\hat{\mathbf{j}}$$

The reader should note that with a high-speed collision occurring in a short period of time, the impulses can be accurately estimated by neglecting the impulses of the weights of the bodies. In this example we would have $0.6v \gg g(t_2 - t_1)$. Other examples of impact problems are treated in Chapter 5.

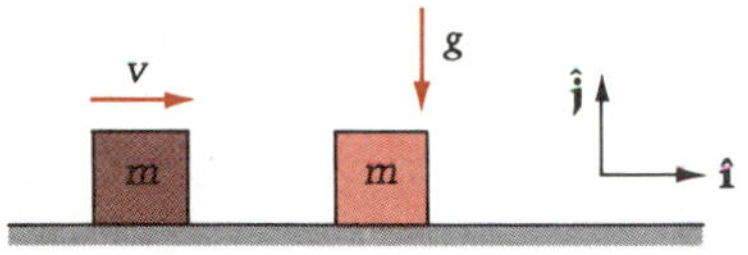

Figure E2.15a

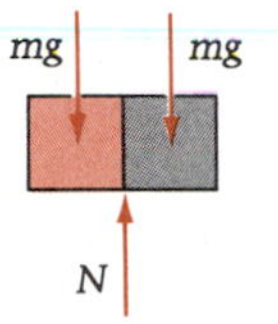

Figure E2.15b

EXAMPLE 2.15

A block is at rest on a smooth horizontal surface before being struck by an identical block sliding at speed v. (See Figure E2.15a.) Find the velocities of the two blocks after the collision assuming (1) that they stick together or (2) that the system experiences no loss in kinetic energy.

Solution

Let v_L and v_R be the speeds of the mass centers of the left and right blocks at the end of the collision; that is, $v_L\hat{\mathbf{i}}$ is the velocity of the left block. The free-body diagram of the system of two blocks during the collision (see Figure E2.15b) shows that there is no external force with a horizontal component. Thus the horizontal component (the only component not zero here) of momentum is conserved and

$$mv_L + mv_R = mv + m(0) \tag{1}$$

or

$$v_L + v_R = v \tag{2}$$

If the blocks remain attached after the collision is completed and they are behaving as rigid bodies, we have

$$v_R = v_L$$

so that

$$v_R = v_L = \frac{v}{2} \tag{3}$$

> **Question 2.8** What would be the common velocity if the right block had 100 times as much mass as the left block? If it had 10,000 times as much mass?

If, however, the blocks do not stick together, the conservation of momentum statement alone is not adequate to determine their subsequent velocities. What we need is some measure of their tendency to bounce off each other — or, to put it another way, a measure of how much energy is expended in permanent deformations or vibrations (or both) of the blocks. The parameter used to describe these effects (the coefficient of restitution) is discussed in the text that follows this example. At this point we simply note that when the blocks stick together the kinetic energy of the system is less after the collision than before. That loss is

$$\frac{1}{2}mv^2 - \left[\frac{1}{2}m\left(\frac{v}{2}\right)^2 + \frac{1}{2}m\left(\frac{v}{2}\right)^2\right] = \frac{1}{4}mv^2 \tag{4}$$

which is to say that one-half of the mechanical energy was dissipated in the collision in this case.

The other extreme case is that in which *no* mechanical energy is expended during the collision process. In this case

---

Answer 2.8 $v/101$; $v/10{,}001$.

$$\frac{1}{2}mv_L^2 + \frac{1}{2}mv_R^2 = \frac{1}{2}mv^2 \tag{5}$$

But since $v_R = v - v_L$ from Equation (2), we have

$$v_L^2 + (v - v_L)^2 = v^2$$

$$v_L^2 + v^2 - 2vv_L + v_L^2 = v^2 \tag{6}$$

or

$$2v_L(v_L - v) = 0 \tag{7}$$

Therefore either $v_L = 0$ and $v_R = v$, or $v_L = v$ and $v_R = 0$. The latter case must be rejected as physically meaningless since it would require the left block to pass through the stationary right block.

An extension of this result to the case of three blocks is shown in Figures E2.15c and E2.15d:

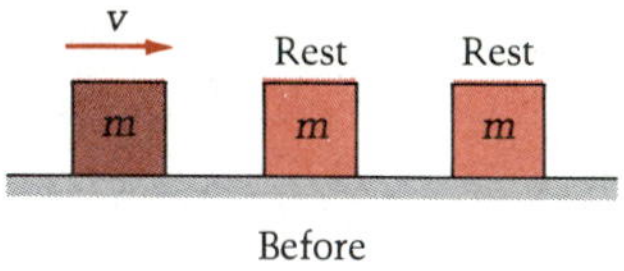

Figure E2.15c

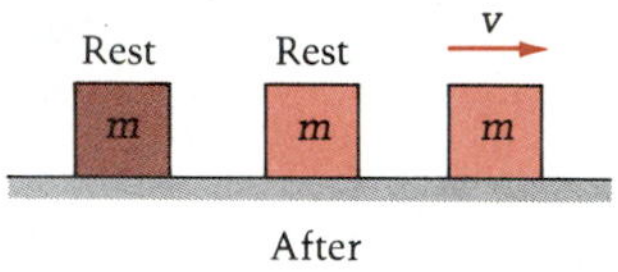

Figure E2.15d

If we let the spacing between the blocks initially at rest approach zero and add more of them, then we have the mechanism for a popular adult toy (see Figures E2.15e,f):

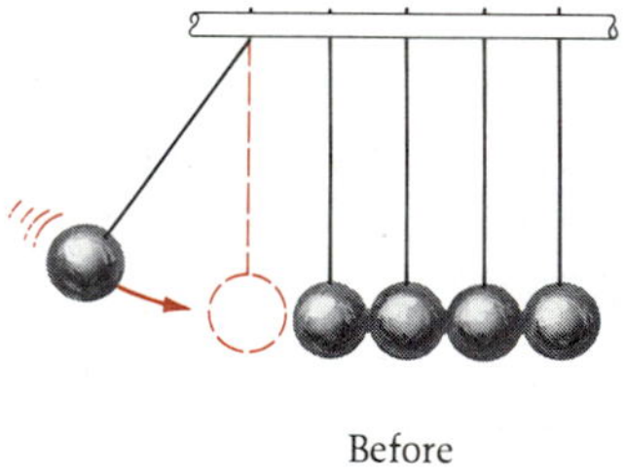

Figure E2.15e

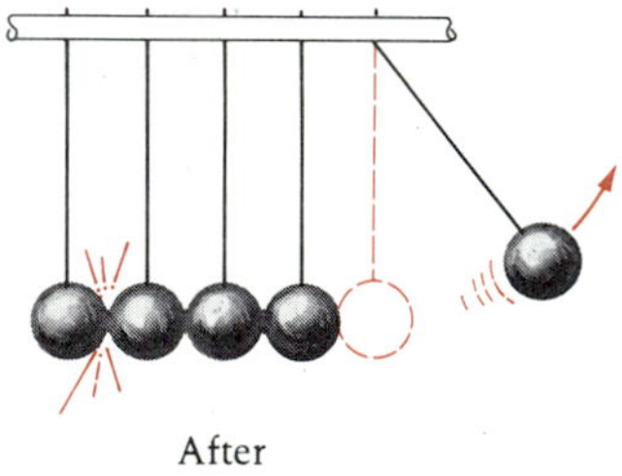

Figure E2.15f

Coefficient of Restitution

In the preceding example we noted the need for some measure of the capacity of colliding bodies to rebound off each other. The introduction of a parameter called the coefficient of restitution which provides this information is most easily accomplished through a simple example.

Suppose that, as depicted in Figure 2.9, two disks are sliding along a smooth floor. The paths are the same straight line and disk $\mathcal{A}$ is just about to overtake and contact disk $\mathcal{B}$ at time t_1. The centers of mass of the disks will approach each other until, at time t_2, they have the common velocity v_C. Then they will recede from each other until at time t_3 contact is broken. The equal and opposite forces of interaction $F(t)$ are shown on the disks in Figure 2.9. Applying the impulse-momentum principle dur-

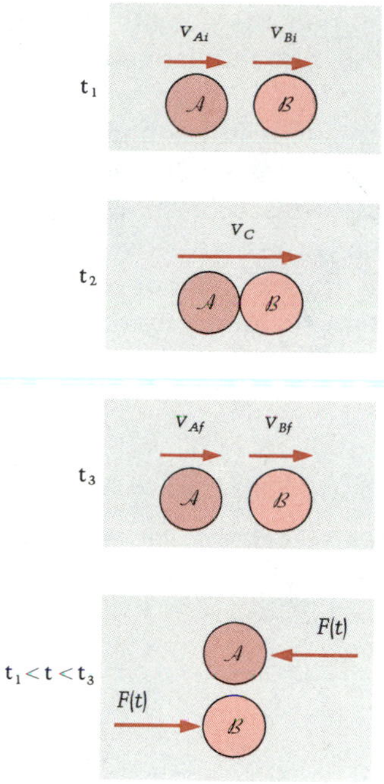

Figure 2.9

ing the intervals of approach and then separation of the centers of mass of the disks,

$$-\int_{t_1}^{t_2} F\,dt = m_{\mathcal{A}}(v_C - v_{Ai})$$

$$\int_{t_1}^{t_2} F\,dt = m_{\mathcal{B}}(v_C - v_{Bi})$$

$$-\int_{t_2}^{t_3} F\,dt = m_{\mathcal{A}}(v_{Af} - v_C) \tag{2.31}$$

and

$$\int_{t_2}^{t_3} F\,dt = m_{\mathcal{B}}(v_{Bf} - v_C)$$

Defining the coefficient of restitution, e, by

$$e = \frac{\int_{t_2}^{t_3} F\,dt}{\int_{t_1}^{t_2} F\,dt} \tag{2.32}$$

we obtain, after using the impulse-momentum equations above,

$$e = \frac{v_{Af} - v_C}{v_C - v_{Ai}} = \frac{v_{Bf} - v_C}{v_C - v_{Bi}}$$

Eliminating v_C, there results

$$e = \frac{v_{Bf} - v_{Af}}{v_{Ai} - v_{Bi}} \tag{2.33}$$

which is seen to be the quotient of the "relative velocity of separation" and the "relative velocity of approach." The coefficient of restitution is inherently nonnegative, and the case $e = 0$ yields $v_{Af} = v_{Bf}$, which means that the disks stick together. In Example 2.15, for the case of no energy dissipation, we had

$$e = \frac{v - 0}{v - 0} = 1$$

Exercise Problem 2.147 provides an outline of proof that, under the conditions of our discussion here, $e \leqq 1$.

The impact just described is called *central,* because the line of action of the equal and opposite forces of interaction is the line joining the mass centers of the bodies. It is also called *direct* because the preimpact velocities are parallel to that line of action. Generalization to the case of indirect, but still central, impact is easily accomplished, assuming the disks are smooth and that the time of contact is so small that there are no significant changes in their positions during the collision. Then, the velocity components perpendicular to the line of action of the impulsive force (called the line of impact) are unchanged by the collision. Equations (2.31) and (2.33) now refer to velocity components along the line of impact. Thus, the coefficient of restitution is the ratio of relative velocity components along the line of impact.

Experiments* indicate that the coefficient of restitution depends upon just about everything involved in an impact: materials, geometry, and initial velocities. Therefore, numerical values must be used with care. Nonetheless, the fact that the coefficient must have a value between zero and unity is valuable information in bounding the behavior of colliding bodies. The use of the coefficient of restitution for other than central impact is discussed in Chapter 5.

EXAMPLE 2.16

Two identical hockey pucks collide, coming into contact in the positions shown. Their velocities before the collision are also shown in Figure E2.16a.

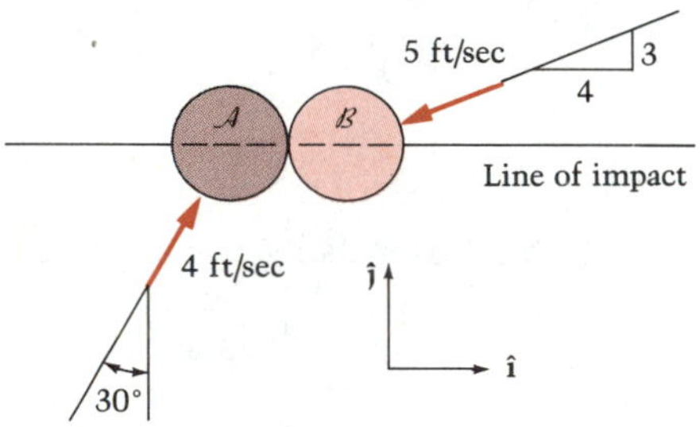

Figure E2.16a

If the coefficient of restitution is 0.8, find the velocities of the pucks after the collision. Then find the impulse of the force of interaction.

Solution

Neglecting friction and assuming insignificant deformation, the forces of interaction will act along the line of impact shown in Figure E2.16a. It is convenient to choose $\hat{\mathbf{i}}$ and $\hat{\mathbf{j}}$ parallel and perpendicular to this line as shown. Let m be the mass of each puck and let $\mathbf{v}_{Af}$ and $\mathbf{v}_{Ai}$ be the final and initial velocities of puck $\mathcal{A}$ and similarly for puck $\mathcal{B}$.

$$\mathbf{v}_{Ai} = 4(\sin 30°\hat{\mathbf{i}} + \cos 30°\hat{\mathbf{j}}) = 2\hat{\mathbf{i}} + 3.46\hat{\mathbf{j}} \text{ ft/sec}$$

Thus

$$\mathbf{v}_{Af} = \mathbf{v}_1\hat{\mathbf{i}} + 3.46\hat{\mathbf{j}}$$

where v_1 is the unknown component along the line of impact. Also

$$\mathbf{v}_{Bi} = 5\left(-\frac{4}{5}\hat{\mathbf{i}} - \frac{3}{5}\hat{\mathbf{j}}\right) = -4\hat{\mathbf{i}} - 3\hat{\mathbf{j}} \text{ ft/sec}$$

and

$$\mathbf{v}_{Bf} = v_2\hat{\mathbf{i}} - 3\hat{\mathbf{j}}$$

* See W. Goldsmith, *Impact* (London: Edward Arnold Publishers, Ltd., 1960).

Since there are no external forces on the system of two pucks, momentum is conserved:

$$m\mathbf{v}_{Af} + m\mathbf{v}_{Bf} = m\mathbf{v}_{Ai} + m\mathbf{v}_{Bi}$$

The component equation for the directions perpendicular to the line of impact is automatically satisfied, and for the $\hat{\mathbf{i}}$-direction

$$mv_1 + mv_2 = m(2) + m(-4)$$

or

$$v_1 + v_2 = -2 \tag{1}$$

By the definition of the coefficient of restitution

$$e = 0.8 = \frac{v_2 - v_1}{2 - (-4)}$$

or

$$v_2 - v_1 = 4.80 \tag{2}$$

Solving (1) and (2),

$$v_1 = -3.40$$

$$v_2 = 1.40$$

so that

$$\mathbf{v}_{Af} = -3.40\hat{\mathbf{i}} + 3.46\hat{\mathbf{j}} \text{ ft/sec}$$

and

$$\mathbf{v}_{Bf} = 1.40\hat{\mathbf{i}} - 3\hat{\mathbf{j}} \text{ ft/sec}$$

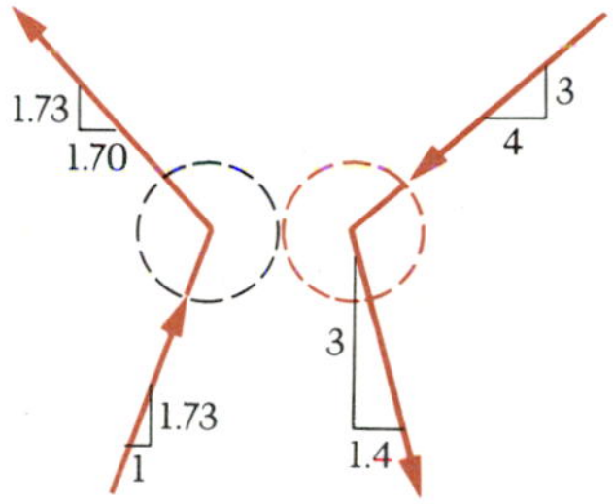

Figure E2.16b

Thus the paths of the pucks are as shown in Figure E2.16b.

To compute the impulse of the force of interaction we apply the impulse-momentum principle to puck $\mathcal{B}$, noting that a hockey puck weighs about 6 ounces. Thus with $m = (6/16)/32.2 = 0.0116$ slug,

$$\int F\,dt = 0.0116[1.40 - (-4)]$$
$$= 0.0626 \text{ lb-sec}$$

EXAMPLE 2.17

In Example 2.9 we found the velocities of the identical Cars 1 and 2, just after they collided in the given position, to be 40.1 and 25.4 ft/sec, respectively, as shown in Figure E2.17. Now, using the principle of impulse and momentum, find the velocities of the cars *prior* to impact. Remember that Cars 1 and 2 were heading west and south, respectively. Assume the collision is instantaneous.

Solution

If the impact occurs over a vanishingly small time Δt, then the impulse from the road (due to the friction force on the tires) during the impact is negligible, so that the linear momentum of the system of two cars may be assumed to be conserved

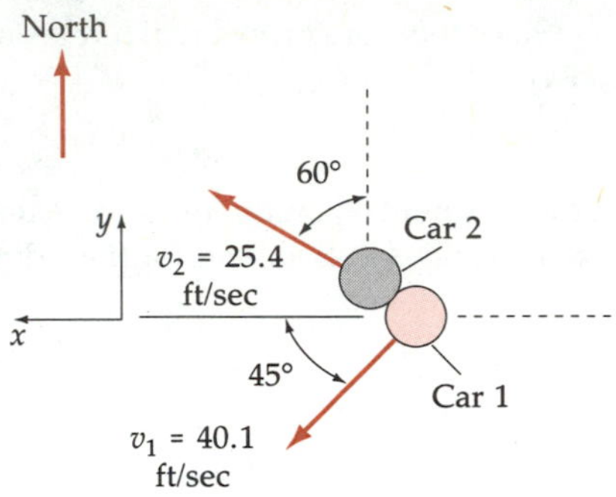

Figure E2.17

during Δt. In expressing this conservation, we shall use "i" for initial (before impact) and "f" for final (after impact):

$$mv_{1_i}\hat{\mathbf{i}} + mv_{2i}(-\hat{\mathbf{j}}) = m(v_{1fx}\hat{\mathbf{i}} + v_{1fy}\hat{\mathbf{j}}) + m(v_{2fx}\hat{\mathbf{i}} + v_{2fy}\hat{\mathbf{j}})$$

or

$$v_{1i} = v_{1fx} + v_{2fx} \tag{1}$$

and

$$-v_{2i} = v_{1fy} + v_{2fy} \tag{2}$$

Using the results of Example 2.9 and the angles in the figure above, we obtain the following post-collision velocity components:

$$v_{1fx} = 40.1 \cos 45° = 28.4 \text{ ft/sec}$$

$$v_{1fy} = -40.1 \sin 45° = -28.4 \text{ ft/sec}$$

$$v_{2fx} = 25.4 \sin 60° = 22.0 \text{ ft/sec}$$

$$v_{2fy} = 25.4 \cos 60° = 12.7 \text{ ft/sec}$$

Using Equations (1,2), we find

$$\begin{aligned} v_{1i} &= v_{1fx} + v_{2fx} \\ &= 28.4 + 22.0 = 50.4 \text{ ft/sec} \end{aligned}$$

$$\begin{aligned} v_{2i} &= -v_{1fy} - v_{2fy} \\ &= +28.4 - 12.7 = 15.7 \text{ ft/sec} \end{aligned}$$

We remark that the energy lost during the collision may now be calculated:

$$\begin{aligned} E_{\text{lost}} = KE_i - KE_f &= [\tfrac{1}{2}m(50.4)^2 + \tfrac{1}{2}m(15.7)^2] - [\tfrac{1}{2}m(40.1)^2 + \tfrac{1}{2}m(25.4)^2] \\ &= 1390m - 1130m = 260m \end{aligned}$$

Note that the work done by the road friction was (see Example 2.9) equal to $0.5(32.2)(50 + 20)m = 1130m$, and that this energy change plus the $260m$ lost in the collision (to deformation, sound, vibration, etc.) gives the total original kinetic energy, 1390 m.

PROBLEMS ► Section 2.5

2.132 Figure P2.132 presents data pertaining to a system of two particles. At the instant shown find the:

- **a.** Position of the mass center
- **b.** Kinetic energy of the system
- **c.** Linear momentum of the system
- **d.** Velocity of the mass center
- **e.** Acceleration of the mass center

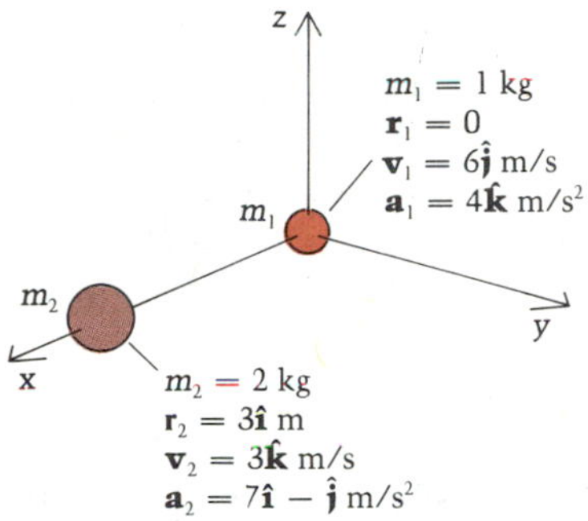

Figure P2.132

Figure P2.133

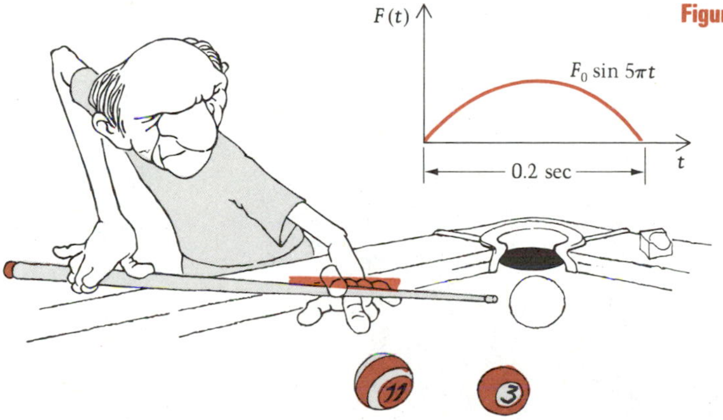

Figure P2.134

2.133 The astronaut in Figure P2.133 is finding it difficult to stop his forward momentum while jogging on the moon. Using a friction coefficient of $\mu = 0.3$ and a gravitational acceleration one-sixth that of earth's, illustrate the difficulty of stopping a forward momentum of mv = (5 slugs)(12 ft/sec). Specifically, use the principle of impulse and momentum to find the time it takes to stop on earth versus on the moon.

2.134 A horizontal force $F(t)$ is applied for 0.2 sec to a cue ball (weighing 0.55 lb) by a cue stick; the form of the force is as shown in Figure P2.134. If the velocity of the center of the ball is 8 ft/sec after contact with the stick is broken, find the peak magnitude F_0 of the force. Neglect friction. Force F is measured in pounds.

2.135 The 50-lb box shown in Figure P2.135 is at rest before the force $F(t) = 5 + 2t$ pounds is applied at $t = 0$. Assume the box to be wide enough not to tip over and suppose the coefficient of friction between box and floor to be 0.2. Find the velocity of (the mass center of) the box at $t = 10$ sec.

2.136 Repeat the preceding problem for the case in which the force $F(t)$ has a vertical component as shown in Figure P2.136.

2.137 A force P applied to $\mathcal{C}$ at $t = 0$ varies with time according to $P = 25 \sin(\pi t/60)$ lb, where t is in seconds. (See Figure P2.137.) How long will it take for $\mathcal{C}$ to begin sliding? What will be its velocity at $t = 30$ sec?

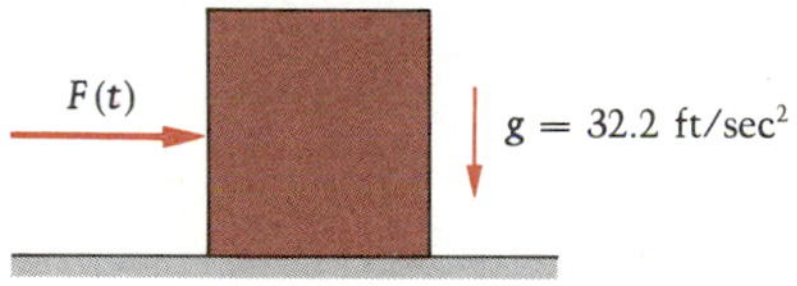

Figure P2.135

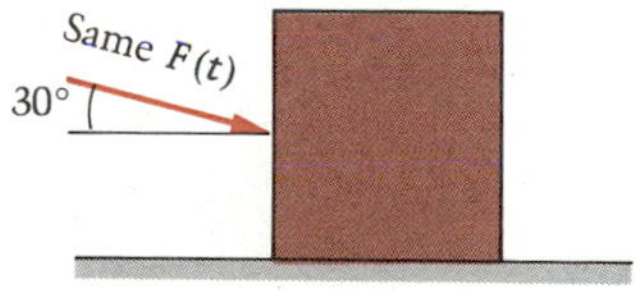

Figure P2.136

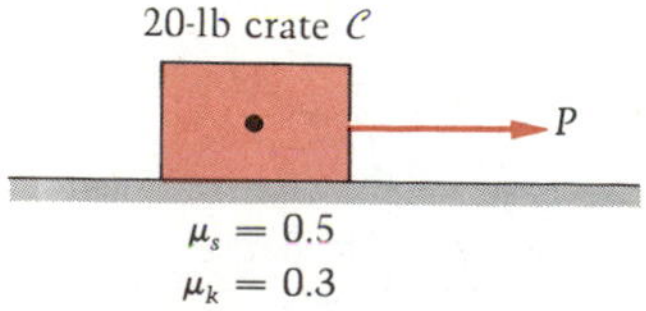

Figure P2.137

2.138 An unattached 2.2-lb roofing shingle slides downward and strikes a gutter. (See Figure P2.138.) The angle at which the shingle would be just on the verge of slipping is 20°. Determine the impulse imparted to the shingle by the gutter if there is no rebound. If the interval of impact is 0.1 sec, find the average force imparted to the gutter by the shingle.

2.139 Two railroad cars are coupled by a collision occurring just after the instant shown in Figure P2.139. Neglecting the impulse caused by friction from the tracks, determine the final velocity of the two cars as they move together.

2.140 In the preceding problem, find the average impulsive force between the cars if the coupling requires 0.6 sec of contact.

2.141 In a rail yard a freight car moving at speed v strikes two identical cars at rest. (See Figure P2.141.) Neglecting any resistance to rolling, find the common velocity of the three-car system after the coupling has been completed and any associated vibrations have died out.

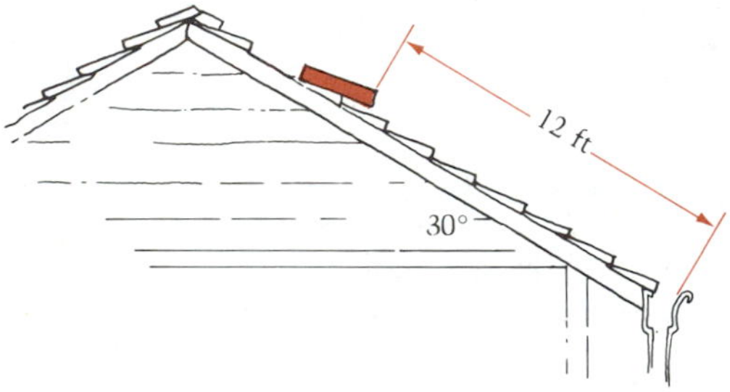

Figure P2.138

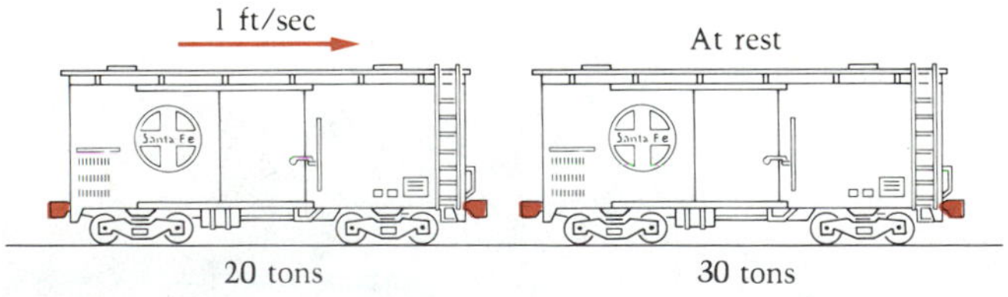

Figure P2.139

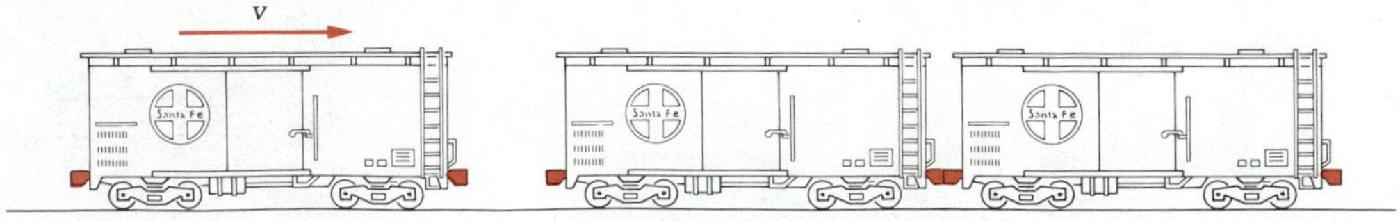

Figure P2.141

Figure P2.142

2.142 A man of mass m and a boat of mass M are at rest as shown in Figure P2.142. If the man walks to the front of the boat, show that his distance from the pier is then $L\mathcal{M}/(1 + \mathcal{M})$, where $\mathcal{M} = m/M$ is the ratio of the masses of man and boat. Explain the answer in the limiting cases in which $m \ll M$ and $M \ll m$. Neglect the resistance of the water to the boat's motion.

* **2.143** Two men each of mass m stand on the ends of a flatcar of mass M. The car is free to move on frictionless level tracks. All is at rest initially. One man runs to the right end of the car and jumps off horizontally, parallel to the tracks with a velocity U relative to the car. Then the other man runs to the left end of the car and jumps off horizontally, parallel to the tracks also with a velocity U relative to the car. Find the final velocity of the car and indicate clearly the direction of its motion.

* **2.144** In Figure P2.144 the man of mass m stands at end A of a 20-ft plank of mass $3m$ that is held at rest on the smooth inclined plane by the cord. The man cuts the cord and runs down to end B of the plank. When he gets there, end B is in the same position on the plane as it was originally. Find the time it takes the man to run down from A to B.

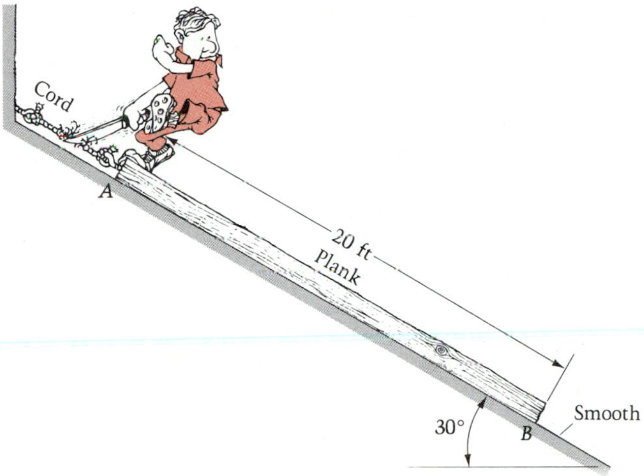

Figure P2.144

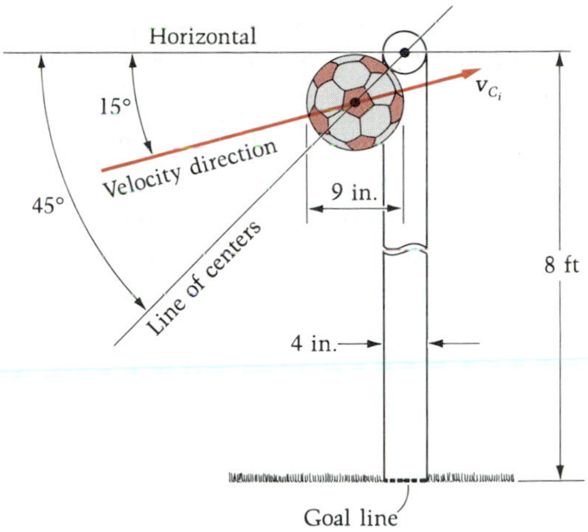

Figure P2.149

2.145 A ball is dropped from a height H and bounces. (See Figure P2.145.) If the coefficient of restitution is e, find the height to which the ball rises after the second bounce.

2.146 Two identical elastic balls $\mathcal{A}$ and $\mathcal{B}$ move toward each other. Find the approach velocity ratio v_{A_i}/v_{B_i} that will result in $\mathcal{A}$ coming to rest following the collision. The coefficient of restitution is e. (See Figure P2.146.)

2.147 Use the two equations

$$m_{\mathcal{A}} v_{A_i} + m_{\mathcal{B}} v_{B_i} = m_{\mathcal{A}} v_{A_f} + m_{\mathcal{B}} v_{B_f}$$

and

$$e = \frac{v_{B_f} - v_{A_f}}{v_{A_i} - v_{B_i}} \qquad (= \text{coefficient of restitution})$$

to prove that the loss in kinetic energy as the bodies $\mathcal{A}$ and $\mathcal{B}$ collide (Figure P2.147) is

$$\Delta T = \frac{m_{\mathcal{A}} m_{\mathcal{B}}(1 - e^2)(v_{A_i} - v_{B_i})^2}{2(m_{\mathcal{A}} + m_{\mathcal{B}})}$$

Deduce from this result that $e \le 1$!

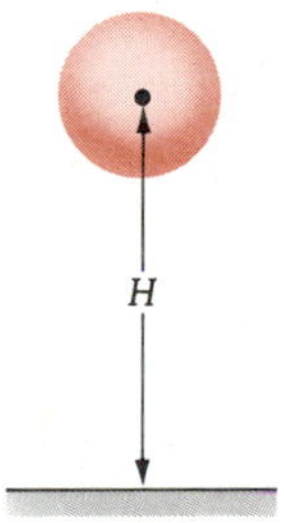

Figure P2.145

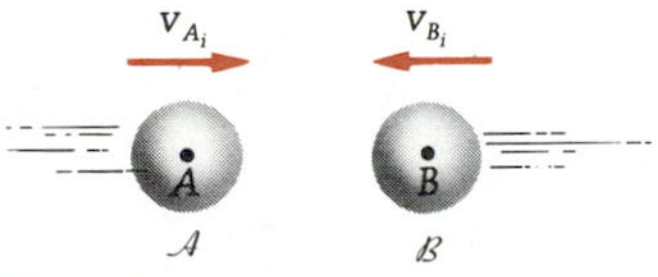

Figure P2.146

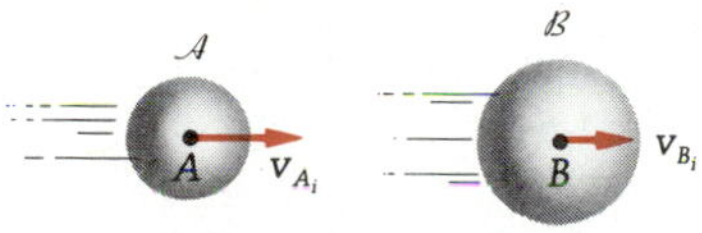

Figure P2.147

2.148 Use the result of the preceding problem to show that for a head-on collision at equal speeds v and equal masses m,

$$\Delta T = 2\frac{mv^2}{2}(1 - e^2)$$

so that if $e = 0$ then *all* of the initial T is lost and if $e = 1$ then *none* of T is lost. Is this true for differing speeds and masses?

2.149 In soccer, a goal is scored only when the *entire* ball is over the *entirety* of the 4-in.-wide goal line. (See Figure P2.149.) Neglecting friction between ball and post, determine the maximum coefficient of restitution for which a goal will be scored before the ball hits the ground. The velocity of the ball's center C makes an angle with the horizontal of 15°. Neglect the deviation caused by gravity on the trajectory between post and ground.

2.150 Repeat Example 2.16 for the line of impact shown in Figure P2.150.

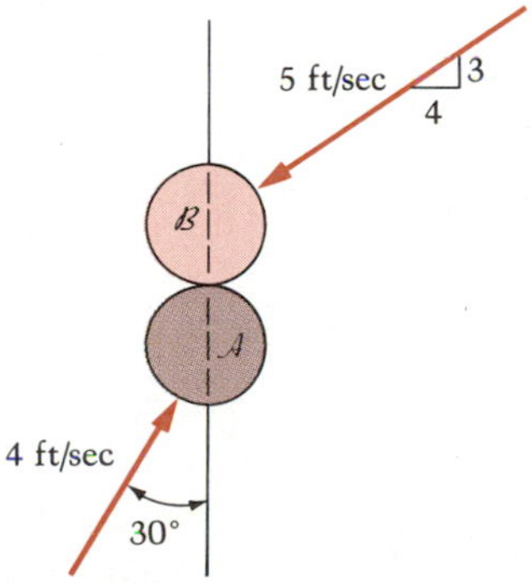

Figure P2.150

2.151 Let disk $\mathcal{A}$ in Problem 2.150 weigh 8 oz (and $\mathcal{B}$ weigh 6 oz as before), and then repeat the problem.

2.152 Repeat Example 2.16 for the case where the line of impact is parallel to the before-collision velocity of $\mathcal{B}$.

2.153 Let disk $\mathcal{A}$ in Problem 2.152 weigh 9 oz (and $\mathcal{B}$ weigh 6 oz as before), and then repeat the problem.

2.154 A 10-kg block swings down as shown in Figure P2.154 and strikes an identical block. Assume that the 6 m rope breaks during impact and the blocks stick together after colliding. How long will it be before they come to rest? How far will they have traveled?

2.155 Using the angle α that will land the cannonball of Problem 2.12 in the cart, find the maximum deflection of the spring. (See Figure P2.155.)

2.156 Find the total time after firing for the cannonball and box to either stop or strike the wall, whichever comes first. (See Figure P2.156.)

2.157 A cannonball is fired as shown in Figure P2.157 with an initial speed of 1600 ft / sec at 60°. Just after the cannon fires, it begins to recoil, and strikes a plate attached to a spring. Find the maximum spring deflection if the plane is smooth and the spring modulus is 500 lb / ft.

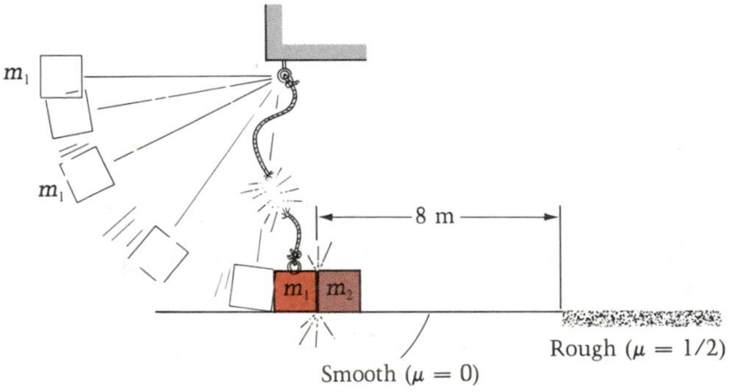

Figure P2.154

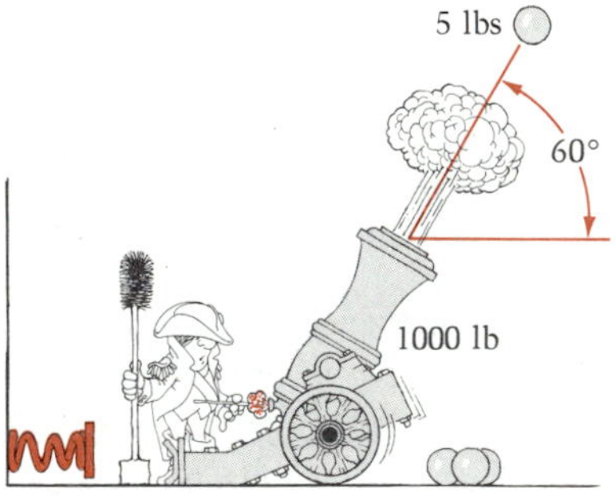

Figure P2.157

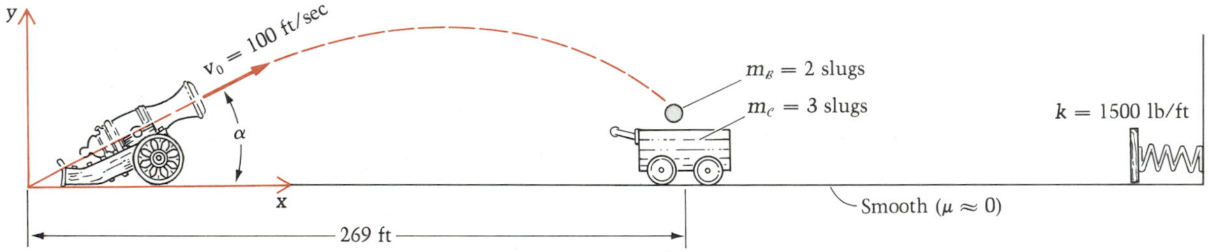

Figure P2.155

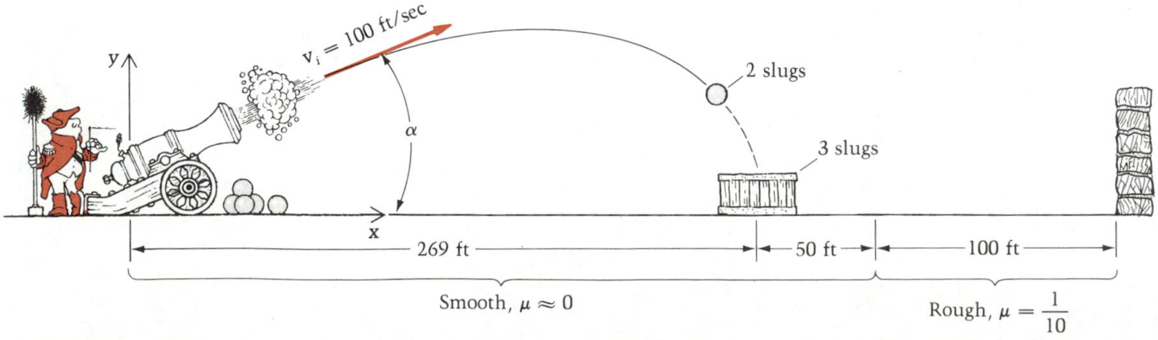

Figure P2.156

2.158 A $\frac{3}{4}$-oz bullet is fired with a speed of 1800 ft/sec into a 10-lb block. (See Figure P2.158.) If the coefficient of friction between block and plane is 0.3, find, neglecting the impulse of friction during the collision:

- **a.** The distance through which the block will slide
- **b.** The percentage of the bullet's loss of initial kinetic energy caused by sliding friction, and the percentage caused by the collision
- **c.** How long it takes block and bullet to come to rest after the impact.

2.159 Weight W_1 falls from rest through a distance H; it lands on another weight W_2, which was in equilibrium atop a spring of modulus k. (See Figure P2.159.) If the coefficient of restitution is zero, find the spring compression when the weights are at their lowest point.

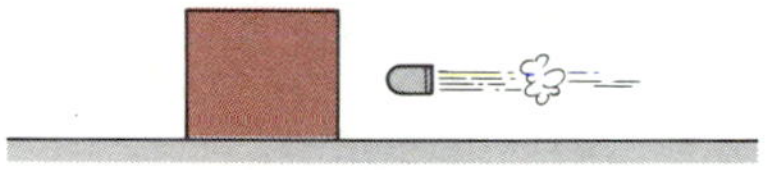

Figure P2.158

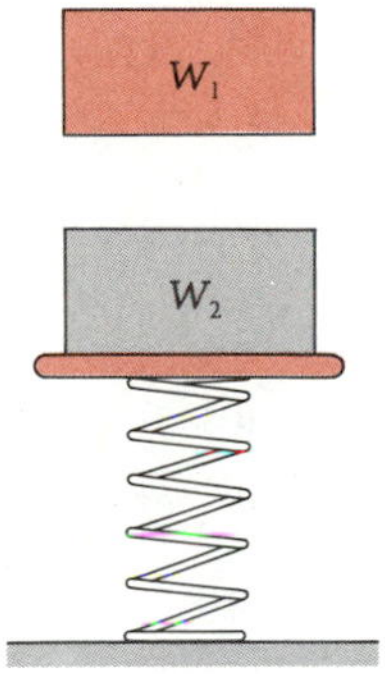

Figure P2.159

2.160 Block $\mathcal{A}$ in Figure P2.160 weighs 16.1 lb and is traveling to the right on the smooth plane at 50 ft/sec. Block $\mathcal{B}$ weighs 8.05 lb and is in equilibrium with the spring barely preventing it from sliding down the rough section of plane. Body $\mathcal{A}$ impacts $\mathcal{B}$; the coefficient of restitution $e = \frac{1}{2}$. Find the maximum spring deflection.

2.161 The 16-kg body $\mathcal{A}$ and the 32-kg body $\mathcal{B}$ shown in Figure P2.161 are connected by a light spring of modulus 12,000 N/m. The unstretched length of the spring is 0.15 m. The blocks are pulled apart on the smooth horizontal plane until the distance between them is 0.3 m and then released from rest. Determine the velocity of each block when the distance between them has decreased to 0.22 m. *Hint*: As in Problem 2.129, form the sum of the work-energy equations for the two blocks.

2.162 The cart and block in Figure P2.162 are initially at rest, when the bullet slams into the block at speed $100\sqrt{gL}$ and sticks inside it. The combined body then starts sliding on the cart.

Find:

- **a.** the speed of the block just after impact;
- **b.** the energy lost during impact;
- ***c.** the time when the block leaves the cart.

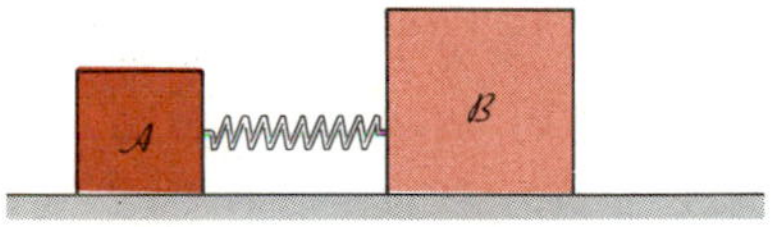

Figure P2.161

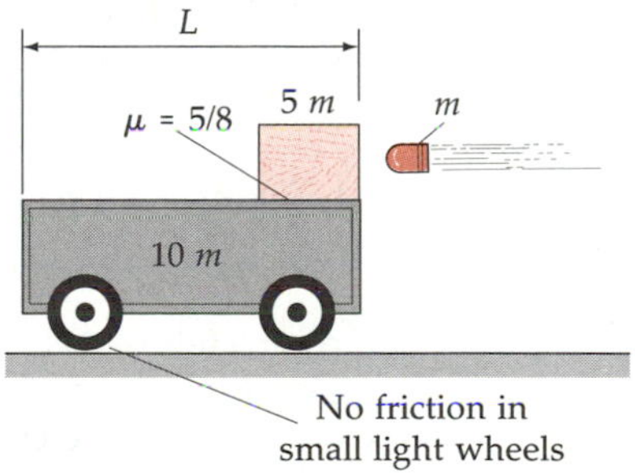

Figure P2.162

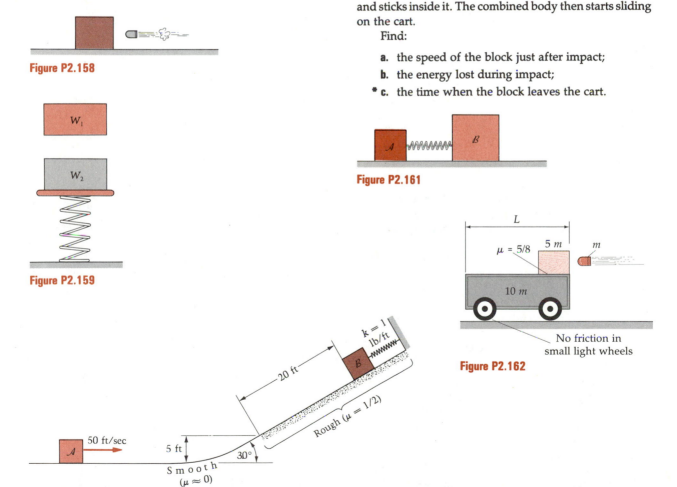

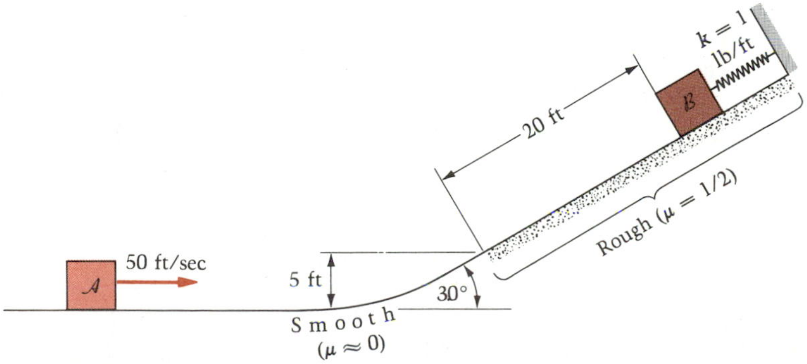

Figure P2.160

2.163 A chain of length L and mass per unit length β is held vertically above the platform scale shown in Figure P2.163 and is released from rest with its lower end just touching the platform. Assume that the links quickly come to rest as they stack up on the platform and that they do not interfere with the links still in free fall above the platform. Draw a free-body diagram of the entire chain and express the momentum as a function of the distance through which the upper end has fallen. Then determine the force read on the scale in terms of this distance.

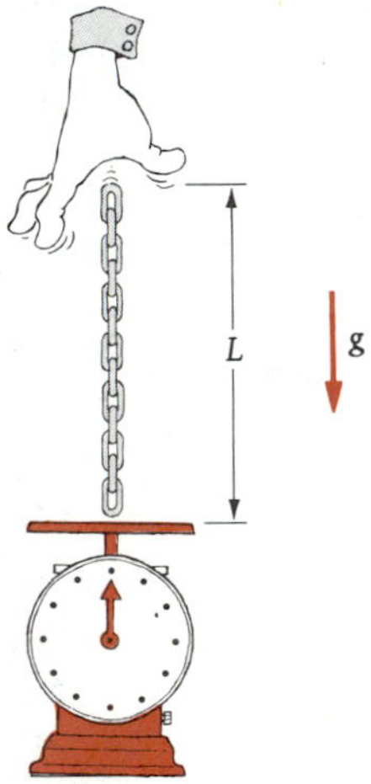

Figure P2.163

* **2.164** A block of mass mL, which can move on a smooth horizontal table, is attached to one end of a uniform chain of mass m per unit length. Initially the block and the chain are at rest, and the chain is completely coiled on the table. A constant horizontal force mLf is then applied to the block so that the chain begins to uncoil. Show that the length x uncoiled after time t is given by

$$(L + x)^2 = Lft^2 + L^2$$

until the chain is completely uncoiled. If the length of the chain is very large compared with L, show that the velocity of the block is approximately equal to $(Lf)^{1/2}$ at the moment when the chain is completely uncoiled.

* **2.165** An important problem in the dynamics of deformable solids is that of describing the motion which ensues when pressure is rapidly applied to the end of a slender, uniform, elastic bar. A useful approximate theory yields the one-dimensional wave equation as the governing equation of motion. This theory predicts that a pressure applied at one end of the bar creates a disturbance (wave) that propagates into the bar at a constant speed c. To be specific, suppose the bar shown in Figure P2.165a is at rest for $t < 0$ and is subjected to the uniform pressure (over the end of area A) shown in Figure P2.165b. If the disturbance has not reached the right end, that is if $t < L/c$, then for $t > t_0$ the particle velocities $\dot{u}\hat{\mathbf{i}}$ and accelerations $\ddot{u}\hat{\mathbf{i}}$, which vary only with x and t, are as shown in Figures P2.165c and d, where ρ is the density of the bar.

The first part of this problem is to evaluate the integral

$$\int_{\mathcal{B}} \mathbf{a}\, dm = \hat{\mathbf{i}} \int_0^L \int_A \rho \ddot{u}\, dA\, dx$$

The value that should be obtained is $p_0 A\hat{\mathbf{i}}$, and, since this equals the external force on the bar, Equation (2.4) is thus confirmed for this case. It is important to recognize that only the interval from $x = ct - ct_0$ to $x = ct$ contributes to the value of the integral; that is, only the particles in that region are accelerating.

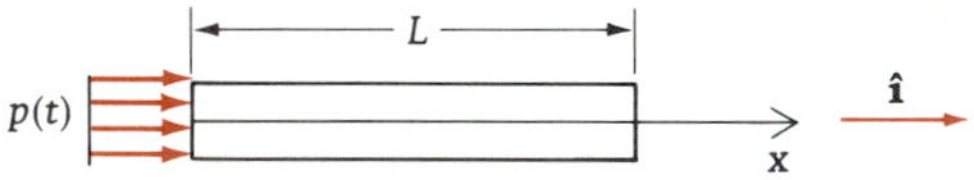

Figure P2.165a

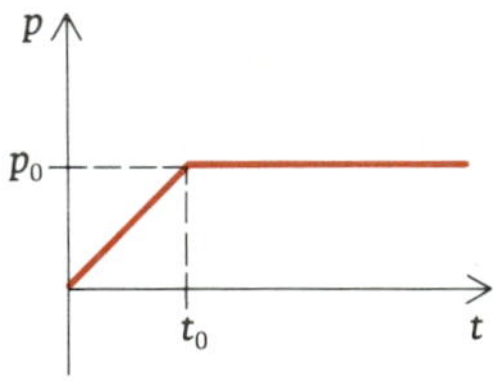

Figure P2.165b

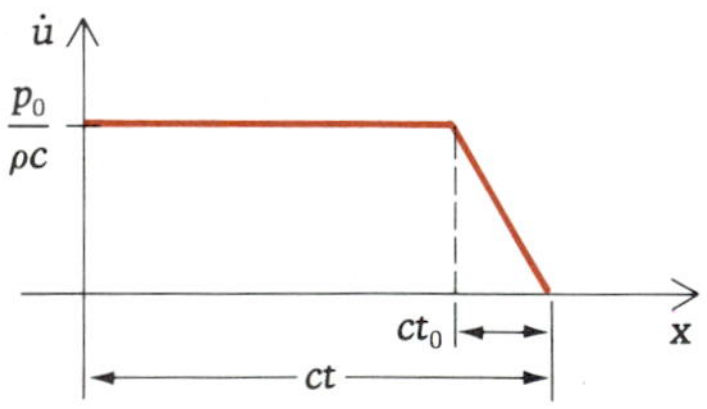

Figure P2.165c

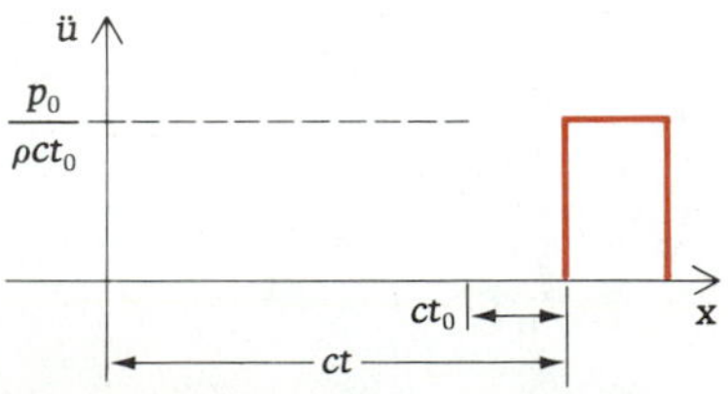

Figure P2.165d

The second element of this problem is to evaluate the momentum

$$\mathbf{L} = \int_B \mathbf{v}\,dm = \hat{\mathbf{i}} \int_0^L \int_A \rho \dot{u}\,dA\,dx$$

The result will be

$$\mathbf{L} = p_0 A \left[(t - t_0) + \frac{t_0}{2} \right] \hat{\mathbf{i}}$$

The second term in the brackets, a constant, is the contribution from integrating over the interval ct_0 where the particles are accelerating. The time dependence of $\mathbf{L}$ appears through the increasing number of particles having velocity $(p_0/\rho c)\hat{\mathbf{i}}$. As expected, we see that $\dot{\mathbf{L}} = p_0 A\hat{\mathbf{i}}$.

In effect we have confirmed Euler's law, $\Sigma\mathbf{F} = \dot{\mathbf{L}}$, in two forms. In the first,

$$\dot{\mathbf{L}} = \int \frac{d\mathbf{v}}{dt}\,dm = \int \mathbf{a}\,dm$$

In the second,

$$\dot{\mathbf{L}} = \frac{d}{dt} \int \mathbf{v}\,dm$$

For the case at hand there is no reason to express a preference for the order of differentiating with respect to time and integrating over the body. If the pressure were suddenly applied at full strength ($t_0 = 0$), however, there would be a discontinuity in particle velocity (shock wave) and a consequent undefined acceleration at the wavefront $x = ct$. Because of this undefined (or infinite) acceleration, $\int \mathbf{a}\,dm$ becomes meaningless and no longer provides $\dot{\mathbf{L}}$. There is no difficulty involved in evaluating $\mathbf{L}$, however, since the particle velocities are $p_0/\rho c$ for $x < ct$ and zero for $x > ct$. Thus

$$\mathbf{L} = \frac{p_0}{\rho c}(\rho A c t)\hat{\mathbf{i}} = p_0 A t \hat{\mathbf{i}}$$

and $\dot{\mathbf{L}} = p_0 A\hat{\mathbf{i}}$.

2.6 Euler's Second Law (The Moment Equation)

A second relationship between the external forces on a particle system or a body is obtained if, referring to Figure 2.10, we take the cross product of $\mathbf{r}_i$ with both sides of Equation (2.1):

$$\mathbf{r}_i \times \mathbf{F}_i + \sum_{j=1}^{N} \mathbf{r}_i \times \mathbf{f}_{ij} = \mathbf{r}_i \times m_i \mathbf{a}_i \tag{2.31}$$

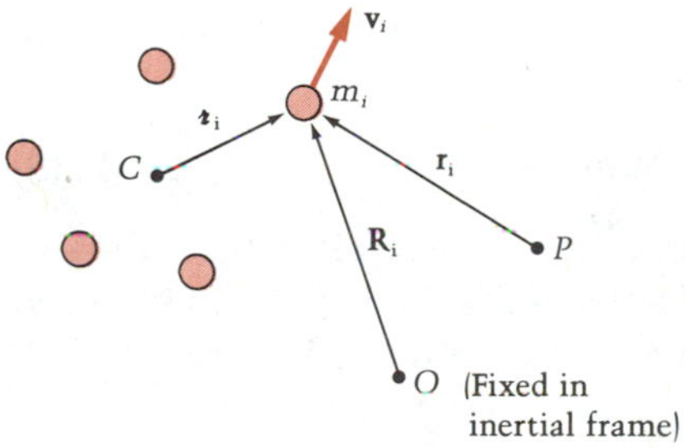

Figure 2.10

The first term on the left is recognized as the moment about point P of the external force $\mathbf{F}_i$. The cross product $\mathbf{r}_i \times \mathbf{f}_{ij}$ is the moment with respect to P of the force exerted on the ith particle by the jth particle. As before, we now sum the N equations typified by Equation (2.31) to obtain

$$\sum_{i=1}^{N} \mathbf{r}_i \times \mathbf{F}_i + \sum_{i=1}^{N} \sum_{j=1}^{N} \mathbf{r}_i \times \mathbf{f}_{ij} = \sum_{i=1}^{N} \mathbf{r}_i \times m_i \mathbf{a}_i \tag{2.32}$$

Terms in the double sum occur in pairs, such as

$$\mathbf{r}_1 \times \mathbf{f}_{12} + \mathbf{r}_2 \times \mathbf{f}_{21}$$

But $\mathbf{r}_2 \times \mathbf{f}_{21} = \mathbf{r}_1 \times \mathbf{f}_{21}$ since $\mathbf{r}_2$ and $\mathbf{r}_1$ both terminate on the line of action of $\mathbf{f}_{21}$. Moreover, $\mathbf{f}_{21} = -\mathbf{f}_{12}$ so that

$$\begin{aligned} \mathbf{r}_1 \times \mathbf{f}_{12} + \mathbf{r}_2 \times \mathbf{f}_{21} &= \mathbf{r}_1 \times (\mathbf{f}_{12} + \mathbf{f}_{21}) \\ &= \mathbf{0} \end{aligned}$$

and similarly for other such pairs. That is, the moments of the internal forces of interaction sum to zero. Thus

$$\Sigma \mathbf{r}_i \times \mathbf{F}_i = \Sigma \mathbf{r}_i \times m_i \mathbf{a}_i$$

or

$$\Sigma \mathbf{M}_P = \Sigma \mathbf{r}_i \times m_i \mathbf{a}_i \tag{2.33}$$

which is the particle-system form of Euler's second law and states that the sum of the moments of the external forces about a point equals the sum of the moments of the $m\mathbf{a}$'s about that point.

For a body whose mass is continuously distributed, the counterpart to Equation (2.33) is

$$\Sigma \mathbf{M}_P = \int \mathbf{r} \times \mathbf{a}\, dm \tag{2.34}$$

Question 2.9 In Equation 2.33 (or 2.34) must point P be fixed in the inertial frame of reference?

Equations (2.4) and (2.34) play the same roles in dynamics as do the equations of equilibrium in statics. And in fact we obtain those equations, $\Sigma \mathbf{F} = \mathbf{0}$ and $\Sigma \mathbf{M} = \mathbf{0}$, from (2.4) and (2.34), if we set to zero the accelerations of all points of a body.

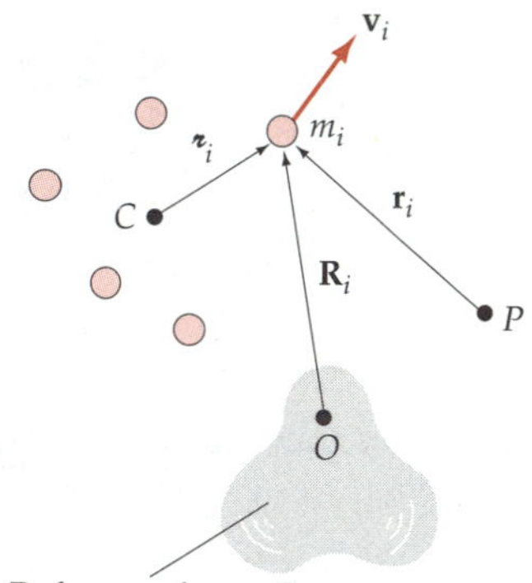

Figure 2.11

Moment of Momentum

Just as Euler's first law can be expressed in terms of the time derivative of momentum of a body, so Euler's second law can be expressed in terms of the time derivative of a quantity called **moment of momentum,** or **angular momentum.*** The moment of momentum with respect to a point P is designated $\mathbf{H}_P$ and is defined to be the sum of the moments (with respect to P) of the momenta of the individual particles making up the body. Referring to Figure 2.11, where v_i is the velocity in reference frame $\mathcal{I}$ of the ith particle, we have

$$\mathbf{H}_P = \Sigma \mathbf{r}_i \times m_i \mathbf{v}_i \tag{2.35}$$

Before proceeding to the development of several of the forms of Euler's second law, we shall develop a very useful relationship between moments of momentum. Noting from the definition, Equation (2.35), and from Figure 2.11 that

$$\mathbf{H}_C = \Sigma \boldsymbol{\mathscr{r}}_i \times m_i \mathbf{v}_i$$

and that

$$\mathbf{r}_i = \mathbf{r}_{PC} + \boldsymbol{\mathscr{r}}_i$$

then for any point P,

$$\begin{aligned} \mathbf{H}_P &= \Sigma(\boldsymbol{\mathscr{r}}_i + \mathbf{r}_{PC}) \times \mathbf{m}_i \mathbf{v}_i = \Sigma \boldsymbol{\mathscr{r}}_i \times m_i \mathbf{v}_i + \mathbf{r}_{PC} \times \Sigma m_i \mathbf{v}_i \\ &= \mathbf{H}_C + \mathbf{r}_{PC} \times \Sigma m_i \mathbf{v}_i \end{aligned}$$

Answer 2.9 No. Nowhere in the development did we need to fix P.

* The term angular momentum stems from the fact that the moment of momentum of a rigid body is related to the angular velocity of the body.

But from Section 2.5, $\Sigma m_i \mathbf{v}_i$ is the momentum $\mathbf{L}$, also expressed as

$$\mathbf{L} = m\mathbf{v}_C$$

so that

$$\mathbf{H}_P = \mathbf{H}_C + \mathbf{r}_{PC} \times \mathbf{L} \tag{2.36}$$

Thus the moment of the momentum of a body about *any* point P (not necessarily fixed in the reference frame) is the sum of the moment of momentum about its mass center, C, and the moment of its (linear) momentum $\mathbf{L}$ about P, where $\mathbf{L}$ is given a "line of action" through C.

Now we can return to the definition of moment of momentum, Equation (2.35), and apply it for the case of a point, O, fixed in the frame of reference. Thus, using definition (2.35) for the third time,

$$\mathbf{H}_O = \Sigma \mathbf{R}_i \times m_i \mathbf{v}_i$$

Now, differentiating with respect to time in $\mathcal{I}$,

$$\frac{d\mathbf{H}_O}{dt} = \Sigma(\dot{\mathbf{R}}_i \times m_i \mathbf{v}_i + \mathbf{R}_i \times m_i \mathbf{a}_i) \tag{2.37}$$

But because O is fixed in $\mathcal{I}$,

$$\dot{\mathbf{R}}_i = \mathbf{v}_i$$

so that

$$\dot{\mathbf{R}}_i \times \mathbf{v}_i = 0$$

Therefore

$$\frac{d\mathbf{H}_O}{dt} = \Sigma \mathbf{R}_i \times m_i \mathbf{a}_i \tag{2.38}$$

Momentum Forms of Euler's Second Law

Now the fundamental form of Euler's second law, Equation (2.33), tells us that if the frame $\mathcal{I}$ in which the $\mathbf{a}_i$ are calculated is an inertial frame, then

$$\Sigma \mathbf{M}_O = \Sigma \mathbf{R}_i \times m_i \mathbf{a}_i \tag{2.39}$$

so that, from the last two equations, we see that

$$\Sigma \mathbf{M}_O = \frac{d\mathbf{H}_O}{dt} \tag{2.40}$$

Another similar form of Euler's second law can be deduced if we first use Equation (2.36) in the case of a fixed point O:

$$\mathbf{H}_O = \mathbf{H}_C + \mathbf{r}_{OC} \times \mathbf{L}$$

Differentiating with respect to time in the reference frame $\mathcal{I}$,

$$\frac{d\mathbf{H}_O}{dt} = \frac{d\mathbf{H}_C}{dt} + \mathbf{v}_C \times \mathbf{L} + \mathbf{r}_{OC} \times \frac{d\mathbf{L}}{dt} \tag{2.41}$$

where we have used the fact that

$$\frac{d\mathbf{r}_{OC}}{dt} = \mathbf{v}_C$$

But we again recall that

$$\mathbf{L} = m\mathbf{v}_C \qquad (\text{so that } \mathbf{v}_C \times \mathbf{L} = \mathbf{0})$$

and therefore

$$\frac{d\mathbf{H}_O}{dt} = \frac{d\mathbf{H}_C}{dt} + \mathbf{r}_{OC} \times \frac{d\mathbf{L}}{dt}$$

Now we know from our study of equipollent force systems in statics that the external forces on the body must produce moments about O and C that are related by

$$\Sigma\mathbf{M}_O = \Sigma\mathbf{M}_C + \mathbf{r}_{OC} \times (\Sigma\mathbf{F}) \tag{2.42}$$

This law of resultants has nothing to do with whether or not the body is in equilibrium. And since $\mathcal{I}$ is an inertial frame, then we also know

$$\Sigma\mathbf{F} = \frac{d\mathbf{L}}{dt} \qquad \text{and} \qquad \Sigma\mathbf{M}_O = \frac{d\mathbf{H}_O}{dt}$$

and thus we may subtract the two Equations (2.41) and (2.42) to obtain

$$\Sigma\mathbf{M}_C = \frac{d\mathbf{H}_C}{dt} \tag{2.43}$$

Neither of the above equations $\Sigma\mathbf{M}_C = \dot{\mathbf{H}}_C$ or $\Sigma\mathbf{M}_O = \dot{\mathbf{H}}_O$ is any more basic or special than the other, as each one can be derived from the other. They are therefore equivalent forms. However, the equation does *not* hold for any arbitrary point P, i.e., in general $\Sigma\mathbf{M}_P \neq \dot{\mathbf{H}}_P$.

Conservation of Moment of Momentum

We next note that — as was the case with linear momentum — there are situations in which a moment of momentum is conserved. In particular, if for an interval of time $\Sigma\mathbf{M}_O = \mathbf{0}$, then during that interval $\dot{\mathbf{H}}_O = \mathbf{0}$, and thus $\mathbf{H}_O$ is constant. For example, let the body of interest be a single spherical planet in its motion around its star. The gravitational force exerted on the planet by the star always passes through the star's mass center O, so $\Sigma\mathbf{M}_O = \mathbf{0}$ and thus $\mathbf{H}_O$ of the planet is a constant. This result is shown in Section 8.4 to lead to the elliptical orbit of the earth around the sun.

For an arbitrary point P, there is a form of Euler's second law that is of particular value in analyzing the motions of rigid bodies, although it remains valid for non-rigid bodies as well. To derive it we again use our knowledge about force systems to write

$$\Sigma\mathbf{M}_P = \Sigma\mathbf{M}_C + \mathbf{r}_{PC} \times (\Sigma\mathbf{F}) \tag{2.44}$$

Thus, using Euler's laws (Equations (2.27 and (2.43))

$$\Sigma \mathbf{M}_P = \frac{d\mathbf{H}_C}{dt} + \mathbf{r}_{PC} \times \frac{d\mathbf{L}}{dt}$$

or

$$\Sigma \mathbf{M}_P = \dot{\mathbf{H}}_C + \mathbf{r}_{PC} \times m\mathbf{a}_C \tag{2.45}$$

which we shall use later.

Question 2.10 Must point P be fixed in the inertial frame of reference, (a) for Equation (2.36) to be true? (b) for Equation (2.45) to be true?

Finally, we remind the reader that all of the relationships of this chapter pertain only to a specific collection of material — that is, a system (or body) of constant mass. However, the momentum forms of Euler's laws provide the natural starting point for developing relationships appropriate to "variable mass" systems such as rockets. If desired, the reader now has the proper background to study that special topic which is found in Section 8.3.

Answer 2.10 (a) No; (b) No. Point P was unrestricted in both derivations.

Figure E2.18a

Example 2.18

Two gymnasts of equal weight (see Figure E2.18a) are hanging in equilibrium at the ends of a rope passing over a relatively light pulley for which the bearing friction can be neglected. Then the gymnast on the right begins to climb the rope, while the gymnast on the left simply holds on. When the right gymnast has raised himself through height h (relative to the floor), what has been the change in position of the left gymnast?

Solution

Constructing a free-body diagram (Figure E2.18b) of the pulley-rope-gymnasts system in which we neglect the weights of the pulley and the rope, we see that

$$\begin{aligned}\dot{\mathbf{H}}_O &= \Sigma \mathbf{M}_O \\ &= mgr\hat{\mathbf{k}} + mgr(-\hat{\mathbf{k}}) \\ &= \mathbf{0}\end{aligned}$$

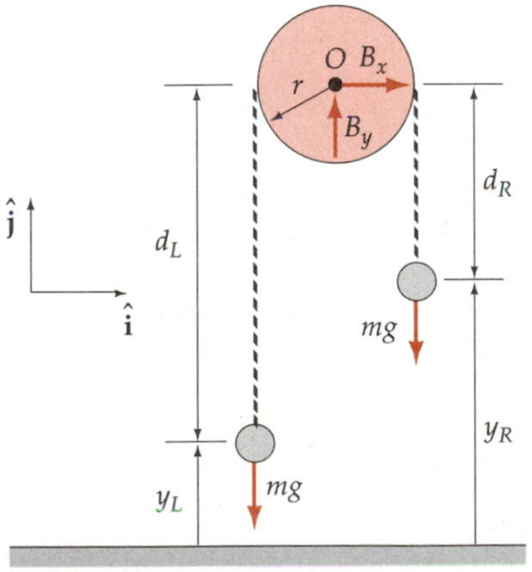

Figure E2.18b

Therefore, $\mathbf{H}_O$ is constant during the motion, and since everything starts from rest,

$$\mathbf{H}_O = \mathbf{0}$$

Treating the gymnasts as particles and neglecting the moment of momentum of the pulley,

$$\mathbf{H}_O = (r\hat{\mathbf{i}} - d_R\hat{\mathbf{j}}) \times (m\dot{y}_R\hat{\mathbf{j}}) + (-r\hat{\mathbf{i}} - d_L\hat{\mathbf{j}}) \times (m\dot{y}_L\hat{\mathbf{j}})$$

or

$$\mathbf{H}_O = rm\dot{y}_R\hat{\mathbf{k}} + rm\dot{y}_L(-\hat{\mathbf{k}})$$

But

$$\mathbf{H}_O = \mathbf{0}$$

so that

$$\dot{y}_R = \dot{y}_L$$

and the left gymnast, "going along for the ride," rises at the same rate as the right one. Thus when the right gymnast has pulled himself up height h, the left one has been pulled up the same height h. Note that if the rope is inextensible, the right gymnast therefore would have climbed $2h$ relative to the rope!

EXAMPLE 2.19

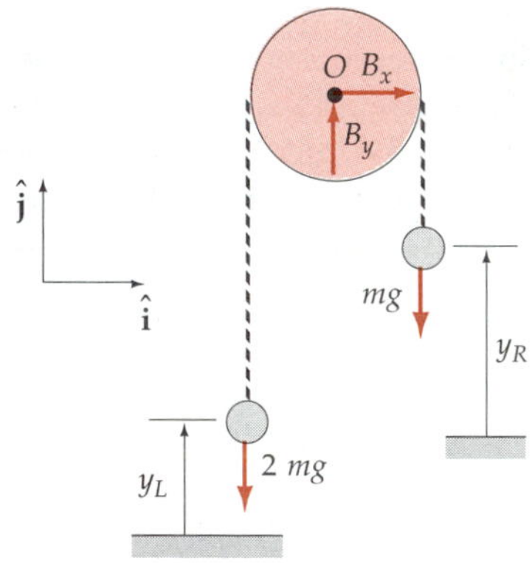

Figure E2.19

Suppose the "counterweight" gymnast on the left in the preceding example were to weigh twice that of the climbing gymnast, as suggested by Figure E2.19. What then would be the relationship between their elevation changes?

Solution

From the free-body diagram

$$\Sigma \mathbf{M}_O = 2mgr\hat{\mathbf{k}} - mgr\hat{\mathbf{k}}$$
$$= mgr\hat{\mathbf{k}}$$

Since $mgr\hat{\mathbf{k}} \neq \mathbf{0}$, the moment of momentum is not conserved this time. Integrating,

$$\int_O^t \Sigma \mathbf{M}_O d\tau = \mathbf{H}_O(t) - \cancel{\mathbf{H}_O(0)}^{\,0}$$

or

$$mgrt\hat{\mathbf{k}} = rm\dot{y}_R\hat{\mathbf{k}} + r(2m)\dot{y}_L(-\hat{\mathbf{k}})$$

so that

$$\dot{y}_L = \frac{1}{2}\dot{y}_R - \frac{gt}{2}$$

If we define y_R and y_L so that $y_R = y_L = 0$ at $t = 0$, then

$$y_L = \frac{1}{2}y_R - \frac{gt^2}{4}$$

Thus we see that it's possible for the lighter gymnast to raise the heavier gymnast by climbing rapidly enough.

For an inextensible rope, the right gymnast climbs, as before, at a rate of $\dot{y}_R + \dot{y}_L$ relative to the rope.

PROBLEMS ► Section 2.6

2.166 The uniform rigid bar $\mathcal{B}$ in Figure P2.166 weighs 60 lb and is pinned at A (and fastened by the cable DB) to the frame $\mathcal{F}$. If the frame is given an acceleration $a = 32.2$ ft/sec^2 as shown, determine the tension T in the cable and the force exerted by the pin at A on the bar.

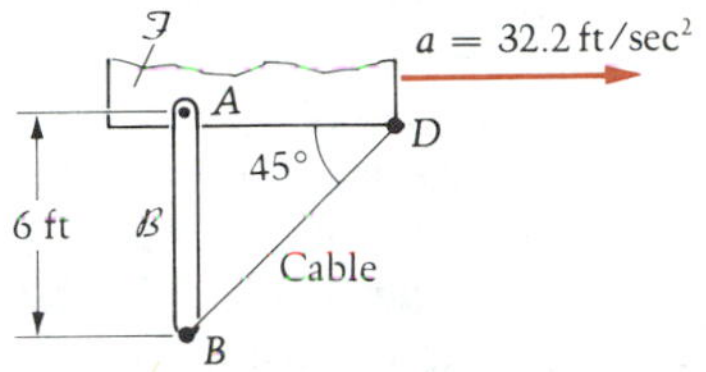

Figure P2.166

2.167 A force F causes the carriage to move with rectilinear horizontal motion defined by a constant acceleration of 20 ft/sec² (see Figure P2.167). A rigid, slender, homogeneous rod of weight 32.2 lb and length 6 ft is welded to the carriage at B and projects vertically upward. Find, in magnitude and direction, the bending moment that the carriage exerts on the rod at B.

2.168 A uniform slender bar of density ρ, cross-sectional area A, and length L undergoes small-amplitude, free transverse vibrations according to $y(x, t) = Y \sin(\pi x / L) \sin \omega t$, where y is the displacement perpendicular to the axis (x) of the bar. (See Figure P2.168.) Neglecting other components of displacement (and hence acceleration), calculate the maximum force generated at one of the supports during the motion.

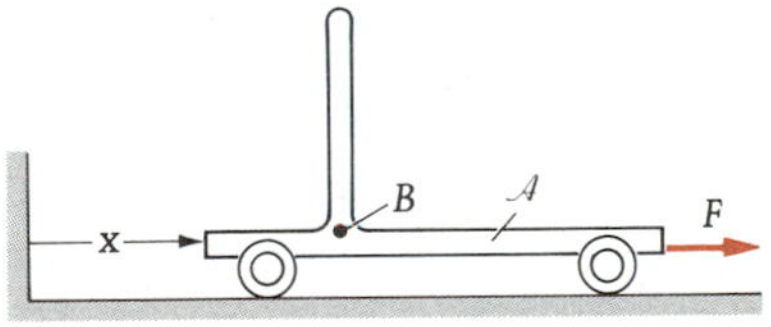

Figure P2.167

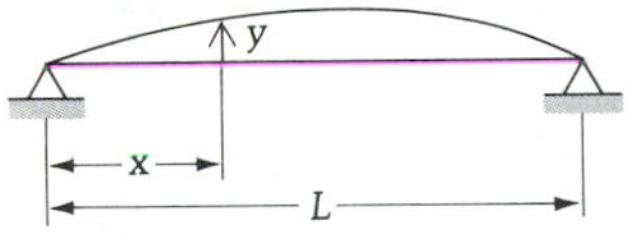

Figure P2.168

2.169 Show that in Equation (2.36), point C need not be the mass center, i.e., if Q is another arbitrary point like P, then show that $\mathbf{H}_P = \mathbf{H}_Q + \mathbf{r}_{PQ} \times \mathbf{L}$ (where $\mathbf{L}$ is of course still $\Sigma m_i \mathbf{v}_i = m\mathbf{v}_C$).

2.170 Let S be a set of vectors $\mathbf{Q}_1, \mathbf{Q}_2, \ldots, \mathbf{Q}_i, \ldots, \mathbf{Q}_N$ of equal dimension. Define the resultant of S as $\mathbf{R} = \Sigma \mathbf{Q}_i$, and place each $\mathbf{Q}_i$ at a point P_i. Define the moment of the set of vectors about a point A by

$$\Sigma \mathbf{M}_A = \Sigma(\mathbf{r}_{AP_i} \times \mathbf{Q}_i)$$

and show that

$$\Sigma \mathbf{M}_P = \Sigma \mathbf{M}_A + \mathbf{r}_{PA} \times \mathbf{R}$$

Now note that (a) if the $\mathbf{Q}_i$ are a set of forces $\mathbf{F}_i$, then Equation (2.42) results with $\mathbf{R}$ being $\Sigma \mathbf{F}_i$; and (b) if the $\mathbf{Q}_i$ are a set of momenta $m_i \mathbf{v}_i$ of a group of particles, then Equation (2.36) results with $\mathbf{R}$ being $\Sigma m_i \mathbf{v}_i$ (or $\mathbf{L}$ or $m\mathbf{v}_c$). Thus we may conclude that both equations, (2.36) and (2.42), are practical examples of the very same law of resultants!

2.171 The angular momentum about point Q is defined as

$$\mathbf{H}_Q = \Sigma(\mathbf{r}_{QP_i} \times m_i \mathbf{v}_i)$$

Differentiate this expression in the inertial reference frame $\mathcal{J}$, and show by the result and that of the preceding problem that, in general, (i.e., not just at an isolated instant of time):

$$\Sigma \mathbf{M}_Q = \dot{\mathbf{H}}_Q$$

only if (a) Q is fixed in $\mathcal{J}$, or (b) Q is the mass center C; or (c) $\mathbf{v}_Q$ is parallel to $\mathbf{v}_C$.

2.172 In Problem 2.132 find: (a) the angular momentum of the system with respect to the origin; (b) the angular momentum of the system with respect to the mass center.

2.173 A massless rope hanging over a massless, frictionless pulley supports two monkeys (one of mass M, the other of mass $2M$). The system is released at rest at $t = 0$, as shown in Figure P2.173. During the following 2 sec, monkey B travels down 15 ft of rope to obtain a massless peanut at end P. Monkey A holds tightly to the rope during these 2 sec. Find the displacement of A during the time interval.

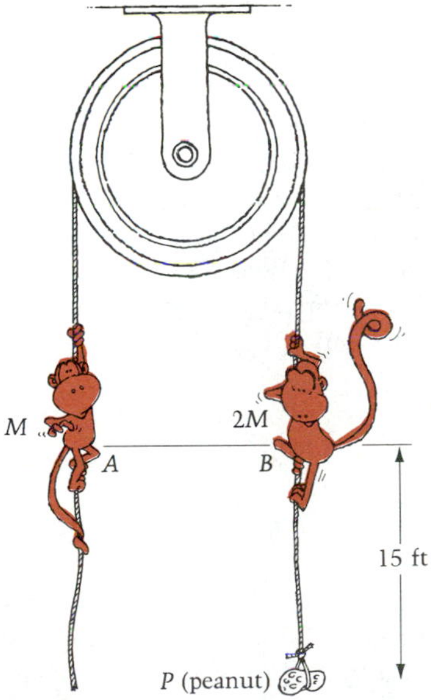

Figure P2.173

2.174 A starving monkey of mass m spies a bunch of delicious bananas of the same mass. (See Figure P2.174.) He climbs at a varying speed relative to the (light) rope. Determine whether the monkey reaches the bananas before they sail over the pulley if the pulley's mass is negligible ($\ll m$).

Figure P2.174

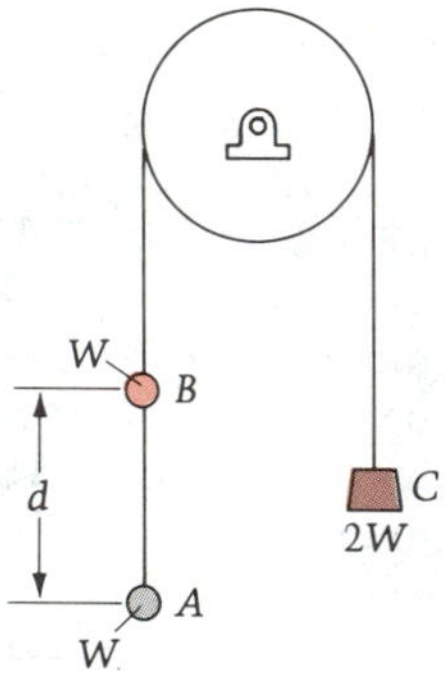

Figure P2.175

2.175 Two gymnasts A and B, each of weight W, hold onto the left side of a rope that passes over a light pulley to a counterweight C of weight $2W$. (See Figure P2.175.) Initially the gymnast A is at depth d below B. He climbs the rope to join gymnast B. Determine the displacement of the counterweight C at the end of the climb.

2.176 Define the angular momentum of a particle about a fixed axis and state the conditions under which the angular momentum remains constant. A man (to be regarded as a particle) stands on a swing. His distance from the smooth horizontal axis of the swing is L when he crouches and $L-H$ when he stands. As the swing falls he crouches; as it rises he stands — the changeover is assumed instantaneous. If the swing falls through an angle α and then rises through an angle β, show that

$$\sin\frac{\beta}{2} = \left(\frac{L}{L-H}\right)^{3/2} \sin\frac{\alpha}{2}$$

The *relative angular momentum* of a body $\mathcal{B}$ with respect to a point P is defined to be

$$\mathbf{H}_{P_{rel}} = \int \mathbf{R} \times (\mathbf{v} - \mathbf{v}_P)\, dm$$

(See Figure P2.177, where the velocity $\mathbf{v}$ of dm is the derivative of $\mathbf{r}$ in inertial frame $\mathcal{I}$.) Note that what makes it "relative" is that the velocity in the integral is the difference between $\mathbf{v}$ (of dm) and $\mathbf{v}_P$. Now solve the following problems.

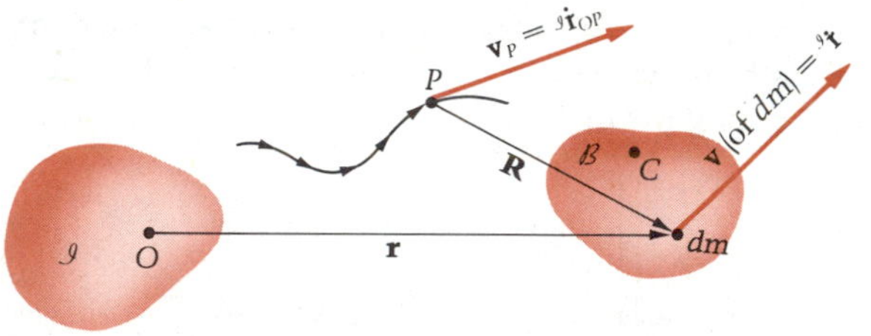

Figure P2.177

2.177 Show that $\mathbf{H}_P = \mathbf{H}_{P_{rel}} + m\mathbf{r}_{PC} \times \mathbf{v}_P$. (Thus $\mathbf{H}_C = \mathbf{H}_{C_{rel}}$ always!)

2.178 Show that $\Sigma\mathbf{M}_P = \dot{\mathbf{H}}_P + \mathbf{v}_P \times m\mathbf{v}_C$. (Thus $\Sigma\mathbf{M}_P$ is not generally equal to $\dot{\mathbf{H}}_P$!)

2.179 Show that $\Sigma\mathbf{M}_P = \dot{\mathbf{H}}_{P_{rel}} + \mathbf{r}_{PC} \times m\mathbf{a}_P$.

2.180 Show that $\dot{\mathbf{H}}_P = \dot{\mathbf{H}}_{P_{rel}}$ if and only if $\mathbf{v}_P \times \mathbf{v}_C = \mathbf{r}_{PC} \times \mathbf{a}_P$.

COMPUTER PROBLEMS ▸ Chapter 2

2.181 In the table on the next page are data of the mass center velocity versus time for a 30-lb crate that was lifted approximately straight up by two people. The "velocity 1" column represents a taller person than the "velocity 2" column. Use the computer to integrate numerically the velocity from $t = 0$ to 4.4 sec, thereby obtaining and comparing the heights to which the crate was lifted by the two people.

Time (sec)	Velocity 1 (in./sec)	Velocity 2 (in./sec)	Time (sec)	Velocity 1 (in./sec)	Velocity 2 (in./sec)
0.0	0.0	0.0	2.3	25.0 (peak)	23.5
0.2	0.0	0.0	2.4	24.0	25.5
0.4	1.5	0.0	2.6	22.0	32.0
0.6	3.5	0.0	2.7	20.5	37.5 (peak)
0.8	5.5	0.0	2.8	19.0	35.0
1.0	8.5	2.0	3.0	16.5	25.0
1.2	12.0	4.0	3.2	14.5	21.5
1.4	16.5	6.5	3.4	13.0	17.0
1.6	19.0	10.0	3.6	12.0	15.0
1.8	21.0	14.0	3.8	10.0	13.0
2.0	22.5	17.0	4.0	8.5	9.5
2.2	24.5	21.0	4.2	7.5	8.5
			4.4	6.5	8.0

SUMMARY ▶ Chapter 2

In this chapter we have set out the fundamental relationships between forces on a body and its motion, and we have illustrated their use for the solution of a variety of problems, many of which are closely associated with our everyday experience.

The starting point here was Newton's second law for a particle,

$$\Sigma \mathbf{F} = m\mathbf{a}$$

where $\Sigma \mathbf{F}$ is the sum of all the forces acting on the particle, m is the mass of the particle and $\mathbf{a}$ is its acceleration relative to an inertial frame of reference. Extending to a system of particles, the ith having mass m_i and acceleration $\mathbf{a}_i$,

$$\Sigma \mathbf{F} = \Sigma m_i \mathbf{a}_i$$

where $\Sigma \mathbf{F}$ is the sum of the *external* forces on the system. Another form of particular value is

$$\Sigma \mathbf{F} = m\mathbf{a}_C$$

where $m = \Sigma m_i$ is the mass of the system of particles, or the body comprising them, and $\mathbf{a}_C$ is the acceleration of the mass center C. In addition, it is sometimes useful to decompose a body into two (or more) parts with masses m_1 and m_2 and mass centers C_1 and C_2, and then we may use:

$$\Sigma \mathbf{F} = m_1\mathbf{a}_{C_1} + m_2\mathbf{a}_{C_2}$$

The preceding equations, for a body of finite size, are forms of what is often called Euler's first law and are counterparts in dynamics to the equilibrium equation, $\Sigma \mathbf{F} = \mathbf{0}$, studied in statics. We have used them to solve a variety of problems such as finding accelerations and constraining forces when some forces and paths were prescribed and also integrating to find the motion of a particle (or mass center of a body) when external forces were prescribed. Central to the problem-solving process was the free-body diagram, the importance of which cannot be overstated.

The Principle of Work and Kinetic Energy is very useful in solving problems in which the speeds of a particle at different locations in space are of interest. The Kinetic Energy, T, of a particle is defined to be

$$T \equiv \tfrac{1}{2}m\,|\mathbf{v}|^2$$

The principle states that the work done by all the forces acting over an interval of time is equal to the change in kinetic energy. Or, in symbols,

$$W = T_2 - T_1$$

This is a derived result following from integrating and assigning the term "work of a force on a particle" to $\Sigma\mathbf{F} = m\mathbf{a}$

$$W = \int_{t_1}^{t_2} \mathbf{F} \cdot \mathbf{v}dt = \int_{\mathbf{r}(t_1)}^{\mathbf{r}(t_2)} \mathbf{F} \cdot d\mathbf{r}$$

Two special forces arise frequently enough in problems to evaluate the work and express it in symbols:

a. Constant force: $W = \mathbf{F} \cdot (\mathbf{r}_2 - \mathbf{r}_1)$, or in words: (magnitude of force) * (magnitude of displacement) * (cosine of angle between force and displacement). For weight (force exerted by gravity near the earth's surface) this means (weight) * (decrease in altitude of mass center).

b. Force exerted by a linear spring:

$$W = -\frac{k}{2}(\delta_2^2 - \delta_1^2)$$

where δ is spring stretch and k is the spring modulus.

A force whose work does not depend upon the path of the point of application is called conservative and a potential energy, φ, is associated with it so that the work done is the negative of the change in that potential energy,

$$W = -[\varphi_2 - \varphi_1]$$

Combining this with $W = \Delta T$, assuming all forces to be conservative,

$$T_2 + \varphi_2 = T_1 + \varphi_1$$

or

$$T + \varphi = \text{constant}$$

which means that in this case, kinetic plus potential energy is conserved. A potential energy for a linear spring is

$$\varphi = \tfrac{1}{2}k\delta^2$$

and for weight (with z being elevation)

$$\varphi = mgz$$

For a system of particles,

$$T = \Sigma\tfrac{1}{2}m_i|\,\mathbf{v}_i\,|^2$$

and the Principle of Work and Kinetic Energy applies so long as one considers the work of all *internal* forces as well as the work of external forces; that is,

$$W_{\text{external}} + W_{\text{internal}} = \Delta T$$

This is of practical value only in special situations where it's possible to readily evaluate the work of internal forces. One example is a rigid body or system of rigidly connected particles, for then $W_{\text{internal}} = 0$. Another case is that of a pair of particles joined by a linear spring, for which the net work of the equal and opposite internal forces is

$$W_{\text{internal}} = -\frac{k}{2}[\delta_2^2 - \delta_1^2].$$

The concept of momentum is particularly useful in problems of impact or collision in which very intense forces of interaction may act for a very brief interval. Momentum is defined to be

$$\mathbf{L} = m\mathbf{v} \qquad \text{(particle)}$$

$$\mathbf{L} = \Sigma m_i \mathbf{v}_i \qquad \text{(system of particles)}$$

from which, for a body in general,

$$\mathbf{L} = m\mathbf{v}_C$$

Euler's first law can be written

$$\Sigma\mathbf{F} = \frac{d\mathbf{L}}{dt} \quad \text{or} \quad \dot{\mathbf{L}}$$

which when integrated yields the impulse-momentum principle

$$\int_{t_1}^{t_2} \Sigma\mathbf{F}\, dt = \mathbf{L}(t_2) - \mathbf{L}(t_1)$$

So if external forces do not act during the interval,

$$\mathbf{L}(t_2) = \mathbf{L}(t_1),$$

and so momentum is conserved. This is quite often (approximately) the case in problems of collision.

Finally in this chapter we have developed the counterpart in dynamics to the second equilibrium equation, $\Sigma\mathbf{M} = \mathbf{0}$, in statics. In dynamics this is, for a system of particles,

$$\Sigma\mathbf{M}_P = \Sigma\mathbf{r}_i \times m_i\mathbf{a}_i$$

when $\Sigma\mathbf{M}_P$ refers, as it did in statics, to the moments of *external* forces. This is often called Euler's second law. There are several different forms in which this law can be expressed, among them expressions involving the moment of momentum, defined as

$$\mathbf{H}_P = \Sigma\mathbf{r}_i \times m_i\mathbf{v}_i$$

for a system of particles. A useful relationship is

$$\mathbf{H}_P = \mathbf{H}_C + \mathbf{r}_{PC} \times \mathbf{L},$$

but the key expressions are the forms that Euler's second law can take,

$$\Sigma \mathbf{M}_C = \dot{\mathbf{H}}_C$$

and

$$\Sigma \mathbf{M}_O = \dot{\mathbf{H}}_O$$

where "O" is a point fixed in the inertial frame of reference.

REVIEW QUESTIONS ▶ Chapter 2

True or False?

1. At a given time, the mass center of a deformable body can be shown to be a unique point.
2. The momentum of any body (or system of bodies) in a frame $\mathcal{I}$ can be shown to be equal to the total mass times the velocity of the mass center in $\mathcal{I}$, even if $\mathcal{I}$ is not an inertial frame.
3. Euler's first law ($\Sigma\mathbf{F} = \dot{\mathbf{L}}$) applies to deformable bodies whether solid, liquid, or gaseous, as well as to rigid bodies and particles.
4. Neither the laws of motion nor the inertial frame is of any value without the other.
5. The mass center of a body $\mathcal{B}$ has to be a physical, or material, point of $\mathcal{B}$.
6. The work done by a linear spring depends on the paths traversed by its endpoints between the initial and final positions.
7. The work done by the friction force upon a block sliding on a fixed plane depends on the path taken by the block.
8. The work done by gravity on a body $\mathcal{B}$ depends on the lateral as well as the vertical displacement of the mass center of $\mathcal{B}$.
9. Since no external work was done on the two bodies of Example 2.14 *during the impact,* their total kinetic energy is the same after the collision as it was before.
10. For all bodies of constant density, the centroid of volume and the center of mass coincide.
11. In studying the motion of the earth around the sun, it is acceptable to treat the earth as a particle; in studying the daily rotation of the earth on its axis, however, it would not make sense to consider the earth as a particle.
12. The external forces acting on a body $\mathcal{B}$, which together form the resultant $\mathbf{F}_r$, must each have a line of action passing through the mass center of $\mathcal{B}$ in order for Euler's first law to apply.
13. Euler's second law can take the form $\Sigma\mathbf{M}_P = \dot{\mathbf{H}}_P$ regardless of the motion of point P in the inertial frame.

Answers: 1. T 2. T 3. T 4. T 5. F 6. F 7. T 8. F 9. F 10. T 11. T 12. F 13. F

3 Kinematics of Plane Motion of a Rigid Body

3.1 Introduction

3.2 Velocity and Angular Velocity Relationship for Two Points of the Same Rigid Body

Development of the Velocity and Angular Velocity Relationship

Important Things to Remember About Equation (3.8)

3.3 Translation

3.4 Instantaneous Center of Zero Velocity

Proof of the Existence of the Instantaneous Center

The Special Case in Which the Normals Do Not Intersect

The Special Case in Which the Normals are Coincident

3.5 Acceleration and Angular Acceleration Relationship for Two Points of the Same Rigid Body

Development of the Acceleration and Angular Acceleration Relationship

3.6 Rolling

Rolling of a Wheel on a Fixed Straight Line

Rolling of a Wheel ($\mathcal{B}_1$) on a Fixed Plane Curve ($\mathcal{B}_2$)

Gears

3.7 Relationship Between the Velocities of a Point with Respect to Two Different Frames of Reference

Relationship Between the Derivatives of a Vector in Two Frames

Velocity Relationship in Two Frames

3.8 Relationship Between the Accelerations of a Point with Respect to Two Different Frames of Reference

Summary

Review Questions

3.1 Introduction

In this chapter our goals are to develop the relationships between velocities, accelerations, angular velocity, and angular acceleration when a rigid body $\mathcal{B}$ moves in plane motion in a reference frame $\mathcal{I}$. Before doing so, however, we shall first explain precisely what we mean by such terms as *rigid body, plane motion, rigid extension, reference plane,* and several other concepts we shall be needing in this chapter and those to follow.

A **rigid body** is taken to be a body in which the distance between each and every pair of its points remains the same throughout the motion.* There is, of course, no such thing as a truly rigid body (since all bodies do *some* deforming); however, the deformations of many bodies are sufficiently small during their motions to allow the bodies to be treated as though they were rigid with good results.

The significance of the rigid-body model is that velocities of different points will be found to differ by something proportional to the rate at which the body turns, what we shall come to call its angular velocity. And accelerations will be found to be related through the angular velocity and its rate of change which we know as the angular acceleration. Thus a very small amount of information will characterize all the accelerations in the body. There are a number of ways in which this is important, but foremost is the fact that the right-hand side of our moment equation in Chapter 2 will in Chapter 4 be seen to take on a compact form involving the angular velocity and the angular acceleration.

Plane motion is treated in this book as motion in the xy plane (fixed in $\mathcal{I}$) or in planes parallel to it. Let a point P be located originally at coordinates (x_p, y_p, z_p). To say that P has plane motion simply means that it stays in the plane $z = z_p$ throughout its motion. Extending this definition, we say that rigid body $\mathcal{B}$ has plane motion whenever *all* its points remain in the same planes (parallel to xy) in which they started.

Question 3.1 How few points of a rigid body must be in plane motion to ensure that they *all* are?

A third concept we need to understand in rigid-body kinematics is that of the **body extended,** also called a **rigid extension** of the body. This idea, briefly mentioned in Chapter 1, says that we sometimes need to imagine points (which are not physical or material points of $\mathcal{B}$) moving with $\mathcal{B}$ as though they were in fact attached to it. An example would be the points on the axis of a pipe that are in the space inside it but of course move rigidly with it. We shall imagine a "rigid extension" of the body to pick up such points whenever it is useful to do so. Note that *any* point Q

* We have already encountered this concept in Chapter 1, where it was seen to be synonymous with the concept of *frame.*

Answer 3.1 Three noncollinear points are needed.

may be considered a point of *any* body $\mathcal{B}$ extended, provided that Q moves with $\mathcal{B}$ as if it were rigidly attached to it.

With these three concepts in mind, we are now prepared to define the **reference plane.** First we must realize that we are faced with the problem of determining where all the points of a body are as functions of time t. These points' locations $(x(t), y(t))$ would take forever to find it we had to do so for each of the infinitely many points of $\mathcal{B}$. Fortunately, for a rigid body in plane motion, if we know the locations of all its points in any one plane of $\mathcal{I}$ (which we shall call the reference plane), then we automatically know the locations of all its *other* points in all *other* planes. The reason for this is as follows. For each point B of $\mathcal{B}$ that does not lie in the reference plane, there is a "companion point" of $\mathcal{B}$ in the reference plane (suggested by A in Figure 3.1) that has the same (x, y) coordinates at the beginning of the motion. It then follows that the (x, y) coordinates of A and B *always* match *throughout* the motion of $\mathcal{B}$!

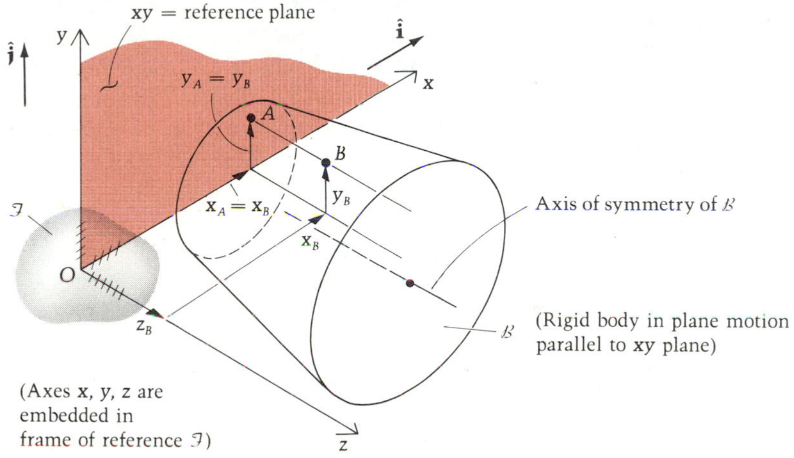

Figure 3.1

Question 3.2 Why is $x_A \equiv x_B$ and $y_A \equiv y_B$ as time passes?

Body $\mathcal{B}$ in Figure 3.1 is a cone pulley, which turns about its axis of symmetry. Note from its varying cross-sectional diameter that a body need not have constant cross section to be in plane motion.

Having laid the necessary groundwork, we now let xy be our reference plane. From Figure 3.1 we see that

$$\mathbf{r}_{OA} = x_A\hat{\mathbf{i}} + y_A\hat{\mathbf{j}} \tag{3.1}$$

Answer 3.2 If ever $x_A \neq x_B$ or $y_A \neq y_B$ (or both), either the rigid body or the plane motion assumptions (or both) will have been violated.

and we may differentiate this equation to obtain

$$\mathbf{v}_A = \dot{x}_A\hat{\mathbf{i}} + \dot{y}_A\hat{\mathbf{j}} = \dot{x}_B\hat{\mathbf{i}} + \dot{y}_B\hat{\mathbf{j}} = \mathbf{v}_B \tag{3.2}$$

$$\mathbf{a}_A = \ddot{x}_A\hat{\mathbf{i}} + \ddot{y}_A\hat{\mathbf{j}} = \ddot{x}_B\hat{\mathbf{i}} + \ddot{y}_B\hat{\mathbf{j}} = \mathbf{a}_B \tag{3.3}$$

In these equations we have used the facts that $x_B \equiv x_A$, $y_B \equiv y_A$, and $z_B =$ constant. Equations (3.2) and (3.3) show clearly that if we completely describe the velocities and accelerations in one reference plane, we then know them for *all* the points of the body. This allows us to focus on one plane of the body throughout this chapter and most of the next two as well.

The reference plane is thus a very important concept, for it allows us to study the motion of an entire body by concerning ourselves only with those of its points that lie in this plane. We say we "know the motion" of $\mathcal{B}$ when we know where all its points are at all times. We have already reduced this task to knowing the locations of the points in the reference plane. (The rest of the body "goes along for the ride.") But in fact if we know the location of just *two* points (say P_1 and P_2) of the reference plane, then we know the whereabouts of *all* points of this plane and thus of the whole body! This is because each point of the reference plane must maintain the same position relative to the points P_1 and P_2. This idea is illustrated in Figure 3.2. Note in the figure that if P_1 and P_2 are correctly located with respect to the reference frame $\mathcal{I}$, all other points of $\mathcal{B}$ are necessarily in their correct positions.

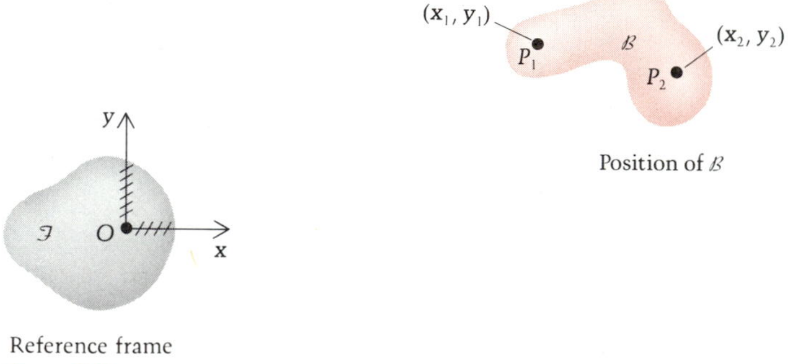

Figure 3.2

Question 3.3 Is knowledge of the locations of two points sufficient for us to know the motion of a body in general (three-dimensional) motion?

Instead of knowing the locations of two points of the body — (x_1, y_1) of P_1 and (x_2, y_2) of P_2 — we may alternatively locate the body if we know where just *one* point, P, is located *plus* the value of the orientation angle θ (about an axis through P_1 and parallel to z); see Figure 3.3.

Answer 3.3 No; the body could rotate around the line joining the two points. In three dimensions it takes three points, not all on the same line!

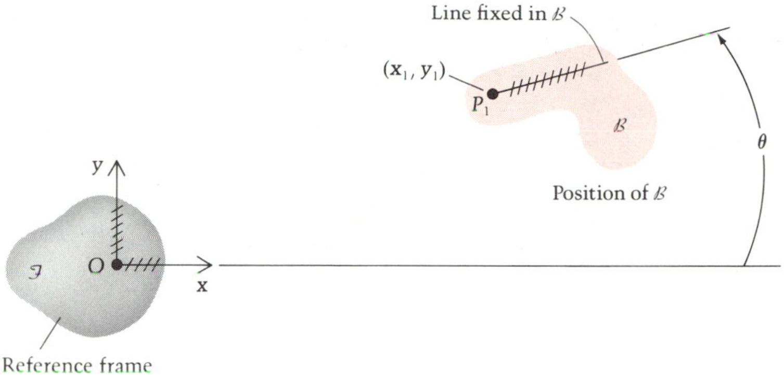

Figure 3.3

> **Question 3.4** Knowing the locations of two points requires four variables (x_1, y_1, x_2, y_2), whereas one point plus the angle takes but three (x_1, y_1, θ). Why do these numbers of variables differ?

The foregoing is intended to suggest quite correctly that the velocities of different points in the reference plane will be linked together because of the rigidity of the body, and similarly for accelerations. In the next section we will turn to the development of the relationship between the velocities of points such as P_1 and P_2 in Figure 3.2.

Answer 3.4 The $2 \times 2 = 4$ coordinates of P_1 and P_2 are not independent. The distance between the points is a constant, so

$$\sqrt{(x_1 - x_2)^2 + (y_1 - y_2)^2} = \text{constant}$$

can be used to find any one of x_1, y_1, x_2, y_2 in terms of the other three.

PROBLEMS ▶ Section 3.1

3.1 Which of the bodies ℬ shown in Figure P3.1(a–f) are in plane motion in frame 𝒥?

(a) A turkey being barbecued by slowly turning on a rotisserie.

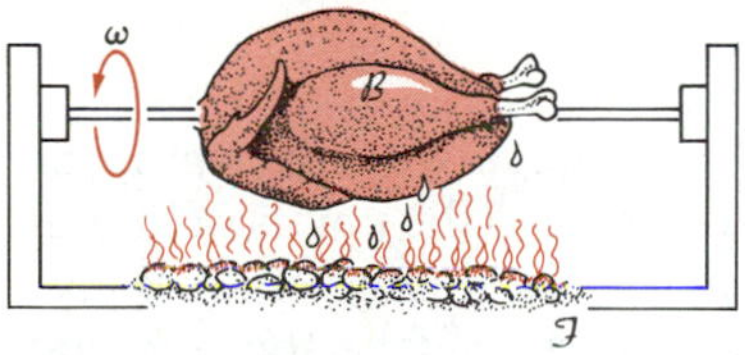

(b) A cone rolling on a tabletop.

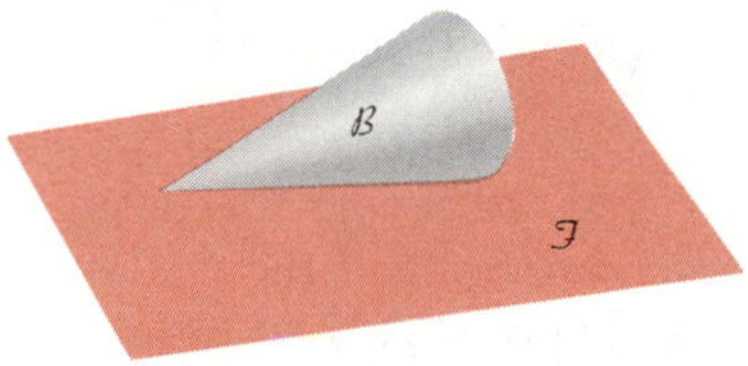

(c) A spinning coin if the base is fixed.

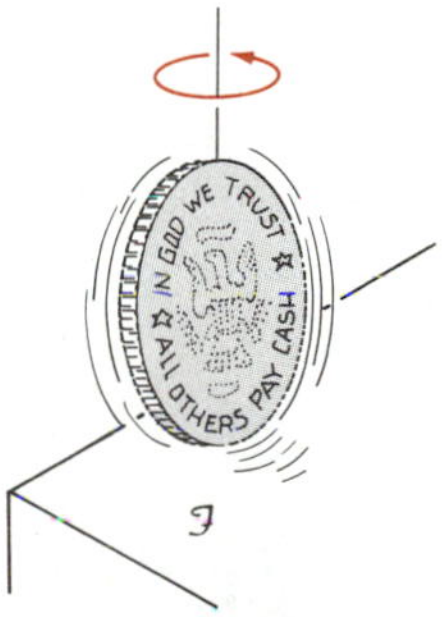

Figure P3.1(a–c) (See next page for d–f)

(d) A can rolling down an inclined plane.

(e) The bevel gear $\mathcal{B}$, which meshes with another bevel gear $\mathcal{A}$.

(f) The (shaded) crosspiece of a universal joint.

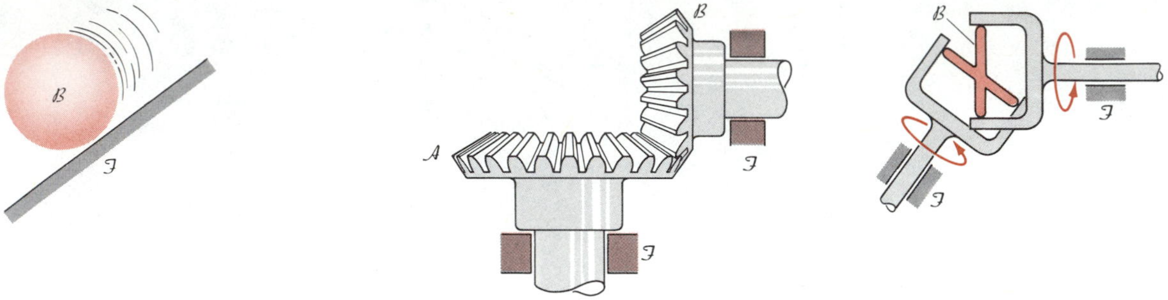

Figure P3.1(d–f)

3.2 Give three examples of plane motion besides those in the previous problem. Then give three examples of motion that is not planar.

3.2 Velocity and Angular Velocity Relationship for Two Points of the Same Rigid Body

In this section we derive a very useful relationship between the velocities in $\mathcal{F}$ of any two points in the reference plane of a rigid body $\mathcal{B}$ in plane motion and the angular velocity vector of $\mathcal{B}$ in $\mathcal{F}$. Let P and Q denote these two points of $\mathcal{B}$, and let us embed the axes (x, y, z) in reference frame $\mathcal{F}$ as shown in Figure 3.4.

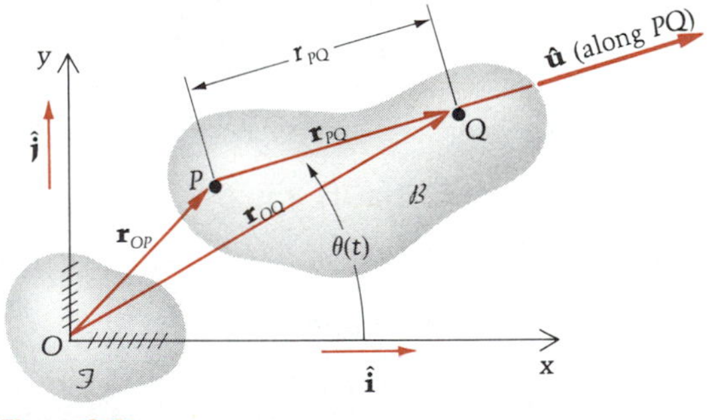

Figure 3.4

We are saying that even though $\mathcal{B}$ may move with respect to the reference frame $\mathcal{F}$, the xy plane of $\mathcal{F}$ always contains the points of interest P and Q of $\mathcal{B}$. A good example is found in the classroom; let the body $\mathcal{B}$ be a blackboard eraser. Letting the blackboard itself be the reference frame $\mathcal{F}$ (so that x and y are fixed in the plane of the blackboard), the eraser undergoes plane motion whenever the professor erases the board. Our points P and Q are any two points of the erasing surface of the eraser. Note how each point of the eraser remains the same z distance from the blackboard (where $z = 0$) during the erasing. The eraser is no longer in plane motion, however, once it *leaves* the surface of the board and its points move with z components of velocity.

Notice from Figure 3.4 that $\hat{\mathbf{u}}$ is a unit vector always directed from P toward Q so that $\mathbf{r}_{PQ} = r_{PQ}\hat{\mathbf{u}}$, where r_{PQ} is the distance PQ (that is, the magnitude of the vector $\mathbf{r}_{PQ}$). Note further that the orientation (angular rotation) of the body is described by the angle θ, measured between any line fixed in the reference frame $\mathcal{I}$ (we shall here use the x axis) and any line fixed in the body (for the moment, we shall use the line segment from P to Q).

Development of the Velocity and Angular Velocity Relationship

We are now ready to develop the velocity and angular velocity relationship for rigid bodies. From Figure 3.4 we see that

$$\mathbf{r}_{OQ} = \mathbf{r}_{OP} + \mathbf{r}_{PQ} \tag{3.4}$$

so that we have, upon differentiation in frame $\mathcal{I}$,

$$\dot{\mathbf{r}}_{OQ} = \dot{\mathbf{r}}_{OP} + \dot{\mathbf{r}}_{PQ}$$

Recognizing the first two vectors as the definitions of the velocities of P and Q (in $\mathcal{I}$, where O is fixed), we may write

$$\mathbf{v}_Q = \mathbf{v}_P + \dot{\mathbf{r}}_{PQ} \tag{3.5}$$

In obtaining Equation (3.5), all derivatives were taken in $\mathcal{I}$, so that, for example, there is no need to write $\mathbf{v}_{Q/\mathcal{I}}$.

In order to write $\dot{\mathbf{r}}_{PQ}$ as a vector we can use, we express $\mathbf{r}_{PQ}$ as a magnitude times a unit vector. With the help of Figure 3.4 we get

$$\mathbf{r}_{PQ} = r_{PQ}\hat{\mathbf{u}} = r_{PQ}(\cos\theta\hat{\mathbf{i}} + \sin\theta\hat{\mathbf{j}}) \tag{3.6}$$

Differentiating this expression as in Section 1.6, we have

$$\begin{aligned}\dot{\mathbf{r}}_{PQ} &= r_{PQ}\dot{\theta}(-\sin\theta\hat{\mathbf{i}} + \cos\theta\hat{\mathbf{j}}) = r_{PQ}\dot{\theta}\hat{\mathbf{k}} \times (\cos\theta\hat{\mathbf{i}} + \sin\theta\hat{\mathbf{j}})^* \\ &= r_{PQ}\dot{\theta}\hat{\mathbf{k}} \times \hat{\mathbf{u}} = \dot{\theta}\hat{\mathbf{k}} \times r_{PQ}\hat{\mathbf{u}}\end{aligned}$$

Therefore we have derived a useful expression for $\dot{\mathbf{r}}_{PQ}$:

$$\dot{\mathbf{r}}_{PQ} = \dot{\theta}\hat{\mathbf{k}} \times \mathbf{r}_{PQ} \tag{3.7}$$

Question 3.5 In the preceding development, why is $\dot{r}_{PQ} = 0$?

Substituting Equation (3.7) into (3.5) yields

$$\mathbf{v}_Q = \mathbf{v}_P + \dot{\theta}\hat{\mathbf{k}} \times \mathbf{r}_{PQ} \tag{3.8}$$

* Throughout this book $\hat{\mathbf{i}}$, $\hat{\mathbf{j}}$, and $\hat{\mathbf{k}}$ constitute a right-handed system so that $\hat{\mathbf{i}} \times \hat{\mathbf{j}} = \hat{\mathbf{k}}$.

Answer 3.5 Because $\mathcal{B}$ is rigid, r_{PQ} is the *constant* distance between points P and Q.

which relates the velocities of the points P and Q and introduces the **angular velocity** of $\mathcal{B}$ in reference frame $\mathcal{I}$. The Greek letter omega is usually used to denote this vector:

$$\boldsymbol{\omega}_{B/\mathcal{I}} = \dot{\theta}\hat{\mathbf{k}} \tag{3.9}$$

When there is no confusion about the body and reference frame involved, we may drop the subscripts and write $\boldsymbol{\omega}_{B/\mathcal{I}}$ as simply $\boldsymbol{\omega}$. Also, some prefer to write $\boldsymbol{\omega}$ as $\omega\hat{\mathbf{k}}$ rather than $\dot{\theta}\hat{\mathbf{k}}$; we shall use both forms, feeling that the latter is a nice reminder that in plane motion, angular velocity is proportional to the time rate of change of an angle.*

The magnitude $|\dot{\theta}|$ (or $|\boldsymbol{\omega}|$ or $|\omega|$) of the angular velocity is called the **angular speed** of $\mathcal{B}$ in frame $\mathcal{I}$. Note that $\dot{\theta}$ (or ω) itself can be negative. Note further that neither the angular velocity vector nor Equation (3.8) depends on which body-fixed line segment (such as PQ above) is chosen to measure θ. The proof of the preceding statement is not difficult and will be given later as an exercise.

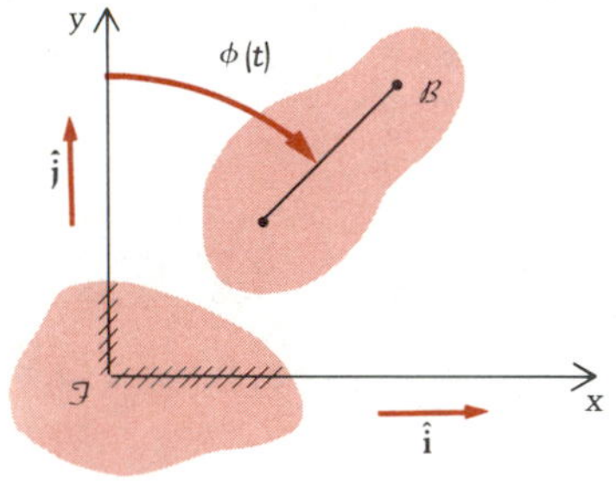

Figure 3.5

Note that if our angle of orientation were chosen as shown in Figure 3.5, then the angular velocity would be given by

$$\boldsymbol{\omega} = \dot{\varphi}(-\hat{\mathbf{k}}) = -\dot{\varphi}\hat{\mathbf{k}}$$

The angular velocity vector is always in the direction given by the right-hand rule when we turn our fingers in the direction of rotation of the body. Referring to both Figures 3.4 and 3.5, $\dot{\varphi} = -\dot{\theta}$, and

$$\boldsymbol{\omega} = \dot{\theta}\hat{\mathbf{k}} = -\dot{\varphi}\hat{\mathbf{k}}$$

and $\boldsymbol{\omega}$ is directed out of the page if the body is turning counterclockwise, and into the page if it's turning clockwise.

Important Things to Remember About Equation (3.8)

We now have the result that the angular velocity vector is a property of the overall body $\mathcal{B}$, and *not* a property of its individual *points*. This idea cannot be overemphasized. Remember:

1. A *point* has position, velocity, and acceleration.
2. A *body* has orientation, angular velocity, and angular acceleration.†

Remember too that a *point* does not have orientation, $\boldsymbol{\omega}$, or $\boldsymbol{\alpha}$,† and a finite-sized *body* does not have a unique **r**, **v**, and **a**.**

We shall emphasize these property differences between points and bodies in the following way: throughout this book, points are denoted by capital italic printed letters while bodies are denoted by ordinary capital

* We remark that in general (three-dimensional) motion, such a simple relationship as $\dot{\theta}\hat{\mathbf{k}}$ between angular velocity and body orientation does *not* exist.

† Angular acceleration ($\boldsymbol{\alpha}$), the derivative of angular velocity, is discussed in Section 3.5.

** Of course the *particle*, being treated as small enough that we need not distinguish between the locations of its points, must be considered as having an **r**, **v**, and **a** — and *not* an $\boldsymbol{\omega}$ or $\boldsymbol{\alpha}$.

cursive letters. Hence, for example, P, A, and B denote points, while $\mathcal{P}$, $\mathcal{A}$, and $\mathcal{B}$ denote bodies. Therefore, we shall simply print the names of points, and write the names of bodies in cursive script.

We now move toward a number of examples of the use of our new Equation (3.8); in each application of this equation, the following three rules must be followed *without exception:*

3. This vector extends *from* the point (P) on *this* (right) side of the equation *to* the point (Q) on the *other* (left) side.

$$\mathbf{v}_Q = \mathbf{v}_P + \dot{\theta}\hat{\mathbf{k}} \times \mathbf{r}_{PQ} \tag{3.10}$$

1. These two points are on the *same* rigid body $\mathcal{B}$.
2. This is the angular velocity vector of $\mathcal{B}$.

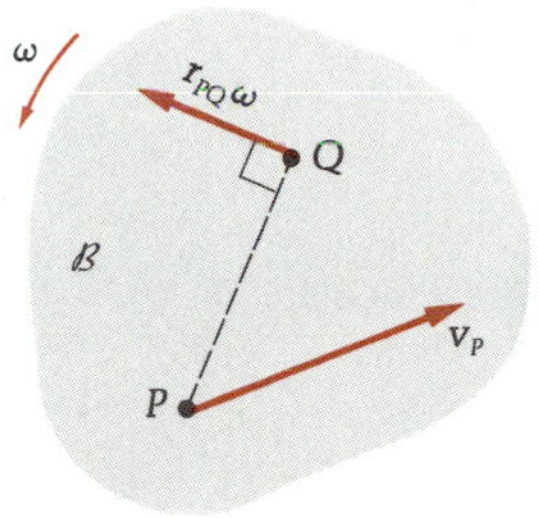

Figure 3.6

Also helpful in using Equation (3.8) is the kinematic diagram presented in Figure 3.6. The velocity of Q, from Equation (3.8), is the sum of the two vectors in Figure 3.6 (see Figure 3.7). Note that depending on the relative sizes of v_P and $r_{PQ}\omega$, the velocity $\mathbf{v}_Q$ could lie on either side, or even along, line PQ. Note further that the difference between the velocities of Q and P — that is, $\mathbf{v}_Q - \mathbf{v}_P$ — is simply $\dot{\theta}\hat{\mathbf{k}} \times \mathbf{r}_{PQ}$. This means that the only way in which the velocities of two points of a rigid body $\mathcal{B}$, in motion in frame $\mathcal{I}$, can differ is by the $r\omega$ term normal to the line joining them. We shall return to this idea following the first three examples of this section.

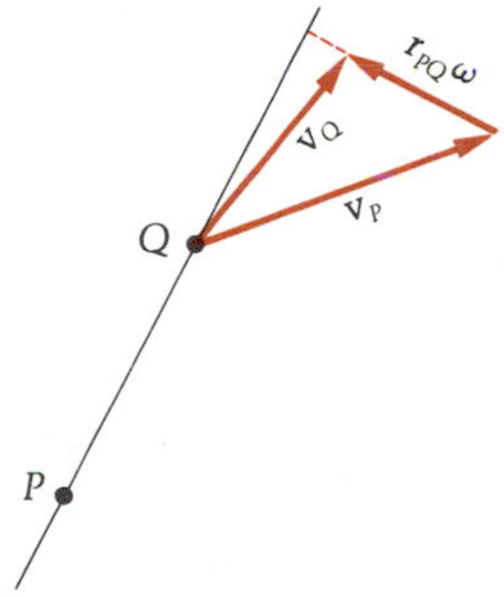

Figure 3.7

Incidentally, some books describe $\mathbf{v}_Q - \mathbf{v}_P$ as "the velocity of point Q relative to point P." We mention this only by way of explanation; our definition of $\mathbf{v}_P$ in Section 1.3 shows that points have velocities relative to frames, *not* relative to other points. If one uses the phrase "the velocity of point Q relative to point P," one means the velocity of Q in a reference frame in which P is fixed and which translates relative to $\mathcal{I}$.*

In each of the examples that follow, note the importance of selecting *and depicting* the unit vectors to be used in the solution. Also, in each of the first three examples, pay careful attention to the way the velocity of a point (say, B) is expressed if the tangent to the path of B is known; using what we learned in Section 1.7, $\mathbf{v}_B$ is expressed as a single unknown scalar (whose absolute value is the speed of B) times a unit vector along the known tangent.

Example 3.1

A 30-ft ladder is slipping down in a warehouse with the upper contact point T moving downward on the wall at a speed of 2 ft/sec in the position shown in Figure E3.1. Find the velocity of point B, which is sliding on the floor.

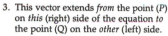

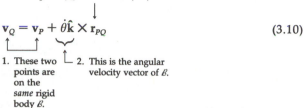

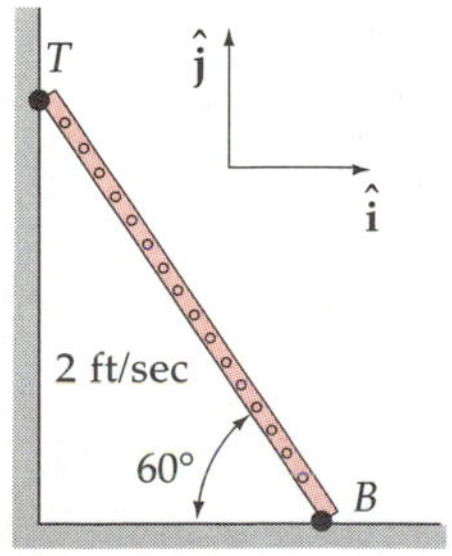

Figure E3.1

* "Translates" means that the frame moves in $\mathcal{I}$ without rotating. Translation is discussed in more detail in Section 3.3.

Solution

We relate $\mathbf{v}_T$ and $\mathbf{v}_B$ by using Equation (3.8):

$$\mathbf{v}_B = \mathbf{v}_T + \dot{\theta}\hat{\mathbf{k}} \times \mathbf{r}_{TB}$$

Noting that $\mathbf{v}_B$ has no $\hat{\mathbf{j}}$-component and that $\mathbf{v}_T$ has no $\hat{\mathbf{i}}$-component, we write:

$$v_B\hat{\mathbf{i}} = -2\hat{\mathbf{j}} + \dot{\theta}\hat{\mathbf{k}} \times 30\left(\frac{1}{2}\hat{\mathbf{i}} - \frac{\sqrt{3}}{2}\hat{\mathbf{j}}\right)$$

$$= \left(30\frac{\sqrt{3}}{2}\dot{\theta}\right)\hat{\mathbf{i}} + (-2 + 15\dot{\theta})\hat{\mathbf{j}} \text{ ft/sec}$$

Matching the $\hat{\mathbf{j}}$ coefficients, we have

$$0 = -2 + 15\dot{\theta} \Rightarrow \dot{\theta} \text{ (or } \omega) = \frac{2}{15} = 0.133 \text{ rad/sec}$$

Matching the $\hat{\mathbf{i}}$ coefficients, we have

$$v_B = 15\sqrt{3}\dot{\theta} = 15\sqrt{3}\left(\frac{2}{15}\right) = 3.46 \text{ ft/sec}$$

Thus the velocity of B is $\mathbf{v}_B = 3.46\hat{\mathbf{i}}$ ft/sec (or 3.46 → ft/sec).

Note that a direction indicator must be attached to $\dot{\theta}$ in order to specify correctly the angular velocity vector of the ladder:

$$\boldsymbol{\omega} = \dot{\theta}\hat{\mathbf{k}} = 0.133\hat{\mathbf{k}} \text{ rad/sec}$$

or, alternatively,

$$\omega = 0.133 \circlearrowleft \text{ rad/sec}$$

Note that the directions of $\mathbf{v}_B(\rightarrow)$ and $\boldsymbol{\omega}$ ($\circlearrowleft$) make sense. Such visual checks on solutions should be made whenever possible.

EXAMPLE 3.2

At the instant shown in Figure E3.2, the velocity of point A is 0.2 m/s to the right. Find the angular velocity of rod $\mathcal{B}$, and determine the velocity of its other end (point B), which is constrained to move in the circular slot.

Solution

We shall use Equation (3.8), featuring the points A and B of the rod:

$$\mathbf{v}_B = \mathbf{v}_A + \dot{\theta}\hat{\mathbf{k}} \times \mathbf{r}_{AB}$$

Noting that the velocity of B has a known direction (tangent to its path), we write $\mathbf{v}_B$ as an unknown scalar times a unit vector in this direction:

$$v_B\left(\frac{\hat{\mathbf{i}} + \hat{\mathbf{j}}}{\sqrt{2}}\right) = 0.2\hat{\mathbf{i}} + \dot{\theta}\hat{\mathbf{k}} \times (0.3\hat{\mathbf{i}} + 0.4\hat{\mathbf{j}})$$

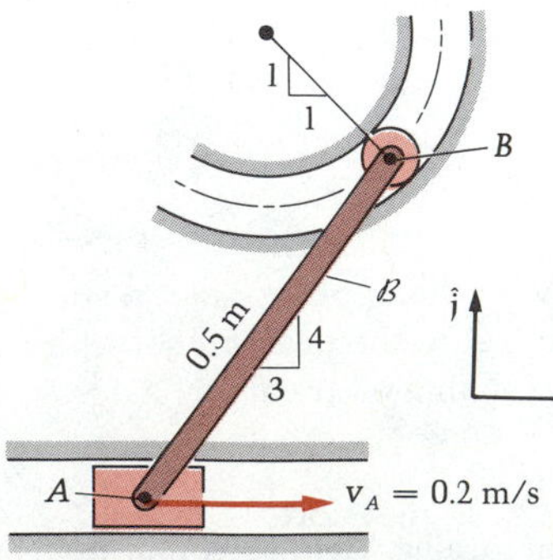

Figure E3.2

The component equations are:

$$\hat{\mathbf{i}} \text{ coefficients:} \qquad \left(\frac{1}{\sqrt{2}}\right)v_B = 0.2 - 0.4\dot{\theta} \qquad (1)$$

$$\hat{\mathbf{j}} \text{ coefficients:} \qquad \left(\frac{1}{\sqrt{2}}\right)v_B = 0.3\dot{\theta} \tag{2}$$

Solving Equations (1) and (2) gives

$$v_B = 0.121 \text{ m/s} \qquad \dot{\theta} = 0.286 \text{ rad/s}$$

and therefore the answers (*vectors* are what are asked for!) are

$$\mathbf{v}_B = 0.121 \measuredangle^{1}_{1} \text{ m/s} \quad \text{and} \quad \omega = 0.286 \circlearrowleft \text{rad/s}$$

or, equivalently,

$$\mathbf{v}_B = 0.0856\hat{\mathbf{i}} + 0.0856\hat{\mathbf{j}} \text{ m/s} \quad \text{and} \quad \dot{\theta}\hat{\mathbf{k}} = \omega = 0.286\hat{\mathbf{k}} \text{ rad/s}$$

In the next example, *two* bodies have angular velocity; thus we shall have to subscript the ω's (or $\dot{\theta}$'s). We shall simply denote by ω_1 (or $\dot{\theta}_1\hat{\mathbf{k}}$) the angular velocity of $\mathcal{B}_1$ and ω_2 (or $\dot{\theta}_2\hat{\mathbf{k}}$) the angular velocity of $\mathcal{B}_2$.

Example 3.3

The crank arm $\mathcal{B}_1$ shown in Figure E3.3a turns about a horizontal z axis, through its pinned end O, with an angular velocity of 10 rad/sec clockwise at the given instant. Find the velocity of the piston pin B.

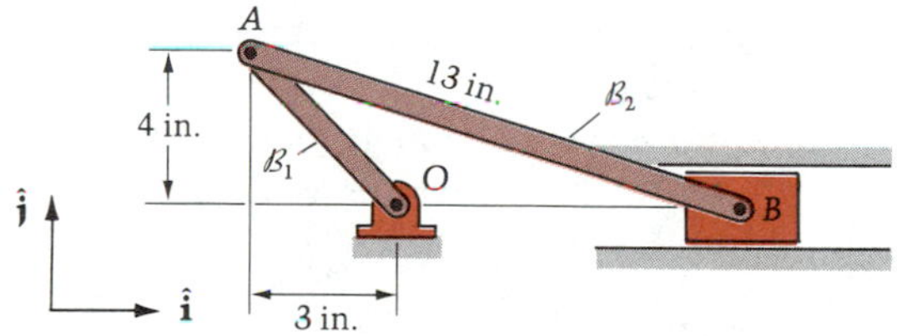

Figure E3.3a

Solution

We apply Equation (3.8) first to relate the velocities of A and O on body $\mathcal{B}_1$ and then to relate $\mathbf{v}_B$ to $\mathbf{v}_A$ on rod $\mathcal{B}_2$. Note that A is a "linking point" of both $\mathcal{B}_1$ and $\mathcal{B}_2$, since it belongs to *both* bodies. On body $\mathcal{B}_1$:

$$\begin{aligned}\mathbf{v}_A &= \mathbf{v}_O + \omega_1\hat{\mathbf{k}} \times \mathbf{r}_{OA} \\ &= 0 + (-10\hat{\mathbf{k}}) \times (-3\hat{\mathbf{i}} + 4\hat{\mathbf{j}}) \\ &= 40\hat{\mathbf{i}} + 30\hat{\mathbf{j}} \text{ in./sec}\end{aligned}$$

On body $\mathcal{B}_2$:

$$\mathbf{v}_B = \mathbf{v}_A + \omega_2\hat{\mathbf{k}} \times \mathbf{r}_{AB}$$

and using the Pythagorean theorem (see Figure E3.3b),

$$\mathbf{v}_B = (40\hat{\mathbf{i}} + 30\hat{\mathbf{j}}) + \omega_2\hat{\mathbf{k}} \times (12.4\hat{\mathbf{i}} - 4\hat{\mathbf{j}})$$

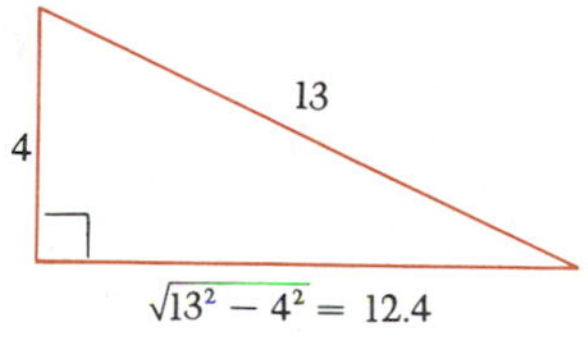

Figure E3.3b

Now point B is constrained to move only horizontally. Therefore,

$$v_B\hat{\mathbf{i}} = (40 + 4\omega_2)\hat{\mathbf{i}} + (30 + 12.4\omega_2)\hat{\mathbf{j}}$$

Equating the $\hat{\mathbf{i}}$ coefficients:

$$v_B = 40 + 4\omega_2 \tag{1}$$

Equating the $\hat{\mathbf{j}}$ coefficients:

$$0 = 30 + 12.4\omega_2$$

$$\omega_2 = -2.42 \text{ rad/sec}$$

Therefore,

$$\omega_2 = 2.42 \circlearrowright \text{ rad/sec} \tag{2}$$

Substituting ω_2 into (1), we have $\mathbf{v}_B = 30.3$ in./sec and $\mathbf{v}_B = 30.3\hat{\mathbf{i}}$ in./sec.

In all three of the preceding examples, we re-emphasize that it is absolutely essential to correctly incorporate the kinematic constraints imposed by slots, walls, floors, and so forth.

It is often helpful in studying the kinematics of rigid bodies to make use of the following result, which is a corollary of Equation (3.8):

> Corollary: If P and Q are two points of a rigid body, their velocity components along the line joining them must be equal.

Intuitively, we see that the difference between these components is the rate of stretching of the line PQ, and this has to vanish. Also, we have seen that $\mathbf{v}_Q$ and $\mathbf{v}_P$ differ only by the term $\dot{\theta}\hat{\mathbf{k}} \times \mathbf{r}_{PQ}$, which is clearly *normal* to the line PQ joining the points. Mathematically, we can see this immediately by dotting Equation (3.8) with the unit vector parallel to $\mathbf{r}_{PQ}$, which is $\mathbf{r}_{PQ}/r_{PQ}$:

$$\underbrace{\frac{\mathbf{r}_{PQ}}{r_{PQ}} \cdot \mathbf{v}_Q}_{\substack{\text{component of} \\ \mathbf{v}_Q \text{ along } PQ}} = \underbrace{\frac{\mathbf{r}_{PQ}}{r_{PQ}} \cdot \mathbf{v}_P}_{\substack{\text{component of} \\ \mathbf{v}_P \text{ along } PQ}} + \underbrace{(\dot{\theta}\hat{\mathbf{k}} \times \mathbf{r}_{PQ}) \cdot \frac{\mathbf{r}_{PQ}}{r_{PQ}}}_{\substack{\text{zero (since } \dot{\theta}\hat{\mathbf{k}} \times \mathbf{r}_{PQ} \\ \text{is } \perp \mathbf{r}_{PQ})}} \tag{3.11}$$

Thus, if we know the velocity of one point of the body, we can find any other without involving the angular velocity by using Equation (3.11). For instance, in Example 3.2, the unit vector $\mathbf{r}_{AB}/r_{AB}$ is simply $(3\hat{\mathbf{i}} + 4\hat{\mathbf{j}})/5$, and dotting this with the equation

$$v_B \frac{\hat{\mathbf{i}} + \hat{\mathbf{j}}}{\sqrt{2}} = 0.2\hat{\mathbf{i}} + \omega\hat{\mathbf{k}} \times (0.3\hat{\mathbf{i}} + 0.4\hat{\mathbf{j}})$$

from that example gives

$$v_B\left(\frac{3(1) + 4(1)}{5\sqrt{2}}\right) = \frac{3}{5}\,(0.2) = 0.120$$

or

$v_B = 0.121 \Rightarrow \mathbf{v}_B = 0.121$ 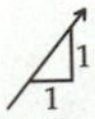 m/s (as before)

The algebra is seen to be simpler; we have worked with one equation in one unknown rather than two in two.

We now return to the vector formulation (Equation 3.8) for two final examples in this section.

Example 3.4

In the linkage shown in Figure E3.4, the velocities of A and C are given to be

$$\mathbf{v}_A = 2 \leftarrow \text{m/s}$$

$$\mathbf{v}_C = 3 \uparrow \text{m/s}$$

at the instant given. Find the velocity of point B at the same instant.

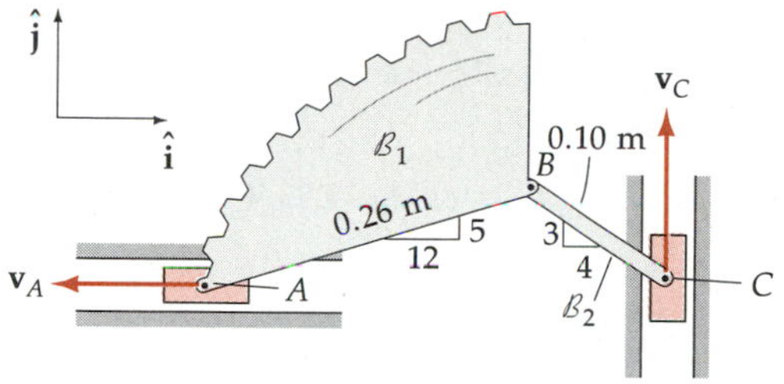

Figure E3.4

Solution

On body $\mathcal{B}_1$:

$$\begin{aligned}\mathbf{v}_B &= \mathbf{v}_A + \omega_1 \hat{\mathbf{k}} \times \mathbf{r}_{AB} \\ &= -2\hat{\mathbf{i}} + \omega_1 \hat{\mathbf{k}} \times (0.24\hat{\mathbf{i}} + 0.10\hat{\mathbf{j}}) \\ &= (-2 - 0.1\omega_1)\hat{\mathbf{i}} + (0.24\omega_1)\hat{\mathbf{j}} \text{ m/s} \qquad (1)\end{aligned}$$

On bar $\mathcal{B}_2$:

$$\begin{aligned}\mathbf{v}_B &= \mathbf{v}_C + \omega_2 \hat{\mathbf{k}} \times \mathbf{r}_{CB} \\ &= 3\hat{\mathbf{j}} + \omega_2 \hat{\mathbf{k}} \times (-0.08\hat{\mathbf{i}} + 0.06\hat{\mathbf{j}}) \\ &= (-0.06\omega_2)\hat{\mathbf{i}} + (3 - 0.08\omega_2)\hat{\mathbf{j}} \text{ m/s} \qquad (2)\end{aligned}$$

Equating the two vector expressions for $\mathbf{v}_B$, we get

$$\hat{\mathbf{i}} \text{ coefficients:} \quad -2 - 0.1\omega_1 = -0.06\omega_2$$

$$\hat{\mathbf{j}} \text{ coefficients:} \quad 0.24\omega_1 = 3 - 0.08\omega_2$$

Solving these two equations,

$$\omega_1 = 0.893 \quad \text{and} \quad \omega_2 = 34.8 \text{ rad/s}$$

From Equation (1), it follows that

$$\mathbf{v}_B = -2.09\hat{\mathbf{i}} + 0.214\hat{\mathbf{j}} \text{ m/s}$$

and the same result follows from (2), as a check.

Question 3.6 If the velocities of A and C were given to be $2 \leftarrow$ m/s and $3 \uparrow$ m/s for an *interval* of time, and not just at the instant shown, would the solution be any different at (a) the same instant? (b) some other instant?

Answer 3.6 (a) No. (b) Yes, because the geometry would be different.

EXAMPLE 3.5

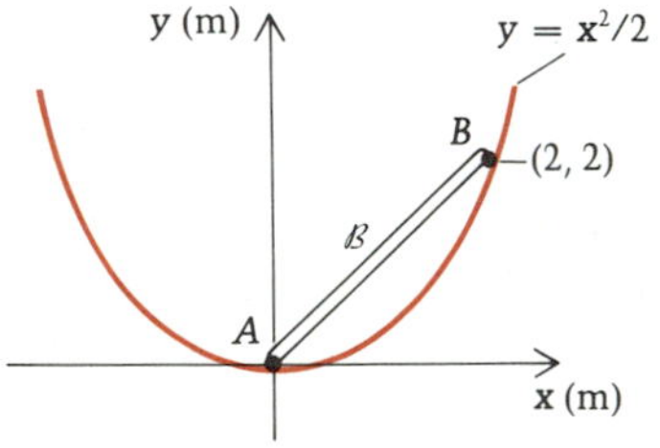

Figure E3.5a

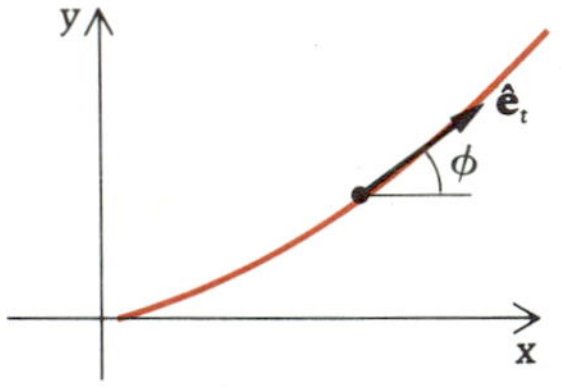

Figure E3.5b

The end B of rod $\mathcal{B}$ travels up the right half of the parabolic incline in Figure E3.5a at the constant speed of 0.3 m/s. Find the angular velocity of $\mathcal{B}$ and the velocity of point A, which is at the origin at the given instant.

Solution

We shall relate $\mathbf{v}_B$ to $\mathbf{v}_A$ using Equation (3.8):

$$\mathbf{v}_B = \mathbf{v}_A + \boldsymbol{\omega} \times \mathbf{r}_{AB} \tag{1}$$

Next we use Equation (1.41) to express $\mathbf{v}_B$:

$$\mathbf{v}_B = 0.3\hat{\mathbf{e}}_t$$

To get the unit tangent $\hat{\mathbf{e}}_t$ for point B, we use Figure E3.5b, noting that $\hat{\mathbf{e}}_t$ is tangent to the parabola at all times:

$$\phi = \tan^{-1}\left(\frac{dy}{dx}\right)$$
$$= \tan^{-1} 2 = 63.4°$$

Therefore, for point B,

$$\hat{\mathbf{e}}_t = \cos\phi\hat{\mathbf{i}} + \sin\phi\hat{\mathbf{j}}$$
$$= 0.447\hat{\mathbf{i}} + 0.894\hat{\mathbf{j}}$$

And thus

$$\mathbf{v}_B = 0.3\hat{\mathbf{e}}_t = 0.134\hat{\mathbf{i}} + 0.268\hat{\mathbf{j}} \text{ m/s}$$

Since point A likewise has a velocity tangent to *its* path, we may write

$$\mathbf{v}_A = v_A\hat{\mathbf{i}}$$

and so Equation (1) gives

$$0.134\hat{\mathbf{i}} + 0.268\hat{\mathbf{j}} = v_A\hat{\mathbf{i}} + \dot{\theta}\hat{\mathbf{k}} \times (2\hat{\mathbf{i}} + 2\hat{\mathbf{j}})$$

Collecting the coefficients of $\hat{\mathbf{i}}$ and $\hat{\mathbf{j}}$, we have

$$\hat{\mathbf{j}} \text{ coefficients:} \qquad 0.268 = 2\dot{\theta} \Rightarrow \dot{\theta} = 0.134 \text{ rad/s}$$

so that

$$\boldsymbol{\omega} = \dot{\theta}\hat{\mathbf{k}} = 0.134\hat{\mathbf{k}} \text{ rad/s or } 0.134 \circlearrowleft \text{rad/s}$$

$$\hat{\mathbf{i}} \text{ coefficients:} \qquad 0.134 = v_A - 2\dot{\theta}$$

Substituting for $\dot{\theta}$ and solving,

$$v_A = 0.402 \text{ m/s}$$

so that

$$\mathbf{v}_A = 0.402\hat{\mathbf{i}} \text{ m/s}$$

Applications of Equation (3.8) to rolling bodies are presented in Section 3.6 after we have examined that topic in detail.

PROBLEMS ▶ Section 3.2

3.3 The angular velocity of the bent bar is indicated in Figure P3.3. Find the velocity of the endpoint B in this position.

3.4 The velocities of the two endpoints A and B of a rigid bar in plane motion are shown in Figure P3.4. Find the velocity of the midpoint of the bar in the given position.

3.5 If $\mathbf{v}_A = 80\hat{\mathbf{i}}$ in./sec, find ω_2 and ω_3. See Figure P3.5.

3.6 At a certain instant, the coordinates of two points A and B of a rigid body $\mathcal{B}$ in plane motion are given in Figure P3.6. Point A has $\mathbf{v}_A = 2\hat{\mathbf{i}}$ m/s, and the velocity of B is vertical. Find $\mathbf{v}_B$ and the angular velocity of $\mathcal{B}$.

3.7–3.11 In the following five problems involving a "four-bar linkage" (the fourth bar in each case is the rigid ground length between fixed pins!), the angular velocity of one of the bars is indicated. Find the angular velocities of the other two bars.

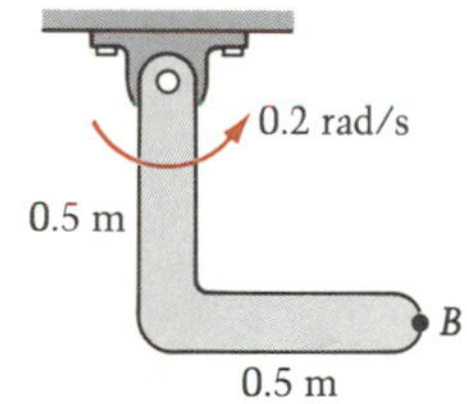

Figure P3.3

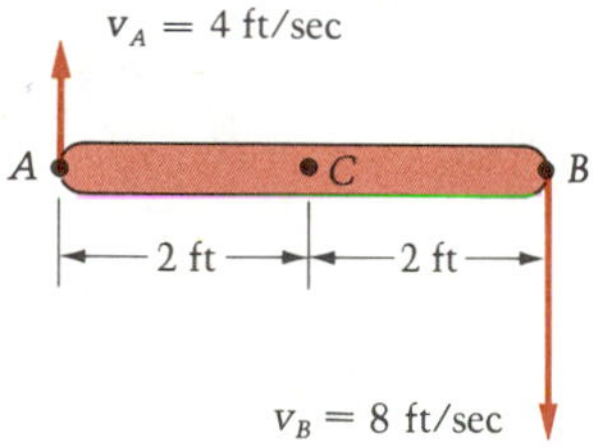

Figure P3.4

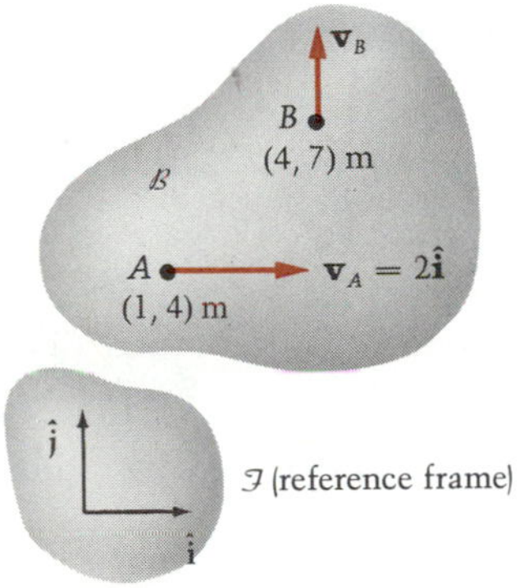

Figure P3.6

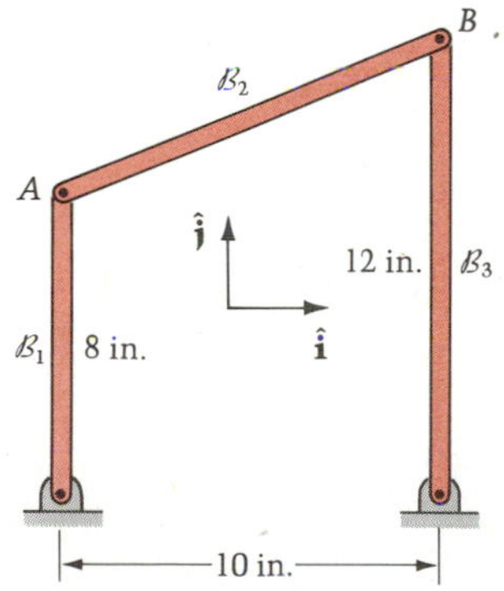

Figure P3.5

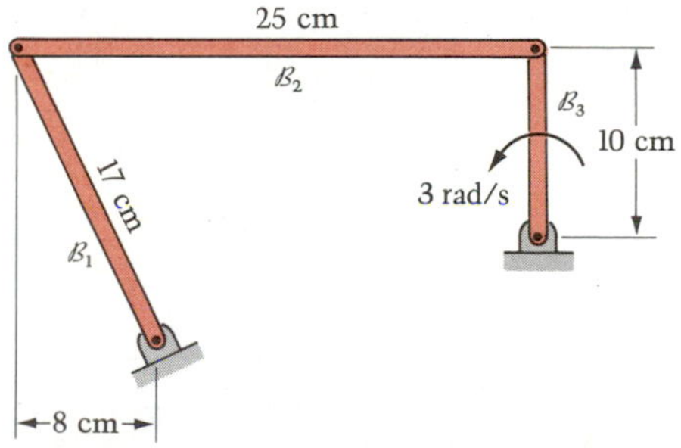

Figure P3.7

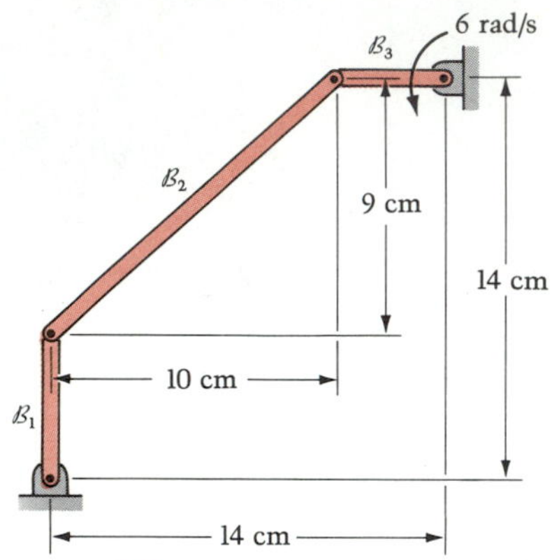

Figure P3.8

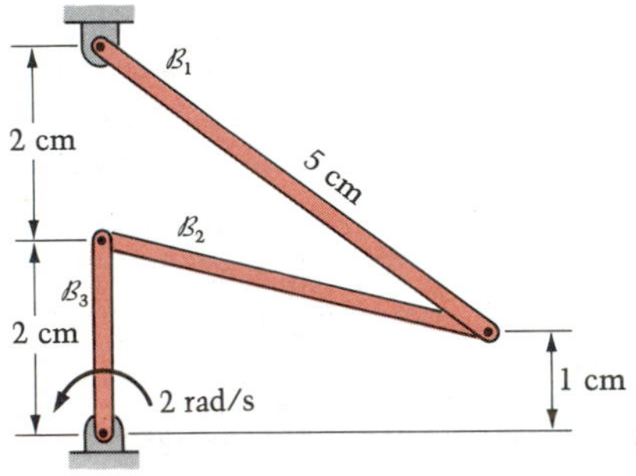

Figure P3.9

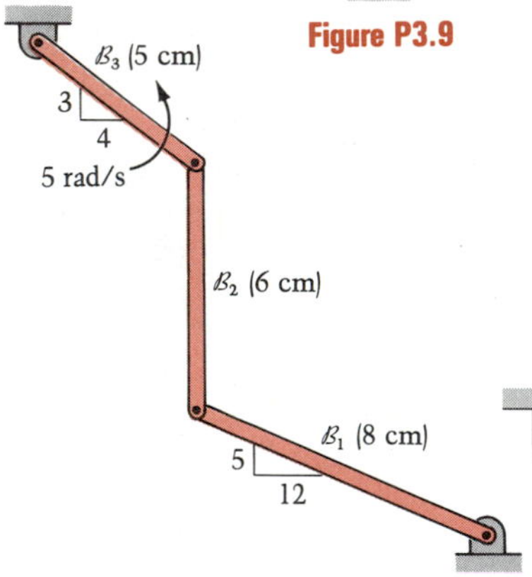

Figure P3.10

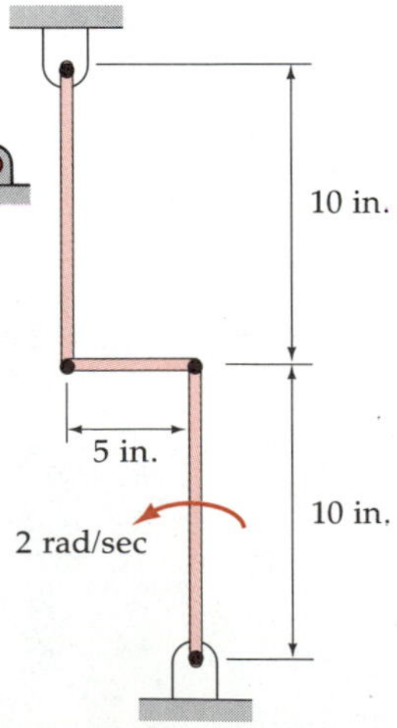

Figure P3.11

3.12 The equilateral triangular plate $\mathcal{B}_1$ shown in Figure P3.12 has three sides of length 0.3 m each. The bar $\mathcal{B}_2$ has an angular velocity $\omega_2 = 2$ rad/s counterclockwise and is pinned to $\mathcal{B}_1$ at A. Body $\mathcal{B}_1$ is also pinned to a block at B, which moves in the indicated slot. At the given time, find the angular velocity of $\mathcal{B}_1$.

3.13 Crank arm $\mathcal{B}_1$ shown in Figure P3.13 turns counterclockwise at a constant rate of 1 rad/s. Rod $\mathcal{B}_2$ is pinned to $\mathcal{B}_1$ at A and to a roller at B that slides in a circular slot. Determine the velocity of B and the angular velocity of $\mathcal{B}_2$ at the given instant.

3.14 The wheel shown in Figure P3.14 turns and slips in such a manner that its angular velocity is 2 rad/s ↺ while the velocity of the center C is 0.3 m/s to the left. Determine the velocity of point A.

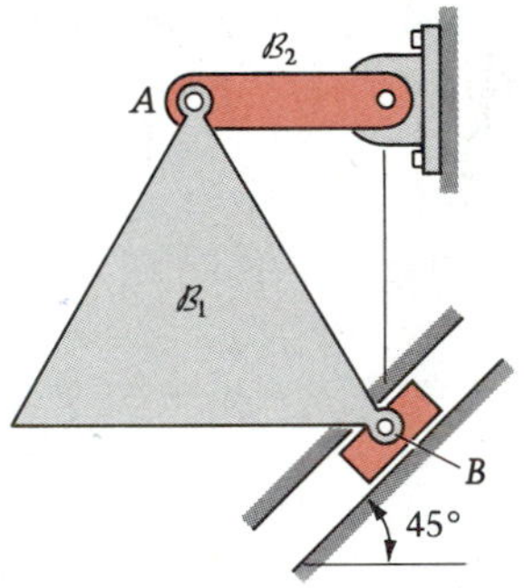

Figure P3.12

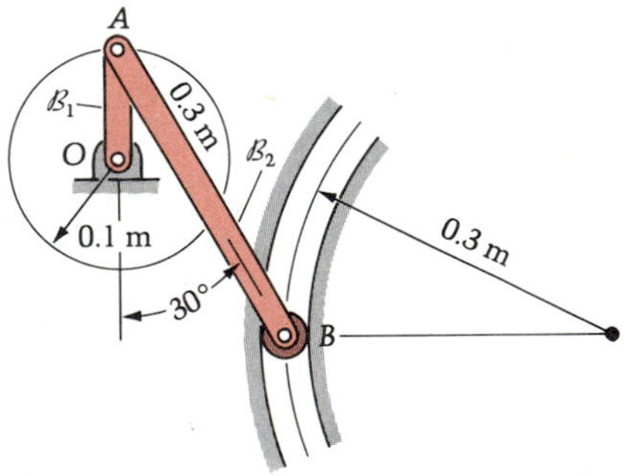

Figure P3.13

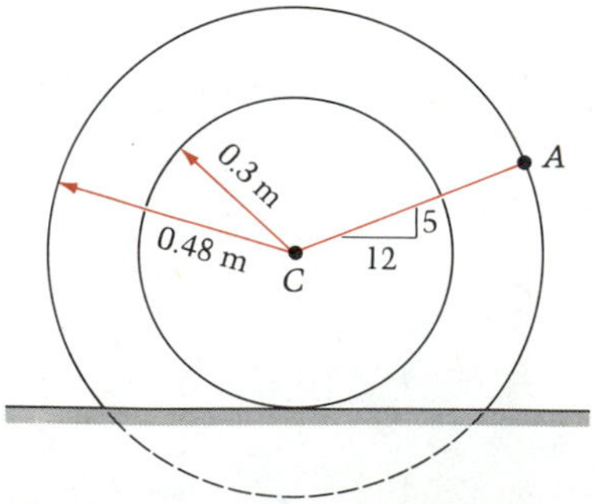

Figure P3.14

3.15 For the configuration shown in Figure P3.15, find the velocity of point P of the disk $\mathcal{B}_3$.

3.16 The speed of block $\mathcal{B}_1$ in Figure P3.16 has the value shown. Find the angular velocity of rod $\mathcal{B}_2$ and determine the velocity of pin A of block $\mathcal{B}_3$, when $\theta = 60°$.

3.17 Wheel $\mathcal{B}_1$ (Figure P3.17) turns and slips in such a way that its angular velocity is 2↻ rad/s while the velocity of C is 0.4 m/s to the left. Determine the velocity of point B, which slides on the plane. Bar $\mathcal{B}_2$ is pinned to $\mathcal{B}_1$ at D.

3.18 Point A of the rod slides along an inclined plane as in Figure P3.18, while the other end, B, slides on the horizontal plane. In the indicated position, $\boldsymbol{\omega} = 1.5\hat{\mathbf{k}}$ rad/sec. Find the velocity of the midpoint of the rod at this instant.

3.19 Wheel $\mathcal{B}_1$ in Figure P3.19 has a counterclockwise angular velocity of 6 rad/s. What is the velocity of point B at the instant shown?

3.20 Block $\mathcal{B}_1$ in Figure P3.20, which slides in a vertical slot, is pinned to bars $\mathcal{B}_2$ and $\mathcal{B}_3$ at A. The other ends of $\mathcal{B}_2$ and $\mathcal{B}_3$ are pinned to blocks that slide in horizontal slots. Block $\mathcal{B}_4$ translates to the left at constant speed 0.2 m/s. Find the velocity of B: (a) at the given instant; (b) when C is at point D; (c) when C is at point E.

3.21 The four links shown in Figure P3.21 each have length 0.4 m, and two of their angular velocities are indicated. Find the velocity of point C and determine the angular velocities of $\mathcal{B}_2$ and $\mathcal{B}_3$ at the indicated instant.

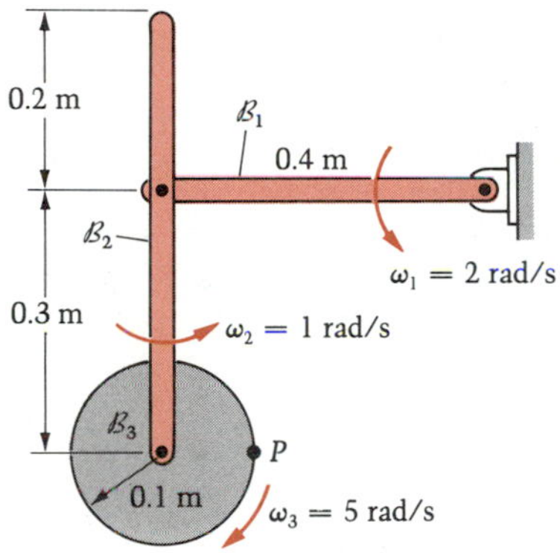

Figure P3.15

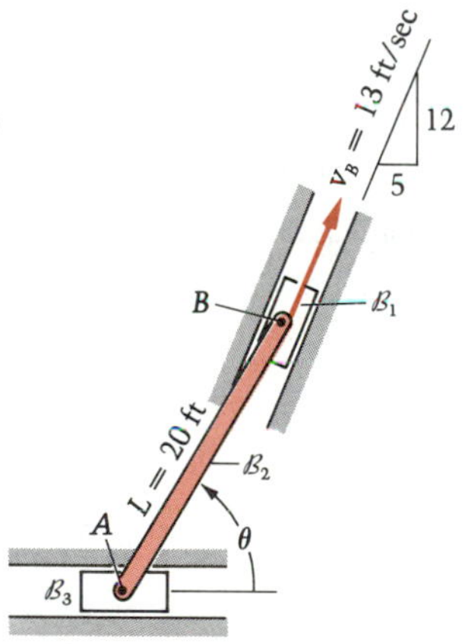

Figure P3.16

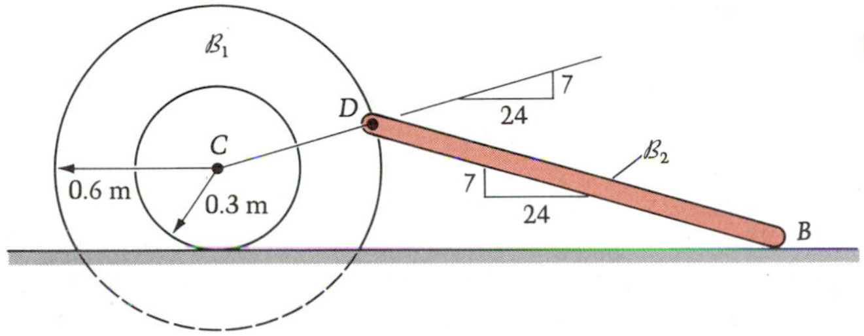

Figure P3.17

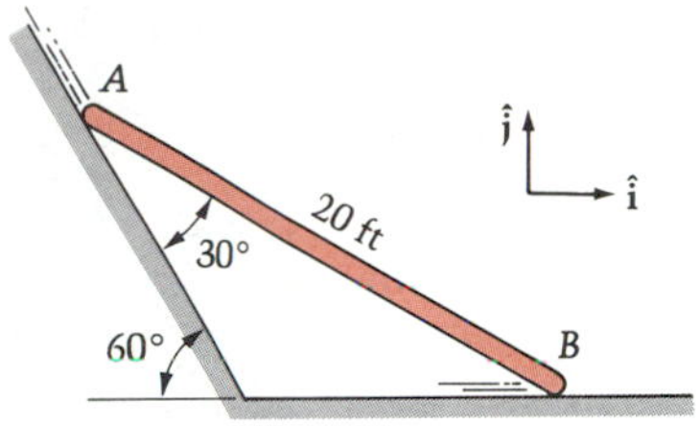

Figure P3.18

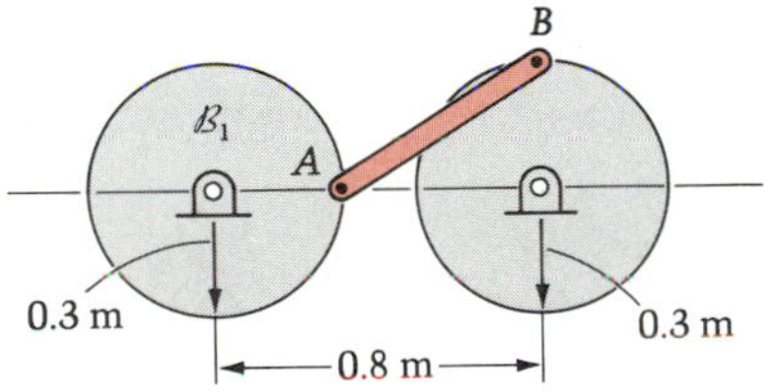

Figure P3.19

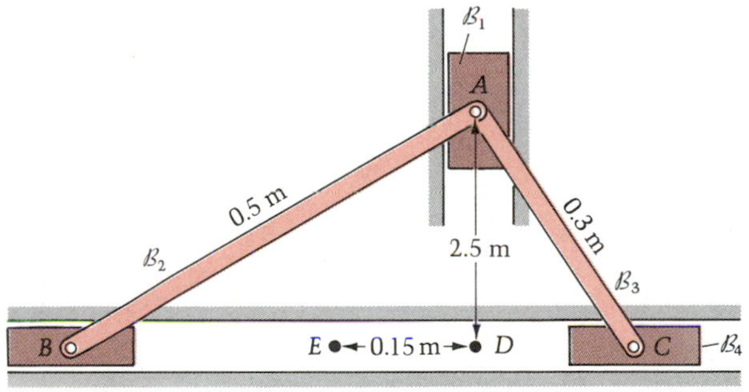

Figure P3.20

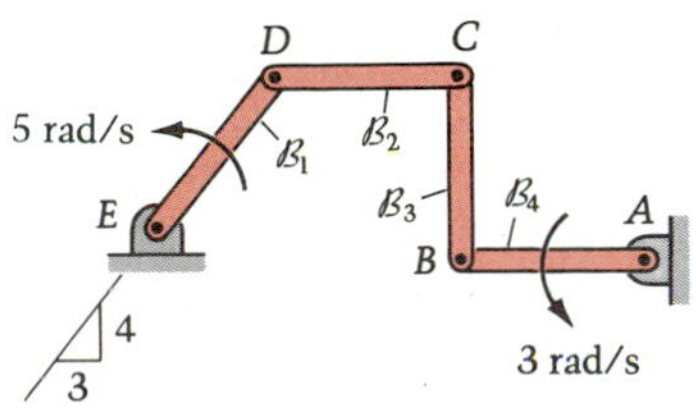

Figure P3.21

3.22 In the mechanism shown in Figure P3.22, the sleeve $\mathcal{B}_1$ is connected to the pivoted bar $\mathcal{B}_2$ by the 15-cm link $\mathcal{B}_3$. Over a certain range of motion of $\mathcal{B}_2$, the angle θ varies according to $\theta = 0.02t^2$ rad, starting at $t = 0$ with $\mathcal{B}_2$ and $\mathcal{B}_3$ horizontal. Find the velocity of pin S and the angular velocities of $\mathcal{B}_2$ and $\mathcal{B}_3$ when $\theta = 30°$. Time t is measured in seconds.

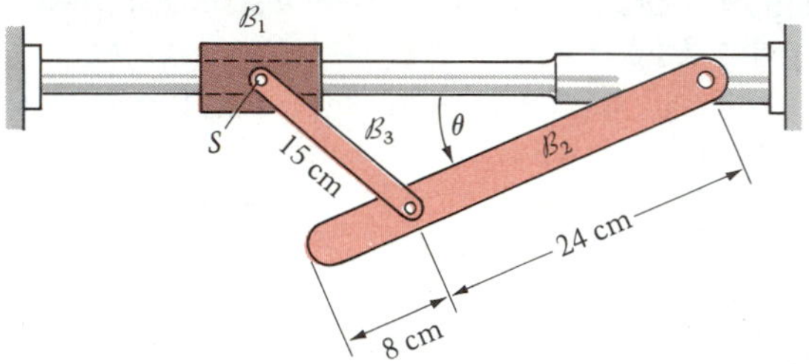

Figure P3.22

* **3.23** Find the velocity of point B of the rod if end A has constant velocity 2 m/s to the right as shown in Figure P3.23. The rollers are small. Compare the use of Equation (3.8) with the procedure used to solve Problem 1.63.

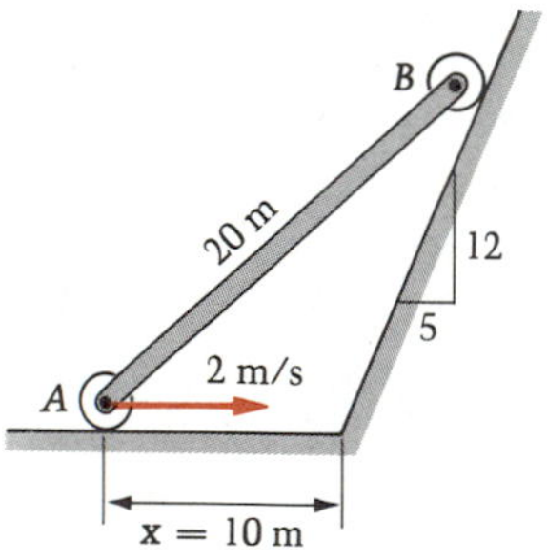

Figure P3.23

3.24 Find the velocity of the guided block at the instant shown in Figure P3.24.

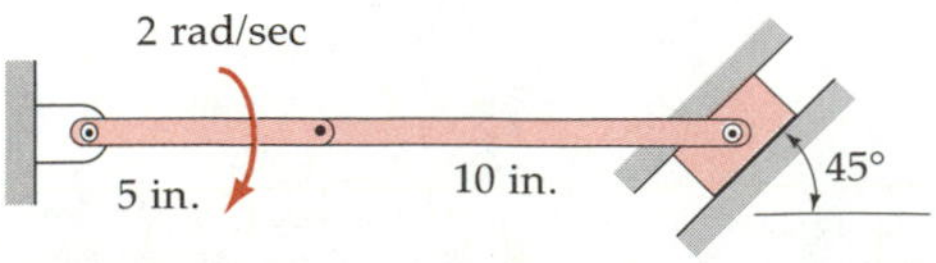

Figure P3.24

* **3.25** Block $\mathcal{B}$ has a controlled position in the slot given by $y = \sqrt{120}\,\sin(\pi t/10)$ in. for $0 \le t \le 10$ sec. (See Figure P3.25.) The time is $t = 0$ sec in the indicated position. Find the angular velocities of the rod and the wheel at (a) $t = 0$ sec and (b) $t = 5$ sec.

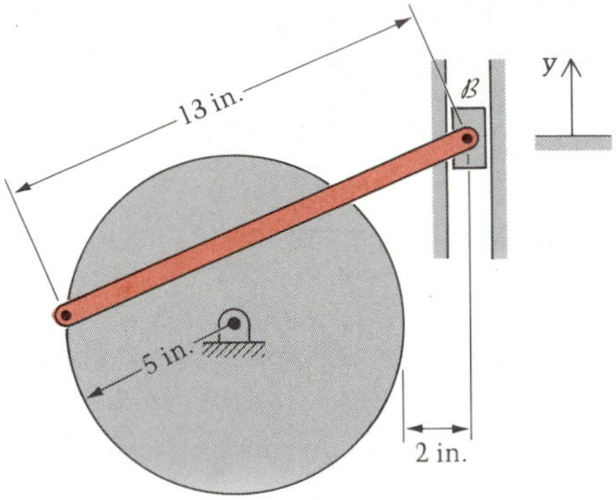

Figure P3.25

3.26 Crank $\mathcal{B}_1$ of the slider-crank mechanism shown in Figure P3.26 has a constant angular speed $\dot{\theta}$. Find the equation for the angular velocity $\dot{\phi}$ of the connecting rod $\mathcal{B}_2$ as a function of r, ℓ, θ, and $\dot{\theta}$.

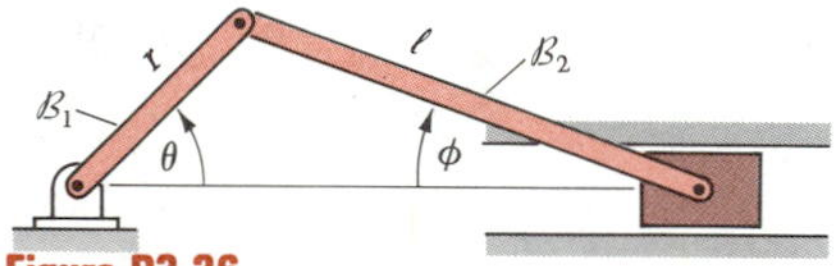

Figure P3.26

3.27 In the preceding problem, plot $\dot{\phi}/\dot{\theta}$ as a function of θ, from $\theta = 0$ to 2π, for $\ell/r = 1$, 2, and 5.

3.28 Referring to Section 3.2, show that neither the angular velocity vector nor Equation (3.8) depends on which body-fixed line segment (such as PQ in the text) is chosen to measure θ. Use two other points P' and Q' and their angle ϕ as suggested in Figure P3.28 for your proof.

* **3.29** Rod $\mathcal{B}_1$ begins moving at $\theta = 0$ (see Figure P3.29) and is made to turn at the constant angular rate $\dot{\theta} = 0.2$ rad/s. The cord is attached to the end of $\mathcal{B}_1$ and passes around a pulley. The other end of the cord is tied to weight $\mathcal{B}_2$ at point B. Observe that $\mathcal{B}_2$ moves downward until $\theta = 90°$, when it reverses direction. Write an equation that gives the velocity of point B as a function of θ for $\pi \ge \theta \ge \pi/2$. *Hint*: Using trigonometry, write y as a function of θ and the length L of the cord. Then differentiate.

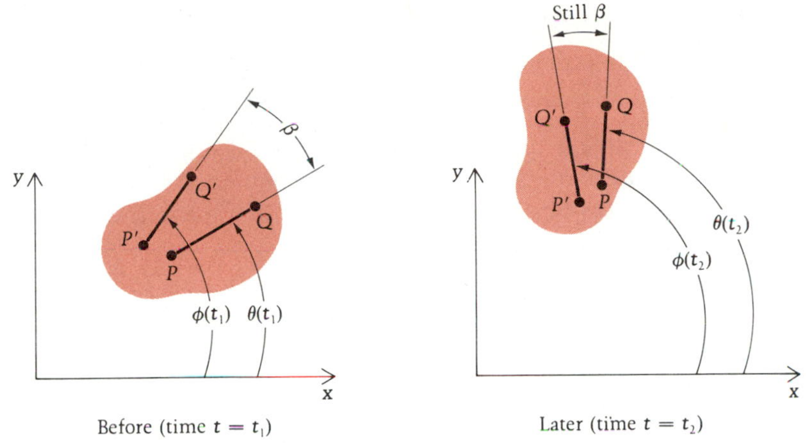

Figure P3.28

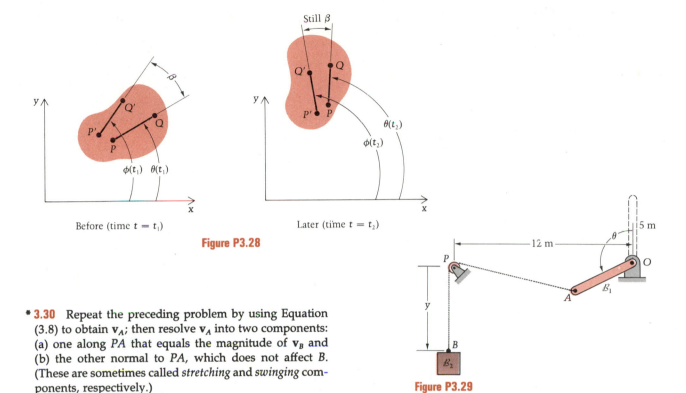

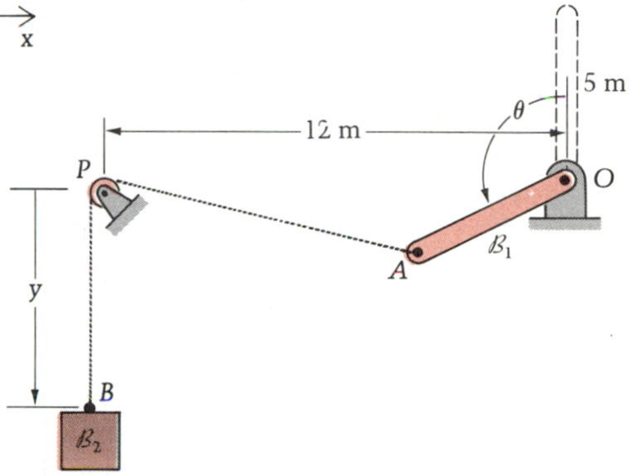

Figure P3.29

* **3.30** Repeat the preceding problem by using Equation (3.8) to obtain $\mathbf{v}_A$; then resolve $\mathbf{v}_A$ into two components: (a) one along PA that equals the magnitude of $\mathbf{v}_B$ and (b) the other normal to PA, which does not affect B. (These are sometimes called *stretching* and *swinging* components, respectively.)

3.3 Translation

When a rigid body $\mathcal{B}$ moves during a certain time interval in such a way that its angular velocity vector remains identically zero, then the body is said to be **translating**, or to be in a state of **translational motion** during that interval. From Equation (3.8) we thus see that for translation

$$\mathbf{v}_Q = \mathbf{v}_P \tag{3.12}$$

That is, all points of the body have the same velocity vector. By differentiating Equation (3.12), we see that the accelerations of all points of $\mathcal{B}$ are also equal for translation. Note that if $\dot{\theta} = 0$ only at an *instant* (that is, at a single value of time rather than over an interval), then all points of the body have equal velocities at that instant but *need not have equal accelerations.*

Question 3.7 Why is this the case?

Answer 3.7 The derivative of $\dot{\theta}\hat{\mathbf{k}} \times \mathbf{r}_{PQ}$ is not zero merely because $\dot{\theta}$ happens to be zero at one instant of time. To be able to differentiate $\mathbf{v}_Q = \mathbf{v}_P$, this equation must be valid for *all* values of t and not just one!

Translation can be either:

1. *Rectilinear:* Each point of $\mathcal{B}$ moves along a straight line in $\mathcal{I}$.
2. *Curvilinear:* Each point of $\mathcal{B}$ moves on a curved path in $\mathcal{I}$.

Examples of translation are shown in Figure 3.8. Part (a) shows an example of rectilinear translation: Body $\mathcal{B}$ is constrained to move in a straight slot. Part (b) shows an example of curvilinear translation: Body $\mathcal{B}$ is constrained by the identical links.

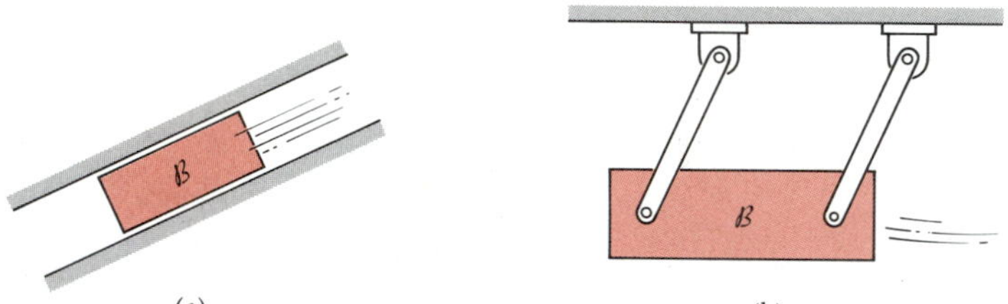

(a) (b)

Figure 3.8 Examples of translation.

Perhaps an even better pair of examples is the blackboard eraser (Figure 3.9), which we used earlier to explain plane motion in Section 3.2. In part (a), the professor moves the eraser so that each of its points stays on a straight line; it is therefore in a state of rectilinear translation. In part (b), the professor moves the eraser on a curve; but if the word *eraser* is always horizontal during the erasing, then $\dot{\theta} \equiv 0$ and the eraser is in a state of curvilinear translation. Even though each of its points moves on a curve, all the velocities (and accelerations) are equal at all times. There is one notable exception to our earlier statement that "points, not bodies, have velocities and accelerations." In this present case of translation, since all the points have the *same* **v**'s and **a**'s, one could loosely refer to "the velocity of the eraser" without ambiguity.

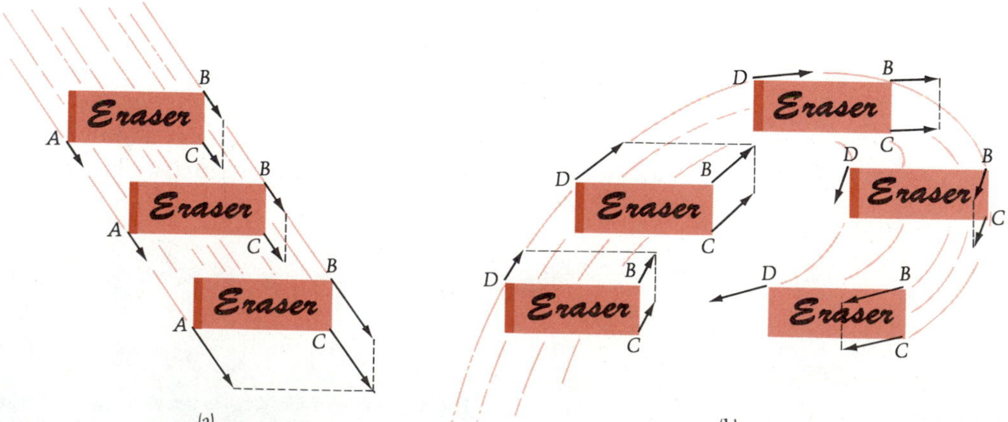

(a) (b)

Figure 3.9 Another example of translation.

There are no examples or problems in this section because translation problems of rigid bodies require no new theory beyond what was developed in Chapter 1.

Summarizing, when a body is translating (either rectilinearly or curvilinearly), its angular velocity $\dot{\theta}\hat{\mathbf{k}}$ is identically zero, and all its points have equal velocities (and accelerations). If $\dot{\theta} = 0$ only at an instant, then all the points of the body have the same velocity at that instant but need not have equal accelerations.

3.4 Instantaneous Center of Zero Velocity

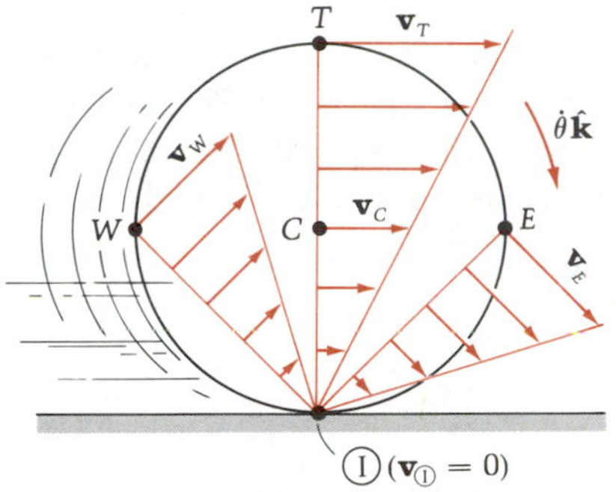

Figure 3.10 Instantaneous center of a rolling wheel.

If P is a point in the reference plane having zero velocity at some instant, then the velocity field of $\mathcal{B}$ is the same as if the body were constrained at that instant to rotate about an axis through P normal to the reference plane. This axis is called the **instantaneous axis of rotation**, and point P is called the **instantaneous center** (abbreviated Ⓘ) of **zero velocity*** of $\mathcal{B}$. Thus if Q is any *other* point of $\mathcal{B}$, then we have

$$\mathbf{v}_Q = \overset{0}{\cancel{\mathbf{v}_Ⓘ}} + \dot{\theta}\hat{\mathbf{k}} \times \mathbf{r}_{ⒾQ} = \dot{\theta}\hat{\mathbf{k}} \times \mathbf{r}_{ⒾQ} \quad \text{or} \quad \boldsymbol{\omega} \times \mathbf{r}_{ⒾQ} \quad (3.13)$$

and since $\mathbf{v}_Q$ is then normal to both of the vectors $\boldsymbol{\omega}$ and $\mathbf{r}_{ⒾQ}$, we see that *each point moves with its velocity perpendicular to the line joining it to* Ⓘ. This concept is illustrated in Figure 3.10 for a rolling† wheel, in which Ⓘ is the contact point.

Proof of the Existence of the Instantaneous Center

We can show that if a body $\mathcal{B}$ has $\dot{\theta} \neq 0$ at a given instant, then it has an instantaneous center.

Question 3.8 Why can there be no point Ⓘ whenever $\dot{\theta}$ is zero?

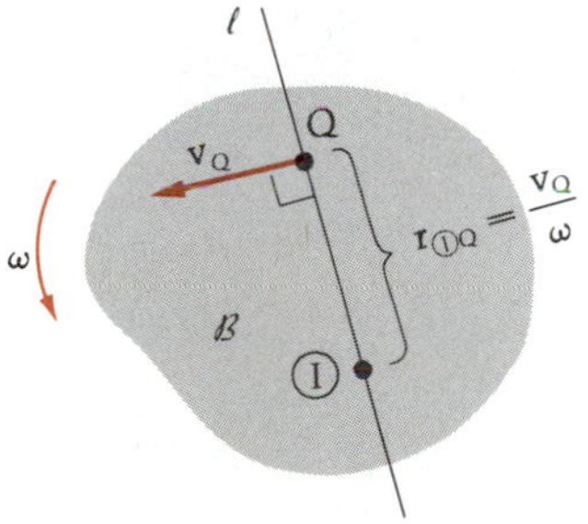

Figure 3.11

To demonstrate the existence of Ⓘ, we shall use Equation (3.13) in conjunction with Figure 3.11. As we have noted above, the vector $\mathbf{v}_Q$, being equal to $\boldsymbol{\omega} \times \mathbf{r}_{ⒾQ}$ for the point Ⓘ having $\mathbf{v}_Ⓘ = \mathbf{0}$, is normal to both $\boldsymbol{\omega}$ and to $\mathbf{r}_{ⒾQ}$. Hence we have these results:

1. The vector $\mathbf{r}_{ⒾQ}$ lies in the reference plane and is normal to $\mathbf{v}_Q$. It thus lies along the line ℓ in Figure 3.11.

* The phrase is admittedly redundant, but it is in common usage. "Instantaneous center of velocity" would perhaps be more concise, and "center of velocity" even more so. "Instantaneous center," however, is inadequate because of the possibility of confusion with points of zero acceleration.

† Rolling means no slipping, according to the definition we adopt in this book (see Section 3.6).

Answer 3.8 If $\dot{\theta} = 0$, Equation (3.8) says that $\mathbf{v}_Q = \mathbf{v}_P$; that is, all points of $\mathcal{B}$ have the *same* velocity vector. This common velocity vector is then zero only if the body is at rest. Incidentally, some think of $\mathcal{B}$ as having an instantaneous center Ⓘ at infinity when $\dot{\theta} = 0$.

2. The point Ⓘ exists (and is unique) because $|\mathbf{r}_{\textcircled{I}Q}|$ is therefore seen to be $|\mathbf{v}_Q/\omega|$ in order that $\mathbf{v}_Q = \boldsymbol{\omega} \times \mathbf{r}_{\textcircled{I}Q}$.

Question 3.9 Why is Ⓘ *below* Q in Figure 3.11 instead of being the same distance *above* Q?

We have thus verified the existence of the instantaneous center (unless $\dot{\theta} = 0$), because we know how to get to it from any arbitrary point Q of the body $\mathcal{B}$ whenever the angular velocity $\omega\hat{\mathbf{k}}$ of $\mathcal{B}$ and the velocity $\mathbf{v}_Q$ of the point Q are known. We also note again that the velocity magnitude of every point of $\mathcal{B}$ (in the reference plane!) equals the product of $|\boldsymbol{\omega}|$ and the distance to the point from Ⓘ.

Sometimes because of a constraint on the motion we know the location of Ⓘ at the outset. This is the case for the rolling wheel of Figure 3.10 in which the bottom point grips the ground and is held at rest (thereby becoming Ⓘ) for the instant of its contact. If the radius of the wheel is 15″ and the velocity of its center C is 88 ft/sec →, then from the above discussion, the angular speed of the wheel is

$$\omega = \frac{v_C}{r_{\textcircled{I}C}} = \frac{88}{15/12} = 70.4 \text{ rad/sec}$$

Note that when viewed from Ⓘ the senses of the velocity direction of any point and the angular velocity direction of the body must always agree. For example, these are possible situations:

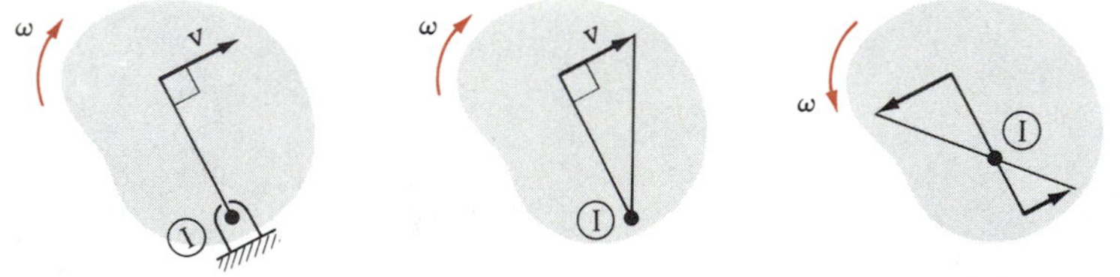

These are not:

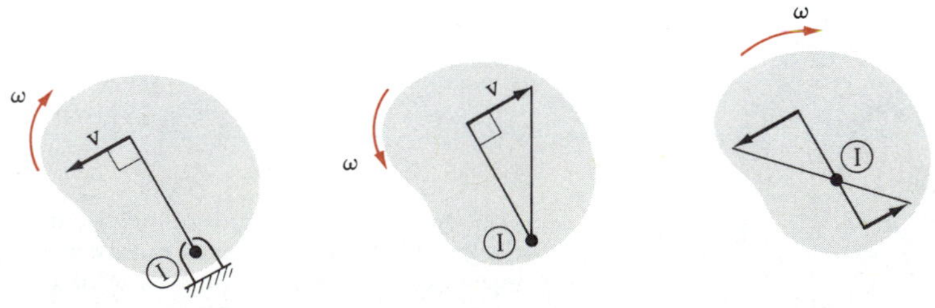

Answer 3.9 If point I were above Q, then $\boldsymbol{\omega} \times \mathbf{r}_{\textcircled{I}Q}$ would give an incorrect direction for the velocity $\mathbf{v}_Q$ — it would be opposite to the actual direction. The senses of $\mathbf{v}_Q$ and $\boldsymbol{\omega}$ when viewed from Ⓘ must agree, as we point out later in this section.

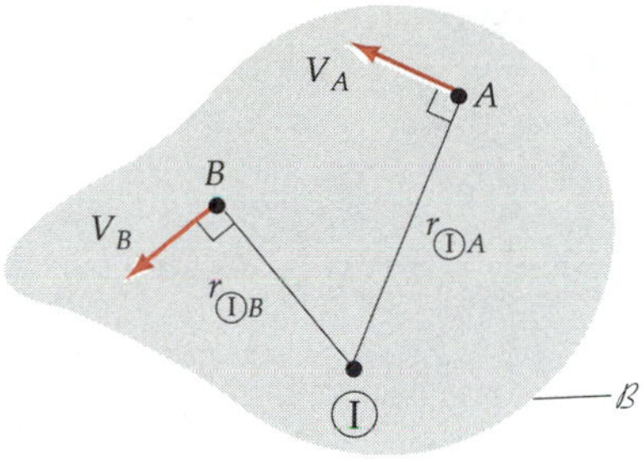

Figure 3.12

Therefore the direction of ω of the wheel in Figure 3.10 is clockwise (↻) in order that the known velocity direction of point C (→) and the angular velocity of the body be in agreement as the body rotates about Ⓘ at the instant shown.

Even if the angular velocity of ℬ is not known, we can still easily find the instantaneous center of zero velocity Ⓘ if we know the velocities — or really, just the directions of the velocities — of *two* points A and B of ℬ.

Constructing perpendicular lines to the velocities of A (at A) and of B (at B) as shown in Figure 3.12, we immediately recognize Ⓘ as the intersection point of the two lines.

Question 3.10 Why?

From Figure 3.12 and the discussion above, we know that:

$$\frac{v_A}{r_{ⒾA}} = \omega = \frac{v_B}{r_{ⒾB}} \qquad \text{(Recall that there is only one } \omega \text{ for the body.)}$$

where we are abbreviating $|\mathbf{v}_A|$ by v_A, $|\mathbf{v}_B|$ by v_B, $|\mathbf{r}_{ⒾA}|$ by $r_{ⒾA}$, $|\mathbf{r}_{ⒾB}|$ by $r_{ⒾB}$, and $|\omega|$ by ω.

We now present three examples of the use of the above procedure for locating Ⓘ when two velocity directions are known in advance.

Answer 3.10 The point Ⓘ is unique. Since there is only one common point on the lines drawn perpendicular to the velocities (to $\mathbf{v}_B$ at B and to $\mathbf{v}_A$ at A), that point is the instantaneous center.

Example 3.6

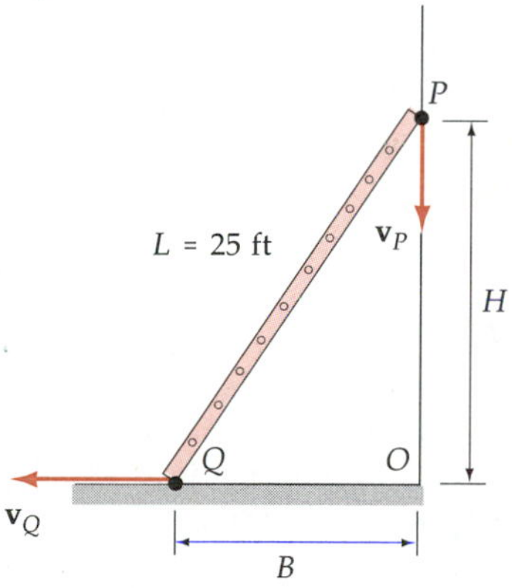

Figure E3.6a

Ladders commonly carry a warning that for safe placement, the distance B in Figure E3.6a should be $\frac{1}{4}$ of the length L (i.e., of $\sqrt{B^2 + H^2}$). Let us suppose that a careless painter temporarily set a ladder against a wall in a dangerous position with $B/L = 0.5$, and went off to get his paint and brushes. Suppose further that the ladder began to slip, with the top of the ladder, point P, sliding down the wall and the bottom, point Q, slipping along the ground as shown. When B is 15 ft, find the instantaneous center of zero velocity Ⓘ of the ladder, and discuss the path of Ⓘ in space as the ladder falls further.

Solution

When $B = 15$ ft, the normals to $\mathbf{v}_P$ at P and to $\mathbf{v}_Q$ at Q intersect at point Ⓘ as shown in Figure E3.6b on the next page.

If we imagine a rigid sheet of very light plastic glued to the ladder as in Figure E3.6c, then the ladder has been "rigidly extended." Note that *only for this instant,* we can think of the extended body as rotating about an imaginary pin at the intersection point Ⓘ of the normals to two velocities as shown. Note further from Figure E3.6c that the velocities of *all* points of the rigid sheet are perpendicular to lines drawn to them from Ⓘ. The velocity magnitude of each point is proportional to the distance from that point to Ⓘ, with the proportionality

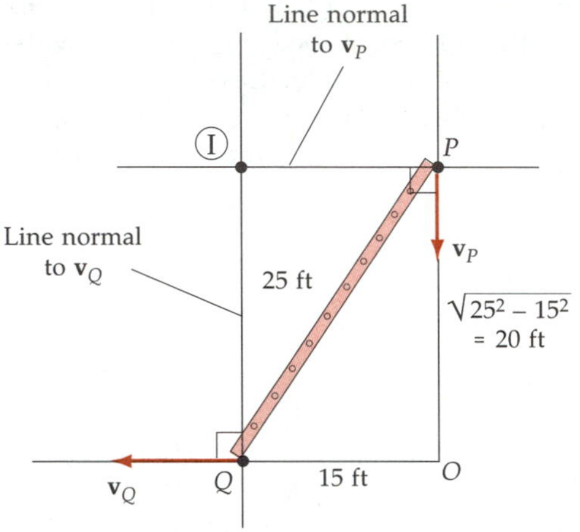

Figure E3.6b

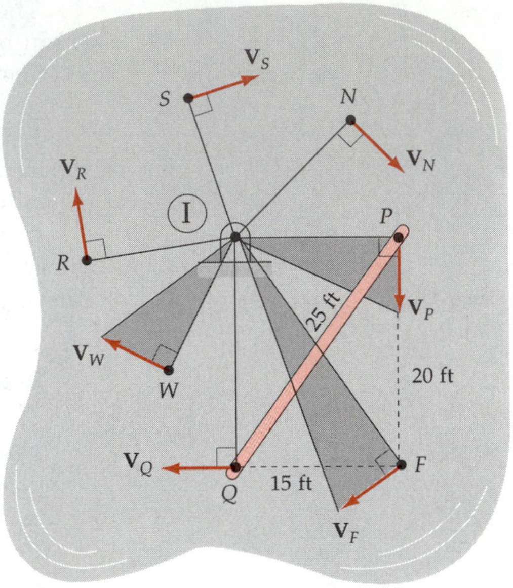

Figure E3.6c

constant being ω of the body at the instant. Hence all triangles like the three shaded in Figure E3.6c are similar.

As the ladder falls, the location of the point Ⓘ changes on the imagined rigid extension (sheet) as time passes, because the perpendiculars to $\mathbf{v}_A$ and $\mathbf{v}_B$ intersect at different points of the sheet, as seen below in Figure E3.6d:

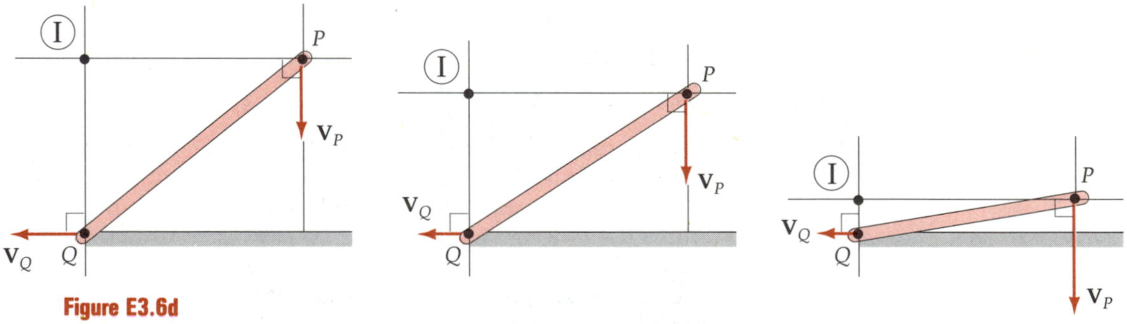

Figure E3.6d

Note that as point P (and thus all of the ladder) gets closer and closer to the ground, point Ⓘ gets closer and closer to point Q. Thus even though ω increases, $r_{\text{Ⓘ}Q}\,\omega$ gets smaller and smaller, until, in the limit (as P contacts the ground) Q *becomes* Ⓘ (the intersection of the perpendiculars) and V_Q is then zero.

When the body $\mathcal{B}$ has a **pivot** (a point that never moves throughout the body's motion, such as a pin), it is clearly *always* Ⓘ; in this case the motion is called *pure rotation.* But otherwise, the point Ⓘ is *not* the same point of $\mathcal{B}$ throughout the motion, as we have already seen with the wheel and ladder. In each of the next two examples, one of the bodies has a pivot.

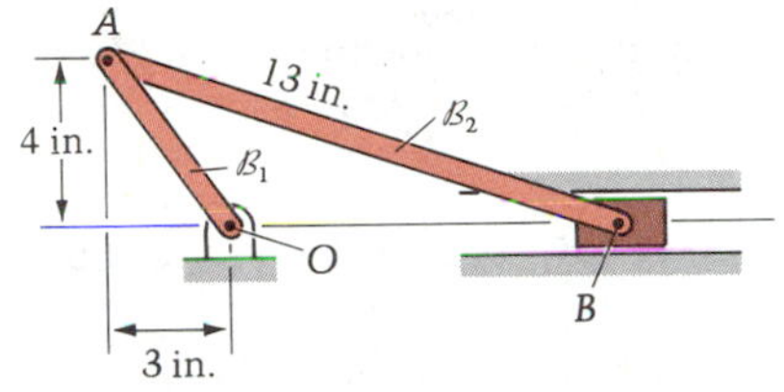

Figure E3.7a

Example 3.7

Rework Example 3.3 using instantaneous centers. (See Figure E3.7a) The crank arm $\mathcal{B}_1$ turns about a horizontal z axis, through its pinned end O, with an angular velocity of 10 rad/sec clockwise at the given instant. Find the velocity of the piston pin B.

Solution

Since O is the point Ⓘ for body $\mathcal{B}_1$, we have

$$\mathbf{v}_A = r\omega \; {}_{4}\!\nearrow^{3} = 5(10) \; {}_{4}\!\nearrow^{3}$$

$$= 50 \; {}_{4}\!\nearrow^{3} \text{ in./sec}$$

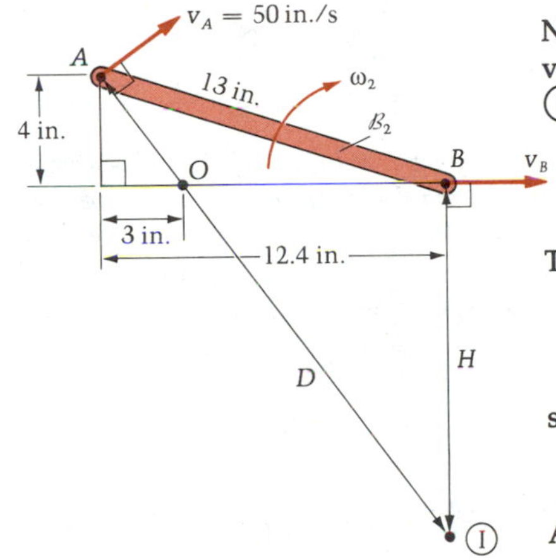

Figure E3.7b

Next we find the Ⓘ of $\mathcal{B}_2$, using the fact that it is on lines perpendicular to the velocities of A and B as shown in Figure 3.7b. If we next find the distance D from Ⓘ to A, then ω_2 will be $v_A/D = 50/D$. By similar triangles:

$$\frac{D}{12.4} = \frac{5}{3} \Rightarrow D = 20.7 \text{ in.}$$

Then

$$\omega_2 = \frac{v_A}{D} = \frac{50}{20.7} = 2.42$$

so

$$\omega_2 = 2.42 \circlearrowright \text{rad/sec}$$

Again by similar triangles:

$$\frac{H+4}{12.4} = \frac{4}{3} \Rightarrow H = 12.5 \text{ in.}$$

so that $v_B = H\omega_2 = 12.5(2.42) = 30.3$ in./sec to the right, as we have seen before.

Example 3.8

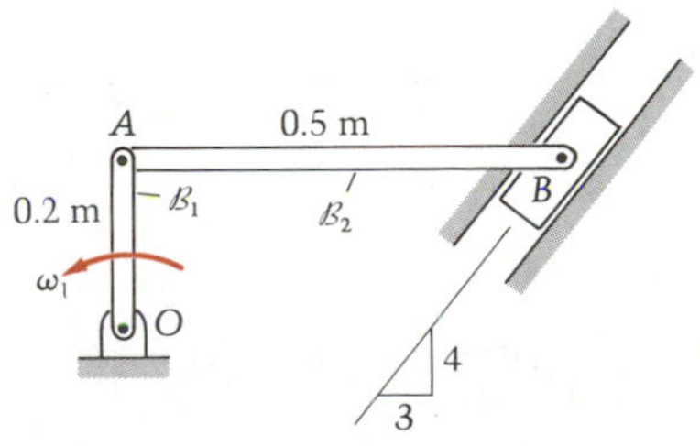

Figure E3.8a

At the instant shown in Figure E3.8a, the angular velocity of bar $\mathcal{B}_1$ is $\omega_1 = 5 \circlearrowleft$ rad/sec. Find the velocity of pin B connecting bar $\mathcal{B}_2$ to the slider block, constrained to slide in the slot as shown.

Solution

As seen in Figure E3.8b on the next page, the point Ⓘ for $\mathcal{B}_1$ is O, since it is pinned to the reference frame. The velocity of A is perpendicular to the line from Ⓘ to A (that is, from O to A) and has a direction in agreement with the angular velocity of $\mathcal{B}_1$ as the body turns about O. Its value is $r_{ⒾA}\omega_1$, or 1 m/s $\leftarrow$.

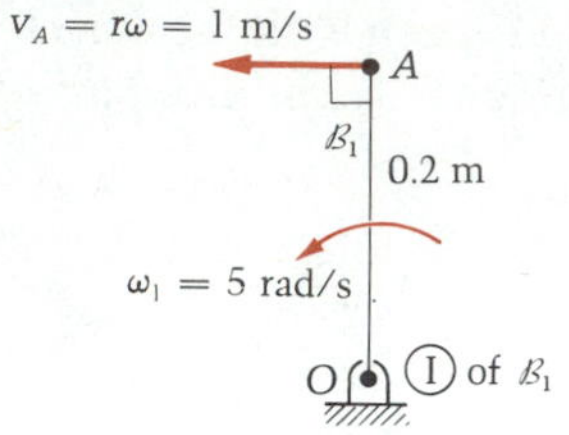

Figure E3.8b

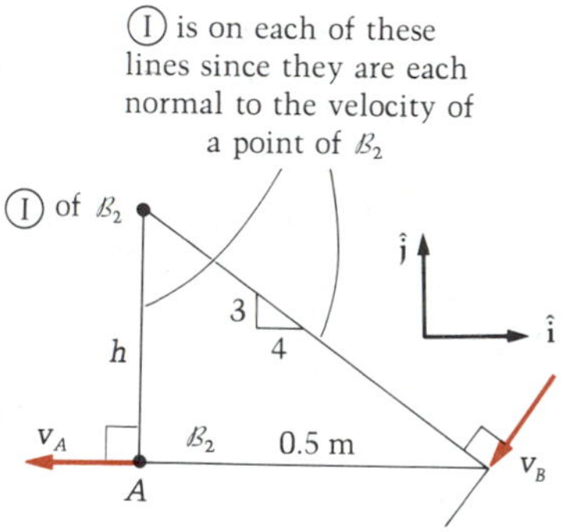

Figure E3.8c

Next we sketch body $\mathcal{B}_2$ and note the position of Ⓘ for $\mathcal{B}_2$, as explained in Figure E3.8c. Similar triangles yield the height H of Ⓘ above A:

$$\frac{H}{0.5} = \frac{3}{4}$$

$$H = 0.375 \text{ m}$$

> **Question 3.11** Why does $\mathbf{v}_B$ have to be "southwest" along the slot and not "northeast"?

We may now use Ⓘ of $\mathcal{B}_2$ to get the angular velocity of $\mathcal{B}_2$; using vectors this time,

$$\mathbf{v}_A = \omega_2 \hat{\mathbf{k}} \times \mathbf{r}_{\text{Ⓘ}A}$$

Substituting, we get

$$-1\hat{\mathbf{i}} = \omega_2 \hat{\mathbf{k}} \times (-0.375\hat{\mathbf{j}})$$

Solving gives

$$\omega_2 = -2.67 \Rightarrow \boldsymbol{\omega}_2 = -2.67\hat{\mathbf{k}} \text{ or } 2.67 \circlearrowright \text{ rad/s}$$

Note that when we write $\boldsymbol{\omega}_2 = \omega_2\hat{\mathbf{k}}$, we are saying that ω_2 is counterclockwise in accordance with the sign convention adopted for the problem in the figure if its value turns out positive. Thus when its value is now found to be negative, we know that $\mathcal{B}_2$ is turning clockwise at the given instant. Of course, as we have seen, we do not have to use vectors on such a simple problem; we can use what we know about the instantaneous center in scalar form to get a quick solution:

$$v_A = H|\omega_2| \Rightarrow |\omega_2| = \frac{v_A}{H} = \frac{1}{0.375} = 2.67 \Rightarrow \boldsymbol{\omega}_2 = 2.67 \circlearrowright \text{ rad/s}$$

where we assign the direction in accordance with the known velocity direction of A and the position of Ⓘ.

Next we use Ⓘ of $\mathcal{B}_2$ to obtain $\mathbf{v}_B$:

$$\begin{aligned} v_B &= r_{\text{Ⓘ}B}|\omega_2| \\ &= \sqrt{0.375^2 + 0.5^2}|\omega_2| \text{ m/s} \\ &= 0.625(2.67) = 1.67 \text{ m/s} \end{aligned}$$

The velocity of B is thus $\mathbf{v}_B = 1.67 \swarrow_{3}^{4}$ m/s.

Note that the arrow in this sketch is just as descriptive of the direction of the vector velocity of B as is the unit vector $-0.6\hat{\mathbf{i}} - 0.8\hat{\mathbf{j}}$.

Answer 3.11 The known velocity direction of A dictates that $\mathcal{B}_2$ is turning clockwise around Ⓘ, so $\mathbf{v}_B$ has to be "southwest" for this to be the case.

The Special Case in Which the Normals Do Not Intersect

Two things can go wrong with the procedure we have been following of intersecting the normals to two points' velocity vectors to find Ⓘ. The first of these is that the two perpendicular lines may be parallel and hence not intersect, as suggested by Figure 3.13 below for the points A and B of bar $\mathcal{B}$. Note that A and B are each at the top of the vertical plane circles on which they move, and since their velocity vectors are tangent to their paths, each is horizontal at this instant.

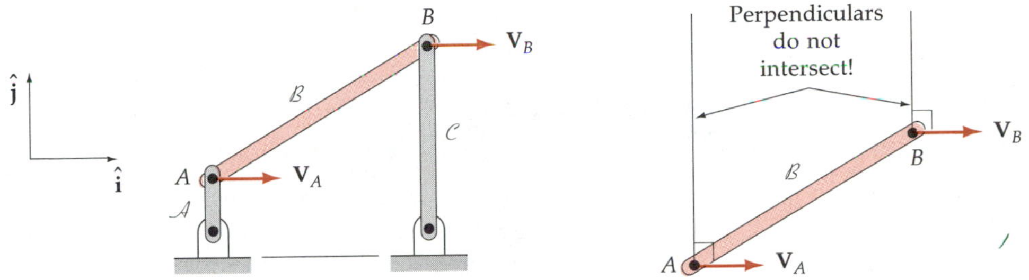

Figure 3.13

Let us examine Equation 3.8 for this case:

$$\mathbf{v}_B = \mathbf{v}_A + \boldsymbol{\omega} \times \mathbf{r}_{AB}$$

Since $\mathbf{v}_B$ and $\mathbf{v}_A$ have only $\hat{\mathbf{i}}$-components while $\boldsymbol{\omega} \times \mathbf{r}_{AB}$ has both $\hat{\mathbf{i}}$ *and* $\hat{\mathbf{j}}$, then $\boldsymbol{\omega}$ must be zero at this instant. This does not mean the body is translating; that occurs when $\boldsymbol{\omega}$ is zero *all the time*. Rather, in this case the body is just stopped for one instant in its angular motion, as its angular velocity is changing from clockwise to counterclockwise (see Figure 3.14).

The equation above also shows that at such an instant when $\boldsymbol{\omega} = \mathbf{0}$, all points of the body have identical velocities. So in Figure 3.13, $\mathbf{v}_A = \mathbf{v}_B = \mathbf{v}_{\textit{any point of } \mathcal{B}}$ at that instant. Conversely, any time two points of a body have equal velocities in plane motion, the body's angular velocity vanishes at that instant.

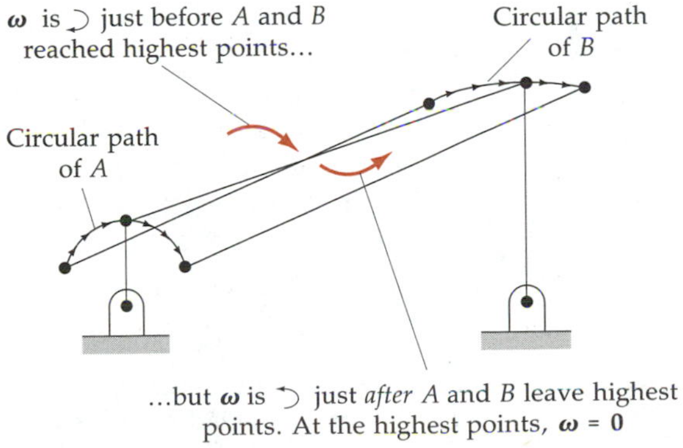

Figure 3.14

The Special Case in Which the Normals are Coincident

The second exceptional case occurs when the perpendiculars to two velocities are one and the same line (see Figures 3.15(a,b)):

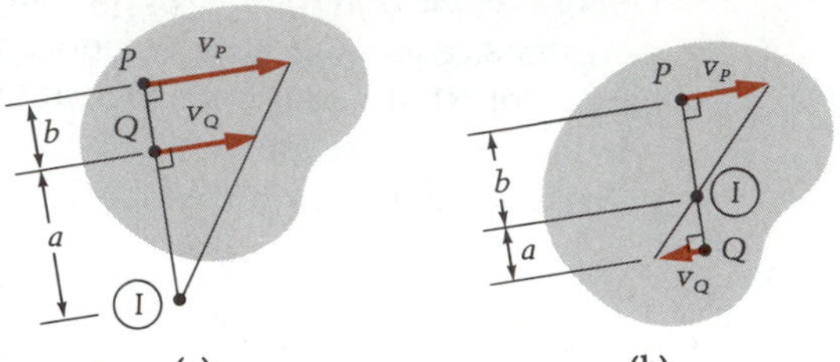

Figure 3.15

In this case, we can find the instantaneous center using similar triangles as shown. This simple procedure works, because, for the case shown in Figure 3.15a,

$$v_Q = a\omega \qquad \text{and} \qquad v_P = (a + b)\omega$$

so that

$$\frac{v_P}{a + b} = \frac{v_Q}{a}$$

If we should get coincident normals with the directions of $\mathbf{v}_P$ and $\mathbf{v}_Q$ *opposite,* as in Figure 3.15b, then Ⓘ lies *between* P and Q, and it may again be found by similar triangles. This time,

$$v_P = b\omega \qquad \text{and} \qquad v_Q = a\omega$$

so that

$$\frac{v_P}{b} = \frac{v_Q}{a}$$

In the examples we have presented in this section, note that use of the instantaneous center may be made with or without vector algebra. Its advantage is in finding and using points of zero velocity in order to simplify the resulting mathematics. Instantaneous centers never *have* to be used to effect a solution. Sometimes they are helpful, but at other times, they may be more trouble to locate than they are worth!

Examples of both the above special cases occur in our last example in this section. It involves four different positions of the same system:

EXAMPLE 3.9

Figure E3.9a shows a rolling wheel $\mathcal{B}_1$ of a large vehicle that travels at a constant speed of 60 mph →. Find the velocity of the piston when θ equals: (a) 0°, (b) 90°, (c) 180°, (d) 270°.

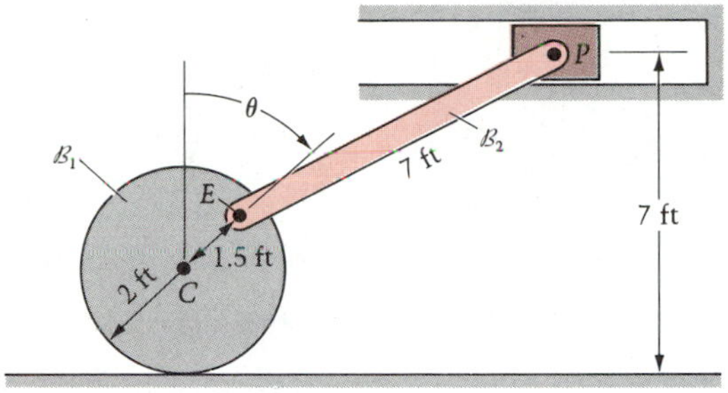

Figure E3.9a

Solution

First we solve for the velocities by using several approaches. In each case, the speed of the wheel's center is the same as the speed of the vehicle: 60 mph or 88 ft/sec. And since the piston translates, all its points have equal velocities and equal accelerations at every instant.

Case (a): As we shall see in detail in Section 3.6, the instantaneous center of a wheel rolling on a fixed track is at the point of contact. Since velocities increase linearly with distance from this point Ⓘ, we have for the point E of $\mathcal{B}_1$:

$$v_E = \frac{2 + 1.5}{2}(88) = 154 \text{ ft/sec}$$

If we draw lines at E and P perpendicular to $\mathbf{v}_E$ and $\mathbf{v}_P$ (P is constrained to move horizontally), they are parallel and thus will not intersect (see Figure E3.9b). Therefore $\omega_2 = 0$ at that instant. Thus $\mathbf{v}_P = \mathbf{v}_E = 154 \rightarrow$ ft/sec.

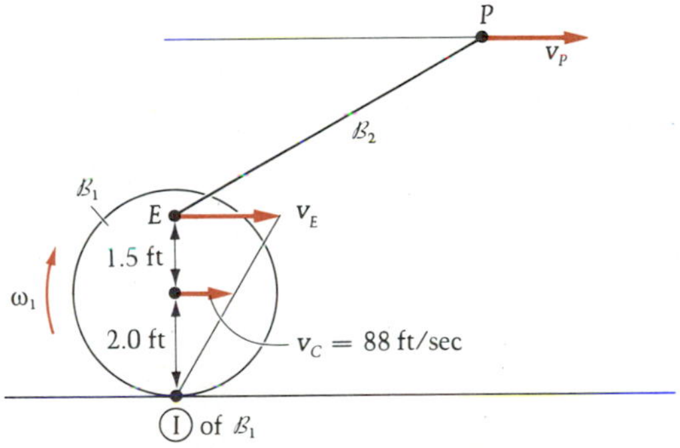

Figure E3.9b

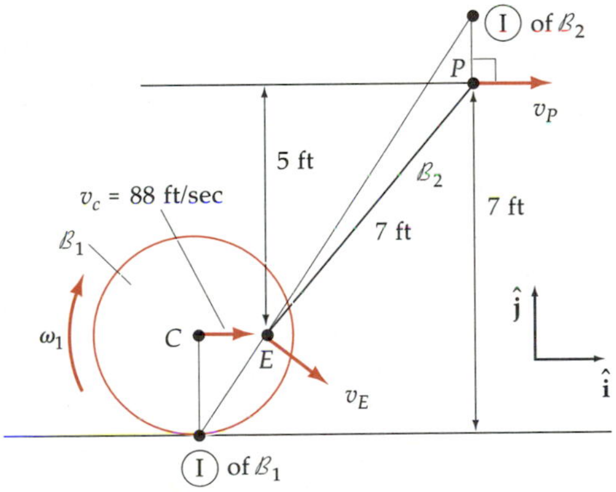

Figure E3.9c

Case (b): This time we shall use vectors; on the wheel (see Figure E3.9c),

$$\mathbf{v}_C = \overset{0}{\cancel{\mathbf{v}_{Ⓘ}}} + \omega_1\hat{\mathbf{k}} \times \overset{2\hat{\mathbf{j}}}{\cancel{\mathbf{r}_{ⒾC}}}$$

$$88\hat{\mathbf{i}} = -2.0\omega_1\hat{\mathbf{i}} \Rightarrow \omega_1 = -44 \Rightarrow \boldsymbol{\omega}_1 = -44\hat{\mathbf{k}}$$

or

$$\omega_1 = 44.0 \circlearrowleft \text{ rad/sec}$$

$$\mathbf{v}_E = -44.0\hat{\mathbf{k}} \times (1.5\hat{\mathbf{i}} + 2\hat{\mathbf{j}}) = 88.0\hat{\mathbf{i}} - 66.0\hat{\mathbf{j}}$$

or

$$\mathbf{v}_E = 110 \overset{4}{\searrow}{}_3 \text{ ft/sec}$$

On $\mathcal{B}_2$ now (after noting the trigonometry results in Figure E3.9d):

$$\mathbf{v}_P = v_P\hat{\mathbf{i}} = \mathbf{v}_E + \omega_2\hat{\mathbf{k}} \times \mathbf{r}_{EP}$$
$$= 88.0\hat{\mathbf{i}} - 66.0\hat{\mathbf{j}} + \omega_2\hat{\mathbf{k}} \times (4.90\hat{\mathbf{i}} + 5.00\hat{\mathbf{j}})$$

Equating coefficients of $\hat{\mathbf{i}}$ and then of $\hat{\mathbf{j}}$, we obtain:

$$\hat{\mathbf{i}}: \quad v_P = 88.0 - 5.00\omega_2$$
$$\hat{\mathbf{j}}: \quad 0 = -66.0 + 4.90\omega_2 \Rightarrow \omega_2 = +13.5 \text{ rad/sec}$$

Therefore

$$v_P = 88.0 - 5.00(+13.5) = 20.5 \text{ ft/sec}$$

so that

$$\mathbf{v}_P = 20.5\hat{\mathbf{i}} \text{ ft/sec}$$

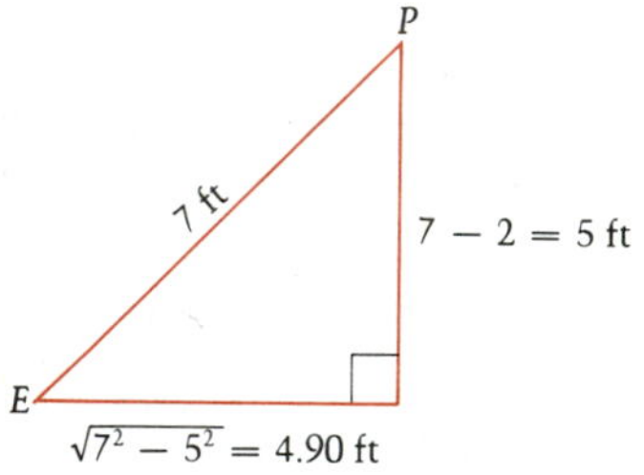

Figure E3.9d

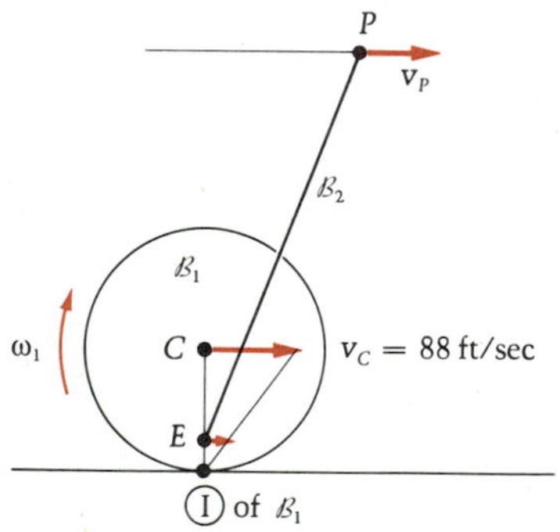

Figure E3.9e

Case (c): In all four cases, $\omega_1 = 44.0 \circlearrowleft$ rad/sec. This time, then (see Figure E3.9e),

$$v_E = (2.0 - 1.5)\omega_1 = 22.0 \text{ ft/sec}$$
$$\mathbf{v}_E = 22.0 \rightarrow \text{ ft/sec}$$

Again, as in Case (a), body $\mathcal{B}_2$ has $\boldsymbol{\omega}_2 = \mathbf{0}$ so that all points of $\mathcal{B}_2$ have the same velocity. Thus

$$\mathbf{v}_P = 22.0\hat{\mathbf{i}} \text{ ft/sec}$$

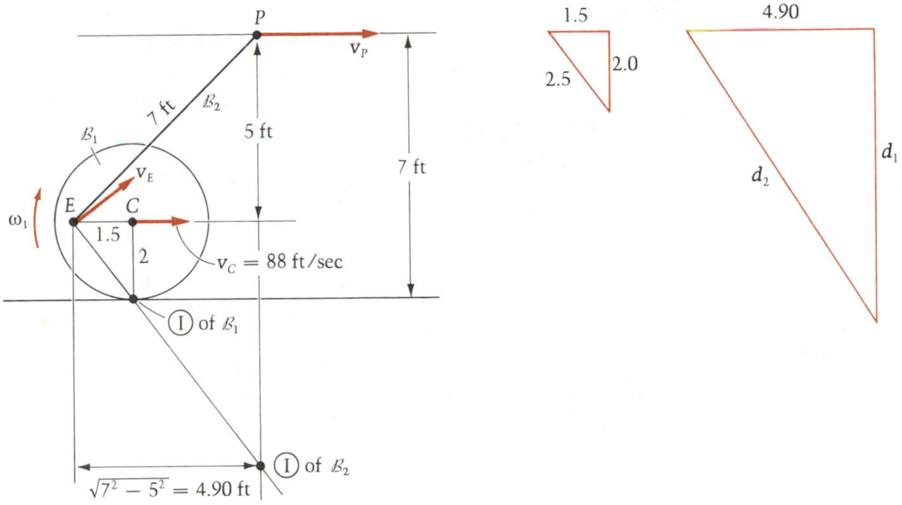

Figure E3.9f

Case (d): Using Figure E3.9f, we see that, on body $\mathcal{B}_1$,

$$\mathbf{v}_E = \underbrace{88\hat{\mathbf{i}}}_{\mathbf{v}_C} + \underbrace{(-44.0\hat{\mathbf{k}})}_{\omega_1\mathbf{k}} \times \underbrace{(-1.5\hat{\mathbf{i}})}_{\mathbf{r}_{CE}}$$

$$= 88.0\hat{\mathbf{i}} + 66.0\hat{\mathbf{j}}$$

We shall now use the instantaneous center of $\mathcal{B}_2$. From the similar triangles in the above figure,

$$d_1 = \frac{2.0}{1.5}(4.90) = 6.53 \text{ ft}$$

$$d_2 = \frac{2.5}{1.5}(4.90) = 8.17 \text{ ft}$$

Therefore, on body $\mathcal{B}_2$,

$$\omega_2 = \frac{v_E}{8.17} = \frac{\sqrt{88.0^2 + 66.0^2}}{8.17} = \frac{110}{8.17} = 13.5 \text{ rad/sec}$$

$$\omega_2 = 13.5 \circlearrowright \text{ rad/sec}$$

and

$$v_P = (6.53 + 5)\omega_2 = 156 \text{ ft/sec} \Rightarrow \mathbf{v}_P = 156 \rightarrow \text{ft/sec}$$

Note from the four answers in the above example that the piston is moving faster during the parts of the wheel's revolution in which $\mathbf{v}_E$ makes small angles with rod $\mathcal{B}_2$; conversely, it moves slower when $\mathbf{v}_E$ is making a large angle with $\mathcal{B}_2$. This is because the components of $\mathbf{v}_E$ and $\mathbf{v}_P$ along $\mathcal{B}_2$ must always be the same, as we saw earlier.

PROBLEMS ► Section 3.4

3.31 The angular velocity of $\mathcal{B}_1$ in Figure P3.31 is $3 \circlearrowright$ rad/s = constant. Trace the five sketches and then show on parts (a) to (d) the position of Ⓘ for the rod $\mathcal{B}_2$. In part (e), using the proper length of $\mathcal{B}_2$, draw the positions of $\mathcal{B}_2$ at the two times when $\mathbf{v}_B = \mathbf{0}$. Rod $\mathcal{B}_2$ has length 0.9 m.

3.32 Solve Problem 3.16 by using instantaneous centers.

3.33 Solve Problem 3.7 by using instantaneous centers.

3.34 Solve Problem 3.8 by using instantaneous centers.

3.35 Solve Problem 3.6 by using instantaneous centers.

3.36 Solve Problem 3.24 by using instantaneous centers.

3.37 Solve Problem 3.11 by using instantaneous centers.

3.38 In Figure P3.38 the crank arm $\mathcal{B}_1$ is 4 in. long and has constant angular velocity $\omega_1 = 2 \circlearrowright$ rad/sec. It is pinned to the triangular plate $\mathcal{B}_2$, which is also pinned to the block in the slot at D. Find the velocity of D at the instant shown.

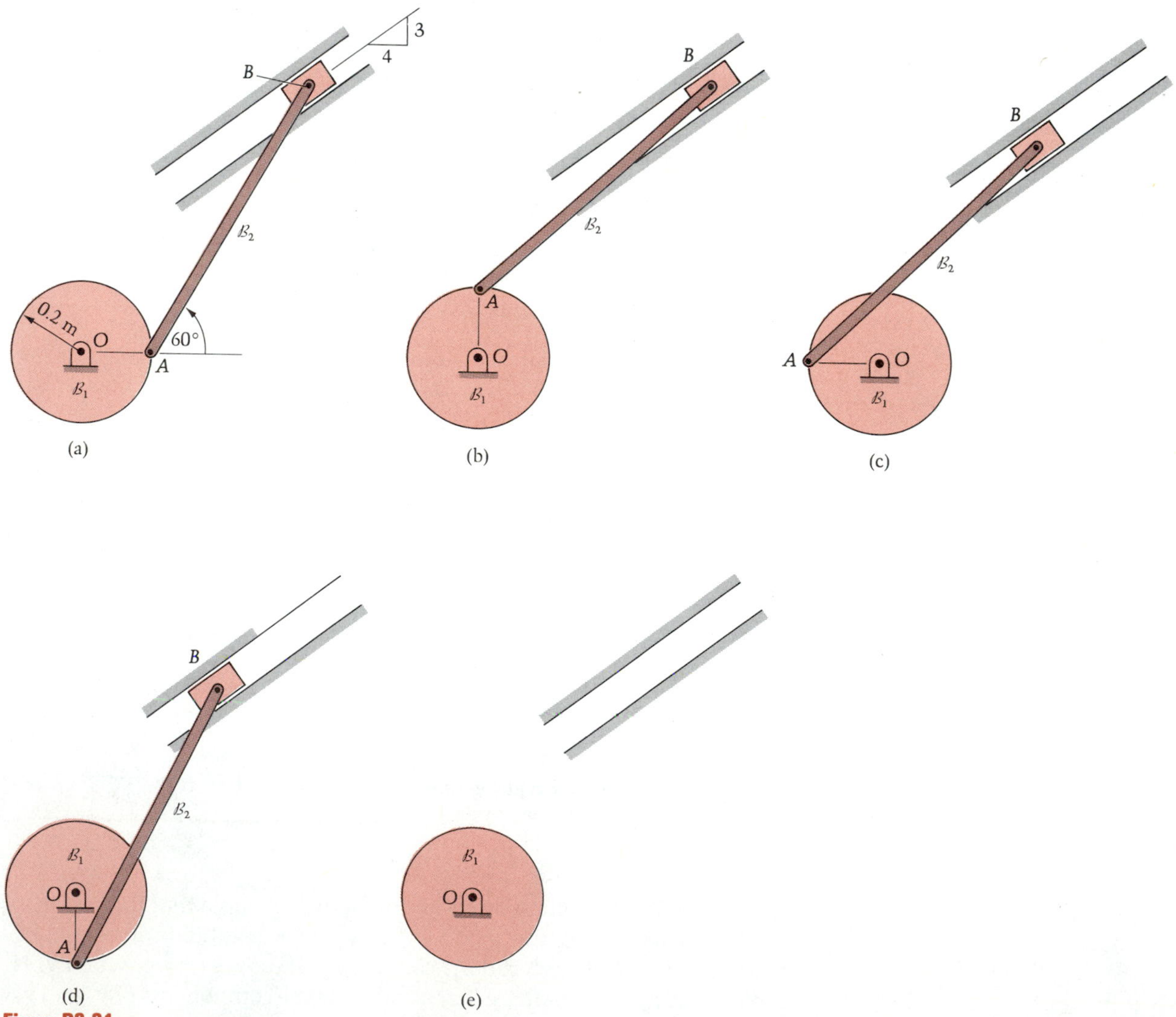

Figure P3.31

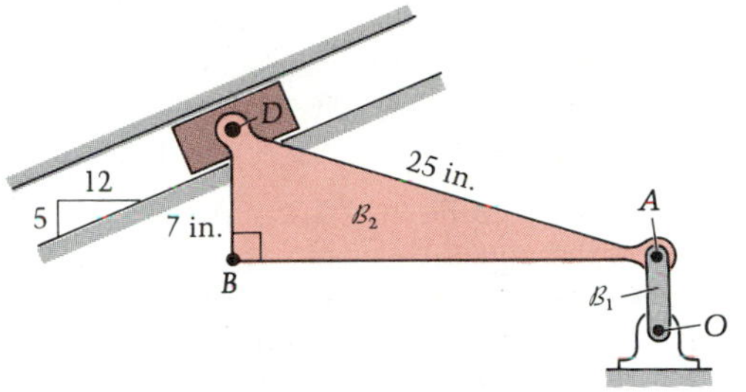

Figure P3.38

3.39 See Figure P3.39. The angular velocity of rod $\mathcal{B}_1$ is a constant: $\boldsymbol{\omega}_1 = 0.3\hat{\mathbf{k}}$ rad/s. Determine the angular velocities of plate $\mathcal{B}_2$ and bar $\mathcal{B}_3$ in the indicated position.

3.40 The pin at B (Figure P3.40) has a constant speed of 51 cm/s and moves in a circle in the clockwise direction. Find the angular velocities of bars $\mathcal{B}_1$ and $\mathcal{B}_2$ in the given position.

3.41 Solve Example 3.5 by using the instantaneous center of the rod.

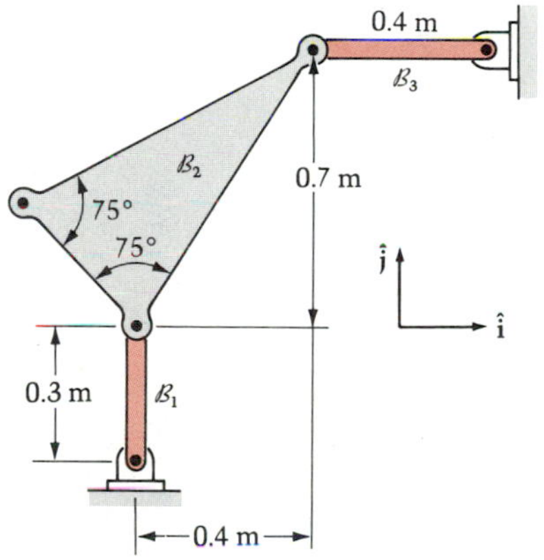

Figure P3.39

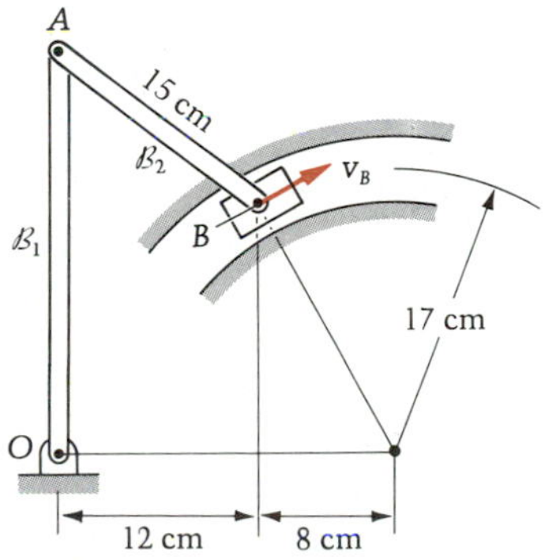

Figure P3.40

3.42 Rods $\mathcal{B}_1$ and $\mathcal{B}_2$ are pinned at B and move in a vertical plane with the constant angular velocities shown in Figure P3.42. Locate the instantaneous center of $\mathcal{B}_2$ for the given position, and use it to find the velocity of point C. Then check by calculating $\mathbf{v}_C$ by relating it (on $\mathcal{B}_2$) to the velocity of B. Note that sometimes Ⓘ is more trouble to locate than it is worth!

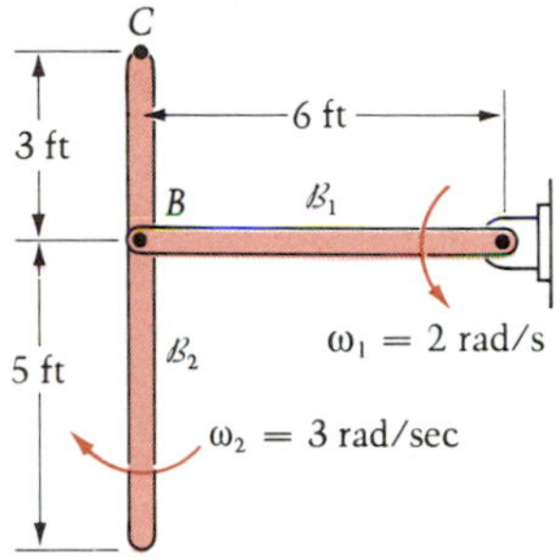

Figure P3.42

3.43 The linkage shown in Figure P3.43 is made up of rods $\mathcal{B}_1$, $\mathcal{B}_2$, and $\mathcal{B}_3$. Rod $\mathcal{B}_1$ has constant angular velocity $\boldsymbol{\omega}_1 = 0.6$ ↺ rad/sec. Determine the angular velocities of $\mathcal{B}_2$ and $\mathcal{B}_3$ when the angle θ is equal to 90° as shown.

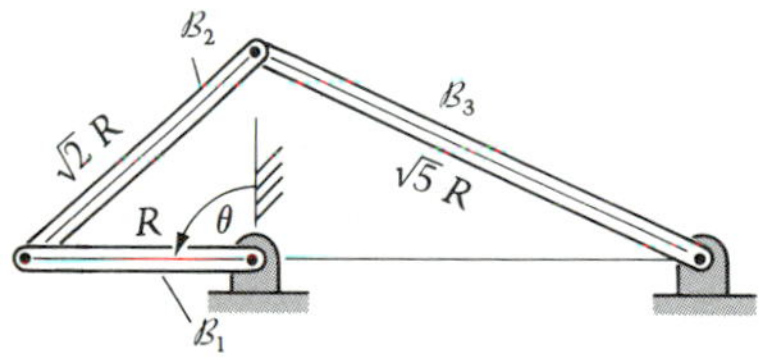

Figure P3.43

3.44 A bar of length $2L$ moves with its ends in contact with the planes shown in Figure P3.44. Find the velocity and acceleration of point C, in terms of θ and its derivatives, by writing and then differentiating the position vector of C. Then check the velocity solution by using the instantaneous center.

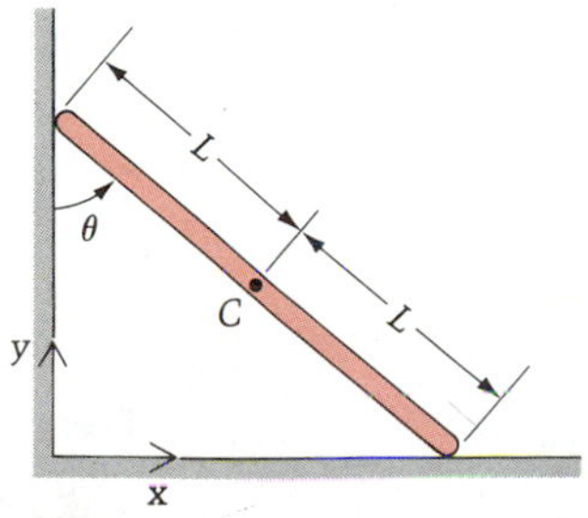

Figure P3.44

3.45 The piston rod of the hydraulic cylinder shown in Figure P3.45 moves outward at the constant speed of 0.13 m/s. Find the angular velocity of $\mathcal{B}_1$ at the instant shown.

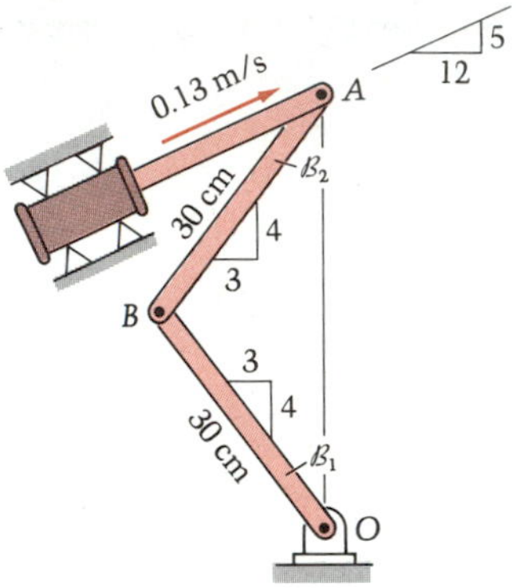

Figure P3.45

3.46 Using the method of instantaneous centers, find the velocity of point B in Figure P3.46, which is constrained to move in the slot as shown. The angular velocity of $\mathcal{B}_1$ is 0.3 ↺ rad/s at the indicated instant.

3.47 The center of block $\mathcal{B}_1$ in Figure P3.47 travels at a constant speed of 30 mph to the right. Disk $\mathcal{B}_2$ is pinned to $\mathcal{B}_1$ at A and spins at 100 rpm counterclockwise. Find: (a) the velocity of P; (b) the instantaneous center Ⓘ of $\mathcal{B}_2$; and (c), using Ⓘ, the velocities of Q, S, and R.

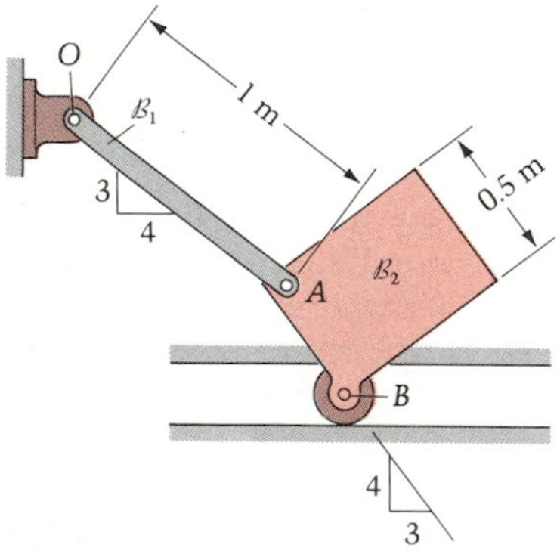

Figure P3.46

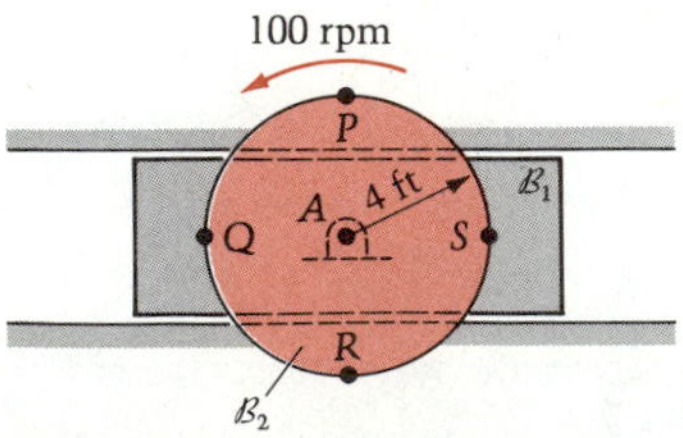

Figure P3.47

3.48 The roller at B, which moves in the parabolic slot, is pinned to bar $\mathcal{B}_1$ as shown in Figure P3.48. Bar $\mathcal{B}_1$ is pinned to $\mathcal{B}_2$ at A. The angular velocity of $\mathcal{B}_2$ at this instant is shown. Find the angular velocity of $\mathcal{B}_1$ at this time.

3.49 Bars $\mathcal{B}_1$ and $\mathcal{B}_2$ (see Figure P3.49) are pinned together at A. Find the angular velocity of bar $\mathcal{B}_1$ and the velocity of point B when the bars are next collinear. *Hint*: To find this configuration, draw a series of rough sketches of $\mathcal{B}_1$ and $\mathcal{B}_2$ as $\mathcal{B}_2$ turns counterclockwise from the position shown, and you will see $\mathcal{B}_1$ and $\mathcal{B}_2$ coming into alignment.

3.50 Repeat the preceding problem at the *second* instant of time when the bars are collinear. Follow the same hint, but this time start just past the first collinear position, found in Problem 3.49.

3.51 The constant angular velocity of wheel $\mathcal{B}_1$ is $\omega_1 = 100$ ↻ rad/sec. It is in rolling (i.e., "no slip") contact with $\mathcal{B}_2$, which means the contacting points have the same velocity. Find the angular velocity of the bar $\mathcal{B}_4$ at the instant shown in Figure P3.51.

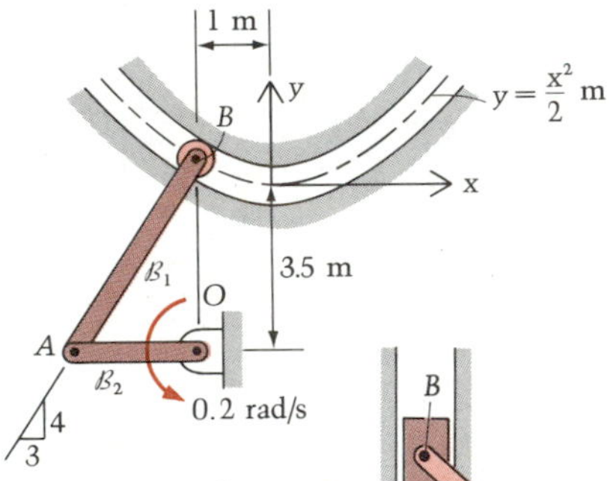

Figure P3.48

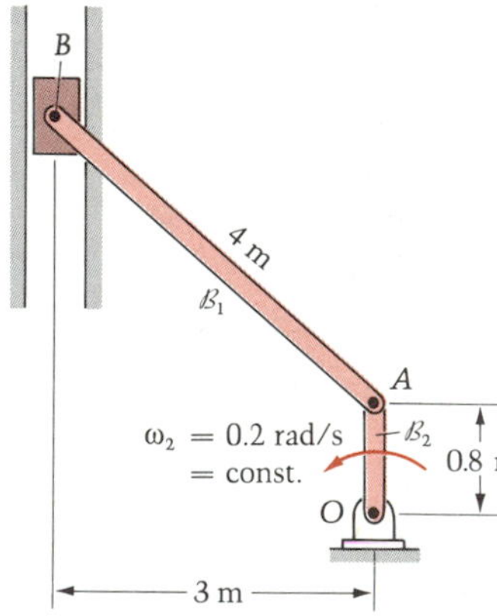

Figure P3.49

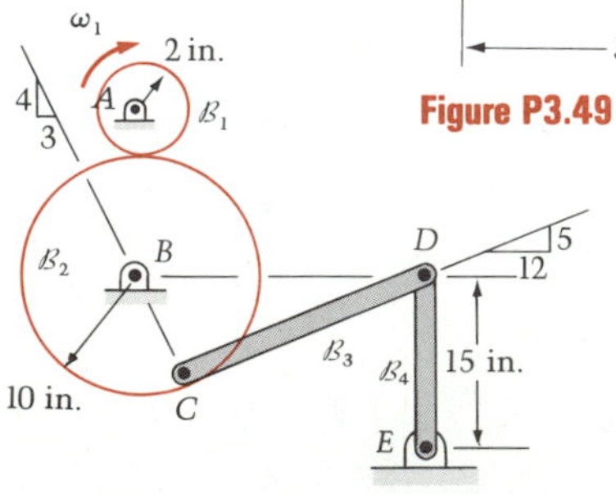

Figure P3.51

3.5 Acceleration and Angular Acceleration Relationship for Two Points of the Same Rigid Body

The **angular acceleration** vector of a rigid body $\mathcal{B}$ in plane motion in a frame $\mathcal{I}$ is defined as the derivative in $\mathcal{I}$ of the angular velocity and is called $\boldsymbol{\alpha}$:

$$\boldsymbol{\alpha}_{\mathcal{B}/\mathcal{I}} = \boldsymbol{\alpha} = \frac{d\boldsymbol{\omega}_{\mathcal{B}/\mathcal{I}}}{dt} = \dot{\boldsymbol{\omega}} = \ddot{\theta}\hat{\mathbf{k}} \tag{3.14}$$

or

$$\boldsymbol{\alpha} = \ddot{\theta}\hat{\mathbf{k}} \quad \text{or} \quad \alpha\hat{\mathbf{k}} \tag{3.15}$$

where $\hat{\mathbf{k}}$ is a constant vector in both $\mathcal{B}$ and $\mathcal{I}$. Note that, as with $\boldsymbol{\omega}$, we delete the subscript when there is no confusion about the frame of reference being used.

Development of the Acceleration and Angular Acceleration Relationship

We now develop the relationship between the accelerations of two points P and Q of a rigid body $\mathcal{B}$. Differentiating both sides of Equation (3.8) yields

$$\dot{\mathbf{v}}_Q = \mathbf{a}_Q = \mathbf{a}_P + \ddot{\theta}\hat{\mathbf{k}} \times \mathbf{r}_{PQ} + \dot{\theta}\hat{\mathbf{k}} \times \dot{\mathbf{r}}_{PQ} \tag{3.16}$$

Now using Equation (3.7),* we may rewrite the last term as

$$\begin{aligned}\dot{\theta}\hat{\mathbf{k}} \times \dot{\mathbf{r}}_{PQ} &= \dot{\theta}\hat{\mathbf{k}} \times (\dot{\theta}\hat{\mathbf{k}} \times \mathbf{r}_{PQ}) \\ &= \dot{\theta}\hat{\mathbf{k}}(\dot{\theta}\hat{\mathbf{k}} \cdot \mathbf{r}_{PQ}) - \mathbf{r}_{PQ}(\dot{\theta}\hat{\mathbf{k}} \cdot \dot{\theta}\hat{\mathbf{k}})\end{aligned} \tag{3.17}$$

or

$$\dot{\theta}\hat{\mathbf{k}} \times \dot{\mathbf{r}}_{PQ} = -\dot{\theta}^2\mathbf{r}_{PQ} \tag{3.18}$$

Question 3.12 Why is the dot product $\dot{\theta}\hat{\mathbf{k}} \cdot \mathbf{r}_{PQ}$ in Equation (3.17) zero?

Equations (3.16) and (3.18) then yield the desired relation between the accelerations of P and Q:

$$\mathbf{a}_Q = \mathbf{a}_P + \ddot{\theta}\hat{\mathbf{k}} \times \mathbf{r}_{PQ} - \dot{\theta}^2\mathbf{r}_{PQ} \tag{3.19}$$

We note that the same three rules spelled out in Equation (3.10) also hold for the use of Equation (3.19). Unlike velocities, however, the accelerations of P and Q do not generally have equal components along the line PQ joining them; these components differ by $r\dot{\theta}^2 = r\omega^2$. Likewise, the components *perpendicular* to PQ differ by $r\ddot{\theta} = r\alpha$ (in the same way as the velocity components normal to PQ differ by $r\dot{\theta} = r\omega$).

Answer 3.12 Because $\mathbf{r}_{PQ}$ lies in the (xy) reference plane and is therefore perpendicular to $\hat{\mathbf{k}}$.

* And the vector identity $\mathbf{A} \times (\mathbf{B} \times \mathbf{C}) = \mathbf{B}(\mathbf{A} \cdot \mathbf{C}) - \mathbf{C}(\mathbf{A} \cdot \mathbf{B})$.

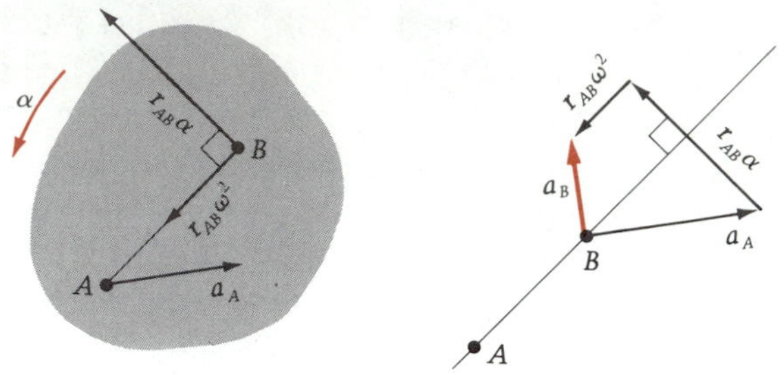

Figure 3.16

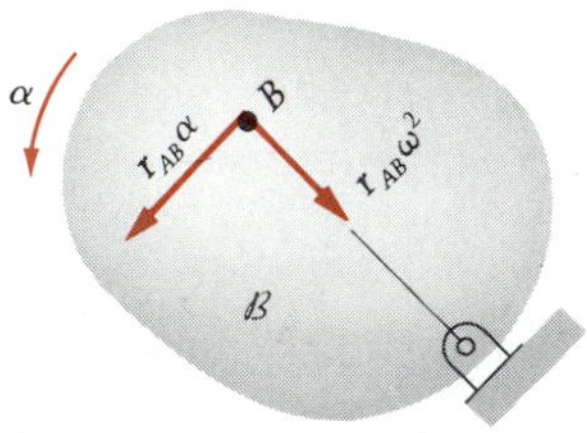

Figure 3.17

If the acceleration of point A of a body $\mathcal{B}$ is $\mathbf{a}_A$, for example, then the acceleration of any point B is the sum of the three vectors shown in Figure 3.16. If A is a pinned point, or pivot,* then $\mathbf{a}_B$ has two components: one along the line from B toward A and the other perpendicular to it and tangent to the circle (on which it necessarily travels when there is a pivot at A). Note that the direction of the tangential component depends on the direction of $\boldsymbol{\alpha}$, but the "radial" component is *always* inward toward the pivot. This case is shown in Figure 3.17. We now illustrate the use of Equation (3.19) with several examples.

EXAMPLE 3.10

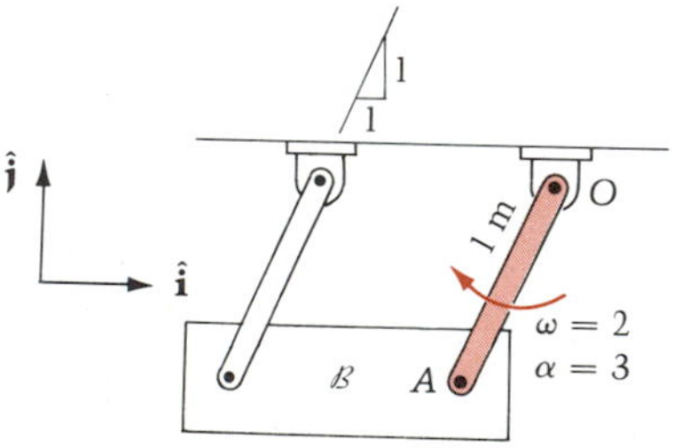

Figure E3.10

In Figure E3.10, let the links have length 1 m and let them each have $\boldsymbol{\omega} = 2 \circlearrowleft$ rad/s and $\boldsymbol{\alpha} = 3 \circlearrowleft$ rad/s² at a time when they make an angle of 45° with the ceiling. Find the acceleration of block $\mathcal{B}$. (That is, find the acceleration of any of its points — they are all the same since $\mathcal{B}$ is translating.)

Solution

All points of $\mathcal{B}$ have the same **v** and **a** as point A. Using Equation (3.19) for the link OA, we get

$$\begin{aligned}\mathbf{a}_A &= \overset{0}{\cancel{\mathbf{a}_O}} + \boldsymbol{\alpha} \times \mathbf{r}_{OA} - \omega^2 \mathbf{r}_{OA} \\ &= 3\hat{\mathbf{k}} \times (-0.707\hat{\mathbf{i}} + 0.707\hat{\mathbf{j}}) - 2^2(-0.707\hat{\mathbf{i}} + 0.707\hat{\mathbf{j}}) \\ &= (-2.12 + 2.83)\hat{\mathbf{i}} + (-2.12 - 2.83)\hat{\mathbf{j}} \\ &= 0.71\hat{\mathbf{i}} - 4.95\hat{\mathbf{j}} \text{ m/s}^2\end{aligned}$$

EXAMPLE 3.11

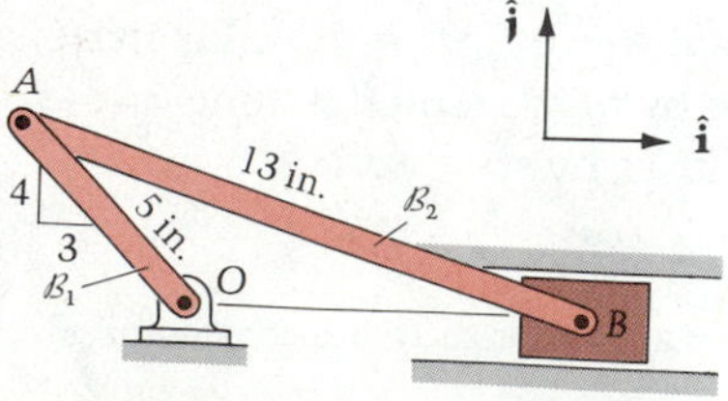

Figure E3.11

In Example 3.3 find the acceleration of the translating piston at the given instant if $\boldsymbol{\omega}_1 = 10 \circlearrowleft$ rad/sec and $\boldsymbol{\alpha}_1 = 5 \circlearrowright$ rad/sec². See Figure E3.11.

* Again, *pivot* means a point of $\mathcal{B}$ that does not move throughout a motion. It includes, but is not limited to, a hinge.

Question 3.13 What does it mean when α is in the opposite direction from that of ω?

Solution

Relating O and A on body $\mathcal{B}_1$ by Equation (3.19), we have

$$\begin{aligned}\mathbf{a}_A &= \overset{0\text{ (pinned)}}{\cancel{\mathbf{a}_O}} + \alpha_1\hat{\mathbf{k}} \times \mathbf{r}_{OA} - \omega_1^2\mathbf{r}_{OA} \\ &= 5\hat{\mathbf{k}} \times (-3\hat{\mathbf{i}} + 4\hat{\mathbf{j}}) - 10^2(-3\hat{\mathbf{i}} + 4\hat{\mathbf{j}}) \\ &= 280\hat{\mathbf{i}} - 415\hat{\mathbf{j}} \text{ in./sec}^2\end{aligned}$$

And relating A and B on body $\mathcal{B}_2$, we have

$$\mathbf{a}_B = \mathbf{a}_A + \alpha_2\hat{\mathbf{k}} \times \mathbf{r}_{AB} - \omega_2^2\mathbf{r}_{AB}$$

In the previous example we found

$$\mathbf{r}_{AB} = 12.4\hat{\mathbf{i}} - 4\hat{\mathbf{j}} \text{ in.} \qquad \text{and} \qquad \omega_2 = 2.42 \circlearrowleft \text{ rad/sec}$$

Noting that the piston translates horizontally, we get

$$\begin{aligned}\mathbf{a}_B = a_B\hat{\mathbf{i}} &= 280\hat{\mathbf{i}} - 415\hat{\mathbf{j}} + 12.4\alpha_2\hat{\mathbf{j}} + 4\alpha_2\hat{\mathbf{i}} - 2.42^2(12.4\hat{\mathbf{i}} - 4\hat{\mathbf{j}}) \\ &= (207 + 4\alpha_2)\hat{\mathbf{i}} + (-392 + 12.4\alpha_2)\hat{\mathbf{j}}\end{aligned}$$

The coefficients of $\hat{\mathbf{j}}$ yield

$$\alpha_2 = \frac{392}{12.4} = 31.6 \text{ rad/sec}^2$$

The coefficients of $\hat{\mathbf{i}}$ then give our answer:

$$\mathbf{a}_B = a_B\hat{\mathbf{i}} = [207 + 4(31.6)]\hat{\mathbf{i}} = 333\hat{\mathbf{i}} \text{ in./sec}^2$$

Let us find the acceleration of the instantaneous center of zero velocity of the rod $\mathcal{B}_2$ in Examples 3.7 and 3.11. Using those examples and Equation (3.19), we have

$$\begin{aligned}\mathbf{a}_{Ⓘ} &= \mathbf{a}_B + \alpha_2\hat{\mathbf{k}} \times \mathbf{r}_{BⒾ} - \omega_2^2\mathbf{r}_{BⒾ} \\ &= 333\hat{\mathbf{i}} + 31.6\hat{\mathbf{k}} \times (-12.5\hat{\mathbf{j}}) - (2.42)^2(-12.5\hat{\mathbf{j}}) \\ &= 728\hat{\mathbf{i}} + 73.2\hat{\mathbf{j}} \text{ in./sec}^2\end{aligned}$$

We see from this result that the instantaneous center of zero *velocity* does not generally have zero acceleration. Unless the point Ⓘ is a pivot (i.e., a permanently fixed point, such as a pin connecting the point to the reference frame), it should *never* be assumed that $\mathbf{a}_{Ⓘ}$ is zero.*

Answer 3.13 It means that the angular speed of the body is decreasing.

* Actually, in nontranslational cases there is a point of zero acceleration, but unless it is a pivot point, or a point of rolling contact at an instant when $\omega = 0$, it is more trouble to find than it is worth. See Problem 3.83.

EXAMPLE 3.12

In Example 3.5, find the angular acceleration of the rod and the acceleration of its endpoint A. See Figure E3.12a.

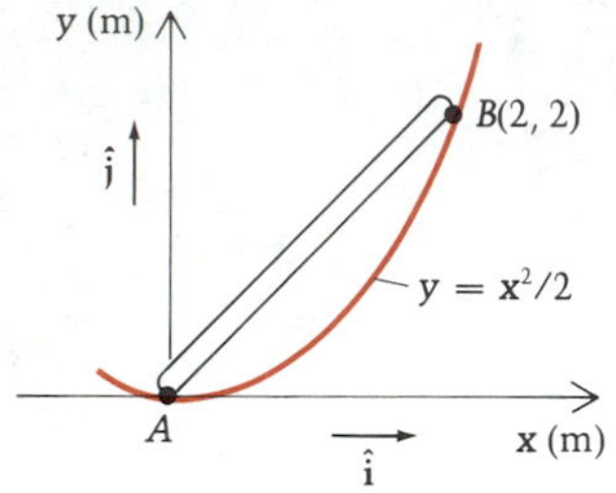

Figure E3.12a

Solution

Relating the accelerations of B and A with Equation (3.19),

$$\mathbf{a}_B = \mathbf{a}_A + \boldsymbol{\alpha} \times \mathbf{r}_{AB} - \omega^2 \mathbf{r}_{AB} \tag{1}$$

The acceleration of point B is

$$\mathbf{a}_B = \ddot{s}\hat{\mathbf{e}}_t + \frac{\dot{s}^2}{\rho}\hat{\mathbf{e}}_n = \frac{\dot{s}^2}{\rho}\hat{\mathbf{e}}_n$$

where $\ddot{s}_B = 0$ since $\dot{s}_B = \text{constant} = 0.3$ m/s. The curvature formula from calculus gives us the radius of curvature at point B:

$$\frac{1}{\rho} = \left| \frac{y''}{(1 + y'^2)^{3/2}} \right|$$
$$= \frac{1}{(1 + x^2)^{3/2}}$$

Therefore

$$\rho_B = (1 + 2^2)^{3/2}$$
$$= 5^{3/2}$$
$$= 11.2 \text{ m}$$

Therefore,

$$\mathbf{a}_B = \frac{(0.3)^2}{11.2}\hat{\mathbf{e}}_n = 0.00804\hat{\mathbf{e}}_n \text{ m/s}^2$$

The unit vector $\hat{\mathbf{e}}_n$ is seen in Figure E3.12b to be

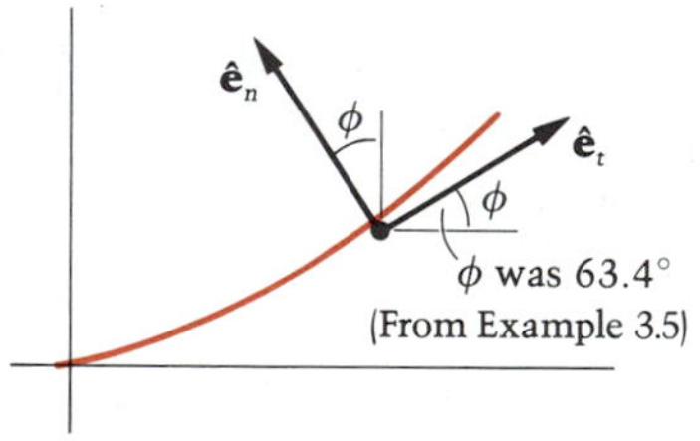

Figure E3.12b

$$\hat{\mathbf{e}}_n = -\sin\phi\hat{\mathbf{i}} + \cos\phi\hat{\mathbf{j}}$$
$$= -0.894\hat{\mathbf{i}} + 0.448\hat{\mathbf{j}}$$

Therefore

$$\mathbf{a}_B = -0.00804\hat{\mathbf{e}}_n$$
$$= -0.00719\hat{\mathbf{i}} + 0.00360\hat{\mathbf{j}} \text{ m/s}^2$$

Substituting a_B into Equation (1) gives

$$-0.00719\hat{\mathbf{i}} + 0.00360\hat{\mathbf{j}} = \ddot{s}_A\hat{\mathbf{e}}_t + \frac{\dot{s}_A^2}{\rho_A}\hat{\mathbf{e}}_n$$
$$+ \alpha\hat{\mathbf{k}} \times (2\hat{\mathbf{i}} + 2\hat{\mathbf{j}}) - \omega^2(2\hat{\mathbf{i}} + 2\hat{\mathbf{j}})$$

where $\mathbf{r}_{AB} = 2\hat{\mathbf{i}} + 2\hat{\mathbf{j}}$ has also been substituted.

Next, the radius of curvature of the path of A at the instant of interest is

$$\rho_A = (1 + x^2)^{3/2} = (1 + 0^2)^{3/2} = 1 \text{ m}$$

Substituting $\rho_A = 1$, $\dot{s}_A = 0.403$ and $\omega = 0.134$ from Example 3.5 (along with $\hat{\mathbf{e}}_t = \hat{\mathbf{i}}$ and $\hat{\mathbf{e}}_n = \hat{\mathbf{j}}$ at point A), we obtain the vector equation:

$$-0.00719\hat{\mathbf{i}} + 0.00360\hat{\mathbf{j}} = \ddot{s}_A\hat{\mathbf{i}} + \frac{(0.403)^2}{1}\hat{\mathbf{j}} - 2\alpha\hat{\mathbf{i}} + 2\alpha\hat{\mathbf{j}} - 2(0.134)^2(\hat{\mathbf{i}} + \hat{\mathbf{j}})$$

Writing the component equations, we have

$$\hat{\mathbf{i}} \text{ coefficients:} \qquad -0.00719 = \ddot{s}_A - 2\alpha - 2(0.134)^2$$

$$\hat{\mathbf{j}} \text{ coefficients:} \qquad 0.00360 = \frac{(0.403)^2}{1} + 2\alpha - 2(0.134)^2$$

The $\hat{\mathbf{j}}$ equation gives

$$\alpha = -0.0612 \Rightarrow \boldsymbol{\alpha} = -0.0612\hat{\mathbf{k}} \text{ rad/s}^2$$

And the $\hat{\mathbf{i}}$ equation then yields

$$\ddot{s}_A = -0.0936 \text{ m/s}^2$$

Therefore

$$\mathbf{a}_A = -0.0936\hat{\mathbf{i}} + \frac{(0.403)^2}{1}\hat{\mathbf{j}}$$

$$= -0.0936\hat{\mathbf{i}} + 0.162\hat{\mathbf{j}} \text{ m/s}^2$$

Many more examples of the use of Equation (3.19) will be found in the next section, after we have discussed the topic of rolling.

PROBLEMS ▶ Section 3.5

3.52 In Figure P3.52 the angular velocity of the bent bar is 0.2 rad/s counterclockwise at an instant when its angular acceleration is 0.3 rad/s² clockwise. Find the acceleration of the endpoint B in the indicated position.

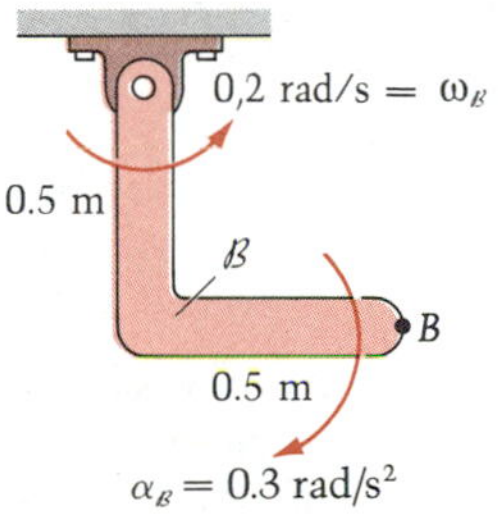

Figure P3.52

3.53 The acceleration of pin B in Figure P3.53 is 9.9 ft/sec² down and to the left, and its velocity is 4 ft/sec up and to the right, when $\mathcal{L}$ passes the horizontal. At this instant, find the angular acceleration of $\mathcal{L}$.

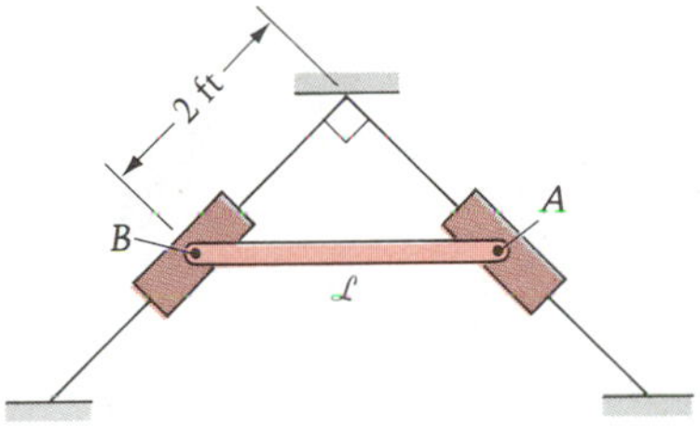

Figure P3.53

3.54 End B of the rod shown in Figure P3.54 has a constant velocity of 10 ft/sec down the plane. For the position shown (rod horizontal) determine the velocity and acceleration of end A of the rod.

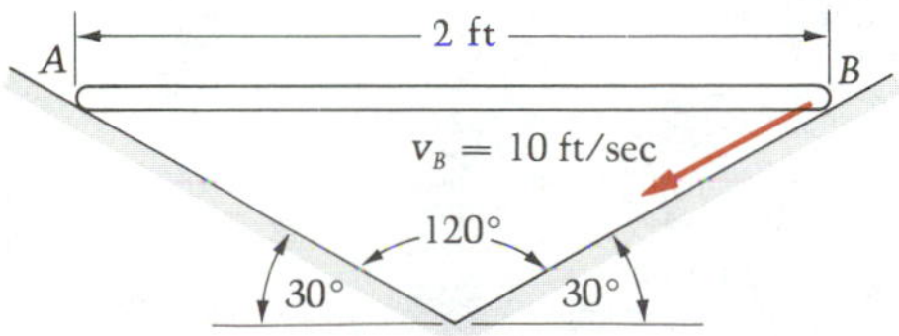

Figure P3.54

3.55 The velocities and accelerations of the two endpoints A and B of a rigid bar in plane motion are as shown in Figure P3.55. Find the acceleration of the midpoint of the bar in the given position.

3.56 In Problem 3.42 find the acceleration of C in the position given in the figure.

3.57 At the instant given, the angular velocity and angular acceleration of bar $\mathcal{B}_1$ are 0.2 ↺ rad/sec and 0.1 ↻ rad/sec². (See Figure P3.57.) Find the angular accelerations of $\mathcal{B}_2$ and $\mathcal{B}_3$ at this instant.

3.58 Bar $\mathcal{B}_1$ rotates with a constant angular velocity of $2\hat{\mathbf{k}}$ rad/sec. Find the angular velocities and angular accelerations of $\mathcal{B}_2$ and $\mathcal{B}_3$ at the instant shown in Figure P3.58.

3.59 In the position indicated in Figure P3.59, the slider block $\mathcal{B}$ has the indicated velocity and acceleration. Find the angular acceleration of the wheel at this instant.

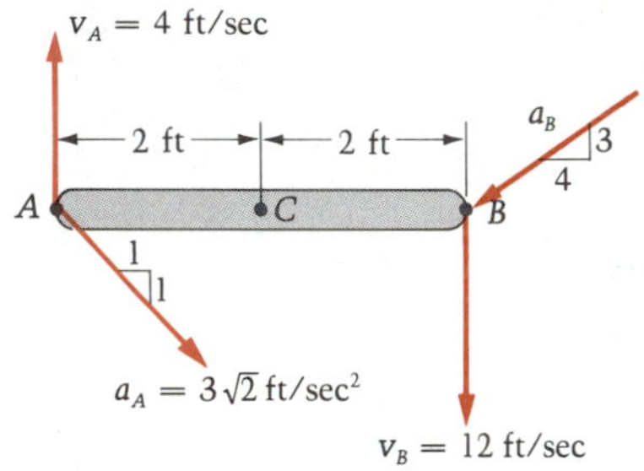

Figure P3.55

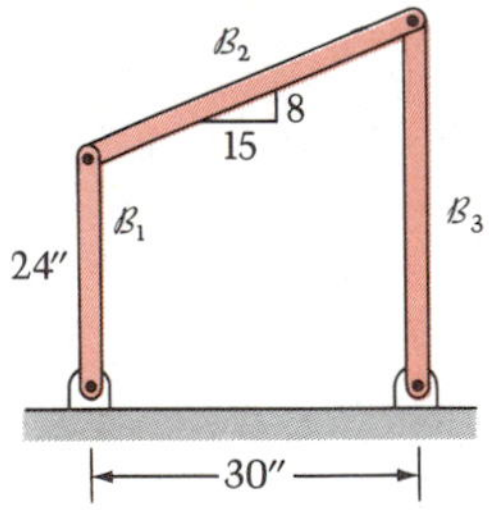

Figure P3.57

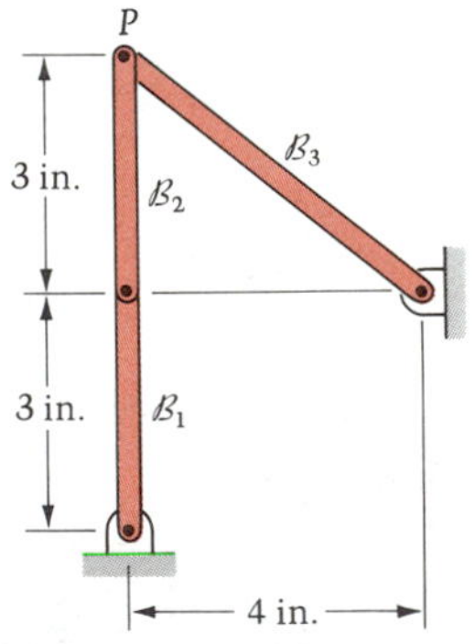

Figure P3.58

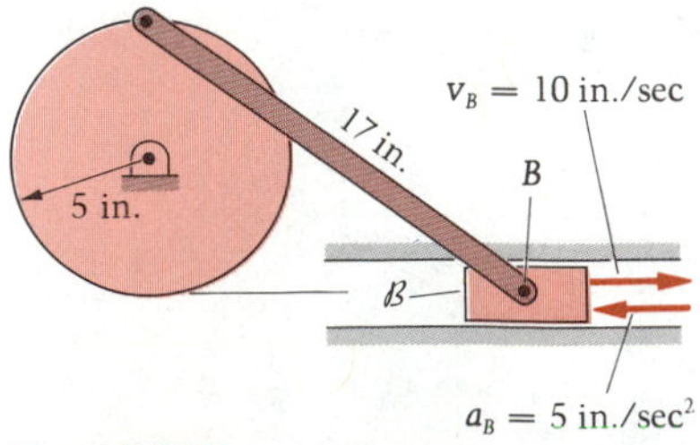

Figure P3.59

3.60 If in Problem 3.11 the bar whose angular velocity is given to be 2 rad/sec has an angular acceleration of zero at that instant, find the angular acceleration of the 5-inch (horizontal) bar.

* **3.61** The angular velocity of $\mathcal{B}_1$ in Figure P3.61 is a constant 3 rad/sec clockwise. Find the velocity and acceleration of point C in the given configuration, and determine the acceleration of point C when $\mathbf{v}_C = \mathbf{0}$.

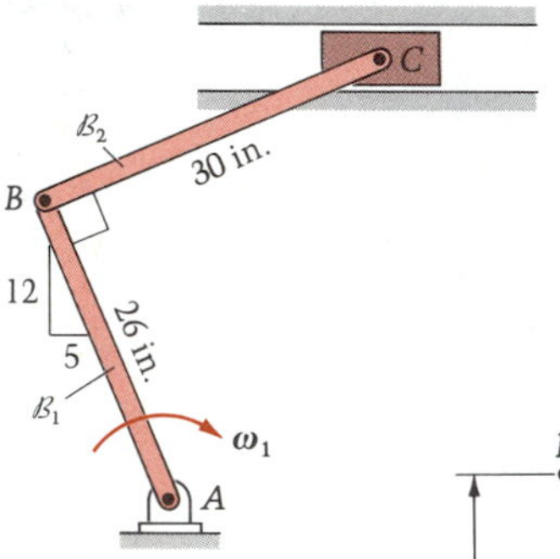

Figure P3.61

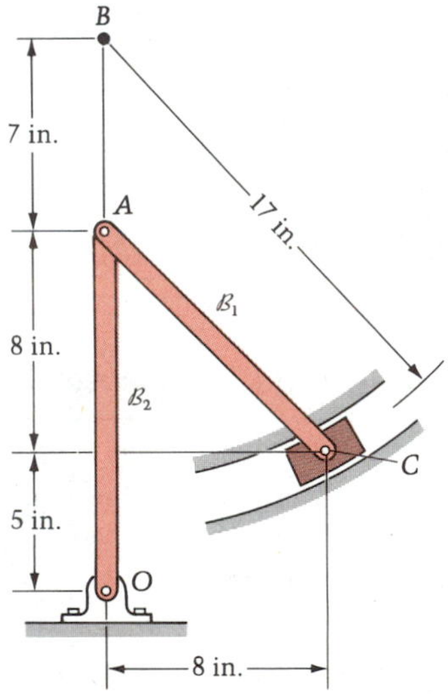

Figure P3.64

3.62 Refer to the preceding problem. At the instant of time when $\omega_2 = 0$, find the acceleration of point C.

3.63 If in Problem 3.24 the angular velocity of the 5-inch bar is constant throughout an interval which includes the instant shown, find at that instant the acceleration of the guided block.

3.64 At the instant of time shown in Figure P3.64, the angular velocity and angular acceleration of rod $\mathcal{B}_1$ are 3 ↺ rad/sec and 2 ↻ rad/sec². At the same time, find the angular acceleration of bar $\mathcal{B}_2$.

3.65 In Problem 3.26 find the equation for the angular acceleration $\ddot{\phi}$ of the connecting rod $\mathcal{B}_2$ a a function of r, ℓ, θ, and $\dot{\theta}$.

3.66 In the preceding problem, plot $\ddot{\phi}/\dot{\theta}^2$ versus θ ($0 \le \theta \le 2\pi$) for $\ell/r = 1, 2$, and 5.

3.67 Crank $\mathcal{B}_1$ in Figure P3.67 is pinned to rod $\mathcal{B}_2$; the other end of $\mathcal{B}_2$ slides on a parabolic incline and is at the origin in the position shown. The angular velocity of $\mathcal{B}_1$ is 3 ⊋ rad/s = constant. Determine the acceleration of A and the angular acceleration of $\mathcal{B}_2$ at the given instant. *Hint*: The radius of curvature ρ of a plane curve $y = y(x)$ can be calculated from

$$\frac{1}{\rho} = \left| \frac{y''}{(1 + y'^2)^{3/2}} \right|$$

Use this result in computing the normal component of $\mathbf{a}_A$ as in Example 3.11.

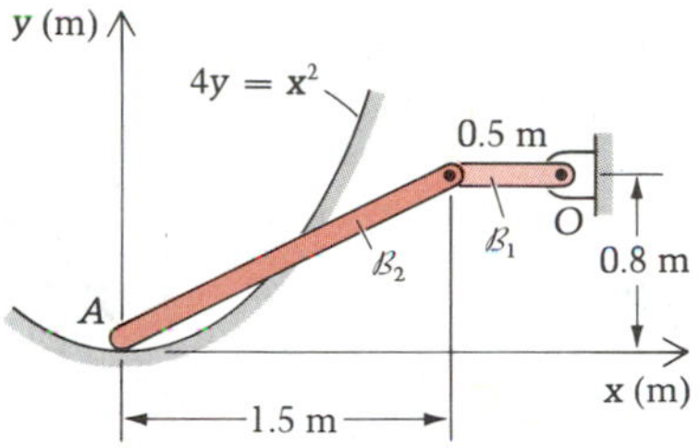

Figure P3.67

3.68 The 10-ft bar in Figure P3.68 is sliding down the 13-ft-radius circle as shown. For the position shown, the bar has an angular velocity of 2 rad/sec and an angular acceleration of 3 rad/sec², both clockwise. Find the x and y components of acceleration of point B for this position.

3.69 In Problem 3.23, find the acceleration of B when $x = 10$ m.

3.70 In Problem 3.31 find the acceleration of the upper end B of rod $\mathcal{B}_2$ in position (a).

3.71 Find the acceleration of P in Problem 3.58 if at the same instant the body $\mathcal{B}_1$ has angular acceleration $\boldsymbol{\alpha}_1 = 3\hat{\mathbf{k}}$ rad/sec² instead of zero.

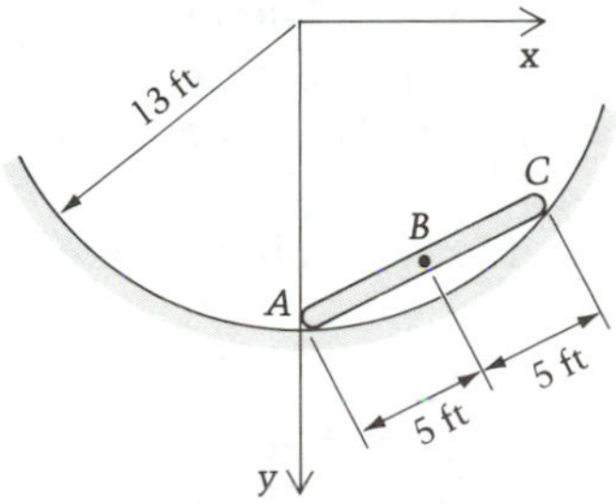

Figure P3.68

3.72 In Problem 3.49 find the acceleration of point B and the angular acceleration of body $\mathcal{B}_1$ at the described instant.

3.73 In Problem 3.50 find the acceleration of point B and the angular acceleration of body $\mathcal{B}_1$ at the described instant.

3.74 In Problem 3.22 find the acceleration of S and the angular accelerations of $\mathcal{B}_2$ and $\mathcal{B}_3$ at the instant when $\theta = 30°$.

3.75 In Problem 3.38 determine the angular acceleration of the plate and the acceleration of pin D in the indicated position.

3.76 In Problem 3.45 determine the angular accelerations of $\mathcal{B}_1$ and $\mathcal{B}_2$ in the indicated position.

3.77 The motion of a rotating element in a mechanism is controlled so that the rate of change of angular speed ω with angular displacement θ is a constant K. If the angular speed is ω_0 when both θ and the time t are zero, determine θ, ω, and the angular acceleration α as *functions of time.*

* **3.78** Rod $\mathcal{B}_1$ in Figure P3.78 is pinned to disk $\mathcal{B}_2$ at A and B. Disk $\mathcal{B}_2$ rotates about a fixed axis through O. The rod makes an angle θ radians with the line $\ell\ell$ as shown, where $\theta = \sin t$. The time is given by t in seconds. Determine the horizontal and vertical components of acceleration of the midpoint P of segment AB when $\theta = -\pi/6$ rad.

3.79 A rod is pinned at A and B to the centers of two small rollers. (See Figure P3.79.) The speed of A is kept constant at v_O even after B encounters the parabolic surface. Find the acceleration of B just after its roller begins to move on the parabola.

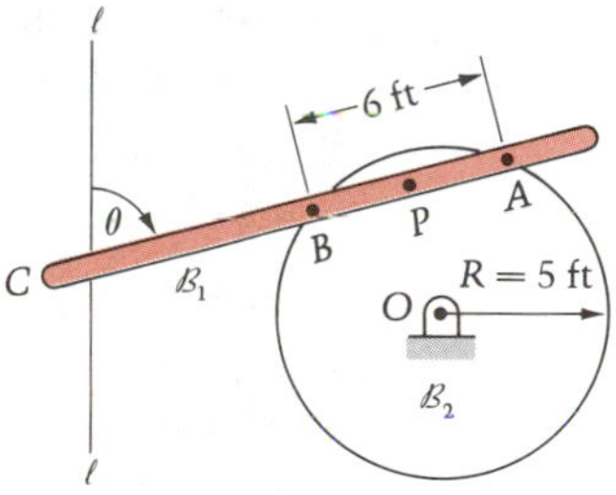

Figure P3.78

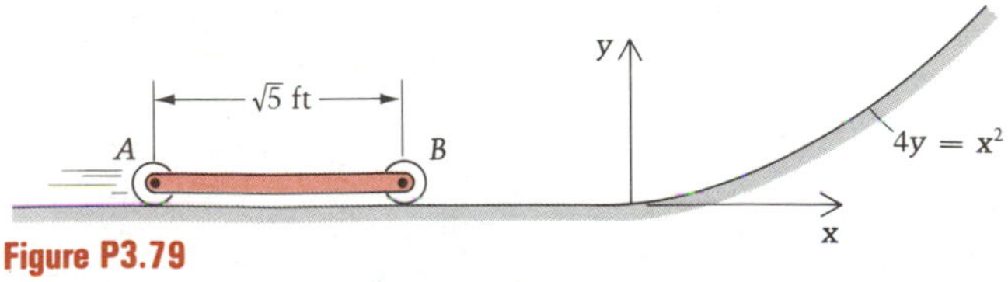

Figure P3.79

* **3.80** In the preceding problem, find $\mathbf{a}_B$ just after the *left* roller has begun to travel on the parabola.

* **3.81** The bent bar shown in Figure P3.81 slides on the vertical and horizontal surfaces. For the position shown, A has an acceleration of 4 ft/sec² to the left, while the bar has an angular velocity of 2 rad/sec clockwise and an angular acceleration of $\boldsymbol{\alpha}$.

   a. Determine $\boldsymbol{\alpha}$ for the position shown.

   b. Find, for the position shown, the angle θ and the distance PA such that point P has zero acceleration.

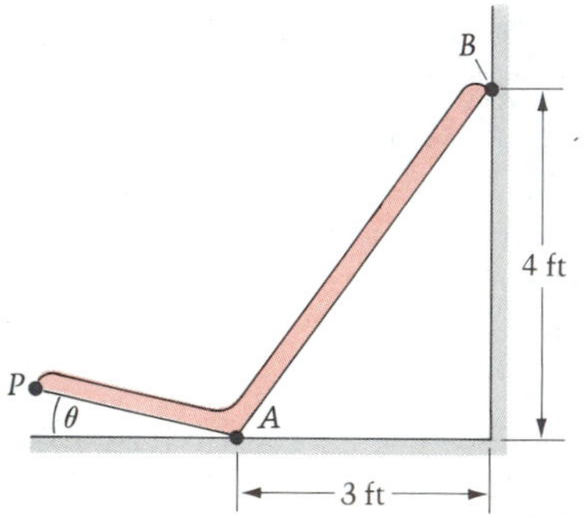

Figure P3.81

* **3.82** The right end P of bar $\mathcal{B}$ is constrained to move to the right on the sine wave shown in Figure P3.82 at the constant speed $\sqrt{10}$ in./sec. The left end A of $\mathcal{B}$ is constrained to slide along the x-axis. At the instant when $x = \pi$ in., find (a) $\omega_{\mathcal{B}}$; (b) $\mathbf{a}_P$.

* **3.83** Show that for a rigid body $\mathcal{B}$ in plane motion, as long as ω and α are not *both* zero there is a point of $\mathcal{B}$ having zero acceleration. *Hint*: Let P be a reference point with acceleration $\mathbf{a}_P = a_{P_x}\hat{\mathbf{i}} + a_{P_y}\hat{\mathbf{j}}$. See if you can find a vector $\mathbf{r}_{PT} = x\hat{\mathbf{i}} + y\hat{\mathbf{j}}$ from P to a point T of zero acceleration. That is, solve

$$\mathbf{a}_T = \mathbf{0} = \mathbf{a}_P + \alpha\hat{\mathbf{k}} \times \mathbf{r}_{PT} - \omega^2\mathbf{r}_{PT}$$

for x and y.

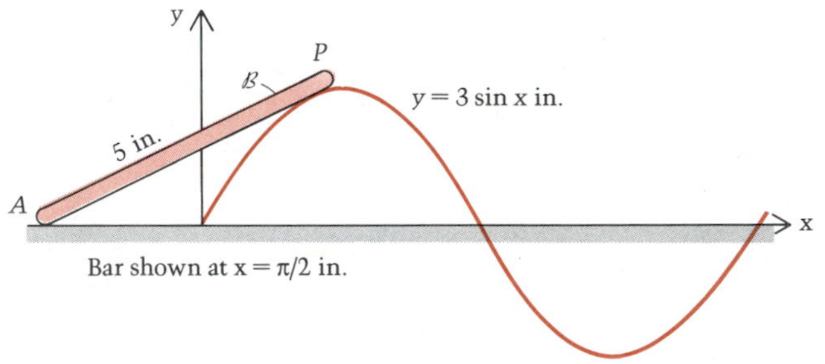

Figure P3.82

3.6 Rolling

Let $\mathcal{B}_1$ and $\mathcal{B}_2$ be two rigid bodies in motion. We define **rolling** to exist between $\mathcal{B}_1$ and $\mathcal{B}_2$ if during their motion:

1. A continuous sequence of points on the surface of $\mathcal{B}_1$ comes into one-to-one contact with a continuous sequence of points on the surface of $\mathcal{B}_2$.
2. At each instant during the interval of the motion, the contacting points have the same velocity vector.

Note that according to this definition there can be no slipping or sliding between the surfaces of $\mathcal{B}_1$ and $\mathcal{B}_2$ if rolling exists. Many authors, however, use the phrase "rolling without slipping" to describe the motion defined here. In their context, "rolling and slipping" would denote turning without the contact points having equal velocities; in our context, rolling *means* no slipping, so we shall say "turning and slipping" in cases of unequal contact-point velocities.

In this section we consider three classes of problems involving rolling contact:

1. Rolling of a wheel on a fixed straight line
2. Rolling of a wheel on a fixed plane curve
3. Gears

Rolling of a Wheel on a Fixed Straight Line

If the wheel ($\mathcal{B}_1$) shown in Figure 3.18 is rolling on the ground ($\mathcal{B}_2$, the reference frame in this case), then the continuous (shaded) sequences of points of $\mathcal{B}_1$ and $\mathcal{B}_2$ are in contact, one pair of points at a time. Since the points of $\mathcal{B}_2$ are all at rest, each point P_1 on the rim of $\mathcal{B}_1$ comes instantaneously to rest as it contacts a point P_2 of $\mathcal{B}_2$ (and is gripped for an instant by the ground). In this case, the velocities of P_1 and P_2 are each zero, although they need not vanish in general for rolling; all that is required is that $\mathbf{v}_{P_1} = \mathbf{v}_{P_2}$.*

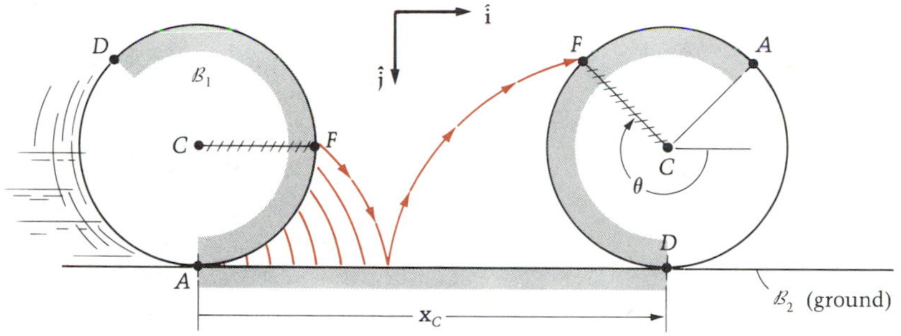

Figure 3.18 Rolling wheel, illustrating contacting sequence of points.

Using Equation (3.8) and noting that the center point C moves only horizontally and that the contact point is always the instantaneous center of $\mathcal{B}_1$,

$$\mathbf{v}_C = \mathbf{v}_{Ⓘ} + \dot{\theta}\hat{\mathbf{k}} \times \mathbf{r}_{ⒾC} \tag{3.20}$$

$$\dot{x}_C\hat{\mathbf{i}} = \mathbf{0} + \dot{\theta}\hat{\mathbf{k}} \times (-r\hat{\mathbf{j}}) \tag{3.21}$$

$$\dot{x}_C\hat{\mathbf{i}} = r\dot{\theta}\hat{\mathbf{i}} \tag{3.22}$$

For the rolling wheel, therefore, the velocity of its center C and the body's angular velocity are related quite simply by

$$\dot{x}_C = r\dot{\theta} \quad \text{or} \quad r\omega \tag{3.23}$$

Question 3.14 How would this expression be different were θ to have been chosen so that it would increase with *counterclockwise* turning of the body?

The relation between the displacement of C and the rotation of $\mathcal{B}_1$ follows from integrating Equation (3.23):

$$x_C = r\theta + C_1 \tag{3.24}$$

where the integration constant is zero if we choose $x_C = 0$ when $\theta = 0$.

* If $\mathcal{B}_2$ is the flatbed of a truck, for example, itself in motion with respect to a ground reference frame $\mathcal{B}_3$, then $\mathbf{v}_{P_1} = \mathbf{v}_{P_2} \neq \mathbf{0}$, but $\mathcal{B}_1$ still rolls on $\mathcal{B}_2$.

Answer 3.14 Then we would have $\dot{x}_C = -r\theta$.

Another approach to rolling is to begin with a small displacement Δx_C while the base point Ⓘ grips the ground. If the angle of rotation produced is $\Delta\theta$, then

$$\Delta x_C = r\,\Delta\theta$$

if we envision the body turning about its instantaneous center. Dividing by a small time increment Δt and taking the limit, we get

$$\lim_{\Delta t \to 0}\left(\frac{\Delta x_C}{\Delta t}\right) = \dot{x}_C = \lim_{\Delta t \to 0}\left(\frac{r\,\Delta\theta}{\Delta t}\right) = r\dot{\theta} = r\omega$$

as was obtained in Equation (3.23).

Next we consider accelerations. From Equation (3.22):

$$\mathbf{a}_C = \dot{\mathbf{v}}_C = \ddot{x}_C\hat{\mathbf{i}} = r\ddot{\theta}\hat{\mathbf{i}} = r\alpha\hat{\mathbf{i}} \tag{3.25}$$

and the center point C accelerates parallel to the plane, as it must since it (alone among all points of the wheel) has rectilinear motion. Now let us compute the acceleration of the instantaneous center Ⓘ of the wheel. Using Equation (3.19), we obtain

$$\begin{aligned}\mathbf{a}_{Ⓘ} &= \mathbf{a}_C + \ddot{\theta}\hat{\mathbf{k}} \times \mathbf{r}_{CⓘI} - \dot{\theta}^2\mathbf{r}_{CⓘI}\\ &= r\ddot{\theta}\hat{\mathbf{i}} + \ddot{\theta}\hat{\mathbf{k}} \times r\hat{\mathbf{j}} - \dot{\theta}^2 r\hat{\mathbf{j}}\\ &= -r\dot{\theta}^2\hat{\mathbf{j}} \quad \text{or} \quad -r\omega^2\hat{\mathbf{j}} \quad \text{or} \quad r\omega^2\uparrow\end{aligned} \tag{3.26}$$

Thus the contact point of a wheel rolling on a flat, fixed plane is accelerated toward its center with a magnitude $r\omega^2$. We see once again that a point of zero velocity need *not* be a point of zero acceleration, although of course it will be such if it is pinned to the reference frame. The point Ⓘ in the example is at rest instantanously, but it has an acceleration — which is why its velocity changes from zero as soon as it moves and a new Ⓘ takes its place in the rolling. We now take up several examples,* each of which deals with a round object rolling on a flat surface.

EXAMPLE 3.13

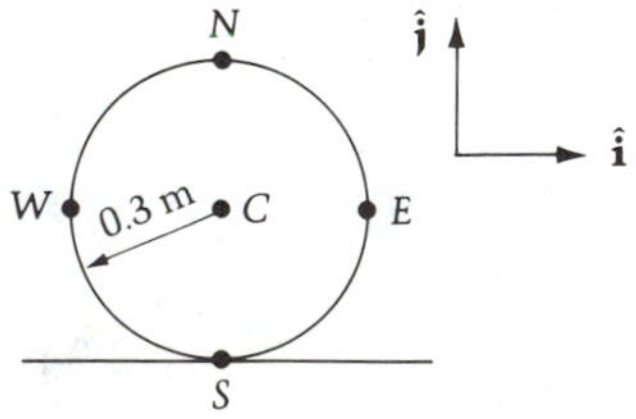

Figure E3.13

At a given instant, the rolling cylinder in Figure E3.13 has $\omega = 2 \circlearrowright$ rad/s and $\alpha = 1.5 \circlearrowleft$ rad/s². Find the velocity and acceleration of points N and E.

Solution

We shall relate the velocities and accelerations of N and E to those of point C. Calculating these, we have, because of the rolling (with S being Ⓘ),

$$\mathbf{v}_C = \dot{x}_C\hat{\mathbf{i}} = \boldsymbol{\omega} \times \mathbf{r}_{SC} = -2\hat{\mathbf{k}} \times 0.3\hat{\mathbf{j}} = 0.3(2)\hat{\mathbf{i}} = 0.6\hat{\mathbf{i}}\ \text{m/s}$$

* The examples of this section make continued use of Equations (3.8), (3.13), and (3.19), with the added feature that a rolling body is involved in each problem.

and

$$\begin{aligned}\mathbf{a}_C = \ddot{x}_C\hat{\mathbf{i}} = \mathbf{a}_S + \boldsymbol{\alpha} \times \mathbf{r}_{SC} - \omega^2\mathbf{r}_{SC} &= 0.3(2)^2\hat{\mathbf{j}} + \alpha\hat{\mathbf{k}} \times \mathbf{r}_{SC} - (2)^2 0.3\hat{\mathbf{j}}\\ &= \alpha\hat{\mathbf{k}} \times \mathbf{r}_{SC}\\ &= 1.5\hat{\mathbf{k}} \times 0.3\hat{\mathbf{j}}\\ &= -0.45\hat{\mathbf{i}}\ \text{m/s}^2\end{aligned}$$

Therefore

$$\begin{aligned}\mathbf{v}_N &= \mathbf{v}_C + \boldsymbol{\omega} \times \mathbf{r}_{CN}\\ &= 0.6\hat{\mathbf{i}} + (-2\hat{\mathbf{k}}) \times 0.3\hat{\mathbf{j}}\\ &= 1.2\hat{\mathbf{i}}\ \text{m/s}\end{aligned}$$

Note the agreement of this result with

$$\begin{aligned}\mathbf{v}_N &= \boldsymbol{\omega} \times \mathbf{r}_{\textcircled{I}N}\\ &= (-2\hat{\mathbf{k}}) \times 0.6\hat{\mathbf{j}}\\ &= 1.2\hat{\mathbf{i}}\ \text{m/s}\end{aligned}$$

Continuing, we get

$$\begin{aligned}\mathbf{a}_N &= \mathbf{a}_C + \boldsymbol{\alpha} \times \mathbf{r}_{CN} - \omega^2\mathbf{r}_{CN}\\ &= -0.45\hat{\mathbf{i}} + 1.5\hat{\mathbf{k}} \times 0.3\hat{\mathbf{j}} - 2^2(0.3\hat{\mathbf{j}})\\ &= -0.90\hat{\mathbf{i}} - 1.2\hat{\mathbf{j}}\ \text{m/s}^2\end{aligned}$$

For point E,

$$\begin{aligned}\mathbf{v}_E &= \mathbf{v}_C + \boldsymbol{\omega} \times \mathbf{r}_{CE}\\ &= 0.6\hat{\mathbf{i}} + (-2\hat{\mathbf{k}}) \times 0.3\hat{\mathbf{i}}\\ &= 0.6\hat{\mathbf{i}} - 0.6\hat{\mathbf{j}}\ \text{m/s}\end{aligned}$$

and

$$\begin{aligned}\mathbf{a}_E &= \mathbf{a}_C + \boldsymbol{\alpha} \times \mathbf{r}_{CE} - \omega^2\mathbf{r}_{CE}\\ &= -0.45\hat{\mathbf{i}} + 1.5\hat{\mathbf{k}} \times 0.3\hat{\mathbf{i}} - 2^2(0.3\hat{\mathbf{i}})\\ &= -1.65\hat{\mathbf{i}} + 0.45\hat{\mathbf{j}}\ \text{m/s}^2\end{aligned}$$

The reader may wish to obtain the results for $\mathbf{a}_N$ and $\mathbf{a}_E$ by relating them instead to $\mathbf{a}_{\textcircled{I}}$, which as we have seen is $r\omega^2\uparrow$, *not* zero.

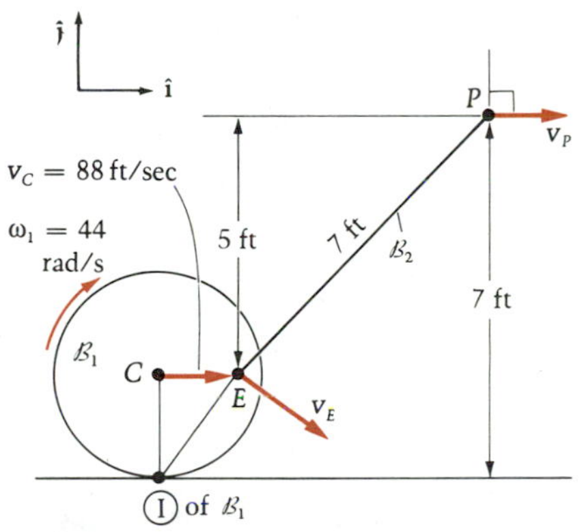

Figure E3.14

Example 3.14

In Example 3.9 find the piston acceleration when $\theta = 90°$. See Figure E3.14.

Solution

In *Case (b)* of Example 3.9 we found $\omega_1 = 44.0 \circlearrowright$ rad/sec, $\mathbf{v}_E = 88.0\hat{\mathbf{i}} + 66.0\hat{\mathbf{j}}$ ft/sec, $\omega_2 = 13.5 \circlearrowright$ rad/sec, and $\mathbf{v}_P = 20.5\hat{\mathbf{i}}$ ft/sec. On body $\mathcal{B}_1$, the velocity of C is constant $(88\hat{\mathbf{i}})$, so that $a_C = 0$. Also, $a_C = r\alpha_1$ so that $\alpha_1 = 0$.

Therefore

$$\mathbf{a}_E = \mathbf{a}_C + \alpha_1\hat{\mathbf{k}} \times \mathbf{r}_{CE} - \omega_1^2\mathbf{r}_{CE}$$
$$= \mathbf{0} + \mathbf{0} - 44.0^2(1.5\hat{\mathbf{i}})$$
$$= -2900\hat{\mathbf{i}}\ \text{ft/sec}^2$$

On $\mathcal{B}_2$, we now relate $\mathbf{a}_E$ to the desired $\mathbf{a}_P$:

$$\underbrace{\mathbf{a}_P = a_P\hat{\mathbf{i}}}_{\text{kinematic constraint}} = \mathbf{a}_E + \alpha_2\hat{\mathbf{k}} \times \mathbf{r}_{EP} - \omega_2^2\mathbf{r}_{EP}$$

$$= -2900\hat{\mathbf{i}} + \underbrace{\alpha_2\hat{\mathbf{k}} \times (4.90\hat{\mathbf{i}} + 5.00\hat{\mathbf{j}})}_{\text{Note that this is a cross product, but ...}} - \underbrace{13.5^2(4.90\hat{\mathbf{i}} + 5.00\hat{\mathbf{j}})}_{\textit{this} \text{ is not!}}$$

$$a_P\hat{\mathbf{i}} = (-2900 - 5.00\alpha_2 - 893)\hat{\mathbf{i}} + (4.90\alpha_2 - 911)\hat{\mathbf{j}}$$

Equating the $\hat{\mathbf{j}}$ coefficients first eliminates the unknown $\mathbf{a}_P$:

$$0 = 4.90\alpha_2 - 911 \Rightarrow \alpha_2 = 186\ \text{rad/sec}^2 \qquad \text{or} \qquad \boldsymbol{\alpha}_2 = 186 \circlearrowleft \text{rad/sec}^2$$

Then the coefficients of $\hat{\mathbf{i}}$ yield our answer:

$$a_P = -4720\ \text{ft/sec}^2 \qquad \text{or} \qquad \mathbf{a}_P = 4720 \leftarrow \text{ft/sec}^2$$

EXAMPLE 3.15

The wheel $\mathcal{B}_1$ in Figure E3.15 rolls to the right on plane $\mathcal{P}$. At the instant shown in the figure, $\mathcal{B}_1$ has angular velocity $\omega_1 = 1.2 \circlearrowright$ rad/s. Rod $\mathcal{B}_2$ is pinned to $\mathcal{B}_1$ at A, and the other end B of $\mathcal{B}_2$ slides along a plane $\mathcal{Q}$ parallel to $\mathcal{P}$. Determine the velocity of B and the angular velocity of $\mathcal{B}_2$ at the given instant.

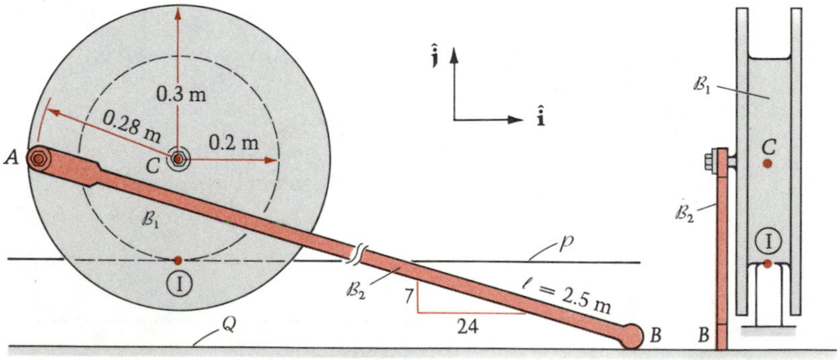

Figure E3.15

Solution

We shall use Equation (3.8) together with the results we have derived for rolling. We first seek the velocity of A; when we have $\mathbf{v}_A$ we shall then relate it to $\mathbf{v}_B$ on rod $\mathcal{B}_2$. We may find $\mathbf{v}_A$ in either of two ways, each on wheel $\mathcal{B}_1$:

$$\mathbf{v}_A = \overset{0}{\cancel{\mathbf{v}_{\text{Ⓘ}}}} + \boldsymbol{\omega}_1 \times \mathbf{r}_{\text{Ⓘ}A} = (-1.2\hat{\mathbf{k}}) \times (-0.28\hat{\mathbf{i}} + 0.2\hat{\mathbf{j}}) = 0.24\hat{\mathbf{i}} + 0.336\hat{\mathbf{j}} \text{ m/s}$$

$$\mathbf{v}_A = \mathbf{v}_C + \boldsymbol{\omega}_1 \times \mathbf{r}_{CA} = 0.2(1.2)\hat{\mathbf{i}} + (-1.2\hat{\mathbf{k}}) \times (-0.28\hat{\mathbf{i}}) = 0.24\hat{\mathbf{i}} + 0.336\hat{\mathbf{j}} \text{ m/s}$$

We note for interest that when point A is beneath Ⓘ, its velocity is to the left — that is, it is going backwards!

Next, on $\mathcal{B}_2$,

$$\mathbf{v}_B = \mathbf{v}_A + \boldsymbol{\omega}_2 \times \mathbf{r}_{AB}$$

Point B is constrained to move horizontally; therefore

$$v_B\hat{\mathbf{i}} = 0.24\hat{\mathbf{i}} + 0.336\hat{\mathbf{j}} + \omega_2\hat{\mathbf{k}} \times (2.4\hat{\mathbf{i}} - 0.7\hat{\mathbf{j}})$$

or

$$v_B\hat{\mathbf{i}} = \hat{\mathbf{i}}(0.24 + 0.7\omega_2) + \hat{\mathbf{j}}(0.336 + 2.4\omega_2)$$

$\hat{\mathbf{j}}$ coefficients: $0 = 0.336 + 2.4\omega_2 \Rightarrow \omega_2 = -0.140$

so that

$$\boldsymbol{\omega}_2 = -0.140\hat{\mathbf{k}} \text{ rad/s or } 0.140 \circlearrowright \text{ rad/s}$$

$\hat{\mathbf{i}}$ coefficients: $v_B = 0.24 + 0.7(-0.140) = 0.142$

so that

$$\mathbf{v}_B = 0.142\hat{\mathbf{i}} \text{ m/s}$$

The reader is encouraged to mentally locate the Ⓘ point for $\mathcal{B}_2$ and from it deduce that the directions of $\boldsymbol{\omega}_2$ and $\mathbf{v}_B$ are correct.

Example 3.16

In the preceding example, at the same instant, $\boldsymbol{\alpha}_1 = 0.8 \circlearrowright \text{ rad/s}^2$. Find the acceleration of point B and the angular acceleration of rod $\mathcal{B}_2$.

Solution

As we did with the velocity of A, we can relate $\mathbf{a}_A$ to the acceleration of either Ⓘ or C:

$$\mathbf{a}_A = \mathbf{a}_{\text{Ⓘ}} + \boldsymbol{\alpha}_1 \times \mathbf{r}_{\text{Ⓘ}A} - \omega_1^2\mathbf{r}_{\text{Ⓘ}A}$$

The terms are

$$\mathbf{a}_{\text{Ⓘ}} = r\omega_1^2\uparrow = 0.2(1.2^2)\hat{\mathbf{j}} = 0.288\hat{\mathbf{j}} \quad \text{(note not zero!)}$$

$$\boldsymbol{\alpha}_1 \times \mathbf{r}_{\text{Ⓘ}A} = -0.8\hat{\mathbf{k}} \times (-0.28\hat{\mathbf{i}} + 0.2\hat{\mathbf{j}}) = 0.16\hat{\mathbf{i}} + 0.224\hat{\mathbf{j}}$$

$$-\omega_1^2\mathbf{r}_{\text{Ⓘ}A} = -1.2^2(-0.28\hat{\mathbf{i}} + 0.2\hat{\mathbf{j}}) = 0.403\hat{\mathbf{i}} - 0.288\hat{\mathbf{j}}$$

Adding the terms, we get

$$\mathbf{a}_A = 0.563\hat{\mathbf{i}} + 0.224\hat{\mathbf{j}} \text{ m/s}^2$$

$$\mathbf{a}_A = \mathbf{a}_C + \boldsymbol{\alpha}_1 \times \mathbf{r}_{CA} - \omega_1^2\mathbf{r}_{CA}$$

The terms are

$$\mathbf{a}_C = r\alpha_1\hat{\mathbf{i}} = 0.2(0.8)\hat{\mathbf{i}} = 0.16\hat{\mathbf{i}}$$

$$\boldsymbol{\alpha}_1 \times \mathbf{r}_{CA} = (-0.8\hat{\mathbf{k}}) \times (-0.28\hat{\mathbf{i}}) = 0.224\hat{\mathbf{j}}$$

$$-\omega_1^2\mathbf{r}_{CA} = -1.2^2(-0.28\hat{\mathbf{i}}) = 0.403\hat{\mathbf{i}}$$

Again adding the terms, we get

$$\mathbf{a}_A = 0.563\hat{\mathbf{i}} + 0.224\hat{\mathbf{j}} \text{ m/s}^2$$

Now we relate $\mathbf{a}_A$ to $\mathbf{a}_B$ on the rod; note that the acceleration of B is constrained by the plane to be horizontal:

$$\mathbf{a}_B = a_B\hat{\mathbf{i}} = \mathbf{a}_A + \boldsymbol{\alpha}_2 \times \mathbf{r}_{AB} - \omega_2^2\mathbf{r}_{AB}$$
$$= 0.563\hat{\mathbf{i}} + 0.224\hat{\mathbf{j}} + \alpha_2\hat{\mathbf{k}} \times (2.4\hat{\mathbf{i}} - 0.7\hat{\mathbf{j}})$$
$$- (-0.140)^2(2.4\hat{\mathbf{i}} - 0.7\hat{\mathbf{j}})$$

$\hat{\mathbf{j}}$ coefficients: $\quad 0 = 0.224 + 2.4\alpha_2 + 0.0137 \Rightarrow \alpha_2 = -0.0991 \text{ rad/s}^2$

so that

$$\boldsymbol{\alpha}_2 = -0.0991\,\hat{\mathbf{k}} \text{ rad/s}^2$$

$\hat{\mathbf{i}}$ coefficients: $\quad a_B = 0.563 + 0.7(-0.0991) - 0.0470 = 0.447$

so that

$$\mathbf{a}_B = 0.447\hat{\mathbf{i}} \text{ m/s}^2$$

Rolling of a Wheel ($\mathcal{B}_1$) on a Fixed Plane Curve ($\mathcal{B}_2$)

In the second class of rolling problems to be considered in this section, the contact surface is curved. Let $\hat{\mathbf{e}}_t$ and $\hat{\mathbf{e}}_n$ be the principal unit tangent and normal vectors for the center point C of the wheel. (See Figure 3.19.) Then, since we are again defining a motion such that the contact point has zero velocity, Equation (3.8) yields

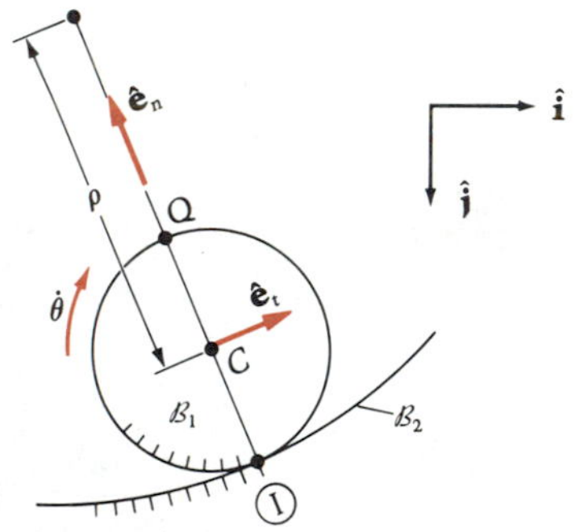

Figure 3.19 Wheel rolling on curved path concave upward.

$$\mathbf{v}_C = \mathbf{v}_{Ⓘ} + \dot{\theta}\hat{\mathbf{k}} \times \mathbf{r}_{ⒾC}$$
$$= \mathbf{0} + \dot{\theta}\hat{\mathbf{k}} \times r\hat{\mathbf{e}}_n$$
$$= r\dot{\theta}\hat{\mathbf{e}}_t \tag{3.27}$$

which gives us the velocity of C. Differentiating Equation (3.27), we get

$$\mathbf{a}_C = r\ddot{\theta}\hat{\mathbf{e}}_t + r\dot{\theta}\dot{\hat{\mathbf{e}}}_t$$
$$= r\ddot{\theta}\hat{\mathbf{e}}_t + r\dot{\theta}\frac{\dot{s}}{\rho}\hat{\mathbf{e}}_n \tag{3.28}$$

in which we have used Equations (1.42) and (1.45), where ρ is the instantaneous radius of curvature of the path on which C moves. If this path is a circle, as is often the case, then ρ = constant = radius of the circle.

Using $\dot{s} = v_C = r\dot{\theta}$ in Equation (3.28), we have

$$\mathbf{a}_C = r\ddot{\theta}\hat{\mathbf{e}}_t + \frac{(r\dot{\theta})^2}{\rho}\hat{\mathbf{e}}_n \tag{3.29}$$

The acceleration of Ⓘ is interesting, and it follows from Equation (3.19):

$$\mathbf{a}_{Ⓘ} = \mathbf{a}_C + \ddot{\theta}\hat{\mathbf{k}} \times \mathbf{r}_{CⒾ} - \dot{\theta}^2\mathbf{r}_{CⒾ} \tag{3.30}$$

or

$$\mathbf{a}_{Ⓘ} = r\ddot{\theta}\hat{\mathbf{e}}_t + \frac{(r\dot{\theta})^2}{\rho}\hat{\mathbf{e}}_n - r\ddot{\theta}\hat{\mathbf{e}}_t + r\dot{\theta}^2\hat{\mathbf{e}}_n = \left(1 + \frac{r}{\rho}\right) r\dot{\theta}^2\hat{\mathbf{e}}_n \tag{3.31}$$

Comparing the accelerations of the contact points Ⓘ of Figures 3.18 and 3.19, we observe from our results (Equations 3.26 and 3.31) that the point contacting the curved track has a greater acceleration than the one touching the flat track, due to the r/ρ term of Equation (3.31). This term represents the normal component of acceleration of C, which was zero on the flat track.

It is also interesting to examine the acceleration of point Q at the top of the wheel in Figure 3.19:

$$\begin{aligned}\mathbf{a}_Q &= \mathbf{a}_C + \ddot{\theta}\hat{\mathbf{k}} \times \mathbf{r}_{CQ} - \dot{\theta}^2\mathbf{r}_{CQ} \\ &= \mathbf{r}\ddot{\theta}\hat{\mathbf{e}}_t + \frac{(r\dot{\theta})^2}{\rho}\hat{\mathbf{e}}_n + r\ddot{\theta}\hat{\mathbf{e}}_t - r\dot{\theta}^2\hat{\mathbf{e}}_n \\ &= 2r\ddot{\theta}\hat{\mathbf{e}}_t - r\dot{\theta}^2\left(1 - \frac{r}{\rho}\right)\hat{\mathbf{e}}_n \end{aligned} \tag{3.32}$$

Note that it is possible for the $\hat{\mathbf{e}}_n$ (normal) component of $\mathbf{a}_Q$ to be either away from or toward C (or even zero), depending on whether $r > \rho$ or $r < \rho$ (or $r = \rho$), respectively (see Figure 3.20).

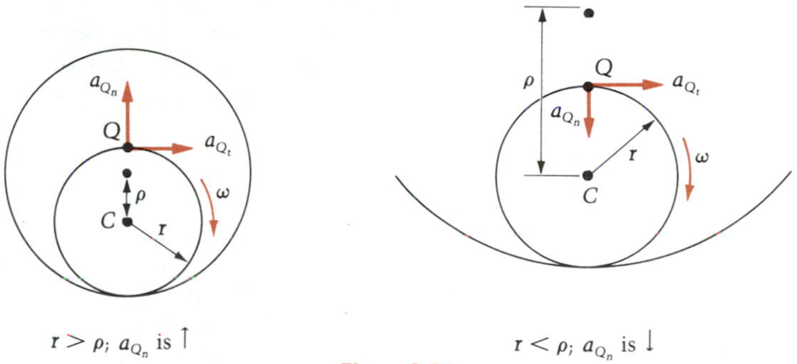

Figure 3.20

If the track is concave *downward* as shown in Figure 3.21, similar results may be obtained for the accelerations of C, Ⓘ, and Q. From Equation (3.29), which still holds:

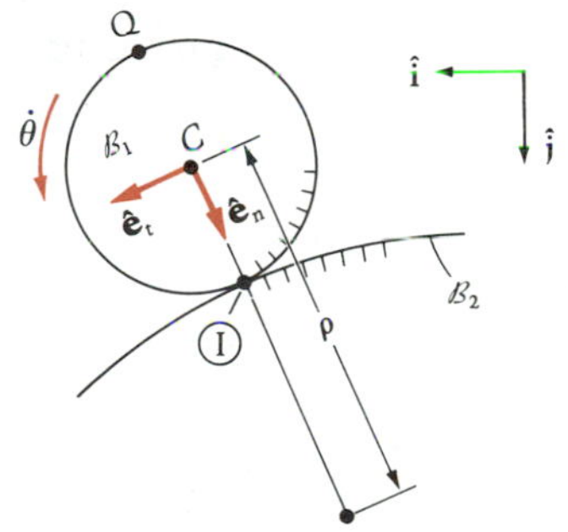

Figure 3.21

$$\mathbf{a}_C = r\ddot{\theta}\hat{\mathbf{e}}_t + \frac{(r\dot{\theta})^2}{\rho}\hat{\mathbf{e}}_n \tag{3.33}$$

After relating $\mathbf{a}_{Ⓘ}$ and $\mathbf{a}_C$ on $\mathcal{B}_1$, we obtain:

$$\mathbf{a}_{Ⓘ} = \left(-1 + \frac{r}{\rho}\right) r\dot{\theta}^2\hat{\mathbf{e}}_n \tag{3.34}$$

And relating Q to either C or Ⓘ gives

$$\mathbf{a}_Q = 2r\ddot{\theta}\hat{\mathbf{e}}_t + \left(1 + \frac{r}{\rho}\right) r\dot{\theta}^2\hat{\mathbf{e}}_n \tag{3.35}$$

Note that this time the radius r cannot exceed ρ, so $\mathbf{a}_{Ⓘ}$ is always outward; the normal component of the acceleration of Q is seen to be always inward in this case.

The reader may find it useful to remember the following form that doesn't depend upon any particular choice of coordinate system:

$$\mathbf{v}_C = \boldsymbol{\omega} \times \mathbf{r}_{\text{Ⓘ}C}$$

$$\mathbf{a}_C = \boldsymbol{\alpha} \times \mathbf{r}_{\text{Ⓘ}C} + \text{(normal part)}$$

If we knew either $\boldsymbol{\alpha}$ or the tangential component of $\mathbf{a}_C$, for example, we could get the other without concern for expressing the normal component of $\mathbf{a}_C$. We now consider several examples that feature a body rolling on a curved surface.

EXAMPLE 3.17

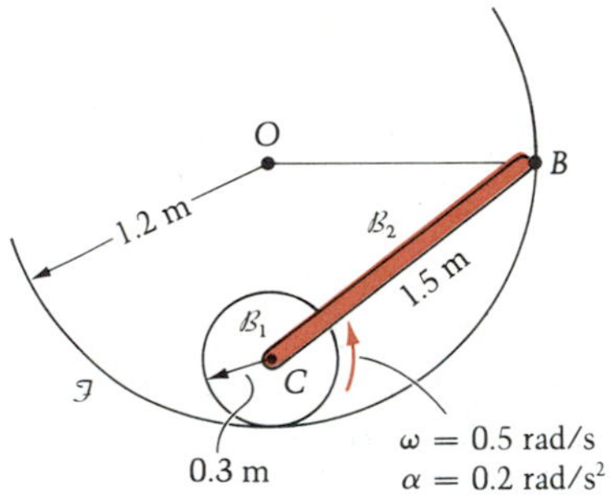

Figure E3.17a

The cylinder $\mathcal{B}_1$ shown in Figure E3.17a is rolling on the fixed, circular track with the indicated angular velocity and acceleration when $\mathcal{B}_1$ is at the bottom of the track. Rod $\mathcal{B}_2$ is pinned to the center C of $\mathcal{B}_1$, and its other end, B, slides on track $\mathcal{T}$. Find the velocity and acceleration of B.

Solution

As we have seen in Section 3.5, problems like this one have two parts. The "velocity part" must be solved before the "acceleration part" because the velocities as well as angular velocities are needed in the expressions for acceleration. We shall use instantaneous centers to get $\mathbf{v}_B$; the steps are:

1. The contact point of $\mathcal{B}_1$ is its Ⓘ point, since the body is rolling (see Figure E3.17b).
2. $\mathbf{v}_C$ is determined as in the diagram from $r_{\text{Ⓘ}C}\,\omega_1$.
3. The velocity of B is vertical (tangent to the path of the point).
4. Thus Ⓘ of $\mathcal{B}_2$ is at the intersection of the normals to $\mathbf{v}_C$ and $\mathbf{v}_B$, namely at point O (see Figure E3.17c).
5. Then $\omega_2 = v_C/d_1 = 0.15/0.9 = 0.167$ or $\omega_2 = 0.167 \circlearrowright$ rad/s
6. Finally, $v_B = d_2\omega_2 = 1.2(0.167) = 0.200$ or $\mathbf{v}_B = 0.200 \downarrow$ m/s

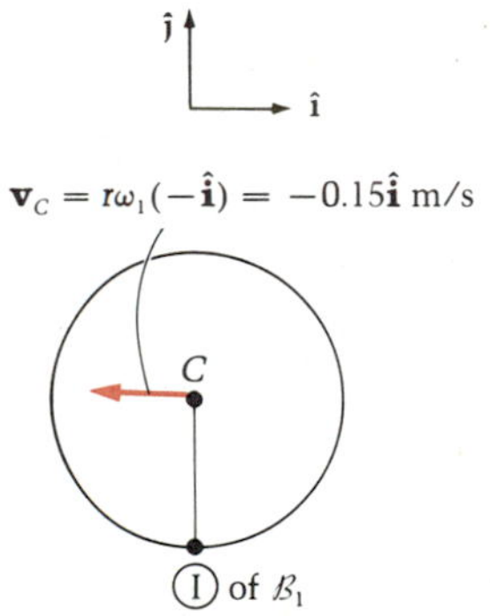

Figure E3.17b

Next, to find $\mathbf{a}_B$ we shall relate it to the acceleration of C, which is, from Equation (3.29),

$$\mathbf{a}_C = r\alpha(-\hat{\mathbf{i}}) + \frac{(r\omega)^2}{\rho}\hat{\mathbf{j}}$$

$$= 0.3(0.2)(-\hat{\mathbf{i}}) + \frac{(0.3 \times 0.5)^2}{0.9}\hat{\mathbf{j}}$$

$$= -0.06\hat{\mathbf{i}} + 0.0250\hat{\mathbf{j}} \text{ m/s}^2$$

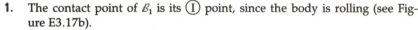

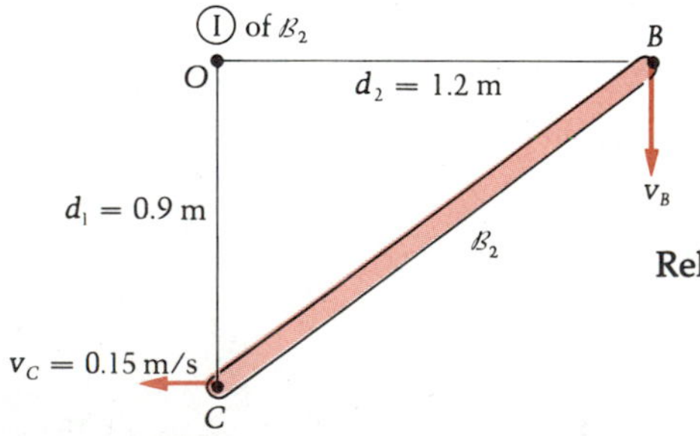

Figure E3.17c

Relating $\mathbf{a}_C$ to $\mathbf{a}_B$, we get, using Figure E3.17c:

$$\mathbf{a}_B = \mathbf{a}_C + \alpha_2\hat{\mathbf{k}} \times \mathbf{r}_{CB} - \omega_2^2\mathbf{r}_{CB}$$

$$\ddot{s}_B(-\hat{\mathbf{j}}) + \frac{\dot{s}_B^2}{\rho_B}(-\hat{\mathbf{i}}) = -0.06\hat{\mathbf{i}} + 0.0250\hat{\mathbf{j}} + \alpha_2\hat{\mathbf{k}} \times (1.2\hat{\mathbf{i}} + 0.9\hat{\mathbf{j}})$$

$$-(0.167)^2(1.2\hat{\mathbf{i}} + 0.9\hat{\mathbf{j}})$$

In substituting for $\mathbf{a}_B$ we have used the facts that for B we have $\hat{\mathbf{e}}_t = -\hat{\mathbf{j}}$ (direction vector of $\mathbf{v}_B$) and $\hat{\mathbf{e}}_n = -\hat{\mathbf{i}}$ (always toward the center of curvature of the path of the point). Substituting $\dot{s}_B = |\mathbf{v}_B| = 0.2$ m/s and $\rho_B = 1.2$ m, we then have two scalar equations in the two unknowns $\ddot{s}_B$ and α_2:

$\hat{\mathbf{i}}$ coefficients:

$$-0.0333 = -0.06 - 0.9\alpha_2 - 0.0333$$
$$\alpha_2 = -0.0667 \text{ rad/s}^2$$

$\hat{\mathbf{j}}$ coefficients:

$$-\ddot{s}_B = 0.0250 + 1.2\alpha_2 - 0.0250 = -0.0800$$
$$\ddot{s}_B = 0.0800 \text{ m/s}^2$$

Therefore

$$\mathbf{a}_B = -0.0333\hat{\mathbf{i}} - 0.0800\hat{\mathbf{j}} \text{ m/s}^2$$

We wish to make a very important point regarding the preceding example. You may have noticed that the values of α_2 and $\ddot{s}_B$ could have been obtained more quickly from

$$|\alpha_2| = \frac{a_C}{r_{OC}} = \frac{0.06}{0.9} = 0.0667 \text{ rad/s}^2$$

and

$$\ddot{s}_B = r_{OB}|\alpha_2| = 1.2(0.0667) = 0.08 \text{ m/s}^2$$

This shortcut is very dangerous because it is not always valid! It is essential to understand when and why this procedure works. (It is *not* because O is the instantaneous center of zero velocity of $\mathcal{B}_2$!) For a counterexample, consider the results of Examples 3.3, 3.7, and 3.11:

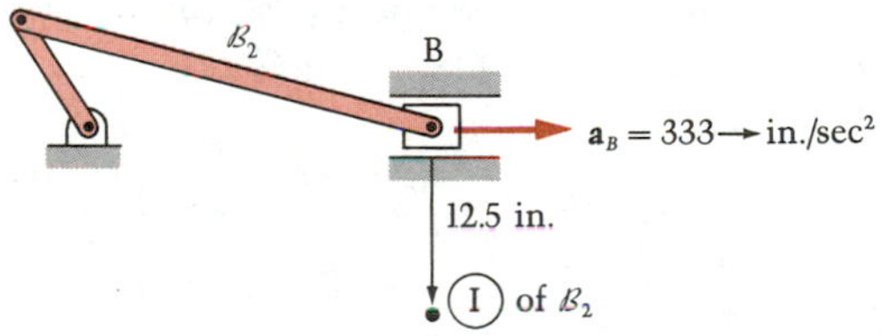

Figure 3.22

If we were to divide $|\mathbf{a}_B|$ by $|\mathbf{r}_{ⒾB}|$ (see Figure 3.22), we would erroneously obtain 26.6 rad/sec² with a direction indicator ↻. But $\boldsymbol{\alpha}_2$ is 31.6 ↺ rad/sec²! The answer to the question of when the procedure is legitimate is covered by the following text question:

Question 3.15 When can we use $r_{ⒾB}\alpha$ to obtain the correct acceleration component of B normal to line ⒾB?

Answer 3.15 From Figure 3.13 we can see exactly when the component of $\mathbf{a}_B$ normal to line AB is given by $r_{AB}\alpha$: It is when $\mathbf{a}_A$ has no component normal to AB. (This result does not require that A be the point Ⓘ, incidentally.)

EXAMPLE 3.18

The velocity magnitude of G in Figure E3.18a is $v_G = t^2$ ft/sec, and G moves on the 8-ft circle in a clockwise direction. The position shown is at $t = 2$ sec. Find the acceleration of B at this instant.

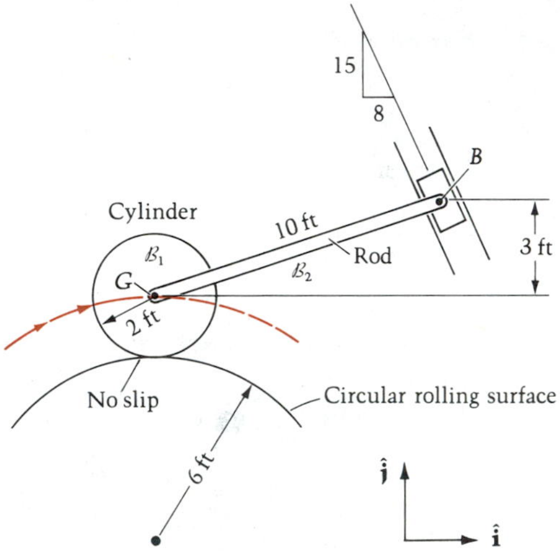

Figure E3.18a

Solution

We shall relate $\mathbf{a}_B$ to $\mathbf{a}_G$ on bar $\mathcal{B}_2$:

$$\mathbf{a}_B = \mathbf{a}_G + \alpha_2\hat{\mathbf{k}} \times \mathbf{r}_{GB} - \omega_2^2\mathbf{r}_{GB} \qquad (1)$$

First, we determine $\mathbf{a}_G$. Since $v_G = \dot{s}_G = t^2$ in this problem, we have $\ddot{s}_G = 2t = 2(2) = 4$ ft/sec². Therefore

$$\mathbf{a}_G = \ddot{s}_G\hat{\mathbf{e}}_t + \frac{\dot{s}_G^2}{\rho_G}\hat{\mathbf{e}}_n = 2t\hat{\mathbf{e}}_t + \frac{t^4}{8}\hat{\mathbf{e}}_n$$

$$= 4\hat{\mathbf{i}} - 2\hat{\mathbf{j}} \text{ ft/sec}^2 \qquad (\text{at } t = 2 \text{ sec})$$

From Equation (1)

$$\mathbf{a}_B = a_B\left(\frac{8\hat{\mathbf{i}} - 15\hat{\mathbf{j}}}{17}\right) = \overset{(4\hat{\mathbf{i}} - 2\hat{\mathbf{j}})}{\mathbf{a}_G} + \alpha_2\hat{\mathbf{k}} \times \overset{(\sqrt{91}\,\hat{\mathbf{i}} + 3\hat{\mathbf{j}})}{\mathbf{r}_{GB}} - \omega_2^2\mathbf{r}_{GB}$$

in which the acceleration of B is an unknown magnitude in a known direction as signified by the unit vector $(8\hat{\mathbf{i}} - 15\hat{\mathbf{j}})/17$ down the slot.

Question 3.16 If the direction of $\mathbf{a}_B$ turns out to be up the slot, will the solution be valid?

Answer 3.16 Yes; this will be manifested by a_B turning out negative. Note that (negative a_B) · $[(8\hat{\mathbf{i}} - 15\hat{\mathbf{j}})/17]$ is the same as (positive a_B) · $[(-8\hat{\mathbf{i}} + 15\hat{\mathbf{j}})/17]$.

We have two equations ($\hat{\mathbf{i}}$ and $\hat{\mathbf{j}}$ coefficients) in the three unknowns a_B, α_2, and ω_2, but the angular speed ω_2 may be found from the instantaneous center of $\mathcal{B}_2$. From the geometry and similar triangles, we use Figure E3.18b to obtain:

$$\frac{\sqrt{91}}{H} = \frac{15}{8} \Rightarrow H = 5.09 \text{ ft}$$

$$D = H - 3 = 2.09 \text{ ft}$$

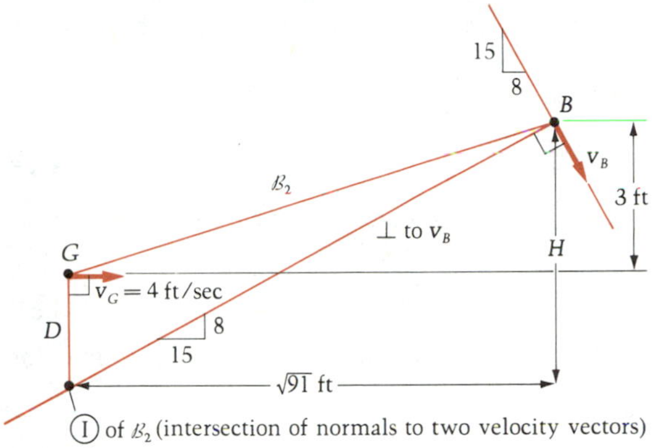

Figure E3.18b

Thus

$$\omega_2 = \frac{v_G}{D} = \frac{4}{2.09} = 1.91 \text{ rad/sec}$$

$$\omega_2 = 1.91 \circlearrowright \text{ rad/sec}$$

Note that until Ⓘ of $\mathcal{B}_2$ is located, we do not know whether $\mathbf{v}_B$ is up or down the slot; however, the *normal* to $\mathbf{v}_B$ is the same line in either case. Once Ⓘ is established, $\mathbf{v}_G$ to the right gives ω_2 clockwise — and then we know that $\mathbf{v}_B$ is *down* the slot. Substituting, we get

$$a_B\left(\frac{8\hat{\mathbf{i}} - 15\hat{\mathbf{j}}}{17}\right) = 4\hat{\mathbf{i}} - 2\hat{\mathbf{j}} + \sqrt{91}\alpha_2\hat{\mathbf{j}} - 3\alpha_2\hat{\mathbf{i}} - 1.91^2(\sqrt{91}\hat{\mathbf{i}} + 3\hat{\mathbf{j}})$$

$\hat{\mathbf{i}}$ coefficients: $\quad \dfrac{8}{17}a_B = 4 - 3\alpha_2 - 1.91^2\sqrt{91}$

$\hat{\mathbf{j}}$ coefficients: $\quad \dfrac{-15}{17}a_B = -2 + \sqrt{91}\,\alpha_2 - 1.91^2(3)$

Eliminating α_2 gives $a_B = -181$, so that

$$\mathbf{a}_B = 181 \; {}^{15}\!\!\nwarrow_{8} \text{ or } (-85.2\hat{\mathbf{i}} + 160\hat{\mathbf{j}}) \text{ ft/sec}^2$$

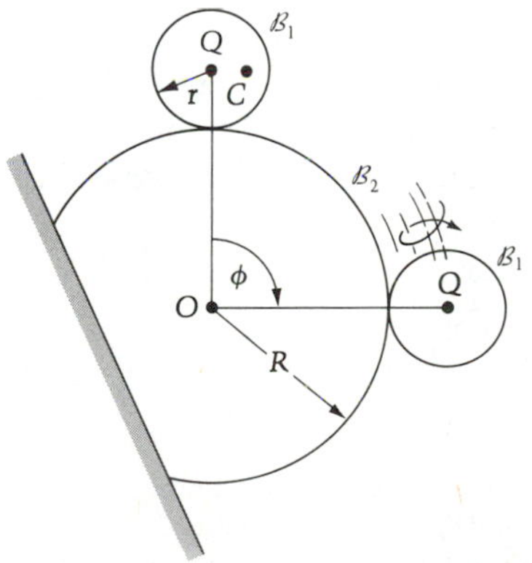

Figure 3.23

The final example in this part of Section 3.6 will be very helpful to us later in situations such as the one shown in Figure 3.23. Suppose we need to know the location of point C of cylinder $\mathcal{B}_1$ after $\mathcal{B}_1$ has rolled on $\mathcal{B}_2$ to the lower position.* The problem is to find the angle θ turned through by $\mathcal{B}_1$ for a given rotation ϕ of the line OQ. The procedure to be followed in solving this problem is illustrated by the next example.

EXAMPLE 3.19

Find the relationship between the angle ϕ (locating the line OC in Figure E3.19) and the angle of rotation θ of the rolling cylinder $\mathcal{B}$.

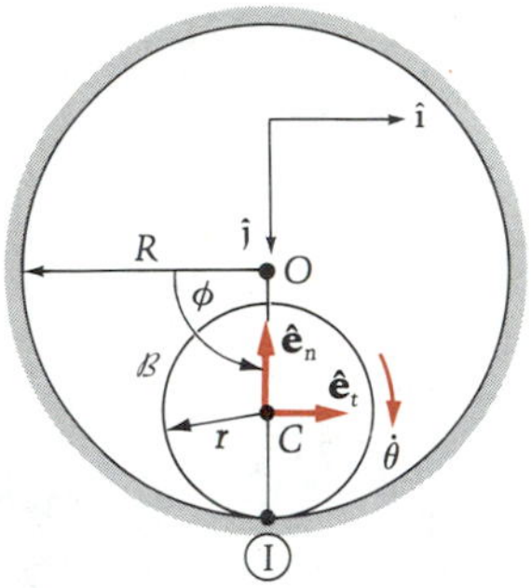

Figure E3.19

Solution

Treating C as a point whose path is a known circle, we get

$$\mathbf{v}_C = \dot{s}\hat{\mathbf{e}}_t = (R - r)\dot{\phi}\hat{\mathbf{i}}$$

Alternatively, we may also treat C as a point on the cylinder with instantaneous center at Ⓘ:

$$\mathbf{v}_C = \omega\hat{\mathbf{k}} \times \mathbf{r}_{\text{Ⓘ}C} = \dot{\theta}\hat{\mathbf{k}} \times (-r\hat{\mathbf{j}}) = r\dot{\theta}\hat{\mathbf{i}}$$

Thus, equating the two expressions for $\mathbf{v}_C$, we have

$$(R - r)\dot{\phi} = r\dot{\theta}$$

Integrating, and letting $\theta = 0$ when $\phi = 0$, we get

$$(R - r)\phi = r\theta + \overset{0}{\cancel{C_1}}$$

or

$$\theta = \left(\frac{R - r}{r}\right)\phi \tag{3.36}$$

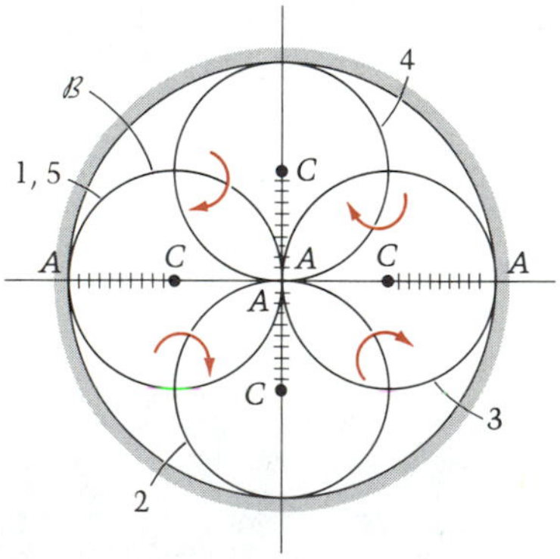

Figure 3.24

From Example 3.19, if we let $R = 2r$, then we see (Figure 3.24) that $\theta \equiv \phi$. Even though the circumferences of cylinder and track are $2\pi r$ and $4\pi r(= 2\pi R)$,† the curvature forces the angular velocities of the line OC and the cylinder to be the same. If the outer track were *straight* and of length $4\pi r$, the cylinder would turn *two* revolutions in space instead of just one in traversing it.

It is seen that the line AC of $\mathcal{B}$ (and hence $\mathcal{B}$ itself) revolves once as C completes its circular path for the case $R = 2r$. When $R > 2r$, then $\theta > \phi$. Such a case is shown in Figure 3.25. If we now let $R = 7r$, then Equation (3.36) gives $\theta = 6\phi$ and $\mathcal{B}$ now revolves once in space for each 60° of turn of line OC.

If the track were *convex*, then for $R = 2r$:

$$\mathbf{v}_C = 3r\dot{\phi}\hat{\mathbf{e}}_t \qquad \text{and} \qquad \mathbf{v}_C = r\dot{\theta}\hat{\mathbf{e}}_t \Rightarrow 3\dot{\phi} = \dot{\theta}$$

* This need will arise, for example, in kinetics problems in which we seek the work done by gravity on $\mathcal{B}_1$ if its mass center is offset from its geometric center.

† Which at first glance might lead one to believe that the cylinder would revolve twice per revolution of OC.

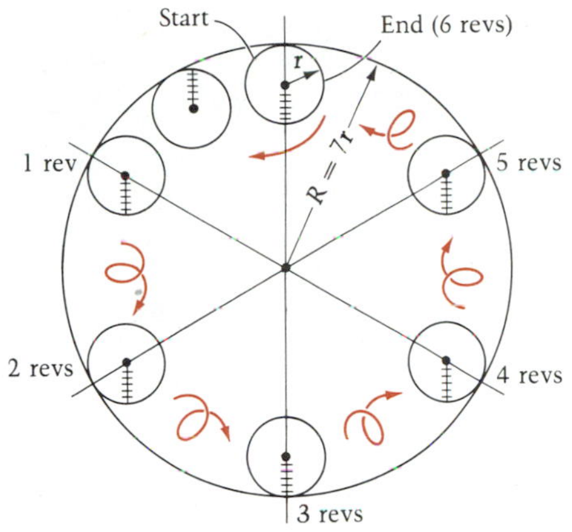

Figure 3.25

and the wheel would turn in space *three times* as fast and as far as line OC (see Figure 3.26).

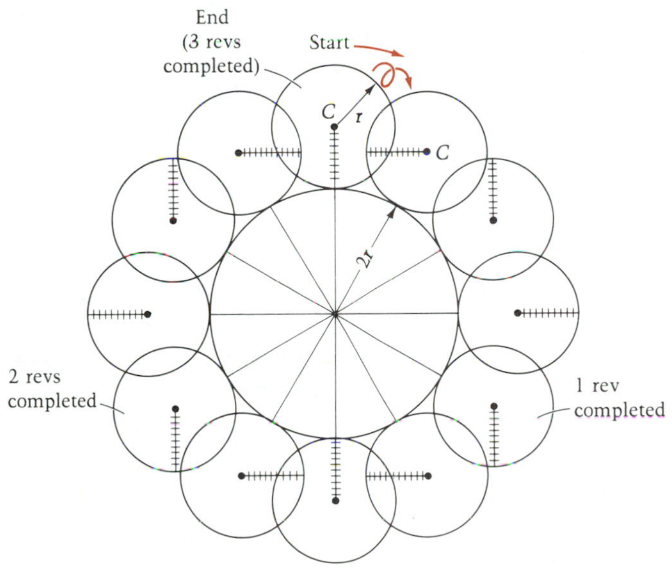

Figure 3.26

Gears

The final class of rolling problems is concerned with gears. Gears are used to transmit power. The teeth of the gears are cut so that they will give constant speed to the driven gear when the driving gear is itself turning at constant angular speed.

However, gears violate the rolling condition; there is necessarily some sliding since the contacting points do not have equal velocities (except at $\theta = 0$), as can be seen in Figure 3.27 for spur gears; nonetheless, the teeth are cut so that we may correctly treat the gears for dynamic

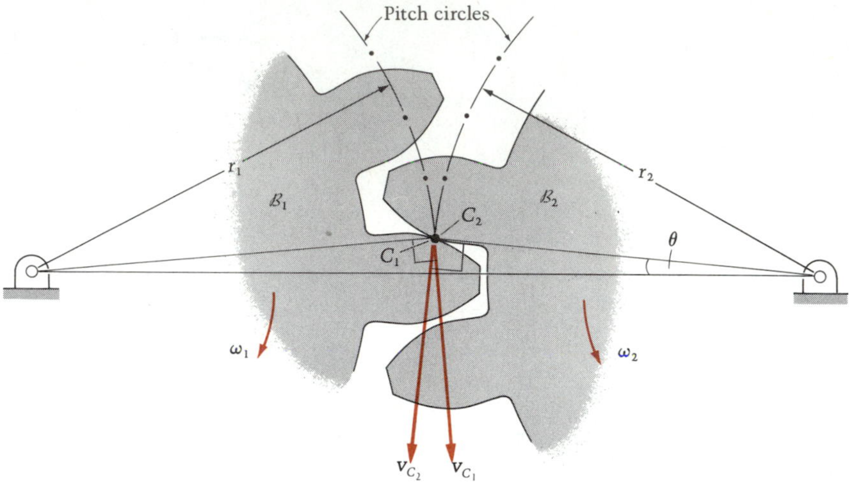

Figure 3.27

purposes as if they were two cylinders rolling on each other at the pitch circles. Thus when the centers are pinned as in Figure 3.27, we may use the relation

$$r_1\omega_1 = r_2\omega_2 \tag{3.37}$$

where r_1 and r_2 are the respective pitch radii of the gears $\mathcal{B}_1$ and $\mathcal{B}_2$. We may also use the derivative of Equation 3.37 (since it is valid for all t):

$$r_1\alpha_1 = r_2\alpha_2 \tag{3.38}$$

We note that the radius ratio is inversely proportional to the ratio of angular speeds (and directly proportional to the ratio of numbers of gear teeth, since the shape and spacing of teeth must match). We now consider several examples.

EXAMPLE 3.20

Find the angular speed of the front sprocket (rigidly fixed to the pedal crank) of the bicycle in Figure E3.20a if the man is traveling at 10 mph. There are 26 teeth on the front sprocket and 9 on the rear sprocket (which turns rigidly with the rear wheel). The wheel diameters are 26 in.

Figure E3.20a

Solution

The velocities of A and B, the two ends of the straight upper length of chain, are equal. (As shown in Figure E3.20b, A is just leaving the rear sprocket $\mathcal{B}_1$; B is just about to enter the front sprocket $\mathcal{B}_2$.) To prove this, we note that the translating section AB of chain is behaving as if rigid, so that, calling this "body" $\mathcal{B}_3$,

$$\mathbf{v}_B = \mathbf{v}_A + \boldsymbol{\omega}_3 \times \mathbf{r}_{AB}$$

But $\boldsymbol{\omega}_3 = \mathbf{0}$, so that

$$\mathbf{v}_B = \mathbf{v}_A$$

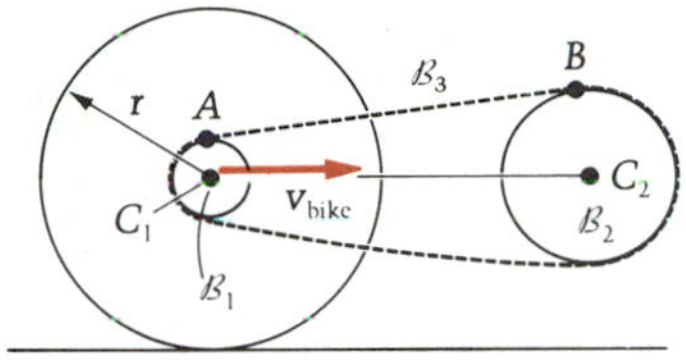

Figure E3.20b

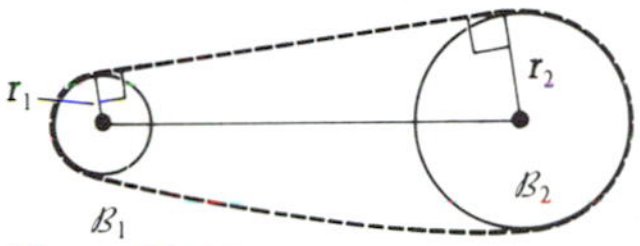

Figure E3.20c

Next we relate the equal velocities of A and B to the respective centers of their sprockets $\mathcal{B}_1$ and $\mathcal{B}_2$:

$$\mathbf{v}_{C_1} + \boldsymbol{\omega}_1 \times \mathbf{r}_{C_1A} = \mathbf{v}_{C_2} + \boldsymbol{\omega}_2 \times \mathbf{r}_{C_2B}$$

Now the velocities of C_1 and C_2 are each equal to the "velocity of the bike," meaning the common velocity of all the points on the translating part of the bike, such as points of the frame and seat. Therefore $\mathbf{v}_{C_1}$ and $\mathbf{v}_{C_2}$ cancel, leaving

$$\boldsymbol{\omega}_1 \times \mathbf{r}_{C_1A} = \boldsymbol{\omega}_2 \times \mathbf{r}_{C_2B}$$

This says simply that (see Figure E3.20c):

$$r_1\omega_1 = r_2\omega_2$$

Now if the speed of the bike is to be 10 mph, we have

$$v_{C_1} = \overset{13\text{ in.}}{\cancel{r}}\omega_1 = 10\text{ mph}\left(\frac{88\text{ ft/sec}}{60\text{ mph}}\right)\frac{12\text{ in.}}{1\text{ ft}} = 176\text{ in./sec}$$

or

$$\omega_1 = 13.5 \circlearrowright \text{ rad/sec}$$

Thus

$$\omega_2 = \frac{r_1}{r_2}\,\omega_1 = \frac{9}{26}(13.5) \qquad \text{(Radii are proportional to number of teeth!)}$$

$$= 4.67\text{ rad/sec}$$

So the rider must turn the pedal crank at $4.67/2\pi = 0.743$ rev/sec.

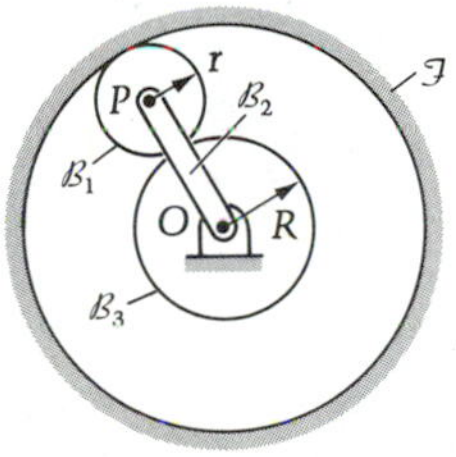

Figure E3.21a

Example 3.21

Frame $\mathcal{I}$ is a fixed ring gear with internal teeth (not shown in Figure E3.21a) that mesh with those of the planetary gear $\mathcal{B}_1$. The teeth of $\mathcal{B}_1$ also mesh with those of the sun gear $\mathcal{B}_3$, which is pinned at its center point O to frame $\mathcal{I}$. The crank arm $\mathcal{B}_2$ is pinned at its ends to O and to the center point P of $\mathcal{B}_1$. The arm $\mathcal{B}_2$ has angular speed $\omega_2(t)$ counterclockwise. Find the angular velocity of $\mathcal{B}_3$ in terms of R, r, and ω_2.

Solution

We take $\mathcal{I}$ to be our reference frame, to which all motions are referred. We work first with the crank $\mathcal{B}_2$, since we know its angular velocity and the velocity of one of its points ($\mathbf{v}_O = \mathbf{0}$). From the sketch of $\mathcal{B}_2$ (Figure E3.21b), we see that we can write

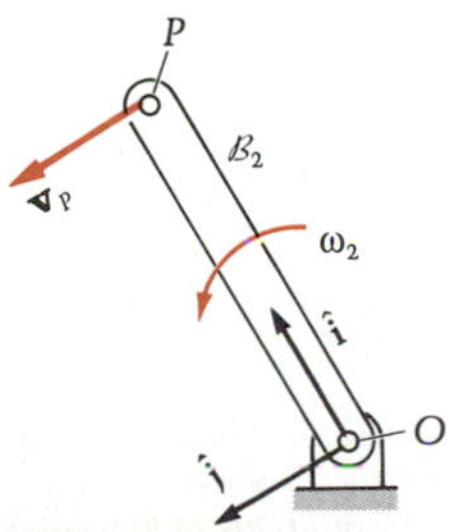

Figure E3.21b

$$\begin{aligned}\mathbf{v}_P &= \overset{0}{\cancel{\mathbf{v}_O}} + \omega_2\hat{\mathbf{k}} \times (R + r)\hat{\mathbf{i}} \\ &= (R + r)\omega_2\hat{\mathbf{j}} \qquad (1)\end{aligned}$$

(Note that we align $\hat{\mathbf{i}}$ parallel to OP for convenience; we need not always draw it to the right.)

Next, the points of $\mathcal{B}_1$ and $\mathcal{B}_2$ that are pinned together at P have the same velocity at all times. Furthermore, the points of $\mathcal{I}$ and $\mathcal{B}_1$ at D (see Figure E3.21c)

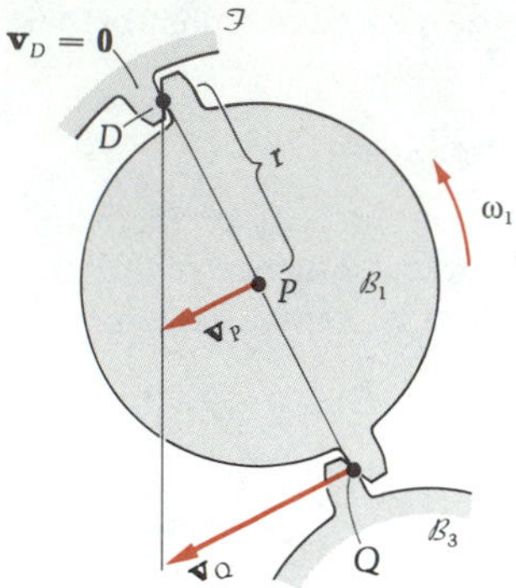

Figure E3.21c

are in contact and each has zero velocity* since D is fixed in $\mathcal{J}$. Thus

$$\mathbf{v}_P = \overset{0}{\cancel{\mathbf{v}_D}} + \omega_1\hat{\mathbf{k}} \times (-r\hat{\mathbf{i}})$$

or

$$(R + r)\omega_2\hat{\mathbf{j}} = -r\omega_1\hat{\mathbf{j}}$$

Solving for ω_1 gives

$$\omega_1 = \frac{-(R + r)\omega_2}{r} \qquad \text{or} \qquad \omega_1 = \frac{R + r}{r}\,\omega_2 \circlearrowright \tag{2}$$

We are now in a position to obtain the velocity of point Q of $\mathcal{B}_1$, the point in contact with the tooth of $\mathcal{B}_3$:

$$\mathbf{v}_Q = \overset{0}{\cancel{\mathbf{v}_D}} + \omega_1\hat{\mathbf{k}} \times (-2r\hat{\mathbf{i}})$$

Substituting for ω_1 in terms of ω_2 from Equation (2), we get

$$\mathbf{v}_Q = 2(R + r)\omega_2\hat{\mathbf{j}} \tag{3}$$

Note that Q has twice the speed of P since it is twice as far from the instantaneous center D of $\mathcal{B}_1$ as is P.

Finally, we come to the body $\mathcal{B}_3$ of interest (see Figure E3.21d). Knowing that the points Q and Q' (the respective tooth points in contact on $\mathcal{B}_1$ and $\mathcal{B}_3$) have equal velocities as they move together tangent to the pitch circle, we obtain

$$\mathbf{v}_{Q'} = \mathbf{v}_O + \omega_3\hat{\mathbf{k}} \times R\hat{\mathbf{i}}$$
$$2(R + r)\omega_2\hat{\mathbf{j}} = \mathbf{0} + R\omega_3\hat{\mathbf{j}}$$

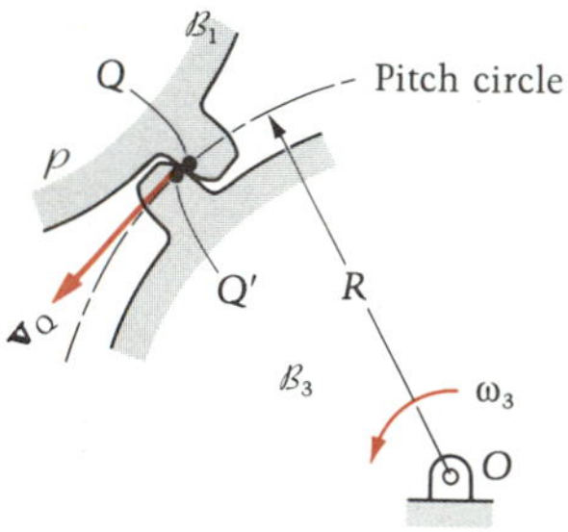

Figure E3.21d

Thus the angular speed of $\mathcal{B}_3$ in $\mathcal{J}$ is

$$\omega_3 = \frac{2(R + r)\omega_2}{R} \tag{4}$$

and the angular velocity of $\mathcal{B}_3$ in $\mathcal{J}$ is

$$\omega_3\hat{\mathbf{k}} = \frac{2(R + r)\omega_2}{R}\,\hat{\mathbf{k}}$$

Note in the previous example that the point of $\mathcal{B}_2$ passing over point Q has velocity $R\omega_2$, which is of course less than the velocity $\mathbf{v}_Q$ of the gear teeth in contact at Q. This velocity is $R\omega_3 = 2(R + r)\omega_2$, which is more than twice as fast as $R\omega_2$.

We also remark that since the answers

$$\omega_1(t) = \frac{R + r}{r}\,\omega_2(t) \qquad \text{and} \qquad \omega_3(t) = \frac{2(R + r)}{R}\,\omega_2(t)$$

* As we have pointed out, the contact points of gear teeth necessarily slide relative to each other. The points used in the analysis are actually not tooth points, however, but imaginary points on the pitch circles of the gears. Furthermore, the radii given in the examples and problems are the radii of these circles.

are completely general functions of time, the angular accelerations are obtainable immediately by differentiation, with $\alpha_2 = \dot{\omega}_2$:

$$\alpha_1(t) = \frac{R+r}{r}\,\alpha_2(t) \qquad \text{and} \qquad \alpha_3(t) = \frac{2(R+r)}{R}\,\alpha_2(t)$$

Rather than differentiating, however, we shall obtain these two results in the following example in another manner: by repeated use of Equation (3.19). The purpose is to gain insight into its use in gearing situations involving several bodies. The procedure in the next example would have to be followed if the previous example had been worked using instantaneous values instead of generally (with symbols).

Example 3.22

Find the angular accelerations α_1 and α_3 of the planetary and sun gears in the previous example in terms of R, r, and ω_2 and α_2, which are given functions of the time t.

Solution

Relating the accelerations of P and O on body $\mathcal{B}_2$ gives (see the bar in Figure E3.22a):

$$\mathbf{a}_P = -(R+r)\omega_2^2\hat{\mathbf{i}} + (R+r)\alpha_2\hat{\mathbf{j}}$$

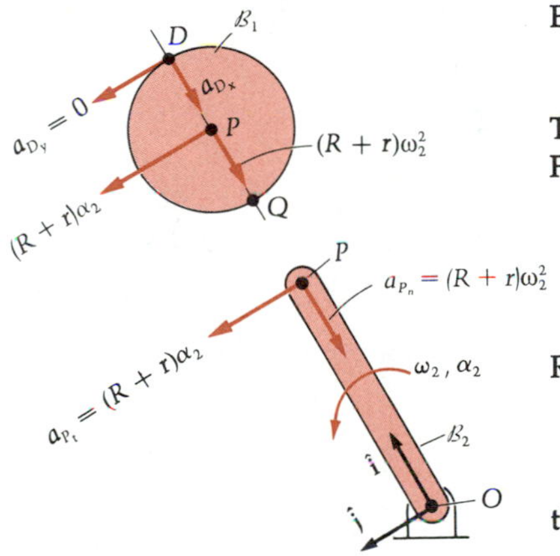

Figure E3.22a

This acceleration is then carried over to the coincident point P on $\mathcal{B}_1$ (again see Figure E3.22a). Relating D and P on the planetary gear $\mathcal{B}_1$, we have

$$\mathbf{a}_D = a_D\hat{\mathbf{i}} = \mathbf{a}_P + \alpha_1\hat{\mathbf{k}} \times \overset{r\hat{\mathbf{i}}}{\mathbf{r}_{PD}} - \omega_1^2\,\overset{r\hat{\mathbf{i}}}{\mathbf{r}_{PD}}$$
$$= -(R+r)\omega_2^2\hat{\mathbf{i}} + (R+r)\alpha_2\hat{\mathbf{j}} + \alpha_1\hat{\mathbf{k}} \times r\hat{\mathbf{i}} - \omega_1^2 r\hat{\mathbf{i}}$$

Recalling that

$$\omega_1 = \frac{R+r}{r}\,\omega_2$$

the coefficients of $\hat{\mathbf{i}}$ then give

$$a_D = \frac{-(R+2r)(R+r)}{r}\,\omega_2^2$$

and the $\hat{\mathbf{j}}$ coefficients yield

$$\alpha_1 = \frac{-(R+r)\alpha_2}{r} \qquad \text{or} \qquad \alpha_1 = \frac{R+r}{r}\,\alpha_2 \circlearrowleft$$

We now need $\mathbf{a}_Q$, where Q is again the tooth point of $\mathcal{B}_1$ in contact with the sun gear $\mathcal{B}_3$:

$$\mathbf{a}_Q = \mathbf{a}_D + \alpha_1\hat{\mathbf{k}} \times \overset{-2r\hat{\mathbf{i}}}{\mathbf{r}_{DQ}} - \omega_1^2\mathbf{r}_{DQ}$$
$$= \frac{-(R+2r)(R+r)}{r}\,\omega_2^2\hat{\mathbf{i}} + 2(R+r)\alpha_2\hat{\mathbf{j}} + \frac{2r(R+r)^2}{r^2}\,\omega_2^2\hat{\mathbf{i}}$$
$$= \frac{(R+r)R}{r}\,\omega_2^2\hat{\mathbf{i}} + 2(R+r)\alpha_2\hat{\mathbf{j}}$$

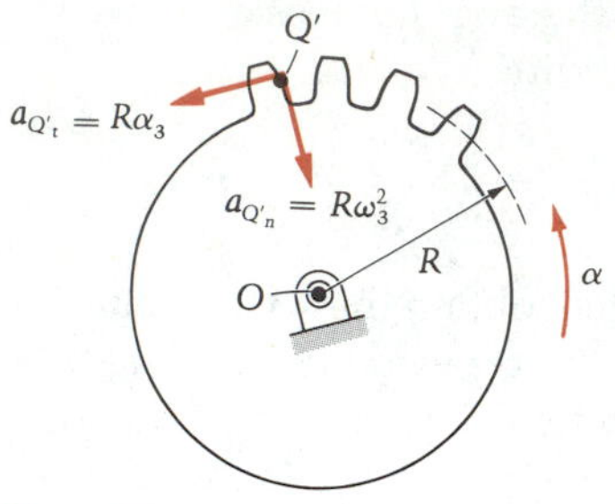

Figure E3.22b

We now go to body $\mathcal{B}_3$ to complete the solution. Relating the tooth point Q' to O on $\mathcal{B}_3$ gives the components of $a_{Q'}$. (see Figure E3.22b). The tangential acceleration components of Q and Q' are equal* as the teeth contact and move together:

$$a_{Q'_t} = a_{Q'_y} = 2(R + r)\alpha_2 = R\alpha_3$$

Thus

$$\boldsymbol{\alpha}_3 = \frac{2(R + r)}{R}\,\alpha_2 \circlearrowleft$$

Note that we may express the normal acceleration component of Q' in terms of ω_2 by using the result for ω_3 from Example 3.21:

$$a_{Q'_n} = -a_{Q'_x} = +R\omega_3^2 = +R\left[\frac{2(R + r)}{R}\,\omega_2\right]^2$$

$$= \frac{4(R + r)^2\omega_2^2}{R}$$

We also note that we could have alternatively obtained the accelerations of D and Q as points on rims of wheels rolling on curved tracks by using Equations (3.31) and (3.32). Noting that ρ for P is $(R + r)$, we present these partial checks on our solution:

$$\mathbf{a}_D = \left(1 + \frac{r}{\rho}\right) r\dot{\theta}^2\hat{\mathbf{e}}_n = \left(1 + \frac{r}{R + r}\right) r\omega_1^2(-\hat{\mathbf{i}})$$

$$= -\left(\frac{R + 2r}{R + r}\right) r\omega_1^2\hat{\mathbf{i}} = -\left(\frac{R + 2r}{R + r}\right) r \left(\frac{R + r}{r}\,\omega_2\right)^2 \hat{\mathbf{i}}$$

$$= \frac{-(R + 2r)(R + r)}{r}\,\omega_2^2\hat{\mathbf{i}} \qquad \text{(as above)}$$

$$\mathbf{a}_Q = 2r\ddot{\theta}\hat{\mathbf{e}}_t - r\dot{\theta}^2\left(1 - \frac{r}{\rho}\right)\hat{\mathbf{e}}_n$$

$$= 2r(-\alpha_1)(+\hat{\mathbf{j}}) - r\omega_1^2\left(1 - \frac{r}{R + r}\right)(-\hat{\mathbf{i}})$$

$$= \frac{Rr}{R + r}\,\omega_1^2\hat{\mathbf{i}} - 2r\alpha_1\hat{\mathbf{j}}$$

$$= \frac{Rr}{R + r}\left(\frac{R + r}{r}\,\omega_2\right)^2\hat{\mathbf{i}} - 2r\left[\frac{-(R + r)}{r}\,\alpha_2\right]\hat{\mathbf{j}}$$

$$= \frac{(R + r)R}{r}\,\omega_2^2\hat{\mathbf{i}} + 2(R + r)\alpha_2\hat{\mathbf{j}} \qquad \text{(as above)}$$

* This is in fact true even when *neither* body's center is fixed and the geometry is irregular. As long as there is rolling, the acceleration components of the contacting points *in the plane tangent to the two bodies* are equal in plane motion at all times. See "Contact Point Accelerations in Rolling Problems," D. J. McGill, *Mechanics Research Communications,* 7(3), 175–179, 1980.

PROBLEMS ▶ Section 3.6

3.84 The wheel in Figure P3.84 rolls on the plane with constant angular velocity 1⟳ rad/sec. Find the velocity of point Q by using the instantaneous center Ⓘ of zero velocity. Then check by using Equation (3.8) to relate $\mathbf{v}_Q$ to $\mathbf{v}_P$.

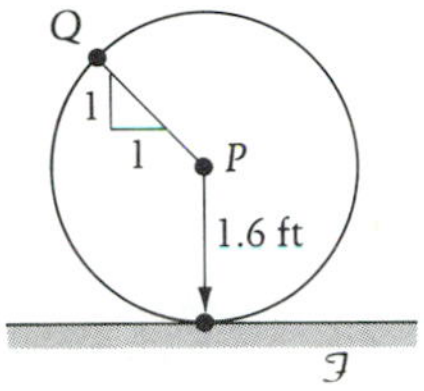

Figure P3.84

3.85 In the preceding problem, suppose that the plane $\mathcal{I}$ on which the wheel rolls is not fixed to the reference frame but instead translates on it (this time the reference frame is $\mathcal{G}$) at constant velocity 3 ft/sec to the left. (See Figure P3.85.)

a. Find the instantaneous center of zero velocity Ⓘ

b. Find $\mathbf{v}_Q$ again.

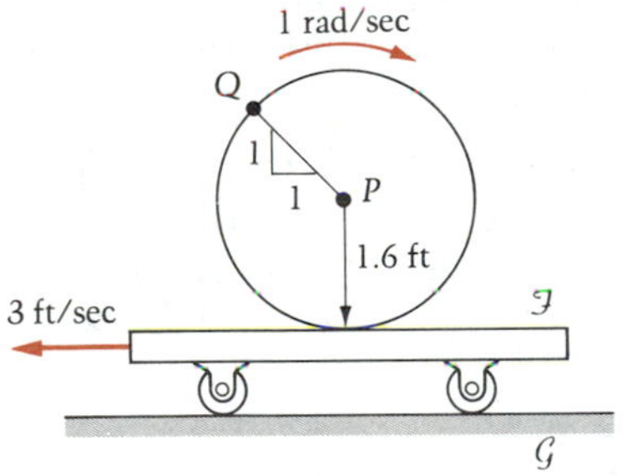

Figure P3.85

3.86 The wheel in Figure P3.86 rolls on the bar. If at a certain instant the bar has a velocity of 2 m/s to the right and the wheel has counterclockwise angular velocity of 0.5 rad/s, determine the velocity of (a) the center of the wheel and (b) point P.

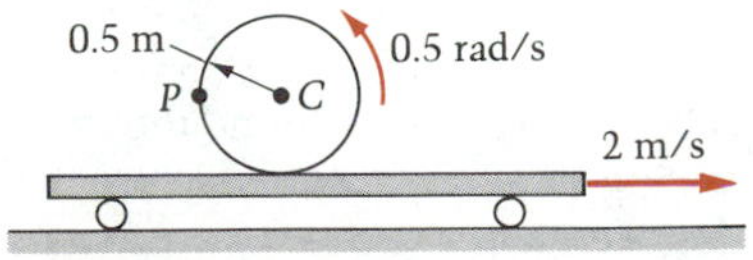

Figure P3.86

3.87 Figure P3.87(a) shows the manner in which a train wheel rests on the track. If the train travels at a constant speed of 80 mph and does not slip on the track, determine the velocities of points A, B, D, and E on the vertical line through the center C in Figure P3.87(b). Which point is traveling backward? Why?

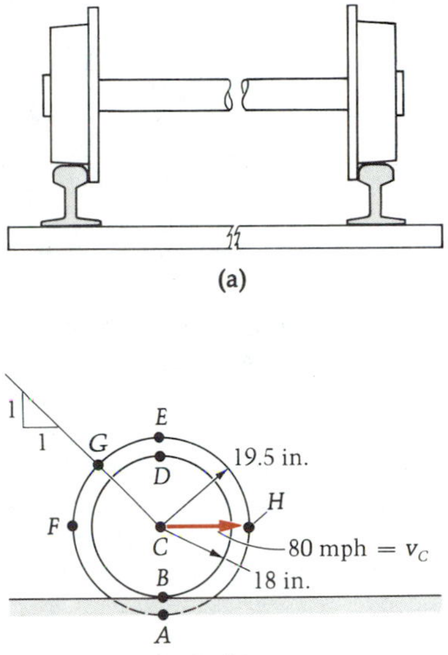

Figure P3.87

3.88 In the preceding problem find the velocities of points F, G, and H.

3.89 Find the velocities of points B and C in Figure P3.89 if the cylinder does not slip on the translating bodies $\mathcal{B}_1$ and $\mathcal{B}_2$.

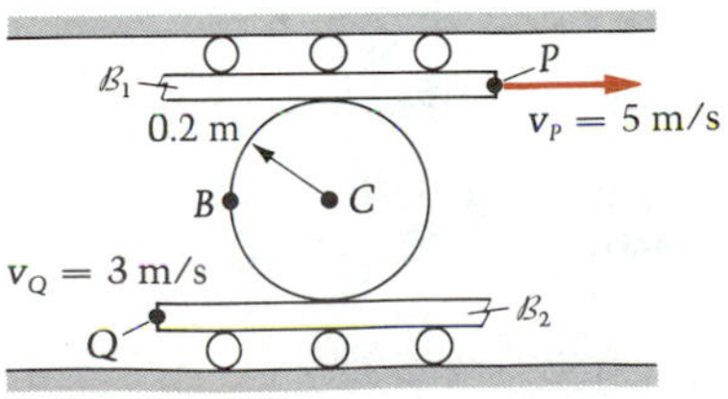

Figure P3.89

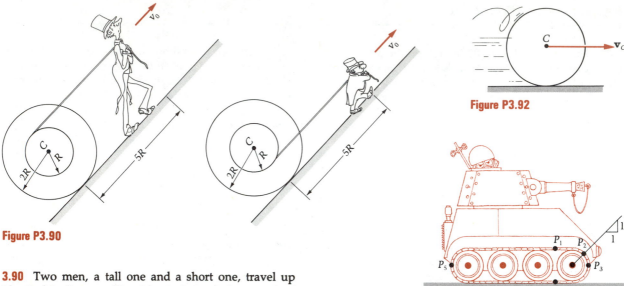

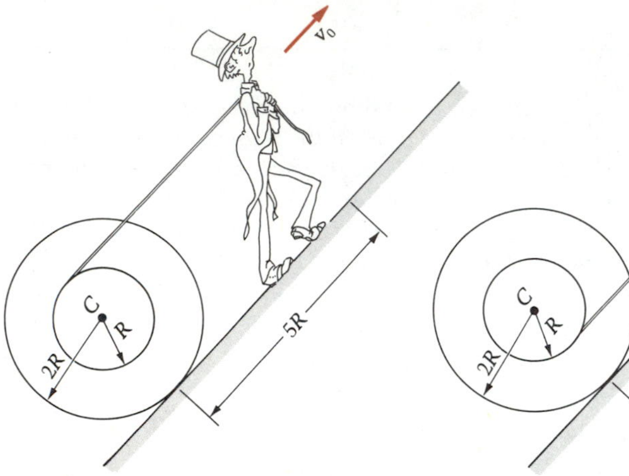

Figure P3.90

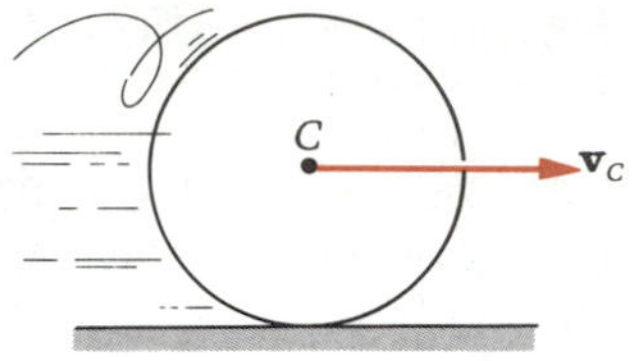

Figure P3.92

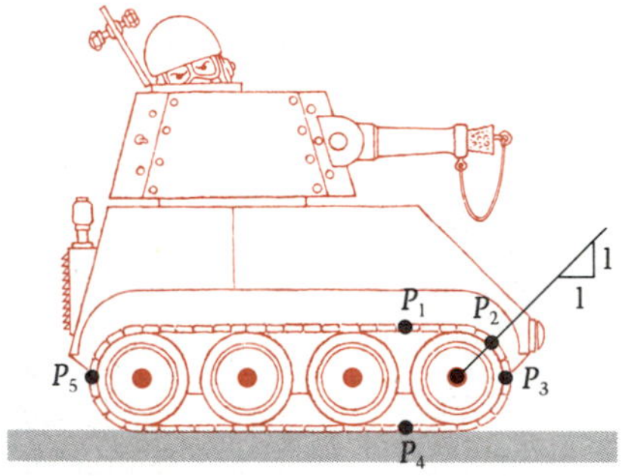

Figure P3.93

3.90 Two men, a tall one and a short one, travel up identical inclines, pulling identical spools by means of ropes wrapped around the hubs. (See Figure P3.90.) The men travel at the same constant speed v_0, and the ropes are wrapped in the opposite directions indicated. If the spools do not slip on the plane, one of the men will be run over by his own spool. Prove which one it is, and show how long it will take, from the instant depicted, for the spool to roll over him.

3.91 A cylinder $\mathcal{B}$ of radius r rolls over a circular arc of constant radius of curvature R (see Figure P3.91). What is the ratio of the angular speed of the cylinder to $\dot{\phi}$?

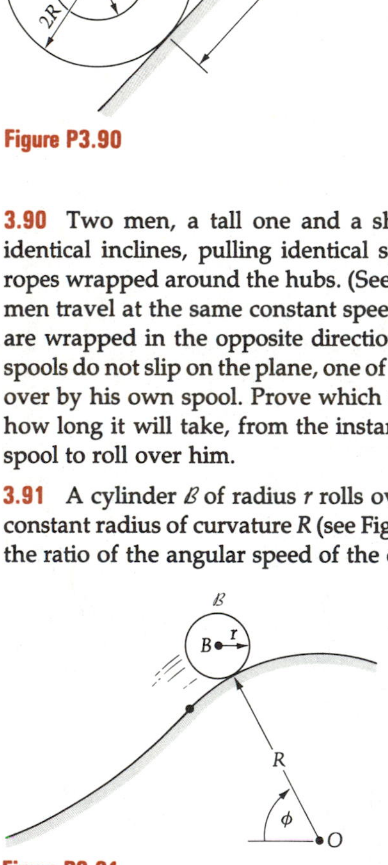

Figure P3.91

3.92 A disk with diameter 1.2 m rolls along the plane as indicated in Figure P3.92. Its center point C has velocity

$$\mathbf{v}_C = (t^2 + 3t + 4)\hat{\mathbf{i}} \text{ m/s}$$

where t is the time in seconds. Find the velocity of the point that lies 0.3 m directly below C when (a) $t = 2$ s and (b) $t = 5$ s.

3.93 The tank shown in Figure P3.93 is translating to the right, and at a certain instant it has velocity $v_0\hat{\mathbf{i}}$ and acceleration $a_0\hat{\mathbf{i}}$. (These values are $\mathbf{v}$ and $\mathbf{a}$ for all points in the body of the tank and for the centers of its wheels.) Find the velocities of the five points P_1, P_2, P_3, P_4, and P_5 if there is no slipping. The wheels have radius R.

3.94 The wheel rolls on both $\mathcal{B}_1$ and $\mathcal{B}_2$ (See Figure P3.94.) The constant angular velocity of the wheel in frame $\mathcal{I}$ is shown in the figure. Find:

a. The velocity of the points of $\mathcal{B}_1$ relative to $\mathcal{B}_2$

b. The constant velocity of C in $\mathcal{I}$ for which the velocities of T (on $\mathcal{B}_1$) and B (on $\mathcal{B}_2$) in $\mathcal{I}$ are equal in magnitude and opposite in direction.

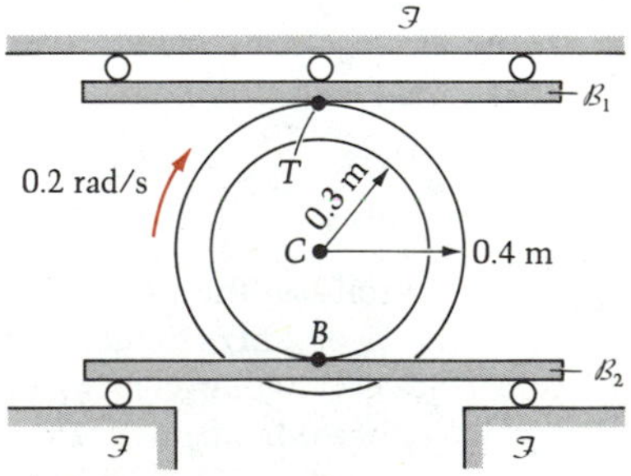

Figure P3.94

3.95 An inextensible string is wrapped around the cylinder in Figure P.3.95, fitting in a small slot near the rim. The center C is moving down the plane at a constant speed of 0.1 m/s. Find the velocities of points A, B, D, and

E. *Hint*: The cylinder is not rolling on the plane, but it *is* rolling on ___?___.

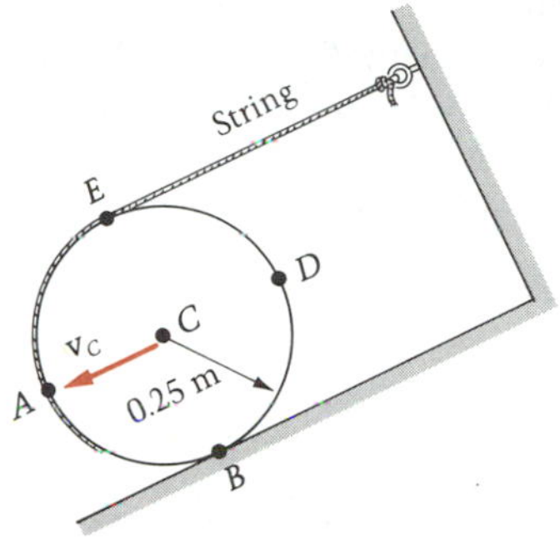

Figure P3.95

3.96 At the instant shown in Figure P3.96, point B of the block (to which rod $\mathcal{B}$ is pinned) has $\mathbf{v}_B = 0.5 \downarrow$ ft/sec. Find the angular velocity of the rolling cylinder.

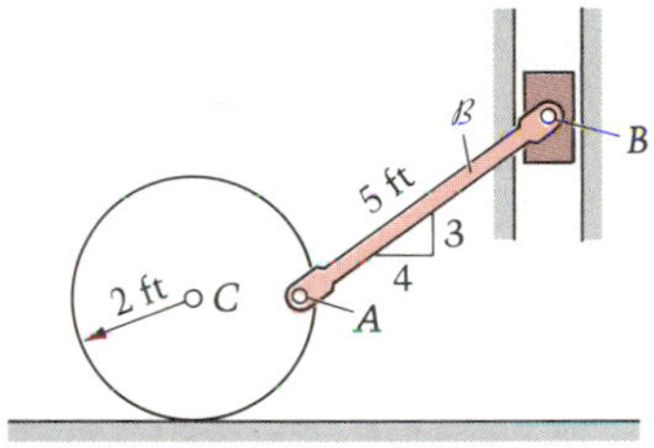

Figure P3.96

3.97 Figure P3.97 shows a circular cam $\mathcal{B}_1$ and an oscillating roller follower consisting of the roller $\mathcal{B}_2$ (which rolls on $\mathcal{B}_1$) and the follower bar $\mathcal{B}_3$. If the cam turns at the constant angular velocity 0.3 ↺ rad/s, find the angular velocity of the follower bar and of the roller at the given instant.

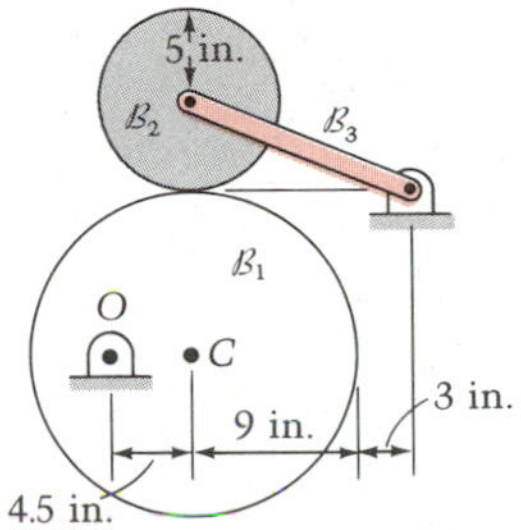

Figure P3.97

3.98 The cylinder in Figure P3.98 is rolling to the left with constant center speed v_C. A stick is pinned to the cylinder at B, and its other end A slides on the plane. Find the velocity of A when $\theta = 0°$, $90°$, $180°$, and $270°$.

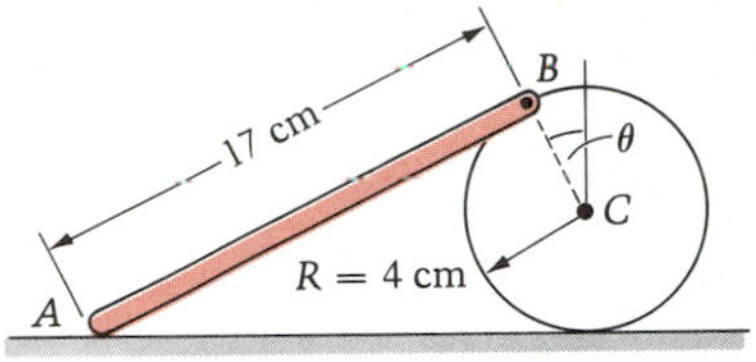

Figure P3.98

3.99 Wheel $\mathcal{B}_1$ in Figure P3.99 has angular velocity 3 ↻ rad/sec. Find the angular velocities of wheel $\mathcal{B}_2$ and the bent bar $\mathcal{B}_3$.

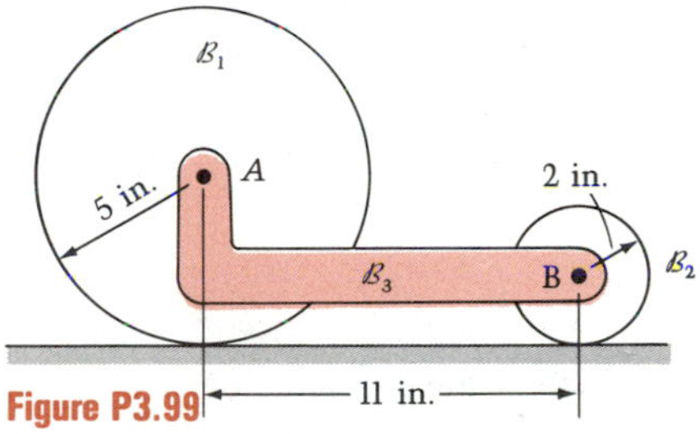

Figure P3.99

* **3.100** The cart $\mathcal{B}_1$ in Figure P3.100 travels from left to right, with its rear wheels $\mathcal{B}_2$ rolling at constant angular velocity 0.2 ↻ rad/sec. The front wheels $\mathcal{B}_3$ are rolling up the parabolic surface shown. The wheels have radius 0.4 m, and are pinned to the cart. Find the angular velocity of cart $\mathcal{B}_1$ at the given instant. *Hint:*

$$\frac{dy}{dx} = \tan\phi = 2$$

$$\phi = 63.43°$$

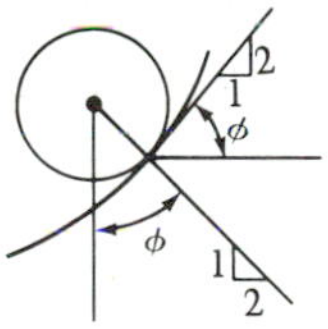

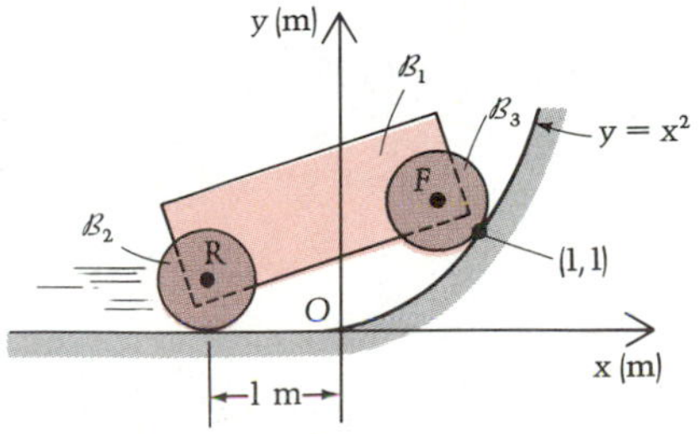

Figure P3.100

3.101 The constant angular velocities of the ring gear $\mathcal{B}_1$ and the spider arm $\mathcal{B}_2$ shown in Figure P3.101 are 2 $\circlearrowleft$ rad/s and 10 $\circlearrowright$ rad/s, respectively. Determine the angular velocity of gear $\mathcal{B}_3$ and the velocity of the point of $\mathcal{B}_4$ having maximum speed in the given position. The centers of $\mathcal{B}_1$ and $\mathcal{B}_2$ are pinned to the reference frame $\mathcal{I}$.

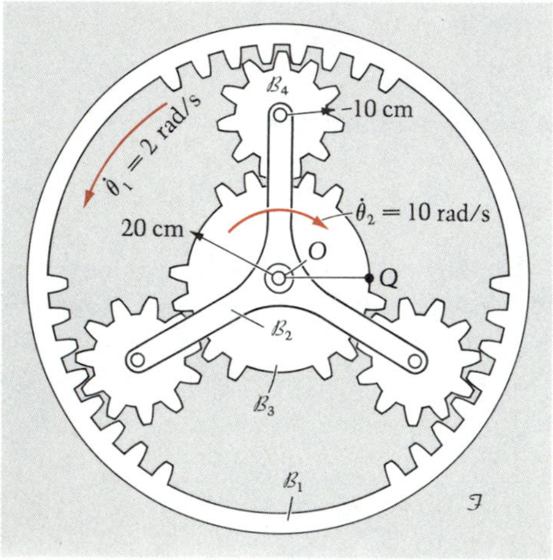

Figure P3.101

3.102 The rod $\mathcal{R}$, which is pinned to cylinder $\mathcal{B}_2$, translates upward in the y-direction at the constant speed 4 ft/sec (see Figure P3.102). The rod $\mathcal{B}_1$ is pinned to the reference frame at O and rests against the rim of $\mathcal{B}_2$ as shown in the figure. There is rolling contact between $\mathcal{B}_1$ and $\mathcal{B}_2$. Find the angular velocities of $\mathcal{B}_1$ and $\mathcal{B}_2$ at the instant shown.

3.103 If the given velocities of P and Q in Problem 3.89 are constant, find the accelerations of C and B.

3.104 If, in Problem 3.89, the respective accelerations of P and Q are $1 \leftarrow$ m/^{2} and $2 \rightarrow$ m/s^2, find the angular acceleration of the cylinder.

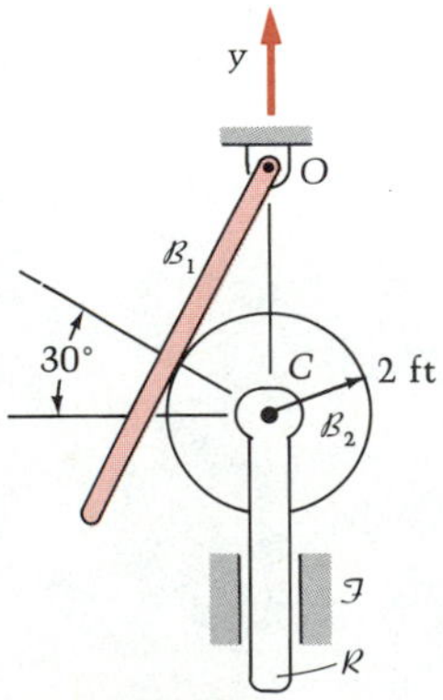

Figure P3.102

3.105 The shaded arcs on $\mathcal{B}_1$ and $\mathcal{B}_2$ (Figure P3.105(a)) are always equal if the two bodies are in rolling contact; however, the converse is not necessarily true. Just because the contacting arclengths are equal does not mean that $\mathcal{B}_1$ rolls on $\mathcal{B}_2$. For the wheel $\mathcal{B}_1$ on the plane $\mathcal{B}_2$ shown in Figure P3.105(b), give constant values $\dot{x}_C$ and $\dot{\theta}$ for which the arclengths of contact are equal but the velocities of the contact points are not. *Hint*: Look at the shaded arcs on $\mathcal{B}_1$ and $\mathcal{B}_2$ in Figure P3.105(b).

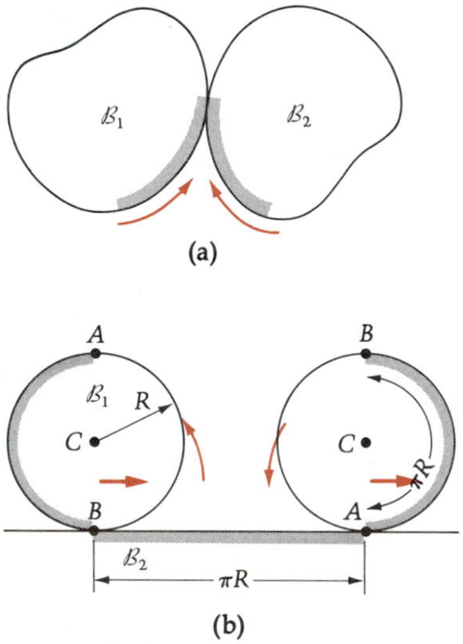

Figure P3.105

3.106 If in Problem 3.99 the angular acceleration of $\mathcal{B}_1$ is $\boldsymbol{\alpha}_1 = 1 \circlearrowleft$ rad/sec^2 at the given instant, find $\boldsymbol{\alpha}_2$ and $\boldsymbol{\alpha}_3$ at that time.

3.107 In Problem 3.93 find the accelerations of the same five points P_1, P_2, P_3, P_4, and P_5. (See Figure P3.107.)

3.108 During startup of the two friction wheels (see Figure P3.108), the angular velocity of $\mathcal{B}_1$ is $\boldsymbol{\omega}_1 = 5t^2 \circlearrowleft$ rad/sec. Assuming rolling contact, compute the acceleration of the point T, which is at the top of $\mathcal{B}_2$ when $t = 3$ sec.

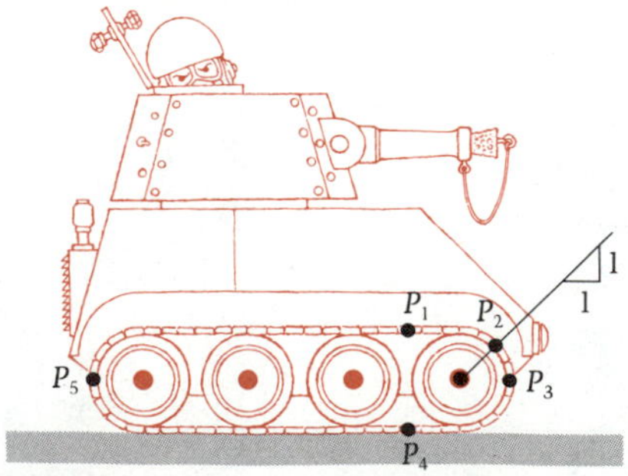

Figure P3.107

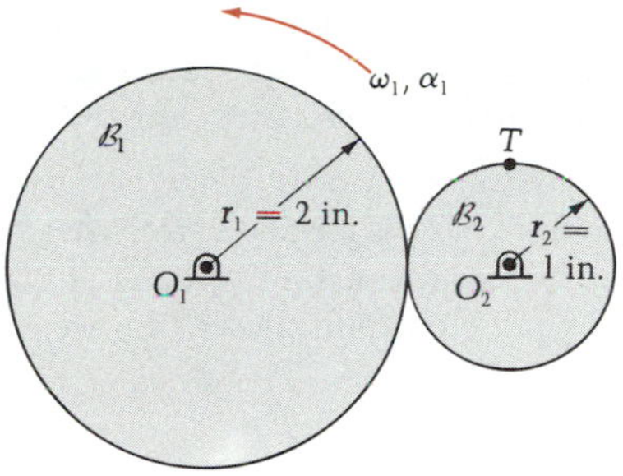

Figure P3.108

3.109 The wheel in Figure P3.109 rolls, its center having a constant velocity of 10 ft/sec to the right. Find $d|\mathbf{v}_A|/dt$ at the instant shown.

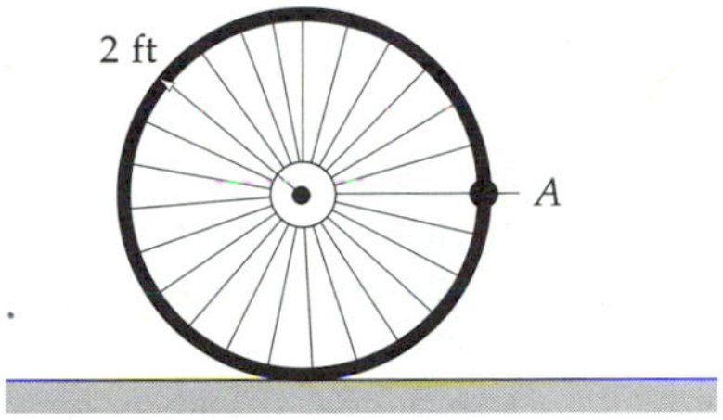

Figure P3.109

3.110 A wheel rolls on a 10-cm-radius track. (See Figure P3.110.) At the instant shown the wheel has an angular velocity of $4\hat{\mathbf{k}}$ rad/s and an angular acceleration of $-3\hat{\mathbf{k}}$ rad/s^2. At the instant shown, find:

- **a.** The velocity and acceleration of C
- **b.** The velocity and acceleration of A
- **c.** $(d/dt)|\mathbf{v}_A|$
- **d.** The center of curvature of the path of point T.

3.111 In Problem 3.87 find the accelerations of points A, B, C, D, and E.

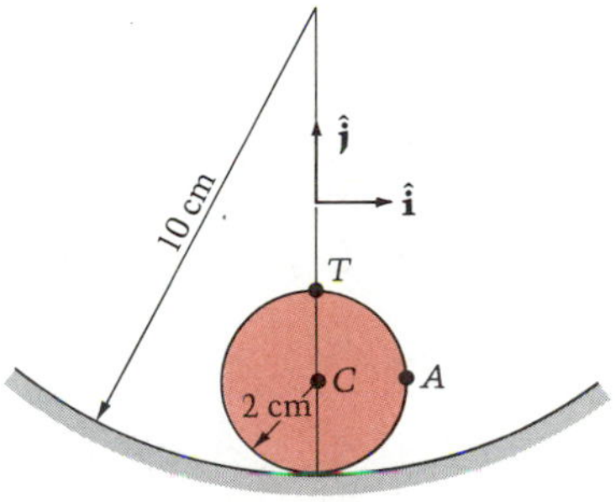

Figure P3.110

3.112 In Problem 3.88 find the accelerations of points F, G, and H.

3.113 A moment applied to gear $\mathcal{B}_1$ in Figure P3.113 results in a constant angular acceleration $\boldsymbol{\alpha}_1 = 2 \circlearrowleft$ rad/sec^2. The other gear, $\mathcal{B}_2$, is fixed in the reference frame. Determine:

- **a.** The time required for C to return to its starting point after one revolution around $\mathcal{B}_2$ from rest
- **b.** The number of revolutions turned through in space by $\mathcal{B}_1$ during the complete revolution.

3.114 The center of the rolling wheel in Figure P3.114 moves to the right at a constant speed of 10 in./sec. The bar is pinned to the wheel at A, and end B always stays in contact with the ground. Find the acceleration of B at the instant shown.

3.115 Calculate the acceleration of the instantaneous center of rod $\mathcal{B}_2$ in Example 3.15.

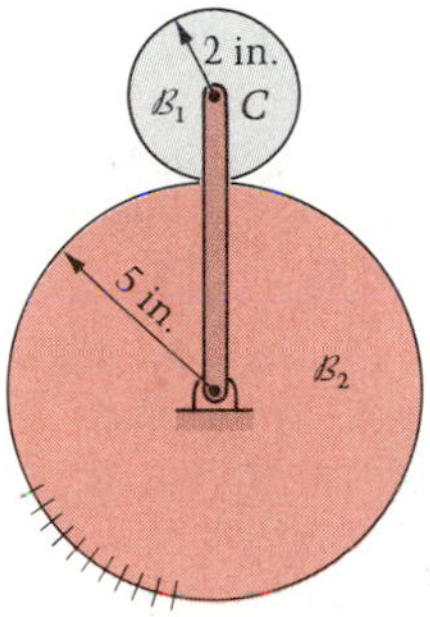

Figure P3.113

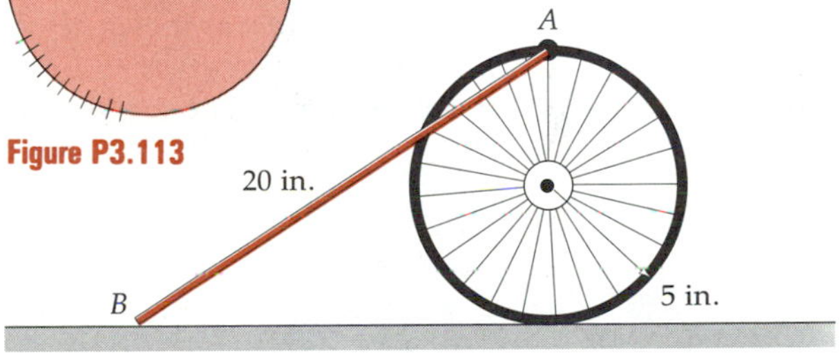

Figure P3.114

3.116 Gear $\mathcal{B}_1$ and crank $\mathcal{B}_2$ have angular speeds ω_0 and angular acceleration magnitudes α_0 at the instant shown in Figure P3.116 in the indicated directions. Find the angular velocities and angular accelerations of gears $\mathcal{B}_3$ and $\mathcal{B}_4$ at the same time, if $\mathcal{B}_2$ is pinned to $\mathcal{B}_1$, $\mathcal{B}_3$ and $\mathcal{B}_4$.

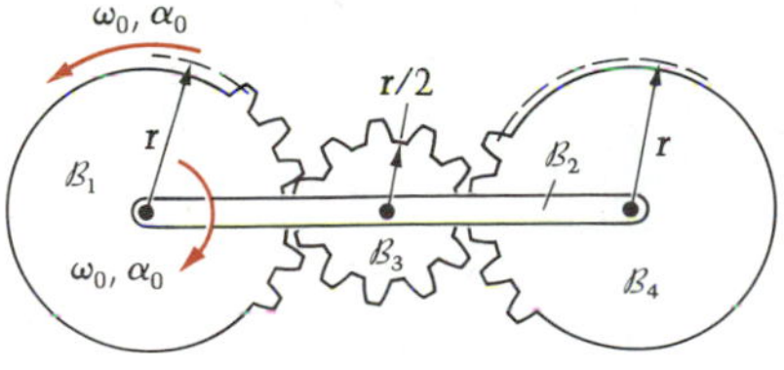

Figure P3.116

* **3.117** The center C of the small cylinder $\mathcal{B}_1$ in Figure P3.117 has a speed of $0.1t^2$ m/s as it moves clockwise on a circle. Body $\mathcal{B}_1$ rolls on the large cylinder $\mathcal{B}_2$. In the position given in the figure, $t = 10$ s. Find the acceleration of point B of the stick $\mathcal{B}_3$ that is in contact with $\mathcal{B}_2$ at the given instant.

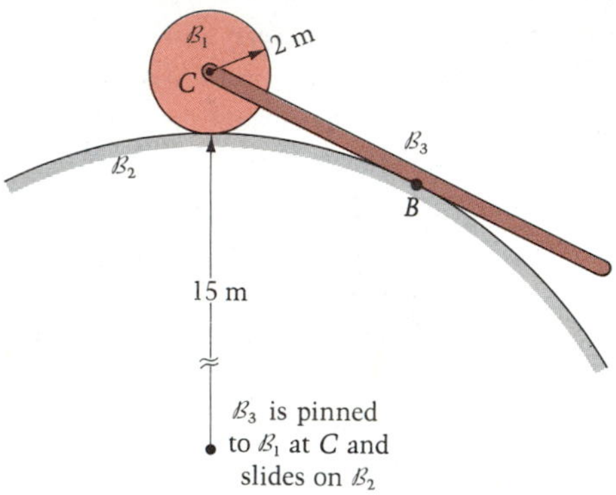

Figure P3.117

3.118 In Problem 3.89 let $\mathbf{a}_P = 0.2 \rightarrow$ m/s^2 and $\mathbf{a}_Q = 0.1 \leftarrow$ m/s^2 at the instant shown. Again assuming no slipping, find the accelerations of B and C.

3.119 The two identical cylinders $\mathcal{B}_1$ and $\mathcal{B}_2$ (Figure P3.119) are connected by bar $\mathcal{B}_3$ (which is pinned to their centers), and they roll on the surface as shown. If the angular velocity of $\mathcal{B}_1$ is $\boldsymbol{\omega}_1 = \omega_0$ ↺ = constant, find the angular accelerations of both $\mathcal{B}_2$ and $\mathcal{B}_3$ at the given instant.

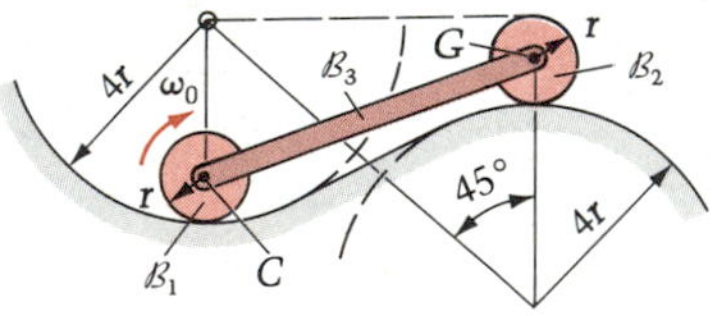

Figure P3.119

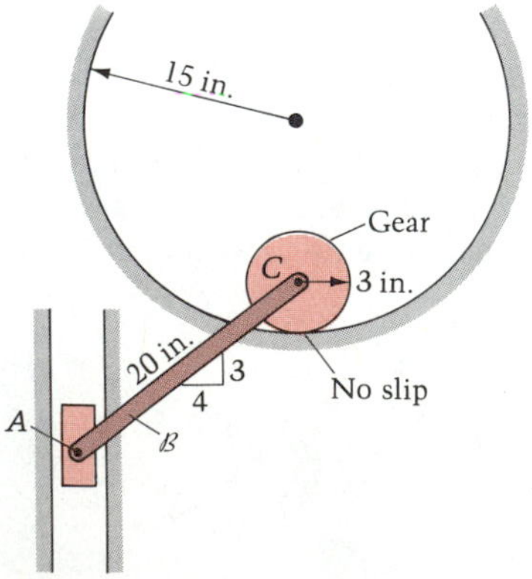

Figure P3.120

3.120 Point A of the slider block has, at the instant shown in Figure P3.120, $\mathbf{v}_A = 12 \uparrow$ in./sec and $\mathbf{a}_A = 6 \downarrow$ in./sec^2. Find the angular acceleration of bar $\mathcal{B}$.

3.121 Two 5-in.-radius wheels roll on a plane surface. (See Figure P3.121.) A 13-in. bar $\mathcal{B}$ is pinned to the wheels at A and B as shown. If C has a constant velocity of 20 ft/sec to the right, find, for the position shown, the acceleration of A.

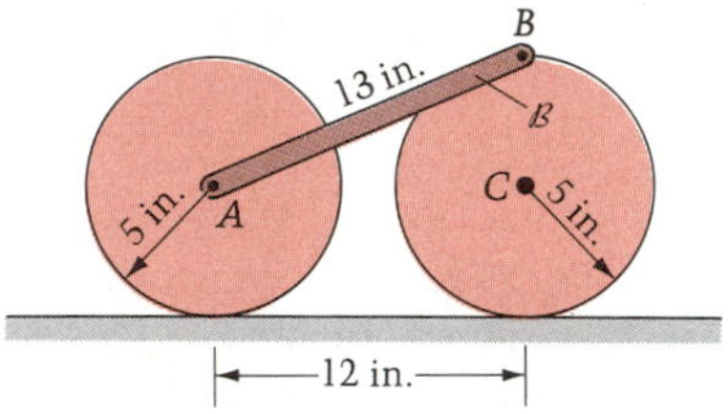

Figure P3.121

3.122 See Figure P3.122. The velocity of the pin in block $\mathcal{B}$ is $0.2 \downarrow$ m/s, and its acceleration is $0.5 \uparrow$ m/s^2 in the given position. Find at this instant the angular velocity and angular acceleration of the cylinder if there is sufficient friction to prevent it from slipping.

3.123 A wheel rolls along a curved surface. In the position shown in Figure P3.123, its angular velocity and angular acceleration are $\boldsymbol{\omega} = 3$ ↺ rad/s and $\boldsymbol{\alpha} = 5$ ↺ rad/s^2. Determine at this instant the angular acceleration of bar $\mathcal{B}$ and the acceleration of pin B of the slider block.

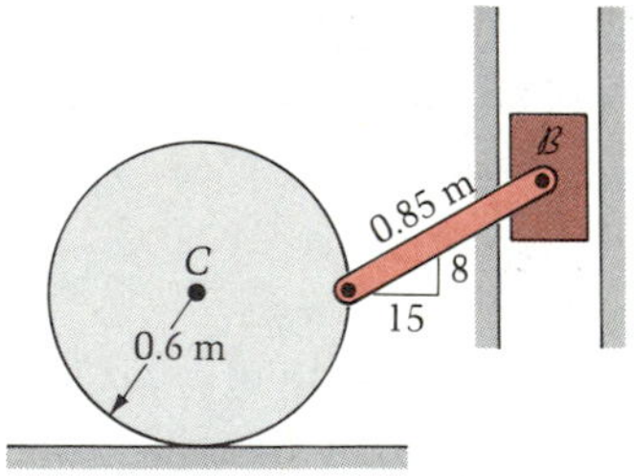

Figure P3.122

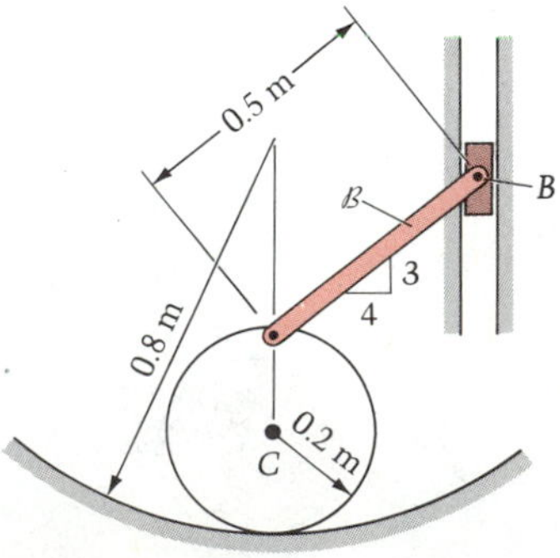

Figure P3.123

3.124 In Example 3.14 find the piston acceleration when $\theta = 0°$, $180°$, and $270°$.

3.125 The center point C of gear $\mathcal{B}_1$ in Figure P3.125 moves in a horizontal plane at constant speed v_0. The ring gear $\mathcal{I}$ is fixed in the reference frame, and the constant angular velocity of $\mathcal{B}_1$ is clockwise. Find the acceleration of point Q of $\mathcal{B}_1$.

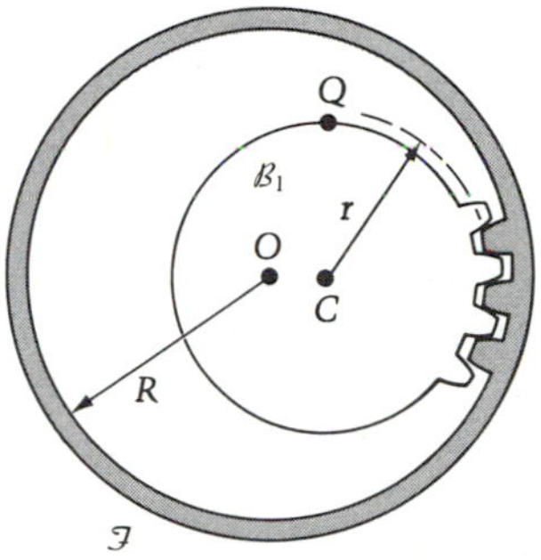

Figure P3.125

3.126 If the ball in a ball bearing assembly (Figure P3.126) neither slips on the shaft nor on the fixed housing, find the velocity and acceleration of the center of the ball in terms of the angular velocity and angular acceleration of the shaft (ω and $\dot{\omega}$).

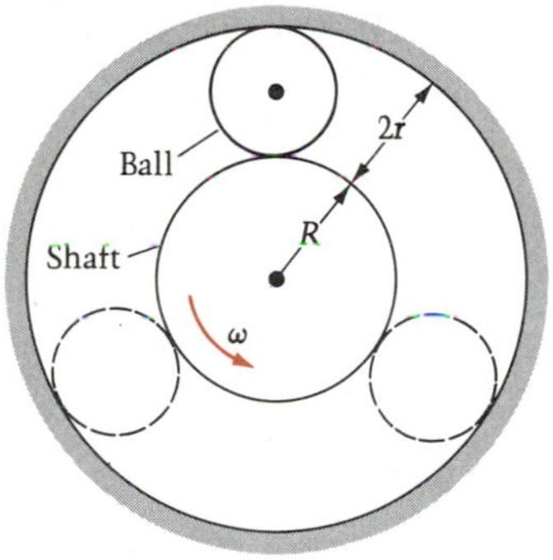

Figure P3.126

3.127 The ball in Figure P3.127 rolls on the fixed surface and at the instant shown has angular velocity $\boldsymbol{\omega} = 3\hat{\mathbf{k}}$ rad/s and angular acceleration $\boldsymbol{\alpha} = -2\hat{\mathbf{k}}$ rad/s². At this instant find:

a. The velocities of A and B

b. The accelerations of A and B

c. $(d/dt)|\mathbf{v}_B|$

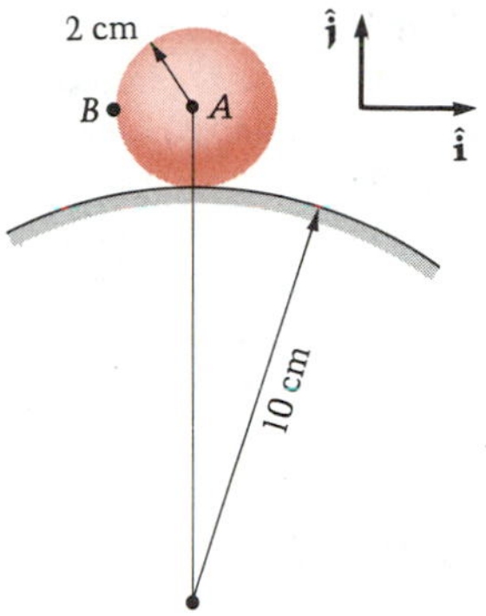

Figure P3.127

3.128 The angular velocity of crank $\mathcal{B}_1$ in Figure P3.128 is a constant 3 ↺ rad/s. In the given position, find the velocity of the center C of wheel $\mathcal{B}_2$ and also determine the angular acceleration of $\mathcal{B}_2$, which rolls on the circular track.

3.129 Cylinders $\mathcal{B}_1$ and $\mathcal{B}_2$ in Figure P3.129 have a radius of 10 in. each and roll on the respective planes. Bar $\mathcal{B}_3$ has length 48 in. and is pinned to the centers of the cylinders. The center G of $\mathcal{B}_1$ has velocity $\mathbf{v}_G = -10t \rightarrow$ in./sec. If the time at the instant shown is $t = 5$ sec, find $\boldsymbol{\alpha}_2$ and $\boldsymbol{\alpha}_3$ at that instant.

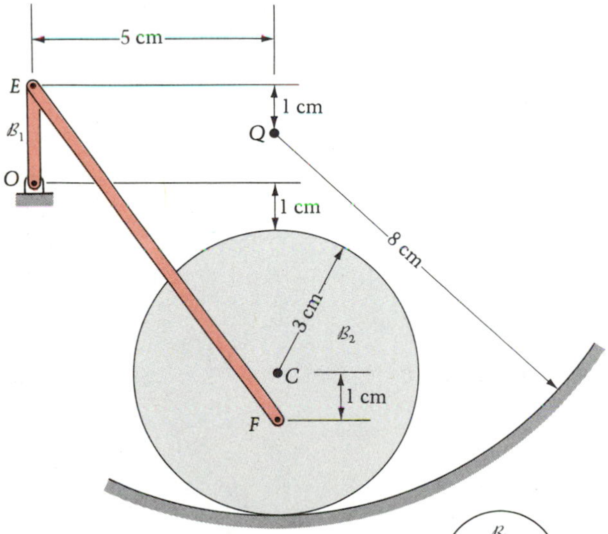

Figure P3.128

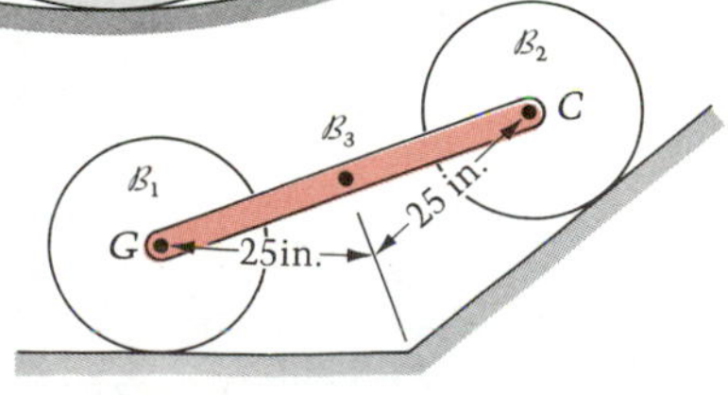

Figure P3.129

3.130 Bar $\mathcal{B}$ in Figure P3.130 is 25 cm long and is pinned to the rolling cylinder at B. The other end of $\mathcal{B}$ is pinned to the roller at A as shown. The center of the cylinder has $v_C = 11.2$ cm/s and $a_C = 16.8$ cm/s^2 down the plane at the given instant; at this time line BⒾ is vertical and AⒾ is horizontal, and BC is parallel to the plane beneath it. Find the acceleration of point A and the angular acceleration of body $\mathcal{B}$ at the given instant.

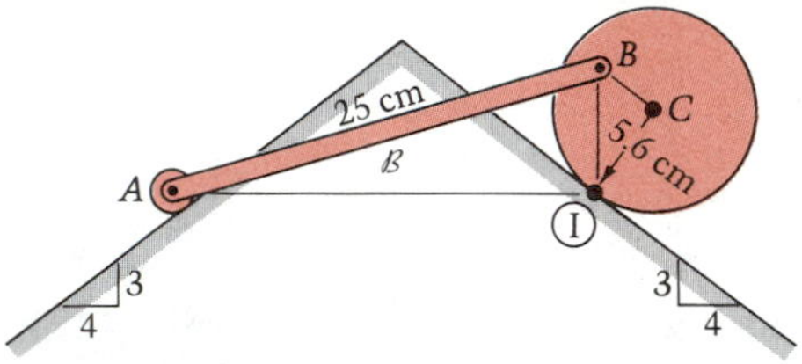

Figure P3.130

3.131 Figure P3.131 shows a 10-ft-radius disk that rolls on a plane surface. It has, at the instant shown, an angular velocity of 2 rad/sec and an angular acceleration of 3 rad/sec^2, both counterclockwise. Find a point on the disk or the disk extended that has zero acceleration at this instant.

3.132 Find the acceleration of point B, the pin connecting rod $\mathcal{B}$ to the block in Figure P3.132, at $t = 1$ sec. The rod is horizontal at $t = 0$, and the velocity of C is $\mathbf{v}_C = 2t^2 \rightarrow$ ft/sec.

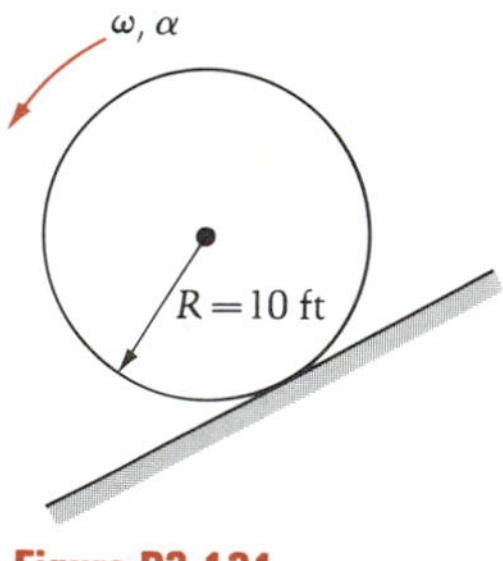

Figure P3.131

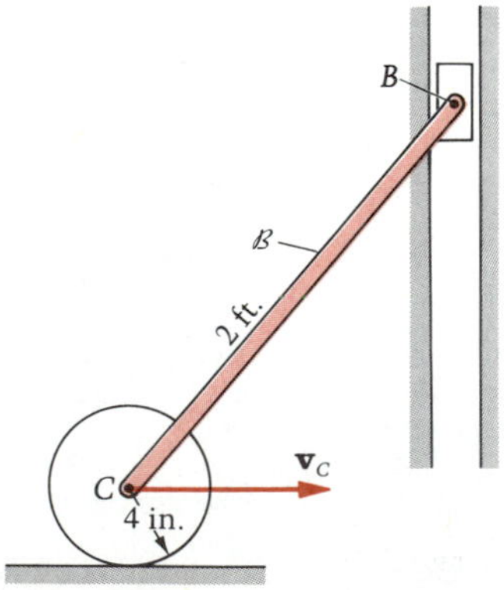

Figure P3.132

3.133 Referring to Problem 3.121, if at the instant shown $\mathbf{v}_C = 20 \rightarrow$ in./sec and $\mathbf{a}_C = 5 \leftarrow$ in./sec^2, find $\mathbf{a}_A$ at this time.

* **3.134** Gears $\mathcal{B}_1$ and $\mathcal{B}_2$ in Figure P3.134 have 25 and 50 teeth, respectively. Rod $\mathcal{B}_3$ is 2 ft long, and the radius of $\mathcal{B}_2$ is 1 ft. Determine the acceleration of point A when $t = 0$ if $x_B = 0.2 \sin \pi t$ ft (positive to the left) with $\theta = 90°$ at $t = 0$.

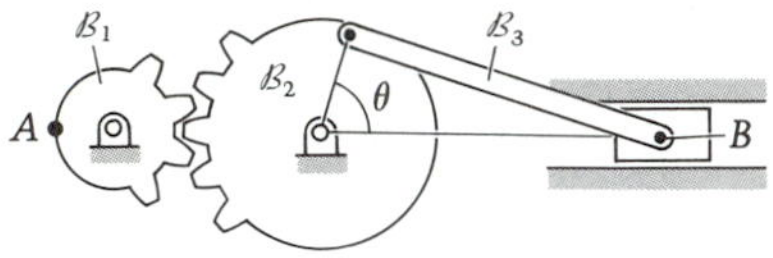

Figure P3.134

* **3.135** At the instant shown in Figure P3.135, bar $\mathcal{B}_1$ has $\omega_1 = \pi$ ↺ rad/sec and $\alpha_1 = (\pi/3)$ ↻ rad/sec^2; and the gear $\mathcal{B}_2$ has $\omega_2 = 2\pi$ rad/sec ↻ and $\alpha_2 = \pi/2$ ↺ rad/sec^2. At this instant, determine the accelerations of each of the two gear tooth contacting points.

* **3.136** The outer gear $\mathcal{I}$ in Figure P3.136 is stationary. Crank $\mathcal{B}_1$ turns at the constant angular velocity of 10 rad/sec counterclockwise and is pinned at its ends to the centers of the sun gear $\mathcal{B}_2$ (at S) and the planetary gear $\mathcal{B}_3$ (at P). Find the accelerations of the points of $\mathcal{B}_2$ and $\mathcal{B}_3$ that are in contact with each other, if the radii of $\mathcal{B}_2$ and $\mathcal{B}_3$ are, respectively, 3 in. and 10 in.

* **3.137** Point O is pinned to the reference frame $\mathcal{I}$. (See Figure P3.137.) The pitch radii of gears $\mathcal{B}_1$ and $\mathcal{B}_2$ are each 0.2 m. The angular velocities of $\mathcal{B}_3$ and $\mathcal{B}_4$ are 2 rad/s, clockwise for $\mathcal{B}_3$ and counterclockwise for $\mathcal{B}_4$, and both constant. Find the maximum acceleration magnitude experienced by any point of $\mathcal{B}_1$.

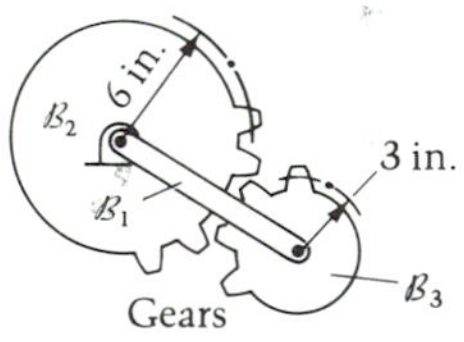

Figure P3.135

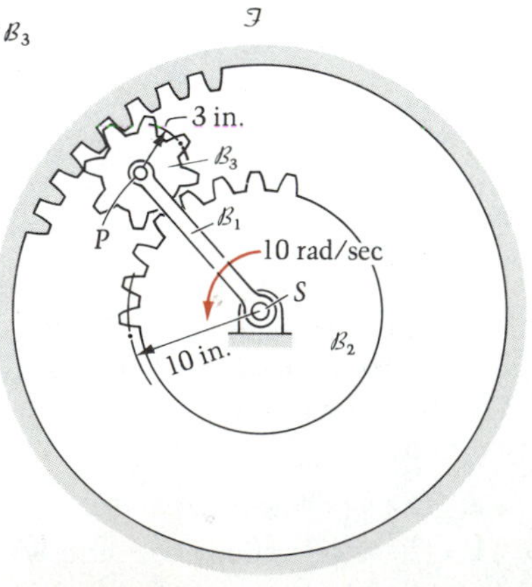

Figure P3.136

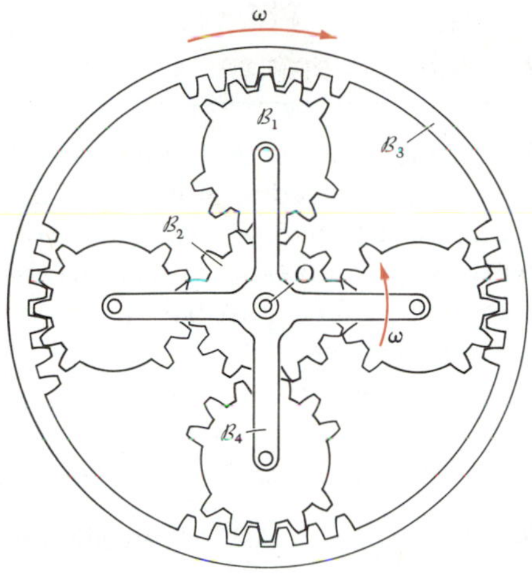

Figure P3.137

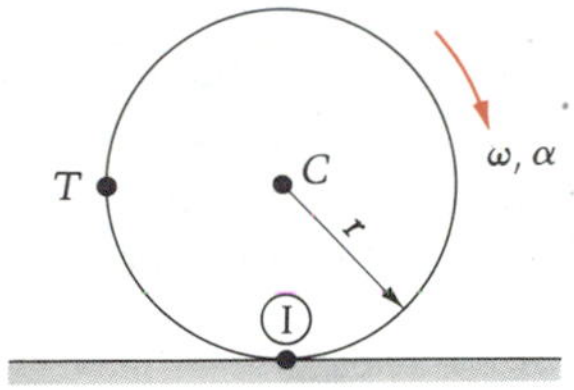

Figure P3.138

* **3.138** The wheel in Figure P3.138 rolls on the plane. Find the radius of curvature and the center of curvature of the path of point T at the given time in terms of r.

* **3.139** A *cycloid* is the curve traced out by a point on the rim of a rolling wheel. The equations for the rectangular coordinates of a point on the cycloid, in terms of the parameter φ (the angle shown in Figure P3.139), are

$$x = a(\varphi - \sin\varphi)$$
$$y = a(1 - \cos\varphi)$$

where a is the wheel's radius. Recall from calculus that the curvature of a plane curve is

$$\frac{1}{\rho} = \left| \frac{y''}{(1 + y'^2)^{3/2}} \right|$$

where ρ is the radius of curvature.

a. Use the chain rule

$$y' = \frac{dy}{dx} = \frac{dy}{d\varphi}\frac{d\varphi}{dx}$$

and show that, for the cycloid,

$$\rho = -2^{3/2}\, a\sqrt{1 - \cos\varphi}$$

b. Explain what the minus sign in the expression for ρ means.

c. Observe that P is at its highest point when $\varphi = \pi$ and that $|\rho| = 4a$ there. In this configuration, show that the following two expressions for the acceleration of P agree:

$$\mathbf{a}_P = \underbrace{a\alpha\hat{\mathbf{i}}}_{\mathbf{a}_C} + \alpha\hat{\mathbf{k}} \times \mathbf{r}_{CP} - \omega^2\mathbf{r}_{CP}$$

$$\mathbf{a}_P = \ddot{s}\hat{\mathbf{e}}_t + \frac{\dot{s}^2}{\rho}\hat{\mathbf{e}}_n$$

* **3.140** Show that for a rigid body $\mathcal{B}$ in the plane motion, as long as $\alpha \neq 0$ there is a circle of points P of $\mathcal{B}$ whose accelerations pass through any point C of $\mathcal{B}$. *Hint*: Write $\mathbf{a}_P = \mathbf{a}_C + \alpha\hat{\mathbf{k}} \times \mathbf{r}_{CP} - \omega^2\mathbf{r}_{CP}$ and dot both sides with the vector $\hat{\mathbf{k}} \times \mathbf{r}_{CP}$. Assume that $\mathbf{a}_P \parallel \mathbf{r}_{CP}$ and see if you can exhibit $\mathbf{r}_{CP}$. For the rolling wheel, show that the points are as shown in Figure P3.140.

* **3.141** Show that for the rolling uniform cylinder in Figure P3.141 there is a point of zero acceleration at the indicated position J if $\boldsymbol{\omega}$ and $\boldsymbol{\alpha}$ are in the given directions and are not both zero. You should find that the coordinates of J are $(x_J, y_J) = [r\alpha\omega^2/(\omega^4 + \alpha^2), r\alpha^2/(\omega^4 + \alpha^2)]$.

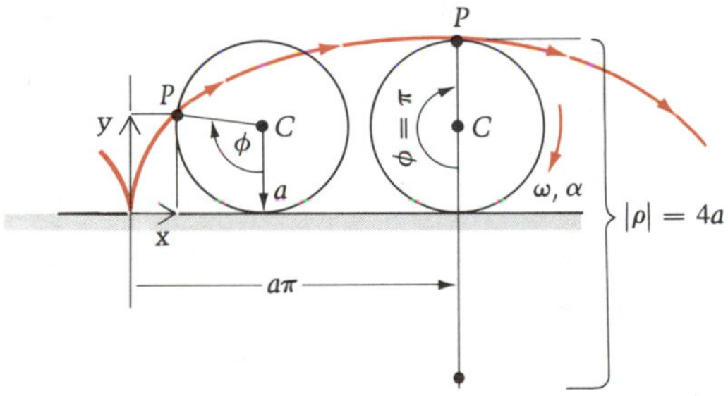

Figure P3.139

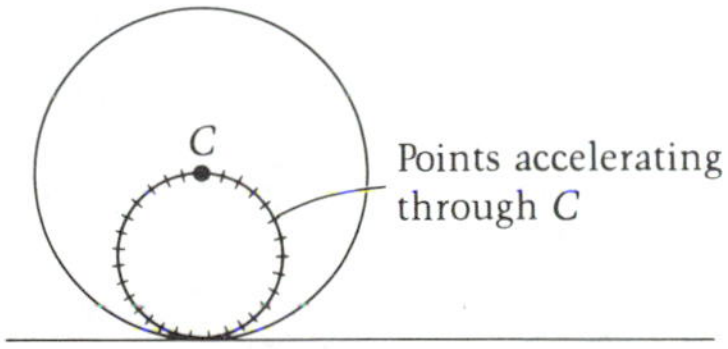

Figure P3.140

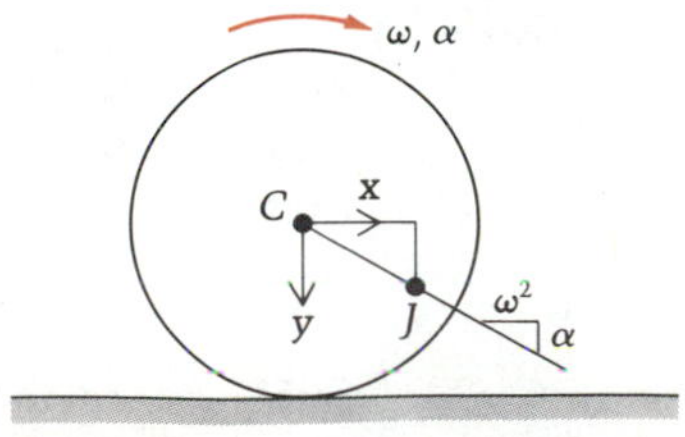

Figure P3.141

3.7 Relationship Between the Velocities of a Point with Respect to Two Different Frames of Reference

While Equation (3.8) gives us the relationship between the velocities in $\mathcal{I}$ of two points of the *same* rigid body, we often need another equation relating the velocities of the *same point* relative to two *different* frames or bodies. This relationship (together with a companion equation for accelerations to be developed in the next section) will be essential in solving kinematics problems involving bodies moving in special ways relative to others (such as a pin of one body sliding in a slot of another).

Relationship Between the Derivatives of a Vector in Two Frames

To develop this equation, we must first find the relationship between the derivatives of an arbitrary vector **A** in two frames $\mathcal{I}$ and $\mathcal{B}$. To do this, we begin by embedding axes X and Y in $\mathcal{I}$, and x and y in $\mathcal{B}$, as suggested by the hatch marks in Figure 3.28. Further, we let $(\hat{\mathbf{I}}, \hat{\mathbf{J}})$ and $(\hat{\mathbf{i}}, \hat{\mathbf{j}})$ be pairs of unit vectors, always respectively parallel to (X, Y) and to (x, y).

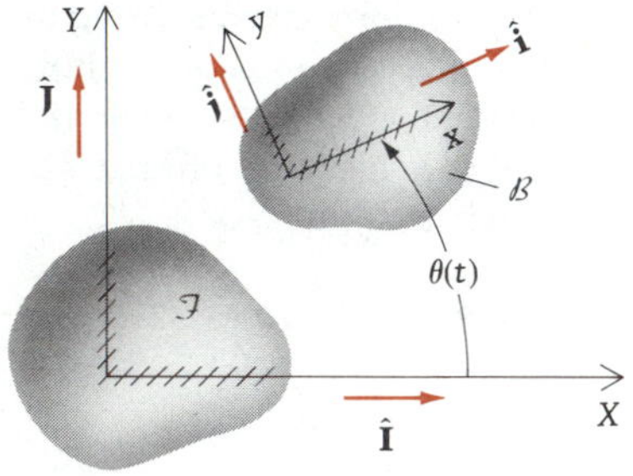

Figure 3.28

We note that if $\theta(t)$ again locates the x axis relative to X as shown, then

$$\hat{\mathbf{i}} = \cos\theta\hat{\mathbf{I}} + \sin\theta\hat{\mathbf{J}} \tag{3.39}$$

and

$$\hat{\mathbf{j}} = -\sin\theta\hat{\mathbf{I}} + \cos\theta\hat{\mathbf{J}} \tag{3.40}$$

Differentiating in frame $\mathcal{I}$ and noting that $\hat{\mathbf{I}}$ and $\hat{\mathbf{J}}$ are constants there, we obtain

$$\frac{d\hat{\mathbf{i}}}{dt} = \dot{\hat{\mathbf{i}}} = (-\sin\theta\hat{\mathbf{I}} + \cos\theta\hat{\mathbf{J}})\dot{\theta} = \dot{\theta}\hat{\mathbf{j}} \tag{3.41}$$

$$\frac{d\hat{\mathbf{j}}}{dt} = \dot{\hat{\mathbf{j}}} = -(\cos\theta\hat{\mathbf{I}} + \sin\theta\hat{\mathbf{J}})\dot{\theta} = -\dot{\theta}\hat{\mathbf{i}} \tag{3.42}$$

Now let the arbitrary vector **A** be written in frame $\mathcal{B}$ (meaning that **A** is expressed in terms of its components there—i.e., in terms of unit vectors fixed in direction in $\mathcal{B}$):

$$\mathbf{A} = A_x\hat{\mathbf{i}} + A_y\hat{\mathbf{j}}$$

Differentiating this vector in $\mathcal{I}$, we get

$${}^{\mathcal{I}}\dot{\mathbf{A}} = \dot{A}_x\hat{\mathbf{i}} + \dot{A}_y\hat{\mathbf{j}} + A_x{}^{\mathcal{I}}\dot{\hat{\mathbf{i}}} + A_y{}^{\mathcal{I}}\dot{\hat{\mathbf{j}}} \tag{3.43}$$

We now note that the first two terms on the right side of Equation (3.43) add up to the derivative of vector **A** in $\mathcal{B}$, because $\hat{\mathbf{i}}$ and $\hat{\mathbf{j}}$ do not change in magnitude *or* direction with time there. Thus

$${}^{\mathcal{I}}\dot{\mathbf{A}} = {}^{\mathcal{B}}\dot{\mathbf{A}} + A_x{}^{\mathcal{I}}\dot{\hat{\mathbf{i}}} + A_y{}^{\mathcal{I}}\dot{\hat{\mathbf{j}}}$$

Substituting the derivatives of $\hat{\mathbf{i}}$ and $\hat{\mathbf{j}}$ in $\mathcal{I}$ from Equations (3.41) and (3.42) yields

$$ {}^{\mathcal{I}}\dot{\mathbf{A}} = {}^{\mathcal{B}}\dot{\mathbf{A}} + \dot{\theta}(A_x\hat{\mathbf{j}} - A_y\hat{\mathbf{i}}) $$
$$ = {}^{\mathcal{B}}\dot{\mathbf{A}} + \dot{\theta}\hat{\mathbf{k}} \times (A_x\hat{\mathbf{i}} + A_y\hat{\mathbf{j}}) $$

or

$$ {}^{\mathcal{I}}\dot{\mathbf{A}} = {}^{\mathcal{B}}\dot{\mathbf{A}} + \dot{\theta}\hat{\mathbf{k}} \times \mathbf{A} \tag{3.44} $$

The angular velocity $\dot{\theta}\hat{\mathbf{k}}$ has reappeared, and Equation (3.44) shows us that this vector has a more general purpose than merely relating velocities in kinematics. It is in fact the link that allows us to relate the derivatives of any vector in two different frames. (This same result is in fact true in general three-dimensional motion, with three-dimensional vectors and a more general expression for angular velocity substituted, as will be seen in Chapter 6.)

Velocity Relationship in Two Frames

We now use Equation (3.44) to relate the velocities of a point P in two frames $\mathcal{B}$ and $\mathcal{I}$. From Figure 3.29, the position vectors of P in these two frames are related by

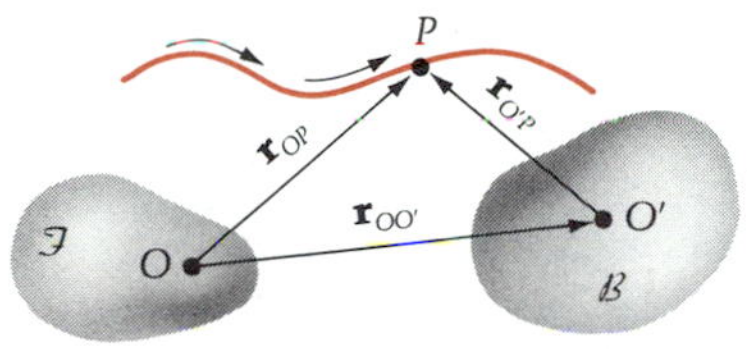

Figure 3.29

$$ \mathbf{r}_{OP} = \mathbf{r}_{OO'} + \mathbf{r}_{O'P} \tag{3.45} $$

Differentiating this equation in $\mathcal{I}$, we have

$$ {}^{\mathcal{I}}\dot{\mathbf{r}}_{OP} = {}^{\mathcal{I}}\dot{\mathbf{r}}_{OO'} + {}^{\mathcal{I}}\dot{\mathbf{r}}_{O'P} \tag{3.46} $$

The first two vectors in Equation (3.46) are the velocities of P and O' in $\mathcal{I}$ by definition:

$$ \mathbf{v}_{P/\mathcal{I}} = \mathbf{v}_{O'/\mathcal{I}} + {}^{\mathcal{I}}\dot{\mathbf{r}}_{O'P} \tag{3.47} $$

Question 3.17 (a) Why is the last vector in Equation (3.47) not the velocity of P in $\mathcal{I}$? (b) Why is it not the velocity of P in $\mathcal{B}$?

To replace ${}^{\mathcal{I}}\dot{\mathbf{r}}_{O'P}$ by a vector that we can operate with, we use Equation (3.44), with $\mathbf{r}_{O'P}$ becoming the vector $\mathbf{A}$:

$$ {}^{\mathcal{I}}\dot{\mathbf{r}}_{O'P} = {}^{\mathcal{B}}\dot{\mathbf{r}}_{O'P} + \dot{\theta}\hat{\mathbf{k}} \times \mathbf{r}_{O'P} \tag{3.48} $$

Therefore, recognizing that ${}^{\mathcal{B}}\dot{\mathbf{r}}_{O'P}$ is $\mathbf{v}_{P/\mathcal{B}}$ and substituting Equation (3.48) into (3.47), we obtain

$$ \mathbf{v}_{P/\mathcal{I}} = \mathbf{v}_{P/\mathcal{B}} + \mathbf{v}_{O'/\mathcal{I}} + \dot{\theta}\hat{\mathbf{k}} \times \mathbf{r}_{O'P} \tag{3.49} $$

Another way of expressing Equation (3.49) is to think of frame $\mathcal{B}$ as a "moving frame" with respect to a "fixed" reference frame $\mathcal{I}$. Then the velocities of P can be written as simply $\mathbf{v}_P$ when the reference is $\mathcal{I}$ (thus $\mathbf{v}_P = \mathbf{v}_{P/\mathcal{I}}$) and as $\mathbf{v}_{\text{rel}}$ when the reference is the "moving frame" $\mathcal{B}$ (thus

Answer 3.17 (a) It is not $\mathbf{v}_{P/\mathcal{I}}$ because the origin of the position vector is not fixed in $\mathcal{I}$. (b) And it is not $\mathbf{v}_{P/\mathcal{B}}$ because the derivative is not taken in $\mathcal{B}$.

$\mathbf{v}_{\text{rel}} = \mathbf{v}_{P/\mathcal{B}}$). Hence we can write Equation (3.49) in abbreviated notation as

$$\mathbf{v}_P = \mathbf{v}_{\text{rel}} + \mathbf{v}_{O'} + \boldsymbol{\omega} \times \mathbf{r} \tag{3.50}$$

where $\mathbf{r} = \mathbf{r}_{O'P}$, the position of P in the moving frame, and $\boldsymbol{\omega}$ is the angular velocity of $\mathcal{B}$ relative to frame $\mathcal{I}$. The reader may find this form of Equation (3.49) easier to use when there is just one "moving frame."

Now let us denote by P' the point of $\mathcal{B}$ (or $\mathcal{B}$ extended) that is coincident with P. Then $\mathbf{r}_{O'P} = \mathbf{r}_{O'P'}$ and the last two terms of Equation (3.49) or (3.50) are seen (by Equation 3.8) to be the velocity of P' in $\mathcal{I}$:

$$\mathbf{v}_{P/\mathcal{I}} = \mathbf{v}_{P/\mathcal{B}} + \mathbf{v}_{P'/\mathcal{I}} \tag{3.51}$$

In words, Equation (3.51) says that at any time, the velocity of P in $\mathcal{I}$ is the sum of the velocity of P in $\mathcal{B}$ plus the velocity in $\mathcal{I}$ of the point of $\mathcal{B}$ coincident with P.*

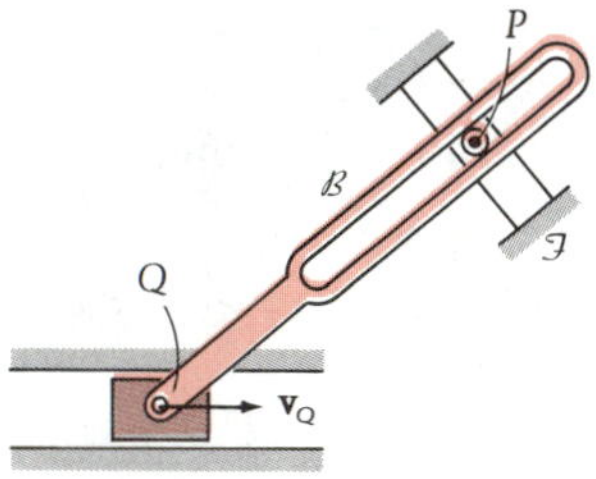

Figure 3.30

As a preliminary example, consider Figure 3.30, in which pin Q is moving to the right. Let P be the center of the other pin, which is attached to frame $\mathcal{I}$. Then we may write

$$\mathbf{v}_{P/\mathcal{I}} = \mathbf{0} = \mathbf{v}_{P/\mathcal{B}} + \mathbf{v}_{P'/\mathcal{I}} \Rightarrow \mathbf{v}_{P'/\mathcal{I}} = -\mathbf{v}_{P/\mathcal{B}}$$

Now the center of the pin's motion in $\mathcal{B}$ is necessarily along a straight line within the slot of $\mathcal{B}$. Therefore the velocity of P' (the point of $\mathcal{B}$ extended coincident with P) is seen to be *also* parallel to the slot and in a direction (↗) opposite to that of $\mathbf{v}_{P/\mathcal{B}}$(↙). We now consider several detailed examples of the use of Equation (3.49).

EXAMPLE 3.23

A yellowjacket walks radially outward at a constant 2 in./sec in a straight line relative to a record turning at $33\frac{1}{3}$ rpm. (See Figure E3.23.) Find the velocity of the yellowjacket in frame $\mathcal{I}$, which is the cabinet on which the stereo rests.

Solution

We shall treat the yellowjacket as a point Y, and we note that the unit vectors in Figure E3.23 are fixed in our "moving frame" $\mathcal{B}$. Also, the "moving" (O') and "fixed" (O) origins are coincident. Then we have, using Equation (3.49):

$$\mathbf{v}_{Y/\mathcal{I}} = \mathbf{v}_{Y/\mathcal{B}} + \mathbf{v}_{O'/\mathcal{I}} + \dot{\theta}\hat{\mathbf{k}} \times \mathbf{r}_{O'Y}$$

$$= 2\hat{\mathbf{i}} + \mathbf{0} + \left(33\frac{1}{3}\right)\left(\frac{2\pi}{60}\right)\hat{\mathbf{k}} \times r\hat{\mathbf{i}}$$

$$= 2\hat{\mathbf{i}} + \frac{10\pi}{9} r\hat{\mathbf{j}} \text{ in./sec}$$

Note how the second term grows linearly with the radius.

Figure E3.23

* This latter term is sometimes called the *vehicle velocity* of P.

In the next example, there is more than one "moving frame." To avoid three levels of subscripts, we shall name the bodies $\mathcal{B}$, $\mathcal{R}$, and $\mathcal{C}$ rather than the usual $\mathcal{B}_1$, $\mathcal{B}_2$, $\mathcal{B}_3$. Thus $\boldsymbol{\omega}_1$ becomes $\boldsymbol{\omega}_{\mathcal{B}}$, and so on.

Example 3.24

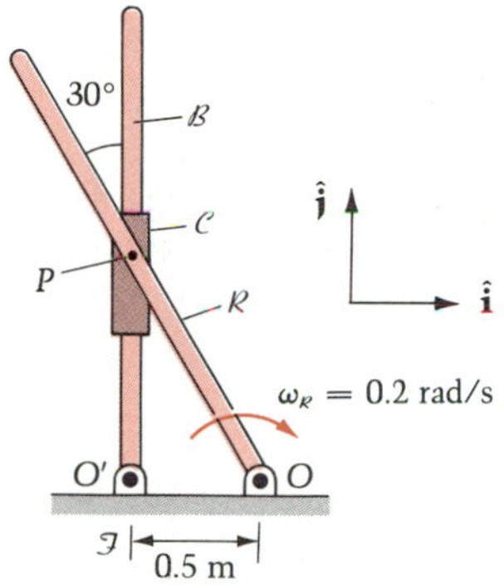

Figure E3.24

Collar $\mathcal{C}$ in Figure E3.24 is pinned to rod $\mathcal{R}$ at P and is free to slide along rod $\mathcal{B}$. The angular velocity of $\mathcal{R}$ is $0.2 \curvearrowright$ rad/s at the instant shown. Find the angular velocity of $\mathcal{B}$ at this time, and determine the velocity of P relative to $\mathcal{B}$.

Solution

We relate the velocities of P in $\mathcal{I}$ and in $\mathcal{B}$:

$$\mathbf{v}_{P/\mathcal{I}} = \mathbf{v}_{P/\mathcal{B}} + \mathbf{v}_{O'/\mathcal{I}} + \boldsymbol{\omega}_{\mathcal{B}} \times \mathbf{r}_{O'P}$$

$$\overbrace{\underbrace{\mathbf{v}_{O/\mathcal{I}}}_{\text{zero}} + \boldsymbol{\omega}_{\mathcal{R}} \times \mathbf{r}_{OP}}^{\text{on } \mathcal{R}} = v_{P/\mathcal{B}}\hat{\mathbf{j}} + (\underbrace{\mathbf{v}_{O'/\mathcal{I}}}_{\text{zero}} + \boldsymbol{\omega}_{\mathcal{B}} \times \mathbf{r}_{O'P})$$

$$-0.2\hat{\mathbf{k}} \times (-0.5\hat{\mathbf{i}} + 0.866\hat{\mathbf{j}}) = v_{P/\mathcal{B}}\hat{\mathbf{j}} + \omega_{\mathcal{B}}\hat{\mathbf{k}} \times 0.866\hat{\mathbf{j}}$$

$\hat{\mathbf{i}}$ coefficients: $\quad 0.173 = -0.866\omega_{\mathcal{B}}$

$$\omega_{\mathcal{B}} = -0.2 \Rightarrow \boldsymbol{\omega}_{\mathcal{B}} = -0.2\hat{\mathbf{k}} \text{ rad/s} \quad \text{or} \quad 0.2 \curvearrowright \text{rad/s}$$

$\hat{\mathbf{j}}$ coefficients: $\quad 0.1 = v_{P/\mathcal{B}}$

$$\mathbf{v}_{P/\mathcal{B}} = 0.1\hat{\mathbf{j}} \text{ m/s}$$

Thus the pin is moving outward on $\mathcal{B}$, which is turning clockwise.

Question 3.18 Will $\mathcal{B}$ *always* have the same $\boldsymbol{\omega}$ as does $\mathcal{R}$?

Note that in the preceding example, we did not need to know the angular velocity of $\mathcal{C}$. Nonetheless, it is important for the reader to realize that $\boldsymbol{\omega}_{\mathcal{C}} \equiv \boldsymbol{\omega}_{\mathcal{B}}$. The reason is that $\mathcal{B}$ and $\mathcal{C}$ can only *translate* relative to each other; thus lines fixed in each will turn at the same time rates in $\mathcal{I}$. This observation will sometimes be needed (as in Problems 3.147 and 3.149).

Example 3.25

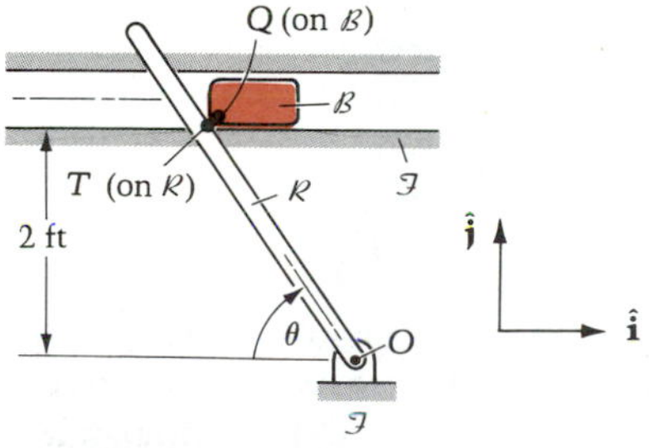

Figure E3.25a

Block $\mathcal{B}$ translates in a horizontal slot (see Figure E3.25a) and is pushed along by a bar $\mathcal{R}$ that turns at angular velocity $\omega_{\mathcal{R}} = 10 \curvearrowright$ rad/sec about the pin at point O. Find $\mathbf{v}_Q$, the velocity of the contact point of $\mathcal{B}$, when $\theta = 60°$.

Answer 3.18 Definitely not! This happened because of the geometry at the given instant.

Solution

Let the ground be the reference frame $\mathcal{I}$, and note that T is the point of $\mathcal{R}$ coincident with Q at the given instant (the point we have been calling P' in the theory). Using Equation (3.51), we obtain

$$\mathbf{v}_{Q/\mathcal{I}} = \mathbf{v}_{Q/\mathcal{R}} + \mathbf{v}_{T/\mathcal{I}}$$

Therefore

$$\mathbf{v}_{Q/\mathcal{I}} = v_{Q/\mathcal{R}}\left(\frac{\hat{\mathbf{i}}}{2} - \frac{\sqrt{3}}{2}\hat{\mathbf{j}}\right) + (-10\hat{\mathbf{k}}) \times \left(\frac{-2}{\sqrt{3}}\hat{\mathbf{i}} + 2\hat{\mathbf{j}}\right) \tag{1}$$

Note that the velocity of Q relative to bar $\mathcal{R}$ has an unknown magnitude $v_{Q/\mathcal{R}}$ but a *known* direction (along $\mathcal{R}$). Now we also know the direction of $\mathbf{v}_{Q/\mathcal{I}}$, so that

$$\mathbf{v}_{Q/\mathcal{I}} = v_{Q/\mathcal{I}}\hat{\mathbf{i}} = \frac{20}{\sqrt{3}}\hat{\mathbf{j}} + 20\hat{\mathbf{i}} + \frac{v_{Q/\mathcal{R}}}{2}\hat{\mathbf{i}} - v_{Q/\mathcal{R}}\frac{\sqrt{3}}{2}\hat{\mathbf{j}} \tag{2}$$

Equating the x components of Equation (2), we get

$$v_{Q/\mathcal{I}} = 20 + \frac{v_{Q/\mathcal{R}}}{2} \tag{3}$$

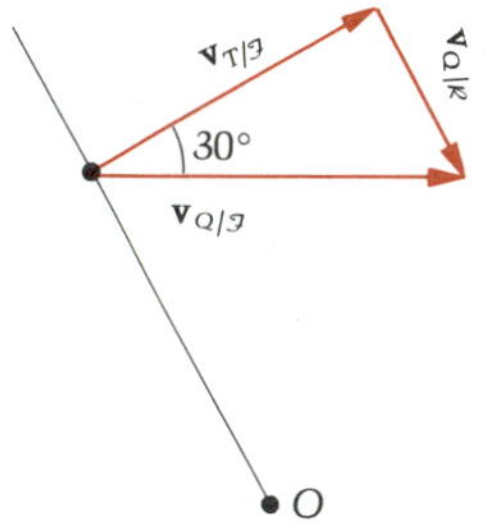

Figure E3.25b

And equating the y components:

$$0 = \frac{20}{\sqrt{3}} - v_{Q/\mathcal{R}}\frac{\sqrt{3}}{2} \tag{4}$$

Equation (4) gives $v_{Q/\mathcal{R}} = 13.3$ ft/sec, from which Equation (3) then yields

$$v_{Q/\mathcal{I}} = v_{Q/\mathcal{I}}\hat{\mathbf{i}} = 26.7\hat{\mathbf{i}} \text{ ft/sec}$$

The correct triangle relating the velocities of Q and T is shown in Figure E3.25b. As a check, $v_Q \cos 30° = (26.7)\sqrt{3}/2 = 23.1$ and $v_r = \sqrt{(20/\sqrt{3})^2 + 20^2} = 23.1$ ft/sec.

Question 3.19 Can the preceding example also be worked by using $\mathbf{v}_{T/\mathcal{I}} = \mathbf{v}_{T/\mathcal{B}} + \mathbf{v}_{Q/\mathcal{I}}$?

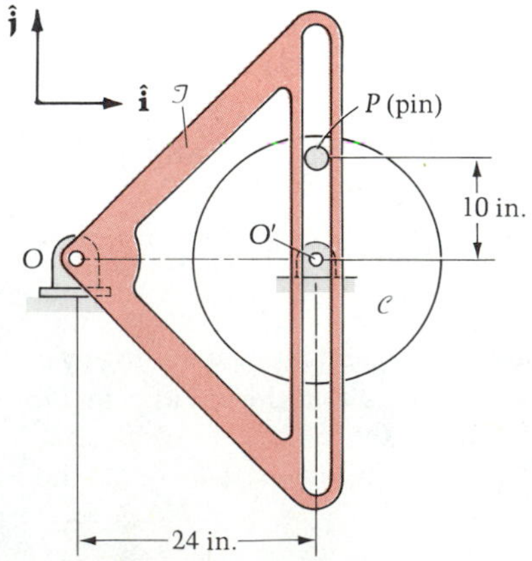

Figure E3.26a

Answer 3.19 Yes, provided we recognize that the direction of $\mathbf{v}_{T/\mathcal{B}}$ is along the axis of the rod (see Problem 3.146). That is, $\mathbf{v}_{T/\mathcal{B}}$ must be tangent to the surface at which $\mathcal{R}$ and $\mathcal{B}$ touch. It is important to realize, however, that while the path of Q in $\mathcal{R}$ is a straight line, the path of T in $\mathcal{B}$ is not. Thus $\mathbf{a}_{T/\mathcal{B}}$ would *not* be in the direction of the axis of $\mathcal{R}$.

EXAMPLE 3.26

Disk $\mathcal{C}$ of Figure E3.26a, with its attached pin P, has limited angular motion. After a 45° clockwise rotation from the original position (see Figure E3.26b), disk $\mathcal{C}$ has angular velocity $\boldsymbol{\omega}_{\mathcal{C}} = 2 \circlearrowright$ rad/sec. At this time, find $\boldsymbol{\omega}_{\mathcal{I}}$ of the slotted triangular body $\mathcal{I}$ and determine the velocity of pin P relative to $\mathcal{I}$.

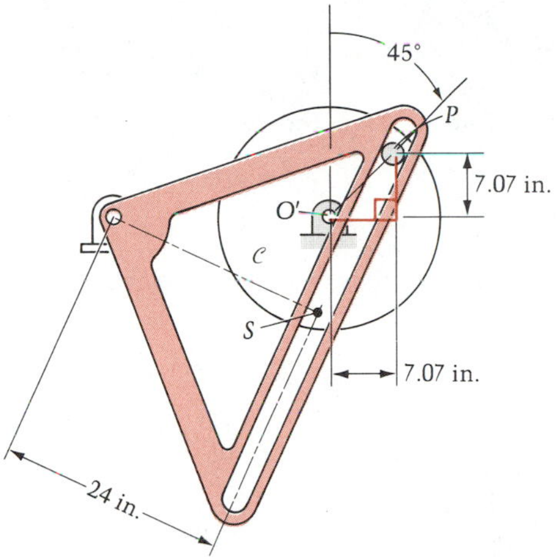

Figure E3.26b

Solution

Calling the ground frame $\mathcal{I}$, we relate the velocities of pin P in $\mathcal{I}$ and in $\mathcal{I}$:

$$\mathbf{v}_{P/\mathcal{I}} = \mathbf{v}_{P/\mathcal{I}} + \mathbf{v}_{P'/\mathcal{I}} \tag{1}$$

where P' is the point of $\mathcal{I}$ coincident with P.

We may find $\mathbf{v}_{P/\mathcal{I}}$ by relating it on body $\mathcal{C}$ to $\mathbf{v}_{O'/\mathcal{I}}$ (which vanishes):

$$\mathbf{v}_{P/\mathcal{I}} = \omega_C \hat{\mathbf{k}} \times \mathbf{r}_{O'P} = -2\hat{\mathbf{k}} \times (7.07\hat{\mathbf{i}} + 7.07\hat{\mathbf{j}})$$
$$= 1.41\hat{\mathbf{i}} - 1.41\hat{\mathbf{j}} \text{ in./sec}$$

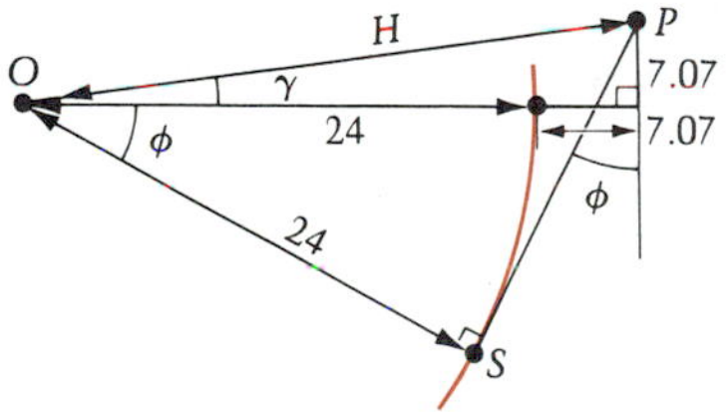

Figure E3.26c

To find $\mathbf{v}_{P/\mathcal{I}}$, we need the orientation of the slot. At 45° ↻, the configuration is as shown in Figure E3.26c. Note that the slot center S moves on a circle about O and that a tangent to this circle at S must pass through P at all times, since P must stay within the slot. From the diagram at the left we get, from geometry and trigonometry,

$$\gamma = \tan^{-1}\left(\frac{7.07}{31.07}\right) = 12.8°$$

$$H = \frac{7.07}{\sin\gamma} = 31.9 \text{ in.}$$

$$\varphi + \gamma = \cos^{-1}\left(\frac{24}{31.9}\right) = 41.2°$$

$$\varphi = 28.4°$$

Using φ to form $\mathbf{v}_{P/\mathcal{I}}$ in Equation (1), we have:

$$1.41\hat{\mathbf{i}} - 1.41\hat{\mathbf{j}} = v_{P/\mathcal{I}}(\sin 28.4°\hat{\mathbf{i}} + \cos 28.4°\hat{\mathbf{j}}) + \mathbf{v}_{P'/\mathcal{I}}$$

in which we have used the fact that we know the direction but not the magnitude of $\mathbf{v}_{P/\mathcal{I}}$. (It moves in the slot at the angle calculated earlier.) Further, relating the velocities of points P' and O on $\mathcal{I}$, we have

$$\mathbf{v}_{P'/\mathcal{I}} = \overset{0}{\cancel{\mathbf{v}_{O/\mathcal{I}}}} + \omega_{\mathcal{I}}\hat{\mathbf{k}} \times \mathbf{r}_{OP'} = \omega_{\mathcal{I}}\hat{\mathbf{k}} \times (31.1\hat{\mathbf{i}} + 7.07\hat{\mathbf{j}})$$

or

$$\mathbf{v}_{P'/\mathcal{J}} = -7.07\omega_{\mathcal{J}}\hat{\mathbf{i}} + 31.1\omega_{\mathcal{J}}\hat{\mathbf{j}}$$

Substituting, and equating the coefficients of $\hat{\mathbf{i}}$ and of $\hat{\mathbf{j}}$, we get

$$\hat{\mathbf{i}} \text{ coefficients:} \quad 0.476v_{P/\mathcal{J}} - 7.07\omega_{\mathcal{J}} = 1.41$$

$$\hat{\mathbf{j}} \text{ coefficients:} \quad 0.880v_{P/\mathcal{J}} + 31.1\omega_{\mathcal{J}} = -1.41$$

Solving these equations gives $v_{P/\mathcal{J}} = 1.61$, so that:

$$\mathbf{v}_{P/\mathcal{J}} = 1.61 \angle 28.4° \text{ in./sec}$$

$$\omega_{\mathcal{J}} = 0.0909 \text{ rad/sec or } \boldsymbol{\omega}_{\mathcal{J}} = 0.0909 \circlearrowright \text{rad/sec}$$

In working out the following problems, the student is urged to begin by thinking carefully about the selection of a point whose velocities in two frames are to be related with Equation (3.49).

PROBLEMS ► Section 3.7

3.142 Boat $\mathcal{B}$ in Figure P3.142 departs from A and is supposed to arrive at point B some 100 ft downstream and on the other side of a river with a current of 5 ft/sec. If $\mathcal{B}$ can move at 10 ft/sec relative to the water, and if it travels on a straight line from A toward B, how long will it take?

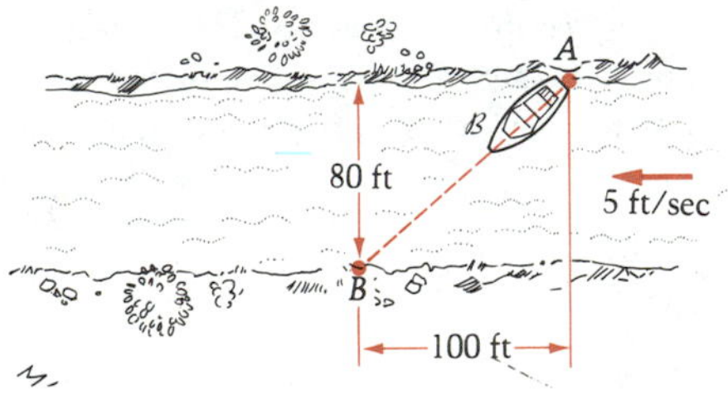

Figure P3.142

3.143 Bar $\mathcal{A}$ in Figure P3.143 is turning clockwise with angular speed 0.25 rad/s, pushing bar $\mathcal{B}$ as it goes. Find $\boldsymbol{\omega}_{\mathcal{B}}$ at the given instant.

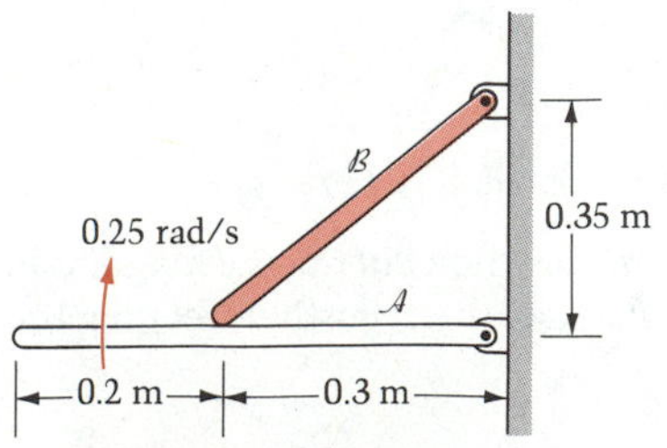

Figure P3.143

3.144 Cylinder $\mathcal{C}$ in Figure P3.144 rolls on a circular surface. When it is at the lowest point of the circle, its angular velocity and acceleration are $\boldsymbol{\omega}_{\mathcal{C}} = 0.2 \circlearrowleft$ rad/s and $\boldsymbol{\alpha}_{\mathcal{C}} = 0.02 \circlearrowleft$ rad/s^2. Rod $\mathcal{B}$ is pinned to $\mathcal{C}$ at E and is also pinned to a block at P that slides in the slot of $\mathcal{S}$. The constant angular velocity of $\mathcal{S}$ is $0.3 \circlearrowleft$ rad/s. Find the velocity of P in $\mathcal{S}$ and the angular velocity of $\mathcal{B}$ at the given instant.

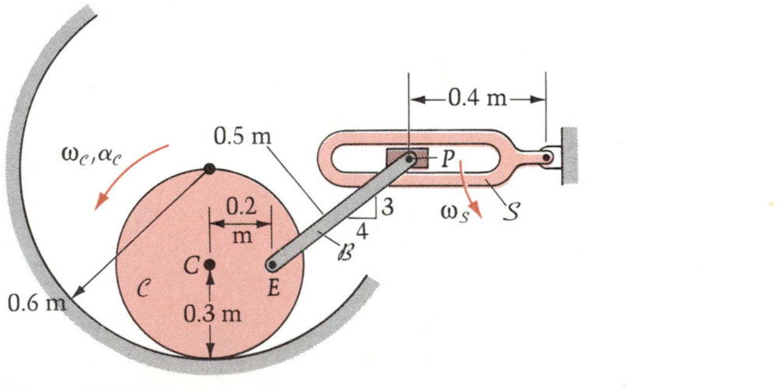

Figure P3.144

3.145 Rod $\mathcal{R}$ is pinned to the ceiling at A and slides on wedge $\mathcal{W}$ at B. (See Figure P3.145.) The wedge moves to the right with constant velocity of 5 ft/sec. Find the angular velocity of the rod in the position shown.

3.146 Referring to Example 3.25 for the meanings of the symbols, show that $\mathbf{v}_{Q/\mathcal{R}} = -\mathbf{v}_{T/\mathcal{B}}$.

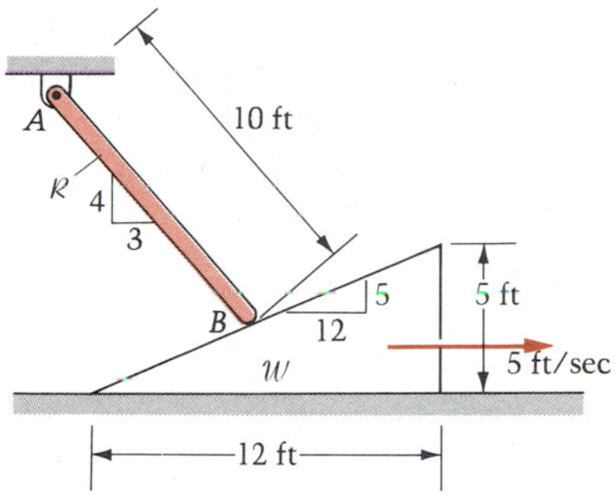

Figure P3.145

3.147 Bar $\mathcal{B}$ slides through a collar in body $\mathcal{C}$ (see Figure P3.147) and is pinned at P to a second bar $\mathcal{R}$. Both $\mathcal{R}$ and $\mathcal{C}$ are pinned to the reference frame as shown, and $\mathcal{R}$ rotates with limited motion at constant angular velocity $\dot{\theta}_{\mathcal{R}} = 1$ rad/s counterclockwise. Find the angular velocity of $\mathcal{C}$ when point P is at the top of the circle on which it travels.

3.148 Plank $\mathcal{P}$ slides on the floor at A and on block $\mathcal{D}$ at Q. Block $\mathcal{D}$ moves to the right with a constant velocity of 6 ft/sec while end A moves to the left with a constant velocity of 4 ft/sec. For the position shown in Figure P3.148, find the angular velocity of the plank.

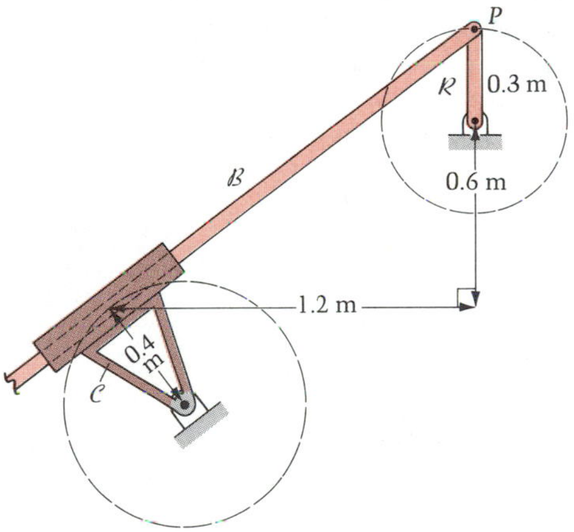

Figure P3.147

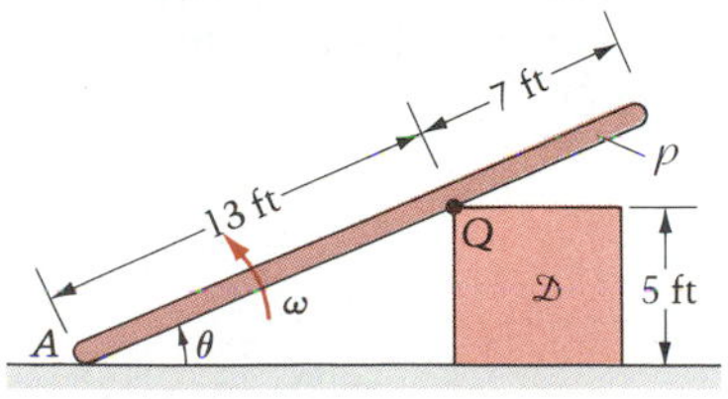

Figure P3.148

3.149 Rod $\mathcal{B}$ in Figure P3.149 has angular velocity 5 ⮌ rad/s. It is pinned to another rod $\mathcal{R}$, which passes through a slot in $\mathcal{A}$ as shown. At the given instant, find the angular velocity of body $\mathcal{A}$ and the velocity of any point of $\mathcal{R}$ relative to $\mathcal{A}$. *Hint*: All points of $\mathcal{R}$ translate in $\mathcal{A}$—what does this mean about the angular velocities of $\mathcal{R}$ and $\mathcal{A}$?

3.150 A mechanism consists of crank $\mathcal{C}$ pinned to O, rocker $\mathcal{R}$ pinned to O', and a small body $\mathcal{B}$ that is pinned to $\mathcal{C}$ and slides in the slot of $\mathcal{R}$. (See Figure P3.150.) The length of $\mathcal{C}$ is $\ell = D\sqrt{2}$, where D is the distance between O and O'. If $\mathcal{C}$ has constant angular velocity $\omega_{\mathcal{C}}$ ⮎ over a range of its motion, find $\omega_{\mathcal{R}}$ when: (a) $\theta_{\mathcal{R}} = \tan^{-1}(1/2)$; (b) $\theta_{\mathcal{R}} = 90°$

3.151 Collars $\mathcal{A}$ and $\mathcal{C}$ in Figure P3.151 are pinned together at C, and they slide on rods $\mathcal{B}$ and $\mathcal{S}$, respectively. Rod $\mathcal{B}$ has a constant angular velocity $\dot{\theta}\hat{\mathbf{k}}$ for $10° \le \theta \le 45°$. Find the velocity of point C relative to $\mathcal{B}$, as a function of D, θ, and $\dot{\theta}$ in this range of angles.

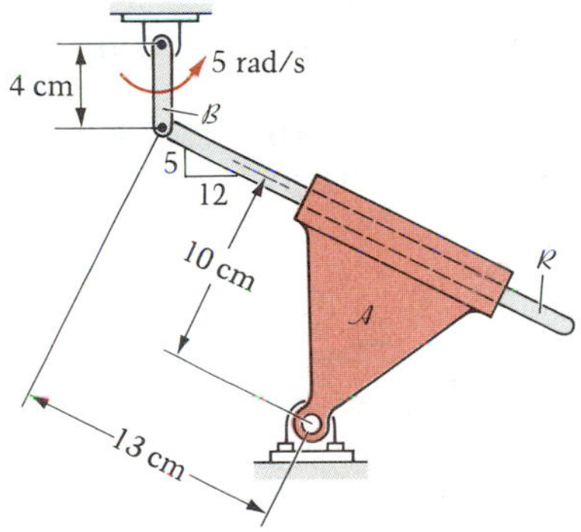

Figure P3.149

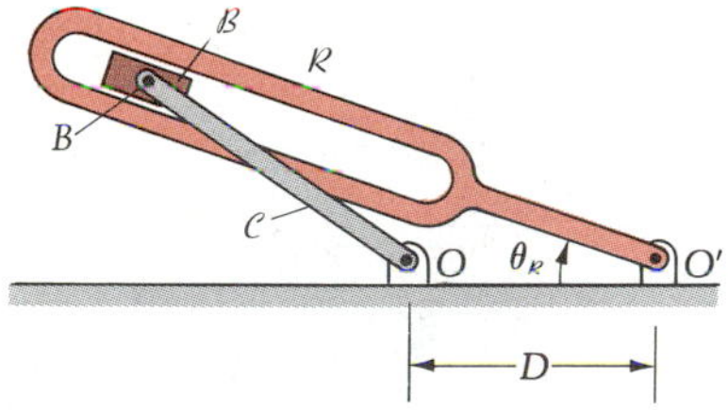

Figure P3.150

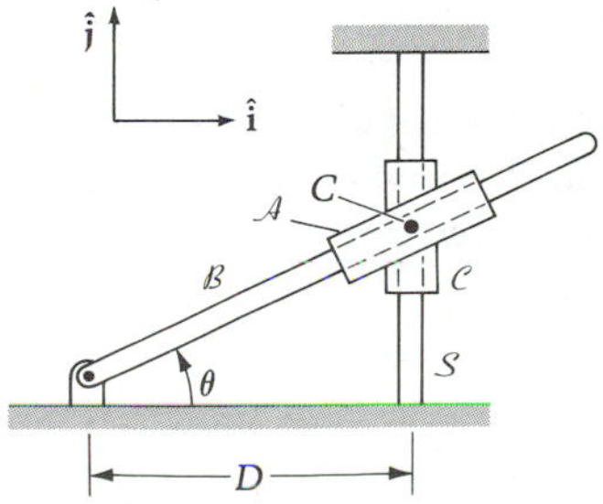

Figure P3.151

3.152 Figure P3.152 shows a circular cam $\mathcal{B}_1$ and a flat-face translating follower $\mathcal{B}_2$. If $\mathcal{B}_1$ rotates with constant angular velocity $\omega_0 \circlearrowleft$, find the maximum velocity in reference frame $\mathcal{I}$ of any point of $\mathcal{B}_2$, in terms of ω_0 and the offset distance δ.

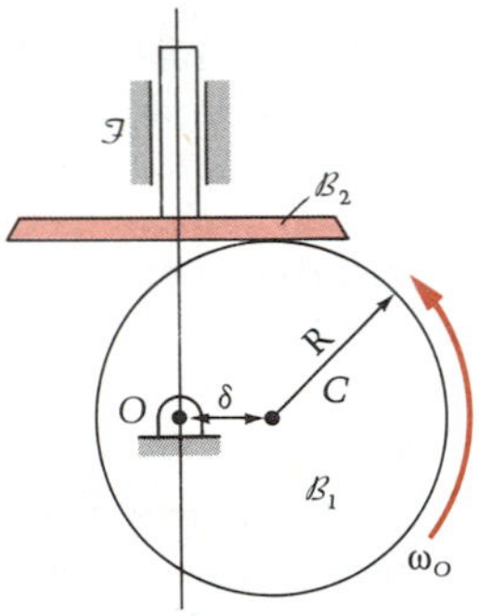

Figure P3.152

• **3.153** Figure P3.153 illustrates a "Geneva mechanism," in which disk $\mathcal{A}$ is driven with a constant counterclockwise angular speed and produces an intermittent (starting and stopping, waiting, then repeating) rotational motion of the slotted disk $\mathcal{B}$. Pin P is fixed to $\mathcal{A}$ and drives disk $\mathcal{B}$ by pressing on the surfaces of the slots. Show with the use of Equation (3.49) that disk $\mathcal{B}$ will have zero angular speed in the two positions shown, a varying angular speed in between these positions, and zero angular speed while P is returning to the $\theta = 135°$ position. Note that the operation of the mechanism requires that the distance between O and O' be $\sqrt{2}R$.

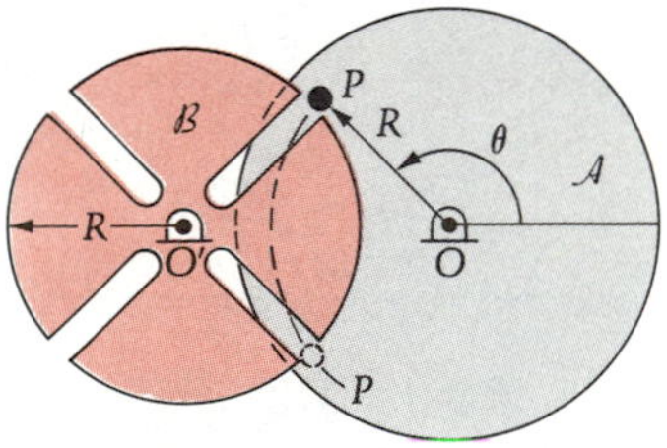

Figure P3.153

• **3.154** In the preceding problem let $R = 0.1$ m and $\dot{\theta}_{\mathcal{A}} = 5$ rad/s = constant. Find the angular velocity of the slotted body $\mathcal{B}$ at the instant when $\theta = 160°$.

• **3.155** Rods $\mathcal{R}$ and $\mathcal{L}$ in Figure P3.155 are pinned at O and O' to a reference frame $\mathcal{I}$. Rod $\mathcal{L}$ is also pinned to the slotted body $\mathcal{B}$ at B. The upper end of $\mathcal{R}$ is pinned at P to a roller that moves freely in the slot of $\mathcal{B}$. The angular velocities of rod $\mathcal{R}$ and link $\mathcal{L}$ are constants:

$$\omega_{\mathcal{R}} = 0.2 \circlearrowright \text{ rad/s}$$
$$\omega_{\mathcal{L}} = 0.4 \circlearrowright \text{ rad/s}$$

Determine the velocity of P in $\mathcal{B}$ and the angular velocity of $\mathcal{B}$ at the given instant.

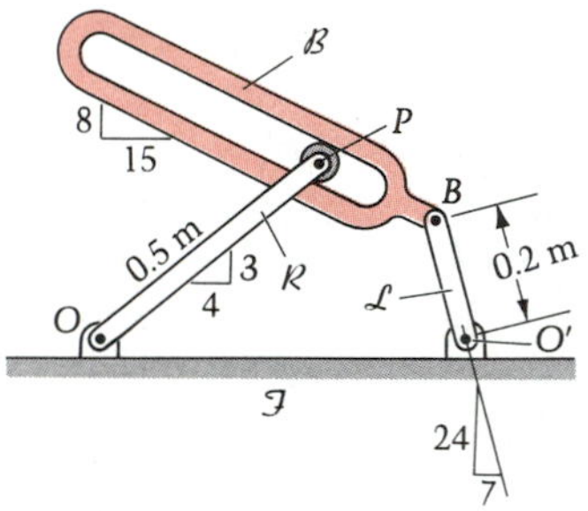

Figure P3.155

• **3.156** In Figure P3.156 collar $\mathcal{C}$ is fixed to arm $\mathcal{A}$ and slides along rod $\mathcal{R}$. Arm $\mathcal{A}$ is pinned to a second collar $\mathcal{K}$ at A; this collar slides on rod $\mathcal{L}$. At the given instant, $\omega_{\mathcal{R}} = 0.2 \circlearrowright$ rad/s, $\omega_{\mathcal{L}} = 0.1 \circlearrowright$ rad/s, and the velocity of all points of $\mathcal{C}$ relative to rod $\mathcal{R}$ is 0.3 m/s outward (along OC). Give the angular velocity of $\mathcal{A}$ by inspection, and find the velocity of the points of $\mathcal{K}$ relative to $\mathcal{L}$.

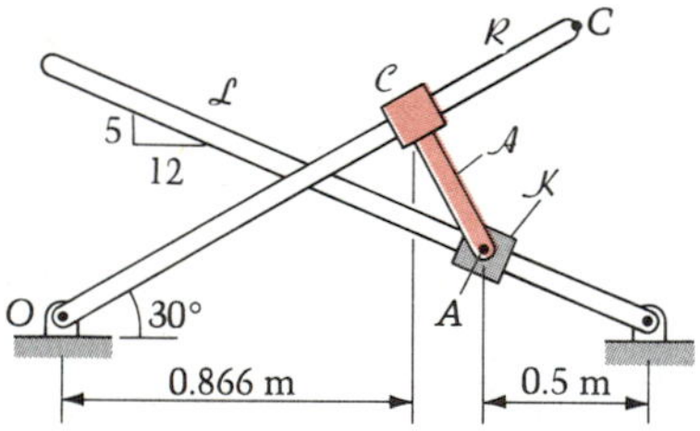

Figure P3.156

3.157 Show that the rigid-body velocity equation (3.8) can be derived from Equation (3.49). *Hint*: Fix P to $\mathcal{B}$ and the equation will relate the velocities in $\mathcal{I}$ of the two points P and O' of $\mathcal{B}$.

3.8 Relationship Between the Accelerations of a Point with Respect to Two Different Frames of Reference

We shall now derive the companion equation to (3.49), this one relating the accelerations of P in $\mathcal{I}$ and $\mathcal{B}$. Differentiating Equation (3.49) in $\mathcal{I}$, we get

$$\mathbf{a}_{P/\mathcal{I}} = {}^{\mathcal{I}}\dot{\mathbf{v}}_{P/\mathcal{B}} + \mathbf{a}_{O'/\mathcal{I}} + \ddot{\theta}\hat{\mathbf{k}} \times \mathbf{r}_{O'P} + \dot{\theta}\hat{\mathbf{k}} \times {}^{\mathcal{I}}\dot{\mathbf{r}}_{O'P} \tag{3.52}$$

Using Equation (3.44) to "move the derivative" in the first and last terms on the right side of Equation (3.52) gives

$$\mathbf{a}_{P/\mathcal{I}} = ({}^{\mathcal{B}}\dot{\mathbf{v}}_{P/\mathcal{B}} + \dot{\theta}\hat{\mathbf{k}} \times \mathbf{v}_{P/\mathcal{B}}) + \mathbf{a}_{O'/\mathcal{I}} + \ddot{\theta}\hat{\mathbf{k}} \times \mathbf{r}_{O'P}$$
$$+ \dot{\theta}\hat{\mathbf{k}} \times ({}^{\mathcal{B}}\dot{\mathbf{r}}_{O'P} + \dot{\theta}\hat{\mathbf{k}} \times \mathbf{r}_{O'P}) \tag{3.53}$$

Recognizing that ${}^{\mathcal{B}}\dot{\mathbf{r}}_{O'P} = \mathbf{v}_{P/\mathcal{B}}$ and that ${}^{\mathcal{B}}\dot{\mathbf{v}}_{P/\mathcal{B}} = \mathbf{a}_{P/\mathcal{B}}$, and rearranging terms, from Equation (3.53) we obtain

$$\mathbf{a}_{P/\mathcal{I}} = \mathbf{a}_{P/\mathcal{B}} + (\mathbf{a}_{O'/\mathcal{I}} + \ddot{\theta}\hat{\mathbf{k}} \times \mathbf{r}_{O'P} - \dot{\theta}^2\mathbf{r}_{O'P}) + 2\dot{\theta}\hat{\mathbf{k}} \times \mathbf{v}_{P/\mathcal{B}} \tag{3.54}$$

in which $\dot{\theta}\hat{\mathbf{k}} \times (\dot{\theta}\hat{\mathbf{k}} \times \mathbf{r}_{O'P}) = -\dot{\theta}^2\mathbf{r}_{O'P}$ as we have already seen in Section 3.5.

The parenthesized term in Equation (3.54) is seen from Equation (3.19) to be $\mathbf{a}_{P'/\mathcal{I}}$, where, as before, P' is the point of $\mathcal{B}$ (or $\mathcal{B}$ extended) that is coincident with P. Therefore we have the following for our result:

$$\mathbf{a}_{P/\mathcal{I}} = \mathbf{a}_{P/\mathcal{B}} + \mathbf{a}_{P'/\mathcal{I}} + 2\dot{\theta}\hat{\mathbf{k}} \times \mathbf{v}_{P/\mathcal{B}} \tag{3.55}$$

In words: The acceleration of P in $\mathcal{I}$ equals its acceleration in $\mathcal{B}$, plus the acceleration in $\mathcal{I}$ of the point of $\mathcal{B}$ coincident with P, *plus* the **Coriolis acceleration**, $2\dot{\theta}\hat{\mathbf{k}} \times \mathbf{v}_{P/\mathcal{B}}$. The Coriolis acceleration is seen to provide an unexpected but essential difference between the forms of Equations (3.51) and (3.55).

Question 3.20 If we differentiate Equation (3.51) instead of (3.49), we might (*erroneously!*) obtain

$$\begin{aligned}\mathbf{a}_{P/\mathcal{I}} &= {}^{\mathcal{I}}\dot{\mathbf{v}}_{P/\mathcal{B}} + \mathbf{a}_{P'/\mathcal{I}} \\ &= {}^{\mathcal{B}}\dot{\mathbf{v}}_{P/\mathcal{B}} + \boldsymbol{\omega}_{\mathcal{B}/\mathcal{I}} \times \mathbf{v}_{P/\mathcal{B}} + \mathbf{a}_{P'/\mathcal{I}} \\ &= \mathbf{a}_{P/\mathcal{B}} + \mathbf{a}_{P'/\mathcal{I}} + \boldsymbol{\omega}_{\mathcal{B}/\mathcal{I}} \times \mathbf{v}_{P/\mathcal{B}}\end{aligned}$$

and we come up (incorrectly) short by half on the Coriolis term. What is wrong with this approach?

In the same procedure we used for velocities, we can simplify the notation of Equation (3.54) if there is but one "moving frame" ($\mathcal{B}$) involved, which is in motion relative to the reference frame ($\mathcal{I}$):

$$\mathbf{a}_P = \mathbf{a}_{\text{rel}} + \mathbf{a}_{O'} + \boldsymbol{\alpha} \times \mathbf{r} - \omega^2\mathbf{r} + 2\boldsymbol{\omega} \times \mathbf{v}_{\text{rel}} \tag{3.56}$$

Answer 3.20 The error is that the derivative ${}^{\mathcal{I}}\dot{\mathbf{v}}_{P'/\mathcal{I}}$ is not equal to $\mathbf{a}_{P'/\mathcal{I}}$, because P' denotes a *succession of points* of $\mathcal{B}$ which are at each instant coincident with P.

In this equation, $\mathbf{a}_P$ and $\mathbf{a}_{\text{rel}}$ are the respective accelerations of P in $\mathcal{I}$ and in $\mathcal{B}$. The vectors $\boldsymbol{\omega}$ and $\boldsymbol{\alpha}$ are the angular velocity and angular acceleration of $\mathcal{B}$ in $\mathcal{I}$ (equal to $\dot{\theta}\hat{\mathbf{k}}$ and $\ddot{\theta}\hat{\mathbf{k}}$) and $\mathbf{r} = \mathbf{r}_{O'P}$ (the position vector of P in the "moving frame"). We now consider several examples showing the use of Equations (3.54) and (3.55).

EXAMPLE 3.27

Find the acceleration in frame $\mathcal{I}$ of the yellowjacket of Example 3.23.

Solution

Using Equation 3.54, we obtain:

$$\mathbf{a}_{Y/\mathcal{I}} = \mathbf{a}_{Y/\mathcal{B}} + (\mathbf{a}_{O'/\mathcal{I}} + \ddot{\theta}\hat{\mathbf{k}} \times \mathbf{r}_{O'Y} - \dot{\theta}^2\mathbf{r}_{O'Y}) + 2\dot{\theta}\hat{\mathbf{k}} \times \mathbf{v}_{Y/\mathcal{B}}$$

$$= 0 + 0 + 0 - \left(\frac{10\pi}{9}\right)^2 (r\hat{\mathbf{i}}) + 2\left(\frac{10\pi}{9}\hat{\mathbf{k}}\right) \times 2\hat{\mathbf{i}}$$

$$= -\frac{100\pi^2}{81} r\hat{\mathbf{i}} + \frac{40\pi}{9}\hat{\mathbf{j}} \text{ in./sec}^2$$

The "$\hat{\mathbf{j}}$-term" in this example is the Coriolis acceleration. Note that the yellowjacket has two nonzero acceleration components, even though both $\ddot{r}$ and $\ddot{\theta}$ are zero in this example.

EXAMPLE 3.28

If in Example 3.24 we have the additional data that $\boldsymbol{\alpha}_{\mathcal{R}} = 0.25\hat{\mathbf{k}}$ rad/s² at the given time, find $\boldsymbol{\alpha}_{\mathcal{B}}$ and $\mathbf{a}_{P/\mathcal{B}}$. (See Figure E3.28.)

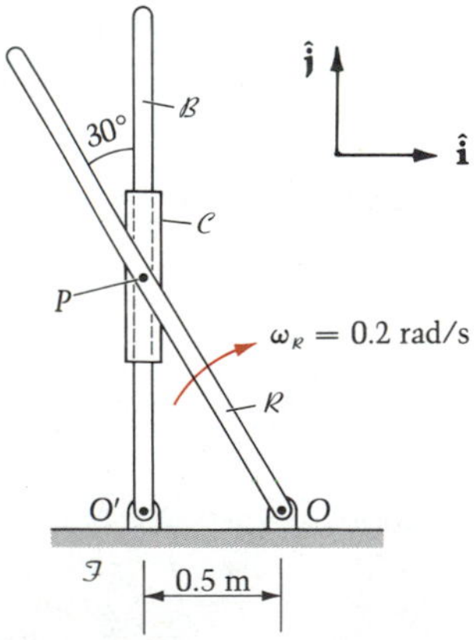

Figure E3.28

Solution

Relating the accelerations of P in $\mathcal{B}$ and $\mathcal{I}$, we have

$$\mathbf{a}_{P/\mathcal{I}} = \mathbf{a}_{P/\mathcal{B}} + \mathbf{a}_{O'/\mathcal{I}} + \boldsymbol{\alpha}_{\mathcal{B}} \times \mathbf{r}_{O'P} - \omega_{\mathcal{B}}^2\mathbf{r}_{O'P} + 2\boldsymbol{\omega}_{\mathcal{B}/\mathcal{I}} \times \mathbf{v}_{P/\mathcal{B}}$$

$$\overbrace{\underbrace{\mathbf{a}_{O/\mathcal{I}}}_{\text{zero}} + \boldsymbol{\alpha}_{\mathcal{R}} \times \mathbf{r}_{OP} - \omega_{\mathcal{R}}^2\mathbf{r}_{OP}}^{\text{on } \mathcal{R}} = a_{P/\mathcal{B}}\hat{\mathbf{j}} + \underbrace{\mathbf{a}_{O'/\mathcal{I}}}_{\text{zero}} + \boldsymbol{\alpha}_{\mathcal{B}} \times \mathbf{r}_{O'P} - \omega_{\mathcal{B}}^2\mathbf{r}_{O'P} + 2(-0.2\hat{\mathbf{k}}) \times (0.1\hat{\mathbf{j}})$$

in which the answers to Example 3.24 are used in the Coriolis term. Filling in the variables, we get

$$0.25\hat{\mathbf{k}} \times (-0.5\hat{\mathbf{i}} + 0.866\hat{\mathbf{j}}) - 0.04(-0.5\hat{\mathbf{i}} + 0.866\hat{\mathbf{j}})$$
$$= a_{P/\mathcal{B}}\hat{\mathbf{j}} + \alpha_{\mathcal{B}}\hat{\mathbf{k}} \times 0.866\hat{\mathbf{j}} - 0.04(0.866\hat{\mathbf{j}}) + 0.04\hat{\mathbf{i}}$$

$\hat{\mathbf{i}}$ coefficients: $\quad -0.217 + 0.02 = -0.866\alpha_{\mathcal{B}} + 0.04$

$$\alpha_{\mathcal{B}} = 0.274 \Rightarrow \boldsymbol{\alpha}_{\mathcal{B}} = 0.274\hat{\mathbf{k}} \text{ rad/s}^2$$

$\hat{\mathbf{j}}$ coefficients: $\quad -0.125 - 0.0346 = a_{P/\mathcal{B}} - 0.0346$

$$\mathbf{a}_{P/\mathcal{B}} = -0.125\hat{\mathbf{j}} \text{ m/s}^2$$

Thus point P is slowing down as it moves outward along $\mathcal{B}$, and $\mathcal{B}$ is slowing down as it rotates clockwise.

The reader should note in the preceding example that $\boldsymbol{\alpha}_{\mathcal{C}}$ and $\boldsymbol{\alpha}_{\mathcal{B}}$ are identical. The sleeve forces the two bodies $\mathcal{B}$ and $\mathcal{C}$ to translate relative to each other, so that, as was pointed out at the end of Example 3.23, $\boldsymbol{\omega}_{\mathcal{B}} \equiv \boldsymbol{\omega}_{\mathcal{C}}$. By differentiating this equation, we see that the angular accelerations are also always equal.

Example 3.29

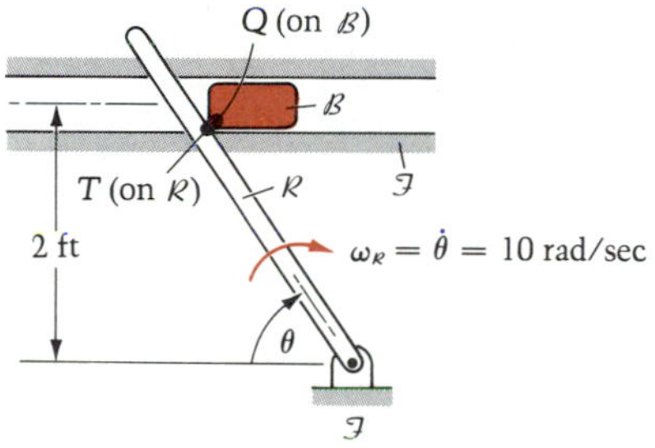

Figure E3.29

In Example 3.25, suppose that at the given instant ($\theta = 60°$) all the data are the same and in addition $\boldsymbol{\alpha}_{\mathcal{R}} = 30 \circlearrowleft$ rad/sec². Find the acceleration of block $\mathcal{B}$ (see Figure E3.29).

Solution

Note that if we again use Q as our point (of the block $\mathcal{B}$) that is moving relative to two bodies ($\mathcal{R}$ and $\mathcal{I}$), we know the direction of $\mathbf{a}_{Q/\mathcal{R}}$. (It is *along* $\mathcal{R}$ since point Q moves on a *straight line* in $\mathcal{R}$.)

Equation (3.55) thus gives

$$\mathbf{a}_{Q/\mathcal{I}} = \mathbf{a}_{Q/\mathcal{R}} + \mathbf{a}_{Q'/\mathcal{I}} + 2\omega_{\mathcal{R}}\hat{\mathbf{k}} \times \mathbf{v}_{Q/\mathcal{R}}$$

Noting that point Q' is the point T, we obtain

$$a_{Q/\mathcal{I}}\hat{\mathbf{i}} = a_{Q/\mathcal{R}}\left(-\frac{1}{2}\hat{\mathbf{i}} + \frac{\sqrt{3}}{2}\hat{\mathbf{j}}\right) + (\overset{0}{\cancel{\mathbf{a}_O}} + \overset{30}{\cancel{\alpha_{\mathcal{R}}}}\hat{\mathbf{k}} \times \overbrace{\cancel{\mathbf{r}_{OP}}}^{\left(\frac{-2}{\sqrt{3}}\hat{\mathbf{i}} + 2\hat{\mathbf{j}}\right)} - \overset{10^2}{\cancel{\omega_{\mathcal{R}}^2}}\mathbf{r}_{OP})$$

$$+ 2(-10\hat{\mathbf{k}}) \times \left(\frac{40}{3}\right)\left(-\frac{1}{2}\hat{\mathbf{i}} + \frac{\sqrt{3}}{2}\hat{\mathbf{j}}\right)$$

This result yields the following scalar component equations:

$\hat{\mathbf{j}}$ coefficients: $$0 = a_{Q/\mathcal{R}}\frac{\sqrt{3}}{2} - \frac{60}{\sqrt{3}} - 200 + \frac{800}{6}$$

$$a_{Q/\mathcal{R}} = 117 \text{ in./sec}^2$$

$\hat{\mathbf{i}}$ coefficients: $$a_{Q/\mathcal{I}} = 117\left(-\frac{1}{2}\right) - 60 + \frac{200}{\sqrt{3}} + \frac{800}{6}\sqrt{3}$$

Thus the acceleration of the translating block is:

$$\mathbf{a}_{Q/\mathcal{I}} = a_{Q/\mathcal{I}}\hat{\mathbf{i}} = 228\hat{\mathbf{i}} \text{ in./sec}^2$$

EXAMPLE 3.30

If in Example 3.26 we add $\boldsymbol{\alpha}_{\mathcal{C}} = 10 \circlearrowleft$ rad/sec² to the data, find the angular acceleration of $\mathcal{T}$ and the acceleration of P relative to $\mathcal{T}$. (See Figures E3.30a,b.)

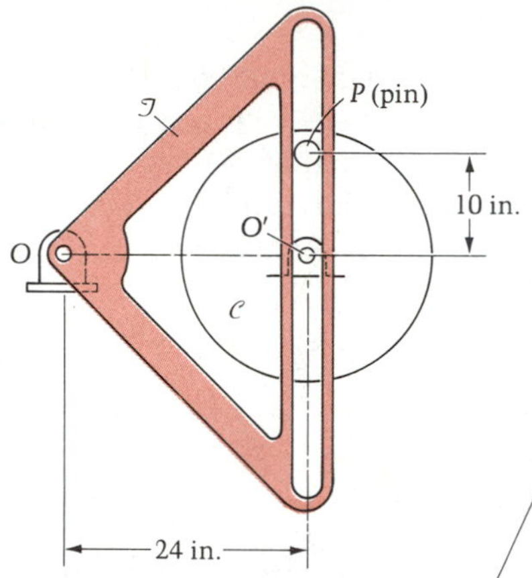

Figure E3.30a

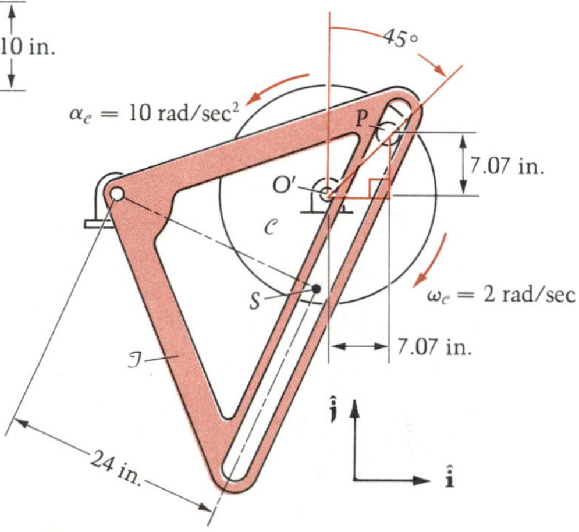

Figure E3.30b

Solution

Again we apply Equation (3.55):

$$\mathbf{a}_{P/\mathcal{I}} = \mathbf{a}_{P/\mathcal{T}} + \mathbf{a}_{P'/\mathcal{I}} + 2\omega_{\mathcal{T}}\hat{\mathbf{k}} \times \mathbf{v}_{P/\mathcal{T}}$$

$$\overset{0}{\cancel{\mathbf{a}}_{O'/\mathcal{I}}} + \overset{10}{\alpha_{\mathcal{C}}}\hat{\mathbf{k}} \times \overset{(7.07\hat{\mathbf{i}} + 7.07\hat{\mathbf{j}})}{\mathbf{r}_{O'P}} - \overset{2^2}{\omega_{\mathcal{C}}^2}\mathbf{r}_{O'P} = a_{P/\mathcal{T}}(\sin 28.4°\hat{\mathbf{i}} + \cos 28.4°\hat{\mathbf{j}}) + \overset{0}{\cancel{\mathbf{a}}_{O'/\mathcal{I}}}$$

$$+ \alpha_{\mathcal{T}}\hat{\mathbf{k}} \times (31.1\hat{\mathbf{i}} + 7.07\hat{\mathbf{j}}) - \overset{0.091^2}{\omega_{\mathcal{T}}^2}(31.1\hat{\mathbf{i}} + 7.07\hat{\mathbf{j}})$$

$$+ 2(-0.091\hat{\mathbf{k}}) \times 1.62(\sin 28.4°\hat{\mathbf{i}} + \cos 28.4°\hat{\mathbf{j}})$$

and again we equate the $\hat{\mathbf{i}}$ coefficients, and then the $\hat{\mathbf{j}}$ coefficients, to obtain the scalar component equations:

$\hat{\mathbf{i}}$ coefficients: $\quad -99.0 = 0.476a_{P/\mathcal{T}} - 7.07\alpha_{\mathcal{T}} - 0.258 + 0.259$

$\hat{\mathbf{j}}$ coefficients: $\quad 42.4 = a_{P/\mathcal{T}}(0.880) + 31.1\alpha_{\mathcal{T}} - 0.0586 - 0.140$

Solving these equations results in $\alpha_{\mathcal{T}} = 5.12$ rad/sec², so that:

$\boldsymbol{\alpha}_{\mathcal{T}} = 5.12 \circlearrowleft$ rad/sec²

$a_{P/\mathcal{T}} = -132$ in./sec² $\quad$ or $\quad \mathbf{a}_{P/\mathcal{T}} = 132$ (arrow at 28.4°) in./sec²

EXAMPLE 3.31

Pin P in Figure E.3.31 is attached to cart $\mathcal{B}$ and slides in the smooth slot cut in wheel $\mathcal{C}$. The wheel rolls on the rough plane $\mathcal{I}$. The cart's position is given by $x_B = 0.3t^2$, with x_B in meters when t is in seconds. Find $\boldsymbol{\omega}_{\mathcal{C}}$ and $\boldsymbol{\alpha}_{\mathcal{C}}$ at the given instant (which is at $t = 3$ s), and determine the acceleration of P in the slot at this time.

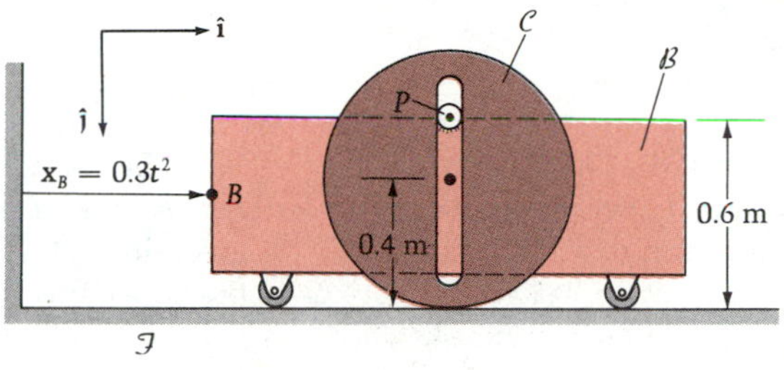

Figure E3.31

Solution

Since $x_B = 0.3t^2$, we have $\mathbf{v}_B = 0.6t\hat{\mathbf{i}}$ and $\mathbf{a}_B = 0.6\hat{\mathbf{i}}$; these are the velocity and acceleration vectors of all points of the cart $\mathcal{B}$, in particular of P. At $t = 3$, we have $\mathbf{v}_B = 1.8\hat{\mathbf{i}}$ m/s and $\mathbf{a}_B = 0.6\hat{\mathbf{i}}$ m/s².

Relating the velocities of P in frames $\mathcal{C}$ and $\mathcal{I}$, we obtain

$$\mathbf{v}_{P/\mathcal{I}} = \mathbf{v}_{P/\mathcal{C}} + \mathbf{v}_{P'/\mathcal{I}}$$

$$v_B\hat{\mathbf{i}} = v_{P/\mathcal{C}}\hat{\mathbf{j}} + 0.6\omega_{\mathcal{C}}\hat{\mathbf{i}}$$

$\hat{\mathbf{i}}$ coefficients: $\quad v_B = 1.8 = 0.6\omega_{\mathcal{C}} \Rightarrow \omega_{\mathcal{C}} = 3 \circlearrowright$ rad/s

$\hat{\mathbf{j}}$ coefficients: $\quad v_{P/\mathcal{C}} = 0$

Relating the accelerations using Equation (3.54), we get

$$\mathbf{a}_{P/\mathcal{I}} = \mathbf{a}_{P/\mathcal{C}} + \underbrace{\mathbf{a}_{P'/\mathcal{I}}} + 2\boldsymbol{\omega}_{\mathcal{C}/\mathcal{I}} \times \mathbf{v}_{P/\mathcal{C}}$$

$$a_B\hat{\mathbf{i}} = a_{P/\mathcal{C}}\hat{\mathbf{j}} + \overbrace{0.6\alpha_{\mathcal{C}}\hat{\mathbf{i}} + 0.2\omega_{\mathcal{C}}^2\hat{\mathbf{j}}} + \mathbf{0}$$

Note that we have related P' (the point of $\mathcal{C}$-extended coincident with P) and the center of $\mathcal{C}$ (call it C) to get

$$\mathbf{a}_{P'/\mathcal{I}} = \mathbf{a}_C + \alpha_{\mathcal{C}}\hat{\mathbf{k}} \times \overset{-0.2\hat{\mathbf{j}}}{\mathbf{r}_{CP'}} - \omega_{\mathcal{C}}^2\mathbf{r}_{CP'}$$

$$= \overset{0.4}{R}\alpha_{\mathcal{C}}\hat{\mathbf{i}} + 0.2\alpha_{\mathcal{C}}\hat{\mathbf{i}} + 0.2\omega_{\mathcal{C}}^2\hat{\mathbf{j}}$$

$$= 0.6\alpha_{\mathcal{C}}\hat{\mathbf{i}} + 0.2\omega_{\mathcal{C}}^2\hat{\mathbf{j}}$$

Solving, we obtain

$\hat{\mathbf{i}}$ coefficients: $\quad a_B = 0.6 = 0.6\alpha_{\mathcal{C}} \Rightarrow \alpha_{\mathcal{C}} = 1 \circlearrowright$ rad/s²

$\hat{\mathbf{j}}$ coefficients: $\quad a_{P/\mathcal{C}} = -0.2\omega_{\mathcal{C}}^2 = -0.2(3^2) = -1.8$ m/s²

Thus the acceleration of P in the slot (which is its acceleration in $\mathcal{C}$) is $1.8 \uparrow$ m/s²; it is upward because we assumed it to be in the positive y direction ($\downarrow$) and got a negative answer. Note that although P is momentarily stopped in the slot ($v_{P/\mathcal{C}} = 0$), it has to be "getting ready" to move outward since $\mathcal{B}$ is translating to the right and $\mathcal{C}$ is rolling that way. This is what is indicated by $\mathbf{a}_{P/\mathcal{C}} = 1.8 \uparrow$ m/s².

PROBLEMS ▶ Section 3.8

3.158 A bug $\mathcal{B}$ is crawling outward at a uniform speed relative to the rotating arm of 3 ft/sec. In the position shown in Figure P3.158, for the arm $\omega = 2$ rad/sec and $\alpha = 4$ rad/sec², both counterclockwise. What is the acceleration of the bug? Indicate the direction in a sketch.

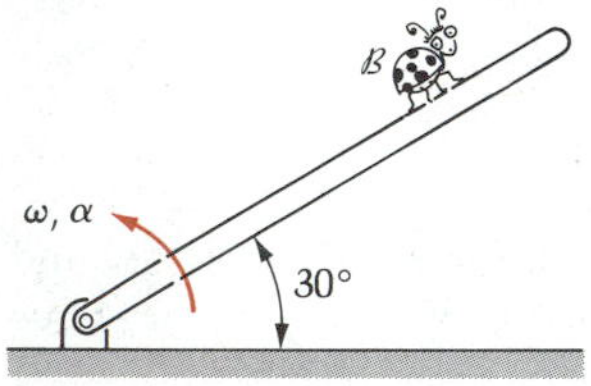

Figure P3.158

3.159 The mechanism shown in Figure P3.159 is used to raise and lower hammer $\mathcal{H}$. The 26-cm crank $\mathcal{C}$ turns clockwise at the constant rate of 30 rpm. It is pinned to block $\mathcal{B}$, which slides in a slot in $\mathcal{H}$. If at $t = 0$ point A is directly above O, find the velocity and acceleration of $\mathcal{H}$ as a function of time. (The block and hammer are slightly offset from $\mathcal{C}$ so they do not interfere with the pin at O.)

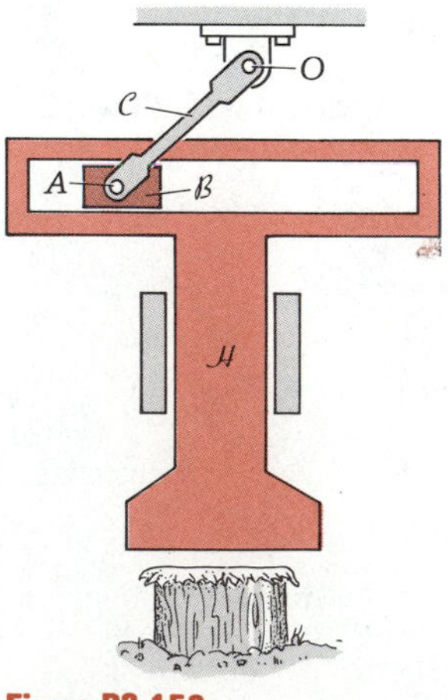

Figure P3.159

3.160 Bar $\mathcal{A}$ in Figure P3.160 has angular velocity 0.25 ↻ rad/s and angular acceleration 0.15 ↺ rad/s² at the given instant. Find the angular acceleration of $\mathcal{B}$ at this time. (See Problem 3.143.)

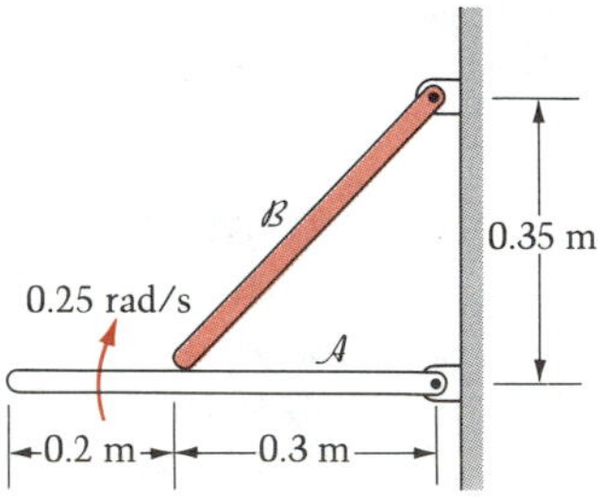

Figure P3.160

3.161 In Problem 3.144 find the acceleration of P in $\mathcal{S}$ and the angular acceleration of $\mathcal{B}$.

3.162 In Problem 3.155 find, at the same instant of time, the acceleration of P in $\mathcal{B}$ and the angular acceleration of $\mathcal{B}$.

3.163 In Figure P3.163,

$$\mathbf{v}_A = 4 \rightarrow \text{in./sec} \quad \text{and} \quad \frac{d}{dt}|\mathbf{v}_A| = 3 \text{ in./sec}^2$$

If the bar stays in contact with both the step and circular

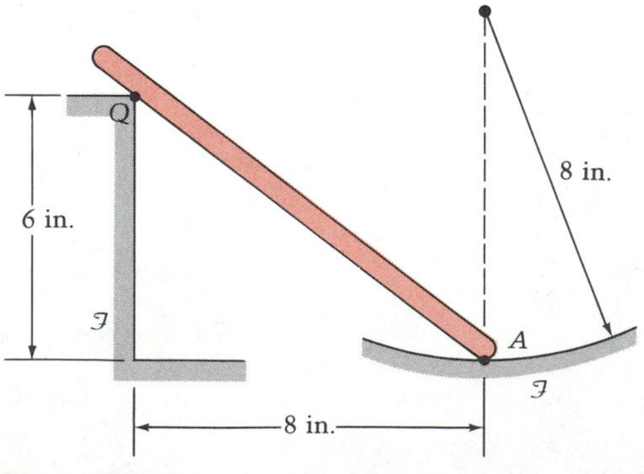

Figure P3.163

trough, find its angular acceleration. *Hint*: Treat point Q (fixed to the step) as the "moving point," and note that Q moves on a straight line relative to the bar.

3.164 In Problem 3.151 determine the acceleration of C relative to $\mathcal{B}$ as a function of D, θ, and $\dot{\theta}$.

3.165 Rods $\mathcal{A}$ and $\mathcal{B}$ (see Figure P3.165) pass smoothly through the short collars, which can turn relative to each other by virtue of the ball-and-socket connection. If bars $\mathcal{A}$ and $\mathcal{B}$ turn with constant angular velocities 0.4 ⮌ rad/sec and 0.2 ⮎ rad/sec, respectively, find the velocity and acceleration of the ball-and-socket connection with respect to $\mathcal{A}$ and to $\mathcal{B}$ in the indicated position.

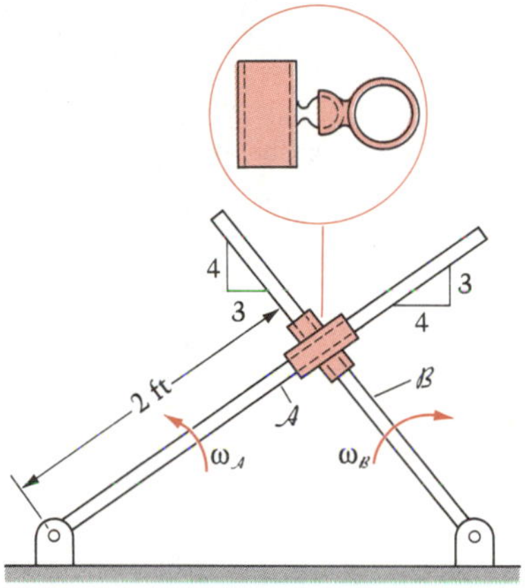

Figure P3.165

• **3.166** Wheel $\mathcal{W}$ in Figure P3.166 has a constant clockwise angular velocity of 2 rad/sec. It is connected by link $\mathcal{L}$ to block $\mathcal{E}$. End B of rod $\mathcal{R}$ slides in a vertical slot in block $\mathcal{E}$. For the position shown, find the angular velocity and angular acceleration of rod $\mathcal{R}$ if block $\mathcal{E}$ translates.

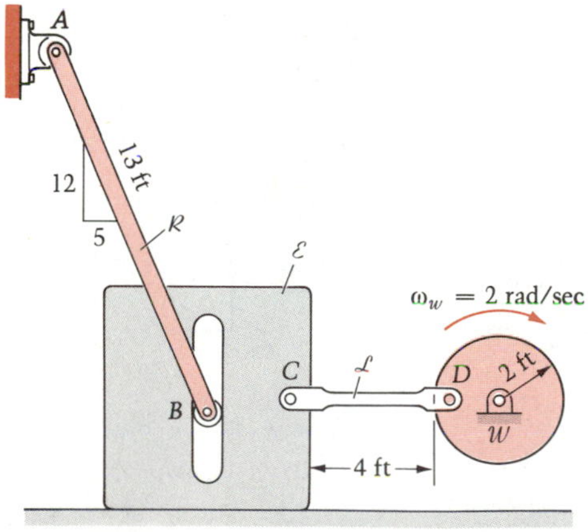

Figure P3.166

• **3.167** Extending Problem 3.149, suppose that the angular velocity of rod $\mathcal{B}$, 5 ⮌ rad/s, is constant in time. (See Figure P3.167.) Find, at the given instant, the angular acceleration of $\mathcal{A}$ and the acceleration of any point of $\mathcal{R}$ relative to $\mathcal{A}$.

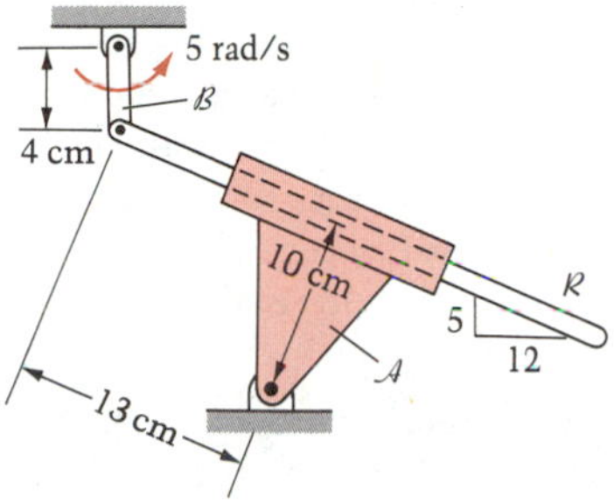

Figure P3.167

3.168 Referring to Example 3.29 for the meanings of the symbols, show that $\mathbf{a}_{Q/\mathcal{R}} \neq -\mathbf{a}_{T/\mathcal{B}}$.

• **3.169** A circular turntable $\mathcal{T}$ (see Figure P3.169) rotates about a vertical axis through O (normal to the plane of the paper) with θ changing at a constant rate ω_0. A block $\mathcal{B}$ rests in a groove cut in the turntable. If the cable is reeled in at a constant velocity v relative to $\mathcal{T}$, find expressions for the radial and transverse components of the block's acceleration. Check your answers using the expression for acceleration in cylindrical coordinates.

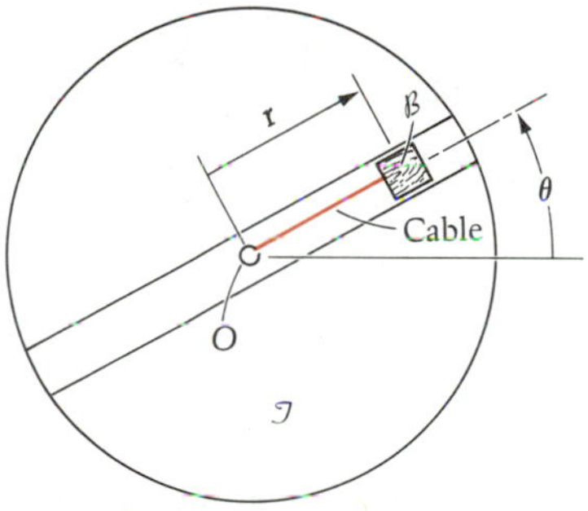

Figure P3.169

3.170 Show that the rigid-body acceleration equation (3.19) can be *derived* from Equation (3.54). *Hint:* If you fix P to $\mathcal{B}$, the equation will then relate the accelerations in $\mathcal{T}$ of the two points P and O' of $\mathcal{B}$.

• **3.171** In Problem 3.153 let $R = 0.1$ m and $\ddot{\theta}_{\mathcal{A}} = 5$ rad/s as in Problem 3.154. This time, again at $\theta = 160°$, find: (a) the angular acceleration of $\mathcal{B}$; (b) the acceleration of P in $\mathcal{B}$ $(= \mathbf{a}_{P/\mathcal{B}})$.

* **3.172** The 26-ft rod $\mathcal{R}$ in Figure P3.172 slides on a plane surface at A and on the fixed half-cylinder at Q as shown below. If end A is moved at a constant velocity of 4 ft/sec to the right along the plane, determine the vertical component of the acceleration of B for the position shown.

* **3.173** Pin P in Figure P3.173 moves along a curved path and is controlled by the motions of the slotted links $\mathcal{A}$ and $\mathcal{B}$. At the instant shown, each point of $\mathcal{A}$ has a velocity of 5 ft/sec and an acceleration of 20 ft/sec², both to the right, while each point of $\mathcal{B}$ has a velocity of 3 ft/sec and an acceleration of 30 ft/sec² in the direction shown in the figure. Find the radius of curvature of the path of P in this position.

* **3.174** Considering the instantaneous center Ⓘ of the bar $\mathcal{B}$ in Figure P3.174 as a point of $\mathcal{B}$ extended, find the acceleration of Ⓘ at the instant shown, if $\mathbf{v}_A = 2\hat{\mathbf{i}}$ in./sec = constant. Also find the acceleration of the point B' of $\mathcal{B}$ passing over the pin, and note that it is not along the slot.

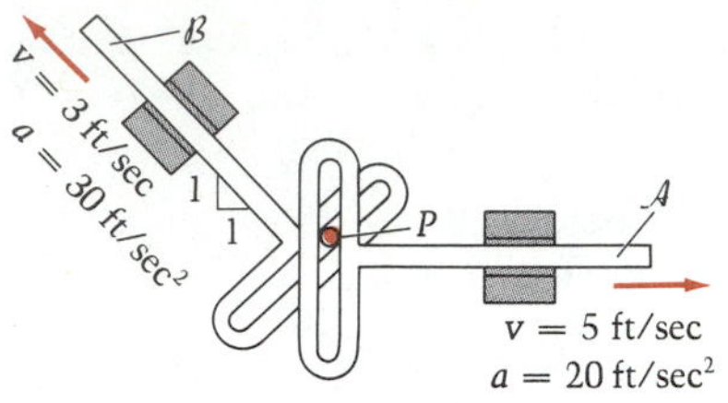

Figure P3.173

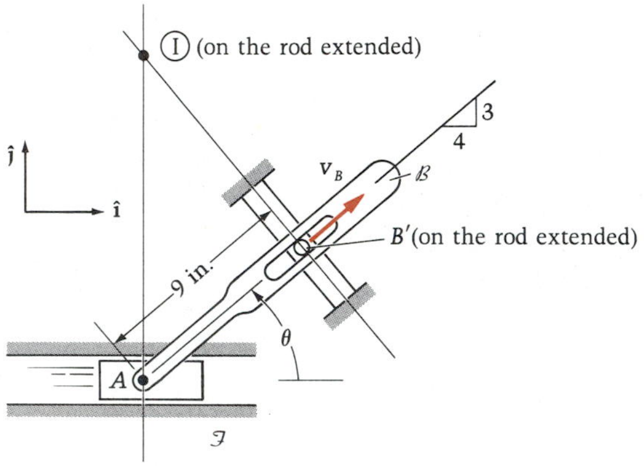

Figure P3.174

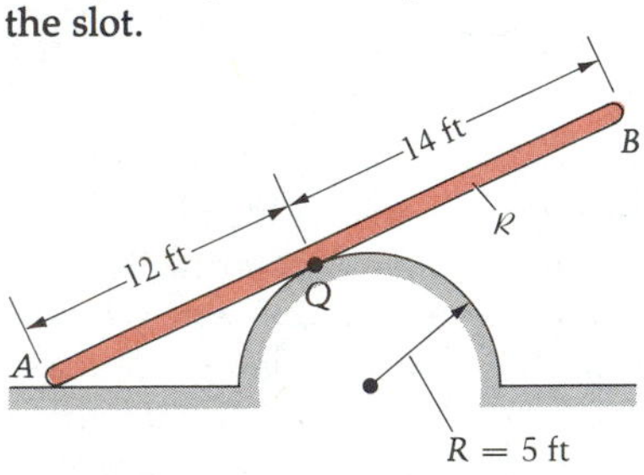

Figure P3.172

COMPUTER PROBLEMS ▸ Chapter 3

* **3.175** Crank $\mathcal{B}_1$ in Figure P3.175 is driven at a constant angular speed ω clockwise. Show that the speed of the piston is maximum when θ satisfies the equation

$$\cos\theta + \frac{\cos 2\theta}{\sqrt{(\ell_2/\ell_1)^2 - \sin^2\theta}} + \frac{\sin^2 2\theta}{4[(\ell_2/\ell_1)^2 - \sin^2\theta]^{3/2}} = 0$$

Solve with a computer for the first root of this equation when the lengths of $\mathcal{B}_1$ and $\mathcal{B}_2$ are 8 cm and 20 cm, respectively. Note that the answer is independent of the value of ω. You may wish to read Appendix B and make use of the Newton-Raphson method described there.

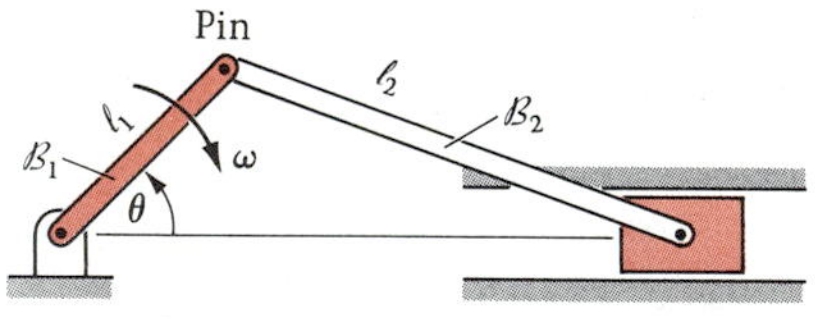

Figure P3.175

* **3.176** Crank $\mathcal{B}_1$ in Figure P3.176 rotates at constant angular velocity $\dot{\theta}\hat{\mathbf{k}}$. Use a computer to generate data for a plot of the following two quantities as functions of θ for the case in which $D = 2\ell$:

  a. The angle φ that locates slider $\mathcal{B}_2$
  b. The ratio $\dot{\varphi}/\dot{\theta}$ of the angular speeds of $\mathcal{B}_2$ and $\mathcal{B}_1$.

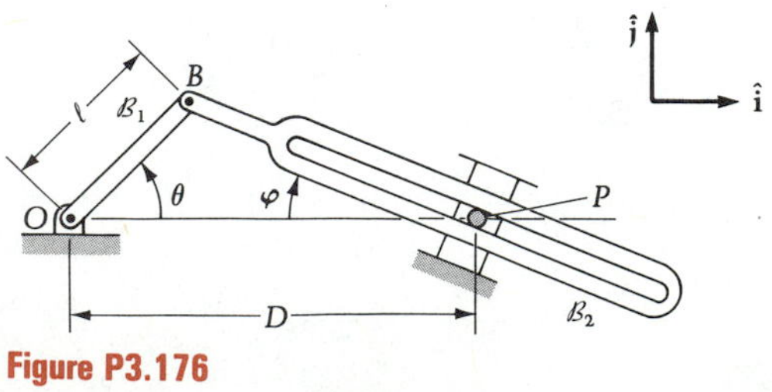

Figure P3.176

SUMMARY ▶ Chapter 3

This chapter has been devoted to presentation of the velocity and acceleration relationships that pertain to a rigid body in plane motion.

If A and B are two points of the body lying in the same plane of motion, then their velocities are linked through the angular velocity $\omega\hat{\mathbf{k}}$ ($\hat{\mathbf{k}}$ is perpendicular to the plane of motion) by

$$\begin{aligned}\mathbf{v}_B &= \mathbf{v}_A + \omega\hat{\mathbf{k}} \times \mathbf{r}_{AB} \\ &= \mathbf{v}_A + r\omega\hat{\mathbf{e}}\end{aligned}$$

If a point instantaneously has zero velocity, call it Ⓘ, then:

$$\begin{aligned}\mathbf{v}_P &= \overset{\mathbf{0}}{\mathbf{v}_{Ⓘ}} + \omega\hat{\mathbf{k}} \times \mathbf{r}_{ⒾP} \\ &= r\omega\hat{\mathbf{e}}\end{aligned}$$

which shows that the speed of any point is the product of the angular speed and the distance between the point and the instantaneous center. For accelerations, where $\alpha\hat{\mathbf{k}}$ is the angular acceleration, we have

$$\mathbf{a}_B = \mathbf{a}_A + \alpha\hat{\mathbf{k}} \times \mathbf{r}_{AB} - \omega^2\, \mathbf{r}_{AB}$$

When one body rolls on another, the points of contact have the same velocity. Rolling of a wheel on a fixed surface means that the point on the wheel in contact with the surface is the instantaneous center of velocity, and with C denoting the center of the wheel,

$$\mathbf{v}_C = \omega\hat{\mathbf{k}} \times \mathbf{r}_{ⒾC}$$

and

$$\mathbf{a}_C = \alpha\hat{\mathbf{k}} \times \mathbf{r}_{ⒾC} + (\text{a part normal to the path of } C)$$

Finally, we have investigated the relationships between velocities and then accelerations of a single point P relative to two reference frames. The results can be expressed compactly if we think of the underlying basic frame of reference as "fixed" and a second body as moving. Then with

$$\begin{aligned}\mathbf{v}_P &= \text{velocity of } P \text{ relative to the fixed frame,} \\ \mathbf{v}_{\text{rel}} &= \text{velocity of } P \text{ relative to the moving body,} \\ \omega\hat{\mathbf{k}} &= \text{angular velocity of the moving body,}\end{aligned}$$

and similarly for accelerations and with O' being a point fixed in the moving body, we have:

$$\mathbf{v}_P = \mathbf{v}_{\text{rel}} + \mathbf{v}_{O'} + \omega\hat{\mathbf{k}} \times \mathbf{r}_{O'P}$$

and

$$\mathbf{a}_P = \mathbf{a}_{\text{rel}} + \mathbf{a}_{O'} + \alpha\hat{\mathbf{k}} \times \mathbf{r}_{O'P} - \omega^2\mathbf{r}_{O'P} + 2\omega\hat{\mathbf{k}} \times \mathbf{v}_{\text{rel}}$$

The last term in the second equation is called the Coriolis acceleration.

REVIEW QUESTIONS ▶ Chapter 3

True or False?

In these questions, P and Q are points in the reference plane of a rigid body $\mathcal{B}$ in plane motion.

1. For a rigid body in plane motion with angular velocity $\dot{\theta}\hat{\mathbf{k}}$, $\dot{\theta}$ depends on a certain choice of points whose velocities are to be related.
2. For a rigid body in plane motion, there is always an instantaneous center Ⓘ of zero velocity located a finite distance from the body.
3. $\mathbf{a}_{Ⓘ}$ is not necessarily zero.
4. $\mathbf{v}_P = \dot{\theta}\hat{\mathbf{k}} \times \mathbf{r}_{ⒾP}$ if the angular velocity of $\mathcal{B}$ is not zero.
5. For a rigid body $\mathcal{B}$ in rectilinear translation, the velocities of all points of $\mathcal{B}$ are equal, and so are the accelerations.
6. For a rigid body $\mathcal{B}$ in curvilinear translation, the velocities of all points of $\mathcal{B}$ are equal, but the accelerations are not.
7. At a given instant, any point can be considered to lie on a rigid extension of any rigid body.
8. The smallest number of scalar parameters required to locate a rigid body in plane motion is four.
9. For a rigid body $\mathcal{B}$ in plane motion,

$$\mathbf{v}_P = \mathbf{v}_Q + \dot{\theta}\hat{\mathbf{k}} \times \mathbf{r}_{QP}$$

10. A point P can have an angular velocity.
11. The components of $\mathbf{v}_P$ and $\mathbf{v}_Q$ along the line PQ are not always equal.
12. The components of $\mathbf{a}_P$ and $\mathbf{a}_Q$ along the line PQ are not always equal.

Answers: 1. F 2. F 3. T 4. T 5. T 6. F 7. T 8. F 9. T 10. F 11. F 12. T

4 Kinetics of a Rigid Body in Plane Motion/ Development and Solution of the Differential Equations Governing the Motion

4.1 Introduction
4.2 Rigid Bodies in Translation
4.3 Moment of Momentum (Angular Momentum)
Inertia Properties
4.4 Moments and Products of Inertia/The Parallel-Axis Theorems
Examples of Moments of Inertia
The Parallel-Axis Theorem for Moments of Inertia
The Radius of Gyration
Products of Inertia
Transfer Theorem for Products of Inertia
4.5 The Mass-Center Form of the Moment Equation of Motion
Development of the Equations of Plane Motion
Helpful Steps to Follow in Generating and Solving the Equations of Motion
4.6 Other Useful Forms of the Moment Equation
Moment Equation in Terms of $\mathbf{a}_C$
Moment Equation in Terms of $\mathbf{a}_P$
Moment Equation for Fixed-Axis Rotation (The "Pivot" Equation)
4.7 Rotation of Unbalanced Bodies
Summary
Review Questions

4.1 Introduction

In this chapter we apply Euler's laws to the plane motions of rigid bodies. Motion of the mass center of any body, rigid or not, is governed by Euler's first law as discussed in Chapter 2. The *rotational* motion of a rigid body is governed by Euler's second law. We saw in Chapter 2 that this law can be expressed in terms of a moment of momentum for any body, rigid or not. However, the moment of momentum for a rigid body can be expressed in a particularly compact way that involves moments and products of inertia of the body and its angular velocity; because of this, the term *angular momentum* is used synonymously with *moment of momentum.*

There is, however, one type of plane motion of a rigid body $\mathcal{B}$ that can be immediately studied, prior to the introduction of angular momentum. This class of motions, called *translation,* is characterized by the angular velocity of $\mathcal{B}$ being always zero. The translation problems treated in the next section (4.2) differ from the particle/mass-center motion problems of Chapter 2 in that a moment equation is required for their solution in addition to the mass-center equation $\Sigma \mathbf{F} = m\mathbf{a}_c$.

In Section 4.3, when we develop the expression for the angular momentum of a rigid body, the moments and products of inertia suddenly appear — in the same way the mass center did, back in Chapter 2. Thus we spend some time in Section 4.4 studying these inertia properties before moving on.

In Sections 4.5 and 4.6 we deduce several especially useful forms of Euler's second law in terms of the inertia properties. After each form of the equation, we illustrate its use with a set of examples. Some of these examples might be termed "snapshot" problems; in these we investigate the relationships between external forces on a body and its accelerations at a single instant. These problems are rather natural extensions of those the student has encountered in statics — that is, we know the geometrical configuration and seek information about forces on the body. Other problems might be called "movie" problems; in one class of these the geometry is of sufficient simplicity that Euler's laws produce differential equations which we can readily integrate so as to predict the motion of the body during some interval of time.

In Section 4.7, we take up the very special problem of rotation of an unbalanced body about a fixed axis. Here we establish the criteria for the technologically important problem of balancing.

In the next chapter, we will continue our study of plane-motion kinetics of rigid bodies by investigating the use of three special solutions (which can be obtained in general) to the differential equations of motion. These special integrals are known as the principles of work and kinetic energy; linear impulse and momentum; and angular impulse and angular momentum.

Finally, we mention to the reader that it is possible to obtain all the results of this chapter on plane motion of rigid bodies from the general three-dimensional results developed in Chapter 7. It is not necessary to travel this complex route in order to learn plane motion, however, and in this chapter we take a simpler path. It is worth noting that while the planar case covers a restricted class of motions, it does in fact contain a large number of problems with important engineering applications.

4.2 Rigid Bodies in Translation

In Chapter 2 we presented the equation

$$\Sigma \mathbf{M}_P = \int \mathbf{r} \times \mathbf{a}\, dm \tag{2.34}$$

which is valid both for *any* body and *any* point P. In particular, if the body is translating, then by definition all its points have the same acceleration — including its mass center — and if we label that common acceleration $\mathbf{a}$, then it may be factored, leaving:

$$\begin{aligned} \Sigma \mathbf{M}_P &= \left(\int \mathbf{r}\, dm\right) \times \mathbf{a} \\ &= m\mathbf{r}_{PC} \times \mathbf{a} \qquad \text{(using the definition of the mass center)} \\ &= \mathbf{r}_{PC} \times (m\mathbf{a}) \end{aligned} \tag{4.1a}$$

and therefore

$$\Sigma \mathbf{M}_C = \mathbf{0} \tag{4.1b}$$

In this section, we shall apply this simple equation to several examples of translating rigid bodies in plane motion. The reader should note, however, that the two forms of Equation (4.1) apply whether the motion is plane or not. It also doesn't even require the body to be "physically rigid," although if it is translating, it is necessarily behaving like a rigid body during that motion.

Question 4.1 If point P is arbitrary in Equation (4.1a), where does the inertial frame come into the equation?

Before wading into the examples, we wish to note what is new here. In Chapter 2, we were concerned with the *mass center motions* of bodies, whether they were rigid or not. In those sections there was no need to sum moments, and *that* is what will distinguish this chapter from those preceding sections. Now the simplest problems by far in which a moment equation is sometimes needed are those involving translation. Translation is simple because it can be studied prior to the introduction of angular momentum forms for rigid bodies with their accompanying inertia properties and angular velocities.

We now examine three examples of translation, and in each, the reader is urged to note two things: (1) how the problem could not be

Answer 4.1 The acceleration $\mathbf{a}$ is the second derivative, taken in an inertial frame, of the position vector from an origin in that frame to any point of the body.

solved without the use of Equation (4.1); and (2) how for translation, the moments generally do *not* sum to zero. They do sum to zero at the mass center, and also at points lying on the line of **a** drawn through C (for then $\mathbf{r}_{PC}$ is parallel to **a**)*, but *not otherwise.* (Equations for translation at constant velocity are trivial and identical to the equilibrium equations; these were studied in statics and are not considered in this book.)

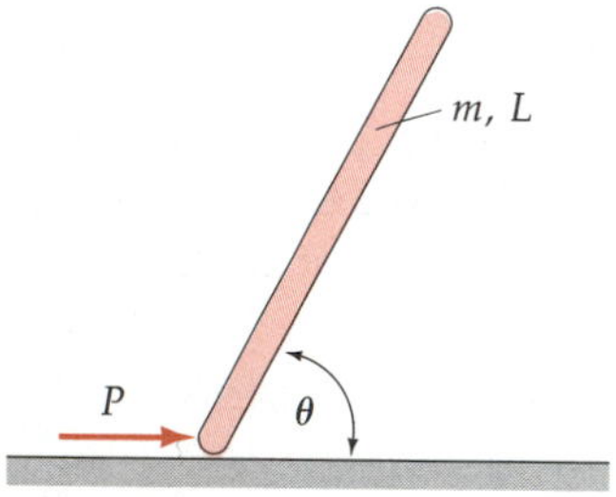

Figure E4.1a

EXAMPLE 4.1

Find the angle θ for which the bar in Figure E4.1a will translate to the right at the given constant acceleration "a." Then find the force P required to produce this motion.

Solution

We sum moments at the contact point A, using the FBD in Figure E4.1b and Equation (4.1):

$$\Sigma \mathbf{M}_A = \mathbf{r}_{AC} \times m\mathbf{a}$$

$$-mg\frac{L}{2}\cos\theta\,\hat{\mathbf{k}} = \frac{L}{2}(\cos\theta\hat{\mathbf{i}} + \sin\theta\hat{\mathbf{j}}) \times ma\hat{\mathbf{i}}$$

$$-mg\frac{L}{2}\cos\theta\,\hat{\mathbf{k}} = -\frac{Lma}{2}\sin\theta\,\hat{\mathbf{k}}$$

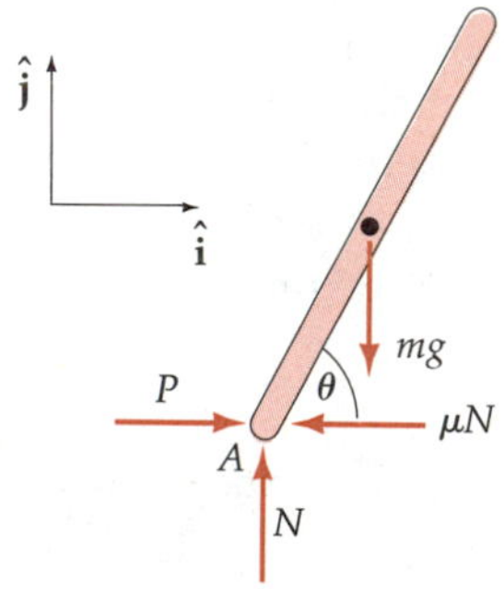

Figure E4.1b

Therefore θ in terms of a is given by:

$$\tan\theta = \frac{g}{a} \Rightarrow \theta = \tan^{-1}\left(\frac{g}{a}\right)$$

To determine P, we write the mass-center equations:

$$\Sigma F_x = P - \mu N = ma \tag{1}$$

and

$$\Sigma F_y = N - mg = m\ddot{y}_c = 0 \Rightarrow N = mg$$

Substituting this value of N into Equation (1) gives

$$P - \mu mg = ma$$

or

$$P = m(\mu g + a)$$

Note that part of P balances the friction and the rest, the unbalanced force in the x-direction, produces the "ma."

* and of course at times when the equal accelerations of all points, **a**, are zero.

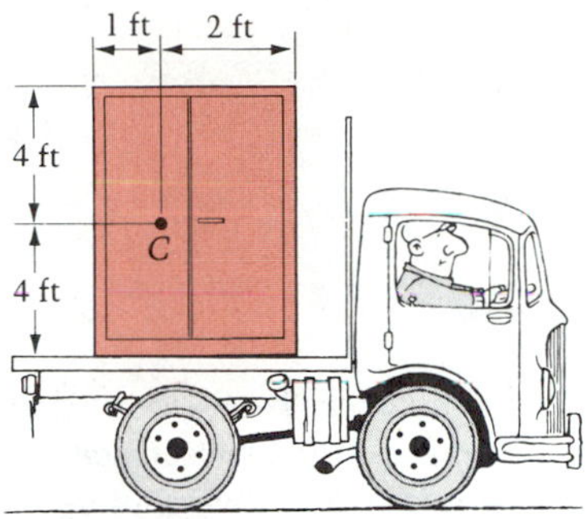

Figure E4.2a

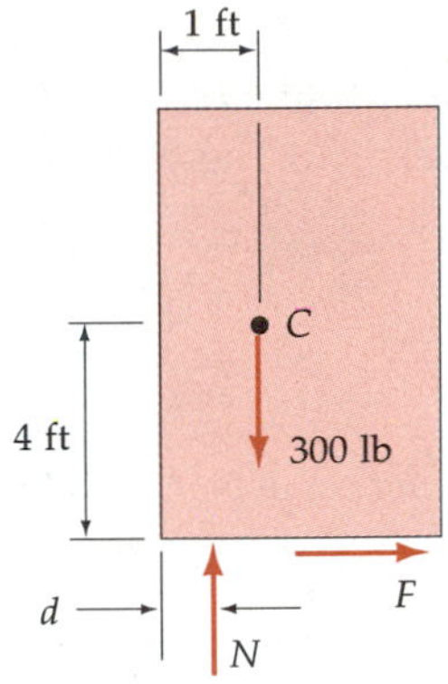

Figure E4.2b

Example 4.2

A 300-lb cabinet is to be transported on a truck as shown in Figure E4.2a. Assuming sufficient friction so that the cabinet will not slide on the truck, for forward acceleration find

a. the forces exerted by the truck bed on the cabinet for 5 ft/sec^2 acceleration;

b. the maximum acceleration for which the cabinet will not tip over.

Solution

a. The acceleration is 5 $ft/sec^2 \rightarrow$, so using the free body diagram (Figure E4.2b) and putting $\Sigma\mathbf{F} = m\mathbf{a}$ into component form,

$$\xrightarrow{+} \quad \Sigma F_x = ma_x$$

$$F = \left(\frac{300}{32.2}\right) 5 = 46.6 \text{ lb}$$

$$+\uparrow \quad \Sigma F_y = ma_y$$

$$N - 300 = 0$$

$$N = 300 \text{ lb}$$

To locate the line of action of N, that is, to find the distance d, we need the moment equation:

$$\Sigma M = 0$$

$$4F - (1 - d)N = 0$$

$$4(46.6) - (1 - d)(300) = 0$$

$$1 - d = 0.621$$

$$d = 0.379 \text{ ft}$$

We might note at this point that the minimum coefficient of friction for which this motion is possible is

$$\mu_{min} = \frac{46.6}{300} = 0.155$$

b. Assuming we still have adequate friction to prevent slip, but treating the acceleration magnitude, a, as an unknown, we can retrace our steps as in part (a) to obtain

$$F = \frac{300}{32.2} a$$

$$N = 300$$

and

$$4F - (1 - d)N = 0$$

When the cabinet is on the verge of tipping, the line of action of N is at the left corner, $d = 0$, so for that condition:

$$4F - (1 - 0)(300) = 0$$

$$F = 75 \text{ lb}$$

and then from

$$F = \frac{300}{32.2} a$$

we find

$$a = \frac{32.2}{300}(75) = 8.05 \text{ ft/sec}^2$$

The reader should note that this tendency to tip over "backwards" is a phenomen uniquely of dynamics; there's nothing quite like it in statics. In addition we sometimes tend to think of friction in oversimplified terms, as perhaps "always opposing motion"; of course it is precisely the friction that here *provides* the motive force to cause the cabinet to accelerate.

In the preceding example we saw how the state of translation could be jeopardized by the tendency of a body to "rock." In the next we *additionally* explore a tendency to slip.

Figure E4.3a

EXAMPLE 4.3

The coefficient of friction at both ends of the uniform slender bar in Figure E4.3a is 0.5. Find the maximum forward acceleration that the truck may have without the bar moving relative to the truck.

Solution

One possibility is that the upper end separates from the truck body. The free-body diagram in Figure E4.3b shows the situation when the end is barely about to break away. Because the bar is in translation, we may write:

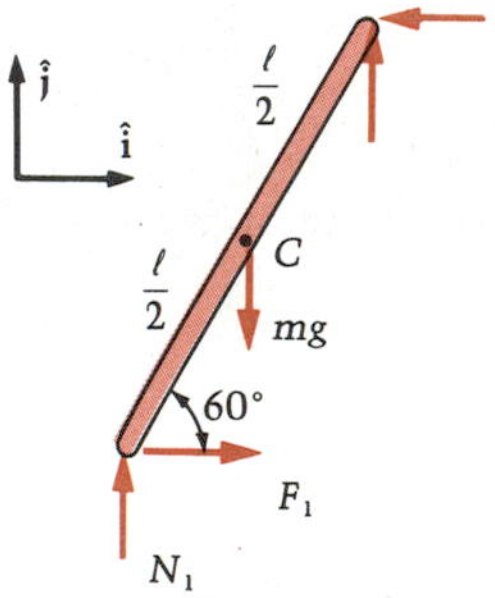

Figure E4.3b

$$\Sigma M_C = 0$$

$$F_1 \frac{\ell}{2} \sin 60° - N_1 \frac{\ell}{2} \cos 60° = 0$$

$$\frac{\sqrt{3}}{2} F_1 - \frac{1}{2} N_1 = 0$$

$$F_1 = N_1/\sqrt{3} = 0.577 N_1$$

But this much friction cannot be generated because $\mu = 0.5 < 0.577$. Therefore motion of the rod relative to the truck will not be initiated by this mechanism.

The other possibility, as shown in the second free-body diagram, Figure E4.3c, is that the bar is on the verge of slipping where it contacts the truck; this must occur at the two surfaces simultaneously. The equations of motion are (with $\mathbf{a} = \ddot{x}\hat{\mathbf{i}}$)

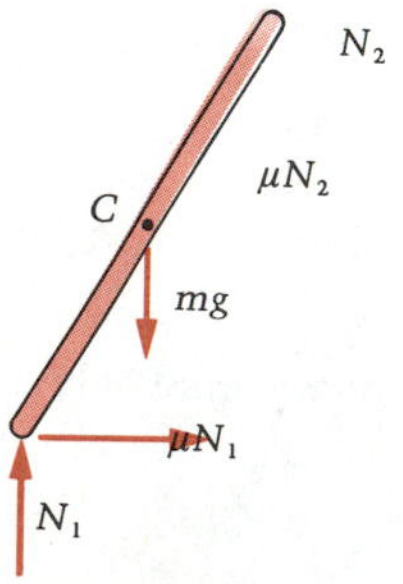

Figure E4.3c

$$\Sigma F_x = m\ddot{x}$$

$$\mu N_1 - N_2 = m\ddot{x} \quad (1)$$

$$\Sigma F_y = m\ddot{y} = 0$$

$$N_1 + \mu N_2 - mg = 0 \quad (2)$$

$$\Sigma M_C = 0$$

$$\frac{\sqrt{3}}{2}\left(\frac{\ell}{2}\right)\mu N_1 - \frac{1}{2}\left(\frac{\ell}{2}\right)N_1 + \frac{1}{2}\left(\frac{\ell}{2}\right)\mu N_2 + \frac{\sqrt{3}}{2}\left(\frac{\ell}{2}\right)N_2 = 0 \tag{3}$$

With $\mu = 0.5$, Equation (3) yields

$$N_1 = 16.7N_2$$

and from (2)

$$N_2 = 0.0581\ mg$$

so that

$$N_1 = 16.7N_2 = 0.971\ mg$$

Equation (1) then gives

$$0.5(0.971\ mg) - 0.0581\ mg = m\ddot{x}$$

$$\ddot{x} = 0.427\ g$$

which is 13.8 ft/sec² (for $g = 32.2$ ft/sec²) or 4.19 m/s² (for $g = 9.81$ m/s²).

The reader is encouraged to rework the previous example using Equation (4.1a) to sum moments about the bottom point of the translating bar.

PROBLEMS ► Section 4.2

4.1 For what force P is it possible for the uniform slender bar (Figure P4.1) to translate across the smooth floor in the position shown? The bar has mass m and length ℓ.

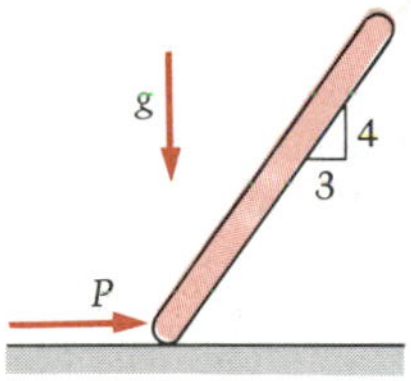

Figure P4.1

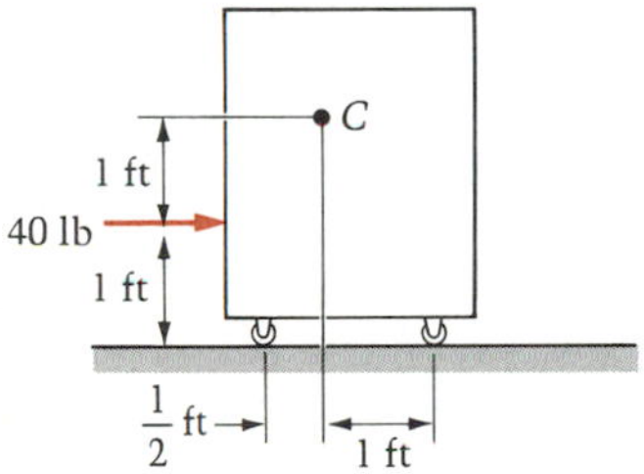

Figure P4.2

4.2 A 100-lb cabinet, rolling on small wheels, is subjected to a 40-lb force as shown in Figure P4.2. Neglecting friction, find (a) the acceleration of the cabinet; (b) the reactions of the floor on the wheels.

4.3 Repeat the preceding problem for the case where the 40-lb force is applied 1 ft *above* C.

4.4 Find the value of F for which one of the wheels of the door in Figure P4.4 lifts out of its track. Which one? Assume negligible friction.

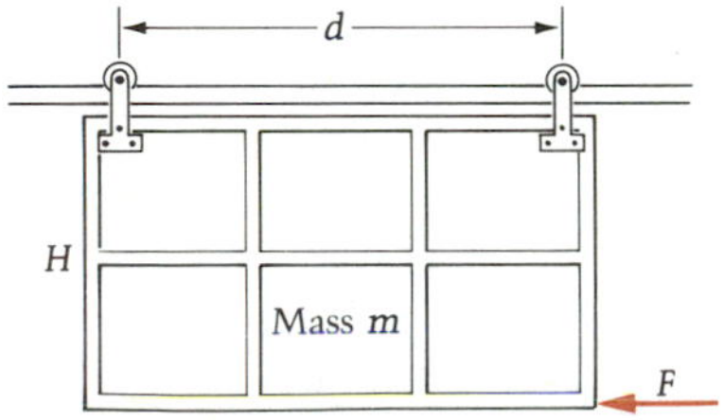

Figure P4.4

4.5 The force P causes the uniform rectangular box of weight W in Figure P4.5 to slide. Find the range of values of H for which the box will not tip about either the front or rear lower corner as it slides on the smooth floor, if $P = W$.

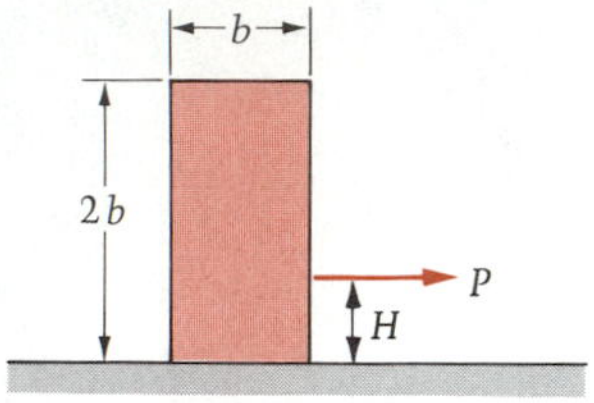

Figure P4.5

4.6 Repeat the preceding problem for a coefficient of sliding friction of 0.2.

4.7 A 400-lb cabinet is to be transported on a truck as shown in Figure P4.7. Assuming sufficient friction so that the cabinet will not slide on the truck, what is the maximum forward acceleration for which the cabinet will not tip over?

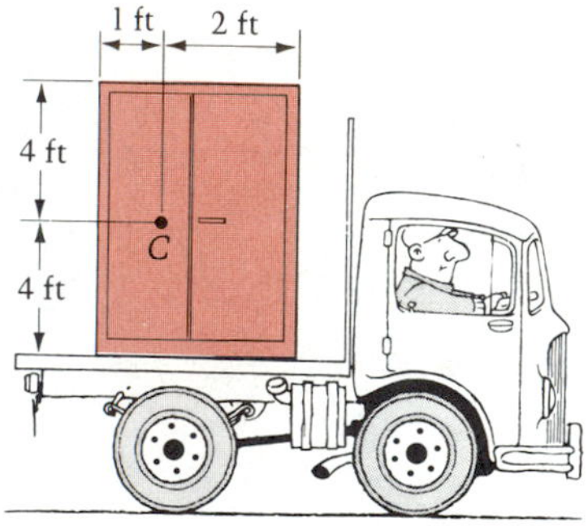

Figure P4.7

4.8 The uniform bar $\mathcal{B}$ in Figure P4.8 weighs 60 lb and is pinned at A (and fastened by the cable DB) to the frame $\mathcal{F}$. If the frame is given an acceleration $a = 32.2$ ft/sec² as shown, determine the tension T in the cable and the force exerted by the pin at A on the bar.

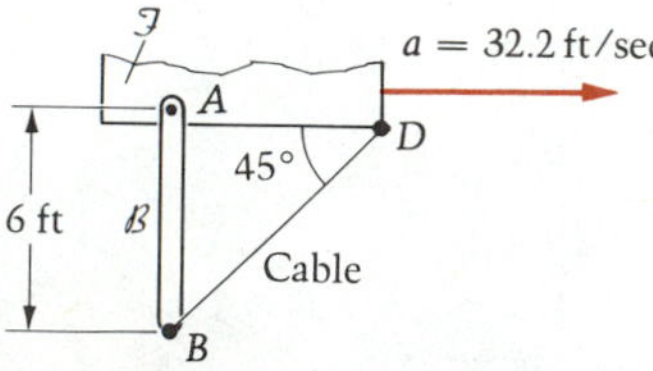

Figure P4.8

4.9 The cords in Figure P4.9 have a tensile strength of 12 N. Cart $\mathcal{C}$ has a mass of 35 kg exclusive of the 10-kg and 1.2-m vertical rod $\mathcal{R}$, which is pinned to it at A. Find the maximum value of P that can be exerted without breaking either cord if: (a) P acts to the right as shown; (b) P acts to the left. Neglect friction, and assume negligible tension in each cord when the cart is at rest.

4.10 A force F, alternating in direction, causes the carriage to move with rectilinear horizontal motion defined by the equation $x = 2 \sin \pi t$ ft, where x is the displacement in feet and t is the time in seconds. (See Figure P4.10.) A rigid, slender, homogeneous rod of weight 32.2 lb and length 6 ft is welded to the carriage at B and projects vertically upward. Find, in magnitude and direction, the bending moment that the carriage exerts on the rod at B when $t = \frac{1}{2}$ sec.

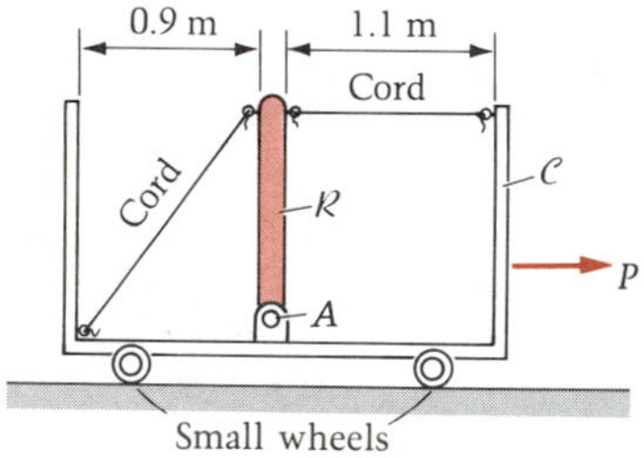

Figure P4.9

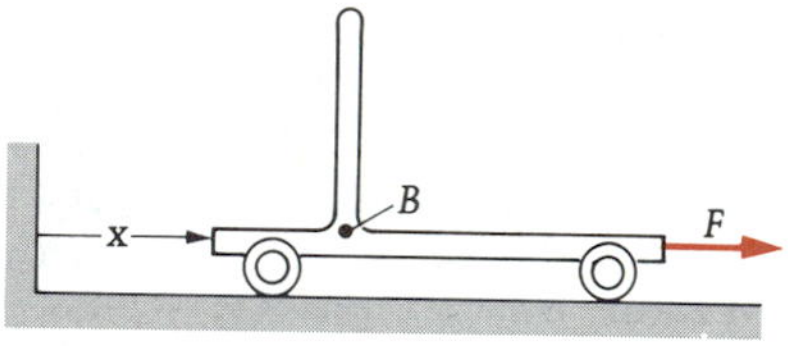

Figure P4.10

4.11 A child notices that sometimes the ball $\mathcal{M}$ does not roll down the inclined surface of toy $\mathcal{T}$ when she pushes it along the floor. (See Figure P4.11.) What is the minimum acceleration a_{min} of $\mathcal{T}$ to prevent this rolling?

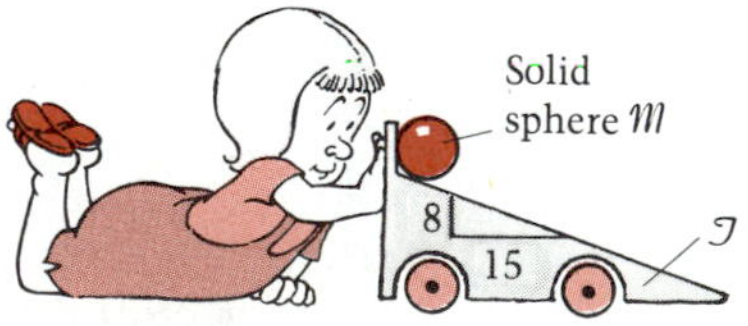

Figure P4.11

4.12 In the preceding problem, suppose the acceleration of $\mathcal{T}$ is $2a_{min}$. What is the normal force between the smooth vertical surface of $\mathcal{T}$ and the ball? The ball's weight is 0.06 lb.

4.13 A can $\mathcal{C}$ that may be considered a uniform solid cylinder (see Figure P4.13) is pushed along a surface $\mathcal{B}$ by a moving arm $\mathcal{A}$. If it is observed that $\mathcal{C}$ *translates* to the right with $\ddot{x}_C = g/10$, what must be the minimum coefficient of friction between $\mathcal{A}$ and $\mathcal{C}$? The coefficient of friction between $\mathcal{C}$ and $\mathcal{B}$ is μ.

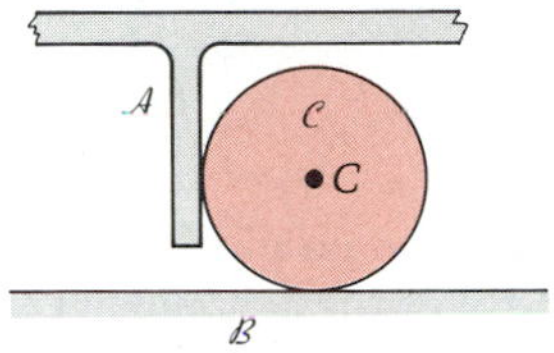

Figure P4.13

4.14 The force P is applied to cart $\mathcal{C}$, and increases slowly from zero, always acting to the right. (See Figure P4.14.) Point C is the mass center of $\mathcal{B}$. At what value of P will the bodies $\mathcal{B}$ and $\mathcal{C}$ no longer move as one?

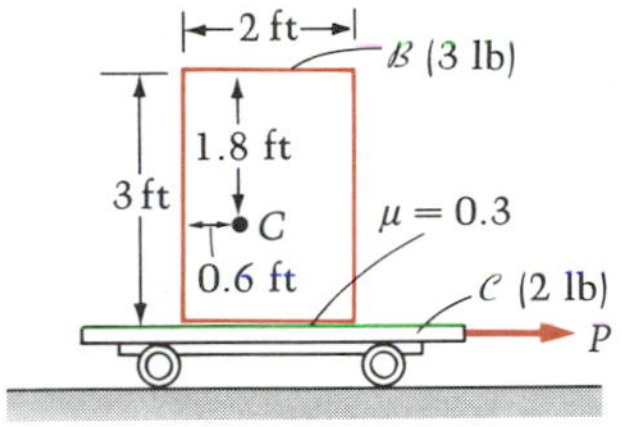

Figure P4.14

4.15 A nonuniform block rests on a flatcar as shown in Figure P4.15. If the coefficient of friction between car and block is 0.40, for what range of accelerations of the car will the block neither tip nor slide?

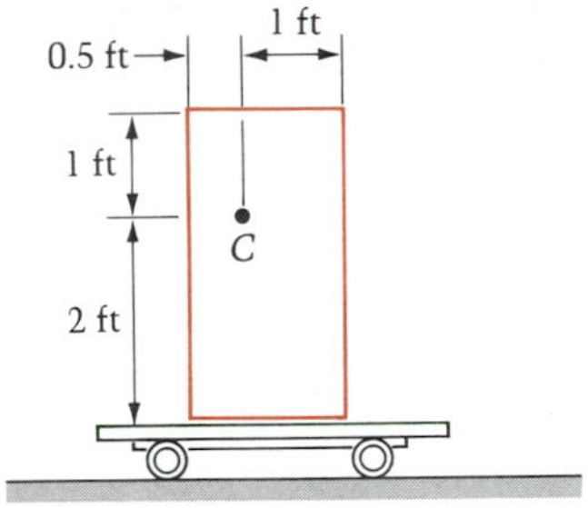

Figure P4.15

4.16 The truck in Figure P4.16 is traveling at 45 mph. Find the minimum stopping distance such that the 250-lb crate will neither slide nor tip over.

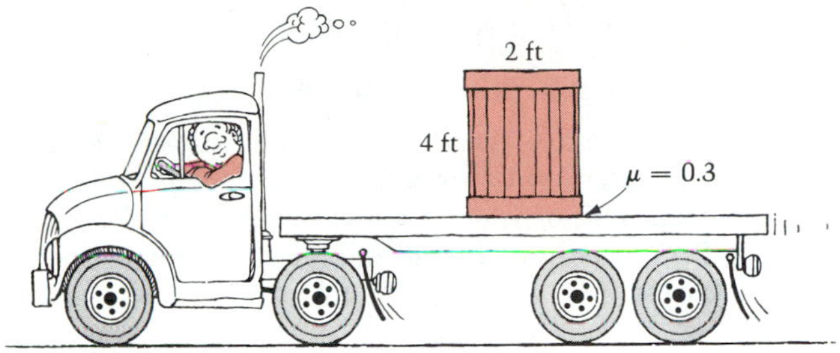

Figure P4.16

4.17 The block $\mathcal{B}$ of mass m in Figure P4.17 is resting on the cart $\mathcal{C}$ of mass M. Force P is applied to the cart, starting it in motion to the right. The wheels are small and frictionless, and the coefficient of friction between $\mathcal{B}$ and $\mathcal{C}$ is $\mu = \frac{1}{4}$. Find the largest value of P for which $\mathcal{B}$ and $\mathcal{C}$ will move together, considering all cases.

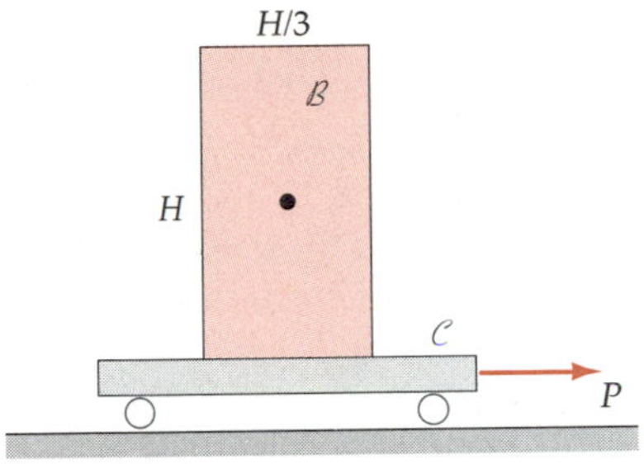

Figure P4.17

4.18 The 25-lb triangular plate is smoothly pinned at vertex A to a small, light wheel (see Figure P4.18). Find the value of force P so that the plate, in theory, will translate along the incline. Also find the acceleration.

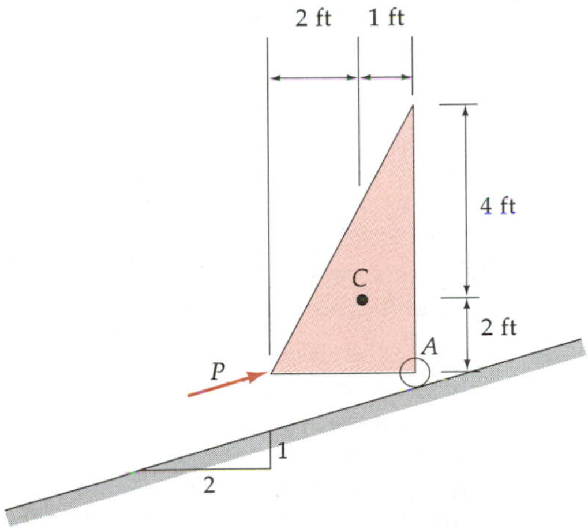

Figure P4.18

4.19 Repeat the preceding problem if the angle of the plane is changed from $\angle^{1}_{2}$ to $\angle^{3}_{4}$.

4.20 The monorail car in Figure P4.20 is driven through its front wheel and moves forward from left to right. If the coefficient of friction between wheels and track is $\mu = 0.55$, determine the maximum acceleration possible for the car.

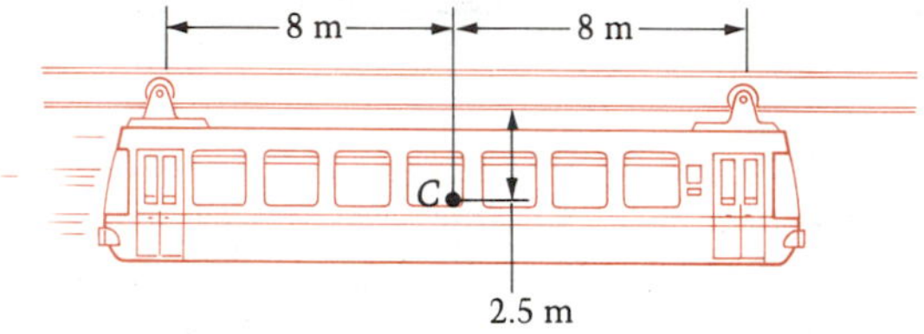

Figure P4.20

4.21 A dragster is all set for the annual neighborhood race. (See Figure P4.21.)

a. In terms of the dimensions b, H, and d and the coefficient of friction μ, find the maximum possible acceleration of the car. Neglect the rotational inertia of the wheels.

b. How would you adjust the four parameters b, H, d, and μ to further increase the driver's acceleration?

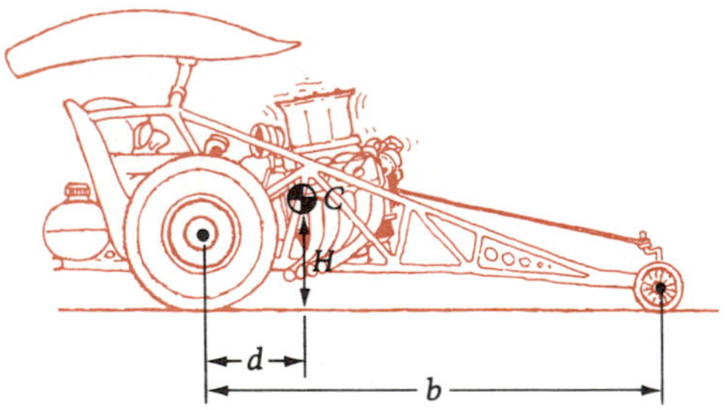

Figure P4.21

4.22 Rework the preceding problem for a car with (a) front-wheel drive and (b) four-wheel drive.

4.23 In an emergency the driver of an automobile applies his brakes; the front brakes fail and the rear wheels are locked. Find the time and distance required to bring the car to rest. Neglect the masses of the wheels, and express the results in terms of the coefficient of sliding friction μ, the initial speed v, the gravitational acceleration g, and the dimensions shown in Figure P4.23.

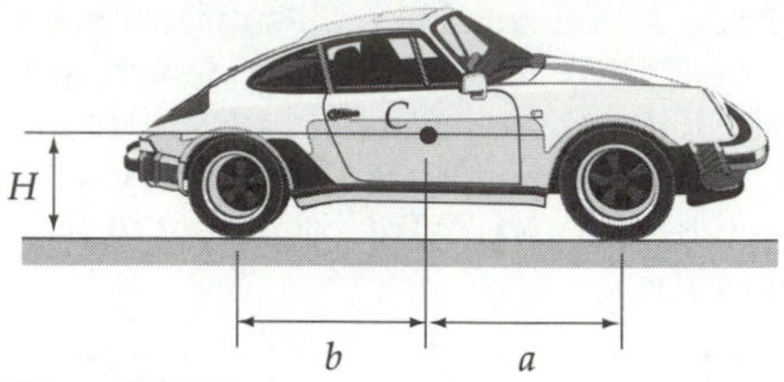

Figure P4.23

4.24 A uniform rod $\mathcal{A}$ of length L and weight W is connected to smooth hinges at E and D by the light members B and $\mathcal{C}$, each of length L. In the position shown in Figure P4.24, $\mathcal{B}$ has an angular velocity of ω rad/sec clockwise. Find the forces in members $\mathcal{B}$ and $\mathcal{C}$, and determine the acceleration of center C of rod $\mathcal{A}$ in terms of the given variables.

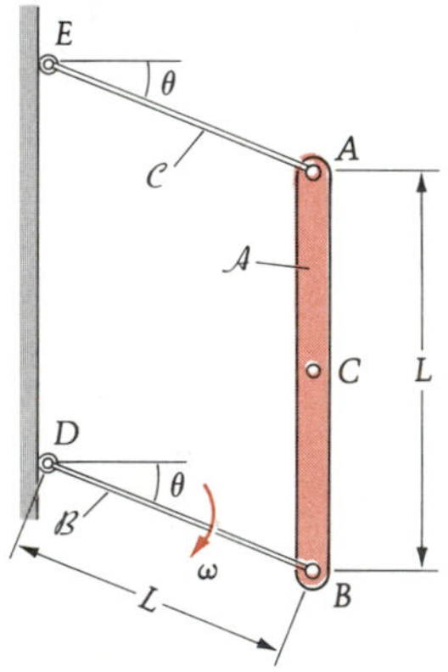

Figure P4.24

• **4.25** Find the range of accelerations of $\mathcal{W}$ that F can produce without $\mathcal{B}$ moving in *any* manner relative to $\mathcal{W}$. (See Figure P4.25.) Note carefully the position of the mass center of $\mathcal{B}$.

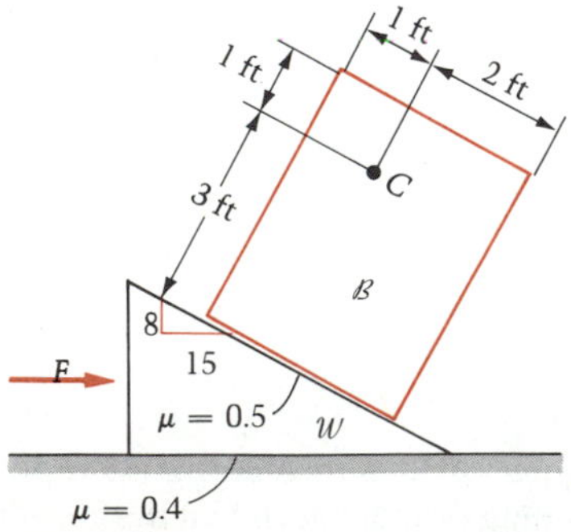

Figure P4.25

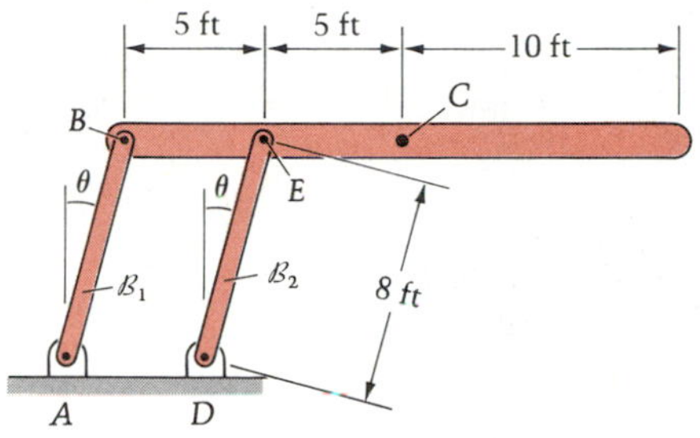

Figure P4.26

* **4.26** A slender homogeneous rod weighing 64.4 lb and 20 ft long is supported as shown in Figure P4.26. Bars $\mathcal{B}_1$ and $\mathcal{B}_2$ are of negligible mass and have frictionless pins at each end. The system is released from rest with $\theta = 0$.

  a. Derive expressions for the angular velocity and acceleration of bars $\mathcal{B}_1$ and $\mathcal{B}_2$ as functions of θ.

  b. Derive expressions for the axial force in bars $\mathcal{B}_1$ and $\mathcal{B}_2$ as a function of θ.

4.3 Moment of Momentum (Angular Momentum)

We recall from Chapter 2 that the moment of momentum of any body $\mathcal{B}$ with respect to a point P (not necessarily fixed in either the body or in the reference frame) was defined by Equation (2.35):

$$\mathbf{H}_P = \int_{\mathcal{B}} \mathbf{R} \times \mathbf{v}\, dm$$

where in this section $\mathbf{R}$ is the vector from P to the element of mass dm of the body, and $\mathbf{v}$ is the velocity of dm in the reference frame $\mathcal{I}$. (See Figure 4.1.) We note that $\mathbf{H}_P$ depends on the location of P as well as the distributions of mass and velocities in the body; it is seen to be the sum of the moments of momenta of all the mass elements of $\mathcal{B}$.

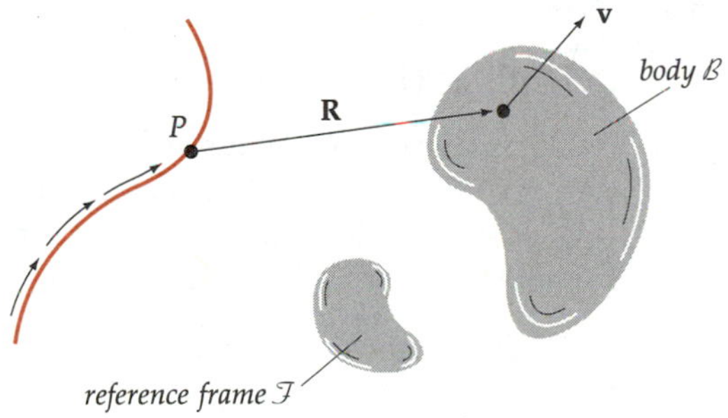

Figure 4.1

We shall now restrict the general body above to be rigid, place it in plane motion, and recall from Chapter 3 that the kinematics of a rigid body in plane motion can be described very simply. If we know the velocity of just one point and the angular velocity of the body, then we know the velocity of *every point* of $\mathcal{B}$—quite a bargain. Because of this simplicity, we shall see that compact and yet completely general expressions can be written for the moment of momentum (or angular momentum, as it is often called for rigid bodies). We shall further restrict the generic point P in the foregoing to be a point of $\mathcal{B}$.

The rectangular axes (x, y, z) have origin at P as shown in Figure 4.2, and the xy-plane is the reference plane, or plane of motion, as described in Sections 3.1, 2.

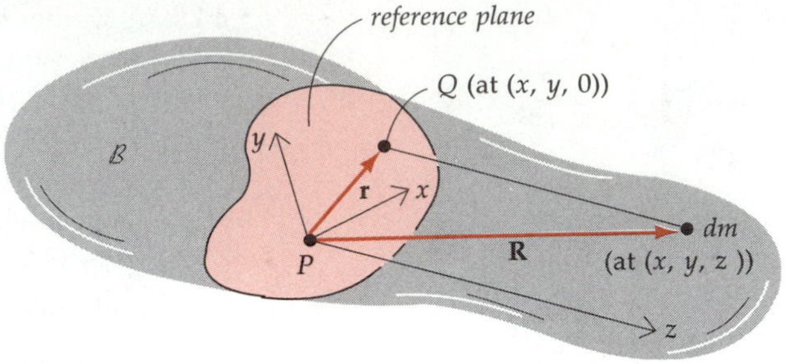

Figure 4.2

Recall also from Section 3.1 that each point of the rigid body has a "companion point" in the reference plane which always has the same x and y as the point, and hence also has the same velocity and acceleration. For the point at "dm" in Figure 4.2, the companion point is Q. Thus, using Equation (3.8),

$$\mathbf{v} = \mathbf{v}_Q = \mathbf{v}_P + \omega\hat{\mathbf{k}} \times \mathbf{r}_{PQ}$$

Substituting this expression for $\mathbf{v}$ and the coordinates of dm into R, $\mathbf{H}_P$ becomes:

$$\mathbf{H}_P = \int \underbrace{(x\hat{\mathbf{i}} + y\hat{\mathbf{j}} + z\hat{\mathbf{k}})}_{\mathbf{R}} \times [\mathbf{v}_P + \omega\hat{\mathbf{k}} \times \underbrace{(x\hat{\mathbf{i}} + y\hat{\mathbf{j}})}_{\mathbf{r}_{PQ}}]\, dm$$

or

$$\mathbf{H}_P = \left(\int \mathbf{R}\, dm\right) \times \mathbf{v}_P + \int (x\hat{\mathbf{i}} + y\hat{\mathbf{j}} + z\hat{\mathbf{k}}) \times [\omega\hat{\mathbf{k}} \times (x\hat{\mathbf{i}} + y\hat{\mathbf{j}})]\, dm$$

The integral $\int \mathbf{R}\, dm$ in the first term above is equal to $m\mathbf{r}_{PC}$ by the definition of the mass center. Making this substitution, and also carrying out the cross products in the second term, gives the moment of momentum vector in terms of $\mathbf{v}_P$, ω, and certain mass distribution integrals:

$$\mathbf{H}_P = \mathbf{r}_{PC} \times m\mathbf{v}_P + \hat{\mathbf{k}}\omega\int (x^2 + y^2)\, dm - \hat{\mathbf{i}}\omega\int xz\, dm - \hat{\mathbf{j}}\omega\int yz\, dm$$

Inertia Properties

We call the integrals in this equation inertia properties. Specifically:

$$\int (x^2 + y^2)\, dm = I_{zz}^P = \text{moment of inertia of mass of } \mathcal{B} \text{ about } z \text{ axis through } P \tag{4.2a}$$

$$-\int xz\, dm = I_{xz}^P = \text{product of inertia of mass of } \mathcal{B} \text{ with respect to } x \text{ and } z \text{ axes through } P^* \tag{4.2b}$$

* If the products of inertia are defined with the minus sign as above, then and only then will the inertia properties transform as a tensor — a topic beyond the scope of this book, however.

$$-\int yz\,dm = I_{yz}^P = \text{product of inertia of mass of } \mathcal{B} \text{ with respect to } y \text{ and } z \text{ axes through } P \tag{4.2c}$$

Thus,

$$\mathbf{H}_P = \mathbf{r}_{PC} \times m\mathbf{v}_P + I_{xz}^P \omega \hat{\mathbf{i}} + I_{yz}^P \omega \hat{\mathbf{j}} + I_{zz}^P \omega \hat{\mathbf{k}} \tag{4.3}$$

We are now at the point in our development where the inertia properties, like the mass center in Chapter 2, have arisen naturally. We shall spend the next section studying the moments and products of inertia; readers already familiar with inertia properties may wish to skip Section 4.4.

Before leaving Equation (4.3), we note for future reference that its first term, $\mathbf{r}_{PC} \times m\mathbf{v}_P$, vanishes if P is the mass center or has zero velocity* In both these cases, which will prove valuable to us, $\mathbf{H}_P$ takes the form

$$\mathbf{H}_P = I_{xz}^P \omega \hat{\mathbf{i}} + I_{yz}^P \omega \hat{\mathbf{j}} + I_{zz}^P \omega \hat{\mathbf{k}} \qquad (\text{if } P \text{ is } C, \text{ or if } \mathbf{v}_P = \mathbf{0}) \tag{4.4}$$

Thus in these cases the moment of momentum can be expressed in terms of the angular velocity of $\mathcal{B}$ (hence its other name: **angular momentum**), along with three measures of its mass distribution.

Question 4.2 In Equation (4.3), does the mass center have to lie in the reference plane with P? How about in Equation (4.4) when P has zero velocity?

4.4 Moments and Products of Inertia / The Parallel-Axis Theorems

Examples of Moments of Inertia

From the definition of moment of inertia,

$$I_{zz}^P = \int (x^2 + y^2)\,dm$$

we see that I_{zz}^P is a measure of "how much mass is located how far" from the z axis through P. In cylindrical coordinates we have $I_{zz}^P = \int r^2\,dm$, and thus I_{zz}^P measures the sum total of mass times distance squared over the body's volume. The quantity I_{zz}^P is thus seen to be always positive. We now compute the mass-center moments of inertia of a number of common shapes. In Examples 4.4–4.12, we are seeking I_{zz}^C.

* Or if $\mathbf{r}_{PC}$ is parallel to $\mathbf{v}_P$, a case we need not consider here.

Answer 4.2 No. No.

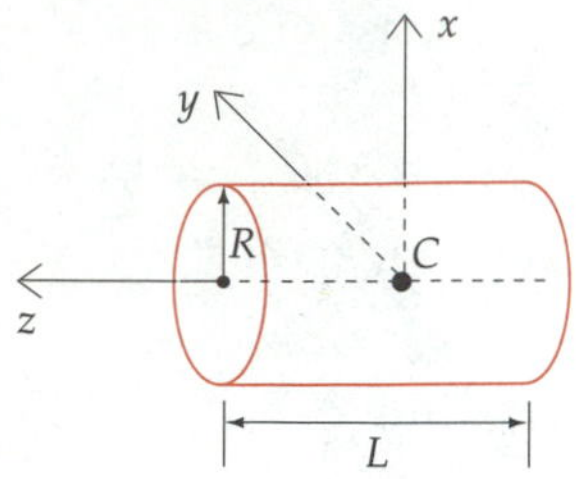

Figure E4.4

EXAMPLE 4.4

Homogeneous solid cylinder if z is its axis (see Figure E4.4).

Solution

Noting that $dm = \rho\, dV$, where ρ = mass density,

$$I_{zz}^{C} = \int_{\text{vol}} (x^2 + y^2)\overbrace{\rho\, dV}^{dm} = \int_{z=-L/2}^{L/2} \int_{\theta=0}^{2\pi} \int_{r=0}^{R} r^2 \rho (r\, dr\, d\theta\, dz)$$

$$I_{zz}^{C} = \rho \frac{r^4}{4}\Big]_0^R \theta\Big]_0^{2\pi} z\Big]_{-L/2}^{L/2} = (\rho\pi R^2 L)\frac{R^2}{2} = \boxed{\frac{mR^2}{2}}$$

EXAMPLE 4.5

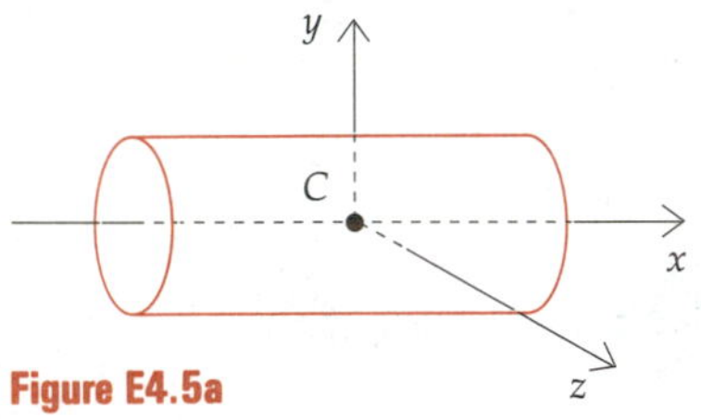

Figure E4.5a

Homogeneous solid cylinder if z is an axis *normal* to the axis of the cylinder (see Figure E4.5a).

Solution

$$I_{zz}^{C} = \int_{\text{vol}} (x^2 + y^2)\rho\, dV = \int_{x=-L/2}^{L/2} \int_{\theta=0}^{2\pi} \int_{r=0}^{R} [\underbrace{(r \sin\theta)^2}_{y = r\sin\theta} + x^2]\rho r\, dr\, d\theta\, dx$$

$$= \int_{x=-L/2}^{L/2} \int_{\theta=0}^{2\pi} \left(\frac{r^4}{4} \underbrace{\sin^2\theta}_{\frac{1-\cos 2\theta}{2}} + \frac{x^2 r^2}{2}\right)\Bigg]_0^R \rho\, d\theta\, dx$$

$$= \int_{-L/2}^{L/2} \left[\frac{R^4}{4}\left(\frac{\theta}{2} - \frac{\sin 2\theta}{4}\right)\Bigg|_0^{2\pi} + \frac{x^2 R^2 \theta}{2}\Bigg|_0^{2\pi}\right] \rho\, dx$$

$$= \rho \frac{\pi R^4}{4} x\Bigg|_{-L/2}^{L/2} + \frac{\rho R^2 \pi x^3}{3}\Bigg|_{-L/2}^{L/2} = (\rho\pi R^2 L)\left(\frac{R^2}{4} + \frac{L^2}{12}\right)$$

$$= \boxed{\frac{mR^2}{4} + \frac{mL^2}{12}}$$

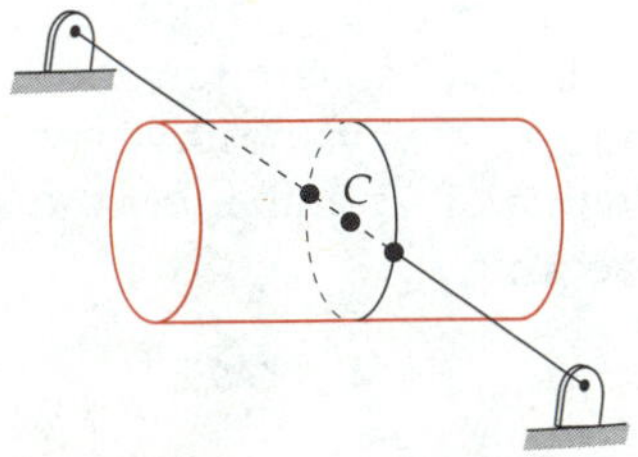

Figure E4.5b

It may be confusing at first to know which moment of inertia to use for a cylinder in a plane kinetics problem; the answer is that it is always the value associated with the axis normal to the xy plane of the motion. If the problem is a rolling cylinder, $I_{zz}^{C} = mR^2/2$. If we have a cylinder turning around a diametral axis (see Figure E4.5b), then $I_{zz}^{C} = mR^2/4 + mL^2/12$.

Example 4.6

Two special cases of Example 4.5: Slender rods and disks (see Figures E4.6a, b).

Solution

1. If the body is "pencil-like" — that is, $L \gg R$ — the moment of inertia for a lateral axis through C is approximated as $mL^2/12$. This is also a correct result even if the cross section is not circular but has a maximum dimension within the cross section much less than L. Such a body is called a *slender bar* or *rod.*
2. If the body is a disk, however, we have $R \gg L$ and the moment of inertia is approximately $mR^2/4$.

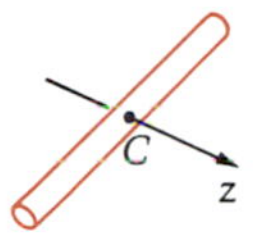

Figure E4.6a

For a cylinder with the dimensions of a pencil, for example, with $R = \frac{1}{8}$ in. and $L = 7$ in., we see that (see Figure E4.6a)

$$I_{zz}^{C} = \frac{mL^2}{12} + \frac{mR^2}{4} = \frac{mL^2}{12}\left[1 + 3\left(\frac{R}{L}\right)^2\right]$$

$$= \frac{mL^2}{12}\left[1 + 3\left(\frac{1/8}{7}\right)^2\right]$$

$$\approx \boxed{\frac{mL^2}{12}}$$

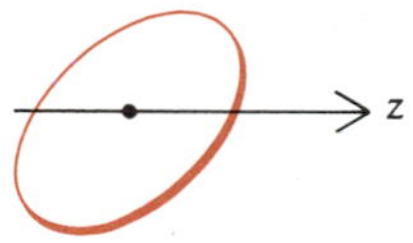

Figure E4.6b

where the second term is less than 0.1 of 1 percent of the retained $mL^2/12$ term. For a typical coin, on the other hand, with $R \approx \frac{15}{16}$ in. and $L \approx \frac{1}{16}$ in., we obtain (see Figure E4.6b).

$$I_{zz}^{C} = \frac{mR^2}{4} + \frac{mL^2}{12} = \frac{mR^2}{4}\left[1 + \frac{1}{3}\left(\frac{L}{R}\right)^2\right]$$

$$= \frac{mR^2}{4}\left[1 + \frac{(1/16)^2}{3(15/16)^2}\right]$$

$$\approx \frac{mR^2}{4}$$

This time it is the $mL^2/12$ term that is negligible; it is less than 0.15 of 1 percent of the $mR^2/4$ term. We emphasize, however, that with respect to the *axis* of any solid homogeneous cylinder (disk, rod, or anything in between), the moment of inertia is $mR^2/2$.

Example 4.7

A uniform rectangular solid (see Figure E4.7).

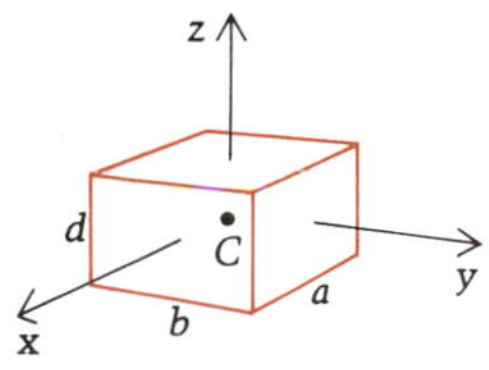

Figure E4.7

Solution

$$I_{zz}^{C} = \int_{-d/2}^{d/2}\int_{-b/2}^{b/2}\int_{-a/2}^{a/2} (x^2 + y^2)\rho \, dx\, dy\, dz$$

This integration yields

$$I_{zz}^{C} = (\rho abd)\,\frac{a^2 + b^2}{12}$$

$$= \boxed{\frac{m}{12}(a^2 + b^2)}$$

EXAMPLE 4.8

Special case of Example 4.7: A rectangular plate.

Solution

If the rectangular solid is a plate — that is, it has one edge much smaller than the other two dimensions — then, referring to Figure E4.8a–c, we have:

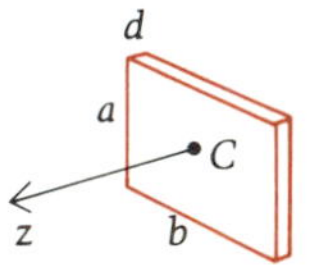

Figure E4.8a

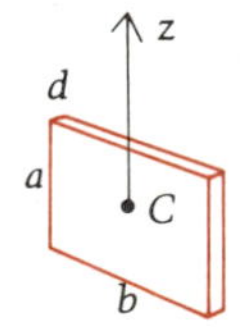

Figure E4.8b

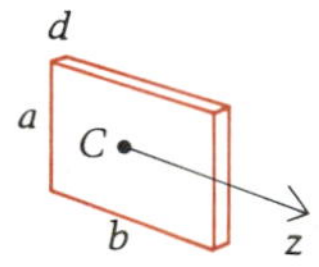

Figure E4.8c

$$I_{zz}^{C} = \boxed{\frac{m}{12}(a^2 + b^2)} \qquad I_{zz}^{C} = \frac{m}{12}(b^2 + d^2) \approx \boxed{\frac{mb^2}{12}} \qquad I_{zz}^{C} = \frac{m}{12}(a^2 + d^2) \approx \boxed{\frac{ma^2}{12}}$$

Again it depends on how the body's plane motion is set up as to which axis is z (normal to the plane of the motion) and hence which formula to use.

EXAMPLE 4.9

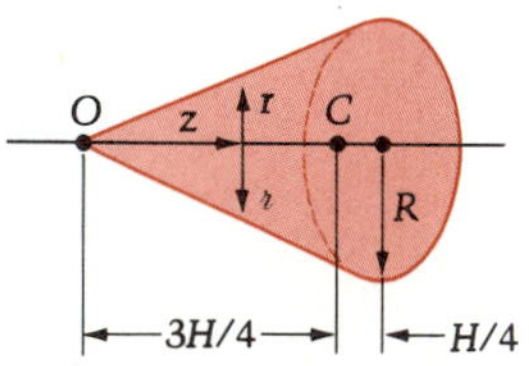

Figure E4.9

Solid, homogeneous, right circular cone about its axis (see Figure E4.9).

Solution

Here we encounter a variable limit, since $r = r(z)$. Noting that C and O are on the same z axis, then (see Figure E4.9):

$$I_{zz}^{C} = I_{zz}^{O} = \int (x^2 + y^2)\,dm = \int_{z=0}^{H} \int_{\theta=0}^{2\pi} \int_{r=0}^{r(z)} r^2 \rho r\,dr\,d\theta\,dz$$

From similar triangles,

$$\frac{r}{z} = \frac{R}{H} \Rightarrow r = \frac{Rz}{H} = r(z)$$

which gives the varying radius in terms of z. Then

$$I_{zz}^{C} = \int_{z=0}^{H} \int_{\theta=0}^{2\pi} \frac{\rho r^4}{4}\Bigg]_0^{Rz/H} \rho\, d\theta\, dz = \int_{z=0}^{H} \frac{\rho R^4 z^4}{4H^4}\,\theta\Bigg]_0^{2\pi} dz$$

$$= \frac{\rho\pi R^4}{2H^4}\frac{z^5}{5}\Bigg]_0^{H} = \left(\frac{\rho\pi R^2 H}{3}\right)\frac{3R^2}{10} = \boxed{\frac{3}{10}mR^2}$$

EXAMPLE 4.10

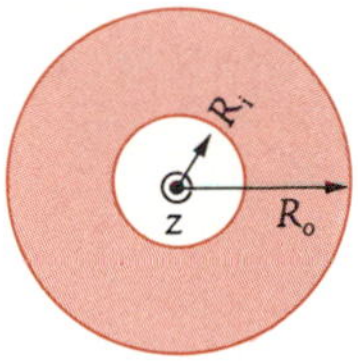

Figure E4.10

Hollow, homogeneous cylinder about its axis (see Figure E4.10).

Solution

$$I_{zz}^{C} = \int_{-L/2}^{L/2} \int_{0}^{2\pi} \int_{R_i}^{R_o} r^2 \rho r\, dr\, d\theta\, dz = \frac{\rho\pi(R_o^4 - R_i^4)L}{2}$$

$$= \underbrace{[\rho\pi(R_o^2 - R_i^2)L]}\,\frac{(R_o^2 + R_i^2)}{2} = \boxed{\frac{m(R_o^2 + R_i^2)}{2}}$$

The same result can be obtained by substracting the moment of inertia of the "hole" (H) from that of the "whole" (W). The basis for this procedure is that we may integrate over *more* than the required region provided we subtract away the integral over the part that is not to be included:

$$I_{zz}^{C} = \frac{m_W R_o^2}{2} - \frac{m_H R_i^2}{2} = \frac{\rho\pi R_o^2 L R_o^2}{2} - \frac{\rho\pi R_i^2 L R_i^2}{2}$$

$$= \underbrace{\rho\pi(R_o^2 - R_i^2)L}_{m}\,\frac{(R_o^2 + R_i^2)}{2} = \frac{m(R_o^2 + R_i^2)}{2}$$

Note that if the wall thickness is small, we have a cylindrical shell (or a hoop if the length is small) for which $R_o \approx R_i$ and

$$I_{zz}^{C} \approx \frac{m(2R^2)}{2} = \boxed{mR^2}$$

(It is obvious that if all the mass is the same distance R from the axis z we should indeed get mR^2.)

EXAMPLE 4.11

A uniform solid sphere about any diameter.

Solution

$$I_{zz}^{C} = \int (x^2 + y^2)\, dm$$

Also:

$$I_{xx}^{C} = \int (y^2 + z^2)\, dm$$

and

$$I_{yy}^{C} = \int (z^2 + x^2)\, dm$$

Adding:

$$\underbrace{I_{xx}^{C} + I_{yy}^{C} + I_{zz}^{C}}_{\substack{= 3I \text{ since they are all} \\ \text{equal by symmetry}}} = \int 2\underbrace{(x^2 + y^2 + z^2)}_{r^2}\, dm$$

Thus (see Figure E4.11 for the spherical coordinates):

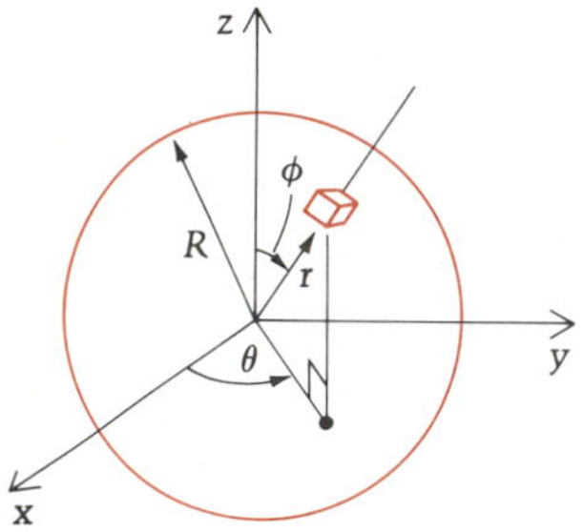

Figure E4.11

$$3I = 2\int r^2 \rho_0\, dV = 2\rho_0 \int_{\theta=0}^{2\pi} \int_{\phi=0}^{\pi} \int_{r=0}^{R} r^2 \underbrace{(r^2 \sin\phi\, dr\, d\phi\, d\theta)}_{dV \text{ in spherical coordinates}}$$

$$I = \frac{2}{3}\rho_0 \frac{r^5}{5}\bigg]_0^R (-\cos\phi)\bigg]_0^{\pi} \theta\bigg]_0^{2\pi} = \frac{8\pi\rho_0 R^5}{15}$$

$$= \underbrace{\left(\frac{4}{3}\pi R^3 \rho_0\right)}_{m} \frac{2}{5} R^2 = \boxed{\frac{2}{5} mR^2}$$

A less tricky way to do the sphere is to use spherical coordinates directly; the integral is

$$I_{zz}^{C} = \int_{\theta=0}^{2\pi} \int_{\phi=0}^{\pi} \int_{r=0}^{R} \rho_0 (r \sin\phi)^2\, dr (r\, d\phi)(r \sin\phi\, d\theta)$$

which yields the same result of $(2/5)mR^2$, as the reader may show by carrying out the integration.

EXAMPLE 4.12

An example in which the density is not constant. Sometimes a body's density varies; if it does, it must stay inside the integral when we calculate the inertia properties. An example is the earth; we now know that the density of the solid central core of the earth is about four times that of the outermost part of its crust and, moreover, that this central density is nearly twice that of steel!

Let us imagine a sphere with the same mass and radius as in the preceding example but with a density that varies linearly and is twice as high at $r = 0$ as at $r = R$. We shall find I about any diameter. (See Figure E4.12(a).)

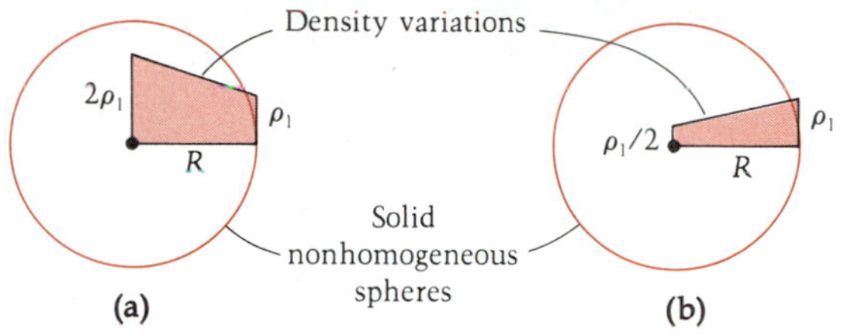

Figure E4.12

Solution

The mass of the body is

$$m = \int \rho \, dV = \int_0^{2\pi} \int_0^{\pi} \int_0^{R} \rho r^2 \sin\phi \, dr \, d\phi \, d\theta$$

Letting the density at R be ρ_1, then

$$\rho = \frac{-\rho_1}{R} r + 2\rho_1$$

Substituting and integrating with the same limits as before, and then equating the new mass to the old, gives

$$\rho_1 = \tfrac{4}{5}\rho_0$$

Thus to have the same mass as the uniform sphere, the density varies from $(8/5)\rho_0$ to $(4/5)\rho_0$ going outward from the center. Then, integrating to find I,

$$3I = 2 \int_0^{2\pi} \int_0^{\pi} \int_0^{R} r^2 \rho(r)(r^2 \sin\phi \, dr \, d\phi \, d\theta)$$

which gives

$$I = \frac{28}{75} mR^2 = 0.373mR^2 \qquad \text{(slightly less than } \tfrac{2}{5}mR^2\text{)}$$

Alternatively, if the density varies linearly but is only half as much $(\rho_1/2)$ at the center as at R, then the results, if m and R are again the same as in the uniform case, are (see Figure E4.12(b)):

$$\rho_1 = \frac{8}{7}\rho_0 \qquad \text{and} \qquad I = \frac{44}{105} mR^2 = 0.419mR^2 \qquad \text{(slightly more than } \tfrac{2}{5}mR^2\text{)}$$

The Parallel-Axis Theorem for Moments of Inertia

In many applications the body consists of a number of different smaller bodies of familiar shapes. In such cases, there is fortunately no need to integrate in order to find the inertia of each part with respect to a common axis of interest, thanks to what is called the **parallel-axis theorem** or **transfer theorem**. If we know the moment of inertia about an axis through the mass center C of any body $\mathcal{B}$, we can then easily find it about any axis parallel to C by a simple calculation. The theorem states that the moment of inertia of the mass of $\mathcal{B}$ about *any* line is the moment of inertia about a parallel line through C plus the mass of $\mathcal{B}$ times the square of the distance between the two axes:

$$I_{zz}^{P} = I_{zz}^{C} + md^2$$

To prove this theorem, we let (x, y, z) and (x_1, y_1, z_1) be rectangular cartesian coordinate axes through P and C with the corresponding axes respectively parallel, as shown in Figure 4.3. Then we have, by definition,

$$I_{zz}^{C} = \int (x_1^2 + y_1^2)\, dm \qquad \text{and} \qquad I_{zz}^{P} = \int (x^2 + y^2)\, dm$$

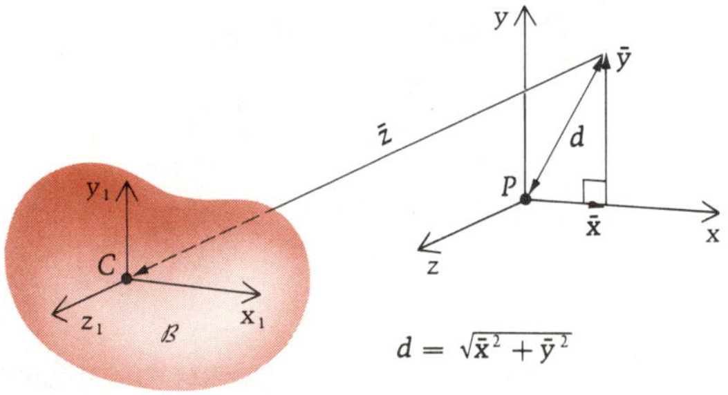

Figure 4.3

Since it is seen that

$$x = x_1 + \bar{x} \qquad \text{and} \qquad y = y_1 + \bar{y},$$

substitution gives

$$\begin{aligned} I_{zz}^{P} &= \int [(x_1 + \bar{x})^2 + (y_1 + \bar{y})^2]\, dm \\ &= \int (x_1^2 + y_1^2)\, dm + (\bar{x}^2 + \bar{y}^2) \int dm \\ &\quad + 2\bar{x} \int x_1\, dm + 2\bar{y} \int y_1\, dm \end{aligned} \tag{4.5}$$

or

$$I_{zz}^{P} = I_{zz}^{C} + md^2 \tag{4.6}$$

in which $\bar{x}^2 + \bar{y}^2 = d^2$, the square of the distance between z axes through C and P.

Question 4.3 Why are the last two integrals in Equation (4.5) zero?

Equation (4.6) is the parallel-axis theorem for moments of inertia. But

Answer 4.3 Any integral such as $\int_{\mathcal{B}} x\, dm$, where x is measured from an origin at, say, Q is equal to $m\bar{x}$; this *is* the definition of the mass center. So if the origin is C, then $\bar{x} = 0$ and $\int_{\mathcal{B}} x\, dm = 0$.

note: *We can only transfer from the mass center C and not from any other point A about which we may happen to know* I_{zz}^{A}.

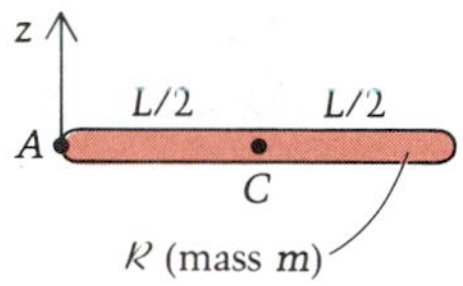

Figure E4.13

Example 4.13

For the uniform slender rod $\mathcal{R}$ shown in Figure E4.13 find I_{zz}^{A}, the moment of inertia of the mass of $\mathcal{R}$ with respect to a lateral axis through one end. (This exercise will be useful in pendulumlike applications in which a rod is pinned at one end.)

Solution

$$I_{zz}^{A} = I_{zz}^{C} + md^2 = \frac{mL^2}{12} + m\left(\frac{L}{2}\right)^2 = \boxed{\frac{mL^2}{3}}$$

We next consider an example of the buildup of the moment of inertia for a composite body.

Example 4.14

Find I_{zz}^{O} for the body shown in Figure E4.14. The mass densities each $= \rho$ = constant, so that the respective masses are:

$$m_{\text{sph}} = m_1 = \rho \frac{4}{3} \pi R^3$$

$$m_{\text{bar}} = m_2 = \rho \frac{\pi\, d^2 L}{4}$$

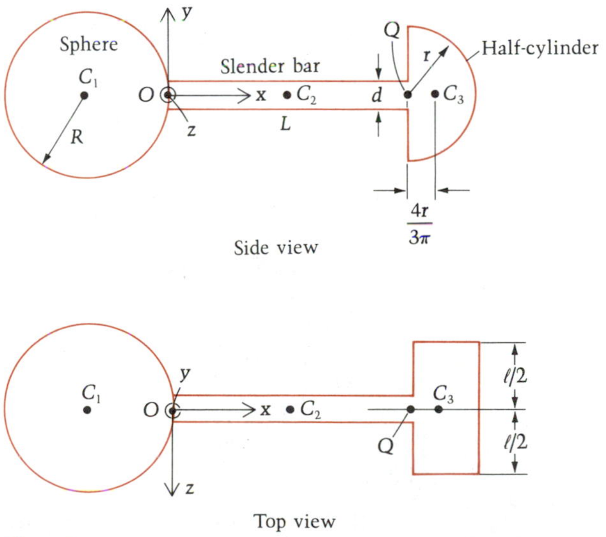

Figure E4.14

and

$$m_{\text{cyl}} = m_3 = \rho\frac{\pi r^2 \ell}{2}$$

Solution

First we observe that $I_{zz}^{O} = I_{zz}^{\text{sph}} = I_{zz}^{\text{bar}} + I_{zz}^{\text{cyl}}$ since the inertia integral may be carried out over the bodies separately, so long as we cover all the elemental masses of the total body. Filling in the separate integrals, we get

$$I_{zz}^{O} = \left(\frac{2}{5}m_1R^2 + m_1R^2\right) + \underbrace{\left[\frac{m_2L^2}{12} + m_2\left(\frac{L}{2}\right)^2\right]}_{\text{or } m_2L^2/3}$$

$$+\left\{\left[\frac{m_3r^2}{2} - m_3\left(\frac{4r}{3\pi}\right)^2\right] + m_3\left(L + \frac{4r}{3\pi}\right)^2\right\}$$

in which, for the half-cylinder,

$$I_{zz}^{C_3} = I_{zz}^{Q} - m_3\left(\frac{4r}{3\pi}\right)^2 = \frac{m_3r^2}{2} - m_3\left(\frac{4r}{3\pi}\right)^2$$

Note that we cannot correctly transfer the inertia of the semicylinder from Q to O; it must be done from the mass center C_3. So first we go "through the back door" to *find* $I_{zz}^{C_3}$ (since we know the moment of inertia already with respect to Q) and only *then* may we transfer to O.

EXAMPLE 4.15

A closed, empty wooden box is 5 ft × 3 ft × 2 ft and weights 124 lb (see Figure E4.15).

a. Find its moment of inertia about an axis through C parallel to the 2-ft dimension.
b. If the box is then filled with homogeneous material weighing 240 lb (excluding the box), how much does the moment of inertia about the axis of part (a) increase?

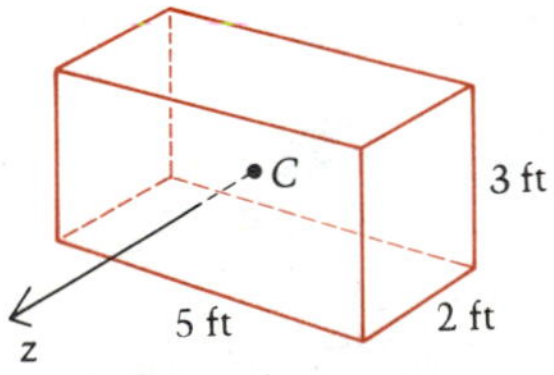

Figure E4.15

Solution

a. The masses of the various sides of the box are proportional to their areas (thickness and density assumed constant):

1. 3 × 5's:
   2 × 15 = 30 ft²
2. 2 × 3's:
   2 × 6 = 12 ft²
3. 2 × 5's:
   2 × 10 = 20 ft²

total area = 62 ft²

$$W_1 = \frac{30}{62}(124) = 60 \text{ lb for two } 3 \times 5\text{'s}$$

$$W_2 = \frac{12}{62}(124) = 24 \text{ lb for two } 2 \times 3\text{'s}$$

$$W_3 = \frac{20}{62}(124) = 40 \text{ lb for two } 2 \times 5\text{'s}$$

Therefore, taking the contributions from the three pairs of sides,

$$I_{zz}^{C} = I_{zz}^{C_1} + I_{zz}^{C_2} + I_{zz}^{C_3}$$

$$= \frac{1}{32.2}\left[\frac{2(30)(3^2+5^2)}{12} + 2(12)\left(\frac{3^2}{12} + 2.5^2\right) + 2(20)\left(\frac{5^2}{12} + 1.5^2\right)\right]$$

$$= 15.9 \text{ slug-ft}^2 \qquad (\text{or lb-ft-sec}^2)$$

b. $$I_{zz_{\text{contents}}}^{C} = \frac{240(5^2+3^2)}{32.2(12)} = 21.1 \text{ slug-ft}^2$$

Even though the box weighs only about half as much as the contents, the position of its mass makes its moment of inertia over three-fourths that of the contents. The *total* moment of inertia is 37.0 slug-ft².

The Radius of Gyration

There is a distance called the **radius of gyration** that is often used in connection with moments of inertia. The radius of gyration of the mass of a body about a line z (through a point P) is called k_{zP}, or just k_P if the axis is understood to be z, and is defined by the equation

$$I_{zz}^{P} = mk_P^2 \tag{4.7}$$

If one insists on a physical interpretation of k_P, it may be thought of as the distance from P, in any direction perpendicular to z, at which a point mass, with the same mass as the body, would have the same resulting moment of inertia that the body itself has about axis z. For example, a solid homogeneous cylinder has a radius of gyration with respect to its axis of $R/\sqrt{2}$, since $I_{zz}^{C} = mk_C^2 = \frac{1}{2}mR^2$. The usefulness of k_C is seen here, since regardless of the mass of a cylinder (and hence of its density) k_C will be the same for all homogeneous cylinders of equal radii.

Note further that (using the parallel-axis theorem)

$$mk_P^2 = I_{zz}^{P} = I_{zz}^{C} + md^2 = m(k_C^2 + d^2)$$

Thus

$$k_P^2 = k_C^2 + d^2 \tag{4.8}$$

and we see from Equation (4.8) that the radius of gyration, like the moment of inertia itself, is a minimum at C.

Products of Inertia

We now turn to the other two measures of mass distribution that have arisen in our study of plane motion of a rigid body — namely I_{yz}^P and I_{xz}^P, taken here to be with respect to axes (x, y, z) through any point P.*

Our first step is to gain insight into the meaning of products of inertia as we show that they in fact vanish for two large classes of commonly occurring symmetry. These classes are defined by the two conditions (with ρ constant in both): (1) z is an axis of symmetry and (2) xy is a plane of symmetry. Let us examine why the two products of inertia I_{xz}^P and I_{yz}^P are zero in these cases. We recall that their definitions are

$$I_{xz}^P = -\int xz\, dm \qquad I_{yz}^P = -\int yz\, dm \tag{4.9}$$

Class 1: z Is an Axis of Symmetry. For each dV at (x, y, z) there is a corresponding dV at $(-x, -y, z)$. Thus the contributions of these two elements cancel in both the I_{xz}^P and I_{yz}^P integrals. Since *each* point of $\mathcal{B}$ has a "canceling point" reflected through the z axis, I_{xz}^P and I_{yz}^P are each zero for this class of bodies. (See Figure 4.4.)
Class 2: xy Is a Plane of Symmetry. In this case each differential volume dV at (x, y, z) necessarily has a mirror image at $(x, y, -z)$. Thus the contributions of these two elements cancel in both integrals, and taken over the whole of $\mathcal{B}$ we see again that I_{xz}^P and I_{yz}^P are each zero. (See Figure 4.5.)

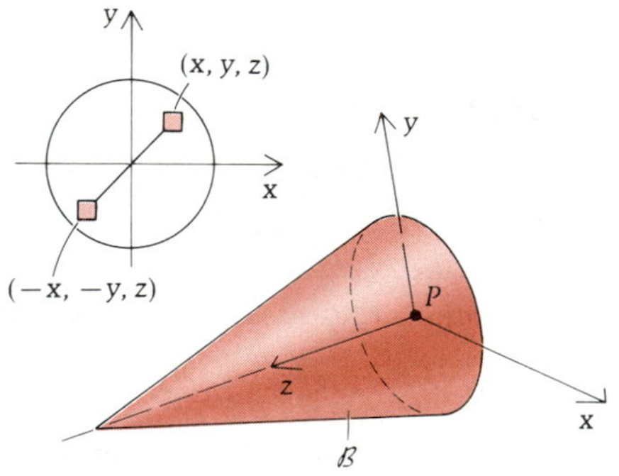

Figure 4.4

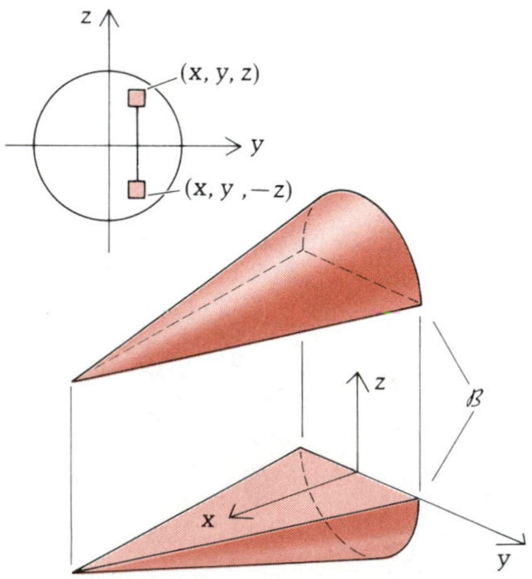

Figure 4.5

* In general (three-dimensional) motion, there arise six distinct inertia properties: three moments of inertia and three products of inertia.

Just because a body does not belong to either of these two classes does not mean it cannot have zero products of inertia. However, these are simply common cases worthy of note.

Question 4.4 Think of a rigid body for which both products of inertia are zero, but which does not fall into either of the two classes.

Transfer Theorem for Products of Inertia

There is a transfer theorem for products of inertia, just as there is one for moments of inertia. To derive it, we write from Figure 4.3:

$$I_{xz}^{P} = -\int xz\, dm = -\int (x_1 + \bar{x})(z_1 + \bar{z})\, dm$$
$$= -\int x_1 z_1\, dm - \bar{x}\bar{z}\int dm - \bar{x}\int z_1\, dm - \bar{z}\int x_1\, dm$$

The last two terms vanish by virtue of the definition of the mass center (for example, $\int z_1\, dm = m$ times the z distance from C to C, which is zero). Thus

$$I_{xz}^{P} = I_{xz}^{C} - m\bar{x}\bar{z} \tag{4.10a}$$

We note that the factor of m in Equation (4.10a) is alternatively the product of the x and z coordinates of P in an axis system with origin at C. Similarly, we have

$$I_{yz}^{P} = I_{yz}^{C} - m\bar{y}\bar{z} \tag{4.10b}$$

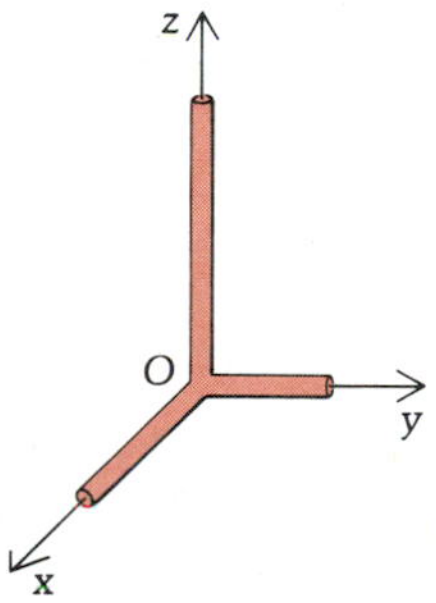

Answer 4.4 Both products of inertia are zero for the body shown in the diagram at the left. Three cylindrical bars of any lengths lying along the (x, y, z) axes are joined at the origin to form a rigid body. Neither the z axis nor the xy plane is one of symmetry, yet I_{xz}^{O} and I_{yz}^{O} are zero.

Example 4.16

Find I_{xz}^{P} and I_{yz}^{P} for the body shown in Figure E4.16; it is composed of eight identical uniform slender rods, each of mass m and length ℓ.

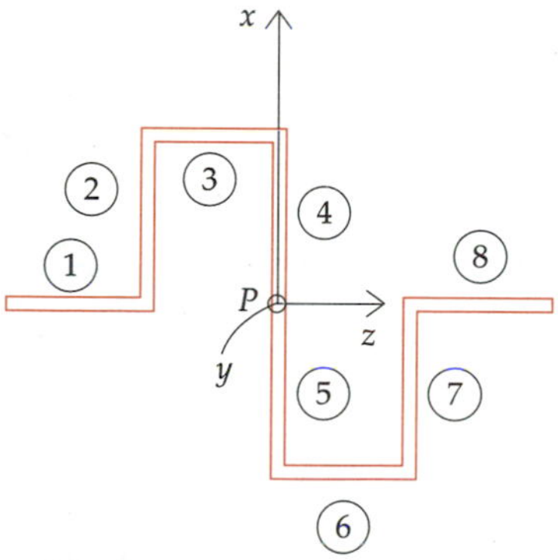

Figure E4.16

Solution

We have $I_{yz}^P = 0$ since xz is a plane of symmetry. Recall that when this happens, the two products of inertia containing (as a subscript) the coordinate normal to the plane are zero.* With superscripts identifying the various rods, we then have the following for the other product of inertia:

$$I_{xz}^P = I_{xz_P}^{①} + I_{xz_P}^{②} + I_{xz_P}^{③} + I_{xz_P}^{④} + I_{xz_P}^{⑤} + I_{xz_P}^{⑥} + I_{xz_P}^{⑦} + I_{xz_P}^{⑧}$$

Note that by symmetry each rod has zero I_{xz} about axes through its *own* center of mass parallel to x and z. Therefore the eight terms listed above will consist only of transfer terms in this problem.

Furthermore, since $\bar{z}$ is zero for rods 4 and 5, and since $\bar{x}$ is zero for rods 1 and 8, only four rods contribute to the overall I_{xz}^P:

$$I_{xz}^P = I_{xz_P}^{②} + I_{xz_P}^{③} + I_{xz_P}^{⑥} + I_{xz_P}^{⑦}$$

$$= -m\left(\frac{\ell}{2}\right)(-\ell) - m(\ell)\left(-\frac{\ell}{2}\right) - m(-\ell)\left(\frac{\ell}{2}\right) - m\left(-\frac{\ell}{2}\right)\ell$$

$$= 4\left(\frac{m\ell^2}{2}\right) = 2m\ell^2$$

Note that the "unbalanced" masses lie in the second and fourth quadrants in this example; hence the sign of I_{xz}^P is positive since its definition carries a minus sign outside the integral. We shall return to this example later in the chapter and examine the reactions caused by the nonzero value of I_{xz}^P when the body is spun up in bearings about the z axis.

PROBLEMS ▶ Section 4.4

4.27 An ellipsoid of revolution is formed by rotating the ellipse about the x axis as in Figure P4.27. Find the moment of inertia of this solid body of density 15 slug/ft³ about the x axis.

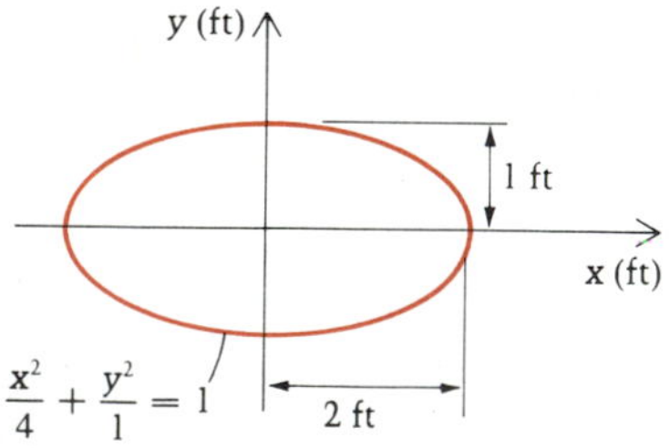

Figure P4.27

4.28 In Figure P4.28, the area bounded by the x and y axes and the parabola $y^2 = 1 - x$ is rotated about the x axis to form a solid of revolution. The density is $\rho = 1000 \cdot (1 - x)^8$ kg/m³. Find the moment of inertia of the solid mass about the x axis.

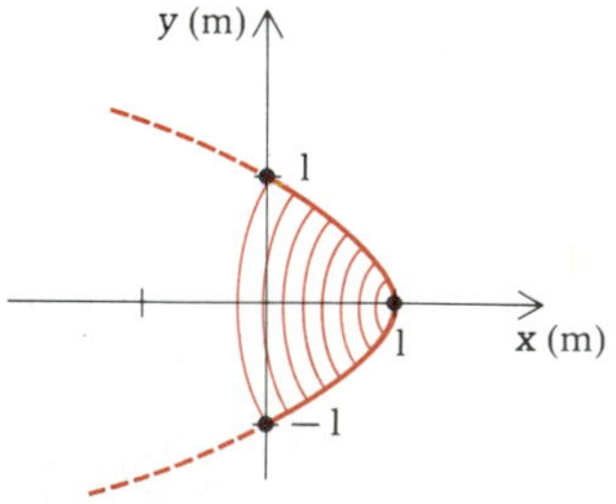

Figure P4.28

* Thus I_{xy}^P is also zero in this example, but I_{xy}^P does not appear anywhere in the plane-motion equations, as we have seen.

4.29 The slender rod in Figure P4.29 has a mass density given by

$$\rho = \rho_1 \left(\frac{x}{L}\right)^2 + \rho_0$$

in which ρ_0 and ρ_1 are constants. The rod has length L. Find its moment of inertia about the line defined by $x = 0$, $y = L/2$.

* **4.30** Find I_{zz}^C for the semielliptical prism shown in Figure P4.30 (density ρ, length normal to plane of paper $= L$).

4.31 The midplane of a uniform triangular plate is shown in Figure P4.31. Find, by integration:

- **a.** I_{xx}^O
- **b.** I_{yy}^O
- **c.** I_{zz}^O
- **d.** I_{xy}^O
- **e.** I_{xz}^O and I_{yz}^O.

What would be good approximations were the plate thin?

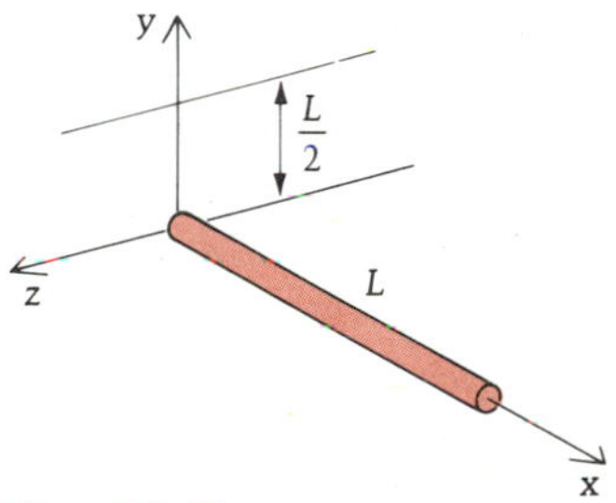

Figure P4.29

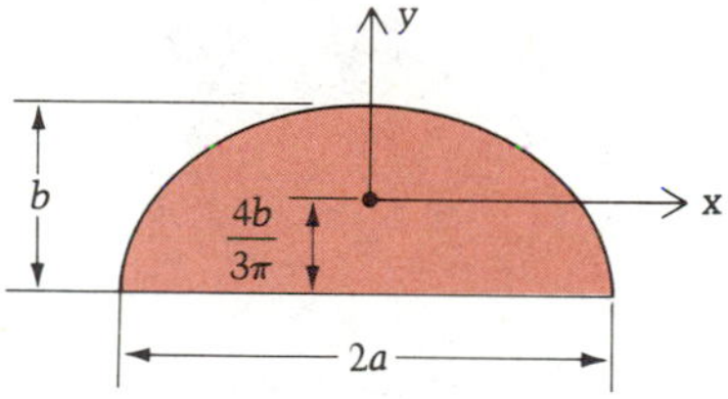

Figure P4.30

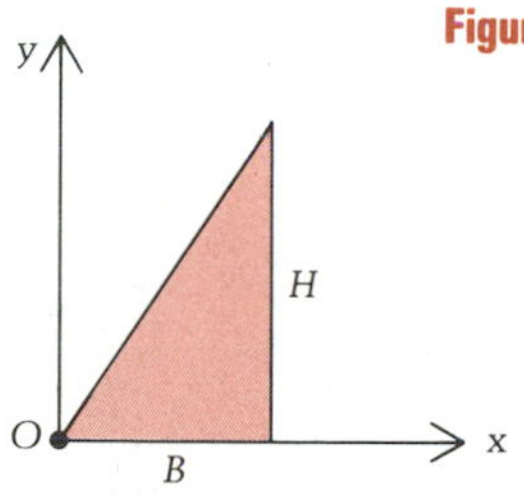

Density $= \rho$, thickness $= t$

Figure P4.31

4.32 Use the parallel-axis theorems and the results of the preceding problem to find, for that plate, the moments and products of inertia at the center of mass.

4.33 For a uniform thin plate with xy axes (and origin O) in the midplane, show that

$$I_{zz}^O \approx I_{xx}^O + I_{yy}^O$$

Confirm this statement with the results of Problem 4.31 for the case when that plate is thin.

4.34 Find I_{xx}^C for a uniform thin plate in the form of a pie-shaped circular sector as shown in Figure P4.34.

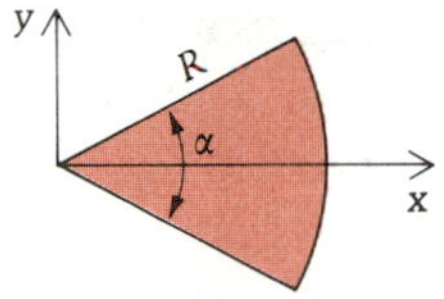

Figure P4.34

4.35 Find I_{yy}^C for the plate in the preceding problem.

4.36 The surface area of a solid of revolution is formed by rotating the curve $y = x^2$ (for $0 \le x \le 2$ m) about the x axis. (See Figure P4.36.) The density of the material varies according to the equation $\rho = 20x$, where ρ is in kg/m^3 when x is in meters. Find I_{xx}^O and tell why your answer is also I_{xx}^C.

4.37 In the preceding problem, find I_{yy}^O and I_{yy}^C.

* **4.38** See Figure P4.38. (a) Show that the moment of inertia (I_{xx}^O) for a solid homogeneous cone about a lateral axis through the base is $I_{xx}^O = (m/20)(3R^2 + 2H^2)$. (b) Using the transfer theorem, find the expression for I_{xx}^C.

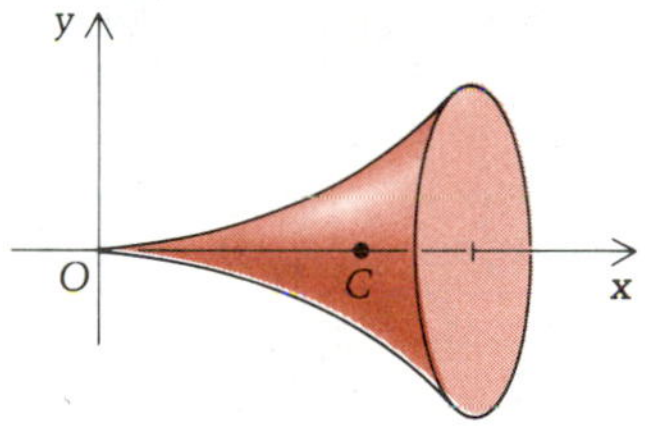

Figure P4.36

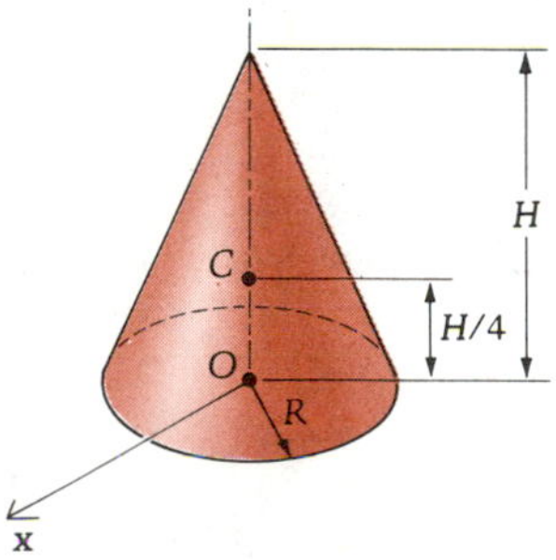

Figure P4.38

* Asterisks identify the more difficult problems.

* **4.39** In the previous problem, find I_{xx}^{O} for the body in the figure if the part above $z = H/2$ is cut away to form a truncated cone.

4.40 The body shown in Figure P4.40 is composed of a slender uniform bar ($m = 4$ slugs) and a uniform sphere ($m = 5$ slugs). Find I_{zz}^{C} for the body, where z is normal to the figure.

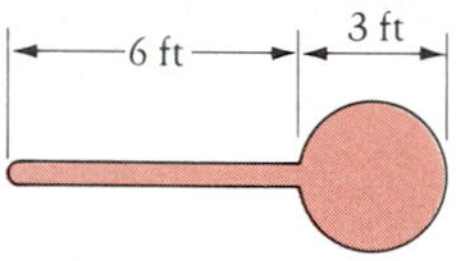

Figure P4.40

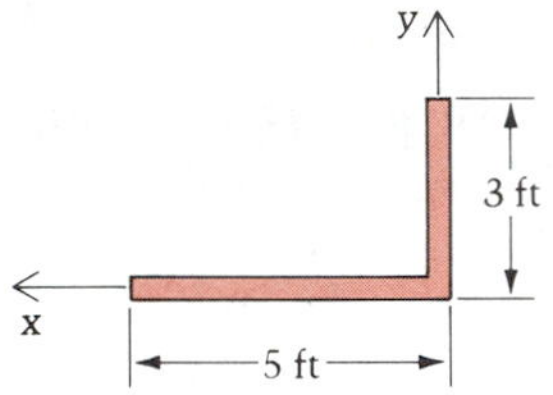

Figure P4.41

4.41 Two bars, each weighing 5 lb per foot, are welded together as shown in Figure P4.41. (a) Locate the center of mass of the body. (b) Find I_{zz}^{C}.

4.42 Find I_{xx} in Problem 4.41.

4.43 Use the result of the preceding problem, together with the transfer (parallel axis) theorem, to find I_{xx}^{C}.

4.44 Find I_{yy} in Problem 4.41.

4.45 Use the result of the preceding problem, together with the transfer theorem, to determine I_{yy}^{C}.

4.46 Find the moment of inertia of the mass of $\mathcal{B}$ about axis z_B if $I_{zz}^{A} = 40 \text{ kg} \cdot \text{m}^2$. (See Figure P4.46.)

4.47 Find the moment of inertia of a uniform hemispherical solid about the lateral axis x_C through its mass center. (See Figure P4.47.)

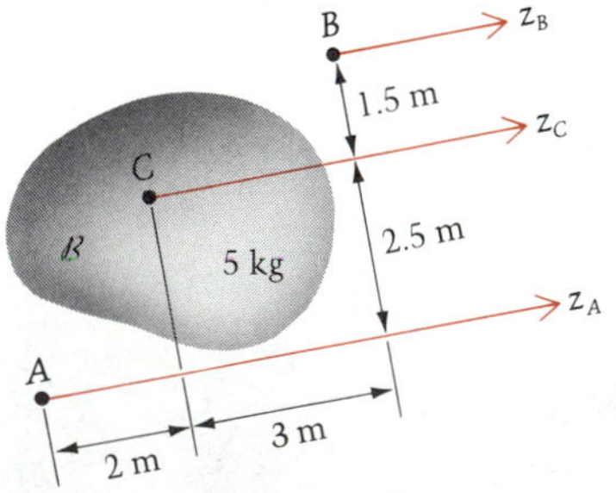

Figure P4.46

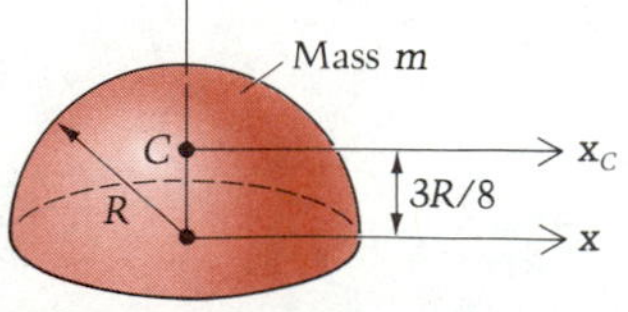

Figure P4.47

4.48 Find I_{zz}^{C} for the semicircular ring $\mathcal{B}$ in Figure P4.48. *Hint*: If the dashed portion were present, I_{zz}^{O} would be $(2m)R^2$; by symmetry, our semicircular ring contributes half of this, so that for $\mathcal{B}$ we have

$$I_{zz}^{O} = mR^2$$

Now use the transfer theorem to complete the solution without integration.

4.49 A rod $\mathcal{R}$ of length 1 m is welded on its ends to a disk $\mathcal{D}$ and a sphere $\mathcal{S}$. (See Figure P4.49.) The uniform bodies have masses $m_{\mathcal{R}} = 10$ kg, $m_{\mathcal{D}} = 5$ kg, and $m_{\mathcal{S}} = 15$ kg. The radii of $\mathcal{D}$ and $\mathcal{S}$ are 0.3 m and 0.1 m, respectively. Find I_{zz}.

4.50 The three bodies shown in Figure P4.50, welded together to form a single body $\mathcal{B}$, have masses of 64 (rectangular plate), 56 (rod), and 48 (disk), all in kilograms. Find the moment of inertia of $\mathcal{B}$ with respect to the z_Q axis.

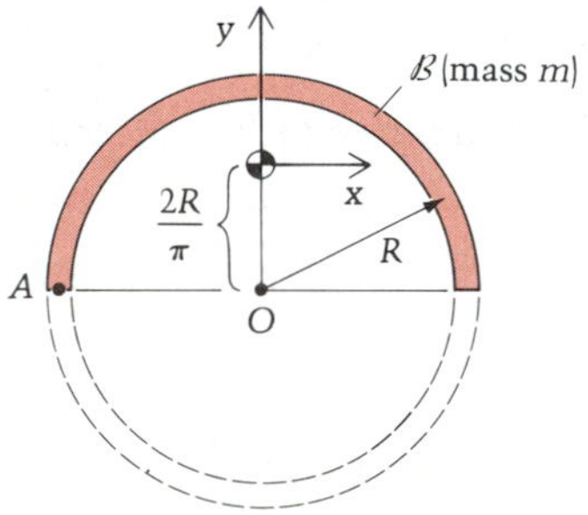

Figure P4.48

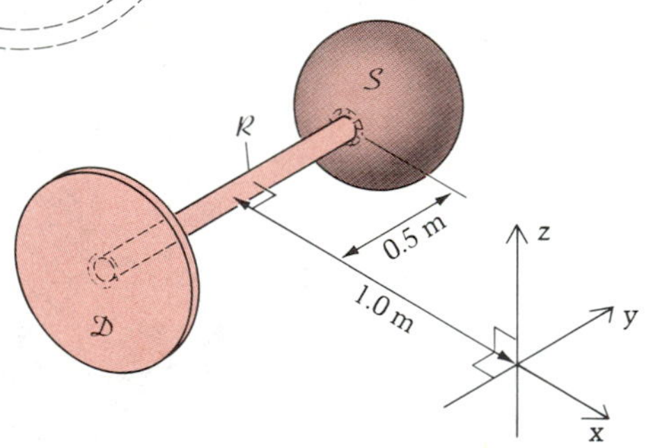

Figure P4.49

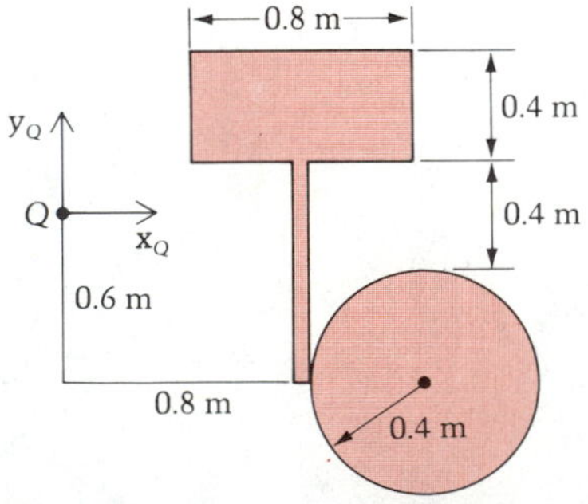

Figure P4.50

4.51 In Figure P4.51, $\mathcal{S}$ is a solid sphere, $\mathcal{C}$ is a solid cylinder, and $\mathcal{R}_1$ and $\mathcal{R}_2$ are slender rods. The center lines of $\mathcal{R}_1$ and $\mathcal{R}_2$ pass through the mass centers of $\mathcal{S}$ and $\mathcal{C}$, respectively. Find I_{zz}^O for the system of four bodies.

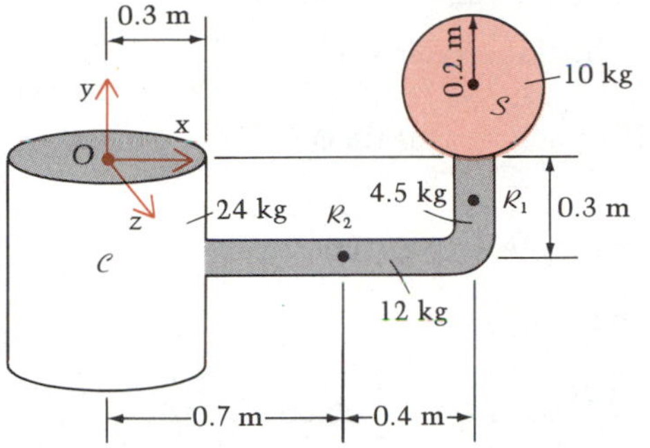

Figure P4.51

4.52 Find I_{xx}^C (which equals I_{yy}^C) for a thin plate (density ρ, thickness t) in the shape of a quarter-circle. (See Figure P4.52.)

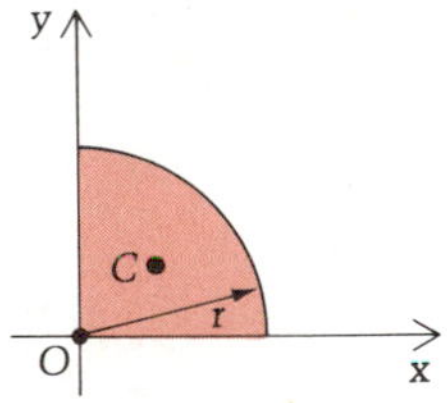

Figure P4.52

4.53 Find the moments of inertia of the pendulum about axes x, y, and z. (See Figure P4.53.) Axes x and y are in the plane of the pendulum, $\mathcal{R}$ is a slender rod, and $\mathcal{D}$ is a semicircular disk, each of constant density.

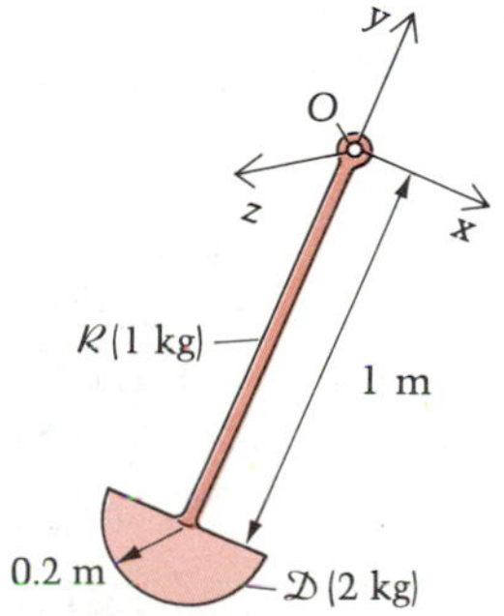

Figure P4.53

4.54 The bent slender rod in Figure P4.54 is located with the axes of the rods parallel to the x- and z-axes, as shown. Find the value of I_{zz}^O.

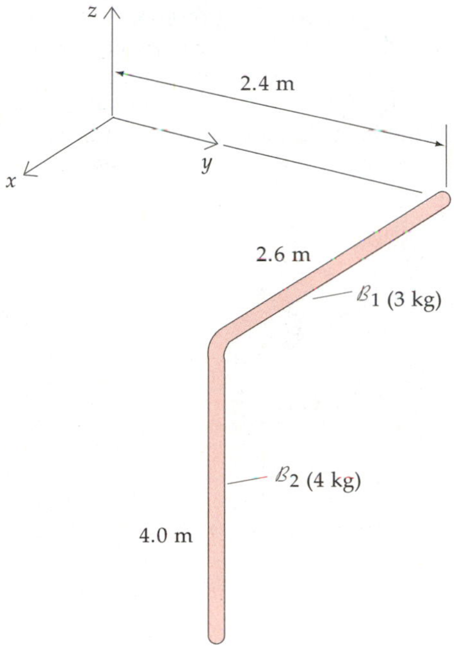

Figure P4.54

4.55 The cylinder $\mathcal{C}$ in Figure P4.55 has a mass of 6 kg and a radius of 0.4 m. Show that the moment of inertia about an axis z_C normal to the page is 0.48 kg · m², and that the corresponding radius of gyration is $k_C = 0.283$ m.

4.56 In the preceding problem, show that it is possible to drill a hole through $\mathcal{C}$ that is below the geometric center Q and satisfy all of the following:

a. the remaining mass is 5.5 kg;

b. the distance between Q and the *new* mass center C (see Figure P4.56) is 0.02 m.

Find the radius r of the hole, the center distance d, and the new value of k_C.

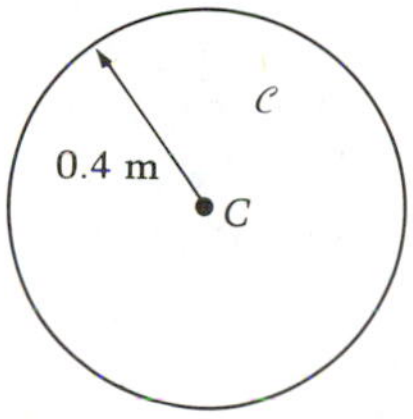

Figure P4.55

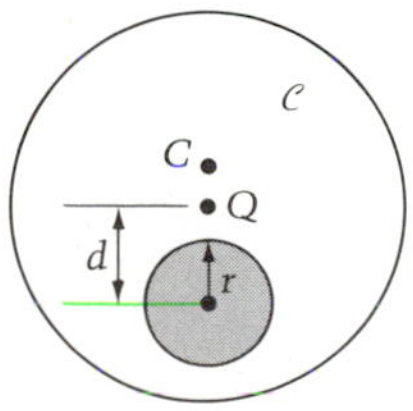

Figure P4.56

4.57 The antenna $\mathcal{A}$ in Figure P4.57 has a moment of inertia about z_C of I_C, and the counterweights $\mathcal{C}$ have a collective moment of inertia about z_G of I_G. Points C and G are the respective mass centers of the antenna and its counterweights. The purpose of the counterweights is to place the combined mass center at O to reduce stresses. Thus $MD = md$, where we neglect the mass of the connecting frame for this problem. Compute the values of M and D that will minimize the total moment of inertia I_O (I of $\mathcal{C}$ about O plus I of $\mathcal{A}$ about O). Use $I_G = Mk_G^2$, where k_G is a constant.

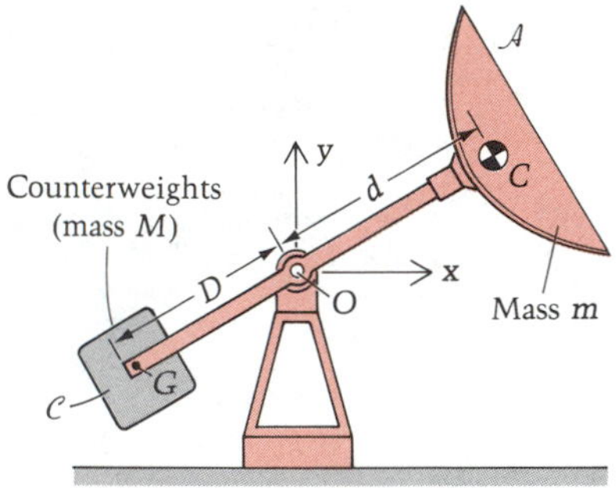

Figure P4.57

4.58 Find the product of inertia I_{xy}^A for the ring of Problem 4.48.

4.59 Find I_{xy}^Q for the welded body $\mathcal{B}$ of Problem 4.50.

4.60 Determine I_{xy} in Problem 4.49. The xy plane contains the centers of $\mathcal{D}$, $\mathcal{R}$, and $\mathcal{S}$.

4.61 Show in the following three ways that the moment of inertia of a uniform, thin spherical shell, about any line through its mass center, is $(2/3)\,mr^2$. (See Figure P4.61.) Which of the three approaches would work if the object were hollow but not thin — that is, if $t \gg R$?

a. Use spherical coordinates:

$$I_{zz}^C = \int_{\theta=0}^{2\pi}\int_{\phi=0}^{\pi}(r\sin\phi)^2\rho t r^2 \sin\phi \, d\phi \, d\theta$$

where r is the average radius, $R - \dfrac{t}{2}$.

b. Use the idea:

$$I_{zz}^C = \frac{2}{5}m_{\text{WHOLE}}R^2 - \frac{2}{5}m_{\text{HOLE}}(R-t)^2$$

c. Use $I_{zz_{\text{sphere}}}^C = \dfrac{2}{5}mR^2 = \dfrac{2}{5}\rho\left(\dfrac{4}{3}\pi R^5\right) = \dfrac{8}{15}\rho\pi R^5$.

Let R increase by ΔR, and compute the change, ΔI_{zz}^C, by using differential calculus with $\Delta R \ll R$. Note that this change *is* the moment of inertia of the shell!

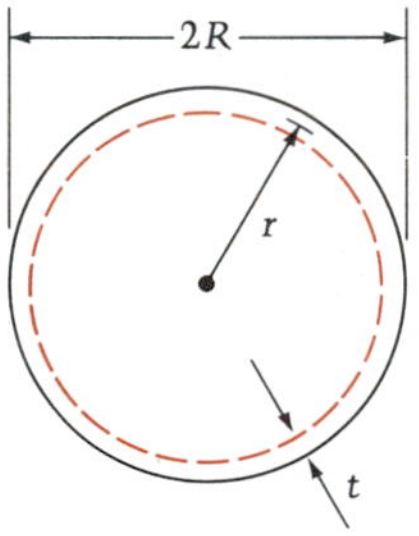

Figure P4.61

4.5 The Mass-Center Form of the Moment Equation of Motion

Development of the Equations of Plane Motion

In Chapter 2 we developed several different forms of Euler's second law. In this section we will continue to study the mass-center form, Equation (2.43):

$$\Sigma \mathbf{M}_C = \dot{\mathbf{H}}_C \tag{2.43}$$

where C is the mass center of an arbitrary body $\mathcal{B}$. If we now restrict the body $\mathcal{B}$ to be rigid and the reference frame to be an inertial frame in Equation (4.4), then we may substitute from that equation for $\mathbf{H}_C$ into Equation (2.43) above and obtain:

$$\Sigma \mathbf{M}_C = \frac{d}{dt}\mathbf{H}_C = \frac{d}{dt}(I_{xz}^C\omega\hat{\mathbf{i}} + I_{yz}^C\omega\hat{\mathbf{j}} + I_{zz}^C\omega\hat{\mathbf{k}}) \tag{4.11}$$

At this stage we must make a decision with regard to how the x and y axes of Equation (4.11), which have their origin at C, will be allowed to change relative to the inertial frame of reference where the derivative in Equation (4.11) is to be taken, and to which $\boldsymbol{\omega}$ of the body is referred. Note that the direction of the z axis has already been fixed perpendicular to the reference plane. If we fix the directions of x and y relative to the inertial frame, then $\hat{\mathbf{i}}$ and $\hat{\mathbf{j}}$ (as well as $\hat{\mathbf{k}}$) are constant relative to that frame but I_{xz}^{C} and I_{yz}^{C} are in general time-dependent. This choice is very difficult to deal with.

Question 4.5 However, I_{zz}^{C} will not change in this case. Why not?

A much more convenient choice is to let the axes x, y, and z all be fixed in the body so that the moments and products of inertia are all constant. Now $\hat{\mathbf{i}}$ and $\hat{\mathbf{j}}$ are time-dependent relative to the inertial frame $\mathcal{I}$. Figure 4.6 shows the unit vectors $\hat{\mathbf{i}}$ and $\hat{\mathbf{j}}$ (fixed to $\mathcal{B}$) expressed in terms of $\hat{\mathbf{I}}$ and $\hat{\mathbf{J}}$ (which are fixed in the inertial reference frame $\mathcal{I}$):

$$\hat{\mathbf{i}} = 1(\cos\theta\hat{\mathbf{I}} + \sin\theta\hat{\mathbf{J}})$$
$$\hat{\mathbf{j}} = 1(-\sin\theta\hat{\mathbf{I}} + \cos\theta\hat{\mathbf{J}})$$

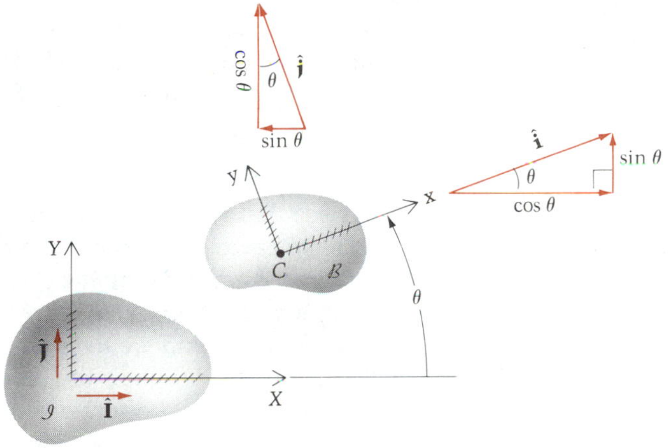

Figure 4.6

The derivatives of these unit vectors, obtained earlier as Equations (3.41) and (3.42), are:

$$\frac{d\hat{\mathbf{i}}}{dt} = (-\sin\theta\hat{\mathbf{I}} + \cos\theta\hat{\mathbf{J}})\dot{\theta} = \dot{\theta}\hat{\mathbf{j}} \text{ or } \omega\hat{\mathbf{j}} \tag{4.12a}$$

Answer 4.5 Let (x, y, z) be fixed in the reference frame $\mathcal{I}$. As the body rotates with respect to frame $\mathcal{I}$, its mass is then distributed differently, at different times, with respect to x and y. But in plane motion the z axis is fixed in direction in *both* the body and in space. Thus since the square of the distance from the z axis is always the same r^2 regardless of the orientation of x and y, we see that I_{zz}^{C} does not change as the body turns.

$$\frac{d\hat{\mathbf{j}}}{dt} = -(\cos\theta\hat{\mathbf{I}} + \sin\theta\hat{\mathbf{J}})\dot{\theta} = -\dot{\theta}\hat{\mathbf{i}} \text{ or } -\omega\hat{\mathbf{i}} \tag{4.12b}$$

Therefore, carrying out the differentiations in Equation (4.11), we have

$$\Sigma\mathbf{M}_C = I_{xz}^C\dot{\omega}\hat{\mathbf{i}} + I_{yz}^C\dot{\omega}\hat{\mathbf{j}} + I_{zz}^C\dot{\omega}\hat{\mathbf{k}} + I_{xz}^C\omega(\omega\hat{\mathbf{j}}) + I_{yz}^C\omega(-\omega\hat{\mathbf{i}})$$

or

$$\Sigma\mathbf{M}_C = (I_{xz}^C\alpha - I_{yz}^C\omega^2)\hat{\mathbf{i}} + (I_{yz}^C\alpha + I_{xz}^C\omega^2)\hat{\mathbf{j}} + I_{zz}^C\alpha\hat{\mathbf{k}} \tag{4.13}$$

This expression represents three scalar equations:

$$\Sigma M_{Cx} = I_{xz}^C\alpha - I_{yz}^C\omega^2 \tag{4.14a}$$

$$\Sigma M_{Cy} = I_{yz}^C\alpha + I_{xz}^C\omega^2 \tag{4.14b}$$

$$\Sigma M_{Cz} = I_{zz}^C\alpha \tag{4.14c}$$

Equations (4.14a,b), along with $\Sigma F_z = 0$, tell us about the nature of reactions necessary to maintain the plane motion. If I_{xz}^C and I_{yz}^C are both zero,* then $\Sigma M_{Cx} = 0 = \Sigma M_{Cy}$ and the system of external forces (loads plus reactions) has a planar resultant, with this plane containing the mass center. Thus, with a coplanar system of external *loads*, the resultant *reactions* must have a resultant in that same plane. This is the basis for "two-dimensionalizing" the analysis of problems for which the two products of inertia above vanish; we shall first work with symmetrical bodies for which this is the case. Then later in this chapter we shall examine some problems in which I_{xz}^C and I_{yz}^C are *not* both zero. In any case, however, the rotational motion of the body is governed by the simple kinetics equation

$$\Sigma M_{Cz} = I_{zz}^C\alpha \qquad (=I_{zz}^C\ddot{\theta}) \tag{4.14d}$$

We see that while force produces acceleration with the "resistance" being the mass, it is also true that moment produces angular acceleration with the "resistance" being the body's moment of inertia. Note also that the moment of the external forces about the z axis through the mass center is $\Sigma\mathbf{M}_C \cdot \hat{\mathbf{k}}$, and we have $\Sigma\mathbf{M}_C \cdot \hat{\mathbf{k}} = I_{zz}^C\alpha$. Thus the resultant moment about this axis equals the moment of inertia about the axis multiplied by the angular acceleration of the body, regardless of whether or not the products of inertia vanish.

We will now restrict our attention in the remainder of this section to a special class of problems of plane, rigid-body motions. This class is defined by the following pair of conditions:

1. $I_{xz}^C = I_{yz}^C = 0$ (usually because the body is either symmetric about the plane of motion of the mass center or else has an axis of physical symmetry which remains normal to the reference plane of motion); and

2. The external *loads* have a planar resultant with the plane containing the mass center.

---

* When I_{xz}^C or I_{yz}^C is *not* zero, there has to be a nonzero ΣM_{Cx} or ΣM_{Cy} present to maintain the motion; these are usually formed by lateral forces (such as bearing reactions) at different positions along the z-axis.

These two conditions necessitate external *reactions* likewise equipollent to a coplanar (in the same plane as the loads) force system. Euler's laws are then effectively reduced to the following three scalar equations:

$$\Sigma F_x = m\ddot{x}_C \tag{4.15a}$$

$$\Sigma F_y = m\ddot{y}_C \tag{4.15b}$$

$$\Sigma M_{Cz} = I_{zz}^C \ddot{\theta} \tag{4.15c}$$

where x_C and y_C are coordinates of the mass center in a rectangular coordinate system fixed in the inertial frame.* Equations (4.15a,b) are of course the x and y components of the mass center equation of motion valid for any body, rigid or not. Since there will be no confusion about the axis in question, we shall often write Equation (4.15c) as simply

$$\Sigma M_C = I_C \ddot{\theta}$$

or

$$\Sigma M_C = I_C \alpha \tag{4.16}$$

Helpful Steps to Follow in Generating and Solving the Equations of Motion

In some instances Equations (4.15a–c) will yield a differential equation(s) that can be readily integrated so that we predict the ongoing motion of a body. More commonly we shall be dealing with what we might call "snapshot" problems where at a specific instant we calculate forces and accelerations. These problems are rather natural extensions of statics, all of the operations in the analysis being algebraic. For problems of either class the following steps are recommended:

1. Draw a free-body diagram (FBD) of each body in the problem.
2. Define a set of unit vectors, or, equivalently, a coordinate system, in terms of which unknown forces and accelerations may be expressed.
3. Substitute into the three equations of motion; some may prefer to do this in vector algebraic form, explicitly displaying previously defined unit vectors, while others may prefer to initiate the analysis using the three component equations (4.15a,b and 4.16).
4. Often the number of scalar unknowns will exceed the number of independent equations (maximum of three), and we must look for supplementary information. This might mean nothing more than applying the Coulomb law of friction, but often the supplementary information will be in the form of a kinematic constraint. For example, for a wheel having its mass center at the geometric center, rolling would imply (with appropriate definitions of variables) $\ddot{x}_C = r\ddot{\theta}$. If some point A has its motion constrained, then the restriction on its acceleration along with $\mathbf{a}_C = \mathbf{a}_A + \alpha\hat{\mathbf{k}} \times \mathbf{r}_{AC} - \omega^2\mathbf{r}_{AC}$ can be used to relate acceleration of

* Which component equations are used and the specific forms they take depend, of course, on what type of coordinate system is used to describe the motion of C. A rectangular coordinate system is the natural choice for most of the problems taken up in this chapter. A polar coordinate system is the natural choice, however, for problems of orbital mechanics (see Section 8.4).

the mass center and angular acceleration of the body (presuming velocities, and thus ω, are known). (Of course, $\mathbf{a}_C$ and α are the kinematical variables which naturally appear in the equations of motion, as we have seen.)

5. Solve for unknown forces and/or accelerations. All problems appearing in this text are "rigid-body dynamically determinate," so this will always be possible. In the case of "movie" problems, integrate accelerations to get velocities and then velocities to get positions as functions of time.

6. Be sure to check on dimensional consistency of results and, in the case of numerical results, check to see that units are correct (especially important in these days of transition from U.S. to SI units). Also, see if your answer seems to make sense.

We now proceed toward a set of examples which will make use of the equations (4.15a,b and 4.16), which we have developed in this section. These examples are designed to illustrate the kinds of problems the student should learn to solve. The first one is a "movie" problem in which one resultant force component and the moment are zero, while the other force component is not.

EXAMPLE 4.17

A horseshoe pitcher releases a horseshoe with $\omega_i = 4.71 \circlearrowleft$ rad/sec in the position shown in Figure E4.17a. If the horseshoe turns exactly once in plane motion and scores a ringer, find the initial velocity of the horseshoe's mass center.

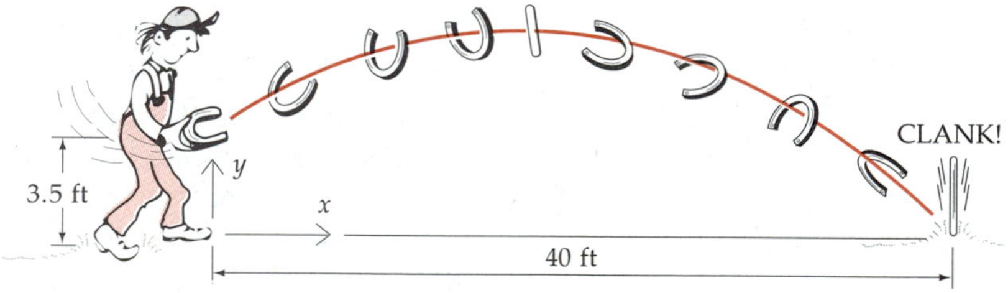

Figure E4.17a

A free-body diagram of the horseshoe in flight is shown in Figure E4.17b. We write and then integrate Equations (4.17a–c):

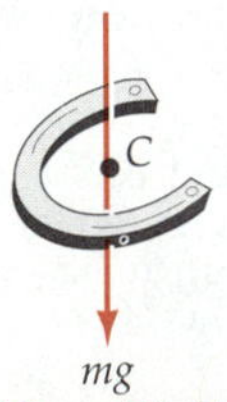

Figure E4.15b

$$\Sigma F_x = 0 = m\ddot{x}_C$$

$$\ddot{x}_C = 0$$

$$\dot{x}_C = \dot{x}_i$$

(the initial velocity of C in x-direction)

$$x_C = \dot{x}_i t + \cancel{x_i}^{\,0}$$

since $x_C = 0$ at $t = 0$

$$\Sigma F_y = -mg = m\ddot{y}_C$$

$$\ddot{y}_C = -g$$

$$\dot{y}_C = -gt + \dot{y}_i$$

(the initial velocity of C in y-direction)

$$y_C = \frac{-gt^2}{2} + \dot{y}_i t + \cancel{y_i}^{\,3.5}$$

since horseshoe is released at $y_C = 3.5$ ft

$$\Sigma M_C = 0 = I_C\ddot{\theta}$$

$$\ddot{\theta} = 0$$

$$\dot{\theta} = \omega_i = 4.71$$

$$\theta = 4.71t + \cancel{\theta_i}^{\,0}$$

Now when the horseshoe lands, $\theta = 2\pi$, so

$$2\pi = 4.71t_f \Rightarrow t_f = 1.33 \text{ sec}$$

Thus

$$x_C = 40 = \dot{x}_i(1.33) \Rightarrow \dot{x}_i = 30.1 \text{ ft/sec}$$

and

$$y_C = 0 = -\frac{32.2}{2}(1.33)^2 + \dot{y}_i(1.33) + 3.5 \Rightarrow \dot{y}_i = 18.8 \text{ ft/sec}$$

Hence the initial velocity of the horseshoe's mass center is

$$\mathbf{v}_i = 30.1\hat{\mathbf{i}} + 18.8\hat{\mathbf{j}} \text{ ft/sec}$$

The next example involves rolling, and is a "movie" (ongoing time) problem:

EXAMPLE 4.18

The cylinder (mass m, radius r) is released from rest on the inclined plane shown in Figure E4.18a. The coefficient of friction between cylinder and plane is μ. Determine the motion of C, assuming that μ is large enough to prevent slipping. (How large must it be?)

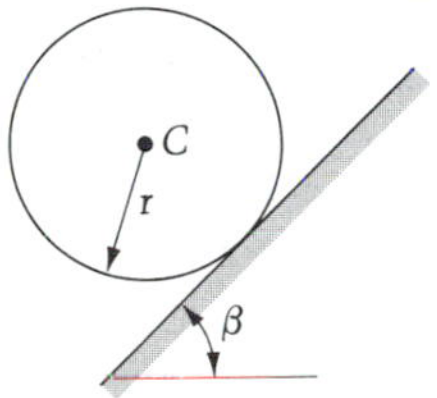

Figure E4.18a

Solution

As in statics, a good first step is to draw a free-body diagram (see Figure E4.18b). We are to assume that the cylinder rolls. In this case the friction force f is an unknown and has a value satisfying

$$0 \le f \le f_{max}$$

where $f_{max} = \mu N$ from the study of Coulomb friction in statics. We shall also use a kinematic equation expressing the rolling. After solving for f, we shall then impose the condition that it be less than μN, since we know the cylinder is not slipping.

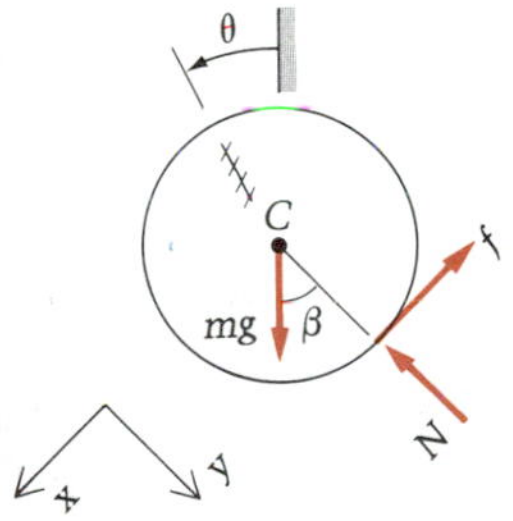

Figure E4.18b

We choose x, y, and θ as shown, motivated by the fact that C will move down the plane and the cylinder will turn counterclockwise. The equations of motion are

$$\Sigma F_x = m\ddot{x}_C \Rightarrow mg \sin\beta - f = m\ddot{x}_C \qquad (1)$$

$$\Sigma F_y = m\ddot{y}_C \Rightarrow mg \cos\beta - N = m\ddot{y}_C = 0 \qquad (2)$$

(Note that, kinematically, y_C is constant so that $\ddot{y}_C$ vanishes.)

$$\Sigma M_C = I_C\alpha \Rightarrow fr = \tfrac{1}{2}mr^2\ddot{\theta} \qquad (3)$$

We can solve (2) for N, getting $N = mg \cos\beta$. There remain two equations in the three unknowns f, $\ddot{x}_C$, and $\ddot{\theta}$. We must therefore supplement our equations of motion with the remaining kinematics result, which comes from the rolling condition:

$$\ddot{x}_C = r\ddot{\theta} \qquad (4)$$

Solving Equations (1), (3), and (4) gives

$$\ddot{x}_C = \frac{2}{3} g \sin \beta \qquad \ddot{\theta} = \frac{2g \sin \beta}{3r} \qquad f = \frac{mg \sin \beta}{3}$$

Integrating twice, and noting that the integration constants vanish, we get

$$x_C = \frac{gt^2}{3} \sin \beta \qquad \theta = \frac{gt^2}{3r} \sin \beta$$

And since we are told that

$$f \leq f_{max} = \mu N$$

that is,

$$\frac{mg \sin \beta}{3} \leq \mu mg \cos \beta$$

then we have

$$\mu \geq \frac{\tan \beta}{3}$$

for the rolling motion to occur.

In the preceding example, things would be quite different if the value of the friction force that satisfies the equations, namely $f = (mg \sin \beta)/3$, were larger than $f_{max} = \mu N = \mu mg \cos \beta$. So if the solution were to yield

$$f > f_{max}, \quad \text{i.e., if} \quad \mu < \frac{\tan \beta}{3},$$

then there would be insufficient friction to permit rolling. We would then need to abandon* the rolling condition, replacing it by the known maximum value of f, i.e.,

$$\ddot{x}_C \neq r\ddot{\theta} \text{ any longer,}$$

but now we know

$$f = \mu N = \mu mg \cos \beta$$

where N still equals $mg \cos \beta$. And since Equation (3) still holds,

$$fr = \mu mgr \cos \beta = \tfrac{1}{2} mr^2 \ddot{\theta}$$

or

$$\ddot{\theta} = \frac{2g\mu \cos \beta}{r}$$

Therefore, integrating twice, we get

$$\theta = \frac{\mu g t^2 \cos \beta}{r}$$

* When you assume something and later arrive at a contradiction of fact, then the logical conclusion is that the assumption was invalid.

where the integration constants are zero since $\theta = \dot{\theta} = 0$ at $t = 0$. Equation (1) now yields, with $f = \mu N$,

$$mg \sin \beta - \mu mg \cos \beta = m\ddot{x}_C$$

or

$$\ddot{x}_C = g(\sin \beta - \mu \cos \beta)$$

Thus

$$x_C = \frac{gt^2}{2}(\sin \beta - \mu \cos \beta)$$

and the solutions for the motion [$x_C(t)$ and $\theta(t)$] are indeed quite different when the cylinder turns and slips than they are when it rolls. If $\mu = 0.5$ and $\beta = 30°$, for example, then

$$\tan \beta = 0.577 \le 3\mu = 1.5$$

and the cylinder rolls. But if $\mu = 0.2$ and $\beta = 60°$, then

$$\tan \beta = 1.73 \nleq 3\mu = 0.6$$

and the cylinder slips. Note that in general (as one would expect) the cylinder rolls for larger μ and smaller β.

Finally, note that if we wish to distinguish between static and kinetic coefficients of friction (μ_s and μ_k), then the rolling assumption would be correct if $\tan \beta \le 3\mu_s$. But if $\tan \beta > 3\mu_s$, we would then use $f = \mu_k N$ in the remainder of the solution, and the μ in the answers for x_C and θ would become μ_k.

Next we take up another rolling problem, only this time the mass center and geometric center are not the same point. As a result, the kinematics equations are more difficult. This example is a "snapshot" (occurring at one instant of time) problem.

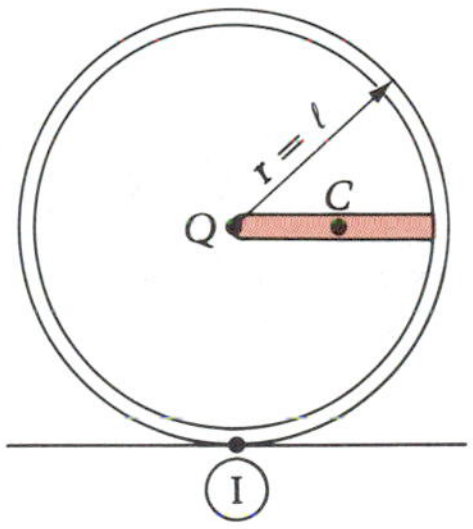

Figure E4.19a

EXAMPLE 4.19

The rigid body $\mathcal{B}$ in Figure E4.19a consists of a heavy bar of mass m welded to a light hoop; the radius of the hoop thus equals the length of the bar. Find the minimum coefficient of friction between the hoop and the ground for which the body will roll when released from rest in the given position.

Solution

The free-body diagram is shown in Figure E4.19b along with the base vectors adopted for the problem.

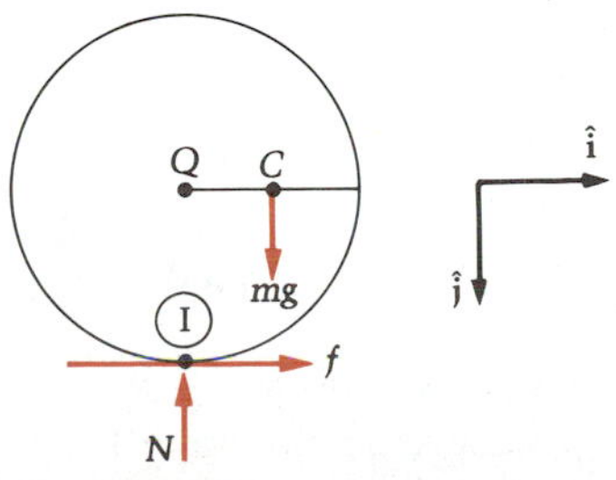

Figure E4.19b

Question 4.6 Why is this a good choice of base vectors to use in this problem?

Answer 4.6 Because the initial acceleration of the mass center C will be to the right and downward (↱) and the initial angular acceleration will be clockwise (↻).

Note that the gravity force resultant acts through the center of the bar since we are neglecting the weight of the hoop.

Next we write the three differential equations of motion, letting $\mathbf{a}_C = \ddot{x}_C\hat{\mathbf{i}} + \ddot{y}_C\hat{\mathbf{j}}$ and $\boldsymbol{\alpha} = \alpha\hat{\mathbf{k}}$.

$$\Sigma F_x = m\ddot{x}_C \Rightarrow f = m\ddot{x}_C \tag{1}$$

$$\Sigma F_y = m\ddot{y}_C \Rightarrow mg - N = m\ddot{y}_C \tag{2}$$

$$\Sigma M_C = I_C\alpha \Rightarrow \frac{N\ell}{2} - f\ell = \frac{m\ell^2}{12}\alpha \tag{3}$$

These equations contain the unknowns f, N, $\ddot{x}_C$, $\ddot{y}_C$, and α. We draw upon kinematics for two more equations. We know that the accleration of the geometric center Q of the hoop is $\mathbf{a}_Q = r\alpha\hat{\mathbf{i}}$, so that

$$\mathbf{a}_C = \underbrace{\mathbf{a}_Q}_{a_Q\hat{\mathbf{i}} = r\alpha\hat{\mathbf{i}}} + \alpha\hat{\mathbf{k}} \times \mathbf{r}_{QC} - \overset{0}{\cancel{\omega^2}}\,\mathbf{r}_{QC}$$

or

$$\ddot{x}_C\hat{\mathbf{i}} + \ddot{y}_C\hat{\mathbf{j}} = r\alpha\hat{\mathbf{i}} + \alpha\hat{\mathbf{k}} \times \left(\frac{\ell}{2}\hat{\mathbf{i}}\right)$$

Therefore, equating the $\hat{\mathbf{i}}$ coefficients and then the $\hat{\mathbf{j}}$ coefficients, we find

$$\ddot{x}_C = \ell\alpha \tag{4}$$

$$\ddot{y}_C = \frac{\ell\alpha}{2} \tag{5}$$

The student may wish to verify that (4) and (5) also result from relating $\mathbf{a}_C$ to $\mathbf{a}_{\text{Ⓘ}} = r\omega^2\uparrow$, true for any round body rolling on a flat, fixed plane. In this problem, $\mathbf{a}_{\text{Ⓘ}}$ is then zero at release because, until time passes, ω is still zero.

Solving Equations (1) to (5) for the five unknowns gives the results:

$$f = \frac{3}{8}mg \qquad N = \frac{13}{16}mg \qquad \ddot{x}_C = \frac{3g}{8} \qquad \ddot{y}_C = \frac{3g}{16} \qquad \alpha = \frac{3g}{8\ell}$$

To complete the solution, we must get the coefficient of friction μ into the picture. We know that for any friction force f,

$$0 \le f \le f_{\max} = \mu N$$

Therefore, in our problem,

$$\frac{3}{8}mg \le \mu\frac{13}{16}mg$$

so that

$$\mu \ge \frac{6}{13}$$

This means that for the body to roll, a friction coefficient of at least $\mu = 6/13$ is required; this is then the desired minimum.

We emphasize that students should always make "eyeball checks" of their answers — glancing over the results to be sure they make sense physically. In this problem, for instance, note that:

1. $N < mg$ as expected, for otherwise the mass center could not begin to move downward as the body rolls.
2. f is positive, and therefore in the correct direction to (since it is the only force in the x-direction) move the mass center to the right.
3. $\ddot{x}_C$, $\ddot{y}_C$, and α are all positive and therefore in the expected directions.

The next example features another "snapshot" problem — one which we are examining only at one instant. It also involves the interesting constraint of a taut string.

Example 4.20

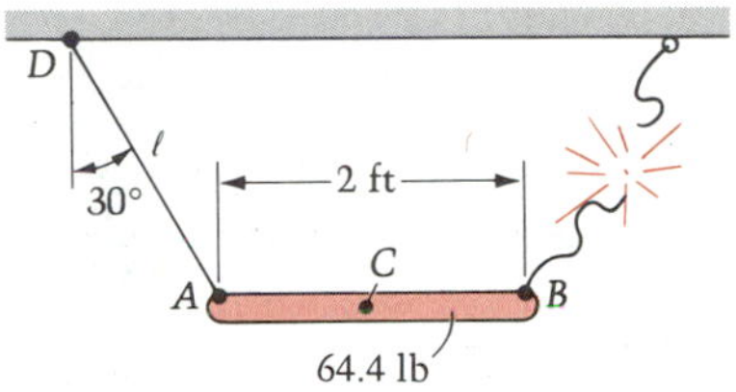

Figure E4.20a

A uniform rod is supported by two cords as shown in Figure E4.20a. If the right-hand cord suddenly breaks, determine the initial tension in the left cord AD. ("Initial" means before the rod has had time to move and before it has had time to generate any velocities.)

Solution

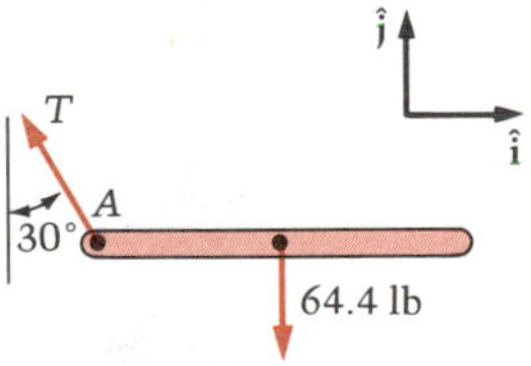

Figure E4.20b

Using the free-body diagram in Figure E4.20b, the equations of motion are, with $\mathbf{a}_C = \ddot{x}_C\hat{\mathbf{i}} + \ddot{y}_C\hat{\mathbf{j}}$ and $\boldsymbol{\alpha} = \alpha\hat{\mathbf{k}}$:

$$\xrightarrow{+} \qquad \Sigma F_x = -T\left(\frac{1}{2}\right) = m\ddot{x}_C = 2\ddot{x}_C \tag{1}$$

$$+\uparrow \qquad \Sigma F_y = -64.4 + T\left(\frac{\sqrt{3}}{2}\right) = m\ddot{y}_C = 2\ddot{y}_C \tag{2}$$

$$\circlearrowleft + \qquad \Sigma M_C = -T\left(\frac{\sqrt{3}}{2}\right)(1) = I_C\alpha = \frac{mL^2}{12}\alpha = \frac{2\alpha}{3} \tag{3}$$

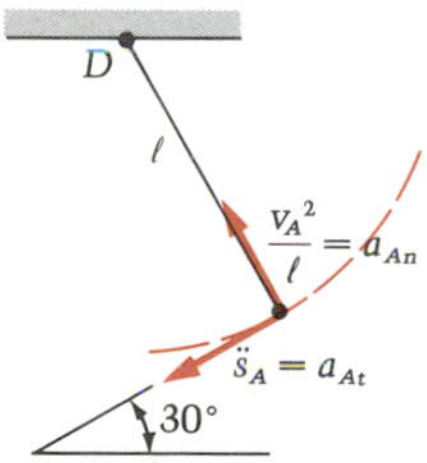

Figure E4.20c

Unfortunately, Equations (1–3) contain four unknowns (T, $\ddot{x}_C$, $\ddot{y}_C$, and α). Thus we seek an additional equation in these unknowns from kinematics. The point A is constrained to move (see Figure E4.20c) on a circle of radius ℓ about D. Thus point A has the tangential and normal components of acceleration shown (see Section 1.7). Furthermore, $v_A = 0$ at the instant of interest (nothing is moving yet!).

We may relate this $\mathbf{a}_A$ to $\mathbf{a}_C$:

$$\mathbf{a}_C = \ddot{x}_C\hat{\mathbf{i}} + \ddot{y}_C\hat{\mathbf{j}} = \underbrace{\mathbf{a}_A}_{\ddot{s}_A} + \underbrace{\alpha\hat{\mathbf{k}} \times \cancelto{1\hat{\mathbf{i}}}{\mathbf{r}_{AC}}}_{\alpha\hat{\mathbf{j}}} - \cancelto{\text{zero at } t=0}{\omega^2}\mathbf{r}_{AC} \tag{4}$$

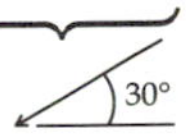

Here now is a nice shortcut:
If we dot Equation (4) with a unit vector in the direction 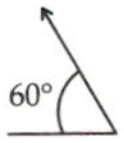 , we will eliminate $\mathbf{a}_A$ (because it is perpendicular to that direction). This is easier than solving the two "$\hat{\mathbf{i}}$ and $\hat{\mathbf{j}}$ equations." Such a unit vector is

$$-\frac{1}{2}\hat{\mathbf{i}} + \frac{\sqrt{3}}{2}\hat{\mathbf{j}}$$

so that, doing the dotting,

$$\ddot{x}_C\left(-\frac{1}{2}\right)+\ddot{y}_C\left(\frac{\sqrt{3}}{2}\right)=0+\alpha\left(\frac{\sqrt{3}}{2}\right)$$

or

$$-\ddot{x}_C+\sqrt{3}\ddot{y}_C=\sqrt{3}\alpha \tag{5}$$

Equations (1), (2), and (3) yield:

$$\ddot{x}_C=-\frac{T}{4} \qquad \ddot{y}_C=-32.2+T\left(\frac{\sqrt{3}}{4}\right) \qquad \alpha=\frac{-3\sqrt{3}T}{4}$$

Substituting these three results into Equation (5) results in:

$$+\frac{T}{4}+\sqrt{3}\left(-32.2+\frac{\sqrt{3}T}{4}\right)=-\sqrt{3}\left(\frac{3\sqrt{3}T}{4}\right)=-\frac{9T}{4}$$

or

$$T=\frac{4}{13}(32.2)\sqrt{3}=17.2 \text{ lb}$$

Note that *before* the right-hand string was cut, the tension, from statics, was:

$$\Sigma F_y=0=2T_{\text{STATIC}}\frac{\sqrt{3}}{2}-64.4 \Rightarrow T_{\text{STATIC}}=37.2 \text{ lb}$$

Forces in *inextensible* strings (ropes, cables, cords) are capable of changing "instantaneously," and indeed we see that this is the case in this problem.

Question 4.7 Can spring forces change instantaneously in this way?

In the preceding example, back-substitution immediately yields

$$\ddot{x}_C=\frac{17.2}{4}=-4.30 \text{ ft/sec}^2$$

$$\ddot{y}_C=\frac{17.2\sqrt{3}}{4}-32.2=-24.8 \text{ ft/sec}^2$$

Thus the mass center will start to move off to the left and down.

$$\alpha=\frac{-17.2(3\sqrt{3})}{4}=-22.3 \text{ rad/sec}^2$$

Thus the body will start to turn clockwise.

The acceleration of A follows from Equation (4):

$$\mathbf{a}_A=-4.30\hat{\mathbf{i}}-24.8\hat{\mathbf{j}}+22.3\hat{\mathbf{j}}$$
$$=-4.30\hat{\mathbf{i}}-2.5\hat{\mathbf{j}}$$

and, as a check, the direction of $\mathbf{a}_A$, $\tan^{-1}\left(\frac{2.5}{4.3}\right)$, is 30°.

Answer 4.7 No, the length must change and that takes time.

The magnitude of $\mathbf{a}_A$ is $\sqrt{(-4.30)^2 + (-2.5)^2} = 4.97$ ft/sec², which is $\ddot{s}_A$ (see Section 1.7) at the initial instant. It is interesting to note that the initial angular acceleration of the *string DA* is $|\mathbf{a}_A|/\ell$ ↻, or $4.97/\ell$ ↻.

Finally, we close the section with an example containing two bodies in rolling contact. The plate is simply translating, but the pipe has a more complicated motion: it rolls on the pipe, but *not* on the inertial frame (ground).

Example 4.21

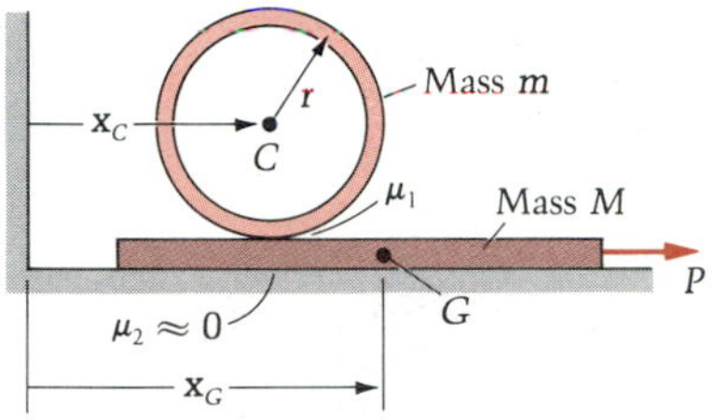

Figure E4.21a

Force P is applied to a plate that rests on a smooth surface. (See Figure E4.21a.) Find the largest force P for which the pipe will not slip on the plate.

Solution

For the pipe (Figure E4.21b), with $\boldsymbol{\alpha} = \alpha\hat{\mathbf{k}}$:

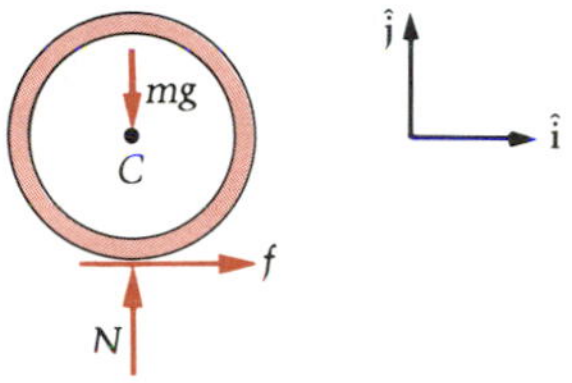

Figure E4.21b

$$\xrightarrow{+} \quad \Sigma F_x = f = m\ddot{x}_C \tag{1}$$

$$+\uparrow \quad \Sigma F_y = N - mg = m\ddot{y}_C = 0 \Rightarrow N = mg \tag{2}$$

$$\circlearrowleft + \quad \Sigma M_C = fr = I_C\alpha = mr^2\alpha \tag{3}$$

(Note that $I_C = (m/2)(r_o^2 + r_i^2) \approx mr^2$ if $r_o \approx r_i$. If the thickness $(r_o - r_i)$ is not given, assume it is small.)

For the plate (Figure E4.21c), we note that only the x equation of motion is of help; $\Sigma F_y = m\ddot{y}_G = 0$ gives $N_2 = N + Mg = (m + M)g$ as expected, and dimensions are not given so moments cannot be taken. (The moment equation would only give us the location of N_2, anyway.) Therefore

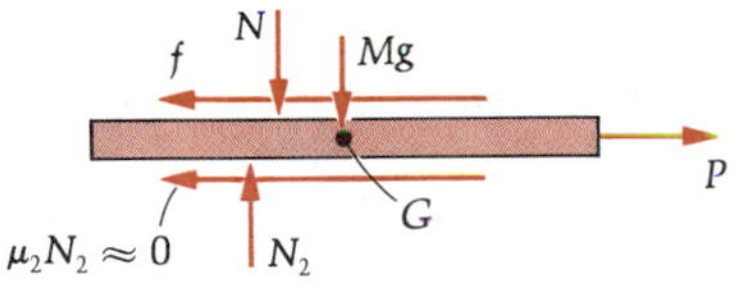

Figure E4.21c

$$\Sigma F_x = P - f = M\ddot{x}_G{}^* \tag{4}$$

Eliminating f between (1) and (3) gives

$$m\ddot{x}_C = mr\alpha \tag{5}$$

And between (1) and (4) gives

$$P = m\ddot{x}_C + M\ddot{x}_G \tag{6}$$

We note that if $m_T = m + M$ and C_T is the mass center of pipe *plus* plate, then Equation (6) could have been written immediately from $\Sigma F_x = m_T\ddot{x}_{C_T}$ for the combined system. Here $\Sigma F_x = P$; the right side follows from two derivatives of the definition of the mass center ($m_T x_{C_T} = m x_C + M x_G$).

The kinematics equation is tricky here. It is a rolling condition, but we must remember that x_G and x_C are necessarily measured relative to an inertial frame, here assumed to be fixed in the ground. Thus it is $\ddot{x}_C - \ddot{x}_G$ that is related to α.† For no slip, $\dot{x}_C + r\omega = \dot{x}_G$ which, when differentiated, yields

$$\ddot{x}_C - \ddot{x}_G = -r\alpha \tag{7}$$

Substituting α from (5) into (7) relates the accelerations of the two mass centers:

$$2\ddot{x}_C = \ddot{x}_G \tag{8}$$

* Sometimes G is used to designate a mass center.

† This difference is just the acceleration of C in the frame consisting of the translating plate.

Then (6) and (8) may be combined to give

$$P = (m + 2M)\ddot{x}_C \tag{9}$$

And combining (9) and (1) gives us the relationship between P and f:

$$P = \frac{m + 2M}{m} f \tag{10}$$

Since $f \le \mu_1 N$ for no slip, (10) gives:

$$\frac{m}{m + 2M} P \le \mu_1 mg \Rightarrow P \le (m + 2M)g\mu_1$$

Any larger P than $(m + 2M)g\mu_1$ will cause the pipe to slip on the plate.

PROBLEMS ▶ Section 4.5

4.62 A uniform sphere (radius r, mass m) rolls on the plane in Figure P4.62. If the sphere is released from rest at $t = 0$ when $x = L$, find $x(t)$.

4.63 A symmetric body $\mathscr{B}$ has mass m and radius R; a cord is wrapped around it as shown in Figure P4.63. Compute the downward acceleration of the center C if $\mathscr{B}$ is (a) a cylinder; (b) a sphere (with a small slot to accommodate the cord); (c) a thin ring. *Hint*: Work the problem just once with radius of gyration k_C; then substitute the three values $R/\sqrt{2}$, $\sqrt{\tfrac{2}{5}}R$, and $1R$ for k_C.

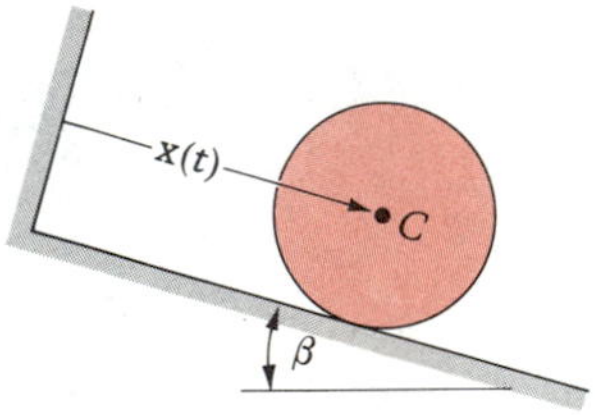

Figure P4.62

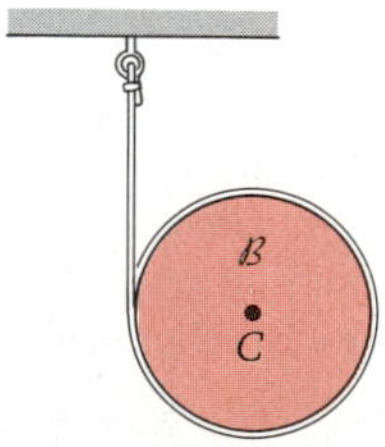

Figure P4.63

4.64 The cord in Figure P4.64 is wrapped around the cylinder, which is released from rest on the 60° incline shown. Find the velocity and position of C as a function of time t.

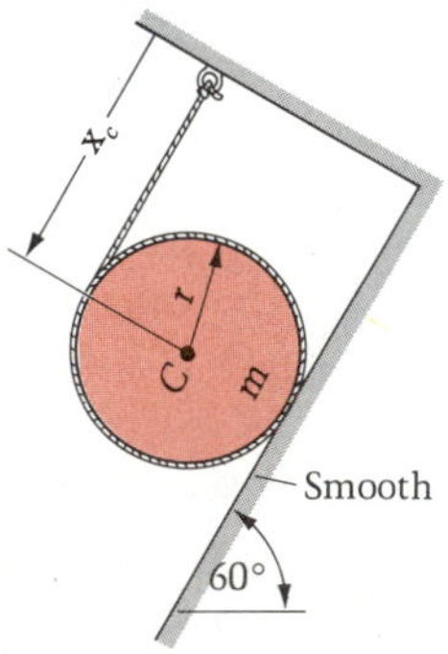

Figure P4.64

4.65 Sally Sphere, Carolyn Cylinder, Harry Hoop, and Wally Wheel each have mass m and radius R. Wally's spokes and rim are very light compared to his hub. (See Figure P4.65.) They are going to have a race by rolling down a rough plane. Give (a) the order in which they finish and (b) the times.

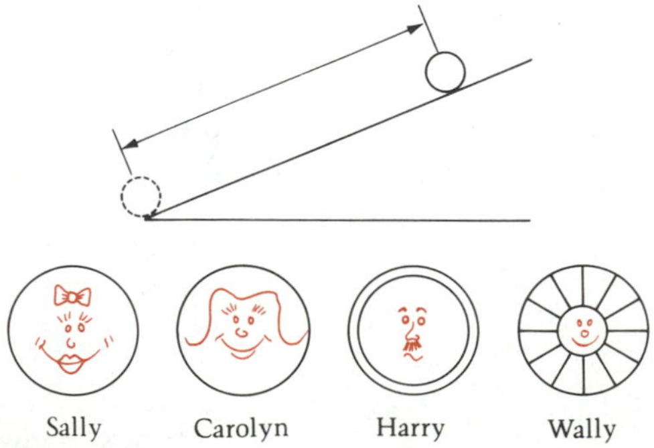

Figure P4.65

4.66 In the preceding problem, Wally and Carolyn are connected by a bar of negligible mass and released from rest on the same incline. (See Figure P4.66.) Determine the force in the bar.

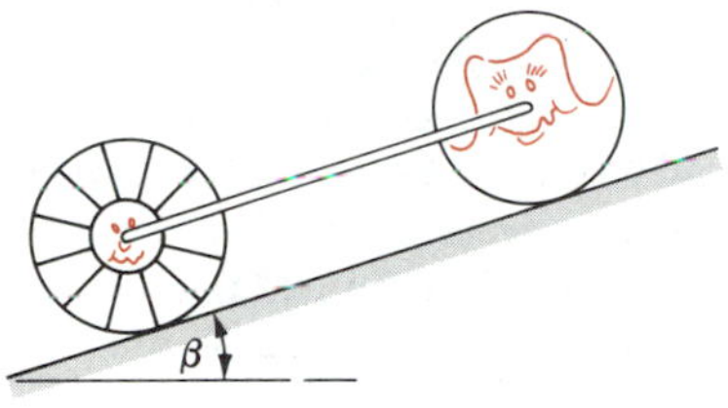

Figure P4.66

4.67 Repeat the preceding problem, but suppose that Wally and Carolyn switch places.

4.68 The two pulleys $\mathcal{P}_1$ and $\mathcal{P}_2$ and the block $\mathcal{B}$ in Figure P4.68 each have mass m and are connected by the cord.

- **a.** Write a brief paragraph explaining in words why the system cannot be in equilibrium. Start with, "If the system were in equilibrium, the tension in the rope above $\mathcal{B}$ would equal mg." Then follow the rope around the pulleys until you reach a contradiction.
- **b.** Find the acceleration of C_1.

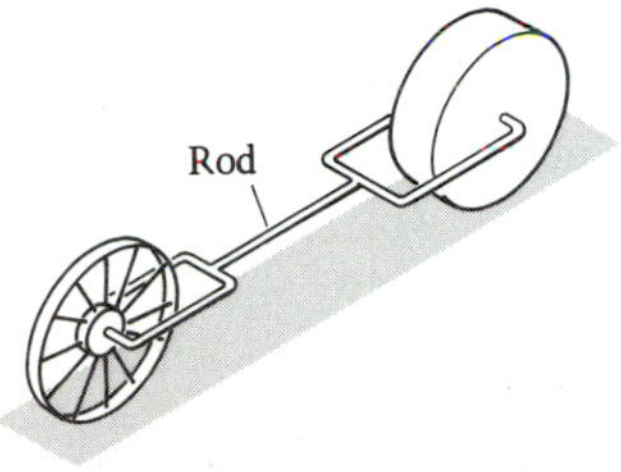

Figure P4.68

4.69 The uniform sphere (mass m, radius r) in Figure P4.69 is at rest before P is applied. If μ is the coefficient of friction between sphere and floor,

- **a.** find the maximum P for there to be no slip;
- **b.** for P twice that found in (a), find $\mathbf{a}_C$ and $\boldsymbol{\alpha}$. Note that $|\mathbf{a}_C|$ does not equal $r|\boldsymbol{\alpha}|$ as it would if there were no slipping, i.e., if the sphere were rolling.

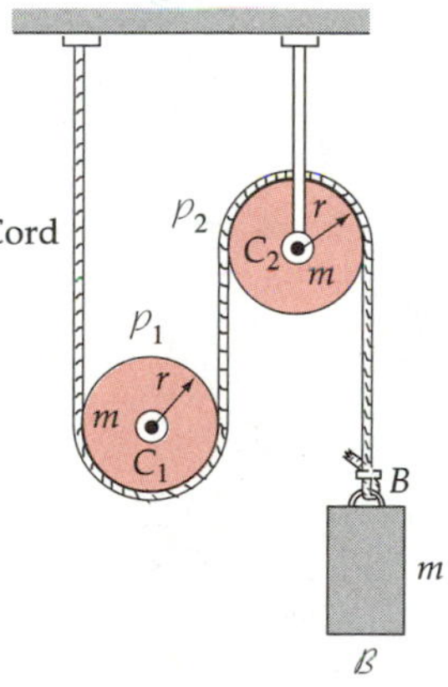

Figure P4.69

4.70 The uniform cylinder in Figure P4.70, of mass m and radius r, is at rest before it is subjected to a couple of moment M_o. The coefficient of friction between cylinder and floor is μ.

- **a.** Find the largest value of M_o for which there is no slip.
- **b.** For M_o twice the value found in (a), find $\mathbf{a}_C$ and $\boldsymbol{\alpha}$.

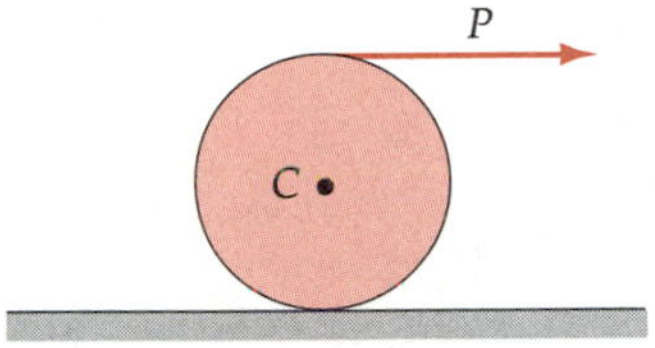

Figure P4.70

4.71 Find the ratio of r to R for which the force T in Figure P4.71 will cause the wheel to roll (no slip) no matter how small the friction. Treat the wheel as a uniform cylinder.

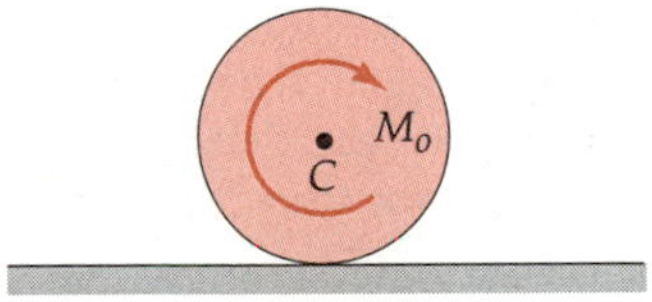

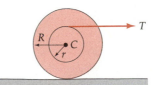

Figure P4.71

4.72 Force T is given to be small enough, and the friction coefficient large enough, that both wheels in Figure P4.72 will roll on the plane.

a. Give arguments why one wheel rolls left and the other right.
b. Find the ratio of r to R for which the accelerations of C are equal in magnitude.

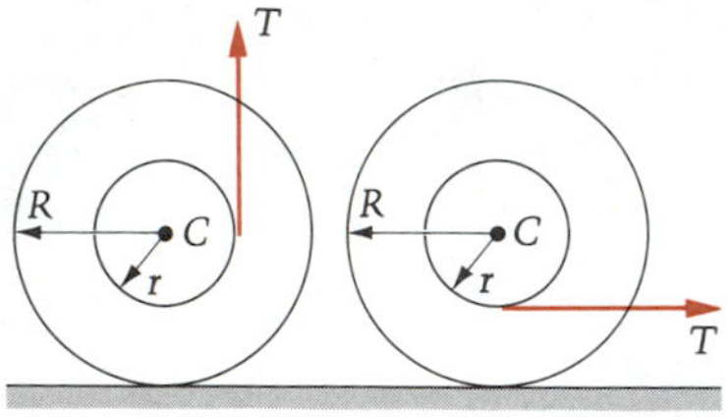

Figure P4.72

4.73 The uniform sphere in Figure P4.73, of mass m and radius r, is at rest when it is subjected to a couple of moment M_0. If there is no slip, find the acceleration of the center of the sphere.

4.74 A cylinder spinning at angular speed ω_0 rad/sec clockwise is placed on an inclined plane. (See Figure P4.74.) Show that the cylinder center will begin moving up the plane if $\mu > \tan \beta$. Why does this result have nothing to do with the size of ω_0?

4.75 The cylinder of weight W and radius r shown in Figure P4.75 has an angular velocity of 100 rad/s clockwise. It is lowered onto the rough incline. If its center C is observed to remain momentarily at rest, determine the coefficient of sliding friction. Find how long the center C remains at rest.

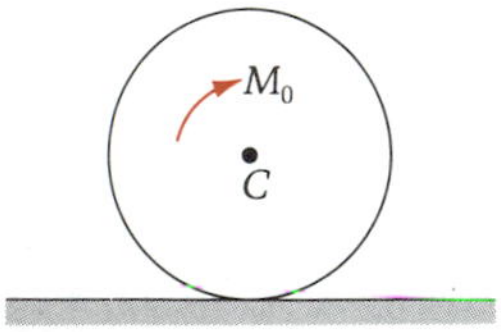

Figure P4.73

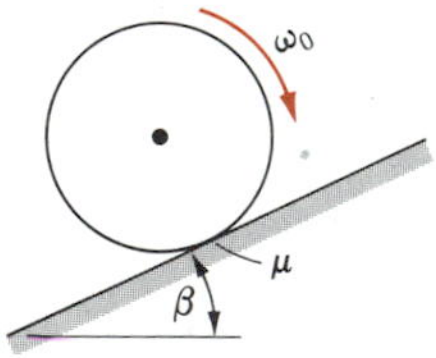

Figure P4.74

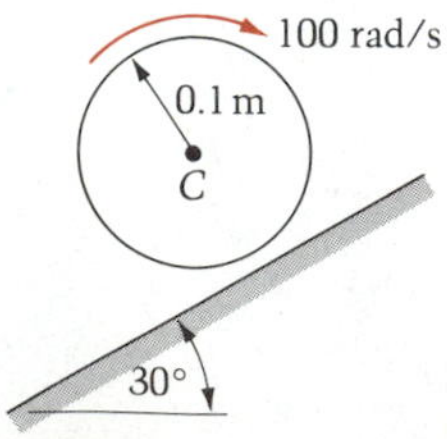

Figure P4.75

4.76 The bowling ball in Figure P4.76 is released with $v_C = 22$ ft/sec and $\omega = 0$ as it contacts the surface of the alley. Neglecting the effect of the three finger holes, and using a coefficient of friction of 0.3, find the distance traveled by the center of the ball before slipping stops.

* **4.77** The force $P = 60$ N is applied as shown in Figure P4.77 to the 10-kg cylinder $\mathcal{C}$, originally at rest beneath the mass center of the thin, 5-kg rectangular plate $\mathcal{P}$. The coefficient of friction between $\mathcal{C}$ and $\mathcal{P}$ is 0.5, and the plane beneath $\mathcal{C}$ is smooth. Determine: (a) the initial acceleration of C; (b) the value of x when $\mathcal{C}$ is slipping on both surfaces. The length of $\mathcal{P}$ is 2 m.

4.78 The constant force F_0 is applied to the cylinder, initially at rest, as shown in the two drawings constituting Figure P4.78. Show in the following two ways that the cylinder will slip provided that

$$\frac{F_0}{mg} > 3\mu$$

a. Assume rolling; then obtain the inequality from $f > \mu N$ after solving for f and N.
b. Assume slipping; then integrate $\ddot{x}_C$ and $\ddot{\theta}$ to obtain $\dot{x}_C$ and $\dot{\theta}$, and then find the velocity of the contact point B; if it is to the right (that is, positive), this is consistent with $f = \mu N$ to the left and we have slipping.

Figure P4.76

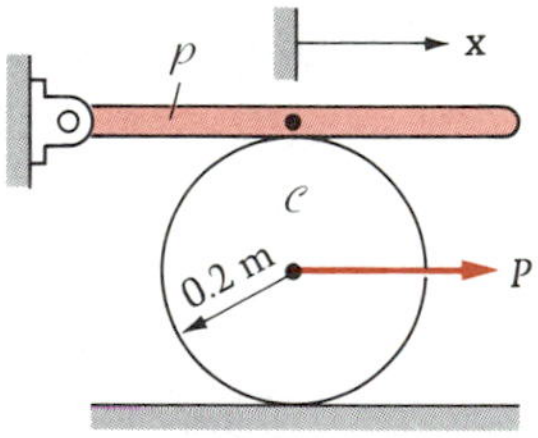

Figure P4.77

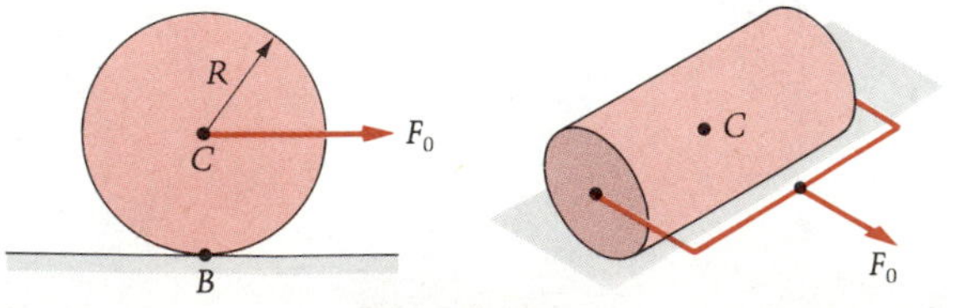

Figure P4.78

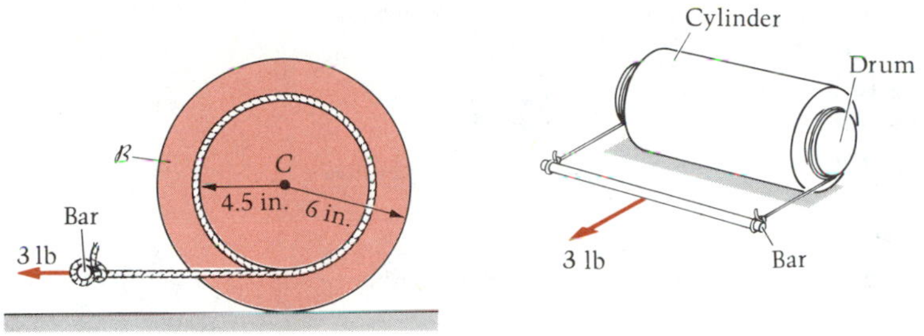

Figure P4.79

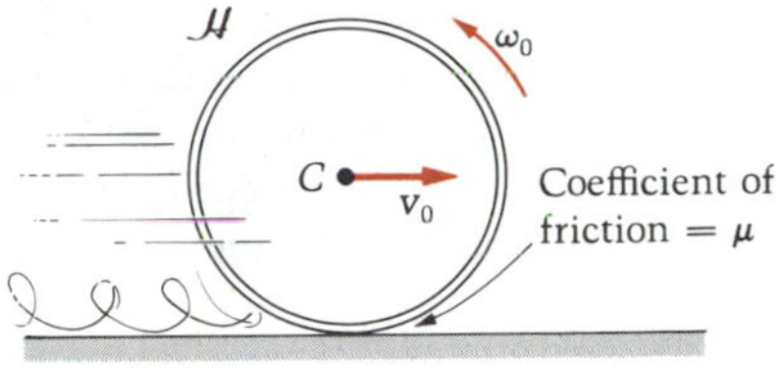

Figure P4.82

4.79 Two drums of radius 4.5 in. are mounted on each end of a cylinder of radius 6 in. to form a 40-lb rigid body $\mathcal{B}$ with radius of gyration $k_{zC} = 5$ in. (See Figure P4.79.) Ropes are wrapped around the drum and tied to a horizontal bar to which a 3-lb force is applied. As $\mathcal{B}$ rolls from rest, tell (a) the number of inches of rope wound or unwound (tell which) in three seconds and (b) the minimum friction coefficient needed for the rolling to take place.

• **4.80** Find the range of possible values of the couple M_0 for which the cylinder in Figure P4.80 will not slip in *either direction* when released from rest on the incline. The mass is 15 kg; the radius is 0.2 m; and the coefficient of friction is $\mu = 0.2$.

4.81 An airplane lands on a level strip at 200 mph. (See Figure P4.81.) Initially, just before the wheels touch the runway, the wheels are not turning. After they touch the runway they will skid for some distance and then roll free. If during this skidding the plane has a constant velocity of 200 mph and the normal force between the wheel and the runway is 10 times the wheel weight, find the length of the skid mark. (The coefficient of friction is $\frac{1}{2}$; the radius of gyration of the wheel is three-fourths of its radius.)

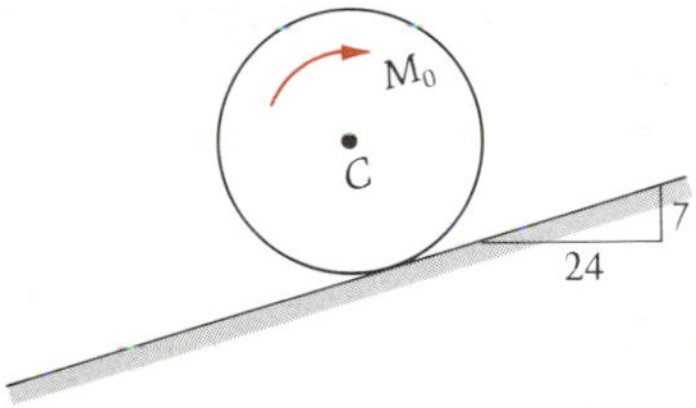

Figure P4.80

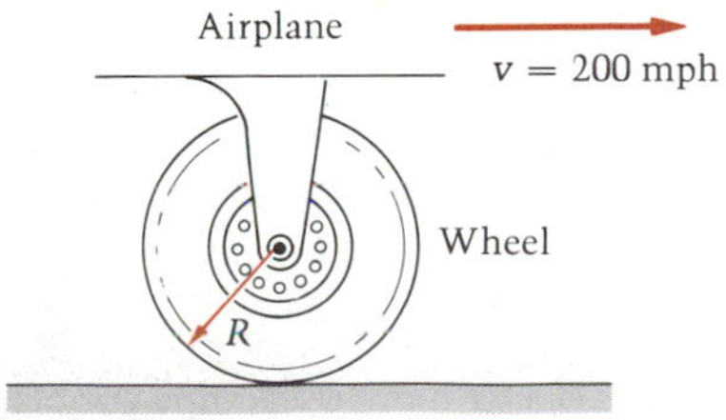

Figure P4.81

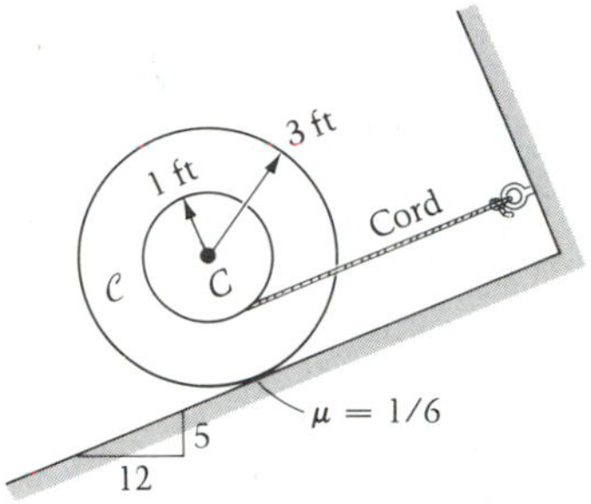

Figure P4.83

• **4.82** A hula hoop $\mathcal{H}$ (mass m, radius r) is thrown forward with backspin; $v_C = v_0$ to the right and $\omega = \omega_0$ counterclockwise. (See Figure P4.82.)

a. How long and how far does the mass center move before $\mathcal{H}$ stops slipping?

b. Find the relationship between v_0 and ω_0 such that when $\mathcal{H}$ stops slipping: (i) it rolls right; (ii) it rolls left; (iii) it stops.

4.83 The strong, flexible cable shown in Figure P4.83 is wrapped around a light hub attached to the 130-lb cylinder $\mathcal{C}$. Find the angular acceleration of $\mathcal{C}$ upon release from rest. Note that it is impossible for the wheel to roll down the plane (meaning without slipping); to do so the cord would have to break.

4.84 Repeat the previous problem for $\mu = 0.25$.

4.85 In Figure P4.85, find how far down the incline C travels in 5 s if the 20-kg cylinder $\mathcal{C}$ is released from rest.

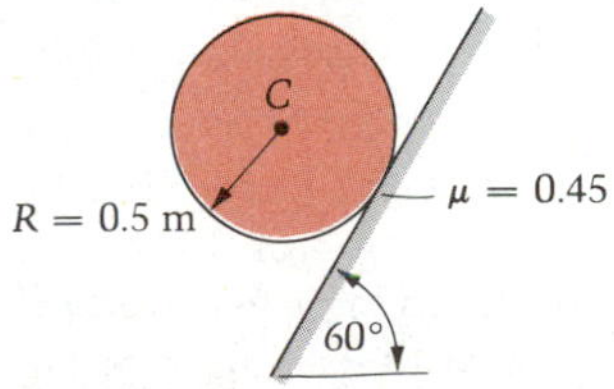

Figure P4.85

4.86 The 50-lb body $\mathcal{C}$ in Figure P4.86 may be treated as a solid cylinder of radius 2 ft. The coefficient of friction between $\mathcal{C}$ and the plane is $\mu = 0.2$, and a force $P = 10$ lb is applied vertically to a cord wrapped around the hub. Find the position of the center C, 10 sec after starting from rest.

4.87 Given that the slot (for the cord) in the cylinder in Figure P4.87 (mass 10 kg) has a negligible effect on I_C, find:

a. The largest θ for which no motion down the plane will occur
b. The time required for C to move 3 m down the incline if $\theta = 60°$.

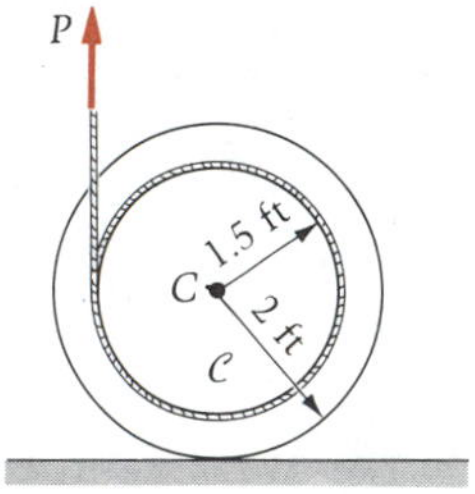

Figure P4.86

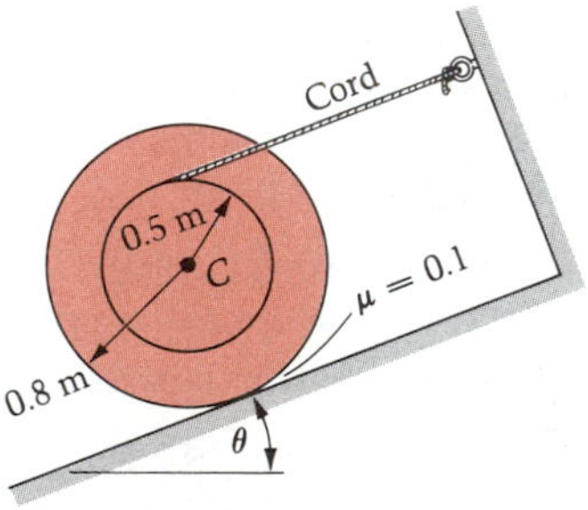

Figure P4.87

4.88 A light 100-ft cord is wrapped around the 32.2-lb spool $\mathcal{S}$, which is pinned at C to the cart $\mathcal{C}$ (see Figure P4.88). The radius of gyration of $\mathcal{S}$ with respect to an axis normal to the figure at C is 1.3 ft. The cart (without $\mathcal{S}$) has weight 64.4 lb. The wheels of $\mathcal{C}$ are small and light, so that friction beneath them is negligible. The 2- and 3-lb forces are applied to the system at rest. If upon complete unwrapping the cord is to end up between points P and Q in the lower figure, where should $\mathcal{C}$ be originally parked along PQ?

4.89 A child pulls on an old wheel with a force of 5 lb by means of a rope looped through the hub of the wheel. (See Figure P4.89.) The friction coefficient between wheel and ground is $\mu = 0.2$. Find I_C for the wheel, and use it to determine the location of C after 3 sec.

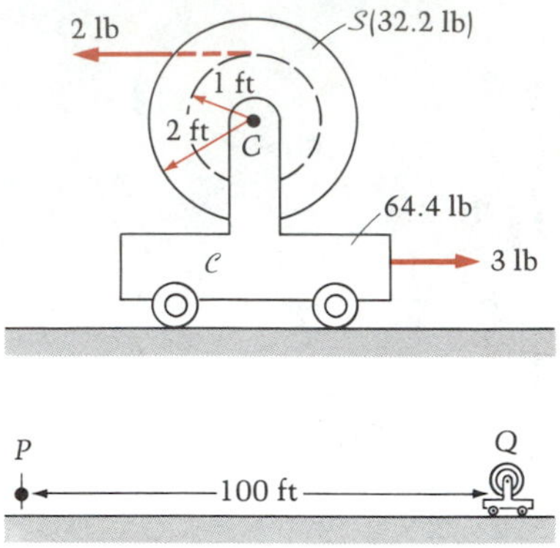

Figure P4.88

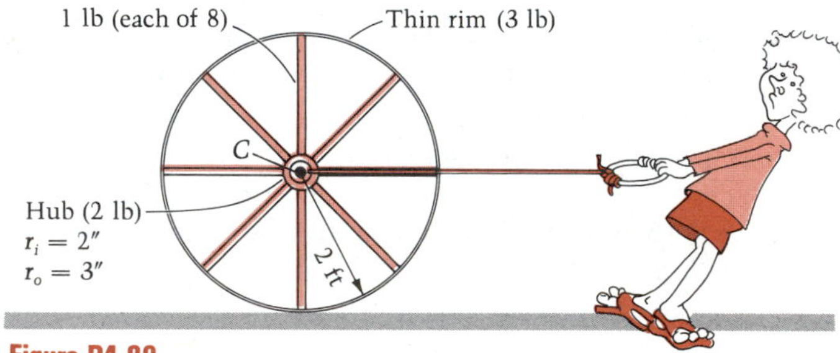

Figure P4.89

4.90 The wheel shown in Figure P4.90 has a mass of 10 kg, a radius of 0.4 m and a radius of gyration with respect to the z-axis through C of 0.3 m. Determine the angular acceleration of the wheel and how far the mass center C moves in 3 seconds if the wheel starts from rest.

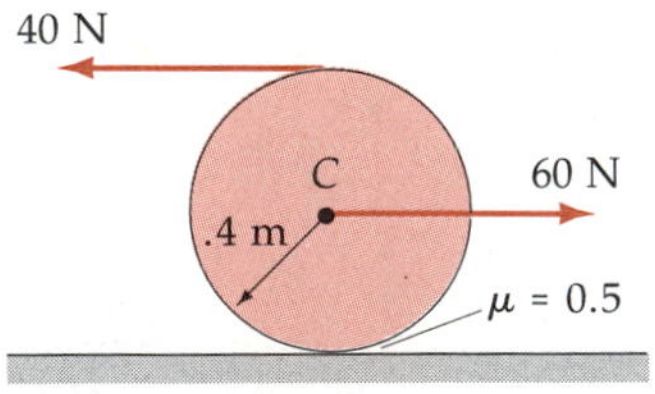

Figure P4.90

4.91 Rework the preceding problem if the friction coefficient is changed to $\mu = 0.2$.

4.92 Two cables are wrapped around the hub of the 10-kg spool shown in Figure P4.92, which has a radius of gyration of 500 mm with respect to its axis. A constant 40-N force is applied to the upper cable as shown. Find the mass center location 5 s after starting from rest if: (a) $\mu = 0.2$; (b) $\mu = 0.5$.

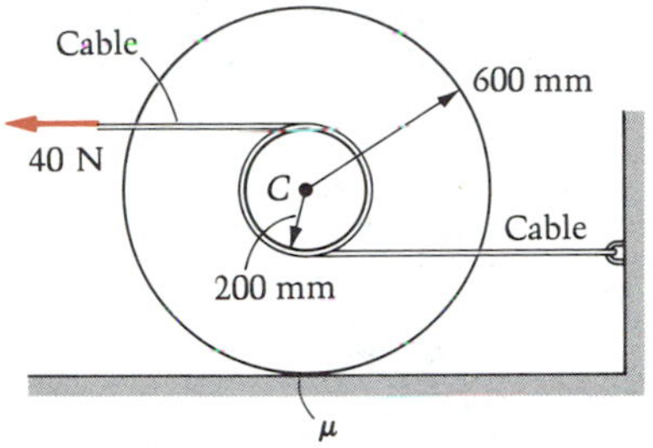

Figure P4.92

4.93 A sphere of radius $\frac{1}{2}$ ft and weight 16.1 lb is projected onto a horizontal plane (Figure P4.93). Its center has initial velocity v_0 at $t = 0$ and the sphere has initial angular velocity ω_0, defined as shown. If the coefficient of sliding friction between the sphere and the plane is 0.15, plot graphs of distance gone (x_C) against time t up to $t = 3$ sec for the following cases:

a. $v_0 = 10$ ft/sec; $\omega_0 = 100$ rad/sec

b. $v_0 = 10$ ft/sec; $\omega_0 = 50$ rad/sec

c. $v_0 = 10$ ft/sec; $\omega_0 = 30$ rad/sec

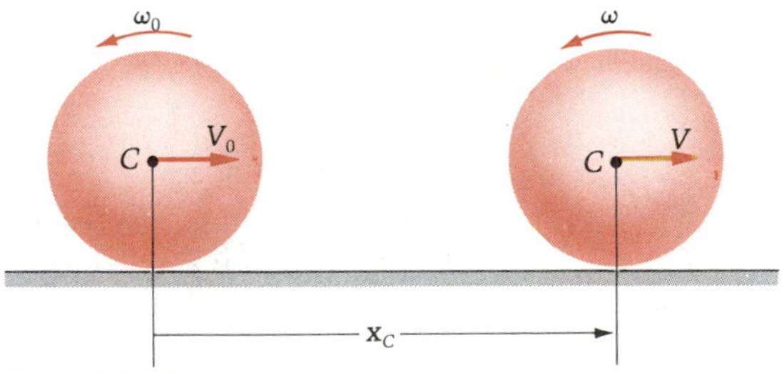

Figure P4.93

4.94 The cylinder $\mathcal{C}$ in Figure P4.94 has a thin slot cut in it which doesn't affect its moment of inertia appreciably. A cord is wrapped in the slot and connects to the cart $\mathcal{B}$, which rests on the plane on small, light wheels. The force of 10 lb is applied to $\mathcal{C}$ with the system initially at rest. Find the length of unwrapped cord after 4 seconds elapse. Assume enough friction to prevent slip of $\mathcal{C}$ on the plane.

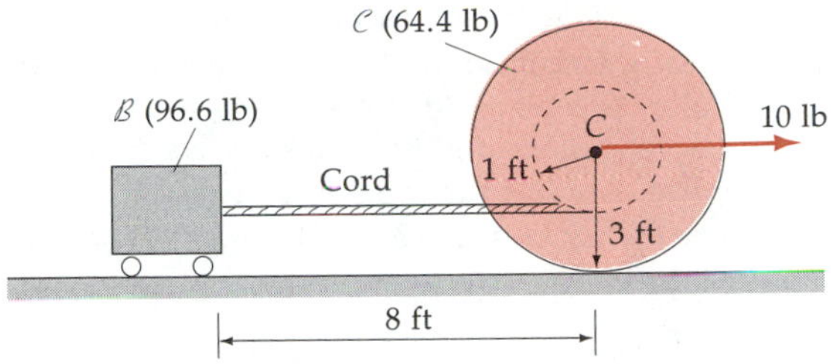

Figure P4.94

4.95 Assume that enough friction is available to prevent the cylinder in Figure P4.95 from slipping.

a. Show that

(i) $\mathcal{A}$ rolls to the right if $\theta < \cos^{-1}(r/R)$.

(ii) $\mathcal{A}$ rolls to the left if $\theta > \cos^{-1}(r/R)$.

(iii) $\mathcal{A}$ is in equilibrium if $\theta = \cos^{-1}(r/R)$ (and will translate if P increases enough to overcome friction).

b. Find $\ddot{x}_C$ and α if $r = 0.2$ m, $R = 0.4$ m, $P = 20$ N, $mg = 40$ N, and $\theta = 45°$.

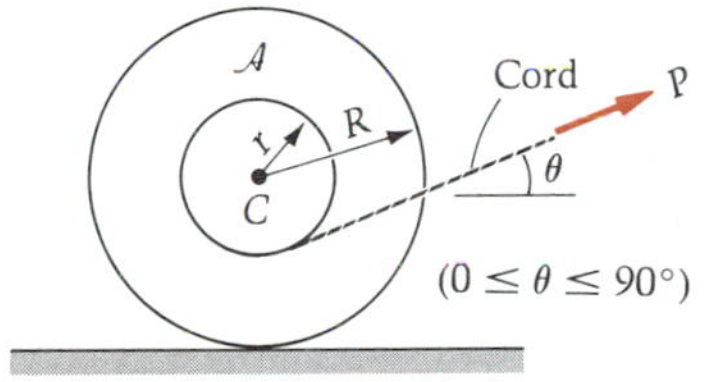

Figure P4.95

4.96 A homogeneous spool of weight W rolls on an inclined plane; a string tension of amount $4W$ acts up the plane as shown in Figure P4.96. With I_C given approximately by $WR^2/2g$, find the acceleration of C. Assume unlimited friction.

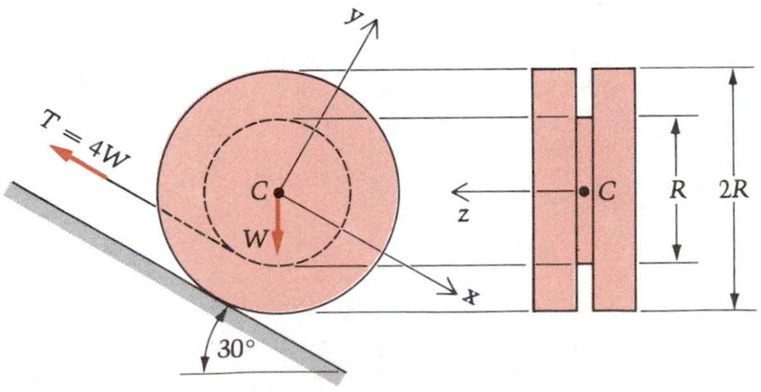

Figure P4.96

• **4.97** Pulley $\mathcal{B}_1$ in Figure P4.97 weighs 100 pounds and has a radius of gyration about the z-axis through O of $k_O = 7$ in. Pulley $\mathcal{B}_2$ weighs 20 lb and has $k_C = 3$ in. Find the angular acceleration of $\mathcal{B}_1$ just after the system is released from rest. Assume the rope doesn't slip on $\mathcal{B}_1$ but that there is *no friction* between $\mathcal{B}_2$ and the rope. Is this the angular acceleration for later times as well?

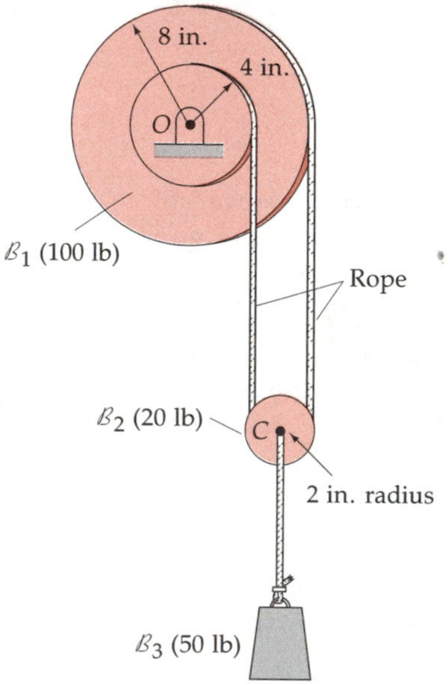

Figure P4.97

4.98 The 32.2-lb body $\mathcal{C}$ in Figure P4.98 is a spool having a radius of gyration $k_C = 6$ in. about its axis. Cords are wrapped around the peripheries; one is connected to a ceiling, the others to the 48.3-lb block $\mathcal{B}$. Find the accelerations of the centers C (of $\mathcal{C}$) and B (of $\mathcal{B}$).

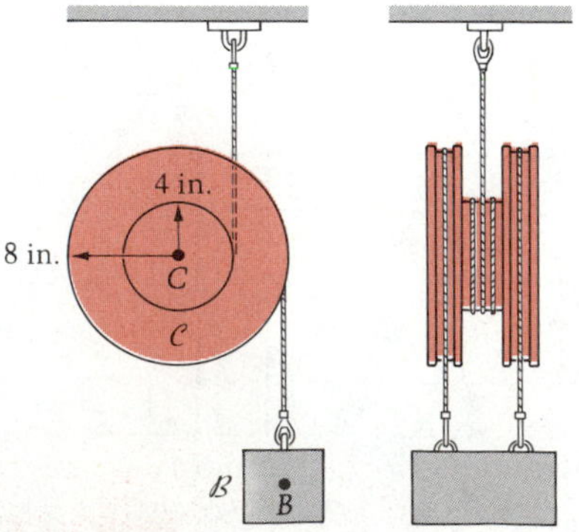

Figure P4.98

4.99 Wheel $\mathcal{C}$ is made up of the solid disk $\mathcal{A}$, rim $\mathcal{R}$, and four spokes $\mathcal{S}$. Masses and radii are given in Figure P4.99 and the table.

a. Compute I_{zz}^C for the wheel.

b. The coefficient of friction between $\mathcal{C}$ and the plane is $\mu = 0.3$. If a cord is wrapped around the disk and connected to the 50-kg body $\mathcal{B}$, determine the acceleration of the mass center C of $\mathcal{C}$.

Part	Mass (kg)
Disk $\mathcal{A}$	20
Spokes $\mathcal{S}$	5 (each)
Rim $\mathcal{R}$	10

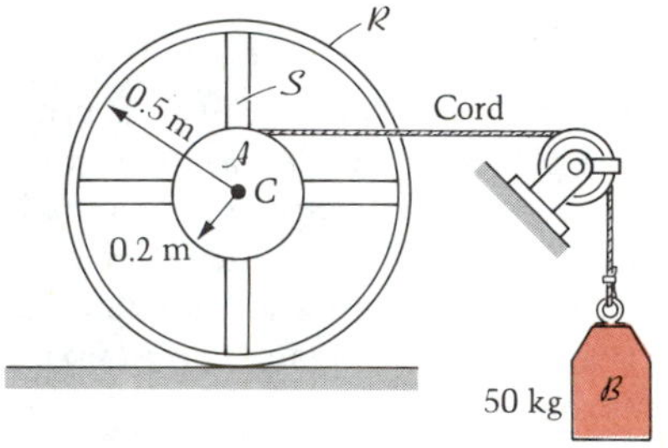

Figure P4.99

4.100 The radius of gyration of the 20-kg wheel in Figure P4.100 with respect to its axis is 0.3 m. Motion starts from rest. Find the acceleration of the mass center C, and determine how far C moves in 5 s.

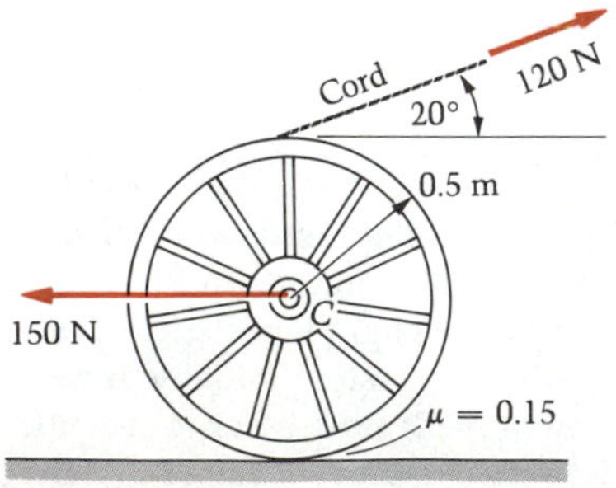

Figure P4.100

4.101 A string is wrapped around the hub of the spool shown in Figure P4.101. There are four indicated string directions. For the direction that will result in the largest displacement of C in 3 s, find this displacement. Assume sufficient friction to prevent slipping. The spool has a mass of 12 kg and a radius of gyration about z_C of 0.6 m. Each force equals 10 N, and the spool starts from rest.

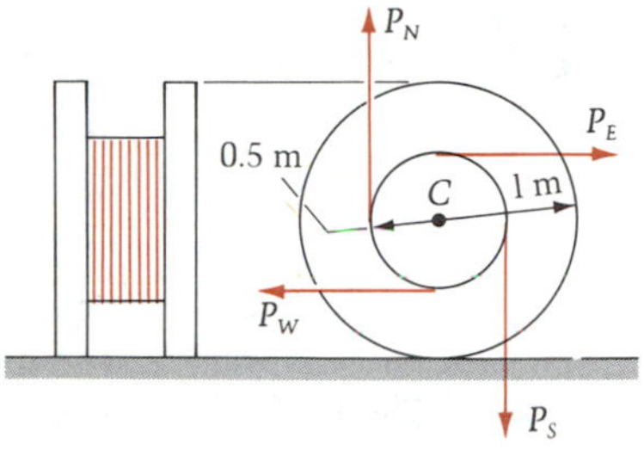

Figure P4.101

4.102 Cylinder $\mathcal{C}$ in Figure P4.102 has a mass of 4 slugs, and the effect of the hub on its moment of inertia is negligible. It is connected by means of a cord to the 1-slug block $\mathcal{B}$. The mass of the pulley is negligible. The coefficient of friction between $\mathcal{C}$ and the plane is $\mu = 0.5$, and the radii of $\mathcal{C}$ are given in the figure. If the system is released from rest, determine the time that will elapse before $\mathcal{B}$ hits the ground.

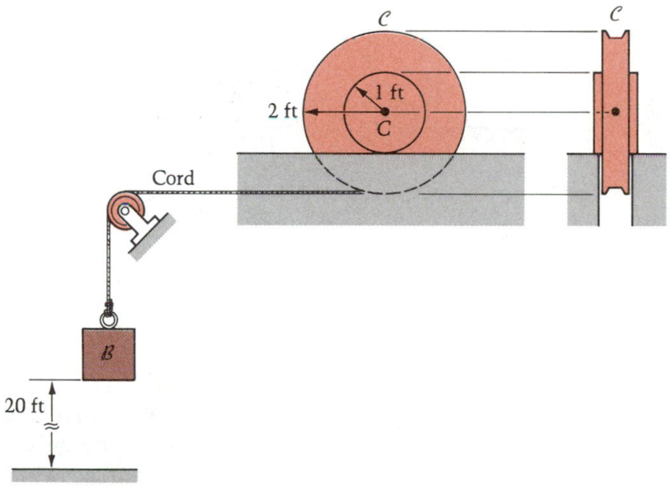

Figure P4.102

4.103 Find how long it takes for $\mathcal{A}$ to roll off the plane in Figure P4.103, assuming sufficient friction to prevent slipping. The system is released from rest.

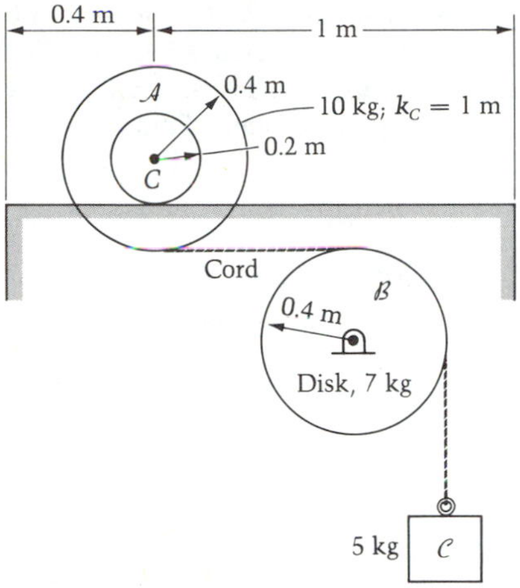

Figure P4.103

4.104 Disks $\mathcal{A}$ and $\mathcal{B}$ each weigh 64.4 lb and are rigidly attached to the light shaft $\mathcal{S}$ that joins their centers. (See Figure P4.104.) A 96.6-lb cylinder $\mathcal{C}$ has a hole drilled along its axis, through which $\mathcal{S}$ passes. A force of 20 lb is applied horizontally to an inextensible string wrapped around $\mathcal{C}$. If friction is negligible between $\mathcal{S}$ and $\mathcal{C}$, and if $\mathcal{A}$ and $\mathcal{B}$ roll on the plane, find:

- **a.** The angular acceleration of the cylinder
- **b.** The angular acceleration of the disks
- **c.** The minimum coefficient of friction between disks and plane for no slipping.

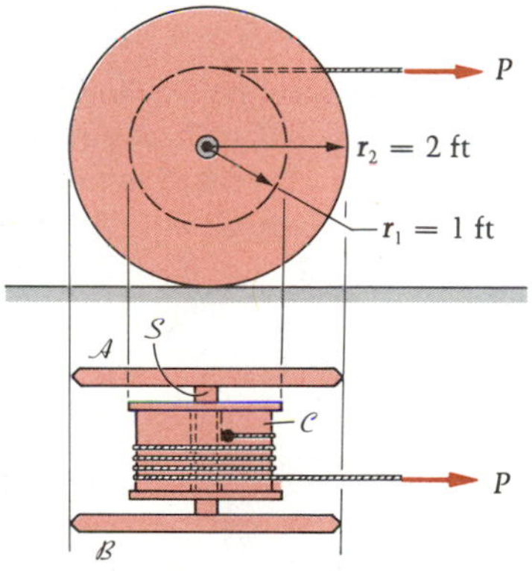

Figure P4.104

4.105 Rework the preceding problem, but this time assume that the string is wrapped so that it comes off the *bottom* of $\mathcal{C}$.

4.106 The two wheels are identical 16.1-lb cylinders with smooth axles at their centers. (See Figure P4.106.) The carriage weighs 32.2 lb and has its mass center at C. The cylinders do not slip on the inclined plane. Find the acceleration of point Q.

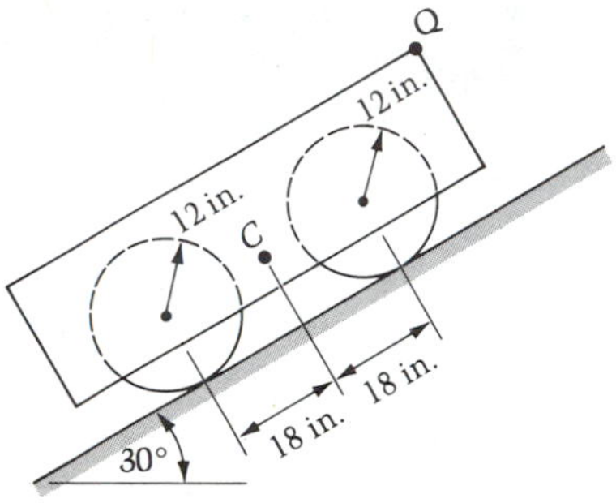

Figure P4.106

* **4.107** Cylinder $\mathcal{C}$ in Figure P4.107 weighs 100 lb; it is rolling on the plane and is pinned at its center C to the 10-lb rod $\mathcal{R}$. If v_C is initially 10 ft/sec to the left, and if the coefficient of kinetic friction between the plane and each body is $\mu = 0.4$, determine how long it will take the system to come to rest.

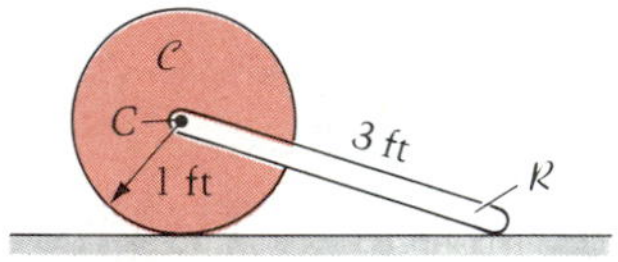

Figure P4.107

* **4.108** A uniform half-cylinder of radius r and mass m is held in the position shown in Figure P4.108 by the string tied to B. Find the reaction of the floor just after the string is cut. There is sufficient friction to prevent slipping.

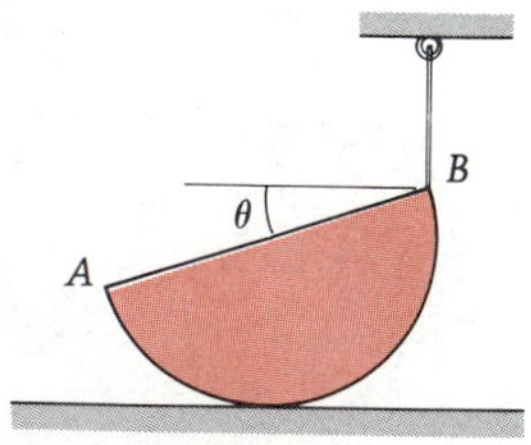

Figure P4.108

* **4.109** In Figure P4.109 the force P is applied to the cord at $t = 0$, when the 25-N cylinder is at rest. Find the position of the mass center when $t = 6$ s.

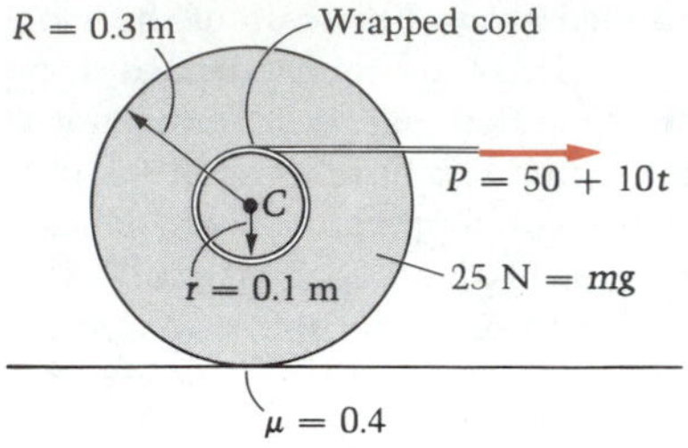

Figure P4.109

* **4.110** The semicylinder in Figure P4.110(a) is released from rest, and there is enough friction to prevent slipping throughout the ensuing motion (Figure P4.110(b)).

a. Find I_{zz}^C.

b. Write the three differential equations of motion of the body (good at any angle θ).

c. Find the two equations relating $\ddot{x}_C$ and $\ddot{y}_C$ to $\ddot{\theta}$, $\dot{\theta}$, and θ.

d. Eliminate f, N, $\ddot{x}_C$, and $\ddot{y}_C$ and obtain the single differential equation in the variable $\theta(t)$. Note the complexity of the equation!

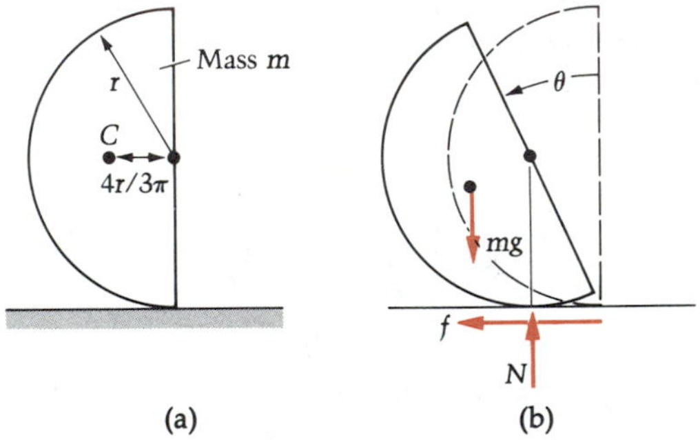

Figure P4.110

4.111 The 15-lb carriage shown in Figure P4.111 is supported by two uniform rollers each of weight 10 lb and radius 3 in. The rollers roll on the ground and on the carriage. Determine the acceleration of the carriage when the 5-lb force is applied to it.

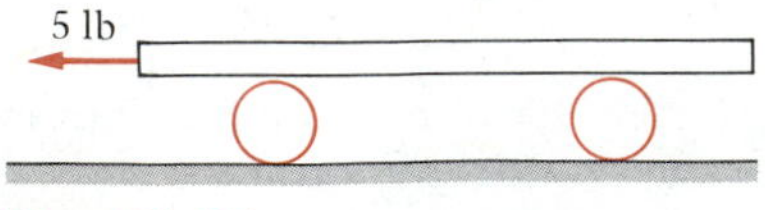

Figure P4.111

4.112 The 128.8-lb homogeneous plank shown in Figure P4.112 is placed on two homogeneous cylindrical rollers, each of weight 32.2 lb. The system is released from rest. Determine the initial acceleration of the plank if no slipping occurs. Is this the acceleration for later times as well?

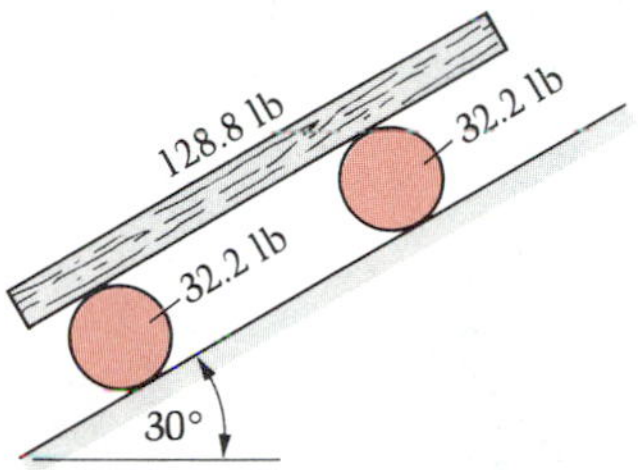

Figure P4.112

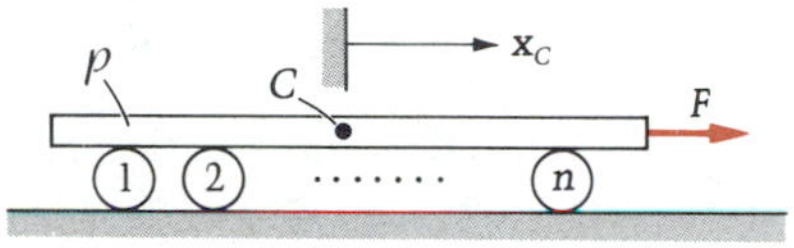

Figure P4.113

4.113 Body $\mathcal{P}$ in Figure P4.113 is a rigid plate of mass M, resting on a number n of cylinders each of mass m and radius R. Force F is constant and starts the system moving from the position shown. If there is no slipping at any surface, find: (a) the acceleration of the plate and (b) its position x_C as a function of M, m, F, n, and time t.

* **4.114** A 6-ft gymnast makes a somersault dive into a net by standing stiff and erect on the edge of a platform and allowing himself to overbalance. He loses foothold (without having slipped) when the platform's reaction on his feet becomes zero; he preserves his rigidity during his fall. Show that he falls flat on his back if the drop from the platform to the net is about 43 ft.

4.115 The homogeneous cylinder $\mathcal{C}$ in Figure P4.115 is at rest on the conveyor belt when the latter is started up with a constant acceleration of 3 ft/sec² to the right. If the cylinder rolls on the belt, find the elapsed time when the cylinder reaches the end A.

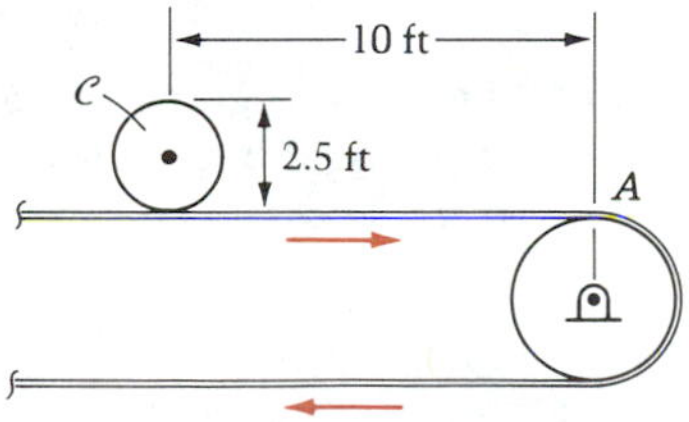

Figure P4.115

4.116 The pipe in Figure P4.116 has a mass of 500 kg and rests on the flatbed of the truck. The coefficient of friction between the pipe and truck bed is $\mu = 0.4$. The truck starts from rest with a constant acceleration a_0.

a. How large can a_0 be without the pipe slipping at any time?

b. For the value of a_0 in part (a), how far has the truck moved when the pipe rolls off the back?

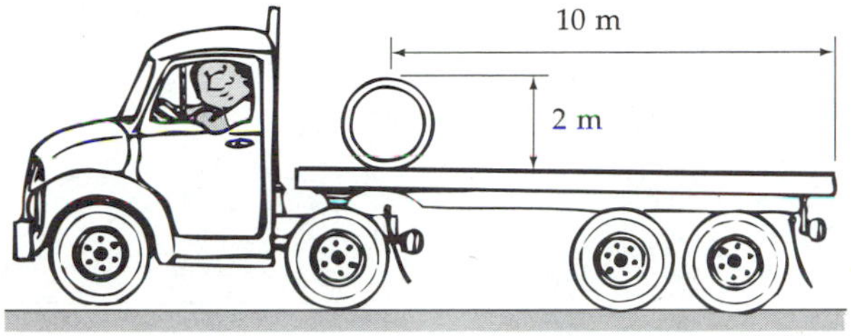

Figure P4.116

4.117 The uniform sphere (mass = 1 slug, radius = 1 ft) and the slab (mass = 2 slugs) shown in Figure P4.117 are at rest before the force $P = 24$ lb is suddenly applied to the slab. The coefficient of friction is 0.2 between the sphere and slab and between the slab and horizontal plane. (a) Does the sphere slip on the slab? (b) What is the acceleration of the center of the sphere?

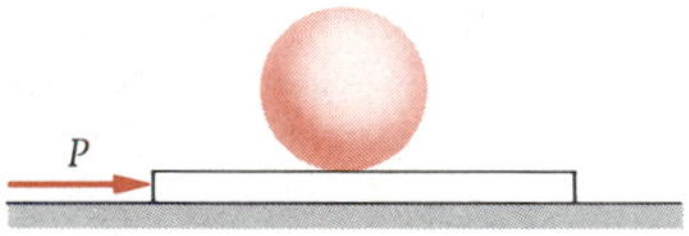

Figure P4.117

4.118 The homogeneous cylinder $\mathcal{C}$ in Figure P4.118 weighs 64.4 lb. The acceleration of the 96.6-lb cart $\mathcal{D}$ is 10 ft/sec² to the right.

a. Determine the acceleration of the center C of the cylinder and the friction force exerted on $\mathcal{C}$ by $\mathcal{D}$ if there is sufficient friction to prevent slipping.

b. How large does the friction coefficient μ have to be for this to occur?

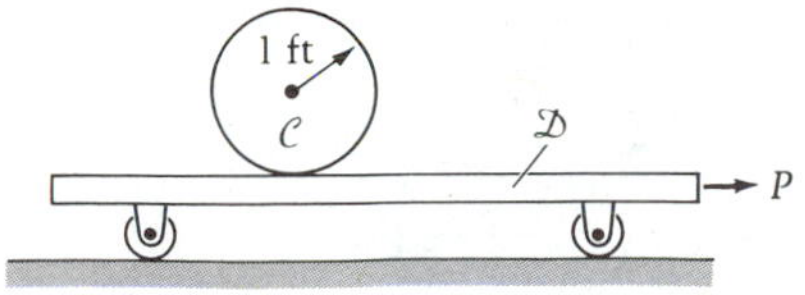

Figure P4.118

4.119 The homogeneous cylinder $\mathcal{A}$ in Figure P4.119 weighs 64.4 lb and rolls on the 96.6-lb truck $\mathcal{B}$. The mass of the truck rollers may be neglected. Find the force P such that C does not move relative to the plane.

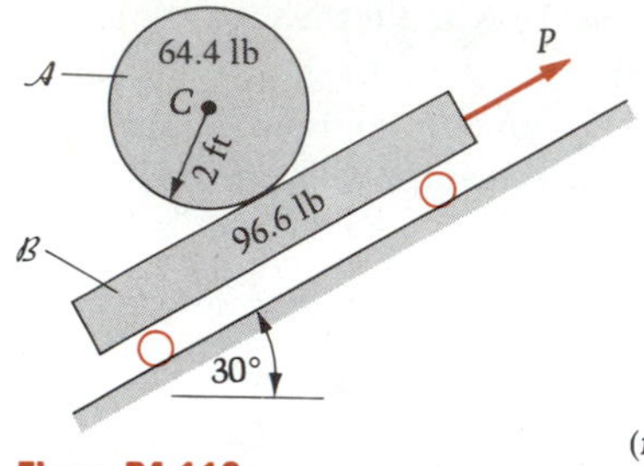

Figure P4.119

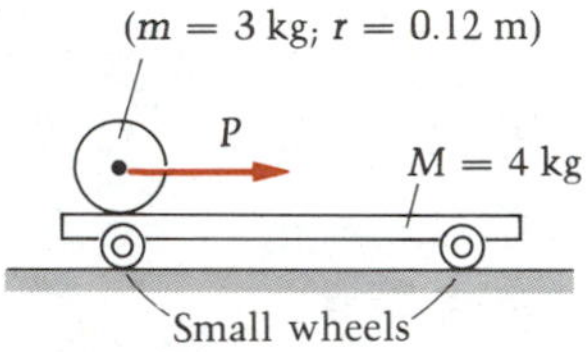

Figure P4.120

4.120 The system shown in Figure P4.120 is initially at rest. A force P is then applied that varies with time according to $P = 7t^2$, where P is in newtons and t in seconds. If the coefficient of friction between cylinder and cart is $\mu = 0.5$, find how much time elapses before the cylinder starts to slip on the cart.

4.121 In the previous problem, determine how much time passes (from $t = 0$) before the cylinder leaves the surface of the cart. Initially, the center of the cylinder is 2 m from the right end of the cart.

4.122 A slender homogeneous bar $\mathcal{B}$ weighing 193 lb has an angular velocity of 2 rad/sec clockwise and an angular acceleration of 8 rad/sec² clockwise when in the position shown in Figure P4.122. The wall at B is smooth; the coefficient of sliding friction at A is 0.10. Find the reactions at A and B on $\mathcal{B}$ in this position. *Hint*: The force P can be found.

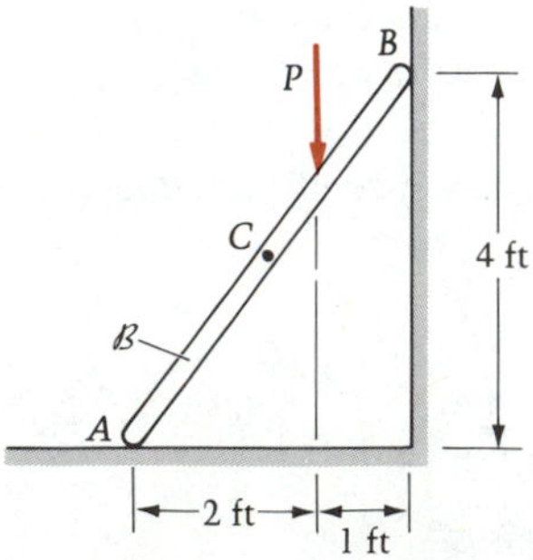

Figure P4.122

4.123 After release from a slightly displaced position, the rod in Figure P4.123 will remain in contact with the floor throughout its fall. Describe the path of C and find the reaction onto the floor just before the rod becomes horizontal.

4.124 The uniform slender bar of mass m and length L is released from rest in the position shown in Figure P4.124. Find the force exerted by the smooth floor at this instant.

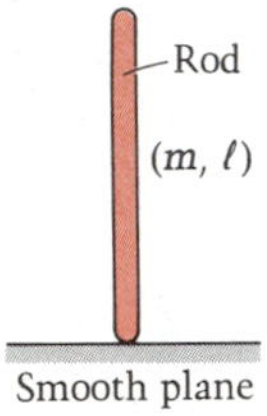

Figure P4.123

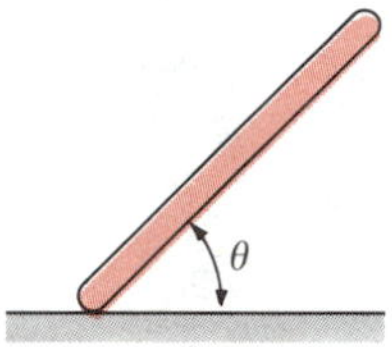

Figure P4.124

4.125 A thin rod AB of length ℓ and mass m is released from rest in the position shown in Figure P4.125. Point A of the rod is in contact with a surface whose coefficient of friction is μ.

a. Determine the minimum value of μ, say $\mu = \mu_{\text{min}}$, required to prevent end A from slipping upon release.

b. Find the acceleration of the mass center of the rod immediately after release for $\mu \geq \mu_{\text{min}}$ and for $\mu < \mu_{\text{min}}$.

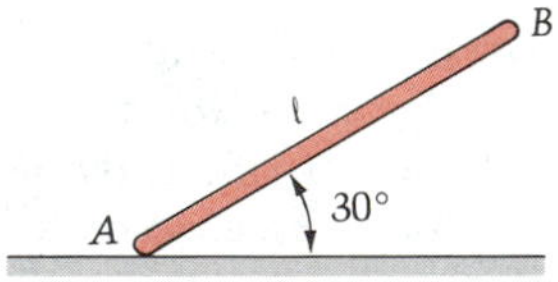

Figure P4.125

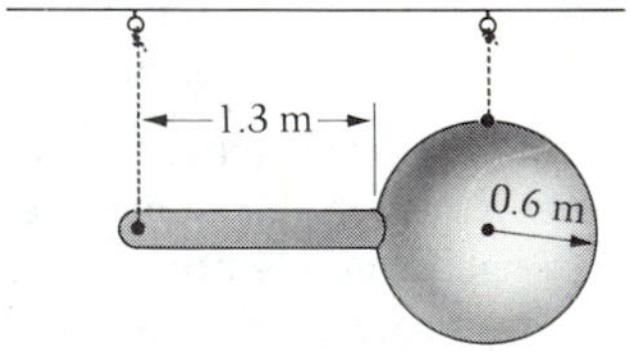

Figure P4.126

4.126 The 30-kg sphere and 15-kg rod in Figure P4.126 are welded together to form a single rigid body. Determine the angular acceleration of the body immediately after the right-hand string is cut.

4.127 If the right-hand string in Figure P4.127 is cut, find the initial tension in the left string. The slender rod has mass m and length L.

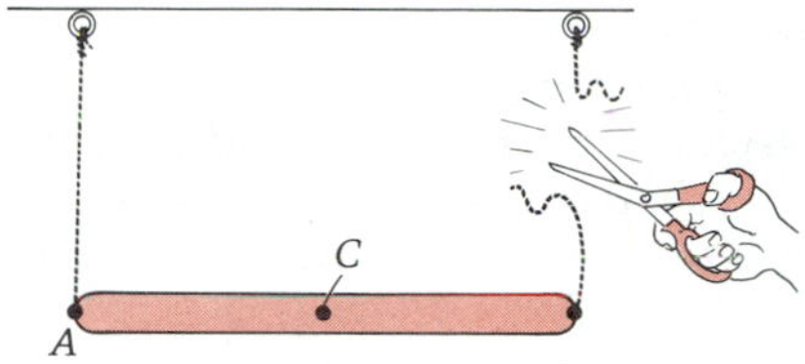

Figure P4.127

4.128 Repeat the preceding problem if the rod is replaced by a rectangular plate suspended from the two upper corners. The width (between strings) is B and the height is H.

4.129 A uniform slender rod, 10 ft long and weighing 90 lb, is supported by wires attached to its ends. (See Figure P4.129.) Find the tension in the right wire just after the left wire is cut. Assume the wires to be inextensible.

4.130 The left end of a slender uniform bar is attached to a light inextensible cable as shown in Figure P4.130. If the bar has mass m and length L and is released from rest in the position shown, find the angular acceleration of the bar at the instant after release.

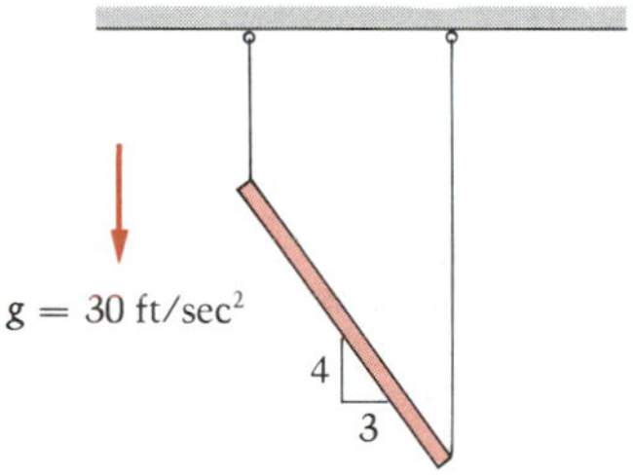

Figure P4.129

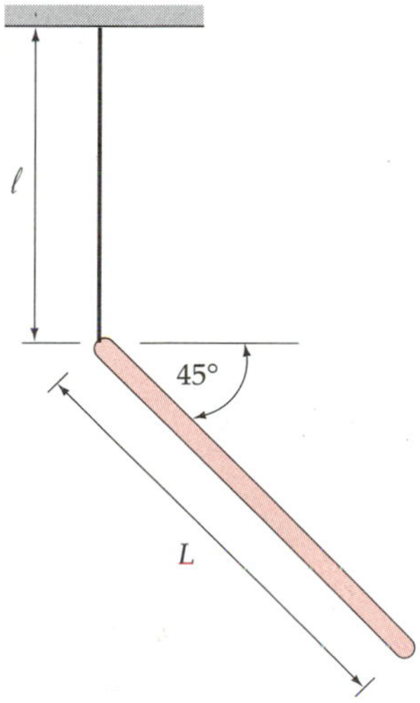

Figure P4.130

4.131 The uniform slender bar of mass m is released from rest in the position shown in Figure P4.131. Find $\mathbf{a}_A$ and the tension in the inextensible cord at the instant after release.

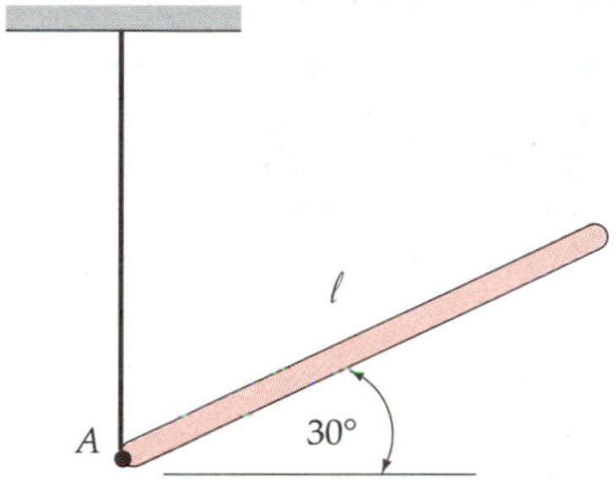

Figure P4.131

4.132 The uniform 20-lb bar is three feet long and has an angular velocity $\omega = 3 \circlearrowleft$ rad/sec with $\mathbf{v}_C = \mathbf{0}$ at the instant shown in Figure P4.132. Neglecting interaction with the air, what is the angular velocity of the bar after its center has dropped 10 feet?

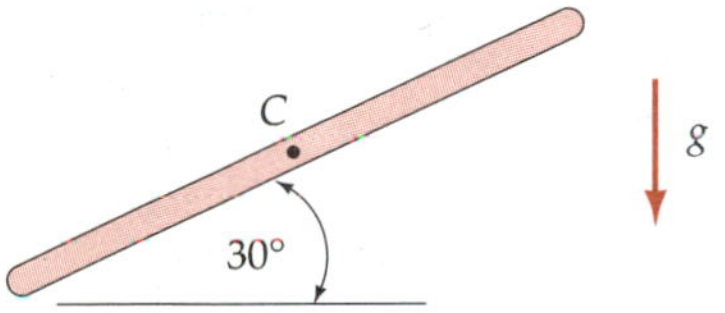

Figure P4.132

4.133 The disk shown in Figure P4.133 has mass m and radius r. Show that at the instant the right-hand string is cut, the tension in the other string changes to $\frac{2}{5}\,mg$, so that the acceleration of the mass center is $\frac{3}{5}\,g \downarrow$.

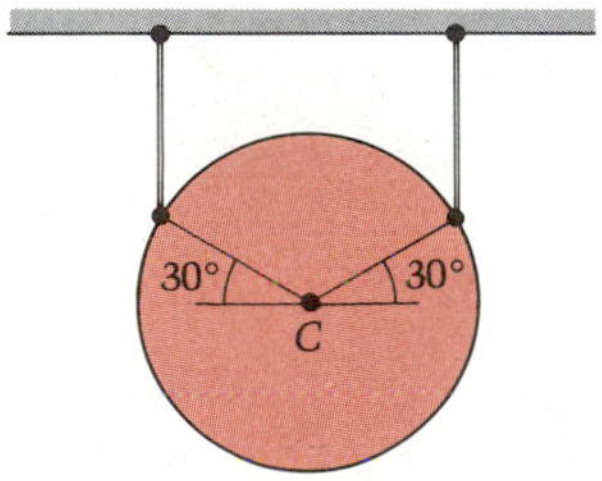

Figure P4.133

4.134 A uniform rod $\mathcal{R}$ is supported by two cords as shown in Figure P4.134. If the right-hand cord suddenly breaks, determine the initial tension in the left cord AD. ("Initial" means before the rod has had time to move and before it has had time to generate any velocities.)

4.135–4.140 The six equilateral triangular plates (Figures P4.135–4.140) are each supported on their rightmost corner B by a string; each has a different support condition at the left corner A. At the instant when the string at B is cut, find $\mathbf{a}_C$ and $\boldsymbol{\alpha}$ in each case. The length of each side is s.

4.141 The uniform 10-lb bar in Figure P4.141 is suspended by two inextensible cables. At the instant shown, when each point in the bar has a velocity of $10\hat{\mathbf{i}}$ ft/sec, the right cable breaks. Find the force in the left cable immediately after the break.

4.142 A slender bar $\mathcal{B}$ weighing 64.4 lb is attached by massless cables to a fixed pivot A as shown in Figure P4.142. The system is swinging about A as a pendulum. At $\theta = 0$ the angular velocity is 2 ↺ rad/sec and cable AD breaks. Find the tension in cable AB just after the break.

* **4.143** A beam of length L and weight W per unit length is supported by two cables at A and B. (See Figure P4.143.) If the cable at B should break, find the shear force V and bending moment M at section xx just after the cable breaks. *Hint*: Euler's laws apply to every part of the body.

* **4.144** See Figure P4.144. Assuming that sufficient friction is present to prevent slipping between $\mathcal{C}$ and the plane, find the angular accelerations of $\mathcal{C}$ and $\mathcal{R}$ just after force P is applied to the bodies at rest. They are connected by a smooth pin.

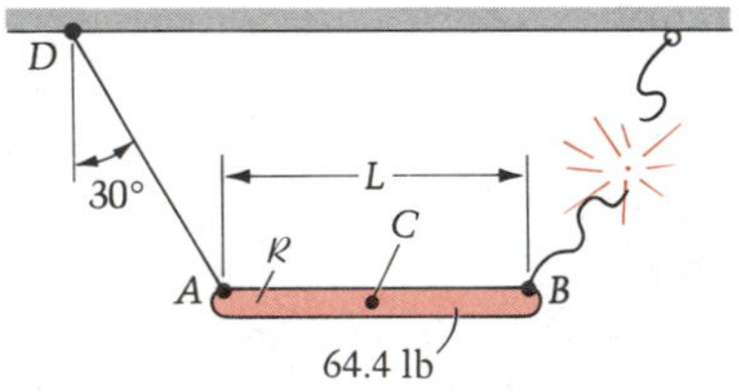

Figure P4.134

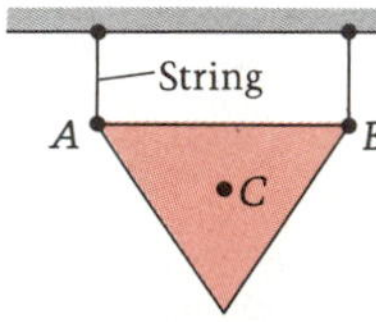

Figure P4.135

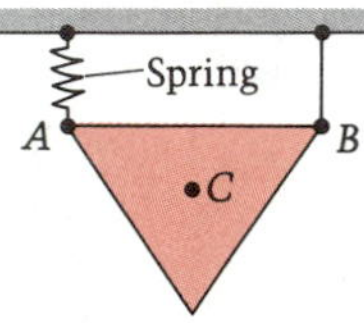

Figure P4.136

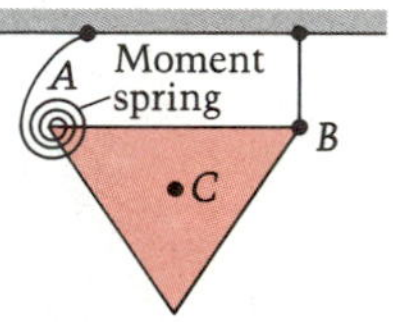

Figure P4.137

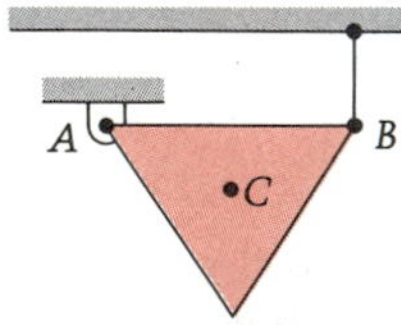

Figure P4.138

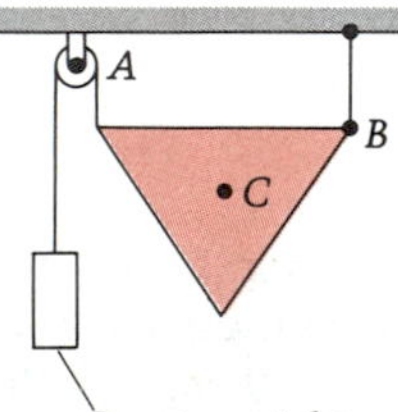

Figure P4.139

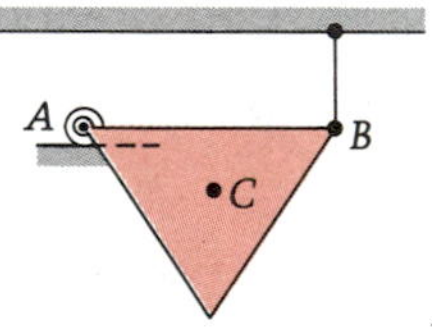

Figure P4.140

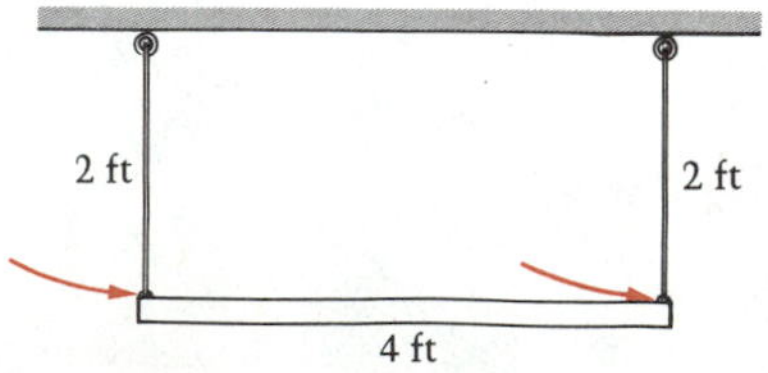

Figure P4.141

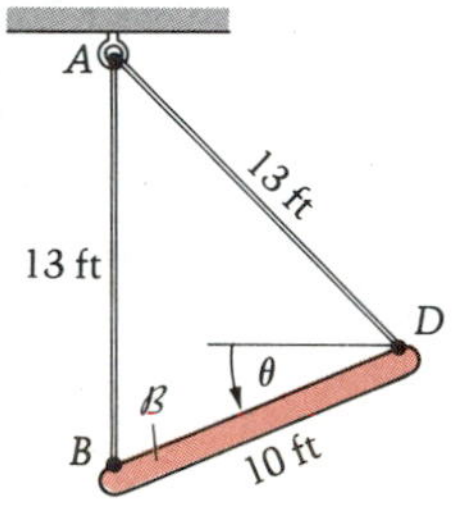

Figure P4.142

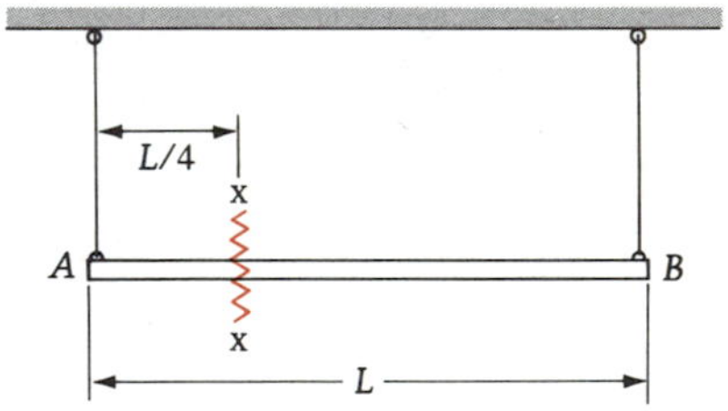

Figure P4.143

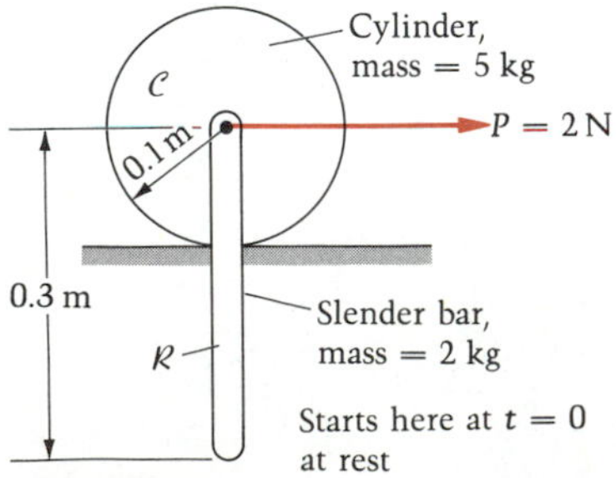

Figure P4.144

* **4.145** Rods $\mathcal{B}_1$ and $\mathcal{B}_2$ each have mass m. (See Figure P4.145.) Upon release from rest in the horizontal position indicated, find the reactions at O, and at A, onto $\mathcal{B}_1$.

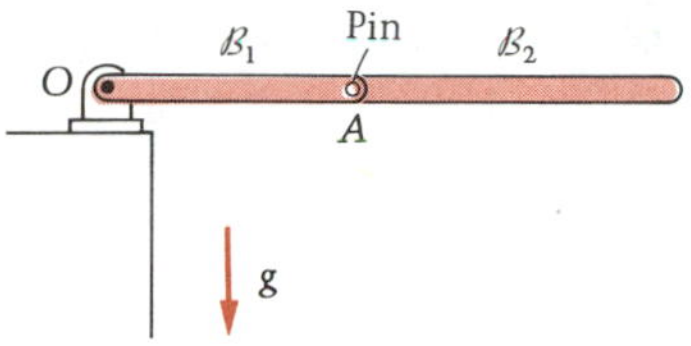

Figure P4.145

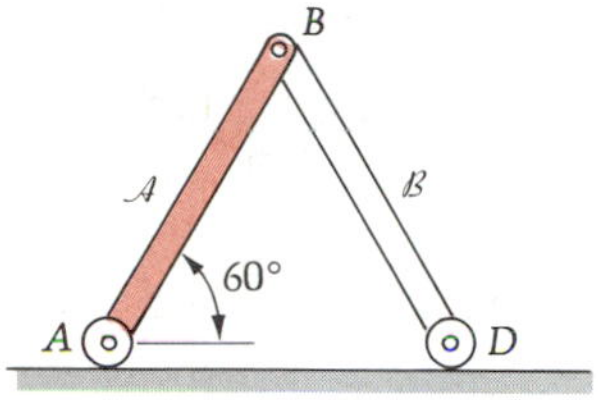

Figure P4.146

* **4.146** Two uniform bars $\mathcal{A}$ and $\mathcal{B}$ are released from rest in the position shown in Figure P4.146. Each bar is 2 ft long and weighs 10 lb. Determine the angular acceleration of each bar and the reactions at A and D immediately after release. The rollers are light and the pins smooth.

4.147 A constant torque T_0 ↺ is applied to the crank arm $\mathcal{C}$ of the planetary mechanism shown in Figure P4.147. The axes of the identical gears $\mathcal{S}$ and $\mathcal{P}$ are vertical, and the ends of the crank are pinned to the centers of $\mathcal{S}$ and $\mathcal{P}$. Determine the angular acceleration of $\mathcal{C}$ if $\mathcal{S}$ is fixed in the inertial frame of reference. Treat the gears as uniform disks. The plane of the page is horizontal.

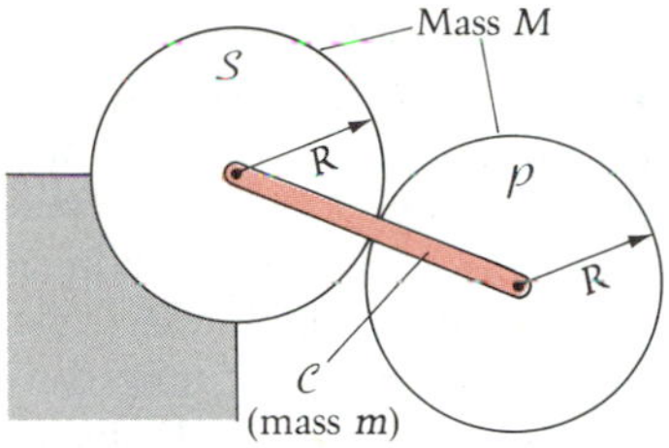

Figure P4.147

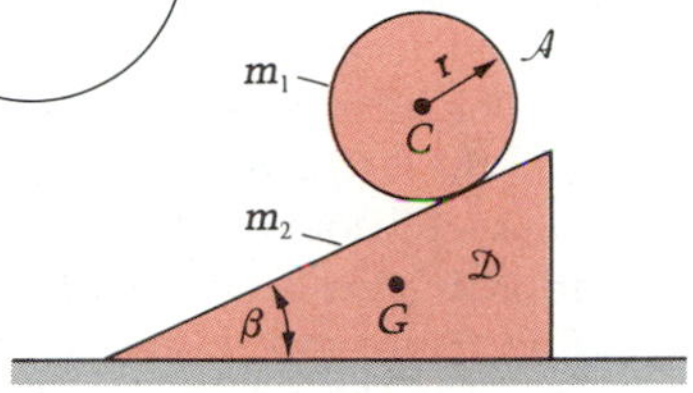

Figure P4.148

* **4.148** Cylinder $\mathcal{A}$ in Figure P4.148 rolls down a wedge that can slide without friction on a smooth floor. Show that the acceleration of wedge $\mathcal{D}$ is a constant given by the equation

$$a_G = \frac{m_1 g \sin 2\beta}{3(m_1 + m_2) - 2m_1 \cos^2 \beta}$$

Extended Problems

4.149 The stick of mass m, shown in Figure P4.149, originally at rest with $\theta = 0$, is disturbed slightly and begins to slide on a smooth wall and floor. Derive the differential equation of motion of the stick. Integrate the equation and find the angle θ at which contact with the wall is lost.

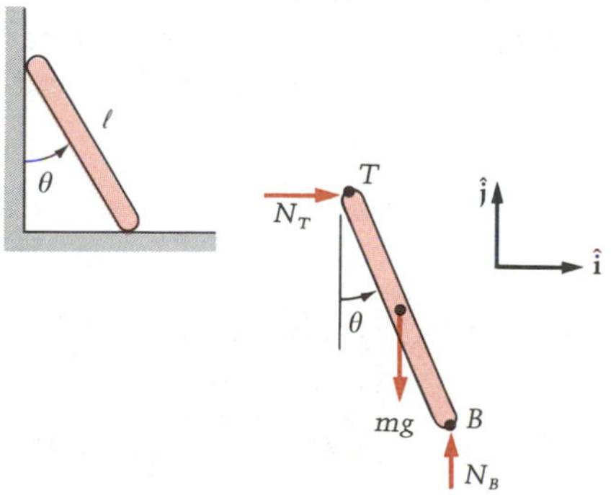

Figure P4.149

Hints: First verify the following equations of motion:

$$\Sigma F_x = N_T = m\ddot{x}_C \quad (1)$$

$$\Sigma F_y = N_B - mg = m\ddot{y}_C \quad (2)$$

$$\Sigma M_C = N_B \frac{\ell}{2} \sin\theta - N_T \frac{\ell}{2} \cos\theta = \frac{m\ell^2}{12}\ddot{\theta} \quad (3)$$

Then note that there are more unknowns than equations. Use kinematics to relate $\mathbf{a}_C$ to $\mathbf{a}_B$, and use the "$\hat{\mathbf{j}}$-component" of that equation to obtain a fourth equation, containing $\ddot{y}_C$. Then relate $\mathbf{a}_C$ to $\mathbf{a}_T$ and get a fifth equation, containing $\ddot{x}_C$. From your five equations, eliminate $\ddot{x}_C$, $\ddot{y}_C$, N_T, and N_B, obtaining the following differential equation governing θ:

$$\ddot{\theta} = \frac{3g}{2\ell} \sin\theta \quad (4)$$

Multiply (4) by $\dot{\theta}$ and integrate, using $\ddot{\theta}\dot{\theta} = d(\dot{\theta}^2)/dt$. Use the initial condition $\dot{\theta} = 0$ at $\theta = 0$ to evaluate the constant of integration. You will now know $\dot{\theta}$ as a function of θ. Then, with N_T expressed as a function of θ, set $N_T = 0$ in (1) to find the angle where contact is lost. Your answer should be $\cos^{-1}(\frac{2}{3})$.

4.150 Disks $\mathcal{B}_3$ and $\mathcal{B}_4$ each weigh 64.4 lb and are rigidly attached to the light shaft $\mathcal{B}_5$ that joins their centers. (See Figure P4.150.) A 96.6-lb cylinder $\mathcal{B}_1$ has a hole drilled along its axis, through which $\mathcal{B}_5$ passes. Let $\mathcal{B}_2$ represent the rigid body comprised of $\mathcal{B}_3$, $\mathcal{B}_4$, and $\mathcal{B}_5$.

While the body $\mathcal{B}_2$ is held fixed on the plane, the cylinder is spun up to an angular velocity 8 ⮌ rad/sec, and the system is then released. Assume that part of the reaction between the axle and the wall of the cylindrical hole in the cylinder is a friction couple proportional to the difference in angular velocities, with proportionality constant k. The friction couple acting on the axle will cause $\mathcal{B}_2$ to roll to the right; the opposite couple on $\mathcal{B}_1$ will slow its angular speed down. As time passes, the bodies $\mathcal{B}_1$ and $\mathcal{B}_2$ will approach the condition of moving as one. Show this, and find the common, limiting-case "terminal" angular velocity shared by $\mathcal{B}_1$ and $\mathcal{B}_2$. There is sufficient friction between $\mathcal{B}_2$ and the ground to prevent slipping there.

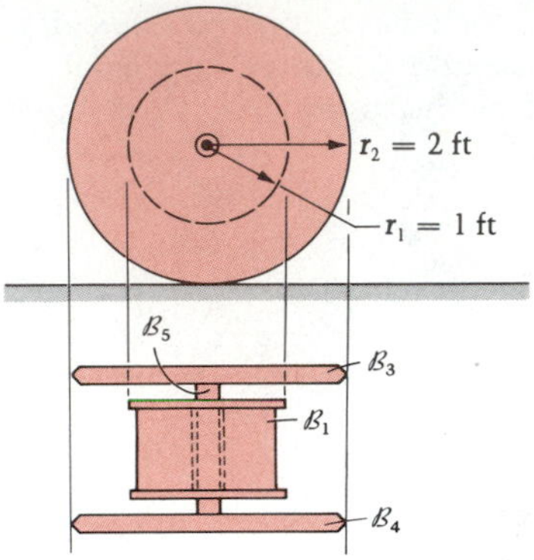

Figure P4.150

4.6 Other Useful Forms of the Moment Equation

For a rigid body in plane motion, there are several other forms of the moment equation of motion [besides the translation equation (4.1) and $\Sigma M_C = I_C\alpha$] which are worthy of special study. The idea is that it is often convenient and helpful to sum the moments about a point other than the mass center C. We will study three of these forms one-by-one and present examples of each as we go along.

Moment Equation in Terms of $\mathbf{a}_C$

To develop this form, we begin with Equation (2.45):

$$\Sigma \mathbf{M}_P = \dot{\mathbf{H}}_C + \mathbf{r}_{PC} \times m\mathbf{a}_C \tag{2.45}$$

where we recall that in this form there is *no restriction at all* on the location of point P, the type of body being studied, or the type of motion. Thus, specializing for a rigid body in plane motion and using the right-hand side of Equation (4.13) to replace $\dot{\mathbf{H}}_C$ for this case,

$$\begin{aligned}\Sigma \mathbf{M}_P = (I_{xz}^C\alpha - I_{yz}^C\omega^2)\hat{\mathbf{i}} + (I_{yz}^C\alpha + I_{xz}^C\omega^2)\hat{\mathbf{j}} \\ + I_{zz}^C\alpha\hat{\mathbf{k}} + \mathbf{r}_{PC} \times m\mathbf{a}_C\end{aligned} \tag{4.17}$$

Whenever the products of inertia in Equation (4.17) vanish, this equation takes the particularly simple form

$$\Sigma \mathbf{M}_P = I_{zz}^C\alpha\hat{\mathbf{k}} + \mathbf{r}_{PC} \times m\mathbf{a}_C \tag{4.18}$$

Note that the translation Equation (4.1) results from Equation (4.18) if $\alpha \equiv 0$.

Note further that if P and C are in the same (reference) plane of motion, then $\mathbf{r}_{PC} \times m\mathbf{a}_C$ is perpendicular to the plane of motion — that is, it is parallel to **k**. When this is the case, we will rewrite Equation (4.18) in scalar form*:

$$\Sigma M_P = I_{zz}^C \alpha + (\mathbf{r}_{PC} \times m\mathbf{a}_C)_z \tag{4.19}$$

in which $(\ )_z$ means the coefficient of the unit vector $\hat{\mathbf{k}}$ within the parenthesis.

We now present two examples of the use of Equation (4.18). In the first one, we eliminate reactions at a pin by summing moments there:

Example 4.22

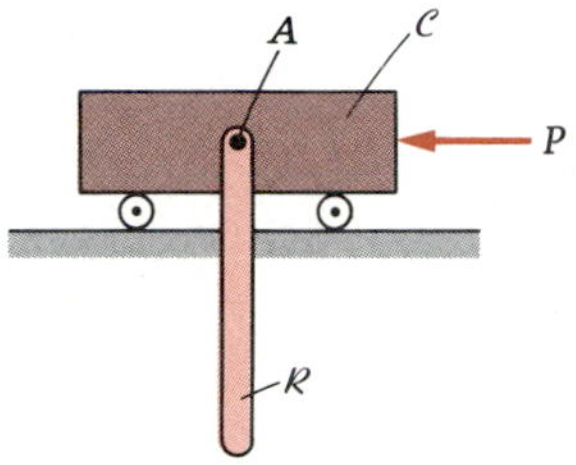

Figure E4.22a

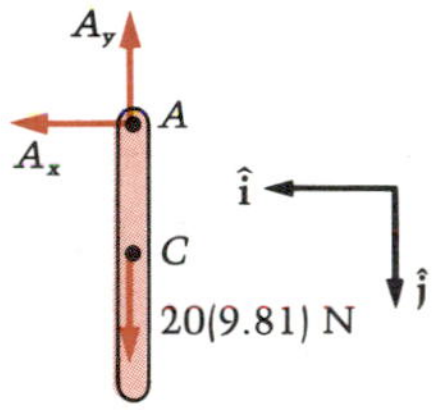

Figure E4.22b

The uniform rod $\mathcal{R}$ in Figure E4.22a (length 80 cm, mass 20 kg) is smoothly pinned to cart $\mathcal{C}$ (50 kg) at point A. Force P, applied to $\mathcal{C}$ with the system initially at rest, causes $\mathcal{C}$ to translate with the acceleration $3 \leftarrow \text{m/s}^2$. Find the initial angular acceleration of the rod.

Solution

By using Equation (4.19), we can sum moments about A of the forces on $\mathcal{R}$ and avoid having to use $\Sigma F_x = m\ddot{x}_C$ in this case. We obtain, using the free-body in Figure E4.22b,

$$\Sigma M_A = I_C\alpha + (\mathbf{r}_{AC} \times m\mathbf{a}_C)_z \tag{1}$$

We note that $I_C = m\ell^2/12 = 20(0.8)^2/12 = 1.07 \text{ kg} \cdot \text{m}^2$, and we get $\mathbf{a}_C$ from kinematics:

$$\mathbf{a}_C = \mathbf{a}_A + \alpha\hat{\mathbf{k}} \times (0.4\hat{\mathbf{j}}) - \overset{0 \text{ at } t=0}{\omega^2}(0.4\hat{\mathbf{j}})$$

$$\mathbf{a}_C = 3\hat{\mathbf{i}} - 0.4\alpha\hat{\mathbf{i}} = (3 - 0.4\alpha)\hat{\mathbf{i}}$$

so that, substituting into (1),

$$0 = 1.07\alpha + [0.4\hat{\mathbf{j}} \times 20(3 - 0.4\alpha)\hat{\mathbf{i}}]_z$$

$$0 = (1.07 + 3.2)\alpha - 24$$

$$\alpha = 5.62 \text{ rad/s}^2$$

Thus the rod starts off with $\alpha = 5.62 \circlearrowleft \text{rad/s}^2$.

Question 4.8 Why would the solution to the preceding example be much more complicated using the mass center form of the moment Equation (4.16)?

Next we use Equation (4.19) to rework an example from the previous section.

---

* When the products of inertia I_{xz}^C and I_{yz}^C both vanish, then by Equations (4.14a,b), $\Sigma M_{Cx} = 0 = \Sigma M_{Cy}$. This leads, when P and C are both in the reference plane, to $\Sigma M_{Px} = 0 = \Sigma M_{Py}$ since $\Sigma \mathbf{M}_P$ always equals $\Sigma \mathbf{M}_C + \mathbf{r}_{PC} \times \Sigma \mathbf{F}$ and $\Sigma \mathbf{F} = m\mathbf{a}_C$.

Answer 4.8 The pin reaction A_x would appear in ΣM_C. Thus we would need to also write Equation (4.15a) to eliminate A_x.

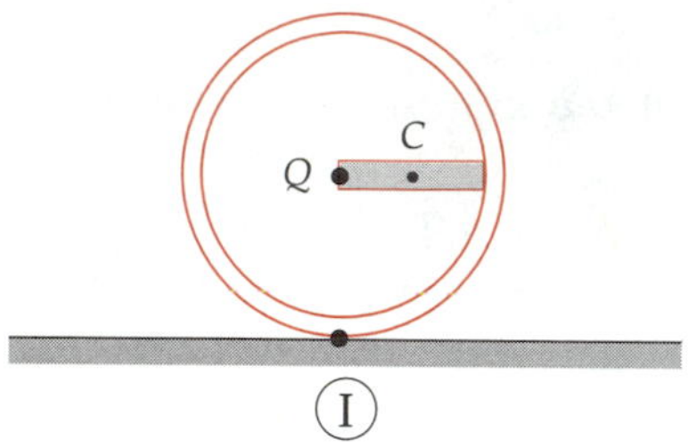

Figure E4.23a

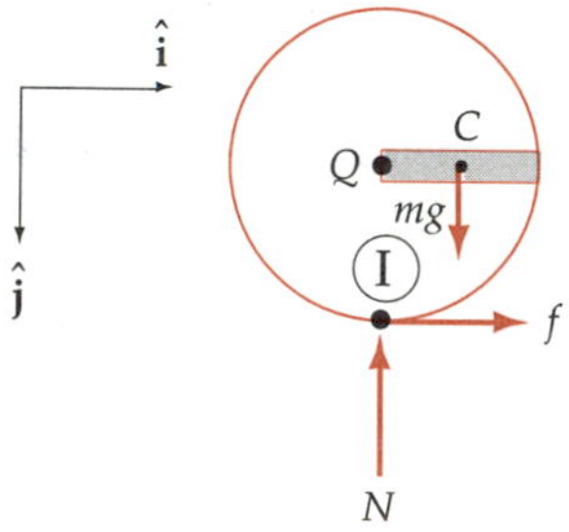

Figure E4.23b

EXAMPLE 4.23

Find the starting angular acceleration for the body of Example 4.19, shown again in Figure E4.23a.

Solution

We sum moments about point Ⓘ, with the help of Equation (4.19) and the free-body diagram in Figure E4.23b:

$$\Sigma M_{Ⓘ} = I_C\alpha + (\mathbf{r}_{ⒾC} \times m\mathbf{a}_C)_z$$

$$\circlearrowright \quad \frac{mg\ell}{2} = \frac{m\ell^2}{12}\alpha + \underbrace{\left[\left(\frac{\ell}{2}\hat{\mathbf{i}} - \ell\hat{\mathbf{j}}\right) \times m(\ddot{x}_C\hat{\mathbf{i}} + \ddot{y}_C\hat{\mathbf{j}})\right]_z}_{m\frac{\ell}{2}\ddot{y}_C + m\ell\ddot{x}_C}$$

From the kinematics in the earlier example, we know $\ddot{x}_C = \ell\alpha$ and $\ddot{y}_C = \frac{\ell}{2}\alpha$. Substituting these into the above equation after cancelling "m",

$$\frac{g\ell}{2} = \frac{\ell^2\alpha}{12} + \frac{\ell}{2}\left(\frac{\ell\alpha}{2}\right) + \ell(\ell\alpha)$$

Therefore,

$$\alpha = \frac{3}{8}\frac{g}{\ell}$$

as before.

Note that the "x" and "y" equations of motion, (4.15a,b), were not needed in this example. They would have been in Example 4.19, however, even if all we had been seeking was α.

Question 4.9 Why?

Moment Equation in Terms of $\mathbf{a}_P$

Another form of the moment equation of motion that is sometimes useful involves the acceleration and inertia properties at some point other than the mass center. Recalling that if P is a point of a rigid body in plane motion, then by Equation (4.3),

$$\mathbf{H}_P = \mathbf{r}_{PC} \times m\mathbf{v}_P + I_{xz}^P\omega\hat{\mathbf{i}} + I_{yz}^P\omega\hat{\mathbf{j}} + I_{zz}^P\omega\hat{\mathbf{k}} \tag{4.3}$$

And for *any* point P, we know from Equation (2.36) that:

$$\mathbf{H}_P = \mathbf{H}_C + \mathbf{r}_{PC} \times m\mathbf{v}_C \tag{2.36}$$

Equating these two expressions for $\mathbf{H}_P$ and then differentiating with respect to time,

Answer 4.9 ΣM_C would have included moments of both f and N.

$$\dot{\mathbf{H}}_C + \mathbf{r}_{PC} \times m\mathbf{a}_C + \dot{\mathbf{r}}_{PC} \times m\mathbf{v}_C$$
$$= \dot{\mathbf{r}}_{PC} \times m\mathbf{v}_P + \mathbf{r}_{PC} \times m\mathbf{a}_P + \frac{d}{dt}(I_{xz}^P\omega\hat{\mathbf{i}} + I_{yz}^P\omega\hat{\mathbf{j}} + I_{zz}^P\omega\hat{\mathbf{k}}) \quad (4.20)$$

By Equations (2.8) and (2.43), we know that $m\mathbf{a}_C = \Sigma\mathbf{F}$ and $\dot{\mathbf{H}}_C = \Sigma\mathbf{M}_C$. Using these results and Equation (2.42), we may replace by $\Sigma\mathbf{M}_P$ the first two terms on the left-hand side of this equation and obtain

$$\Sigma\mathbf{M}_P = \dot{\mathbf{r}}_{PC} \times m(\mathbf{v}_P - \mathbf{v}_C) + \mathbf{r}_{PC} \times m\mathbf{a}_P$$
$$+ \frac{d}{dt}(I_{xz}^P\omega\hat{\mathbf{i}} + I_{yz}^P\omega\hat{\mathbf{j}} + I_{zz}^P\omega\hat{\mathbf{k}}) \quad (4.21)$$

On the right side of Equation (4.21) the first term vanishes since $\mathbf{v}_P - \mathbf{v}_C = -\dot{\mathbf{r}}_{PC}$, and the third term is of the same form as $\dot{\mathbf{H}}_C$, except that here the inertia properties are with respect to axes with origin at P. Thus retracing the steps between Equations (4.11) and (4.13), we obtain

$$\Sigma\mathbf{M}_P = (I_{xz}^P\alpha - I_{yz}^P\omega^2)\hat{\mathbf{i}} + (I_{yz}^P\alpha + I_{xz}^P\omega^2)\hat{\mathbf{j}}$$
$$+ I_{zz}^P\alpha\hat{\mathbf{k}} + \mathbf{r}_{PC} \times m\mathbf{a}_P \quad (4.22)$$

Question 4.10 Why can we say, as was done above, that $\mathbf{v}_P - \mathbf{v}_C = -\dot{\mathbf{r}}_{PC}$ and why does that cause the first term on the right side of Equation (4.16) to vanish?

When the products of inertia I_{xz}^P and I_{yz}^P vanish, Equation (4.22) simplifies to

$$\Sigma\mathbf{M}_P = I_{zz}^P\alpha\hat{\mathbf{k}} + \mathbf{r}_{PC} \times m\mathbf{a}_P \quad (4.23)$$

As in Section 4.5, when P and C are in the same (reference) plane the $\hat{\mathbf{i}}$ and $\hat{\mathbf{j}}$ components of this equation vanish. Thus the scalar form of the equation is

$$\Sigma M_P = I_{zz}^P\alpha + (\mathbf{r}_{PC} \times m\mathbf{a}_P)_z \quad (4.24)$$

where $(\quad)_z$ again means the coefficient of $\hat{\mathbf{k}}$ within the parentheses.

This equation is very useful if we happen to know the acceleration of a point other than the mass center, as in the following examples:

Answer 4.10 Let $\mathbf{r}_{OP}$ and $\mathbf{r}_{OC}$ be position vectors for P and C; differentiating $\mathbf{r}_{OC} = \mathbf{r}_{OP} + \mathbf{r}_{PC}$ gives $\mathbf{v}_C = \mathbf{v}_P + \dot{\mathbf{r}}_{PC}$. The cross product of parallel vectors always vanishes.

Example 4.24

Solve the problem of Example 4.22 by using Equation 4.24.

Solution

Equation (4.24) makes a problem such as this even simpler than did Equation (4.19):

$$\circlearrowleft\!+\quad \Sigma M_A = I_A\alpha + (\mathbf{r}_{AC} \times m\mathbf{a}_A)_z$$

$$0 = \frac{m\ell^2}{3}\alpha - \frac{\ell}{2}ma_A = \frac{20(0.8)^2}{3}\alpha - 0.4(20)3$$

$$\alpha = 5.62 \text{ rad/sec}^2\text{, as before.}$$

To determine other information in the preceding example, for instance the starting value of force P, we can write the x-component of the mass center equation of motion for the cart and then for the bar (see Figure 4.7):

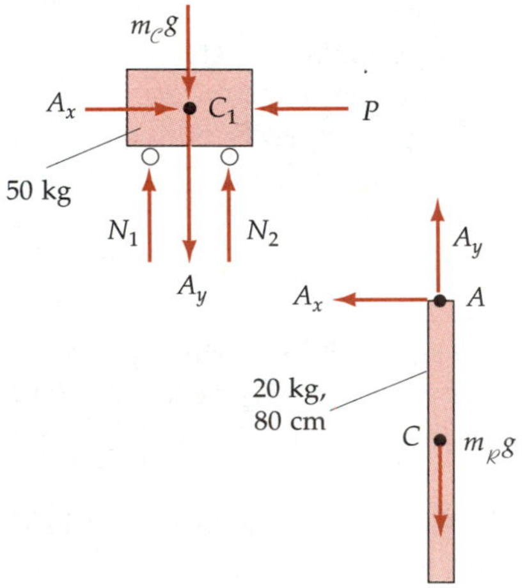

Figure 4.7

For $\mathcal{C}$:

$$\overset{+}{\longleftarrow}\quad \Sigma F_x = P - A_x = m_{\mathcal{C}}\ddot{x}_{C_1} = 50\,\ddot{x}_{C_1}$$

For $\mathcal{R}$:

$$\overset{+}{\longleftarrow}\quad \Sigma F_x = A_x = m_{\mathcal{R}}\ddot{x}_C = 20\,\ddot{x}_C$$

where

$$\mathbf{a}_C = \mathbf{a}_A + \alpha \times \mathbf{r}_{AC} - \underbrace{\cancel{\omega^2}}_{0 \text{ at } t = 0}\mathbf{r}_{AC}$$

$$= \ddot{x}_{C_1}\hat{\mathbf{i}} + \alpha\hat{\mathbf{k}} \times \frac{\ell}{2}\hat{\mathbf{j}}$$

$$\mathbf{a}_C = (\ddot{x}_{C_1}\hat{\mathbf{i}} - \frac{\ell}{2}\alpha)\hat{\mathbf{i}}$$

or

$$\ddot{x}_C = \ddot{x}_{C_1} - \frac{\ell}{2}\alpha$$

Thus

$$A_x = 20\left(3 - \frac{0.8}{2}5.62\right) = 15.0 \text{ N}$$

and

$$P - A_x = P - 15.0 = 50(3)$$
$$P = 165 \text{ N}$$

Example 4.25

Repeat Examples 4.19,23, using Equation (4.24) this time to find the starting angular acceleration.

Solution

We again sum moments about point Ⓘ:

$$\Sigma M_{Ⓘ} = I_{Ⓘ}\alpha + (\mathbf{r}_{ⒾC} \times m\mathbf{a}_{Ⓘ})_z$$

$$\circlearrowleft^{+} \quad \frac{mg\ell}{2} = \left[\frac{m\ell^2}{12} + m\left(\left(\frac{\ell}{2}\right)^2 + \ell^2\right)\right]\alpha + \left[\left(\frac{\ell}{2}\hat{\mathbf{i}} - \ell\hat{\mathbf{j}}\right) \times m(-\ell\overset{0 \text{ initially}}{\cancel{\omega^2}}\hat{\mathbf{j}})\right]_z$$

$$\frac{mg\ell}{2} = \frac{4}{3}m\ell^2\alpha$$

$$\alpha = \frac{3}{8}\frac{g}{\ell}$$

once again. Notice that the cross-product term is simpler this time, but the moment of inertia at Ⓘ has to be calculated by the parallel axis theorem. In Problems 4.153, 154, this problem is to be reworked one last time using the geometric center of the hoop as point P in Equations 4.19 and 4.24, respectively.

In the preceding example, it *accidentally* happened that $\Sigma M_{Ⓘ} = I_{Ⓘ}\alpha$. *This is not generally true.* For example, an instant later ω would not be zero and the cross-product term would *not* be zero.

Moment Equation for Fixed-Axis Rotation (The "Pivot" Equation)

Consider now the case in which the acceleration of a point P of the rigid body in plane motion is identically zero. Equation (4.22) tells us that for such a point,

$$\Sigma \mathbf{M}_P = (I_{xz}^P\alpha - I_{yz}^P\omega^2)\hat{\mathbf{i}} + (I_{yz}^P\alpha + I_{xz}^P\omega^2)\hat{\mathbf{j}} + I_{zz}^P\alpha\hat{\mathbf{k}} \qquad (4.25)$$

If we take the dot product of this equation with $\hat{\mathbf{k}}$, we see that whether or not the products of inertia I_{xz}^P and I_{yz}^P vanish, we always have:

$$\Sigma \mathbf{M}_P \cdot \hat{\mathbf{k}} = (\Sigma \mathbf{M}_P)_z = I_{zz}^P\alpha \qquad (4.26)$$

When the point P is fixed in the inertial frame as well as in the body, we will then usually label the point as O for emphasis. In this case, the only way the body can move is to rotate about the z-axis through O, a motion which we call *fixed-axis rotation.* The point O, which does not move during the motion of interest, is called a *pivot,* and we abbreviate Equation (4.26) as

$$(\Sigma M_O)_z = I_O \alpha \tag{4.27}$$

or as

$$\Sigma M_{axis} = I_{axis} \alpha \tag{4.28}$$

When the products of inertia vanish at a pivot O, as they will for all problems in this section, then of course $(\Sigma M_O)_z$ can be further abbreviated to simply ΣM_O, for then only the z-component of $\Sigma \mathbf{M}_O$ is non-zero. This is *not* always the case, as we shall later see in the final section 4.7 of this chapter. When it is, however, we simply write

$$\Sigma M_O = I_O \alpha \tag{4.29}$$

Question 4.11 Since Equation (4.4) applies for any point P having zero velocity, why can we not use equations such as (4.29) for the instantaneous center Ⓘ of $\mathcal{B}$ when Ⓘ is *not* a pivot?

Because of the importance of fixed-axis rotation in engineering, we present more examples of it than we did for the earlier forms in this section. In the first example, we examine a composite body rotating about a pivot:

Answer 4.11 Although $\mathbf{H}_O$ may always be written as $I_{xz}^O \omega \hat{\mathbf{i}} + I_{yz}^O \omega \hat{\mathbf{j}} + I_{zz}^O \omega \hat{\mathbf{k}}$ whenever $\mathbf{v}_O$ is zero, its derivative is only equal to $\Sigma \mathbf{M}_O$ when $\mathbf{v}_O$ is *identically* zero — in other words, *zero all the time.*

EXAMPLE 4.26

The rod $\mathcal{R}$ and sphere $\mathcal{S}$ in Figure E4.26a are welded together to form a combined rigid body which is attached to the ground at O by means of a smooth pin. Find the force exerted by the pin onto the body, upon release of the system from rest.

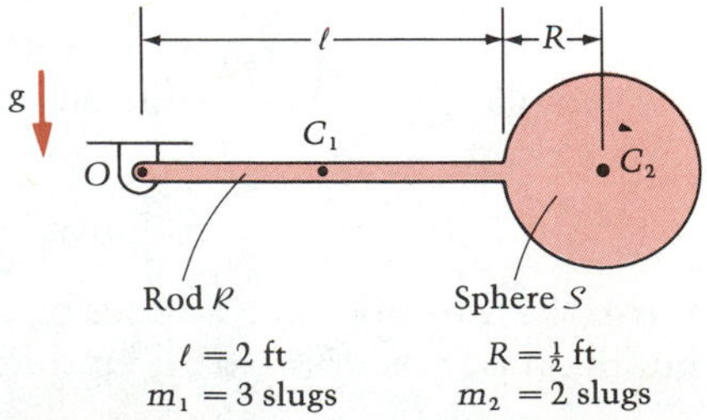

Figure E4.26a

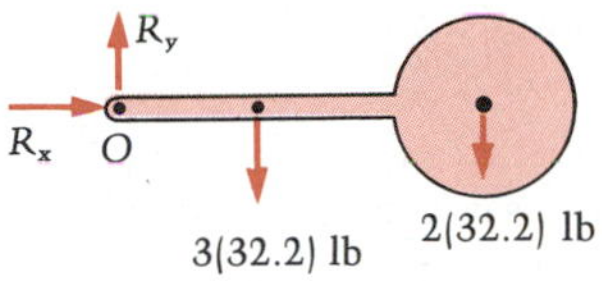

Figure E4.26b

Solution

Because O is a pivot of the combined body, we use Equation 4.29:

$$\Sigma \mathbf{M}_O = I_O \alpha \hat{\mathbf{k}}$$

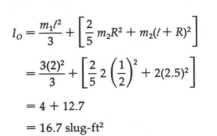

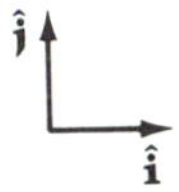

where

$$\begin{aligned} I_O &= \frac{m_1 \ell^2}{3} + \left[\frac{2}{5} m_2 R^2 + m_2(\ell + R)^2\right] \\ &= \frac{3(2)^2}{3} + \left[\frac{2}{5}2\left(\frac{1}{2}\right)^2 + 2(2.5)^2\right] \\ &= 4 + 12.7 \\ &= 16.7 \text{ slug-ft}^2 \end{aligned}$$

so that, at release, using the FBD in Figure E4.26b,

$$-3(32.2)(1) - 2(32.2)(2.5) = 16.7\alpha$$

or

$$\alpha = -15.4 \text{ rad/sec}^2$$

To calculate the pin reaction we shall use $\Sigma \mathbf{F} = m\mathbf{a}_C$. We first locate the mass center, C, of the body. The distance from O to C is

$$d = \frac{3(1) + 2(2.5)}{3 + 2} = \frac{8}{5} = 1.6 \text{ ft}$$

At this instant we have

$$\begin{aligned} \mathbf{a}_C &= \overset{0}{\cancel{\mathbf{a}_O}} + \boldsymbol{\alpha} \times \mathbf{r}_{OC} - \overset{0}{\cancel{\omega^2}}\,\mathbf{r}_{OC} \\ \mathbf{a}_C &= (-15.4\hat{\mathbf{k}}) \times (1.6\hat{\mathbf{i}}) \\ &= -24.6\hat{\mathbf{j}} \text{ ft/sec}^2 \end{aligned}$$

Thus

$$(R_x\hat{\mathbf{i}} + R_y\hat{\mathbf{j}}) - 96.6\hat{\mathbf{j}} - 64.4\hat{\mathbf{j}} = 5(-24.6\hat{\mathbf{j}})$$

So

$$\begin{aligned} R_x &= 0 \\ R_y &= 161 - 123 = 38 \text{ lb} \end{aligned}$$

An alternate approach to the mass center calculation in the preceding example would *not* require that we explicitly locate C, because

$$\Sigma \mathbf{F} = m\mathbf{a}_C = m_1\mathbf{a}_{C_1} + m_2\mathbf{a}_{C_2}$$

or

$$(R_x\hat{\mathbf{i}} + R_y\hat{\mathbf{j}}) - 96.6\hat{\mathbf{j}} - 64.4\hat{\mathbf{j}} = 3[(-15.4\hat{\mathbf{k}}) \times \hat{\mathbf{i}}] + 2[(-15.4\hat{\mathbf{k}}) \times 2.5\hat{\mathbf{i}}]$$

from which

$$R_x = 0$$

and

$$R_y = 161 - 46.2 - 77 = 38 \text{ lb}$$

as above.

In our second example, we feature distinct bodies connected by an unwinding rope. One body has a pivot and the others don't.

EXAMPLE 4.27

A rope is wrapped around the 10-lb cylinder $\mathcal{C}$ as indicated in Figure E4.27a. The rope passes through a hole in the 5-lb annular disk $\mathcal{A}$ and is then tied to the 15-lb block $\mathcal{B}$. When the system is let go from rest with the rope just taut, what is the reaction exerted on $\mathcal{A}$ by $\mathcal{B}$?

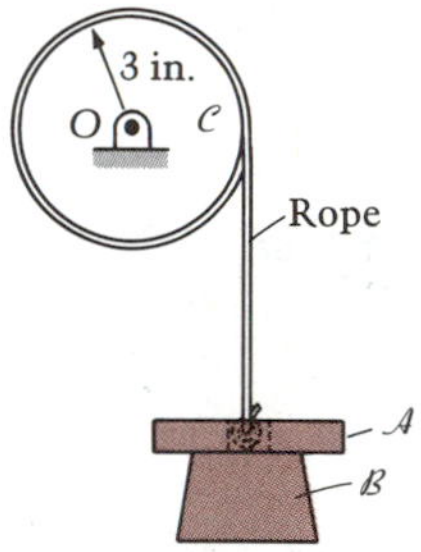

Figure E4.27a

Solution

Using the free-body diagrams in Figures E4.27b, 4.27c, and 4.27d, we write the following equations of motion of the respective bodies. For $\mathcal{C}$, using the "pivot equation" (4.29):

$$\circlearrowleft^{+} \quad \Sigma M_O = I_O \alpha$$

$$T\left(\frac{3}{12}\right) = \left[\frac{1}{2}\left(\frac{10}{32.2}\right)\left(\frac{3}{12}\right)^2\right]\alpha$$

which gives us the tension T in terms of α:

$$T = 0.0388\alpha \tag{1}$$

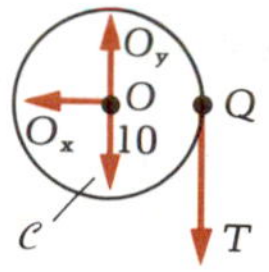

Figure E4.27b

For the block $\mathcal{B}$ by itself,

$$\downarrow + \quad \Sigma F_y = m\ddot{y}_B$$

$$15 - T + R = \frac{15}{32.2}\ddot{y}_B \tag{2}$$

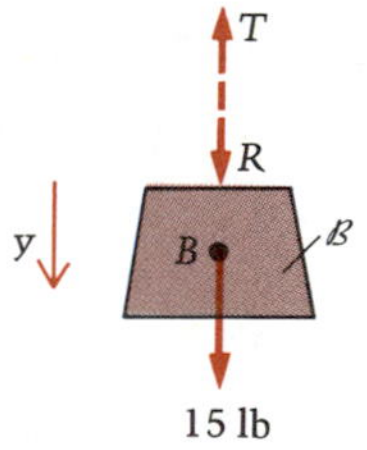

Figure E4.27c

where R is the force exerted by $\mathcal{A}$ onto $\mathcal{B}$. Now, on the disk $\mathcal{A}$,

$$\downarrow + \quad \Sigma F_y = m\ddot{y}_A$$

$$5 - R = \frac{5}{32.2}\ddot{y}_A \tag{3}$$

At this point we have three equations in the five unknowns T, α, R, $\ddot{y}_B$, and $\ddot{y}_A$.

One constraint is that the vertical component of $\mathbf{a}_Q$ (see Figure E4.27b) is the same as the acceleration of the points of the straight portion of rope, and these accelerations are each $\ddot{y}_B$:

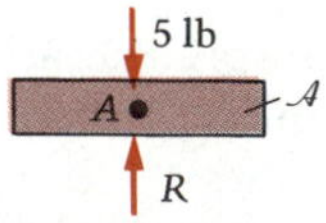

Figure E4.27d

$$\ddot{y}_B = (\mathbf{a}_Q)_y = \frac{3}{12}\alpha \tag{4}$$

Also, the accelerations of $\mathcal{B}$ and $\mathcal{A}$ are equal. Without any rope tension, they would each be $g\downarrow$; with this tension, the acceleration of $\mathcal{B}$ is slowed, guaranteeing continuing contact of the two bodies. Therefore:

$$\ddot{y}_A = \ddot{y}_B \tag{5}$$

Adding Equations (1) and (2) and using (4),

$$15 + R = \left[0.0388(4) + \frac{15}{32.2}\right]\ddot{y}_B$$

or

$$0.621\ddot{y}_B - R = 15 \tag{6}$$

Equations (5) and (3) give:

$$0.155\ddot{y}_B + R = 5 \tag{7}$$

Adding Equations (6) and (7),

$$\ddot{y}_B = \frac{20}{0.776} = 25.8 \text{ ft/sec}^2$$

so that, by (6),

$$R = -15 + 0.621(25.8) = 1.02 \text{ lb}$$

Note that the acceleration of $\mathcal{B}$ is less than "g" (32.2 ft/sec^2), as it must be, and that R is less than the weight of $\mathcal{A}$, thereby allowing it to fall, but not freely.

In our third example, we again have two bodies, but this time *both* have pivots and one is in fact in equilibrium. The contact between these bodies involves sliding friction:

Example 4.28

Just after the brake arm in Figure E4.28a contacts the top of the cylinder $\mathcal{A}$, the cylinder is turning at 1000 rpm ↻. The coefficient of kinetic friction between $\mathcal{A}$ and $\mathcal{B}$ is $\mu = 0.3$. Find (a) how long it takes for $\mathcal{A}$ to come to rest under the constant force $P = 40$ N; and (b) the pin reactions exerted onto $\mathcal{A}$ at the pin O.

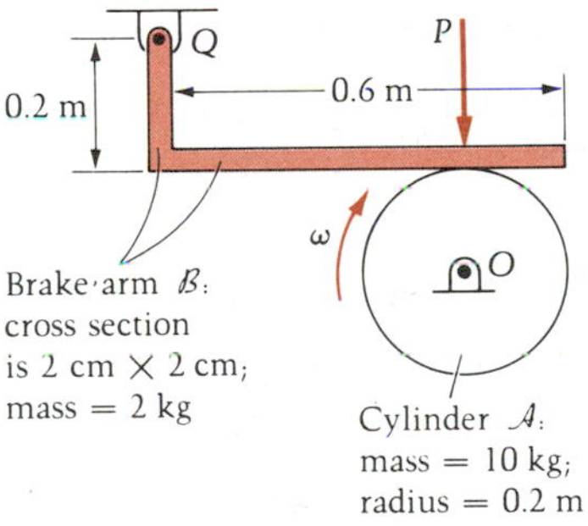

Figure E4.28a

Solution

Since body $\mathcal{B}$ is in equilibrium, we may find the normal force between it and the cylinder by statics. The bar's weight is proportioned between its horizontal and vertical parts as shown in Figure E4.28b. Note that equilibrium requires that $f = Q_x$, where $f = f_{max} = \mu N$ since slipping is taking place.

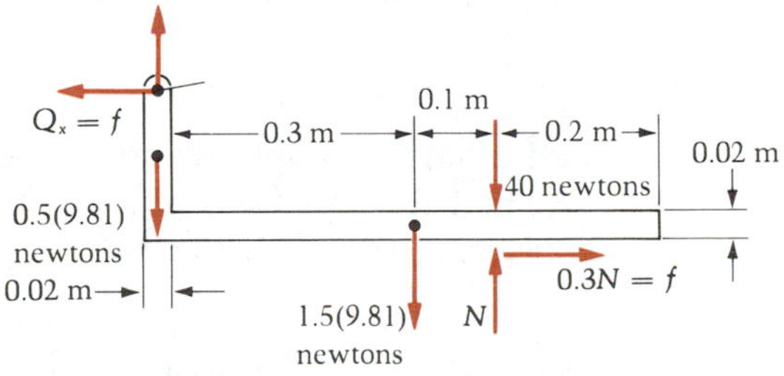

Figure E4.28b

Summing moments about Q, we have

$$0.401N + 0.2(0.3N) - 40(0.401) - 1.5(9.81)(0.301) = 0$$

from which we get

$$N = \frac{20.5}{0.461} = 44.6 \text{ newtons}$$

and

$$\mu N = 0.3(44.6) = 13.4 \text{ newtons}$$

Question 4.12 Would the solution for the normal force N be any different if ω were counterclockwise?

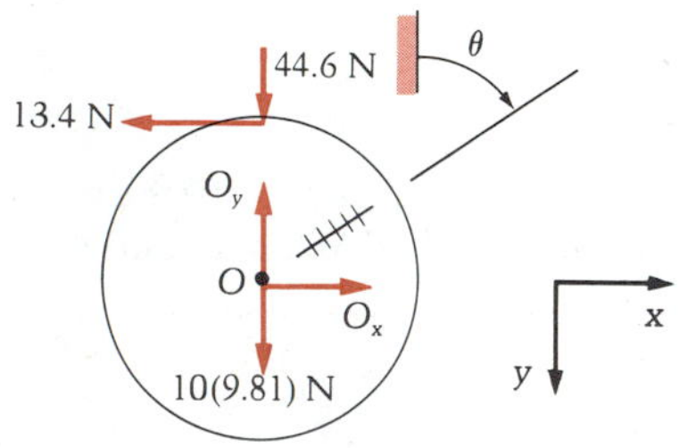

Figure E4.28c

The motion of body $\mathcal{A}$ is one of pure rotation. Its free-body diagram is shown in Figure E4.28c. We use $\ddot{\theta}$ for α since we are ultimately interested in equating $\dot{\theta}$ to zero.

$$\circlearrowleft\!+ \quad \Sigma M_O = I_O\ddot{\theta}$$

$$-13.4(0.2) = \left[\frac{1}{2}\,10(0.2)^2\right]\ddot{\theta}$$

$$\ddot{\theta} = \frac{-2.68}{0.20} = -13.4 \text{ rad/s}^2$$

Integrating, we get

$$\dot{\theta} = -13.4t + C_1$$

$$= -13.4t + 1000\left(\frac{2\pi}{60}\right)$$

where the initial condition on ω allows us to calculate the integration constant C_1.

Body $\mathcal{A}$ stops when $\dot{\theta} = 0$ at a time t_s that we are now in a position to calculate:

$$0 = -13.4t_s + 105$$

$$t_s = 7.84 \text{ s}$$

Note that the mass center $O = C$ of $\mathcal{A}$ is fixed in the inertial frame, so that the pin reactions follow from the mass center equations:

$$\xrightarrow{+} \quad \Sigma F_x = -13.4 + O_x = m\ddot{x}_C = 0 \Rightarrow O_x = 13.4 \text{ newtons}$$

$$+\!\downarrow \quad \Sigma F_y = +44.6 + 98.1 - O_y = m\ddot{y}_C = 0 \Rightarrow O_y = 143 \text{ newtons}$$

In the fourth example, we are concerned with *internal* forces and with the fact that the equations of motion can be applied to a *part* of a body, considered as a body in itself:

Answer 4.12 Yes, for then the friction force would be in the opposite direction and ΣM_Q would give a larger N. The normal force N would then have to balance the moments about Q of all three of P, f, and the weight.

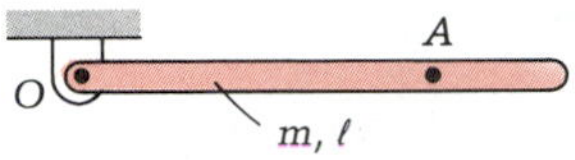

Figure E4.29a

Example 4.29

The slender bar, pinned smoothly at O, is released from rest in the position shown in Figure E4.29a. After a rotation of 45°↻, its angular speed is $\omega = \sqrt{(3g)/(\sqrt{2}\ell)}$. Find at that instant the axial force, shear force, and bending moment in the bar at point A, which is one-fourth the length of the bar from its free end.

Solution

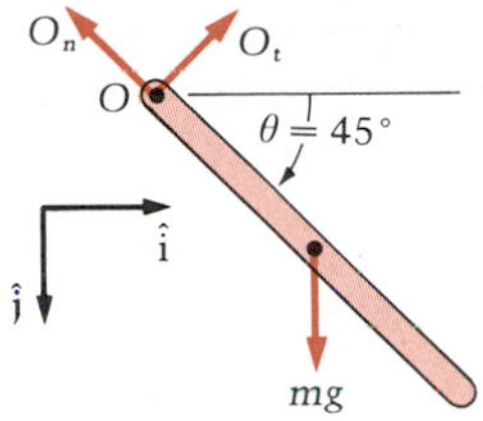

Figure E4.29b

First, we find the angular acceleration, making use of the FBD in Figure E4.29b:

$$\Sigma M_O = I_O \alpha$$

$$\frac{mg}{\sqrt{2}}\frac{\ell}{2} = \left(\frac{1}{3}m\ell^2\right)\alpha$$

$$\alpha = \frac{3g}{2\sqrt{2}\ell}$$

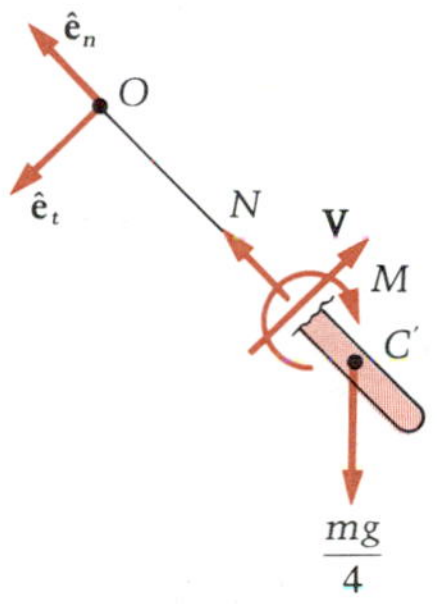

Figure E4.29c

Next we expose the desired forces and moment by drawing a free-body diagram (Figure E4.29c) of the lower fourth of the bar, and writing the equations of motion for just that body (C' is its mass center):

$$\Sigma F_t = m a_{C't}$$

$$\frac{mg}{4}\frac{1}{\sqrt{2}} - V = \frac{m}{4}\left(\frac{7}{8}\ell\alpha\right) = \frac{7}{32}m\ell\left(\frac{3g}{2\sqrt{2}\ell}\right)$$

so that

$$V = \frac{-5}{64\sqrt{2}}mg$$

where we note that ω and α are the same for this "sub-body" as they were for the whole bar, and that O is a pivot of the sub-body extended. The other mass center equation is:

$$\Sigma F_n = m a_{C'n}$$

$$N - \frac{mg}{4}\frac{1}{\sqrt{2}} = \frac{m}{4}\left(\frac{7}{8}\ell\omega^2\right) = \frac{7\,m\ell}{32}\frac{3g}{\sqrt{2}\ell} = \frac{21mg}{32\sqrt{2}}$$

$$N = \frac{29mg}{32\sqrt{2}}$$

Finally, the "moment equation of motion," written for the sub-body this time, is

$$\Sigma M_{C'} = I_{C'}\alpha$$

$$M + V\left(\frac{\ell}{8}\right) = \left(\frac{\frac{m}{4}\left(\frac{\ell}{4}\right)^2}{12}\right)\left(\frac{3g}{2\sqrt{2}\ell}\right)$$

$$M = \frac{3}{256\sqrt{2}}mg\ell$$

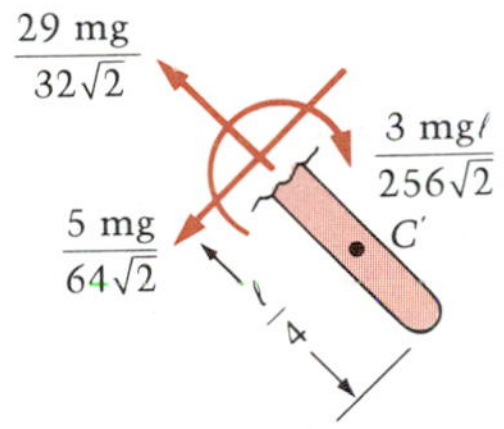

Figure E4.29d

The final three results are summarized pictorially on the cut section in Figure E4.29d.

We now examine a pivot problem in which the mass center of the body is offset from the fixed axis of rotation:

EXAMPLE 4.30

How must the applied couple C in Figure E4.30a vary with time in order to turn the unbalanced (but round) wheel $\mathcal{C}$ at constant angular velocity $\omega_O \circlearrowright$? What is the initial angular acceleration if the couple is absent and the wheel is released in the position shown in the figure? The moment of inertia of the mass of $\mathcal{C}$ with respect to its axis of rotation is I_O, and the mass center of $\mathcal{C}$ is located at G.

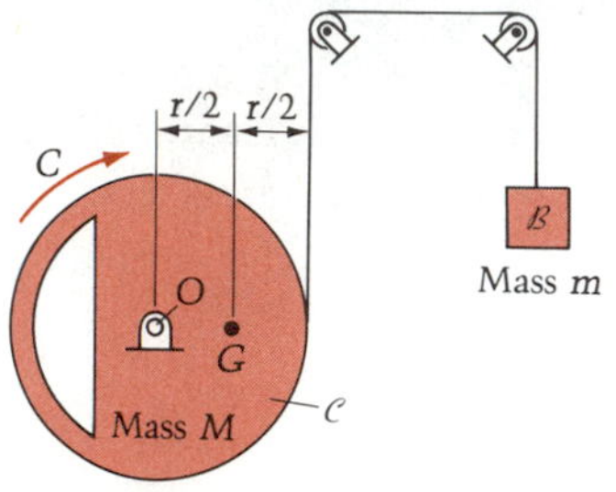

($\theta = 0$ at $t = 0$)
Mass center $= G$

Figure E4.30a

Solution

Free-body diagrams of $\mathcal{C}$ and $\mathcal{B}$ are shown in Figure E4.30b. Since $\Sigma M_O = I_O\alpha$ for body $\mathcal{C}$, we have:

$$\stackrel{+}{\circlearrowright} \qquad -Tr + C + Mg\frac{r}{2}\cos\theta = I_O\ddot{\theta} \tag{1}$$

And for the translating block $\mathcal{B}$ we may write*

$$\uparrow + \qquad T - mg = m\ddot{y}_B \tag{2}$$

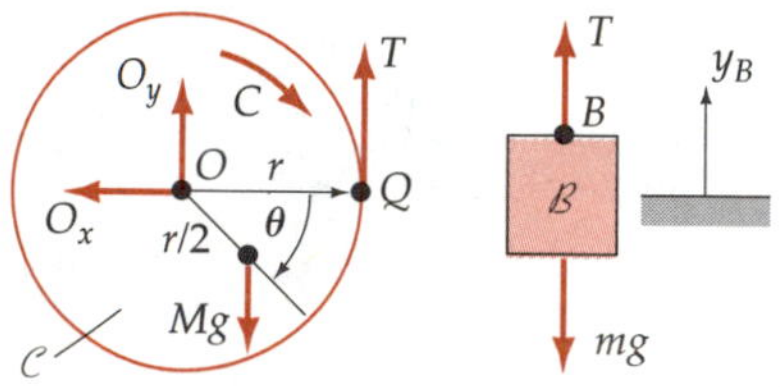

Figure E4.30b

The point Q of $\mathcal{C}$ located where the rope leaves the rim has the same velocity $(r\dot{\theta}\downarrow)$ and tangential acceleration component $(r\ddot{\theta}\downarrow)$ as does the rope itself at that point. Since the rope is assumed inextensible, this acceleration has the same magnitude as $\ddot{y}_B$. Therefore our kinematics gives us the following additional equation for the translating block:

$$\ddot{y}_B = r\ddot{\theta} \tag{3}$$

Substituting (3) into (2) gives

$$T = mg + mr\ddot{\theta}$$

And substituting T into the moment equation (1) for $\mathcal{C}$ then yields

$$C = mgr - Mg\frac{r}{2}\cos\theta + (I_O + mr^2)\ddot{\theta}$$

Since $\dot{\theta} = \omega_0 =$ constant, we have $\theta = \omega_0 t$ and $\ddot{\theta} = 0$, so that

$$C = mgr - \frac{Mgr}{2}\cos\omega_0 t$$

and the required couple varies harmonically.

If there is *no* couple C and the system is released from rest in the position shown, then the equations are still valid and, with couple $C = 0$,

$$(I_O + mr^2)\ddot{\theta} = Mg\frac{r}{2}\cos\theta - mgr$$

The initial angular acceleration of $\mathcal{C}$ (with $\theta = 0$) is thus seen to be

$$\ddot{\theta}_0 = \frac{(Mgr/2) - mgr}{I_O + mr^2}$$

which is positive ($\circlearrowright$) if $M > 2m$.

---

* Note that since $\mathcal{B}$ translates, the acceleration of its mass center is $\ddot{y}_B\downarrow$.

Question 4.13 What happens if $M < 2m$? If $M = 2m$?

In the preceding example, it is interesting to write the equations of motion in the following form for the wheel:

$$\overset{+}{\leftarrow} \quad \Sigma F_x = O_x = M\ddot{x}_G \tag{4}$$

$$+\uparrow \quad \Sigma F_y = O_y + T - Mg = M\ddot{y}_G \tag{5}$$

$$\circlearrowleft + \quad \Sigma M_G = O_y \frac{r}{2} \cos\theta - O_x \frac{r}{2} \sin\theta - T\left(r - \frac{r}{2}\cos\theta\right) = I_G\ddot{\theta} \tag{6}$$

Equations (4) and (5) are useful if the pin reactions are desired,* but Equation (6) is nowhere near as handy to use as $\Sigma M_O = I_O\,\alpha$, which we have used earlier in the example since the body has a pin. The student may wish to eliminate O_x, O_y and T from (6) by using (4), (5), and the previous equation for $\mathcal{B}$ $(T = mg + mr\ddot{\theta})$ and to show that the same result is obtained (after a good deal more work than in the example) for $\ddot{\theta}$. (Kinematics must also be used to relate $\ddot{x}_G$ and $\ddot{y}_G$ to θ, $\dot{\theta}$ and $\ddot{\theta}$!)

In our last "pivot-equation" example, we take up a much longer problem, involving two bodies neither of which is translating. Only one of the bodies has a pivot, and so we shall have to use both the pivot equation (4.29) *and* the mass center form of the moment equation (4.16) before finally getting the problem solved:

Answer 4.13 If $M = 2m$, then (when the couple is not present) $T = mg$ and $\ddot{\theta} = 0$ are solutions to the problem and there is *no* motion. If $M < 2m$, the block moves *downward* and $\ddot{\theta}$ is negative ($\circlearrowleft$).

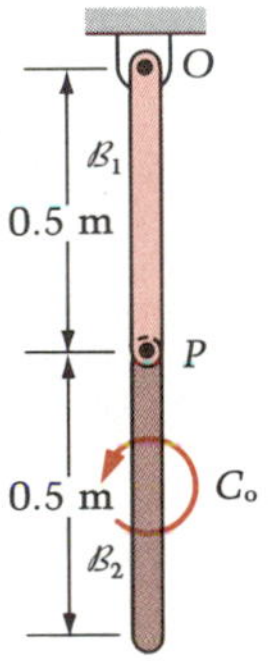

Figure E4.31a

EXAMPLE 4.31

The two uniform, slender rods $\mathcal{B}_1$ and $\mathcal{B}_2$ in Figure E4.31a, each of mass 2 kg, are pinned together at P, and then $\mathcal{B}_1$ is suspended from a pin at O. (This arrangement is called a *double pendulum*.) The counterclockwise couple C_O, having moment 150 N · m, is applied to $\mathcal{B}_2$ beginning at $t = 0$. Find the angular accelerations of $\mathcal{B}_1$ and of $\mathcal{B}_2$ upon application of the couple, and the force exerted on $\mathcal{B}_2$ at P.

Solution

The equations of motion for $\mathcal{B}_1$ and $\mathcal{B}_2$, using the respective free-body diagrams in Figures E4.31b and 4.31c, are:

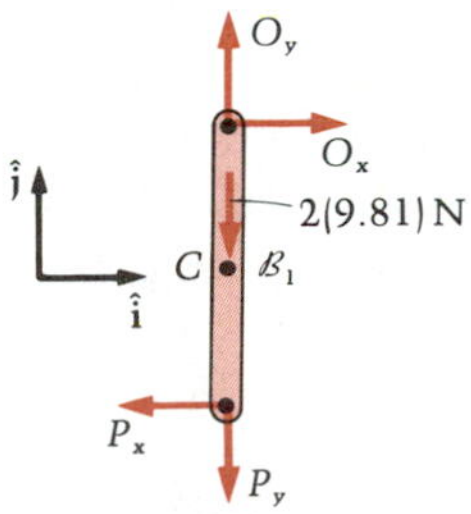

Figure E4.31b

$$\Sigma F_x = O_x - P_x = 2\ddot{x}_C \tag{1}$$

$$\Sigma F_y = O_y - P_y - 19.6 = 2\ddot{y}_C \tag{2}$$

$$\Sigma M_O = -0.5P_x = \frac{2(0.5)^2}{3}\,\alpha_1 \tag{3}$$

* For instance, the pins must be designed strong enough to take the forces caused by the accelerations.

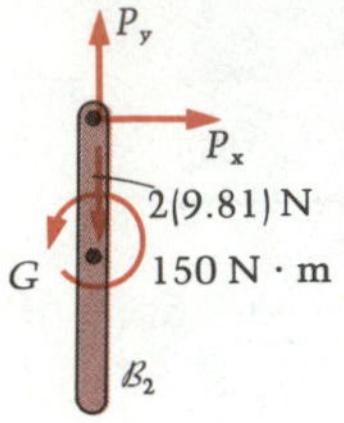

Figure E4.31c

(Note in Equation (3) that O is a pivot of $\mathcal{B}_1$.)

$$\Sigma F_x = P_x = 2\ddot{x}_G \tag{4}$$

$$\Sigma F_y = P_y - 19.6 = 2\ddot{y}_G \tag{5}$$

$$\Sigma M_G = -P_x(0.25) + 150 = \frac{2(0.5)^2}{12}\alpha_2 \tag{6}$$

Thus far we have six equations in the *ten* unknowns $O_x, O_y, P_x, P_y, \ddot{x}_C, \ddot{y}_C, \alpha_1, \ddot{x}_G, \ddot{y}_G$, and α_2. Kinematics gives the four additional equations we will need:

$$\underbrace{\ddot{x}_C\hat{\mathbf{i}} + \ddot{y}_C\hat{\mathbf{j}}}_{\mathbf{a}_C} = \overset{\mathbf{0}}{\cancel{\mathbf{a}_O}} + \alpha_1\hat{\mathbf{k}} \times \overset{-0.25\hat{\mathbf{j}}}{\cancel{\mathbf{r}_{OC}}} - \overset{0 \text{ at } t=0}{\cancel{\omega_1^2}}\mathbf{r}_{OC}$$

$$\hat{\mathbf{i}}\text{-coefficients} \Rightarrow \ddot{x}_C = 0.25\alpha_1 \tag{7}$$

$$\hat{\mathbf{j}}\text{-coefficients} \Rightarrow \ddot{y}_C = 0 \tag{8}$$

Also,

$$\overbrace{\ddot{x}_G\hat{\mathbf{i}} + \ddot{y}_G\hat{\mathbf{j}}}^{\mathbf{a}_G} = \mathbf{a}_P + \alpha_2\hat{\mathbf{k}} \times \overset{-0.25\hat{\mathbf{j}}}{\cancel{\mathbf{r}_{PG}}} - \overset{0 \text{ at } t=0}{\cancel{\omega_2^2}}\mathbf{r}_{PG}$$

But

$$\mathbf{a}_P = \overset{\mathbf{0}}{\cancel{\mathbf{a}_O}} + \alpha_1\hat{\mathbf{k}} \times (-0.5\hat{\mathbf{j}}) - \overset{0}{\cancel{\omega_1^2}}(-0.5\hat{\mathbf{j}})$$
$$= 0.5\alpha\hat{\mathbf{i}}$$

Thus

$$\ddot{x}_G = 0.5\alpha_1 + 0.25\alpha_2 \tag{9}$$

$$\ddot{y}_G = 0 \tag{10}$$

Solving these equations gives:

$\boldsymbol{\alpha}_1 = -771\hat{\mathbf{k}}\ \text{rad/s}^2$
$\boldsymbol{\alpha}_2 = 2060\hat{\mathbf{k}}\ \text{rad/s}^2$ and
Force on $\mathcal{B}_2$ at $P = P_x\hat{\mathbf{i}} + P_y\hat{\mathbf{j}} = 257\hat{\mathbf{i}} + 19.6\hat{\mathbf{j}}$ N

As a point of interest, the values of the other four variables (besides $\ddot{y}_C = \ddot{y}_G = 0$) are: $O_x = -129$ N, $O_y = 39.2$ N, $\ddot{x}_C = -193\ \text{m/s}^2$, and $\ddot{x}_G = 129\ \text{m/s}^2$.

Finally, we remark that any of the alternative moment equations discussed in the preceding two sections may be used. However, the student is cautioned to realize that, just as in statics, once the vector "force equation" and "moment equation" have been written, no new (independent) information will arise from summing moments at a different point.

PROBLEMS ► Section 4.6

4.151 The rod is pinned to the light roller, which moves in the smooth slot, and with the system at rest as shown in Figure P4.151, the 6-N force is applied. Find the angular acceleration of the rod at the given instant, by using Equation (4.19) together with $\Sigma F_x = m\ddot{x}_C$. Then check your solution using only $\Sigma M_C = I_C\alpha$.

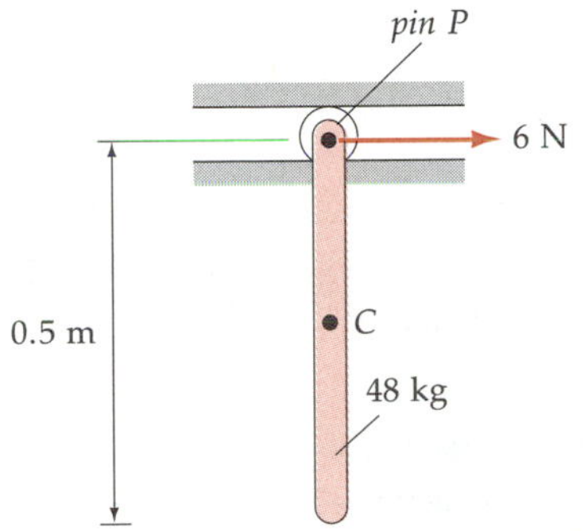

Figure P4.151

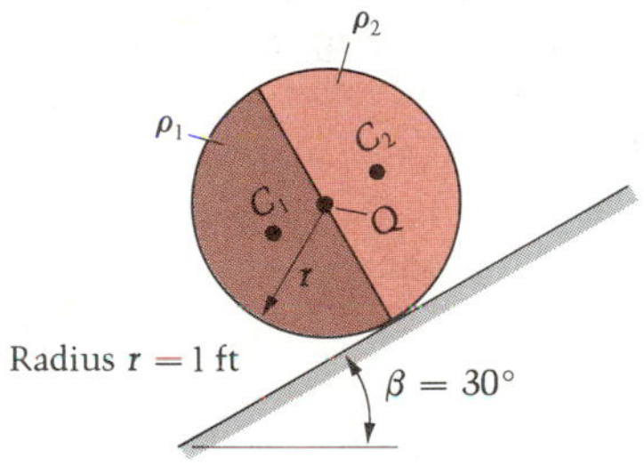

Figure P4.155

4.152 If in the preceding problem we replace the 6-N force by a force F which produces a constant acceleration of pin P of 0.5 m/s², again find the initial value of α, this time using only Equation (4.24). Explain why the answers are the same. (*Hint*: Solve for F at the given instant.)

4.153 Solve the problem of Examples 4.19, 23, 25 using Equation (4.19), with the point P (about which moments are taken) being the geometric center Q of the hoop.

4.154 Solve the problem of Examples 4.19, 23, 25 using Equation (4.24), with the point P again (see the preceding problem) being the geometric center Q of the hoop.

* **4.155** The cylinder shown in Figure P4.155 is made of two halves of different densities. The left half is steel, with mass density $\rho_1 = 15.2$ slug/ft³; the right half is wood with $\rho_2 = 1.31$ slug/ft³. Recalling that the mass center of each half is located $4r/3\pi$ from the geometric center Q, find the acceleration of Q when the cylinder is released from rest. Assume enough friction to prevent slipping. *Hint*: Use Equation (4.17) with Q as point P.

4.156 (a) Use Equation (4.24) to categorize the restrictions on point P for which we may correctly write $\Sigma M_P = I_P\alpha$. Show that there are only three cases, and that the mass center form is one of them, while the "fixed-axis-of-rotation form" is but a special case of one of the other two. (b) Note that the instantaneous center of zero velocity Ⓘ is *not* a point P for which, in general, $\Sigma M_P = I_P\alpha$. (c) Finally, determine in which of the problems in Figure P4.156 (a–e) it is true that $\Sigma M_Q = I_Q\alpha$.

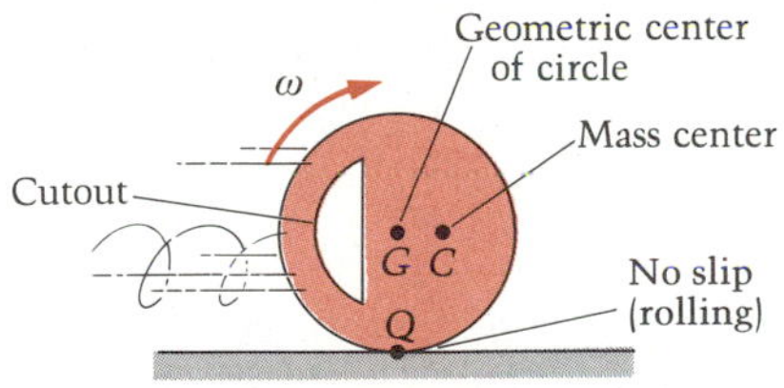

Figure P4.156(a)

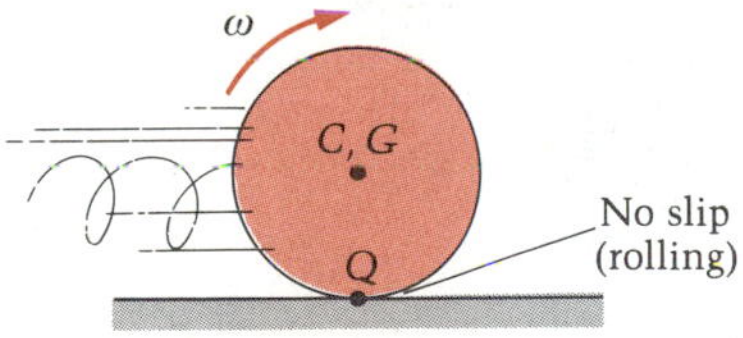

Figure P4.156(b)

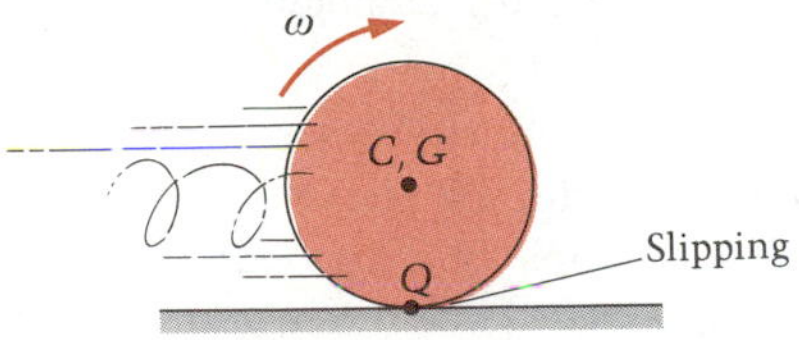

Figure P4.156(c)

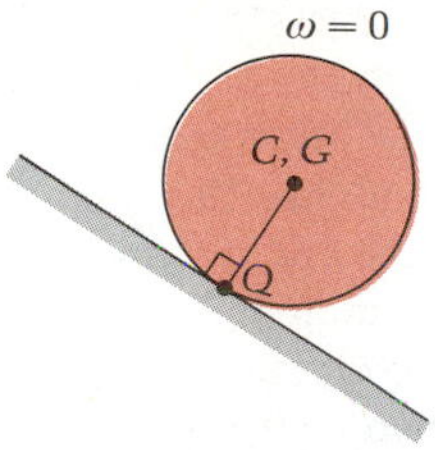

Figure P4.156(d)

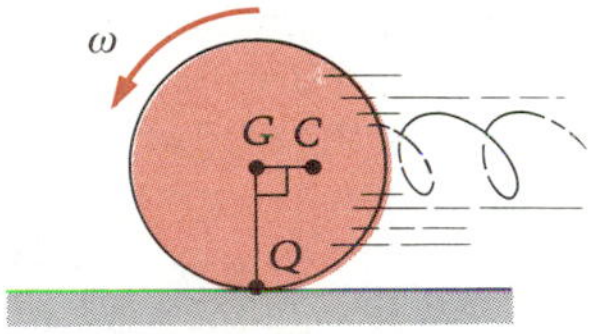

Figure P4.156(e)

4.157 Let $\mathcal{B}$ be a rigid body in plane motion, in constant contact with a surface $\mathcal{S}$. Let Q be the point of $\mathcal{B}$ in contact with $\mathcal{S}$ (Q can be different points of $\mathcal{B}$ at different times). Use the result of Problem 4.156 to prove that if the following four conditions hold, then $\Sigma M_Q = I_Q\alpha$ at all times: (1) $\mathcal{S}$ is fixed in an inertial frame; (2) $\mathcal{B}$ is rolling on $\mathcal{S}$; (3) $\mathcal{B}$ is round; and (4) the mass center of $\mathcal{B}$ is at its geometric center.

* **4.158** In the preceding problem, suppose that conditions (1) and (2) hold, and that at a certain instant, ω of $\mathcal{B}$ is zero. Using the result of Problem 4.156, show that, *at that instant,* $\Sigma M_Q = I_Q\alpha$ regardless of whether conditions (3) or (4) hold.

4.159 The thin-walled hollow sphere of Figure P4.159 (moment of inertia about any diameter $= \frac{2}{3}mr^2$) has average radius $r = 0.5$ m and mass 50 kg. It is pinned at the bottom of the cart. The force F applied to the cart at rest produces a constant acceleration of all points of the cart of $a\hat{\mathbf{i}}$.

a. Find the maximum value of a if the sphere is to translate and the breaking strength of each of the cords is 100 N.

b. Suppose a is twice the result of (a), so that one cord breaks at $t = 0$. Find the angular acceleration of the sphere at the instant it has turned through 90°, using Equation (4.24). Does the answer to (b) depend on the value of a?

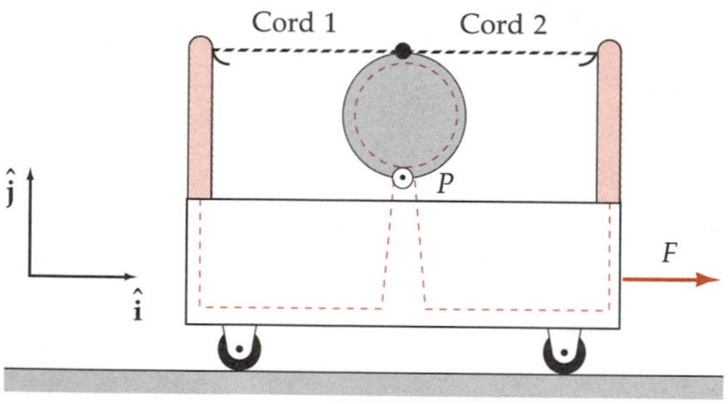

Figure P4.159

4.160 The rectangular door of a railroad car has mass m (Figure P4.160); it is of uniform width 2ℓ and has its hinges on the side of the doorway closest to the engine. Initially the door makes an angle β with the train, which begins to move forward from rest at constant acceleration a_o. Find the initial resultant horizontal reaction component that the hinges exert on the door.

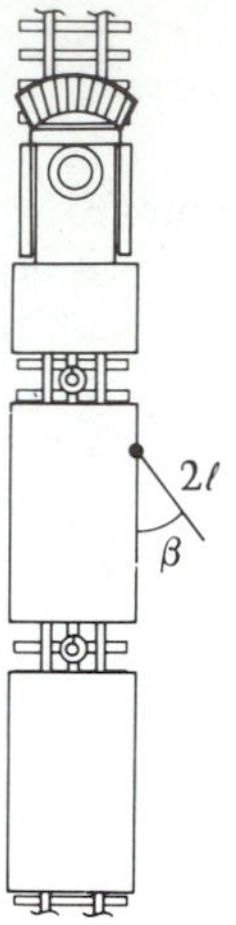

Figure P4.160

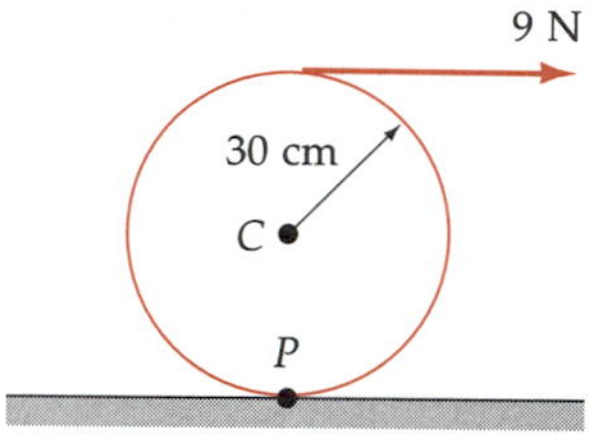

Figure P4.161

4.161 The cylinder in Figure P4.161 has mass 10 kg. The 9-N force is applied to a string wrapped around a thin slot near the surface of the cylinder. Find the angular acceleration of the cylinder, assuming enough friction to prevent slip, using (a) Equation (4.19); (b) Equation (4.24). Observe how both right-hand-sides add up to $\frac{3}{2}mr^2\alpha$ even though the individual terms are different.

4.162 In Problem 4.152 at a later time, let θ be the angle between the vertical and the rod (see Figure P4.162). Using Equation (4.24), find the angular acceleration of the rod as a function of θ.

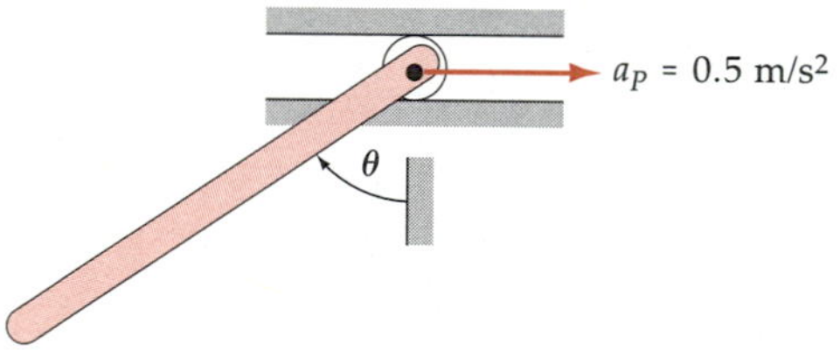

Figure P4.162

4.163 The crank arm OP is turned by the couple M_o at constant angular velocity 1.0 rad/s ⟲. In the position shown in Figure P4.163, determine the reaction of the smooth plane onto the 20-kg slender bar PQ. *Hints*: First solve for $\mathbf{a}_P$ and for $\boldsymbol{\alpha}$ of PQ using kinematics, then use Equation (4.24).

4.164 A body $\mathcal{B}$ weighing 805 lb with radius of gyration 0.8 ft about its z_C axis (see Figure P4.164) is pinned at its mass center. A clockwise couple of magnitude e^t lb-ft is applied to $\mathcal{B}$ starting at $t = 0$. Find the angle through which $\mathcal{B}$ has turned during the interval $0 \le t \le 3$ sec.

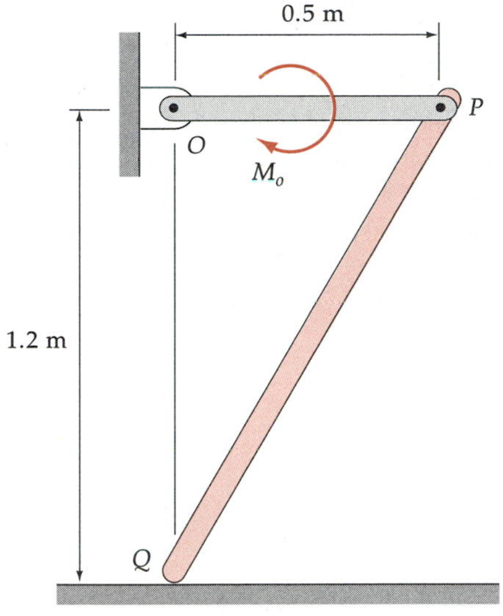

Figure P4.163

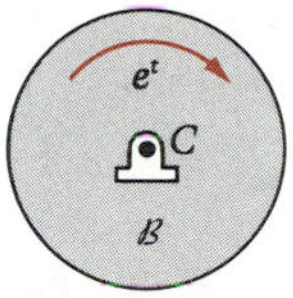

Figure P4.164

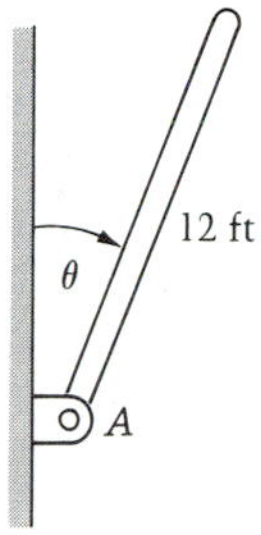

Figure P4.165

4.165 The slender homogeneous rod in Figure P4.165 is 12 ft long and weighs 5 lb; it is connected by a rusty hinge to a support at A. Because of friction in the hinge, the hinge exerts a couple of 9 lb-ft on the rod when it rotates. If the rod is released from rest with $\theta = 30°$, find: (a) the angular acceleration of the rod when $\theta = 30°$, 60° and 90°; (b) the angle θ at which the angular acceleration of the rod is zero.

4.166 A wagon wheel spinning counterclockwise is placed in a corner and contacts the wall and floor. (See Figure P4.166.)

- **a.** Show with a free-body diagram that the wheel cannot climb the wall.
- **b.** Show with a free-body diagram that the wheel cannot move to the right along the floor either.
- **c.** Therefore the wheel stays in the corner. Treat it as a ring of mass m and radius R, with friction coefficient μ at both surfaces of contact. Determine how long it takes for the wheel to stop, and find how many radians it has turned through since first contacting the surfaces.

4.167 The cylinder in Figure P4.167 has a mass of 30 kg and rotates about an axis normal to the clevis at O. At the instant shown, $\omega = 5$ ↻ rad/s and $\dot{\omega} = 10$ ↺ rad/s^2. Find the force P that acts on the cylinder, and determine the reactions exerted by the pin onto the clevis at O, all at the given instant.

Figure P4.166

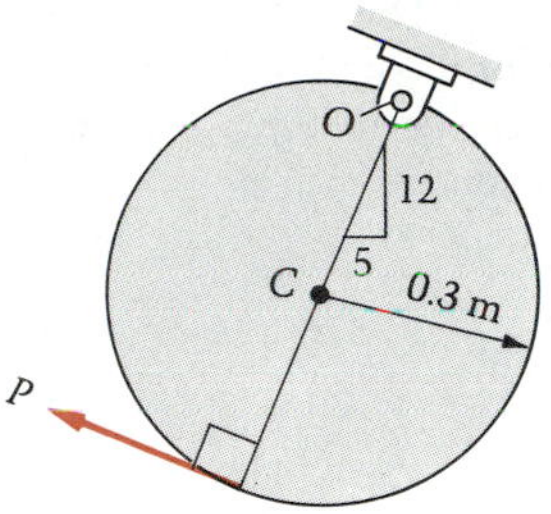

Figure P4.167

4.168 Figure P4.168 shows a scene from Edgar Allen Poe's "The Pit and the Pendulum." Find the reaction of the pin onto the bar if the pendulum is instantaneously at rest in a horizontal position.

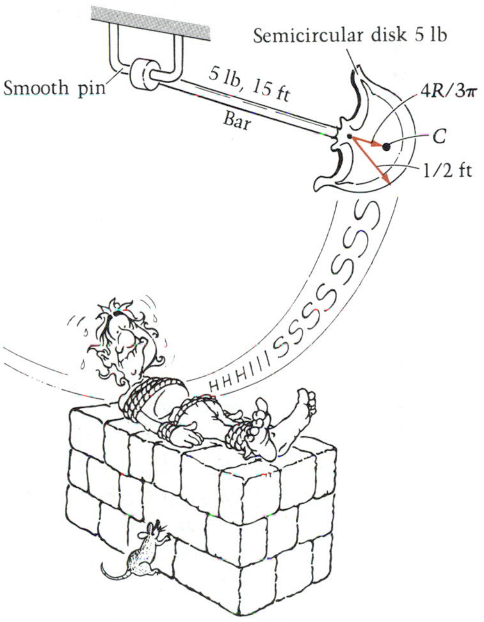

Figure P4.168

4.169 The uniform slender bar of mass m is released from rest in the position shown in Figure P4.169. Find the angular acceleration when the bar has turned through 45°

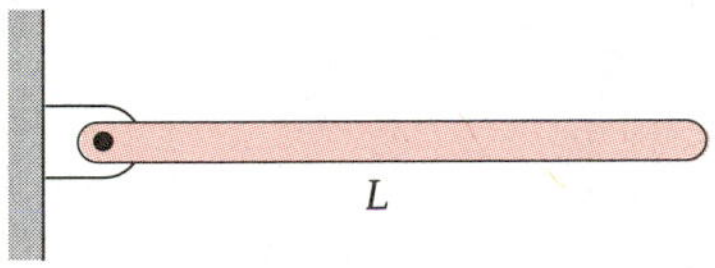

Figure P4.169

* **4.170** The uniform slender bar in Figure P4.170, of mass m and length ℓ, is released from rest at $\theta =$ zero plus a tiny increment. Find the magnitude of the bearing reaction when $\theta = \pi/2$.

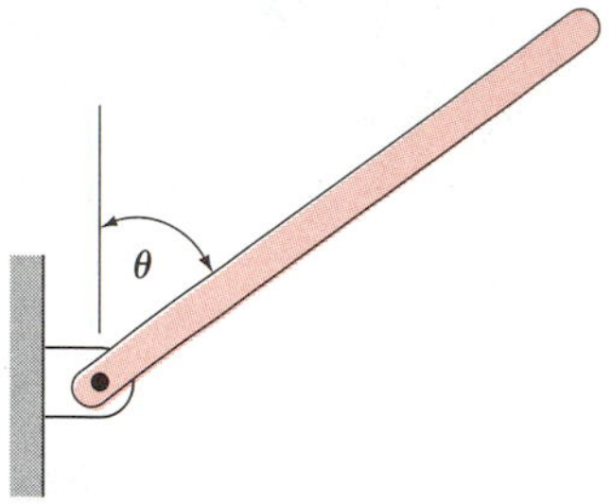

Figure P4.170

4.171 Body $\mathcal{B}$ is a slender bar bent into the shape of a quarter-circle (Figure P4.171). Find the tensions in strings OA and OB when the system is released from rest.

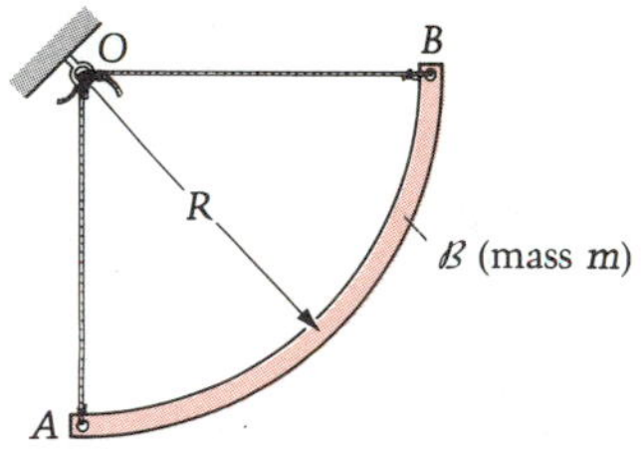

Figure P4.171

4.172 Two weights W_1 and W_2 in Figure P4.172 are connected by an inextensible cord that passes over a pulley. The pulley weighs W and its mass is concentrated at the rim of radius R. Show that if the system is released and the cord does not slip on the pulley, the acceleration magnitude of W_1 and W_2 is

$$\left|\frac{W_2 - W_1}{W_1 + W_2 + W}\,g\right|$$

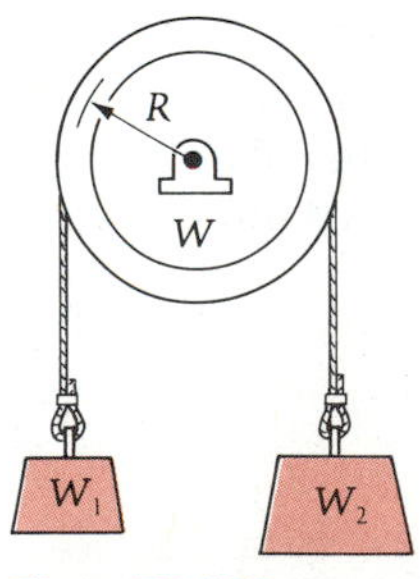

Figure P4.172

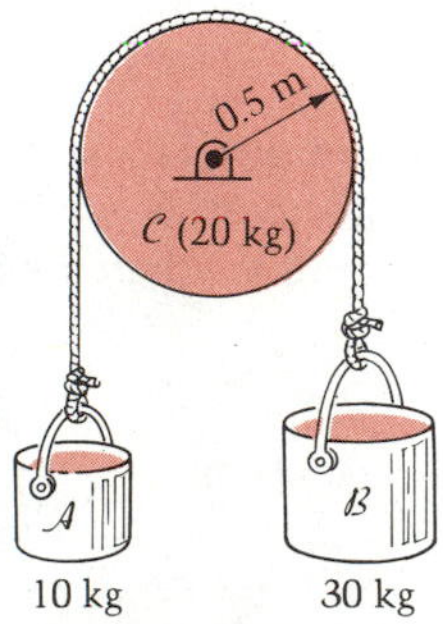

Figure P4.173

4.173 Find the angular acceleration of cylinder $\mathcal{C}$ in Figure P4.173. The rope passes over it without slipping and ties to $\mathcal{A}$ and $\mathcal{B}$ as shown.

4.174 Body $\mathcal{C}$ is a pulley made of the cylinders $\mathcal{D}$ and $\mathcal{E}$, which are butted together and rigidly attached. (See Figure P4.174.) The combined body $\mathcal{C}$ is smoothly pinned to the ground through its axis of symmetry (which passes through its mass center). Ropes wrapped around $\mathcal{E}$ and $\mathcal{D}$ are tied to bodies $\mathcal{A}$ and $\mathcal{B}$, respectively. If the system is released from rest, what will be the angular acceleration of $\mathcal{C}$?

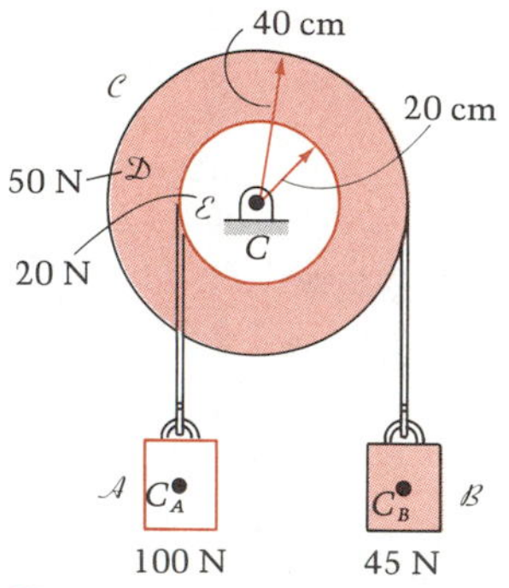

Figure P4.174

4.175 The 32.2-lb particle P rests on the 128.8-lb plank as shown in Figure P4.175. If the cord at B suddenly breaks, find the initial acceleration of the particle, and the force exerted on it by the bar.

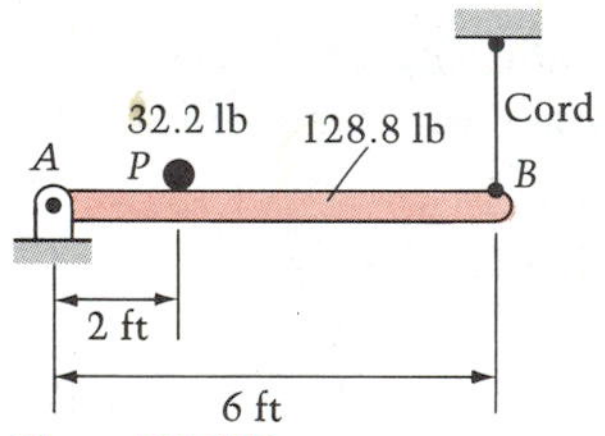

Figure P4.175

4.176 In Example 4.29, note that pivot point O is also a point of the "subbody" used in that example. Thus for that body, $\Sigma M_O = I_O\alpha$. Separately compute ΣM_O and $I_O\alpha$ for the subbody and show that their values are the same.

* **4.177** The uniform slender bar of weight W and length L in Figure P4.177 is released from rest at $\theta = 0$ and pivots on its square end about corner O.

a. If the bar is observed to slip at $\theta = 30°$, find the coefficient of limiting static friction μ_s.

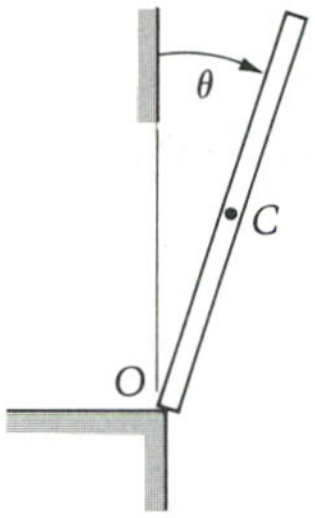

Figure P4.177

b. If the end of the bar is notched so that it cannot slip, find the angle θ at which contact between bar and corner ceases. *Hint*: Write the moment equation of motion about the pivot O, multiply it by $\dot{\theta}$, and integrate, obtaining $\dot{\theta}$ as a function of θ. Use this relation together with the component equation of $\Sigma \mathbf{F} = m\mathbf{a}_C$ in the $\hat{\mathbf{e}}_n$ direction.

4.178 The chain drive in Figure P4.178 may be considered as two disks with equal density and thickness. The larger sprocket has a mass of 2 kg and a radius of 0.2 m. If the couple is applied starting from rest at $t = 0$, find the angular speed of the smaller sprocket at $t = 10$ s. *Hint*: What does a dentist look at?

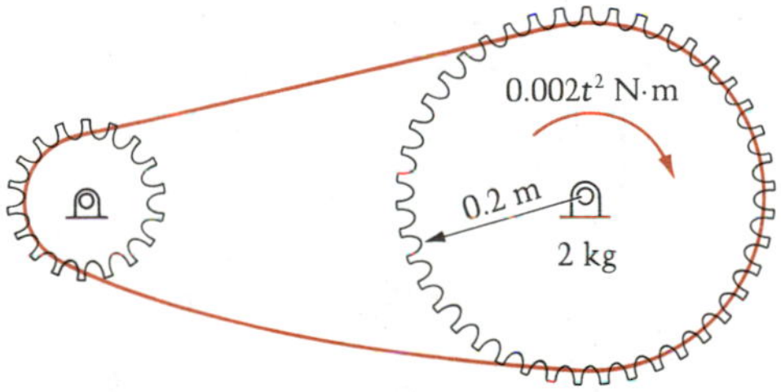

Figure P4.178

• **4.179** Cylinder $\mathcal{C}$ in Figure P4.179 with four cutouts is rotating at 200 rpm initially. A uniform 100-lb cylinder $\mathcal{D}$ is placed in the position shown, and the friction produces a braking moment that will stop $\mathcal{C}$. The friction coefficient is $\mu = \frac{1}{3}$, and before the four holes were drilled the uniform body $\mathcal{C}$ weighed 200 lb. For whichever rotation direction of $\mathcal{C}$ results in a quicker stop, find the stopping time.

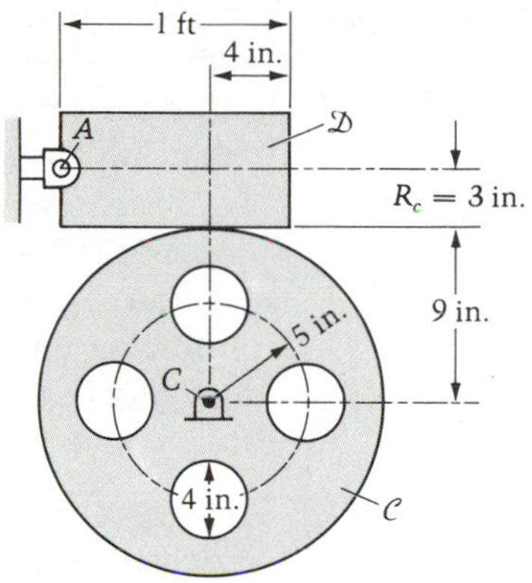

Figure P4.179

• **4.180** The slender, homogeneous rod in Figure P4.180 is supported by a cord at A and a horizontal pin at B. The cord is cut. Determine, at that instant, the location of pin B that will result in the maximum initial angular acceleration of the rod.

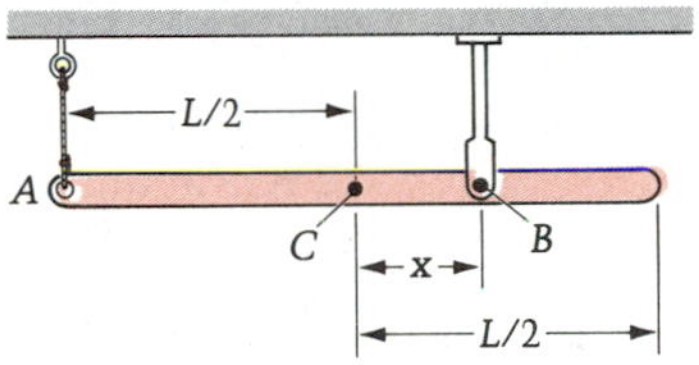

Figure P4.180

4.181 The uniform rod of mass m is released from rest in the horizontal position indicated in Figure P4.181. Consider the force exerted by the smooth pin.

a. How does the magnitude of the force vary with the angle θ through which it has turned?

b. What is the maximum value of this magnitude?

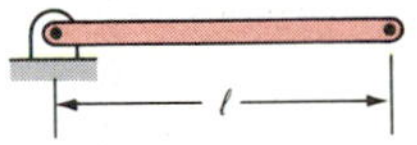

Figure P4.181

4.7 Rotation of Unbalanced Bodies

When a rigid body is mounted in bearings and made to rotate by means of a moment about the axis of the bearings, it is said to be balanced (for rotation about that axis) if the external reactions exerted by the bearings onto the body are only what are required to support the weight of the body. The bearing reactions accompanying *im*balance result in vibration and wear of rotating machinery and are the reason, for example, for balancing automobile tires.

There are two distinct causes for a rotating body to be out of balance. The first is if the mass center is located (a distance "d") off the axis of rotation. Then as the body turns, there will be forces at the bearings producing and equaling $m\mathbf{a}_C$. Clearly, these forces will be constantly changing in direction (relative to the inertial frame) if not also in magnitude.

Question 4.14 What would cause them to change in magnitude?

Moving the mass center onto the axis of rotation by addition or deletion of mass is called static balancing. It carries that name because only then will the body remain in equilibrium when turned to any position and released, this being true regardless of the orientation of the axis.

The second cause of imbalance is nonzero products of inertia I_{xz}^P and/or I_{yz}^P, where z is the axis of rotation and P is a point on that axis. In the same way that an off-axis mass center causes bearing forces which produce $m\mathbf{a}_C$, these products of inertia likewise cause bearing forces to exist; for a system which has been statically balanced, they produce the "bearing moments" (see Equations 4.14 a,b) ΣM_{Cx} and ΣM_{Cy}. And similar to removing the offset "d," we can also add or delete material to force the values of I_{xz}^P and I_{yz}^P to be zero. When this is done, in addition to having ensured that C lies on the axis, the body is then said to be dynamically (and of course also statically) balanced. We shall develop the equations to accomplish this in what follows.

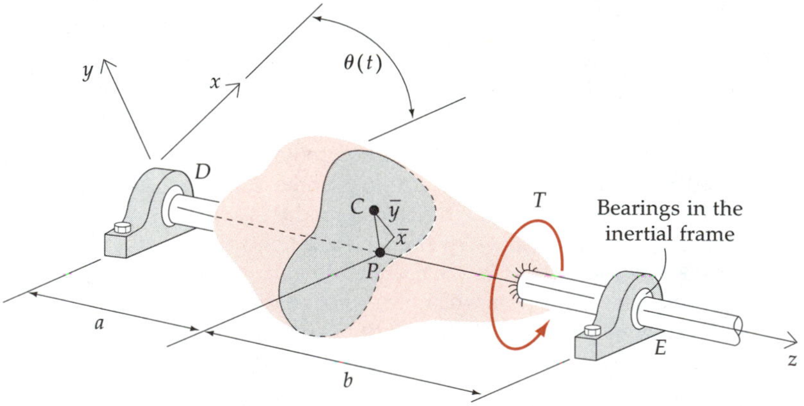

Figure 4.8

Answer 4.14 Changes in rotational speed.

We show body $\mathcal{B}$ in Figure 4.8, set in ball bearings at D and E. T is an externally applied torque about z, the axis of rotation. Let us say that T is the driving torque less any frictional resistance moments from the air or the bearings. Finally, note that the x and y axes are also fixed in $\mathcal{B}$, and $(\bar{x}, \bar{y}, a)$ are the coordinates of C in this system.

For an unbalanced body rotating about a horizontal axis, the bearing reactions required to support the weight of the body (when it is not rotating) may be simply added to the dynamic reactions that would be generated were there no gravity. Therefore, for the sake of simplicity we shall ignore the effects of gravity in our discussion here.

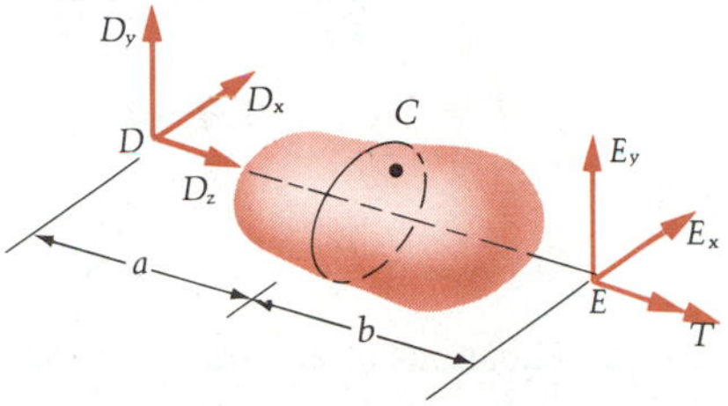

Figure 4.9

Now consider the free-body diagram, Figure 4.9, and let the bearing-reaction components be referred to the body-fixed axes (x, y, z).

Then $\Sigma \mathbf{F} = m\mathbf{a}_C$ yields the component equations

$$D_x + E_x = m(-\omega^2\bar{x} - \alpha\bar{y}) \tag{4.24a}$$

$$D_y + E_y = m(-\omega^2\bar{y} + \alpha\bar{x}) \tag{4.24b}$$

$$D_z = 0 \tag{4.24c}$$

Using Euler's second law in the form

$$\Sigma \mathbf{M}_D = (I_{xz}^D\alpha - I_{yz}^D\omega^2)\hat{\mathbf{i}} + (I_{yz}^D\alpha + I_{xz}^D\omega^2)\hat{\mathbf{j}} + I_{zz}^D\alpha\hat{\mathbf{k}} \tag{4.20}$$

we obtain the following component equations:

$$-(a + b)E_y = I_{xz}^D\alpha - I_{yz}^D\omega^2 \tag{4.25a}$$

$$(a + b)E_x = I_{yz}^D\alpha + I_{xz}^D\omega^2 \tag{4.25b}$$

$$T = I_{zz}^D\alpha \tag{4.25c}$$

We observe, then, that if we know ω, α, and the geometric and inertia properties of the body, we can solve Equations (4.24a,b) and (4.25a,b) for the bearing reactions D_x, D_y, E_x, and E_y. We now illustrate such a calculation with an example before taking up the issue of how to balance a body.

EXAMPLE 4.32

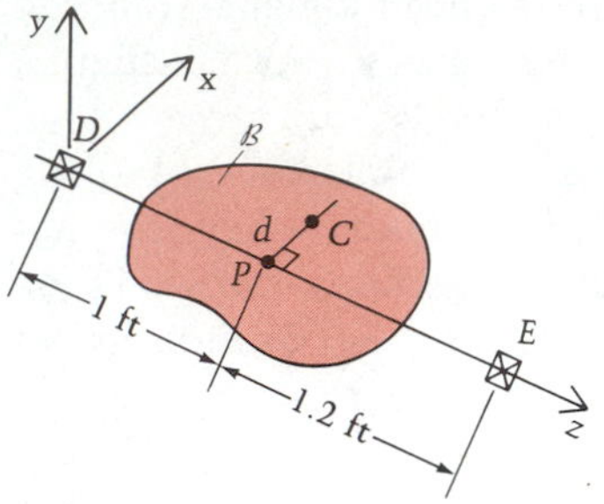

Figure E4.32

The body in Figure E4.32 has mass $m = 2$ slugs, and its mass center is off-axis by the amount $d = 1/64$ in. in the x-z plane so that $\bar{x} = 0.0156$ in. and $\bar{y} = 0$. Its products of inertia are $I_{xz}^C = I_{yz}^C = 0.000380$ slug-ft^2. If the body is spun up to a constant angular speed of 3000 rpm, what then are the dynamic reactions at bearings D and E?

Solution

By the parallel-axis theorem

$$I_{xz}^D = 0.000380 - 2\left(\frac{.0156}{12}\right)(1) = -0.00222 \text{ slug-ft}^2$$

$$I_{yz}^D = 0.000380 - 2(0)(1) = 0.000380 \text{ slug-ft}^2$$

Also,

$$\alpha = 0 \qquad \text{and} \qquad \omega = 3000(2\pi)/60 = 100\pi \text{ rad/sec}$$

Thus by Equations (4.24) and (4.25) we obtain

$$E_y = \frac{0.000380}{2.2}(100\pi)^2 = 17.0 \text{ lb}$$

$$D_y = 0 - E_y = -17.0 \text{ lb}$$

$$E_x = -\frac{0.00222(100\pi)^2}{2.2} = -99.6 \text{ lb}$$

and

$$D_x = -2\left(\frac{.0156}{12}\right)(100\pi)^2 + 99.6 = -157 \text{ lb}$$

Let us now assume that we have a rotating body (ω and α not both zero) mounted in bearings. For such a body, we can show that the bearing reactions vanish if and only if $\bar{x} = 0 = \bar{y}$ and $I_{xz}^D = 0 = I_{yz}^D$. The "if" proof is simple, for if $I_{xz}^D = 0 = I_{yz}^D$, Equations (4.25 a,b) yield $E_x = 0 = E_y$, and substituting these zero values along with $\bar{x} = 0 = \bar{y}$ into Equations (4.24 a,b), we find $D_x = 0 = D_y$.

For the "only if" part of the proof, if the bearing reactions are all zero, Kramer's rule applied to Equations (4.24 a,b) gives $\bar{x} = 0 = \bar{y}$; and to Equations (4.25 a,b) yields $I_{xz}^D = 0 = I_{yz}^D$. When these two products of inertia are zero at a point D, then z is called a *principal axis of inertia* at D; this concept is discussed in considerable detail in Chapter 7.

In summary, then, we can say that the bearing reactions vanish, and hence the body is balanced, if and only if the axis of rotation is a principal axis of inertia containing the mass center of the body.

Now let us see what can be done about *correcting* imbalance. Suppose values of m, $\bar{x}$, $\bar{y}$, I_{xz}^P and I_{yz}^P of a body are known, where P, lying on the axis of rotation, is the origin of the coordinates. We can, for example, determine the coordinates (x_A, y_A) and (x_B, y_B) and masses (m_A and m_B) of a pair of weights which, when placed in two "correction planes" A (at $z = z_A$) and B (at $z = z_B$), will ensure that the shaft is dynamically balanced. All we have to do is (a) force the mass center C^* of the combined system (m plus m_A and m_B) to lie on the axis of the shaft, and (b) force the products of inertia of the combined system to vanish:

$$x\text{-coordinate of } C^* = 0: \qquad m_A x_A + m_B x_B + m\bar{x} = 0 \qquad (4.26a)$$

$$y\text{-coordinate of } C^* = 0: \qquad m_A y_A + m_B y_B + m\bar{y} = 0 \qquad (4.26b)$$

$$^*I_{xz}^P = 0: \qquad -m_A x_A z_A - m_B x_B z_B + I_{xz}^P = 0 \qquad (4.26c)$$

$$^*I_{yz}^P = 0: \qquad -m_A y_A z_A - m_B y_B z_B + I_{yz}^P = 0 \qquad (4.26d)$$

Note that we assume that the "balance weights" are small enough to be treated as particles.

These four equations (4.26) may be solved for the four quantities $m_A x_A$, $m_B x_B$, $m_A y_A$, and $m_B y_B$. Thus there is some freedom to select two of the six quantities m_A, m_B, x_A, x_B, y_A, and y_B, provided there is no other condition linking them; for an example of such a constraint, the weights might have to be placed on a circle of given radius (such as when tires are balanced and weights are clamped to a rim). In this case we would additionally have, for example,

$$x_A^2 + y_A^2 = R_A^2$$

and

$$x_B^2 + y_B^2 = R_B^2$$

and now there are six equations in six unknowns. Let us illustrate the use of these equations in the following example.

Example 4.33

In Example 4.32, suppose that we are to balance the body by adding weights in two correction planes midway between C and the two bearings. Furthermore, the weights are each to be placed on a circle of radius ½ ft. Find the masses and coordinates of the weights.

Solution

We had $m = 2$ slugs, $\bar{x} = 1/64$ in., $\bar{y} = 0$, and $I_{xz}^C = I_{yz}^C = 0.000380$ slug-ft². If we choose P to have the same axial position as C, then $z_A = -0.5$ ft, $z_B = 0.6$ ft, and $\bar{z} = 0$. Also, $I_{xz}^P = I_{xz}^C - 0 = 0.000380$ slug-ft² and $I_{yz}^P = I_{yz}^C - 0 = 0.000380$ slug-ft², and:

$$(4.26a) \Rightarrow m_A x_A + m_B x_B = -2\left(\frac{1}{64(12)}\right) = -0.00260$$

$$(4.26c) \Rightarrow m_A x_A(-0.5) - m_B x_B(0.6) = -0.000380$$

Solving these we get

$$m_B x_B = -0.000836$$

$$m_A x_A = -0.00176$$

Similarly,

$$(4.26\text{b}) \Rightarrow m_A y_A + m_B y_B = 0$$
$$(4.26\text{d}) \Rightarrow -m_A y_A(-0.5) - m_B y_B(0.6) = -0.000380$$

from which we obtain

$$m_B y_B = 0.000345$$
$$m_A y_A = -0.000345$$

Squaring and adding,

$$m_B^2 x_B^2 + m_B^2 y_B^2 = 0.818 \times 10^{-6}$$
$$m_B^2 \underbrace{(x_B^2 + y_B^2)}_{\left(\frac{1}{2}\right)^2} = 0.818 \times 10^{-6}$$

$$m_B = 1.81 \times 10^{-3} \text{ slug}$$

So the weight of B is $W_B = 1.81 \times 10^{-3}(32.2) = 0.0582$ lb, or 0.932 oz. For the coordinates,

$$x_B = \frac{-0.000836}{1.81 \times 10^{-3}} = -0.462 \text{ ft}$$

and

$$y_B = \frac{0.000345}{1.81 \times 10^{-3}} = 0.191 \text{ ft}$$

(These add vectorially to $\sqrt{(0.462)^2 + (0.191)^2} = 0.500$ ft, as a check.)

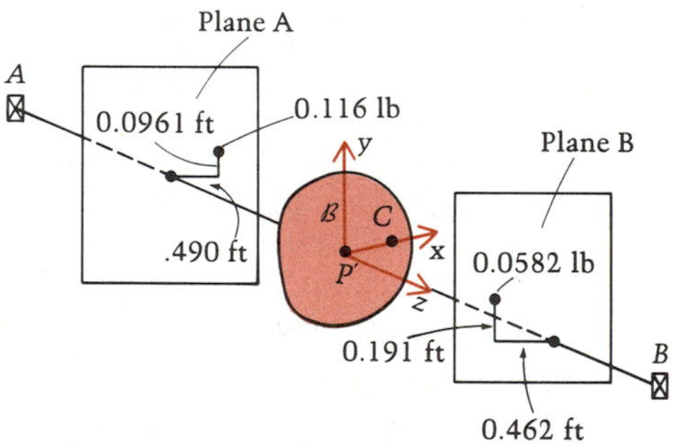

Figure E4.33

Now, for the mass and coordinates of m_A in plane A, we have, by again squaring and adding,

$$m_A^2 \underbrace{(x_A^2 + y_A^2)}_{\left(\frac{1}{2}\right)^2} = 3.22 \times 10^{-6}$$

$$m_A = 3.59 \times 10^{-3} \text{ slug}$$

Thus the weight of A is $W_A = (3.59 \times 10^{-3})\ 32.3 = 0.116$ lb, or 1.85 oz. The coordinates are

$$x_A = \frac{-0.00176}{3.59 \times 10^{-3}} = 0.490 \text{ ft}$$

$$y_A = \frac{-0.000345}{3.59 \times 10^{-3}} = 0.0961 \text{ ft}$$

Again checking,

$$\sqrt{x_A^2 + y_A^2} = 0.499 \text{ ft}$$

The above results are all shown in Figure E.4.33.

In our last example, we shall add mass in the form of two rods to balance the body in Example 4.16.

Example 4.34

For the body of Example 4.16, find the length L of the pair of rods, each of mass $3m$, that will dynamically balance the shaft when attached to it as shown in Figure E4.34.

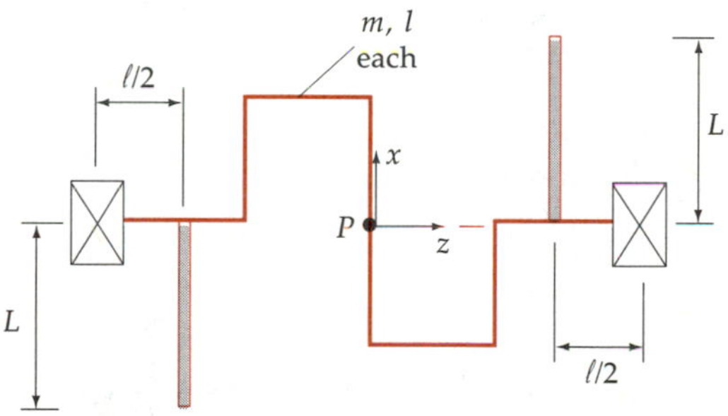

Figure E4.34

Solution

In the previous example, I_{xz}^P was computed to be $2m\ell^2$. Note that the mass center of the original and modified systems is at P, so the system is already statically balanced. Thus since $I_{yz}^P = 0$ (all the mass is still in the xz-plane), all we need for dynamic balance is:

$$(I_{xz}^P)_{\text{TOTAL}} = (I_{xz}^P)_{\text{RODS}} + 2m\ell^2 = 0$$

$$2\left[0 - 3m\left(-\frac{L}{2}\right)\left(-\frac{3}{2}\ell\right)\right] = -2m\ell^2$$

$$\frac{9}{2}mL\ell = 2m\ell^2$$

$$L = \frac{4}{9}\ell$$

PROBLEMS ▶ Section 4.7

4.182 Explain why the uniform plate in Figure P4.182 is dynamically balanced.

4.183 A light rod of length ℓ, with a concentrated end mass M, is welded to a vertical shaft turning at constant ω. (See Figure P4.183.) Find the force and moment exerted by the rod onto the shaft. Include the effect of gravity.

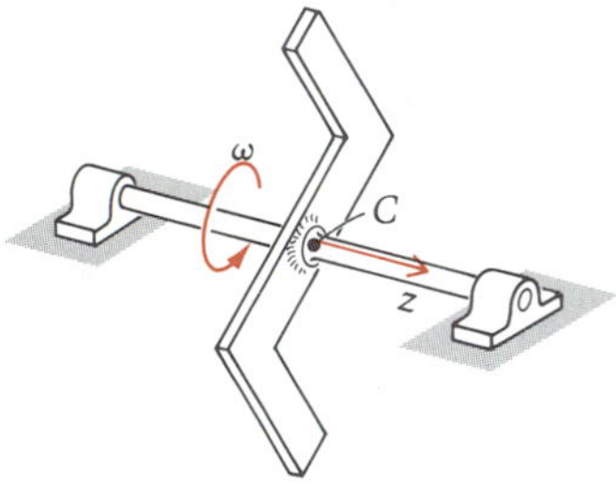

Figure P4.182

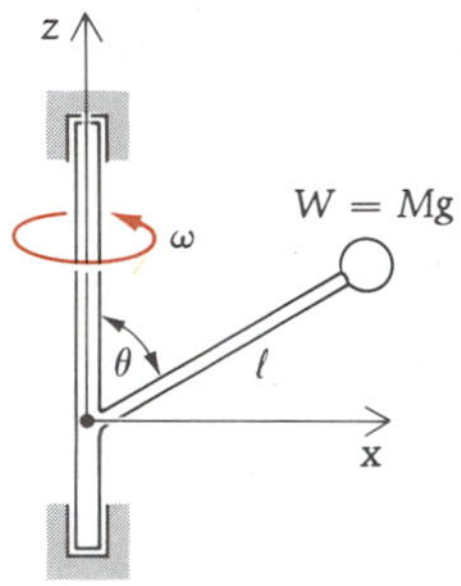

Figure P4.183

4.184 The shaft in Figure P4.184 turns at constant angular velocity 10 rad/sec. If the bars are light compared with the two weights, determine the bending moment exerted on $\mathcal{S}_2$ (length 2ℓ) by $\mathcal{S}_1$ at the point where they are welded together. Sketch the way the shaft will deform in reality under the action of this couple. Ignore gravity.

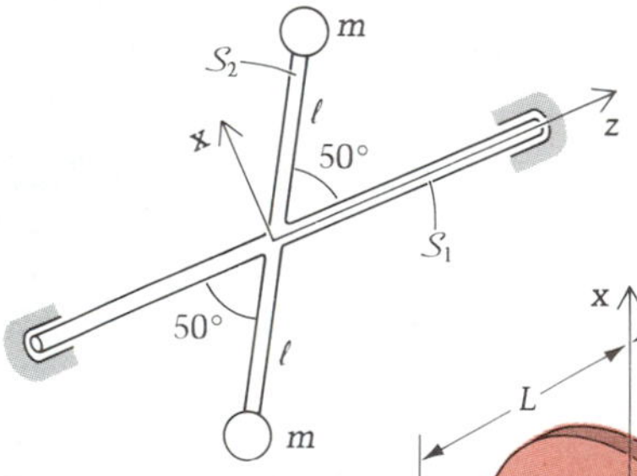

Figure P4.184

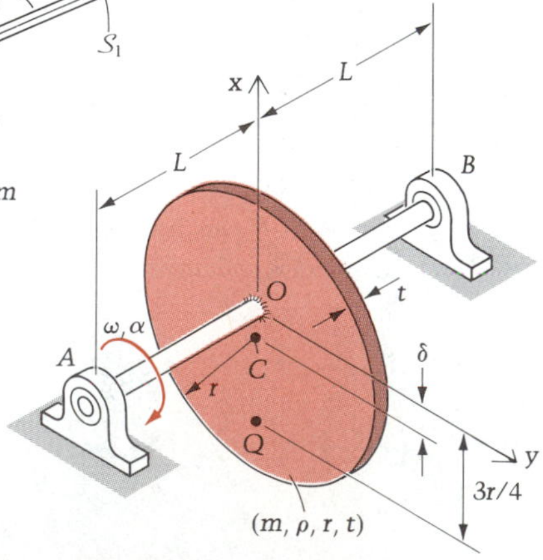

Figure P4.185

4.185 The circular disk in Figure P4.185 is offset by the amount δ from the shaft to which it is attached.

a. Find the dynamic bearing reactions at A and B in terms of the system parameters shown in the figure.

b. If $\delta = r/20$, find the radius of a hole (in terms of r) at Q that will eliminate these bearing reactions.

4.186 Two thin disks are mounted on a shaft, each midway between the center and one of the bearings, as indicated in Figure P4.186. The disks are each mounted off center by the amount $\delta = 0.05$ in. as shown. Determine the x and y locations of two small 4-oz magnetic weights (one for each disk), which when stuck to the disks will balance the shaft. Neglect the thicknesses of the disks, and treat the weights as particles.

4.187 In Figure P4.187, $\mathcal{A}$ is the axle of a bicycle, mounted in bearings $2d$ apart. The cranks $\mathcal{C}$ are rigidly connected to the axle and also to the pedal shafts $\mathcal{P}_1$ and $\mathcal{P}_2$. If the rigid body consisting of axle $\mathcal{A}$, the cranks $\mathcal{C}$, and the pedal shafts is turning freely about axis z_C at constant angular speed ω, find the forces exerted on the bearings in the given configuration.

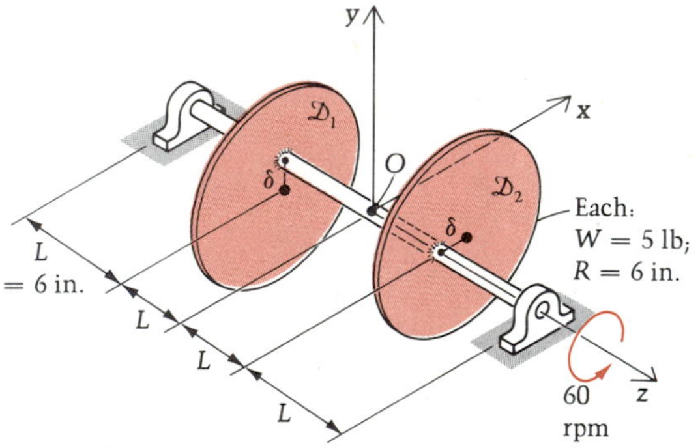

Figure P4.186

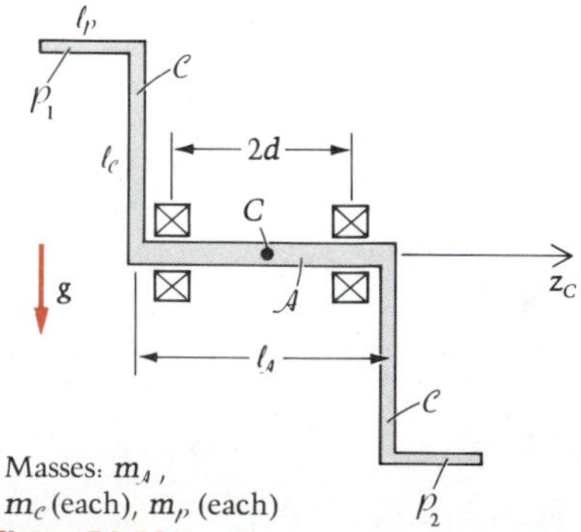

Masses: $m_\mathcal{A}$, $m_\mathcal{C}$ (each), $m_\mathcal{P}$ (each)

Figure P4.187

4.188 A solid cylinder (of mass m, radius r, and length $\ell = 4r$) and a light rod are welded together at angle ψ as shown in Figure P4.188. The rigid assembly is spun up to ω_0 rad/sec and then maintained at that speed. Find the dynamic bearing reactions after $\omega = \omega_0$.

$I_{xz} = (-6.5 \sin 2\psi)(mr^2/12)$

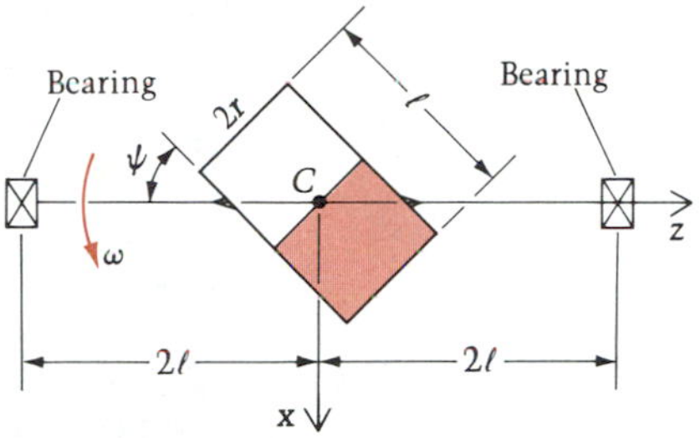

Figure P4.188

4.189 Repeat the preceding problem if the lower half (shaded) of the cylinder is missing. (The mass is now $m/2$.)

4.190 The S-shaped shaft in Figure P4.190(a) is made of two half-rings, each of radius R and mass $m/2$. Find the dynamic bearing reactions for the instant given. *Hint*: For a half-ring, the mass center is located as shown in Figure P4.190(b).

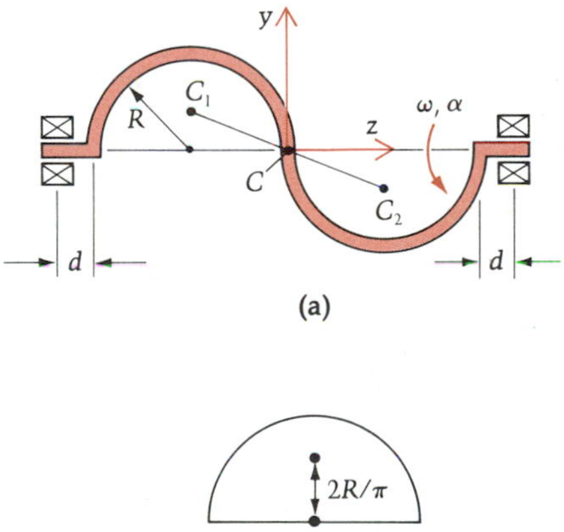

Figure P4.190

4.191 Show that, if a body mounted on a shaft is statically balanced, and if I_{xz}^D and I_{yz}^D are zero for any point D on the shaft, it follows that I_{xz}^Q and I_{yz}^Q are zero for any *other* point Q on the shaft.

4.192 The shaft in Figure P4.192 supports the eccentrically located weights W_1 (0.1 lb) and W_2 (0.2 lb) as shown. It is desired to add a 0.3-lb weight in plane A and a 0.4-lb weight in plane B to balance the shaft dynamically. Determine the x and y coordinates of the added weights.

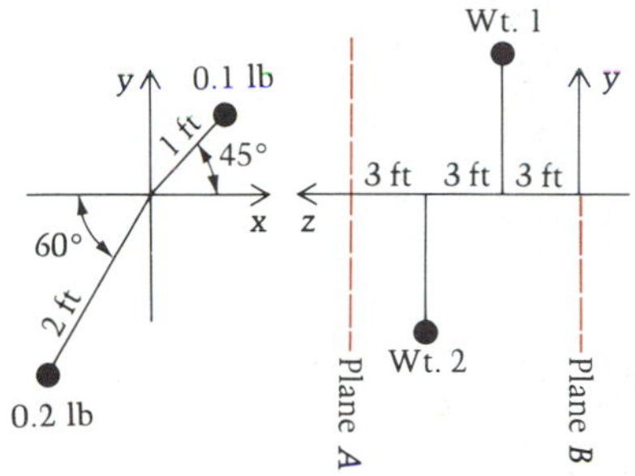

Figure P4.192

4.193 Rotor $\mathcal{R}$ in Figure P4.193 has a mass of 2 slugs, and its mass center C is at a 5-in. offset from its shaft as shown ($x = 3$ in., $y = 4$ in., and $z = 10$ in. in the coordinate system fixed in the shaft at the point B). The products of inertia of $\mathcal{R}$ with respect to the center of mass axes (x_C, y_C, z_C) are $I_{xz}^C = -\frac{5}{3}$ lb-in.-sec^2 and $I_{yz}^C = -\frac{5}{6}$ lb-in.-sec^2. By adding a $\frac{1}{7}$-slug mass in each of the two correction planes $\mathcal{A}$ and $\mathcal{B}$, balance the rotor. That is, determine the x and y coordinates of each of the two added masses by ensuring that the mass center of the final system is on the shaft and that the products of inertia vanish.

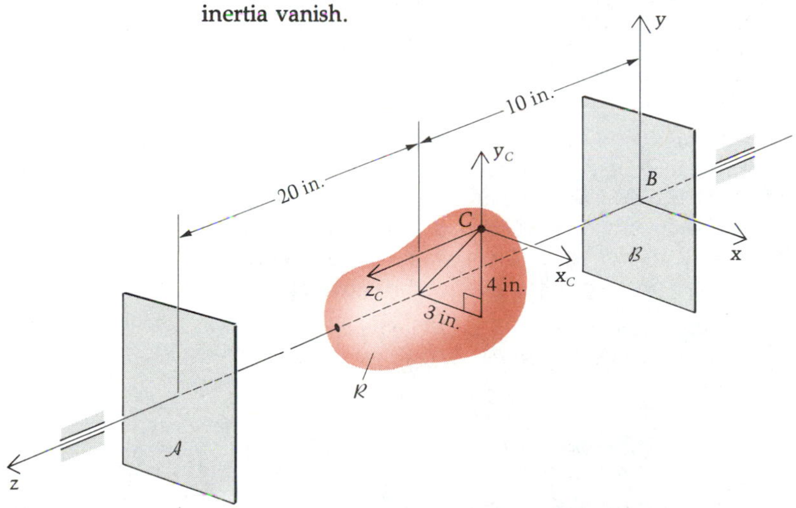

Figure P4.193

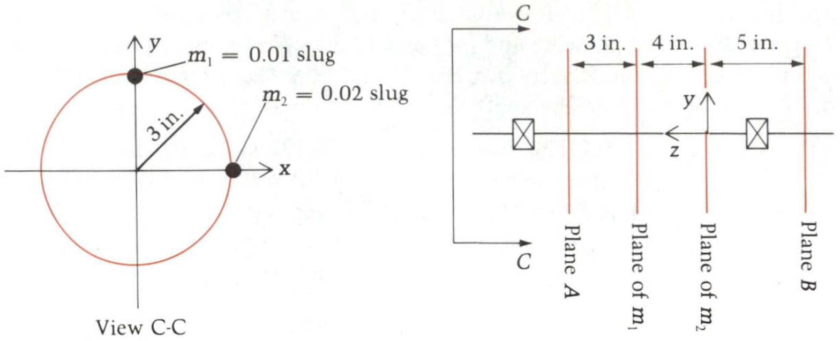

Figure P4.194

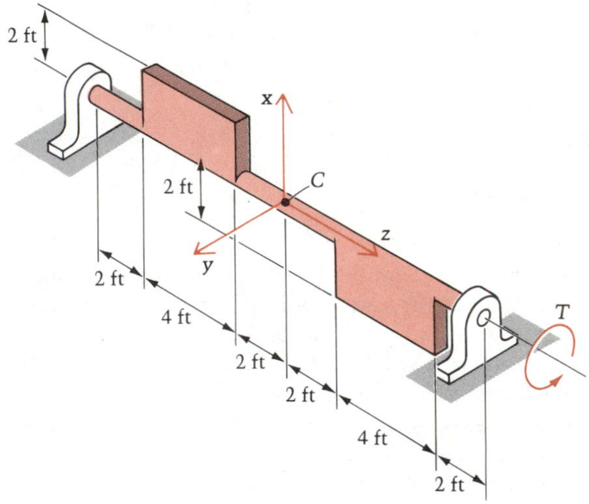

Figure P4.195

4.194 Balance the shaft in Figure P4.194 by adding a mass of 0.003 slug in plane A and a mass of 0.004 slug in plane B.

* **4.195** Two plates, each weighing 32.2 lb, are welded to a light shaft as shown in Figure P4.195. A torque T of 10 lb-ft is applied about the z axis until the assembly is turning at angular speed ω_0, then T is removed. If the bearings can hold a force perpendicular to the shaft of no more than 320 lb, find the maximum value that ω_0 can be without failure. Note that xz is the plane of the plates and (x, y, z) are fixed to the assembly.

COMPUTER PROBLEM ▶ Chapter 4

* **4.196** A cylinder of mass m and radius R is rolling to the left and encounters a pothole of length s, as shown in Figure P4.196(a). The angular velocity when the mass center C is directly above O is ω_i↺. We are interested in the condition(s) for which there will be no slip at O while the cylinder pivots prior to striking the corner at A.

a. Show that for no slip at O, the equations of motion are (see Figure P4.196(b)):

1. $mR\alpha = mg \sin \theta - f$
2. $mR\omega^2 = mg \cos \theta - N$
3. $mgR \sin \theta = \dfrac{3}{2} mR^2\alpha$

where $a_{Ct} = R\alpha$ and $a_{Cn} = R\omega^2$ have been substituted.

b. Multiply Equation (3) by $\dot{\theta}$ and integrate, obtaining

4. $mgR\,(1 - \cos \theta) = \dfrac{3}{4} mR^2\omega^2 - \dfrac{3}{4} mR^2\omega_i^2$

Solve these equations for f and N, and show that the no-slip condition $f \leqq \mu N$ requires that

$$\sin \theta \leqq \mu(7 \cos \theta - 4 - 3r\omega_i^2/g)$$

Note that for very low ω_i, this is easily satisfied if μ is not too small and s (and therefore θ) is not too large. But, for example, if $\omega_i^2 = g/R$, then the cyl-

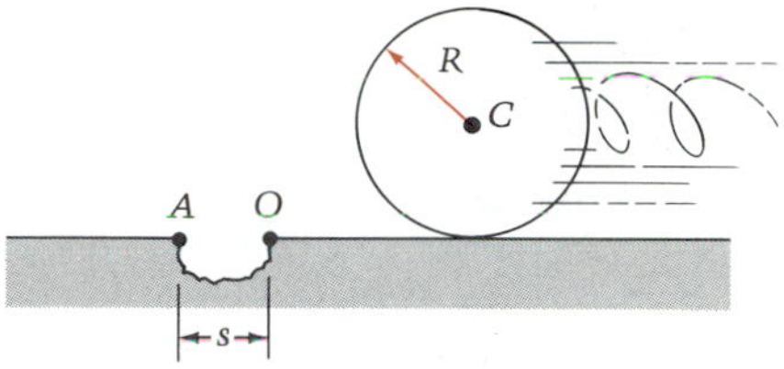

Figure P4.196(a)

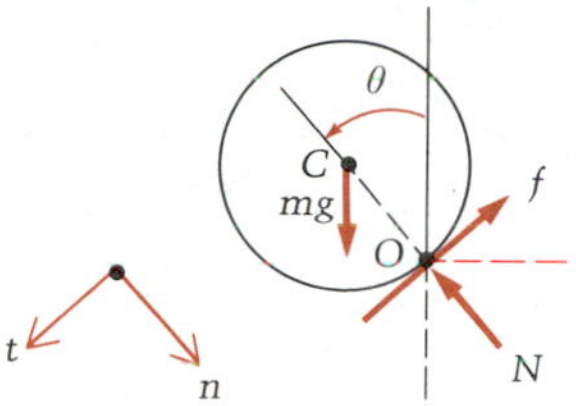

Figure P4.196(b)

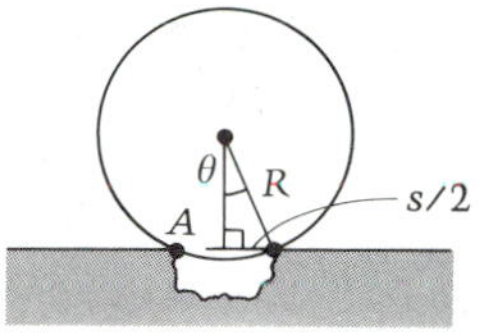

Figure P4.196(c)

inder will slip regardless of the value of μ because then the inequality cannot hold. Note too that if the cylinder is not slipping just prior to impact, it has not slipped at all.

c. Next, use Figure P4.196(c) to compute the angle θ for which the cylinder will strike the left corner A of the depression, and show that no slip will have occurred at any time during the pivoting if

$$\frac{s}{2R} \leqq \mu\left(7\sqrt{1-\left(\frac{s}{2R}\right)^2} - 4 - 3R\omega_i^2/g\right)$$

Finally, use the computer to create data for plots of the minimum μ required for no slip at O versus $R\omega_i^2/g$ for three values of $s/2R$: 0.1, 0.2, 0.5. Draw the three curves on the same graph.

Extra Credit Project Problem ► Chapter 4

4.197 Construct a round object with a challenging I_{zz}^{C} to calculate (as an example, see Figure P4.197(a)). On an inclined plane (see Figure P4.197(b)), roll your object down a 5-ft, 15° grade and with a stop-watch measure the descent time. Do this twice and average the times. Then explain the experiment, calculate the expected time, compare with the actual time, and give possible reasons for the difference in a brief report. (Note: It is fun to do all the students' experiments in the same session.)

Figure P4.197(a)

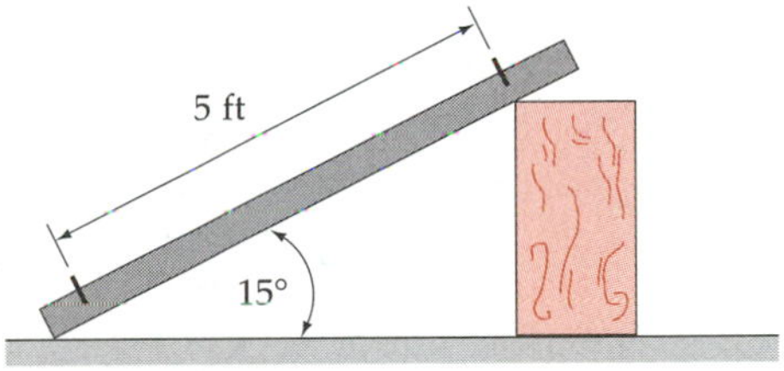

Figure P4.197(b)

Summary ► Chapter 4

In this chapter we have developed compact forms for the right-hand sides of moment equations for plane motion of a rigid body. The most general forms we have studied are, for an arbitrary point P,

$$\Sigma \mathbf{M}_P = (I_{xz}^{C}\,\alpha - I_{yz}^{C}\,\omega^2)\hat{\mathbf{i}} + (I_{yz}^{C}\,\alpha + I_{xz}^{C}\,\omega^2)\hat{\mathbf{j}} + I_{zz}^{C}\,\alpha\hat{\mathbf{k}} + \mathbf{r}_{PC} \times m\mathbf{a}_C$$

and

$$\Sigma \mathbf{M}_P = (I_{xz}^P\,\alpha - I_{yz}^P\,\omega^2)\hat{\mathbf{i}} + (I_{yz}^P\,\alpha + I_{xz}^P\,\omega^2)\hat{\mathbf{j}} + I_{zz}^P\,\alpha\hat{\mathbf{k}} + \mathbf{r}_{PC} \times m\mathbf{a}_P$$

where moments and products of inertia are defined by

$$I_{zz} = \int (x^2 + y^2)\, dm$$

$$I_{xz} = -\int xz\, dm \qquad \text{and} \qquad I_{yz} = -\int yz\, dm$$

and for which there are the very useful parallel-axis theorems

$$I_{zz}^P = I_{zz}^C + m(\bar{x}^2 + \bar{y}^2)$$

and

$$I_{xz}^P = I_{xz}^C - m\bar{x}\bar{z} \qquad \text{and} \qquad I_{yz}^P = I_{yz}^C - m\bar{y}\bar{z}$$

Except for the topic of balancing of rotating bodies (Section 4.7) we have restricted our attention to situations in which the products of inertia vanish, usually because of the body having an xy plane of symmetry. In those cases,

$$\Sigma \mathbf{M}_P = I_{zz}^C\,\alpha\,\hat{\mathbf{k}} + \mathbf{r}_{PC} \times m\mathbf{a}_C$$

and

$$\Sigma \mathbf{M}_P = I_{zz}^P\,\alpha\,\hat{\mathbf{k}} + \mathbf{r}_{PC} \times m\mathbf{a}_P$$

Important special cases are

a. Translation (in which every point has the same acceleration, $\mathbf{a}$, and of course $\alpha = 0$), for which:

$$\Sigma \mathbf{M}_P = \mathbf{r}_{PC} \times m\mathbf{a}$$

and so

$$\Sigma \mathbf{M}_C = \mathbf{0}$$

b. Summing moments at the mass center C:

$$\Sigma \mathbf{M}_C = I_{zz}^C\alpha\hat{\mathbf{k}}$$

or, more simply,

$$\Sigma M_C = I_C\alpha$$

c. P is a pivot (body rotates about a fixed axis), so that $\mathbf{a}_P = \mathbf{0}$:

$$\Sigma \mathbf{M}_P = I_{zz}^P\alpha\hat{\mathbf{k}}$$

or, more simply,

$$\Sigma M_P = I_P\alpha$$

It is more important to realize that while we have a number of options as to form for the moment equation, one moment equation plus the force equation, $\Sigma \mathbf{F} = m\mathbf{a}_C$, are all we can bring to bear independently for a given (free) body. That is, the situation is the same as in statics: we may sum moments wherever we like, but the two vector equations — one force and one moment — give us all the independent relationships involving external forces on the body. Many practical problems are solved by augmenting these equations with kinematic constraint conditions that can be invoked to generate relationships between $\mathbf{a}_C$ and α.

A body rotating about a fixed axis is said to be statically balanced if the mass center is located on the axis. It is said to be dynamically balanced (no bearing reactions induced by the rotation) if in addition the products of inertia associated with the rotation axis all vanish. Industrial equipment and automobile tires are modified by the addition of "balance weights" so as to ensure these conditions.

REVIEW QUESTIONS ▶ Chapter 4

True or False?

These questions all refer to rigid bodies in plane motion.

1. Euler's second law enables us to study the rotational motion of rigid bodies.
2. The moment of inertia is always positive, whereas the products of inertia can have either sign.
3. The formula $m\ell^2/12$ gives the exact value of the moment of inertia of a slender rod about a lateral axis through its mass center.
4. Euler's second law, $\Sigma\mathbf{M}_O = \dot{\mathbf{H}}_O$, is valid only in an inertial frame (meaning that the position vectors and velocities inherent in $\mathbf{H}_O$, the origin O, and the time derivative are all taken in an inertial frame).
5. In $I_{zz}^P = I_{zz}^C + md^2$, the quantity d is the distance between the points P and C. (C is in the reference plane, whereas P is *any* point of the body.)
6. Euler's second law, $\Sigma\mathbf{M}_O = \dot{\mathbf{H}}_O$, applies to deformable bodies, liquids, and gases, as well as to rigid bodies.
7. If Ⓘ represents the instantaneous center of zero velocity, then $\Sigma M_{Ⓘ} \neq I_{Ⓘ}\alpha$ in general.
8. Products of inertia are not found in the equations of plane motion.
9. $\Sigma\mathbf{M}_C = \dot{\mathbf{H}}_C$ is just as general as $\Sigma\mathbf{M}_O = \dot{\mathbf{H}}_O$, where O is fixed in an inertial frame.
10. In translation problems, the moments of external forces and couples taken about any point add to zero.
11. Suppose you buy a new set of automobile tires and a dynamic balance is performed on each wheel by adding weights in two planes (inner and outer rims). The products of inertia I_{xz} and I_{yz} have thus been eliminated, which otherwise would have caused bearing reactions and vibration.
12. $\Sigma M_C = I_C\alpha$ applies to deformable as well as to rigid bodies, as long as they are in plane motion.
13. For two bodies $\mathcal{B}_1$ and $\mathcal{B}_2$, the sum of the equations $\Sigma\mathbf{F} = m_i\mathbf{a}_{C_i}$ written for each will be $\Sigma\mathbf{F} = m\mathbf{a}_C$ for the combined body.
14. If the bodies of Question (13) are turning relative to each other, it makes no sense to talk about a combined $\Sigma M_C = I_C\alpha$ equation.

Answers: 1. T 2. T 3. F 4. T 5. F 6. T 7. T 8. F 9. T 10. F 11. T 12. F 13. T 14. T

5 SPECIAL INTEGRALS OF THE EQUATIONS OF PLANE MOTION OF RIGID BODIES: WORK-ENERGY AND IMPULSE-MOMENTUM METHODS

5.1 **Introduction**

5.2 **The Principle(s) of Work and Kinetic Energy**

Kinetic Energy of a Rigid Body in Plane Motion
An Alternative Form for Kinetic Energy
Derivation of the Principle $W = \Delta T$; Power and Work of Systems of Forces and Couples
Restriction of $W = \Delta T$ to a Rigid Body; A Notable Exception
Computing the Work Done by Various Types of Forces and Moments
Examples Solved by the Principle $W = \Delta T$
Two Subcases of the Work and Kinetic Energy Principle
Potential Energy, Conservative Forces, and Conservation of Mechanical Energy

5.3 **The Principles of Impulse and Momentum**

The Equations of Impulse and Momentum for the Rigid Body in Plane Motion
Conservation of Momentum
Impact
The Center of Percussion

SUMMARY
REVIEW QUESTIONS

5.1 Introduction

Just as in Chapter 4, the framework here is rigid bodies in plane motion. But we shall focus our attention now on problems which most efficiently can be attacked by using work-and-kinetic-energy and/or impulse-and-momentum principles. We shall employ these principles, rather broadly stated in Chapter 2, taking advantage of the simple forms that kinetic energy and angular momentum take when the body is rigid and constrained to plane motion.

In Chapter 2 we defined the kinetic energy of a body to be the sum of the kinetic energies of the particles making up the body; that is, $T = \frac{1}{2}\Sigma m_i v_i^2$. Because we have found in Chapter 3 that velocities of different points in a rigid body are related through the body's angular velocity, the reader should not be surprised to find kinetic energy for such a body to be expressible in terms of the velocity of one point and the angular velocity. Moreover, in Chapter 2 we observed that, for a body in general, change in kinetic energy equals work of external *and* internal forces. But for a rigid body the net work of internal forces vanishes, so that the work W in $W = \Delta T$ is the work only of external forces. We shall derive this work-and-kinetic-energy relationship directly from the force and moment equations (Euler's laws) as studied in Chapter 4, but it is helpful to recall the discussion of Chapter 2 and note the consistency of that material with the result we shall develop here.

The relationship between angular impulse and angular momentum developed in Section 5.3 takes on a quite useful form for a rigid body in plane motion, owing to the fact, as shown in Chapter 4, that the angular momentum can be expressed then in terms of inertia properties and angular velocity. Thus we shall find ourselves in a position to evaluate sudden changes in rates of turning for colliding bodies and to study quantitatively the relationship between the spin rate and the arm-trunk configuration of a skater.

It is very important for the reader to always keep in mind that the principle of work and kinetic energy and the principles of impulse and momentum do not stand as principles somehow separate from Newton's laws or their extensions to bodies of finite size, Euler's laws. Rather, here it will be seen, as was observed before in Chapter 2, that these relationships, which involve velocities, are really just special first integrals of the more fundamental second-order expressions relating forces and accelerations. Thus the principles of this chapter allow us to begin our solutions halfway between accelerations and positions. They therefore involve velocities but not accelerations.

5.2 The Principle(s) of Work and Kinetic Energy

Kinetic Energy of a Rigid Body in Plane Motion

There is a principle, derived from the equations of motion, that will help us to solve for unknowns of interest in kinetics problems. In this section we shall see that this principle arises from first deriving and then differentiating the kinetic energy of the body.

Kinetic energy, which we have examined in Chapter 2, is usually denoted by the letter T; for any body or system of bodies, it is defined as the summation of $\frac{1}{2}(dm)v^2$ over all its elements of mass:

$$T = \frac{1}{2} \int (\mathbf{v} \cdot \mathbf{v})\, dm \tag{5.1}$$

In this section we need to specialize Definition (5.1) for a rigid body $\mathcal{B}$ in plane motion. To this end we kinematically relate the velocity $\mathbf{v}$ of the differential mass to the velocity of the mass center C. Using the fact that $\mathbf{v}$ is at all times equal to the velocity of its companion point in the reference plane containing C (see Figure 5.1), we may write

$$\mathbf{v} = \mathbf{v}_C + \omega\hat{\mathbf{k}} \times (x\hat{\mathbf{i}} + y\hat{\mathbf{j}}) = \mathbf{v}_C + \omega(-y\hat{\mathbf{i}} + x\hat{\mathbf{j}})$$

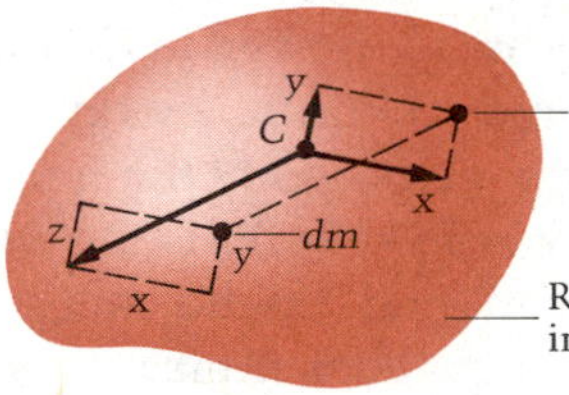

Figure 5.1

We note that the x and y axes are fixed in $\mathcal{B}$ with their origin at C. Forming v^2, that is, $\mathbf{v} \cdot \mathbf{v}$, we have

$$\mathbf{v} \cdot \mathbf{v} = \mathbf{v}_C \cdot \mathbf{v}_C + \omega^2(x^2 + y^2) + 2\omega\mathbf{v}_C \cdot (x\hat{\mathbf{j}} - y\hat{\mathbf{i}}) \tag{5.2}$$

Thus the kinetic energy becomes, substituting (5.2) into (5.1),

$$T = \frac{1}{2}\mathbf{v}_C \cdot \mathbf{v}_C \int dm + \frac{\omega^2}{2} \int (x^2 + y^2)\, dm$$
$$+ \omega(\mathbf{v}_C \cdot \hat{\mathbf{j}}) \int x\, dm - \omega(\mathbf{v}_C \cdot \hat{\mathbf{i}}) \int y\, dm$$

Recognizing the moment of inertia integral in the second term, we obtain

$$T = \frac{1}{2} m v_C^2 + \frac{1}{2} I_{zz}^C \omega^{2*} \tag{5.3}$$

in which $\mathbf{v}_C \cdot \mathbf{v}_C = |\mathbf{v}_C|^2 = v_C^2$, the square of the magnitude of the velocity of C.

Question 5.1 Why do $\int x\, dm$ and $\int y\, dm$ both vanish?

We note that the (scalar) kinetic energy has two identifiable *parts (not components!)*: one relating to the motion of the mass center C ($T_v = \frac{1}{2}mv_C^2$)

* Henceforth, we shall use the abbreviation I_C for I_{zz}^C throughout this chapter.

Answer 5.1 By the definition of the mass center.

and the other to the motion of the body relative to C ($T_\omega = \frac{1}{2}I_C\omega^2$). This clear division of T even exists in general motion of rigid bodies (that is, in three dimensions), though there are more terms in T_ω then.

Example 5.1

Calculate the kinetic energy of the round rolling body $\mathcal{B}$ in Figure E5.1, which has mass m, radius R, and radius of gyration k_C with respect to the z_C axis. The mass center C (see Figure E5.1) lies at the geometric center.

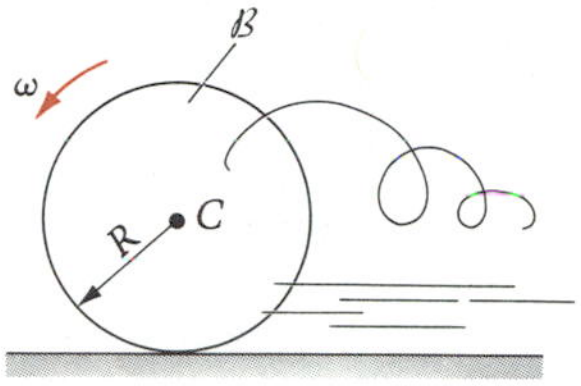

Figure E5.1

Solution

$$T = \frac{1}{2} m v_C^2 + \frac{1}{2} \overset{mk_C^2}{\cancel{I_C}} \omega^2$$

$$= \frac{1}{2} m[(R\omega)^2 + k_C^2\omega^2]$$

$$= \frac{mR^2\omega^2}{2}\left[1 + \left(\frac{k_C}{R}\right)^2\right]$$

Note that if $\mathcal{B}$ is a solid cylinder, then $k_C = R/\sqrt{2}$ and

$$T = \frac{mR^2\omega^2}{2}\left(1 + \frac{1}{2}\right)$$

In this case, two-thirds of the kinetic energy rests in the translation term of T ($\frac{1}{2}mv_C^2$).

If $\mathcal{B}$ is a ring (or hoop), however, $I_C = mR^2$ so that $k_C = R$ and

$$T = \frac{mR^2\omega^2}{2}(1 + 1)$$

and this time half the kinetic energy is in each of the translational and rotational terms.

Example 5.2

Work Example 5.1 for the case when the mass center C is offset by a distance r from the geometric center Q of a round rolling body $\mathcal{B}$. (See Figure E5.2.)

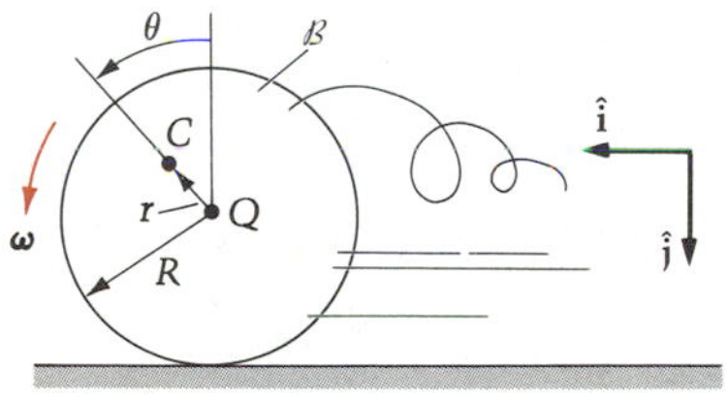

Figure E5.2

Solution

In order to use our equation for kinetic energy,

$$T = \frac{1}{2} m v_C^2 + \frac{1}{2} I_C\omega^2$$

we must first calculate v_C^2:

$$\mathbf{v}_C = \mathbf{v}_Q + \dot{\theta}\hat{\mathbf{k}} \times \overset{r \sin\theta\hat{\mathbf{i}} - r\cos\theta\hat{\mathbf{j}}}{\cancel{\mathbf{r}_{QC}}}$$

$$= (R\dot{\theta} + r\dot{\theta}\cos\theta)\hat{\mathbf{i}} + r\dot{\theta}\sin\theta\hat{\mathbf{j}}$$

Therefore

$$v_C^2 = R^2\dot{\theta}^2 + r^2\dot{\theta}^2 + 2Rr\dot{\theta}^2\cos\theta$$

Substituting, we get

$$T = \frac{mR^2\dot{\theta}^2}{2}\left[1 + \frac{2r}{R}\cos\theta + \left(\frac{r}{R}\right)^2 + \left(\frac{k_C}{R}\right)^2\right]$$

Note that if $r = 0$, the answer agrees as it should with Example 5.1.

An Alternative Form for Kinetic Energy

There is an alternative means of writing the kinetic energy T of a rigid body in plane motion by making use of the instantaneous center of zero velocity Ⓘ (see Figure 5.2):

$$T = \overbrace{\frac{1}{2}mv_C^2}^{T_v} + \overbrace{\frac{1}{2}I_C\omega^2}^{T_\omega}$$

$$= \frac{1}{2}m(d\omega)^2 + \frac{1}{2}I_C\omega^2$$

$$= \frac{1}{2}(I_C + md^2)\omega^2$$

Thus by using the parallel-axis theorem we obtain

$$T = \frac{1}{2}I_{Ⓘ}\omega^2 \tag{5.4}$$

The translational (T_v) and rotational (T_ω) terms composing the scalar T are thus seen to collapse into the one term $\frac{1}{2}I_{Ⓘ}\omega^2$ if we choose to work with Ⓘ instead of C.

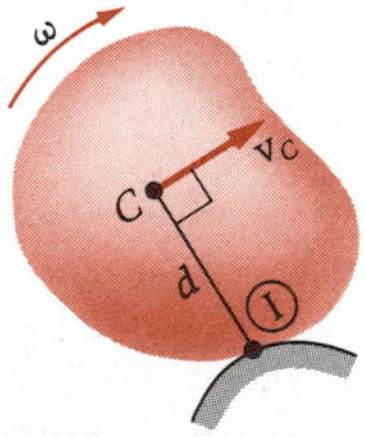

Figure 5.2

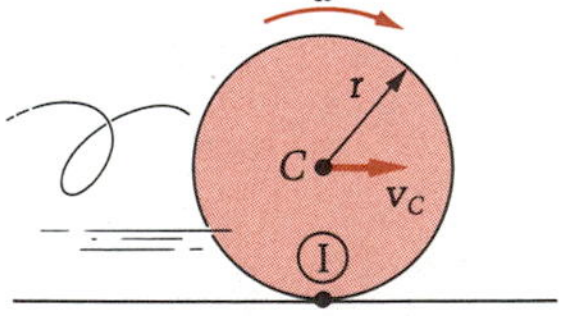

Figure 5.3

As an example, we consider a rolling cylinder again (see Figure 5.3):

$$T = T_v + T_\omega = \frac{1}{2}mv_C^2 + \frac{1}{2}I_C\omega^2$$

$$= \frac{1}{2}m(r\omega)^2 + \frac{1}{2}\left(\frac{1}{2}mr^2\right)\omega^2$$

$$= \frac{1}{2}mr^2\omega^2\left(1 + \frac{1}{2}\right) = \frac{3}{4}mr^2\omega^2$$

We have noted that two-thirds of the cylinder's kinetic energy is associated with the "translational part" of T and one-third with the "rotational part." If we now use Ⓘ, we get *all* of T at once:

$$T = \frac{1}{2} I_{Ⓘ}\omega^2 = \frac{1}{2}\underbrace{\Bigg[\underbrace{\frac{1}{2} mr^2}_{I_C} + \underbrace{mr^2}_{\text{transfer term}}\Bigg]}_{I_{Ⓘ}} \omega^2 = \frac{3}{4} mr^2\omega^2 \qquad \text{(as above)}$$

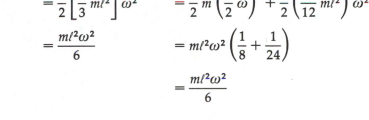

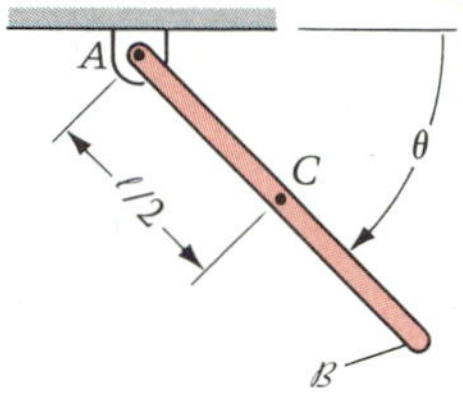

Figure 5.4

As a second example of the use of Equation (5.4), consider the slender rod $\mathcal{B}$ swinging about a pivot at A as shown in Figure 5.4. The kinetic energy of $\mathcal{B}$ may be found in either of two ways:

$$T = \frac{1}{2} I_{Ⓘ}\omega^2 \qquad\qquad T = \frac{1}{2} mv_C^2 + \frac{1}{2} I_C\omega^2$$

$$= \frac{1}{2}\left[\frac{1}{3} m\ell^2\right]\omega^2 \qquad\qquad = \frac{1}{2} m\left(\frac{\ell}{2}\omega\right)^2 + \frac{1}{2}\left(\frac{1}{12} m\ell^2\right)\omega^2$$

$$= \frac{m\ell^2\omega^2}{6} \qquad\qquad = m\ell^2\omega^2\left(\frac{1}{8} + \frac{1}{24}\right)$$

$$= \frac{m\ell^2\omega^2}{6}$$

Derivation of the Principle $W = \Delta T$; Power and Work of Systems of Forces and Couples

Returning now to the derivation of our principle, we next compute the rate of change of kinetic energy:

$$\frac{dT}{dt} = \frac{d}{dt}\left(\frac{1}{2} m\mathbf{v}_C \cdot \mathbf{v}_C + \frac{1}{2} I_C\omega\hat{\mathbf{k}} \cdot \omega\hat{\mathbf{k}}\right)$$

$$= \frac{1}{2} m\underbrace{(\mathbf{a}_C \cdot \mathbf{v}_C + \mathbf{v}_C \cdot \mathbf{a}_C)}_{2\mathbf{a}_C \cdot \mathbf{v}_C} + \frac{1}{2} I_C\underbrace{(\alpha\hat{\mathbf{k}} \cdot \omega\hat{\mathbf{k}} + \omega\hat{\mathbf{k}} \cdot \alpha\hat{\mathbf{k}})}_{2\alpha\hat{\mathbf{k}} \cdot \omega\hat{\mathbf{k}}}$$

Therefore

$$\frac{dT}{dt} = m\mathbf{a}_C \cdot \mathbf{v}_C + (I_C\alpha\hat{\mathbf{k}}) \cdot \omega\hat{\mathbf{k}} \tag{5.5}$$

Recalling that $\Sigma\mathbf{F} = m\mathbf{a}_C$ and that the z component of $\Sigma\mathbf{M}_C$ is $I_C\alpha$ for rigid bodies in plane motion, we may write

$$\frac{dT}{dt} = \Sigma\mathbf{F} \cdot \mathbf{v}_C + \Sigma\mathbf{M}_C \cdot \omega\hat{\mathbf{k}} \tag{5.6}$$

Question 5.2 Since $\Sigma \mathbf{M}_C$ can contain x and y components (see Equation 4.13) why may we substitute the *total* vector $\Sigma \mathbf{M}_C$ for just the z component $I_C \alpha \hat{\mathbf{k}}$ in Equation 5.5?

Our next goal is to get the individual external forces and couples acting on the body $\mathcal{B}$ into the equation. (See Figure 5.5.) Note the abbreviations $\mathbf{r}_{CP_1} = \mathbf{r}_1$, $\mathbf{r}_{CP_2} = \mathbf{r}_2$, and so on, of the vectors to the points of application of $\mathbf{F}_1$, $\mathbf{F}_2$, and so on. We assume that the external mechanical actions on the body arise from a system of forces ($\mathbf{F}_1, \mathbf{F}_2 \ldots$) and couples with moment vectors ($\mathbf{C}_1, \mathbf{C}_2, \ldots$), as shown in Figure 5.5. Further, we let ($\mathbf{v}_1, \mathbf{v}_2, \ldots$) be the velocities of the material points ($P_1, P_2, \ldots$) on which the forces act instantaneously.

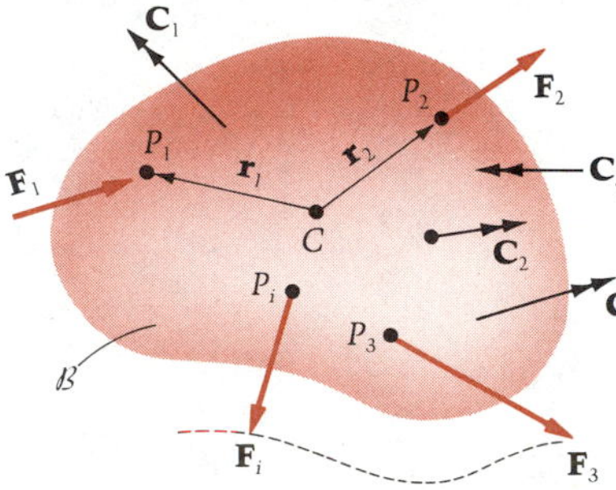

Figure 5.5

Clearly, then, the resultant of the external forces is

$$\Sigma \mathbf{F} = \mathbf{F}_1 + \mathbf{F}_2 + \cdots \tag{5.7}$$

and the moment of the $\mathbf{F}_i$'s and $\mathbf{C}_j$'s about C is

$$\Sigma \mathbf{M}_C = \mathbf{r}_1 \times \mathbf{F}_1 + \mathbf{r}_2 \times \mathbf{F}_2 + \cdots + \mathbf{C}_1 + \mathbf{C}_2 + \cdots \tag{5.8}$$

Substituting Equations (5.7) and (5.8) into (5.6) gives

$$\begin{aligned} \frac{dT}{dt} &= (\mathbf{F}_1 + \mathbf{F}_2 + \cdots) \cdot \mathbf{v}_C + (\mathbf{r}_1 \times \mathbf{F}_1 + \mathbf{r}_2 \times \mathbf{F}_2 + \cdots) \cdot \omega \hat{\mathbf{k}} \\ &\quad + (\mathbf{C}_1 + \mathbf{C}_2 + \cdots) \cdot \omega \hat{\mathbf{k}} \\ &= (\mathbf{F}_1 + \mathbf{F}_2 + \cdots) \cdot \mathbf{v}_C + \omega \hat{\mathbf{k}} \cdot (\mathbf{r}_1 \times \mathbf{F}_1 + \mathbf{r}_2 \times \mathbf{F}_2 + \cdots) \\ &\quad + (\mathbf{C}_1 + \mathbf{C}_2 + \cdots) \cdot \omega \hat{\mathbf{k}} \end{aligned} \tag{5.9}$$

Now since the dot and cross may be interchanged without altering the value of a scalar triple product,

$$\begin{aligned} \frac{dT}{dt} &= (\mathbf{F}_1 + \mathbf{F}_2 + \cdots) \cdot \mathbf{v}_C + (\omega \hat{\mathbf{k}} \times \mathbf{r}_1) \cdot \mathbf{F}_1 + (\omega \hat{\mathbf{k}} \times \mathbf{r}_2) \cdot \mathbf{F}_2 \\ &\quad + \cdots + (\mathbf{C}_1 + \mathbf{C}_2 + \cdots) \cdot \omega \hat{\mathbf{k}} \end{aligned} \tag{5.10}$$

Answer 5.2 Because $(\Sigma M_{Cx} \hat{\mathbf{i}} + \Sigma M_{Cy} \hat{\mathbf{j}}) \cdot \omega \hat{\mathbf{k}}$ is zero!

But the velocities of P_1 and C are related:

$$\mathbf{v}_{P_1} = \mathbf{v}_1 = \mathbf{v}_C + \omega\hat{\mathbf{k}} \times \mathbf{r}_1$$

so that

$$\frac{dT}{dt} = \mathbf{F}_1 \cdot \mathbf{v}_1 + \mathbf{F}_2 \cdot \mathbf{v}_2 + \cdots + (\mathbf{C}_1 + \mathbf{C}_2 + \cdots) \cdot \omega\hat{\mathbf{k}} \quad (5.11)$$

The right-hand side of Equation (5.11) is called the **power**, or **rate of work**, of the external system of forces and couples acting on the body. The power of a force is its dot product with the velocity of the point on which it acts; the power of a couple is its dot product with the angular velocity of the body on which it acts:

$$\text{Rate of work of force } \mathbf{F}_1 = \mathbf{F}_1 \cdot \mathbf{v}_1 = \text{power of } \mathbf{F}_1 \quad (5.12)$$

$$\text{Rate of work of couple } \mathbf{C}_1 = \mathbf{C}_1 \cdot \omega\hat{\mathbf{k}} = \text{power of } \mathbf{C}_1 \quad (5.13)$$

Hence one form of the principle of this section is

$$\text{Power} = \frac{dT}{dt} \quad (5.14)$$

or

$$P = \dot{T}$$

Integrating, we obtain another principle.*

$$\int_{t_1}^{t_2} P\,dt = T(t_2) - T(t_1) = T_2 - T_1$$

or

$$W = \Delta T = \left(\frac{1}{2}\,mv_C^2 + \frac{1}{2}\,I_C\omega^2\right)\Bigg]_{t_1}^{t_2} \quad (5.15)$$

where the integral of the power is called the **work** W of the external forces and couples. It is the work done by the $\mathbf{F}_i$'s and $\mathbf{C}_i$'s on the body between the two times t_1 and t_2. Hence we have a principle that can be stated in words:

Work done by external forces and couples on $\mathcal{B}$ = Change of kinetic energy of $\mathcal{B}$

Restriction of $W = \Delta T$ to a Rigid Body; A Notable Exception

It is essential to recognize that our derivation of the principle of work and kinetic energy *depends crucially on the body being rigid.* In fact the work of external forces on a deformable body is *not* in general equal to the change in its kinetic energy. That is the case even when the "deformable" body is

* Sometimes (t_i, t_f) is used to denote the time interval, rather than (t_1, t_2); the subscripts stand for "initial" and "final" values.

composed of several individually rigid parts. However, there are a number of special circumstances, usually easy to recognize, for which the principle is valid for such a system of rigid bodies. To give an example for which this is true, suppose we have two rigid bodies, $\mathcal{B}_1$ and $\mathcal{B}_2$, making up the system, and suppose the bodies are connected by a pin (or hinge) with negligible friction as shown in Figure 5.6.

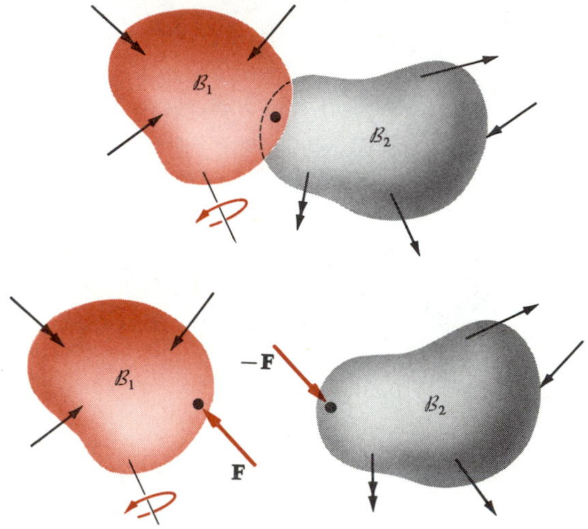

Figure 5.6

Let **F** be the force exerted by $\mathcal{B}_1$ on $\mathcal{B}_2$ at the pin, and consequently $-\mathbf{F}$ is the force exerted by $\mathcal{B}_2$ on $\mathcal{B}_1$. Furthermore, let

$\mathbf{v}$ = common velocity of attachment points in the two bodies

P_{E_1} = power (rate of work) of forces acting on $\mathcal{B}_1$ that are also external to system

P_{E_2} = power of forces acting on $\mathcal{B}_2$ that are also external to system

$T_{\mathcal{B}_1}$ = kinetic energy of $\mathcal{B}_1$

$T_{\mathcal{B}_2}$ = kinetic energy of $\mathcal{B}_2$

Now if we apply Equations (5.11) and (5.14) to each of the bodies, we obtain

$$P_{E_1} + (-\mathbf{F}) \cdot \mathbf{v} = \frac{dT_{\mathcal{B}_1}}{dt}$$

and

$$P_{E_2} + \mathbf{F} \cdot \mathbf{v} = \frac{dT_{\mathcal{B}_2}}{dt}$$

which may be added to yield

$$P_{E_1} + P_{E_2} = \frac{d}{dt}\left(T_{\mathcal{B}_1} + T_{\mathcal{B}_2}\right)$$

or

$$P = \frac{dT}{dt}$$

where P is the power of the external forces on the system and T is the kinetic energy of the system.

With friction in the pin, however, we also would have interactive couples $\mathbf{C}$ and $-\mathbf{C}$, and the sum of their work rates would be

$$\mathbf{C} \cdot (\omega_{\mathcal{B}_2} - \omega_{\mathcal{B}_1})\hat{\mathbf{k}}$$

which in general would *not* vanish.* This net rate of work of friction couples would be negative, reflecting the fact that the friction will reduce the kinetic energy of the system. We can expect the principle of work (of external forces) and kinetic energy to be valid for a system of rigid bodies whenever the interaction of the bodies leads neither to dissipation of mechanical energy by friction nor to a storing of energy as in a spring. When in doubt, follow the procedure we have just been through — that is, apply Equation (5.14) to each of the bodies, add the equations, and see whether the rates of work of interactive forces cancel out.

Computing the Work Done by Various Types of Forces and Moments

Before we can put Equation (5.15) to use, it is essential to demonstrate how to compute the work W done on $\mathcal{B}$ by a number of common types of forces and moments:

Type 1: $\mathbf{F}_1$ is constant. In this case, as in Chapter 2,

$$W = \int \mathbf{F}_1 \cdot \mathbf{v}_1 \, dt = \mathbf{F}_1 \cdot \int \mathbf{v}_1 \, dt \tag{5.16}$$

Type 2: $\mathbf{F}_1$ acts on the same point P_1 of $\mathcal{B}$ throughout its motion.† In this case,

$$W = \int_{t_i}^{t_f} \mathbf{F}_1 \cdot \mathbf{v}_1 \, dt = \int_{t_i}^{t_f} \mathbf{F}_1 \cdot \frac{d\mathbf{r}_1}{dt} \, dt = \int_{\mathbf{r}(t_i)}^{\mathbf{r}(t_f)} \mathbf{F}_1 \cdot d\mathbf{r}_1 \tag{5.17}$$

where $\mathbf{r}_{OP_1} = \mathbf{r}_1$ and i and f denote starting (initial) and ending (final) times and positions. It is true, of course, that the velocity $\mathbf{v}_1$, which combines with $\mathbf{F}_1$ to produce its power, is at each instant the derivative of *some* position vector. If the force acts on *different* material points of $\mathcal{B}$ at different times throughout a motion (such as friction from a brake), however, the path integral $\int \mathbf{F}_1 \cdot d\mathbf{r}_{OP_1}$ has no real functional utility and the general $\int \mathbf{F}_1 \cdot \mathbf{v}_1 \, dt$ must be used.

Type 3: $\mathbf{F}_1$ is due to gravity. This is an example of *both* Types 1 *and* 2. Thus, letting z be positive downward, we get

$$W = \int mg\hat{\mathbf{k}} \cdot d\mathbf{r}_{OC} = mg\hat{\mathbf{k}} \cdot \int d\mathbf{r}_{OC}{}^{**}$$

* It would vanish, of course, if the friction were enough to prevent relative rotation so that $\omega_{\mathcal{B}_1} = \omega_{\mathcal{B}_2}$; then the system would behave as a single rigid body!

† Which was *necessarily* the case in Chapter 2!

** The work done by *any* constant force $\mathbf{F}$ always acting on the same point with position vector $\mathbf{r}$ is thus $\mathbf{F} \cdot [\mathbf{r}(t_f) - \mathbf{r}(t_i)]$.

Expressing the differential of the position vector in terms of rectangular cartesian coordinates, we get

$$d\mathbf{r}_{OC} = dx_C\hat{\mathbf{i}} + dy_C\hat{\mathbf{j}} + dz_C\hat{\mathbf{k}}$$

and substituting we obtain a simple result for the work of gravity:

$$W = mg\int_{z_{C_1}}^{z_{C_2}} dz_C = mg(z_{C_2} - z_{C_1}) = mgh \tag{5.18}$$

as we observed in Chapter 2. Note that gravity does positive work if the body moves downward. (Indeed, a good rule of thumb to remember is that a force does positive work if it "gets to move in the direction it wants to" — that is, it has a component in the direction of the motion of the point on which it acts. If it does not, it does negative work during the motion of that point.)

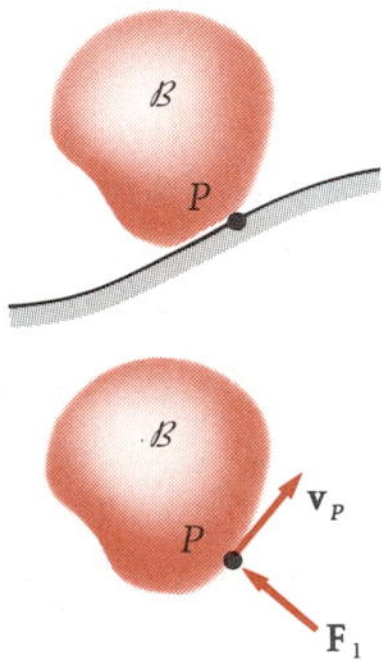

Figure 5.7

Type 4: $\mathbf{F}_1$ is the normal force exerted at the point of contact on a rigid body that is maintaining contact with a fixed surface, whether rolling or slipping. Note in the lower portion of Figure 5.7 that the normal force $\mathbf{F}_1$ is always perpendicular to the velocity of P. That is,

$$W = \int \mathbf{F}_1 \cdot \mathbf{v}_P\, dt = 0$$

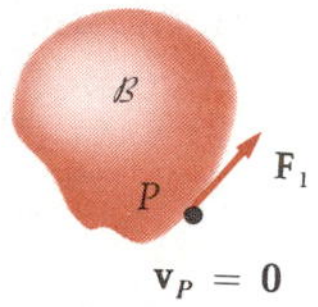

Figure 5.8

Type 5: $\mathbf{F}_1$ is the friction force exerted at the point of contact when a rigid body rolls on a fixed surface (Figure 5.8). This time, the force $\mathbf{F}_1$ (which may or may not be zero) does zero work because it always acts on a point of zero velocity:

$$W = \int \mathbf{F}_1 \cdot \overset{0}{\cancel{\mathbf{v}_P}}\, dt = 0$$

Type 6: $\mathbf{F}_1$ is the force in a linear spring connected to the same two points P and Q of bodies $\mathcal{B}$ and $\mathcal{R}$ during an interval of their motions. (See Figure 5.9.) We denote:

k = spring modulus (which when multiplied by the stretch yields the force in the linear spring)

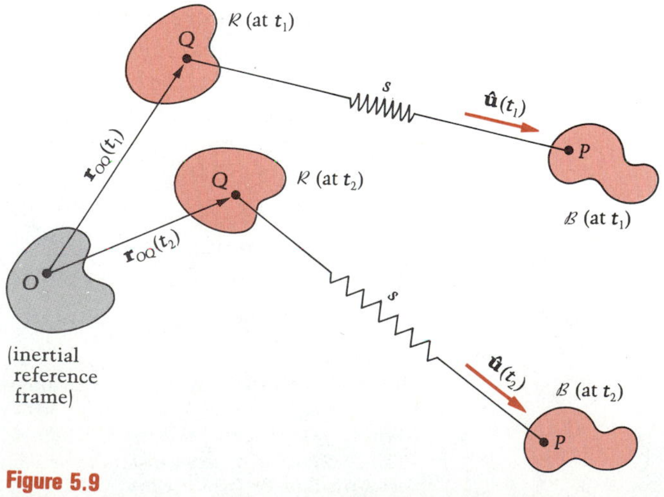

Figure 5.9

ℓ_u = unstretched length

δ = stretch ($\delta < 0$ if compressed)

$\hat{\mathbf{u}}$ = unit vector along spring toward body $\mathcal{B}$

We first note that the work of spring S on body $\mathcal{B}$ is

$$W_{s \text{ on } \mathcal{B}} = \int_{t_1}^{t_2} \mathbf{F} \cdot \mathbf{v}_P \, dt = \int_{t_1}^{t_2} - k\delta \, \hat{\mathbf{u}} \cdot \mathbf{v}_P \, dt$$

Using

$$\mathbf{r}_{OP} = \mathbf{r}_{OQ} + (\ell_u + \delta)\hat{\mathbf{u}}$$

we may differentiate and obtain

$$\mathbf{v}_P = \mathbf{v}_Q + \dot{\delta}\hat{\mathbf{u}} + (\ell_u + \delta)\dot{\hat{\mathbf{u}}}$$

Therefore, substituting for $\mathbf{v}_P$, we get

$$W_{s \text{ on } \mathcal{B}} = \int_{t_1}^{t_2} - k\,\delta\hat{\mathbf{u}} \cdot [\mathbf{v}_Q + \dot{\delta}\hat{\mathbf{u}} + (\ell_u + \delta)\dot{\hat{\mathbf{u}}}] \, dt$$

$$= - \int_{t_1}^{t_2} k\,\delta\hat{\mathbf{u}} \cdot \mathbf{v}_Q \, dt - k \int_{t_1}^{t_2} \delta\dot{\delta} \, dt - k \int_{t_1}^{t_2} \delta(\ell_u + \delta)\hat{\mathbf{u}} \cdot \dot{\hat{\mathbf{u}}} \, dt$$

Since the derivative of a unit vector is perpendicular to the unit vector, the last integral vanishes and we obtain

$$W_{s \text{ on } \mathcal{B}} = -W_{s \text{ on } \mathcal{R}} - k \int_{\delta_1}^{\delta_2} \delta \, d\delta$$

Thus

$$W_{s \text{ on } \mathcal{B}} + W_{s \text{ on } \mathcal{R}} = W_{s \text{ on system of } (\mathcal{B}+\mathcal{R})} = \frac{k}{2}(\delta_1^2 - \delta_2^2) \qquad (5.19)$$

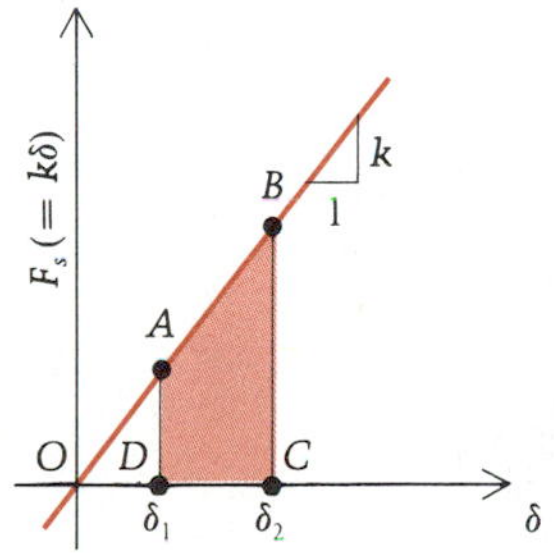

Figure 5.10

If Q is fixed in the inertial reference frame, the work of S on $\mathcal{B}$ alone is given by the right side of (5.19);* if Q moves, however, we can only say that the total work on *both* bodies by S is given by $(k/2)(\delta_1^2 - \delta_2^2)$.

We note from the spring's force-stretch diagram (Figure 5.10) that the work done by the spring is in fact the negative of the change in energy E stored in it; namely, in stretching from δ_1 to δ_2,

$$E = (\text{area of triangle } OCB) - (\text{are of } ODA)$$

$$= \frac{k}{2}(\delta_2^2 - \delta_1^2)$$

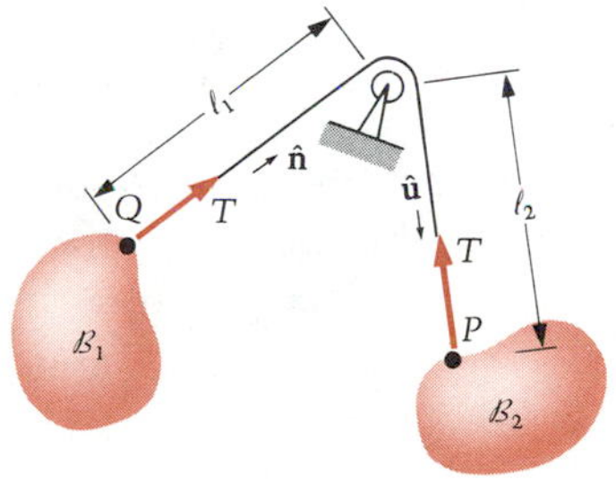

$$W_{\text{by cable on } \mathcal{B}_1} = \int T\hat{\mathbf{n}} \cdot \mathbf{v}_Q dt$$

$$W_{\text{by cable on } \mathcal{B}_2} = \int (-T\hat{\mathbf{u}}) \cdot \mathbf{v}_P dt$$

Figure 5.11

Type 7: We now consider the work done by the force in an inextensible cable (or rope, string, cord) connected to two points P and Q of bodies $\mathcal{B}_1$ and $\mathcal{B}_2$ during an interval of their motions (Figure 5.11). The cable under consideration may pass over one or more light, frictionless pulleys

* And its work on $\mathcal{R}$ is of course then zero.

between the bodies, but it is assumed to remain taut throughout the motion.

The work done by the cable tension on the system of $\mathcal{B}_1$ *plus* $\mathcal{B}_2$ is zero, which we proceed to prove as follows:

We write $\mathbf{v}_Q$ and $\mathbf{v}_P$ in terms of their components parallel and perpendicular to the cord:

$$\mathbf{v}_Q = \mathbf{v}_{Q\parallel} + \mathbf{v}_{Q\perp}$$
$$\mathbf{v}_P = \mathbf{v}_{P\parallel} + \mathbf{v}_{P\perp}$$

Noting that the perpendicular components $\mathbf{v}_{Q\perp}$ and $\mathbf{v}_{P\perp}$ have zero dot products with the unit vectors $\hat{\mathbf{n}}$ and $\hat{\mathbf{u}}$ (see Figure 5.11), we obtain for the works of the tensions:

$$W_{\text{by cable on } \mathcal{B}_1} = \int T\hat{\mathbf{n}} \cdot \mathbf{v}_{Q\parallel}\, dt = \int T\hat{\mathbf{n}} \cdot \frac{d\ell_1}{dt}(-\hat{\mathbf{n}})\, dt$$
$$= \int (-T)\, d\ell_1$$

and

$$W_{\text{by cable on } \mathcal{B}_2} = \int (-T\hat{\mathbf{u}}) \cdot \mathbf{v}_{P\parallel}\, dt = \int -T\hat{\mathbf{u}} \cdot \frac{d\ell_2}{dt}\hat{\mathbf{u}}\, dt$$
$$= \int (-T)\, d\ell_2$$

But by the cable's inextensibility, $d(\ell_1 + \ell_2) = 0$ so that $d\ell_2 = -d\ell_1$ and

$$W_{\text{by cable on } \mathcal{B}_2} = \int T\, d\ell_1 = -W_{\text{by cable on } \mathcal{B}_1}$$

so that

$$W_{\text{by cable on both bodies}} = 0$$

Type 8: We have a couple $\mathbf{C}$. In this case, the work of the couple in plane motion is given by

$$W = \int_{t_1}^{t_2} \mathbf{C} \cdot \boldsymbol{\omega}\, dt = \int_{t_1}^{t_2} C\hat{\mathbf{k}} \cdot \dot{\theta}\hat{\mathbf{k}}\, dt$$
$$= \int_{t_1}^{t_2} C\dot{\theta}\, dt \quad \text{or} \quad \int_{\theta_1}^{\theta_2} C\, d\theta \tag{5.20}$$

Thus if C is constant, the work of the couple is given by

$$W = C(\theta_2 - \theta_1) \tag{5.21}$$

That is, the work of C is the strength of the couple times the angle through which the body turns. As with the work of forces, the couple's work is positive if it "gets to move" in the direction in which it acts (or turns, in this case).

Examples Solved by the Principle $W = \Delta T$

We are now in a position to solve some problems by using the principle of work and kinetic energy. A number of examples follow. In the first, work is done only by gravity, and $W = \Delta T$ is used to supplement the equations of motion.

Example 5.3

Figure E5.3a

Find the pin reaction at O when the uniform bar in Figure E5.3a has fallen through 45° from rest.

Solution

We first find the angular speed ω_2 in the final (45°) position by using the principle of work and kinetic energy $W = \Delta T$.

Letting T_2 be the kinetic energy in the final position, and noting that the work done by gravity $= mgh = mg[\ell/(2\sqrt{2})]$, we find:

$$mg\frac{\ell}{2\sqrt{2}} = \frac{1}{2}\overbrace{\left(\frac{1}{3}m\ell^2\right)}^{I_{①} = I_O}\omega_2^2 \Rightarrow \omega_2^2 = \frac{3g}{\sqrt{2}\ell} \tag{1}$$

We must now return to the differential equations to obtain equations in the desired reaction. (Note that $W = \Delta T$ alone can only give us the solution to one scalar unknown!) In the final position, we have:

$$\Sigma \mathbf{F} = m\mathbf{a}_C$$

Expressing this equation in its tangential and normal components with the help of the free-body diagram (Figure E5.3b),

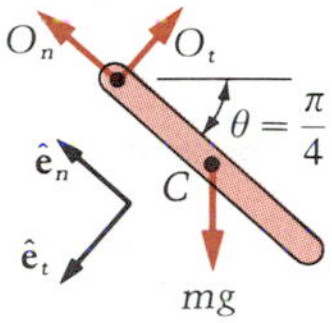

Figure E5.3b

$$\left(O_n - mg\frac{1}{\sqrt{2}}\right)\hat{\mathbf{e}}_n + \left(mg\frac{1}{\sqrt{2}} - O_t\right)\hat{\mathbf{e}}_t = m\mathbf{a}_C \tag{2}$$

But

$$\mathbf{a}_C = \overset{0}{\cancel{\mathbf{a}_O}} + \alpha\hat{\mathbf{k}} \times \mathbf{r}_{OC} - \omega^2\mathbf{r}_{OC}$$

and with $\hat{\mathbf{k}}$ defined as $\hat{\mathbf{e}}_t \times \hat{\mathbf{e}}_n$,

$$\mathbf{a}_C = \frac{\ell}{2}\alpha\hat{\mathbf{e}}_t + \frac{\ell}{2}\omega^2\hat{\mathbf{e}}_n$$

so that, from the $\hat{\mathbf{e}}_n$ component of Equation (2),

$$O_n - mg\frac{1}{\sqrt{2}} = m\frac{\ell}{2}\omega_2^2 = m\frac{\ell}{2}\frac{3g}{\sqrt{2}\ell}$$

where we have substituted ω_2^2 from Equation (1). Therefore, the normal component of the reaction is

$$O_n = \frac{5}{2\sqrt{2}}mg \tag{3}$$

Next, from the $\hat{\mathbf{e}}_t$ component of Equation (2),

$$mg\frac{1}{\sqrt{2}} - O_t = m\frac{\ell}{2}\alpha \tag{4}$$

Also, since point O is a pivot of the rod, we know:

$$\Sigma M_O = I_O\alpha$$

$$\frac{mg}{\sqrt{2}}\frac{\ell}{2} = \left(\frac{1}{3}m\ell^2\right)\alpha \Rightarrow \alpha = \frac{3g}{2\sqrt{2}\ell} \tag{5}$$

Substituting α from Equation (5) into (4) gives the tangential component of the reaction:

$$O_t = \frac{mg}{\sqrt{2}} - m\frac{\ell}{2}\frac{3g}{2\sqrt{2}\ell} = \frac{mg}{4\sqrt{2}}$$

Thus the pin reaction is

$$O_n\hat{\mathbf{e}}_n + O_t\hat{\mathbf{e}}_t = \frac{5mg}{2\sqrt{2}}\hat{\mathbf{e}}_n + \frac{mg}{4\sqrt{2}}\hat{\mathbf{e}}_t$$

$$= (1.77\hat{\mathbf{e}}_n + 0.177\hat{\mathbf{e}}_t)mg$$

In the second example, work is done only by a spring; however, calculating the final stretch is tricky.

EXAMPLE 5.4

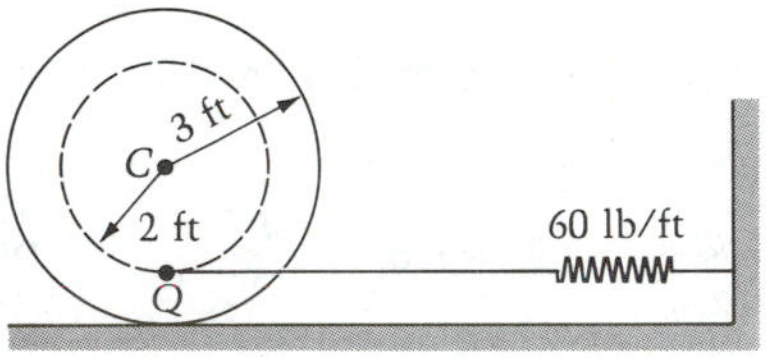

Figure E5.4a

One end of the linear spring in Figure E5.4a is attached to a thin inextensible cord that is lightly wrapped around a narrow groove in the wheel (mass = 1 slug, radius of gyration about center = 1.5 ft). If the wheel rolls, and starts from rest when the spring is stretched 1 ft, find the velocity of the center of the wheel when the center has moved 2 ft. The mass center of the wheel coincides with the geometric center.

Solution

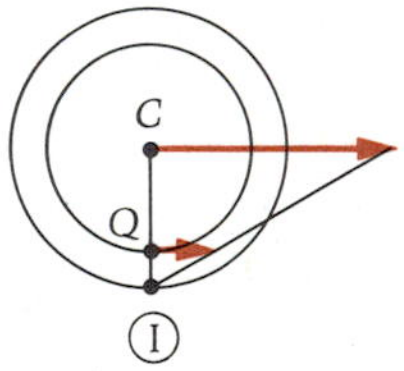

Figure E5.4b

We first note that the cord is not attached to a specific material point on the wheel. However, as time passes, the various "wrapping points" on the end of the straight portion of the cord (such as Q in Figure E5.4b) have, at every instant the cord is taut, the same velocity as the coincident point of the wheel at Q. Thus Equation (5.19) gives the work done on the wheel by the spring.

We have at all times, by kinematics (see the figure),

$$\dot{x}_C = 3\omega \qquad \text{and} \qquad \dot{x}_Q = 1\omega$$

so that

$$\dot{x}_C = 3\dot{x}_Q$$

or

$$x_C = 3x_Q$$

Thus the net shortening of the spring when C has moved 2 ft to the right is $\frac{2}{3}$ ft. Another way to see this is to let C move to the right the amount x_C. This compresses the spring (if it were able to do so!) the same amount, x_C. Then turn the wheel clockwise about C through the angle $\theta = x_C/R = x_C/3$ radians, until the correct point is on the ground. ("Correct" means the point that would be on the ground had the wheel rolled normally over to the final position.) The rotation wraps $r\theta = 2\theta = 2x_C/3$ of string around the inner radius and "takes back" $\frac{2}{3}x_C$ of the compression. Thus $\frac{1}{3}x_C = \frac{1}{3}(2)$ is the reduction in the original 1 ft of stretch, leaving $\frac{1}{3}$ ft, as before. Hence

$$\delta_1 = 1 \text{ ft} \qquad \text{and} \qquad \delta_2 = 1 - \frac{2}{3} = \frac{1}{3} \text{ ft}$$

We then find, noting that gravity does no work here,

$$W = \Delta T = T_2 - \cancelto{0}{T_1}$$

$$\frac{60}{2}\left[(1)^2 - \left(\frac{1}{3}\right)^2\right] = \frac{1}{2}(1)v_C^2 + \frac{1}{2}[1(1.5)^2](\cancelto{v_C/3}{\omega})^2$$

$$v_C = \sqrt{42.7} = 6.53$$

$$\mathbf{v}_C = 6.53 \rightarrow \text{ft/sec}$$

We note that when the center has moved 3 ft, then $\frac{3}{3} = 1 = x_Q$ and all the stretch is gone. At this time, the spring would simply drop out of the problem.

The next example is an actual application of $W = \Delta T$ from industry.

Example 5.5

This example involves a practical application in the antenna industry of the work and kinetic energy principle. The antenna positioner in Figure E5.5 is equipped with a mechanical stop spring so that if the elevation drive overruns its lower limit, the antenna motion (a pure rotation about the horizontal elevation rotation axis) will be arrested before the reflector strikes another part and is damaged.

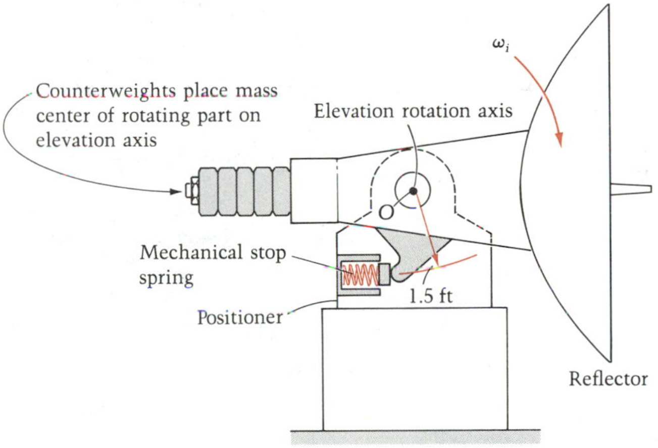

Figure E5.5

The elevation motor has an armature rotational mass moment of inertia of 0.01 lb-ft-sec² (or slug-ft²) and drives the reflector through a gear reducer with a 700 : 1 gear ratio. The combined moment of inertia of the reflector, its counterweights, and the supporting structure is 12,000 slug-ft² $= I_O$.

It is desired to arrest a rotational speed of 30°/sec during a rotation from contact to full stop of 3°. The radius from the elevation rotation axis to the stop spring is 1.5 ft. The spring is unstretched at initial contact and may be assumed to have linear load-deflection behavior. It is further assumed that the motor is switched off but remains mechanically coupled while the rotation is being arrested. Find:

a. The required stiffness of the spring.
b. The maximum force induced in it.
c. The rotational position when it sustains its maximum force.
d. The angular accelerations of the reflector and motor armature at the position of maximum force. (Are these the maximum accelerations?)

Solution

Since the spring is linear, its greatest force is the spring stiffness times the maximum deflection. This is also the position for which motion is completely arrested. At this position the kinetic energy has been brought to zero with the stop spring storing the energy; the principle $W = \Delta T$ gives

$$W = \frac{1}{2} k(\overset{0}{\cancel{\delta_i^2}} - \delta_f^2) = \Delta T = \frac{1}{2} I_O \overset{0}{\cancel{\omega_f^2}} - \frac{1}{2} I_O \omega_i^2$$

Note that point O is Ⓘ for the rotating body and that gravity does no work between contact and stop.

Question 5.3 Why does gravity do no work?

The values of δ_f, I_O, and ω_i needed in the equation are calculated as follows:

$$\begin{aligned} I_O &= \text{total moment of inertia at axis of rotation} \\ &= I_{\text{motor}_O} + I_{(\text{reflector, counterweights, structure})_O} \\ &= 0.01 \times 700^2 * + 12{,}000 = 16{,}900 \text{ slug-ft}^2 \end{aligned}$$

$$\omega_i = 30 \times \frac{\pi}{180} = 0.524 \text{ rad/sec}$$

$$\delta_f = 3 \times \frac{\pi}{180} \times 1.5 = 0.0785 \text{ ft}$$

(Note that over the very small angle of 3° the spring compression is approximately the arclength $R\theta$.)

Solving for the spring's stiffness, we get

$$k = \frac{I_O \omega_i^2}{\delta_f^2} = \frac{16{,}900 \times 0.524^2}{0.0785^2}$$

$$= 753{,}000 \text{ lb/ft}$$

The maximum spring force $= k\delta_f = 753{,}000 \times 0.0785 = 59{,}100$ lb. The rotational position is 3° beyond contact — that is, the position at full stop. The

Answer 5.3 Since the counterweights place the mass center on the elevation axis, the mass center does not move.

* As the reader may wish to prove, moments of inertia reflect through gear trains from input to output with the gear ratio squared as a factor; also, the torque increases (while the speed decreases) with the gear ratio as the multiplying factor.

angular acceleration of the reflector is

$$\alpha = \frac{\Sigma M_O}{I_O} = \frac{1.5 \times 59{,}100}{16{,}900} = 5.25 \text{ rad/sec}^2$$

and that of the motor armature is $5.25 \times 700^* = 3680$ rad/sec².

These are the maximum accelerations, since here the force (and torque) are greatest. In closing, we note that motor torque and friction, omitted in this problem for simplicity, limit the rebound in the actual case.

The next example illustrates work done by a force acting on different points of the body as time passes. A shortcut for calculating this work is presented.

Example 5.6

This example illustrates the work done by forces and couples belonging to Types 1, 2, and 8 on the preceding pages. The force **F** (52 lb) is applied to the uniform cylinder $\mathcal{C}$ at rest in Figure E5.6a at the left. (This type of force might be applied by a cord on a hub, as is suggested by Figure E5.6b.) If force **F** continues to act with the same magnitude and direction as the cylinder rolls, find:

a. The work done by **F** during transit to the dashed position

b. The velocity of C and the angular velocity of the cylinder in the dashed position

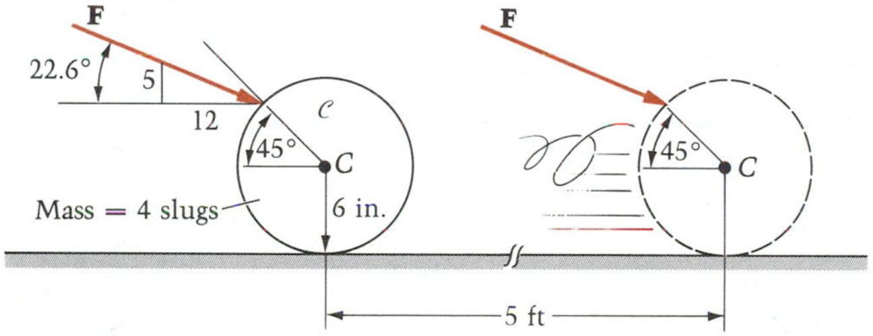

Figure E5.6a

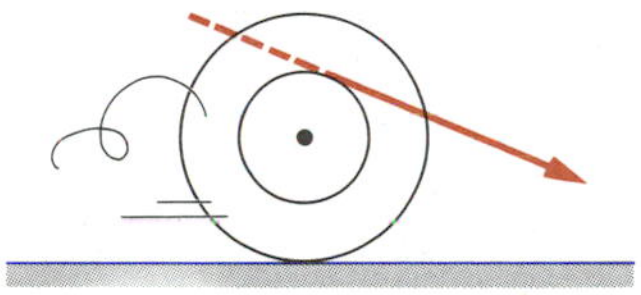

Figure E5.6b

Solution

We shall work part (a) in two ways. First, the definition of the work of **F** is

$$W = \int \mathbf{F} \cdot \mathbf{v}_Q \, dt$$

where Q is the point of $\mathcal{C}$ in contact with **F** at any time. The geometry in Figure E5.6c gives an angle of 45.1° between **F** and $\mathbf{v}_Q$, since

$$\alpha = \tan^{-1}\left(\frac{r/\sqrt{2}}{r + r/\sqrt{2}}\right) = 22.5^\circ$$

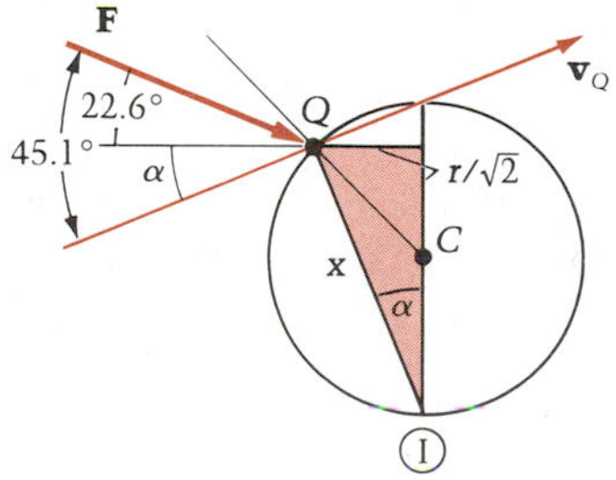

Figure E5.6c

* See preceding footnote.

Also,

$$r_{①Q} = \sqrt{\left(\frac{1}{\sqrt{2}}\right)^2 + \left(1 + \frac{1}{\sqrt{2}}\right)^2}\, r = 1.85r$$

Therefore

$$\mathbf{F} \cdot \mathbf{v}_Q = F(r_{①Q}\dot{\theta}) \cos 45.1°$$

and

$$W = \int (52 \cos 45.1°) 1.85r \frac{d\theta}{dt}\, dt$$

$$= (52 \cos 45.1°)(1.85) \left.\frac{\theta}{2}\right]_0^{5/0.5}$$

$$= 340 \text{ ft-lb}$$

A second and simpler approach is to note that **F** at Q may be moved to C, using the idea of resultants, as in Figure E5.6d. The force at Q is replaced by the force and couple at C which produce the same effect on the rigid body. Since, at C, force **F** always acts on the *same point of the body* (it didn't at Q!) we may write

$$W = \text{work of } \mathbf{F} \text{ at } Q = (\text{work of } \mathbf{F} \text{ at } C) + (\text{work of couple on } \mathcal{C})$$

$$= (F \cos 22.6°)x_C + (Fr \sin 22.4°)\theta = (0.923F)5 + (0.191F)10$$

$$= 6.53F = 340 \text{ ft-lb} \qquad \text{(as before)}$$

Note that the work of a constant couple in plane motion is simply the moment of the couple times the angle through which the body turns.

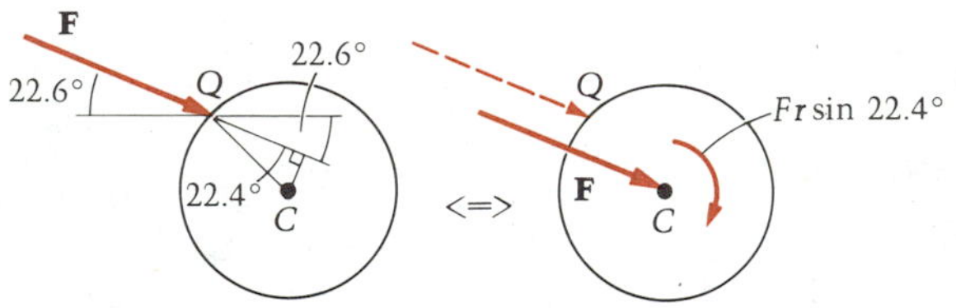

Figure E5.6d

For part (b) we equate the work to the change in the kinetic energy of $\mathcal{C}$:

$$W = \Delta T = \frac{1}{2} m v_{C_f}^2 + \frac{1}{2} I_C \omega_f^2 - 0 \qquad (\text{initial } T = 0)$$

$$340 = \frac{1}{2} 4 v_{C_f}^2 + \frac{1}{2}\left(\frac{1}{2} \cdot 4 \cdot \frac{1}{4}\right)\left(\frac{v_{C_f}}{1/2}\right)^2$$

$$v_{C_f} = \sqrt{\frac{340}{2+1}} = 10.6 \text{ ft/sec}$$

Note that the gravity, friction, and normal forces do no work in this problem, for the reasons given in Types 3, 4, and 5 of the text preceding the examples.

The hardest part of the next example is finding where the mass center is in the final position!

Example 5.7

The unstretched length of the spring in Figure E5.7a is $\ell_u = 0.3$ m. The initial angular velocity of body $\mathcal{A}$ in the top position is $\omega_i = 2.5$ ⤾ rad/s. There is enough friction to prevent slipping of $\mathcal{A}$ on $\mathcal{B}$ at all times. Determine the modulus of the spring that will cause $\mathcal{A}$ to stop in the $\varphi = 90°$ position.

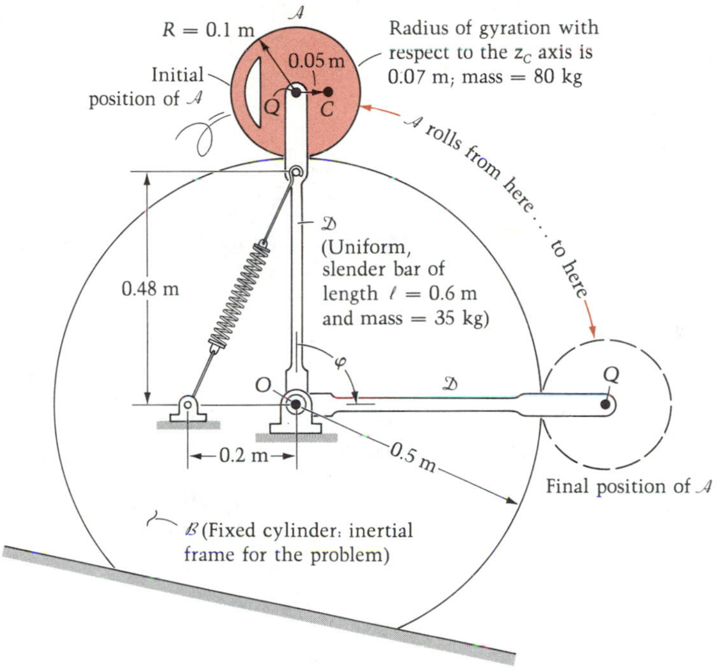

Figure E5.7a

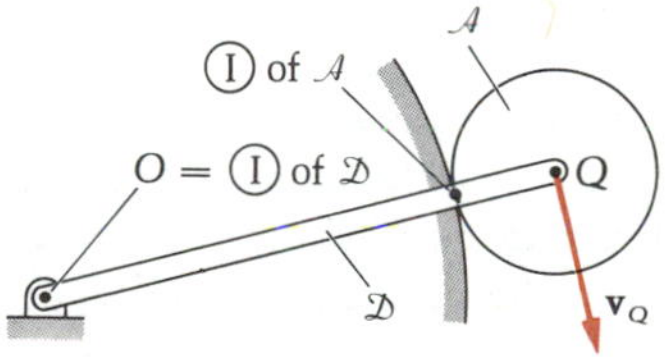

Figure E5.7b

Solution

Part of the work W in this problem is done by gravity. To express this work, we must determine where the mass center C is located when $\mathcal{A}$ reaches the final position. Reviewing the kinematics, we find that the velocity of the geometric center Q of $\mathcal{A}$ (see Figure E5.7b) is expressible in two ways:

1. As a point of $\mathcal{A}$, $\mathbf{v}_Q = \overset{0}{\cancel{\mathbf{v}_{Ⓘ}}} + R\dot{\theta}\hat{\mathbf{e}}_t$.
2. As a point of $\mathcal{D}$, $\mathbf{v}_Q = \overset{0}{\cancel{\mathbf{v}_{O}}} + \ell\dot{\varphi}\hat{\mathbf{e}}_t$.

Thus we see that $R\dot{\theta} = \ell\dot{\varphi}$. Integrating, we get

$$R\theta = \ell\varphi$$

in which the constant of integration is zero if we select $\theta = 0$ when $\varphi = 0$. Therefore when $\varphi = \pi/2$, we may find the orientation of body $\mathcal{A}$:

θ = angle that body $\mathcal{A}$ turns through in reference frame (angle seen by *stationary* observer in body $\mathcal{B}$)

$$= \frac{\ell}{R}\varphi = \frac{0.6}{0.1} \times \frac{\pi}{2} = 3\pi$$

And so the final position of C is to the left of Q (see Figure E5.7c). We can now write the work of gravity W_g because we now know the h moved through by C:

$$W_g = (mgh)_{\mathcal{D}} + (mgh)_{\mathcal{A}} = 35(9.81)(0.3) + 80(9.81)(0.6) = 574 \text{ J}$$

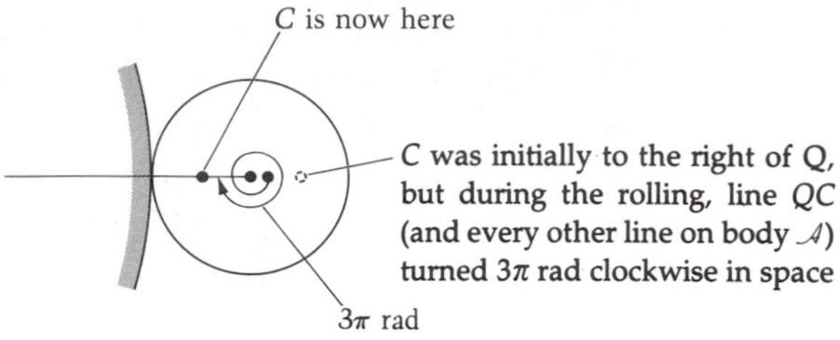
C is now here
C was initially to the right of Q, but during the rolling, line QC (and every other line on body 𝒜) turned 3π rad clockwise in space
3π rad

Figure E5.7c

The work done by the linear spring is *always* given by $(k/2)(\delta_i^2 - \delta_f^2)$:

$$W_s = \frac{k}{2}(\delta_i^2 - \delta_f^2) = \frac{k}{2}(0.220^2 - 0.380^2)$$

$$= -0.0480k \text{ J}$$

where k is our unknown and the initial and final stretches are computed as follows:

$$\ell_i = \ell_u + \delta_i \qquad \text{(unstretched length plus initial stretch = initial length of spring)}$$

and

$$\ell_f = \ell_u + \delta_f$$

so that the stretches are

$$\delta_i = \sqrt{0.2^2 + 0.48^2} - 0.3 = 0.520 - 0.300 = 0.220 \text{ m}$$

$$\delta_f = (0.48 + 0.2) - 0.3 = 0.680 - 0.300 = 0.380 \text{ m}$$

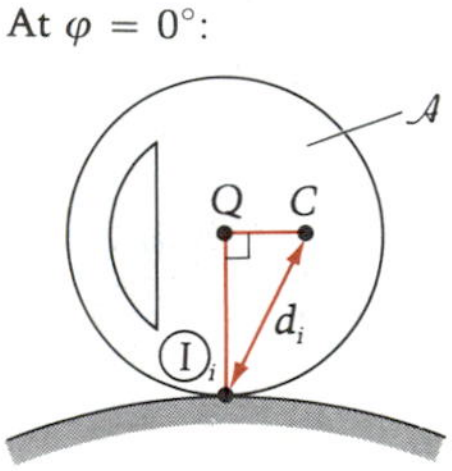
At φ = 0°:
𝒜
Q
C
d_i
I_i

Figure E5.7d

For the kinetic energy side of $W = \Delta T$, we need the moments of inertia; first we consider body $\mathcal{A}$:

$$I_C = mk_C^2 = 80(0.07^2) = 0.392 \text{ kg-m}^2$$

We shall use the "short form" of T — namely $T = \frac{1}{2}I_{\textcircled{I}}\omega^2$ (always valid whenever $\omega \neq 0$ in plane motion). Thus we need $I_{\textcircled{I}i}$ and $I_{\textcircled{I}f}$.* Note that when Ⓘ is a different point of a body in the initial and final positions, the value of $I_{\textcircled{I}}$ is generally different in the two configurations, as is the case in this problem. Using Figures E5.7d and E5.7e, we find:

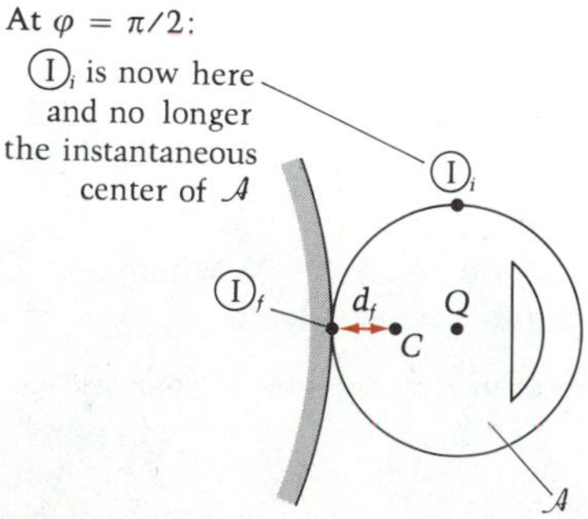
At φ = π/2:
I_i is now here and no longer the instantaneous center of 𝒜
I_i
I_f
d_f
Q
C
𝒜

Figure E5.7e

$I_{\textcircled{I}i} = I_C + md_i^2$	$I_{\textcircled{I}f} = I_C + md_f^2$
$= 0.392 + 80(0.1^2 + 0.05^2)$	$= 0.392 + 80(0.1 - 0.05)^2$
$= 1.39 \text{ kg} \cdot \text{m}^2$	$= 0.592 \text{ kg} \cdot \text{m}^2$

* Since $\omega_f = 0$, we do not have to calculate $I_{\textcircled{I}f}$ here, but we do so to illustrate the procedure in general.

Therefore the kinetic energies of $\mathcal{A}$ we need are

$$T_i^{\mathcal{A}} = \frac{1}{2} I_{①i}\omega_i^2 = \frac{1}{2}(1.39)2.5^2 = 4.34 \text{ J}$$

$$T_f^{\mathcal{A}} = \frac{1}{2} I_{①f}\omega_f^2 = \frac{1}{2}(0.592)0^2 = 0 \text{ J} \qquad \text{(since the final angular speed is to be zero)}$$

For the bar $\mathcal{D}$, $I_{①}$ is the same in any position since ① is point O, which is pinned to the reference frame. Therefore, using $v_Q = R\omega_{\mathcal{A}}$, we have

$$T_i^{\mathcal{D}} = \frac{1}{2} I_{①i}\omega_i^2 = \frac{1}{2}\left(\frac{m\ell^2}{3}\right)\left(\frac{v_{Qi}}{r_{①Q}}\right)^2 = \frac{1}{2}\left(\frac{35 \times 0.6^2}{3}\right)\left(\frac{0.1 \times 2.5}{0.6}\right)^2$$

$$= 0.365 \text{ J}$$

$$T_f^{\mathcal{D}} = \frac{1}{2} I_{①f}\omega_f^2 = \frac{1}{2}\left(\frac{m\ell^2}{3}\right)\left(\frac{v_{Qf}}{r_{①Q}}\right)^2 = 0 \qquad (\text{since } v_{Qf} = 0.1\omega_{\mathcal{A}f} = 0)$$

Applying the work and kinetic energy principle, we get

$$W = \Delta T$$

$$W_g + W_s = T_f - T_i = -T_i$$

$$574 - 0.0480k = 0 - (4.34 + 0.365)$$

$$k = 12{,}100 \text{ N/m}$$

This is equivalent to 829 lb/ft of stiffness in the U.S. system of units, since 1 lb/ft is the same stiffness as 14.6 N/m.

Question 5.4 What happens if k is larger than the calculated value? What happens if it is smaller?

In the next example work and kinetic energy is used to help determine the point where rolling stops and slipping starts.

Answer 5.4 Larger: the rod will not reach the horizontal position; smaller: the system passes through this position without stopping.

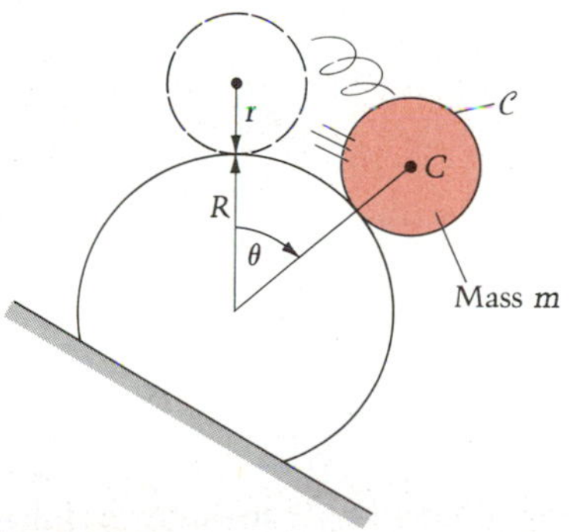

Figure E5.8a

Example 5.8

As we found in Example 5.3, sometimes it is useful to combine the work and kinetic energy principle with one or more of the differential equations of motion in order to obtain a desired solution. This example involves such a combination. The small cylinder $\mathcal{C}$ starts from rest at $\theta = 0$ in the dotted position (see Figure E5.8a) and begins to roll down the large cylinder. Find the angle θ_s at which slipping starts, and show that the small cylinder will *always* slip before it leaves the surface for a finite coefficient of friction.

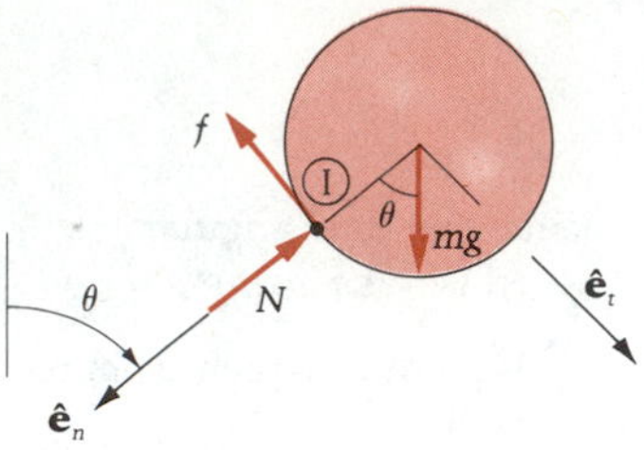

Figure E5.8b

Solution

Using the free-body diagram in Figure E5.8b, the equations of motion are

$$\Sigma F_n = mg\cos\theta - N = ma_{C_n} = \frac{mv_C^2}{R+r} = \frac{mr^2\omega^2}{R+r} \tag{1}$$

$$\Sigma F_t = mg\sin\theta - f = ma_{C_t} = m\ddot{s}_C = mr\alpha \tag{2}$$

$$\Sigma M_C = fr = I_C\alpha = \frac{mr^2}{2}\alpha \tag{3}$$

Just prior to slipping, the friction force $f \approx \mu N$ while a_C is still equal to $r\alpha$ and v_C is still equal to $r\omega$. Therefore the equations can be rewritten as

$$mg\cos\theta_s - N = \frac{mr^2\omega_s^2}{R+r} \tag{1a}$$

$$mg\sin\theta_s - \mu N = mr\alpha_s \tag{2a}$$

$$\mu N r = \frac{mr^2}{2}\alpha_s \tag{3a}$$

These equations may be supplemented with the work and kinetic energy equation for body $\mathcal{C}$, written between $\theta = 0$ and $\theta = \theta_s$:

$$W_g = mg(R+r)(1-\cos\theta_s) = \frac{1}{2}I_{①}\omega_s^2 = \frac{1}{2}\left(\frac{3}{2}mr^2\right)\omega_s^2 \tag{4}$$

Equations (1) to (4) may now be treated as four equations in the unknowns N, θ_s, ω_s^2, and α_s, where the last three quantities are the angle, angular velocity, and angular acceleration at slip. Solving them for θ_s yields the equation

$$7\mu\cos\theta_s - 4\mu = \sin\theta_s$$

Writing $\sqrt{1-\cos^2\theta_s}$ for $\sin\theta_s$, and then squaring and solving the resulting quadratic for $\cos\theta_s$, gives

$$\theta_s = \cos^{-1}\left(\frac{28\mu^2 + \sqrt{33\mu^2+1}}{1+49\mu^2}\right)$$

which plots as shown in Figure E5.8c.

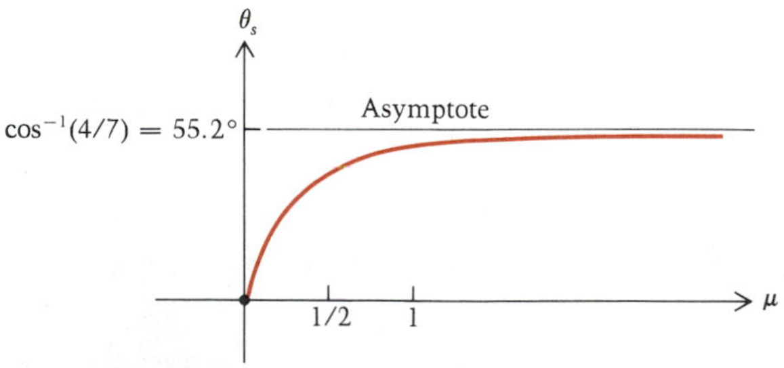

Figure E5.8c

The curve in the diagram gives the slipping angle as a function of the friction coefficient; this is *not* the angle at which body $\mathcal{C}$ leaves the surface. We note that if we were to look for the angle at which the small cylinder leaves the surface of the large cylinder, assuming no slipping has occurred, we would be trying to solve a problem with no solution if the friction coefficient is finite. The curve clearly shows that for $\mathcal{C}$ to reach the angle $\cos^{-1}(4/7)$, an infinite coefficient of friction is required. Since the solution to the "leaving without slip" problem is precisely $\cos^{-1}(4/7)$, shown below, then regardless of the friction coefficient (so long as it is finite) $\mathcal{C}$ will have to slip before it leaves.

Assuming now that the cylinder $\mathcal{C}$ leaves without having slipped, we obtain the (simpler) solution:

$$\text{Equation (1)} \Rightarrow mg\cos\theta_L - \underbrace{\cancel{N}}_{\text{0 at leaving!}} = \frac{mr^2\omega_L^2}{R+r}$$

$$\text{Equation (4)} \Rightarrow mg(R+r)(1-\cos\theta_L) = \frac{1}{2}\left(\frac{3}{2}mr^2\right)\omega_L^2$$

Eliminating ω_L gives

$$\theta_L = \cos^{-1}(4/7) = 55.2°$$

As we have noted, this solution is valid only for an infinite coefficient of friction between the cylinders. If $\mathcal{C}$ were a *particle* (no rotational kinetic energy) with a smooth surface, we would obtain (Example 2.13) $\theta_L = \cos^{-1}(2/3) = 48.2°$. Note the differences between these solutions.

Two Subcases of the Work and Kinetic Energy Principle

There is an important subcase of the principle of work and kinetic energy that we have already seen in Chapter 2. Using

$$\Sigma\mathbf{F} = \frac{d}{dt}(m\mathbf{v}_C)$$

we obtained

$$\int_{t_1}^{t_2} \Sigma\mathbf{F}\cdot\mathbf{v}_C\,dt = m\int_{t_1}^{t_2}\dot{\mathbf{v}}_C\cdot\mathbf{v}_C\,dt = \frac{1}{2}mv_C^2\bigg]_{v_C(t_1)}^{v_C(t_2)} \tag{5.22}$$

This principle states again that the work done by the external force resultant, when considered to act on the mass center, equals the change in the translational part of the kinetic energy:

$$\int \Sigma\mathbf{F}\cdot\mathbf{v}_C\,dt = \Delta T_v \tag{5.23}$$

The integral of Equation (5.6) is

$$\int \Sigma\mathbf{F}\cdot\mathbf{v}_C\,dt + \int \Sigma\mathbf{M}_C\cdot\omega\hat{\mathbf{k}}\,dt = \int\frac{dT}{dt}\,dt = \Delta T = \Delta T_v + \Delta T_\omega \tag{5.24}$$

If we subtract (5.23) from (5.24), we obtain yet another result:

$$\int \Sigma\mathbf{M}_C\cdot\omega\hat{\mathbf{k}}\,dt = \Delta T_\omega = \Delta\left(\frac{1}{2}I_C\omega^2\right)$$

or

$$\int \Sigma M_C\,d\theta = \Delta T_\omega \tag{5.25}$$

This second subprinciple says that the work done by the external *moments* (about C) on the body, as it turns in the inertial frame, is equal to the change in the *rotational part* of the kinetic energy. We may use the "total" $W = \Delta T$ principle or either of its two "subparts" (Figure 5.12).

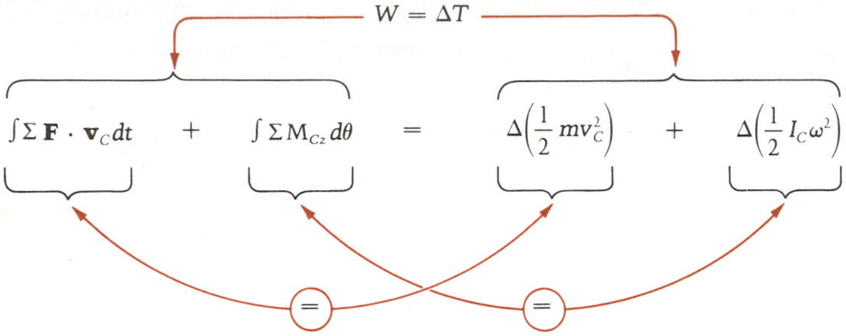

Figure 5.12

Let us now examine an example in which these subparts may be seen to add to the "total" $W = \Delta T$ equation:

EXAMPLE 5.9

As an illustration of the two subcases of the principle of work and kinetic energy, we consider the cylinder of mass m rolling down the inclined plane shown in Figure E5.9a. If the cylinder is released from rest, find the velocity v_C of its mass center as a function of the distance x_C traveled by C.

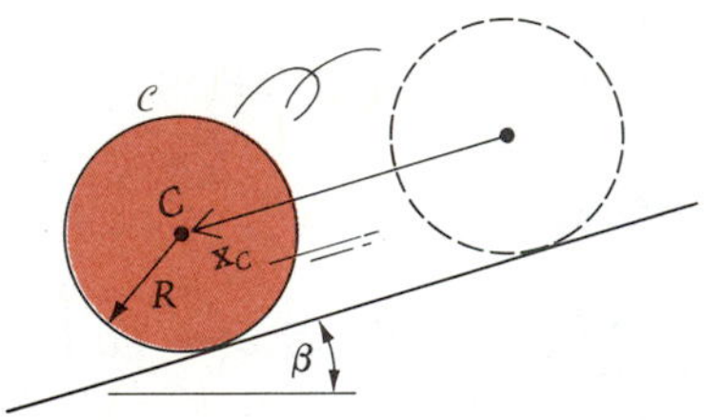

Figure E5.9a

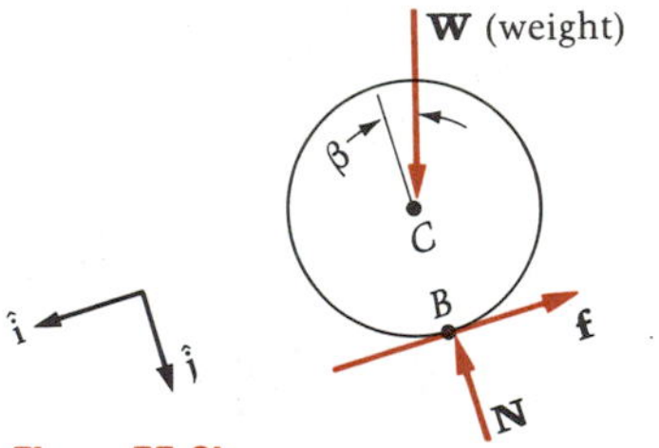

Figure E5.9b

Solution

Referring to the free-body diagram in Figure E5.9b, we see that the normal and friction forces do no work because, as the cylinder rolls on the incline, they always act on a point at rest. That is,

$$\int \mathbf{N} \cdot \mathbf{v}_B \, dt = 0 \qquad \text{and} \qquad \int \mathbf{f} \cdot \mathbf{v}_B \, dt = 0$$

Applying the principle that $W = \Delta T$, only the component of the gravity force $\mathbf{W}$ that acts parallel to the plane does any work:

$$\int (mg \sin \beta \hat{\mathbf{i}}) \cdot v_C \hat{\mathbf{i}} \; dt = \frac{1}{2} m v_C^2 + \frac{1}{2} I_C \omega^2 \tag{1}$$

Since $\mathbf{W}$ always acts on the same point (C) of $\mathcal{C}$, and since $dx_C/dt = v_C = R\omega$,

$$\int mg \sin \beta \, dx_C = \frac{1}{2} m v_C^2 + \frac{1}{2}\left(\frac{1}{2} mR^2\right) \frac{v_C^2}{R^2} \tag{2}$$

The mass center's velocity is therefore

$$\mathbf{v}_C = v_C\hat{\mathbf{i}} = \sqrt{\frac{4g \sin\beta\, x_C}{3}}\,\hat{\mathbf{i}} \tag{3}$$

Now suppose we apply Equation (5.23):

$$\int \Sigma\mathbf{F} \cdot \mathbf{v}_C\, dt = \Delta\left(\frac{1}{2}\, m v_C^2\right) \tag{4}$$

In this problem the resultant force acting on $\mathcal{C}$ is

$$\Sigma\mathbf{F} = \mathbf{W} + \mathbf{f} + \mathbf{N} = mg(\sin\beta\hat{\mathbf{i}} + \cos\beta\hat{\mathbf{j}}) - f\hat{\mathbf{i}} - N\hat{\mathbf{j}}$$

and Equation (4) becomes

$$\int mg \sin\beta\, dx_C - \int f v_C\, dt = \frac{1}{2}\, m v_C^2 \tag{5}$$

We see that, as expected, the friction force (though it does no *net* work) retards the motion of the mass center C while *turning* the cylinder, as can be seen from the *other* subcase of $W = \Delta T$:

$$\int \underbrace{\Sigma\mathbf{M}_C \cdot \boldsymbol{\omega}\, dt}_{\Sigma M_C\, d\theta} = \Delta\,\frac{1}{2}\, I_C\omega^2 \tag{6}$$

$$\int fR\, d\theta = \frac{1}{2}\left(\frac{1}{2}\, mR^2\right)\omega^2 \tag{7}$$

or

$$\int f\,\frac{d(R\theta)}{dt}\, dt = \int f v_C\, dt = \frac{1}{2}\left(\frac{1}{2}\, mR^2\right)\frac{v_C^2}{R^2} \tag{8}$$

And the sum of Equations (5) and (8) indeed gives Equation (2): the total $W = \Delta T$ equation!

Potential Energy, Conservative Forces, and Conservation of Mechanical Energy

In Section 2.4 we introduced the concept of **potential energy**, or the potential of a force. When the work done by a force on a body is independent of the path taken as the body moves from one configuration to another, the force is said to be **conservative** and the work is expressible as the decrease in a scalar function φ, the potential (energy). Thus as a body moves from a configuration at time t_1 to a second configuration at time t_2, the work done by an external conservative force is

$$W = \varphi(t_1) - \varphi(t_2)$$

or simply

$$W = \varphi_1 - \varphi_2$$

If all the external forces that do work on a rigid body are conservative and

φ is now the *sum* of the potentials of those forces, Equation (5.15) yields

$$\varphi_1 - \varphi_2 = W = \Delta T = T_2 - T_1$$

or

$$T_2 + \varphi_2 = T_1 + \varphi_1$$

or

$$T + \varphi = \text{constant}$$

which expresses the **conservation of mechanical energy**.

From Chapter 2 and earlier in this section we can easily identify two common conservative forces: (1) the constant force acting always on the same material point in the body and (2) the force exerted on a body by a linear spring attached at one end to the body and at the other to a point fixed in the inertial frame of reference.

In the case of the constant force, a potential is $\varphi = -\mathbf{F} \cdot \mathbf{r}$, where $\mathbf{r}$ is a position vector for the point of application. When the force is that exerted by gravity (weight) on a body near the surface of the earth,

$$\varphi = mgz$$

where h is the altitude of the mass center of the body.

For the linear spring, we recall that $\varphi = (k/2)\delta^2$, where k is the spring modulus, or stiffness, and δ is the stretch. It is important to recognize that when a spring is attached to, or between, two bodies that are both moving (relative to the inertial frame), then $(k/2)\delta^2$ is a potential for the two spring forces *taken together* (see Equation 5.19). That is, while neither of the forces acting on the bodies can be judged by itself to be conservative, the net work done on the two bodies by the two forces is expressible as a decrease in the potential, $\varphi = (k/2)\delta^2$. This is helpful in the analysis of problems in which we have two or more interacting rigid bodies. We have already noted earlier in this section that the work of the *external* forces on a system of rigid bodies is not in general equal to the change in kinetic energy of the system; this is because there may be net work done on the rigid bodies by the equal and opposite forces of interaction. Suppose now that our system is made up of two bodies joined by a spring, and suppose the spring forces are the *only* internal ones that produce net work on the system. We may then write $W = \Delta T$ for each rigid body. Upon adding these equations there results

(Work of forces external to system) + (Work of pair of spring forces)
= (Change in kinetic energy of system)

If the forces external to the system that do work are conservative, we may add the various potential energies associated with them to that for the pair of spring forces and conclude that

$$T + \varphi = \text{constant}$$

That is to say, in this case the mechanical energy of the *system* is conserved.

An example of a *nonconservative* force is sliding friction. A potential cannot be found for friction, since the work it does depends on the path taken by the body on which it acts. In this case, $W = \Delta T$ must be used, and it is seen to be more general than the principle of conservation of mechanical energy.

Example 5.10

Show that the same equation for the spring modulus in Example 5.7 is obtained by conservation of mechanical energy. (See Figure E5.10.)

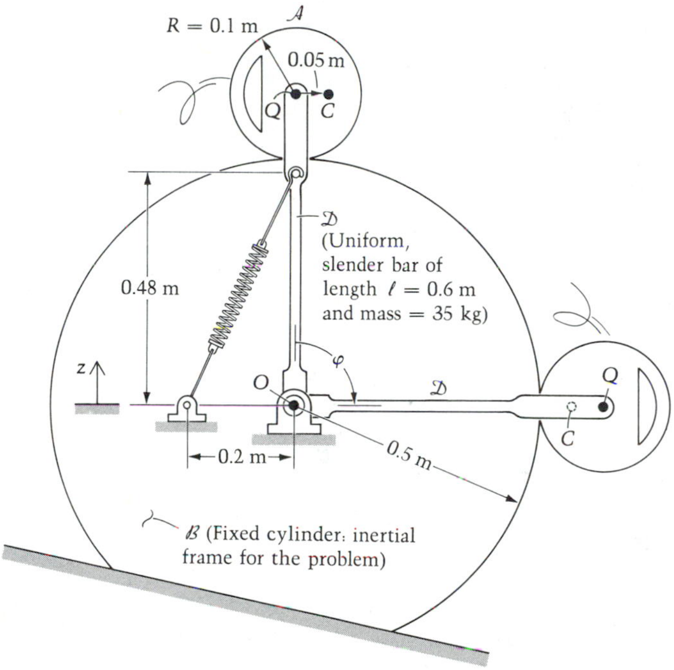

Figure E5.10

Solution

The potentials for gravity and for the spring are

$$\varphi_g = +m_{\mathcal{A}} g z_{C_{\mathcal{A}}} + m_{\mathcal{D}} g z_{C_{\mathcal{D}}} \qquad \varphi_{\text{spr}} = \frac{k\delta^2}{2}$$

Therefore, measuring z_C from O, we have

$$\begin{aligned} \varphi_{g_i} &= +80(9.81)(+0.6) + 35(9.81)(+0.3) \qquad & \varphi_{g_f} &= -mg(0) \\ &= +471 + 103 & & \\ &= +574 \text{ J} & &= 0 \end{aligned}$$

For the spring, using i for the initial and f for the final configuration, we have

$$\varphi_{\text{spr}_i} = \frac{k(0.22)^2}{2} = 0.0242k \text{ J} \qquad \varphi_{\text{spr}_f} = \frac{k(0.38)^2}{2} = 0.0722k \text{ J}$$

Thus, adding the potentials ($\varphi = \varphi_g + \varphi_{spr}$), we get

$$\varphi_i = 574 + 0.0242k \text{ J} \qquad \varphi_f = 0 + 0.0722k \text{ J}$$

The kinetic energies were $T_i = 4.34 + 0.365 = 4.71$ J and $T_f = 0$. Therefore

$$\varphi_i + T_i = \varphi_f + T_f$$

$$574 + 0.0242k + 4.71 = 0.0722k + 0$$

or, rearranging,

$$574 - 0.0480k = -4.71$$

This is the same final equation that resulted from $W = \Delta T$ in the earlier Example 5.7.

PROBLEMS ► Section 5.2

5.1 Find the kinetic energy of the system of bodies $\mathcal{B}_1$, $\mathcal{B}_2$, and $\mathcal{B}_3$ at an instant when the speed of $\mathcal{B}_1$ is 5 ft/sec. (See Figure P5.1.)

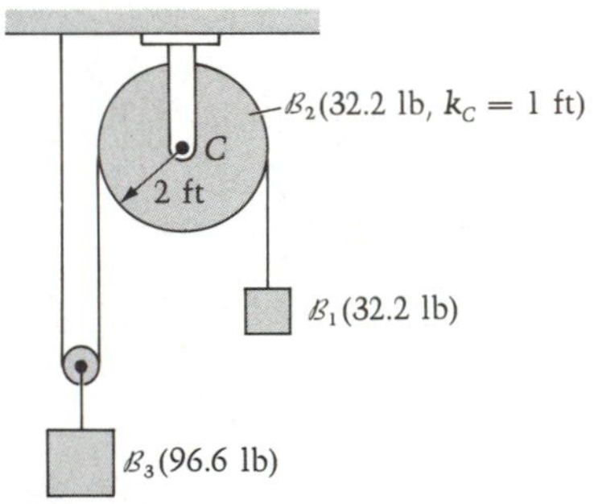

Figure P5.1

5.2 See Figure P5.2. (a) Explain why the friction force f does no work on the rolling cylinder $\mathcal{B}$ if the plane $\mathcal{I}$ is the reference frame. (b) If, however, $\mathcal{I}$ is the top surface of a moving block (dotted lines) and the reference frame is now the ground $\mathcal{G}$, does f then do work on $\mathcal{B}$? Why or why not?

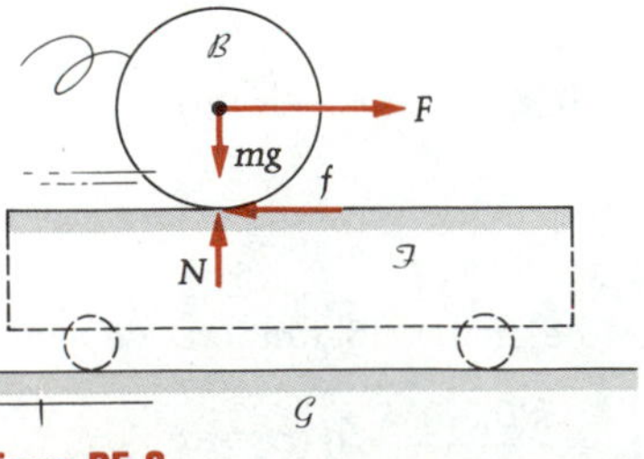

Figure P5.2

5.3 Upon application of the 10-N force F to the cord in Figure P5.3, the cylinder begins to roll to the right. After C has moved 5 m, how much work has been done by F?

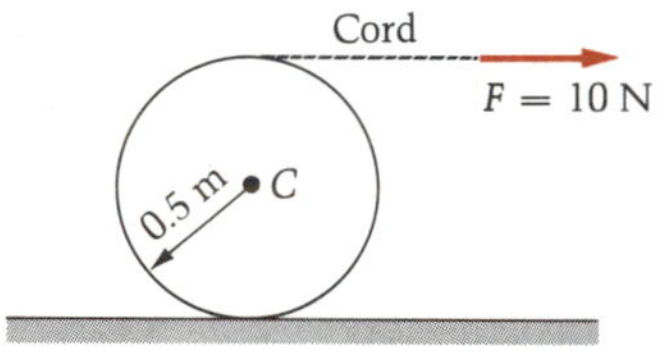

Figure P5.3

5.4 The suspended log shown in Figure P5.4 is to be used as a battering ram. At what angle θ should the ruffian release the log from rest so that it strikes the door at $\theta = 0$ with a velocity of 20 ft/sec?

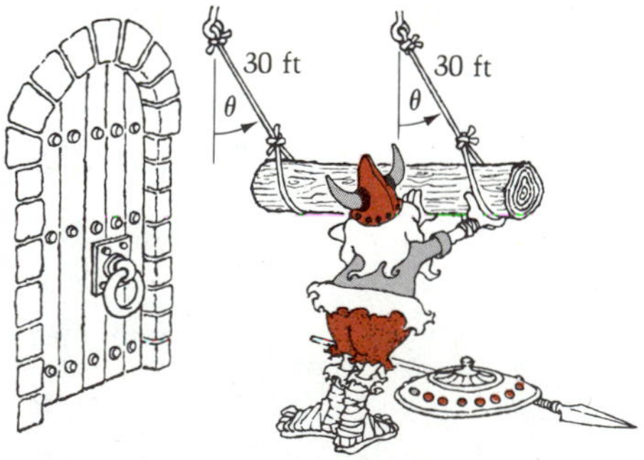

Figure P5.4

5.5 The 20-kg bar in Figure P5.5 has an angular velocity of 3 rad/s clockwise in the horizontal configuration shown. In that position the tensile force in the spring is

30 N. After a 90° clockwise rotation the angular velocity has increased to 4 rad/s. Determine the spring modulus k.

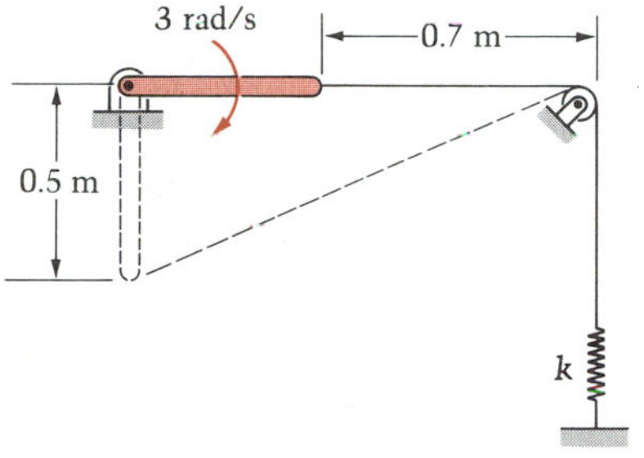

Figure P5.5

5.6 A uniform 40-lb sphere (radius = 1 ft) is released from rest in the position shown in Figure P5.6. If the sphere rolls (no slip), find its maximum angular speed.

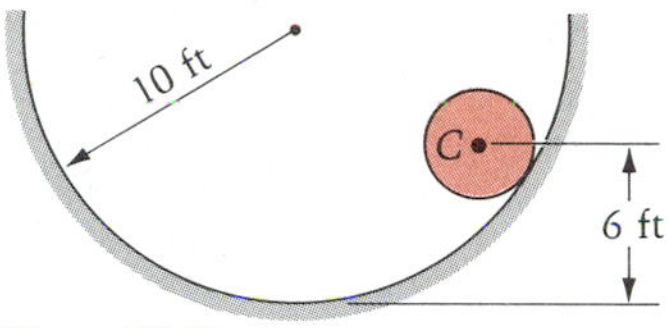

Figure P5.6

5.7 The unbalanced wheel of radius 2 ft and weight 64.4 lb shown in Figure P5.7 has a mass center moment of inertia of 6 slug-ft². In position 1, with C above O, the wheel has a clockwise angular velocity of 2 rad/sec. The wheel then rolls to position 2, where OC is horizontal. Determine the angular velocity of the wheel in position 2.

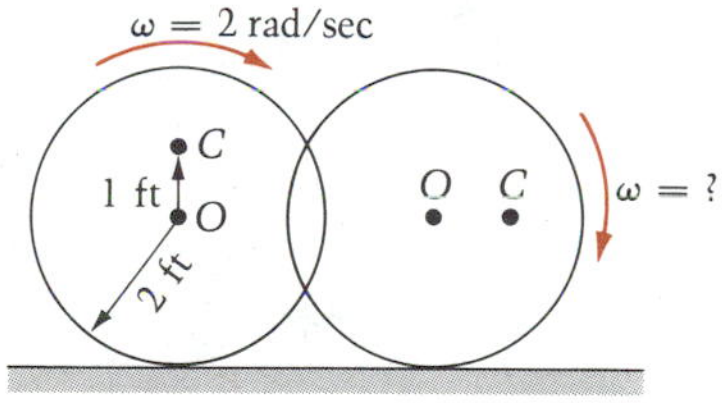

Figure P5.7

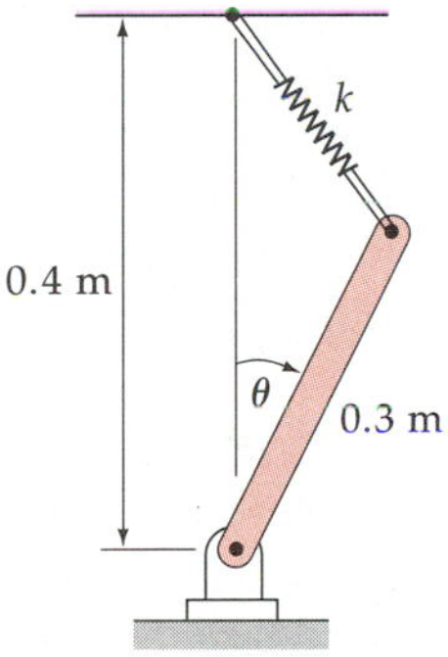

Figure P5.8

5.8 Determine the spring modulus that will allow the 2-kg bar in Figure P5.8 to arrive at the position $\theta = 90°$ at zero angular velocity if it passed through the vertical (where the spring is compressed 0.1 m) at 8 rad/s ⮌.

5.9 Bar $\mathcal{B}_1$ is smoothly pinned to the support at A and smoothly pin-jointed to $\mathcal{B}_2$ at B. (See Figure P5.9.) End D slides on a smooth horizontal surface. If D starts from rest at $\theta = \theta_0$, determine the angular velocities of the rods just before they become horizontal.

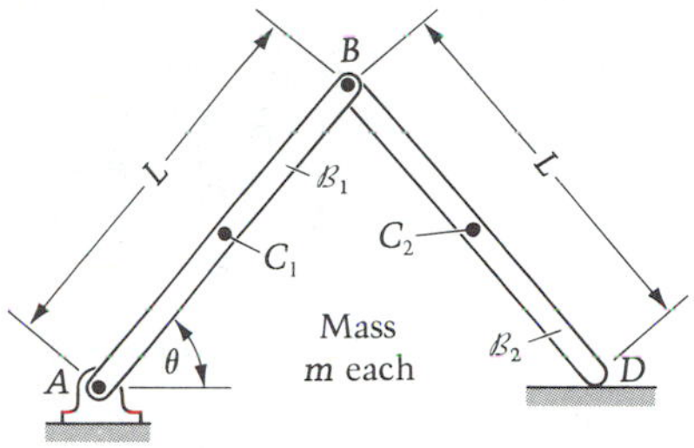

Figure P5.9

5.10 The prehistoric car shown in Figure P5.10 is powered by the falling rock m, connected to the main wheel (a cylinder of mass M) by a vine as shown. If the weights of the frame, pulley, and front wheel are small compared with Mg, find the velocity v_C of the car as a function of y if there is no slipping and it starts from rest with $y = 0$ at $t = 0$. Assume that m moves only vertically relative to the car's frame.

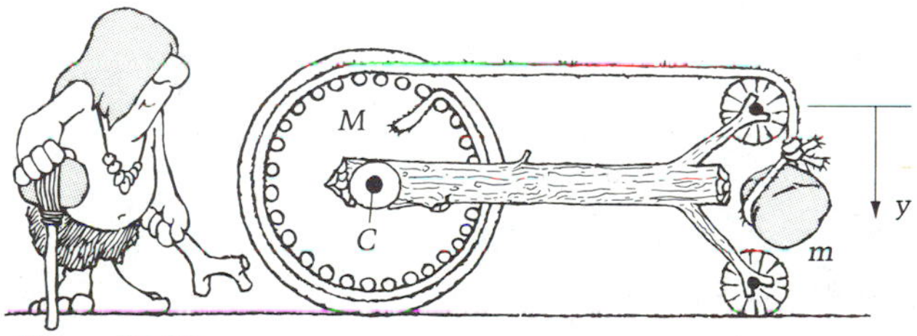

Figure P5.10

5.11 A truck body weighing 4000 lb is carried by four solid disk wheels that roll on the sloping surface. (See Figure P5.11.) Each wheel weighs 322 lb and is 3 ft in diameter. The truck has a velocity of 5 ft/sec in the position shown. Determine the modulus of the spring if the truck is brought to rest by compressing the spring 6 in.

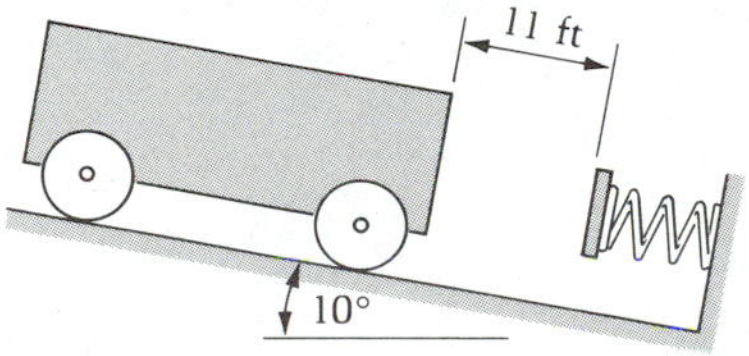

Figure P5.11

5.12 For the cylinder of Problem 4.70, assuming no slip and that the cylinder starts from rest, use work and kinetic energy to find the speed of its center in terms of the displacement of the center.

Ideally, the following five problems should be worked sequentially:

5.13 A cylinder with mass 6 kg has a 20-N force applied to it as shown in Figure P5.13. Find the angular velocity of the cylinder after it has rolled through 90° from rest.

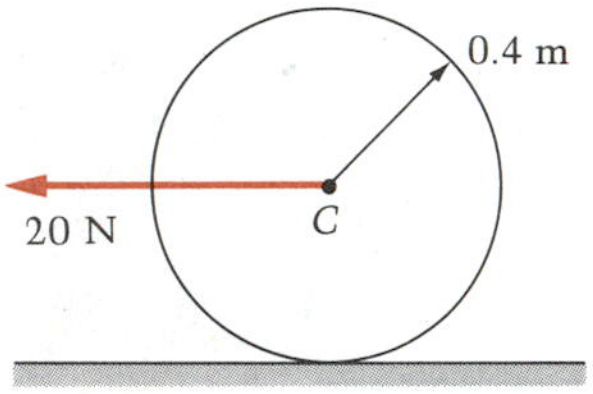

Figure P5.13

5.14 Rework Problem 5.13 if a slot is cut in the cylinder and a cord is wrapped around the slot, with the 20-N force now applied to the end of the cord as shown in Figure P5.14. Neglect the effect of the thin slot on the moment of inertia of the cylinder.

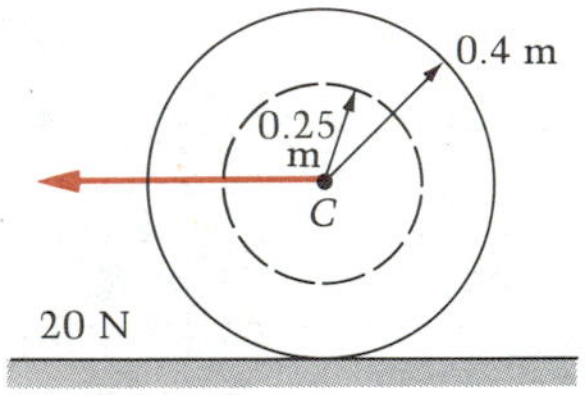

Figure P5.14

5.15 Suppose in Problem 5.14 we remove some material from the cylinder so as to offset the mass center C from the geometric center Q as shown in Figure P5.15. The removal reduces the mass to 5.5 kg and makes the radius of gyration with respect to the axis through C normal to the plane of the figure $k_C = 0.286$ m. Repeat the problem.

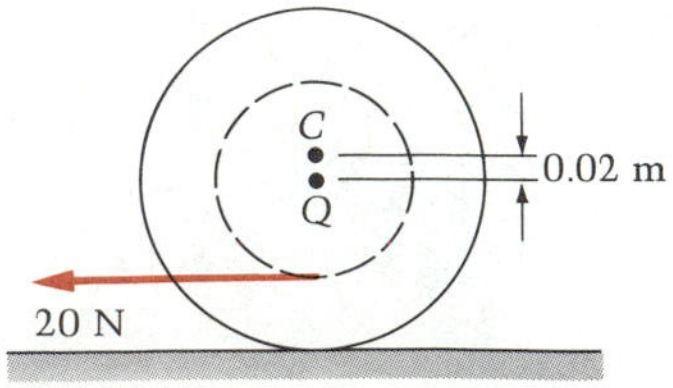

Figure P5.15

5.16 To the data of Problem 5.15 we add a constant counterclockwise couple of moment 2 N · m acting as shown in Figure P5.16. Repeat the problem.

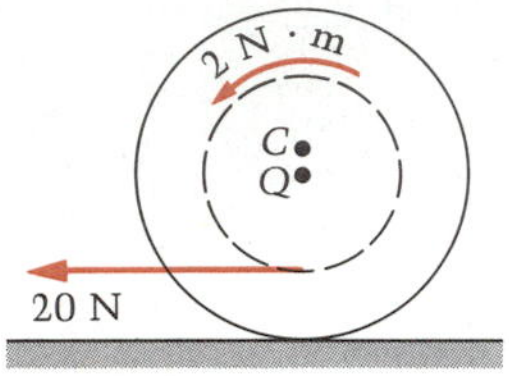

Figure P5.16

5.17 To the data of Problem 5.16, we add a spring, attached to a cord wrapped around a second slot in the cylinder near its outer rim as shown in Figure P5.17. The spring has modulus 6 N/m and is initially stretched 0.2 m. Repeat the problem.

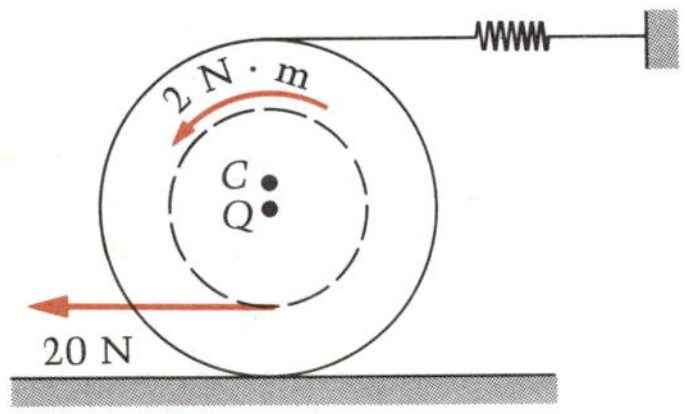

Figure P5.17

5.18 The 5-lb cylinder in Figure P5.18 rolls on the incline. If the velocity of the mass center C is 5 ft/sec down the plane in the upper (starting) position, find v_C in the bottom position.

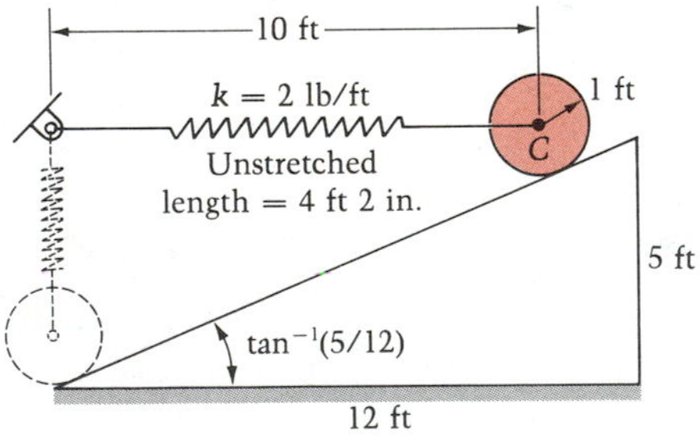

Figure P5.18

5.19 The spring in Figure P5.19 has an unstretched length of 0.8 m and a modulus of 60 N/m. The 20-kg wheel is released from rest in the upper position. Find its angular velocity when it passes through the lower (dashed) position if its radius of gyration is $k_C = 0.2$ m.

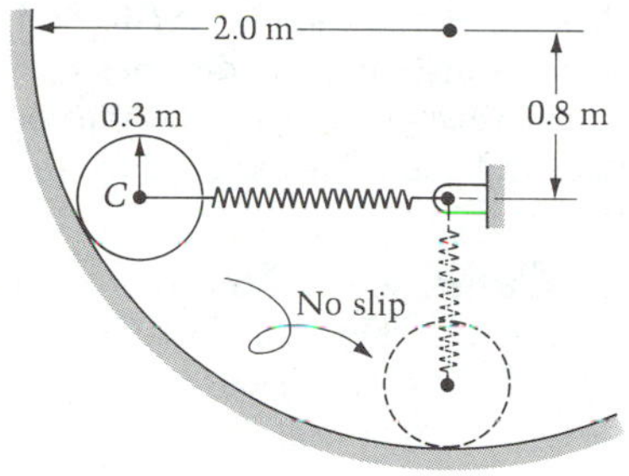

Figure P5.19

5.20 Bar $\mathcal{B}$ in Figure P5.20 is initially at rest in the vertical position, where the spring is unstretched. The wall and floor are smooth. Point B is then given a very slight displacement to the right, opening up a small angle $\Delta\theta$.

a. Draw a free-body diagram of the slightly displaced bar and use it to show that the bar will start to slide downward if $k < mg/2\ell$.

b. Find the angular velocity of $\mathcal{B}$ as a function of θ for such a spring.

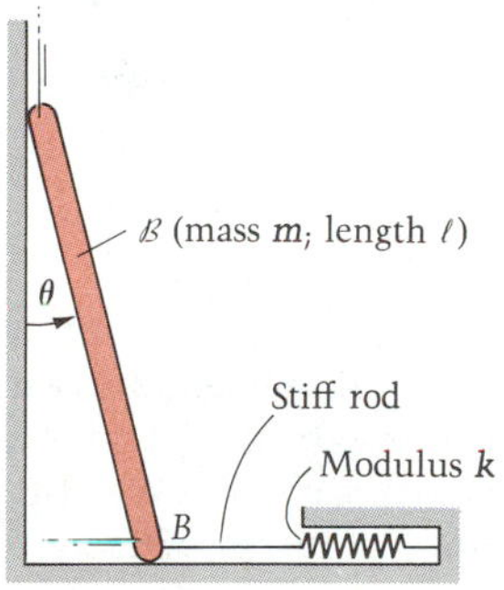

Figure P5.20

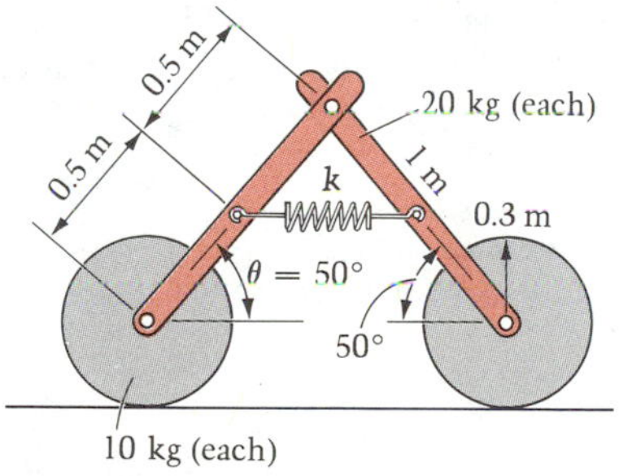

Figure P5.21

5.21 Find the spring modulus k that will result in the system momentarily stopping at $\theta = 35°$ after being released from rest at $\theta = 50°$ if the initial stretch δ_i in the spring is zero. (See Figure P5.21.) *Hint*: Use symmetry!

5.22 The wheel in Figure P5.22 has a mass of 5 slugs and a radius of gyration for the z axis through C of 0.7 ft. The spring has modulus 20 lb/ft and natural length 4 ft. The wheel is released from rest, and it rolls without slipping on the plane. Find how far down the plane the mass center C will move.

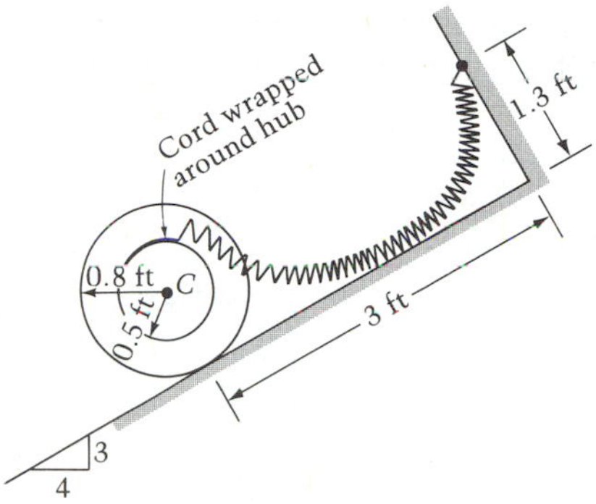

Figure P5.22

5.23 The wheel in Figure P5.23 weighs 200 N and has a radius of gyration 0.3 m with respect to the z_C axis. It is released from rest with the spring stretched $\frac{1}{2}$ m. If there is no slipping, find how far the cylinder center C moves

a. up, and

b. down the plane in the subsequent motion.

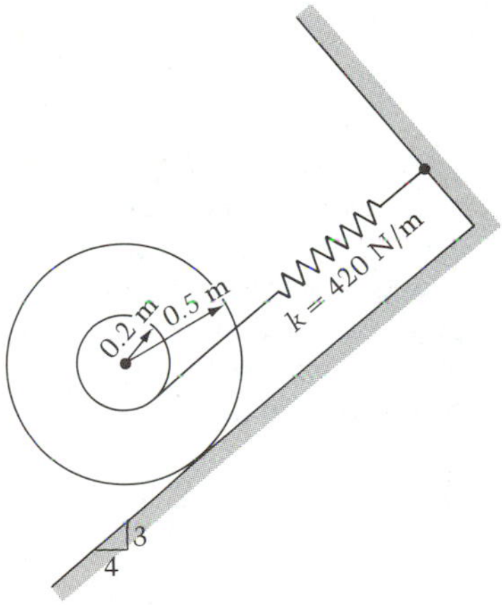

Figure P5.23

5.24 Show that if the rolling body in Example 5.8 is a sphere instead of a cylinder, it will slip at the angle θ_s satisfying the equation

$$\mu = \frac{2 \sin \theta_s}{17 \cos \theta_s - 10}$$

5.25 Use $W = \Delta T$ in Problem 4.98(b) to find the velocity of C when it has moved 3 m down the incline.

* **5.26** In Problem 4.134 use the principle of work and energy to obtain an upper bound on the rod's angular speed in its subsequent motion after the right-hand string is cut.

5.27 A thin disc of mass m and radius a is pinned smoothly at A to a thin rod of mass $m/2$ and length $3a$ (see Figure P5.27). The rod is then pinned at B. If the body is held in equilibrium in the configuration shown, then released from rest, find the velocity of point A as the system passes through the vertical.

5.28 Repeat the preceding problem if the pin at A is replaced by a weld.

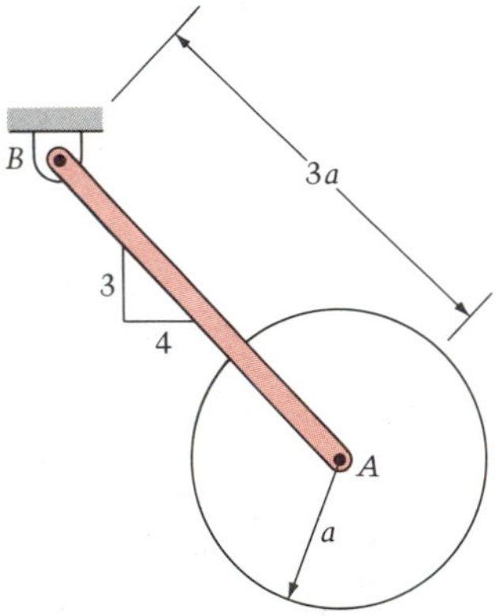

Figure P5.27

5.29 The 10-lb wheel shown in Figure P5.29 is attached at its center to a spring of modulus 20 lb/in. The radius of gyration of the wheel about the center is 2.5 in. The wheel rolls (no slip) after being released from rest with the spring stretched 1 in. Find: (a) the maximum magnitude of force in the spring; (b) the maximum speed of the center of the wheel during the ensuing motion.

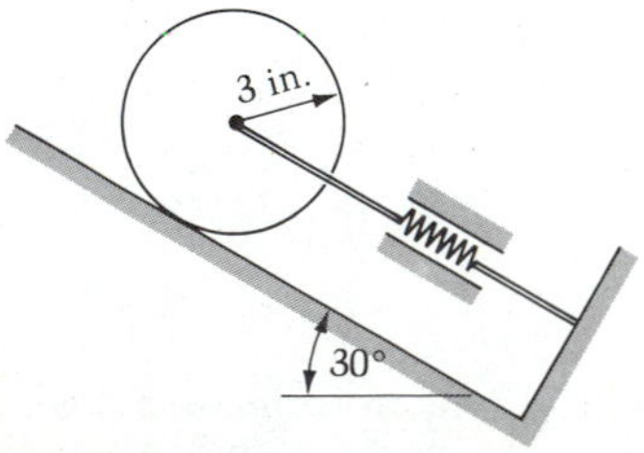

Figure P5.29

* Asterisks identify the more difficult problems.

5.30 For the data of Problem 4.113 use $W = \Delta T$ to find the speed $\dot{x}_C$ of the plate as a function of the distance x_C it has traveled to the right. Use the $x_C = x_C(t)$ result to check your answer; differentiate and eliminate t to produce the same $\dot{x}_C = \dot{x}_C(x_C)$ result.

5.31 Figure P5.31 shows a fire door on the roof of a building. The door $\mathcal{B}_1$, 4 ft wide, 6 ft long, and 4 in. thick, is wooden (at 30 lb/ft^3) and can rotate about a frictionless hinge at O. A cantilever arm $\mathcal{B}_2$ of negligible weight is rigidly fastened to the door and carries a 150-lb weight at its free end. During a fire the link $\mathcal{B}_3$ melts and the door swings open 45°. Find the angular velocity of the door just before the 150-lb weight hits the roof: (a) with no snow on the roof; (b) with snow at 1 lb/ft^2 on the roof.

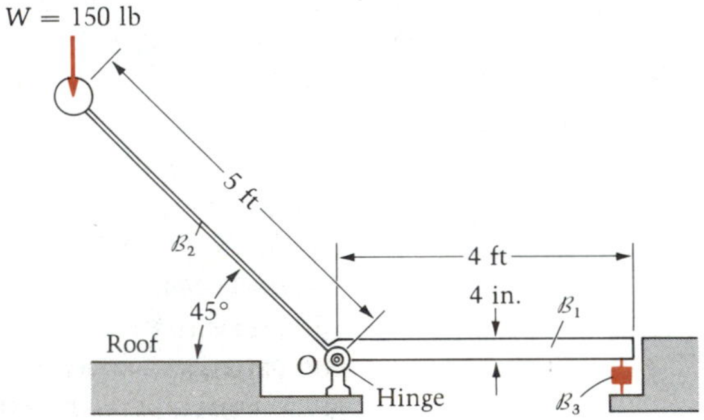

Figure P5.31

5.32 Block $\mathcal{B}_1$ in Figure P5.32 is moving downward at 5 ft/sec at a certain time when the spring is compressed 1 ft. The coefficient of friction between block $\mathcal{B}_2$ and the plane is 0.2, and the radius of cylinder $\mathcal{B}_3$ is 0.5 ft. Weights of $\mathcal{B}_1$, $\mathcal{B}_2$, and $\mathcal{B}_3$ are 161, 193, and 322 lb, respectively.

a. Find the distance that $\mathcal{B}_1$ falls from its initial position before coming to zero speed.

b. Determine whether or not body $\mathcal{B}_1$ will start to move back upward.

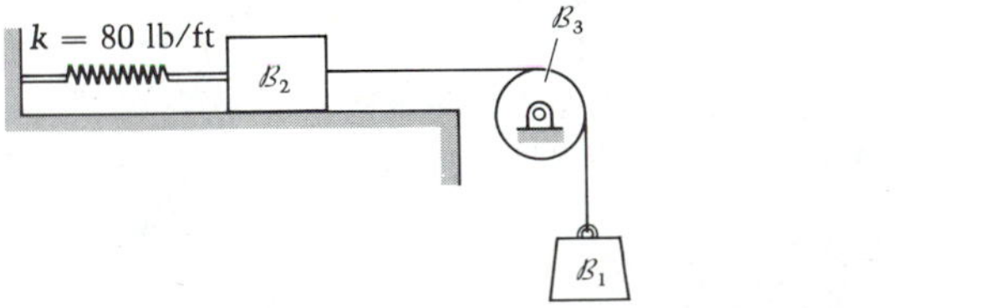

Figure P5.32

5.33 The system in Figure P5.33 consists of a cylinder $\mathcal{B}_1$ (100 kg) and (equilateral) triangular plate $\mathcal{B}_2$ (20 kg) pinned together at the mass center C_1 of the cylinder. The other two vertices of the plate are connected to springs,

The study of classical mechanics is a profound experience. The deeper one delves into it, the more he appreciates the contributions of the great masters. Y. C. FUNG

INTRODUCTION

COMPREHENDING MECHANICS GOES beyond reading the textbook and working problems. Knowing dynamics means that you understand the physics embodied in the laws of mechanics, recognize their limitations and assumptions, and can correctly apply them to situations you encounter in practicing engineering. Knowing mechanics requires that you also develop a reasonable sense of the physical consequences of the fundamental principles along with the mathematical consequences.

GREAT MASTERS OF MECHANICS such as Galileo, Leonardo da Vinci, Hooke, Kepler, and Newton formulated the laws of statics and dynamics from the results of numerous observations and experiments. They devised simple experiments to test and clarify their ideas. Using empirical findings, they developed theories for predicting the behavior of mechanical systems. The principles they discovered and the mathematical expressions that describe these principles are the cornerstones of engineering mechanics.

STUDENTS (AND TEACHERS) of dynamics often overemphasize analysis and pay too little attention to the relationship between theory and the actual physical behavior of mechanical systems. Understanding both aspects of mechanics is essential. Engineers cannot successfully model the behavior of a mechanical system if they are unsure of the physics of the system. And the ability to predict successfully the behavior of physical systems is fundamental to the process of engineering design.

OVERVIEW

THIS SECTION PRESENTS experiments not unlike those used in early empirical studies of mechanics. These experiments demonstrate actual behaviors of simple mechanical systems and are intended to strengthen your understanding of the basic laws of dynamics. These exercises emphasize physical reality to help you develop qualitative intuitive skills that are essential in the practice of engineering. In addition, the demonstrations provide a way to check the soundness of certain mathematical models that are used to describe real-world problems.

THESE EXPERIMENTS PROVIDE only a starting point for your explorations and will raise additional questions as you conduct them. Do not leave those questions unanswered. To answer them you may need to modify a demonstration, design new experiments, or simply concentrate on interpreting a mathematical model. The important point is that you should pursue the answers. Along the way you will develop new insights into dynamics, you will become more proficient, and your ability to explain and predict the physical world will improve.

THE EXERCISES IN dynamics are keyed to specific sections and problems in the text. Be sure to review the text material before attempting these exercises. Each demonstration requires that you compare your observations with behavior predicted from a mathematical model. In most cases, experimental and theoretical results should be reasonably close. Remember, however, that models are only approximate and that experiments are never perfect. So, if your results disagree, find out why. To do so, verify your measurements and, if necessary, repeat or redesign the demonstration. Review the assumptions and limitations of the theory and check your analysis or computer program for mistakes. If you still find a disparity between the results, you may be applying the wrong principles or using incorrect equations.

MATERIALS

THE EXPERIMENTAL setups are simple and easy to construct. To conduct them, however, you will need some materials, all of which are readily available and can be obtained at little or no cost. These materials can be found at hardware stores, hobby shops, toy stores, and in the engineering shop at your college. We encourage the use of scrap materials and creative scrounging!

FOR MECHANICAL PARTS you should collect an assortment of cylinders, tubes, spheres, wheels, and rectangular blocks. The only requirement is that the parts be homogeneous and reasonably uniform. For example, if you need a cylindrical tube, select one that is straight and has a constant diameter and thickness. Manufactured tubes such as the following are excellent:

- Wood dowel
- PVC pipe, copper pipe, steel pipe
- Conduit tubing
- Aluminum rod, steel rod
- Empty coffee can with ends removed, tennis ball can
- Cardboard tube from roll of paper towels or toilet tissue
- Cardboard mailing tube
- Hockey puck
- Thread spool, metal adhesive tape spool

Be creative and resourceful when selecting materials.

IN ADDITION, you will need some laboratory supplies. They include string, duct tape, protractor, graph paper, stopwatch, tape measure or ruler, scissors, inexpensive calipers, and a scale or access to a scale for weighing parts.

ROLLING CYLINDERS AND MOMENTS OF INERTIA

Lay out a length L on a flat board (or table), raise one end between 15° and 30°, and measure the angle. Select an assortment of cylinders, tubes, rods, wheels, and disks. Measure the time required for each object to roll distance L (see the figure below). Take several measurements for each object and obtain an average time. Work the theoretical problem of a round object starting from rest and rolling down an incline. Compare the results and explain any differences. Note that the results are independent of mass and radius. Repeat the procedure for different angles of inclination.

REFERENCE: *Dynamics* Sections 4.4-4.6

A "JUMPY" CYLINDER

Build an unbalanced but round cylinder. One approach, shown in the figure to the right, is to tape or glue a rod or small cylinder to the inside of a cylindrical tube. (Be sure that the two axes are parallel.) A cardboard tube from a roll of paper towels and a length of 5/8" diameter PVC pipe work well for this demonstration. Place the cylinder on an inclined surface and release it from rest, as in the figure below. The slope of the surface should be great enough that the statically unbalanced cylinder rolls without stopping and does not slip. The "herky-jerky" motion of the cylinder should be obvious. Increase the inclination of the surface and repeat the test. The cylinder may begin to slip as it rolls. If so, increase the friction between the cylinder and the surface and continue your observations. You can increase the coefficient of friction by placing flat, wide rubber bands on the large cylinder to serve as treads or by wrapping the cylinder with several turns of duct tape. Experiment with different angles of inclination and different coefficients of friction until the composite body jumps up and loses contact with the incline. Did you anticipate this behavior? Can you explain it? The equations of motion predict this behavior and can help you interpret what you observed.

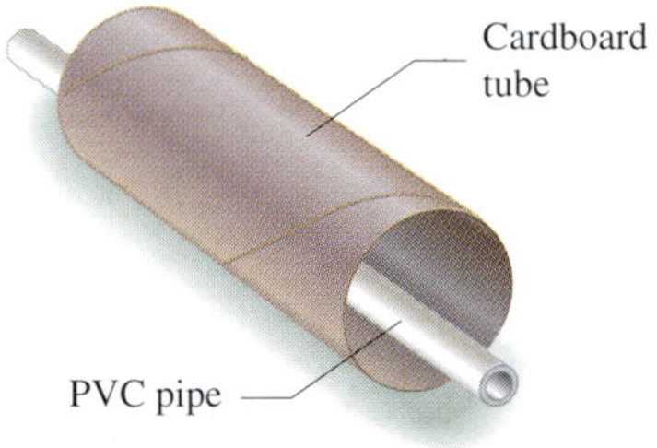

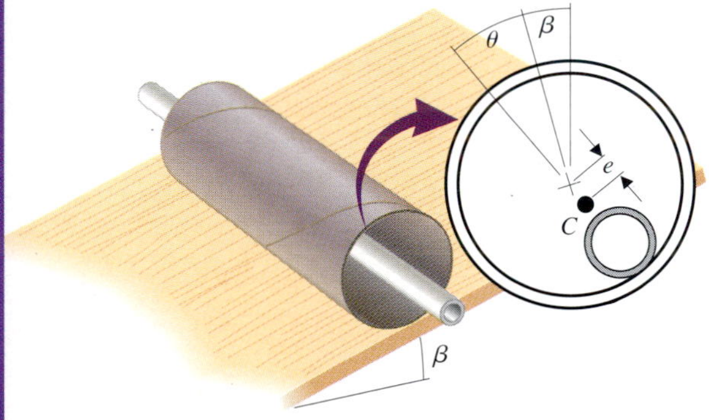

Derive the governing equations of motion for the unbalanced cylinder. Let m denote the mass of the body, e the eccentricity of the mass center C, I_G the body's moment of inertia about the central axis of the cylinder, R the cylinder radius, and g the acceleration of gravity. Assume that the cylinder rolls down the incline without slipping and show that

$$\ddot{\theta} = \frac{mg[e\cos\beta\sin\theta + (R + e\cos\theta)\sin\beta] + mRe\dot{\theta}^2\sin\theta}{I_G + mR^2 - 2m\,Re\cos\theta}$$

The equations also show that the normal force exerted on the body by the incline is

$$N = mg\cos\beta - me\dot{\theta}^2\cos\theta - me\ddot{\theta}\sin\theta$$

and that the upslope frictional force on the body is

$$F = mg\sin\beta + me\dot{\theta}^2\sin\theta - m(R + e\cos\theta)\ddot{\theta}$$

Note that N may vanish, which corresponds to the cylinder losing contact with the surface. Hence this equation predicts the possibility of the cylinder jumping up from the incline if certain conditions are met.

Develop the expression

$$\dot{\theta}^2 = \frac{2mgR\sin\beta(\theta - \theta_0) + 2mge[\cos(\theta_0 + \beta) - \cos(\theta + \beta)]}{I_G + mR^2 - 2m\,Re\cos\theta}$$

for the square of the body's angular speed if it starts from rest with $\theta = \theta_0$.

Construct a computer program that evaluates $\dot{\theta}$, $\ddot{\theta}$, N, F, and $|F|/N$ as functions of θ for a statically unbalanced cylinder that is released from rest and rolls down an incline without slipping. Your program should accept values of I_G, m, e, R, θ_0, and β as input and evaluate $\dot{\theta}$, $\ddot{\theta}$, N, F, and $|F|/N$ for equally spaced values of θ over several revolutions. A computer speadsheet works well for this analysis. Does the behavior predicted by your analysis match the actual motion of the unbalanced cylinder (at least qualitatively)?

REFERENCE: *Dynamics* Sections 4.4 - 4.6

ROLLING A WHEEL ON ITS AXLE

Assemble a single wheel and axle as shown in (a). You may use a pulley wheel (sheave) or a wheel with a grooved rim. The wheel should not slip around the axle. Tape a string to the wheel and wind it around the wheel at the center of the rim. The axle stubs need to ride on parallel rails that are high enough for the wheel to turn freely. The rails may be two boards on edge, two books of equal thickness, or two cylinders or tubes of equal diameters. Place the model on the rails, as in (b) and (c). Before pulling the string in the direction suggested in (c), decide which way the wheel will roll. Then pull the string and check your intuition. Solve the equation of motion for the angular acceleration α and confirm that the direction of α agrees with your observation. Repeat the experiment when the string force is vertical, as in (d), and when the force is horizontal and comes off the top of the wheel, as in (e). Verify your results using the equations of motion.

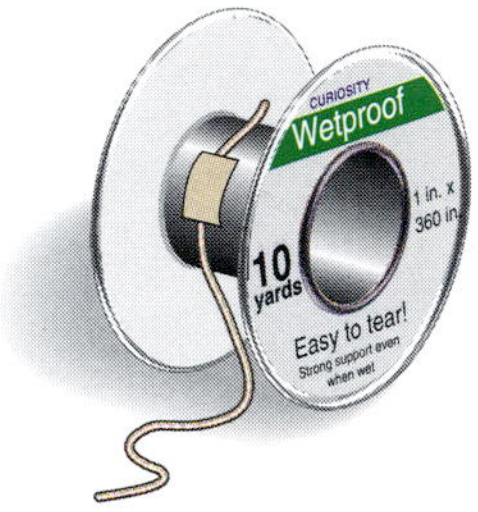

(a)

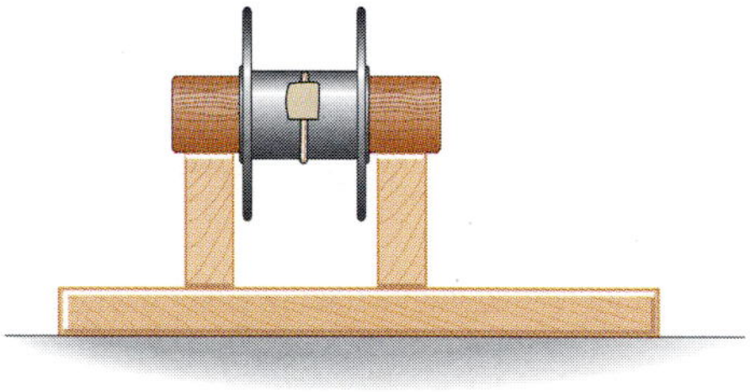

(b)

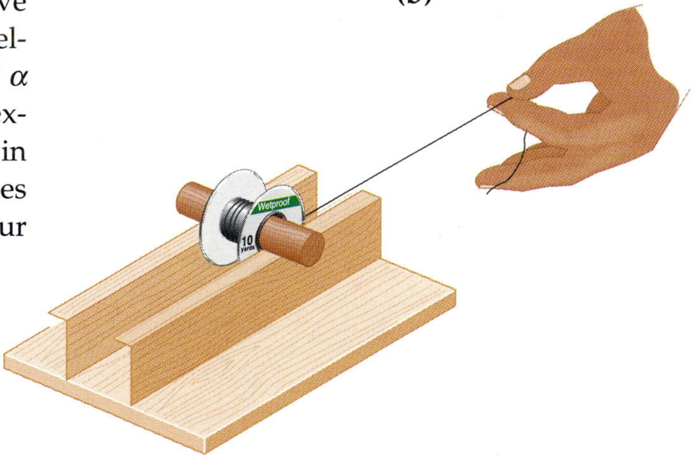

(c)

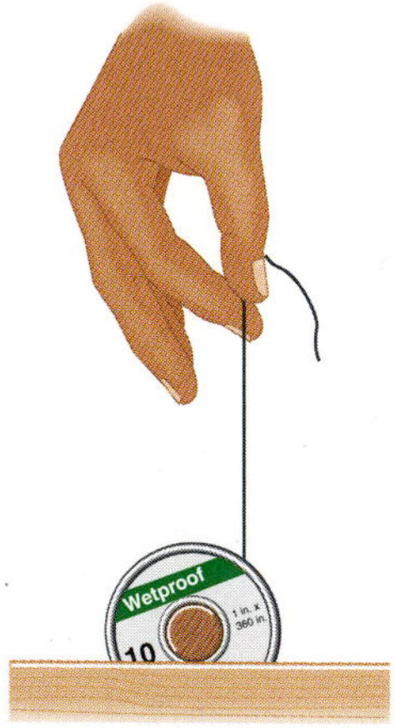

(d)

(e)

REFERENCE: *Dynamics* Sections 4.5 and 4.6

UNWINDING A SPOOL ON AN INCLINE

Assemble a spool from two identical disks or cylinders and a single axle, as shown in the figure here. A metal adhesive tape spool or a yo-yo also works well for this demonstration. Tape a string to the center of the axle and wind it up. Bring the string from the bottom of the axle and hold it parallel to the surface of an inclined board, as shown in the top figure. Be sure that the string is attached midway between the disks. For small values of θ the spool remains in static equilibrium. Slowly increase the inclination of the board and observe when the spool begins to move. Note that the spool cannot roll down the plane (without slipping); to do so the string would break. The spool moves by slipping upward against the board at the point of contact while it actually rolls downward on the string. Derive the equations of motion for the mass center of the spool as it moves down the incline. Integrate twice and determine the time required for the center to move length L along the board. Mark off L on the incline and then measure the time for the center to traverse this distance when the spool is released from rest. Compare your predictions and observations. Repeat the experiment by pulling on the string from the top of the axle, as shown in the bottom figure.

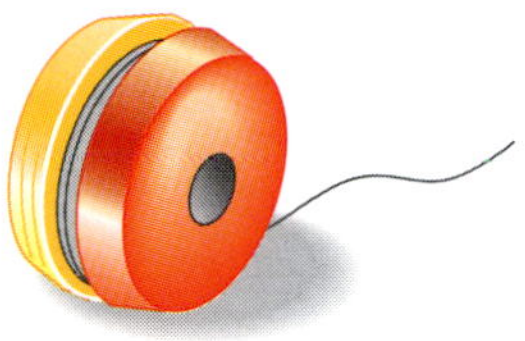

REFERENCE : *Dynamics* Sections 4.5 and 4.6

WALKING A YO-YO

Assemble a spool from two identical tubes or cylinders and a single axle, as shown in the top figure. Tape a string to the center of the axle and wind it up. Now you have a yo-yo! You may use a metal adhesive tape spool or an actual yo-yo for this demonstration. Be sure that the string is attached midway between the disks. If necessary, build up edges on both sides of the center of the axle with strips of tape or rubber bands. How will the spool move if you pull the string horizontally from the bottom of the axle, as shown in the middle figure? Vertically, as in the bottom figure? Gently pull the string from these positions and check your intuition. Use the equations of motion to compute α and compare its direction with your observations.

The interesting and somewhat unexpected behavior of the spool for these two situations should intrigue you and lead you to ask whether a pull angle between horizontal and vertical exists for which the spool will not roll either way. If so, what is that angle? Is the result the same if the string comes off the top of the axle?

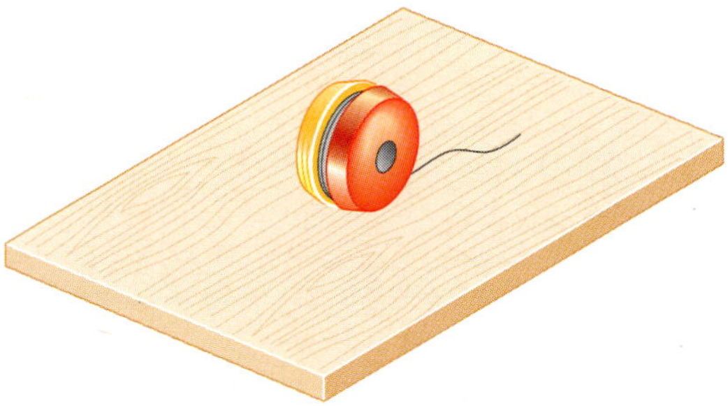

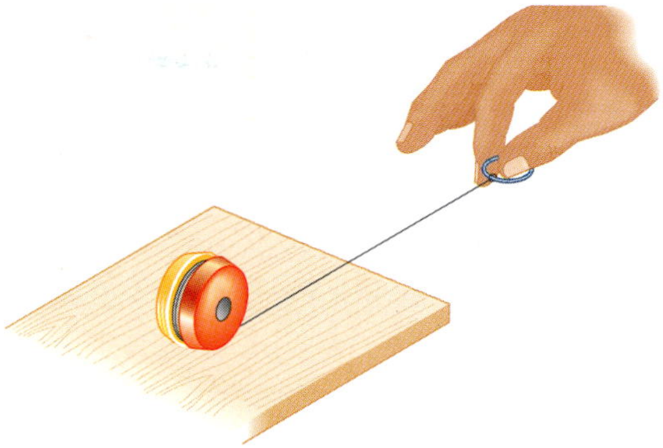

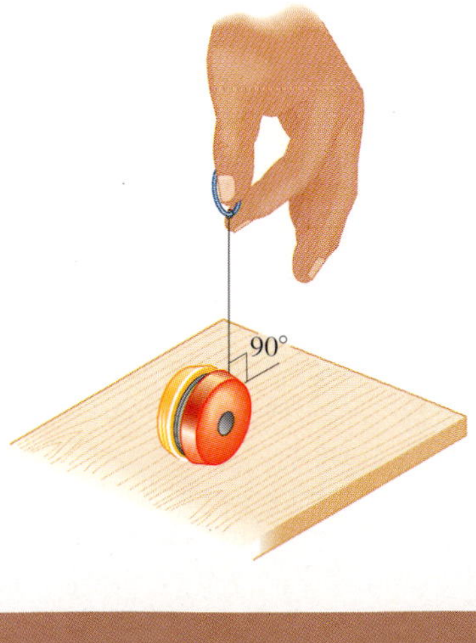

REFERENCE : *Dynamics* Sections 4.5 and 4.6

the left one of which (S_1) remains vertical in the slot. (Spring S_1 is shown only in its initial position.) The initial stretches of the two springs (in the position shown) are 0.2 m for S_1 and 0.04 m for S_2. The moduli are 40 N/m for S_1 and 10 N/m for S_2. If the system is released from rest in the given position, find the velocity of C_1 when vertex B reaches its lowest point in the slot. Assume sufficient friction to prevent $\mathcal{B}_1$ from slipping on the plane. The moment of inertia of an equilateral triangular plate of side s about its z_C axis is $ms^2/12$.

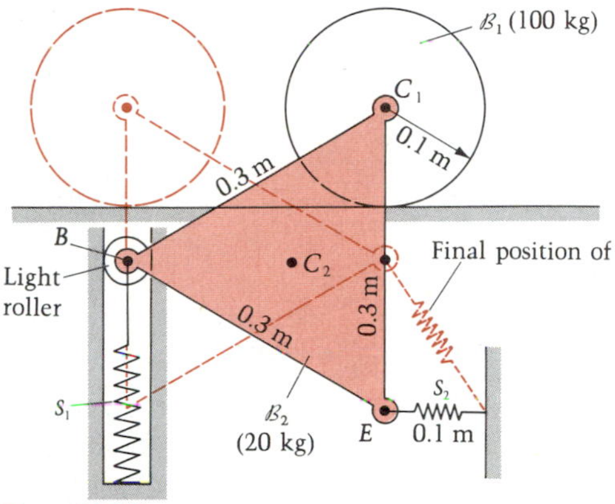

Figure P5.33

5.34 The mass center C of a rolling 2-kg wheel of radius $R = 15$ cm is located 5 cm from its geometric center Q. (See Figure P5.34.) The spring is attached at C and is not shown in position 2; its unstretched length is 0.3 m, and its modulus is 3 N/m. The radius of gyration is $k_C = 0.09$ m. Find the angular speed in position 2 (one-quarter turn from position 1).

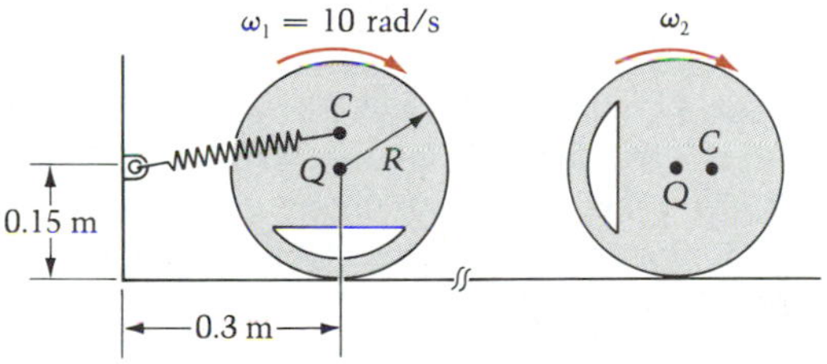

Figure P5.34

* **5.35** The three rods shown in Figure P5.35 are pinned together with one vertex also pinned to the ground. The length of the bar labeled $\mathcal{B}$ is given by $2b = 0.4$ m, and the density of the material of all bars is 7850 kg/m³. Their cross-sectional area is 0.002 m². Find the angular velocity of the combined body after it swings 90° from rest if: (a) $H = 2b$; (b) $H = \sqrt{3}\, b$; (c) $H = b$.

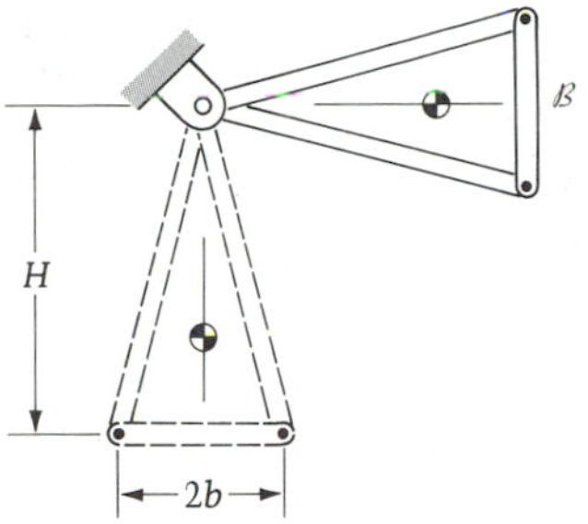

Figure P5.35

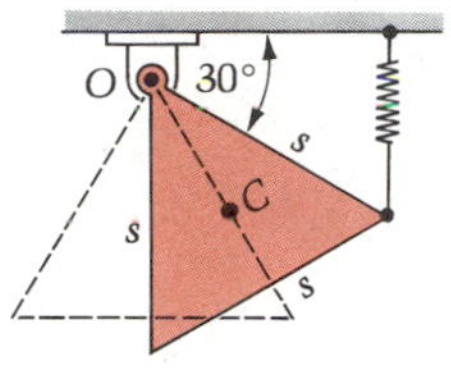

Figure P5.36

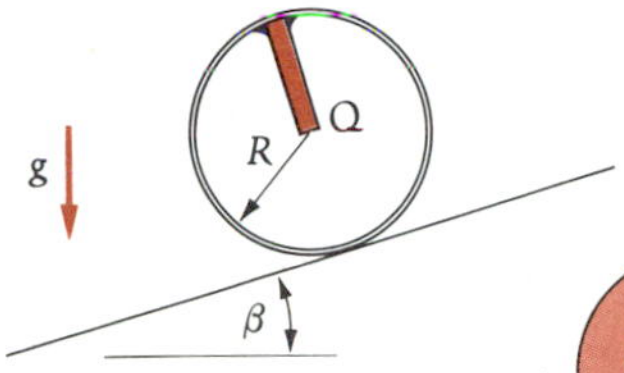

Figure P5.37

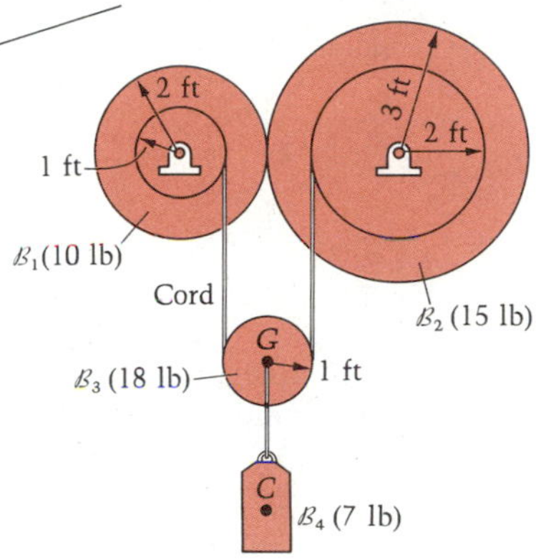

Figure P5.38

5.36 The center of mass of a uniform triangular plate is two-thirds of the distance from any vertex to the opposite side. The moment of inertia of an equilateral triangular plate is $ms^2/12$ with respect to the z axis through C. For the plate shown in Figure P5.36, with mass 30 kg and side 2 m, find its angular velocity when it reaches the dotted position where C is beneath O. The spring has unstretched length 0.5 m and modulus 20 N/m, and the plate is released from rest.

5.37 The bar in Figure P5.37 weighs the same (W) as the hoop to which it is welded. The combined body is released from rest on the incline in the position shown. If there is no slipping, determine the velocity of Q after one revolution of the hoop.

5.38 Cylinders $\mathcal{B}_1$ and $\mathcal{B}_2$ in Figure P5.38 are released from rest and turn without slip at the contact point. A cord is wrapped around an attached hub of each, which has negligible effect on the moment of inertia. There is enough friction to prevent the rope from slipping on the pulley. Find the velocity of the mass center of the pulley $\mathcal{B}_3$ after body $\mathcal{B}_4$ has fallen 20 ft.

5.39 A 5-lb cylinder is raised from rest by a force $P = 20$ lb. (See Figure P5.39.) Find the modulus of the spring that will cause the cylinder to stop after its center has been raised 2 ft. Will it then start back down? The spring is initially unstretched.

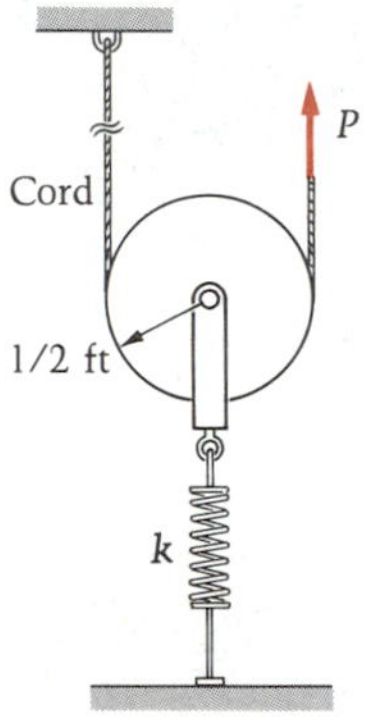

Figure P5.39

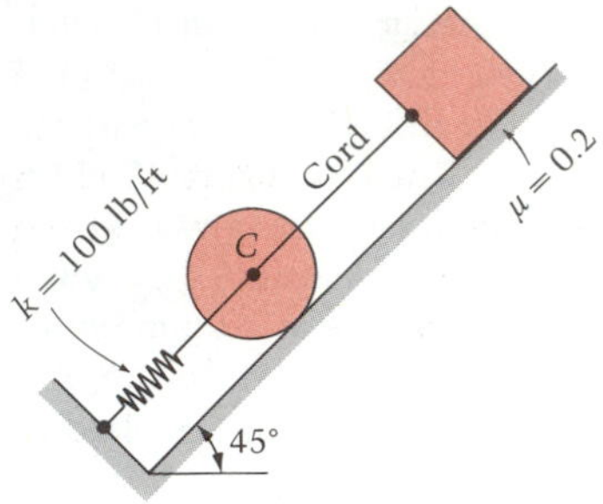

Figure P5.41

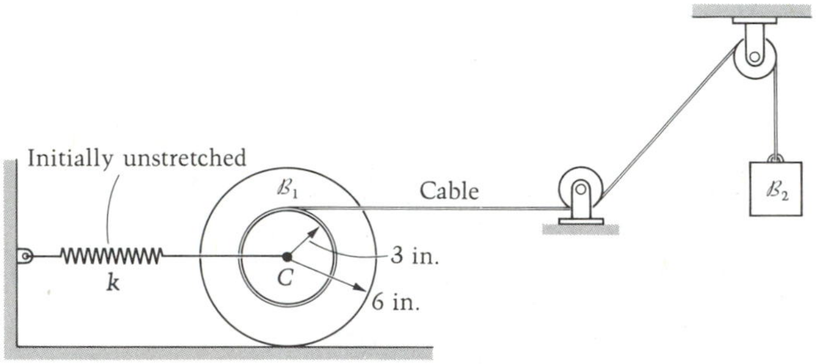

Figure P5.42

5.40 Body $\mathcal{B}_1$ in Figure P5.40 rolls to the right along the plane and has a radius of gyration with respect to its axis of symmetry of $k_C = 0.5$ m. The corresponding radius of gyration for $\mathcal{B}_2$ is 0.12 m. The spring is stretched 0.6 m at an instant when $\omega_1 = 5\circlearrowright$ rad/s. Find ω_1 after C has traveled 1 m to the right. (C_0 is an externally applied couple acting on $\mathcal{B}_1$.)

5.41 The cylinder and the block each weigh 100 lb. They are connected by a cord and released from rest on the inclined plane as shown in Figure P5.41. The spring, connected to the center C of the cylinder, is initially stretched 6 in. Find the velocity of the block at the instant the spring becomes unstretched, if there is sufficient friction between the plane and the cylinder to prevent it from slipping.

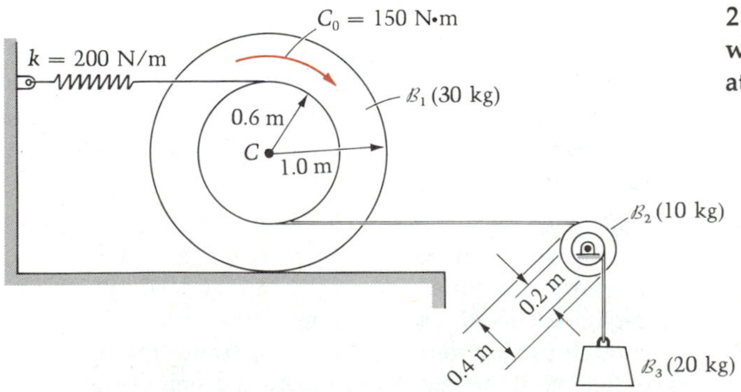

Figure P5.40

5.42 The 20-lb wheel $\mathcal{B}_1$ in Figure P5.42 has a radius of gyration of 4 in. with respect to its (z_C) axis. A cable wrapped around its inner radius passes under and over two small pulleys and is then tied to the 50-lb block $\mathcal{B}_2$. The spring has a modulus of 90 lb/ft and is constrained to remain horizontal. There is sufficient friction to prevent $\mathcal{B}_1$ from slipping on the plane. (a) If the system is released from rest, find the angular speed of $\mathcal{B}_1$ after the block then falls 1 ft. (b) Would the answer be different if block $\mathcal{B}_2$ were replaced by a device that keeps the cable force constant at 50 lb? Why or why not?

* **5.43** Rod $\mathcal{B}_1$ and disk $\mathcal{B}_2$ in Figure P5.43 have weights $W_1 = 5$ lb and $W_2 = 6$ lb. The rod's length is 8 in., the disk's radius is 4 in., the mass center offset of the disk is 2 in. from Q, and the radius of gyration of the mass of $\mathcal{B}_2$ with respect to the z axis through C is 3 in. It is desired to attach a spring between point Q and a fixed point so that

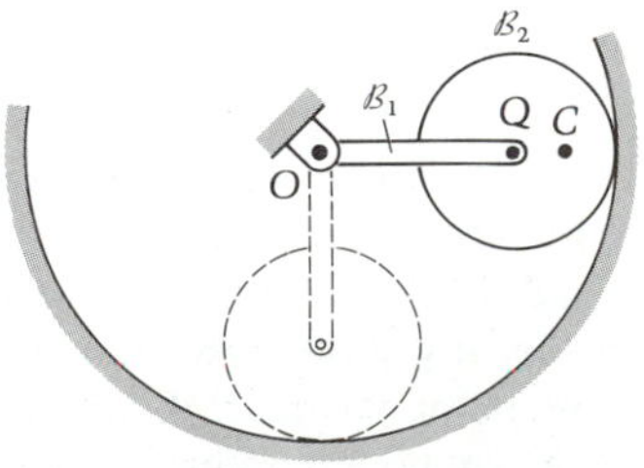

Figure P5.43

the disk and rod come to a stop (in the dotted position) after $\mathcal{B}_1$ turns 90° clockwise from rest. The spring has a modulus of 25.5 lb/ft and an unstretched length of 4 in.; it is to be unstretched initially. Find the final spring stretch, and from this result determine where to attach the fixed end of the spring. (There are two possible points!)

5.44 The bodies in Figure P5.44 have masses $m_1 = 0.3$ slug, $m_2 = 0.5$ slug, and m_3 negligible. A spring is attached to A that is stretched 25 in. in the dotted position when everything is at rest. Find the spring modulus if $\omega_{\mathcal{B}_1} = 2 \circlearrowright$ rad/sec when $\mathcal{B}_1$ is horizontal.

5.45 A vertical rod is resting in unstable equilibrium when it begins to fall over. (See Figure P5.45.) End A slides along a smooth floor. Find the velocity of the mass center C as a function of L, g, and its height H above the floor.

5.46 Pulley $\mathcal{B}_1$ weighs 100 lb and has a centroidal radius of gyration $k_C = 7$ in. (See Figure P5.46.) The disk pulley $\mathcal{B}_2$ weighs 20 lb. Find the velocity of weight $\mathcal{B}_3$ (50 lb) after it falls 2 ft from rest. (Assume that the rope does not slip on the pulleys.)

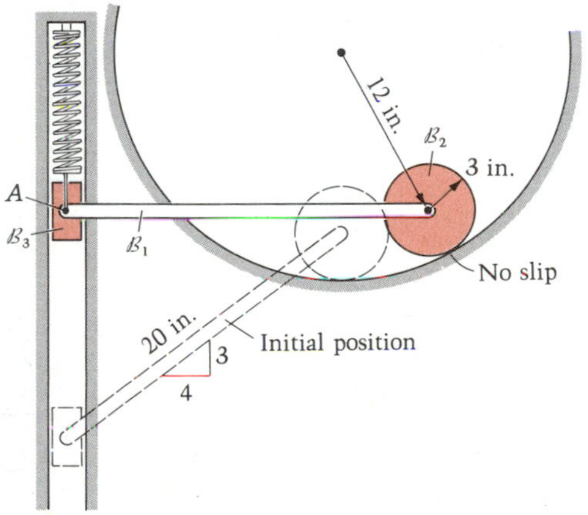

Figure P5.44

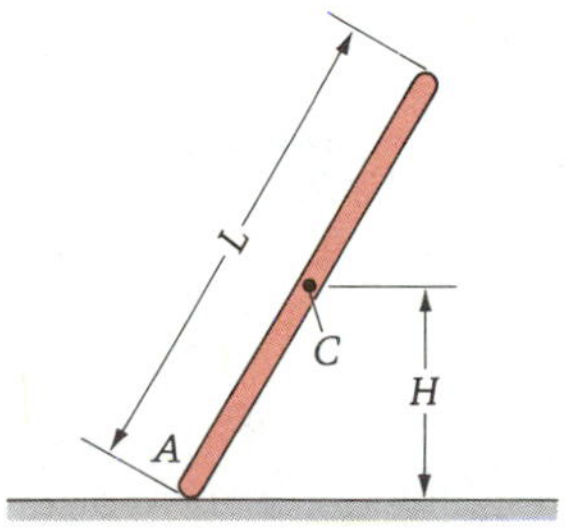

Figure P5.45

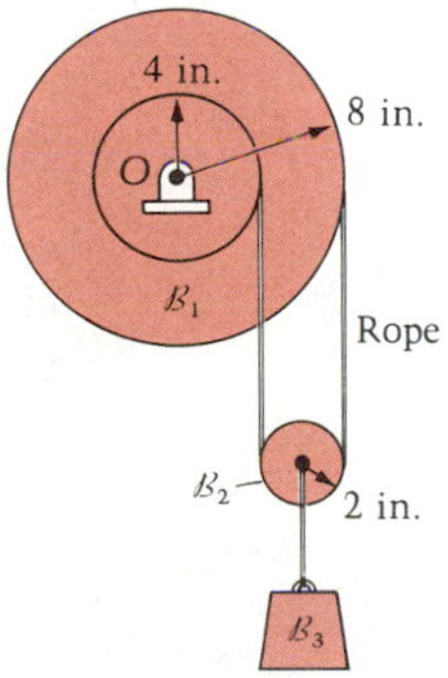

Figure P5.46

5.47 In Problem 4.108 determine the velocity of corner B of the half-cylinder when the diameter AB becomes horizontal for the first time.

5.48 Body $\mathcal{B}_1$ translates in the slot without friction. (See Figure P5.48.) Disk $\mathcal{B}_2$ (radius R) is pinned to block $\mathcal{B}_1$ through their mass centers at G. Body $\mathcal{B}_1$ and body $\mathcal{B}_2$ *each* has mass m; body $\mathcal{B}_3$ has mass $2m$. The system is released from rest a distance D above the floor. Find: (a) the starting accelerations of $\mathcal{B}_3$ and G; (b) the velocity of $\mathcal{B}_3$ when it hits the floor, using $W = \Delta T$.

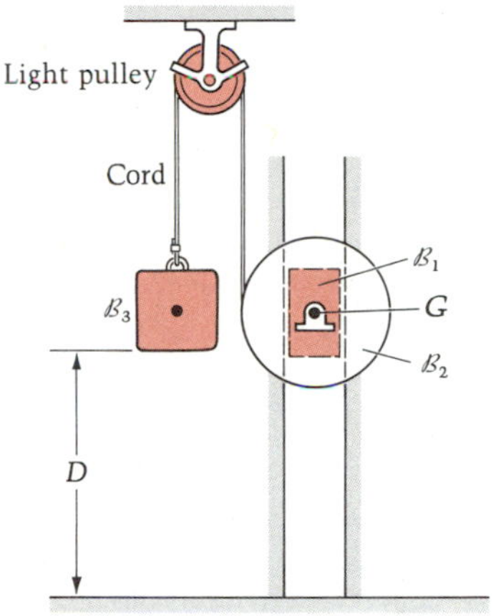

Figure P5.48

5.49 Link $\mathcal{B}_1$ weighs 10 lb and may be treated as a uniform slender rod (Figure P5.49). The 15-lb wheel is a circular disk with sufficient friction on the horizontal surface to prevent slipping. The spring is unstretched as shown. Link $\mathcal{B}_1$ is released from rest, and the light block $\mathcal{B}_2$ slides down the smooth slot. Neglecting friction in the pins, determine: (a) the angular velocity of the link as A strikes the spring with $\mathcal{B}_1$ horizontal; (b) the maximum deflection of the spring. (The modulus k of the spring is 10 lb/in.)

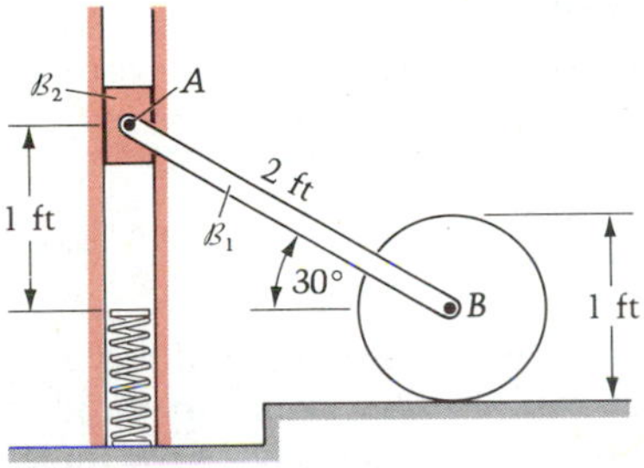

Figure P5.49

5.50 The masses of four bodies are shown in Figure P5.50. The radius of gyration of wheel $\mathcal{B}_4$ with respect to its axis is $k_C = 0.4$ m. Initially there is 0.6 m of slack in the cord between $\mathcal{B}_1$ and the linear spring. (Modulus $k = 1000$ N/m, and the spring is initially unstretched.) Determine how far downward body $\mathcal{B}_2$ will move.

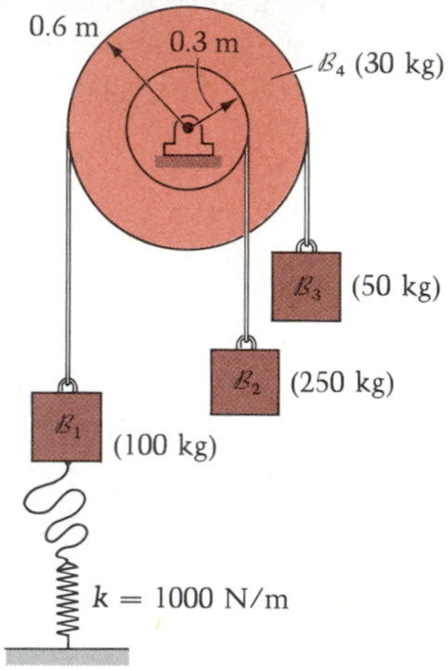

Figure P5.50

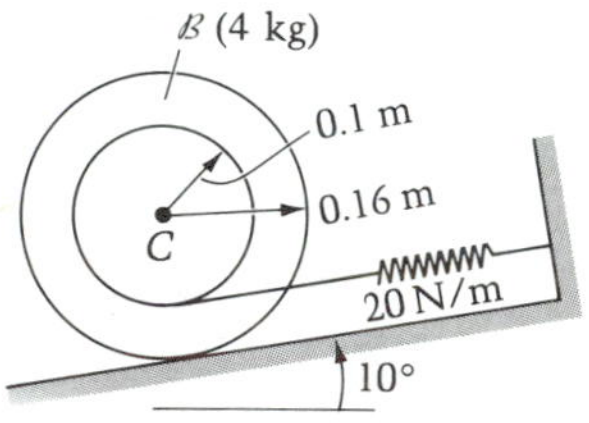

Figure P5.51

5.51 Cylinder $\mathcal{B}$ in Figure P5.51 is moving up the plane with $v_C = 0.3$ m/s at an initial instant when the spring is stretched 0.2 m. If $\mathcal{B}$ does not slip at any time, determine how far *down* the plane the point C will move in the subsequent motion. *Note*: The spring, connected to the cord, cannot be in compression.

5.52 Solve Problem 4.64 for v_C as a function of x_C using $W = \Delta T$.

5.53 In Problem 4.165, find the angular velocity of the rod when $\theta = 90°$.

5.54 Solve Problem 4.65 (a) by $W = \Delta T$.

5.55 For each of the wheels in Problem 4.72, solve for v_C as a function of x_C using $W = \Delta T$. The wheels start from rest.

• **5.56** Solve Problem 4.177 with the help of $W = \Delta T$. Ignore the hint.

5.57 The radius of gyration of the wheel and hub $\mathcal{B}$ in Figure P5.57, with respect to its axis of symmetry through C, is $k_C = 2.5$ m. The springs are unstretched at an initial position of rest, when the 50-N force is applied.

a. Find how far to the right the mass center moves in the ensuing motion, assuming sufficient friction to prevent slipping.

b. When $\mathcal{B}$ stops instantaneously at its farthest right point, what increase in the 50-N force and what minimum friction coefficient are needed to keep it there?

5.58 The slender nonuniform bar in Figure P5.58 (the mass is m and the radius of gyration with respect to the mass center C is $L/2$) is supported by two inextensible wires. If the bar is released from rest with $\theta = 0$, find the tension in each wire as a function of θ.

5.59 The system depicted in Figure P5.59 is released from rest with 2 ft of initial stretch in the spring. There is sufficient friction to prevent slipping at all times. Determine whether $\mathcal{B}_1$ will leave the horizontal surface during the subsequent motion. Note that the string $\mathcal{S}$ goes slack if the stretch tries to become negative.

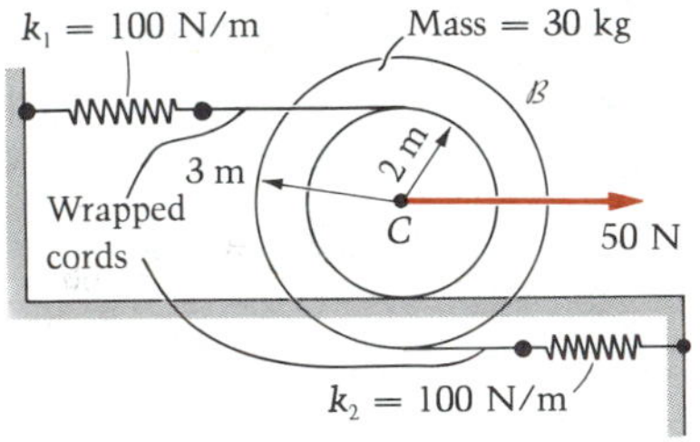

Figure P5.57

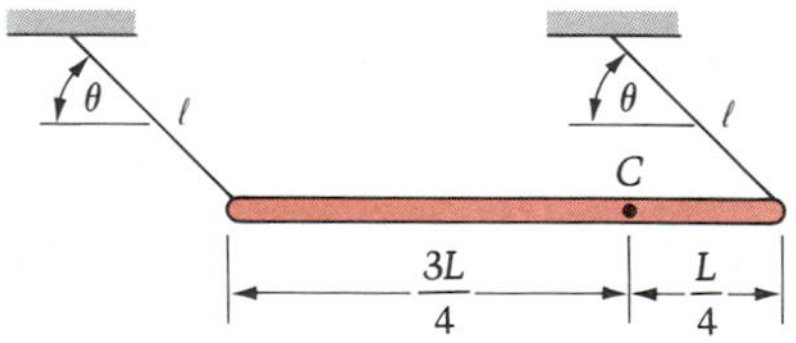

Figure P5.58

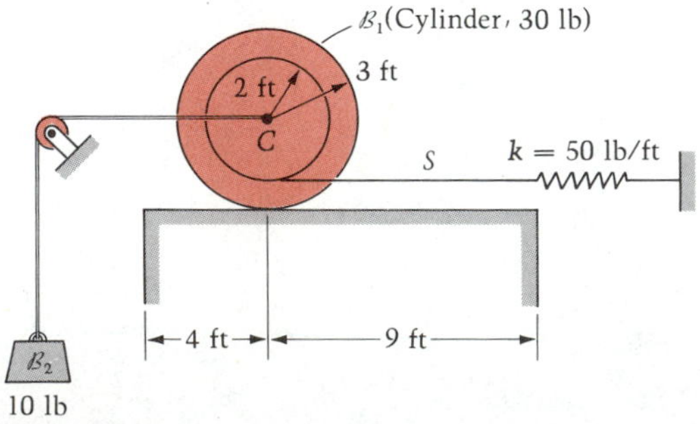

Figure P5.59

5.60 The system is released from rest in the position shown in Figure P5.60. Force P is constant, 60 lb, and the cord is wrapped around the inner radius of $\mathcal{B}_1$. Note the mass center of $\mathcal{B}_1$ is at C. Find the normal force exerted onto $\mathcal{B}_1$ by the plane (after using $W = \Delta T$ to get ω_1) at the instant when $\mathcal{B}_1$ has rotated 90° ⥁. The spring is initially unstretched, and there is enough friction to prevent slipping.

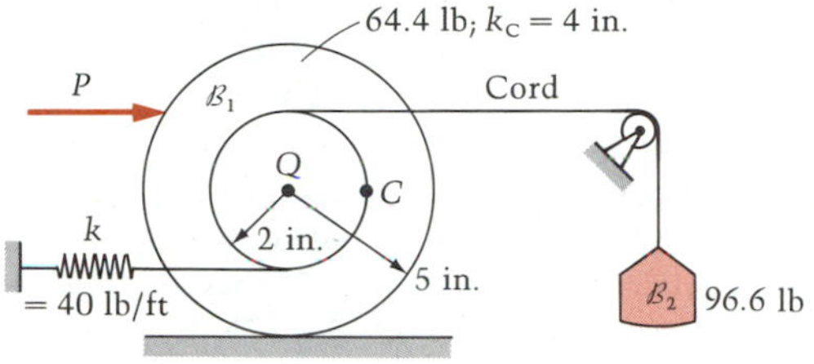

Figure P5.60

5.61 The two identical links $\mathcal{B}_1$ and $\mathcal{B}_2$ in Figure P5.61, each of mass m and length ℓ, are pinned together at A, and $\mathcal{B}_1$ is pinned to the ground at B. The end C of $\mathcal{B}_2$ slides in the vertical slot. Friction is negligible, and the system is released from rest. Find the velocity of point C just before point A reaches its lowest point.

5.62 The body of mass $2m$ in Figure P5.62 is composed of two identical uniform slender rods welded together. If friction in the bearing at O is neglected and the body is released from rest in the position shown, find the *magnitude* of the force exerted on the rod by the bearing after the body has rotated through 90°.

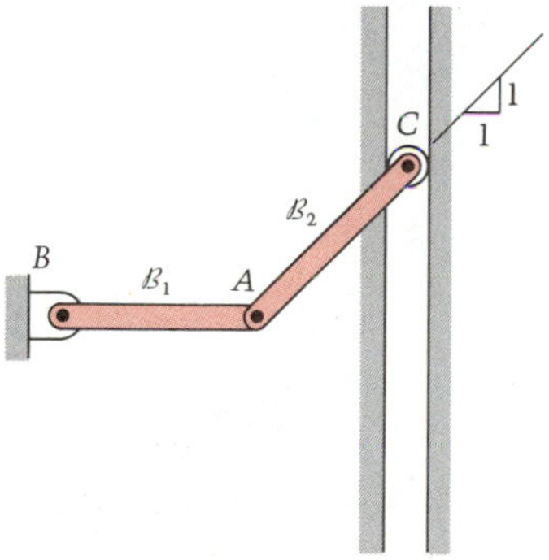

Figure P5.61

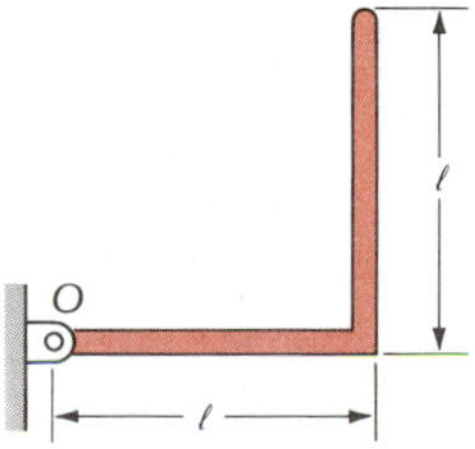

Figure P5.62

5.63 The cord connects the slotted cylinder $\mathcal{B}_1$ to the cylinder $\mathcal{B}_2$ as shown in Figure P5.63. Assume that neither body slips after the system is released from rest. The spring is initially unstretched, and is stiff and guided so it can take compression. Find the angular velocity of $\mathcal{B}_2$ after its center C has moved 1 m.

5.64 The 12-ft, 32.2-lb homogeneous rod $\mathcal{B}_1$ shown in Figure P5.64 is free to move on the smooth horizontal and vertical guides as shown. The modulus of the spring is 15 lb/ft and the spring is unstretched when in the position shown. Rod $\mathcal{B}_1$ is released from rest with $\theta = \pi/2$ and nudged to the right to begin motion. (a) Determine the angular velocity of the rod when it becomes horizontal. (b) What is the angular acceleration of the rod in this position ($\theta = 0$)?

5.65 The 50-kg wheel in Figure P5.65 is to be treated as a cylinder of radius $R = 0.2$ m. If it is rolling to the left with $v_C = 0.07$ m/s at an initial instant when the spring is unstretched, find: (a) the distance moved by C before v_C is instantaneously zero; (b) the minimum coefficient of friction that will prevent slip.

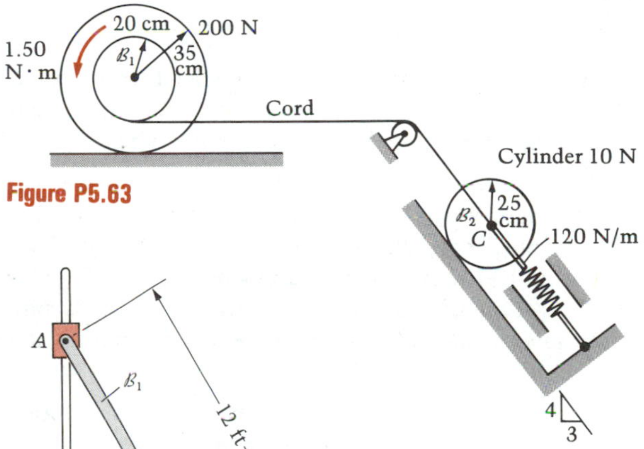

Figure P5.63

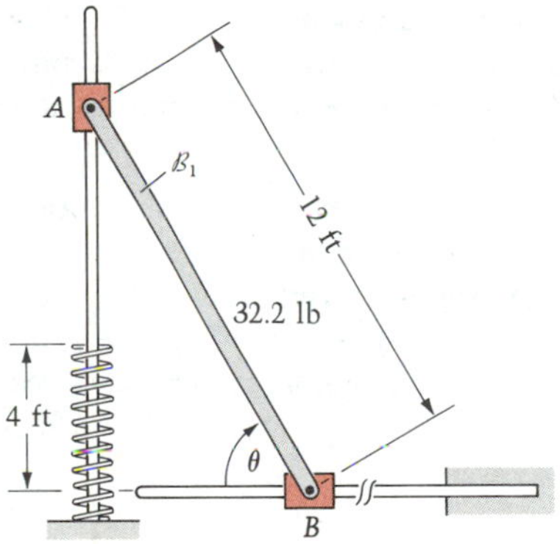

Figure P5.64

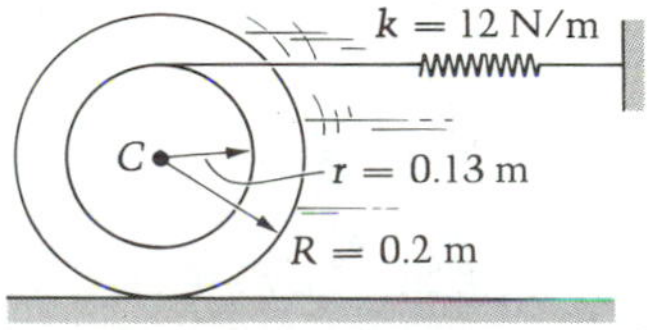

Figure P5.65

5.66 The cylinder in Figure P5.66 is rolling at $\omega = 2$ ↺ rad/sec in the initial (i) position, where the spring is unstretched. Other data are:

$$m = 2 \text{ slugs}$$
$$r = 3 \text{ ft}$$
$$k = 3 \text{ lb/ft}$$
$$\mu = 0.2$$
$$\ell_u = \text{unstretched spring length} = 9 \text{ ft}$$

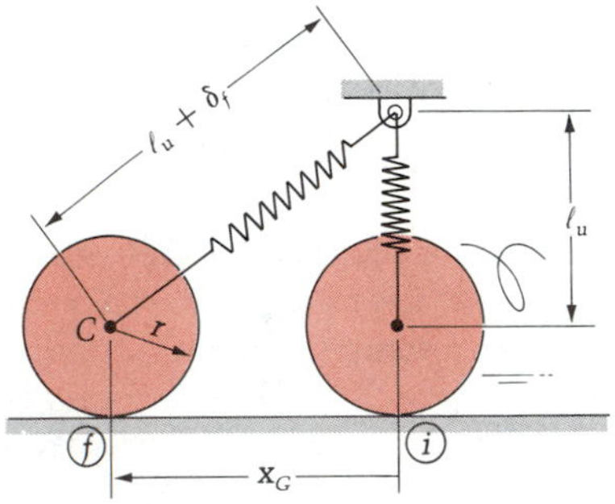

Figure P5.66

Find the final position of C (x_C) at which either the cylinder has stopped (for an instant) or started to slip, whichever comes first. *Hint*: Try one, check the other!

5.67 The uniform slender rod in Figure P5.67 (mass = 5 slugs, length = 10 ft) is released from rest in the position shown. Neglecting friction, find the force that the floor exerts on the lower end of the rod when the upper end is 6 ft above the floor. *Hint*: First use a free-body diagram and the equations of motion to deduce the path of the mass center.

5.68 In Figure P5.68, the ends of the bar are constrained to vertical and horizontal paths by the smooth rollers in the slots shown. The bar, originally vertical, is very gently nudged at its lower end to initiate motion. Find the reactions onto the bar at A and B just before the bar becomes horizontal.

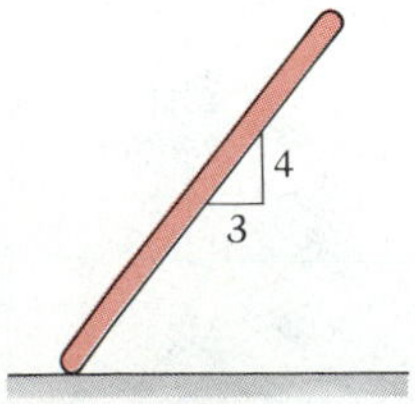

Figure P5.67

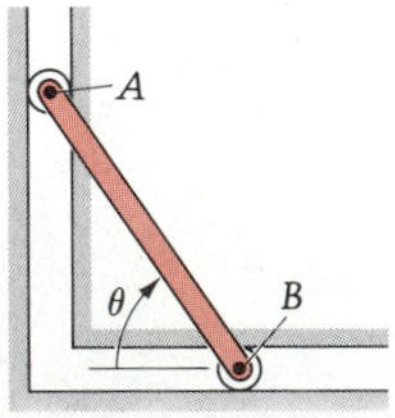

Figure P5.68

5.69 A slender uniform rod of weight W is smoothly hinged to a fixed support at A and rests on a block at B. (See Figure P5.69.) The block is suddenly removed. Find: (a) the initial angular acceleration and components of reaction at A; (b) the components of reaction at A when the rod becomes horizontal.

5.70 Two quarter-rings are pinned together at P and released from rest in the indicated position (Figure P5.70) on a smooth plane. Find the angular velocities of the rings when their mass centers are passing through their lowest points. *Hint*: By symmetry, point P always has only a vertical velocity component; this means that no work is done on either ring by the other, because (again by symmetry) the force between the rings has only a horizontal component normal to the velocity of P. More generally, as long as the pin is smooth, the work done by two pinned bodies in motion on each other will be the negative of each other because the velocities will be equal whereas the forces will be opposites.

* **5.71** A slender rod is placed on a table as shown in Figure P5.71. It will begin to pivot about the edge E and, at some angle θ_s, it will begin to slip. Find this angle, which will depend on the coefficient of friction μ and on k. *Hint*: Use all three equations of motion together with $W = \Delta T$. Eliminate α and ω^2, obtaining expressions for f and N. Setting $f = \mu N$ then permits a solution for θ_s. Solve the resulting equation when $\mu = 0.3$ and $k = 0.25$.

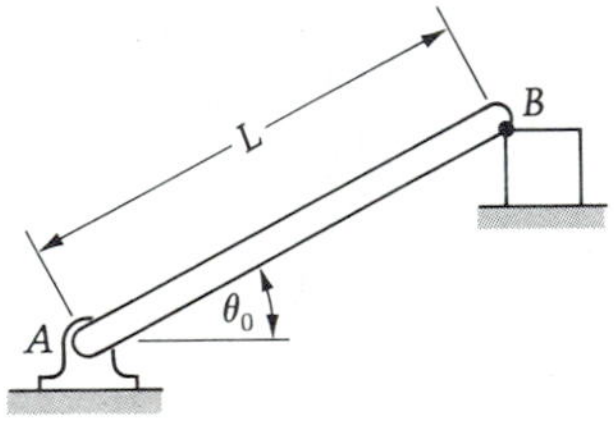

Figure P5.69

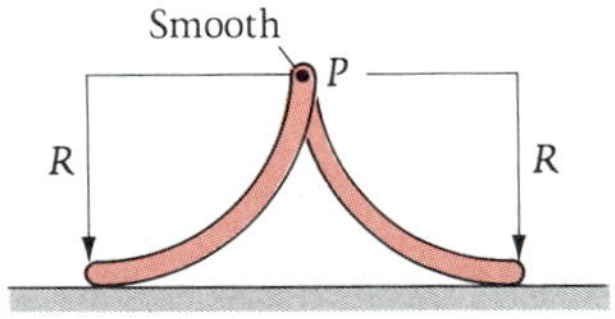

Figure P5.70

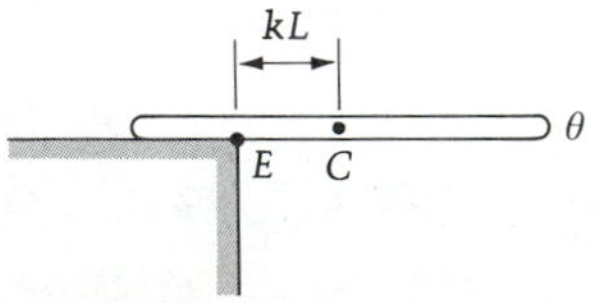

Figure P5.71

***5.72** The uniform equilateral triangular plate *ABE* in Figure P5.72 weighs *W* and is pinned to a fixed point at *A* and to a rope at *E*. The rope passes over a small, frictionless pulley at *D* and is then tied to the (equal) weight *W* which is constrained to move vertically. If the system is released in the given position with the angular velocity of *ABE* being 3 ↺ rad/sec, find the angular velocity of *ABE* in the dashed position (i.e., when side *AE* becomes horizontal).

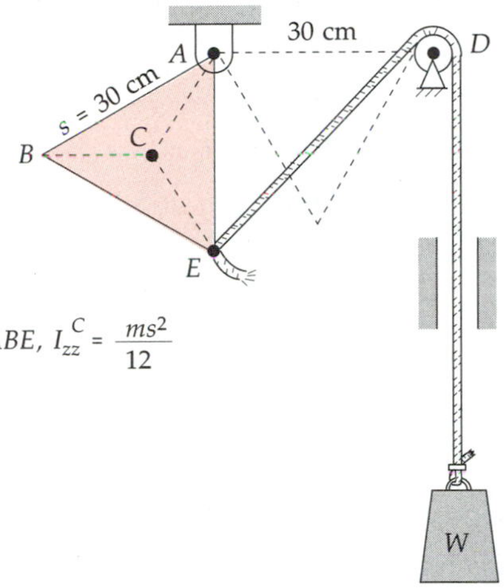

Figure P5.72

***5.73** This problem, continuing and using the results of Problem 4.150, is to verify that $W = \Delta T$ for that solution. First, note that only the friction couple does net work on the system of $\mathcal{B}_1$ plus $\mathcal{B}_2$, and verify:

$$\begin{aligned} W &= W_{\text{on } \mathcal{B}_1} + W_{\text{on } \mathcal{B}_2} \\ &= \int -k(\omega_1 - \omega_2)\omega_1\, dt + \int k(\omega_1 - \omega_2)\omega_2\, dt \\ &= -k \int (\omega_1 - \omega_2)^2\, dt \\ &= -46.1 \text{ lb-ft} \end{aligned}$$

Next, show that an identical result is obtained (as it must be) for ΔT:

$$\Delta T = T_f - T_i = T_f^{\mathcal{B}_1} + T_f^{\mathcal{B}_2} - T_i^{\mathcal{B}_1} - \overset{0}{\cancel{T_i^{\mathcal{B}_2}}}$$

***5.74** The cylinder $\mathcal{B}_1$ in Figure P5.74 has a spring attached and is released from rest. Assume that there is sufficient friction between $\mathcal{B}_1$ and the plane to prevent slip throughout the motion, and that the slot around which the cord is wrapped has a negligible effect on the cylinder's moment of inertia. Find the velocity of *B* when the unwrapped length of rope is completely vertical (that is, when *C* has 0.3 m left to travel before it would be directly above *E*). Assume also that the weight $\mathcal{B}_2$ moves only vertically — that is, that the rope does not start to sway.

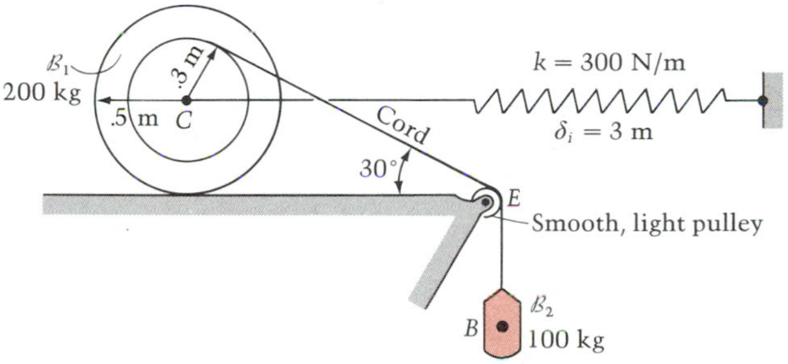

Figure P5.74

5.3 The Principles of Impulse and Momentum

The Equations of Impulse and Momentum for the Rigid Body in Plane Motion

The principle of work and kinetic energy is very helpful when the problem is posed in terms of positions and velocities. When time, rather than position, is the main concern, we often draw on a pair of principles concerned with impulse and momentum vectors. Just like $W = \Delta T$, these principles are obtained by general integrations of the equations of motion, but now the integration is directly with respect to time, without first dotting the equations with velocity. Thus they leave us with a set of vector equations instead of a single scalar result.

We have encountered one of the principles in Section 2.5 in our study of mass center motion: The impulse of the external forces imparted to any system equals its change of momentum over the same time interval. In Chapter 2 the system was general, so this principle holds for the rigid bodies we are now studying. From Equation (2.27), $\Sigma \mathbf{F} = \dot{\mathbf{L}}$ so that

$$\int \Sigma\mathbf{F}\,dt = \int \dot{\mathbf{L}}\,dt = \int d\mathbf{L} = \mathbf{L}\Big]_i^f = \mathbf{L}_f - \mathbf{L}_i = \Delta\mathbf{L} = m\mathbf{v}_{C_f} - m\mathbf{v}_{C_i} \tag{5.26}$$

Reviewing, we note that the integral $\int \Sigma\mathbf{F}\,dt$ is called the **impulse** (or **linear impulse**) imparted to the system by external forces. The vector $\Delta\mathbf{L}$ is the change in the system's momentum (or linear momentum) from the initial to the final time.

We need only the x and y components of Equation (5.26) for the rigid body in plane motion:

$$\int \Sigma F_x\,dt = \Delta L_x = \Delta(m\dot{x}_C) = m\dot{x}_{C_f} - m\dot{x}_{C_i} \tag{5.27}$$

$$\int \Sigma F_y\,dt = \Delta L_y = \Delta(m\dot{y}_C) = m\dot{y}_{C_f} - m\dot{y}_{C_i} \tag{5.28}$$

There is also a corresponding principle of *angular* impulse and momentum. From Equation (2.43),

$$\Sigma\mathbf{M}_C = \dot{\mathbf{H}}_C$$

so that

$$\int \Sigma\mathbf{M}_C\,dt = \int \dot{\mathbf{H}}_C\,dt = \int d\mathbf{H}_C = \mathbf{H}_C\Big]_i^f = \mathbf{H}_{C_f} - \mathbf{H}_{C_i} = \Delta\mathbf{H}_C \tag{5.29}$$

This equation may be put into a convenient form for rigid bodies in plane motion by recalling Equation (4.4) for the angular momentum:

$$\mathbf{H}_C = I_{xz}^C\omega\hat{\mathbf{i}} + I_{yz}^C\omega\hat{\mathbf{j}} + I_{zz}^C\omega\hat{\mathbf{k}}$$

For symmetric bodies in which the products of inertia vanish and $\Sigma\mathbf{M}_C = \Sigma M_C\hat{\mathbf{k}}$, this equation becomes

$$\mathbf{H}_C = I_{zz}^C\omega\hat{\mathbf{k}}$$

Therefore

$$\int \Sigma M_C\hat{\mathbf{k}}\,dt = \Delta\mathbf{H}_C = \Delta(I_{zz}^C\omega\hat{\mathbf{k}})$$

or

$$\int \Sigma M_C\,dt = \Delta(I_{zz}^C\omega) = (I_{zz}^C\omega)_f - (I_{zz}^C\omega)_i \tag{5.30}$$

The integral $\int \Sigma M_C\,dt$ is called the **angular impulse** imparted to the system by the external forces and couples, and the quantity $\Delta(I_{zz}^C\omega)$ is the change in angular momentum, both taken about C.

A subtle but important point regarding Equation (5.30) must be understood here. We note from Equation (5.29) that angular impulse equals the change in angular momentum for any body (deformable as well as rigid); therefore the use of Equation (5.30) only requires that the body of interest behave rigidly at the start (t_i) and end (t_f) of the time interval (t_i, t_f). At those times the moment of momentum is $\mathbf{H}_C = I_{zz}^C\omega\hat{\mathbf{k}}$, even though this simple expression for $\mathbf{H}_C$ may not apply *between* t_i and t_f. A good example is an ice skater drawing in her arms to increase angular speed, as we shall see later in Example 5.14.

In summary, for the rigid body in plane motion we have the following two principles at our disposal:

1. Linear impulse and momentum:

$$\int \Sigma \mathbf{F}\, dt = \Delta(m\mathbf{v}_C)$$

from which we get

$$\hat{\mathbf{i}} \text{ coefficients:} \quad \int_{t_i}^{t_f} \Sigma F_x\, dt = m(\dot{x}_{C_f} - \dot{x}_{C_i}) \tag{5.31}$$

$$\hat{\mathbf{j}} \text{ coefficients:} \quad \int_{t_i}^{t_f} \Sigma F_y\, dt = m(\dot{y}_{C_f} - \dot{y}_{C_i}) \tag{5.32}$$

where the directions of x, y, and z are fixed in the inertial frame.

2. Angular impulse and momentum:

$$\int_{t_i}^{t_f} \Sigma M_{Cz}\, dt = I_{zz_f}^{C}\omega_f - I_{zz_i}^{C}\omega_i \tag{5.33}$$

We note also that if the products of inertia are not zero, we have

$$\int \Sigma M_{Cx}\, dt = (I_{xz}^{C}\omega)_f - (I_{xz}^{C}\omega)_i \tag{5.34}$$

and

$$\int \Sigma M_{Cy}\, dt = (I_{yz}^{C}\omega)_f - (I_{yz}^{C}\omega)_i \tag{5.35}$$

Question 5.5 Are the coordinate axes associated with Equations (5.34) and (5.35) the same as those of (5.31) and (5.32)?

In the remainder of this section we treat only examples of symmetric bodies, for which Equations (5.31) to (5.33) are the impulse and momentum equations. The first example deals with both linear and angular impulse and momentum for a single body.

Answer 5.5 Yes.

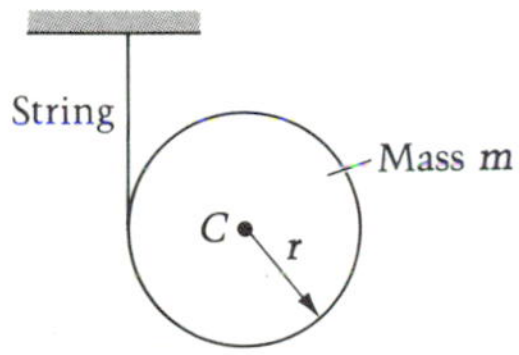

Figure E5.11a

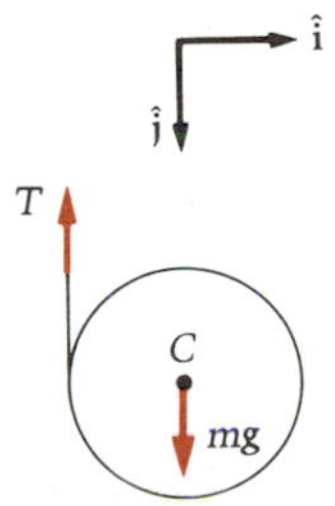

Figure E5.11b

EXAMPLE 5.11

A cylinder has a string wrapped around it (see Figure E5.11a) and is released from rest. Determine the velocity of C as a function of time.

Solution

We choose the sign convention to be as shown in Figure E5.11b, since the cylinder turns clockwise as C moves downward. Applying the impulse and momentum equations in the y and θ directions (note that $\Sigma F_x = 0$ means $\ddot{x}_C = 0$ so that $\dot{x}_C = \text{constant} = 0$) gives

$$\int_0^t \Sigma F_y \, dt = m(\dot{y}_{C_f} - \dot{y}_{C_i})$$

$$\int_0^t (mg - T) \, dt = m(\dot{y}_C - 0)$$

$$mgt - \int_0^t T \, dt = m\dot{y}_C \qquad (1)$$

and

$$\int_0^t \Sigma M_C \, dt = I_C(\omega_f - \omega_i)$$

$$\int_0^t Tr \, dt = \frac{1}{2} mr^2\omega$$

$$\int_0^t T \, dt = \frac{1}{2} mr\omega \qquad (2)$$

We note that $\int T\,dt$ is itself an unknown and should be treated as such. (In this problem, use of the equations of motion would separately tell us that $T = mg/3$, so that the integral is in fact Tt. But sometimes T is time-dependent, in which case Tt would be incorrect for the value of the integral.)

Eliminating $\int_0^t T \, dt$ gives

$$mgt - m\dot{y}_C = \frac{1}{2} mr\omega$$

However, $\dot{y}_C = r\omega$ because the cylinder rolls on the rope, so that

$$gt = \left(1 + \frac{1}{2}\right)\dot{y}_C$$

which gives our result:

$$\dot{y}_C = \frac{2}{3} gt$$

The cylinder falls with "$\frac{2}{3}$ of a g" because of the retarding force of the rope.

In the next example, several bodies are involved, making the solution a little more difficult. Two limiting case checks are discussed at the end.

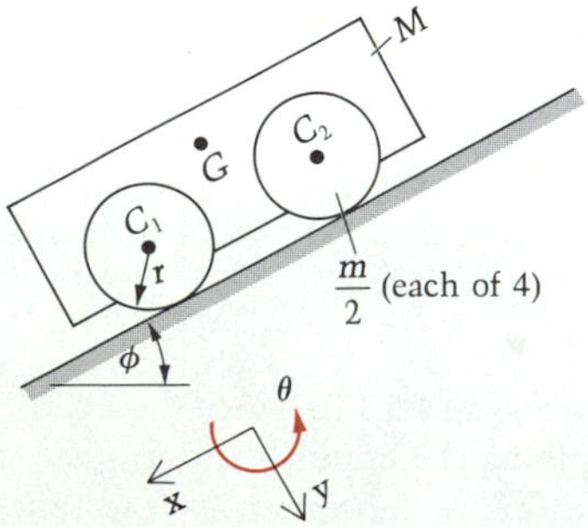

Figure E5.12a

EXAMPLE 5.12

The cart shown in Figure E5.12a has mass M exclusive of its four wheels, each of which is a disk of mass $m/2$. The front wheels and their axle are rigidly connected, and the same is true for the rear wheels. If the axles are smooth, find the velocity of G (the cart's mass center) as a function of time. The system starts from rest. Assume that there is enough friction to prevent the wheels from slipping.

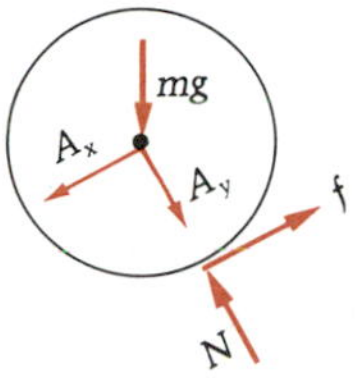

Figure E5.12b

Solution

We first consider the free-body diagram (Figure E5.12b) of a wheel pair (either front or back). Since the front and rear wheels are constrained to have identical angular velocities at all times, the *fr*'s (front and back) must produce identical $I\alpha$'s; hence the friction force is the same for the rear wheels as for the front. And since the wheels' mass centers must always have identical velocities, the forces acting down the plane on each pair (front and back) must also be equal. These resultants are $A_x + mg \sin \phi - f$, so the reaction A_x is also the same on each pair of wheels.

Question 5.6 Are A_y and N also the same for front and rear wheels?

From the free-body diagram, we may write the following linear and angular equations of impulse and momentum:

$$\int_0^t \Sigma F_x \, dt = m(\dot{x}_{C_f} - \dot{x}_{C_i})$$

$$\int_0^t (A_x + mg \sin \phi - f) \, dt = m\dot{x}_{C_{1 \text{ or } 2}} \tag{1}$$

$$\int_0^t \Sigma M_C \, dt = I_C(\omega_f - \omega_i) = I_C(\omega - 0)$$

$$\int_0^t fr \, dt = \frac{1}{2} mr^2\omega \tag{2}$$

We may also isolate a free-body diagram of the translating cart (Figure E5.12c) and write its equation of impulse and momentum in the x direction down the plane:

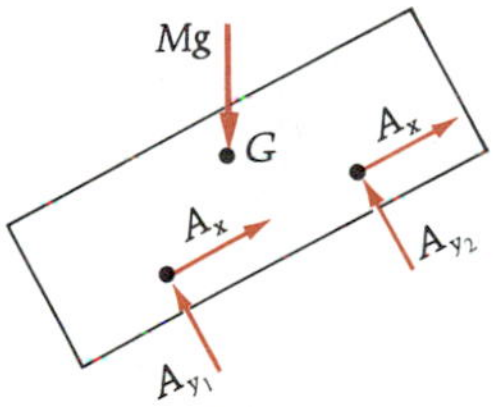

Figure E5.12c

$$\int_0^t \Sigma F_x \, dt = M(\dot{x}_{G_f} - \dot{x}_{G_i})$$

$$\int_0^t (-2A_x + Mg \sin \phi) \, dt = M(\dot{x}_G - 0) \tag{3}$$

Next we note that C_1 and C_2, the mass centers of the front and back wheel pairs, are also points of the cart; thus $\dot{x}_{C_1} \equiv \dot{x}_{C_2} \equiv \dot{x}_G$. Using this relation, and adding Equation (3) to twice Equation (1), eliminates the unknown impulse of the reaction A_x:

$$\int_0^t [(M + 2m)g \sin \phi - 2f] \, dt = (M + 2m)\dot{x}_G{}^* \tag{4}$$

Similarly, adding Equation (4) to twice Equation (2) results in an equation free of the unknown friction force:

$$\int_0^t (M + 2m)g \sin \phi \, dt = (M + 2m)\dot{x}_G + m \overbrace{(r\omega)}^{\dot{x}_G}$$

Answer 5.6 Not in general. They depend on the position of G relative to C_1 and C_2.

* This is the equation of linear impulse and momentum for the total (nonrigid) system of cart plus wheels.

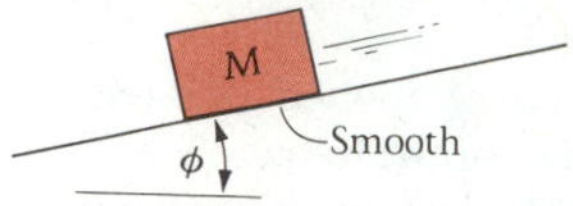

Figure E5.12d

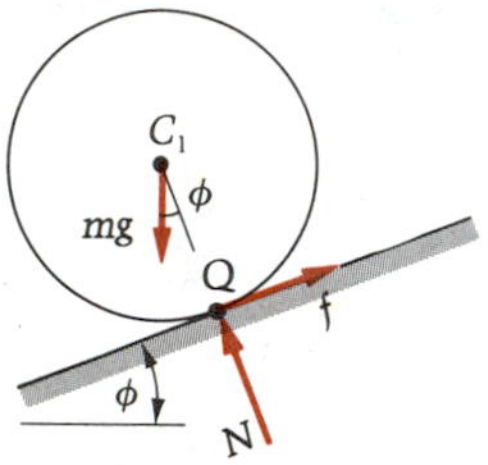

Figure E5.12e

Carrying out the integration and solving for $\dot{x}_G$, we get

$$\dot{x}_G = \frac{(M + 2m)\, gt \sin \phi}{M + 3m}$$

We note from this result that if the wheels are very light compared to the weight of the cart ($m \ll M$), then $\dot{x}_G = gt \sin \phi$, which is the answer for the problem of Figure E5.12d. Thus light wheels on smooth axles makes the cart move as if it were on a smooth plane, as expected.

The reader may wish to examine the other limiting case, that of the cart being light compared to heavy wheels ($M \ll m$). In this case the result, using the free-body diagram in Figure E5.12e, is $\dot{x}_G = \frac{2}{3}\, gt \sin \phi$.

In the next example, the two bodies — one rolling and the other translating — are connected by an inextensible cord.

EXAMPLE 5.13

Force P acts on the rolling cylinder $\mathcal{C}$ beginning at $t = 0$ with $\mathcal{C}$ at rest. (See Figure E5.13a.) Force P varies with the time t in seconds according to

$$P = 5 \sin \frac{\pi t}{10} \text{ N} \qquad \text{(positive to the left as shown)}$$

Cylinder $\mathcal{C}$ and body $\mathcal{B}$ respectively weigh 100 and 40 N. Find the velocity of G (the mass center of $\mathcal{B}$) when $t = 10$ s. Neglect the effect of the hubs in Figure E5.13b (and the drilled hole to accommodate force P) on the moment of inertia of $\mathcal{C}$.

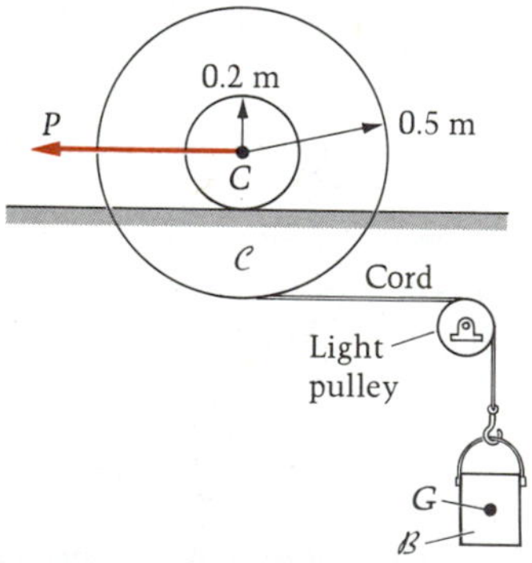

Figure E5.13a

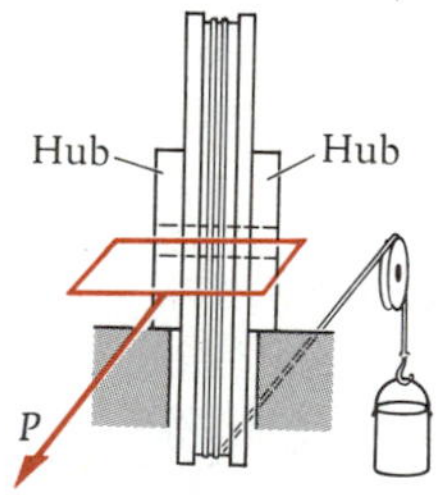

Figure E5.13b

Solution

Using the free-body diagrams (Figures E5.13c and E5.13d), we may write the equations of impulse and momentum. On $\mathcal{C}$, using Figure E5.13c,

$$\int \Sigma F_x \, dt = m\dot{x}_{C_f} - m\cancelto{0}{\dot{x}_{C_i}}$$

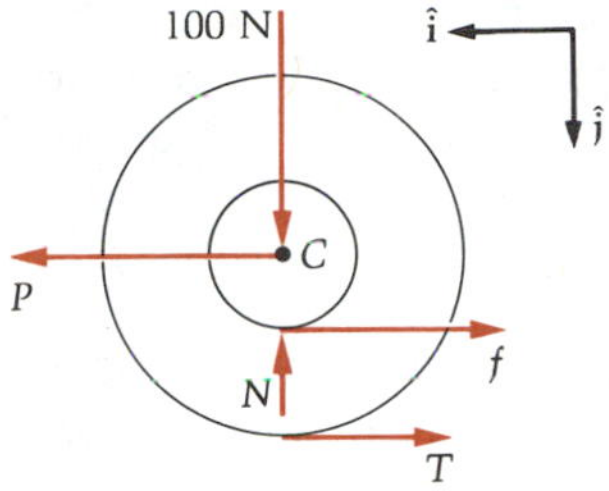

Figure E5.13c

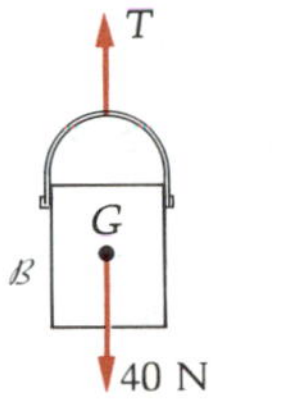

Figure E5.13d

or

$$\int_0^{10}\left(5\sin\frac{\pi t}{10} - f - T\right)dt = \frac{100}{9.81}\,\dot{x}_{C_f} = 10.2\dot{x}_{C_f} \tag{1}$$

Also on $\mathcal{C}$:

$$\int \Sigma M_C\,dt = I_C\omega_f - \cancelto{0}{I_C\omega_i}$$

or

$$\int_0^{10}(0.2f + 0.5T)\,dt = \left[\frac{1}{2}\,\frac{100}{9.81}\,(0.5)^2\right]\omega_f = 1.27\omega_f \tag{2}$$

On $\mathcal{B}$, using Figure E5.13d,

$$\int \Sigma F_y\,dt = m\dot{y}_{G_f} - \cancelto{0}{m\dot{y}_{G_i}}$$

or

$$\int_0^{10}(40 - T)\,dt = \frac{40}{9.81}\,\dot{y}_{G_f} = 4.08\dot{y}_{G_f} \tag{3}$$

Subtracting Equation (3) from (1), after integrating the sine function, we obtain

$$\frac{100}{\pi} - \int_0^{10} f\,dt - 40(10) = 10.2\dot{x}_{C_f} - 4.08\dot{y}_{G_f} \tag{4}$$

To obtain a second, independent equation that is also free of the integral of the unknown tension T, we add Equation (1) to twice Equation (2):

$$\frac{100}{\pi} - 0.6\int_0^{10} f\,dt = 10.2\dot{x}_{C_f} + 2.54\omega_f \tag{5}$$

Multiplying Equation (4) by 0.6 and subtracting from Equation (5) gives

$$\frac{40}{\pi} + 240 = 4.08\dot{x}_{C_f} + 2.54\omega_f + 2.45\dot{y}_{G_f} \tag{6}$$

Kinematics now relates $\dot{x}_{C_f}$, $\dot{y}_{G_f}$, and ω_f; the lowest point of $\mathcal{C}$ has the same velocity magnitude as does G because of the inextensibility of the cord (see Figure E5.13e):

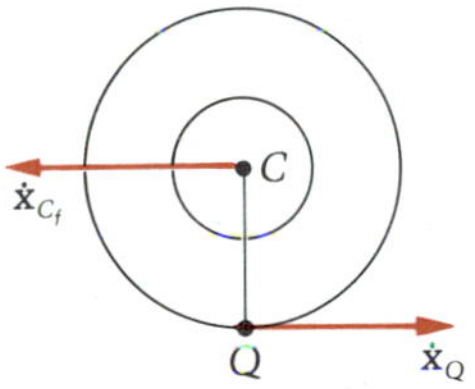

Figure E5.13e

$$\text{Kinematic conditions:}\qquad \left.\begin{aligned}\dot{x}_{C_f} &= 0.2\omega_f\\ \dot{x}_Q &= 0.3\omega_f\\ \dot{x}_Q &= \dot{y}_{G_f}\end{aligned}\right\} \Rightarrow \left\{\begin{aligned}\omega_f &= 3.33\dot{y}_{G_f}\\ \dot{x}_{C_f} &= 0.2\omega_f = 0.667\dot{y}_{G_f}\end{aligned}\right.$$

Substituting these expressions for $\dot{x}_{C_f}$ and ω_f into Equation (6) gives

$$253 = \dot{y}_{G_f}\,[4.08(0.667) + 2.54(3.33) + 2.45]$$

$$\dot{y}_{G_f} = \frac{253}{13.6} = 18.6 \text{ m/s}$$

Hence the velocity of the mass center of $\mathcal{B}$ at $t = 10$ s (when the force changes direction) is

$$\mathbf{v}_G = 18.6\downarrow \text{ m/s}$$

We emphasize that Equations (1), (2), and (3) in the preceding example are merely first integrals of the equations of motion studied in Chapter 4.

Conservation of Momentum

As we saw in Section 2.5, if the force in any direction (let us use x, for example) vanishes over a time interval, then the impulse in that direction vanishes also:

$$\int_{t_i}^{t_f} \Sigma F_x \, dt = 0$$

Since this impulse equals the change in momentum in the x direction, we have zero change when $\Sigma F_x \equiv 0$ and thus the momentum is *conserved* in that direction between t_i and t_f:

$$0 = m(\dot{x}_{C_f} - \dot{x}_{C_f})$$

or

$$m\dot{x}_{C_i} = m\dot{x}_{C_f} \tag{5.36}$$

We would of course also have **conservation of momentum** in the y (or any other) direction in which the force resultant vanished.

Finally, if the z component of $\Sigma \mathbf{M}_C$ is zero between t_i and t_f, then the *angular* impulse vanishes and we have **conservation of angular momentum:**

$$\int_{t_i}^{t_f} \Sigma M_{Cz} \, dt = 0 = H_{C_f} - H_{C_i}$$

or

$$H_{C_i} = H_{C_f} \tag{5.37}$$

For plane motion of symmetric bodies (I_{xz}^C and $I_{yz}^C = 0$) we have $\mathbf{H}_C = I_{zz}^C \omega \hat{\mathbf{k}}$; there is then no need for the z subscript, and Equation (5.37) may be rewritten

$$(I_C \omega)_i = (I_C \omega)_f \tag{5.38}$$

We now consider a well-known example of conservation of angular momentum.

Figure E5.14

EXAMPLE 5.14

A skater spinning about a point on the ice (see Figure E5.14) draws in her arms and her angular speed increases.

a. Is angular momentum conserved?
b. Is kinetic energy conserved?
c. Account for any gains or losses if either answer is no.

Solution

We begin with part (a). *Before* the skater draws in her arms, we may treat her as a rigid body and thus $H_{Ci} = I_1\omega_1$. The same is true *after* the arms are drawn in, so that $H_{Cf} = I_2\omega_2$. If we neglect the small friction couple at the skates and the small drag moments caused by air resistance, then the answer to part (a) is yes because $\Sigma \mathbf{M}_C$ is then zero. Thus

$$I_1\omega_1 = I_2\omega_2$$

Therefore

$$\omega_2 = \frac{I_1}{I_2}\omega_1 > \omega_1 \qquad \text{(showing an angular speed increase since } I_1 > I_2\text{)}$$

For part (b) the kinetic energies are

$$T_1 = \frac{1}{2} I_1\omega_1^2$$

$$T_2 = \frac{1}{2} I_2\omega_2^2 = \frac{1}{2} I_1\omega_1^2\left(\frac{I_1}{I_2}\right) > T_1$$

Thus kinetic energy is *not* conserved.

For part (c) the change in kinetic energy is seen to be positive:

$$\Delta T = T_2 - T_1 = \frac{1}{2} I_1\omega_1^2 \underbrace{\left(\frac{I_1}{I_2} - 1\right)}_{>0 \text{ since } I_1 > I_2}$$

Since there is no work done by the external forces and couples,* it is clear that this kinetic energy increase is accompanied by an internal energy decrease within the skater's body as her muscles do (nonexternal) work on her (nonrigid) arms in drawing them inward. Since *total* energy is always conserved (first law of thermodynamics), the skater has lost *internal* energy in the process.

The next example is similar to Examples 2.18 and 2.19 except that now the pulley has mass.

Example 5.15

Two identical twin gymnasts, *L* and *R*, of mass *m* are in equilibrium holding onto a stationary rope in the position shown in Figure E5.15a. The rope passes over the pulley $\mathcal{B}$, which has moment of inertia *I* with respect to the axis through *O* normal to the figure. The gymnasts then begin to move on the rope at speeds relative to it of $\dot{y}_{L\,\text{rel}}$ upward and $\dot{y}_{R\,\text{rel}}$ downward. When gymnast *R* reaches the end of the rope, he discovers he is in the same spot in space at which he began. How far up or down (tell which) has *L* moved (a) relative to the rope? (b) in space?

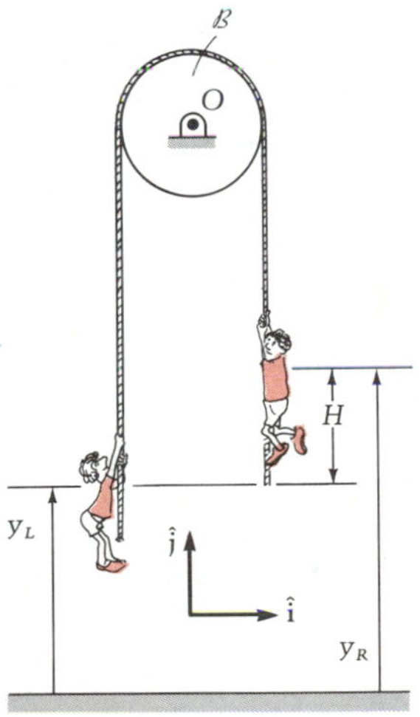

Figure E5.15a

* Assuming her arms are drawn in at the same level.

Solution

If we select our system to be "everything": L, R, the rope, and $\mathcal{B}$, then $\Sigma \mathbf{M}_O = \mathbf{0}$, and thus angular momentum about O is conserved. (Note that the gymnasts' gravity forces' moments about O cancel, and the pin reactions and weight of $\mathcal{B}$ pass through O.) Therefore,

$$\mathbf{H}_{O_f} = \mathbf{H}_{O_i} = \mathbf{0} \qquad \text{(since all bodies are at rest initially)} \tag{1}$$

After motion begins, the angular momentum about O of $\mathcal{B}$ is simply $I_O^{\mathcal{B}} \omega_{\mathcal{B}} \hat{\mathbf{k}}$, since O is a pivot of $\mathcal{B}$. For L and R, however, the situation is different and needs some discussion. The gymnasts, of course, are not rigid bodies, and we shall resort to Equation (2.38) to write their angular momenta with respect to O. For L, with velocity $\mathbf{v}_L$ in the inertial frame (the ground), and mass center C_L,

$$\mathbf{H}_O^L = \mathbf{H}_{C_L}^L + \mathbf{r}_{OC_L} \times m\mathbf{v}_{C_L}$$

We now assume that the gymnast is a particle; this is equivalent to neglecting $\mathbf{H}_{C_L}^L$, the gymnast's angular momentum about his mass center, in comparison with the "$\mathbf{r} \times m\mathbf{v}$" term. This is a good assumption because whatever body motions are not translatory are caused by parts (arms, mostly) in relative motion fairly close to the mass center. Thus,

$$\mathbf{H}_O^L \approx \mathbf{r}_{OC_L} \times m\mathbf{v}_L = \mathbf{r}_{OP} \times m\mathbf{v}_L$$

where P is shown in Figure E5.15b. Similarly,

$$\mathbf{H}_O^R \approx \mathbf{r}_{OC_R} \times m\mathbf{v}_R = \mathbf{r}_{OQ} \times m\mathbf{v}_R$$

Question 5.7 Why is $\mathbf{r}_{OC_R} \times m\mathbf{v}_R = \mathbf{r}_{OQ} \times m\mathbf{v}_R$?

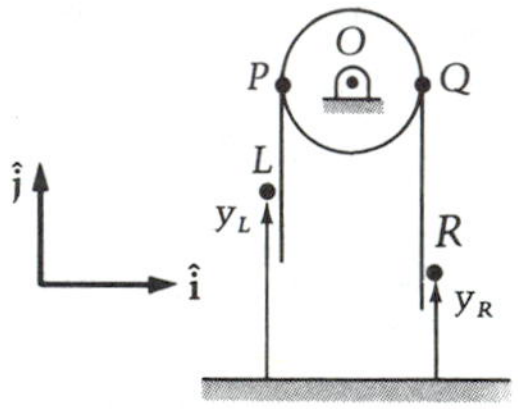

Figure E5.15b

Now, if we locate (see Figure E5.15b) the gymnasts' position *in space* with the coordinates y_L and y_R, then

$$\mathbf{H}_O^L = \mathbf{r}_{OP} \times m\mathbf{v}_L = -r\hat{\mathbf{i}} \times m\dot{y}_L\hat{\mathbf{j}} = -mr\dot{y}_L\hat{\mathbf{k}}$$
$$\mathbf{H}_O^R = \mathbf{r}_{OQ} \times m\mathbf{v}_R = r\hat{\mathbf{i}} \times m\dot{y}_R\hat{\mathbf{j}} = mr\dot{y}_R\hat{\mathbf{k}}$$

Note that if the gymnasts move in *opposite* vertical directions on opposite sides of the pulley, their angular momenta about O will be in the *same direction.* Note further that even though the "particles" L and R are in rectilinear motion, they still have angular momenta about points such as O that do not lie on their lines of motion.

Substituting the three angular momenta into Equation (1), we find:

$$\mathbf{H}_{O_f} = \mathbf{0}$$
$$\mathbf{H}_O^{\mathcal{B}} + \mathbf{H}_O^L + \mathbf{H}_O^R = 0$$
$$I\omega_{\mathcal{B}}\hat{\mathbf{k}} - mr\dot{y}_L\hat{\mathbf{k}} + mr\dot{y}_R\hat{\mathbf{k}} = 0 \tag{2}$$

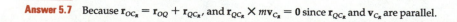

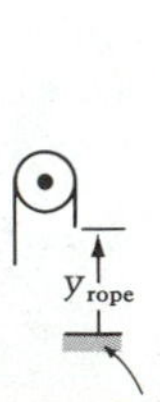

Starting postion of end of rope at $t = 0$

Figure E5.15c

Next, we know from the data that the rope moves counterclockwise around the pulley. Therefore, calling its speed $\dot{y}_{\text{rope}}$, we have (see Figure E5.15c)

$$\omega_B = \frac{\dot{y}_{\text{rope}}}{r}; \qquad \dot{y}_L = \dot{y}_{L\,\text{rel}} - \dot{y}_{\text{rope}}; \qquad \text{and} \qquad \dot{y}_R = -\dot{y}_{R\,\text{rel}} + \dot{y}_{\text{rope}}$$

Answer 5.7 Because $\mathbf{r}_{OC_R} = \mathbf{r}_{OQ} + \mathbf{r}_{QC_R}$, and $\mathbf{r}_{QC_R} \times m\mathbf{v}_{C_R} = \mathbf{0}$ since $\mathbf{r}_{QC_R}$ and $\mathbf{v}_{C_R}$ are parallel.

Substituting these into Equation (2), and simplifying the result, we obtain the equation

$$I\ddot{y}_{\text{rope}} + mr^2(-\dot{y}_{L\,\text{rel}} + \dot{y}_{\text{rope}} - \dot{y}_{R\,\text{rel}} + \dot{y}_{\text{rope}}) = 0 \tag{3}$$

Integrating,

$$(I + 2mr^2)y_{\text{rope}} - mr^2 y_{L\,\text{rel}} - mr^2 y_{R\,\text{rel}} = C_1$$

But $C_1 = 0$ since $y_{\text{rope}} = y_{L\,\text{rel}} = y_{R\,\text{rel}} = 0$ at $t = 0$. Now at the time when R is at the end of the rope, $y_{R\,\text{rel}} = H$, and $y_{\text{rope}} = H$ also, so that

$$mr^2 y_{L\,\text{rel}} = (I + 2mr^2)H - mr^2 H = (I + mr^2)H \Rightarrow y_{L\,\text{rel}} = \left(\frac{I + mr^2}{mr^2}\right)H$$

Therefore

$$y_L = y_{L\,\text{rel}} - y_{\text{rope}} = \left(\frac{I + mr^2}{mr^2}\right)H - H = \frac{I}{mr^2}H$$

and gymnast L moves up $\left(\dfrac{I + mr^2}{mr^2}\right)H$ relative to the rope and $\dfrac{IH}{mr^2}$ in space from his original position.

We can check Example 2.18 by going back to Equation (2) of the preceding example, and letting $I \to 0$ (note that at this point nothing has been said about the relative motions):

$$\overset{0}{\cancel{I}}\,\omega_{\mathcal{B}} - mr\dot{y}_L + mr\dot{y}_R = 0$$

or

$$\dot{y}_L = \dot{y}_R$$

Therefore, regardless of relative motions (with respect to the rope), the two gymnasts rise in space equal amounts. The reader may wish to show that for the problem of Example 2.18 but with $I > 0$, the left gymnast is pulled up less than the height of climb in space of the right gymnast.

We now consider an example in which (unusually) both angular momentum *and* kinetic energy are conserved.

EXAMPLE 5.16

The 2-kg collar $\mathcal{C}$ in Figures E5.16a,b turns along with the smooth rod $\mathcal{R}$ (see Figure E5.16a), which is 1 m long, has a mass of 3 kg, and is mounted in bearings with negligible friction. The angular speed is increased until the cord breaks (its tensile strength is 60 N), and at that instant the external moment is removed. Determine the angular velocity of $\mathcal{R}$ and the velocity of C (the mass center of $\mathcal{C}$) when the collar leaves the rod.

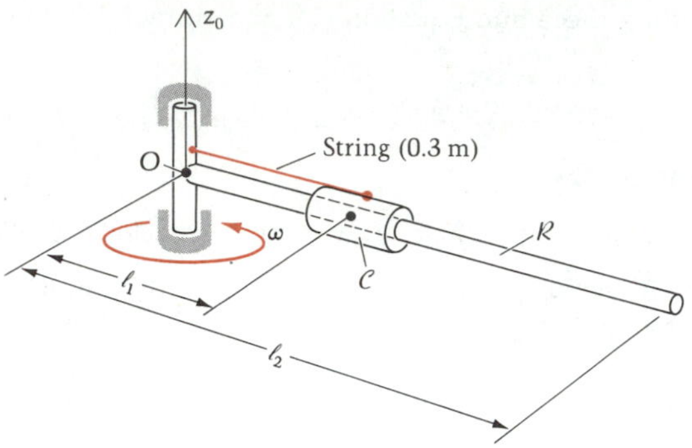

Figure E5.16a

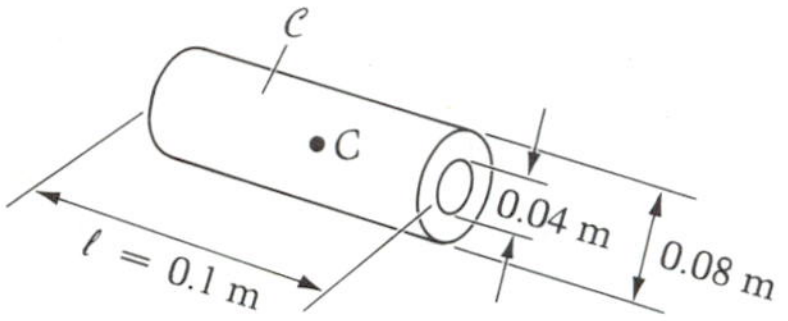

Figure E5.16b

Solution

The string provides the force causing the centripetal (inward) acceleration until it breaks. At that instant, we may solve for the angular velocity of $\mathcal{C}$:

$$\Sigma F_n = ma_{C_n}$$
$$T = m\ell_1\omega_1^2$$
$$60 = 2(0.3)\omega_1^2$$
$$\omega_1 = 10 \text{ rad/s}$$

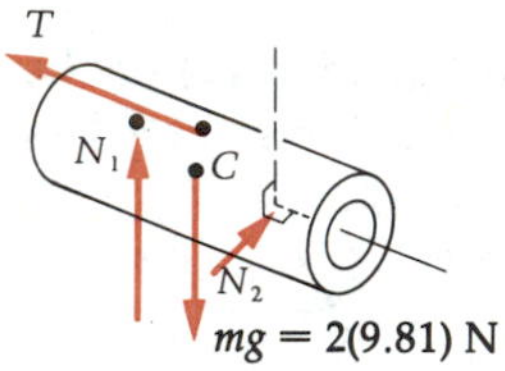

Figure E5.16c

In the accompanying free-body diagram, Figure E5.16c, N_1 and N_2 are the vertical and horizontal resultants of the pressures of the inside wall of $\mathcal{C}$ exerted by $\mathcal{R}$.

After the rope breaks at time t_1, collar $\mathcal{C}$ moves outward in addition to turning with $\mathcal{R}$; this is because there is no longer any inward force to keep it from "flying off on a tangent." Between times t_1 and t_2 (when it leaves $\mathcal{R}$), we have the following for the system ($\mathcal{C}$ plus $\mathcal{R}$):

1. Conservation of angular momentum H_O about z_O (because the external forces have no moment about z_O).
2. Conservation of kinetic energy T (since no net work is done on the system). Note that the normal forces between rod and collar, being equal in magnitude but opposite in direction, act on points with equal velocity components in the direction of either force; hence their net work vanishes.

Condition 1 gives

$$H_{O_i} = H_{O_f}$$

$$\underbrace{I_O^{\mathcal{C}}\omega_i + I_O^{\mathcal{R}}\omega_i}_{\substack{\text{Until the string breaks,} \\ O \text{ is a point of both bodies!}}} = I_O^{\mathcal{R}}\omega_f + \overbrace{I_C^{\mathcal{C}}\omega_f + \left[\left(\ell_2 + \frac{\ell}{2}\right)\hat{\mathbf{i}} \times m_{\mathcal{C}}\mathbf{v}_C\right]_z}^{H_{O_f}^{\mathcal{C}} = H_{C_f}^{\mathcal{C}} + (\mathbf{r}_{OC} \times m_{\mathcal{C}}\mathbf{v}_{C_f})_z}$$

Thus

$$\left[\underbrace{m_{\mathcal{C}}\left(\frac{r_i^2 + r_o^2}{4} + \frac{\ell^2}{12}\right)}_{I_C^{\mathcal{C}} = 0.00267} + m_{\mathcal{C}}\ell_1^2\right]10 + \frac{m_R\ell_2^2}{3}\,10 = \frac{m_R\ell_2^2}{3}\,\omega_f + I_C^{\mathcal{C}}\omega_f + \left(\ell_2 + \frac{\ell}{2}\right)^2 \omega_f m_c$$

$$(0.00267 + 0.180)10 + \frac{3(1^2)}{3}\,10 = \omega_f + 0.00267\omega_f + (1.05)^2 2\omega_f$$

$$1.83 + 10 = \omega_f(1 + 0.00267 + 2.21)$$

$$\omega_f = 3.69 \text{ rad/s}$$

The component of $\mathbf{v}_C$ perpendicular to the rod $\mathcal{R}$ is thus $v_C = 1.05\omega_f = 3.87$ m/s. We can now obtain the radial component by conservation of T (condition 2):

$$\underbrace{\frac{1}{2} I_O \omega_i^2}_{\substack{O \text{ is (I) for both } \mathcal{C} \\ \text{and } \mathcal{R} \text{ initially}}} = \frac{1}{2} I_O^{\mathcal{R}} \omega_f^2 + \frac{1}{2} m_{\mathcal{C}} (\underbrace{v_{C_\parallel}^2 + v_{C_\perp}^2}_{\substack{\text{components of } \mathbf{v}_C \text{ parallel} \\ \text{and perpendicular to } \mathcal{R}}}) + \frac{1}{2} I_C^{\mathcal{C}} \omega_f^2$$

$$\tfrac{1}{2}(0.183 + 1)10^2 = \tfrac{1}{2}(1)3.69^2 + \tfrac{1}{2}(2)(v_{C_\parallel}^2 + 3.87^2) + \tfrac{1}{2}(0.00267)3.69^2$$

$$59.2 = 6.81 + v^2_{C_\parallel} + 15.0 + 0.0182$$

$$v_{C_\parallel} = 6.11 \text{ m/s}$$

Thus since the initial kinetic energy was 59.2 J and since

$$\frac{\tfrac{1}{2}(2)(6.11^2)}{59.2} = 0.631$$

we see that 63 percent of the original energy has gone into the outward motion of the collar.

Impact

We studied the impact of a pair of particles in Section 2.5. In this section we shall extend this study to two bodies colliding in plane motion.

The large forces occurring during an impact between two bodies $\mathcal{B}_1$ and $\mathcal{B}_2$ obviously deform the bodies. Because of vibrations and permanent deformations that are produced, some of the mechanical energy will be dissipated in the collision. However, it is often possible to treat a body as rigid *before,* and then again *after,* the impact in order to gain information of value. In impact problems we assume that:

1. Velocities and angular velocities may change greatly over the short impact interval Δt.
2. Positions of the bodies do not change appreciably.
3. Forces (and moments) that do not grow large over the interval Δt are neglected (such as gravity and spring forces). Such forces are called **nonimpulsive**; the large contact forces are called **impulsive**. It is the impulsive forces and moments that produce the sudden changes in velocities and angular velocities.

In Chapter 2 we introduced the coefficient of restitution as a measure of the capacity for colliding bodies to rebound off each other. We shall continue to use this parameter in this section, where now the relative velocities of separation and approach are of the impacting points of $\mathcal{B}_1$ and $\mathcal{B}_2$. Thus rigid-body kinematics will be needed to relate these velocities to those of the mass centers of the bodies. We emphasize again that the coefficient of restitution "e" is not the best of physical properties to measure; it depends upon the materials, geometry, and initial velocities. But as long as we take "e" with a grain of salt and remain aware of the limiting values $e = 0$ (bodies stick together) and $e = 1$ (no loss of energy), the definition of e does provide an approximate, much-needed equation that allows us to solve many problems of impact. We now consider two forms of the angular impulse and angular momentum equation that are applicable at the beginning and end of impacts involving the plane motion of bodies that may be regarded as rigid except during the collision phase of the motion.

If the body has a pivot O, we recall that

$$\mathbf{H}_O = I_{xz}^O \omega \hat{\mathbf{i}} + I_{yz}^O \omega \hat{\mathbf{j}} + I_{zz}^O \omega \hat{\mathbf{k}}$$

With O fixed in the inertial frame we have

$$\Sigma \mathbf{M}_O = \dot{\mathbf{H}}_O$$

Thus we may replace the C by an O in the angular impulse and momentum equation (5.33) for such pivot cases. The resulting equation about the axis of rotation is

$$\int_{t_i}^{t_f} \Sigma M_O \, dt = H_{Of} - H_{Oi} = I_{zz}^O(\omega_f - \omega_i)^* \tag{5.39}$$

This formula is of considerable value in impact problems because impulsive pivot reactions have no moment about O and thus do not appear in the equation. Note that if O is C, then Equation (5.39) is the same as our previous Equation (5.33) written about the mass center.

Another useful equation follows from

$$\Sigma \mathbf{M}_P = \dot{\mathbf{H}}_C + (\mathbf{r}_{PC} \times m\mathbf{a}_C) \tag{5.40}$$

In scalar form, for rigid bodies in plane motion this equation is

$$\Sigma M_P = I_{zz}^C \alpha + (\mathbf{r}_{PC} \times m\mathbf{a}_C)_z$$

in which P is an arbitrary point. If we state that P is now a fixed point O of the inertial frame $\mathcal{I}$, we may integrate this equation, getting

$$\int_{t_i}^{t_f} \Sigma M_O \, dt = (H_{Cf} - H_{Ci}) + (\mathbf{r}_{OC} \times m\mathbf{v}_C)_z \Big]_i^f$$

$$= \{I_{zz}^C \omega^* + (\mathbf{r}_{OC} \times m\mathbf{v}_C)_z\} \Big]_i^f \tag{5.41}$$

* We emphasize again that the angular momentum $\mathbf{H}_O$ (or $\mathbf{H}_C$) is not equal to its rigid body form $I_{zz}^O \omega \hat{\mathbf{k}}$ (or $I_{zz}^C \omega \hat{\mathbf{k}}$) *during* the impact, but these substitutions may be made at t_i before the collision and at t_f afterward.

Question 5.8 Why is the right-hand side not the integral of the right side of Equation (5.40) if O is moving?

Answer 5.8 If O is not fixed in $\mathcal{I}$, then $(d/dt)(\mathbf{r}_{OC} \times m\mathbf{v}_C) = \mathbf{r}_{OC} \times m\mathbf{a}_C - \mathbf{v}_O \times m\mathbf{v}_C$ and the second term is not zero then!

We shall now use these principles to solve a pair of example problems.

Example 5.17

An arrow of length L traveling with speed v_0 strikes a smooth hard wall obliquely as shown in the figure. End A does not penetrate but slides downward along the wall without friction or rebound. Find the angular velocity of the arrow after impact.

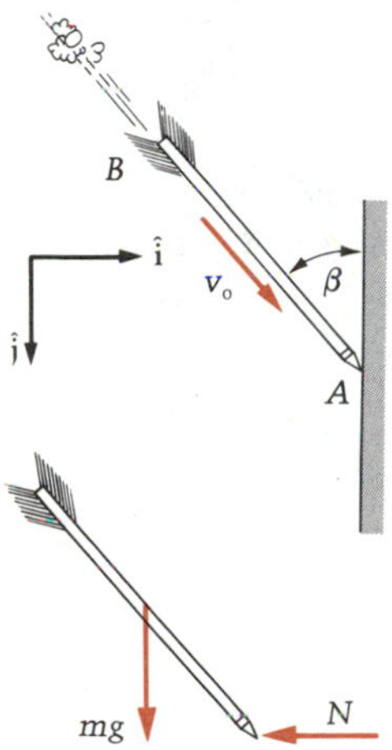

Figure E5.17

Solution

The only impulsive force acting on the arrow during its impact with the wall is the normal force N shown in the free-body diagram in Figure E5.17. We note that the gravity force over the short time interval is nonimpulsive:

$$\int_{t=0}^{\Delta t} mg\, dt\hat{\mathbf{j}} = mg\, \Delta t\hat{\mathbf{j}}$$

This is negligible in magnitude if Δt is very small, since mg does not grow large during impact. But for the normal force, that is not the case, as

$$\int_{t=0}^{\Delta t} N\, dt(-\hat{\mathbf{i}}) = -N_{\text{average}}\, \Delta t\hat{\mathbf{i}}$$

and this is non-negligible since N grows large "impulsively" during the short interval Δt, with average value N_{average}. In what follows, we shall delete the subscript and simply denote the impulse of $-N\hat{\mathbf{i}}$ as $-N\,\Delta t\hat{\mathbf{i}}$.

The impulse and momentum equation is then:

$$-N\,\Delta t\hat{\mathbf{i}} = m(v_{Cx_f} - v_0 \sin\beta)\hat{\mathbf{i}} + m(v_{Cy_f} - v_0\cos\beta)\hat{\mathbf{j}}$$

or

$$\frac{-N\,\Delta t}{m} = v_{Cx_f} - v_0 \sin\beta \tag{1}$$

and

$$v_{Cy_f} = v_0 \cos\beta \tag{2}$$

where we note that momentum is conserved in the y-direction during impact.

The angular impulse and angular momentum principle yields:

$$N\,\Delta t\,\frac{L}{2}\cos\beta = \frac{mL^2}{12}\,\omega_f \tag{3}$$

"No rebound" means the x-component of $\mathbf{v}_{A_f}$ is zero; thus:

$$\mathbf{v}_C = v_{Cx_f}\hat{\mathbf{i}} + v_{Cy_f}\hat{\mathbf{j}} = v_{Ay_f}\hat{\mathbf{j}} + \omega_f\hat{\mathbf{k}} \times \mathbf{r}_{AC}$$

Using $\mathbf{r}_{AC} = L/2\,(-\sin\beta\hat{\mathbf{i}} - \cos\beta\hat{\mathbf{j}})$, we find that the x-component of this equation is:

$$v_{Cx_f} = \frac{L\omega_f}{2}\cos\beta \tag{4}$$

We have four equations in the unknowns v_{Cx_f}, v_{Cy_f}, ω_f, and $N\,\Delta t$. Solving, we find

$$\boldsymbol{\omega}_f = \omega_f\hat{\mathbf{k}} = \frac{6v_0\sin\beta\cos\beta}{L(1+3\cos^2\beta)}\,\hat{\mathbf{k}}$$

Note the obvious, that a head-on impact ($\beta = \pi/2$) brings the arrow to a dead stop with $\omega_f = 0$ and, from Equation (4), $v_{Cx_f} = 0$ also.

The next example, and the comments following it, constituted the solution to an actual engineering problem.

EXAMPLE 5.18

A 770-ton steel nuclear reactor vessel is being transported down a 6.5 percent grade using a specially designed suspended hauling platform together with crawler transporters. (See Figure E5.18a.) Determine the maximum velocity at which the reactor can be transported without tipping over if it should strike, and pivot about, a rigid obstacle at the front edge of the vessel's base ring.

Figure E5.18a *(Courtesy American Rigging Co.)*

Solution

The reactor vessel $\mathcal{B}$ will tip over if there is any kinetic energy left after it pivots about the front edge at O (see Figure E5.18b) and the mass center $\mathcal{C}$ reaches its highest point B, directly above O. Thus we solve for the velocity that will cause C to reach B; any higher velocity will cause overturning.

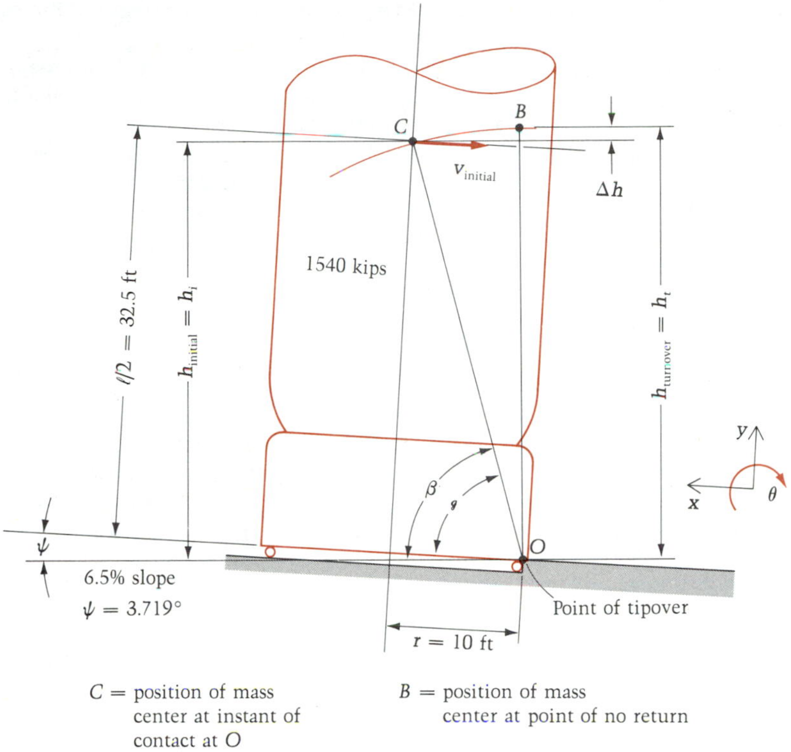

Figure E5.18b

We begin the solution with some preliminary geometric and trigonometric calculations based on the diagram. The 6.5 percent slope means $\psi = \tan^{-1}(0.065) = 3.719°$.* The turnover height h_t is given by the distance OC:

$$h_t = OC = OB = \sqrt{10^2 + 32.5^2} = 34.00 \text{ ft}$$

Also

$$\delta = \tan^{-1}\left(\frac{32.5}{10}\right) = 72.90°$$

and

$$\beta = \delta + \psi = 72.90° + 3.72° = 76.62°$$

The initial height h_i of C, above the horizontal line through O, is

$$h_i = (OC) \sin \beta$$
$$= (34.00) \sin 76.62° = 33.08 \text{ ft}$$

and thus the vertical distance through which C will move in reaching B is

$$\Delta h = h_t - h_i = 34.00 - 33.08$$
$$= 0.92 \text{ ft}$$

* We use four significant digits in this example.

(If we had adhered to three significant digits, the subtraction would have reduced us to just one good digit.)

There now remain two separate main parts to the solution of this problem. We first have to consider that mechanical energy is lost during the impact of $\mathcal{B}$ with the obstacle at O. Thus we are prevented from using the principle of work and kinetic energy over the short period of impact. What does apply, however, is conservation of angular momentum about O. This is because the impulsive forces (in both the x *and* y directions!) causing the sudden changes in the mass center velocity $\mathbf{v}_C$ and in the angular velocity ω are acting at O, so that $\Sigma\mathbf{M}_O = \mathbf{0}$. Therefore

$$\Sigma\mathbf{M}_O = \mathbf{0} = {}^{\mathcal{I}}\dot{\mathbf{H}}_O \Rightarrow \mathbf{H}_O = \text{constant vector}$$

or

$$\mathbf{H}_{Oi} = \mathbf{H}_{Of} \tag{1}$$

We shall use Equation (2.36) to express $\mathbf{H}_{Oi}$; this is the best formula for $\mathbf{H}_{Oi}$ for translation problems because $\mathbf{H}_{Ci} = \mathbf{0}$ in that case. For $\mathbf{H}_{Of}$, however, we have a nonzero ω. Thus we draw on the fact that since the vessel does not bounce at O, we may consider O a fixed point of both $\mathcal{B}$ and the inertial frame during and following the short period of impact. This in turn means that $\mathbf{H}_{Of}$ is simply $I_O\omega_f\hat{\mathbf{k}}$ after impact. Therefore Equation (1) becomes

$$\overset{\mathbf{0}}{\cancel{\mathbf{H}_{Ci}}} + \mathbf{r}_{OC} \times m\mathbf{v}_{Ci} = I_O\omega_f\hat{\mathbf{k}} \tag{2}$$

Now since

$$\mathbf{r}_{OC} = +r\hat{\mathbf{i}} + \frac{\ell}{2}\hat{\mathbf{j}} \qquad \text{and} \qquad \mathbf{v}_{Ci} = -v_{Ci}\hat{\mathbf{i}}$$

the left side of Equation (2) is simply

$$\frac{\ell}{2}mv_{Ci}\hat{\mathbf{k}} \tag{3}$$

By the parallel-axis (transfer) theorem for moments of inertia we have

$$I_O = I_C + md^2 = m\left[k_C^2 + \left(\frac{\ell}{2}\right)^2 + r^2\right]$$

The vessel is essentially a thick shell; considering then that

$$I_C = mk_C^2 \approx \frac{mr^2}{2} + \frac{m\ell^2}{12} \tag{4}$$

we see that

$$I_O = \frac{m\ell^2}{3} + \frac{3}{2}mr^2 \tag{5}$$

By substituting Equations (3) and (5) into (2), we thus obtain

$$m\frac{\ell}{2}v_{Ci} = m\left(\frac{\ell^2}{3} + \frac{3}{2}r^2\right)\omega_f$$

and the angular velocity after impact is then

$$\omega_f = \frac{3\ell v_{Ci}}{2\ell^2 + 9r^2} \circlearrowleft \tag{6}$$

We are now ready to proceed to the second part of the solution.

Between the start of pivoting (immediately following impact) and the arrival at point B, the system is easily analyzed by work and kinetic energy:

$$W = \Delta T = T_f - T_i$$

To obtain the least possible value of v_{Ci} for no overturning, we set $T_f = 0$. The only work done in this phase of the vessel's motion is by gravity, so that

$$W = -mg\,\Delta h = 0 - T_i = -\frac{1}{2} I_O \omega_f^2$$

where ω_f is now an *initial* angular speed for this final stage of the problem and O is still a pivot point for $\mathcal{B}$. Thus

$$-mg\,\Delta h = \frac{1}{2} m \left(\frac{\ell^2}{3} + \frac{3}{2} r^2\right)\left(\frac{3\ell v_{C_i}}{2\ell^2 + 9r^2}\right)^2 \tag{7}$$

or

$$v_{Ci} = \sqrt{\frac{4g\,\Delta h(2\ell^2 + 9r^2)}{3\ell^2}} = \sqrt{\frac{4(32.17)(0.92)(2 \times 65^2 + 9 \times 10^2)}{3 \times 65^2}}$$

$$= 9.4 \text{ ft/sec}$$

There are several important follow-on remarks to be made about the preceding example. The first is that it can be shown (with a coefficient of restitution analysis) that more energy is lost with no bounce at O than if rebounding takes place. This energy loss for the $e = 0$ case just studied is

$$\Delta E = T_i - T_f$$

where i and f refer to the instants just before and after the impact. Substituting (for the case of no bounce), we get

$$\Delta E = \frac{1}{2} m v_{Ci}^2 - \frac{1}{2} I_O \omega_f^2$$

$$= \frac{1}{2} m\left[v_{Ci}^2 - \left(\frac{\ell^2}{3} + \frac{3}{2} r^2\right)\left(\frac{3\ell v_{Ci}}{2\ell^2 + 9r^2}\right)^2\right]$$

$$= \frac{1}{2} m v_{Ci}^2 \left(1 - \frac{9\ell^2/6}{2\ell^2 + 9r^2}\right)$$

For $\ell = 65$ ft and $r = 10$ ft, we obtain

$$\Delta E = \frac{1}{2} m v_{Ci}^2 (0.322)$$

Thus 32.2%, or nearly a third of the original mechanical energy, is lost during impact if the lower front corner of $\mathcal{B}$ sticks to, and pivots about, point O. Of great importance here is the fact that we do not *know* how much rebounding would actually occur in the physical situation and hence how much energy would be lost. This means that 9.4 ft/sec *may not be a conservative engineering answer* for the safe speed. If the plane is flat ($\psi = 0$), for example, it can be shown that:

1. The speed corresponding to pivoting as in this example is 11.9 ft/sec. (It has farther to pivot so it can be going faster prior to impact.)
2. At this speed initially, and with a no-energy-lost rebound, the vessel will easily overturn even though the striking corner backs up.

A conservative safe speed of the vessel in the inclined plane case can be obtained by assuming that no energy is dissipated during the impact and that all the vessel's initial kinetic energy goes into tilting it up about O. This approach gives

$$-mg\,\Delta h = -\frac{1}{2}mv_{Ci}^2$$

$$v_{Ci} = \sqrt{2g\,\Delta h} = 7.7 \text{ ft/sec}$$

In practice, the engineers in this case decided not to exceed 3 ft/sec, in view of the importance of the work and the danger involved.

The Center of Percussion

We now turn our attention to a new topic. Besides the mass center C (but of much lesser importance), there is another special point of interest associated with a rigid body $\mathcal{B}$ in plane motion, a point that differs from C in that it depends not only on the mass distribution of $\mathcal{B}$ but also on the motion of the body. This point lies along the resultant of the $m\mathbf{a}$ vectors of all the body's mass elements. The point is called the **center of percussion**, and it has value in certain applications such as impact testing.

Before getting into the theory behind the center of percussion, we first illustrate its existence and demonstrate its value by means of an example. If a youngster hits a baseball with a stick and does not translate his hands too much, we may model the situation as shown in Figure 5.13 and ask where the ball should hit the stick in order to eliminate the "sting" (transverse reaction R_y of the stick onto the boy's hands). If the boy hits the ball at just the right place, called the "sweet spot," he hits it a long way while hardly feeling it and is said to have gotten "good wood" on the ball. Assuming the stick to be rigid at the beginning ($t = 0$) and end ($t = \Delta t$) of the short impact interval, the two principles of impulse and momentum are used as follows.

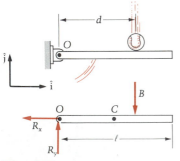

Figure 5.13

The impulse-momentum equation for the stick in the y direction is

$$\int_0^{\Delta t} (R_y - B)\,dt = m(\dot{y}_{C_f} - \dot{y}_{C_i}) \tag{5.42}$$

The angular impulse-angular momentum equation is

$$\int_0^{\Delta t} -Bd\,dt = I_O(\omega_f - \omega_i) = \frac{m\ell^2}{3}(\omega_f - \omega_i) \tag{5.43}$$

Using $R_y = 0$ for "no sting," setting $\dot{y}_C = (\ell/2)\omega$ by kinematics at $t = 0$ and Δt, and multiplying Equation (5.42) by d gives

$$-\int_0^{\Delta t} Bd\, dt = \frac{md\ell}{2}(\omega_f - \omega_i) \tag{5.44}$$

Dividing Equation (5.44) by (5.43) gives

$$1 = \frac{d}{2} \div \frac{\ell}{3} \Rightarrow d = \frac{2}{3}\ell$$

Thus if the ball is struck two-thirds of the way from O to the end of the stick, the transverse reaction R_y will be zero.

In this example the point at which the ball is struck ($d = 2\ell/3$) is the center of percussion of the stick. To show this, at least for the case when the bat is rigid, we first recall that in Chapter 2 we saw that for *any* point P (moving or not, fixed to $\mathcal{B}$ or not), we can always write

$$\Sigma \mathbf{M}_P = \int \mathbf{R} \times \mathbf{a}\, dm$$

in which $\mathbf{R}$ is the position vector from point P to a generic differential mass element. Therefore, since the integral of the vectors ($\mathbf{R} \times \mathbf{a}\, dm$) over the body in fact represents the resultant moment about P of the $m\mathbf{a}$ vectors over the mass of $\mathcal{B}$, this integral vanishes for all points P^* on the line along which the resultant of the $m\mathbf{a}$ vectors lies and hence $\Sigma \mathbf{M}_{P^*} = \mathbf{0}$ for these points.

Armed with the fact that $\Sigma \mathbf{M}_{P^*} = \mathbf{0}$ for the center of percussion, we can now derive the general equation for the distance from a pivot O to the center of percussion P^* in plane motion (Figure 5.14):

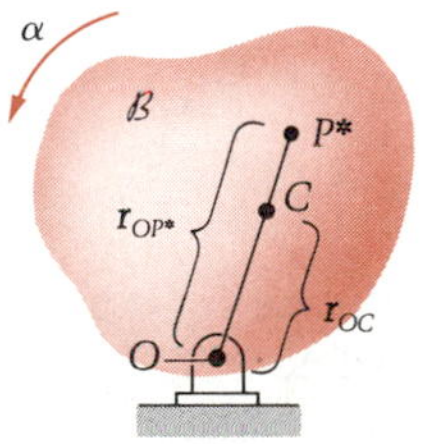

Figure 5.14

$$\Sigma M_O = \overset{0}{\cancel{\Sigma M_{P^*}}} + (\mathbf{r}_{OP^*} \times \overset{m\mathbf{a}_C}{\cancel{\Sigma \mathbf{F}}})_z$$

$$\alpha(I_C + mr_{OC}^2) = r_{OP^*} m(r_{OC}\alpha)$$

$$mk_C^2 + mr_{OC}^2 = mr_{OP^*} r_{OC}$$

Therefore

$$r_{OP^*} = r_{OC} + \frac{k_C^2}{r_{OC}} \tag{5.45}$$

and we see that, for the stick,

$$r_{OP^*} = \frac{\ell}{2} + \frac{\ell^2/12}{\ell/2} = \frac{2}{3}\ell \qquad \text{(as before)}$$

Note from Equation (5.45) that the center of percussion is always farther from the pivot than is the mass center. We make one final remark about the center of percussion. If we treat the "**a** dm's" of $\mathcal{B}$ as a collection of vectors, its resultant may be expressed (for a rigid body in plane motion) at the mass center (Figure 5.15), where

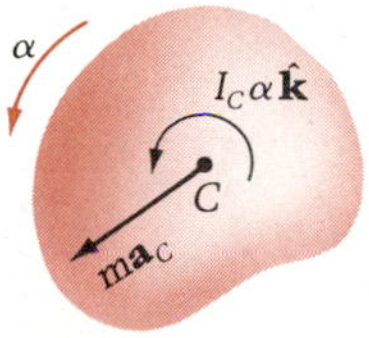

Figure 5.15

$$\int \mathbf{a}\, dm = m\mathbf{a}_C$$

$$\int \mathbf{r} \times \mathbf{a}\, dm = \Sigma \mathbf{M}_C = I_C \alpha \hat{\mathbf{k}}$$

In a manner identical to reducing a force and couple to its simplest form, we may reduce this resultant of the $m\mathbf{a}$ vectors as shown in Figure 5.16, where the distance D is $I_C\alpha/(ma_C)$. We note that there is in fact a *line* ℓ of points along the resultant of the $m\mathbf{a}$ vectors, making the location of a single point P^* ambiguous. However, the concept of the center of percussion is usually used in conjunction with problems in which the body has a pivot O (as in the previous example). In these problems P^* is the well-defined single point at the intersection of lines ℓ and OC as in Figure 5.16.

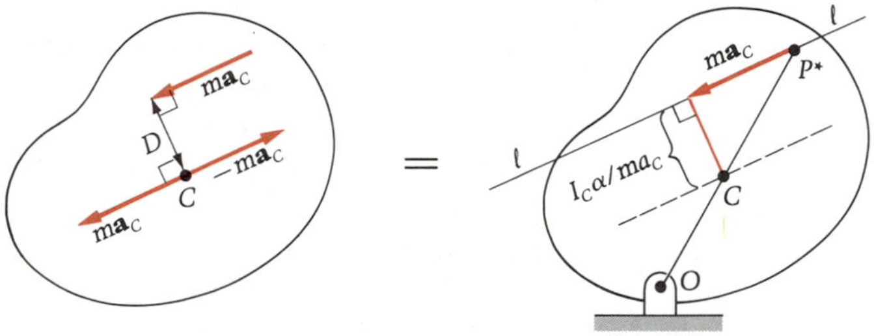

Figure 5.16

EXAMPLE 5.19

Find the center of percussion P^* for a pendulum $\mathcal{P}$ consisting of a rod plus disk, each of which has equal mass m. (See Figure E5.19.)

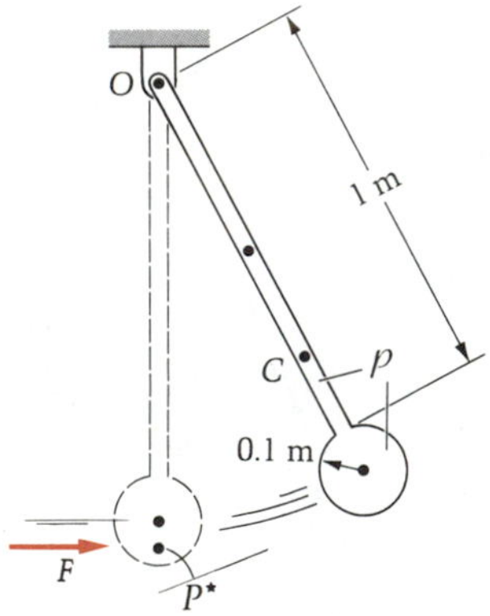

Figure E5.19

Solution

The mass center of $\mathcal{P}$ is located at a distance r_{OC} from O given by

$$m(0.5) + m(1.1) = 2mr_{OC}$$

$$r_{OC} = 0.8 \text{ m}$$

(Note that with equal masses C lies halfway between the mass centers of the rod and disk.) The radius of gyration with respect to C is calculated next:

$$I_{z_C} = \frac{m(1^2)}{12} + m(0.3^2) + \frac{m(0.1^2)}{2} + m(0.3^2)$$

$$= 0.268m = 2mk_C^2$$

Thus $k_C = 0.366$ m and Equation (5.45) then gives us the location of P^*:

$$r_{OP^*} = r_{OC} + \frac{k_C^2}{r_{OC}} = 0.8 + \frac{(0.366)^2}{0.8}$$

$$= 0.968 \text{ m}$$

Striking the pendulum at P^* eliminates the horizontal pin reaction at O, as we have seen.

PROBLEMS ▶ Section 5.3

5.75 Drum $\mathscr{B}_1$ has a radius of gyration of mass with respect to a horizontal axis through O of 1 m and a mass of 800 kg. Body $\mathscr{B}_2$ has a mass of 600 kg and a velocity of 20 m/s upward when in the position shown in Figure P5.75. Find the velocity of $\mathscr{B}_2$ 3 s later.

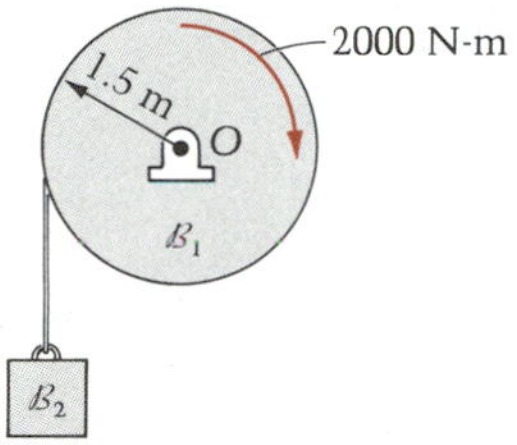

Figure P5.75

5.76 The hollow drum shown in Figure P5.76 weighs 161 lb and rotates about a fixed horizontal axis through O. The diameter of the drum is 2.4 ft, and the radius of gyration of the mass with respect to the axis through O is 0.8 ft. The angular speed changes from 30 rpm ↻ to 90 rpm ↺ during a certain time interval. Find the time interval.

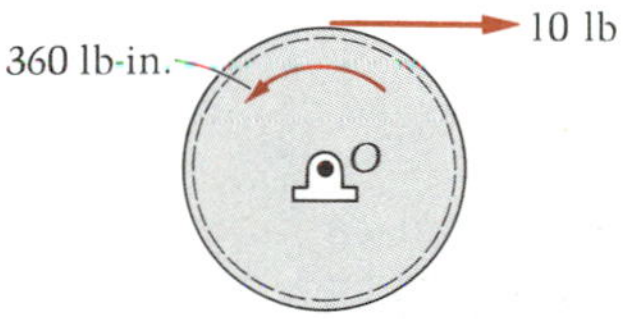

Figure P5.76

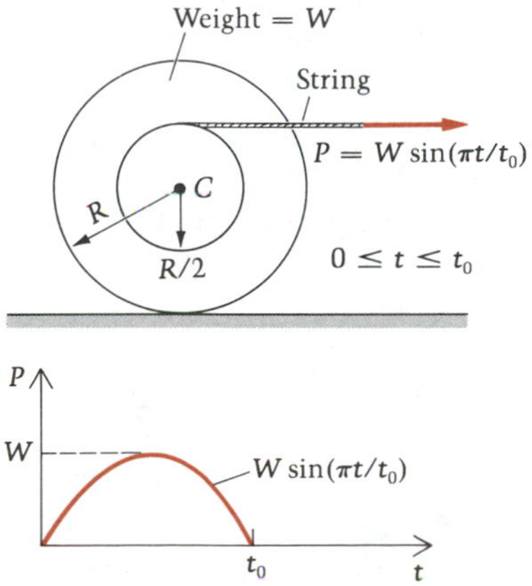

Figure P5.77

5.77 The sinusoidal force P is applied to the string in Figure P5.77 for a half-cycle. If the cylinder (initially at rest) does not slip, find its angular velocity at the end of the load application (at $t = t_0$).

5.78 A massless rope hanging over a frictionless pulley of mass M supports two monkeys (one of mass M, the other of mass $2M$). The system is released at rest at $t = 0$, as shown in Figure P5.78. During the following 2 sec, monkey B travels down 15 ft of rope to obtain a massless peanut at end P. Monkey A holds tightly to the rope during these 2 sec. Find the displacement of A during the time interval. Treat the pulley as a uniform cylinder of radius R.

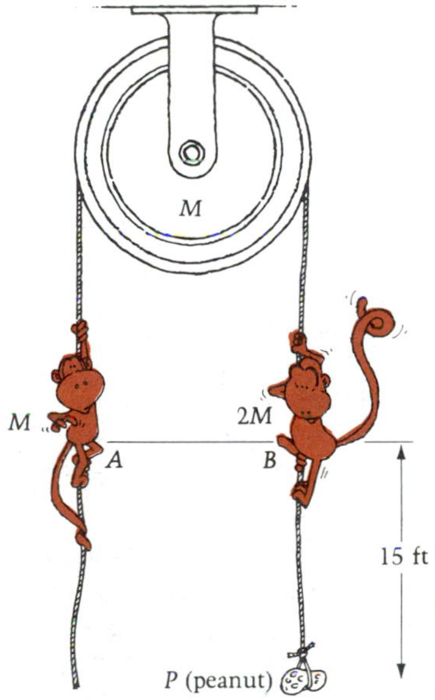

Figure P5.78

5.79 Force F in Figure P5.79 varies with time according to $F = 0.02t^2$ newtons, where t is measured in seconds. If there is enough friction to prevent slipping of the cylinder on the plane, find the velocity of C at: (a) $t = 3$ s; (b) $t = 10$ s. The cylinder starts from rest at $t = 0$.

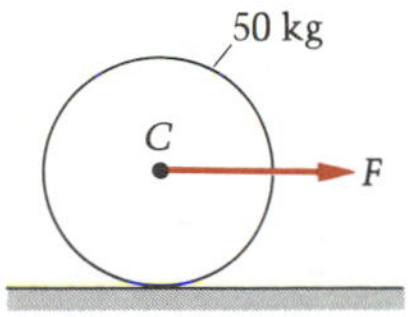

Figure P5.79

5.80 The cylinder in Figure P5.80 has mass $m = 3$ slugs and radius of gyration $k_C = 1.5$ ft with respect to C. There is sufficient friction to prevent slipping on the plane. A rope is wrapped around the inner radius, and a tension $T = 40$ lb is applied parallel to the plane as shown. Use impulse/momentum principles to find the velocity of C after 3 sec if motion starts from rest.

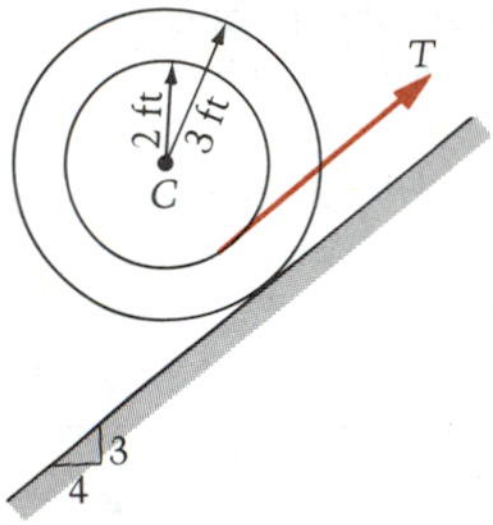

Figure P5.80

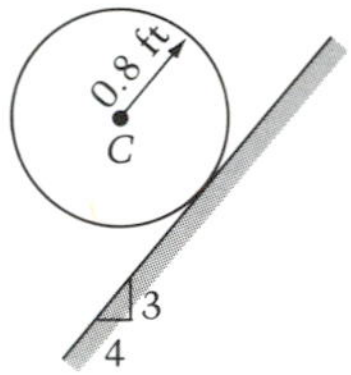

Figure P5.81

5.81 The 161-lb round body is rolling up the plane with $\omega = 5$ ↻ rad/sec at the instant shown in Figure P5.81. The radius of gyration of the mass of the body with respect to the axis through the mass center C normal to the page is 0.7 ft. Find the time required for the mass center to reach its highest point.

5.82 The cylinder $\mathcal{B}_1$, turning at 200 rpm ↻, is brought to rest by applying the 50-lb force to the light brake arm $\mathcal{B}_2$ as shown in Figure P5.82. Friction in the bearings at O produces a constant resistance torque of 7 lb-ft, and the coefficient of friction at the contact point A between $\mathcal{B}_1$ and $\mathcal{B}_2$ is $\mu = 0.3$. (a) Find the stopping time, and (b) find the number of revolutions turned by $\mathcal{B}_1$ during the braking.

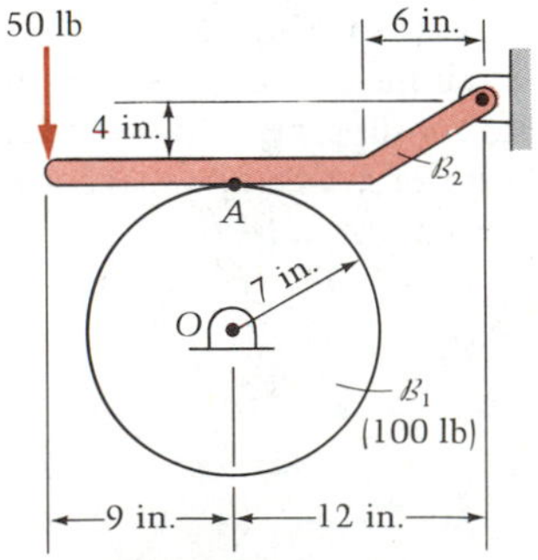

Figure P5.82

5.83 Acting on the gear is a couple C with a time-dependent strength given by $C = (6 + 0.8t)$ N-m, where t is measured in seconds. (See Figure P5.83.) If the system is released from rest at $t = 0$, find the velocity of block $\mathcal{B}$ when (a) $t = 3$ s; (b) $t = 10$ s. The centroidal radius of gyration of the gear is 0.25 m.

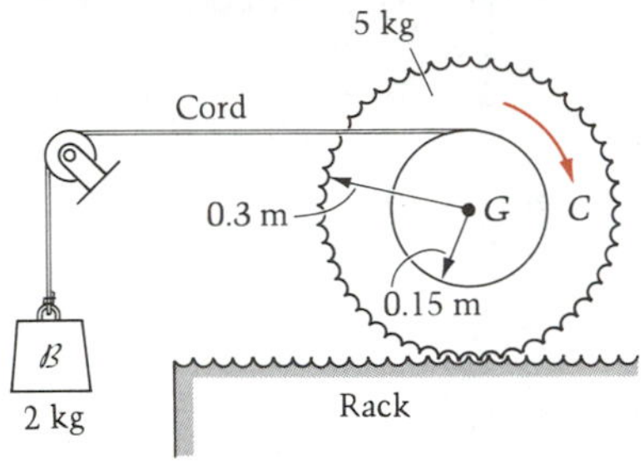

Figure P5.83

Solve the following problems by making use of the impulse and momentum, and/or angular impulse and angular momentum, methods.

5.84 Problem 4.107

5.85 Problem 4.166(c)

5.86 Problem 4.179

5.87 Problem 4.102

5.88 Problem 4.101

5.89 Problem 4.87(b)

5.90 A pipe rolls (from rest) down an incline (Figure P5.90). Using the *equations of motion*, find:

a. $\dot{x}_C$ at time t

b. $\dot{x}_C$ after C moves the distance x_C.

Then use work and energy to verify the answer to part (b) and impulse and momentum to verify part (a). Finally, give the minimum μ to prevent slipping.

5.91 A body $\mathcal{B}$ weighing 805 lb with radius of gyration 0.8 ft about its z_C axis (see Figure P5.91) is pinned at its mass center. A clockwise couple of magnitude e^t lb-ft is

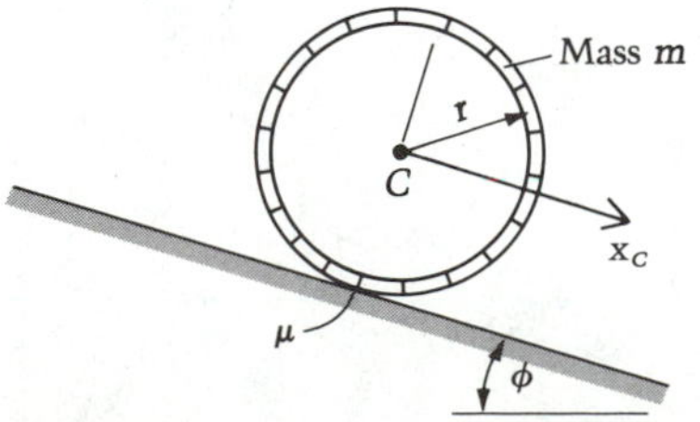

Figure P5.90

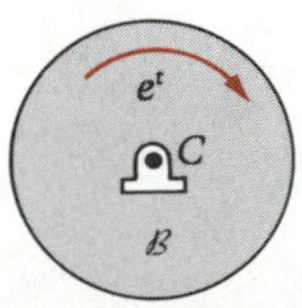

Figure P5.91

applied to $\mathcal{B}$ starting at $t = 0$. Find the angular velocity of $\mathcal{B}$ when $t = 3$ seconds.

5.92 Given that the slot (for the cord) in the cylinder in Figure P5.92 (mass 10 kg) has a negligible effect on I_C, find the velocity of the mass center C as a function of time, if $\theta = 60°$.

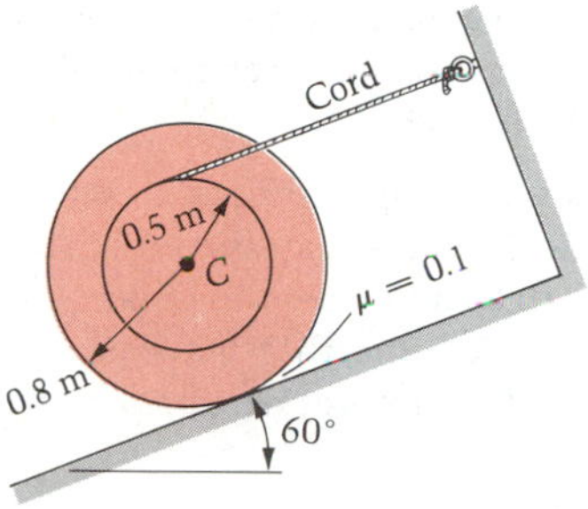

Figure P5.92

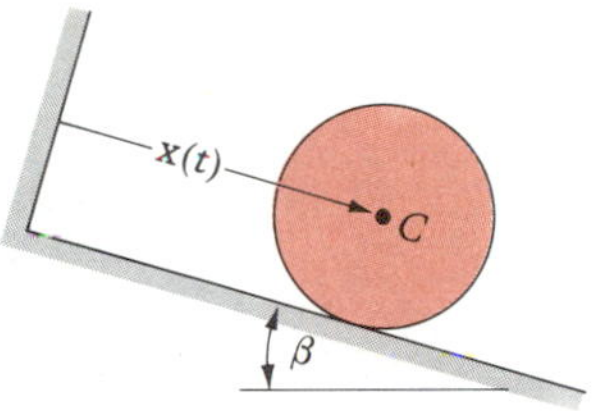

Figure P5.93

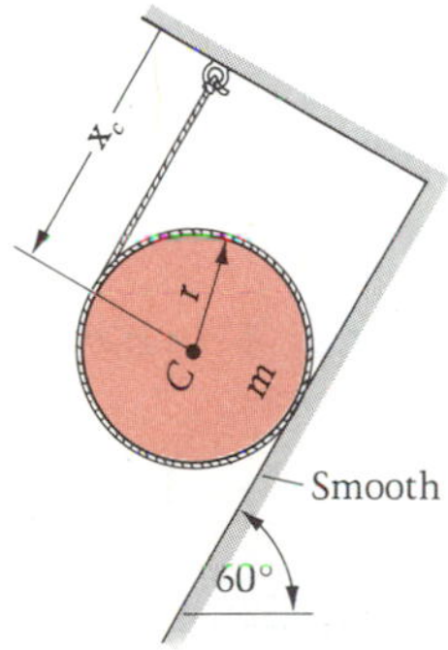

Figure P5.94

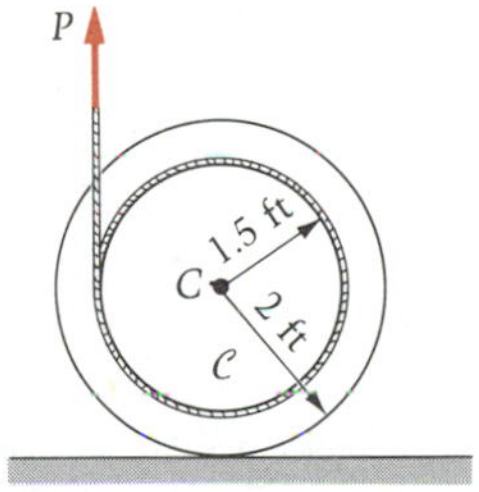

Figure P5.95

5.93 A uniform sphere (radius r, mass m) rolls on the plane in Figure P5.93. If the sphere is released from rest at $t = 0$ when $x = L$, find $\dot{x}(t)$.

5.94 The cord in Figure P5.94 is wrapped around the cylinder, which is released from rest on the 60° incline shown. Find the velocity of C as a function of time t.

* **5.95** The 50-lb body $\mathcal{C}$ in Figure P5.95 may be treated as a solid cylinder of radius 2 ft. The coefficient of friction between $\mathcal{C}$ and the plane is $\mu = 0.2$, and a force $P = 10$ lb is applied vertically to a cord wrapped around the hub. Find the velocity of the center C 10 sec after starting from rest.

* **5.96** A child pulls on an old wheel with a force of 5 lb by means of a rope looped through the hub of the wheel. (See Figure P5.96.) The friction coefficient between wheel and ground is $\mu = 0.2$. Find I_C for the wheel, and use it to determine the velocity of C 3 sec after starting from rest.

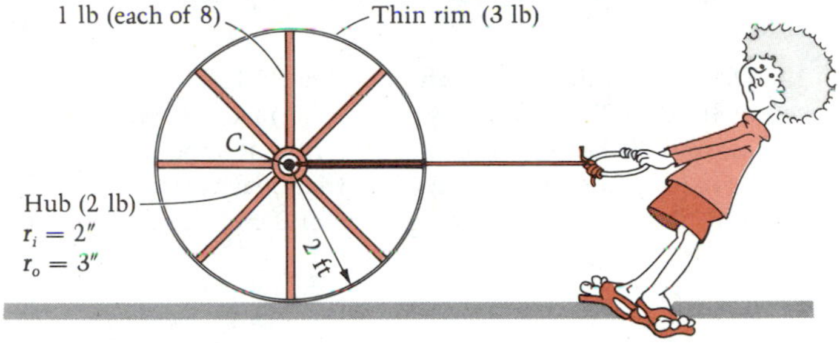

Figure P5.96

* **5.97** Two cables are wrapped around the hub of the 10-kg spool shown in Figure P5.97, which has a radius of gyration of 500 mm with respect to its axis. A constant 40-N force is applied to the upper cable as shown. Find the velocity of the mass center C 5 sec after starting from rest if: (a) $\mu = 0.2$; (b) $\mu = 0.5$.

* **5.98** The cart $\mathcal{B}_1$ is given an initial velocity v_i to the right at $t = 0$. The rod $\mathcal{B}_2$ is pinned to $\mathcal{B}_1$ at its mass center G, as shown in Figure P5.98(a). At $t = 0$, the mass center C of $\mathcal{B}_2$ is held fixed at the instant the cart starts off, then immediately released. At a later time (see Figure P5.98(b)), it is observed that $\mathcal{B}_2$ has $\omega_2 = 0$ at an instant when $\mathcal{B}_2$ has turned 90° clockwise. If $M = m$, find the velocity of G at that instant. Use $W = \Delta T$ *and* an impulse and momentum principle.

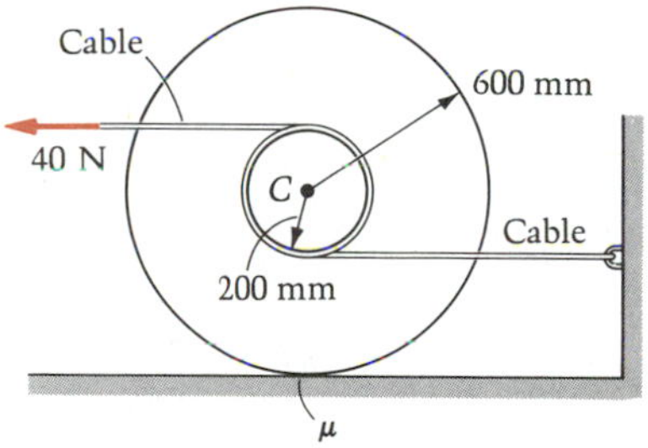

Figure P5.97

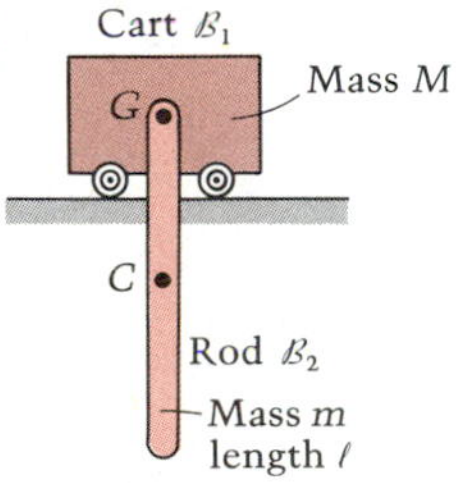

Figure P5.98(a)

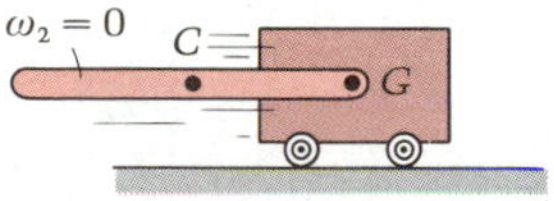

Figure P5.98(b)

5.99 Two gymnasts at A and B, each of weight W, hold onto the left side of a rope that passes over a cylindrical pulley (weight W, radius R) to a counterweight C of weight $2W$. (See Figure P5.99.) Initially the gymnast A is at depth d below B. He climbs the rope to join gymnast B. Determine the displacement of the counterweight C at the end of the climb.

5.100 Disk $\mathcal{B}_1$ and the light shaft in Figure P5.100 rotate freely at 40 rpm. Disk $\mathcal{B}_2$ (initially not turning) slides down the shaft and strikes $\mathcal{B}_1$; after a brief period of slipping, they move together. Find the average frictional moment exerted on $\mathcal{B}_1$ by $\mathcal{B}_2$ if the slipping lasts for 3 sec.

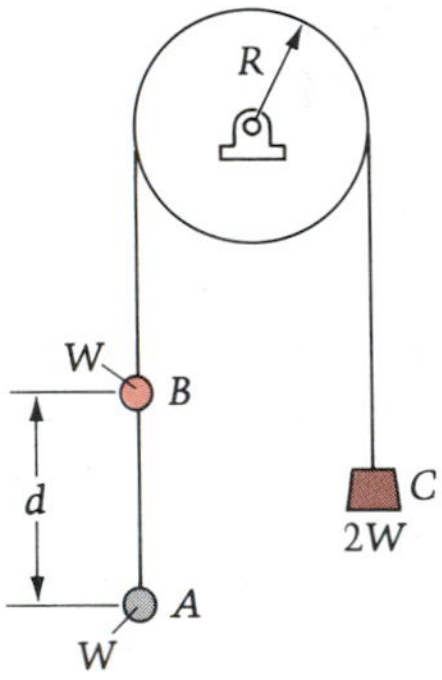

Figure P5.99

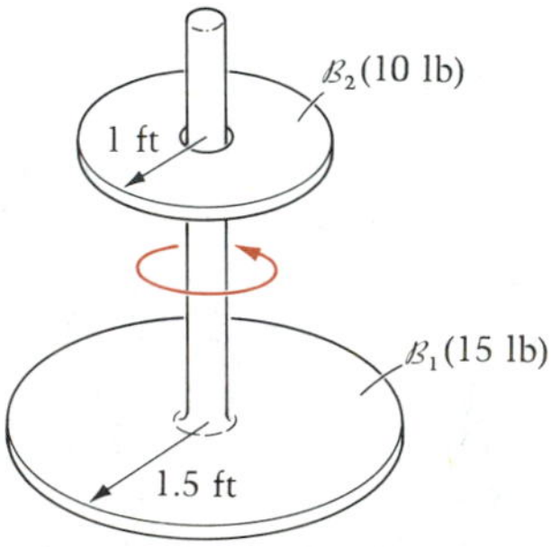

Figure P5.100

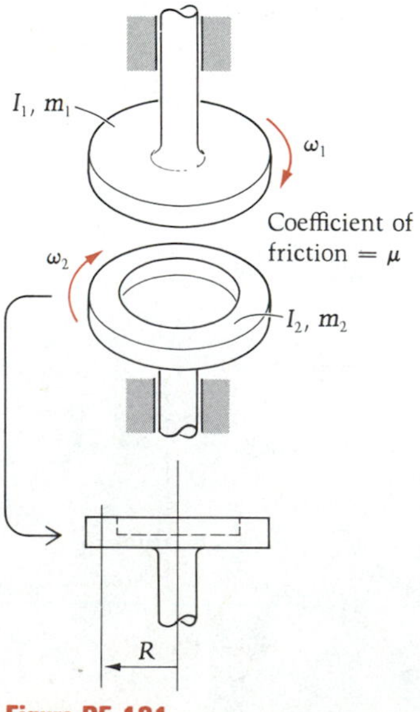

Figure P5.101

5.101 Two disks are spinning in the directions shown in Figure P5.101. The upper disk is lowered until it contacts the bottom disk (around the rim). Find how long it takes for the two disks to reach a common angular velocity, and determine its value. Finally, determine the energy lost. Show that if $I_1 = I_2$ and $\omega_1 = -\omega_2$, your solution predicts that 100 percent of the energy is lost (as it should). Determine which of the three answers (time, ω_f, energy loss) are the same if the two disks are instantaneously locked together instead of slipping.

5.102 Figure P5.102(a) shows a rough guess at a skater's mass distribution. Calculate the percentage increase in his angular speed about the vertical if he draws in his arms as shown in Figure P5.102(b). Assume that his arms are wrapped around the 6-in. radius circle of his upper body.

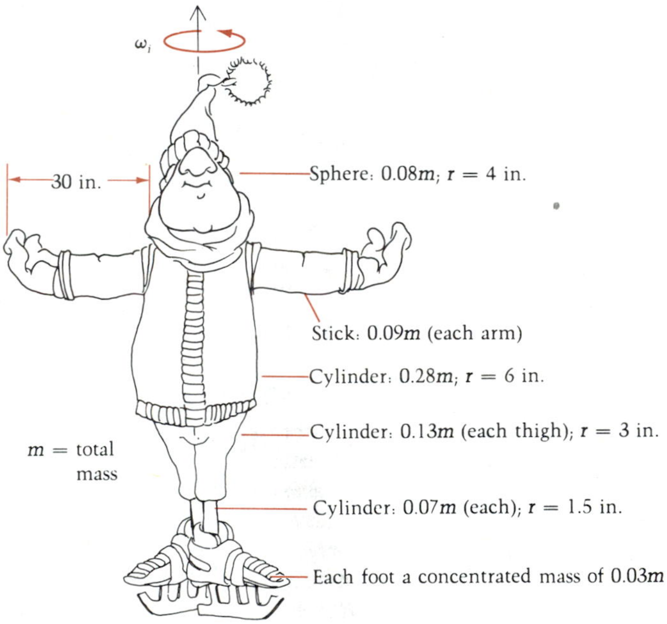

Figure P5.102(a)

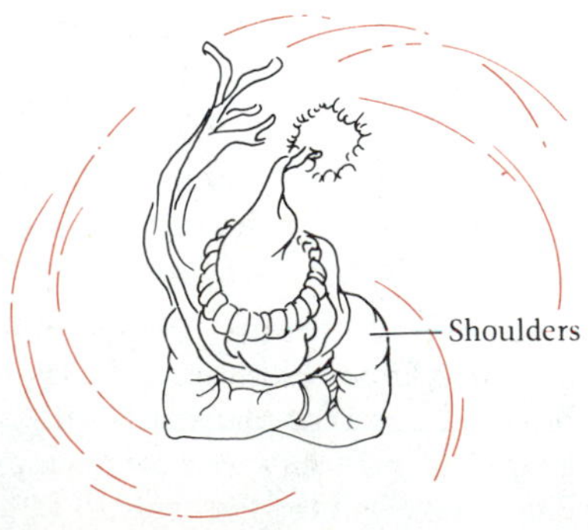

Figure P5.102(b)

* **5.103** A starving monkey of mass m spies a bunch of delicious bananas of the same mass. (See Figure P5.103.) He climbs at a varying speed relative to the (light) rope. Determine whether the monkey reaches the bananas before they sail over the pulley of radius R if:

   a. The pulley's mass is negligible ($\ll m$).
   b. The pulley's mass is fm, where $f > 0$ and the radius of gyration of the pulley with respect to its axis is k.

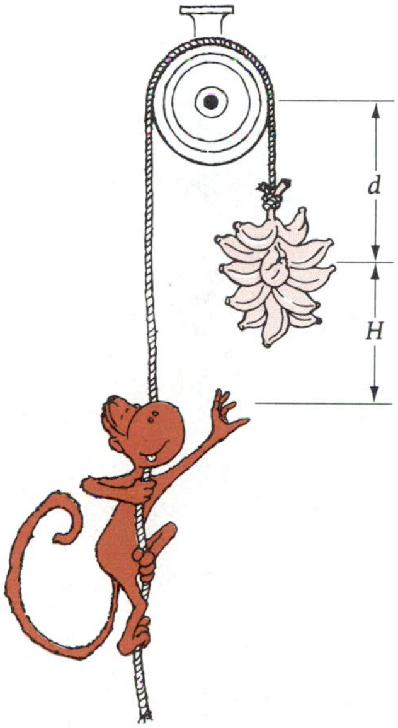

Figure P5.103

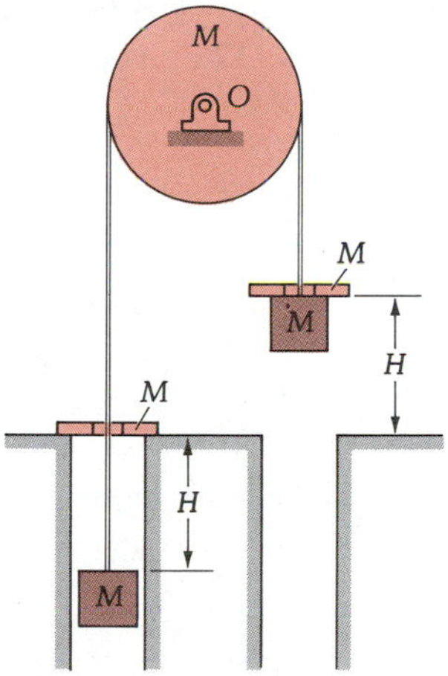

Figure P5.104

If either answer is yes, give the relationship between d and H in the figure for which overtaking the bananas is possible.

* **5.104** A circular disk of mass M rotates without friction about O. (See Figure P5.104.) A string passed over the disk (and not slipping on it) carries a mass M at each end. The system is released at rest as shown with the right-hand mass carrying a washer of mass M. As the system moves, the left-hand weight picks up a washer of mass M at the same instant the right mass deposits its washer. Find the velocity of the right-hand weight just after this exchange of washers.

* **5.105** A bird of mass m, flying horizontally at speed v_0 perpendicular to a stick, lands on the stick and holds fast to it. (See Figure P5.105.) The stick (mass M, length ℓ) is lying on a frozen pond. Find the angular velocity of the bird and stick as they move together. (Answer in terms of m, M, ℓ, and v_0.) Assume that the bird lands on the end of the stick.

Figure P5.105

* **5.106** A hemispherical block of mass M and radius a whose surfaces are smooth rests with its plane face in contact with a *smooth* horizontal table. A particle of mass m is placed at the highest point of the block and is slightly disturbed. Show that as long as the particle remains in contact with the block, the radius to the particle makes an angle θ with the upward vertical where

$$a\dot{\theta}^2(M + m\sin^2\theta) = 2g(M + m)(1 - \cos\theta)$$

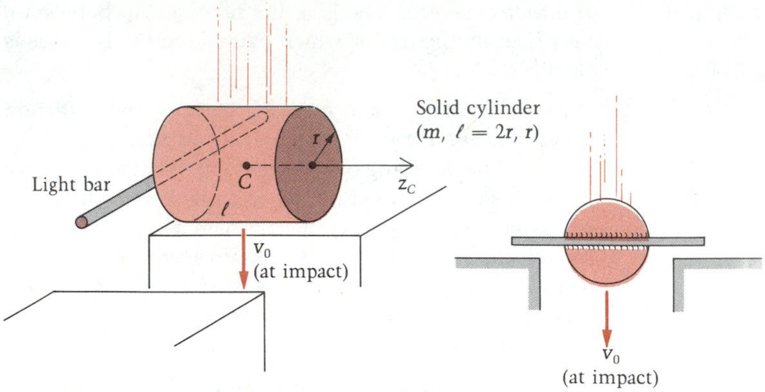

Figure P5.107

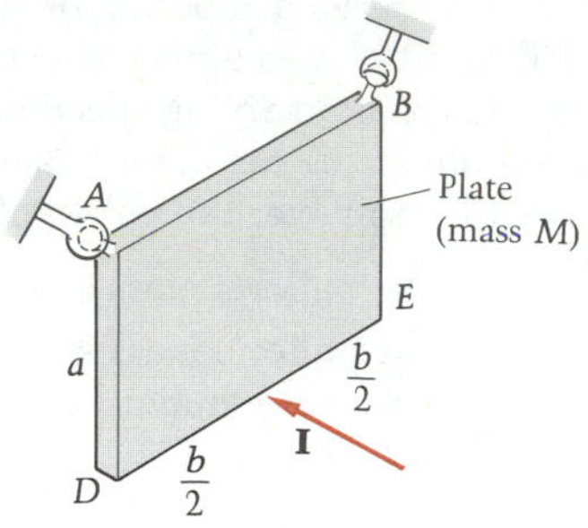

Figure P5.109

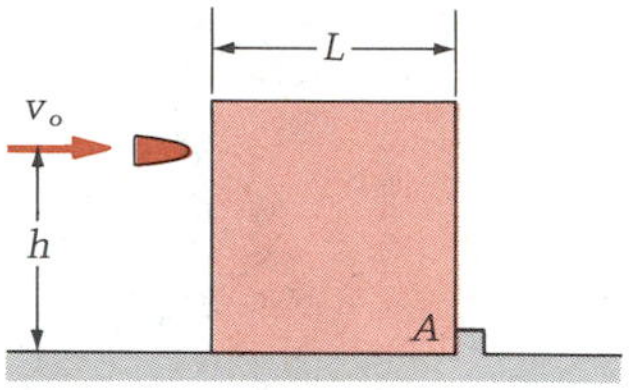

Figure P5.111

5.107 The bar in Figure P5.107 is welded to the end of the cylinder, which is traveling downward in translation. The bar strikes the tables at speed v_0, and the cylinder begins to rotate about the bar without rebound.

- **a.** Find the angular velocity of the cylinder when C is at its lowest point.
- **b.** Find the percentage of energy lost during the impact; that is

$$\left(1 - \frac{\text{new energy}}{\text{old energy}}\right) \times 100$$

5.108 A CARE package (Figure P5.108) consists of the box plus contents described in Example 4.12. At impact the crate has $\mathbf{v}_{G_i} = 50 \downarrow$ ft/sec and is translating. If there is no rebound, find the angular velocity of the box and the velocity of its mass center G just after the impact.

5.109 The plate in Figure P5.109, supported by ball joints at its top corners, is suddenly struck as shown with a force that produces the impulse I normal to the plate. Find the kinetic energy produced by the impact.

5.110 Work the preceding problem but assume that the plate is initially free. If you work both problems, show that the difference in the energies is $I^2/(2M)$.

5.111 A bullet (see Figure P5.111) of mass m_1 strikes a square homogeneous block of mass m_2, where $m_2 \gg m_1$. The bullet is traveling with initial velocity $\mathbf{v}_O$ and becomes embedded in the block. After impact, the block is observed to be pivoting about corner A. What is the maximum speed $|\mathbf{v}_O|$ of the bullet such that the block will not tip all the way over?

5.112 The cylinder $\mathcal{B}$ (radius 10 cm, length 40 cm) swings down from a position of rest where $\theta = 0$, and strikes the particle P of mass 5 kg. (See Figure P5.112.) The coeffi-

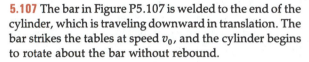

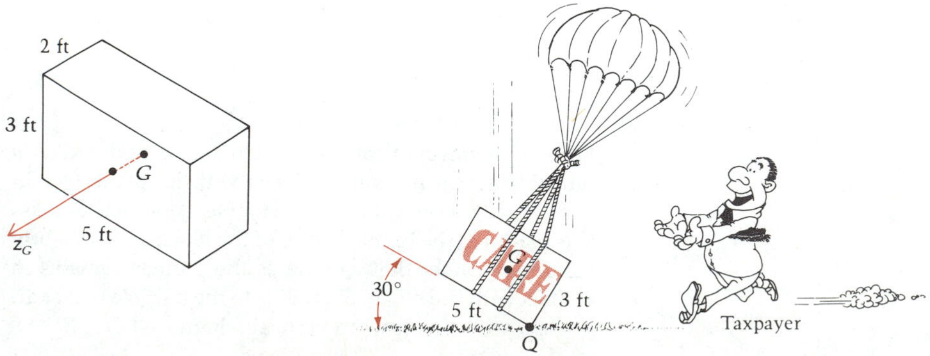

Figure P5.108

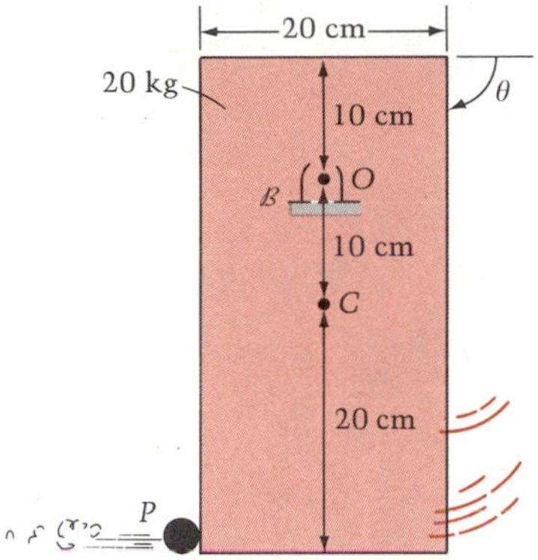

Figure P5.112

cient of restitution is $e = 0.5$, and at impact the particle has $\mathbf{v}_P = 2 \rightarrow$ m/s. Find the angle through which $\mathcal{B}$ will turn about its pivot O after impact.

5.113 An equilateral triangular plate of mass 2 slugs and side 2 ft is released in the upper position from rest (see Figure P5.113). It swings down and strikes the stationary cylinder. The coefficient of restitution for the impact is $e = 1/2$. Find elapsed time after impact until the cylinder no longer slips on the plane.

5.114 A block slides to the right and strikes a small obstruction at a speed of 20 ft/sec. (See Figure P5.114.)

- **a.** If the coefficient of restitution is zero, find the energy loss caused by the impact.
- **b.** What is the minimum striking velocity required to overturn the block after collision?

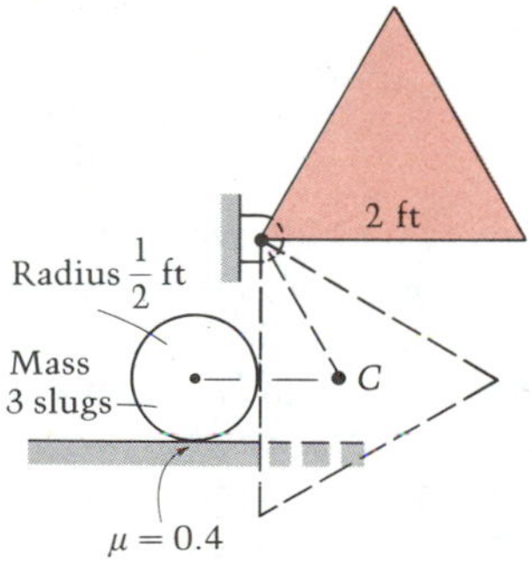

Figure P5.113

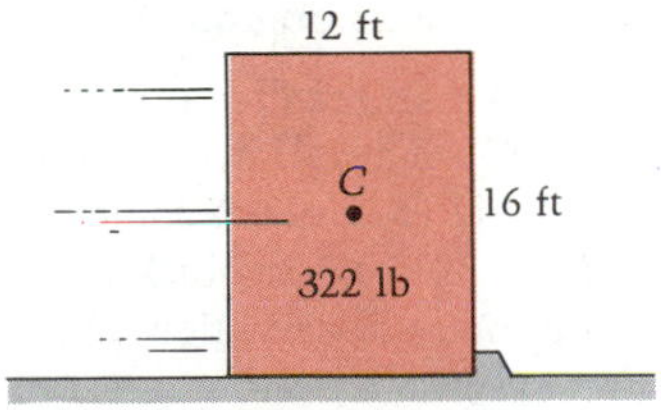

Figure P5.114

5.115 A homogeneous cube of side a and mass M slides on a level, frictionless table with velocity v_0. See Figure P5.115.) It strikes a small lip on the table at A of negligible height. Find the velocity of the center of mass just after impact if the coefficient of restitution is unity. (The centroidal moment of inertia of a cube about an axis parallel to an edge is $Ma^2/6$.)

5.116 There is only one height H above a pool table at which a cue ball may be struck by the stick without the ball slipping for a while after the impact. (See Figure P5.116.) Find this value of H, in terms of R, for which the ball immediately rolls.

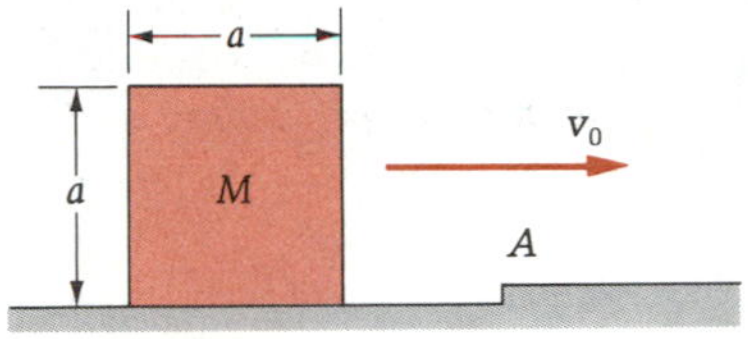

Figure P5.115

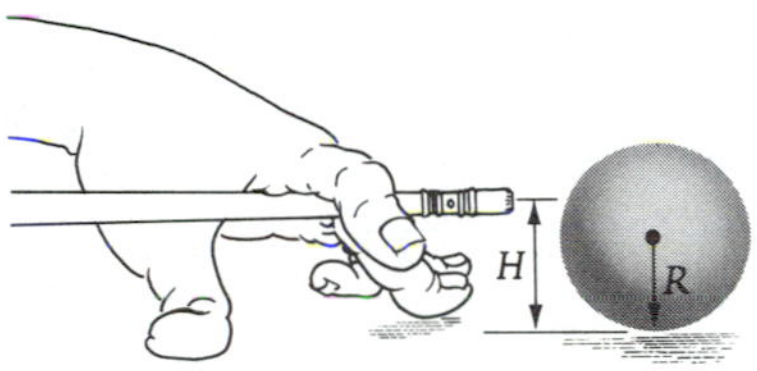

Figure P5.116

5.117 Compute the error in I_{zz}^C in Example 5.18 that was incurred in assuming the vessel to be a shell (so that I_{zz}^C was $m\ell^2/12 + mr^2/2$). Use the weight, height, outer radius, and density to compute the thickness of the vessel; then calculate a more accurate I_{zz}^C and compare.

5.118 A uniform rod of length L is dropped and translates downward at an angle θ with the vertical as shown in Figure P5.118. If end A does not rebound after striking the ground at speed v_0, find: (a) the energy lost during the impact of A with the ground; (b) the speed at which the other end B then hits the ground.

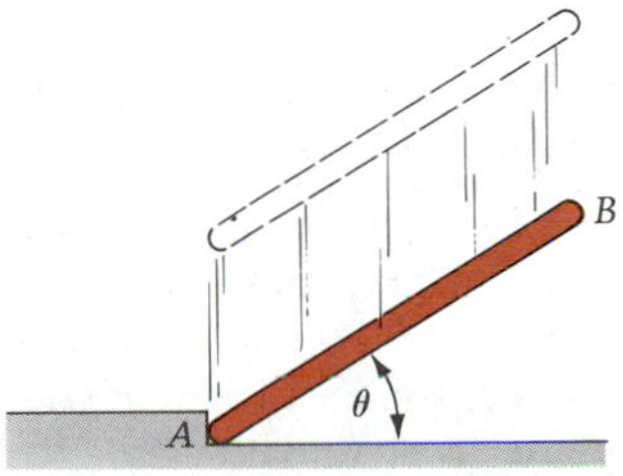

Figure P5.118

5.119 A uniform bar AB of length L and mass M is moving on a smooth horizontal plane with $\mathbf{v}_C = v_0\hat{\mathbf{i}}$ and $\boldsymbol{\omega} = \omega_0\hat{\mathbf{k}}$, when end B strikes a peg P (see Figure P5.119). If $\omega_0 = 2v_0/L$ and the coefficient of restitution $e = \frac{1}{6}$, find the loss of kinetic energy.

5.120 The 80-lb solid block hits a smooth, rigid wall (see Figure P5.120) and rebounds with a coefficient of restitution of $e = 0.2$. Prior to impact the block had: $\boldsymbol{\omega}_i = 1.2 \circlearrowleft$ rad/sec and $\mathbf{v}_C = 0.8\hat{\mathbf{i}} + 0.6\hat{\mathbf{j}}$ ft/sec. Find the angular velocity of the block immediately following the impact.

5.121 The rod in Figure P5.121 is freely falling in a vertical plane. At a certain instant it is horizontal with its ends A and B having the velocities shown. If end A is suddenly fixed, prove that the rod will start to rise around end A provided that $v_1 < 2v_2$.

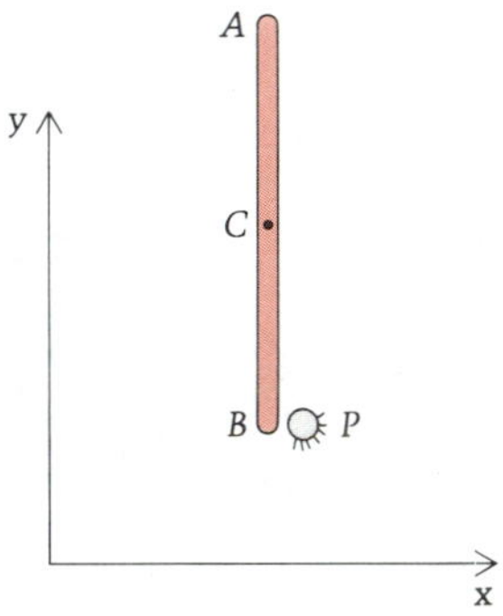

Figure P5.119

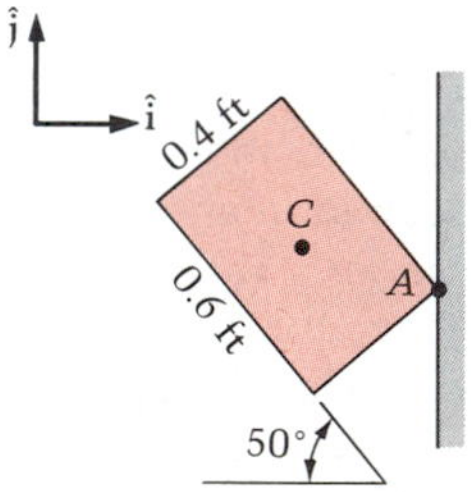

Figure P5.120

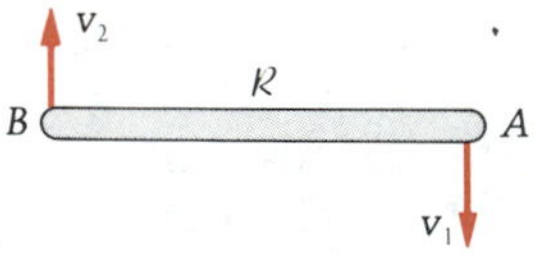

Figure P5.121

5.122 In the preceding problem show that the energy loss in instantaneously stopping point A is independent of v_2.

5.123 The bar in Figure P5.123 swings downward from the dotted horizontal position and strikes mass m. The bar has mass M and length ℓ. The collision takes place with a coefficient of restitution of zero. If the coefficient of friction between m and the plane is μ, find the distance moved by m before stopping. Treat m as a particle.

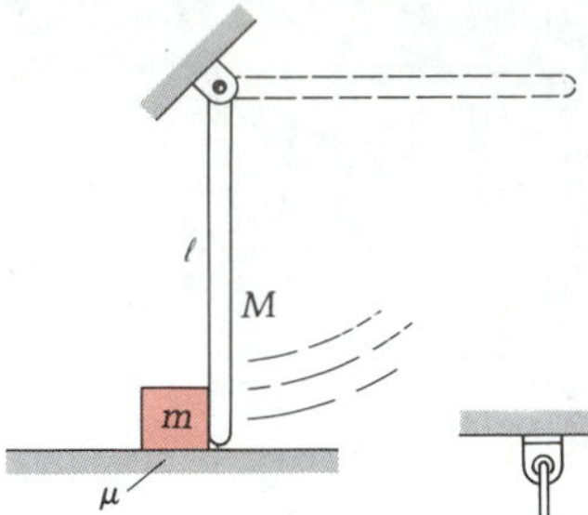

Figure P5.123

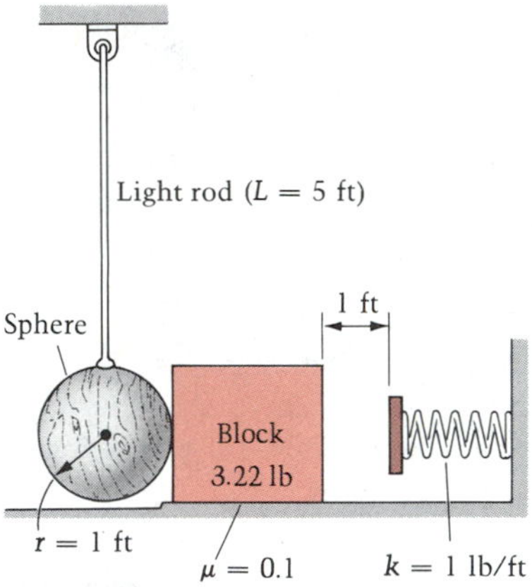

Figure P5.124

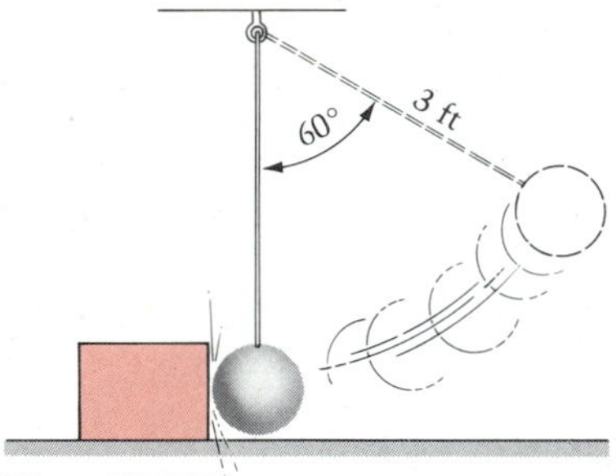

Figure P5.125

5.124 A wooden sphere weighing 0.644 lb swings down from a position where the rod is horizontal, and it impacts the block. (See Figure P5.124.) The coefficient of restitution is $e = 0.6$. The block weighs 3.22 lb and is initially at rest. Find the position of the block when it comes *permanently* to rest. (Assume that the sphere is removed from the problem after impact and that the spring cannot rebound past its original unstretched position.)

5.125 A 4-lb sphere is released from rest in the position shown in Figure P5.125, and two observations are then made: (1) The sphere comes immediately to rest after the impact; (2) the 5-lb block slides 3 ft before coming to rest. Using these observations, find the coefficients of restitution (between sphere and block) and friction (between block and floor).

5.126 The rod-sphere rigid body in Figure P5.126 is released from rest in the horizontal position. It swings down and at its lowest point strikes the box. Find how far the box slides before coming to rest if the coefficient of restitution is $e = 0.5$. The data are:

1. Rod: length = 1 m; mass = 3 kg
2. Sphere: radius = 0.2 m; mass = 10 kg
3. Block: $b = 0.3$ m; $H = 0.35$ m; mass = 5 kg
4. Coefficient of friction between block and plane = 0.3

Assume the sphere hits the block just once.

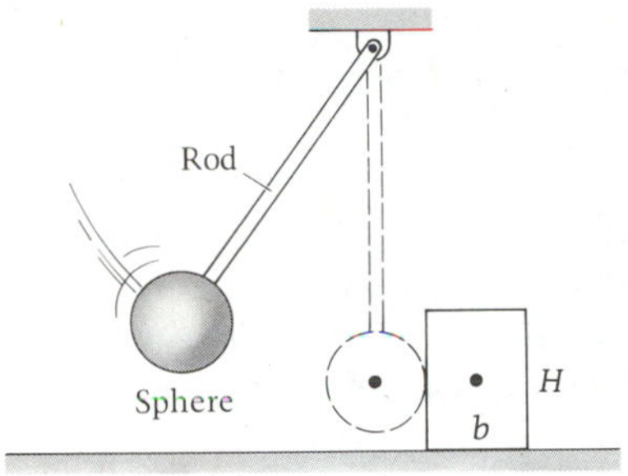

Figure P5.126

5.127 The assembly in Figure P5.127 is turning at $\omega_i = 2$ rad/sec when the collar is released. All surfaces are smooth, and the disk is fixed to the bar; the light vertical shaft ends in bearings and turns freely. The data are:

$$m_3 = \tfrac{1}{4}\text{ slug} \qquad \ell_2 = 2\text{ ft}$$
$$m_1 = 1\text{ slug} \qquad \ell_1 = 4\text{ ft}$$
$$m_2 = \tfrac{1}{4}\text{ slug}$$

The collar moves outward and impacts the disk without rebound. Find: (a) the angular speed of the bar just before and just after impact; (b) the percentage of energy lost during impact. The radii of $\mathcal{B}_1$ and $\mathcal{B}_2$ are small compared to their lengths. Treat $\mathcal{B}_3$ as a particle.

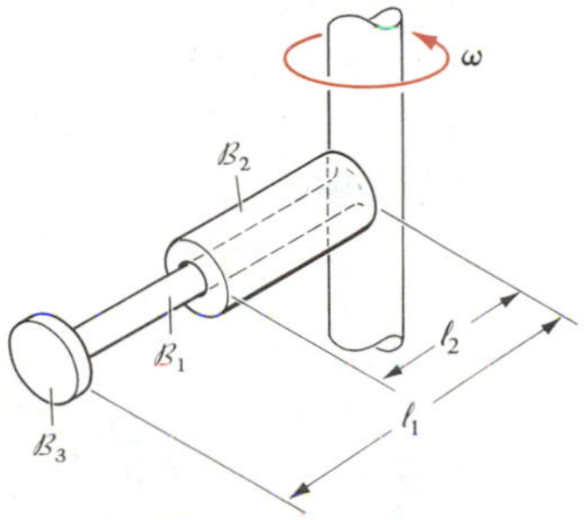

Figure P5.127

5.128 Two toothed gear wheels, which may be treated as uniform disks of radii a and b and masses M and m, respectively, are rotating in the same plane. They are not quite in contact and have angular velocities ω_1 and ω_2 about fixed axes through their centers. Their axes are then slightly moved so that the wheels engage. Prove that the loss of energy is

$$\frac{Mm}{4(M+m)}(a\omega_1 + b\omega_2)^2$$

5.129 Sphere $\mathcal{B}_1$ has mass m and radius r, and it rolls with mass center velocity $v_0 \rightarrow$ on a horizontal plane. (See Figure P5.129.) It hits squarely an identical sphere $\mathcal{B}_2$ that is at rest. The coefficient of friction between a sphere and the plane is μ, and between spheres it is negligible. The impact is nearly elastic ($e \approx 1$).

a. Find v_{C_f} and ω_f of each sphere right after impact.
b. Find v_C of each sphere after it has started rolling uniformly.
c. Discuss the special case when $\mu = 0$.

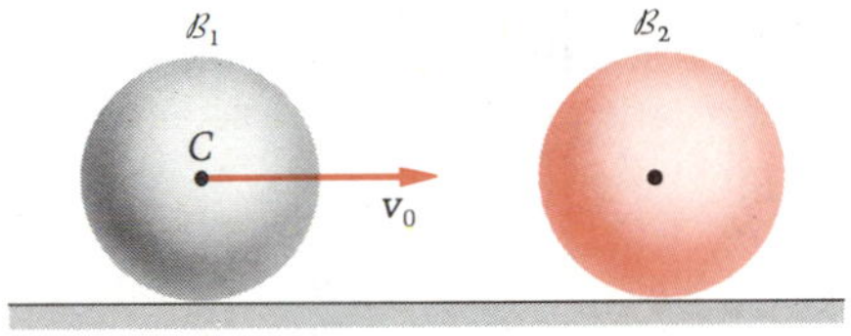

Figure P5.129

* **5.130** The 30-kg bent bar in Figure P5.130 falls from the dashed position onto the spinning cylinder, which was initially turning at 3000 rad/s ↻. If the bar does not bounce (coefficient of restitution is zero), find the stopping time for the cylinder following the impact.

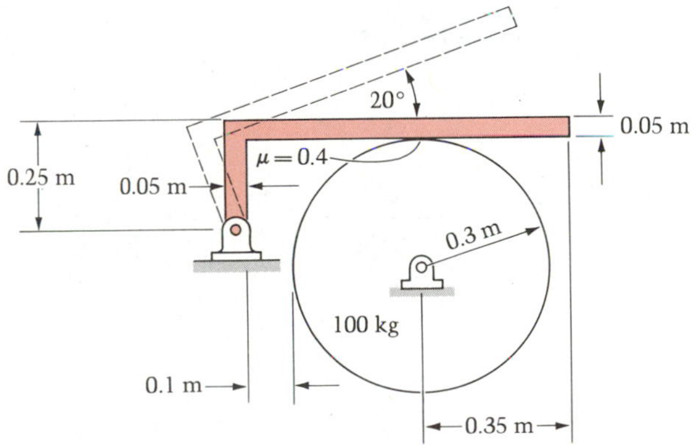

Figure P5.130

* **5.131** In Example 5.18 assume that the lower front striking corner of the vessel $\mathcal{B}$ rebounds back up the plane with velocity ev_{C_i}, where e is the coefficient of restitution ($0 < e \le 1$). Use the equations of impulse and momentum in the x and y directions, and the equation of angular impulse and angular momentum, to find the two mass-center velocity components and the angular velocity of $\mathcal{B}$ after impact. Compare the results of $\dot{x}_{C_f}$, $\dot{y}_{C_f}$, and $\dot{\theta}_f$ for $e = 1$ with those at $e = 0$. Show that no energy is lost when $e = 1$; that is, show

$$\frac{1}{2}mv_{C_i}^2 = \frac{1}{2}m(\dot{x}_{C_f}^2 + \dot{y}_{C_f}^2) + \frac{1}{2}I_C\dot{\theta}_f^2$$

* **5.132** *After* the impact in the preceding problem, show that the equations of motion of the vessel are

$$m\ddot{x}_C = -mg \sin \psi \qquad (1)$$

$$m\ddot{y}_C = N - mg \cos \psi \qquad (2)$$

$$I_C\ddot{\theta} = \frac{\ell}{2} N \sin \theta - rN \cos \theta \qquad (3)$$

where N is the normal reaction at the corner Q (see Figure P5.132). Observe that until there is another impact, the mass center has a constant x component of acceleration. Use kinematics to prove that

$$\ddot{y}_C = (r \cos \theta - \frac{\ell}{2} \sin \theta)\ddot{\theta} - \dot{\theta}^2(r \sin \theta + \frac{\ell}{2} \cos \theta) \qquad (4)$$

Use Equations (2) and (4) to eliminate N from (3), thus obtaining a single differential equation in θ governing the rotational motion of $\mathcal{B}$, and note its complexity.

5.133 Prove statement 1 near the end of Example 5.18 for the case when the plane is level ($\psi = 0$) and $e = 0$. *Hint*: Note carefully that the angle of the plane does not affect Equation (6), so you only need to alter the Δh in Equation (7) to obtain the new result.

5.134 Prove statement 2 near the end of Example 5.18. Again the plane is to be level in this problem, but now $e = 1$. *Hint*: The x_C-component of velocity is constant after impact, since with $\psi = 0$ all external forces (mg and N) are vertical. To find this velocity $\dot{x}_{C_f}$, use ω_f and the velocity of the striking corner Q just after impact (ω_f is the same as with $\psi = 3.719°$ in Problem 5.131 with $e = 1$; v_Q is v_{C_i} back to the left). Then use

$$W = \Delta T = \frac{1}{2}m\dot{x}_{C_f}^2 + \frac{1}{2}m\dot{y}_{C_f}^2 + \frac{1}{2}I_C\omega_f^2 - \frac{1}{2}mv_{C_i}^2$$

to show that C reaches the top with energy to spare.

* **5.135** Show that the cylinder in Figure P5.135, following release from rest, will reach the lower wall. Find the velocity with which C will rebound up the plane following impact, and determine the amount of energy lost. All data are shown on the figure.

* **5.136** A sphere rolling with speed v_C on a horizontal surface strikes an obstacle of height H. What is the largest value that H can have if the sphere is able to make it over the obstacle? Consider the coefficient of restitution to be zero during the sphere's impact with the corner point O. The answer will be a function of g, r, and v_C—in fact, H/r may be solved for as a function of the single nondimensional parameter gr/v_C^2. (See Figure P5.136.)

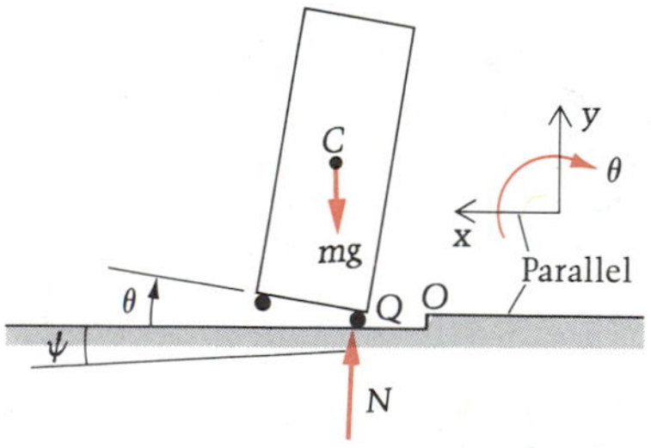

Figure P5.132

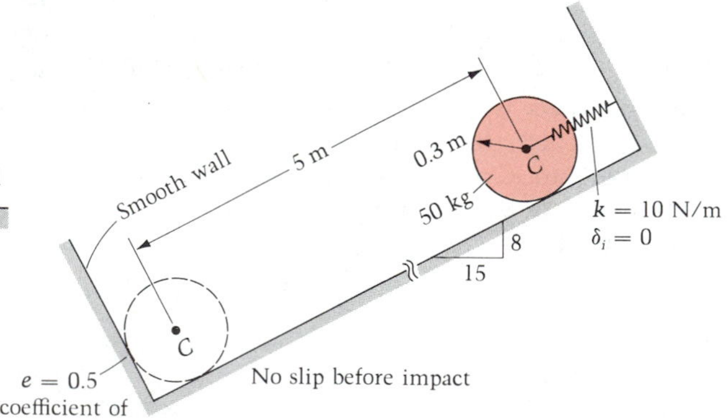

Figure P5.135

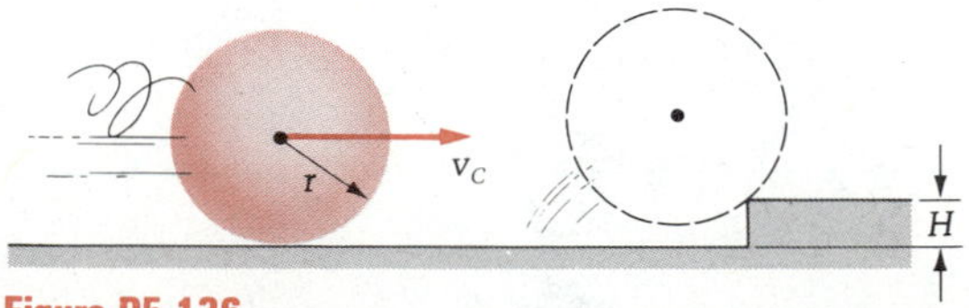

Figure P5.136

* **5.137** Using a density of wood of 673 kg/m^3, find the mass center C of the baseball bat in Figure P5.137 and then determine its moment of inertia with respect to the z_0 axis, perpendicular to the axis of symmetry of the bat. Use the parallel-axis theorem to obtain I_{zz}^{C}, and find the bat's center of percussion if it is swinging about a fixed point O.

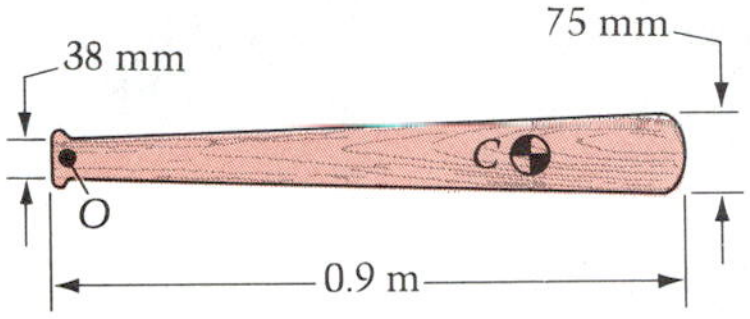

Figure P5.137

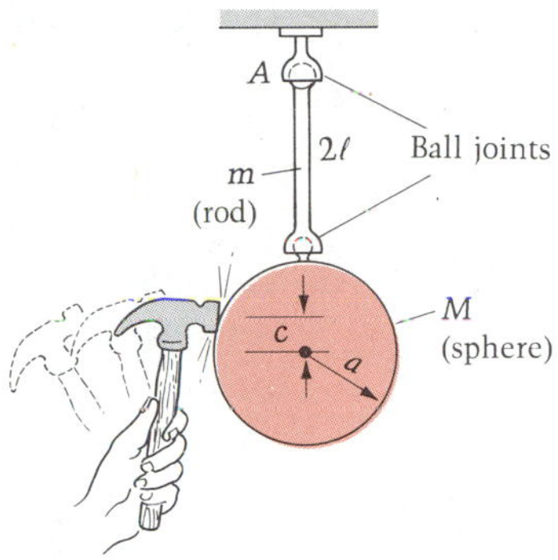

Figure P5.138

* **5.138** The hammer in Figure P5.138 strikes the sphere and imparts a horizontal impulse I to it. Determine the initial angular velocity of the sphere.

5.139 Repeat the preceding problem but suppose that the sphere and rod are welded together to form one rigid body.

COMPUTER PROBLEM ▶ Chapter 5

* **5.140** The system in Figure P5.140 is released from rest in the given position. With the help of a computer, generate data for a plot of the angle θ turned through by wheel $\mathcal{B}_1$ before first stopping, as a function of the mass ratio M/m. *Hint*: First show using $W = \Delta T$ that the equation governing θ is

$$\sin\theta = \frac{2m}{M}\theta$$

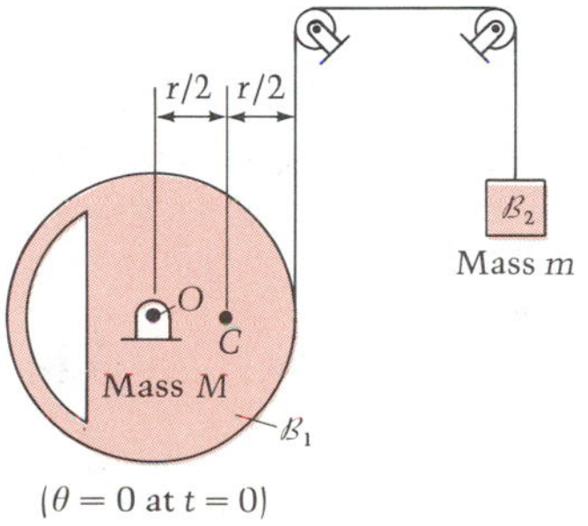

Figure P5.140

SUMMARY ▶ Chapter 5

For a rigid body in plane motion, the kinetic energy, T, can be expressed as

$$T = \tfrac{1}{2}mv_C^2 + \tfrac{1}{2}I_C\omega^2$$

or as

$$T = \tfrac{1}{2}I_{①}\omega^2$$

With W standing for the net work over a time interval (t_1, t_2) of all the *external* forces on the body, the principle of work and kinetic energy states

$$W = T_2 - T_1$$

or more compactly

$$W = \Delta T$$

The work done by a force, **F**, is defined as

$$\int_{t_1}^{t_2} \mathbf{F} \cdot \mathbf{v}\, dt$$

where $\mathbf{F} \cdot \mathbf{v}$ is called the power, or rate of work of force **F**, with **v** being the velocity of the material point being acted upon instantaneously by the force. This definition is necessary to accommodate the possibility that the force moves around on the body. When the force always acts on the same material point for which a position vector is **r**,

$$\int_{t_1}^{t_2} \mathbf{F} \cdot \mathbf{v}\, dt = \int_{t_1}^{t_2} \mathbf{F} \cdot \frac{d\mathbf{r}}{dt}\, dt$$

$$= \int_{\mathbf{r}_1}^{\mathbf{r}_2} \mathbf{F} \cdot d\mathbf{r}$$

The work done by a couple of moment **C** is

$$\int_{t_1}^{t_2} \mathbf{C} \cdot \omega\hat{\mathbf{k}}\, dt$$

which, with $\mathbf{C} = C\hat{\mathbf{k}}$, is

$$\int_{t_1}^{t_2} C\omega dt = \int_{t_1}^{t_2} C d\theta$$

For some special cases:

(a) **F** is constant:

$$W = \mathbf{F} \cdot (\mathbf{r}_2 - \mathbf{r}_1)$$

and for weight mg, with y being elevation,

$$W = -mg(y_2 - y_1)$$

(b) Spring forces (on two bodies):

$$W = -(k/2)(\delta_2^2 - \delta_1^2)$$

(c) Workless force:

$$W = \int_{t_1}^{t_2} \mathbf{F} \cdot \mathbf{v}\, dt = 0$$

if $\mathbf{F} \perp \mathbf{v}$ as with a normal force acting on a sliding body, or if $\mathbf{v} = \mathbf{0}$ at each instant as for the contact point of a rolling body.

(d) Constant couple:

$$W = C(\theta_2 - \theta_1)$$

A conservative force does work independent of path and can have associated with it a potential φ which we define so that its change is the negative of the work done by the force. Examples are weight, for which $\varphi = mgy$, and a linear spring, $\varphi = (k/2)\delta^2$. The minus sign is used for the convenience that follows when all forces acting on a body are conservative, so that

$$W = T_2 - T_1$$
$$-(\varphi_2 - \varphi_1) = T_2 - T_1$$

or

$$T_2 + \varphi_2 = T_1 + \varphi_1$$

which expresses the conservation of mechanical energy.

For a rigid body in plane motion, we found in Chapter 4 that

$$\mathbf{H}_C = I_{xz}^C\, \omega\hat{\mathbf{i}} + I_{yz}^C\, \omega\hat{\mathbf{j}} + I_{zz}^C\, \omega\hat{\mathbf{k}},$$

and similarly for a pivot. Concerning ourselves only with the case when the products of inertia vanish,

$$\mathbf{H}_C = I_{zz}^C\, \omega\hat{\mathbf{k}}$$

As long as a body is behaving rigidly before and after an interval of time of interest, the principle of angular impulse and angular momentum from Chapter 2 gives

$$\int_{t_1}^{t_2} \Sigma M_C\, dt = (I_C\omega)_2 - (I_C\omega)_1$$

even though it may be that $I_C(t_2) \neq I_C(t_1)$ as in the example of the spinning ice skater. The above is, of course, paired with the principle of linear impulse and momentum,

$$\int_{t_1}^{t_2} \Sigma\mathbf{F}\, dt = m\mathbf{v}_C\,(t_2) - m\mathbf{v}_C\,(t_1)$$

to effect solution of collision problems.

REVIEW QUESTIONS ▶ Chapter 5

True or False?

These questions all refer to rigid bodies in plane motion.

1. If you raise a 2-lb object 3 ft from rest and stop it there, gravity has done −6 ft-lb of work and you have done +6 ft-lb on the object.
2. The work of a constant couple $C\hat{\mathbf{k}}$ on body $\mathcal{B}$ is always $C\hat{\mathbf{k}} \cdot \theta\hat{\mathbf{k}} = C\theta$, where θ is the angle through which the body turns.
3. The work done by a linear spring is always $k(\delta_i^2 - \delta_f^2)/2$, where δ_i and δ_f are the amounts of initial and final stretch. (If negative, they represent compression.)
4. There are actually three separate work and kinetic energy principles; two of the equations add to give the third.
5. The principles of work and kinetic energy, and (linear and angular) impulse and momentum, result from general integrations of the equations of motion, and thus they are free of accelerations.

6. Not all forces acting on a body have to do non-zero work on it in general.
7. The friction force beneath a rolling wheel does work on it if the surface of contact is curved and fixed.
8. The normal force exerted on a rolling wheel by a surface, whether fixed or in motion, never does work on the wheel.
9. The principle $\int \Sigma \mathbf{F}\, dt = \int \dot{\mathbf{L}}\, dt = m\mathbf{v}_{C_f} - m\mathbf{v}_{C_i}$ is valid for deformable bodies.
10. The principle $\int \Sigma M_C \; dt = I_{zz}^C \omega_f - I_{zz}^C \omega_i$ is valid for deformable bodies.
11. Any problem that can be solved by $W = \Delta T$ can likewise be solved by using "kinetic + potential energy = constant."
12. The formula $T = \frac{1}{2} I_{①} \omega^2$ gives *all* the kinetic energy of the rigid body in plane motion, assuming the body is not translating.

Answers: 1. T 2. T 3. T 4. T 5. T 6. T 7. F 8. F 9. T 10. F 11. F 12. T

6 Kinematics of a Rigid Body in Three-Dimensional Motion

6.1 Introduction
6.2 Relation Between Derivatives / The Angular Velocity Vector
6.3 Properties of Angular Velocity
The Derivative Formula
Uniqueness of the Angular Velocity Vector
The Addition Theorem
Simple Angular Velocity
Summary of Properties of Angular Velocity
6.4 The Angular Acceleration Vector
6.5 Velocity and Acceleration in Moving Frames of Reference
The Velocity Relationship in Moving Frames
The Acceleration Relationship in Moving Frames / Coriolis Acceleration
6.6 The Earth as a Moving Frame
6.7 Velocity and Acceleration Equations for Two Points of the Same Rigid Body
Does An Instantaneous Axis of Rotation Exist in General?
6.8 Describing the Orientation of a Rigid Body
The Eulerian Angles
6.9 Rotation Matrices
Summary
Review Questions

6.1 Introduction

In this chapter we study the kinematics of a rigid body in general motion — we now do for general motion what we did in Chapter 3 for plane motion. There we found that at any instant the velocities of different points are linked together because of the rigidity of the body, and the connecting link is the angular velocity of the body. We also found there that it is angular velocity that links the derivatives of the *same* vector relative to two different frames of reference. We now wish to remove the plane-motion restriction.

The principal difficulty encountered in the study of general motion of a rigid body is that the angular velocity $\boldsymbol{\omega}$ does not always take the $\dot{\theta}\hat{\mathbf{k}}$ form of plane motion. The fact that $\boldsymbol{\omega}$ can be changing in direction as the body moves causes it to be difficult to visualize. An efficient way to deal with this abstract concept is to start with derivatives of the same vector in different frames of reference (rigid bodies). Angular velocities will naturally arise out of connecting these derivatives, and properties of relative angular velocities amongst several bodies, almost self-evident for the case of plane motion, will surface for the general case.

Application of the derivative/angular-velocity relationship to position vectors, in two frames, of a point leads us to the velocities of the point as observed in those frames, connected, in part, by the relative angular velocity of the frames. Subsequent mathematical analysis leads to: a relationship between accelerations of a point in two frames; a relationship between velocities, relative to a frame, of two points fixed in the same rigid body; and a relationship between accelerations, relative to a frame, of two such body-fixed points.

We close this chapter with development of methods to describe the orientation of a rigid body (relative, of course, to some reference frame). This was easy to do in plane motion — an angle θ is all that was required. In the general case we shall see the need for three angles, called Euler angles in one popular scheme.

The reflective reader will notice that the sequence of coverages in this chapter is almost precisely the reverse of its counterpart in Chapter 3, which is after all just a special case. There we were able to capitalize on an ease of visualization not available to us here.

6.2 Relation Between Derivatives/The Angular Velocity Vector

In this section we consider the relationship between the derivatives of a vector taken in two different frames. In the process we shall arrive at a concise and useful definition of angular velocity. The reader is strongly encouraged to persevere until this section and the next are fully understood. Even though the angular velocity vector in three dimensions is a difficult subject at first, it *must* be comprehended before we can consider the kinematics and kinetics of general rigid-body motion. The angular velocity vector is the key to the subject. It will either make life easier for students of three-dimensional motion (if they work hard at understanding it) or much more difficult (if they do not).

Let **Q** be an arbitrary vector. We may express **Q** in terms of its components (Q_x, Q_y, Q_z) associated with directions fixed in a frame $\mathcal{B}$ by

$$\begin{aligned}\mathbf{Q} &= (\mathbf{Q}\cdot\hat{\mathbf{i}})\hat{\mathbf{i}} + (\mathbf{Q}\cdot\hat{\mathbf{j}})\hat{\mathbf{j}} + (\mathbf{Q}\cdot\hat{\mathbf{k}})\hat{\mathbf{k}} \\ &= Q_x\hat{\mathbf{i}} + Q_y\hat{\mathbf{j}} + Q_z\hat{\mathbf{k}} \end{aligned} \tag{6.1}$$

in which the unit vectors ($\hat{\mathbf{i}}$, $\hat{\mathbf{j}}$, $\hat{\mathbf{k}}$) are parallel at all times to the respective axes of a Cartesian coordinate system fixed in $\mathcal{B}$. Now consider another reference frame $\mathcal{A}$, in which we wish to differentiate vector **Q** (see Figure 6.1). As an example, we may wish to find the velocity of a point in frame $\mathcal{A}$ even though the point's location may be defined in $\mathcal{B}$ (say by the vector **Q**). In this case, part of the solution will require that we be able to differentiate **Q** in $\mathcal{A}$ even though it is expressed in terms of its components in $\mathcal{B}$.

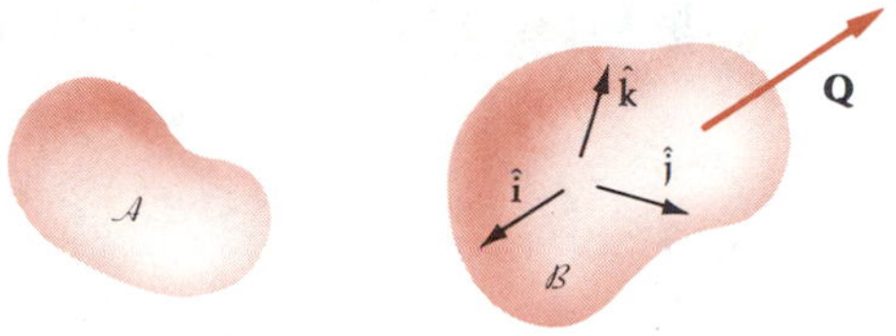

Figure 6.1 Vector Q and frames $\mathcal{A}$ and $\mathcal{B}$.

Therefore it is now time to learn how to relate derivatives of a vector taken in two different frames. We emphasize at the outset that these vectors are completely arbitrary — they need not even be related to dynamics! Nor does the derivative have to be with respect to time, although this is the independent variable of interest to us in dynamics and thus the one we shall use in the development to follow.

Letting ${}^{\mathcal{A}}\dot{\mathbf{Q}}$ represent (see Equation 1.8) the derivative of **Q** with respect to time in $\mathcal{A}$, we have

$${}^{\mathcal{A}}\dot{\mathbf{Q}} = \dot{Q}_x\hat{\mathbf{i}} + \dot{Q}_y\hat{\mathbf{j}} + \dot{Q}_z\hat{\mathbf{k}} + Q_x{}^{\mathcal{A}}\dot{\hat{\mathbf{i}}} + Q_y{}^{\mathcal{A}}\dot{\hat{\mathbf{j}}} + Q_z{}^{\mathcal{A}}\dot{\hat{\mathbf{k}}} \tag{6.2}$$

Recognizing the first three terms on the right of Equation (6.2) as the derivative of **Q** in $\mathcal{B}$, we have

$${}^{\mathcal{A}}\dot{\mathbf{Q}} = {}^{\mathcal{B}}\dot{\mathbf{Q}} + (Q_x{}^{\mathcal{A}}\dot{\hat{\mathbf{i}}} + Q_y{}^{\mathcal{A}}\dot{\hat{\mathbf{j}}} + Q_z{}^{\mathcal{A}}\dot{\hat{\mathbf{k}}}) \tag{6.3}$$

Clearly the last three (parenthesized) terms in Equation (6.3) represent a vector depending upon both **Q** *and* the change of orientation of frame $\mathcal{B}$ with respect to $\mathcal{A}$. We now proceed to obtain a useful and compact expression for this vector; in the process, the angular velocity vector will arise.

Since

$$\hat{\mathbf{i}}\cdot\hat{\mathbf{i}} = \hat{\mathbf{j}}\cdot\hat{\mathbf{j}} = \hat{\mathbf{k}}\cdot\hat{\mathbf{k}} = 1 \tag{6.4}$$

it follows that

$$\hat{\mathbf{i}}\cdot{}^{\mathcal{A}}\dot{\hat{\mathbf{i}}} = 0 = \hat{\mathbf{j}}\cdot{}^{\mathcal{A}}\dot{\hat{\mathbf{j}}} = \hat{\mathbf{k}}\cdot{}^{\mathcal{A}}\dot{\hat{\mathbf{k}}} \tag{6.5}$$

so that the three derivatives of the unit vectors in Equation (6.3) are each perpendicular to the respective unit vectors themselves.*

Question 6.1 Will this be true for *any* vector of constant magnitude (not necessarily a unit vector)?

This means that there are three vectors $\boldsymbol{\alpha}$, $\boldsymbol{\beta}$, and $\boldsymbol{\gamma}$ for which

$$\begin{aligned} {}^{\mathcal{A}}\dot{\hat{\mathbf{i}}} &= \boldsymbol{\alpha} \times \hat{\mathbf{i}} \\ {}^{\mathcal{A}}\dot{\hat{\mathbf{j}}} &= \boldsymbol{\beta} \times \hat{\mathbf{j}} \\ {}^{\mathcal{A}}\dot{\hat{\mathbf{k}}} &= \boldsymbol{\gamma} \times \hat{\mathbf{k}} \end{aligned} \tag{6.6}$$

The cross products ensure that $\hat{\mathbf{i}}$, $\hat{\mathbf{j}}$, and $\hat{\mathbf{k}}$ are each perpendicular to their derivatives ($\hat{\mathbf{i}} \perp {}^{\mathcal{A}}\dot{\hat{\mathbf{i}}}$ and so on) and the magnitudes of $\boldsymbol{\alpha}$, $\boldsymbol{\beta}$, and $\boldsymbol{\gamma}$ give to ${}^{\mathcal{A}}\dot{\hat{\mathbf{i}}}$, ${}^{\mathcal{A}}\dot{\hat{\mathbf{j}}}$, and ${}^{\mathcal{A}}\dot{\hat{\mathbf{k}}}$ their correct magnitudes.

In terms of their components in $\mathcal{B}$, we can write $\boldsymbol{\alpha}$, $\boldsymbol{\beta}$, and $\boldsymbol{\gamma}$ as

$$\begin{aligned} \boldsymbol{\alpha} &= \alpha_x\hat{\mathbf{i}} + \alpha_y\hat{\mathbf{j}} + \alpha_z\hat{\mathbf{k}} \\ \boldsymbol{\beta} &= \beta_x\hat{\mathbf{i}} + \beta_y\hat{\mathbf{j}} + \beta_z\hat{\mathbf{k}} \\ \boldsymbol{\gamma} &= \gamma_x\hat{\mathbf{i}} + \gamma_y\hat{\mathbf{j}} + \gamma_z\hat{\mathbf{k}} \end{aligned} \tag{6.7}$$

Substituting these component expressions into Equations (6.6) results in

$$\begin{aligned} {}^{\mathcal{A}}\dot{\hat{\mathbf{i}}} &= \alpha_z\hat{\mathbf{j}} - \alpha_y\hat{\mathbf{k}} \\ {}^{\mathcal{A}}\dot{\hat{\mathbf{j}}} &= \beta_x\hat{\mathbf{k}} - \beta_z\hat{\mathbf{i}} \\ {}^{\mathcal{A}}\dot{\hat{\mathbf{k}}} &= \gamma_y\hat{\mathbf{i}} - \gamma_x\hat{\mathbf{j}} \end{aligned} \tag{6.8}$$

and we see that α_x, β_y, and γ_z, at this point, remain arbitrary.

Question 6.2 Why do they remain arbitrary?

Here we are seeking to relate the components of the vectors $\boldsymbol{\alpha}$, $\boldsymbol{\beta}$, and $\boldsymbol{\gamma}$ in the hope of finding a way to express the last three terms of Equation (6.3). To this end we note that, for all time t,

$$\hat{\mathbf{i}} \cdot \hat{\mathbf{j}} = \hat{\mathbf{j}} \cdot \hat{\mathbf{k}} = \hat{\mathbf{k}} \cdot \hat{\mathbf{i}} = 0$$

from the first of which, differentiation yields

$${}^{\mathcal{A}}\dot{\hat{\mathbf{i}}} \cdot \hat{\mathbf{j}} + {}^{\mathcal{A}}\dot{\hat{\mathbf{j}}} \cdot \hat{\mathbf{i}} = 0 \tag{6.9}$$

* This assumes that the unit vectors are not constant in frame $\mathcal{A}$. If two of them are constant in $\mathcal{A}$, then all three are and the angular velocity vanishes, if only *one* is constant in $\mathcal{A}$, we have a simple special case to be considered later.

Answer 6.1 Sure, as we have seen in Section 1.6.

Answer 6.2 Since $\alpha_x\hat{\mathbf{i}} \times \hat{\mathbf{i}} = \mathbf{0}$, then α_x can be anything and not affect the first of Equations (6.6).

Substitution of the first two equations of (6.6) into (6.9) yields

$$(\boldsymbol{\alpha} \times \hat{\mathbf{i}}) \cdot \hat{\mathbf{j}} + (\boldsymbol{\beta} \times \hat{\mathbf{j}}) \cdot \hat{\mathbf{i}} = 0 \tag{6.10}$$

Interchanging the dot and cross in each term (which leaves the scalar triple product unchanged) results in

$$\boldsymbol{\alpha} \cdot \hat{\mathbf{k}} - \boldsymbol{\beta} \cdot \hat{\mathbf{k}} = 0 \tag{6.11}$$

so that

$$\alpha_z = \beta_z \tag{6.12}$$

Similarly from

$$\hat{\mathbf{j}} \cdot \hat{\mathbf{k}} = 0 \quad \text{and} \quad \hat{\mathbf{k}} \cdot \hat{\mathbf{i}} = 0 \tag{6.13}$$

we respectively obtain (as the student should verify)

$$\beta_x = \gamma_x \quad \text{and} \quad \gamma_y = \alpha_y \tag{6.14}$$

The only components not involved in Equations (6.12) and (6.14) are α_x, β_y, and γ_z, which were arbitrary. If we now select them as follows,

$$\begin{aligned} \alpha_x &= \beta_x = \gamma_x \\ \beta_y &= \gamma_y = \alpha_y \\ \gamma_z &= \alpha_z = \beta_z \end{aligned} \tag{6.15}$$

then *all three vectors are identical,* and we call the resulting common vector $\boldsymbol{\omega}_{\mathcal{B}/\mathcal{A}}$:

$$\boldsymbol{\alpha} = \boldsymbol{\beta} = \boldsymbol{\gamma} = \boldsymbol{\omega}_{\mathcal{B}/\mathcal{A}} \tag{6.16}$$

If we now dot the three equations (6.8) respectively with $\hat{\mathbf{j}}$, $\hat{\mathbf{k}}$, and $\hat{\mathbf{i}}$, we get the three components of $\boldsymbol{\omega}_{\mathcal{B}/\mathcal{A}}$:

$$\begin{aligned} {}^{\mathcal{A}}\dot{\hat{\mathbf{i}}} \cdot \hat{\mathbf{j}} &= \alpha_z = \omega_{\mathcal{B}/\mathcal{A}_z} \\ {}^{\mathcal{A}}\dot{\hat{\mathbf{j}}} \cdot \hat{\mathbf{k}} &= \beta_x = \omega_{\mathcal{B}/\mathcal{A}_x} \\ {}^{\mathcal{A}}\dot{\hat{\mathbf{k}}} \cdot \hat{\mathbf{i}} &= \gamma_y = \omega_{\mathcal{B}/\mathcal{A}_y} \end{aligned} \tag{6.17}$$

Thus the vector $\boldsymbol{\omega}_{\mathcal{B}/\mathcal{A}}$ may be expressed as*

$$\boldsymbol{\omega}_{\mathcal{B}/\mathcal{A}} = ({}^{\mathcal{A}}\dot{\hat{\mathbf{j}}} \cdot \hat{\mathbf{k}})\hat{\mathbf{i}} + ({}^{\mathcal{A}}\dot{\hat{\mathbf{k}}} \cdot \hat{\mathbf{i}})\hat{\mathbf{j}} + ({}^{\mathcal{A}}\dot{\hat{\mathbf{i}}} \cdot \hat{\mathbf{j}})\hat{\mathbf{k}} \tag{6.18}$$

We call the vector $\boldsymbol{\omega}_{\mathcal{B}/\mathcal{A}}$ defined by Equation (6.18) the **angular velocity of frame $\mathcal{B}$ with respect to frame $\mathcal{A}$**, or more briefly, the **angular velocity of $\mathcal{B}$ in $\mathcal{A}$**. It is clear that the angular velocity vector depends intimately on the way frame $\mathcal{B}$ is changing its orientation with respect to $\mathcal{A}$. In the next section we examine some special properties of this vector. We shall see that $\boldsymbol{\omega}_{\mathcal{B}/\mathcal{A}}$ is *unique,* which means that we lost no generality when we let $\alpha_x = \beta_x = \gamma_x$, $\beta_y = \gamma_y = \alpha_y$, and $\gamma_z = \alpha_z = \beta_z$ in our development above of angular velocity.

---

* This is the *definition* of angular velocity set forth by the dynamicist T. R. Kane. See his books *Dynamics: Theory and Applications* (New York: McGraw-Hill, 1985), p. 16 and *Spacecraft Dynamics* (New York: McGraw-Hill, 1983), p. 49.

6.3 Properties of Angular Velocity

The Derivative Formula

We now return to Equation (6.3). Substituting from Equations (6.6) for ${}^{\mathcal{A}}\dot{\hat{\mathbf{i}}}$, ${}^{\mathcal{A}}\dot{\hat{\mathbf{j}}}$, and ${}^{\mathcal{A}}\dot{\hat{\mathbf{k}}}$ and using Equation (6.16) to replace $\boldsymbol{\alpha}$, $\boldsymbol{\beta}$, and $\boldsymbol{\gamma}$ by $\boldsymbol{\omega}_{\mathcal{B}/\mathcal{A}}$, we obtain

$$ {}^{\mathcal{A}}\dot{\mathbf{Q}} = {}^{\mathcal{B}}\dot{\mathbf{Q}} + Q_x(\boldsymbol{\omega}_{\mathcal{B}/\mathcal{A}} \times \hat{\mathbf{i}}) + Q_y(\boldsymbol{\omega}_{\mathcal{B}/\mathcal{A}} \times \hat{\mathbf{j}}) + Q_z(\boldsymbol{\omega}_{\mathcal{B}/\mathcal{A}} \times \hat{\mathbf{k}}) \quad (6.19) $$

We call this the **derivative formula**, which may be expressed, using Equation (6.1), as

$$ {}^{\mathcal{A}}\dot{\mathbf{Q}} = {}^{\mathcal{B}}\dot{\mathbf{Q}} + \boldsymbol{\omega}_{\mathcal{B}/\mathcal{A}} \times \mathbf{Q} \quad (6.20) $$

Equation (6.20) will turn out to be of vital importance in this chapter and, moreover, to be equally invaluable in our later study of the kinetics of rigid bodies in general motion in Chapter 7. It permits us easily to calculate the derivative of a vector in one frame if it is expressed in terms of base vectors fixed in another; the only price we have to pay is to add the cross product $\boldsymbol{\omega}_{\mathcal{B}/\mathcal{A}} \times \mathbf{Q}$. Thus the first property of $\boldsymbol{\omega}_{\mathcal{B}/\mathcal{A}}$ is that it allows us to relate (by Equation 6.20) the derivatives of any vector in two different frames. We have already encountered this for the special case of plane motion in Section 3.7, where Equation (3.44) may be seen to be the plane motion counterpart of Equation (6.20).

Uniqueness of the Angular Velocity Vector

There remains the nagging question of whether there might be more than one vector satisfying Equation (6.20); remember that we arbitrarily selected the components α_x, β_y, and γ_z in the preceding section in order to make $\boldsymbol{\alpha} = \boldsymbol{\beta} = \boldsymbol{\gamma} = \boldsymbol{\omega}_{\mathcal{B}/\mathcal{A}}$. We now proceed to show that the angular velocity vector is indeed unique. We do so by postulating that *two* vectors $\boldsymbol{\omega}_{\mathcal{B}/\mathcal{A}_1}$ and $\boldsymbol{\omega}_{\mathcal{B}/\mathcal{A}_2}$ *both* satisfy Equation (6.20) and then showing that they are necessarily equal.* We have

$$ {}^{\mathcal{A}}\dot{\mathbf{Q}} = {}^{\mathcal{B}}\dot{\mathbf{Q}} + \boldsymbol{\omega}_{\mathcal{B}/\mathcal{A}_1} \times \mathbf{Q} \quad (6.21) $$

$$ {}^{\mathcal{A}}\dot{\mathbf{Q}} = {}^{\mathcal{B}}\dot{\mathbf{Q}} + \boldsymbol{\omega}_{\mathcal{B}/\mathcal{A}_2} \times \mathbf{Q} \quad (6.22) $$

so that, subtracting, we get

$$ (\boldsymbol{\omega}_{\mathcal{B}/\mathcal{A}_1} - \boldsymbol{\omega}_{\mathcal{B}/\mathcal{A}_2}) \times \mathbf{Q} = \mathbf{0} \quad (6.23) $$

Finally, since $\mathbf{Q}$ is arbitrary, the parenthesized expression of Equation (6.23) must vanish, and thus the angular velocity vector has been shown to be unique as the two $\boldsymbol{\omega}$'s are one and the same.

Nothing has yet been said about dynamics in this section or in the preceding one; thus it is clear that angular velocity is a far more general

* Let $\boldsymbol{\omega}_{\mathcal{B}/\mathcal{A}_1}$ be calculated with $\hat{\mathbf{i}}$, $\hat{\mathbf{j}}$, and $\hat{\mathbf{k}}$ as described in Section 6.2, for example, and let $\boldsymbol{\omega}_{\mathcal{B}/\mathcal{A}_2}$ be computed with another triad of unit vectors fixed in $\mathcal{B}$. The question of uniqueness is whether the resulting $\boldsymbol{\omega}_{\mathcal{B}/\mathcal{A}}$'s are the same.

vector than one that is simply useful in describing rotational motions of rigid bodies. We have seen that angular velocity is in fact the vector that may be used in relating the derivatives in two frames of any arbitrary vector. Furthermore, even though we have used time as the independent variable, these derivatives may be taken with respect to *any* scalar variable. Finally, we note from the defining equation (6.18) for $\boldsymbol{\omega}_{\mathcal{B}/\mathcal{A}}$ that angular velocity is a vector relating two *frames*; thus it is meaningless to talk about the angular velocity of a point.

Now let us consider several additional properties of $\boldsymbol{\omega}_{\mathcal{B}/\mathcal{A}}$ that will prove useful in what is to follow. First we note for emphasis that if two frames $\mathcal{A}$ and $\mathcal{B}$ maintain a constant orientation (even if they are each in motion in a third frame $\mathcal{C}$), then $\boldsymbol{\omega}_{\mathcal{B}/\mathcal{A}} \equiv \mathbf{0}$.* The proof is simply to observe that no unit vector fixed in direction in $\mathcal{B}$ can change with time in $\mathcal{A}$ if there is no change in orientation between $\mathcal{A}$ and $\mathcal{B}$. Thus from Equation (6.18), $\boldsymbol{\omega}_{\mathcal{B}/\mathcal{A}} \equiv \mathbf{0}$.

Next we shall prove that the angular velocity of $\mathcal{B}$ in $\mathcal{A}$ is the negative of the angular velocity of $\mathcal{A}$ in $\mathcal{B}$. If we add Equation (6.20) to the equation

$$^{\mathcal{B}}\dot{\mathbf{Q}} = {}^{\mathcal{A}}\dot{\mathbf{Q}} + \boldsymbol{\omega}_{\mathcal{A}/\mathcal{B}} \times \mathbf{Q} \tag{6.24}$$

we obtain

$$(\boldsymbol{\omega}_{\mathcal{B}/\mathcal{A}} + \boldsymbol{\omega}_{\mathcal{A}/\mathcal{B}}) \times \mathbf{Q} = \mathbf{0} \tag{6.25}$$

Again, since $\mathbf{Q}$ is arbitrary, we have the expected result:

$$\boldsymbol{\omega}_{\mathcal{B}/\mathcal{A}} = -\boldsymbol{\omega}_{\mathcal{A}/\mathcal{B}} \tag{6.26}$$

The Addition Theorem

We now prove the **addition theorem**, which states that

$$\boldsymbol{\omega}_{\mathcal{C}/\mathcal{A}} = \boldsymbol{\omega}_{\mathcal{C}/\mathcal{B}} + \boldsymbol{\omega}_{\mathcal{B}/\mathcal{A}} \tag{6.27}$$

For the proof, we know from the first property of $\boldsymbol{\omega}_{\mathcal{B}/\mathcal{A}}$ that

$$^{\mathcal{A}}\dot{\mathbf{Q}} = {}^{\mathcal{C}}\dot{\mathbf{Q}} + \boldsymbol{\omega}_{\mathcal{C}/\mathcal{A}} \times \mathbf{Q} \tag{6.28}$$

$$^{\mathcal{B}}\dot{\mathbf{Q}} = {}^{\mathcal{C}}\dot{\mathbf{Q}} + \boldsymbol{\omega}_{\mathcal{C}/\mathcal{B}} \times \mathbf{Q} \tag{6.29}$$

$$^{\mathcal{A}}\dot{\mathbf{Q}} = {}^{\mathcal{B}}\dot{\mathbf{Q}} + \boldsymbol{\omega}_{\mathcal{B}/\mathcal{A}} \times \mathbf{Q} \tag{6.30}$$

Adding Equations (6.29) and (6.30) yields

$$^{\mathcal{A}}\dot{\mathbf{Q}} = {}^{\mathcal{C}}\dot{\mathbf{Q}} + (\boldsymbol{\omega}_{\mathcal{C}/\mathcal{B}} + \boldsymbol{\omega}_{\mathcal{B}/\mathcal{A}}) \times \mathbf{Q} \tag{6.31}$$

and subtracting Equation (6.31) from (6.28) gives

$$(\boldsymbol{\omega}_{\mathcal{C}/\mathcal{A}} - \boldsymbol{\omega}_{\mathcal{C}/\mathcal{B}} - \boldsymbol{\omega}_{\mathcal{B}/\mathcal{A}}) \times \mathbf{Q} = \mathbf{0} \tag{6.32}$$

* *Constant orientation* means that $\mathcal{A}$ and $\mathcal{B}$ move as if they were rigidly attached except for a possible translation of one with respect to the other.

Therefore, again since **Q** is arbitrary,

$$\boldsymbol{\omega}_{\mathcal{C}/\mathcal{A}} = \boldsymbol{\omega}_{\mathcal{C}/\mathcal{B}} + \boldsymbol{\omega}_{\mathcal{B}/\mathcal{A}} \tag{6.33}$$

and the theorem is proved. It may seem intuitively obvious to the reader that Equation (6.33) is true, but in the next section we show that such a relationship *does not exist* for angular acceleration!

The addition theorem is an extremely powerful result. With it we are able to build up the angular velocity, one pair of frames at a time, of a body turning in complicated ways relative to a reference frame. This theorem makes it possible for us to avoid using the definition (6.18), which has served us well but in practice is normally supplanted by the properties described in this section.

We recall for emphasis that if $\mathcal{A}$ and $\mathcal{B}$ both move in $\mathcal{C}$ and maintain constant orientation with respect to each other, then $\boldsymbol{\omega}_{\mathcal{B}/\mathcal{A}} = \mathbf{0}$. Thus, by the addition theorem,

$$\boldsymbol{\omega}_{\mathcal{B}/\mathcal{C}} = \boldsymbol{\omega}_{\mathcal{B}/\mathcal{A}} + \boldsymbol{\omega}_{\mathcal{A}/\mathcal{C}} = \boldsymbol{\omega}_{\mathcal{A}/\mathcal{C}}$$

so that, as expected, the angular velocities in $\mathcal{C}$ of two frames $\mathcal{A}$ and $\mathcal{B}$ maintaining constant orientation with each other are identical.

Conversely, if two frames $\mathcal{A}$ and $\mathcal{B}$ have equal angular velocities in $\mathcal{C}$, we may show that their relative orientation is constant. Using the addition theorem, $\boldsymbol{\omega}_{\mathcal{B}/\mathcal{A}} = \mathbf{0}$; thus with $\hat{\mathbf{i}}$, $\hat{\mathbf{j}}$, and $\hat{\mathbf{k}}$ still fixed in direction in $\mathcal{B}$, then from Equations (6.6) and (6.16)

$$\begin{aligned} {}^{\mathcal{A}}\dot{\hat{\mathbf{i}}} &= \boldsymbol{\omega}_{\mathcal{B}/\mathcal{A}} \times \hat{\mathbf{i}} = \mathbf{0} \\ {}^{\mathcal{A}}\dot{\hat{\mathbf{j}}} &= \boldsymbol{\omega}_{\mathcal{B}/\mathcal{A}} \times \hat{\mathbf{j}} = \mathbf{0} \\ {}^{\mathcal{A}}\dot{\hat{\mathbf{k}}} &= \boldsymbol{\omega}_{\mathcal{B}/\mathcal{A}} \times \hat{\mathbf{k}} = \mathbf{0} \end{aligned}$$

Therefore, $\hat{\mathbf{i}}$, $\hat{\mathbf{j}}$, and $\hat{\mathbf{k}}$ are constant in $\mathcal{A}$, so the orientation of $\mathcal{B}$ in $\mathcal{A}$ is constant.

Thus, for two frames $\mathcal{A}$ and $\mathcal{B}$, the descriptions "constant orientation" and "$\boldsymbol{\omega}_{\mathcal{B}/\mathcal{A}} \equiv \mathbf{0}$" are completely equivalent. Note also that the addition theorem can be extended to *any number of frames* by repetition of the following procedure, two frames at a time:

$$\boldsymbol{\omega}_{\mathcal{A}/\mathcal{D}} = \boldsymbol{\omega}_{\mathcal{A}/\mathcal{B}} + \boldsymbol{\omega}_{\mathcal{B}/\mathcal{D}} = \boldsymbol{\omega}_{\mathcal{A}/\mathcal{B}} + \boldsymbol{\omega}_{\mathcal{B}/\mathcal{C}} + \boldsymbol{\omega}_{\mathcal{C}/\mathcal{D}}$$

Simple Angular Velocity

Next we show that when there exists a unit vector $\hat{\mathbf{k}}$ whose time derivative in each of two frames $\mathcal{A}$ and $\mathcal{B}$ vanishes (that is, ${}^{\mathcal{A}}\dot{\hat{\mathbf{k}}} = {}^{\mathcal{B}}\dot{\hat{\mathbf{k}}} = \mathbf{0}$), then

$$\boldsymbol{\omega}_{\mathcal{B}/\mathcal{A}} = \dot{\theta}\hat{\mathbf{k}} \tag{6.34}$$

in which θ is the angle between a pair of directed line segments $\ell_{\mathcal{A}}$ and $\ell_{\mathcal{B}}$ fixed respectively in $\mathcal{A}$ and $\mathcal{B}$, each perpendicular to $\hat{\mathbf{k}}$. The angle is measured in a reference plane containing projections of the two lines intersecting at point P as shown in Figure 6.2. The sign of the angle θ is given by the right-hand rule: If the right thumb is placed in the positive $\hat{\mathbf{k}}$ direction at P, then the direction of positive θ is that of the right hand's fingers when they curl from $\ell_{\mathcal{A}}$ into $\ell_{\mathcal{B}}$ as shown.

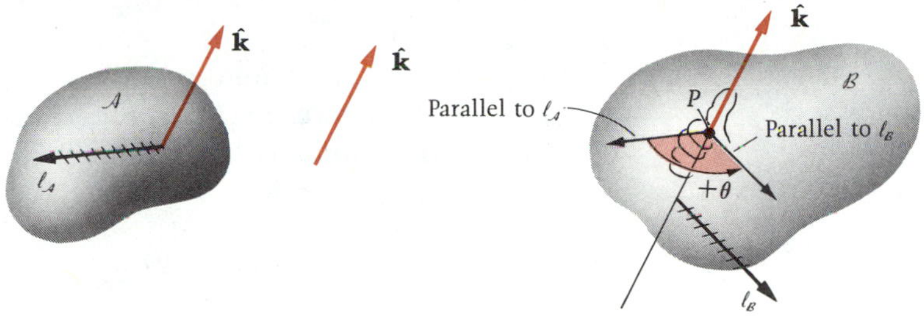

Figure 6.2 Simple angular velocity.

The type of rotational motion given by Equation (6.34) is called **simple angular velocity**. One case in which Equation (6.34) holds is that of plane motion; note, however, that there are more general cases of simple angular velocity in which the body $\mathcal{B}$ may also have a translational motion in $\mathcal{A}$ parallel to $\hat{\mathbf{k}}$ that would prevent the plane motion designation.

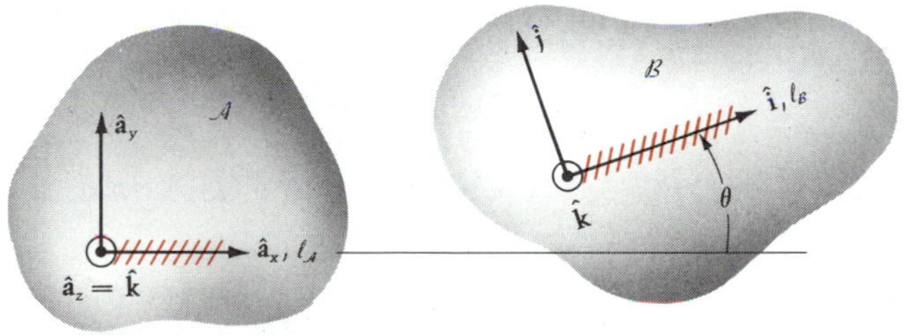

Figure 6.3 Unit vectors drawn in the reference plane for simple angular velocity.

To prove Equation (6.34), we make use of Figure 6.3. (The reference plane is the plane of the paper, and the unit vector that is constant in $\mathcal{A}$ and in $\mathcal{B}$ is $\hat{\mathbf{k}}$, perpendicular to the paper.) From this figure we may write

$$\begin{aligned} \hat{\mathbf{i}} &= \hat{\mathbf{a}}_x \cos\theta + \hat{\mathbf{a}}_y \sin\theta \\ \hat{\mathbf{j}} &= -\hat{\mathbf{a}}_x \sin\theta + \hat{\mathbf{a}}_y \cos\theta \\ \hat{\mathbf{k}} &= \hat{\mathbf{a}}_z \end{aligned} \tag{6.35}$$

Therefore, differentiating Equation (6.35), we get

$$\begin{aligned} {}^{\mathcal{A}}\dot{\hat{\mathbf{i}}} &= (-\hat{\mathbf{a}}_x \sin\theta + \hat{\mathbf{a}}_y \cos\theta)\dot{\theta} = \dot{\theta}\hat{\mathbf{j}} \\ {}^{\mathcal{A}}\dot{\hat{\mathbf{j}}} &= (-\hat{\mathbf{a}}_x \cos\theta - \hat{\mathbf{a}}_y \sin\theta)\dot{\theta} = -\dot{\theta}\hat{\mathbf{i}} \\ {}^{\mathcal{A}}\dot{\hat{\mathbf{k}}} &= \mathbf{0} \end{aligned} \tag{6.36}$$

and Equation (6.18) yields, upon direct substitution of Equation (6.36),

$$\omega_{\mathcal{B}/\mathcal{A}} = 0\hat{\mathbf{i}} + 0\hat{\mathbf{j}} + \dot{\theta}\hat{\mathbf{k}} \tag{6.37}$$

which is the desired result.

One interesting final property of $\omega_{\mathcal{B}/\mathcal{A}}$ is that its derivative is the same whether computed in $\mathcal{A}$ or in $\mathcal{B}$. Using Equation (6.20) and letting $\mathbf{Q}$ be $\omega_{\mathcal{B}/\mathcal{A}}$ itself, we get

$$^{\mathcal{A}}\dot{\omega}_{\mathcal{B}/\mathcal{A}} = {}^{\mathcal{B}}\dot{\omega}_{\mathcal{B}/\mathcal{A}} + \omega_{\mathcal{B}/\mathcal{A}} \times \omega_{\mathcal{B}/\mathcal{A}} = {}^{\mathcal{B}}\dot{\omega}_{\mathcal{B}/\mathcal{A}}$$

This result is not true for any other nonvanishing vector, unless it happens to be parallel to $\omega_{\mathcal{B}/\mathcal{A}}$.

Summary of Properties of Angular Velocity

The properties of $\omega_{\mathcal{B}/\mathcal{A}}$ that we have examined are summarized here:

1. It is a unique vector that satisfies

$$^{\mathcal{A}}\dot{\mathbf{Q}} = {}^{\mathcal{B}}\dot{\mathbf{Q}} + \omega_{\mathcal{B}/\mathcal{A}} \times \mathbf{Q}$$

which is called "The Derivative Formula."

2. $\omega_{\mathcal{B}/\mathcal{A}} = \mathbf{0}$ is synonymous with "the orientations of $\mathcal{B}$ and $\mathcal{A}$ do not change." And if $\mathcal{A}$ and $\mathcal{B}$ maintain constant orientation, their angular velocities in any third frame are equal.
3. $\omega_{\mathcal{B}/\mathcal{A}} = -\omega_{\mathcal{A}/\mathcal{B}}$.
4. "The Addition Theorem": $\omega_{\mathcal{D}/\mathcal{A}} = \omega_{\mathcal{D}/\mathcal{B}} + \omega_{\mathcal{B}/\mathcal{A}} = (\omega_{\mathcal{D}/\mathcal{C}} + \omega_{\mathcal{C}/\mathcal{B}}) + \omega_{\mathcal{B}/\mathcal{A}}$, which can be further extended to any number of frames.
5. "Simple Angular Velocity": If $\hat{\mathbf{k}}$ is constant in both $\mathcal{A}$ and $\mathcal{B}$, then

$$\omega_{\mathcal{B}/\mathcal{A}} = \dot{\theta}\hat{\mathbf{k}}$$

where θ was defined earlier.

6. $^{\mathcal{A}}\dot{\omega}_{\mathcal{B}/\mathcal{A}} = {}^{\mathcal{B}}\dot{\omega}_{\mathcal{B}/\mathcal{A}}$.

In the following example we use the addition theorem to write an angular velocity, and then we express it in three different frames.

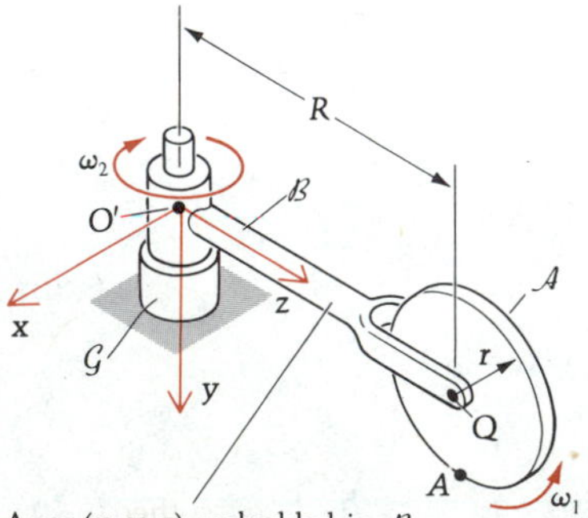

Axes (x, y, z) embedded in $\mathcal{B}$

Figure E6.1a

EXAMPLE 6.1

Body $\mathcal{B}$ in Figure E6.1a rotates in frame $\mathcal{G}$ about the vertical at constant angular speed ω_2; in $\mathcal{B}$, disk $\mathcal{A}$ rotates about its pinned axis at constant angular speed ω_1 relative to $\mathcal{B}$. (The directions of rotation are as shown. Determine the angular velocity of $\mathcal{A}$ in $\mathcal{G}$.

Solution

The coordinate axes shown are fixed in $\mathcal{B}$. By the addition theorem,*

$$\boldsymbol{\omega}_{\mathcal{A}/\mathcal{G}} = \boldsymbol{\omega}_{\mathcal{A}/\mathcal{B}} + \boldsymbol{\omega}_{\mathcal{B}/\mathcal{G}}$$
$$= \omega_1\hat{\mathbf{i}} + \omega_2\hat{\mathbf{j}} \qquad (1)$$

We see from this answer that expressing $\boldsymbol{\omega}_{\mathcal{A}/\mathcal{G}}$ in terms of its components in the intermediate ("between" $\mathcal{A}$ and $\mathcal{G}$) frame $\mathcal{B}$ has yielded a neat, simple result. If we had chosen instead to write $\boldsymbol{\omega}_{\mathcal{A}/\mathcal{G}}$ in terms of its components in $\mathcal{G}$, then (see Figure E6.1b) with $\hat{\mathbf{I}}, \hat{\mathbf{J}}, \hat{\mathbf{K}}$ fixed in $\mathcal{G}$,

$$\hat{\mathbf{i}} = (\cos\omega_2 t)\hat{\mathbf{I}} - (\sin\omega_2 t)\hat{\mathbf{K}}$$
$$\hat{\mathbf{j}} = \hat{\mathbf{J}}$$

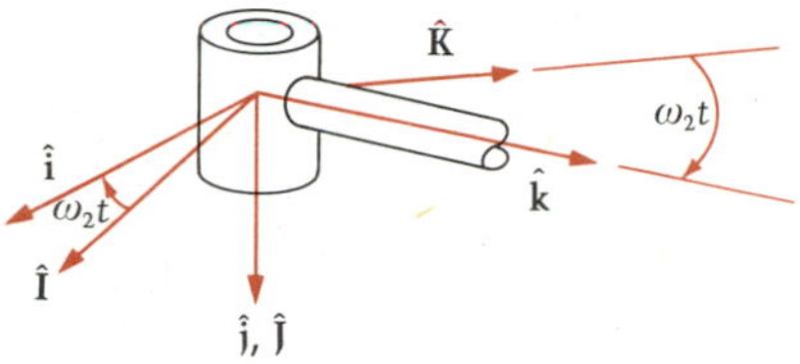

Figure E6.1b

so that, substituting into Equation (1),

$$\boldsymbol{\omega}_{\mathcal{A}/\mathcal{G}} = (\omega_1\cos\omega_2 t)\hat{\mathbf{I}} + \omega_2\hat{\mathbf{J}} - (\omega_1\sin\omega_2 t)\hat{\mathbf{K}}$$

And if we had written $\boldsymbol{\omega}_{\mathcal{A}/\mathcal{G}}$ in terms of its components along directions ($\hat{\mathbf{i}}_1, \hat{\mathbf{j}}_1, \hat{\mathbf{k}}_1$ in Figure E6.1c) fixed in $\mathcal{A}$, then with

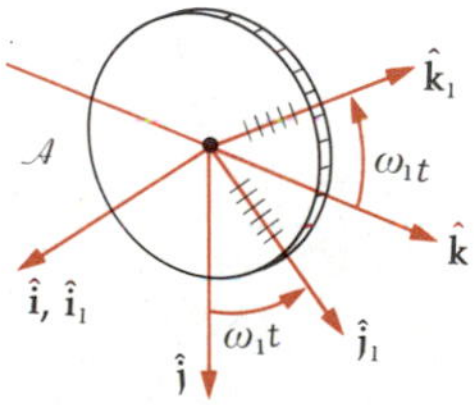

Figure E6.1c

$$\hat{\mathbf{i}} = \hat{\mathbf{i}}_1$$
$$\hat{\mathbf{j}} = (\cos\omega_1 t)\hat{\mathbf{j}}_1 - (\sin\omega_1 t)\hat{\mathbf{k}}_1$$

we obtain, again using Equation (1),

$$\boldsymbol{\omega}_{\mathcal{A}/\mathcal{G}} = \omega_1\hat{\mathbf{i}}_1 + (\omega_2\cos\omega_1 t)\hat{\mathbf{j}}_1 - (\omega_2\sin\omega_1 t)\hat{\mathbf{k}}_1$$

We see that expressing $\boldsymbol{\omega}_{\mathcal{A}/\mathcal{G}}$ in terms of its components in either $\mathcal{A}$ or $\mathcal{G}$ gives a lengthier expression than in $\mathcal{B}$; moreover, these expressions become even more complicated if ω_1 or ω_2 vary with time.

Question 6.3 Why?

The reader should note, however, that even though each of the three above representations of $\boldsymbol{\omega}_{\mathcal{A}/\mathcal{G}}$ appear to be different, they all yield the same vector.

---

* While the defining equation (6.18) is always available for directly computing the angular velocity, it is usually easier to build up the $\boldsymbol{\omega}$ vector by using the addition theorem.

Answer 6.3 Because then the angles (arguments of the sines and cosines) are not simply $\omega_1 t$ or $\omega_2 t$, but integrals of ω_1 or ω_2 with respect to time.

In the next example we illustrate the use of the "Derivative Formula" (Equation 6.20) three times.

EXAMPLE 6.2

Two children are playing in the park on a seesaw mounted on a merry-go-round as shown in Figure E6.2. The merry-go-round rotates about the ground-fixed (frame $\mathcal{G}$) vertical at $\omega_v = 3$ rad/sec, and at the instant shown the seesaw turns at $\omega_H = 2$ rad/sec relative to the merry-go-round $\mathcal{F}$. The vector from the girl to the boy is always $\mathbf{Q} = -10\,\hat{\mathbf{j}}$ ft, $(\hat{\mathbf{i}}, \hat{\mathbf{j}}, \hat{\mathbf{k}})$ being fixed in the seesaw board $\mathcal{B}$. Find ${}^{\mathcal{B}}\dot{\mathbf{Q}}$, ${}^{\mathcal{F}}\dot{\mathbf{Q}}$ and ${}^{\mathcal{G}}\dot{\mathbf{Q}}$ at the given instant.

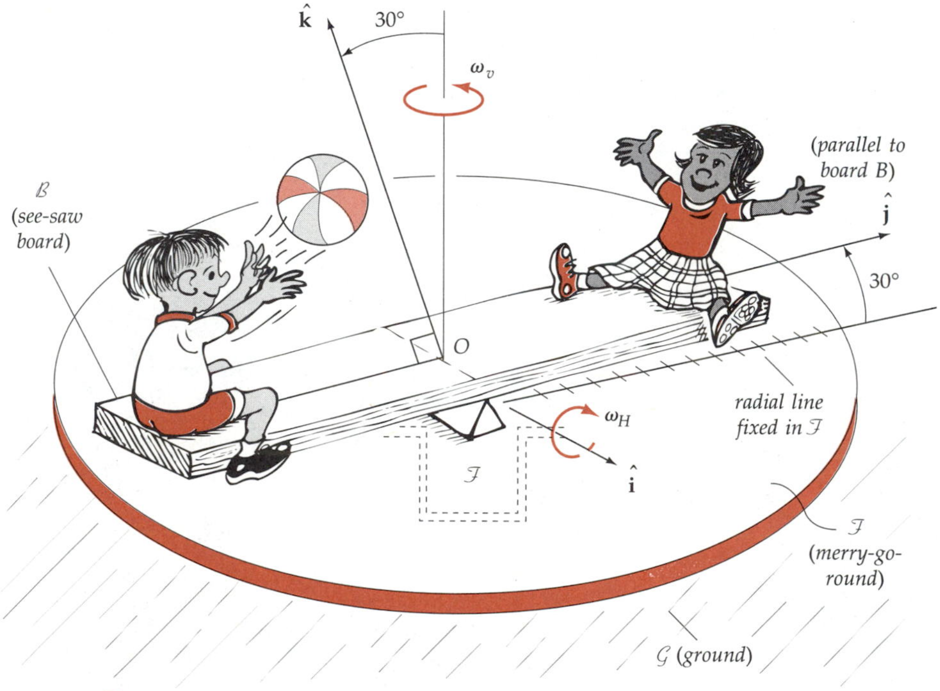

Figure E6.2

Solution

$\mathbf{Q} = -10\hat{\mathbf{j}}$ is constant relative to $\mathcal{B}$ so

$$ {}^{\mathcal{B}}\dot{\mathbf{Q}} = \mathbf{0} $$

To find ${}^{\mathcal{F}}\dot{\mathbf{Q}}$ we shall use the derivative formula:

$$ {}^{\mathcal{F}}\dot{\mathbf{Q}} = {}^{\mathcal{B}}\dot{\mathbf{Q}} + \boldsymbol{\omega}_{\mathcal{B}/\mathcal{F}} \times \mathbf{Q} $$

where

$$ \boldsymbol{\omega}_{\mathcal{B}/\mathcal{F}} = \omega_H(-\hat{\mathbf{i}}) = -2\hat{\mathbf{i}} \text{ rad/sec}, $$

so

$$\begin{aligned}{}^{\mathcal{I}}\dot{\mathbf{Q}} &= 0 + (-2\hat{\mathbf{i}}) \times (-10\hat{\mathbf{j}}) \\ &= 20\hat{\mathbf{k}} \text{ ft/sec}\end{aligned}$$

One way to find ${}^{\mathcal{G}}\dot{\mathbf{Q}}$ is to use

$${}^{\mathcal{G}}\dot{\mathbf{Q}} = {}^{\mathcal{I}}\dot{\mathbf{Q}} + \boldsymbol{\omega}_{\mathcal{I}/\mathcal{G}} \times \mathbf{Q}$$

where

$$\begin{aligned}\boldsymbol{\omega}_{\mathcal{I}/\mathcal{G}} &= \omega_v(\sin 30°\hat{\mathbf{j}} + \cos 30°\hat{\mathbf{k}}) \\ &= 3(0.500\hat{\mathbf{j}} + 0.866\hat{\mathbf{k}}) \\ &= 1.50\hat{\mathbf{j}} + 2.60\hat{\mathbf{k}}\end{aligned}$$

so

$$\begin{aligned}{}^{\mathcal{G}}\dot{\mathbf{Q}} &= 20\hat{\mathbf{k}} + (1.50\hat{\mathbf{j}} + 2.60\hat{\mathbf{k}}) \times (-10\hat{\mathbf{j}}) \\ &= 20\hat{\mathbf{k}} + 26\hat{\mathbf{i}} \text{ ft/sec}\end{aligned}$$

Had we not desired to obtain ${}^{\mathcal{I}}\dot{\mathbf{Q}}$ we might have used

$${}^{\mathcal{G}}\dot{\mathbf{Q}} = {}^{\mathcal{B}}\dot{\mathbf{Q}} + \boldsymbol{\omega}_{\mathcal{B}/\mathcal{G}} \times \mathbf{Q}$$

where, by the addition theorem,

$$\begin{aligned}\boldsymbol{\omega}_{\mathcal{B}/\mathcal{G}} &= \boldsymbol{\omega}_{\mathcal{B}/\mathcal{I}} + \boldsymbol{\omega}_{\mathcal{I}/\mathcal{G}} \\ &= -2\hat{\mathbf{i}} + (1.50\hat{\mathbf{j}} + 2.60\hat{\mathbf{k}})\end{aligned}$$

so that

$$\begin{aligned}{}^{\mathcal{G}}\dot{\mathbf{Q}} &= 0 + (-2\hat{\mathbf{i}} + 1.50\hat{\mathbf{j}} + 2.60\hat{\mathbf{k}}) \times (-10\hat{\mathbf{j}}) \\ &= 20\hat{\mathbf{k}} + 26\hat{\mathbf{i}} \text{ ft/sec}\end{aligned}$$

as before.

We now present an extended practical example of the use of the $\boldsymbol{\omega}$ properties. In this example three separate bodies are in motion in a reference frame, and their angular velocities are related by using the simple angular velocity and addition theorem properties.

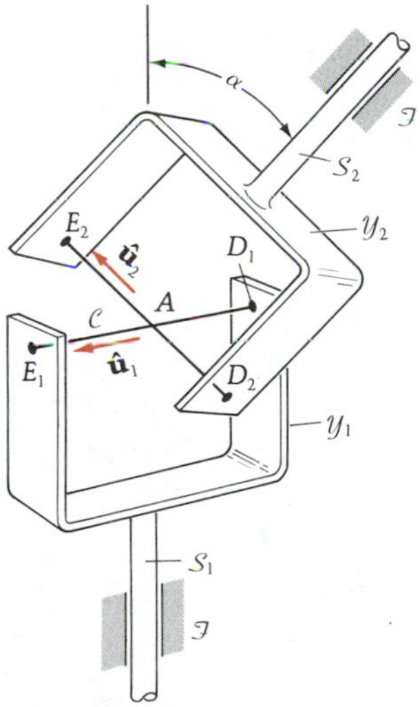

Figure E6.3a

Example 6.3

The *Hooke's joint,* or universal joint, is a device used to transmit power between two shafts that are not collinear. Figure E6.3a shows a Hooke's joint in which the shafts $\mathcal{S}_1$ and $\mathcal{S}_2$ are out of alignment by the angle α.

Each shaft is mounted in a bearing fixed to the reference frame $\mathcal{I}$. The shafts, whose axes intersect at point A, are rigidly attached to the yokes $\mathcal{Y}_1$ and $\mathcal{Y}_2$. A rigid cross $\mathcal{C}$ is the connecting body between the yokes. One leg of the cross (indicated by the unit vector $\hat{\mathbf{u}}_1$) turns in bearings fixed in $\mathcal{Y}_1$ at D_1 and E_1, while the other leg (unit vector $\hat{\mathbf{u}}_2$) turns in bearings fixed in $\mathcal{Y}_2$ at D_2 and E_2. The arms of cross $\mathcal{C}$ are identical; they form a right angle with each other, and each is perpendicular to its respective shaft.

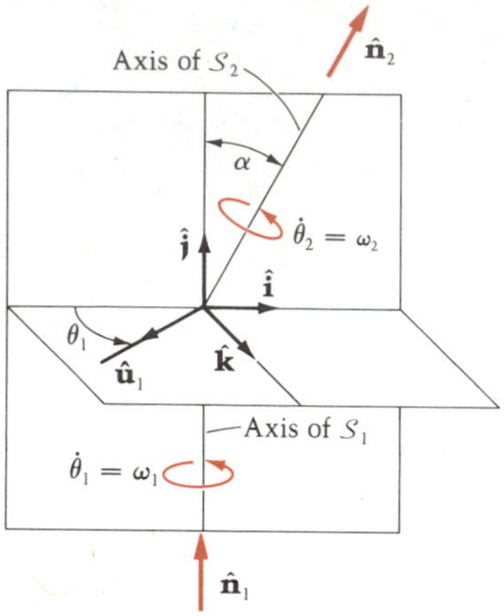

Figure E6.3b

Figure E6.3b shows that θ_1 measures the angular position of $\mathcal{S}_1$ in $\mathcal{I}$. If $\mathcal{S}_1$ (considered to be the drive shaft) has angular velocity $\boldsymbol{\omega}_{\mathcal{S}_1/\mathcal{I}} = \omega_1\hat{\mathbf{n}}_1$ and the resulting angular velocity of $\mathcal{S}_2$ is $\boldsymbol{\omega}_{\mathcal{S}_2/\mathcal{I}} = \omega_2\hat{\mathbf{n}}_2$, find the ratio of ω_2 to ω_1 in terms of θ_1 and α, and plot ω_2/ω_1 versus θ_1 for $\alpha = 0, 20, 40, 60$, and $80°$. Letting θ_2 be the rotation angle of $\mathcal{S}_2$, further investigate θ_2 versus θ_1 for the same five α values.

Question 6.4 In Figure E6.3b note that $\hat{\mathbf{n}}_2$ has no $\hat{\mathbf{k}}$ component. Why does this not represent a loss of generality?

Solution

Using the addition theorem, we may relate the angular velocities of the four rigid bodies $\mathcal{B}_2$ (shaft $\mathcal{S}_2$ plus its yoke $\mathcal{Y}_2$), $\mathcal{C}$, $\mathcal{B}_1$ (shaft $\mathcal{S}_1$ plus its yoke $\mathcal{Y}_1$), and $\mathcal{I}$:

$$\boldsymbol{\omega}_{\mathcal{B}_2/\mathcal{I}} = \boldsymbol{\omega}_{\mathcal{B}_2/\mathcal{C}} + \boldsymbol{\omega}_{\mathcal{C}/\mathcal{B}_1} + \boldsymbol{\omega}_{\mathcal{B}_1/\mathcal{I}} \tag{1}$$

Since $\mathcal{B}_1$ and $\mathcal{B}_2$ both have simple angular velocity in $\mathcal{I}$, we may write

$$\boldsymbol{\omega}_{\mathcal{B}_1/\mathcal{I}} = \omega_1\hat{\mathbf{n}}_1 \quad \text{or} \quad \dot\theta_1\hat{\mathbf{n}}_1 \qquad \boldsymbol{\omega}_{\mathcal{B}_2/\mathcal{I}} = \omega_2\hat{\mathbf{n}}_2 \quad \text{or} \quad \dot\theta_2\hat{\mathbf{n}}_2 \tag{2}$$

We also know from Figure E6.3a that the cross $\mathcal{C}$ has a simple angular velocity in *each* of $\mathcal{B}_1$ and $\mathcal{B}_2$. For example, the only motion that $\mathcal{C}$ can have with respect to $\mathcal{B}_1$ is a rotation about D_1E_1, the line fixed in *both* bodies. The same is true for the motion of $\mathcal{C}$ in $\mathcal{B}_2$. Thus

$$\boldsymbol{\omega}_{\mathcal{C}/\mathcal{B}_1} = \omega_{\mathcal{C}/\mathcal{B}_1}\hat{\mathbf{u}}_1 \qquad \boldsymbol{\omega}_{\mathcal{C}/\mathcal{B}_2} = -\boldsymbol{\omega}_{\mathcal{B}_2/\mathcal{C}} = -\omega_{\mathcal{B}_2/\mathcal{C}}\hat{\mathbf{u}}_2 \tag{3}$$

in which $\omega_{\mathcal{C}/\mathcal{B}_1}$ and $\omega_{\mathcal{B}_2/\mathcal{C}}$ are the unknown magnitudes of the respective vectors.

Next we must express all of $\hat{\mathbf{n}}_1$, $\hat{\mathbf{n}}_2$, $\hat{\mathbf{u}}_1$, and $\hat{\mathbf{u}}_2$ in terms of a common set of unit vectors. Then we shall be able to obtain three scalar equations from (1) and hence solve for ω_2 in terms of ω_1. From Figure E6.3b, three of the unit vectors are obvious:

$$\begin{aligned}\hat{\mathbf{n}}_1 &= \hat{\mathbf{j}}\\ \hat{\mathbf{n}}_2 &= \sin\alpha\hat{\mathbf{i}} + \cos\alpha\hat{\mathbf{j}}\\ \hat{\mathbf{u}}_1 &= -\cos\theta_1\hat{\mathbf{i}} + \sin\theta_1\hat{\mathbf{k}}\end{aligned} \tag{4}$$

To obtain $\hat{\mathbf{u}}_2$, we note that it is perpendicular to both $\hat{\mathbf{n}}_2$ and $\hat{\mathbf{u}}_1$. Crossing $\hat{\mathbf{u}}_1$ into $\hat{\mathbf{n}}_2$ then gives the assigned direction of $\hat{\mathbf{u}}_2$ (note that $\hat{\mathbf{n}}_2 \times \hat{\mathbf{u}}_1$ is opposite!); $\hat{\mathbf{u}}_1 \times \hat{\mathbf{n}}_2$ is not generally a unit vector, however, so to get $\hat{\mathbf{u}}_2$ we divide this vector by its magnitude:

$$\hat{\mathbf{u}}_2 = \frac{\hat{\mathbf{u}}_1 \times \hat{\mathbf{n}}_2}{|\hat{\mathbf{u}}_1 \times \hat{\mathbf{n}}_2|} = \frac{-\cos\alpha\sin\theta_1\hat{\mathbf{i}} + \sin\alpha\sin\theta_1\hat{\mathbf{j}} - \cos\alpha\cos\theta_1\hat{\mathbf{k}}}{\sqrt{\cos^2\alpha + \sin^2\alpha\sin^2\theta_1}} \tag{5}$$

Substituting Equations (4) and (5) for the four unit vectors into Equations (2) and (3), and then substituting the resulting angular velocity expressions into Equation (1), we get a vector equation that has the following three scalar component equations:

$$\hat{\mathbf{i}} \text{ coefficients:} \qquad \begin{aligned}\omega_2\sin\alpha = {}& R\omega_{\mathcal{B}_2/\mathcal{C}}(-\cos\alpha\sin\theta_1)\\ &+ \omega_{\mathcal{C}/\mathcal{B}_1}(-\cos\theta_1)\end{aligned} \tag{6}$$

Answer 6.4 The (xy) plane of the paper can be chosen to be the plane containing $\hat{\mathbf{n}}_1$ and $\hat{\mathbf{n}}_2$ without loss of generality.

$$\hat{\mathbf{j}} \text{ coefficients:} \qquad \omega_2 \cos\alpha = R\omega_{\mathcal{B}_2/\mathcal{C}}(\sin\alpha \sin\theta_1) + \omega_1 \tag{7}$$

$$\hat{\mathbf{k}} \text{ coefficients:} \qquad 0 = R\omega_{\mathcal{B}_2/\mathcal{C}}(-\cos\alpha\cos\theta_1) + \omega_{\mathcal{C}/\mathcal{B}_1}(\sin\theta_1) \tag{8}$$

in which $R = 1/\sqrt{\cos^2\alpha + \sin^2\alpha \sin^2\theta_1}$. Eliminating $\omega_{\mathcal{C}/\mathcal{B}_1}$ between (6) and (8) gives

$$\omega_{\mathcal{B}_2/\mathcal{C}} = \frac{-\omega_2 \tan\alpha \sin\theta_1}{R} \tag{9}$$

and substitution of (9) into (7) yields

$$\omega_2 = \left(\frac{\cos\alpha}{\cos^2\alpha + \sin^2\alpha \sin^2\theta_1}\right)\omega_1$$

so that

$$\frac{\omega_2}{\omega_1} = \frac{\cos\alpha}{1 - \sin^2\alpha\cos^2\theta_1} \tag{10}$$

A plot of this expression (Figure E6.3c) shows the manner in which ω_2 changes over a quarter-turn of $\mathcal{S}_1$ in space. Note that since $\cos\theta_1$ is squared, the curves reflect around the vertical line at $\theta_1 = 90°$ for $90° \le \theta_1 \le 180°$; between 180° and 360° we again have a mirror image, this time of the curves between 0° and 180°. Note that for large misalignment angles α, shaft $\mathcal{S}_2$ must turn very rapidly at and near $\theta_1 = 0$; in fact, when $\alpha = 90°$ the bodies reach a configuration in which they cannot turn at all. This is called *gimbal lock.* Note further that a misalignment of as much as 10° results in an output speed variation (ω_2) over a revolution of only about 3 percent.

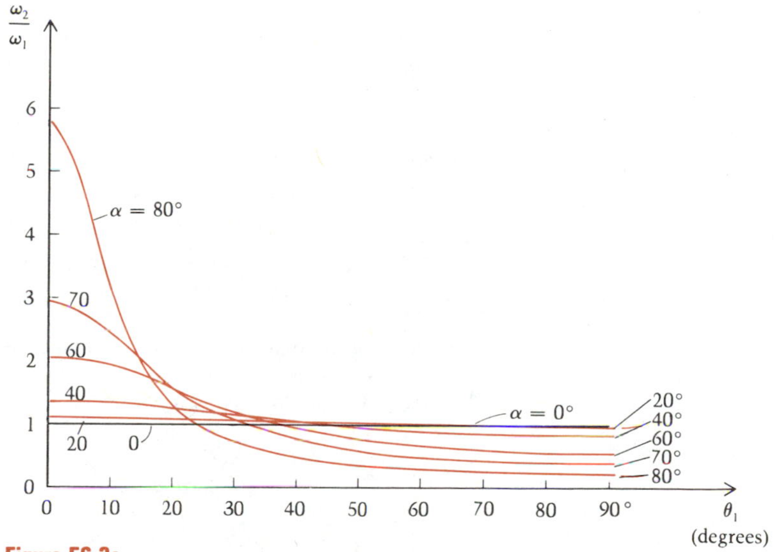

Figure E6.3c

Next we examine the angles of rotation θ_1 (of $\mathcal{S}_1$) and θ_2 (of $\mathcal{S}_2$). Since $\mathcal{S}_1$ and $\mathcal{S}_2$ both have simple angular velocity in $\mathcal{I}$, we have $\omega_1 = \dot{\theta}_1$ and $\omega_2 = \dot{\theta}_2$, so that (from Equation 10):

$$\dot{\theta}_2 = \frac{\dot{\theta}_1 \cos\alpha}{1 - \sin^2\alpha\cos^2\theta_1} \tag{11}$$

Integrating (11) gives

$$\theta_2 + \text{constant} = \int \frac{\cos\alpha\, d\theta_1}{1 - \sin^2\alpha \cos^2\theta_1}$$

$$= \tan^{-1}\left(\frac{\tan\theta_1}{\cos\alpha}\right) \quad (12)$$

The constant of integration is zero if we define $\theta_2 = 0$ when $\theta_1 = 0$. Then θ_2 may be plotted as a function of θ_1 for the same representative values of α (see Figure E6.3d). If $\alpha = 0$, then $\theta_1 \equiv \theta_2$ and the curve is a 45° line. If there is misalignment ($\alpha \neq 0$), then note from the curves that for $0 \le \theta_1 \le 90°$, shaft $\mathcal{S}_2$ is always turned *more* than $\mathcal{S}_1$. Then $\mathcal{S}_1$ catches up at $\theta = 90°$, and from $\theta_1 = 90°$ to 180° the angle θ_2 of $\mathcal{S}_2$ lags *behind* θ_1.

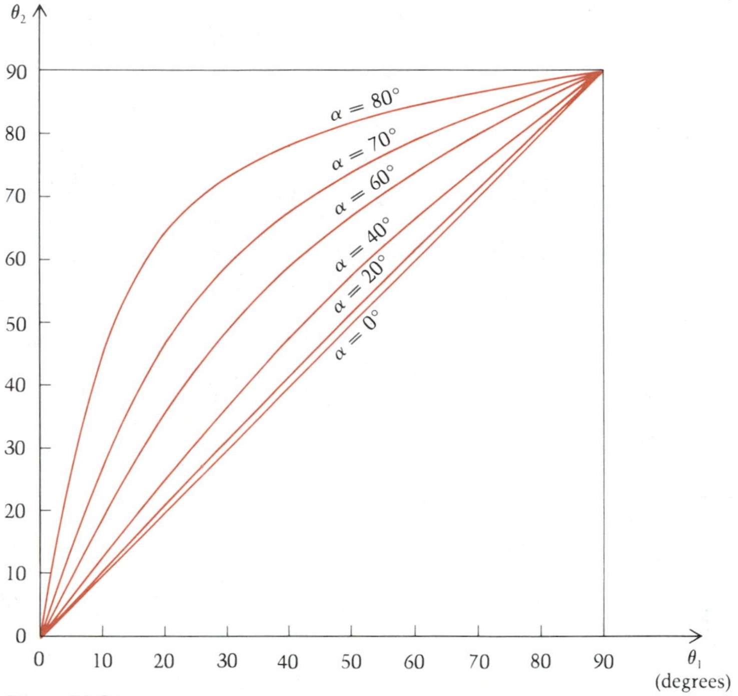

Figure E6.3d

Question 6.5 Explain why this is so by using Equation (12).

From 180° to 360°, the cycle repeats and everything returns to the same starting position ($\theta_1 = \theta_2 = 360°$) at the same time.

Answer 6.5 For θ_1 in the second quadrant, and with $0 < \cos\alpha \le 1$, we see from Equation (12) that $\tan\theta_2$ is a more negative number than $\tan\theta_1$ (unless $\alpha = 0$, in which case $\theta_2 = \theta_1$). Therefore, θ_1 and θ_2 are angles between 90° and 180°, and $\theta_2 < \theta_1$.

PROBLEMS ▶ Section 6.3

(See also the Project Problem 6.93 at the end of this chapter.)

6.1 Verify, in Example 1.1, that ${}^{\mathcal{I}}\dot{\mathbf{A}}$ is indeed $\boldsymbol{\omega}_{\mathcal{B}/\mathcal{I}} \times \mathbf{A}$, where $\boldsymbol{\omega}_{\mathcal{B}/\mathcal{I}} = \dot{\theta}(-\hat{\mathbf{k}})$. (See Figure P6.1.)

6.2 The angular velocities of $\mathcal{A}$ and $\mathcal{B}$ in a reference frame $\mathcal{I}$ are, respectively, $10\hat{\mathbf{n}}_1$ rad/sec and $7\hat{\mathbf{n}}_2$ rad/sec. Find the angular velocity of $\mathcal{B}$ in $\mathcal{A}$ expressed in terms of $\hat{\mathbf{i}}$ and $\hat{\mathbf{j}}$. (See Figure P6.2.)

6.3 A vector $\mathbf{v}$ is given as a function of time t by $\mathbf{v} = t^3\hat{\mathbf{i}} + t^2\hat{\mathbf{j}} + t\hat{\mathbf{k}}$ m/s, where $(\hat{\mathbf{i}}, \hat{\mathbf{j}}, \hat{\mathbf{k}})$ are unit vectors whose directions are fixed in a frame $\mathcal{H}$. The angular velocity of $\mathcal{H}$ in frame $\mathcal{R}$ is $\boldsymbol{\omega}_{\mathcal{H}/\mathcal{R}} = t\hat{\mathbf{i}} + t^2\hat{\mathbf{j}} + t^3\hat{\mathbf{k}}$ rad/s. Find the derivative of $\mathbf{v}$ in frame $\mathcal{R}$, that is, ${}^{\mathcal{R}}\dot{\mathbf{v}}$: (a) as a function of t; (b) at $t = 1$ s; (c) at $t = 2$ s.

6.4 In the preceding problem, find ${}^{\mathcal{R}}\ddot{\mathbf{v}}$.

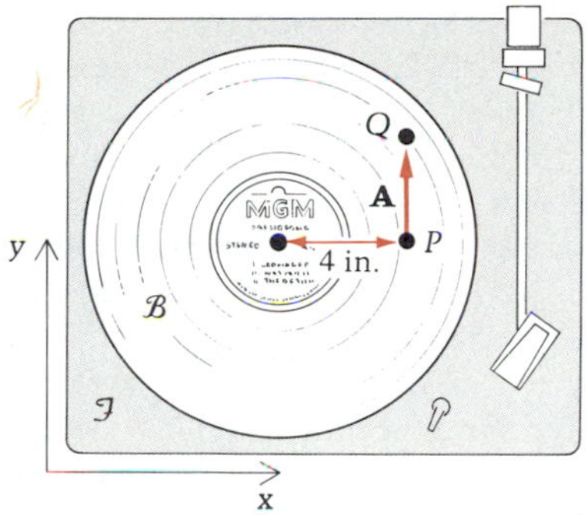

Figure P6.1

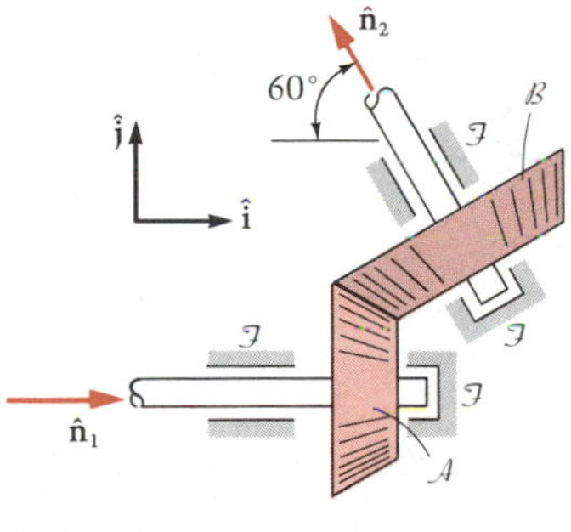

Figure P6.2

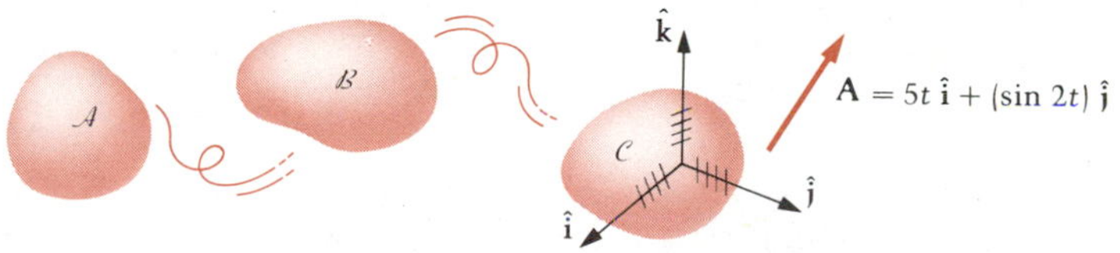

Figure P6.5

6.5 Note the three frames $\mathcal{A}$, $\mathcal{B}$, and $\mathcal{C}$, and the vector $\mathbf{A}$ defined in Figure P6.5 expressed in terms of its components in $\mathcal{C}$. Also, $\boldsymbol{\omega}_{\mathcal{C}/\mathcal{B}} = t^2\hat{\mathbf{i}} + t^3\hat{\mathbf{j}}$ and $\boldsymbol{\omega}_{\mathcal{B}/\mathcal{A}} = 5\hat{\mathbf{k}}$. Find ${}^{\mathcal{A}}\dot{\mathbf{A}}$ at $t = \pi/4$ sec.

6.6 Review Problem 1.155 in which the unit tangent, normal, and binormal of a curve in space $\mathcal{S}$ are defined. Let $\mathcal{B}$ be a frame moving relative to $\mathcal{S}$ in such a way that $\hat{\mathbf{e}}_t$, $\hat{\mathbf{e}}_n$, and $\hat{\mathbf{e}}_b$ are always fixed in $\mathcal{B}$. Use the definition (Equation (6.18)) of angular velocity to find the angular velocity of $\mathcal{B}$ in $\mathcal{S}$. Note that

$$\frac{d}{dt}(\;) = \left[\frac{d}{ds}(\;)\right]\frac{ds}{dt} = \dot{s}\,\frac{d(\;)}{ds}$$

6.7 The antenna $\mathcal{A}$ in Figure P6.7 is oriented with the following three rotations:

1. Azimuth, about y fixed in $\mathcal{G}$, at the rate $\omega_{\mathcal{I}_1/\mathcal{G}} = 3t^2$ rad/sec
2. Elevation, about z_1 fixed in a first intermediate frame $\mathcal{I}_1$, at the constant rate $\omega_{\mathcal{I}_2/\mathcal{I}_1} = 1$ rad/sec
3. A polarization rotation about the antenna axis x_2 (fixed in both the second intermediate frame $\mathcal{I}_2$ and in $\mathcal{A}$) at $\omega_{\mathcal{A}/\mathcal{I}_2} = 4t$ rad/sec

If the structure is in the $\phi = 0$ position at $t = 0$, find $\boldsymbol{\omega}_{\mathcal{A}/\mathcal{G}}$ at the time $t = \pi/2$ sec. Use these unit vectors fixed in direction in $\mathcal{I}_1$: $\hat{\mathbf{j}}$ parallel to y, $\hat{\mathbf{k}}$ parallel to z_1, and $\hat{\mathbf{i}} = \hat{\mathbf{j}} \times \hat{\mathbf{k}}$.

6.8 Show that the output angle θ_2 of the Hooke's joint in Example 6.3 can be alternatively obtained by dotting $\hat{\mathbf{k}}$ with $\hat{\mathbf{u}}_2$.

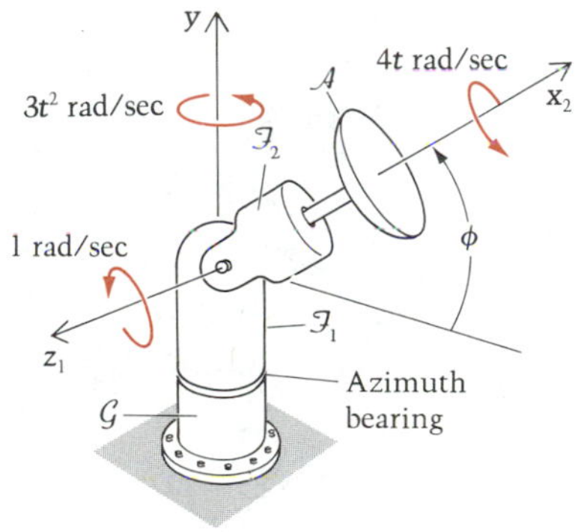

Figure P6.7

6.9 A device for simulating conditions in space allows rotations about orthogonal axes as shown in Figure P6.9. Determine the angular velocity in frame $\mathcal{G}$ of the capsule $\mathcal{C}$ containing the astronaut. Express the result in terms of unit vectors $(\hat{\mathbf{i}}, \hat{\mathbf{j}}, \hat{\mathbf{k}})$ fixed in the beam $\mathcal{B}$. Note that the ω_y rotation is about an axis fixed in $\mathcal{C}$ and in $\mathcal{A}$ but *not* in $\mathcal{B}$; this axis is parallel to y at $t = 0$, and ω_x is a constant.

** **6.10** The outer cone $\mathcal{B}$ in Figure P6.10 has the following prescribed motion with respect to the fixed inner cone $\mathcal{C}$:

1. The vertices remain together.
2. The line AB (a base radius fixed in $\mathcal{B}$) always lies in some vertical plane parallel to XY.
3. Cone $\mathcal{B}$ slides on $\mathcal{C}$; that is, there is always a line of contact between O and a point of the base circle of $\mathcal{C}$.
4. Point A of $\mathcal{B}$ revolves around the x axis in a vertical circle at constant speed $H\dot{\theta}$.

Use the addition theorem to show that the angular velocity of $\mathcal{B}$ in $\mathcal{C}$ is given by

$$\boldsymbol{\omega}_{\mathcal{B}/\mathcal{C}} = \frac{\dot{\theta}(\tan^2 \gamma \cos^2 \theta \hat{\mathbf{i}} - \tan \gamma \cos \theta \hat{\mathbf{j}} - \sin \theta \tan \gamma \hat{\mathbf{k}})}{1 + \tan^2 \gamma \cos^2 \theta}$$

6.11 In the preceding problem find $\boldsymbol{\omega}_{\mathcal{B}/\mathcal{C}}$ if the projection of AB into the YZ plane through A is always aligned with the radius. (See Figure P6.11.)

6.12 Find $\boldsymbol{\omega}_{\mathcal{B}/\mathcal{C}}$ in the preceding problem if the outer cone *rolls* on the fixed inner cone. (See Figure P6.12.)

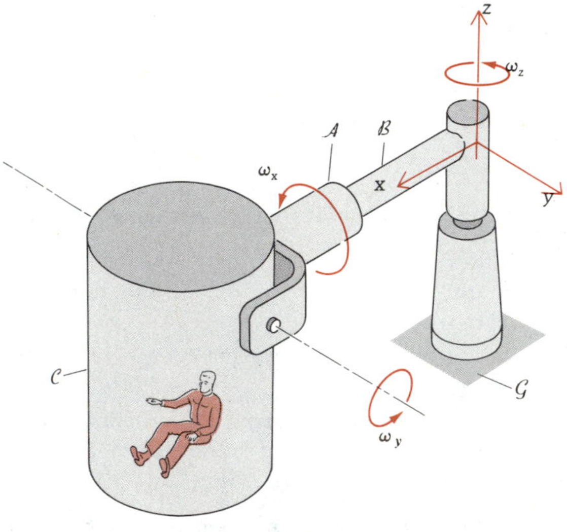

Figure P6.9

A popular method of stabilizing shipboard antennas is by means of pendulous masses together with the gyroscopic effect of spinning flywheels. In Figure P6.13 the ship (frame $\mathcal{S}$) pitches (about x), rolls (about y), and yaws (about z) in the sea (frame $\mathcal{E}$). Frame $\mathcal{I}$, just above a Hooke's joint, is to form a stable platform on which the antenna can then be easily positioned in azimuth (angle A) and elevation (angle E). The frame $\mathcal{I}$ remains level by a "depitching" rotation P above the "derolling" rotation R. The following three problems are based on this system.

The INMARSAT communications satellite system required that shipboard antenna systems remain operational up to the following oscillatory limits:

Pitch:	±10° in 6 sec
Roll:	±30° in 8 sec
Yaw:	±8° in 50 sec

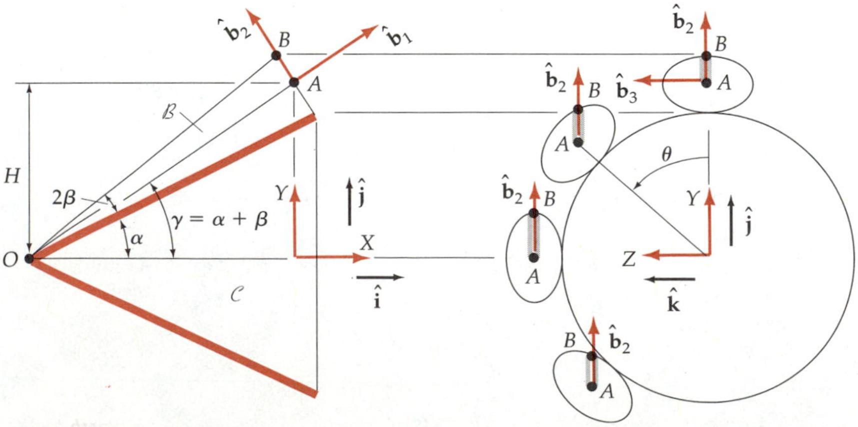

Figure P6.10

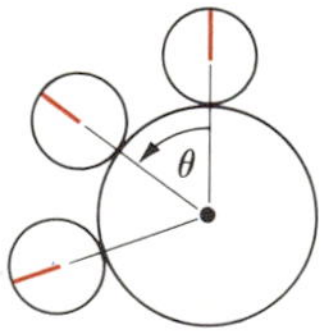

Figure P6.11

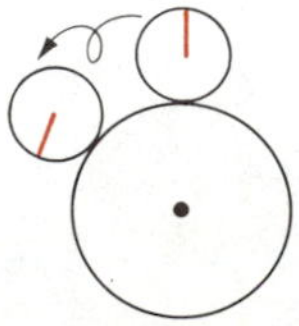

Figure P6.12

*** 6.13** Assume sine waves for each of these three motions and assume yaw over roll over pitch — that is, the assumed order of ship rotations is (1) pitch, from frame $\mathcal{E}$ to an intermediate frame $\mathcal{F}_1$; and (2) roll, from $\mathcal{F}_1$ to a second

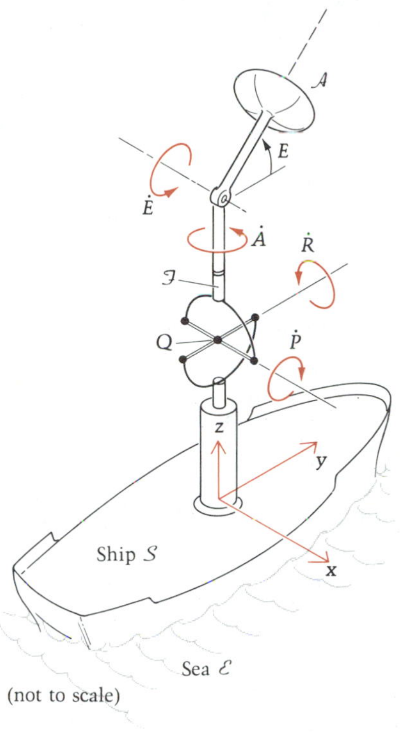

Figure P6.13

intermediate frame $\mathcal{F}_2$; and (3) yaw, from $\mathcal{F}_2$ to frame $\mathcal{S}$. Write the angular velocity of the ship $\mathcal{S}$ in the sea (earth-fixed frame $\mathcal{E}$), expressed in the ship-fixed axes (x, y, z). *Hint*: For example, θ_{pitch} will be $10\pi/180 \sin 2\pi t/6$ rad.

6.14 Write the angular velocity of $\mathcal{F}$ in $\mathcal{S}$, expressed in the axes (x, y, z).

6.15 Write the angular velocity of $\mathcal{F}$ in $\mathcal{E}$, using the results of the preceding two problems together with the addition theorem.

*** 6.16** A robot manufactured by the Heath Company has the mechanical arm shown in Figure P6.16(a) and accompanying photograph. Its shoulder $\mathcal{S}$ extends from the head $\mathcal{H}$, which can itself rotate 350° about z in 30 sec relative to the reference frame $\mathcal{F}$. The arm is able to travel 150° about axis y in 26 sec. (The axes (x, y, z) are fixed in the head $\mathcal{H}$.) The part of the arm to the left of point E is able to extend (and retract) up to 5 in. in 30 sec. The wrist $\mathcal{W}$ has two motions: It is able to pivot up to 180° about y' in 10 sec and to rotate (about x') up to 350° in 4 sec. Axes (x', y', z') are fixed in $\mathcal{W}$. Finally, the gripper $\mathcal{G}$ is able to open (and close) 3.5 in. in 3 sec, but is assumed here to be a closed circle with a 2.5-inch diameter. Approximate dimensions are shown in the figure.

For this problem, assume that all the robot's motions (except the gripper opening) are occurring simultaneously about positive axes with their respective average speeds. Find the angular velocity of the gripper $\mathcal{G}$, relative to $\mathcal{F}$ and

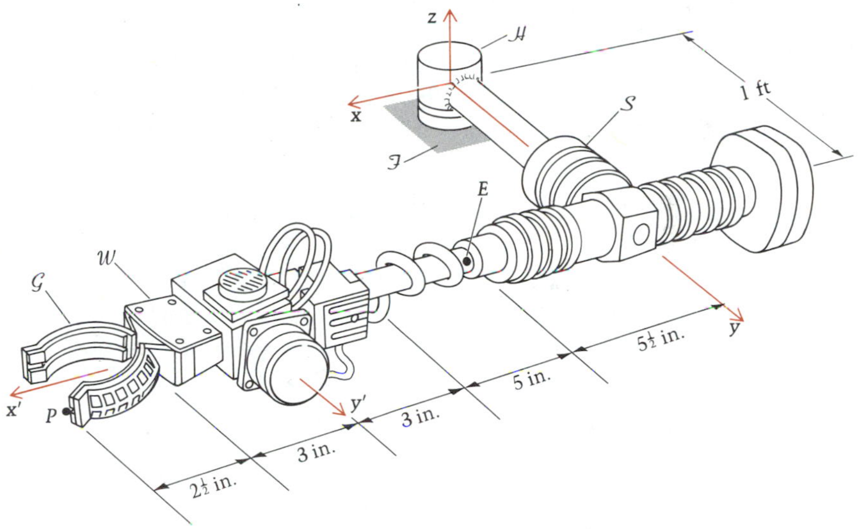

Figure P6.16(a)

Figure P6.16(b) *(Courtesy of the Heath Company.)*

* Asterisks identify the more difficult problems.

expressed in terms of unit vectors in $\mathcal{A}$, at an instant when these two conditions hold:

1. The shoulder rotation angle θ is $-60°$:

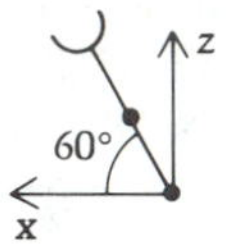

2. The wrist is pivoted 30°:

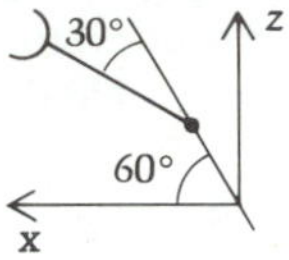

6.4 The Angular Acceleration Vector

In applying what we have learned about angular velocity to the kinematics of rigid bodies, we also need to understand its derivative. The **angular acceleration** of frame $\mathcal{B}$ relative to frame $\mathcal{A}$ is defined to be

$$\boldsymbol{\alpha}_{\mathcal{B}/\mathcal{A}} = {}^{\mathcal{A}}\dot{\boldsymbol{\omega}}_{\mathcal{B}/\mathcal{A}} \tag{6.38}$$

(Note from the last property of $\boldsymbol{\omega}_{\mathcal{B}/\mathcal{A}}$ in the previous section that the derivative could equally well be taken in $\mathcal{B}$, but generally not in any other frame.)

It is important to note that the addition theorem (Equation 6.27) *does not hold* for angular acceleration. Watch:

$$\begin{aligned}\boldsymbol{\alpha}_{\mathcal{C}/\mathcal{A}} &= {}^{\mathcal{A}}\dot{\boldsymbol{\omega}}_{\mathcal{C}/\mathcal{A}} = {}^{\mathcal{A}}\overset{\cdot}{\overline{\boldsymbol{\omega}_{\mathcal{C}/\mathcal{B}} + \boldsymbol{\omega}_{\mathcal{B}/\mathcal{A}}}} \\ &= {}^{\mathcal{A}}\dot{\boldsymbol{\omega}}_{\mathcal{C}/\mathcal{B}} + {}^{\mathcal{A}}\dot{\boldsymbol{\omega}}_{\mathcal{B}/\mathcal{A}} \\ &= ({}^{\mathcal{B}}\dot{\boldsymbol{\omega}}_{\mathcal{C}/\mathcal{B}} + \boldsymbol{\omega}_{\mathcal{B}/\mathcal{A}} \times \boldsymbol{\omega}_{\mathcal{C}/\mathcal{B}}) + \boldsymbol{\alpha}_{\mathcal{B}/\mathcal{A}}\end{aligned}$$

$$\boldsymbol{\alpha}_{\mathcal{C}/\mathcal{A}} = \boldsymbol{\alpha}_{\mathcal{C}/\mathcal{B}} + \boldsymbol{\alpha}_{\mathcal{B}/\mathcal{A}} + \boldsymbol{\omega}_{\mathcal{B}/\mathcal{A}} \times \boldsymbol{\omega}_{\mathcal{C}/\mathcal{B}} \tag{6.39}$$

We see that there is an extra term (the cross product of two angular velocity vectors) that prevents the simple theorem we have derived for $\boldsymbol{\omega}$'s from working for $\boldsymbol{\alpha}$'s. This term is sometimes called a *gyroscopic term*; note that it vanishes for plane motion, in which case we do have an addition theorem for the $\boldsymbol{\alpha}$'s (which are then of the form $\ddot{\theta}\hat{\mathbf{k}}$).

In each of the two examples to follow, the reader should notice how the various properties of $\boldsymbol{\omega}$ — simple angular velocity (Equation (6.34)), the addition theorem (6.27), and the derivative formula (6.20) — are used to great advantage.

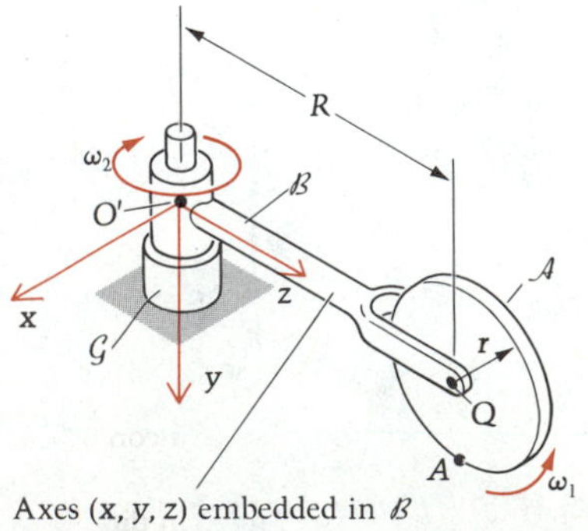

Axes (x, y, z) embedded in $\mathcal{B}$

Figure E6.4

EXAMPLE 6.4

Body $\mathcal{B}$ in Figure E6.4 rotates in frame $\mathcal{G}$ about the vertical at constant angular speed ω_2; in $\mathcal{B}$, disk $\mathcal{A}$ rotates about its pinned axis at constant angular speed ω_1. (The directions of rotation are as shown.) Determine the angular acceleration of $\mathcal{A}$ in $\mathcal{G}$.

Solution

As we saw in Example 6.1,

$$\boldsymbol{\omega}_{\mathcal{A}/\mathcal{G}} = \boldsymbol{\omega}_{\mathcal{A}/\mathcal{B}} + \boldsymbol{\omega}_{\mathcal{B}/\mathcal{G}} = \omega_1\hat{\mathbf{i}} + \omega_2\hat{\mathbf{j}}$$

Next, using Equation (6.20) and noting that $\boldsymbol{\omega}_{\mathcal{A}/\mathcal{G}}$ is expressed in terms of axes embedded in $\mathcal{B}$, we "move the derivative" using the derivative formula and obtain*

$$\begin{aligned}\boldsymbol{\alpha}_{\mathcal{A}/\mathcal{G}} &= {}^{\mathcal{G}}\dot{\boldsymbol{\omega}}_{\mathcal{A}/\mathcal{G}} \\ &= {}^{\mathcal{B}}\dot{\boldsymbol{\omega}}_{\mathcal{A}/\mathcal{G}} + \boldsymbol{\omega}_{\mathcal{B}/\mathcal{G}} \times \boldsymbol{\omega}_{\mathcal{A}/\mathcal{G}} \\ &= \mathbf{0} + \omega_2\hat{\mathbf{j}} \times (\omega_1\hat{\mathbf{i}} + \omega_2\hat{\mathbf{j}}) \\ &= -\omega_1\omega_2\hat{\mathbf{k}}\end{aligned}$$

Note that the same result is obtained by using Equation (6.39) with frames $\mathcal{C}$, $\mathcal{B}$, and $\mathcal{A}$ replaced by $\mathcal{A}$, $\mathcal{B}$, and $\mathcal{G}$, respectively.

Question 6.6 Why was ${}^{\mathcal{B}}\dot{\boldsymbol{\omega}}_{\mathcal{A}/\mathcal{G}} = \mathbf{0}$ in the above example?

EXAMPLE 6.5

Determine the angular acceleration of the cross $\mathcal{C}$ relative to frame $\mathcal{I}$ in Example 6.3, for the case $\dot{\theta}_1$ = constant. Express the result in terms of unit vectors $\hat{\mathbf{n}}_2$, $\hat{\mathbf{u}}_2$, and $\hat{\mathbf{v}}_2 = \hat{\mathbf{n}}_2 \times \hat{\mathbf{u}}_2$ fixed in $\mathcal{B}_2$.

Solution

Using the definition of angular acceleration [Equation (6.38)], the addition theorem, and the derivative formula,

$$\begin{aligned}\boldsymbol{\alpha}_{\mathcal{C}/\mathcal{I}} &= {}^{\mathcal{I}}\dot{\boldsymbol{\omega}}_{\mathcal{C}/\mathcal{I}} \\ &= {}^{\mathcal{I}}\dot{\boldsymbol{\omega}}_{\mathcal{C}/\mathcal{B}_2} + {}^{\mathcal{I}}\dot{\boldsymbol{\omega}}_{\mathcal{B}_2/\mathcal{I}} \\ &= ({}^{\mathcal{B}_2}\dot{\boldsymbol{\omega}}_{\mathcal{C}/\mathcal{B}_2} + \boldsymbol{\omega}_{\mathcal{B}_2/\mathcal{I}} \times \boldsymbol{\omega}_{\mathcal{C}/\mathcal{B}_2}) + {}^{\mathcal{I}}\dot{\boldsymbol{\omega}}_{\mathcal{B}_2/\mathcal{I}} \qquad (1)\end{aligned}$$

Let us first concentrate on the first term on the right side of Equation (1). In Example 6.3 we had

$$\boldsymbol{\omega}_{\mathcal{C}/\mathcal{B}_2} = \dot{\theta}_2 T_\alpha s_1 \sqrt{1 - s_\alpha^2 c_1^2}\,\hat{\mathbf{u}}_2 \qquad (2)$$

where we are using the notation $s_\alpha = \sin\alpha$, $c_\alpha = \cos\alpha$, $T_\alpha = \tan\alpha$, $s_1 = \sin\theta_1$, and $c_1 = \cos\theta_1$. We also know from Example 6.3 that

$$\dot{\theta}_2 = \frac{\dot{\theta}_1 c_\alpha}{1 - s_\alpha^2 c_1^2} \qquad (3)$$

Substituting $\dot{\theta}_2$ from Equation (3) into (2) and differentiating the result in $\mathcal{B}_2$ (and noting $\hat{\mathbf{u}}_2$ is constant there) yields, after simplifying,

$${}^{\mathcal{B}_2}\dot{\boldsymbol{\omega}}_{\mathcal{C}/\mathcal{B}_2} = \frac{c_1 s_\alpha c_\alpha^2 \dot{\theta}_1^2}{(1 - s_\alpha^2 c_1^2)^{3/2}}\,\hat{\mathbf{u}}_2 \qquad (4)$$

* By "moving the derivative," we mean shifting it *from* a frame in which it is inconvenient to differentiate, *to* a frame in which we desire to differentiate.

Answer 6.6 ω_1 and ω_2 are constant scalars, and $\hat{\mathbf{i}}$ and $\hat{\mathbf{j}}$ are unit vectors fixed in direction in $\mathcal{B}$.

The second term in Equation (1) is

$$\dot{\theta}_2 \hat{\mathbf{n}}_2 \times \dot{\theta}_2 T_\alpha s_1 \sqrt{1 - s_\alpha^2 c_1^2}\, \hat{\mathbf{u}}_2$$

which upon simplification is

$$\frac{\dot{\theta}_1^2 s_\alpha c_\alpha s_1}{(1 - s_\alpha^2 c_1^2)^{3/2}}\, \hat{\mathbf{v}}_2 \qquad (5)$$

The last term in Equation (1) is:

$${}^{\mathcal{I}}\dot{\boldsymbol{\omega}}_{\mathcal{B}_2/\mathcal{I}} = \ddot{\theta}_2 \hat{\mathbf{n}}_2 = \frac{-2\dot{\theta}_1^2 s_\alpha^2 c_\alpha s_1 c_1}{(1 - s_\alpha^2 c_1^2)^2}\, \hat{\mathbf{n}}_2 \qquad (6)$$

where we have differentiated Equation (3).

The solution for $\boldsymbol{\alpha}_{\mathcal{C}/\mathcal{I}}$ is therefore the sum of the vectors in Equations (4), (5), and (6):

$$\boldsymbol{\alpha}_{\mathcal{C}/\mathcal{I}} = \left[c_\alpha c_1 \hat{\mathbf{u}}_2 + s_1 \hat{\mathbf{v}}_2 - \frac{2 s_\alpha s_1 c_1}{\sqrt{1 - s_\alpha^2 c_1^2}}\, \hat{\mathbf{n}}_2 \right] \frac{s_\alpha c_\alpha \dot{\theta}_1^2}{(1 - s_\alpha^2 c_1^2)^{3/2}}$$

We note that up to this point in Chapter 6 we have not mentioned velocities or accelerations. As long as the angular velocities from one body to another are all simple, we can do a considerable amount of angular velocity and angular acceleration computation merely by using the definitions and properties of $\boldsymbol{\omega}$ and $\boldsymbol{\alpha}$.

PROBLEMS ▶ Section 6.4

6.17 We know that one of the properties of the angular velocity vector is that ${}^{\mathcal{I}}\dot{\boldsymbol{\omega}}_{\mathcal{B}/\mathcal{I}} = {}^{\mathcal{B}}\dot{\boldsymbol{\omega}}_{\mathcal{B}/\mathcal{I}}$. Show that this is *not* a property of the angular acceleration vector $\boldsymbol{\alpha}_{\mathcal{B}/\mathcal{I}}$.

6.18 In Problem 6.6, find the angular acceleration of $\mathcal{B}$ in $\mathcal{S}$.

6.19 The components of two angular velocity vectors are shown in the following table as functions of time. The orthogonal unit vectors $(\hat{\mathbf{i}}, \hat{\mathbf{j}}, \hat{\mathbf{k}})$ are fixed in direction in frame $\mathcal{B}$. Find the angular acceleration of $\mathcal{C}$ in $\mathcal{A}$: (a) as a function of time; (b) at $t = 0$ sec; (c) at $t = 0.5$ sec. (See Figure P6.19.)

	$\hat{\mathbf{i}}$	$\hat{\mathbf{j}}$	$\hat{\mathbf{k}}$
$\omega_{\mathcal{C}/\mathcal{B}}$	$4t^2$	$2t$	6
$\omega_{\mathcal{B}/\mathcal{A}}$	$\sin t$	$\cos t$	$7t$

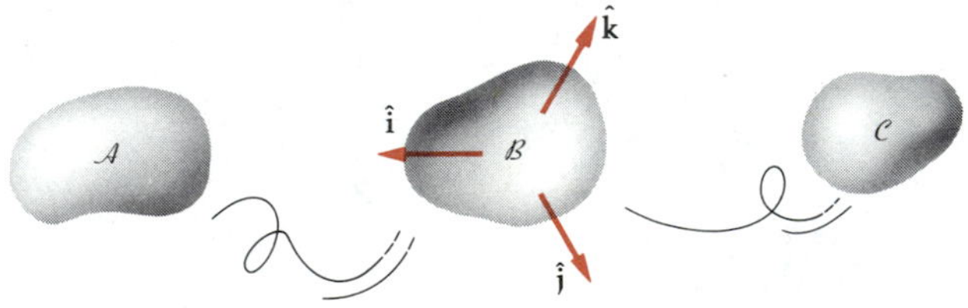

Figure P6.19

6.20 Let the angular velocity and angular acceleration vectors $\boldsymbol{\omega}_{\mathcal{C}/\mathcal{A}}$ and $\boldsymbol{\alpha}_{\mathcal{C}/\mathcal{A}}$ be expressed in terms of their components in a third frame $\mathcal{B}$:

$$\left.\begin{aligned} \boldsymbol{\omega}_{\mathcal{C}/\mathcal{A}} &= \omega_1 \hat{\mathbf{i}} + \omega_2 \hat{\mathbf{j}} + \omega_3 \hat{\mathbf{k}} \\ \boldsymbol{\alpha}_{\mathcal{C}/\mathcal{A}} &= \alpha_1 \hat{\mathbf{i}} + \alpha_2 \hat{\mathbf{j}} + \alpha_3 \hat{\mathbf{k}} \end{aligned}\right\} \quad \text{(The unit vectors are fixed in } \mathcal{B}\text{)}$$

Find the restriction on frame $\mathcal{B}$ for which $\alpha_i = \dot{\omega}_i$ ($i = 1, 2, 3$).

6.21 The antenna $\mathcal{A}$ in Figure P6.21 (see Problem 6.7) is oriented with the following three rotations:

1. Azimuth, about y fixed in $\mathcal{G}$, at the rate $\omega_{\mathcal{I}_1/\mathcal{G}} = 3t^2$ rad/sec
2. Elevation, about z_1 fixed in a first intermediate frame $\mathcal{I}_1$, at the constant rate $\omega_{\mathcal{I}_2/\mathcal{I}_1} = 1$ rad/sec

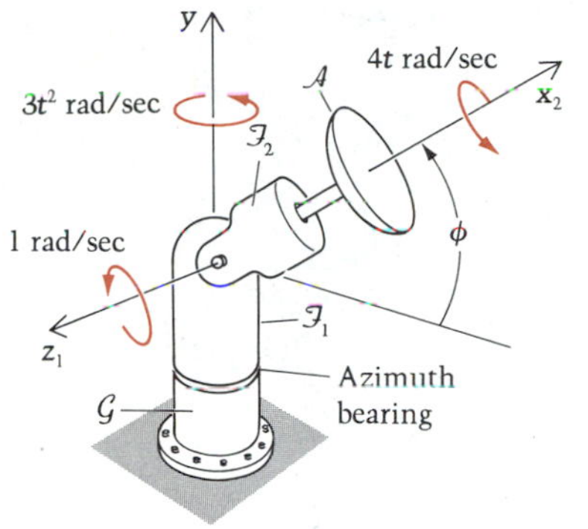

Figure P6.21

3. A polarization rotation about the antenna axis x_2 (fixed in both the second intermediate frame $\mathcal{I}_2$ and in $\mathcal{A}$) at $\omega_{\mathcal{A}/\mathcal{I}_2} = 4t$ rad/sec

If the structure is in the $\phi = 0$ position at $t = 0$, find $\boldsymbol{\alpha}_{\mathcal{A}/\mathcal{G}}$ at the time $t = \pi/2$ sec. Use unit vectors fixed in direction in $\mathcal{I}_1$.

6.22 In problem 6.9 find the angular acceleration of the capsule $\mathcal{C}$ in $\mathcal{G}$. Take ω_z and ω_x to be constants.

6.23 In Example 6.2 find the angular acceleration of the see-saw board $\mathcal{B}$ in the ground $\mathcal{G}$, if in addition to the given data $\dot{\omega}_V = 2$ rad/sec² and $\dot{\omega}_H = 1.5$ rad/sec².

6.24 See Figure P6.24. Axes x, y, and z are fixed in body $\mathcal{A}$, which rotates in $\mathcal{I}$ about the z axis with angular velocity $\omega_1\hat{\mathbf{k}}$. The arm $\mathcal{B}$, attached rigidly to $\mathcal{A}$, supports a bearing about which $\mathcal{C}$ turns with angular velocity $\omega_2\hat{\mathbf{j}}$ relative to $\mathcal{A}$. Finally, body $\mathcal{D}$ turns about the direction $\hat{\mathbf{u}}$ (which lies along axes of symmetry of both $\mathcal{C}$ and $\mathcal{D}$) with $\omega_3\hat{\mathbf{u}}$ relative to $\mathcal{C}$. If ω_1, ω_2, and ω_3 are all functions of time, find the angular acceleration of $\mathcal{D}$ in $\mathcal{I}$ at an instant when $\hat{\mathbf{u}}$ makes angles with x and z of 135° and 45°, respectively.

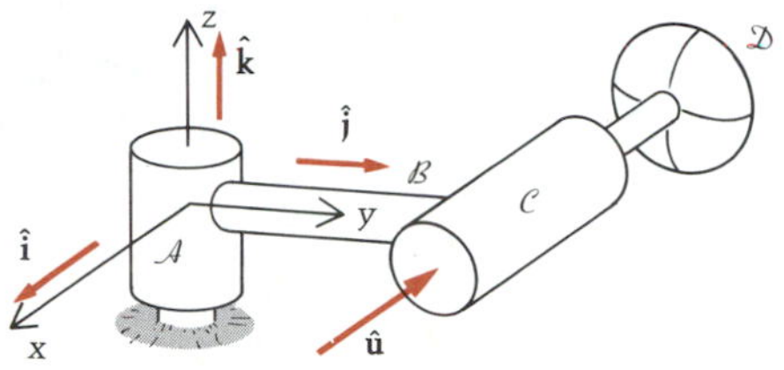

Figure P6.24

6.5 Velocity and Acceleration in Moving Frames of Reference

In certain practical situations a point is moving relative to *two* frames (or bodies) of interest. For example, a pin P may be sliding in a slot of a body $\mathcal{B}$ that is itself in motion in another frame $\mathcal{I}$ (see Figure 6.4). In problems such as these, we are often interested in the relationship between the velocities (and also the accelerations) of P in the two frames $\mathcal{B}$ and $\mathcal{I}$. We studied this problem in plane motion in Sections 3.7 and 3.8, and we now wish to expand the treatment to three dimensions.

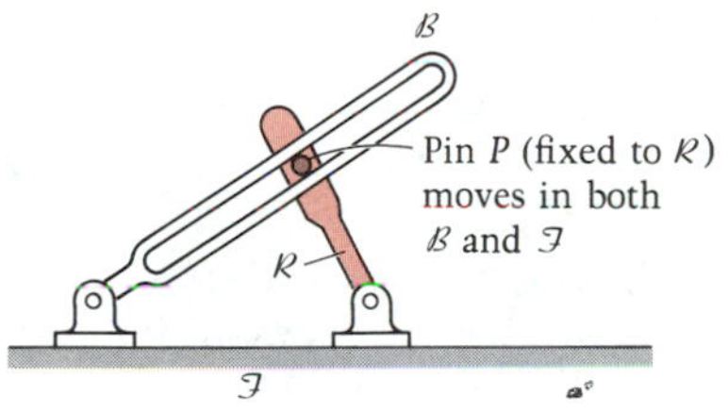

Figure 6.4 Example of a point moving relative to two frames.

The Velocity Relationship in Moving Frames

We shall arbitrarily choose $\mathcal{I}$ as a reference frame for the moving body $\mathcal{B}$, but we emphasize that both are frames and both are bodies; as long as they are considered rigid, the terms mean the same. Figure 6.5 shows the general picture.

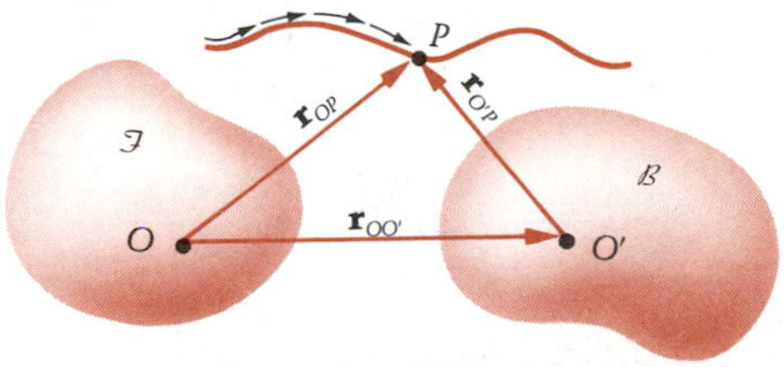

Figure 6.5 Point P moving with respect to frames $\mathcal{B}$ and $\mathcal{I}$.

Letting O and O' be fixed points of $\mathcal{I}$ and $\mathcal{B}$, respectively, we differentiate the connecting relation

$$\mathbf{r}_{OP} = \mathbf{r}_{OO'} + \mathbf{r}_{O'P} \tag{6.40}$$

in $\mathcal{I}$ and obtain*

$${}^{\mathcal{I}}\dot{\mathbf{r}}_{OP} = {}^{\mathcal{I}}\dot{\mathbf{r}}_{OO'} + {}^{\mathcal{I}}\dot{\mathbf{r}}_{O'P} \tag{6.41}$$

Because O is fixed in $\mathcal{I}$, the first two of the three vectors in Equation (6.41) are the velocities in $\mathcal{I}$ of P and O':

$$\mathbf{v}_{P/\mathcal{I}} = \mathbf{v}_{O'/\mathcal{I}} + {}^{\mathcal{I}}\dot{\mathbf{r}}_{O'P} \tag{6.42}$$

The last vector in Equation (6.42) presents a problem. It is not the velocity of P in $\mathcal{I}$, because the point O' is not fixed in $\mathcal{I}$; nor is it the velocity of P in $\mathcal{B}$, because the derivative is not taken there. To overcome this dilemma, we shall rewrite the term by moving the derivative from $\mathcal{I}$ to $\mathcal{B}$ using Equation (6.20):

$$\mathbf{v}_{P/\mathcal{I}} = \mathbf{v}_{O'/\mathcal{I}} + ({}^{\mathcal{B}}\dot{\mathbf{r}}_{O'P} + \boldsymbol{\omega}_{\mathcal{B}/\mathcal{I}} \times \mathbf{r}_{O'P} \tag{6.43}$$

$$= \mathbf{v}_{O'/\mathcal{I}} + \mathbf{v}_{P/\mathcal{B}} + \boldsymbol{\omega}_{\mathcal{B}/\mathcal{I}} \times \mathbf{r}_{O'P}$$

$$= \mathbf{v}_{P/\mathcal{B}} + (\mathbf{v}_{O'/\mathcal{I}} + \boldsymbol{\omega}_{\mathcal{B}/\mathcal{I}} \times \mathbf{r}_{O'P}) \tag{6.44}$$

We are now in a position to derive the equation relating the velocities of two points of the same rigid body from the (therefore more general) equation (6.44). Temporarily let P be a fixed point of $\mathcal{B}$; then $\mathbf{v}_{P/\mathcal{B}} = \mathbf{0}$ and

$$\mathbf{v}_{P/\mathcal{I}} = \mathbf{v}_{O'/\mathcal{I}} + \boldsymbol{\omega}_{\mathcal{B}/\mathcal{I}} \times \mathbf{r}_{O'P} \tag{6.45}$$

or

$$\mathbf{v}_P = \mathbf{v}_{O'} + \boldsymbol{\omega} \times \mathbf{r}_{O'P} \tag{6.46}$$

which is the same as the plane motion equation (3.5), except that now the **r** and **v** vectors may also have z components and the $\boldsymbol{\omega}$ vector can have x and y components in addition to z.

Returning to the general case in which P is not necessarily attached to either $\mathcal{B}$ or $\mathcal{I}$, let us denote the point of $\mathcal{B}$ (or $\mathcal{B}$ extended) coincident with P by $P_{\mathcal{B}}$. Then Equation (6.44) becomes

$$\mathbf{v}_{P/\mathcal{I}} = \mathbf{v}_{P/\mathcal{B}} + \mathbf{v}_{P_{\mathcal{B}}/\mathcal{I}} \tag{6.47}$$

In words, we may restate Equation (6.47) as follows:

$$\begin{bmatrix}\text{Velocity of}\\ P \text{ in } \mathcal{I}\end{bmatrix} = \begin{bmatrix}\text{Velocity of}\\ P \text{ in } \mathcal{B}\end{bmatrix} + \begin{bmatrix}\text{Velocity in } \mathcal{I} \text{ of}\\ \text{the fixed point of}\\ \mathcal{B} \text{ coincident with } P\end{bmatrix}$$

$$\mathbf{v}_{P/\mathcal{I}} \quad = \quad \mathbf{v}_{P/\mathcal{B}} \quad + \quad \mathbf{v}_{P_{\mathcal{B}}/\mathcal{I}}$$

Equation (6.47) has the virtue of compactness. However, it is a less con-

* The superscripts in Equation (6.41) are now necessary to denote the frame in which the derivative is taken.

venient form than is (6.44) for differentiation to produce a corresponding relationship of accelerations.

Question 6.7 Why?

Another common alternative means of stating Equation (6.44) is to view body $\mathcal{B}$ as a moving frame — that is, as a body moving relative to another frame $\mathcal{I}$. (See Figure 6.6.) We may then rewrite Equation (6.44) as

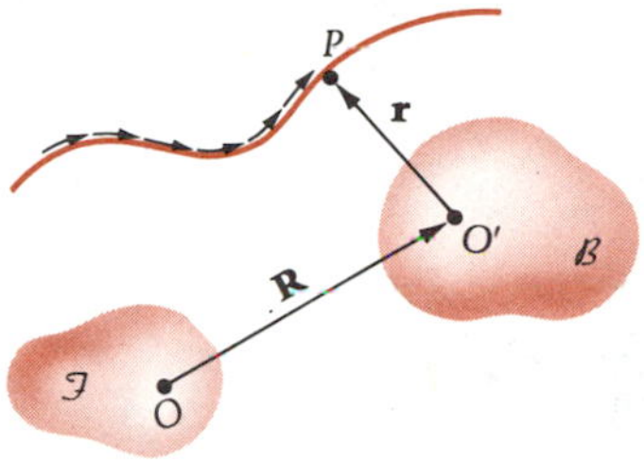

Figure 6.6

$$\mathbf{v}_P = \dot{\mathbf{R}} + \mathbf{v}_{\text{rel}} + \boldsymbol{\omega} \times \mathbf{r} \tag{6.48}$$

in which

$$\begin{aligned}
\mathbf{v}_P &= \text{velocity of } P \text{ in reference frame } \mathcal{I} \\
\dot{\mathbf{R}} &= \text{velocity of moving origin} = \mathbf{v}_{O'/\mathcal{I}} \\
\mathbf{v}_{\text{rel}} &= \text{velocity of } P \text{ in moving frame} = \mathbf{v}_{P/\mathcal{B}} \\
\boldsymbol{\omega} &= \text{angular velocity of moving frame} = \boldsymbol{\omega}_{\mathcal{B}/\mathcal{I}} \\
\mathbf{r} &= \text{position vector of } P \text{ in moving frame} = \mathbf{r}_{O'P}
\end{aligned}$$

We now consider two examples involving the use of the velocity equation for moving frames (Equation 6.44, 6.47, or 6.48).

Example 6.6

Find the velocity in $\mathcal{G}$ of point A at the bottom of the disk in Example 6.1. (See Figure E6.6.)

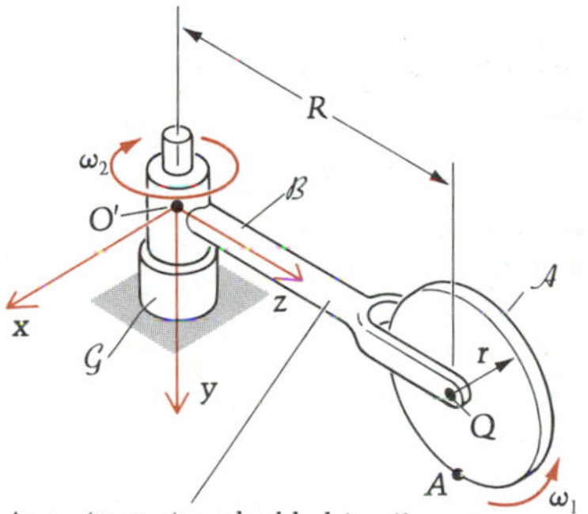

Axes (x, y, z) embedded in $\mathcal{B}$

Figure E6.6

Solution

We select $\mathcal{B}$ as the moving frame, and Equation (6.44) gives the following (where $\mathcal{G}$ is $\mathcal{I}$ and A is P):

$$\mathbf{v}_{A/\mathcal{G}} = \mathbf{v}_{A/\mathcal{B}} + \mathbf{v}_{O'/\mathcal{G}} + \boldsymbol{\omega}_{\mathcal{B}/\mathcal{G}} \times \mathbf{r}_{O'A}$$

In this case $\mathbf{v}_{O'/\mathcal{G}} = \mathbf{0}$; also $\boldsymbol{\omega}_{\mathcal{B}/\mathcal{G}} = \omega_2\hat{\mathbf{j}}$ from the previous example. The velocity of A in $\mathcal{B}$ is given by

$$\mathbf{v}_{A/\mathcal{B}} = \overset{\mathbf{0}}{\cancel{\mathbf{v}_{Q/\mathcal{B}}}} + \omega_1\hat{\mathbf{i}} \times r\hat{\mathbf{j}} = r\omega_1\hat{\mathbf{k}}$$

Therefore

$$\mathbf{v}_{A/\mathcal{G}} = r\omega_1\hat{\mathbf{k}} + \omega_2\hat{\mathbf{j}} \times (r\hat{\mathbf{j}} + R\hat{\mathbf{k}}) = R\omega_2\hat{\mathbf{i}} + r\omega_1\hat{\mathbf{k}}$$

Answer 6.7 If we differentiate Equation (6.47) in $\mathcal{I}$, we encounter the term ${}^{\mathcal{I}}\dot{\mathbf{v}}_{P_{\mathcal{B}}/\mathcal{I}} (\neq \mathbf{a}_{P_{\mathcal{B}}/\mathcal{I}})$, which is not recognizable as a standard kinematic quantity; this is because $P_{\mathcal{B}}$ denotes a *succession* of points of $\mathcal{B}$ which are at each instant coincident with P.

EXAMPLE 6.7

Crank $\mathcal{C}$ in Figure E6.7 rotates about axis z through point O. Its other end, Q, is attached to a ball and socket joint as shown. The ball forms the end of rod $\mathcal{B}$, which passes through a hole in the ceiling $\mathcal{I}$. find the velocity of point P of the bar, which is passing through the hole when $\theta = 90°$ as shown.

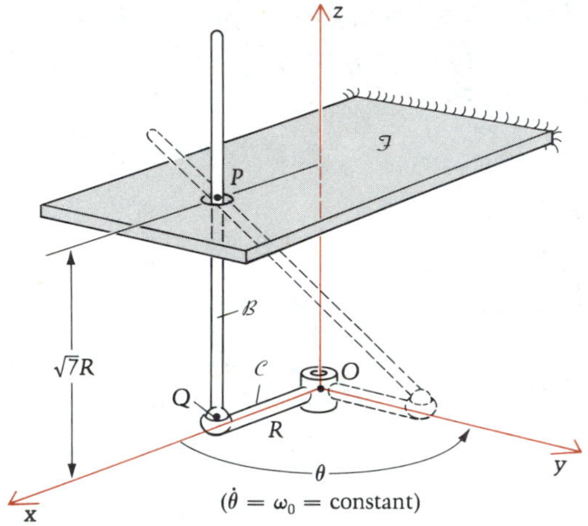

Figure E6.7

Solution

We denote by H the point of $\mathcal{I}$ at the center of the hole; then:

$$\overset{0}{\mathbf{v}_{H/\mathcal{I}}} = \mathbf{v}_{H/\mathcal{B}} + \mathbf{v}_{P/\mathcal{I}} \tag{1}$$

in which P is the point of $\mathcal{B}$ coincident with point H. Knowing that the motion of H in $\mathcal{B}$ must be *along* the axis of $\mathcal{B}$, we have:

$$\mathbf{0} = v_{H/\mathcal{B}}\left(\frac{R\hat{\mathbf{i}} - R\hat{\mathbf{j}} + \sqrt{7}\,R\hat{\mathbf{k}}}{3R}\right) + \overbrace{(\mathbf{v}_{Q/\mathcal{I}} + \boldsymbol{\omega}_{\mathcal{B}/\mathcal{I}} \times \mathbf{r}_{QP})}^{\mathbf{v}_{P/\mathcal{I}}} \tag{2}$$

or

$$\mathbf{0} = v_{H/\mathcal{B}}\left(\frac{\hat{\mathbf{i}} - \hat{\mathbf{j}} + \sqrt{7}\hat{\mathbf{k}}}{3}\right)$$

$$+ [-R\omega_0\hat{\mathbf{i}} + (\omega_x\hat{\mathbf{i}} + \omega_y\hat{\mathbf{j}} + \omega_z\hat{\mathbf{k}}) \times (R\hat{\mathbf{i}} - R\hat{\mathbf{j}} + \sqrt{7}R\hat{\mathbf{k}})] \tag{3}$$

Collecting the coefficients of $\hat{\mathbf{i}}$, $\hat{\mathbf{j}}$, and $\hat{\mathbf{k}}$, respectively, we find

$$\frac{1}{3}v_{H/\mathcal{B}} + \sqrt{7}R\omega_y + R\omega_z = R\omega_O \tag{4}$$

$$-\frac{1}{3}v_{H/\mathcal{B}} + R\omega_z - \sqrt{7}R\omega_x = 0 \tag{5}$$

$$\frac{\sqrt{7}}{3}v_{H/\mathcal{B}} - R\omega_x - R\omega_y = 0 \tag{6}$$

These three equations (4–6) obviously cannot be solved for unique values of all four unknowns $\omega_x, \omega_y, \omega_z, v_{H/\mathcal{B}}$. However, an answer for $v_{H/\mathcal{B}}$ is obtainable by subtracting Equation (5) from (4) and then adding $\sqrt{7}$ times Equation (6). The result is

$$v_{H/\mathcal{B}} = \frac{R\omega_0}{3}$$

Substituting this result back into Equations (4–6) gives three equations in ω_x, ω_y, and ω_z whose coefficient matrix is singular (has a zero determinant). Thus they cannot be solved for the angular velocity components. A more physical reason for this is that the component of $\boldsymbol{\omega}_{\mathcal{B}/\mathcal{I}}$ along bar $\mathcal{B}$ cannot affect the answer for $\mathbf{v}_{P/\mathcal{I}}$ because $\mathcal{B}$ can turn freely in its socket about its axis without altering $\mathbf{v}_{P/\mathcal{I}}$. Mathematically this is manifested by this "axial" component of $\boldsymbol{\omega}_{\mathcal{B}/\mathcal{I}}$ being parallel to $\mathbf{r}_{QP}$ and thus canceling out of Equation (2). These $\boldsymbol{\omega}$ components are not needed for a solution, however, because from Equation (1) we can obtain our desired result:

$$\mathbf{v}_{P/\mathcal{I}} = -\mathbf{v}_{H/\mathcal{B}} = \frac{-R\omega_0}{3}\left(\frac{\hat{\mathbf{i}} - \hat{\mathbf{j}} + \sqrt{7}\hat{\mathbf{k}}}{3}\right)$$

An added note is that although we cannot solve for the component of $\boldsymbol{\omega}_{\mathcal{B}/\mathcal{I}}$ along $\mathcal{B}$ with the given information, we *are* able to calculate the component of $\boldsymbol{\omega}_{\mathcal{B}/\mathcal{I}}$ normal to $\mathcal{B}$. It will be made up of values ω_x', ω_y', and ω_z' that enforce the relationship

$$\boldsymbol{\omega}' \cdot \mathbf{r}_{QP} = 0$$

That is,

$$\cancel{R}(\omega_x' - \omega_y' + \sqrt{7}\omega_z') = 0 \tag{7}$$

This equation states that these $\boldsymbol{\omega}'$ components form a vector normal to the line QP; again, this vector is the only part of $\boldsymbol{\omega}$ that can affect $v_{P/\mathcal{I}}$.

After adding primes to $\omega_x, \omega_y, \omega_z$ in Equations (4–6), the solution of Equations (4–7), as the reader may verify, is

$$\omega_x' = 0 \qquad \omega_y' = \frac{\sqrt{7}}{9}\omega_0 \qquad \omega_z' = \frac{\omega_0}{9} \qquad v_{H/\mathcal{B}} = \frac{R\omega_0}{3}$$

And we may now calculate the velocity of P from Equation (1) as before or from Equation (2) as follows:

$$\begin{aligned}
\mathbf{v}_{P/\mathcal{I}} &= \mathbf{v}_{Q/\mathcal{I}} + \boldsymbol{\omega}_{\mathcal{B}/\mathcal{I}} \times \mathbf{r}_{QP} \\
&= -R\omega_0\hat{\mathbf{i}} + \omega_0\left(\frac{\sqrt{7}\hat{\mathbf{j}} + \hat{\mathbf{k}}}{9}\right) \times R(\hat{\mathbf{i}} - \hat{\mathbf{j}} + \sqrt{7}\hat{\mathbf{k}}) \\
&= \frac{R\omega_0}{9}(-\hat{\mathbf{i}} + \hat{\mathbf{j}} - \sqrt{7}\hat{\mathbf{k}}) \qquad \text{(as before)}
\end{aligned}$$

The Acceleration Relationship in Moving Frames/Coriolis Acceleration

After these examples we are now ready to derive the corresponding relationship between the *accelerations* of P in two frames $\mathcal{B}$ and $\mathcal{I}$. Differentiating Equation (6.44) gives

$${}^{\mathcal{I}}\dot{\mathbf{v}}_{P/\mathcal{I}} = {}^{\mathcal{I}}\dot{\mathbf{v}}_{O'/\mathcal{I}} + {}^{\mathcal{I}}\dot{\boldsymbol{\omega}}_{\mathcal{B}/\mathcal{I}} \times \mathbf{r}_{O'P} + \boldsymbol{\omega}_{\mathcal{B}/\mathcal{I}} \times {}^{\mathcal{I}}\dot{\mathbf{r}}_{O'P} + {}^{\mathcal{I}}\dot{\mathbf{v}}_{P/\mathcal{B}} \tag{6.49}$$

or, again using Equation (6.20) (once in each of the last two terms), we get

$$\mathbf{a}_{P/\mathcal{I}} = \mathbf{a}_{O'/\mathcal{I}} + \boldsymbol{\alpha}_{\mathcal{B}/\mathcal{I}} \times \mathbf{r}_{O'P} + \boldsymbol{\omega}_{\mathcal{B}/\mathcal{I}} \times ({}^{\mathcal{B}}\dot{\mathbf{r}}_{O'P} + \boldsymbol{\omega}_{\mathcal{B}/\mathcal{I}} \times \mathbf{r}_{O'P}) + ({}^{\mathcal{B}}\dot{\mathbf{v}}_{P/\mathcal{B}} + \boldsymbol{\omega}_{\mathcal{B}/\mathcal{I}} \times {}^{\mathcal{B}}\dot{\mathbf{r}}_{O'P}) \tag{6.50}$$

Rearranging the terms, we have

$$\mathbf{a}_{P/\mathcal{I}} = \underbrace{{}^{\mathcal{B}}\dot{\mathbf{v}}_{P/\mathcal{B}}}_{\mathbf{a}_{P/\mathcal{B}}} + \underbrace{\mathbf{a}_{O'/\mathcal{I}} + \boldsymbol{\alpha}_{\mathcal{B}/\mathcal{I}} \times \mathbf{r}_{O'P} + \boldsymbol{\omega}_{\mathcal{B}/\mathcal{I}} \times (\boldsymbol{\omega}_{\mathcal{B}/\mathcal{I}} \times \mathbf{r}_{O'P})}_{\mathbf{a}_{P_{\mathcal{B}}/\mathcal{I}}} + 2\boldsymbol{\omega}_{\mathcal{B}/\mathcal{I}} \times {}^{\mathcal{B}}\dot{\mathbf{r}}_{O'P} \tag{6.51}$$

$$\mathbf{a}_{P/\mathcal{I}} = \mathbf{a}_{P/\mathcal{B}} + \mathbf{a}_{P_{\mathcal{B}}/\mathcal{I}} + 2\boldsymbol{\omega}_{\mathcal{B}/\mathcal{I}} \times \mathbf{v}_{P/\mathcal{B}} \tag{6.52}$$

The middle three terms on the right side of Equation (6.51) make up the acceleration of the point $P_{\mathcal{B}}$ of $\mathcal{B}$ (or $\mathcal{B}$ extended) coincident with P. (The proof is brief: If P is *fixed* to $\mathcal{B}$ at point $P_{\mathcal{B}}$, then the other two terms vanish since $\mathbf{r}_{O'P}$ becomes a constant vector in $\mathcal{B}$, and what remains is necessarily $\mathbf{a}_{P_{\mathcal{B}}/\mathcal{I}}$.) The term ${}^{\mathcal{B}}\dot{\mathbf{v}}_{P/\mathcal{B}}$ (which is ${}^{\mathcal{B}}\ddot{\mathbf{r}}_{O'P}$, with both derivatives taken in $\mathcal{B}$) is clearly the acceleration of P in $\mathcal{B}$. The last term, $2\boldsymbol{\omega}_{\mathcal{B}/\mathcal{I}} \times \mathbf{v}_{P/\mathcal{B}}$, is called the **Coriolis acceleration** of P. Note that due to the presence of the Coriolis acceleration it is *not true* that the acceleration of P in frame $\mathcal{I}$ is its acceleration in $\mathcal{B}$ plus the acceleration of the point of $\mathcal{B}$ with which it is coincident (as was in fact the case with the velocity of P). This result is interestingly analogous to the fact that the addition theorem for angular velocity is not true for angular acceleration.

As we did for the velocity equation, we now restate Equation (6.52) in words:

$$\begin{bmatrix}\text{Acceleration}\\ \text{of } P \text{ in } \mathcal{I}\end{bmatrix} = \begin{bmatrix}\text{Acceleration}\\ \text{of } P \text{ in } \mathcal{B}\end{bmatrix} + \begin{bmatrix}\text{Acceleration in } \mathcal{I} \text{ of}\\ \text{the fixed point of } \mathcal{B}\\ \text{coincident with } P\end{bmatrix} + \begin{bmatrix}\text{Coriolis}\\ \text{acceleration}\end{bmatrix}$$

$$\mathbf{a}_{P/\mathcal{I}} \quad = \quad \mathbf{a}_{P/\mathcal{B}} \quad + \quad \mathbf{a}_{P_{\mathcal{B}}/\mathcal{I}} \quad + \quad 2\boldsymbol{\omega}_{\mathcal{B}/\mathcal{I}} \times \mathbf{v}_{P/\mathcal{B}}$$

In the abbreviated notation of Equation (6.48) we have

$$\mathbf{a}_P = \ddot{\mathbf{R}} + \boldsymbol{\alpha} \times \mathbf{r} + \boldsymbol{\omega} \times (\boldsymbol{\omega} \times \mathbf{r}) + 2\boldsymbol{\omega} \times \mathbf{v}_{\text{rel}} + \mathbf{a}_{\text{rel}} \tag{6.53}$$

in which

$\mathbf{a}_P$ = acceleration of P in reference frame $\mathcal{I}$
$\ddot{\mathbf{R}}$ = acceleration of moving origin = $\mathbf{a}_{O'/\mathcal{I}}$
$\boldsymbol{\alpha}$ = angular acceleration of moving frame = $\boldsymbol{\alpha}_{\mathcal{B}/\mathcal{I}}$
$\mathbf{a}_{\text{rel}}$ = acceleration of P in moving frame = $\mathbf{a}_{P/\mathcal{B}}$

and in which all other terms in Equation (6.53) are defined directly after Equation (6.48).

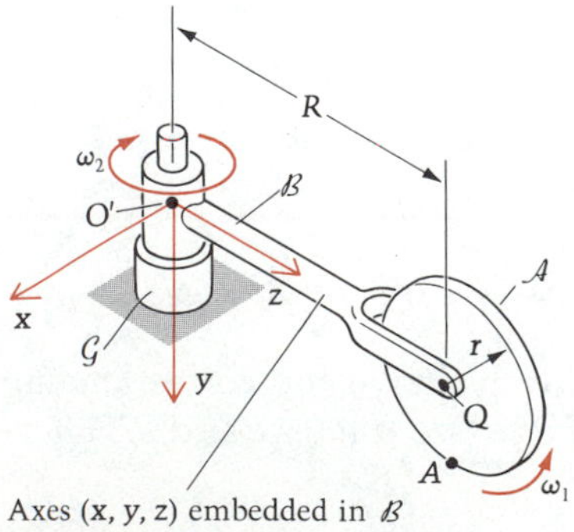

Axes (x, y, z) embedded in $\mathcal{B}$

Figure E6.8

EXAMPLE 6.8

Compute the acceleration in $\mathcal{G}$ of point A in Examples 6.1, 6.4, and 6.6 (see Figure E6.8).

Solution

We shall use Equation (6.51): $\mathcal{B}$ is again the moving frame; the reference frame $\mathcal{I}$ is $\mathcal{G}$, and the moving point P is A:

$$\mathbf{a}_{A/\mathcal{G}} = \mathbf{a}_{A/\mathcal{B}} + \mathbf{a}_{O'/\mathcal{G}} + \boldsymbol{\alpha}_{\mathcal{B}/\mathcal{G}} \times \mathbf{r}_{O'A} + \boldsymbol{\omega}_{\mathcal{B}/\mathcal{G}} \times (\boldsymbol{\omega}_{\mathcal{B}/\mathcal{G}} \times \mathbf{r}_{O'A}) + 2\boldsymbol{\omega}_{\mathcal{B}/\mathcal{G}} \times \mathbf{v}_{A/\mathcal{B}}$$

The various terms on the right side are calculated as follows:

1. $\mathbf{a}_{A/\mathcal{B}} = \cancelto{0}{\mathbf{a}_{Q/\mathcal{B}}} + \cancelto{0}{\boldsymbol{\alpha}_{\mathcal{A}/\mathcal{B}}} \times \mathbf{r}_{QA} - \underbrace{\omega_1^2 \mathbf{r}_{QA}}$

   Note that $\mathcal{A}$ is in plane motion relative to $\mathcal{B}$, so that this short form is acceptable.

   $= -r\omega_1^2\hat{\mathbf{j}}$ (A moves on a circle at constant speed in $\mathcal{B}$)
2. $\mathbf{a}_{O'/\mathcal{G}} = \mathbf{0}$. Note that O' is fixed in $\mathcal{G}$ in this example.
3. $\boldsymbol{\alpha}_{\mathcal{B}/\mathcal{G}} \times \mathbf{r}_{O'A} = \mathbf{0}$. Note that $\boldsymbol{\alpha}_{\mathcal{B}/\mathcal{G}} = {}^{\mathcal{G}}(d/dt)(\omega_2\hat{\mathbf{j}}) = \mathbf{0}$ since ω_2 is a constant, and $\hat{\mathbf{j}}$ does not change in direction in $\mathcal{G}$.
4. $\boldsymbol{\omega}_{\mathcal{B}/\mathcal{G}} \times (\boldsymbol{\omega}_{\mathcal{B}/\mathcal{G}} \times \mathbf{r}_{O'A}) = \omega_2\hat{\mathbf{j}} \times [\omega_2\hat{\mathbf{j}} \times (r\hat{\mathbf{j}} + R\hat{\mathbf{k}})] = -R\omega_2^2\hat{\mathbf{k}}$.
5. $2\boldsymbol{\omega}_{\mathcal{B}/\mathcal{G}} \times \mathbf{v}_{A/\mathcal{B}} = 2\omega_2\hat{\mathbf{j}} \times (r\omega_1\mathbf{k}) = 2r\omega_1\omega_2\hat{\mathbf{i}}$.

Thus the answer for the acceleration of point A is

$$\mathbf{a}_{A/\mathcal{G}} = 2r\omega_1\omega_2\hat{\mathbf{i}} - r\omega_1^2\hat{\mathbf{j}} - R\omega_2^2\hat{\mathbf{k}}$$

In Section 6.7, we shall rework the above example using another approach.

PROBLEMS ▶ Section 6.5

6.25 In Example 6.2, the children are each 5 ft from the fulcrum O. At a later time the girl slides a stone P along the see-saw board toward the boy. The velocity of the stone relative to the board is $-4\hat{\mathbf{j}}$ ft/sec at an instant when (a) the stone is 2 ft from the girl; (b) $\theta = 0$; (c) $\omega_V = 3$ rad/sec; and (d) $\omega_H = 2$ rad/sec. Find $\mathbf{v}_{P/\mathcal{I}}$ and $\mathbf{v}_{P/\mathcal{G}}$ at this instant.

6.26 In Problem 6.25, if at the given instant $\dot{\omega}_V = 0$, $\dot{\omega}_H = 1.5$ rad/sec², and the stone's acceleration relative to the board is $0.8\,\hat{\mathbf{j}}$ ft/sec², find $\mathbf{a}_{P/\mathcal{I}}$ and $\mathbf{a}_{P/\mathcal{G}}$.

6.27 The large disk $\mathcal{A}$ in Figure P6.27 rotates at 10 rad/sec counterclockwise (looking down on its horizontal surface). A small disk $\mathcal{B}$ rolls radially outward along a radius OD of $\mathcal{A}$. At the instant shown, the center C of $\mathcal{B}$ is 4 ft from the axis of rotation of $\mathcal{A}$, and this distance is increasing at the constant rate of 2 ft/sec. Determine the velocity and acceleration of point E, which is at the top of $\mathcal{B}$ at the given instant.

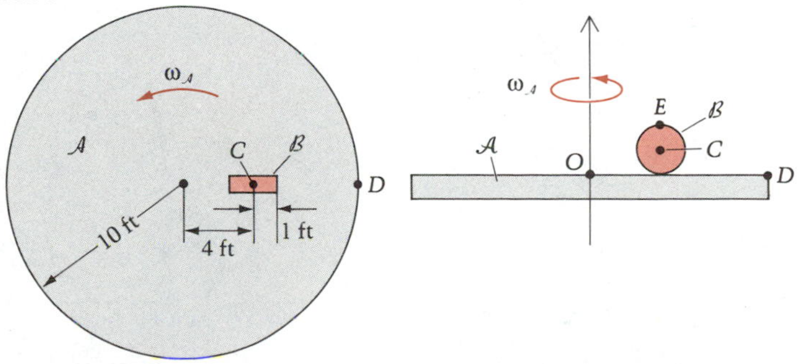

Figure P6.27

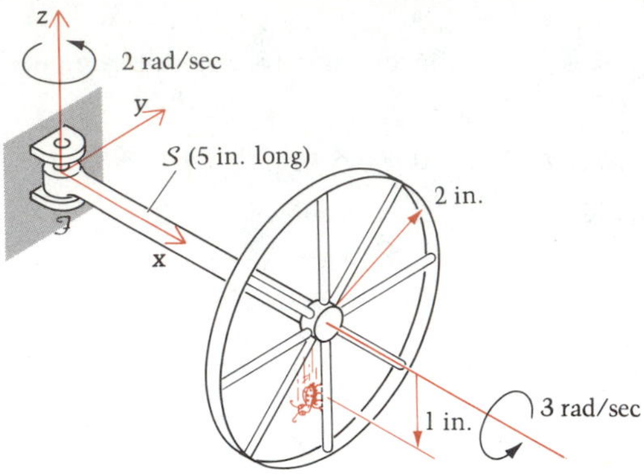

Figure P6.28

6.28 Shaft 𝒮 in Figure P6.28 turns in the clevis 𝒥 at 2 rad/sec in the direction shown. The wheel simultaneously rotates at 3 rad/sec about its axis as indicated. Both rates are constants. The bug is crawling outward on a spoke at 0.2 in./sec with an acceleration of 0.1 in./sec^2 both relative to the spoke. At the instant shown, find: (a) the angular velocity of the wheel; (b) the velocity of the bug.

6.29 Find the angular acceleration of the wheel and the acceleration of the bug in Problem 6.28.

* **6.30** The crane $\mathcal{C}$ in Figure P6.30 turns about the vertical at $\omega_v = 0.2$ rad/sec = constant, and simultaneously its boom $\mathcal{B}$ is being elevated at the increasing rate $\omega_H = 0.1t$ rad/sec. The (x, y, z) axes are fixed to the crane $\mathcal{C}$ at O, and the boom was along the y axis when t was zero. When the boom makes a 60° angle with the horizontal, find: (a) $\boldsymbol{\omega}_{\mathcal{B}/\mathcal{G}}$; (b) $\boldsymbol{\alpha}_{\mathcal{B}/\mathcal{G}}$, (c) $\mathbf{v}_{P/\mathcal{G}}$; (d) $\mathbf{a}_{P/\mathcal{G}}$.

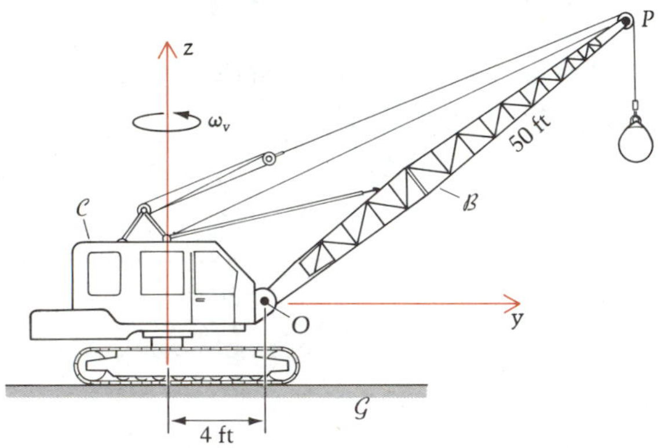

Figure P6.30

6.31 A platform translates past a turntable at 6 mph. (See Figure P6.31.) People step onto the turntable and walk straight toward the center where they exit onto stairs. There is rolling contact between platform and turntable. Suppose the people walk at the constant rate of approximately 3 mph relative to the turntable. If it is desired that they do not experience more than 3 ft/sec^2 of lateral acceleration, find the required turntable radius.

6.32 The truck in Figure P6.32 moves to the left at a constant speed of 7.07 ft/sec. At the instant shown, the loading compartment has an angular speed $\dot{\theta} = \frac{1}{3}$ rad/sec and an angular acceleration of $\ddot{\theta} = -\frac{1}{15}$ rad/sec^2. The cylinder shown on the truck bed comes loose and rolls

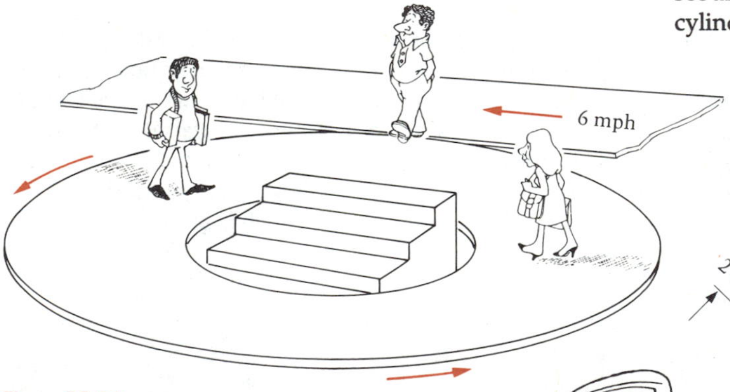

Figure P6.31

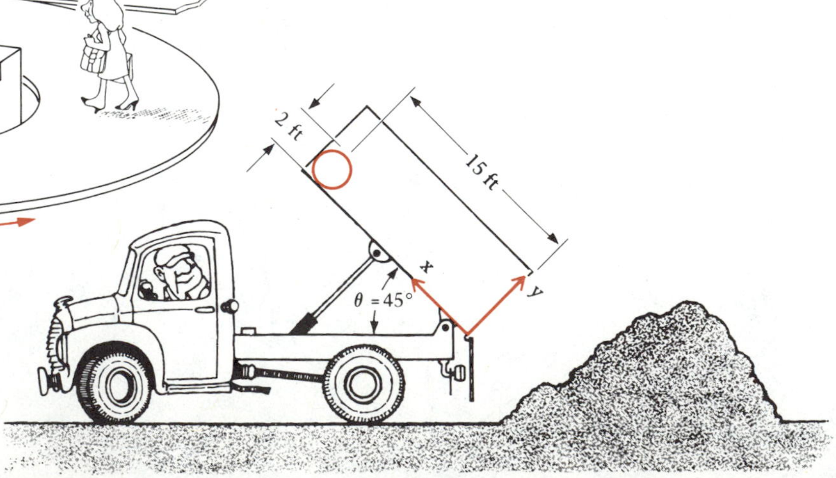

Figure P6.32

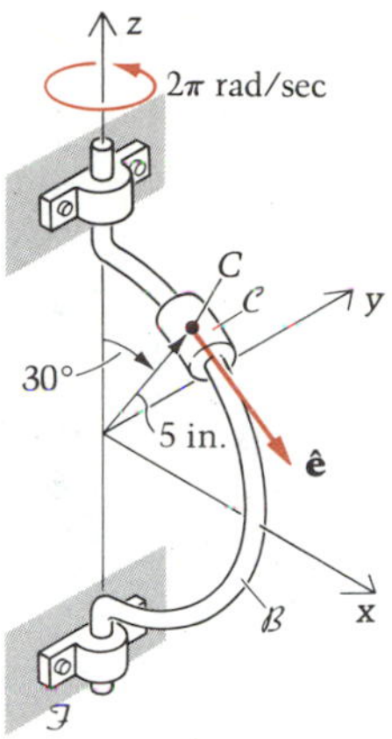

Figure P6.33

toward the ground at an angular velocity, relative to the compartment, of 1 ⥀ rad/sec at this instant; it is speeding up at a rate of $\frac{1}{2}$ rad/sec², also relative to the compartment. Find the velocity and acceleration of the center of the cylinder relative to the ground. Use the rotating reference frame shown.

6.33 The bent bar $\mathcal{B}$ in Figure P6.33 revolves about the vertical at $2\pi\hat{\mathbf{k}}$ rad/sec. The center C of the collar $\mathcal{C}$ has velocity and acceleration relative to $\mathcal{B}$ of $20\hat{\mathbf{e}}$ in./sec and $-10\hat{\mathbf{e}}$ in./sec², respectively, where $\hat{\mathbf{e}}$ is in the direction of the velocity of C in $\mathcal{B}$. At the given instant, find the velocity and acceleration of C in the frame $\mathcal{I}$ in which $\mathcal{B}$ turns.

6.34 A centrifugal pump $\mathcal{P}$ turns at 500 rpm in $\mathcal{I}$, and the water particles have respective tangential velocity and acceleration components *relative to the blades* of 120 ft/sec and 80 ft/sec² outward when they reach the outermost point of their blades. (See Figure P6.34.) At the instant before exit determine the velocity and acceleration vectors of the water particle at P with respect to the ground $\mathcal{I}$.

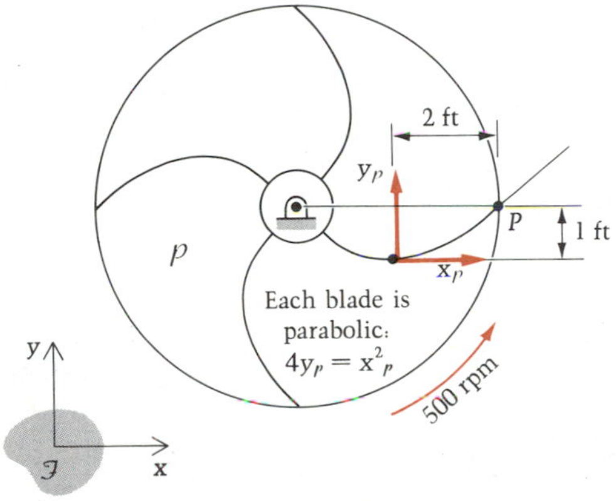

Figure P6.34

6.35 A man walks dizzily outward along a sine wave fixed to a merry go-round that is 40 ft in diameter and turns at 10 rpm. (See Figure P6.35.) If the man's speed relative to the turntable is a constant 2 ft/sec, what is the magnitude of his acceleration when he is 10 ft from the center?

6.36 Disk $\mathcal{D}$ rotates about its axis at the constant angular speed $\omega_0 = 0.1$ rad/s. The round wire is rigidly affixed to $\mathcal{D}$ at points A and B as shown in Figure P6.36. A bug crawls around the wire from A to B; its speed relative to the wire (initially zero) is always increasing at the constant rate of 0.001 m/s². Find the velocity and acceleration of the bug, relative to the reference frame $\mathcal{I}$ in which the disk turns, when it arrives at B.

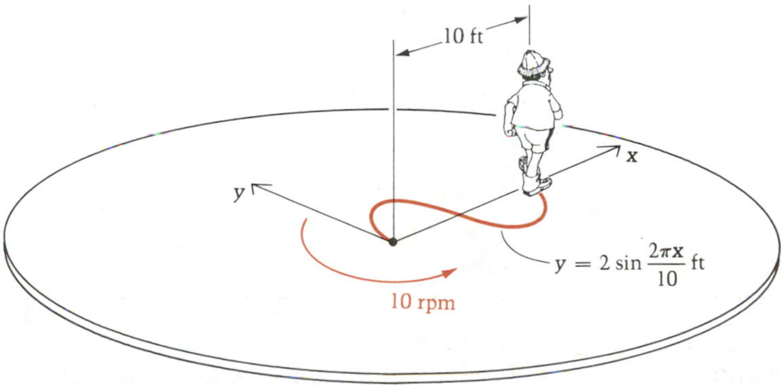

Figure P6.35

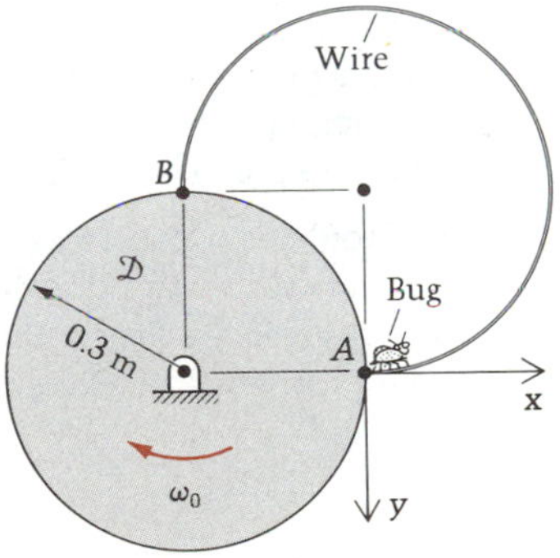

Figure P6.36

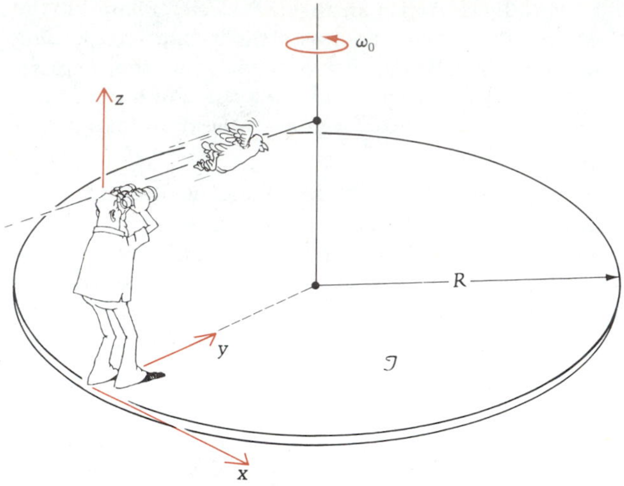

Figure P6.37

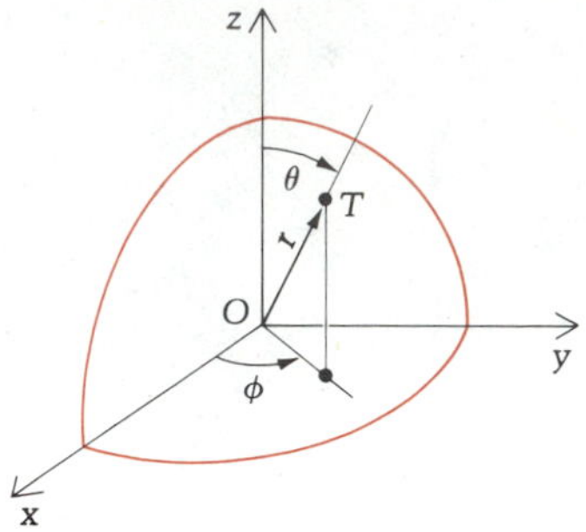

Figure P6.38

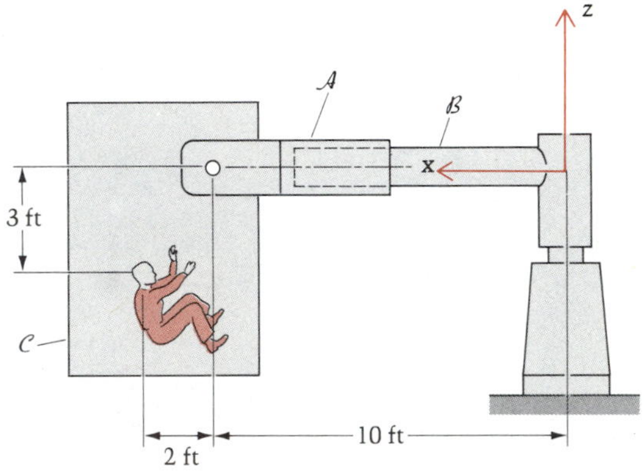

Figure P6.39

6.37 A bird flies horizontally past a man's head in a straight line at constant speed v_0 toward the axis of a turntable on which the man is standing. The axes xy have origin at the man's feet; the z axis is vertical and y always points toward the center of the turntable, which has radius R and angular velocity $\boldsymbol{\omega} = \omega_0\hat{\mathbf{k}}$. Derive $x(t)$ and $y(t)$ of the bird in terms of R, v_0, ω_0, and the time t. (See Figure P6.37.) Use Equation (6.47), integrate, then check your results by inspection.

* **6.38** In Figure P6.38 the axes x and y, and the origin O, are fixed on the deck of a ship. The ship $\mathcal{S}$ has an angular velocity relative to the earth $\mathcal{E}$ of

$$\boldsymbol{\omega}_{\mathcal{S}/\mathcal{E}} = \omega_r\hat{\mathbf{i}} + \omega_P\hat{\mathbf{j}}$$

where x and y are fore-and-aft and athwartships axes, respectively; thus ω_r is a rolling component and ω_P a pitching component of angular velocity. Point T is a target fixed relative to earth, such as a geosynchronous satellite. Find the angular rates $\dot{\theta}$ and $\dot{\phi}$, in terms of θ, ϕ, ω_r, and ω_P, that are required to track point T.

* **6.39** In Problems 6.9 and 6.22, let $\omega_z = 0.5$ rad/sec and $\omega_x = 0.7$ rad/sec in the directions indicated. With the dimensions given in Figure P6.39, find the maximum value of ω_y for which the acceleration magnitude of the astronaut's head will not exceed 5g at the given instant. Let $\omega_y =$ constant.

* **6.40** In Problem 6.16 find the velocity of point P at the tip of the gripper $\mathcal{G}$ at the instant given.

* **6.41** Prove that if two bodies are in rolling contact, the shaded arcs in Figure P6.41, representing the loci of former contact points, are equal in length. (Note from Problem 3.105 that the converse is not true.)

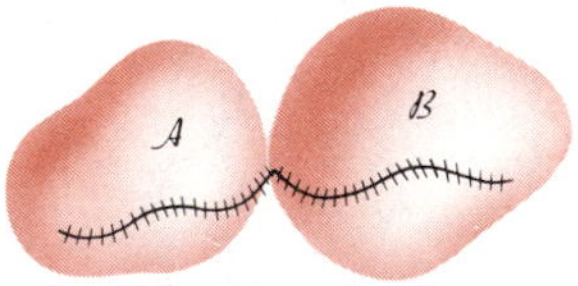

Figure P6.41

6.6 The Earth as a Moving Frame

In this section, we shall make use of Equation (6.53) to set up the differential equation governing the position of the mass center C of a body $\mathcal{B}$ that is in motion near the earth. This equation will allow us to measure the position of C with respect to a desired site O' at latitude λ, which is itself in motion as the earth turns on its axis from west to east. We assume that for describing certain motions near the earth, a frame $\mathcal{I}$ with origin at the earth's mass center O is "sufficiently fixed" to be justifiably called inertial. The frame $\mathcal{I}$ moves as does the (assumed rigid) earth, except that it does not share the earth's daily spin. Thus the site O' has an acceleration $\ddot{\mathbf{R}}$ in $\mathcal{I}$, directed toward the earth's north-south polar axis.

We set up the moving frame $\mathcal{F}$ as shown as Figure 6.7. The frame $\mathcal{F}$ is the turning earth, and the (x, y, z) axes are embedded in it at O' with x pointing east, y north, and z in the direction of local vertical. The acceleration of the mass center C of a body $\mathcal{B}$, moving near the earth and whose position is desired relative to O', is known from Equation (6.53) to be

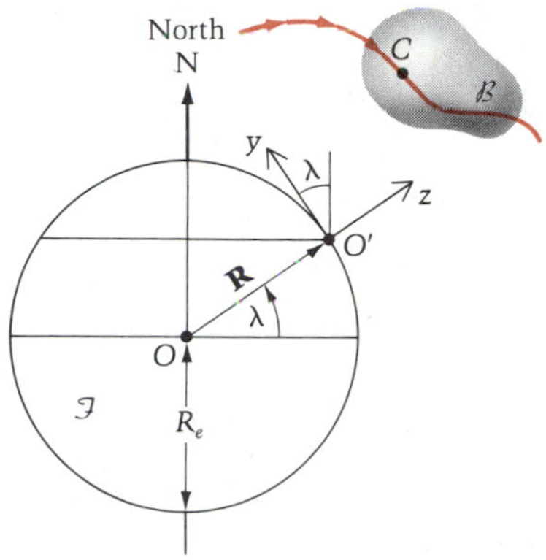

Figure 6.7

$$\mathbf{a}_{C/\mathcal{I}} = \ddot{\mathbf{R}} + \boldsymbol{\alpha} \times \mathbf{r} + \boldsymbol{\omega} \times (\boldsymbol{\omega} \times \mathbf{r}) + 2\boldsymbol{\omega} \times \mathbf{v}_{\text{rel}} + \mathbf{a}_{\text{rel}}$$

where $\mathbf{r}$ is the position vector of C in $\mathcal{F}$, and $\mathbf{v}_{\text{rel}}$ and $\mathbf{a}_{\text{rel}}$ are the velocity and acceleration vectors of C in $\mathcal{F}$. Further, $\boldsymbol{\omega} = \boldsymbol{\omega}_{\mathcal{F}/\mathcal{I}}$ and $\boldsymbol{\alpha} = \boldsymbol{\alpha}_{\mathcal{F}/\mathcal{I}}$. We now use the mass-center equation of motion from Chapter 2:

$$\Sigma\mathbf{F} = m\mathbf{a}_{C/\mathcal{I}}$$

to obtain

$$\mathbf{F} - mg\hat{\mathbf{k}} = m[\ddot{\mathbf{R}} + \boldsymbol{\alpha} \times \mathbf{r} + \boldsymbol{\omega} \times (\boldsymbol{\omega} \times \mathbf{r}) + 2\boldsymbol{\omega} \times \mathbf{v}_{\text{rel}} + \mathbf{a}_{\text{rel}}] \quad (6.54)$$

where $\mathbf{F}$ represents all external forces on $\mathcal{B}$ besides gravity, which is written separately.

Question 6.8 This is a good place to ask: Why is Equation (6.54) restricted to bodies in motion near the earth?

We now proceed to compute the various terms in Equation (6.54). First we note that

$$\mathbf{r} = x\hat{\mathbf{i}} + y\hat{\mathbf{j}} + z\hat{\mathbf{k}}$$
$$\mathbf{v}_{\text{rel}} = \dot{x}\hat{\mathbf{i}} + \dot{y}\hat{\mathbf{j}} + \dot{z}\hat{\mathbf{k}}$$
$$\mathbf{a}_{\text{rel}} = \ddot{x}\hat{\mathbf{i}} + \ddot{y}\hat{\mathbf{j}} + \ddot{z}\hat{\mathbf{k}}$$
$$\boldsymbol{\omega} = \omega_e(\cos\lambda\hat{\mathbf{j}} + \sin\lambda\hat{\mathbf{k}}) \qquad (\text{where } \omega_e = 2\pi \text{ rad/day} \approx 0.0000727 \text{ rad/sec})$$
$$\boldsymbol{\alpha} = \mathbf{0}$$
$$\mathbf{R} = R_e\hat{\mathbf{k}}$$

Next we compute the acceleration $\ddot{\mathbf{R}}$ of the site (O'). We utilize Equation (6.53) again, this time with the "moving point" being O' and the origin in the moving frame being O:

Answer 6.8 It has been assumed that the strength of the gravitational field is constant, but of course gravity decreases as C moves farther and farther from the earth's surface.

$$\mathbf{a}_{O'} = \ddot{\mathbf{R}} = \cancelto{0}{\mathbf{a}_O} + \cancelto{0}{\dot{\boldsymbol{\omega}}} \times \mathbf{R} + \boldsymbol{\omega} \times (\boldsymbol{\omega} \times \mathbf{R}) + 2\boldsymbol{\omega} \times \mathbf{v}_{\text{rel}} + \mathbf{a}_{\text{rel}}$$

This time, however, note that $\mathbf{v}_{\text{rel}}$ and $\mathbf{a}_{\text{rel}}$ are zero; this follows from the fact that O' has *no* motion relative to the moving frame (the earth). Thus the only term surviving is:

$$\ddot{\mathbf{R}} = \boldsymbol{\omega} \times (\boldsymbol{\omega} \times \mathbf{R})$$

Therefore, from Equation (6.54),

$$\mathbf{F} - mg\hat{\mathbf{k}} = m\{\boldsymbol{\omega} \times [\boldsymbol{\omega} \times (\mathbf{R} + \mathbf{r})] + 2\boldsymbol{\omega} \times (\dot{x}\hat{\mathbf{i}} + \dot{y}\hat{\mathbf{i}} + \dot{z}\hat{\mathbf{k}}) + (\ddot{x}\hat{\mathbf{i}} + \ddot{y}\hat{\mathbf{j}} + \ddot{z}\hat{\mathbf{k}})\}$$

Neglecting $|\mathbf{r}|$ with respect to $|\mathbf{R}|$ and expressing $\mathbf{F}$ in terms of its components (F_x, F_y, F_z), we arrive at a set of differential equations governing the motion of C:

$$\ddot{x} = 2\omega_e(\dot{y}\sin\lambda - \dot{z}\cos\lambda) + \frac{F_x}{m}$$

$$\ddot{y} = -2\omega_e\dot{x}\sin\lambda - R_e\omega_e^2\cos\lambda\sin\lambda + \frac{F_y}{m}$$

$$\ddot{z} = 2\omega_e\dot{x}\cos\lambda + R_e\omega_e^2\cos^2\lambda - g + \frac{F_z}{m} \tag{6.55}$$

We complete this brief section with an example illustrating the use of Equations (6.55).

EXAMPLE 6.9

Due to the earth's rotation, the resultant force it exerts on a particle P at rest on its surface is not quite directed toward its center of mass. Use Equations (6.55) to find this deviation, assuming a spherical earth.

Solution

We have $\dot{x}$, $\dot{y}$, $\dot{z}$, $\ddot{x}$, $\ddot{y}$, and $\ddot{z}$ all zero, so that the equations of motion of P (which moves!) in the inertial frame $\mathcal{I}$ are:

$$F_x = 0$$

$$F_y = mR_e\omega_e^2\cos\lambda\sin\lambda = k\sin\lambda$$

$$F_z = -mR_e\omega_e^2\cos^2\lambda + mg = mg - k\cos\lambda$$

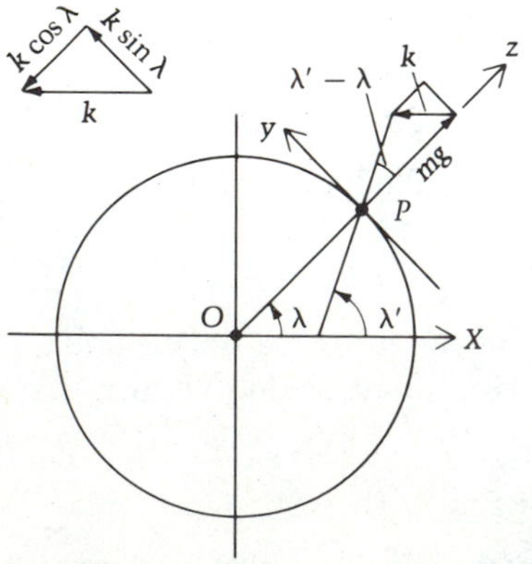

Figure E6.9a

Calling $mR_e\omega_e^2\cos\lambda = k$, we see (Figure E6.9a) that the earth must push on P at a small angle $(\lambda' - \lambda)$ with the *"geometric vertical"* in order for P to remain at rest in the "moving frame" $\mathcal{F}$, rigidly attached to the surface of the planet. It is the angle λ', *not* the angle λ, that defines *"local vertical"*; this is because λ' is the angle that a plumb bob string makes with the axis X in the equatorial plane.

Figure E6.9b shows that (with $\phi = \lambda' - \lambda$)

$$\tan\phi \approx \frac{k\sin\lambda}{mg - k\cos\lambda}$$

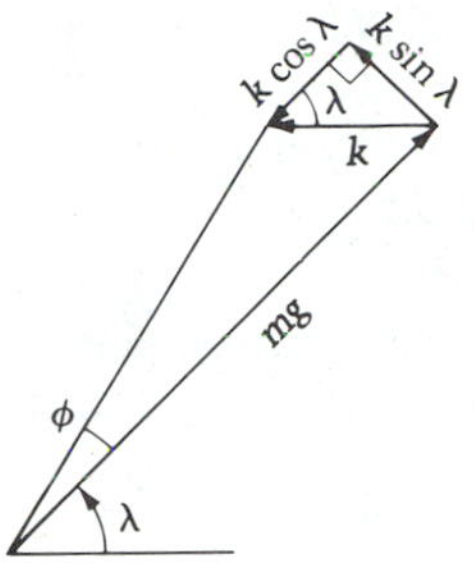

Figure E6.9b

We note that k, in comparison with mg, is quite small:

$$\frac{k}{mg} = \frac{R_e\omega_e^2 \cos\lambda}{g} \approx \frac{6370(0.0000727)^2 \cos\lambda}{(9.81/1000)}$$

$$= \frac{6370(1000)(0.0000727)^2 \cos\lambda}{9.81}$$

$$\approx 0.00343 \cos\lambda$$

This means that we can use small angle approximations on the angle ϕ:

$$\phi \approx \tan\phi \approx \frac{k \sin\lambda}{mg} = \frac{R_e\omega_e^2 \sin\lambda \cos\lambda}{g} = 0.00343 \sin\lambda \cos\lambda$$

so that the deviation of the local "plumb-line" vertical from the "geometric vertical" is

$$\lambda' - \lambda = \phi \approx 0.00343 \sin\lambda \cos\lambda$$

Of course there is *no* deviation in direction at the poles (where $\cos\lambda = 0$) or the equator (where $\sin\lambda = 0$). The maximum, at 45° latitude, is 0.0017 rad, or about 0.1°.

PROBLEMS ▶ Section 6.6

6.42 If it were possible for a train to travel continuously around the world on a meridional track as shown in Figure P6.42(a), one side of the track would wear out in time due to the Coriolis acceleration. Explain which side will wear out in each of the four numbered quadrants of the circular path. Figure P6.42(b) shows how the train's wheels rest on the track.

6.43 Explain in detail how the Coriolis acceleration is related to the deflection of the air rushing toward a low-pressure area, thereby forming a hurricane. (See Figure P6.43.) Do the problem for each hemisphere!

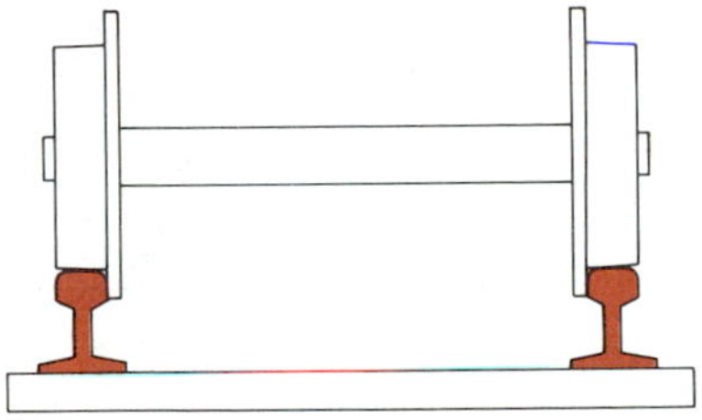
Figure P6.42(b)

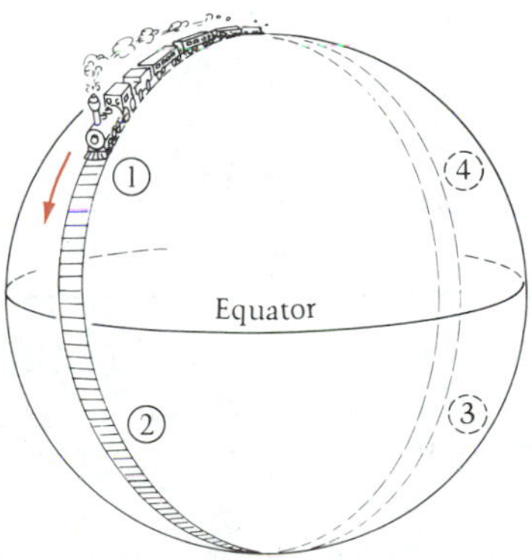

Figure P6.42(a)

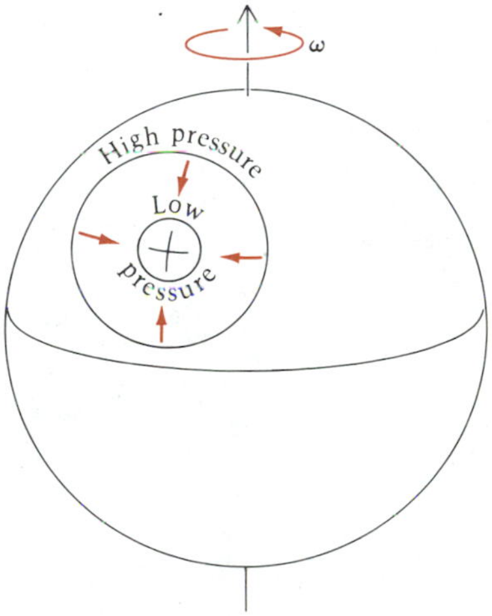

Figure P6.43

6.44 A car travels south along a meridian at a certain time; its speed relative to the earth is 60 mph, increasing at the rate of 2 ft/sec². (See Figure P6.44.) Find the acceleration of the car in a frame having origin always at the center of the earth and z axis along the polar axis of rotation, but not rotating about the axis with the earth.

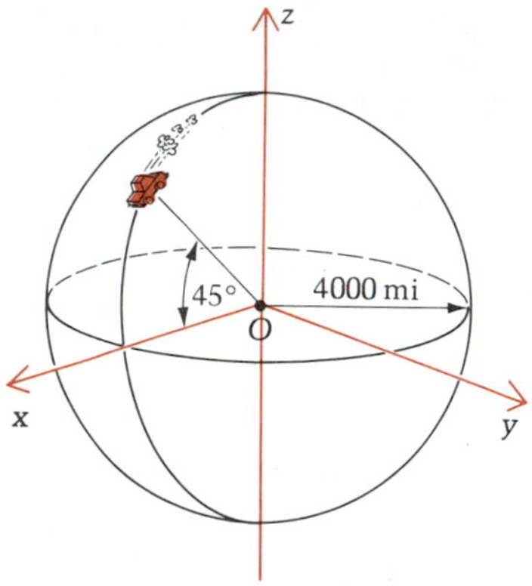

Figure P6.44

6.45 If in the preceding problem the car is traveling from west to east at 45°N latitude instead of along a meridian, find the car's acceleration in the same frame. (See Figure P6.45.)

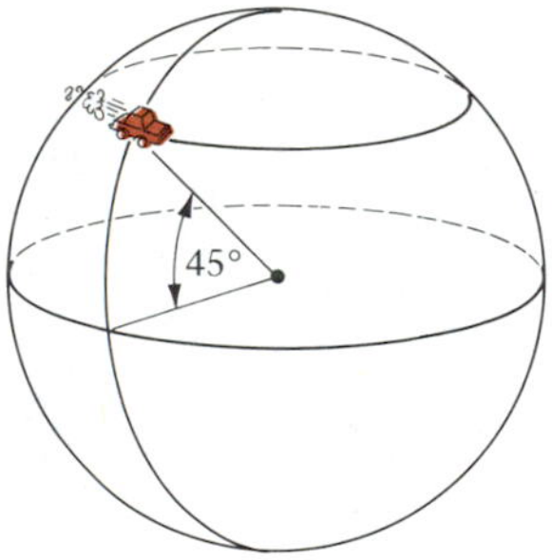

Figure P6.45

* **6.46** A projectile is fired from a site at latitude λ, with the initial velocity components (at time $t = 0$) $= \dot{x}_i, \dot{y}_i, \dot{z}_i$. Determine the maximum height reached by the projectile, neglecting air resistance and the $R_e\omega_e^2$ terms in Equations (6.55).

* **6.47** Refer to Example 6.9. Show that if the component equations of (6.54) are written in terms of axes (x', y', z') (see Figure P6.47) instead of (x, y, z), they will have the same form as do Equations (6.55) without their $R_e\omega_e^2$ terms, provided that

$$G = \sqrt{(g - R_e\omega_e^2 C_\lambda^2)^2 + (R_e\omega_e^2 C_\lambda S_\lambda)^2}$$

replaces g; λ' replaces λ; and y' and z', respectively, replace y and z.

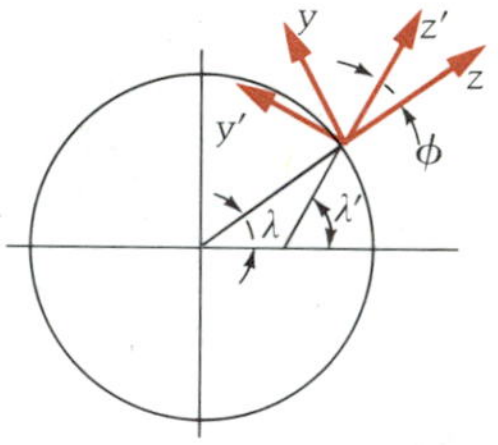

Figure P6.47

* **6.48** Using equations from the preceding problem, find the location at which a falling rock will strike the earth if dropped from rest on the z' axis from a height H. Neglect air resistance and assume the rock strikes the earth when $z' = 0$.

6.7 Velocity and Acceleration Equations for Two Points of the Same Rigid Body

We next apply the concepts of position, velocity, acceleration, angular velocity, and angular acceleration (developed in Chapter 1 and Sections 6.2 to 6.5) to the kinematics of a rigid body $\mathcal{B}$ in general motion in a frame $\mathcal{I}$. The equation relating the velocities in $\mathcal{I}$ of two points of a rigid body $\mathcal{B}$ to its angular velocity $\boldsymbol{\omega}_{\mathcal{B}/\mathcal{I}}$ is a special case of Equation (6.44). We have seen in the text following that equation

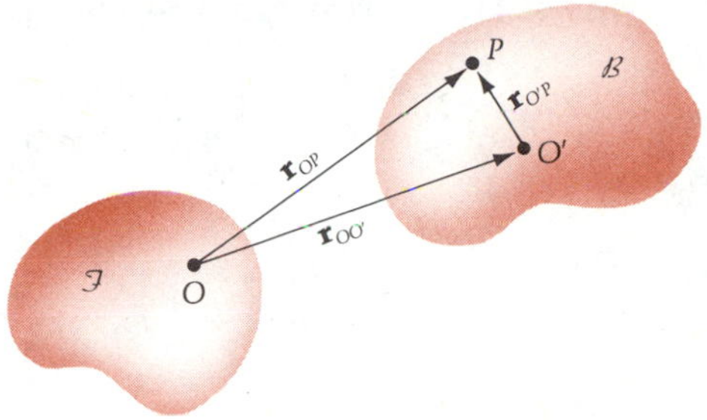

Figure 6.8 Points P and O' of a rigid body $\mathcal{B}$ in general motion.

that if P joins O' as a fixed point of $\mathcal{B}$ (see Figure 6.8), the relationship between the velocities of these two points is given by the equation

$$\mathbf{v}_{P/\mathcal{I}} = \mathbf{v}_{O'/\mathcal{I}} + \boldsymbol{\omega}_{\mathcal{B}/\mathcal{I}} \times \mathbf{r}_{O'P} \tag{6.56}$$

From the text following Equation (6.52), we also know the relationship between the accelerations of these two points of $\mathcal{B}$:

$$\mathbf{a}_{P/\mathcal{I}} = \mathbf{a}_{O'/\mathcal{I}} + \boldsymbol{\alpha}_{\mathcal{B}/\mathcal{I}} \times \mathbf{r}_{O'P} + \boldsymbol{\omega}_{\mathcal{B}/\mathcal{I}} \times (\boldsymbol{\omega}_{\mathcal{B}/\mathcal{I}} \times \mathbf{r}_{O'P}) \tag{6.57}$$

We emphasize that Equations (6.56) and (6.57) follow from the general equations (6.44) and (6.51) when and if $\mathbf{r}_{O'P}$ is a constant in $\mathcal{B}$; in that case, $\mathbf{v}_{P/\mathcal{B}} = {}^{\mathcal{B}}\dot{\mathbf{r}}_{O'P} = \mathbf{0}$ and $\mathbf{a}_{P/\mathcal{B}} = {}^{\mathcal{B}}\ddot{\mathbf{r}}_{O'P} = \mathbf{0}$. Furthermore, the two points P and O' in Equations (6.56) and (6.57) may be replaced by *any* pair of points fixed to $\mathcal{B}$, since being fixed to $\mathcal{B}$ is the only restriction on either of them. Speaking loosely, with Equations (6.56, 6.57) we are interested in *two points* on *one body*, whereas with Equations (6.44, 6.51) we were studying *one point* in motion relative to *two bodies*. We now consider examples involving the use of the two rigid-body equations (6.56) and (6.57).

Example 6.10

Rework Examples 6.6 and 6.8 by treating point A as a point of body $\mathcal{A}$ instead of as a point moving with a known motion in $\mathcal{B}$.

Solution

From Equation (6.56), and Figure E6.10, we have

$$\mathbf{v}_{A/\mathcal{G}} = \mathbf{v}_{Q/\mathcal{G}} + \boldsymbol{\omega}_{\mathcal{A}/\mathcal{G}} \times \mathbf{r}_{QA}$$

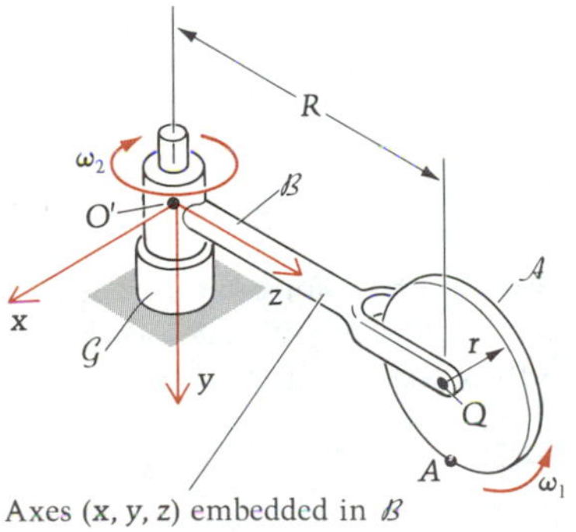

Axes (x, y, z) embedded in $\mathcal{B}$

Figure E6.10

Now recognizing that Q is also a point of $\mathcal{B}$, we obtain

$$\begin{aligned}\mathbf{v}_{A/\mathcal{G}} &= (\mathbf{v}_{O'/\mathcal{G}} + \boldsymbol{\omega}_{\mathcal{B}/\mathcal{G}} \times \mathbf{r}_{O'Q}) + \boldsymbol{\omega}_{\mathcal{A}/\mathcal{G}} \times \mathbf{r}_{QA} \\ &= \mathbf{0} + \omega_2\hat{\mathbf{j}} \times R\hat{\mathbf{k}} + (\omega_1\hat{\mathbf{i}} + \omega_2\hat{\mathbf{j}}) \times r\hat{\mathbf{j}} \\ &= R\omega_2\hat{\mathbf{i}} + r\omega_1\hat{\mathbf{k}} \qquad \text{(as we obtained in Example 6.6)}\end{aligned}$$

Next we relate the accelerations of A and Q with Equation (6.57):

$$\mathbf{a}_{A/\mathcal{G}} = \mathbf{a}_{Q/\mathcal{G}} + \boldsymbol{\alpha}_{\mathcal{A}/\mathcal{G}} \times \mathbf{r}_{QA} + \boldsymbol{\omega}_{\mathcal{A}/\mathcal{G}} \times (\boldsymbol{\omega}_{\mathcal{A}/\mathcal{G}} \times \mathbf{r}_{QA})$$

The first term on the right side, using Q and O' as points of body $\mathcal{B}$, is

$$\begin{aligned}\mathbf{a}_{Q/\mathcal{G}} &= \mathbf{a}_{O'/\mathcal{G}} + \boldsymbol{\alpha}_{\mathcal{B}/\mathcal{G}} \times \mathbf{r}_{O'Q} + \boldsymbol{\omega}_{\mathcal{B}/\mathcal{G}} \times (\boldsymbol{\omega}_{\mathcal{B}/\mathcal{G}} \times \mathbf{r}_{O'Q}) \\ &= 0 + 0 + \omega_2\hat{\mathbf{j}} \times (R\omega_2\hat{\mathbf{i}}) \\ &= -R\omega_2^2\hat{\mathbf{k}}\end{aligned}$$

Thus

$$\mathbf{a}_{A/\mathcal{G}} = -R\omega_2^2\hat{\mathbf{k}} + (-\omega_1\omega_2\hat{\mathbf{k}}) \times (r\hat{\mathbf{j}}) + (\omega_1\hat{\mathbf{i}} + \omega_2\hat{\mathbf{j}}) \times (r\omega_1\hat{\mathbf{k}})$$

or

$$\mathbf{a}_{A/\mathcal{G}} = 2r\omega_1\omega_2\hat{\mathbf{i}} - r\omega_1^2\hat{\mathbf{j}} - R\omega_2^2\hat{\mathbf{k}}$$

which we previously obtained in Example 6.8 by another approach.

In the next example we will see that sometimes there is an indeterminate component of angular velocity.

EXAMPLE 6.11

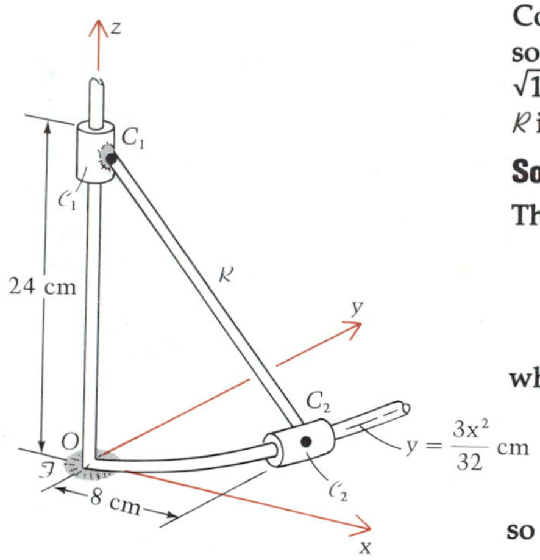

Figure E6.11a

Collars $\mathcal{C}_1$ and $\mathcal{C}_2$ in Figure E6.11a are attached at C_1 and C_2 to rod $\mathcal{R}$ by ball and socket joints. At the instant shown, C_2 is moving away from the origin at speed $\sqrt{13}$ cm/s. Find the velocity of C_1 at the same instant. Can the angular velocity of $\mathcal{R}$ in $\mathcal{I}$ be found?

Solution

The velocity of C_2 is determined by (see Figure E6.11b):

$$\mathbf{v}_{C_2} = \sqrt{13}\hat{\mathbf{e}}_t = \sqrt{13}\left(\frac{2}{\sqrt{13}}\hat{\mathbf{i}} + \frac{3}{\sqrt{13}}\hat{\mathbf{j}}\right) = 2\hat{\mathbf{i}} + 3\hat{\mathbf{j}}$$

where

$$\frac{dy}{dx} = \frac{6x}{32}$$

so

$$\left.\frac{dy}{dx}\right|_{x=8} = \frac{3}{2}$$

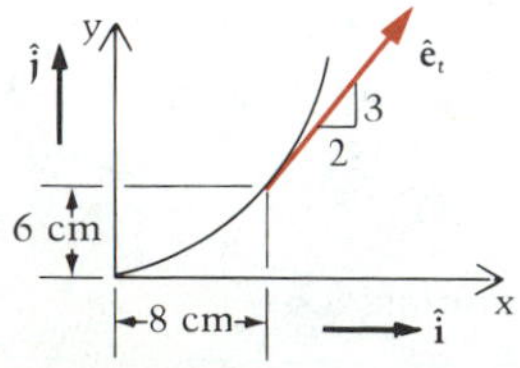

Figure E6.11b

Now we can calculate the velocities of C_1 and C_2 using Equation (6.56):

$$\overbrace{\mathbf{v}_{C_1}}^{v_{C_1}\hat{\mathbf{k}}} = \overbrace{\mathbf{v}_{C_2}}^{(2\hat{\mathbf{i}} + 3\hat{\mathbf{j}})} + \boldsymbol{\omega}_{\mathcal{R}/\mathcal{I}} \times \mathbf{r}_{C_2C_1} \tag{1}$$

We note that the component of the angular velocity $\boldsymbol{\omega}_{\mathcal{R}/\mathcal{I}}$ along the line C_1C_2 cannot be determined from the given information, because any value of it whatsoever will not affect Equation (1). However, dotting this equation with

$$\mathbf{r}_{C_2C_1} = -8\hat{\mathbf{i}} - 6\hat{\mathbf{j}} + 24\hat{\mathbf{k}} \text{ yields:}$$

$$24v_{C_1} = -8(2) - 6(3) + 0 = -34$$

so that

$$v_{C_1} = -1.42$$

and

$$\mathbf{v}_{C_1} = -1.42\hat{\mathbf{k}} \text{ cm/s}$$

Example 6.12

The cone $\mathcal{B}_1$ in Figures E6.12a,b rolls on the floor in such a way that the center Q of the base of the cone travels on a horizontal circle at constant velocity $v_Q\hat{\mathbf{j}}$. Let $\mathcal{B}_2$ denote an intermediate frame ("between" cone $\mathcal{B}_1$ and the ground $\mathcal{B}_3$) in which $\hat{\mathbf{i}}$, $\hat{\mathbf{j}}$, and $\hat{\mathbf{k}}$ are fixed. The unit vector $\hat{\mathbf{i}}$ is always directed along OQ, and $\hat{\mathbf{j}}$ is normal to the plane of $\hat{\mathbf{i}}$ and the contact line, in a direction parallel to $\mathbf{v}_Q$; finally, $\hat{\mathbf{k}} = \hat{\mathbf{i}} \times \hat{\mathbf{j}}$. Find the angular velocity of $\mathcal{B}_1$ in $\mathcal{B}_3$.

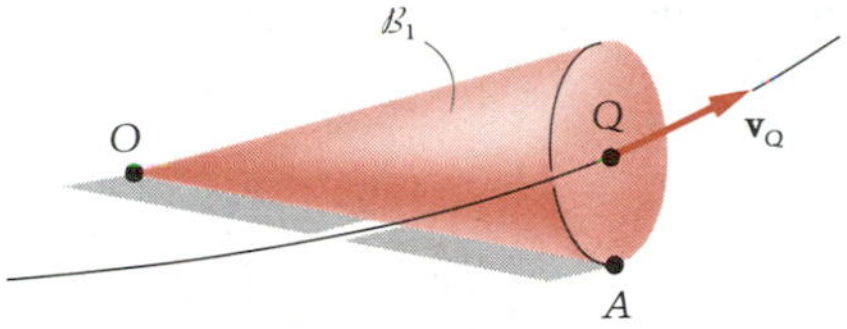

Figure E6.12a

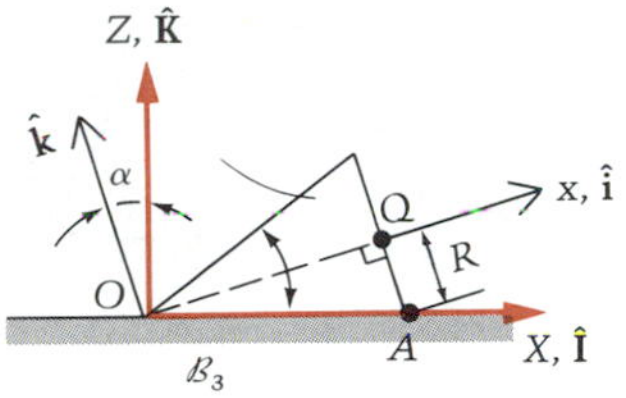

Figure E6.12b

Solution

We shall denote $\boldsymbol{\omega}_{\mathcal{B}_1/\mathcal{B}_3}$ by simply $\boldsymbol{\omega}_{1/3}$. Since $\mathbf{v}_A = \mathbf{v}_O = \mathbf{0}$, then their difference, $\boldsymbol{\omega}_{1/3} \times \mathbf{r}_{OA}$, must also vanish. This requires $\boldsymbol{\omega}_{1/3}$ to be parallel at all times to the line of contact of $\mathcal{B}_1$ with the ground. Thus

$$\boldsymbol{\omega}_{1/3} = \omega(-\cos\alpha\hat{\mathbf{i}} + \sin\alpha\hat{\mathbf{k}})$$

Next, using Equation (6.56),

$$\mathbf{v}_Q = \mathbf{v}_A + \boldsymbol{\omega}_{1/3} \times \mathbf{r}_{AQ}$$
$$v_Q\hat{\mathbf{j}} = \mathbf{0} + \omega(-\cos\alpha\hat{\mathbf{i}} + \sin\alpha\hat{\mathbf{k}}) \times R\hat{\mathbf{k}}$$

Therefore

$$v_Q = \omega R \cos\alpha$$

so that

$$\omega = v_Q/(R\cos\alpha)$$

Thus

$$\boldsymbol{\omega}_{1/3} = \frac{v_Q}{R}(-\hat{\mathbf{i}} + \tan\alpha\,\hat{\mathbf{k}})$$

We wish to make some further remarks about the preceding example. The addition theorem gives:

$$\boldsymbol{\omega}_{1/3} = \boldsymbol{\omega}_{1/2} + \boldsymbol{\omega}_{2/3} \tag{1}$$

where $\boldsymbol{\omega}_{2/3}$ is given by

$$\boldsymbol{\omega}_{2/3} = \underbrace{\left[\frac{v_Q}{\text{horizontal projection of } OQ}\right]}_{(R/\tan\alpha)\cos\alpha} \underbrace{(\sin\alpha\hat{\mathbf{i}} + \cos\alpha\hat{\mathbf{k}})}_{\hat{\mathbf{K}}}$$

Substituting $\boldsymbol{\omega}_{2/3}$ and $\boldsymbol{\omega}_{1/3}$ into the addition theorem equation (1) above, we find

$$\boldsymbol{\omega}_{1/2} = \frac{v_Q}{R\cos^2\alpha}(-\hat{\mathbf{i}})$$

Note the check on the direction $(-\hat{\mathbf{i}})$, since the only way $\mathcal{B}_1$ can move relative to $\mathcal{B}_2$ is to rotate around OQ.

Consider finally two ways of depicting the components of $\boldsymbol{\omega}_{1/3}$. From Figure 6.9a, note that the vector sum of the two components of $\boldsymbol{\omega}_{1/3}$ is parallel to the contact line since

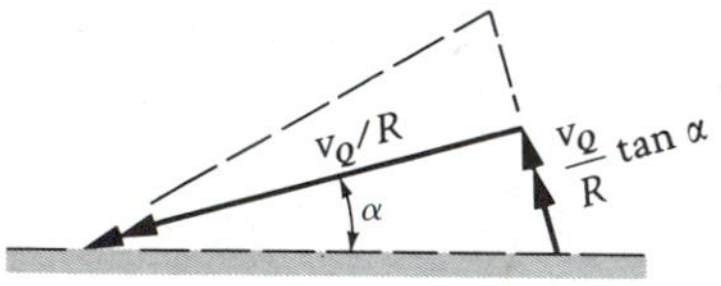

Figure 6.9a

$$\frac{\dfrac{v_Q}{R}\tan\alpha}{v_Q/R} = \tan\alpha$$

Figure 6.9b illustrates the addition theorem. Note that the "direction check" again results in $\boldsymbol{\omega}_{1/3}$ being along the contact line, this time because

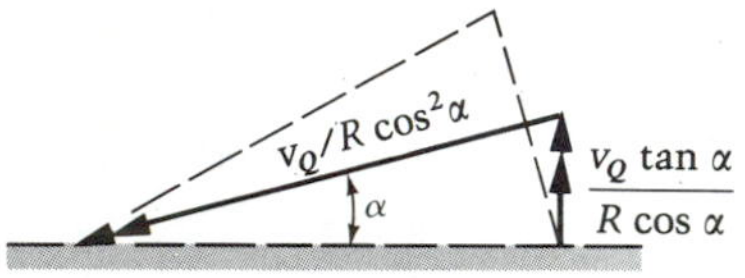

Figure 6.9b

$$\frac{\left(\dfrac{v_Q\tan\alpha}{R\cos\alpha}\right)}{\left(\dfrac{v_Q}{R\cos^2\alpha}\right)} = \sin\alpha$$

EXAMPLE 6.13

In Example 6.12, find the angular acceleration of $\mathcal{B}_1$ in $\mathcal{B}_3$.

Solution

Differentiating $\boldsymbol{\omega}_{1/3}$,

$$\boldsymbol{\alpha}_{1/3} = {}^{\mathcal{B}_3}\dot{\boldsymbol{\omega}}_{1/3} = {}^{\mathcal{B}_2}\dot{\boldsymbol{\omega}}_{1/3} + \boldsymbol{\omega}_{2/3} \times \boldsymbol{\omega}_{1/3}$$

We had

$$\boldsymbol{\omega}_{1/3} = \frac{v_Q}{R}(-\hat{\mathbf{i}} + \tan\alpha\hat{\mathbf{k}})$$

and

$$\boldsymbol{\omega}_{2/3} = \frac{v_Q}{R}(\tan^2\alpha\hat{\mathbf{i}} + \tan\alpha\hat{\mathbf{k}})$$

Thus

$${}^{\mathcal{B}_2}\dot{\boldsymbol{\omega}}_{1/3} = \mathbf{0}$$

and we obtain

$$\boldsymbol{\alpha}_{1/3} = \left(\frac{v_Q}{R}\right)^2 \begin{vmatrix} \hat{\mathbf{i}} & \hat{\mathbf{j}} & \hat{\mathbf{k}} \\ \tan^2\alpha & 0 & \tan\alpha \\ -1 & 0 & \tan\alpha \end{vmatrix}$$

$$= \frac{v_Q^2}{R^2}(-\hat{\mathbf{j}})(\tan^3\alpha + \tan\alpha)$$

or

$$\boldsymbol{\alpha}_{1/3} = \frac{\sin\alpha}{\cos^3\alpha}\frac{v_Q^2}{R^2}(-\hat{\mathbf{j}})$$

We see in the above example that acceleration equations need not be used to compute $\boldsymbol{\alpha}$ if $\boldsymbol{\omega}$ is known.

The final example in this section illustrates the workings of a complicated three-dimensional gear train.

Example 6.14

Very large alterations in speed along a given direction may be obtained by using the gear arrangement shown in the diagram. Gears $\mathcal{A}$, $\mathcal{B}$, and $\mathcal{D}$ all rotate about the x axis in $\mathcal{I}$, but $\mathcal{C}$ has a more complicated motion:

1. It rolls on the fixed (to $\mathcal{I}$) gear $\mathcal{G}$, currently contacting it at P.
2. It rolls on $\mathcal{A}$. (The contacting teeth are at A in Figure E6.14a.)
3. It rolls on $\mathcal{B}$ (at B in the figure).
4. It turns with respect to $\mathcal{D}$ about the line ℓ, which is fixed in both $\mathcal{C}$ and $\mathcal{D}$.

Considering $\mathcal{B}$ to be the driven gear, find the ratio of $\omega_{\mathcal{A}}$ to $\omega_{\mathcal{B}}$.

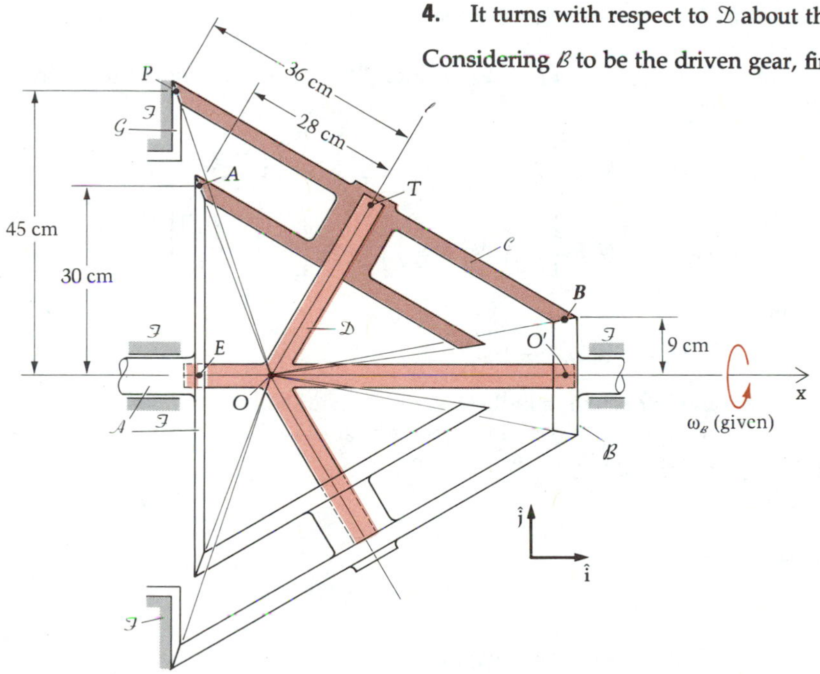

Figure E6.14a

Solution

The velocity of the contact points of $\mathcal{B}$ and $\mathcal{C}$ is, using body $\mathcal{B}$,

$$\mathbf{v}_B = \cancelto{0}{\mathbf{v}_{O'}} + \boldsymbol{\omega}_{\mathcal{B}/\mathcal{I}} \times \mathbf{r}_{O'B}$$

$$\mathbf{v}_B = \omega_{\mathcal{B}}\hat{\mathbf{i}} \times 9\hat{\mathbf{j}} = 9\omega_{\mathcal{B}}\hat{\mathbf{k}} \text{ cm/s} \tag{1}$$

in which we use just one subscript on ω when it is the angular velocity of a body with respect to $\mathcal{I}$.

Next we find another expression for $\mathbf{v}_B$, this time by relating the velocity of the tooth point of $\mathcal{C}$ at B to that of the point of $\mathcal{C}$ that contacts the reference frame ($\mathcal{G}$ is fixed to $\mathcal{I}$) at P.

$$\mathbf{v}_B = \cancelto{0}{\mathbf{v}_P} + \boldsymbol{\omega}_{\mathcal{C}} \times \mathbf{r}_{PB} \tag{2}$$

To obtain $\mathbf{r}_{PB}$, we use Figure E6.14b and see that

$$\phi = \cos^{-1}\left(\frac{36}{72}\right) = 60°$$

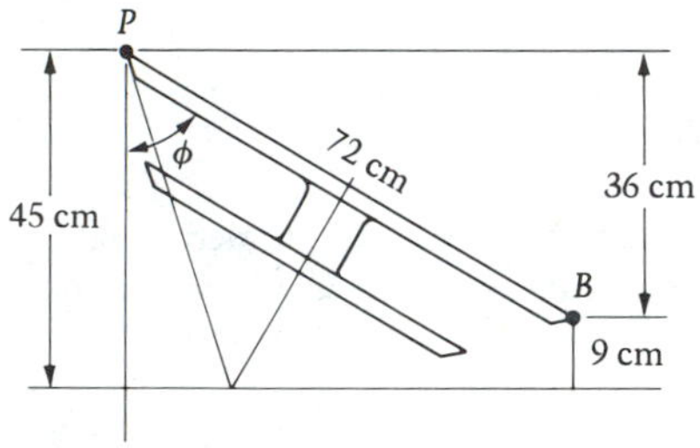

Figure E6.14b

Therefore

$$\mathbf{r}_{PB} = 72\left(\frac{\sqrt{3}}{2}\hat{\mathbf{i}} - \frac{1}{2}\hat{\mathbf{j}}\right) \text{ cm} \tag{3}$$

Also, by the addition theorem,

$$\boldsymbol{\omega}_{\mathcal{C}} = \boldsymbol{\omega}_{\mathcal{C}/\mathcal{I}} = \boldsymbol{\omega}_{\mathcal{C}/\mathcal{D}} + \boldsymbol{\omega}_{\mathcal{D}/\mathcal{I}}$$

$$= \omega_{\mathcal{C}/\mathcal{D}}\left(\frac{1}{2}\hat{\mathbf{i}} + \frac{\sqrt{3}}{2}\hat{\mathbf{j}}\right) + \omega_{\mathcal{D}}\hat{\mathbf{i}} \tag{4}$$

in which we have used the fact that we know the directions (but not the magnitudes yet) of $\boldsymbol{\omega}_{\mathcal{C}/\mathcal{D}}$ and $\boldsymbol{\omega}_{\mathcal{D}/\mathcal{I}}$.

Substituting Equations (3) and (4) into (2) and substituting the result into Equation (1) gives us

$$9\omega_{\mathcal{B}}\hat{\mathbf{k}} = \left[\omega_{\mathcal{C}/\mathcal{D}}\left(\frac{1}{2}\hat{\mathbf{i}} + \frac{\sqrt{3}}{2}\hat{\mathbf{j}}\right) + \omega_{\mathcal{D}}\hat{\mathbf{i}}\right] \times 72\left(\frac{\sqrt{3}}{2}\hat{\mathbf{i}} - \frac{1}{2}\hat{\mathbf{j}}\right)$$

$$= \left\{\omega_{\mathcal{C}/\mathcal{D}}\left[72\left(-\frac{1}{4} - \frac{3}{4}\right)\right] + \omega_{\mathcal{D}}\left[72\left(-\frac{1}{2}\right)\right]\right\}\hat{\mathbf{k}}$$

or, simplifying,

$$\omega_{\mathcal{B}} = -8\omega_{\mathcal{C}/\mathcal{D}} - 4\omega_{\mathcal{D}} \tag{5}$$

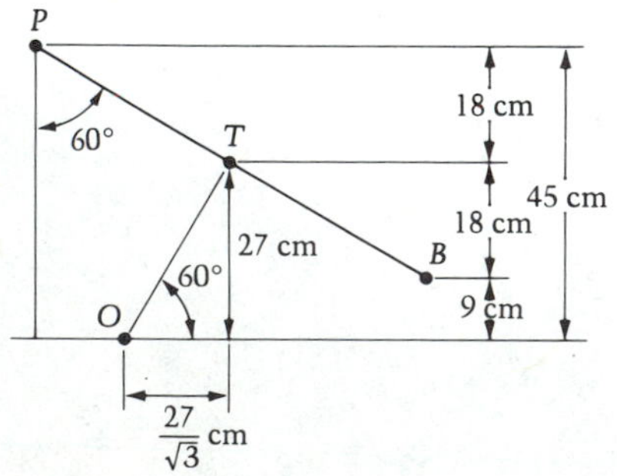

Figure E6.14c

To get another equation in these variables, we shall use the point T, which belongs to both $\mathcal{C}$ and $\mathcal{D}$ and is shown in Figure E6.14c along with some essential geometry. First, as a point of $\mathcal{D}$, we have

$$\mathbf{v}_T = \cancelto{0}{\mathbf{v}_O} + \boldsymbol{\omega}_{\mathcal{D}} \times \mathbf{r}_{OT}$$

$$= \omega_{\mathcal{D}}\hat{\mathbf{i}} \times \left(\frac{27}{\sqrt{3}}\hat{\mathbf{i}} + 27\hat{\mathbf{j}}\right)$$

$$= 27\omega_{\mathcal{D}}\hat{\mathbf{k}} \text{ cm/s} \tag{6}$$

Next, as a point of $\mathcal{C}$,

$$\mathbf{v}_T = \cancelto{0}{\mathbf{v}_P} + \boldsymbol{\omega}_{\mathcal{C}} \times \mathbf{r}_{PT} \tag{7}$$

The cross product in Equation (7) is one-half the cross product in Equation (2).

Question 6.9 Why is this so?

Therefore

$$\mathbf{v}_T = (-36\omega_{\mathcal{C}/\mathcal{D}} - 18\omega_{\mathcal{D}})\hat{\mathbf{k}} \tag{8}$$

Equating the right sides of Equations (6) and (8) gives

$$27\omega_{\mathcal{D}} = -36\omega_{\mathcal{C}/\mathcal{D}} - 18\omega_{\mathcal{D}}$$

or

$$5\omega_{\mathcal{D}} = -4\omega_{\mathcal{C}/\mathcal{D}} \tag{9}$$

Substituting Equation (9) into (5) leads to

$$\omega_{\mathcal{D}} = \frac{\omega_{\mathcal{B}}}{6} \text{ rad/s} \tag{10}$$

and

$$\omega_{\mathcal{C}/\mathcal{D}} = -\frac{5}{24}\,\omega_{\mathcal{B}} \text{ rad/s} \tag{11}$$

Substituting Equations (10) and (11) into (4) gives us $\boldsymbol{\omega}_{\mathcal{C}}$, which we shall need in the last step of the problem. The result is

$$\boldsymbol{\omega}_{\mathcal{C}} = \left(-\frac{5}{24}\,\omega_{\mathcal{B}}\right)\left(\frac{1}{2}\hat{\mathbf{i}} + \frac{\sqrt{3}}{2}\hat{\mathbf{j}}\right) + \left(\frac{\omega_{\mathcal{B}}}{6}\right)\hat{\mathbf{i}}$$

$$\boldsymbol{\omega}_{\mathcal{C}} = \frac{\omega_{\mathcal{B}}}{16}\hat{\mathbf{i}} - \frac{5\sqrt{3}}{48}\,\omega_{\mathcal{B}}\hat{\mathbf{j}} \tag{12}$$

We can now relate the velocities of the contacting points of bodies $\mathcal{C}$ and $\mathcal{A}$ at point A; first, using points A and E of $\mathcal{A}$, we get

$$\mathbf{v}_A = \cancelto{0}{\mathbf{v}_E} + \omega_{\mathcal{A}}\hat{\mathbf{i}} \times 30\hat{\mathbf{j}}$$

$$\mathbf{v}_A = 30\omega_{\mathcal{A}}\hat{\mathbf{k}} \text{ cm/s} \tag{13}$$

To get another expression for $\mathbf{v}_A$, we relate the velocities of the two points A and P on body $\mathcal{C}$:

$$\mathbf{v}_A = \cancelto{0}{\mathbf{v}_P} + \boldsymbol{\omega}_{\mathcal{C}} \times \mathbf{r}_{PA} \tag{14}$$

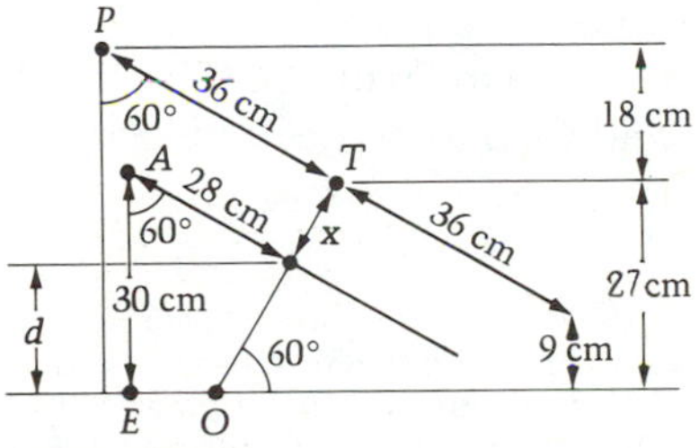

Figure E6.14d

To obtain the position vector $\mathbf{r}_{PA}$, we use the geometry shown in Figure E6.14d. The distance x, needed in forming $\mathbf{r}_{PA}$, is equal to $(27 - d)/\sin 60°$:

$$x = \frac{27 - d}{\sqrt{3}/2} = \frac{27 - [30 - 28(\frac{1}{2})]}{\sqrt{3}/2} = \frac{22}{\sqrt{3}} \tag{15}$$

Answer 6.9 Because $\mathbf{r}_{PT} = \mathbf{r}_{PB}/2$.

Therefore

$$\mathbf{r}_{PA} = (36 - 28)\left(\frac{\sqrt{3}}{2}\hat{\mathbf{i}} - \frac{1}{2}\hat{\mathbf{j}}\right) + \frac{22}{\sqrt{3}}\left(-\frac{1}{2}\hat{\mathbf{i}} - \frac{\sqrt{3}}{2}\hat{\mathbf{j}}\right)$$

$$= 0.577\hat{\mathbf{i}} - 15.0\hat{\mathbf{j}} \text{ cm} \qquad (16)$$

Substituting Equations (12) and (16) into (14), we get

$$\mathbf{v}_A = \omega_{\mathcal{B}}\left(\frac{1}{16}\hat{\mathbf{i}} - \frac{5\sqrt{3}}{48}\hat{\mathbf{j}}\right) \times (0.577\hat{\mathbf{i}} - 15.0\hat{\mathbf{j}})$$

$$\mathbf{v}_A = -0.833\omega_{\mathcal{B}}\hat{\mathbf{k}} \text{ cm/s} \qquad (17)$$

Equating the two expressions for $\mathbf{v}_A$ in Equations (13) and (17) gives

$$\omega_{\mathcal{A}} = -0.0278\omega_{\mathcal{B}} \text{ rad/s} \qquad (18)$$

It is seen that the angular speed of gear $\mathcal{B}$ is 36 times that of $\mathcal{A}$, and in the opposite direction.

Question 6.10 Give an argument why $\mathcal{A}$ has to be turning in the opposite direction from that of $\mathcal{B}$. (*Hint:* Use the original figure and focus your attention on points P and O of $\mathcal{C}$.)

Does An Instantaneous Axis of Rotation Exist in General?

We recall that in Chapter 3 we were able to show that in plane motion a rigid body $\mathcal{B}$, except when its angular velocity vanishes, always has a point of zero velocity (the instantaneous center Ⓘ, and hence a line of points of zero velocity exists which we may call the instantaneous axis of rotation). We now show that in general (three-dimensional) motion, such an axis does not always exist. We start with an arbitrary point P with velocity $\mathbf{v}_P$, and sketch its velocity along with the angular velocity vector $\boldsymbol{\omega}_{\mathcal{B}/\mathcal{I}} = \boldsymbol{\omega}$ of $\mathcal{B}$ in the reference frame $\mathcal{I}$. (See Figure 6.10(a).)

Note that in Figure 6.10a there is a plane $\mathcal{P}$ defined by the vectors $\mathbf{v}_P$ and $\boldsymbol{\omega}$ drawn through P, unless the two vectors are parallel. If they are, then the motion of $\mathcal{B}$ in $\mathcal{I}$ is like that of a screwdriver — the body turns around a line that translates along its axis. The general case ($\mathbf{v}_P$ not parallel to $\boldsymbol{\omega}$) may also be reduced to a screwdriver motion as follows. First we replace $\mathbf{v}_P$ by its components parallel ($v_{P\parallel}$) and perpendicular ($v_{P\perp}$) to $\boldsymbol{\omega}$. (See Figure 6.10b.) Next we consider a plane $\mathcal{R}$ parallel to $\mathcal{P}$ and separated from $\mathcal{P}$ by the distance d as shown in Figure 6.10(b). Point Q is the projection of P into the plane $\mathcal{R}$, and we may write its velocity in

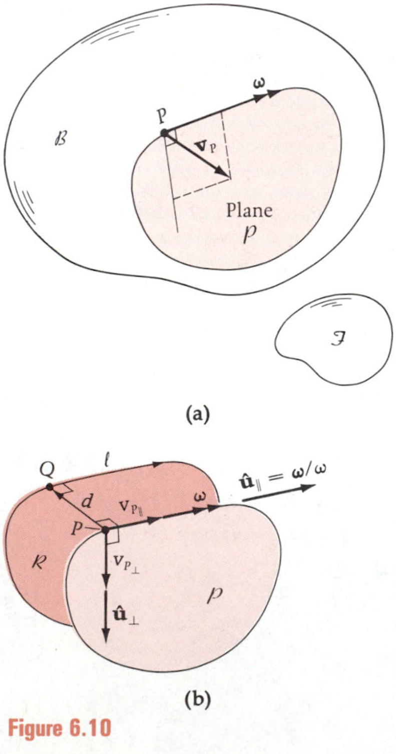

Figure 6.10

Answer 6.10 Each point of $\mathcal{C}$ (extended) which lies on line $\overline{PO}$ has zero velocity. Thus the velocity of B is its distance from $\overline{PO}$ times $|\boldsymbol{\omega}_{\mathcal{C}}|$, and the same is true for point A. The former is seen to be coming out of the paper, and the latter going into it, both about $\overline{PO}$. Therefore $\boldsymbol{\omega}_{\mathcal{A}}$, as determined by the direction of $\mathbf{v}_A$, is in the negative x direction, opposite to $\boldsymbol{\omega}_{\mathcal{B}}$.

of P by Equation (6.46):

$$\mathbf{v}_Q = \mathbf{v}_P + \boldsymbol{\omega} \times \mathbf{r}_{PQ}$$
$$= v_{P\|}\hat{\mathbf{u}}_{\|} + v_{P\perp}\hat{\mathbf{u}}_{\perp} = \boldsymbol{\omega} \times d\underbrace{(\hat{\mathbf{u}}_{\|} \times \hat{\mathbf{u}}_{\perp})}$$

Note from Figure 6.10b that this is the unit vector directed from P to Q.

The vector triple product is equal to

$$d\hat{\mathbf{u}}_{\|}(\boldsymbol{\omega} \cdot \hat{\mathbf{u}}_{\perp}) - d\hat{\mathbf{u}}_{\perp}(\boldsymbol{\omega} \cdot \hat{\mathbf{u}}_{\|}) = \mathbf{0} - d\hat{\mathbf{u}}_{\perp}(\omega)$$

so that

$$\mathbf{v}_Q = v_{P\|}\hat{\mathbf{u}}_{\|} + \hat{\mathbf{u}}_{\perp}(v_{P\perp} - d\omega)$$

Therefore if d is chosen equal to $v_{P\perp}/\omega$, the line ℓ will be a "screwdriver line" and the motion of $\mathcal{B}$ will be as shown in Figure 6.10(c).

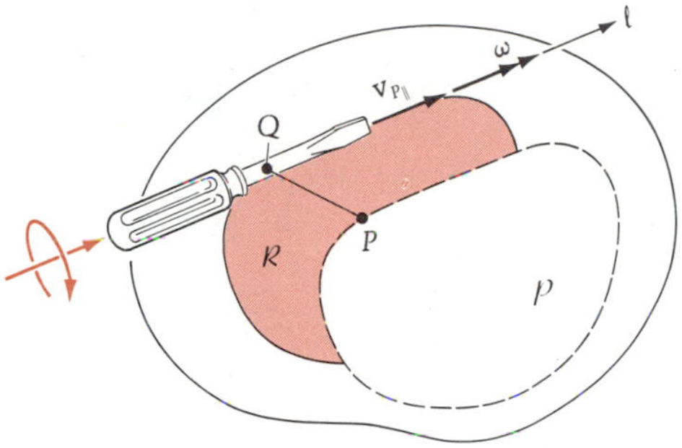

Figure 6.10(c)

All points on ℓ have velocities *along* ℓ at the given instant, while those off the line have the same velocity component parallel to ℓ; in addition, they rotate around it. This is the simplest reduction possible for the motion of $\mathcal{B}$, and it is clear that unless $v_{P\|} = 0$, *no* points can have zero velocity. Thus in three dimensions we are no longer assured of having an instantaneous axis as we were in dealing with plane motion.

There are, however, special cases in three dimensions in which an instantaneous axis exists; a good example is a cone rolling on a plane (Figure 6.11). Note that in this case the *entire line* of contact is at each

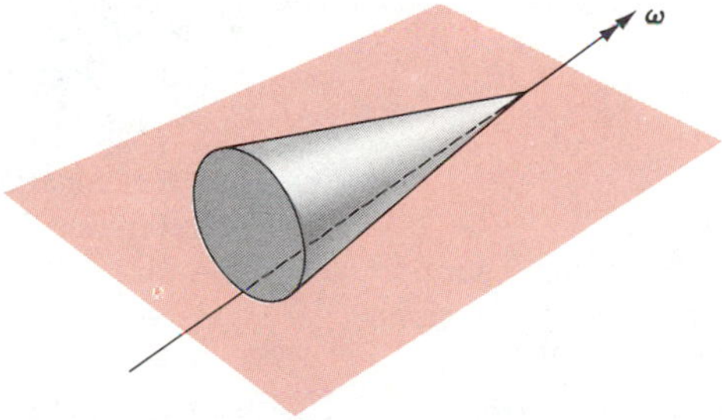

Figure 6.11

instant at rest on the plane. Since the angular velocity of the cone is parallel to this line, *all* the contact points have $v_{P\|} = 0$, which we have shown must be true if the instantaneous axis is to exist in three dimensions.

PROBLEMS ▶ Section 6.7

6.49 A youngster finds an old wagon wheel $\mathcal{W}$ and pushes it around with the center C moving at constant speed in a horizontal circle. (See Figure P6.49(a).) If the end O of the axle stays fixed while C returns to its starting point in T seconds, compute the angular velocity vector of the wheel $\boldsymbol{\omega}_{\mathcal{W}/\mathcal{I}}$, where $\mathcal{I}$ is the ground frame (Figure P6.49(b)). Give the result in terms of b, β, and T.

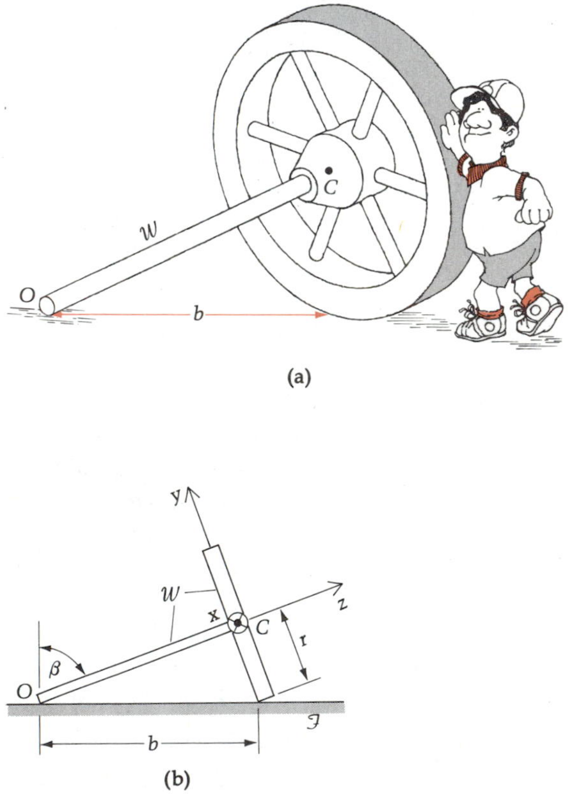

Figure P6.49

6.50 The bevel gears $\mathcal{B}_1$ and $\mathcal{B}_2$ in Figure P6.50 support the turning shaft $\mathcal{S}$, whose angular velocity is given by $\boldsymbol{\omega}_{\mathcal{S}/\mathcal{I}} = 30\hat{\mathbf{k}}$ rad/sec, in which $\hat{\mathbf{k}}$ is parallel to the z direction in both frames $\mathcal{I}$ and $\mathcal{S}$. (Gears $\mathcal{A}$ and $\mathcal{C}$ are seen to be part of $\mathcal{I}$.) Find $\boldsymbol{\omega}_{\mathcal{B}_1/\mathcal{I}}$ and $\boldsymbol{\omega}_{\mathcal{B}_2/\mathcal{I}}$.

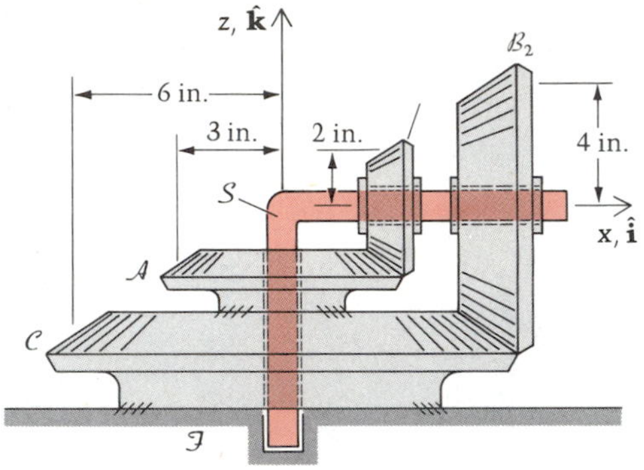

Figure P6.50

6.51 In the preceding problem, find $\boldsymbol{\alpha}_{\mathcal{B}_1/\mathcal{I}}$ and $\boldsymbol{\alpha}_{\mathcal{B}_2/\mathcal{I}}$.

6.52 Find the angular acceleration of the wagon wheel of Problem 6.49.

6.53 Using the data of Problems 6.25 and 6.26, find the velocity and acceleration in $\mathcal{G}$ of the point Q at the end of board $\mathcal{B}$ (beneath the little girl) having position vector $\mathbf{r}_{OQ} = 5\hat{\mathbf{j}}$ ft. Do this using Equations (6.56, 6.57) of this section, then check your answers using Equations (6.44, 6.51).

6.54 The angular velocity of a rigid body $\mathcal{B}$, in motion in frame $\mathcal{I}$, is $\boldsymbol{\omega} = 3\hat{\mathbf{i}} + 2\hat{\mathbf{j}}$ rad/sec. If possible, locate (from P) a point Q of $\mathcal{B}$ with zero velocity when:

- **a.** $\mathbf{v}_{P/\mathcal{I}} = 5\hat{\mathbf{i}}$ in./sec
- **b.** $\mathbf{v}_{P/\mathcal{I}} = 6\hat{\mathbf{i}} - 9\hat{\mathbf{j}} + 3\hat{\mathbf{k}}$ in./sec

In both cases explain why Q exists or not in light of the discussion about points of zero velocity at the end of the preceding section.

6.55 Disk $\mathcal{D}$ in Figure P6.55 spins relative to the bent shaft $\mathcal{B}$ at constant angular speed Ω_1 rad/sec; $\mathcal{B}$ rotates in the reference frame $\mathcal{I}$ at the constant rate Ω_2 rad/sec. (The directions are indicated in the figures.) Using the rigid-body equations (6.56 and 6.57), find $\mathbf{v}_{Q/\mathcal{I}}$ and $\mathbf{a}_{Q/\mathcal{I}}$ for point Q on the periphery of the disk. Express the result in terms of components along the (x, y, z) axes, which are fixed in $\mathcal{B}$.

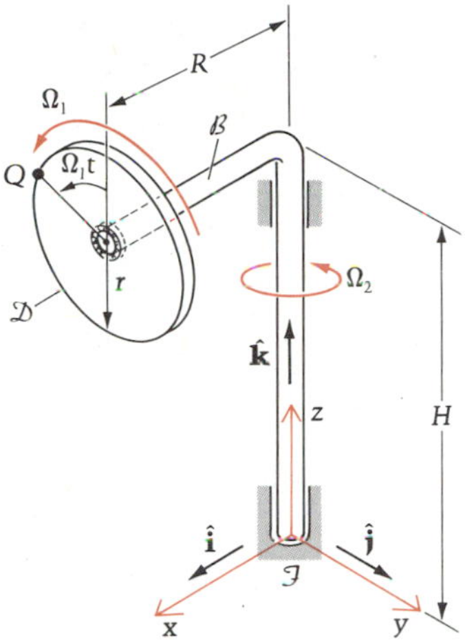

Figure P6.55

6.56 Rework the preceding problem, this time using the moving-frame concept of Section 6.5. Let bar $\mathcal{B}$ be the frame in motion with respect to $\mathcal{I}$, and let Q be the point moving relative to both $\mathcal{B}$ and $\mathcal{I}$.

6.57 A wheel of radius r turns on an axle that rotates with angular velocity $\omega_1\hat{\mathbf{k}}$ about a vertical axis (z) fixed relative to ground (Figure P6.57). If the wheel rolls on the horizontal plane and ω_1 is constant, find:

a. The angular velocity and angular acceleration of the wheel relative to the ground

b. The acceleration, relative to the ground, of the point on the wheel in contact with the horizontal plane.

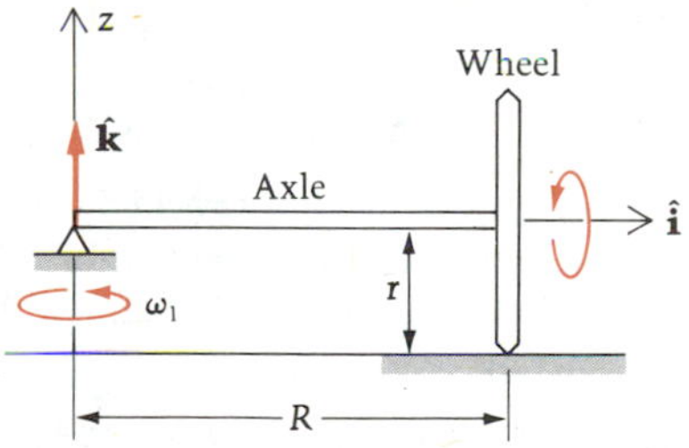

Figure P6.57

6.58 Rework Example 6.14, but this time suppose that the radius of gear $\mathcal{A}$ is 44 cm instead of 30 cm (and that its center is at the same point of $\mathcal{I}$). Explain why gears $\mathcal{A}$ and $\mathcal{B}$ are now moving in the same direction.

6.59 The bevel gear $\mathcal{A}$ in Figure P6.59 is fixed to a reference frame in which the mating gear $\mathcal{B}$ moves. The axis OC of $\mathcal{B}$ turns about the z axis at the constant rate $\Omega = 0.2$ rad/sec, and the angle OQC is 30°. Find the angular velocity of $\mathcal{B}$ in $\mathcal{A}$.

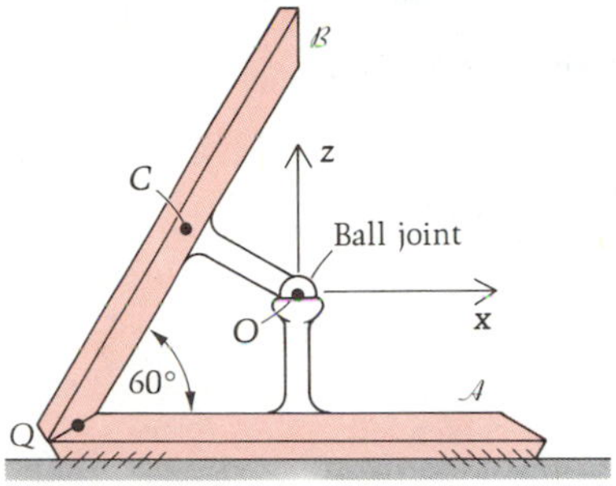

Figure P6.59

6.60 In the preceding problem, find the angular acceleration of $\mathcal{B}$ in $\mathcal{A}$ for the same defined motion.

6.61 A differential friction gear can be made with either bevel gears, as shown at the top of Figure P6.61, or friction disks, as shown at the bottom. In each case, body $\mathcal{D}$ rolls on $\mathcal{A}$ and $\mathcal{B}$ and may turn without resistance on the crank arm $\mathcal{C}$. Find $\boldsymbol{\omega}_{\mathcal{C}/\mathcal{I}}$, $\boldsymbol{\omega}_{\mathcal{D}/\mathcal{C}}$, and the velocities of points A and B of $\mathcal{D}$.

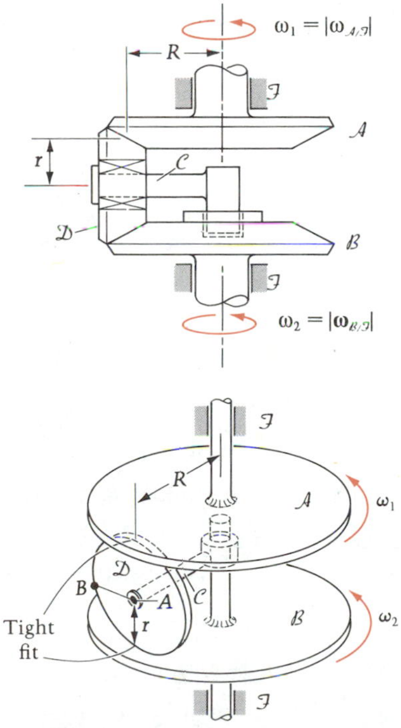

Figure P6.61

6.62 The uniform, solid, right circular cone $\mathcal{C}$ in Figure P6.62 rolls on the horizontal plane $\mathcal{I}$. Let $\mathcal{P}$ represent a frame in which the vertical axis z and the cone's axis ℓ are fixed. (Hence $\mathcal{P}$ has a simple angular velocity in $\mathcal{I}$ about the vertical.) Show that the cone can roll so that $|\boldsymbol{\omega}_{\mathcal{C}/\mathcal{I}}| = \omega$ and $|\boldsymbol{\omega}_{\mathcal{P}/\mathcal{I}}| = n$ are constants and

$$\omega \sin \alpha = n \cos \alpha$$

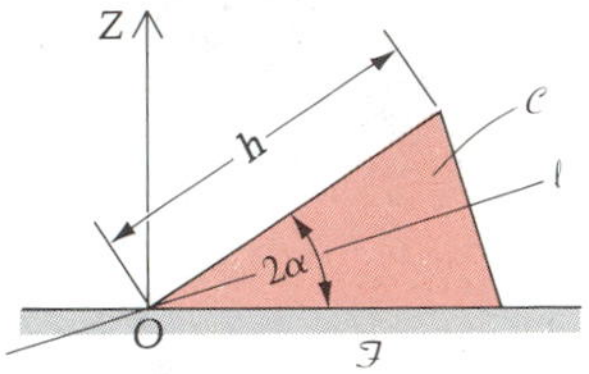

Figure P6.62

6.63 The center C of the bevel gear $\mathcal{C}$ in Figure P6.63 rotates in a horizontal circle at a constant speed of 40 mm/s (clockwise when viewed from above). The mating gear $\mathcal{D}$ is fixed to the reference frame $\mathcal{I}$; shaft $\mathcal{S}$, rigidly attached to $\mathcal{C}$, is connected to $\mathcal{I}$ through a ball and socket joint at O. Find the angular velocity vector of $\mathcal{C}$ in $\mathcal{I}$.

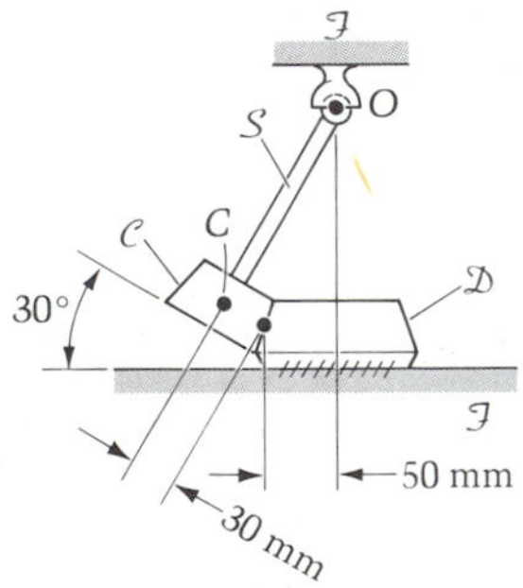

Figure P6.63

6.64 Find $\boldsymbol{\alpha}_{\mathcal{C}/\mathcal{I}}$ for the gear in Problem 6.63.

* **6.65** Plate $\mathcal{P}$ in Figure P6.65 has the following motion:

1. Corner A moves on the x axis.
2. Corner B moves on the y axis with constant velocity $6\hat{\mathbf{j}}$ in./sec.
3. Some point of the top edge of $\mathcal{P}$ (point Q at the instant shown) is always in contact with the z axis.

Find the angular velocity vector of the plate when $x_A = 3$ in.

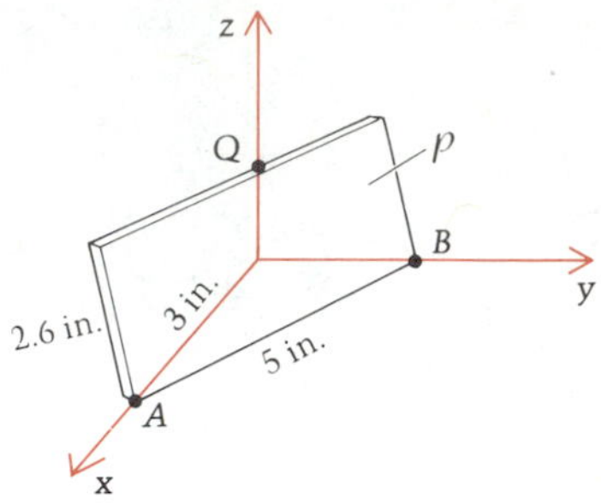

Figure P6.65

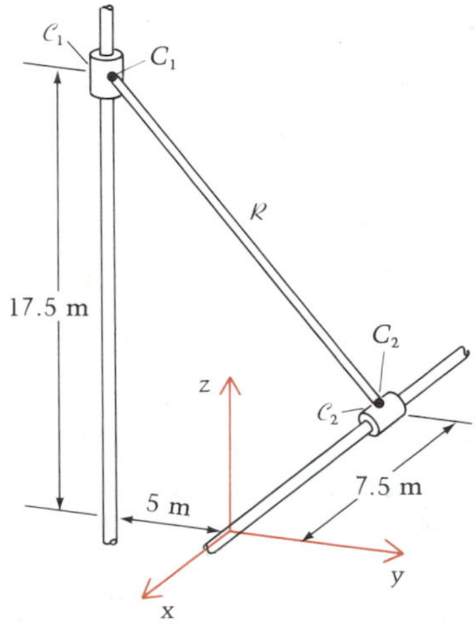

Figure P6.69

6.66 In Example 6.14 find the radius of gear $\mathcal{A}$ for which it will remain stationary in $\mathcal{I}$ as $\mathcal{B}$, $\mathcal{C}$, and $\mathcal{D}$ turn.

6.67 In Example 6.14 label the radius of gear $\mathcal{A}$ as H and call the radius of the 28-cm gear R. Show that the relationship between $\omega_{\mathcal{A}}$ and $\omega_{\mathcal{B}}$ is given by $96\omega_{\mathcal{A}}H = (16H - 20R)\omega_{\mathcal{B}}$. (The center of $\mathcal{A}$ is fixed in $\mathcal{I}$.)

6.68 If the speed of point C_2 is constant in Example 6.11, find the acceleration of C_1 at the instant given.

6.69 Collars $\mathcal{C}_1$ and $\mathcal{C}_2$ in Figure P6.69 are attached at C_1 and C_2 to rod $\mathcal{R}$ by ball and socket joints. Point C_2 has a motion along the x axis given by $x_2 = -0.012t^3$ m. Find the velocity of C_1 as a function of time.

* **6.70** Cone $\mathcal{C}_1$ rolls on cone $\mathcal{C}_2$ so that its axis of symmetry (x) moves in a horizontal plane through O, turning about z at rate Ω_2 rad/sec. (See Figure P6.70.) Cone $\mathcal{C}_2$ is rotating about $(-z)$ at Ω_1 rad/sec. Find, with respect to the frame $\mathcal{I}$ in which $\mathcal{C}_2$ turns, the angular velocity and angular acceleration of $\mathcal{C}_1$, and the acceleration of point A. (Axes (x, y, z) are fixed in frame $\mathcal{B}$, which turns so that x is always along the axis of symmetry of $\mathcal{C}_1$ and z is always vertical. Also, Ω_1 and Ω_2 are constants.)

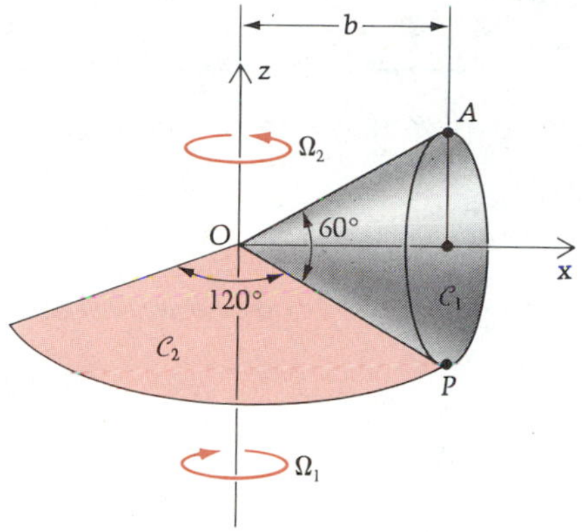

Figure P6.70

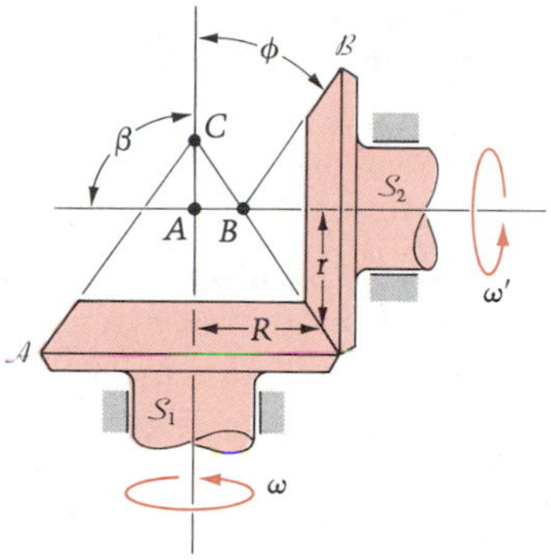

Figure P6.71

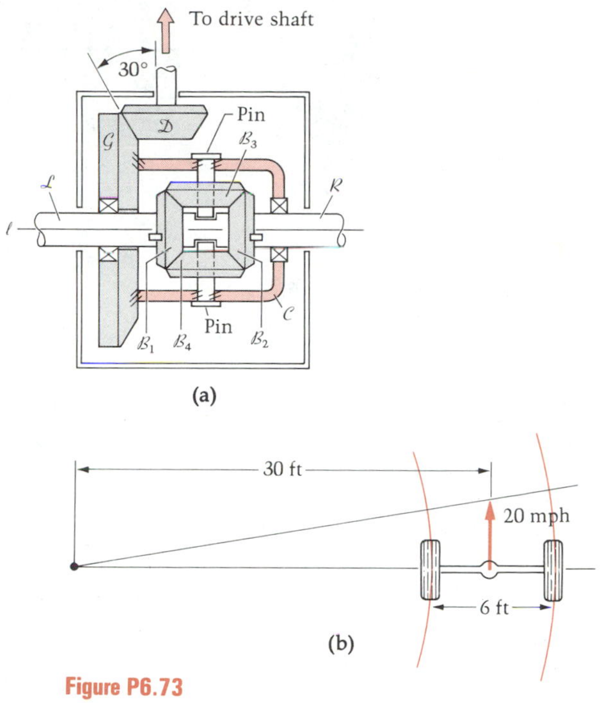

Figure P6.73

6.71 The two shafts $\mathcal{S}_1$ and $\mathcal{S}_2$ are fixed to bevel gears $\mathcal{A}$ and $\mathcal{B}$ as shown in Figure P6.71. (a) Prove that if the velocities of each pair of contacting points are to match along the line of contact of the gears, then points A, B, and C must coincide. (b) Let A, B, and C coincide and find the ratio of ω to ω'.

6.72 (a) In the preceding problem show that if $0 < \beta < 90°$, the result is still true about A, B, and C coinciding (b) Find ω/ω' for this case.

6.73 Depicted in Figure P6.73(a) are the main features of an automobile differential. The left and right axles, $\mathcal{L}$ and $\mathcal{R}$, are keyed to the bevel gears $\mathcal{B}_1$ and $\mathcal{B}_2$. Gear $\mathcal{G}$ is fixed to the case $\mathcal{C}_1$ and the combination is free to turn in bearings around line ℓ. Gear $\mathcal{G}$ meshes with gear $\mathcal{D}$ attached to the car's drive shaft. As the casing turns about the common axis ℓ of $\mathcal{L}$ and $\mathcal{R}$, its pins bear against the other two bevel gears within $\mathcal{C}$, which are $\mathcal{B}_3$ and $\mathcal{B}_4$. (Observe that these two gears do not turn about their axes at all on a straight road.)

Suppose the car makes a 30-ft turn at a speed of 20 mph (Figure P6.73(b)). If the tire radius is 14 in., find the angular velocities of $\mathcal{L}$ and $\mathcal{R}$ and use them to compute the angular velocity of $\mathcal{D}$. In the process note that the differential allows the driven wheels to turn at different angular speeds.

It is assumed in the following three problems that not all points of body $\mathcal{B}$ have zero acceleration.

6.74 Show that for a rigid body $\mathcal{B}$ in general motion, there is a point Q of zero acceleration if $\boldsymbol{\omega} \neq \mathbf{0}, \boldsymbol{\alpha} \neq \mathbf{0}$, and $\boldsymbol{\alpha}$ is not parallel to $\boldsymbol{\omega}$. *Hint:* Let P be an arbitrary point, and let $\boldsymbol{\omega} = \omega\hat{\mathbf{i}}$, $\boldsymbol{\alpha} = \alpha_1\hat{\mathbf{i}} + \alpha_2\hat{\mathbf{j}}$, and $\mathbf{a}_P = a_{P_1}\hat{\mathbf{i}} + a_{P_2}\hat{\mathbf{j}} + a_{P_3}\hat{\mathbf{k}}$, noting that there is no loss in generality in these assumptions. Set $\mathbf{a}_Q = \mathbf{0} = \mathbf{a}_P + \boldsymbol{\alpha} \times \mathbf{r}_{PQ} + \boldsymbol{\omega} \times (\boldsymbol{\omega} \times \mathbf{r}_{PQ})$, set $\mathbf{r}_{PQ} = x\hat{\mathbf{i}} + y\hat{\mathbf{j}} + z\hat{\mathbf{k}}$, and solve for x, y, and z.

6.75 (a) Following up the previous problem, show that if $\boldsymbol{\omega} = \mathbf{0}$ and $\boldsymbol{\alpha} \neq \mathbf{0}$ at a given instant, then at this instant there is a point Q of zero acceleration if and only if the accelerations of all points of B are perpendicular to $\boldsymbol{\alpha}$. (b) Investigate the case $\boldsymbol{\omega} \neq \mathbf{0}$ and $\boldsymbol{\alpha} = \mathbf{0}$.

6.76 Here is another follow-up on Problem 6.74: Show that at any instant when the two vectors $\boldsymbol{\omega}$ and $\boldsymbol{\alpha}$ are parallel, there is a point of $\mathcal{B}$ with zero acceleration if and only if the accelerations of all its points are perpendicular to $\boldsymbol{\omega}$ and $\boldsymbol{\alpha}$.

* **6.77** Find the angular acceleration of the plate of Problem 6.65 at the same instant of time.

6.8 Describing the Orientation of a Rigid Body

In the case of plane motion of a rigid body $\mathcal{B}$, if we know the angular velocity

$$\omega = \omega\hat{\mathbf{k}} = \dot{\theta}\hat{\mathbf{k}}$$

as a function of time, we may clearly integrate to find the orientation of $\mathcal{B}$ at any time t:

$$\int_0^t \dot{\theta}(\xi)\, d\xi = \theta(t) - \theta(0)$$

Thus in plane motion we may completely specify the position of $\mathcal{B}$ by giving the xy coordinates of a point (usually the mass center C is chosen to be the point) and the orientation angle θ.

The Eulerian Angles

A major difference between planar and general motion is that the angles which yield a body's orientation in space in three-dimensional motion are *not* the integrals by simple quadratures of the angular velocity components of the body. In fact, finding (in general) the orientation of a body $\mathcal{B}$ in closed form, given $\omega(t)$ and the orientation of $\mathcal{B}$ at $t = 0$, is an unsolved fundamental problem in rigid-body kinematics. We now introduce the **Eulerian angles** in order to show the difficulty of determining a body's orientation in space when the motion is nonplanar.

We begin with the body $\mathcal{B}$ oriented so that the body-fixed axes (x, y, z) initially coincide respectively with axes (X, Y, Z) embedded in the reference frame $\mathcal{I}$. Let $(\hat{\mathbf{i}}, \hat{\mathbf{j}}, \hat{\mathbf{k}})$ and $(\hat{\mathbf{I}}, \hat{\mathbf{J}}, \hat{\mathbf{K}})$ be sets of unit vectors respectively parallel to (x, y, z) and (X, Y, Z). Three successive rotations about specific axes will now be described that will orient $\mathcal{B}$ in $\mathcal{I}$. (See Figure 6.12(a))

In Figure 6.12(b) the first rotation is through the angle ϕ about the Z axis. Let the new positions of (x, y, z) after this first rotation be denoted by (x_1, y_1, z_1) as shown; these positions are embedded in an intermediate frame $\mathcal{I}_1$. Note that axes Z and z_1 are identical and that $\mathcal{I}_1$ has the simple angular velocity $\dot{\phi}\hat{\mathbf{K}}$ in $\mathcal{I}$. Note also from Figure 6.12 that $\hat{\mathbf{n}}_{11}$, $\hat{\mathbf{n}}_{12}$, and $\hat{\mathbf{n}}_{13}$ are unit vectors that are respectively and always parallel to x_1, y_1, z_1. Next (Figure 6.12(c)) a rotation through the angle θ about axis y_1 moves the body axes into the coordinate directions (x_2, y_2, z_2) of a second intermediate frame $\mathcal{I}_2$ having unit vectors $(\hat{\mathbf{n}}_{21}, \hat{\mathbf{n}}_{22}, \hat{\mathbf{n}}_{23})$. A final rotation, this time of amount ψ about z_2 (Figure 6.12(d)), turns the body axes into their final positions in $\mathcal{B}$, indicated by (x, y, z).

It is clear that we may use the addition theorem to express the angular velocity of $\mathcal{B}$ in $\mathcal{I}$ as follows:

$$\omega_{\mathcal{B}/\mathcal{I}} = \omega_{\mathcal{B}/\mathcal{I}_2} + \omega_{\mathcal{I}_2/\mathcal{I}_1} + \omega_{\mathcal{I}_1/\mathcal{I}}$$

$$\omega_{\mathcal{B}/\mathcal{I}} = \dot{\psi}\hat{\mathbf{k}} + \dot{\theta}\hat{\mathbf{n}}_{12} + \dot{\phi}\hat{\mathbf{K}} \qquad (6.58)$$

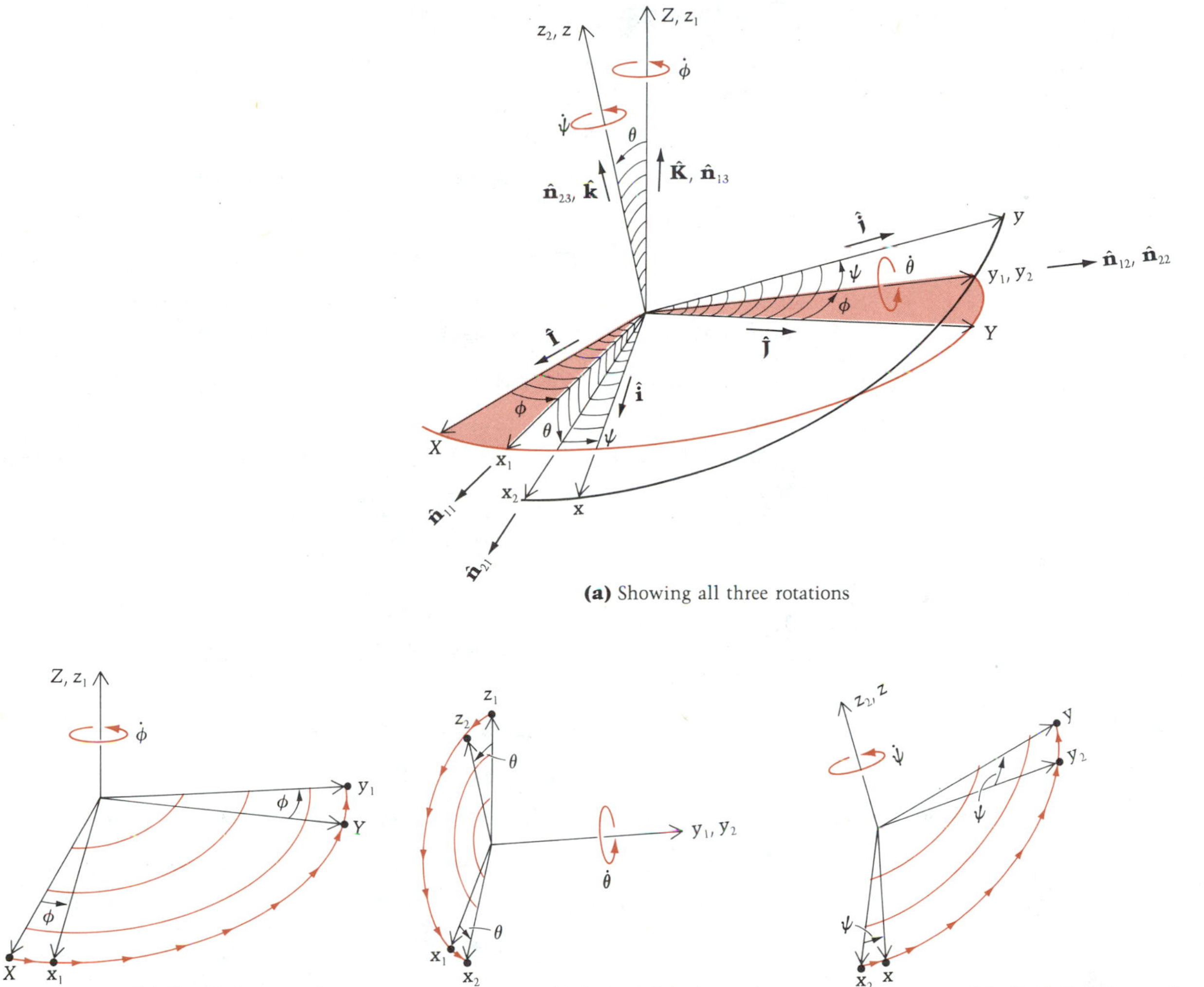

(a) Showing all three rotations

(b) First Eulerian angle rotation (ϕ), about Z

(c) Second Eulerian angle rotation (θ), about y_1

(d) Third Eulerian angle rotation (ψ), about z_2

Figure 6.12 Eulerian angles.

in which we again remark that $\hat{\mathbf{n}}_{ij}$ is a unit vector in the jth coordinate direction of frame $\mathcal{F}_i (i = 1, 2)$.

If the preceding expression for $\boldsymbol{\omega}_{\mathcal{B}/\mathcal{I}}$ is to be functionally useful, its three terms should all be expressed in the *same frame* — that is, in terms of components associated with the directions of base vectors all fixed in direction in a common frame. Now in practice, as we shall see, they are sometimes expressed by components associated with body-fixed directions, at other times with space-fixed directions, and at still other times with directions fixed in intermediate frames such as $\mathcal{F}_1$ or $\mathcal{F}_2$. For example, let us write $\boldsymbol{\omega}_{\mathcal{B}/\mathcal{I}}$ in body components. This means, looking at Equation (6.58), that we must express $\hat{\mathbf{n}}_{12}$ and $\hat{\mathbf{K}}$ in terms of $\hat{\mathbf{i}}$, $\hat{\mathbf{j}}$, and $\hat{\mathbf{k}}$.

We see from Figure 6.12 that

$$\hat{\mathbf{n}}_{12} = \sin\psi\hat{\mathbf{i}} + \cos\psi\hat{\mathbf{j}} \tag{6.59}$$

and that

$$\begin{aligned}\hat{\mathbf{K}} &= -\sin\theta\hat{\mathbf{n}}_{21} + \cos\theta\hat{\mathbf{k}} \\ &= -\sin\theta(\cos\psi\hat{\mathbf{i}} - \sin\psi\hat{\mathbf{j}}) + \cos\theta\hat{\mathbf{k}}\end{aligned}$$

or

$$\hat{\mathbf{K}} = -s_\theta c_\psi\hat{\mathbf{i}} + s_\theta s_\psi\hat{\mathbf{j}} + c_\theta\hat{\mathbf{k}} \tag{6.60}$$

where $s_\theta = \sin\theta$, $c_\psi = \cos\psi$, and so forth. Substituting these expressions into Equation (6.58) then gives $\boldsymbol{\omega}_{\mathcal{B}/\mathcal{I}}$ in body components:

$$\boldsymbol{\omega}_{\mathcal{B}/\mathcal{I}} = (s_\psi\dot{\theta} - s_\theta c_\psi\dot{\phi})\hat{\mathbf{i}} + (c_\psi\dot{\theta} + s_\theta s_\psi\dot{\phi})\hat{\mathbf{j}} + (\dot{\psi} + c_\theta\dot{\phi})\hat{\mathbf{k}} \tag{6.61}$$

Alternatively, we may express $\boldsymbol{\omega}_{\mathcal{B}/\mathcal{I}}$ in terms of its components in frame $\mathcal{I}$ by writing $\hat{\mathbf{k}}$ and $\hat{\mathbf{n}}_{12}$ in terms of $\hat{\mathbf{I}}$, $\hat{\mathbf{J}}$, and $\hat{\mathbf{K}}$. Again referring to the figure, we see that

$$\hat{\mathbf{n}}_{12} = -s_\phi\hat{\mathbf{I}} + c_\phi\hat{\mathbf{J}} \tag{6.62}$$

and

$$\begin{aligned}\hat{\mathbf{k}} &= \sin\theta\hat{\mathbf{n}}_{11} + \cos\theta\hat{\mathbf{K}} \\ &= \sin\theta(\cos\phi\hat{\mathbf{I}} + \sin\phi\hat{\mathbf{J}}) + \cos\theta\hat{\mathbf{K}}\end{aligned}$$

or

$$\hat{\mathbf{k}} = s_\theta c_\phi\hat{\mathbf{I}} + s_\theta s_\phi\hat{\mathbf{J}} + c_\theta\hat{\mathbf{K}} \tag{6.63}$$

so that, substituting into Equation (6.58), we have $\boldsymbol{\omega}_{\mathcal{B}/\mathcal{I}}$ written in $\mathcal{I}$:

$$\boldsymbol{\omega}_{\mathcal{B}/\mathcal{I}} = (-s_\phi\dot{\theta} + s_\theta c_\phi\dot{\psi})\hat{\mathbf{I}} + (c_\phi\dot{\theta} + s_\theta s_\phi\dot{\psi})\hat{\mathbf{J}} + (\dot{\phi} + c_\theta\dot{\psi})\hat{\mathbf{K}} \tag{6.64}$$

Finally, we notice that expressing $\boldsymbol{\omega}_{\mathcal{B}/\mathcal{I}}$ in the *intermediate* frames is easier still. To write it in terms of its components in $\mathcal{I}_1$, we note that

$$\hat{\mathbf{K}} \equiv \hat{\mathbf{n}}_{13} \tag{6.65}$$

and

$$\hat{\mathbf{k}} = c_\theta\hat{\mathbf{n}}_{13} + s_\theta\hat{\mathbf{n}}_{11} \tag{6.66}$$

so that from Equation (6.58) we get

$$\boldsymbol{\omega}_{\mathcal{B}/\mathcal{I}} = s_\theta\dot{\psi}\hat{\mathbf{n}}_{11} + \dot{\theta}\hat{\mathbf{n}}_{12} + (c_\theta\dot{\psi} + \dot{\phi})\hat{\mathbf{n}}_{13} \tag{6.67}$$

All these expressions for $\boldsymbol{\omega}_{\mathcal{B}/\mathcal{I}}$ appear different, but of course they all represent the *same vector* written in different frames. The components may vary from frame to frame, but the vector is the same. In the first exercise at the end of this section, the reader will be asked to write $\boldsymbol{\omega}_{\mathcal{B}/\mathcal{I}}$ in yet another frame, $\mathcal{I}_2$.

The angles (ϕ, θ, ψ) are known as the Eulerian angles. They represent one way of orientating a rigid body in space. Unfortunately the Eulerian angles (ϕ, θ, ψ) do not carry the same symbol from one book to the next; worse still, the order and even the directions of the rotations vary from writer to writer. Obviously, then, it is important to choose a set to work with and then be consistent.

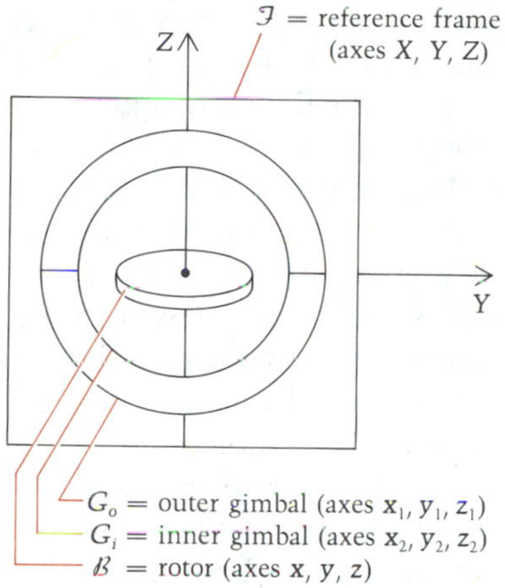

Figure 6.13

A physical feel for the Eulerian angles may be gained by considering a gyroscope $\mathcal{B}$ spinning in a Cardan suspension as shown in Figure 6.13. In this system the Eulerian angles (ϕ, θ, ψ) may be used as follows to pinpoint the orientation of the rotor $\mathcal{B}$ in space:

1. First Rotation: With respect to frame $\mathcal{J}$, we rotate the plane of the outer gimbal G_o (frame $\mathcal{J}_1$ in the earlier theory) about axis Z through angle ϕ. Axis X is turned into x_1 and axis Y into y_1; frames (bodies) G_i and $\mathcal{B}$ are not shown yet, but they move rigidly with G_o in this first rotation. (See Figure 6.14.)

2. Second Rotation: Next we turn the plane of the inner gimbal $G_i(\mathcal{J}_2)$ about axis y_1, through angle θ, thereby tilting G_i with respect to G_o. Axis z_1 is thereby rotated into z_2 and axis x_1 into x_2. Body $\mathcal{B}$ (not shown in Figure 6.15) goes along for the ride.

3. Third Rotation: The third and last rotation turns the rotor $\mathcal{B}$ about axis z_2 through angle ψ. (See Figure 6.16.) This allows $\mathcal{B}$ to spin relative to G_i. Axis x_2 is turned into x and axis y_2 into y.

Through these three Eulerian angle rotations, the body ($\mathcal{B}$) can be positioned in any desired orientation in space ($\mathcal{J}$). We are now able to see the difficulty of solving for the orientation of a rigid body in general motion. If we write $\boldsymbol{\omega}_{\mathcal{B}/\mathcal{J}}$ as $\omega_1\hat{\mathbf{i}} + \omega_2\hat{\mathbf{j}} + \omega_3\hat{\mathbf{k}}$, then Equation (6.61) is equivalent to

$$\begin{aligned} s_\psi\dot{\theta} - s_\theta c_\psi\dot{\phi} &= \omega_1 \\ c_\psi\dot{\theta} + s_\theta s_\psi\dot{\phi} &= \omega_2 \\ \dot{\psi} + c_\theta\dot{\phi} &= \omega_3 \end{aligned} \tag{6.68}$$

Solving for the rates of change of the Eulerian angles gives

$$\begin{aligned} \dot{\theta} &= \omega_1 s_\psi + \omega_2 c_\psi \\ \dot{\phi} &= \frac{\omega_2 s_\psi - \omega_1 c_\psi}{s_\theta} \\ \dot{\psi} &= \frac{\omega_3 s_\theta - \omega_2 c_\theta s_\psi + \omega_1 c_\theta c_\psi}{s_\theta} \end{aligned} \tag{6.69}$$

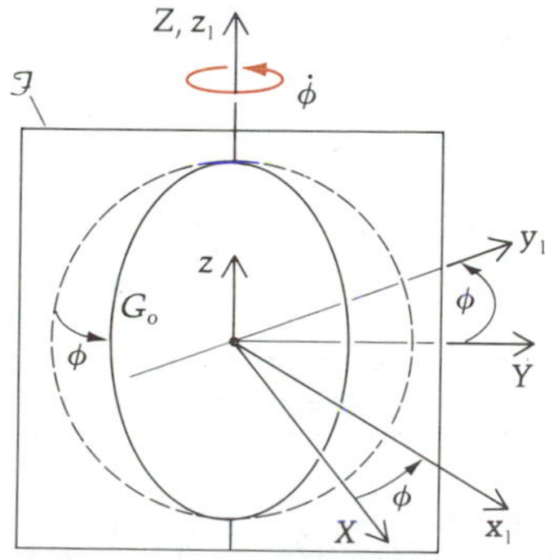

Figure 6.14 First rotation.

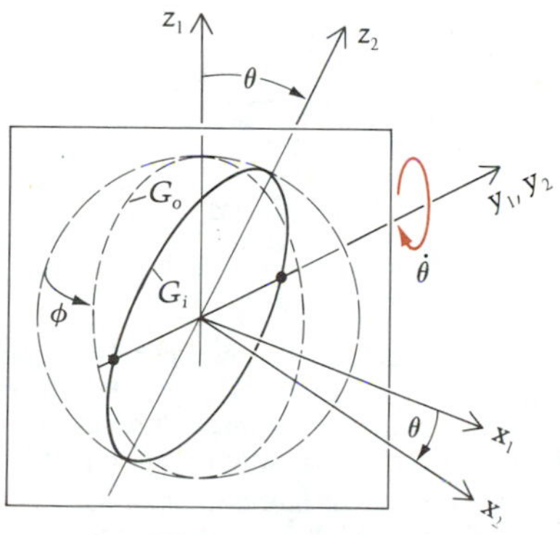

Figure 6.15 Second rotation.

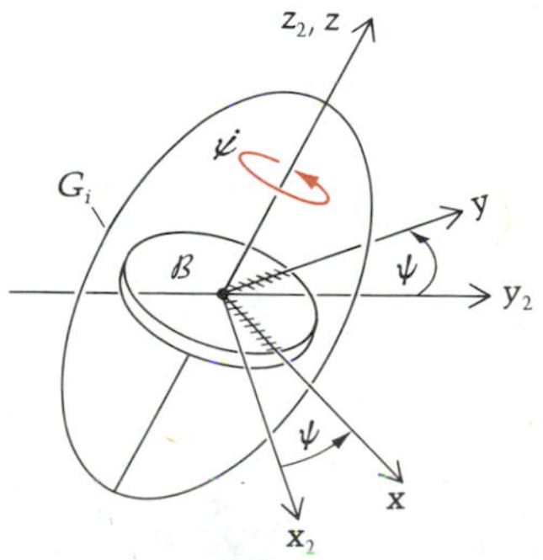

Figure 6.16 Third rotation.

We see from these equations that even if we knew the ω components as functions of time in closed form,* it would still be a formidable task to integrate Equations (6.69) analytically to obtain the Eulerian angles and thus to know the body's orientation in space. This is usually not even possible, and resort is made to computers that can numerically carry out integrations with a step-by-step scheme such as Runge-Kutta.

Incidentally, the $\sin\theta$ denominators in Equations (6.69) present serious obstacles in the dynamics of space vehicles; whenever θ is zero or a multiple of π, the equations develop a singularity. Sophisticated programming or, in some cases, completely different mathematical schemes for orienting the body are required to overcome such difficulties.

We mention that use of the preceding set of Eulerian angles as defined requires that we maintain the *order* of rotation. To illustrate the importance of rotation order, we remark that if this book is rotated through two $\pi/2$ rotations about the space axes Y and Z, in opposite orders as suggested by Figure 6.17, it will end up in a different position. We should point out, however, that there are ways of setting up the axes and angles which make a body's final orientation independent of order. For instance, just as in our Eulerian angle development, let Z be fixed in $\mathcal{I}$, let z be fixed in $\mathcal{B}$, and let y be always perpendicular to both Z and z. But now restrict Z and z to be nonparallel. (Let z, the axis of $\mathcal{B}$, lie along X initially, for example.) In this case, the angles (ϕ, θ, ψ) as defined earlier may be performed in any of the six possible orders and the body's resulting orientation in $\mathcal{I}$ will be the same each time!†

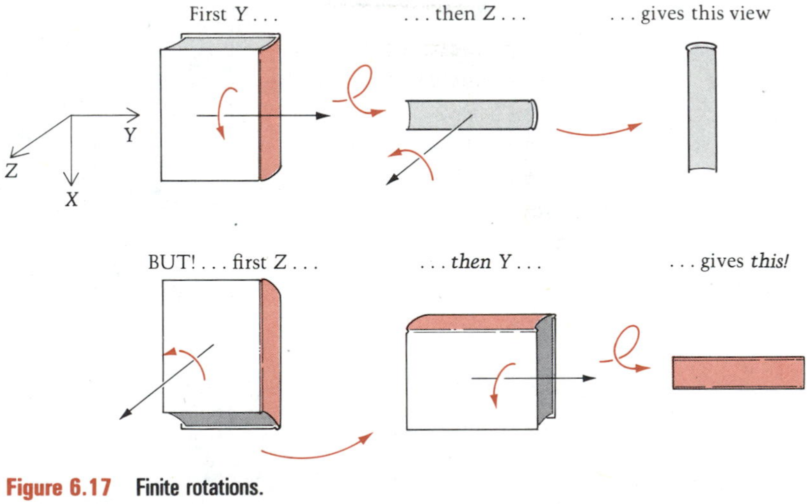

Figure 6.17 Finite rotations.

* The equations governing these three components of angular velocity are the Euler equations of rigid-body kinetics. They themselves are also nonlinear and unsolvable in closed form in general three-dimensional motion except for a few special cases. We shall be studying these equations in Chapter 7.

† Also, subsequent reorientations will likewise be order-independent; see "Successive Finite Rotations," by T. R. Kane and D. A. Levinson, *Journal of Applied Mechanics*, Dec. 1978, Vol. 45, pp. 945–946.

An alternative set of three rotations has become popular in the literature in recent years:

1. Rotate through θ_1 about X.
2. Then rotate through θ_2 about y_1.
3. Then rotate through θ_3 about z_2.

This sequence results in angular velocity components in $\mathcal{B}$ — that is, in the body-axis system (x, y, z) — of

$$\begin{aligned}\omega_1 &= \dot{\theta}_1 \cos\theta_2 \cos\theta_3 + \dot{\theta}_2 \sin\theta_3 \\ \omega_2 &= -\dot{\theta}_1 \cos\theta_2 \sin\theta_3 + \dot{\theta}_2 \cos\theta_3 \\ \omega_3 &= \dot{\theta}_1 \sin\theta_2 + \dot{\theta}_3\end{aligned} \tag{6.70}$$

Although (6.70) is not made up of classic Eulerian angles, the result is an equally valid set of relations between the $\boldsymbol{\omega}_{\mathcal{B}/\mathcal{I}}$ components and the angles θ_i of rotation. The order and axes of rotations of the θ_i in this system are quite simple to remember.

There are alternatives to using the Eulerian angles or the angles in Equation (6.70) to orient a body in space. One such alternative is to use the quaternions of Hamilton, which are free of the singular points caused by zero denominators at certain values of the Eulerian angles. (For this reason, quaternions were used in the Skylab Orbital Assembly's attitude control system.) Another approach to the orientation of a body in space is to determine the direction cosines of a unit vector, fixed in direction in space $\mathcal{I}$, with respect to a set of axes fixed in the body $\mathcal{B}$. Let this unit vector (call it $\hat{\mathbf{u}}$) have direction cosines (p, q, r) with respect to axes (x, y, z) fixed in $\mathcal{B}$. Further, let $(\hat{\mathbf{i}}, \hat{\mathbf{j}}, \hat{\mathbf{k}})$ be unit vectors, always respectively parallel to (x, y, z). Then

$$\hat{\mathbf{u}} = p\hat{\mathbf{i}} + q\hat{\mathbf{j}} + r\hat{\mathbf{k}}$$

and, differentiating in $\mathcal{I}$, we obtain

$$\begin{aligned}{}^{\mathcal{I}}\dot{\hat{\mathbf{u}}} = 0 &= {}^{\mathcal{B}}\dot{\hat{\mathbf{u}}} + \boldsymbol{\omega}_{\mathcal{B}/\mathcal{I}} \times \hat{\mathbf{u}} \\ &= (\dot{p}\hat{\mathbf{i}} + \dot{q}\hat{\mathbf{j}} + \dot{r}\hat{\mathbf{k}}) + (\omega_x\hat{\mathbf{i}} + \omega_y\hat{\mathbf{j}} + \omega_z\hat{\mathbf{k}}) \times (p\hat{\mathbf{i}} + q\hat{\mathbf{j}} + r\hat{\mathbf{k}}) \\ &= (\dot{p} + \omega_y r - \omega_z q)\hat{\mathbf{i}} + (\dot{q} + \omega_z p - \omega_x r)\hat{\mathbf{j}} + (\dot{r} + \omega_x q - \omega_y p)\hat{\mathbf{k}}\end{aligned}$$

This vector equation has the following scalar component equations:

$$\begin{aligned}\dot{p} &= \omega_z q - \omega_y r \\ \dot{q} &= \omega_x r - \omega_z p \\ \dot{r} &= \omega_y p - \omega_x q\end{aligned} \tag{6.71}$$

If the $\boldsymbol{\omega}$ components are either prescribed or else found from kinetics equations (to be studied in Chapter 7), then Equations (6.71) may be solved for the direction cosines, thereby orienting $\mathcal{B}$ in $\mathcal{I}$. Equations (6.71) are known as the Poisson equations. They are alternatives to equations such as (6.68) and (6.70).

PROBLEMS ► Section 6.8

6.78 Using the Eulerian angles (ϕ, θ, ψ) discussed in this section, express $\boldsymbol{\omega}_{\mathcal{B}/\mathcal{I}}$ in (terms of its components in) the frame $\mathcal{I}_2$.

6.79 Show that the magnitudes of $\boldsymbol{\omega}_{\mathcal{B}/\mathcal{I}}$ are all the same as expressed (a) in $\mathcal{B}$ in Equation (6.61); (b) in $\mathcal{I}$ in Equation (6.64); and (c) in $\mathcal{I}_1$ in Equation (6.67).

6.80 Derive Equations (6.70).

6.81 Write $\boldsymbol{\omega}_{\mathcal{B}/\mathcal{I}}$ in $\mathcal{I}$ by using the successive rotations θ_1, θ_2, and θ_3 that resulted in Equation (6.70) when expressed in $\mathcal{B}$.

6.82 Euler's theorem for finite rotations is stated as follows: The most general rotation of a rigid body $\mathcal{B}$ with respect to a point A is equivalent to a rotation about some axis through A. Prove the theorem. *Hint*: Let A be considered fixed in the reference frame in which $\mathcal{B}$ moves. (See Figure P6.82.) Let point P be at P_1 prior to the rotation and at P_2 afterward; assume the same for point Q (Q_1 before, Q_2 after). Bisect angle P_1AP_2 with a plane normal to the plane of the angle. Do the same for Q_1AQ_2 and consider the intersection of the two planes.

6.83 Chasle's theorem states: The most general displacement of a rigid body is equivalent to the translation of some point A followed by a rotation about an axis through A. Show that this result follows immediately from the previous problem.

6.84 The circular drum of radius R in Figure P6.84 is pivoted to a support at O, where O is a distance $R/2$ from the center C of the drum. A weight $\mathcal{W}$ (particle) hangs from a cord wrapped around the drum. The drum is slowly rotated $\pi/2$ rad clockwise about O. Find the displacement of $\mathcal{W}$. *Hint*: Use the result of the preceding problem, with C being the point A and with a $\pi/2$ ↻ rotation following the translation. Add the displacements of $\mathcal{W}$ during each part.

6.85 In Figure P6.85, the sphere $\mathcal{S}$ rolls on the plane, and its angular velocity in the reference frame $\mathcal{I}$, in which (x, y, z) are fixed, is given by Equation (6.64). Noting that $\mathbf{v}_C = \dot{x}\hat{\mathbf{i}} + \dot{y}\hat{\mathbf{j}}$, write the constraint equations (the "no slip" conditions) relating $\dot{x}$ and $\dot{y}$ to the Eulerian angles ϕ, θ, ψ and their derivatives.

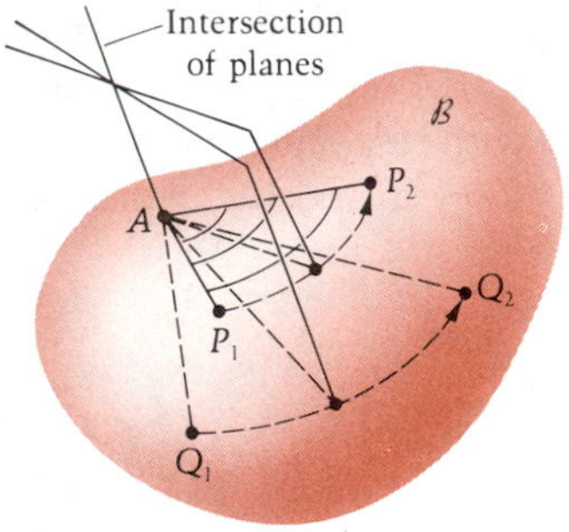

Figure P6.82

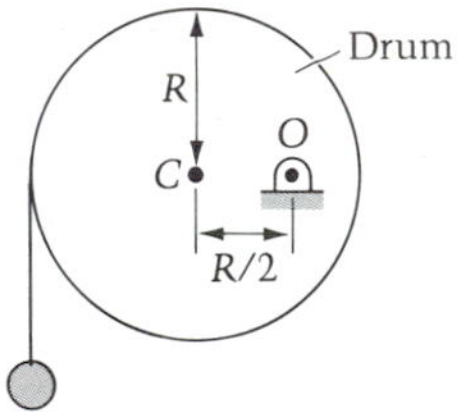

Figure P6.84

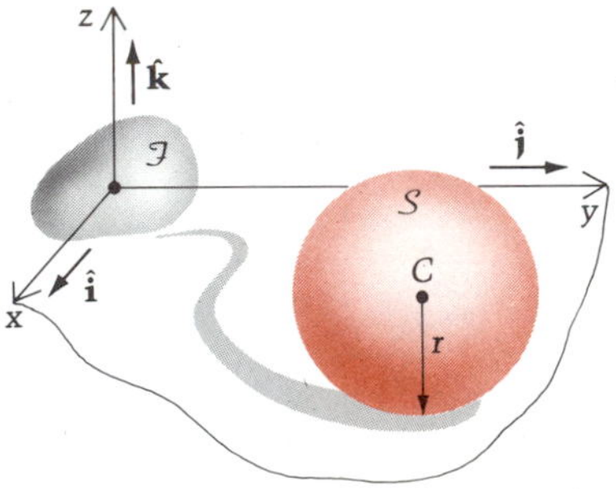

Figure P6.85

6.9 Rotation Matrices

Suppose that a vector $\mathbf{Q}$ is written in a frame $\mathcal{I}$ and that its components are the elements of a column matrix $\{Q\}_{\mathcal{I}}$. It is possible to develop a set of 3×3 matrices $[T_x]$, $[T_y]$, and $[T_z]$ each of which, when postmultiplied by $\{Q\}_{\mathcal{I}}$, gives the components of $\mathbf{Q}$ in a new frame rotated about an x, y, or z axis, respectively, of $\mathcal{I}$. These matrices are handy work-savers. For example, they reduce the work involved, in going from Equation (6.58) to (6.61) or (6.64), to a pair of simple matrix multiplications.

We shall develop $[T_z]$ and then state the results for $[T_x]$ and $[T_y]$, which are derived similarly. Let $Q = Q_x\hat{\mathbf{i}} + Q_y\hat{\mathbf{j}} + Q_z\hat{\mathbf{k}}$, in which ($\hat{\mathbf{i}}$, $\hat{\mathbf{j}}$, $\hat{\mathbf{k}}$) are a triad of unit vectors having fixed directions along axes (x, y, z) of frame $\mathcal{I}$. Suppose further that $\mathcal{B}$ is a frame whose orientation may be obtained from that of $\mathcal{I}$ by a rotation through the angle θ_z about z. The rotated axes, which were aligned with (x, y, z) prior to the rotation, will be denoted (x_1, y_1, z_1) with associated unit vectors $(\hat{\mathbf{i}}_1, \hat{\mathbf{j}}_1, \hat{\mathbf{k}}_1)$.

We note that in the "new" frame $\mathcal{B}$, we may write $\mathbf{Q} = Q_{x_1}\hat{\mathbf{i}}_1 + Q_{y_1}\hat{\mathbf{j}}_1 + Q_{z_1}\hat{\mathbf{k}}_1$, where $Q_{z_1} = Q_z$ and $\hat{\mathbf{k}}_1 = \hat{\mathbf{k}}$ since the rotation is about this axis, common to both frames. To get Q_{x_1} and Q_{y_1} in terms of Q_x, Q_y, and θ_z, we use (see Figure 6.18):

$$\begin{aligned} Q_{x_1} &= \mathbf{Q} \cdot \hat{\mathbf{i}}_1 = (Q_x\hat{\mathbf{i}} + Q_y\hat{\mathbf{j}} + Q_z\hat{\mathbf{k}}) \cdot \hat{\mathbf{i}}_1 \\ &= Q_x(\hat{\mathbf{i}} \cdot \hat{\mathbf{i}}_1) + Q_y(\hat{\mathbf{j}} \cdot \hat{\mathbf{i}}_1) + Q_z\cancelto{0}{(\hat{\mathbf{k}} \cdot \hat{\mathbf{i}}_1)} \\ &= Q_x \cos\theta_z + Q_y \sin\theta_z \end{aligned} \tag{6.72}$$

and similarly,

$$\begin{aligned} Q_{y_1} &= \mathbf{Q} \cdot \hat{\mathbf{j}}_1 = Q_x(\hat{\mathbf{i}} \cdot \hat{\mathbf{j}}_1) + Q_y(\hat{\mathbf{j}} \cdot \hat{\mathbf{j}}_1) + Q_z\cancelto{0}{(\hat{\mathbf{k}} \cdot \hat{\mathbf{j}}_1)} \\ &= Q_x(-\sin\theta_z) + Q_y \cos\theta_z \end{aligned} \tag{6.73}$$

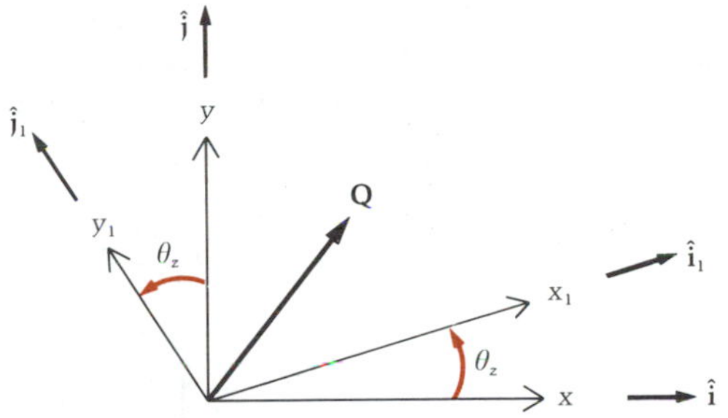

Figure 6.18

Together with $Q_{z_1} = Q_z$, the Equations (6.72, 6.73) give the components of $\mathbf{Q}$ in the rotated frame $\mathcal{B}$, in terms of its components Q_x, Q_y, Q_z) back in $\mathcal{I}$. Now we are ready to observe that if the matrix $[T_z]$ is defined as

$$[T_z] = \begin{bmatrix} \cos\theta_z & \sin\theta_z & 0 \\ -\sin\theta_z & \cos\theta_z & 0 \\ 0 & 0 & 1 \end{bmatrix}$$

then the same results for the components of $\mathbf{Q}$ in the rotated frame $\mathcal{B}$ are obtained from the matrix product $[T_z]\{Q\}_{\mathcal{I}}$:

$$\{Q\}_{\mathcal{B}} = [T_z]\{Q\}_{\mathcal{I}} = \begin{bmatrix} \cos\theta_z & \sin\theta_z & 0 \\ -\sin\theta_z & \cos\theta_z & 0 \\ 0 & 0 & 1 \end{bmatrix} \begin{Bmatrix} Q_x \\ Q_y \\ Q_z \end{Bmatrix}$$

$$= \begin{Bmatrix} Q_x\cos\theta_z + Q_y\sin\theta_z \\ -Q_x\sin\theta_z + Q_y\cos\theta_z \\ Q_z \end{Bmatrix}$$

The rotation matrices for rotations of θ_x about x, and θ_y about y, are respectively given by $[T_x]$ and $[T_y]$:

$$[T_x] = \begin{bmatrix} 1 & 0 & 0 \\ 0 & \cos\theta_x & \sin\theta_x \\ 0 & -\sin\theta_x & \cos\theta_x \end{bmatrix} \qquad [T_y] = \begin{bmatrix} \cos\theta_y & 0 & -\sin\theta_y \\ 0 & 1 & 0 \\ \sin\theta_y & 0 & \cos\theta_y \end{bmatrix}$$

The student may wish to verify one or both of these matrices, as we did for $[T_z]$. Note the change in the "sign of sine" in the $[T_y]$ matrix. Moreover, if we must turn about an axis through a negative angle, we need only change the signs of both sine terms; this follows from the fact that $\cos(-\theta) = \cos\theta$, while $\sin(-\theta) = -\sin\theta$. We now consider examples of the use of the rotation matrices. We shall use some shorthand common in the literature of kinematics: s_θ for $\sin\theta$, c_ϕ for $\cos\phi$, etc.

EXAMPLE 6.15

Use rotation matrices to obtain the components of $\boldsymbol{\omega}_{\mathcal{B}/\mathcal{I}}$ in body coordinates, given its representation (Equation 6.64) in the reference or space frame $\mathcal{I}$.

Solution

We premultiply $\boldsymbol{\omega}_{\mathcal{B}/\mathcal{I}}$, expressed in $\mathcal{I}$ in matrix form, with rotation matrices of ϕ about the 3-axis, then θ about the new 2-axis, and then ψ about the new and final 3-axis:

$$\{\omega_{\mathcal{B}/\mathcal{I}}\}_\mathcal{B} = \underset{(\text{angle } \psi)}{[T_z]} \; \underset{(\text{angle } \theta)}{[T_y]} \; \underset{(\text{angle } \phi)}{[T_z]} \; \{\omega_{\mathcal{B}/\mathcal{I}}\}_\mathcal{I}$$

$$= \begin{bmatrix} c_\psi & s_\psi & 0 \\ -s_\psi & c_\psi & 0 \\ 0 & 0 & 1 \end{bmatrix} \begin{bmatrix} c_\theta & 0 & -s_\theta \\ 0 & 1 & 0 \\ s_\theta & 0 & c_\theta \end{bmatrix} \underbrace{\begin{bmatrix} c_\phi & s_\phi & 0 \\ -s_\phi & c_\phi & 0 \\ 0 & 0 & 1 \end{bmatrix} \overbrace{\begin{Bmatrix} s_\theta c_\phi \dot{\psi} - s_\phi \dot{\theta} \\ c_\phi \dot{\theta} + s_\theta s_\phi \dot{\psi} \\ \dot{\phi} + c_\theta \dot{\psi} \end{Bmatrix}}^{\text{components of } \omega_{\mathcal{B}/\mathcal{I}} \text{ in } \mathcal{I}}}_{\text{This gives the components of } \omega_{\mathcal{B}/\mathcal{I}} \text{ in } \mathcal{I}_1}$$

$$= \begin{bmatrix} c_\psi & s_\psi & 0 \\ -s_\psi & c_\psi & 0 \\ 0 & 0 & 1 \end{bmatrix} \underbrace{\begin{bmatrix} c_\theta & 0 & -s_\theta \\ 0 & 1 & 0 \\ s_\theta & 0 & c_\theta \end{bmatrix} \begin{Bmatrix} s_\theta \dot{\psi} \\ \dot{\theta} \\ \dot{\phi} + c_\theta \dot{\psi} \end{Bmatrix}}_{\text{This yields the components of } \omega_{\mathcal{B}/\mathcal{I}} \text{ in } \mathcal{I}_2}$$

$$= \begin{bmatrix} c_\psi & s_\psi & 0 \\ -s_\psi & c_\psi & 0 \\ 0 & 0 & 1 \end{bmatrix} \begin{Bmatrix} -s_\theta \dot{\phi} \\ \dot{\theta} \\ \dot{\psi} + c_\theta \dot{\phi} \end{Bmatrix}$$

Finally, we obtain the components of $\boldsymbol{\omega}_{\mathcal{B}/\mathcal{I}}$ in $\mathcal{B}$:

$$= \begin{Bmatrix} -s_\theta c_\psi \dot{\phi} + s_\psi \dot{\theta} \\ s_\psi s_\theta \dot{\phi} + c_\psi \dot{\theta} \\ \dot{\psi} + c_\theta \dot{\phi} \end{Bmatrix}$$

Comparing the elements of this matrix with the components in Equation (6.61), we see that rotation matrices indeed furnish us with a rapid means of "converting" a vector from one frame to another. Note also that the bracket in the second line above contains $\boldsymbol{\omega}_{\mathcal{B}/\mathcal{I}}$ expressed in $\mathcal{I}_1$, previously derived as Equation (6.67). The bracket in the third line gives the components of $\boldsymbol{\omega}_{\mathcal{B}/\mathcal{I}}$ in $\mathcal{I}_2$.

The next example illustrates a use that was made of rotation matrices by one of the authors in the development of earth stations.

Example 6.16

Using rotation matrices, compute the angles A (azimuth) and E (elevation) through which an antenna must respectively turn about (a) the negative of local vertical and then (b) the new, rotated position of the elevation axis in order to sight a satellite in geosynchronous orbit (see Problems 2.64,65). Angles A and E are called *look angles,* and an antenna that performs azimuth followed by elevation in this manner is said to have "el over az" positioning. The azimuth angle A is a function of the local latitude λ and the relative west longitude δ of the satellite; the elevation angle E depends additionally on R_e/R, the ratio of the earth and orbit radii. (See Figure E6.16a.)

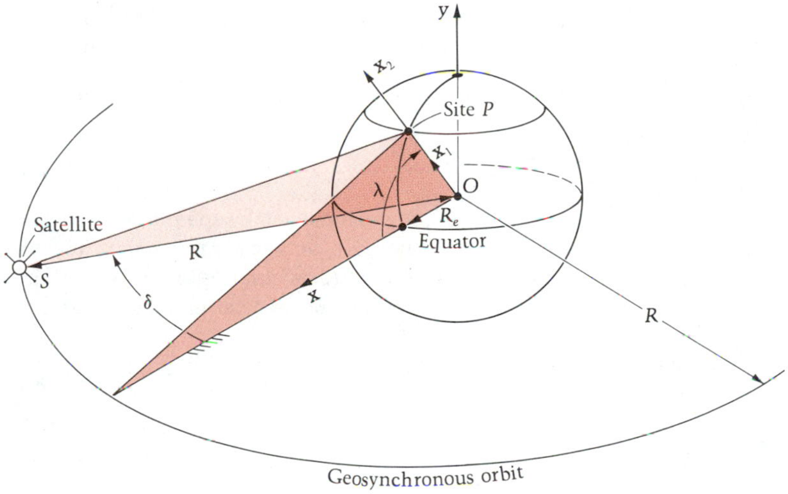

Figure E6.16a

Solution

The frame $\mathcal{I}(x, y, z)$ has origin at the center of the earth as shown; the xy plane contains the site P and its meridian. The coordinates of the satellite in this frame are seen to be given by the position vector

$$\mathbf{r}_{OS} = R \cos \delta \hat{\mathbf{i}} + R \sin \delta \hat{\mathbf{k}}$$

First we rotate the frame $\mathcal{I}$ through the latitude angle λ about z in order to line up the new axis x_1 with the local vertical at the site P. We call the resulting rotated frame $\mathcal{I}_1$ and obtain the following for the new components of $\mathbf{r}_{OS}$ ($c_\lambda = \cos\lambda$, $s_\delta = \sin\delta$, and so forth):

$$\{r_{OS}\}_{\mathcal{I}_1} = \underset{(\text{angle } \lambda)}{[T_z]} \{r_{OS}\}_{\mathcal{I}}$$

$$\{r_{OS}\}_{\mathcal{I}_1} = \begin{bmatrix} c_\lambda & s_\lambda & 0 \\ -s_\lambda & c_\lambda & 0 \\ 0 & 0 & 1 \end{bmatrix} \begin{Bmatrix} Rc_\delta \\ 0 \\ Rs_\delta \end{Bmatrix} = \begin{Bmatrix} Rc_\lambda c_\delta \\ -Rs_\lambda c_\delta \\ Rs_\delta \end{Bmatrix}$$

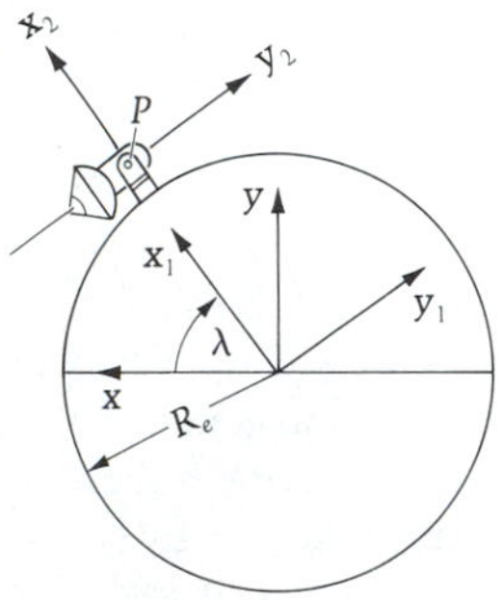

Figure E6.16b

Next we translate the axes to the site, as shown in Figure E6.16b. The only component of $\{r_{OS}\}_{\mathcal{I}_1}$ that changes is x_1, and we see by inspection that

$$\{r_{PS}\}_{\mathcal{I}_2} = \begin{Bmatrix} Rc_\lambda c_\delta - R_e \\ -Rs_\lambda c_\delta \\ Rs_\delta \end{Bmatrix}$$

Note that we must subtract the earth radius R_e from x_1 to get the proper x_2 coordinate of the satellite.

The second rotation is the azimuth rotation about $-x_2$; if we call the rotated frame $\mathcal{I}_3$, the coordinates (x_3, y_3, z_3) of S in this frame are given by

$$\{r_{PS}\}_{\mathcal{I}_3} = \underset{(\text{angle } -A)}{[T_x]} \{r_{PS}\}_{\mathcal{I}_2}$$

$$= \begin{bmatrix} 1 & 0 & 0 \\ 0 & c_A & -s_A \\ 0 & s_A & c_A \end{bmatrix} \begin{Bmatrix} Rc_\lambda c_\delta - R_e \\ -Rs_\lambda c_\delta \\ Rs_\delta \end{Bmatrix}$$

$$= \begin{Bmatrix} Rc_\lambda c_\delta - R_e \\ -Rs_\lambda c_\delta c_A - Rs_\delta s_A \\ -Rs_\lambda c_\delta s_A + Rs_\delta c_A \end{Bmatrix}$$

Here we take an important step. We want the z_3 component of $\mathbf{r}_{PS}$ to be zero because we wish to rotate next in elevation about z_3 and end up with the "boresight" (axis) of the antenna aiming at the satellite. Thus angle A (see Figure E6.16c) is determined by setting the third element of the preceding matrix to zero:

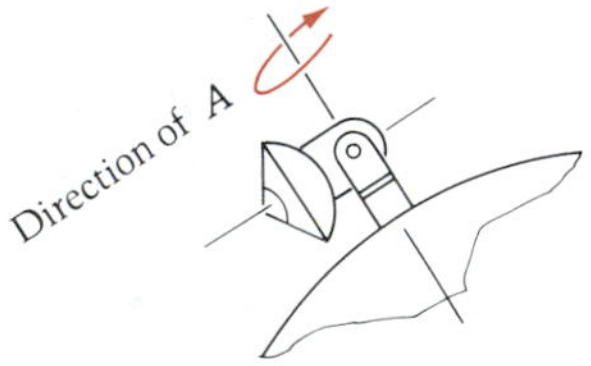

Figure E6.16c

$$-Rs_\lambda c_\delta s_A + Rs_\delta c_A = 0$$

$$\tan A = \tan\delta \csc\lambda$$

$$A = \tan^{-1}(\tan\delta \csc\lambda) \tag{1}$$

Finally we rotate through angle E about the z_3 axis:

$$\{r_{PS}\}_{\mathcal{I}_4} = \underset{(\text{angle } E)}{[T_z]} \{r_{PS}\}_{\mathcal{I}_3}$$

$$= \begin{bmatrix} c_E & s_E & 0 \\ -s_E & c_E & 0 \\ 0 & 0 & 1 \end{bmatrix} \begin{Bmatrix} Rc_\lambda c_\delta - R_e \\ -Rs_\lambda c_\delta c_A - Rs_\delta s_A \\ 0 \end{Bmatrix}$$

$$= \begin{Bmatrix} c_E(Rc_\lambda c_\delta - R_e) - s_E(Rs_\lambda c_\delta c_A + Rs_\delta s_A) \\ -s_E(Rc_\lambda c_\delta - R_e) - c_E(Rs_\lambda c_\delta c_A + Rs_\delta s_A) \\ 0 \end{Bmatrix}$$

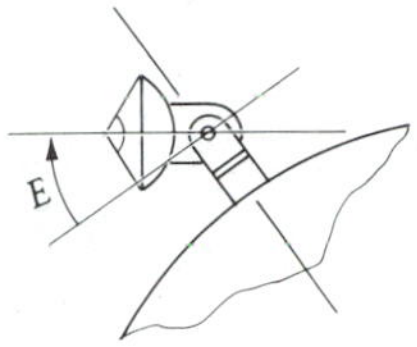

Figure E6.16d

Now we come to the condition that will allow us to determine the value of angle E (see Figure E6.16d): We wish the antenna to aim directly at the satellite. Since the antenna boresight is now in the $-y_3$ direction, we wish the elevation rotation to stop when the x_3 coordinate is zero:

$$c_E(Rc_\lambda c_\delta - R_e) - s_E(Rs_\lambda c_\delta c_A + Rs_\delta s_A) = 0$$

$$E = \tan^{-1}\left(\frac{Rc_\lambda c_\delta - R_e}{Rs_\lambda c_\delta c_A + Rs_\delta s_A}\right) \quad (2)$$

If $r = R_e/R(\approx 1/6.61)$, then (2) becomes

$$E = \tan^{-1}\left(\frac{c_\lambda c_\delta - r}{s_\lambda c_\delta c_A + s_\delta s_A}\right)$$

in which the azimuth angle A is given by Equation (1), so that

$$E = \tan^{-1}\left(\frac{c_\lambda c_\delta - r}{\sqrt{1 - c_\delta^2 c_\lambda^2}}\right) \quad (3)$$

There is a single circle in the sky in which geosynchronous satellites can exist. This circle, which was examined in Problem 2.64, has rapidly become very crowded, however. As of the summer of 1994, there were nearly 500 satellites in geosynchronous orbit. Six years earlier, there were just over 100, and in early 1982, only around 30.

PROBLEMS ▶ Section 6.9

6.86 For the United States the eastern and western limits of usable satellite positions in the geosynchronous arc (see Problem 2.64) are about 70° and 143°W longitude, respectively. Find the ranges in azimuth and elevation that are required if the antenna in Example 6.16 is to sweep from the eastern to the western limit for a site at: (a) 34°N latitude and 84°W longitude; (b) your home town. (Select a city in the contiguous United States if you are from another country.)

6.87 Use Equations (6.70) together with rotation matrices to compute the angular velocity components in space-fixed axes.

6.88 An antenna $\mathcal{P}$ has three rotational degrees of freedom (see Figure P6.88):

1. Azimuth angle A about local vertical z,
2. Elevation angle E about an axis originally parallel to x,
3. Polarization angle P about the axis of symmetry of the dish (originally parallel to y).

Use the addition theorem together with rotation matrices to calculate $\boldsymbol{\omega}_{\mathcal{P}/\mathcal{I}}$ in terms of its components in $\mathcal{I}$.

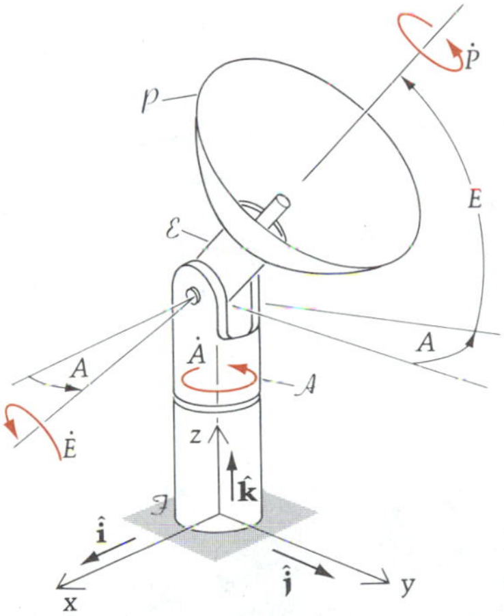

Figure P6.88

6.89 In the preceding problem calculate $\boldsymbol{\omega}_{\mathcal{P}/\mathcal{I}}$ in terms of its components in $\mathcal{P}$.

* **6.90** Plot the elevation angle E versus the satellite angle δ (see Example 6.16) for the following values of λ (on the same graph): $\lambda = 0°$, $20°$, $40°$, $60°$, and $80°$. What do the crossings of the δ axis of these curves physically represent?

* **6.91** Calculate the look angles (see Example 6.16) for the case in which the elevation rotation is performed *prior* to azimuth ("az over el" positioning).

6.92 It takes six *orbital parameters* to establish the location of a planet with respect to a frame fixed in space. (See Figure P6.92.) To find its orbital path, we first turn through the angle Ω in the ecliptic plane (the plane containing the path of the sun as we see it from earth) to the *ascending node* (the intersection of the ecliptic plane with the planet's path when going north). Next we turn in inclination through the angle i about x_1' to obtain the tilt of the planet's plane. (Thus earth's inclination is defined as zero.) Finally, the angle θ_o locates the perihelion of the planet's orbit. This is the closest point of the orbit to the sun's center S. Two other quantities give the orbit's shape, and a sixth one locates the planet in its orbit with respect to the perihelion. If $\mathbf{v} = (v_x, v_y, v_z)$ is a vector defined in the space frame $\mathcal{I}$, use rotation matrices to obtain the components of $\mathbf{v}$ in the frame $\mathcal{B}\,(x, y, z)$ located as shown in the orbital path at P, in terms of Ω, i, and θ_o.

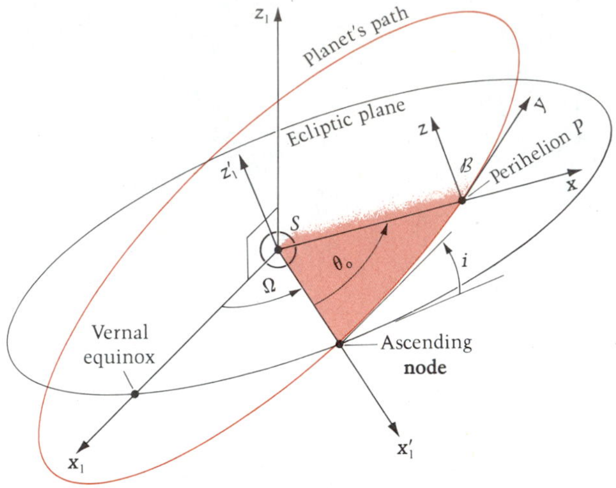

Figure P6.92

PROJECT PROBLEM ▸ Chapter 6

6.93 After reading Example 6.3, construct a simple model that illustrates the workings of a universal joint (see Figure P6.93 for some examples).

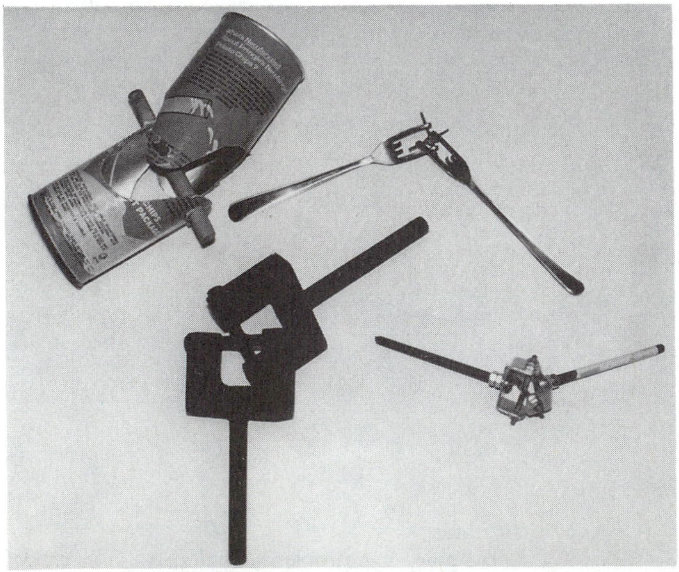

Figure P6.93

SUMMARY ▶ Chapter 6

The key concept of this chapter is angular velocity. If $\mathcal{A}$ and $\mathcal{B}$ are two rigid bodies (or frames of reference), the angular velocity of $\mathcal{B}$ relative to $\mathcal{A}$ (or of $\mathcal{B}$ *in* $\mathcal{A}$), $\boldsymbol{\omega}_{\mathcal{B}/\mathcal{A}}$, is the unique vector that connects the derivatives relative to $\mathcal{A}$ and $\mathcal{B}$ of any vector, say $\mathbf{Q}$, by

$$^{\mathcal{A}}\dot{\mathbf{Q}} = {}^{\mathcal{B}}\dot{\mathbf{Q}} + \boldsymbol{\omega}_{\mathcal{B}/\mathcal{A}} \times \mathbf{Q}$$

The angular velocity describes the rate at which the orientation of one body changes relative to another. Among the important properties are:

$$\boldsymbol{\omega}_{\mathcal{B}/\mathcal{A}} = -\boldsymbol{\omega}_{\mathcal{A}/\mathcal{B}}$$

and the addition theorem (a chain rule):

$$\boldsymbol{\omega}_{\mathcal{C}/\mathcal{A}} = \boldsymbol{\omega}_{\mathcal{C}/\mathcal{B}} + \boldsymbol{\omega}_{\mathcal{B}/\mathcal{A}}$$

which can be extended by repetition to

$$\boldsymbol{\omega}_{\mathcal{D}/\mathcal{A}} = \boldsymbol{\omega}_{\mathcal{D}/\mathcal{C}} + \boldsymbol{\omega}_{\mathcal{C}/\mathcal{B}} + \boldsymbol{\omega}_{\mathcal{B}/\mathcal{A}}$$

and in fact to any number of bodies.

This rather abstract-seeming concept of angular velocity reduces to the familiar form of Chapter 3 when we have "plane-motion" situations in which a direction is fixed in both of two bodies. So if $\hat{\mathbf{k}}$ is the constant vector (denotes a fixed direction) in both of the bodies $\mathcal{A}$ and $\mathcal{B}$, we have

$$\boldsymbol{\omega}_{\mathcal{B}/\mathcal{A}} = \dot{\theta}\hat{\mathbf{k}}$$

This equation, along with the addition theorem, gives us an important tool to deal with systems of bodies that appear complex but where each body moves relative to a neighbor by rotating about an axis fixed in the neighbor.

Angular acceleration is the derivative of angular velocity, and the derivative may be calculated in either of the two bodies (frames) involved.

$$\boldsymbol{\alpha}_{\mathcal{B}/\mathcal{A}} = {}^{\mathcal{A}}\dot{\boldsymbol{\omega}}_{\mathcal{B}/\mathcal{A}} = {}^{\mathcal{B}}\dot{\boldsymbol{\omega}}_{\mathcal{B}/\mathcal{A}}$$

Angular accelerations cannot be added with the simplicity of an addition theorem such as (6.27), but instead obey the formula:

$$\boldsymbol{\alpha}_{\mathcal{C}/\mathcal{A}} = \boldsymbol{\alpha}_{\mathcal{C}/\mathcal{B}} + \boldsymbol{\alpha}_{\mathcal{B}/\mathcal{A}} + \boldsymbol{\omega}_{\mathcal{B}/\mathcal{A}} \times \boldsymbol{\omega}_{\mathcal{C}/\mathcal{B}}$$

Velocities of a point P relative to two frames $\mathcal{B}$ and $\mathcal{I}$ are related by

$$\mathbf{v}_{P/\mathcal{I}} = \mathbf{v}_{P/\mathcal{B}} + \mathbf{v}_{O'/\mathcal{I}} + \boldsymbol{\omega}_{\mathcal{B}/\mathcal{I}} \times \mathbf{r}_{O'P}$$

where O' is a point fixed in $\mathcal{B}$. Similarly for accelerations, we have

$$\mathbf{a}_{P/\mathcal{I}} = \mathbf{a}_{P/\mathcal{B}} + \mathbf{a}_{O'/\mathcal{I}} + \boldsymbol{\alpha}_{\mathcal{B}/\mathcal{I}} \times \mathbf{r}_{O'P} + \boldsymbol{\omega}_{\mathcal{B}/\mathcal{I}} \times (\boldsymbol{\omega}_{\mathcal{B}/\mathcal{I}} \times \mathbf{r}_{O'P}) + 2\boldsymbol{\omega}_{\mathcal{B}/\mathcal{I}} \times \mathbf{v}_{P/\mathcal{B}}$$

By letting P be fixed in $\mathcal{B}$ (as is O') we can deduce from the two previous equations a pair of expressions relating the velocities and then the accelerations (relative to a frame $\mathcal{I}$) of two points such as P and O', both fixed in $\mathcal{B}$:

$$\mathbf{v}_{P/\mathcal{I}} = \mathbf{v}_{O'/\mathcal{I}} + \boldsymbol{\omega}_{\mathcal{B}/\mathcal{I}} \times \mathbf{r}_{O'P}$$
$$\mathbf{a}_{P/\mathcal{I}} = \mathbf{a}_{O'/\mathcal{I}} + \boldsymbol{\alpha}_{\mathcal{B}/\mathcal{I}} \times \mathbf{r}_{O'P} + \boldsymbol{\omega}_{\mathcal{B}/\mathcal{I}} \times (\boldsymbol{\omega}_{\mathcal{B}/\mathcal{I}} \times \mathbf{r}_{O'P})$$

These are the counterparts in three dimensions of Equations (3.8) and (3.19).

Finally, we have discussed a system that is often used to describe the orientation of a body relative to a frame of reference — Euler angles. Angular velocity in terms of rates of change of Euler angles has been developed.

REVIEW QUESTIONS ▶ Chapter 6

True or False?

1. The angular velocity of body $\mathcal{B}$ in frame $\mathcal{I}$ depends only upon the changes in orientation of $\mathcal{B}$ with respect to $\mathcal{I}$.
2. The addition theorem for angular velocity applies equally well to angular acceleration.
3. The formula relating the velocities of two points of a rigid body in plane motion, $\mathbf{v}_B = \mathbf{v}_A + \boldsymbol{\omega} \times \mathbf{r}_{AB}$, applies to three-dimensional problems provided that $\boldsymbol{\omega}$, $\mathbf{r}_{AB}$, and the $\mathbf{v}$'s become three-dimensional vectors.
4. The formula relating the accelerations of two points of a rigid body in plane motion, $\mathbf{a}_B = \mathbf{a}_A + \boldsymbol{\alpha} \times \mathbf{r}_{AB} - \omega^2\mathbf{r}_{AB}$, applies to three-dimensional problems provided that $\boldsymbol{\alpha}$, $\mathbf{r}_{AB}$, $\boldsymbol{\omega}$, and the $\mathbf{a}$'s become three-dimensional vectors.
5. The equation $\omega_z = \dot{\theta}$ in plane motion extends to three similar linear equations in general motion for determining the orientation angles.
6. For a point P to have a nonvanishing Coriolis acceleration, there must be both a relative velocity of P with respect to the "moving frame" and an angular velocity of the moving frame relative to the reference frame.
7. If we premultiply a vector $\{v\}$ by a rotation matrix $[T]$, the 3×1 vector we get contains the "new" components of $\mathbf{v}$ in the rotated frame.
8. The Eulerian angles are used to orient a body in three-dimensional space.
9. The Eulerian angles are three rotations ϕ, θ, and ψ about axes which were originally distinct and orthogonal.
10. It is possible, for any moving point P, to choose a moving frame such that the Coriolis acceleration of P vanishes identically.
11. The angular velocity vector is used to relate the derivatives of a vector in two frames.
12. If one yoke of a misaligned universal (Hooke's) joint turns at constant angular speed, so does the other.

13. In general motion of a rigid body $\mathcal{B}$, as long as $\boldsymbol{\omega} \neq \mathbf{0}$ there is a point of zero velocity of $\mathcal{B}$ or $\mathcal{B}$-extended.

14. The order of rotations is important in orienting a body if the Eulerian angles are used as defined in this chapter (in conjunction with the Cardan suspension of the gyroscope).

Answers: 1. T 2. F 3. T 4. F 5. F 6. T 7. T 8. T 9. F 10. T 11. T 12. F 13. F 14. T

7 Kinetics of a Rigid Body in General Motion

7.1 Introduction
7.2 Moment of Momentum (Angular Momentum) in Three Dimensions
7.3 Transformations of Inertia Properties
Transformation of Inertia Properties at a Point
7.4 Principal Axes and Principal Moments of Inertia
Principal Axes at *C*
Calculation of Principal Moments of Inertia
Calculation of Principal Directions
Principal Axes at Any Point
Orthogonality of Principal Axes
Equal Moments of Inertia
Maximum and Minimum Moments of Inertia
7.5 The Moment Equation Governing Rotational Motion
The Euler Equations
Use of Non-Principal Axes
Use of an Intermediate Frame
7.6 Gyroscopes
Steady Precession
Torque-Free Motion
7.7 Impulse and Momentum
7.8 Work and Kinetic Energy
Summary
Review Questions

7.1 Introduction

In Chapter 2 we found that, relative to an inertial frame of reference, motion of a body is governed by

$$\Sigma \mathbf{F} = \frac{d\mathbf{L}}{dt} = m\mathbf{a}_C \tag{7.1}$$

and a moment equation

$$\Sigma \mathbf{M}_C = \frac{d\mathbf{H}_C}{dt} \tag{7.2}$$

or

$$\Sigma \mathbf{M}_O = \frac{d\mathbf{H}_O}{dt} \tag{7.3}$$

with O being fixed in the inertial frame. These general equations were specialized to *plane* motion of a rigid body $\mathcal{B}$ in Chapter 4 and will now be used to study the *general* motion of $\mathcal{B}$ in three dimensions.

As we indicated in Chapter 2, the first of the two vector equations given above describes the mass center motion of *any* system.* It is applicable, for example, to rigid or deformable solids, systems of small masses, liquids, and gases. For a body in general (three-dimensional) motion, Equation (7.1) now possesses three nontrivial scalar component equations whose solutions allow us to locate the mass center C.

We note that, as was the case with plane motion, the mass center can move independently of the body's changing orientation (provided that the external forces do not themselves depend on the body's angular motion, which is frequently the case). We saw such an example in Section 6.6 when we examined the motion of (a particle or) the mass center of a body near the rotating earth. We emphasize that such a simple and natural extension from two to three dimensions will *not* occur with the *orientation* (or angular) motion of $\mathcal{B}$, as we shall see in Section 7.5. The reason is that $\dot{\mathbf{H}}_C$ in Equation (7.2) *cannot* be written as the sum of three terms of the form $I_{zz}^C \ddot{\theta}\hat{\mathbf{k}}$.

In the remainder of the chapter we shall first develop the expression for the moment of momentum of a rigid body $\mathcal{B}$ in general motion. This will then lead us into a study of the inertia properties of $\mathcal{B}$. Having partially examined the concept of inertia in Chapter 4, we shall extend this study to include transformations at a point as well as principal moments and axes of inertia. Then and only then shall we be fully prepared to derive the Euler equations that govern the rotational motion of a rigid body in general motion. We shall also examine, as we did for the plane

* Excluding throughout, of course, relativistic effects occurring when velocities are not small compared to the speed of light.

motion in Chapter 5, some special integrals of the equations of motion, which are known as the principles of impulse and momentum, angular impulse and angular momentum, and work and kinetic energy.

7.2 Moment of Momentum (Angular Momentum) in Three Dimensions

We saw in Chapter 4 that when it is reasonable to treat a body $\mathcal{B}$ as rigid, the equations of motion of $\mathcal{B}$ are greatly simplified. The mass center becomes fixed in the body, and the moment of momentum is expressible in terms of the angular velocity and the inertia properties of $\mathcal{B}$—hence the other name of moment of momentum: **angular momentum**. We shall proceed now to study the angular momentum $\mathbf{H}_P$ of $\mathcal{B}$ about a point P in general (three-dimensional) motion. We shall see that the equations that result are much more complicated than their plane motion counterparts.

Let us begin by introducing a system of rectangular axes (x, y, z) which have their origin at P. The angular velocity of $\mathcal{B}$ in reference frame $\mathcal{I}$ may then be expressed in terms of its components along these axes by

$$\begin{aligned}\boldsymbol{\omega}_{\mathcal{B}/\mathcal{I}} = \boldsymbol{\omega} &= (\boldsymbol{\omega} \cdot \hat{\mathbf{i}})\hat{\mathbf{i}} + (\boldsymbol{\omega} \cdot \hat{\mathbf{j}})\hat{\mathbf{j}} + (\boldsymbol{\omega} \cdot \hat{\mathbf{k}})\hat{\mathbf{k}} \\ &= \omega_x\hat{\mathbf{i}} + \omega_y\hat{\mathbf{j}} + \omega_z\hat{\mathbf{k}} \end{aligned} \tag{7.4}$$

The location, relative to P, of a typical point in the body is given by

$$\mathbf{r} = x\hat{\mathbf{i}} + y\hat{\mathbf{j}} + z\hat{\mathbf{k}} \tag{7.5}$$

The moment of momentum of $\mathcal{B}$ relative to P is now defined (see Sections 2.6 and 4.3) to be

$$\mathbf{H}_P = \int \mathbf{r} \times \mathbf{v}\, dm \tag{7.6}$$

in which $\mathbf{v}$, the velocity of the mass element dm, is not the derivative of $\mathbf{r}$ but rather of the position vector to the element from a point fixed in the reference frame $\mathcal{I}$, as shown in Figure 7.1.

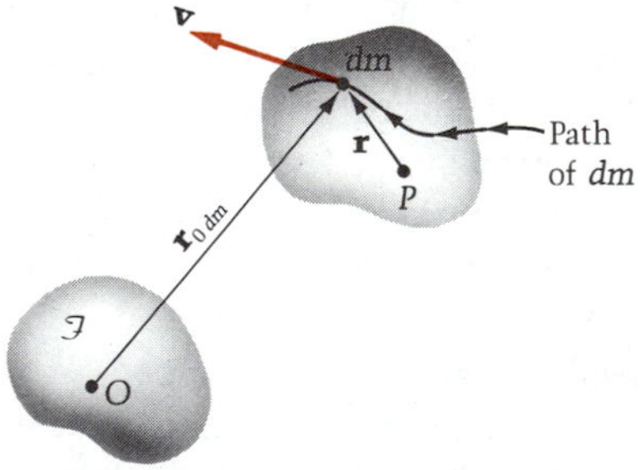

Figure 7.1

Since $\mathcal{B}$ is a rigid body, we know from Equation (6.56) that $\mathbf{v} = \mathbf{v}_P + \boldsymbol{\omega} \times \mathbf{r}$, and we may substitute this expression into Equation (7.6) to obtain

$$\begin{aligned}\mathbf{H}_P &= \int \mathbf{r} \times \mathbf{v}_P\, dm + \int \mathbf{r} \times (\boldsymbol{\omega} \times \mathbf{r})\, dm \\ &= \left(\int \mathbf{r}\, dm\right) \times \mathbf{v}_P + \int \mathbf{r} \times (\boldsymbol{\omega} \times \mathbf{r})\, dm\end{aligned}$$

Using the mass center definition, the integral in the first term on the right-hand side is $m\mathbf{r}_{PC}$. Therefore

$$\mathbf{H}_P = m\mathbf{r}_{PC} \times \mathbf{v}_P + \int \mathbf{r} \times (\boldsymbol{\omega} \times \mathbf{r})\, dm \tag{7.7}$$

In the cases where either (a) P is chosen to be the mass center C, or (b) $\mathbf{v}_P = \mathbf{0}$, or (c) $\mathbf{r}_{PC} \parallel \mathbf{v}_P$, the first term on the right side of Equation (7.7) vanishes. For these cases,

$$\mathbf{H}_P = \int \mathbf{r} \times (\boldsymbol{\omega} \times \mathbf{r})\, dm \tag{7.8}$$

Substituting $\boldsymbol{\omega}$ and $\mathbf{r}$ from Equations (7.4) and (7.5) and using the identity

$$\mathbf{A} \times (\mathbf{B} \times \mathbf{C}) \equiv (\mathbf{A} \cdot \mathbf{C})\mathbf{B} - (\mathbf{A} \cdot \mathbf{B})\mathbf{C} \tag{7.9}$$

we obtain the result

$$\begin{aligned}\mathbf{H}_P = {} & [\omega_x \textstyle\int (y^2 + z^2)\, dm - \omega_y \int xy\, dm - \omega_z \int xz\, dm]\hat{\mathbf{i}} \\ & + [-\omega_x \textstyle\int xy\, dm + \omega_y \int (x^2 + z^2)\, dm - \omega_z \int yz\, dm]\hat{\mathbf{j}} \\ & + [-\omega_x \textstyle\int xz\, dm - \omega_y \int yz\, dm + \omega_z \int (x^2 + y^2)\, dm]\hat{\mathbf{k}}\end{aligned} \tag{7.10}$$

Question 7.1 Why may the $\boldsymbol{\omega}$ components be brought outside the various integrals in Equation (7.10)?

Once we recognize the inertia properties (see Sections 4.3 and 4.4), this angular momentum expression becomes, for the case when P is C,

$$\begin{aligned}\mathbf{H}_C = {} & (I_{xx}^C\omega_x + I_{xy}^C\omega_y + I_{xz}^C\omega_z)\hat{\mathbf{i}} \\ & + (I_{xy}^C\omega_x + I_{yy}^C\omega_y + I_{yz}^C\omega_z)\hat{\mathbf{j}} \\ & + (I_{xz}^C\omega_x + I_{yz}^C\omega_y + I_{zz}^C\omega_z)\hat{\mathbf{k}}\end{aligned} \tag{7.11}$$

The form of the equation is identical if point P is not C, but rather either $\mathbf{v}_P = \mathbf{0}$ or $\mathbf{r}_{PC} \parallel \mathbf{v}_P$; the only difference is that the inertia properties are calculated with respect to axes at P instead of at the mass center:

$$\begin{aligned}\mathbf{H}_P = {} & (I_{xx}^P\omega_x + I_{xy}^P\omega_y + I_{xz}^P\omega_z)\hat{\mathbf{i}} \\ & + (I_{xy}^P\omega_x + I_{yy}^P\omega_y + I_{yz}^P\omega_z)\hat{\mathbf{j}} \\ & + (I_{xz}^P\omega_x + I_{yz}^P\omega_y + I_{zz}^P\omega_z)\hat{\mathbf{k}}\end{aligned} \tag{7.12}$$

Both of these forms for the angular momentum vector (Equations 7.11 and 7.12) will prove important to us in the sections to follow.

Answer 7.1 As we have seen in Chapter 6, $\boldsymbol{\omega}_{\mathcal{B}/\mathcal{I}}$ depends only on how a set of unit vectors, locked into $\mathcal{B}$, change their directions in $\mathcal{I}$. The angular velocity is a constant with regard to integration at a particular instant over the body's volume.

PROBLEMS ► Section 7.2

7.1 Find the angular momentum vector $\mathbf{H}_O$ of the wagon wheel of Problem 6.49.

7.2 Find the angular momentum vector of the disk $\mathcal{B}$ in Problem 6.27 about (a) C and (b) O.

7.3 Find the angular momentum vector for the bent bar of Example 4.16 about the mass center, if it turns about the z axis at angular speed ω.

7.4 A thin homogeneous disk $\mathcal{D}$ of mass M and radius r rotates with constant angular speed ω_2 about the shaft $\mathcal{S}$ (Figure P7.4). This shaft is cantilevered from the vertical shaft $\mathcal{R}$ and rotates with constant angular speed ω_1 about the axis of $\mathcal{R}$. Find the angular momentum of the disk about point Q, and show the direction of the vector in a sketch.

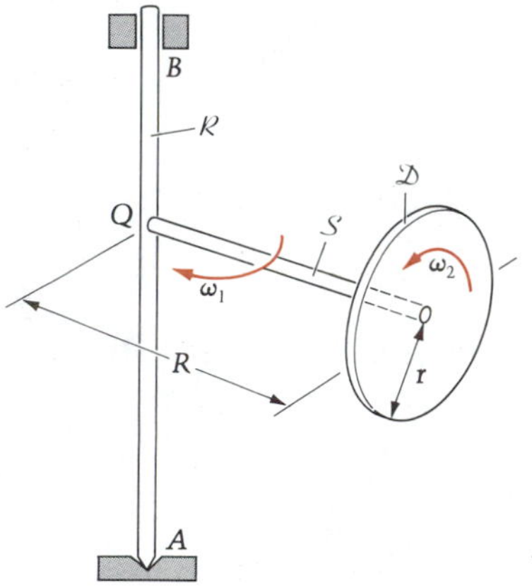

Figure P7.4

7.5 Depicted in Figure P7.5 is a grinder in a grinding mill that is composed of three main parts:

1. The vertical shaft $\mathcal{S}$, which rotates at constant angular speed Ω
2. The slanted shaft $\mathcal{B}$ of length ℓ, which is pinned to $\mathcal{S}$ and turns with it
3. The grinder $\mathcal{D}$ of radius r, turning in bearings at C about $\mathcal{B}$, and rolling on the inner surface of $\mathcal{I}$.

As shaft $\mathcal{S}$ gets up to speed Ω, body $\mathcal{B}$ swings outward, and then the angle ϕ remains constant during operation. Treat the grinder $\mathcal{D}$ as a disk and find its angular momentum vector $\mathbf{H}_C$ in convenient coordinates. (A suggested set is shown.)

7.6 In the preceding problem note that point O is a fixed point of all three bodies $\mathcal{S}$, $\mathcal{B}$, and $\mathcal{D}$ extended. Compute the angular momentum $\mathbf{H}_O$ of $\mathcal{D}$, and verify that $\mathbf{H}_O = \mathbf{H}_C + \mathbf{r}_{OC} \times \mathbf{L}$.

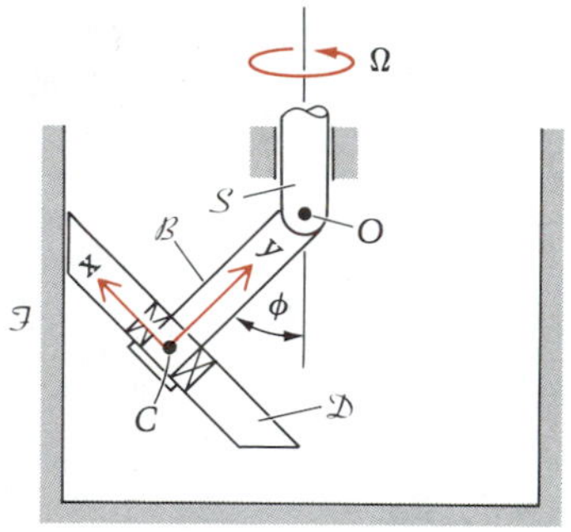

Figure P7.5

7.3 Transformations of Inertia Properties

Sometimes we need the moments and products of inertia at points other than the mass center C of a rigid body $\mathcal{B}$. These properties can be found without further integration by using the parallel-axis theorems, which were derived in Chapter 4. These are restated below, where $(\bar{x}, \bar{y}, \bar{z})$ are the coordinates of the mass center C relative to axes at P. For the moments of inertia at P:

$$I_{xx}^P = I_{xx}^C + m(\bar{y}^2 + \bar{z}^2) \quad (7.13a)$$

$$I_{yy}^P = I_{yy}^C + m(\bar{z}^2 + \bar{x}^2) \quad (7.13b)$$

$$I_{zz}^P = I_{zz}^C + m(\bar{x}^2 + \bar{y}^2) \quad (7.13c)$$

And for the products of inertia at P:

$$I^P_{xy} = I^C_{xy} - m\bar{x}\bar{y} \tag{7.14a}$$

$$I^P_{yz} = I^C_{yz} - m\bar{y}\bar{z} \tag{7.14b}$$

$$I^P_{zx} = I^C_{zx} - m\bar{z}\bar{x} \tag{7.14c}$$

Example 7.1

As a review example, compute the inertia properties at the corner B of the uniform rectangular solid of mass m shown in Figure E7.1.

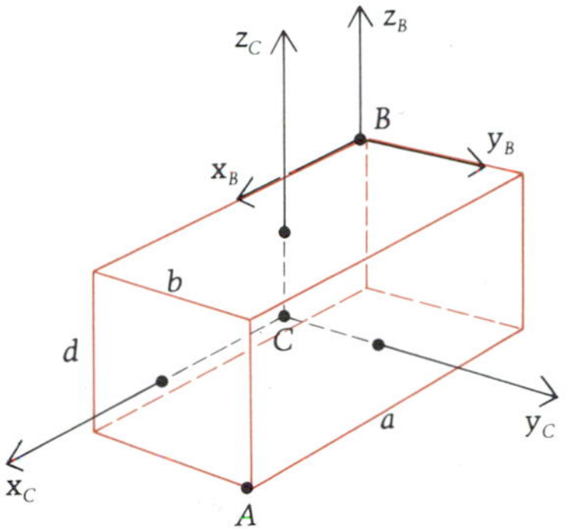

Figure E7.1

Solution

For the moments of intertia we obtain

$$I^B_{xx} = I^C_{xx} + m(\bar{y}^2 + \bar{z}^2) = \frac{m}{12}(b^2 + d^2) + m\left[\left(\frac{b}{2}\right)^2 + \left(\frac{d}{2}\right)^2\right] = \frac{m}{3}(b^2 + d^2)$$

Note that the distance between x_B and x_C is $\sqrt{(b/2)^2 + (d/2)^2}$. In the same way,

$$I^B_{yy} = \frac{m}{3}(d^2 + a^2) \qquad \text{and} \qquad I^B_{zz} = \frac{m}{3}(a^2 + b^2)$$

For the products of inertia,

$$I^B_{xy} = I^C_{xy} - m\bar{x}\bar{y} = 0 - m\left(\frac{a}{2}\right)\left(\frac{b}{2}\right) = \frac{-mab}{4}$$

$$I^B_{yz} = I^C_{yz} - m\bar{y}\bar{z} = 0 - m\left(\frac{b}{2}\right)\left(-\frac{d}{2}\right) = \frac{mbd}{4}$$

$$I^B_{xz} = I^C_{xz} - m\bar{x}\bar{z} = 0 - m\left(\frac{a}{2}\right)\left(-\frac{d}{2}\right) = \frac{mad}{4}$$

Transformation of Inertia Properties at a Point

We now consider a second and equally important transformation, which will demonstrate that if we know the moments and products of inertia associated with a set of orthogonal axes through a point P, we can easily compute the moments and products of inertia associated with any *other* set of axes having the same origin. Consider two sets of axes with a common origin at P. (See Figure 7.2.) Let ℓ_x, ℓ_y, ℓ_z be the direction cosines of x' relative to x, y, and z, respectively. Then the rectangular coordinate x' of a point Q in the body is related to the rectangular coordinates x, y, z by

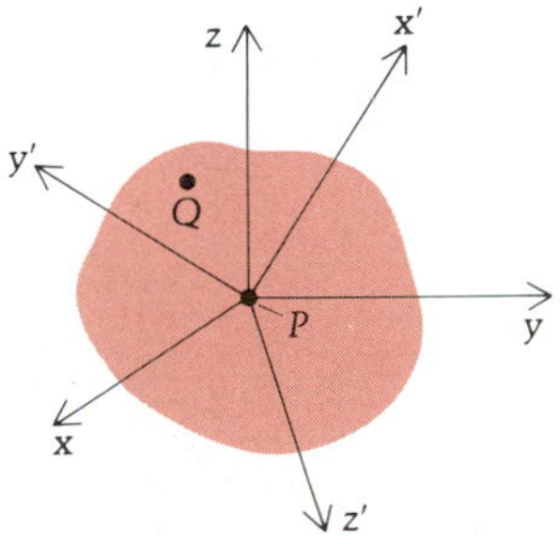

Figure 7.2

$$x' = \mathbf{r}_{PQ} \cdot \underbrace{(\ell_x\hat{\mathbf{i}} + \ell_y\hat{\mathbf{j}} + \ell_z\hat{\mathbf{k}})}_{\text{unit vector along } x' \text{ axis}}$$

$$= (x\hat{\mathbf{i}} + y\hat{\mathbf{j}} + z\hat{\mathbf{k}}) \cdot (\ell_x\hat{\mathbf{i}} + \ell_y\hat{\mathbf{j}} + \ell_z\hat{\mathbf{k}})$$

$$x' = x\ell_x + y\ell_y + z\ell_z \tag{7.15}$$

We seek a formula for $I^P_{x'x'}$ in terms of the inertia properties written with respect to the (x, y, z) axes. The definition of $I^P_{x'x'}$ is

$$I^P_{x'x'} = \int (y'^2 + z'^2)\, dm \tag{7.16}$$

Since $x'^2 + y'^2 + z'^2 = x^2 + y^2 + z^2$ (each is the square of the length of $\mathbf{r}_{PQ}$), we may add and subtract x'^2 to produce this quantity in Equation (7.16):

$$\begin{aligned} I^P_{x'x'} &= \int [(x'^2 + y'^2 + z'^2) - x'^2]\, dm \\ &= \int [(x^2 + y^2 + z^2) - x'^2]\, dm \end{aligned} \tag{7.17}$$

Substituting x' from Equation (7.15) into (7.17) gives

$$I^P_{x'x'} = \int [x^2 + y^2 + z^2 - (x\ell_x + y\ell_y + z\ell_z)^2]\, dm$$

Expanding the trinomial and rearranging, we get

$$\begin{aligned} I^P_{x'x'} = \int [&(1 - \ell_x^2)x^2 + (1 - \ell_y^2)\, y^2 + (1 - \ell_z^2)z^2 \\ &- 2xy\ell_x\ell_y - 2xz\ell_x\ell_z - 2yz\ell_y\ell_z]\, dm \end{aligned} \tag{7.18}$$

Since the ℓ's are the direction cosines of the vector in the direction of the x' axis, we know that

$$\ell_x^2 + \ell_y^2 + \ell_z^2 = 1$$

Using this relation in the first three terms of the integrand in Equation (7.18) gives

$$\begin{aligned} I^P_{x'x'} = \int [&(\ell_y^2 + \ell_z^2)\, x^2 + (\ell_x^2 + \ell_z^2)\, y^2 + (\ell_x^2 + \ell_y^2)\, z^2 \\ &- 2xy\ell_x\ell_y - 2xz\ell_x\ell_z - 2yz\ell_y\ell_z)]\, dm \end{aligned}$$

Rearranging, we have

$$\begin{aligned} I^P_{x'x'} = {}&\ell_x^2 \int (y^2 + z^2)\, dm + \ell_y^2 \int (x^2 + z^2)\, dm + \ell_z^2 \int (x^2 + y^2)\, dm \\ &+ 2\ell_x\ell_y\left(-\int xy\, dm\right) + 2\ell_x\ell_z\left(-\int zx\, dm\right) + 2\ell_y\ell_z\left(-\int yz\, dm\right) \end{aligned} \tag{7.19}$$

Recognizing the six integrals in Equation (7.19) as the inertia properties associated with the (x, y, z) directions at P, we arrive at our goal:

$$I^P_{x'x'} = \ell_x^2 I^P_{xx} + \ell_y^2 I^P_{yy} + \ell_z^2 I^P_{zz} + 2\ell_x\ell_y I^P_{xy} + 2\ell_x\ell_z I^P_{xz} + 2\ell_y\ell_z I^P_{yz} \tag{7.20}$$

This formula allows us to compute the moment of inertia of the mass of $\mathcal{B}$ about any line through P if we know the properties at P for any set of orthogonal axes. We now illustrate its use with an example.

EXAMPLE 7.2

Compute the moment of inertia about the diagonal BA of the rectangular solid of Example 7.1.

Solution

We define the axis x' to emanate from B, pointing toward A as in Figure E7.2. The inertia properties at B were computed in the prior example. The direction cosines

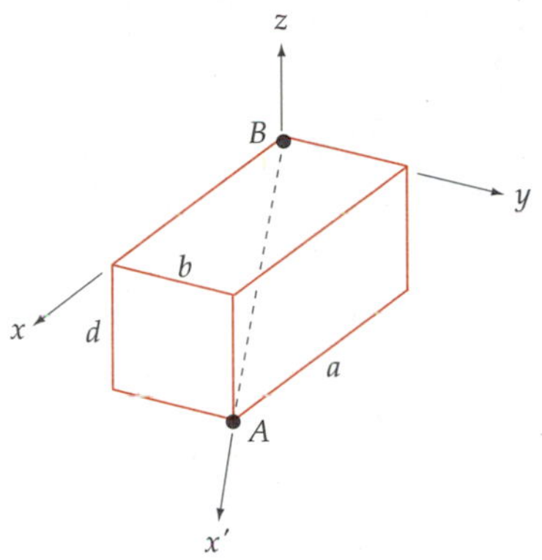

Figure E7.2

of x' are seen by inspection to be

$$\ell_x = \frac{a}{H} \qquad \ell_y = \frac{b}{H} \qquad \ell_z = \frac{-d}{H}$$

in which

$$H = \sqrt{a^2 + b^2 + d^2} = |\mathbf{r}_{BA}|$$

Substituting the ℓ_i's and the inertia properties at B into Equation (7.20) gives

$$\begin{aligned} I^B_{x'x'} &= \frac{m}{3H^2}[(b^2+d^2)a^2 + (d^2+a^2)b^2 + (a^2+b^2)d^2] \\ &\quad + \frac{2m}{H^2}\left[(ab)\left(\frac{-ab}{4}\right) + (-ad)\left(\frac{ad}{4}\right) + (-bd)\left(\frac{bd}{4}\right)\right] \\ &= \frac{m(a^2b^2 + b^2d^2 + d^2a^2)}{6(a^2+b^2+d^2)} \end{aligned}$$

In the preceding example we observe that the line BA also passes through the mass center C. The moment of inertia about line BA is of course the same no matter which point on the line one uses to make the calculation. (It would actually be easier to compute at C in this case, because the ℓ_i's are the same while the products of inertia vanish!)

A result similar to Equation (7.20) for products of inertia will now be derived. Let n_x, n_y, and n_z be the direction cosines of axis y'. Note that the rectangular coordinate y' may then be written, in the same way as Equation (7.15), as follows:

$$y' = xn_x + yn_y + zn_z \tag{7.21}$$

By definition,

$$\begin{aligned} I^P_{x'y'} &= -\int x'y'\,dm \\ &= -\int (x\ell_x + y\ell_y + z\ell_z)(xn_x + yn_y + zn_z)\,dm \end{aligned}$$

Therefore, expanding and recognizing the product of inertia integrals,

$$\begin{aligned} I^P_{x'y'} &= \int -(x^2\ell_x n_x + y^2\ell_y n_y + z^2\ell_z n_z)\,dm + (\ell_x n_y + \ell_y n_x)I^P_{xy} \\ &\quad + (\ell_x n_z + \ell_z n_x)I^P_{xz} + (\ell_y n_z + \ell_z n_y)I^P_{yz} \end{aligned} \tag{7.22}$$

Since (ℓ_x, ℓ_y, ℓ_z) and (n_x, n_y, n_z) are components of unit vectors along the mutually perpendicular axes x' and y', we may dot these vectors together and obtain

$$\ell_x n_x + \ell_y n_y + \ell_z n_z = 0$$

The integrand in Equation (7.22) may therefore be written as

$$\begin{aligned} &-(x^2\ell_x n_x + y^2\ell_y n_y + z^2\ell_z n_z) \\ &\qquad = x^2(\ell_y n_y + \ell_z n_z) + y^2(\ell_x n_x + \ell_z n_z) + z^2(\ell_x n_x + \ell_y n_y) \\ &\qquad = \ell_x n_x(y^2+z^2) + \ell_y n_y(x^2+z^2) + \ell_z n_z(x^2+y^2) \end{aligned} \tag{7.23}$$

Substituting Equation (7.23) into (7.22) then yields the desired transformation equation for the products of inertia:

$$I^P_{x'y'} = \ell_x n I^P_{xx} + \ell_y n_y I^P_{yy} + \ell_z n_z I^P_{zz} + (\ell_x n_y + \ell_y n_x) I^P_{xy} + (\ell_x n_z + \ell_z n_x) I^P_{xz} + (\ell_y n_z + \ell_z n_y) I^P_{yz} \tag{7.24}$$

EXAMPLE 7.3

In Examples 7.1 and 7.2 let the solid be a cube ($a = b = d$) and let y' be defined as follows (see Figure E7.3):

1. y' is perpendicular to x'.
2. y' is the same plane as z and x'.

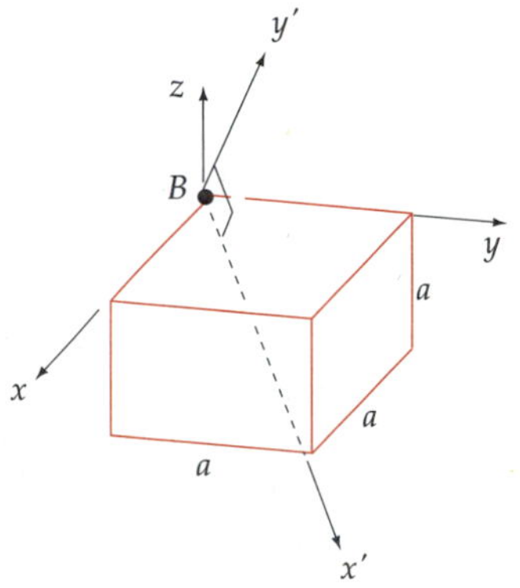

Figure E7.3

Find $I^B_{x'y'}$.

Solution

From the equations in Example 7.2 we have $\hat{\ell} = (\ell_x, \ell_y, \ell_z) = (1/\sqrt{3}, 1/\sqrt{3}, -1/\sqrt{3})$ for the unit vector along x'; we now force the components of $\hat{\mathbf{n}}$ — that is, (n_x, n_y, n_z) — to be such that conditions 1 and 2 are satisfied:

1. $\hat{\ell} \perp \hat{\mathbf{n}} \Rightarrow \dfrac{1}{\sqrt{3}} n_x + \dfrac{1}{\sqrt{3}} n_y - \dfrac{1}{\sqrt{3}} n_z = 0 \Rightarrow n_x + n_y - n_z = 0$

2. $\underbrace{(\hat{\ell} \times \hat{\mathbf{n}})} \cdot \hat{\mathbf{k}} = 0 \Rightarrow \dfrac{1}{\sqrt{3}} n_y - \dfrac{1}{\sqrt{3}} n_x = 0 \Rightarrow n_y - n_x = 0$

(a vector perpendicular to the plane of x' and y')

These two equations give $n_y = n_x$ and $n_z = 2n_x$. We must also ensure that $\hat{\mathbf{n}}$ is a unit vector:

$$1 = n_x^2 + n_y^2 + n_z^2 = n_x^2(1 + 1 + 4) = 6n_x^2$$

Thus $n_x = 1/\sqrt{6}$, so

$$n_x = \frac{1}{\sqrt{6}} \qquad n_y = \frac{1}{\sqrt{6}} \qquad n_z = \frac{2}{\sqrt{6}}$$

Substituting the components of $\hat{\ell}$ and $\hat{\mathbf{n}}$ and the inertia properties at B (letting the general point P in (7.24) be B in this problem) into Equation (7.24) gives

$$I^B_{x'y'} = \frac{2ma^2}{3}\left[\frac{1}{\sqrt{3}}\frac{1}{\sqrt{6}} + \frac{1}{\sqrt{3}}\frac{1}{\sqrt{6}} - \frac{1}{\sqrt{3}}\left(\frac{2}{\sqrt{6}}\right)\right] + \frac{ma^2}{4}\left[-\left(\frac{1}{\sqrt{3}}\frac{1}{\sqrt{6}} + \frac{1}{\sqrt{3}}\frac{1}{\sqrt{6}}\right) + \left(\frac{1}{\sqrt{3}}\frac{2}{\sqrt{6}} - \frac{1}{\sqrt{3}}\frac{1}{\sqrt{6}}\right) + \left(\frac{1}{\sqrt{3}}\frac{2}{\sqrt{6}} - \frac{1}{\sqrt{3}}\frac{1}{\sqrt{6}}\right)\right] = 0$$

We note that in Example 7.3 the zero result is not obvious at this point in our study. While it is true that, for this case of $a = b = d$, the $x'y'$

plane is a plane of symmetry, this guarantees (see Section 4.4) that $I^B_{x'z'}$ and $I^B_{y'z'}$ are zero but not necessarily $I^B_{x'y'}$. Note also that there are two directions (180° apart) for y' that both satisfy conditions 1 and 2 in the preceding example.

Question 7.2 Where in the solution did we choose one of these directions? (And does it matter?)

We close this section by noting that Equations (7.20) and (7.24) are the transformation equations satisfied by a symmetric second-order tensor; thus the inertia properties do indeed form such a tensor. We also note from these two equations that only if the products of inertia are defined with the minus sign (see Equations 4.2) do we get the correct tensor transformation equations.

Answer 7.2 When we said that $n_x = +\sqrt{1/6}$, that is, took the positive square root, we chose y' to be in the direction making an acute angle with x_B; had we chosen $-\sqrt{1/6}$, we would have gotten the opposite direction for y'. And had $I_{x'y'}$ been nonzero, the sign of the answer would have been opposite also.

PROBLEMS ▶ Section 7.3

7.7 The three homogeneous rods in Figure P7.7 are welded together at O to form a rigid body. Find the mass moments and products of inertia at point Q with respect to axes there that are parallel to x, y, and z.

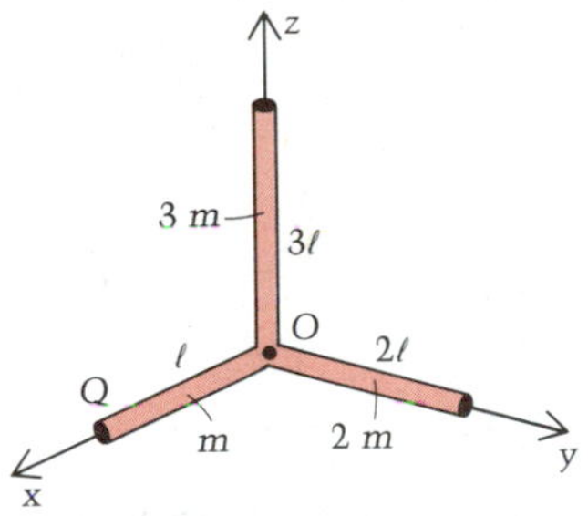

Figure P7.7

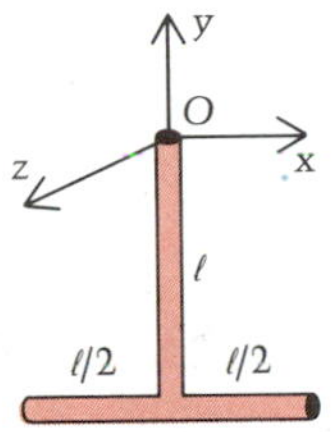

Figure P7.8

7.8 Find the mass moments and products of inertia of the body in Figure P7.8, with respect to a set of axes through the point P at $\left(\frac{\ell}{2}, \frac{\ell}{2}, \frac{\ell}{2}\right)$, respectively, parallel to x, y, and z. Each of the two perpendicular rods of the "T" has mass m and length ℓ.

7.9 In the preceding problem, find the moment of inertia about the line OP.

7.10 Compute the moment of inertia with respect to line AB for the bent bar in Figure P7.10. The bar lies in a plane and has mass 4 m.

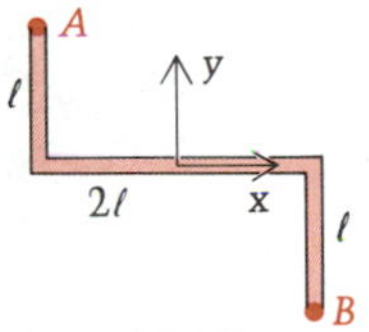

Figure P7.10

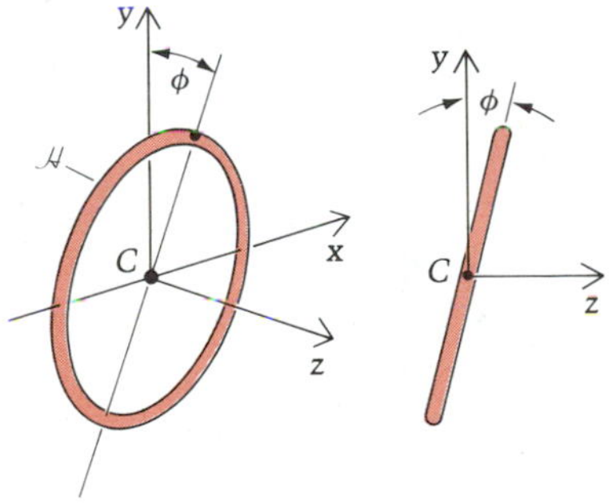

Figure P7.11

7.11 Find the product of inertia I^C_{yz} for the hoop $\mathcal{H}$ of mass m and radius R in Figure P7.11. The plane of $\mathcal{H}$ is misaligned with the xy plane by angle ϕ.

7.12 Compute the moment of inertia about line *BA* in Example 7.2 by using Equation (7.20) at *C* instead of *B*. Show that if $a = b = d$, the answer becomes equal to I_{xx}^C (which equals $I_{yy}^C = I_{zz}^C$ in this case).

7.13 The centroidal moments of inertia for the solid ellipsoid $\mathcal{E}$ in Figure P7.13 are

$$I_{xx}^C = \frac{m}{5}(b^2 + c^2)$$

$$I_{yy}^C = \frac{m}{5}(a^2 + c^2)$$

$$I_{zz}^C = \frac{m}{5}(a^2 + b^2)$$

Further, the mass of $\mathcal{E}$ is $(4\pi/3)\,\rho abc$, where ρ is the mass density. Find the moment of inertia of the mass of $\mathcal{E}$ about the line making equal angles with x, y, and z.

7.14 Show that the sum of any two of I_{xx}^P, I_{yy}^P, and I_{zz}^P always exceeds the third.

7.15 Part of a special-purpose, dual-driven antenna system consists of an octagonal rotator as shown in Figure P7.15. Each of the eight equal sections is a square steel tube with the indicated dimensions and thickness $\frac{1}{8}$ in. Find the moment of inertia of the rotator about the axis of rotation. (Consider each section to have squared-off ends at the average 18-in. length and ignore the small overlaps. Use a density of 15.2 slug/ft³.)

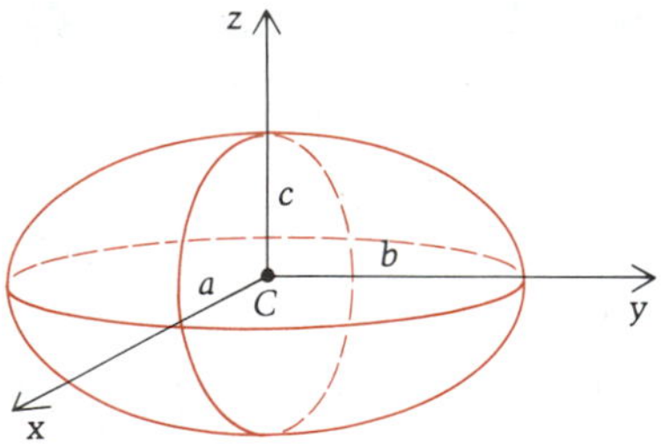

Figure P7.13

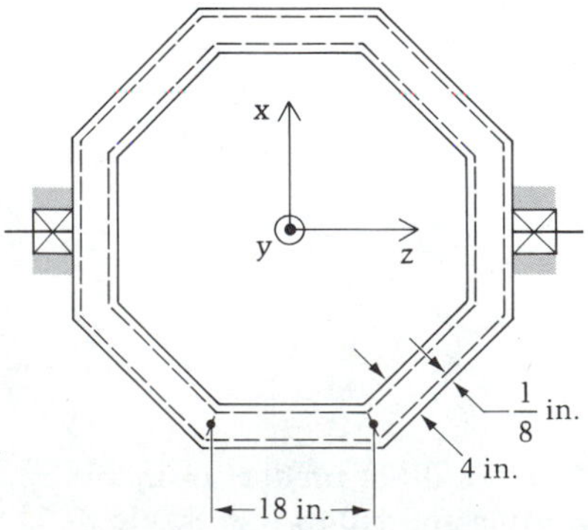

Figure P7.15

7.16 Find the inertia properties at *O* for the body shown in Figure P7.16, which is composed of a rod and ring that have equal cross sections and densities. The rod is perpendicular to the plane of the ring.

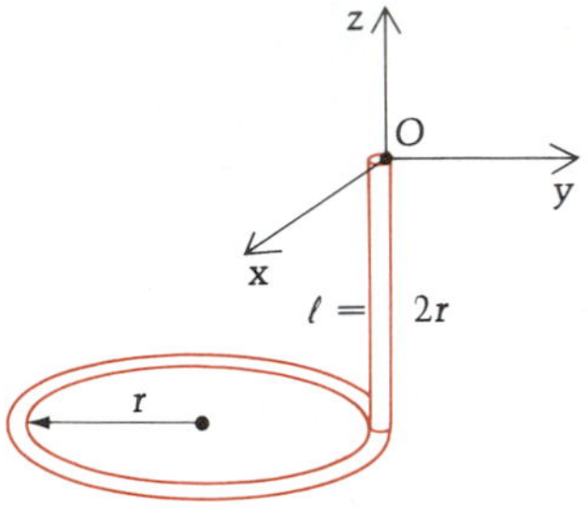

Figure P7.16

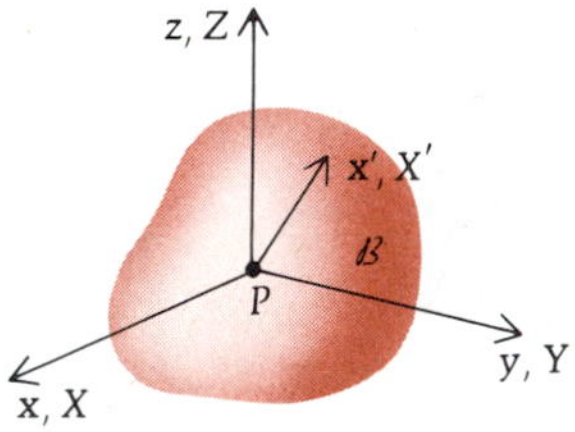

Figure P7.18

* **7.17** Find $I_{x'y'}^B$ for the problem of Example 7.3 if $b = d = a/2$. The axes have their origin at *B* just as in the example.

* **7.18** For the rigid body $\mathcal{B}$ in Figure P7.18 let the inertia properties at *P* be known ($I_{xx}^P, \ldots, I_{yz}^P$). Further, let there be measured along the same axes the quantities

$$X = \frac{\ell_x}{\sqrt{I_{x'x'}^P}}$$

$$Y = \frac{\ell_y}{\sqrt{I_{x'x'}^P}}$$

$$Z = \frac{\ell_z}{\sqrt{I_{x'x'}^P}}$$

where $I_{x'x'}^P$ and (ℓ_x, ℓ_y, ℓ_z) are as defined in Section 7.3. Show that Equation (7.20) then implies

$$I_{xx}^P X^2 + I_{yy}^P Y^2 + I_{zz}^P Z^2 + 2XYI_{xy}^P + 2XZI_{xz}^P + 2YZI_{yz}^P = 1$$

This is the equation of an ellipsoid centered at *P*. Developed by Cauchy in 1827, it is called the *ellipsoid of inertia*. Show that the moment of inertia about any line x' through *P* equals the reciprocal of the square of the distance from *P* to the point where X' intersects the ellipsoid.

* Asterisks identify the more difficult problems.

7.19 In the preceding problem, if the products of inertia vanish, the equation of the ellipsoid of inertia written in terms of the resulting moments of inertia (I_1^P, I_2^P, I_3^P) is

$$I_1^P X^2 + I_2^P Y^2 + I_3^P Z^2 = 1$$

Show that not all ellipsoids of the form $ax^2 + by^2 + cz^2 = 1$ can be ellipsoids of inertia. *Hint:* The sum of any two moments of inertia must always exceed the third, as stated in Problem 7.14.

** **7.20** Calculate the inertia properties at O of the three circular fan blades connected by light rods in Figure P7.20. The blades are tilted 30° with respect to the axes OC_1, OC_2, and OC_3; the shaded halves of each are behind the plane of the drawing and the unshaded halves are in front of it.

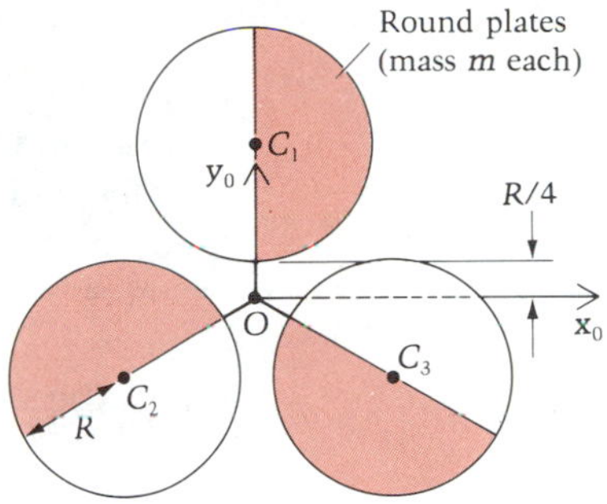

Figure P7.20

7.4 Principal Axes and Principal Moments of Inertia

In this section we demonstrate a particularly useful way of describing the inertia characteristics of a rigid body $\mathcal{B}$. It happens that, at any point P of $\mathcal{B}$, it is always possible to find a set of rectangular axes so that the products of inertia at P with respect to these axes all vanish. These axes are called the **principal axes of inertia** at P, and the moments of inertia with respect to them are known as the **principal moments of inertia** for the point.

Specifically, an axis x is a principal axis at P if $I_{x\beta}^P = 0$, where β is any axis through P that is perpendicular to x. (See Figure 7.3.) We can show that if $I_{xy}^P = I_{xz}^P = 0$, in which (x, y, z) form a triad of rectangular axes at P, then x is a principal axis at P. For the proof, we shall show that $I_{xy}^P = 0 = I_{xz}^P$ implies $I_{x\beta}^P = 0$, where β is the arbitrary axis through P normal to x. Using Equation (7.24) we may write

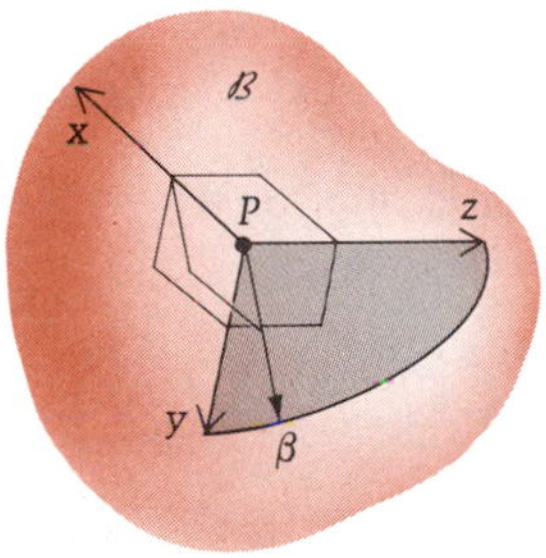

Figure 7.3

$$I_{x\beta}^P = \ell_x n_x I_{xx}^P + \ell_y n_y I_{yy}^P + \ell_z n_z I_{zz}^P + (\ell_x n_y + \ell_y n_x) I_{xy}^P$$
$$+ (\ell_x n_z + \ell_z n_x) I_{xz}^P + (\ell_y n_z + \ell_z n_y) I_{yz}^P$$

In this equation $(\ell_x, \ell_y, \ell_z) = (1, 0, 0)$ are the direction cosines of x (the first subscript in $I_{x\beta}$) with respect to x, y, and z, respectively, while (n_x, n_y, n_z) are the direction cosines of β (the second subscript in $I_{x\beta}$), also with respect to x, y, and z. Since $\beta \perp x$, we have $(n_x, n_y, n_z) = (0, n_y, n_z)$. Substituting these ℓ's and n's, we get

$$I_{x\beta}^P = 0 + 0 + 0 + (\ell_x n_y + 0) I_{xy}^P + (\ell_x n_z + 0) I_{xz}^P + 0$$

But $I_{xy}^P = I_{xz}^P = 0$, so that $I_{x\beta}^P = 0$ and we see that all it takes to make an axis, such as x, a principal axis at a point is to show that $I_{xy}^P = 0$ and $I_{xz}^P = 0$, where (x, y, z) form an orthogonal triad at P. We shall need this result in what is to follow.

Principal Axes at C

We now proceed to a computational procedure for finding the principal axes and moments of inertia. We shall do the derivation when P is C and then later explain how it applies equally well to *all* points of $\mathcal{B}$.

For the moment let (x, y, z) be centered at C and at some instant let x be parallel to the angular velocity $\boldsymbol{\omega}_{\mathcal{B}/\mathcal{I}} = \boldsymbol{\omega}$ of $\mathcal{B}$ in a reference frame $\mathcal{I}$. Then $\boldsymbol{\omega} = \omega\hat{\mathbf{i}}$ and Equation (7.11) gives, for the angular momentum of $\mathcal{B}$ about C, the simplified expression

$$\mathbf{H}_C = I_{xx}^C\omega\hat{\mathbf{i}} + I_{xy}^C\omega\hat{\mathbf{j}} + I_{xz}^C\omega\hat{\mathbf{k}} \tag{7.25}$$

We note from Equation (7.25) that $\mathbf{H}_C$ is parallel to $\boldsymbol{\omega}$ if and only if $I_{xy}^C = I_{xz}^C = 0$ — that is, if x is a principal axis at C. Thus we see, as did Leonhard Euler himself in the middle of the eighteenth century, that the principal axes have the property that when the angular velocity lies along one of them, so will the angular momentum. Euler was seeking an axis through C for which, when $\mathcal{B}$ was set spinning about it, the motion would continue about this axis without any need for external moments to maintain it. Note further that when $\mathbf{H}_C \parallel \boldsymbol{\omega}$ ("$\mathbf{H}_C$ is parallel to $\boldsymbol{\omega}$"), the proportionality constant is necessarily the moment of inertia I about their common axis.

Now we are ready for the big step. We let $\hat{\mathbf{n}} = n_x\hat{\mathbf{i}} + n_y\hat{\mathbf{j}} + n_z\hat{\mathbf{k}}$ be a unit vector and we seek the direction of $\hat{\mathbf{n}}$ that will ensure its being a principal axis. In other words, we want to find the values of the direction cosines of $\hat{\mathbf{n}}$ (n_x, n_y, n_z) such that if $\boldsymbol{\omega} = \omega\hat{\mathbf{n}}$, then $\mathbf{H} = I\boldsymbol{\omega}$. Writing $\boldsymbol{\omega}$ in component form gives

$$\boldsymbol{\omega} = \omega n_x\hat{\mathbf{i}} + \omega n_y\hat{\mathbf{j}} + \omega n_z\hat{\mathbf{k}} \tag{7.26}$$

Substituting from Equation (7.11) for $\mathbf{H}_C$, the vector relation $\mathbf{H}_C = I\boldsymbol{\omega}$ then gives the following three scalar component equations:

$$\begin{aligned}
I_{xx}^C\omega_x + I_{xy}^C\omega_y + I_{xz}^C\omega_z &= I\omega n_x \\
I_{xy}^C\omega_x + I_{yy}^C\omega_y + I_{yz}^C\omega_z &= I\omega n_y \\
I_{xz}^C\omega_x + I_{yz}^C\omega_y + I_{zz}^C\omega_z &= I\omega n_z
\end{aligned} \tag{7.27}$$

If we divide all three of Equations (7.27) by ω and note that since $\boldsymbol{\omega} = \omega\hat{\mathbf{n}}$ we have $n_x = \omega_x/\omega$, $n_y = \omega_y/\omega$, and $n_z = \omega_z/\omega$, then we may rearrange the equations as follows:

$$\begin{aligned}
(I_{xx}^C - I)n_x + I_{xy}^C n_y + I_{xz}^C n_z &= 0 \\
I_{xy}^C n_x + (I_{yy}^C - I)n_y + I_{yz}^C n_z &= 0 \\
I_{xz}^C n_x + I_{yz}^C n_y + (I_{zz}^C - I)n_z &= 0
\end{aligned} \tag{7.28}$$

We now have a set of three equations that are algebraic, linear, and homogeneous in the three variables n_x, n_y, n_z. Such a system is known to have a nontrivial solution if and only if the determinant of the coefficients of the variables is zero.* In this case, we may drop the "nontrivial" adjective because the trivial solution ($n_x = n_y = n_z = 0$) fails to satisfy the side condition

$$n_x^2 + n_y^2 + n_z^2 = 1 \tag{7.29}$$

which must always be true for the direction cosines of a vector.

Calculation of Principal Moments of Inertia

Setting the determinant of the coefficients in (7.28) equal to zero will lead first to the special values of I for which the three equations have a solution. Each special value I is called an *eigenvalue,* or *characteristic value,* and will be a principal moment of inertia; the corresponding $\hat{\mathbf{n}}$ (with components n_x, n_y, n_z) is called the *eigenvector* associated with this eigenvalue I. The unit vector $\hat{\mathbf{n}}$ points in the direction of a principal axis of inertia at C. The determinant is equated to zero below:

$$\begin{vmatrix} I_{xx}^C - I & I_{xy}^C & I_{xz}^C \\ I_{xy}^C & I_{yy}^C - I & I_{yz}^C \\ I_{xz}^C & I_{yz}^C & I_{zz}^C - I \end{vmatrix} = 0 \tag{7.30}$$

If we expand this characteristic determinant, we clearly get a cubic polynomial in I:

$$I^3 + a_1 I^2 + a_2 I + a_3 = 0 \tag{7.31}$$

The a_i are of course functions of the inertia properties. Now we know from algebra that if polynomials with real coefficients have any complex roots, they must occur in conjugate pairs. Thus the polynomial derived above has at this point at least one real root I_1. (It is positive by the definition of the quantity "moment of inertia" that it represents.) We now are guaranteed at least one principal moment of inertia and corresponding principal axis of inertia.

In order to show that there are two others, we next reorient our orthogonal triad of reference axes so that one of them (x) coincides with the already identified principal axis; this then allows us to write $I_{xy}^C = 0$, $I_{xz}^C = 0$, and $I_{xx}^C = I_1$, where y and z are now a new pair of axes normal to our new (principal) x axis. Equations (7.28) now appear as

$$\begin{aligned} (I_1 - I)n_x &+ 0n_y &&+ 0n_z &&= 0 \\ 0n_x &+ (I_{yy}^C - I)n_y &&+ I_{yz}^C n_z &&= 0 \\ 0n_x &+ I_{yz}^C n_y &&+ (I_{zz}^C - I)n_z &&= 0 \end{aligned} \tag{7.32}$$

and the determinantal equation becomes:

$$\begin{vmatrix} I_1 - I & 0 & 0 \\ 0 & I_{yy}^C - I & I_{yz}^C \\ 0 & I_{yz}^C & I_{zz}^C - I \end{vmatrix} = 0 \tag{7.33}$$

* Cramer's rule clearly gives $n_x = n_y = n_z = 0$ as the only solution if the equations are independent, in which case the determinant D of the coefficients is not zero. If $D = 0$, then Cramer's rule yields the indeterminate form $0/0$ for the n's and there is a chance for other solutions, as the equations are then dependent.

This time the resulting cubic becomes factorable. Expanding the determinant, we have

$$(I_1 - I)[(I_{yy}^C - I)(I_{zz}^C - I) - I_{yz}^{C^2}] = 0 \tag{7.34}$$

The principal moments of inertia at C are the roots of Equation (7.34). The first root (the one we already knew) is reaffirmed by setting the first factor to zero:

$$I = I_1 \qquad (= I_{xx}^C)$$

The others will be seen to come from equating the second factor to zero:

$$I^2 - (I_{yy}^C + I_{zz}^C)I + (I_{yy}^C I_{zz}^C - I_{yz}^{C^2}) = 0 \tag{7.35}$$

This is, of course, a quadratic equation in I. Recalling that the two roots to

$$aI^2 + bI + c = 0$$

are

$$\frac{-b \pm \sqrt{b^2 - 4ac}}{2a}$$

we see that we shall have two (more) real roots I_2 and I_3 if the discriminant is positive or zero:

$$\begin{aligned} b^2 - 4ac &= (I_{yy}^C + I_{zz}^C)^2 - 4(1)(I_{yy}^C I_{zz}^C - I_{yz}^{C^2}) \\ &= I_{yy}^{C^2} - 2I_{yy}^C I_{zz}^C + I_{zz}^{C^2} + 4I_{yz}^{C^2} \\ &= (I_{yy}^C - I_{zz}^C)^2 + 4I_{yz}^{C^2} \geq 0 \end{aligned} \tag{7.36}$$

Therefore all three roots of the characteristic cubic equation are real (and positive), and so we always have three principal moments of inertia at C, each with its own corresponding principal axis.*

Calculation of Principal Directions

We mention at this point the procedure for obtaining the principal direction, given by n_x, n_y, and n_z, for *each* of the principal moments of inertia (I_1, I_2, or I_3). Equations (7.28), being dependent, may not be solved for the three components of each $\hat{\mathbf{n}}$ in themselves; however, together with the identity $n_x^2 + n_y^2 + n_z^2 = 1$, a solution may be found. The idea is to solve for, say, n_y and n_z in terms of n_x from two of Equations (7.28); then we substitute into Equation (7.29) and solve for n_x. Either sign may be used in taking the final square root, because there are obviously two legitimate sets of direction cosines. These two sets are negatives of each other, and each yields the correct principal axis. In Figure 7.4, either $\hat{\mathbf{n}}$ or $-\hat{\mathbf{n}}$ defines a principal axis through C. The principal axis is an undirected line.

In the first of two examples to follow, we shall again see (as in the preceding discussion) that if at least two of the products of inertia vanish,

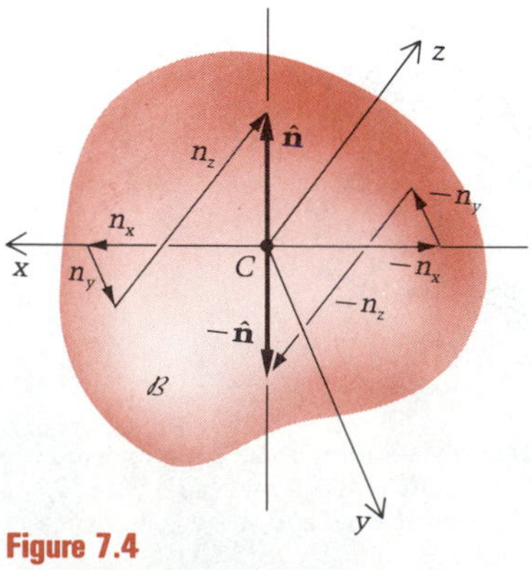

Figure 7.4

* This was proved in 1755 for the first time by Segner, a contemporary of Euler. Segner also showed that the principal axes (for distinct principal moments of inertia) are orthogonal.

then the cubic equation (7.31) becomes factorable. In this case, we do not have to solve it numerically.

EXAMPLE 7.4

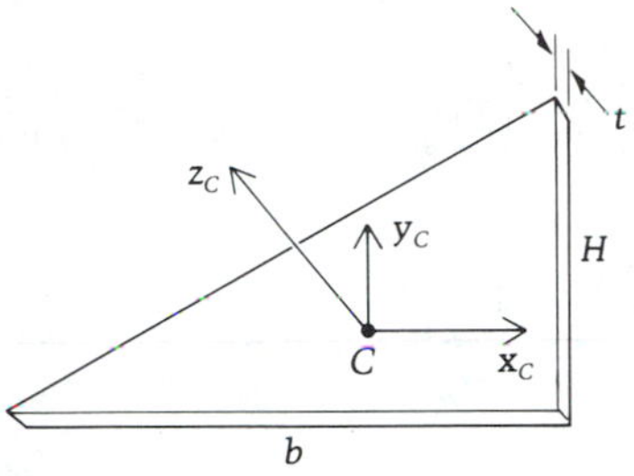

Figure E7.4a

The inertia properties of a right triangular plate (see Figure E7.4a) are

$$I_{xx}^C = \frac{mH^2}{18} \qquad I_{xy}^C = \frac{-mbH}{36}$$

$$I_{yy}^C = \frac{mb^2}{18} \qquad I_{xz}^C = 0$$

$$I_{zz}^C = \frac{m(H^2 + b^2)}{18} \qquad I_{yz}^C = 0$$

Find the principal moments of inertia of the plate. Then find their associated principal axes when $b = H$.

Solution

Equations (7.28), which lead to the principal moments and axes of inertia, become, for the plate,

$$\left(\frac{mH^2}{18} - I\right) n_x - \frac{mbH}{36} n_y + 0n_z = 0 \tag{1}$$

$$\frac{-mbH}{36} n_x + \left(\frac{mb^2}{18} - I\right) n_y + 0n_z = 0 \tag{2}$$

$$0n_x + 0n_y + \left(\frac{m(b^2 + H^2)}{18} - I\right) n_z = 0 \tag{3}$$

The algebra is simplified by dividing by $mH^2/36$ and defining

$$B = \frac{b}{H} \quad \text{and} \quad I^* = \frac{I}{(mH^2/36)} \tag{4}$$

This gives, in terms of the nondimensional parameters B and I^*,

$$(2 - I^*)n_x + (-B)n_y + 0n_z = 0 \tag{5}$$

$$(-B)n_x + (2B^2 - I^*)n_y + 0n_z = 0 \tag{6}$$

$$0n_x + 0n_y + [2(B^2 + 1) - I^*]n_z = 0 \tag{7}$$

Therefore the determinantal equation becomes

$$\begin{vmatrix} 2 - I^* & -B & 0 \\ -B & 2B^2 - I^* & 0 \\ 0 & 0 & 2(B^2 + 1) - I^* \end{vmatrix} = 0 \tag{8}$$

Expanding across the third row (or down the third column), we get

$$[2(B^2 + 1) - I^*][(2 - I^*)(2B^2 - I^*) - B^2] = 0 \tag{9}$$

Therefore one of the brackets must vanish and the roots come from

$$I^* = 2(B^2 + 1) \tag{10}$$

and

$$I^{*2} - 2I^*(B^2 + 1) + 3B^2 = 0 \tag{11}$$

Equation (10) gives, using Equations (4),

$$I_3 = \frac{m(b^2 + H^2)}{18} \tag{12}$$

Equation (11) gives, by the quadratic formula,

$$I^*_{1,2} = \frac{2(B^2 + 1) \pm \sqrt{4(B^2 + 1)^2 - 12B^2}}{2} \tag{13}$$

Thus

$$I_{1,2} = \frac{mH^2}{36}\left(\frac{b^2}{H^2} + 1 \pm \sqrt{\frac{b^4}{H^4} - \frac{b^2}{H^2} + 1}\right)$$

$$= \frac{m(b^2 + H^2)}{36} \pm \frac{m}{36}\sqrt{b^4 - b^2H^2 + H^4} \tag{14}$$

The three principal moments of inertia of the plate are given by Equations (12) and (14). In the case when the right triangle is isosceles ($b = H$), we have $B = 1$ so that, from Equation (13),

$$I^*_{1,2} = 2 \pm \sqrt{4 - 3} = 2 \pm 1$$

or

$$I^*_1 = 3 \qquad \text{and} \qquad I^*_2 = 1 \tag{15}$$

Also, from Equation (10),

$$I^*_3 = 2(1^2 + 1) = 4$$

Changing back to dimensional inertias by Equation (4), we see that for a right isosceles triangular plate,

$$I_1 = \frac{mH^2}{12} \qquad I_2 = \frac{mH^2}{36} \qquad I_3 = \frac{mH^2}{9} \tag{16}$$

We shall now determine the principal axis associated with each of these principal moments of inertia. We first substitute $I^*_1 = 3$ (with $B = 1$) in each of Equations (5) to (7) and get

$$-n_x - n_y = 0$$

$$-n_x - n_y = 0$$

$$n_z = 0 \tag{17}$$

The third of these equations says that the principal axis for I_1 is in the plane of the plate (xy); the other two equations both give

$$n_x = -n_y \tag{18}$$

Substituting this result into

$$n_x^2 + n_y^2 + n_z^2 = 1 \tag{19}$$

gives

$$(-n_y)^2 + n_y^2 = 1 \tag{20}$$

$$n_y^2 = \frac{1}{2}$$

$$n_y = \pm\frac{1}{\sqrt{2}} \tag{21}$$

Thus, by Equation (18),

$$n_x = \mp\frac{1}{\sqrt{2}} \tag{22}$$

so that either

$$(n_x, n_y, n_z) = \left(-\frac{1}{\sqrt{2}}, \frac{1}{\sqrt{2}}, 0\right) \tag{23}$$

or

$$(n_x, n_y, n_z) = \left(\frac{1}{\sqrt{2}}, -\frac{1}{\sqrt{2}}, 0\right) \tag{24}$$

The lines ℓ_1 defined by these two sets of direction cosines are shown in Figure E7.4b and Figure E7.4c. It is seen that the two preceding results represent the *same line;* the positive directions are opposite but unimportant. The inertia value, being the integral of $r^2\,dm$, is independent of the directivity of the line.

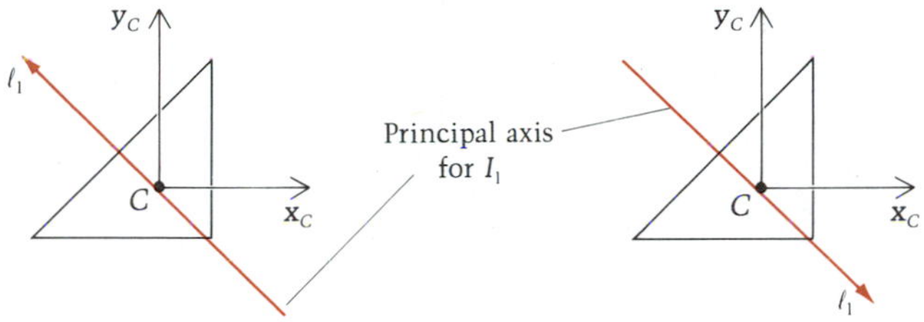

(a) Line of Eq. (7.12)

Figure E7.4b

(b) Line of Eq. (7.13)

Figure E7.4c

For I_2^*, Equations (5) to (7) become

$$n_x - n_y = 0$$
$$-n_x + n_y = 0$$
$$n_z = 0 \tag{25}$$

This time $n_x = n_y$ with n_z again equaling zero, so this principal axis makes equal angles with x_C and y_C. (See Figure E7.4d and Figure E7.4e.) Note from the pre-

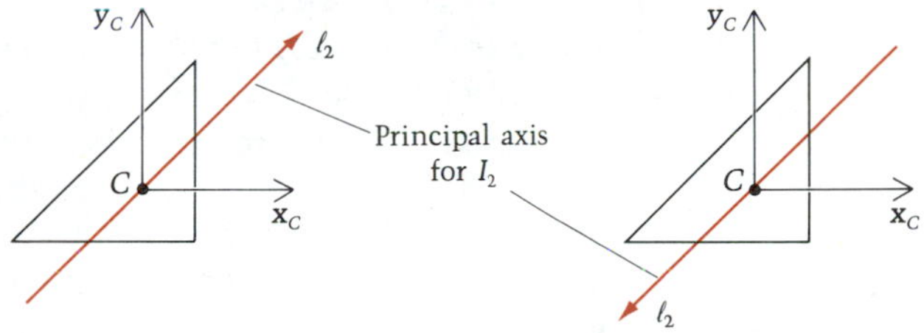

$n_x = n_y = 1/\sqrt{2}$

Figure E7.4d

$n_x = n_y = -1/\sqrt{2}$

Figure E7.4e

ceding diagrams that there is much more inertia about the principal axis for I_1 than there is for I_2; the mass is more closely clustered about ℓ_2 than it is about ℓ_1.

The third principal axis is found from Equations (5) to (7) when, for $B = 1$,

$$I^* = I_3^* = 2(B^2 + 1) = 4 \tag{26}$$

These equations are

$$\begin{aligned} -2n_x - n_y &= 0 \\ -n_x - 2n_y &= 0 \\ [2(2) - 4]n_z &= 0 \end{aligned} \tag{27}$$

The first two of these equations have the solution $n_x = n_y = 0$. The third leaves n_z indeterminate. But from Equation (19) we have $n_z = 1$ or -1. Thus the principal axis for I_3 is the line normal to the plate at C. This will in fact always be true: When a body is a *plate* (that is, flat with negligible thickness compared to its other dimensions), the moment of inertia about the axis normal to the plate at any point is principal for that point. It is even true that it is the sum of the other two and therefore the largest.

Principal Axes at Any Point

We now wish to remark that a set of principal axes exists at *every* point of $\mathcal{B}$, not just at the mass center C. To show this we first recall Equation (7.7), which with Equation (7.12) gives the angular momentum $\mathbf{H}_P$ about any point P of body $\mathcal{B}$:

$$\begin{aligned} \mathbf{H}_P = m\mathbf{r}_{PC} \times \mathbf{v}_P &+ [I_{xx}^P\omega_x + I_{xy}^P\omega_y + I_{xz}^P\omega_z]\hat{\mathbf{i}} \\ &+ [I_{xy}^P\omega_x + I_{yy}^P\omega_y + I_{yz}^P\omega_z]\hat{\mathbf{j}} \\ &+ [I_{xz}^P\omega_x + I_{yz}^P\omega_y + I_{zz}^P\omega_z]\hat{\mathbf{k}} \end{aligned}$$

Whenever the cross-product term in $\mathbf{H}_P$ vanishes, the terms that remain are identical to those of Equation (7.11) if P replaces C. Therefore, for cases in which $\mathbf{r}_{PC} \times \mathbf{v}_P = \mathbf{0}$, we need only recall our arguments made for C and we shall be led, through an identical procedure, to the principal axes and moments of inertia for *any* point P of $\mathcal{B}$.

Therefore we only need to imagine that at some instant our point P of interest has either $\mathbf{v}_P = \mathbf{0}$ or $\mathbf{v}_P \parallel \mathbf{r}_{PC}$. In either case, $\mathbf{r}_{PC} \times \mathbf{v}_P = \mathbf{0}$ and $\mathbf{H}_P$ then has the form of Equation (7.12). Retracing our steps from that point, we arrive at the three axes (through P this time) for which all three products of inertia are zero; these are the same three axes through P for which $\mathbf{H}_P \parallel \boldsymbol{\omega}$ whenever $\boldsymbol{\omega}$ is aligned with one of them and $\mathbf{v}_P$ either vanishes or is parallel to $\mathbf{r}_{PC}$. Of course, all references to the motion conditions that led to the determinantal equation are again lost (as they were for C), so that the principal moments and axes of inertia depend only on the body's mass distribution.

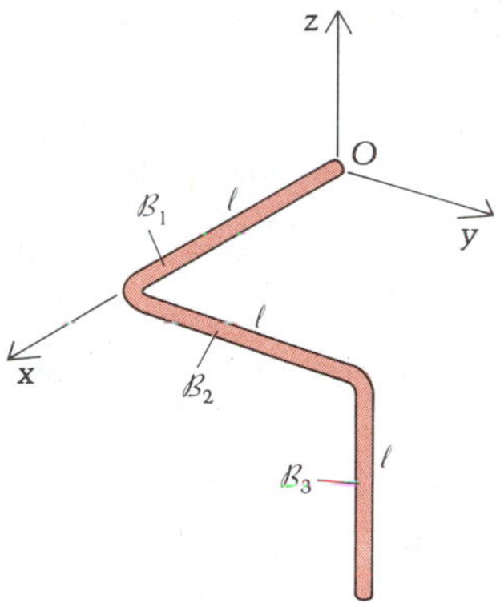

Figure E7.5a

Example 7.5

Find the principal moments of inertia at O and the directions of their associated principal axes for the body shown in Figure E7.5a. It is made up of three rigid, identical slender rods welded together at right angles to form a single rigid body $\mathcal{B}$.

Solution

Using the moments of inertia shown for one rod in Figure E7.5b, plus the parallel-axis (transfer) theorem, the six inertia properties are calculated below. The reader should verify each of the entries.

$$I_{xx}^O = \underbrace{\frac{m\ell^2}{3}}_{\mathcal{B}_2} + \underbrace{\frac{m\ell^2}{12} + m\left[\ell^2 + \left(\frac{\ell}{2}\right)^2\right]}_{\mathcal{B}_3} = m\ell^2\left(\frac{4+1+15}{12}\right) = \frac{10}{6}m\ell^2$$

$$I_{yy}^O = \underbrace{\frac{m\ell^2}{3}}_{\mathcal{B}_1} + \underbrace{m\ell^2}_{\mathcal{B}_2} + \underbrace{\frac{m\ell^2}{12} + \frac{5}{4}m\ell^2}_{\mathcal{B}_3} = m\ell^2\left(\frac{4+12+1+15}{12}\right) = \frac{16}{6}m\ell^2$$

$$I_{zz}^O = \underbrace{\frac{m\ell^2}{3}}_{\mathcal{B}_1} + \underbrace{\frac{m\ell^2}{12} + \frac{5}{4}m\ell^2}_{\mathcal{B}_2} + \underbrace{2m\ell^2}_{\mathcal{B}_3} = m\ell^2\left(\frac{4+1+15+24}{12}\right) = \frac{22}{6}m\ell^2$$

$$I_{xy}^O = \underbrace{-m\ell\frac{\ell}{2}}_{\mathcal{B}_2} \underbrace{- m\ell^2}_{\mathcal{B}_3} = -\frac{3}{2}m\ell^2 = -\frac{9}{6}m\ell^2$$

$$I_{yz}^O = \underbrace{-\left(-m\frac{\ell^2}{2}\right)}_{\mathcal{B}_3} = +\frac{m\ell^2}{2} + \frac{3m\ell}{6}$$

$$I_{xz}^O = \underbrace{-\left(-m\frac{\ell^2}{2}\right)}_{\mathcal{B}_3} = +\frac{m\ell^2}{2} = \frac{3m\ell^2}{6}$$

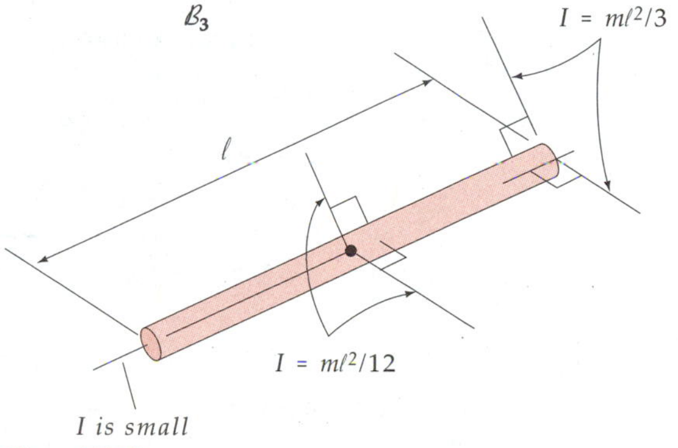

Figure E7.5b

For this problem, then, Equations (7.28) may be expressed as

$$\left(10 - \frac{6I}{m\ell^2}\right) n_x - 9n_y + 3n_z = 0$$

$$-9n_x + \left(16 - \frac{6I}{m\ell^2}\right) n_y + 3n_z = 0$$

$$3n_x + 3n_y + \left(22 - \frac{6I}{m\ell^2}\right) n_z = 0 \qquad (1)$$

in which we have multiplied the three equations by $(6/m\ell^2)$; we may replace $6I/m\ell^2$ by $\mathcal{I}$ and write the determinant as

$$\begin{vmatrix} 10 - \mathcal{I} & -9 & 3 \\ -9 & 16 - \mathcal{I} & 3 \\ 3 & 3 & 22 - \mathcal{I} \end{vmatrix} = 0$$

Expanding gives the characteristic cubic equation:

$$f(\mathcal{I}) = -\mathcal{I}^3 + 48\mathcal{I}^2 - 633\mathcal{I} + 1342 = 0$$

If a computer or programmable calculator is not available,* we can always solve a cubic by trial and error rather quickly. Noting that $f(\mathcal{I})$ is 1342 at $\mathcal{I} = 0$ and is negative at $\mathcal{I} = 3$, for example, on a calculator we may proceed and within a few minutes obtain the root between these values.† The procedure is as follows:

$\mathcal{I}$	$f(\mathcal{I})$	
3	−206	(thus a root is probably just past $\mathcal{I} = 2.5$)
2	260	
2.6	3.10	(still positive)
2.61	−0.93	(so it is >2.6 and <2.61, closer to the latter)
2.608	−0.123	(back up slightly!)
2.607	0.2802	(so it is about two-thirds of the way from 2.607 to 2.608)
2.6076	0.03835	(so just a little farther)
2.6077	−0.00195	(the root is close to this number!)
2.60769	0.00208	(should be halfway between the last one and this one . . .)
2.607695	0.00006	(now double-check)
2.607696	−0.00034	(so $\mathcal{I}_1 = 2.607695$ to seven figures)

Next we use synthetic division to obtain the reduced quadratic:

2.607695	−1	48	−633	1342
		−2.607695	118.369287	−1341.999938
	−1	45.392305	−514.630713	0.000062 ≈ 0

$$-\mathcal{I}^2 + 45.392305\mathcal{I} - 514.630713 = 0$$

* See Appendix B for a numerical solution to this problem using the Newton-Raphson method.

† We abandon our three-digit consistency in numerical analyses like this one in order to illustrate the speed of convergence.

Using the quadratic formula, we obtain

$$\mathcal{I}_2 = 22.000002$$
$$\mathcal{I}_3 = 23.392303$$

The value of $\mathcal{I}_2$ strongly hints that 22 might be a rational root. Synthetic division shows that it is, and refined values from the reduced quadratic are then

$$\mathcal{I}_1 = 2.607695$$
$$\mathcal{I}_2 = 22$$
$$\mathcal{I}_3 = 23.392305$$

Since $\mathcal{I} = 6I/m\ell^2$, our dimensional principal moments of inertia are, to six significant figures,

$$I_1 = 0.434623m\ell^2$$
$$I_2 = 3.66667m\ell^2$$
$$I_3 = 3.89872m\ell^2$$

We next illustrate the computation of the direction cosines, which locate for us the principal axes of inertia. We find them from Equations (1), for which $\mathcal{I}_1$, $\mathcal{I}_2$, and $\mathcal{I}_3$ are the only special values (eigenvalues) of $\mathcal{I}$ for which these equations have a solution. First we seek the principal axis associated with $\mathcal{I}_1 = 2.607695$. The first of Equations (1) becomes

$$7.392305n_x - 9n_y + 3n_z = 0$$

Solving for n_z in terms of n_x and n_y and substituting the result into the second of Equations (1) gives

$$n_y = 0.732051n_x$$

Thus

$$n_z = -0.267950n_x$$

Substituting these expressions for n_y and n_z into $n_x^2 + n_y^2 + n_z^2 = 1$ yields

$$\text{Unit vector } \hat{\mathbf{n}}_1 \begin{cases} n_x = 0.788675 \\ n_y = 0.577350 \\ n_z = -0.211325 \end{cases}$$

Thus the angles that the principal axis of minimum moment of inertia makes with x, y, and z are, respectively, 37.94°, 54.74°, and 102.20°. This axis has to be the one to which, loosely speaking, the mass finds itself closest. Examination of Figure E7.5c at the left, together with these angles, shows that this makes sense.

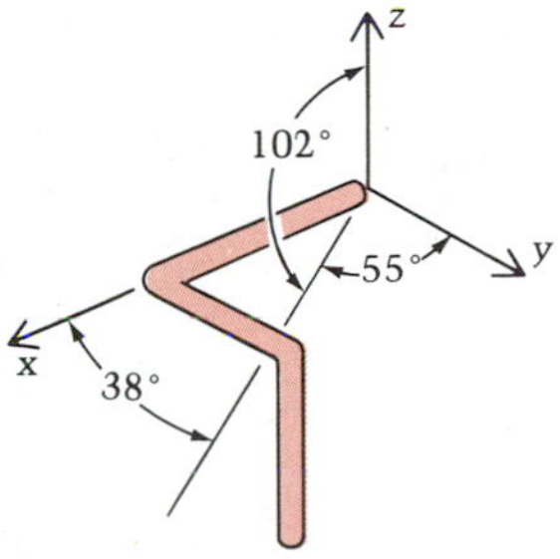

Figure E7.5c

Next we may follow the same procedure for the principal axis of maximum moment of inertia ($I_3 = \mathcal{I}_3 m\ell^2/6 = 3.898718m\ell^2$). The results are, as the reader may verify,

$$\text{Unit vector } \hat{\mathbf{n}}_3 \begin{cases} n_x = 0.211325 & (\theta_x = \cos^{-1} n_x = 77.80°) \\ n_y = -0.577350 & (\theta_y = 125.26°) \\ n_z = -0.788675 & (\theta_z = 142.06°) \end{cases}$$

And for the intermediate moment of inertia ($I_2 = \mathcal{I}_2 m\ell^2/6 = 3.666667\ m\ell^2$), the principal axis is defined by

$$\text{Unit vector } \hat{\mathbf{n}}_2 \begin{cases} n_x = \ \ 0.577350 & (\theta_x = 54.74°) \\ n_y = -0.577350 & (\theta_y = 125.26°) \\ n_z = \ \ 0.577350 & (\theta_z = 54.74°) \end{cases}$$

Orthogonality of Principal Axes

Note in the preceding example that $\hat{\mathbf{n}}_1 \cdot \hat{\mathbf{n}}_2 = \hat{\mathbf{n}}_2 \cdot \hat{\mathbf{n}}_3 = \hat{\mathbf{n}}_3 \cdot \hat{\mathbf{n}}_1 = 0$ and that $\hat{\mathbf{n}}_1 \times \hat{\mathbf{n}}_2 = \hat{\mathbf{n}}_3$.* These are very good checks on the solution since the principal axes, when the principal moments of inertia are distinct ($I_1 \neq I_2 \neq I_3 \neq I_1$), are orthogonal. To prove this in general, let $I_1 \neq I_2$ or I_3, and let x lie along the axis of I_1. Then from the first of Equations (7.32), which is

$$(I_1 - I)n_x = 0$$

we see that if I is either I_2 or I_3, then $n_x = 0$; that is, the cosine of the angle between x and the corresponding principal axis has to be zero. This means that x is perpendicular to the other two principal axes. In turn, these two axes are normal to each other; this follows from reorienting the axes once more so that x still lies along the axis of I_1 but now y lies along the axis of I_2. This time I_{yz}^C is also zero, so the new second equation becomes

$$(I_2 - I)n_y = 0$$

This shows that for $I = I_3(I_2 \neq I_3)$, the value of n_y for the third principal axis vanishes and it is then normal not only to x (which it still is, since we have only rotated it about x) but also to y. Thus the three principal axes are orthogonal if the principal moments of inertia are all different. We now turn our attention to what happens if they are not.

Equal Moments of Inertia

A commonly occurring case is for two of the principal moments of inertia at a point P to be equal to each other but different from the third. When this happens, we can show that *every* line through P in the plane of the two axes (call them x and y) with equal moments of inertia (say $I_1 = I_2$) is a principal axis having this same value for its moment of inertia.

To do this, we first show that if $I_1 = I_2 \neq I_3$, then the axis associated with I_3 (call it z) is perpendicular to those (x, y) of I_1 and I_2. The third equation of (7.32) gives

$$(I_3 - I)n_z = 0$$

Thus when I is I_1 or I_2, then $n_z = 0$. Hence the angle between z and x (and

* This could have been $\hat{\mathbf{n}}_1 \times \hat{\mathbf{n}}_2 = -\hat{\mathbf{n}}_3$—equally correct!

between z and y) is also 90°. It does not follow in like manner from the equations, however, that axes x and y are perpendicular. To handle this case, we begin by showing that if the axes of the other two principal moments of inertia ($I_1 = I_2$) are *not* presumed to be orthogonal (say they are q and x in Figure 7.5 with $I_{qq}^P = I_1 = I_{xx}^P$), then all is well because I_{yy} is *also* equal to I_1.

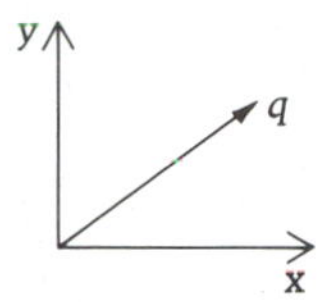

Figure 7.5

To prove this, we use Equation (7.20):

$$I_{qq}^P = I_{xx}^P \ell_x^2 + I_{yy}^P \ell_y^2 + \underbrace{I_{zz}^P \ell_z^2}_{0} + \overbrace{0 + 0 + 0}^{x \text{ and } z \text{ are principal!}}$$

$$I_1 = I_1 \ell_x^2 + I_{yy}^P \ell_y^2$$

$$I_1 \underbrace{(1 - \ell_x^2)}_{\ell_y^2} = I_{yy}^P \ell_y^2$$

Thus we see that

$$I_{yy} = I_1$$

Next we let $\hat{\ell}$ be a unit vector in the direction of an arbitrary axis x' in the plane of the perpendicular axes x and y. Then using Equation (7.20) again gives

$$I_{x'x'}^P = \ell_x^2 I_1 + \ell_y^2 \underbrace{I_2}_{= I_1} + \ell_z^2 I_3 \,(\ell_z = 0) + \underbrace{0 + 0 + 0}_{I_{xy}^P = I_{xz}^P = I_{yz}^P = 0}$$

$$= I_1(\ell_x^2 + \ell_y^2) = I_1$$

And equation (7.24) yields (with y' perpendicular to x' and lying in the plane of x', x, and y as in Figure 7.6:

$$I_{x'y'}^P = \ell_x n_x I_1 + \ell_y n_y \underbrace{I_2}_{= I_1} + \ell_z n_z I_3 \,(\ell_z = 0,\ n_z = 0) + 0 + 0 + 0$$

$$= I_1(\ell_x n_x + \ell_y n_y)$$

$$= I_1(\hat{\ell} \cdot \hat{\mathbf{n}}) = 0$$

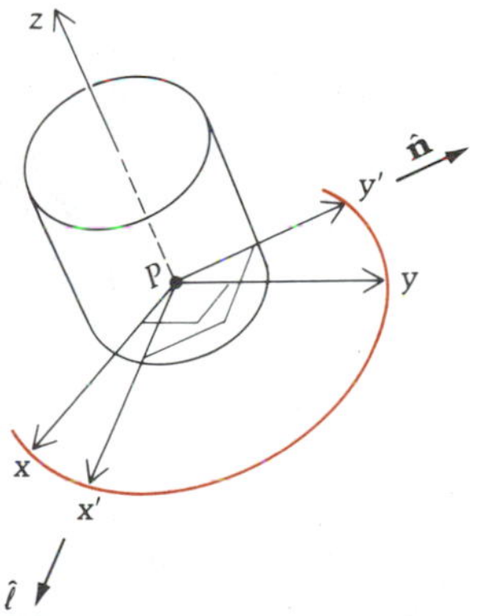

Figure 7.6

Noting that z is perpendicular to x' and that $I_{x'z}^P = 0$ since z is principal, we then have the result that x' is principal with the same moment of inertia as x and y, and this is what we wanted to prove. Note then, for a body with an axis of symmetry, for example, that any axis which passes through and is normal to the symmetry axis of $\mathcal{B}$ is always principal; this is true *even if the axis is not fixed in the body*, which will prove useful to us later.

Finally, if all three principal axes (x, y, z) through P have the same corresponding principal moment of inertia I_1, then *every* axis through P is principal with principal moment of inertia I_1. Let $\hat{\ell}$ be the unit vector along an arbitrary axis x' through P in this case of three equal principal moments of inertia. Also let y' and z' complete an orthogonal triad with x', with $\hat{\mathbf{m}}$ and $\hat{\mathbf{n}}$ being unit vectors in the respective y' and z' directions. Thus (ℓ_x, ℓ_y, ℓ_z), (m_x, m_y, m_z), and (n_x, n_y, n_z) are the respective sets of direction cosines of x', y', and z' with respect to (x, y, z). Equation (7.24) then gives

$$I^P_{x'y'} = (\ell_x m_x + \ell_y m_y + \ell_z m_z)I_1 + \underbrace{0 + 0 + 0}_{\substack{I^P_{xy} = I^P_{yz} = I^P_{xz} = 0 \text{ since} \\ (x,\, y,\, z) \text{ are principal axes}}}$$

$$= 0 \quad (\text{since } \hat{\ell} \perp \hat{\mathbf{m}})$$

In the same way, $I^P_{x'z'} = 0$ since $\hat{\ell} \perp \hat{\mathbf{n}}$ as well. Thus the arbitrary axis x' through P is principal, and Equation (7.20) shows that its moment of inertia is also I_1:

$$I^P_{x'x'} = \underbrace{(\ell_x^2 + \ell_y^2 + \ell_z^2)}_{= 1 \text{ (since } \hat{\ell} \text{ is a unit vector)}} I_1 = I_1$$

Examples of the preceding results regarding two and three equal principal moments of inertia are given in the table on the next page.

Maximum and Minimum Moments of Inertia

An important property of principal moments of inertia is that the largest and smallest of these are the largest and smallest moments of inertia associated with *any* axis through the point in question. To show that this is true, let $I_1 \le I_2 \le I_3$ be the principal moments of inertia at P and let the corresponding principal axes be x, y, and z. The moment of inertia about some other axis, x', is given by Equation (7.20):

$$I^P_{x'x'} = I_1 \ell_x^2 + I_2 \ell_y^2 + I_3 \ell_z^2$$

or

$$\frac{I^P_{x'x'}}{I_1} = \ell_x^2 + \left(\frac{I_2}{I_1}\right)\ell_y^2 + \left(\frac{I_3}{I_1}\right)\ell_z^2 \ge 1$$

since

$$\frac{I_3}{I_1} \ge \frac{I_2}{I_1} \ge 1 \qquad \text{and} \qquad \ell_x^2 + \ell_y^2 + \ell_z^2 = 1$$

Thus $I^P{}_{x'x'} \ge I_1$, so that *no* line through P has a smaller moment of inertia than the smallest *principal* moment of inertia. By a similar argument $I^P_{x'x'}/I_3 \le 1$, which demonstrates that $I^P_{x'x'} \le I_3$ so that no line through P has a larger associated moment of inertia than the largest *principal* moment of inertia. Thus the largest and smallest moments of inertia at a point P are found among the principal moments of inertia for P. It may now be shown quite easily that the smallest moment of inertia at the mass center is the minimum I for *any* line through *any point* of $\mathcal{B}$ or $\mathcal{B}$ extended.

Two Equal I's

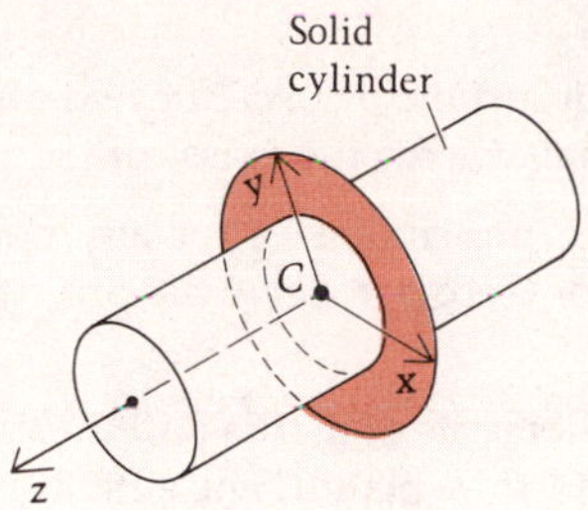

$$I_{xx}^{C} = I_{yy}^{C} = \frac{mR^2}{4} + \frac{m\ell^2}{12}$$

All axes through C in the shaded plane have this same inertia and are principal. Note that $I_{zz}^{C} = mR^2/2$ and is generally not equal to I_{xx} and I_{yy}; if, however, $\ell = \sqrt{3}R$, then *every* axis through C is principal with the same principal moment of inertia!

Three Equal I's

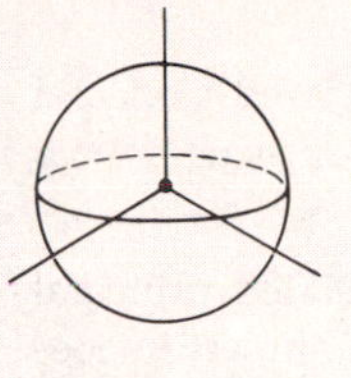

Solid sphere:

$$I = \frac{2}{5}mR^2$$

in any direction through C.

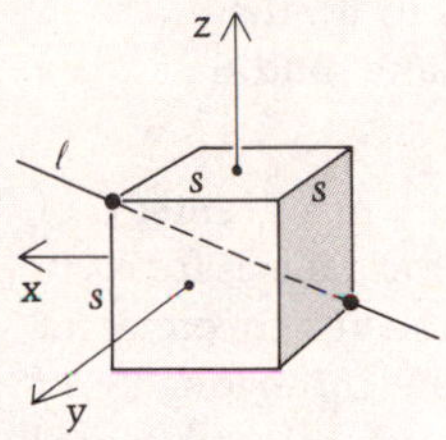

Solid cube:

$$I_x = I_y = I_z = \frac{m}{12}(s^2 + s^2) = \frac{ms^2}{6}$$

in any direction through C. Therefore if line ℓ is a diagonal of the cube, then $I_{\ell\ell} = ms^2/6$ even though it would be formidable to obtain this result by integration. Note that the direction cosines of ℓ are $(\ell_x, \ell_y, \ell_z) = (1/\sqrt{3}, 1/\sqrt{3}, 1/\sqrt{3})$ since it makes equal angles with x, y, and z. Thus Equation (7.20) gives

$$I_{\ell\ell}^{C} = \frac{1}{3}I_x + \frac{1}{3}I_y + \frac{1}{3}I_z + 0 + 0 + 0$$

$$= \left(\frac{1}{3} + \frac{1}{3} + \frac{1}{3}\right)\frac{ms^2}{6} = \frac{ms^2}{6}$$

Question 7.3 Write a one-sentence proof of this statement by using the preceding results together with the parallel-axis theorem.

Answer 7.3 The moment of inertia about any line ℓ through any point P other than C is larger than the moment of inertia about the line through C parallel to ℓ, by the transfer term md^2, and thus the smallest I at C is the smallest of all.

PROBLEMS ▶ Section 7.4

7.21 Find the principal axes and associated principal moments of inertia at O for the semicircular plate of mass m and radius R shown in Figure P7.21.

7.22 Find the principal axes and associated principal moments of inertia for the planar wire shown in Figure P7.22, at the mass center.

7.23 Find the vector from O to the mass center of the bent bar in Example 7.5. Observe that it does not lie along any of the three principal axes at O.

7.24 Use the definitions of the moments of inertia to prove that if a body lies essentially in the xy plane (that is, it has very small dimensions normal to it), then $I_{zz}^{P} \approx I_{xx}^{P} + I_{yy}^{P}$, where P is *any point* in the plane, and that z is a principal axis at P.

7.25 Show that if an axis through the mass center C of body $\mathcal{B}$ is principal at C, then it is a principal axis for *every point* on that axis. *Hint*: Use the transfer theorem for products of inertia, together with the orthogonality of principal axes. (See Figure P7.25.) Transfer I_{yz}^{C} and I_{xz}^{C} to P!

7.26 Show that if a principal axis for a point (such as P in the preceding problem) passes through C, then it is also principal for C. (Same hint!)

7.27 Show that if a line is a principal axis for two of its points, then it is a principal axis for the mass center.

7.28 Show that the three principal axes for any point lying on a principal axis for C are parallel to the principal axes for C.

7.29 Find the principal moments of inertia and their associated axes at O for the thin plate (Figure P7.29) in terms of its density ρ and thickness t.

* **7.30** Find the moments of inertia I_{xx}^{P}, I_{yy}^{P}, and I_{zz}^{P} for the body depicted in Example 4.16. Then find the principal moments of inertia and corresponding principal axes at P.

* **7.31** In Problem 7.16 extend the problem and find the principal moments of inertia, and their principal axes, at point O.

* **7.32** Calculate the principal moments of inertia at O, and the direction cosines of their respective principal axes, for body $\mathcal{B}$ in Figure P7.32. It is made up of three bent bars welded together; all legs are either along, or parallel to, the coordinate axes.

* **7.33** Find the principal moments of inertia and related principal axes at the origin for the body in Figure P7.33.

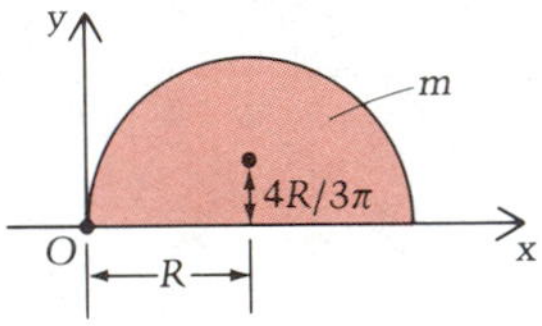

Figure P7.21

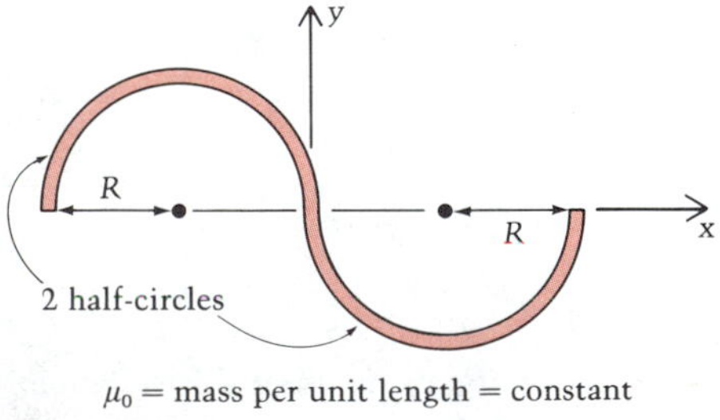

Figure P7.22

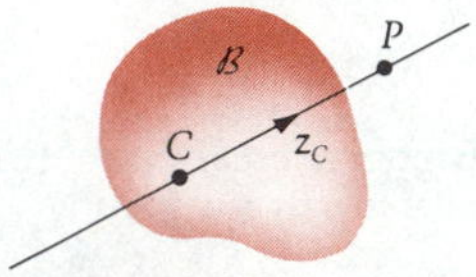

Figure P7.25

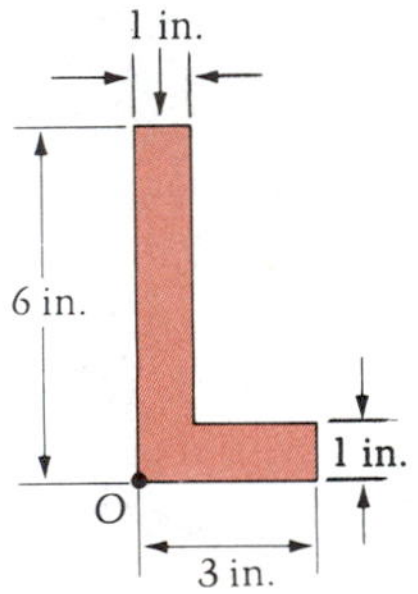

Figure P7.29

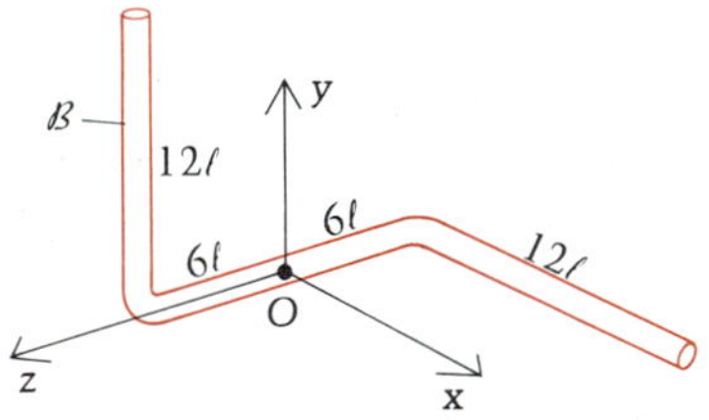

Figure P7.32

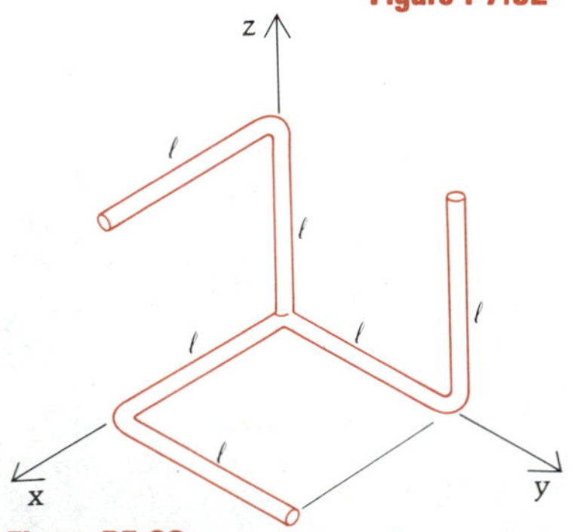

Figure P7.33

* **7.34** In Example 7.5 find the principal moments and axes of inertia at the mass center *C*.

* **7.35** Find the principal moments of inertia and their associated principal axes for the plate of Example 7.4 when $b = 2H$.

7.36 Find the principal moments and axes of inertia if the body of Example 7.4 has depth *L* instead of being a thin plate. (See Figure P7.36.) *Hint*: The *xy* plane is still one of symmetry, so $I_{xz}^{C} = I_{yz}^{C} = 0$ again!

* **7.37** For the homogeneous rectangular solid shown in Figure P7.37, find the smallest of the three angles between line *AB* and the principal axes of inertia at *A*.

* **7.38** At the origin, find the principal moments of inertia and associated principal axes for a body consisting of three square plates welded along their edges as shown in Figure P7.38. (The axis of the smallest value of *I* should be the one that the mass lies closest to in an overall sense. Make this rough check on your solution.) Mass $= 3m$, side $= a$.

* **7.39** Four slender bars, each of mass *m* and length *ℓ*, are welded together to form the body shown in Figure P7.39. Find: (a) the inertia properties at the mass center *C*; (b) the principal axes and principal moments of inertia at *C*.

* **7.40** In Figure P7.40, the axis of symmetry, y_C, of the disk is parallel to *y*; the plane of the disk is parallel to *xz*. Find the principal moments of inertia of $\mathcal{D}$ at the origin *O*, and for the smallest one, determine the angles that its associated principal axis forms with *x*, *y*, and *z*.

** **7.41** Figure P7.41 shows part of a space station being constructed in orbit. Find the principal axes and moments of inertia at C_3. The modules have 33-ft diameters, but due to the material within they are not hollow. For the purposes of this problem, treat each as a uniform hollow shell with a radius of gyration about its axis of 12 ft.

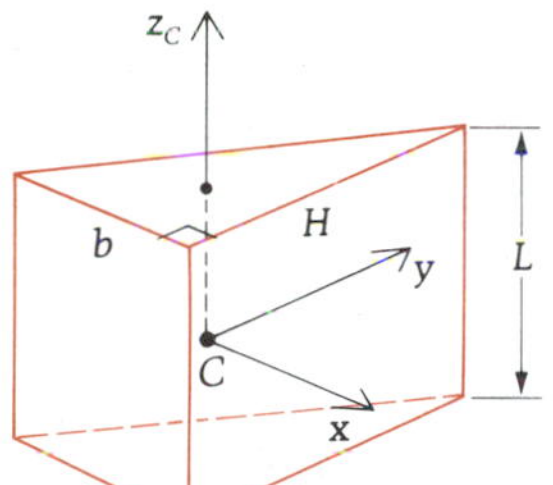

Figure P7.36

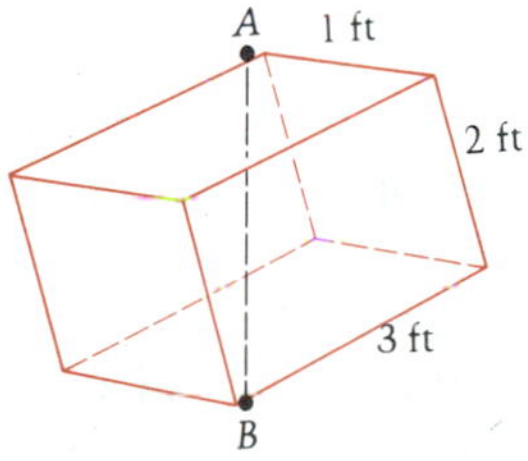

Figure P7.37

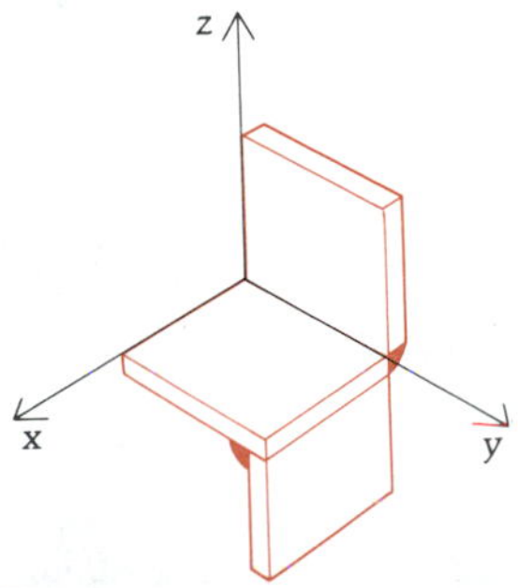

Figure P7.38

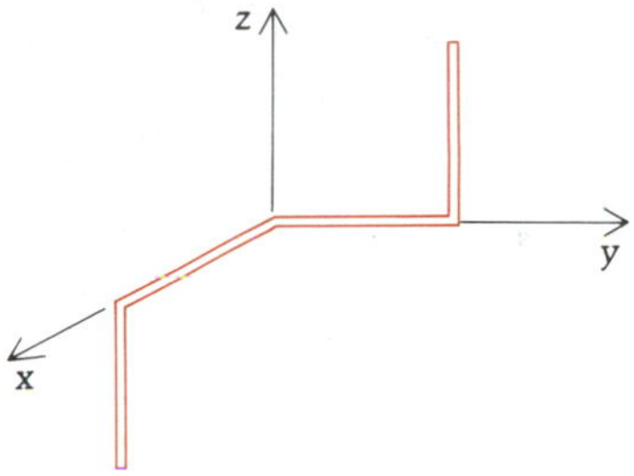

Figure P7.39

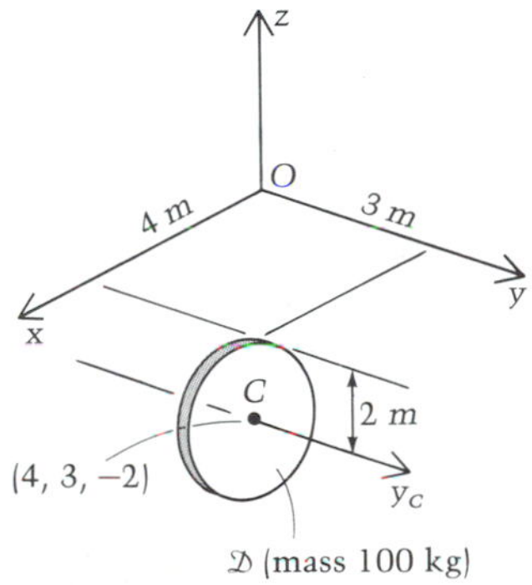

Figure P7.40

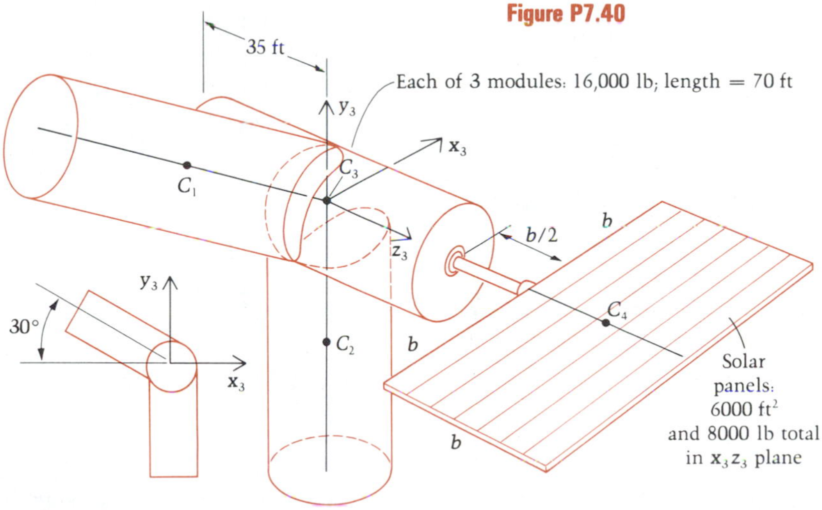

Figure P7.41

7.5 The Moment Equation Governing Rotational Motion

The Euler Equations

In this section we derive the three differential equations governing the angular motion of a rigid body $\mathcal{B}$. Their solution, which is difficult to obtain in closed form in most cases, yields the three components of the angular velocity of $\mathcal{B}$ in an inertial frame $\mathcal{I}$. We begin with Equation (7.2), noting that the time derivative is taken in $\mathcal{I}$. However, the angular momentum $\mathbf{H}_C$ was most conveniently expressed "in"* body $\mathcal{B}$ in Equation (7.11). Thus we shall use Equation (6.20) to move the derivative in (7.2) from $\mathcal{I}$ to $\mathcal{B}$:

$$\Sigma\mathbf{M}_C = {}^{\mathcal{I}}\dot{\mathbf{H}}_C = {}^{\mathcal{B}}\dot{\mathbf{H}}_C + \boldsymbol{\omega}_{\mathcal{B}/\mathcal{I}} \times \mathbf{H}_C \tag{7.37}$$

We now fix the axes (x, y, z) to body $\mathcal{B}$ so that, relative to $\mathcal{B}$, the inertia properties are constant. Using Equation (7.11), the first term on the right side of (7.37) is

$$\begin{aligned}{}^{\mathcal{B}}\dot{\mathbf{H}}_C = &(I_{xx}^C\dot{\omega}_x + I_{xy}^C\dot{\omega}_y + I_{xz}^C\dot{\omega}_z)\hat{\mathbf{i}}\\ &+ (I_{xy}^C\dot{\omega}_x + I_{yy}^C\dot{\omega}_y + I_{yz}^C\dot{\omega}_z)\hat{\mathbf{j}}\\ &+ (I_{xz}^C\dot{\omega}_x + I_{yz}^C\dot{\omega}_y + I_{zz}^C\dot{\omega}_z)\hat{\mathbf{k}}\end{aligned} \tag{7.38}$$

where the unit vectors $\hat{\mathbf{i}}$, $\hat{\mathbf{j}}$, and $\hat{\mathbf{k}}$ are respectively parallel to x, y, and z, and therefore now are fixed in direction in $\mathcal{B}$. The second term in Equation (7.37), after computing the cross product, is

$$\begin{aligned}\boldsymbol{\omega}_{\mathcal{B}/\mathcal{I}} \times \mathbf{H}_C = &[(I_{zz}^C - I_{yy}^C)\omega_y\omega_z + I_{yz}^C(\omega_y^2 - \omega_z^2) + \omega_x(\omega_y I_{xz}^C - \omega_z I_{xy}^C)]\hat{\mathbf{i}}\\ &+ [(I_{xx}^C - I_{zz}^C)\omega_z\omega_x + I_{xz}^C(\omega_z^2 - \omega_x^2) + \omega_y(\omega_z I_{xy}^C - \omega_x I_{yz}^C)]\hat{\mathbf{j}}\\ &+ [(I_{yy}^C - I_{xx}^C)\omega_x\omega_y + I_{xy}^C(\omega_x^2 - \omega_y^2) + \omega_z(\omega_x I_{yz}^C - \omega_y I_{xz}^C)]\hat{\mathbf{k}}\end{aligned} \tag{7.39}$$

The sum of Equations (7.38) and (7.39) yields the right side of Equation (7.37), which in turn equals the moment about the mass center C of all the external forces and couples acting on $\mathcal{B}$.

It is clear that this equation is extremely lengthy and complicated. If we select the body-fixed axes (x, y, z) to be the *principal* axes through C, however, then all product-of-inertia terms vanish and we obtain

$${}^{\mathcal{B}}\dot{\mathbf{H}}_C = I_{xx}^C\dot{\omega}_x\hat{\mathbf{i}} + I_{yy}^C\dot{\omega}_y\hat{\mathbf{j}} + I_{zz}^C\dot{\omega}_z\hat{\mathbf{k}}$$

and

$$\begin{aligned}\boldsymbol{\omega}_{\mathcal{B}/\mathcal{I}} \times \mathbf{H}_C = &(I_{zz}^C - I_{yy}^C)\omega_y\omega_z\hat{\mathbf{i}} + (I_{xx}^C - I_{zz}^C)\omega_z\omega_x\hat{\mathbf{j}}\\ &+ (I_{yy}^C - I_{xx}^C)\omega_x\omega_y\hat{\mathbf{k}}\end{aligned}$$

so that, substituting into Equation (7.37) and equating the respective

* "Expressing a vector in a frame" simply means the vector is expressed in terms of unit vectors fixed in that frame.

coefficients of $\hat{\mathbf{i}}$, $\hat{\mathbf{j}}$, and $\hat{\mathbf{k}}$, we obtain the **Euler equations:**

$$\Sigma M_{Cx} = I_{xx}^{C}\dot{\omega}_x - (I_{yy}^{C} - I_{zz}^{C})\omega_y\omega_z$$
$$\Sigma M_{Cy} = I_{yy}^{C}\dot{\omega}_y - (I_{zz}^{C} - I_{xx}^{C})\omega_z\omega_x$$
$$\Sigma M_{Cz} = I_{zz}^{C}\dot{\omega}_z - (I_{xx}^{C} - I_{yy}^{C})\omega_x\omega_y \tag{7.40}$$

We note that the Euler equations are nonlinear in the $\boldsymbol{\omega}$ components and that the plane-motion equation $\Sigma M_{Cz} = I_{zz}^{C}\dot{\omega}$ does not extend simply to general motion.

It is very important to realize that, if a body has a pivot (permanently fixed point), equations analogous to the preceding pertain. And in fact these are merely what is obtained by substituting O (the pivot) for C in all equations from (7.37) through (7.40).

We can use Equations (7.40) to make an important observation about the special case of "torque-free" motion, meaning $\Sigma \mathbf{M}_C = \mathbf{0}$. Suppose at an instant $\omega_x \neq 0$, $\omega_y = \omega_z = 0$, where x, y, and z are principal axes. Then, with $\Sigma \mathbf{M}_C = \mathbf{0}$, Equations (7.40) tell us that $\dot{\omega}_x = \dot{\omega}_y = \dot{\omega}_z = 0$; that is, if the body were initially to be spun about a principal axis it would continue to spin about that axis and at constant rate. Conversely, if we seek conditions for which $\dot{\omega}_x = \dot{\omega}_y = \dot{\omega}_z = 0$, we find from Equations (7.40) that two of ω_x, ω_y and ω_z must vanish. Thus the spin will persist in the absence of external moment if and only if the axis of initial spin is a principal axis. This investigation is what led Euler to discover the principal-axis concept in 1750.

Question 7.4 Can a similar conclusion be drawn for spinning about an axis through a pivot?

Use of Non-Principal Axes

Sometimes other forms of moment equations are more advantageous to apply in particular problems than Equations (7.40). Firstly, we may find it convenient to use reference axes that are body-fixed but not principal. Often it is less trouble to simply deal with nonzero products of inertia and axes that are convenient to the body (or its angular velocity) than to compute principal directions and associated inertia properties. The component equations are formed by combining Equations (7.37)–(7.39). When specialized to (x, y) plane motion, we recover the following equations developed in Chapter 4:

$$\Sigma M_{Cx} = I_{xz}^{C}\dot{\omega}_z - I_{yz}^{C}\omega_z^2$$
$$\Sigma M_{Cy} = I_{yz}^{C}\dot{\omega}_z + I_{xz}^{C}\omega_z^2$$
$$\Sigma M_{Cz} = I_{zz}^{C}\dot{\omega}_z \tag{7.41}$$

Answer 7.4 Yes, provided there is no net moment about the pivot.

Use of an Intermediate Frame

Secondly, we often find it convenient to express external moments and/or angular momentum in terms of components associated with directions fixed neither in the body nor in the inertial frame. That is, we may choose to involve an intermediate frame, say $\mathcal{I}$, and use

$$\Sigma \mathbf{M}_C = {}^{\mathcal{J}}\dot{\mathbf{H}}_C = {}^{\mathcal{I}}\dot{\mathbf{H}}_C + \boldsymbol{\omega}_{\mathcal{I}/\mathcal{J}} \times \mathbf{H}_C \qquad (7.42)$$

or, when there is a pivot O,

$$\Sigma \mathbf{M}_O = {}^{\mathcal{J}}\dot{\mathbf{H}}_O = {}^{\mathcal{I}}\dot{\mathbf{H}}_O + \boldsymbol{\omega}_{\mathcal{I}/\mathcal{J}} \times \mathbf{H}_O \qquad (7.43)$$

This approach is particularly useful when, usually because of symmetries, moments and products of inertia of the body remain constant relative to axes fixed in the intermediate frame.

We close this section with five examples, the first employing Euler's equations to study torque-free motion when the initial spin is not about a principal axis. In the second we use body-fixed axes that are not principal in a practical problem of a satellite dish antenna. The final three examples illustrate the use of intermediate frames of reference.

EXAMPLE 7.6

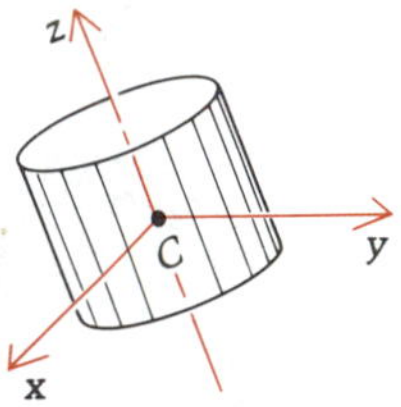

Figure E7.6

A satellite* is moving through deep space far from the influence of atmospheric drag and gravity. (See Figure E7.6.) If the z axis is one of symmetry and if at some instant called $t = 0$ we have $\boldsymbol{\omega} = (\omega_{x_i}, \omega_{y_i}, \omega_{z_i})$ along the body-fixed mass-center axes, find $\boldsymbol{\omega}(t)$. Assume the satellite to be a rigid body.

Solution

The Euler equations, if $I_{xx}^C = I_{yy}^C = I$ and $I_{zz}^C = J$, are

$$I\dot{\omega}_x - (I - J)\omega_y\omega_z = 0 \qquad (1)$$

$$I\dot{\omega}_y - (J - I)\omega_z\omega_x = 0 \qquad (2)$$

$$J\dot{\omega}_z - (I - I)\omega_x\omega_y = 0 \qquad (3)$$

in which the moment components are zero in the absence of external forces and couples. Equation (3) gives

$$\omega_z = \text{constant} = \omega_{z_i} \qquad (4)$$

so that Equations (1) and (2) become linear and are:

$$I\dot{\omega}_x - (I - J)\omega_y\omega_{z_i} = 0 \qquad (5)$$

$$I\dot{\omega}_y - (J - I)\omega_{z_i}\omega_x = 0 \qquad (6)$$

Differentiating Equation (6) and solving for $\dot{\omega}_x$, we get

$$\dot{\omega}_x = \frac{I\ddot{\omega}_y}{(J - I)\omega_{z_i}} \qquad (7)$$

* Such as the *Voyager* spacecraft, which left our solar system in 1983.

This expression may be substituted into (5) to yield an equation free of ω_x:

$$\ddot{\omega}_y + \left(\frac{(J-I)\omega_{z_i}}{I}\right)^2 \omega_y = 0$$

or

$$\ddot{\omega}_y + p^2\omega_y = 0$$

in which $p = (J-I)\omega_{z_i}/I$. The solution to this equation is harmonic:

$$\omega_y = A\cos pt + B\sin pt$$

Since $\omega_y = \omega_{y_i}$ at $t = 0$, we see that $A = \omega_{y_i}$. Finally, Equation (6) gives

$$\omega_x = \frac{\dot{\omega}_y}{p} = \frac{-\omega_{y_i}p\sin pt + Bp\cos pt}{p}$$

or

$$\omega_x = -\omega_{y_i}\sin pt + B\cos pt$$

The initial condition for ω_x gives us

$$\omega_{x_i} = 0 + B \Rightarrow B = \omega_{x_i}$$

so that the other two components (besides $\omega_2 = \omega_{z_i}$) of $\boldsymbol{\omega}(t)$ are

$$\omega_x = \omega_{x_i}\cos pt - \omega_{y_i}\sin pt$$
$$\omega_y = \omega_{y_i}\cos pt + \omega_{x_i}\sin pt$$

Example 7.7

For reasons of interference with other bodies, an antenna was recently designed and built with an offset axis as shown in Figure E7.7a and Figure E7.7b. The antenna is composed of a 12-ft, 1200-lb parabolic reflector $\mathcal{R}$, a counterweight $\mathcal{W}$, a reflector support structure $\mathcal{S}$, and a positioner. The positioner consists of (1) a pedestal $\mathcal{P}$ that is fixed to the (inertial) reference frame; (2) an azimuth bearing at

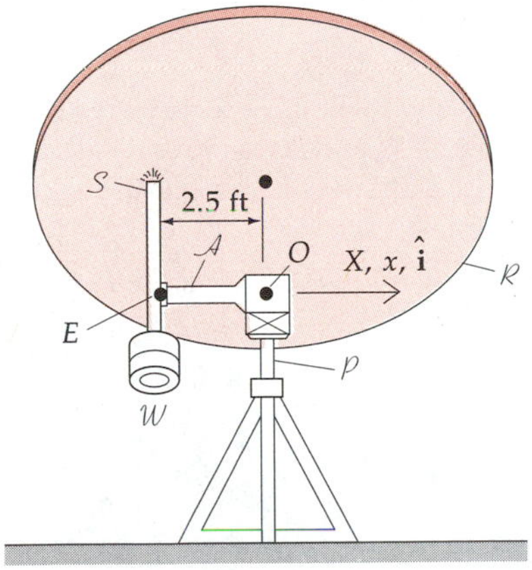

Figure E7.7a Back View

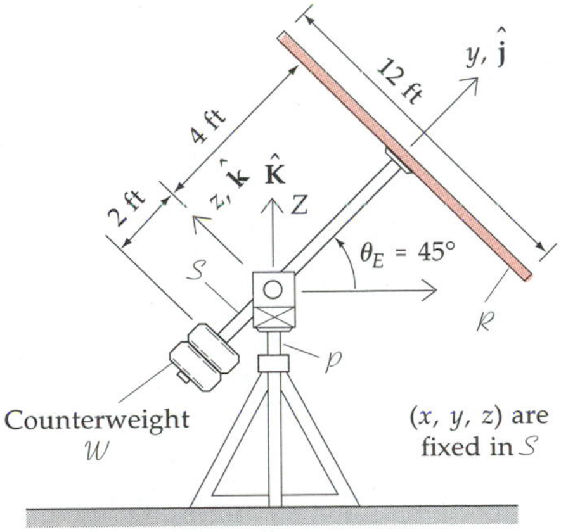

Figure E7.7b Side View

O and ring gear by means of which the arm $\mathcal{A}$ is made to rotate about the vertical; and (3) an elevation torque motor at E that rotates the support structure $\mathcal{S}$ with respect to the arm $\mathcal{A}$. Rotations of the reflector consist of an azimuth angle θ_A about Z, and an elevation angle θ_E about x.

a. Find the weight W of the counterweight $\mathcal{W}$.

b. Find the inertia properties at point O of the (assumed rigid) body $\mathcal{B}$ composed of $\mathcal{R} + \mathcal{W} + \mathcal{S}$.

c. Write the equations of rotational motion of $\mathcal{B}$.

d. Determine the net moments that must be exerted about axes through O for the position shown if first (in units of radians and seconds) $\dot{\theta}_A = \pi/6$, $\ddot{\theta}_E = \pi/6$, $\dot{\theta}_E = \ddot{\theta}_A = 0$; and then $\dot{\theta}_E = \pi/6$, $\ddot{\theta}_A = \pi/6$, $\dot{\theta}_A = \ddot{\theta}_E = 0$.

Solution

Let us model the antenna as follows. The reflector is treated as a thin disk, the counterweight as a point mass, and the support structure as being rigid but light. Axes (X, Y, Z) are fixed in the inertial frame $\mathcal{I}$ while (x, y, z) are attached to $\mathcal{B}$.

In part (a) the purpose of the counterweight is to place the mass center of $\mathcal{B}$ on its elevation axis. Thus

$$1200(4) = W(2)$$

$$W = 2400 \text{ lb}$$

For part (b) we generate the inertia properties of $\mathcal{B}$ about the point O. (See the table on the next page.)

Therefore the *inertia matrix* may be written as

$$\begin{bmatrix} 1230 & -373 & 0 \\ -373 & 1140 & 0 \\ 0 & 0 & 1700 \end{bmatrix} \quad \text{slug-ft}^2 \tag{1}$$

For part (c), one way to proceed is to compute the principal axes and moments of inertia from this matrix and then use the Euler equations (7.40). This would be a very unwise approach in this case, however. Not only is it tedious to locate the principal axes, but following this we would have to break up the angular velocity into its components along these directions; we would then obtain not-so-useful moment components about axes skewed with respect to the rotation axes. It is much simpler to use the second of Euler's laws:*

$$\Sigma \mathbf{M}_O = {}^{\mathcal{I}}\dot{\mathbf{H}}_O$$

The only prices we will have to pay are (1) to retain and deal with the nonzero product of inertia I^O_{xy} and (2) to move the derivative from frame $\mathcal{I}$ to $\mathcal{B}$. The angular momentum of $\mathcal{B}$ about O is, for our problem,

$$\begin{aligned} \mathbf{H}_O = {} & (I^O_{xx}\omega_x + I^O_{xy}\omega_y + \cancelto{0}{I^O_{xz}}\omega_z)\hat{\mathbf{i}} \\ & + (I^O_{yx}\omega_x + I^O_{yy}\omega_y + \cancelto{0}{I^O_{yz}}\omega_z)\hat{\mathbf{j}} \\ & + (\cancelto{0}{I^O_{zx}}\omega_x + \cancelto{0}{I^O_{zy}}\omega_y + I^O_{zz}\omega_z)\hat{\mathbf{k}} \end{aligned} \tag{2}$$

* All that is required of O for this equation to be valid is that it be a point of the inertial frame; in what follows, however, it also needs to be, and is, a pivot of $\mathcal{B}$.

	Reflector	+	Counterweight	=	Total
I^O_{xx}	$= m\left(\frac{r^2}{4} + d^2\right)$	+	$\frac{2400}{32.2} d^2$		
	$= \frac{1200}{32.2}\left(\frac{6^2}{4} + 4^2\right)$	+	$\frac{2400}{32.2} 2^2$		
	+ 932	+	298		= 1230 slug-ft²
I^O_{yy}	$= m\frac{r^2}{2}$	+	md^2		
	$= \frac{1200}{32.2}\frac{6^2}{2}$	+	$\frac{2400}{32.2} 2.5^2$		
	= 671	+	466		= 1140 slug-ft²
I^O_{zz}	$= m\left(\frac{r^2}{4} + d^2\right)$	+	md^2		
	$= \frac{1200}{32.2}\left(\frac{6^2}{4} + 4^2\right)$	+	$\frac{2400}{32.2}(2^2 + 2.5^2)$		
	= 932	+	764		= 1700 slug-ft²
I^O_{xy}	= 0*	+	$(-m\overline{xy})$		
	= 0	+	$-\frac{2400}{32.2}(2.5)(2)$		
	= 0	+	−373		= −373 slug-ft²
I^O_{yz}	= 0	+	$(-m\overline{yz})$		
	= 0	+	$-\frac{2400}{32.2}(2)(0)$		
	= 0	+	0	=	0
I^O_{xz}	= 0	+	$(-m\overline{xz})$		
	= 0	+	$-\frac{2400}{32.2}(2.5)(0)$		
	= 0	+	0	=	0

* The y axis through O is an axis of symmetry of $\mathcal{R}$; hence $I^O_{xy} = 0 = I^O_{yz}$. And the third product of inertia of $\mathcal{R}$, I^O_{xz}, vanishes because zy is a plane of symmetry for body $\mathcal{R}$.

The angular velocity of $\mathcal{B}$ in frame $\mathcal{I}$ is found by the addition theorem. The reflector and its supporting structure rotate in elevation with a simple angular velocity $\dot{\theta}_E\hat{\mathbf{i}}$ with respect to the housing $\mathcal{A}$; likewise, $\mathcal{A}$ rotates in azimuth with a simple angular velocity $\dot{\theta}_A\hat{\mathbf{K}}$ with respect to the pedestal (which is rigidly fixed to the reference frame $\mathcal{I}$). Therefore

$$\begin{aligned}\boldsymbol{\omega}_{\mathcal{B}/\mathcal{I}} &= \boldsymbol{\omega}_{\mathcal{B}/\mathcal{A}} + \boldsymbol{\omega}_{\mathcal{A}/\mathcal{I}} \\ &= \omega_{\mathcal{B}/\mathcal{A}}\hat{\mathbf{i}} + \omega_{\mathcal{A}/\mathcal{I}}\hat{\mathbf{K}} \\ &= \dot{\theta}_E\hat{\mathbf{i}} + \dot{\theta}_A(\sin\theta_E\hat{\mathbf{j}} + \cos\theta_E\hat{\mathbf{k}}) \qquad (3)\end{aligned}$$

Substituting the $\boldsymbol{\omega}$ components into Equation (2) gives the angular momentum of $\mathcal{B}$ in $\mathcal{I}$, expressed in (terms of its components in) $\mathcal{B}$:

$$\mathbf{H}_O = (I_{xx}^O\dot{\theta}_E + I_{xy}^O\dot{\theta}_A \sin\theta_E)\hat{\mathbf{i}} + (I_{xy}^O\dot{\theta}_E + I_{yy}^O\dot{\theta}_A \sin\theta_E)\hat{\mathbf{j}} + I_{zz}^O\dot{\theta}_A \cos\theta_E\hat{\mathbf{k}} \quad (4)$$

Next we use Equation (6.20) to differentiate $\mathbf{H}_O$ (note that point O is fixed in $\mathcal{B}$ and $\mathcal{I}$):

$$\Sigma\mathbf{M}_O = {}^{\mathcal{I}}\dot{\mathbf{H}}_O = {}^{\mathcal{B}}\dot{\mathbf{H}}_O + \boldsymbol{\omega}_{\mathcal{B}/\mathcal{I}} \times \mathbf{H}_O \quad (5)$$

Taking the derivative and performing the cross product, we find that the three component equations are as follows. (Note that the inertia properties do not change in $\mathcal{B}$.)

$$\begin{aligned}\Sigma M_{Ox} &= I_{xx}^O\ddot{\theta}_E + I_{xy}^O(\ddot{\theta}_A \sin\theta_E + \dot{\theta}_A\dot{\theta}_E\cos\theta_E) + \dot{\theta}_A \sin\theta_E(I_{zz}^O\dot{\theta}_A\cos\theta_E) \\ &\quad - \dot{\theta}_A\cos\theta_E(I_{xy}^O\dot{\theta}_E + I_{yy}^O\dot{\theta}_A\sin\theta_E) \\ &= I_{xx}^O\ddot{\theta}_E + I_{yy}^O(-\dot{\theta}_A^2\sin\theta_E\cos\theta_E) + I_{zz}^O(\dot{\theta}_A^2\sin\theta_E\cos\theta_E) \\ &\quad + I_{xy}^O(\ddot{\theta}_A\sin\theta_E) \end{aligned} \quad (6a)$$

$$\begin{aligned}\Sigma M_{Oy} &= I_{xy}^O\ddot{\theta}_E + I_{yy}^O(\ddot{\theta}_A \sin\theta_E + \dot{\theta}_A\dot{\theta}_E\cos\theta_E) + \dot{\theta}_A\cos\theta_E(I_{xx}^O\dot{\theta}_E + I_{xy}^O\dot{\theta}_A\sin\theta_E) \\ &\quad - \dot{\theta}_E(I_{zz}^O\dot{\theta}_A\cos\theta_E)\end{aligned} \quad (6b)$$

$$\begin{aligned}\Sigma M_{Oz} &= I_{zz}^O(\ddot{\theta}_A\cos\theta_E - \dot{\theta}_A\dot{\theta}_E\sin\theta_E) + \dot{\theta}_E(I_{xy}^O\dot{\theta}_E + I_{yy}^O\dot{\theta}_A\sin\theta_E) \\ &\quad - \dot{\theta}_A\sin\theta_E(I_{xx}^O\dot{\theta}_E + I_{xy}^O\dot{\theta}_A\sin\theta_E)\end{aligned} \quad (6c)$$

As an indication of the increased difficulty of three-dimensional dynamics problems, note that *all four* of the nonvanishing inertia properties contribute to *each component* of the external moment acting on $\mathcal{B}$ at O!

In the indicated position, $\theta_E = 45°$. Therefore

$$\Sigma M_{Ox} = I_{xx}^O\ddot{\theta}_E + I_{yy}^O\left(\frac{-\dot{\theta}_A^2}{2}\right) + I_{zz}^O\left(\frac{\dot{\theta}_A^2}{2}\right) + I_{xy}^O\left(\frac{\ddot{\theta}_A}{\sqrt{2}}\right) \quad (7a)$$

$$\Sigma M_{Oy} = I_{xx}^O\left(\frac{\dot{\theta}_A\dot{\theta}_E}{\sqrt{2}}\right) + I_{yy}^O\left(\frac{\ddot{\theta}_A + \dot{\theta}_A\dot{\theta}_E}{\sqrt{2}}\right) + I_{zz}^O\left(\frac{-\dot{\theta}_A\dot{\theta}_E}{\sqrt{2}}\right) + I_{xy}^O\left(\ddot{\theta}_E + \frac{\dot{\theta}_A^2}{2}\right) \quad (7b)$$

$$\Sigma M_{Oz} = I_{xx}^O\left(\frac{-\dot{\theta}_A\dot{\theta}_E}{\sqrt{2}}\right) + I_{yy}^O\left(\frac{\dot{\theta}_A\dot{\theta}_E}{\sqrt{2}}\right) + I_{zz}^O\left(\frac{\ddot{\theta}_A - \dot{\theta}_A\dot{\theta}_E}{\sqrt{2}}\right) + I_{xy}^O\left(\dot{\theta}_E^2 - \frac{\dot{\theta}_A^2}{2}\right) \quad (7c)$$

In part (d), for the case specified, $\ddot{\theta}_E = \pi/6$ while $\ddot{\theta}_A = 0$. Also, $\dot{\theta}_A = \pi/6$ while $\dot{\theta}_E = 0$. This case physically corresponds to the antenna, at 45° elevation, swinging around the vertical at 30°/sec and suddenly sensing an object traveling toward zenith; the controls activate a motor whose torque produces an angular acceleration that will send the antenna upward in elevation. The angular accelerations are large because the need is to get there quickly.

Substituting these values of $\dot{\theta}_E$, $\ddot{\theta}_E$, $\dot{\theta}_A$, and $\ddot{\theta}_A$ into Equations (7), along with the inertia values, gives our answer:

$$\Sigma M_{Ox} = 1230\left(\frac{\pi}{6}\right) + (-1140 + 1700)\frac{(\pi/6)^2}{2} + (-373)(0)$$

$$= 721 \text{ lb-ft}$$

$$\Sigma M_{Oy} = 1230(0) + 1140(0) + 1700(0) + (-373)\left[\frac{\pi}{6} + \frac{(\pi/6)^2}{2}\right]$$

$$= -246 \text{ lb-ft}$$

$$\Sigma M_{Oz} = 1230(0) + 1140(0) + 1700(0) + (-373)\left(\frac{-(\pi/6)^2}{2}\right)$$

$$= 51 \text{ lb-ft} \tag{8}$$

These are the moments exerted by $\mathcal{P}$ onto $\mathcal{A}$, excluding that required to balance the dead weight of the antenna. In the inertial reference frame, the moments are

$$\Sigma M_{OX} = \Sigma M_{Ox} = 721 \text{ lb-ft}$$
$$\Sigma M_{OY} = \Sigma M_{Oy} \cos\theta_E + \Sigma M_{Oz}(-\sin\theta_E) = -210 \text{ lb-ft}$$
$$\Sigma M_{OZ} = \Sigma M_{Oy} \sin\theta_E + \Sigma M_{Oz}(\cos\theta_E) = -138 \text{ lb-ft} \tag{9}$$

In the opposite case when the antenna is tracking in elevation, say $\dot{\theta}_E = \pi/6$ rad/sec, and receives a sudden command resulting in $\ddot{\theta}_A = \pi/6$ rad/sec^2 at $\theta_E = 45°$, the moment components become (here $\ddot{\theta}_E = 0 = \dot{\theta}_A$):

$$\Sigma M_{Ox} = 0 + 0 + 0 + (-373)\frac{\pi}{6}\frac{1}{\sqrt{2}}$$

$$= -138 \text{ lb-ft}$$

$$\Sigma M_{Oy} = 0 + 1140\left(\frac{\pi/6 + 0}{\sqrt{2}}\right) + 0 + 0$$

$$= 422 \text{ lb-ft}$$

$$\Sigma M_{Oz} = 0 + 0 + 1700\left(\frac{\pi/6 - 0}{\sqrt{2}}\right) + (-373)\left[\left(\frac{\pi}{6}\right)^2 - 0\right]$$

$$= 629 - 102$$

$$= 527 \text{ lb-ft} \tag{10}$$

Once again we see the considerable effect of the product of inertia term.

The negatives of the X, Y, Z components respectively bend, bend, and twist the pedestal and are considerations in its design; far larger and more important moments, however, arise from the wind blowing against the "dish" and from gravity. There are also forces exerted on $\mathcal{A}$ at O due to gravity and the mass center acceleration.

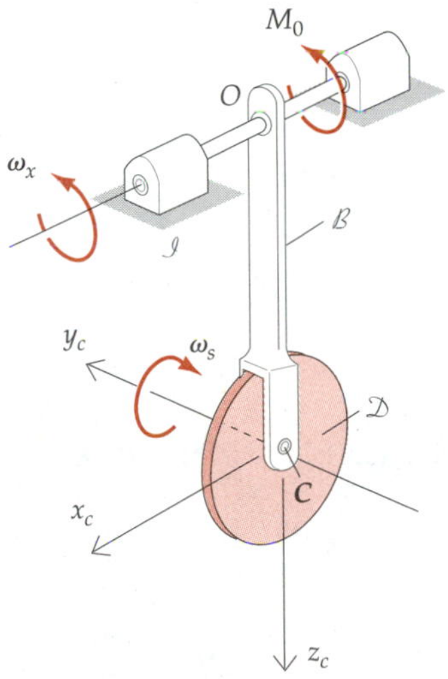

Figure E7.8a

EXAMPLE 7.8

A symmetric wheel $\mathcal{D}$ spins at angular speed ω_s about its axis, which is a line fixed in body $\mathcal{B}$ as well as in $\mathcal{D}$. (See Figure E7.8a.) The moments of inertia of $\mathcal{D}$ at C are $I_{yy}^C = J$ and $I_{xx}^C = I_{zz}^C = I$. Body $\mathcal{B}$ has negligible mass and rotates at angular speed

ω_x about the x axis at O. If a moment M_0 is applied to $\mathcal{B}$ parallel to the x_C axis, find, at the instant shown, the rates of change of ω_s and ω_x and the force and moment exerted on $\mathcal{D}$ at C by the pin. Neglect friction.

Solution

Let us denote the inertial reference frame (in which $\mathcal{B}$ moves) by $\mathcal{I}$. The addition theorem for angular velocity then gives, for the wheel,

$$\begin{aligned}\boldsymbol{\omega}_{\mathcal{D}/\mathcal{I}} &= \boldsymbol{\omega}_{\mathcal{D}/\mathcal{B}} + \boldsymbol{\omega}_{\mathcal{B}/\mathcal{I}} \\ &= \omega_s\hat{\mathbf{j}} + \omega_x\hat{\mathbf{i}}\end{aligned}$$

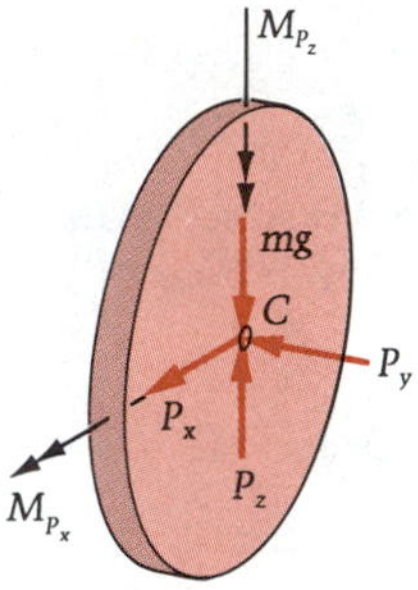

Figure E7.8b

where the axes (x_C, y_C, z_C) are fixed in $\mathcal{B}$ and their associated unit vectors $(\hat{\mathbf{i}}, \hat{\mathbf{j}}, \hat{\mathbf{k}})$ are fixed in direction in $\mathcal{B}$. Note that we may use the equation

$$\mathbf{H}_C = I_{xx}^C\omega_x\hat{\mathbf{i}} + I_{yy}^C\omega_y\hat{\mathbf{j}} + I_{zz}^C\omega_z\hat{\mathbf{k}}$$

for the angular momentum of $\mathcal{D}$ because even though (x_C, y_C, z_C) are not fixed in body $\mathcal{D}$, they are nonetheless permanently principal. Therefore

$$\mathbf{H}_C = I\omega_x\hat{\mathbf{i}} + J\omega_s\hat{\mathbf{j}}$$

The second law of Euler then yields

$$\Sigma\mathbf{M}_C = {}^{\mathcal{I}}\dot{\mathbf{H}}_C = {}^{\mathcal{B}}\dot{\mathbf{H}}_C + \boldsymbol{\omega}_{\mathcal{B}/\mathcal{I}} \times \mathbf{H}_C$$

where for convenience we differentiate $\mathbf{H}_C$ in frame $\mathcal{B}$ since the vector has been written in terms of its components there. Continuing,

$$\begin{aligned}\Sigma\mathbf{M}_C &= I\dot{\omega}_x\hat{\mathbf{i}} + J\dot{\omega}_s\hat{\mathbf{j}} + (\omega_x\hat{\mathbf{i}}) \times (I\omega_x\hat{\mathbf{i}} + J\omega_s\hat{\mathbf{j}}) \\ &= I\dot{\omega}_x\hat{\mathbf{i}} + J\dot{\omega}_s\hat{\mathbf{j}} + J\omega_x\omega_s\hat{\mathbf{k}}\end{aligned}$$

From Figure E7.8b, a free-body diagram of $\mathcal{D}$, we get the components of $\Sigma\mathbf{M}_C$ so that

$$M_{P_x}\hat{\mathbf{i}} + M_{P_z}\hat{\mathbf{k}} = I\dot{\omega}_x\hat{\mathbf{i}} + J\dot{\omega}_s\hat{\mathbf{j}} + J\omega_x\omega_s\hat{\mathbf{k}}$$

Thus we see that

$$M_{P_x} = I\dot{\omega}_x \tag{1}$$

$$0 = J\dot{\omega}_s \Rightarrow \omega_s = \text{constant} \tag{2}$$

$$M_{P_z} = J\omega_x\omega_s \tag{3}$$

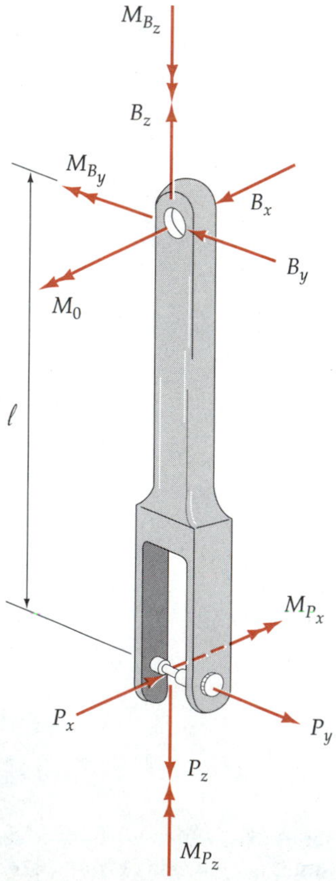

Figure E7.8c

In addition, $\Sigma\mathbf{F} = m\mathbf{a}_C$ for the disk yields

$$P_x\hat{\mathbf{i}} + P_y\hat{\mathbf{j}} - P_z\hat{\mathbf{k}} + mg\hat{\mathbf{k}} = m(-\ell\dot{\omega}_x\hat{\mathbf{j}} - \ell\omega_x^2\hat{\mathbf{k}})$$

so that

$$P_x = 0 \tag{4}$$

$$P_y = -m\ell\dot{\omega}_x \tag{5}$$

$$P_z = mg + m\ell\omega_x^2 \tag{6}$$

Turning now to Figure E7.8c, a free-body diagram of the *light* body $\mathcal{B}$, we have $\Sigma M_{Ox} \approx 0$ so that

$$\ell P_y + M_0 - M_{P_x} = 0 \tag{7}$$

Combining Equations (1), (5), and (7),

$$\dot{\omega}_x = \frac{M_0}{I + m\ell^2} \tag{8}$$

and

$$P_y = \frac{-m\ell M_0}{I + m\ell^2} \tag{9}$$

and

$$M_{P_x} = \frac{IM_0}{I + m\ell^2} \tag{10}$$

The results (4), (6), (8), (9), and (10) are recognizable as what would have been obtained had the disk been frozen in its bearings, in which case $\mathcal{B}$ and $\mathcal{D}$ would constitute a single rigid body in plane motion.

Equation (3), however, does not follow intuitively from the study of plane motion. The term $J\omega_x\omega_s$ is sometimes called a gyroscopic moment, and the equation says that a moment of this magnitude must act on $\mathcal{D}$ about z_C if the given motion is to occur. Note that in this case the body $\mathcal{D}$ is *not allowed* to turn about z_C as it spins (about y_C). If it were, say by means of a bearing between C and O, then the moment component M_{P_z} would become zero, and a third ω component (about z_C) would appear. Note also from Figure E7.8c that the gyroscopic moment *twists* the shaft of $\mathcal{B}$. This, for example, is a consideration in the retraction of the wheels of some airplanes.

EXAMPLE 7.9

The thin disk $\mathcal{D}$ turns on the light arm $\mathcal{B}$ by way of a smooth bearing that keeps the axes of $\mathcal{D}$ and $\mathcal{B}$ aligned. The arm is hinged to a shaft driven at constant angular speed Ω by a motor. The system is set up so that the arm is horizontal when the disk contacts the ground as shown in Figure E7.9a. Assuming the disk, of mass m, to roll on the ground, find the forces exerted by the ground on the disk, the forces and moments exerted by the arm on the disk, and the forces and moments exerted by the shaft on the arm.

Solution

We shall begin the analysis by applying to the disk the equations of motion:

$$\Sigma \mathbf{F} = m\mathbf{a}_C$$

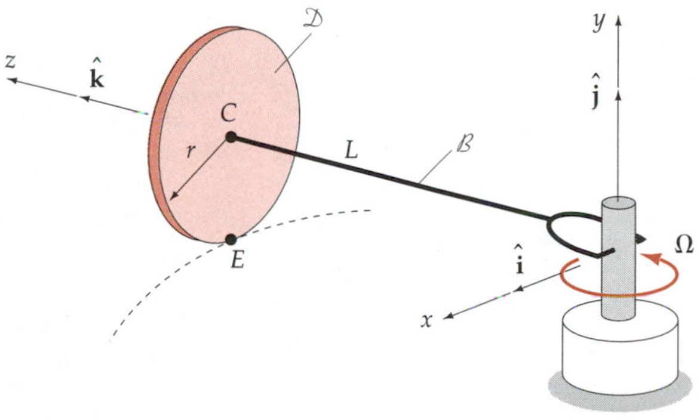

Figure E7.9a

and

$$\Sigma \mathbf{M}_C = {}^{\mathcal{I}}\dot{\mathbf{H}}_C$$

Because C is on the axis of the arm which turns at constant rate Ω, we know that

$$\mathbf{a}_C = -L\Omega^2\hat{\mathbf{k}}$$

where we are using the axes and unit vectors shown in Figure E7.9a and these are *fixed in the arm* $\mathcal{B}$. To obtain an expression for $\mathbf{H}_C$ we need to determine $\boldsymbol{\omega}_{\mathcal{D}/\mathcal{I}}$. So, using the addition theorem,

$$\begin{aligned}\boldsymbol{\omega}_{\mathcal{D}/\mathcal{I}} &= \boldsymbol{\omega}_{\mathcal{D}/\mathcal{B}} + \boldsymbol{\omega}_{\mathcal{B}/\mathcal{I}} \\ &= \omega\hat{\mathbf{k}} + \Omega\hat{\mathbf{j}}\end{aligned}$$

where the only motion $\mathcal{D}$ can have in $\mathcal{B}$ is to turn (in simple angular velocity) about their common axis.

Because the disk rolls, its point E in contact with the ground has zero velocity; using Equation (6.56), we obtain

$$\begin{aligned}\mathbf{v}_E &= \mathbf{v}_C + \boldsymbol{\omega}_{\mathcal{D}/\mathcal{I}} \times \mathbf{r}_{CE} \\ 0 &= \mathbf{v}_C + (\omega\hat{\mathbf{k}} + \Omega\hat{\mathbf{j}}) \times (-r\hat{\mathbf{j}}) \\ 0 &= L\Omega\hat{\mathbf{i}} + r\omega\hat{\mathbf{i}}\end{aligned}$$

or

$$\omega = -\frac{L}{r}\Omega$$

and

$$\boldsymbol{\omega}_{\mathcal{D}/\mathcal{I}} = -\frac{L}{r}\Omega\hat{\mathbf{k}} + \Omega\hat{\mathbf{j}}$$

Therefore, since we assume the disk to be relatively thin so that to good approximation

$$I_{xx}^C = I_{yy}^C = \frac{mr^2}{4},$$

and since $I_{zz}^C = \dfrac{mr^2}{2}$, we have

$$\mathbf{H}_C = \frac{mr^2}{4}\Omega\hat{\mathbf{j}} - \frac{mrL}{2}\Omega\hat{\mathbf{k}}$$

Note that since Ω is constant,

$${}^{\mathcal{B}}\dot{\mathbf{H}}_C = \mathbf{0}$$

and so

$$\begin{aligned}{}^{\mathcal{I}}\dot{\mathbf{H}}_C &= \overset{0}{\cancel{{}^{\mathcal{B}}\dot{\mathbf{H}}_C}} + \boldsymbol{\omega}_{\mathcal{B}/\mathcal{I}} \times \mathbf{H}_C \\ &= \mathbf{0} + \Omega\hat{\mathbf{j}} \times \left(\frac{mr^2}{4}\Omega\hat{\mathbf{j}} - \frac{mrL}{2}\Omega\hat{\mathbf{k}}\right) \\ &= -\frac{mrL}{2}\Omega^2\hat{\mathbf{i}}\end{aligned}$$

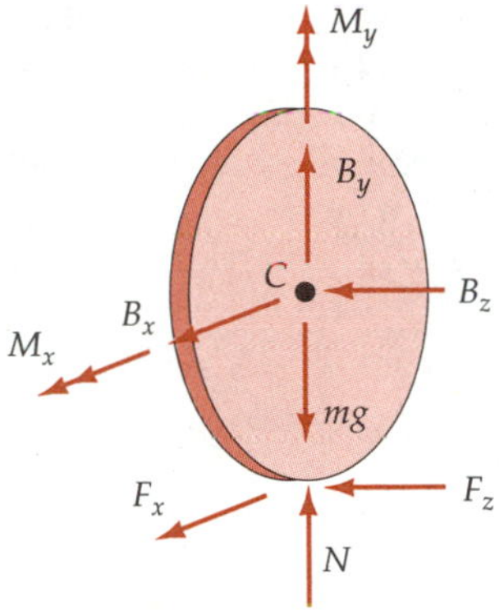

Figure E7.9b

Now using Figure E7.9b, the free-body diagram of the disk, the equation

$$\Sigma \mathbf{F} = m\mathbf{a}_C$$

yields

$$B_x + F_x = 0 \tag{1}$$

$$N + B_y - mg = 0 \tag{2}$$

$$B_z + F_z = -mL\Omega^2 \tag{3}$$

and the equation

$$\Sigma \mathbf{M}_C = {}^{\mathscr{I}}\dot{\mathbf{H}}_C$$

likewise yields

$$M_x - rF_z = -\frac{mrL}{2}\Omega^2 \tag{4}$$

$$M_y = 0 \tag{5}$$

$$rF_x = 0 \tag{6}$$

Using Figure E7.9c, the free-body diagram of the massless bar, we have

$$\Sigma M_{Ax} = 0$$

$$LB_y - M_x = 0 \tag{7}$$

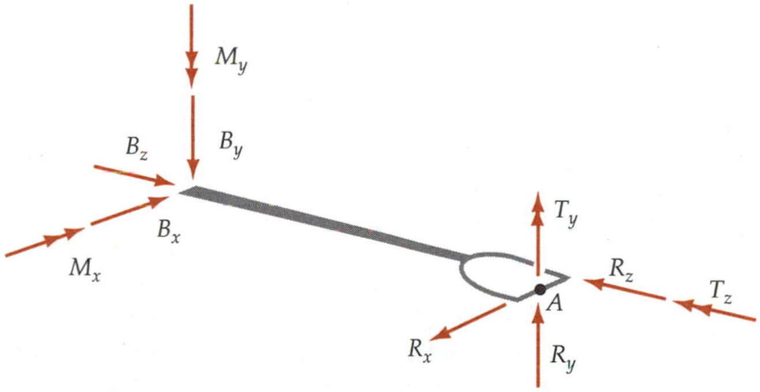

Figure E7.9c

We shall delay using $\Sigma M_{Ay} = 0$, $\Sigma M_{Az} = 0$ and $\Sigma \mathbf{F} = \mathbf{0}$, which will ultimately yield R_x, R_y, R_z, T_y and T_z, and focus on Equations (1)–(7) from which we obtain

$$M_y = F_x = B_x = 0$$

$$M_x = \frac{-mLr}{2}\Omega^2 + rF_z$$

$$B_y = \frac{-mr\Omega^2}{2} + \frac{r}{L}F_z$$

$$B_z = -mL\Omega^2 - F_z$$

$$N = mg + \frac{mr}{2}\Omega^2 - \frac{r}{L}F_z$$

which is as far as we can go because this problem is actually "dynamically" indeterminate. However, if we assume friction to be small (just enough to produce rolling as this system is slowly brought up to speed) so that $F_z \approx 0$, then we have M_x, B_y, B_z and N uniquely determined. With that condition we may now use the remaining "equilibrium" equations for the light rod to obtain

$$R_x = 0$$

$$R_y = \frac{-mr}{2}\Omega^2$$

$$R_z = -mL\Omega^2$$

$$T_y = T_z = 0$$

Let us make a couple of observations: first, note that $T_y = 0$ means that the motor doesn't have to supply any driving torque in order to maintain the constant Ω — remember we have smooth bearings; and secondly note that if Ω is large, so is N:

$$N = mg + \frac{mr\Omega^2}{2}$$

and that extra part of N, over and above the weight mg, is often said to be due to gyroscopic action.

EXAMPLE 7.10

It is possible for the homogeneous cone $\mathcal{B}_1$ of base radius R in the Figure E7.10a to roll steadily around on a flat horizontal table $\mathcal{B}_3$ in such a way that its unconstrained vertex O remains fixed and the center point Q of its base travels on a horizontal circle at constant speed. Let this motion be begun by forces which are then released. Assume that there is sufficient friction between the cone and the table to prevent slipping. Besides having enough friction, there is yet another special condition that must be satisfied in order that the motion occur. Find the friction and normal force resultants, their lines of action, and the special condition. Refer to Example 6.12 for some related kinematics.

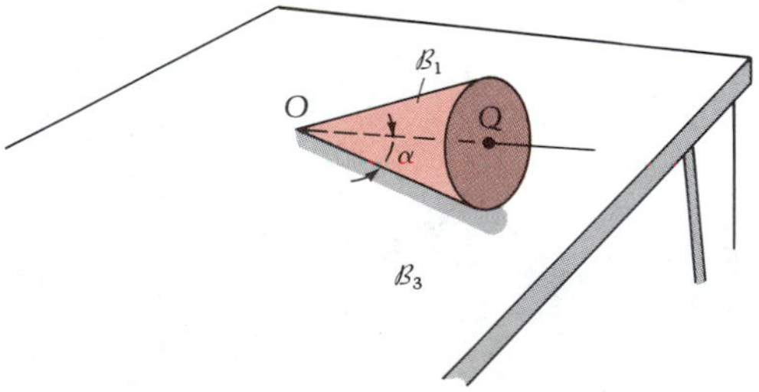

Figure E7.10a

Solution

Since the mass center C moves on a horizontal circle at the constant speed $\frac{3}{4}v_Q$ (see Figure E7.10b), the friction force f is given by:

$$\overset{+}{\longleftarrow} \qquad \Sigma F_x = ma_{CX} = \frac{m\dot{s}_C^2}{\rho_C}$$

or

$$f = \frac{m\left(\dfrac{3}{4}v_Q\right)^2}{\left(\dfrac{3}{4}H\cos\alpha\right)} = \frac{3}{4}\frac{mv_Q^2\sin\alpha}{R\cos^2\alpha}$$

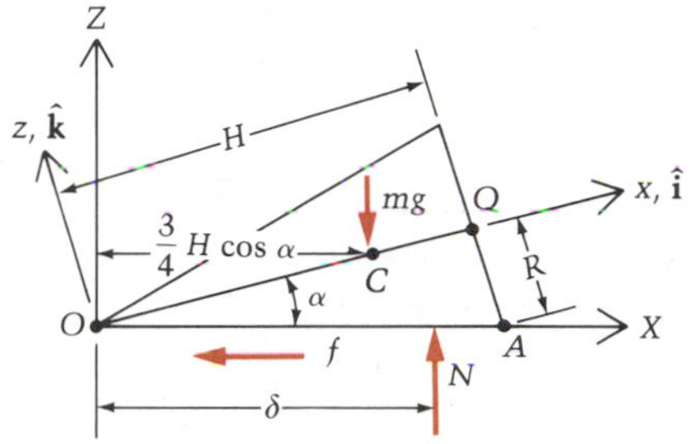

Figure E7.10b

since $H = R/\tan\alpha$. The normal force N is simply mg since $a_{CZ} = 0$. There is no friction force component normal to the plane of the paper, since a_{CY} also vanishes. The directed line of action of f is as shown above, along AO. For the line of action of N, we use the moment equation of motion; note that $\mathcal{B}_3$ is our inertial frame:

$$\Sigma \mathbf{M}_O = {}^{\mathcal{B}_3}\dot{\mathbf{H}}_O = {}^{\mathcal{B}_2}\dot{\mathbf{H}}_O + \boldsymbol{\omega}_{2/3} \times \mathbf{H}_O$$

where frame $\mathcal{B}_2$ contains the Z axis and the axis of the cone. Now the angular momentum $\mathbf{H}_O$ is, since $(\hat{\mathbf{i}}, \hat{\mathbf{j}}, \hat{\mathbf{k}})$ are in principal directions at the fixed point O:

$$\mathbf{H}_O = J\omega_x\hat{\mathbf{i}} + I\omega_y\hat{\mathbf{j}} + I\omega_z\hat{\mathbf{k}}$$

In the preceding equation, the $\boldsymbol{\omega}$-components are those of the body $\mathcal{B}_1$ in $\mathcal{B}_3$. From the previous examples, $\omega_y = 0$ and, substituting for ω_x, ω_z, J, and I,

$$\mathbf{H}_O = \frac{mR^2v_Q}{R}\left[\frac{-3\hat{\mathbf{i}}}{10} + \left(\frac{3}{20} + \frac{3H^2}{5R^2}\right)\tan\alpha\hat{\mathbf{k}}\right]$$

$$= \frac{3mR^2v_Q}{20R}[-2\hat{\mathbf{i}} + (1 + 4/\tan^2\alpha)\tan\alpha\hat{\mathbf{k}}]$$

Noting that ${}^{\mathcal{B}_2}\dot{\mathbf{H}}_O = \mathbf{0}$, we find

$$\Sigma\mathbf{M}_O = \frac{v_Q\sin\alpha}{R\cos^2\alpha}\frac{3mR^2v_Q}{20R}\begin{vmatrix} \hat{\mathbf{i}} & \hat{\mathbf{j}} & \hat{\mathbf{k}} \\ \sin\alpha & 0 & \cos\alpha \\ -2 & 0 & \left(1 + \dfrac{4}{\tan^2\alpha}\right)\tan\alpha \end{vmatrix}$$

$$= \frac{3mv_Q^2\sin\alpha}{20\cos^3\alpha}(1 + 5\cos^2\alpha)(-\hat{\mathbf{j}})$$

and from the free-body diagram, using $N = mg$,

$$-mg\,\delta + mg\frac{3R\cos\alpha}{4\tan\alpha} = \frac{-3mv_Q^2\sin\alpha}{20\cos^3\alpha}(1 + 5\cos^2\alpha)$$

or

$$\delta = \frac{3R\cos\alpha}{4\tan\alpha} + \frac{3v_Q^2\sin\alpha}{20g\cos^3\alpha}(1 + 5\cos^2\alpha)$$

which gives the line of action of the normal force resultant.

The "special condition" mentioned in the problem statement is that the normal force must intersect the plane at a point of physical contact with the cone, i.e.:

$$\delta \le R/\sin\alpha$$

Thus

$$\frac{3R\cos^2\alpha}{4\sin\alpha} + \frac{3v_Q^2\sin\alpha}{20g\cos^3\alpha}(1 + 5\cos^2\alpha) \le \frac{R}{\sin\alpha}$$

which when simplified is

$$3\frac{v_Q^2}{gR}\sin^2\alpha(1 + 5\cos^2\alpha) \le 5\cos^3\alpha(4 - 3\cos^2\alpha)$$

Holding α constant, we see that what happens if v_Q is too large (or g or R too small) is that the normal force needs to act beyond the point A. Since that cannot physically happen, the specified motion will not occur but will give way to another, that of tipping outward.

PROBLEMS ▶ Section 7.5

7.42 The rod $\mathcal{L}$ is rigidly attached to shaft $\mathcal{S}$ which is free to turn in the two bearings as indicated in Figure P7.42. The y and y' axes point into the page at C. Show that the moment with respect to C that must be supplied by the bearings to shaft $\mathcal{S}$ to sustain the motion must have the components:

$$\Sigma M_{Cx'} = -\frac{m\ell^2}{12}(\sin\beta\cos\beta)\alpha$$

$$\Sigma M_{Cy'} = -\frac{m\ell^2}{12}(\sin\beta\cos\beta)\omega^2$$

Do this in two ways: (1) Use the Euler equations (7.40) with the principal axes (x, y, z); (2) use the Equations (7.41) with the axes (x', y', z') in the figure.

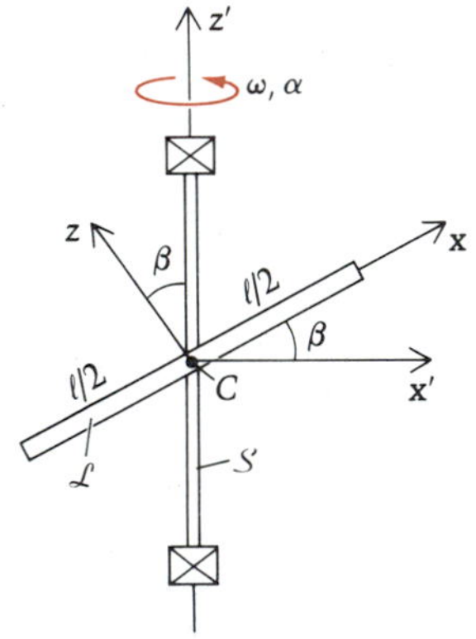

Figure P7.42

7.43 Find the reactions exerted by the bearings on the shaft to which a 30-kg thin plate $\mathcal{P}$ is welded. (See Figure P7.43.) The assembly is turning at the constant angular speed of 30 rad/s. Work the problem by using

$$\Sigma \mathbf{M}_C = {}^{\mathcal{I}}\dot{\mathbf{H}}_C = {}^{\mathcal{P}}\dot{\mathbf{H}}_C + \boldsymbol{\omega}_{\mathcal{P}/\mathcal{I}} \times \mathbf{H}_C$$

that is, by expressing $\mathbf{H}_C$ using principal directions in $\mathcal{P}$ (which omits the need for computing the nonzero product of inertia I_{xz}).

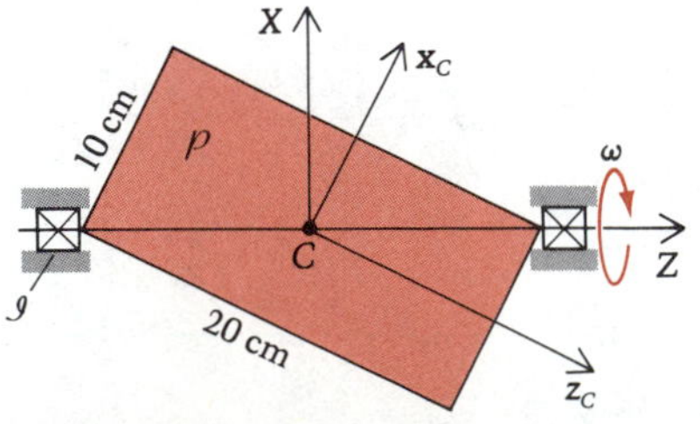

Figure P7.43

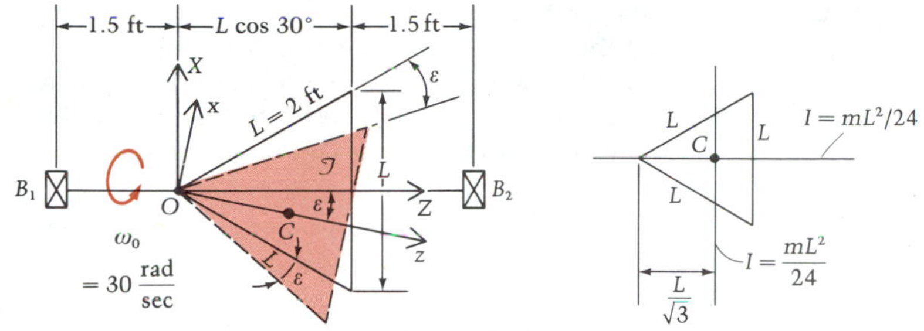

Figure P7.45

7.44 Rework and check the results of the preceding problem by using Equations (7.24) to calculate I_{xz}; then use Equations (7.41) to obtain the bearing reactions.

7.45 The equilateral triangular plate $\mathcal{T}$ was to be mounted onto the rotating shaft as shown in Figure P7.45 in the solid figure. The installation resulted in the misalignment angle ϵ (see the dashed position). The plate has mass 2 slugs, and the shaft is light. Determine the dynamic bearing reactions at B_1 and B_2 caused by the misalignment, in terms of ϵ, by: (a) using principal axes; (b) without using principal axes (i.e., calculate and utilize the products of inertia).

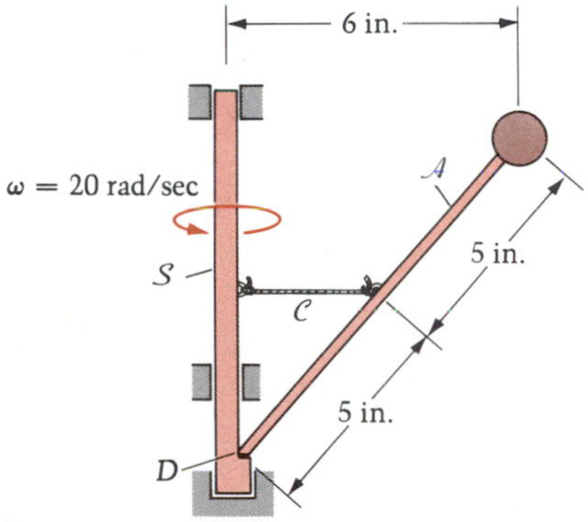

Figure P7.46

7.46 A slender rod $\mathcal{A}$ and a small ball each weigh 0.3 lb. (See Figure P7.46.) The bodies rotate about the vertical along with the slender shaft $\mathcal{S}$ and are supported by a smooth step or thrust bearing at D and by the cord $\mathcal{C}$. Find the tension in the cord if the angular speed of the system is 20 rad/sec.

7.47 The disk (mass m, radius r) in Figure P7.47 is rigidly attached to the shaft, and the assembly is spun up to angular speed Ω about the z_C axis. Determine the bearing reactions at A and B in terms of m, r, g, β, Ω, and L.

7.48 A bicycle wheel weighing 5 lb and having a 14-in. radius is misaligned by 1° with the vertical. The top of the wheel tilts toward the right with the bike moving forward. If the bike is driven along a straight path at 15 mph, find the moment $\Sigma \mathbf{M}_C$ exerted on the wheel. Use the result of Problem 7.11, neglecting the spokes and hub.

7.49 In the preceding problem, suppose instead that the bicycle is in a 50-ft-radius turn to the (a) right and (b) left. Determine the new values of the moment exerted on the misaligned wheel. Neglect the lean angle.

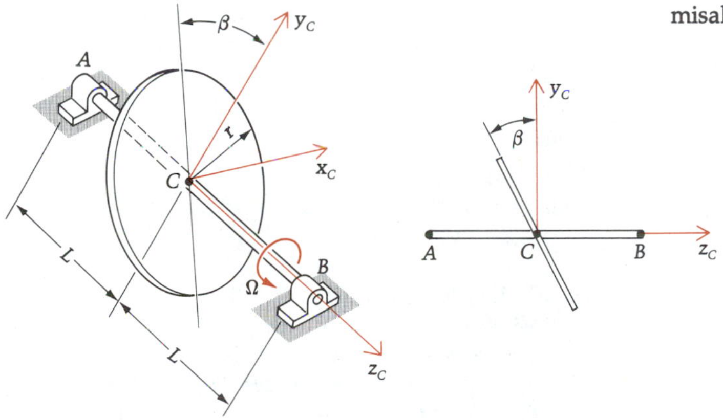

Figure P7.47

7.50 Writing Euler's laws as $\Sigma\mathbf{F} + (-m\mathbf{a}_C) = \mathbf{0}$ and $\Sigma\mathbf{M}_C + (-\dot{\mathbf{H}}_C) = \mathbf{0}$ results in what is known as the *reversed effective force* $(-m\mathbf{a}_C)$ and the *inertia torque* $(-\dot{\mathbf{H}}_C)$. If these quantities are added to the free-body diagram, the object may be treated as though it were in equilibrium. For a car at the instant of overturning while traveling at speed v on a curve of radius R and bank angle θ, such a diagram would appear as shown in Figure P7.50.

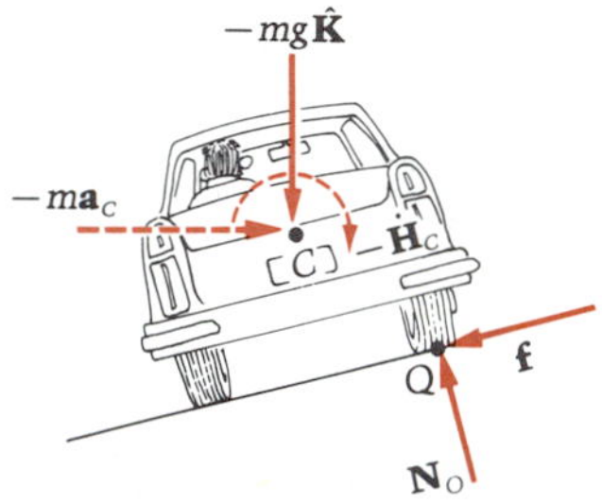

Figure P7.50

Calculate the moment about the point Q and, noting that $\Sigma\mathbf{M}_Q = \mathbf{0}$ for this free-body diagram, compare the effects of the reversed effective force (or inertia force) $-m\mathbf{a}_C$ with that of the inertia torque $-\dot{\mathbf{H}}_C$ on the overturning tendency. What factors make the car more likely to turn over?

7.51 Suppose that the components of the mass center velocity $\mathbf{v}_C$ are written in body $\mathcal{B}$ instead of in an inertial frame $\mathcal{I}$. Use the property (6.20) of $\boldsymbol{\omega}_{\mathcal{B}/\mathcal{I}}$ to derive the scalar equations of motion of the mass center from Euler's first law: $\Sigma\mathbf{F} = {}^{\mathcal{I}}\dot{\mathbf{L}}$.

7.52 Show that if a rigid body $\mathcal{B}$ undergoing torque-free motion in an inertial frame $\mathcal{I}$ has three equal principal moments of inertia at its mass center, then its angular velocity is constant in $\mathcal{I}$.

7.53 In Example 7.8 define the (x, y, z) axes at O as principal for $\mathcal{B}$, with associated respective moments of inertia $\bar{I}, \bar{J}$, and $\bar{K}$. Rework the problem without assuming that $\mathcal{B}$ has negligible mass. The two sets of axes are respectively parallel prior to the application of M_O.

7.54 Show that if the solution $\boldsymbol{\omega}(t)$ of Example 7.6 is projected into the xy plane, the tip of the projection vector travels on a circle of radius $\sqrt{\omega_{x_i}^2 + \omega_{y_i}^2}$ at the frequency $(J - I)\omega_{z_i}/I$.

7.55 A result of the earth's bulge is that $I/J = 0.997$. Use this result to compute the period of one revolution of the earth's angular velocity vector (North Pole!) about its axis of symmetry. (The answer, obtained by Euler in 1752, is about 4 months less than the actual period first observed by S. Chandler in 1891. The difference is attributed to the nonrigidity of the earth. Although energy dissipation should damp out this "wobble," in fact it does not. The ongoing cause of the wobble is an unsolved problem in geodynamics at this time. See *Science News*, 24 October 1981.)

7.56 A screwdriver-like motion between the planar case and general (three-dimensional) motion is defined as follows. All points of the body $\mathcal{B}$ have, at any time, identical z components of velocity in a reference frame $\mathcal{I}$. The unit vector $\hat{\mathbf{k}}$ of this $\dot{z}\hat{\mathbf{k}}$ component is constant in both $\mathcal{B}$ and $\mathcal{I}$, though $\dot{z}$ can vary with time. Thus the angular velocity vector is still expressible as $\boldsymbol{\omega} = \dot{\theta}\hat{\mathbf{k}}$. Derive a moment equation for $\mathcal{B}$ that is valid for this motion.

7.57 The disk in Figure P7.57 is spinning about the light axle, and the axle is precessing at the constant rate of $32\hat{\mathbf{i}}$ rad/s. If the axle is observed to remain horizontal, find the magnitude and direction of the spin of the disk.

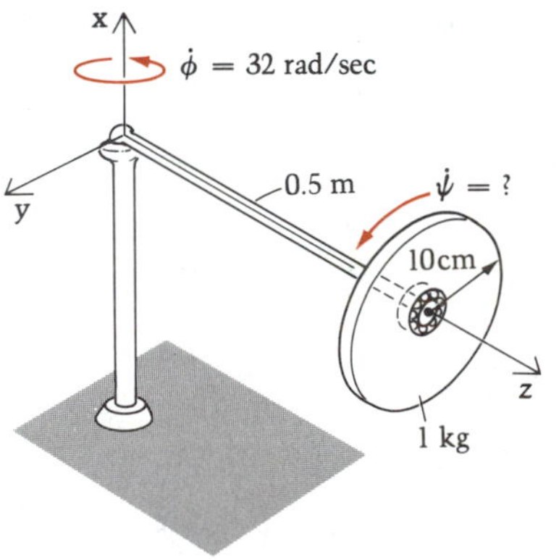

Figure P7.57

7.58 A single-engine aircraft has a four-bladed propeller weighing 128 lb with a radius of gyration about its center of mass of 3 ft. It rotates counterclockwise at 2000 rpm when viewed from the rear. Find the gyroscopic moment on the propeller shaft when the plane is at the bottom of a vertical loop of 2000-ft radius with a speed of 500 mph. In which direction will the tail of the plane tend to move because of this moment?

7.59 The blades of a fan turn at 1750 rpm, and the fan oscillates about the vertical axis z (Figure P7.59(a)) at the rate of one cycle every 10 sec. Assuming that the fan travels at the constant angular velocity of 0.2 rad/sec except when it is reversing direction (Figure P7.59(b)), find the moment exerted by the base on the arm section $\mathcal{A}$ at the $\frac{1}{4}$-cycle point due to gyroscopic action. For the calculation (*only*!) consider the blades (Figure P7.59(c)) to be 4-in.-diameter circular aluminum plates all in the same plane and $\frac{1}{32}$ in. thick. Use a density of 0.1 lb/in.^{3}.

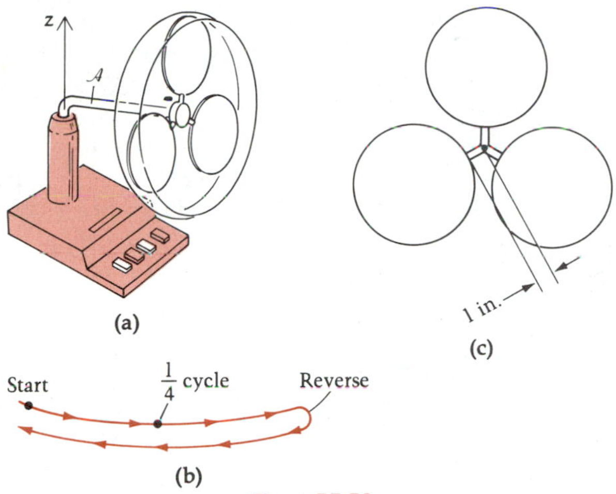

Figure P7.59

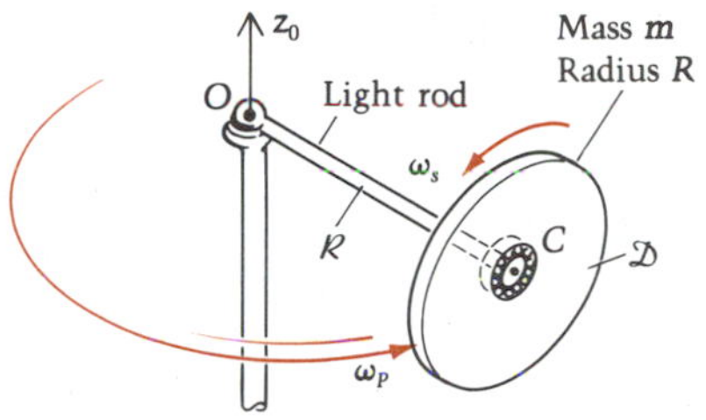

Figure P7.60

7.60 Disk $\mathcal{D}$ in Figure P7.60 turns in bearings at C at angular rate ω_s about the light rod $\mathcal{R}$, and both precess about axis z_O at angular rate ω_P as shown. Show with a free-body diagram how it is possible for the mass center C to remain in a horizontal plane. Then find the reactions exerted onto $\mathcal{R}$ by the socket at O. Is there any difference in the solution if $\mathcal{D}$ and $\mathcal{R}$ are rigidly connected?

7.61 In the grinding mill of Problem 7.5, suppose that the wall is absent. (See Figure P7.61.) Find, for a given Ω (constant angular speed of $\mathcal{S}$), the angle ϕ that the axis of the grinder $\mathcal{D}$ will make with the vertical. Observe that with the wall present and ϕ fixed, larger speeds than this Ω will allow the grinder to work. In particular, show that the following set of parameters is satisfactory: $r = 2.5$ ft, $\ell = 6$ ft, $\Omega = 2\pi$ rad/sec, and $\phi = 60°$. Neglect the mass of body $\mathcal{B}$ in comparison with the heavy grinding disk $\mathcal{D}$.

7.62 Find the grinding force N produced at the wall of the grinding mill of Problems 7.5 and 7.61 for the given parameters.

7.63 Find the magnitude and direction of the force and/or couple exerted on disk $\mathcal{D}$ by the shaft $\mathcal{S}$ in Problem 7.4.

7.64 A disk $\mathcal{D}$ rolls around in a circle with its plane vertical and its center traveling at constant speed v_C. Find the tension in the string, and the friction force exerted on $\mathcal{D}$ by the floor. (See Figure P7.64.)

7.65 There is a relationship among v_C, g, r, R, and θ such that the disk can roll around in a circle as shown in Figures P7.65(a) and (b), with v_C and θ remaining constant. Find this relationship.

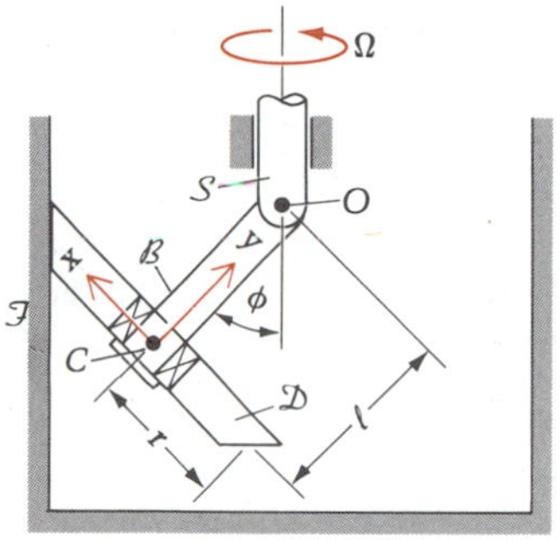

Figure P7.61

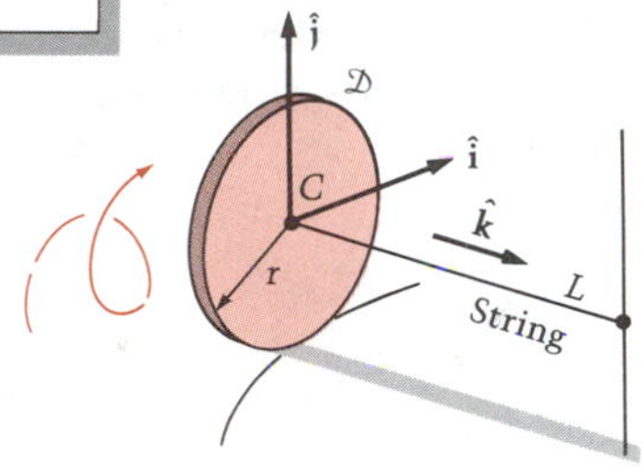

Figure P7.64

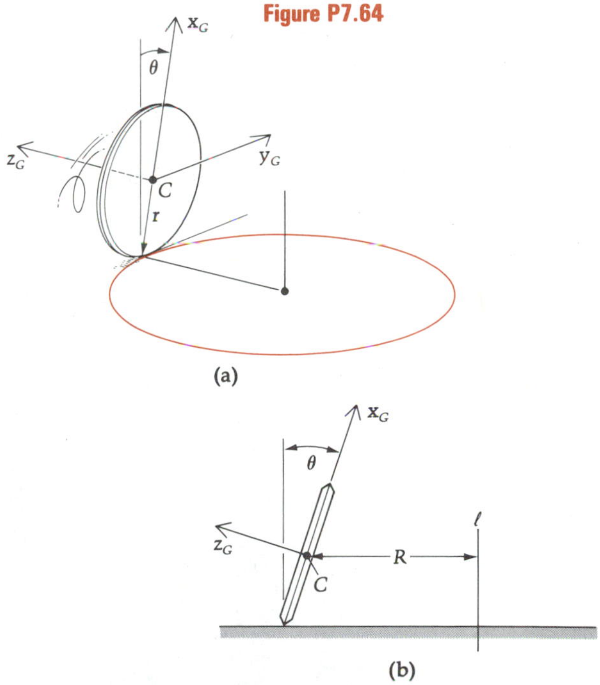

Figure P7.65

7.66 Obtain the results of Example 7.9 by using the Euler equations (7.42). *Hint*: This time the axes are body-fixed, and the Ω part of $\boldsymbol{\omega}$ changes direction in $\mathcal{D}$; therefore the components ω_x and ω_y have derivatives that were formerly zero in $\mathcal{B}$. The Ω component of $\boldsymbol{\omega}_{\mathcal{D}/\mathcal{I}}$ is

$$\Omega\,(\cos\theta_r\hat{\mathbf{j}} - \sin\theta_r\hat{\mathbf{i}})$$

where θ_r is the angle of roll as shown in Figure P7.66. Differentiate this expression, substitute $\dot{\theta}_r = v_C/r$, and *then* take $\theta_r = 0$. Finally, go to Equations (7.42) and substitute your results.

7.67 A ship's turbine has a mass of 2500 kg and a radius of gyration about its axis (y_c in Figure P7.67) of 0.45 m. It is mounted on bearings as indicated and turns at 5000 rpm clockwise when viewed from the stern (rear) of the boat.

a. If the ship is in a steady turn to the right of radius 500 m and is traveling at 15 knots, what are the reactions exerted on the shaft by the bearings? (1 knot = 1.15 mph = 1.85 km/hr)

b. If the ship on a straight course in rough seas pitches sinusoidally at ±12° amplitude with a 6-s period, what are the maximum bearing reactions then?

7.68 A heavy disk $\mathcal{D}$ of mass m and radius r spins at the angular rate ω_3 $(=|\boldsymbol{\omega}_{\mathcal{D}/\mathcal{B}}|)$ with respect to the rigid, but light, bent bar $\mathcal{B}$. (See Figure P7.68.) Body $\mathcal{B}$ turns at rate ω_2 $(=|\boldsymbol{\omega}_{\mathcal{B}/\mathcal{G}}|)$ about a vertical axis through O, a point of both $\mathcal{B}$ and the inertial frame $\mathcal{G}$. Find the force and couple

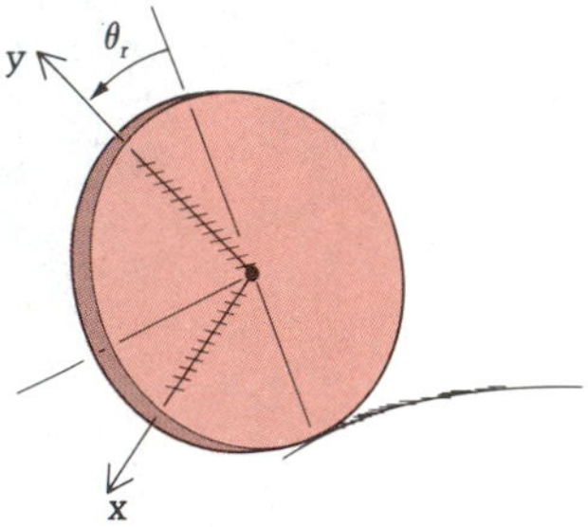

Figure P7.66

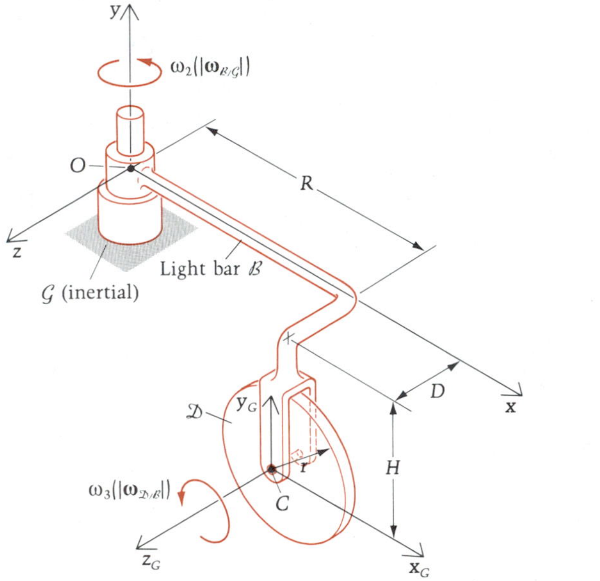

Figure P7.68

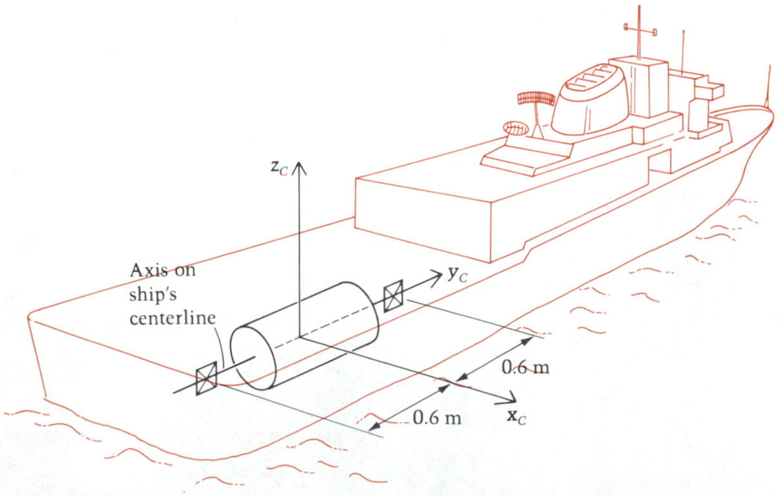

Figure P7.67

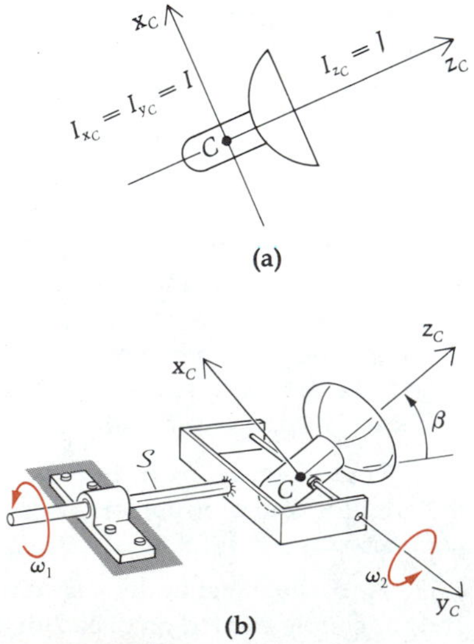

Figure P7.69

that must be acting on $\mathcal{B}$ at O to produce a motion of the system for which ω_2 and ω_3 are constants. Both sets of axes in the figure are fixed in $\mathcal{B}$, and note that (x_C, y_C, z_C) are always principal axes for $\mathcal{D}$ at C even though they are not fixed in $\mathcal{D}$.

7.69 Compute the moment M applied to the shaft $\mathcal{S}$ in Figures P7.69(a) and (b), on the preceding page, as a function of the angle $\mathcal{B}$ if ω_1 and ω_2 $(=\dot{\beta})$ are constants.

7.70 A bike rider enters a turn of radius R at a constant speed of v_C. (See Figure P7.70.) Other quantities are defined below:

r = radius of wheel

d = distance between axle and C

I_1, I_2 = principal moments of inertia of entire bike plus rider with respect to $\hat{\mathbf{n}}_1$ and $\hat{\mathbf{n}}_2$ directions through C

i = moment of inertia, with respect to $\hat{\mathbf{n}}_2$ direction, of one wheel about its axis of symmetry

m = total mass

ϕ = angle shown

Solve for the resultant force $\Sigma\mathbf{F}$ and moment $\Sigma\mathbf{M}_C$ in terms of these quantities. Compare the effects of the *D'Alembert force* $(-m\mathbf{a}_C)$ and the *inertia torque* $(-\dot{\mathbf{H}}_C)$ in righting the bike when ϕ is small. Neglect the products of inertia.

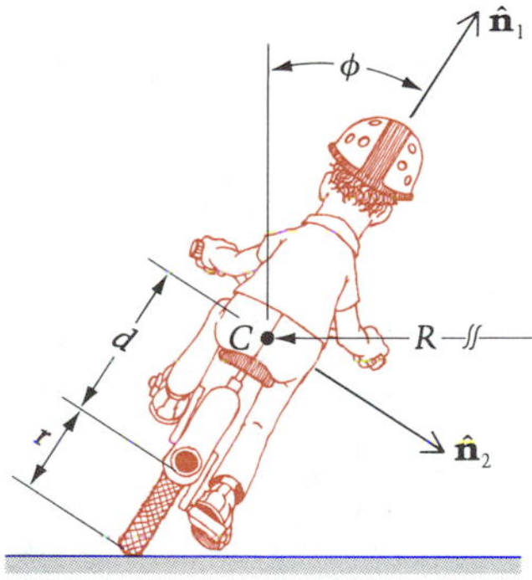

Figure P7.70

* **7.71** Disk $\mathcal{D}$ in Figure P7.71 turns in bearings around rod $\mathcal{R}$ as it rolls on the ground. Find all the forces acting on $\mathcal{D}$, and contrast the solution with that of Example 7.9. The rod is free to slide along the vertical post about which it turns at constant rate.

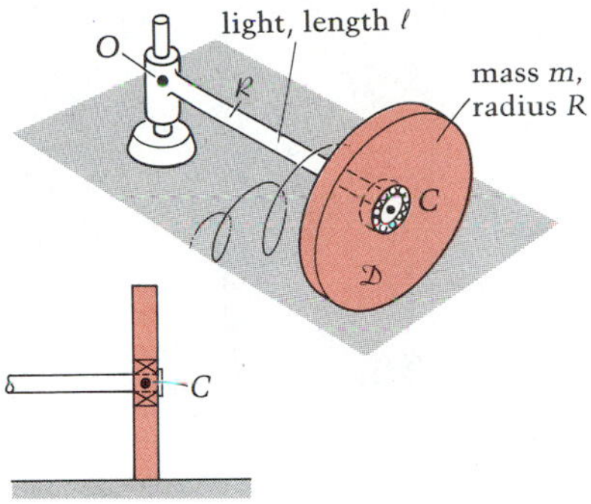

Figure P7.71

7.72 In Problem 6.24, if ω_1, ω_2, and ω_3 are constants, find the resultant moment exerted on $\mathcal{D}$ at its mass center, when $\hat{\mathbf{u}}$ is pointing straight up. The body $\mathcal{D}$ is symmetrical, with centroidal principal moments of inertia J along $\hat{\mathbf{u}}$, and I normal to $\hat{\mathbf{u}}$.

* **7.73** The body $\mathcal{B}$ in Figure P7.73 is an ellipsoid of revolution with mass = 1 slug, and semiminor and semimajor axis lengths a and $2a$, with a = 1 ft. Thus

$$I_{xx}^C = \frac{m}{5}(a^2 + a^2) = 0.4 \text{ slug-ft}^2$$

and

$$I_{yy}^C = I_{zz}^C = \frac{m}{5}[a^2 + (2a)^2] = 1.0 \text{ sl-ft}^2$$

The shaded light frame $\mathcal{A}$ is driven around the fixed post $\mathcal{P}$, with angular velocity $\omega_2\hat{\mathbf{j}}$, by a motor torque T_2 applied at P. Another motor (neither is shown) between $\mathcal{A}$ and $\mathcal{B}$ applies a torque $T_1\hat{\mathbf{i}}$ which causes the body $\mathcal{B}$ to spin in the frame. The axes and unit vectors shown are fixed in $\mathcal{A}$. During an interval of motion, $\omega_1 = 3t^2$ rad/sec and $\omega_2 = 2t$ rad/sec. Find all forces and couples applied onto $\mathcal{A}$ at P when t = 1 sec. The distance from P to the x axis is 2 ft.

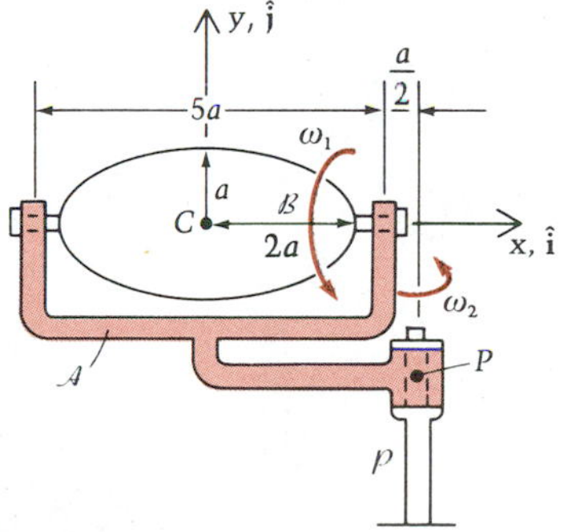

Figure P7.73

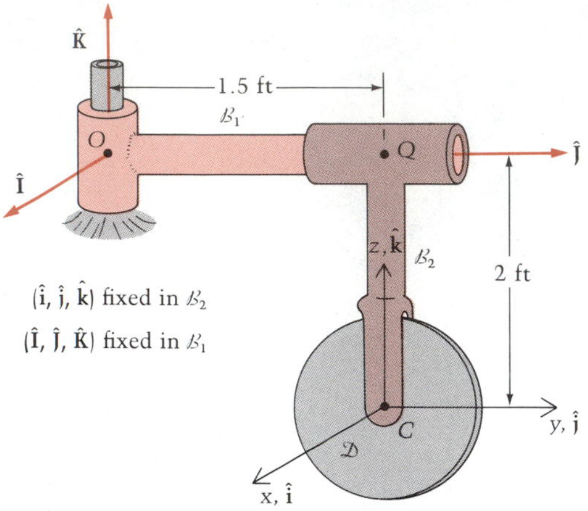

Figure P7.74

7.74 By means of appropriately mounted control motors (not shown), the disk $\mathcal{D}$ in Figure P7.74 is made to turn about its axis of symmetry relative to the arm $\mathcal{B}_2$; $\mathcal{B}_2$ is made to turn relative to arm $\mathcal{B}_1$ about the axis of $\mathcal{B}_1$; and $\mathcal{B}_1$ is made to turn with respect to the ground (inertial frame) $\mathcal{I}$ about the vertical through point O. At the given instant, the angular velocity of $\mathcal{D}$ in $\mathcal{I}$, expressed in terms of unit vectors $(\hat{\mathbf{i}}, \hat{\mathbf{j}}, \hat{\mathbf{k}})$ fixed in $\mathcal{B}_2$, is $\omega_1\hat{\mathbf{i}} + \omega_2\hat{\mathbf{j}} + \omega_3\hat{\mathbf{k}} = \hat{\mathbf{i}} + 2\hat{\mathbf{j}} + 3\hat{\mathbf{k}}$ rad/sec. In addition, at this instant we are given $\dot{\omega}_1 = 4$, $\dot{\omega}_2 = 5$, and $\dot{\omega}_3 = 6$ rad/sec^2. For the disk, $m = 10$ slugs, $I_{xx}^C = 1.4$ slug-ft^2, and $I_{yy}^C = 0.7$ slug-ft^2. Find all forces and couples that are acting on $\mathcal{D}$ at C at the given instant.

* **7.75** In the preceding problem, use $\omega_1, \omega_2, \omega_3$, and their derivatives to compute the right-hand sides of the Euler equations (7.40). Explain why these results are not the components of the moments of external forces acting on $\mathcal{D}$ at C.

* **7.76** In the preceding two problems, at the given instant find $\dot{\theta}_i$ and $\ddot{\theta}_i$, for $i = 1, 2$, and 3, where θ_1 is the angle of rotation of $\mathcal{D}$ with respect to $\mathcal{B}_2$, θ_2 is the angle of rotation of $\mathcal{B}_2$ with respect to $\mathcal{B}_1$, and θ_3 is the angle of rotation of $\mathcal{B}_1$ with respect to $\mathcal{I}$.

7.6 Gyroscopes

We now return our attention to the gyroscope whose orientation was examined in Section 6.8. First we shall derive the equations of rotational motion of such a gyroscope $\mathcal{G}$.* We begin by expressing its angular velocity $\boldsymbol{\omega}_{\mathcal{G}/\mathcal{I}}$ in the frame $\mathcal{F}_2$; see Figures 7.7, 7.8 and 7.9, repeated from Section 6.8.

$$\boldsymbol{\omega}_{\mathcal{G}/\mathcal{I}} = -\dot{\phi} \sin \theta \hat{\mathbf{i}}_2 + \dot{\theta}\hat{\mathbf{j}}_2 + (\dot{\psi} + \dot{\phi} \cos \theta)\hat{\mathbf{k}}_2 \tag{7.44}$$

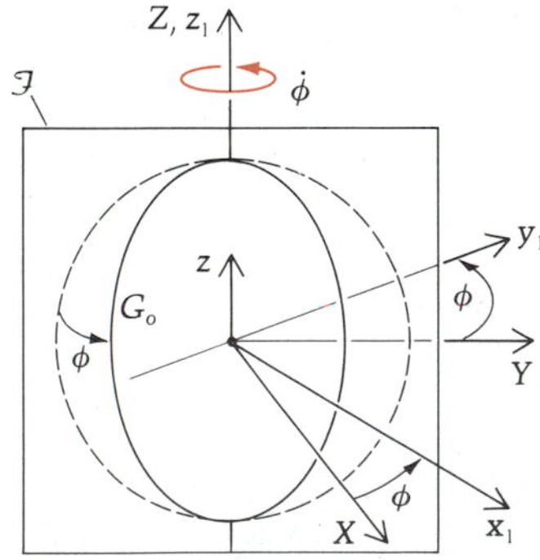

Figure 7.7 First rotation.

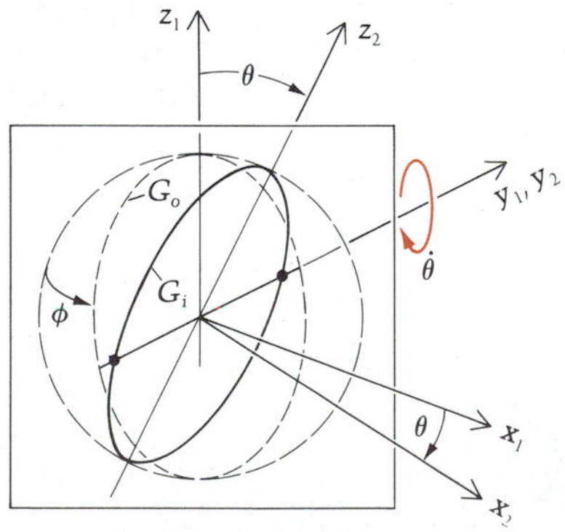

Figure 7.8 Second rotation.

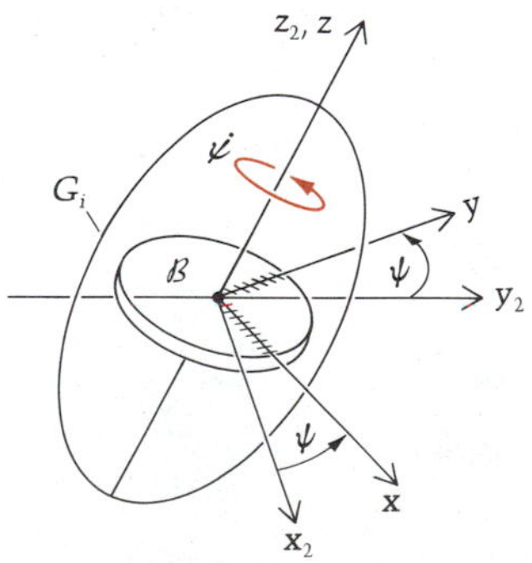

Figure 7.9 Third rotation.

* We are taking $\mathcal{G}$ to be the rotor and are considering it as heavy with respect to the inner and outer gimbals, whose mass we then neglect. We also assume $\mathcal{G}$ to be symmetric about its axis. Of course, a gyroscope does not have to possess *any* gimbals; the earth is a massive gyro, as will be seen in an example to follow.

The axes (x_2, y_2, z_2) of the inner gimbal are not fixed in $\mathcal{G}$ because of its spin $\dot{\psi}$ — *but they are nonetheless permanently principal.* This important fact allows us to write the angular momentum of the gyroscope in the frame $\mathcal{F}_2$ as

$$\mathbf{H}_C = -I\dot{\phi}\sin\theta\hat{\mathbf{i}}_2 + I\dot{\theta}\hat{\mathbf{j}}_2 + J(\dot{\psi} + \dot{\phi}\cos\theta)\hat{\mathbf{k}}_2 \tag{7.45}$$

Euler's second law, together with Property (6.20) of the angular velocity vector, then gives the equations of motion of the gyroscope as follows:

$$\begin{aligned}\Sigma\mathbf{M}_C &= {}^{\mathcal{I}}\dot{\mathbf{H}}_C = {}^{\mathcal{F}_2}\dot{\mathbf{H}}_C + \boldsymbol{\omega}_{\mathcal{F}_2/\mathcal{I}} \times \mathbf{H}_C \\ &= [-I(\ddot{\phi}\sin\theta + \dot{\phi}\dot{\theta}\cos\theta) + \dot{\theta}\dot{\phi}\cos\theta(J - I) + \dot{\theta}\dot{\psi}J]\hat{\mathbf{i}}_2 \\ &\quad + [I\ddot{\theta} + (J - I)\dot{\phi}^2\sin\theta\cos\theta + J\dot{\phi}\dot{\psi}\sin\theta]\hat{\mathbf{j}}_2 \\ &\quad + \left[J\frac{d}{dt}(\dot{\psi} + \dot{\phi}\cos\theta)\right]\hat{\mathbf{k}}_2\end{aligned} \tag{7.46}$$

In the preceding calculations we have used the addition theorem to observe the following:

$$\begin{aligned}\boldsymbol{\omega}_{\mathcal{G}/\mathcal{I}} &= \boldsymbol{\omega}_{\mathcal{G}/\mathcal{F}_2} + \boldsymbol{\omega}_{\mathcal{F}_2/\mathcal{I}} \\ &= \dot{\psi}\hat{\mathbf{k}}_2 + \boldsymbol{\omega}_{\mathcal{F}_2/\mathcal{I}}\end{aligned} \tag{7.47}$$

so that $\boldsymbol{\omega}_{\mathcal{F}_2/\mathcal{I}}$ is the same vector as in Equation (7.44) if $\dot{\psi}$ is omitted. The equations of motion of $\mathcal{G}$ are therefore

$$\begin{aligned}\Sigma M_{Cx_2} &= -I(\ddot{\phi}\sin\theta + 2\dot{\phi}\dot{\theta}\cos\theta) + J\dot{\theta}(\dot{\psi} + \dot{\phi}\cos\theta) \\ \Sigma M_{Cy_2} &= I(\ddot{\theta} - \dot{\phi}^2\sin\theta\cos\theta) + J\dot{\phi}\sin\theta(\dot{\psi} + \dot{\phi}\cos\theta) \\ \Sigma M_{Cz} &= J\frac{d}{dt}(\dot{\psi} + \dot{\phi}\cos\theta) = J\frac{d\omega_z}{dt}\end{aligned} \tag{7.48}$$

where ω_z is the component of $\boldsymbol{\omega}_{\mathcal{G}/\mathcal{I}}$ about the spin axis of symmetry of the gyroscope. Note that it is made up of part of the precession speed as well as all of the spin.

The gyroscope equations are seen to be nonlinear, including not only products of the angles' derivatives but also trigonometric functions of them. Their general solution is an unsolved problem; however, there are two special solutions that are quite worthy of study. The first of these is steady precession; the second is torque-free motion. We shall have a look at each in turn.

Steady Precession

Steady precession is defined by the nutation angle θ, the precession speed $\dot{\phi}$, and the spin speed $\dot{\psi}$ each being constant throughout the motion. Let us call these constants θ_0, $\dot{\phi}_0$, and $\dot{\psi}_0$, and substitute them into Equation (7.48) to obtain:

$$\begin{aligned}\Sigma M_{Cx_2} &= 0 \\ \Sigma M_{Cy_2} &= -I\dot{\phi}_0^2\sin\theta_0\cos\theta_0 + J\dot{\phi}_0\sin\theta_0(\dot{\psi}_0 + \dot{\phi}_0\cos\theta_0) \\ \Sigma M_{Cz} &= 0\end{aligned} \tag{7.49}$$

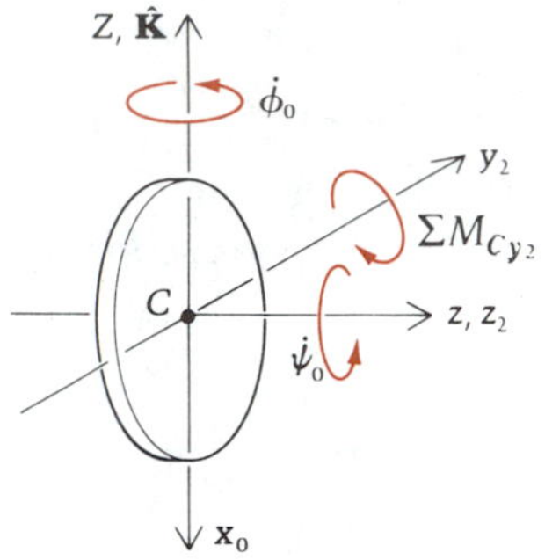

Figure 7.10

We see that only a moment about the y_2 axis is needed to sustain steady precession. Also $\Sigma M_{Cz} = 0 = J(d\omega_z/dt)$ means that ω_z is a constant.*

In the case in which $\theta = 90°$, we have the precession and spin axes orthogonal; the situation is shown in Figure 7.10. If the gyro is spinning and the torque is applied, there will simultaneously occur a precession that tends to turn the spin vector toward the torque vector. This is sometimes called the law of gyroscopic precession. The torque in this case, from Equation (7.49), is

$$\Sigma M_{Cy_2} = J\dot{\psi}_0\dot{\phi}_0 \tag{7.50}$$

and it is seen to be the product of the spin momentum $J\dot{\psi}_0$ and the precessional angular speed $\dot{\phi}_0$.

We turn now to an illustration of the law of gyroscopic precession. We have just seen that when a freely spinning body is torqued about an axis normal to the spin axis, it precesses about a third axis that forms an orthogonal triad with the spin and torque vectors. The direction of the precession is such that it turns the spin vector toward the torque vector. This law of gyroscopic precession is responsible for the lunisolar precession of the equinoxes.

What is the lunisolar precession? Because of the billions of years of gravitational pull from the sun and moon, the earth is slightly bulged instead of round. It is in fact about 27 miles shorter across the poles than it is across the equator. This bulge, plus the fact that its axis is tilted $23\frac{1}{2}°$ to the ecliptic, causes the sun (and moon) to torque the earth in addition to the gravity pull, as shown in Figure 7.11.

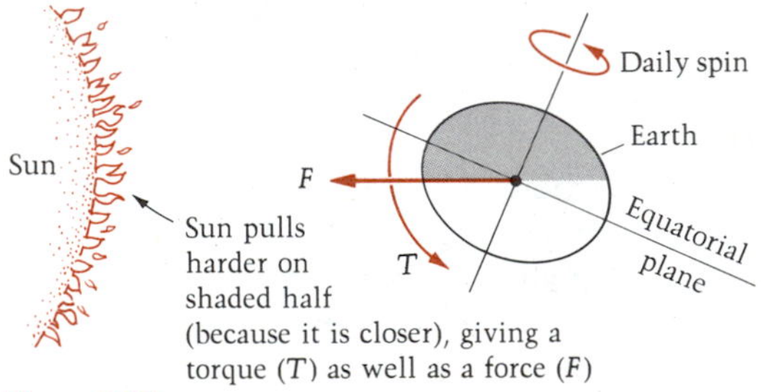

Figure 7.11

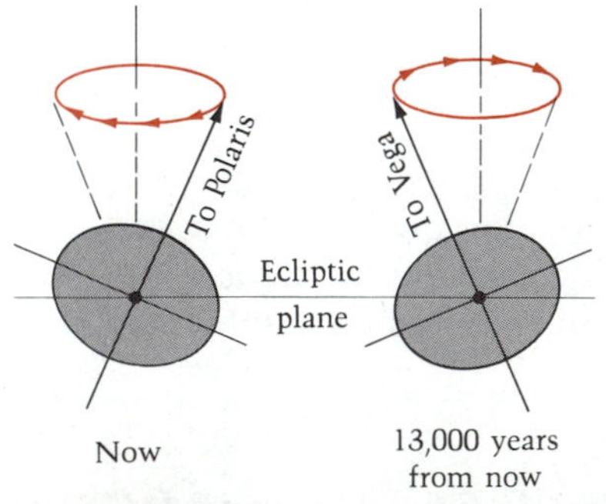

Figure 7.12

We see, then, that we live on the surface of a spinning gyroscope that is constantly being acted on by an external torque, and a precession is thus ongoing. This precession, shown in Figure 7.12, turns the spin axis of the earth out of the plane of the paper toward the torque vector.† This motion results in counterclockwise movement, on the celestial sphere, of the celestial pole to which the earth's rotation axis is directed. (This point

* Note that z_2 and z are the same axis.

† The spin axis of the earth is aligned with its angular velocity vector ω. The point where the ω vector, placed at C, cuts the surface of the earth is the real meaning of the North Pole. The North Pole wanders about the geometric pole (on the symmetry axis) as time passes; it has remained within a few feet of it in this century.

is currently close to Polaris, the North Star.) The period of this rotation is about 26,000 years, and it is interesting that the moon's effect is 2.2 times that of the sun's because it is much closer.

Torque-Free Motion

We now take up the other example of a solution to the gyroscope equations: the case of **torque-free motion**. "Torque-free" means that $\Sigma \mathbf{M}_C$ vanishes, so that $\mathbf{H}_C$ is a constant. This follows from Euler's second law:

$$\Sigma \mathbf{M}_C = \mathbf{0} = {}^{\mathcal{I}}\dot{\mathbf{H}}_C \Rightarrow \mathbf{H}_C = \text{constant vector in the inertial frame } \mathcal{I}$$

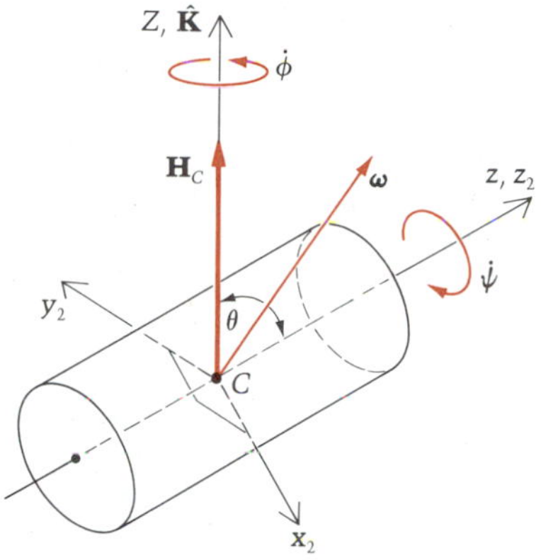

Figure 7.13

We shall conveniently let the direction of the Z axis (see Figure 7.13), which is arbitrary, coincide with the constant direction of $\mathbf{H}_C$. Then Z becomes the precession axis of the motion, and the (x_2, y_2, z_2) axes appear as shown in Figure 7.13. it is seen that since

$$\begin{aligned}\mathbf{H}_C &= H_{C_{x_2}}\hat{\mathbf{i}}_2 + H_{C_{y_2}}\hat{\mathbf{j}}_2 + H_{C_{z_2}}\hat{\mathbf{k}}_2 = \text{constant} \\ &= -I\dot{\phi}\sin\theta\hat{\mathbf{i}}_2 + I\dot{\theta}\hat{\mathbf{j}}_2 + J(\dot{\psi} + \dot{\phi}\cos\theta)\hat{\mathbf{k}}_2 \end{aligned} \tag{7.51}$$

and since $\mathbf{H}_C$ is seen always to lie in the x_2z_2 plane, then $H_{C_{y_2}}$ must vanish:

$$H_{C_{y_2}} = 0 = I\dot{\theta} \Rightarrow \theta = \text{constant} \tag{7.52}$$

Note that $\boldsymbol{\omega}$ lies in the x_2z_2 plane along with $\mathbf{H}_C$, since its y_2 component, $\dot{\theta}$, vanishes. Furthermore, we see that

$$\begin{aligned} H_{C_{x_2}} &= -I\dot{\phi}\sin\theta \quad \text{and} \quad & H_{C_{z_2}} &= J(\dot{\psi} + \dot{\phi}\cos\theta) \\ &= I\omega_{x_2}{}^{*} & &= J\omega_{z_2} \end{aligned} \tag{7.53}$$

But also, as we can see from Figure 7.13,

$$H_{C_{x_2}} = -H_C \sin\theta \quad \text{and} \quad H_{C_{z_2}} = H_C\cos\theta \tag{7.54}$$

Therefore, equating the first of Equations (7.53) and (7.54) for $H_{C_{x_2}}$, we obtain

$$\omega_{x_2} = \frac{-H_C \sin\theta}{I} = \text{constant} \quad \text{(Since } H_C, \theta, \text{ and } I \text{ are constants)} \tag{7.55}$$

$$= -\dot{\phi}\sin\theta \tag{7.56}$$

and we see that

$$\dot{\phi} = \frac{H_C}{I} \quad \text{(a constant)} \tag{7.57}$$

Similarly, equating the two preceding values of $H_{C_{z_2}}$ gives

$$\omega_{z_2} = \frac{H_C\cos\theta}{J} = \text{constant} \tag{7.58}$$

$$= \dot{\psi} + \dot{\phi}\cos\theta \tag{7.59}$$

* Note again that the (x_2, y_2, z_2) axes are permanently principal, even though x_2 and y_2 are not body-fixed, and this lets us write $\mathbf{H}_C$ in terms of the $I\omega$'s along these axes.

so that

$$\dot{\psi} = \frac{H_C \cos\theta}{J} - \left(\frac{H_C}{I}\right)\cos\theta = H_C\cos\theta\left(\frac{I-J}{IJ}\right)$$
$$= \text{constant} \tag{7.60}$$

Therefore all conditions are satisfied for the torque-free body to be in a state of steady precession about the z axis fixed in $\mathcal{I}$!

Dividing Equation (7.55) by (7.58) leads to

$$\frac{\omega_{x_2}}{\omega_{z_2}} = -\frac{J}{I}\tan\theta \tag{7.61}$$

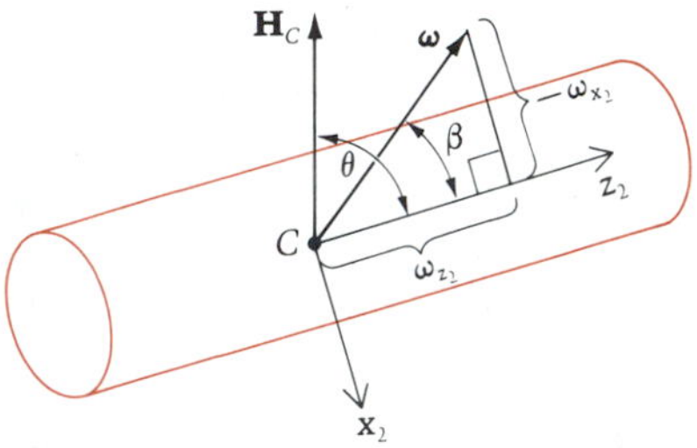

Figure 7.14

and Figure 7.14 shows that

$$\frac{-\omega_{x_2}}{\omega_{z_2}} = \tan\beta \tag{7.62}$$

where β is the angle between z_2 and ω. Therefore

$$\tan\beta = \frac{J}{I}\tan\theta$$

and we see that the answer to whether β is larger or smaller than θ depends on the ratio of J to I. If $J < I$, as in the elongated shape in Figure 7.14, then $\beta < \theta$ and the angular velocity vector lies inside of $\mathbf{H}_C$ and z_2, making a constant angle with each. Two cones may be imagined — one fixed to the body, the other in space ($\mathcal{I}$). The body cone is seen to roll on the fixed space cone as its spin and precession vectorially add to the vector $\boldsymbol{\omega}$, which changes only in direction.

This precession (Figure 7.15) is called *direct* because $\dot{\phi}$ and $\dot{\psi}$ have the same counterclockwise sense when observed from the $\boldsymbol{\omega}$ vector outside the cones. If, however, $J > I$, then $\beta > \theta$ and $\boldsymbol{\omega}$ lies *outside* the angle

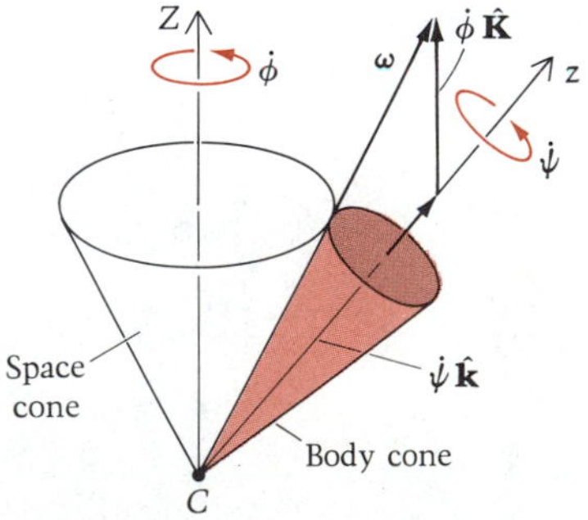

Figure 7.15

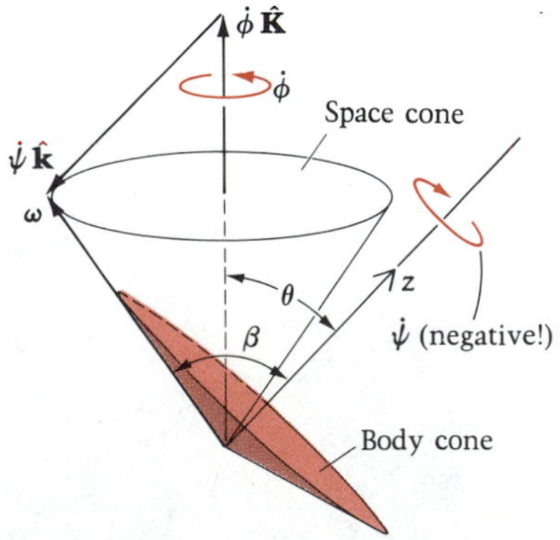

Figure 7.16

ZCz. This situation is harder to depict, but just as important. This time the body cone rolls around the *outside* of the fixed space cone (Figure 7.16), and the two rotations $\dot{\phi}$ and $\dot{\psi}$ have opposite senses. This precession is called *retrograde.*

A final note on the theory of the torque-free body: If the constant value of θ is either 0 or 90°, there is *no* precession and the gyroscope is simply in a state of pure rotation, planar motion:

$\theta = 0$:

Here $z = Z$, so that $\omega_{x_2} = 0$ and $\omega_{z_2} = H_C/J$. If $\theta = 0°$, the body simply spins about its axis. In this case, the rates $\dot{\phi}$ and $\dot{\psi}$ cannot be distinguished.

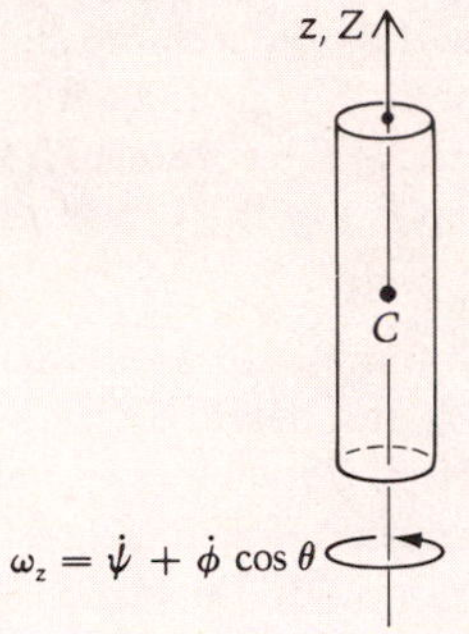

Question 7.5 Why can we not use Equations (7.57) and (7.60) to get $\dot{\phi}$ and $\dot{\psi}$ in this case?

$\theta = 90°$:

Here we have $\omega_{x_2} = -H_C/I$ and $\omega_{z_2} = 0$. Thus:

$$\dot{\phi} = \frac{H_C}{I}$$

$$\dot{\psi} = 0$$

If $\theta = 90°$, the body spins about a transverse axis without precessing.

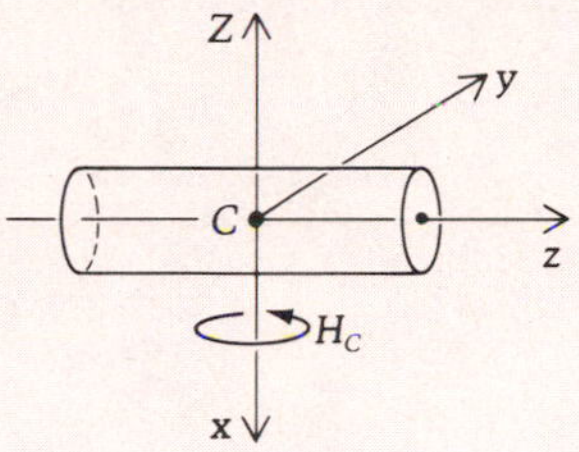

It requires a great many terms to describe the complex motion of the earth. We have seen one example of this in the lunisolar precession caused by the gravity torque exerted on the earth by the sun and moon. This motion is analogous to a differential equation's particular solution, which it has whenever the equation has a nonzero right-hand side. The complementary, or homogeneous, solution is analogous to the torque-free part of the solution to the earth's rotational motion. This part, called the free precession of the earth, is in fact retrograde. Both the space and body cones are very thin as the ω, $\mathbf{H}_C$, and z axes are all quite close together; each lies about $23\frac{1}{2}°$ off the normal to the ecliptic plane.

Answer 7.5 In deriving (7.57), if $\theta = 0$ we have divided both sides of an equation by zero. This result is then used in getting $\dot{\psi}$ in (7.60).

PROBLEMS ▶ Section 7.6

7.77 Find the angular acceleration in $\mathcal{I}$ of the gyroscope for the case of steady precession.

7.78 The spinning top (Figure P7.78) is another example of a gyroscope. Show that if the top's peg is not moving across the floor, the condition for steady precession is given by

$$mgd = J\dot{\psi}\dot{\phi} + (J - I)\dot{\phi}^2 \cos\theta$$

7.79 A top steadily precesses about the fixed direction Z at 60 rpm. (See Figure P7.79.) Treating the top as a cone of radius 1.2 in. and height 2.0 in., find the rate of spin $\dot{\psi}$ of the top about its axis of symmetry.

7.80 Cone $\mathcal{C}$ in Figure P7.80 has radius 0.2 m and height 0.5 m. It is precessing about the vertical axis through the ball joint, in the direction shown, at the rate of $\dot{\phi} = 0.5$ rad/s. If the angle θ is observed to be 20° and unchanging, what must be the rate of spin $\dot{\psi}$ of the cone?

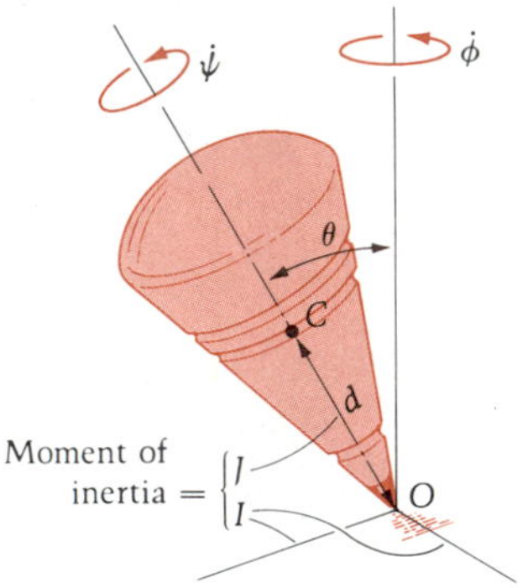

Figure P7.78

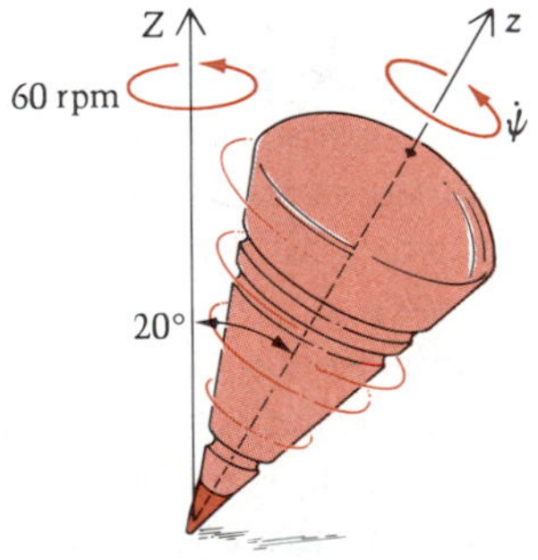

Figure P7.79

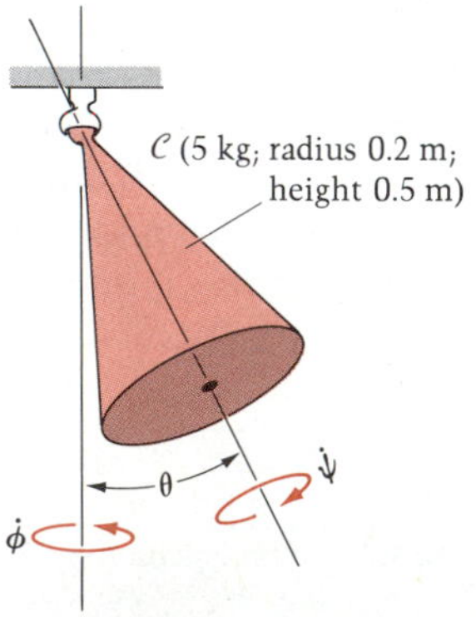

Figure P7.80

7.81 In the preceding problem, suppose that $\dot{\psi}$ is given to be 400 rad/s in the same direction as given in the figure and that the cone's height H is not given. Find the value of H for which this steady precession will occur.

7.82 Using the fact that the sum of any two moments of inertia at a point is always larger than the third (Problem 7.14), show that for a torque-free axisymmetric body undergoing retrograde precession, $\dot{\phi} \geq 2|\dot{\psi}|$ and that the z axis of the body is always outside the space cone.

7.83 The graph in Figure P7.83a depicts the stability of symmetrical satellites spinning about the axis z_C normal to the orbital plane. The abscissa is the ratio of I_{z_C} to the moment of inertia I_ℓ about any lateral axis (they are all the same for what is called a "symmetrical" satellite—it need not be *physically* symmetric about z_C). The ordinate is the ratio of the spin speed ω_s (about z_C) *in* the orbit to the orbital angular speed ω_0.

a. For a satellite equivalent to four solid cylinders each of mass m, radius R, and height $3R$, find I_{z_C} and I_ℓ. The distance from C to any cylinder's center is $2R$, and the connecting cross is light. The cylinders' axes are normal to the orbital plane. (See Figure 7.83b.)

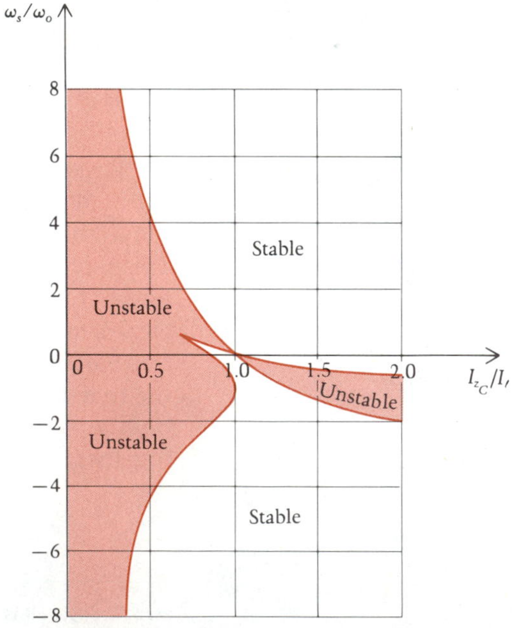

Figure P7.83a

b. Determine whether the station is stable for the following cases:
  i. The station's orientation is fixed in inertial space.
  ii. The station travels around the earth as the moon does.
  iii. The station has twice the angular velocity of the orbiting frame.
  iv. The same as (iii), but the spin is opposite in direction to the orbital angular speed.

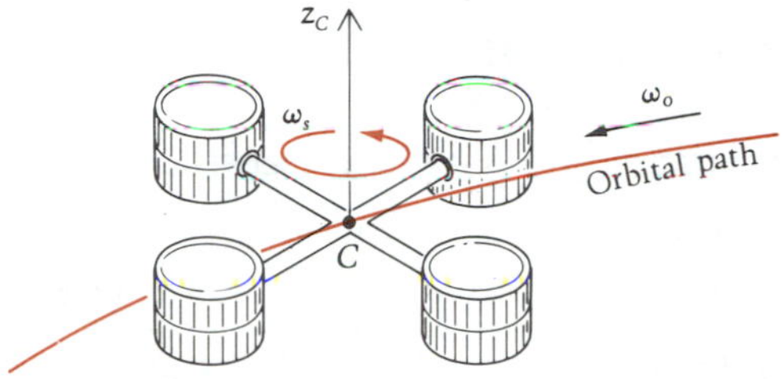

Figure P7.83b

The following five problems are advanced looks at statics of rigid bodies that depend on our study of dynamics.

7.84 It is possible for a spinning top to "sleep," meaning that its axis remains vertical and its peg stationary as it spins on a floor. (See Figure P7.84.) In the absence of a friction couple about the axis of the top, note that the spin speed ω_z is constant and that the equations of motion then reduce to $\Sigma \mathbf{M}_C = \mathbf{0}$ and $\Sigma \mathbf{F} = \mathbf{0}$. It thus follows that

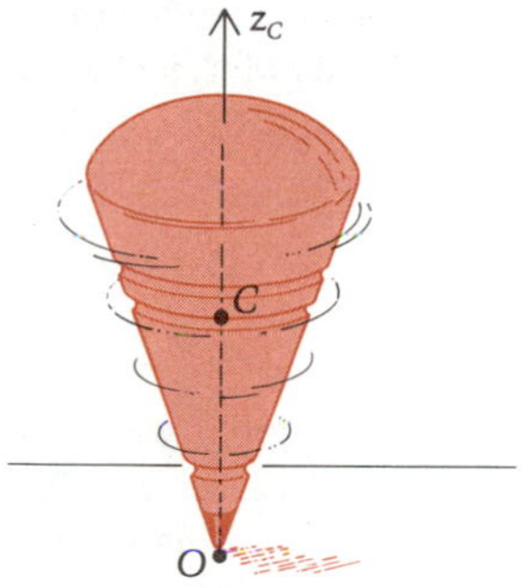

Figure P7.84

$\Sigma \mathbf{M}_O$ is also zero. None of the particles of the top not on the z_C axis are in equilibrium, however, because they all have (inward) accelerations $r\omega^2$. A body is in equilibrium if and only if all its particles are in equilibrium, so the sleeping top cannot be in equilibrium. Explain this statement in light of $\Sigma \mathbf{F} = \mathbf{0}$ and $\Sigma \mathbf{M}_O = \mathbf{0}$, which were the equilibrium equations for a body in statics. *Hint*: If O is a fixed point of rigid body $\mathcal{B}$ in an inertial frame $\mathcal{I}$, then

$$\Sigma \mathbf{M}_O = {}^{\mathcal{I}}\dot{\mathbf{H}}_O = {}^{\mathcal{I}}\frac{d}{dt}(\mathbf{H}_C + \mathbf{r}_{OC} \times \mathbf{L})$$

$$= {}^{\mathcal{B}}\dot{\mathbf{H}}_C + \boldsymbol{\omega}_{\mathcal{B}/\mathcal{I}} \times \mathbf{H}_C + \mathbf{r}_{OC} \times \underbrace{m\mathbf{a}_C}_{\Sigma \mathbf{F}}$$

Therefore show that just because $\Sigma \mathbf{F} = \mathbf{0}$ and $\Sigma \mathbf{M}_O = \mathbf{0}$, $\boldsymbol{\omega}_{\mathcal{B}/\mathcal{I}}$ need not be zero. Use the top as a counterexample and explain why the first two terms on the right side of the preceding equation vanish. Thus $\Sigma \mathbf{F} = \Sigma \mathbf{M}_O = \mathbf{0}$ are necessary but not *sufficient* conditions for equilibrium of a rigid body.

7.85 In the sleeping top counterexample of Problem 7.84, the terms ${}^{\mathcal{B}}\dot{\mathbf{H}}_C$ and $\boldsymbol{\omega}_{\mathcal{B}/\mathcal{I}} \times \mathbf{H}_C$ both vanish independently. Show that there are more complicated counterexamples in which $\boldsymbol{\omega}_{\mathcal{B}/\mathcal{I}}$ is not constant in direction in $\mathcal{B}$ and $\mathcal{I}$ and in which the two terms *add* to zero. *Hint*: $\Sigma \mathbf{M}_O - \mathbf{r}_{OC} \times \Sigma \mathbf{F} = \Sigma \mathbf{M}_C$. What is $\Sigma \mathbf{M}_C$ for the torque-free body?

7.86 Show that if $\boldsymbol{\omega}_{\mathcal{B}/\mathcal{I}} = \mathbf{0}$ at all times, then so is $\Sigma \mathbf{M}_C$. Is the converse true?

7.87 If a frame $\mathcal{B}$ is moving relative to an inertial frame $\mathcal{I}$, it can be shown that $\mathcal{B}$ is also an inertial frame if and only if $\boldsymbol{\omega}_{\mathcal{B}/\mathcal{I}} = \mathbf{0}$ at all times *and* the acceleration in $\mathcal{I}$ of at least one point of $\mathcal{B}$ is zero at all times. Use this theorem to show that if a rigid body $\mathcal{B}$ is in equilibrium in an inertial frame $\mathcal{I}$, then $\mathcal{B}$ is *itself* an inertial frame. Is the converse true?

7.88 Show that a rigid body $\mathcal{B}$ is in equilibrium in an inertial frame $\mathcal{I}$ if and only if (a) at least one point of $\mathcal{B}$ is fixed in $\mathcal{I}$ and (b) $\boldsymbol{\omega}_{\mathcal{B}/\mathcal{I}} = \mathbf{0}$ at all times. What is the minimum number of constraints on $\mathcal{B}$ that will satisfy (a) and (b)? Describe one set of physical constraints that will assure equilibrium.

7.7 Impulse and Momentum

As we did in Chapter 5 for the case of plane motion, we could apply the principles of impulse and momentum and those of angular impulse and angular momentum to the three-dimensional motion of a rigid body $\mathcal{B}$. As we saw in Section 5.3, however, these applications are really nothing more than time integrations of the equations of motion.

There is one type of problem, however, in which these two principles furnish us with a means of solution—problems involving impact. Some three-dimensional aspects are sufficiently different from the planar case to warrant an example. But first we use the integrals of the Euler laws to derive the needed relations:

$$\int_{t_i}^{t_f} \Sigma \mathbf{F}\, dt = \mathbf{L}_f - \mathbf{L}_i \tag{7.63}$$

$$\int_{t_i}^{t_f} \Sigma \mathbf{M}_C\, dt = \mathbf{H}_{C_f} - \mathbf{H}_{C_i} \tag{7.64}$$

An alternative to the rotational equation (7.64) is to integrate the equally general equation

$$\Sigma \mathbf{M}_O = \dot{\mathbf{H}}_O \tag{7.65}$$

where O is now a fixed point of the inertial frame $\mathscr{I}$:

$$\int_{t_i}^{t_f} \Sigma \mathbf{M}_O\, dt = \mathbf{H}_{O_f} - \mathbf{H}_{O_i} = (\mathbf{H}_{C_f} - \mathbf{H}_{C_i}) + (\mathbf{r}_{PC} \times m\mathbf{v}_C)\Big|_i^f \tag{7.66}$$

To use either Equation (7.64) or (7.66) in an impact situation, we use Equation (7.11) for the body's angular momentum *before the deformation starts* (at t_i) and then again *after it ends* (at t_f). The following example illustrates the procedure.

EXAMPLE 7.11

The bent bar $\mathscr{B}$ of Example 4.16 is dropped from a height H and strikes a rigid, smooth surface on one end of $\mathscr{B}$ as shown in Figure E7.11. If the coefficient of restitution is e, find the angular velocity of $\mathscr{B}$, as well as the velocity of C, just after the collision.

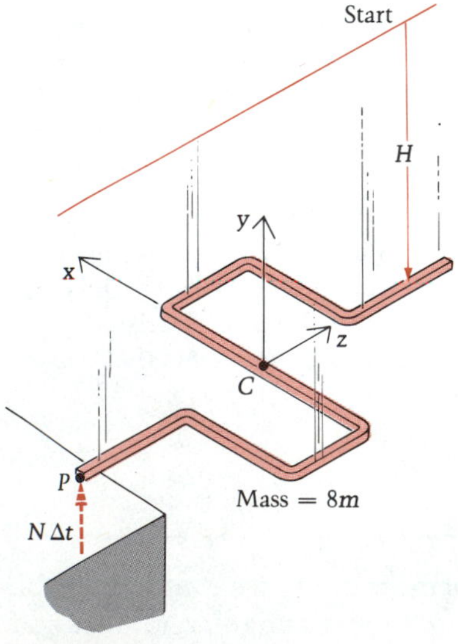

Figure E7.11

Solution

Using the y component equation of (7.63) yields

$$N\,\Delta t = 8m\dot{y}_{C_f} - 8m(-\sqrt{2gH}) \tag{1}$$

in which the impulse of the gravity force is neglected as small in comparison with the impulsive upward force exerted by the surface over the short time interval Δt.

Next we write the component equations of (7.64); we first need the inertia properties of the body, which can be computed to be

$$I_{xx}^C = \frac{22}{3}ml^2 \qquad I_{xy}^C = 0$$

$$I_{yy}^C = \frac{32}{3}ml^2 \qquad I_{yz}^C = 0$$

$$I_{zz}^C = \frac{10}{3}ml^2 \qquad I_{zx}^C = -2ml^2 \tag{2}$$

We then obtain, from Equation (7.64),

$$2\ell N\,\Delta t\hat{\mathbf{i}} = m\ell^2\left[\frac{22}{3}\,\omega_x + 0\omega_y - 2\omega_z\right]\hat{\mathbf{i}} + m\ell^2\left[0\omega_x + \frac{32}{3}\,\omega_y + 0\omega_z\right]\hat{\mathbf{j}} + m\ell^2\left[-2\omega_x + 0\omega_y + \frac{10}{3}\,\omega_z\right]\hat{\mathbf{k}} \tag{3}$$

in which the initial angular velocity components vanish and the desired final components are $(\omega_x, \omega_y, \omega_z)$. The component equations of (3) are

$$\frac{22}{3}\,\omega_x - 2\omega_z = \frac{+2N\,\Delta t}{m\ell} \tag{4}$$

$$\frac{32}{3}\,\omega_y = 0 \tag{5}$$

$$-2\omega_x + \frac{10}{3}\,\omega_z = 0 \tag{6}$$

At this point we have four equations in the five unknowns $\dot{y}_C$, ω_x, ω_y, ω_z, and the impulse $N\,\Delta t$. We get a fifth equation from the definition of the coefficient of restitution together with the y component of the rigid-body velocity relationship between P and C:

$$e = \frac{\dot{y}_{P_f} - 0}{0 - (-\sqrt{2gH})} \Rightarrow \dot{y}_{P_f} = \sqrt{2gH}\,e \tag{7}$$

and

$$\mathbf{v}_C = \mathbf{v}_P + \boldsymbol{\omega} \times \underbrace{\mathbf{r}_{PC}}_{2\ell\hat{\mathbf{k}}} \tag{8}$$

which has the y-component equation

$$\dot{y}_{C_f} = \dot{y}_{P_f} - 2\ell\omega_x \tag{9}$$

Using Equation (7), we obtain

$$\dot{y}_{Cf} = \sqrt{2gH}\,e - 2\ell\omega_x \tag{10}$$

The solution to the five equations (1, 4, 5, 6, 10) is

$$\omega_x = \frac{60(1+e)\sqrt{2gH}}{143\ell} \qquad \dot{y}_{C_f} = \frac{(23e - 120)\sqrt{2gH}}{143}$$

$$\omega_y = 0 \tag{11}$$

$$\omega_z = \frac{36(1+e)\sqrt{2gH}}{143\ell} \qquad N\,\Delta t = \frac{184m(1+e)\sqrt{2gH}}{143}$$

Returning to Equation (8), we find that the x and z components of $\mathbf{v}_{C_f}$ vanish:

$$x \text{ components} \Rightarrow \dot{x}_{C_f} = 0 + 2\omega_y\ell = 0$$
$$z \text{ components} \Rightarrow \dot{z}_{C_f} = 0 + \quad 0 \quad = 0 \tag{12}$$

The results in Equations (12) are obvious, since if there is no friction at the point of contact there can be no impulsive forces in the horizontal plane to change the momentum (from zero) in the x or z directions.

It is seen that the single nonzero product of inertia causes a coupling between ω_x and ω_z (see Equations (4) and (6)), which prevents ω_z from vanishing—even though the only moment component with respect to C is about the x axis!

We shall now see with another example the advantages of Equation (7.66), which may be used to eliminate undesired forces from moment equations, just as was done in our study of statics.

EXAMPLE 7.12

Rework the preceding example by using Equation (7.66) instead of the combination of Equations (7.63) and (7.64). Find the value of $\boldsymbol{\omega}$ after impact.

Solution

Equation (7.66) allows us to eliminate the impulse $N\,\Delta t$ by summing moments about the point (P) of impact:

$$\int_{t_i}^{t_f} \Sigma \mathbf{M}_P\,dt = \mathbf{0} = (\mathbf{H}_{C_f} - \overset{0}{\cancel{\mathbf{H}_{C_i}}}) + (\mathbf{r}_{PC} \times m\mathbf{v}_C)\Big|_i^f$$

$$= \left[\left(\frac{22}{3}\,\omega_x - 2\omega_z\right) m\ell^2\hat{\mathbf{i}} + \frac{32}{3}\,\omega_y m\ell^2\hat{\mathbf{j}} + \left(-2\omega_x + \frac{10}{3}\,\omega_z\right) m\ell^2\hat{\mathbf{k}}\right]$$

$$+ 2\ell\mathbf{k} \times [8m\dot{y}_{C_f}\hat{\mathbf{j}} - 8m\sqrt{2gH}(-\hat{\mathbf{j}})]$$

We still have to use the coefficient of restitution and relate $\mathbf{v}_P$ and $\mathbf{v}_C$ exactly as before; making this substitution for $\dot{y}_{C_f}$ leads to the following three scalar component equations:

$$\frac{118}{3}\,\omega_x - 2\omega_z = \frac{16\sqrt{2gH}}{\ell}\,(1+e)$$

$$\frac{32}{3}\,\omega_y = 0$$

$$-2\omega_x + \frac{10}{3}\,\omega_z = 0$$

These equations, of course, have the same solution as $(\omega_x, \omega_y, \omega_z)$ in the preceding example.

PROBLEMS ▶ Section 7.7

7.89 Bend a coat hanger or pipe cleaner into the shape of the bent bar of Example 7.11. Drop it onto the edge of a table as in the example and observe that the angular velocity direction following impact agrees with the results of the example.

* **7.90** The equilateral triangular dinner bell in Figure P7.90 is struck with a horizontal force in the y direction that imparts an impulse $F\,\Delta t\hat{\mathbf{j}}$ to the bell. Find the angular velocity of the bell immediately after the blow is struck. Is the answer the same if the bell is an equilateral triangular *plate* of the same mass? Why or why not?

* **7.91** In the preceding problem, suppose the hammer is replaced by a bullet of mass m and speed v_b that rebounds straight back with a coefficient of restitution $e = 0.1$. Determine the resulting angular velocity of the bell.

7.92 Repeat Problem 7.90, but this time suppose the bell hangs from a string instead of from a ball and socket joint.

7.93 The bent bar $\mathcal{B}$ of Figure P7.93 has the inertia properties listed below. It is in motion in an inertial frame $\mathcal{I}$, and at a certain instant has angular velocity $\boldsymbol{\omega}_{\mathcal{B}/\mathcal{I}} = \omega(4\hat{\mathbf{i}} + 2\hat{\mathbf{j}} + 7\hat{\mathbf{k}})$ rad/sec. Use the angular impulse and angular momentum principle to answer the following question: Is it possible to strike $\mathcal{B}$ at point Q with an impulse $\mathbf{F} = F_x\,\Delta t\hat{\mathbf{i}} + F_y\,\Delta t\hat{\mathbf{j}} + F_z\,\Delta t\hat{\mathbf{k}}$ that reduces $\boldsymbol{\omega}_{\mathcal{B}/\mathcal{I}}$ to zero after the impulse? If so, find the components of the impulse in terms of m, ℓ, ω, and Δt. If not, show why not.

$$I^C_{xx} = \frac{22}{3}m\ell^2 \qquad I^C_{zz} = \frac{10}{3}m\ell^2$$

$$I^C_{yy} = \frac{32}{3}m\ell^2 \qquad I^C_{xz} = -2\,m\ell^2$$

$$I^C_{yz} = I^C_{xy} = 0$$

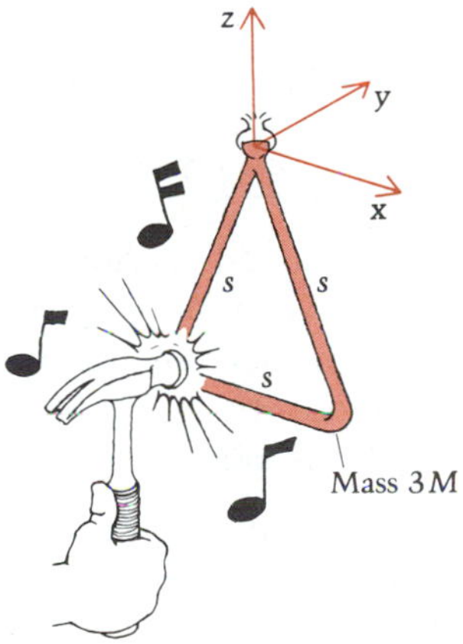

Figure P7.90

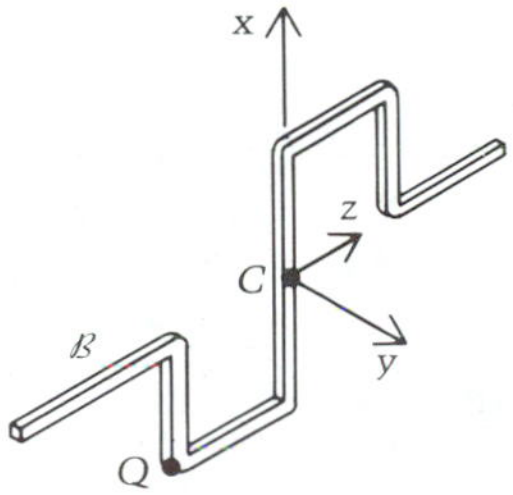

Figure P7.93

* **7.94** A diver $\mathcal{D}$ leaves a diving board in a straight, symmetric position with angular velocity and angular momentum vectors each in the x direction as indicated in Figure P7.94a. Since $\Sigma\mathbf{M}_C$ is zero, there will be no change in the angular momentum $\mathbf{H}_C$ in the inertial frame (the swimming pool) as long as the diver is in the air. Therefore, as long as he remains in the straight position, his constant angular momentum is expressed by

$$I^C_{xx}\omega_x + \overset{0}{\cancel{I^C_{xy}}}\overset{0}{\cancel{\omega_y}} + \overset{0}{\cancel{I^C_{xz}}}\overset{0}{\cancel{\omega_z}} = H_{C_1} = \text{constant} \qquad (1)$$

$$\overset{0}{\cancel{I^C_{yx}}}\omega_x + I^C_{yy}\overset{0}{\cancel{\omega_y}} + \overset{0}{\cancel{I^C_{yz}}}\overset{0}{\cancel{\omega_z}} = 0 \qquad (2)$$

$$\overset{0}{\cancel{I^C_{zx}}}\omega_x + \overset{0}{\cancel{I^C_{zy}}}\overset{0}{\cancel{\omega_y}} + I^C_{zz}\overset{0}{\cancel{\omega_z}} = 0 \qquad (3)$$

where we assume the body to be sufficiently internally symmetric so that the products of inertia all vanish. Now suppose the diver instantaneously moves his arms as shown in Figure P7.94b to initiate a twist. Following the maneuver, he may again be treated as a rigid body and we may use the same body-fixed axes as before. (Note that the mass center changes very little.)

a. From Figure P7.94c argue that the indicated changes in the products of inertia occur. (Only the shaded arms contribute to the products of inertia.) Argue also that I^C_{yz} is smaller than I^C_{xz} and also less than $|I^C_{xy}|$. Observe that all three

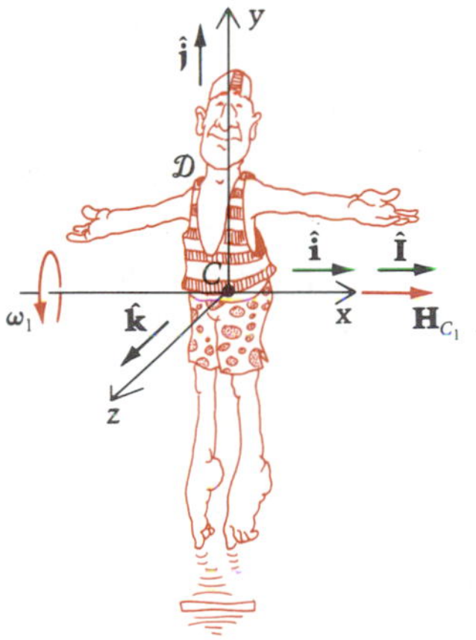

Figure P7.94a

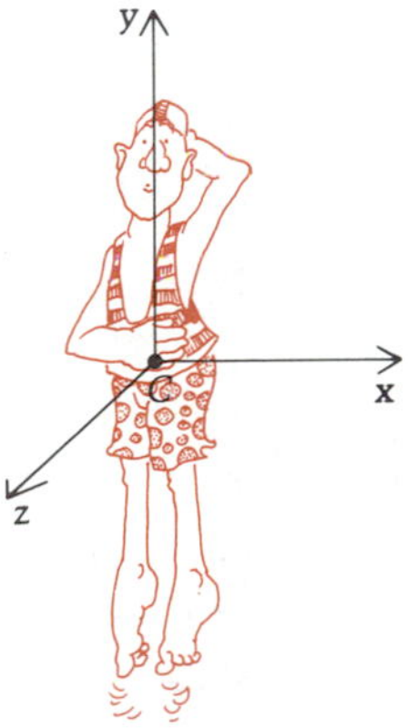

Figure P7.94b

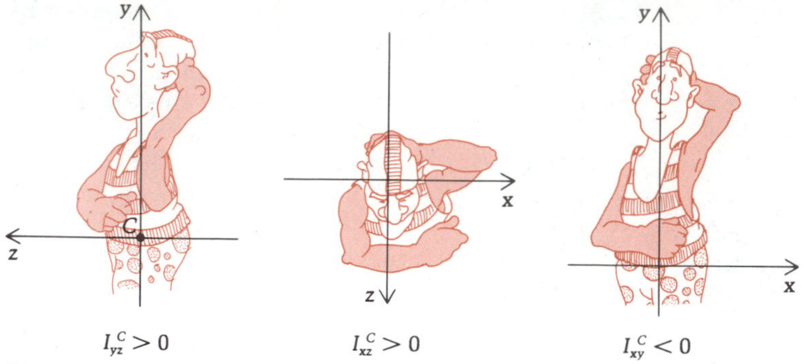

Figure P7.94c

products of inertia are small compared with the three moments of inertia and that $I_{yy}^C < I_{xx}^C < I_{zz}^C$, with I_{yy}^C being very much smaller than the other two moments of inertia. Note that (x, y, z) are no longer principal, but this does not matter since we are not making use of principal axes here.

As the diver's body begins to twist and turn, the right sides of Equations (1) to (3) will change and *none* of the quantities on the left will remain zero. But the right sides will constitute the components *in the body frame* $\mathcal{D}$ of the vector $\mathbf{H}_C$, which will still vectorially add to $H_{C_1}\hat{\mathbf{I}}$, where $\hat{\mathbf{I}}$ is the original direction in $\mathcal{I}$ of $\mathbf{H}_C$ after the diver leaves the diving board (to the right in the first sketch).

b. After the rapid twist maneuver, but *before* the diver begins to twist, his axes are still instantaneously aligned with those of the frame $\mathcal{I}$. Use Equations (1) to (3), with the right sides $(H_{C_1}, 0, 0)$ and the now nonzero products of inertia, to show that:

   i. There will be a small (compared to the original ω_x) angular velocity developed about the $-z_C$ direction (negative ω_z).

   ii. There will be an angular velocity of twist developed about y_C (positive ω_y).

   iii. There will be an increase in the somersaulting angular velocity component ω_x.

In arguing statements (i) to (iii), assume nothing about the ω's following the maneuver except that ω_x is still in the same direction as before.

7.8 Work and Kinetic Energy

A special integral of the equations of motion of a rigid body $\mathcal{B}$ yields a relationship between the work of the external forces (and/or couples) and the change in the kinetic energy of $\mathcal{B}$. To develop this relation, we must first explore expressions for the kinetic energy of the rigid body. **Kinetic energy** is usually denoted by the letter T and is defined by (see Section 5.2)

$$T = \frac{1}{2}\int \mathbf{v} \cdot \mathbf{v}\, dm \tag{7.67}$$

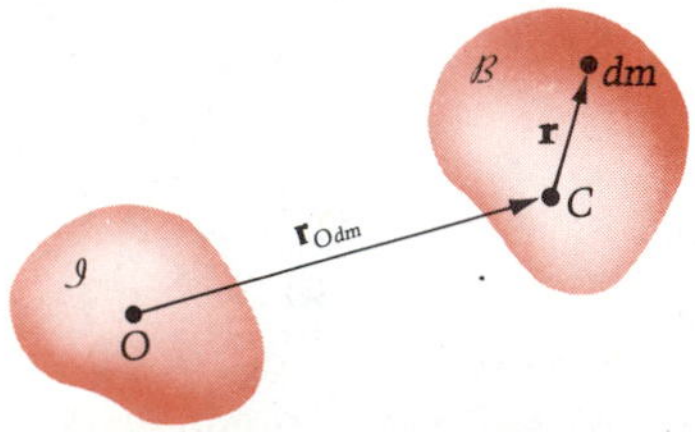

Figure 7.17

in which $\mathbf{v}$ is the derivative of the position vector from O (fixed point in the inertial frame $\mathcal{I}$ in Figure 7.17) to the differential mass element dm. In this section all time derivatives, velocities, and angular velocities are taken in $\mathcal{I}$ unless otherwise specified.

Since $\mathcal{B}$ is a rigid body, we may relate $\mathbf{v}$ to the velocity $\mathbf{v}_C$ of the mass center C of $\mathcal{B}$:

$$\mathbf{v} = \mathbf{v}_C + \boldsymbol{\omega} \times \mathbf{r} \tag{7.68}$$

in which $\boldsymbol{\omega}$ is $\boldsymbol{\omega}_{\mathcal{B}/\mathcal{I}}$ and $\mathbf{r}$ is the position vector from C to dm as shown in Figure 7.17. Substituting Equation (7.68) into (7.67), we get

$$T = \frac{1}{2}\mathbf{v}_C \cdot \mathbf{v}_C \int_{\mathcal{B}} dm + \frac{1}{2}\int_{\mathcal{B}} (\boldsymbol{\omega} \times \mathbf{r}) \cdot (\boldsymbol{\omega} \times \mathbf{r})\, dm + \mathbf{v}_C \cdot \left[\boldsymbol{\omega} \times \int_{\mathcal{B}} \mathbf{r}\, dm\right] \tag{7.69}$$

where $\mathbf{v}_C$ and $\boldsymbol{\omega}$ do not vary over the body's volume and can thus be taken outside the integrals. The integral in the last term is zero by virtue of the definition of the mass center:

$$\int_{\mathcal{B}} \mathbf{r}\, dm = m\mathbf{r}_{CC} = \mathbf{0} \tag{7.70}$$

The integral in the first term on the right side of Equation (7.69) is of course the mass m of $\mathcal{B}$. The integrand of the remaining term may be simplified by the vector identity:*

$$(\mathbf{A} \times \mathbf{B}) \cdot (\mathbf{C} \times \mathbf{D}) = \mathbf{A} \cdot [\mathbf{B} \times (\mathbf{C} \times \mathbf{D})] \tag{7.71}$$

Therefore Equation (7.69) becomes

$$T = \frac{m}{2}(\mathbf{v}_C \cdot \mathbf{v}_C) + \frac{1}{2}\boldsymbol{\omega} \cdot \int [\mathbf{r} \times (\boldsymbol{\omega} \times \mathbf{r})]\, dm \tag{7.72}$$

As we have already seen in Section 7.2, the integral in Equation (7.72) is the angular momentum (moment of momentum) of the body with respect to C, and thus we can write

$$T = \overbrace{\frac{m}{2}\mathbf{v}_C \cdot \mathbf{v}_C}^{T_v} + \overbrace{\frac{1}{2}\boldsymbol{\omega} \cdot \mathbf{H}_C}^{T_\omega} \tag{7.73}$$

It is seen that the kinetic energy can be represented as the sum of two terms:

1. A part $T_v = (m/2)\mathbf{v}_C \cdot \mathbf{v}_C$ that the body possesses if its mass center is in motion.
2. A part $T_\omega = \frac{1}{2}\boldsymbol{\omega} \cdot \mathbf{H}_C$ that is due to the difference between the velocities of the points of $\mathcal{B}$ and the velocity of its mass center.

The term T_ω can be interpreted quite simply if at an instant we let $\boldsymbol{\omega} = \omega\hat{\mathbf{i}}$—that is, if we align the reference axis x with the angular velocity vector at that instant. In this case, using Equations (7.11), we obtain

$$\mathbf{H}_C = I_{xx}^C\omega\hat{\mathbf{i}} + I_{xy}^C\omega\hat{\mathbf{j}} + I_{xz}^C\omega\hat{\mathbf{k}} \tag{7.74}$$

* Which is nothing more than interchanging the dot and cross of the scalar triple product $(\mathbf{A} \times \mathbf{B}) \cdot \mathbf{E}$, where $\mathbf{E}$ is the vector $\mathbf{C} \times \mathbf{D}$.

so that

$$\frac{1}{2}\,\boldsymbol{\omega}\cdot\mathbf{H}_C=\frac{1}{2}\,I_{xx}^{C}\omega^2 \tag{7.75}$$

This means that the "rotational part" of T is *instantaneously* of the same form as it was for the plane case in Chapter 4. The difference, of course, is that the direction of the angular velocity vector $\boldsymbol{\omega}$ changes in the general (three-dimensional) case.

Suppose the body $\mathcal{B}$ has a point P with zero velocity. (This is not always the case in general motion as we have already seen in Chapter 6.) Then if $\mathbf{v}$ in Equation (7.67) is replaced by $\mathbf{v}_P+\boldsymbol{\omega}\times\mathbf{r}'=\boldsymbol{\omega}\times\mathbf{r}'$, where $\mathbf{r}'$ extends from P to the mass element dm, we obtain

$$T=\int(\boldsymbol{\omega}\times\mathbf{r}')\cdot(\boldsymbol{\omega}\times\mathbf{r}')\,dm \tag{7.76}$$

The identical steps that produced the second term of Equation (7.73) from the middle term of (7.69) then give

$$T=\frac{1}{2}\,\boldsymbol{\omega}\cdot\mathbf{H}_P \tag{7.77}$$

and the two terms of Equation (7.73) have collapsed into one if $\mathbf{H}$ is expressed relative to a point of zero velocity instead of C.

In Sections 2.4 and 5.2 we demonstrated one work and kinetic energy principle that remains true for the general case. This result came from integrating $\Sigma\mathbf{F}=m\mathbf{a}_C$:

$$\int_{t_1}^{t_2}\Sigma\mathbf{F}\cdot\mathbf{v}_C\,dt=\frac{1}{2}\,m|\mathbf{v}_C(t_2)|^2-\frac{1}{2}\,m|\mathbf{v}_C(t_1)|^2$$

$$=\frac{1}{2}\,m(v_{C_2}^2-v_{C_1}^2) \tag{7.78}$$

A second principle will now be deduced from the moment equation*

$$\Sigma\mathbf{M}_C=\dot{\mathbf{H}}_C \tag{7.79}$$

but first we need to prove the non-obvious result that:

$$\dot{\boldsymbol{\omega}}\cdot\mathbf{H}_C=\boldsymbol{\omega}\cdot\dot{\mathbf{H}}_C$$

To do this, we first recall that

$$\mathbf{H}_C=\int[\mathbf{r}\times(\boldsymbol{\omega}\times\mathbf{r})]\,dm \tag{7.80}$$

If ${}^{\mathcal{B}}\dot{\mathbf{H}}_C$ is the derivative of $\mathbf{H}_C$ taken in the body $\mathcal{B}$, then the derivative relative to the inertial frame can be written

$$\dot{\mathbf{H}}_C={}^{\mathcal{B}}\dot{\mathbf{H}}_C+\boldsymbol{\omega}\times\mathbf{H}_C \tag{7.81}$$

* Derivatives such as $\dot{\boldsymbol{\omega}}$ are taken in the inertial frame $\mathcal{I}$ in this section unless the letter $\mathcal{B}$ appears to the left of the dot, in which case the derivative is taken in the body.

Dotting $\boldsymbol{\omega}$ with both sides of Equation (7.81) shows that

$$\boldsymbol{\omega} \cdot \dot{\mathbf{H}}_C = \boldsymbol{\omega} \cdot {}^{\mathcal{B}}\dot{\mathbf{H}}_C \tag{7.82}$$

and since $\mathbf{r}$ is constant in time relative to body $\mathcal{B}$, we can differentiate Equation (7.80) there and obtain

$$ {}^{\mathcal{B}}\dot{\mathbf{H}}_C = \int \mathbf{r} \times (\dot{\boldsymbol{\omega}} \times \mathbf{r})\, dm \tag{7.83}$$

In Equation (7.83) we have used the property of $\boldsymbol{\omega}$ that its derivatives in $\mathcal{I}$ and $\mathcal{B}$ are the same; that is,

$$ {}^{\mathcal{I}}\dot{\boldsymbol{\omega}} = {}^{\mathcal{I}}\dot{\boldsymbol{\omega}}_{\mathcal{B}/\mathcal{I}} = {}^{\mathcal{B}}\dot{\boldsymbol{\omega}}_{\mathcal{B}/\mathcal{I}} + \boldsymbol{\omega}_{\mathcal{B}/\mathcal{I}} \times \boldsymbol{\omega}_{\mathcal{B}/\mathcal{I}} = {}^{\mathcal{B}}\dot{\boldsymbol{\omega}}_{\mathcal{B}/\mathcal{I}} = {}^{\mathcal{B}}\dot{\boldsymbol{\omega}}$$

Substituting Equation (7.83) into (7.82) then gives

$$\begin{aligned}
\boldsymbol{\omega} \cdot \dot{\mathbf{H}}_C &= \boldsymbol{\omega} \cdot \int \mathbf{r} \times (\dot{\boldsymbol{\omega}} \times \mathbf{r})\, dm = \int \boldsymbol{\omega} \cdot [\mathbf{r} \times (\dot{\boldsymbol{\omega}} \times \mathbf{r})]\, dm \\
&= \int (\boldsymbol{\omega} \times \mathbf{r}) \cdot (\dot{\boldsymbol{\omega}} \times \mathbf{r})\, dm = \int [(\dot{\boldsymbol{\omega}} \times \mathbf{r}) \cdot (\boldsymbol{\omega} \times \mathbf{r})]\, dm \\
&= \dot{\boldsymbol{\omega}} \cdot \int \mathbf{r} \times (\boldsymbol{\omega} \times \mathbf{r})\, dm
\end{aligned}$$

Hence

$$\boldsymbol{\omega} \cdot \dot{\mathbf{H}}_C = \dot{\boldsymbol{\omega}} \cdot \mathbf{H}_C \tag{7.84}$$

We are now in a position to observe that

$$\boldsymbol{\omega} \cdot \Sigma \mathbf{M}_C = \boldsymbol{\omega} \cdot \dot{\mathbf{H}}_C = \frac{d}{dt}\left(\frac{\boldsymbol{\omega} \cdot \mathbf{H}_C}{2}\right) \tag{7.85}$$

Integrating Equation (7.85), we have

$$\int_{t_1}^{t_2} \Sigma \mathbf{M}_C \cdot \boldsymbol{\omega}\, dt = \frac{1}{2}\boldsymbol{\omega}(t_2) \cdot \mathbf{H}_C(t_2) - \frac{1}{2}\boldsymbol{\omega}(t_1) \cdot \mathbf{H}_C(t_1) \tag{7.86}$$

Note that the right sides of Equations (7.78) and (7.86) each represents the change, occurring in the time interval $t_1 \le t \le t_2$, of part of the kinetic energy of the body. The left sides of these equations are usually called a form of *work*.

While the relationships between work and kinetic energy that have been developed are important, another relationship that combines them is often more useful. We can differentiate Equation (7.73) and get

$$\frac{dT}{dt} = m\mathbf{v}_C \cdot \mathbf{a}_C + \frac{1}{2}\dot{\boldsymbol{\omega}} \cdot \mathbf{H}_C + \frac{1}{2}\boldsymbol{\omega} \cdot \dot{\mathbf{H}}_C$$

Using Euler's laws and Equation (7.84), this may be put into the form

$$\frac{dT}{dt} = \Sigma \mathbf{F} \cdot \mathbf{v}_C + \Sigma \mathbf{M}_C \cdot \boldsymbol{\omega} \tag{7.87}$$

If we now let $\mathbf{F}_1, \mathbf{F}_2, \ldots$ represent the external forces acting on the body, and $\mathbf{C}_1, \mathbf{C}_2, \ldots$ represent the moments of the external couples, then

$$\Sigma \mathbf{F} = \mathbf{F}_1 + \mathbf{F}_2 + \cdots \tag{7.88a}$$

$$\Sigma \mathbf{M}_C = \mathbf{r}_1 \times \mathbf{F}_1 + \mathbf{r}_2 \times \mathbf{F}_2 + \cdots + \mathbf{C}_1 + \mathbf{C}_2 + \cdots \tag{7.88b}$$

where $P_1, P_2, \ldots$ are the points of $\mathcal{B}$ where $\mathbf{F}_1, \mathbf{F}_2, \ldots$ are respectively applied and where $\mathbf{r}_1 = \mathbf{r}_{CP_1}$, $\mathbf{r}_2 = \mathbf{r}_{CP_2}$, and so forth, as shown in Figure 7.18. Recall from statics that a couple has the same moment about any point in space, so that the $\mathbf{C}_i$'s are simply added into the moment equation (7.88b).

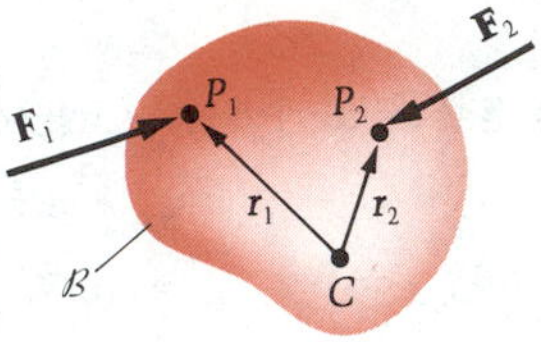

Figure 7.18

Substituting Equation (7.88) into (7.87), we obtain

$$\frac{dT}{dt} = \mathbf{F}_1 \cdot \mathbf{v}_C + \mathbf{F}_2 \cdot \mathbf{v}_C + \cdots + \boldsymbol{\omega} \cdot (\mathbf{r}_1 \times \mathbf{F}_1)$$

$$+ \boldsymbol{\omega} \cdot (\mathbf{r} \times \mathbf{F}_2) + \cdots + \boldsymbol{\omega} \cdot \mathbf{C}_1 + \boldsymbol{\omega} \cdot \mathbf{C}_2 + \cdots \tag{7.89}$$

However,

$$\boldsymbol{\omega} \cdot (\mathbf{r}_i \times \mathbf{F}_i) = \mathbf{F}_i \cdot (\boldsymbol{\omega} \times \mathbf{r}_i) \qquad (i = 1, 2, \ldots)$$

so that

$$\frac{dT}{dt} = \mathbf{F}_1 \cdot (\mathbf{v}_C + \boldsymbol{\omega} \times \mathbf{r}_1) + \mathbf{F}_2 \cdot (\mathbf{v}_C + \boldsymbol{\omega} \times \mathbf{r}_2) + \cdots$$

$$+ \boldsymbol{\omega} \cdot \mathbf{C}_1 + \boldsymbol{\omega} \cdot \mathbf{C}_2 + \cdots \tag{7.90}$$

We note that $\mathbf{v}_C + \boldsymbol{\omega} \times \mathbf{r}_1$ is the velocity $\mathbf{v}_1$ of point P_1, the point of application of $\mathbf{F}_1$. Therefore

$$\frac{dT}{dt} = \dot{T} = \Sigma \mathbf{F}_i \cdot \mathbf{v}_i + \boldsymbol{\omega} \cdot \Sigma \mathbf{C}_i \tag{7.91}$$

Equation (7.91) leads us to define the **power**, or **rate of work**, as follows:

$$\text{Power (rate of work) of a force } (\mathbf{F}_1) = \mathbf{F}_1 \cdot \mathbf{v}_1$$

$$\text{Power (rate of work) of a couple } (\mathbf{C}_1) = \boldsymbol{\omega} \cdot \mathbf{C}_1 \tag{7.92}$$

Therefore

$$\dot{T} = \text{rate of work of external forces and couples}$$

Integrating Equation (7.91), we get

$$\int_{t_1}^{t_2} (\text{rate of work})\, dt = T(t_2) - T(t_1) = \Delta T \tag{7.93}$$

The integral on the left side of Equation (7.93) is called the **work** done on $\mathcal{B}$ between t_1 and t_2 by the external forces and couples. Hence

$$\text{Work} = \int (\Sigma \mathbf{F}_i \cdot \mathbf{v}_i + \boldsymbol{\omega} \cdot \Sigma \mathbf{C}_i)\, dt = \Delta T \tag{7.94}$$

That is, the work done on $\mathcal{B}$ equals its change in kinetic energy. It is left as an exercise for the reader to show that Equation (7.94) is in fact the sum of the two "subequations" (7.78) and (7.86).

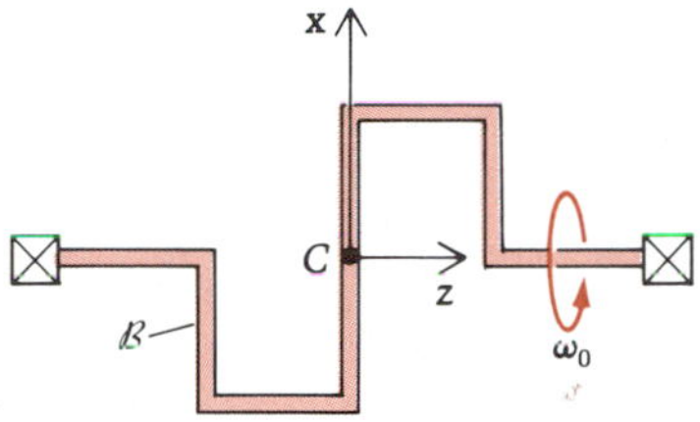

Figure E7.13

Example 7.13

Find the work done on the bent bar of Example 7.11 by a motor that brings it up to speed ω_0 from rest. (See Figure E7.13.)

Solution

The mass center C does not move, so that Equation (7.73) gives, in this case,

$$T_f = \frac{1}{2}\,\boldsymbol{\omega}\cdot\mathbf{H}_C \tag{1}$$

Since $\boldsymbol{\omega}$ has only a $\hat{\mathbf{k}}$ component, $\omega_0\hat{\mathbf{k}}$, we may substitute Equation (7.11) into (1) and get

$$T_f = \frac{1}{2}\,\omega_0(I_{xz}^C \overset{0}{\cancel{\omega_x}} + I_{yz}^C \overset{0}{\cancel{\omega_y}} + I_{zz}^C \overset{\omega_0}{\cancel{\omega_z}}) \tag{2}$$

We note that even though I_{xz}^C is not zero, it has no effect on the kinetic energy of ℬ since it is multiplied by ω_x, which is forced to vanish by the bearings aligned with z.

Thus the work done by the motor on ℬ is given simply by Equation (7.94):

$$\begin{aligned} W = \Delta T &= T_f - \overset{0}{\cancel{T_i}} \\ &= \frac{1}{2}\, I_{zz}^C\omega_0^2 \\ &= \frac{5}{3}\, m\ell^2\omega_0^2 \end{aligned} \tag{3}$$

where $I_{zz}^C = (10/3)m\ell^2$ from Example 7.11. The motor would, of course, have to do additional work besides that given by (3) to overcome its own armature inertia, bearing and belt friction, and air resistance.

We now consider an example in three dimensions in which the products of inertia do play a role in the kinetic energy calculation.

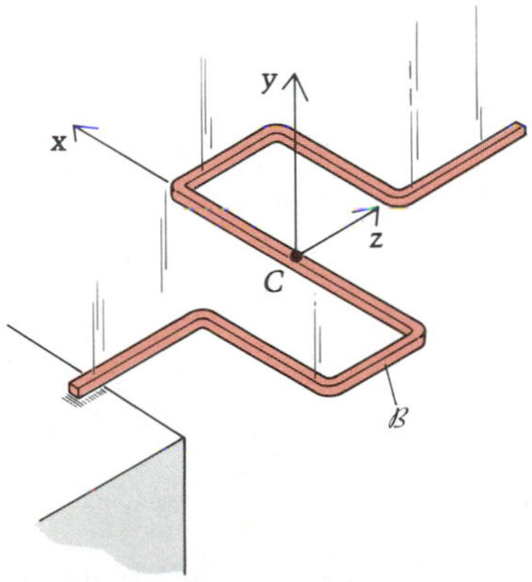

Figure E7.14a

Example 7.14

Find the kinetic energy lost by the bent bar of Example 7.10 when it strikes the table top as shown in Figure E7.14a.

Solution

During the impact with the table top, the bodies do not behave rigidly. The kinetic energy lost by bar ℬ is transformed into noise, heat, vibration, and both elastic and permanent deformation. In Example 7.11 we found $\mathbf{v}_C$ and $\boldsymbol{\omega}_i$ just before and after impact; we now use these vectors to find the kinetic energy lost by ℬ. Just after impact we have

$$T_f = \frac{1}{2}\, m v_{C_f}^2 + \frac{1}{2}\,\boldsymbol{\omega}_f\cdot\mathbf{H}_{C_f}$$

The term $\boldsymbol{\omega}_f \cdot \mathbf{H}_{C_f}$ can be written just after impact, using Equation (7.11), as follows. (Note that ω_y and two of the products of inertia are zero here.)

$$\begin{aligned}\boldsymbol{\omega}_f \cdot \mathbf{H}_{C_f} &= \omega_x(I_{xx}^C\omega_x + \overset{0}{\cancel{I_{xy}^C}}\overset{0}{\cancel{\omega_y}} + I_{xz}^C\omega_z) \\ &\quad + \overset{0}{\cancel{\omega_y}}(\overset{0}{\cancel{I_{yx}^C}}\omega_x + I_{yy}^C\overset{0}{\cancel{\omega_y}} + \overset{0}{\cancel{I_{yz}^C}}\omega_z) \\ &\quad + \omega_z(I_{zx}^C\omega_x + \overset{0}{\cancel{I_{zy}^C}}\overset{0}{\cancel{\omega_y}} + I_{zz}^C\omega_z) \\ &= I_{xx}^C\omega_x^2 + 2I_{xz}\omega_x\omega_z + I_{zz}^C\omega_z^2\end{aligned}$$

Using this result and Equations (11) and (12) from Example 7.14, we obtain

$$\begin{aligned}T_f = \frac{1}{2}(8m)\left[\frac{(23e-120)\sqrt{2gh}}{143}\right]^2 &+ \frac{1}{2}\left\{\left[\frac{60(1+e)\sqrt{2gh}}{143\ell}\right]^2 \frac{22}{3}m\ell^2\right. \\ &+ 2\left[\frac{60(1+e)\sqrt{2gh}}{143\ell}\right]\left[\frac{36(1+e)\sqrt{2gh}}{143\ell}\right](-2m\ell^2) \\ &+ \left.\left[\frac{36(1+e)\sqrt{2gh}}{143\ell}\right]^2 \frac{10}{3}m\ell^2\right\}\end{aligned}$$

which, after simplification, equals

$$T_f = mgh(1.29e^2 + 6.71)$$

The initial kinetic energy (just prior to the collision) was

$$T_i = \frac{1}{2}(8m)(\sqrt{2gh})^2 = 8mgh$$

Thus the change in kinetic energy of the bent bar is given by

$$\Delta T = T_f - T_i = mgh(1.29e^2 - 1.29)$$

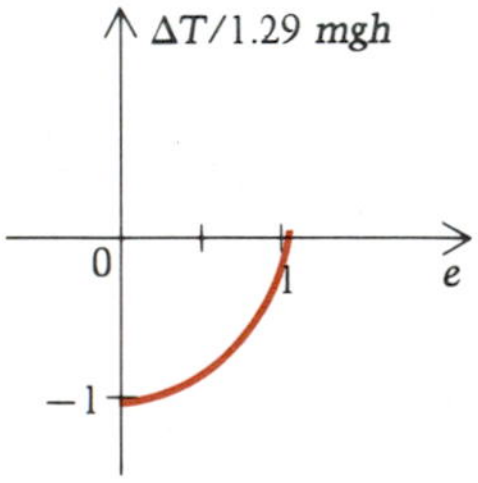

Figure E7.14b

We see that if $e = 1$ (purely elastic collision), no loss in kinetic energy occurs and hence no work is done in changing T. The energy lost is seen in Figure E7.14b to vary quadratically, with a maximum percentage loss (when $e = 0$) of

$$\frac{1.29mgh}{8mgh} \cdot 100 = 16.1\%$$

in this case. Because the point of striking is the end of the bar, 83.9 percent of the kinetic energy is retained. If the mass center of the bar were the point that struck the table, however, *all* the kinetic energy would have been lost if $e = 0$.

PROBLEMS ▶ Section 7.8

7.95 Find the kinetic energy of disk $\mathcal{B}$ in Problem 6.27.

7.96 The center of mass C of a gyroscope $\mathcal{G}$ is fixed. Show that the kinetic energy of $\mathcal{G}$ is

$$\tfrac{1}{2}A(\dot\theta^2 + \dot\phi^2\sin^2\theta) + \tfrac{1}{2}C(\dot\phi\cos\theta + \dot\psi)^2$$

where ϕ, θ, ψ are the Eulerian angles and A, A, C are the principal moments of inertia of $\mathcal{G}$ at C.

7.97 Find the kinetic energy of the wagon wheel in Problem 6.49 and use it to deduce the work done by the boy in getting it up to its final speed from rest.

7.98 A disk $\mathcal{D}$ of mass 10 kg and radius 25 cm is welded at a 45° angle to a vertical shaft $\mathcal{S}$. (See Figure P7.98.) The shaft is then spun up from rest to a constant angular speed $\omega_f = 10$ rad/s.

a. How much work is done in bringing the assembly up to speed?
b. Find the force and couple system acting on the plate at C after it is turning at the constant speed ω_f.

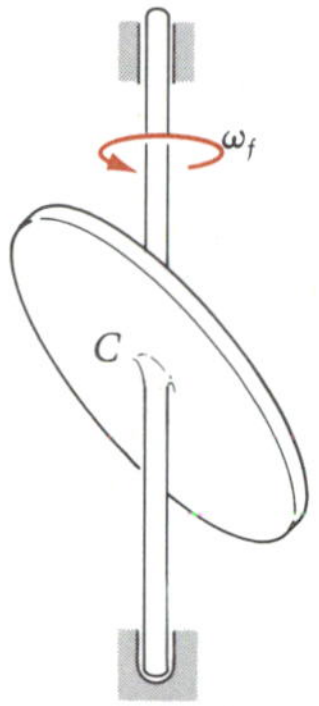

Figure P7.98

7.99 A thin rectangular plate (Figure P7.99) is brought up from rest to speed ω_0 about a horizontal axis Y.

a. Find the work that is done.
b. If two concentrated masses of $m/2$ each are added on the x_C axis, one on each side of the mass center, find their distances d from the mass center that will eliminate the bearing reactions.

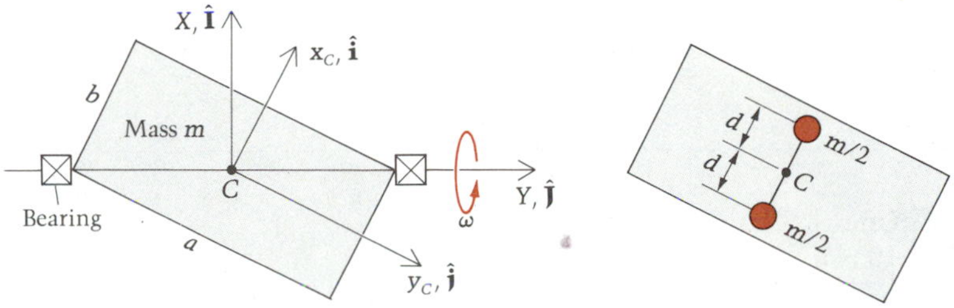

Figure P7.99

*** 7.100** The rigid body in Figure P7.100 consists of a disk $\mathcal{D}$ and rod $\mathcal{R}$, welded together perpendicularly as shown in the figure. If the body is spun up to angular speed ω_o about the z axis, how much work was done on it (excluding the overcoming of frictional resistance)?

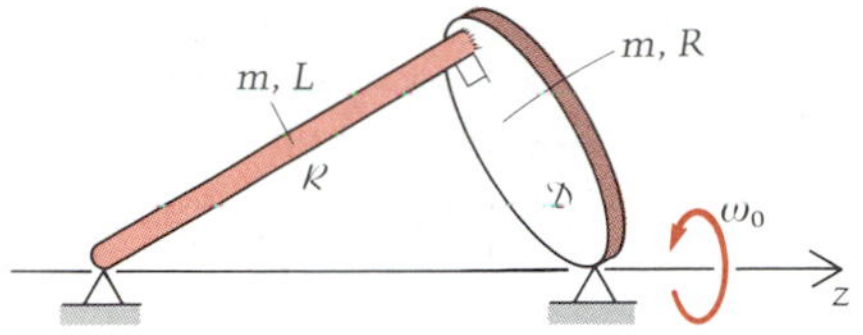

Figure P7.100

*** 7.101** Figure P7.101 shows a thin homogeneous triangular plate of mass m, base a, and height $2a$. It is welded to a light axle that can turn freely in bearings at A and B. Given:

$$I_{xx}^A = \frac{2ma^2}{3} \qquad I_{yy}^A = \frac{ma^2}{6} \qquad I_{xz}^A = I_{yz}^A = 0$$

$$I_{zz}^A = \frac{5ma^2}{6} \qquad I_{xy}^A = \frac{-ma^2}{6}$$

a. If the plate is turning at constant angular speed ω, find the torque that must be applied to the axle, and find the dynamic bearing reactions.
b. Find the principal axes at A and the principal moments of inertia there. Draw the axes on a sketch.
c. If possible, give the radius of a hole that, when drilled at C, will eliminate the bearing reactions. Give the answer in terms of m and ρt (density times thickness) of the plate.
d. Find the work done in bringing the plate up to speed ω from rest.

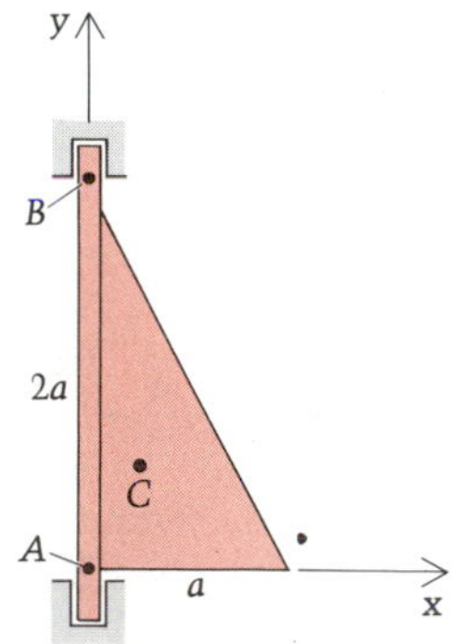

Figure P7.101

* **7.102** A thin equilateral triangular plate $\mathcal{P}$ of side s is welded to the vertical shaft at A in Figure P7.102. The shaft is brought up to speed ω_0 from rest by a motor.

a. How much work is done in bringing the system up to speed?

b. Find the force and couple system acting on the plate at A after it is turning at the speed ω_0 and the motor is turned off.

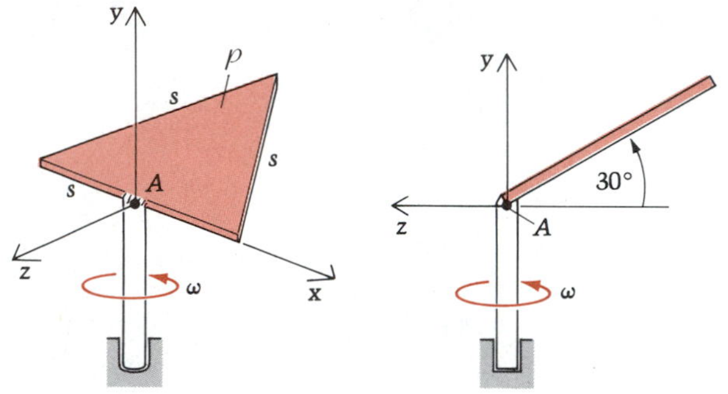

Figure P7.102

7.103 Two concentrated masses $m_1 = 10$ kg and $m_2 = 20$ kg are connected by a 15-kg slender rod m_3 of length 1.5 m. As shown in Figure P7.103, $(\hat{\mathbf{I}}, \hat{\mathbf{J}}, \hat{\mathbf{K}})$ are unit vectors fixed in direction in the inertial frame $\mathcal{I}$ and $(\hat{\mathbf{i}}, \hat{\mathbf{j}}, \hat{\mathbf{k}})$ are parallel to principal axes fixed at C in the combined body. At two times t_1 and t_2, the velocities of C and the angular velocities of the combined body are

$$\mathbf{v}_C(t_1) = \hat{\mathbf{I}} + 2\hat{\mathbf{J}} \text{ m/s} \qquad \boldsymbol{\omega}(t_1) = \hat{\mathbf{i}} + 2\hat{\mathbf{j}} - 4\hat{\mathbf{k}} \text{ rad/s}$$
$$\mathbf{v}_C(t_2) = 3\hat{\mathbf{J}} - 4\hat{\mathbf{K}} \text{ m/s} \qquad \boldsymbol{\omega}(t_2) = 3\hat{\mathbf{j}} - \hat{\mathbf{k}} \text{ rad/s}$$

Find the total work done on the system between t_1 and t_2.

7.104 Find the kinetic energy of the grinder in Problem 7.62. Is this equal to the work done by a motor on $\mathcal{S}$ which brings the system up to speed? (Neglect the masses of $\mathcal{S}$ and $\mathcal{B}$.)

* **7.105** A ring is welded to a rod at a point A as shown in Figure P7.105. The cross sections and densities of the rod and ring are the same. The combined body is released with a gentle nudge with end B of the rod connected to the smooth plane by a ball joint and with point A at its highest point as shown. At the instant when A reaches its lowest point, find the relationship between the horizontal and vertical angular velocity components of the body.

7.106 If in the preceding problem the plane is rough enough to prevent slipping, find the magnitude of the angular velocity when A reaches the floor.

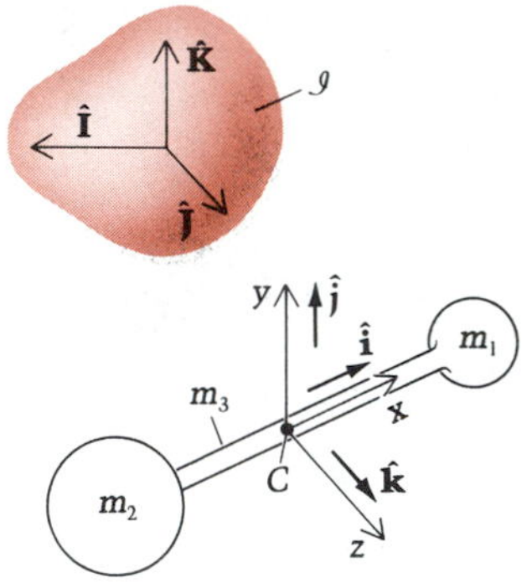

Figure P7.103

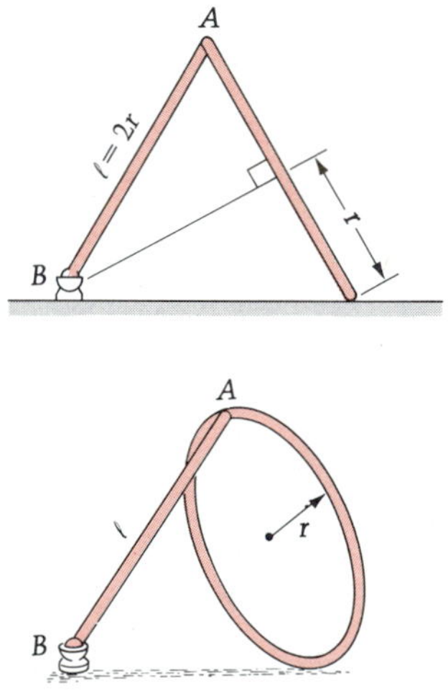

Figure P7.105

COMPUTER PROBLEM ▸ Chapter 7

* **7.107** Use a computer to generate data for a plot of maximum values of $V_Q^2/(gR)$ versus α in Example 7.10, for $0 < \alpha \leqq \pi/2$.

SUMMARY ▶ Chapter 7

In this chapter we have developed expressions for angular momentum of a rigid body in general three-dimensional motion. With respect to the mass center, it is

$$\begin{aligned}\mathbf{H}_C = &(I^C_{xx}\omega_x + I^C_{xy}\omega_y + I^C_{xz}\omega_z)\hat{\mathbf{i}}\\ &+ (I^C_{xy}\omega_x + I^C_{yy}\omega_y + I^C_{yz}\omega_z)\hat{\mathbf{j}}\\ &+ (I^C_{xz}\omega_x + I^C_{yz}\omega_y + I^C_{zz}\omega_z)\hat{\mathbf{k}}\end{aligned}$$

And if P is the location of a point of the body with zero velocity,

$$\begin{aligned}\mathbf{H}_P = &(I^P_{xx}\omega_x + I^P_{xy}\omega_y + I^P_{xz}\omega_z)\hat{\mathbf{i}}\\ &+ (I^P_{xy}\omega_x + I^P_{yy}\omega_y + I^P_{yz}\omega_z)\hat{\mathbf{j}}\\ &+ (I^P_{xz}\omega_x + I^P_{yz}\omega_y + I^P_{zz}\omega_z)\hat{\mathbf{k}}\end{aligned}$$

Transformation properties of moments and products of inertia include the parallel-axis theorems

$$\begin{aligned}I^P_{xx} &= I^C_{xx} + m(\bar{y}^2 + \bar{z}^2)\\ I^P_{yy} &= I^C_{yy} + m(\bar{z}^2 + \bar{x}^2)\\ I^P_{zz} &= I^C_{zz} + m(\bar{x}^2 + \bar{y}^2)\\ I^P_{xy} &= I^C_{xy} - m\bar{x}\bar{y}\\ I^P_{yz} &= I^C_{yz} - m\bar{y}\bar{z}\\ I^P_{zx} &= I^C_{zx} - m\bar{z}\bar{x}\end{aligned}$$

together with formulas for obtaining the moments and products of inertia associated with axes through a point when those properties are known for other axes through the point:

$$\begin{aligned}I^P_{x'x'} &= \ell_x^2 I^P_{xx} + \ell_y^2 I^P_{yy} + \ell_z^2 I^P_{zz} + 2\ell_x\ell_y I^P_{xy} + 2\ell_x\ell_z I^P_{xz} + 2\ell_y\ell_z I^P_{yz}\\ I^P_{x'y'} &= \ell_x n_x I^P_{xx} + \ell_y n_y I^P_{yy} + \ell_z n_z I^P_{zz} + (\ell_x n_y + \ell_y n_x) I^P_{xy}\\ &\quad + (\ell_x n_z + \ell_z n_x) I^P_{xz} + (\ell_y n_z + \ell_z n_y) I^P_{yz}\end{aligned}$$

In these two equations ℓ_x, ℓ_y, ℓ_z and n_x, n_y, n_z are respectively the direction cosines of x' and y' relative to axes x, y, and z.

Principal axes of inertia are very important and have the key property that were a body to rotate about a principal axis at a point P, then the angular momentum with respect to P would be in the same direction as the angular velocity, or

$$\mathbf{H}_P = I\boldsymbol{\omega}$$

where I is the moment of inertia about the principal axis, and is called a principal moment of inertia.

All products of inertia associated with a principal axis vanish, and at any point there are three mutually perpendicular principal axes. The largest and smallest of the principal moments of inertia are the largest and smallest of all the moments of inertia about axes through the point.

Some important special cases are:

1. If P lies in a plane of symmetry of the body, then the axis through P and perpendicular to the plane is a principal axis.
2. If P lies on an axis of symmetry of the body, then that axis and every line through P and perpendicular to it is a principal axis. Furthermore, the moments of inertia about these transverse axes through a given point are all the same.
3. If P is a point of spherical symmetry, e.g., the center of a uniform sphere, then every line through P is a principal axis and all of the corresponding principal moments of inertia are equal.

The most convenient form of Euler's second law, $\Sigma \mathbf{M}_C = \dot{\mathbf{H}}_C$, to use in a particular problem is often dependent on the problem. When body-fixed principal axes are used for reference, then we have what are usually referred to as the Euler equations:

$$\Sigma M_{C_x} = I_{xx}^C \, \dot{\omega}_x - (I_{yy}^C - I_{zz}^C) \, \omega_y \omega_z$$
$$\Sigma M_{C_y} = I_{yy}^C \, \dot{\omega}_y - (I_{zz}^C - I_{xx}^C) \, \omega_z \omega_x$$
$$\Sigma M_{C_z} = I_{zz}^C \, \dot{\omega}_z - (I_{xx}^C - I_{yy}^C) \, \omega_x \omega_y$$

However, it is very often more convenient to express the angular momentum in terms of its components parallel to reference axes associated with some intermediate frame of reference, say $\mathcal{J}$, which is neither the body itself nor the inertial frame $\mathcal{I}$, so that

$$\Sigma \mathbf{M}_C = {}^{\mathcal{J}}\dot{\mathbf{H}}_C + \boldsymbol{\omega}_{\mathcal{J}/\mathcal{I}} \times \mathbf{H}_C$$

Just as in the case of plane motion (Chapter 5), the work of external forces equals the change in kinetic energy for rigid bodies in general motion. In three-dimensional motion the kinetic energy, T, can be written in general as

$$T = \frac{m}{2} \mathbf{v}_C \cdot \mathbf{v}_C + \frac{1}{2} \boldsymbol{\omega} \cdot \mathbf{H}_C$$

The second term may be compactly written as

$$\frac{1}{2} I \omega^2$$

where I is the moment of inertia about the axis, through C, that is instantaneously aligned with $\boldsymbol{\omega}$.

REVIEW QUESTIONS ► Chapter 7

True or False?

1. Products of inertia associated with principal axes always vanish, but only at the mass center.
2. If the principal moments of inertia at a point are distinct, then the principal axes of inertia associated with them are orthogonal.

3. The maximum moment of inertia about any line through point P of rigid body $\mathcal{B}$ is the largest principal moment of inertia at P.
4. General motion is a much more difficult subject than plane motion. A major reason for this is that neither the kinematics nor kinetics differential equations governing the orientation motion of the body are linear.
5. If we solve the Euler equations (7.40), we immediately know the orientation of the rigid body in space.
6. The sun and the moon exert gravity torques on the earth, and they cause the axis of our planet to precess.
7. If at a certain instant the moment of inertia of the mass of body $\mathcal{B}$ about an axis through C parallel to the angular velocity vector is I, then the kinetic energy of $\mathcal{B}$ at that instant is $\frac{1}{2}mv_C^2 + \frac{1}{2}I\omega^2$.
8. The earth's lunisolar precession is the result of *both* the bulge at the equator *and* the tilt of the axis.
9. The kinetic energy lost during a collision of two bodies does not depend on the angular velocities of the bodies prior to impact.
10. The work-energy and impulse-momentum principles are general integrals of the equations of motion for a rigid body.
11. Sometimes it is better to use the products of inertia in $\Sigma \mathbf{M}_C = {}^{\mathcal{I}}\dot{\mathbf{H}}_C$ than to take the time to compute principal moments and axes of inertia so as to be able to utilize Euler's equations (7.40).
12. In steady precession with the nutation angle θ equaling 90°, the spin vector always precesses away from the torque vector.

Answers: 1. F 2. T 3. T 4. T 5. F 6. T 7. T 8. T 9. F 10. T 11. T 12. F

8 Special Topics

8.1 Introduction

8.2 Introduction to Vibrations

Free Vibration

Damped Vibration

Forced Vibration

8.3 Euler's Laws for a Control Volume

8.4 Central Force Motion

Review Questions

8.1 Introduction

In this chapter, we examine three subjects which are of considerable practical importance in Dynamics. In the first of these special topics, we introduce the reader to the subject of vibrations, limiting the presentation to a single degree of freedom (in which the oscillatory motion can be described by just one coordinate).

The second special topic deals with problems in which mass is continuously leaving and / or entering a region of space known as a control volume. A rocket is a good example: as the fuel is burned and the combustion products are ejected from a control volume enveloping the rocket, its momentum changes and it is propelled through the atmosphere. Euler's laws still apply, though the resulting equations are a bit more complicated than they were for the "constant mass" particles and bodies of earlier chapters.

The final topic in the chapter is central force motion, the most common example of which is that of orbits—such as the motions of planets around the sun, and of the moon and of man-made satellites around the earth.

The topics of Sections 8.2, 3, and 4 all stand alone, and can be read and understood after the reader has mastered Chapters 1 and 2, except for some of the problems in Section 8.2 in which the moment equation for rigid bodies in plane motion from Chapter 4 is also needed.

8.2 Introduction to Vibrations

Vibration is a term used to describe oscillatory motions of a body or system of bodies. These motions may be caused by isolated disturbances as when the wheel of an automobile strikes a bump or by fluctuating forces as in the case of the fuselage panels in an airplane vibrating in response to engine noise. Similarly, the oscillatory ground motions resulting from an earthquake cause vibrations of buildings. In each of these cases the undesirable motion may cause discomfort to occupants; moreover, the oscillating stresses induced within the body may lead to a fatigue failure of the structure, vehicle, or machine.

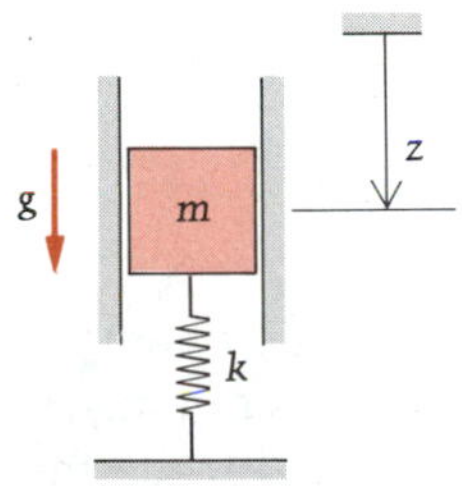

Figure 8.1

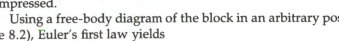

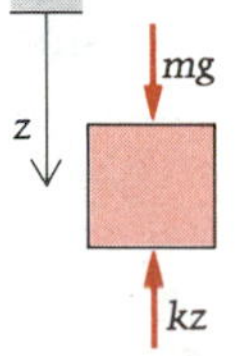

Figure 8.2

Free Vibration

For perhaps the simplest example of a mechanical oscillator consider the rigid block and linear spring shown in Figure 8.1. The block is constrained to translate vertically; thus a single parameter (scalar) is sufficient to establish position and hence the system is called a **single-degree-of-freedom system**. We choose z to be the parameter and let $z = 0$ correspond to the configuration in which the spring is neither stretched nor compressed.

Using a free-body diagram of the block in an arbitrary position (Figure 8.2), Euler's first law yields

$$m\ddot{z} = mg - kz$$

or

$$m\ddot{z} + kz = mg \tag{8.1}$$

which is a second-order linear differential equation with constant coefficients describing the motion of the block. The fact that the differential equation is nonhomogeneous (the right-hand side is not zero) is a consequence of our choice of datum for the displacement parameter z. For if we make the substitution $y = z - mg/k$, the governing equation (8.1) becomes

$$m\ddot{y} + ky = 0 \tag{8.2}$$

which is a *homogeneous* differential equation. It is not a coincidence that this occurs when the displacement variable is chosen so that it vanishes when the block is in the equilibrium configuration—that is, when the spring is compressed mg/k.

Motion described by an equation such as (8.2) is called a **free vibration** since there is no external force (external, that is, to the spring-mass system) stimulating it.

Rewriting Equation (8.2), we obtain

$$\ddot{y} + \frac{k}{m} y = 0$$

or, defining $\omega_n = \sqrt{k/m}$,

$$\ddot{y} + \omega_n^2 y = 0 \tag{8.3}$$

which has as its general solution

$$y = A \sin \omega_n t + B \cos \omega_n t \tag{8.4}$$

or

$$y = C \sin(\omega_n t + \varphi) \tag{8.5}$$

where

$$C = \sqrt{A^2 + B^2} \qquad \text{and} \qquad \tan \varphi = \frac{B}{A}$$

Whether expressed in the form of (8.4) or (8.5), y is called a **simple harmonic** function of time, ω_n is called the **natural circular frequency**, C is called the **amplitude** of the displacement y, and φ is said to be the **phase angle** by which y *leads* the reference function, $\sin \omega_n t$. The simple harmonic function is *periodic* and its **period** is $\tau_n = 2\pi/\omega_n$. Another quantity called **frequency** is $f_n = 1/\tau_n = \omega_n/2\pi$, which gives the number of cycles in a unit of time. When the unit of time is the second, the unit for f_n is the hertz (Hz); 1 Hz is 1 cycle per second.

The constants A and B in (8.4), or equivalently C and φ in (8.5), are determined from initial conditions of position and velocity. Thus if

$$y(0) = y_0$$

and

$$\dot{y}(0) = v_0$$

then

$$B = y_0$$

and

$$A = \frac{v_0}{\omega_n}$$

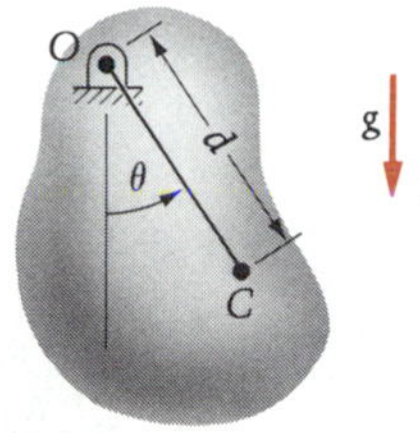

Figure 8.3

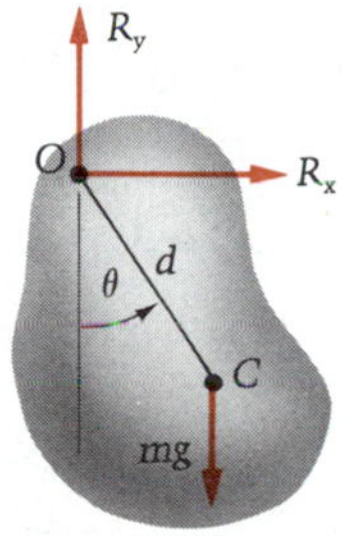

Figure 8.4

Now let us investigate what might seem an entirely different situation—that of a rigid body constrained to rotate about a fixed horizontal axis (through O as in Figure 8.3). Since the only kinematic freedom the body has is that of rotation, a single angle is sufficient to describe a configuration of the body. Let the angle be θ as shown, where we note that when $\theta = 0$ the mass center C is located directly below the pivot O.

Neglecting any friction at the axis of rotation, the free-body diagram appropriate to an arbitrary instant during the motion is shown in Figure 8.4. Summing moments about the axis of rotation, we get

$$-mg\, d \sin\theta = I_0 \ddot{\theta} \tag{8.6}$$

where I_0 is the mass moment of inertia about the axis of rotation. Equation (8.6) is a nonlinear differential equation because $\sin\theta$ is a nonlinear function of θ, but if we restrict our attention to sufficiently small angles so that $\sin\theta \approx \theta$, Equation (8.6) becomes

$$I_0 \ddot{\theta} + (mg\, d)\theta = 0 \tag{8.7}$$

That is, θ is a simple harmonic function:

$$\theta = A \sin \omega_n t + B \cos \omega_n t$$

where now

$$\omega_n^2 = \frac{mg\, d}{I_0}$$

The two preceding examples have an important feature in common: Motion near the equilibrium configuration is governed by a homogeneous, second-order, linear differential equation with constant coefficients, and in each case the motion is simple harmonic. A point of difference is that in the block-spring case the gravitational field plays no role other than establishing the equilibrium configuration; in particular the natural frequency does not depend on the strength (g) of the field. In the second case where the body is basically behaving as a pendulum, the gravitational field provides the "restoring action" and the natural frequency is proportional to $\sqrt{g}$.

EXAMPLE 8.1

Find the natural frequency of small oscillations about the equilibrium position of a uniform ball (sphere) rolling on a cylindrical surface.

Solution

Let m be the mass of the ball, let R be the radius of the path of its center, and let θ be the polar coordinate angle locating the center as shown in Figure E8.1a. Thus

$$\mathbf{a}_C = -R\dot{\theta}^2\hat{\mathbf{e}}_R + R\ddot{\theta}\hat{\mathbf{e}}_\theta$$

and the angular acceleration of the ball is $\boldsymbol{\alpha} = -(R\ddot{\theta}/r)\hat{\mathbf{k}}$ because of the no-slip condition. We shall now use $\mathbf{a}_C$ and $\boldsymbol{\alpha}$ in the equations of motion:

$$\Sigma\mathbf{F} = m\mathbf{a}_C \tag{1}$$

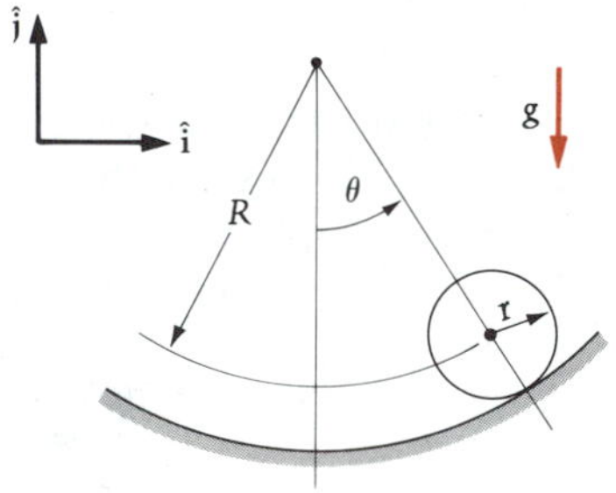

Figure E8.1a

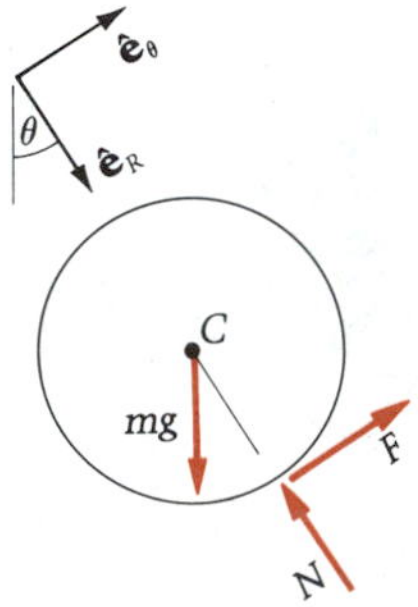

Figure E8.1b

Hence, from the free-body diagram shown in Figure E8.1b, the $\hat{\mathbf{e}}_R$ and $\hat{\mathbf{e}}_\theta$ component equations of (1) are:

$$F - mg\sin\theta = mR\ddot{\theta} \tag{2}$$

and

$$N - mg\cos\theta = mR\dot{\theta}^2 \tag{3}$$

Also, from summing moments about C, we have

$$Fr = \left(\frac{2}{5}mr^2\right)\left(-\frac{R\ddot{\theta}}{r}\right)$$

or

$$F = -\frac{2}{5}mR\ddot{\theta} \tag{4}$$

Eliminating the friction force F between Equations (2) and (4), we obtain the differential equation

$$\frac{7}{5}mR\ddot{\theta} + mg\sin\theta = 0$$

For small θ so that $\sin\theta \approx \theta$,

$$\frac{7}{5}R\ddot{\theta} + g\theta = 0$$

from which we see that

$$\omega_n^2 = \frac{5g}{7R}$$

or

$$\omega_n = 0.845\sqrt{\frac{g}{R}}$$

Damped Vibration

The simple harmonic motion in our examples of free vibration has a feature that conflicts with our experience in the real world; that is, the motion calculated persists forever unabated. Intuition would suggest decaying oscillations and finally the body coming to rest. Of course the problem here is that we have not incorporated any mechanism for energy dissipation in the analytical model. To do that, we shall return to the simple block-spring system and introduce a new element: a viscous damper (Figure 8.5). The *rate* of extension of this element is proportional to the force applied, through a damping constant c, so that the force is c times the rate of extension.

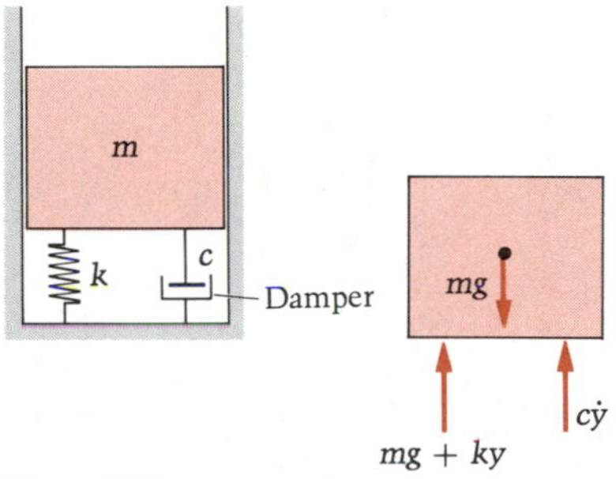

Figure 8.5

Referring to the free-body diagram in Figure 8.5 and letting $y = 0$ designate the equilibrium position as before, we have

$$m\ddot{y} = mg - (mg + ky) - c\dot{y}$$

or

$$m\ddot{y} + c\dot{y} + ky = 0 \tag{8.8}$$

The appearance of the $c\dot{y}$ term in (8.8) has a profound effect on the solution to the differential equation and hence on the motion being described. Solutions to (8.8) may be found from

$$y = Ae^{rt} \tag{8.9}$$

where A is an arbitrary constant and r is a characteristic parameter. Substituting (8.9) into (8.8), we obtain

$$(mr^2 + cr + k)Ae^{rt} = 0 \tag{8.10}$$

which is satisfied nontrivially (i.e., for $A \neq 0$) with

$$mr^2 + cr + k = 0 \tag{8.11}$$

This characteristic equation has two roots given by

$$r = -\frac{c}{2m} \pm \sqrt{\left(\frac{c}{2m}\right)^2 - \frac{k}{m}} \tag{8.12}$$

Except for the case in which $(c/2m)^2 = k/m$, the roots are distinct; if we call them r_1 and r_2, then the general solution to (8.8) is

$$y = A_1e^{r_1t} + A_2e^{r_2t}$$

In the exceptional case $(c/2m)^2 = k/m$, there is only the one repeated root $r = -c/2m$, but direct substitution will verify that there is a solution to (8.8) of the form $te^{-(c/2m)t}$ so that the general solution in that case is

$$y = A_1 e^{-(c/2m)t} + A_2 t e^{-(c/2m)t} \tag{8.13}$$

With initial conditions

$$y(0) = y_0$$

and

$$\dot{y}(0) = v_0$$

we find that

$$A_1 = y_0$$

and

$$A_2 = v_0 + \left(\frac{c}{2m}\right) y_0$$

Since

$$\left(\frac{c}{2m}\right)^2 = \frac{k}{m}$$
$$= \omega_n^2$$

the solution is

$$y = e^{-\omega_n t}[y_0 + (v_0 + \omega_n y_0)t] \tag{8.14}$$

Displacements given by (8.14) are plotted in Figure 8.6 for several representative sets of initial conditions (positive y_0 but positive and negative v_0). Two features of the motion are apparent:

1. $y \to 0$ (the equilibrium position) as $t \to \infty$.
2. The motion is not oscillatory; the equilibrium position is "overshot" at most once and only then when the initial speed is sufficiently large and in the direction opposite to that of the initial displacement.

In the case we have just studied, the damping is called **critical damping**, because it separates two quite different mathematical solu-

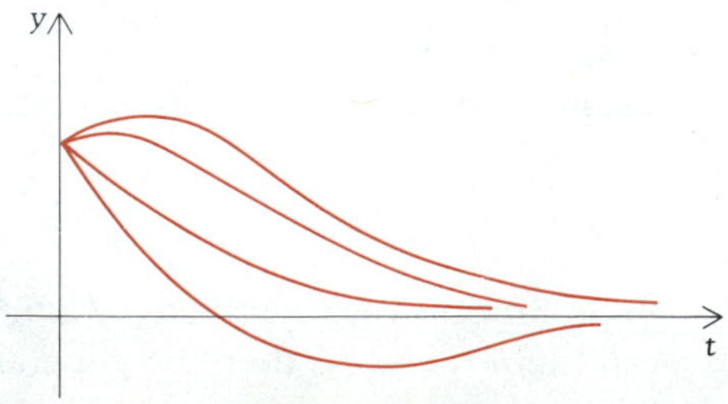

Figure 8.6 Motion of a critically damped system.

tions: For greater damping the roots of the characteristic equation (8.11) are both real and negative, and for small damping the roots are complex conjugates. If we let the critical damping be denoted by c_{crit} then we have seen that

$$\begin{aligned} c_{crit} &= 2\sqrt{km} \\ &= 2m\omega_n \end{aligned} \tag{8.15}$$

Now let us consider the case for which $c > c_{crit}$; the mechanical system is then said to be **overdamped** or the damping is said to be **supercritical**. In this case the roots given by (8.12) are both real and negative since $(c/2m)^2 > k/m$; if we call these roots $-a_1$ and $-a_2$, with $a_2 > a_1 > 0$, then the general solution to the differential equation of motion is

$$y = A_1 e^{-a_1 t} + A_2 e^{-a_2 t} \tag{8.16}$$

The motion described here is in no way qualitatively different from that for the case of critical damping, which we have just discussed. For a given set of initial conditions, Equation (8.16) yields a slower approach to $y = 0$ than does (8.13). That is, the overdamped motion is more "sluggish" than the critically damped motion as we would anticipate because of the greater damping.

Finally we consider the case in which the system is said to be **underdamped** or **subcritically** damped; that is, $c < c_{crit}$. The roots given by (8.12) are the complex conjugates

$$-\frac{c}{2m} \pm i\sqrt{\frac{k}{m} - \left(\frac{c}{2m}\right)^2}$$

where $i = \sqrt{-1}$. It is possible to express the general solution to the governing differential equation as

$$y = e^{-(c/2m)t}(A_1 \sin \omega_d t + A_2 \cos \omega_d t) \tag{8.17}$$

where $\omega_d = \sqrt{k/m - (c/2m)^2}$. A typical displacement history corresponding to (8.17) is shown in Figure 8.7. We note that, just as in the preceding cases, $y \to 0$ as $t \to \infty$; however, here the motion is oscillatory. We see that the simple harmonic motion obtained for the model without

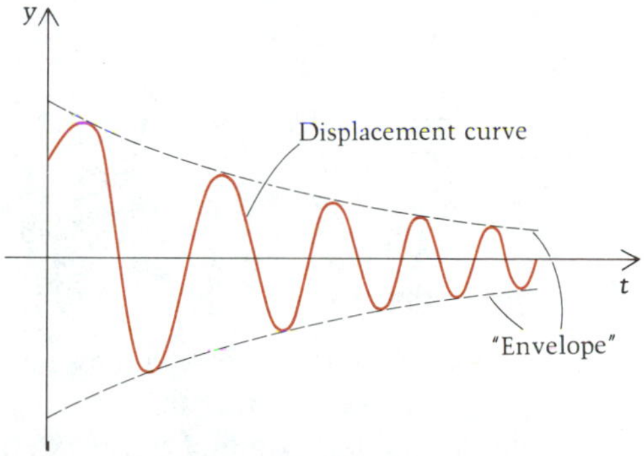

Figure 8.7 Motion of an underdamped system.

damping is given by (8.17) with $c = 0$. Moreover, we see that with light damping (small c) the analytical model that does not include damping adequately describes the motion during the first several oscillations. It is this case—subcritical damping—that is of greatest practical importance in studies of vibration.

EXAMPLE 8.2

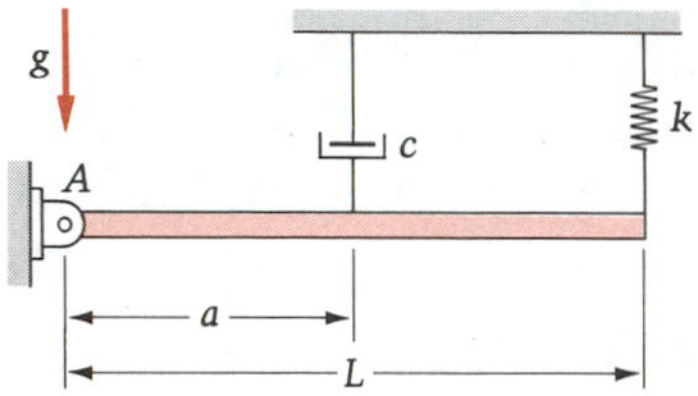

Figure E8.2a

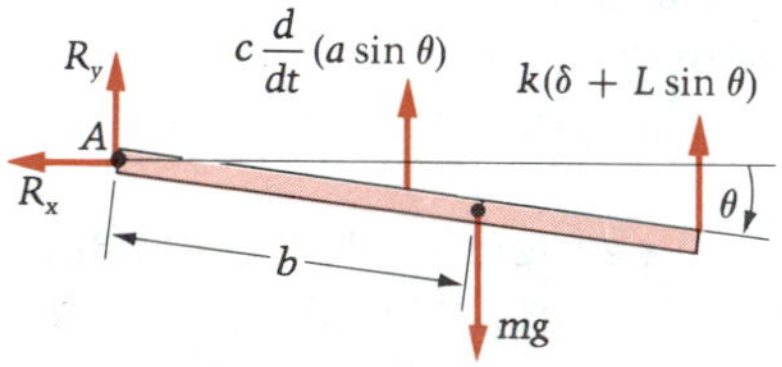

Figure E8.2b

Find the damping constant c that gives critical damping of the rigid bar executing motions near the equilibrium position shown in Figure E8.2a.

Solution

We are going to restrict our attention to small angles θ, and thus we may ignore any tilting of the damper or the spring. However, it may help us develop the equation of motion in an orderly way if we assume that the upper ends of the spring and damper slide along so that each remains vertical as the bar rotates through the angle θ. Without further restriction, if we sum moments about A with the help of the FBD in Figure E8.2b, we obtain, with I_A the moment of inertia of the mass of the bar about the axis of rotation at A,

$$I_A\ddot{\theta} = mg(b \cos \theta) - \left[c \frac{d}{dt}(a \sin \theta)\right](a \cos \theta) - k(\delta + L \sin \theta)(L \cos \theta) \quad (1)$$

where δ is the spring stretch at equilibrium. Thus for small θ (that is, $\sin \theta \approx \theta$, $\cos \theta \approx 1$) we linearize Equation (1) and obtain

$$I_A\ddot{\theta} = mgb - ca^2\dot{\theta} - kL\delta - kL^2\theta$$

Of course, $\theta = 0$ is the equilibrium configuration so that

$$mgb = kL\delta$$

The linear governing differential equation is then

$$I_A\ddot{\theta} + ca^2\dot{\theta} + kL^2\theta = 0$$

and for critical damping we get, associating the coefficients of $\theta, \dot{\theta}, \ddot{\theta}$ with those of $y, \dot{y}, \ddot{y}$ in Equation (8.8),

$$c_{crit}a^2 = 2\sqrt{(kL^2)I_A}$$

or

$$c_{crit} = 2\frac{L}{a^2}\sqrt{kI_A}$$

Any c less than this critical value will result in oscillations of decreasing amplitude.

Forced Vibration

Fluctuating external forces may have destructive effects on mechanical systems; this is perhaps the primary motivation for studying mechanical vibration. It is common for the external loading to be a periodic function of time, in which case the loading may be expressed as a series of simple

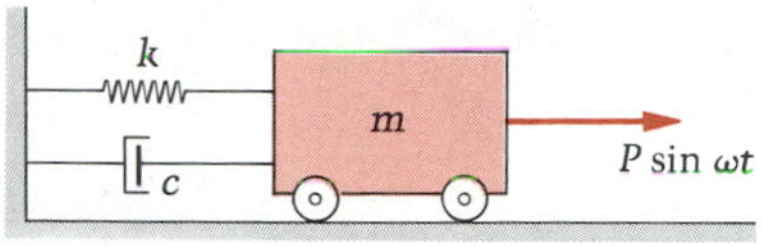

Figure 8.8

harmonic functions (Fourier series). Consequently it is instructive to consider the case in which the loading is simple-harmonic. For the mass-spring-damper system shown in Figure 8.8, the differential equation of motion is

$$m\ddot{x} + c\dot{x} + kx = P \sin \omega t \tag{8.18}$$

The general solution is composed of two parts: a *particular* solution (anything that satisfies the differential equation) and what is called the *complementary* solution (the general solution to the homogeneous differential equation). A particular solution of the form $x = X \sin(\omega t - \varphi)$ may be found. If we substitute this expression in (8.18) we obtain

$$-m\omega^2 X \sin(\omega t - \varphi) + c\omega X \cos(\omega t - \varphi) + kX \sin(\omega t - \varphi) = P \sin \omega t$$

or

$$(k - m\omega^2)X(\sin \omega t \cos \varphi - \sin \varphi \cos \omega t) + c\omega X(\cos \omega t \cos \varphi + \sin \omega t \sin \varphi) = P \sin \omega t$$

or

$$[(k - m\omega^2) \cos \varphi + c\omega \sin \varphi]X \sin \omega t - [(k - m\omega^2) \sin \varphi - c\omega \cos \varphi]X \cos \omega t = P \sin \omega t$$

For this to be satisfied at every instant of time,

$$[(k - m\omega^2) \cos \varphi + c\omega \sin \varphi]X = P \tag{8.19}$$

and

$$-c\omega \cos \varphi + (k - m\omega^2) \sin \varphi = 0 \tag{8.20}$$

From (8.20) we get

$$\tan \varphi = \frac{c\omega}{k - m\omega^2} \tag{8.21}$$

so that

$$\sin \varphi = \frac{c\omega}{\sqrt{(k - m\omega^2)^2 + (c\omega)^2}}$$

and

$$\cos \varphi = \frac{k - m\omega^2}{\sqrt{(k - m\omega^2)^2 + (c\omega)^2}}$$

Substituting these expressions for $\sin \varphi$ and $\cos \varphi$ into (8.19), we obtain

$$\left[\frac{(k - m\omega^2)^2}{\sqrt{(k - m\omega^2)^2 + (c\omega)^2}} + \frac{(c\omega)^2}{\sqrt{(k - m\omega^2)^2 + (c\omega)^2}}\right] X = P$$

so that

$$X = \frac{P}{\sqrt{(k - m\omega^2)^2 + (c\omega)^2}} \tag{8.22}$$

We may now write the complete solution to the differential equation (8.18):

$$x = x_c(t) + X\sin(\omega t - \varphi) \tag{8.23}$$

where x_c is the complementary solution and is one of the three cases enumerated in the preceding section. That is, the form of x_c depends on whether the system is overdamped, critically damped, or underdamped. However, in each of these cases the negative exponent causes the function to approach zero as time becomes large. Thus for large time x_c tends to zero and $x(t)$ tends to the particular solution. For this reason the simple-harmonic particular solution is called the **steady-state displacement**, since it represents the long-term behavior of the system.

We note that the steady-state motion is a simple harmonic function having amplitude X and lagging the excitation (force) function by the phase angle φ. We may put these in a convenient form by dividing numerator and denominator of (8.21) and (8.22) by k, so that

$$\tan\varphi = \frac{c\omega/k}{1 - \omega^2/\omega_n^2} \qquad \left(\text{where } \omega_n^2 = \frac{k}{m}\right) \tag{8.24}$$

and

$$X = \frac{P/k}{\sqrt{(1 - \omega^2/\omega_n^2)^2 + (c\omega/k)^2}} \tag{8.25}$$

Investigating the dimensionless quantity $c\omega/k$, we find

$$\frac{c\omega}{k} = \frac{c\omega}{m\omega_n^2} = \frac{2c}{2m\omega_n}\left(\frac{\omega}{\omega_n}\right)$$

But we know that $2m\omega_n = c_{crit}$, the critical damping, so if we let ζ be the damping ratio (c/c_{crit}),

$$\frac{c\omega}{k} = 2\zeta\frac{\omega}{\omega_n} \tag{8.26}$$

and

$$\tan\varphi = \frac{2\zeta(\omega/\omega_n)}{1 - \omega^2/\omega_n^2} \tag{8.27}$$

and

$$X = \frac{P/k}{\sqrt{(1 - \omega^2/\omega_n^2)^2 + [2\zeta(\omega/\omega_n)]^2}} \tag{8.28}$$

The phase angle φ and the dimensionless displacement amplitude kX/P are plotted against the frequency ratio ω/ω_n in Figures 8.9 and 8.10, respectively, for various values of the damping ratio ζ. We see that, with small damping, large amplitudes of displacement occur when the excitation frequency ω is near the natural frequency ω_n. This phenomenon is called **resonance**, and the desire to avoid it has led to the development of methods for estimating natural frequencies of mechanical sys-

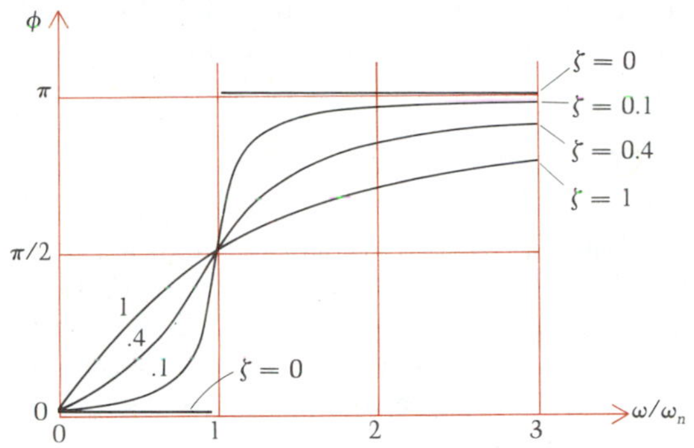

Figure 8.9

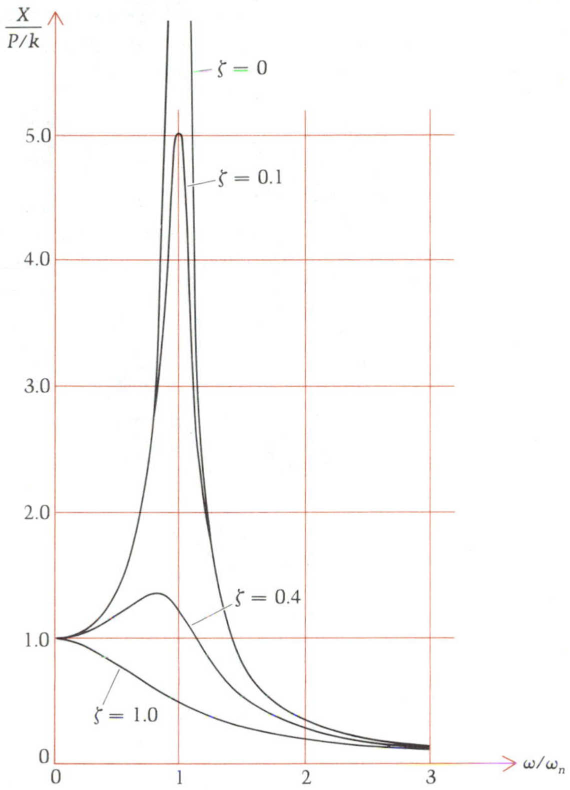

Figure 8.10

tems. Note that the steady-state response curves are insensitive to the damping for sufficiently small damping (say $\zeta < 0.1$) provided that we are not in the near vicinity of $\omega/\omega_n = 1$. This is an important observation because often in engineering practice we have reason to believe that the damping is small but we do not have accurate quantitative information about it.

We close this section by discussing the usual source of a simple harmonic external loading—an imbalance in a piece of rotating machinery. Let the machine be made up of two parts. The first, of mass m_1, is a rigid body constrained to rotate about an axis fixed in the second body (mass m_2), which translates relative to the inertial frame of reference. Let the mass center of the rotating body lie off the axis of rotation a distance e and let the body rotate at constant angular speed ω. (See Figure 8.11(a).)

Referring to the free-body diagram in Figure 8.11(b), we have

$$-kx - c\dot{x} = m_2\ddot{x} + m_1 \frac{d^2}{dt^2}(x + e \sin \omega t)$$

$$= (m_1 + m_2)\ddot{x} - m_1 e\omega^2 \sin \omega t$$

If we denote the total mass of the machine by m, then $m = m_1 + m_2$ and

$$m\ddot{x} + c\dot{x} + kx = m_1 e\omega^2 \sin \omega t$$

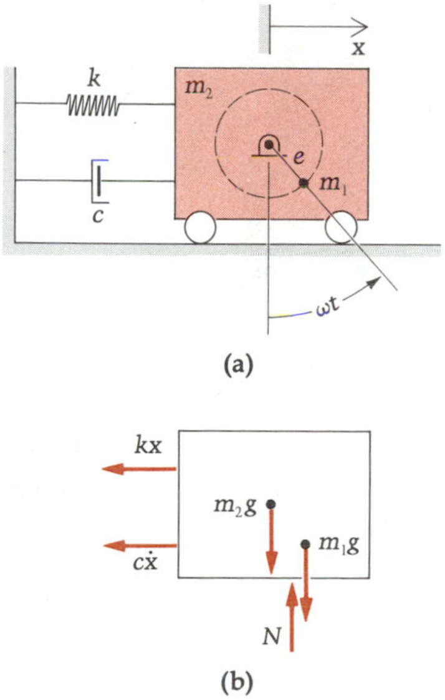

Figure 8.11

Thus the amplitude of the apparent "external" sinusoidal loading is $m_1e\omega^2$ and its frequency is the angular speed of the rotating element.

EXAMPLE 8.3

A piece of machinery weighing 200 lb has a rotating element with imbalance (m_1e times the acceleration of gravity, which is $32.2 \times 12 = 386$ in./sec²) 5 lb-in. and an operating speed of 1200 rpm. There are four springs, each of stiffness 1500 lb/in., supporting the machine whose frame is constrained to translate vertically. The damping ratio is $\zeta = 0.3$. Find the steady-state displacement of the frame.

Solution

The effective spring stiffness is

$$k = 4(1500) = 6000 \text{ lb/in.}$$

so that

$$\begin{aligned}\omega_n &= \sqrt{\frac{6000}{200/386}} \\ &= 108 \text{ rad/sec} \\ \omega &= \frac{1200}{60}(2\pi) \\ &= 126 \text{ rad/sec}\end{aligned}$$

The effective external force amplitude is

$$P = m_1e\omega^2 = \left(\frac{5}{386}\right)(126)^2 = 206 \text{ lb}$$

From Equation (8.28) we get

$$\begin{aligned}X &= \frac{P/k}{\sqrt{(1-\omega^2/\omega_n^2)^2 + (2\zeta\omega/\omega_n)^2}} \\ &= \frac{206/6000}{\sqrt{[1-(126/108)^2]^2 + [2(0.3)(126/108)]^2}} \\ &= 0.0436 \text{ in.}\end{aligned}$$

The phase angle φ is given by

$$\begin{aligned}\tan\varphi &= \frac{2\zeta\omega/\omega_n}{1-\omega^2/\omega_n^2} \\ &= \frac{2(0.3)(126/108)}{1-(126/108)^2} = -1.94\end{aligned}$$

so that $\varphi = 205$ rad (117°).

Example 8.4

The machine of Example 8.3 (weight = 200 lb, imbalance = 5 lb-in., operating speed = 1200 rpm) is to be supported by springs with negligible damping. If the machine were bolted directly to the floor, the amplitude of force transmitted to the floor would be

$$(m_1 e)\omega^2 = 206 \text{ lb}$$

What should the stiffness of the support system be so that the amplitude of the force transmitted to the floor is less than 20 lb?

Solution

The force exerted on the floor is transmitted through the supporting springs and is of amplitude kX, where X is the amplitude of displacement of the machine. From Equation (8.28) we have

$$kX = \frac{m_1 e\omega^2}{\sqrt{(1-\omega^2/\omega_n^2)^2 + (2\zeta\omega/\omega_n)^2}}$$

or with negligible damping (i.e., $\zeta \approx 0$)

$$kX = \frac{m_1 e\omega^2}{\sqrt{(1-\omega^2/\omega_n^2)^2}} = \frac{m_1 e\omega^2}{|1-\omega^2/\omega_n^2|}$$

Thus for

$$\frac{1}{|1-\omega^2/\omega_n^2|} = \frac{kX}{m_1 e\omega^2} < \frac{20}{206} = 0.0971$$

it is clear that $1-\omega^2/\omega_n^2$ is negative. Note that only when $\omega^2/\omega_n^2 > 2$ is

$$\frac{1}{|1-\omega^2/\omega_n^2|} < 1$$

Therefore we inquire into the condition for which

$$\frac{1}{1-\omega^2/\omega_n^2} < -0.0971$$

or

$$\frac{\omega^2}{\omega_n^2} > 1 + \frac{1}{0.0971} = 11.3$$

or

$$\omega_n^2 < \frac{(126)^2}{11.3} = 1400$$

since $\omega = 126$ rad / sec. But

$$k = m\omega_n^2 < \frac{200}{386}(1400) = 725 \text{ lb/in.}$$

Thus to satisfy the given conditions the support stiffness must be *less than* 725 lb / in.

If the only springs available give a greater stiffness, the problem may be solved by increasing the mass; particularly we might mount the machine on a block of material, say concrete, and then support the machine and block by springs. For example, if the only springs available were those of Example 8.3 for which $k = 6000$ lb/in., then we need m to be *at least* that given by

$$m = \frac{k}{\omega_n^2} = \frac{6000}{1400} = 4.29 \text{ lb-sec}^2/\text{in.}$$

for which the weight is

$$(4.29)(386) = 1660 \text{ lb}$$

Therefore we need a slab or block weighing

$$1660 - 200 = 1460 \text{ lb}$$

EXAMPLE 8.5

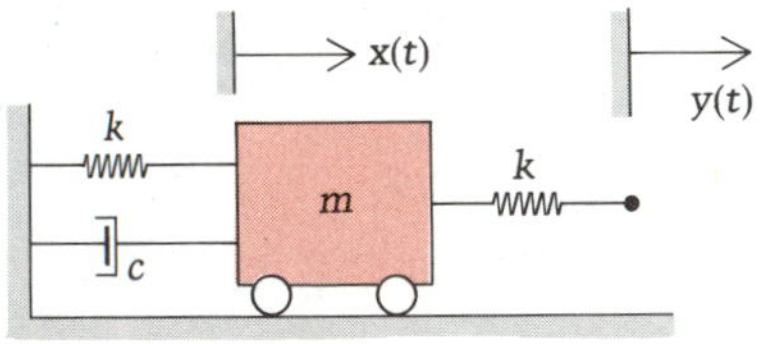

Figure E8.5

Find the steady-state displacement $x(t)$ of the mass in Figure E8.5 if $y(t) = 0.1 \cos 120t$ inch, where t is in seconds, $m = 0.01$ lb-sec²/in., $k = 100$ lb/in., and $c = 2$ lb-sec/in. In particular: (a) What is the amplitude of $x(t)$? (b) What is the angle by which $x(t)$ leads or lags $y(t)$?

Solution

The differential equation of motion of the mass is seen to be

$$m\ddot{x} = -kx - c\dot{x} - k(x - y)$$

or

$$m\ddot{x} + c\dot{x} + 2kx = kY \cos \omega t$$

where $Y = 0.1$ in. and $\omega = 120$ rad/sec. Using Equation (8.18), we see that kY is playing the same role as the oscillating force P, so that the steady-state amplitude is

$$X = \frac{kY}{\sqrt{(2k - m\omega^2)^2 + (c\omega)^2}} = \frac{100(0.1)}{\sqrt{[200 - 0.01(120)^2]^2 + [2(120)]^2}}$$

or

$$X = 0.0406 \text{ in.}$$

The phase angle is

$$\varphi = \tan^{-1}\left(\frac{c\omega}{2k - m\omega^2}\right) = \tan^{-1}\left[\frac{2(120)}{200 - 0.01(120^2)}\right] = 76.9^\circ \text{ or } 1.34 \text{ rad} \quad \text{(lagging)}$$

Thus the steady-state motion is

$$x_{ss} = 0.0406 \cos(120t - 1.34) \text{ in.}$$

PROBLEMS ► Section 8.2

8.1 Find the frequency of small vibrations of the round wheel $\mathscr{C}$ as it rolls back and forth on the cylindrical surface in Figure P8.1. The radius of gyration of $\mathscr{C}$ with respect to the axis through C normal to the plane of the figure is k_C. Verify the result of Example 8.1 with your answer.

8.2–8.4 Find the equations of motion and periods of vibration of the systems shown in Figure P8.2 to P8.4. In each case, neglect the mass of the rigid bar to which the ball (particle) is attached.

8.5 The cylinder in Figure P8.5 is in equilibrium in the position shown. For no slipping, find the natural frequency of free vibration about this equilibrium position.

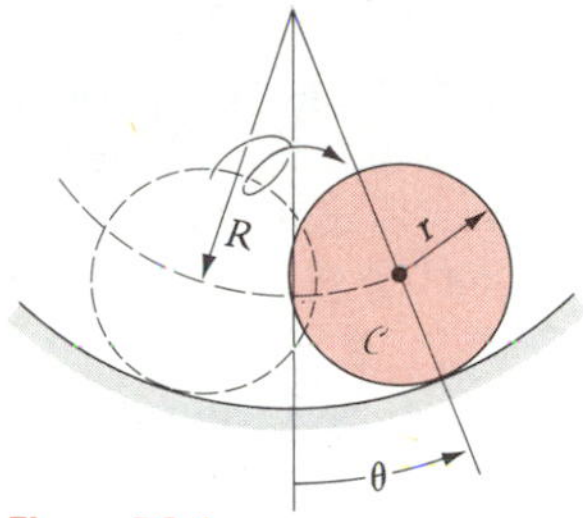

Figure P8.1

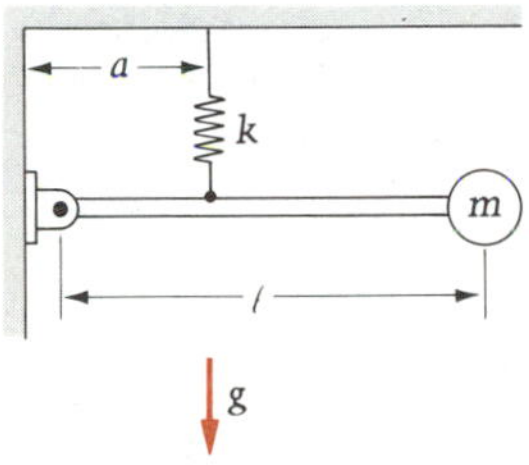

Figure P8.2

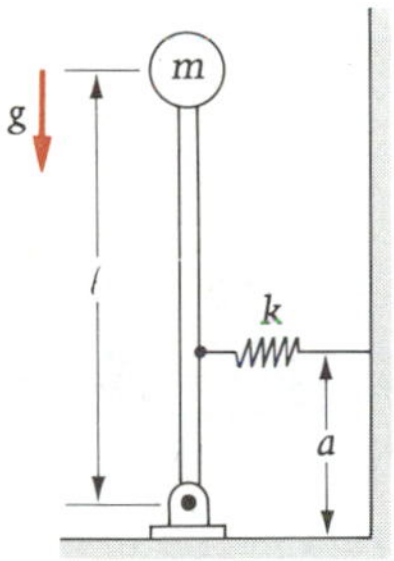

Figure P8.3

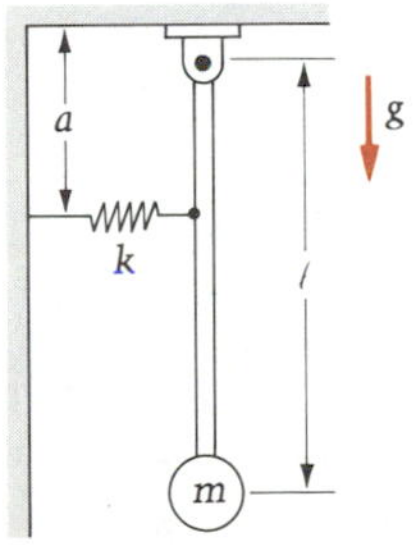

Figure P8.4

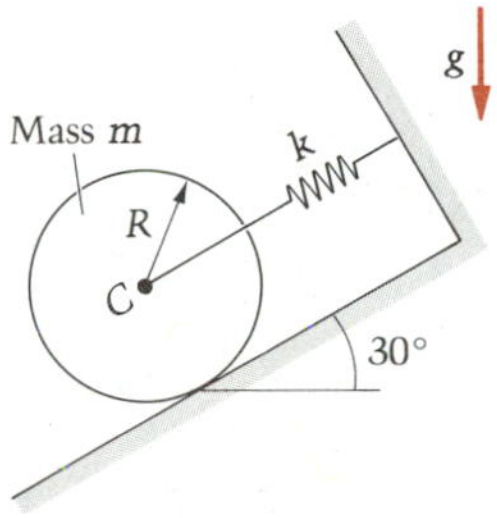

Figure P8.5

8.6 A uniform cylinder of mass m and radius R is floating in water. (See Figure P8.6.) The cylinder has a spring of modulus k attached to its top center point. If the specific weight of the water is γ, find the frequency of the vertical bobbing motion of the cylinder. *Hint*: The upward (buoyant) force on the bottom of the cylinder equals the weight of water displaced at any time (Archimedes' principle).

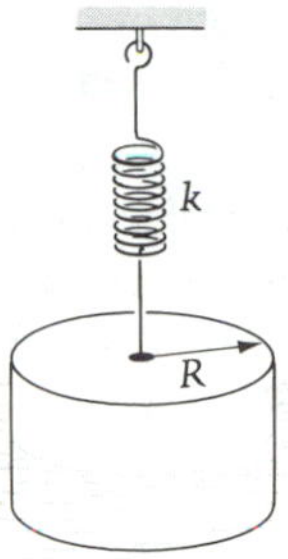

Figure P8.6

8.7 It is possible to determine experimentally the moments of inertia of large objects, such as the rocket shown in Figure P8.7. If the rocket is turned through a slight angle about z_C and released, for example, it oscillates with a period of 2.8 sec. Find the radius of gyration k_{z_C}.

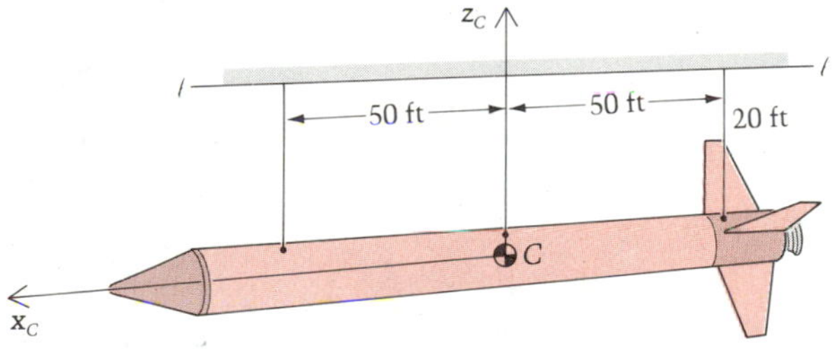

Figure P8.7

8.8 In the preceding problem, when the rocket is caused to swing with small angles about axis $\ell\ell$ as shown, the period is observed to be 8 sec. Find from this information the value of k_{x_C}.

8.9 Prove statements (1) and (2) on page 522.

* **8.10** Find the frequency of small amplitude oscillations of the uniform half-cylinder near the equilibrium position shown in Figure P8.10. Assume that the cylinder rolls on the horizontal plane.

8.11 A particle of mass m is attached to a light, taut string. The string is under tension, T, sufficiently large that the string is, for all practical purposes, straight when the system is in equilibrium as shown in Figure P8.11. Find the natural frequency of small transverse oscillations of the particle.

* **8.12** The masses in Figure P8.12 are connected by an inextensible string. Find the frequency of small oscillations if mass m is lowered slightly and released.

* **8.13** The solid homogeneous cylinder in Figure P8.13 weighs 200 lb and rolls on the horizontal plane. When the cylinder is at rest, the springs are each stretched 2 ft. The modulus of each spring is 15 lb/ft. The mass center C is given an initial velocity of $\frac{1}{2}$ ft/sec to the right.

a. How far to the right will C go?
b. How long will it take to get there?
c. How long will it take to go halfway to the extreme position?

8.14 A sack of cement of mass m is to be dropped on the center of a simply supported beam as shown in Figure P8.14. Assume that the mass of the beam may be neglected, so that it may be treated as a simple linear spring of stiffness k. Estimate the maximum deflection at the center of the beam.

* **8.15** A particle $\mathcal{P}$ of mass m moves on a rough, horizontal rail with friction coefficient μ. (See Figure P8.15.) It is attached to a fixed point on the rail by a linear spring of modulus k. The initial stretch of the spring is $7\,\mu gm/k$. Describe the subsequent motion if it is known that the particle starts from rest. Show that the mass stops for good when $t = 3\pi/\sqrt{k/m}$. *Hint*: The differential equation doesn't have quite the form found in the text; also, every time the particle reverses direction, so does the friction force—thus the equation needs rewriting with each stop.

8.16 A spring with modulus 120 lb/in. supports a 200-lb block. (See Figure P8.16.) The block is fastened to the spring. A 400-lb downward force is applied to the top of the block at $t = 0$ when the block is at rest. Find the maximum deflection of the spring in the ensuing time.

* **8.17** A block weighing 1 lb is dropped from height $H = 0.1$ in. (See Figure P8.17.) If $k = 2.5$ lb/in., find the time interval for which the ends of the springs are in contact with the ground.

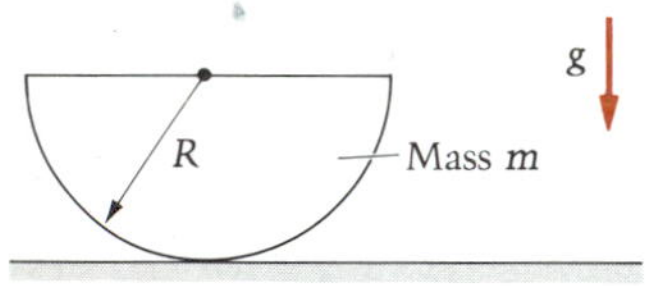

Figure P8.10

Figure P8.11

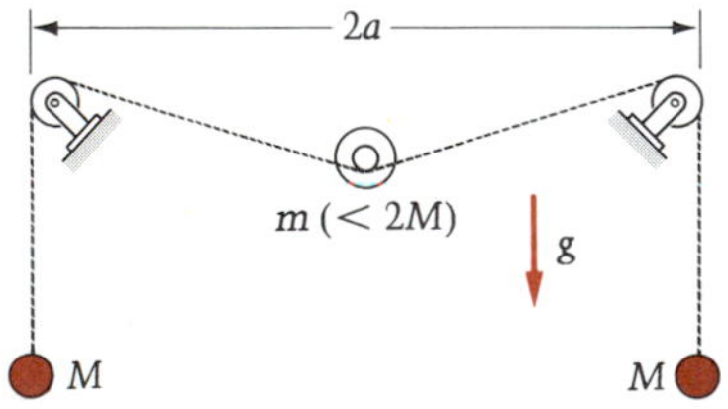

Figure P8.12

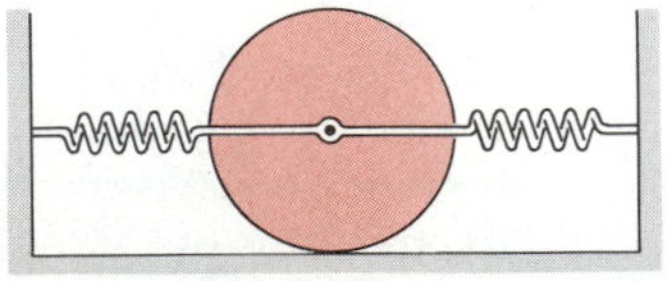

Figure P8.13

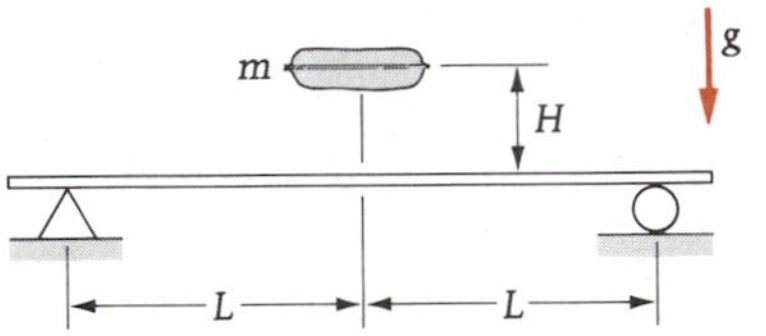

Figure P8.14

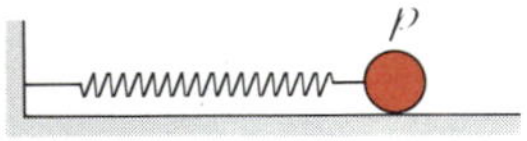

Figure P8.15

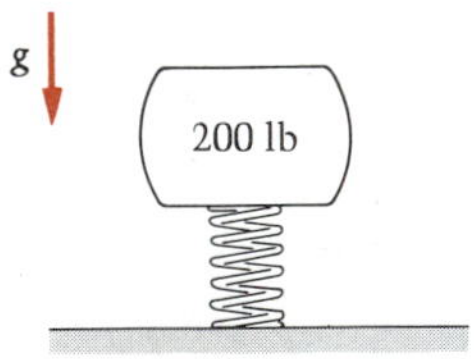

Figure P8.16

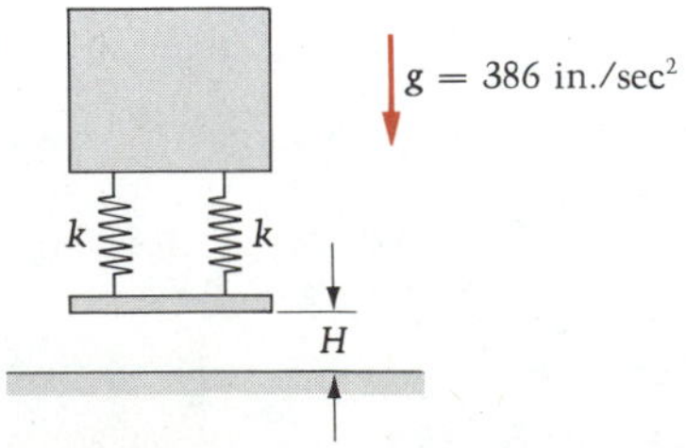

Figure P8.17

8.18 Assume that the slender rigid bar $\mathcal{B}$ in Figure P8.18 undergoes only small angles of rotation. Find the angle of rotation $\theta(t)$ if the bar is in equilibrium prior to $t = 0$, at which time the constant force P begins to traverse the bar at constant speed v.

* **8.19** Refer to the preceding problem: (a) Find the work done by P in traversing the bar $\mathcal{B}$; (b) show that this work equals the change in mechanical energy (which is the kinetic energy of $\mathcal{B}$ plus the potential energy stored in the spring).

* **8.20** The turntable in Figure P8.20 rotates in a horizontal plane at a *constant* angular speed ω. The particle P (mass $= m$) moves in the frictionless slot and is attached to the spring (modulus k, free length ℓ) as shown.

   a. Derive the differential equation describing the motion $y(t)$ of the particle relative to the slot.
   b. What is the extension of the spring such that P does not accelerate relative to the slot?
   c. Suppose the motion is initiated with the spring unstretched and the particle at rest relative to the slot. Find the ensuing motion $y(t)$.

8.21 Find the value of c to give critical damping of the pendulum in Figure P8.21. Neglect the mass of the rigid bar to which the particle of mass m is attached.

8.22 If $k = 100$ lb / in. and the mass of the uniform, slender, rigid bar in Figure P8.22 is 0.03 lb-sec^2 / in., what damping constant c results in critical damping?

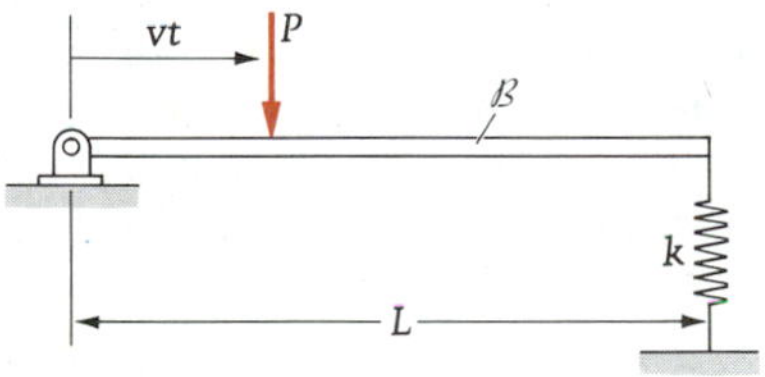

Figure P8.18

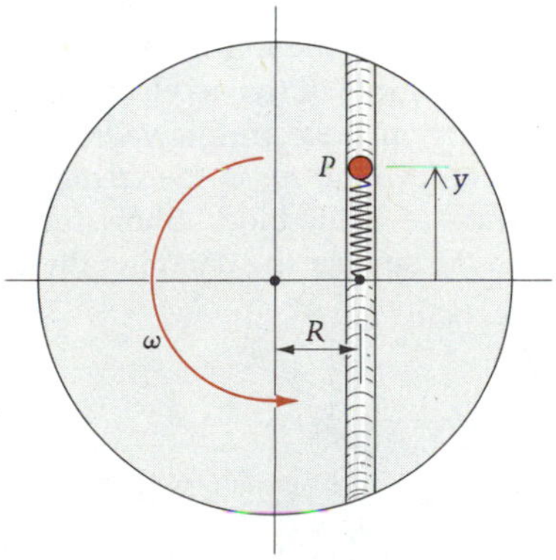

Figure P8.20

8.23 If $k = 100$ lb / in. and the mass of the uniform, slender, rigid bar in Figure P8.23 is 0.03 lb-sec^2 / in., what damping modulus c results in critical damping? Compare with the c from Problem 8.22. For this damping, find $\theta(t)$ if the bar is turned through a small angle θ_0 and then released from rest. If the dashpot were removed, what would be the period of free vibration?

8.24 A cannon weighing 1200 lb shoots a 100-lb cannonball at a velocity of 600 ft / sec. (See Figure P8.24.) It then immediately comes into contact with a spring of stiffness 149 lb / ft and a dashpot that is set up to critically damp the system. Assuming that there is no friction between the wheels and the plane, find the displacement toward the wall after $\frac{1}{2}$ sec has elapsed.

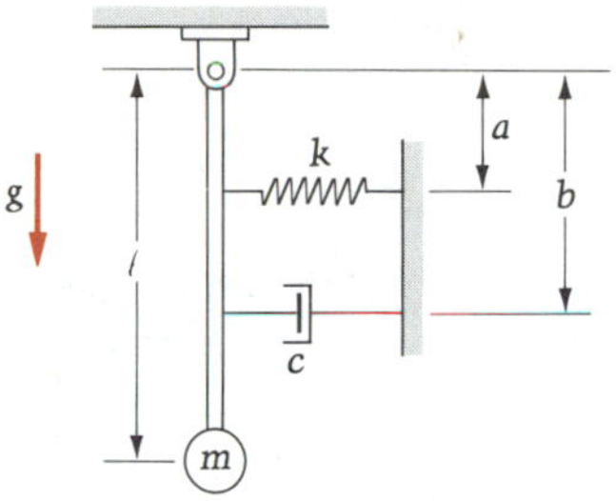

Figure P8.21

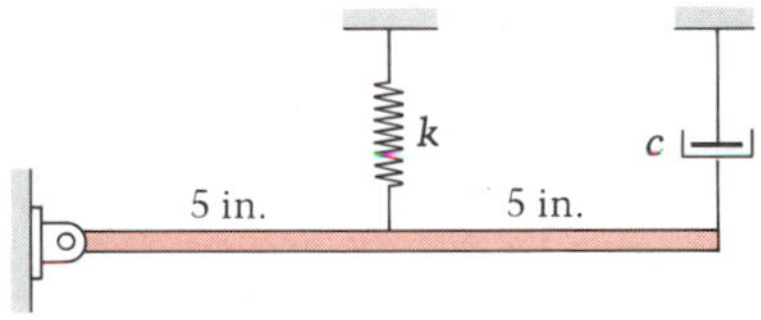

Figure P8.22

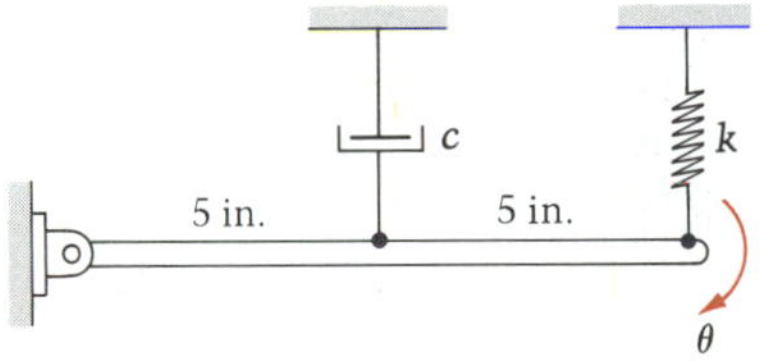

Figure P8.23

Figure P8.24

8.25 Consider free oscillations of a subcritically damped oscillator. Do local maxima in the response occur periodically?

8.26 In Figure P8.26, find the steady-state displacement $x(t)$ if $y(t) = 0.1 \sin 100t$ inch, where t is in seconds, $m = 0.01$ lb-sec^2/in., $k = 100$ lb/in., and $c = 2$ lb-sec/in. In particular:

- **a.** What is the amplitude of $x(t)$?
- **b.** What is the angle by which $x(t)$ leads or lags $y(t)$?

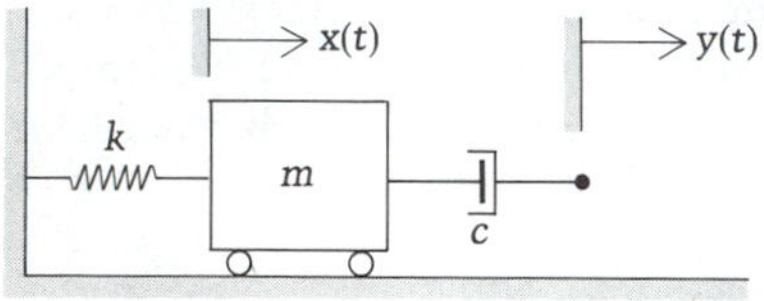

Figure P8.26

8.27 In Figure P8.27 find the steady-state displacement $x(t)$ if $y(t) = 0.2 \sin 90t$ inch, where t is in seconds, $m = 0.01$ lb-sec^2/in., $k = 50$ lb/in., and $c = 1$ lb-sec/in. In particular:

- **a.** What is the amplitude of $x(t)$?
- **b.** What is the phase angle by which $x(t)$ leads or lags $y(t)$?

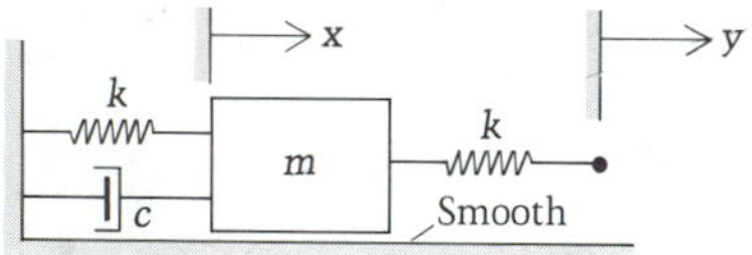

Figure P8.27

8.28 The cart in Figure P8.28 is at rest prior to $t = 0$, at which time the right end of the spring is given the motion $y = vt$, where v is a constant. Find $x(t)$.

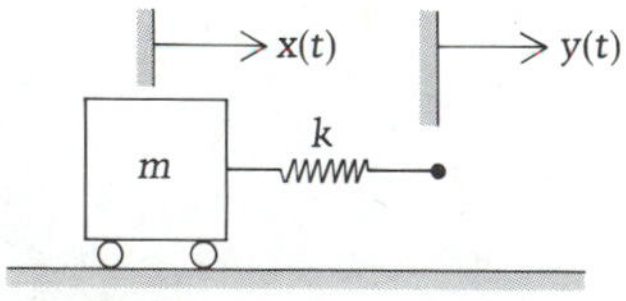

Figure P8.28

8.29 The block in Figure P8.29 is at rest in equilibrium prior to the application of the constant force $P = 50$ lb at $t = 0$. If $k = 100$ lb/in., $m = 0.01$ lb-sec^2/in., and the system is critically damped, find $x(t)$.

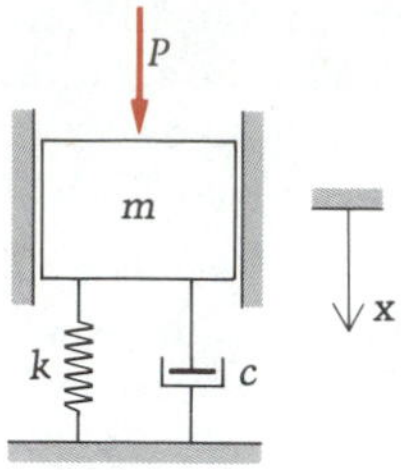

Figure P8.29

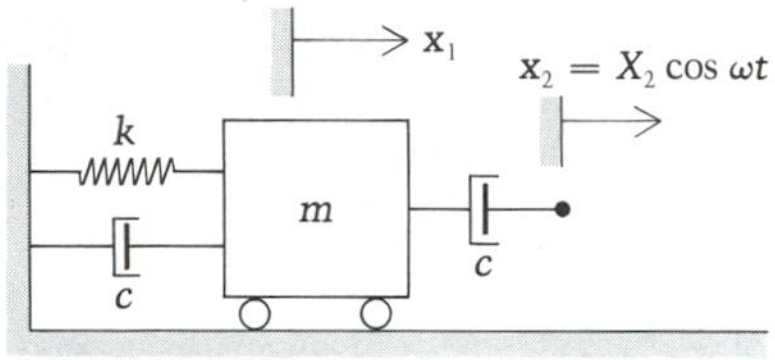

Figure P8.30

* **8.30** In Figure P8.30 find the response $x_1(t)$ for the initial conditions $x_1(0) = \dot{x}_1(0) = 0$ if

$$k = 100 \text{ lb/in.}$$
$$m = 0.01 \text{ lb-sec}^2\text{/in.}$$
$$c = 1.0 \text{ lb-sec/in.}$$
$$X_2 = 0.05 \text{ in.}$$
$$\omega = 100 \text{ rad/sec}$$

* **8.31** Repeat the preceding problem if (a) $c = 2.0$ lb-sec/in.; (b) $c = 0.5$ lb-sec/in.

8.32 Optical equipment is mounted on a table whose four legs are pneumatic springs. If the table and equipment together weigh 700 lb, what should be the stiffness of each spring so that the amplitude of steady, simple-harmonic, vertical displacement of the table will not be greater than 5 percent of a corresponding motion of the floor? The forcing frequency is 30 rad/sec. Neglect damping in your calculations.

* **8.33** The block of mass m in Figure P8.33 is mounted through springs k and damper c on a vibrating floor. Derive an expression for the steady-state acceleration of the block (whose motion is vertical translation). Show that the amplitude of the acceleration is less than that of the floor, regardless of the value of c, provided that $\omega > \sqrt{2}\omega_n$, where ω_n is the frequency of free undamped vibrations of the block. Show further that if $\omega > \sqrt{2}\omega_n$, then the smaller the damping the better the isolation.

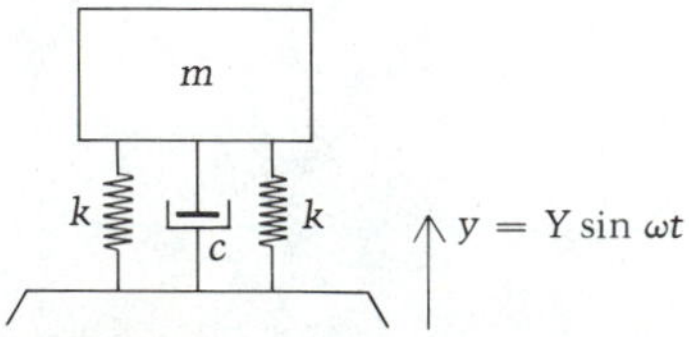

Figure P8.33

8.3 Euler's Laws for a Control Volume

Euler's laws describe the relationship between external forces and the motion of any body whether it be a solid, liquid, or gas. Sometimes, however, it is desirable to focus attention on some region of space (control volume) through which material may flow rather than on the fixed collection of particles that constitute a body. Examples of this sort are abundant in the field of fluid mechanics and include the important problem of describing and analyzing rocket-powered flight. Our purpose in this section is to discuss the forms taken by Euler's laws when the focus of attention is the control volume rather than the body.

We take as self-evident what might be called the "law of accumulation, production, and transport" — that is, the rate of accumulation of something within a region of space is equal to the rate of its production within the region plus the rate at which it is transported into the region.* Thus, for example, the rate of accumulation of peaches in Georgia equals the rate of production of peaches in the state plus the net rate at which they are shipped in. This idea can be applied in mechanics whenever we are dealing with a quantity whose measure for a body is the sum of the measures for the particles making up the body. Thus we can apply this principle to things such as mass, momentum, moment of momentum, and kinetic energy.

Suppose that at an instant a closed region V (control volume) contains material (particles) making up body $\mathcal{B}$. Let $m_{\mathcal{B}}$ denote the mass of body $\mathcal{B}$ and m_V denote the mass associated with V (that is, the mass of whatever particles happen to be in V at some time). Instantaneously $m_V = m_{\mathcal{B}}$, but because some of the material of $\mathcal{B}$ is flowing out of V and some other material is flowing in, $\dot{m}_V \neq \dot{m}_{\mathcal{B}}$. In fact by the accumulation principle stated above

$$\dot{m}_V = \dot{m}_{\mathcal{B}} + (\text{net rate of mass flow into } V) \tag{8.29}$$

since clearly $\dot{m}_V$ represents the rate of buildup (accumulation) of mass in V and since $\dot{m}_{\mathcal{B}}$, the rate of change of mass of the material instantaneously within V, represents the production term. Of course a body, being a specific collection of particles, has constant mass; thus $\dot{m}_{\mathcal{B}} = 0$ and (8.29) becomes $\dot{m}_V =$ (rate of mass flow into V), which is often called the **continuity equation**.

For momentum **L**, the statement corresponding to Equation (8.29) is

$$\dot{\mathbf{L}}_V = \dot{\mathbf{L}}_{\mathcal{B}} + (\textbf{net rate of flow of momentum into } V) \tag{8.30}$$

But Euler's first law applies to a body (such as $\mathcal{B}$) so that $\Sigma\mathbf{F} = \dot{\mathbf{L}}_{\mathcal{B}}$, where $\Sigma\mathbf{F}$ is the resultant of the external forces on $\mathcal{B}$ — or, in other words, the resultant of the external forces acting on the material instantaneously in V. Thus Equation (8.30) becomes

$$\dot{\mathbf{L}}_V = \Sigma\mathbf{F} + (\textbf{net rate of flow of momentum into } V) \tag{8.31}$$

* The mathematical statement of this is known as the Reynolds Transport Theorem.

which is the **control-volume form of Euler's first law.** The momentum flow rate on the right of (8.31) is calculated by summing up (or integrating) the momentum flow rates across infinitesimal elements of the boundary of V, where the momentum flow rate per unit of boundary area is the product of the mass flow rate per unit of area and the instantaneous velocity of the material as it crosses the boundary.

A similar derivation produces a **control-volume form of Euler's second law,** for which the result is

$$(\dot{\mathbf{H}}_O)_V = \Sigma \mathbf{M}_O + \begin{array}{c}\textbf{(net rate of flow into } V \textbf{ of moment}\\ \textbf{of momentum with respect to } O\textbf{)}\end{array} \tag{8.32}$$

where O is a point fixed in the inertial frame of reference.

It is important to realize that nothing in our derivations here has restricted the control volume except that it be a closed region in space. It may be moving relative to the frame of reference in almost any imaginable way, and it may be changing in shape or volume with time. We conclude this section with examples of two of the most common applications of Equation (8.31).

EXAMPLE 8.6

A fluid undergoes steady flow in a pipeline and encounters a bend at which the cross-sectional area of pipe changes from A_1 to A_2. At inlet 1 the density is ρ_1 and the velocity (approximately uniform over the cross section) is $v_1\hat{\mathbf{i}}$. At outlet 2 the density is ρ_2. Find the resultant force exerted on the pipe bend by the fluid. (See Figure E8.6a.)

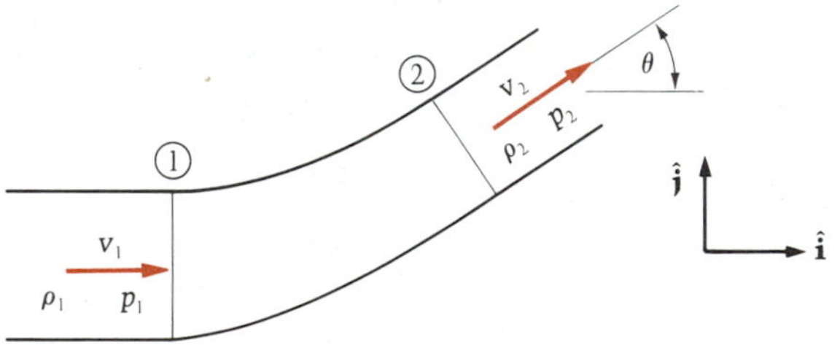

Figure E8.6a

Solution

Let the velocity of flow at the outlet be given by $v_2(\cos\theta\hat{\mathbf{i}} + \sin\theta\hat{\mathbf{j}})$. Then for steady flow the rate of mass flow at the inlet section is the same as that at the outlet section:

$$\rho_1 A_1 v_1 = \rho_2 A_2 v_2$$

so that

$$v_2 = \frac{\rho_1 A_1}{\rho_2 A_2} v_1$$

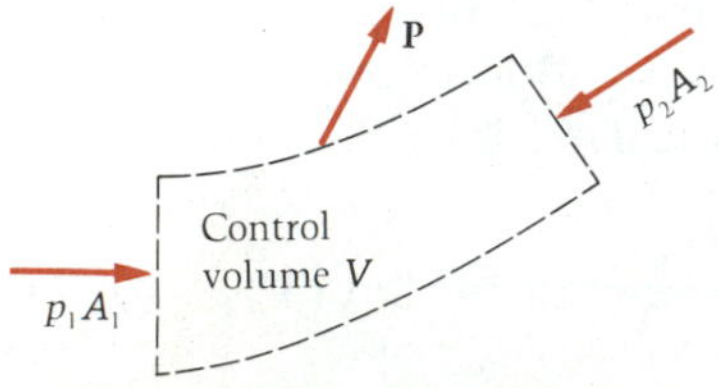

Figure E8.6b

Let the control volume be the region bounded by the inner surface of the pipe bend and the inlet and outlet cross sections. (See Figure 8.6b.) A consequence of the condition of steady flow is that within the control volume the distributions of velocity and density are independent of time. Thus the total momentum associated with V is a constant and

$$\frac{d\mathbf{L}_V}{dt} = 0$$

But

$$\frac{d\mathbf{L}_V}{dt} = \Sigma\mathbf{F} + (\text{net rate of momentum flow into } V)$$

Therefore

$$\begin{aligned}
\Sigma\mathbf{F} &= (\text{net rate of momentum flow out of } V) \\
&= \rho_2 A_2 v_2 (v_2 \cos\theta\hat{\mathbf{i}} + v_2 \sin\theta\hat{\mathbf{j}}) - \rho_1 A_1 v_1 (v_1 \hat{\mathbf{i}}) \\
&= \rho_1 A_1 v_1 (v_2 \cos\theta\hat{\mathbf{i}} + v_2 \sin\theta\hat{\mathbf{j}} - v_1 \hat{\mathbf{i}}) \\
&= \rho_1 A_1 v_1 \left[\left(\frac{\rho_1 A_1}{\rho_2 A_2} v_1 \cos\theta - v_1\right)\hat{\mathbf{i}} + \left(\frac{\rho_1 A_1}{\rho_2 A_2} v_1 \sin\theta\right)\hat{\mathbf{j}}\right] \\
&= \rho_1 A_1 v_1^2 \left[\left(\frac{\rho_1 A_1}{\rho_2 A_2} \cos\theta - 1\right)\hat{\mathbf{i}} + \frac{\rho_1 A_1}{\rho_2 A_2} \sin\theta\hat{\mathbf{j}}\right]
\end{aligned}$$

And if $\mathbf{P}$ is the force exerted on the fluid by the bend, then

$$\Sigma\mathbf{F} = \mathbf{P} + p_1 A_1 \hat{\mathbf{i}} + p_2 A_2 (-\cos\theta\hat{\mathbf{i}} - \sin\theta\hat{\mathbf{j}})$$

where p_1 and p_2 are the inlet and outlet fluid pressures respectively. Therefore

$$\mathbf{P} = \Sigma\mathbf{F} - p_1 A_1 \hat{\mathbf{i}} + p_2 A_2 (\cos\theta\hat{\mathbf{i}} + \sin\theta\hat{\mathbf{j}})$$

and the force exerted on the bend by the fluid is $-\mathbf{P}$ with

$$\begin{aligned}
-\mathbf{P} = &\left[p_1 A_1 - p_2 A_2 \cos\theta + \rho_1 A_1 v_1^2 \left(1 - \frac{\rho_1 A_1}{\rho_2 A_2}\cos\theta\right)\right]\hat{\mathbf{i}} \\
&- \left[p_2 A_2 \sin\theta + \rho_1 A_1 v_1^2 \left(\frac{\rho_1 A_1}{\rho_2 A_2}\right)\sin\theta\right]\hat{\mathbf{j}}
\end{aligned}$$

Example 8.7

To illustrate how the control volume form of Euler's first law is used to describe the motion of a rocket vehicle, consider such a vehicle climbing in a vertical rectilinear flight. Let $v\hat{\mathbf{j}}$ be the velocity of the vehicle from which combustion products are being expelled at velocity $-v_e\hat{\mathbf{j}}$ relative to the rocket. Further let $M(t)$ be the mass at time t of the vehicle and its contents, let μ be the rate of mass flow of the ejected gases, and let p be the gas pressure at the nozzle exit of cross section A.

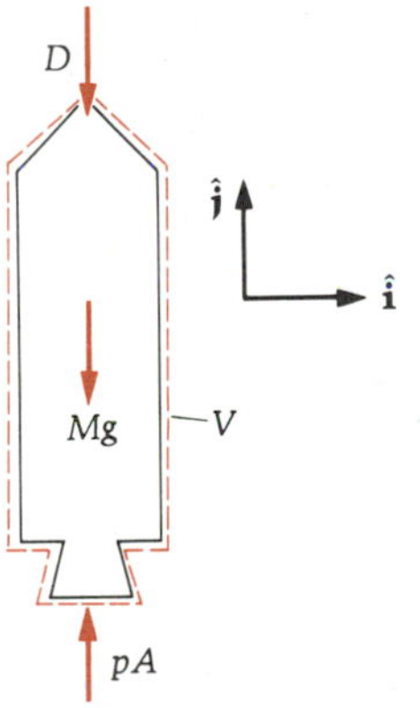

Figure E8.7 Free-body diagram of rocket.

Solution

Force D in the free-body diagram (Figure E8.7), representing the drag or resistance to motion, is the resultant of (1) all the shear stresses acting on the surface of the vehicle and (2) all the pressure on the surface. Thus $(pA - D)\hat{\mathbf{j}}$ represents the resultant of all the surface-distributed forces on the control volume. The two terms have been separated so that we may point out that the force pA remains even after the rocket has cleared the atmosphere. That is, p is pressure exerted by the gas particles *about to pass across* the nozzle exit plane* on the particles that have *just passed across,* and vice-versa.

If we let the control volume V surround the vehicle, Equation (8.31) becomes

$$\frac{d}{dt}\mathbf{L}_V = (pA - D)\hat{\mathbf{j}} - Mg\hat{\mathbf{j}} - \mu(v - v_e)\hat{\mathbf{j}}$$

At this point we must approximate $\mathbf{L}_V$ by $Mv\hat{\mathbf{j}}$; this is an approximation because some of the products of combustion inside the rocket are of course moving relative to the vehicle. Therefore

$$\frac{d}{dt}(Mv\hat{\mathbf{j}}) = (pA - D)\hat{\mathbf{j}} - Mg\hat{\mathbf{j}} - \mu(v - v_e)\hat{\mathbf{j}}$$

or

$$v\frac{dM}{dt} + M\frac{dv}{dt} = pA - D - Mg - \mu v + \mu v_e$$

But of course

$$\frac{dM}{dt} = -\mu$$

so that

$$M\frac{dv}{dt} = pA - D - Mg + \mu v_e$$

which is of the form of force = mass × acceleration, where one of the "forces" is the "thrust" μv_e.

* This term may be neglected if exhaust gases have expanded to atmospheric pressure or nearly so.

PROBLEMS ▶ Section 8.3

8.34 Let dm_i/dt and dm_o/dt be the respective rates at which mass enters and leaves a system. Show that Equation (8.31) may be expressed in terms of these rates as

$$\Sigma\mathbf{F} = m\frac{dv}{dt} + \left(\frac{dm_i}{dt} - \frac{dm_o}{dt}\right)\mathbf{v} + \frac{dm_o}{dt}\mathbf{v}_o - \frac{dm_i}{dt}\mathbf{v}_i$$

where $m\mathbf{v} = \mathbf{L}_V$ and it is assumed that all the incoming particles have a common velocity $\mathbf{v}_i$ (in an inertial frame) and that all the exiting particles have a common velocity $\mathbf{v}_o$.

8.35 Liquid of specific weight w flows out of a hole in the side of a tank in a jet of cross section A. If the velocity of the jet is $\mathbf{v}$, determine the force exerted on the tank by the supporting structure that holds the tank at rest. Note that the pressure in the jet will be atmospheric pressure.

8.36 A child aims a garden hose at the back of a friend. (See Figure P8.36.) If the water (specific weight 62.4 lb/ft^3) stream has a diameter of $\frac{1}{4}$ in. and a speed of 50 ft/sec, estimate the force exerted on the "target" if: (a) he is stationary; (b) he is running away from the stream at a speed of 10 ft/sec. Assume the flow in contact with the

Figure P8.36

boy's back to be vertical relative to him; that is, neglect any splashback.

8.37 A steady jet of liquid is directed against a smooth rigid surface and the jet splits as shown in Figure P8.37. Assume that each fluid particle moves in a plane parallel to that of the figure and ignore gravity. Ignoring gravity and friction, it can be shown that the particle speed after the split is still v as depicted. Estimate the fraction of the flow rate occurring in each of the upper and lower branches. *Hint*: Use the fact that no external force tangent to the surface acts on the liquid.

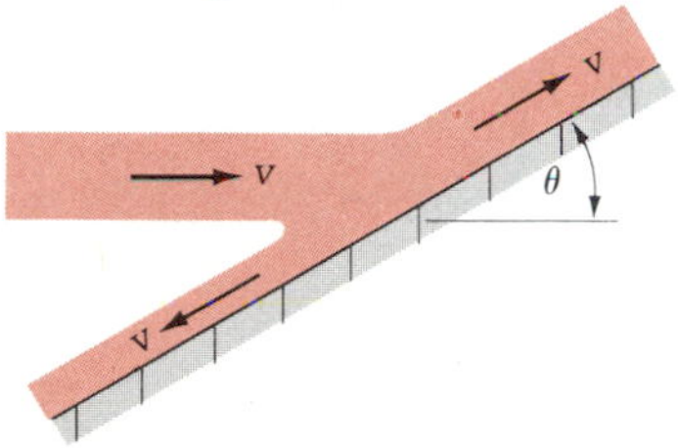

Figure P8.37

8.38 Air flows into the intake of a jet engine at mass flow rate q (slug / sec or kg / s). If v is the speed of the airplane flying through still air and u is the speed of engine exhaust relative to the plane, derive an expression for the force (thrust) of the flowing fluid on the engine. Neglect the fact that the rate of exhaust is slightly greater than q because of the addition of fuel in the engine.

8.39 Revise the analysis of the preceding problem to account for the mass of fuel injected into the engine. Let f be the mass flow rate of the fuel, and assume that the fuel is injected with no velocity relative to the engine housing.

8.40 In a quarry, rocks slide onto a conveyor belt at the constant mass flow rate k, and at speed v_{rel} relative to the ground. (See Figure P8.40.) The belt is driven by a motor (with torque M applied to the drum on the right) at constant speed v_B. Find the power that the motor must deliver, neglecting friction in the shaft bearings and assuming the belt does not slip. *Hint:* Use the control volume indicated by the dashed lines to compute the difference in belt tensions, neglecting any sag of the belt due to the weight of the rocks.

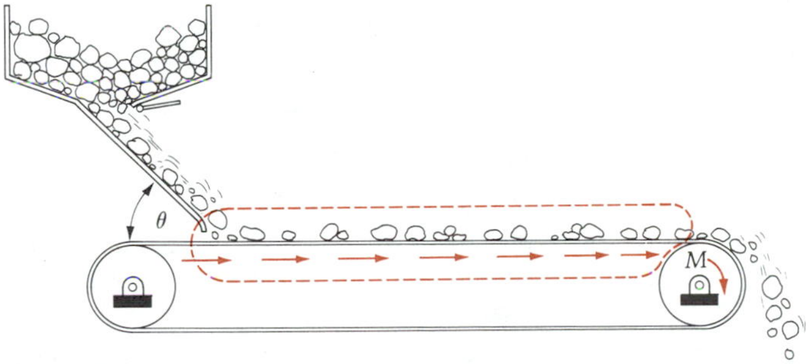

Figure P8.40

8.41 Sand is being dumped on a flatcar of mass M at the constant mass flow rate of q. (See Figure P8.41.) The car is being pulled by a constant force P, and friction is negligible. The car was at rest at $t = 0$. Determine the car's acceleration as a function of P, M, q, and t.

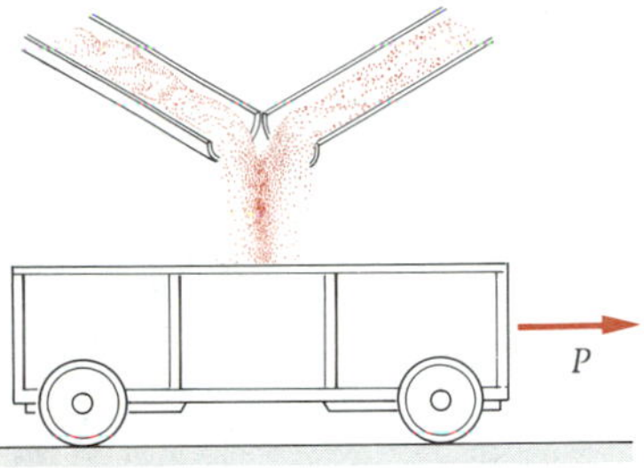

Figure P8.41

8.42 The pressure in a 90° bend of a water pipe is 2 psi (gauge). The inside diameter of the pipe is 6 in. and water flows steadily at the rate of 1.5 ft^3 / sec. Find the magnitude of the force exerted on the bend by the fluid. The specific weight of water is 62.4 lb / ft^3.

8.43 The reducing section in Figure P8.43 connects a 36-in. inside-diameter pipe to a 24-in. inside-diameter pipe. Water enters the reducer at 10 ft / sec and 5 psi (gauge) and leaves at 2 psi. Find the force exerted on the reducer by the steadily flowing water.

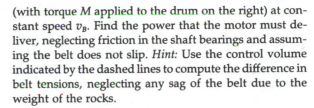

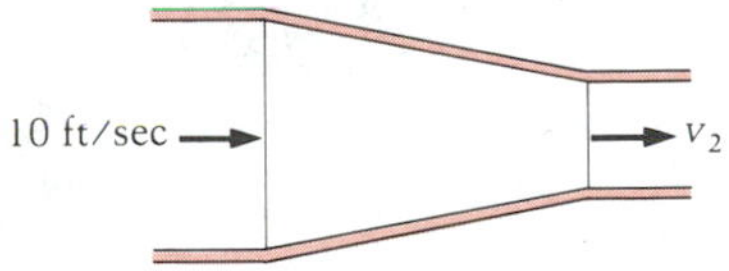

Figure P8.43

8.44 If the plane of Figure P8.44 is horizontal, find the force and moment at O that will allow the body $\mathcal{B}$ to remain in equilibrium when the open stream of water impinges steadily on it as shown. The stream's velocity is 60 ft/sec, and its constant area is 12 in.^{2}; its specific weight is 62.4 lb/ft^3.

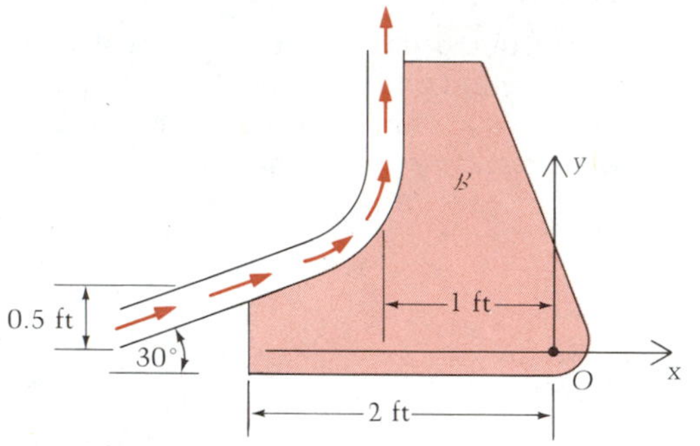

Figure P8.44

8.45 A coal truck weighs 5 tons when empty. It is pushed under a loading chute by a constant force of 500 lb. The chute, inclined at 60° as shown in Figure P8.45, delivers 100 lb of coal per second to the truck at a velocity of 30 ft/sec. When the truck contains 10 tons of coal, its velocity is 10 ft/sec to the right. (a) What is its acceleration at this instant? The wheels are light and all horizontal frictional forces may be taken as included in the 500-lb force. (b) What is the horizontal component of force on the truck from the coal at this instant?

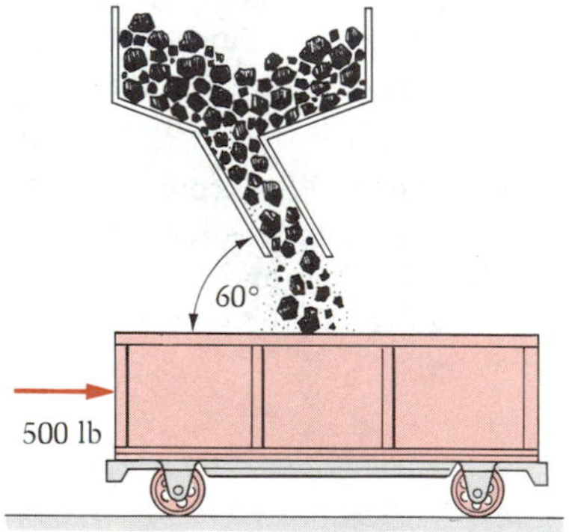

Figure P8.45

8.46 Bonnie and Clyde are making a getaway in a cart with negligible friction beneath its wheels. (See Figure P8.46.) Clyde is killing two birds with one stone by using his machine gun to propel the car as well as to ward off pursuers. He fires 500 rounds (shots) per minute with each bullet weighing 1 oz and exiting the muzzle with a speed relative to the car of 2500 ft/sec. The bullets originally comprised 2 percent of an initial total mass of $m_0 = 20$ slugs. If the system starts from rest at $t = 0$, find: (a) the maximum speed of Bonnie and Clyde; (b) how long it takes to attain this speed.

Figure P8.46

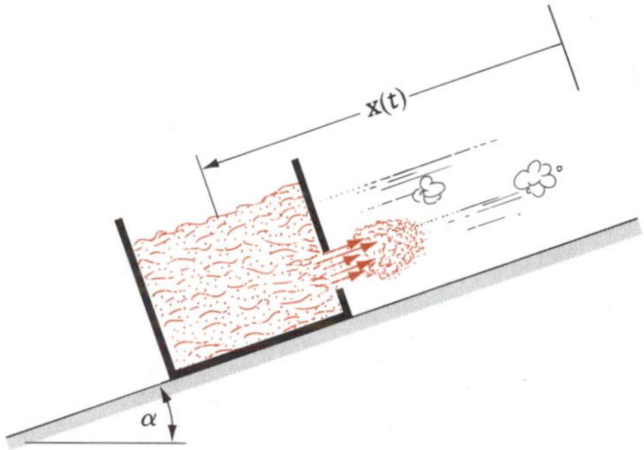

Figure P8.47

8.47 A black box with an initial mass of m_0 (of which 10 percent is box and 90 percent is fuel) is released from rest on the inclined plane in Figure P8.47. The coefficient of friction is μ between the box and the plane.

a. Show that with $\tan\alpha > \mu$, the box will begin to slide down the plane.
b. Assume now that $\tan\alpha > \mu$ and that a mechanism in the box is able to sense its velocity and eject particles rearward (up the plane) at a constant mass flow rate of k_0, and at a relative velocity always equal to the negative of the velocity of the box. Find the velocity of the box at the time t_f when the last of the fuel leaves.
c. Show that the box is going 5.5 times faster at $t = t_f$ than it would have gone if no fuel had been ejected.

8.48 Santa Claus weighs 450 lb and drops down a 20-ft chimney (Figure P8.48). He gains mass in the form of ashes and soot at the rate of 3 slugs / sec from a *very* dirty chimney.

a. Find Santa's velocity as a function of time. (Neglect friction.)

b. Calculate the velocity v_b and the time t_b at which he would hit bottom *without* adding mass and then compare v_b with his "ashes and soot velocity" at the same t_b.

Figure P8.48

8.49 A small rocket is fired vertically upward. Air resistance is neglected. Show that for the rocket to have constant acceleration upward, its mass m must vary with time t according to the equation

$$\frac{dm}{dt} = -\frac{a+g}{u}m$$

where a is the acceleration of the rocket and u is the velocity of the escaping gas relative to the rocket.

8.50 The end of a chain of length L and weight per unit length w, which is piled on a platform, is lifted vertically by a variable force P so that it has a constant velocity v. (See Figure P8.50.) Find P as a function of x. *Hint*: Choose a control-volume boundary so that material crosses the boundary (with negligible velocity) just *before* it is acted on by the moving material already in the control volume. That is, there is no force transmitted across the boundary of the control volume. The solution will be an approximation to reality because of assuming arbitrarily small individual links; but the more links having the common velocity of the fully engaged links within the volume, the better the approximation will be.

8.51 Solve the final equation of Example 8.7 for the velocity $v\hat{\jmath}$ of the rocket as a function of time in the case in which the pressure force pA and the drag D are negligible, the initial mass of the rocket is m_0, and the gravitational acceleration g, the rate μ, and the relative velocity v_e are all constants. Initially, the rocket is at rest.

8.52 Repeat the preceding problem, but this time include a drag force of $-kv$, where k is a positive constant.

8.53 (a) Extending Problem 8.51, find the height of the rocket as a function of time. (b) If the fraction of m_0 which is fuel is f, find the rocket's "burnout" velocity and position when all the fuel is spent.

8.54 Spherical raindrops produced by condensation are precipitated form a cloud when their radius is a. They fall freely from rest, and their radii increase by accretion of moisture at a uniform rate k. Find the velocity of a raindrop at time t, and show that the distance fallen in that time is

$$\frac{gt^2}{8}\left(\frac{2a+kt}{a+kt}\right)^2$$

* **8.55** A chain of length L weighing γ per unit length begins to fall through a hole in a ceiling. (See Figure P8.55.) Referring to the hint in Problem 8.50:

a. Find $v(x)$ if $v = 0$ when x and t are zero.

b. Show that the acceleration of the end of the falling chain is the constant $g/3$.

c. Show that when the last link has left the ceiling, the chain has lost more potential energy than it has gained in kinetic energy, the difference being $\gamma L^2/6$. Give the reason for this loss.

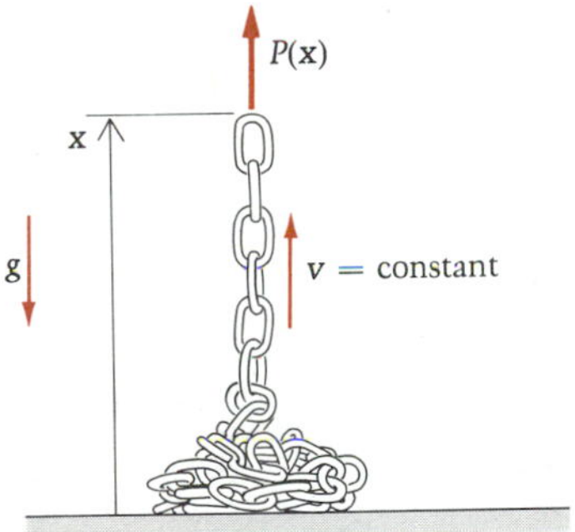

Figure P8.50

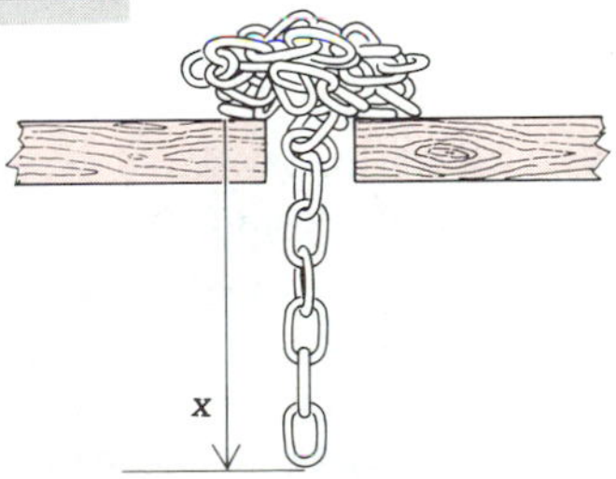

Figure P8.55

8.56 The machine gun in Figure P8.56 has mass M exclusive of its bullets, which have mass M' in total. The bullets are fired at the mass rate of K_0 "slugs" per second, with velocity u_0 relative to the ground. If the coefficient of friction between the gun's frame and the ground is μ, find the velocity of the gun at the instant the last bullet is fired.

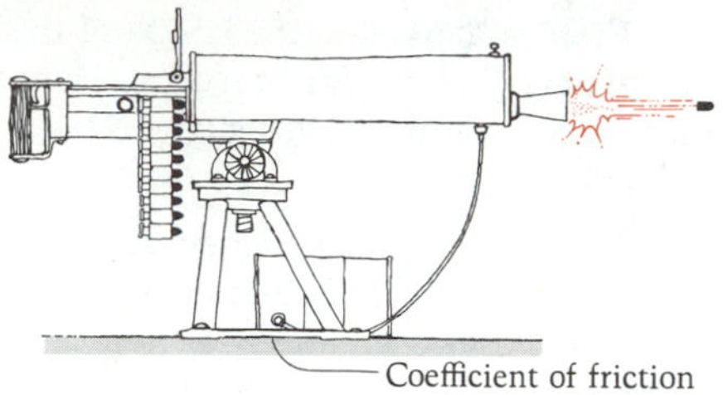

Figure P8.56

* **8.57** A particle of mass m, initially at rest, is projected with velocity $\mathbf{v}_0$ at an angle α to the horizontal and moves under gravity. (See Figure P8.57.) During its flight, it gains mass at the uniform rate k. If air resistance is neglected, show that its equation of motion is

$$(m + kt)\ddot{\mathbf{r}} + k\dot{\mathbf{r}} = (m + kt)g\hat{\mathbf{k}}$$

and that the equation of its path is

$$\mathbf{r} = \frac{m^2}{4k^2}\left[\left(1 + \frac{kt}{m}\right)^2 - 1 - 2\log\left(1 + \frac{kt}{m}\right)\right]g\hat{\mathbf{k}} + \frac{m}{k}\log\left(1 + \frac{kt}{m}\right)\mathbf{v}_0$$

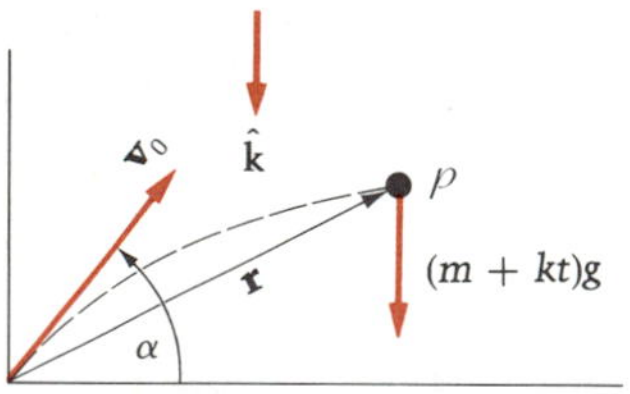

Figure P8.57

* **8.58** If in the preceding problem the air offers a resistance $-c\dot{\mathbf{r}}$, determine the equation of the path.

* **8.59** From a rocket that is free to move vertically upward, matter is ejected downward with a constant relative velocity gT at a constant rate $2M/T$. Initially the rocket is at rest and has mass $2M$, half of which is available for ejection. Neglecting air resistance and variations in the gravitational attraction, (a) show that the greatest upward speed is attained when the mass of the rocket is reduced to M, and determine this speed. (b) Show also that the rocket rises to a height

$$\tfrac{1}{2}gT^2(1 - \ln 2)^2$$

8.60 With the same notation and conditions as in Problem 8.34, show that Equation (8.32) may be written as

$$\Sigma\mathbf{M}_O = (\dot{\mathbf{H}}_O)_V + \frac{dm_o}{dt}(\mathbf{r}_o \times \mathbf{v}_o) - \frac{dm_i}{dt}(\mathbf{r}_i \times \mathbf{v}_i)$$

where $\mathbf{r}_i$ and $\mathbf{r}_o$ are position vectors for the mass centers of the incoming and exiting particles.

* **8.61** A pinwheel of radius a, which can turn freely about a horizontal axis, is initially of mass M and moment of inertia I about its center. A charge is spread along the rim and ignited at time $t = 0$. While the charge is burning, the rim of the wheel loses mass at a constant rate m_1 mass units per second, and at the rim a mass m_2 of gas is taken up per second from the atmosphere, which is at rest. The total mass $m_1 + m_2$ is discharged per second tangentially from the rim, with velocity v relative to the rim. Prove that if θ is the angle through which the wheel has turned after t sec, then

$$\theta = \frac{v}{a(\mu - \lambda)}\left[\mu t - 1 + (1 - \lambda t)^{\mu/\lambda}\right]$$

where

$$\lambda = \frac{m_1 a^2}{I} \qquad \mu = \frac{(m_1 + m_2)a^2}{I}$$

* **8.62** A wheel of radius a starts from rest and fires out matter at a uniform rate from all points on the rim (Figure P8.62). The matter leaves tangentially with relative speed v and at such a rate that the mass decreases at the rim by m mass units per second. Show that the angle θ turned through by the wheel is given by

$$\theta = \frac{vI_0}{ma^3}\left[\left(1 - \frac{ma^2t}{I_0}\right)\ln\left(1 - \frac{ma^2t}{I_0}\right) + \frac{ma^2t}{I_0}\right]$$

in which I_0 is the initial moment of inertia of the wheel about its axis.

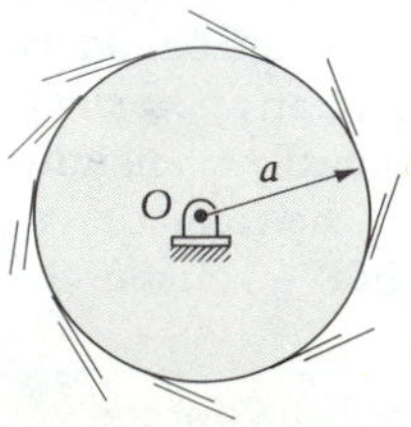

Figure P8.62

8.4 Central Force Motion

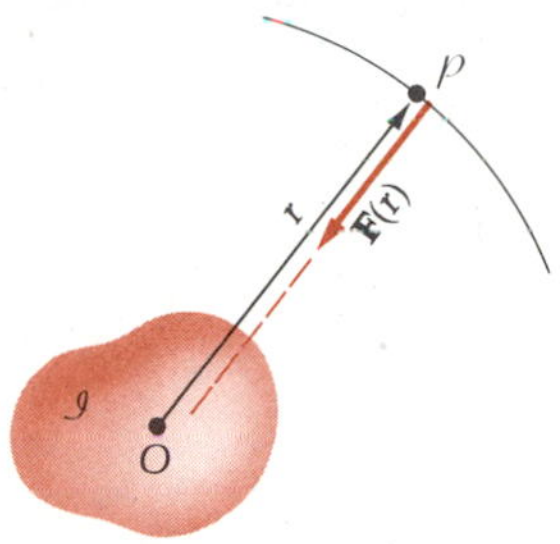

Figure 8.12

In Chapter 2 we defined a **central force** acting on a particle $\mathcal{P}$ as one which (1) always passes through a certain point O fixed in the inertial reference frame $\mathcal{I}$ and (2) depends only on the distance r between O and $\mathcal{P}$. (See Figure 8.12.) In this section we are going to treat the central force in more detail. We shall go as far as we can without specializing $\mathbf{F}(r)$ — that is, without saying how $\mathbf{F}$ depends on r. In the second part of the section we shall study the most important of central forces: gravitational attraction.

If the central force $\mathbf{F}$ is the only force acting on the particle, then $\mathbf{F} = m\mathbf{a}$; and since the central force always passes through point O, $\mathbf{r} \times \mathbf{F}$ is identically zero. These two facts allow us to write:

$$\mathbf{r} \times \mathbf{F} = \mathbf{r} \times m\mathbf{a} = \mathbf{0}$$

or, since $\dot{\mathbf{r}} = \mathbf{v}$,

$$\frac{d}{dt}(\mathbf{r} \times m\mathbf{v}) = \mathbf{0}$$

Therefore for a particle acted on only by a central force,

$$\mathbf{r} \times \mathbf{v} = \text{constant vector in } \mathcal{I} = \mathbf{h}_O \tag{8.33}$$

Dotting this equation with $\mathbf{r}$, we find, since $\mathbf{r} \times \mathbf{v}$ is perpendicular to $\mathbf{r}$,

$$\mathbf{r} \cdot (\mathbf{r} \times \mathbf{v}) = 0 = \mathbf{r} \cdot \mathbf{h}_O$$

and we see that $\mathbf{r}$ is always perpendicular to a vector that is constant in $\mathcal{I}$; therefore $\mathcal{P}$ moves in a plane in $\mathcal{I}$. Using polar coordinates to then describe the motion of $\mathcal{P}$ in this plane, the governing equations are:

$$\Sigma F_r = -F(r) = m(\ddot{r} - r\dot{\theta}^2) \tag{8.34}$$

and

$$\Sigma F_\theta = 0 = m(r\ddot{\theta} + 2\dot{r}\dot{\theta}) = \frac{m}{r}\frac{d}{dt}(r^2\dot{\theta}) \tag{8.35}$$

From Equation (8.35) we see immediately that

$$r^2\dot{\theta} = \text{constant} = h_O \tag{8.36}$$

where h_O is the magnitude of the constant vector $\mathbf{h}_O$ of Equation (8.33) because, expressing $\mathbf{r}$ and $\mathbf{v}$ in polar coordinates, we find

$$\mathbf{h}_O = \mathbf{r} \times \mathbf{v} = \text{constant} = r\hat{\mathbf{e}}_r \times (\dot{r}\hat{\mathbf{e}}_r + r\dot{\theta}\hat{\mathbf{e}}_\theta) = r^2\dot{\theta}\hat{\mathbf{k}}$$

so that

$$|\mathbf{h}_O| = h_O = r^2\dot{\theta} = \text{constant} \tag{8.37}$$

Equation (8.37) is a statement of the conservation of the moment of momentum, or angular momentum of $\mathcal{P}$; the constant h_O is the magnitude of the angular momentum $\mathbf{H}_O$ of $\mathcal{P}$ divided by its mass m. Thus we shall call h_O the angular momentum (magnitude) per unit mass.

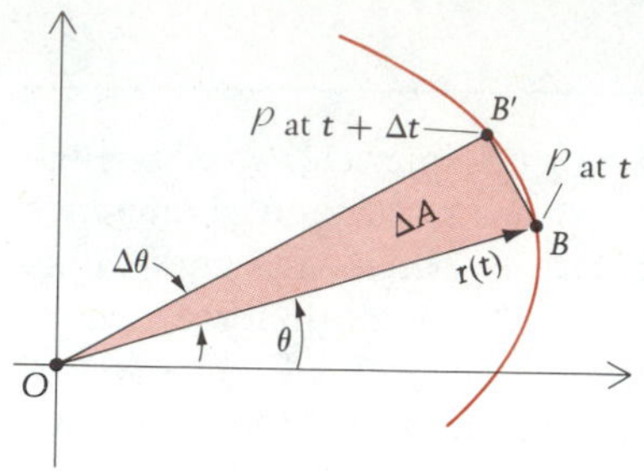

Figure 8.13

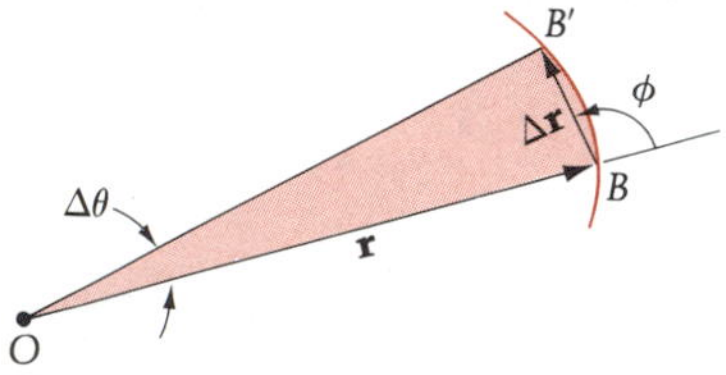

Figure 8.14

We can use the previous pair of results to show that the second of Kepler's three laws of planetary motion is in fact valid for any central force. This law states that the radius vector from the sun to a planet sweeps out equal areas in equal time intervals. From Figure 8.13 the incremental planar area ΔA swept out by $\mathcal{P}$ between θ (at t) and $\theta + \Delta\theta$ (at $t + \Delta t$) is approximately given by the area of the triangle OBB'* (see Figure 8.14):

$$\Delta A \approx \tfrac{1}{2} \times \text{base} \times \text{height}$$
$$\approx \tfrac{1}{2} r(\Delta r \sin \phi) \approx \tfrac{1}{2} |\mathbf{r} \times \Delta\mathbf{r}|$$

Dividing by the time increment Δt and taking the limit as $\Delta t \to 0$, we have

$$\lim \frac{\Delta A}{\Delta t} = \frac{dA}{dt} = \frac{1}{2} \lim_{\Delta t \to 0} \left| \mathbf{r} \times \frac{\Delta \mathbf{r}}{\Delta t} \right| = \frac{1}{2} \overbrace{|\mathbf{r} \times \mathbf{v}|}^{\mathbf{h}_O}$$

or

$$\frac{dA}{dt} = \frac{h_O}{2} \qquad \text{(a constant that is } r^2\dot{\theta}/2\text{)} \tag{8.38}$$

Thus the rate of sweeping out area is a constant. This is why a satellite or a planet in elliptical orbit (Figure 8.15) has to travel faster when it is near the *perigee* than the *apogee*—the same area must be swept out in the same period of time.† We emphasize again that this result is valid for *all* central force trajectories, not just elliptical orbits and not just if the central force is gravity.

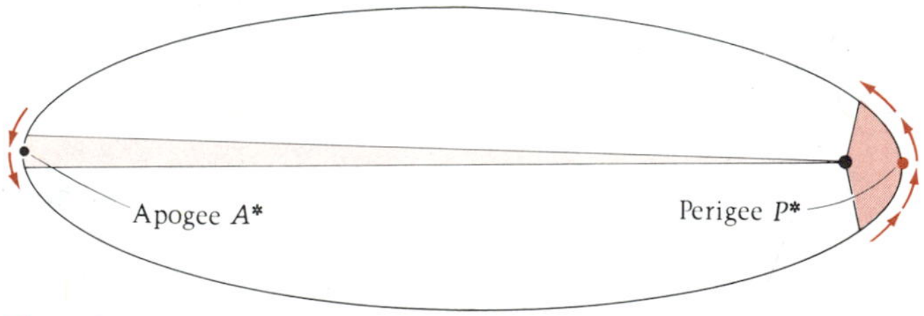

Figure 8.15

Next we focus our attention on the most important central force: gravitational attraction. If G is the universal gravitation constant and M and m are the masses of what we are considering to be the attracting and attracted bodies,‡ then the central force acting on m for this case is

* "Approximately" because the area between arc and chord is outside the triangle.

† We use *perigee* and *apogee* in a general sense; technically these words refer to the nearest and farthest points, respectively, for the moon and artificial satellites. For the orbits of planets, the proper terms are *perihelion* and *aphelion*.

‡ Actually, of course, both are attracting and both are attracted — each to the other! The constant GM is, for the sun, 4.68×10^{21} ft³/sec².

$$F(r) = \frac{GMm}{r^2} \tag{8.39}$$

and Equation (8.34) becomes

$$m(\ddot{r} - r\dot{\theta}^2) = -\frac{GMm}{r^2} \tag{8.40}$$

Canceling m and inserting h_O for $r^2\dot{\theta}$ gives

$$\ddot{r} - \frac{h_O^2}{r^3} = -\frac{GM}{r^2} \tag{8.41}$$

Multiplying Equation (8.41) by $\dot{r}$ will allow us to integrate it:

$$\dot{r}\ddot{r} - h_O^2 r^{-3}\dot{r} = -GMr^{-2}\dot{r} \tag{8.42}$$

Integrating, we get

$$\frac{\dot{r}^2}{2} + h_O^2\frac{r^{-2}}{2} = \frac{-GMr^{-1}}{-1} + C_1 \tag{8.43}$$

If we multiply Equation (8.43) by m and replace h_O by $r^2\dot{\theta}$, we see that

$$\frac{m}{2}[\dot{r}^2 + (r\dot{\theta})^2] - \frac{GMm}{r} = C_1 m \tag{8.44}$$

and the left side of Equation (8.44) is seen to be the total energy of $\mathcal{P}$, kinetic plus potential. Thus we shall replace C_1 by E, the energy of $\mathcal{P}$ per unit mass, and obtain

$$\dot{r}^2 + h_O^2 r^{-2} = +2GMr^{-1} + 2E \tag{8.45}$$

This equation will be helpful to us later. But now we are interested in studying the trajectory of particle $\mathcal{P}$ — that is, in finding r as a function of θ. By the chain rule,

$$\frac{dr}{dt} = \frac{dr}{d\theta}\frac{d\theta}{dt} = \dot{\theta}\frac{dr}{d\theta}$$

and since $\dot{\theta} = h_O/r^2$ from Equation (8.36),

$$\frac{dr}{dt} = \frac{h_O}{r^2}\frac{dr}{d\theta} \tag{8.46}$$

We need the second derivative of r in Equation (8.41), so we apply the chain rule once more:

$$\begin{aligned}\frac{d^2r}{dt^2} &= \left[\frac{d}{d\theta}\left(\frac{h_O}{r^2}\frac{dr}{d\theta}\right)\right]\frac{d\theta}{dt} \\ &= \frac{h_O}{r^2}\left[-\frac{2h_O}{r^3}\left(\frac{dr}{d\theta}\right)^2 + \frac{h_O}{r^2}\frac{d^2r}{d\theta^2}\right] \\ &= \frac{h_O^2}{r^4}\frac{d^2r}{d\theta^2} - \frac{2h_O^2}{r^5}\left(\frac{dr}{d\theta}\right)^2 = \frac{h_O^2}{r^2}\frac{d}{d\theta}\left(\frac{1}{r^2}\frac{dr}{d\theta}\right)\end{aligned} \tag{8.47}$$

Substituting into Equation (8.41), we get

$$\frac{h_O^2}{r^2}\frac{d}{d\theta}\left(\frac{1}{r^2}\frac{dr}{d\theta}\right) - \frac{h_O^2}{r^3} = -\frac{GM}{r^2}$$

or

$$\frac{d}{d\theta}\left(\frac{1}{r^2}\frac{dr}{d\theta}\right) - \frac{1}{r} = -\frac{GM}{h_O^2} \tag{8.48}$$

The following simple change of variables will make the solution to this differential equation immediately recognizable:

$$u = \frac{1}{r} \tag{8.49}$$

Substituting Equation (8.49) into (8.48) along with

$$\frac{dr}{d\theta} = \frac{dr}{du}\frac{du}{d\theta} = \frac{-1}{u^2}\frac{du}{d\theta} \tag{8.50}$$

gives

$$\frac{d}{d\theta}\left[u^2\left(\frac{-1}{u^2}\frac{du}{d\theta}\right)\right] - u = \frac{-GM}{h_O^2} \tag{8.51}$$

or

$$\frac{d^2u}{d\theta^2} + u = \frac{GM}{h_O^2} \tag{8.52}$$

The solution to Equation (8.52), from elementary differential equations, consists of a homogeneous (or complementary) part plus a particular part:

$$u = \overbrace{A_1 \cos\theta + B_1 \sin\theta}^{u_H} + \overbrace{\frac{GM}{h_O^2}}^{u_P} \tag{8.53}$$

Switching variables back from u to r by Equation (8.49), we obtain

$$r = \frac{h_O^2/GM}{1 + (h_O^2/GM)(A_1\cos\theta + B_1\sin\theta)} \tag{8.54}$$

This solution for $r(\theta)$ is the equation of a **conic**; it can be put into a more recognizable form after a brief review of conic sections. For every point P on a conic, the ratio of the distances from P to a fixed point (O: the focus) and to a fixed line (ℓ: the directrix) is a constant called the **eccentricity** of the conic:

$$e = \frac{OP}{L} \tag{8.55}$$

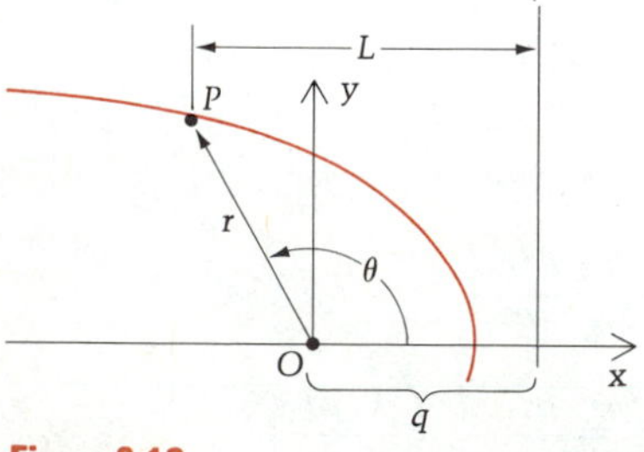

Figure 8.16

Therefore, in terms of the parameters in Figure 8.16,

$$e = \frac{r}{q - r\cos\theta} \tag{8.56}$$

or, solving for r,

$$r = \frac{eq}{1 + e \cos \theta} \tag{8.57}$$

The conic specified by Equation (8.57) is a:

$$\begin{array}{l} \text{Hyperbola if } |e| > 1 \\ \text{Parabola if } e = 1 \\ \text{Ellipse if } -1 < e < 1 \end{array} \tag{8.58}$$

The ellipse becomes a circle if $e = 0$. It is an ellipse with perigee (closest point to O) at $\theta = 0$ if $0 < e < 1$ and an ellipse with apogee (farthest point from O) at $\theta = 0$ if $-1 < e < 0$; this latter type is called a subcircular ellipse.

Returning to our solution (8.54) for $r(\theta)$, it is customary to select one of the constants A_1 and B_1 so that, as suggested by Figure 8.16, $dr/d\theta = 0$ when $\theta = 0$. This condition easily gives $B_1 = 0$, as the reader may wish to demonstrate using calculus. The result simply means that we are measuring θ from the perigee of the conic. At this point we should compare Equations (8.57) and (8.54) with $B_1 = 0$:

$$r = \frac{h_O^2/GM}{1 + (h_O^2/GM)A_1 \cos \theta} \tag{8.59}$$

and

$$r = \frac{eq}{1 + e \cos \theta} \tag{8.60}$$

By direct comparison of these two expressions for r, we see that

$$A_1 = \frac{eGM}{h_O^2} \qquad \text{and} \qquad eq = \frac{h_O^2}{GM}$$

It is more customary, however, to express the constant A_1 (as well as the eccentricity) in terms of the energy E of the orbit. To do this, Equations (8.45) and (8.46) give

$$\frac{h_O^2}{r^4}\left(\frac{dr}{d\theta}\right)^2 + \frac{h_O^2}{r^2} - \frac{2GM}{r} = 2E \tag{8.61}$$

At the point r_P where $\theta = 0$ and $dr/d\theta = 0$, we see that

$$\frac{h_O^2}{r_P^2} - \frac{2GM}{r_P} = 2E \tag{8.62}$$

Thus not all of h_O, r_P, and E are independent. We shall eliminate r_P. Multiplying Equation (8.62) by r_P^2, we get

$$2Er_P^2 + 2GMr_P - h_O^2 = 0 \tag{8.63}$$

Solving via the quadratic formula, we have

$$r_P = \frac{-2GM + \sqrt{4G^2M^2 + 8Eh_O^2}}{4E} \tag{8.64}$$

in which we use the plus sign since we need the smaller root for closed conics ($E < 0$). The positive sign also ensures a positive r_P for open conics ($E > 0$).

Returning to our solution (8.59), when $\theta = 0$ then

$$r_P = \frac{h_O^2/GM}{1 + (h_O^2/GM)A_1} \tag{8.65}$$

Equating the two expressions for r_P, Equations (8.64) and (8.65), we can solve for A_1.

We see by comparing Equations (8.59) and (8.60) that the eccentricity e of our conic will be $(h_O^2/GM)A_1$. Equating the right sides of Equations (8.64) and (8.65) and solving for this quantity, we get

$$\frac{h_O^2}{GM}A_1 = e = \sqrt{1 + \frac{2Eh_O^2}{G^2M^2}} \tag{8.66}$$

Therefore

$$r = \frac{h_O^2/GM}{1 + \sqrt{1 + (2Eh_O^2/G^2M^2)}\cos\theta} \tag{8.67}$$

which expresses r as a function of θ, the constant GM, the energy E, and the angular momentum per unit mass h_O. Note that by again comparing Equations (8.59) and (8.60) we can obtain the distance q between the focus O and the directrix ℓ:

$$\frac{h_O^2}{GM} = eq \Rightarrow q = \frac{h_O^2/GM}{\sqrt{1 + 2Eh_O^2/G^2M^2}} \tag{8.68}$$

The first of **Kepler's three laws of planetary motion** states that the planets travel in elliptical orbits with the sun at one focus.* These ellipses are very nearly circular for most of the planets; the eccentricity of earth is $e = 0.017$. To obtain the third of Kepler's laws, we return once more to our equations and obtain for elliptic orbits, from (8.67), the distance r when $\theta = 90°$:

$$r_{90} = \frac{h_O^2/GM}{1 + 0} = \ell \tag{8.69}$$

This distance, the *semilatus rectum*, may be used to express the distance r_A between the focus O and apogee A^*, and the distance r_P between O and the perigee P^*. (See Figure 8.17.) At apogee, $\theta = \pi$ and Equations (8.59), (8.60), and (8.69) give

$$r_A = \frac{h_O^2/GM}{1 - e} = \frac{\ell}{1 - e} \tag{8.70}$$

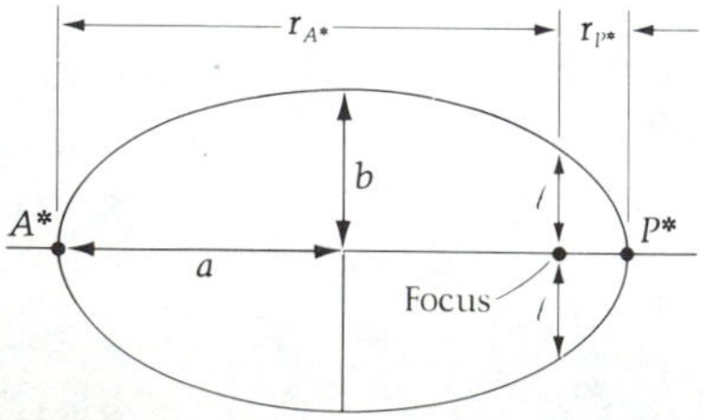

Figure 8.17

* Kepler's laws, based on his astronomical observations and set forth in 1609 and 1619, were studied by Newton before the Englishman published the *Principia*, which contained his own laws of motion.

and at perigee ($\theta = 0$),

$$r_{P*} = \frac{\ell}{1+e} \tag{8.71}$$

The semimajor axis length of the ellipse is

$$a = \frac{r_{A*} + r_{P*}}{2} = \frac{\ell}{1-e^2} \tag{8.72}$$

and the semiminor axis length is, from analytic geometry,

$$b = a\sqrt{1-e^2} = \frac{\ell}{\sqrt{1-e^2}} \tag{8.73}$$

An ellipse has area

$$A_T = \pi ab = \pi \left(\frac{\ell}{1-e^2}\right)\left(\frac{\ell}{\sqrt{1-e^2}}\right)$$

or

$$A_T = \frac{\pi \ell^2}{(1-e^2)^{3/2}} = \pi a^2\sqrt{1-e^2} \tag{8.74}$$

With these results in hand, we shall now prove Kepler's third law. Since dA/dt is constant,

$$\frac{dA}{dt} = \frac{h_O}{2} \Rightarrow A = \frac{h_O t}{2} \tag{8.75}$$

where we take $A = 0$ when $t = 0$, say at the perigee. Over one orbit we have, with T being the orbit period,

$$\begin{aligned} A_T &= \text{area of ellipse} \\ &= \pi a^2\sqrt{1-e^2} = \frac{h_O T}{2} \end{aligned} \tag{8.76}$$

Since (from Equations (8.69) and (8.72))

$$h_O = \sqrt{GM\ell} = \sqrt{GMa(1-e^2)} \tag{8.77}$$

we obtain the following from Equation (8.76):

$$\pi a^2\sqrt{1-e^2} = \frac{\sqrt{GMa(1-e^2)}}{2}\,T \tag{8.78}$$

so that

$$T = \frac{2\pi a^{3/2}}{\sqrt{GM}}$$

or

$$T^2 = \frac{4\pi^2}{GM}\,a^3 \tag{8.79}$$

Equation (8.79) states the third of Kepler's laws: The squares of the planets' orbital periods are proportional to the cubes of the semimajor axes of their orbits.

EXAMPLE 8.8

Calculate the semimajor axis length of an earth satellite with a period of 90 min.

Solution

We can solve this problem using Kepler's third law. The weight of a particle (mass m) on the earth's (mass M) surface is both mg and GMm/r_e^2; thus

$$mg = \frac{GMm}{r_e^2} \Rightarrow GM = gr_e^2$$

and we see that the product of the unwieldy constants G and M is

$$GM = gr_e^2 = \frac{32.2}{5280}(3960)^2 = 95{,}600 \text{ mi}^3/\text{sec}^2$$

Therefore

$$T^2 = (90 \times 60)^2 = \frac{4\pi^2}{95{,}600}a^3$$

$$a = 4130 \text{ mi.}$$

which is about 170 mi above the earth's surface.

We shall present one more example on elliptical orbits under gravity, but first we need two equations relating the velocities v_1 and v_2 at any two points P_1 and P_2 with radii r_1 and r_2 on the orbit. The first of these comes from Equation (8.37), which states that $r^2\dot{\theta} = \text{constant}$. From Figure 8.18, since the velocity $\mathbf{v}$ is always tangent to the path, we see that if ϕ is the angle between $\mathbf{r}$ and $\mathbf{v}$, then in cylindrical coordinates

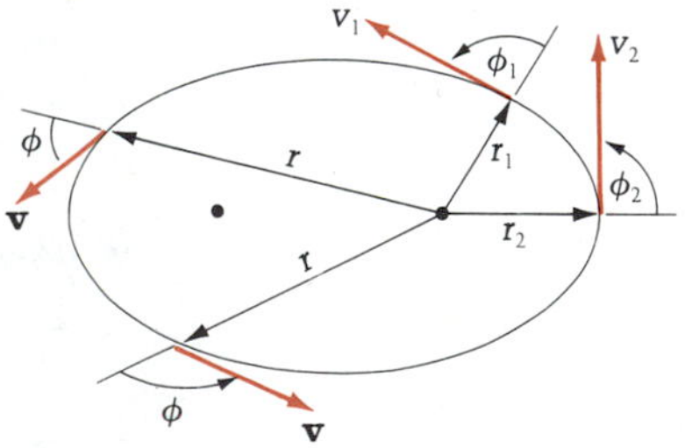

Figure 8.18

$$v \sin\phi = (\text{transverse component of } \mathbf{v}) = r\dot{\theta}$$

so that

$$r(r\dot{\theta}) = rv\sin\phi = \text{constant}$$

or, for two points P_1 and P_2 on the orbital path,

$$r_1 v_1 \sin\phi_1 = r_2 v_2 \sin\phi_2 \qquad (8.80)$$

Note that at apogee and perigee, $\phi = 90°$. Thus letting P_1 and P_2 be these two points, we get from Equation (8.80)

$$r_A v_A = r_p v_p \qquad (8.81)$$

and the two velocities are inversely proportional to the radii, with v being faster at perigee as we have already seen from Kepler's second law.

The other equation relating v_1 and v_2 comes from the potential for gravity, which from Equation (2.27) and Example 8.8 is

$$\varphi = -\frac{gr_e^2 m}{r} = -\frac{GMm}{r}$$

Using conservation of energy between P_1 and P_2,

$$T_1 + \varphi_1 = T_2 + \varphi_2$$

$$\frac{mv_1^2}{2} - \frac{GMm}{r_1} = \frac{mv_2^2}{2} - \frac{GMm}{r_2}$$

$$v_2^2 - v_1^2 = 2GM\left(\frac{1}{r_2} - \frac{1}{r_1}\right) \tag{8.82}$$

If we let point P_2 represent the perigee P^*, as suggested in Figure 8.18, then Equation (8.80) becomes

$$v_2 = v_{P^*} = \frac{r_1 v_1 \sin\phi_1}{r_{P^*}} \tag{8.83}$$

where $\sin\phi_2 = \sin 90° = 1$. Now if r_1, v_1, and ϕ_1 are initial (launch) values of r, v, and ϕ, then we may consider these as given quantities. Substituting Equation (8.83) into (8.82), we can obtain an equation for the perigee radius $r_2 (= r_{P^*})$:

$$\frac{r_1^2 v_1^2 \sin^2\phi_1}{r_{P^*}^2} - v_1^2 = 2GM\left(\frac{1}{r_{P^*}} - \frac{1}{r_1}\right)$$

Multiplying through by $-r_{P^*}^2/(r_1^2 v_1^2)$ and rearranging, we get

$$\left(\frac{r_{P^*}}{r_1}\right)^2\left(1 - \frac{2GM}{r_1 v_1^2}\right) + \left(\frac{r_{P^*}}{r_1}\right)\frac{2GM}{r_1 v_1^2} - \sin^2\phi_1 = 0 \tag{8.84}$$

We see that Equation (8.84) is simply a quadratic equation in the ratio (r_{P^*}/r_1) and that $[2GM/(r_1 v_1^2)]$ is a nondimensional parameter of the orbit. We now illustrate the use of this important equation in an example.

Example 8.9

A satellite is put into an orbit with the following launch parameters: $H = 1000$ mi, $v_1 = 17{,}000$ mph, and $\phi = 100°$ (see Figure E8.9). Find the apogee and perigee radii of the resulting orbit.

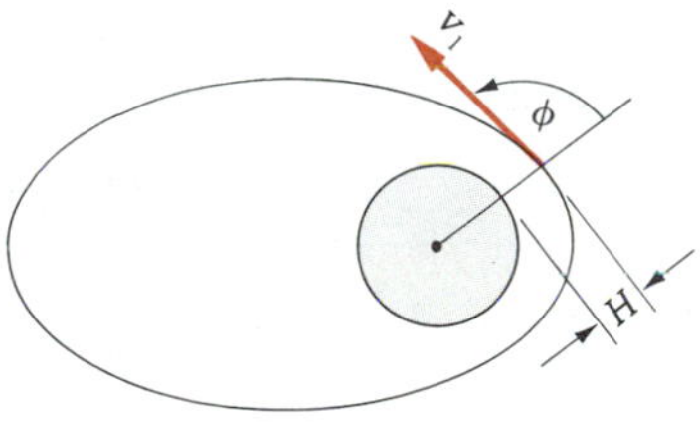

Figure E8.9

Solution

We need GM in mi³/hr²; therefore

$$GM = gr_e^2 = 32.2(3960)^2 \text{ ft-mi}^2/\text{sec}^2$$

$$= 32.2(3960)^2 \times \frac{1}{5280} \times 3600^2 \text{ mi}^3/\text{hr}^2$$

$$= 124 \times 10^{10} \text{ mi}^3/\text{hr}^2$$

The parameter $2GM/(r_1 v_1^2)$ in Equation (8.84) is therefore

$$\frac{2GM}{r_1 v_1^2} = \frac{2(124 \times 10^{10})}{(1000 + 3960)17{,}000^2} = 1.73$$

Equation (8.84) becomes

$$\left(\frac{r_{P^\bullet}}{r_1}\right)^2(-0.73) + 1.73\left(\frac{r_{P^\bullet}}{r_1}\right) - 0.970 = 0$$

The quadratic formula gives

$$\left(\frac{r_{P^\bullet}}{r_1}\right)_{1,2} = \frac{-1.73 \pm \sqrt{1.73^2 - 4(-0.73)(-0.970)}}{2(-0.73)}$$

$$= 0.911 \text{ and } 1.46$$

Therefore

$$r_{P_1} = r_{P^\bullet} = 0.911(4960) = 4520 \text{ mi}$$

The other root corresponds to the apogee. (Since the starting condition of $\sin\phi = 1$ is the same for apogee and perigee, both answers are produced by the quadratic formula!)

$$r_{P_2} = r_{A^\bullet} = 1.46(4960) = 7240 \text{ mi}$$

The altitudes are

$$\text{Perigee height} = 4520 - 3960 = 560 \text{ mi}$$

$$\text{Apogee height} = 7240 - 3960 = 3280 \text{ mi}$$

To pin down the orbit in space, we need to know the angle to the perigee point from the launch point and also the orbit's eccentricity. Problems 8.81 and 8.82 will be concerned with finding these two quantities given initial values of r, v, and ϕ.

PROBLEMS ▸ Section 8.4

8.63 Show that a satellite in orbit has a period T given by

$$T = \frac{2\pi ab}{r_{A^\bullet}v_{A^\bullet}(\text{or } r_{P^\bullet}v_{P^\bullet})}$$

8.64 Show that, for a body in elliptical orbit (Figure P8.64), $b = \sqrt{r_{A^\bullet}r_{P^\bullet}}$.

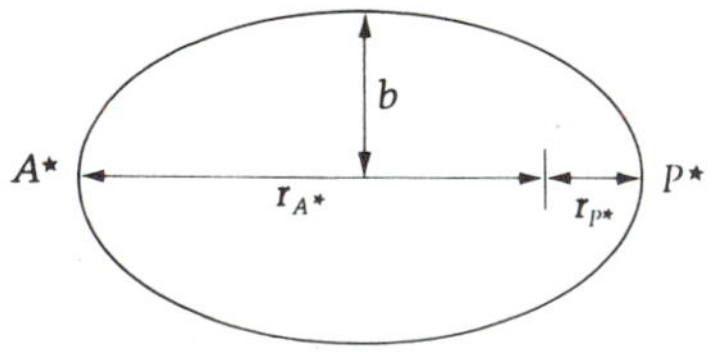

Figure P8.64

8.65 Halley's comet's latest return to Earth was in 1986. The comet orbits the sun in an elongated ellipse every 74 to 79 years; the period varies due to perturbations in its orbit caused by the four largest (Jovian) planets. (Its passages have been recorded since 240 B.C.!) What is the approximate semimajor axis length of Halley's comet? (Use 76 years as the period.)

8.66 Find the minimum period of a satellite in circular orbit about the earth. Upon what assumption is your answer based?

8.67 Repeat the preceding problem if the satellite orbits the moon. Assume

$$g_{\text{moon}} = \tfrac{1}{6}g_{\text{earth}}$$

$$r_{\text{moon}} = 0.27r_{\text{earth}}$$

8.68 Show that for circular orbits around an attracting body of mass M, $rv^2 = GM$. Then use the 93×10^6 mi average orbital radius of earth, and the fact that its orbit is nearly circular, to find the constant GM_s for the sun as attracting center (heliocentric system).

8.69 The first artificial satellite to orbit the earth was the Russians' Sputnik I. Following insertion into orbit it had a period of 96.2 min. Find the semimajor axis length. If the initial eccentricity was 0.0517, find the maximum and minimum distances from earth following its injection into orbit.

8.70 Show that if a satellite is in a circular orbit at radius r around a planet of mass M, the velocity to which it must increase to *escape* the planet's gravitational attraction is given by

$$v_{escape} = \sqrt{\frac{2GM}{r}}$$

Find v_{escape} if "to which" is replaced by "by which."

8.71 Show that if the launch velocity in Example 8.9 is 15,000 mi/hr, the satellite will fail to orbit the earth.

8.72 Using Equation (8.54), show that $B_1 = 0$ follows from the condition $dr/d\theta = 0$ when $\theta = 0$.

* **8.73** Prove that Equation (8.66) follows from (8.64) and (8.65).

8.74 Find the form of the central force $\mathbf{F}(r)$ for which all circular orbits of a particle about an attracting center O have the same angular momentum (and the same rate of sweeping out area).

8.75 Show that, in terms of the radius $r_{P\bullet}$ and speed $v_{P\bullet}$ at perigee, the energy and eccentricity of the orbit may be expressed as

$$\frac{Er_{P\bullet}}{GM} = \frac{r_{P\bullet}v_{P\bullet}^2}{2GM} - 1$$

and

$$e = \frac{r_{P\bullet}v_{P\bullet}^2}{2GM} - 1$$

8.76 Use Equations (8.81) and (8.82) to show that the velocities at apogee and perigee, in terms of the known radii $r_{A\bullet}$ and $r_{P\bullet}$, are:

$$v_{A\bullet} = \sqrt{\frac{2GMr_{P\bullet}}{r_{A\bullet}(r_{A\bullet} + r_{P\bullet})}}$$

and

$$v_{P\bullet} = \sqrt{\frac{2GMr_{A\bullet}}{r_{P\bullet}(r_{A\bullet} + r_{P\bullet})}}$$

* **8.77** A rocket is in a 200-mi-high circular parking orbit above a planet. What velocity boost at point P will result in the new, elliptical orbit shown in Figure P8.77? *Hint:* Use the results of Problem 8.75.

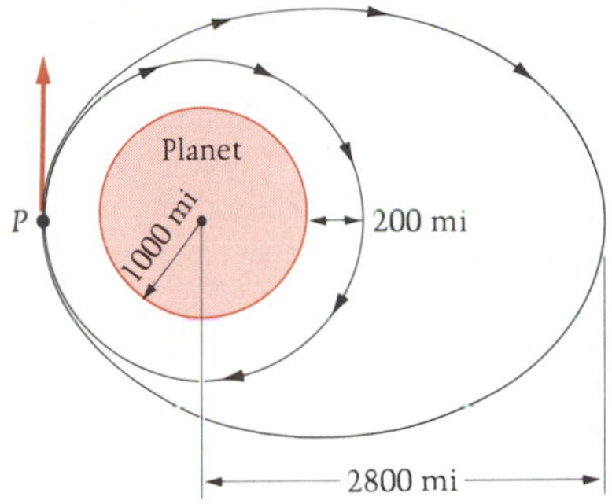

Figure P8.77

* **8.78** A satellite is in a circular orbit of radius R_1. (See Figure P8.78.) Find the (negative) velocity increment that will send the satellite to position A, at radius R_2 ($<R_1$), 180° away. Then find the second negative velocity increment, this time applied at A, that will put the satellite in a circular orbit of radius R_2. *Hint*: Use Problem 8.76.

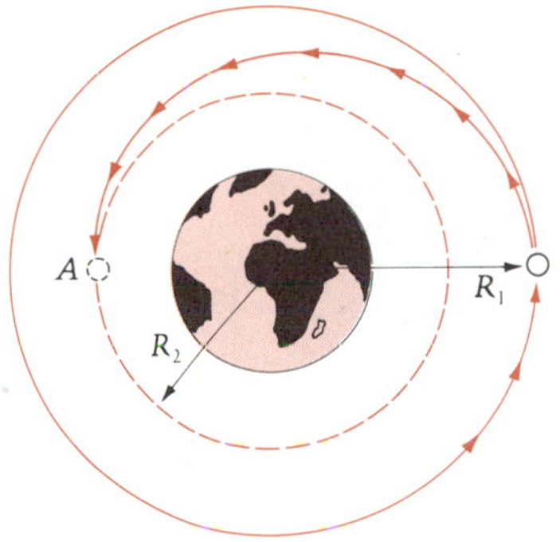

Figure P8.78

8.79 A satellite has $r_{A\bullet} = 8000$ mi and $r_{P\bullet} = 5000$ mi. If it was launched with a velocity of 15,000 mi/hr, what was its launch radius? What was the angle ϕ between $\mathbf{r}$ and $\mathbf{v}$ at launch? *Hint*: Use Problem 8.76.

8.80 What is the period of the satellite in the preceding problem?

* **8.81** Show that if the launch parameters, r_1, v_1, and ϕ_1 are known, then the angle θ from perigee to the launch point of a satellite in orbit is given by

$$\tan\theta = \frac{r_1v_1^2}{GM}\sin\phi_1\cos\phi_1 \Bigg/ \left[\frac{r_1v_1^2}{GM}\sin^2\phi_1 - 1\right]$$

* **8.82** Show that if the launch parameters r_1, v_1, and ϕ_1 are known, then the eccentricity of the resulting conic is:

$$e = \sqrt{\left(\frac{r_1v_1^2}{GM} - 1\right)^2 \sin^2\phi_1 + \cos^2\phi_1}$$

* **8.83** A large meteorite approaches the earth. (See Figure P8.83.) Measurements indicate that at a given time it has a speed of 8000 mph at a radius of 100,000 mi. Will it orbit the earth? If so, what is the period? If not, what is the maximum velocity v that would have resulted in an orbit? *Hint*: Use the result of the preceding problem.

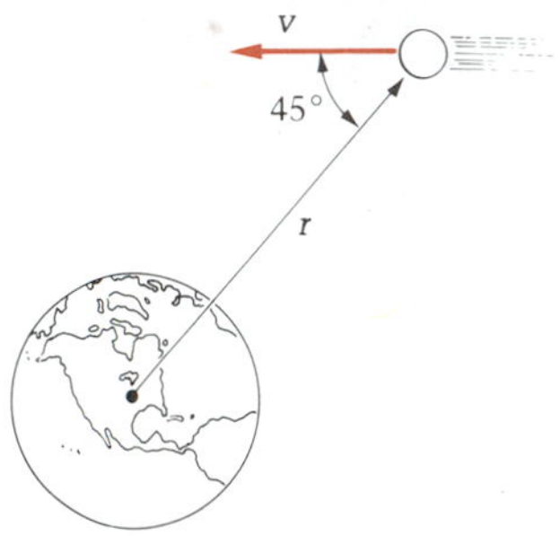

Figure P8.83

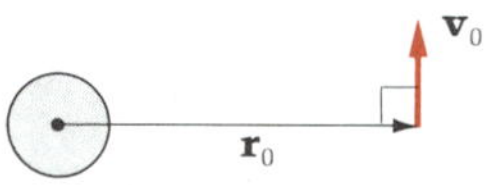

Figure P8.84

8.84 Classify the various orbits according to values of the dimensionless parameter $GM/(r_0v_0^2)$ for a satellite launched with the conditions of Figure P8.84.

* **8.85** Find the kinetic energy increase needed to move a satellite from radius R to $nR(n > 1)$ in circular orbits. *Hint*: Use Problem 8.76.

* **8.86** A particle of mass m moves in the xy plane under the influence of an attractive central force that is proportional to its distance from the origin ($F(r) = kr$). It has the same initial conditions as Problem 8.84. Find the largest and smallest values of r in the ensuing motion.

8.87 A satellite has apogee and perigee points 1000 and 180 mi, respectively, above the earth's surface. Compute the satellite's period.

8.88 In the preceding problem find the speeds of the satellite at perigee and at apogee. *Hint*: Use Problem 8.76.

REVIEW QUESTIONS ▶ Chapter 8

True or False?

1. Frequencies of vibration of periodic motion are associated only with small amplitudes.
2. Natural frequencies of vibration are associated only with translational motion.
3. Natural frequencies of vibration of bodies in a gravitational field do not have to depend on "g."
4. Free vibrations will always decrease in time due to "real world damping."
5. There are three types of damped vibrations, and the subcritical case has the greatest practical importance.
6. In forced vibration, the steady-state part of the response dies out due to damping inherent in the physical system.
7. In general, the rate of accumulation of momentum within a region of space equals the rate at which it is transported into that region.

In using the control volume form of Euler's Laws, the control volume:

8. has to be fixed in the inertial frame of reference;
9. may change in shape or volume with time;
10. has to be a closed region of space.
11. In a one-dimensional control-volume problem, Euler's first law becomes, in general,

$$\Sigma F = \frac{d}{dt}(mv_C) = \dot{m}v_C + m\dot{v}_C$$

12. The law of accumulation, production, and transport applies only to scalar quantities.
13. A central force depends only on the distance *r* between the attracted and attracting particles.
14. There are central forces besides gravity.
15. In every central force problem the angular momentum about the attracting point *O* is a constant.
16. All three of Kepler's laws apply to motions of a particle under the influence of any type of central force.
17. All central force problems result in paths which are conics.
18. In a gravitational central force problem, the type of conic is determined by the eccentricity.

Answers

1. F 2. F 3. T 4. T 5. T 6. F 7. F 8. F 9. T 10. T 11. F 12. F 13. T 14. T 15. T 16. F 17. F 18. T

Appendices Contents

Appendix A	Units
Appendix B	Examples of Numerical Analysis/The Newton-Raphson Method
Appendix C	Moments of Inertia of Masses
Appendix D	Answers to Odd-Numbered Problems
Appendix E	Additional Model-Based Problems for Engineering Mechanics

A UNITS

The numerical value assigned to a physical entity expresses the relationship of that entity to certain standards of measurement called **units**. There is currently an international set of standards called the International System (SI) of Units, a descendant of the meter-kilogram-second (mks) metric system. In the SI system the unit of time is the **second** (s), the unit of length is the **meter** (m), and the unit of mass is the **kilogram** (kg). These independent (or *basic*) units are defined by physical entities or phenomena. The second is defined by the period of a radiation occurring in atomic physics; the meter is defined by the wavelength of a different radiation; the kilogram is defined to be the mass of a certain body of material stored in France. Any other SI units we shall need are *derived* from these three basic units. For instance, the unit of force, the **newton** (N), is a derived quantity in the SI system, as we shall see.

Until recently almost all engineers in the United States used the system (sometimes called British gravitational or U.S.) in which the basic units are the second (sec) for time, the **foot** (ft) for length,* and the **pound** (lb) for force. The pound is the weight, at a standard gravitational condition (location), of a certain body of material stored in the United States. In this system the unit of mass, the **slug**, is a derived quantity. It is a source of some confusion that sometimes there is used a unit of mass called the pound (the mass whose weight is one pound of force at standard gravitational conditions); also, particularly in Europe, the term *kilogram* is also sometimes used for a unit of force.† Grocery shoppers in the United States are exposed to this confusion by the fact that packages are marked by weight (or is it mass?) both in pounds and in kilograms. Throughout this text, without exception, *the pound is a unit of force* and *the kilogram is a unit of mass.*

The United States is currently in the painful process of gradual changeover to the metric system of units after more than 200 years of attachment to the U.S. system. The new engineers who begin practicing their profession in the 1990s will doubtless encounter both systems, and thus it is crucial to master both (including *thinking* in terms of the units of either) and to be able to convert from one to the other. The units mentioned here are summarized in Table A.1 for the SI and the U.S. systems.

---

* Sometimes, particularly in the field of mechanical vibrations, the inch is used as the unit of length; in that case the unit of mass is 1 $\text{lb-sec}^2/\text{in.}$, which equals 12 slugs.

† A kilogram *was* a force unit in one of two mks systems, compounding the misunderstanding.

Table A.1

Quantity	SI (Standard International or "Metric") Unit	U.S. Unit
force	newton (N)	pound (lb)
mass	kilogram (kg)	slug
length	meter (m)	foot (ft)
time	second (s)	second (sec)

We now examine how the newton of force is derived in SI units and the slug of mass is derived in U.S. units. Let the dimensions of the four basic dimensional quantities be labeled as F (force), M (mass), L (length), and T (time). From the first law of motion (discussed in detail in Chapter 2), $\mathbf{F} = m\mathbf{a}$, we observe that the four basic units are always related as follows:

$$F = \frac{ML}{T^2}$$

This means, of course, that we may select three of the units as basic and derive the fourth. Two ways in which this has been done are the *gravitational* and the *absolute* systems. The former describes the U.S. system; the latter describes SI. (See Table A.2.) Therefore, in U.S. units the mass of an object weighing W lb is $W/32.2$ slugs. Similarly, in SI units the weight of an object having a mass of M kg is $9.81M$ newtons.

Table A.2

Gravitational System	Absolute System
The basic units are force, length, and time, and mass is derived: $M = \frac{FT^2}{L}$ This system has traditionally been more popular with engineers. As an example, in the U.S. system of units the pound, foot, and second are basic. Thus the mass unit, the slug, is derived: $1 \text{ slug} = 1 \frac{\text{lb-sec}^2}{\text{ft}}$ This is summed up by: A slug is the quantity of mass that will be accelerated at 1 ft/sec^2 when acted upon by a force of 1 lb.	The basic units are mass, length, and time, and force is derived: $F = \frac{ML}{T^2}$ This system has traditionally been more popular with physicists. As an example, in the SI (metric) system of units the kilogram, meter, and second are basic. Thus the force unit, the newton, is derived: $1 \text{ newton} = 1 \frac{\text{kg} \cdot \text{m}}{\text{s}^2}$ This is summed up by: A newton is the amount of force that will accelerate a mass of 1 kg at 1 m/s^2.

In the SI system the unit of moment of force is the newton · meter (N · m); in the U.S. system it is the pound-foot (lb-ft). Work and energy have this same dimension; the U.S. unit is the ft-lb whereas the SI unit is the joule (J), which equals 1 N · m. In the SI system the unit of power is called the watt (W) and equals one joule per second (J/s); in the U.S. system it is the ft-lb/sec. The unit of pressure or stress in the SI system is called the pascal (Pa) and equals 1 N/m^2; in the U.S. system it is the lb/ft^2, although often the inch is used as the unit of length so that the unit of pressure is the $lb/in.^2$ (or psi). In both systems the unit of frequency is called the hertz (Hz), which is one cycle per second. Other units of interest in dynamics include those in Table A.3.

Table A.3

Quantity	SI Unit	U.S. Unit
velocity	m/s	ft/sec
angular velocity	rad/s	rad/sec
acceleration	m/s^2	ft/sec^2
angular acceleration	rad/s^2	rad/sec^2
mass moment of inertia	$kg \cdot m^2$	$slug\text{-}ft^2$
momentum	$kg \cdot m/s$	slug-ft/sec
moment of momentum	$kg \cdot m^2/s$	$slug\text{-}ft^2/sec$
impulse	$N \cdot s(= kg \cdot m/s)$	lb-sec
angular impulse	$N \cdot m \cdot s(= kg \cdot m^2/s)$	lb-ft-sec
mass density	kg/m^3	$slug/ft^3$
specific weight	N/m^3	lb/ft^3

Moreover, in the SI system there are standard prefixes to indicate multiplication by powers of 10. For example, kilo (k) is used to indicate multiplication by 1000, or 10^3; thus 5 kilonewtons, written 5 kN, stands for 5×10^3 N. Other prefixes that commonly appear in engineering are shown in Table A.4. We reemphasize that for the foreseeable future American engineers will find it desirable to know both the U.S. and SI systems well; for that reason we have used both sets of units in examples and problems throughout this book.

Table A.4

tera	T	10^{12}	centi	c	10^{-2}
giga	G	10^9	milli	m	10^{-3}
mega	M	10^6	micro	μ	10^{-6}
kilo	k	10^3	nano	n	10^{-9}
hecto	h	10^2	pico	p	10^{-12}
deka	da	10^1	femto	f	10^{-15}
deci	d	10^{-1}	atto	a	10^{-18}

We turn now to the question of unit conversion. The conversion of units is quickly and efficiently accomplished by multiplying by equivalent fractions until the desired units are achieved. Suppose we wish to know how many newton-meters ($N \cdot m$) of torque are equivalent to 1 lb-ft. Since we know there to be 3.281 ft per meter and 4.448 N per pound,

$$1 \text{ lb-ft} = 1 \cancel{\text{lb-ft}} \left(\frac{1 \text{ m}}{3.281 \cancel{\text{ft}}}\right)\left(\frac{4.448 \text{ N}}{1 \cancel{\text{lb}}}\right) = 1.356 \text{ N} \cdot \text{m}$$

Note that if the undesired unit (such as lb in this example) does not cancel, the conversion fraction is upside-down!

For a second example, let us find how many slugs of mass there are in a kilogram:

$$1\text{ kg} = 1\,\frac{\not{N}\cdot s^2}{\not{m}}\cdot\left(\frac{1\text{ lb}}{4.448\,\not{N}}\right)\cdot\left(\frac{1\,\not{m}}{3.281\text{ ft}}\right) = \frac{1}{14.59}\,\frac{\text{lb-sec}^2}{\text{ft}} = 0.06852\text{ slug}$$

Inversely, 1 slug = 14.59 kg. A set of conversion factors to use in going back and forth between SI and U.S. units is given in Table A.5.*

Table A.5

To Convert From	To	Multiply By	Reciprocal (to Get from SI to U.S. Units)
Length, area, volume			
foot (ft)	meter (m)	0.30480	3.2808
inch (in.)	m	0.025400	39.370
statute mile (mi)	m	1609.3	6.2137×10^{-4}
foot² (ft²)	meter² (m²)	0.092903	10.764
inch² (in.²)	m²	6.4516×10^{-4}	1550.0
foot³ (ft³)	meter³ (m³)	0.028317	35.315
inch³ (in.³)	m³	1.6387×10^{-5}	61024
Velocity			
feet/second (ft/sec)	meter/second (m/s)	0.30480	3.2808
feet/minute (ft/min)	m/s	0.0050800	196.85
knot (nautical mi/hr)	m/s	0.51444	1.9438
mile/hour (mi/hr)	m/s	0.44704	2.2369
mile/hour (mi/hr)	kilometer/hour (km/h)	1.6093	0.62137
Acceleration			
feet/second² (ft/sec²)	meter/second² (m/s²)	0.30480	3.2808
inch/second² (in./sec²)	m/s²	0.025400	39.370
Mass			
pound-mass (lbm)	kilogram (kg)	0.45359	2.20462
slug (lb-sec²/ft)	kg	14.594	0.068522
Force			
pound (lb) or pound-force (lbf)	newton (N)	4.4482	0.22481
Density			
pound-mass/inch³ (lbm/in.³)	kg/m³	2.7680×10^{4}	3.6127×10^{-5}
pound-mass/foot³ (lbm/ft³)	kg/m³	16.018	0.062428
slug/foot³ (slug/ft³)	kg/m³	515.38	0.0019403
Energy, work, or moment of force			
foot-pound or pound-foot (ft-lb) (lb-ft)	joule (J) or newton · meter (N · m)	1.3558	0.73757
Power			
foot-pound/minute (ft-lb/min)	watt (W)	0.022597	44.254
horsepower (hp) (550 ft-lb/sec)	W	745.70	0.0013410
Stress, pressure			
pound/inch² (lb/in.² or psi)	N/m² (or Pa)	6894.8	1.4504×10^{-4}
pound/foot² (lb/ft²)	N/m² (or Pa)	47.880	0.020886
Mass moment of inertia			
slug-foot² (slug-ft² or lb-ft-sec²)	kg · m²	1.3558	0.73756

* Rounded to the five digits cited. Note, for example, that 1 ft = 0.30480 m, so that

$$(\text{Number of feet}) \times \left(\frac{0.30480\text{ m}}{1\text{ ft}}\right) = \text{number of meters}$$

Table A.5 Continued

To Convert From	To	Multiply By	Reciprocal (to Get from SI to U.S. Units)
Momentum (or linear momentum)			
slug-foot/second (slug-ft/sec)	kg · m/s	4.4482	0.22481
Impulse (or linear impulse)			
pound-second (lb-sec)	N · s (or kg · m/s)	4.4482	0.22481
Moment of momentum (or angular momentum)			
slug-foot²/second (slug-ft²/sec)	kg · m²/s	1.3558	0.73756
Angular impulse			
pound-foot-second (lb-ft-sec)	N · m · s (or kg · m²/s)	1.3558	0.73756

Note that the units for time (s or sec), angular velocity (rad/s or 1/s), and angular acceleration (rad/s² or 1/s²) are the same for the two systems. To five digits, the acceleration of gravity at sea level is 32.174 ft/s² in the U.S. system and 9.8067 m/s² in SI units.

We wish to remind the reader of the care that must be exercised in numerical calculations involving different units. For example, if two lengths are to be summed in which one length is 2 ft and the other is 6 in., the simple sum of these measures, $2 + 6 = 8$, does not of course provide a measure of the desired length. It is also true that we may not add or equate the numerical measures of different types of entities; thus it makes no sense to attempt to add a mass to a length. These are said to have different dimensions. A dimension is the name assigned to the *kind* of measurement standard involved as contrasted with the choice of a particular measurement standard (unit). In science and engineering we attempt to develop equations expressing the relationships among various physical entities in a physical phenomenon. We express these equations in symbolic form so that they are valid regardless of the choice of a system of units, but nonetheless they must be *dimensionally consistent.* In the following equation, for example, we may check that the units on the left and right sides agree; r is a radial distance, P is a force, and dots denote time derivatives:

$$P - mg\cos\theta = m(\ddot{r} - r\dot{\theta}^2)$$

Dimensions of . . .	. . . are
P	F
$mg\cos\theta$	$M\left(\frac{L}{T^2}\right)(1) = F$
$m\ddot{r}$	$M\frac{L}{T^2} = F$
$-mr\dot{\theta}^2$	$ML\left(\frac{1}{T}\right)^2 = F$

Therefore the units of (every term in) the equation are those of force. If such a check is made prior to the substitution of numerical values, much time can be saved if an error has been made.

PROBLEMS ▶ Appendix A

A.1 Find the units of the universal gravitational constant G, defined by

$$F = \frac{GMm}{r^2}$$

in (a) the SI system and (b) the U.S. system.

A.2 Find the weight in pounds of 1 kg of mass.

A.3 Find the weight in newtons of 1 slug of mass.

A.4 One pound-mass (lbm) is the mass of a substance that is acted on by 1 lb of gravitational force at sea level. Find the relationship between (a) 1 lbm and 1 slug; (b) 1 lbm and 1 kg.

A.5 The momentum of a body is the product of its mass m and the velocity v_C of its mass center. A child throws an 8-oz ball into the air with an initial speed of 20 mph. Find the magnitude of the momentum of the ball in (a) slug-ft/sec; (b) kg · m/s.

A.6 Is the following equation dimensionally correct?

$$\int_0^5 Fv\, dt = \frac{mv^2}{2} + ma \qquad (v = \text{velocity};\ a = \text{acceleration})$$

A.7 The equation for the distance r_s from the center of the earth to the geosynchronous satellite orbit is

$$gr_e^2 = r_s^3\omega^2 \qquad (\omega = \text{angular speed of earth};\ r_e = \text{earth radius})$$

a. Show that the equation is dimensionally correct.

b. Use the equation to find the ratio of the orbit radius to earth radius.

A.8 The universal gravitational constant is $G = 6.67 \times 10^{-11}\ \text{N} \cdot \text{m}^2/\text{kg}^2$. Express G in units of lb-ft²/slug².

B Examples of Numerical Analysis/The Newton-Raphson Method

There are a few places in this book where equations arise whose solutions are not easily found by elementary algebra; they are either polynomials of degree higher than 2 or else transcendental equations. In this appendix we explain in brief the fundamental idea behind the Newton-Raphson numerical method for solving such equations. We shall first do this while applying the method to the solution for one of the roots of a cubic polynomial equation that occurs in Chapter 7.

To solve the cubic equation of Example 7.5,

$$f(\mathcal{J}) = -\mathcal{J}^3 + 48\mathcal{J}^2 - 633\mathcal{J} + 1342 = 0$$

we could, alternatively, use the Newton-Raphson algorithm. This procedure finds a root of the equation $f(\mathcal{J}) = 0$ (it need not be a polynomial equation, however) by using the slope of the curve. The algorithm, found in more detail in any book on numerical analysis, works as follows. If $\mathcal{J}_{1_0}$ is an initial estimate of a root $\mathcal{J}_1$, then a better approximation is

$$\mathcal{J}_{1_1} = \mathcal{J}_{1_0} - \frac{f(\mathcal{J}_{1_0})}{f'(\mathcal{J}_{1_0})}$$

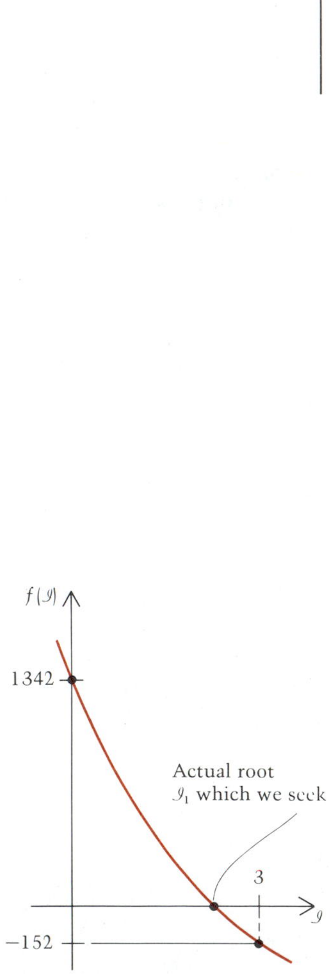

Figure B.1

Figures B.1 and B.2 indicate what is happening. The quantity $f(\mathcal{J}_{1_0})/f'(\mathcal{J}_{1_0})$ causes a backup in the $\mathcal{J}_1$ approximation — in our case from the *initial* value of 3 to the *improved* estimate $\mathcal{J}_{1_1}$:

$$\mathcal{J}_{1_1} = 3 - \frac{f(3)}{f'(3)} = 3 - \frac{-152}{-372}$$

$$= 3 - 0.408602150$$

$$= 2.591397850$$

where

$$f'(\mathcal{J}) = -3\mathcal{J}^2 + 96\mathcal{J} - 633$$

so that $f'(3) = -372$. Repeating the algorithm, we get

$$\mathcal{J}_{1_2} = \mathcal{J}_{1_1} - \frac{f(\mathcal{J}_{1_1})}{f'(\mathcal{J}_{1_1})}$$

$$= 2.591397850 - \frac{6.579491260}{-404.3718348}$$

$$= 2.591397850 + 0.016270894$$

$$= 2.607668744$$

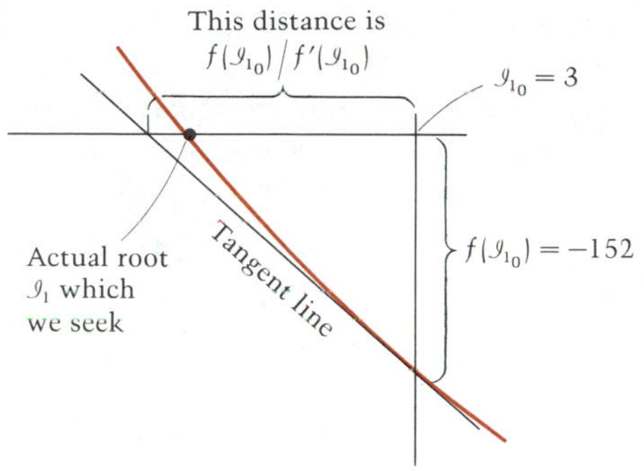

Figure B.2

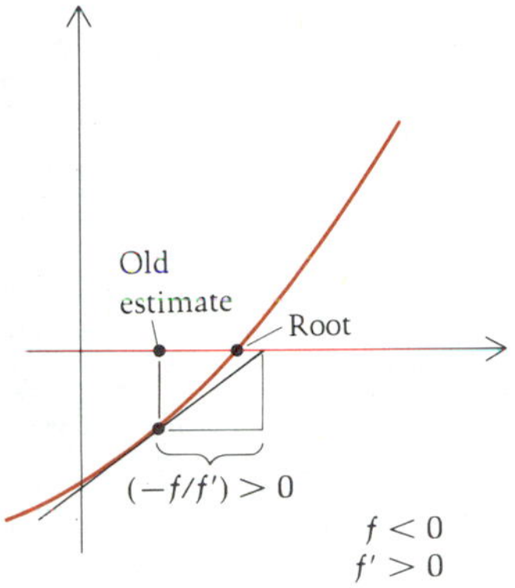

Figure B.3

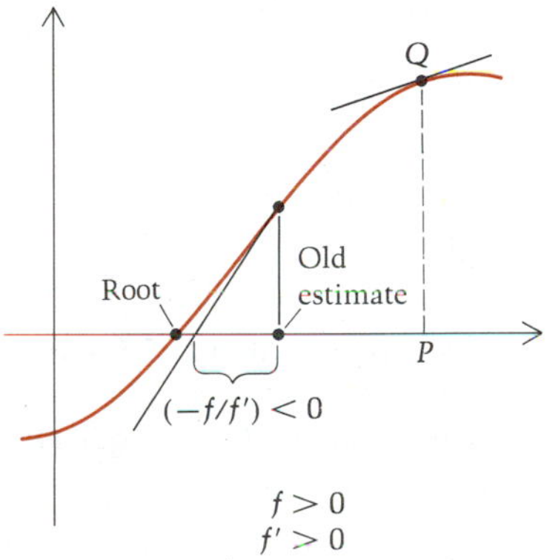

Figure B.4

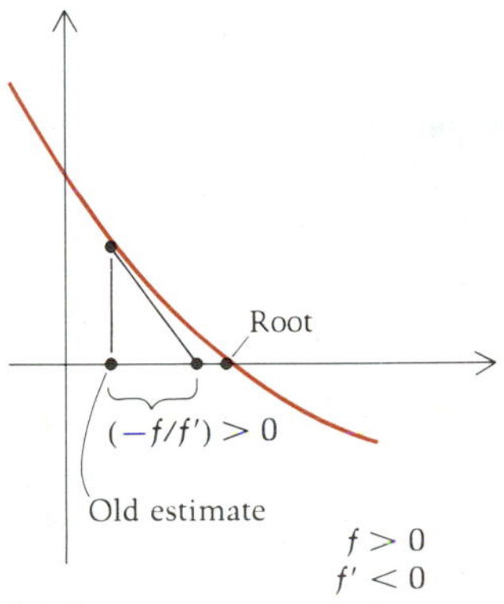

Figure B.5

And one more time:

$$\mathcal{J}_{1_3} = \mathcal{J}_{1_2} - \frac{f(\mathcal{J}_{1_2})}{f'(\mathcal{J}_{1_2})}$$

$$= 2.607668744 - \frac{0.010645000}{-403.0636094}$$

$$= 2.607668744 + 0.000026410$$
$$= 2.607695154$$

This algorithm is easily programmed on a computer. After doing this, the results (with the same initial guess $\mathcal{J}_{1_0} = 3$) are:

$$\mathcal{J}_{1_0} = 3$$
$$\mathcal{J}_{1_1} = 2.591397850$$
$$\mathcal{J}_{1_2} = 2.607668744$$
$$\mathcal{J}_{1_3} = 2.607695154$$
$$\mathcal{J}_{1_4} = 2.607695156$$
$$\left.\begin{aligned}\mathcal{J}_{1_5} &= 2.607695153\\ \mathcal{J}_{1_6} &= 2.607695153\\ \mathcal{J}_{1_7} &= 2.607695153\end{aligned}\right\} \quad \text{convergence!}$$

which is in agreement with the results in Example 7.5.

Incidentally, note from Figures B.3 to B.5 that adding $(-f/f')$ to form the new estimate works equally well for the three other sign combinations of f and f'. Note also that if the estimate is *too far* from the root, such as P in Figure B.4, the procedure might not converge; the tangent at Q in this case would send us far from the desired root.

We next consider the equation from Problem 5.140 when $M = 4m$:

$$f(\theta) = \sin\theta - \frac{\theta}{2} = 0 \tag{B.1}$$

with the derivative of f being

$$f'(\theta) = \cos\theta - \frac{1}{2}$$

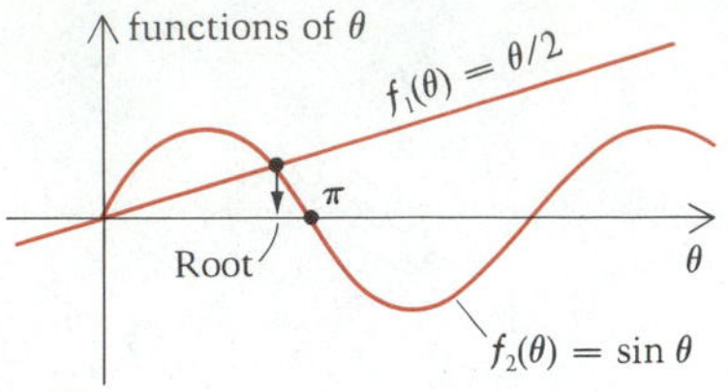

Figure B.6

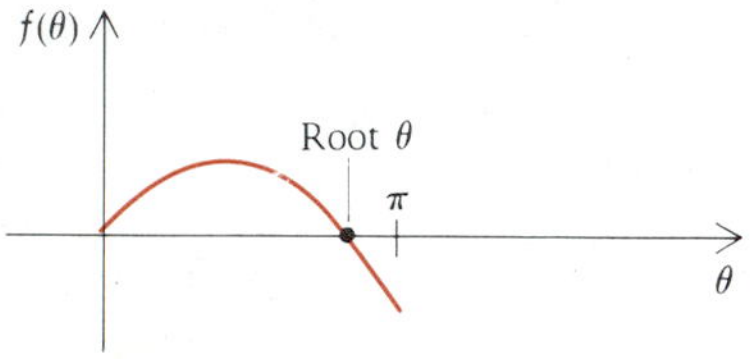

Figure B.7

There is but one root of Equation (B.1) for $\theta > 0$, as can be seen from Figure B.6, which shows the two functions making up $f(\theta)$. To find this root, we can use Newton-Raphson as previously described. Figure B.7 suggests that π might serve as a good first guess at the root. A Newton-Raphson program shows that it is, and yields the answer below very quickly:

$$\theta_0 = 3.141592654$$
$$\theta_1 = 2.094395103$$
$$\theta_2 = 1.913222955$$
$$\theta_3 = 1.895671752$$
$$\theta_4 = 1.895494285$$
$$\left.\begin{aligned}\theta_5 &= 1.895494267\\ \theta_6 &= 1.895494267\\ \theta_7 &= 1.895494267\end{aligned}\right\} \quad \text{convergence!}$$

The last example in this appendix will be to solve the equation

$$\cos\left(\frac{\pi}{4} - q\right) = 0.373q$$

from Example 2.6. We write this equation as

$$f(q) = \cos\left(\frac{\pi}{4} - q\right) - 0.373q = 0$$

with

$$f'(q) = \sin\left(\frac{\pi}{4} - q\right) - 0.373$$

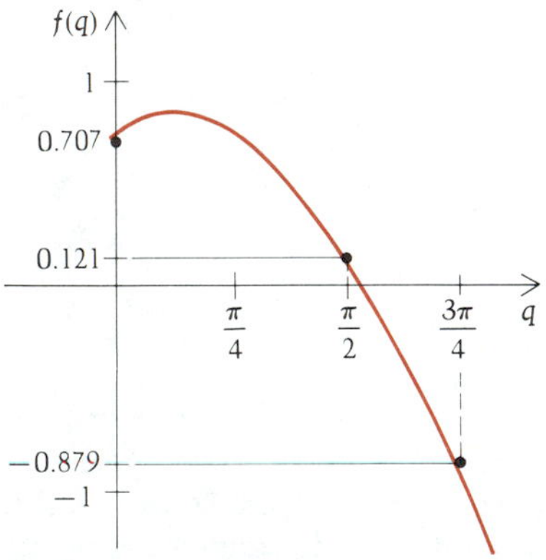

Figure B.8

The rough plot in Figure B.8 shows a few points which indicate that $\pi/2$ is fairly close to the root. Here are the results of a program, which uses the Newton-Raphson method as in the first two examples, to narrow down on the root quickly and accurately:

$$q_0 = 1.570796327$$
$$q_1 = 1.683007224$$
$$q_2 = 1.679300543$$
$$\left.\begin{aligned}q_3 &= 1.679296821\\ q_4 &= 1.679296821\\ q_5 &= 1.679296821\end{aligned}\right\} \quad \text{convergence!}$$

C

Moments of Inertia of Masses (See also Section 4.3)

	Object	Mass Center Coordinates and Volume V	Moments of Inertia About Indicated Axes
slender rod	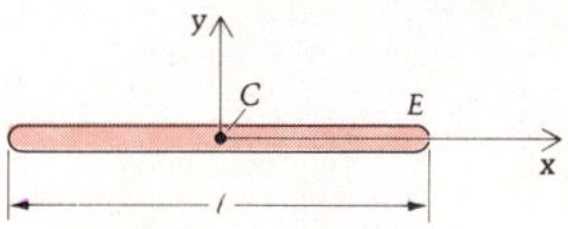	$(0, 0, 0)$ $V = A\ell$ (A = area of cross section)	$I^C_{xx} \approx 0,\ I^C_{yy} = I^C_{zz} \approx \dfrac{m\ell^2}{12}$ (Note: $I^E_{yy} = I^E_{zz} \approx m\ell^2/3$.)
slender circular rod	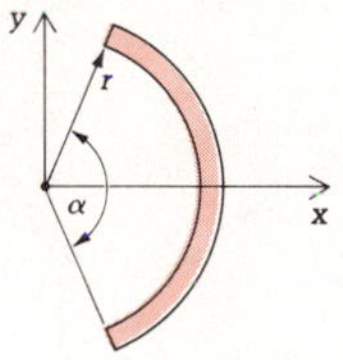	$\left(\dfrac{r \sin(\alpha/2)}{\alpha/2}, 0, 0\right)$ $V = Ar\alpha$ (A = area of cross section)	$I_{xx} = \dfrac{mr^2}{2}\left(1 - \dfrac{\sin\alpha}{\alpha}\right)$ $I_{yy} = \dfrac{mr^2}{2}\left(1 + \dfrac{\sin\alpha}{\alpha}\right)$ $I_{zz} = mr^2$ (Note special cases for $\alpha = \pi$ and 2π.)
bent slender rod	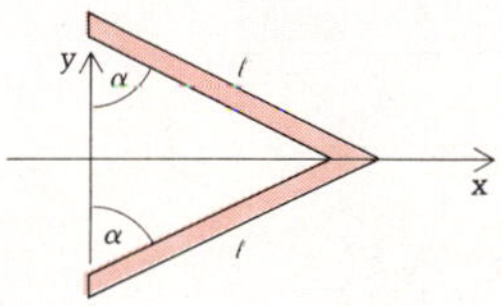	$\left(\dfrac{\ell \sin\alpha}{2}, 0, 0\right)$ $V = 2A\ell$ (A = area of cross section)	$I_{xx} = \dfrac{m\ell^2 \cos^2\alpha}{3}$ $I_{yy} = \dfrac{m\ell^2 \sin^2\alpha}{3}$ $I_{zz} = \dfrac{m\ell^2}{3}$

	Object	Mass Center Coordinates and Volume V	Moments of Inertia About Indicated Axes
rectangular solid	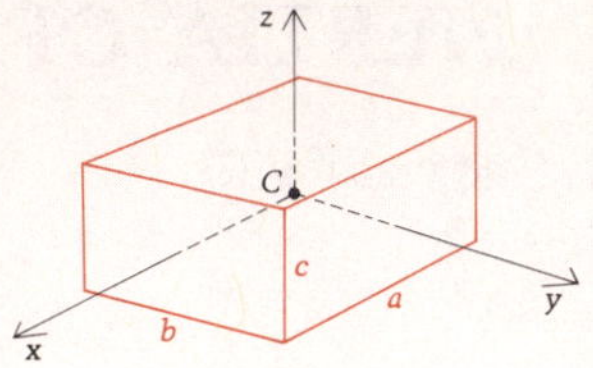	(0, 0, 0) $V = abc$	$I_{xx} = \frac{m}{12}(b^2 + c^2)$ $I_{yy} = \frac{m}{12}(a^2 + c^2)$ $I_{zz} = \frac{m}{12}(a^2 + b^2)$ (Note: If it is a rectangular plate, the thickness dimension is neglected.)
hollow cylinder	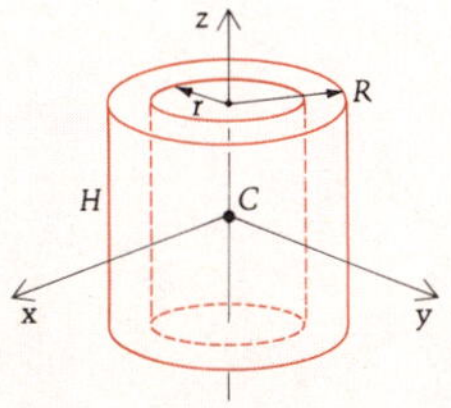	(0, 0, 0) $V = \pi(R^2 - r^2)H$	$I_{xx} = I_{yy} = \frac{m}{4}\left(R^2 + r^2 + \frac{H^2}{3}\right)$ $I_{zz} = \frac{m}{2}(R^2 + r^2)$ (Note special cases when $r = 0$, $r \approx R$, and $H \ll R$, below.) For the case $R \ll H$, see the slender rod above.
Four Special Cases 1. If $r = 0$: solid cylinder	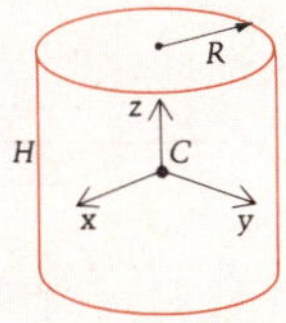	(0, 0, 0) $V = \pi R^2 H$	$I_{xx} = I_{yy} = \frac{mR^2}{4} + \frac{mH^2}{12}$ $I_{zz} = \frac{mR^2}{2}$
2. If $r \approx R$: cylindrical shell	R = average radius 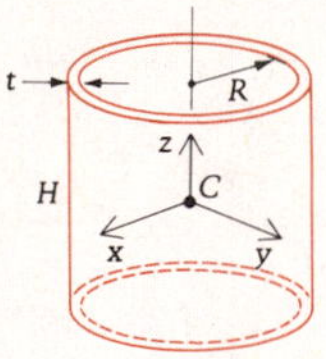	(0, 0, 0) $V = 2\pi RtH$	$I_{xx} = I_{yy} \approx \frac{mR^2}{2} + \frac{mH^2}{12}$ $I_{zz} \approx mR^2$
3. If $H \approx r$: annular disk	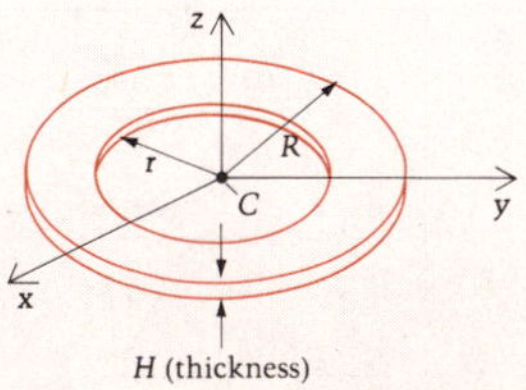	(0, 0, 0) $V = r(R^2 - r^2)H$	$I_{xx} = I_{yy} \approx \frac{m}{4}(R^2 + r^2)$ $I_{zz} = \frac{m(R^2 + r^2)}{2}$

	Object	Mass Center Coordinates and Volume V	Moments of Inertia About Indicated Axes
4. If $H \ll r \approx R$: thin ring with rectangular cross section (area)	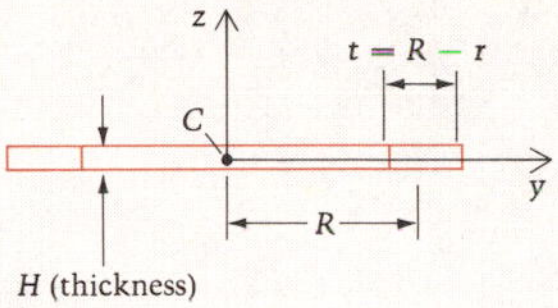	$(0, 0, 0)$ $V = 2\pi RtH$ $= 2\pi R(A)$ $(A = tH$ $=$ area of cross section)	$I_{xx} = I_{yy} \approx \dfrac{mR^2}{2}$ $I_{zz} \approx mR^2$ (Note: These apply for *any* cross-sectional shape as long as the ring is thin.)
thin right triangular plate	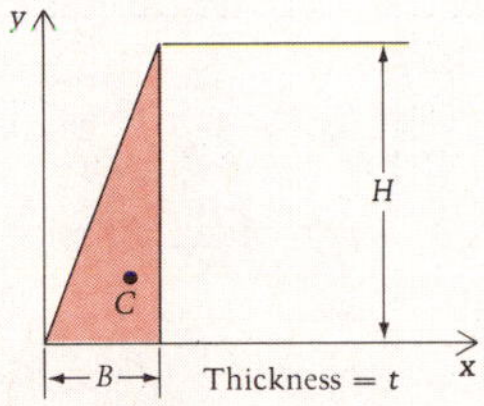	(We are interested here in properties for any base axis, such as x in the figure.) $\left(\dfrac{2B}{3}, \dfrac{H}{3}, 0\right)$ $V = \frac{1}{2}BHt$	$I_{xx} = \dfrac{mH^2}{6}$ $I_{yy} = \dfrac{mB^2}{2}$ $I_{zz} = \dfrac{mH^2}{6} + \dfrac{mB^2}{2}$ $I_{xy} = -\dfrac{mBH}{4}$
thin elliptical plate	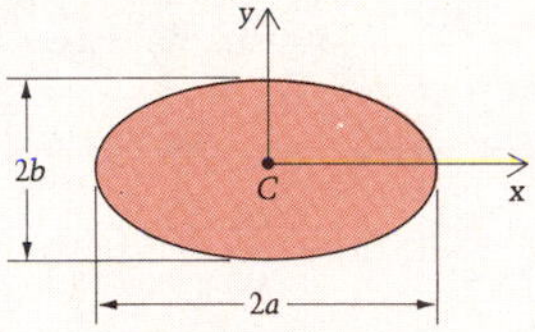 Thickness = t in z direction	$(0, 0, 0)$ $V = \pi abt$	$I_{xx} = \dfrac{mb^2}{4}$ $I_{yy} = \dfrac{ma^2}{4}$ $I_{zz} = \dfrac{m(a^2 + b^2)}{4}$
thin paraboloidal plate	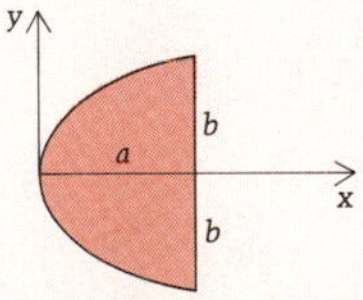 Thickness = t in z direction	$\left(\dfrac{3a}{5}, 0, 0\right)$ $V = \dfrac{4}{3}abt$	$I_{xx} = \dfrac{mb^2}{5}$ $I_{yy} = \dfrac{3ma^2}{7}$ $I_{zz} = \dfrac{m(15a^2 + 7b^2)}{35}$
thin circular sector plate	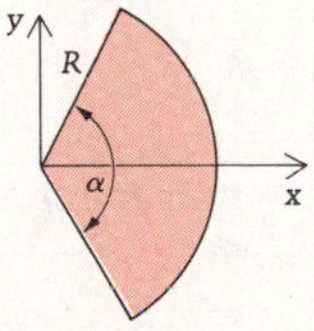 Thickness = t	$\left(\dfrac{2R\sin(\alpha/2)}{3\alpha/2}, 0, 0\right)$ $V = \dfrac{\alpha R^2 t}{2}$	$I_{xx} = \dfrac{mR^2}{4}\left(1 - \dfrac{\sin\alpha}{\alpha}\right)$ $I_{yy} = \dfrac{mR^2}{4}\left(1 + \dfrac{\sin\alpha}{\alpha}\right)$ $I_{zz} = \dfrac{mR^2}{2}$
Two Special Cases 1. If $\alpha = \pi$: semicircular plate	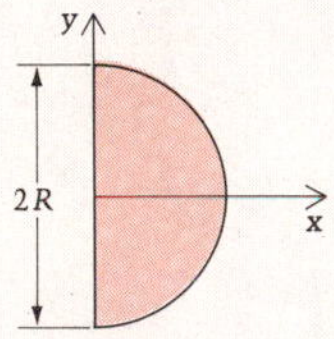 Thickness = t	$\left(\dfrac{4R}{3\pi}, 0, 0\right)$ $V = \dfrac{\pi R^2 t}{2}$	$I_{xx} = I_{yy} \approx \dfrac{mR^2}{4}$ $I_{zz} = \dfrac{mR^2}{2}$

	Object	Mass Center Coordinates and Volume V	Moments of Inertia About Indicated Axes
2. If $\alpha = 2\pi$: circular plate	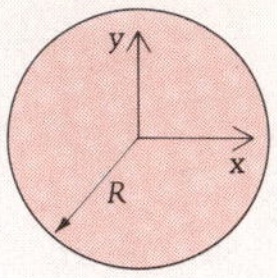 Thickness = t	$(0, 0, 0)$ $V = \pi R^2 t$	$I_{xx} = I_{yy} \approx \dfrac{mR^2}{4}$ $I_{zz} = \dfrac{mR^2}{2}$ (Note: The results appear to be the same as those of the semicircular plate, but the masses differ by a factor of 2.)
thin circular segment plate	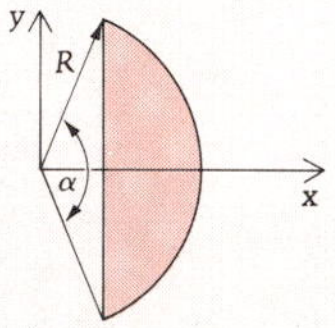	$\left(\dfrac{4R \sin^3(\alpha/2)}{3(\alpha - \sin\alpha)}\right)$ $V = \dfrac{R^2 t}{2}(\alpha - \sin\alpha)$	$I_{xx} = \dfrac{mR^2}{12}(3 - k)$ $I_{yy} = \dfrac{mR^2}{4}(1 + k)$ $I_{zz} = \dfrac{mR^2}{6}(3 + k)$ where $k = \dfrac{(1 - \cos\alpha)\sin\alpha}{\alpha - \sin\alpha}$ (Note special cases for $\alpha = \pi, 2\pi$.)
rectangular tetrahedron	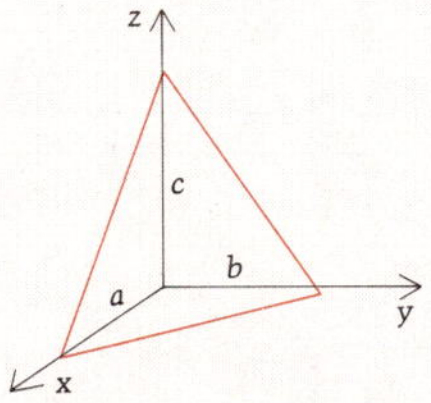	$\left(\dfrac{a}{4}, \dfrac{b}{4}, \dfrac{c}{4}\right)$ $V = \dfrac{abc}{6}$	$I_{xx} = \dfrac{m}{10}(b^2 + c^2)$ $I_{yy} = \dfrac{m}{10}(a^2 + c^2)$ $I_{zz} = \dfrac{m}{10}(a^2 + b^2)$
hollow sphere	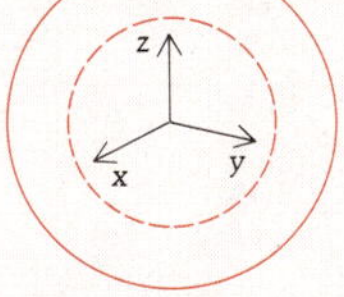	$(0, 0, 0)$ $V = \frac{4}{3}\pi(R^3 - r^3)$	$I_{xx} = I_{yy} = I_{zz} = \dfrac{2}{5} m \left(\dfrac{R^5 - r^5}{R^3 - r^3}\right)$ (Note: If $r = 0$, get $\frac{2}{5}mR^2$, and if $r \approx R$ (spherical shell), get $\frac{2}{3}mR^2$ with $V \approx 4\pi R^2 t$, where $t = R - r =$ thickness of shell. For the shell result, $R - r$ divides both numerator and denominator evenly.)
solid ellipsoid	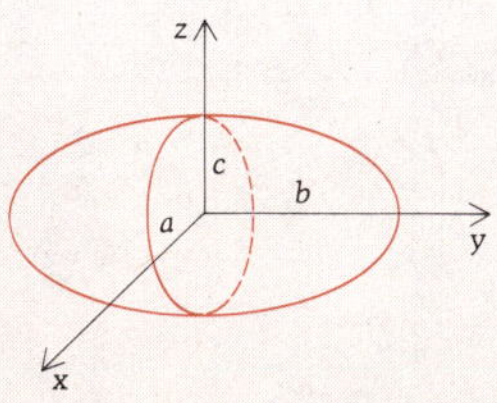	$(0, 0, 0)$ $V = \frac{4}{3}\pi abc$	$I_{xx} = \dfrac{m}{5}(b^2 + c^2)$ $I_{yy} = \dfrac{m}{5}(a^2 + c^2)$ $I_{zz} = \dfrac{m}{5}(a^2 + b^2)$

	Object	Mass Center Coordinates and Volume V	Moments of Inertia About Indicated Axes
solid spherical cap	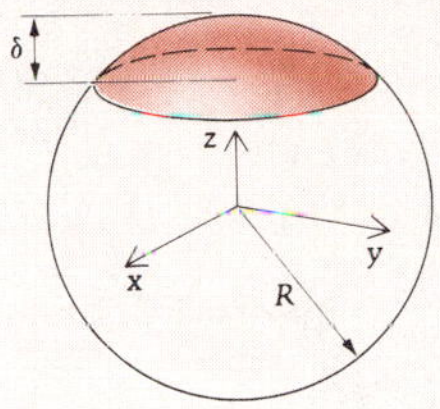	$\left(0, 0, \dfrac{3(2R-\delta)^2}{4(3R-\delta)}\right)$ $V = \dfrac{\pi}{3}\delta^2(3R-\delta)$	$I_{xx} = I_{yy}$ $= \dfrac{m}{2}\left[2R^2 - \dfrac{3(10R^2-\delta^2)\delta}{5(3R-\delta)} + \dfrac{3\delta^2}{2}\right]$ $I_{zz} = \dfrac{m\delta}{10}\left[\dfrac{20R^2 - 15R\delta + 3\delta^2}{3R-\delta}\right]$ (Note: If $\delta = R$, we have a hemisphere and $I_{xx} = I_{yy} = I_{zz} = \frac{2}{5}mR^2$, with $V = \frac{2}{3}\pi R^3$ and $\bar{z} = \frac{3}{8}R$.)
paraboloid of revolution	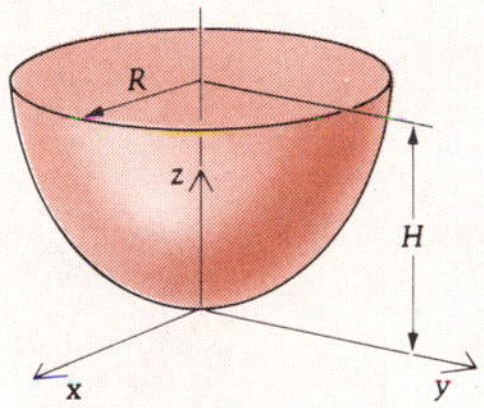	$(0, 0, \frac{2}{3}H)$ $V = \dfrac{\pi R^2 H}{2}$	$I_{xx} = I_{yy} = \dfrac{m}{6}(R^2 + 3H^2)$ $I_{zz} = \dfrac{mR^2}{3}$
elliptic paraboloid	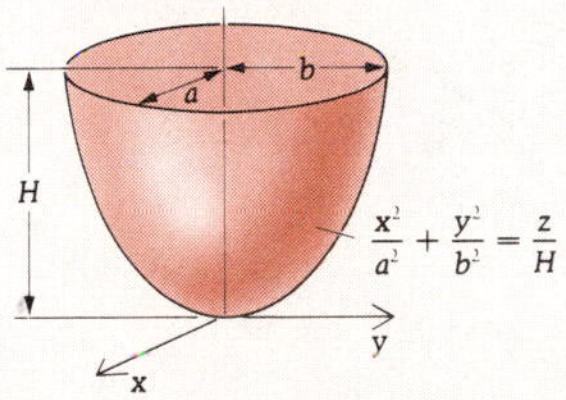	$\left(0, 0, \dfrac{2H}{3}\right)$ $V = \dfrac{\pi abH}{2}$	$I_{xx} = \dfrac{m}{6}(b^2 + 3H^2)$ $I_{yy} = \dfrac{m}{6}(a^2 + 3H^2)$ $I_{zz} = \dfrac{m}{6}(a^2 + b^2)$
solid cone	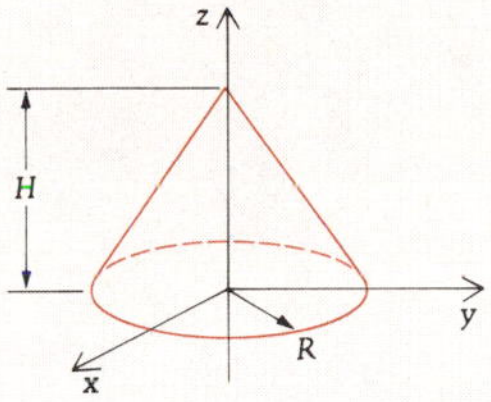	$\left(0, 0, \dfrac{H}{4}\right)$ $V = \dfrac{\pi R^2 H}{3}$	$I_{xx} = I_{yy} = \dfrac{m}{20}(3R^2 + 2H^2)$ $I_{zz} = \dfrac{3}{10}mR^2$
solid right rectangular prism	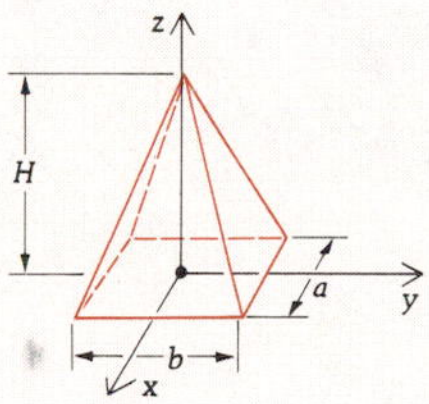	$\left(0, 0, \dfrac{H}{4}\right)$ $V = \dfrac{abH}{3}$	$I_{xx} = \dfrac{m}{80}(4b^2 + 8H^2)$ $I_{yy} = \dfrac{m}{80}(4a^2 + 8H^2)$ $I_{zz} = \dfrac{m}{20}(a^2 + b^2)$

	Object	Mass Center Coordinates and Volume V	Moments of Inertia About Indicated Axes
solid toroid	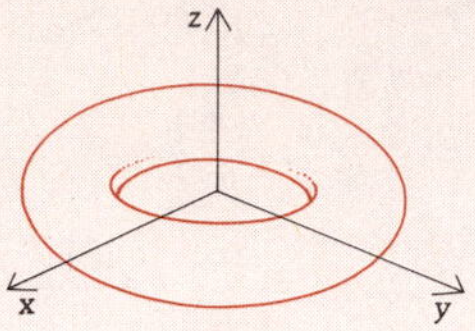	$(0, 0, 0)$ $V = 2\pi^2 R r^2$	$I_{xx} = I_{yy} = \frac{m}{8}(4R^2 + 5r^2)$ $I_{zz} = \frac{m}{4}(4R^2 + 3r^2)$ (Note: If $R \gg r$, we have a hoop for which $I_{xx} = I_{yy} = mR^2/2$ and $I_{zz} = mR^2$.)
frustum of cone	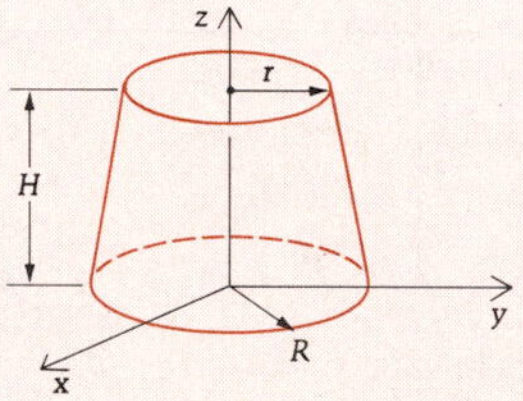	$\left(0, 0, \frac{H(R^2 + 2Rr + 3r^2)}{4(R^2 + Rr + r^2)}\right)$ $V = \frac{\pi H}{3}(R^2 + Rr + r^2)$	$I_{xx} = I_{yy} = \frac{m}{20}\left[3(R^2 + r^2) + \frac{(2R^2 + 6Rr + 12r^2)H^2 - 3r^2R^2}{R^2 + Rr + r^2}\right]$ $I_{zz} = \frac{3m(R^5 - r^5)}{10(R^3 - r^3)}$

D ANSWERS TO ODD-NUMBERED PROBLEMS

In the solutions to problems in Chapters 1–5, unless identified otherwise below, $\hat{\mathbf{i}}$, $\hat{\mathbf{j}}$, and $\hat{\mathbf{k}}$ are unit vectors in the respective directions →, ↑, and out of the page. In Chapters 6–8, the unit vectors are respectively parallel to axes defined in the problems.

CHAPTER 1

1.1 $18\hat{\mathbf{j}} - 8\hat{\mathbf{k}}$ kg · m/s²
1.3 $11.2\hat{\mathbf{i}} + 20\hat{\mathbf{j}} - 30\hat{\mathbf{k}}$ slug-ft/sec²
1.5 $5\hat{\mathbf{i}} + 0.889\hat{\mathbf{j}} - 0.222\hat{\mathbf{k}}$ slug-ft/sec²
1.7 $0.00420\hat{\mathbf{i}} + 29.7\hat{\mathbf{j}}$ kg · m/s²
1.9 $6\hat{\mathbf{i}} + 117\hat{\mathbf{j}} - 84\hat{\mathbf{k}}$ N · s
1.11 $-57.2\hat{\mathbf{i}} + 210\hat{\mathbf{j}} - 195\hat{\mathbf{k}}$ lb-sec
1.13 $52.5\hat{\mathbf{i}} - 3.60\hat{\mathbf{j}} + 0.630\hat{\mathbf{k}}$ lb-sec
1.15 $-0.0183\hat{\mathbf{i}} + 213\hat{\mathbf{j}}$ N · s
1.17 Answer given in problem.
1.19 $-1.64\hat{\mathbf{i}} + 12.9\hat{\mathbf{j}}$ ft/sec² **1.21** $\hat{\mathbf{i}} - (\pi/2)\hat{\mathbf{j}}$ ft/sec²
1.23 $120\hat{\mathbf{i}} + 3\hat{\mathbf{j}} - 152\hat{\mathbf{k}}$ m; 194 m
1.25 $\dfrac{-80}{\pi}\hat{\mathbf{i}} + \dfrac{80}{\pi}\hat{\mathbf{j}}$ m; 36.0 m
1.27 $6.08\hat{\mathbf{i}} - 1.11\hat{\mathbf{k}}$ m; 6.18 m **1.29** 387 ft
1.31 $32.3\hat{\mathbf{i}}$ ft/sec²
1.33 15 sec to *return* (21 seconds total elapsed time)
1.35 $v_P = -10t + 150$

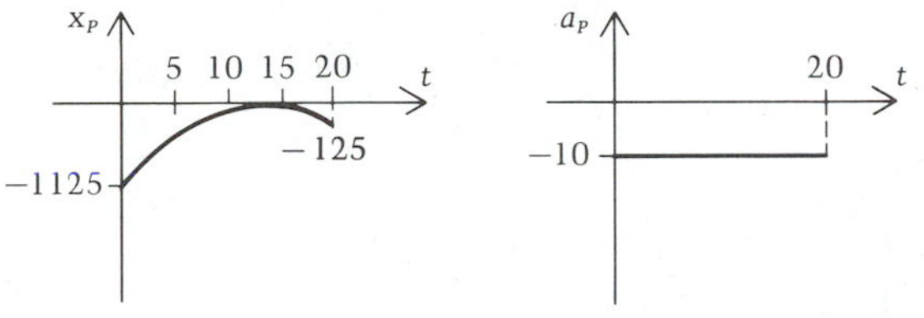

$x_P = -5t^2 + 150t - 1125$ $a_P = -10$

1.37 $v_P = t^2/2$, $a_P = t$, $x_P = \dfrac{t^3}{6} + 10$

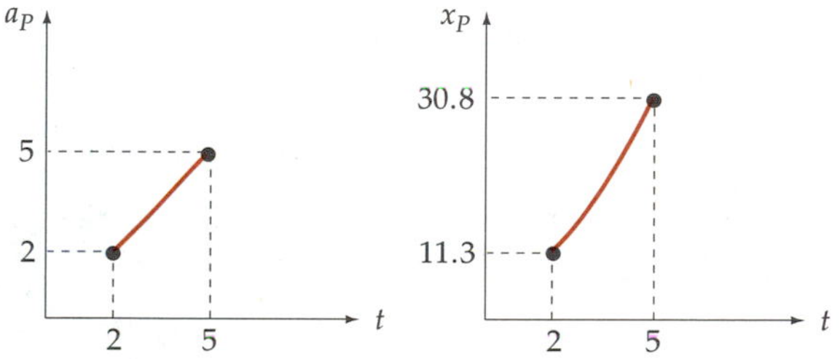

1.39 (a) 2.57 s; (b) 11.7 m/s; (c) 15.0 m
1.41 2 → ft/sec **1.43** 512 ft
1.45 (a) $230\hat{\mathbf{i}}$ m; (b) 234 m **1.47** 8.46 m
1.49 0.469 s **1.51** $\dot{x} = 1/(5t + 1.67)$ m/s
1.53 52.6 sec
1.55 (a) $\mathscr{B}$ wins by 198 ft; (b) 660 ft; (c) 2.4 miles
1.57 $\mathbf{v}_A = 10$ ↓ m/s and $\mathbf{v}_B = 20$ ↑ m/s (at $t = 2$ s)
1.59 11.3 m **1.61** 6.21 ↑ m/s
1.63 $1.73\left(\dfrac{5\hat{\mathbf{i}} + 12\hat{\mathbf{j}}}{13}\right)$ m/s **1.65** $(8a_0/15)\hat{\mathbf{j}}$
1.67 $20\pi\hat{\mathbf{j}}; -20\pi\hat{\mathbf{i}}; 20\pi\hat{\mathbf{i}}, 20\pi\hat{\mathbf{j}}$ m/s
1.69 $\mathbf{v}_P = -10 \sin 5t\hat{\mathbf{i}} + 10 \cos 5t\hat{\mathbf{j}}$ m/s;
$\mathbf{a}_P = -50 \cos 5t\hat{\mathbf{i}} - 50 \sin 5t\hat{\mathbf{j}}$ m/s²
1.71 $3\hat{\mathbf{i}} + 6\hat{\mathbf{j}}$ ft/sec; $18\hat{\mathbf{j}}$ ft/sec²
1.73 (a) $1.5\hat{\mathbf{i}} + 11.8\hat{\mathbf{j}}$ m/s²; (b) 0 and 2.52 s

1.75 (a) $\mathbf{v}_P = -6\pi \sin \frac{\pi t}{2}\hat{\mathbf{i}} + 4\pi \cos \frac{\pi t}{2}\hat{\mathbf{j}}$ m/s;

$\mathbf{a}_P = -3\pi^2 \cos \frac{\pi t}{2}\hat{\mathbf{i}} - 2\pi^2 \sin \frac{\pi t}{2}\hat{\mathbf{j}}$ m/s^2
(b) $\mathbf{r}_{OP} = 12\hat{\mathbf{i}}$ m; $\mathbf{v}_P = 4\pi\hat{\mathbf{j}}$ m/s; $\mathbf{a}_P = -3\pi^2\hat{\mathbf{i}}$ m/s^2
(c) $(x/12)^2 + (y/8)^2 = 1$ (an ellipse) **1.77** kr
1.79 $y = \frac{2}{3}x - 11$, a straight line
1.81 $\frac{(x-2)^2}{3^2} + \frac{y^2}{4^2} = 1$, an ellipse
1.83 $3\hat{\mathbf{i}} + 0.0968\hat{\mathbf{j}}$ in./sec
1.85 $3\hat{\mathbf{i}} - 0.00312\hat{\mathbf{j}}$ in./sec **1.87** $-0.00937\hat{\mathbf{j}}$ in./sec^2
1.89 $0.000604\hat{\mathbf{j}}$ in./s^2 **1.91** 14.9 m at 0.690 s
1.93 $0.671\hat{\mathbf{i}} + 2.30\hat{\mathbf{j}} + 0.500\hat{\mathbf{k}}$ m/s
1.95 (a) 7 m/s; (b) $4.46 \times 10^5\hat{\mathbf{k}}$ m/s^2; (c) 3 m/s
1.97 For x within the intervals (290, 1200) ft and (1800, 2700) ft
1.99 0.000219 ft/sec^2; at $x = 0$, 1500 and 3000 ft
1.101 $(\dot\theta, \ddot\theta) = (1.29$ rad/sec, -3.44 rad/sec$^2)$ or $(-1.29$ rad/sec, 3.44 rad/sec$^2)$
1.103 $2\sqrt{2}\hat{\mathbf{j}}$ ft/sec; $-12\sqrt{2}\hat{\mathbf{i}} + 6\sqrt{2}\hat{\mathbf{j}}$ ft/sec^2
1.105 In the order $\theta = 0, \pi/2, \pi$, and $-\pi/2$: $\mathbf{v}_P = 2aK\hat{\mathbf{j}}, -aK(\hat{\mathbf{i}} + \hat{\mathbf{j}}), \mathbf{0}$, and $aK(\hat{\mathbf{i}} - \hat{\mathbf{j}})$. Note respectively that $\hat{\mathbf{e}}_r = \hat{\mathbf{i}}, \hat{\mathbf{j}}, -\hat{\mathbf{i}}, -\hat{\mathbf{j}}$ and $\hat{\mathbf{e}}_\theta = \hat{\mathbf{j}}, -\hat{\mathbf{i}}, -\hat{\mathbf{j}}, \hat{\mathbf{i}}$.
1.107 $\mathbf{v}_P = \frac{9}{\sqrt{2}}\hat{\mathbf{e}}_r + \frac{3}{\sqrt{2}}\hat{\mathbf{e}}_\theta$; $\mathbf{a}_P = \frac{18}{\sqrt{2}}(\hat{\mathbf{e}}_r + \hat{\mathbf{e}}_\theta)$
1.109 (a) 0.577 ft/sec; (b) $-0.136\hat{\mathbf{i}} - 0.385\hat{\mathbf{j}}$ ft/sec^2
1.111 (a) $-535\hat{\mathbf{e}}_r + 560\hat{\mathbf{e}}_\theta$ m/s^2 (at $\theta = 114°$, $t = 0.446$ s) (b) $-1380\hat{\mathbf{e}}_r + 160\hat{\mathbf{e}}_\theta$ m/s^2
1.113 (a) yes; (b) yes
1.115 $\frac{2.93x}{\sqrt{361 + x^2}}$ ↑ ft/sec; measuring r from pulley to bumper, $\dot r$ is the velocity of the shingles; it is also the component of $\mathbf{v}_A$ along the rope.
1.117 First,

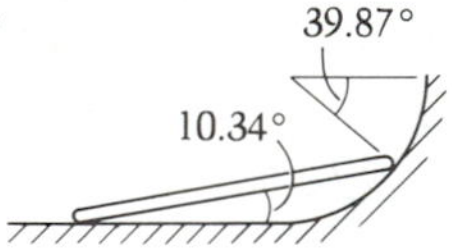

Then, $\mathbf{v}_B = 0.640$ 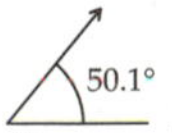

ft/sec.
1.119 $1.65\hat{\mathbf{i}} + 1.13\hat{\mathbf{j}} + 0.600\hat{\mathbf{k}}$ m; $-0.0226\hat{\mathbf{i}} + 0.0330\hat{\mathbf{j}} + 0.0200\hat{\mathbf{k}}$ m/s
1.121 $R\hat{\mathbf{e}}_r + H\hat{\mathbf{k}}$; $k_1\hat{\mathbf{e}}_r + \frac{2k_2R^2}{k_1}\hat{\mathbf{e}}_\theta + \frac{2Hk_1}{R}\hat{\mathbf{k}}$;
$-\frac{4k_2^2R^3}{k_1^2}\hat{\mathbf{e}}_r + 6Rk_2\hat{\mathbf{e}}_\theta + \frac{2Hk_1^2}{R^2}\hat{\mathbf{k}}$
1.123 largest: 73.3 ft/sec, at the top; smallest: 1.17 ft/sec, at the bottom.
1.125 Answer given in problem.
1.127 Answer given in problem.
1.129 $6.5t^2 + C$, where C is a constant of integration
1.131 16 ft **1.133** -34.6 m/s^2; 0.2 m
1.135 4 m/s^2; 1.11 m **1.137** 0.471 ft
1.139 $(-\dot x_0^2/a)\hat{\mathbf{j}}$ **1.141** $\frac{-aK^2}{\sqrt{2}}\hat{\mathbf{e}}_t + \frac{3aK^2}{\sqrt{2}}\hat{\mathbf{e}}_n$; $\frac{2\sqrt{2}}{3}a$
1.143 $(x, y) = (2.88, -0.584)$ m
1.145 $-1.85\hat{\mathbf{i}} - 1.91\hat{\mathbf{j}} - 3.41\hat{\mathbf{k}}$ m; 28.9 m/s^2
1.147 $11.3\hat{\mathbf{i}} - 6.24\hat{\mathbf{k}}$ m
1.149 Use $\mathbf{v} = \dot x\hat{\mathbf{i}} + \dot y\hat{\mathbf{j}}$ and $\mathbf{a} = \ddot x\hat{\mathbf{i}} + \ddot y\hat{\mathbf{j}}$ for the proof; 9680 ft
1.151 $a_t = \frac{1}{Ka}$; $a_n = \frac{1}{a}$ m/s^2

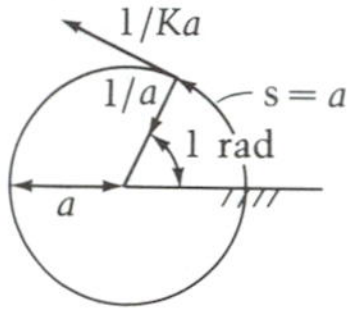

1.153 $a_t = 0$; $a_n = 0.708$ ft/sec^2; $\ddot x = 0.0637$ and $\ddot y = -0.701$ with $\sqrt{\ddot x^2 + \ddot y^2} = 0.704$ ft/sec^2
1.155 Answers given in problem.
1.157 $y_M = 2D/3$ at $t = 2D/(3V_0)$

CHAPTER 2

2.1 Answer given in problem.
2.3 With $x \rightarrow$ and $y \uparrow$ in center of sphere, $(\bar x, \bar y) = (-1.55, -0.20)$ m **2.5** $\left(-\frac{\ell}{5}, \frac{\ell}{5}, -\frac{\ell}{2}\right)$
2.7 $1.34R$ **2.9** $\sqrt{6}R$ **2.11** By inspection, $g \downarrow$
2.13 Home run, clears fence by 2 ft
2.15 88.1 ft/sec at $\theta = 14.9°$; 0.705 sec
2.17 3 ft **2.19** 63.8 ft/sec **2.21** 35.2 ft/sec
2.23 $\phi = \tan^{-1}\left[\frac{H}{d} + \frac{g(D-d)^2}{2dv_o^2}\right]$; time $= \frac{D-d}{v_o}$;
$v_i = \sqrt{\left(\frac{dv_o}{D-d}\right)^2 + \left(\frac{Hv_o}{D-d}\right)^2 + gH + \frac{g^2(D-d)^2}{4v_o^2}}$
2.25 43.4° **2.27** 540 m
2.29 5.45 m/s^2; Apollo: 1.94 m/s^2 **2.31** W
2.33 time $= v_i/(\mu g)$; distance $= v_i^2/(2\mu g)$
2.35 225 ft **2.37** 3.35 → ft/sec^2
2.39 No slip, and the force is 40 → N
2.41 164 N **2.43** $\mu < 0.10$ **2.45** 43.7 lb
2.47 Answer given in problem. **2.49** 0.788 sec
2.51 (a) 13.1 m (b) smaller because now μN resists the motion of $\mathcal{A}$. **2.53** 0.032 lb **2.55** 136 N
2.57 $\mathbf{a}_{\mathcal{A}} = 0.0204g \downarrow$, $\mathbf{a}_{\mathcal{B}} = 0.224g \uparrow$, $\mathbf{a}_{\mathcal{C}} = 0.184g \downarrow$; $\mathcal{C}$; 0.941 s ($g = 9.81$ m/s^2)
2.59 Answer given in problem.
2.61 (a) a (vertical) component of the string tension;
(b) $\sqrt{(mg)^2 + \left(\frac{mv_0^2}{R}\right)^2}$;

(c) another component of the string tension, this one in the $\hat{\mathbf{e}}_t$ direction.
2.63 170 miles **2.65** Answer given in problem.
2.67 3.13 rad/sec **2.69** $\rho = a^2/b$; 52.0 mph, so yes.
2.71 0.5, at $\theta = 0$
2.73 $\ddot{r} = 1$ ft/sec²; $\mathbf{F} = 0.00207\hat{\mathbf{e}}_\theta + 0.0500\hat{\mathbf{k}}$ lb
2.75 $m\ell\omega_0^2 \cos\phi = (m + M)g$ **2.77** $mr_0^4\,\dot{\theta}_0^2/(r_0 - v_C t)^3$
2.79 Answer given in problem.
2.81 1.85 sec; 24.6 ↑ ft/sec **2.83** $\sqrt{2gR}$
2.85 $\dfrac{v_0 m}{K}(1 - e^{-Kt/m})$ **2.87** 1.16 m/D
2.89 191 ↓ ft/sec, compared to 200 ↓ ft/sec without air resistance. **2.91** 15.9 ↓ ft/sec **2.93** 0.736 sec
2.95 $m\ddot{x} = -kx + mg$, x measured down from point of contact; $x(0) = 0 = \dot{x}(0)$; $2mg$; $\pi\sqrt{m/k}$
2.97 Answer given in problem.
2.99 Answer given in problem. (Set $N = 0$ to find the leaving point.) **2.101** 76,300 lb/ft
2.103 20.7 ft/sec down the plane.
2.105 (a) 12 lb (b) 2.84 ft/sec **2.107** 72 lb/ft
2.109 2.56 ← m/s **2.111** 4.0 ft **2.113** $5d/4$
2.115 1020 ↓ N; 9.80 ↑ m/s² **2.117** $\sqrt{2gR}$
2.119 $\sqrt{2K/(3mS)}$ **2.121** W and ΔT are each 400 J.
2.123 $-0.849\, mgr_e$
2.125 (a) 4.78 ft (b) $\mathcal{A}$ will start back upward.
2.127 $v_A = 5.3$ ↓ ft/sec; $v_B = 10.6$ ↑ ft/sec
2.129 Distance between them is $(\ell_u - 0.02)$ m, where ℓ_u = unstretched length. They are 0.22 m closer together. Final spring force = 1 lb (compressive).
2.131 1.41 miles/sec **2.133** 7.45 sec
2.135 36.2 → ft/sec **2.137** 7.86 sec; 491 ft/sec
2.139 0.400 → ft/sec **2.141** $v/3$ →
2.143 $\dfrac{m^2U}{(M + 2m)(M + m)}$ → **2.145** e^4H
2.147 Answer given in problem. (If $e > 1$, there would be an energy *gain*!) **2.149** 0.446
2.151 $\mathbf{v}_A = 2\hat{\mathbf{i}} - 1.53\hat{\mathbf{j}}$ ft/sec; $\mathbf{v}_B = -4\hat{\mathbf{i}} + 3.65\hat{\mathbf{j}}$ ft/sec; Impulse on $\mathcal{B} = 0.077$ ↑ lb-sec = −(Impulse on $\mathcal{A}$)
2.153 With 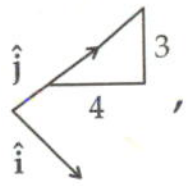

, $\mathbf{v}_A = -0.479\hat{\mathbf{i}} - 2.49\hat{\mathbf{j}}$ ft/sec;
$\mathbf{v}_B = 4.69\hat{\mathbf{j}}$ ft/sec; impulse on $B = 0.113\hat{\mathbf{j}}$ lb-sec
2.155 2.00 ft **2.157** 0.997 ft
2.159 $\dfrac{W_1 + W_2}{k} + \sqrt{\dfrac{W_1^2(W_1 + W_2) + 2HkW_1^2}{k^2(W_1 + W_2)}}$
2.161 $\mathbf{v}_A = 2.97$ → m/s; $\mathbf{v}_B = 1.48$ ← m/s
2.163 $3\beta gy$ **2.165** Answers given in problem.
2.167 60 ↻ lb-ft
2.169 The derivation of Equation (2.36) nowhere requires that the point be the mass center.
2.171 Answer given in problem.
2.173 11.5 ft upward **2.175** $d/4$ upward
2.177 Answer given in problem.
2.179 Answer given in problem.
2.181 $H_1 = 57.0$ in.; $H_2 = 56.5$ in.

CHAPTER 3

3.1 a,c,d,e **3.3** $0.1\hat{\mathbf{i}} + 0.1\hat{\mathbf{j}}$ m/s
3.5 0; $-6.67\hat{\mathbf{k}}$ rad/sec
3.7 $\omega_1 = 2$ ↺ rad/s; $\omega_2 = 0.640$ ↺ rad/s
3.9 $\omega_1 = 2$ ↻ rad/s; $\omega_2 = 2$ ↻ rad/s
3.11 $\omega_2 = 0$; $\omega_3 = -2\hat{\mathbf{k}}$ rad/sec
3.13 $0.058\hat{\mathbf{j}}$ m/s; 0.385 ↺ rad/s
3.15 $0.3\hat{\mathbf{i}} - 1.3\hat{\mathbf{j}}$ m/s **3.17** $0.272\hat{\mathbf{i}}$ m/s
3.19 1.08 → m/s
3.21 $-1.6\hat{\mathbf{i}} - 1.2\hat{\mathbf{j}}$ m/s; $-6\hat{\mathbf{k}}$ rad/s; $4\hat{\mathbf{k}}$ rad/s
3.23 $1.73\left(\dfrac{5\hat{\mathbf{i}} + 12\hat{\mathbf{j}}}{13}\right)$ m/s
3.25 (a) 0, 0.688 ↻ rad/sec; 0, 0
3.27 The plots can be constructed from the answer to 3.26, which is $[(r\cos\theta)/\sqrt{\ell^2 - r^2\sin^2\theta}]\,\dot{\theta}$
3.29 $\dfrac{-12\cos\theta\hat{\mathbf{j}}}{\sqrt{169 - 120\sin\theta}}$ m/s
3.31 In each case, Ⓘ is at the intersection of the radial line OA and the normal to the slot at B; $\mathbf{v}_B = \mathbf{0}$ when $\overline{OAB}$ and $\overline{AOB}$ are straight lines.
3.33 See 3.7. **3.35** $2\hat{\mathbf{j}}$ m/s; $\frac{2}{3}\hat{\mathbf{k}}$ rad/s **3.37** See 3.11.
3.39 0.129 ↻ rad/s; 0.129 ↺ rad/s
3.41 See Example 3.5. **3.43** $\omega_2 = 0.2$ ↺ rad/sec $= \omega_3$
3.45 $-0.389\hat{\mathbf{k}}$ rad/sec **3.47** 2.11 → ft/sec; 0.201 ft above P;

$\mathbf{v}_Q = 60.7$ 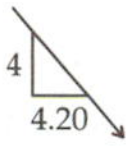

ft/sec;

$\mathbf{v}_S = 60.7$ 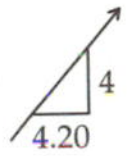

ft/sec;

$\mathbf{v}_R = 85.9$ → ft/sec
3.49 0.0400 ↻ rad/s; 0 **3.51** 14 ↻ rad/sec
3.53 8.95 ↺ rad/sec² **3.55** $-29\hat{\mathbf{i}} - 24.4\hat{\mathbf{j}}$ ft/sec²
3.57 $\boldsymbol{\alpha}_2 = 0.0128\hat{\mathbf{k}}$ rad/sec²; $\boldsymbol{\alpha}_3 = -0.055\hat{\mathbf{k}}$ rad/sec²
3.59 3.91 ↺ rad/sec²
3.61 84.5 → in./sec; 38.1 ← in./sec²; 565 ← in./sec²
3.63 $-30\hat{\mathbf{i}} - 30\hat{\mathbf{j}}$ in./sec² **3.65** $\ddot{\varphi} = \dfrac{r\dot{\theta}^2\sin\theta(r^2 - \ell^2)}{(\ell^2 - r^2\sin^2\theta)^{3/2}}$
3.67 $6.26\hat{\mathbf{i}} + 0.320\hat{\mathbf{j}}$ m/s²; 0.320 ↺ rad/s²
3.69 $-0.0938\hat{\mathbf{i}} - 0.225\hat{\mathbf{j}}$ m/s²
3.71 $-18\hat{\mathbf{i}} - 24\hat{\mathbf{j}}$ in./sec²
3.73 0.0735 ↑ m/s²; 0.0172 ↻ rad/s²
3.75 0.779 ↻ rad/sec²; 6.56 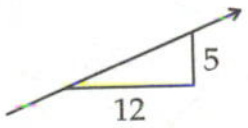

in./sec²

3.77 $\frac{\omega_o}{K}(e^{Kt}-1)$; $\omega_o e^{Kt}$; $\omega_o K e^{Kt}$ **3.79** $(v_0^2/2)\hat{\mathbf{j}}$ ft/sec^2
3.81 (a) 4 ↻ rad/sec^2; (b) $\theta = 45°$; $\overline{PA} = \sqrt{2}/2$ ft
3.83 $x = \dfrac{a_{P_x}\omega^2 - a_{P_y}\alpha}{\omega^4+\alpha^2}$; $y = \dfrac{a_{P_x}\alpha - a_{P_y}\omega^2}{\omega^4+\alpha^2}$
3.85 (a) 1.4 ft above P; (b) $-0.27\hat{\mathbf{i}} + 1.13\hat{\mathbf{j}}$ ft/sec
3.87 $\mathbf{v}_A = 117 \leftarrow$ in./sec; $\mathbf{v}_B = \mathbf{0}$; $\mathbf{v}_D = 2820 \rightarrow$ in./sec; $\mathbf{v}_E = 2930 \rightarrow$ in./sec; A is going backwards (to the left) since it's below Ⓘ, and the wheel turns ↻.
3.89 $\mathbf{v}_C = 1\hat{\mathbf{i}}$ m/s; $\mathbf{v}_B = 1\hat{\mathbf{i}} + 4\hat{\mathbf{j}}$ m/s
(Note: radius superfluous.)
3.91 $\dfrac{\omega}{\dot{\phi}} = \dfrac{R+r}{r}$
3.93 In order, $2v_0\hat{\mathbf{i}}$; $1.71v_0\hat{\mathbf{i}} - 0.707v_0\hat{\mathbf{j}}$; $v_0\hat{\mathbf{i}} - v_0\hat{\mathbf{j}}$; $\mathbf{0}$; $v_0\hat{\mathbf{i}} + v_0\hat{\mathbf{j}}$
3.95 $\mathbf{v}_A = 0.1\hat{\mathbf{i}} + 0.1\hat{\mathbf{j}}$ m/s; $\mathbf{v}_B = 0.2\hat{\mathbf{i}}$ m/s;
$\mathbf{v}_D = 0.1\hat{\mathbf{i}} - 0.1\hat{\mathbf{j}}$ m/s; $\mathbf{v}_E = \mathbf{0}$; 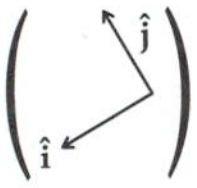

3.97 $\omega_3 = 0.113$ ↻ rad/sec; $\omega_2 = 0.653$ ↻ rad/sec
3.99 $\omega_2 = 7.5$ ↻ rad/sec; $\omega_3 = 0$
3.101 34 ↻ rad/s; $680\hat{\mathbf{i}}$ cm/s, the speed of the tooth point of $\mathcal{B}_4$ in contact with $\mathcal{B}_3$ **3.103** 0; 80 → m/s^2
3.105 $\dot{x}_C$ = any positive constant k; $\dot{\theta} = k/R$
(Directions are → and ↺.)
3.107 $2a_0\hat{\mathbf{i}}$; $(1.71a_0 - 0.707v_0^2/R)\hat{\mathbf{i}} + (0.707a_0 + 0.707v_0^2/R)\hat{\mathbf{j}}$; $(a_0 - v_0^2/R)\hat{\mathbf{i}} + a_0\hat{\mathbf{j}}$; 0; $(a_0 + v_0^2/R)\hat{\mathbf{i}} - a_0\hat{\mathbf{j}}$ **3.109** -35.4 ft/sec^2
3.111 A: $9890\hat{\mathbf{j}}$; B: $9130\hat{\mathbf{j}}$; C: $\mathbf{0}$; D: $-9130\hat{\mathbf{j}}$; E: $-9890\hat{\mathbf{j}}$, all in ft/sec^2. **3.113** (a) 4.69 sec (b) 3.5 revs
3.115 $0.347\hat{\mathbf{i}} + 0.0198\hat{\mathbf{j}}$ m/s^2
3.117 $-0.889\hat{\mathbf{i}} - 5.41\hat{\mathbf{j}}$ m/s^2
3.119 $\alpha_3 = 0.0889\omega_0^2$ ↻; $\alpha_2 = 0.178\ \omega_0^2$ ↻
3.121 $-400\hat{\mathbf{i}}$ ft/sec^2
3.123 14.7 ↻ rad/s^2; $-11.9\hat{\mathbf{j}}$ m/s^2
3.125 $-\dfrac{v_0^2}{R-r}\hat{\mathbf{i}} - \dfrac{v_0^2}{r}\hat{\mathbf{j}}$
3.127 (a) $-6\hat{\mathbf{i}}$ and $-6\hat{\mathbf{i}} - 6\hat{\mathbf{j}}$ cm/s;
(b) $4\hat{\mathbf{i}} - 3\hat{\mathbf{j}}$ and $22\hat{\mathbf{i}} + \hat{\mathbf{j}}$ cm/s^2; -16.3 cm/s^2
3.129 2.70 ↺ rad/sec^2; 0.216 ↻ rad/sec^2
3.131 Let x and y respectively be directed down and toward the plane, with origin at the center of the disk. Then the point has $(x, y) = (4.80, 3.60)$ ft.
3.133 $-43.3\hat{\mathbf{i}}$ in./sec^2
3.135 Let $\hat{\mathbf{i}}$ be from the center of $\mathcal{B}_2$ along $\mathcal{B}_1$, and $\hat{\mathbf{k}}$ be out of the page. Answers are then $-24\pi^2\hat{\mathbf{i}} + 3\pi\hat{\mathbf{j}}$ in./sec^2 ($\mathcal{B}_2$), and $138\pi^2\hat{\mathbf{i}} + 3\pi\hat{\mathbf{j}}$ in./sec^2 ($\mathcal{B}_3$).
3.137 21.6 m/s^2 (It is the highest point of $\mathcal{P}$.)
3.139 (a) Answer given in problem.
(b) Curve is concave downward.
(c) First is $a\alpha\hat{\mathbf{i}} + a\alpha\hat{\mathbf{i}} - \omega^2 a\hat{\mathbf{j}}$, and second is $(2a\alpha)\hat{\mathbf{i}} - [(2a\omega)^2/(4a)]\hat{\mathbf{j}}$, the same.
3.141 Answer given in problem. **3.143** 0.25 ↻ rad/s
3.145 0.781 ↻ rad/sec **3.147** 0.120 ↺ rad/s
3.149 0.592 ↻ rad/s; $11.6\hat{\mathbf{i}} - 4.85\hat{\mathbf{j}}$ cm/s
3.151 $(D\dot{\theta}\sin\theta/\cos^2\theta)(\cos\theta\hat{\mathbf{i}} + \sin\theta\hat{\mathbf{j}})$
3.153 Answer given in problem.
3.155 0.0334 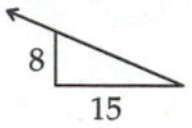

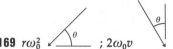

m/s; 0.428 ↻ rad/s
3.157 Answer given in problem.
3.159 $26\pi \sin \pi t \downarrow$; $26\pi^2 \cos \pi t \downarrow$
3.161 $-0.165\hat{\mathbf{i}}$ m/s^2; 0.170 ↺ rad/s^2
3.163 $0.186\hat{\mathbf{k}}$ rad/sec^2
3.165 $\mathbf{v}_{A/\mathcal{A}} = 0.240\hat{\mathbf{i}} + 0.180\hat{\mathbf{j}}$ ft/sec;
$\mathbf{v}_{A/\mathcal{B}} = -0.480\hat{\mathbf{i}} + 0.640\hat{\mathbf{j}}$ ft/sec;
$\mathbf{a}_{A/\mathcal{A}} = 0.512\hat{\mathbf{i}} + 0.384\hat{\mathbf{j}}$ ft/sec^2;
$\mathbf{a}_{A/\mathcal{B}} = -0.180\hat{\mathbf{i}} + 0.240\hat{\mathbf{j}}$ ft/sec^2
3.167 $-8.51\hat{\mathbf{k}}$ rad/s^2; $-118\hat{\mathbf{i}} + 49.2\hat{\mathbf{j}}$ cm/sec^2
3.169 $r\omega_0^2$ 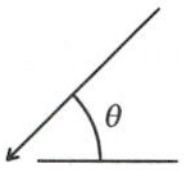

; $2\omega_0 v$ 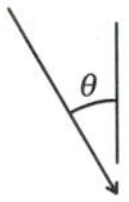

3.171 (a) 103 ↻ rad/s^2; (b) $2.25\hat{\mathbf{i}} + 1.62\hat{\mathbf{j}}$ m/s^2
3.173 9.11 ft
3.175 $\theta = 70.7°$ for the maximum piston speed

CHAPTER 4

4.1 $\frac{3}{4}mg$
4.3 (a) 12.9 → ft/sec^2; (b) left: 40 ↑ lb; right: 60 ↑ lb **4.5** $\dfrac{b}{2} \le H \le \dfrac{3}{2}b$ **4.7** 8.05 ft/sec^2
4.9 (a) 108 N; (b) 64.8 N **4.11** $8g/15 \rightarrow$
4.13 $10\mu/(1+9\mu)$ **4.15** $-12.9 \le a \le 8.05$ ft/sec^2
4.17 $(m+M)g/4$ **4.19** $g/5$ up the plane
4.21 (a) $(b-d)g\mu/(b-\mu H)$
(b) $d = \mu H$ gives $\ddot{x}_{C_{MAX}} = \mu g$
4.23 time $= (a+b+\mu H)v/(\mu a g)$;
distance $= (a+b+\mu H)v^2/(2\mu g a)$
4.25 $0.847 \le \ddot{x}_C \le 33.9$ ft/sec^2
4.27 16π slug-ft^2 **4.29** $\pi R^2 L^3(17\rho_1 + 35\rho_0)/60$
4.31 (a) $\dfrac{mH^2}{6} + \dfrac{mt^2}{12}$; (b) $\dfrac{mB^2}{2} + \dfrac{mt^2}{12}$;
(c) $\dfrac{mH^2}{6} + \dfrac{mB^2}{2}$; (d) $-\dfrac{mBH}{4}$;
(e) 0, 0; neglect t^2 terms in (a) and (b).
4.33 Answer given in problem. (I_{zz}^O exceeds $I_{xx}^O + I_{yy}^O$ by $mt^2/6$.) **4.35** $mR^2\left(\dfrac{1}{4} + \dfrac{\sin\alpha}{4\alpha} - \dfrac{16}{9\alpha^2}\sin^2\dfrac{\alpha}{2}\right)$
4.37 3619 kg · m^2; 1660 kg · m^2
4.39 $0.145\ mR^2 + 0.0481\ mH^2$

4.41 From the corner, 1.56 ft ← and 0.563 ft ↑; $I_{z_C} = 4.45$ slug-ft². **4.43** 1.00 slug-ft²
4.45 3.44 slug-ft² **4.47** 0.259 mR^2
4.49 37.7 kg · m² **4.51** 29.3 kg · m²
4.53 2.69, 0.0200, and 2.71 kg · m², respectively.
4.55 Answer given in problem
4.57 $M = md/k_G$; $D = k_G$ **4.59** 23.1 kg · m²
4.61 Answer given in problem; only (b) starts without approximation. **4.63** $2g/3$; $5g/7$; $g/2$ ↓
4.65 (a) Wally, Sally, Carolyn, Harry;
(b) 1.41K, 1.67K, 1.73K, 2K where $K = \sqrt{\dfrac{D}{g \sin \beta}}$
4.67 0.200 $mg \sin \beta$ (compression)
4.69 (a) 0.428 μmg (b) 1.86 $\mu g\hat{\mathbf{i}}$; $(0.360\ \mu g/r)\hat{\mathbf{k}}$
4.71 1/2 **4.73** $\dfrac{0.714\ M_O}{mr}$ → **4.75** 0.577; 1.02 s
4.77 (a) 4.00 → m/s² (b) 0.23 m
4.79 (a) 14.4 in. wrapped; 0.064 **4.81** 300 ft
4.83 0.450 ⟲ rad/sec² **4.85** 78.6 m
4.87 (a) 14.6°; (b) 1.38 s **4.89** 0.753 slug-ft²; 38.3 → ft
4.91 2.88$\hat{\mathbf{k}}$ rad/s²; 0.180 m to the right

4.93 (a) 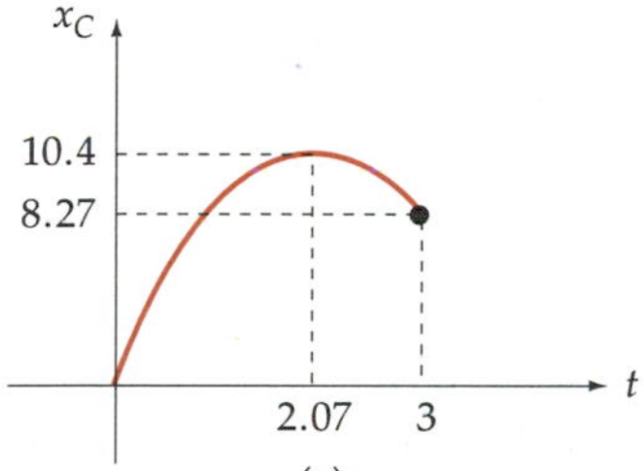

(a) ;

(b) 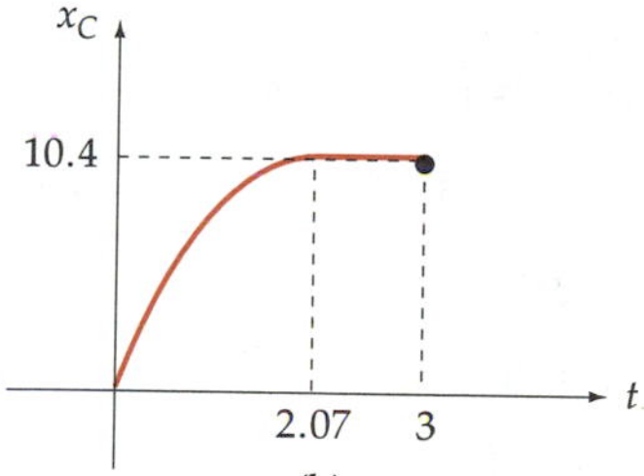

(b) ;

(c) 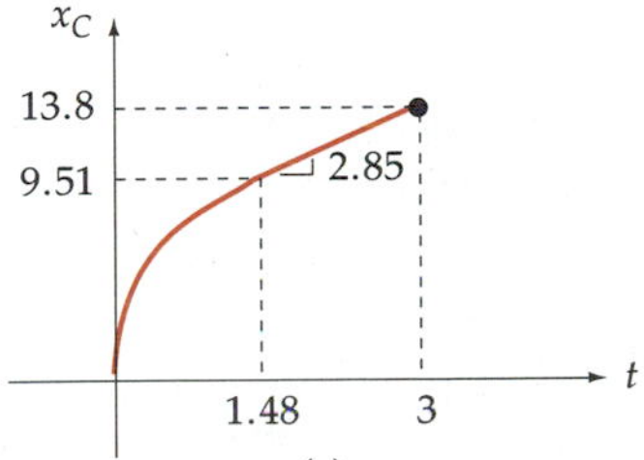

(c)

4.95 (a) Answer given in problem.
(b) 0.676 → m/s², 1.69 ⟳ rad/s²
4.97 21.8 ⟳ rad/sec²; Yes, because the geometry doesn't change. **4.99** 5.50 kg · m²; 4.04 → m/s²
4.101 Largest for P_E, and is 4.14 → m
4.103 2.09 s (rolls off left corner)
4.105 (a) 13.3 ⟲ rad/sec² (b) 1.11 ⟳ rad/sec² (c) 0.0197
4.107 28.2 sec **4.109** 440 m **4.111** 7.16 ← ft/sec²
4.113 (a) $\dfrac{F}{M + 3mn/8}$ (b) $\dfrac{Ft^2}{2M + 3mn/4}$
4.115 4.47 sec **4.117** (a) No (b) 0.585 → ft/sec²
4.119 177 lb up the plane **4.121** 2.10 s
4.123 straight down; $mg/4$ ↓
4.125 (a) $3\sqrt{3}/7$ (b) For $\mu \geq \mu_{min}$, $\dfrac{3\sqrt{3}}{16} g\hat{\mathbf{i}} - \dfrac{9}{16} g\hat{\mathbf{j}}$;
for $\mu < \mu_{min}$, $\dfrac{g}{13 - 3\sqrt{3}\,\mu}[4\mu\hat{\mathbf{i}} - (9 - 3\sqrt{3}\,\mu)\hat{\mathbf{j}}]$
4.127 $mg/4$ **4.129** 43.3 lb **4.131** $-0.400\ g\hat{\mathbf{i}}$; 0.308 mg
4.133 Answers given in problem.
4.135 $\mathbf{a}_C = 3g/4$ ↓; $\boldsymbol{\alpha} = 3g/(2s)$ ⟳
4.137 $\mathbf{a}_C = g$ ↓; $\boldsymbol{\alpha} = 6g/s$ ⟳
4.139 $\mathbf{a}_C = 2g/3$ ↓; $\boldsymbol{\alpha} = 2g/s$ ⟳ **4.141** 6.38 lb
4.143 On the section to the left of the cut, $V = 3WL/64$ ↓ and $M = 9WL^2/256$ ⟲.
4.145 At O, $\dfrac{2}{7}mg$ ↑; at A, $\dfrac{mg}{14}$ ↑
4.147 $3T_0/[(2R^2)(9M + 2m)]$ ⟲
4.149 Answer given in problem. **4.151** 1.5 ⟳ rad/s²
4.153 $\alpha = \dfrac{3}{8}\dfrac{g}{\ell}$ ⟳ **4.155** 17.4 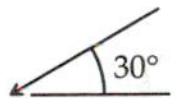

ft/sec²
4.157 For rolling on fixed surface, $\mathbf{a}_Q$ is normal to the surface, hence toward geometric center of round body; thus $\mathbf{r}_{QC} \times \mathbf{a}_Q = \mathbf{0}$ since geometric center is mass center.
4.159 (a) 4 m/s² (b) 11.8 ⟲ rad/s²; no
4.161 0.400 ⟳ rad/s² **4.163** 55.8 ↑ N
4.165 (a) 30°: 0.805 ⟳ rad/sec²; 60°: 2.28 ⟳ rad/sec²; 90°: 2.82 ⟳ rad/sec² (b) 163°
4.167 10.8 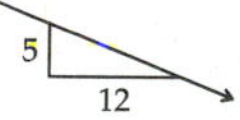

N;

497 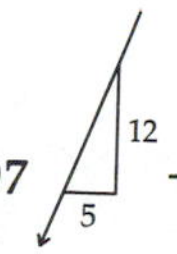

+ 34.0 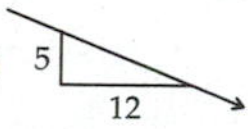

N

4.169 1.06 g/L ⟳ **4.171** OB: 0.405 mg; OA: 0.595 mg
4.173 7.85 ⟳ rad/s² **4.175** 17.3 ↓ ft/sec²; 14.9 ↑ lb
4.177 (a) 0.188 (b) 53.1°
4.179 counterclockwise, 1.48 sec
4.181 (a) $mg\sqrt{\cos^2\theta + (10 \sin \theta)^2}/4$ (b) $5mg/2$

4.183 $M\ell\omega^2 \sin\theta\hat{\mathbf{i}} - Mg\hat{\mathbf{j}}$; $(Mg\ell \sin\theta + M\ell^2\omega^2 \sin\theta\cos\theta)$ ↻

4.185 (a) $\frac{m\delta\omega^2}{2}\hat{\mathbf{i}} - \frac{m\delta\alpha}{2}\hat{\mathbf{j}}$ each, with $(\hat{\mathbf{i}}, \hat{\mathbf{j}})$ along (x, y) in figure; (b) 0.258 r

4.187 $\left(m_p + m_c + \frac{m_{\mathcal{A}}}{2}\right) g \downarrow$ each, plus $\left[\frac{m_c\ell_{\mathcal{A}}\ell_c}{2} + m_p(\ell_{\mathcal{A}} + \ell_p)\ell_c\right]\frac{\omega^2}{2d}$, up on left bearing and down on right bearing.

4.189 $\frac{13\, mr^2\omega_0^2 \sin\psi\cos\psi}{96\ell}$, down on left and up on right, onto shaft and turning with it.

4.191 By parallel-axis theorem, $I_{xz}^{D} = 0 \Rightarrow I_{xz}^{C} = 0$ since C is on z-axis. Thus using the theorem again, $I_{xz}^{Q} = 0$. Same arguments for I_{yz}.

4.193 In $\mathcal{A}$, $(x, y) = (-56/3, -21)$ in.; in $\mathcal{B}$, $(x, y) = (-70/3, -35)$ in. **4.195** 24.0 rad/sec

CHAPTER 5

5.1 25 ft-lb **5.3** 100 J **5.5** 140 N/m

5.7 3.75 ↻ rad/sec

5.9 $\sqrt{\frac{3g\sin\theta_0}{L}}$ (↻ for $\mathcal{A}$ and ↺ for $\mathcal{B}$)

5.11 103,000 lb/ft **5.13** 4.18 ↺ rad/s

5.15 2.95 ↺ rad/s **5.17** 2.00 ↺ rad/s

5.19 12.1 ↻ rad/s **5.21** (a) 2440 N/m

5.23 (a) 0.0794 m; (b) back to the starting point

5.25 4.36 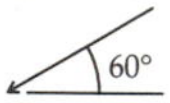

m/s **5.27** $\sqrt{\frac{18ag}{7}} \leftarrow$

5.29 (a) 30 lb; (b) 26.7 in./sec

5.31 (a) 1.56 ↺ rad/sec; (b) 1.40 ↺ rad/sec

5.33 0.0931 ← m/s

5.35 (a) 7.19 ↻ rad/s; (b) 7.53 ↻ rad/s; (c) 8.58 ↻ rad/s

5.37 $\sqrt{24\pi gr\sin\beta/13}$ 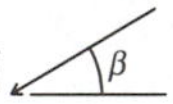

5.39 35 lb/ft; yes

5.41 4.43 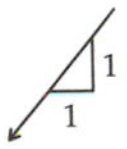

ft/sec

5.43 8 in.; the two points are the intersections of the perimeter of $\mathcal{B}_2$ (in the starting position) with a circle of radius 12 in. and center at Q (in the final position)

5.45 $\frac{L\omega}{2}\sqrt{1 - \left(\frac{2H}{L}\right)^2} \downarrow$

5.47 $4\sqrt{2}R\sqrt{\frac{g(1-\cos\theta)}{R(9\pi - 16)}}$ 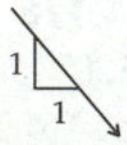

5.49 4.91 ↺ rad/sec; $\frac{1}{3}$ ft **5.51** 3.82 m

5.53 2.11 ↻ rad/sec

5.55 For $T\uparrow$, $v_C = \sqrt{\frac{2RrTx_C}{I_C + mR^2}}$ (moves to the left); For $T\rightarrow$, $v_C = \sqrt{\frac{2R(R-r)Tx_C}{I_C + mR^2}}$ (moves to the right)

5.57 (a) 4/17 m; (b) 50 N *increase*, $\mu_{\min} = 0.220$

5.59 It starts out to the right, the spring goes slack, and then it leaves on the right. (It would need one more foot of plane to stay on.)

5.61 $\sqrt{6\sqrt{2}g\ell}\downarrow$ **5.63** 1.33 ↻ rad/s

5.65 0.106 m to the left; 0.000428 **5.67** $41.5\hat{\mathbf{j}}$ lb

5.69 (a) $\frac{3g\cos\theta_o}{2L}$ ↻; $W\sin\theta_o$ 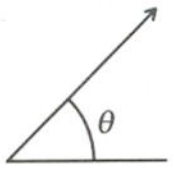

$+ \frac{W\cos\theta}{4}$ 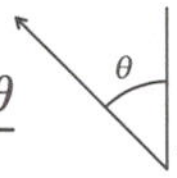

; (b) $\frac{3W\sin\theta_o}{2}\leftarrow + \frac{W}{4}\uparrow$

5.71 $\tan^{-1}[\mu/(1 + 36k^2)]$; 5.27°

5.73 Answer given in problem.

5.75 5.66 ↑ m/s **5.77** $2gt_0/(R\pi)$

5.79 (a) 0.00240 → m/s; (b) 0.0889 → m/s

5.81 0.366 sec

5.83 (a) 1.88 ↓ m/s; (b) 4.51 ↑ m/s **5.85** See 4.166(c). **5.87** 4.02 sec **5.89** 1.38 sec

5.91 1.19 ↻ rad/sec **5.93** $\frac{5\,gt\sin\beta}{7}$

5.95 32.2 → ft/sec **5.97** (a) 2.82 ← m/s; (b) zero

5.99 $2d/9$ upward

5.101 $\Delta t = \frac{I_1I_2(\omega_2 - \omega_1)}{\mu m_1 gR(I_1 + I_2)}$; $\omega_f = \frac{I_1\omega_1 + I_2\omega_2}{I_1 + I_2}$; $\Delta T = \frac{I_1I_2(\omega_1 - \omega_2)^2}{2(I_1 + I_2)}$; ω_f and ΔT are the same

5.103 (a) no; (b) yes, if $R^2H \le dfk^2$

5.105 $6v_0m/[\ell(4m + M)]$, counterclockwise looking down

5.107 (a) $\sqrt{1.26g/r + 0.399\, v_0^2/r^2}$; (b) 36.8%

5.109 $3I^2/(2M)$ **5.111** $0.526\frac{m_2L}{m_1h}\sqrt{gL}$

5.113 0.159 sec **5.115** $-0.25v_0\hat{\mathbf{i}} + 0.75v_0\hat{\mathbf{j}}$

5.117 0.8 ft; $19.0(10^6)$ slug-ft²; 1.05% difference

5.119 $\Delta T = \frac{35}{72}mv_0^2$, or 72.9% loss

5.121 Answer given in problem. **5.123** $\frac{3M^2\ell}{2\mu(M + 3m)^2}$

5.125 $e = 0.8$; $\mu = 0.32$

5.127 (a) 1.66 rad/sec for both; (b) 17%

5.129 (a) $\mathbf{v}_{G_{\mathcal{A}}} = 0$; $\mathbf{v}_{G_{\mathcal{B}}} = v_0 \rightarrow$; $\omega_{\mathcal{A}} = v_0/r$ ↻; $\omega_{\mathcal{B}} = 0$; (b) $\mathbf{v}_{G_{\mathcal{A}}} = 2v_0/7 \rightarrow$; $\mathbf{v}_{G_{\mathcal{B}}} = 5v_0/7 \rightarrow$; (c) If $\mu = 0$, final motion is given by (a).

5.131 $\omega_f = \dfrac{3\ell v_{C_i}(1+e)}{2\ell^2 + 9r^2}$ clockwise;
$\mathbf{v}_{C_f} = (-\ell\omega_f/2 + ev_{C_i})\hat{\mathbf{i}} + r\omega_f\hat{\mathbf{j}}$
(see Example 5.18 for $\hat{\mathbf{i}}$, $\hat{\mathbf{j}}$); $\omega_f(e=0) = \frac{1}{2}\omega_f(e=1)$;
$\dot{x}_{C_f}(e=0) = 1.91\dot{x}_{C_f}(e=1)$; $\dot{y}_{C_f}(e=0) = \frac{1}{2}\dot{y}_{C_f}(e=1)$
5.133 Answer given in problem.

5.135 $v_{C_{\text{rebound}}} = 2.63$ 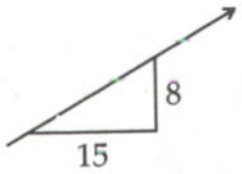 m/s; $\Delta T = 512$ ft-lb

5.137 0.545 m from left end; 0.562 kg-m²; 0.0957 kg-m²; 0.657 m from left end
5.139 $\dfrac{F\,\Delta t(2\ell + a - c)}{\dfrac{4m\ell^2}{3} + M\left[\dfrac{2a^2}{5} + (a+2\ell)^2\right]}$ ↶

CHAPTER 6

6.1 Answer given in problem.
6.3 (a) $(-t^5 + t^3 + 3t^2)\hat{\mathbf{i}} + (t^6 - t^2 + 2t)\hat{\mathbf{j}} + (-t^5 + t^3 + 1)\hat{\mathbf{k}}$ m/s²;
(b) $3\hat{\mathbf{i}} + 2\hat{\mathbf{j}} + \hat{\mathbf{k}}$ m/s²; (c) $-12\hat{\mathbf{i}} + 64\hat{\mathbf{j}} - 23\hat{\mathbf{k}}$ m/s²
6.5 $19.6\hat{\mathbf{j}} - 1.29\hat{\mathbf{k}}$ **6.7** $\boldsymbol{\omega}_{\mathcal{A}/\mathcal{G}} = 13.7\hat{\mathbf{j}}_1 + \hat{\mathbf{k}}_1$
6.9 $\omega_x\hat{\mathbf{i}} + \omega_y \cos\omega_x t\,\hat{\mathbf{j}} + (\omega_z + \omega_y \sin\omega_x t)\hat{\mathbf{k}}$ **6.11** $\dot{\theta}\hat{\mathbf{i}}$
6.13 The solution is Equation (6.70) with
$\theta_1 = \theta_{\text{pitch}} = \dfrac{\pi}{18}\sin\dfrac{2\pi t}{6}$; $\theta_2 = \theta_{\text{roll}} = \dfrac{\pi}{6}\sin\dfrac{2\pi t}{8}$;
and $\theta_3 = \theta_{\text{yaw}} = \dfrac{\pi}{22.5}\sin\dfrac{2\pi t}{50}$.
6.15 To the components in 6.13, respectively, add $-\dot{P}\cos R$, $-\dot{R}$, and $-\dot{P}\sin R$.
6.17 ${}^{\mathcal{I}}\dot{\boldsymbol{\alpha}}_{\mathcal{B}/\mathcal{I}} = {}^{\mathcal{B}}\dot{\boldsymbol{\alpha}}_{\mathcal{B}/\mathcal{I}} + \boldsymbol{\omega}_{\mathcal{B}/\mathcal{I}} \times \boldsymbol{\alpha}_{\mathcal{B}/\mathcal{I}}$, and the cross-product is not generally zero this time.
6.19 (a) $(7\cos t + 8t - 14t^2)\hat{\mathbf{i}} + (-7\sin t + 2 + 28t^3)\hat{\mathbf{j}} + (2t\sin t - 4t^2\cos t + 7)\hat{\mathbf{k}}$; (b) $7\hat{\mathbf{i}} + 2\hat{\mathbf{j}} + 7\hat{\mathbf{k}}$;
(c) $6.64\hat{\mathbf{i}} + 2.14\hat{\mathbf{j}} + 6.60\hat{\mathbf{k}}$ rad/sec²
6.21 $\boldsymbol{\alpha}_{\mathcal{A}/\mathcal{G}} = 1.12\hat{\mathbf{i}}_1 + 13.4\hat{\mathbf{j}}_1$
6.23 $-1.5\hat{\mathbf{i}} - 4.20\hat{\mathbf{j}} + 4.73\hat{\mathbf{k}}$ rad/sec²
6.25 $\mathbf{v}_{P/\mathcal{I}} = -4\hat{\mathbf{j}} - 6\hat{\mathbf{k}}$ ft/sec;
$\mathbf{v}_{P/\mathcal{G}} = -9\hat{\mathbf{i}} - 4\hat{\mathbf{j}} - 6\hat{\mathbf{k}}$ ft/sec
6.27 $4\hat{\mathbf{i}} + 40\hat{\mathbf{j}}$ ft/sec; $-400\hat{\mathbf{i}} + 80\hat{\mathbf{j}} - 4\hat{\mathbf{k}}$ ft/sec²
6.29 $6\hat{\mathbf{j}}$ rad/sec²; $-32.0\hat{\mathbf{i}} + 1.20\hat{\mathbf{j}} + 8.90\hat{\mathbf{k}}$ in./sec²
6.31 36.5 ft
6.33 $17.3\hat{\mathbf{i}} + 15.7\hat{\mathbf{j}} - 10\hat{\mathbf{k}}$ in./sec;
$-147\hat{\mathbf{i}} + 218\hat{\mathbf{j}} - 64.3\hat{\mathbf{k}}$ in./sec²
6.35 $\mathbf{a} = -14.3\hat{\mathbf{i}} + 2.61\hat{\mathbf{j}}$ ft/sec²;
magnitude = 14.5 ft/sec²
6.37 $x = (v_o t - R)\sin\omega_o t$; $y = (v_o t - R)\cos\omega_o t + R$
6.39 6.60 rad/sec **6.41** Answer given in problem.

6.43

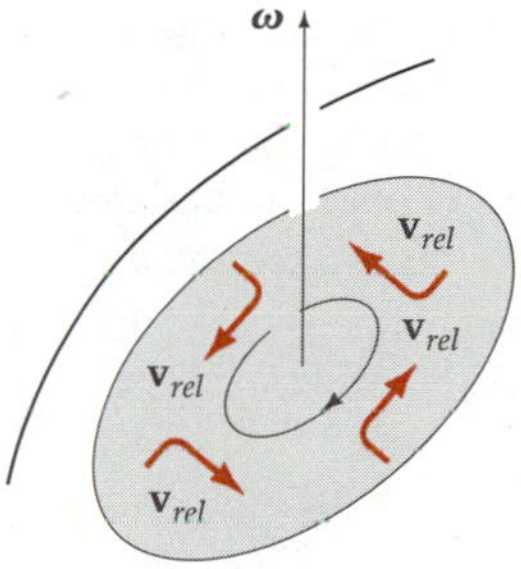

"$-2m\boldsymbol{\omega} \times \mathbf{v}_{rel}$" is a "force" that will change the particles' velocity directions relative to the earth so as to produce ccw rotation in the northern hemisphere. The effect is opposite in the southern hemisphere.
6.45 $-0.0923\hat{\mathbf{i}}$ ft/sec² **6.47** Answer given in problem.
6.49 $\dfrac{2\pi}{T}\left(\sin\beta\hat{\mathbf{j}} - \dfrac{\sin^2\beta}{\cos\beta}\hat{\mathbf{k}}\right)$
6.51 $-1350\hat{\mathbf{j}}$ rad/sec² (each)
6.53 $\mathbf{v}_{P/\mathcal{G}} = -15\hat{\mathbf{i}} - 10\hat{\mathbf{k}}$ ft/sec;
$\mathbf{a}_{P/\mathcal{G}} = -65\hat{\mathbf{j}} - 7.5\hat{\mathbf{k}}$ ft/sec²
6.55 $\mathbf{v}_Q = r\Omega_2 s_1\hat{\mathbf{i}} + (R\Omega_2 - r\Omega_1 C_1)\hat{\mathbf{j}} - r\Omega_1 s_1\hat{\mathbf{k}}$;
$\mathbf{a}_Q = (-R\Omega_2^2 + 2r\Omega_1\Omega_2 C_1)\hat{\mathbf{i}} + (r\Omega_1^2 s_1 + r\Omega_2^2 s_1)\hat{\mathbf{j}} - r\Omega_1^2 C_1\hat{\mathbf{k}}$, where $s_1 = \sin\Omega_1 t$ and $C_1 = \cos\Omega_1 t$
6.57 (a) $\boldsymbol{\omega} = -(R\omega_1/r)\hat{\mathbf{i}} + \omega_1\hat{\mathbf{k}}$; $\boldsymbol{\alpha} = -(R\omega_1^2/r)\hat{\mathbf{j}}$
(b) $R\omega_1^2\hat{\mathbf{i}} + (R^2\omega_1^2/r)\hat{\mathbf{k}}$
6.59 $0.2\left(\dfrac{\sqrt{3}}{2}\hat{\mathbf{i}} + \dfrac{1}{2}\hat{\mathbf{k}}\right)$ rad/sec
6.61 $\dfrac{\omega_1 + \omega_2}{2}\hat{\mathbf{k}}$; $\dfrac{R(\omega_1 - \omega_2)}{2r}\hat{\mathbf{j}}$; $\dfrac{R(\omega_1 + \omega_2)}{2}\hat{\mathbf{i}}$;
$\dfrac{R(\omega_1 + \omega_2)}{2}\hat{\mathbf{i}} - \dfrac{r(\omega_1 + \omega_2)}{2}\hat{\mathbf{j}} - \dfrac{R(\omega_2 - \omega_1)}{2}\hat{\mathbf{k}}$
6.63 1.36 rad/s, directed from O through the line of contact between $\mathcal{C}$ and $\mathcal{D}$.
6.65 $-1.68\hat{\mathbf{i}} + 2.24\hat{\mathbf{j}} - 2\hat{\mathbf{k}}$ rad/sec
6.67 Answer given in problem.
Also, $R = \dfrac{H}{2} + 13$ so right side is $(6H - 260)\omega_{\mathcal{B}}$
6.69 $-0.000432t^5/\sqrt{363 - 0.000144t^6}\,\hat{\mathbf{k}}$ m/s
6.71 (a) Answer given in problem; (b) r/R
6.73 $\boldsymbol{\omega}_{\mathcal{L}} = -22.6\hat{\mathbf{i}}$ rad/sec; $\boldsymbol{\omega}_{\mathcal{R}} = -27.7\hat{\mathbf{i}}$ rad/sec;
$\boldsymbol{\omega}_{\mathcal{D}} = -43.6\hat{\mathbf{j}}$ rad/sec with 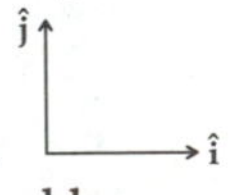

6.75 (a) Answer given in problem.
(b) Answer same if $\boldsymbol{\alpha}$ is replaced by $\boldsymbol{\omega}$.
6.77 $-11.8\hat{\mathbf{i}} + 25.0\hat{\mathbf{j}} - 5.33\hat{\mathbf{k}}$ rad/sec²
6.79 $|\boldsymbol{\omega}_{\mathcal{B}/\mathcal{I}}| = \sqrt{\dot{\phi}^2 + \dot{\theta}^2 + \dot{\psi}^2 + 2\dot{\phi}\dot{\psi}\cos\theta}$
6.81 Components are:
$(\dot{\theta}_1 + s_2\dot{\theta}_3,\ c_1\dot{\theta}_2 - s_1c_2\dot{\theta}_3,\ s_1\dot{\theta}_2 + c_1c_2\dot{\theta}_3)$

6.83 Let point A be displaced from its original to its final position. Then, using Euler's Theorem, all other points of the body may be placed in their final positions via a single rotation about an axis through A.
6.85 $\dot{x} = r(\dot{\psi}s_\phi s_\theta + \dot{\theta}c_\phi)$; $\dot{y} = -r(\dot{\psi}c_\phi s_\theta - \dot{\theta}s_\phi)$
6.87 Components are:
$(\dot{\theta}_1 + \dot{\theta}_3 s_2,\ \dot{\theta}_2 c_1 - \dot{\theta}_3 s_1 c_2,\ \dot{\theta}_2 s_1 + \dot{\theta}_3 c_1 c_2)$,
the same as Problem 6.73.
6.89 $\dot{E}\hat{\mathbf{n}}_{21} + (S_E\dot{A} + \dot{P})\hat{\mathbf{n}}_{22} + \dot{A}C_E\hat{\mathbf{n}}_{23}$
6.91 $E = \tan^{-1}[(c_\lambda c_\delta - r)/(s_\lambda c_\delta)]$;
$A = \tan^{-1}[s_\delta/\sqrt{c_\delta^2 - 2rc_\lambda c_\delta + r^2}]$, where $r = R_e/R$
6.93 Be sure the cross is rigid and planar!

CHAPTER 7

7.1 With $\hat{\mathbf{i}}$ along the axle from O through the wheel center, and $\hat{\mathbf{k}}$ out of the page,
$$\mathbf{H}_O = \frac{2\pi mb^2}{T}\left[-\cos\beta\sin^2\beta\hat{\mathbf{i}} + \left(1 - \frac{\cos^2\beta}{2}\right)\sin\beta\hat{\mathbf{j}}\right]$$
7.3 $m\ell^2\omega_0[2\hat{\mathbf{i}} + (10/3)\hat{\mathbf{k}}]$
7.5 With $\hat{\mathbf{i}}$ and $\hat{\mathbf{j}}$ parallel to x and y of the figure,
$$\mathbf{H}_C = \frac{mr^2}{4}\Omega\sin\phi\left(\hat{\mathbf{i}} - \frac{2\ell}{r}\hat{\mathbf{j}}\right)$$
7.7 $I^Q_{xx} = 11.7\ m\ell^2$; $I^Q_{yy} = 14.3m\ell^2$; $I^Q_{zz} = 8.00m\ell^2$;
$I^Q_{xy} = 2.00m\ell^2$; $I^Q_{xz} = 4.50m\ell^2$; $I^Q_{yz} = 0$ **7.9** $0.944m\ell^2$
7.11 $\dfrac{-mR^2}{2}\sin\phi\cos\phi$ **7.13** $2m(a^2 + b^2 + c^2)/15$
7.15 4.33 slug-ft² **7.17** $-0.0186ma^2$
7.19 If $c > a + b$, the ellipsoid $ax^2 + by^2 + cz^2 = 1$ cannot be an ellipsoid of inertia, for then it would represent a body having one moment of inertia > sum of other two, a physical impossibility.
7.21 $I_1 = 1.5mR^2$, d.c.'s (0,0,1);
$I_2 = 1.41mR^2$, $(-0.345, 0.939, 0)$;
$I_3 = 0.0942mR^2$, $(0.939, 0.345, 0)$
7.23 $\frac{5}{6}\ell\hat{\mathbf{i}} + \frac{1}{2}\ell\hat{\mathbf{j}} - \frac{1}{6}\ell\hat{\mathbf{k}}$; the direction is different.
7.25 $I^P_{xz} = \overset{0}{\cancel{I^C_{xz}}} - m\overset{0}{\cancel{a}}c = 0$ and $I^P_{yz} = \overset{0}{\cancel{I^C_{xz}}} - m\overset{0}{\cancel{b}}c = 0$,
so that z is a principal axis for every point on that axis

7.27 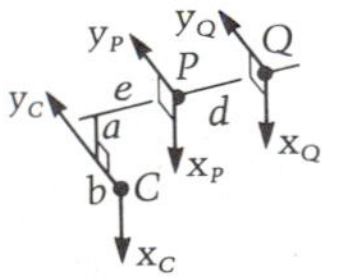

$$I^C_{xz} = \overset{0}{\cancel{I^P_{xz}}} + m(-a)e = mae$$
$$I^C_{xz} = \overset{0}{\cancel{I^Q_{xz}}} + m(-a)(e + d)$$
$$= ma(e + d)$$
$\therefore a = 0$. From I_{yz} equations, get $b = 0$, so C is on the z axis and by Problem 7.26, z is a principal axis for C.
7.29 $I_1 = 83.4\rho t$, d.c.'s (0, 0, 1);
$I_2 = 74.6\rho t$, $(0.985, -0.170, 0)$;
$I_3 = 8.80\rho t$, $(0.170, 0.985, 0)$
7.31 $I_1 = 37.2\rho Ar^3$, d.c.'s (1, 0, 0);
$I_2 = 6.18\rho Ar^3$, (0, 0.454, 0.891);
$I_3 = 37.3\rho Ar^3$, $(0, -0.891, 0.454)$
7.33 $I_1 = \dfrac{7}{3}m\ell^2$ with $\hat{\mathbf{n}}_1 = \left(\dfrac{1}{\sqrt{3}}, \dfrac{1}{\sqrt{3}}, \dfrac{1}{\sqrt{3}}\right)$;
$I_2 = I_3 = \dfrac{23}{6}m\ell^2$ with $\hat{\mathbf{n}}$ in any direction normal to $\hat{\mathbf{n}}_1$
7.35 $I_1 = 0.278mH^2$, d.c.'s (0, 0, 1);
$I_2 = 0.239mH^2$, $(0.290, -0.957, 0)$;
$I_3 = 0.0387mH^2$, $(0.957, 0.290, 0)$
7.37 3.58°, working with six digits and rounding at the end
7.39 (a) $\begin{bmatrix} 1.44m\ell^2 & 0.563m\ell^2 & 0.500m\ell^2 \\ & 1.44\ m\ell^2 & -0.500m\ell^2 \\ & \text{Symmetric} & 1.54\ m\ell^2 \end{bmatrix}$;
(b) $I_1 = 2.00m\ell^2$, d.c.'s (0.707, 0.707, 0);
$I_2 = 1.99m\ell^2$, $(0.379, -0.379, 0.844)$;
$I_3 = 0.427m\ell^2$, $(0.597, -0.597, -0.535)$
7.41 $I_1 = 34.3 \times 10^5$ slug-ft², $\hat{\mathbf{n}} = (0, 0, 1)$
$I_2 = 33.7 \times 10^5$ slug-ft², $\hat{\mathbf{n}} = (0.566, -0.824, 0)$
$I_3 = 47.5 \times 10^5$ slug-ft², $\hat{\mathbf{n}} = (0.824, 0.566, 0)$
Note: There is a precision problem here because (1) I_{xx} and I_{yy} are so much larger than I_{xy}, and (2) I_{xx} and I_{yy} are nearly equal.
7.43 Left: 121 ↓ N; right: 121 ↑ N
7.45 At B_1: $(1419 - 507\cos\epsilon)\sin\epsilon$ ↑;
at B_2: $(659 + 507\cos\epsilon)\sin\epsilon$ ↑
7.47 At A: $\dfrac{mr^2\Omega^2\sin 2\beta}{16L}$ ↓;
at B: same magnitude but ↑
7.49 (a) 2.41 ⮌ lb-ft; (b) 1.09 ⮌ lb-ft
7.51 $\Sigma F_x = m\dot{v}_{Cx} + \omega_y mv_{Cz} - \omega_z mv_{Cy}$;
$\Sigma F_y = m\dot{v}_{Cy} + \omega_z mv_{Cx} - \omega_x mv_{Cz}$;
$\Sigma F_z = m\dot{v}_{Cz} + \omega_x mv_{Cy} - \omega_y mv_{Cx}$;
7.53 $\dot{\omega}_s = 0$, $\dot{\omega}_x = \dfrac{M_o}{I + \bar{I} + m\ell^2}$,
$$\text{Force} = \left(0, \frac{-m\ell M_o}{I + \bar{I} + m\ell^2}, mg + m\ell\omega_x^2\right),$$
$$\text{Couple} = \left(\frac{IM_o}{I + \bar{I} + m\ell^2}, 0, J\omega_x\omega_s\right)$$
7.55 332 days
7.57 $\dot{\psi} = 30.7$ rad/s, about the $+z$ axis
7.59 0.123 lb-in.
7.61 $\phi = \cos^{-1}\{g\ell/[\Omega^2(\ell^2 - r^2/4)]\}$; for $\phi = 60°$,
$\Omega_{\min} = 3.35$ rad/sec $< 2\pi$ rad/sec
7.63 With x along $\mathcal{S}$ from Q and y upward,
$$\mathbf{F} = (-mR\omega_1^2, mg, 0) \text{ and } \mathbf{M}_C = \left(0, 0, \frac{mr^2}{2}\omega_1\omega_2\right)$$
7.65 $\sin\theta - \dfrac{v_C^2}{gR}\left(\dfrac{3}{2}\cos\theta + \dfrac{r\sin\theta\cos\theta}{4R}\right) = 0$

7.67 (a) stern bearing: $149\hat{\mathbf{i}} + R_y\hat{\mathbf{j}} + 8850\hat{\mathbf{k}}$ N;
bow bearing: $149\hat{\mathbf{i}} - R_y\hat{\mathbf{j}} + 15650\hat{\mathbf{k}}$ N;
(b) stern: $48460\hat{\mathbf{i}} + R_y'\hat{\mathbf{j}} + 12250\hat{\mathbf{k}}$ N; bow: $-48460\hat{\mathbf{i}} - R_y'\hat{\mathbf{j}} + 12250\hat{\mathbf{k}}$ N (R_y and R_y' are indeterminate).

7.69

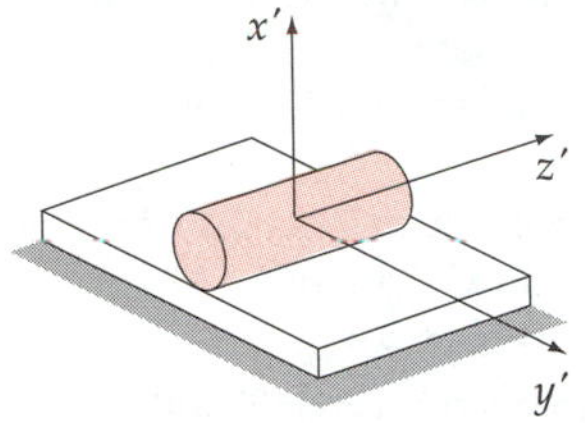

$$M_{x'} = \omega_1\dot{\beta}\,(2IC_\beta^2 - J\cos 2\beta);$$
$$M_{y'} = (J - I)\omega_1^2 S_\beta C_\beta;$$
$$M_{z'} = 2(J - I)\omega_1\dot{\beta} S_\beta C_\beta$$

7.71 With $\hat{\mathbf{i}}$ from O toward C, and $\hat{\mathbf{j}}$ in the direction of $\mathbf{v}_c$, $\mathbf{F}$ on $\mathcal{D} = \dfrac{-mv_C^2}{R}\hat{\mathbf{i}}$ by the bearing at C, $-mg\hat{\mathbf{k}}$ from gravity, and $mg\hat{\mathbf{k}}$ from the ground;
and $\mathbf{C}$ on $\mathcal{D}$ by bearing $= \dfrac{-2IR\Omega^2}{r}\hat{\mathbf{j}}$.
The difference is that in this problem the normal force from the ground is just the weight.

7.73 $12\hat{\mathbf{i}} + 32.2\hat{\mathbf{j}} + 6\hat{\mathbf{k}}$ lb; $14.4\hat{\mathbf{i}} + 20\hat{\mathbf{j}} - 123\hat{\mathbf{k}}$ lb-ft

7.75 RHS of x-eqn $= 5.6$ (the same);
RHS of y-eqn $= 5.6 \neq 7.7$; RHS of z-eqn $= 2.8 \neq 1.4$;
Axes used for Euler Equations must be *body*-fixed, not just permanently principal.

7.77 $(-\ddot{\phi}s_\theta - \dot{\phi}\dot{\theta}C_\theta + \dot{\psi}\dot{\theta})\hat{\mathbf{i}}_2 + (\ddot{\theta} + \dot{\phi}\dot{\psi} - s_\theta)\hat{\mathbf{j}}_2 + (\ddot{\psi} + \ddot{\phi}C_\theta - \dot{\theta}\dot{\phi}s_\theta)\hat{\mathbf{k}}_2$ **7.79** $\dot{\psi} = 244$ rad/sec

7.81 0.328 m (Also, 51.9 m is a solution!)

7.83 $I_{zz}^C = 18mR^2$; $I_t = 12mR^2$; ratio $= 1.5$;
(i) is unstable and the other three are stable.

7.85 For a torque-free body $\mathcal{B}$ in general motion in an inertial frame $\mathcal{I}$, with $\boldsymbol{\omega}_{\mathcal{B}/\mathcal{I}}$ not parallel to $\mathbf{H}_C$, we have $\Sigma \mathbf{M}_C = \mathbf{0} = {}^{\mathcal{B}}\dot{\mathbf{H}}_C + \boldsymbol{\omega}_{\mathcal{B}/\mathcal{I}} \times \mathbf{H}_C$, and the two terms *add* to zero.

7.87 If $\mathcal{B}$ is in equilibrium in $\mathcal{I}$, all its points are stationary there; thus $\mathbf{a} = \mathbf{0}$ for all these points, and also $\boldsymbol{\omega}_{\mathcal{B}/\mathcal{I}} = \mathbf{0}$. Hence $\mathcal{B}$ is an inertial frame. But if $\mathcal{B}$ is an inertial frame, it can at most translate at constant velocity with respect to another inertial frame $\mathcal{I}$. Thus it need not be stationary in $\mathcal{I}$, i.e., need not be in equilibrium in $\mathcal{I}$ even though none of its points accelerates in $\mathcal{I}$!

7.89 Answer agrees with Example 7.11.

7.91 $\dfrac{11.4\,v_b m}{s(15M + 24m)}\hat{\mathbf{i}} - \dfrac{33v_b m}{s(15M + 24m)}\hat{\mathbf{k}}$

7.93 No. Two different results are obtained for F_y.

7.95 $815m$ ft-lb, where m is the mass in slugs

7.97 $W = T_f = \dfrac{4\pi^2 m(b^2 - r^2)(b^2 - 0.75r^2)}{b^2T^2}$

7.99 (a) $\dfrac{a^2b^2m\omega^2}{12(a^2 + b^2)}$; (b) $\sqrt{\dfrac{a^2 - b^2}{12}}$

7.101 (a) Torque $= 0$, At B: $\dfrac{ma\omega^2}{12} \leftarrow$, At A: $\dfrac{ma\omega^2}{4} \leftarrow$;
(b) $I_1 = \dfrac{5}{6}ma^2$, z axis;
$I_2 = 0.116ma^2$, d.c.'s (0.290, 0.957, 0);
$I_3 = 0.717ma^2$, d.c.'s (0.957, −0.290, 0);
(c) no hole is physically possible; (d) $ma^2\omega^2/12$

7.103 358 J

7.105 $0.339\,\dfrac{g}{r} = \omega_x^2 - 1.33\omega_x\omega_y + 2.03\,\omega_y^2$

7.107 Some check values: $\left(\alpha, \dfrac{v_Q^2}{gR}\right) = (1°, 913)$, $(45°, 0.842)$, and $(80°, 0.031)$

CHAPTER 8

8.1 $\omega_n = \sqrt{\dfrac{gr^2/R}{k_C^2 + r^2}}$; for $k_C = \sqrt{\dfrac{2}{5}}\,r$, $\omega_n = \sqrt{\dfrac{5g}{7R}}$

8.3 $m\ell^2\ddot{\theta} + ka^2\theta - mg\ell\theta = 0$; $2\pi\sqrt{\dfrac{m\ell^2}{ka^2 - mg\ell}}$

8.5 $\sqrt{2k/(12m\pi^2)}$ **8.7** 40.0 ft

8.9 Answer given in problem. **8.11** $\omega_n = 3\sqrt{\dfrac{T}{2m\ell}}$

8.13 (a) 0.278 ft (b) 0.876 sec (c) 0.291 sec

8.15 (1) moves to left with $x = \dfrac{6\mu gm}{k}\left(1 - \cos\sqrt{\dfrac{k}{m}}\,t\right)$
(2) moves to right with $x = \dfrac{4\mu gm}{k}\left(1 - \cos\sqrt{\dfrac{k}{m}}\,t\right)$
(3) moves to left with $x = \dfrac{2\mu gm}{k}\left(1 - \cos\sqrt{\dfrac{k}{m}}\,t\right)$ and stops for good $\mu gm/k$ to left of unstretched position. Time in *each* interval is $\pi\sqrt{m/k}$, and total distance traveled $= 24\,\mu gm/k$. **8.17** 0.107 sec

8.19 (a) Work $= \displaystyle\int_0^{L/v} Pvt[\dot{\theta}(t)]\,dt =$
$(P^2/k)\left[\dfrac{1}{2} - \dfrac{v}{\omega L}\sin\dfrac{\omega L}{v} - \dfrac{v^2}{\omega^2L^2}\cos\dfrac{\omega L}{v} + \dfrac{v^2}{\omega^2L^2}\right]$,
(b) which agrees with $T + \phi$.

8.21 $\dfrac{2\ell}{b^2}\sqrt{(mg\ell + ka^2)m}$

8.23 8 lb-sec/in.; 8 times as great;
$\theta_o e^{-100t}(1 + 100t)$; 0.0628 sec

8.25 Yes, they do. (Start with $x = Ae^{-\zeta\omega_n t}\sin(\omega_d t - \varphi)$ and investigate when $\dot{x} = 0$!)

8.27 (a) 0.109 in.;
(b) 1.36 rad or 78.1°, with x lagging y
8.29 $x(t) = 0.5[1 - (1 + 100t)e^{-100t}]$ in.
8.31 (a) $x = 0.025 \cos 100t - 0.0269e^{-26.8t} + 0.0019e^{-373t}$ in. (b) $x = 0.025 \cos 100t - e^{-50t} \cdot (0.0144 \sin 86.6t + 0.025 \cos 86.6t)$ in.
8.33 $-Y\omega^2\sqrt{\dfrac{(c\omega)^2 + (2k)^2}{(2k - m\omega^2)^2 + (c\omega)^2}} \sin(\omega t - \phi)$,
where $\phi = \tan^{-1}\left[\dfrac{cm\omega^3}{2k(2k - m\omega^2) + (c\omega)^2}\right]$;
the radicand is < 1 if $\omega > \sqrt{2}\,\omega_n$, and for these ω's, the radicand is smaller for smaller values of c.
8.35 $wA|\mathbf{v}|\mathbf{v}/g - \mathbf{W}$,
where $\mathbf{W}$ is the weight of tank plus fluid.
8.37 $Q_u/Q = (1 + \cos\theta)/2$; $Q_l/Q = (1 - \cos\theta)/2$
8.39 $q(u - v) + fu$ **8.41** $MP/(M + qt)^2$
8.43 $2480\hat{\mathbf{i}}$ lb **8.45** (a) 0.553 ft/sec²; (b) 328 ← lb
8.47 (a, c) Answer given in problem;
(b) $4.95 \dfrac{m_o g}{k_o}(\sin\alpha - \mu\cos\alpha)$
8.49 Answer given in problem.
8.51 $v(t) = -gt + v_e \ln[m_0/(m_0 - \mu t)]$, valid until fuel gone
8.53 $x(t) = -gt^2/2 + v_e\left[\left(\dfrac{m_0}{\mu} - t\right)\ln\left(1 - \dfrac{\mu t}{m_0}\right) + t\right]$;

$$v_{\text{burnout}} = \frac{-gfm_0}{\mu} - v_e \ln(1 - f);$$

$$x_{\text{burnout}} = \frac{-gf^2m_0^2}{2\mu^2} + \frac{v_e m_0}{\mu}[(1 - f)\ln(1 - f) + f]$$

8.55 (a) $v = \sqrt{2gx/3}$; (b) $a = g/3$;
(c) Mechanical energy is lost (to heat, deformation, vibration, etc.) as the links suddenly join the falling part of the chain. **8.57** Answers given in problem.
8.59 (a) $0.193gT$. (b) Answer given in problem.
8.61 Answer given in problem.
8.63 Answer given in problem. (Use Kepler's laws!)
8.65 1.67×10^9 mi
8.67 107 min ("No air resistance" needn't be assumed this time!) **8.69** 4320 mi; 583 mi and 137 mi
8.71 r_{p*} is 3620 mi, and this is $< r_{\text{earth}}$
8.73 Answer given in problem. (Isolate the radical and square both sides.) **8.75** Answer given in problem.
8.77 $0.00530\sqrt{GM}$ mi/hr **8.79** 5960 mi; 77.5°
8.81 Answer given in problem.
8.83 No; 4980 mph **8.85** $GMm(n - 1)/(2Rn)$
8.87 104 min

APPENDIX A

A.1 (a) lb-ft²/slug²; (b) N · m²/kg² **A.3** 143 N
A.5 (a) 0.456 slug-ft/sec; (b) 2.03 kg · m/s
A.7 (a) $\dfrac{L}{T^2}L^2 = L^3\left(\dfrac{1}{T}\right)^2$;
(b) 6.61, using $g = 32.17$ ft/sec²,
$\omega = 2\pi\left(1 + \dfrac{1}{365}\right)$ rad/day, and $r_e = 3960$ mi

E

Additional Model-Based Problems for Engineering Mechanics

Kinematics of Rolling

Draw radial lines on the ends of a large and a small cylinder or tube, measure their outer diameters, and lay them on a table as shown in (a) below. Predict the number of times the small cylinder will turn in space if it rolls completely around the large cylinder. Hold the large cylinder still and slowly and carefully turn the small cylinder clockwise, maintaining contact and not allowing them to slip. If slipping is a problem, increase the friction between the surfaces by placing a rubber band or strip of tape around the small cylinder. Record the number of times the small cylinder turns in space before it contacts the starting point on the large cylinder. Did you correctly predict this number? Dividing the circumferences to predict the number of revolutions is incorrect! Place a small cylinder inside a large tube as shown in (b). Carefully rotate the small cylinder, not allowing it to slip, on the inside of the tube and measure the number of turns before it returns to the starting point. Use the appropriate kinematic relation, the inside diameter of the large tube, and the outside diameter of the small cylinder to calculate the same result.

REFERENCE — *Dynamics,* Section 3.6.

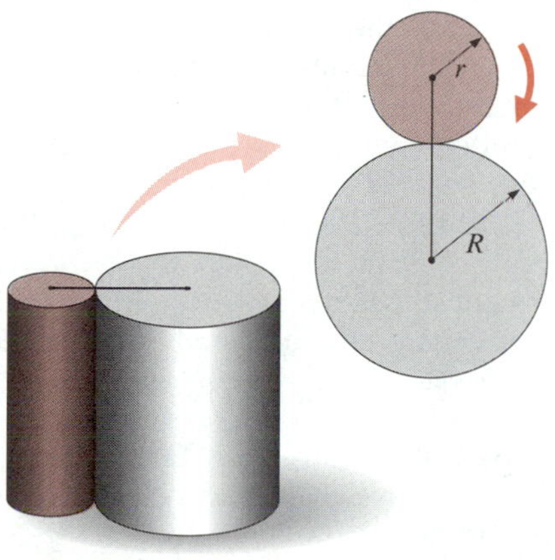

(a)

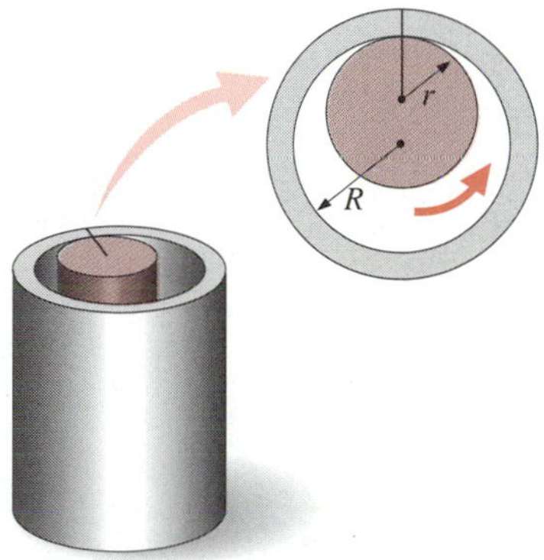

(b)

IMPACT AND TIPPING OF A SLIDING BLOCK

Install a small lip or bump on an inclined straight board and place a rectangular block on the board. Hold the board at an angle $\theta > \tan^{-1}\mu_s$ and release the block from rest behind the lip, as depicted in (a) below. Observe the motion of the block as it slides toward and impacts the lip. Release the block from different positions until you find the length L for which the block slides to the lip and barely tips over after impact, as in (b). Use the equations of motion and the impulse–momentum and work–energy principles to calculate length L in terms of μ_k, θ, and the dimensions of the block. Compare the experimental and theoretical results and explain any differences. Repeat this demonstration for several different values of θ and sizes of blocks. Graph the predicted and actual values of L and θ.

REFERENCES — *Dynamics,* Sections 5.2 and 5.3.

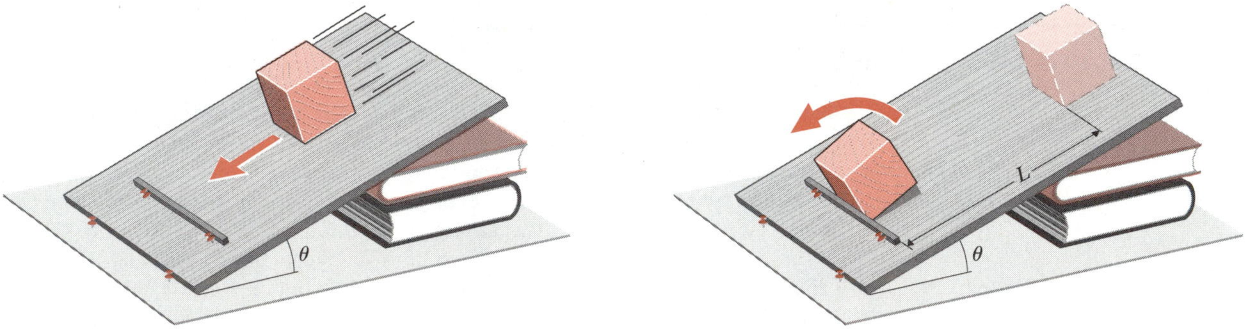

(a) (b)

ROLLING A CYLINDER OVER A STEP

Select a cylinder of radius R and install a step of height $H \cong 0.1R$ on an inclined straight board (a). Hold the board at some convenient angle θ between 10° and 20°. Release the cylinder from rest behind the step as depicted in (b) below. Observe the motion of the cylinder as it rolls toward and impacts the step. Release the cylinder from different positions until you find the length L for which the cylinder just rolls over the step after impact. Repeat the experiment for several values of step height $H \leq R$ and measure the minimum distance L the cylinder must roll to barely climb the step. (An easy way to vary the height is to build the steps from layers of stiff cardboard cut from the backing of a note pad.)

Using the equations of motion and the principles of impulse–momentum and work–energy, calculate the length L in terms of the ratio H/R and the slope angle θ. Plot the predicted and measured values of L versus H/R for different slopes and compare the results. Show that your results are independent of the friction coefficient and weight of the cylinder. Do your results depend upon the coefficient of restitution?

REFERENCES — *Dynamics,* Sections 5.2 and 5.3.

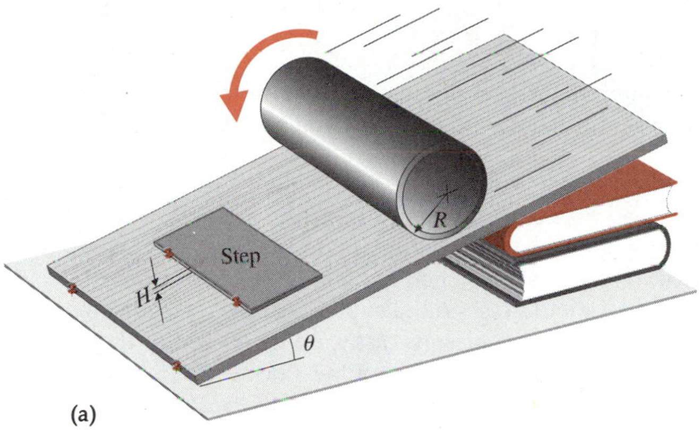

(a)

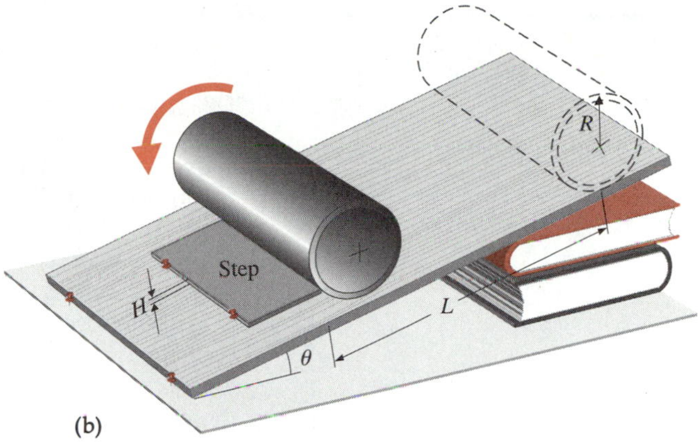

(b)

Motion, Momentum, and Energy of a Simple Pendulum

Build a simple pendulum of length L using string and a small sphere as shown in (a) below. Release the pendulum from rest at angle θ_0 and measure its period of vibration. Repeat the experiment for different lengths and release angles. Compare your values with those predicted from the linear theory of vibrations. Indicate where the actual behavior of your pendulum begins to deviate from the theoretical. Explain any differences.

Place a small peg a distance d below the point of attachment of the pendulum (b). Predict what will happen if the mass is released from rest at $\theta_0 = 90°$. Calculate the minimum distance d_{min} for which the ball will make at least one **complete** revolution around the peg. Determine d_{min} experimentally and explain any difference in the two values. Repeat the calculation and experiment for different release angles. Do you find better agreement between the values?

REFERENCES — *Dynamics,* Sections 2.4 and 8.2.

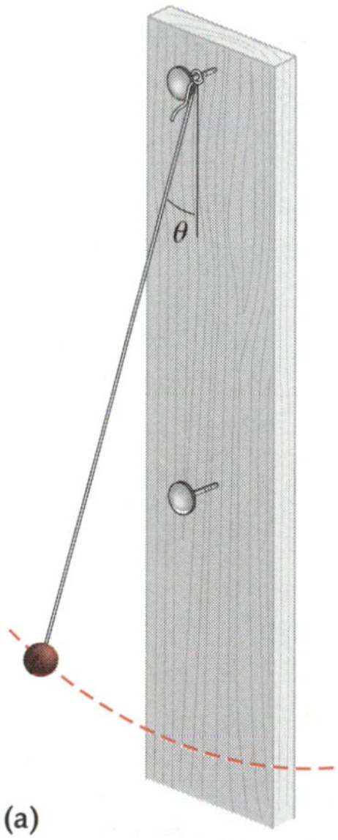

(a)

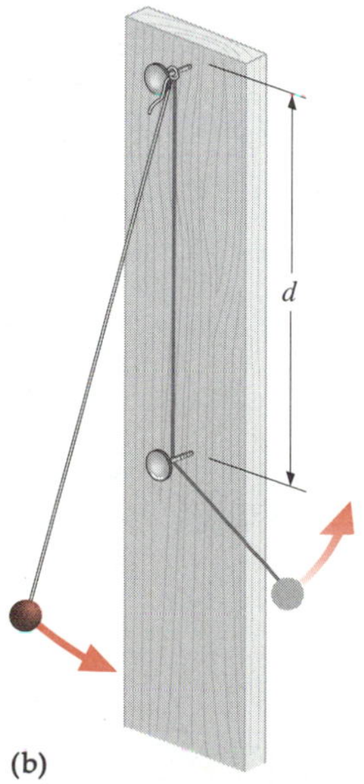

(b)

USING VIBRATIONS TO DETERMINE MOMENT OF INERTIA

This experiment explores a method for determining the mass moment of inertia of a rigid body about an axis of rotation and about a parallel axis through the body's center of mass.

Construct a triangle from a piece of stiff wire (steel coathanger wire is suitable). The sides should be several inches long and of unequal length. Pass a small-diameter rod or thin rigid support through the frame and hold it horizontally, as shown in (a) below (an ordinary notched key works well as a support). One of the corners of the frame should rest on the support. Allow the frame to oscillate about the fixed support in its own plane. Measure the period of these oscillations and the mass and location of the center of mass of the frame. From these values you can compute the moment of inertia I_0 about the axis of rotation. You can also calculate I_0 from the definition of mass moment of inertia if you know the density of the material and the geometry of the frame. Compare the experimental and theoretical values of I_0.

Devise a scheme to determine the mass moment of inertia I_C about the centroidal axis perpendicular to the plane of a steel wire coathanger, as shown in (b). *Hints*: Measure the periods of oscillation about axes through points A and B and the distance AB. Do not attempt to locate the center of mass C. Use the parallel axis theorem. Generalize this approach for irregular rigid bodies with a plane of symmetry.

REFERENCES — *Dynamics,* Sections 4.4, 8.2, and Appendix C.

(a)

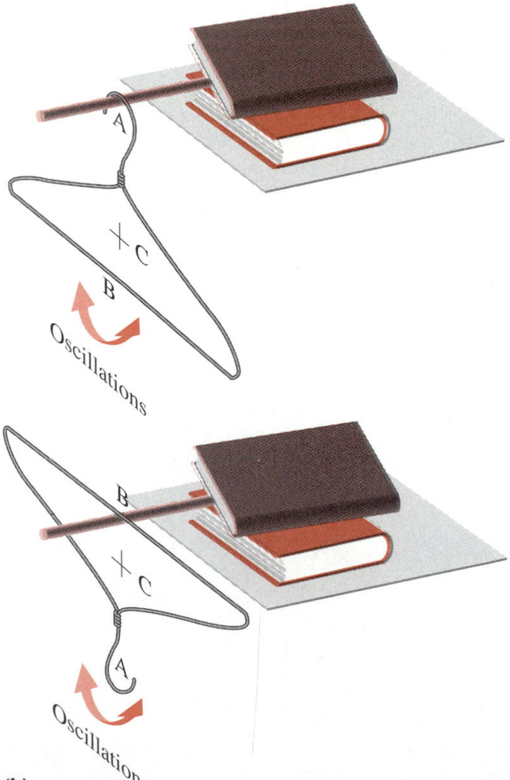

(b)

INDEX

Absolute system, 559
Acceleration, 6, 8
    angular, 163–164, 398–399
    Coriolis, 207, 405–406
    in different frames, 207–208
    equations of, 414–415
    in moving frames of reference, 401, 405–406
    radial and transverse components of, 32
    tangential and normal components of, 43
Action-reaction principle, 64
Addition theorem for angular velocities, 385–386, 428
Amplitude, 518
Angle
    Eulerian, 428–433
    friction, 66
    nutation, 493
    phase, 518
Angular acceleration, 163–164
Angular acceleration vector, 398–399
Angular impulse, 344
Angular momentum, 118, 227–228
    conservation of, 350
    relative, 124
    in three dimensions, 446
Angular speed, 136, 184
Angular velocity, 134–135, 136, 380–381, 383, 432
    addition theorem and, 385–386
    differentiation of, 387
    properties of, 384–388
    simple, 386–388
Apogee, 544
Arclength, 43
Axes of inertia
    non-principal, 473
    principal, 455, 456, 462, 466
Axis of rotation, 277–278, 422

Balance, static, 292
Balancing a rotating body, 295
Bearing reactions, 291–294
Binormal vector, unit, 51
Body, 56
    cone, 496
    mass center of, 62–64
    rigid. *See* Rigid body
    unbalanced, rotation of, 291–293
Body extended, 130

Cardan suspension, 431
Cartesian coordinates, 24, 381
Center
    of curvature, 49
    of mass, 58–60, 62, 219
    of percussion, 362
    of zero velocity, 149–151
Central force, 89–90
Central force motion, 543
Central impact, 107
Chasle's theorem, 434
Circular frequency, 518
Circular orbit, 547
Coefficient of friction, 65
Coefficient of restitution, 106–108, 355, 361
Cones, body and space, 496
Conic, 546
Conservation
    of angular momentum, 350
    of energy, 92
    of mechanical energy, 92, 330–331
    of moment of momentum, 120–121
    of momentum, 102–103, 350
Conservative forces, 91–92, 329–331
Constant force, 89
Continuity equation, 535

Control volume
    Euler's laws for, 535–536
Coordinates
    Cartesian, 24, 381
    cylindrical, 31
    polar, 31
    spherical, 43
Coriolis acceleration, 207, 405–406
Coulomb law of friction, 249
Couple, 309, 311
Cramer's rule, 457
Critical damping, 522
Curvature, 45, 149
    center of, 49
    radius of, 45
Curvilinear translation, 148
Cylindrical coordinates, 31

Damped vibration, 521
Damping
    critical, 522
    subcritical, 523
    supercritical, 523
Deformation, 60, 116
Derivative, 399
    of a vector, 3
        in different frames, 198–200
Derivative formula, 384
Differential equations
    complementary solution, 526
    energy integral, 68
    particular solution, 526
    of plane motion, 246
Direct impact, 107
Direct precession, 496
Displacement, 522
    steady-state, 526
Drag, 74, 85

Earth
    as a moving frame, 410–411
    gravity and, 91
    lunisolar precession of, 494, 497
    motion near the surface, 410
    orbit around sun, 440
    satellites, geosynchronous, 81, 439
Eccentricity, 546
Eigenvalue, 457
Eigenvector, 457
Ellipsoid of inertia, 454
Energy
    conservation of, 92
    kinetic. *See* Kinetic energy
    mechanical, 330–331
    potential, 329–331
Energy integral, 68
Equations of motion, 249–250, 272–273, 274
    for fixed-axis rotation, 277–278
Escape velocity, 85, 553
Euler equations, 432, 472–473
Eulerian angles, 428–433
Euler's first law, 56–58, 62
    control-volume form of, 536
    momentum forms of, 101–102
Euler's second law, 117–118, 495
    control-volume form of, 536
    momentum forms of, 119–120
External forces and couples, 58

Fixed-axis rotation, 277–278
Fluid flow, 535
Focus of conic, 546
Foot, 558
Force, 355
    conservative, 91–92, 329–331
    nonconservative, 331
    work and, 89–90, 309, 311, 313
Forced vibration, 524–525
Frame
    earth as moving, 410–411
    intermediate, 474
    of reference, 3, 7, 401, 405–406
        acceleration and, 207–208, 401, 405–406
        inertial, 57
        velocity and, 198–200, 401
Free vibration, 517–518
Free-body diagram, 62–64
Frequency, 518
Friction, 249

Gears, 170, 183–184
Geosynchronous orbit, 81, 439
Gravitation, 57
Gravitational system, 559
Gravity, 91, 351
Gyration, 239–240
Gyroscopes, 492–495
Gyroscopic moment, 481
Gyroscopic precession, 494

Harmonic motion, simple, 518
Helix, 35
Hertz, 560
Hooke's joint, 391

Impact, 103, 355–356
    central, 107
    direct, 107
    energy loss, 113, 356
    restitution coefficient for, 106–108
Impulse, 102–103, 344, 499–500
Impulsive force, 355
Inertia
    ellipsoid of, 455

moments of, 229, 235–236, 455, 457, 458, 466
maximum and minimum, 468
principal axes of, 455, 456, 462, 466
products of, 240–241
torque (moment), 488
Inertia properties, 228–229
transformations of, 448–449
at a point, 449
Inertial frame of reference, 57
Instantaneous axis of rotation, 149–151, 422
Instantaneous center of zero velocity, 149–151
Intermediate frame, 474
Intrinsic components of velocity and acceleration, 43

Jerk, 52
Joule, 88, 359

Kane, T. R., 383, 432
Kepler's laws of planetary motion, 544, 548, 550
Kilogram, 558
Kinematics, 2, 7
of a point, 8
Kinetic energy, 504
alternative form of, 308
mass center, 87, 306
of a particle, 88
rate of change of, 309
of rigid body, 305–306
rotational, 308, 506
translational, 308, 506
work and, 87–89, 97
principle of, and, 305–306, 316, 327–328
Kinetics, 2

Levinson, D. A., 432
Linear impulse, 102–103, 344, 499–500
Linear momentum. *See* Momentum
Linear spring, 90
Local vertical, 411
Lunisolar precession, 494, 497

Mass center, 58–60, 62, 219, 246–247
Matrices, 434–435
Maximum moments of inertia, 468
Mechanical action, 56
Mechanical energy, 92, 330–331
Meter, 558
Minimum moments of inertia, 468
Moment
of inertia, 229, 235–236, 455, 457, 458, 466
of masses, 567–572
maximum and minimum, 468
of momentum, 118–119, 120–121, 227–228
conservation of, 120–121
in three dimensions, 446
work of, 313
Moment equations of motion, 249–250, 272–273, 274
for fixed-axis rotation, 277–278
governing rotational motion, 472–473
Momentum, 343, 499–500
angular, 118, 227–228, 350, 446
conservation of, 102–103, 350
and Euler's second law, 119–120
linear, 101
moment of, 118–119, 120–121, 227–228, 446
net rate of flow of, 535
of a particle, 101–102
Motion, 176, 305–306
central force, 543
equations of, 249–250, 272–273, 274, 277–278
of mass centers, 58–60, 62
of particles, 62
planetary, Kepler's laws of, 544, 548, 550
rectilinear, 6
rotational, 492
simple harmonic, 518
torque-free, 473, 495, 497
translational, 147
Moving a derivative, 399
Moving frames of reference, 401, 405–406

Natural circular frequency, 518
Net rate of flow of momentum, 535
Newton, 558
Newton, I., 56
Newton frame of reference, 57
Newton-Raphson method, 564
Newton's laws, 56–58
Nonconservative force, 331
Nonimpulsive force, 355
Non-principal axes, 473
Normal component of acceleration, 46
Normal, principal unit, 46
Normal and tangential components of velocity and acceleration, 43
Nutation angle, 493

Orbit
apogee of, 544
eccentricity of, 546
perigee of, 544
period of, 549
Orientation of a rigid body, 132, 428
Orthogonal components, 4

Orthogonality of principal axes, 466
Overdamped system, 523

Parallel-axis theorem
for moments of inertia, 235–236
for products of inertia, 241
Particle, 2, 136
momentum of, 101–102
motion of, 62
work and kinetic energy for, 87–89, 97
Particular solution, 526
Path, 6
Percussion, 362
Perigee, 544
Period, natural, of vibration, 518
orbit, 549
Phase angle, 518
Pivot, 152–153, 277–278, 473
Plane, reference, 131–133
Plane motion, 130
equations of
impulse and momentum for rigid body in, 343–345
motion of, 246–247
kinetic energy of rigid body in, 305–306
Point, 8
Poisson equations, 433
Polar coordinates, 31, 249
Position vector, 6
Potential energy, 92, 329–331
of a central force, 90, 544
of a constant force, 90, 330
of a linear spring, 90, 330
Pound, 558
Power, 309, 311, 508
Precession, 91, 493–495, 496, 497
direct, 496
gyroscopic, 494
lunisolar, 494, 497
retrograde, 497
Primitive, 56
Principal axes of inertia, 294, 455, 456, 462, 466
Principal directions, 458–459
Principal moments of inertia, 455, 457, 458, 466
Principal unit normal vector, 46
Principle of
action and reaction, 64
impulse and momentum, 343–345, 500
Principle of
work and kinetic energy, 305–306, 316, 327–328
Products of inertia, 240–241
parallel-axis theorem for, 241
Projectile motion, 65

Quantity of matter, 56

Radial components of velocity and acceleration, 32
Radius
of curvature, 45
of gyration, 239–240
Rate
of change of kinetic energy, 309
of work of a couple, 311, 508
of work of a force, 81, 311, 508
Rectangular Cartesian coordinates, 24
Rectilinear motion, 8
Rectilinear translation, 148
Reference, frame of. *See* Frame of reference
Reference plane, 131–133
Relative
acceleration, 207, 405
angular momentum, 124
velocity, 198, 402
Resonance, 526
Restitution, coefficient of, 106–108, 355, 361
Resultant force, 57, 248
Resultant moment, 248
Retrograde precession, 497
Reversed effective forces, 488
Reynolds transport theorem, 535
Rigid body(ies), 3, 130, 198
angular acceleration of, 163–164
kinetic energy of, 305–306
orientation of, 132, 428
points on
acceleration of, 163–164
velocities of, 134–135
principle of impulse and momentum for, 343–345
in translation, 219–220
translational motion of, 147, 348
velocity and acceleration equations for, 414–415
work and kinetic energy principles for, 311–312
Rigid extension of the body, 130
Robot, 397
Rocket, 537
Rolling, 170–172, 176
Rotation
of axes (transformation of inertia properties), 448–449
axis of, 277–278
instantaneous, 149–151, 422
of unbalanced bodies, 291–293
Rotation matrices, 434–435

Rotational motion, 492

Satellite, 81, 439, 474, 498
Second, 558
Serret-Frenet formulas, 52
SI units, 559
Simple angular velocity, 386–388
Simple harmonic motion, 518
Single-degree-of-freedom system, 517
Sliding friction, 66
Slug, 558
Space cone, 496
Speed, 6
  angular, 136
Spring, linear
  modulus, 90
  work done by, 90, 314–315
Steady-state displacement, 526
Subcritical damping, 523
Supercritical damping, 523

Tangent vector, unit, 44
Tangential and normal components of velocity and acceleration, 43
Thrust, rocket, 537
Torque-free motion, 473, 495, 497
Total energy, 92
Trajectory, ballistic, 64
Transfer theorem. *See* Parallel-axis theorem
Transformation of inertia properties, 449
Translation, 147–149
  rigid bodies in, 219–220
Translational motion, 147, 348
Transverse components of velocity and acceleration, 32
Truesdell, C., 57

Unbalanced bodies, 291–293
Underdamped system, 523
Unit, 558–562
Unit conversion, 560–562
Unit tangent, 44
Universal gravitation, 57
Universal joint, 391

Vector, 472
  angular acceleration, 398–399
  angular velocity, 384–385
  derivatives of, 3, 384
  position, 6
  principal unit normal, 46
Velocity, 6, 107
  angular. *See* Angular velocity
  in different frames, 198–200, 401
  of points in a rigid body, 134–135
  radial and transverse components of, 32
  tangential and normal components of, 43
  zero, 149–151, 176, 422–423
Velocity equations for rigid bodies, 414–415
Vibration, 517
  damped, 521
  forced, 524–525
  free, 517–518
v-t diagram, 11

Wave propagation, 116
Work, 309, 311, 313
  of a couple, 311
  of a force, 89–90, 309, 311, 313
  of gravity, 91
  and kinetic energy, 87–89, 97, 305–306, 316, 327–328, 504
  rate of, 508
  of a spring, 90

Zero velocity, 149–151, 176, 422–423